## Abbreviations for Units

| | | | | | |
|---|---|---|---|---|---|
| A | ampere | H | henry | | |
| Å | angstrom ($10^{-10}$ m) | h | hour | pt | pint |
| atm | atmosphere | Hz | hertz | qt | quart |
| Btu | British thermal unit | in | inch | rev | revolution |
| Bq | becquerel | J | joule | R | roentgen |
| C | coulomb | K | kelvin | Sv | seivert |
| °C | degree Celsius | kg | kilogram | s | second |
| cal | calorie | km | kilometer | T | tesla |
| Ci | curie | keV | kilo-electron volt | u | unified mass unit |
| cm | centimeter | lb | pound | V | volt |
| dyn | dyne | L | liter | W | watt |
| eV | electron volt | m | meter | Wb | weber |
| °F | degree Fahrenheit | MeV | mega-electron volt | y | year |
| fm | femtometer, fermi ($10^{-15}$ m) | Mm | megameter ($10^6$ m) | yd | yard |
| ft | foot | mi | mile | $\mu$m | micrometer ($10^{-6}$ m) |
| Gm | gigameter ($10^9$ m) | min | minute | $\mu$s | microsecond |
| G | gauss | mm | millimeter | $\mu$C | microcoulomb |
| Gy | gray | ms | millisecond | $\Omega$ | ohm |
| g | gram | N | newton | | |

## Some Conversion Factors

*Length*

1 m = 39.37 in = 3.281 ft = 1.094 yd
1 m = $10^{15}$ fm = $10^{10}$ Å = $10^9$ nm
1 km = 0.6215 mi
1 mi = 5280 ft = 1.609 km
1 lightyear = 1 $c\cdot$y = $9.461 \times 10^{15}$ m
1 in = 2.540 cm

*Volume*

1 L = $10^3$ cm$^3$ = $10^{-3}$ m$^3$ = 1.057 qt

*Time*

1 h = 3600 s = 3.6 ks
1 y = 365.24 d = $3.156 \times 10^7$ s

*Speed*

1 km/h = 0.278 m/s = 0.6215 mi/h
1 ft/s = 0.3048 m/s = 0.6818 mi/h

*Angle–angular speed*

1 rev = $2\pi$ rad = 360°
1 rad = 57.30°
1 rev/min = 0.1047 rad/s

*Force–pressure*

1 N = $10^5$ dyn = 0.2248 lb
1 lb = 4.448 N
1 atm = 101.3 kPa = 1.013 bar = 76.00 cmHg = 14.70 lb/in$^2$

*Mass*

1 u = $[(10^{-3}\ \text{mol}^{-1})/N_A]$ kg = $1.661 \times 10^{-27}$ kg
1 tonne = $10^3$ kg = 1 Mg
1 slug = 14.59 kg
1 kg weighs about 2.205 lb

*Energy–power*

1 J = $10^7$ erg = 0.7373 ft·lb = $9.869 \times 10^{-3}$ L·atm
1 kW·h = 3.6 MJ
1 cal = 4.184 J = $4.129 \times 10^{-2}$ L·atm
1 L·atm = 101.325 J = 24.22 cal
1 eV = $1.602 \times 10^{-19}$ J
1 Btu = 778 ft·lb = 252 cal = 1054 J
1 horsepower = 550 ft·lb/s = 746 W

*Thermal conductivity*

1 W/(m·K) = 6.938 Btu·in/(h·ft$^2$·F°)

*Magnetic field*

1 T = $10^4$ G

*Viscosity*

1 Pa·s = 10 poise

*fifth edition*

# PHYSICS

## FOR SCIENTISTS AND ENGINEERS

*Extended*

W. H. Freeman and Company
*New York*

| | |
|---:|:---|
| *Publisher:* | Susan Finnemore Brennan |
| *Senior Development Editors:* | Kathleen Civetta/Jennifer Van Hove |
| *Assistant Editors:* | Rebecca Pearce/Amanda McCorquodale/Eileen McGinnis |
| *Marketing Manager:* | Mark Santee |
| *Project Editors:* | Georgia L. Hadler/Cathy Townsend, PreMediaONE, A Black Dot Group Company |
| *Text Designer:* | Marsha Cohen |
| *Cover Designer:* | Blake Logan |
| *Illustrations:* | Network Graphics/PreMediaONE, A Black Dot Group Company |
| *Photo Editors:* | Patricia Marx/Dena Betz |
| *Production Manager:* | Julia DeRosa |
| *Media and Supplements Editor:* | Brian Donnellan |
| *Composition:* | PreMediaONE, A Black Dot Group Company |
| *Manufacturing:* | RR Donnelley & Sons Company |

Cover image: Digital Vision

ISBN: 0-7167-4389-2 (EAN: 9780716743897)

**Library of Congress Cataloging-in-Publication Data**
Physics for Scientists and Engineers. - 5th ed.
    p. cm.
    By Paul A. Tipler and Gene Mosca
    Includes index.
    ISBN: 0-7167-0809-4 (Vol. 1 Hardback Ch. 1-20, R)
    ISBN: 0-7167-0900-7 (Vol. 1A Softcover Ch. 1-13, R)
    ISBN: 0-7167-0903-1 (Vol. 1B Softcover Ch. 14-20)
    ISBN: 0-7167-0810-8 (Vol. 2 Hardback Ch. 21-41)
    ISBN: 0-7167-0902-3 (Vol. 2A Softcover Ch. 21-25)
    ISBN: 0-7167-0901-5 (Vol. 2B Softcover Ch. 26-33)
    ISBN: 0-7167-0906-6 (Vol. 2C Softcover Ch. 34-41)
    ISBN: 0-7167-8339-8 (Standard Hardback Ch. 1-33, R)
    ISBN: 0-7167-4389-2 (Extended Hardback Ch. 1-41)

Printed in the United States of America

Second printing

**PT:** For Claudia

**GM:** For Vivian

# CONTENTS IN BRIEF

**VOLUME 1**

1    SYSTEMS OF MEASUREMENT /1

## PART I    MECHANICS

2    MOTION IN ONE DIMENSION /17
3    MOTION IN TWO AND THREE DIMENSIONS /53
4    NEWTON'S LAWS /85
5    APPLICATIONS OF NEWTON'S LAWS /117
6    WORK AND ENERGY /151
7    CONSERVATION OF ENERGY /183
8    SYSTEMS OF PARTICLES AND CONSERVATION OF LINEAR MOMENTUM /217
9    ROTATION /267
10   CONSERVATION OF ANGULAR MOMENTUM /309
R    SPECIAL RELATIVITY /R-1
11   GRAVITY /339
12   STATIC EQUILIBRIUM AND ELASTICITY /370
13   FLUIDS /395

## PART II    OSCILLATIONS AND WAVES

14   OSCILLATIONS /425
15   TRAVELING WAVES /465
16   SUPERPOSITION AND STANDING WAVES /503

## PART III    THERMODYNAMICS

17   TEMPERATURE AND THE KINETIC THEORY OF GASES /532
18   HEAT AND THE FIRST LAW OF THERMODYNAMICS /558
19   THE SECOND LAW OF THERMODYNAMICS /595
20   THERMAL PROPERTIES AND PROCESSES /628

## PART IV    ELECTRICITY AND MAGNETISM

21   THE ELECTRIC FIELD I: DISCRETE CHARGE DISTRIBUTIONS /651
22   THE ELECTRIC FIELD II: CONTINUOUS CHARGE DISTRIBUTIONS /682
23   ELECTRIC POTENTIAL /717
24   ELECTROSTATIC ENERGY AND CAPACITANCE /748

**VOLUME 2**

25 ELECTRIC CURRENT AND DIRECT-CURRENT CIRCUITS /768

26 THE MAGNETIC FIELD /829

27 SOURCES OF THE MAGNETIC FIELD /856

28 MAGNETIC INDUCTION /897

29 ALTERNATING-CURRENT CIRCUITS /934

30 MAXWELL'S EQUATIONS AND ELECTROMAGNETIC WAVES /971

**PART V   LIGHT**

31 PROPERTIES OF LIGHT /997

32 OPTICAL IMAGES /1038

33 INTERFERENCE AND DIFFRACTION /1084

**PART VI   MODERN PHYSICS: QUANTUM MECHANICS, RELATIVITY, AND THE STRUCTURE OF MATTER**

34 WAVE–PARTICLE DUALITY AND QUANTUM PHYSICS /1117

35 APPLICATIONS OF THE SCHRÖDINGER EQUATION /1149

36 ATOMS /1171

37 MOLECULES /1208

38 SOLIDS /1228

39 RELATIVITY /1267

40 NUCLEAR PHYSICS /1306

41 ELEMENTARY PARTICLES AND THE BEGINNING OF THE UNIVERSE /1335

**APPENDIX**

A SI UNITS AND CONVERSION FACTORS /AP-1

B NUMERICAL DATA /AP-3

C PERIODIC TABLE OF ELEMENTS /AP-6

D REVIEW OF MATHEMATICS /AP-8

ILLUSTRATION CREDITS /IL-1

ANSWERS TO ODD-NUMBERED PROBLEMS /A-1

INDEX /I-1

# CONTENTS

## VOLUME 1

PREFACE
ABOUT THE AUTHORS      **\*** = optional material

### CHAPTER 1

**SYSTEMS OF MEASUREMENT / 1**

Classical and Modern Physics /2
1-1      Units /3
International System of Units /3
Other Systems of Units /5
1-2      Conversion of Units /6
1-3      Dimensions of Physical Quantities   7
1-4      Scientific Notation /8
1-5      Significant Figures and Order of Magnitude /10
Summary /13
Problems /14

### PART I      MECHANICS/17

### CHAPTER 2

**MOTION IN ONE DIMENSION / 17**

2-1      Displacement, Velocity, and Speed /17
Instantaneous Velocity /21
Relative Velocity /23
2-2      Acceleration /24
2-3      Motion With Constant Acceleration /27
Problems With One Object /28
Problems With Two Objects /33
2-4      Integration /36
Summary /41
Problems /42

### CHAPTER 3

**MOTION IN TWO AND THREE DIMENSIONS / 53**

3-1      The Displacement Vector /53
Addition of Displacement Vectors /54
3-2      General Properties of Vectors /55
Multiplying a Vector by a Scalar /55

Subtracting Vectors /55
Components of Vectors /55
Unit Vectors /58
3-3      Position, Velocity, and Acceleration /59
Position and Velocity Vectors /59
Relative Velocity /61
The Acceleration Vector /62
3-4      Special Case 1: Projectile Motion /65
3-5      Special Case 2: Circular Motion /72
Uniform Circular Motion /73
Summary /75
Problems /76

### CHAPTER 4

**NEWTON'S LAWS / 85**

4-1      Newton's First Law: The Law of Inertia /86
Inertial Reference Frames /86
4-2      Force, Mass, and Newton's Second Law /87
4-3      The Force Due to Gravity: Weight /90
Units of Force and Mass /91
4-4      Forces in Nature /92
The Fundamental Forces /92
Action at a Distance /92
Contact Forces /93
4-5      Problem Solving: Free Body Diagrams /95
4-6      Newton's Third Law /101
4-7      Problems With Two or More Objects /103
Summary /105
Problems /107

# CHAPTER 5

## APPLICATIONS OF NEWTON'S LAWS / 117

5-1     Friction /117
            Static Friction /117
            Kinetic Friction /118
            Rolling Friction /118
            Friction Explained /119
5-2     Motion Along a Curved Path /129
            ＊ Banked Curves /132
5-3     ＊ Drag Forces /134
5-4     ＊ Numerical Integration: Euler's Method /136
Summary /139
Problems /140

# CHAPTER 6

## WORK AND ENERGY / 151

6-1     Work and Kinetic Energy /152
            Motion in One Dimension With Constant Forces /152
            The Work–Kinetic Energy Theorem /153
            Work Done by a Variable Force /156
6-2     The Dot Product /159
            Power /163
6-3     Work and Energy in Three Dimensions /165
6-4     Potential Energy /167
            Conservative Forces /168
            Potential-Energy Functions /168
            Nonconservative Forces /172
            Potential Energy and Equilibrium /172
Summary /174
Problems /175

# CHAPTER 7

## CONSERVATION OF ENERGY / 183

7-1     The Conservation of Mechanical Energy /184
            Applications /185
7-2     The Conservation of Energy /191
            The Work–Energy Theorem /192
            Problems Involving Kinetic Friction /194
            Systems With Chemical Energy /199
7-3     Mass and Energy /201
            Nuclear Energy /202
            Newtonian Mechanics and Relativity /204
7-4     Quantization of Energy /204
Summary /205
Problems /207

# CHAPTER 8

## SYSTEMS OF PARTICLES AND CONSERVATION OF LINEAR MOMENTUM / 217

8-1     The Center of Mass /218
            Gravitational Potential Energy of a System /221
8-2     ＊ Finding the Center of Mass by Integration /222
            Uniform Rod /222
            Semicircular Hoop /222
8-3     Motion of the Center of Mass /223
8-4     Conservation of Linear Momentum /227
8-5     Kinetic Energy of a System /232
8-6     Collisions /233
            Impulse and Average Force /234
            Collisions in One Dimension (Head-on Collisions) /237
            Collisions in Three Dimensions /244
8-7     ＊ The Center-of-Mass Reference Frame /247
8-8     ＊ Systems With Continuously Varying Mass: Rocket Propulsion /248
Summary /252
Problems /254

# CHAPTER 9

## ROTATION / 267

9-1     Rotational Kinematics: Angular Velocity and Angular Acceleration /267
9-2     Rotational Kinetic Energy /271
9-3     Calculating the Moment of Inertia /272
            Systems of Discrete Particles /273

Continuous Objects /273

The Parallel-Axis Theorem /275

* Proof of the Parallel-Axis Theorem /276

9-4     Newton's Second Law for Rotation /280

Calculating Torques /281

Torque Due to Gravity /281

9-5     Applications of Newton's Second Law for Rotation /282

Problem-Solving Guidelines for Applying Newton's Second Law for Rotation /282

Nonslip Conditions /283

Problem-Solving Guidelines for Applying Newton's Second Law for Rotation /284

Power /286

9-6     Rolling Objects /288

Rolling Without Slipping /288

* Rolling With Slipping /292

Summary /294

Problems /296

# CHAPTER 10

## CONSERVATION OF ANGULAR MOMENTUM / 309

10-1     The Vector Nature of Rotation /309

The Cross Product /310

10-2     Torque and Angular Momentum /311

The Gyroscope /316

10-3     Conservation of Angular Momentum /317

Proofs of Equations 10-10, 10-12, 10-13, 10-14, and 10-15 /324

10-4    * Quantization of Angular Momentum /326

Summary /328

Problems /329

# CHAPTER R

## SPECIAL RELATIVITY / R-1

R-1     The Principle of Relativity and the Constancy of the Speed of Light /R-2

R-2     Moving Sticks /R-4

R-3     Moving Clocks /R-4

R-4     Moving Sticks Again /R-7

R-5     Distant Clocks and Simultaneity /R-8

R-6     Applying the Rules /R-10

R-7     Relativistic Momentum, Mass, and Energy /R-12

Summary /R-13

Problems /R-14

# CHAPTER 11

## GRAVITY / 339

11-1     Kepler's Laws /340

11-2     Newton's Law of Gravity /342

Measurement of $G$ /345

Gravitational and Inertial Mass /345

Derivation of Kepler's Laws /346

11-3     Gravitational Potential Energy /349

Escape Speed /350

Classification of Orbits by Energy /351

11-4     The Gravitational Field $\vec{g}$ / 353

$\vec{g}$ of Spherical Shell and of a Solid Sphere /355

$\vec{g}$ Inside a Solid Sphere /356

11-5    * Finding the Gravitational Field of a Spherical Shell by Integration /358

Summary /361

Problems /362

# CHAPTER 12 *

## STATIC EQUILIBRIUM AND ELASTICITY / 370

12-1     Conditions for Equilibrium /371

12-2     The Center of Gravity /371

12-3     Some Examples of Static Equilibrium /372

12-4     Couples /377

12-5     Static Equilibrium in an Accelerated Frame /377

12-6     Stability of Rotational Equilibrium /378

12-7    Indeterminate Problems /379
12-8    Stress and Strain /380
Summary /383
Problems /384

# CHAPTER 13 *

## FLUIDS / 395

13-1    Density /396
13-2    Pressure in a Fluid /397
13-3    Buoyancy and Archimedes' Principle /402
13-4    Fluids in Motion /407
          Bernoulli's Equation /408
        * Viscous Flow /413
Summary /416
Problems /417

# PART II    OSCILLATIONS AND WAVES /425

# CHAPTER 14

## OSCILLATIONS / 425

14-1    Simple Harmonic Motion /426
          Simple Harmonic Motion and Circular Motion /433
14-2    Energy in Simple Harmonic Motion /434
        * General Motion Near Equilibrium /436
14-3    Some Oscillating Systems /437
          Object on a Vertical Spring /437
          The Simple Pendulum /440
        * The Physical Pendulum /443

14-4    Damped Oscillations /445
14-5    Driven Oscillations and Resonance /449
        * Mathematical Treatment of Resonance /451
Summary /454
Problems /455

# CHAPTER 15

## TRAVELING WAVES / 465

15-1    Simple Wave Motion /465
          Transverse and Longitudinal Waves /465
          Wave Pulses /466
          Speed of Waves /467
        * The Wave Equation /470
15-2    Periodic Waves /473
          Harmonic Waves /473
          Harmonic Sound Waves /477
          Electromagnetic Waves /478
15-3    Waves in Three Dimensions /478
          Wave Intensity /479
15-4    Waves Encountering Barriers /482
          Reflection and Refraction /482
          Diffraction /484
15-5    The Doppler Effect /486
          Shock Waves /490
Summary /492
Problems /495

# CHAPTER 16

## SUPERPOSITION AND STANDING WAVES / 503

16-1    Superposition of Waves /504
        * Superposition and the Wave Equation /504
          Interference of Harmonic Waves /505
16-2    Standing Waves /511
          Standing Waves on Strings /511
          Standing Sound Waves /517
16-3    * The Superposition of Standing Waves /519
16-4    * Harmonic Analysis and Synthesis /520
16-5    * Wave Packets and Dispersion /521
Summary /522
Problems /524

## PART III  THERMODYNAMICS /532

### CHAPTER 17

**TEMPERATURE AND THE KINETIC THEORY OF GASES / 532**

17-1   Thermal Equilibrium and Temperature /532

17-2   The Celsius and Fahrenheit Temperature Scales /533

17-3   Gas Thermometers and the Absolute Temperature Scale /535

17-4   The Ideal-Gas Law /537

17-5   The Kinetic Theory of Gases /541

    Calculating the Pressure Exerted by a Gas /542

    The Molecular Interpretation of Temperature /542

    The Equipartition Theorem /544

    Mean Free Path /545

    * The Distribution of Molecular Speeds /546

Summary /551

Problems /552

### CHAPTER 18

**HEAT AND THE FIRST LAW OF THERMODYNAMICS / 558**

18-1   Heat Capacity and Specific Heat /559

    Calorimetry /561

18-2   Change of Phase and Latent Heat /562

18-3   Joule's Experiment and the First Law of Thermodynamics /565

18-4   The Internal Energy of an Ideal Gas /568

18-5   Work and the PV Diagram for a Gas /569

    Quasi-Static Processes /569

    PV Diagrams /570

18-6   Heat Capacities of Gases /572

    Heat Capacities and the Equipartition Theorem /576

18-7   Heat Capacities of Solids /577

18-8   Failure of the Equipartition Theorem /578

18-9   The Quasi-Static Adiabatic Compression of a Gas /581

    Speed of Sound Waves /584

Summary /585

Problems /587

### CHAPTER 19

**THE SECOND LAW OF THERMODYNAMICS / 595**

19-1   Heat Engines and the Second Law of Thermodynamics /596

19-2   Refrigerators and the Second Law of Thermodynamics /600

19-3   Equivalence of the Heat-Engine and Refrigerator Statements /602

19-4   The Carnot Engine /602

    The Thermodynamic or Absolute Temperature Scale /608

19-5   * Heat Pumps /609

19-6   Irreversibility and Disorder /610

19-7   Entropy /610

    Entropy of an Ideal Gas /611

    Entropy Changes for Various Processes /612

19-8   Entropy and the Availability of Energy /617

19-9   Entropy and Probability /618

Summary /620

Problems /621

### CHAPTER 20 *

**THERMAL PROPERTIES AND PROCESSES / 628**

20-1   Thermal Expansion /628

20-2   The van der Waals Equation and Liquid-Vapor Isotherms /632

20-3   Phase Diagrams /634

20-4 The Transfer of Thermal Energy /635

Conduction /635

Convection /641

Radiation /642

Summary /645

Problems /646

# VOLUME 2

## PART IV ELECTRICITY AND MAGNETISM/651

### CHAPTER 21

**THE ELECTRIC FIELD I: DISCRETE CHARGE DISTRIBUTIONS / 651**

21-1 Electric Charge /652

Charge Quantization /652

Charge Conservation /653

21-2 Conductors and Insulators /654

Charging by Induction /655

21-3 Coulomb's Law /656

Force Exerted by a System of Charges /658

21-4 The Electric Field /661

Electric Dipoles /665

21-5 Electric Field Lines /666

21-6 Motion of Point Charges in Electric Fields /668

21-7 Electric Dipoles in Electric Fields /671

Summary /673

Problems /674

### CHAPTER 22

**THE ELECTRIC FIELD II: CONTINUOUS CHARGE DISTRIBUTIONS / 682**

22-1 Calculating $\vec{E}$ From Coulomb's Law /683

$\vec{E}$ on the Axis of a Finite Line Charge /683

$\vec{E}$ off the Axis of a Finite Line Charge /684

$\vec{E}$ due to an Infinite Line Charge /685

$\vec{E}$ on the Axis of a Ring Charge /688

$\vec{E}$ on the Axis of a Uniformly Charged Disk /688

$\vec{E}$ Due to an Infinite Plane of Charge /689

22-2 Gauss's Law /690

Electric Flux /691

Quantitative Statement of Gauss's Law /692

22-3 Calculating $\vec{E}$ From Gauss's Law /694

Plane Symmetry /695

Spherical Symmetry /696

$\vec{E}$ due to a Spherical Shell of Charge /696

$\vec{E}$ Due to a Uniformly Charged Sphere /699

Cylindrical Symmetry /700

22-4 Discontinuity of $\vec{E}_n$ /701

22-5 Charge and Field at Conductor Surfaces /702

22-6 * Derivation of Gauss's Law From Coulomb's Law /707

Summary /708

Problems /710

### CHAPTER 23

**ELECTRIC POTENTIAL / 717**

23-1 Potential Difference /717

Continuity of V /718

Units /719

Potential and Electric Field Lines /719

23-2 Potential Due to a System of Point Charges /720

23-3 Computing the Electric Field From the Potential /724

* General Relation Between $\vec{E}$ and V /725

23-4 Calculations of $V$ for Continuous Charge Distributions /726

    $V$ on the Axis of a Charged Ring /726

    $V$ on the Axis of a Uniformly Charged Disk /727

    $V$ due to an Infinite Plane of Charge /729

    $V$ Inside and Outside a Spherical Shell of Charge /731

    $V$ due to an Infinite Line Charge /733

23-5 Equipotential Surfaces /733

    The Van de Graaff Generator /736

    Dielectric Breakdown /737

Summary /739

Problems /741

## CHAPTER 24

### ELECTROSTATIC ENERGY AND CAPACITANCE / 748

24-1 Electrostatic Potential Energy /749

24-2 Capacitance /752

    Capacitors /753

    Parallel-Plate Capacitors /753

    Cylindrical Capacitors /754

24-3 The Storage of Electrical Energy /756

    Electrostatic Field Energy /759

24-4 Capacitors, Batteries, and Circuits /760

    Combinations of Capacitors /761

24-5 Dielectrics /767

    Energy Stored in the Presence of a Dielectric /770

24-6 Molecular View of a Dielectric /772

    Magnitude of the Bound Charge /775

    * The Piezoelectric Effect /775

Summary /776

Problems /777

## CHAPTER 25

### ELECTRIC CURRENT AND DIRECT-CURRENT CIRCUITS / 786

25-1 Current and the Motion of Charges /787

25-2 Resistance and Ohm's Law /790

25-3 Energy in Electric Circuits /794

    EMF and Batteries /795

25-4 Combinations of Resistors /798

    Resistors in Series /798

    Resistors in Parallel /799

25-5 Kirchhoff's Rules /803

    Single-Loop Circuits /804

    Multiloop Circuits /806

    Ammeters, Voltmeters, and Ohmmeters /809

25-6 $RC$ Circuits /811

    Discharging a Capacitor /811

    Charging a Capacitor /813

    Energy Conservation in Charging a Capacitor /815

Summary /816

Problems /818

## CHAPTER 26

### THE MAGNETIC FIELD / 829

26-1 The Force Exerted by a Magnetic Field /830

26-2 Motion of a Point Charge in a Magnetic Field /834

    * The Velocity Selector /836

    * Thomson's Measurement of $q/m$ for Electrons /837

    * The Mass Spectrometer /838

    The Cyclotron /839

26-3 Torques on Current Loops and Magnets /841

    Potential Energy of a Magnetic Dipole in a Magnetic Field /843

26-4 The Hall Effect /845

    * The Quantum Hall Effects /847

Summary /848

Problems /849

# CHAPTER 27

## SOURCES OF THE MAGNETIC FIELD / 856

27-1    The Magnetic Field of Moving Point Charges /857
27-2    The Magnetic Field of Currents: The Biot–Savart Law /858
        $\vec{B}$ Due to a Current Loop /859
        $\vec{B}$ Due to a Current in a Solenoid /863
        $\vec{B}$ Due to a Current in a Straight Wire /865
        Magnetic Force Between Parallel Wires /868
27-3    Gauss's Law for Magnetism /870
27-4    Ampère's Law /871
        Limitations of Ampère's Law /874
27-5    Magnetism in Matter /874
        Magnetization and Magnetic Susceptibility /875
        Atomic Magnetic Moments /876
        * Paramagnetism /879
        * Ferromagnetism /880
        * Diamagnetism /884
Summary /886
Problems /888

# CHAPTER 28

## MAGNETIC INDUCTION / 897

28-1    Magnetic Flux /898
28-2    Induced EMF and Faraday's Law /899
28-3    Lenz's Law /903
28-4    Motional EMF /907
28-5    Eddy Currents /912
28-6    Inductance /912
        Self-Inductance /912
        Mutual Inductance /914
28-7    Magnetic Energy /915
28-8    * *RL* Circuits /917
28-9    * Magnetic Properties of Superconductors /922
        * Meissner Effect /922
        * Flux Quantization /923
Summary /923
Problems /925

# CHAPTER 29

## ALTERNATING-CURRENT CIRCUITS / 934

29-1    Alternating Current Generators /935
29-2    Alternating Current in a Resistor /936
        Root-Mean-Square Values /937
29-3    Alternating-Current Circuits /939
        Inductors in Alternating Current Circuits /939
        Capacitors in Alternating Current Circuits /941
29-4    * Phasors /943
29-5    * *LC* and *RLC* Circuits Without a Generator /944
29-6    * Driven *RLC* Circuit /948
        Series *RLC* Circuit /948
        Resonance /950
        Parallel *RLC* Circuit /956
29-7    * The Transformer /956
Summary /960
Problems /962

# CHAPTER 30

## MAXWELL'S EQUATIONS AND ELECTROMAGNETIC WAVES / 971

30-1    Maxwell's Displacement Current /972
30-2    Maxwell's Equations /975
30-3    Electromagnetic Waves /976
        The Electromagnetic Spectrum /976

Production of Electromagnetic Waves /978

Electric Dipole Radiation /978

Energy and Momentum in an Electromagnetic Wave /981

30-4 * The Wave Equation for Electromagnetic Waves /985

* Derivation of the Wave Equation /986

Summary /991

Problems /992

PART V    LIGHT/997

CHAPTER 31

PROPERTIES OF LIGHT / 997

31-1    Wave–Particle Duality /998

31-2    Light Spectra /998

31-3    Sources of Light /999

Line Spectra /999

Absorption, Scattering, Spontaneous Emission, and Stimulated Emission /1001

Lasers /1003

31-4    The Speed of Light /1005

31-5    The Propagation of Light /1010

Huygen's Principle /1010

Fermat's Principle /1011

31-6    Reflection and Refraction /1011

Physical Mechanisms for Reflection and Refraction /1013

Specular Reflection and Diffuse Reflection /1013

Relative Intensity of Reflected and Transmitted Light /1014

Total Internal Reflection /1015

Mirages /1017

Dispersion /1018

31-7    Polarization /1021

Polarization by Absorption /1022

Polarization by Reflection /1023

Polarization by Scattering /1024

Polarization by Birefringence /1025

31-8    Derivation of the Laws of Reflection and Refraction /1027

Huygen's Principle /1027

Fermat's Principle /1028

Summary /1030

Problems /1032

CHAPTER 32

OPTICAL IMAGES / 1038

32-1    Mirrors /1038

Plane Mirrors /1038

Spherical Mirrors /1041

Ray Diagrams for Mirrors /1045

32-2    Lenses /1049

Images Formed by Refraction /1049

Thin Lenses /1052

Ray Diagrams for Lenses /1057

Combinations of Lenses /1059

Compound Lenses /1061

32-3    * Aberrations /1062

32-4    * Optical Instruments /1063

* The Eye /1063

* The Simple Magnifier /1066

* The Compound Microscope /1068

* The Telescope /1070

Summary /1074

Problems /1076

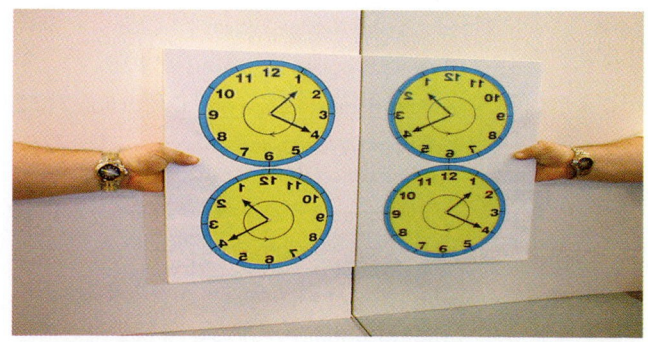

# CHAPTER 33

## INTERFERENCE AND DIFFRACTION / 1084

33-1 Phase Difference and Coherence /1084
33-2 Interference in Thin Films /1086
33-3 Two-Slit Interference Pattern /1088
 Calculation of Intensity /1090
33-4 Diffraction Pattern of a Single Slit /1091
 Interference–Diffraction Pattern of Two Slits /1093
33-5 * Using Phasors to Add Harmonic Waves /1094
 * The Interference Pattern of Three or More Equally Spaced Sources /1096
 * Calculating the Single-Slit Diffraction Pattern /1098
 * Calculating the Interference-Diffraction Pattern of Multiple Slits /1100
33-6 Fraunhofer and Fresnel Diffraction /1101
33-7 Diffraction and Resolution /1103
33-8 * Diffraction Gratings /1105
 * Holograms /1108
Summary /1109
Problems /1110

# PART VI MODERN PHYSICS: QUANTUM MECHANICS, RELATIVITY, AND THE STRUCTURE OF MATTER

# CHAPTER 34

## WAVE-PARTICLE DUALITY AND QUANTUM PHYSICS / 1117

34-1 Light /1118
34-2 The Particle Nature of Light: Photons /1119
 The Photoelectric Effect /1119
 Compton Scattering /1122
34-3 Energy Quantization in Atoms /1124
34-4 Electrons and Matter Waves /1125
 The de Broglie Hypothesis /1125
 Electron Interference and Diffraction /1127
 Standing Waves and Energy Quantization /1129
34-5 The Interpretation of the Wave Function /1129
34-6 Wave–Particle Duality /1131
 The Two-Slit Experiment Revisited /1131
 The Uncertainty Principle /1132

34-7 A Particle in a Box /1133
 Standing-Wave Functions /1134
34-8 Expectation Values /1138
 * Calculating Probabilities and Expectation Values /1138
34-9 Energy Quantization in Other Systems /1140
 The Harmonic Oscillator /1140
 The Hydrogen Atom /1141
Summary /1142
Problems /1144

# CHAPTER 35

## APPLICATIONS OF THE SCHRÖDINGER EQUATION / 1149

35-1 The Schrödinger Equation /1150
 A Particle in an Infinite Square-Well Potential /1151
35-2 A Particle in a Finite Square Well /1152
35-3 The Harmonic Oscillator /1155
 Wave Functions and Energy Levels /1155
35-4 Reflection and Transmission of Electron Waves: Barrier Penetration /1157
 Step Potential /1157
 Barrier Penetration /1159
35-5 The Schrödinger Equation in Three Dimensions /1161
35-6 The Schrödinger Equation for Two Identical Particles /1164
Summary /1166
Problems /1168

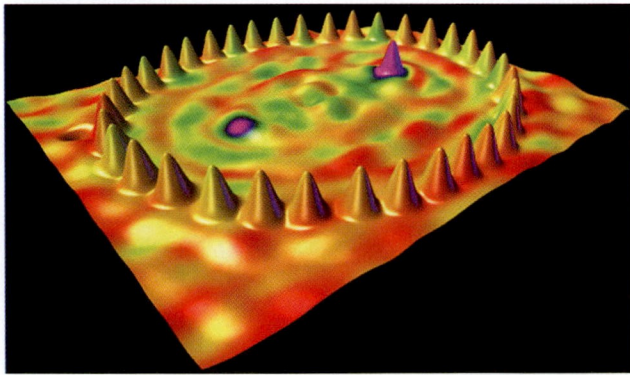

# CHAPTER 36

## ATOMS / 1171

36-1 The Nuclear Atom /1172
 Atomic Spectra /1172
36-2 The Bohr Model of the Hydrogen Atom /1173
 Energy for a Circular Orbit /1173

Bohr's Postulates /1174

Energy Levels /1177

36-3 Quantum Theory of Atoms /1178

The Schrödinger Equation in Spherical Coordinates /1178

Quantum Numbers in Spherical Coordinates /1179

36-4 Quantum Theory of the Hydrogen Atom /1181

Energy Levels /1181

Wave Functions and Probability Densities /1182

36-5 The Spin–Orbit Effect and Fine Structure /1186

36-6 The Periodic Table /1189

Helium ($Z = 2$) /1190

Lithium ($Z = 3$) /1191

Beryllium ($Z = 4$) /1193

Boron to Neon ($Z = 5$ to $Z = 10$) /1193

Sodium to Argon ($Z = 11$ to $Z = 18$) /1193

Elements With $Z > 18$ /1193

36-7 Optical Spectra and X-Ray Spectra /1197

Optical Spectra /1197

X-Ray Spectra /1199

Summary /1201

Problems /1203

# CHAPTER 37

## MOLECULES / 1208

37-1 Molecular Bonding /1208

The Ionic Bond /1209

The Covalent Bond /1210

Other Bonding Types /1213

37-2 * Polyatomic Molecules /1215

37-3 Energy Levels and Spectra of Diatomic Molecules /1217

Rotational Energy Levels /1217

Vibrational Energy Levels /1219

Emission Spectra /1220

Absorption Spectra /1222

Summary /1223

Problems /1225

# CHAPTER 38

## SOLIDS / 1228

38-1 The Structure of Solids /1229

38-2 A Microscopic Picture of Conduction /1233

Classical Interpretation of $v_{av}$ and $\lambda$ /1235

Successes and Failures of the Classical Model /1236

38-3 The Fermi Electron Gas /1236

Energy Quantization in a Box /1237

The Exclusion Principle /1237

The Fermi Energy /1238

The Fermi Factor at $T = 0$ /1240

The Fermi Factor for $T > 0$ /1240

Contact Potential /1241

Heat Capacity due to Electrons in a Metal /1242

38-4 Quantum Theory of Electrical Conduction /1243

The Scattering of Electron Waves /1243

38-5 Band Theory of Solids /1244

38-6 Semiconductors /1246

38-2 * Semiconductor Junctions and Devices /1248

 * Diodes /1248

 * Transistors /1250

38-8 Superconductivity /1252

The BCS Theory /1254

The Josephson Effect /1255

38-9 The Fermi–Dirac Distribution /1256

Summary /1259

Problems /1262

# CHAPTER 39

## RELATIVITY / 1267

39-1 Newtonian Relativity /1268

Ether and the Speed of Light /1268

39-2 Einstein's Postulates /1269

39-3 The Lorentz Transformation /1270

Time Dilation /1272

Length Contraction /1274

The Relativistic Doppler Effect /1276

39-4 Clock Synchronization and Simultaneity /1278

The Twin Paradox /1282

39-5 The Velocity Transformation /1284

39-6 Relativistic Momentum /1287

Illustration of Conservation of the Relativistic Momentum /1288

39-7 Relativistic Energy /1289

Mass and Energy /1292

39-8 General Relativity /1296

Summary /1299

Problems /1300

# CHAPTER 40

## NUCLEAR PHYSICS / 1306

40-1      Properties of Nuclei /1306
         Size and Shape /1307
         N and Z Numbers /1308
         Mass and Binding Energy /1308

40-2      Radioactivity /1310
         Beta Decay /1314
         Gamma Decay /1316
         Alpha Decay /1316

40-3      Nuclear Reactions /1317
         Reactions With Neutrons /1319

40-4      Fission and Fusion /1319
         Fission /1321
         Nuclear Fission Reactors /1322
         Fusion /1326

Summary /1328
Problems /1330

# CHAPTER 41

## ELEMENTARY PARTICLES AND THE BEGINNING OF THE UNIVERSE / 1335

41-1      Hadrons and Leptons /1336
41-2      Spin and Antiparticles /1339
41-3      The Conservation Laws /1342

41-4      Quarks /1345
         Quark Confinement /1347
41-5      Field Particles /1347
41-6      The Electroweak Theory /1348
41-7      The Standard Model /1349
         Grand Unification Theories /1350
41-8      The Evolution of the Universe /1350
         The 2.7-K Background Radiation /1351
         The Big Bang /1352

Summary /1353
Problems /1354

# APPENDIX A

## SI UNITS AND CONVERSION FACTORS / AP-1

# APPENDIX B

## NUMERICAL DATA / AP-3

# APPENDIX C

## PERIODIC TABLE OF ELEMENTS / AP-6

# APPENDIX D

## REVIEW OF MATHEMATICS / AP-8

Illustration Credits /IL-1
Answers /A-1
Index /I-1
Glossary/Inside front and back covers

# PREFACE

We are exceptionally pleased to present the fifth edition of *Physics for Scientists and Engineers*. Over the course of this revision, we have built upon the strengths of the fourth edition so that the new text is an even more reliable, engaging and motivating learning tool for the calculus-based introductory physics course. With the help of reviewers and the many users of the fourth edition we have carefully scrutinized and refined every aspect of the book, with an eye toward improving student comprehension and success. Our goals included helping students to increase their problem-solving ability, making the text more accessible and fun to read, and keeping the text flexible for the instructor.

## Examples

One of the most important ways we've addressed our goals was to add some new features to the side-by-side worked examples that were introduced in the fourth edition. These examples juxtapose the problem-solving steps with the necessary equations so that it's easier for students to watch the problem unfold.

### The side-by-side format for the worked examples came from a student suggestion; we've just added a few finishing touches:

• After each problem statement, students are asked to *Picture the Problem*. Here, the problem is analyzed both conceptually and visually, with students frequently directed to draw a free-body diagram. Each step of the solution is then presented with a written statement in the left-hand column and the corresponding mathematical equations in the right-hand column.

• *Remarks* at the end of the example point out the importance or relevance of the example, or suggest a different way to approach it.

• NEW *Plausibility Checks* remind students to check their results for mathematical accuracy, and for reasonableness as well.

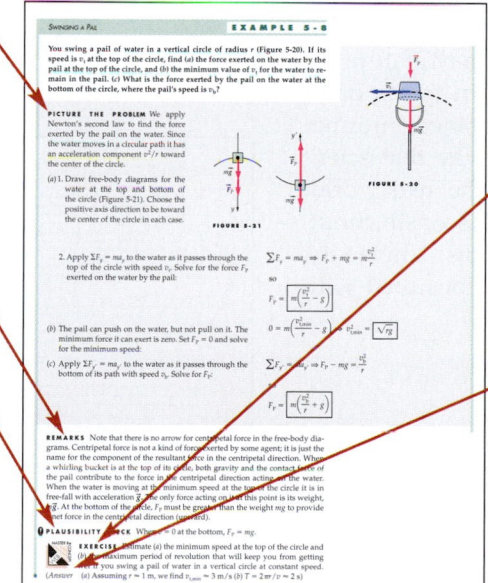

• An *Exercise* often follows the solution of the example, allowing students to check their understanding by solving a similar problem without help. Answers are included with the Exercise to provide immediate feedback and alternative solutions.

• NEW *Master the Concept Exercises* appear at least once in each chapter and help build students' problem-solving skills online.

Every example has been scrutinized, with additional steps added wherever an assumption might have been made, new Remarks included, and new follow-up exercises, free-body diagrams added where appropriate. The answers are now boxed to make them easier to find. Our new features include the Plausibility Check, which offers quick tests that help students learn to evaluate their answers with logic. We've also added interactive Master the Concept exercises to help students work through key problems. The exercises follow examples in the textbook and are marked with a Master the Concept icon that directs students to our Web site. There, the exercise is set up with algorithmically generated variables and students work the problem with step-by-step guidance and immediate feedback.

This edition also includes two types of specialized examples that provide unique problem-solving opportunities for students. The Try it Yourself examples prompt students to take an active role in solving the problem, and the Put It in Context examples approximate the real life scenarios they might encounter as scientists.

## Try It Yourself examples

Like the regular worked example, these use the side-by-side format, but here the Picture the Problem section is sometimes missing, and the descriptions in the left-hand column are more terse. These examples take students step-by-step through the solution without doing the math for them. Students find it helpful to cover the right-hand column and attempt to perform the calculations on their own before looking at the equations. In this way, students can think through the steps as they fill in the answers.

## New Put It in Context examples

Each chapter now identifies at least one worked example as "context rich." These examples may include information not needed to solve the problem, or may require the student to find additional information in tables or to draw from experience or previously obtained information. Context-rich examples reflect the way that scientists and engineers solve problems in the real world. Laura McCullough of the University of Wisconsin, Stout, and Thomas Foster of Southern Illinois University, Edwardsville, initiated this feature and consulted with us in creating many of these examples.

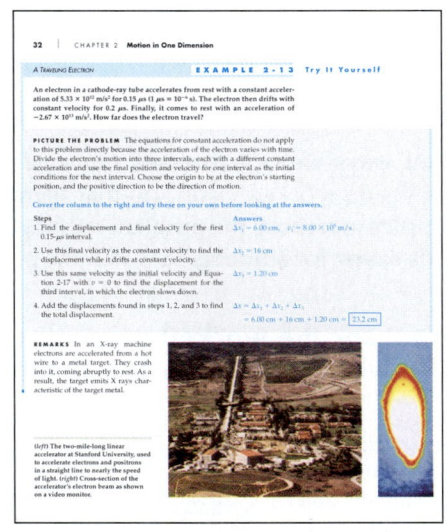

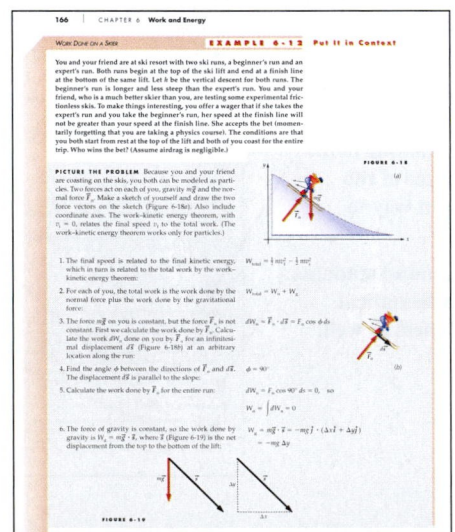

# Practice Problems

Care has been taken to improve the quality and clarity of the end-of-chapter problems. About twenty percent of the 4,500 problems are new, written by Charles Adler of St. Mary's College of Maryland. Conceptual problems have been grouped together at the beginning of each problem set, and a new category of Estimation and Approximation problems have been added to encourage students to think more like scientists or engineers. Answers to odd-numbered problems appear at the back of the text. Solutions to approximately twenty-five percent of the problems appear in the newly revised Student Solutions Manual. This was written by David Mills of the College of the Redwoods to provide detailed solutions and to mirror the popular side-by-side format of the textbook examples.

About 1,100 of the text's problems are included in the new iSOLVE homework service. These problems can be accessed at www.whfreeman.com/tipler5e. About a third of the iSOLVE problems are Checkpoint Problems, which ask students to note the key principles and equations they're using and indicate their confidence level.

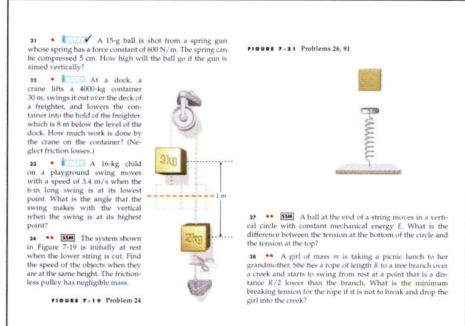

**Each problem is marked with:**

• a series of one, two, or three bullets, which identify its level of difficulty

• a SSM icon if the answer is in the Student Solutions Manual

• an iSOLVE icon if the problem is part of the isolve homework service and a iSOLVE ✓ icon if the problem is a Checkpoint problem.

# Features

This new edition of *Physics* has a number of textual features that make the book a valuable teaching tool. Key aspects of the last edition have been revised, and some new features have been added to make the book more engaging, inviting and up to date.

## New chapter-opening pedagogy

• Each chapter now begins with a photograph and a question that is answered in a worked example within the chapter. These draw students into the material and provide motivation for problem solving.

• Chapter outlines list the major section headings, giving students a "road map" to the chapter.

• Chapter goal statements highlight the main ideas of the chapter.

PART I **MECHANICS**

CHAPTER **2**

**Motion in One Dimension**

MOTION IN ONE DIMENSION IS MOTION ALONG A STRAIGHT LINE, LIKE THAT OF A CAR ON A STRAIGHT ROAD. THIS DRIVER ENCOUNTERS STOPLIGHTS AND DIFFERENT SPEED LIMITS ON HER COMMUTE ALONG A STRAIGHT HIGHWAY TO SCHOOL.

How can she estimate her arrival time? (See Example 2-2.)

2-1 Displacement, Velocity, and Speed
2-2 Acceleration
2-3 Motion With Constant Acceleration
2-4 Integration

## Content improvements

**Chapter R**, an optional "mini" chapter in Volume 1, brief enough to be covered in a lecture or two, allows instructors to include this popular modern topic early in the course. The chapter avoids the abstraction associated with the Lorentz transformations and focuses on the basic concepts of length contraction, time dilation, and simultaneity, using thought experiments involving meter sticks and light clocks. The relation between relativistic momentum and relativistic energy is also developed.

**Quantum Theory:** Chapters 17, "Wave-Particle Duality and Quantum Physics," and 27, "The Microscopic Theory of Electrical Conduction" of the fourth edition have been moved to their more traditional location in Volume II of the fifth edition as Chapters 37 and 38. Should instructors wish to include these chapters earlier in the course, both chapters are available on the web at www.whfreeman.com/tipler5e.

**Changes in Approach:** Dozens of smaller, yet significant improvements in content have been made throughout the book. For example:

- Motion-diagrams are introduced in Section 3-3 and used to estimate the direction of the acceleration vector using the definition of acceleration.

- In Section 4-4, frictional forces are now introduced qualitatively, allowing for free-body diagrams that include frictional forces. A quantitative treatment of frictional forces appears in Section 5-1.

- Section 4-7 introduces problems with two or more objects. Selecting a separate set of coordinate axes for each object is a robust problem-solving practice when using Newton's laws with systems consisting of two or more objects. The value of this practice is revealed in the example where Steve is sliding down the glacier while Paul has already fallen over its edge.

- In Section 8-8, "Systems With Variable Mass," the basic equation of motion for an object with continuously varying mass (the rocket equation) is developed using an object that is acquiring mass—like an open boxcar in the rain—rather than one that is losing mass—like a rocket spewing exhaust gasses. This approach facilitates both the development of the basic equation of motion and the application of it to certain situations.

- In Chapter 9, "Rotation," there is a new section that provides problem-solving guidelines for applying Newton's Second Law to rotation.

- In Section 13-3 the discussion of buoyancy now includes the buoyant force on objects supported by a submerged surface.

- In Chapter 18, work-energy relations are expressed in terms of the work done on the system. The first law of thermodynamics is now expressed in terms of the work done on the system also. (The Educational Testing Service has adopted the convention that the work term in first law of thermodynamics be the work done on the system. This will be adhered to on all Advanced Placement physics exams.)

## More engineering and biological applications

Additional applications emphasize the relevance of physics to students' experiences, further studies, and future careers.

## New focus on common pitfalls

Topics that commonly cause confusion are identified with a new ❶ icon where the difficulty is addressed. For example, in Section 3-4 the icon is used to identify the discussion pointing out that the horizontal and vertical motions are independent in projectile motion.

For instructor and student convenience, the fifth edition of *Physics for Scientists and Engineers* is available in five paperback volumes—

***Vol. 1A Mechanics*** (Ch. 1-13, plus a mini-chapter on relativity, Ch. R) 0-7167-0900-7

***Vol. 1B Oscillations & Waves; Thermodynamics*** (Ch. 14-20) 0-7167-0903-1

***Vol. 2A Electricity*** (Ch. 21-25) 0-7167-0902-3

***Vol. 2B Electrodynamics, Light*** (Ch. 26-33) 0-7167-0901-5

***Vol. 2C Elementary Modern Physics*** (Ch. 34-41) 0-7167-0906-6

or in four hardcover versions—

***Vol. 1 Mechanics, Oscillations and Waves; Thermodynamics*** (Ch. 1-20, R) 0-7167-0809-4

***Vol. 2 Electricity, Magnetism, Light & Modern Physics*** (Ch. 21-41) 0-7167-0810-8

**Standard Version** (Vol 1A-2B) 0-7167-8339-8

**Extended Version** (Vol 1A-2C) 0-7167-4389-2

## New design and improved illustrations

The book has a warmer, more colorful look. Each piece of art has been carefully considered and many have been revised to increase clarity. Approximately 245 new figures have been added, including many new free-body diagrams within the worked examples. New photos bring to life the many real-world applications of physics.

## Optional sections

The book was designed to allow professors to be flexible by designating certain sections "optional." These sections are marked with an *, and professors who choose to skip this section can do so knowing that their students won't be missing any material they will need in later chapters.

## Summary

End of chapter summaries are organized with important topics on the left and relevant remarks and equations on the right. Here the key equations from the chapter appear together for easy reference.

## Exploring essays

Students are invited to examine interesting extensions of the chapter concepts in Exploring sections, which are now found on the Web. These short pieces relate the chapter concepts to everything from the weather to transducers.

# Media and Print Supplements

The supplements package has been updated and improved in response to reviewer suggestions and those from users of the fourth edition.

## For the Student:

**Student Solutions Manual:** *Vol. 1, 0-7167-8333-9; Vol. 2, 0-7167-8334-7.* The new manual prepared by David Mills of College of the Redwoods, Charles Adler of St. Mary's College of Maryland, Ed Whittaker of Stevens Institute of Technology, George Zober of Yough Senior High School and Patricia Zober of Ringgold High School provides solutions for about twenty-five percent of the problems in the textbook, using the same side-by-side format and level of detail as the textbook's worked examples.

**Study Guide:** *Vol. 1, 0-7167-8332-0; Vol. 2, 0-7167-8331-2.* Prepared by Gene Mosca of the United States Naval Academy and Todd Ruskell of Colorado School of Mines, the Study Guide describes the key ideas and potential pitfalls of each chapter, and also includes true and false questions that test essential definitions and relations, questions and answers that require qualitative reasoning, and problems and solutions.

**Student Web Site:** Robin Jordan of Florida Atlantic University has put together a site designed to make studying and testing easier for both students and professors. The Web site includes:

- **On-line quizzing:** Multiple choice quizzes are available for each chapter. Students will receive immediate feedback, and the quiz results are collected for the instructor in a grade book.

- **iSOLVE homework service**: *0-7167-5802-4.* About one-fourth of the book's end-of-chapter problems, 1,100 altogether, are available on-line in W.H. Freeman's iSOLVE homework service. This service will offer each student a different version of every problem similar to CAPA and WebAssign, and the iSOLVE problems will be marked with an icon in the textbook. Homework scores can be collected in a grade book. Students may purchase access to iSOLVE for three semesters at a time.

• **iSOLVE Checkpoint problems:** A third of our iSOLVE questions are Checkpoint problems, which prompt students to describe how they arrived at their answer and to indicate their confidence level. All student responses will be gathered and included in the instructor's grade book report. Rolf Enger of the U.S. Air Force Academy inspired the development of Checkpoints to help professors gauge their student's understanding of the material.

• **Master the Concept exercises:** For each chapter, one or more exercises from the book will be available on-line so students can practice working the problem with randomized variables and step-by-step guidance. The on-line exercise will walk the student slowly through the problem-solving process and use interactive animations, simulations, video, and other graphic aids to help students visualize the problem. Teachers can collect grade book information on their progress. These premium examples are called out in the book with a Master the Concept icon.

**Homework services:** In addition to the iSOLVE network, there are three other homework services that are compatible with this textbook. End of chapter problems are available in WebAssign as well as CAPA: A Computer-Assisted Personalized Approach. A list of all the fifth edition problems included in WebAssign and CAPA is posted on the instructor's section of the *Physics* Web site. Our text is also compatible with the University of Texas Interactive Homework Service.

**The iSolve homework service is available at**
www.whfreeman.com/tipler5e

**For more information about WebAssign, CAPA or UTX homework services, find their Web sites at:**
http://webassign.net/info
http://www.pa.msu.edu/educ/CAPA/
http://hw.utexas.edu/hw.html

## For the Instructor:

**Instructor's Resource CD-ROM:** *0-7167-9839-5*. This multi-faceted resource will give instructors the tools to make their own Web sites and presentations. The CD contains illustrations from the text in .jpg format, Powerpoint Lecture Slides for each chapter of the book, Lab Demonstration Videos, and Applied Physics videos in QuickTime format, and Presentation Manager Pro v.2.0, as well as all of the solutions to the end-of-chapter problems in editable Microsoft Word format.

**Instructor's Resource Manual:** The updated IRM contains Classroom Demonstrations for each chapter, a film and video guide with suggestions for each chapter, links to valuable Web sites, and links to free sources for Physlets, animations, and other teaching tools. This manual will be available on the book's Web site at www.whfreeman.com/tipler5e.

**Instructor's Solutions Manual:** *Vol. 1, 0-7167-9640-6; Vol. 2, 0-7167-9639-2*. This guide contains fully worked solutions for all of the problems in the textbook, using the side-by-side format wherever possible. It is available in print and is also included in editable Word files on the Instructor's CD-ROM.

**Test Bank:** *In print, 0-7167-9652-X; CD-ROM, 0-7167-9653-8*. Prepared by Mark Riley of Florida State University and David Mills of College of the Redwoods, this set of more than 4,000 multiple choice questions is available both in print and on a CD-ROM for Windows and Macintosh users. All questions refer to specific sections in the book. The CD-ROM version of the Test Bank makes it easy to add, edit and re-sequence questions to suit your needs.

**Transparencies:** *0-7167-9664-3*. Approximately 150 full color acetates of figures and tables from the text are included, with type enlarged for projection.

## Acknowledgments

We are grateful to the many instructors, students, colleagues, and friends who have contributed to this, and to earlier editions.

Charles Adler of St. Mary's College of Maryland authored the excellent new problems. David Mills of the College of the Redwoods extensively revised the solutions manual. Robin Jordan of Florida Atlantic University created the innovative Master the Concept exercises and iSOLVE Checkpoint problems. Laura McCullough of the University of Wisconsin, Stout, and Thomas Foster of Southern Illinois University, Edwardsville, drawing from their background in Physics Education Research, were instrumental in providing context-rich examples in every chapter as well as our new Estimation and Approximation problems. We received invaluable help in accuracy checking of text and problems from professors:

**Karamjeet Arya,**
San Jose State University

**Michael Crivello,**
San Diego Mesa College

**David Faust,**
Mt. Hood Community College

**Jerome Licini,**
Lehigh University

**Dan Lucas,**
University of Wisconsin

**Jeannette Myers,**
Clemson University

**Marian Peters,**
Appalachian State University

**Paul Quinn,**
Kutztown University

**Michael G. Strauss,**
University of Oklahoma

**George Zober,**
Yough Senior High School

**Patricia Zober,**
Ringgold High School

Many instructors and students have provided extensive and helpful reviews of one or more chapters. They have each made a fundamental contribution to the quality of this revision, and deserve our gratitude. We would like to thank the following reviewers:

**Edward Adelson,**
The Ohio State University

**Todd Averett,**
The College of William and Mary

**Yildirim M. Aktas,**
University of North Carolina at Charlotte

**Karamjeet Arya,**
San Jose State University

**Alison Baski,**
Virginia Commonwealth University

**Gary Stephen Blanpied,**
University of South Carolina

**Ronald Brown,**
California Polytechnic State University

**Robert Coakley,**
University of Southern Maine

**Robert Coleman,**
Emory University

**Andrew Cornelius,**
University of Nevada at Las Vegas

**Peter P. Crooker,**
University of Hawaii

**N. John DiNardo,**
Drexel University

**William Ellis,**
University of Technology - Sydney

**John W. Farley,**
University of Nevada at Las Vegas

**David Flammer,**
Colorado School of Mines

**Tom Furtak,**
Colorado School of Mines

**Patrick C. Gibbons,**
Washington University

**John B. Gruber,**
San Jose State University

**Christopher Gould,**
University of Southern California

**Phuoc Ha,**
Creighton University

**Theresa Peggy Hartsell,**
Clark College

**James W. Johnson,**
Tallahassee Community College

**Thomas O. Krause,**
Towson University

**Donald C. Larson,**
Drexel University

**Paul L. Lee,**
California State University, Northridge

**Peter M. Levy,**
New York University

**Jerome Licini,**
Lehigh University

**Edward McCliment,**
University of Iowa

**Robert R. Marchini,**
The University of Memphis

**Pete E.C. Markowitz,**
Florida International University

Fernando Medina,
Florida Atlantic University

Laura McCullough,
University of Wisconsin at Stout

John W. Norbury,
University of Wisconsin at Milwaukee

Melvyn Jay Oremland,
Pace University

Antonio Pagnamenta,
University of Illinois at Chicago

John Parsons,
Columbia University

Dinko Pocanic,
University of Virginia

Bernard G. Pope,
Michigan State University

Yong-Zhong Qian,
University of Minnesota

Ajit S. Rupaal,
Western Washington University

Todd G. Ruskell,
Colorado School of Mines

Mesgun Sebhatu,
Winthrop University

Marllin L. Simon,
Auburn University

Zbigniew M. Stadnik,
University of Ottawa

G. R. Stewart,
University of Florida

Michael G. Strauss,
University of Oklahoma

Chin-Che Tin,
Auburn University

Stephen Weppner,
Eckerd College

Suzanne E. Willis,
Northern Illinois University

Ron Zammit,
California Polytechnic State University

## Problems/solutions reviewers

Lay Nam Chang,
Virginia Polytechnic Institute

Mark W. Coffey,
Colorado School of Mines

Brent A. Corbin,
UCLA

Alan Cresswell,
Shippensburg University

Ricardo S. Decca,
Indiana University-Purdue University

Michael Dubson,
University of Colorado at Boulder

David Faust,
Mount Hood Community College

Philip Fraundorf,
University of Missouri, Saint Louis

Clint Harper,
Moorpark College

Kristi R.G. Hendrickson,
University of Puget Sound

Michael Hildreth,
University of Notre Dame

David Ingram,
Ohio University

James J. Kolata,
University of Notre Dame

Eric Lane,
University of Tennessee, Chattanooga

Jerome Licini,
Lehigh University

Daniel Marlow,
Princeton University

Laura McCullough,
University of Wisconsin at Stout

Carl Mungan,
United States Naval Academy

Jeffry S. Olafsen,
University of Kansas

Robert Pompi,
The State University of New York
at Binghamton

R. J. Rollefson,
Wesleyan University

Andrew Scherbakov,
Georgia Institute of Technology

Bruce A. Schumm,
University of California, Santa Cruz

Dan Styer,
Oberlin College

Jeffrey Sundquist,
Palm Beach Community College - South

Cyrus Taylor,
Case Western Reserve University

Fulin Zuo,
University of Miami

## Study Guide & Test Bank reviewers

Anthony J. Buffa,
California Polytechnic State University

Mirela S. Fetea,
University of Richmond

James Garner,
University of North Florida

Tina Harriott,
Mount Saint Vincent, Canada

Roger King,
City College of San Francisco

John A. McClelland,
University of Richmond

Chun Fu Su,
Mississippi State University

John A. Underwood,
Austin Community College

## Media reviewers

Mick Arnett,
Kirkwood Community College

**Colonel Rolf Enger,**
U.S. Air Force Academy

**John W. Farley,**
The University of Nevada at Las Vegas

**David Ingram,**
Ohio University

**Shawn Jackson,**
The University of Tulsa

**Dan MacIsaac,**
Northern Arizona University

**Peter E.C. Markowitz,**
Florida International University

**Dean Zollman,**
Kansas State University

**Media focus group participants**

**Edwin R. Jones,**
University of South Carolina

**William C. Kerr,**
Wake Forest University

**Taha Mzoughi,**
Mississippi State University

**Charles Niederriter,**
Gustavus Adolphus College

**Cindy Schwarz,**
Vassar College

**Dave Smith,**
University of the Virgin Islands

**D.J. Wagner,**
Grove City College

**George Watson,**
University of Delaware

**Frank Wolfs,**
University of Rochester

We also remain indebted to the reviewers of past editions. We would therefore like to thank the following reviewers, who provided immeasurable support as we developed the fourth edition:

**Michael Arnett,**
Iowa State University

**William Bassichis,**
Texas A&M

**Joel C. Berlinghieri,**
The Citadel

**Frank Blatt,**
Michigan State University

**John E. Byrne,**
Gonzaga University

**Wayne Carr,**
Stevens Institute of Technology

**George Cassidy,**
University of Utah

**I.V. Chivets,**
Trinity College, University of Dublin

**Harry T. Chu,**
University of Akron

**Jeff Culbert,**
London, Ontario

**Paul Debevec,**
University of Illinois

**Robert W. Detenbeck,**
University of Vermont

**Bruce Doak,**
Arizona State University

**John Elliott,**
University of Manchester, England

**James Garland,**
Retired

**Ian Gatland,**
Georgia Institute of Technology

**Ron Gautreau,**
New Jersey Institute of Technology

**David Gavenda,**
University of Texas at Austin

**Newton Greenberg,**
SUNY Binghamton

**Huidong Guo,**
Columbia University

**Richard Haracz,**
Drexel University

**Michael Harris,**
University of Washington

**Randy Harris,**
University of California at Davis

**Dieter Hartmann,**
Clemson University

**Robert Hollebeek,**
University of Pennsylvania

**Madya Jalil,**
University of Malaya

**Monwhea Jeng,**
University of California – Santa Barbara

**Ilon Joseph,**
Columbia University

**David Kaplan,**
University of California – Santa Barbara

**John Kidder,**
Dartmouth College

**Boris Korsunsky,**
Northfield Mt. Hermon School

**Andrew Lang (graduate student),**
University of Missouri

**David Lange,**
University of California – Santa Barbara

**Isaac Leichter,**
Jerusalem College of Technology

**William Lichten,**
Yale University

**Robert Lieberman,**
Cornell University

**Fred Lipschultz,**
University of Connecticut

**Graeme Luke,**
Columbia University

**Howard McAllister,**
University of Hawaii

**M. Howard Miles,**
Washington State University

**Matthew Moelter,**
University of Puget Sound

**Eugene Mosca,**
United States Naval Academy

**Aileen O'Donughue,**
St. Lawrence University

**Jack Ord,**
University of Waterloo

**Richard Packard,**
University of California

**George W. Parker,**
North Carolina State University

**Edward Pollack,**
University of Connecticut

**John M. Pratte,**
Clayton College and State University

**Brooke Pridmore,**
Clayton State College

**David Roberts,**
Brandeis University

**Lyle D. Roelofs,**
Haverford College

**Larry Rowan,**
University of North Carolina
at Chapel Hill

**Lewis H. Ryder,**
University of Kent, Canterbury

**Bernd Schuttler,**
University of Georgia

**Cindy Schwarz,**
Vassar College

**Murray Scureman,**
Amdahl Corporation

**Scott Sinawi,**
Columbia University

**Wesley H. Smith,**
University of Wisconsin

**Kevork Spartalian,**
University of Vermont

**Kaare Stegavik,**
University of Trondheim, Norway

**Jay D. Strieb,**
Villanova University

**Martin Tiersten,**
City College of New York

**Oscar Vilches,**
University of Washington

**Fred Watts,**
College of Charleston

**John Weinstein,**
University of Mississippi

**David Gordon,**
Wilson, MIT

**David Winter,**
Columbia University

**Frank L.H. Wolfe,**
University of Rochester

**Roy C. Wood,**
New Mexico State University

**Yuriy Zhestkov,**
Columbia University

Of course, our work is never done. We hope to continue to receive comments and suggestions from our readers so that we can improve the text and correct any errors. If you believe you have found an error, or have any other comments, suggestions, or questions, send us a note at asktipler@whfreeman.com. We will incorporate corrections into the text during subsequent reprinting.

Finally, we would like to thank our friends at W. H. Freeman and Company for their help and encouragement. Susan Brennan, Kathleen Civetta, Georgia Lee Hadler, Julia DeRosa, Margaret Comaskey, Dena Betz, Rebecca Pearce, Brian Donnellan, Jennifer Van Hove, Patricia Marx, and Mark Santee were extremely generous with their creativity and hard work at every stage of the process. We are also grateful for the contributions of Cathy Townsend and Denise Kadlubowski at PreMediaONE and the help of our colleagues Larry Tankersley, John Ertel, Steve Montgomery, and Don Treacy.

Paul Tipler
Alameda, California

Gene Mosca
Annapolis, Maryland

# ABOUT THE AUTHORS

PAUL A TIPLER

**Paul Tipler** was born in the small farming town of Antigo, Wisconsin, in 1933. He graduated from high school in Oshkosh, Wisconsin, where his father was superintendent of the Public Schools. He received his B.S. from Purdue University in 1955 and his Ph.D. at the University of Illinois in 1962, where he studied the structure of nuclei. He taught for one year at Wesleyan University in Connecticut while writing his thesis, then moved to Oakland University in Michigan, where he was one of the original members of the Physics department, playing a major role in developing the physics curriculum. During the next 20 years, he taught nearly all the physics courses and wrote the first and second editions of his widely used textbooks *Modern Physics* (1969, 1978) and *Physics* (1976, 1982). In 1982, he moved to Berkeley, California, where he now resides, and where he wrote *College Physics* (1987) and the third edition of *Physics* (1991). In addition to physics, his interests include music, hiking, and camping, and he is an accomplished jazz pianist and poker player.

GENE MOSCA

**Gene Mosca** was born in New York City and grew up on Shelter Island, New York. His undergraduate studies were at Villanova University and his graduate studies were at the University of Michigan and the University of Vermont, where he received his Ph.D. in 1974. He taught at Southampton High School, the University of South Dakota, and Emporia State University. Since 1986 Gene has been teaching at the U.S. Naval Academy. There he coordinated the core physics course for 16 semesters, and instituted numerous enhancements to both the laboratory and classroom. Proclaimed by Paul Tipler as, "the best reviewer I ever had," Mosca authored the popular Study Guide for the third and fourth editions of the text.

# Systems of Measurement

THE NUMBER OF GRAINS OF SAND ON A BEACH MAY BE TOO GREAT TO COUNT, BUT WE CAN ESTIMATE THE NUMBER BY USING REASONABLE ASSUMPTIONS AND SIMPLE CALCULATIONS.

**?** **How many grains of sand are on your favorite beach? (See Example 1-6.)**

1-1  Units

1-2  Conversion of Units

1-3  Dimensions of Physical Quantities

1-4  Scientific Notation

1-5  Significant Figures and Order of Magnitude

**W**e have always been curious about the world around us. Since the beginnings of recorded thought, we have sought ways to impose order on the bewildering diversity of events that we observe. Science is a process of searching for fundamental and universal principles that govern causes and effects in the universe. The process of science is the building, testing, and connecting of falsifiable models in an effort to describe, explain, and predict reality. The process involves hypotheses, repeatable experiments and observations, and new hypotheses. The prime criteria for determining the value of a scientific model is its simplicity and its usefulness in correctly making predictions or explaining observations concerning a broad spectrum of phenomena.

Today we think of science as divided into separate fields, although this division occurred only in the last century or so. The separation of complex systems into smaller categories that can be more easily studied is one of the great successes of science in general. Biology, for example, is the study of living organisms. Chemistry deals with the interaction of elements and compounds. Geology is the study of the earth. Astronomy is the study of the solar system, the stars and galaxies, and the universe as a whole. Physics is the science of matter and energy, space and time. It includes the principles that govern the motion of particles and waves, the interactions of particles, and the properties of molecules, atoms, and atomic nuclei, as well as larger-scale systems such as gases, liquids, and solids.

Some consider physics the most fundamental science because its principles supply the foundation of the other scientific fields.

Physics is the science of the exotic and the science of everyday life. At the exotic extreme, black holes boggle the imagination. In everyday life, engineers, musicians, architects, chemists, biologists, doctors, and many others routinely command such subjects as heat transfer, fluid flow, sound waves, radioactivity, and stresses in buildings or bones to perform their daily work. Countless questions about our world can be answered with a basic knowledge of physics. Why must a helicopter have two rotors? Why do astronauts float in space? Why do moving clocks run slow? Why does sound travel around corners while light appears to travel in straight lines? Why does an oboe sound different from a flute? How do CD players work? Why is there no hydrogen in the atmosphere? Why do metal objects feel colder than wood objects at the same temperature? Why is copper an electrical conductor while wood is an insulator? Why is lithium, with its three electrons, extremely reactive, whereas helium, with two electrons, is chemically inert?

➤ **In this chapter, we will begin to prepare ourselves to answer some of these questions by examining the units and dimensions used. Any time a measurement is made, the accuracy of the measurement should be stated. If a fuel-gauge reading is 100 gallons, it does not mean that there are exactly 100 gallons of fuel left. So, what exactly does it mean, and how do I express it?**

## Classical and Modern Physics

The earliest recorded efforts to systematically assemble knowledge concerning motion came from ancient Greece. In Aristotle's (384–322 B.C.) system of natural philosophy, explanations of physical phenomena were deduced from assumptions about the world, rather than derived from experimentation. For example, it was a fundamental assumption that every substance had a "natural place" in the universe. Motion was thought to be the result of a substance trying to reach its natural place. Because of the agreement between the deductions of Aristotelian physics and the motions observed throughout the physical universe, and the lack of experimentation that could overturn the ancient physical ideas, the Greek view was accepted for nearly two thousand years. It was the Italian scientist Galileo Galilei (1564–1642), whose brilliant experiments on motion established once and for all the absolute necessity of experimentation in physics, who initiated the disintegration of Aristotelian physics. Within a hundred years, Isaac Newton had generalized the results of Galileo's experiments into his three spectacularly successful laws of motion, and the reign of the natural philosophy of Aristotle was over.

Experimentation during the next two hundred years brought a flood of discoveries—and raised a flood of new questions. These inspired the development of new models to explain them. By the end of the nineteenth century, Newton's laws for the motions of mechanical systems had been joined by equally impressive laws from James Maxwell, James Joule, Sadi Carnot, and others to describe electromagnetism and thermodynamics. The subjects that occupied physical scientists through the end of the nineteenth century—mechanics, light, heat, sound, electricity and magnetism—-are usually referred to as *classical physics*. Because classical physics is what we need to understand the macroscopic world we live in, it dominates Parts I through V of this text.

The remarkable success of classical physics led many scientists to believe that the description of the physical universe was complete. However, the discoveries of X rays by Wilhelm Röntgen in 1895, and of radioactivity by Antoine Becquerel and Marie and Pierre Curie in the next few years seemed to be outside the framework of classical physics. The theory of special relativity proposed by Albert Einstein in 1905 contradicted the ideas of space and time of Galileo and Newton. In the same year, Einstein suggested that light energy is quantized; that is, that light comes in discrete packets rather than being wavelike and continuous as was thought in classical physics. The generalization of this insight to the quantization of all types of energy is a central idea of quantum mechanics, one that has many amazing and important consequences. The application of special relativity, and particularly quantum theory, to such microscopic systems as atoms, molecules, and nuclei, which has led to a detailed understanding of solids, liquids, and gases, is often referred to as *modern physics*. Modern physics is the subject of Part VI of this text.

While classical physics is the main subject of this book, from time to time in the earlier parts of the text we will note the relationship between classical and modern physics. For example, when we discuss velocity in Chapter 2, we will take a moment to consider velocities near the speed of light and briefly cross over to the relativistic universe first imagined by Einstein. After discussing the conservation of energy in Chapter 7, we will discuss the quantization of energy and Einstein's famous relation between mass and energy, $E = mc^2$. Just a few chapters later, in Chapter R, we will study the nature of space and time as revealed by Einstein in 1903.

# 1-1 Units

We all know of things that cannot be measured—the beauty of a flower, or of a Bach fugue. As certain as our knowledge of these things may be, we readily admit that this knowledge is not science. The ability not only to define but also to measure is a requisite of science, and in physics, more than in any other field of knowledge, the precise definition of terms and the accurate measurement of quantities have led to great discoveries. We begin with some preliminary chores, such as establishing some basic definitions and learning about units and how to deal with them in equations. The fun comes later.

Measurement of any quantity involves comparison with some precisely defined unit value of the quantity. For example, to measure the distance between two points, we need a standard unit, such as a meter. The statement that a certain distance is 25 meters means that it is 25 times the length of the unit meter. That is, a standard meterstick fits into that distance 25 times. It is important to include the unit, in this case meters, along with the number 25 in expressing this distance, because there are other units of distance such as kilometers or miles that are in common use. To say that a distance is 25 is meaningless. The magnitude of any physical quantity must include both a number and a unit.

## The International System of Units

A small number of fundamental units are sufficient to express all physical quantities. Many of the quantities that we shall be studying, such as velocity, force, momentum, work, energy, and power can be expressed in terms of three fundamental measures: length, time, and mass. The choice of standard units for these fundamental quantities determines a system of units. The system used universally in the scientific community is called SI (for *Système International*). The standard SI unit for length is the meter, the standard unit of time is the second, and the standard unit of mass is the kilogram. Complete definitions of the SI units are given in Appendix B.

**Length** The standard unit of length, the **meter** (abbreviated m), was originally defined by two scratches on a bar made of a platinum-iridium alloy kept at the International Bureau of Weights and Measures in Sèvres, France. This length was chosen so that the distance between the equator and the North Pole along the meridian through Paris would be 10 million meters (Figure 1-1). The meter is

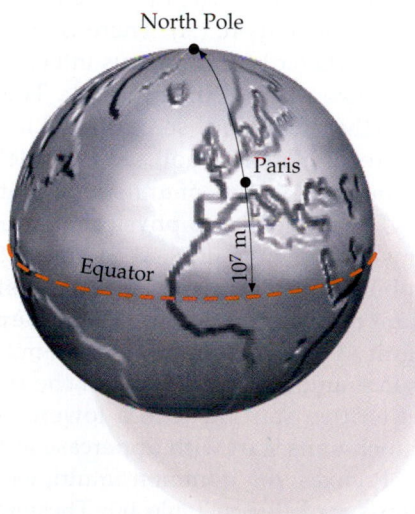

**FIGURE 1-1** The meter was originally chosen so that the distance from the equator to the North Pole along the meridian through Paris would be $10^7$ m.

now defined in terms of the speed of light—the meter is the distance light travels through empty space in 1/299,729,458 second. (This makes the speed of light exactly 299,792,458 m/s.)

**EXERCISE** What is the circumference of the earth in meters? (*Answer* About $4 \times 10^7$ m)

**Time** The unit of time, the **second** (s), was originally defined in terms of the rotation of the earth and was equal to (1/60)(1/60)(1/24) of the mean solar day. The second is now defined in terms of a characteristic frequency associated with the cesium atom. All atoms, after absorbing energy, emit light with wavelengths and frequencies characteristic of the particular element. There is a set of wavelengths and frequencies for each element, with a particular frequency and wavelength associated with each energy transition within the atom. As far as we know, these frequencies remain constant. The second is defined so that the frequency of the light from a certain transition in cesium is exactly 9,192,631,770 cycles per second. With these definitions, the fundamental units of length and time are accessible to laboratories throughout the world.

(a)

**Mass** The unit of mass, the **kilogram** (kg), which equals 1000 grams (g), is defined to be the mass of a standard body, also kept at Sèvres. A duplicate of the standard 1-kg body is kept at the National Institute of Standards and Technology (NIST) in Gaithersburg, Maryland. We shall discuss the concept of mass in detail in Chapter 4, where we will see that the weight of an object at a given point on earth is proportional to its mass. Thus the masses of objects of ordinary size can be compared by weighing them.

In our study of thermodynamics and electricity, we shall need three more fundamental physical units: one for temperature, the kelvin (K) (formerly the degree Kelvin); one for the amount of a substance, the mole (mol); and one for electrical current, the ampere (A). There is another fundamental unit, the candela (cd) for luminous intensity, which we shall have no occasion to use in this book. These seven fundamental units, the meter (m), second (s), kilogram (kg), kelvin (K), ampere (A), mole (mol), and candela (cd), constitute the international system of units or SI units.

The unit of every physical quantity can be expressed in terms of the fundamental SI units. Some frequently used combinations are given special names. For example, the SI unit of force, kg·m/s² is called a newton (N). Similarly, the SI unit of power, 1 kg·m²/s³ = N·m/s, is called a watt (W). When a unit like the newton or the watt is someone's name, it is written starting with a lowercase letter. Abbreviations for such units start with uppercase letters.

Prefixes for common multiples and submultiples of SI units are listed in Table 1-1. These multiples are all powers of 10. Such a system is called a decimal system. The decimal system based on the meter is called the metric system. The prefixes can be applied to any SI unit; for example, 0.001 second is 1 millisecond (ms); 1,000,000 watts is 1 megawatt (MW).

(b)

(*a*) Water clock used to measure time intervals in the thirteenth century. (*b*) Cesium fountain clock with developers Jefferts & Meekhof.

## TABLE 1-1

**Prefixes for Powers of 10†**

| Multiple | Prefix | Abbreviation |
|----------|--------|--------------|
| $10^{18}$ | exa | E |
| $10^{15}$ | peta | P |
| $10^{12}$ | tera | T |
| $10^{9}$ | giga | G |
| $10^{6}$ | mega | M |
| $10^{3}$ | kilo | k |
| $10^{2}$ | hecto | h |
| $10^{1}$ | deka | da |
| $10^{-1}$ | deci | d |
| $10^{-2}$ | centi | c |
| $10^{-3}$ | milli | m |
| $10^{-6}$ | micro | $\mu$ |
| $10^{-9}$ | nano | n |
| $10^{-12}$ | pico | p |
| $10^{-15}$ | femto | f |
| $10^{-18}$ | atto | a |

† The prefixes hecto (h), deka (da), and deci (d) are not multiples of $10^3$ or $10^{-3}$ and are rarely used. The other prefix that is not a multiple of $10^3$ or $10^{-3}$ is centi (c). The prefixes frequently used in this book are printed in red. Note that all prefix abbreviations for multiples $10^6$ and higher are uppercase letters; all others are lowercase letters.

## Other Systems of Units

Another decimal system still in use but gradually being replaced by SI units is the cgs system, based on the centimeter, gram, and second. The centimeter is defined as 0.01 m. The gram is now defined as 0.001 kg. Originally the gram was defined as the mass of one cubic centimeter ($cm^3$) of water at 4°C. The kilogram is then the mass of 1 liter (1000 $cm^3$) of water.

In another system of units, the U.S. customary system, a unit of force, the pound, is chosen to be a fundamental unit. In this system, the unit of mass is then defined in terms of the fundamental unit of force. The pound is defined in terms of the gravitational attraction of the earth at a particular place for a standard body. The fundamental unit of length in this system is the foot and the unit of time is the second, which has the same definition as the SI unit. The foot is defined as exactly one-third of a yard, which is now defined in terms of the meter:

$$1 \text{ yd} = 0.9144 \text{ m} \tag{1-1}$$

$$1 \text{ ft} = \tfrac{1}{3}\text{yd} = 0.3048 \text{ m} \tag{1-2}$$

making the inch exactly 2.54 cm. This scheme is not a decimal system. It is less convenient than the SI or other decimal systems because common multiples of the unit are not powers of 10. For example, 1 yd = 3 ft and 1 ft = 12 in. We will see in Chapter 4 that mass is a better choice for a fundamental unit than force because mass is an intrinsic property of an object, independent of its location. Relations between the U.S. customary system and SI units are given in Appendix A.

## 1-2 Conversion of Units

All physical quantities contain both a number and a unit. When such quantities are added, subtracted, multiplied, or divided in an algebraic equation, the unit can be treated like any other algebraic quantity. For example, suppose you want to find the distance traveled in 3 hours (h) by a car moving at a constant rate of 80 kilometers per hour (km/h). The distance is the product of the speed $v$ and the time $t$:

$$x = vt = \frac{80 \text{ km}}{\text{h}} \times 3 \text{ h} = 240 \text{ km}$$

We cancel the unit of time, the hours, just as we would any algebraic quantity to obtain the distance in the proper unit of length, the kilometer. This method of treating units makes it easy to convert from one unit of distance to another. Suppose we want to convert our answer of 240 km to miles (mi). Using the fact that 1 mi = 1.61 km, we divide each side of this equality by 1.61 km to obtain

$$\frac{1 \text{ mi}}{1.61 \text{ km}} = 1$$

Because any quantity can be multiplied by 1 without changing its value, we can now change 240 km to miles by multiplying by the factor (1 mi)/(1.61 km):

$$240 \text{ km} = 240 \text{ km} \times \frac{1 \text{ mi}}{1.61 \text{ km}} = 149 \text{ mi}$$

The factor (1 mi)/(1.61 km) is called a **conversion factor.** All conversion factors have a value of 1 and are used to convert a quantity expressed in one unit of measure into its equivalent in another unit of measure. By writing out the units explicitly and canceling them, we do not need to think about whether we multiply by 1.61 or divide by 1.61 to change kilometers to miles, because the units tell us whether we have chosen the correct or incorrect factor.

(*a*) Laser beam from the Macdonald Observatory used to measure the distance to the moon. The distance can be measured within a few centimeters by measuring the time required for the beam to go to the moon and back after reflecting off a mirror (*b*) placed on the moon by the Apollo 14 astronauts.

(a)

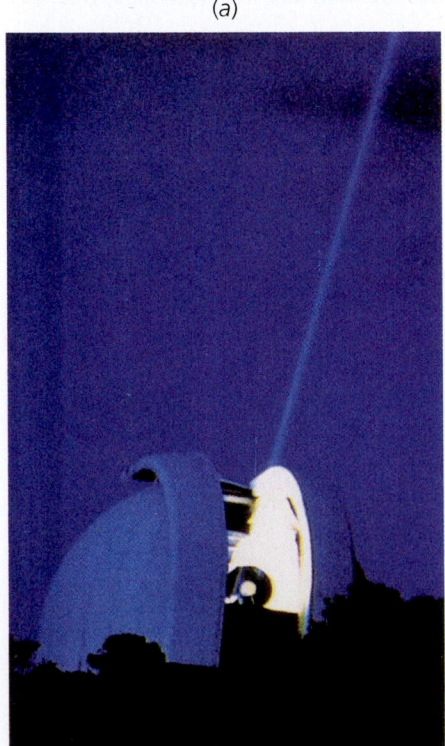

(b)

## EXAMPLE 1-1

**Your employer sends you on a trip to a foreign country where the road signs give distances in kilometers and the automobile speedometers are calibrated in kilometers per hour. If you drive 90 km/h, how fast are you going in meters per second and in miles per hour?**

**PICTURE THE PROBLEM** We use the facts that 1000 m = 1 km, 60 s = 1 min, and 60 min = 1 h to convert to meters per second. The quantity 90 km/h is multiplied by a set of conversion factors each having the value 1, so the value of the speed is not changed. To convert to miles per hour, we use the conversion factor (1 mi)/(1.61 km) = 1.

1. Multiply 90 km/h by a set of conversion factors that convert km to m and h to s:

$$\frac{90 \text{ km}}{\text{h}} \times \frac{1000 \text{ m}}{1 \text{ km}} \times \frac{1 \text{ h}}{60 \text{ min}} \times \frac{1 \text{ min}}{60 \text{ s}} = \boxed{25 \text{ m/s}}$$

2. Multiply 90 km/h by 1 mi/1.61 km:

$$\frac{90 \text{ km}}{\text{h}} \times \frac{1 \text{ mi}}{1.61 \text{ km}} = \boxed{55.9 \text{ mi/h}}$$

**EXERCISE** What is the equivalent of 65 mi/h in m/s? (*Answer* 29.1 m/s)

# 1-3 Dimensions of Physical Quantities

The area of a surface is found by multiplying one length by another. For example, the area of a rectangle of sides 2 m and 3 m is $A = (2 \text{ m})(3 \text{ m}) = 6 \text{ m}^2$. The units of this area are square meters. Because area is the product of two lengths, it is said to have the dimensions of length times length, or length squared, often written $L^2$. The idea of dimensions is easily extended to other nongeometric quantities. For example, speed is said to have the dimensions of length divided by time, or $L/T$. The dimensions of other quantities such as force or energy are written in terms of the fundamental quantities of length, time, and mass. Adding two physical quantities makes sense only if the quantities have the same dimensions. For example, we cannot add an area to a speed to obtain a meaningful sum. If we have an equation like

$$A = B + C$$

the quantities $A$, $B$, and $C$ must all have the same dimensions. Addition of $B$ and $C$ also requires that these quantities be in the same units. For example, if $B$ is an area of 500 in.$^2$ and $C$ is 4 ft$^2$, we must either convert $B$ into square feet or $C$ into square inches in order to find the sum of the two areas.

We can often find mistakes in a calculation by checking the dimensions or units of the quantities in our result. Suppose, for example, that we mistakenly use the formula $A = 2\pi r$ for the area of a circle. We can see immediately that this cannot be correct because $2\pi r$ has dimensions of length whereas area must have dimensions of length squared. Dimensional consistency is a necessary but not a sufficient condition for an equation to be correct. An equation can have the correct dimensions in each term without describing any physical situation. Table 1-2 gives the dimensions of some quantities we will encounter in physics.

## TABLE 1-2

**Dimensions of Physical Quantities**

| Quantity | Symbol | Dimension |
|---|---|---|
| Area | $A$ | $L^2$ |
| Volume | $V$ | $L^3$ |
| Speed | $v$ | $L/T$ |
| Acceleration | $a$ | $L/T^2$ |
| Force | $F$ | $ML/T^2$ |
| Pressure ($F/A$) | $p$ | $M/LT^2$ |
| Density ($M/V$) | $\rho$ | $M/L^3$ |
| Energy | $E$ | $ML^2/T^2$ |
| Power ($E/T$) | $P$ | $ML^2/T^3$ |

**EXAMPLE 1-2**

The pressure in a fluid in motion depends on its density $\rho$ and its speed $v$. Find a simple combination of density and speed that gives the correct dimensions of pressure.

**PICTURE THE PROBLEM** We note from Table 1-2 that both pressure and density have units of mass in the numerator, whereas speed does not contain $M$. We therefore divide the units of pressure by those of density and inspect the result.

1. Divide the units of pressure by those of density:

$$\frac{[p]}{[\rho]} = \frac{M(L/T^2)}{M/L^3} = \frac{L^2}{T^2}$$

2. By inspection, we note that the result has dimensions of $v^2$. The dimensions of pressure are thus the same as the dimensions of density times speed squared:

$$[p] = [\rho][v^2] = \frac{M}{L^3}\left(\frac{L}{T}\right)^2 = \boxed{\frac{M}{LT^2}}$$

**REMARKS** When we study fluids in motion in Chapter 13, we will see from Bernoulli's law that for a fluid moving at a constant height, $p + \frac{1}{2}\rho v^2$ is constant where $p$ is the pressure in the fluid.

## 1-4 Scientific Notation

Handling very large or very small numbers is simplified by using scientific notation. In this notation, the number is written as a product of a number between 1 and 10 and a power of 10, such as $10^2$ (= 100) or $10^3$ (= 1000). For example, the number 12,000,000 is written $1.2 \times 10^7$; the distance from the earth to the sun, about 150,000,000,000 m, is written $1.5 \times 10^{11}$ m. The number 11 in $10^{11}$ is called the **exponent.** For numbers smaller than 1, the exponent is negative. For example, $0.1 = 10^{-1}$, and $0.000\ 1 = 10^{-4}$. The diameter of a virus, which is about 0.00000001 m, is written $1 \times 10^{-8}$ m.

When numbers in scientific notation are multiplied, the exponents are added; when divided, the exponents are subtracted. These rules can be seen from some simple examples:

$$10^2 \times 10^3 = 100 \times 1,000 = 100,000 = 10^5$$

Similarly,

$$\frac{10^2}{10^3} = \frac{100}{1000} = \frac{1}{10} = 10^{2-3} = 10^{-1}$$

In this notation, $10^0$ is defined to be 1. To see why, suppose we divide 1000 by itself. We have

$$\frac{1000}{1000} = \frac{10^3}{10^3} = 10^{3-3} = 10^0 = 1$$

**EXAMPLE 1-3**

In 12 g of carbon, there are $N_A = 6.02 \times 10^{23}$ carbon atoms (Avogadro's number). If you could count 1 atom per second, how long would it take to count the atoms in 1 g of carbon? Express your answer in years.

**PICTURE THE PROBLEM** We need to find the total number of atoms to be counted, $N$, and then use the fact that the number counted equals the counting rate $R$ multiplied by the time $t$.

1. The time is related to the total number of atoms $N$, and the rate of counting $R = 1$ atom/s:

$$t = \frac{N}{R}$$

2. Find the number of carbon atoms in 1 gram:

$$N = \frac{6.02 \times 10^{23} \text{ atoms}}{12 \text{ g}} \times 1 \text{ g} = 5.02 \times 10^{22} \text{ atoms}$$

3. Calculate the number of seconds it takes to count these at 1 per second:

$$t = \frac{N}{R} = \frac{5.02 \times 10^{22} \text{ atoms}}{1 \text{ atom/s}} = 5.02 \times 10^{22} \text{ s}$$

4. Calculate the number of seconds in a year:

$$n = \frac{365 \text{ d}}{1 \text{ y}} \times \frac{24 \text{ h}}{1 \text{ d}} \times \frac{3600 \text{ s}}{1 \text{ h}} = 3.15 \times 10^7 \text{ s/y}$$

5. Use the conversion factor $3.15 \times 10^7$ s/y (a handy quantity to remember) to convert the answer in step 3 to years:

$$t = 5.02 \times 10^{22} \text{ s} \times \frac{1 \text{ y}}{3.15 \times 10^7 \text{ s}}$$

$$= \frac{5.02}{3.15} \times 10^{22-7} \text{ y} = \boxed{1.59 \times 10^{15} \text{ y}}$$

**REMARKS** The time required is about 100,000 times the age of the universe.

**EXERCISE** If you divided the task so that each person counted different atoms, how long would it take for 5 billion ($5 \times 10^9$) people to count the atoms in 1 g of carbon? (*Answer* $3.19 \times 10^5$ y)

---

*HOW MUCH WATER?*                         **EXAMPLE 1-4**

**A liter (L) is the volume of a cube that is 10 cm by 10 cm by 10 cm. If you drink 1 liter of water, how much volume in cubic centimeters and in cubic meters would it occupy in your stomach?**

**PICTURE THE PROBLEM** The volume of a cube of side $\ell$ is $V = \ell^3$. The volume in cubic centimeters is found directly from $\ell = 10$ cm. To find the volume in cubic meters, convert $cm^3$ to $m^3$ using the conversion factor $1 \text{ cm} = 10^{-2}$ m.

1. Calculate the volume in $cm^3$:

$$V = \ell^3 = (10 \text{ cm})^3 = 10^3 \text{ cm}^3$$

2. Convert to $m^3$:

$$10^3 \text{ cm}^3 = 10^3 \text{ cm}^3 \times \left(\frac{10^{-2} \text{ m}}{1 \text{ cm}}\right)^3 = 10^3 \text{ cm}^3 \times \left(\frac{10^{-6} \text{ m}^3}{1 \text{ cm}^3}\right) = \boxed{10^{-3} \text{ m}^3}$$

**REMARKS** Note that the conversion factor (which equals 1) can be raised to the third power without changing its value, enabling us to cancel units.

Care is required when adding or subtracting numbers written in scientific notation when their exponents don't match. Consider, for example,

$$(1.200 \times 10^2) + (8 \times 10^{-1}) = 120.0 + 0.8 = 120.8$$

To find the sum without converting both numbers into ordinary decimal form, it is sufficient to rewrite either of the numbers so that its power of 10 is the same as that of the other. For example, we can find the sum by writing $1.200 \times 10^2 = 1200 \times 10^{-1}$ and then adding:

$$(1200 \times 10^{-1}) + (8 \times 10^{-1}) = 1208 \times 10^{-1} = 120.8$$

When the exponents are very different, one of the numbers is much smaller than the other. The smaller number can often be neglected in addition or subtraction.

For example,

$$(2 \times 10^6) + (9 \times 10^{-3}) = 2,000,000 + 0.009$$
$$= 2,000,000.009 \approx 2 \times 10^6$$

where the symbol $\approx$ means "is approximately equal to."

When raising a power to another power, the exponents are multiplied. For example,

$$(10^2)^4 = 10^2 \times 10^2 \times 10^2 \times 10^2 = 10^8$$

# 1-5 Significant Figures and Order of Magnitude

Many of the numbers in science are the result of measurement and are therefore known only to within some degree of experimental uncertainty. The magnitude of the uncertainty depends on the skill of the experimenter and the apparatus used, and often can only be estimated. A rough indication of the uncertainty in a measurement is inferred by the number of digits used. For example, if we say that a table is 2.50 m long, we are often saying that its length is between 2.495 m and 2.505 m. That is, we know the length to about $\pm$ 0.005 m = $\pm$ 0.5 cm. If we used a meterstick with millimeter markings and measured the table length carefully, we might estimate that we could measure the length to $\pm$ 0.5 mm rather than $\pm$ 0.5 cm. We would indicate this precision when giving the length by using four digits, such as 2.503 m. A reliably known digit (other than a zero used to locate the decimal point) is called a **significant figure.** The number 2.50 has three significant figures; 2.503 m has four. The number 0.001 03 has three significant figures. (The first three zeroes are not significant figures but merely locate the decimal point.) In scientific notation, the number 0.001 03 is written $1.03 \times 10^{-3}$. A common student error is to carry more digits than the certainty of measurement warrants. Suppose, for example, that you measure the area of a circular playing field by pacing off the radius and using the formula for the area of a circle, $A = \pi r^2$. If you estimate the radius to be 8 m and use a 10-digit calculator to compute the area, you obtain $\pi(8 \text{ m})^2 = 201.0619298 \text{ m}^2$. The digits after the decimal point give a false indication of the accuracy with which you know the area. If you found the radius by pacing, you might expect that your measurement was accurate to only about 0.5 m. That is, the radius could be as great as 8.5 m or as small as 7.5 m. If the radius is 8.5 m, the area is $\pi(8.5 \text{ m})^2 = 226.9800692 \text{ m}^2$, whereas if it is 7.5 m, the area is $\pi(7.5 \text{ m})^2 = 176.714587 \text{ m}^2$. There is a general rule to guide you when combining several numbers in multiplication or division:

> The number of significant figures in the result of multiplication or division is no greater than the least number of significant figures in any of the factors.

In the previous example, the radius is known to only one significant figure, so the area is also known only to one significant figure. It should be written as $2 \times 10^2 \text{ m}^2$, which says that the area is probably between 150 m$^2$ and 250 m$^2$.

The precision of the sum or difference of two measurements is only as good as the precision of the least precise of the two measurements. A general rule is:

> The result of addition or subtraction of two numbers has no significant figures beyond the last decimal place where both of the original numbers had significant figures.

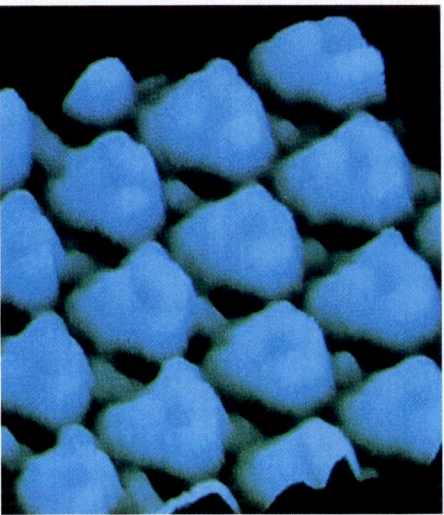

Benzene molecules of the order of $10^{-10}$ m in diameter as seen in a scanning electron microscope.

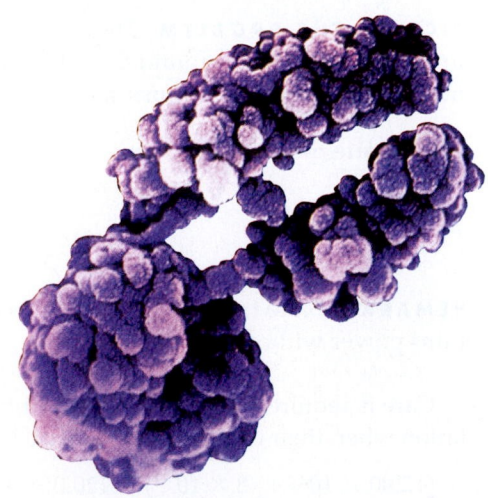

Chromosomes measuring on the order of $10^{-6}$ m across as seen in a scanning electron microscope.

**E X A M P L E   1 - 5**

Find the sum of 1.040 and 0.21342.

**PICTURE THE PROBLEM** The first number, 1.040, has only three significant figures beyond the decimal point, whereas the second, 0.21342 has five. According to the rule stated above, the sum can have only three significant figures beyond the decimal point.

Sum the numbers, keeping only three digits beyond the decimal point:

$$1.040 + 0.21342 = \boxed{1.253}$$

**EXERCISE** Apply the appropriate rule for significant figures to calculate the following: (*a*) $1.58 \times 0.03$, (*b*) $1.4 + 2.53$, (*c*) $2.34 \times 10^2 + 4.93$ (*Answer* (*a*) 0.05, (*b*) 3.9, (*c*) $2.39 \times 10^2$)

Most examples and exercises in this book will be done with data to three (or sometimes four) significant figures, but occasionally we will say, for example, that a table top is 3 ft by 8 ft rather than taking the time and space to say it is 3.00 ft by 8.00 ft. Any data you see in an example or exercise can be assumed to be known to three significant figures unless otherwise indicated. The same assumption holds for the data in the end-of-chapter problems. In doing rough calculations or comparisons, we sometimes round off a number to the nearest power of 10. Such a number is called an **order of magnitude.** For example, the height of a small insect, say an ant, might be $8 \times 10^{-4}$ m $\approx 10^{-3}$ m. We would say that the order of magnitude of the height of an ant is $10^{-3}$ m. Similarly, though the height of most people is about 2 m, we might round that off and say that the order of magnitude of the height of a person is $10^0$ m. By this we do not mean to imply that a typical height is really 1 m but that it is closer to 1 m than to 10 m or to $10^{-1} = 0.1$ m. We might say that a typical human being is three orders of magnitude taller than a typical ant, meaning that the ratio of heights is about 1000 to 1. An order of magnitude does not provide any digits that are reliably known.

Distances familiar in our everyday world. The height of the woman is of the order of $10^0$ m and that of the mountain is of the order of $10^4$ m.

Earth, with a diameter of the order of $10^7$ m, as seen from space.

The diameter of the Andromeda galaxy is of the order of $10^{21}$ m.

## TABLE 1-3

**The Universe by Orders of Magnitude**

| Size or Distance | (m) | Mass | (kg) | Time Interval | (s) |
|---|---|---|---|---|---|
| Proton | $10^{-15}$ | Electron | $10^{-30}$ | Time for light to cross nucleus | $10^{-23}$ |
| Atom | $10^{-10}$ | Proton | $10^{-27}$ | Period of visible light radiation | $10^{-15}$ |
| Virus | $10^{-7}$ | Amino acid | $10^{-25}$ | Period of microwaves | $10^{-10}$ |
| Giant amoeba | $10^{-4}$ | Hemoglobin | $10^{-22}$ | Half-life of muon | $10^{-6}$ |
| Walnut | $10^{-2}$ | Flu virus | $10^{-19}$ | Period of highest audible sound | $10^{-4}$ |
| Human being | $10^{0}$ | Giant amoeba | $10^{-8}$ | Period of human heartbeat | $10^{0}$ |
| Highest mountain | $10^{4}$ | Raindrop | $10^{-6}$ | Half-life of free neutron | $10^{3}$ |
| Earth | $10^{7}$ | Ant | $10^{-4}$ | Period of earth's rotation | $10^{5}$ |
| Sun | $10^{9}$ | Human being | $10^{2}$ | Period of earth's revolution around sun | $10^{7}$ |
| Distance from earth to sun | $10^{11}$ | Saturn V rocket | $10^{6}$ | | |
| | | Pyramid | $10^{10}$ | Lifetime of human being | $10^{9}$ |
| Solar system | $10^{13}$ | Earth | $10^{24}$ | Half-life of plutonium-239 | $10^{12}$ |
| Distance to nearest star | $10^{16}$ | Sun | $10^{30}$ | Lifetime of mountain range | $10^{15}$ |
| Milky Way galaxy | $10^{21}$ | Milky Way galaxy | $10^{41}$ | Age of earth | $10^{17}$ |
| Visible universe | $10^{26}$ | Universe | $10^{52}$ | Age of universe | $10^{18}$ |

It may be thought of as having no significant figures. Table 1-3 gives some typical order-of-magnitude values for a variety of sizes, masses, and time intervals encountered in physics.

In many cases the order of magnitude of a quantity can be estimated using reasonable assumptions and simple calculations. The physicist Enrico Fermi was a master at using cunning order-of-magnitude estimations to generate answers for questions that seemed impossible to calculate because of lack of information. Problems like these are often called **Fermi questions.** The following is an example of a Fermi question.

### EXPLORING

*How many piano tuners are there in Chicago? Find out this, and more, at* www.whfreeman.com/tipler5e.

---

*BURNING RUBBER*                                **EXAMPLE 1-6**

**What thickness of rubber tread is worn off the tire of an automobile as it travels 1 km (0.6 mi)?**

**PICTURE THE PROBLEM** We assume the tread thickness of a new tire is 1 cm. This may be off by a factor of two or so, but 1 mm is certainly too small and 10 cm is too large. Since tires have to be replaced after about 60,000 km (about 37,000 mi), we will assume that the tread is completely worn off after 60,000 km. In other words, the rate of wear is 1 cm of tire per 60,000 km of travel.

Use 1 cm wear per 60,000 km travel to compute the thickness worn after 1 km of travel:

$$\frac{1 \text{ cm wear}}{60{,}000 \text{ km travel}} = \frac{1.7 \times 10^{-5} \text{ cm wear}}{1 \text{ km travel}}$$

$$\boxed{\approx 0.2 \ \mu\text{m wear per km of travel}}$$

**EXERCISE** How many grains of sand are on a 0.50 km stretch of beach that is 100 m wide? *Hint: Assume that the sand is 3 m deep. Estimate that the diameter of one grain of sand is 1.00 mm. (Answer $\approx 2 \times 10^{14}$)*

## SUMMARY

The fundamental units in the SI system are the meter (m), the second (s), the kilogram (kg), the kelvin (K), the ampere (A), the mole (mol), and the candela (cd). The unit(s) of every physical quantity can be expressed in terms of these fundamental units.

| Topic | Relevant Equations and Remarks |
|---|---|
| **1. Units** | The magnitude of physical quantities (for example, length, time, force, and energy) are expressed as a number times a unit. |
| Fundamental units | The fundamental units in the SI system (short for *Système International*) are the meter (m), the second (s), the kilogram (kg), the kelvin (K), the ampere (A), the mole (mol), and the candela (cd). The unit(s) of every physical quantity can be expressed in terms of these fundamental units. |
| Units in equations | Units in equations are treated just like any other algebraic quantity. |
| Conversion | Conversion factors, which are always equal to 1, provide a convenient method for converting from one kind of unit to another. |
| **2. Dimensions** | The two sides of an equation must have the same dimensions. |
| **3. Scientific Notation** | For convenience, very small and very large numbers are generally written as a factor times a power of 10. |
| **4. Exponents** | |
| Multiplication | When multiplying two numbers, the exponents are added. |
| Division | When dividing two numbers, the exponents are subtracted. |
| Raising to a power | When a number containing an exponent is itself raised to a power, the exponents are multiplied. |
| **5. Significant Figures** | |
| Multiplication and division | The number of significant figures in the result of multiplication or division is *no greater than* the least number of significant figures in any of the numbers. |
| Addition and subtraction | The result of addition or subtraction of two numbers has no significant figures beyond the last decimal place where both of the original numbers had significant figures. |
| **6. Order of Magnitude** | A number rounded to the nearest power of 10 is called an order of magnitude. The order of magnitude of a quantity can often be estimated using reasonable assumptions and simple calculations. |

## Conceptual Problems

**1** • [SSM] [iSOLVE] Which of the following is *not* one of the fundamental physical quantities in the SI system? (*a*) Mass. (*b*) Length. (*c*) Force. (*d*) Time. (*e*) All of the above are fundamental physical quantities.

**2** • [iSOLVE] In doing a calculation, you end up with m/s in the numerator and m/s² in the denominator. What are your final units? (*a*) $m^2/s^3$. (*b*) 1/s. (*c*) $s^3/m^2$. (*d*) s. (*e*) m/s.

**3** • [iSOLVE] The prefix giga means (*a*) $10^3$, (*b*) $10^6$, (*c*) $10^9$, (*d*) $10^{12}$, (*e*) $10^{15}$.

**4** • [iSOLVE] The prefix mega means (*a*) $10^{-9}$, (*b*) $10^{-6}$, (*c*) $10^{-3}$, (*d*) $10^6$, (*e*) $10^9$.

**5** • [SSM] [iSOLVE] The prefix pico means (*a*) $10^{-12}$, (*b*) $10^{-6}$, (*c*) $10^{-3}$, (*d*) $10^6$, (*e*) $10^9$.

**6** • [iSOLVE] The number 0.0005130 has _____ significant figures. (*a*) one, (*b*) three, (*c*) four, (*d*) seven, (*e*) eight.

**7** • [iSOLVE] The number 23.0040 has _____ significant figures. (*a*) two, (*b*) three, (*c*) four, (*d*) five, (*e*) six.

**8** • What are the advantages and disadvantages of using the length of your arm for a standard length?

**9** • True or false:

(*a*) Two quantities must have the same dimensions in order to be added.
(*b*) Two quantities must have the same dimensions in order to be multiplied.
(*c*) All conversion factors have the value 1.

## Estimation and Approximation

**10** •• [SSM]
The angle subtended by the moon's diameter at a point on the earth is about 0.524° (Figure 1-2). Use this and the fact that the moon is about 384 Mm away to find the

**FIGURE 1-2** Problem 10

diameter of the moon. (The angle $\theta$ subtended by the moon is approximately $D/r_m$, where $D$ is the diameter of the moon and $r_m$ is the distance to the moon.)

**11** •• [SSM] [iSOLVE✓] The sun has a mass of $1.99 \times 10^{30}$ kg and is composed mostly of hydrogen, with only a small fraction being heavier elements. The hydrogen atom has a mass of $1.67 \times 10^{-27}$ kg. Estimate the number of hydrogen atoms in the sun.

**12** •• Most soft drinks are sold in aluminum cans. The mass of a typical can is about 0.018 kg. (*a*) Estimate the number of aluminum cans used in the United States in one year. (*b*) Estimate the total mass of aluminum in a year's consumption from these cans. (*c*) If aluminum returns $1/kg at a recycling center, how much is a year's accumulation of aluminum cans worth?

**13** •• In his essay "There's plenty of room at the bottom," Richard Feynman proposed writing the entire *Encyclopaedia Brittanica* on the head of a pin. (*a*) Estimate the size of the letters needed if we assume that a pinhead is 1/16 in across (the value that Feynman used). (*b*) If the atomic spacing in a metal is about 0.5 nm ($5 \times 10^{-10}$ m), about how many atoms across is each letter?

**14** •• [SSM] (*a*) Estimate the number of gallons of gasoline used per day by automobiles in the United States and the total amount of money spent on it. (*b*) If 19.4 gal of gasoline can be made from one barrel of crude oil, estimate the total number of barrels of oil imported into the United States per year to make gasoline. How many barrels per day is this?

**15** •• There is an environmental debate over the use of cloth versus disposable diapers. (*a*) If we assume that between birth and 2.5 y of age, a child uses 3 diapers per day, estimate the total number of disposable diapers used in the United States per year. (*b*) Estimate the total landfill volume due to these diapers, assuming that 1000 kg of waste fills about 1 $m^3$ of landfill volume. (*c*) How many square miles of landfill area at an average height of 10 m is needed for the disposal of diapers each year?

**16** ••• Each binary digit is termed a bit. A series of bits grouped together is called a word. An 8-bit word is called a byte. Suppose a computer hard disk has a capacity of 20 gigabytes. (*a*) How many bits can be stored on the disk? (*b*) Estimate the number of typical books that can be stored on the disk assuming each character requires one 8-bit word.

**17** •• [SSM] Estimate the yearly toll revenue of the George Washington Bridge in New York. At last glance, the toll is $6 to go into New York from New Jersey; going from New York into New Jersey is free. There are a total of 14 lanes.

## Units

**18** • Express the following quantities using the prefixes listed in Table 1-1 and the abbreviations listed on page EP-1; for example, 10,000 meters = 10 km. (a) 1,000,000 watts, (b) 0.002 gram, (c) $3 \times 10^{-6}$ meter, (d) 30,000 seconds.

**19** • Write each of the following without using prefixes: (a) 40 $\mu$W, (b) 4 ns, (c) 3 MW, (d) 25 km.

**20** • [SSM] Write the following (which are not SI units) without using abbreviations. For example, $10^3$ meters = 1 kilometer: (a) $10^{-12}$ boo, (b) $10^9$ low, (c) $10^{-6}$ phone, (d) $10^{-18}$ boy, (e) $10^6$ phone, (f) $10^{-9}$ goat, (g) $10^{12}$ bull.

**21** •• [SOLVE] In the following equations, the distance $x$ is in meters, the time $t$ is in seconds, and the velocity $v$ is in meters per second. What are the SI units of the constants $C_1$ and $C_2$? (a) $x = C_1 + C_2 t$, (b) $x = \frac{1}{2}C_1 t^2$, (c) $v^2 = 2C_1 x$, (d) $x = C_1 \cos C_2 t$, (e) $v^2 = 2C_1 v - (C_2 x)^2$.

**22** •• [SOLVE] If $x$ is in feet, $t$ is in seconds, and $v$ is in feet per second, what are the units of the constants $C_1$ and $C_2$ in each part of Problem 21?

## Conversion of Units

**23** • From the original definition of the meter in terms of the distance from the equator to the North Pole, find in meters (a) the circumference of the earth and (b) the radius of the earth. (c) Convert your answers for (a) and (b) from meters into miles.

**24** • [SOLVE] The speed of sound in air is 340 m/s. What is the speed of a supersonic plane that travels at twice the speed of sound? Give your answer in kilometers per hour and miles per hour.

**25** • [SSM] [SOLVE] A basketball player is 6 ft $10\frac{1}{2}$ in. tall. What is his height in centimeters?

**26** • Complete the following: (a) 100 km/h = _____ mi/h, (b) 60 cm = _____ in., (c) 100 yd = _____ m.

**27** • The main span of the Golden Gate Bridge is 4200 ft. Express this distance in kilometers.

**28** • [SSM] Find the conversion factor to convert from miles per hour into kilometers per hour.

**29** • Complete the following: (a) $1.296 \times 10^5$ km/h$^2$ = _____ km/(h·s), (b) $1.296 \times 10^5$ km/h$^2$ = _____ m/s$^2$, (c) 60 mi/h = _____ ft/s, (d) 60 mi/h = _____ m/s.

**30** • There are 1.057 quarts in a liter and 4 quarts in a gallon. (a) How many liters are there in a gallon? (b) A barrel equals 42 gallons. How many cubic meters are there in a barrel?

**31** • [SOLVE] There are 640 acres in a square mile. How many square meters are there in one acre?

**32** •• [SOLVE] A right circular cylinder has a diameter of 6.8 in and a height of 2 ft. What is the volume of the cylinder in (a) cubic feet, (b) cubic meters, (c) liters?

**33** •• [SSM] In the following, $x$ is in meters, $t$ is in seconds, $v$ is in meters per second, and the acceleration $a$ is in meters per second squared. Find the SI units of each combination: (a) $v^2/x$, (b) $\sqrt{x/a}$, (c) $\frac{1}{2}at^2$.

## Dimensions of Physical Quantities

**34** • What are the dimensions of the constants in each part of Problem 21?

**35** •• The law of radioactive decay is $N(t) = N_0 e^{-\lambda t}$, where $N_0$ is the number of radioactive nuclei at $t = 0$, $N(t)$ is the number remaining at time $t$, and $\lambda$ is a quantity known as the decay constant. What is the dimension of $\lambda$?

**36** •• [SSM] The SI unit of force, the kilogram-meter per second squared (kg·m/s$^2$) is called the newton (N). Find the dimensions and the SI units of the constant $G$ in Newton's law of gravitation $F = Gm_1 m_2/r^2$.

**37** •• An object on the end of a string moves in a circle. The force exerted by the string has units of $ML/T^2$ and depends on the mass of the object, its speed, and the radius of the circle. What combination of these variables gives the correct dimensions?

**38** •• [SOLVE✓] Show that the product of mass, acceleration, and speed has the dimensions of power.

**39** •• The momentum of an object is the product of its velocity and mass. Show that momentum has the dimensions of force multiplied by time.

**40** •• What combination of force and one other physical quantity has the dimensions of power?

**41** •• [SSM] [SOLVE✓] When an object falls through air, there is a drag force that depends on the product of the surface area of the object and the square of its velocity, that is, $F_{air} = CAv^2$, where $C$ is a constant. Determine the dimensions of $C$.

**42** •• Kepler's third law relates the period of a planet to its orbital radius $r$, the constant $G$ in Newton's law of gravitation ($F = Gm_1 m_2/r^2$), and the mass of the sun $M_s$. What combination of these factors gives the correct dimensions for the period of a planet?

## Scientific Notation and Significant Figures

**43** • [SSM] Express as a decimal number without using powers of 10 notation: (a) $3 \times 10^4$, (b) $6.2 \times 10^{-3}$, (c) $4 \times 10^{-6}$, (d) $2.17 \times 10^5$.

**44** • Write the following in scientific notation: (a) 3.1 GW = _____ W, (b) 10 pm = _____ m, (c) 2.3 fs = _____ s, (d) 4 $\mu$s = _____ s.

**45** • [SOLVE] Calculate the following, round off to the correct number of significant figures, and express your result in scientific notation: (a) $(1.14)(9.99 \times 10^4)$, (b) $(2.78 \times 10^{-8}) - (5.31 \times 10^{-9})$, (c) $12\pi/(4.56 \times 10^{-3})$, (d) $27.6 + (5.99 \times 10^2)$.

**46** • Calculate the following, round off to the correct number of significant figures, and express your result in scientific notation: (a) (200.9)(569.3), (b) (0.000000513)(62.3 × 10⁷), (c) 28,401 + (5.78 × 10⁴), (d) 63.25/(4.17 × 10⁻³).

**47** • **SSM** **ISOLVE** A cell membrane has a thickness of about 7 nm. How many cell membranes would it take to make a stack 1 in high?

**48** • Calculate the following, round off to the correct number of significant figures, and express your result in scientific notation: (a) (2.00 × 10⁴)(6.10 × 10⁻²), (b) (3.141592) (4.00 × 10⁵), (c) (2.32 × 10³)/(1.16 × 10⁸), (d) (5.14 × 10³) + (2.78 × 10²), (e) (1.99 × 10²) + (9.99 × 10⁻⁵).

**49** • **SSM** Perform the following calculations and round off the answers to the correct number of significant figures: (a) 3.141592654 × (23.2)², (b) 2 × 3.141592654 × 0.76, (c) 4/(3 π) × (1.1)³, (d) (2.0)⁵/3.141592654.

## General Problems

**50** • On many of the roads in Canada the speed limit is 100 km/h. What is the speed limit in miles per hour?

**51** • **SSM** If you could count $1 per second, how many years would it take to count 1 billion dollars (1 billion = 10⁹)?

**52** • Sometimes a conversion factor can be derived from the knowledge of a constant in two different systems. (a) The speed of light in vacuum is 186,000 mi/s = 3 × 10⁸ m/s. Use this fact to find the number of kilometers in a mile. (b) The weight of 1 ft³ of water is 62.4 lb. Use this and the fact that 1 cm³ of water has a mass of 1 g to find the weight in pounds of a 1-kg mass.

**53** •• **ISOLVE** The mass of one uranium atom is 4.0 × 10⁻²⁶ kg. How many uranium atoms are there in 8 g of pure uranium?

**54** •• **ISOLVE✔** During a thunderstorm, a total of 1.4 in. of rain falls. How much water falls on one acre of land? (1 mi² = 640 acres.)

**55** •• An iron nucleus has a radius of 5.4 × 10⁻¹⁵ m and a mass of 9.3 × 10⁻²⁶ kg. (a) What is its mass per unit volume in kg/m³? (b) If the earth had the same mass per unit volume, what would be its radius? (The mass of the earth is 5.98 × 10²⁴ kg.)

**56** •• Evaluate the following expressions: (a) (5.6 × 10⁻⁵) (0.0000075)/(2.4 × 10⁻¹²), (b) (14.2)(6.4 × 10⁷)(8.2 × 10⁻⁹) − 4.06, (c) (6.1 × 10⁻⁶)²(3.6 × 10⁴)³/(3.6 × 10⁻¹¹)¹ᐟ², (d) (0.000064)¹ᐟ³/[(12.8 × 10⁻³)(490 × 10⁻¹)¹ᐟ²].

**57** •• **SSM** The astronomical unit (AU) is defined in terms of the distance from the earth to the sun, namely 1.496 × 10¹¹ m. The parsec is the radius of a circle for which a central angle of 1 s intercepts an arc of length 1 AU. The light-year is the distance that light travels in 1 y. (a) How many parsecs are there in one astronomical unit? (b) How many meters are in a parsec? (c) How many meters in a light-year? (d) How many astronomical units in a light-year? (e) How many light-years in a parsec?

**58** •• If the average density of the universe is at least 6 × 10⁻²⁷ kg/m³, then the universe will eventually stop expanding and begin contracting. (a) How many electrons are needed in a cubic meter to produce the critical density? (b) How many protons per cubic meter would produce the critical density? ($m_e = 9.11 × 10^{-31}$ kg; $m_p = 1.67 × 10^{-27}$ kg.)

**59** •• **SSM** The Super-Kamiokande neutrino detector in Japan is a large transparent cylinder filled with ultrapure water. The height of the cylinder is 41.4 m and the diameter is 39.3 m. Calculate the mass of the water in the cylinder. Does this match the claim posted on the official Super-K Web site that the detector uses 50,000 tons of water? The density of water is 1,000 kg/m³.

**60** ••• The table below gives experimental results for a measurement of the period of motion $T$ of an object of mass $m$ suspended on a spring versus the mass of the object. These data are consistent with a simple equation expressing $T$ as a function of $m$ of the form $T = Cm^n$, where $C$ and $n$ are constants and $n$ is not necessarily an integer. (a) Find $n$ and $C$. (There are several ways to do this. One is to guess the value of $n$ and check by plotting $T$ versus $m^n$ on graph paper. If your guess is right, the plot will be a straight line. Another is to plot log $T$ versus log $m$. The slope of the straight line on this plot is $n$.) (b) Which data points deviate the most from a straight-line plot of $T$ versus $m^n$?

| Mass $m$, kg | 0.10 | 0.20 | 0.40 | 0.50 | 0.75 | 1.00 | 1.50 |
|---|---|---|---|---|---|---|---|
| Period $T$, s | 0.56 | 0.83 | 1.05 | 1.28 | 1.55 | 1.75 | 2.22 |

**61** ••• The table below gives the period $T$ and orbit radius $r$ for the motions of four satellites orbiting a dense, heavy asteroid. (a) These data can be fitted by the formula $T = Cr^n$. Find $C$ and $n$. (b) A fifth satellite is discovered to have a period of 6.20 y. Find the radius for the orbit of this satellite, which fits the same formula.

| Period $T$, y | 0.44 | 1.61 | 3.88 | 7.89 |
|---|---|---|---|---|
| Radius $r$, Gm | 0.088 | 0.208 | 0.374 | 0.600 |

**62** ••• **SSM** The period $T$ of a simple pendulum depends on the length $L$ of the pendulum and the acceleration of gravity $g$ (dimensions $L/T^2$). (a) Find a simple combination of $L$ and $g$ that has the dimensions of time. (b) Check the dependence of the period $T$ on the length $L$ by measuring the period (time for a complete swing back and forth) of a pendulum for two different values of $L$. (c) The correct formula relating $T$ to $L$ and $g$ involves a constant that is a multiple of $\pi$, and cannot be obtained by the dimensional analysis of Part (a). It can be found by experiment as in Part (b) if $g$ is known. Using the value $g = 9.81$ m/s² and your experimental results from Part (b), find the formula relating $T$ to $L$ and $g$.

**63** ••• **ISOLVE✔** The weight of the earth's atmosphere pushes down on the surface of the earth with a force of 14.7 lb for each square inch of the earth's surface. What is the weight in pounds of the earth's atmosphere? (The radius of the earth is about 6370 km.)

# Motion in One Dimension

MOTION IN ONE DIMENSION IS MOTION ALONG A STRAIGHT LINE, LIKE THAT OF A CAR ON A STRAIGHT ROAD. THIS DRIVER ENCOUNTERS STOPLIGHTS AND DIFFERENT SPEED LIMITS ON HER COMMUTE ALONG A STRAIGHT HIGHWAY TO SCHOOL.

**?** **How can she estimate her arrival time? (See Example 2-2.)**

2-1   Displacement, Velocity, and Speed

2-2   Acceleration

2-3   Motion With Constant Acceleration

2-4   Integration

**W**e begin our study of the physical universe by examining objects in motion. The study of motion, whose measurement, more than 400 years ago gave birth to physics, is called **kinematics.**

➤ In this chapter, we start with the simplest case of kinematics, the motion of a particle along a straight line, like the motion of a car moving along a flat, straight, narrow road. A particle is an object whose position can be described by a single point. Anything can be considered to be a particle—a molecule, a person, or a galaxy—as long as we can reasonably ignore its internal structure.

## 2-1 Displacement, Velocity, and Speed

Figure 2-1 shows a car at position $x_i$ at time $t_i$ and at position $x_f$ at a later time $t_f$. The change in the car's position, called the **displacement,** is given by $x_f - x_i$. We use the Greek letter $\Delta$ (uppercase delta) to indicate the change in a quantity; thus, the change in $x$ is written $\Delta x$:

$$\Delta x = x_f - x_i \qquad \qquad 2\text{-}1$$

DEFINITION—DISPLACEMENT

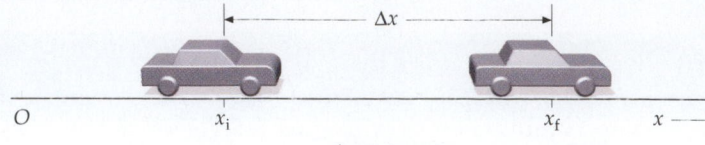

$$\Delta x = x_f - x_i$$

The notation $\Delta x$ (read "delta x") stands for a single quantity, the change in $x$. It is not a product of $\Delta$ and $x$ any more than $\cos \theta$ is a product of cos and $\theta$. By convention, the change in a quantity is always its final value minus its initial value.

Velocity is the rate at which the position changes. The **average velocity** of the particle is defined as the ratio of the displacement $\Delta x$ to the time interval $\Delta t = t_f - t_i$:

$$v_{av} = \frac{\Delta x}{\Delta t} = \frac{x_f - x_i}{t_f - t_i} \qquad \qquad 2\text{-}2$$

DEFINITION—AVERAGE VELOCITY

**FIGURE 2-1** A car moving in a straight line. A coordinate axis consists of a line along the path of the car. A point on this line is chosen to be the origin $O$. Other points on it are assigned a number $x$, the value of $x$ being proportional to its distance from $O$. Points to the right of $O$ are positive as shown, and points to the left are negative. When the car travels from point $x_i$ to point $x_f$, its displacement is $\Delta x = x_f - x_i$.

Displacement and average velocity may be positive or negative. A positive value indicates motion in the positive $x$ direction. The SI unit of velocity is meters per second (m/s), and the U.S. customary unit is feet per second (ft/s).

---

DISPLACEMENT AND VELOCITY OF A COMET | **EXAMPLE 2-1**

A comet moving toward the sun is first seen at $x_i = 3.0 \times 10^{12}$ m relative to the sun (see Figure 2-2). Exactly one year later, it is seen at $x_f = 2.1 \times 10^{12}$ m. Find its displacement and average velocity.

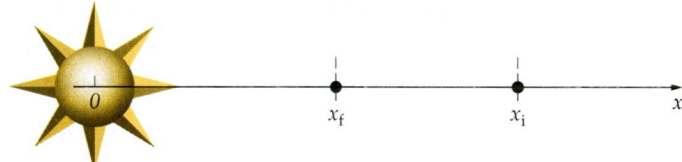

**FIGURE 2-2**

**PICTURE THE PROBLEM** Comets move in elliptical orbits around the sun. Here we consider just the distance from the sun as if the comet moved in one dimension. We are given $x_i$ and $x_f$. If we choose $t_i = 0$, then $t_f = 1$ y $= 3.16 \times 10^7$ s. The average velocity is $\Delta x / \Delta t$.

1. The displacement is found from its definition:

$$\Delta x = x_f - x_i = (2.1 \times 10^{12} \text{ m}) - (3.0 \times 10^{12} \text{ m})$$
$$= -9 \times 10^{11} \text{ m}$$

2. The average velocity is the displacement divided by the time interval:

$$v_{av} = \frac{\Delta x}{\Delta t} = \frac{-9 \times 10^{11} \text{ m}}{3.16 \times 10^7 \text{ s}}$$
$$= -2.85 \times 10^4 \text{ m/s} = \boxed{-28.5 \text{ km/s}}$$

**REMARKS** Both displacement and average velocity are negative, because the comet moved toward smaller values of $x$. Note that the units, m for $\Delta x$, and m/s or km/s for $v_{av}$, are essential parts of the answers. It is meaningless to say "the displacement is $-9 \times 10^{11}$" or "the average velocity of a particle is $-28.5$."

**EXERCISE** A jet plane leaves the gate in Detroit at 2:15 P.M. Its average velocity is 500 km/h for the trip to Chicago, which is 483 km away. When does it arrive at the gate in Chicago? (*Answer* 3:13 P.M. Detroit time, which is actually 2:13 P.M. Chicago time)

*DRIVING TO SCHOOL*          **EXAMPLE 2-2**   **Put It in Context**

It normally takes you 10 min to travel 5 mi to school along a straight road. You leave home 15 min before class begins. Delays caused by a broken traffic light slow down traffic to 20 mph for the first 2 mi of the trip. Will you be late for class?

**PICTURE THE PROBLEM** You need to find the total time that it will take you to travel to class. To do so, you must find the time $\Delta t_{2\,mi}$ that you will be driving at 20 mph, and the time $\Delta t_{3\,mi}$ for the remainder of the trip, during which you are driving at your usual speed.

1. The total time equals the time to travel the first 2 mi plus the time to travel the remaining 3 mi:

$$\Delta t_{tot} = \Delta t_{2\,mi} + \Delta t_{3\,mi}$$

2. Using $\Delta x = v_{av}\Delta t$, solve for the time taken to travel 2 mi at 20 mi/h:

$$\Delta t_{2\,mi} = \frac{\Delta x}{v_{av}} = \frac{2\,mi}{20\,mi/h} = 0.1\,h = 6\,min$$

3. Using $\Delta x = v_{av}\Delta t$, relate the time to travel the last 3 mi at the usual speed:

$$\Delta t_{3\,mi} = \frac{\Delta x}{v_{av}} = \frac{3\,mi}{v_{usual}}$$

4. Using $\Delta x = v_{av}\Delta t$, solve for the $v_{usual}$, the speed needed for you to travel the 5 mi in 10 min:

$$v_{usual} = \frac{\Delta x_{tot}}{\Delta t_{usual}} = \frac{5\,mi}{10\,min} = 0.5\,mi/min$$

5. Solve for $t_{3\,mi}$:

$$\Delta t_{3\,mi} = \frac{3\,mi}{0.5\,mi/min} = 6\,min$$

6. Solve for the total time:

$$\Delta t_{tot} = \Delta t_{2\,mi} + \Delta t_{3\,mi} = 12\,min$$

7. The trip takes 12 min with the delay, compared to the usual 10 minutes. Because you wisely allowed yourself 15 min for the trip, *you will not be late for class.*

The **average speed** of a particle is the ratio of the total distance traveled to the total time from start to finish:

$$\text{Average speed} = \frac{\text{total distance}}{\text{total time}} = \frac{s}{t} \qquad\qquad 2\text{-}3$$

Since the total distance and total time are both always positive, the average speed is always positive.

*AVERAGE SPEED IN A FOOT RACE*          **EXAMPLE 2-3**

You run 100 m in 12 s then turn around and jog 50 m back toward the starting point in 30 s (see Figure 2-3). Calculate (*a*) your average speed, and (*b*) your average velocity for the total trip.

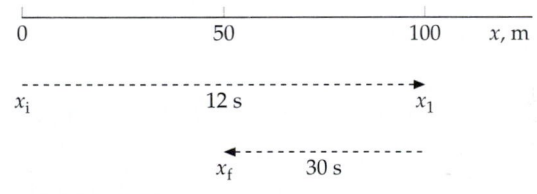

**FIGURE 2-3**

**PICTURE THE PROBLEM** We use the definitions of average speed and average velocity, noting that average *speed* is the total *distance* divided by $\Delta t$, whereas the average *velocity* is the net *displacement* divided by $\Delta t$:

(*a*) 1. Your average speed equals the total distance divided by the total time:

$$\text{Average speed} = \frac{s}{t}$$

2. Calculate the total distance traveled and the total time:

$$s = s_1 + s_2 = 100\,m + 50\,m = 150\,m$$

$$t = 12\,s + 30\,s = 42\,s$$

3. Use $s$ and $t$ to find your average speed:

$$\text{Average speed} = \frac{150 \text{ m}}{42 \text{ s}} = \boxed{3.57 \text{ m/s}}$$

(b) 1. Your average velocity is the ratio of the net displacement $\Delta x$ to the time interval $\Delta t$:

$$v_{av} = \frac{\Delta x}{\Delta t}$$

2. Your net displacement is $x_f - x_i$, where $x_i = 0$ is your initial position and $x_f = 50$ m is your final position:

$$\Delta x = x_f - x_i = 50 \text{ m} - 0 = 50 \text{ m}$$

3. Use $\Delta x$ and $\Delta t$ to find your average velocity:

$$v_{av} = \frac{\Delta x}{\Delta t} = \frac{50 \text{ m}}{42 \text{ s}} = \boxed{1.19 \text{ m/s}}$$

**PLAUSIBILITY CHECK** The world record for a 100-m race is just under 10 s, so 10 m/s is about the maximum possible speed obtainable. The result of 3.57 m/s for the average speed in (*a*) is reasonable, given that you merely jogged for one-third of the distance. If you had obtained 35.7 m/s for the average speed, that would have been a clue that something was wrong with the calculation.

**REMARKS** Note that your average speed is greater than your average velocity because the total distance traveled is greater than the total displacement. Also, note that the net displacement is the sum of the individual displacements. That is, $\Delta x = \Delta x_1 + \Delta x_2 = (100 \text{ m}) + (-50 \text{ m}) = 50 \text{ m}$, which is the Part (*b*), step 2 result.

---

*A Train-Hopping Bird*                         **EXAMPLE 2-4**

**Two trains 75 km apart approach each other on parallel tracks, each moving at 15 km/h. A bird flies back and forth between the trains at 20 km/h until the trains pass each other. How far does the bird fly?**

**PICTURE THE PROBLEM** This problem seems difficult at first, but viewed in the right way it is actually quite simple. We approach it by first writing an equation for the quantity to be found, the total distance $s$ flown by the bird.

1. The total distance traveled by the bird equals its speed times the time:

$$s = (\text{average speed})_{bird} \times t$$
$$= (\text{speed})_{av\,bird} \times t$$

2. The time that the bird is in the air is the time taken for the trains to meet. The sum of the distances traveled by the two trains is $D = 75$ km. Find the time it will take the two trains to travel a total distance $D$:

$$s_1 + s_2 = (\text{speed})_{av\,1} \times t + (\text{speed})_{av\,2} \times t = D$$

so

$$t = \frac{D}{(\text{speed})_{av\,1} + (\text{speed})_{av\,2}}$$

3. The total distance traveled by the bird is therefore:

$$s = (\text{speed})_{av\,bird} \times t$$
$$= (\text{speed})_{av\,bird} \frac{D}{(\text{speed})_{av\,1} + (\text{speed})_{av\,2}}$$
$$= 20 \text{ km/h} \frac{75 \text{ km}}{15 \text{ km/h} + 15 \text{ km/h}}$$
$$= \boxed{50 \text{ km}}$$

**REMARKS** Some try to solve this problem by finding and summing the distances flown by the bird each time it moves from one train to the other. This makes a relatively easy problem quite difficult. It is important to develop a thoughtful, systematic approach to solving problems. Begin by writing an equation for the unknown quantity in terms of other quantities. Then proceed by determining the values for each of the other quantities in the equation.

Figure 2-4 depicts average velocity graphically. A straight line connects points $P_1$ and $P_2$ and forms the hypotenuse of the triangle having sides $\Delta x$ and $\Delta t$. The ratio $\Delta x / \Delta t$ is the line's **slope,** which gives us a geometric interpretation of average velocity:

> The average velocity is the slope of the straight line connecting the points $(t_1, x_1)$ and $(t_2, x_2)$.

GEOMETRIC INTERPRETATION OF AVERAGE VELOCITY

Generally, average velocity depends on the time interval on which it is based. In Figure 2-4, for example, the smaller time interval indicated by $t_2'$ and $P_2'$ gives a larger average velocity, as shown by the greater steepness of the line connecting points $P_1$ and $P_2'$.

## Instantaneous Velocity

On first consideration, defining the velocity of a particle at a single instant seems impossible. At a given instant, a particle is at a single point. If it is at a single point, how can it be moving? If it is not moving, how can it have a velocity? This age-old paradox is resolved when we realize that observing and defining motion requires that we look at the position of the object at more than one time. Consider Figure 2-5. As we consider successively shorter time intervals beginning at $t_1$, the average velocity for the interval approaches the slope of the tangent at $t_1$. We define the slope of this tangent as the **instantaneous velocity** at $t_1$. This tangent is the limit of the ratio $\Delta x / \Delta t$ as $\Delta t$, and therefore $\Delta x$, approaches zero. So we can say:

> The instantaneous velocity is the limit of the ratio $\Delta x / \Delta t$ as $\Delta t$ approaches zero.
>
> $$v(t) = \lim_{\Delta t \to 0} \frac{\Delta x}{\Delta t}$$
>
> $$= \text{slope of the line tangent to the } x\text{-versus-}t \text{ curve}^\dagger \qquad 2\text{-}4$$

DEFINITION—INSTANTANEOUS VELOCITY

This limit is called the **derivative** of $x$ with respect to $t$. In the usual calculus notation, the derivative is written $dx/dt$

$$v(t) = \lim_{\Delta t \to 0} \frac{\Delta x}{\Delta t} = \frac{dx}{dt} \qquad 2\text{-}5$$

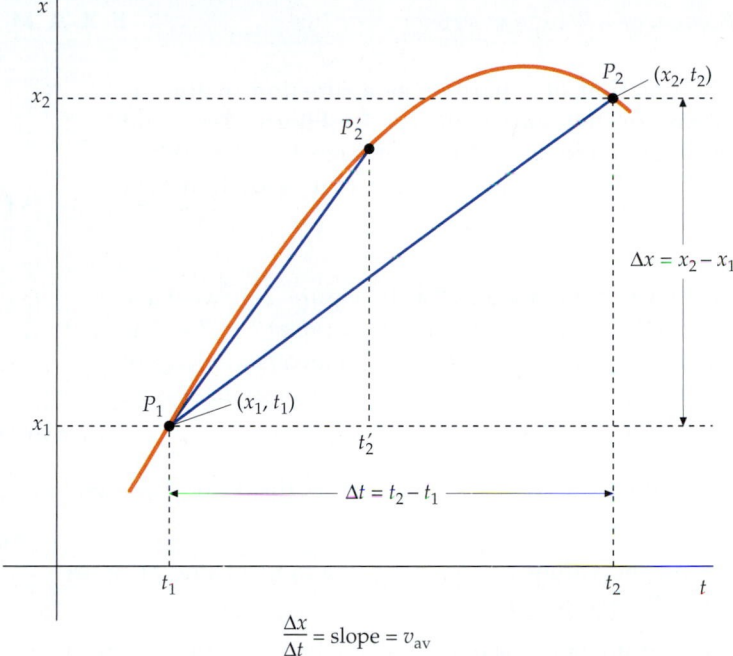

$$\frac{\Delta x}{\Delta t} = \text{slope} = v_{\text{av}}$$

**FIGURE 2-4** Graph of $x$ versus $t$ for a particle moving in one dimension. Each point on the curve represents the location $x$ at a particular time $t$. We have drawn a straight line between positions $P_1$ and $P_2$. The displacement $\Delta x = x_2 - x_1$ and the time interval $\Delta t = t_2 - t_1$ between these points are indicated. The straight line between $P_1$ and $P_2$ is the hypotenuse of the triangle having sides $\Delta x$ and $\Delta t$, and the ratio $\Delta x/\Delta t$ is its slope. In geometric terms, the slope is a measure of the line's steepness.

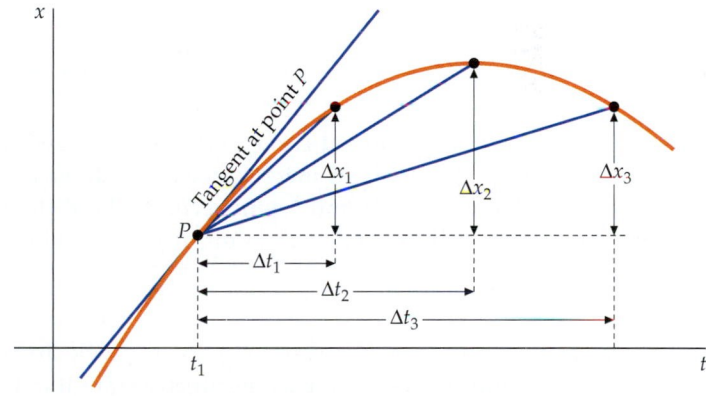

**FIGURE 2-5** Graph of $x$ versus $t$. Note the sequence of successively smaller time intervals, $\Delta t_1, \Delta t_2, \Delta t_3, \ldots$. The average velocity of each interval is the slope of the straight line for that interval. As the time intervals become smaller, these slopes approach the slope of the tangent to the curve at point $t_1$. The slope of this line is defined as the instantaneous velocity at time $t_1$.

A line's slope may be positive, negative, or zero; consequently, instantaneous velocity (in one-dimensional motion) may be positive ($x$ increasing), negative ($x$ decreasing), or zero (no motion). The magnitude of the instantaneous velocity is the **instantaneous speed.**

---

† The slope of the line tangent to a curve is often referred to more simply as the "slope of the curve."

**EXAMPLE   2 - 5**   **Try It Yourself**

The position of a particle as a function of time is given by the curve shown in Figure 2-6. Find the instantaneous velocity at time $t = 2$ s. When is the velocity greatest? When is it zero? Is it ever negative?

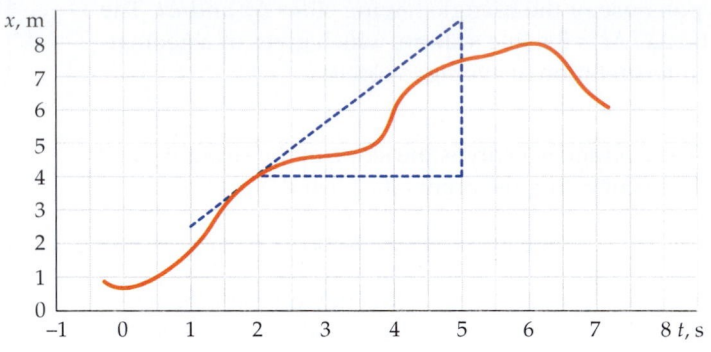

**PICTURE THE PROBLEM** In Figure 2-6, we have sketched the line tangent to the curve at $t = 2$ s. The tangent line's slope is the instantaneous velocity of the particle at the given time. You can use this figure to measure the slope of the tangent line.

**FIGURE 2-6**

**Cover the column to the right and try these on your own before looking at the answers.**

**Steps**

1. Find the values $x_1$ and $x_2$ on the tangent line at times $t_1 = 2$ s and $t_2 = 5$ s.

2. Compute the slope of the tangent line from these values. This slope equals the instantaneous velocity at $t = 2$ s.

3. From the figure, the slope (and therefore velocity) is greatest at about $t = 4$ s. The slope and velocity are zero at $t = 0$ and $t = 6$ s and are negative before 0 and after 6 s.

**Answers**

$x_1 \approx 4$ m,     $x_2 \approx 8.5$ m

$v = \text{slope} \approx \dfrac{8.5 \text{ m} - 4 \text{ m}}{5 \text{ s} - 2 \text{ s}} = \boxed{1.5 \text{ m/s}}$

**EXERCISE** What is the average velocity of this particle between $t = 2$ s and $t = 5$ s? (*Answer*   1.17 m/s)

---

*A STONE DROPPED FROM A CLIFF*    **EXAMPLE   2 - 6**

The position of a stone dropped from a cliff is described approximately by $x = 5t^2$, where $x$ is in meters measured downward from the original position at $t = 0$, and $t$ is in seconds. Find the velocity at any time $t$. (We omit explicit indication of units to simplify the notation.)

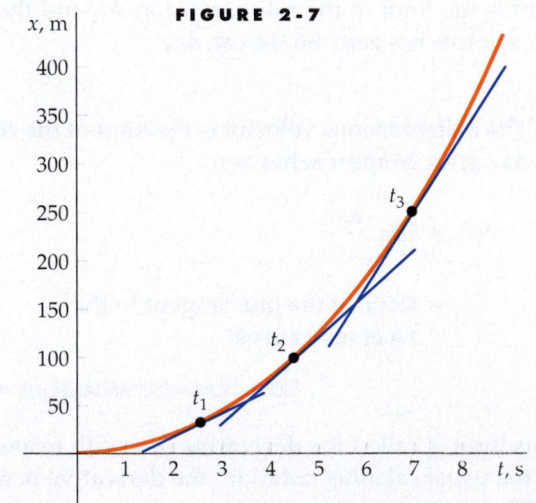

**FIGURE 2-7**

**PICTURE THE PROBLEM** We can compute the velocity at some time $t$ by computing the derivative $dx/dt$ directly from the definition in Equation 2-4. The corresponding curve giving $x$ versus $t$ is shown in Figure 2-7. Tangent lines are drawn at times $t_1$, $t_2$, and $t_3$. The slopes of these tangent lines increase steadily, indicating that the instantaneous velocity increases steadily with time.

1. By definition the instantaneous velocity is:

$$v(t) = \lim_{\Delta t \to 0} \frac{\Delta x}{\Delta t} = \lim_{\Delta t \to 0} \frac{x(t + \Delta t) - x(t)}{\Delta t}$$

2. We compute the displacement $\Delta x$ from the position function $x(t)$:

$$x(t) = 5t^2$$

3. At a later time $t + \Delta t$, the position is $x(t + \Delta t)$, given by:

$$x(t + \Delta t) = 5(t + \Delta t)^2 = 5\left[t^2 + 2t\Delta t + (\Delta t)^2\right]$$
$$= 5t^2 + 10t\Delta t + 5(\Delta t)^2$$

4. The displacement for this time interval is thus:

$$\Delta x = x(t + \Delta t) - x(t)$$
$$= \left[5t^2 + 10t\Delta t + 5(\Delta t)^2\right] - 5t^2$$
$$= 10t\Delta t + 5(\Delta t)^2$$

5. Divide $\Delta x$ by $\Delta t$ to find the average velocity for this time interval:

$$v_{av} = \frac{\Delta x}{\Delta t} = \frac{10t\Delta t + 5(\Delta t)^2}{\Delta t} = 10t + 5\Delta t$$

6. As we consider shorter and shorter time intervals, $\Delta t$ approaches zero and the second term $5\Delta t$ approaches zero, though the first term, $10t$, remains unchanged:

$$v(t) = \lim_{\Delta t \to 0} \frac{\Delta x}{\Delta t} = \lim_{\Delta t \to 0} (10t + 5\Delta t) = \boxed{10t}$$

**REMARKS** If we had set $\Delta t = 0$ in steps 4 and 5, the displacement would be $\Delta x = 0$, in which case the ratio $\Delta x/\Delta t$ would be undefined. Instead, we leave $\Delta t$ as a variable until the final step, when the limit $\Delta t \to 0$ is well defined.

To find derivatives quickly, we use rules based on the limiting process above (see Appendix Table A-4). A particularly useful rule is

$$\text{If } x = Ct^n, \quad \text{then} \quad \frac{dx}{dt} = Cnt^{n-1} \qquad\qquad 2\text{-}6$$

where $C$ and $n$ are any constants. Using this rule in Example 2-6, we have $x = 5t^2$, and $v = dx/dt = 10t$, in agreement with our previous results.

## Relative Velocity

If you are sitting in an airplane moving with a velocity of 500 mi/h toward the east, your velocity is also 500 mi/h toward the east. However, 500 mi/h toward the east might be your velocity relative to the surface of the earth, or it might be your velocity relative to the air outside the airplane. (If the plane is flying in a jet stream, these two velocities would be very different.) Furthermore, your velocity is zero relative to the airplane itself. To specify the velocity of a particle you must also specify the **frame of reference.** In this discussion three different frames of reference are specified, the surface of the earth, the air outside the airplane, and the airplane itself.

Midair refueling. Each plane is nearly at rest relative to the other, though both are moving with very large velocities relative to the earth.

A frame of reference is an extended object whose parts are at rest relative to each other.

DEFINITION—FRAME OF REFERENCE

To make position measurements we use coordinate axes that are attached to reference frames. For a horizontal coordinate axis attached to the plane your position remains constant, at least it does if you remain in your seat. However, for a horizontal coordinate axis attached to the surface of the earth, and for a horizontal coordinate axis attached to a balloon floating in the air outside the plane, your position keeps changing. If you have trouble imagining a coordinate axis attached to the air outside the plane, instead imagine a coordinate axis attached to a balloon that is suspended in the air. The air and the balloon are at rest relative to each other, so together they form a single reference frame.

If a particle moves with velocity $v_{pA}$ relative to reference frame A, which is in turn moving with velocity $v_{AB}$ relative to reference frame B, the velocity of the particle relative to B is

$$v_{pB} = v_{pA} + v_{AB} \qquad\qquad 2\text{-}7a$$

For example, if you swim in a river parallel to the direction of the flow, your velocity relative to the shore $v_{ys}$ equals your velocity relative to the water $v_{yw}$ plus the velocity of the water relative to the shore $v_{ws}$:

$$v_{ys} = v_{yw} + v_{ws}$$

If you are swimming upstream at 2 m/s relative to the water, and the water is moving at 1.2 m/s relative to the shore, then your velocity relative to the shore is $v_{ys} = -2$ m/s $+ 1.2$ m/s $= -0.8$ m/s, where we have chosen downstream to be the positive direction.

A great surprise of twentieth-century physics was the discovery that Equation 2-7a is only an approximation. When we study the theory of relativity in Chapter 39, we will see that the exact expression for relative velocities is

$$v_{pB} = \frac{v_{pA} + v_{AB}}{1 + v_{pA}v_{AB}/c^2} \qquad\qquad 2\text{-}7b$$

where $c = 3 \times 10^8$ m/s is the speed of light in a vacuum. In all everyday cases with macroscopic objects, $v_{pA}$ and $v_{AB}$ are both so much smaller than $c$ that Equations 2-7a and b give essentially the same result, but for high-speed objects, such as electrons or distant galaxies, the difference between these two equations becomes significant. Equation 2-7b has the interesting property that if $v_{pA} = c$, then $v_{pB}$ also equals $c$, which is a core tenet of relativity, namely that the speed of light is the same in all reference frames.

**EXERCISE** Using Equation 2-7b, substitute $c$ for $v_{pA}$ and solve for $v_{pB}$, verifying that Equation 2-7b is in agreement with the result "the speed of light is the same in all reference frames."

# 2-2 Acceleration

Acceleration is the rate of change of the instantaneous velocity. When you step on your car's accelerator, for example, you expect to change your velocity. The **average acceleration** for a particular time interval $\Delta t = t_2 - t_1$ is defined as the ratio $\Delta v/\Delta t$, where $\Delta v = v_2 - v_1$:

$$a_{av} = \frac{\Delta v}{\Delta t} \qquad\qquad 2\text{-}8$$

DEFINITION—AVERAGE ACCELERATION

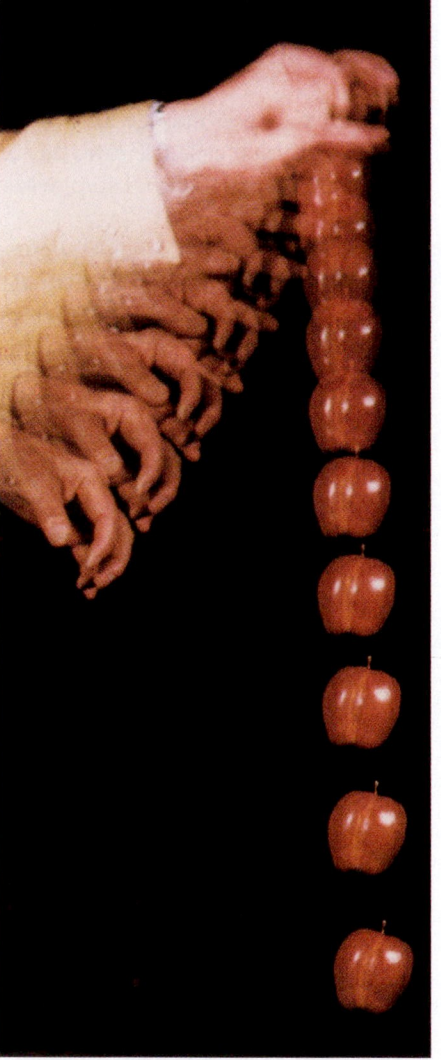

A falling apple captured by strobe photography at 60 flashes per second. The acceleration of the apple is indicated by the widening spaces between the images.

Acceleration has dimensions of length divided by time squared. The SI unit is meters per second squared, m/s$^2$. (In Equation 2-8, if the numerator is in m/s and the denominator is in s, then $\Delta v/\Delta t$ is in units of (m/s)/s. Multiplying the numerator and the denominator by 1 s, we find the units of $\Delta v/\Delta t$ to equal $m/s^2$.) We can write Equation 2-8 as $\Delta v = a_{av}\Delta t$. Thus, if a particle at rest accelerates at 5.1 $m/s^2$, its velocity after 1 s is 5.1 m/s, its velocity after 2 s is 10.2 m/s, and so on. **Instantaneous acceleration** is the limit of the ratio $\Delta v/\Delta t$ as $\Delta t$ approaches zero. On a plot of velocity versus time, the instantaneous acceleration at time $t$ is the slope of the line tangent to the curve at that time:

$$a = \lim_{\Delta t \to 0} \frac{\Delta v}{\Delta t}$$

$$= \text{slope of the line tangent to the } v\text{-versus-}t \text{ curve} \qquad 2\text{-}9$$

DEFINITION—INSTANTANEOUS ACCELERATION

Thus, acceleration is the derivative of velocity with respect to time, $dv/dt$. Since velocity is the derivative of the position $x$ with respect to $t$, acceleration is the second derivative of $x$ with respect to $t$, $d^2x/dt^2$. We can see the reason for this notation when we write the acceleration as $dv/dt$ and replace $v$ with $dx/dt$:

$$a = \frac{dv}{dt} = \frac{d(dx/dt)}{dt} = \frac{d^2x}{dt^2} \qquad 2\text{-}10$$

If acceleration remains zero, there is no change in velocity over time—velocity is constant. In this case, the curve of $x$ versus $t$ is a straight line. If acceleration is nonzero and constant, as in Example 2-9, then velocity varies linearly with time and $x$ varies quadratically with time.

---

*A Fast Cat*                                **EXAMPLE   2 - 7**

A cheetah can accelerate from 0 to 96 km/h (60 mi/h) in 2 s, whereas a Corvette requires 4.5 s. Compute the average accelerations for the cheetah and Corvette and compare them with the free-fall acceleration due to gravity, $g = 9.81$ m/s².

1. Find the average acceleration from the information given: cat $a_{av} = \dfrac{\Delta v}{\Delta t} = \dfrac{96 \text{ km/h} - 0}{2 \text{ s}}$

$= 48 \text{ km/(h·s)}$

car $a_{av} = \dfrac{\Delta v}{\Delta t} = \dfrac{96 \text{ km/h} - 0}{4.5 \text{ s}}$

$= 21.3 \text{ km/(h·s)}$

2. Convert to m/s$^2$ using 1 h = 3600 s = 3.6 ks: cat $\dfrac{48 \text{ km}}{\text{h·s}} \times \dfrac{1 \text{ h}}{3.6 \text{ ks}} = 13.3 \text{ m/s}^2$

car $\dfrac{21.3 \text{ km}}{\text{h·s}} \times \dfrac{1 \text{ h}}{3.6 \text{ ks}} = 5.92 \text{ m/s}^2$

3. To compare the result with the acceleration due to gravity, multiply each by the conversion factor $1g/9.81 \text{ m/s}^2$: cat $13.3 \text{ m/s}^2 \times \dfrac{1g}{9.81 \text{ m/s}^2} = \boxed{1.36g}$

car $5.92 \text{ m/s}^2 \times \dfrac{1g}{9.81 \text{ m/s}^2} = \boxed{0.60g}$

**REMARKS** Note that by expressing the time in kiloseconds in step 2, the kilo prefixes in km and ks cancel.

**EXERCISE** A car is traveling at 45 km/h at time $t = 0$. It accelerates at a constant rate of 10 km/(h·s). (*a*) How fast is it traveling at $t = 2$ s? (*b*) At what time is the car traveling at 70 km/h?

(*Answer* (*a*) 65 km/h (*b*) 2.5 s)

**EXERCISE IN DIMENSIONAL ANALYSIS** If a car starts from rest at $x = 0$ with constant acceleration $a$, its velocity $v$ depends on $a$ and the distance traveled $x$. Which of the following equations has the correct dimensions and therefore could be a possible equation relating $x$, $a$, and $v$?

(*a*) $v = 2ax$  (*c*) $v = 2ax^2$

(*b*) $v^2 = 2a/x$  (*d*) $v^2 = 2ax$

(*Answer* Only (*d*) has the same dimensions on both sides of the equation. Although we cannot obtain the exact equation from dimensional analysis, we can often obtain the functional dependence.)

---

*VELOCITY AND ACCELERATION AS FUNCTIONS OF TIME*  **EXAMPLE 2-8**

**The position of a particle is given by $x = Ct^3$, where $C$ is a constant having units of m/s$^3$. Find the velocity and acceleration as functions of time.**

1. We find the velocity by applying $dx/dt = Cnt^{n-1}$ (Equation 2-6): $x = Ct^3$

$v = \dfrac{dx}{dt} = \boxed{3Ct^2}$

2. The time derivative of velocity gives the acceleration: $a = \dfrac{dv}{dt} = \boxed{6Ct}$

**ⓘ PLAUSIBILITY CHECK** We can check the units of our answers. For velocity, $[v] = [C][t^2] = (\text{m/s}^3)(\text{s}^2) = \text{m/s}$. For acceleration, $[a] = [C][t] = (\text{m/s}^3)(\text{s}) = \text{m/s}^2$.

# 2-3 Motion With Constant Acceleration

The motion of a particle that has constant acceleration is common in nature. For example, near the earth's surface all unsupported objects fall vertically with constant acceleration (provided air resistance is negligible). If a particle has a constant acceleration $a$, it follows that the average acceleration for any time interval is also $a$. Thus,

$$a_{av} = \frac{\Delta v}{\Delta t} = a \qquad\qquad 2\text{-}11$$

If the velocity is $v_0$ at time $t = 0$, and $v$ at some later time $t$, the corresponding acceleration is

$$a = \frac{\Delta v}{\Delta t} = \frac{v - v_0}{t - 0} = \frac{v - v_0}{t}$$

Rearranging yields $v$ as a function of time:

$$v = v_0 + at \qquad\qquad 2\text{-}12$$

CONSTANT ACCELERATION, **v** VERSUS **t**

This is the equation for a straight line in a $v$-versus-$t$ plot (Figure 2-8). The line's slope is the acceleration $a$ and its $v$ intercept is the initial velocity $v_0$.

The displacement $\Delta x = x - x_0$ in the time interval $\Delta t = t - 0$ is

$$\Delta x = v_{av}\Delta t = v_{av}t \qquad\qquad 2\text{-}13$$

For constant acceleration, the velocity varies linearly with time, and the average velocity is the mean value of the initial and final velocities. (This relation holds only if the acceleration is constant.) If $v_0$ is the initial velocity and $v$ is the final velocity, the average velocity is

$$v_{av} = \tfrac{1}{2}(v_0 + v) \qquad\qquad 2\text{-}14$$

CONSTANT ACCELERATION, **v**$_{av}$

The displacement is then

$$\Delta x = x - x_0 = v_{av}t = \tfrac{1}{2}(v_0 + v)t \qquad\qquad 2\text{-}15$$

We can eliminate $v$ by substituting $v = v_0 + at$ from Equation 2-12:

$$\Delta x = \tfrac{1}{2}(v_0 + v)t = \tfrac{1}{2}(v_0 + v_0 + at)t = v_0t + \tfrac{1}{2}at^2$$

The displacement is thus

$$\Delta x = x - x_0 = v_0t + \tfrac{1}{2}at^2 \qquad\qquad 2\text{-}16$$

CONSTANT ACCELERATION, **x**(**t**)

The first term on the right, $v_0t$, is the displacement that would occur if $a$ were zero, and the second term, $\tfrac{1}{2}at^2$, is the additional displacement due to the constant acceleration.

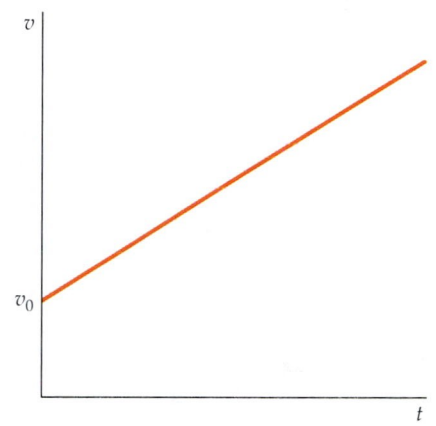

**FIGURE 2-8** Graph of velocity versus time for constant acceleration.

*"It goes from zero to 60 in about 3 seconds."*
© **Sydney Harris**

Let's eliminate $t$ from Equations 2-12 and 2-14 and find a relation between $\Delta x$, $a$, $v$, and $v_0$. From Equation 2-12, $t = (v - v_0)/a$. Substituting this into Equation 2-14,

$$\Delta x = v_{av}t = \frac{1}{2}(v_0 + v)t = \frac{1}{2}(v_0 + v)\left(\frac{v - v_0}{a}\right) = \frac{v^2 - v_0^2}{2a}$$

or

$$v^2 = v_0^2 + 2a\Delta x \qquad\qquad 2\text{-}17$$

CONSTANT ACCELERATION

Equation 2-17 is useful, for example, if we want to find the final velocity of a ball dropped from rest at some height $x$ and we are not interested in the time the fall takes.

## Problems with One Object

Many practical problems deal with objects in free-fall, that is, falling freely under the influence of gravity only. All objects in free-fall with the same initial velocity move identically. As shown in Figure 2-9, a feather and an apple, simultaneously released from rest in a large vacuum chamber, fall with identical motions. They have the same acceleration. The magnitude of this acceleration, designated by $g$, has the approximate value

$$g \approx 9.81 \text{ m/s}^2 = 32.2 \text{ ft/s}^2$$

Because $g$ is the *magnitude* of the acceleration, it is always positive. If downward is the positive direction, then the free-fall acceleration is $a = g$; if upward is positive, then $a = -g$.

**FIGURE 2-9**

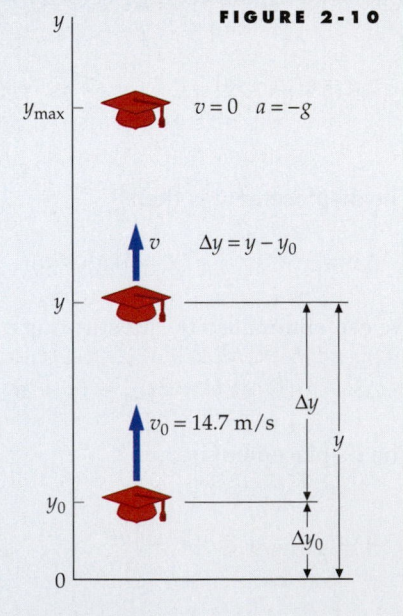

**FIGURE 2-10**

*THE FLYING CAP*

**EXAMPLE 2-9**

**Upon graduation, a joyful physics student throws her cap straight upward with an initial speed of 14.7 m/s. Given that its acceleration is 9.81 m/s² downward (we neglect air resistance), (a) how long does it take to reach its highest point? (b) What is the distance to the highest point? (c) Assuming the cap is caught at the same height from which it was released, what is the total time the cap is in flight?**

**PICTURE THE PROBLEM** When the cap is at its highest point, its instantaneous velocity is zero. **Thus, we translate the phrase "at its highest point" into the mathematical condition $v = 0$.**

(a) 1. Make a sketch of the cap in its initial position and again at its highest point. Include a coordinate axis and label the origin and the two positions of the cap.

2. The time is related to the velocity and acceleration: $\qquad v = v_0 + at$

3. To find the time at which the cap reaches its greatest height, set $v = 0$ and solve for $t$: $\qquad t = \dfrac{0 - v_0}{a} = \dfrac{-14.7 \text{ m/s}}{-9.81 \text{ m/s}^2} = \boxed{1.50 \text{ s}}$

(b) We can find the displacement from the time $t$ and the average velocity:

$$\Delta y = v_{av}t = \tfrac{1}{2}(v_0 + v)t$$

$$= \tfrac{1}{2}(14.7 \text{ m/s} + 0)(1.50 \text{ s}) = \boxed{11.0 \text{ m}}$$

(c) 1. Set $\Delta y = 0$ in Equation 2-16 and solve for $t$:

$$\Delta y = v_0 t + \tfrac{1}{2}at^2$$

$$0 = (v_0 + \tfrac{1}{2}at)t$$

2. There are two solutions for $t$ if $\Delta y = 0$. The first corresponds to the time at which the cap is released, the second to the time at which the cap is caught:

$$t = 0 \qquad \text{(first solution)}$$

$$t = -\frac{2v_0}{a} = -\frac{2(14.7 \text{ m/s})}{-9.81 \text{ m/s}^2} = \boxed{3 \text{ s}} \qquad \text{(second solution)}$$

**REMARKS** The $t = 3$ s solution also follows from a symmetry in the system—it takes the same time for the cap to fall from its greatest height as to rise to that height (see Figure 2-11). In reality the cap will not have a constant acceleration because air resistance has a significant effect on a light object like a cap. If air resistance is not negligible, the fall time will exceed the rise time.

**EXERCISE** Find $y_{max} - y_0$ using (a) Equation 2-15 and (b) Equation 2-16. (c) Find the velocity of the cap when it returns to its starting point. (*Answer* (a) and (b) $y_{max} - y_0 = 11.0$ m (c) $-14.7$ m/s; notice that the final speed is the same as the initial speed)

**EXERCISE** What is the velocity of the cap at the following points in time? (a) 0.1 s before it reaches its highest point; (b) 0.1 s after it reaches its highest point. (c) Compute $\Delta v / \Delta t$ for this 0.2-s-long time interval. (*Answer* (a) $+ 0.981$ m/s (b) $- 0.981$ m/s (c) $[(-0.981 \text{ m/s}) - (+0.981 \text{ m/s})]/(0.2 \text{ s}) = -9.81$ m/s$^2$)

**EXERCISE** A car starting from rest gains speed at a constant rate of 8 m/s$^2$. (a) How fast is it going after 10 s? (b) How far has it gone after 10 s? (c) What is its average velocity for the interval $t = 0$ to $t = 10$ s? (*Answer* (a) 80 m/s (b) 400 m (c) 40 m/s)

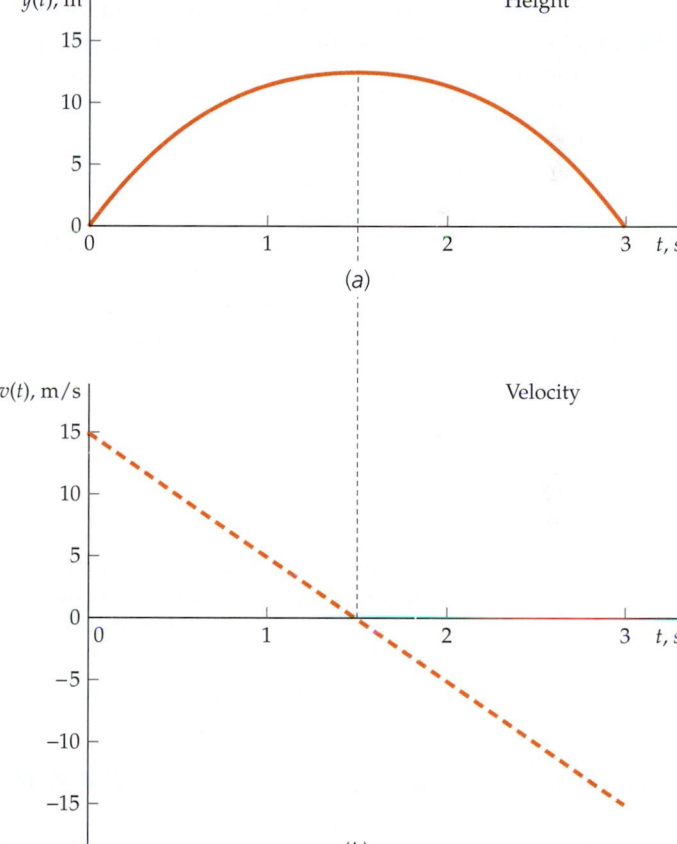

(a)

(b)

**FIGURE 2-11**

The next example concerns a car's **stopping distance**—how far it travels while coming to a halt.

*A CAR'S STOPPING DISTANCE*    **EXAMPLE 2-10**

**On a highway at night you see a stalled vehicle and brake your car to a stop with an acceleration of magnitude 5 m/s$^2$. (An acceleration that reduces the speed is often called a deceleration.) What is the car's stopping distance if its initial speed is (a) 15 m/s (about 34 mi/h) or (b) 30 m/s?**

**PICTURE THE PROBLEM** If we choose the direction of motion to be positive, the stopping distance and the initial velocity are positive, but the acceleration is negative. Thus, the initial velocity is $v_0 = 15$ m/s, the final velocity is $v = 0$, and the acceleration is $a = -5$ m/s². We seek the distance traveled, $\Delta x$. We do not need to know the time it takes for the car to stop, so Equation 2-17 is the most convenient formula to use.

(a) Using Equation 2-17 with $v = 0$, calculate the displacement $\Delta x$:

$$v^2 = v_0^2 + 2a\Delta x$$

so

$$\Delta x = \frac{v^2 - v_0^2}{2a} = \frac{0^2 - v_0^2}{2a} = -\frac{v_0^2}{2a}$$

$$= -\frac{(15 \text{ m/s})^2}{2(-5 \text{ m/s}^2)} = \boxed{22.5 \text{ m}}$$

(b) From the previous step we see that if $v = 0$, then $\Delta x = -v_0^2/(2a)$. Thus, $\Delta x$ is proportional to the square of the initial speed. Using this observation and the Part (a) result, find the stopping distance for an initial speed equal to twice that in Part (a):

$$\Delta x = 2^2(22.5 \text{ m}) = \boxed{90 \text{ m}}$$

**REMARKS** The answer to (b) can also be gotten by directly substituting the initial speed of 30 m/s into the expression for $\Delta x$ obtained in Part (a). Ninety meters is a considerable distance, roughly the length of a football field. Changing $v_0$ by a factor of 2 changes the stopping distance by a factor $2^2 = 4$ (see Figure 2-12). The practical implication of this squared dependence is that even modest increases in speed cause significant increases in stopping distance.

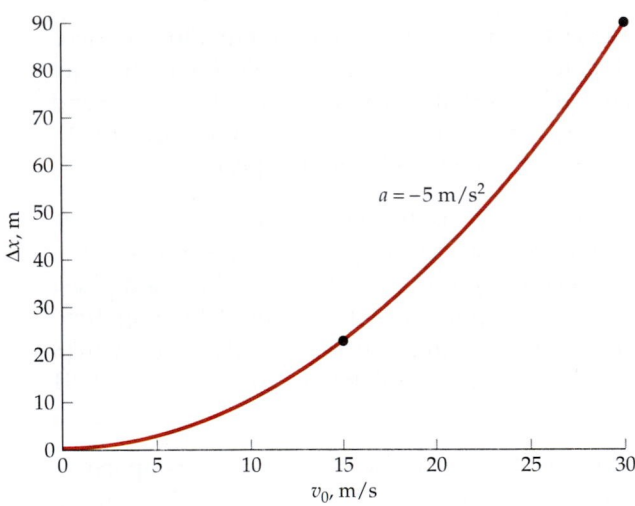

**FIGURE 2-12** Stopping distance as a function of the initial velocity. The curve shows the case for Example 2-10, where the acceleration is $a = -5.0$ m/s²; the points shown on the red curve are the solutions to parts (a) and (b).

---

*STOPPING DISTANCE*                    **EXAMPLE 2-11**   **Try It Yourself**

In Example 2-10, (a) how much time does it take for the car to stop if its initial velocity is 30 m/s? (b) How far does the car travel in the last second?

**PICTURE THE PROBLEM** (a) Except for the numbers, this is the same as Part (a) of Example 2-9. Use the same procedure shown in Example 2-10. (b) Since the speed decreases by 5 m/s each second, the velocity 1 s before the car stops must be 5 m/s. Find the average velocity during the last second and use that to find the distance traveled.

**Cover the column to the right and try these on your own before looking at the answers.**

| Steps | Answers |
|---|---|
| (a) Find the total stopping time $t$. | $t = 6$ s |

(b) 1. Find the average velocity during the last second.     $v_{av} = \boxed{2.5 \text{ m/s}}$

2. Compute the distance traveled from $\Delta x = v_{av}\Delta t$.     $\Delta x_1 = v_{av}\Delta t = \boxed{2.5 \text{ m}}$

**REMARKS** If Part (b) asked for the average velocity during the last 1.3 seconds (rather than during the last second), we can find the initial velocity $v_1$ during this interval using $\Delta v = a\Delta t$ (Equation 2-11).

Sometimes valuable insight can be gained about the motion of an object by assuming that the constant-acceleration formulas still apply even when the acceleration is not constant. This is the case in the following example.

---

*THE CRASH TEST*                                    **EXAMPLE 2-12**

In a crash test, a car traveling 100 km/h (about 62 mi/h) hits an immovable concrete wall. What is its acceleration?

**PICTURE THE PROBLEM** In this example, it is not appropriate to treat the car as a particle because different parts of the vehicle will have different accelerations as the car crumples to a halt. Moreover, the accelerations are not constant. However, we *can* treat a bolt in the center of the car as a particle. We need additional information to solve this problem—either the stopping distance or the time to stop. We can estimate the stopping distance using common sense. Upon impact, the center of the car will certainly move forward less than half the length of the car. We'll choose 0.75 m as a reasonable estimate of the stopping distance. Since the problem neither asks for nor provides the time we will use the relation $v^2 = v_0^2 + 2a\Delta x$.

1. Using $v^2 = v_0^2 + 2a\Delta x$, solve for the acceleration:

$$v^2 = v_0^2 + 2a\Delta x$$

so

$$a = \frac{v^2 - v_0^2}{2\Delta x} = \frac{0^2 - (100 \text{ km/h})^2}{2(0.75 \text{ m})}$$

2. Convert the velocity from km/h to m/s. In one hour there are $60^2$ s = 3.6 ks:

$$(100 \text{ km/h}) \times \left(\frac{1 \text{ h}}{3.6 \text{ ks}}\right) = 27.8 \text{ m/s}$$

3. Complete the calculation of the acceleration:

$$a = \frac{0^2 - (100 \text{ km/h})^2}{2(0.75 \text{ m})} = -\frac{(27.8 \text{ m/s})^2}{1.5 \text{ s}}$$

$$= -514 \text{ m/s}^2 \approx \boxed{-500 \text{ m/s}^2}$$

**REMARKS** Note that the magnitude of this acceleration is greater than $50g$. This estimate is what the magnitude of the acceleration would be both if the displacement of the center of the car were actually 0.75 m and if the acceleration were constant.

*A Traveling Electron*                  **EXAMPLE 2-13** Try It Yourself

An electron in a cathode-ray tube accelerates from rest with a constant acceleration of $5.33 \times 10^{12}$ m/s$^2$ for 0.15 $\mu$s (1 $\mu$s = $10^{-6}$ s). The electron then drifts with constant velocity for 0.2 $\mu$s. Finally, it comes to rest with an acceleration of $-2.67 \times 10^{13}$ m/s$^2$. How far does the electron travel?

**PICTURE THE PROBLEM** The equations for constant acceleration do not apply to this problem directly because the acceleration of the electron varies with time. Divide the electron's motion into three intervals, each with a different constant acceleration and use the final position and velocity for one interval as the initial conditions for the next interval. Choose the origin to be at the electron's starting position, and the positive direction to be the direction of motion.

**Cover the column to the right and try these on your own before looking at the answers.**

**Steps**

1. Find the displacement and final velocity for the first 0.15-$\mu$s interval.

2. Use this final velocity as the constant velocity to find the displacement while it drifts at constant velocity.

3. Use this same velocity as the initial velocity and Equation 2-17 with $v = 0$ to find the displacement for the third interval, in which the electron slows down.

4. Add the displacements found in steps 1, 2, and 3 to find the total displacement.

**Answers**

$\Delta x_1 = 6.00$ cm, $\quad v_1 = 8.00 \times 10^5$ m/s

$\Delta x_2 = 16$ cm

$\Delta x_3 = 1.20$ cm

$\Delta x = \Delta x_1 + \Delta x_2 + \Delta x_3$

$= 6.00$ cm $+ 16$ cm $+ 1.20$ cm $= \boxed{23.2 \text{ cm}}$

**REMARKS** In an X-ray machine electrons are accelerated from a hot wire to a metal target. They crash into it, coming abruptly to rest. As a result, the target emits X rays characteristic of the target metal.

 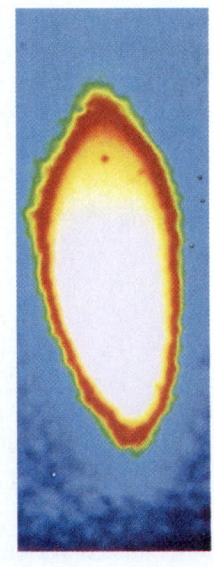

(*left*) The two-mile-long linear accelerator at Stanford University, used to accelerate electrons and positrons in a straight line to nearly the speed of light. (*right*) Cross-section of the accelerator's electron beam as shown on a video monitor.

*Tossing Binoculars*                  **EXAMPLE 2-14** Try It Yourself

John climbs a tree to get a better view of the speaker at an outdoor graduation ceremony. Unfortunately, he leaves his binoculars behind. Marsha throws them up to John, but her strength is greater than her accuracy. The binoculars pass John's outstretched hand after 0.69 s and again 1.68 s later. How high is John?

**PICTURE THE PROBLEM** There are two unknowns in this problem, John's height $h$ and the initial velocity of the binoculars $v_0$. We know that $y = h$ at $t_1 = 0.69$ s and $y = h$ at $t_2 = 0.69$ s $+ 1.68$ s $= 2.37$ s. Expressing $h$ as a function of time $t$ gives us two equations from which the two unknowns can be determined.

**Cover the column to the right and try these on your own before looking at the answers.**

**Steps**

**Answers**

1. Using $\Delta y = v_0 t + \frac{1}{2} a t^2$, equate $y$ for times $t_1$ and $t_2$, noting that $y = h$, and $a = -g$ in each case.

$h = v_0 t_1 - \frac{1}{2} g t_1^2$ and $h = v_0 t_2 - \frac{1}{2} g t_2^2$

2. Eliminate $v_0$ from these two equations and solve for $h$ in terms of the times $t_1$ and $t_2$. This can be done by solving the first equation for $v_0$ and substituting into the second equation.

$h = \left( \dfrac{h + \frac{1}{2} g t_1^2}{t_1} \right) t_2 - \frac{1}{2} g t_2^2$

so

$h = \boxed{8.02 \text{ m}}$

**REMARKS** We have two unknowns, $h$ and $v_0$, but are given two times, $t_1$ and $t_2$, so we can write two equations and solve them for either or both of the two unknowns.

**EXERCISE** Find the initial velocity of the binoculars and the velocity of the binoculars as they pass John on the way down. (*Answer* $v_0 = 15.0$ m/s; $v_2 = -8.24$ m/s)

## Problems with Two Objects

We now give some examples of problems involving two objects moving with constant acceleration.

CATCHING A SPEEDING CAR    **EXAMPLE 2-15**

A car is speeding at 25 m/s (~90 km/h; ~56 mi/h) in a school zone. A police car starts from rest just as the speeder passes and accelerates at a constant rate of 5 m/s². (*a*) When does the police car catch the speeding car? (*b*) How fast is the police car traveling when it catches up with the speeder?

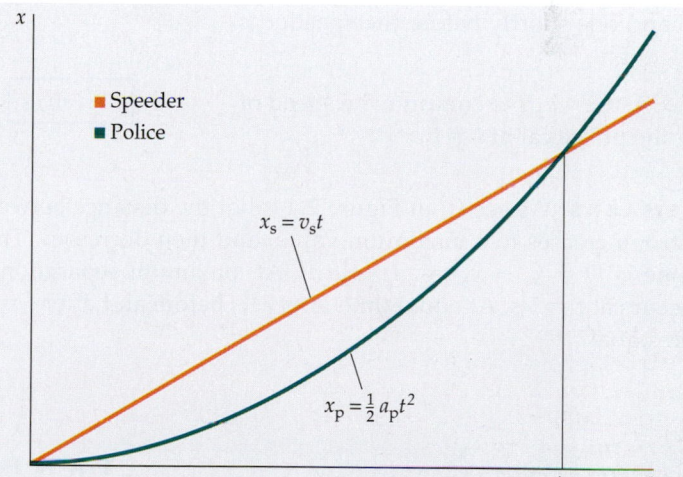

**PICTURE THE PROBLEM** To determine when the two cars will be at the same position, we write the positions of the speeder $x_s$ and of the police car $x_p$ as functions of time and solve for the time $t$ when $x_s = x_p$.

**FIGURE 2-13** The two curves depict the positions of the speeder and the police car. They have the same position at $t = 0$ and again at $t = t_c$.

(*a*) 1. Write the position functions for the speeder and the police car:

$x_s = v_s t$ and $x_p = \frac{1}{2} a_p t^2$

2. Set $x_s = x_p$ and solve for the time $t_c$, for $t_c > 0$:

$v_s t_c = \frac{1}{2} a_p t_c^2 \Rightarrow v_s = \frac{1}{2} a_p t_c \qquad t_c \neq 0$

$t_c = \dfrac{2 v_s}{a_p} = \dfrac{2(25 \text{ m/s})}{5 \text{ m/s}^2} = \boxed{10 \text{ s}}$

(*b*) 1. The velocity of the police car is given by $v = v_0 + a t_c$, with $v_0 = 0$:

$v_p = a_p t_c = (5 \text{ m/s}^2)(10 \text{ s}) = \boxed{50 \text{ m/s}}$

**REMARKS** Notice that the final speed of the police car in (b) is exactly twice that of the speeder. Since the two cars covered the same distance in the same time, they must have had the same average speed. The speeder's average speed, of course, is 25 m/s. For the police car to start from rest and have an average speed of 25 m/s, it must reach a final speed of 50 m/s.

**EXERCISE** How far have the cars traveled when the police car catches the speeder? (*Answer* 250 m)

---

*THE POLICE CAR*                  **EXAMPLE 2-16**  **Try It Yourself**

**How fast is the police car in Example 2-15 traveling when it is 25 m behind the speeding car?**

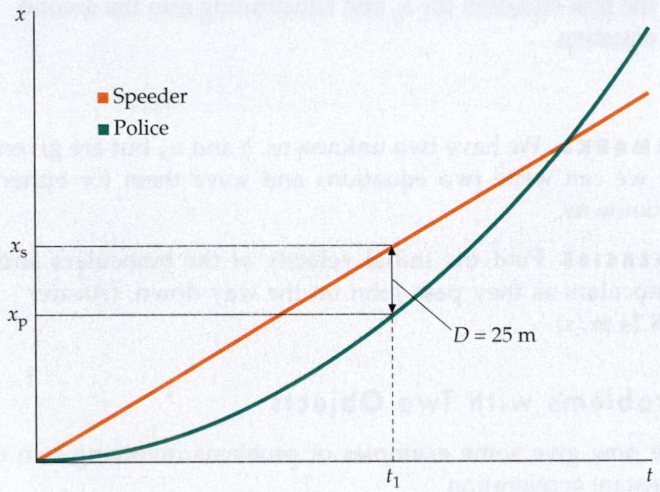

**PICTURE THE PROBLEM** The speed is given by $v_p = at_1$, where $t_1$ is the time at which $D = x_s - x_p = 25$ m.

**Cover the column to the right and try these on your own before looking at the answers.**

| Steps | Answers |
|---|---|
| 1. Sketch an x-versus-t curve showing the positions of the two cars at time $t_1$ (Figure 2-14). | |
| 2. Using the equations for $x_p$ and $x_s$ from Example 2-15, solve for $t_1$ when $x_s - x_p = 25$ m. We expect two solutions, one shortly after the start time and one shortly before the speeder is caught. | $t_1 = (5 \pm \sqrt{15})$ s |
| 3. Use $v_{p1} = a_p t_1$ to compute the speed of the police car at $t = t_1$. | $v_{p1} = \boxed{5.64 \text{ m/s}}$ and $\boxed{44.4 \text{ m/s}}$ |

**FIGURE 2-14**

**REMARKS** We see from Figure 2-14 that the distance between the cars starts at zero, increases to a maximum value, and then decreases. The separation at any time is $D = x_s - x_p = v_s t - \frac{1}{2}a_p t^2$. At maximum separation, $dD/dt = 0$, which occurs at $t = 5$ s. At equal time intervals before and after $t = 5$ s, the separations are equal.

---

*A MOVING ELEVATOR*                  **EXAMPLE 2-17**

**While standing in an elevator, you see a screw fall from the ceiling. The ceiling is 3 m above the floor. How long does it take for the screw to hit the floor if the elevator is moving upward and gaining speed at a constant rate of $a_f = 4.0$ m/s²?**

**PICTURE THE PROBLEM** Write the position as a function of time for both the screw, $y_s$, and the floor, $y_f$. When the screw hits the floor $y_s = y_f$. Choose the origin to be the initial position of the floor, and designate "upward" as the positive direction.

1. Make a sketch of the elevator and the screw as shown in Figure 2-15. Include a coordinate axis indicating the positions of the screw and the floor:

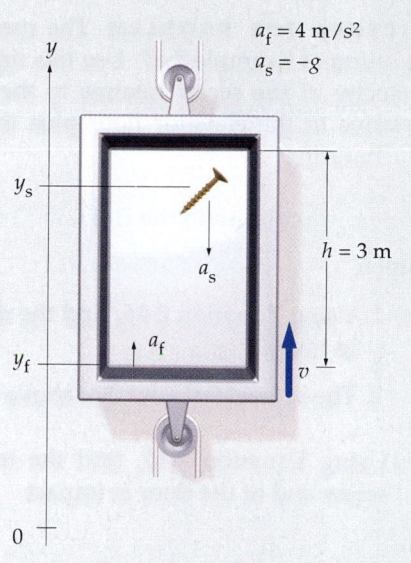

$a_f = 4 \text{ m/s}^2$
$a_s = -g$

2. Write the position function for the elevator floor and for the screw:

$$y_f - y_{0f} = v_{0f}t + \tfrac{1}{2}a_f t^2$$

$$y_s - y_{0s} = v_{0s}t + \tfrac{1}{2}a_s t^2$$

3. At $t = t_1$ the screw hits the floor. At this time these positions are equal:

$$y_s = y_f$$

$$y_{0s} + v_{0s}t_1 + \tfrac{1}{2}a_s t_1^2 = y_{0f} + v_{0f}t_1 + \tfrac{1}{2}a_f t_1^2$$

4. At $t = 0$, the floor and the screw have the same velocity. Use this to simplify the step 3 result:

$$v_{0f} = v_{0s}$$

so

$$y_{0s} + v_{0s}t_1 + \tfrac{1}{2}a_s t_1^2 = y_{0f} + v_{0s}t_1 + \tfrac{1}{2}a_f t_1^2$$

$$y_{0s} + \tfrac{1}{2}a_s t_1^2 = y_{0f} + \tfrac{1}{2}a_f t_1^2$$

5. Use the given information to further simplify:

$$y_{0f} = 0, \quad a_f = 4.0 \text{ m/s}^2$$

$$y_{0s} = h = 3 \text{ m}, \quad a_s = -g$$

so

$$h - \tfrac{1}{2}g t_1^2 = 0 + \tfrac{1}{2}a_f t_1^2$$

or

$$h = \tfrac{1}{2}(g + a_f)t_1^2$$

6. Solve for the time:

$$t_1 = \sqrt{\frac{2h}{g + a_f}} = \sqrt{\frac{2(3 \text{ m})}{9.81 \text{ m/s}^2 + 4.0 \text{ m/s}^2}}$$

$$= \boxed{0.659 \text{ s}}$$

**FIGURE 2-15** The coordinate axis is fixed to the building.

**REMARKS** The time of fall depends on the acceleration of the elevator, but not on its velocity. There is an "effective gravity" $g' = g + a_f$ in the frame of reference of the elevator. In the case (presumably hypothetical) in which the elevator itself is in free-fall, that is, $a_f = -g'$, the time of fall becomes infinite and the screw appears "weightless."

**EXERCISE** The speed of a good base runner is 9.5 m/s. The distance between bases is 26 m, and the pitcher is about 18.5 m from home plate. If a runner on first base edges 2 m off the base and then begins running at the instant the ball leaves the pitchers's hand, what is the likelihood that the runner will steal second base safely?

---

*THE MOVING ELEVATOR*                     **EXAMPLE 2-18** Try It Yourself

Consider the elevator and screw in Example 2-17. Assume the velocity of the elevator is 16 m/s upward when the screw separates from the ceiling. (*a*) How far does the elevator rise while the screw is falling? How far does the screw fall? (*b*) What is the velocity of the screw and the velocity of the elevator at impact? (*c*) What is the velocity of the screw relative to the floor at impact?

**PICTURE THE PROBLEM** The time of flight of the screw is obtained in the solution of Example 2-17. Use this time to solve parts (*a*) and (*b*). For part (*c*), the velocity of the screw relative to the building equals the velocity of the screw relative to the elevator floor plus the velocity of the elevator floor relative to the building.

**Cover the column to the right and try these on your own before looking at the answers.**

| Steps | Answers |
|---|---|
| (*a*) 1. Using Equation 2-16, find the distance the floor rises in time $t_1$. | $\Delta y_f = v_{f0}t_1 + \frac{1}{2}a_f t_1^2 = \boxed{11.4 \text{ m}}$ |
| 2. The screw starts out 3 m above the floor. | $\Delta y_s = \boxed{8.42 \text{ m}}$ |
| (*b*) Using Equation 2-12, find the impact velocity of the screw and of the floor at impact. | $v = v_0 + at$, so |
| | $v_s = v_{s0} - gt_1 = \boxed{9.53 \text{ m/s}}$ |
| | $v_f = v_{f0} + a_f t_1 = \boxed{18.6 \text{ m/s}}$ |
| (*c*) Using Equation 2-7*a*, find the velocity of the screw relative to the elevator floor. | $v_{sb} = v_{sf} + v_{fb}$ |
| | so |
| | $v_{sf} = v_{sb} - v_{fb} = 9.53 \text{ m/s} - 18.6 \text{ m/s}$ |
| | $= \boxed{-9.10 \text{ m/s}}$ |

**REMARKS** The screw strikes the floor 8.4 m above its position when it leaves the ceiling. Relative to the building it is still rising when it strikes the floor. Note that at impact the velocity of the screw relative to the building is positive.

# 2-4 Integration

To find the velocity from a given acceleration, we note that the velocity is the function $v(t)$ whose derivative is the acceleration $a(t)$:

$$\frac{dv(t)}{dt} = a(t)$$

If the acceleration is constant, the velocity is that function of time which, when differentiated, equals this constant. One such function is

$$v = at, \qquad a = \text{constant}$$

More generally, we can add any constant to $at$ without changing the time derivative. Calling this constant $c$, we have

$$v = at + c$$

When $t = 0$, $v = c$. Thus, $c$ is the initial velocity $v_0$.

Similarly, the position function $x(t)$ is that function whose derivative is the velocity:

$$\frac{dx}{dt} = v = v_0 + at$$

We can treat each term separately. The function whose derivative is a constant $v_0$ is $v_0 t$ plus any constant. The function whose derivative is $at$ is $\frac{1}{2}at^2$ plus any constant. Writing $x_0$ for the combined arbitrary constants, we have

$$x = x_0 + v_0 t + \tfrac{1}{2}at^2$$

When $t = 0$, $x = x_0$. Thus, $x_0$ is the initial position.

Whenever we find a function from its derivative, we must include an arbitrary constant in the general function. Since we go through the integration process twice to find $x(t)$ from the acceleration, two constants arise. These constants are usually determined from the velocity and position at some given time, which is usually chosen to be $t = 0$. They are therefore called the **initial conditions.** A common problem, called the **initial-value problem,** takes the form "given $a(t)$ and the initial values of $x$ and $v$, find $x(t)$." This problem is particularly important in physics because the acceleration of a particle is determined by the forces acting on it. Thus, if we know the forces acting on a particle and the position and velocity of the particle at some particular time, we can find its position at all other times.

A function $F(t)$ whose derivative (with respect to $t$) equals the function $f(t)$ is called the **antiderivative** of $f(t)$. Finding the antiderivative of a function is related to the problem of finding the area under a curve. Consider motion with a constant velocity $v_0$. The change in position $\Delta x$ during an interval $\Delta t$ is

$$\Delta x = v_0 \Delta t$$

This is the area under the $v$-versus-$t$ curve (Figure 2-16). If $v_0$ is negative, both the displacement and the area under the curve are negative. Normally we think of area as a quantity that cannot be negative. However, in this context that is not the case. In this case the "area under the curve" (the area between the curve and the time axis) is a negative quantity.

The geometric interpretation of the displacement as the area under the $v$-versus-$t$ curve is true not only for constant velocity, but it is true in general, as illustrated in Figure 2-17. To show this we first divide the time interval into numerous small intervals, $\Delta t_1$, $\Delta t_2$, and so on. Then we draw a set of rectangles as shown. The area of the rectangle corresponding to the $i$th time interval $\Delta t_i$ (shaded in the figure) is $v_i \Delta t_i$, which is approximately equal to the displacement $\Delta x_i$ during the interval $\Delta t_i$. The sum of the rectangular areas is therefore approximately the sum of the displacements during the time intervals and is approximately equal to the total displacement from time $t_1$ to $t_2$. Mathematically, we write this as

$$\Delta x \approx \sum_i v_i \Delta t_i$$

where the Greek letter $\Sigma$ (uppercase sigma) stands for sum. We can make the approximation as accurate as we wish by putting enough rectangles under the curve, each rectangle having a sufficiently small value for $\Delta t$. For the limit of smaller and smaller time intervals (and more and more rectangles), the resulting sum approaches the area under the curve, which in turn equals the displacement. The limit as $\Delta t$ approaches zero (and the number of rectangles approaches infinity) is called an **integral** and is written

$$\Delta x = x(t_2) - x(t_1) = \lim_{\Delta t \to 0}\left( \sum_i v_i \Delta t_i \right) = \int_{t_1}^{t_2} v\, dt \qquad 2\text{-}18$$

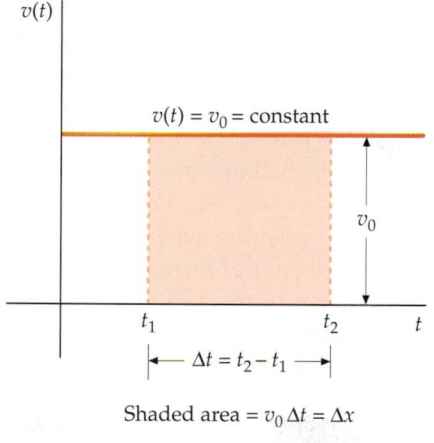

Shaded area = $v_0 \Delta t = \Delta x$

**FIGURE 2-16** The displacement for the interval $\Delta t$ equals the area under the velocity-versus-time curve for that interval. For $v(t) = v_0 =$ constant, the displacement equals the area of the shaded rectangle.

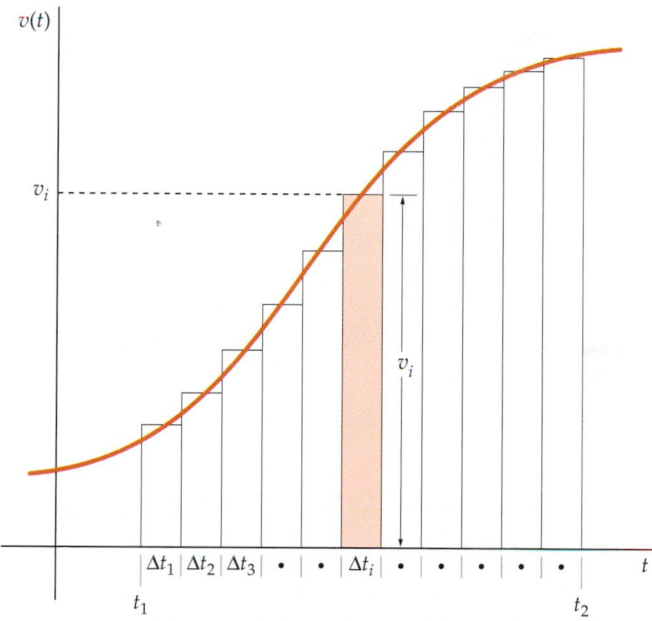

**FIGURE 2-17** Graph of a general $v(t)$-versus-$t$ curve. The total displacement from $t_1$ to $t_2$ is the area under the curve for this interval, which can be approximated by summing the areas of the rectangles.

It is helpful to think of the integral sign $\int$ as an elongated $S$ indicating a sum. The limits $t_1$ and $t_2$ indicate the initial and final values of the variable $t$. The displacement is thus the area under the $v$-versus-$t$ curve. Figure 2-18 demonstrates that the average velocity has a simple geometric interpretation in terms of the "area under a curve."

To illustrate that the displacement equals the area under the $v$-versus-$t$ curve, consider what happens when you throw a golf ball straight up. The ball rises a meter or so, reverses direction, and then descends, gaining speed until you catch it. Assuming air resistance is negligible, the velocity of the ball is given by $v = v_0 + at$ (Equation 2-12), where the up direction is taken as positive and $a = -g$. Figure 2-19 is a plot of this velocity during the time that the ball is in free-fall. As shown, the velocity, which is initially positive, equals zero half way through the flight. As the ball descends, the velocity remains negative and, just before the ball is caught, reaches $-v_0$. During the rising portion of the motion, the area under the curve is positive, whereas during the descending portion it is negative. Thus, the total area under the curve for the entire flight is zero. It is easy to see that the displacement of the ball is also zero. Because the ball is thrown from the same place at which it is caught, the change in position is zero. Therefore, the displacement and the area under the $v$-versus-$t$ curve are equal because they both are zero.

The process of computing an integral is called **integration.** In Equation 2-18, $v$ is the derivative of $x$, and $x$ is the antiderivative of $v$. This is an example of the fundamental theorem of calculus, whose formulation in the seventeenth century greatly accelerated the mathematical development of physics.

$$\text{If } f(t) = \frac{dF(t)}{dt}, \quad \text{then} \quad F(t_2) - F(t_1) = \int_{t_1}^{t_2} f(t)\, dt \qquad \text{2-19}$$

FUNDAMENTAL THEOREM OF CALCULUS

The antiderivative of a function is also called the indefinite integral of the function and is written without limits on the integral sign, as in

$$x = \int v\, dt$$

Finding the function $x$ from the derivative $v$ (that is, finding the antiderivative) is also called integration. For example, if $v = v_0$, a constant, then

$$x = \int v_0\, dt = v_0 t + x_0$$

where $x_0$ is the arbitrary constant of integration. We can find a general rule for the integration of a power of $t$ from Equation 2-6, which gives the general rule for the derivative of a power. The result is

$$\int t^n\, dt = \frac{t^{n+1}}{n+1} + C, \qquad n \neq -1 \qquad \text{2-20}$$

where $C$ is an arbitrary constant. This can be checked by differentiating the right side using the rule of Equation 2-6. (For the special case $n = -1$, $\int t^{-1}\, dt = \ln t + C$, where $\ln t$ is the natural logarithm of $t$.)

The change in velocity for some time interval can similarly be interpreted as the area under the $a$-versus-$t$ curve for that interval. This is written

$$\Delta v = \lim_{\Delta t \to 0} \left( \sum_i a_i \Delta t_i \right) = \int_{t_1}^{t_2} a\, dt \qquad \text{2-21}$$

We can now derive the constant-acceleration equations by computing the indefinite integrals of the acceleration and velocity. If $a$ is constant, we have

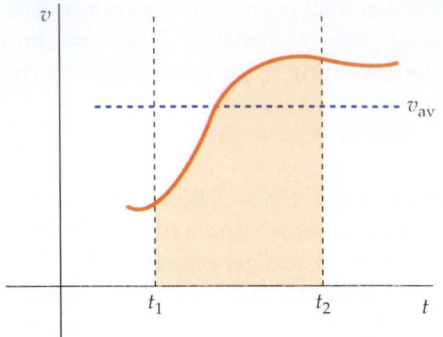

**FIGURE 2-18** The displacement $\Delta x$ during the time interval $\Delta t = t_2 - t_1$ is equal to the area of the shaded region. We know from the definition of average velocity that $\Delta x = v_{av}\, \Delta t$. This is just the area of a rectangle of height $v_{av}$ and width $\Delta t$. Thus, the rectangular area $v_{av}\, \Delta t$ and the area under the $v$-versus-$t$ curve must be equal.

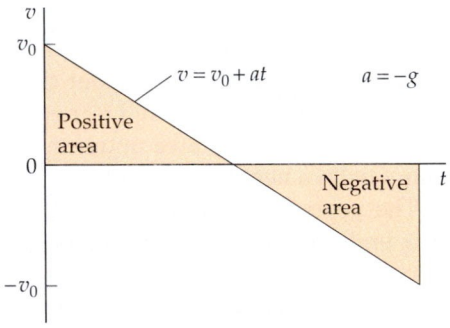

**FIGURE 2-19** A $v$-versus-$t$ curve for a golf ball thrown straight up that is caught by the thrower. The area under the curve is positive for the rising portion of the motion and negative for the descending portion. The area under the curve for the entire flight is zero.

$$v = \int a\, dt = a \int dt = v_0 + at \qquad\qquad 2\text{-}22$$

where we have expressed $a$ times the constant of integration as $v_0$. Integrating again, and writing $x_0$ for the constant of integration, gives

$$x = \int (v_0 + at)dt = x_0 + v_0 t + \tfrac{1}{2}at^2 \qquad\qquad 2\text{-}23$$

It is instructive to derive Equations 2-22 and 2-23 using definite integrals instead of indefinite ones. For constant acceleration, Equation 2-21, with $t_1 = 0$, gives

$$v(t_2) - v(0) = a \int_0^{t_2} dt = a(t_2 - 0)$$

where the time $t_2$ is arbitrary. Because it is arbitrary, we can set $t_2 = t$ to obtain

$$v = v_0 + at$$

where $v = v(t)$ and $v_0 = v(0)$. To derive Equation 2-23, we substitute $v_0 + at$ for $v$ in Equation 2-18 with $t_1 = 0$. This gives

$$x(t_2) - x(0) = \int_0^{t_2} (v_0 + at)dt$$

This integral is equal to the area under the $v$-versus-$t$ curve (Figure 2-20). Evaluating the integral and solving for $x$ gives

$$x(t_2) - x(0) = \int_0^{t_2} (v_0 + at)dt = v_0 t + \tfrac{1}{2}at^2 \Big|_0^{t_2} = v_0 t_2 + \tfrac{1}{2}at_2^2$$

where $t_2$ is arbitrary. Setting $t_2 = t$ we obtain

$$x = x_0 + v_0 t + \tfrac{1}{2}at^2$$

where $x = x(t)$ and $x_0 = x(0)$.

Having derived the constant-acceleration kinematic equations without any reference to average velocity, we can now show that the average velocity for the special case of constant acceleration is the mean value between the initial and final velocities as given by Equation 2-14. Let $v_0$ be the initial velocity at $t = 0$, and $v$ be the final velocity at time $t$. According to the definition of average velocity, the displacement is

$$\Delta x = v_{av}\Delta t = v_{av}(t - 0) = v_{av}t \qquad\qquad 2\text{-}24$$

Also, from Equation 2-23, we have

$$\Delta x = v_0 t + \tfrac{1}{2}at^2$$

We eliminate the acceleration using $a = (v - v_0)/t$ from Equation 2-12. This gives

$$\Delta x = v_0 t + \frac{1}{2}\left(\frac{v - v_0}{t}\right)t^2 = v_0 t + \frac{1}{2}vt - \frac{1}{2}v_0 t = \frac{1}{2}(v + v_0)t \qquad\qquad 2\text{-}25$$

Comparing this with $\Delta x = v_{av}t$ (Equation 2-24), we have

$$v_{av} = \tfrac{1}{2}(v_0 + v_f)$$

which is Equation 2-14.

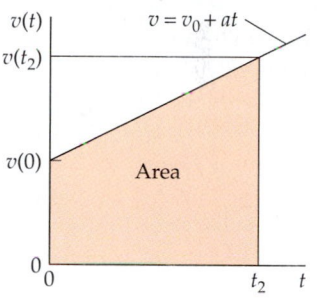

**FIGURE 2-20** The area under the $v$-versus-$t$ curve equals the displacement $\Delta x = x(t_2) - x(0)$.

The average velocity can be visualized using a $v$-versus-$t$ curve (Figure 2-21). The displacement $\Delta x$ equals the area under this curve. However, the average velocity is the area under the curve $v = v_{av}$ for the same time interval. Thus, the height of the $v = v_{av}$ curve is such that the areas under the two curves are equal. This implies that the areas of the two gray-shaded triangles are equal and that $v_{av} = \frac{1}{2}(v_1 + v_2)$.

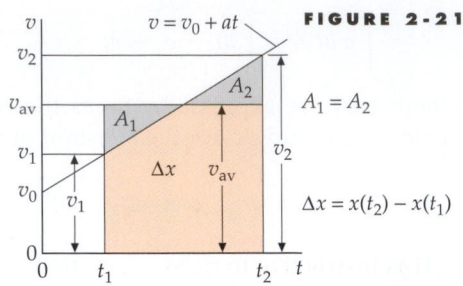

**FIGURE 2-21**

---

*A COASTING BOAT*

**EXAMPLE 2-19**

A Shelter Island ferryboat moves with constant velocity $v_0 = 8$ m/s for 60 s. It then shuts off its engines and coasts. Its coasting velocity is given by $v = v_0 t_1^2 / t^2$, where $t_1 = 60$ s. What is the displacement of the boat for the interval $0 < t < \infty$?

**PICTURE THE PROBLEM** This velocity function is shown in Figure 2-22. The total displacement is calculated as the sum of the displacement $\Delta x_1$ during the interval $0 < t < t_1 = 60$ s and the displacement $\Delta x_2$ during the interval $t_1 < t < \infty$.

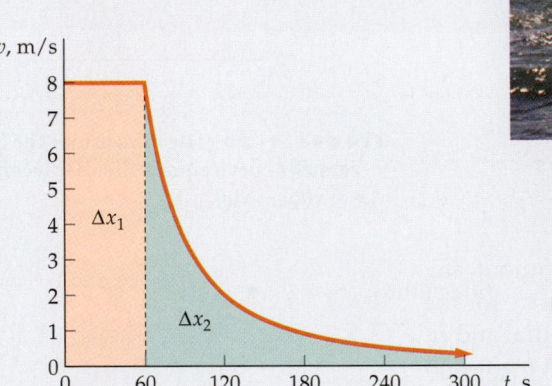

**FIGURE 2-22**

1. The velocity of the boat is constant during the first 60 s; thus the displacement is simply the velocity times the elapsed time:

$$\Delta x_1 = v_0 \Delta t = v_0 t_1 = (8 \text{ m/s})(60 \text{ s}) = 480 \text{ m}$$

2. The remaining displacement is given by the integral of the velocity from $t = t_1$ to $t = \infty$. We use Equation 2-18 to calculate the integral:

$$\Delta x_2 = \int_{t_1}^{\infty} v \, dt = \int_{t_1}^{\infty} \frac{v_0 t_1^2}{t^2} \, dt = v_0 t_1^2 \int_{t_1}^{\infty} t^{-2} \, dt = v_0 t_1^2 \left. \frac{t^{-1}}{-1} \right|_{t_1}^{\infty}$$

$$= -v_0 t_1^2 \left( \frac{1}{\infty} - \frac{1}{t_1} \right) = v_0 t_1 = (8 \text{ m/s})(60 \text{ s}) = 480 \text{ m}$$

3. The total displacement is the sum of the displacements found above:

$$\Delta x = \Delta x_1 + \Delta x_2 = 480 \text{ m} + 480 \text{ m} = \boxed{960 \text{ m}}$$

**REMARKS** Note that the area under the $v$-versus-$t$ curve is finite. Thus, even though the boat never stops moving, it travels only a finite distance. A better representation of the velocity of a coasting boat might be the exponentially decreasing function $v = v_0 e^{-b(t-t_1)}$, where $b$ is a positive constant. In that case, the boat would also coast a finite distance in the interval $t_1 \leq t \leq \infty$.

# SUMMARY

Displacement, velocity, and acceleration are important *defined* kinematics quantities.

| Topic | Relevant Equations and Remarks | |
|---|---|---|
| **1. Displacement** | $\Delta x = x_2 - x_1$ | 2-1 |
| Graphical interpretation | Displacement is the area under the $v$-versus-$t$ curve. | |
| **2. Velocity** | | |
| Average velocity | $v_{av} = \dfrac{\Delta x}{\Delta t}$ | 2-2 |
| Instantaneous velocity | $v(t) = \lim\limits_{\Delta t \to 0} \dfrac{\Delta x}{\Delta t} = \dfrac{dx}{dt}$ | 2-5 |
| Graphical interpretation | The instantaneous velocity is represented graphically as the slope of the $x$-versus-$t$ curve. | |
| Relative velocity | If a particle moves with velocity $v_{pA}$ relative to reference frame A, which is in turn moving with velocity $v_{AB}$ relative to a second reference frame B, the velocity of the particle relative to B is $$v_{pB} = v_{pA} + v_{AB}$$ | 2-7 |
| **3. Speed** | | |
| Average speed | $\text{Average speed} = \dfrac{\text{total distance}}{\text{total time}} = \dfrac{s}{t}$ | 2-3 |
| Instantaneous speed | Instantaneous speed is the magnitude of the instantaneous velocity. | |
| **4. Acceleration** | | |
| Average acceleration | $a_{av} = \dfrac{\Delta v}{\Delta t}$ | 2-8 |
| Instantaneous acceleration | $a = \dfrac{dv}{dt} = \dfrac{d^2x}{dt^2}$ | 2-10 |
| Graphical interpretation | The instantaneous acceleration is represented graphically as the slope of the $v$-versus-$t$ curve. | |
| Acceleration due to gravity | The acceleration of an object near the surface of the earth in free-fall under the influence of gravity is directed downward and has the magnitude $$g = 9.81 \text{ m/s}^2 = 32.2 \text{ ft/s}^2$$ | |
| **5. Displacement and velocity as integrals** | Displacement is represented graphically as the area under the $v$-versus-$t$ curve. This area is the integral of $v$ over time from some initial time $t_1$ to some final time $t_2$ and is written $$\Delta x = \lim_{\Delta t \to 0} \sum_i v_i \Delta t_i = \int_{t_1}^{t_2} v\, dt$$ | 2-18 |

Similarly, change in velocity is represented graphically as the area under the *a*-versus-*t* curve:

$$\Delta v = \lim_{\Delta t \to 0} \sum_i a_i \Delta t_i = \int_{t_1}^{t_2} a \, dt \qquad \text{2-21}$$

| | | |
|---|---|---|
| Velocity | $v = v_0 + at$ | 2-12 |
| Displacement in terms of $v_{av}$ | $\Delta x = x - x_0 = v_{av}t = \frac{1}{2}(v_0 + v)t$ | 2-15 |
| Displacement in terms of $a$ | $\Delta x = x - x_0 = v_0 t + \frac{1}{2}at^2$ | 2-16 |
| $v$ in terms of $a$ and $\Delta x$ | $v^2 = v_0^2 + 2a\,\Delta x$ | 2-17 |

# PROBLEMS

- Single-concept, single-step, relatively easy
- •• Intermediate-level, may require synthesis of concepts
- ••• Challenging
- **SSM** Solution is in the *Student Solutions Manual*
- **iSOLVE** Problems available on iSOLVE online homework service
- **iSOLVE✓** These "Checkpoint" online homework service problems ask students additional questions about their confidence level, and how they arrived at their answer

**In a few problems, you are given more data than you actually need; in a few other problems, you are required to supply data from your general knowledge, outside sources, or informed estimates.**

**For all problems, use $g = 9.81$ m/s² for the acceleration due to gravity and neglect friction and air resistance unless instructed to do otherwise.**

## Conceptual Problems

**1** • What is the average velocity over the "round trip" of an object that is launched straight up from the ground and falls straight back down to the ground?

**2** • **SSM** An object thrown straight up falls back to the ground. Its time of flight is $T$, its maximum height is $H$, and its height at release is negligible. Its average speed for the entire flight is (*a*) $H/T$, (*b*) 0, (*c*) $H/(2T)$, (*d*) $2H/T$.

**3** • **iSOLVE** To avoid falling too fast during a landing, an airplane must maintain a minimum airspeed (the speed of the plane relative to the air). However, the slower the ground speed (speed relative to the ground) during a landing, the safer the landing. Is it safer for an airplane to land with the wind or against the wind?

**4** • Give an example of one-dimensional motion where (*a*) the velocity is positive and acceleration is negative, and (*b*) the velocity is negative and the acceleration is positive.

**5** • **SSM** Stand in the center of a large room. Call movement to your right "positive," and movement to your left "negative." Walk across the room along a straight line in such a way that, after getting started, your velocity is negative but your acceleration is positive. (*a*) Is your displacement initially positive or negative? Explain. (*b*) Describe how you vary your speed as you walk. (*c*) Sketch a graph of *v* versus *t* for your motion.

**6** • True/false; explain: The displacement *always* equals the product of the average velocity and the time interval.

**7** • True/false; explain:
(*a*) For the velocity to remain constant, the acceleration *must* remain zero.
(*b*) For the speed to remain constant, the acceleration *must* remain zero.

**8** •• **SSM** Draw careful graphs of the position and velocity and acceleration over the time period $0 \le t \le 25$ s for a cart that

(*a*) moves away from the origin at a slow and steady (constant) velocity for the first 5 s;
(*b*) moves away at a medium-fast, steady (constant) velocity for the next 5 s;
(*c*) stands still for the next 5 s;
(*d*) moves toward the origin at a slow and steady (constant) velocity for the next 5 s;
(*e*) stands still for the last 5 s.

**9** • True/false; explain: The average velocity *always* equals one-half the sum of initial plus final velocities.

**10** • Identical twin brothers standing on a horizontal bridge each throw a rock straight down into the water below. They throw rocks at exactly the same time, but one hits the water before the other. How can this occur if the rocks have the same starting time?

**11** •• [SSM] Dr. Josiah S. Carberry stands at the top of the Sears Tower in Chicago. Wanting to emulate Galileo, and ignoring the safety of the pedestrians below, he drops a bowling ball from the top of the tower. One second later, he drops a second bowling ball. While the balls are in the air, does their separation (*a*) increase over time, (*b*) decrease, or (*c*) stay the same? Ignore any effects that air resistance may have.

**12** •• Which of the position-versus-time curves in Figure 2-23 best shows the motion of an object with constant positive acceleration?

**FIGURE 2-23** Problem 12

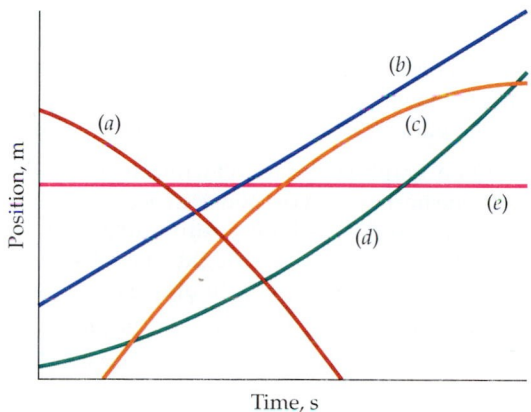

**13** • [SSM] Which of the velocity-versus-time curves in Figure 2-24 best describes the motion of an object with constant positive acceleration?

**FIGURE 2-24** Problem 13

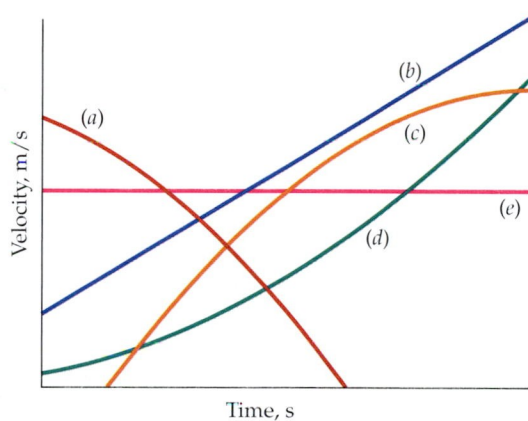

**14** • Does the following statement make sense? "The average velocity of the car at 9 A.M. was 60 km/h."

**15** • [SSM] Is it possible for the average velocity of an object to be zero during some interval, even though its average velocity for the first half of the interval is not zero? Explain.

**16** • The diagram in Figure 2-25 tracks the path of an object moving in a straight line along the *x* axis. Assuming that the object is at the origin ($x_o = 0$) at $t_o = 0$, which point on the position-versus-time graph represents the instant the object is farthest from the origin? (*a*) A (*b*) B (*c*) C (*d*) D (*e*) E.

**FIGURE 2-25** Problem 16

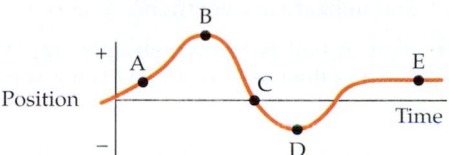

**17** • [iSOLVE] If the instantaneous velocity does not change, will the average velocities for different intervals differ?

**18** • If $v_{av} = 0$ for some time interval $\Delta t$, must the instantaneous velocity $v$ be zero at some point in the interval? Support your answer by sketching a possible *x*-versus-*t* curve that has $\Delta x = 0$ for some interval $\Delta t$.

**19** •• An object moves along a line as shown in Figure 2-26. At which point or points is its speed at a minimum? (*a*) A and E. (*b*) B, D, and E. (*c*) C only. (*d*) E only. (*e*) None of these is correct.

**FIGURE 2-26** Problem 19

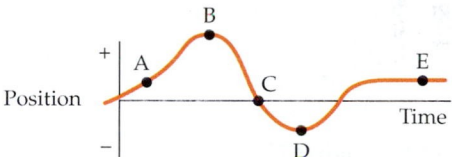

**20** •• [SSM] [iSOLVE] For each of the four graphs of *x* versus *t* in Figure 2-27, answer the following questions. (*a*) Is the velocity at time $t_2$ greater than, less than, or equal to the velocity at time $t_1$? (*b*) Is the speed at time $t_2$ greater than, less than, or equal to the speed at time $t_1$?

**FIGURE 2-27** Problem 20

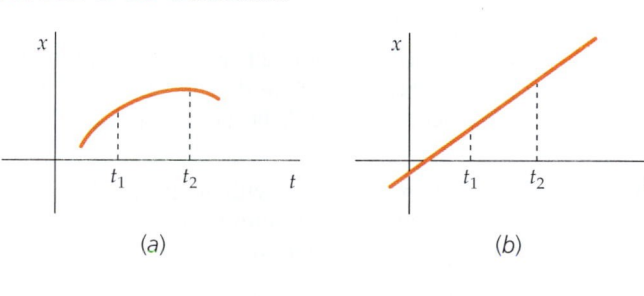

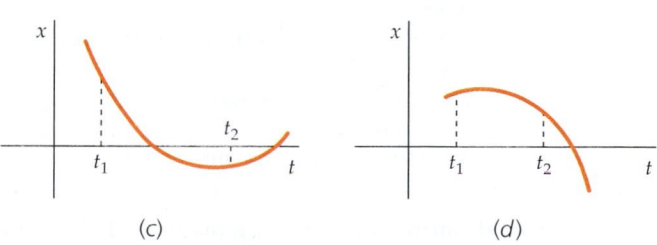

**21** • True/false; explain:

(*a*) If the acceleration remains zero, the body cannot be moving.
(*b*) If the acceleration remains zero, the *x*-versus-*t* curve must be a straight line.

**22** • Is it possible for a body to simultaneously have zero velocity and nonzero acceleration?

**23** • **ISOLVE** A ball is thrown straight up. What is the velocity of the ball at the top of its flight? What is its acceleration at that point?

**24** • Find the average speed over the "round trip" of an object that is launched straight up from the ground, reaches a height $H$, and falls straight back down to the ground, hitting it after $T$ seconds have elapsed. Express this in terms of the initial launch speed $v_0$.

**25** • A bowling ball is thrown upward. While it is in flight, its acceleration is (a) decreasing, (b) constant, (c) zero, (d) increasing.

**26** • At $t = 0$, object A is dropped from the roof of a building. At the same instant, object B is dropped from a window 10 m below the roof. During their descent to the ground, the distance between the two objects (a) is proportional to $t$, (b) is proportional to $t^2$, (c) decreases, (d) remains 10 m throughout.

**27** •• **SSM** Assume that the Porsche accelerates uniformly from 80.5 km/h (50 mi/h) at $t = 0$ to 113 km/h (70 mi/h) at $t = 9$ s. Which graph in Figure 2-28 best describes the motion of the car?

**FIGURE 2-28** Problem 27

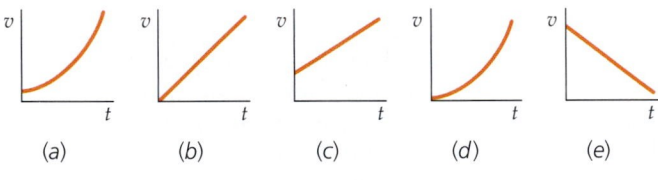

(a)    (b)    (c)    (d)    (e)

**28** •• **SSM** An object is dropped from rest and falls a distance $D$ in a given time. If the time during which it falls is doubled, the distance it falls will be (a) $4D$, (b) $2D$, (c) $D$, (d) $D/2$, (e) $D/4$.

**29** •• A ball is thrown upward with an initial velocity $v_0$. Its velocity halfway to its highest point is (a) $0.25v_0$, (b) $0.5v_0$, (c) $0.707v_0$, (d) $v_0$, (e) cannot be determined from the information given.

**30** • True or false:

(a) The equation $\Delta x = v_0 t + \frac{1}{2}at^2$ is valid for all particle motion in one dimension.
(b) If the velocity at a given instant is zero, the acceleration at that instant must also be zero.
(c) The equation $\Delta x = v_{av}\Delta t$ holds for all particle motion in one dimension.

**31** • **SSM** If an object is moving at constant acceleration in a straight line, its instantaneous velocity halfway through any time interval is (a) greater than its average velocity, (b) less than its average velocity, (c) equal to its average velocity, (d) half its average velocity, (e) twice its average velocity.

**32** • On a graph showing position on the vertical axis and time on the horizontal axis, a straight line with a negative slope represents motion with (a) constant positive acceleration, (b) constant negative acceleration, (c) zero velocity, (d) constant positive velocity, (e) constant negative velocity.

**33** •• On a graph showing position on the vertical axis and time on the horizontal axis, a parabola that opens upward represents (a) a positive acceleration, (b) a negative acceleration, (c) no acceleration, (d) a positive followed by a negative acceleration, (e) a negative followed by a positive acceleration.

**34** •• On a graph showing velocity on the vertical axis and time on the horizontal axis, a constant acceleration of zero is represented by (a) a straight line with positive slope, (b) a straight line with negative slope, (c) a straight line with zero slope, (d) either (a), (b), or (c), (e) none of the above.

**35** •• On a graph showing velocity on the vertical axis and time on the horizontal axis, constant acceleration is represented by (a) a straight line with positive slope, (b) a straight line with negative slope, (c) a straight line with zero slope, (d) either (a), (b), or (c), (e) none of the above.

**36** •• Which graph of $v$ versus $t$ in Figure 2-29 best describes the motion of a particle with positive velocity and negative acceleration?

**FIGURE 2-29** Problem 36

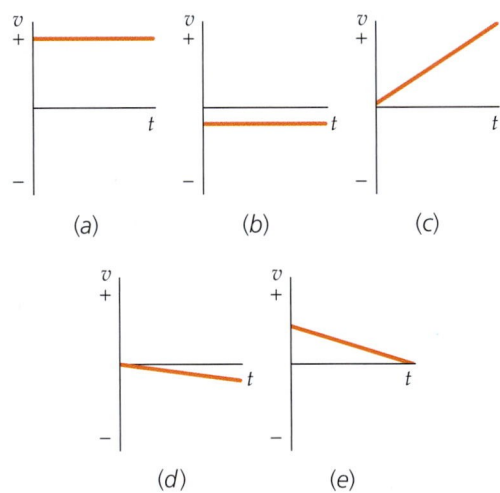

(a)    (b)    (c)

(d)    (e)

**37** •• **ISOLVE✔** Which graph of $v$ versus $t$ in Figure 2-29 best describes the motion of a particle with negative velocity and negative acceleration?

**38** •• A graph of the motion of an object is plotted with the velocity on the vertical axis and time on the horizontal axis. The graph is a straight line. Which of these quantities *can* be determined from this graph? (a) The displacement from time $t = 0$ to any other time shown. (b) The initial velocity at $t = 0$. (c) The acceleration as a function of time. (d) The average velocity for any specified time interval. (e) All of the above.

**39** •• **SSM** Figure 2-30 shows the position of a car plotted as a function of time. At which times $t_0$ to $t_7$ is the velocity (*a*) negative? (*b*) positive? (*c*) zero? At which times is the acceleration (*a*) negative? (*b*) positive? (*c*) zero?

**FIGURE 2-30** Problem 39

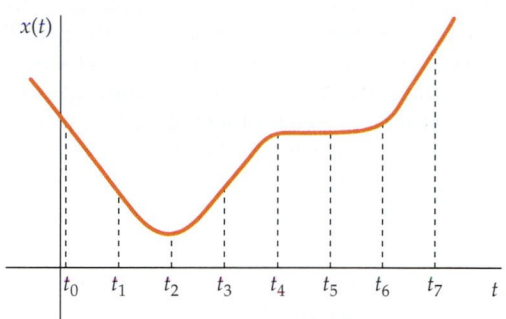

**40** •• Sketch *v*-versus-*t* curves for each of the following conditions: (*a*) Acceleration is zero and constant while velocity is not zero. (*b*) Acceleration is constant but not zero. (*c*) Velocity and acceleration are both positive. (*d*) Velocity and acceleration are both negative. (*e*) Velocity is positive and acceleration is negative. (*f*) Velocity is negative and acceleration is positive. (*g*) Velocity is momentarily zero but the acceleration is not zero.

**41** •• Figure 2-31 shows nine graphs of position, velocity, and acceleration for objects in motion along a straight line. Indicate the graphs that meet the following conditions: (*a*) velocity is constant, (*b*) velocity reverses its direction, (*c*) acceleration is constant, (*d*) acceleration is not constant. (*e*) Which graphs of position, velocity, and acceleration are mutually consistent?

**FIGURE 2-31** Problem 41

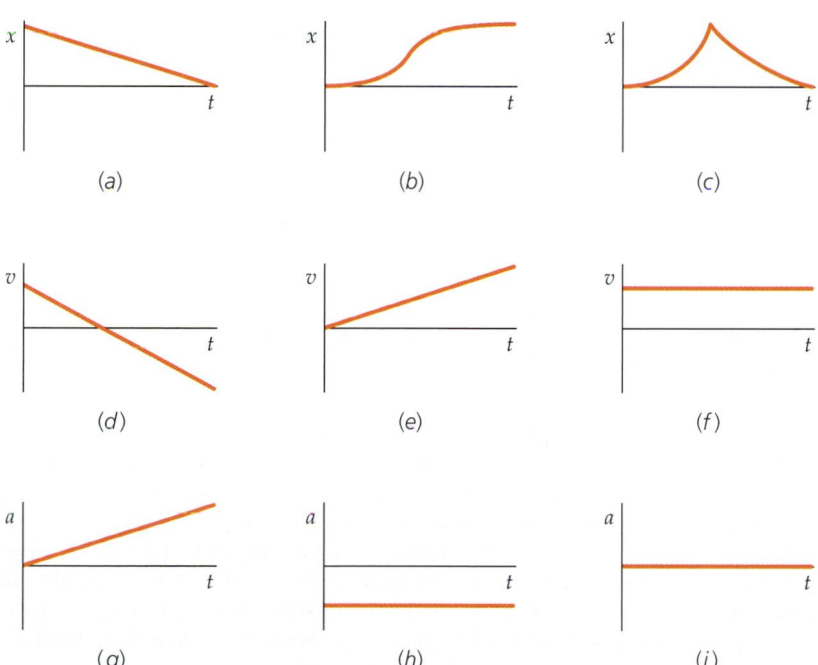

## Estimation and Approximation

**42** • Measure your own pulse rate (the number of heart beats per minute). Typical adult "resting rates" fall between 60 bpm (beats per min) and 80 bpm. (*a*) How many times will your heart beat during the time that it takes you to drive 1 mi at 60 mph? (*b*) How many times will your heart beat during your lifetime? (Assume a lifetime of 95 y.)

**43** •• **SSM** **iSOLVE** Occasionally, people can survive after falling large distances if the surface they fall on is soft enough. During a traverse of the Eiger's infamous Nordvand, mountaineer Carlos Ragone's rock anchor pulled out and he plummeted 500 ft to land in snow. Amazingly, he suffered only a few bruises and a wrenched shoulder. (*a*) What final speed did he reach before impact? Ignore air resistance. (*b*) Assuming that his impact left a hole in the snow 4 ft deep, estimate his acceleration as he slowed to a stop. Assume that the acceleration was constant. Express this as a multiple of *g* (the magnitude of free-fall acceleration at the surface of the earth).

**44** •• When we solve problems involving free-fall above the surface of the earth, it's important to remember that air resistance always exists; if we naively assume that objects always fall with constant acceleration, we may get answers that are wrong by orders of magnitude. How can we tell when it is valid to assume that a body is falling with (almost) constant acceleration? As a real body falls from rest through the air, as its speed increases, its acceleration downward decreases. The velocity will approach, but never quite reach, a *terminal velocity* that depends on the mass and cross-sectional area of the body; at the terminal velocity, the forces of gravity and air resistance exactly balance. For a "typical" skydiver falling through the air, a reasonable estimate for the terminal velocity is about 50 m/s (roughly 120 mph). At a speed of half the terminal velocity, the skydiver's acceleration will be $\frac{3}{4}g$. (*a*) Let's take half the terminal velocity as a reasonable "upper bound" beyond which we shouldn't use the constant acceleration formulas to calculate velocities and displacements. Roughly how far, and for how long, will the skydiver fall before we can't use these formulas anymore? (*b*) Repeat the analysis for a mouse, which has a terminal velocity of about 1 m/s.

**45** •• On June 16, 1999 Maurice Greene of the United States set a new world's record for the 100-m dash with a time $t = 9.79$ s. Suppose that he started from rest at constant acceleration *a* and reached his maximum velocity in 3.00 s, which he then kept until the finish line. What was the acceleration *a*?

**46** •• [SSM] The photograph in Figure 2-32 is a short-time exposure (1/30 s) of a juggler with two tennis balls in the air. The tennis ball near the top of its trajectory is less blurred than the lower one. Why is that? Can you estimate the speed of the lower ball from the picture?

**FIGURE 2-32** Problem 46

**47** •• Look up the speed at which a nerve impulse travels through the body. Estimate the time between stubbing your toe on a rock and feeling the pain due to this.

## Speed, Displacement, and Velocity

**48** • (a) An electron in a television tube travels the 16-cm distance from the grid to the screen at an average speed of $4 \times 10^7$ m/s. How long does the trip take? (b) An electron in a current-carrying wire travels at an average speed of $4 \times 10^{-5}$ m/s. How long does it take to travel 16 cm?

**49** • [SSM] A runner runs 2.5 km, in a straight line, in 9 min and then takes 30 min to walk back to the starting point. (a) What is the runner's average velocity for the first 9 min? (b) What is the average velocity for the time spent walking? (c) What is the average velocity for the whole trip? (d) What is the average speed for the whole trip?

**50** • [iSOLVE] A car travels in a straight line with an average velocity of 80 km/h for 2.5 h and then with an average velocity of 40 km/h for 1.5 h. (a) What is the total displacement for the 4-h trip? (b) What is the average velocity for the total trip?

**51** • One busy air route across the Atlantic Ocean is about 5500 km. (a) How long does it take for a supersonic jet flying at 2 times the speed of sound to make the trip? Use 340 m/s for the speed of sound. (b) How long does it take a subsonic jet flying at 0.9 times the speed of sound to make the same trip? (c) Allowing 2 h at each end of the trip for ground travel, check-in, and baggage handling, what is your average speed, door to door, when traveling on the supersonic jet? (d) What is your average speed taking the subsonic jet?

**52** • [SSM] The speed of light, c, is $3 \times 10^8$ m/s. (a) How long does it take for light to travel from the sun to the earth, a distance of $1.5 \times 10^{11}$ m? (b) How long does it take light to travel from the moon to the earth, a distance of $3.84 \times 10^8$ m? (c) A light-year is a unit of distance equal to that traveled by light in 1 year. Convert 1 light-year into kilometers and miles.

**53** • Proxima Centauri, a dim companion to Alpha Centauri, is $4.1 \times 10^{13}$ km away. From the vicinity of this star, Gregor places an order at Tony's Pizza in Hoboken, New Jersey, communicating via light signals. Tony's fastest delivery craft travels at $10^{-4}c$ (see Problem 52). (a) How long does it take Gregor's order to reach Tony's Pizza? (b) How long does Gregor wait between sending the signal and receiving the pizza? If Tony's has a 1000-years-or-it's-free delivery policy, does Gregor have to pay for the pizza?

**54** • A car making a 100-km journey travels 40 km/h for the first 50 km. How fast must it go during the second 50 km to average 50 km/h?

**55** •• [SSM] An archer fires an arrow, which produces a muffled "thwok" as it hits a target. If the archer hears the "thwok" exactly 1 s after firing the arrow and the average speed of the arrow was 40 m/s, what was the distance separating the archer and the target? Use 340 m/s for the speed of sound.

**56** •• John can run 6 m/s. Marcia can run 15% faster than John. (a) By what distance does Marcia beat John in a 100-m race? (b) By what time does Marcia beat John in a 100-m race?

**57** • [iSOLVE] Figure 2-33 shows the position of a particle as a function of time. Find the average velocities for the time intervals a, b, c, and d indicated in the figure.

**FIGURE 2-33** Problem 57

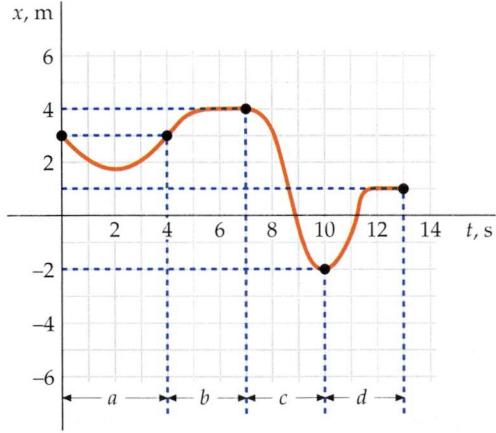

**58** •• It has been found that galaxies are moving away from the earth at a speed that is proportional to their distance from the earth. This discovery is known as Hubble's law. The speed of a galaxy at a distance r from the earth is given by $v = Hr$, where H is the Hubble constant, equal to $1.58 \times 10^{-18}$ s$^{-1}$. What is the speed of a galaxy (a) $5 \times 10^{22}$ m from earth and (b) $2 \times 10^{25}$ m from earth? (c) If each of these galaxies has traveled with constant speed, how long ago were they both located at the same place as the earth?

**59** •• SSM iSOLVE The cheetah can run as fast as $v_1 = 113$ km/h, the falcon can fly as fast as $v_2 = 161$ km/h, and the sailfish can swim as fast as $v_3 = 105$ km/h. The three of them run a relay with each covering a distance $L$ at maximum speed. What is the average speed $v$ of this relay team? Compare this with the average of the three speeds.

**60** •• Two cars are traveling along a straight road. Car A maintains a constant speed of 80 km/h; car B maintains a constant speed of 110 km/h. At $t = 0$, car B is 45 km behind car A. How much farther will car A travel before it is overtaken by car B?

**61** •• SSM A car traveling at a constant speed of 20 m/s passes an intersection at time $t = 0$, and 5 s later another car traveling at a constant speed of 30 m/s passes the same intersection in the same direction. (a) Sketch the position functions $x_1(t)$ and $x_2(t)$ for the two cars. (b) Determine when the second car will overtake the first. (c) How far from the intersection will the two cars be when they pull even? (d) Where is the first car when the second car passes the intersection?

**62** • Joe and Sally tend to argue when they travel. Just as they reached the moving sidewalk at the airport, their tempers flared to a point where neither was talking to the other. Though they stepped on the moving belt at the same time, Joe chose to stand and ride, while Sally opted to keep walking. Sally reached the end in 1 min, while Joe took 2 min. How long would it have taken Sally if she had walked twice as fast relative to the moving belt?

**63** •• Margaret has just enough gas in her speedboat to get to the marina, an upstream journey that takes 4.0 h. Finding it closed for the season, she spends the next 8.0 h floating back downstream (out of gas) to her shack. The entire trip took 12.0 h. How long would it have taken if she had bought gas at the marina? Assume that the effect of the wind is negligible.

## Acceleration

**64** • iSOLVE A BMW-M3 sports car can accelerate in third gear from 48.3 km/h (30 mi/h) to 80.5 km/h (50 mi/h) in 3.7 s. (a) What is the average acceleration of this car in m/s²? (b) If the car continued at this acceleration for another second, how fast would it be moving?

**65** • At $t = 5$ s, an object at $x = 3$ m has a velocity of $+5$ m/s. At $t = 8$ s, it is at $x = 9$ m and its velocity is $-1$ m/s. Find the average acceleration for this interval.

**66** •• A particle moves with velocity $v = (8$ m/s²$) t - 7$ m/s. (a) Find the average acceleration for two 1-s intervals, one beginning at $t = 3$ s and the other beginning at $t = 4$ s. (b) Sketch $v$ versus $t$. What is the instantaneous acceleration at any time?

**67** •• iSOLVE ✔ The position of a certain particle depends on time according to the equation $x(t) = t^2 - 5t + 1$, where $x$ is in meters if $t$ is in seconds. (a) Find the displacement and average velocity for the interval $3$ s $\leq t \leq 4$ s. (b) Find the general formula for the displacement for the time interval from $t$ to $t + \Delta t$. (c) Use the limiting process to obtain the instantaneous velocity for any time $t$.

**68** •• SSM iSOLVE The position of an object is related to time by $x = At^2 - Bt + C$, where $A = 8$ m/s², $B = 6$ m/s, and $C = 4$ m. Find the instantaneous velocity and acceleration as functions of time.

**69** •• The one-dimensional motion of a particle is plotted in Figure 2-34. (a) What is the average acceleration in the intervals AB, BC, and CE? (b) How far is the particle from its starting point after 10 s? (c) Sketch the displacement of the particle as a function of time; label the instants A, B, C, D, and E on your figure. (d) At what time is the particle traveling most slowly?

**FIGURE 2-34** Problem 69

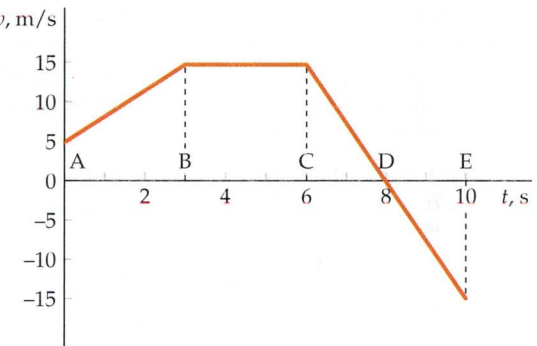

## Constant Acceleration and Free-Fall

**70** • SSM An object projected upward with initial velocity $v_0$ attains a height $h$. Another object projected up with initial velocity $2v_0$ will attain a height of (a) $4h$, (b) $3h$, (c) $2h$, (d) $h$.

**71** • A car starting at $x = 50$ m accelerates from rest at a constant rate of 8 m/s². (a) How fast is it going after 10 s? (b) How far has it gone after 10 s? (c) What is its average velocity for the interval $0 \leq t \leq 10$ s?

**72** • An object with an initial velocity of 5 m/s has a constant acceleration of 2 m/s². When its speed is 15 m/s, how far has it traveled?

**73** • SSM An object with constant acceleration has a velocity of 10 m/s when it is at $x = 6$ m and of 15 m/s when it is at $x = 10$ m. What is its acceleration?

**74** • The speed of an object increases at a constant rate of 4 m/s each second. At $t = 0$, its velocity is 1 m/s and its position is $x = 7$ m. How fast is it moving when it is at $x = 8$ m and what does $t$ equal then?

**75** •• iSOLVE ✔ A ball is thrown upward with an initial velocity of 20 m/s. (a) How long is the ball in the air? (Neglect the height of the release point.) (b) What is the greatest height reached by the ball? (c) How long after release is the ball 15 m above the release point?

**76** •• [ISOLVE] ✔ In the Blackhawk landslide in California, a mass of rock and mud fell 460 m down a mountain and then traveled 8 km across a level plain. It has been theorized that the rock and mud moved on a cushion of water vapor. Assume that the mass dropped with the free-fall acceleration and then slid horizontally, losing speed at a constant rate. (a) How long did the mud take to drop the 460 m? (b) How fast was it traveling when it reached the bottom? (c) How long did the mud take to slide the 8 km horizontally?

**77** •• [SSM] A load of bricks is being lifted by a crane at a steady velocity of 5 m/s when one brick falls off 6 m above the ground. (a) Sketch $x(t)$ to show the motion of the free brick. (b) What is the greatest height the brick reaches above the ground? (c) How long does it take to reach the ground? (d) What is its speed just before it hits the ground?

**78** •• A bolt comes loose from underneath an elevator that is moving upward at a speed of 6 m/s. The bolt reaches the bottom of the elevator shaft in 3 s. (a) How high up is the elevator when the bolt comes loose? (b) What is the speed of the bolt when it hits the bottom of the shaft?

**79** •• [SSM] An object is dropped from rest at a height of 120 m. Find the distance it falls during its final second in the air.

**80** •• An object is released from rest at a height $h$. During the final second of its fall, it traverses a distance of 38 m. What is $h$?

**81** • [SSM] A stone is thrown vertically from a 200-m-tall cliff. During the last half second of its flight the stone travels a distance of 45 m. Find the initial speed of the stone.

**82** •• An object is released from rest at a height $h$. It travels 0.4$h$ during the first second of its descent. Determine the average velocity of the object during its entire descent.

**83** •• A bus accelerates at 1.5 m/s² from rest for 12 s. It then travels at constant velocity for 25 s, after which it slows to a stop with an acceleration of −1.5 m/s². (a) How far does the bus travel? (b) What is its average velocity?

**84** •• [SSM] It is relatively easy to use a spreadsheet program such as Microsoft Excel to solve certain types of physics problems. For example, you probably solved Problem 75 using algebra. Let's solve Problem 75 in a different way, this time using a spreadsheet program. While we can solve this problem using algebra, there are many places in physics where we can't get an alternative solution so easily, and have to rely on numerical methods like the one shown here. (a) Using Microsoft Excel or some other spreadsheet program, generate a graph of the height versus time for the ball in Problem 75 (thrown upward with an initial velocity of 20 m/s). Determine the maximum height, the time it was in the air, and the time(s) when the ball is 15 m above the ground by inspection (i.e., look at the graph and find them.) (b) Now change the initial velocity to 10 m/s and find the maximum height the ball reaches and the time the ball spends in the air.

**85** •• [SSM] [ISOLVE] Al and Bert are jogging side-by-side on a trail in the woods at a speed of 0.75 m/s. Suddenly Al

sees the end of the trail 35 m ahead and decides to speed up to reach it. He accelerates at a constant rate of 0.5 m/s², leaving Bert behind, who continues on at a constant speed. (a) How long does it take Al to reach the end of the trail? (b) Once he reaches the end of the trail, he immediately turns around and heads back along the trail with a constant speed of 0.85 m/s. How long does it take him to reach Bert? (c) How far are they from the end of the trail when they meet?

**86** •• Solve Problem 85, parts (b) and (c), using a spreadsheet program.

**87** •• A rocket is fired vertically with an upward acceleration of 20 m/s². After 25 s, the engine shuts off and the rocket then continues rising (while in free-fall). The rocket eventually stops rising and then falls back to the ground. Calculate (a) the highest point the rocket reaches, (b) the total time the rocket is in the air, (c) the speed of the rocket just before it hits the ground.

**88** •• [ISOLVE] A flowerpot falls from the ledge of an apartment building. A person in an apartment below, coincidentally holding a stopwatch, notices that it takes 0.2 s for the pot to fall past his window, which is 4 m high. How far above the top of the window is the ledge from which the pot fell?

**89** •• [SSM] In a classroom demonstration, a glider moves along an inclined air track with constant acceleration. It is projected from the start of the track with an initial velocity. After 8 s have elapsed, it is 100 cm from its starting point and is moving along the track at a velocity of −15 cm/s. Find the initial velocity and the acceleration.

**90** •• A rock dropped from a cliff falls one-third of its total distance to the ground in the last second of its fall. How high is the cliff?

**91** •• A typical automobile under hard braking loses speed at a rate of about 7 m/s²; the typical reaction time to engage the brakes is 0.50 s. A school board sets the speed limit in a school zone to meet the condition that all cars should be able to stop in a distance of 4 m. (a) What maximum speed should be allowed for a typical automobile? (b) What fraction of the 4 m is due to the reaction time?

**92** •• Two trains face each other on adjacent tracks. They are initially at rest, and their front ends are 40 m apart. The train on the left accelerates rightward at 1.4 m/s². The train on the right accelerates leftward at 2.2 m/s². How far does the train on the left travel before the front ends of the trains pass?

**93** •• Two stones are dropped from the edge of a 60-m cliff, the second stone 1.6 s after the first. How far below the top of the cliff is the second stone when the separation between the two stones is 36 m?

**94** •• [SSM] A motorcycle officer hidden at an intersection observes a car that ignores a stop sign, crosses the intersection, and continues on at constant speed. The police officer takes off in pursuit 2.0 s after the car has passed the stop sign, accelerates at 6.2 m/s² until her speed is 110 km/h, and then continues at this speed until she catches the car. At that instant, the car is 1.4 km from the intersection. How fast was the car traveling?

**95** •• ▌**SOLVE** At $t = 0$, a stone is dropped from the top of a cliff above a lake. Another stone is thrown downward 1.6 s later from the same point with an initial speed of 32 m/s. Both stones hit the water at the same instant. Find the height of the cliff.

**96** ••• ▌**SOLVE** A passenger train is traveling at 29 m/s when the engineer sees a freight train 360 m ahead of his train traveling in the same direction on the same track. The freight train is moving at a speed of 6 m/s. (*a*) If the reaction time of the engineer is 0.4 s, what is the minimum rate at which the passenger train must lose speed if a collision is to be avoided? (*b*) If the engineer's reaction time is 0.8 s and the train loses speed at the minimum rate described in part (*a*), what is the relative speed with which the two trains collide? How far will the passenger train have traveled in the time between the sighting of the freight train and the collision?

**97** • Intent on studying the effects of gravity near the surface of the earth, a student launches a small projectile straight up with a speed of 300 m/s. How high will the projectile rise? (Ignore air resistance.)

**98** • ▌**SSM** At the end of *Charlie and the Chocolate Factory*, Willie Wonka presses a button that shoots the great glass elevator through the roof of his chocolate factory. (*a*) If the elevator reaches a maximum height of 10 km above the roof, what was its speed immediately after crashing through the roof? Ignore air resistance, even though in this case it makes little sense to ignore it. (*b*) Assume that the elevator's speed just after it crashes through the roof was half of what it was just before its impact with the roof. Assuming that it started from rest on the ground floor of the chocolate factory, and that the height of the roof is 150 m above the ground floor, what uniform acceleration is needed for it to reach this high speed?

**99** •• The click beetle can project itself vertically with an acceleration of about $a = 400g$ (an order of magnitude more than a human could stand). The beetle jumps by "unfolding" its legs, which are about $d = 0.6$ cm long. How high can the click beetle jump? How long is the beetle in the air? (Assume constant acceleration while in contact with the ground, and neglect air resistance.)

**100** • ▌**SOLVE**✓ A test of the prototype of a new automobile shows that the minimum distance for a controlled stop from 98 km/h to zero is 50 m. Find the acceleration, assuming it to be constant, and express your answer as a fraction of the free-fall acceleration. How much time does the car take to stop?

**101** •• ▌**SSM** Consider the motion of a particle that experiences free-fall with a constant acceleration. Before the advent of computer-driven data-logging software, we used to do a free-fall experiment in which a coated tape was placed vertically next to the path of a dropped conducting puck. A high-voltage spark generator would cause an arc to jump between two vertical wires through the falling puck and through the tape, thereby marking the tape at fixed time intervals $\Delta t$. Show that the change in height in successive time intervals for an object falling from rest follows *Galileo's Rule of Odd Numbers*: $\Delta y_{21} = 3\Delta y_{10}$, $\Delta y_{32} = 5\Delta y_{10}$, . . . , where $\Delta y_{10}$ is the change in $y$ during the first interval of duration $\Delta t$, $\Delta y_{21}$ is the change in $y$ during the second interval of duration $\Delta t$, etc.

**102** •• A particle moves with a constant acceleration of 3 m/s². At a time of 4 s, it is at a position of 100 m with respect to some coordinate system; at a time of 6 s, it has a velocity of 15 m/s. Find its position at a time of 6 s.

**103** •• ▌**SSM** ▌**SOLVE**✓ A plane landing on a small tropical island has just 70 m of runway on which to stop. If its initial speed is 60 m/s, (*a*) what is the minimum acceleration of the plane during landing, assuming it to be constant? (*b*) How long does it take for the plane to stop with this acceleration?

**104** •• ▌**SOLVE**✓ An automobile accelerates from rest at 2 m/s² for 20 s. The speed is then held constant for 20 s, after which there is an acceleration of $-3$ m/s² until the automobile stops. What is the total distance traveled?

**105** •• ▌**SSM** ▌**SOLVE** If it were possible for a spacecraft to maintain a constant acceleration indefinitely, trips to the planets of the Solar System could be undertaken in days or weeks, while voyages to the nearer stars would only take a few years. (*a*) Show that $g$, the magnitude of free-fall acceleration on earth, is approximately $1 \ c \cdot y/y^2$. (See Problem 52 for the definition of a light-year.) (*b*) Using data from the tables at the back of the book, find the time it would take for a one-way trip from Earth to Mars (at Mars' closest approach to Earth). Assume that the spacecraft starts from rest, travels along a straight line, accelerates halfway at $1g$, and then flips around and decelerates at $1g$ for the rest of the trip.

**106** • ▌**SOLVE**✓ The Stratosphere Tower in Las Vegas is 1137 ft high. It takes 1 min, 20 s to ascend from the ground floor to the top of the tower using the high-speed elevator. Assuming that the elevator starts and ends at rest, and its acceleration has a constant *magnitude* when moving, find the acceleration of the elevator. Express it in terms of a multiple of the acceleration of gravity.

**107** •• ▌**SOLVE**✓ A train pulls away from a station with a constant acceleration of 0.4 m/s². A passenger arrives at the track 6.0 s after the end of the train has passed the very same point. What is the slowest constant speed at which she can run and catch the train? Sketch curves for the motion of the passenger and the train as functions of time.

**108** ••• Ball A is dropped from the top of a building at the same instant that ball B is thrown vertically upward from the ground. When the balls collide, they are moving in opposite directions, and the speed of A is twice the speed of B. At what fraction of the height of the building does the collision occur?

**109** ••• Solve Problem 108 if the collision occurs when the balls are moving in the same direction and the speed of A is 4 times that of B.

**110** •• ▌**SSM** Starting at one station, a subway train accelerates from rest at a constant rate of 1.0 m/s² for half the distance to the next station, then slows down at the same rate for the second half of the journey. The total distance between stations is 900 m. (*a*) Sketch a graph of the velocity $v$ as a function of time over the full journey. (*b*) Sketch a graph of the distance covered as a function of time over the full journey. Place appropriate numerical values on both axes.

**111** •• **ISOLVE** A speeder traveling at a constant speed of 125 km/h races past a billboard. A patrol car pursues from rest with constant acceleration of (8 km/h)/s until it reaches its maximum speed of 190 km/h, which it maintains until it catches up with the speeder. (*a*) How long does it take the patrol car to catch the speeder if it starts moving just as the speeder passes? (*b*) How far does each car travel? (*c*) Sketch $x(t)$ for each car.

**112** •• When the patrol car in Problem 111 (traveling at 190 km/h), pulls within 100 m behind the speeder (traveling at 125 km/h), the speeder sees the police car and slams on his brakes, locking the wheels. (*a*) Assuming that each car can brake at 6 m/s² and that the driver of the police car brakes instantly as she sees the brake lights of the speeder (reaction time = 0 s), show that the cars collide. (*b*) At what time after the speeder applies his brakes do the two cars collide? (*c*) Discuss how reaction time affects this problem.

**113** •• Urgently needing the cash prize, Lou enters the Rest-to-Rest auto competition, in which each contestant's car begins and ends at rest, covering a distance $L$ in as short a time as possible. The intention is to demonstrate mechanical and driving skills, and to consume the largest amount of fossil fuels in the shortest time possible. The course is designed so that maximum speeds of the cars are never reached. (*a*) If Lou's car has a maximum acceleration of $a$ and a maximum deceleration of $2a$, then at what fraction of $L$ should Lou move his foot from the gas pedal to the brake? (*b*) What fraction of the time for the trip has elapsed at that point?

**114** •• **ISOLVE✓** A physics professor demonstrates her new "anti-gravity parachute" by exiting from a helicopter at an altitude of 575 m with zero initial velocity. For 8 s, she falls freely. Then she switches on the "parachute" and her rate of descent slows at a constant rate of 15 m/s² until her downward speed reaches 5 m/s, whereupon she adjusts her controls to maintain that speed until she reaches the ground. (*a*) On a single graph, sketch her acceleration and velocity as functions of time. (Take upward to be positive.) (*b*) What is her speed at the end of the first 8 s? (*c*) For how long is she losing speed? (*d*) How far does she travel while losing speed? (*e*) How much time is required for the entire trip from the helicopter to the ground? (*f*) What is her average velocity for the entire trip?

## Integration of the Equations of Motion

**115** • **SSM** The velocity of a particle is given by $v(t) =$ (6 m/s²)$t$ + (3 m/s). (*a*) Sketch $v$ versus $t$ and find the area under the curve for the interval $t = 0$ to $t = 5$ s. (*b*) Find the position function $x(t)$. Use it to calculate the displacement during the interval $t = 0$ to $t = 5$ s.

**116** • Figure 2-35 shows the velocity of a particle versus time. (*a*) What is the magnitude in meters represented by the area of the shaded box? (*b*) Estimate the displacement of the particle for the two 1-s intervals beginning at $t = 1$ s and at $t = 2$ s. (*c*) Estimate the average velocity for the interval

1 s $\leq t \leq$ 3 s. (*d*) The equation of the curve is $v = (0.5$ m/s³)$t^2$. Find the displacement of the particle for the interval 1 s $\leq t \leq$ 3 s by integration and compare this answer with your answer for part (*b*). Is the average velocity equal to the mean of the initial and final velocities for this case?

**FIGURE 2-35** Problem 116

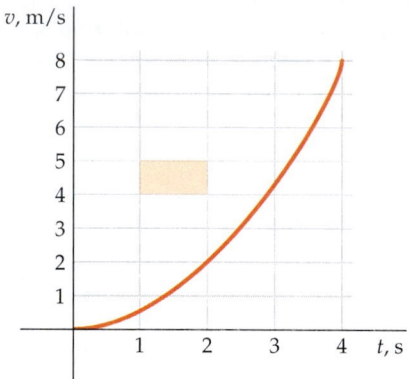

**117** •• **SSM** The velocity of a particle is given by $v = (7$ m/s³)$t^2$ − 5 m/s, where $t$ is in seconds and $v$ is in meters per second. If the particle starts from the origin, $x_0 = 0$, at $t_0 = 0$, find the general position function $x(t)$.

**118** •• Consider the velocity graph in Figure 2-36. Assuming $x = 0$ at $t = 0$, write correct algebraic expressions for $x(t)$, $v(t)$, and $a(t)$ with appropriate numerical values inserted for all constants.

**FIGURE 2-36** Problem 118

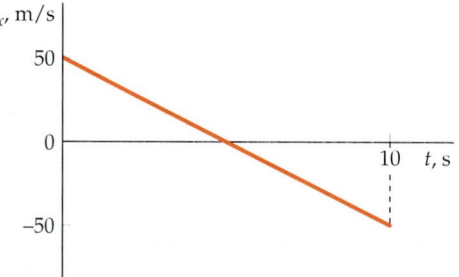

**119** ••• Figure 2-37 shows the acceleration of a particle versus time. (*a*) What is the magnitude, in m/s, of the area of the shaded box? (*b*) The particle starts from rest at $t = 0$. Estimate the velocity at $t = 1$ s, 2 s, and 3 s by counting the boxes under the curve. (*c*) Sketch the curve $v$ versus $t$ from your results for part (*b*), and using it, estimate how far the particle travels in the interval $t = 0$ to $t = 3$ s.

**FIGURE 2-37** Problem 119

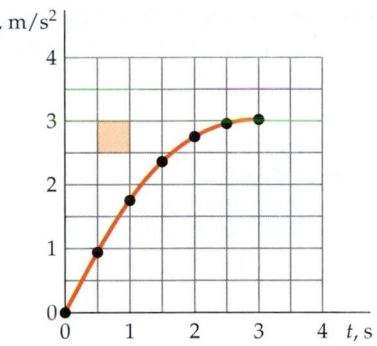

**120** •• Figure 2-38 is a graph of $v$ versus $t$ for a particle moving along a straight line. The position of the particle at time $t_0 = 0$ is $x_0 = 5$ m. (a) Find $x$ for various times $t$ by counting boxes, and sketch $x$ as a function of $t$. (b) Sketch a graph of the acceleration $a$ as a function of the time $t$.

**FIGURE 2-38** Problem 120

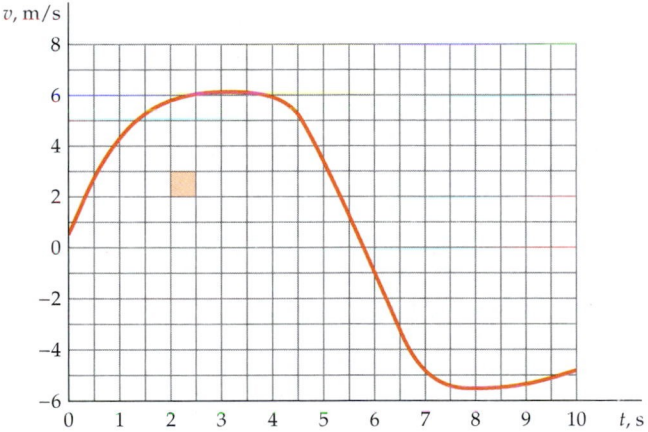

**121** •• [SSM] Figure 2-39 shows a plot of $x$ versus $t$ for a body moving along a straight line. Sketch rough graphs of $v$ as a function of $t$ and $a$ as a function of $t$ for this motion.

**FIGURE 2-39** Problem 121

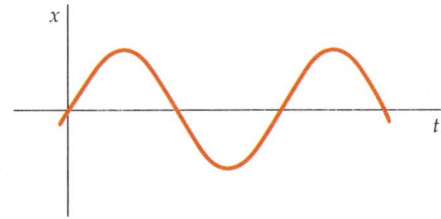

**122** •• The acceleration of a certain rocket is given by $a = bt$, where $b$ is a positive constant. (a) Find the general position function $x(t)$. (b) Find the position and velocity at $t = 5$ s if $x = 0$ and $v = 0$ at $t = 0$ and $b = 3$ m/s³.

**123** •• [iSOLVE] In the time interval from 0.0 s to 10.0 s, the acceleration of a particle is given by $a = (0.20$ m/s³)$t$ for one-dimensional motion. If a particle starts from rest at the origin, (a) calculate first its *instantaneous velocity* at any time during the interval, then (b) calculate its *average velocity* over the time interval from 2.0 s to 7.0 s.

**124** • Consider the motion of a particle that experiences a nonconstant acceleration $a$ given by $a = a_0 + bt$, where $a_0$ and $b$ are constant. (a) Find the instantaneous velocity as a function of time. (b) Find the position as a function of time. (c) Find the average velocity over the time interval with an initial time of zero and arbitrary final time $t$.

## General Problems

**125** ••• [iSOLVE] The following apparatus is used in a science class to determine the free-fall acceleration. Two photogates are set up, one at the edge of a table exactly 1.0 m high, the second one directly below it at a height of 0.5 m. A marble is dropped through the photogates from the height of 1 m above the floor (i.e., exactly the same as the height of the table), starting a timer when it enters the first photogate and stopping it when it enters the second one. The magnitude of free-fall acceleration is determined by $g_{exp} = (1$ m)$/(\Delta t)^2$, where $\Delta t$ is the time measured by the timer. A careless student sets up the first photogate 0.5 cm below the edge of the table. (Assume that the second photogate is properly placed at a height of 0.5 m.) What value of $g_{exp}$ will he determine? What percentage difference does this represent from the accepted value at sea level?

**126** ••• [SSM] The position of a body oscillating on a spring is given by $x = A \sin \omega t$, where $A$ and $\omega$ are constants with values $A = 5$ cm and $\omega = 0.175$ s⁻¹. (a) Plot $x$ as a function of $t$ for $0 \le t \le 36$ s. (b) Measure the slope of your graph at $t = 0$ to find the velocity at this time. (c) Calculate the average velocity for a series of intervals beginning at $t = 0$ and ending at $t = 6, 3, 2, 1, 0.5$, and 0.25 s. (d) Compute $dx/dt$ and find the velocity at time $t = 0$. (e) Compare your results in parts (c) and (d).

**127** ••• [iSOLVE] Consider an object that is attached to a driving motor so that the object moves with a velocity given by $v = v_{max} \sin(\omega t)$, where $\omega$ is in radians/s. (a) What is the acceleration of the object and is it constant? (b) At $t = 0$, the position is $x_0$. What is the position as a function of time?

**128** ••• Suppose the acceleration of a particle is a function of $x$, where $a(x) = (2$ s⁻²$)$ $x$. (a) If the velocity at $x = 1$ m is zero, what is the speed of the particle at $x = 3$ m? (b) How long does it take the particle to travel from $x = 1$ m to $x = 3$ m?

**129 •••** Suppose that a particle moves in a straight line such that, at each instant of time, its position and velocity have the same numerical value if expressed in SI units. (*a*) Express the position $x$ as a function of time $t$. (*b*) Show that at each instant of time the acceleration has the same numerical value as the position and velocity.

**130 •••** A small rock sinking through water experiences an exponentially decreasing acceleration as a function of time given by $a(t) = ge^{-bt}$, where $b$ is a positive constant that depends on the shape and size of the rock and the physical properties of the water. Based upon this result, derive an expression for the position of the rock as a function of time. Assume that its initial velocity is 0.

**131 •••** **SSM** In Problem 130, a rock falls through water with a continuously decreasing acceleration of the form $a(t) = ge^{-bt}$, where $b$ is a positive constant. In physics, we are not often given acceleration directly as a function of time, but usually either as a function of position or of velocity. Assume that the rock's acceleration as a function of *velocity* has the form $a = g - bv$ where $g$ is the magnitude of free-fall acceleration and $v$ is the rock's speed. Prove that if the rock has an initial velocity $v = 0$ at time $t = 0$, it will have the dependence on *time* given above.

**132 •••** The acceleration of a skydiver jumping from an airplane is given by the formula $a = g - cv^2$, where $c$ is a constant depending on the skydiver's cross-sectional area and the density of the surrounding atmosphere she is diving through. (*a*) If her initial speed is 0 when jumping from the plane, show that her speed as a function of time is given by the formula $v(t) = v_r \tanh(t/T)$, where $v_T$ is the terminal velocity $(v_T = \sqrt{g/c})$ and $T = v_T/g$ is a time scale determining very roughly the time it takes for her speed to approach $v_T$. (*b*) Use a spreadsheet program such as Microsoft Excel to graph $v(t)$ as a function of time, using a terminal velocity of 56 m/s (use this to calculate $c$ and $T$). Does the resulting curve make sense?

# Motion in Two and Three Dimensions

SAILBOATS DON'T TRAVEL IN STRAIGHT LINES TO THEIR DESTINATIONS, BUT INSTEAD MUST "TACK" BACK AND FORTH ACROSS THE WIND. THIS BOAT MUST SAIL EAST, THEN SOUTH, AND THEN EAST AGAIN, IN ITS JOURNEY TO A SOUTHEASTERN PORT.

 **How can we calculate its displacement and its average velocity? (See Example 3-3.)**

3-1   The Displacement Vector

3-2   General Properties of Vectors

3-3   Position, Velocity, and Acceleration

3-4   Special Case 1: Projectile Motion

3-5   Special Case 2: Circular Motion

**H**ere we extend the ideas of Chapter 2 to two and three dimensions. To do this, we introduce vectors and show how they are used to analyze and describe motion.

➤ Our major goal in this chapter is to develop the concept of the acceleration vector, a concept that is central to the development of Newton's laws in Chapters 4 and 5.

## 3-1 The Displacement Vector

When motion occurs, the displacement of a particle has a direction in space as well as a magnitude. The quantity that gives the straight-line distance between two points in space as well as the direction is a line segment called the **displacement vector.** It is represented graphically by an arrow whose direction is the same as that of the displacement vector and whose length is proportional to the magnitude of the displacement vector. We denote vectors by italic boldface letters with an overhead arrow, $\vec{A}$. The magnitude of $\vec{A}$ is written $|\vec{A}|$, $\|\vec{A}\|$, or simply $A$, and has dimensions of length. The magnitude of a vector is never negative.

## Addition of Displacement Vectors

Figure 3-1 shows the path of a particle that moves from point $P_1$ to a second point $P_2$ and then to a third point $P_3$. The vector $\vec{A}$ represents the displacement from $P_1$ to $P_2$, while $\vec{B}$ represents the displacement from $P_2$ to $P_3$. Note that the displacement vectors depend only on the end points and not on the actual path of the particle. The *resultant* displacement from $P_1$ to $P_3$, labeled $\vec{C}$, is called the sum of the two successive displacements $\vec{A}$ and $\vec{B}$:

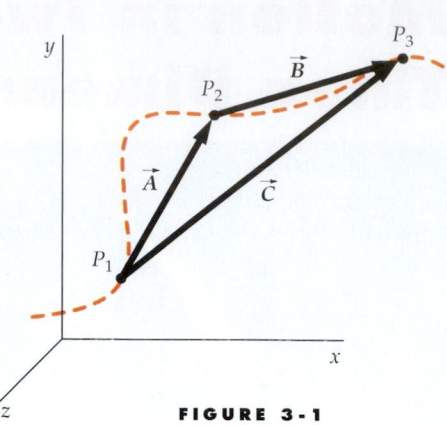

**FIGURE 3-1**

$$\vec{C} = \vec{A} + \vec{B} \qquad\qquad 3\text{-}1$$

Two displacement vectors are added graphically by placing the tail of one at the head of the other (Figure 3-2). The resultant vector extends from the tail of the first to the head of the second. Note that $C$ does not equal $A + B$ unless $\vec{A}$ and $\vec{B}$ are in the same direction. That is, $\vec{C} = \vec{A} + \vec{B}$ does not imply that $C = A + B$.

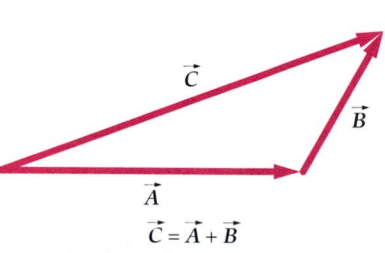

**FIGURE 3-2**
**Head-to-tail method of vector addition.**

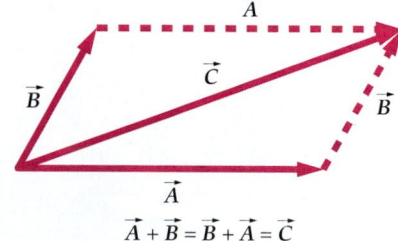

**FIGURE 3-3**
**Parallelogram method for adding vectors.**

An equivalent way of adding vectors, called the **parallelogram method,** is to move $\vec{B}$ so that it is tail-to-tail with $\vec{A}$. The diagonal of the parallelogram formed by $\vec{A}$ and $\vec{B}$ then equals $\vec{C}$. From Figure 3-3, we can see that it makes no difference in which order we add two vectors; that is, $\vec{A} + \vec{B} = \vec{B} + \vec{A}$.

---

*YOUR DISPLACEMENT*           **EXAMPLE 3-1**

**You walk 3 km east and then 4 km north. What is your resultant displacement?**

**PICTURE THE PROBLEM** Draw and label the two displacements $\vec{A}$ and $\vec{B}$ as well as the resultant displacement $\vec{C}$ (Figure 3-4). Include axes indicating the directions north and east. $\vec{A}$ and $\vec{B}$ are at right angles to each other, and $\vec{C} = \vec{A} + \vec{B}$ is the hypotenuse of the corresponding right triangle. The magnitude of $\vec{C}$ can be obtained using the Pythagorean theorem, and the direction of $\vec{C}$ is found using trigonometry.

**FIGURE 3-4**

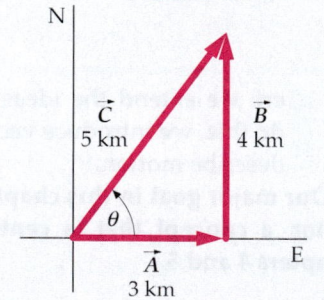

1. The magnitude of the resultant displacement is related to the magnitudes of the two sequential displacements by the Pythagorean theorem:

$$C^2 = A^2 + B^2$$
$$= (3\ \text{km})^2 + (4\ \text{km})^2$$
$$= 25\ \text{km}^2$$

so

$$C = \sqrt{25\ \text{km}^2} = \boxed{5\ \text{km}}$$

2. Let $\theta$ be the angle from the east axis to the resultant displacement $\vec{C}$. From the figure we find $\tan\theta$; using a calculator with trigonometric functions yields $\theta$:

$$\tan\theta = \frac{B}{A}$$

so

$$\theta = \tan^{-1}\frac{B}{A} = \tan^{-1}\frac{4\ \text{km}}{3\ \text{km}} = \boxed{53.1°}$$

**REMARKS** A vector is described by its magnitude and its direction. Your resultant displacement is a vector of length 5 km in a direction 53.1° north of east.

# 3-2 General Properties of Vectors

Many quantities in physics have magnitude and direction, and add like displacements. Examples include velocity, acceleration, momentum, and force. Such quantities are called **vectors.** Quantities with magnitude but no associated direction—for example, distance and speed—are called **scalars.**

> Vectors are quantities with magnitude and direction that add like displacements.
>
> DEFINITION—VECTORS

A vector is represented graphically by an arrow drawn in the same direction as that of the vector, and with a length that is proportional to the magnitude of the vector. When the magnitude of a vector is given, its units must also be given. The magnitude of a velocity vector, for example, requires units such as m/s. Two vectors are defined to be equal if they have the same magnitude and the same direction. Graphically, this means that they have the same length and are parallel to each other. A consequence of this definition is that moving a vector so that it remains parallel to itself does not change it. Thus, all the vectors in Figure 3-5 are equal. If we translate or rotate the coordinate system, all the vectors in Figure 3-5 remain equal. Vectors do not depend on the coordinate system used to represent them (except for position vectors, which are introduced in Section 3.3).

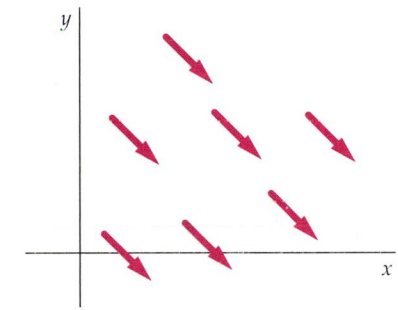

**FIGURE 3-5** Vectors are equal if their magnitudes and directions are the same. All vectors in this figure are equal.

## Multiplying a Vector by a Scalar

A vector $\vec{A}$ multiplied by a scalar $s$ is the vector $\vec{B} = s\vec{A}$, which has magnitude $|s|A$ and is parallel to $\vec{A}$ if $s$ is positive and antiparallel if $s$ is negative. The vector $-\vec{A}$ is the vector you add to $\vec{A}$ to get zero. Thus, $-\vec{A}$ has the same magnitude as $\vec{A}$ but points in the opposite direction. The dimensions of $s\vec{A}$ are those of $s$ multiplied by those of $A$.

## Subtracting Vectors

We subtract vector $\vec{B}$ from vector $\vec{A}$ by adding $-\vec{B}$ to $\vec{A}$. The result is $\vec{C} = \vec{A} + (-\vec{B}) = \vec{A} - \vec{B}$ (Figure 3-6a). An equivalent way of subtracting $\vec{B}$ from $\vec{A}$ is to draw them tail-to-tail (Figure 3-6b) and noting that $\vec{C}$ is the vector that must be added to $\vec{B}$ to obtain the resultant vector $\vec{A}$. The rules for adding or subtracting any two vectors, such as two velocity vectors or two acceleration vectors, are the same as for adding or subtracting displacement vectors.

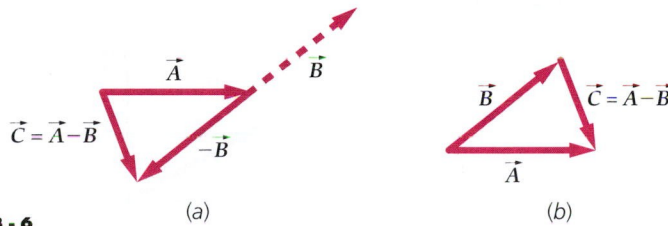

**FIGURE 3-6**

(a)                (b)

## Components of Vectors

The component of a vector in a given direction is a number whose absolute value is the length of the projection of the vector onto an axis (directed line) in that direction. It is found by dropping a perpendicular from the head of the vector to the axis, as shown in Figure 3-7. The sign of the component is positive if the projection of the head of the vector is in the positive direction relative to the tail of the vector. Thus, $A_S$ is positive and $B_S$ is negative. The components of a vector along the $x$, $y$, and $z$ directions, illustrated in Figure 3-8 for a vector in the $xy$ plane, are called rectangular components. Note that the components of a vector *do* depend on the coordinate system used, although the vector itself does not.

Rectangular components are useful for the addition or subtraction of vectors. If $\theta$ is the angle between $\vec{A}$ and the $x$ axis, then

$$A_x = A \cos \theta \qquad 3\text{-}2$$

X COMPONENT OF A VECTOR

and

$$A_y = A \sin \theta \qquad 3\text{-}3$$

Y COMPONENT OF A VECTOR

where $A$ is the magnitude of $\vec{A}$.

If we know $A_x$ and $A_y$, we can find the angle $\theta$ from

$$\tan \theta = \frac{A_y}{A_x}, \qquad \theta = \tan^{-1} \frac{A_y}{A_x} \qquad 3\text{-}4$$

and the magnitude $A$ from the Pythagorean theorem:

$$A = \sqrt{A_x^2 + A_y^2} \qquad 3\text{-}5a$$

In three dimensions,

$$A = \sqrt{A_x^2 + A_y^2 + A_z^2} \qquad 3\text{-}5b$$

Components can be positive or negative. For example, if $\vec{A}$ points in the negative $x$ direction, then $A_x$ is negative.

Consider two vectors $\vec{A}$ and $\vec{B}$ that lie in the $xy$ plane. The rectangular components of each vector and those of the sum $\vec{C} = \vec{A} + \vec{B}$ are shown in Figure 3-9. We see that the vector equation $\vec{C} = \vec{A} + \vec{B}$ is equivalent to the two component equations

$$C_x = A_x + B_x \qquad 3\text{-}6a$$

and

$$C_y = A_y + B_y \qquad 3\text{-}6b$$

**EXERCISE** A car travels 20 km in a direction 30° north of west. Let the $x$ axis point east and the $y$ axis point north as in Figure 3-10. Find the $x$ and $y$ components of the displacement vector of the car. (*Answer* $A_x = -17.3$ km, $A_y = +10$ km)

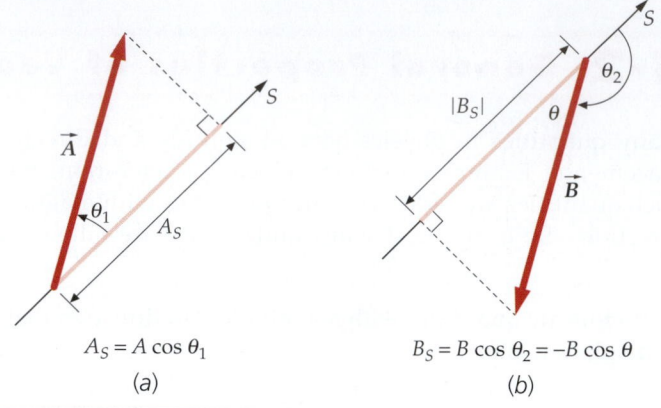

$A_S = A \cos \theta_1$      $B_S = B \cos \theta_2 = -B \cos \theta$

(a)      (b)

**FIGURE 3-7** Definition of the component of a vector. The component of the vector $\vec{A}$ in the positive $S$ direction is $A_S$, and $A_S$ is positive. The component of the vector $\vec{B}$ in the positive $S$ direction is $B_S$, and $B_S$ is negative.

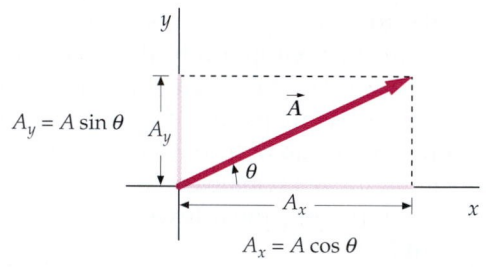

**FIGURE 3-8** The rectangular components of a vector. $A_x = A \cos \theta$, $A_y = A \sin \theta$.

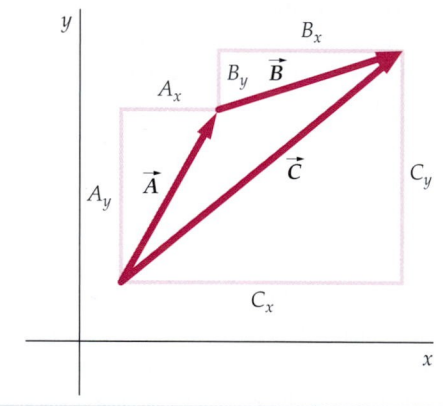

**FIGURE 3-9**

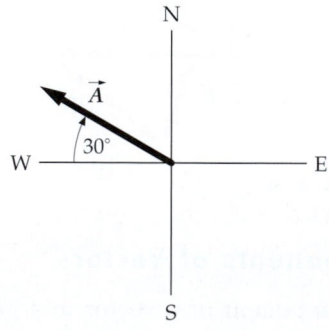

**FIGURE 3-10**

*A TREASURE MAP*                          **EXAMPLE  3 - 2**    **Put  It  in  Context**

**You are working as a counselor at a tropical resort. You've been given a map and told to follow its directions in order to bury the "treasure" at a specific location. You don't want to waste time walking around the island, because you want to finish early and go surfing. The directions are to walk 3 km headed due west and then 4 km headed 60° north of east. Where should you head and how far must you walk to get the job done quickly? Find your answer (*a*) graphically and (*b*) using vector components.**

**PICTURE THE PROBLEM**  You need to find your resultant displacement, which is $\vec{C}$ in Figure 3-11. The triangle formed by the three vectors is not a right triangle, so the magnitudes of the vectors are not related by the Pythagorean theorem. We find the resultant graphically by drawing each of the displacements to scale and measuring the resultant displacement.

**FIGURE 3-11**

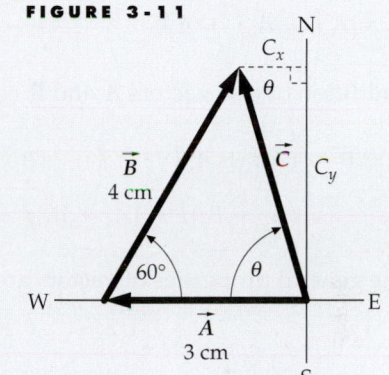

(*a*) If you draw the first displacement vector $\vec{A}$ 3 cm long and the second one $\vec{B}$ 4 cm long, you find the resultant vector $\vec{C}$ to be about 3.5 cm long. Thus, the magnitude of the resultant displacement is $\boxed{3.5 \text{ km}}$. The angle $\theta$ made between the resultant displacement and the west direction can then be measured with a protractor. It is about 75°. Therefore, you should walk 3.5 km headed $\boxed{75° \text{ north of west}}$.

(*b*) 1. To solve using vector components, let $\vec{A}$ be the first displacement and choose the direction of increasing $x$ to be east and the direction of increasing $y$ to be north. Compute $A_x$ and $A_y$ from Equations 3-2 and 3-3:

$$A_x = -3 \text{ km} \quad \text{and} \quad A_y = 0$$

2. Similarly, compute the components of the second displacement $\vec{B}$:

$$B_x = (4 \text{ km}) \cos 60° = 2 \text{ km}$$

$$B_y = (4 \text{ km}) \sin 60° = 2\sqrt{3} \text{ km} = 3.46 \text{ km}$$

3. The components of the resultant displacement $\vec{C} = \vec{A} + \vec{B}$ are found by addition:

$$C_x = A_x + B_x = -3 \text{ km} + 2 \text{ km} = -1 \text{ km}$$

$$C_y = A_y + B_y = 0 + 2\sqrt{3} \text{ km} = 3.46 \text{ km}$$

4. The Pythagorean theorem gives the magnitude of $\vec{C}$:

$$C^2 = C_x^2 + C_y^2$$

$$= (-1 \text{ km})^2 + (2\sqrt{3} \text{ km})^2 = 13 \text{ km}^2$$

$$C = \sqrt{13} \text{ km} = \boxed{3.61 \text{ km}}$$

5. The ratio of $C_y$ to $|C_x|$ equals the tangent of the angle $\theta$ between $\vec{C}$ and the negative $x$ direction:

$$\tan \theta = \frac{C_y}{|C_x|}$$

so

$$\theta = \tan^{-1} \frac{2\sqrt{3} \text{ km}}{1 \text{ km}} = \tan^{-1} (2\sqrt{3})$$

$$= \boxed{74° \text{ north of west}}$$

**REMARKS**  Since the displacement (which is a vector) was asked for, the answer must include either both the magnitude *and* direction, or both components. In (*b*) we could have stopped at step 3 because the *x* and *y* components completely define the displacement vector. We converted to the magnitude and direction to compare with the answer to Part (*a*). Note that step 5 of Part (*b*) gives the angle as 74°. This agrees with the results in (*a*) within the accuracy of our measurement.

## Unit Vectors

A **unit vector** is a *dimensionless* vector with unit magnitude. The vector $A^{-1}\vec{A}$ is an example of a unit vector that points in the direction of $\vec{A}$. Unit vectors are often written with an overhead caret as in $\hat{A} = A^{-1}\vec{A}$. Unit vectors that point in the positive $x$, $y$, and $z$ directions are convenient for expressing vectors in terms of their rectangular components. They are usually written $\hat{i}$, $\hat{j}$, and $\hat{k}$, respectively. For example, the vector $A_x\hat{i}$ has a magnitude $|A_x|$ and points in the positive $x$ direction if $A_x$ is positive (or the negative $x$ direction if $A_x$ is negative). A general vector $\vec{A}$ can be written as the sum of three vectors, each of which is parallel to a coordinate axis (Figure 3-12):

$$\vec{A} = A_x\hat{i} + A_y\hat{j} + A_z\hat{k}$$ 
3-7

The addition of two vectors $\vec{A}$ and $\vec{B}$ can be written in terms of unit vectors as

$$\vec{A} + \vec{B} = (A_x\hat{i} + A_y\hat{j} + A_z\hat{k}) + (B_x\hat{i} + B_y\hat{j} + B_z\hat{k})$$

$$= (A_x + B_x)\hat{i} + (A_y + B_y)\hat{j} + (A_z + B_z)\hat{k}$$
3-8

The general properties of vectors are summarized in Table 3-1.

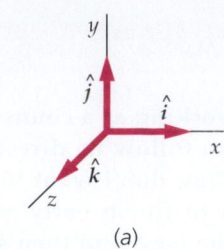

(a)

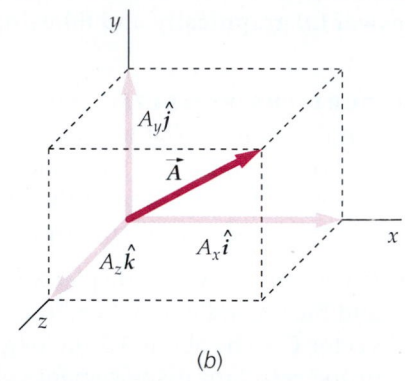

(b)

**FIGURE 3-12** (a) The unit vectors $\hat{i}$, $\hat{j}$, and $\hat{k}$ in a rectangular coordinate system. (b) The vector $\vec{A}$ in terms of unit vectors: $\vec{A} = A_x\hat{i} + A_y\hat{j} + A_z\hat{k}$.

## TABLE 3-1

**Properties of Vectors**

| Property | Explanation | Figure | Component representation |
|---|---|---|---|
| Equality | $\vec{A} = \vec{B}$ if $|\vec{A}| = |\vec{B}|$ and their directions are the same | | $A_x = B_x$ $A_y = B_y$ $A_z = B_z$ |
| Addition | $\vec{C} = \vec{A} + \vec{B}$ | | $C_x = A_x + B_x$ $C_y = A_y + B_y$ $C_z = A_z + B_z$ |
| Negative of a vector | $\vec{A} = -\vec{B}$ if $|\vec{B}| = |\vec{A}|$ and their directions are opposite | | $A_x = -B_x$ $A_y = -B_y$ $A_z = -B_z$ |
| Subtraction | $\vec{C} = \vec{A} - \vec{B}$ | | $C_x = A_x - B_x$ $C_y = A_y - B_y$ $C_z = A_z - B_z$ |
| Multiplication by a scalar | $\vec{B} = s\vec{A}$ has magnitude $|\vec{B}| = |s||\vec{A}|$ and has the same direction as $\vec{A}$ if $s$ is positive or as $-\vec{A}$ if $s$ is negative | | $B_x = sA_x$ $B_y = sA_y$ $B_z = sA_z$ |

**EXERCISE** Given two vectors

$$\vec{A} = (4\ \text{m})\hat{i} + (3\ \text{m})\hat{j} \qquad \text{and} \qquad \vec{B} = (2\ \text{m})\hat{i} - (3\ \text{m})\hat{j}$$

find (a) $A$, (b) $B$, (c) $\vec{A} + \vec{B}$, and (d) $\vec{A} - \vec{B}$. (Answers (a) $A = 5$ m, (b) $B = 3.61$ m, (c) $\vec{A} + \vec{B} = (6\ \text{m})\hat{i}$, (d) $\vec{A} - \vec{B} = (2\ \text{m})\hat{i} + (6\ \text{m})\hat{j}$ )

# 3-3 Position, Velocity, and Acceleration

## Position and Velocity Vectors

The **position vector** of a particle is a vector drawn from the origin of a coordinate system to the position of the particle. For a particle at the point $(x, y)$, the position vector $\vec{r}$ is

$$\vec{r} = x\hat{i} + y\hat{j} \qquad\qquad 3\text{-}9$$

DEFINITION—POSITION VECTOR

Figure 3-13 shows the actual path or trajectory of the particle. (Don't confuse the trajectory with the $x$-versus-$t$ plots of Chapter 2.) At time $t_1$, the particle is at $P_1$, with position vector $\vec{r}_1$; by $t_2$, the particle has moved to $P_2$, with position vector $\vec{r}_2$. The particle's change in position is the displacement vector $\Delta\vec{r}$:

$$\Delta\vec{r} = \vec{r}_2 - \vec{r}_1 \qquad\qquad 3\text{-}10$$

DEFINITION—DISPLACEMENT VECTOR

The ratio of the displacement vector to the time interval $\Delta t = t_2 - t_1$ is the **average-velocity vector**:

$$\vec{v}_{av} = \frac{\Delta\vec{r}}{\Delta t} \qquad\qquad 3\text{-}11$$

DEFINITION—AVERAGE-VELOCITY VECTOR

This vector points in the direction of the displacement.

The magnitude of the displacement vector is less than the distance traveled along the curve unless the particle moves in a straight line. However, if we consider smaller and smaller time intervals (Figure 3-14), the magnitude of the displacement approaches the distance along the curve, and the direction of $\Delta\vec{r}$ approaches the tangent to the curve at the beginning of the interval. We define the **instantaneous-velocity vector** as the limit of the average-velocity vector as $\Delta t$ approaches zero:

$$\vec{v} = \lim_{\Delta t \to 0}\frac{\Delta\vec{r}}{\Delta t} = \frac{d\vec{r}}{dt} \qquad\qquad 3\text{-}12$$

DEFINITION—INSTANTANEOUS-VELOCITY VECTOR

The instantaneous-velocity vector is the derivative of the position vector with respect to time. Its magnitude is the speed and its direction is the direction of motion of the particle along the line tangent to the curve.

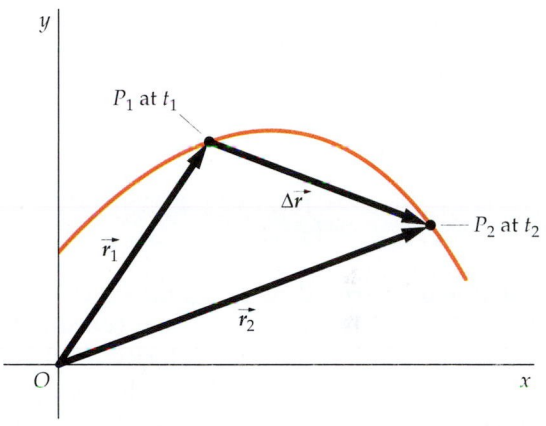

**FIGURE 3-13** The displacement vector $\Delta\vec{r}$ is the difference in the position vectors, $\Delta\vec{r} = \vec{r}_2 - \vec{r}_1$. Equivalently, $\Delta\vec{r}$ is the vector that, when added to $\vec{r}_1$, yields the new position vector $\vec{r}_2$.

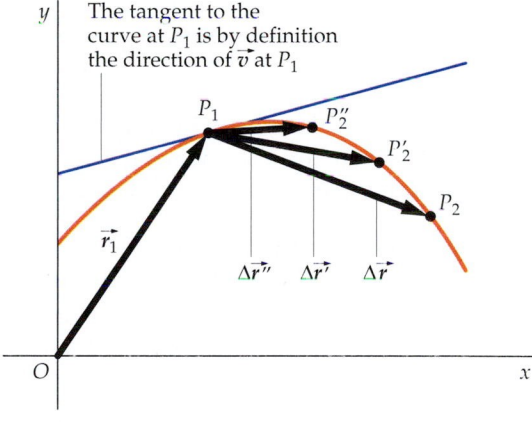

**FIGURE 3-14** As the time interval is made smaller, the direction of the displacement approaches the tangent to the curve.

To calculate the derivative in Equation 3-12, we write the position vectors in terms of their components:

$$\Delta\vec{r} = \vec{r}_2 - \vec{r}_1 = (x_2 - x_1)\hat{i} + (y_2 - y_1)\hat{j} = \Delta x\hat{i} + \Delta y\hat{j}$$

Then

$$\vec{v} = \lim_{\Delta t \to 0}\frac{\Delta\vec{r}}{\Delta t} = \lim_{\Delta t \to 0}\frac{\Delta x\hat{i} + \Delta y\hat{j}}{\Delta t} = \lim_{\Delta t \to 0}\left(\frac{\Delta x}{\Delta t}\right)\hat{i} + \lim_{\Delta t \to 0}\left(\frac{\Delta y}{\Delta t}\right)\hat{j}$$

or

$$\vec{v} = \frac{dx}{dt}\hat{i} + \frac{dy}{dt}\hat{j} = v_x\hat{i} + v_y\hat{j} \qquad\qquad 3\text{-}13$$

---

*THE VELOCITY OF A SAILBOAT* **EXAMPLE 3-3**

A sailboat has coordinates $(x_1, y_1) = (110\text{ m}, 218\text{ m})$ at $t_1 = 60$ s. Two minutes later, at time $t_2$, it has coordinates $(x_2, y_2) = (130\text{ m}, 205\text{ m})$. (a) Find the average velocity for this 2-min interval. Express $\vec{v}_{av}$ in terms of its rectangular components. (b) Find the magnitude and direction of this average velocity. (c) For $t \geq 20$ s, the position of the sailboat as a function of time is $x(t) = b_1 + b_2 t$ and $y(t) = c_1 + c_2/t$, where $b_1 = 100$ m, $b_2 = \frac{1}{6}$ m/s, $c_1 = 200$ m, and $c_2 = 1080$ m·s. Find the instantaneous velocity at a general time $t \geq 20$ s.

**PICTURE THE PROBLEM** The initial and final positions of the sailboat are given. (a) The average-velocity vector points from the initial toward the final position. (b) The instantaneous-velocity components are calculated from Equation 3-13: $v_x = dx/dt$ and $v_y = dy/dt$.

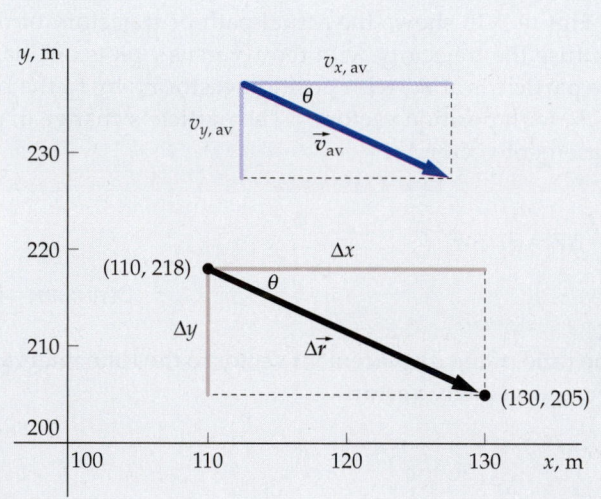

**FIGURE 3-15**

(a) 1. Draw a coordinate system (Figure 3-15) showing the displacement of the sailboat. The average-velocity vector and the displacement are in the same direction:

    2. The $x$ and $y$ components of the average velocity $\vec{v}_{av}$ are calculated directly from their definitions:

$$\vec{v}_{av} = v_{x,av}\,\hat{i} + v_{y,av}\,\hat{j}$$

where

$$v_{x,av} = \frac{\Delta x}{\Delta t} = \frac{130\text{ m} - 110\text{ m}}{120\text{ s}} = 0.167\text{ m/s}$$

$$v_{y,av} = \frac{\Delta y}{\Delta t} = \frac{205\text{ m} - 218\text{ m}}{120\text{ s}} = -0.108\text{ m/s}$$

so

$$\boxed{\vec{v}_{av} = (0.167\text{ m/s})\hat{i} - (0.108\text{ m/s})\hat{j}}$$

(b) 1. The magnitude of $\vec{v}_{av}$ is found from the Pythagorean theorem:

$$v_{av} = \sqrt{(v_{x,av})^2 + (v_{y,av})^2} = \boxed{0.199\text{ m/s}}$$

2. The ratio of $v_{y,\,av}$ to $v_{x,\,av}$ gives the tangent of the angle $\theta$ between $\vec{v}_{av}$ and the $x$ axis:

$$\tan\theta = \frac{v_{y,av}}{v_{x,av}}$$

so

$$\theta = \tan^{-1}\frac{v_{y,av}}{v_{x,av}} = \tan^{-1}\frac{-0.108\ \text{m/s}}{0.167\ \text{m/s}} = \boxed{-33.0°}$$

(c) We find the instantaneous velocity $\vec{v}$ by calculating $dx/dt$ and $dy/dt$:

$$\vec{v} = \frac{dx}{dt}\hat{i} + \frac{dy}{dt}\hat{j} = b_2\hat{i} - c_2t^{-2}\hat{j} = \boxed{\left(\frac{1}{6}\,\text{m/s}\right)\hat{i} - \frac{1080\ \text{m}\cdot\text{s}}{t^2}\hat{j}}$$

**REMARKS** The magnitude of $\vec{v}$ can be found from $v = \sqrt{v_x^2 + v_y^2}$ and its direction can be found from $\tan\theta = v_y/v_x$.

**EXERCISE** Find the $x$ and $y$ components and the magnitude and direction of the instantaneous velocity of the sailboat at time $t_1 = 60$ s. (*Answer* $\vec{v}_1 = (\frac{1}{6}\,\text{m/s})\hat{i} - (0.30\ \text{m/s})\hat{j}$, $v_1 = 0.34$ m/s, $\theta_1 = -60.9°$)

## Relative Velocity

Relative velocities in two and three dimensions combine just as they do in one dimension, except that the velocity vectors are not necessarily along the same line. If a particle moves with velocity $\vec{v}_{pA}$ relative to reference frame A, which is in turn moving with velocity $\vec{v}_{AB}$ relative to reference frame B, the velocity of the particle relative to reference frame B is

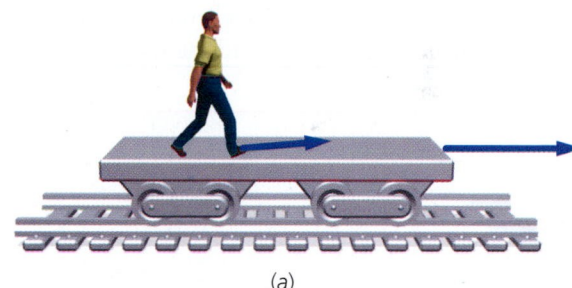

(a)

$$\boxed{\vec{v}_{pB} = \vec{v}_{pA} + \vec{v}_{AB}} \qquad\qquad 3\text{-}14$$

RELATIVE VELOCITY

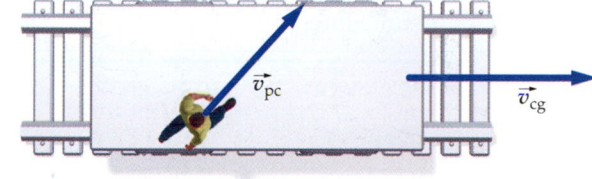

(b)

For example, if you are on a railroad flatcar that is moving with velocity $\vec{v}_{cg}$ relative to the ground (Figure 3-16a), and you are walking with velocity $\vec{v}_{pc}$ (Figure 3-16b) relative to the car, then your velocity relative to the ground is the vector sum of these two velocities: $\vec{v}_{pg} = \vec{v}_{pc} + \vec{v}_{cg}$ (Figure 3-16c).

The velocity of object A relative to object B is equal in magnitude and opposite in direction to the velocity of object B relative to object A. For example, $\vec{v}_{pc} = -\vec{v}_{cp}$ where $\vec{v}_{cp}$ is the velocity of the car relative to the person. The addition of relative velocities is done in the same way as the addition of displacements, either graphically, using either the head-to-tail or the parallelogram method, or analytically, using vector components.

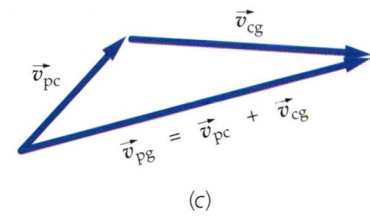

(c)

**FIGURE 3-16**

*A FLYING PLANE*

# EXAMPLE 3-4

A pilot wants to fly a plane due north. The speed of the plane relative to the air is 200 km/h and the wind is blowing from west to east at 90 km/h. (*a*) In which direction should the plane head? (*b*) How fast does the plane travel relative to the ground?

**PICTURE THE PROBLEM** Because the wind is blowing toward the east, a plane headed due north will drift off course toward the east. To compensate for the crosswind, the plane must head west of due north. The velocity of the plane relative to the ground $\vec{v}_{pg}$ will be the sum of the velocity of the plane relative to the air $\vec{v}_{pa}$ and the velocity of the air relative to the ground $\vec{v}_{ag}$.

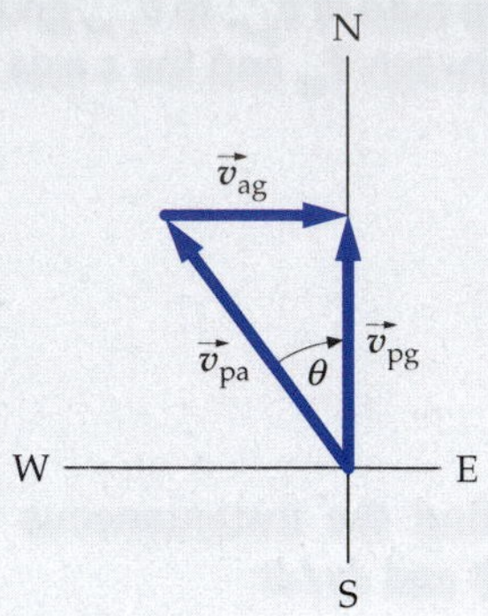

**FIGURE 3-17**

(a) 1. The velocity of the plane relative to the ground is given by Equation 3-14:

$$\vec{v}_{pg} = \vec{v}_{pa} + \vec{v}_{ag}$$

2. Make a velocity addition diagram showing the addition of the vectors in step 1. Include direction axes as shown in Figure 3-17.

3. The sine of the angle $\theta$ between the velocity of the plane and north equals the ratio of $v_{ag}$ and $v_{pa}$:

$$\sin \theta = \frac{v_{ag}}{v_{pa}}$$

so

$$\theta = \sin^{-1} \frac{v_{ag}}{v_{pa}} = \sin^{-1} \frac{90 \text{ km/h}}{200 \text{ km/h}} = \boxed{26.7°}$$

(b) Because $\vec{v}_{ag}$ and $\vec{v}_{pg}$ are perpendicular, we can use the Pythagorean theorem to find the magnitude of $\vec{v}_{pg}$:

$$v_{pa}^2 = v_{pg}^2 + v_{ag}^2$$

so

$$v_{pg} = \sqrt{v_{pa}^2 - v_{ag}^2}$$

$$= \sqrt{(200 \text{ km/h})^2 - (90 \text{ km/h})^2} = \boxed{179 \text{ km/h}}$$

## The Acceleration Vector

The **average-acceleration** is the ratio of the change in the instantaneous-velocity vector $\Delta \vec{v}$ to the time interval $\Delta t$:

$$\vec{a}_{av} = \frac{\Delta \vec{v}}{\Delta t} \qquad \text{3-15}$$

DEFINITION—AVERAGE-ACCELERATION VECTOR

The **instantaneous-acceleration** is the limit of this ratio as $\Delta t$ approaches zero; in other words, it is the derivative of the velocity vector with respect to time:

$$\vec{a} = \lim_{\Delta t \to 0} \frac{\Delta \vec{v}}{\Delta t} = \frac{d\vec{v}}{dt} \qquad \text{3-16}$$

DEFINITION—INSTANTANEOUS-ACCELERATION VECTOR

To calculate the instantaneous acceleration, we express $\vec{v}$ in rectangular coordinates:

$$\vec{v} = v_x \hat{i} + v_y \hat{j} + v_z \hat{k} = \frac{dx}{dt} \hat{i} + \frac{dy}{dt} \hat{j} + \frac{dz}{dt} \hat{k}$$

Then

$$\vec{a} = \frac{dv_x}{dt} \hat{i} + \frac{dv_y}{dt} \hat{j} + \frac{dv_z}{dt} \hat{k} = \frac{d^2x}{dt^2} \hat{i} + \frac{d^2y}{dt^2} \hat{j} + \frac{d^2z}{dt^2} \hat{k}$$

$$= a_x \hat{i} + a_y \hat{j} + a_z \hat{k} \qquad \text{3-17}$$

*A Thrown Baseball*

**EXAMPLE 3-5**

The position of a thrown baseball is given by $\vec{r} = 1.5\text{ m }\hat{i} + (12\text{ m/s }\hat{i} + 16\text{ m/s }\hat{j})t - 4.9\text{ m/s}^2\,\hat{j}\,t^2$. Find its velocity and acceleration.

**PICTURE THE PROBLEM** Because $\vec{r} = x\hat{i} + y\hat{j}$, we have $x = 1.5\text{ m} + (12\text{ m/s})t$ and $y = (16\text{ m/s})t - (4.9\text{ m/s}^2)t^2$. We can find the $x$ and $y$ components of the velocity and acceleration by differentiating $x$ and $y$.

1. The $x$ and $y$ components of the velocity are found by differentiating $x$ and $y$:

$$v_x = \frac{dx}{dt} = \frac{d}{dt}[1.5\text{ m} + (12\text{ m/s})t] = 12\text{ m/s}$$

$$v_y = \frac{dy}{dt} = \frac{d}{dt}[(16\text{ m/s})t - (4.9\text{ m/s}^2)t^2]$$

$$= 16\text{ m/s} - (9.8\text{ m/s}^2)t$$

2. We differentiate again to obtain the components of the acceleration:

$$a_x = \frac{dv_x}{dt} = 0$$

$$a_y = \frac{dv_y}{dt} = -9.8\text{ m/s}^2$$

$$\vec{v} = \boxed{(12\text{ m/s})\hat{i} + [16\text{ m/s} - (9.8\text{ m/s}^2)t]\hat{j}}$$

3. In vector notation, the velocity and acceleration are:

$$\vec{a} = \boxed{(-9.8\text{ m/s}^2)\hat{j}}$$

**REMARKS** This is an example of projectile motion, a topic we will study in Section 3-4.

For a vector to be constant, both its magnitude and direction must remain constant. If either changes, the vector changes. Thus, if a car rounds a curve in the road at constant speed, it is accelerating because the velocity is changing due to the change in direction of the velocity vector.

**FIGURE 3-18**

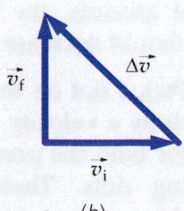

(a)

*Cornering a Turn*

**EXAMPLE 3-6**

A car is traveling east at 60 km/h. It rounds a curve, and 5 s later it is traveling north at 60 km/h. Find the average acceleration of the car.

**PICTURE THE PROBLEM** The initial and final velocity vectors are shown in Figure 3-18. We choose the unit vector $\hat{i}$ to be east and $\hat{j}$ to be north, and calculate the average acceleration from its definition, $\vec{a}_{av} = \Delta\vec{v}/\Delta t$. Note that $\Delta\vec{v}$ is the vector that, when added to $\vec{v}_i$, results in $\vec{v}_f$.

1. The average acceleration is the ratio of the velocity change to the time interval:

$$\vec{a}_{av} = \frac{\Delta\vec{v}}{\Delta t}$$

2. The change in velocity is related to the initial and final velocities:

$$\Delta\vec{v} = \vec{v}_f - \vec{v}_i$$

3. Express the initial and final velocities as vectors:

$$\vec{v}_i = (60\text{ km/h})\hat{i}, \qquad \vec{v}_f = (60\text{ km/h})\hat{j}$$

4. Substitute these results to find the average acceleration:

$$\vec{a}_{av} = \frac{\vec{v}_f - \vec{v}_i}{\Delta t} = \frac{(60\text{ km/h})\hat{j} - (60\text{ km/h})\hat{i}}{5\text{ s}}$$

$$= \boxed{-[(12\text{ km/h})/s]\hat{i} + [(12\text{ km/h})/s]\hat{j}}$$

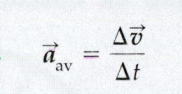

(b)

**REMARKS** Note that the car accelerates, even though its speed does not change.

**EXERCISE** Find the magnitude and direction of the average acceleration vector. (*Answer* $a_{av} = (17.0 \text{ km/h})/\text{s}$ at 45° west of north)

The motion of an object traveling in a circle is a common example of motion in which the velocity of an object changes even though its speed remains constant.

**The Direction of the Acceleration Vector** In the next few chapters we will want to determine the direction of the acceleration vector from a description of the motion. For example, consider a bungee jumper as she slows down prior to reversing direction at the lowest point of her jump. To find the direction of her acceleration as she loses speed we make a series of dots representing her position at successive ticks of a clock, as shown in Figure 3-19a. The faster she moves, the greater the distance she travels between ticks, and the greater the space between the dots in the diagram. Next we number the dots, starting with zero and increasing in the direction of her motion. At time $t_0$ she is at dot 0, at time $t_1$ she is at dot 1, and so forth. To determine the direction of the acceleration at time $t_3$ we draw vectors representing the jumper's velocities at times $t_2$ and $t_4$. The average acceleration during the interval from $t_2$ to $t_4$ equals $\Delta \vec{v}/\Delta t$, where $\Delta \vec{v} = \vec{v}_4 - \vec{v}_2$ and $\Delta t = t_4 - t_2$. We use this as an estimate of her acceleration at time $t_3$. That is, $\vec{a}_3 \approx \Delta \vec{v}/\Delta t$. Since $\vec{a}_3$ and $\Delta \vec{v}$ are in the same direction, by finding the direction of $\Delta \vec{v}$ we also find the direction of $\vec{a}_3$. The direction of $\Delta \vec{v}$ is obtained by using the relation $\vec{v}_2 + \Delta \vec{v} = \vec{v}_4$ and drawing the corresponding vector addition diagram (Figure 3-19b). Because the jumper is moving faster (the dots are farther apart) at $t_2$ than at $t_4$, we draw $\vec{v}_2$ longer than $\vec{v}_4$. From this figure we get the direction of $\Delta \vec{v}$, and thus the direction of $\vec{a}_3$.

**EXERCISE** Figure 3-20 is a motion diagram of the bungee jumper before, during, and after time $t_6$, when she momentarily comes to rest at the lowest point in her descent. During the part of her ascent shown she is moving upward with increasing speed. Use this diagram to determine the direction of the jumper's acceleration (*a*) at time $t_6$ and (*b*) at time $t_9$. (*Answer* (*a*) upward, (*b*) upward)

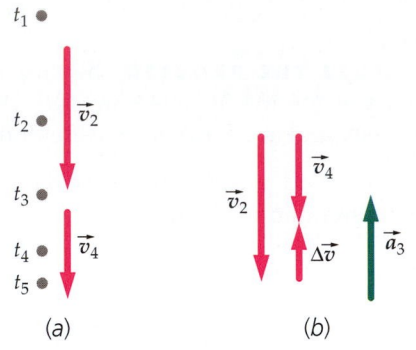

**FIGURE 3-19** (*a*) A motion diagram of a bungee jumper losing speed as she descends. The dots are drawn at successive ticks of a clock. (*b*) We draw vectors $\vec{v}_2$ and $\vec{v}_4$ starting from the same point. Then we draw $\Delta \vec{v}$ from the head of $\vec{v}_2$ to the head of $\vec{v}_4$ to obtain the graphical expression of the relation $\vec{v}_2 + \Delta \vec{v} = \vec{v}_4$. The acceleration $\vec{a}_3$ is in the same direction as $\Delta \vec{v}$.

**FIGURE 3-20** The dots for the bungee jumper's ascent are drawn to the right of those for her descent so that they do not overlap each other. Her motion, however, is straight down and then straight up.

---

**EXAMPLE 3-7**

**A graduating physics student throws his cap into the air with an initial angle of 60° above the horizontal. Using a motion diagram, find the direction of the acceleration of the cap during the ascending portion of its flight.**

**PICTURE THE PROBLEM** As the cap rises it both loses speed and changes direction. We should draw a motion diagram and then make a sketch of the relation $\vec{v}_i + \Delta \vec{v} = \vec{v}_f$ to find the direction of $\Delta \vec{v}$ and thus the direction of the acceleration.

1. Make a motion diagram (Figure 3-21a) of the cap's motion for the ascending portion of its flight. Because it slows as it ascends, the spacing between dots should decrease as it rises.

2. Pick a dot on the motion diagram and draw a velocity vector on the diagram for both the preceding and the following dots. These vectors should be drawn tangent to the path of the cap.

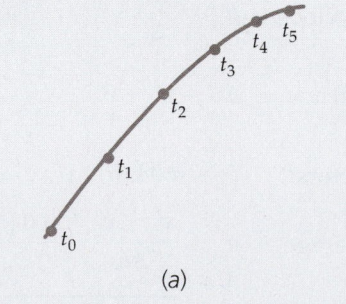

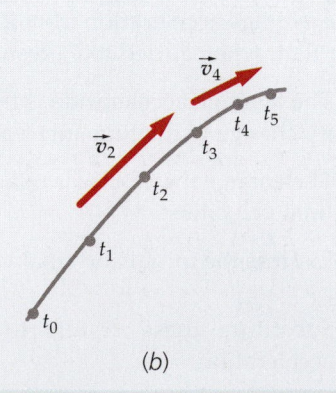

**FIGURE 3-21**

3. Draw the graphical expression of the relation $\vec{v}_i + \Delta\vec{v} = \Delta\vec{v}_f$. Begin by drawing the two velocity vectors from the same point. These vectors should have the same magnitude and direction as the vectors drawn for step 2. Then draw the $\Delta\vec{v}$ vector connecting their heads.

4. Draw the acceleration vector in the same direction as $\Delta\vec{v}$, but not necessarily the same length since $\vec{a} = \Delta\vec{v}/\Delta t$.

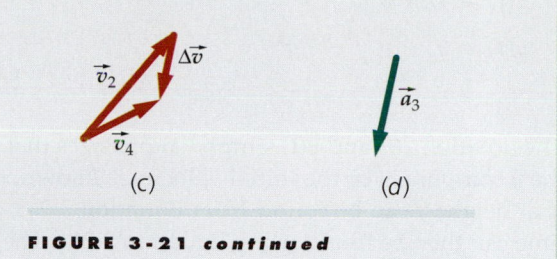

(c)                    (d)

**FIGURE 3-21** *continued*

**REMARKS** The process of finding the direction of the acceleration using a motion diagram is not precise. Therefore the result is an estimate of the direction of the acceleration, as opposed to a precise determination.

**EXERCISE** Use a motion diagram to find the direction of the acceleration of the cap in Example 3-7 during the descending portion of its flight. (*Answer* directly downward)

# 3-4 Special Case 1: Projectile Motion

Figure 3-22 shows a particle launched with initial speed $v_0$ at angle $\theta_0$ with the horizontal axis. Let the launch point be at $(x_0, y_0)$; $y$ is positive upward and $x$ is positive to the right.
   The initial velocity then has components

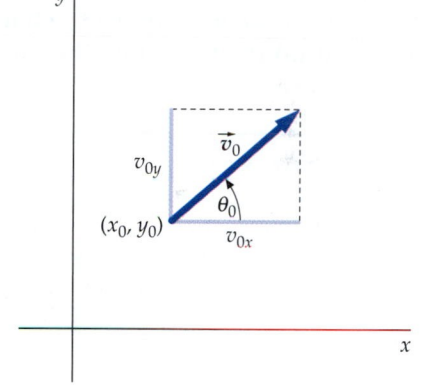

$$v_{0x} = v_0 \cos \theta_0 \qquad\qquad 3\text{-}18a$$

$$v_{0y} = v_0 \sin \theta_0 \qquad\qquad 3\text{-}18b$$

**FIGURE 3-22**

In the absence of air resistance, the acceleration is free-fall acceleration, vertically downward:

$$a_x = 0 \qquad\qquad 3\text{-}19a$$

and

$$a_y = -g \qquad\qquad 3\text{-}19b$$

Since the acceleration is constant, we can use the kinematics equations discussed in Chapter 2. The $x$ component of the velocity is constant because there is no horizontal acceleration:

$$v_x = v_{0x} \qquad\qquad 3\text{-}20a$$

The $y$ component of the velocity varies with time according to Equation 2-12, with $a = -g$:

$$v_y = v_{0y} - gt \qquad\qquad 3\text{-}20b$$

❶ Notice that $v_x$ does not depend on $v_y$ and vice versa: *The horizontal and vertical components of projectile motion are independent.* (Dropping a ball from a desktop and projecting a second ball horizontally at the same time can demonstrate this. The two balls strike the floor simultaneously.) The displacements $x$ and $y$ are given by (see Equation 2-16)

$$x(t) = x_0 + v_{0x}t \qquad\qquad\qquad\qquad \text{3-21}a$$

$$y(t) = y_0 + v_{0y}t - \tfrac{1}{2}gt^2 \qquad\qquad\qquad \text{3-21}b$$

EQUATIONS OF MOTION FOR A PROJECTILE

The notation $x(t)$ and $y(t)$ simply emphasizes that $x$ and $y$ are functions of time. If the $y$ component of the initial velocity is known, the time $t$ for which the particle is at height $y$ can be found from Equation 3-21$b$. The horizontal position at that time can then be found using Equation 3-21$a$. The total horizontal distance a projectile travels is called its **range.** (The vector forms of equations 3-19 to 3-21 are given on page 71.)

---

A *CAP IN THE AIR*                      **E X A M P L E   3 - 8**

**A delighted physics graduate throws her cap into the air with an initial velocity of 24.5 m/s at 36.9° above the horizontal. It is later caught by another student. Find (a) the total time the cap is in the air and (b) the total horizontal distance traveled.**

**PICTURE THE PROBLEM** We choose the origin to be the initial position of the cap so that $x_0 = y_0 = 0$. We assume it is caught at the same height. The total time the cap is in the air is found by setting $y = 0$ in Equation 3-21$b$. We can then use this result in Equation 3-21$a$ to find the total horizontal distance traveled.

(a) 1. Set $y = 0$ in Equation 3-21$b$ and solve for $t$:
$$y = v_{0y}t - \tfrac{1}{2}gt^2 = t(v_{0y} - \tfrac{1}{2}gt) = 0$$

2. There are two solutions for $t$:
$$t = 0 \qquad \text{(initial conditions)}$$
$$t = \frac{2v_{0y}}{g}$$

3. Compute the vertical component of the initial velocity vector:
$$v_{0y} = v_0 \sin \theta_0 = (24.5 \text{ m/s}) \sin 36.9° = 14.7 \text{ m/s}$$

4. Substitute for $v_{0y}$ in the Step 2 result to find the total time $t$:
$$t = \frac{2v_{0y}}{g} = \frac{2v_0 \sin \theta_0}{g}$$
$$= \frac{2(24.5 \text{ m/s})\sin 36.9°}{9.81 \text{ m/s}^2} = \boxed{3.00 \text{ s}}$$

(b) Use this value for the time to calculate the total horizontal distance traveled:
$$x = v_{0x}t = (v_0 \cos \theta_0)t$$
$$= (24.5 \text{ m/s}) \cos 36.9°(3.00 \text{ s}) = \boxed{58.8 \text{ m}}$$

**REMARKS** The vertical component of the initial velocity of the cap is 14.7 m/s, the same as that of the cap in Example 2-9 (Chapter 2), where the cap was thrown straight up with $v_0 = 14.7$ m/s. The time the cap is in the air is also the same as in Example 2-9. Figure 3-23 shows the height $y$ versus $t$ for the cap. This curve is identical to Figure 2-11 (Example 2-9) because the caps each have the same vertical acceleration and vertical velocity. Figure 3-23 can be reinterpreted as a graph of $y$ versus $x$ if its time scale is converted to a distance scale, as shown in the figure. This can be done by multiplying the time values by 19.6 m/s, because the cap moves at $(24.5 \text{ m/s}) \cos 36.9° = 19.6$ m/s horizontally. The curve $y$ versus $x$ is a parabola.

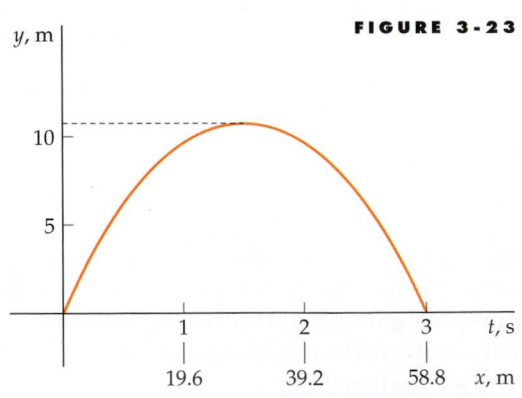

FIGURE 3-23

Figure 3-24 shows graphs of the vertical heights versus the horizontal distances for projectiles with an initial speed of 24.5 m/s and several different initial angles. The angles drawn are 45°, which has the maximum range, and pairs of angles at equal amounts above and below 45°. Notice that the paired angles have the same range. The green curve has an initial angle of 36.9° (0.64 rad), as in this example.

FIGURE 3-24

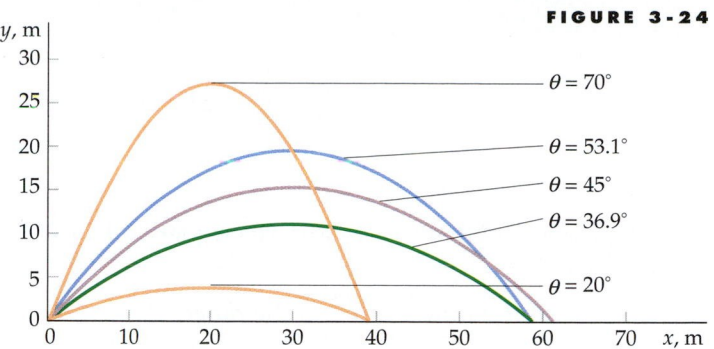

The general equation for the path $y(x)$ of a projectile can be obtained from Equations 3-21 by eliminating the variable $t$. Choosing $x_0 = 0$ and $y_0 = 0$, we obtain $t = x/v_{0x}$ from Equation 3-21a. Substituting this into Equation 3-21b gives

$$y(x) = v_{0y}\left(\frac{x}{v_{0x}}\right) - \frac{1}{2}g\left(\frac{x}{v_{0x}}\right)^2 = \left(\frac{v_{0y}}{v_{0x}}\right)x - \left(\frac{g}{2v_{0x}^2}\right)x^2$$

Writing out the velocity components yields

$$y(x) = (\tan \theta_0)x - \left(\frac{g}{2v_0^2 \cos^2 \theta_0}\right)x^2 \qquad 3\text{-}22$$

PATH OF A PROJECTILE

for the projectile's path. This is of the form $y = ax + bx^2$, the equation for a parabola passing through the origin. Figure 3-25 shows the path of a projectile with its velocity vector and components at several points. The path is for a projectile that impacts the ground at $P$. The horizontal distance between launch and impact is the range $R$.

If the initial and final elevations are equal, the range of a projectile can be written in terms of its initial speed and the angle of projection. As in the preceding examples, we find the range by multiplying the $x$ component of the velocity by the total time that the projectile is in the air. The total flight time $T$ is obtained by setting $y = 0$ in Equation 3-21b.

$$y(T) = v_{0y}T - \tfrac{1}{2}gT^2 = 0, \qquad T > 0$$

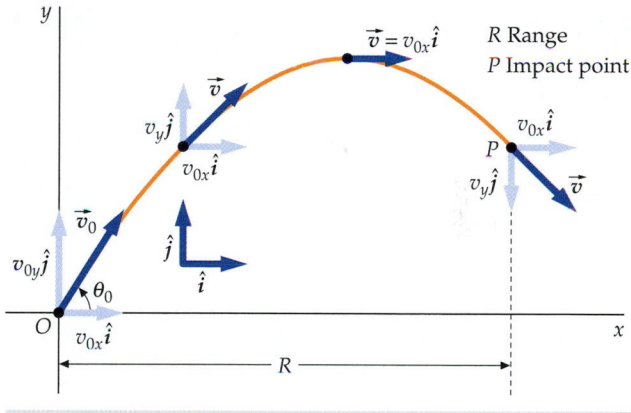

**FIGURE 3-25** Path of a projectile showing velocity vectors.

Dividing through by $T$ gives

$$v_{0y} - \tfrac{1}{2}gT = 0$$

The flight time of the projectile is thus

$$T = \frac{2v_{0y}}{g} = \frac{2v_0}{g}\sin \theta_0$$

and the range is

$$R = v_{0x}T = (v_0 \cos \theta_0)\left(\frac{2v_0}{g}\sin \theta_0\right) = \frac{2v_0^2}{g}\sin \theta_0 \cos \theta_0$$

This can be further simplified by using the following trigonometric identity:

$$\sin 2\theta = 2 \sin \theta \cos \theta$$

Thus,

$$R = \frac{v_0^2}{g} \sin 2\theta_0 \qquad\qquad\qquad 3\text{-}23$$

RANGE OF A PROJECTILE FOR EQUAL INITIAL AND FINAL ELEVATIONS

**EXERCISE** Use Equation 3-22 for the path to derive Equation 3-23. (*Answer* Set $y(x) = 0$ and solve for $x$)

Equation 3-23 is useful if you want to find the range for many projectiles with equal initial and final elevations. More importantly, this equation shows how the range depends on $\theta$. Since the maximum value of $\sin 2\theta$ is 1 when $2\theta = 90°$ or $\theta = 45°$, the range is greatest when $\theta = 45°$. In many practical applications, the initial and final elevations may not be equal, and other considerations are important. For example, in the shot put, the ball ends its flight when it hits the ground, but it is projected from an initial height of about 2 m above the ground. This causes the range to be maximum at an angle somewhat lower than 45°, as shown in Figure 3-26. Studies of the best shot-putters show that maximum range occurs at an initial angle of about 42°.

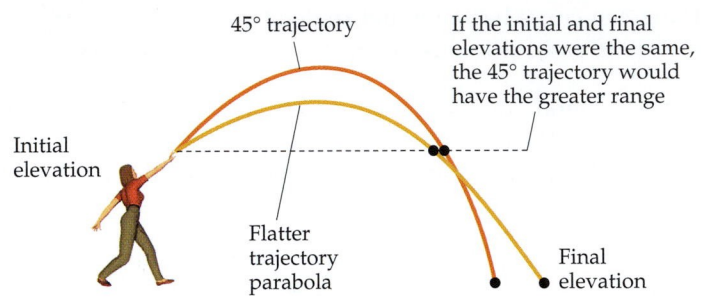

**FIGURE 3-26** If a projectile lands at an elevation lower than the elevation of projection, the maximum range is achieved when the projection angle is somewhat lower than 45°.

---

*A Supply Drop*                    **EXAMPLE 3-9**

**A helicopter drops a supply package to flood victims on a raft on a swollen lake. When the package is released, the helicopter is 100 m above the raft and flying at 25 m/s at an angle $\theta_0 = 36.9°$ above the horizontal. (*a*) How long is the package in the air? (*b*) How far from the raft does the package land? (*c*) If the helicopter flies at constant velocity, where is the helicopter when the package lands?**

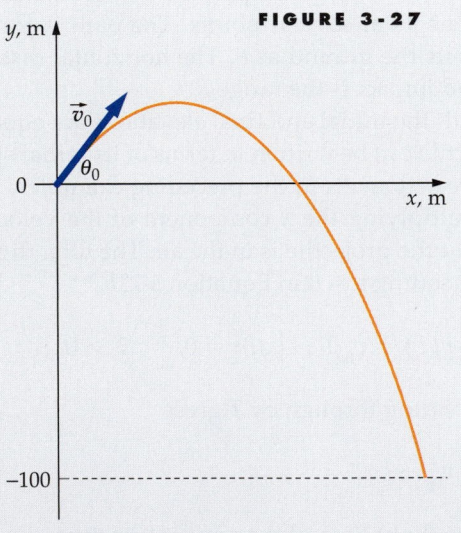

**FIGURE 3-27**

**PICTURE THE PROBLEM** The time in the air depends only on the vertical aspects of the motion. Using Equation 3-21*b*, solve for the time. Choose the origin to be at the location of the package when it is released. The initial velocity of the package is the initial velocity of the helicopter. The horizontal distance traveled by the package is given by Equation 3-21*a*, where $t$ is the time the package is in the air.

(*a*) 1. Sketch the trajectory of the package during the time it is in the air. Include coordinate axes as shown in Figure 3-27.

2. To find the time of flight, write $y(t)$:

$$y(t) = y_0 + v_{0y}t - \tfrac{1}{2}gt^2$$

3. Set $y_0 = 0$ and apply the quadratic formula to solve for $t$:

$$y(t) = v_{0y}t - \tfrac{1}{2}gt^2 \quad \text{or} \quad 0 = \tfrac{1}{2}gt^2 - v_{0y}t + y(t)$$

so

$$t = \frac{v_{0y} \pm \sqrt{v_{0y}^2 - 2gy(t)}}{g}$$

4. Solve for the time when $y(t) = -100$ m. First solve for $v_{0y}$:

$$v_{0y} = v_0 \sin \theta_0 = (25 \text{ m/s})\sin 36.9° = 15 \text{ m/s}$$

so

$$t = \frac{15\text{ m/s} \pm \sqrt{(15\text{ m/s})^2 - 2(9.81\text{ m/s}^2)(-100\text{ m})}}{9.81\text{ m/s}^2}$$

so

Because the package is released at $t = 0$, the time of impact cannot be negative. Hence:

$t = -3.24\text{ s}$ or $t = 6.30\text{ s}$

$t = \boxed{6.30\text{ s}}$

(b) At impact the package has traveled a horizontal distance $x$, where $x$ is the horizontal velocity times the time of flight. First solve for the horizontal velocity:

$v_{0x} = v_0 \cos \theta_0 = (25\text{ m/s}) \cos 36.9° = 20\text{ m/s}$

so

$x = v_{0x}t = (20\text{ m/s})(6.3\text{ s}) = \boxed{126\text{ m}}$

(c) The coordinates of the helicopter at the time of impact are:

$x_h = v_{0x}t = (20\text{ m/s})(6.3\text{ s}) = \boxed{126\text{ m}}$

$y_h = y_{h0} + v_{h0}t = 0 + (15\text{ m/s})(6.30\text{ s}) = \boxed{94.5\text{ m}}$

The height of the helicopter is 194.5 m directly above the package.

**REMARKS** The positive time is appropriate because it corresponds to a time after the package is dropped (which occurs at $t = 0$). The negative time is when the package would have been at $y = -100$ m if its motion had started earlier as shown in Figure 3-28. Note that the helicopter is directly above the package when the package hits the water (and at all other times before then). Figure 3-29 shows a graph of $y$ versus $x$ for various initial angles and an initial speed of 25 m/s. The curve with an initial angle of 36.9° is the one given in this example. Note that the maximum range occurs at an angle less than 45°.

**FIGURE 3-29**

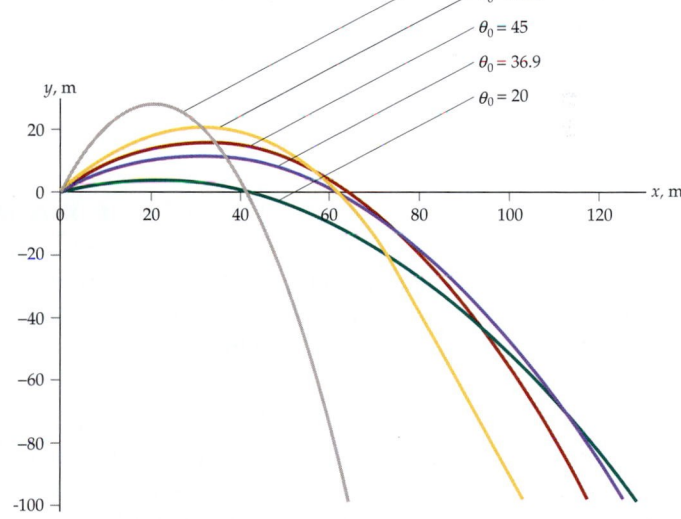

**FIGURE 3-28**

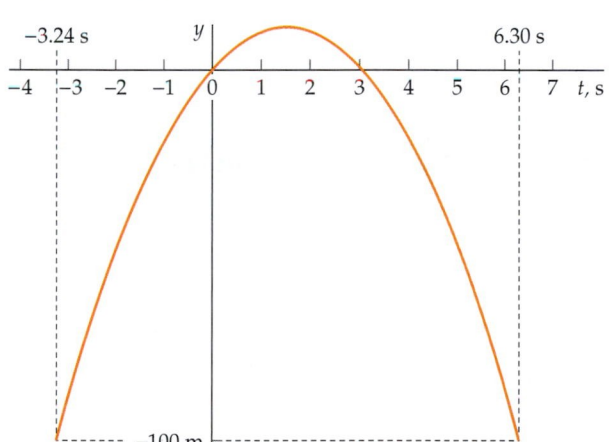

TO CATCH A THIEF

**EXAMPLE 3-10** **Try It Yourself**

**FIGURE 3-30**

A police officer chases a master jewel thief across city rooftops. They are both running when they come to a gap between buildings that is 4 m wide and has a drop of 3 m (Figure 3-30). The thief, having studied a little physics, leaps at 5 m/s and at 45° and clears the gap easily. The police officer did not study physics and thinks he should maximize his horizontal velocity, so he leaps at 5 m/s horizontally. (a) Does he clear the gap? (b) By how much does the thief clear the gap?

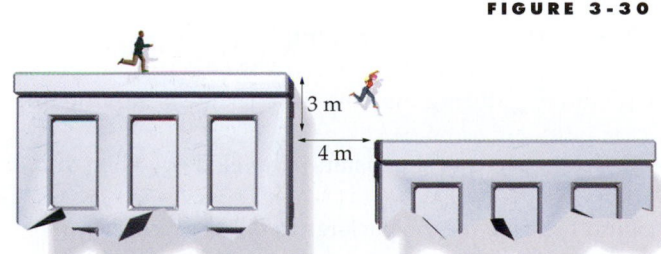

**PICTURE THE PROBLEM** Assuming they both clear the gap, the total time in the air depends only on the vertical aspects of the motion. Choose the origin at the launch point, with upward positive so that Equations 3-21 apply. Use Equation 3-21b for $y(t)$ and solve for the time when $y = -3$ m for $\theta_0 = 0°$ and again for $\theta_0 = 45°$. The horizontal distances traveled are the values of $x$ at these times.

**Cover the column to the right and try these on your own before looking at the answers.**

| Steps | Answers |
|---|---|
| (a) 1. Write $y(t)$ for the police officer and solve for $t$ when $y = -3$ m. | $t = 0.782$ s |
| 2. Substitute this into the equation for $x(t)$ and find the horizontal distance traveled during this time. | $x = \boxed{3.91 \text{ m}}$ |
| | Since this is less than 4 m, it appears the police officer fails to make it across the gap between buildings. |
| (b) 1. Write $y(t)$ for the thief and solve for $t$ when $y = -3$ m. $y(t)$ is a quadratic equation with two solutions, but only one of its solutions is acceptable. | $t = -0.5$ s or $t = 1.22$ s <br><br> She must land after she leaps, so <br><br> $t = 1.22$ s |
| 2. Find the horizontal distance covered for the positive value of $t$. | $x = v_{0x}t = 4.31$ m |
| 3. Subtract 4.0 m from this distance. | $4.31$ m $- 4.0$ m $= \boxed{0.31 \text{ m}}$ |

**REMARKS** The thief probably knew that she should jump at slightly less than 45°, but she didn't have time to solve the problem exactly. The police officer actually did make it across by tightening his abdominal muscles before impact. This raised his feet more than the 9 cm needed for him to complete the leap.

---

*DROPPING SUPPLIES*        **EXAMPLE 3-11**    **Try It Yourself**

In Example 3-9, find (a) the time $t_1$ for the package to reach its greatest height $h$ above the water, (b) its greatest height $h$, and (c) the time $t_2$ for the package to fall to the water from its greatest height.

**PICTURE THE PROBLEM** The time $t_1$ is the time at which the vertical component of the velocity is zero. Using Equation 3-20b solve for $t_1$.

**Cover the column to the right and try these on your own before looking at the answers.**

| Steps | Answers |
|---|---|
| (a) 1. Write $v_y(t)$ for the package. | $v_y(t) = v_{0y} - gt$ |
| 2. Set $v_y(t_1) = 0$ and solve for $t_1$. | $t_1 = \boxed{1.53 \text{ s}}$ |
| (b) 1. Find $v_{y\,av}$ during the time the package is moving up. | $v_{y\,av} = 7.5$ m/s |
| 2. Use $v_{y\,av}$ to find the distance traveled up. Then find $h$. | $\Delta y = 11.5$ m, $h = \boxed{111.5 \text{ m}}$ |
| (c) Find the time for the package to fall a distance $h$. | $t_1 = \boxed{4.77 \text{ s}}$ |

**REMARKS** Note that $t_1 + t_2 = 6.3$ s, in agreement with Example 3-9.

**EXERCISE** Solve Part (b) of Example 3-11 using $y(t)$ (Equation 3-21b) instead of finding $v_{y\,av}$.

Equations 3-19a and b can be expressed in vector form. Multiplying through Equation 3-19a by $\hat{i}$, multiplying through Equation 3-19b by $\hat{j}$, and then adding the two equations gives $a_x\hat{i} + a_y\hat{j} = -g\hat{j}$, or

$$\vec{a} = \vec{g}$$  3-19c

where $\vec{g}$ is the free-fall acceleration vector. At the surface of the earth the magnitude of $\vec{g}$ is $g = 9.81$ m/s². Equations 3-20a and b can also be expressed in vector form. Multiplying through Equation 3-20a by $\hat{i}$, multiplying through Equation 3-20b by $\hat{j}$, and then adding the two equations gives $(v_x\hat{i} + v_y\hat{j}) = (v_{0x}\hat{i} + v_{0y}\hat{j}) - gt\hat{j}$ or

$$\vec{v} = \vec{v}_0 + \vec{g}t \qquad \text{or} \qquad \Delta\vec{v} = \vec{g}t$$  3-20c

where $\vec{v} = v_x\hat{i} + v_y\hat{j}$, $\vec{v}_0 = v_{0x}\hat{i} + v_{0y}\hat{j}$, and $\vec{g} = -g\hat{j}$. Repeating the process for Equations 3-21a and b we obtain

$$\vec{r} = \vec{r}_0 + \vec{v}_0 t + \tfrac{1}{2}\vec{g}t^2 \qquad \text{or} \qquad \Delta\vec{r} = \vec{v}_0 t + \tfrac{1}{2}\vec{g}t^2$$  3-21c

where $\vec{r} = x\hat{i} + y\hat{j}$ and $\vec{r}_0 = x_0\hat{i} + y_0\hat{j}$. For a number of problems the vector forms of the kinematic equations (Equations 3-20c and 3-21c) are more suitable. This is the case in the following example.

---

*THE RANGER AND THE MONKEY*                 **EXAMPLE 3-12**

A park ranger with a tranquilizer dart gun intends to shoot a monkey hanging from a branch. The ranger aims directly at the monkey, not realizing that the dart will follow a parabolic path that will pass below the present position of the creature. The monkey, seeing the gun discharge, lets go of the branch and drops out of the tree, expecting to avoid the dart. (a) Show that the monkey will be hit regardless of the initial speed of the dart so long as it is great enough for the dart to travel the horizontal distance to the tree. Assume the reaction time of the monkey is negligible. (b) Let $\vec{v}_{d0}$ be the velocity of the dart as it leaves the gun. Find the velocity of the dart relative to the monkey at an arbitrary time $t$ during the dart's flight.

**PICTURE THE PROBLEM** We apply Equation 3-21c to both the monkey and the gun.

(a) 1. Apply Equation 3-21c to the monkey at time $t$:

$\Delta\vec{r}_m = \tfrac{1}{2}\vec{g}t^2$

where the initial velocity of the monkey is zero.

2. Apply Equation 3-21c to the dart at time $t$:

$\Delta\vec{r}_d = \vec{v}_{d0}t + \tfrac{1}{2}\vec{g}t^2$

where $\vec{v}_{d0}$ is the velocity of the dart as it leaves the gun.

3. Make a sketch of the monkey, the dart, and the gun, as shown in Figure 3-31. Show the dart and the monkey at their initial locations and at their locations a time $t$ later. On the figure draw a vector representing each term in the step 1 and step 2 results. Note that at time $t$ the dart and the monkey both are a distance $\tfrac{1}{2}gt^2$ below the line of sight of the gun. The dart will strike the monkey when it reaches the monkey's line of fall:

**FIGURE 3-31**

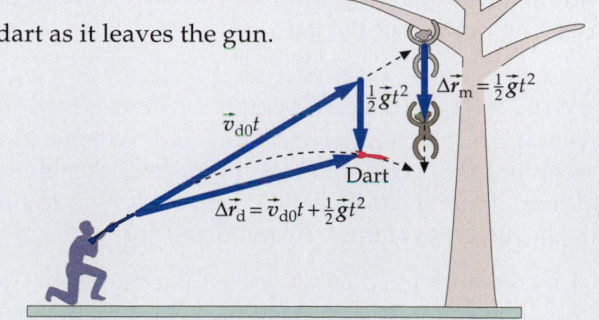

(b) 1. The velocity of the dart relative to the monkey equals the velocity of the dart relative to the gun plus the velocity of the gun relative to the monkey:

$$\vec{v}_{dm} = \vec{v}_{dg} + \vec{v}_{gm}$$

2. The velocity of the gun relative to the monkey is the negative of the velocity of the monkey relative to the gun:

$$\vec{v}_{dm} = \vec{v}_{dg} - \vec{v}_{mg}$$

3. Using Equation 3-20c, express the velocity of the dart relative to the gun and the velocity of the monkey relative to the gun:

$$\vec{v}_{dg} = \vec{v}_{d0} + \vec{g}t$$
$$\vec{v}_{mg} = \vec{g}t$$

4. Substitute these expressions into the Part (b) step 2 result:

$$\vec{v}_{dm} = (\vec{v}_{d0} + \vec{g}t) - (\vec{g}t) = \boxed{\vec{v}_{d0}}$$

**REMARKS** Relative to the monkey, the dart moves with constant speed $v_{d0}$ in a straight line. The dart strikes the monkey at time $t = L/v_{d0}$, where $L$ is the distance from the muzzle of the gun to the initial position of the monkey.

**REMARKS** In a familiar lecture demonstration, a target is suspended by an electromagnet. When the dart leaves the gun, the circuit to the magnet is broken and the target falls. The initial velocity of the dart is varied so that for large $v_{d0}$ the target is hit very near its original height and for small $v_{d0}$ it is hit just before it reaches the floor.

**EXERCISE** A hockey puck at ice level is struck such that it misses the net and clears the top of the Plexiglass wall of height $h = 2.80$ m. The flight time at the moment the puck clears the wall is $t_1 = 0.650$ s, and the horizontal distance is $x_1 = 12.0$ m. (a) Find the initial speed and direction of the puck. (b) When does the puck reach its maximum height? (c) What is the maximum height of the puck? (*Answer* (a) $\vec{v} = 20.0$ m/s, $\theta_0 = 22.0°$, (b) $t = 0.764$ s, (c) $v_{y \, av} t = 2.86$ m)

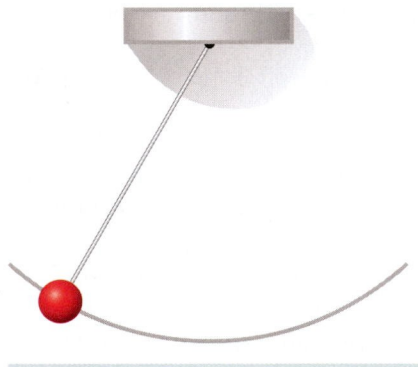

**FIGURE 3-32**

## 3-5 Special Case 2: Circular Motion

Figure 3-32 shows a pendulum bob swinging back and forth in a vertical plane. The path of the bob is a segment of a circular path. Motion along a circular path, or a segment of a circular path, is called **circular motion.**

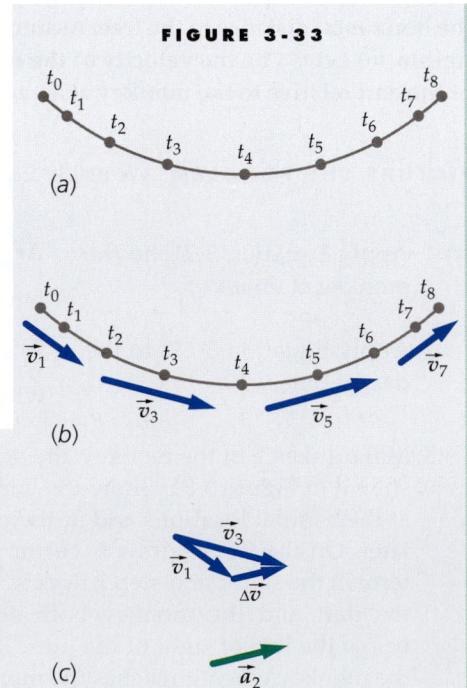

**FIGURE 3-33**

### A SWINGING PENDULUM

**EXAMPLE 3-13**

Consider the motion of the pendulum bob shown in Figure 3-32. Using a motion diagram (Figure 3-33), find the direction of the acceleration vector when the bob is swinging from left to right and (a) on the descending portion of the path, (b) passing through the lowest point on the path, and (c) on the ascending portion of the path.

**PICTURE THE PROBLEM** As the bob descends it both gains speed and changes direction. The acceleration is related to the change in velocity by $\vec{a} \approx \Delta\vec{v}/\Delta t$. The direction of the acceleration at a point can be estimated by constructing a vector addition diagram for the relation $\vec{v}_i + \Delta\vec{v} = \vec{v}_f$ to find the direction of $\Delta\vec{v}$, and thus the direction of the acceleration vector.

(a) 1. Make a motion diagram for a full left-to-right swing of the bob. The spacing between dots is greatest at the lowest point where the speed is greatest.

2. Pick a dot on the descending portion of the motion diagram and draw a velocity vector on the diagram for both the preceding and the following dot. The velocity vectors should be drawn tangent to the path and with lengths proportional to the speed.

**FIGURE 3-33** *continued*

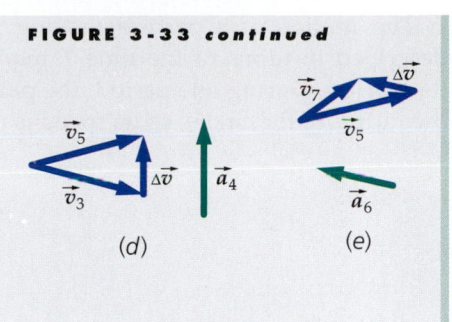

(d)          (e)

3. Draw the graphical expression of the relation $\vec{v}_i + \Delta\vec{v} = \vec{v}_f$. On this figure draw the acceleration vector. Since $\vec{a} \approx \Delta\vec{v}/\Delta t$, $\vec{a}$ is in the same direction as $\Delta\vec{v}$.

(b) Repeat steps 2 and 3 for the lowest point on the path.

(c) Repeat steps 2 and 3 for a point on the ascending portion of the path.

**REMARKS** At the lowest point the acceleration vector is directed straight upward (Figure 3-34)—toward point $P$ at the center of the circle. Where the speed is increasing (on the descending portion), the acceleration has a component in the forward direction as well as toward $P$. Where the speed is decreasing, the acceleration has a rearward component as well as a component toward $P$.

If a particle moves along a circular arc, the direction from the particle toward the center of the circle is called the **centripetal direction.** In Example 3-13 the acceleration at the lowest point of the pendulum bob's path was found to be in the centripetal direction. At other points the acceleration was found to have a tangential component as well as a centripetal component.

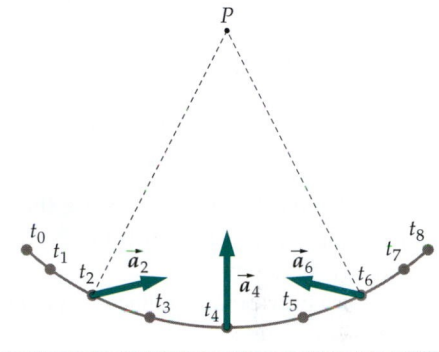

**FIGURE 3-34**

## Uniform Circular Motion

Motion in a circle at constant speed is called **uniform circular motion.** To find an expression for the acceleration of a particle moving in a circle at constant speed we will extend the method used in Example 3-13 to find the direction of the acceleration. The position and velocity vectors for a particle moving in a circle at constant speed are shown in Figure 3-35. The angle $\Delta\theta$ between $\vec{v}(t)$ and $\vec{v}(t + \Delta t)$ is the same as between $\vec{r}(t)$ and $\vec{r}(t + \Delta t)$ because the position and velocity vectors must both rotate through equal angles to remain mutually perpendicular. An isosceles triangle is formed by the two velocity vectors and $\Delta\vec{v}$, and a second isosceles triangle is formed by the two position vectors and $\Delta\vec{r}$. To find the direction of the acceleration vector we examine the triangle formed by the two velocity vectors and $\Delta\vec{v}$. The sum of the angles of any triangle is 180° and the base angles of any isosceles triangle are equal. In the limit that $\Delta t$ approaches zero, $\Delta\theta$ also approaches zero, so in this limit the two base angles must each approach 90°. This means $\Delta\vec{v}$ is perpendicular to the velocity. If $\Delta\vec{v}$ is drawn from the position of the particle then it points in the centripetal direction.

The two triangles are similar, so $|\Delta\vec{v}|/v = |\Delta\vec{r}|/r$ (corresponding lengths of similar shapes are proportional). Dividing both sides by $\Delta t$ and rearranging gives

$$\frac{|\Delta\vec{v}|}{\Delta t} = \frac{v|\Delta\vec{r}|}{r\ \Delta t}$$

In the limit as $\Delta t$ approaches zero, the term $|\Delta\vec{v}|/\Delta t$ approaches $a$, the magnitude of the instantaneous acceleration, and the term $|\Delta\vec{r}|/\Delta t$ approaches $v$, the magnitude of the instantaneous velocity (the speed). With these substitutions we have

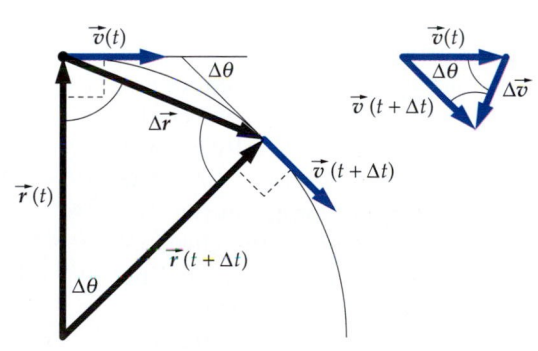

**FIGURE 3-35** Position and velocity vectors for a particle moving in a circle at constant speed.

$$a = a_c = \frac{v^2}{r} \qquad\qquad 3\text{-}24$$

CENTRIPETAL ACCELERATION

The motion of a particle moving in a circle with constant speed is often described in terms of the time $T$ required for one complete revolution, called the **period.** During one period, the particle travels a distance of $2\pi r$ (where $r$ is the radius of the circle), so its speed is related to $r$ and $T$ by

$$v = \frac{2\pi r}{T}$$ 3-25

---

*A SATELLITE'S MOTION* **EXAMPLE 3-14**

**A satellite moves at constant speed in a circular orbit about the center of the earth and near the surface of the earth. If its acceleration is $g = 9.81$ m/s², find (a) its speed and (b) the time for one complete revolution.**

**PICTURE THE PROBLEM** Since the satellite orbits near the surface of the earth, we take the radius of the orbit to be the radius of the earth, $r = 6370$ km.

(a) Set the centripetal acceleration $v^2/r$ equal to $g$ and solve for the speed $v$:

$$a = \frac{v^2}{r} = g \quad \text{or}$$

$$v = \sqrt{rg} = \sqrt{(6370 \text{ km})(9.81 \text{ m/s}^2)}$$

$$= \boxed{7.91 \text{ km/s} = 17{,}700 \text{ mi/h}}$$

(b) Use Equation 3-25 to get the period $T$:

$$T = \frac{2\pi r}{v} = \frac{2\pi(6370 \text{ km})}{7.91 \text{ km/s}} = \boxed{5060 \text{ s} = 84.3 \text{ min}}$$

**REMARKS** For actual satellites in orbit a few hundred kilometers above the earth's surface, the orbital radius $r$ is slightly greater than 6370 km. As a result, the centripetal acceleration is slightly less than 9.81 m/s² because of the decrease in the gravitational force with distance from the center of the earth. Many satellites are launched into such orbits, and their periods are roughly 90 min.

**EXERCISE** A car rounds a curve of radius 40 m at 48 km/h. What is its centripetal acceleration? (*Answer* 4.44 m/s²)

A particle moving in a circle with *varying speed* has a component of acceleration tangent to the circle, $dv/dt$, as well as the radially inward centripetal acceleration, $v^2/r$. For general motion along a curve, we can treat a portion of the curve as an arc of a circle (Figure 3-36). The particle then has acceleration $v^2/r$ toward the center of curvature, and if the speed $v$ is changing, it has tangential acceleration

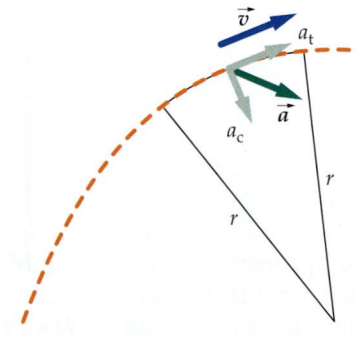

**FIGURE 3-36** A particle moving along an arbitrary curve can be considered to be moving in a circular arc during a small time interval. Its instantaneous acceleration vector has a component $a_c = v^2/r$ toward the center of curvature of the arc and a component $a_t = dv/dt$ that is tangential to the curve.

$$a_t = \frac{dv}{dt}$$ 3-26

TANGENTIAL ACCELERATION

| Topic | Relevant Equations and Remarks |
|---|---|
| **1. Vectors** | |
| Definition | Vectors are quantities that have both magnitude and direction. Vectors add like consecutive displacements. |
| Components | The component of a vector in a direction in space is its projection on an axis in that direction. If $\vec{A}$ makes an angle $\theta$ with the positive $x$ direction, its $x$ and $y$ components are |
| | $A_x = A \cos \theta$      3-2 |
| | $A_y = A \sin \theta$      3-3 |
| Magnitude | $A = \sqrt{A_x^2 + A_y^2}$      3-5a |
| Adding vectors graphically | Any two vectors whose magnitudes have the same units may be added graphically by placing the tail of one arrow at the head of the other. |
| Adding vectors using components | If $\vec{C} = \vec{A} + \vec{B}$, then |
| | $C_x = A_x + B_x$      3-6a |
| | and |
| | $C_y = A_y + B_y$      3-6b |
| Unit vectors | A vector $\vec{A}$ can be written in terms of unit vectors $\hat{i}$, $\hat{j}$, and $\hat{k}$, which have unit magnitude and lie along the $x$, $y$, and $z$ axes, respectively |
| | $\vec{A} = A_x \hat{i} + A_y \hat{j} + A_z \hat{k}$      3-7 |
| Position vector | The position vector $\vec{r}$ points from the origin of the coordinate system to the particle's position. |
| Instantaneous-velocity vector | The velocity vector $\vec{v}$ is the rate of change of the position vector. Its magnitude is the speed and it points in the direction of motion. |
| | $\vec{v} = \lim\limits_{\Delta t \to 0} \dfrac{\Delta \vec{r}}{\Delta t} = \dfrac{d\vec{r}}{dt}$      3-12 |
| Instantaneous-acceleration vector | $\vec{a} = \lim\limits_{\Delta t \to 0} \dfrac{\Delta \vec{v}}{\Delta t} = \dfrac{d\vec{v}}{dt}$      3-16 |
| **2. Relative Velocity** | If a particle moves with velocity $\vec{v}_{pA}$ relative to reference frame A, which is in turn moving with velocity $\vec{v}_{AB}$ relative to reference frame B, the velocity of the particle relative to B is |
| | $\vec{v}_{pB} = \vec{v}_{pA} + \vec{v}_{AB}$      3-14 |
| **3. Projectile Motion** | The positive $x$ direction is horizontal and the positive $y$ direction is upward for the equations in this section. |
| Independence of motion | In projectile motion, the horizontal and vertical motions are independent. Thus, |
| | $a_x = 0$    and    $a_y = -g$. |

| Dependence on time | $v_x(t) = v_{0x} + a_x t, \qquad v_y(t) = v_{0y} + a_y t$ | 2-14 |
| | $x(t) = x_0 + v_{0x}t + \frac{1}{2}a_x t^2, \qquad y(t) = y_0 + v_{0y}t + \frac{1}{2}a_y t^2$ | 2-16 |
| | where $a_x = 0$, $a_y = -g$, $v_{0x} = v_0 \cos\theta_0$, and $v_{0y} = v_0 \sin\theta_0$. Alternatively, | |
| | $\Delta\vec{v} = \vec{g}t, \qquad \Delta\vec{r} = \vec{v}_0 t + \frac{1}{2}\vec{g}\,t^2$ | 3-20c, 3-21c |
| | where $\vec{g} = -g\hat{j}$ | |
| Range | The range is found by multiplying $v_x$ by the total time the projectile is in the air. | |

## 3. Circular Motion

| Centripetal acceleration | $a_c = \dfrac{v^2}{r}$ | 3-24 |
| Tangential acceleration | $a_t = \dfrac{dv}{dt}$ | |
| | where $v$ is the speed. | 3-26 |
| Period | $v = \dfrac{2\pi r}{T}$ | 3-25 |

# PROBLEMS

- • Single-concept, single-step, relatively easy
- •• Intermediate-level, may require synthesis of concepts
- ••• Challenging
- **SSM** Solution is in the *Student Solutions Manual*
-  Problems available on iSOLVE online homework service
- ✓ These "Checkpoint" online homework service problems ask students additional questions about their confidence level, and how they arrived at their answer

In a few problems, you are given more data than you actually need; in a few other problems, you are required to supply data from your general knowledge, outside sources, or informed estimates.

## Conceptual Problems

**1** • **SSM** Can the magnitude of the displacement of a particle be less than the distance traveled by the particle along its path? Can its magnitude be more than the distance traveled? Explain.

**2** • Give an example in which the distance traveled is a significant amount, yet the corresponding displacement is zero.

**3** • What is the approximate average velocity of the race cars during the Indianapolis 500?

**4** • **iSOLVE** True or false: The magnitude of the sum of two vectors *must be* greater than the magnitude of either vector.

**5** • Can a component of a vector have a magnitude greater than the magnitude of the vector? Under what circumstances can a component of a vector have a magnitude equal to the magnitude of the vector?

**6** • **SSM** Can a vector be equal to zero and still have one or more components not equal to zero?

**7** • **iSOLVE** Are the components of $\vec{C}$, where $\vec{C} = \vec{A} + \vec{B}$, necessarily larger than the corresponding components of either $\vec{A}$ or $\vec{B}$?

**8** • **SSM** True or false: The instantaneous-acceleration vector is *always* in the direction of motion.

**9** • If an object is moving toward the west at some instant, in what direction is its acceleration? (*a*) North (*b*) East (*c*) West (*d*) South (*e*) May be any direction.

**10** • **iSOLVE** A golfer drives the ball from the tee down the fairway in a high arcing shot. When the ball is at the highest point of its flight, (*a*) its velocity and acceleration are both zero, (*b*) its velocity is zero but its acceleration is nonzero, (*c*) its velocity is nonzero but its acceleration is zero, (*d*) its velocity and acceleration are both nonzero, (*e*) insufficient information is given to answer correctly.

**11** • The velocity of a particle is in the eastward direction while the acceleration is directed toward the northwest as shown in Figure 3-37. The particle is (a) speeding up and turning toward the north, (b) speeding up and turning toward the south, (c) slowing down and turning toward the north, (d) slowing down and turning toward the south, (e) maintaining constant speed and turning toward the south.

**FIGURE 3-37**
**Problem 11**

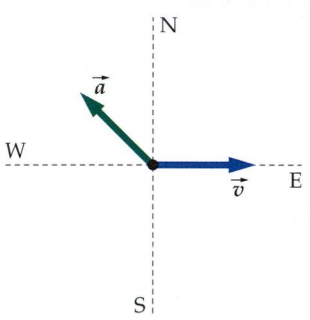

**12** • **SSM** Assuming constant acceleration, if you know the position vectors of a particle at two points on its path and also know the time it took to move from one point to the other, you can then compute (a) the particle's average velocity, (b) the particle's average acceleration, (c) the particle's instantaneous velocity, (d) the particle's instantaneous acceleration, (e) insufficient information is given to describe the particle's motion.

**13** •• Consider the path of a particle as it moves in space. (a) How is the velocity vector related geometrically to the path of the particle? (b) Sketch a curved path and draw the velocity vector for the particle for several positions along the path.

**14** • The acceleration of a car is zero when it is (a) turning right at a constant speed, (b) driving up a long straight incline at constant speed, (c) topping the crest of a hill at constant speed, (d) bottoming out at the lowest point of a valley at constant speed, (e) speeding up as it descends a long straight decline.

**15** • **SSM** Give examples of motion in which the directions of the velocity and acceleration vectors are (a) opposite, (b) the same, and (c) mutually perpendicular.

**16** • **iSOLVE** How is it possible for a particle moving at constant speed to be accelerating? Can a particle with constant velocity be accelerating at the same time?

**17** •• Imagine throwing a dart straight upward so that it sticks into the ceiling. After it leaves your hand, it steadily slows down as it rises before it sticks. (a) Draw the dart's velocity vector at times $t_1$ and $t_2$, where $\Delta t = t_2 - t_1$ is small. From your drawing find the direction of the change in velocity $\Delta \vec{v} = \vec{v}_2 - \vec{v}_1$, and thus the direction of the acceleration vector. (b) After it has stuck in the ceiling for a few seconds, the dart falls down to the floor. As it falls it speeds up, of course, until it hits the floor. Repeat part (a) to find the direction of its acceleration vector as it falls. (c) Now imagine tossing the dart horizontally. What is the direction of its acceleration vector after it leaves your hand, but before it strikes the floor?

**18** •• **SSM** As a bungee jumper approaches the lowest point in her descent, the rubber band holding her stretches and she loses speed as she continues to move downward. Assuming that she is dropping straight down, make a motion diagram to find the direction of her acceleration vector as she slows down by drawing her velocity vectors at times $t_1$ and $t_2$, where $\Delta t = t_2 - t_1$ is small. From your drawing find the direction of the change in velocity $\Delta \vec{v} = \vec{v}_2 - \vec{v}_1$, and thus the direction of the acceleration vector.

**19** •• After reaching the lowest point in her jump at time $t_{low}$, the bungee jumper in Problem 18 moves upward, gaining speed for a short time until gravity again dominates her motion. Draw her velocity vectors at times $t_1$ and $t_2$, where $\Delta t = t_2 - t_1$ is small and $t_1 < t_{low} < t_2$. From your drawing find the direction of the change in velocity $\Delta \vec{v} = \vec{v}_2 - \vec{v}_1$, and thus the direction of the acceleration vector.

**20** • A river is 0.76 km wide. The banks are straight and parallel (Figure 3-38). The current is 4.0 km/h and is parallel to the banks. A boat has a maximum speed of 4 km/h in still water. The pilot of the boat wishes to go on a straight line from A to B, where AB is perpendicular to the banks. The pilot should (a) head directly across the river, (b) head 53° upstream from the line AB, (c) head 37° upstream from the line AB, (d) give up—the trip from A to B is not possible with a boat of this limited speed, (e) do none of the above.

**FIGURE 3-38**
**Problem 20**

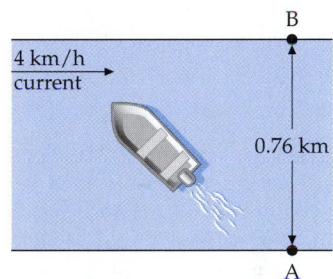

**21** • **SSM** True or false: When a projectile is fired horizontally, it takes the same amount of time to reach the ground as an identical projectile dropped from rest from the same height. Ignore the effects of air resistance.

**22** • **iSOLVE** A projectile is fired at 35° above the horizontal. At the highest point in its trajectory, its speed is 200 m/s. The initial velocity had a horizontal component of (a) 0; (b) (200 m/s) cos 35°; (c) (200 m/s) sin 35°; (d) (200 m/s)/cos 35°; (e) 200 m/s. Neglect the effects of air resistance.

**23** • Figure 3-39 represents the parabolic trajectory of a ball going from A to E. What is the direction of the acceleration at point B? (a) Up and to the right (b) Down and to the left (c) Straight up (d) Straight down (e) The acceleration of the ball is zero.

**FIGURE 3-39**
**Problems 23 and 24**

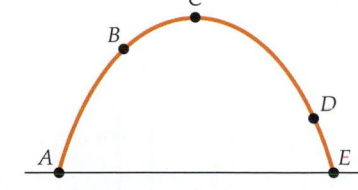

**24** • Referring to the motion described in Problem 23, (a) at which point(s) is the speed the greatest? (b) At which point(s) is the speed the lowest? (c) At which two points is the speed the same? Is the velocity the same at those points?

**25** • **iSOLVE** True or false:

(a) If the speed is constant, the acceleration must be zero.
(b) If the acceleration is zero, the speed must be constant.

**26** • The initial and final velocities of an object are as shown in Figure 3-40. Indicate the direction of the average acceleration.

**FIGURE 3-40** Problem 26

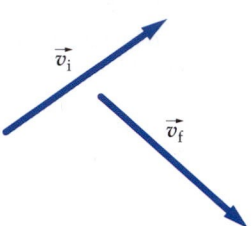

**27** • The velocities of objects A and B are shown in Figure 3-41. Draw a vector that represents the velocity of B relative to A.

**FIGURE 3-41** Problem 27

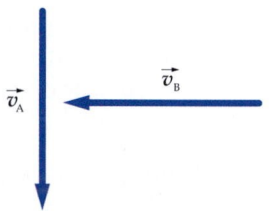

**28** •• **SSM** A vector $\vec{A}(t)$ has a constant magnitude but is changing direction. (a) Find $d\vec{A}/dt$ in the following manner: Draw the vectors $\vec{A}(t + \Delta t)$ and $\vec{A}(t)$ for a small time interval $\Delta t$, and find the difference $\Delta\vec{A} = \vec{A}(t + \Delta t) - \vec{A}(t)$ graphically. How is the direction of $\Delta\vec{A}$ related to $\vec{A}$ for small time intervals? (b) Interpret this result for the special cases where $\vec{A}$ represents the position of a particle with respect to some coordinate system. (c) Could $\vec{A}$ represent a velocity vector? Explain.

**29** •• The automobile path shown in Figure 3-42 is made up of straight lines and arcs of circles. The automobile starts from rest at point A. After it reaches point B, it travels at constant speed until it reaches point E. It comes to rest at point F. (a) At the middle of each segment (AB, BC, CD, DE, and EF), what is the direction of the velocity vector? (b) At which of these points does the automobile have an acceleration? In those cases, what is the direction of the acceleration? (c) How do the magnitudes of the acceleration compare for segments BC and DE?

**FIGURE 3-42**
**Problem 29**

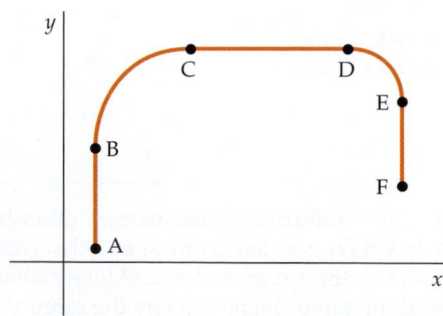

**30** •• **SSM** Two cannons are pointed directly toward each other as shown in Figure 3-43. When fired, the cannonballs will follow the trajectories shown—P is the point where the trajectories cross each other. If we want the cannonballs to hit each other, should the gun crews fire cannon A first, cannon B first, or should they fire simultaneously? Ignore the effects of air resistance.

**FIGURE 3-43** Problem 30

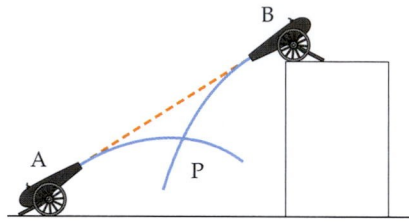

**31** •• Galileo wrote the following in his *Dialogue concerning the two world systems:* "Shut yourself up . . . in the main cabin below decks on some large ship, and . . . hang up a bottle that empties drop by drop into a wide vessel beneath it. When you have observed [this] carefully . . . have the ship proceed with any speed you like, so long as the motion is uniform and not fluctuating this way and that. . . . The droplets will fall as before into the vessel beneath without dropping towards the stern, although while the drops are in the air the ship runs many spans." Explain this quotation.

**32** • A man swings a stone attached to a rope in a horizontal circle at constant speed. Figure 3-44 represents the path of the rock looking down from above. (a) Which of the vectors $\vec{A}$ to $\vec{E}$ could represent the velocity of the stone? (b) Which could represent the acceleration?

**FIGURE 3-44** Problem 32

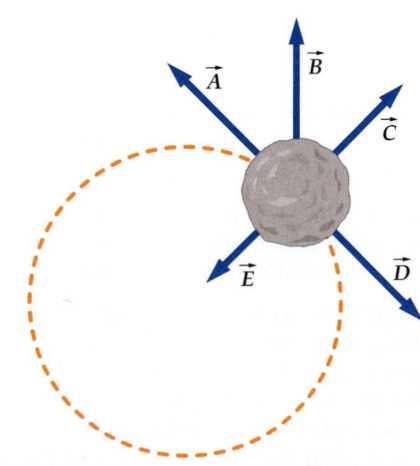

**33** • True or false: An object cannot move in a circle unless it is accelerating.

**34** •• Using a motion diagram, find the direction of the acceleration of the bob of a pendulum when the bob is at a point where it is just reversing its direction.

**35** • The speed of a batted baseball immediately after being struck can reach 110 mph. Let's say that the baseball has a "launch angle" of 35°, which is fairly typical for the sport. Naively using the range equation (Equation 3-23) to calculate the distance the ball will travel, we find a range of 760 ft (232 m)! In reality, it will only travel about 400 ft. Can you give a reason why the range equation breaks down so badly here? Be specific: If you can, look up the *terminal speed* for a baseball.

## Estimation and Approximation

**36** •• **SSM** Estimate how far you can throw a ball if you throw it (*a*) horizontally while standing on level ground, (*b*) at $\theta = 45°$ while standing on level ground, (*c*) horizontally from the top of a building 12 m high, (*d*) at $\theta = 45°$ from the top of a building 12 m high.

**37** •• **ISOLVE** In 1978, Geoff Capes of Great Britain threw a heavy brick a horizontal distance of 44.5 m. Find the approximate velocity of the brick at the highest point of its flight, neglecting the effects of air resistance.

## Vectors, Vector Addition, and Coordinate Systems

**38** • A wall clock has a minute hand that has a length of 0.5 m and an hour hand with a length of 0.25 m. Taking the center of the clock as the origin, and choosing an appropriate coordinate system, write the position of the hour and minute hands as vectors when the time reads (*a*) 12:00, (*b*) 3:30, (*c*) 6:30, (*d*) 7:15. (*e*) Call the position of the tip of the minute hand $\vec{A}$ and the position of the tip of the hour hand $\vec{B}$. Find $\vec{A} - \vec{B}$ for the times given in (*a*)–(*d*) above.

**39** • **SSM** A bear walks northeast for 12 m and then east for 12 m. Show each displacement graphically and find the resultant displacement vector graphically, as in Example 3-2(*a*).

**40** • **ISOLVE✓** A circular arc is centered at $x = 0$, $y = 0$. (*a*) A student walks along the circular arc from the position $x = 5$ m, $y = 0$ to a final position $x = 0$, $y = 5$ m. What is her displacement? (*b*) A second student walks from the same initial position along the *x* axis to the origin and then along the *y* axis to $y = 5$ m and $x = 0$. What is his displacement?

**41** • **SSM** **ISOLVE✓** For the two vectors $\vec{A}$ and $\vec{B}$ of Figure 3-45, find the following graphically as in Example 3-2(*a*): (*a*) $\vec{A} + \vec{B}$, (*b*) $\vec{A} - \vec{B}$, (*c*) $2\vec{A} + \vec{B}$, (*d*) $\vec{B} - \vec{A}$, (*e*) $2\vec{B} - \vec{A}$.

**FIGURE 3-45** Problem 41

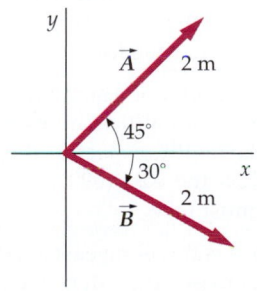

**42** • A Scout walks 2.4 km due east from camp, then turns left and walks 2.4 km along the arc of a circle centered at the campsite, and finally walks 1.5 km directly toward the camp. (*a*) How far is the Scout from camp at the end of his walk? (*b*) In what direction is the Scout's position relative to the campsite? (*c*) What is the ratio of the final magnitude of the displacement to the total distance walked?

**43** • A velocity vector has an *x* component of +5.5 m/s and a *y* component of −3.5 m/s. Which diagram in Figure 3-46 shows the direction of the vector correctly?

**FIGURE 3-46** Problem 43

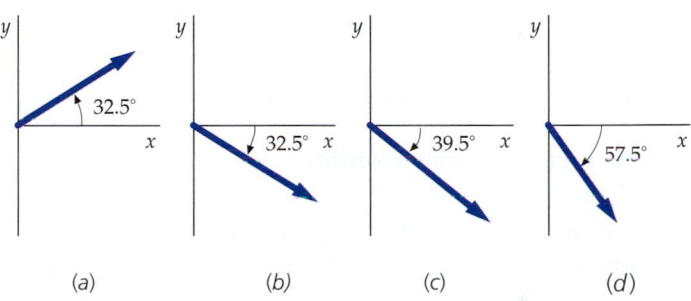

(*a*)      (*b*)      (*c*)      (*d*)

(*e*) None of the above.

**44** • **ISOLVE** Three vectors $\vec{A}$, $\vec{B}$, and $\vec{C}$ have the following *x* and *y* components: $A_x = 6$, $A_y = -3$; $B_x = -3$, $B_y = 4$; $C_x = 2$, $C_y = 5$. The magnitude of $\vec{A} + \vec{B} + \vec{C}$ is (*a*) 3.3, (*b*) 5.0, (*c*) 11, (*d*) 7.8, (*e*) 14.

**45** • Find the rectangular components of the following vectors $\vec{A}$ which lie in the *xy* plane and make an angle $\theta$ with the *x* axis (Figure 3-47) if (*a*) $A = 10$ m, $\theta = 30°$, (*b*) $A = 5$ m, $\theta = 45°$, (*c*) $A = 7$ km, $\theta = 60°$, (*d*) $A = 5$ km, $\theta = 90°$, (*e*) $A = 15$ km/s, $\theta = 150°$, (*f*) $A = 10$ m/s, $\theta = 240°$, (*g*) $A = 8$ m/s², $\theta = 270°$.

**FIGURE 3-47**
**Problem 45**

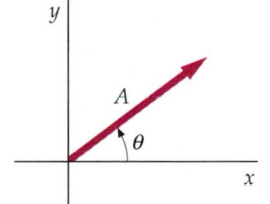

**46** • **SSM** Vector $\vec{A}$ has a magnitude of 8 m at an angle of 37° with the positive *x* axis; vector $\vec{B} = (3 \text{ m})\hat{i} - (5 \text{ m})\hat{j}$; vector $\vec{C} = (-6 \text{ m})\hat{i} + (3 \text{ m})\hat{j}$. Find the following vectors: (*a*) $\vec{D} = \vec{A} + \vec{C}$, (*b*) $\vec{E} = \vec{B} - \vec{A}$, (*c*) $\vec{F} = \vec{A} - 2\vec{B} + 3\vec{C}$, (*d*) a vector $\vec{G}$ such that $\vec{G} - \vec{B} = \vec{A} + 2\vec{C} + 3\vec{G}$.

**47** •• Find the magnitude and direction of the following vectors: (*a*) $\vec{A} = 5\hat{i} + 3\hat{j}$, (*b*) $\vec{B} = 10\hat{i} - 7\hat{j}$, (*c*) $\vec{C} = -2\hat{i} - 3\hat{j} + 4\hat{k}$.

**48** • Find the magnitude and direction of $\vec{A}$, $\vec{B}$, and $\vec{C} = \vec{A} + \vec{B}$ for (*a*) $\vec{A} = -4\hat{i} - 7\hat{j}$, $\vec{B} = 3\hat{i} - 2\hat{j}$, and (*b*) $\vec{A} = 1\hat{i} - 4\hat{j}$, $\vec{B} = 2\hat{i} + 6\hat{j}$.

**49** • Describe the following vectors using the unit vectors $\hat{i}$ and $\hat{j}$: (*a*) a velocity of 10 m/s at an angle of elevation of 60°, (*b*) a vector $\vec{A}$ of magnitude $A = 5$ m and $\theta = 225°$, (*c*) a displacement from the origin to the point $x = 14$ m, $y = -6$ m.

**50** • For the vector $\vec{A} = 3\hat{i} + 4\hat{j}$, find any three other vectors $\vec{B}$ that also lie in the $xy$ plane and have the property that $A = B$ but $\vec{A} \neq \vec{B}$. Write these vectors in terms of their components and show them graphically.

**51** •• **SSM** The faces of a cube with 3-m-long edges are parallel to the coordinate planes. The cube has one corner at the origin. A fly begins at the origin and walks along three edges until it is at the far corner. Write the displacement vector of the fly using the unit vectors $\hat{i}$, $\hat{j}$, and $\hat{k}$, and find the magnitude of this displacement.

**52** • **SSM** A ship at sea receives radio signals from two transmitters A and B, which are 100 km apart, one due south of the other. The direction finder shows that transmitter A is $\theta = 30°$ south of east, while transmitter B is due east. Calculate the distance between the ship and transmitter B.

## Velocity and Acceleration Vectors

**53** • A stationary radar operator determines that a ship is 10 km south of him. An hour later the same ship is 20 km southeast. If the ship moved at constant speed and always in the same direction, what was its velocity during this time?

**54** • A particle's position coordinates $(x, y)$ are (2 m, 3 m) at $t = 0$; (6 m, 7 m) at $t = 2$ s; and (13 m, 14 m) at $t = 5$ s. (a) Find the average velocity $v_{av}$ from $t = 0$ to $t = 2$ s. (b) Find $v_{av}$ from $t = 0$ to $t = 5$ s.

**55** • **SSM** **ISOLVE** ✓ A particle moving at a velocity of 4.0 m/s in the positive $x$ direction is given an acceleration of 3.0 m/s² in the positive $y$ direction for 2.0 s. The final speed of the particle is (a) $-2.0$ m/s, (b) 7.2 m/s, (c) 6.0 m/s, (d) 10 m/s, (e) none of the above.

**56** • Initially, a particle is moving due west with a speed of 40 m/s; 5 s later it is moving due north with a speed of 30 m/s. (a) What was the change in the magnitude of the particle's velocity during this time? (b) What was the change in the direction of the velocity? (c) What are the magnitude and direction of $\Delta\vec{v}$ for this interval? (d) What are the magnitude and direction of $\vec{a}_{av}$ for this interval?

**57** • At $t = 0$, a particle located at the origin has a velocity of 40 m/s at $\theta = 45°$. At $t = 3$ s, the particle is at $x = 100$ m and $y = 80$ m with a velocity of 30 m/s at $\theta = 50°$. Calculate (a) the average velocity and (b) the average acceleration of the particle during this interval.

**58** •• **SSM** **ISOLVE** ✓ A particle moves in the $xy$ plane with constant acceleration. At time zero, the particle is at $x = 4$ m, $y = 3$ m and has velocity $\vec{v} = (2 \text{ m/s})\hat{i} + (-9 \text{ m/s})\hat{j}$. The acceleration is given by $\vec{a} = (4 \text{ m/s}^2)\hat{i} + (3 \text{ m/s}^2)\hat{j}$. (a) Find the velocity at $t = 2$ s. (b) Find the position at $t = 4$ s. Give the magnitude and direction of the position vector.

**59** •• **ISOLVE** A particle has a position vector given by by $\vec{r} = (30t)\hat{i} + (40t - 5t^2)\hat{j}$, where $r$ is in meters and $t$ is in seconds. Find the instantaneous-velocity and instantaneous-acceleration vectors as functions of time $t$.

**60** •• A particle has a constant acceleration of $\vec{a} = (6 \text{ m/s}^2)\hat{i} + (4 \text{ m/s}^2)\hat{j}$. At time $t = 0$, the velocity is zero and the position vector is $\vec{r}_0 = (10 \text{ m})\hat{i}$. (a) Find the velocity and position vectors at any time $t$. (b) Find the equation of the particle's path in the $xy$ plane and sketch the path.

**61** ••• **ISOLVE** Starting from rest at a dock, a motor boat heads *north* while accelerating at a constant 3 m/s² for 20 s. The boat then turns *west* at the speed that it had at 20 s and travels west at this constant speed for 10 s. (a) What was the average velocity of the boat during the 30-s trip? (b) What was the average acceleration of the boat during the 30-s trip? (c) What is the displacement of the boat from the dock at the end of the 30-s trip?

**62** ••• **SSM** Mary and Robert decide to rendezvous on Lake Michigan. Mary departs in her boat from Petoskey at 9:00 A.M. and travels due north at 8 mi/h. Robert leaves from his home on the shore of Beaver Island, 26 mi, 30° west of north of Petoskey, at 10:00 A.M. and travels at a constant speed of 6 mi/h. In what direction should Robert be heading to intercept Mary, and where and when will they meet?

## Relative Velocity

**63** •• A plane flies at an airspeed of 250 km/h. There is a wind blowing at 80 km/h in the northeast direction at exactly 45° to the east of north. (a) In what direction should the plane head in order to fly due north? (b) What is the speed of the plane relative to the ground?

**64** •• **ISOLVE** ✓ A swimmer heads directly across a river, swimming at 1.6 m/s relative to the water. She arrives at a point 40 m downstream from the point directly across the river, which is 80 m wide. (a) What is the speed of the river current? (b) What is the swimmer's speed relative to the shore? (c) In what direction should the swimmer head to arrive at the point directly opposite her starting point?

**65** •• **SSM** A small plane departs from point A heading for an airport 520 km due north at point B. The airspeed of the plane is 240 km/h and there is a steady wind of 50 km/h blowing northwest to southeast. Determine the proper heading for the plane and the time of flight.

**66** •• **ISOLVE** ✓ Two boat landings are 2.0 km apart on the same bank of a stream that flows at 1.4 km/h. A motorboat makes the round trip between the two landings in 50 min. What is the speed of the boat relative to the water?

**67** •• In radio-controlled model airplane competition each plane must fly from the center of a 1-km-radius circle to any point on the circle and back to the center. The winner is the plane with the shortest round-trip time. The contestants are free to fly their planes along any route so long as the plane begins at the center, travels to the circle, and then returns to the center. On the day of the race, a steady wind blows out of the north at 5 m/s. Your plane can maintain an airspeed of 15 m/s. Strategy is paramount. Should you fly your plane upwind on the first leg and downwind on the trip back, or across the wind flying east and then west? Optimize your chances by calculating the round-trip time for both routes.

**68** • **ISOLVE** The pilot of a small plane maintains an air speed of 150 kts (knots, or nautical miles per hour) and wants to fly due north (000°) with respect to the earth. If a wind of 30 kts is blowing *from* the east (090°), calculate the heading (azimuth) the pilot must take.

**69** •• **SSM** Car A is traveling east at 20 m/s toward an intersection. As car A crosses the intersection, car B starts from rest 40 m north of the intersection and moves south with a constant acceleration of 2 m/s². Six seconds after A crosses the intersection find (a) the position of B relative to A, (b) the velocity of B relative to A, (c) the acceleration of B relative to A.

**70** ••• SSM A tennis racket is held horizontally and a tennis ball is held above the racket. When the ball is dropped from rest, and bounces off the strings of the racket, the ball always rebounds to 64% of its initial height. (a) Express the speed of the tennis ball just after it bounces as some fraction of the speed of the ball just before the bounce. (b) The tennis ball is now thrown up into the air and served using the same racket. Assuming the ball's pre-impact speed is zero, and that the racket's speed through impact is 25 m/s, with what speed does the tennis ball come off the racket strings? *Hint: Using the results of part (a), solve for the post-impact speed of the ball in the reference frame of the racket, and then calculate the ball's speed in the reference frame of the earth.* (c) From some well-established laws of physics, we never see a ball bounce higher than the point from which it was released. From this, can you give an upper bound on the speed of a served tennis ball in relation to the speed of the racket, no matter how well the racket is designed? (We will see later that these results can be explained in a different context: the idea of *conservation of momentum.*)

## Circular Motion and Centripetal Acceleration

**71** • What is the acceleration of the extreme tip of the minute hand of the clock in Problem 38? Express it as a fraction of the magnitude of free-fall acceleration $g$.

**72** • A centrifuge spins at a rate of 15,000 rev/min. (a) Calculate the centripetal acceleration of a test-tube sample held in the centrifuge arm 15 cm from the rotation axis. (b) It takes 1 min, 15 s for the centrifuge to spin up to its maximum rate of revolution from rest. Calculate the *magnitude* of the tangential acceleration of the centrifuge while it is spinning up, assuming that the tangential acceleration is constant.

**73** • An object resting on the equator has an acceleration toward the center of the earth due to the earth's rotational motion about its axis, and an acceleration toward the sun due to the earth's orbital motion. Calculate the magnitudes of both of these accelerations, and express them as a fraction of the magnitude of free-fall acceleration $g$. Use values from the physical-data table in the textbook.

**74** •• SSM Determine the acceleration of the moon toward the earth, using values for its mean distance and orbital period from the physical-data table in the textbook. Assume a circular orbit. Express the acceleration as a fraction of the magnitude of free-fall acceleration $g$.

**75** • A boy whirls a ball on a string in a horizontal circle of radius 0.8 m. How many revolutions per minute does the ball make if the magnitude of its centripetal acceleration is $g$ (the magnitude of the free-fall acceleration)?

## Projectile Motion and Projectile Range

**76** • A pitcher throws a fastball at 140 km/h (about 87 mi/h) toward home plate, which is 18.4 m away. Neglecting air resistance (not a good idea if you are the batter), how far does the ball drop because of gravity by the time it reaches home plate?

**77** • A projectile is launched with speed $v_0$ at an angle of $\theta_0$ with the horizontal. Find an expression for the maximum height it reaches above its starting point in terms of $v_0$, $\theta_0$, and $g$.

**78** •• SSM A cannonball is fired with initial speed $v_0$ at an angle 30° above the horizontal from a height of 40 m above the ground. The projectile strikes the ground with a speed of $1.2v_0$. Find $v_0$.

**79** •• In Figure 3-48, if $x$ is 50 m and $h = 10$ m, what is the minimum initial speed of the dart if it is to hit the monkey before hitting the ground, which is 11.2 m below the initial position of the monkey?

**FIGURE 3-48** Problem 79

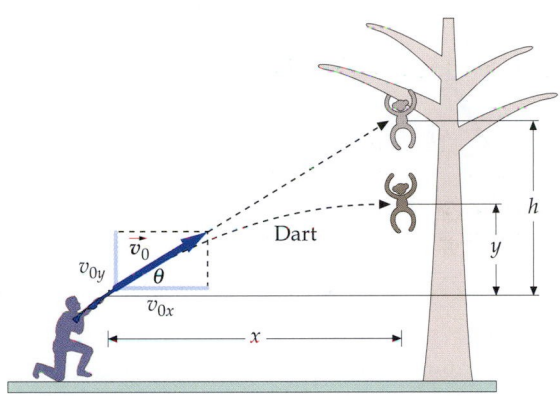

**80** •• A projectile is fired with an initial speed of 53 m/s. Find the angle of projection such that the maximum height of the projectile is equal to its horizontal range.

**81** •• A ball thrown into the air lands 40 m away 2.44 s later. Find the direction and magnitude of the initial velocity.

**82** •• SSM iSOLVE Consider a ball that is thrown with initial speed $v_0$ at an angle $\theta$ above the horizontal. If we consider its speed $v$ at some height $h$ above the ground, show that $v(h)$ is independent of $\theta$.

**83** •• At $\frac{1}{2}$ of its maximum height, the speed of a projectile is $\frac{3}{4}$ of its initial speed. What was its launch angle?

**84** • iSOLVE✓ A cargo plane is flying horizontally at an altitude of 12 km with a speed of 900 km/h when a large crate falls out of the rear loading ramp. (a) How long does it take the crate to hit the ground? (b) How far horizontally is the crate from the point where it fell off when it hits the ground? (c) How far is the crate from the aircraft when the crate hits the ground, assuming that the plane continues to fly with constant velocity?

**85** •• SSM iSOLVE Wile E. Coyote (*Carnivorous hungribilous*) is chasing the Roadrunner (*Speedibus cantcatchmi*) yet again. While running down the road, they come to a deep gorge, 15 m straight across and 100 m deep. The Roadrunner launches itself across the gorge at a launch angle of 15° above the horizontal, and lands with 1.5 m to spare. (a) What was the Roadrunner's launch speed? Ignore air resistance. (b) Wiley Coyote launches himself across the gorge with the same initial speed, but at a different launch angle. To his horror, he is short the other lip by 0.5 m. What was his launch angle? (Assume that it was lower than 15°.)

**86** • A cannon is elevated at an angle of 45°. It fires a ball with a speed of 300 m/s. (*a*) What height does the ball reach? (*b*) How long is the ball in the air? (*c*) What is the horizontal range of the cannon?

**87** •• A stone thrown horizontally from the top of a 24-m tower hits the ground at a point 18 m from the base of the tower. (*a*) Find the speed with which the stone was thrown. (*b*) Find the speed of the stone just before it hits the ground.

**88** •• **ISOLVE** A projectile is fired into the air from the top of a 200-m cliff above a valley (Figure 3-49). Its initial velocity is 60 m/s at 60° above the horizontal. Where does the projectile land?

**FIGURE 3-49** Problem 88

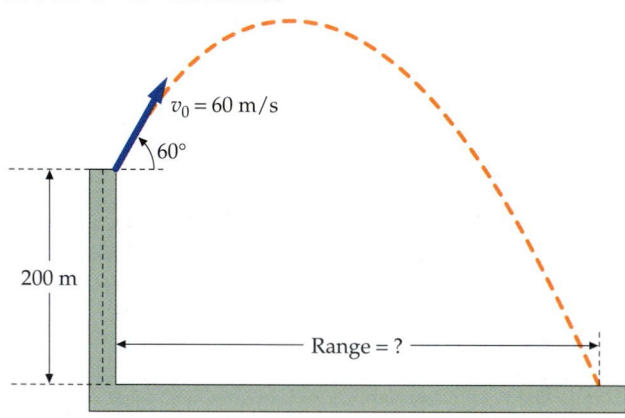

**89** •• The range of a cannonball fired horizontally from a cliff is equal to the height of the cliff. What is the direction of the velocity vector when the projectile strikes the ground?

**90** • A projectile is fired at an angle of 60° above the horizontal with an initial speed of 300 m/s. Calculate (*a*) the horizontal distance traveled and (*b*) the vertical height attained in the first 6 s.

**91** •• A cannonball is fired with an initial speed of 42.2 m/s at an angle of 30° above the horizontal from an initial height of 40 m. Find the range of the cannonball.

**92** •• **SSM** **ISOLVE** The speed of an arrow fired from a compound bow is about 45 m/s. (*a*) A Tartar archer sits astride his horse and launches an arrow into the air, elevating the bow at an angle of 10° above the horizontal. If the bow is 2.25 m above the ground, what is the arrow's range? Assume that the ground is level, and ignore air resistance. (*b*) Now assume that his horse is at full gallop, moving in the same direction as he will fire the arrow, and that he elevates the bow in the same way as in part (*a*) and fires. If the horse's speed is 12 m/s, what is the arrow's range now?

**93** • **ISOLVE✓** Using a potato cannon, Chuck launches a spud-plug horizontally with an initial velocity of 50.0 m/s (about 112 mi/h). (*a*) If Chuck holds the potato cannon so that it is 1.00 m above ground, how long is the spud-plug in the air? (*b*) How far does the plug travel before hitting the ground? (You can find more about potato cannons on the Internet.)

**94** •• Compute $dR/d\theta$ from $R = (v_0^2/g)\sin(2\theta_0)$ and show that setting $dR/d\theta = 0$ gives $\theta = 45°$ for the maximum range.

**95** • **SSM** In a science fiction short story written in the 1970s, Ben Bova described a conflict between two hypothetical colonies on the moon—one founded by the United States and the other by the USSR. In the story, colonists from each side started firing bullets at each other, only to find to their horror that their rifles had a high enough muzzle velocity that the bullets went into orbit. (*a*) If the magnitude of free-fall acceleration on the moon is 1.67 m/s², what is the maximum range of a rifle bullet with a muzzle velocity of 900 m/s? (Assume the curvature at the surface of the moon is negligible.) (*b*) What would the muzzle velocity have to be to send the bullet into a circular orbit just above the surface of the moon? (You will first need to look up the radius of the moon.)

**96** ••• In the text, we calculated the range for a projectile that lands at the same elevation from which it is fired as $R = (v_0^2/g)\sin 2\theta_0$. Show that the change in the range for a small change in free-fall acceleration is given by $\Delta R/R = -\Delta g/g$.

**97** ••• In the text, we calculated the range for a projectile that lands at the same elevation from which it is fired as $R = (v_0^2/g)\sin 2\theta_0$. Show that the change in the range for a small change in launch velocity is given by $\Delta R/R = 2\Delta v_0/v_0$.

**98** ••• In the text, we calculated the range for a projectile that lands at the same elevation from which it is fired as $R = (v_0^2/g)\sin 2\theta_0$. Show that the range for the more general problem (Figure 3-50) where $\Delta y \neq 0$ is given by

**FIGURE 3-50** Problem 98

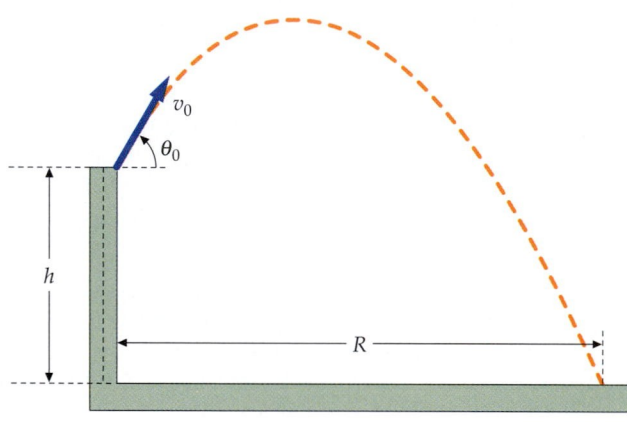

$$R = \left(1 + \sqrt{1 + \frac{2gh}{v_0^2 \sin^2\theta_0}}\right)\frac{v_0^2}{2g}\sin 2\theta_0$$

**99** •• **SSM** A projectile is launched over level ground at an elevation angle of $\theta$. An observer standing at the launch site sights the projectile at the point of its highest elevation, and measures the angle $\phi$ shown in Figure 3-51. Show that $\tan \phi = \frac{1}{2}\tan\theta$.

**FIGURE 3-51** Problem 99

**100** • A projectile, fired with unknown initial velocity, lands 20 s later on the side of a hill, 3000 m away horizontally and 450 m vertically above its starting point. (*a*) What is the vertical component of its initial velocity? (*b*) What is the horizontal component of its initial velocity?

**101** •• **SSM** A stone is thrown horizontally from the top of an incline that makes an angle $\theta$ with the horizontal. If the stone's initial speed is $v_0$, how far down the incline will it land?

**102** ••• A toy cannon is placed on a ramp that has a slope of angle $\phi$. (*a*) If the cannonball is projected up the hill at an angle of $\theta_0$ above the horizontal (Figure 3-52) and has a muzzle speed of $v_0$, show that the range $R$ of the cannonball (as measured along the ramp) is given by

$$R = \frac{2v_0^2 \cos^2\theta_0 \, (\tan\theta_0 - \tan\phi)}{g \cos\phi}$$

Ignore air resistance.

**FIGURE 3-52** Problem 102

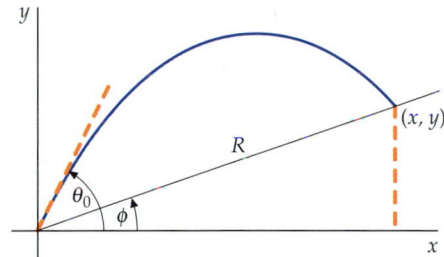

**103** •• A rock is thrown from the top of a 20-m building at an angle of 53° above the horizontal. If the horizontal range of the throw is equal to the height of the building, with what speed was the rock thrown? What is the velocity of the rock just before it strikes the ground?

**104** •• A girl throws a ball at a vertical wall 4 m away (Figure 3-53). The ball is 2 m above ground when it leaves the girl's hand with an initial velocity of $\vec{v}_0 = (10 \text{ m/s})(\hat{i} + \hat{j})$ or $10\sqrt{2}$ m/s at 45°. When the ball hits the wall, the horizontal component of its velocity is reversed; the vertical component remains unchanged. Where does the ball hit the ground? *Hint: The wall can be thought of as a mirror. Determine the range, neglecting the wall, and then consider the mirror-like reflection.*

**FIGURE 3-53** Problem 104

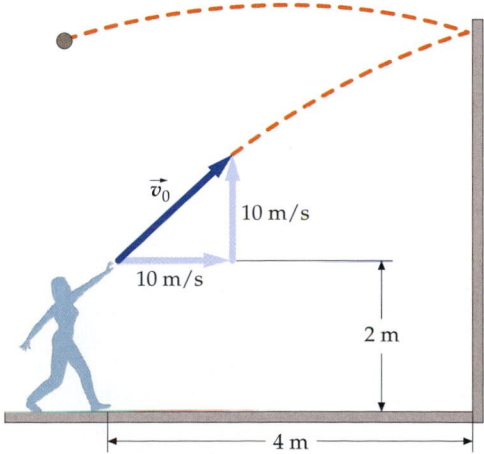

### Hitting Targets and Related Problems

**105** • A boy uses a slingshot to project a pebble at a shoulder-height target 40 m away. He finds that to hit the target he must aim 4.85 m above the target. Determine the velocity of the pebble on leaving the slingshot and the time of flight of the pebble.

**106** •• **SSM** The distance from the pitcher's mound to home plate is 18.4 m. The mound is 0.2 m above the level of the field. A pitcher throws a fastball with an initial speed of 37.5 m/s. At the moment the ball leaves the pitcher's hand, it is 2.3 m above the mound. What should the angle between $\vec{v}_0$ and the horizontal be so that the ball crosses the plate 0.7 m above ground? (Neglect interaction with air.)

**107** •• Suppose that a hockey puck is struck in such a way that, when it is at its highest point, it just clears a Plexiglass wall of height $h = 2.80$ m. Find $v_{0y}$, the time $t$ to reach the wall, and $v_{0x}$, $v_0$, and $\theta_0$ for this case. Assume that the horizontal distance is $x_1 = 12.0$ m.

**108** •• Carlos is on his trail bike, approaching a creek bed that is 7 m wide. A ramp with an incline of 10° has been built for daring people who try to jump the creek. Carlos is traveling at his bike's maximum speed, 40 km/h. (*a*) Should Carlos attempt the jump or emphatically hit the brakes? (*b*) What is the minimum speed a bike must have to make this jump? Assume equal elevations on either side of the creek.

**109** •• If a bullet that leaves the muzzle of a gun at 250 m/s is to hit a target 100 m away at the level of the muzzle, the gun must be aimed at a point above the target. How far above the target is that point?

### General Problems

**110** • The displacement vectors $\vec{A}$ and $\vec{B}$ in Figure 3-54 both have a magnitude of 1 m. (*a*) Find their $x$ and $y$ components. (*b*) Find the components, magnitude, and direction of the sum $\vec{S} = \vec{A} + \vec{B}$. (*c*) Find the components, magnitude, and direction of the difference $\vec{D} = \vec{A} - \vec{B}$.

**FIGURE 3-54** Problem 110

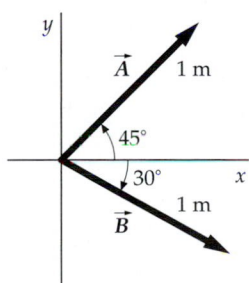

**111** • **SSM** A plane is inclined at an angle of 30° from the horizontal. Choose the $x$ axis pointing down the slope of the plane and the $y$ axis perpendicular to the plane. Find the $x$ and $y$ components of the acceleration of gravity, which has the magnitude 9.81 m/s² and points vertically down.

**112** • Two vectors $\vec{A}$ and $\vec{B}$ lie in the $xy$ plane. Under what conditions does the ratio $A/B$ equal $A_x/B_x$?

**113** • The position vector of a particle is given by $\vec{r} = (5 \text{ m/s})t\hat{i} + (10 \text{ m/s})t\hat{j}$, where $t$ is in seconds and $\vec{r}$ is in meters. (a) Draw the path of the particle in the $xy$ plane. (b) Find $\vec{v}$ in component form and then find its magnitude.

**114** •• **iSOLVE** A worker on the roof of a house drops her hammer, which slides down the roof at constant speed of 4 m/s. The roof makes an angle of 30° with the horizontal, and its lowest point is 10 m from the ground. What is the horizontal distance traveled by the hammer between the time it leaves the roof of the house and the time it hits the ground?

**115** •• In 1940, Emanuel Zacchini flew about 53 m as a human cannonball, a record that remains unbroken. His initial velocity was 24.2 m/s at an angle $\theta$. Find $\theta$ and the maximum height $h$ Emanuel achieved during the record flight. Ignore the effects of air resistance.

**116** •• A particle moves in the $xy$ plane with constant acceleration. At $t = 0$ the particle is at $\vec{r}_1 = (4 \text{ m})\hat{i} + (3 \text{ m})\hat{j}$, with velocity $\vec{v}_1$. At $t = 2$ s the particle has moved to $\vec{r}_2 = (10 \text{ m})\hat{i} - (2 \text{ m})\hat{j}$ and its velocity has changed to $\vec{v}_2 = (5 \text{ m/s})\hat{i} - (6 \text{ m/s})\hat{j}$. (a) Find $\vec{v}_1$. (b) What is the acceleration of the particle? (c) What is the velocity of the particle as a function of time? (d) What is the position vector of the particle as a function of time?

**117** •• **SSM** A small steel ball is projected horizontally off the top landing of a long, rectangular staircase. The initial speed of the ball is 3 m/s. Each step is 0.18 m high and 0.3 m wide. Which step does the ball strike first?

**118** •• Suppose you can throw a ball a distance $x_0$ when standing on level ground. How far can you throw it from a building of height $h = x_0$ if you throw it at (a) 0°? (b) 30°? (c) 45°?

**119** ••• Darlene is a stunt motorcyclist in a traveling circus. For the climax of her show, she takes off from the ramp at angle $\theta$, clears a fiery ditch of width $x$, and lands on an elevated platform (height $h$) on the other side (Figure 3-55). (a) For a given height $h$, find the minimum necessary takeoff speed $v_{min}$ needed to make the jump successfully. (b) What is $v_{min}$ for a launch angle $\theta = 30°$ with a pit width of 8 m and a platform height $h = 4$ m? (c) Show that no matter what her takeoff speed is, the maximum height of the platform is $h_{max} < x \tan \theta$. Interpret this result physically. (Neglect the effects of air resistance and treat the bike as if it were a particle.)

**FIGURE 3-55** Problem 119

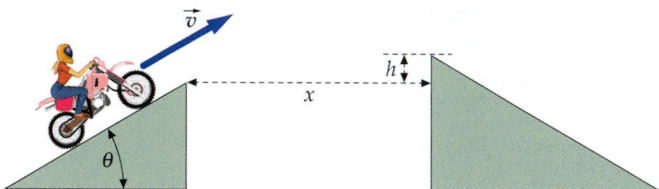

**120** ••• A small boat is headed for a harbor 32 km northwest of its current position when it is suddenly engulfed in heavy fog. The captain maintains a compass bearing of northwest and a speed of 10 km/h relative to the water. The fog lifts 3 h later and the captain notes that he is now exactly 4.0 km south of the harbor. (a) What was the average velocity of the current during those 3 h? (b) In what direction should the boat have been heading to reach its destination along a straight course? (c) What would its travel time have been if it had followed a straight course?

**121** •• **SSM** Galileo showed that, if air resistance is neglected, the ranges for projectiles whose angles of projection exceed or fall short of 45° by the same amount are equal. Prove Galileo's result.

**122** •• Two balls are thrown with equal speeds from the top of a cliff of height $h$. One ball is thrown at an angle of $\alpha$ above the horizontal. The other ball is thrown at an angle of $\beta$ below the horizontal. Show that each ball strikes the ground with the same speed, and find that speed in terms of $h$ and the initial speed $v_0$.

# Newton's Laws

THIS AIRPLANE IS ACCELERATING AS IT HEADS DOWN THE RUNWAY BEFORE TAKEOFF. NEWTON'S LAWS RELATE AN OBJECT'S ACCELERATION TO ITS MASS AND THE FORCES ACTING ON IT.

 **If you were a passenger on this plane, how might you use Newton's laws to determine the plane's acceleration? (See Example 4-9.)**

4-1   Newton's First Law: The Law of Inertia

4-2   Force, Mass, and Newton's Second Law

4-3   The Force Due to Gravity: Weight

4-4   Forces in Nature

4-5   Problem Solving: Free-Body Diagrams

4-6   Newton's Third Law

4-7   Problems With Two or More Objects

**N**ow that we have studied how objects move in one, two, and three dimensions, we can ask the question, "why do objects start to move?" What causes a moving object to gain speed or change direction?

Classical mechanics relates the forces objects exert on each other, and relates changes in the motion of an object to the forces that act on it. It describes phenomena using Newton's three laws of motion. While we may already have an intuitive idea of a force as a push or a pull, like that exerted by our muscles or by stretched rubber bands and springs, Newton's laws allow us to refine our understanding of forces.

➤ **In this chapter, we describe Newton's three laws of motion and begin using them to solve problems involving objects in motion and at rest.**

A modern wording of Newton's laws is:

**First law.** An object at rest stays at rest *unless* acted on by an external force. An object in motion continues to travel with constant velocity *unless* acted on by an external force.

**Second law.** The direction of the acceleration of an object is in the direction of the net external force acting on it. The acceleration is proportional to the net external force $\vec{F}_{net}$, in accordance with $\vec{F}_{net} = m\vec{a}$, where $m$ is the mass of the object. The net force acting on an object, also called the resultant force, is the vector sum of all the forces acting on it: $\vec{F}_{net} = \Sigma\vec{F}$. Thus,

$$\Sigma\vec{F} = m\vec{a} \qquad\qquad 4\text{-}1$$

NEWTON'S SECOND LAW

**Third law.** Forces always occur in equal and opposite pairs. If object A exerts a force $\vec{F}_{A,B}$ on object B, an equal but opposite force $\vec{F}_{B,A}$ is exerted by object B on object A. Thus,

$$\vec{F}_{B,A} = -\vec{F}_{A,B} \qquad\qquad 4\text{-}2$$

NEWTON'S THIRD LAW

# 4-1 Newton's First Law: The Law of Inertia

Push a piece of ice on a counter top: It slides, then stops. If the counter is wet, the ice will travel farther before stopping. A piece of dry ice (frozen carbon dioxide) riding on a cushion of carbon dioxide vapor slides quite far with little change in velocity. Before Galileo it was thought that a force, such as a push or pull, was always needed to keep an object moving with constant velocity. But Galileo, and later Newton, recognized that the slowing of objects in everyday experience is due to the force of friction. If friction is reduced, the rate of slowing is reduced. A water slick or a cushion of gas is especially effective at reducing friction, allowing the object to slide a great distance with little change in velocity. Galileo reasoned that, if we could remove from an object all external forces including friction, then the velocity of the object would never change—a property of matter he described as **inertia.** This conclusion, restated by Newton as his first law, is also called the **law of inertia.**

Friction is greatly reduced by a cushion of air that supports the hovercraft.

## Inertial Reference Frames

Newton's first law makes no distinction between an object at rest and an object moving with constant (nonzero) velocity. Whether an object remains at rest or remains moving with constant velocity depends on the reference frame in which the object is observed. Suppose you are a passenger on an airplane that is flying along a straight path at constant altitude and you carefully place a small ball on your seat tray (which is horizontal). Relative to the plane, the ball will remain at rest as long as the plane continues to fly at constant velocity relative to the ground. Relative to the ground, the ball remains moving with the same velocity as the plane.

Now, suppose that the pilot opens the throttle and the plane suddenly accelerates forward (relative to the ground). You will then observe that the ball on your tray suddenly starts to roll backward, accelerating (relative to the plane) even though there is no horizontal force acting on it.

A reference frame accelerating relative to an inertial reference frame is not an inertial reference frame. *Newton's first law thus gives us the criterion for determining if a reference frame is an inertial frame.* In fact, it is useful to think of Newton's first law as a statement that defines inertial reference frames.

> If no forces act on an object, any reference frame with respect to which the acceleration of the object remains zero is an **inertial reference frame.**
>
> DEFINITION—INERTIAL REFERENCE FRAME

Both the plane, when cruising at constant velocity, and the ground are, to a good approximation, inertial reference frames. Any reference frame moving with constant velocity relative to an inertial reference frame is also an inertial reference frame.

A reference frame attached to the surface of the earth is not quite an inertial reference frame because of the small acceleration of the surface of the earth due to the rotation of the earth and the small acceleration of the earth itself due to its revolution around the sun. However, these accelerations are of the order of $0.01 \text{ m/s}^2$ or less, so to a good approximation, a reference frame attached to the surface of the earth is an inertial reference frame.

The concept of inertial reference frame is of central importance because *Newton's first, second, and third law statements are valid only in inertial reference frames.*

# 4-2 Force, Mass, and Newton's Second Law

Newton's first and second laws allow us to define force. A **force** is an external influence on an object that causes it to accelerate relative to an inertial reference frame. (We assume there are no other forces acting.) The direction of the force is the direction of the acceleration it causes. The magnitude of the force is the product of the mass of the object and the magnitude of its acceleration. This definition is given in Equation 4-1.

Forces can be compared by stretching identical rubber bands. For example, if two identical rubber bands are stretched by the same amount, then they exert forces of equal magnitudes.

Objects intrinsically resist being accelerated. Imagine kicking a soccer ball or a bowling ball. The bowling ball resists being accelerated much more than does the soccer ball, as would be evidenced by your sore toes. This intrinsic property is called the object's **mass.** It is a measure of the object's inertia. The ratio of two masses is defined quantitatively by applying the same force to each and comparing their accelerations. If a force $F$ produces acceleration $a_1$ when applied to an object of mass $m_1$, and an equal force produces acceleration $a_2$ when applied to an object of mass $m_2$, then the ratio of the two masses is defined by

$$\frac{m_2}{m_1} = \frac{a_1}{a_2} \qquad\qquad 4\text{-}3$$

DEFINITION—MASS

This definition agrees with our intuitive idea of mass. If a force is applied to an object and a force of equal magnitude is applied to a second object, then the object with more mass will accelerate less. The ratio $a_1/a_2$ produced by forces of equal magnitude acting on two objects is found experimentally to be independent of the magnitude, direction, or type of force used. Mass is an intrinsic property of an object that does not depend on its location—it remains the same whether the object is on the earth, on the moon, or in outer space.

If a direct comparison shows that $m_2/m_1 = 2$ and $m_3/m_1 = 4$, then $m_3$ will be twice $m_2$ when objects 2 and 3 are compared with each other. We can therefore establish a mass scale by choosing a standard object and assigning it a mass of 1 unit. As we noted in Chapter 1, the object chosen as the international standard for mass is a platinum-iridium alloy cylinder carefully preserved at the International Bureau of Weights and Measures at Sèvres, France. The mass of the standard object is 1 **kilogram** (kg), the SI unit of mass. The force required to produce an acceleration of 1 m/s² on the standard object is defined to be 1 **newton** (N). The force that produces an acceleration of 2 m/s² on the standard object is 2 N, and so on.

---

*A SLIDING ICE CREAM CARTON*                    **E X A M P L E   4 - 1**

A given force produces an acceleration of 5 m/s² on the standard object of mass $m_1$. When an equal force is applied to a carton of ice cream of mass $m_2$, it produces an acceleration of 11 m/s². (*a*) What is the mass of the carton of ice cream? (*b*) What is the magnitude of the force?

**PICTURE THE PROBLEM** Apply $\Sigma\vec{F} = m\vec{a}$ to each object and solve for the mass of the ice-cream carton and the magnitude of the force.

(*a*) 1. Apply $\Sigma\vec{F} = m\vec{a}$ to each object. There is only one force and we only need to consider magnitudes of the vector quantities:

$$F_1 = m_1 a_1 \quad \text{and} \quad F_2 = m_2 a_2$$

2. The ratio of the masses varies inversely as the ratio of the accelerations under applied forces of equal magnitude:

$$F_1 = F_2 = F, \quad \text{so} \quad m_1 a_1 = m_2 a_2$$

and

$$\frac{m_2}{m_1} = \frac{a_1}{a_2} = \frac{5\ \text{m/s}^2}{11\ \text{m/s}^2}$$

3. Solve for $m_2$ in terms of $m_1$, which is 1 kg:

$$m_2 = \frac{5}{11} m_1 = \frac{5}{11}(1\ \text{kg}) = \boxed{0.45\ \text{kg}}$$

(*b*) The magnitude $F$ is found by using the mass and acceleration of either object:

$$F = m_1 a_1 = (1\ \text{kg})(5\ \text{m/s}^2) = \boxed{5\ \text{N}}$$

**EXERCISE** A force of 3 N produces an acceleration of 2 m/s² on an object of unknown mass. (*a*) What is the mass of the object? (*b*) If the force is increased to 4 N, what is the acceleration? (*Answer* (*a*) 1.5 kg (*b*) 2.67 m/s²)

It is found experimentally that two or more forces acting on an object accelerate it as if the object were acted upon by a single force equal to the vector sum of the individual forces. That is, forces combine as vectors. Newton's second law is thus

$$\Sigma\vec{F} = \vec{F}_{\text{net}} = m\vec{a}$$

---

*A WALK IN SPACE*                    **E X A M P L E   4 - 2**

You're stranded in space away from your spaceship. Fortunately, you have a propulsion unit that provides a constant force $\vec{F}$ for 3 s. After 3 s you have moved 2.25 m. If your mass is 68 kg, find $\vec{F}$.

The propulsion unit (not shown) is pushing the astronaut to the right.

**PICTURE THE PROBLEM**   The force acting on you is constant, so your acceleration $\vec{a}$ is also constant. We can use the kinematic equations of Chapter 2 to find $\vec{a}$, and then obtain the force from $\Sigma\vec{F} = m\vec{a}$. Choose $\vec{F}$ to be along the $x$ axis, so that $\vec{F} = F_x\hat{i}$ (Figure 4-1). The component of Newton's second law along the $x$ axis is then $F_x = ma_x$.

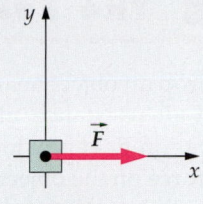

**FIGURE 4-1**

1. Apply $\Sigma\vec{F} = m\vec{a}$ to relate the net force to the mass and the acceleration:

$$F_x = ma_x$$

2. To find the acceleration, we use Equation 2-15 with $v_0 = 0$:

$$\Delta x = v_0 t + \tfrac{1}{2}a_x t^2 = \tfrac{1}{2}a_x t^2$$

$$a_x = \frac{2\Delta x}{t^2} = \frac{2(2.25\text{ m})}{(3\text{ s})^2} = 0.500\text{ m/s}^2$$

3. Substitute $a_x = 0.500\text{ m/s}^2$ and $m = 68\text{ kg}$ to find the force:

$$F_x = ma_x = (68\text{ kg})(0.500\text{ m/s}^2) = \boxed{34.0\text{ N}}$$

---

*A PARTICLE SUBJECTED TO FORCES*                  **EXAMPLE 4-3**   **Try It Yourself**

A particle of mass 0.4 kg is subjected simultaneously to two forces $\vec{F}_1 = -2\text{ N }\hat{i} - 4\text{ N }\hat{j}$ and $\vec{F}_2 = -2.6\text{ N }\hat{i} + 5\text{ N }\hat{j}$. If the particle is at the origin and starts from rest at $t = 0$, find (*a*) its position vector $\vec{r}$ and (*b*) its velocity $\vec{v}$ at $t = 1.6$ s.

**PICTURE THE PROBLEM**   Since $\vec{F}_1$ and $\vec{F}_2$ are constant, the acceleration of the particle is constant. Hence, we use the kinematic equations of Chapter 2 to determine the particle's position and velocity as functions of time.

**Cover the column to the right and try these on your own before looking at the answers.**

**Steps**

**Answers**

(*a*) 1. Write the general equation for the position vector $\vec{r}$ as a function of time $t$ for constant acceleration $\vec{a}$ in terms of $\vec{r}_0$, $\vec{v}_0$, and $\vec{a}$, and substitute $\vec{r}_0 = \vec{v}_0 = 0$.

$\vec{r} = \vec{r}_0 + \vec{v}_0 t + \tfrac{1}{2}\vec{a}t^2 = \tfrac{1}{2}\vec{a}t^2$

2. Use $\Sigma\vec{F} = m\vec{a}$ to write the acceleration $\vec{a}$ in terms of the resultant force $\Sigma\vec{F}$ and the mass $m$.

$\vec{a} = \dfrac{\Sigma\vec{F}}{m}$

3. Compute $\Sigma\vec{F}$ from the given forces.

$\Sigma\vec{F} = \vec{F}_1 + \vec{F}_2 = -4.6\text{ N }\hat{i} + 1.0\text{ N }\hat{j}$

4. Find the acceleration vector $\vec{a}$.

$\vec{a} = \dfrac{\Sigma\vec{F}}{m} = -11.5\text{ m/s}^2\,\hat{i} + 2.5\text{ m/s}^2\,\hat{j}$

5. Find the position vector $\vec{r}$ for a general time $t$.

$\vec{r} = \tfrac{1}{2}\vec{a}t^2 = \tfrac{1}{2}a_x t^2\hat{i} + \tfrac{1}{2}a_y t^2\hat{j}$

$\quad = -5.75\text{ m/s}^2\,t^2\hat{i} + 1.25\text{ m/s}^2\,t^2\hat{j}$

6. Find $\vec{r}$ at $t = 1.6$ s.

$\vec{r} = \boxed{-14.7\text{ m }\hat{i} + 3.20\text{ m }\hat{j}}$

(*b*) Write the velocity vector $\vec{v}$ in terms of the acceleration and time and compute its components for the time $t = 1.6$ s.

$\vec{v} = \vec{a}t = (-11.5\text{ m/s}^2\,\hat{i} + 2.5\text{ m/s}^2\,\hat{j})t$

$\quad = \boxed{-18.4\text{ m/s }\hat{i} + 4.00\text{ m/s }\hat{j}}$

# 4-3 The Force Due to Gravity: Weight

If we drop an object near the earth's surface, it accelerates toward the earth. If we neglect air resistance, all objects have the same acceleration, called the free-fall acceleration $\vec{g}$, at any location. The force causing this acceleration is the gravitational force on the object, called its weight.[†] If its weight $\vec{w}$ is the *only* force acting on an object, the object is said to be in **free-fall.** If its mass is $m$, Newton's second law ($\Sigma\vec{F} = m\vec{a}$) defines the weight $\vec{w}$:

$$\vec{w} = m\vec{g} \qquad\qquad 4\text{-}4$$

WEIGHT

Since $\vec{g}$ is the same for all objects, it follows that the weight of an object is proportional to its mass. The vector $\vec{g}$ is the force per unit mass exerted by the earth on any object and is called the **gravitational field** of the earth. It is equal to the free-fall acceleration.[‡] Near the surface of the earth, $g$ has the value

$$g = 9.81 \text{ N/kg} = 9.81 \text{ m/s}^2$$

Careful measurements show that $\vec{g}$ varies with location. In particular, at points above the surface of the earth, $\vec{g}$ points toward the center of the earth and varies inversely with the square of the distance to the center of the earth. Thus an object weighs slightly less at very high altitudes than it does at sea level. The gravitational field also varies slightly with latitude because the earth is not exactly spherical but is slightly flattened at the poles. **Thus weight, unlike mass, is not an intrinsic property of an object.** Although the weight of an object varies from place to place because of changes in $g$, this variation is too small to be noticed in most practical applications on or near the surface of the earth.

An example should help clarify the difference between mass and weight. Consider a bowling ball near the moon. Its weight is the gravitational force exerted on it by the moon, but that force is a mere sixth of the force exerted on the bowling ball when it is similarly positioned on earth. The ball weighs about one-sixth as much on the moon, and lifting the ball on the moon requires one-sixth the force. However, because the mass of the ball is the same on the moon as on the earth, throwing the ball with some horizontal acceleration requires the same force on the moon as on the earth or in free space.

Although the weight of an object may vary from one place to another, at any particular location the weight of the object is proportional to its mass. Thus we can conveniently compare the masses of two objects at a given location by comparing their weights.

Our sensation of our own weight comes from other forces that balance it. When you sit on a chair, you feel a force exerted by the chair that balances your weight and prevents you from falling to the floor. When you stand on a spring scale, your feet feel the force exerted by the scale. The scale is calibrated to read the force it must exert (by the compression of its springs) to balance your weight. This force is called your **apparent weight.** If there is no force to balance your weight, as in free-fall, your apparent weight is zero. This condition, called **weightlessness,** is experienced by astronauts in orbiting satellites. A satellite in a circular orbit near the surface of the earth is accelerating toward the earth. The only force acting on the satellite is that of gravity (its weight), so it is in free-fall.

---

† Referring to the gravitation force as "its weight" is unfortunate because it appears to imply that "its weight" is a property of the object rather than an external force acting on it. To avoid buying into the apparent implication of the wording, every time you read "its weight," mentally translate it to read "the gravitational force acting on it."

‡ $\vec{g}$ refers to the free-fall acceleration, which is the acceleration (of an object in free-fall) relative to the ground. It is not entirely correct to attribute it to the gravitational attraction of the earth. This distinction is discussed further in Chapter 11.

Astronauts in the satellite are also in free-fall. The only force on them is their weight, which produces the acceleration $g$. Because there is no force balancing the force of gravity, the astronauts have zero apparent weight.

## Units of Force and Mass

Like the second and the meter, the SI unit of mass, the kilogram, is a fundamental unit. The unit of force, the newton, and the units for other quantities that we will study such as momentum and energy are derived from the three fundamental units s, m, and kg.

As noted in Section 4-2, the newton is defined as the force that produces an acceleration of 1 m/s² when it acts on an object with a mass of 1 kg. Then Newton's second law gives

$$1 \, \text{N} = (1 \, \text{kg})(1 \, \text{m/s}^2) = 1 \, \text{kg·m/s}^2 \qquad 4\text{-}5$$

A convenient standard unit for mass in atomic and nuclear physics is the **unified mass unit** (u), which is defined as one-twelfth of the mass of the neutral carbon-12 ($^{12}$C) atom. The unified mass unit is related to the kilogram by

$$1 \, \text{u} = 1.660\,540 \times 10^{-27} \, \text{kg} \qquad 4\text{-}6$$

The mass of a hydrogen atom is approximately 1 u.

Although we will generally use SI units in this book, we need to know another scheme, the U.S. customary system, still used in the United States, which is based on the second, the foot, and the pound. The U.S. customary system differs from SI in that a unit of force, the pound, has been chosen as a fundamental unit rather than a unit of mass. The **pound** was originally defined as the weight of a particular standard object at a particular location. It is now defined as 4.448222 N. Rounding to three places, we have 1 lb ≈ 4.45 N. Since 1 kg weighs 9.81 N, its weight in pounds is

$$9.81 \, \text{N} = 2.20 \, \text{lb} \qquad 4\text{-}7$$

WEIGHT OF 1 KG

The unit of mass in the U.S. customary system is the rarely encountered slug, defined as the mass of an object that weighs 32.2 lb. When working problems in the U.S. customary system, we substitute $w/g$ for mass $m$, where $w$ is the weight in pounds and $g$ is the acceleration due to gravity in feet per second per second:

$$g = 32.2 \, \text{ft/s}^2 \qquad 4\text{-}8$$

---

*AN ACCELERATING STUDENT* **EXAMPLE 4 - 4**

**The net force acting on a 130-lb student is 25 lb. What is her acceleration?**

**PICTURE THE PROBLEM** Apply $\Sigma F = ma$ and solve for the acceleration. The mass can be found from the student's weight.

According to Newton's second law, the student's acceleration is the force divided by her mass:

$$a = \frac{F}{m} = \frac{F}{w/g} = \frac{25 \, \text{lb}}{(130 \, \text{lb})/(32.2 \, \text{ft/s}^2)}$$

$$= \boxed{6.19 \, \text{ft/s}^2}$$

**EXERCISE**  What force is needed to give an acceleration of 3 ft/s² to a 5-lb block? (*Answer*  0.466 lb)

## 4-4 Forces in Nature

The full power of Newton's second law emerges when it is combined with the force laws that describe the interactions of objects. For example, Newton's law for gravitation, which we study in Chapter 11, gives the gravitational force exerted by one object on another in terms of the distance between the objects and the masses of each. This, combined with Newton's second law, enables us to calculate the orbits of planets around the sun, the motion of the moon, and variations with altitude of $g$, the gravitational field strength.

(a)

### The Fundamental Forces

All the different forces observed in nature can be explained in terms of four basic interactions that occur between elementary particles (see Figure 4.2):

1. The gravitational force—the force of mutual attraction between objects
2. The electromagnetic force—the force between electric charges
3. The strong nuclear force—the force between subatomic particles
4. The weak nuclear force—the force between subatomic particles during certain radioactive decay processes

The everyday forces that we observe between macroscopic objects are due to either the gravitational force or the electromagnetic force.

(b)

### Action at a Distance

The fundamental forces of gravity and electromagnetism act between particles that are separated in space. This creates a philosophical problem referred to as **action at a distance.** Newton perceived action at

(c)

**FIGURE 4-2** (a) The gravitational force between the earth and an object near the earth's surface is the weight of the object. The gravitational force exerted by the sun on the earth and the other planets is responsible for keeping the planets in their orbits around the sun. Similarly, the gravitational force exerted by the earth on the moon keeps the moon in its nearly circular orbit around the earth. The gravitational forces exerted by the moon and the sun on the oceans of the earth are responsible for the tides. Mont-Saint-Michel, France, shown in the photo, is an island when the tide is in. (b) The electromagnetic force includes both the electric and the magnetic forces. A familiar example of the electric force is the attraction between small bits of paper and a comb that is electrified after being run through hair. The lightning bolts above the Kitt Peak National Observatory, shown in the photo, are the result of the electromagnetic force. (c) The strong nuclear force occurs between elementary particles called hadrons, which include protons and neutrons, the constituents of atomic nuclei. This force results from the interaction of quarks, which are the building blocks of hadrons, and is responsible for holding nuclei together. The hydrogen bomb explosion shown here illustrates the strong nuclear force. (d) The weak nuclear force occurs between leptons (which include electrons and muons) and between hadrons (which include protons and neutrons). This false-color cloud chamber photograph illustrates the weak interaction between a cosmic ray muon (green) and an electron (red) knocked out of an atom.

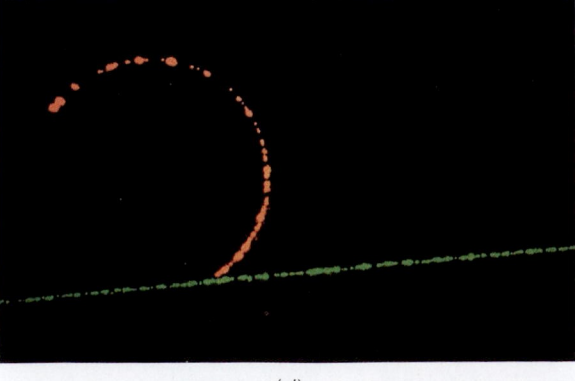

(d)

a distance as a flaw in his theory of gravitation but avoided giving any other hypothesis. Today the problem is avoided by introducing the concept of a field, which acts as an intermediary agent. For example, we consider the attraction of the earth by the sun in two steps. The sun creates a condition in space that we call the gravitational field. This field then exerts a force on the earth. Similarly, the earth produces a gravitational field that exerts a force on the sun. Your weight is the force exerted by the gravitational field of the earth on you. When we study electricity and magnetism (Chapters 21–30) we will study electric fields, which are produced by electrical charges, and magnetic fields, which are produced by electrical charges in motion.

## Contact Forces

Many forces we encounter are exerted by objects in direct contact—they touch. These forces are electromagnetic in origin. They are exerted between the surface molecules of the objects in contact.

**Solids**  If a surface is pushed against, it pushes back. Consider the ladder leaning against a wall shown in Figure 4-3. At the region of contact, the ladder pushes against the wall with a horizontal force, compressing the molecules in the surface of the wall. Like mattress springs, the compressed molecules in the wall push back on the ladder with a horizontal force. Such a force, *perpendicular* to the contacting surfaces, is called a **normal force** (the word *normal* means perpendicular). The wall bends slightly in response to a load, though this is rarely noticeable to the naked eye.

Normal forces can vary over a wide range of magnitudes. A table, for instance, will exert an upward normal force on any object resting on it. As long as the table doesn't break, this normal force will balance the downward weight force on the object. Furthermore, if you press down on the object, the upward normal force exerted by the table will increase, countering the extra force, thus preventing the object from accelerating downward.

Surfaces in contact can also exert forces on each other that are *parallel* to the contacting surfaces. Consider the large block on the floor shown in Figure 4-4. If the block is pushed sideways with a gentle enough force, it will not slide. The surface of the floor exerts a force back on the block, opposing its tendency to slide in the direction of the push. However, if the block is pushed sideways with a sufficiently hard force, it will start to slide. To keep it sliding it is necessary to continue to push it. If the sideways push is not sustained, the contact force will slow the motion of the box until it stops. A component of a contact force that opposes sliding, or the tendency to slide, is called a **frictional force;** it acts parallel to the contacting surfaces.

Although the frictional and normal forces are shown in diagrams as if they act at a single point, they are, in reality, distributed over the entire region of contact. Frictional forces are treated in more depth in Chapter 5.

**Springs**  When a spring is compressed or extended by a small amount $\Delta x$, the force it exerts is found experimentally to be

$$F_x = -k \, \Delta x \qquad 4\text{-}9$$

HOOKE'S LAW

where $k$ is the force constant, a measure of the stiffness of the spring (Figure 4-5). The negative sign in Equation 4-9 signifies that when the spring is stretched or compressed, the force it exerts back is in the opposite direction. This relation, known as Hooke's law, turns out to be quite important. An object at rest under the influence of forces that balance is said to be in static equilibrium. If a small displacement results in a net restoring force toward

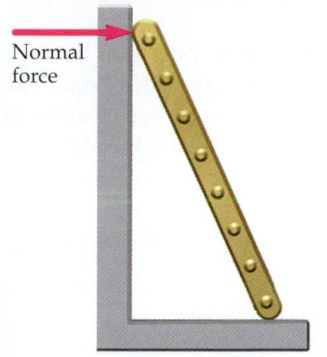

**FIGURE 4-3**  The wall supports the ladder by pushing on the ladder with a force normal to the wall.

**FIGURE 4-4**  The frictional force exerted by the floor on the block opposes its sliding motion or its tendency to slide.

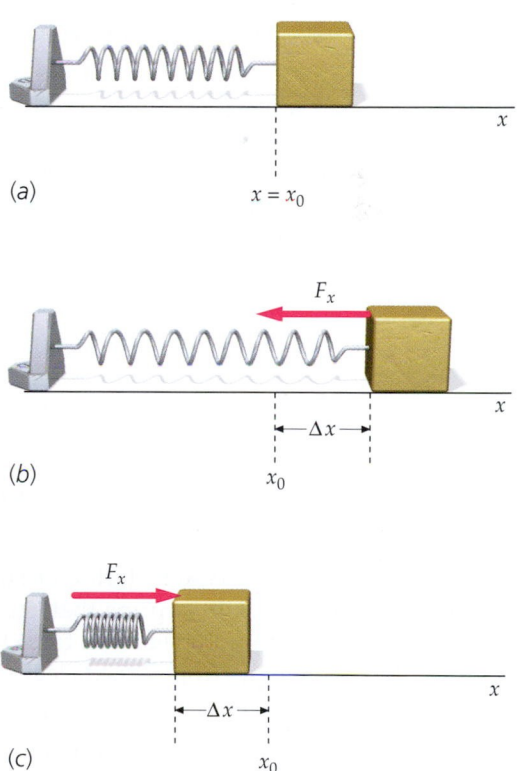

**FIGURE 4-5**  A horizontal spring. (*a*) When the spring is unstretched, it exerts no force on the block. (*b*) When the spring is stretched so that $\Delta x$ is positive, it exerts a force of magnitude $k \, \Delta x$ in the negative $x$ direction. (*c*) When the spring is compressed so that $\Delta x$ is negative, the spring exerts a force of magnitude $k \, |\Delta x|$ in the positive direction.

the equilibrium position, the equilibrium is called stable equilibrium. For small displacements, nearly all restoring forces obey Hooke's law.

The molecular force of attraction between atoms in a molecule or solid varies approximately linearly with the change in separation (for small changes); the force varies much like that of a spring. We can therefore use two masses on a spring to model a diatomic molecule, or a set of masses connected by springs to model a solid as shown in Figure 4-6.

(a)

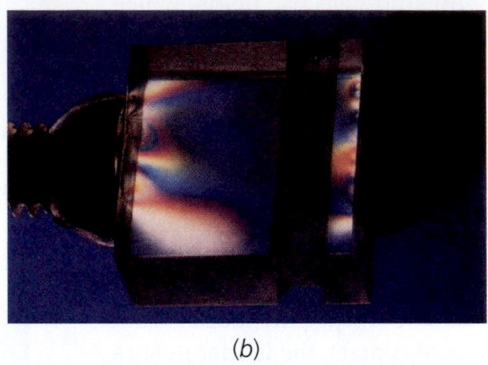

(b)

**FIGURE 4-6** (a) Model of a solid consisting of atoms connected to each other by springs. The springs are very stiff (large force constant) so that when a weight is placed on the solid its deformation is not visible. However, compression such as that produced by the clamp on the plastic block in (b) leads to stress patterns that are visible when viewed with polarized light.

*THE SLAM DUNK*                                    **EXAMPLE  4 - 5**

**A 110-kg basketball player hangs on the rim following a slam dunk (Figure 4-7). Prior to dropping to the floor, he hangs motionless with the front of the rim deflected down a distance of 15 cm. Assume the rim can be approximated by a spring and calculate the force constant $k$.**

**PICTURE THE PROBLEM** Since the acceleration of the player is zero, the net force exerted on him must also be zero. The upward force exerted by the rim balances his weight. Let $y = 0$ be the original position of the rim and choose down to be the positive $y$ direction. Then $\Delta y$ is positive, the weight $mg$ is positive, and the force exerted by the rim, $-k\,\Delta y$, is negative.

**FIGURE 4-7**

$\vec{F} = -k\Delta y \hat{j}$

$\vec{w} = m\vec{g}$

Apply $\Sigma\vec{F} = m\vec{a}$ to the player, and solve for $k$:

$$\Sigma F_y = w_y + F_y = ma_y$$
$$mg + (-k\Delta y) = 0$$

$$k = \frac{mg}{\Delta y} = \frac{(110\ \text{kg})(9.81\ \text{N/kg})}{0.15\ \text{m}}$$

$$= \boxed{7.19 \times 10^3\ \text{N/m}}$$

**REMARKS** Although a basketball rim doesn't look much like a spring, the rim is suspended by a hinge with a spring that is distorted when the front of the rim is pulled down. As a result, the upward force the rim exerts on the player's hands is proportional to the rim front's displacement and oppositely directed. Note that we used N/kg for the units of $g$ so that kg cancels, giving N/m for the units of $k$. We can use either 9.81 N/kg or 9.81 m/s² for $g$, whichever is more convenient, because 1 N/kg = 1 m/s².

**EXERCISE** A 4-kg bunch of bananas is suspended motionless from a spring balance whose force constant is $k = 300$ N/m. By how much is the spring stretched? (*Answer* 13.1 cm)

**EXERCISE** A spring of force constant 400 N/m is attached to a 3-kg block that rests on a horizontal air track that renders friction negligible. What extension of the spring is needed to give the block an acceleration of 4 m/s² upon release? (*Answer* 3.0 cm)

**EXERCISE IN DIMENSIONAL ANALYSIS** An object of mass $m$ oscillates at the end of an ideal spring of force constant $k$. The time for one complete oscillation is the period $T$. Assuming that $T$ depends on $m$ and $k$, use dimensional analysis to find the form of the relationship $T = f(m, k)$, ignoring numerical constants. This is most easily found by looking at the units. Note that the units of $k$ are $N/m = (kg \cdot m/s^2)/m = kg/s^2$, and the units of $m$ are kg. (*Answer* $T = C\sqrt{m/k}$ where $C$ is some dimensionless constant. The correct expression for the period, as we will see in Chapter 14, is $T = 2\pi\sqrt{m/k}$.

**Strings** Strings (ropes) are used to pull things. We can think of a string as a spring with such a large force constant that the extension of the string is negligible. Strings are flexible, however, so unlike springs, they cannot push things. Instead they flex or bend. The magnitude of the force that one segment of a string exerts on an adjacent segment is called **tension.** It follows that if a string pulls on an object, the magnitude of the force equals the tension. The concept of tension in a string or rope is further developed in Section 4-7.

**Constraints** A railroad car moves along a track. A wooden pony on a merry-go-round moves in a circle. A sled sliding along the surface of a frozen pond moves in a horizontal plane. Such conditions on the motion of objects are called **constraints.**

# 4-5 Problem Solving: Free-Body Diagrams

Imagine a dogsled being pulled across icy ground. The dog in front pulls on a rope attached to the sled (Figure 4-8*a*) with a horizontal force causing the sled to gain speed. We can think of the sled and rope together as a single object. What forces act on the sled-rope object? Both the dog and the ice touch the object, so we know that the dog and the ice exert contact forces on it. We also know that the earth exerts a gravitational force on the sled-rope (the object's weight). Thus, a total of three forces act on the object (assuming that friction is negligible):

1. The weight of the sled-rope $\vec{w}$.
2. The contact force $\vec{F}_n$ exerted by the ice (without friction, the contact force is directed normal to the ice.)
3. The contact force $\vec{F}$ exerted by the dog.

A diagram that shows schematically all the forces acting on a system, such as Figure 4-8*b*, is called a **free-body diagram.** It is called a free-body diagram because the body (object) is drawn free from (without) its surroundings. Drawing the force vectors on a free-body diagram to scale requires that we first determine the direction of the acceleration vector using kinematic methods. We know the object is moving to the right with increasing speed. It follows from kinematics that its acceleration vector is in the direction of its motion—to the right. Note that $\vec{F}_n$ and $\vec{w}$ in the diagram have equal magnitudes. The magnitudes must be equal because the vertical component of the acceleration is zero. As a check on the correctness of our free-body diagram, we draw a vector-addition diagram (Figure 4-9) verifying that the sum of the force vectors is in the direction of the acceleration vector.

The $x$ component of Newton's second law gives

$$\Sigma F_x = F_{n,x} + w_x + F_x = ma_x$$
$$0 + 0 + F = ma_x$$

or

$$a_x = \frac{F}{m}$$

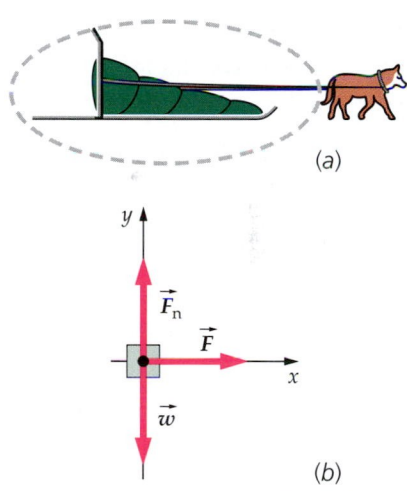

(a)

(b)

**FIGURE 4-8** (*a*) A dog pulling a sled. The first step in problem solving is to isolate the system to be analyzed. In this case, the closed dashed curve represents the boundary between the sled-rope object and its surroundings. (*b*) The forces acting on the sled in Figure 4.8*a*.

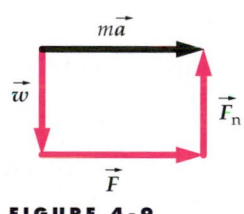

**FIGURE 4-9**

The $y$ component of Newton's second law gives

$$\Sigma F_y = F_{n,y} + w_y + F_y = ma_y$$
$$F_n - w + 0 = 0$$

or

$$F_n = w$$

In this simple example, we found two things: the horizontal acceleration ($a_x = F/m$), and the vertical force $\vec{F}_n$ exerted by the ice ($F_n = w$).

---

*A DOGSLED RACE*                    **EXAMPLE 4-6**

During your winter break, you enter a dogsled race in which students replace the dogs. Wearing cleats for traction, you begin the race by pulling on a rope attached to the sled with a force of 150 N at 25° with the horizontal. The mass of the sled-passenger-rope object is 80 kg and there is negligible friction between the sled runners and the ice. Find (*a*) the acceleration of the sled and (*b*) the normal force $\vec{F}_n$ exerted by the surface on the sled.

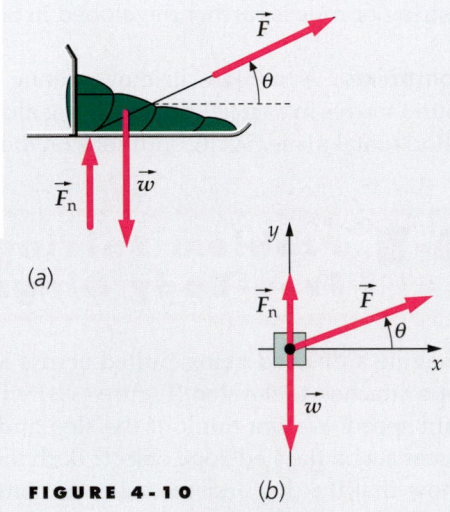

**PICTURE THE PROBLEM** Three forces act on the object: its weight $\vec{w}$, which acts downward; the normal force $\vec{F}_n$, which acts upward; and the force with which you pull the rope $\vec{F}$, directed 25° above the horizontal. Since the forces are not all parallel to a single line, we study the system by applying Newton's second law to the $x$ and $y$ directions separately.

(*a*) 1. Sketch a free-body diagram (Figure 4-10*b*) of the sled. Include a coordinate system with one of the coordinate axes in the direction of the sled's acceleration. The object moves to the right with increasing speed, so we know the acceleration is in that direction:

**FIGURE 4-10**     (*b*)

2. *Note:* Add the force vectors on the free-body diagram (Figure 4-11) to verify that their sum can be in the direction of the acceleration:

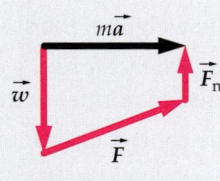

3. Apply Newton's second law to the object. Write out the equation in both vector and component form:

$$\vec{F}_n + \vec{w} + \vec{F} = m\vec{a}$$

or

**FIGURE 4-11**

$$F_{n,x} + w_x + F_x = ma_x$$
$$F_{n,y} + w_y + F_y = ma_y$$

4. Express the $x$ components of $\vec{F}_n$, $\vec{w}$, and $\vec{F}$:

$$F_{n,x} = 0, \quad w_x = 0, \quad \text{and} \quad F_x = F\cos\theta$$

5. Substitute the step 4 results into the $x$ component equation in step 3. Then solve for the acceleration $a_x$:

$$0 + 0 + F\cos\theta = ma_x$$
$$a_x = \frac{F\cos\theta}{m} = \frac{(150\ \text{N})\cos 25°}{80\ \text{kg}} = \boxed{1.70\ \text{m/s}^2}$$

(*b*) 1. Express the $y$ component of $\vec{a}$:

$$a_y = 0$$

2. Express the $y$ components of $\vec{F}_n$, $\vec{w}$, and $\vec{F}$:

$$F_{n,y} = F_n, \quad w_y = -mg, \quad \text{and} \quad F_y = F\sin\theta$$

3. Substitute the step *b*1 and *b*2 results into the $y$ component equation in step *a*3. Then solve for $F_n$:

$$\Sigma F_y = F_n - mg + F\sin\theta = 0$$
$$F_n = mg - F\sin\theta$$
$$= (80\ \text{kg})(9.81\ \text{N/kg}) - (150\ \text{N})\sin 25° = \boxed{721\ \text{N}}$$

**REMARKS** Note that only the $x$ component of $\vec{F}$, which is $F \cos \theta$, causes the object to accelerate. Also note that the ice supports less than the full weight of the object since part of the weight, $F \sin \theta$, is supported by the rope.

**⓿ PLAUSIBILITY CHECK** If $\theta = 0$, the object is accelerated by a force $F$ and the ice supports all the object's weight. Our results agree, giving $a_x = F/m$ and $F_n = mg$.

**EXERCISE** If $\theta = 25°$, what is the maximum force $F$ that can be applied to the rope without lifting the sled off the surface? (*Answer*    $F = 1.86$ kN)

Example 4-6 illustrates a general method for solving problems using Newton's laws:

1. Draw a neat diagram that includes the important features of the problem.

2. Isolate the object (particle) of interest, and draw a free-body diagram showing each external force that acts on it. If there is more than one object of interest in the problem, draw a separate free-body diagram for each. Choose a convenient coordinate system for each object and include it on that object's free-body diagram. If the direction of the acceleration is known, choose a coordinate axis that is parallel to it. For objects sliding along a surface, choose one coordinate axis parallel to the surface and the other perpendicular to it.

3. Apply Newton's second law, $\Sigma \vec{F} = m\vec{a}$, usually in component form.

4. For problems involving two or more objects, make use of Newton's third law, $\vec{F}_{A,B} = -\vec{F}_{B,A}$, and any constraints to simplify the equations obtained from applying $\Sigma \vec{F} = m\vec{a}$.

5. Solve the resulting equations for the unknowns.

6. Check to see whether your results have the correct units and seem plausible. Substituting extreme values into your solution is a good way to check your work for errors.

SOLVING PROBLEMS USING NEWTON'S LAWS

---

*UNLOADING A TRUCK*

**EXAMPLE 4-7**    **Put It in Context**

You are working for a big delivery company, and must unload a large, fragile package from your truck, using a delivery ramp (Figure 4-12). If the downward component of the velocity of the package when it reaches the bottom of the ramp is greater than 2.5 m/s (the speed an object would have if it were dropped from a height of about 1 ft), the package will break. What is the largest angle at which you can safely unload? The ramp is 1-m high, has rollers (i.e., the ramp is approximately frictionless), and is inclined at an angle $\theta$ to the horizontal.

**PICTURE THE PROBLEM** Two forces act on the box, the weight $\vec{w}$ and the normal force $\vec{F}_n$. Since these forces are not parallel to a single line, they cannot sum to zero, hence there is a net force on the box causing it to accelerate. The ramp constrains the box to move parallel to its surface, so we choose down the incline as the positive $x$ direction. To determine the acceleration we apply Newton's second law to the box. Once the acceleration is known, we can use kinematics to determine the largest safe angle.

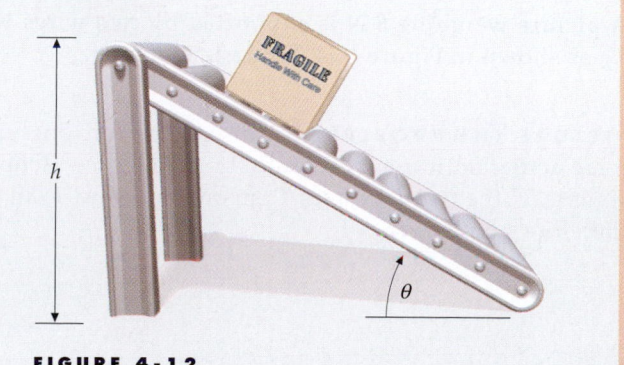

**FIGURE 4-12**

1. Relate the downward component of the velocity of the box to its speed $v$ along the ramp:

$$v_d = v \sin \theta$$

2. The speed $v$ is related to the displacement $\Delta x$ along the ramp by the kinematic equation:

$$v^2 = v_0^2 + 2a_x \, \Delta x$$

3. To find $a_x$ we apply Newton's second law ($\Sigma F_x = ma_x$) to the package. First we draw a free-body diagram (Figure 4-13). Two forces act on the package, the weight force and the normal force. We choose the direction of the acceleration, down the ramp, as the $+x$ direction.

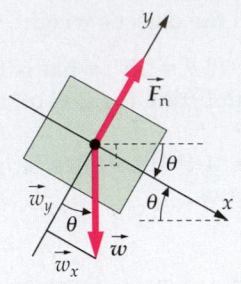

*Note:* The angle between $\vec{w}$ and the negative $y$ axis is the same as the angle between the incline and the horizontal as we see from the free-body diagram. We can also see that $w_x = w \sin \theta$.

**FIGURE 4-13**

4. Applying Newton's second law gives:

$$F_{n,x} + w_x = ma_x$$
where

*Note:* $F_n$ is perpendicular to the $x$ axis and $w = mg$.

$$F_{n,x} = 0 \quad \text{and} \quad w_x = w \sin \theta = mg \sin \theta$$

5. Substituting and solving for the acceleration gives:

$$0 + mg \sin \theta = ma_x$$
so
$$a_x = g \sin \theta$$

6. Substituting for $a_x$ in the kinematic equation (step 2) and setting $v_0$ to zero gives:

$$v^2 = 2g \sin \theta \, \Delta x$$

7. From Figure 4-12 we can see that when $\Delta x$ equals the length of the ramp, $\Delta x \sin \theta = h$, where $h$ is the height of the ramp:

$$v^2 = 2gh$$

8. Using $v_d = v \sin \theta$, solve for $v_d$:

$$v_d = \sqrt{2gh} \sin \theta$$

9. Solve for the maximum angle:

$$2.5 \text{ m/s} = \sqrt{2(9.81 \text{ m/s}^2)(1.0 \text{ m})} \sin \theta_{max}$$

$$\boxed{\theta_{max} = 34.4°}$$

**REMARKS** The acceleration down the incline is constant and equal to $g \sin \theta$. Also, the speed $v$ at the bottom ($v = \sqrt{2gh}$) does not depend on the angle $\theta$.

**EXERCISE** Apply $\Sigma F_y = ma_y$ to the package and show that $F_n = mg \cos \theta$.

---

*PICTURE HANGING*　　　　　　　**EXAMPLE 4-8 Try It Yourself**

**A picture weighing 8 N is supported by two wires with tensions $T_1$ and $T_2$, as shown in Figure 4-14. Find each tension.**

**PICTURE THE PROBLEM** Because the picture does not accelerate, the net force acting on it must be zero. The three forces acting on the picture (its weight $\vec{w}$, the tension forces $\vec{T}_1$ in one wire and $\vec{T}_2$ in the other wire) must therefore sum to zero.

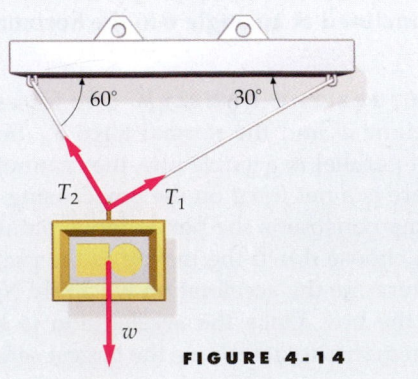

**FIGURE 4-14**

**Cover the column to the right and try these on your own before looking at the answers.**

**Steps:**

1. Draw a free-body diagram for the picture (Figure 4-15). On your diagram show the $x$ and $y$ components of the two tension forces.

2. Apply $\Sigma \vec{F} = m\vec{a}$ in vector form to the picture.

3. Resolve each force into its $x$ and $y$ components. This gives you two equations for the two unknowns $T_1$ and $T_2$.

4. Solve the $x$ component equation for $T_2$ in terms of $T_1$.

5. Substitute your result for $T_2$ (from step 4) into the $y$ component equation and solve for $T_1$.

6. Use your result for $T_1$ to find $T_2$.

**Answers:**

$\vec{T}_1 + \vec{T}_2 + \vec{w} = m\vec{a}$

$T_{1,x} + T_{2,x} + w_x = 0$

$T_{1,y} + T_{2,y} + w_y = 0$

$T_1 \cos 30° - T_2 \cos 60° + 0 = 0$

$T_1 \sin 30° + T_2 \sin 60° - w = 0$

$T_2 = T_1 \dfrac{\cos 30°}{\cos 60°} = T_1 \sqrt{3}$

$T_1 \sin 30° + (T_1 \sqrt{3})\sin 60° - w = 0$

$T_1 = \tfrac{1}{2}w = \boxed{4\text{N}}$

$T_2 = T_1 \sqrt{3} = \boxed{6.93 \text{ N}}$

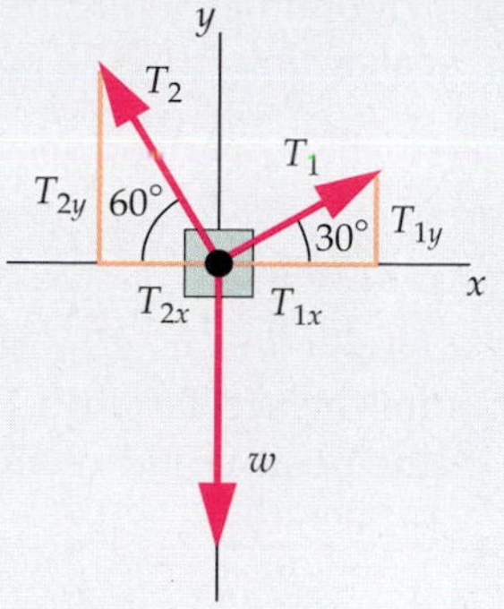

**FIGURE 4-15**

**REMARKS** Note that the more vertical of the two wires supports the greater share of the load, as you might expect. Also, we see that $T_1 + T_2 > 8$ N. The "extra" force is due to the wires pulling to the right and left.

---

*AN ACCELERATING JET PLANE*                          **E X A M P L E    4 - 9**

**As your jet plane speeds down the runway on takeoff, you decide to determine its acceleration, so you take out your yo-yo and note that when you suspend it, the string makes an angle of 22° with the vertical (Figure 4-16a). (a) What is the acceleration of the plane? (b) If the mass of the yo-yo is 40 g, what is the tension in the string?**

**PICTURE THE PROBLEM** The yo-yo and plane have the same acceleration to the right. The net force on the yo-yo is in the direction of its acceleration. This force is supplied by the horizontal component of the tension force $\vec{T}$. The vertical component of $\vec{T}$ balances the weight of the yo-yo. We choose a coordinate system in which the $x$ direction is parallel to the acceleration vector $\vec{a}$ and the $y$ direction is vertical. Writing Newton's second law for both the $x$ and $y$ directions gives two equations to determine the two unknowns, $a$ and $T$.

(*a*) 1. Draw a free-body diagram for the yo-yo (Figure 4-16*b*). Choose the postive $x$ direction to be the direction of the acceleration.

    2. Apply $\Sigma F_x = ma_x$ to the yo-yo. Then simplify using trigonometry:

$T_x + w_x = ma_x$

$T \sin \theta + 0 = ma_x$

or

$T \sin \theta = ma_x$

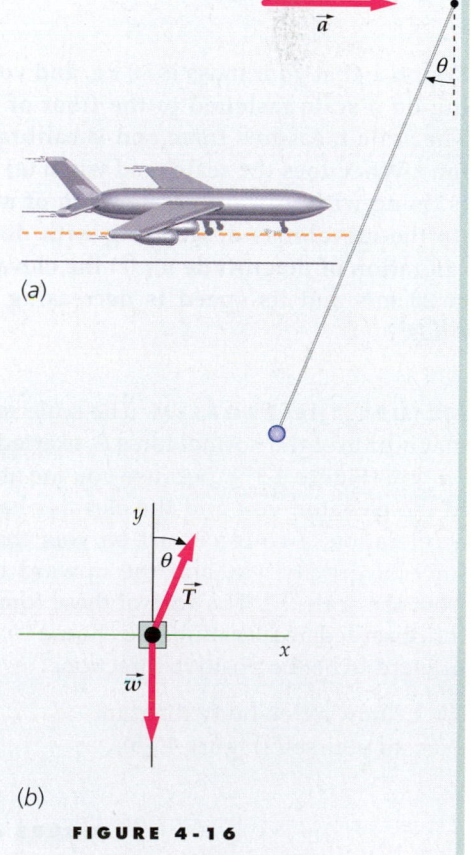

(a)

(b)

**FIGURE 4-16**

3. Apply $\Sigma F_y = ma_y$ to the yo-yo. Then simplify using trigonometry (Figure 4-16c) and $w = mg$. Since the acceleration is in the positive $x$ direction, $a_y = 0$:

$$T_y + w_y = ma_y$$

$$T\cos\theta - mg = 0$$

or

$$T\cos\theta = mg$$

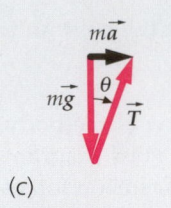

(c)

**FIGURE 4-16**

4. Divide the step 2 result by the step 3 result and solve for the acceleration. Since the acceleration vector is in the positive $x$ direction, $a = a_x$:

$$\frac{T\sin\theta}{T\cos\theta} = \frac{ma_x}{mg}, \quad \text{so} \quad \tan\theta = \frac{a_x}{g}$$

and

$$a = g\tan\theta = (9.81\ \text{m/s}^2)\tan 22° = \boxed{3.96\ \text{m/s}^2}$$

(b) Using the step 3 result, solve for the tension:

$$T = \frac{mg}{\cos\theta} = \frac{(0.04\ \text{kg})(9.81\ \text{m/s}^2)}{\cos 22°} = \boxed{0.423\ \text{N}}$$

**REMARKS** Notice that $T$ is greater than the weight of the yo-yo ($mg = 0.392$ N) because the cord not only keeps the yo-yo from falling but also accelerates it in the horizontal direction. Here we use the units m/s² for $g$ because we are calculating acceleration.

**PLAUSIBILITY CHECK** At $\theta = 0$, we find that $T = mg$ and $a = 0$.

**EXERCISE** For what acceleration magnitude $a$ would the tension in the string be equal to $3mg$? What is $\theta$ in this case? (*Answer* $a = 27.8$ m/s², $\theta = 70.5°$)

Our next example is the application of Newton's second law to objects that are at rest relative to a reference frame that is itself accelerating.

---

*YOUR WEIGHT IN AN ELEVATOR*                    **EXAMPLE 4-10**

Suppose that your mass is 80 kg, and you are standing on a scale fastened to the floor of an elevator. The scale measures force and is calibrated in newtons. What does the scale read when (*a*) the elevator is rising with upward acceleration of magnitude $a$; (*b*) the elevator is descending with downward acceleration of magnitude $a'$; (*c*) the elevator is rising at 20 m/s and its speed is decreasing at a rate of 8 m/s²?

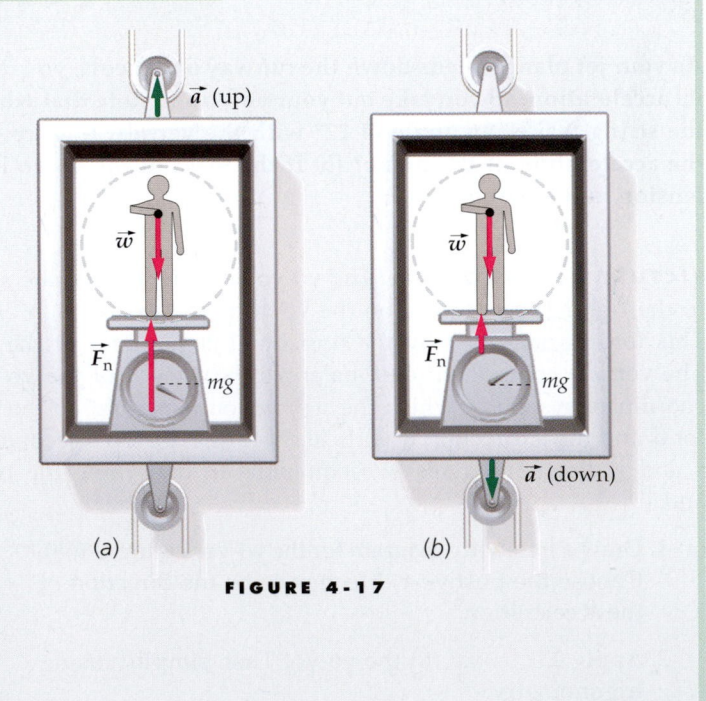

(a)          (b)

**FIGURE 4-17**

**PICTURE THE PROBLEM** The scale reading is the magnitude of the normal force $F_n$ exerted by the scale on you (Figure 4-17). Because you are at rest relative to the elevator, you and the elevator have the same acceleration. Two forces act on you: the downward force of gravity, $mg$, and the upward normal force from the scale, $F_n$. The sum of these forces gives you the observed acceleration. We choose upward to be the positive direction.

(*a*) 1. Draw a free-body diagram of yourself (Figure 4-18):

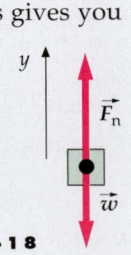

**FIGURE 4-18**

2. Apply $\Sigma\vec{F} = m\vec{a}$ in the $y$ direction:

$$F_{n,y} + w_y = ma_y$$
$$F_n - mg = ma$$

3. Solve for $F_n$. This is the reading on the scale (your apparent weight):

$$F_n = mg + ma = \boxed{m(g + a)}$$

(b) 1. Apply $\Sigma\vec{F} = m\vec{a}$ in the $y$ direction for the case in which the elevator accelerates downward with magnitude $a'$:

$$F_{n,y} + w_y = ma_y$$
$$F_n - mg = m(-a')$$

2. Solve for $F_n$:

$$F_n = mg - ma' = \boxed{m(g - a')}$$

(c) 1. Apply $\Sigma\vec{F} = m\vec{a}$ in the $y$ direction. Note that the acceleration is downward. (Why is that?) It follows that $a_y$ is negative:

$$F_{n,y} + w_y = ma_y$$

2. Solve for $F_n$:

$$F_n - mg = ma_y$$
$$F_n = m(g + a_y) = (80 \text{ kg})(9.81 \text{ m/s}^2 - 8.00 \text{ m/s}^2) = \boxed{145 \text{ N}}$$

**REMARKS** Whether the elevator is ascending or descending, if it accelerates upward, then your apparent weight is greater than $mg$ by $ma$. For you, it is as if gravity were increased from $g$ to $g + a$. If it accelerates downward, then your apparent weight is less than $mg$ by the amount $ma'$. You feel lighter, as if gravity were $g - a'$. If $a' = g$, the elevator is in free-fall, and you experience weightlessness.

**EXERCISE** An elevator descending comes to a stop with an acceleration of magnitude 4 m/s². If your mass is 70 kg and you are standing on a scale in the elevator, what does the scale read as the elevator is stopping? (*Answer*   967 N)

**EXERCISE** A man stands on a scale in an elevator that has an upward acceleration $a$. The scale reads 960 N. When he picks up a 20 kg box, the scale reads 1200 N. Find the mass of the man, his weight, and the acceleration $a$.

# 4-6   Newton's Third Law

When two objects interact, they exert forces on each other. Newton's third law states that these forces are equal in magnitude and opposite in direction. That is, if object A exerts a force on object B, then object B exerts a force on A that is equal in magnitude and opposite in direction. Thus forces always occur in pairs. It is common to refer to one force in the pair as an action and the other as a reaction. This terminology is unfortunate because it sounds like one force "reacts" to the other, which is not the case. The two forces occur simultaneously. **Either can be called the action and the other the reaction.** If we refer to an external force acting on a particular object as an action force, then the corresponding reaction force must act on a different object. Thus no two external forces acting on a single object can ever constitute an action–reaction pair.

In Figure 4-19, a block rests on a table. The force acting downward on the block is the weight $\vec{w}$ due to the attraction of the earth. An equal and opposite force $\vec{w}' = -\vec{w}$ is exerted by the block on the earth. These forces form an action–reaction pair. If they were the only forces present, the block would accelerate downward because it would have only a single force acting on it (and the earth would accelerate upward toward the block). However, the table exerts an

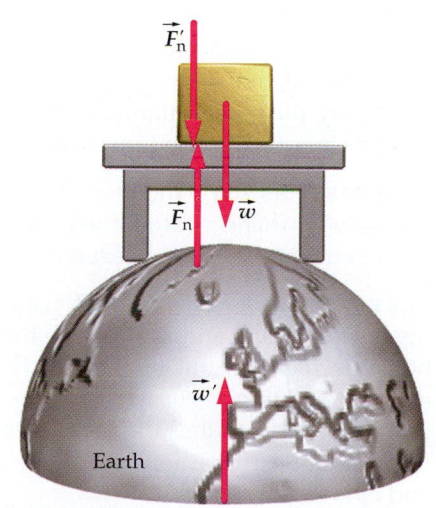

Earth

**FIGURE 4-19**

upward force $\vec{F}_n$ on the block that balances the block's weight. The block also exerts a force $\vec{F}'_n = -\vec{F}_n$ downward on the table. The forces $\vec{F}_n$ and $\vec{F}'_n$ also form an action–reaction pair.

**EXERCISE** Do the forces $\vec{w}$ and $\vec{F}_n$ in Figure 4-19 form an action–reaction pair? (*Answer*   No, they do not. These forces are both external forces and they both act on the same object, the block. Thus, they cannot constitute an action–reaction pair.)

---

*THE HORSE BEFORE THE CART*     **EXAMPLE 4-11**

**A horse refuses to pull a cart (Figure 4-20a). The horse reasons, "according to Newton's third law, whatever force I exert on the cart, the cart will exert an equal and opposite force on me, so the net force will be zero and I will have no chance of accelerating the cart." What is wrong with this reasoning?**

**PICTURE THE PROBLEM** Because we are interested in the motion of the cart, we draw a simple diagram for it (Figure 4-20b). The force exerted by the horse on the harness is labeled $\vec{F}$. (The harness is attached to the cart, and so we can consider it to be part of the cart.) Other forces acting on the cart are its weight $\vec{w}$, the vertical support force of the ground $\vec{F}_n$, and the horizontal force exerted by the pavement, labeled $\vec{f}$ (for friction).

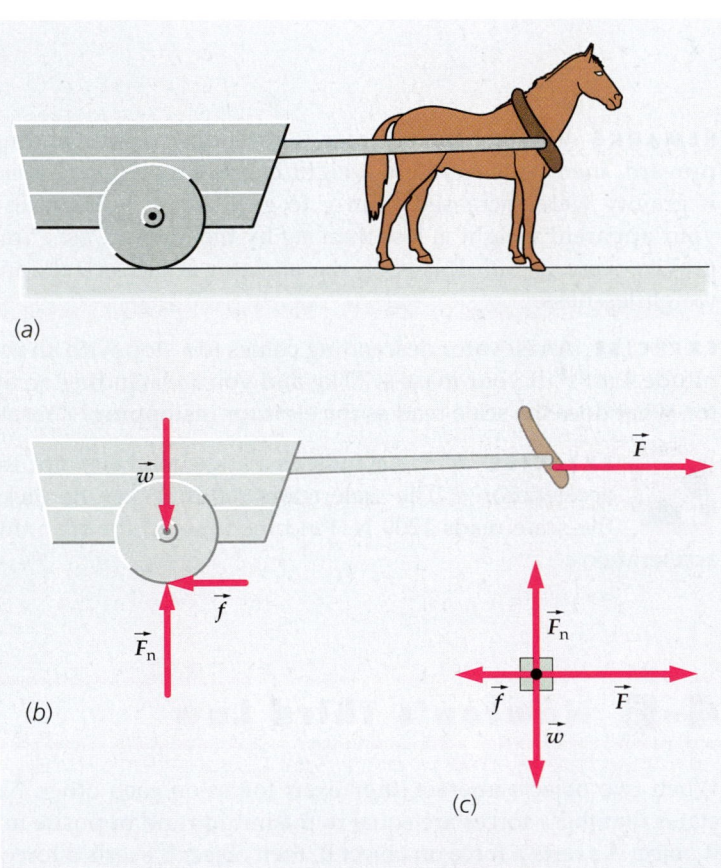

1. Draw a free-body diagram for the cart (see Figure 4-20c). Because the cart does not accelerate vertically, the verticle forces must sum to zero. The horizontal forces are $\vec{F}$ to the right and $\vec{f}$ to the left. The cart will accelerate to the right if $\vec{F}$ is greater than $\vec{f}$.

2. Note that the reaction force to $\vec{F}$, which we call $\vec{F}'$, is exerted on the horse, not on the cart (Figure 4-20d). It has no effect on the motion of the cart, but it does affect the motion of the horse. If the horse is to accelerate to the right, there must be a force $\vec{F}_P$ (to the right) exerted by the pavement on the horse's hooves that is greater than $\vec{F}'$.

**REMARKS** This example illustrates the importance of drawing a simple diagram when solving mechanics problems. Had the horse done so, he would have seen that he need only push back hard against the pavement so that the pavement will push him forward.

**EXERCISE** As you stand facing a friend, place your palms against your friend's palms and push. Can your friend exert a force on you if you do not exert a force back? Try it.

**EXERCISE** True or false: The force exerted by the cart on the horse is equal and opposite to the force exerted by the horse on the cart, but only when the horse and cart are not accelerating. (*Answer* False! An action–reaction pair of forces describes the interaction of two objects. One force cannot exist without the other. They are *always* equal and opposite.)

**FIGURE 4-20**

# 4-7 Problems With Two or More Objects

In some problems, two or more objects are in contact or are connected by a string or spring. Such problems are solved by drawing a separate free-body diagram for each object and then applying Newton's second law to each object. The resultant equations, together with any equations describing interactions and constraints, are solved simultaneously for the unknown quantities. If the objects are in direct contact, the forces they exert on each other must be equal and opposite, as stated in Newton's third law. For two objects, each moving in a straight line, that are connected by a taut nonstretching string, the acceleration components parallel to the string are the same for both objects. This is so because, for each object, its motion parallel to the string is identical with that of the other object. If the string passes over a pulley or peg, the phrase "parallel to the string" means parallel with that segment of the string attached to the object.

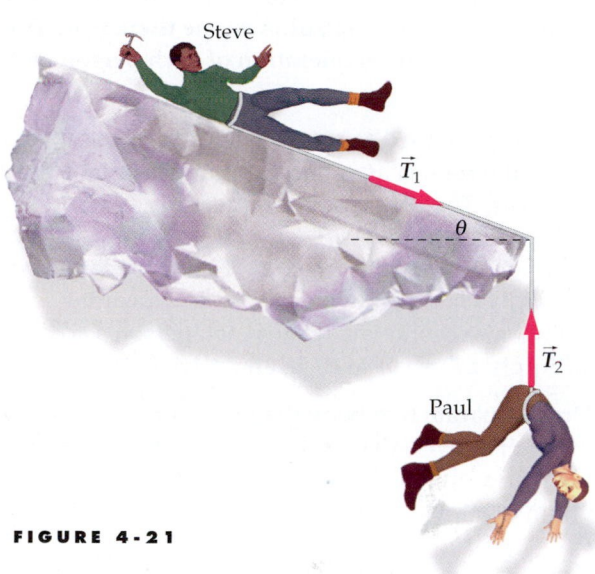

**FIGURE 4-21**

Consider the motion of Steve and Paul in Figure 4-21. The rate at which Paul descends equals the rate at which Steve slides along the glacier. That is, Paul's velocity component parallel with the segment of the rope attached to him equals Steve's velocity component parallel with the segment attached to him. These two velocity components must remain equal. If Steve and Paul are changing speed, they do so at the same rate. Their acceleration components parallel with the rope must be equal.

The tension in a string or rope is the magnitude of the force that one segment of the rope exerts on a neighboring segment. The tension can vary throughout the rope. For a rope dangling from a girder at the ceiling of a gymnasium, the tension is greatest near the top because the segment at the top has to support the weight of all the rope below it. However, for the problems in this book, the masses of strings and ropes are normally assumed to be so small that variations in tension due to the weight of a string or rope can be neglected. Conveniently, this also means that variations in the tension due to any acceleration of the rope can also be neglected. To see that this is so, consider the free-body diagram of a segment of the rope attached to Steve, where $\Delta m_s$ is the segment's mass (Figure 4-22).

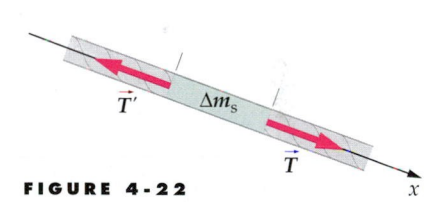

**FIGURE 4-22**

Applying Newton's second law to the segment gives $T - T' = \Delta m_s a_x$. If the mass of the segment is negligible, then $T = T'$. No net force is needed to give the segment an acceleration. (That is, only a negligible difference in tension is needed to give a rope segment of negligible mass any finite acceleration.)

Next we consider the entire rope connecting Steve and Paul. Neglecting gravity, there are three forces acting on the rope. Steve and Paul each exert a force, as does the ice at the edge of the glacier. Neglecting any friction between the ice and the rope means that the force exerted by the ice is always a normal force (Figure 4-23). A normal force has no component along the rope, so it cannot produce a change in the tension. Thus the tension is the same throughout the entire length of the rope. To summarize, if a taut rope of negligible mass changes direction by passing over a frictionless surface, the tension is the same throughout the rope.

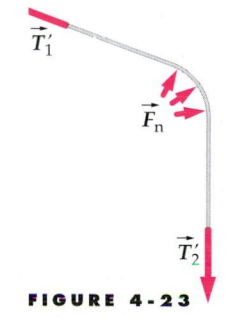

**FIGURE 4-23**

**EXERCISE** Suppose that instead of passing over the edge of a glacier, the rope passed around a pulley with frictionless bearings as shown in Figure 4-24. Would the tension then be the same throughout the length of the rope? (*Answer* No. Doing away with friction in the bearing is one thing, but the pulley still has inertia (mass). A difference in tension is needed to change the rate of rotation of the pulley.)

**FIGURE 4-24**

**EXAMPLE 4-12**

Paul (mass $m_P$) accidentally falls off the edge of a glacier as shown in Figure 4-21. Fortunately he is tied by a long rope to Steve (mass $m_S$), who has a climbing ax. Before Steve sets his ax to stop them, he slides without friction along the ice, attached by the rope to Paul. Assume there is no friction between the rope and the glacier. Find the acceleration of each person and the tension in the rope.

**PICTURE THE PROBLEM** The tension forces $\vec{T}_1$ and $\vec{T}_2$ have equal magnitudes because the rope is assumed to be massless and the glacier ice is assumed to be frictionless. The rope does not stretch or become slack, so Paul and Steve have equal speeds at all times. Their accelerations $\vec{a}_S$ and $\vec{a}_P$ must therefore be equal in magnitude, but not in direction. Steve accelerates down the face of the glacier whereas Paul accelerates vertically downward.

Newton's second law relates each person's acceleration to the forces acting on him. Apply $\Sigma\vec{F} = m\vec{a}$ to each, and solve for the accelerations and the tension.

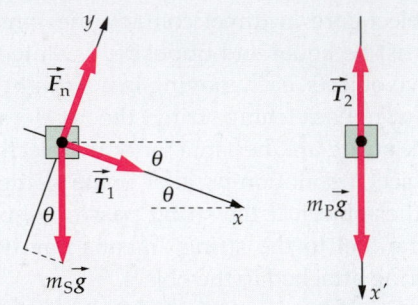

**FIGURE 4-25**

1. Draw separate free-body diagrams for Steve and Paul (Figure 4-25). Put axes $x$ and $y$ on Steve's diagram, choosing the direction of Steve's acceleration as the positive $x$ direction. Choose the direction of Paul's acceleration as the positive $x'$ direction.

2. Apply $\Sigma\vec{F} = m\vec{a}$ in the $x$ direction to Steve:

$$F_{n,x} + T_{1,x} + m_S g_x = m_S a_{S,x}$$

3. Apply $\Sigma\vec{F} = m\vec{a}$ in the $x'$ direction to Paul:

$$T_{2,x'} + m_P g_{x'} = m_P a_{P,x'}$$

4. Because they are each moving in a straight line and are connected by a taut length of rope that does not stretch, the accelerations of Paul and Steve are related. Express this relation:

$$a_{P,x'} = a_{S,x} = a_x$$

5. Because the rope is of negligible mass and slides over the ice with negligible friction, the forces $\vec{T}_1$ and $\vec{T}_2$ are related. Express this relation:

$$T_2 = T_1 = T$$

6. Substitute the steps 4 and 5 results into the step 2 and step 3 equations:

$$T + m_S g \sin\theta = m_S a_x$$
$$-T + m_P g = m_P a_x$$

7. Solve the step 6 equations for the acceleration by eliminating $T$ and solving for $a_x$:

$$\boxed{a_x = \frac{m_S \sin\theta + m_P}{m_S + m_P} g}$$

8. Substitute the step 7 result into either step 6 equation and solve for $T$:

$$\boxed{T = \frac{m_S m_P}{m_S + m_P}(1 - \sin\theta)g}$$

**REMARKS** In Step 3 we chose downward to be positive to keep the solution as simple as possible. With this choice, when Steve moves in the positive $x$ direction (down the glacier), Paul moves in the positive $x'$ direction (downward).

**PLAUSIBILITY CHECK** If $m_P$ is very much greater than $m_S$, we expect the acceleration to be approximately $g$ and the tension to be approximately zero. Substituting $m_S = 0$ does indeed give $a = g$ and $T = 0$ for this case. If $m_P$ is much less than $m_S$, we expect the acceleration to be approximately $g \sin\theta$ (see Example 4-8) and the tension to be zero. Substituting $m_P = 0$ in steps 7 and 8, we indeed obtain $a_x = g \sin\theta$ and $T = 0$. At the extreme value of the inclination ($\theta = 90°$) we check our answers. Substituting $\theta = 90°$ in steps 7 and 8, we obtain $a_x = g$ and $T = 0$. This seems right since Steve and Paul would be in free-fall for $\theta = 90°$.

**EXERCISE** (*a*) Find the acceleration if $\theta = 15°$ and if the masses are $m_S = 78$ kg and $m_P = 92$ kg. (*b*) Find the acceleration if these two masses are interchanged. (*Answer* (*a*) $a_x = 0.660g$ (*b*) $a_x = 0.599g$)

---

BUILDING A SPACE STATION                          **EXAMPLE 4-13**   **Try It Yourself**

You are an astronaut constructing a space station, and you push on a box of mass $m_1$ with a force of $\vec{F}_A$. The box is in direct contact with a second box of mass $m_2$ (Figure 4-26). (*a*) What is the acceleration of the boxes? (*b*) What is the magnitude of the force exerted by one box on the other?

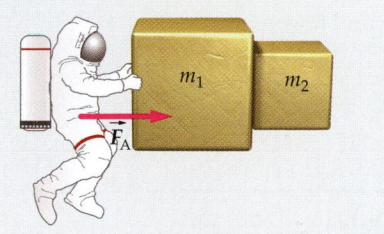

**PICTURE THE PROBLEM** Let $\vec{F}_{2,1}$ be the force exerted by box 2 on box 1, and $\vec{F}_{1,2}$ be the force exerted by box 1 on box 2. In accord with Newton's third law, these forces are equal and opposite ($\vec{F}_{2,1} = -\vec{F}_{1,2}$), so $F_{2,1} = F_{1,2}$. Apply Newton's second law to each box separately. The motions of the two boxes are identical, so the accelerations $\vec{a}_1$ and $\vec{a}_2$ are equal.

**FIGURE 4-26**

**Cover the column to the right and try these on your own before looking at the answers.**

Steps

(*a*) 1. Draw free-body diagrams for the two boxes (Figure 4-27).

2. Apply $\Sigma\vec{F} = m\vec{a}$ to box 1.

3. Apply $\Sigma\vec{F} = m\vec{a}$ to box 2.

4. Express both the relation between the two accelerations and the relation between the magnitudes of the forces the blocks exert on each other.

5. Substitute these back into the step 2 and step 3 results and solve for $a_x$.

(*b*) Substitute your expression for $a_x$ into either the step 2 or the step 3 result and solve for $F$.

Answers

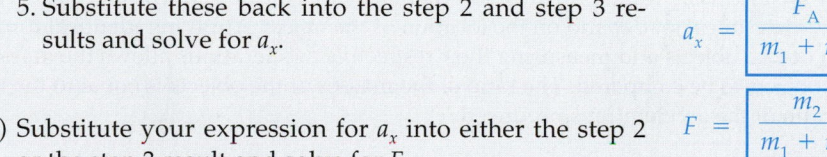

$F_A - F_{2,1} = m_1 a_{1,x}$

$F_{1,2} = m_2 a_{2,x}$

$a_{2,x} = a_{1,x} = a_x$

$F_{2,1} = F_{1,2} = F$

**FIGURE 4-27**

$$a_x = \frac{F_A}{m_1 + m_2}$$

$$F = \frac{m_2}{m_1 + m_2} F_A$$

**REMARKS** Note that the result in step 5 is the same as if the force $\vec{F}_A$ had acted on a single mass equal to the sum of the masses of the two boxes. In fact, since the two boxes have the same acceleration, we can consider them to be a single object with mass $m_1 + m_2$.

**EXERCISE** (*a*) Find the acceleration and the contact force if $m_1 = 2$ kg, $m_2 = 3$ kg, and $F_A = 12$ N. (*b*) Find the contact force if the two boxes are interchanged so that the first block has a mass of 3 kg and the second block has a mass of 2 kg. (*Answer* (*a*) $a_x = 2.4$ m/s², $F = 7.2$ N (*b*) $F = 4.8$ N)

---

# SUMMARY

1. Newton's laws of motion are fundamental laws of nature that serve as the basis for our understanding of mechanics.

2. Mass is an *intrinsic* property of an object.

3. Force is an important *derived* dynamic quantity.

| Topic | Relevant Equations and Remarks |
|-------|-------------------------------|
| **1. Newton's Laws** | |
| First law | An object at rest stays at rest unless acted on by an external force. An object in motion continues to travel with constant velocity unless acted on by an external force. (Reference frames in which this occurs are called inertial reference frames.) |
| Second law | The magnitude of the acceleration is proportional to the magnitude of the net external force $\vec{F}_{net}$, in accordance with $\vec{F}_{net} = m\vec{a}$, where $m$ is the mass of the object. The net force acting on an object, also called the resultant force, is the vector sum of all the forces acting on it: $\vec{F}_{net} = \Sigma\vec{F}$. Thus $$\Sigma\vec{F} = m\vec{a} \qquad \text{4-1}$$ |
| Third law | Forces always occur in equal and opposite pairs. If object A exerts a force on object B, an equal but opposite force is exerted by object B on object A: $$\vec{F}_{A,B} = -\vec{F}_{B,A} \qquad \text{4-2}$$ |
| **2. Inertial Reference Frames** | Our statements of Newton's laws are valid only in an inertial reference frame—a reference frame for which an object at rest remains at rest if no force acts on the object. Any reference frame that is moving with constant velocity relative to an inertial reference frame is itself an inertial reference frame, and any reference frame that is accelerating relative to an inertial frame is not an inertial reference frame. The earth's surface is, to a good approximation, an inertial reference frame. |
| **3. Force, Mass, and Weight** | |
| Force | Force is defined in terms of the acceleration it produces on a given object. A force of 1 newton (N) is that force which produces an acceleration of $1\ \text{m/s}^2$ on a mass of 1 kilogram (kg). |
| Mass | Mass is the intrinsic property of an object that measures its inertial resistance to acceleration. Mass does not depend on the location of the object. Applying identical forces to each of two objects and measuring their respective accelerations allows the masses of two objects to be compared. The ratio of the masses of the objects is equal to the inverse ratio of the accelerations produced: $$\frac{m_2}{m_1} = \frac{a_1}{a_2} \qquad \text{4-3}$$ |
| Weight | The weight $\vec{w}$ of an object is the force of gravitational attraction exerted by the earth on the object. It is proportional to the mass $m$ of the object and the gravitational field $\vec{g}$, which equals the free-fall acceleration: $$\vec{w} = m\vec{g} \qquad \text{4-4}$$ Weight is not an intrinsic property of an object; it depends on the location of the object. |
| **4. Fundamental Forces** | All the forces observed in nature can be explained in terms of four basic interactions: 1. The gravitational force 2. The electromagnetic force 3. The strong nuclear force (also called the hadronic force) 4. The weak nuclear force |
| **5. Contact Forces** | Contact forces of support and friction and those exerted by springs and strings are due to molecular forces that arise from the basic electromagnetic force. |
| Hooke's law | When a relaxed spring is compressed or extended by a small amount $\Delta x$, the force it exerts is proportional to $\Delta x$: $$F_x = -k\,\Delta x \qquad \text{4-9}$$ |

# PROBLEMS

- Single-concept, single-step, relatively easy
- •• Intermediate-level, may require synthesis of concepts
- ••• Challenging
- SSM Solution is in the *Student Solutions Manual*
-  Problems available on iSOLVE online homework service
- ✓ These "Checkpoint" online homework service problems ask students additional questions about their confidence level, and how they arrived at their answer

In a few problems, you are given more data than you actually need; in a few other problems, you are required to supply data from your general knowledge, outside sources, or informed estimates.

**For all problems, use $g = 9.81$ m/s² for the free-fall acceleration and neglect friction and air resistance unless instructed to do otherwise.**

## Conceptual Problems

**1** •• SSM How can you tell if a particular reference frame is an inertial reference frame?

**2** •• Suppose you observe an object from a reference frame and find that it has an acceleration $\vec{a}$ when there are no forces acting on it. How can you use this information to find an inertial frame?

**3** • If an object has no acceleration when observed from an inertial reference frame, can you conclude that no forces are acting on it?

**4** • SSM If only a single nonzero force acts on an object, must the object have an acceleration relative to any inertial reference frame? Can it ever have zero velocity?

**5** • If an object is acted upon by a single known force, can you tell in which direction the object will move, using no other information?

**6** • An object is observed to be moving at constant velocity in an inertial reference frame. It follows that (*a*) no forces act on the object, (*b*) a constant force acts on the object in the direction of motion, (*c*) the net force acting on the object is zero, (*d*) the net force acting on the object is equal and opposite to its weight.

**7** • Suppose an object was sent far out in space, away from galaxies, stars, or other bodies. How would its mass change? Its weight?

**8** • SSM How would an astronaut in apparent weightlessness be aware of her mass?

**9** • SSM Under what circumstances would your apparent weight be greater than your true weight?

**10** •• It is often said that Newton's first and second laws imply that it is impossible to use the laws of mechanics to tell if you are standing still or moving with a constant velocity. Explain.

**11** • Suppose a block of mass $m_1$ rests on a block of mass $m_2$ and the combination rests on a table as shown in Figure 4-28. Find the force exerted (*a*) by $m_1$ on $m_2$, (*b*) by $m_2$ on $m_1$, (*c*) by $m_2$ on the table, (*d*) by the table on $m_2$.

**FIGURE 4-28** Problem 11

**12** • SSM True or false:
(*a*) If two external forces that are both equal in magnitude and opposite in direction act on the same object, the two forces can never be an action-reaction force pair.
(*b*) Action equals reaction only if the objects are not accelerating.

**13** • An 80-kg man on ice skates pushes his 40-kg son, also on skates, with a force of 100 N. The force exerted by the boy on his father is (*a*) 200 N, (*b*) 100 N, (*c*) 50 N, (*d*) 40 N.

**14** • A girl holds a bird in her hand. The reaction force to the weight of the bird is (*a*) the gravitational force of the earth on the bird, (*b*) the gravitational force of the bird on the earth, (*c*) the contact force of the hand on the bird, (*d*) the contact force of the bird on the hand, (*e*) the gravitational force of the earth on the hand.

**15** • A baseball player hits a ball with a bat. If the force with which the bat hits the ball is considered the action force, what is the reaction force? (*a*) The force the bat exerts on the batter's hands. (*b*) The force on the ball exerted by the glove of the person who catches it. (*c*) The force the ball exerts on the bat. (*d*) The force the pitcher exerts on the ball while throwing it. (*e*) Friction, as the ball rolls to a stop.

**16** • Consider any situation in which an external force, say a push, is applied to an object. If Newton's third law requires that *for every action there is an equal and opposite reaction,* why doesn't the reaction force always cancel out the applied force, leaving no acceleration at all?

**17** • **SSM** A 2.5-kg object hangs at rest from a string attached to the ceiling. (*a*) Draw a diagram showing all the forces on the object and indicate the reaction force to each. (*b*) Do the same for each force acting on the string.

**18** • Which of the free-body diagrams in Figure 4-29 represents a block sliding down a frictionless inclined surface?

**FIGURE 4-29** Problem 18

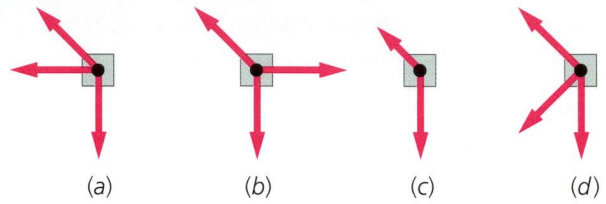

(*a*)        (*b*)        (*c*)        (*d*)

**19** • For an observer in an inertial reference frame, identify which (if any) of the following statements are true and which (if any) are false.

(*a*) If there are no forces acting on an object, it will not accelerate.
(*b*) If an object is not accelerating, there must be no forces acting on it.
(*c*) The motion of an object is always in the direction of the resultant force.
(*d*) The mass of an object depends on its location.

**20** • A sky diver of weight $w$ is descending near the surface of the earth. What is the magnitude of the force exerted by her body *on the earth*? (*a*) $w$. (*b*) Greater than $w$. (*c*) Less than $w$. (*d*) $9.8w$. (*e*) 0. (*f*) It depends on the air resistance.

**21** • **SSM** The net force on a moving object is suddenly reduced to zero and remains zero. As a consequence, the object (*a*) stops abruptly, (*b*) stops during a short time interval, (*c*) changes direction, (*d*) continues at constant velocity, (*e*) changes velocity in an unknown manner.

**22** • A clothesline is stretched taut between two poles. Then a wet towel is hung at the center of the line. Can the line remain horizontal? Explain.

**23** • What effect does the velocity of an elevator have on the apparent weight of a person in the elevator?

## Estimation and Approximation

**24** •• A car traveling 90 km/h crashes into the rear end of an unoccupied stalled vehicle. Fortunately, the driver is wearing a seat belt. Using reasonable values for the mass of the driver and the stopping distance, estimate the force (assuming it to be constant) exerted on the driver by the seat belt.

**25** ••• **SSM** Making any necessary assumptions, find the normal force and the tangential force exerted by the road on the wheels of your bicycle (*a*) as you climb an 8% grade at constant speed, and (*b*) as you descend the 8% grade at constant speed. (An 8% grade means that the angle of inclination $\theta$ is given by $\tan \theta = 0.08$.)

## Newton's First and Second Laws: Mass, Inertia, and Force

**26** • A particle of mass $m$ is traveling at an initial speed $v_0 = 25.0$ m/s. When a net force of 15.0 N acts on it, it comes to a stop in a distance of 62.5 m. What is $m$? (*a*) 37.5 kg. (*b*) 3.00 kg. (*c*) 1.50 kg. (*d*) 6.00 kg. (*e*) 3.75 kg.

**27** • (*a*) An object has an acceleration of 3 m/s² when the only force acting on it is $F_0$. What is its acceleration when this force is doubled? (*b*) A second object has an acceleration of 9 m/s² under the influence of the force $F_0$. What is the ratio of the masses of the two objects? (*c*) If the two objects are glued together, what acceleration will the force $F_0$ produce?

**28** • **iSOLVE** ✓ A tugboat tows a ship with a constant force $F_1$. The increase in the ship's speed in a 10-s interval is 4 km/h. When a second tugboat applies an additional constant force $F_2$ in the same direction, the speed increases by 16 km/h in a 10-s interval. How do the magnitudes of the two forces compare? (Neglect water resistance.)

**29** •• **SSM** **iSOLVE** A bullet of mass $1.8 \times 10^{-3}$ kg moving at 500 m/s impacts a large fixed block of wood and travels 6 cm before coming to rest. Assuming that the acceleration of the bullet is constant, find the force exerted by the wood on the bullet.

**30** •• **SSM** A cart on a horizontal, linear track has a fan attached to it. The cart is positioned at one end of the track, and the fan is turned on. Starting from rest, the cart takes 4.55 s to travel a distance of 1.5 m. The mass of the cart plus fan is 355 g. Assume that the cart travels with constant acceleration. (*a*) What is the net force exerted on the cart? (*b*) Weights are added to the cart until its mass is 722 g, and the experiment is repeated. How long does it take for the cart to travel 1.5 m now? Ignore the effects of friction.

**31** • A horizontal force $F_0$ causes an acceleration of 3 m/s² when it acts on an object of mass $m$ sliding on a frictionless surface. Find the acceleration of the same object in the circumstances shown in Figure 4-30*a* and *b*.

**FIGURE 4-30** Problem 31

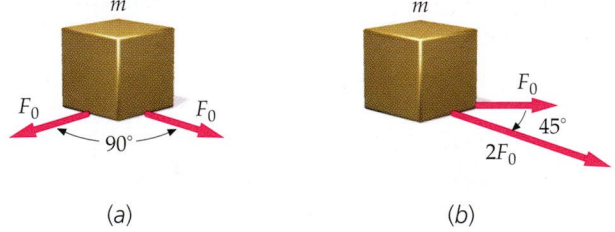

(*a*)        (*b*)

**32** • **iSOLVE** A force $\vec{F} = (6 \text{ N})\hat{i} - (3 \text{ N})\hat{j}$ acts on an object of mass 1.5 kg. Find the acceleration $\vec{a}$. What is the magnitude $a$?

**33** • A single force of 12 N acts on a particle of mass $m$. The particle starts from rest and travels in a straight line a distance of 18 m in 6 s. Find $m$.

**34** • **SSM** Al and Bert stand in the middle of a large frozen lake. Al pushes on Bert with a force of 20 N for a period of 1.5 s. Bert's mass is 100 kg. Assume that both are at rest before Al pushes Bert. (*a*) What is the speed that Bert reaches as he is pushed away from Al? Treat the ice as frictionless. (*b*) What speed does Al reach if his mass is 80 kg?

**35** • If I push a block whose mass is $m_1$ across a frictionless floor with a force of a given magnitude, the block has acceleration 12 m/s². If I push on a different block whose mass is $m_2$ with a force of the same magnitude, its acceleration is 3 m/s². (Both forces are applied horizontally.) (*a*) What acceleration will this force give to a block with mass $m_2 - m_1$? The force is still applied horizontally. (*b*) What acceleration will this force give to a block with mass $m_2 + m_1$?

**36** • To drag a 75-kg log along the ground at constant velocity, you have to pull on it with a horizontal force of 250 N. (*a*) What is the resistive force exerted by the ground? (*b*) What horizontal force must you exert if you want to give the log an acceleration of 2 m/s²?

**37** • **ISOLVE** A 4-kg object is subjected to two forces, $\vec{F}_1 = (2\,\text{N})\hat{i} + (-3\,\text{N})\hat{j}$ and $\vec{F}_2 = (4\,\text{N})\hat{i} - (11\,\text{N})\hat{j}$. The object is at rest at the origin at time $t = 0$. (*a*) What is the object's acceleration? (*b*) What is its velocity at time $t = 3$ s? (*c*) Where is the object at time $t = 3$ s?

## Mass and Weight

**38** • **SSM** On the moon, the acceleration due to gravity is only about 1/6 of that on earth. An astronaut whose weight on earth is 600 N travels to the lunar surface. His mass as measured on the moon will be (*a*) 600 kg, (*b*) 100 kg, (*c*) 61.2 kg, (*d*) 9.81 kg, (*e*) 360 kg.

**39** • **ISOLVE** Find the weight of a 54-kg student in (*a*) newtons and (*b*) pounds.

**40** • Find the mass of a 165-lb engineer in kilograms.

## Contact Forces

**41** • **SSM** **ISOLVE** ✓ A vertical spring of force constant 600 N/m has one end attached to the ceiling and the other to a 12-kg block resting on a horizontal surface so that the spring exerts an upward force on the block. The spring stretches by 10 cm. (*a*) What force does the spring exert on the block? (*b*) What is the force that the surface exerts on the block?

**42** • A 6-kg box on a frictionless horizontal surface is attached to a horizontal spring with a force constant of 800 N/m. If the spring is stretched 4 cm from its equilibrium length, what is the acceleration of the box?

## Free-Body Diagrams: Static Equilibrium

**43** • A traffic light is supported by two wires as in Figure 4-31. Is the tension in the wire that is more nearly vertical greater than or less than the tension in the other wire?

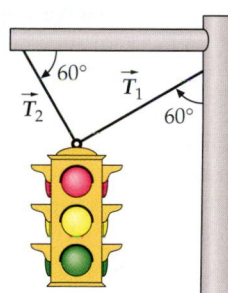

**FIGURE 4-31** Problem 43

**44** • **ISOLVE** A lamp with mass $m = 42.6$ kg is hanging from wires as shown in Figure 4-32. The ring has negligible mass. The tension $T_1$ in the vertical wire is (*a*) 209 N, (*b*) 418 N, (*c*) 570 N, (*d*) 360 N, (*e*) 730 N.

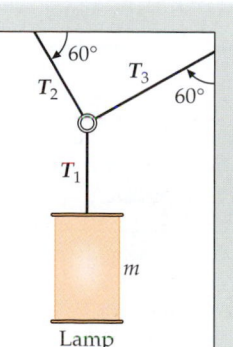

**FIGURE 4-32** Problem 44

**45** •• **SSM** **ISOLVE** ✓ In Figure 4-33*a*, a 0.500-kg block is suspended from a 1.25-m-long string. The ends of the string are attached to the ceiling at points separated by 1.00 m. (*a*) What angle does the string make with the ceiling? (*b*) What is the tension in the string? (*c*) The 0.500-kg block is removed and two 0.250-kg blocks are attached to the string such that the lengths of the three string segments are equal (Figure 4-33*b*). What is the tension in each segment of the string?

**FIGURE 4-33** Problem 45

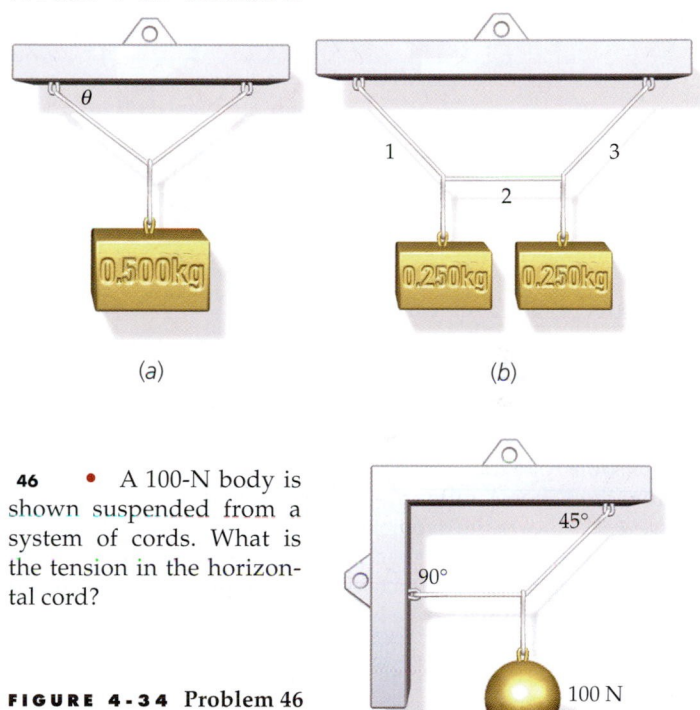

(*a*)                    (*b*)

**46** • A 100-N body is shown suspended from a system of cords. What is the tension in the horizontal cord?

**FIGURE 4-34** Problem 46

**47** • **ISOLVE** A 10-kg object on a frictionless table is subjected to two horizontal forces, $\vec{F}_1$ and $\vec{F}_2$, with magnitudes $F_1 = 20$ N and $F_2 = 30$ N, as shown in Figure 4-35. (*a*) Find the acceleration $\vec{a}$ of the object. (*b*) A third force $\vec{F}_3$ is applied so that the object is in static equilibrium. Find $\vec{F}_3$.

**FIGURE 4-35** Problem 47

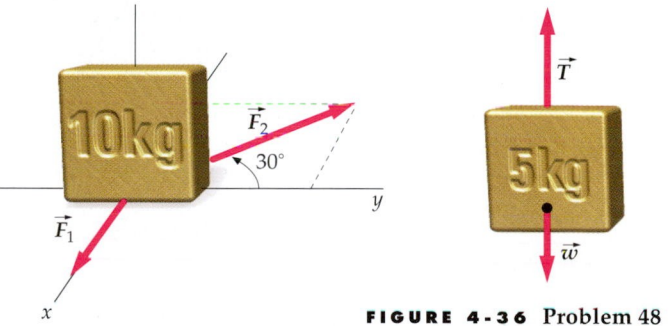

**FIGURE 4-36** Problem 48

**48** • **SSM** **ISOLVE** ✓ A vertical force $\vec{T}$ is exerted on a 5-kg object near the surface of the earth, as shown in Figure 4-36. Find the acceleration of the object if (*a*) $T = 5$ N, (*b*) $T = 10$ N, and (*c*) $T = 100$ N.

**49** •• A 2-kg picture is hung by two wires of equal length. Each makes an angle $\theta$ with the horizontal, as shown in Figure 4-37. (*a*) Find the general equation for the tension $T$, given $\theta$ and the weight $w$ for the picture. For what angle $\theta$ is $T$ the least? The greatest? (*b*) If $\theta = 30°$, what is the tension in the wires?

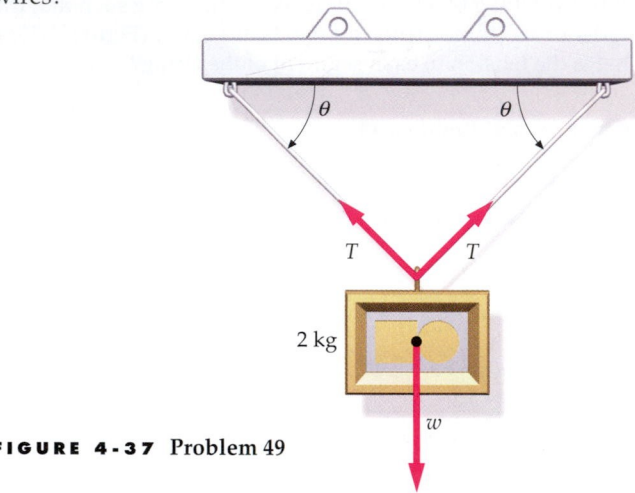

**FIGURE 4-37** Problem 49

**50** ••• **SSM** Balloon arches are often seen at festivals or celebrations; they are made by attaching helium-filled balloons to a rope that is fixed to the ground at each end. The lift from the balloons raises the structure into the arch shape. Figure 4-38*a* shows the geometry of such a structure: $N$ balloons are attached at equally spaced intervals along a massless rope of length $L$, which is attached to two supports. Each balloon provides a lift force $F$. The horizontal and vertical coordinates of the point on the rope where the *i*th balloon is attached are $x_i$ and $y_i$, and $T_i$ is the tension in the *i*th segment (with segment 0 being the segment between the point of attachment and the first balloon, and segment $N$ being the segment between the last balloon and the other point of attachment).

**FIGURE 4-38** Problem 50

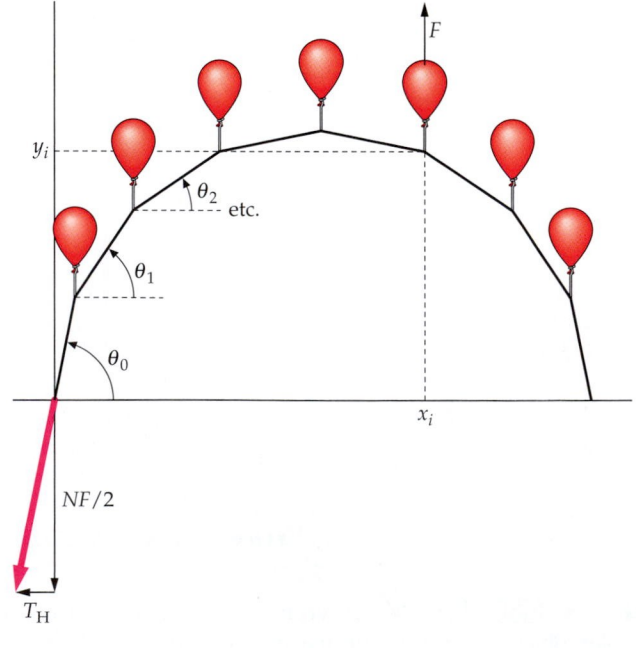

(*a*)

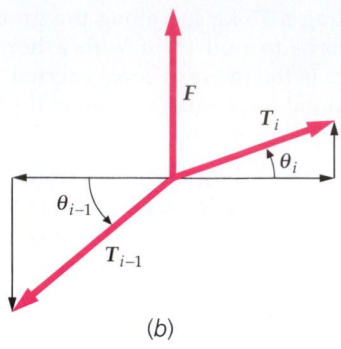

(*b*)

(*a*) Figure 4-38*b* shows a free-body diagram for the *i*th balloon. From this diagram, show that the horizontal component of the force $T_i$ (call it $T_H$) is the same for all the balloons, and that by considering the vertical component of the force, one can derive the following equation relating the tension in the *i*th and $(i - 1)$th segments:

$$T_{i-1} \sin \theta_{i-1} - T_i \sin \theta_i = F$$

(*b*) Show that $\tan \theta_0 = -\tan \theta_{N+1} = NF/2T_H$.
(*c*) From the diagram and the two expressions above, show that

$$\tan \theta_i = (N - 2i)F/2T_H$$

and that

$$x_i = \frac{L}{N+1} \sum_{j=0}^{i-1} \cos \theta_j, \qquad y_i = \frac{L}{N+1} \sum_{j=0}^{i-1} \sin \theta_j$$

(*d*) Write a spreadsheet program to make a graph of the shape of a balloon arch with the following parameters: $N = 10$ balloons giving a lift force $F = 1$ N each attached to a rope length $L = 10$ m, with a horizontal component of tension $T_H = 10$ N. How far apart are the two points of attachment? How high is the arch at its highest point?
(*e*) Note that we haven't specified the spacing between the supports—it is determined by the other parameters. Vary $T_H$ while keeping the other parameters the same until you create an arch that has a spacing of 8 m between the supports. What is $T_H$ then? As you increase $T_H$, the arch should get flatter and more spread out. Does your spreadsheet model show this?

**51** •• A 1000-kg load is being moved by a crane. Find the tension in the cable that supports the load as (*a*) it moves upward with a speed increasing by 2 m/s each second, (*b*) it is lifted at constant speed, and (*c*) it moves upward with speed decreasing by 2 m/s each second.

**52** •• **i·SOLVE** For the systems in equilibrium in Figures 4-39*a*, 4-39*b*, and 4-39*c*, find the unknown tensions and masses.

**FIGURE 4-39** Problem 52

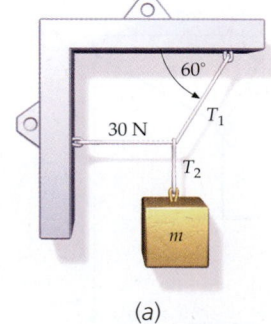

(*a*)

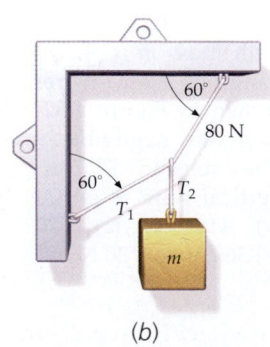

(*b*)

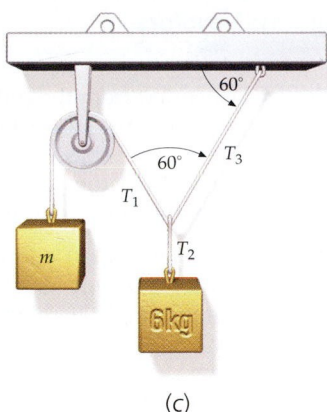

(c)

**53** •• [SOLVE] ✓ Your car is stuck in a mud hole. You are alone, but you have a long, strong rope. Having studied physics, you tie the rope tautly to a telephone pole and pull on it sideways, as shown in Figure 4-40. (a) Find the force exerted by the rope on the car when the angle $\theta$ is 3° and you are pulling with a force of 400 N but the car does not move. (b) How strong must the rope be if it takes a force of 600 N to move the car when $\theta$ is 4°?

**FIGURE 4-40** Problem 53

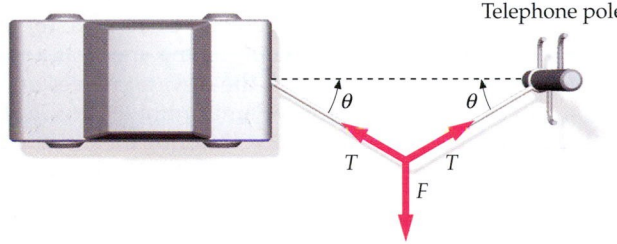

### Free-Body Diagrams: Inclined Planes and the Normal Force

**54** • [SSM] [SOLVE] A large box whose mass is 20 kg rests on a frictionless floor. A mover pushes on the box with a force of 250 N at an angle 35° below the horizontal. What is the acceleration of the box across the floor?

**55** • [SOLVE] The box from Problem 54 now rests on a frictionless ramp with a 15° slope. The mover pulls up on a rope attached to the box to pull it up the incline (see Figure 4-41). If the rope makes an angle of 40° with the horizontal, what is the smallest force $F$ the mover will have to exert to move the box up the ramp?

**FIGURE 4-41** Problem 55

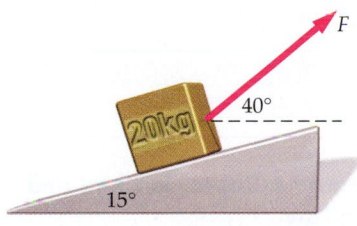

**56** • A box slides down a frictionless inclined plane. Draw a diagram showing the forces acting on the box. For each force in your diagram, indicate the reaction force.

**57** • The system shown in Figure 4-42 is in equilibrium and the incline is frictionless. It follows that the mass $m$ is (a) 3.5 kg, (b) 3.5 sin 40° kg, (c) 3.5 tan 40° kg, (d) none of these answers.

**FIGURE 4-42** Problem 57

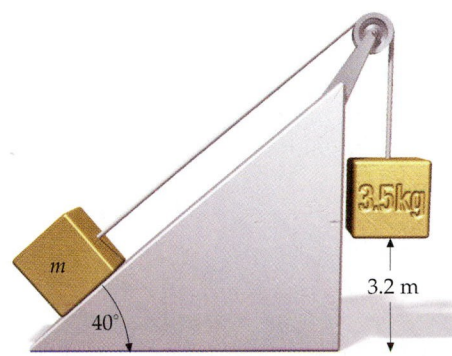

**58** • [SSM] In Figure 4-43, the objects are attached to spring balances calibrated in newtons. Give the reading of the balance(s) in each case, assuming that the strings are massless.

**FIGURE 4-43** Problem 58

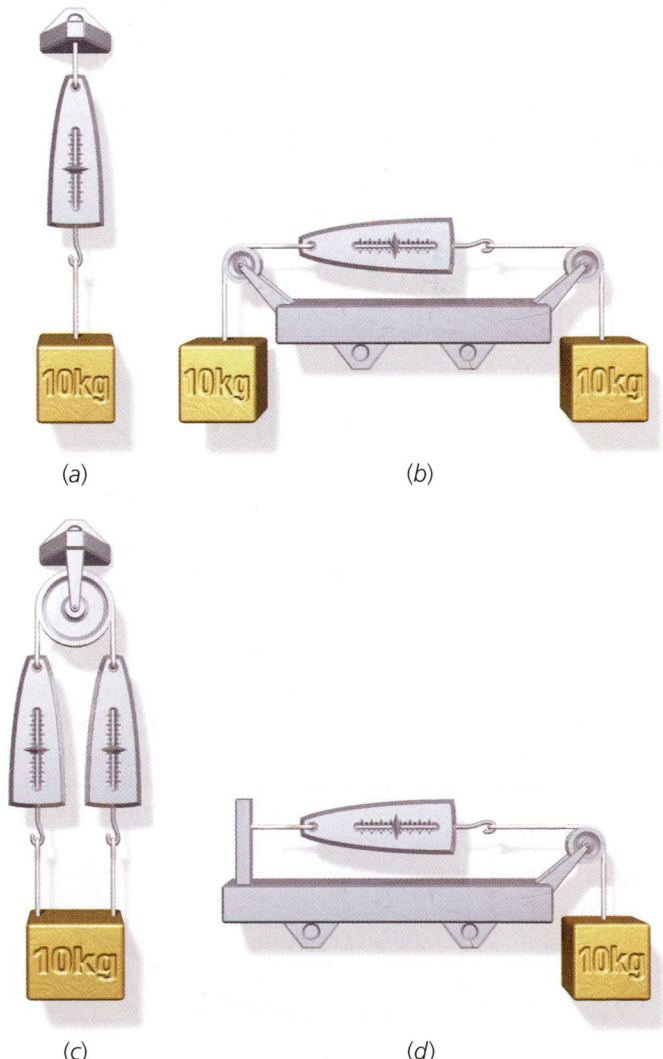

(a)                    (b)

(c)                    (d)

**59** •• A box is held in position by a cable along a frictionless incline (Figure 4-44). (*a*) If $\theta = 60°$ and $m = 50$ kg, find the tension in the cable and the normal force exerted by the incline. (*b*) Find the tension as a function of $\theta$ and $m$, and check your result for $\theta = 0°$ and $\theta = 90°$.

**FIGURE 4-44** Problem 59

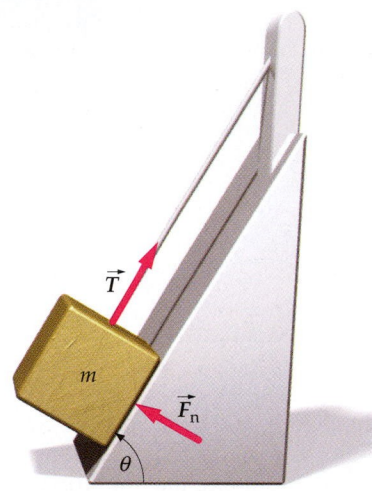

**60** •• A horizontal force of 100 N pushes a 12-kg block up a frictionless incline that makes an angle of 25° with the horizontal. (*a*) What is the normal force that the incline exerts on the block? (*b*) What is the acceleration of the block?

**61** •• SSM ISOLVE A 65-kg student weighs himself by standing on a scale mounted on a skateboard that is rolling down an incline, as shown in Figure 4-45. Assume there is no friction so that the force exerted by the incline on the skateboard is normal to the incline. What is the reading on the scale if $\theta = 30°$?

**FIGURE 4-45** Problem 61

**62** •• A block of mass $m$ slides across a frictionless floor and then up a frictionless ramp (see Figure 4-46). The angle of the ramp is $\theta$ and the speed of the block before it starts up the ramp is $v_0$. The block will slide up to some maximum height $h$ above the floor before starting to slide back down. Show that $h$ is independent of $\theta$.

**FIGURE 4-46** Problem 62

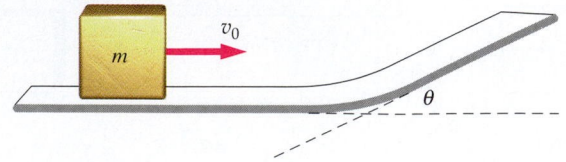

### Free-Body Diagrams: Elevators

**63** • An object is suspended from the ceiling of an elevator that is descending at a constant speed of 9.81 m/s. The tension in the string holding the object is (*a*) equal to the weight of the object, (*b*) less than the weight of the object, (*c*) greater than the weight of the object, (*d*) zero.

**64** • Suppose you are standing on a force scale in a descending elevator as it comes to a stop on the ground floor. Will the scale's report of your weight be high, low, or correct as the elevator slows down?

**65** • SSM A person of weight $w$ is in an elevator going up when the cable suddenly breaks. What is the person's apparent weight immediately after the cable breaks? (*a*) $w$. (*b*) Greater than $w$. (*c*) Less than $w$. (*d*) $9.8w$. (*e*) Zero.

**66** • ISOLVE A person in an elevator is holding a 10-kg block by a cord rated to withstand a tension of 150 N. When the elevator starts up, the cord breaks. What was the minimum acceleration of the elevator?

**67** •• A 2-kg block hangs from a spring scale calibrated in newtons that is attached to the ceiling of an elevator (Figure 4-47). What does the scale read when (*a*) the elevator is moving up with a constant velocity of 30 m/s, (*b*) the elevator is moving down with a constant velocity of 30 m/s, (*c*) the elevator is ascending at 20 m/s and gaining speed at a rate of 3 m/s²? (*d*) From $t = 0$ to $t = 5$ s, the elevator moves up at 10 m/s. Its velocity is then reduced uniformly to zero in the next 4 s, so that it is at rest at $t = 9$ s. Describe the reading of the scale during the interval $0 < t < 9$ s.

**FIGURE 4-47** Problem 67

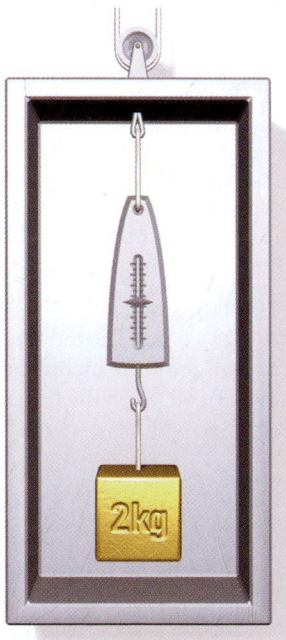

### Free-Body Diagrams: Ropes, Tension, and Newton's Third Law

**68** • Two boxes of mass $m_1$ and $m_2$ connected by a massless string are accelerated uniformly on a frictionless surface,

**79** •• **iSOLVE** An 8-kg block and a 10-kg block connected by a rope that passes over a frictionless peg slide on frictionless inclines, as shown in Figure 4-54. (a) Find the acceleration of the blocks and the tension in the rope. (b) The two blocks are replaced by two others of masses $m_1$ and $m_2$ such that there is no acceleration. Find whatever information you can about the masses of these two new blocks.

**FIGURE 4-54** Problem 79

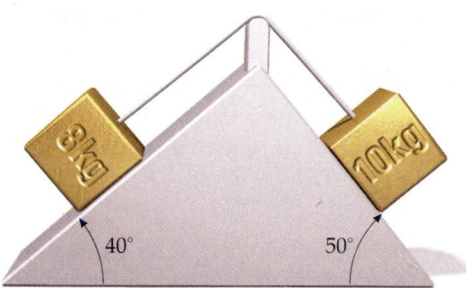

**80** •• A heavy rope of length 5 m and mass 4 kg lies on a frictionless horizontal table. One end is attached to a 6-kg block. At the other end of the rope, a constant horizontal force of 100 N is applied. (a) What is the acceleration of the system? (b) Give the tension in the rope as a function of position along the rope.

**81** •• **SSM** A 60-kg house-painter stands on a 15-kg aluminum platform. The platform is attached to a rope that passes through an overhead pulley, which allows the painter to raise herself and the platform (Figure 4-55). (a) To accelerate herself and the platform at a rate of 0.8 m/s², with what force $F$ must she pull on the rope? (b) When her speed reaches 1 m/s, she pulls in such a way that she and the platform go up at a constant speed. What force is she exerting on the rope? (Ignore the mass of the rope.)

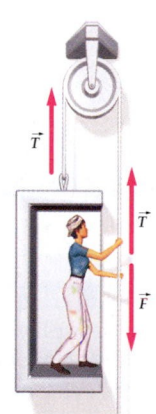

**FIGURE 4-55** Problem 81

**82** ••• Figure 4-56 shows a 20-kg block sliding on a 10-kg block. All surfaces are frictionless. Find the acceleration of each block and the tension in the string that connects the blocks.

**FIGURE 4-56** Problem 82

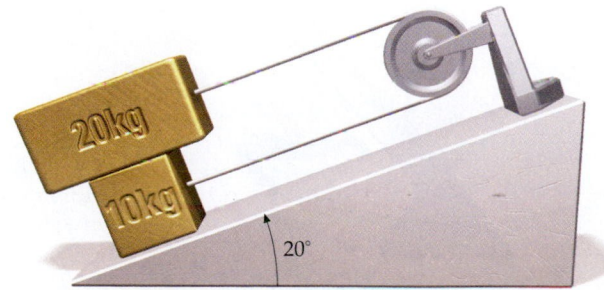

**83** ••• **iSOLVE** ✓ A 20-kg block with a pulley attached slides along a frictionless ledge. It is connected by a massless string to a 5-kg block via the arrangement shown in Figure 4-57. (a) Find the horizontal distance the 20-kg block moves when the 5-kg block descends a distance of 10 cm. (b) Find the acceleration of each block and the tension in the connecting string.

**FIGURE 4-57** Problem 83

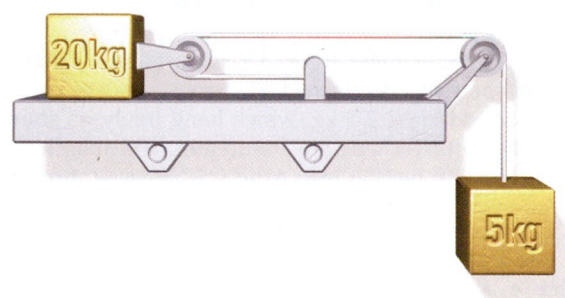

## Free-Body Diagrams: The Atwood's Machine

**84** •• **SSM** The apparatus in Figure 4-58 is called an *Atwood's machine* and is used to measure the free-fall acceleration $g$ by measuring the acceleration of the two blocks. Assuming a massless, frictionless pulley and a massless string, show that the magnitude of the acceleration of either body and the tension in the string are

$$a = \frac{m_1 - m_2}{m_1 + m_2} g \quad \text{and} \quad T = \frac{2m_1 m_2 g}{m_1 + m_2}$$

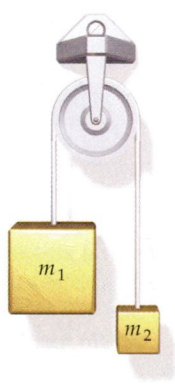

**FIGURE 4-58**
**Problems 84–87**

**85** •• **iSOLVE** If one of the masses of the Atwood's machine in Figure 4-58 is 1.2 kg, what should be the other mass so that the displacement of either mass during the first second following release is 0.3 m?

**86** •• A very small pebble of mass $m$ rests on the block of mass $m_2$ of the Atwood's machine in Figure 4-58. Find the force exerted by the pebble on $m_2$.

**87** •• Find the force exerted by the Atwood's machine on the hanger to which the pulley is attached, as shown in Figure 4-58, while the blocks accelerate. Neglect the mass of the pulley. Check your answer by considering limiting values for $m_1$ and/or $m_2$ for which you can determine the answer by qualitative reasoning.

**88** ••• The acceleration of gravity $g$ can be determined by measuring the time $t$ it takes for a mass $m_2$ in an Atwood's machine to fall a distance $L$, starting from rest. (a) Find an expression for $g$ in terms of $m_1$, $m_2$, $L$, and $t$. (b) Show that if there is a small error in the time measurement $dt$, it will lead to an error in the determination of $g$ by an amount $dg$ given by $dg/g = -2dt/t$. (c) If $L = 3$ m and $m_1$ is 1 kg, find the value of $m_2$ such that $g$ can be measured with an accuracy of ±5% with a time measurement that is accurate to 0.1 s. Assume that the only significant uncertainty in the measurement is the time of fall.

as shown in Figure 4-48. The ratio of the tensions $T_1/T_2$ is given by (a) $m_1/m_2$, (b) $m_2/m_1$, (c) $(m_1 + m_2)/m_2$, (d) $m_1/(m_1 + m_2)$, (e) $m_2/(m_1 + m_2)$.

**FIGURE 4-48** Problem 68

**69** •• A box of mass $m_2 = 3.5$ kg rests on a frictionless horizontal shelf and is attached by strings to boxes of masses $m_1 = 1.5$ kg and $m_3 = 2.5$ kg, which hang freely, as shown in Figure 4-49. Both pulleys are frictionless and massless. The system is initially held at rest. After it is released, find (a) the acceleration of each of the boxes and (b) the tension in each string.

**FIGURE 4-49** Problem 69

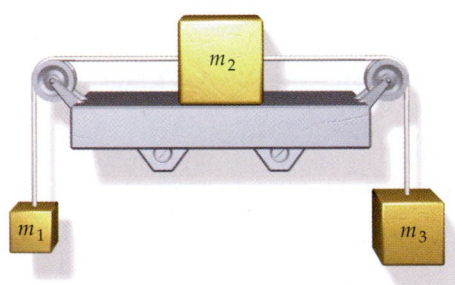

**70** •• **SSM** Two blocks are in contact on a frictionless horizontal surface. The blocks are accelerated by a horizontal force $\vec{F}$ applied to one of them (Figure 4-50). Find the acceleration and the contact force for (a) general values of $F$, $m_1$, and $m_2$, and (b) for $F = 3.2$ N, $m_1 = 2$ kg, and $m_2 = 6$ kg.

**FIGURE 4-50** Problems 70 and 71

**71** •• Repeat Problem 70, but with the two blocks interchanged.

**72** •• **iSOLVE**✔ Two 100-kg boxes are dragged along a frictionless surface with a constant acceleration of 1.0 m/s², as shown in Figure 4-51. Each rope has a mass of 1 kg. Find the force $F$ and the tension in the ropes at points A, B, and C.

**FIGURE 4-51** Problem 72

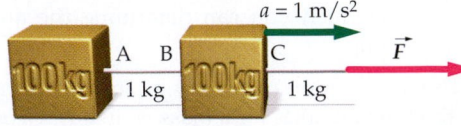

**73** •• A block of mass $m$ is being lifted vertically by a rope of mass $M$ and length $L$. The rope is being held at its top end, and the rope and block are accelerating upward with acceleration $a$. The distribution of mass in the rope is uniform. Show that the tension in the rope at a distance $x$ ($< L$) above the block is $(a + g)[m + (x/L)M]$.

**74** •• **SSM** **iSOLVE** A chain consists of 5 links, each having a mass of 0.1 kg. The chain is lifted vertically with an upward acceleration of 2.5 m/s². The chain is held at the top link; no point of the chain touches the floor. Find (a) the force $F$ exerted on the top of the chain, (b) the net force on each link, and (c) the force each link exerts on the link below it.

**75** • A 40.0-kg object supported by a vertical rope is initially at rest. The object is then accelerated upward. The tension in the rope needed to give the object an upward speed of 3.50 m/s in 0.700 s is (a) 590 N, (b) 390 N, (c) 200 N, (d) 980 N, (e) 720 N.

**76** • **iSOLVE**✔ A hovering helicopter of mass $m_h$ is lowering a truck of mass $m_t$. If the truck's downward speed is increasing at the rate of $0.1g$, what is the tension in the supporting cable? (a) $1.1m_tg$. (b) $m_tg$. (c) $0.9m_tg$. (d) $1.1(m_h + m_t)g$. (e) $0.9(m_h + m_t)g$.

**77** •• Two objects are connected by a massless string, as shown in Figure 4-52. The incline and pulley are frictionless. Find the acceleration of the objects and the tension in the string for (a) general values of $\theta$, $m_1$, and $m_2$, and (b) $\theta = 30°$ and $m_1 = m_2 = 5$ kg.

**FIGURE 4-52** Problem 77

**78** • **iSOLVE** In a stage production of Peter Pan, the 50-kg actress playing Peter has to fly in vertically, and to be in time with the music, she must be lowered, starting from rest, a distance of 3.2 m in 2.2 s. Backstage, a smooth surface sloped at 50° supports a counterweight of mass $m$, as shown in Figure 4-53. Show the calculations that the stage manager must perform to find (a) the mass of the counterweight that must be used and (b) the tension in the wire.

**FIGURE 4-53** Problem 78

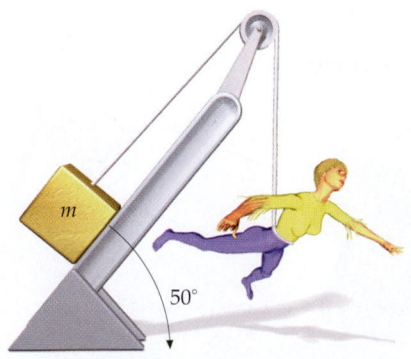

**89** •• **SSM** You are given an Atwood's machine and a set of weights whose total mass is $M$. You are told to attach some of the weights to one side of the machine, and the rest to the other side. If $m_1$ represents the mass attached to the left side and $m_2$ is the mass attached to the right side, the tension in the rope is given by the expression

$$T = \frac{2m_1 m_2}{m_1 + m_2} g$$

as was shown in Problem 85. Show that the tension will be greatest when $m_1 = m_2 = M/2$.

**90** ••• An Atwood's machine has a fixed mass $m_1$ attached on one side and variable mass $m_2$ ($> m_1$) on the other side. (a) Show that the largest possible magnitude of the tension in the rope is $2m_1 g$. (b) Interpret this result physically, without the use of calculus.

## General Problems

**91** • A redheaded woodpecker hits the bark of a tree extremely hard—the speed of its head reaches approximately $v = 3.5$ m/s before impact. If the mass of the bird's head is 0.060 kg, and the average force acting on the head during impact is $F = 6.0$ N, find (a) the acceleration of its head (assuming it is constant), (b) the depth of penetration into the bark, and (c) the time $t$ it takes the woodpecker's head to stop.

**92** •• **SSM** A simple accelerometer can be made by suspending a small object from a string attached to a fixed point on an accelerating object. Suppose such an accelerometer is attached to the ceiling of an automobile traveling on a large flat surface. When there is acceleration, the object will deflect and the string will make some angle with the vertical. (a) How is the direction in which the suspended object is deflected related to the direction of the acceleration? (b) Show that the acceleration $a$ is related to the angle $\theta$ that the string makes by $a = g \tan \theta$. (c) Suppose the automobile brakes to rest from 50 km/h in a distance of 60 m. What angle will the accelerometer make? Will the object swing forward or backward?

**93** •• **iSOLVE** The mast of a sailboat is supported at bow and stern by stainless steel wires, the forestay and backstay, anchored 10 m apart (Figure 4-59). The 12-m long mast weighs 800 N and stands vertically on the deck of the boat. The mast is positioned 3.6 m behind where the forestay is attached. The tension in the forestay is 500 N. Find the tension in the backstay and the force that the mast exerts on the deck.

**FIGURE 4-59** Problem 93

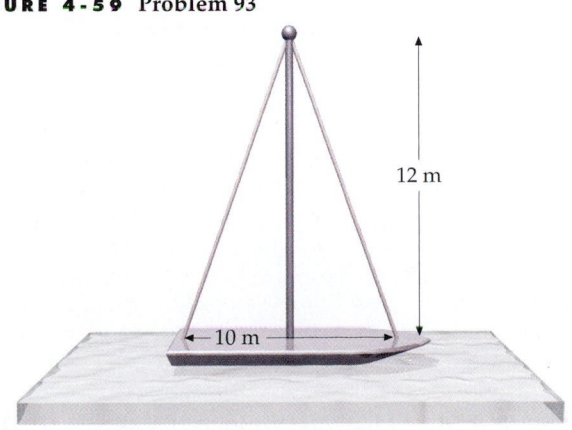

**94** •• **iSOLVE** A large uniform chain is hanging from the ceiling, supporting a block of mass 50 kg. The mass of the chain itself is 20 kg, and the length of the chain is 1.5 m. Determine the tension in the chain (a) at the point where the chain is supporting the block, (b) midway up the chain, and (c) at the top of the chain where it is attached to the ceiling.

**95** ••• **SSM** **iSOLVE** A man pushes a 24-kg box across a frictionless floor. The box begins moving from rest. He initially pushes on the box gently, but gradually increases his force so that the force he exerts on the box varies in time as $F = (8$ N/s$)t$. After 3 s, he stops pushing the box. The force is always exerted in the same direction. (a) What is the velocity of the box after 3 s? (b) How far has the man pushed the box in 3 s? (c) What is the average velocity of the box between 0 s and 3 s? (d) What is the average force that the man exerts on the box while he is pushing it?

**96** •• Suppose that a frictionless surface is inclined at an angle of 30° to the horizontal. The 270-g block is attached to a 75-g hanging weight using a pulley, as shown in Figure 4.60. (a) Draw two free-body diagrams, one for the block and the other for the hanging weight. (b) Find the tension in the string and the acceleration of the block. (c) The block is released from rest. How long does it take for it to slide a distance of 1.00 m down the surface?

**FIGURE 4-60** Problem 96

**97** •• A box of mass $m_1$ is pulled along a smooth horizontal surface by a force $F$ exerted at the end of a rope that has a much smaller mass $m_2$, as shown in Figure 4-61. (a) Find the acceleration of the rope and block, assuming them to be one object. (b) What is the net force acting on the rope? (c) Find the tension in the rope at the point where it is attached to the block. (d) The diagram, with the rope perfectly horizontal along its length, is not quite accurate. Correct the diagram and state how this correction affects your solution.

**FIGURE 4-61** Problem 97

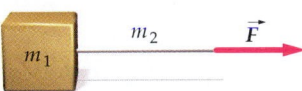

**98** •• **SSM** **iSOLVE** A 2-kg block rests on a frictionless wedge that has an inclination of 60° and an acceleration $a$ to the right such that the mass remains stationary relative to the wedge (Figure 4-62). (a) Find $a$. (b) What would happen if the wedge were given a greater acceleration?

**FIGURE 4-62** Problem 98

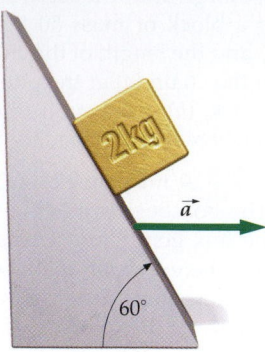

**99** •• ⓘSOLVE✓ The masses attached to each side of an Atwood's machine consist of a stack of five washers, each of mass $m$, as shown in Figure 4-63. The tension in the string is $T_0$. When one of the washers is removed from the left side, the remaining washers accelerate and the tension decreases by 0.3 N. (*a*) Find $m$. (*b*) Find the new tension and the acceleration of each mass when a second washer is removed from the left side.

**FIGURE 4-63** Problems 99 and 100

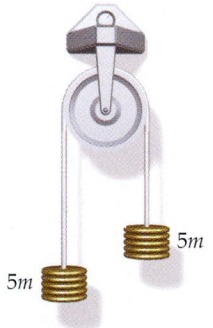

**100** •• Consider the Atwood's machine in Figure 4-63. When $N$ washers are transferred from the left side to the right side, the right side drops 47.1 cm in 0.40 s. Find $N$.

**101** •• Blocks of mass $m$ and $2m$ are connected by a string (Figure 4-64). (*a*) If the forces are constant, find the tension in the connecting string. (*b*) If the forces vary with time as $F_1 = Ct$ and $F_2 = 2Ct$, where $C$ is a constant and $t$ is time, find the time $t_0$ at which the tension in the string is $T_0$.

**FIGURE 4-64** Problem 101

**102** ••• SSM ⓘSOLVE The pulley in an Atwood's machine is given an upward acceleration $a$, as shown in Figure 4-65. Find the acceleration of each mass and the tension in the string that connects them. *Hint: a constant upward acceleration has the same effect as an increase in the acceleration due to gravity.*

**FIGURE 4-65** Problem 102

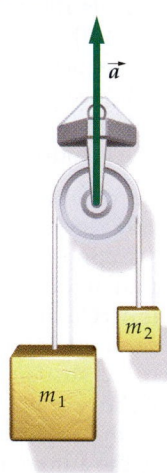

# Applications of Newton's Laws

AS THIS CAR ROUNDS THE CURVE, IT IS PREVENTED FROM SLIDING RADIALLY BY THE STATIC FRICTIONAL FORCE EXERTED ON THE TIRES BY THE ROAD.

 **What factors determine how fast the car can go around the corner without skidding? (See Example 5-10.)**

5-1    Friction

5-2    Motion Along a Curved Path

*5-3    Drag Forces

*5-4    Numerical Integration: Euler's Method

In Chapter 4 we introduced Newton's laws and applied them to situations where action was restricted to straight-line motion, and the quantitative effects of friction were excluded.

➤ **In this chapter we will extend the application of Newton's laws to motion along curved paths, and we will include the quantitative effects of friction.**

## 5-1 Friction

Without friction our ground-based transportation system, from walking to automobiles, could not function. To start walking on a horizontal surface requires friction, and once you are already walking, friction is required if you are to change either speed or direction. Friction is required to hold a nut on a screw or a nail in wood. As important as friction is, it is often not desirable. Lubricants, such as motor oil in an automobile engine, or synovial fluid in our joints, can reduce friction.

### Static Friction

When you apply a small horizontal force $\vec{F}$ (see Figure 5-1) to a large box resting on the floor, the box may not move because the force of **static friction** $\vec{f}_s$ exerted by the floor on the box, balances the force you are applying. The force of static friction, which opposes the applied force on the box, can vary from zero to some maximum force $f_{s,max}$, depending on how hard you push. Data show that $f_{s,max}$ is proportional to the normal force exerted by one surface on the other:

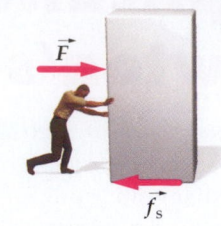

$$f_{s,max} = \mu_s F_n \qquad\qquad\qquad 5\text{-}1$$

DEFINITION—COEFFICIENT OF STATIC FRICTION

FIGURE 5-1

where the proportionality constant $\mu_s$, called the **coefficient of static friction,** depends on the nature of the surfaces in contact. If you exert a horizontal force smaller than $f_{s,max}$ on the box, the frictional force will just balance this horizontal force. In general, we can write

$$f_s \leq \mu_s F_n \qquad\qquad\qquad 5\text{-}2$$

## Kinetic Friction

If you push the box in Figure 5-1 hard enough, it will slide across the floor. As it slides, the floor exerts a force of **kinetic friction** $\vec{f}_k$ (also called sliding friction) that opposes the motion. To keep the box sliding with constant velocity, you must exert a force on the box that is equal in magnitude and opposite in direction to the force of kinetic friction exerted by the floor.

The **coefficient of kinetic friction** $\mu_k$ is the ratio of the magnitudes of the kinetic frictional force $f_k$ and the normal force $F_n$:

$$f_k = \mu_k F_n \qquad\qquad\qquad 5\text{-}3$$

DEFINITION—COEFFICIENT OF KINETIC FRICTION

where $\mu_k$ depends on the nature of the surfaces in contact. Experimentally, it is found that $\mu_k$ is less than $\mu_s$, and is approximately constant for speeds ranging from about 1 cm/s to several meters per second—the only situations we will consider.

## Rolling Friction

When an ideal, rigid wheel rolls *at constant speed* along an ideal, rigid horizontal road without slipping, no frictional force slows its motion. However, because real tires and roads continually deform and because the tread and the road are continually peeled apart, the road exerts a force of **rolling friction** $\vec{f}_r$ that opposes the motion. To keep the wheel rolling with constant velocity, you must exert a force on the wheel that is equal in magnitude and opposite in direction to the force of rolling friction exerted by the road.

The **coefficient of rolling friction** $\mu_r$ is the ratio of the magnitudes of the rolling frictional force $f_r$ and the normal force $F_n$:

$$f_r = \mu_r F_n \qquad\qquad\qquad 5\text{-}4$$

DEFINITION—COEFFICIENT OF ROLLING FRICTION

where $\mu_r$ depends on the nature of the surfaces in contact and the composition of the wheel and road. Typical values of $\mu_r$ are 0.01 to 0.02 for rubber tires on concrete and 0.001 to 0.002 for steel wheels on steel rails.

## Friction Explained

Friction is a complex, incompletely understood phenomenon that arises from the attraction of molecules between two surfaces that are in close contact. The nature of this attraction is electromagnetic—the same as the molecular bonding that holds an object together. This attractive force is short ranged and becomes negligible at distances of only a few atomic diameters.

As shown in Figure 5-2, ordinary objects that look smooth and feel smooth are rough and bumpy at the microscopic (atomic) scale. This is the case even if the surfaces are highly polished. When surfaces come into contact, they touch only at widely spaced prominences, called asperities, shown in Figure 5-2. The normal force exerted by a surface is exerted at the tips of these asperities where the force per unit area is very large, large enough to flatten the tips of the asperities. As the normal force increases, so does this flattening, resulting in a larger microscopic contact area. Under a wide range of conditions the microscopic area of contact is proportional to the normal force. The frictional force is proportional to the microscopic contact area, so it is also proportional to the normal force.

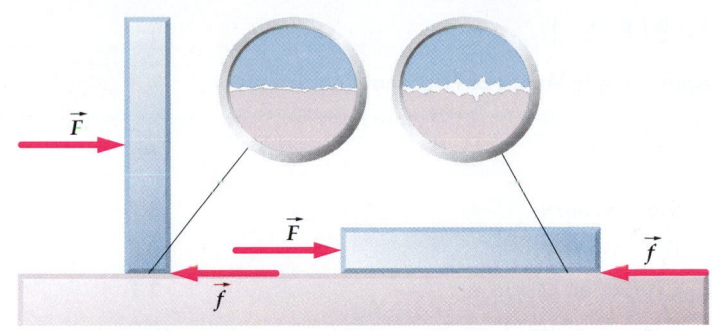

**FIGURE 5-2** The microscopic area of contact between box and floor is only a small fraction of the macroscopic area of the box's bottom surface. This fraction is proportional to the normal force exerted between the surfaces. If the box rests on its side, the macroscopic area is increased, but the force per unit area is decreased so the microscopic area of contact is unchanged. Whether the box is upright or on its side, the same horizontal applied force *F* is required to keep it sliding at constant speed.

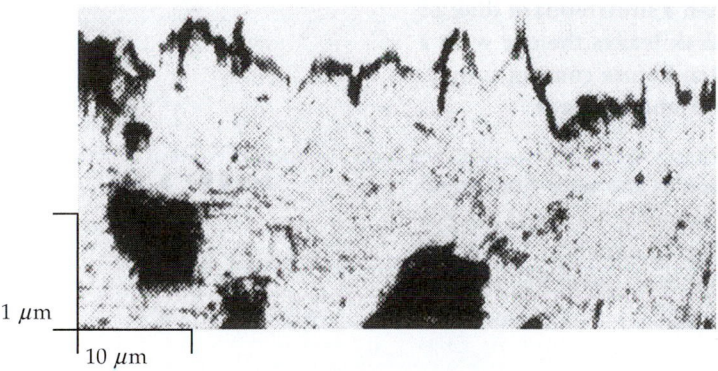

Magnified section of a polished steel surface showing surface irregularities. The irregularities are about $5 \times 10^{-5}$ cm high, corresponding to several thousand atomic diameters.

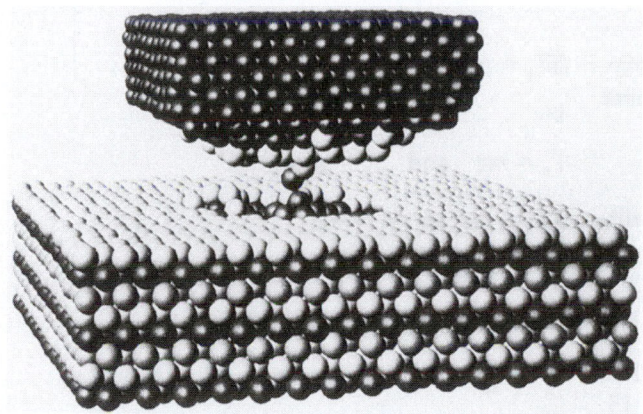

The computer graphic shows gold atoms (bottom) adhering to the fine point of a nickel probe (top) that has been in contact with the gold surface.

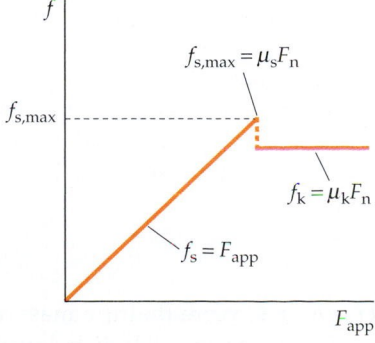

Figure 5-3 shows a plot of the frictional force exerted on the box by the floor as a function of the applied force. The force of friction balances the applied force until the applied force equals $\mu_s F_n$, at which point the box begins to slide. Then the frictional force is constant and equal to $\mu_k F_n$. Table 5-1 lists some approximate values of $\mu_s$ and $\mu_k$ for various surfaces.

**FIGURE 5-3**

## TABLE 5-1

**Approximate Values of Frictional Coefficients**

| Materials | $\mu_S$ | $\mu_k$ |
|---|---|---|
| Steel on steel | 0.7 | 0.6 |
| Brass on steel | 0.5 | 0.4 |
| Copper on cast iron | 1.1 | 0.3 |
| Glass on glass | 0.9 | 0.4 |
| Teflon on Teflon | 0.04 | 0.04 |
| Teflon on steel | 0.04 | 0.04 |
| Rubber on concrete (dry) | 1.0 | 0.80 |
| Rubber on concrete (wet) | 0.30 | 0.25 |
| Waxed ski on snow (0°C) | 0.10 | 0.05 |

---

*A GAME OF SHUFFLEBOARD*   **EXAMPLE 5-1**

**A cruise-ship passenger uses a shuffleboard cue to push a shuffleboard disk of mass 0.40 kg horizontally along the deck so that the disk leaves the cue with a speed of 5.5 m/s. The disk then slides a distance of 8 m before coming to rest. Find the coefficient of kinetic friction between the disk and the deck.**

**PICTURE THE PROBLEM** The force of kinetic friction is the only horizontal force acting on the disk after it separates from the cue. Since the frictional force is constant, the acceleration is constant. We can find the acceleration using the constant-acceleration equations of Chapter 2 and relate it to $\mu_k$ using $\Sigma F_x = ma_x$.

**FIGURE 5-4**

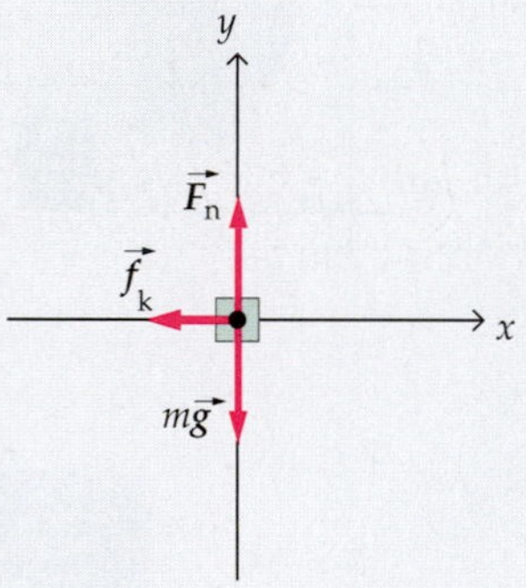

1. Draw a free-body diagram for the disk after it leaves the cue (Figure 5-4). The arrows on the $x$ and $y$ axes indicate the positive $x$ and $y$ directions.

2. The coefficient of friction relates the frictional and normal forces:

$$f_k = \mu_k F_n$$

3. Apply $\Sigma F_y = ma_y$ to the disk. Solve for the normal force. Then, using the step 2 result, solve for the frictional force:

$$\Sigma F_y = ma_y \Rightarrow F_n - mg = 0$$

so

$$F_n = mg \quad \text{and} \quad f_k = \mu_k mg$$

4. Apply $\Sigma F_x = ma_x$ to the disk. Using the step 3 result, solve for the acceleration:

$$\Sigma F_x = ma_x \Rightarrow -f_k = ma_x$$

so

$$-\mu_k mg = ma_x \quad \text{and} \quad a_x = -\mu_k g$$

5. The acceleration is constant. Relate it to the total distance traveled and the initial velocity using Equation 2-15. Using the step 4 result, solve for $\mu_k$:

$$v^2 = v_0^2 + 2a_x \Delta x \Rightarrow 0 = v_0^2 - 2\mu_k g\, \Delta x$$

so

$$\mu_k = \frac{v_0^2}{2g\,\Delta x} = \frac{(5.5\text{ m/s})^2}{2(9.81\text{ m/s}^2)(8\text{ m})} = \boxed{0.193}$$

**REMARKS** Note that the mass $m$ of the disk cancels. The greater the mass, the harder it is to stop the disk, but the greater mass is also accompanied by greater friction. The net result is that mass has no effect.

Example 5-1 illustrates guidelines for solving problems involving friction. They are as follows:

1. Choose the $y$ axis normal to the contacting surfaces. Choose the $x$ axis parallel with the surface and either parallel or antiparallel with the frictional force.
2. Apply $\Sigma F_y = ma_y$ and solve for the normal force $F_n$.
   - If the friction is *kinetic*, solve for the frictional force using $f_k = \mu_k F_n$.
   - If the friction is *static*, relate the maximum frictional force to the normal force using $f_{s,max} = \mu_s F_n$.
   - If the friction is *rolling*, solve for the frictional forces using $f_r = \mu_r F_n$.
3. Apply $\Sigma F_x = ma_x$ to the object and solve for the desired quantity.

SOLVING PROBLEMS INVOLVING FRICTION

---

*A SLIDING COIN*                          **E X A M P L E    5 - 2**

**A hardcover book is resting on a table top with its front cover facing upward. You place a coin on this cover and very slowly open the book until the coin starts to slide. The angle $\theta_{max}$ is the angle the front cover makes with the horizontal just as the coin starts to slide. Find the coefficient of static friction $\mu_s$ between the book cover and the coin in terms of $\theta_{max}$.**

**PICTURE THE PROBLEM** The forces acting on the coin are its weight $mg$, the normal force $F_n$ exerted by the plane, and the force of friction $f$. We should follow the guidelines for solving problems with static friction.

**FIGURE 5-5**

1. Draw a free-body diagram for the coin with the book cover inclined at angle $\theta$, where $\theta \leq \theta_{max}$ (Figure 5-6):

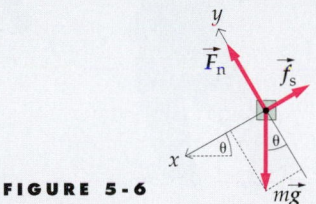

**FIGURE 5-6**

2. Apply $\Sigma F_y = ma_y$ to the coin and solve for the normal force. Then determine the maximum static frictional force:

$$\Sigma F_y = ma_y \Rightarrow F_n - mg \cos \theta = 0$$

so

$$F_n = mg \cos \theta$$

$$f_{s,max} = \mu_s F_n \quad \text{so} \quad f_{s,max} = \mu_s mg \cos \theta_{max}$$

3. Apply $\Sigma F_x = ma_x$ to the coin and solve for the frictional force. Then substitute into the step 2 result:

$$\Sigma F_x = ma_x \Rightarrow -f_{s,max} + mg \sin \theta_{max} = 0$$

so

$$f_{s,max} = mg \sin \theta_{max} \quad \text{and} \quad mg \sin \theta_{max} = \mu_s mg \cos \theta_{max}$$

4. Solve the step 3 result for $\mu_s$:

$$\mu_s = \frac{mg \sin \theta_{max}}{mg \cos \theta_{max}} = \boxed{\tan \theta_{max}}$$

**EXERCISE** The coefficient of static friction between a car's tires and the road on a particular day is 0.7. What is the steepest angle of inclination of the road for which the car can be parked with its wheels locked and not slide down the hill? (*Answer* 35°)

From Example 5-2 we see that the coefficient of static friction is related to the angle $\theta_{max}$ at which an object begins to slip by

$$\mu_s = \tan \theta_{max}$$
5-5

ANGLE OF REPOSE

and $\theta_{max}$ is called the angle of repose.

---

*PULLING A SLED*                                 **EXAMPLE 5-3**

**Two children sitting on a sled at rest in the snow ask you to pull them. You oblige by pulling on the sled's rope, which makes an angle of 40° with the horizontal (Figure 5-7). The children have a combined mass of 45 kg and the sled has a mass of 5 kg. The coefficients of static and kinetic friction are $\mu_s = 0.2$ and $\mu_k = 0.15$. Find the frictional force exerted by the ground on the sled and the acceleration of the children and sled, starting from rest, if the tension in the rope is (*a*) 100 N and (*b*) 140 N.**

**PICTURE THE PROBLEM** First we need to find out whether the frictional force is static or kinetic. To do this we solve for the maximum tension in the rope without the sled sliding.

FIGURE 5-7

(*a*) 1. Draw a free-body diagram for the sled (Figure 5-8):

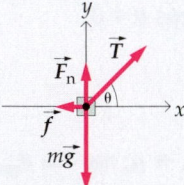

FIGURE 5-8

2. Apply $\Sigma F_y = ma_y$ to the sled and solve for the normal force. Then solve for the maximum static frictional force:

$$\Sigma F_y = ma_y \Rightarrow F_n + T \sin \theta - mg = 0$$

so

$$F_n = mg - T \sin \theta$$

$$f_{s,max} = \mu_s F_n \quad \text{so} \quad f_{s,max} = \mu_s(mg - T_{max} \sin \theta)$$

3. Apply $\Sigma F_x = ma_x$ to the sled and solve for the frictional force. Then substitute into the step 2 result:

$$\Sigma F_x = ma_x \Rightarrow -f_{s,max} + T_{max} \cos \theta = 0$$

so

$$f_{s,max} = T_{max} \cos \theta \quad \text{and} \quad T_{max} \cos \theta = \mu_s(mg - T_{max} \sin \theta)$$

$$T_{max} \cos \theta = \mu_s(mg - T_{max} \sin \theta)$$

4. Solve the step 3 result for the maximum tension without slipping:

$$T_{max}(\mu_s \sin \theta + \cos \theta) = \mu_s mg$$

so

$$T_{max} = \frac{\mu_s mg}{\mu_s \sin \theta + \cos \theta}$$

$$= \frac{0.2(50 \text{ kg})(9.81 \text{ m/s}^2)}{0.2 \sin 40° + \cos 40°} = 110 \text{ N}$$

5. The tension is 100 N, which is less than 110 N. The sled is *not* sliding. To find the frictional force, use the step 3 expression for $f_s$:

$$a_x = \boxed{0}$$

$$f_s = T \cos \theta = (100 \text{ N}) \cos 40° = \boxed{76.6 \text{ N}}$$

(b) 1. The tension is 140 N, which is greater than $T_{max} = 110$ N, so the sled is sliding. Relate the kinetic frictional force $f_k$ to the normal force:

$$f_k = \mu_k F_n$$

2. In step (a)2 we applied $\Sigma F_y = ma_y$ to the sled and found $F_n = mg - T \sin \theta$. Use this result along with the step (b)1 result to solve for the kinetic frictional force:

$$f_k = \mu_k(mg - T \sin \theta)$$

$$= 0.15[(50 \text{ kg})(9.81 \text{ N/kg}) - (140 \text{ N}) \sin 40°]$$

$$= \boxed{60.1 \text{ N}}$$

3. Apply $\Sigma F_x = ma_x$ to the sled and solve for the frictional force. Then substitute the step 2 result for $f_k$ and solve for the acceleration:

$$\Sigma F_x = ma_x \Rightarrow -f_k + T \cos \theta = ma_x$$

so

$$a_x = \frac{-f_k + T \cos \theta}{m}$$

$$= \frac{(-60.1 \text{ N}) + (140 \text{ N}) \cos 40°}{50 \text{ kg}}$$

$$= \boxed{0.943 \text{ m/s}^2}$$

**REMARKS** There are two important points to note about this example: (1) the normal force is not equal to the weight of the children and the sled. That is because the vertical component of the tension helps lift the sled off the ground. (2) In Part (a), the force of static friction is less than $\mu_s F_n$.

---

*A SLIDING BLOCK* | **EXAMPLE 5-4** **Try It Yourself**

The block of mass $m_2$ in Figure 5-9 has been adjusted so that the block of mass $m_1$ is on the verge of sliding. (a) If $m_1 = 7$ kg and $m_2 = 5$ kg, what is the coefficient of static friction between the table and the block? (b) With a slight nudge, the blocks move with acceleration $a$. Find $a$ if the coefficient of kinetic friction between the table and the block is $\mu_k = 0.54$.

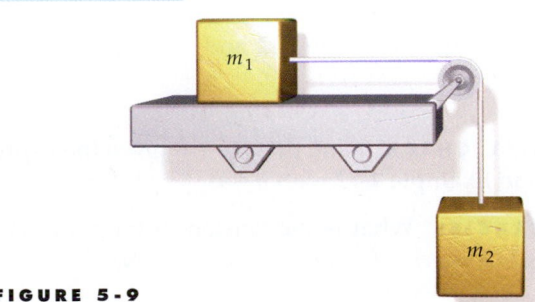

**FIGURE 5-9**

**PICTURE THE PROBLEM** Apply Newton's second law to each block. By neglecting the masses of both the rope and the pulley, and by neglecting friction in the pulley bearing, the tension has the same magnitude throughout the rope, so $T_1 = T_2 = T$, and, because the rope remains taut but does not stretch, the accelerations have the same magnitude, so $a_1 = a_2 = a$.

To find the coefficient of static friction $\mu_s$, as required in Part ($a$), set the force of static friction on $m_1$ equal to its maximum value $f_{max} = \mu_s F_n$ and set the acceleration equal to zero.

**Cover the column to the right and try these on your own before looking at the answers.**

**FIGURE 5-10**

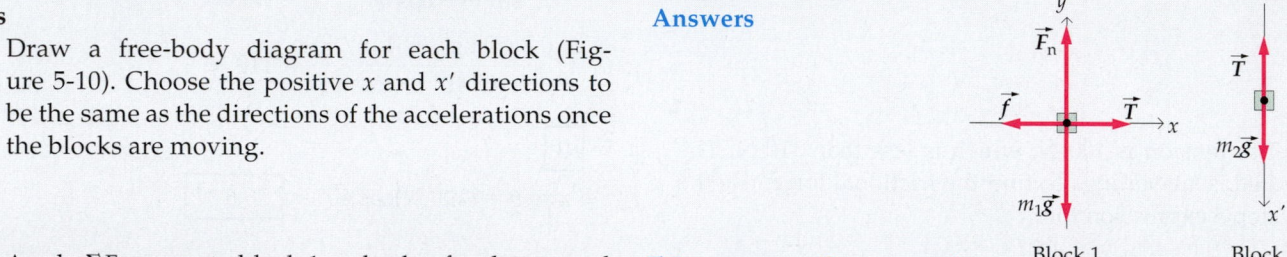

Block 1          Block 2

| **Steps** | **Answers** |
|---|---|
| ($a$) 1. Draw a free-body diagram for each block (Figure 5-10). Choose the positive $x$ and $x'$ directions to be the same as the directions of the accelerations once the blocks are moving. | |
| 2. Apply $\Sigma F_y = ma_y$ to block 1 and solve for the normal force. Then solve for the static frictional force. | $\Sigma F_y = m_1 a_{1y} \Rightarrow F_n - m_1 g = 0$ <br> so <br> $F_n = m_1 g$ <br> $f_{s,max} = \mu_s F_n$ so $f_{s,max} = \mu_s m_1 g$ |
| 3. Apply $\Sigma F_x = ma_x$ to block 1 and solve for the frictional force. Then substitute into the step 2 result. | $\Sigma F_x = m_1 a_{1x} \Rightarrow T - f_{s,max} = 0$ <br> so <br> $f_{s,max} = T$ and $T = \mu_s m_1 g$ |
| 4. Apply $\Sigma F_x = ma_x$ to block 2 and solve for the tension. Then substitute into the step 3 result. | $\Sigma F_{x'} = m_2 a_{2x'} \Rightarrow m_2 g - T = 0$ <br> so <br> $T = m_2 g$ and $m_2 g = \mu_s m_1 g$ |
| 5. Solve the step 4 result for $\mu_s$ | $\mu_s = \dfrac{m_2}{m_1} = \dfrac{5\,\text{kg}}{7\,\text{kg}} = \boxed{0.714}$ |
| ($b$) 1. During sliding the frictional force is kinetic. Relate the kinetic frictional force $f_k$ to the normal force. The normal force was found in step 2 of Part ($a$). | $f_k = \mu_k F_n$ <br> so <br> $f_k = \mu_k m_1 g$ |
| 2. Apply $\Sigma F_x = ma_x$ to block 1. Then substitute for the frictional force using the result from step 1 of Part ($b$). | $\Sigma F_x = m_1 a_{1x} \Rightarrow T - f_k = m_1 a$ <br> so <br> $T - \mu_k m_1 g = m_1 a$ |
| 3. Apply $\Sigma F_{x'} = ma_{x'}$ to block 2. | $\Sigma F_{x'} = m_2 a_{2x'} \Rightarrow m_2 g - T = m_2 a$ |
| 4. Add the equations in steps 2 and 3 of Part ($b$) and solve for $a$. | $a = \dfrac{m_2 - \mu_k m_1}{m_1 + m_2} g = \boxed{0.997\ \text{m/s}^2}$ |

**⟁ PLAUSIBILITY CHECK** Note that $\mu_k = 0$ gives the expression for the acceleration derived in Example 4-12 with $\theta = 0$.

**EXERCISE** What is the tension in the rope when the blocks are sliding? (*Answer* $T = m_2(g - a) = 44.1$ N)

*THE RUNAWAY BUGGY*                    **E X A M P L E   5 - 5**

**A runaway baby buggy is sliding without friction across a frozen pond toward a hole in the ice (Figure 5-11). You race after the buggy on skates. As you grab it, you and the buggy are moving toward the hole at speed $v_0$. The coefficient of friction between your skates and the ice as you turn out the blades to brake is $\mu_k$. $D$ is the distance to the hole when you reach the buggy, $M$ is the total mass of the buggy, and $m$ is your mass. (a) What is the lowest value of $D$ such that you stop the buggy before it reaches the hole in the ice? (b) What force do you exert on the buggy?**

**PICTURE THE PROBLEM** Initially, you and the buggy are moving toward the hole with speed $v_0$, which we take to be in the positive $x$ direction. If you exert a force $\vec{F} = -F\hat{i}$ on the buggy, the buggy, in accord with Newton's third law, exerts a force $\vec{F}' = F\hat{i}$ on you. Apply Newton's second law to determine the acceleration. After finding the acceleration, find the distance $D$ the buggy travels while slowing to a stop. The lowest value of $D$ is that for which your speed reaches zero just as the buggy reaches the hole.

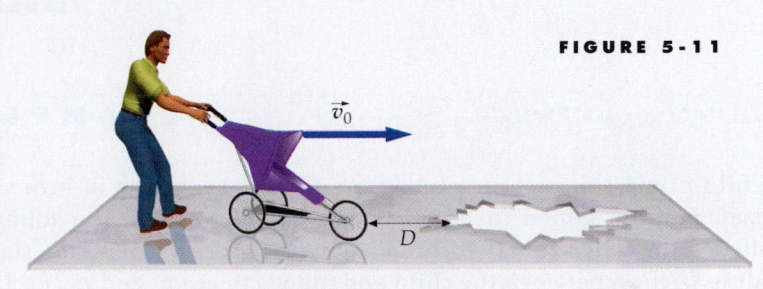

**FIGURE 5-11**

(a) 1. Draw separate free-body diagrams for yourself and the buggy.

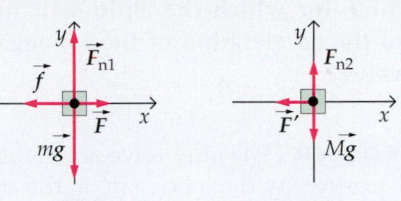

Yourself              Buggy    **FIGURE 5-12**

2. Apply $\Sigma F_y = ma_y$ to yourself and solve first for the normal force and then for the frictional force:

$$\Sigma F_y = ma_y \Rightarrow F_n - mg = 0$$

and

$$f_k = \mu_k F_n \quad \text{so} \quad f_k = \mu_k mg$$

3. Apply $\Sigma F_x = ma_x$ to yourself. Then substitute in the step 2 result:

$$\Sigma F_x = ma_x \Rightarrow F - f_k = ma_x$$

so

$$F - \mu_k mg = ma_x$$

4. Apply $\Sigma F_x = ma_x$ to the buggy. Then substitute for $F$ in the step 3 result:

$$\Sigma F_x = Ma_x \Rightarrow -F = Ma_x$$

so

$$-Ma_x - \mu_k mg = ma_x$$

5. Solve the step 4 result for $a_x$:

$$a_x = -\frac{\mu_k}{1 + M/m}g$$

(The acceleration is negative, as expected.)

6. Substitute the step 5 result into a kinematic equation and solve for $D$:

$$v_x^2 = v_{0x}^2 + 2a_x\Delta x \Rightarrow 0 = v_0^2 + 2a_x D$$

so

$$D = -\frac{v_0^2}{2a_x} = \boxed{\left(1 + \frac{M}{m}\right)\frac{v_0^2}{2\mu_k g}}$$

(b) $F$ can be found from Newton's second law applied to the buggy:

$$F = -Ma_x \Rightarrow F = \boxed{\frac{\mu_k M}{1 + M/m}g}$$

**REMARKS** The minimum value of $D$ is proportional to $v_0^2$ and inversely proportional to $\mu_k$. Figure 5-13 shows the stopping distance $D$ versus initial velocity squared for values of $M/m$ equal to 0.1, 0.3, and 1.0, with $\mu_k = 0.5$. Note that the larger the mass ratio $M/m$, the greater the distance $D$ needed to stop for a given initial velocity. This is akin to stopping a car that is pulling a trailer that does not have its own brakes. The mass of the trailer increases the stopping distance for a given speed.

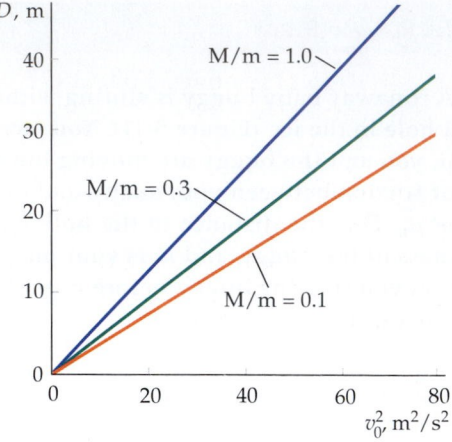

**FIGURE 5-13**

---

*PULLING A CHILD ON A TOBOGGAN*                **EXAMPLE 5-6** **Try It Yourself**

A child of mass $m_c$ sits on a toboggan of mass $m_t$, which in turn sits on a frozen pond assumed to be frictionless (Figure 5-14). The toboggan is pulled with a horizontal force as shown. The coefficients of static and sliding friction between the child and toboggan are $\mu_s$ and $\mu_k$. (*a*) Find the maximum value of $F$ for which the child will not slide relative to the toboggan. (*b*) Find the acceleration of the toboggan and child when $F$ is greater than this value.

**FIGURE 5-14**

**PICTURE THE PROBLEM** The only force accelerating the child forward is the frictional force exerted by the toboggan on the child. In Part (*a*) the challenge is to find $F$ when this frictional force is static and maximum. To do this, apply $\Sigma \vec{F} = m\vec{a}$ to the child and solve for the acceleration when static frictional force is maximum. Then apply $\Sigma \vec{F} = m\vec{a}$ to the toboggan and solve for $F$. In Part (*b*), we follow a parallel procedure. However, in this part $F$ is given and we solve for the acceleration of the toboggan.

**Cover the column to the right and try these on your own before looking at the answers.**

**Steps**

**Answers**

(*a*) 1. Draw a free-body diagram for each object (Figure 5-15).

Two alternative styles for the free-body diagram are shown for the toboggan. In the first, the two downward forces are displaced slightly so they can be seen. In the second, they are drawn head to tail.

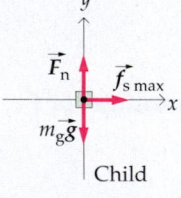 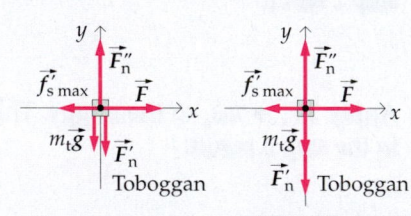

**FIGURE 5-15**

2. Equate the magnitudes of the forces in each action–reaction force pair appearing in the two free-body diagrams. Express the relation between the accelerations due to the nonslipping constraint.

$F_n' = F_n$   and   $f_{s,max}' = f_{s,max}$

and

$a_{cx} = a_{tx} = a_x$

3. Apply $\Sigma F_y = ma_y$ to the child. Solve first for the normal force, then the frictional force.

$\Sigma F_{cy} = m_c a_y \Rightarrow F_n - m_c g = 0$

$F_n = m_c g$   and

$f_{s,max} = \mu_s F_n$   so   $f_{s,max} = \mu_s m_c g$

4. Apply $\Sigma F_x = ma_x$ to the child and solve for the acceleration.

$\Sigma F_{cx} = m_c a_x \Rightarrow f_{s,max} = m_c a_x$

and

$\mu_s m_c g = m_c a_x$,   so   $a_x = \mu_s g$

5. Apply $\Sigma F_x = ma_x$ to the toboggan and, using the acceleration relations from step 2 and the result from step 3, solve for $F$.

$\Sigma F_{tx} = m_t a_x \Rightarrow F - f_{s,max} = m_t a_x$

and

$F - \mu_s m_c g = m_t \mu_s g$    so    $\boxed{F = (m_c + m_t) \mu_s g}$

(b) 1. Equate the magnitudes of each action–reaction force pair and express the change in the relation between the accelerations due to the removal of the nonslipping constraint.

$F'_n = F_n$   and   $f'_k = f_k$

but

$a_{cx} < a_{tx}$

2. Solve for the kinetic frictional force using the result from step 3 of Part (a) for the normal force.

$f_k = \mu_k F_n$   so   $f_k = \mu_k m_c g$

3. Apply $\Sigma F_x = ma_x$ to the child and solve for *her acceleration*.

$\Sigma F_{cx} = m_c a_{cx} \Rightarrow f_k = m_c a_{cx}$

and

$\mu_k m_c g = m_c a_{cx}$   so   $a_{cx} = \mu_k g$

4. Apply $\Sigma F_x = ma_x$ to the toboggan. Using the result from step 2 of Part (b), solve for *its acceleration*.

$\Sigma F_{tx} = m_t a_{tx} \Rightarrow F - f_k = m_t a_{tx}$

and

$F - \mu_k m_c g = m_t a_{tx}$   so   $\boxed{a_{tx} = \dfrac{F - \mu_k m_c g}{m_t}}$

**REMARKS** Friction is a force between two surfaces in contact, and it is not correct that friction always opposes motion, or the tendency to motion. In this example, friction does not oppose the motion of the child, it causes it. It is correct that friction always opposes motion, or the tendency to motion, of each surface *relative to the other surface*. For example, even though the child moves forward *relative to the ice*, she moves, or tends to move, backward (leftward) *relative to the toboggan*. Friction opposes this relative motion or tendency to motion.

Figure 5-16 shows the forces acting on a front-wheel-drive car that is just starting to move from rest on a horizontal road. The weight of the car is balanced by the normal force $F_n$ exerted on the tires. To start the car moving, the engine delivers power to the axle that makes the wheels rotate (we discuss power in Chapter 6). If the road were perfectly frictionless, the wheels would merely spin. When friction is present, the frictional force exerted by the road on the tires is in the forward direction, opposing the tendency of the tire surface to slip backward. This frictional force provides the acceleration needed for the car to start moving forward. If the power delivered by the engine is small enough so that the frictional force exerted by the tire surface on the road surface is sufficiently small, then the two surfaces do not slip. The wheels roll without slipping and the tire tread touching the road is at rest relative to the road. The friction between the road and the tire tread is then static friction. The largest frictional force that the tire can exert on the road (and that the road can exert on the tire) is $\mu_s F_n$.

For a car moving in a straight line with speed $v$ relative to the road, the center of each of its wheels also moves with speed $v$, as shown in Figure 5-17. If a wheel is rolling without slipping, its top is moving faster than $v$ whereas its bottom is moving slower than $v$. However, *relative to the car*, each point on the perimeter of the wheel moves in a circle with the same speed $v$. Moreover, the speed of the point on the tire momentarily in contact with the ground is zero *relative to the ground*. (Otherwise, the tire would be skidding along the road.)

If the engine power is great enough, the tire will slip and the wheels will spin. Then the force that accelerates the car is the force of kinetic friction, which is less than the force of static friction. If we are stuck on ice or snow, our chances of getting free are better if we use a light touch on the accelerator pedal. Similarly,

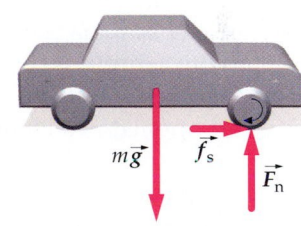

**FIGURE 5-16** Forces acting on a car (front-wheel drive). The normal forces $\vec{F}_n$ are usually larger on the front tires because typically the engine of the car is mounted at the front of the car.

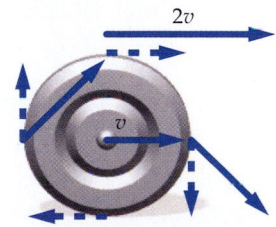

**FIGURE 5-17** In this figure, dashed lines represent velocities relative to the body of the car, solid lines represent velocities relative to the ground.

when braking a car to a stop, the force exerted by the road on the tires may be either static friction or kinetic friction, depending on how the brakes are applied. If the brakes are applied so hard that the wheels lock, the tires will skid along the road and the stopping force will be that of kinetic friction. If the brakes are applied less forcefully, so that no slipping occurs between the tires and the road, the stopping force will be that of static friction. Antilock braking systems in cars have wheel-speed sensors. If the control unit senses that a wheel is about to lock, the module signals the brake pressure modulator to drop, hold, and then restore the pressure to that wheel up to 15 times per second. This varying of pressure is much like "pumping" the brake, but with the ABS system, the wheel that is locking is the only one being pumped. This is called threshold braking. With threshold braking, maximum friction for stopping is maintained.

When wheels do lock and tires skid, two undesirable things happen. The minimum stopping distance is increased *and* the ability to control the direction of the car's motion is greatly diminished. Obviously, this loss of directional control can have dire consequences.

---

*The Effect of Antilock Brakes*　　　　　　　　　　　　　**EXAMPLE 5-7**

**A car is traveling at 30 m/s along a horizontal road. The coefficients of friction between the road and the tires are $\mu_s = 0.5$ and $\mu_k = 0.3$. How far does the car travel before stopping if (a) the car is braked with an antilock braking system so that threshold braking is sustained, and (b) the car is braked hard with no antilock braking system so that the wheels lock?**

**PICTURE THE PROBLEM** The force that stops a car when it brakes without skidding is the force of static friction exerted by the road on the tires (Figure 5-18). We use Newton's second law to solve for the frictional force and the car's acceleration. Kinematics is then used to find the stopping distance.

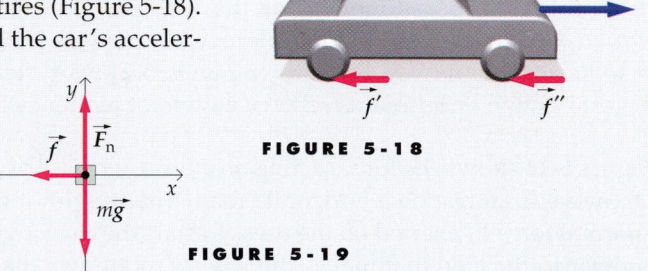

**FIGURE 5-18**

(a) 1. Draw a free-body diagram for the car (Figure 5-19). Treat all four wheels as if they were a single point of contact with the ground. Assume further that the brakes are applied to all four wheels. Let $\vec{f} = \vec{f}' + \vec{f}''$.

**FIGURE 5-19**

2. Assuming that the acceleration is constant, we use Equation 2-15 to relate the stopping distance $\Delta x$ to the initial speed $v_0$. The coefficients of friction change with temperature. Since skidding heats up the tires, these coefficients can be expected to change. Such changes are neglected here:

$$v^2 = v_0^2 + 2a_x\Delta x$$

When $v = 0$,

$$\Delta x = -\frac{v_0^2}{2a_x}$$

3. Apply $\Sigma F_y = ma_y$ to the car. Solve first for the normal force, then for the frictional force:

$$\Sigma F_y = ma_y \Rightarrow F_n - mg = 0, \quad \text{so}$$

$$F_n = mg \quad \text{and}$$

$$f_{s,max} = \mu_s F_n, \quad \text{so} \quad f_{s,max} = \mu_s mg$$

4. Apply $\Sigma F_x = ma_x$ to the car and solve for the acceleration:

$$\Sigma F_x = ma_x \Rightarrow -f_{s,max} = ma_x$$

and

$$-\mu_s mg = ma_x \quad \text{so} \quad a_x = -\mu_s g$$

5. Substituting these results in the equation for $\Delta x$ in step 2 gives the stopping distance:

$$\Delta x = -\frac{v_0^2}{2a_x} = \frac{v_0^2}{2\mu_s g}$$

$$= \frac{(30 \text{ m/s})^2}{2(0.5)(9.81 \text{ m/s}^2)} = \boxed{91.8 \text{ m}}$$

(b) 1. When the wheels lock, the force exerted by the road on the car is that of kinetic friction. Using reasoning similar to that in Part (a), we obtain for the acceleration:

$$a_x = -\mu_k g$$

2. The stopping distance is then:

$$\Delta x = -\frac{v_0^2}{2a_x} = \frac{v_0^2}{2\mu_k g}$$

$$= \frac{(30 \text{ m/s})^2}{2(0.3)(9.81 \text{ m/s}^2)} = \boxed{153 \text{ m}}$$

**REMARKS** Notice that the stopping distance is more than 50% greater when the wheels are locked. Also note that the stopping distance is independent of the car's mass—the stopping distance is the same for a subcompact car as for a large truck—provided the coefficients of friction are the same.

**EXERCISE** What must be the coefficient of static friction between the road and the tires of a four-wheel-drive car if the car is to accelerate from rest to 25 m/s in 8 s? (*Answer* 0.319)

# 5-2 Motion Along a Curved Path

In Chapter 3 we established that a particle moving with speed $v$ along a curved path with a radius of curvature $r$ has an acceleration component $a_c = v^2/r$ in the centripetal direction (toward the center of curvature), and an acceleration component $a_t = dv/dt$ in the tangential direction.

As with any acceleration, the net force is in the direction of the acceleration. The component of the net force in the centripetal direction is called the **centripetal force**. The centripetal force is not a new force, or a new kind of force. It is merely the name for the net-force component perpendicular to the direction of motion. It may be due to a string, spring, or other contact force such as a normal or frictional force; it may be an action-at-a-distance type of force such as a gravitational force; or it may be any combination of these. It is always directed inward—toward the center of curvature of the path.

*SWINGING A PAIL* **EXAMPLE 5-8**

**You swing a pail of water in a vertical circle of radius $r$ (Figure 5-20). If its speed is $v_t$ at the top of the circle, find (a) the force exerted on the water by the pail at the top of the circle, and (b) the minimum value of $v_t$ for the water to remain in the pail. (c) What is the force exerted by the pail on the water at the bottom of the circle, where the pail's speed is $v_b$?**

**PICTURE THE PROBLEM** We apply Newton's second law to find the force exerted by the pail on the water. Since the water moves in a circular path it has an acceleration component $v^2/r$ toward the center of the circle.

(a) 1. Draw free-body diagrams for the water at the top and bottom of the circle (Figure 5-21). Choose the positive axis direction to be toward the center of the circle in each case.

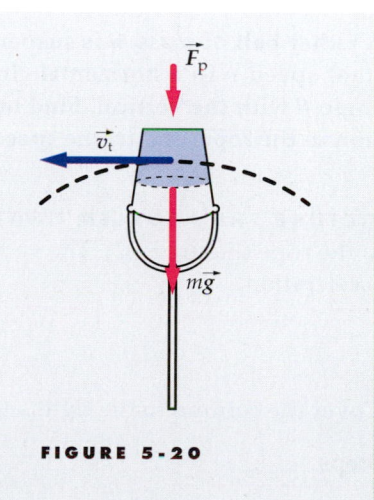

**FIGURE 5-20**

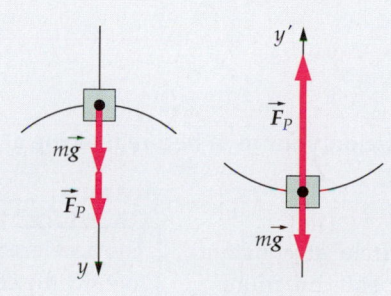

**FIGURE 5-21**

2. Apply $\Sigma F_y = ma_y$ to the water as it passes through the top of the circle with speed $v_t$. Solve for the force $F_P$ exerted on the water by the pail:

$$\Sigma F_y = ma_y \Rightarrow F_P + mg = m\frac{v_t^2}{r}$$

so

$$\boxed{F_P = m\left(\frac{v_t^2}{r} - g\right)}$$

(b) The pail can push on the water, but not pull on it. The minimum force it can exert is zero. Set $F_P = 0$ and solve for the minimum speed:

$$0 = m\left(\frac{v_{t,min}^2}{r} - g\right) \Rightarrow v_{t,min} = \boxed{\sqrt{rg}}$$

(c) Apply $\Sigma F_{y'} = ma_{y'}$ to the water as it passes through the bottom of its path with speed $v_b$. Solve for $F_P$:

$$\Sigma F_{y'} = ma_{y'} \Rightarrow F_P - mg = \frac{mv_b^2}{r}$$

so

$$\boxed{F_P = m\left(\frac{v_b^2}{r} + g\right)}$$

**REMARKS** Note that there is no arrow for centripetal force in the free-body diagrams. Centripetal force is not a kind of force exerted by some agent; it is just the name for the component of the resultant force in the centripetal direction. When a whirling bucket is at the top of its circle, both gravity and the contact force of the pail contribute to the force in the centripetal direction acting on the water. When the water is moving at the minimum speed at the top of the circle it is in free-fall with acceleration $\vec{g}$. The only force acting on it at this point is its weight, $m\vec{g}$. At the bottom of the circle, $F_P$ must be greater than the weight $mg$ to provide a net force in the centripetal direction (upward).

**PLAUSIBILITY CHECK** When $v = 0$ at the bottom, $F_P = mg$.

**EXERCISE** Estimate (a) the minimum speed at the top of the circle and (b) the maximum period of revolution that will keep you from getting wet if you swing a pail of water in a vertical circle at constant speed. (*Answer* (a) Assuming $r \approx 1$ m, we find $v_{t,min} \approx 3$ m/s (b) $T = 2\pi r/v \approx 2$ s)

---

*TETHER BALL*                    **EXAMPLE 5-9**  **Try It Yourself**

A tether ball of mass $m$ is suspended from a length of rope and travels at constant speed $v$ in a horizontal circle of radius $r$ as shown. The rope makes an angle $\theta$ with the vertical. Find (a) the direction of the acceleration, (b) the tension in the rope, and (c) the speed of the ball.

**FIGURE 5-22**

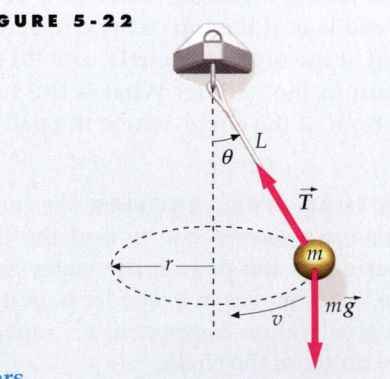

**PICTURE THE PROBLEM** Two forces act on the ball; its weight and the tension in the rope (Figure 5-22). The vector sum of these forces is in the direction of the acceleration.

**Cover the column to the right and try these on your own before looking at the answers.**

**Steps**

(a) The ball is moving in a horizontal circle at constant speed. The acceleration is in the centripetal direction.

**Answers**

The acceleration is horizontal and directed from the ball toward the center of the circle it is moving in.

(b) 1. Draw a free-body diagram for the ball. Choose the positive $x$ direction as the direction of the ball's acceleration.

FIGURE 5-23

2. Apply $\Sigma F_y = ma_y$ to the ball and solve for the tension $T$.

$$\Sigma F_y = ma_y \Rightarrow T\cos\theta - mg = 0$$

so

$$\boxed{T = \frac{mg}{\cos\theta}}$$

(c) 1. Apply $\Sigma F_x = ma_x$ to the ball.

$$\Sigma F_x = ma_x \Rightarrow T\sin\theta = m\frac{v^2}{r}$$

2. Substitute $mg/\cos\theta$ for $T$ and solve for $v$.

$$\frac{mg}{\cos\theta}\sin\theta = m\frac{v^2}{r} \Rightarrow \tan\theta = \frac{v^2}{rg}$$

so

$$v = \boxed{\sqrt{rg\tan\theta}}$$

**REMARKS** An object attached to a string and moving in a horizontal circle so that the string makes an angle $\theta$ with the vertical is called a *conical pendulum*.

When a car rounds a curve on a horizontal road, both the centripetal force and the forward forces are provided by the force of static friction exerted by the road on the tires of the car. If the car is traveling at constant speed, then the forward component of the frictional force is balanced by the rearward-directed forces of air drag and rolling friction. If air drag is negligible, then the forward component of the frictional force is zero.

---

*A ROAD TEST*    **EXAMPLE  5-10**    **Put It in Context**

**You have a summer job as part of an automobile tire design team. You are testing new prototype tires to see whether or not the tires perform as well as predicted. In a skid test, a new BMW 530i was able to travel at constant speed in a circle of radius 45.7 m in 15.2 s without skidding. (a) What was its speed $v$? (b) What is the acceleration? (c) Assuming air drag and rolling friction to be negligible, what is the minimum value for the coefficient of static friction between the tires and the road?**

**PICTURE THE PROBLEM** Figure 5-24 shows the forces acting on the car. The normal force $\vec{F}_n$ balances the downward force due to gravity $m\vec{g}$. The horizontal force is the force of static friction, which provides the centripetal acceleration. The faster the car travels, the greater the centripetal acceleration. The speed can be found from the circumference of the circle and the period $T$. This speed puts a lower limit on the maximum value of the coefficient of static friction.

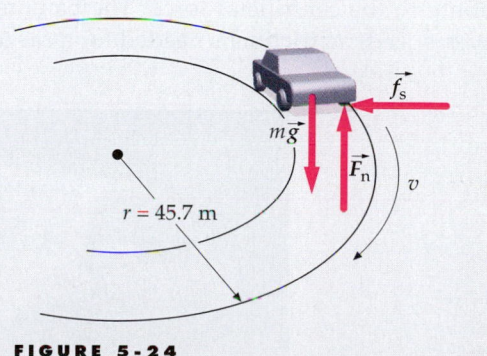

FIGURE 5-24

(a) 1. Draw a free-body diagram for the car (Figure 5-25). The positive $r$ direction is away from the center of curvature.

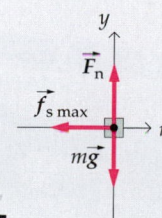

FIGURE 5-25

2. Use "speed equals distance divided by the time" to determine the speed $v$:

$$v = \frac{2\pi r}{T} = \frac{2\pi(45.7 \text{ m})}{15.2 \text{ s}} = \boxed{18.9 \text{ m/s}}$$

(b) Use $v$ to calculate the acceleration:

$$a_c = \frac{v^2}{r} = \frac{(18.9 \text{ m/s})^2}{(45.7 \text{ m})} = \boxed{7.81 \text{ m/s}^2}$$

$$a_t = \frac{dv}{dt} = \boxed{0}$$

The acceleration is 7.81 m/s² in the centripetal direction.

(c) 1. Apply $\Sigma F_y = ma_y$ to the car. Solve for the normal and maximum frictional force:

$$\Sigma F_y = ma_y \Rightarrow F_n - mg = 0$$

so

$$F_n = mg \quad \text{and} \quad f_{s,max} = \mu_s mg$$

2. Apply $\Sigma F_r = ma_r$ to the car. Substituting from step (c), solve 1, for $\mu_s$:

$$\Sigma F_r = ma_r \Rightarrow -f_{s,max} = m\left(-\frac{v^2}{r}\right)$$

so

$$\mu_s mg = m\frac{v^2}{r} \quad \text{and} \quad \mu_s = \frac{v^2}{rg}$$

$$\mu_s = \frac{(18.9 \text{ m/s})^2}{(45.7 \text{ m})(9.81 \text{ m/s}^2)} = \boxed{0.796}$$

**REMARKS** A good practice when calculating values to three figures is to calculate all intermediate values to at least four figures. For example, if you use the values shown in Part (b) you obtain $a_c = 7.816$ m/s². This should not be rounded to 7.82 m/s² because in step 2 of Part (a), substituting the exact values gives, to four figures, $v = 18.89$ m/s. Calculating $a_c$ using $v = 18.89$ m/s (rather than 18.9 m/s) gives $a_c = 7.808$ m/s², which rounds up to $a_c = 7.81$ m/s². Storing intermediate values in the memory of your calculator and recalling them for later calculations facilitates this process.

**❶PLAUSIBILITY CHECK** If $\mu_s$ were equal to 1, the inward force would be equal to $mg$ and the centripetal acceleration would be $g$. Here $\mu_s$ is about 0.8 and the centripetal acceleration is about 0.8 $g$.

### *Banked Curves

If a curved road is not horizontal but banked, the normal force of the road will have a component directed inward toward the center of the circle that will contribute to the centripetal force. The banking angle can be chosen so that, for a given speed, no friction is needed for a car to complete the curve.

As a car rounds the curve, the tire is distorted by the frictional force exerted by the road.

Large centrifuge used for research at
Sandia National Laboratories.

FIGURE 5-26

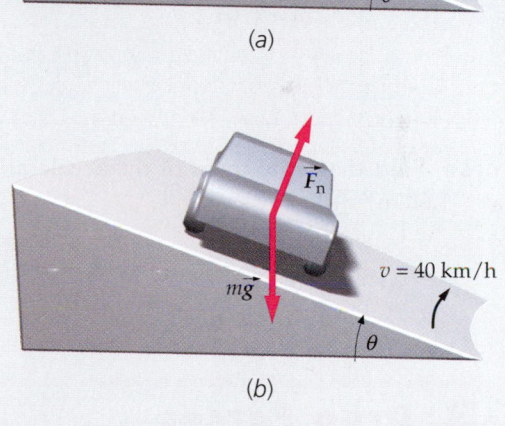

*ROUNDING A BANKED CURVE*    **EXAMPLE 5-11**

A curve of radius 30 m is banked at an angle $\theta$. Find $\theta$ for which a car can round the curve at 40 km/h even if the road is covered with ice so that friction is negligible.

**PICTURE THE PROBLEM** In this case only two forces act on the car: gravity and the normal force. Since the car is traveling in a circle at constant speed, the acceleration is in the centripetal direction. The vector sum of the two forces is in the direction of the acceleration.

**FIGURE 5-27**

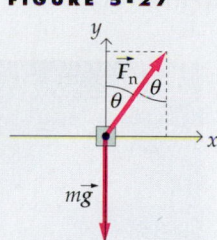

1. In Figure 5-26a the forces exerted by the road on the car are represented by $\vec{F}_{n1}$ and $\vec{F}_{n2}$. These forces are combined into $\vec{F}_n$ in Figure 5-26b. The angle between the normal force $\vec{F}_n$ and the vertical is $\theta$, the same as the banking angle. Draw a free-body diagram for the car.

2. Apply $\Sigma F_y = ma_y$ to the car:

$$\Sigma F_y = ma_y \Rightarrow F_n \cos\theta - mg = 0$$

so

$$F_n = \frac{mg}{\cos\theta}$$

3. Apply $\Sigma F_x = ma_x$ to the car. Substitute for $F_n$ using the step 2 result, then solve for $\theta$:

$$\Sigma F_x = ma_x \Rightarrow F_n \sin\theta = m\frac{v^2}{r}$$

and

$$\frac{mg}{\cos\theta}\sin\theta = m\frac{v^2}{r} \quad so \quad \theta = \tan^{-1}\frac{v^2}{rg}$$

$$\theta = \tan^{-1}\frac{[(40{,}000 \text{ m})/(3600 \text{ s})]^2}{(30 \text{ m})(9.81 \text{ m/s}^2)} = \boxed{22.8°}$$

**REMARKS** The banking angle $\theta$ depends on $v$ and $r$, but not the mass $m$; $\theta$ increases with increasing $v$, and decreases with increasing $r$. When the banking angle, speed, and radius satisfy $\tan\theta = v^2/rg$, the car rounds the curve smoothly, with no tendency to slide either inward or outward. If the car speed is greater than $\sqrt{rg\tan\theta}$, the road will exert a frictional force down the incline. This force has an inward horizontal component, which provides the additional centripetal force needed to keep the car from moving outward (sliding up the incline). If the car speed is less than this amount, the road must exert a frictional force up the incline.

**ALTERNATIVE SOLUTION** In the preceding solution we followed the guideline to choose one of the coordinate axis directions to be the direction of the acceleration vector, the centripetal direction. However, the solution is no more difficult if we choose one of the axis directions to be down the incline. This choice is taken in the following solution.

**FIGURE 5-28**

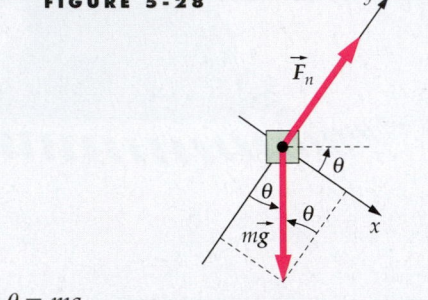

1. Draw a free-body diagram for the car (Figure 5-28). The $x$ axis direction is down the incline and the $y$ axis direction is the normal direction.

2. Apply $\Sigma F_x = ma_x$ to the car:

$$\Sigma F_x = ma_x \Rightarrow mg\sin\theta = ma_x$$

3. Draw a sketch and use trigonometry to obtain an expression for $a_x$ in terms of $a$ and $\theta$ (Figure 5-29):

$$a_x = a\cos\theta$$

**FIGURE 5-29**

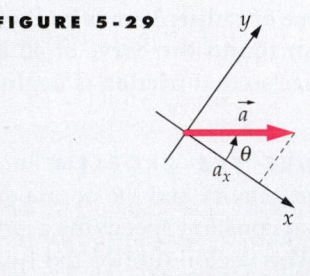

4. Substitute the step 3 result into the step 2 result. Then substitute $v^2/r$ for $a$ and solve for $\theta$:

$$mg\sin\theta = ma\cos\theta$$

$$g\sin\theta = \frac{v^2}{r}\cos\theta$$

$$\tan\theta = \frac{v^2}{rg} \Rightarrow \theta = \tan^{-1}\frac{v^2}{rg}$$

**EXERCISE** Find the component of the acceleration normal to the road surface. (*Answer* 1.60 m/s²)

# *5-3 Drag Forces

When an object moves through a fluid such as air or water, the fluid exerts a **drag force** or retarding force that opposes the motion of the object. The drag force depends on the shape of the object, the properties of the fluid, and the speed of the object relative to the fluid. Unlike ordinary friction, the drag force increases as the speed of the object increases. At low speeds, the drag force is approximately proportional to the speed of the object; at higher speeds, it is more nearly proportional to the square of the speed.

Consider an object dropped from rest and falling under the influence of the force of gravity, which we assume to be constant. Now add a drag force of magnitude $bv^n$, where $b$ and $n$ are constants. We then have

a constant downward force $mg$ and an upward force $bv^n$ (Figure 5-30).

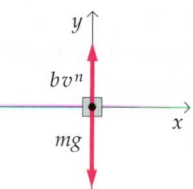

If we take the downward direction to be positive, we obtain from Newton's second law

$$\Sigma F_y = mg - bv^n = ma_y \qquad 5\text{-}6$$

At $t = 0$, the instant when the object is dropped, the speed is zero, so the drag force is zero and the acceleration is $g$ downward. As the speed of the object increases, the drag force increases and the acceleration becomes less than $g$. Eventually, the speed is great enough for the magnitude of the drag force $bv^n$ to approach the force of gravity $mg$. As this happens, the acceleration approaches zero and the speed approaches the **terminal speed** $v_t$. At terminal speed the drag force balances the weight force and the acceleration is zero. Setting the acceleration $a_y$ in Equation 5-6 equal to zero, we obtain

$$bv_t^n = mg$$

Solving for the terminal speed, we get

$$v_t = \left(\frac{mg}{b}\right)^{1/n} \qquad 5\text{-}7$$

The larger the constant $b$, the smaller the terminal speed. A parachute is designed to maximize $b$ so that the terminal speed will be small. On the other hand, cars are designed to minimize $b$ to reduce the effect of wind resistance.

The terminal speed of a sky diver before release of the parachute is about 200 km/h, which is about 60 m/s. When the parachute is opened, the drag force rapidly increases, becoming greater than the force of gravity, and the sky diver experiences an upward acceleration while falling; that is, the speed of the descending sky diver decreases. As the speed of the sky diver decreases, the drag force decreases and the speed approaches a new terminal speed of about 20 km/h.

---

*TERMINAL SPEED*                                    **EXAMPLE 5-12**

**A sky diver of mass 64 kg falls with a terminal speed of 180 km/h with her arms and legs outspread. (a) What is the magnitude of the upward drag force $F_d$ on the sky diver? (b) If the drag force is equal to $bv^2$, what is the value of $b$?**

(a) 1. Draw a free-body diagram.

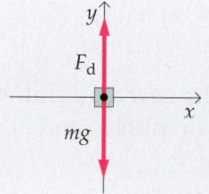

**FIGURE 5-31**

2. Apply $\Sigma F_y = ma_y$. Since the sky diver is moving with constant velocity, the acceleration is zero:

$$\Sigma F_y = ma_y \Rightarrow F_d - mg = 0$$

so

$$F_d = mg = (64 \text{ kg})(9.81 \text{ N/kg}) = \boxed{628 \text{ N}}$$

(b) 1. To find $b$ we set $F_d = bv^2$:

$$F_d = mg = bv^2$$

so

$$b = \frac{mg}{v^2}$$

2. Find the speed in m/s, then calculate $b$:

$$180 \text{ km/h} = \frac{180 \text{ km}}{1 \text{ h}}\left(\frac{1000 \text{ m}}{1 \text{ km}}\right)\left(\frac{1 \text{ h}}{3600 \text{ s}}\right)$$

$$= \frac{180 \text{ km}}{1 \text{ h}}\left(\frac{1 \text{ m/s}}{3.6 \text{ km/h}}\right) = 50 \text{ m/s}$$

$$b = \frac{(64 \text{ kg})(9.81 \text{ N/kg})}{(50 \text{ m/s})^2} = \boxed{0.251 \text{ kg/m}}$$

# *5-4 Numerical Integration: Euler's Method

If a particle moves under the influence of a *constant* force, its acceleration is constant and we can find its velocity and position from the constant-acceleration kinematic formulas in Chapter 2. But consider a particle moving through space where the force on it, and therefore its acceleration, depends on its position and its velocity. The position, velocity, and acceleration of the particle at one instant determine the position and velocity the next instant, which then determine its acceleration at that instant. The actual position, velocity, and acceleration of an object all change continuously with time. We can approximate this by replacing the continuous time variations with small time steps of duration $\Delta t$. The simplest approximation is to assume constant acceleration during each step. This approximation is called **Euler's method.** If the time interval is sufficiently short, the change in acceleration during the interval will be small and can be neglected.

Let $x_0$, $v_0$, and $a_0$ be the known position, velocity, and acceleration of a particle at some initial time $t_0$. If we assume constant acceleration during $\Delta t$, the velocity at time $t_1 = t_0 + \Delta t$ is given by

$$v_1 = v_0 + a_0 \Delta t$$

Similarly, if we neglect any change in velocity during the time interval, the new position is given by

$$x_1 = x_0 + v_0 \Delta t$$

We can use the values $v_1$ and $x_1$ to compute the new acceleration $a_1$ using Newton's second law, and then use $x_1$, $v_1$, and $a_1$ to compute $x_2$ and $v_2$.

$$x_2 = x_1 + v_1 \Delta t \qquad\qquad\qquad 5\text{-}8$$

$$v_2 = v_1 + a_1 \Delta t \qquad\qquad\qquad 5\text{-}9$$

The connection between the position and velocity at time $t_n$ and time $t_{n+1} = t_n + \Delta t$ is given by

$$x_{n+1} = x_n + v_n \Delta t \qquad \text{5-10}$$

and

$$v_{n+1} = v_n + a_n \Delta t \qquad \text{5-11}$$

To find the velocity and position at some time $t$, we therefore divide the time interval $t - t_0$ into a large number of smaller intervals $\Delta t$ and apply Equations 5-10 and 5-11, beginning at the initial time $t_0$. This involves a large number of simple, repetitive calculations that are most easily done on a computer. The technique of breaking the time interval into small steps and computing the acceleration, velocity, and position at each step using the values from the previous step is called numerical integration.

**Drag Forces** To illustrate the use of numerical integration, let us consider a problem in which a sky diver is dropped from rest at some height under the influences of both gravity and a drag force that is proportional to the square of the speed. We will find the velocity $v$ and the distance traveled $x$ as functions of time.

The equation describing the motion of an object of mass $m$ dropped from rest is Equation 5-6 with $n = 2$:

$$mg - bv^2 = ma$$

where down is the positive direction. The acceleration is thus

$$a = g - \frac{b}{m}v^2 \qquad \text{5-12}$$

It is convenient to write the constant $b/m$ in terms of the terminal speed $v_t$. Setting $a = 0$ in Equation 5-12 we obtain

$$0 = g - \frac{b}{m}v_t^2$$

$$\frac{b}{m} = \frac{g}{v_t^2}$$

Substituting $g/v_t^2$ for $b/m$ in Equation 5-12 gives

$$a = g\left(1 - \frac{v^2}{v_t^2}\right) \qquad \text{5-13}$$

The acceleration at time $t_n$ is calculated using the values $x_n$ and $v_n$.

To solve Equation 5-13 numerically, we need to use numerical values for $g$ and $v_t$. A reasonable terminal speed for a sky diver is 60 m/s. If we choose $x_0 = 0$ for the initial position, the initial values are $x_0 = 0$, $v_0 = 0$, and $a_0 = g = 9.81$ m/s$^2$. To find the velocity $v$ and position $x$ after some time, say $t = 20$ s, we divide the time interval $0 < t < 20$ s into many small intervals $\Delta t$ and apply Equations 5-10, 5-11, and 5-13. We do this by writing a computer program or by using a computer spreadsheet, shown in Figure 5-32. This spreadsheet has $\Delta t = 0.5$ s and at $t = 20$ s, the computed values are $x = 59.89$ m and $v = 939.9$ m/s.

**FIGURE 5-32** (a) Spreadsheet to compute the position and speed of a sky diver with air drag proportional to $v^2$. (b) The same Excel spreadsheet displaying the formulas rather than the values.

(a)

| | A | B | C | D |
|---|---|---|---|---|
| 1 | Δt = | 0.5 | s | |
| 2 | x0 = | 0 | m | |
| 3 | v0 = | 0 | m/s | |
| 4 | a0 = | 9.81 | m/s^2 | |
| 5 | vt = | 60 | m/s | |
| 6 | | | | |
| 7 | t | x | v | a |
| 8 | (s) | (m) | (m/s) | (m/s^2) |
| 9 | 0.00 | 0.0 | 0.00 | 9.81 |
| 10 | 0.50 | 0.0 | 4.91 | 9.74 |
| 11 | 1.00 | 2.5 | 9.78 | 9.55 |
| 12 | 1.50 | 7.3 | 14.55 | 9.23 |
| 13 | 2.00 | 14.6 | 19.17 | 8.81 |
| 14 | 2.50 | 24.2 | 23.57 | 8.30 |
| 15 | 3.00 | 36.0 | 27.72 | 7.72 |
| 41 | 16.00 | 701.0 | 59.55 | 0.15 |
| 42 | 16.50 | 730.7 | 59.62 | 0.16 |
| 43 | 17.00 | 760.6 | 59.68 | 0.10 |
| 44 | 17.50 | 790.4 | 59.74 | 0.09 |
| 45 | 18.00 | 820.3 | 59.78 | 0.07 |
| 46 | 18.50 | 850.2 | 59.82 | 0.06 |
| 47 | 19.00 | 880.1 | 59.85 | 0.05 |
| 48 | 19.50 | 910.0 | 59.87 | 0.04 |
| 49 | 20.00 | 939.9 | 59.89 | 0.04 |
| 50 | | | | |

(b)

| | A | B | C | D |
|---|---|---|---|---|
| 1 | Δt = | 0.5 | s | |
| 2 | x0 = | 0 | m | |
| 3 | v0 = | 0 | m/s | |
| 4 | a0 = | 9.81 | m/s^2 | |
| 5 | vt = | 60 | m/s | |
| 6 | | | | |
| 7 | t | x | v | a |
| 8 | (s) | (m) | (m/s) | (m/s^2) |
| 9 | 0 | =B2 | =B3 | =$B$4*(1−C9^2/$B$5^2) |
| 10 | =A9+$B$1 | =B9+C9*$B$1 | =C9+D9*$B$1 | =$B$4*(1−C10^2/$B$5^2) |
| 11 | =A10+$B$1 | =B10+C10*$B$1 | =C10+D10*$B$1 | =$B$4*(1−C11^2/$B$5^2) |
| 12 | =A11+$B$1 | =B11+C11*$B$1 | =C11+D11*$B$1 | =$B$4*(1−C12^2/$B$5^2) |
| 13 | =A12+$B$1 | =B12+C12*$B$1 | =C12+D12*$B$1 | =$B$4*(1−C13^2/$B$5^2) |
| 14 | =A13+$B$1 | =B13+C13*$B$1 | =C13+D13*$B$1 | =$B$4*(1−C14^2/$B$5^2) |
| 15 | =A14+$B$1 | =B14+C14*$B$1 | =C14+D14*$B$1 | =$B$4*(1−C15^2/$B$5^2) |
| 41 | =A40+$B$1 | =B40+C40*$B$1 | =C40+D40*$B$1 | =$B$4*(1−C41^2/$B$5^2) |
| 42 | =A41+$B$1 | =B41+C41*$B$1 | =C41+D41*$B$1 | =$B$4*(1−C42^2/$B$5^2) |
| 43 | =A42+$B$1 | =B42+C42*$B$1 | =C42+D42*$B$1 | =$B$4*(1−C43^2/$B$5^2) |
| 44 | =A43+$B$1 | =B43+C43*$B$1 | =C43+D43*$B$1 | =$B$4*(1−C44^2/$B$5^2) |
| 45 | =A44+$B$1 | =B44+C44*$B$1 | =C44+D44*$B$1 | =$B$4*(1−C45^2/$B$5^2) |
| 46 | =A45+$B$1 | =B45+C45*$B$1 | =C45+D45*$B$1 | =$B$4*(1−C46^2/$B$5^2) |
| 47 | =A46+$B$1 | =B46+C46*$B$1 | =C46+D46*$B$1 | =$B$4*(1−C47^2/$B$5^2) |
| 48 | =A47+$B$1 | =B47+C47*$B$1 | =C47+D47*$B$1 | =$B$4*(1−C48^2/$B$5^2) |
| 49 | =A48+$B$1 | =B48+C48*$B$1 | =C48+D48*$B$1 | =$B$4*(1+C49^2/$B$5^2) |
| 50 | | | | |

Figure 5-33 shows graphs of $v$ versus $t$ and $x$ versus $t$ plotted from this data.

But how accurate are our computations? We can estimate the accuracy by running the program again using a smaller time interval. If we use $\Delta t = 0.25$ s, one-half of the value we originally used, we obtain $v = 59.86$ m/s and $x = 943.1$ m at $t = 20$ s. The difference in $v$ is about 0.05 percent and that in $x$ is about 0.4 percent. These are our estimates of the accuracy of the original computations.

Since the difference between the value of $a_{av}$ for some time interval $\Delta t$ and the value of $a$ at the beginning of the interval becomes smaller as the time interval becomes smaller, we might expect that it would be better to use very small time intervals, say $\Delta t = 0.000000001$ s. But there are two reasons for not using very small time intervals. First, the smaller the time interval, the larger the number of calculations that are required and the longer the program takes to run. Second, the computer keeps only a fixed number of digits at each step of the calculation, so that at each step there is a round-off error. These round-off errors add up. The larger the number of calculations, the more significant the total round-off errors become. When we first decrease the time interval, the accuracy improves because $a_i$ more nearly approximates $a_{av}$ for the interval. However, as the time interval is decreased further, the round-off errors build up and the accuracy of the computation decreases. A good rule of thumb to follow is to use no more than about $10^4$ or $10^5$ time intervals for the typical numerical integration.

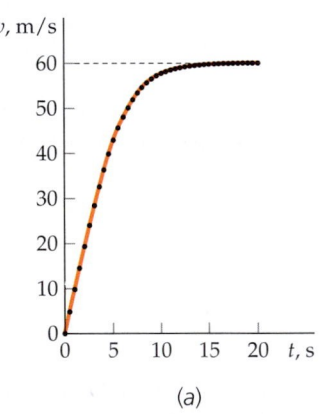

(a)

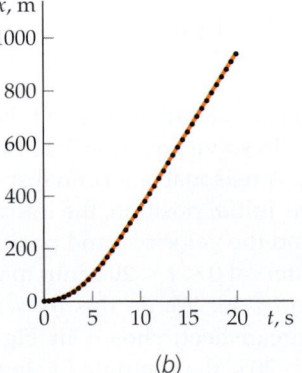

(b)

**FIGURE 5-33** (a) Graph of $v$ versus $t$ for a sky diver, found by numerical integration using $\Delta t = 0.5$ s. The horizontal dashed line is the terminal speed $v_t = 60$ m/s. (b) Graph of $x$ versus $t$ using $\Delta t = 0.5$ s.

## SUMMARY

Friction and drag forces are complex phenomena empirically approximated by simple equations.

| Topic | Relevant Equations and Remarks |
|---|---|
| **1. Friction** | Two objects in contact exert frictional forces on each other. These forces are parallel to the surfaces of the objects at the points of contact and directed opposite to the direction of sliding or tendency to slide. |
| Static friction | $f_s \leq \mu_s F_n$      5-2 <br><br> where $F_n$ is the normal force of contact and $\mu_s$ is the coefficient of static friction. |
| Kinetic friction | $f_k = \mu_k F_n$      5-3 <br><br> where $\mu_k$ is the coefficient of kinetic friction. The coefficient of kinetic friction is slightly less than the coefficient of static friction. |
| **2. Motion Along a Curved Path** | A particle moving along an arbitrary curve can be considered to be moving along a circular arc during a small time interval. Its instantaneous acceleration vector has a component $a_c = v^2/r$ toward the center of curvature of the arc and a component $a_t = dv/dt$ that is tangential to the curve. If the particle is moving along a circular path of radius $r$ at constant speed $v$, $a_t = 0$ and the speed, radius, and period $T$ are related by $2\pi r = vT$. |
| **3. *Drag Forces** | When an object moves through a fluid it experiences a drag force that opposes its motion. The drag force increases with increasing speed. If the body is dropped from rest, its speed increases. As it does, the magnitude of the drag force comes closer and closer to the magnitude of the force of gravity, so the net force, and thus the acceleration, approaches zero. As the acceleration approaches zero, the speed approaches a constant value called its terminal speed. The terminal speed depends on the shape of the body and on the medium through which it falls. |
| **4. *Numerical Integration: Euler's Method** | To estimate the position $x$ and velocity $v$ at some time $t$, we first divide $t$ into a large number of small intervals, each of length $\Delta t$. The initial acceleration $a_0$ is then calculated from the initial position $x_0$ and velocity $v_0$. The position $x_1$ and velocity $v_1$ a time $\Delta t$ later are estimated using the relations <br><br> $x_{n+1} = x_n + v_n \Delta t$      5-10 <br><br> and <br><br> $v_{n+1} = v_n + a_n \Delta t$      5-11 <br><br> with $n = 0$. The acceleration $a_{n+1}$ is calculated using the values for $x_{n+1}$ and $v_{n+1}$ and the process is repeated. This continues until estimations for the position and velocity at time $t$ are calculated. |

- Single-concept, single-step, relatively easy
- • Intermediate-level, may require synthesis of concepts
- • • Challenging
- **SSM** Solution is in the *Student Solutions Manual*
- **iSOLVE** Problems available on iSOLVE online homework service
- **iSOLVE✓** These "Checkpoint" online homework service problems ask students additional questions about their confidence level, and how they arrived at their answer

In a few problems, you are given more data than you actually need; in a few other problems, you are required to supply data from your general knowledge, outside sources, or informed estimates.

## Conceptual Problems

**1** • Various objects lie on the floor of a truck moving along a straight horizontal road. If the truck gradually speeds up, what force acts on the objects to cause them to speed up too?

**2** • **SSM** Any object resting on the floor of a truck will slide if the truck's acceleration is too great. How does the critical acceleration at which a light object slips compare with that at which a much heavier object slips?

**3** • A block of mass $m$ rests on a plane inclined at an angle $\theta$ with the horizontal. It follows that the coefficient of static friction between the block and plane is (a) $\mu_s \geq g$, (b) $\mu_s = \tan \theta$, (c) $\mu_s \leq \tan \theta$, (d) $\mu_s \geq \tan \theta$.

**4** • **SSM** A block of mass $m$ is at rest on a plane inclined at an angle of 30° with the horizontal, as shown in Figure 5-34. Which of the following statements about the force of static friction is necessarily true? (a) $f_s > mg$. (b) $f_s > mg \cos 30°$. (c) $f_s = mg \cos 30°$. (d) $f_s = mg \sin 30°$. (e) None of these statements are true.

**FIGURE 5-34** Problem 4

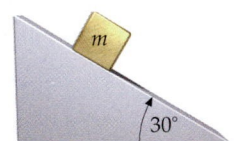

**5** • • On an icy winter day, the coefficient of friction between the tires of a car and a roadway is reduced to one-half of its value on a dry day. As a result, the maximum speed at which a curve of radius $R$ can be safely negotiated is (a) the same as on a dry day, (b) reduced to 71% of its value on a dry day, (c) reduced to 50% of its value on a dry day, (d) reduced to 25% of its value on a dry day, (e) reduced by an unknown amount depending on the car's mass.

**6** • • **SSM** Show with a force diagram how a motorcycle can travel in a circle on the inside vertical wall of a hollow cylinder. Assume reasonable parameters (coefficient of friction, radius of the circle, mass of the motorcycle, or whatever is required), and calculate the minimum speed needed.

**7** • • This is an interesting experiment that you can perform at home: take a wooden block and rest it on the floor or some other flat surface. Attach a spring (a slinky will work well) to the block and pull gently and steadily on the spring in the horizontal direction. At some point, the block will start moving, but it will not move smoothly. Instead, it will start moving, stop again, start moving again, stop again, etc. Explain why the motion of the block (called "stick–slip" motion) behaves this way.

**8** • True or false: Viewed from an inertial reference frame, an object cannot move in a circle unless there is a net force acting on it.

**9** • • A particle is traveling in a vertical circle at constant speed. One can conclude that the _____ is constant.
(a) velocity, (b) acceleration, (c) net force, (d) apparent weight, (e) none of the properties listed.

**10** • **SSM** You place a lightweight piece of iron on a table and hold a small kitchen magnet above the iron at a distance of 1 cm. You find that the magnet cannot lift the iron, even though there is obviously a force between the iron and the magnet. Next, you again hold the piece of iron and the magnet 1 cm apart with the magnet above the iron, but this time you drop them from arm's length. As they fall, the magnet and the piece of iron are pulled together before hitting the floor. (a) Draw free-body diagrams illustrating all of the forces on the magnet and the iron for each demonstration. (b) Explain why the magnet and iron are pulled together when they are dropped, even though the magnet cannot pull up the piece of iron when it is sitting on the table.

**11** • • • **SSM** The following question is an excellent "brain-twister" invented by Boris Korsunsky[†]: Two identical blocks are attached by a massless string running over a pulley as shown in Figure 5-35. The rope initially runs over the pulley at its (the rope's) midpoint, and the surface that block 1 rests on is frictionless. Blocks 1 and 2 are initially at rest when block 2 is released with the string taut and horizontal. Will block 1 hit the pulley before or after block 2 hits the wall? (Assume that the initial distance from block 1 to the pulley is the same as the initial distance from block 2 to the wall.) There is a very simple solution.

---

[†] Boris Korsunsky, "Braintwisters for Physics Students," *The Physics Teacher*, **33**, 550 (1995).

**FIGURE 5-35**
**Problem 11**

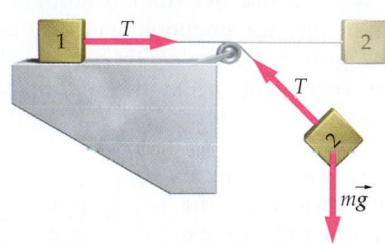

**12** • True or false: The terminal speed of a falling object depends on its shape.

**13** • SSM As a sky diver falls through the air, her terminal speed (a) depends on her mass, (b) depends on her orientation as she falls, (c) depends on the density of the air, (d) depends on all of the above.

**14** •• You are sitting in the passenger seat in a car driving around a circular, horizontal, flat racetrack at a high speed. As you sit there, you "feel" a "force" pushing you toward the outside of the track. What is the true direction of the force acting on you, and where does it come from? (Assume that you don't slide across the seat.)

**15** • SSM The mass of the moon is about 1% of that of the earth. The centripetal force that keeps the moon in its orbit around the earth (a) is much smaller than the gravitational force exerted on the moon by the earth, (b) depends on the phase of the moon, (c) is much greater than the gravitational force exerted on the moon by the earth, (d) is the gravitational force exerted on the moon by the earth, (e) cannot be answered; we haven't studied Newton's law of gravity yet.

**16** • A block is sliding on a frictionless surface along a loop-the-loop as in Figure 5-36. The block is moving fast enough that it never loses contact with the track. Match the points along the track to the appropriate free-body diagrams in Figure 5-37.

**Problem 16   FIGURE 5-36**

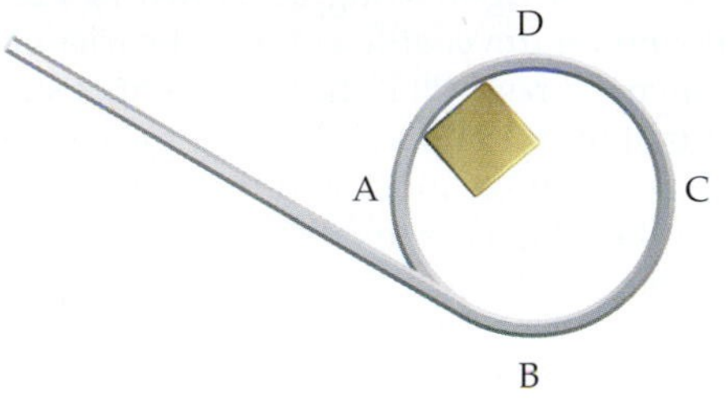

**FIGURE 5-37**

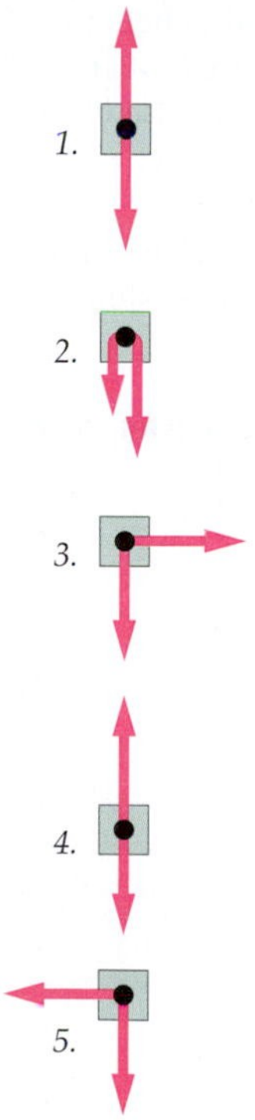

1.

2.

3.

4.

5.

**17** •• If you hold a rock and a feather at the same height above the ground and drop them, the rock will hit the ground first. From this you might conclude that the average aerodynamic drag force on the feather as it falls is larger than that on the rock, but in fact the opposite is true. Explain in detail why the rock hits the ground first. Assume that the drag force is given by the Newtonian formula $F_D = (1/2) CA\rho v^2$ (see Problem 18(c) for an explanation of this formula).

## Estimation and Approximation

**18** • SSM To determine the aerodynamic drag on a car, the "coast-down" method is often used. The car is driven on a long, flat road at some convenient speed (60 mph is typical), shifted into neutral, and allowed to coast to a stop. The time that it takes for the speed to drop by successive 5-mph intervals is measured and used to compute the net force slowing the car down. (a) It was found that a Toyota Tercel with a mass of 1020 kg coasted down from 60 mph to 55 mph in 3.92 s. What is the average force slowing the car down? (b) If the coefficient of rolling friction for the car is 0.02, what is the force of rolling friction that is acting to slow the car down? If we assume that the only two forces acting on the car are rolling friction and aerodynamic drag, what is the average drag force acting on the car? (c) The drag force will have the form $\frac{1}{2}C\rho Av^2$, where $A$ is the cross-sectional area of the car facing the wind, $v$ is the car's speed, $\rho$ is the density of air, and $C$ is a dimensionless constant of order 1. If the cross-sectional area of the car is 1.91 m², determine $C$ from the data given. (The density of air is 1.21 kg/m³; use 55 mph for the speed of the car in this computation.)

**19** • Using dimensional analysis, determine the units and dimensions of the constant $b$ in the retarding force $bv^n$ if (a) $n = 1$ and (b) $n = 2$. (c) Newton showed that the air resistance of a falling object with a circular cross section should be approximately $\frac{1}{2}\rho\pi r^2 v^2$, where $\rho = 1.2$ kg/m³, the density of air. Show that this is consistent with your dimensional analysis for part (b). (d) Find the terminal speed for a 56-kg sky diver; approximate his or her cross-sectional area as a disk of radius 0.30 m. The density of air near the surface of the earth is 1.20 kg/m³. (e) The density of the atmosphere decreases with height above the surface of the earth; at a height of 8 km, the density is only 0.514 kg/m³. What is the terminal velocity at this height?

**20** •• While hailstones the size of golf balls are a bit unusual, hailstones are generally substantially larger than raindrops. Estimate the terminal velocity of a raindrop and a golf-ball sized hailstone.

## Friction

**21** • SSM A block of mass $m$ slides at constant speed down a plane inclined at an angle of $\theta$ with the horizontal. It follows that (a) $\mu_k = mg \sin \theta$, (b) $\mu_k = \tan \theta$, (c) $\mu_k = 1 - \cos \theta$, (d) $\mu_k = \cos \theta - \sin \theta$.

**22** • A block of wood is pulled by a horizontal string across a horizontal surface at a constant velocity with a force of 20 N. The coefficient of kinetic friction between the surfaces is 0.3. The force of friction is (*a*) impossible to determine without knowing the mass of the block, (*b*) impossible to determine without knowing the speed of the block, (*c*) 0.3 N, (*d*) 6 N, (*e*) 20 N.

**23** • **SSM** **iSOLVE** A 20-N block rests on a horizontal surface. The coefficients of static and kinetic friction between the surface and the block are $\mu_s = 0.8$ and $\mu_k = 0.6$. A horizontal string is attached to the block and a constant tension $T$ is maintained in the string. What is the force of friction acting on the block if (*a*) $T = 15$ N or (*b*) $T = 20$ N.

**24** • A block of mass $m$ is pulled at a constant velocity across a horizontal surface by a string as shown in Figure 5-38. The magnitude of the frictional force is (*a*) $\mu_k mg$, (*b*) $T \cos \theta$, (*c*) $\mu_k (T - mg)$, (*d*) $\mu_k T \sin \theta$, (*e*) $\mu_k (mg - T \sin \theta)$.

**FIGURE 5-38** Problem 24

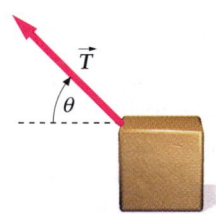

**25** • **iSOLVE** A tired worker pushes with a horizontal force of 500 N on a 100-kg crate initially resting on a thick pile carpet. The coefficients of static and kinetic friction are 0.6 and 0.4, respectively. Find the frictional force exerted by the carpet on the crate.

**26** • **iSOLVE** A box weighing 600 N is pushed along a horizontal floor at constant velocity with a force of 250 N parallel to the floor. What is the coefficient of kinetic friction between the box and the floor?

**27** • **iSOLVE** The coefficient of static friction between the tires of a car and a horizontal road is $\mu_s = 0.6$. If the net force on the car is the force of static friction exerted by the road, (*a*) what is the magnitude of the maximum acceleration of the car when it is braked? (*b*) What is the shortest distance in which the car can stop if it is initially traveling at 30 m/s?

**28** • **SSM** The force that accelerates a car along a flat road is the frictional force exerted by the road on the car's tires. (*a*) Explain why the acceleration can be greater when the wheels do not slip. (*b*) If a car is to accelerate from 0 to 90 km/h in 12 s, what is the minimum coefficient of friction needed between the road and tires? Assume that half the weight of the car is supported by the drive wheels.

**29** • A 5-kg block is held at rest against a vertical wall by a horizontal force of 100 N. (*a*) What is the frictional force exerted by the wall on the block? (*b*) What is the minimum horizontal force needed to prevent the block from falling if the coefficient of friction between the wall and the block is $\mu_s = 0.40$?

**30** • A tired and overloaded student is attempting to hold a large physics textbook wedged under his arm as shown in Figure 5-39. The textbook has a mass of 10.2 kg, while the coefficient of static friction of the textbook against the student's underarm is 0.32 and the coefficient of static friction of the book against the student's shirt is 0.16. (*a*) What is the minimum horizontal force that the student must apply to the textbook to prevent it from falling? (*b*) If the student can only exert a force of 195 N, what is the acceleration of the textbook as it slides from under his arm? The coefficient of kinetic friction of arm against textbook is 0.20, while that of shirt against textbook is 0.09.

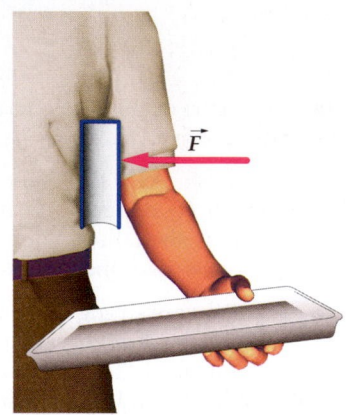

**FIGURE 5-39** Problem 30

**31** • On a snowy day with the temperature near the freezing point, the coefficient of static friction between a car's tires and an icy road is 0.08. What is the maximum incline that a vehicle with four-wheel drive can climb at constant speed?

**32** • **SSM** A 50-kg box that is resting on a level floor must be moved. The coefficient of static friction between the box and the floor is 0.6. One way to move the box is to push down on the box at an angle $\theta$ with the horizontal. Another method is to pull up on the box at an angle $\theta$ with the horizontal. (*a*) Explain why one method is better than the other. (*b*) Calculate the force necessary to move the box by each method if $\theta = 30°$ and compare the answer with the results when $\theta = 0°$.

**33** • **iSOLVE** A 3-kg block rests on a horizontal table, attached to a 2-kg block by a light string as shown in Figure 5-40. (*a*) What is the minimum coefficient of static friction such that the objects remain at rest? (*b*) If the coefficient of static friction is less than that found in part (*a*), and if the coefficient of kinetic friction between the block and the table is 0.3, find the time it takes for the 2-kg mass to fall 2 m to the floor if the system starts from rest.

**FIGURE 5-40** Problem 33

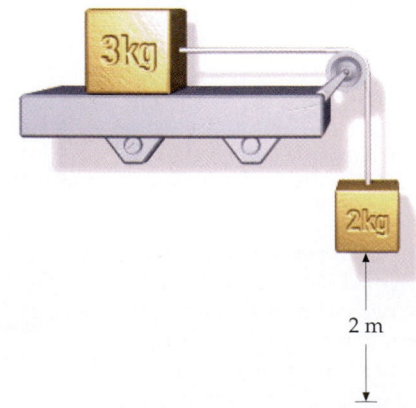

**34** •• A block on a horizontal plane is given an initial speed $v$. Traveling in a straight line, it comes to rest after sliding a distance $d$. Show that the coefficient of kinetic friction is $\mu_k = v^2/2gd$.

**35** •• $\boxed{\text{SSM}}$ A block of mass $m_1 = 250$ g is at rest on a plane that makes an angle of $\theta = 30°$ above the horizontal. The coefficient of kinetic friction between the block and the plane is $\mu_k = 0.100$. The block is attached to a second block of mass $m_2 = 200$ g that hangs freely by a string that passes over a frictionless, massless pulley (Figure 5-41). When the second block has fallen 30.0 cm, its speed is (a) 83 cm/s, (b) 48 cm/s, (c) 160 cm/s, (d) 59 cm/s, (e) 72 cm/s.

**FIGURE 5-41** Problems 35–37

**36** •• Returning to Figure 5-41, this time $m_1 = 4$ kg. The coefficient of static friction between the block and the incline is 0.4. (a) Find the range of possible values for $m_2$ for which the system will be in static equilibrium. (b) What is the frictional force on the 4-kg block if $m_1 = 1$ kg?

**37** •• Returning once again to Figure 5-41, this time $m_1 = 4$ kg, $m_2 = 5$ kg, and the coefficient of kinetic friction between the inclined plane and the 4-kg block is $\mu_k = 0.24$. Find the acceleration of the masses and the tension in the cord.

**38** •• $\boxed{\text{SSM}}$ $\boxed{\text{iSOLVE}}$✓ The coefficient of static friction between the bed of a truck and a box resting on it is 0.30. The truck is traveling at 80 km/h along a horizontal road. What is the shortest distance in which the truck can stop if the box is not to slide?

**39** •• A 4.5-kg block is given an initial velocity of 14 m/s up an incline that makes an angle of 37° with the horizontal. When its displacement is 8.0 m, its velocity up the incline has diminished to 5.2 m/s. Find (a) the coefficient of kinetic friction between the block and plane, (b) the displacement of the block from its starting point at the time when it momentarily comes to rest, and (c) the speed of the block when it again reaches its starting point.

**40** •• An automobile is going up a grade of 15° at a speed of 30 m/s. The coefficient of static friction between the tires and the road is 0.7. (a) What minimum distance does it take to stop the car? (b) What minimum distance would it take if the car were going down the grade?

**41** •• A rear-wheel-drive car supports 40 percent of its weight on its two drive wheels and has a coefficient of static friction of 0.7 with a horizontal straight road. (a) Find the vehicle's maximum acceleration. (b) What is the shortest possible time in which this car can achieve a speed of 100 km/h? (Assume the engine has unlimited power.)

**42** •• $\boxed{\text{SSM}}$ Lou bets an innocent stranger that he can place a 2-kg box against the side of a cart, as in Figure 5-42, and that the box will not fall to the ground, even though Lou will use no hooks, ropes, fasteners, magnets, glue, or adhesives of any kind. When the stranger accepts the bet, Lou begins to push the cart in the direction shown. The coefficient of static friction between the box and the cart is 0.6. (a) Find the minimum acceleration for which Lou will win the bet. (b) What is the magnitude of the frictional force in this case? (c) Find the force of friction on the box if the acceleration is twice the minimum needed for the box not to fall. (d) Show that, for a box of any mass, the box will not fall if the acceleration is $a \geq g/\mu_s$, where $\mu_s$ is the coefficient of static friction.

**FIGURE 5-42** Problem 42

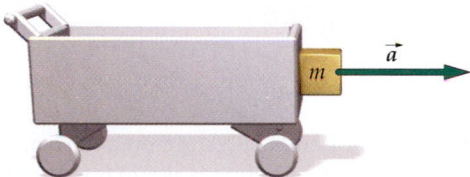

**43** •• $\boxed{\text{iSOLVE}}$ Two blocks attached by a string (Figure 5-43) slide down a 10° incline. The lower block has a mass of $m_2 = 0.25$ kg and a coefficient of kinetic friction $\mu_k = 0.2$. For the upper block, $m_1 = 0.8$ kg and $\mu_k = 0.3$. Find (a) the magnitude and direction of the acceleration of the blocks and (b) the tension in the string.

**44** •• $\boxed{\text{SSM}}$ As in Problem 43, two blocks of masses $m_1$ and $m_2$ are sliding down an incline as shown in Figure 5-43. They are connected by a massless *rod* this time; the rod behaves in exactly the same way as a string, except that the force can be compressive as well as tensile. The coefficient of kinetic friction for block 1 is $\mu_1$ and the coefficient of kinetic friction for block 2 is $\mu_2$. (a) Determine the acceleration of the two blocks. (b) Determine the force that the rod exerts on the two blocks. Show that the force is 0 when $\mu_1 = \mu_2$ and give a simple, nonmathematical argument why this is true.

**FIGURE 5-43** Problems 43 and 44

**45** •• Two blocks attached by a string are at rest on an inclined surface. The lower block has a mass of $m_1 = 0.2$ kg and a coefficient of static friction $\mu_s = 0.4$. The upper block has a mass $m_2 = 0.1$ kg and $\mu_s = 0.6$. The angle $\theta$ is slowly increased. (a) At what angle $\theta_c$ do the blocks begin to slide? (b) What is the tension in the string just before sliding begins?

**46** •• Two blocks connected by a massless, rigid rod slide on a surface inclined at an angle of 20°. The lower block has a mass $m_1 = 1.2$ kg, and the upper block's mass is $m_2 = 0.75$ kg. (a) If the coefficient of kinetic friction is $\mu_k = 0.3$ for the lower block and $\mu_k = 0.2$ for the upper block, what is the acceleration of the blocks? (b) Determine the force exerted by the rod on either block.

**47** •• **SSM** A block of mass $m$ rests on a horizontal table (Figure 5-44). The block is pulled by a massless rope with a force $\vec{F}$ at an angle $\theta$. The coefficient of static friction is 0.6. The minimum value of the force needed to move the block depends on the angle $\theta$. (*a*) Discuss qualitatively how you would expect this force to depend on $\theta$. (*b*) Compute the force for the angles $\theta = 0°$, $10°$, $20°$, $30°$, $40°$, $50°$, and $60°$, and make a plot of $F$ versus $\theta$ for $mg = 400$ N. From your plot, at what angle is it most efficient to apply the force to move the block?

**FIGURE 5-44** Problem 47

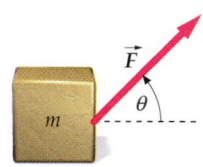

**48** •• For the block in Problem 47, show that, in the general case for a block of mass $m$ resting on a horizontal surface whose coefficient of static friction is $\mu_s$: (*a*) If we want to apply the minimum possible force to move the block, it should be applied with the force pulling upward at an angle $\theta_{min} = \tan^{-1} \mu_s$, and (*b*) the minimum force necessary to start the block moving is $F_{min} = \left(\mu_s / \sqrt{1 + \mu_s^2}\right) mg$. (*c*) Once you start the block moving, if you want to apply the least possible force to *keep* it moving, should you keep the angle at which you are pulling the same, increase it, or decrease it?

**49** •• Answer the questions in Problem 48, but for force $\vec{F}$ that pushes down on the block at an angle $\theta$ below the horizontal (Figure 5-45).

**FIGURE 5-45** Problem 49

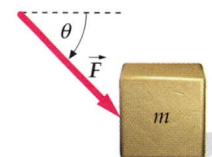

**50** •• A 100-kg mass is pulled along a frictionless surface by a horizontal force $\vec{F}$ such that its acceleration is $a_1 = 6$ m/s² (Figure 5-46). A 20-kg mass slides along the top of the 100-kg mass and has an acceleration of $a_2 = 4$ m/s². (It thus slides backward relative to the 100-kg mass.) (*a*) What is the frictional force exerted by the 100-kg mass on the 20-kg mass? (*b*) What is the net force acting on the 100-kg mass? What is the force $F$? (*c*) After the 20-kg mass falls off the 100-kg mass, what is the acceleration of the 100-kg mass? (Assume that the force $F$ does not change.)

**FIGURE 5-46** Problem 50

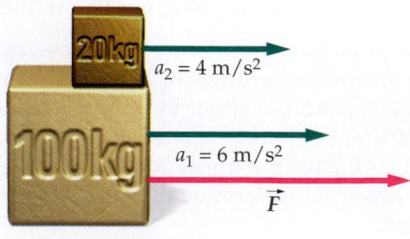

**51** •• **iSOLVE** A 60-kg block slides along the top of a 100-kg block. The 60-kg block has an acceleration of 3 m/s² when a horizontal force $\vec{F}$ of 320 N is applied, as in Figure 5-47. There is no friction between the 100-kg block and a horizontal frictionless surface, but there is friction between the two blocks. (*a*) Find the coefficient of kinetic friction between the blocks. (*b*) Find the acceleration of the 100-kg block during the time that the 60-kg block remains in contact.

**FIGURE 5-47** Problem 51

**52** •• **SSM** **iSOLVE✓** The coefficient of static friction between a rubber tire and the road surface is 0.85. What is the maximum acceleration of a 1000-kg four-wheel-drive truck if the road makes an angle of 12° with the horizontal and the truck is (*a*) climbing and (*b*) descending?

**53** •• A 2-kg block sits on a 4-kg block that is on a frictionless table (Figure 5-48). The coefficients of friction between the blocks are $\mu_s = 0.3$ and $\mu_k = 0.2$. (*a*) What is the maximum horizontal force $F$ that can be applied to the 4-kg block if the 2-kg block is not to slip? (*b*) If $F$ has half this value, find the acceleration of each block and the force of friction acting on each block. (*c*) If $F$ is twice the value found in (*a*), find the acceleration of each block.

**FIGURE 5-48** Problem 53

**54** •• In Figure 5-49, the mass $m_2 = 10$ kg slides on a frictionless table. The coefficients of static and kinetic friction between $m_2$ and $m_1 = 5$ kg are $\mu_s = 0.6$ and $\mu_k = 0.4$. (*a*) What is the maximum acceleration of $m_1$? (*b*) What is the maximum value of $m_3$ if $m_1$ moves with $m_2$ without slipping? (*c*) If $m_3 = 30$ kg, find the acceleration of each body and the tension in the string.

**FIGURE 5-49** Problem 54

**55** • A massive sandstone block is being dragged up a ramp by a counterweight suspended over a pulley, as shown in Figure 5-50. The mass of the block is 1600 kg, while the mass of the counterweight is 550 kg. The coefficient of kinetic friction of the block against the ramp is 0.15, and the ramp is sloped at an angle of 10°. (a) What is the acceleration of the block up the ramp? (b) Three seconds after the block begins moving up the ramp, the rope holding the counterweight breaks. How much farther will the block continue to slide up the ramp before stopping? (c) Unfortunately, the block begins to slide back down the ramp once it reaches its highest point. What is the acceleration of the block as it slides down the ramp?

**FIGURE 5-50** Problem 55

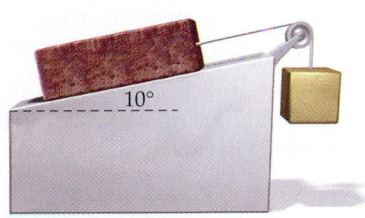

**56** ••• **iSOLVE** A 10-kg block rests on a 5-kg bracket shown in Figure 5-51. The 5-kg bracket sits on a frictionless surface. The coefficients of friction between the 10-kg block and the bracket on which it rests are $\mu_s = 0.40$ and $\mu_k = 0.30$. (a) What is the maximum force $F$ that can be applied if the 10-kg block is not to slide on the bracket? (b) What is the corresponding acceleration of the 5-kg bracket?

**FIGURE 5-51** Problem 56

**57** •• **SSM** On planet Vulcan, an introductory physics class performs several experiments involving friction. In one of these experiments the acceleration of a block is measured both when it is sliding up an incline and when it is sliding down the same incline. You copy the following data and diagram (Figure 5-52) out of one of the lab notebooks, but can't find any translations into metric units (Negative sign indicates that the acceleration is pointing down the incline.):

Acceleration of block
Going up inclined plane      $-1.73$ glapp/plip²
Going down plane             $-1.42$ glapp/plip²

**FIGURE 5-52** Problem 57

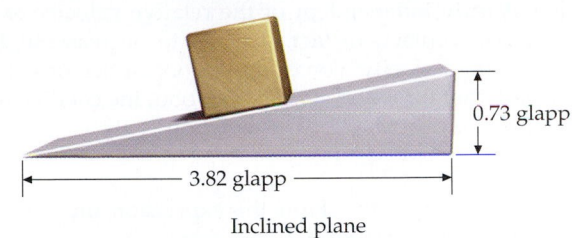

0.73 glapp

3.82 glapp

Inclined plane

From these data, determine the acceleration of gravity on Vulcan (in glapps/plip²) and the coefficient of kinetic friction between the block and the incline.

**58** •• **SSM** A 100-kg block on an inclined plane is attached to another block of mass $m$ via a string, as in Figure 5-53. The coefficients of static and kinetic friction of the block and the incline are $\mu_s = 0.4$ and $\mu_k = 0.2$. The angle of the incline is 18° above horizontal. (a) Determine the range of values for $m$, the mass of the hanging block, for which the block on the incline will not move unless disturbed, but if nudged, will slide *down* the incline. (b) Determine a range of values for $m$ for which the block on the incline will not move unless nudged, but if nudged will slide *up* the incline.

**FIGURE 5-53** Problem 58

**59** ••• **iSOLVE ✓** A block of mass 0.5 kg rests on the inclined surface of a wedge of mass 2 kg, as in Figure 5-54. The wedge is acted on by a horizontal force $\vec{F}$ and slides on a frictionless surface. (a) If the coefficient of static friction between the wedge and the block is $\mu_s = 0.8$ and the angle of the incline is 35°, find the maximum and minimum values of $F$ for which the block does not slip. (b) Repeat part (a) with $\mu_s = 0.4$.

**FIGURE 5-54** Problem 59

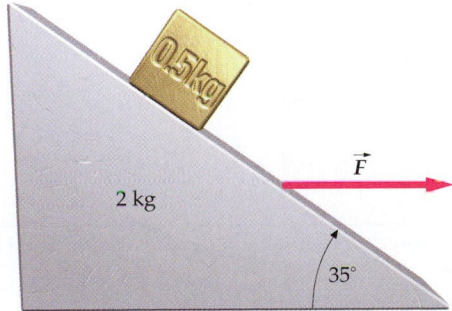

**60** • The coefficient of kinetic friction between two surfaces is not truly independent of the relative velocity of the two objects in contact. In fact, it tends to decrease slightly with increasing velocity. For example, one series of experiments found that for wood sliding on wood, the coefficient of friction could be described by the equation

$$\mu_k = 0.11/(1 + 2.3\times10^{-4}\,v^2)^2$$

where $v$ is measured in m/s. From this expression, find the force of kinetic friction acting on a 100-kg block of wood moving across a horizontal wooden surface at (a) 10 m/s and (b) 20 m/s.

**61** •• In some experiments, it has been found that the coefficient of kinetic friction is not independent of the normal force acting on the object. Vaclav Konecny[†] found, in a set of experiments for wood sliding over wood, that the force of static friction between the two wood surfaces was given by the expression $f_k = 0.4\,F_n{}^{0.91}$, where $f_k$ is the force of friction and $F_n$ is the normal force acting on the object (both forces measured in newtons). Because of this, the acceleration of a block down an inclined plane will not be independent of the mass of the block. Using this formula, calculate the acceleration and stopping distance for (a) a 10-kg block and (b) a 100-kg block sliding along a horizontal surface with an initial speed of 10 m/s.

**62** ••• 〔SSM〕 A block of wood with a mass of 10 kg is pushed, starting from rest, with a constant horizontal force of 70 N across a wooden floor. Assuming that the coefficient of kinetic friction varies with particle speed as $\mu_k = 0.11/(1 + 2.3\times10^{-4}\,v^2)^2$ (see Problem 60), write a spreadsheet program using Euler's method to calculate and graph the speed of the block and its displacement as a function of time from 0 to 10 s. Compare this to the case where the coefficient of kinetic friction is equal to 0.11, independent of $v$.

**63** •• To determine the coefficient of kinetic friction of a block of wood on a horizontal table surface, you are given the following assignment: Take the block of wood and shove it across the surface of the table. Using a stopwatch, measure the time it takes for the block to come to a stop ($\Delta t$) and the total distance that the block travels after the push ($\Delta x$). (a) Show that from these measurements, $\mu_k = 2\Delta x/[g(\Delta t)^2]$. If the block slides a distance of 1.37 m in 0.97 s, calculate $\mu_k$. (b) What was the initial speed of the block?

**64** •• 〔SSM〕 The following data show the acceleration of a block down an inclined plane as a function of the angle of incline $\theta$[‡]:

| $\theta$ (degrees) | Acceleration (m/s²) |
|---|---|
| 25 | 1.6909 |
| 27 | 2.1043 |
| 29 | 2.4064 |
| 31 | 2.8883 |
| 33 | 3.1750 |
| 35 | 3.4886 |
| 37 | 3.7812 |
| 39 | 4.1486 |
| 41 | 4.3257 |
| 43 | 4.7178 |
| 45 | 5.1056 |

(a) Show that, for a block sliding down an incline, graphing $a/\cos\theta$ versus $\tan\theta$ should give a straight line with slope $g$ and y-intercept $-\mu_k g$. (b) Using a spreadsheet program, graph these data and fit a straight line to them to determine $\mu_k$ and $g$. What is the percentage error in $g$ from the commonly accepted value of 9.81 m/s²?

## Motion Along a Curved Path

**65** •• A stone with a mass $m = 95$ g is being whirled in a horizontal circle on the end of a string that is 85 cm long. The length of time required for the stone to make one complete revolution is 1.22 s. The angle that the string makes with the horizontal is (a) 52°, (b) 46°, (c) 26°, (d) 23°, (e) 3°.

**66** •• A 0.20-kg stone attached to a 0.8-m string is rotated in the horizontal plane. The string makes an angle of 20° with the horizontal. Determine the speed of the stone.

**67** •• A 0.75-kg stone attached to a string is whirled in a horizontal circle of radius 35 cm as in the conical pendulum of Example 5-9. The string makes an angle of 30° with the vertical. (a) Find the speed of the stone. (b) Find the tension in the string.

**68** •• 〔SSM〕 〔iSOLVE〕 A 50-kg pilot comes out of a vertical dive in a circular arc such that at the bottom of the arc her upward acceleration is 8.5g. (a) What is the magnitude of the force exerted by the airplane seat on the pilot at the bottom of the arc? (b) If the speed of the plane is 345 km/h, what is the radius of the circular arc?

**69** •• 〔iSOLVE✓〕 A 65-kg airplane pilot pulls out of a dive by following, at constant speed, the arc of a circle whose radius is 300 m. At the bottom of the circle, where her speed is 180 km/h, (a) what are the direction and magnitude of her acceleration? (b) What is the net force acting on her at the bottom of the circle? (c) What is the force exerted on the pilot by the airplane seat?

**70** •• 〔iSOLVE✓〕 Mass $m_1$ moves in a circular path of radius $r$ on a frictionless horizontal table (Figure 5-55). It is attached to a string that passes through a frictionless hole in the center of the table. A second mass $m_2$ is attached to the other end of the string. Derive an expression for $r$ in terms of $m_1$, $m_2$, and the time $T$ for one revolution.

**FIGURE 5-55** Problem 70

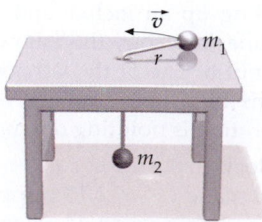

† Vaclav Konecny, "On the first law of friction," *American Journal of Physics*, **41**, 588 (1973).

‡ Data taken from Dennis W. Phillips, "Science Friction Adventure—Part II," *The Physics Teacher*, 553 (Nov. 1990).

**71** •• **SSM** A block of mass $m_1$ is attached to a cord of length $L_1$, which is fixed at one end. The block moves in a horizontal circle on a frictionless table. A second block of mass $m_2$ is attached to the first by a cord of length $L_2$ and also moves in a circle, as shown in Figure 5-56. If the period of the motion is $T$, find the tension in each cord.

**FIGURE 5-56** Problem 71

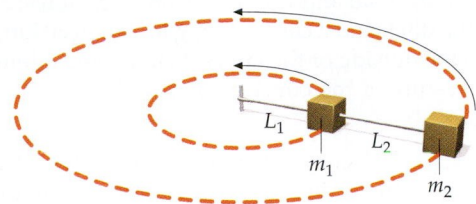

**72** •• **SSM** A particle moves with constant speed in a circle of radius 4 cm. It takes 8 s to complete each revolution. Draw the path of the particle to scale, and indicate the particle's position at 1-s intervals. Draw displacement vectors for each interval. These vectors also indicate the directions for the average-velocity vectors for each interval. Find graphically the magnitude of the change in the average velocity $|\Delta \vec{v}|$ for two consecutive 1-s intervals. Compare $|\Delta \vec{v}|/\Delta t$, measured in this way, with the instantaneous acceleration computed from $a = v^2/r$.

**73** •• A man swings his child in a circle of radius 0.75 m as shown in Figure 5-57. If the mass of the child is 25 kg and the child makes one revolution in 1.5 s, what is the magnitude and direction of the force that must be exerted by the man on the child? (Assume the child to be a point particle.)

**FIGURE 5-57** Problem 73

**74** •• The string of a conical pendulum is 50 cm long and the mass of the bob is 0.25 kg. Find the angle between the string and the horizontal when the tension in the string is six times the weight of the bob. Under those conditions, what is the period of the pendulum?

**75** •• **ISOLVE** ✓ A 100-g coin sits on a horizontally rotating turntable. The turntable makes one revolution each second. The coin is located 10 cm from the axis of rotation of the turntable. (a) What is the frictional force acting on the coin?

(b) The coin will slide off the turntable if it is located more than 16 cm from the axis of rotation. What is the coefficient of static friction?

**76** •• A 0.25-kg tether ball is attached to a vertical pole by a 1.2-m cord. Assume that the cord is attached to the center of the ball. If the ball moves in a horizontal circle with the cord making an angle of 20° with the vertical, (a) what is the tension in the cord? (b) What is the speed of the ball?

**77** •• **SSM** An object on the equator has both an acceleration toward the center of the earth because of the earth's rotation, and an acceleration toward the sun because of the earth's motion along its orbit. Calculate the magnitudes of both accelerations, and express them as fractions of the free-fall acceleration $g$ at the earth's surface.

**78** • (a) Using data available from the textbook, calculate the net force acting on the earth that holds it in orbit around the sun. Assume that the orbit is circular. (b) Using data available from the textbook, calculate the force acting on the moon that holds it in orbit around the earth. Assume that the orbit is circular.

**79** •• **ISOLVE** ✓ A small bead with a mass of 100 g slides without friction along a semicircular wire with a radius of 10 cm that rotates about a vertical axis at a rate of 2 revolutions per second, as in Figure 5-58. Find the value of $\theta$ for which the bead will remain stationary relative to the rotating wire.

**FIGURE 5-58** Problem 79

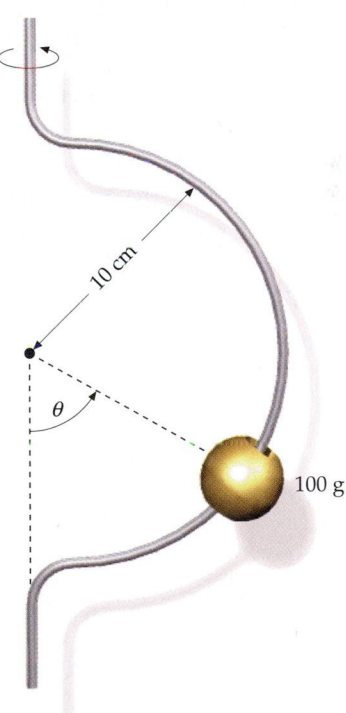

**80** ••• Consider a bead of mass $m$ that is free to move on a thin, circular wire of radius $r$. The bead is given an initial speed $v_0$, and there is a coefficient of kinetic friction $\mu_k$. The experiment is performed in a spacecraft drifting in space. Find the speed of the bead at any subsequent time $t$.

**81 •••** Revisiting Problem 80, (a) find the centripetal acceleration of the bead. (b) Find the tangential acceleration of the bead. (c) What is the magnitude of the resultant acceleration?

## Concepts of Centripetal Force

**82 •** SSM A ride at an amusement park carries people in a vertical circle at constant speed such that the normal forces exerted by the seats are always inward—toward the center of the circle. At the top, the normal force exerted by a seat equals the person's weight, $mg$. At the bottom of the loop, the force exerted by the seat will be (a) 0, (b) $mg$, (c) $2mg$, (d) $3mg$, (e) greater than $mg$, but it cannot be calculated from the information given.

**83 •** The radius of curvature of a loop-the-loop roller coaster is 12.0 m. At the top of the loop, the force that the seat exerts on a passenger of mass $m$ is $0.4mg$. Find the speed of the roller coaster at the top of the loop.

**84 •** A car speeds along the curved exit ramp of a freeway. The radius of the curve is 80 m. A 70-kg passenger holds the arm rest of the car door with a 220-N force in order to keep from sliding across the front seat of the car. (Assume the exit ramp is not banked and ignore friction with the car seat.) What is the car's speed? (a) 16 m/s. (b) 57 m/s. (c) 18 m/s. (d) 50 m/s. (e) 28 m/s.

**85 •••** SSM Suppose you ride a bicycle on a horizontal surface in a circle with a radius of 20 m. The resultant force exerted by the road on the bicycle (normal force plus frictional force) makes an angle of 15° with the vertical. (a) What is your speed? (b) If the frictional force is half its maximum possible value, what is the coefficient of static friction?

**86 ••** An airplane is flying in a horizontal circle at a speed of 480 km/hr. Banked for this turn, the wings of the plane are tilted at an angle of 40° from the horizontal (Figure 5-59). Assume that a lift force acting perpendicular to the wings holds the aircraft in the sky. What is the radius of the circle in which the plane is flying?

**FIGURE 5-59**
**Problem 86**

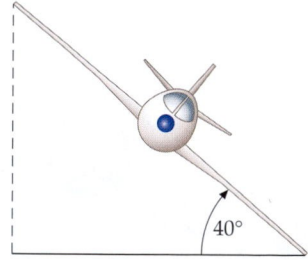

40°

**87 •** ISOLVE A 750-kg car travels around a curve of 160-m radius at 90 km/h. What should the banking angle of the curve be so that the force of the pavement on the tires of the car is in the normal direction?

**88 ••** SSM A curve of radius 150 m is banked at an angle of 10°. An 800-kg car negotiates the curve at 85 km/h without skidding. Find (a) the normal force exerted by the pavement on the tires, (b) the frictional force exerted by the pavement on the tires, (c) the minimum coefficient of static friction between the pavement and the tires.

**89 ••** On another occasion, the car in Problem 88 negotiates the curve at 38 km/h. Find (a) the normal force exerted on the tires by the pavement, and (b) the frictional force exerted on the tires by the pavement.

**90 •••** SSM ISOLVE A civil engineer is asked to design a curved section of roadway that meets the following conditions: With ice on the road, when the coefficient of static friction between the road and rubber is 0.08, a car at rest must not slide into the ditch and a car traveling less than 60 km/h must not skid to the outside of the curve. What is the minimum radius of curvature of the curve and at what angle should the road be banked?

**91 •••** ISOLVE A curve of radius 30 m is banked so that a 950-kg car traveling at 40 km/h can round it even if the road is so icy that the coefficient of static friction is approximately zero. Find the range of speeds at which a car can travel around this curve without skidding if the coefficient of static friction between the road and the tires is 0.3.

## Drag Forces

**92 •** A small pollution particle settles toward the earth in still air with a terminal speed of 0.3 mm/s. The particle has a mass of $10^{-10}$ g and a retarding force of the form $bv$. What is the value of $b$?

**93 •** A Ping-Pong ball has a mass of 2.3 g and a terminal speed of 9 m/s. The retarding force is of the form $bv^2$. What is the value of $b$?

**94 •** SSM A sky diver of mass 60 kg can slow herself to a constant speed of 90 km/h by adjusting her form. (a) What is the magnitude of the upward drag force on the sky diver? (b) If the drag force is equal to $bv^2$, what is the value of $b$?

**95 ••** An 800-kg car rolls down a very long 6° grade. The drag force for motion of the car has the form $F_d = 100$ N $+ (1.2$ N·s²/m²$)v^2$. Neglect rolling friction. What is the terminal velocity for the car rolling down this grade?

**96 •••** Small spherical particles experience a viscous drag force given by Stokes' law: $F_d = 6\pi\eta rv$, where $r$ is the radius of the particle, $v$ is its speed, and $\eta$ is the viscosity of the fluid medium. (a) Estimate the terminal speed of a spherical pollution particle of radius $10^{-5}$ m and density of 2000 kg/m³. (b) Assuming that the air is still and that $\eta$ is $1.8 \times 10^{-5}$ N·s/m², estimate the time it takes for such a particle to fall from a height of 100 m.

**97 •••** SSM A sample of air containing pollution particles of the size and density given in Problem 96 is captured in a test tube 8.0 cm long. The test tube is then placed in a centrifuge with the midpoint of the test tube 12 cm from the center of the centrifuge. The centrifuge spins at 800 revolutions per minute. Estimate the time required for nearly all of the pollution particles to settle at the end of the test tube and compare this to the time required for a pollution particle to fall 8.0 cm under the action of gravity and subject to the viscous drag of air.

## Euler's Method

**98** •• You are riding in a hot air balloon when you throw a baseball straight down with an initial speed of 35 km/h. A baseball falls with a terminal speed of 150 km/h. Assuming air drag is proportional to the speed squared, use Euler's method to estimate the speed of the ball after 10 s. What is the uncertainty in this estimate? You drop a second baseball, this one released from rest. How long does it take for it to reach 99% of its terminal speed? How far does it fall during this time?

**99** •• **SSM** You throw a baseball straight up with an initial speed of 150 km/h. Its terminal speed when falling is also 150 km/h. Use Euler's method to estimate its height 3.5 s after release. What is the maximum height it reaches? How long after release does it reach its maximum height? How much later does it return to the ground? Is the time the ball spends on the way up less than, the same as, or greater than the time it spends on the way down?

**100** •• A 0.80-kg block on a horizontal frictionless surface is held against a massless spring, compressing it 30 cm. The force constant of the spring is 50 N/m. The block is released and the spring pushes it 30 cm. Use Euler's method with $\Delta t = 0.005$ s to estimate the time it takes for the spring to push the block the 30 cm. How fast is the block moving at this time? What is the uncertainty in this speed?

## General Problems

**101** • **iSOLVE** A 4.5-kg block slides down an inclined plane that makes an angle of 28° with the horizontal. Starting from rest, the block slides a distance of 2.4 m in 5.2 s. Find the coefficient of kinetic friction between the block and plane.

**102** • **iSOLVE** A model airplane of mass 0.4 kg is attached to a horizontal string and flies in a horizontal circle of radius 5.7 m. (The weight of the plane is balanced by the upward "lift" force of the air on the wings of the plane.) The plane makes 1.2 revolutions every 4 s. (a) Find the speed $v$ of the plane. (b) Find the tension in the string.

**103** •• **SSM** An 800-N box rests on an incline making a 30° angle with the horizontal. A physics student finds that she can prevent the box from sliding if she pushes on it with a force of at least 200 N parallel to the incline. (a) What is the coefficient of static friction between the box and the incline? (b) What is the greatest force that can be applied to the box parallel to the incline before the box slides up the incline?

**104** •• The position of a particle is given by the vector $\vec{r} = -10$ m cos $\omega t \hat{i} + 10$ m sin $\omega t \hat{j}$, where $\omega = 2$ s$^{-1}$. (a) Show that the path of the particle is a circle. (b) What is the radius of the circle? (c) Does the particle move clockwise or counterclockwise around the circle? (d) What is the speed of the particle? (e) What is the time for one complete revolution?

**105** •• **iSOLVE** A crate of books is to be put on a truck with the help of some planks sloping up at 30°. The mass of the crate is 100 kg, and the coefficient of sliding friction between it and the planks is 0.5. You and your friends push *horizontally* with a force $\vec{F}$. Once the crate has started to move, how large must $F$ be in order to keep the crate moving at constant speed?

**106** •• An object with a mass of 5.5 kg is allowed to slide from rest down an inclined plane. The plane makes an angle of 30° with the horizontal and is 72 m long. The coefficient of kinetic friction between the plane and the object is 0.35. The speed of the object at the bottom of the plane is (a) 5.3 m/s, (b) 15 m/s, (c) 24 m/s, (d) 17 m/s, (e) 11 m/s.

**107** •• **SSM** A brick slides down an inclined plank at constant speed when the plank is inclined at an angle $\theta_0$. If the angle is increased to $\theta_1$, the block accelerates down the plank with acceleration $a$. The coefficient of kinetic friction is the same in both cases. Given $\theta_0$ and $\theta_1$, calculate $a$.

**108** •• Three forces act on an object, shown in Figure 5-60, that is in static equilibrium. (a) If $F_1$, $F_2$, and $F_3$ represent the magnitudes of the forces acting on the object, show that $F_1/\sin\theta_{23} = F_2/\sin\theta_{31} = F_3/\sin\theta_{12}$. (b) Show that $F_1^2 = F_2^2 + F_3^2 + 2F_2F_3\cos\theta_{23}$.

**FIGURE 5-60** Problem 108

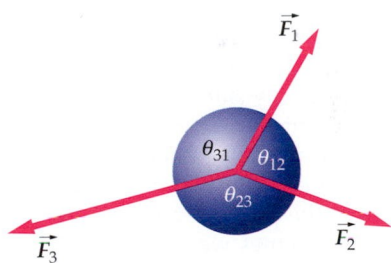

**109** •• In a carnival ride, the passenger sits on a seat in a compartment that rotates with constant speed in a vertical circle of radius $r = 5$ m. The heads of the seated passengers always point toward the center of the circle. (a) If the carnival ride completes one full circle in 2 s, find the acceleration of the passenger. (b) Find the slowest rate of rotation (in other words, the longest time $T_m$ to complete one full circle) if the seat belt is to exert no force on the passenger at the top of the ride.

**110** •• **SSM** A flat-topped toy cart moves on frictionless wheels, pulled by a rope under tension $T$. The mass of the cart is $m_1$. A load of mass $m_2$ rests on top of the cart with a coefficient of static friction $\mu_s$. The cart is pulled up a ramp that is inclined at angle $\theta$ above the horizontal. The rope is parallel to the ramp. What is the maximum tension $T$ that can be applied without making the load slip?

**111** •• A sled weighing 200 N rests on a 15° incline, held in place by static friction (Figure 5-61). The coefficient of static friction is 0.5. (*a*) What is the magnitude of the normal force on the sled? (*b*) What is the magnitude of the static frictional force on the sled? (*c*) The sled is now pulled up the incline at constant speed by a child. The child weighs 500 N and pulls on the rope with a constant force of 100 N. The rope makes an angle of 30° with the incline and has negligible mass. What is the magnitude of the kinetic frictional force on the sled? (*d*) What is the coefficient of kinetic friction between the sled and the incline? (*e*) What is the magnitude of the force exerted on the child by the incline?

**FIGURE 5-61** Problem 111

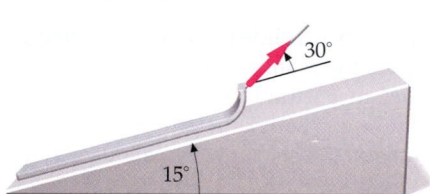

**112** • In 1976, Gerard O'Neill proposed the building of large space stations for human habitation in orbit around the earth and the moon. Because prolonged free-fall has adverse medical effects, he proposed making the stations in the form of long cylinders and spinning them around the cylinder axis to provide the inhabitants inside with the sensation of gravity. This idea has found its way into mainstream science fiction; for one example, the TV show *Babylon 5* was set on an O'Neill colony 5 mi long, with a diameter of 0.6 mi. Because of the rotation, someone on the inside of the colony would experience a sense of "gravity," because he or she would be in an accelerated frame of reference. (*a*) Show that the "acceleration of gravity" experienced by someone in an O'Neill colony is equal to his or her centripetal acceleration. *Hint: consider someone "looking in" from outside the colony.* (*b*) If we assume that the space station is composed of several decks which are at varying distances from the axis of rotation, show that the "acceleration of gravity" becomes weaker the closer one gets to the axis. (*c*) How many revolutions per minute would Babylon 5 have to make to give an "acceleration of gravity" of 9.8 m/s² at the outermost edge of the station?

**113** •• A child slides down a slide inclined at 30° in time $t_1$. The coefficient of kinetic friction between her and the slide is $\mu_k$. She finds that if she sits on a small cart with frictionless wheels, she slides down the same slide in time $\frac{1}{2}t_1$. Find $\mu_k$.

**114** •• **SSM** The position of a particle of mass $m = 0.8$ kg as a function of time is

$$\vec{r} = x\hat{i} + y\hat{j} = R\sin\omega t\,\hat{i} + R\cos\omega t\,\hat{j}$$

where $R = 4.0$ m and $\omega = 2\pi$ s⁻¹. (*a*) Show that the path of this particle is a circle of radius $R$ with its center at the origin. (*b*) Compute the velocity vector. Show that $v_x/v_y = -y/x$. (*c*) Compute the acceleration vector and show that it is in the radial direction and has the magnitude $v^2/r$. (*d*) Find the magnitude and direction of the net force acting on the particle.

**115** •• In an amusement-park ride, riders stand with their backs against the wall of a spinning vertical cylinder. The floor falls away and the riders are held up by friction. If the

radius of the cylinder is 4 m, find the minimum number of revolutions per minute necessary when the coefficient of static friction between a rider and the wall is 0.4.

**116** •• **SOLVE** A mass $m_1$ on a horizontal table is attached by a thin string that passes over a frictionless, massless pulley to a 2.5-kg mass $m_2$ that hangs over the side of the table 1.5 m above the ground (Figure 5-62). The system is released from rest at $t = 0$ and the 2.5-kg mass strikes the ground at $t = 0.82$ s. The system is now placed in its initial position and a 1.2-kg mass is placed on top of the block of mass $m_1$. Released from rest, the 2.5-kg mass now strikes the ground 1.3 s later. Determine the mass $m_1$ and the coefficient of kinetic friction between $m_1$ and the table.

**FIGURE 5-62** Problem 116

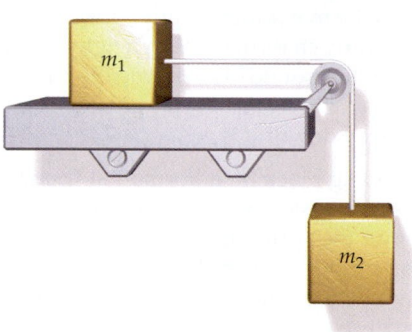

**117** ••• **SSM** (*a*) Show that a point on the surface of the earth at latitude $\theta$, shown in Figure 5-63, has an acceleration of magnitude (3.37 cm/s²)cos $\theta$ relative to a reference frame not rotating with the earth. What is the direction of this acceleration? (*b*) Discuss the effect of this acceleration on the apparent weight of an object near the surface of the earth. (*c*) The free-fall acceleration of an object at sea level measured *relative to the earth's surface* is 9.78 m/s² at the equator and 9.81 m/s² at latitude $\theta = 45°$. What are the values of the gravitational field $g$ at these points?

**FIGURE 5-63** Problem 117

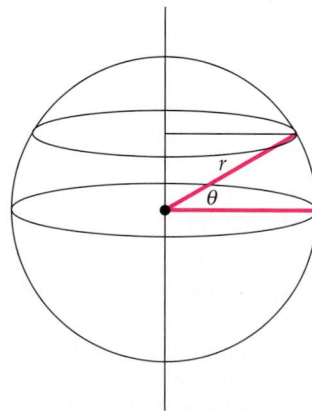

**118** ••• **SSM** A small block of mass 0.01 kg is at rest atop a smooth (frictionless) sphere of radius 0.8 m. The block is given a tiny nudge and starts to slide down the sphere. The mass loses contact with the sphere when the angle between the vertical and the line from the center of the sphere to the position of the mass is $\theta$. Find the angle $\theta$.

# Work and Energy

DOWNHILL RACING CAN LOOK LIKE A
LOT OF FUN, OR A LOT OF WORK,
DEPENDING UPON YOUR PERSPECTIVE.

**?** **Did you know that, in fact,
work is being done on the skier?
(See Example 6-12.)**

6-1   Work and Kinetic Energy

6-2   The Dot Product

6-3   Work and Energy in Three Dimensions

6-4   Potential Energy

**W**ork and energy are important concepts in physics as well as in our everyday lives. In physics, a force does **work** if its point of application moves through a distance and there is a component of the force in the direction of the velocity of the force's point of application. For a constant force in one dimension, the work done equals the force component in the direction of the displacement times the displacement. (This differs somewhat from the everyday use of the word work. When you study hard for an exam, the only work you do according to the use of the word in physics, is in pushing your pencil on the paper, or turning the pages of your book.)

**Energy** is closely associated with work. When work is done by one system on another, energy is transferred between the two systems. For example, when you do work pushing a swing, chemical energy of your body is transferred to the swing and appears as kinetic energy of motion or as gravitational potential energy of the earth–swing system. There are many forms of energy. Kinetic energy is associated with the motion of an object. Potential energy is associated with the configuration of a system, such as the separation distance between two objects that attract each other. Thermal energy is associated with the random motion of the molecules within a system and is closely connected with the temperature of the system.

➤ **In this chapter we study the concepts of work, kinetic energy, and potential energy. Because these concepts arise from Newton's laws, the concepts introduced in this chapter are a continuation of those introduced in previous chapters.**

These new concepts provide powerful methods of solving a wide class of problems. Many of the end-of-chapter problems for this chapter can be solved using the concepts and methods of the previous chapters. However, it is important that you resist any temptation to solve them in that way. The concepts and methods developed in this chapter are developed further in Chapter 7.

# 6-1 Work and Kinetic Energy

## Motion in One Dimension With Constant Forces

The work $W$ done by a constant force $\vec{F}$ moving through a displacement $\Delta x \hat{i}$ is given by

$$W = F_x \Delta x = F \cos \theta \, \Delta x \qquad \text{6-1}$$

WORK BY A CONSTANT FORCE

where $\theta$ is the angle between the directions of $\vec{F}$ and $\hat{i}$, and $\Delta x \hat{i}$ is the displacement of the point of application of the force as shown in Figure 6-1.

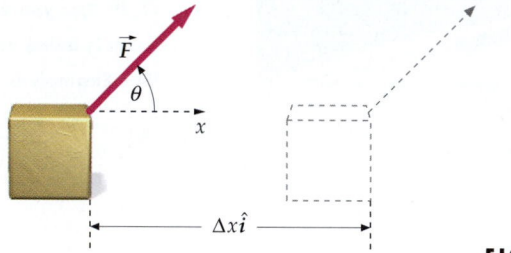

**FIGURE 6-1**

Work is a scalar quantity that is positive if $\Delta x$ and $F_x$ have the same signs and negative if they have opposite signs. The dimensions of work are those of force times distance. The SI unit of work and energy is the **joule** (J), which equals the product of a newton and a meter:

$$1 \, J = 1 \, N \cdot m \qquad \text{6-2}$$

(In the U.S. customary system, the unit of work is the foot-pound: 1 ft·lb = 1.356 J.) A convenient unit of work and energy in atomic and nuclear physics is the electron volt (eV):

$$1 \, eV = 1.6 \times 10^{-19} \, J \qquad \text{6-3}$$

Commonly used multiples are keV ($10^3$ eV) and MeV ($10^6$ eV). The work required to remove an electron from an atom is of the order of a few eV, whereas the work needed to remove a proton or a neutron from an atomic nucleus is of the order of several MeV.

**EXERCISE** A force of 12 N is exerted on a box at an angle of $\theta = 20°$, as in Figure 6-1. How much work is done by the force on the box as the box moves along the table a distance of 3 m? (*Answer* 33.8 J)

If there are several forces that do work, the total work is found by computing the work done by each force and summing.

$$W_{\text{total}} = F_{1x} \Delta x_1 + F_{2x} \Delta x_2 + F_{3x} \Delta x_3 + \dots$$

A **particle** is any object that moves so that all of its parts undergo identical displacements during any interval of time. That is, an object can be modeled as a particle as long as it remains perfectly rigid and moves without rotating.

When several forces do work on a *particle,* the displacements of the points of application of each these forces are equal. Let the displacement of the point of application of any one of the forces be $\Delta x$. Then

$$W_{\text{total}} = F_{1x}\Delta x + F_{2x}\Delta x + \cdots = (F_{1x} + F_{2x} + \cdots)\Delta x = F_{\text{net }x}\Delta x \qquad 6\text{-}4$$

For a particle constrained to move along the x axis, the net force has only an x component. That is, $\vec{F}_{\text{net}} = F_{\text{net }x}\hat{i}$. Thus, for a particle the total work can be found by first finding the net force, and then multiplying the net force by the displacement.

## The Work–Kinetic Energy Theorem

There is an important relation between the total work done on a particle and the initial and final speeds of the particle. If $F_{\text{net }x}$ is the net force acting on a particle, Newton's second law gives

$$F_{\text{net }x} = ma_x$$

For a constant force, the acceleration is constant, and we can relate the displacement to the initial speed $v_i$ and final speed $v_f$ by using the constant-acceleration formula $v_f^2 = v_i^2 + 2a_x\Delta x$ (Equation 2-16). Solving this for $a_x$ gives

$$a_x = \frac{1}{2\Delta x}(v_f^2 - v_i^2)$$

Substituting for $a_x$ in $F_{\text{net }x} = ma_x$ and then multiplying both sides by $\Delta x$ gives

$$F_{\text{net }x}\Delta x = \tfrac{1}{2}mv_f^2 - \tfrac{1}{2}mv_i^2$$

We recognize the term on the left as the total work done on the particle. Thus

$$W_{\text{total}} = \tfrac{1}{2}mv_f^2 - \tfrac{1}{2}mv_i^2 \qquad 6\text{-}5$$

The quantity $\tfrac{1}{2}mv^2$ is a scalar quantity called the **kinetic energy** $K$ of the particle:

$$K = \tfrac{1}{2}mv^2 \qquad 6\text{-}6$$

DEFINITION—KINETIC ENERGY

The quantity on the right side of Equation 6-5 is the change in the kinetic energy of the particle. Thus,

The total work done on a particle is equal to the change in its kinetic energy:

$$W_{\text{total}} = \Delta K \qquad 6\text{-}7$$

WORK–KINETIC ENERGY THEOREM

This result is known as the **work–kinetic energy theorem.** The derivation presented here is valid only if the net force remains constant. However, as we will see later in this chapter, this theorem is valid even when the net force varies and the motion is not along a straight line.

**EXERCISE** A woman of mass 50 kg is running at 3.5 m/s. What is her kinetic energy? (*Answer* 306 J)

**EXAMPLE 6-1**

A truck of mass 3000 kg is to be loaded onto a ship by a crane that exerts an upward force of 31 kN on the truck. This force, which is strong enough to overcome the gravitational force and get the truck started upward, is applied over a distance of 2 m. Find (*a*) the work done by the crane, (*b*) the work done by gravity, and (*c*) the upward speed of the truck after the 2 m.

**PICTURE THE PROBLEM** Sketch the truck at its initial and final positions and choose the positive *y* direction to be the direction of the displacement (Figure 6-2). Use the work–kinetic energy theorem to find the truck's final kinetic energy. The final speed of the truck can be obtained from the final kinetic energy. The total work is the sum of the results for (*a*) and (*b*).

**FIGURE 6-2**

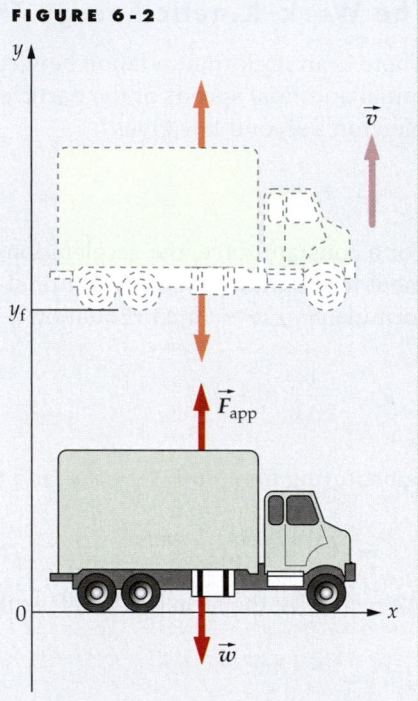

(*a*) Calculate the work done by the applied force:

$$W_{app} = F_{app\,y}\,\Delta y = F_{app}\cos 0°\,\Delta y$$

$$= (31\text{ kN})(1)(2\text{ m}) = \boxed{62.0\text{ kJ}}$$

(*b*) Calculate the work done by the force of gravity:

$$W_g = mg_y\,\Delta y = mg\cos 180°\,\Delta y$$

$$= (3000\text{ kg})(9.81\text{ N/kg})(-1)(2\text{ m})$$

$$= \boxed{-58.9\text{ kJ}}$$

(*c*) Apply the work–kinetic energy theorem and solve for $v_f$:

$$W_{total} = \Delta K$$

$$W_{app} + W_g = K_f - K_i$$

$$= \tfrac{1}{2}mv_f^2 - \tfrac{1}{2}mv_i^2$$

$$v_f^2 = v_i^2 + \frac{2(W_{app} + W_g)}{m}$$

$$= 0 + \frac{2(62{,}000\text{ J} - 58{,}900\text{ J})}{3000\text{ kg}}$$

$$= 2.09\text{ m}^2/\text{s}^2$$

$$v_f = \boxed{1.45\text{ m/s}}$$

**REMARKS** Notice that we treat each force separately when calculating the work done. We could also find the total work by first adding the forces to obtain the net force, then applying $W_{total} = F_{net\,y}\,\Delta y$. In either case, the work–kinetic energy theorem applies only to the total work.

**EXERCISE** Find the final speed of the truck if the same upward force were applied for 2 m after it was already moving upward at 1 m/s. (*Answer* 1.73 m/s. Note that the answer is *not* 1.45 m/s + 1.00 m/s. Why not?)

**EXAMPLE 6-2**

In a television tube, an electron is accelerated from rest to a kinetic energy of 2.5 keV over a distance of 80 cm. (The force that accelerates the electron is an electric force due to the electric field in the tube.) Find the force on the electron, assuming it to be constant and in the direction of motion.

**PICTURE THE PROBLEM** Because the electron starts from rest, the work done equals the final kinetic energy.

Set the work done to be equal to the change in kinetic energy and solve for the force. The initial and final kinetic energies are both given:

$$W_{total} = \Delta K$$

$$F_x \, \Delta x = K_f - K_i$$

$$F_x = \frac{K_f - K_i}{\Delta x}$$

$$= \frac{2500 \text{ eV} - 0}{0.8 \text{ m}} \times \frac{1.6 \times 10^{-19} \text{ J}}{1 \text{ eV}}$$

$$= \boxed{5.0 \times 10^{-16} \text{ N}}$$

**REMARKS** When we discuss electricity we will see that the work done per unit charge is called the potential difference and is measured in volts. Thus, 1 eV is the energy acquired or lost by a particle of charge $e$ (an electron or proton, for example) when its potential difference changes by 1 V.

---

*A DOGSLED RACE*　　　　　　　**EXAMPLE 6-3**

During your winter break you enter a dogsled race across a frozen lake. To get started you pull the sled (total mass 80 kg) with a force of 180 N at 20° above the horizontal. Find (a) the work you do and (b) the final speed of the sled after it moves $\Delta x = 5$ m, assuming that it starts from rest and there is no friction.

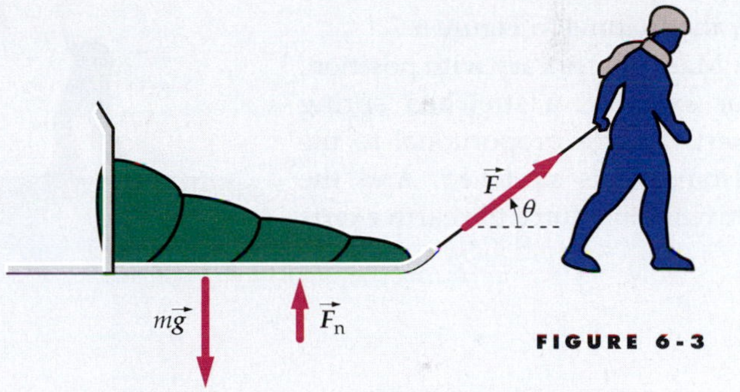

**FIGURE 6-3**

**PICTURE THE PROBLEM** The work done by you is $F_x \, \Delta x$, where we choose the direction of the displacement as the positive $x$ direction. This is also the *total* work done on the sled because the other forces, $mg$ and $F_n$, have no $x$ components (Figure 6-3). The final speed of the sled is found by applying the work–kinetic energy theorem to the sled.

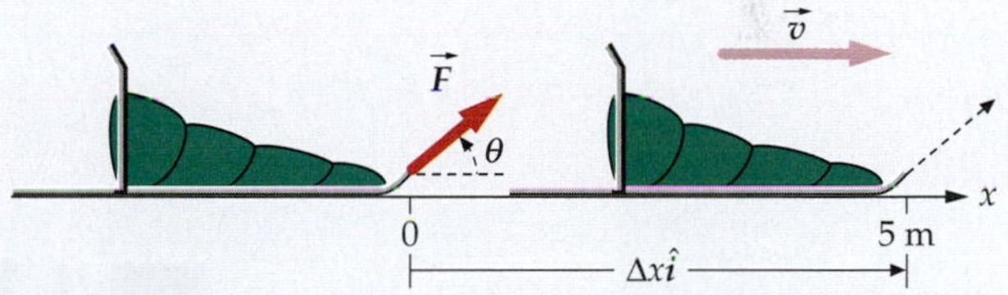

(a) 1. Sketch the sled both in its initial position and in its position after moving the 5 m. Draw the $x$ axis in the direction of motion (Figure 6-4).

**FIGURE 6-4**

2. The work done by you on the sled is $F_x \, \Delta x$. (This is also the total work done on the sled since the other forces act perpendicular to the $x$ direction):

$$W = F_x \, \Delta x = F \cos \theta \, \Delta x$$

$$= (180 \text{ N})(\cos 20°)(5 \text{ m}) = \boxed{846 \text{ J}}$$

(b) Apply the work–kinetic energy theorem to the sled and solve for the final speed:

$$W_{total} = \tfrac{1}{2} mv_f^2 - \tfrac{1}{2} mv_i^2$$

$$v_f^2 = v_i^2 + \frac{2W_{total}}{m}$$

$$= 0 + \frac{2(846 \text{ J})}{80 \text{ kg}}$$

$$= 21.1 \text{ m}^2/\text{s}^2$$

$$v_f = \boxed{4.60 \text{ m/s}}$$

**REMARKS** We do not need to work out the units. If we have a correct equation, and all quantities are in SI units, the result will be in the correct SI units. However, as a check on the equation, we can show that 1 J/kg = 1 m²/s². We have 1 J/kg = 1 N·m/kg = (1 kg·m/s²)·m/kg = 1 m²/s².

**EXERCISE** What is the magnitude of the force you exert if the sled starts with a speed of 2 m/s and its final speed is 4.5 m/s after you pull it through a distance of 5 m? (*Answer* 138 N)

What if you hold a massive object in a fixed position? You are expending energy, but are you doing work? According to the definition of work, you are not doing work *on the object* because the object does not move in the direction of the force you exert (Figure 6-5). But your muscles are continually contracting and relaxing as you hold the weight. In this process internal chemical energy in your body is converted to thermal energy (Figure 6-6).

## Work Done by a Variable Force

In Figure 6-7 we plot a constant force $F_x$ as a function of position $x$. The work done on a particle whose displacement is $\Delta x$ is represented by the area under the force-versus-position curve, indicated by the shading in Figure 6-7.

Many forces vary with position. For example, a stretched spring exerts a force proportional to the distance it is stretched. And the gravitational force the earth exerts

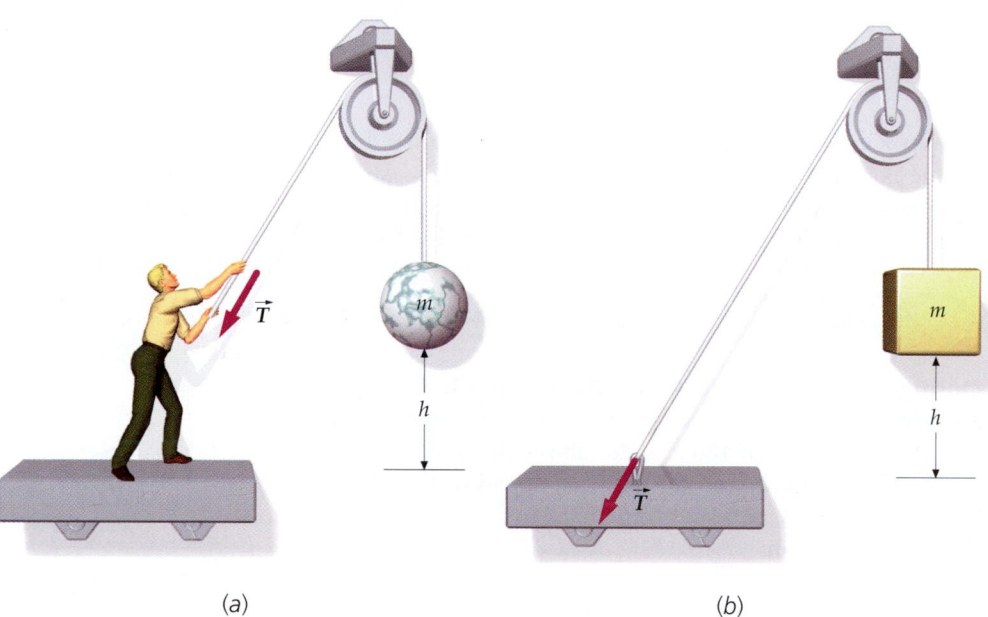

(a)　　　　　　　　　(b)

**FIGURE 6-5** (*a*) The man standing on a ledge does not do work on the weight when holding it at a fixed position. (*b*) The same task could be accomplished by tying the rope to a fixed point.

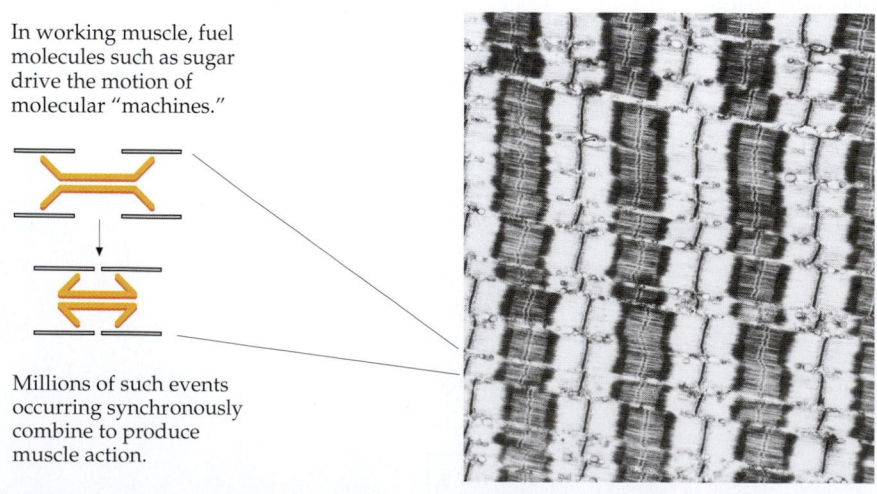

In working muscle, fuel molecules such as sugar drive the motion of molecular "machines."

Millions of such events occurring synchronously combine to produce muscle action.

**FIGURE 6-6** Muscle work. While the man holding the weight in Figure 6-5 may be doing no work on the weight, his body is expending energy on the molecular level, as structures within the muscle slide over each other during muscular extension and contraction.

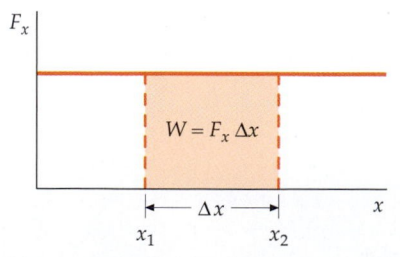

$$W = F_x \, \Delta x$$

**FIGURE 6-7** The work done by a constant force is represented graphically as the area under the $F_x$-versus-$x$ curve.

on a spaceship varies inversely with the square of the distance between the two bodies. We can approximate a variable force by a series of constant forces (Figure 6-8). The work done by a variable force is then

$$W = \lim_{\Delta x_i \to 0} \sum_i F_x \, \Delta x_i = \text{area under the } F_x\text{-versus-}x \text{ curve} \qquad 6\text{-}8$$

This limit is the integral of $F_x$ over $x$. So the work done by a variable force $F_x$ acting on a particle as it moves from $x_1$ to $x_2$ is

$$W = \int_{x_1}^{x_2} F_x \, dx = \text{area under the } F_x\text{-versus-}x \text{ curve} \qquad 6\text{-}9$$

WORK BY A VARIABLE FORCE

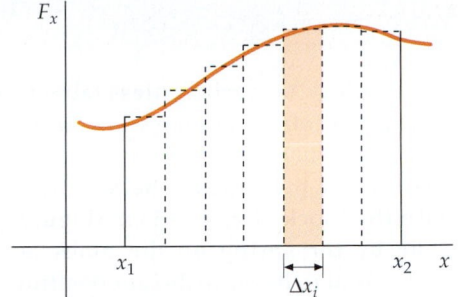

For each displacement interval $\Delta x_i$, the force is essentially constant. Therefore the work done equals the area of the rectangle of height $F_{x\,i}$ and width $\Delta x_i$. As was shown earlier in Section 6-1, this work equals the change in kinetic energy for this displacement interval (if the force is the net force). The total work done is the sum of the areas over all displacement intervals. It follows that the total work equals the change in kinetic energy for the entire displacement. Thus, $W_{\text{total}} = \Delta K$ holds for variable forces as well as for constant forces.

**FIGURE 6-8** A variable force can be approximated by a series of constant forces over small intervals. The work done by the constant force in each interval is the area of the rectangle beneath the force curve. The sum of these rectangular areas is the sum of the work done by the set of constant forces that approximates the varying force. In the limit of infinitesimally small $\Delta x_i$, the sum of the areas of the rectangles equals the area under the complete force curve.

**EXERCISE IN DIMENSIONAL ANALYSIS** A spring is characterized by its force constant $k$, which has dimensions[†] N/m. How does the work required to stretch a spring by an amount $x_0$ depend on $k$ and $x_0$? (*Answer* Because work has dimensions of N·m, the work must depend on $k$ and $x_0$ in the combination $kx_0^2$. We will see in Example 6-5 that the actual expression is $W = \frac{1}{2}kx_0^2$. The factor $\frac{1}{2}$ arises because the force varies from 0 to a maximum value of $kx_0$, and has the average[‡] value $\frac{1}{2}kx_0$.)

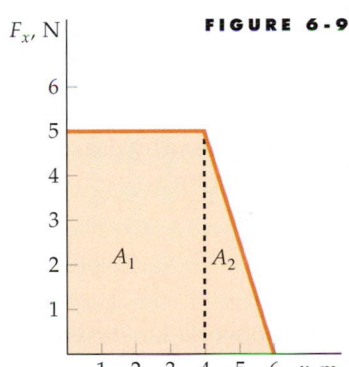

**FIGURE 6-9**

---

WORK DONE ON A PARTICLE                    **E X A M P L E   6 - 4**

A force $F_x$ varies with $x$ as shown in Figure 6-9. Find the work done by the force on a particle as the particle moves from $x = 0$ to $x = 6$ m.

**PICTURE THE PROBLEM** The work is the area under the curve. Because the curve consists of straight-line segments, the easiest approach is to use the geometric formulas for area. (The alternative approach is to set up and execute an integration, which is done in Example 6-5.)

1. We find the work done by calculating the area under the $F_x$-versus-$x$ curve:

$$W = A_{\text{total}}$$

2. This area is the sum of the two areas shown. The area of a triangle is one half the altitude times the base:

$$W = A_{\text{total}} = A_1 + A_2$$
$$= (5\,\text{N})(4\,\text{m}) + \tfrac{1}{2}(5\,\text{N})(2\,\text{m})$$
$$= 20\,\text{J} + 5\,\text{J} = \boxed{25\,\text{J}}$$

**EXERCISE** The force shown is the only force that acts on a particle of mass 3 kg. If the particle starts from rest at $x = 0$, how fast is it moving when it reaches $x = 6$ m? (*Answer* 4.08 m/s)

---

[†] The SI units of $k$ are N/m, but the dimensions of $k$ are $[M][T]^{-2}$.
[‡] Here average refers not to an average over time but to an average over distance.

WORK DONE ON A BLOCK BY A SPRING  **EXAMPLE 6-5**  **FIGURE 6-10**

A 4-kg block on a frictionless table is attached to a horizontal spring that obeys Hooke's law and exerts a force $\vec{F} = -kx\hat{i}$, where $k = 400$ N/m and $x$ is measured from the equilibrium position of the block. The spring is originally compressed with the block at $x_1 = -5$ cm (Figure 6-10). Find (a) the work done by the spring on the block as the block moves from $x_1 = -5$ cm to its equilibrium position $x_2 = 0$ and (b) the speed of the block at $x_2 = 0$.

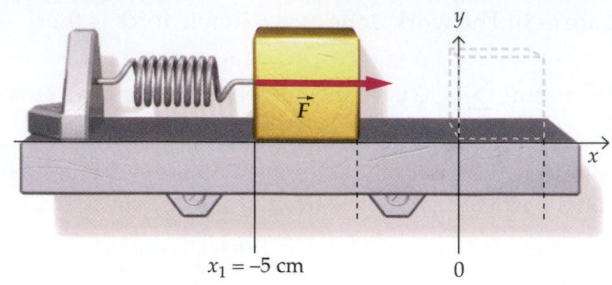

$x_1 = -5$ cm         0

**PICTURE THE PROBLEM** Make a graph of $F_x$ versus $x$. The work done on the block as it moves from $x_1$ to $x_2 = 0$ equals the area under the $F_x$-versus-$x$ curve between these limits, shaded in Figure 6-11, which can be calculated by integrating the force over the distance. The work done equals the change in kinetic energy, which is simply its final kinetic energy because the initial kinetic energy is zero. The speed of the block at $x = 0$ is found from the kinetic energy of the block.

**FIGURE 6-11**

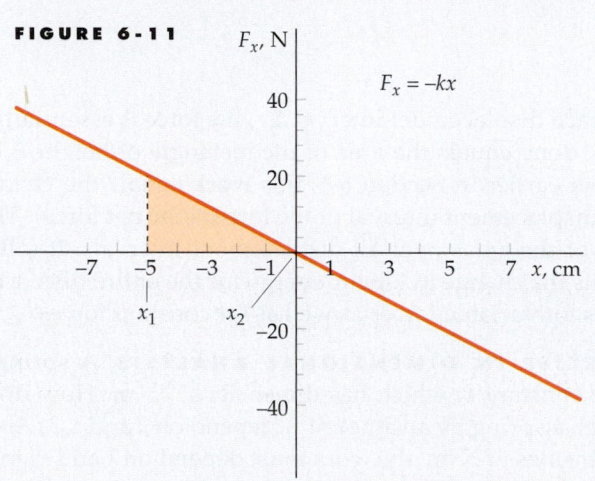

(a) The work $W$ done by the spring on the block is the integral of $F_x\, dx$ from $x_1 = -5$ cm to $x_2 = 0$:

$$W = \int_{x_1}^{x_2} F_x\, dx = \int_{x_1}^{x_2} -kx\, dx = -k \int_{x_1}^{x_2} x\, dx$$

$$= -\tfrac{1}{2} kx^2 \big|_{x_1}^{x_2} = -\tfrac{1}{2} k(x_2^2 - x_1^2)$$

$$= -\tfrac{1}{2}(400 \text{ N/m})\big[0 - (0.05 \text{ m})^2\big] = \boxed{0.500 \text{ J}}$$

(b) Apply the work–kinetic energy theorem and solve for $v_2$:

$$W_{\text{total}} = \tfrac{1}{2} mv_2^2 - \tfrac{1}{2} mv_1^2$$

$$v_2^2 = v_1^2 + \frac{2W_{\text{total}}}{m} = 0 + \frac{2(0.500 \text{ J})}{4 \text{ kg}}$$

$$= 0.250 \text{ m}^2/\text{s}^2$$

$$v_2 = \boxed{0.500 \text{ m/s}}$$

**REMARKS** Besides the spring force, two other forces act on the block; the force of gravity, $m\vec{g}$, and the normal force of the table, $\vec{F}_n$. These forces do no work because they have no component in the direction of the displacement. Only the spring does work on the block. (The force it exerts does have a component in the direction of the displacement.)

**EXERCISE** Find the speed of the block when it reaches $x = 3$ cm if it starts from $x = 0$ with velocity $v_x = 0.5$ m/s. (*Answer* 0.4 m/s)

Note that we could *not* have solved Example 6-5 by first applying Newton's second law to find the acceleration, and then using the constant-acceleration kinematic equations. Because the force exerted by the spring on the block, $F_x = -kx$, varies with position, the acceleration also varies, and thus the constant-acceleration condition is not met.

## 6-2 The Dot Product

The component $F_s$ in Figure 6-12 is related to the angle $\phi$ between the directions of $\vec{F}$ and $d\vec{s}$ by $F_s = F \cos \phi$, so the work done by $\vec{F}$ for the displacement $d\vec{s}$ is

$$W = F_s ds = F \cos \phi \, ds$$

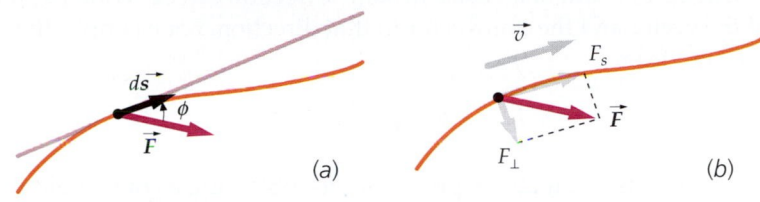

**FIGURE 6-12** (*a*) A particle moving along an arbitrary curve in space. (*b*) The perpendicular component of the force $F_\perp$ changes the direction of the particle's motion but not its speed. The tangential component $F_s$ changes the particle's speed but not its direction. $F_s$ equals the mass $m$ times the tangential acceleration $dv/dt$. Only this component does work.

This combination of two vectors and the cosine of the angle between their directions is called the **dot product** (or **scalar product**) of the vectors. The dot product of two general vectors $\vec{A}$ and $\vec{B}$ is written $\vec{A} \cdot \vec{B}$ and defined by

$$\vec{A} \cdot \vec{B} = AB \cos \phi \qquad\qquad 6\text{-}10$$

DEFINITION—DOT PRODUCT

where $\phi$ is the angle between $\vec{A}$ and $\vec{B}$. (The angle between two vectors is defined as the angle between their directions in space.) The dot product $\vec{A} \cdot \vec{B}$ can be thought of either as $A$ times the component of $\vec{B}$ in the direction of $\vec{A}$ ($A[B \cos \phi]$), or as $B$ times the component of $\vec{A}$ in the direction of $\vec{B}$ ($B[A \cos \phi]$) (see Figure 6-13). Properties of the dot product are summarized in Table 6-1.

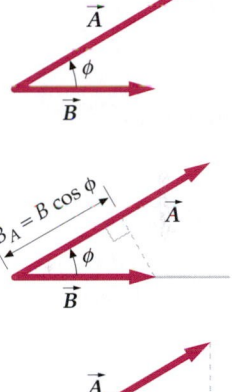

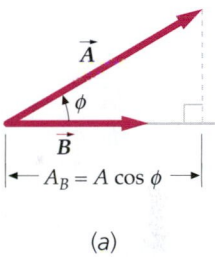

(a)

**FIGURE 6-13** (*a*) The dot product $\vec{A} \cdot \vec{B}$ is the product of $A$ and the projection of $\vec{B}$ on $\vec{A}$ or the product of $B$ and the projection of $\vec{A}$ on $\vec{B}$. That is, $\vec{A} \cdot \vec{B} = AB \cos \phi = AB_A = BA_B$.

## TABLE 6-1

**Properties of Dot Products**

| If | then |
|---|---|
| $\vec{A}$ and $\vec{B}$ are perpendicular, | $\vec{A} \cdot \vec{B} = 0$ (since $\phi = 90°$, $\cos 90° = 0$) |
| $\vec{A}$ and $\vec{B}$ are parallel, | $\vec{A} \cdot \vec{B} = AB$ (since $\phi = 0°$, $\cos 0° = 1$) |
| $\vec{A} \cdot \vec{B} = 0$, | Either $\vec{A} = 0$ or $\vec{B} = 0$ or $\vec{A}$ and $\vec{B}$ are perpendicular |
| *Furthermore,* | |
| $\vec{A} \cdot \vec{A} = A^2$ | Since $\vec{A}$ is parallel to itself |
| $\vec{A} \cdot \vec{B} = \vec{B} \cdot \vec{A}$ | Commutative rule of multiplication |
| $(\vec{A} + \vec{B}) \cdot \vec{C} = \vec{A} \cdot \vec{C} + \vec{B} \cdot \vec{C}$ | Distributive rule of multiplication |

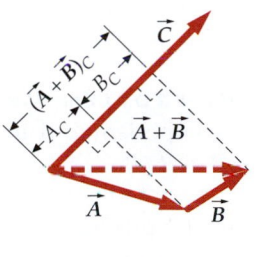

(b)

We can use unit vectors to write the dot product in terms of the rectangular components of the two vectors:

$$\vec{A} \cdot \vec{B} = (A_x\hat{i} + A_y\hat{j} + A_z\hat{k}) \cdot (B_x\hat{i} + B_y\hat{j} + B_z\hat{k})$$

The dot product of any unit vector with itself, like $\hat{i} \cdot \hat{i}$, is 1, so a term like $A_x\hat{i} \cdot B_x\hat{i}$ equals $A_xB_x$. Also, because the unit vectors $\hat{i}$, $\hat{j}$, and $\hat{k}$ are mutually perpendicular, the dot product of one of them with any other one, like $\hat{i} \cdot \hat{j}$, is zero. Thus, terms like $A_x\hat{i} \cdot B_y\hat{j}$ (called cross terms) each equal zero. The result is

$$\vec{A} \cdot \vec{B} = A_xB_x + A_yB_y + A_zB_z \qquad\qquad 6\text{-}11$$

(b) $(\vec{A} + \vec{B}) \cdot \vec{C}$ equals $(\vec{A} + \vec{B})_C C$ (the projection of $\vec{A} + \vec{B}$ in the direction of $\vec{C}$ times C). However, $(\vec{A} + \vec{B})_C = A_C + B_C$, so $(\vec{A} + \vec{B}) \cdot \vec{C} = (A_C + B_C)C = A_C C + B_C C = \vec{A} \cdot \vec{C} + \vec{B} \cdot \vec{C}$. That is, for the dot product multiplication is distributive over addition.

The component of a vector in some direction can be written as the dot product of the vector and the unit vector in that direction. For example, the component $A_x$ is found from

$$\vec{A} \cdot \hat{i} = (A_x\hat{i} + A_y\hat{j} + A_z\hat{k}) \cdot \hat{i} = A_x \qquad \text{6-12}$$

This suggests an algebraic procedure for obtaining a component equation, given a vector equation. That is, multiplying both sides of the vector equation $\vec{A} + \vec{B} = \vec{C}$ by $\hat{i}$ gives $\vec{A} \cdot \hat{i} + \vec{B} \cdot \hat{i} = \vec{C} \cdot \hat{i}$, which in turn gives $A_x + B_x = C_x$.

To establish the product rule for differentiating dot products we differentiate both sides of Equation 6-11. For brevity we do this for vectors with two dimensions.

$$\frac{d}{dt}(\vec{A} \cdot \vec{B}) = \frac{d}{dt}(A_xB_x + A_yB_y)$$

$$= \frac{dA_x}{dt}B_x + A_x\frac{dB_x}{dt} + \frac{dA_y}{dt}B_y + A_y\frac{dB_y}{dt}$$

Rearranging gives

$$\frac{d}{dt}(\vec{A} \cdot \vec{B}) = \frac{dA_x}{dt}B_x + \frac{dA_y}{dt}B_y + A_x\frac{dB_x}{dt} + A_y\frac{dB_y}{dt}$$

$$= \left(\frac{dA_x}{dt}\hat{i} + \frac{dA_y}{dt}\hat{j}\right) \cdot (B_x\hat{i} + B_y\hat{j}) + (A_x\hat{i} + A_y\hat{j}) \cdot \left(\frac{dB_x}{dt}\hat{i} + \frac{dB_y}{dt}\hat{j}\right)$$

so

$$\frac{d}{dt}(\vec{A} \cdot \vec{B}) = \frac{d\vec{A}}{dt} \cdot \vec{B} + \vec{A} \cdot \frac{d\vec{B}}{dt} \qquad \text{6-13}$$

---

USING THE DOT PRODUCT **EXAMPLE 6-6**

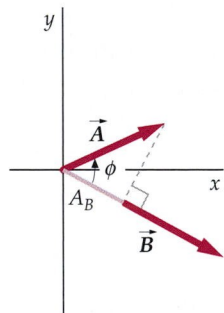

**FIGURE 6-14**

(a) Find the angle between the vectors $\vec{A} = 3$ m$\hat{i}$ + 2 m$\hat{j}$ and $\vec{B} = 4$ m$\hat{i}$ − 3 m$\hat{j}$ (Figure 6-14). (b) Find the component of $\vec{A}$ in the direction of $\vec{B}$.

**PICTURE THE PROBLEM** We find the angle $\phi$ from the definition of the dot product. The component of $\vec{A}$ in the direction of $\vec{B}$ is found from the dot product $\vec{A} \cdot \hat{B}$, where $\hat{B} = \vec{B}/B$.

(a) 1. Write the dot product of $\vec{A}$ and $\vec{B}$ in terms of $A$, $B$, and $\cos\phi$ and solve for $\cos\phi$:

$$\vec{A} \cdot \vec{B} = AB\cos\phi, \quad \text{so}$$

$$\cos\phi = \frac{\vec{A} \cdot \vec{B}}{AB}$$

2. Find $\vec{A} \cdot \vec{B}$ from their components:

$$\vec{A} \cdot \vec{B} = A_xB_x + A_yB_y$$

$$= (3\,\text{m})(4\,\text{m}) + (2\,\text{m})(-3\,\text{m})$$

$$= 12\,\text{m}^2 - 6\,\text{m}^2 = 6\,\text{m}^2$$

3. The magnitudes of the vectors are obtained from the dot product of the vector with itself:

$$\vec{A} \cdot \vec{A} = A^2 = A_x^2 + A_y^2$$

$$= (3\text{ m})^2 + (2\text{ m})^2 = 13\text{ m}^2, \quad \text{so}$$

$$A = \sqrt{13}\text{ m}$$

and

$$\vec{B} \cdot \vec{B} = B^2 = B_x^2 + B_y^2$$

$$= (4\text{ m})^2 + (-3\text{ m})^2 = 25\text{ m}^2, \quad \text{so}$$

$$B = 5\text{ m}$$

4. Substitute these values into the equation in step 1 for $\cos\phi$ to find $\phi$:

$$\cos\phi = \frac{\vec{A} \cdot \vec{B}}{AB} = \frac{6\text{ m}^2}{(\sqrt{13}\text{ m})(5\text{ m})} = 0.333$$

$$\phi = \boxed{70.6°}$$

(b) The component of $\vec{A}$ in the direction of $\vec{B}$ is the dot product of $\vec{A}$ with the unit vector $\hat{B} = \vec{B}/B$:

$$A_B = \vec{A} \cdot \hat{B} = \vec{A} \cdot \frac{\vec{B}}{B} = \frac{\vec{A} \cdot \vec{B}}{B}$$

$$= \frac{6\text{ m}^2}{5\text{ m}} = \boxed{1.2\text{ m}}$$

**PLAUSIBILITY CHECK**  The component of $A$ along $B$ is $A\cos\phi = (\sqrt{13}\text{ m})$ $\cos 70.6° = 1.2$ m.

**EXERCISE**  (a) Find $\vec{A} \cdot \vec{B}$ for $\vec{A} = 3\text{ m}\hat{i} + 4\text{ m}\hat{j}$ and $\vec{B} = 2\text{ m}\hat{i} + 8\text{ m}\hat{j}$. (b) Find $A$, $B$, and the angle between $\vec{A}$ and $\vec{B}$ for these vectors. (*Answer*  (a) 38 m² (b) $A = 5$ m, $B = 8.25$ m, $\phi = 23°$)

In dot-product notation, the work $dW$ done by a force $\vec{F}$ on a particle over a displacement $d\vec{s}$ is

$$dW = F\cos\phi\, ds = \vec{F} \cdot d\vec{s} \qquad\qquad 6\text{-}14$$

INCREMENTAL WORK

where $ds = |d\vec{s}|$ (the magnitude of $d\vec{s}$). The work done on the particle as it moves from point 1 to point 2 is

$$W = \int_{s_1}^{s_2} \vec{F} \cdot d\vec{s} \qquad\qquad 6\text{-}15$$

THE DEFINITION OF WORK

(If the force remains constant, the work can be expressed $W = \vec{F} \cdot \vec{s}$, where $\vec{s}$ is the net displacement.)

When several forces $\vec{F}_i$ act on a particle whose displacement is $d\vec{s}$, the total work done on it is

$$dW_{\text{total}} = \vec{F}_1 \cdot d\vec{s} + \vec{F}_2 \cdot d\vec{s} + \cdots = (\vec{F}_1 + \vec{F}_2 + \cdots) \cdot d\vec{s} = (\Sigma\vec{F}_i) \cdot d\vec{s} \qquad 6\text{-}16$$

**E X A M P L E   6 - 7**

You push a box up a ramp using a horizontal 100-N force $\vec{F}$. For each 5 m of distance along the ramp the box gains 3 m of height. Find the work done by $\vec{F}$ for each 5 m it moves along the ramp (*a*) by directly computing the dot product from the components of $\vec{F}$ and the displacement $\vec{s}$, (*b*) by multiplying the product of the magnitudes of $\vec{F}$ and $\vec{s}$ with the cosine of the angle between their directions, (*c*) by finding $F_s$ (the component of the force in the direction of the displacement) and multiplying it by the magnitude of the displacement, and (*d*) by finding the component of the displacement in the direction of the force and multiplying it by the magnitude of the force.

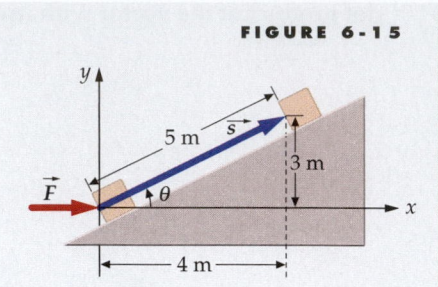

**FIGURE 6-15**

**PICTURE THE PROBLEM**  Draw a sketch of the box in its initial and final positions. Place coordinate axes on the sketch with the *x* axis horizontal. Express the force and displacement vectors in component form and take the dot product. Then find the component of the force in the direction of the displacement, and vice versa.

(*a*) Express $\vec{F}$ and $\vec{s}$ in component form and take the dot product:

$$\vec{F} = 100\ \text{N}\hat{i} + 0\hat{j}$$

$$\vec{s} = 4\ \text{m}\hat{i} + 3\ \text{m}\hat{j}$$

$$W = \vec{F} \cdot \vec{s} = F_x\Delta x + F_y\Delta y$$

$$= (100\ \text{N})(4\ \text{m}) + 0(3\ \text{m}) = \boxed{400\ \text{J}}$$

(*b*) Calculate $Fs\cos\phi$, where $\phi$ is the angle between the directions of the two vectors as shown. Equate this expression with the Part (*a*) result and solve for $\cos\phi$; then solve for the work:

$$\vec{F} \cdot \vec{s} = Fs\cos\phi \quad \text{and} \quad \vec{F} \cdot \vec{s} = F_x\Delta x + F_y\Delta y$$

so

$$\cos\phi = \frac{F_x\Delta x + F_y\Delta y}{Fs} = \frac{(100\ \text{N})(4\ \text{m}) + 0}{(100\ \text{N})(5\ \text{m})} = 0.8$$

and

$$W = Fs\cos\phi$$

$$= (100\ \text{N})(5\ \text{m})0.8 = \boxed{400\ \text{J}}$$

(*c*) Find $F_s$ and multiply it by *s*:

$$F_s = F\cos\phi = (100\ \text{N})0.8 = 80\ \text{N}$$

$$W = F_s s = (80\ \text{N})(5\ \text{m}) = \boxed{400\ \text{J}}$$

(*d*) Multiply *F* and $s_F$, where $s_F$ is the component of $\vec{s}$ in the direction of $\vec{F}$. First calculate $s_F$:

$$s_F = s\cos\phi = (5\ \text{m})0.8 = 4\text{m}$$

$$W = Fs_F = (100\ \text{N})(4\ \text{m}) = \boxed{400\ \text{J}}$$

**REMARKS**  For this problem, computing the work is easiest using the procedure in Part (*d*). For other problems, the procedure in Part (*a*), Part (*b*), or Part (*c*) may be the easiest. You should be comfortable with all four procedures. That way, for a given problem, you can choose the easiest procedure.

---

**E X A M P L E   6 - 8**  **Try It Yourself**

A particle is given a displacement $\vec{s} = 2\ \text{m}\hat{i} - 5\ \text{m}\hat{j}$ along a straight line. During the displacement, a constant force $\vec{F} = 3\ \text{N}\hat{i} + 4\ \text{N}\hat{j}$ acts on the particle. Find (*a*) the work done by the force and (*b*) the component of the force in the direction of the displacement.

FIGURE 6-16

**PICTURE THE PROBLEM** The work $W$ is found by computing $W = \vec{F} \cdot \vec{s} = F_x \Delta x + F_y \Delta y$. Combining this with the relation $\vec{F} \cdot \vec{s} = F_s s$, we can find the component of $\vec{F}$ in the direction of the displacement. Make a sketch showing $\vec{F}, \vec{s}$, and $|F_s|$ (Figure 6-16).

$\vec{F} = 3\,\text{N}\hat{i} + 4\,\text{N}\hat{j}$

$\vec{s} = 2\,\text{m}\hat{i} - 5\,\text{m}\hat{j}$

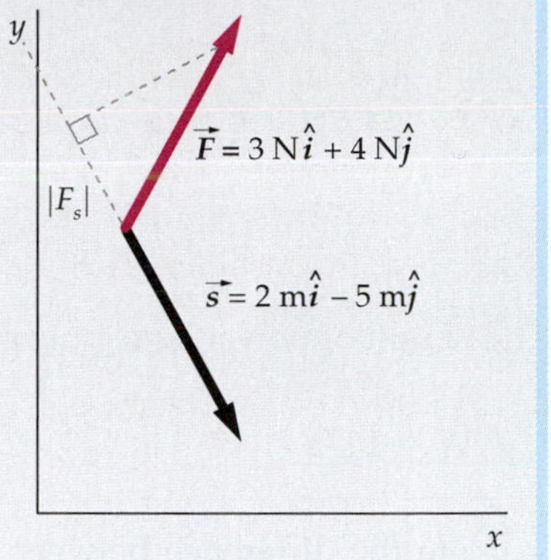

**Cover the column to the right and try these on your own before looking at the answers.**

| Steps | Answers |
|---|---|
| (a) Compute the work done $W$. | $W = \vec{F} \cdot \vec{s} = \boxed{-14\,\text{N} \cdot \text{m}}$ |
| (b) 1. Compute $\vec{s} \cdot \vec{s}$ and use your result to find the distance $|\vec{s}|$. | $\lvert\vec{s}\rvert^2 = \vec{s} \cdot \vec{s} = 29\,\text{m}^2,\quad$ so $\lvert\vec{s}\rvert = \sqrt{29}\,\text{m}$ |
| 2. Using $\vec{F} \cdot \vec{s} = F_s s$, solve for $F_s$. | $F_s = \boxed{-2.60\,\text{N}}$ |

**REMARKS** The component of the force in the direction of the displacement is negative, so the work done is negative.

**EXERCISE** Find the magnitude of $\vec{F}$ and the angle $\phi$ between $\vec{F}$ and $\vec{s}$. (*Answer* $F = 5\,\text{N}, \phi = 121°$)

---

*DIFFERENTIATING A DOT PRODUCT*      **EXAMPLE 6-9**

**Show that $d(v^2)/dt = 2\vec{a} \cdot \vec{v}$, where $v$ is the speed, $\vec{v}$ the velocity, and $\vec{a}$ the acceleration.**

**PICTURE THE PROBLEM** Note that $v^2 = \vec{v} \cdot \vec{v}$, so the rule for differentiating dot products can be used here.

Apply the rule for differentiating dot products to the dot product $\vec{v} \cdot \vec{v}$:

$$\frac{d}{dt}(v^2) = \frac{d}{dt}(\vec{v} \cdot \vec{v})$$

$$= \frac{d\vec{v}}{dt} \cdot \vec{v} + \vec{v} \cdot \frac{d\vec{v}}{dt} = 2\frac{d\vec{v}}{dt} \cdot \vec{v}$$

so

$$\boxed{\frac{d}{dt}(v^2) = 2\vec{a} \cdot \vec{v}}$$

**REMARKS** This example involves only kinematic parameters, so the resulting relation is a strictly kinematic relation. Also, the result is unconditionally valid because the computation involved only the definition of acceleration and the rule for differentiating a dot product.

## Power

The **power** $P$ supplied by a force is the rate at which the force does work. Consider a particle moving with instantaneous velocity $\vec{v}$. In a short time interval $dt$, the particle has a displacement $d\vec{s} = \vec{v}\,dt$. The work done by a force $\vec{F}$ acting on the particle during this time interval is

$$dW = \vec{F} \cdot d\vec{s} = \vec{F} \cdot \vec{v}\,dt$$

The power delivered to the particle is then

$$P = \frac{dW}{dt} = \vec{F} \cdot \vec{v} \qquad\qquad 6\text{-}17$$

DEFINITION—POWER

The SI unit of power, one joule per second, is called a watt (W):

$$1\,W = 1\,J/s$$

Note the difference between power and work. Two motors that lift a given load a given distance expend the same amount of energy, but the one that does it in the least time is more powerful. Gas and electric companies charge for energy, not power, usually by the kilowatt-hour (kW·h). A kilowatt-hour of energy is

$$1\,kW\cdot h = (10^3\,W)(3600\,s) = 3.6 \times 10^6\,W\cdot s = 3.6\,MJ$$

In the U.S. customary system, the unit of energy is the foot-pound and the unit of power is the foot-pound per second. A commonly used multiple of this unit, called a horsepower (hp) is defined as

$$1\,hp = 550\,ft\cdot lb/s = 746\,W$$

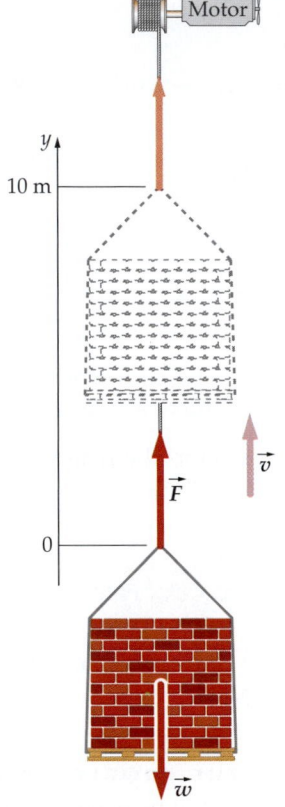

**FIGURE 6-17**

---

*THE POWER OF A MOTOR*                **EXAMPLE 6-10**

**A small motor is used to operate a lift that raises a load of bricks weighing 800 N to a height of 10 m in 20 s (Figure 6-17). What is the minimum power the motor must deliver?**

**PICTURE THE PROBLEM** To find the *minimum* power we assume that the bricks are lifted at constant speed. Because the acceleration is zero, the magnitude of the upward force $\vec{F}$ exerted by the motor is equal to the weight of the bricks, 800 N. The power delivered by the motor is the power supplied by $\vec{F}$.

The power is given by $\vec{F} \cdot \vec{v}$:

$$P = \vec{F} \cdot \vec{v} = Fv \cos \theta = Fv \cos 0$$

$$= (800\,N)\frac{10\,m}{20\,s}(1) = \boxed{400\,W}$$

**REMARKS** This minimum power output of 400 W is slightly more than $\frac{1}{2}$ hp.

**EXERCISE** (*a*) Find the total work done by the force. (*b*) Calculate the power by dividing the total work by the total time. (*Answer* (*a*) 8000 J (*b*) 400 W)

---

*POWER AND KINETIC ENERGY*              **EXAMPLE 6-11**

**Show that the power delivered to a particle by the net force acting on it equals the rate at which the kinetic energy of the particle is changing.**

**PICTURE THE PROBLEM** The power delivered by the net force equals $\vec{F}_{net} \cdot \vec{v}$. Show that $\vec{F}_{net} \cdot \vec{v} = dK/dt$, where $K = \frac{1}{2} mv^2$.

1. Apply the result of Example 6-9 together with $\vec{a} = \vec{F}_{net}/m$ and solve for $\vec{F}_{net} \cdot \vec{v}$:

$$2\vec{a} \cdot \vec{v} = \frac{d}{dt}(v^2)$$

$$2\frac{\vec{F}_{net}}{m} \cdot \vec{v} = \frac{d}{dt}(v^2)$$

$$\vec{F}_{net} \cdot \vec{v} = \frac{m}{2}\frac{d}{dt}(v^2)$$

2. The mass is constant so it can be moved inside the argument of the derivative:

$$\vec{F}_{net} \cdot \vec{v} = \frac{d}{dt}\left(\frac{1}{2} mv^2\right)$$

$$P_{net} = \frac{d}{dt}\left(\frac{1}{2} mv^2\right) = \boxed{\frac{dK}{dt}}$$

**REMARKS** In the next section we use the results of this example to obtain the work–kinetic energy theorem in three dimensions.

In Example 6-10, the power delivered to the bricks by the lower end of the rope was calculated. In that example the rate of change in kinetic energy of the rope is negligible, so the power delivered to the rope by the motor is equal to the power the rope delivers to the bricks.

# 6-3 Work and Energy in Three Dimensions

From Example 6-11 we have $\vec{F}_{net} \cdot \vec{v} = dK/dt$, where $K = \frac{1}{2} mv^2$. The work–kinetic energy theorem for three dimensions can be established by integrating both sides of this equation with respect to time. Integrating both sides gives

$$\int_{t_1}^{t_2} \vec{F}_{net} \cdot \vec{v}\, dt = \int_{t_1}^{t_2} \frac{dK}{dt}\, dt \qquad\qquad 6\text{-}18$$

Because $d\vec{s} = \vec{v}\, dt$, where $d\vec{s}$ is the displacement during time $dt$, and because $(dK/dt)\, dt = dK$, Equation 6-18 can be expressed

$$\int_1^2 \vec{F}_{net} \cdot d\vec{s} = \int_1^2 dK$$

where the integral on the left is the total work $W_{total}$ done on the particle. (In Chapter 7 work–energy relations for objects that cannot be modeled as a particle are presented.) The integral on the right can be integrated giving

$$W_{total} = \int_1^2 \vec{F}_{net} \cdot d\vec{s} = K_2 - K_1 = \Delta K \qquad\qquad 6\text{-}19$$

WORK–KINETIC ENERGY EQUATION IN THREE DIMENSIONS

Equation 6-19 follows directly from Newton's second law of motion.

**EXAMPLE 6-12** **Put It in Context**

You and your friend are at ski resort with two ski runs, a beginner's run and an expert's run. Both runs begin at the top of the ski lift and end at a finish line at the bottom of the same lift. Let $h$ be the vertical descent for both runs. The beginner's run is longer and less steep than the expert's run. You and your friend, who is a much better skier than you, are testing some experimental frictionless skis. To make things interesting, you offer a wager that if she takes the expert's run and you take the beginner's run, her speed at the finish line will not be greater than your speed at the finish line. She accepts the bet (momentarily forgetting that you are taking a physics course). The conditions are that you both start from rest at the top of the lift and both of you coast for the entire trip. Who wins the bet? (Assume airdrag is negligible.)

**PICTURE THE PROBLEM** Because you and your friend are coasting on the skis, you both can be modeled as particles. Two forces act on each of you, gravity $m\vec{g}$ and the normal force $\vec{F}_n$. Make a sketch of yourself and draw the two force vectors on the sketch (Figure 6-18a). Also include coordinate axes. The work–kinetic energy theorem, with $v_i = 0$, relates the final speed $v_f$ to the total work. (The work–kinetic energy theorem works only for particles.)

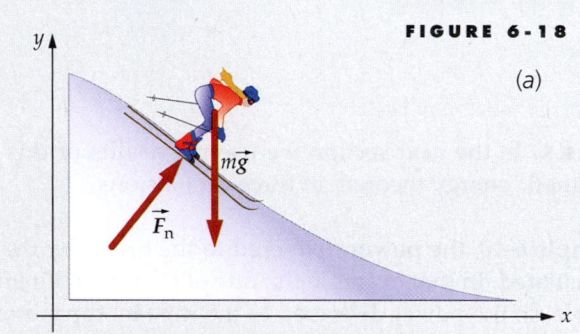

**FIGURE 6-18**

(a)

1. The final speed is related to the final kinetic energy, which in turn is related to the total work by the work–kinetic energy theorem:

$$W_{total} = \tfrac{1}{2} mv_f^2 - \tfrac{1}{2} mv_i^2$$

2. For each of you, the total work is the work done by the normal force plus the work done by the gravitational force:

$$W_{total} = W_n + W_g$$

3. The force $m\vec{g}$ on you is constant, but the force $\vec{F}_n$ is not constant. First we calculate the work done by $\vec{F}_n$. Calculate the work $dW_n$ done on you by $\vec{F}_n$ for an infinitesimal displacement $d\vec{s}$ (Figure 6-18b) at an arbitrary location along the run:

$$dW_n = \vec{F}_n \cdot d\vec{s} = F_n \cos \phi \, ds$$

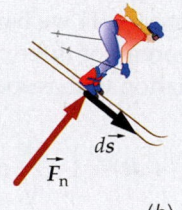

(b)

4. Find the angle $\phi$ between the directions of $\vec{F}_n$ and $d\vec{s}$. The displacement $d\vec{s}$ is parallel to the slope:

$$\phi = 90°$$

5. Calculate the work done by $\vec{F}_n$ for the entire run:

$$dW_n = F_n \cos 90° \, ds = 0, \quad \text{so}$$

$$W_n = \int dW_n = 0$$

6. The force of gravity is constant, so the work done by gravity is $W_g = m\vec{g} \cdot \vec{s}$, where $\vec{s}$ (Figure 6-19) is the net displacement from the top to the bottom of the lift:

$$W_g = m\vec{g} \cdot \vec{s} = -mg\hat{j} \cdot (\Delta x \hat{i} + \Delta y \hat{j})$$

$$= -mg \, \Delta y$$

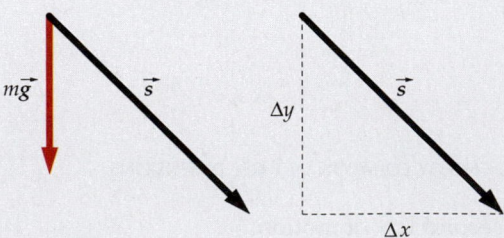

**FIGURE 6-19**

7. The skier is descending the hill, so $\Delta y$ is negative. From Figure 6-18a, we see that $\Delta y = -h$:

$$\Delta y = -h$$

8. Substituting gives:

$$W_g = mgh$$

9. Apply the work–kinetic energy theorem to find $v_f$:

$$W_n + W_g = \Delta K$$

10. The final speed depends only on $h$, which is the same for both runs. Both of you will have the same final speeds.

$$0 + mgh = \tfrac{1}{2} mv_f^2 - 0$$

so

$$v_f = \sqrt{2gh}$$

$\boxed{\text{YOU WIN!}}$ (The bet was that she would not be going faster than you.)

**REMARKS** Your friend on the steeper trail will cross the finish line in less time, but that was not the bet. What was shown here is that the work done by the gravitational force equals $mgh$. It does not depend upon the shape of the hill or upon the length of the path taken. It depends only upon the vertical drop $h$ between the starting point and the finishing point.

In Example 6-12 we found that the work done by the gravitational force is independent of the path taken. This leads us to the concept of potential energy, which is the topic of the next section.

# 6-4 Potential Energy

The total work done on a particle equals the change in its kinetic energy. But we are often interested in the work done on a *system* consisting of two or more particles.[†] Often, the work done by external forces on a *system* does not increase the total kinetic energy *of the system*, but instead is stored as **potential energy**— energy associated with the configuration of the system.

Consider lifting a barbell of mass $m$ to a height $h$. The work done by the gravitational force on the barbell is $-mgh$. The barbell starts at rest and ends at rest. Because the kinetic energy of the barbell does not change, we know the total work done on the barbell is zero. That means the work done by the force of your hands on the barbell is $+mgh$. Now consider the barbell and the earth to be a *system* of two particles. (You are not part of this system.) The external forces *on the earth–barbell system* are the gravitational attraction you exert *on the earth*, the force your feet exert *on the earth*, and the force $mg$ exerted by your hands *on the barbell* (Figure 6-20). (We can neglect the gravitational force you exert on the barbell.) The barbell moves, but the motion of the earth is negligible, so only the force exerted on the barbell by your hands does work on the system. The total work done on the earth–barbell system by all forces *external* to the system is $mgh$. This work is stored as potential energy, which is energy associated with the position of the barbell relative to the earth. That is, it is energy associated with the configuration of the earth–barbell system. This kind of energy is called gravitational potential energy.

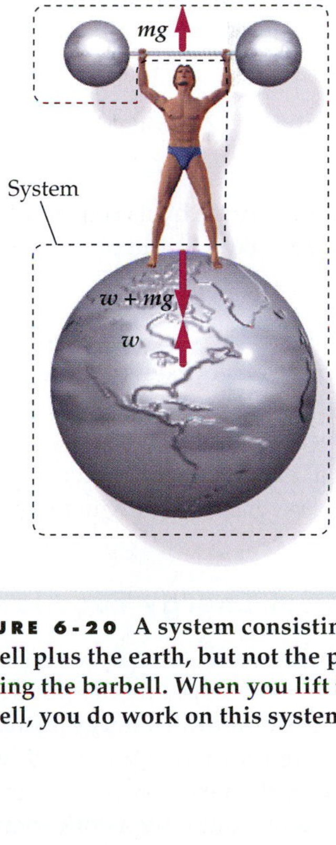

Another system that stores energy associated with its configuration is a spring. If you stretch or compress a spring, energy associated with the length of the spring is stored as potential energy. Consider the spring shown in Figure 6-21 as the system. You compress the spring, pushing it with equal and opposite forces $\vec{F}_1$ and $\vec{F}_2$. These forces sum to zero, so the net force on the spring remains

---

† Systems of particles are discussed more thoroughly in Chapter 8.

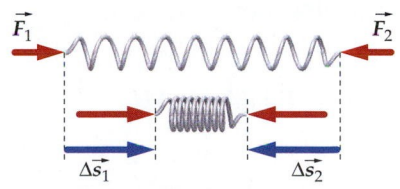

**FIGURE 6-20** A system consisting of a barbell plus the earth, but not the person holding the barbell. When you lift the barbell, you do work on this system.

**FIGURE 6-21** The spring is compressed by external forces $\vec{F}_1$ and $\vec{F}_2$. Both forces do positive work on the spring as they compress it, so the elastic potential energy of the spring increases as it is compressed.

zero. Thus, there is no change in the kinetic energy of the spring. The work you do on this system is stored not as kinetic energy, but as elastic potential energy. The configuration of this system has been changed, as evidenced by the change in the length of the spring. The total work done on the spring is positive because both $\vec{F}_1$ and $\vec{F}_2$ do positive work. (The work done by $\vec{F}_1$ is positive because both $\vec{F}_1$ and the displacement $\Delta\vec{s}_1$ of its point of application are in the same direction. The same can be said for $\vec{F}_2$ and $\Delta\vec{s}_2$.)

## Conservative Forces

When you ride a ski lift to the top of a hill of height $h$, the work done by gravity on you is $-mgh$ and the work done by the lift on you is $+mgh$. When you ski down the hill to the bottom, the work done by gravity is $+mgh$ independent of the shape of the hill (as we saw in Example 6-12). The total work done by gravity on you during the round trip up and down the hill is zero, independent of the path you take. The force of gravity, exerted by the earth on you, is a **conservative force.**

> A force is conservative if the total work it does on a particle is zero when the particle moves around *any* closed path, returning to its initial position.

DEFINITION—CONSERVATIVE FORCE

From Figure 6-22 we see that this definition implies that:

> The work done by a conservative force on a particle is independent of the path taken as the particle moves from one point to another.

ALTERNATIVE DEFINITION—CONSERVATIVE FORCE

Now consider yourself and the earth to be a *two-particle system*. (The ski lift is not part of this system.) When a ski lift raises you to the top of the hill, it does work $mgh$ on the you–earth system. This work is stored as the gravitational potential energy of the system. When you ski down the hill, this potential energy is converted to the kinetic energy of your motion.

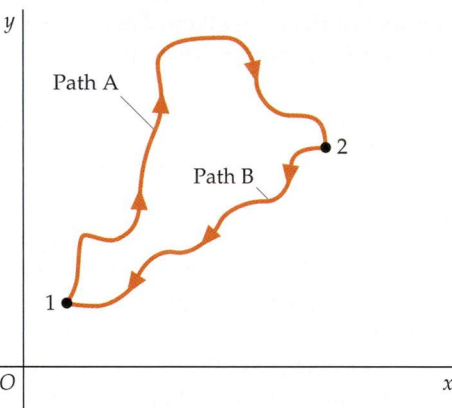

## Potential-Energy Functions

Because the work done by a conservative force on a particle does not depend on the path, it can depend only on the endpoints 1 and 2. We can use this property to define the **potential-energy function** $U$ that is associated with a conservative force. Note that when you ski down the hill, the work done by gravity *decreases* the potential energy of the system. We define the potential-energy function such that the work done by a conservative force equals the decrease in the potential-energy function:

$$W = \int_1^2 \vec{F} \cdot d\vec{s} = -\Delta U$$

or

$$\Delta U = U_2 - U_1 = -\int_1^2 \vec{F} \cdot d\vec{s} \qquad 6\text{-}20a$$

**FIGURE 6-22** Two paths in space connecting the points 1 and 2. If the work done by a conservative force along path A from 1 to 2 is $W$, the work done on the return trip along path B must be $-W$ because the roundtrip work is zero. When traversing path B from 1 to 2, the force is the same at each point, but the displacement is opposite to that when going from 2 to 1. Then the work done along path B from 1 to 2 must also be $W$. It follows that the work done as a particle goes from point 1 to 2 is the same along any path connecting the two points.

DEFINITION—POTENTIAL-ENERGY FUNCTION

For an infinitesimal displacement, we have

$$dU = -\vec{F} \cdot d\vec{s} \qquad\qquad 6\text{-}20b$$

We can calculate the potential-energy function associated with the gravitational force near the surface of the earth from Equation 6-20b. For the force $\vec{F} = -mg\hat{j}$, we have

$$dU = -\vec{F} \cdot d\vec{s} = -(-mg\hat{j}) \cdot (dx\hat{i} + dy\hat{j} + dz\hat{k}) = +mg\,dy$$

Integrating, we obtain

$$U = \int mg\,dy = mgy + U_0$$

$$U = U_0 + mgy \qquad\qquad 6\text{-}21$$

GRAVITATIONAL POTENTIAL ENERGY NEAR THE EARTH'S SURFACE

where $U_0$, the arbitrary constant of integration, is the value of the potential energy at $y = 0$. Because only a change in the potential energy is defined, the actual value of $U$ is not important. We are free to choose $U$ to be zero at any convenient reference point. For example, if the gravitational potential energy of the earth–skier system is chosen to be zero when the skier is at the bottom of the hill, its value when the skier is at a height $h$ above that level is $mgh$. Or we could choose the potential energy to be zero when the skier is at point $P$ half way down the ski slope, in which case its value at any other point would be $mgy$, where $y$ is the height of the skier above point $p$. On the lower half of the slope the potential energy would then be negative.

**EXERCISE** A 55-kg window washer stands on a platform 8 m above the ground. What is the potential energy $U$ of the window-washer–earth system if (a) $U$ is chosen to be zero on the ground, (b) $U$ is chosen to be zero 4 m above the ground, and (c) $U$ is chosen to be zero 10 m above the ground? (*Answer* (a) 4.32 kJ (b) 2.16 kJ (c) −1.08 kJ)

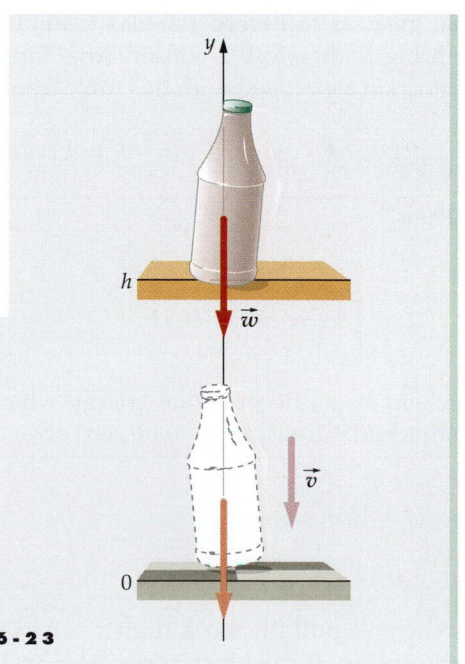

*A FALLING BOTTLE*                     **EXAMPLE 6-13**

**A bottle of mass 0.350 kg falls from rest from a shelf that is 1.75 m above the floor. Find the potential energy of the bottle–earth system when the bottle is on the shelf and just before impact with the floor. Find the kinetic energy of the bottle just before impact.**

**PICTURE THE PROBLEM** The work done by the earth on the bottle as it falls equals the negative of the change in the potential energy of the bottle–earth system. Knowing the work, we can use the work–kinetic energy theorem to find the kinetic energy.

1. Make a sketch showing the bottle on the shelf and again when it is about to impact the floor (Figure 6-23). Choose the potential energy of the bottle–earth system to be zero when the bottle is on the floor, and place a $y$ axis on the sketch with the origin at the floor:

**FIGURE 6-23**

2. The only force doing work on the falling bottle is the force of gravity, so $W_{total} = W_g$. Apply the work–kinetic energy theorem to the falling bottle:

$$W_{total} = W_g = \Delta K$$

3. The force exerted by the earth on the falling bottle is internal to the bottle–earth system. It is also a conservative force, so the work done by it equals the change in the potential energy of the system:

$$W_g = -\Delta U = -(U_f - U_i) = -(mgy_f - mgy_i)$$
$$= mg(y_i - y_f) = mg(h - 0) = mgh$$

4. Substitute the step 3 result into the step 2 result and solve for the final kinetic energy. The original kinetic energy is zero:

$$mgh = \Delta K$$
$$mgh = K_f - K_i$$
$$K_f = K_i + mgh$$
$$= 0 + (0.350 \text{ kg})(9.81 \text{ N/kg})(1.75 \text{ m})$$
$$= \boxed{6.01 \text{ J}}$$

**REMARKS** In this example, the potential energy lost by the bottle–earth system is converted entirely to kinetic energy of the bottle as it falls. Note in step 4 that we have used the definition 1 J = 1 N·m.

Potential energy is associated with the configuration of a *system of particles,* but we sometimes have systems such as the bottle–earth system, in which only one particle moves (the earth's motion is negligible). For brevity, then, we sometimes refer to the potential energy of the bottle–earth system as simply the potential energy of the bottle.

**Potential Energy of a Spring** Another example of a conservative force is that of a stretched (or compressed) spring. Suppose we pull a block attached to a spring from a position $x = 0$ (equilibrium) to $x_1$ (Figure 6-24). The spring does negative work because the force it exerts on the block is opposite to the block's displacement. If we then release the block, the force of the spring does positive work on the block as it accelerates toward its initial position. The total work done by the spring on the block as it moves from its original position to $x = x_1$, and then back, is zero. This result is independent of the size of $x_1$ (as long as the stretching is not so great as to exceed the elastic limit of the spring). The force exerted by the spring is therefore a conservative force. We can calculate the potential-energy function associated with this force from Equation 6-20*b*:

$$dU = -\vec{F} \cdot d\vec{s} = -F_x \, dx = -(-kx) \, dx = +kx \, dx$$

Then

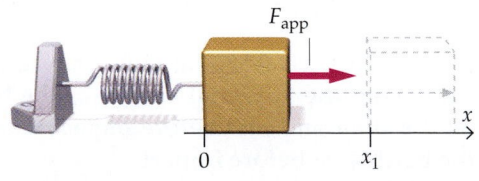

$$U = \int kx \, dx = \tfrac{1}{2}kx^2 + U_0$$

**FIGURE 6-24** The applied force $F_{app}$ pulls the block to the right, stretching the spring by $x_1$.

where $U_0$ is the potential energy when $x = 0$, that is, when the spring is unstretched. Choosing $U_0$ to be zero gives

$$U = \tfrac{1}{2}kx^2 \qquad\qquad 6\text{-}22$$

POTENTIAL ENERGY OF A SPRING

When we pull the block from $x = 0$ to $x = x_1$, we must exert an applied force to the spring. If the block starts from rest at $x = 0$ and ends at rest at $x = x_1$, the change in its kinetic energy is zero. The work–energy theorem then implies that the total work done on the block is zero. That is, $W_{app} + W_{spring} = 0$, or

$$W_{app} = -W_{spring} = \Delta U_{spring} = \tfrac{1}{2}kx_1^2 - 0 = \tfrac{1}{2}kx_1^2$$

This work is stored as potential energy in the spring–block system.

---

*POTENTIAL ENERGY OF A BASKETBALL PLAYER*                **EXAMPLE 6-14**

A system consists of a basketball player, the rim of a basket-ball hoop, and the earth. Assume that the potential energy of this system is zero when the player is standing on the floor and the rim is horizontal. Find the total potential energy of this system when the player is hanging on the rim (as in Figure 6-25). Also assume that the player can be described as a point mass of 110 kg at 0.8 m above the floor when standing and at 1.3 m above the floor when hanging. The force constant of the rim is 7.2 kN/m and the front of the rim is displaced a distance $s = 15$ cm.

**FIGURE 6-25**

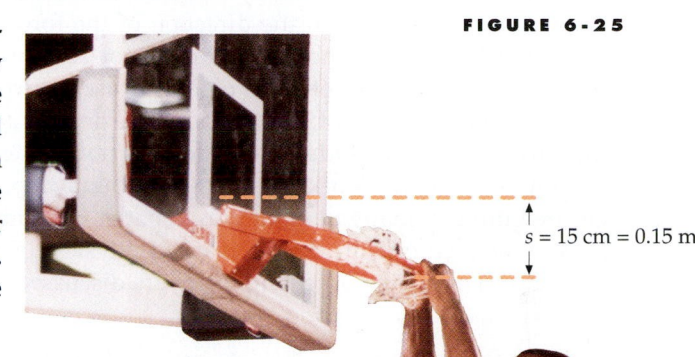

$s = 15\ \text{cm} = 0.15\ \text{m}$

**PICTURE THE PROBLEM** In the player's change in position from standing on the floor to hanging on the rim, the total change in potential energy consists of gravitational potential energy, $U_g = mgy$, and energy stored in the deformed rim, whose potential energy can be measured just as if it were a spring: $U_s = \tfrac{1}{2}ks^2$. Choose $y = 0$ at 0.8 m above the floor for the gravitational potential-energy reference point.

The total potential en-ergy is the sum of gravita-tional potential energy and the elastic potential energy of the rim (see Figure 6-26):

$$U = U_g + U_s = mgy + \tfrac{1}{2}ks^2$$
$$= (110\ \text{kg})(9.81\ \text{N/kg})(0.5\ \text{m}) + \tfrac{1}{2}(7.2\ \text{kN/m})(0.15\ \text{m})^2$$
$$= 540\ \text{J} + 81\ \text{J} = \boxed{621\ \text{J}}$$

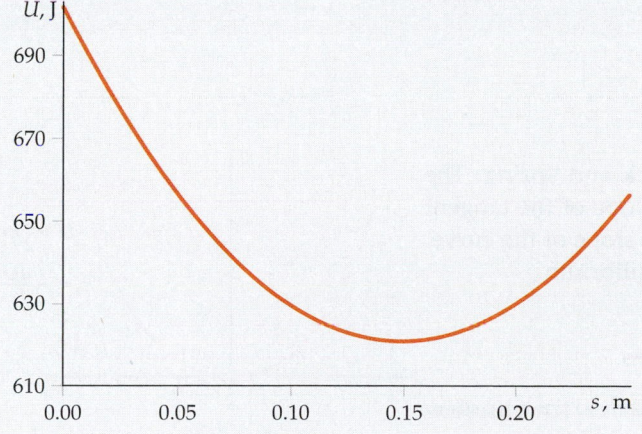

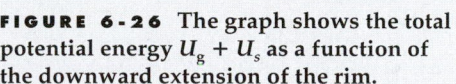

**FIGURE 6-26** The graph shows the total potential energy $U_g + U_s$ as a function of the downward extension of the rim.

**REMARKS** Most of the potential energy is gravitational in this case, because of the choice of the potential-energy reference point.

**EXERCISE** A 3-kg block is hung vertically from a spring with a force constant of 600 N/m. (*a*) By how much is the spring stretched? (*b*) How much potential energy is stored in the spring? (*Answer* (*a*) 4.9 cm (*b*) 0.72 J)

## Nonconservative Forces

Not all forces are conservative. Suppose that you push a box across a table along a straight line from point A to point B and back, so that the box ends up at its original position. Friction opposes the block's motion, so the force you push with is in the direction of motion and does positive work on both legs of the round trip. The total work done by the push does not equal zero. Thus, the push is an example of a **nonconservative force** and no potential-energy function can be defined for it.

Sometimes we can show that a given force is not conservative by computing the work done by the force around some chosen closed curve and showing that it is not zero. Consider the force $\vec{F} = F_0\hat{\phi}$, where $\hat{\phi}$ is a unit vector directed tangent to a circle of radius $r$. The work done by this force as we move around a circle of radius $r$ is $+F_0 2\pi r$ if we move in the direction of the force (and $-F_0 2\pi r$ if we move opposite to the force). Since this work is not zero, we conclude that the force is not conservative. This method is of limited use in investigating whether a given force is conservative or not. If the work done around *any* particular closed path is not zero, we may conclude that the force is not conservative. However, if the force is conservative, the work must be zero around *all* possible closed paths. Since there are infinitely many possible closed paths, it is impossible to calculate the work done for each one. In more advanced physics courses, more sophisticated mathematical methods for testing forces are discussed.

## Potential Energy and Equilibrium

For a general conservative force in one dimension, $\vec{F} = F_x\hat{i}$, Equation 6-20b is

$$dU = -\vec{F} \cdot d\vec{s} = -F_x\, dx$$

The force is therefore the negative derivative of the potential-energy function:

$$F_x = -\frac{dU}{dx} \qquad\qquad\qquad 6\text{-}23$$

We can illustrate this general relation for a block–spring system by differentiating the function $U = \frac{1}{2}kx^2$. We obtain

$$F_x = -\frac{dU}{dx} = -\frac{d}{dx}\left(\frac{1}{2}kx^2\right) = -kx$$

Figure 6-27 shows a plot of $U = \frac{1}{2}kx^2$ versus $x$ for a block and spring. The derivative of this function is represented graphically as the slope of the tangent line to the curve. The force is thus equal to the negative of the slope of the curve. At $x = 0$, the force $F_x = -dU/dx$ is zero and the block is in equilibrium.

A particle is in equilibrium if the net force acting on it is zero.

CONDITION FOR EQUILIBRIUM

When $x$ is positive in Figure 6-27, the slope is positive and the force $F_x$ is negative. When $x$ is negative, the slope is negative and the force $F_x$ is positive. In either case, the force is in the direction that will accelerate the block toward lower potential energy. If the block is displaced slightly from $x = 0$, the force is directed back toward $x = 0$. The equilibrium at $x = 0$ is thus **stable equilibrium** because a small displacement results in a restoring force that accelerates the particle back toward its equilibrium position.

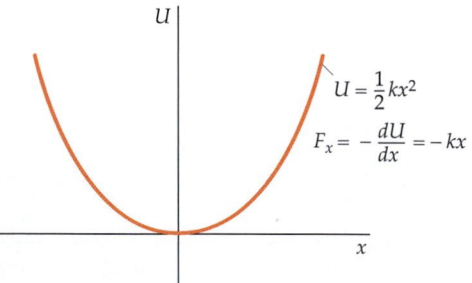

**FIGURE 6-27** Plot of the potential-energy function $U$ versus $x$ for an object on a spring. A minimum in a potential-energy curve is a point of stable equilibrium. Displacement in either direction results in a force directed toward the equilibrium position.

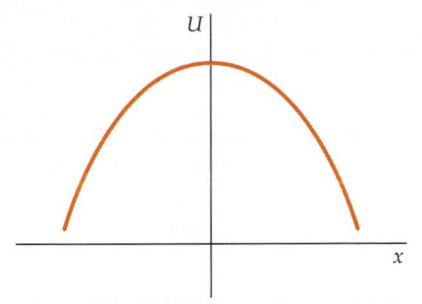

In stable equilibrium, a small displacement in any direction results in a restoring force that accelerates the particle back toward its equilibrium position.

Figure 6-28 shows a potential-energy curve with a maximum rather than a minimum at the equilibrium point $x = 0$. Such a curve could represent the potential energy of a skier at the top of a hill. For this curve, when $x$ is positive, the slope is negative and the force $F_x$ is positive, and when $x$ is negative, the slope is positive and the force $F_x$ is negative. Again, the force is in the direction that will accelerate the particle toward lower potential energy, but this time the force is away from the equilibrium position. The maximum at $x = 0$ in Figure 6-28 is a point of **unstable equilibrium** because a small displacement results in a force that accelerates the particle away from its equilibrium position.

**FIGURE 6-28** A particle at $x = 0$ on this potential-energy curve will be in unstable equilibrium because a displacement in either direction results in a force directed away from the equilibrium position.

In unstable equilibrium, a small displacement results in a force that accelerates the particle away from its equilibrium position.

Figure 6-29 shows a potential-energy curve that is flat in the region near $x = 0$. No force acts on a particle at $x = 0$, and hence the particle is at equilibrium; furthermore, there will be no resulting force if the particle is displaced slightly in either direction. This is an example of **neutral equilibrium.**

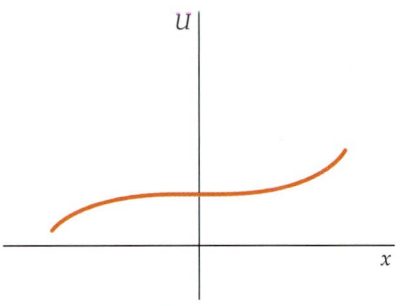

In neutral equilibrium, a small displacement results in zero force and the particle remains in equilibrium.

**FIGURE 6-29** Neutral equilibrium. The force $F_x = -dU/dx$ is zero at $x = 0$ and at neighboring points, so displacement away from $x = 0$ results in no force, and the system remains in equilibrium.

*FORCE AND THE POTENTIAL-ENERGY FUNCTION*    **EXAMPLE 6-15**

**Try It Yourself**

In the region $-a < x < a$ the force on a particle is represented by the potential energy function

$$U = -b\left(\frac{1}{a + x} + \frac{1}{a - x}\right)$$

where $a$ and $b$ are positive constants. (*a*) Find the force $F_x$ in the region $-a < x < a$. (*b*) At what value of $x$ is the force zero? (*c*) At the location where the force equals zero, is the equilibrium stable or unstable?

**PICTURE THE PROBLEM** The force is the negative of the derivative of the potential-energy function. The equilibrium is stable where the potential-energy function is a minimum and it is unstable where the potential-energy function is a maximum.

**Cover the column to the right and try these on your own before looking at the answers.**

| Steps | Answers |
|---|---|
| (*a*) Compute $F_x = -dU/dx$. | $F_x = -\dfrac{d}{dx}\left[-b\left(\dfrac{1}{a + x} + \dfrac{1}{a - x}\right)\right]$ |
| | $= \boxed{-b\left(\dfrac{1}{(x + a)^2} - \dfrac{1}{(x - a)^2}\right)}$ |

(b) Set $F_x$ equal to zero and solve for $x$.

$$F_x = 0 \quad \text{at} \quad \boxed{x = 0}$$

(c) Compute $d^2U/dx^2$. If it is positive at the equilibrium position, then $U$ is a minimum and the equilibrium is stable. If it is negative, then $U$ is a maximum and the equilibrium is unstable.

At $x = 0$, $\dfrac{d^2U}{dx^2} = \dfrac{-4b}{a^3}$

Thus, $\boxed{\text{unstable}}$ equilibrium.

**REMARKS** This potential-energy function is that for a particle under the influence of the gravitational forces exerted by two identical fixed masses, one at $x = -a$ and the other at $x = +a$. The particle is located on the line joining the masses. Midway between the two masses the net force on the particle is zero. Otherwise, it is toward the closest mass.

# SUMMARY

1. Work, kinetic energy, potential energy, and power are important derived dynamic quantities.

2. The work–kinetic energy theorem is an important relation derived from Newton's laws applied to a particle. (In this context, a particle is a perfectly rigid object that moves without rotating.)

3. The dot product of vectors is a mathematical definition that is useful throughout physics.

| Topic | Relevant Equations and Remarks | |
|---|---|---|
| 1. **Work (definition)** | $W = \displaystyle\int_1^2 \vec{F} \cdot d\vec{s}$ | 6-15 |
| Constant force | $W = \vec{F} \cdot \vec{s}$ | |
| In one dimension | | |
| Constant force | $W = F_x \, \Delta x = F \cos\theta \, \Delta x$ | 6-1 |
| Variable force | $W = \displaystyle\int_{x_1}^{x_2} F_x \, dx = \text{area under the } F_x\text{-versus-}x \text{ curve}$ | 6-9 |
| 2. **Kinetic Energy (definition)** | $K = \frac{1}{2}mv^2$ | 6-6 |
| 3. **Work–Kinetic Energy Theorem** | $W_{\text{total}} = \Delta K = \frac{1}{2}mv_f^2 - \frac{1}{2}mv_i^2$ | 6-7 |
| 4. **Dot Product (definition)** | $\vec{A} \cdot \vec{B} = AB \cos\phi$ | 6-10 |
| In terms of components | $\vec{A} \cdot \vec{B} = A_x B_x + A_y B_y + A_z B_z$ | 6-11 |
| Component of vector | $\vec{A} \cdot \hat{i} = A_x$ | 6-12 |
| Derivative | $\dfrac{d}{dt}(\vec{A} \cdot \vec{B}) = \dfrac{d\vec{A}}{dt} \cdot \vec{B} + \vec{A} \cdot \dfrac{d\vec{B}}{dt}$ | 6-13 |
| 5. **Power** | $P = \dfrac{dW}{dt} = \vec{F} \cdot \vec{v}$ | 6-17 |

**6. Conservative Force**

A force is conservative if the total work it does on a particle is zero when the particle moves along any path that returns it to its initial position. Alternatively, the work done by a conservative force on a particle is independent of the path taken by the particle as it moves from one point to another.

**7. Potential Energy**

The potential energy of a system is the energy associated with the configuration of the system. The change in the potential energy of a system is defined as the negative of the work done by all internal conservative forces acting on the system.

| | | |
|---|---|---|
| Definition | $\Delta U = U_2 - U_1 = -\int_1^2 \vec{F} \cdot d\vec{s}$ | 6-20 |
| | $dU = -\vec{F} \cdot d\vec{s}$ | 6-20 |
| Gravitational | $U = U_0 + mgy$ | 6-21 |
| Elastic (spring) | $U = \frac{1}{2}kx^2$ | 6-22 |
| Conservative force | $F_x = -\dfrac{dU}{dx}$ | 6-23 |

| | |
|---|---|
| Potential-energy curve | At a minimum on the curve of the potential-energy function versus the displacement, the force is zero and the system is in stable equilibrium. At a maximum, the force is zero and the system is in unstable equilibrium. A conservative force always tends to accelerate a particle toward a position of lower potential energy. |

# PROBLEMS

• Single-concept, single-step, relatively easy

•• Intermediate-level, may require synthesis of concepts

••• Challenging

**SSM** Solution is in the *Student Solutions Manual*

**iSOLVE** Problems available on iSOLVE online homework service

**iSOLVE✓** These "Checkpoint" online homework service problems ask students additional questions about their confidence level, and how they arrived at their answer

In a few problems, you are given more data than you actually need; in a few other problems, you are required to supply data from your general knowledge, outside sources, or informed estimates.

**Take $g$ = 9.81 N/kg = 9.81 m/s² and neglect friction in all problems unless otherwise stated.**

## Conceptual Problems

**1** • **SSM** True or false:

(a) Only the net force acting on an object can do work.
(b) No work is done on a particle that remains at rest.
(c) A force that is always perpendicular to the velocity of a particle never does work on the particle.

**2** • You are to move a heavy box from the top of one table to the top of another table of the same height on the other side of the room. What is the minimum amount of work you must do on the box to accomplish the move? Explain.

**3** • True or False: A person on a Ferris wheel is moving in a circle at constant speed. Thus, no force is doing work on the person.

**4** • **SSM** By what factor does the kinetic energy of a car change when its speed is doubled?

**5** • A particle moves in a circle at constant speed. Only one of the forces acting on the particle is in the centripetal direction. Does the net force on the particle do work on it? Explain.

**6** • An object initially has kinetic energy $K$. The object then moves in the opposite direction with three times its initial speed. What is the kinetic energy now? (a) $K$. (b) $3K$. (c) $-3K$. (d) $9K$. (e) $-9K$.

**7** • **SSM** How does the work required to stretch a spring 2 cm from its natural length compare with that required to stretch it 1 cm from its natural length?

**8** • Suppose there is a net force acting on a particle but it does no work. Can the particle be moving in a straight line?

**9** • The dimension of power is (a) $[M][L]^2[T]^2$, (b) $[M][L]^2/[T]$, (c) $[M][L]^2/[T]^2$, (d) $[M][L]^2/[T]^3$.

**10** • Two knowledge seekers decide to ascend a mountain. Sal chooses a short, steep trail, while Joe, who weighs the same as Sal, goes up via a long, gently sloped trail. At the top, they get into an argument about who gained more potential energy. Which of the following is true:

(a) Sal gains more gravitational potential energy than Joe.
(b) Sal gains less gravitational potential energy than Joe.
(c) Sal gains the same gravitational potential energy as Joe.
(d) To compare energies, we must know the height of the mountain.
(e) To compare the energies, we must know the length of the two trails.

**11** • True or false:

(a) Only conservative forces can do work.
(b) If only conservative forces act, the kinetic energy of a particle does not change.
(c) The work done by a conservative force equals the decrease in the potential energy associated with that force.

**12** •• [SSM] Figure 6-30 shows the plot of a potential-energy function $U$ versus $x$. (a) At each point indicated, state whether the force $F_x$ is positive, negative, or zero. (b) At which point does the force have the greatest magnitude? (c) Identify any equilibrium points, and state whether the equilibrium is stable, unstable, or neutral.

**FIGURE 6-30** Problem 12

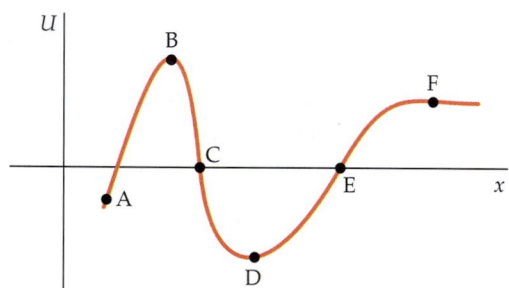

**13** • True or false:

(a) The gravitational force cannot do work because it acts at a distance.
(b) Work is the area under the force-versus-time curve.

**14** • Negative work means (a) the kinetic energy of the object increases, (b) the applied force is variable, (c) the angle between applied force is perpendicular to the displacement, (d) the applied force and the displacement is greater than 90°, (e) nothing; there is no such thing as negative work.

## Estimation and Approximation

**15** •• [SSM] A tightrope walker whose mass is 50 kg walks across a tightrope held between two supports 10 m apart; the tension in the rope is 5000 N. The height of the rope is 10 m above the ground. Estimate: (a) the sag in the tightrope when she stands in the exact center and (b) the change in her gravitational potential energy from just before stepping onto the tightrope to when she stands at its dead center.

**16** • Estimate (a) the change in your potential energy on taking an elevator from the ground floor to the top of the Empire State building, (b) the average force acting on you by the elevator to bring you to the top, and (c) the average power due to that force. The building is 102 stories high.

**17** • The nearest stars, apart from the sun, are at light-years away from the earth. (One light-year is the distance that light travels during a year: $9.47 \times 10^{15}$ m.) If we ever want to send spacecraft to investigate the stars, they will have to have velocities that are an appreciable fraction of the speed of light. Calculate the kinetic energy of a 10,000-kg spacecraft traveling at a speed of 10 percent of the speed of light, and compare that to the amount of energy that the United States uses in a year (about $5 \times 10^{20}$ J). NOTE At velocities approaching the speed of light, the theory of relativity tells us that the formula $\frac{1}{2}mv^2$ for kinetic energy is not correct. However, the correction is only about 1 percent for this particular speed ($0.1c$).

**18** •• [SSM] The mass of the Space Shuttle orbiter is about $8 \times 10^4$ kg and the period of its orbit is 90 min. Estimate (a) the kinetic energy of the orbiter and (b) the change in its potential energy between resting on the surface of the earth and in its orbit, 200 mi above the surface of the earth. (c) Why is the change in potential energy much smaller than the shuttle's kinetic energy? Shouldn't they be equal?

**19** • Ten inches of snow have fallen during the night, and you must shovel out your 50-ft-long driveway (Figure 6-31). Estimate how much work you do on the snow by completing this task. Make a plausible guess of any value(s) needed (the width of the driveway, for example), and state the basis for each guess.

**FIGURE 6-31** Problem 19

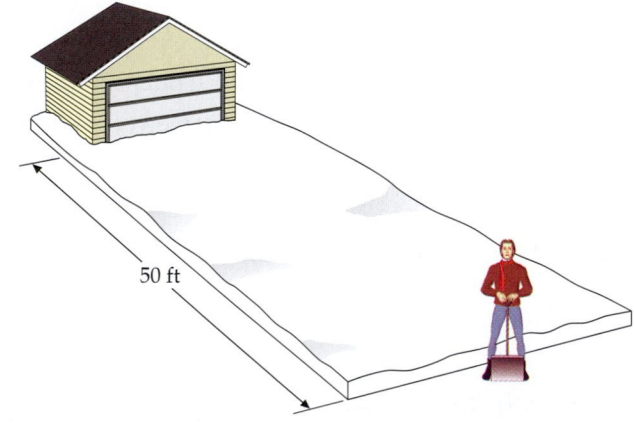

50 ft

## Work and Kinetic Energy

**20** • [SSM] A 15-g bullet has a speed of 1.2 km/s. (a) What is its kinetic energy in joules? (b) What is its kinetic energy if its speed is halved? (c) What is its kinetic energy if its speed is doubled?

**21** • Find the kinetic energy in joules of (a) a 0.145-kg baseball moving with a speed of 45 m/s and (b) a 60-kg jogger running at a steady pace of 9 min/mi.

**22** • A 6-kg box is raised a distance of 3 m from rest by a vertical applied force of 80 N. Find (*a*) the work done by the force, (*b*) the work done by gravity, and (*c*) the final kinetic energy of the box.

**23** • A constant force of 80 N acts on a box of mass 5.0 kg that is moving in the direction of the force with a speed of 20 m/s. A few seconds later the box is moving with a speed of 68 m/s. Determine the work done by this force.

**24** •• **SSM** You run a race with a friend. At first you each have the same kinetic energy, but you find that she is beating you. When you increase your speed by 25 percent, you are running at the same speed she is. If your mass is 85 kg, what is her mass?

## Work Done by a Variable Force

**25** •• A 3-kg particle is moving along the *x* axis with a velocity of 2 m/s as it passes through $x = 0$. It is subjected to a single force $F_x$ that varies with position as shown in Figure 6-32. (*a*) What is the kinetic energy of the particle as it passes through $x = 0$? (*b*) How much work is done by the force as the particle moves from $x = 0$ to $x = 4$ m? (*c*) What is the speed of the particle when it is at $x = 4$ m?

**FIGURE 6-32** Problem 25

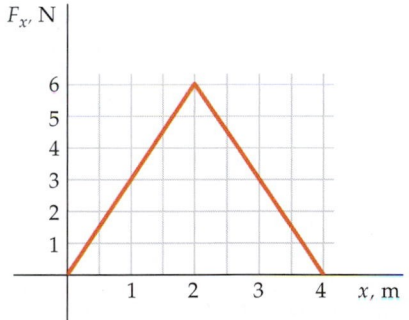

**26** •• **SSM** **iSOLVE** ✓ A force $F_x$ acts on a particle. The force is related to the position of the particle by the formula $F_x = Cx^3$, where *C* is a constant. Find the work done by this force on the particle as the particle moves from $x = 1.5$ m to $x = 3$ m.

**27** •• **iSOLVE** Lou's latest invention, aimed at urban dog owners, is the X-R-Leash. It is made of a rubber-like material that exerts a force $F_x = -kx - ax^2$ when it is stretched a distance *x*, where *k* and *a* are constants. The ad claims, "You'll never go back to your old dog leash after you've had the thrill of an X-R-Leash experience. And you'll see a new look of respect in the eyes of your proud pooch." Find the work done on a dog by the leash if the person remains stationary, and the dog bounds off, stretching the X-R-Leash from $x = 0$ to $x = x_1$.

**28** •• A 3-kg object moving along the *x* axis has a velocity of 2.40 m/s as it passes through the origin. It is acted on by a single force $F_x$ that varies with *x* as shown in Figure 6-33. (*a*) Find the work done by the force from $x = 0$ to $x = 2$ m. (*b*) What is the kinetic energy of the object at $x = 2$ m? (*c*) What is the speed of the object at $x = 2$ m? (*d*) Find the work done on the object from $x = 0$ to $x = 4$ m. (*e*) What is the speed of the object at $x = 4$ m?

**FIGURE 6-33** Problem 28

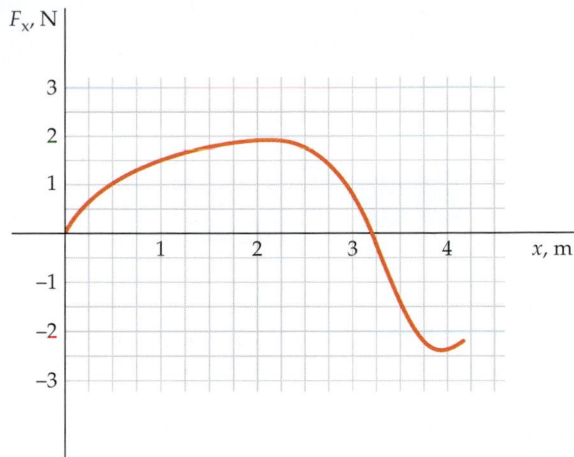

**29** •• **SSM** Near Margaret's cabin is a 20-m water tower that attracts many birds during the summer months. During a hot spell, the tower went dry, and Margaret had to have her water hauled in. She was lonesome without the birds visiting, so she decided to carry some water up the tower to attract them back. Her bucket has a mass of 10 kg and holds 30 kg of water when it is full. However, the bucket has a hole, and as Margaret climbed at a constant speed, water leaked out at a constant rate. When she reached the top, only 10 kg of water remained for the birdbath. (*a*) Write an expression for the mass of the bucket plus water as a function of the height (*y*) climbed. (*b*) Find the work done by Margaret on the bucket.

## Work, Energy, and Simple Machines

**30** • **iSOLVE** ✓ A 6-kg block slides down a frictionless incline making an angle of 60° with the horizontal. (*a*) List all the forces acting on the block, and find the work done by each force when the block slides 2 m (measured along the incline). (*b*) What is the total work done on the block? (*c*) What is the speed of the block after it has slid 1.5 m if it starts from rest? (*d*) What is its speed after 1.5 m if it starts with an initial speed of 2 m/s?

**31** • A 2-kg object attached to a horizontal string moves with a speed of 2.5 m/s in a circle of radius 3 m on a frictionless horizontal surface. (*a*) Find the tension in the string. (*b*) List the forces acting on the object, and find the work done by each force during one revolution.

**32** • **SSM** *Simple machines* are used for reducing the amount of force that must be supplied to perform a task such as lifting a heavy weight. Such machines include the screw, block-and-tackle systems, and levers, but the simplest of the simple machines is the inclined plane. In Figure 6-34 you are raising the heavy box to the height of the truck bed by pushing it up an inclined plane (a ramp). (*a*) We define the *mechanical advantage M* of the inclined plane as the ratio of the force it would take to lift the block into the truck directly from the ground (at constant speed) to the force it takes to push it up the ramp (at constant speed). If the plane is frictionless, show that $M = 1/\sin \theta = L/H$, where *H* is the height of the truck bed and *L* is the length of the ramp. (*b*) Show that the work you do by moving the block into the truck is the same whether you lift it directly into the truck or push it up the frictionless ramp.

**FIGURE 6-34** Problem 32

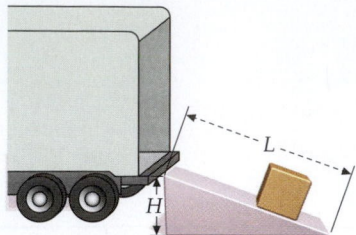

**33** •• The screw is a form of the inclined plane. Figure 6-35 shows a type of car jack used for changing flat tires. The screw on the jack has a pitch (distance between threads) of $p$ and a handle of length $R$. As the handle is turned through a full circle, the jack's height increases, raising the weight $w$ by $p$. Assuming no friction, the work done turning the jack handle is the same as the increase in the potential energy of whatever the jack is lifting. Show that the jack has a mechanical advantage (defined in Problem 32) of $2\pi R/p$.

**FIGURE 6-35** Problem 33

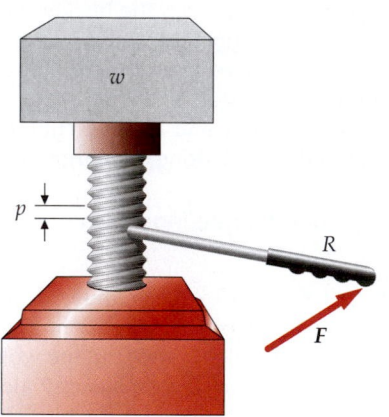

**34** • Figure 6-36 shows two pulleys arranged to help lift a heavy load: a rope runs around two massless, frictionless pulleys and the weight $\vec{w}$ hangs from one pulley. You exert a force of magnitude $F$ on the free end of the cord. (a) If the weight is to move up a distance $h$, through what distance must the force move? (b) How much work is done by the rope on the weight? (c) How much work do you do? (d) What is the mechanical advantage (defined in Problem 32) of this system?

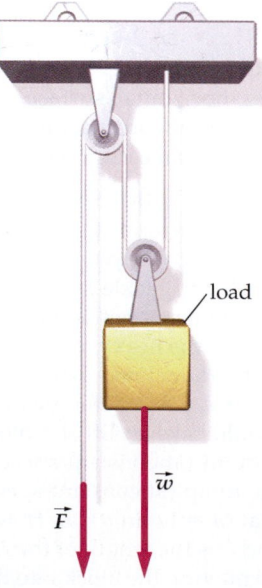

**FIGURE 6-36** Problem 34

## Dot Products

**35** • **SSM** What is the angle between the vectors $\vec{A}$ and $\vec{B}$ if $\vec{A} \cdot \vec{B} = -AB$?

**36** • Two vectors $\vec{A}$ and $\vec{B}$ each have magnitudes of 6 m and make an angle of 60° with each other. Find $\vec{A} \cdot \vec{B}$.

**37** • Find $\vec{A} \cdot \vec{B}$ for the following vectors: (a) $\vec{A} = 3\hat{i} - 6\hat{j}$, $\vec{B} = -4\hat{i} + 2\hat{j}$; (b) $\vec{A} = 5\hat{i} + 5\hat{j}$, $\vec{B} = 2\hat{i} - 4\hat{j}$; and (c) $\vec{A} = 6\hat{i} + 4\hat{j}$, $\vec{B} = 4\hat{i} - 6\hat{j}$.

**38** • Find the angles between the vectors $\vec{A}$ and $\vec{B}$ in Problem 37.

**39** • **ISOLVE**✓ A 2-kg object is given a displacement $\Delta\vec{s} = 3 \text{ m}\hat{i} + 3 \text{ m}\hat{j} - 2 \text{ m}\hat{k}$ along a straight line. During the displacement, a constant force $\vec{F} = 2 \text{ N}\hat{i} - 1 \text{ N}\hat{j} + 1 \text{ N}\hat{k}$ acts on the object. (a) Find the work done by $\vec{F}$ for this displacement. (b) Find the component of $\vec{F}$ in the direction of the displacement.

**40** •• (a) Find the unit vector that is parallel to vector $\vec{A} = A_x\hat{i} + A_y\hat{j} + A_z\hat{k}$. (b) Find the component of the vector $\vec{A} = 2\hat{i} - \hat{j} - \hat{k}$ in the direction of the vector $\vec{B} = 3\hat{i} + 4\hat{j}$.

**41** •• **SSM** Given two vectors $\vec{A}$ and $\vec{B}$, show that if $|\vec{A} + \vec{B}| = |\vec{A} - \vec{B}|$, then $\vec{A} \perp \vec{B}$.

**42** •• $\hat{A}$ and $\hat{B}$ are two unit vectors in the $xy$ plane. They make angles of $\theta_1$ and $\theta_2$ with the positive $x$-axis, respectively. (a) Find the $x$ and $y$ components of the two vectors. (b) By considering the dot product of $\hat{A}$ and $\hat{B}$, show that $\cos(\theta_1 - \theta_2) = \cos\theta_1 \cos\theta_2 + \sin\theta_1 \sin\theta_2$.

**43** • If $\vec{A} \cdot \vec{B} = \vec{A} \cdot \vec{C}$, must $\vec{B} = \vec{C}$? If no, give a counterexample, if yes, explain why.

**44** •• (a) Let $\vec{A}$ be a constant vector in the $xy$ plane with its tail at the origin. Let $\vec{r} = x\hat{i} + y\hat{j}$ be a vector in the $xy$ plane that satisfies the relation $\vec{A} \cdot \vec{r} = 1$. Show that the points $(x,y)$ lie on a straight line. (b) If $\vec{A} = 2\hat{i} - 3\hat{j}$, find the slope and $y$ intercept of the line. (c) If we now let $\vec{A}$ and $\vec{r}$ be vectors in three-dimensional space, show that the relation $\vec{A} \cdot \vec{r} = 1$ specifies a plane.

**45** •• **SSM** When a particle moves in a circle centered at the origin with constant speed, the magnitudes of its position vector and velocity vectors are constant. (a) Differentiate $\vec{r} \cdot \vec{r} = r^2 = $ constant with respect to time to show that $\vec{v} \cdot \vec{r} = 0$ and therefore $\vec{v} \perp \vec{r}$. (b) Differentiate $\vec{v} \cdot \vec{v} = v^2 = $ constant with respect to time to show that $\vec{a} \cdot \vec{v} = 0$ and therefore $\vec{a} \perp \vec{v}$. What do the results of (a) and (b) imply about the direction of $\vec{a}$? (c) Differentiate $\vec{v} \cdot \vec{r} = 0$ with respect to time and show that $\vec{a} \cdot \vec{r} + v^2 = 0$ and therefore $a_r = -v^2/r$.

## Power

**46** •• Force $A$ does 5 J of work in 10 s. Force $B$ does 3 J of work in 5 s. Which force delivers greater power?

**47** • A 5-kg box is being lifted upward at a constant velocity of 2 m/s by a force equal to the weight of the box. (a) What is the power input of the force? (b) How much work is done by the force in 4 s?

**48** • Fluffy has just caught a mouse and decides to bring it to the bedroom so that his human roommate can admire it when she wakes up. A constant horizontal force of 3 N is enough to drag the mouse across the rug at a constant speed $v$. If Fluffy's force does work at the rate of 6 W, (a) what is his speed $v$? (b) How much work does Fluffy do in 4 s?

**49** • A single force of 5 N acts in the $x$ direction on an 8-kg object. (a) If the object starts from rest at $x = 0$ at time $t = 0$, find its velocity $v$ as a function of time. (b) Write an expression for the power input as a function of time. (c) What is the power input of the force at time $t = 3$ s?

**50** • **ISOLVE** Find the power input of a force $\vec{F}$ acting on a particle that moves with a velocity $\vec{v}$ for (a) $\vec{F} = 4\,N\hat{i} + 3\,N\hat{k}$, $\vec{v} = 6\,m/s\hat{i}$; (b) $\vec{F} = 6\,N\hat{i} - 5\,N\hat{j}$, $\vec{v} = -5\,m/s\hat{i} + 4\,m/s\hat{j}$; and (c) $\vec{F} = 3\,N\hat{i} + 6\,N\hat{j}$, $\vec{v} = 2\,m/s\hat{i} + 3\,m/s\hat{j}$.

**51** • **SSM** A small food service elevator (dumbwaiter) in a cafeteria is connected over a pulley system to a motor as shown in Figure 6-37; the motor raises and lowers the dumbwaiter. The mass of the dumbwaiter is 35 kg. In operation, it moves at a speed of 0.35 m/s upward, without accelerating (except for a brief initial period just after the motor is turned on). If the output power from the motor is 27 percent of its input power, what is the input power to the motor? Assume that the pulleys are frictionless.

**FIGURE 6-37** Problem 51

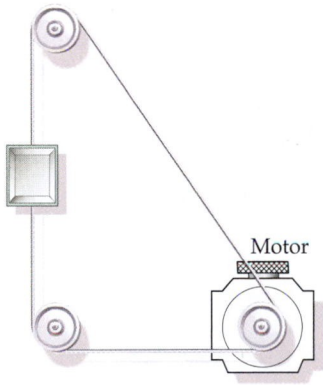

Motor

**52** •• A sky diver falls though the air toward the ground at a constant speed of 120 mph, her terminal velocity, before opening her parachute. (a) If her mass is 55 kg, calculate the magnitude of the power due to the drag force. (b) After she opens her parachute, her speed slows to 15 mph. What is the magnitude of the power due to the drag force now?

**53** •• **SSM** A cannon placed at the top of a cliff of height $H$ fires a cannonball into the air with an initial speed $v_0$, shooting directly upward. The cannonball rises, falls back down, missing the cannon by a little bit, and lands at the foot of the cliff. Neglecting air resistance, calculate the velocity $\vec{v}(t)$ for all times while the cannonball is in the air, and show explicitly that the integral of $\vec{F} \cdot \vec{v}$ over the time that the cannonball spends in the air is equal to the change in the kinetic energy of the cannonball.

**54** •• A particle of mass $m$ moves from rest at $t = 0$ under the influence of a single constant force $\vec{F}$. Show that the power delivered by the force at time $t$ is $P = F^2 t/m$.

## Potential Energy

**55** • **ISOLVE** ✓ An 80-kg man climbs up a 6-m high flight of stairs. What is the increase in the gravitational potential energy of the man–earth system?

**56** • **ISOLVE** Water flows over Victoria Falls, which is 128 m high, at an average rate of $1.4 \times 10^6$ kg/s. If half the potential energy of this water were converted into electric energy, how much power would be produced by these falls?

**57** • **ISOLVE** A 2-kg box slides down a long, frictionless incline of angle 30°. It starts from rest at time $t = 0$ at the top of the incline at a height of 20 m above the ground. (a) What is the original potential energy of the box relative to the ground? (b) From Newton's laws, find the distance the box travels during the interval $0 < t < 1$ s and its speed at $t = 1$ s. (c) Find the potential energy and the kinetic energy of the box at $t = 1$ s. (d) Find the kinetic energy and the speed of the box just as it reaches the bottom of the incline.

**58** • A force $F_x = 6$ N is constant. (a) Find the potential-energy function $U(x)$ associated with this force for an arbitrary reference position $x_0$ at which $U = 0$. (b) Find $U(x)$ such that $U = 0$ at $x = 4$ m. (c) Find $U(x)$ such that $U = 14$ J at $x = 6$ m.

**59** • **ISOLVE** ✓ A spring has a force constant $k = 10^4$ N/m. How far must it be stretched for its potential energy to be (a) 50 J and (b) 100 J?

**60** •• **SSM** **ISOLVE** A simple Atwood's machine (Figure 6-38) uses two masses $m_1$ and $m_2$. Starting from rest, the speed of the two masses is 4.0 m/s at the end of 3.0 s. At that time, the kinetic energy of the system is 80 J and each mass has moved a distance of 6.0 m. Determine the values of $m_1$ and $m_2$.

**FIGURE 6-38** Problem 60

$m_1$

$m_2$

**61** •• A straight rod of negligible mass is mounted on a frictionless pivot as shown in Figure 6-39. Masses $m_1$ and $m_2$ are attached to the rod at distances $\ell_1$ and $\ell_2$. (a) Write an expression for the gravitational potential energy of the masses as a function of the angle $\theta$ made by the rod and the horizontal. (b) For what angle $\theta$ is the potential energy a minimum? Is the statement "systems tend to move toward a configuration of minimum potential energy" consistent with your result? (c) Show that if $m_1\ell_1 = m_2\ell_2$, the potential energy is the same for all values of $\theta$. (When this holds, the system will balance at any angle $\theta$. This result is known as *Archimedes' law of the lever*.)

**FIGURE 6-39** Problem 61

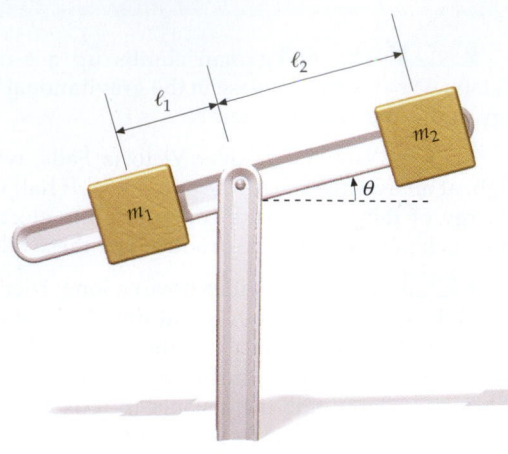

## Force, Potential Energy, and Equilibrium

**62** • (*a*) Find the force $F_x$ associated with the potential-energy function $U = Ax^4$, where $A$ is a constant. (*b*) At what point(s) is the force zero?

**63** •• A potential-energy function is given by $U = C/x$, where $C$ is a positive constant. (*a*) Find the force $F_x$ as a function of $x$. (*b*) Is this force directed toward the origin or away from it? (*c*) Does the potential energy increase or decrease as $x$ increases? (*d*) Answer parts (*b*) and (*c*) where $C$ is a negative constant.

**64** •• **SSM** On the potential-energy curve for $U$ versus $y$ shown in Figure 6-40, the segments AB and CD are straight lines. Sketch a plot of the force $F_y$ versus $y$.

**FIGURE 6-40** Problem 64

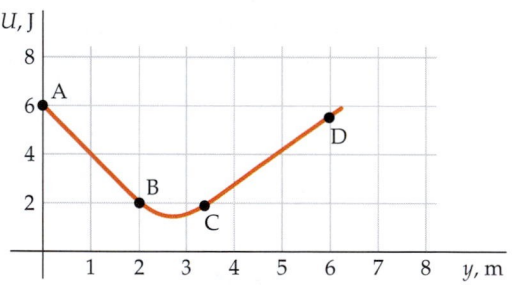

**65** •• The force acting on an object is given by $F_x = a/x^2$. Determine the potential energy of the object as a function of $x$.

**66** •• The potential energy of an object is given by $U(x) = 3x^2 - 2x^3$, where $U$ is in joules and $x$ is in meters. (*a*) Determine the force acting on this object. (*b*) At what positions is this object in equilibrium? (*c*) Which of these equilibrium positions are stable and which are unstable?

**67** •• The potential energy of an object is given by $U(x) = 8x^2 - x^4$, where $U$ is in joules and $x$ is in meters. (*a*) Determine the force acting on this object. (*b*) At what positions is this object in equilibrium? (*c*) Which of these equilibrium positions are stable and which are unstable?

**68** •• The force acting on an object is given by $F(x) = x^3 - 4x$. Locate the positions of unstable and stable equilibrium and show that at these points $U(x)$ is a local maximum or minimum, respectively.

**69** •• The potential energy of a 4-kg object is given by $U = 3x^2 - x^3$ for $x \le 3$ m and $U = 0$ for $x \ge 3$ m, where $U$ is in joules and $x$ is in meters. (*a*) At what positions is this object in equilibrium? (*b*) Sketch a plot of $U$ versus $x$. (*c*) Discuss the stability of the equilibrium for the values of $x$ found in (*a*). (*d*) If the total energy of the particle is 12 J, what is its speed at $x = 2$ m?

**70** •• A force is given by $F_x = Ax^{-3}$, where $A = 8$ N·m³. (*a*) For positive values of $x$, does the potential energy associated with this force increase or decrease with increasing $x$? (You can determine the answer to this question by imagining what happens to a particle that is placed at rest at some point $x$ and is then released.) (*b*) Find the potential-energy function $U$ associated with this force such that $U$ approaches zero as $x$ approaches infinity. (*c*) Sketch $U$ versus $x$.

**71** ••• **SSM** A novelty desk clock is shown in Figure 6-41: The clock (which has mass $m$) is supported by two light cables running over the two pulleys, which are attached to counter-weights that each have mass $M$. (*a*) Find the potential energy of the system as a function of the distance $y$. (*b*) Find the value of $y$ for which the potential energy of the system is a minimum. (*c*) If the potential energy is a minimum, then the system is in equilibrium. Apply Newton's second law to the clock and show that it is in equilibrium (the forces on it sum to zero) for the value of $y$ obtained for part (*b*). Is this a point of stable or unstable equilibrium?

**FIGURE 6-41** Problem 71

## General Problems

**72** • **SSM** In February 2002, a total of 60.7 billion kW·h of electrical energy was generated by nuclear power plants in the United States. At that time, the population of the United States was about 287 million people. If the average American has a mass of 60 kg, and if the entire energy output of all nuclear power plants was diverted to supplying energy for a single giant elevator, estimate the height $h$ to which the entire population of the country could be lifted by the elevator. In your calculations, assume that 25 percent of the energy goes into lifting the people; assume also that $g$ is constant over the entire height $h$.

**73** • **SOLVE** ✔ One of the most powerful cranes in the world, operating in Switzerland, can slowly raise a load of $M = 6000$ tonne to a height of $h = 12.0$ m (1 tonne = 1000 kg). (a) How much work is done by the crane? (b) If it takes 1.00 min to lift the load at constant velocity to this height, find the power developed by the crane.

**74** • **SOLVE** In Austria, there once was a ski lift of length 5.6 km. It took about 60 min for a gondola to travel all the way up. If there were 12 gondolas going up, each with a cargo of mass 550 kg, and if there were 12 empty gondolas going down, and the angle of ascent was 30°, estimate the power $P$ of the engine needed to operate the ski lift.

**75** • **SOLVE** ✔ A 2.4-kg object attached to a horizontal string moves with constant speed in a circle of radius $R$ on a frictionless horizontal surface. The kinetic energy of the object is 90 J and the tension in the string is 360 N. Find $R$.

**76** • **SSM** The movie crew arrives in the Badlands ready to shoot a scene. The script calls for a car to crash into a vertical rock face at 100 km/h. Unfortunately, the car won't start, and there is no mechanic is sight. The crew are about to skulk back to the studio to face the producer's wrath when the cameraman gets an idea. They use a crane to lift the car by its rear end and then drop it vertically, filming at an angle that makes the car appear to be traveling horizontally. How high should the 800-kg car be lifted so that it reaches a speed of 100 km/h in the fall?

**77** ••• The four strings pass over the bridge of a violin as shown in Figure 6-42. The angle that the strings make with the normal to the plane of the instrument is 72° on either side. The total normal force pressing the bridge into the violin is 103 N. The length of the strings from the bridge to the peg to which each is attached is 32.6 cm. (a) Determine the tension in the violin strings, assuming that the tension is the same for each string. (b) One of the strings is plucked out a distance of 4 mm, as shown in the figure. Make a free-body diagram showing all of the forces acting on the string at that point, and determine the force pulling the string back to its equilibrium position. Assume that the tension in the string remains constant. (c) Determine the work done on the string in plucking it out that distance. Remember that the net force pulling the string back to its equilibrium position is changing as the string is being pulled out, but assume that the magnitude of the tension in the string remains constant.

**78** •• The force acting on a particle that is moving along the $x$ axis is given by $F_x = -ax^2$, where $a$ is a constant. Calculate the potential-energy function $U$ relative to $U = 0$ at $x = 0$, and sketch a graph of $U$ versus $x$.

**79** •• **SSM** **SOLVE** A horizontal force acts on a cart of mass $m$ such that the speed $v$ of the cart increases with distance $x$ as $v = Cx$, where $C$ is a constant. (a) Find the force acting on the cart as a function of position. (b) What is the work done by the force in moving the cart from $x = 0$ to $x = x_1$?

**80** •• **SOLVE** ✔ A force $\vec{F} = (2 \text{ N/m}^2)x^2\hat{i}$ is applied to a particle. Find the work done on the particle as it moves a total distance of 5 m (a) parallel to the $y$ axis from point (2 m, 2 m) to point (2 m, 7 m) and (b) in a straight line from (2 m, 2 m) to (5 m, 6 m).

**81** •• A particle of mass $m$ moves along the $x$ axis. Its position varies with time according to $x = 2t^3 - 4t^2$, where $x$ is in meters and $t$ is in seconds. Find (a) the velocity and acceleration of the particle at any time $t$, (b) the power delivered to the particle at any time $t$, and (c) the work done by the force from $t = 0$ to $t = t_1$.

**82** •• **SOLVE** A 3-kg particle starts from rest at $x = 0$ and moves under the influence of a single force $F_x = 6 + 4x - 3x^2$, where $F_x$ is in newtons and $x$ is in meters. (a) Find the work done by the force as the particle moves from $x = 0$ to $x = 3$ m. (b) Find the power delivered to the particle when it is at $x = 3$ m.

**83** •• **SSM** **SOLVE** ✔ The initial kinetic energy imparted to a 0.020-kg bullet is 1200 J. Neglecting air resistance, find the range of this projectile when it is fired at an angle such that the range equals the maximum height attained.

**84** •• **SOLVE** ✔ A force $F_x$ acting on a particle is shown as a function of $x$ in Figure 6-43. (a) From the graph, calculate the work done by the force when the particle moves from $x = 0$ to the following values of $x$: $-4, -3, -2, -1, 0, 1, 2, 3$, and 4 m. (b) Plot the potential energy $U$ versus $x$ for the range of values of $x$ from $-4$ m to $+4$ m, assuming that $U = 0$ at $x = 0$.

**FIGURE 6-43** Problem 84

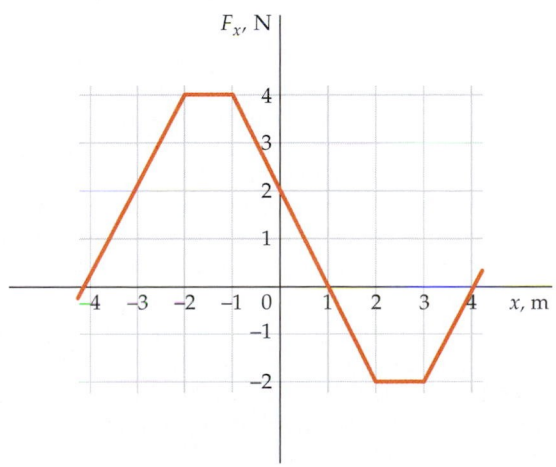

**FIGURE 6-42** Problem 77

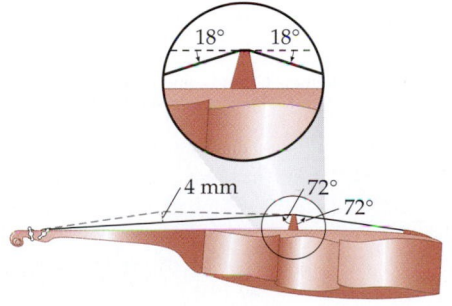

**85** •• Repeat Problem 84 for the force $F_x$ shown in Figure 6-44.

**FIGURE 6-44** Problem 85

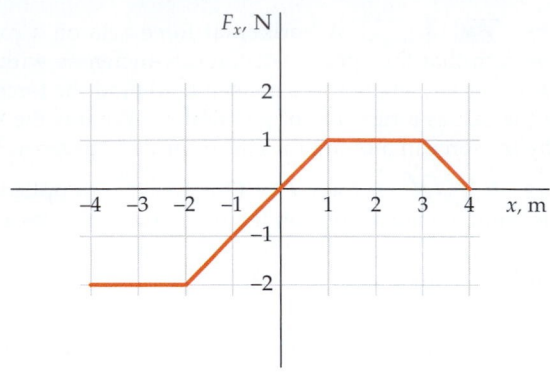

**86** •• **iSOLVE** A box of mass $M$ is at rest at the bottom of a frictionless inclined plane (Figure 6-45). The box is attached to a string that pulls with a constant tension $T$. (a) Find the work done by the tension $T$ as the box moves through a distance $x$ along the plane. (b) Find the speed of the box as a function of $x$ and $\theta$. (c) Determine the power produced by the tension in the string as a function of $x$ and $\theta$.

**FIGURE 6-45** Problem 86

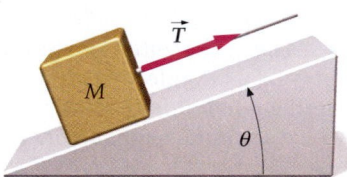

**87** ••• A force in the $xy$ plane is given by $\vec{F} = (F_0/r)(y\hat{i} - x\hat{j})$, where $F_0$ is a constant and $r = \sqrt{x^2 + y^2}$. (a) Show that the magnitude of this force is $F_0$ and that its direction is perpendicular to $\vec{r} = x\hat{i} + y\hat{j}$. (b) Find the work done by this force on a particle that moves once around a circle of radius 5 m centered at the origin. Is this force conservative?

**88** ••• **SSM** A force in the $xy$ plane is given by:

$F = -(b/r^3)(x\hat{i} + y\hat{j})$, where $b$ is a positive constant and $r = \sqrt{x^2 + y^2}$. (a) Show that the magnitude of the force varies as the inverse of the square of the distance to the origin, and that its direction is antiparallel (opposite) to the radius vector $\vec{r} = x\hat{i} + y\hat{j}$. (b) If $b = 3$ N·m$^2$, find the work done by this force on a particle moving along a straight-line path between an initial position $x = 2$ m, $y = 0$ m and a final position $x = 5$ m, $y = 0$ m. (c) Find the work done by this force on a particle moving once around a circle of radius $r = 7$ m centered around the origin. (d) If this force is the only force acting on the particle, what is the particle's speed as it moves along this circular path? Assume that the particle's mass is $m = 2$ kg.

**89** ••• In the words of Richard Feynman, one of the greatest physicists of the twentieth century, "If, in some cataclysm, all of scientific knowledge were to be destroyed, and only one sentence passed on to the next generation, . . . what statement would contain the most information in the fewest words? I believe it is . . . that all things are made of atoms—little particles . . . attracting each other when they are a little distance apart, but repelling upon being squeezed into each other."[†] When looking at the forces between atoms, modern physicists and chemists often model their interactions by the so-called "6–12" potential, where the potential energy function between two atoms is given by the functional form

$$U(r) = \frac{a}{r^{12}} - \frac{b}{r^6}$$

where $r$ is the distance between atomic nuclei and $a$ and $b$ are constants that can be determined spectroscopically. Because they do not form atomic bonds, noble gas atoms have potential-energy functions that can be modeled well by a 6–12 potential and measured reasonably accurately. For argon, these parameters are $a = 1.09 \times 10^{-7}$ and $b = 6.84 \times 10^{-5}$, where $r$ is measured in nm (1 nm $= 10^{-9}$ m) and $U$ is measured in eV (1 eV $= 1.6 \times 10^{-19}$ J.) (a) Using a spreadsheet program, make a graph of the interatomic potential energy function for two argon atoms as a function of separation, for values of $r$ between 0.3 nm and 0.7 nm. Does the shape of the potential energy function support Feynman's claim? Explain. (b) What is the minimum value of the potential energy (compared to where the atoms are separated by a very large distance)? At what separation does this occur? Is this minimum a point of stable or unstable equilibrium? (c) From either the graph or the formula given above, estimate the force of attraction between two argon atoms separated by a distance of 5 Å and the force of repulsion for a separation of 3.5 Å. Make sure that you convert to MKS units!

**90** ••• **SSM** A theoretical formula for the potential energy associated with the nuclear force between two protons, two neutrons, or a neutron and a proton is the Yukawa potential $U(r) = -U_0(a/r)e^{-r/a}$, where $U_0$ and $a$ are constants, and $r$ is the separation between the two nucleons. (a) Using a spreadsheet program such as Microsoft Excel™, make a graph of $U$ versus $r$, using $U_0 = 4$ pJ (a picojoule, pJ, is $1 \times 10^{-12}$ J) and $a = 2.5$ fm (a femtometer, fm, is $1 \times 10^{-15}$ m). (b) Find the force $F(r)$ as a function of the separation of the two nucleons. (b) Find the force $F(r)$ as a function of the separation of the two nucleons. (c) Compare the magnitude of the force at the separation $r = 2a$ to that at $r = a$. (d) Compare the magnitude of the force at the separation $r = 5a$ to that at $r = a$.

† Richard P. Feynman, Matthew L. Sands, and Robert B. Leighton, *The Feynman Lectures on Physics*, Vol. 1, p. 1.1. Boston: Addison-Wesley (1970).

# Conservation of Energy

7-1   The Conservation of Mechanical Energy

7-2   The Conservation of Energy

7-3   Mass and Energy

7-4   Quantization of Energy

ENERGY CAN BE CONVERTED FROM ONE FORM INTO ANOTHER. THIS ROLLER COASTER CONVERTS ELECTRICAL ENERGY PURCHASED FROM THE POWER COMPANY INTO GRAVITATIONAL POTENTIAL ENERGY OF THE PASSENGERS AND THEIR VEHICLES. THEN GRAVITATIONAL POTENTIAL ENERGY IS CONVERTED INTO KINETIC ENERGY, SOME OF WHICH IS CONVERTED BACK TO POTENTIAL ENERGY. EVENTUALLY FRICTION TRANSFORMS THE KINETIC AND POTENTIAL ENERGY INTO THERMAL ENERGY. CHEMICAL ENERGY STORED IN THE MUSCLES OF THE PASSENGERS IS CONVERTED INTO SOUND ENERGY EACH TIME SOMEONE SCREAMS.

 **How does conservation of mechanical energy allow us to determine how high up the cars must be when they start their descent, in order to complete the loop-the-loop? (See Example 7-5.)**

A system is a collection of particles. External forces are any forces exerted by particles not in the system on particles that are in the system, and internal forces are those exerted by particles in the system on other particles in the system. The many ways there are to change the total energy of the system can be divided into just two categories, work and heat. Total energy can change if external forces do work on the system or if energy is transferred because of a temperature difference between the system and its surroundings. (Energy transferred due to a temperature difference is called heat.) Because we do not analyze systems where heat plays a significant role until Chapter 18, here we will consider the change in energy of the system to be equal to the total work done on it by external forces. We define the potential energy of a system so that the change in the potential energy of the system equals the negative of the total work done by all internal conservative forces. If no external forces do work on the system, and if the internal conservative forces are the only internal forces that do work, then the work they do equals the change in the system's kinetic energy. Because the change in the system's potential energy equals the negative of the change in its kinetic energy, the sum of the potential and kinetic energies does not change. This is known as the conservation of mechanical energy. It follows from Newton's laws and is a useful alternative to them for solving many problems in mechanics.

The usefulness of mechanical energy conservation is limited by the presence of nonconservative forces such as friction. When kinetic friction is present in a system, the mechanical energy of the system does not stay the same but instead decreases.[†] Because mechanical energy is often not conserved, the importance of energy conservation was not realized until the nineteenth century, when it was discovered that the disappearance of macroscopic mechanical energy is always accompanied by the appearance of some other kind of energy, often thermal energy, which is usually indicated by an increase in temperature or a change in phase (like the melting of ice). We now know that, on the microscopic scale, this thermal energy is associated with the kinetic and potential energies at the molecular level.

There are a number of different forms of energy, such as chemical energy, sound energy, electromagnetic energy, and nuclear energy. Whenever the energy of a system changes, we can account for the change by the appearance or disappearance of energy somewhere else. This experimental observation is the **law of conservation of energy,** one of the most fundamental and important laws in all of science. Although energy changes from one form to another, it is never created or destroyed.

➤ **In this chapter, we continue the study of energy begun in Chapter 6 by describing and applying the law of conservation of energy and examining different types of energy, including thermal energy. We also discuss Einstein's famous relation between mass and energy, and discover that energy changes for a system are not continuous, but occur in discrete "bundles" or "lumps" called quanta. Although in a macroscopic system the quantum of energy typically is so small that it goes unnoticed, its presence has profound consequences for microscopic systems such as atoms and molecules.**

# 7-1 The Conservation of Mechanical Energy

The total work done on each particle in a system equals the change in the kinetic energy $\Delta K_i$ of that particle, so the total work done by all the forces $W_{total}$ equals the change in the total kinetic energy of the system $\Delta K_{sys}$:

$$W_{total} = \sum \Delta K_i = \Delta K_{sys} \qquad\qquad 7\text{-}1$$

Each internal force is either conservative or nonconservative. The negative of the total work done by all the conservative internal forces $-W_c$ equals the change in the potential energy of the system $\Delta U_{sys}$:

$$-W_c = \Delta U_{sys} \qquad\qquad 7\text{-}2$$

The total work done by all forces equals the work done by all external forces $W_{ext}$, plus the work done by all internal nonconservative forces $W_{nc}$, plus that done by all internal conservative forces $W_c$:

$$W_{total} = W_{ext} + W_{nc} + W_c$$

Rearranging gives $W_{ext} + W_{nc} = W_{total} - W_c$. Substituting from Equations 7-1 and 7-2 we have

$$W_{ext} + W_{nc} = \Delta K_{sys} + \Delta U_{sys} \qquad\qquad 7\text{-}3$$

The right side of this equation can be simplified as

$$\Delta K_{sys} + \Delta U_{sys} = \Delta(K_{sys} + U_{sys}) \qquad\qquad 7\text{-}4$$

The sum of the kinetic energy $K_{sys}$ and the potential energy $U_{sys}$ is called the **total mechanical energy** $E_{mech}$:

---

† Other factors in the system, such as a battery that runs a motor, can cause the mechanical energy to increase, so the presence of kinetic friction does not necessarily mean that the mechanical energy of the system will decrease.

$$E_{mech} = K_{sys} + U_{sys} \qquad\qquad\qquad 7\text{-}5$$

DEFINITION—TOTAL MECHANICAL ENERGY

Combining Equations 7-4 and 7-5, and then substituting into Equation 7-3 gives

$$W_{ext} = \Delta E_{mech} - W_{nc}$$

The mechanical energy of a system is conserved ($E_{mech}$ = constant) if the total work done by all external forces and by all internal nonconservative forces is zero.

$$E_{mech} = K_{sys} + U_{sys} = \text{constant} \qquad\qquad 7\text{-}6$$

CONSERVATION OF MECHANICAL ENERGY

This is **conservation of mechanical energy** and is the origin of the expression "conservative force."

If $E_{mech\,i} = K_i + U_i$ is the initial mechanical energy of the system and $E_{mech\,f} = K_f + U_f$ is the final mechanical energy of the system, conservation of mechanical energy implies that

$$E_{mech\,f} = E_{mech\,i} \qquad (\text{or } K_f + U_f = K_i + U_i) \qquad 7\text{-}7$$

CONSERVATION OF MECHANICAL ENERGY

Many mechanics problems can be solved by setting the final mechanical energy of a system equal to its initial mechanical energy.

## Applications

Suppose that you are wearing skis and, starting at rest from a height $h_0$ above the bottom of a hill, you coast down the hill. Assuming that the skis are frictionless, what is your speed as you pass through a gate at height $h$ above the bottom of the hill? The mechanical energy of the earth–skier system is conserved because the only force doing work is the internal, conservative force of gravity. If we choose $U = 0$ at the bottom of the hill, the initial potential energy is $mgh_0$. This is also the total mechanical energy because the initial kinetic energy is zero. Thus,

$$E_{mech\,i} = K_i + U_i = 0 + mgh_0$$

At height $h$, the potential energy is $mgh$ and the speed is $v$. Hence,

$$E_{mech\,f} = K_f + U_f = \tfrac{1}{2}mv^2 + mgh$$

Setting $E_{mech\,f} = E_{mech\,i}$ we find

$$\tfrac{1}{2}mv^2 + mgh = mgh_0$$

or

$$v = \sqrt{2g(h_0 - h)}$$

Your speed is the same as if you had undergone free-fall through a distance $h_0 - h$. However, by skiing down the hill, you travel farther and take more time than you would if you were in free-fall and falling straight down.

Multiflash photograph of a simple pendulum. As the bob descends, gravitational potential energy is converted into kinetic energy, and the speed increases as indicated by the increased spacing of the recorded positions. The speed decreases as the bob moves up, and the kinetic energy is changed into potential energy.

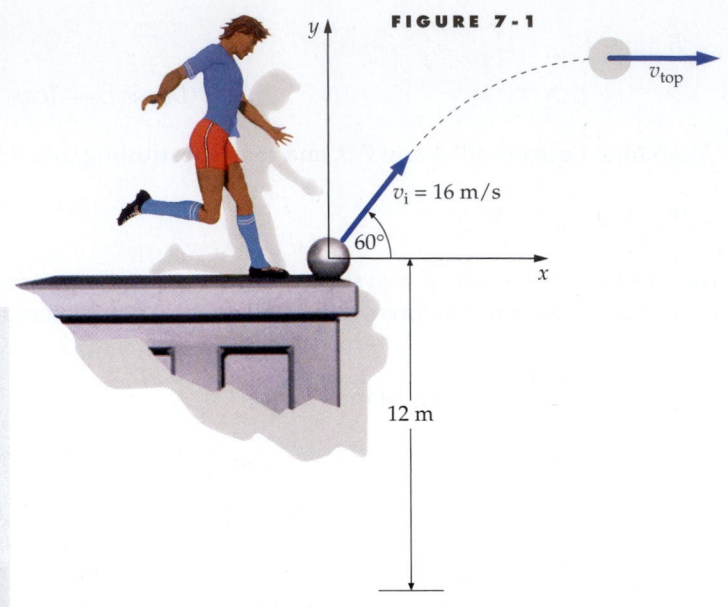

FIGURE 7-1

*KICKING A BALL* **EXAMPLE 7-1**

Standing near the edge of the roof of a 12-m-high building, you kick a ball with an initial speed of $v_i = 16$ m/s at an angle of 60° above the horizontal (Figure 7-1). Neglecting air resistance, find (a) how high above the height of the building the ball rises and (b) its speed just before it hits the ground.

**PICTURE THE PROBLEM** We choose the ball and the earth as the system. We consider this system during the interval from just after the kick to just before impact with the ground. There are no external forces doing work on the system, and there are no internal nonconservative forces doing work, so the mechanical energy of the system is conserved. At the top of its flight, the ball is moving horizontally with a speed $v_{top}$, equal to the horizontal component of its initial horizontal velocity $v_{ix}$. We choose $y = 0$ at the roof of the building.

(a) 1. Conservation of mechanical energy relates the height $h$ above the top of the building to the initial speed $v_i$ and the speed at the top of its flight $v_{top}$:

$$E_{\text{mech top}} = E_{\text{mech i}}$$
$$\tfrac{1}{2}mv_{top}^2 + mgy_{top} = \tfrac{1}{2}mv_i^2 + mgy_i$$
$$\tfrac{1}{2}mv_{top}^2 + mgh_{top} = \tfrac{1}{2}mv_i^2 + 0$$

2. Solve for $h_{top}$:

$$h_{top} = \frac{v_i^2 - v_{top}^2}{2g}$$

3. The velocity at the top of its flight equals its initial horizontal velocity:

$$v_{top} = v_{ix} = v_i \cos\theta$$

4. Substitute the step 3 result into the step 2 result and solve for $h_{top}$:

$$h_{top} = \frac{v_i^2 - v_{top}^2}{2g} = \frac{v_i^2 - v_i^2 \cos^2\theta}{2g} = \frac{v_i^2(1 - \cos^2\theta)}{2g}$$
$$= \frac{(16 \text{ m/s})^2 (1 - \cos^2 60°)}{2(9.81 \text{ m/s}^2)} = \boxed{9.79 \text{ m}}$$

(b) 1. If $v_f$ is the speed of the ball just before it hits the ground, where $y = y_f = -12$ m, its energy is:

$$E_{\text{mech f}} = \tfrac{1}{2}mv_f^2 + mgy_f$$

2. Set the final mechanical energy equal to the initial mechanical energy:

$$\tfrac{1}{2}mv_f^2 + mgy_f = \tfrac{1}{2}mv_i^2 + 0$$

3. Solve for $v_f$, and set $y = -12$ m to find the final velocity:

$$v_f = \sqrt{v_i^2 - 2gy_f}$$
$$= \sqrt{(16 \text{ m/s})^2 - 2(9.81 \text{ m/s}^2)(-12 \text{ m})} = \boxed{22.2 \text{ m/s}}$$

*A PENDULUM* **EXAMPLE 7-2**

A pendulum consists of a bob of mass $m$ attached to a string of length $L$. The bob is pulled aside so that the string makes an angle $\theta_0$ with the vertical, and is released from rest. As it passes through the bottom of the arc, find expressions for (a) the speed and (b) the tension. Air resistance is negligible.

**PICTURE THE PROBLEM** The system consists of the pendulum, the earth, and the pendulum mount and its supports. There are no external forces on this system. The two internal forces acting on the bob (air resistance being negligible) are the force of gravity $m\vec{g}$, which is conservative, and the tension force $\vec{T}$. The rate at which $\vec{T}$ does work is $\vec{T} \cdot \vec{v}$. Because $\vec{T}$ is perpendicular to $\vec{v}$, we know that $\vec{T} \cdot \vec{v} = 0$. Since only $m\vec{g}$ does work, we know that the mechanical energy of the bob–earth system is conserved. To find the speed of the bob, we equate the initial and final mechanical energies. The tension in the string is obtained using Newton's second law.

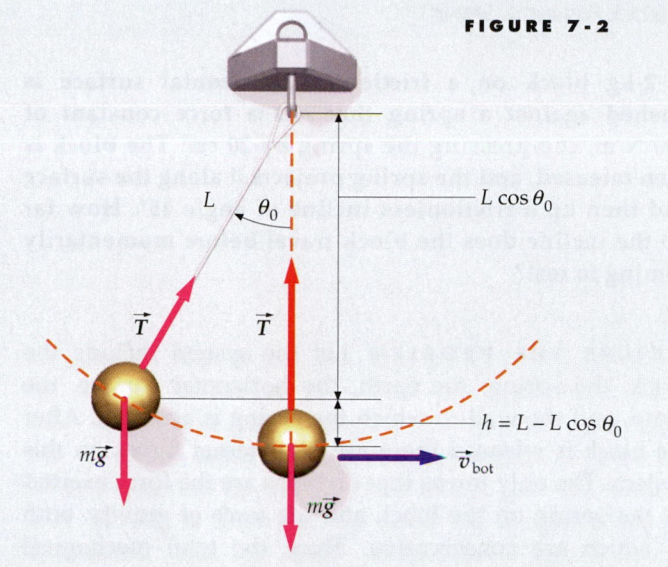

FIGURE 7-2

(a) 1. Make a sketch of the system in its initial and final configurations (Figure 7-2). We choose $y = 0$ at the bottom of the swing and $y = h$ at the initial position:

2. Apply conservation of mechanical energy. Initially the bob is at rest:

$$E_{\text{mech f}} = E_{\text{mech i}}$$

$$\tfrac{1}{2}mv_f^2 + mgy_f = \tfrac{1}{2}mv_i^2 + mgy_i$$

$$\tfrac{1}{2}mv_{\text{bot}}^2 + 0 = 0 + mgh$$

3. Conservation of mechanical energy thus relates the speed $v_{\text{bot}}$ to the height $h$:

$$\tfrac{1}{2}mv_{\text{bot}}^2 = mgh$$

4. Solve for the speed $v_{\text{bot}}$:

$$v_{\text{bot}} = \sqrt{2gh}$$

5. To express speed in terms of the initial angle $\theta_0$, we need to relate $h$ to $\theta_0$. This relation is illustrated in Figure 7-2:

$$L = L \cos \theta_0 + h$$

so

$$h = L - L \cos \theta_0 = L(1 - \cos \theta_0)$$

6. Substitute this value for $h$ to express the speed at the bottom in terms of $\theta_0$:

$$v_{\text{bot}} = \boxed{\sqrt{2gL(1 - \cos \theta_0)}}$$

(b) 1. When the bob is at the bottom of the circle, the forces on it are $m\vec{g}$ and $\vec{T}$. Apply $\Sigma F_y = ma_y$:

$$T - mg = ma_y$$

2. At the bottom, the bob has an acceleration $v_{\text{bot}}^2/L$ toward the center of the circle, which is upward:

$$a_y = \frac{v_{\text{bot}}^2}{L} = \frac{2gL(1 - \cos \theta_0)}{L}$$

$$= 2g(1 - \cos \theta_0)$$

3. Substitute for $a_y$ the Part (b) step 1 result and solve for $T$:

$$T = mg + ma_y = m(g + a_y)$$

$$= m[g + 2g(1 - \cos \theta_0)]$$

$$= \boxed{(3 - 2 \cos \theta_0)mg}$$

**REMARKS** 1. The tension at the bottom is greater than the weight of the bob because the bob is accelerating upward. 2. Step 3 in Part (b) shows that for $\theta_0 = 0$, $T = mg$, the expected result for a stationary bob hanging from a string. 3. Step 4 in Part (a) shows that the speed at the bottom is the same as if the bob had dropped in free-fall from a height $h$. The speed of the bob at the bottom of the arc can also be found using Newton's laws directly (see Problem 7-92), but such a solution is more challenging because the acceleration component tangential to the circle varies with position, and therefore with time, so the constant-acceleration formulas do not apply.

**EXAMPLE 7-3** **Try It Yourself**

A 2-kg block on a frictionless horizontal surface is pushed against a spring that has a force constant of 500 N/m, compressing the spring by 20 cm. The block is then released, and the spring projects it along the surface and then up a frictionless incline of angle 45°. How far up the incline does the block travel before momentarily coming to rest?

**PICTURE THE PROBLEM** Let the system include the block, the spring, the earth, the horizontal surface, the ramp, and the wall to which the spring is attached. After the block is released there are no external forces on this system. The only forces that do work are the force exerted by the spring on the block and the force of gravity, both of which are conservative. Thus, the total mechanical energy of the system is conserved. Find the maximum height $h$ from the conservation of mechanical energy, and then the distance up the incline $s$ from $\sin 45° = h/s$.

Cover the column to the right and try these on your own before looking at the answers.

**Answers**

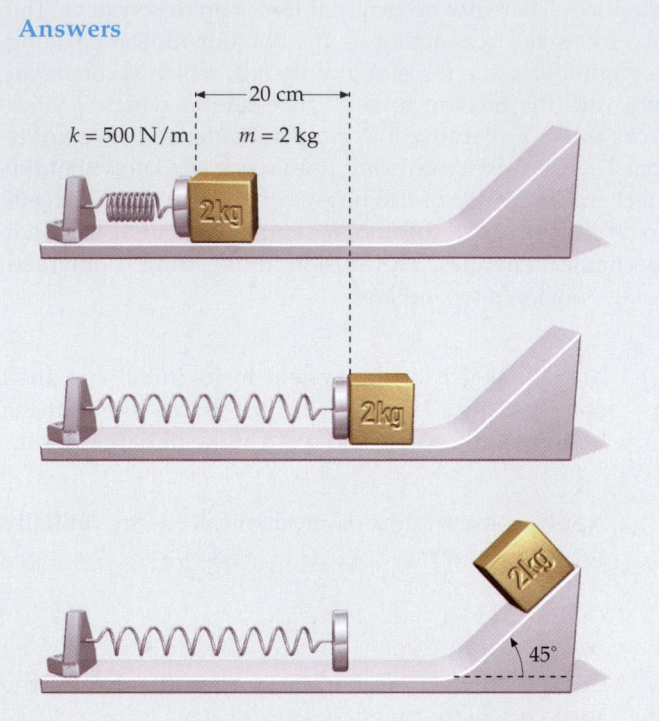

**FIGURE 7-3**

**Steps**

1. Sketch the system with both the initial and final configurations (Figure 7-3).

2. Write the initial mechanical energy in terms of the compression distance $x$.

$$E_{\text{mech i}} = U_{\text{s i}} + U_{\text{g i}} + K_{\text{i}} = \tfrac{1}{2}kx^2 + 0 + 0$$

3. Write the final mechanical energy in terms of the height $h$.

$$E_{\text{mech f}} = U_{\text{s f}} + U_{\text{g f}} + K_{\text{f}} = 0 + mgh + 0$$

4. Apply conservation of mechanical energy and solve for $h$.

$$mgh = \tfrac{1}{2}kx^2$$

$$h = \frac{kx^2}{2mg} = 0.51 \text{ m}$$

5. Find the distance $s$ from $h$ and the angle of inclination.

$$h = s \sin \theta$$

$$s = \boxed{0.72 \text{ m}}$$

**REMARKS** In this problem, the initial mechanical energy of the system is the potential energy of the spring. This energy is converted first into kinetic energy and then into gravitational potential energy.

**EXERCISE** Find the speed of the block just after it leaves the spring. (*Answer* 3.16 m/s)

**EXAMPLE 7-4**

A spring with a force constant of $k$ hangs vertically. A block of mass $m$ is attached to the unstretched spring and allowed to fall from rest. Find an expression for the maximum distance the block falls before it begins moving upward.

**PICTURE THE PROBLEM** As the block drops, its speed first increases, then reaches some maximum value, and then decreases until it is again zero when the block is at its lowest point. There are only conservative forces present, so we apply the conservation of mechanical energy to the earth–spring–block system.

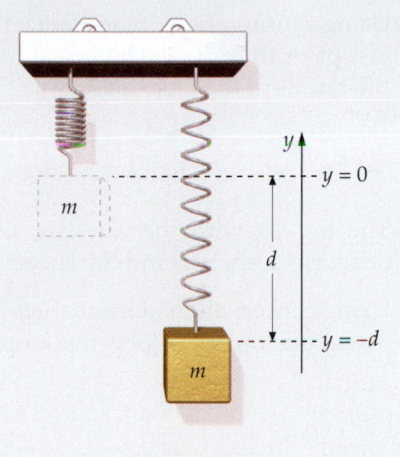

**FIGURE 7-4**

1. Sketch the system showing the initial and final positions of the block and spring (Figure 7-4). Include a $y$ axis with up as the positive $y$ direction. Choose the gravitational potential energy of the system to be zero at the original position $y = 0$ when the spring is unstretched. The potential energy of the spring is zero when the spring is unstretched. Let $d$ be the distance the block falls.

2. Apply conservation of mechanical energy to relate the initial ($y = 0$) and final ($y = -d$) positions:

$$mgy_f + \tfrac{1}{2}ky_f^2 + \tfrac{1}{2}mv_f^2 = mgy_i + \tfrac{1}{2}ky_i^2 + \tfrac{1}{2}mv_i^2$$

$$mg(-d) + \tfrac{1}{2}k(-d)^2 + 0 = 0 + 0 + 0$$

3. Solve for $d$. There are two solutions. One gives $d = 0$ and the other is the solution we want—the maximum distance the block falls before it begins moving upward:

$$\tfrac{1}{2}kd^2 - mgd = 0$$

$$(\tfrac{1}{2}kd - mg)d = 0$$

Since $d \neq 0$, $\boxed{d = \dfrac{2mg}{k}}$

**REMARKS** Gravitational potential energy is converted into the kinetic energy of the block plus the potential energy of the spring. At the lowest point, where the block is momentarily at rest, the elastic potential energy of the spring equals the decrease in the gravitational potential energy of the system.

*BACK TO THE FUTURE*                **EXAMPLE 7-5    Put It in Context**

Imagine that you have time-traveled back to the late 1800s and are watching your great-great-great-grandparents on their honeymoon taking a ride on the **Flip Flap Railway**, a Coney Island roller coaster with a circular loop-the-loop. The car they are in is about to enter the loop-the-loop when a 100-lb sack of sand falls from a construction-site platform and lands in the back seat of the car. No one is hurt, but the impact causes the car to lose 25 percent of its speed. The car started from rest at a high point 2 times as high as the top of the loop-the-loop. Neglect any losses due to either friction or air drag. Will their car make it over the top of the loop-the-loop without falling off?

**PICTURE THE PROBLEM** The car has to have enough speed at the top of the loop-the-loop to maintain contact with the track. We can use conservation of mechanical energy to determine the speed just before the sandbag hits the car, and we can use it again to determine the speed it has at the top of the track. Then we can use Newton's second law to determine the normal force exerted on the car by the track.

**FIGURE 7-5**

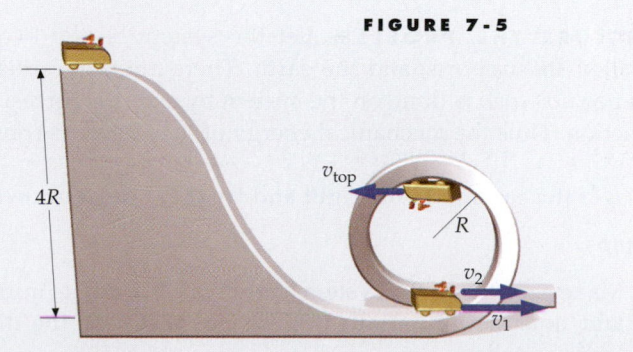

1. Draw a picture of the car and track, with the car at the starting point, at the bottom of the track, and at the top of the loop-the-loop (Figure 7-5):

2. Apply Newton's second law to relate the speed at the top of the loop-the-loop to the normal force:

$$F_n + mg = m\frac{v_{top}^2}{R}$$

3. Using conservation of mechanical energy, find the speed just prior to impact. The initial height is $4R$, where $R$ is the radius of the loop-the-loop:

$$U_1 + K_1 = U_0 + K_0$$
$$0 + \tfrac{1}{2}mv_1^2 = mg4R + 0$$

so

$$v_1 = \sqrt{8Rg}$$

4. The impact with the sandbag results in a 25 percent decrease in speed. Find the speed after impact:

$$v_2 = 0.75v_1 = 0.75\sqrt{8Rg}$$

5. Using conservation of mechanical energy, find the speed just at the top of the loop-the-loop:

$$U_{top} + K_{top} = U_2 + K_2$$
$$mg\,2R + \tfrac{1}{2}mv_{top}^2 = 0 + \tfrac{1}{2}m(0.75^2\,8Rg)$$

so

$$v_{top}^2 = (0.75^2\,8 - 4)Rg = 0.5Rg$$

6. Substituting for $v_{top}^2$ in the step 2 result gives:

$$F_n + mg = 0.5mg$$

7. Solve for $F_n$:

$$F_n = -0.5mg$$

8. $F_n$ is the magnitude of the normal force. It cannot be negative:

Oops! The car has left the track.

**REMARKS** A loss of 25 percent of your speed means losing almost 44 percent of your kinetic energy. Fortunately, there were safety devices to prevent the cars from falling, so your ancestors likely would have survived. The biggest concern for riders on the Flip Flap Railway was a broken neck. In the loop-the-loop, the riders were subjected to up to $12gs$. This was the last of the circular loop-the-loop roller coasters. Loop-the-loops on more recent rides are oval shaped, higher than they are wide.

---

*Two Blocks on a String*     **EXAMPLE 7-6**  **Try It Yourself**

**Two blocks are attached to a light string that passes over a massless, frictionless pulley. The two blocks have masses $m_1$ and $m_2$, where $m_2 > m_1$, and are initially at rest. Find the speed of either block when $m_2$ falls a distance $h$. Find the magnitude of the acceleration of the blocks as $m_2$ falls.**

**PICTURE THE PROBLEM** Let the system be the two blocks, the string, the pulley, its supports, and the earth. There are no external forces on this system, hence no work is done on the system by external forces. Also, there is no kinetic friction. Thus the mechanical energy of the system is conserved.

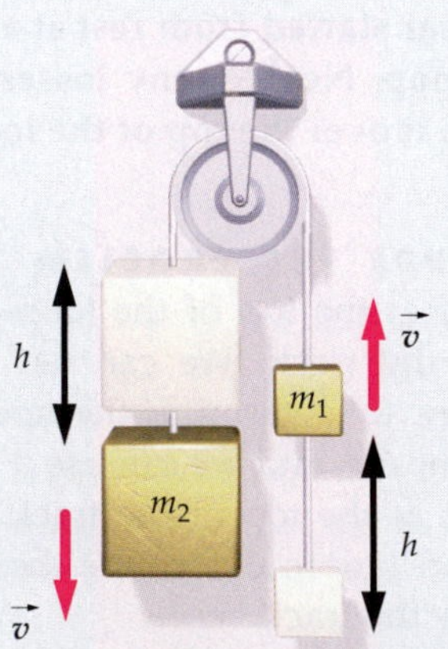

**Cover the column to the right and try these on your own before looking at the answers.**

**Steps**

1. Make a sketch of the system showing it in both its initial and final configurations (Figure 7-6). Let $h$ be the distance $m_2$ falls.

2. Initially, the total kinetic energy of the system is zero. Write the total kinetic energy $K$ of the system when the blocks are moving with speed $v$.

**Answers**

$$K = \tfrac{1}{2}m_1v^2 + \tfrac{1}{2}m_2v^2 = \tfrac{1}{2}(m_1 + m_2)v^2$$

**FIGURE 7-6**

3. Initially, choose the total gravitational potential energy of the system $U$ to be zero. Write an expression for $U$ when $m_1$ has moved up a distance $h$ (and $m_2$ has moved down the same distance).

$$U = m_1 gh - m_2 gh = (m_1 - m_2)gh$$

4. Add $U$ and $K$ to obtain the total mechanical energy $E$.

$$E = K + U$$
$$= \tfrac{1}{2}(m_1 + m_2)v^2 + (m_1 - m_2)gh$$

5. Apply conservation of mechanical energy.

$$E = E_i$$
$$\tfrac{1}{2}(m_1 + m_2)v^2 + (m_1 - m_2)gh = 0$$

6. Solve for $v$.

$$v = \sqrt{\frac{2(m_2 - m_1)}{(m_1 + m_2)}gh}$$

7. Find the acceleration $a$ by $dv/dt$. Use the step 6 result and the chain rule for differentiation. Note that $v = dh/dt$.

$$a = \frac{dv}{dt} = \frac{dv}{dh}\frac{dh}{dt} = \frac{(m_2 - m_1)}{(m_1 + m_2)}g$$

**◉ PLAUSIBILITY CHECK** This device, called an *Atwood's machine*, is analyzed in terms of forces in Problems 84–90 in Chapter 4. There we see that the acceleration is given by the step 7 result.

**EXERCISE** What is the magnitude of the acceleration of either block if the masses are $m_1 = 3$ kg and $m_2 = 5$ kg? (*Answer* $a = 0.25g = 2.45$ m/s²)

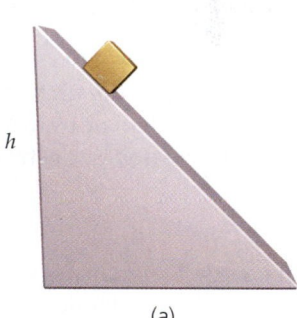

(a)

We've seen that mechanical energy conservation can be used as an alternative to Newton's laws for solving certain problems in mechanics. When we are not interested in the time $t$, the conservation of mechanical energy is often much easier to use than Newton's second law (Figure 7-7). Because mechanical energy conservation is derived from Newton's laws, any problem that can be solved using it can also be solved directly from Newton's laws, though often with much more difficulty.

# 7-2 The Conservation of Energy

In the macroscopic world, dissipative nonconservative forces, like kinetic friction, are always present to some extent. Such forces tend to decrease the mechanical energy of a system. However, any such decrease in mechanical energy is accompanied by a corresponding increase in thermal energy. (Consider how the tires of a car feel warm after a long trip.) Another type of nonconservative force is that involved in the deformations of objects. When you bend a metal coat hanger back and forth, you do work on the coat hanger but the work you do does not appear as mechanical energy. Instead the coat hanger becomes warm. The work done in deforming the hanger is dissipated as thermal energy. Similarly, when a falling ball of putty lands on the floor, it becomes warm as it deforms. The dissipated kinetic energy appears as thermal energy. For the ball–floor–earth system, the total energy is the sum of the thermal energy and the mechanical energy. The total energy of the system is conserved even though neither the total mechanical energy nor the total thermal energy are individually conserved.

A third type of nonconservative force is associated with chemical reactions. When we include systems in which chemical reactions take place, the sum of

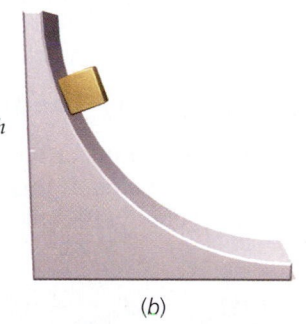

(b)

**FIGURE 7-7** (a) One can easily find the speed of a block sliding down a frictionless incline of constant slope by applying Newton's second law or by using conservation of mechanical energy. However, if the incline is frictionless but not of constant slope, as in (b), the problem can still be solved easily using conservation of mechanical energy, whereas it can be solved using Newton's second law only if the slope of the incline is known at each point, and then the calculation is quite tedious.

mechanical energy plus thermal energy is not conserved. For example, suppose that you begin running from rest. Initially you have no kinetic energy. When you begin to run, internal chemical energy in your muscles is converted to kinetic energy of your body and thermal energy is produced. It is possible to identify and measure the chemical energy that is used. In this case, the sum of mechanical, thermal, and chemical energy is conserved.

Even when thermal energy and chemical energy are included, the total energy of the system does not always remain constant, because energy can be converted to radiation energy, such as sound waves or electromagnetic waves. But *the increase or decrease in the total energy of a system can always be accounted for by the appearance or disappearance of energy somewhere else.* This experimental result, known as the **law of conservation of energy,** is one of the most important laws in all of science. Let $E_{sys}$ be the total energy of a given system, $E_{in}$ be the energy that enters the system, and $E_{out}$ be the energy that leaves the system. The law of conservation of energy then states

$$E_{in} - E_{out} = \Delta E_{sys} \qquad 7\text{-}8$$

LAW OF CONSERVATION OF ENERGY

Alternatively,

The total energy of the universe is constant. Energy can be converted from one form to another, or transmitted from one region to another, but energy can never be created or destroyed.

LAW OF CONSERVATION OF ENERGY

The total energy $E$ of many systems familiar from everyday life can be accounted for completely by mechanical energy $E_{mech}$, thermal energy $E_{therm}$, and chemical energy $E_{chem}$. To be comprehensive and include other possible forms of energy, such as electromagnetic or nuclear energy, we include $E_{other}$, and write generally

$$E_{sys} = E_{mech} + E_{therm} + E_{chem} + E_{other} \qquad 7\text{-}9$$

**EXPLORING**

*Devices that convert one form of energy to another are called transducers. To find out what your inner ear, a strain gauge, a microphone, and your sense of touch have in common, go to* www.whfreeman.com/tipler5e.

## The Work–Energy Theorem

A common way to transfer energy into or out of a system is to do work on the system from the outside. If this is the only method of energy transfer,[†] the law of conservation of energy becomes

$$W_{ext} = \Delta E_{sys} = \Delta E_{mech} + \Delta E_{therm} + \Delta E_{chem} + \Delta E_{other} \qquad 7\text{-}10$$

WORK–ENERGY THEOREM

where $W_{ext}$ is the work done on the system by external forces and $\Delta E_{sys}$ is the change in the system's total energy. This work–energy theorem for systems, which we will call simply the work–energy theorem, is a powerful tool for studying a wide variety of systems. Note that if the system is just a single particle its energy can only be kinetic, so Equation 7-10 is equivalent to the work–kinetic energy theorem studied in Chapter 6.

---

† Energy can also be transferred when heat is exchanged between a system and its surroundings. Exchanges of heat energy, which occur when there is a temperature difference between a system and its surroundings, are discussed in Chapter 18.

*FALLING CLAY*

**EXAMPLE 7-7**

**A ball of modeling clay with mass $m$ is released from rest from a height $h$ and falls to the perfectly rigid floor. Discuss the application of the law of conservation of energy to (a) the system consisting of the ball alone, and (b) the system consisting of the earth, the floor, and the ball.**

**PICTURE THE PROBLEM** There are two forces that act on the ball: gravity and the contact force of the floor. Since the floor does not move, the contact force it exerts on the ball does no work. There are no chemical or other energy changes so we can neglect $\Delta E_{chem}$ and $\Delta E_{other}$. We also neglect the sound energy radiated when the ball hits the floor. Thus, the only energy transferred to or from the ball is the work done by the force of gravity. Apply the work–energy theorem.

(a) 1. Write the work–energy theorem for the ball:

$$W_{ext} = \Delta E_{sys} = \Delta E_{mech} + \Delta E_{therm}$$

2. The two external forces on the system are gravity and the force exerted by the floor. The floor does not move and therefore does no work. The only work done on the ball is by gravity:

$$W_{ext} = mgh$$

3. Since the ball alone is our system, its mechanical energy is entirely kinetic, which is zero both initially and finally. Thus, the change in mechanical energy is zero:

$$\Delta E_{mech} = 0$$

4. Substitute $mgh$ for $W_{ext}$ and 0 for $\Delta E_{mech}$ in step 1:

$$W_{ext} = \Delta E_{mech} + \Delta E_{therm}$$
$$mgh = 0 + \Delta E_{therm}$$

so

$$\boxed{\Delta E_{therm} = mgh}$$

(b) 1. There are no external forces acting on the ball–earth–floor system (the force of gravity and the force of the floor are now internal to the system), so there is no external work done:

$$W_{ext} = 0$$

2. Write the work–energy theorem with $W_{ext} = 0$:

$$W_{ext} = \Delta E_{sys} = \Delta E_{mech} + \Delta E_{therm}$$
$$0 = \Delta E_{mech} + \Delta E_{therm}$$

3. The original mechanical energy of the ball–earth system is the original gravitational potential energy, and the final mechanical energy is zero:

$$E_{mech\,i} = mgh$$
$$E_{mech\,f} = 0$$

4. The change in mechanical energy of the ball–earth system is thus:

$$\Delta E_{mech} = 0 - mgh = -mgh$$

5. The work–energy theorem thus gives the same result as in Part (a):

$$\Delta E_{therm} = \boxed{-\Delta E_{mech} = mgh}$$

**REMARKS** In (a), energy is transferred to the ball by the work done on it by gravity. This energy appears as the kinetic energy of the ball before it hits the floor and as thermal energy after. The ball warms slightly and the energy is eventually transferred to the surroundings via heat. In (b), no energy is transferred to the ball–earth–floor system. The original potential energy of the system is converted to kinetic energy of the ball just before it hits, and then into thermal energy.

In this power plant in Kansas, energy stored in the fossil fuel coal (black area at the right) is released by burning the coal to produce steam; the steam is then used to drive turbines to produce electricity. The excess heat is dissipated by the cooling towers.

The potential energy of the water at the top of Niagara Falls is used to produce electrical energy.

## Problems Involving Kinetic Friction

When surfaces slide across each other kinetic friction decreases the mechanical energy of the system and increases the thermal energy. Consider a block that begins with initial velocity $v_i$ and slides along a table until it stops (Figure 7-8). We choose the block and table to be our system. Then $\Delta E_{chem} = \Delta E_{other} = 0$ and there is no external work done on this system. The work–energy theorem gives

$$0 = \Delta E_{mech} + \Delta E_{therm}$$

The mechanical energy lost is the initial kinetic energy of the block

$$\Delta E_{mech} = -\tfrac{1}{2}mv_i^2 \qquad\qquad 7\text{-}11$$

We can relate the loss in mechanical energy to frictional force. If $f$ is the magnitude of the frictional force, Newton's second law gives

$$-f = ma$$

Multiplying both sides of this equation by $\Delta s$, we find

$$-f\Delta s = ma\,\Delta s = m(\tfrac{1}{2}v_f^2 - \tfrac{1}{2}v_i^2) = -\tfrac{1}{2}mv_i^2 \qquad\qquad 7\text{-}12$$

where $\Delta s$ is the displacement of the block, and we have used the constant-acceleration formula $2a\,\Delta s = v_f^2 - v_i^2$ (with $v_f = 0$). Equating the left sides of Equations 7-11 and 7-12 we obtain

$$f\,\Delta s = -\Delta E_{mech} \qquad\qquad 7\text{-}13$$

Note that the quantity $-f\,\Delta s$ is *not* the work done by friction on the sliding block, because careful analysis shows the actual displacement of the kinetic frictional

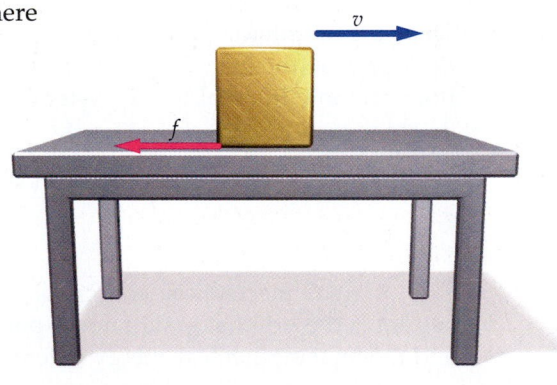

**FIGURE 7-8** A block sliding on a table. The force of friction reduces the mechanical energy of the block–table system.

force on the block is not in general equal to the displacement of the block.[†] However, it can be shown that $f\Delta s$ does equal the increase in thermal energy of the block–table system due to the dissipation of the system's mechanical energy. This thermal energy is produced both on the bottom surface of the block and on the upper surface of the tabletop as the block slides across the tabletop. Thus,

$$f \, \Delta s = \Delta E_{\text{therm}} \qquad\qquad 7\text{-}14$$

ENERGY DISSIPATED BY FRICTION

Substituting this result into the work–energy theorem (with $E_{\text{chem}} = E_{\text{other}} = 0$), we obtain

$$W_{\text{ext}} = \Delta E_{\text{mech}} + \Delta E_{\text{therm}} = \Delta E_{\text{mech}} + f \, \Delta s \qquad\qquad 7\text{-}15$$

WORK–ENERGY THEOREM FOR PROBLEMS WITH KINETIC FRICTION

where $\Delta s$ is the distance one contacting surface slides relative to the other. When there is no external work done on the system, the energy dissipated by friction equals the decrease in mechanical energy:

$$\Delta E_{\text{therm}} = f \, \Delta s = -\Delta E_{\text{mech}} \qquad (W_{\text{ext}} = 0) \qquad\qquad 7\text{-}16$$

Irish wind farm now under construction. The Arklow Banks project in the Irish Sea will have 200 turbines and produce 520 MW of electrical power—three times the combined capacity of all offshore wind farms currently in production in the world.

---

*PUSHING A BOX*

**EXAMPLE 7-8**

You push a 4-kg box, which is initially at rest on a horizontal table, a distance of 3 m with a horizontal force of 25 N. The coefficient of kinetic friction between the box and table is 0.35. Find (*a*) the external work done on the block–table system, (*b*) the energy dissipated by friction, (*c*) the final kinetic energy of the box, and (*d*) the speed of the box.

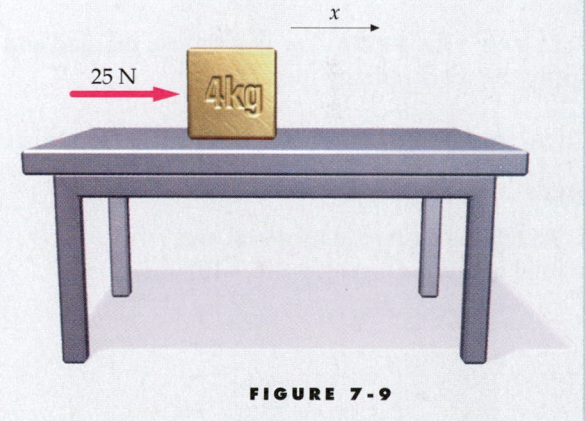

**FIGURE 7-9**

**PICTURE THE PROBLEM** The box plus table is the system (Figure 7-9). You are external to this system, so the force you push with is an external force. The final speed of the box is found from its final kinetic energy, which we find using the work–energy theorem with $\Delta E_{\text{chem}} = 0$ and $\Delta E_{\text{therm}} = f \, \Delta s$. The energy of the system is increased by the external work. Some of the energy increase is kinetic energy and some is thermal energy.

(*a*) There are four external forces acting on the system. However, only one of them does work. The total external work done is the product of the push force and the distance traveled:

$$\begin{aligned} \Sigma W_{\text{ext}} &= W_{\text{by push on block}} + W_{\text{by gravity on block}} \\ &\quad + W_{\text{by gravity on table}} + W_{\text{by floor on table}} \\ &= F_{\text{push}} \, \Delta x + 0 + 0 + 0 = (25\,\text{N})(3\,\text{m}) \\ &= \boxed{75\,\text{J}} \end{aligned}$$

(*b*) The energy dissipated by friction is $f\,\Delta x$ (the magnitude of the normal force equals $mg$):

$$\begin{aligned} \Delta E_{\text{therm}} &= f \, \Delta x = \mu_k F_n \, \Delta x = \mu_k mg \, \Delta x \\ &= (0.35)(4\,\text{kg})(9.81\,\text{N/kg})(3\,\text{m}) = \boxed{41.2\,\text{J}} \end{aligned}$$

---

[†] The work done by kinetic friction is examined in detail in B. A. Sherwood and W. H. Bernard, "Work and Heat Transfer in the Presence of Sliding Friction," *American Journal of Physics,* **52,** 1001 (1984).

(c) 1. Apply the work–energy theorem to find the final kinetic energy:

$$W_{ext} = \Delta E_{mech} + \Delta E_{therm}$$

2. There are no internal conservative forces doing work, so the change in potential energy is zero. The change in kinetic energy equals the change in mechanical energy:

$$\Delta E_{mech} = \Delta U + \Delta K$$
$$= 0 + (K_f - 0) = K_f$$

3. Substitute this result into the work–energy theorem:

$$W_{ext} = K_f + \Delta E_{therm}$$

so

$$K_f = W_{ext} - \Delta E_{therm}$$
$$= 75.0 \text{ J} - 41.2 \text{ J} = \boxed{33.8 \text{ J}}$$

(d) The final speed of the box is related to its kinetic energy. Solve for the final speed of the box:

$$K_f = \tfrac{1}{2}mv_f^2$$

so

$$v_f = \sqrt{\frac{2K_f}{m}} = \sqrt{\frac{2(33.8 \text{ J})}{4 \text{ kg}}} = \boxed{4.11 \text{ m/s}}$$

---

*A MOVING SLED*                    **EXAMPLE 7-9**   **Try It Yourself**

**A sled is coasting on a horizontal snow-covered surface with an initial speed of 4 m/s. If the coefficient of friction between the sled and the snow is 0.14, how far will the sled go before coming to rest?**

**PICTURE THE PROBLEM**  We choose the sled and snow as our system and then apply the work–energy theorem.

**Cover the column to the right and try these on your own before looking at the answers.**

**Steps**                                          **Answers**

1. Sketch the system in its initial and final configurations (Figure 7-10).

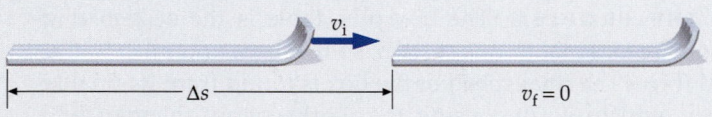

**FIGURE 7-10**

2. Apply the work–energy theorem.

$$W_{ext} = \Delta E_{mech} + \Delta E_{therm}$$
$$= (\Delta U + \Delta K) + f\,\Delta s$$

3. Solve for $f$. The normal force is equal to the weight.

$$f = \mu_k F_n = \mu_k mg$$

4. There are no conservative forces doing work and there are no external forces acting on the system. Use these observations to eliminate two terms from the step 2 result.

$$W_{ext} = \Delta U = 0$$

and

$$W_{ext} = \Delta U + \Delta K + f\,\Delta s$$
$$0 = 0 + \Delta K + \mu_k mg\,\Delta s$$

5. Express the change in kinetic energy in terms of the mass and the initial speed, and solve for $\Delta s$.

$$\Delta s = \frac{v_i^2}{2\mu_k g} = \boxed{5.82 \text{ m}}$$

**EXAMPLE 7-10**

**A child of mass 40 kg goes down an 8.0-m-long slide inclined at 30° with the horizontal. The coefficient of kinetic friction between the child and the slide is 0.35. If the child starts from rest at the top of the slide, how fast is she traveling when she reaches the bottom?**

**PICTURE THE PROBLEM** As the child slides down, some of her potential energy is converted into kinetic energy and some into thermal energy because of friction. We choose the child–slide–earth as our system and apply the conservation of energy theorem.

1. Make a sketch of the system showing both its initial and final configurations (Figure 7-11).

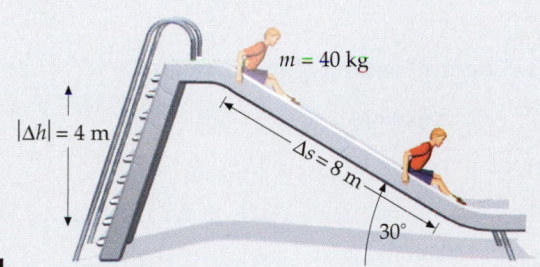

$m = 40$ kg

$|\Delta h| = 4$ m

$\Delta s = 8$ m

30°

**FIGURE 7-11**

2. Write out the conservation of energy equation:

$$W_{ext} = \Delta E_{mech} + \Delta E_{therm}$$
$$= (\Delta U + \Delta K) + f\,\Delta s$$

3. The initial kinetic energy is zero. The speed at the bottom is related to the final kinetic energy:

$$\Delta K = K_f - 0 = \tfrac{1}{2}mv_f^2$$

4. There are no external forces acting on the system:

$$W_{ext} = 0$$

5. The change in potential energy is related to the change in height $\Delta h$ (which is negative).

$$\Delta U = mg\,\Delta h$$

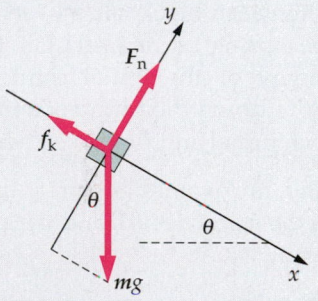

$y$

$F_n$

$f_k$

$\theta$

$\theta$

$mg$

$x$

**FIGURE 7-12**

6. To find $f_k$ we apply Newton's second law to the child. First we draw a free-body diagram (Figure 7-12):

7. Next we apply Newton's second law. The normal component of the acceleration is zero. To find $F_n$ we take components in the normal direction. Then we solve for $f_k$ using $f_k = \mu_k F_n$:

$$F_n - mg\cos\theta = 0$$
so
$$f_k = \mu_k F_n = \mu_k mg\cos\theta$$

8. We use trigonometry to relate $\Delta s$ to $\Delta h$:

$$|\Delta h| = \Delta s\sin\theta$$

9. Substituting into the step 2 result gives:

$$0 = mg\,\Delta h + \tfrac{1}{2}mv_f^2 + \mu_k mg\cos\theta\,\Delta s$$
$$= -mg\,\Delta s\sin\theta + \tfrac{1}{2}mv_f^2 + \mu_k mg\cos\theta\,\Delta s$$

10. Solving for $v_f$ gives:

$$v_f^2 = 2g\,\Delta s(\sin\theta - \mu_k\cos\theta)$$
$$= 2(9.81\ \text{m/s}^2)(8\ \text{m})(\sin 30° - 0.35\cos 30°)$$
$$= 30.9\ \text{m}^2/\text{s}^2$$

so

$$v_f = \boxed{5.60\ \text{m/s}}$$

**REMARKS** Note that the result is independent of the mass of the child.

**EXERCISE** For the earth–child–slide system, calculate (*a*) the initial mechanical energy, (*b*) the final mechanical energy, and (*c*) the energy dissipated by friction. (*Answer* (*a*) 1570 J, (*b*) 618 J, (*c*) 952 J)

TWO BLOCKS AND A SPRING **EXAMPLE 7-11** **Try It Yourself**

A 4-kg block hangs by a light string that passes over a massless, frictionless pulley and is connected to a 6-kg block that rests on a shelf. The coefficient of kinetic friction is 0.2. The 6-kg block is pushed against a spring, compressing it 30 cm. The spring has a force constant of 180 N/m. Find the speed of the blocks after the 6-kg block is released and the 4-kg block has fallen a distance of 40 cm.

**PICTURE THE PROBLEM** The speed of the blocks is obtained from their final kinetic energy. Consider the system to be everything shown in Figure 7-13 plus the earth. This system has both gravitational and elastic potential energy. Apply the work–energy theorem. Knowing the kinetic energy of the blocks means you can solve for their speed.

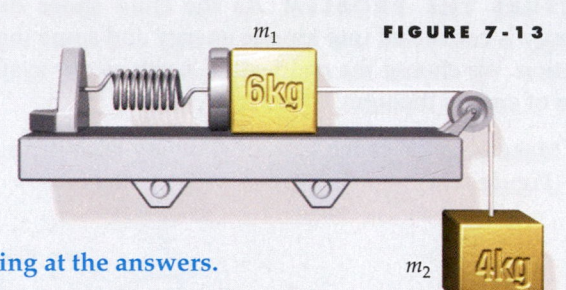

$m_1$   **FIGURE 7-13**

$m_2$

**Cover the column to the right and try these on your own before looking at the answers.**

**Steps**

**Answers**

1. Write out the equation for the conservation of energy of the system.

$W_{ext} = \Delta E_{mech} + \Delta E_{therm}$

$= (\Delta U_s + \Delta U_g + \Delta K) + f \Delta s$

2. There are no external forces on the system.

$W_{ext} = 0$

3. Make a table of the mechanical energy terms both initially, when the spring is compressed 30 cm, and finally, when each block has moved a distance $\Delta s = 40$ cm and the spring is relaxed. Let the gravitational potential energy of the initial configuration equal zero. Also, write down the difference (final minus initial) between each initial and final expression.

|  | $U_s$ | $U_g$ | $K$ |
|---|---|---|---|
| Final | 0 | $-m_2 g \, \Delta s$ | $\frac{1}{2}(m_1 + m_2)v_f^2$ |
| Initial | $\frac{1}{2}kx_i^2$ | 0 | 0 |
| Difference | $-\frac{1}{2}kx_i^2$ | $-m_2 g \, \Delta s$ | $\frac{1}{2}(m_1 + m_2)v_f^2$ |

4. Find an expression for $f_k$ that includes $\mu_k$. Substitute it, the step 2 result, and the step 3 results into the step 1 result.

$0 = -\frac{1}{2}kx_i^2 - m_2 g \, \Delta s + \frac{1}{2}(m_1 + m_2)v_f^2 + \mu_k m_1 g \, \Delta s$

5. Solve the step 4 result for $v_f^2$. Then substitute numerical values and solve for $v_f$:

$v_f^2 = \dfrac{kx_i^2 + 2m_2 g \, \Delta s - 2\mu_k m_1 g \, \Delta s}{m_1 + m_2}$

so

$v_f = \boxed{1.95 \text{ m/s}}$

**REMARKS** This solution assumes that the string remains taut at all times. This will be true if the acceleration of block 1 remains less than $g$, that is, if the net force on block 1 is less than $m_1 g = (6 \text{ kg})(9.81 \text{ N/kg}) = 58.9$ N. Initially the force exerted by the spring on block 1 has the magnitude $kx_1 = (180 \text{ N/m})(0.3 \text{ m}) = 54.0$ N and the frictional force has magnitude $f_k = \mu_k m_1 g = 0.2(58.9 \text{ N}) = 11.8$ N. These forces combine to produce a net force of 42.2 N directed to the right. Since the spring force decreases as block 1 moves following release, the acceleration of the 6-kg block will never exceed $g$, so the string will remain taut.

This pizza contains about 16 MJ of energy, about half the energy in 1 L (0.26 U.S. gal) of gasoline.

## Systems With Chemical Energy

Sometimes a system's internal chemical energy is converted into mechanical energy and thermal energy with no work being done on the system by external forces. For example, at the beginning of this section we described the energy conversions that take place when you start running. In order to move forward, you push back on the floor and the floor pushes forward on you with a static frictional force. This force causes you to accelerate, but it does not do work. It does no work because the displacement of the point of application of the force is zero (assuming your shoes do not slip on the floor). Because no work is done, no energy is transferred from the floor to your body. The kinetic-energy increase of your body comes from the conversion of internal chemical energy derived from the food you eat. Consider the following example.

---

*CLIMBING STAIRS*                    **EXAMPLE 7-12**

**Suppose that you have mass $m$ and you walk up a flight of stairs of height $h$. Discuss the application of energy conservation to a system consisting of you alone.**

**PICTURE THE PROBLEM**  There are two forces that act on you: gravity and the force of the stair treads on your feet. The force of gravity does negative work on you. To determine the work done by the force of the stair treads on your feet, consider the force of one stair tread on one of your feet. This force is a contact force. It does no work because the sole of your foot does not move while it is in contact with the tread. Thus, no work is done by the force of the stair treads on your feet.

1. Write the work–energy equation for the you-alone system:

$$W_{ext} = \Delta E_{sys} = \Delta E_{mech} + \Delta E_{therm} + \Delta E_{chem}$$

2. There are two external forces acting on this system, the force of gravity and the force of the stair treads on your feet. The force of gravity does negative work because this force is directed opposite to your displacement. The force of the stair treads does no work because the point of application, the soles of your feet, do not move while this force is applied:

$$W_{ext} = -mgh$$

3. As stated in the problem, the system consists of you alone. Because your configuration does not change, any change in your mechanical energy is entirely a change in your kinetic energy, which is the same initially and finally:

$$\Delta E_{mech} = 0$$

4. Substitute these results into the work–energy equation:

$$-mgh = \boxed{\Delta E_{therm} + \Delta E_{chem}}$$

**REMARKS**  If there were no change in thermal energy, then your chemical energy would decrease by $mgh$. Because the human body is relatively inefficient, the increase in the amount of thermal energy will be considerably greater than $mgh$. The decrease in stored chemical energy equals $mgh$ plus any thermal energy, which eventually is transferred from your body to your surroundings via heat.

**EXERCISE**  Discuss the energy conservation for a system consisting of both you and the earth. (*Answer*   For the you–earth system no external work is done, so the total energy, which now includes gravitational potential energy, is conserved. The change in mechanical energy is $mgh$, so the work–energy theorem gives $0 = mgh + \Delta E_{therm} + \Delta E_{chem}$.)

*AN UPHILL DRIVE*                          **EXAMPLE 7 - 1 3**

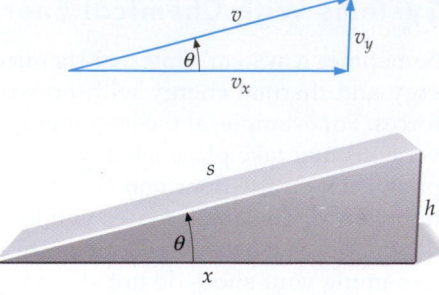

You are driving a 1000-kg gasoline-powered car at a constant speed of 100 km/h (= 27.8 m/s = 62.2 mi/h) up a 10-percent grade (Figure 7-14). (A 10-percent grade means that the road rises 1 m for each 10 m of horizontal distance—that is, the angle of inclination $\theta$ is given by $\tan \theta = 0.1$.) (a) If the efficiency is 15 percent, what is the rate at which the chemical energy of the car–earth–atmosphere system changes? (The efficiency is the fraction of the chemical energy consumed that appears as mechanical energy.) (b) What is the rate at which thermal energy is created?

**FIGURE 7-14**    $\tan \theta = h/x \sim \sin \theta = h/s$

**PICTURE THE PROBLEM** Some of the chemical energy goes into increasing the potential energy of the car as it climbs the hill, and some goes into an increase in thermal energy, much of which is expelled by the car as exhaust. Apply the work–energy theorem to a system consisting of the car, the hill, the air, and the earth.

(a) 1. The rate of loss of chemical energy equals the negative of change in chemical energy per unit time:

$$\text{chemical energy loss rate} = \frac{-\Delta E_{\text{chem}}}{\Delta t}$$

2. The increase in mechanical energy equals 15% of the decrease in chemical energy:

$$\Delta E_{\text{mech}} = 0.15 |\Delta E_{\text{chem}}| = -0.15 \, \Delta E_{\text{chem}}$$

3. Solve for the loss rate of chemical energy:

$$\frac{-\Delta E_{\text{chem}}}{\Delta t} = \frac{1}{0.15} \frac{\Delta E_{\text{mech}}}{\Delta t}$$

4. The car moves at constant speed, so $\Delta K = 0$ and $\Delta E_{\text{mech}} = \Delta U$. Relate the change in mechanical energy to the change in height $\Delta h$ and substitute it into the step 3 result:

$$\Delta E_{\text{mech}} = mg \, \Delta h$$

so

$$\frac{-\Delta E_{\text{chem}}}{\Delta t} = \frac{mg}{0.15} \frac{\Delta h}{\Delta t}$$

5. Convert the changes to time derivatives. That is, take the limit of both sides as $\Delta t$ approaches zero:

$$-\frac{dE_{\text{chem}}}{dt} = \frac{mg}{0.15} \frac{dh}{dt}$$

6. The rate of change of $h$ equals $v_y$, which is related to the speed $v$ as shown in Figure 7-14:

$$\frac{dh}{dt} = v_y = v \sin \theta$$

7. We can approximate $\sin \theta$ by $\tan \theta$ because the angle is small:

$$\sin \theta \approx \tan \theta$$

8. Solve for the loss rate of chemical energy:

$$-\frac{dE_{\text{chem}}}{dt} = \frac{mg}{0.15} v \sin \theta \approx \frac{mg}{0.15} v \tan \theta$$

$$= \frac{(1000 \text{ kg})(9.81 \text{ N/kg})}{0.15} (27.8 \text{ m/s})0.1$$

$$= \boxed{182 \text{ kW}}$$

(b) 1. Write out the work–energy relation:

$$W_{\text{ext}} = \Delta E_{\text{mech}} + \Delta E_{\text{therm}} + \Delta E_{\text{chem}}$$

2. Convert to derivatives and solve for $dE_{\text{therm}}/dt$:

$$0 = \frac{dE_{\text{mech}}}{dt} + \frac{dE_{\text{therm}}}{dt} + \frac{dE_{\text{chem}}}{dt}$$

so

$$\frac{dE_{\text{therm}}}{dt} = -\frac{dE_{\text{mech}}}{dt} - \frac{dE_{\text{chem}}}{dt} = 0.15 \frac{dE_{\text{chem}}}{dt} - \frac{dE_{\text{chem}}}{dt}$$

$$= -0.85 \frac{dE_{\text{chem}}}{dt} = 0.85(182 \text{ kW}) = \boxed{155 \text{ kW}}$$

**REMARKS** Gasoline-powered cars are typically only about 15 percent efficient. About 85 percent of the chemical energy of the gasoline goes to thermal energy, most of which is expelled as heat exhaust. Additional thermal energy is created by rolling friction and wind resistance. The energy content of gasoline is about 31.8 MJ/L.

# 7-3 Mass and Energy

In 1905, Albert Einstein published his special theory of relativity, a result of which is the famous equation

$$E_0 = mc^2 \qquad\qquad 7\text{-}17$$

where $c = 3 \times 10^8$ m/s is the speed of light in vacuum. We will study this theory in some detail in later chapters.

According to Equation 7-17, a particle or system of mass $m$ has "rest" energy $mc^2$. This energy is intrinsic to the particle. Consider the positron, a particle emitted in a nuclear process called beta decay. Positrons and electrons have identical masses, but equal and opposite electrical charge. When a positron encounters an electron in matter, electron–positron annihilation occurs, a process in which the electron and positron disappear and their energy appears as electromagnetic radiation. If the two particles are initially at rest, the energy of the electromagnetic radiation equals the rest energy of the electron plus that of the positron.

Energies in atomic and nuclear physics are usually expressed in units of electron volts (eV) or mega-electron-volts (1 MeV = $10^6$ eV). A convenient unit for the masses of atomic particles is eV/$c^2$ or MeV/$c^2$. Table 7-1 lists rest energies (and therefore the masses) of some elementary particles and light nuclei. The total rest energy of a positron plus an electron is 2(0.511 MeV), which is the radiation energy emitted upon annihilation.

## TABLE 7-1

**Rest Energies of Some Elementary Particles and Light Nuclei**

| Particle | Symbol | Rest Energy (MeV) | |
|----------|--------|-------------------|---|
| Electron | $e^-$ | 0.5110 | |
| Positron | $e^+$ | 0.5110 | |
| Proton | p | 938.272 | |
| Neutron | n | 939.565 | |
| Deuteron | d | 1875.613 | |
| Triton | t | 2808.410 | |
| Helium-3 | $^3$He | 2808.39 | |
| Alpha particle | $\alpha$ | 3727.379 | |

The rest energy of a *system* can consist of the potential energy of the system or other internal energies of the system, in addition to the intrinsic rest energies of the particles in the system. If the system at rest absorbs energy $\Delta E$ and remains at rest, its rest energy increases by $\Delta E$ and its mass increases by

$$\Delta M = \frac{\Delta E}{c^2} \qquad\qquad 7\text{-}18$$

Consider two 1-kg blocks connected by a spring of force constant $k$. If we stretch the spring a distance $x$, the potential energy of the system increases by $\Delta U = \frac{1}{2}kx^2$. According to Equation 7-18, the mass of the system has also increased by $\Delta M = \Delta U/c^2$. Because $c$ is such a large number, this increase in mass cannot be observed in macroscopic systems. For example, suppose $k = 800$ N/m and $x = 10$ cm $= 0.1$ m. The potential energy of the spring system is then $\frac{1}{2}kx^2 = \frac{1}{2}(800 \text{ N/m})(0.1 \text{ m})^2 = 4$ J. The increase in mass of the system is

$$\Delta M = \frac{\Delta U}{c^2} = \frac{4 \text{ J}}{(3 \times 10^8 \text{ m/s})^2} = 4.44 \times 10^{-17} \text{ kg}$$

The relative mass increase $\Delta M/M \approx 2 \times 10^{-17}$ is much too small to be observed.

## Nuclear Energy

In nuclear reactions, the energy changes are often an appreciable fraction of the rest energy of the system. Consider the deuteron, which is the nucleus of deuterium, an isotope of hydrogen also called heavy hydrogen. The deuteron consists of a proton and neutron bound together. From Table 7-1 we see that the mass of the proton is 938.28 MeV/$c^2$ and the mass of the neutron is 939.57 MeV/$c^2$. The sum of these two masses is 1877.85 MeV/$c^2$. But the mass of the deuteron is 1875.63 MeV/$c^2$, which is less than the sum of the masses of the proton and neutron by 2.22 MeV/$c^2$. Note that this mass difference $\Delta M/M \approx 1.2 \times 10^{-3}$, much greater than any uncertainties in the measurement of these masses, and almost 14 orders of magnitude greater than the $2 \times 10^{-17}$ discussed above for the spring–blocks system.

Heavy water (deuterium oxide) molecules are produced in the primary cooling water of a nuclear reactor when neutrons collide with the hydrogen nuclei (protons) of the water molecules. If a slow moving neutron is captured by a proton, 2.22 MeV of energy are released in the form of electromagnetic radiation. Thus, the mass of a deuterium atom is 2.22 MeV/$c^2$ less than the sum of the masses of an isolated hydrogen atom and an isolated neutron.

This process can be reversed by breaking a deuteron into its constituent parts if at least 2.22 MeV of energy is transferred to the deuteron with electromagnetic radiation or by collisions with other energetic particles. Any transferred energy in excess of 2.22 MeV appears as kinetic energy of the resulting proton and neutron.

The energy needed to completely separate a nucleus into individual neutrons and protons is called the **binding energy.** The binding energy of a deuteron is 2.22 MeV.

The deuteron is an example of a bound system. Its rest energy is less than the rest energy of its parts, so energy must be put into the system to break it apart. If the rest energy of a system is greater than the rest energy of its parts, the system is unbound. An example is uranium-236, which breaks apart or **fissions** into two smaller nuclei.[†] The sum of the masses of the resultant parts is less than the mass of the original nucleus. Thus the mass of the system decreases, and energy is released.

In nuclear fusion, two very light nuclei such as a deuteron and a triton (the nucleus of the hydrogen isotope tritium) fuse together. The rest mass of the resultant nucleus is less than that of the original parts and again energy is released. In a chemical reaction that produces energy, such as burning coal, the mass decrease is of the order of 1 eV/$c^2$ per atom. This is more than a million times smaller than the mass changes per nucleus in many nuclear reactions, and is not readily observable.

---

† Uranium-236, written $^{236}$U, is made in a nuclear reactor when the stable isotope uranium-235 absorbs a neutron. This reaction will be discussed in Chapter 34.

*BINDING ENERGY*                    **EXAMPLE 7-14**

**A hydrogen atom consisting of a proton and an electron has a binding energy of 13.6 eV. By what percentage is the mass of a proton plus the mass of an electron greater than that of the hydrogen atom?**

1. The fractional difference between the mass of the hydrogen atom and the masses of its parts is the ratio of the binding energy $E_b$ divided by $c^2$ to $m_e + m_p$:

$$\text{Fractional difference} = \frac{E_b/c^2}{m_e + m_p} = \frac{13.6 \text{ eV}/c^2}{m_e + m_p}$$

2. Obtain the rest masses of the proton and electron from Table 7-1:

$$m_p = 938.28 \text{ MeV}/c^2; \qquad m_e = 0.511 \text{ MeV}/c^2$$

3. Add to find the sum of these masses:

$$m_p + m_e = 938.79 \text{ MeV}/c^2$$

4. The rest mass of the hydrogen atom is less than this by $13.6 \text{ eV}/c^2$. The percentage difference is:

$$\text{Fractional difference} = \frac{13.6 \text{ eV}/c^2}{938.79 \times 10^6 \text{ eV}/c^2}$$

$$= 1.45 \times 10^{-8}$$

$$= \boxed{1.45 \times 10^{-6} \%}$$

**REMARKS** This mass difference is too small to be measured directly. However, binding energies can be accurately measured, so the mass difference can be found from $E_b = (\Delta m)c^2$.

*NUCLEAR FUSION*               **EXAMPLE 7-15**   **Try It Yourself**

**In a typical nuclear fusion reaction, a tritium nucleus ($^3$H) and a deuterium nucleus ($^2$H) fuse together to form a helium nucleus ($^4$He) plus a neutron. The reaction is written $^2$H + $^3$H → $^4$He + n. If the initial kinetic energy of the particles is negligible, how much energy is released in this fusion reaction?**

**PICTURE THE PROBLEM** Because energy is released, the total rest energy of the initial particles must be greater than that of the final particles. This difference equals the energy released.

**Cover the column to the right and try these on your own before looking at the answers.**

| Steps | Answers |
|---|---|
| 1. Write down the rest energies of $^2$H and $^3$H from Table 7-1 and add to find the total initial rest energy. | $E_0$ (initial) = 1875.628 MeV + 2808.944 MeV<br>= 4684.572 MeV |
| 2. Do the same for $^4$He and n to find the final rest energy. | $E_0$ (final) = 3727.409 MeV + 939.573 MeV<br>= 4666.982 MeV |
| 3. Find the energy released from $E_{\text{released}} = E_0$ (initial) − $E_0$ (final). | $E_{\text{released}}$ = 4684.572 MeV − 4666.982 MeV<br>= $\boxed{17.59 \text{ MeV} \approx 17.6 \text{ MeV}}$ |

**REMARKS** This and other fusion reactions occur in the sun. The energy that is released bathes the earth and is ultimately responsible for all life on the planet. The energy emitted by the sun is accompanied by a continuous decrease in the sun's rest mass.

## Newtonian Mechanics and Relativity

When the speed of a particle approaches the speed of light, Newton's second law breaks down, and we must modify Newtonian mechanics according to Einstein's theory of relativity. The criterion for the validity of Newtonian mechanics can also be stated in terms of the energy of a particle. In Newtonian mechanics, the kinetic energy of a particle moving with speed $v$ is

$$K = \tfrac{1}{2}mv^2 = \tfrac{1}{2}mc^2\frac{v^2}{c^2} = \tfrac{1}{2}E_0\frac{v^2}{c^2}$$

where $E_0 = mc^2$ is the rest energy of the particle. Then

$$\frac{v}{c} = \sqrt{\frac{2K}{E_0}}$$

Newtonian mechanics is valid if the speed of the particle is much less than the speed of light, or alternatively, the kinetic energy of a particle is much less than its rest energy.

# 7-4 Quantization of Energy

When energy is put into a system that remains at rest, the internal energy of the system increases. While it might seem that we could put any amount of energy into a system, this is found not to be true for microscopic systems such as atoms or molecules. The internal energy of a microscopic system can increase only by discrete increments.

If we have two blocks attached to a spring and we pull the blocks apart, we do work on the block–spring system, and its potential energy increases. If we then release the blocks, they oscillate back and forth. The energy of oscillation $E$—the kinetic energy of motion of the blocks plus the potential energy due to the stretching of the spring—equals the initial potential energy. In time, the energy of the system decreases because of various damping effects such as friction and air resistance. As closely as we can measure, the energy decreases continuously. Eventually all the energy is dissipated and the energy of oscillation is zero.

Now consider a diatomic molecule such as molecular oxygen, $O_2$. The force of attraction between the two oxygen atoms varies approximately linearly with the change in separation (for small changes), like that of two blocks connected by a spring. If a diatomic molecule is set oscillating with some energy $E$, the energy decreases with time as the molecule radiates or interacts with its surroundings, but the decrease is *not continuous*. The energy decreases in finite steps, and the lowest energy state, called the ground state, is not zero. The vibrational energy of a diatomic molecule is said to be **quantized;** that is, the molecule can absorb or release energies only in certain amounts, known as quanta.

When either blocks on a spring or diatomic molecules oscillate, the time for one oscillation is called the period $T$. The reciprocal of the period is the frequency of oscillation $f = 1/T$. We will see in Chapter 14 that the period and frequency of an oscillator do not depend on the energy of oscillation. As the energy decreases, the frequency remains the same. Figure 7-15 shows an **energy-level diagram** for an oscillator. The allowed energies are approximately equally spaced, and are given by[†]

$$E_n = (n + \tfrac{1}{2})hf, \qquad n = 0, 1, 2, 3, \ldots \tag{7-19}$$

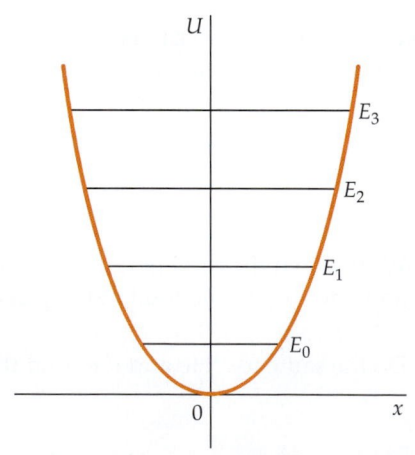

**FIGURE 7-15** Energy-level diagram for an oscillator.

---

[†] A diatomic molecule can also have rotational energy. The rotational energy is also quantized, but the energy levels are not equally spaced, and the lowest possible energy is zero. We will study rotational energy in Chapters 9 and 10.

where $f$ is the frequency of oscillation and $h$ is a fundamental constant of nature called Planck's constant:[†]

$$h = 6.626 \times 10^{-34} \text{ J·s} \qquad 7\text{-}20$$

The integer $n$ is called a **quantum number.** The lowest possible energy is the **ground state energy** $E_0 = \frac{1}{2}hf$.

Microscopic systems often gain or lose energy by absorbing or emitting electromagnetic radiation. By conservation of energy, if $E_i$ and $E_f$ are the initial and final energies of a system, the energy of the radiation emitted or absorbed is

$$E_{rad} = E_i - E_f$$

Since the system energies $E_i$ and $E_f$ are quantized, the radiated energy is also quantized.[‡] The quantum of radiation energy is called a **photon.** The energy of a photon is given by

$$E_{photon} = hf \qquad 7\text{-}21$$

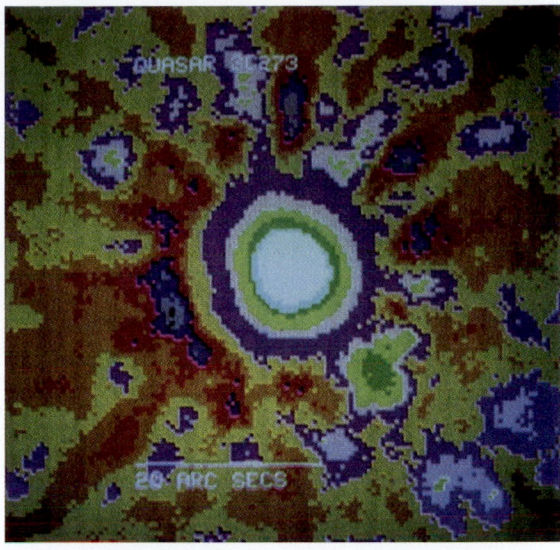

The quasar 3C 273 is shown imaged via X-ray energy. The X-ray energy emitted by this quasar is more than a million times that emitted by the entire Milky Way galaxy.

where $f$ is the frequency of the electromagnetic radiation.[§]

As far as we know, all bound systems exhibit energy quantization. For macroscopic bound systems, the steps between energy levels are so small that they are unobservable. For example, typical oscillation frequencies for two blocks on a spring are 1 to 10 times per second. If $f = 10$ oscillations per second, the spacing between allowed levels is $hf = (6.626 \times 10^{-34} \text{ J·s}) \times (10/\text{s}) \approx 6 \times 10^{-33} \text{ J}$. Since the energy of a macroscopic system is of the order of joules, a quantum step of $10^{-33}$ J is too small to be noticed. To put it another way, if the energy of a system is 1 J, the value of $n$ is of the order of $10^{32}$ and changes of one or two quantum units will not be observable.

For a diatomic molecule, a typical frequency of vibration is $10^{14}$ vibrations per second, and a typical energy is $10^{-19}$ J. The spacing between allowed levels is then

$$E_{n+1} - E_n = hf \approx (6.63 \times 10^{-34} \text{ J·s})(10^{14} \text{ s}) \approx 6 \times 10^{-20} \text{ J}$$

Thus, changes in the energy of oscillation are of the same order of magnitude as the energy of the molecule, and quantization is definitely not negligible.

## SUMMARY

1. The conservation of mechanical energy is an important relation derived from Newton's laws for conservative forces. It is useful in solving many problems.

2. The work–energy theorem and the conservation of energy are fundamental laws of nature that have applications in all areas of physics.

3. Einstein's equation $E_0 = mc^2$ is a fundamental relation between mass and energy.

4. Quantization is a fundamental property of the energy in bound systems.

---

[†] In 1900, the German physicist Max Planck introduced this constant as a calculational device to explain discrepancies between the theoretical curves and experimental data on the spectrum of blackbody radiation. The significance of Planck's constant was not appreciated by Planck or anyone else until Einstein postulated in 1905 that the energy of electromagnetic radiation is not continuous, but occurs in packets of size $hf$ where $f$ is the frequency of the radiation.

[‡] Historically, the quantization of electromagnetic radiation, as proposed by Max Planck and Albert Einstein, was the first "discovery" of energy quantization.

[§] Electromagnetic radiation includes light, microwaves, radio waves, television waves, X rays, and gamma rays. These differ from one another in their frequencies.

| Topic | Relevant Equations and Remarks |
|---|---|
| **1. Mechanical Energy** | The sum of the kinetic and potential energy of a system is called the total mechanical energy |
| | $$E_{mech} = K_{sys} + U_{sys} \qquad \text{7-5}$$ |
| Conservation of mechanical energy | If no external forces do work on a system, and if the internal forces that do work are all conservative, the total mechanical energy of the system remains constant: |
| | $$E_{mech} = K_{sys} + U_{sys} = \text{constant} \qquad \text{7-6}$$ |
| | $$K_f + U_f = K_i + U_i \qquad \text{7-7}$$ |
| **2. Total Energy of a System** | The energy of a system consists of mechanical energy $E_{mech}$, thermal energy $E_{therm}$, chemical energy $E_{chem}$, and other types of energy $E_{other}$, such as sound radiation and electromagnetic radiation. |
| | $$E_{sys} = E_{mech} + E_{therm} + E_{chem} + E_{other} \qquad \text{7-9}$$ |
| **3. Conservation of Energy** | |
| Universe | The total energy of the universe is constant. Energy can be converted from one form to another, or transmitted from one region to another, but energy can never be created or destroyed. |
| System | The energy of a system can be changed by work being done on the system and by energy transfer via heat. (These include the emission or absorption of radiation.) The increase or decrease in the energy of the system can always be accounted for by the appearance or disappearance of some kind of energy somewhere else: |
| | $$E_{in} - E_{out} = \Delta E_{sys} \qquad \text{7-8}$$ |
| Work–energy theorem | $$W_{ext} = \Delta E_{sys} = \Delta E_{mech} + \Delta E_{therm} + \Delta E_{chem} + \Delta E_{other} \qquad \text{7-10}$$ |
| **4. Energy Dissipated by Friction** | For a system that involves a pair of sliding surfaces, the total energy dissipated by friction on both surfaces equals the increase in thermal energy of the system and is given by |
| | $$f\,\Delta s = \Delta E_{therm} \qquad \text{7-14}$$ |
| | where $\Delta s$ is the distance one surface slides relative to the other. |
| **5. Problem Solving** | The conservation of mechanical energy and the work–energy theorem can be used as an alternative to Newton's laws to solve mechanics problems that require the determination of the speed of a particle as a function of its position. |
| **6. Mass and Energy** | A particle with mass $m$ has an intrinsic rest energy $E_0$ given by |
| | $$E_0 = mc^2 \qquad \text{7-17}$$ |
| | where $c = 3 \times 10^8$ m/s is the speed of light in vacuum. A system with mass $M$ also has a rest energy $E_0 = Mc^2$. If a system gains or loses internal energy $\Delta E$, it simultaneously gains or loses mass $\Delta M = \Delta E/c^2$. |
| Binding energy | The energy required to separate a bound system into its constituent parts is called its binding energy. The binding energy is $\Delta Mc^2$, where $\Delta M$ is the sum of the masses of the constituent parts, less the mass of the bound system. |
| **8. Newtonian Mechanics and Special Relativity** | When the speed of a particle approaches the speed of light $c$ (when the kinetic energy of the particle approaches its rest energy), Newtonian mechanics breaks down, and must be replaced by Einstein's special theory of relativity. |

9. **Energy Quantization**

The internal energy of a system is found to have only a discrete set of possible values. For a system oscillating with frequency $f$, the allowed energy values are separated by an amount $hf$, where $h$ is Planck's constant:

$$h = 6.626 \times 10^{-34} \text{ J·s}$$

**7-20**

Photons

Microscopic systems often exchange energy with their surroundings by emitting or absorbing electromagnetic radiation, which is also quantized. The quantum of energy of radiation is called the photon:

$$E_{photon} = hf$$

**7-21**

where $f$ is the frequency of the electromagnetic radiation.

# PROBLEMS

- • Single-concept, single-step, relatively easy
- •• Intermediate-level, may require synthesis of concepts
- ••• Challenging
- SSM Solution is in the *Student Solutions Manual*
- iSOLVE Problems available on iSOLVE online homework service
- iSOLVE ✓ These "Checkpoint" online homework service problems ask students additional questions about their confidence level, and how they arrived at their answer

In a few problems, you are given more data than you actually need; in a few other problems, you are required to supply data from your general knowledge, outside sources, or informed estimates.

**Take $g = 9.81$ N/kg $= 9.81$ m/s² and neglect friction in all problems unless otherwise stated.**

## Conceptual Problems

1 • SSM Two objects of unequal mass are connected by a massless cord passing over a frictionless peg. After the objects are released from rest, which of the following statements are true? ($U$ = gravitational potential energy, $K$ = kinetic energy of the system.) (a) $\Delta U < 0$ and $\Delta K > 0$. (b) $\Delta U = 0$ and $\Delta K > 0$. (c) $\Delta U < 0$ and $\Delta K = 0$. (d) $\Delta U = 0$ and $\Delta K = 0$. (e) $\Delta U > 0$ and $\Delta K < 0$.

2 • Two stones are thrown with the same initial speed at the same instant from the roof of a building. One stone is thrown at an angle of 30° above the horizontal, the other is thrown horizontally. (Neglect air resistance.) Which statement below is true?

(a) The stones strike the ground at the same time and with equal speeds.
(b) The stones strike the ground at the same time with different speeds.
(c) The stones strike the ground at different times with equal speeds.
(d) The stones strike the ground at different times with different speeds.

3 • True or false:

(a) The total energy of a system cannot change.
(b) When you jump into the air, the floor does work on you increasing your mechanical energy.

4 • You stand on roller skates next to a rigid wall. To get started, you push off against the wall. Discuss the energy changes pertinent to this situation.

5 • SSM In *Surely You're Joking, Mr. Feynman,*[†] Richard Feynman described his annoyance at how the concept of energy was portrayed in a children's textbook in the following way: "There was a book which started out with four pictures: first, there was a wind-up toy; then there was an automobile; then there was a boy riding a bicycle; then there was something else. And underneath each picture it said 'What makes it go?' . . . I turned the page. The answer was . . . for everything, 'Energy makes it go' . . . It's also not even true that 'energy makes it go' because if it stops, you could say 'energy makes it stop' just as well. . . . Energy is neither increased or decreased in these examples; it's just changed from one form to another." Describe how energy changes from one form to another when a little girl pedals her bike up a hill, then freewheels down the hill, and brakes to a stop.

6 • SSM A body falling through the atmosphere (air resistance is present) gains 20 J of kinetic energy. The amount of gravitational potential energy that it lost is (a) 20 J, (b) more than 20 J, (c) less than 20 J, (d) impossible to tell without knowing the mass of the body, (e) impossible to tell without knowing how far the body falls.

---

† Richard P. Feynman and Ralph Leighton, *Surely You're Joking, Mr. Feynman,* New York: Bantam Books (1985).

**7** •• Assume that, on applying the brakes, a constant frictional force acts on the wheels of a car. If that is so, it follows that (a) the distance the car travels before coming to rest is proportional to the speed of the car before the brakes are applied, (b) the car's kinetic energy diminishes at a constant rate, (c) the kinetic energy of the car is inversely proportional to the time that has elapsed since the application of the brakes, (d) none of the above are true.

**8** • You are given two frictionless ramps and a block to slide down them (Figure 7-16). One ramp is in the form of an inclined plane with height H and length L. The other has a ramp cut in the form of a partial arc of a circle, but also has height H and length L. You slide the block down each ramp, releasing it from rest, and measure the time it takes to reach the bottom and the speed of the block upon getting there. You find that (a) the block takes the same time to slide down each ramp, (b) the block has the same speed on reaching the bottom of each ramp, (c) both (a) and (b) are correct, (d) neither (a) nor (b) is correct.

**FIGURE 7-16** Problem 8

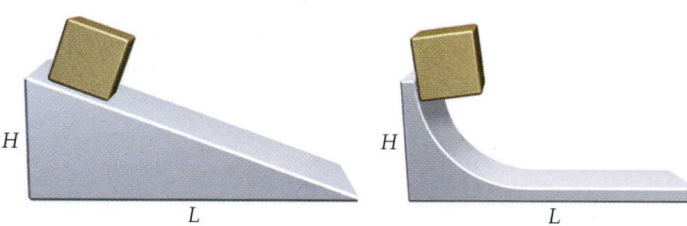

**9** •• If a rock is attached to a massless, rigid rod and swung in a vertical circle at a constant speed, it will not have a constant total energy, as the kinetic energy of the rock will be constant, but the potential energy will be continually changing. Is any total work being done on the rock? Does the rod exert a tangential force on the rock?

### Estimation and Approximation

**10** •• **SSM** The *metabolic rate* is the rate at which the body uses chemical energy to sustain its life functions. Experimentally, the average metabolic rate is proportional to the total skin surface area of the body. The surface area for a 5-ft, 10-in. male weighing 175 lb is just about 2.0 m² and for a 5-ft, 4-in. female weighing 110 lb is approximately 1.5 m². There is about a 1 percent change in surface area for every 3 lb above or below the weights quoted here and a 1 percent change for every inch above or below the heights quoted. (a) Estimate your average metabolic rate over the course of a day using the following guide for physical activity: sleeping, metabolic rate = 40 W/m²; sitting, 60 W/m²; walking, 160 W/m²; moderate physical activity, 175 W/m²; and moderate aerobic exercise, 300 W/m². How does it compare to the power of a 100-W light bulb? (b) Express your average metabolic rate in terms of kcal/day (1 kcal = 4190 J). (A kcal is the "food calorie" used by nutritionists.) (c) An estimate used by nutritionists is that the "average person" must eat roughly 12–15 kcal/lb of body weight a day to maintain his or her weight. From the calculations in part (b), are these estimates reasonable?

**11** • Assume that the maximum rate at which your body can expend energy is 250 W. Assuming a 20 percent efficiency for the conversion of chemical energy into mechanical energy, estimate how quickly you can run up four flights of stairs, with each flight 3.5 m high.

**12** • How much rest mass is consumed in the core of a nuclear-fueled electric generating plant in producing (a) 1 J of thermal energy? (b) enough electrical energy to keep a 100-W light bulb burning for 10 y? (For each joule of electrical energy produced by the generator, the reactor core must produce 3 J of nuclear energy.)

**13** • **SSM** The chemical energy released by burning a gallon of gasoline is approximately $2.6 \times 10^5$ kJ. Estimate the total energy used by all of the cars in the United States during the course of 1 y. What fraction does this represent of the total energy use by the United States in 1 y (about $5 \times 10^{20}$ J)?

**14** • The maximum efficiency of a solar energy panel in converting solar energy into useful electrical energy is about 12 percent. Using the known value of the solar intensity reaching the earth's surface (1.0 kW/m²), what area would have to be covered by solar panels in order to supply the energy requirements of the United States (approximately $5 \times 10^{20}$ J/y)? Assume cloudless skies.

**15** • Hydroelectric power plants convert gravitational potential energy into more useful forms by flowing water downhill through a turbine system to generate electrical energy. The Hoover Dam on the Colorado River is 211 m high and generates 4 billion kW·h/y (1 W · h = $3.6 \times 10^3$ J.) At what flow rate (in L/s) must water be flowing through the turbines to generate this power? The density of water is 1 kg/L. Assume a total efficiency of 20 percent in converting the water's potential energy into electrical energy.

### The Conservation of Mechanical Energy

**16** • A block of mass m is pushed against a spring, compressing it a distance x, and the block is then released. The spring projects the block along a frictionless horizontal surface, giving a speed v. The same spring projects a second block of mass 4m, giving it a speed 3v. What distance was the spring compressed in the second case?

**17** • A bicyclist traveling at 10 m/s on a horizontal road stops pedaling as she starts up a hill inclined at 3.0° to the horizontal. Ignoring frictional forces, how far up the hill will she travel before stopping? (a) 5.1 m. (b) 30 m. (c) 97 m. (d) 10.2 m. (e) The answer depends on the mass of the person.

**18** • **SSM** A pendulum of length L with a bob of mass m is pulled aside until the bob is at a height L/4 above its equilibrium position. The bob is then released. Find the speed of the bob as it passes the equilibrium position.

**19** • **ISOLVE** A 3-kg block slides along a frictionless horizontal surface with a speed of 7 m/s (Figure 7-17). After sliding a distance of 2 m, the block makes a smooth transition to a frictionless ramp inclined at an angle of 40° to the horizontal. How far up the ramp does the block slide before coming momentarily to rest?

**FIGURE 7-17** Problems 19, 46

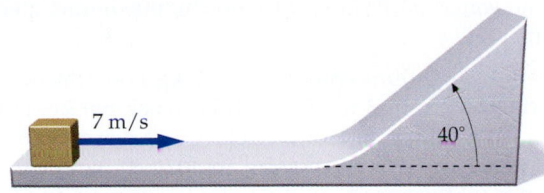

**20** • **iSOLVE** ✓ The 3-kg object in Figure 7-18 is released from rest at a height of 5 m on a curved frictionless ramp. At the foot of the ramp is a spring of force constant $k = 400$ N/m. The object slides down the ramp and into the spring, compressing it a distance $x$ before coming momentarily to rest. (a) Find $x$. (b) What happens to the object after it comes to rest?

**FIGURE 7-18** Problem 20

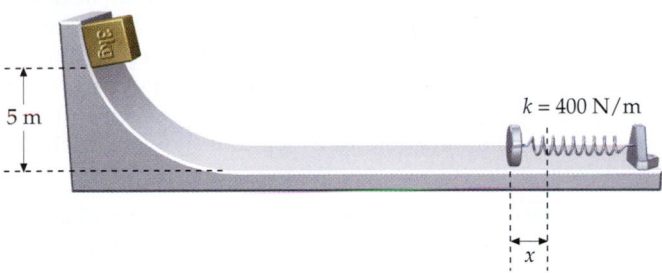

**21** • **iSOLVE** ✓ A 15-g ball is shot from a spring gun whose spring has a force constant of 600 N/m. The spring can be compressed 5 cm. How high will the ball go if the gun is aimed vertically?

**22** • **iSOLVE** At a dock, a crane lifts a 4000-kg container 30 m, swings it out over the deck of a freighter, and lowers the container into the hold of the freighter, which is 8 m below the level of the dock. How much work is done by the crane on the container? (Neglect friction losses.)

**23** • **iSOLVE** A 16-kg child on a playground swing moves with a speed of 3.4 m/s when the 6-m long swing is at its lowest point. What is the angle that the swing makes with the vertical when the swing is at its highest point?

**24** •• **SSM** The system shown in Figure 7-19 is initially at rest when the lower string is cut. Find the speed of the objects when they are at the same height. The frictionless pulley has negligible mass.

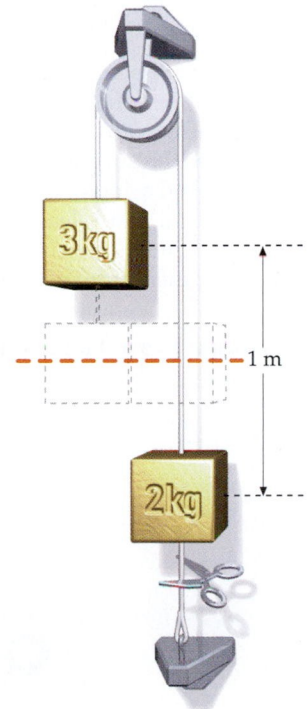

**FIGURE 7-19** Problem 24

**25** •• A block rests on an inclined plane as shown in Figure 7-20. A spring to which it is attached via a pulley is being pulled downward with gradually increasing force. The value of $\mu_s$ is known. Find the potential energy $U$ of the spring at the moment when the block begins to move.

**FIGURE 7-20** Problem 25

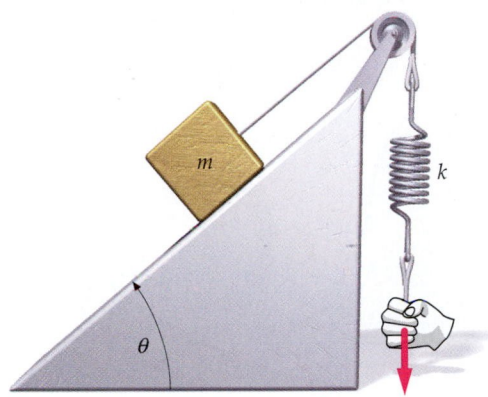

**26** •• **iSOLVE** A 2.4-kg block is dropped onto a spring of spring constant 3955 N/m from a height of 5.0 m (Figure 7-21). When the block is momentarily at rest, the spring has been compressed by 25 cm. Find the speed of the block when the compression of the spring is 15 cm.

**FIGURE 7-21** Problems 26, 91

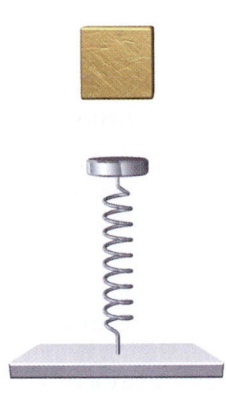

**27** •• **SSM** A ball at the end of a string moves in a vertical circle with constant mechanical energy $E$. What is the difference between the tension at the bottom of the circle and the tension at the top?

**28** •• A girl of mass $m$ is taking a picnic lunch to her grandmother. She ties a rope of length $R$ to a tree branch over a creek and starts to swing from rest at a point that is a distance $R/2$ lower than the branch. What is the minimum breaking tension for the rope if it is not to break and drop the girl into the creek?

**29** •• [SOLVE] A roller coaster car of mass 1500 kg starts at a distance $H = 23$ m above the bottom of a loop 15 m in diameter (Figure 7-22). If friction is negligible, the downward force of the rails on the car when it is upside down at the top of the loop is (a) $4.6 \times 10^4$ N, (b) $3.1 \times 10^4$ N, (c) $1.7 \times 10^4$ N, (d) 980 N, (e) $1.6 \times 10^3$ N.

**FIGURE 7-22** Problem 29

**30** • A single-car roller coaster pushes off, and on the first section of track, descends a 5-m-deep valley, then climbs to the top of a hill that is 9.5 m above the valley floor. (a) What is the minimum initial speed required to carry the coaster beyond the first hill? Assume that the track is frictionless. (b) Can we affect this speed by changing the depth of the valley to make the coaster pick up more speed at the bottom?

**31** •• The Gravitron single-car roller coaster has a loop-the-loop that is constructed so that riders on the coaster will feel perfectly weightless when they reach the top of the circular arc. How heavy will they feel when they reach the bottom of the arc (that is, what is the normal force pressing up into their seats at the bottom of the loop)? Express the answer as a multiple of their normal weight. Assume that the arc is perfectly circular and no frictional forces act on the car.

**32** •• [SSM] [SOLVE] A stone is thrown upward at an angle of 53° above the horizontal. Its maximum height during the trajectory is 24 m. What was the stone's initial speed?

**33** •• [SOLVE]✓ A baseball of mass 0.17 kg is thrown from the roof of a building 12 m above the ground. Its initial velocity is 30 m/s at an angle of 40° above the horizontal. (a) What is the maximum height the ball reaches? (b) What is the work done by gravity as the ball moves from the roof to its maximum height? (c) What is the speed of the ball as it strikes the ground?

**34** •• An 80-cm-long pendulum with a 0.6-kg bob is released from rest at initial angle of $\theta_0$ with the vertical. At the bottom of the swing, the speed of the bob is 2.8 m/s. (a) What was the initial angle of the pendulum? (b) What angle does the pendulum make with the vertical when the speed of the bob is 1.4 m/s?

**35** •• [SSM] [SOLVE]✓ The Royal Gorge bridge over the Arkansas River is $L = 310$ m above the river. A bungee jumper of mass 60 kg has an elastic cord of length $d = 50$ m attached to her feet. Assume the cord acts like a spring of force constant $k$. The jumper leaps, barely touches the water,

and after numerous ups and downs comes to rest at a height $h$ above the water. (a) Find $h$. (b) Find the maximum speed of the jumper.

**36** •• A pendulum consists of a 2-kg bob attached to a light string of length 3 m. The bob is struck horizontally so that it has an initial horizontal velocity of 4.5 m/s. For the point at which the string makes an angle of 30° with the vertical, what is (a) the speed, (b) the potential energy, and (c) the tension in the string? (d) What is the angle of the string with the vertical when the bob reaches its greatest height?

**37** •• A pendulum consists of a string of length $L$ and a bob of mass $m$. The string is brought to a horizontal position and the bob is given the minimum initial speed enabling the pendulum to make a full turn in the vertical plane. (a) What is the maximum kinetic energy $K$ of the bob? (b) What is the tension in the string when the kinetic energy is maximum?

**38** •• [SOLVE]✓ A child whose weight is 360 N swings out over a pool of water using a rope attached to the branch of a tree at the edge of the pool. The branch is 12 m above ground level and the surface of the water is 1.8 m below ground level. The child holds on to the rope at a point 10.6 m from the branch and moves back until the angle between the rope and the vertical is 23°. When the rope is in the vertical position, the child lets go and drops into the pool. Find the speed of the child at the surface of the water.

**39** •• [SSM] [SOLVE] Walking by a pond, you find a rope attached to a stout tree limb 5.2 m off the ground. You decide to use the rope to swing out over the pond. The rope is a bit frayed but supports your weight, which is about 650 N. You estimate that the rope might break if the tension is 80 N greater than your weight. You grab the rope at a point 4.6 m from the limb and move back to swing out over the pond. (a) What is the maximum safe initial angle between the rope and the vertical at which it will not break during the swing? (b) If you begin at this maximum angle, and the surface of the pond is 1.2 m below the level of the ground, with what speed will you enter the water if you let go of the rope when the rope is vertical?

**40** •• A pendulum bob of mass $m$ is attached to a light string of length $L$ and is also attached to a spring of force constant $k$. With the pendulum in the position shown in Figure 7-23, the spring is at its unstretched length. If the bob is now pulled aside so that the string makes a *small* angle $\theta$ with the vertical and released, what is the speed of the bob as it passes through the equilibrium position? For $\theta$ in radians, if $|\theta| \ll 1$, then $\sin\theta \approx \tan\theta \approx \theta$ and $\cos\theta \approx 1$.

**FIGURE 7-23** Problem 40

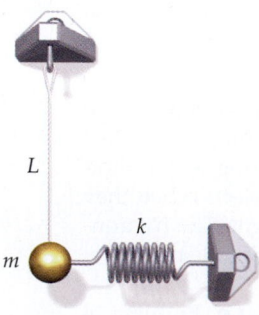

**41 •••** A pendulum is suspended from the ceiling and attached to a spring fixed to the floor directly below the pendulum support (Figure 7-24). The mass of the pendulum bob is $m$, the length of the pendulum is $L$, and the spring constant is $k$. The unstretched length of the spring is $L/2$ and the distance between the floor and ceiling is $1.5L$. The pendulum is pulled aside so that it makes an angle $\theta$ with the vertical and is then released from rest. Obtain an expression for the speed of the pendulum bob when $\theta = 0$.

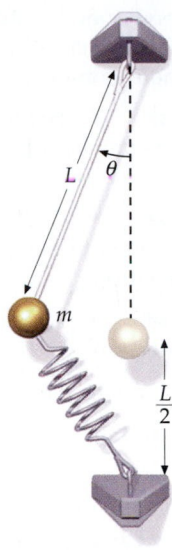

**FIGURE 7-24** Problem 41

## The Conservation of Energy

**42 •** **ISOLVE** In a volcanic eruption, 4 km³ of mountain with a density of 1600 kg/m³ was lifted an average height of 500 m. (a) How much energy in joules was released in this eruption? (b) The energy released by thermonuclear bombs is measured in megatons of TNT, where 1 megaton of TNT $= 4.2 \times 10^{15}$ J. Convert your answer for (a) to megatons of TNT.

**43 ••** **ISOLVE** An 80-kg physics student climbs a 120-m-high hill. (a) What is the increase in the gravitational potential energy of the student? (b) Where does this energy come from? (c) The student's body is 20 percent efficient; that is, for every 20 J that are converted to mechanical energy, 100 J of chemical energy are expended, with 80 J going into thermal energy. How much chemical energy is expended by the student during the climb?

## Kinetic Friction

**44 •** **ISOLVE** A 2000-kg car moving at an initial speed of 25 m/s along a horizontal road skids to a stop in 60 m. (a) Find the energy dissipated by friction. (b) Find the coefficient of kinetic friction between the tires and the road.

**45 •** An 8-kg sled is initially at rest on a horizontal road. The coefficient of kinetic friction between the sled and the road is 0.4. The sled is pulled a distance of 3 m by a force of 40 N applied to the sled at an angle of 30° above the horizontal. (a) Find the work done by the applied force. (b) Find the energy dissipated by friction. (c) Find the change in the kinetic energy of the sled. (d) Find the speed of the sled after it has traveled 3 m.

**46 •** **SSM** Returning to Problem 19 and Figure 7-17, suppose that the surfaces described are not frictionless and that the coefficient of kinetic friction between the block and the surfaces is 0.30. Find (a) the speed of the block when it reaches the ramp, and (b) the distance that the block slides up the ramp before coming momentarily to rest. (Neglect any energy dissipated along the transition curve.)

**47 •** The 2-kg block in Figure 7-25 slides down a frictionless curved ramp, starting from rest at a height of 3 m. The block then slides 9 m on a rough horizontal surface before coming to rest. (a) What is the speed of the block at the bottom of the ramp? (b) What is the energy dissipated by friction? (c) What is the coefficient of friction between the block and the horizontal surface?

**FIGURE 7-25** Problem 47

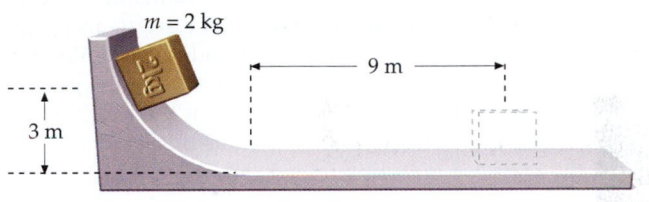

**48 ••** **ISOLVE** A 20-kg girl slides down a playground slide that is 3.2 m high. When she reaches the bottom of the slide, her speed is 1.3 m/s. (a) How much energy was dissipated by friction? (b) If the slide is inclined at 20°, what is the coefficient of friction between the girl and the slide?

**49 ••** In Figure 7-26, the coefficient of kinetic friction between the 4-kg block and the table is 0.35. (a) Find the energy dissipated by friction when the 2-kg block falls a distance $y$. (b) Find the total mechanical energy $E$ of the two-block–earth system after the 2-kg block falls a distance $y$, assuming that $E = 0$ initially. (c) Use your result for (b) to find the speed of either block after the 2-kg block falls 2 m.

**FIGURE 7-26** Problem 49

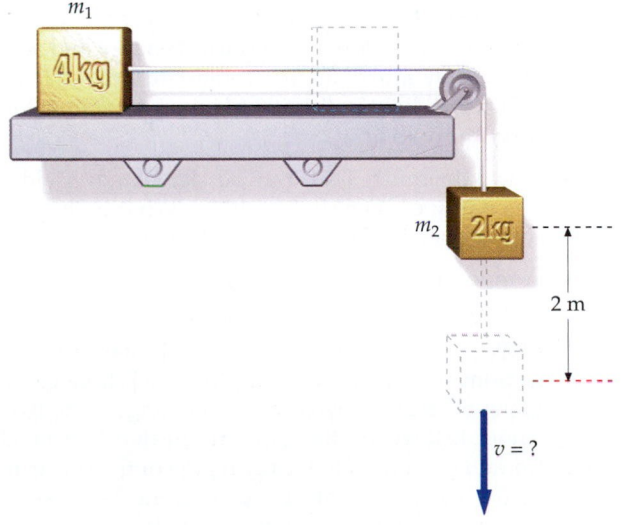

**50** •• [SSM] A compact object of mass $m$ moves in a horizontal circle of radius $r$ on a rough table. It is attached to a horizontal string fixed at the center of the circle. The speed of the object is initially $v_0$. After completing one full trip around the circle, the speed of the object is $\frac{1}{2}v_0$. (a) Find the energy dissipated by friction during that one revolution in terms of $m$, $v_0$, and $r$. (b) What is the coefficient of kinetic friction? (c) How many more revolutions will the object make before coming to rest?

**51** •• [ISOLVE✓] A 2.4-kg box has an initial velocity of 3.8 m/s upward along a plane inclined at 37° to the horizontal. The coefficient of kinetic friction between the box and the plane is 0.30. How far up the incline does the box travel? What is its speed when it passes its starting point on its way down the incline?

**52** ••• A block of mass $m$ rests on a plane inclined at $\theta$ with the horizontal. The block is attached to a spring of constant $k$ as shown in Figure 7-27. The coefficients of static and kinetic friction between the block and plane are $\mu_s$ and $\mu_k$, respectively. Very slowly, the spring is pulled upward along the plane until the block starts to move. (a) Obtain an expression for the extension $d$ of the spring the instant the block moves. (b) Determine the value of $\mu_k$ such that the block comes to rest just as the spring is in its unstressed condition, that is, neither extended nor compressed.

**FIGURE 7-27** Problem 52

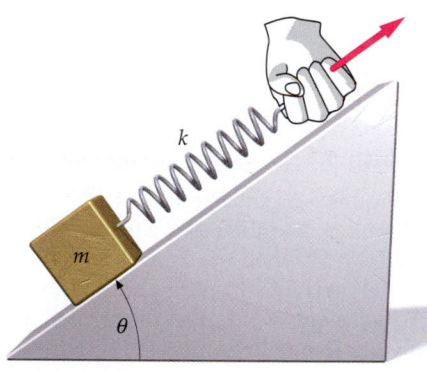

### Mass and Energy

**53** • (a) Calculate the rest energy in 1 g of dirt. (b) If you could convert this energy into electrical energy and sell it for $0.10/kW·h, how much money would you get? (c) If you could power a 100-W light bulb with this energy, for how long could you keep the bulb lit?

**54** • One kiloton of TNT can be detonated, yielding an explosive energy of roughly $5 \times 10^{12}$ J. What is the mass equivalent of the energy of the explosion?

**55** • A muon has a rest energy of 105.7 MeV. Calculate its rest mass in kilograms.

**56** • [SSM] If a black hole and a "normal" star orbit each other, gases from the star falling into the black hole can be heated millions of degrees by frictional heating in the black hole's accretion disk. When the gases are heated that much, they begin to radiate light in the X-ray region of the spectrum. Cygnus X-1, the second brightest source in the X-ray sky, is thought to be one such binary system; it radiates an estimated

power of $4 \times 10^{31}$ W. If we assume that 1 percent of the infalling mass-energy escapes as X rays, at what rate is the black hole gaining mass?

**57** • For the fusion reaction in Example 7-15, calculate the number of reactions per second that are necessary to generate 1 kW of power.

**58** • Use Table 7-1 to calculate how much energy is needed to remove one neutron from $^4$He, leaving $^3$He plus a neutron.

**59** • A free neutron at rest decays into a proton plus an electron:

$$n \rightarrow p + e$$

Use Table 7-1 to calculate the energy released in this reaction.

**60** •• In one nuclear fusion reaction, two $^2$H nuclei combine to produce $^4$He. (a) How much energy is released in this reaction? (b) How many such reactions must take place per second to produce 1 kW of power?

**61** •• A large nuclear power plant produces 3000 MW of electrical power by nuclear fission, which converts matter into energy. (a) How many kilograms of matter does the plant consume in one year? (Assume an efficiency of 33 percent for a nuclear power plant.) (b) In a coal-burning power plant, each kilogram of coal releases 31 MJ of thermal energy when burned. How many kilograms of coal are needed each year for a 3000-MW plant? (Assume an efficiency of 38 percent for a coal-burning power plant.)

### General Problems

**62** •• [SSM] A block of mass $m$, starting from rest, is pulled up a frictionless inclined plane that makes an angle $\theta$ with the horizontal by a string parallel to the plane. The tension in the string is $T$. After traveling a distance $L$, the speed of the block is $v$. The work done by the tension $T$ is (a) $mgL \sin \theta$, (b) $mgL \cos \theta + \frac{1}{2}mv^2$, (c) $mgL \sin \theta + \frac{1}{2}mv^2$, (d) $mgL \cos \theta$, (e) $TL \cos \theta$.

**63** •• A block of mass $m$ slides with constant velocity $v$ down a plane inclined at $\theta$ with the horizontal. During the time interval $\Delta t$, the energy dissipated by friction is (a) $mgv \Delta t \tan \theta$, (b) $mgv \Delta t \sin \theta$, (c) $\frac{1}{2}mv^3 \Delta t$, (d) cannot be determined without knowing the coefficient of kinetic friction.

**64** • [ISOLVE] A 3.5-kg box rests on a horizontal frictionless surface in contact with a spring of spring constant 6800 N/m. The spring is fixed at its other end and is initially at its uncompressed length. A constant horizontal force of 70 N is applied to the box so that the spring compresses. Determine the distance the spring is compressed when the box is momentarily at rest.

**65** • [SSM] [ISOLVE] The average energy per unit time per unit area that reaches the upper atmosphere of the earth from the sun, called the solar constant, is 1.35 kW/m². Because of absorption and reflection by the atmosphere, about 1 kW/m² reaches the surface of the earth on a clear day. How much energy is collected during 8 h of daylight by a window 1 m by 2 m? The window is on a mount that rotates, keeping the window facing the sun so the sun's rays remain perpendicular to the window.

**66** •• (a) Using the value of the solar constant given in Problem 65 and the known distance from the earth to the sun, find the total energy that the sun radiates every second. (b) This energy is due to fusion reactions such as the one detailed in Example 7-15. There are several such reactions taking place in the "Phoenix cycle" in the sun, but the overall result is that 4 hydrogen nuclei fuse together to form 1 helium nucleus, liberating 26.7 MeV. If we assume that the sun is primarily made up of hydrogen, and that it will continue burning until roughly 10% of its hydrogen fuel is used up, use the known mass of the sun to determine approximately how long it can keep burning (that is, determine the solar lifetime).

**67** • **iSOLVE** ✔ After the 1250-kg jet-powered car *Spirit of America* went out of control during a test drive at Bonneville Salt Flats, Utah, it left skid marks about 9.5 km long. (a) If the car was moving initially at a speed of 708 km/h, estimate the coefficient of kinetic friction $\mu_k$. (b) What was the kinetic energy $K$ of the car 60 s after the brakes were applied?

**68** •• **iSOLVE** A T-bar tow is required to pull 80 skiers up a 600-m slope inclined at 15° above the horizontal at a speed of 2.5 m/s. The coefficient of kinetic friction is 0.06. Find the motor power required if the mass of the average skier is 75 kg.

**69** •• **iSOLVE** A 2-kg box is projected with an initial speed of 3 m/s up a rough plane inclined at 60° above the horizontal. The coefficient of kinetic friction is 0.3. (a) List all the forces acting on the box. (b) How far up the plane does the box slide before it stops momentarily? (c) What is the energy dissipated by friction as the box slides up the plane? (d) What is the speed of the box when it again reaches its initial position?

**70** • **SSM** **iSOLVE** A 1200-kg elevator can safely carry a maximum load of 800 kg. What is the power provided by the electric motor powering the elevator when the elevator ascends with a full load at a speed of 2.3 m/s?

**71** •• **iSOLVE** To reduce the power requirement of elevator motors, elevators are counterbalanced with weights connected to the elevator by a cable that runs over a pulley at the top of the elevator shaft. If the elevator in Problem 70 is counterbalanced with a mass of 1500 kg, what is the power provided by the motor when the elevator ascends fully loaded at a speed of 2.3 m/s? How much power is provided by the motor when the elevator ascends without a load at 2.3 m/s?

**72** •• **iSOLVE** ✔ The spring constant of a toy dart gun is 5000 N/m. To cock the gun the spring is compressed 3 cm. The 7-g dart, fired straight upward, reaches a maximum height of 24 m. Determine the energy dissipated by air friction during the dart's ascent. Estimate the speed of the projectile when it returns to its starting point.

**73** •• **SSM** **iSOLVE** ✔ In a volcanic eruption, a 2-kg piece of porous volcanic rock is thrown vertically upward with an initial speed of 40 m/s. It travels upward a distance of 50 m before it begins to fall back to the earth. (a) What is the initial kinetic energy of the rock? (b) What is the increase in thermal energy due to air friction during ascent? (c) If the increase in thermal energy due to air friction on the way down is 70% of that on the way up, what is the speed of the rock when it returns to its initial position?

**74** •• A block of mass $m$ starts from rest at a height $h$ and slides down a frictionless plane inclined at $\theta$ with the horizontal as shown in Figure 7-28. The block strikes a spring of force constant $k$. Find the compression of the spring when the block is momentarily at rest.

**FIGURE 7-28** Problem 74

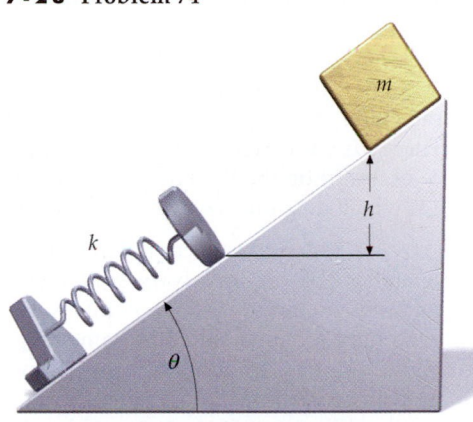

**75** • **SSM** A $1.5 \times 10^4$-kg stone slab rests on a steel girder. On a very hot day, you find that the girder has expanded, lifting the slab by 0.1 cm. (a) What work does the girder do on the slab? (b) Where does the energy come from to lift the slab? Give a microscopic picture of what is happening in the steel.

**76** •• **iSOLVE** ✔ A 1500-kg car traveling at 24 m/s is at the foot of a hill that rises 120 m in 2.0 km. At the top of the hill, the speed of the car is 10 m/s. Assuming constant acceleration, find the average power delivered by the car's engine, neglecting any internal frictional losses.

**77** •• **SSM** A block of mass $m$ is suspended from the ceiling by a spring and is free to move vertically in the $y$ direction as indicated (Figure 7-29). We are given that the potential energy as a function of position is $U = \frac{1}{2}ky^2 - mgy$. (a) Using a spreadsheet program or graphing calculator, make a graph of $U$ as a function of $y$. For what value of $y$ is the spring *unstretched*? (b) From the expression given for $U$, find the net force acting on $m$ at any position $y$. (c) The block is released from rest at $y = 0$; if there is no friction, what is the maximum value of $y$, $y_{max}$, that will be reached by the mass? Indicate $y_{max}$ on your graph. (d) Now consider the effect of friction. The mass ultimately settles into an equilibrium position $y_{eq}$. Find this point on your sketch. (e) Find the amount of thermal energy produced by friction from the start of the operation to the final equilibrium.

**FIGURE 7-29** Problem 77

**78** •• A spring-loaded gun is cocked by compressing a short, strong spring by a distance $d$. It fires a signal flare of mass $m$ directly upward. The flare has speed $v_0$ as it leaves the spring and is observed to rise to a maximum height $h$ above the point where it leaves the spring. After it leaves the spring, effects of drag force by the air on the flare are significant. (Express answers in terms of $m$, $v_0$, $d$, $h$, and $g$.) (a) How much work is done on the spring in the course of the compression? (b) What is the value of the spring constant $k$? (c) How much mechanical energy is converted to thermal energy because of the drag force of the air on the flare between the time of firing and the time at which maximum elevation is reached?

**79** •• **SOLVE** A roller-coaster car having a total mass (including passengers) of 500 kg travels freely along the winding frictionless track shown in Figure 7-30. Points A, E, and G are on horizontal straight sections, all at the same height of 10 m above the ground. Point C is at a height of 10 m above the ground on an inclined section of slope angle 30°. Point B is at the crest of a hill, while point D is at ground level at the bottom of a valley; the radius of curvature at both of these points is 20 m. Point F is at the middle of a banked horizontal curve with a radius of curvature of 30 m, and at the same height of 10 m above the ground as points A, E, and G. At point A the speed of the car is 12 m/s. (a) If the car is just barely able to make it over the hill at point B, what is the height of that point above the ground? (b) If the car is just barely able to make it over the hill at point B, what is the magnitude of the total force exerted on the car by the track at that point? (c) What is the acceleration of the car at point C? (d) What are the magnitude and direction of the total force exerted on the car by the track at point D? (e) What are the magnitude and direction of the total force exerted on the car by the track at point F? (f) At point G a constant braking force is applied to the car, bringing it to a halt in a distance of 25 m. What is the magnitude of the braking force?

**FIGURE 7-30** Problem 79

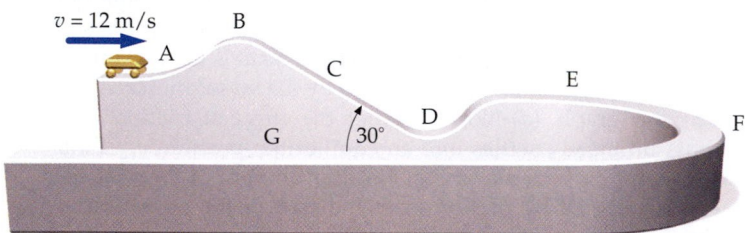

**80** • **SSM** **SOLVE** An elevator (mass $M$ = 2000 kg) is moving downward at $v_0$ = 1.5 m/s. A braking system prevents the downward speed from increasing. (a) At what rate (in J/s) is the braking system converting mechanical energy to thermal energy? (b) While the elevator is moving downward at $v_0$ = 1.5 m/s, the braking system fails and the elevator is in free-fall for a distance $d$ = 5 m before hitting the top of a large safety spring with force constant $k$ = 1.5 × 10$^4$ N/m. After the elevator hits the top of the spring, we want to know the distance $\Delta y$ that the spring is compressed before the elevator is brought to rest. Write an algebraic expression for the value of $\Delta y$ in terms of the known quantities $M$, $v_0$, $g$, $k$, and $d$, and substitute the given values to find $\Delta y$.

**81** • To measure the force of friction (rolling friction plus air drag) on a moving car, engineers turn off the engine and allow the car to coast down hills of known steepness. The engineers collect the following data:

1. On a 2.87° hill, the car can coast at a steady 20 m/s.

2. On a 5.74° hill, the steady coasting speed is 30 m/s.

The total mass of the car is 1000 kg. (a) What is the force of friction at 20 m/s ($F_{20}$) and at 30 m/s ($F_{30}$)? (b) How much useful power must the engine deliver to drive the car on a level road at steady speeds of 20 m/s ($P_{20}$) and 30 m/s ($P_{30}$)? (c) At full throttle, the engine delivers 40 kW. What is the angle of the steepest incline up which the car can maintain a steady 20 m/s? (d) Assume that the engine delivers the same total useful work from each liter of gas, no matter what the speed. At 20 m/s on a level road, the car gets 12.7 km/L. How many kilometers per liter does it get if it goes 30 m/s instead?

**82** •• A 2-kg block is released 4 m from a massless spring with a force constant $k$ = 100 N/m that is along a plane inclined at 30°, as shown in Figure 7-31. (a) If the plane is frictionless, find the maximum compression of the spring. (b) If the coefficient of kinetic friction between the plane and the block is 0.2, find the maximum compression. (c) For the plane in Part (b), how far up the incline will the block travel after leaving the spring?

**FIGURE 7-31** Problem 82

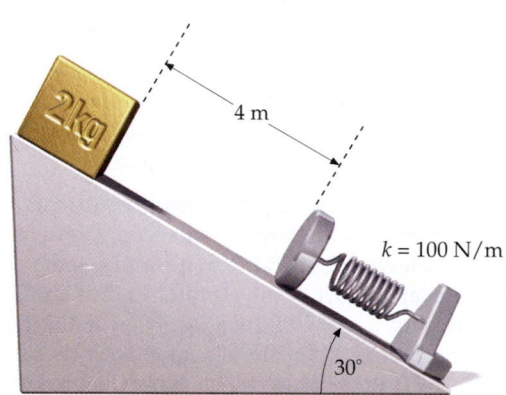

**83** •• **SOLVE** A train with a total mass of 2 × 10$^6$ kg rises 707 m while traveling a distance of 62 km at an average speed of 15.0 km/h. The frictional force is 0.8 percent of the weight. Find (a) the kinetic energy of the train, (b) the total change in its potential energy, (c) the energy dissipated by friction, and (d) the power output of the train's engines.

**84** •• **SSM** While driving, one expects to expend more power when accelerating than when driving at a constant speed. (a) Neglecting friction, calculate the energy required to give a 1200-kg car a speed of 50 km/h. (b) If friction (rolling friction and air drag) results in a retarding force of 300 N at a speed of 50 km/h, what is the energy needed to move the car a distance of 300 m at a constant speed of 50 km/h? (c) Assuming the energy losses due to friction in Part (a) are 75 percent of those found in Part (b), estimate the ratio of the energy consumption for the two cases considered.

**85** ••• **SSM** A pendulum consisting of a string of length $L$ with a small bob on the end (mass $M$) is pulled horizontally, then released (Figure 7-32). At the lowest point of the swing, the string catches on a thin peg a distance $R$ above the lowest point. Show that $R$ must be smaller than $2L/5$ if the bob is to swing around the peg in a full circle.

**FIGURE 7-32** Problem 85

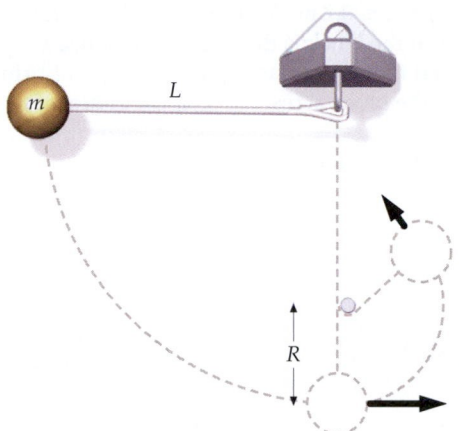

**86** •• A rifle bullet is fired into a massive, stationary block of wood to determine its penetrating power. It is found that it penetrates the wood to a distance $D$. Another bullet that has the same mass, but double the velocity of the first bullet, is fired into an identical block. Assume that the average force acting on the bullet from the wood doesn't depend on the bullet's speed. The penetration depth is now found to be (a) $D$, (b) $2D$, (c) $4D$, (d) undeterminable from the information given.

**87** •• A standard introductory physics lab to examine the conservation of energy and Newton's laws is shown in Figure 7-33. A glider is set up on a linear air track and is attached by a string over a massless pulley to a hanging weight. The mass of the glider is $M$, while the mass of the hanging weight is $m$. When the air supply to the air track is turned on, the track is essentially frictionless; you then release the hanging weight and measure the speed of the glider after the weight has fallen a given distance ($Y$). To show that the laws of physics are consistent, work out the speed of the glider in two different ways: (a) by conservation of mechanical energy; (b) by using Newton's second and third laws directly—make a free-body diagram for the two masses, find their acceleration, and use that to determine the speed of the glider.

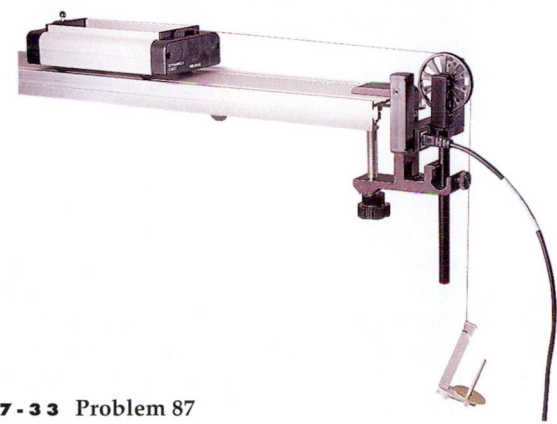

**FIGURE 7-33** Problem 87

**88** •• **SSM** In one model of jogging, the energy expended is assumed to go into accelerating and decelerating the legs. If the mass of the leg is $m$ and the running speed is $v$, the energy needed to accelerate the leg from rest to $v$ is $\frac{1}{2}mv^2$, and the same energy is needed to decelerate the leg back to rest for the next stride. Thus the energy required for each stride is $mv^2$. Assume that the mass of a man's leg is 10 kg and that he jogs at a speed of 3 m/s with 1 m between one footfall and the next. Therefore, the energy he must provide to his legs in each second is $3 \times mv^2$. Calculate the rate of the man's energy expenditure using this model and assuming that his muscles have an efficiency of 20 percent.

**89** •• A high school chemistry teacher once suggested measuring the magnitude of free-fall acceleration by the following method: Hang a weight on a very fine thread (length $L$) to make a pendulum, which is then put on the edge of a table so that the weight is a height $H$ above the floor when at its lowest point. Pull the pendulum back so that the thread makes an angle $\theta_0$ with the vertical. At the bottom point of the pendulum, place a razor blade that is positioned to cut through the thread at its lowest point. Once the thread is cut, the weight is projected horizontally, and is thrown a distance $D$ from the edge of the table. The idea was that the measurement of $D$ as a function of $\theta_0$ should somehow determine $g$. Apart from some obvious experimental difficulties, the experimental had one fatal flaw: $D$ doesn't depend on $g$! Show that this is true, and that $D$ depends only on the angle $\theta_0$.

**90** •• A 5-kg block is held against a spring of force constant 20 N/cm, compressing it 3 cm. The block is released and the spring extends, pushing the block along a horizontal surface. The coefficient of friction between the surface and the block is 0.2. (a) Find the work done on the block by the spring as it extends from its compressed position to its equilibrium position. (b) Find the energy dissipated by friction while the block moves the 3 cm to the equilibrium position of the spring. (c) What is the speed of the block when the spring is at its equilibrium position? (d) If the block is not attached to the spring, how far will it slide along the surface before coming to rest?

**91** •• A block of mass $m$ is dropped onto the top of a vertical spring whose force constant is $k$ (Figure 7-21, Problem 26). If the block is released from a height $h$ above the top of the spring, (a) what is the maximum kinetic energy of the block? (b) What is the maximum compression of the spring? (c) At what compression is the block's kinetic energy half its maximum value?

**92** ••• The bob of a pendulum of length $L$ is pulled aside so that the string makes an angle $\theta_0$ with the vertical, and the bob is then released. In Example 7-2 the conservation of energy was used to obtain the speed of the bob at the bottom of its swing. In this problem, you are to obtain the same result using Newton's second law. (a) Show that the tangential component of Newton's second law gives $dv/dt = -g \sin \theta$, where $v$ is the speed and $\theta$ is the angle between the string and the vertical. (b) Show that $v$ can be written $v = L \, d\theta/dt$ (c) Use this result and the chain rule for derivatives to obtain

$$\frac{dv}{dt} = \frac{dv}{d\theta}\frac{d\theta}{dt} = \frac{dv}{d\theta}\frac{v}{L}$$

(d) Combine the results of (a) and (c) to obtain $v \, dv = -gL \sin \theta \, d\theta$. (e) Integrate the left side of the equation in Part (d) from

$v = 0$ to the final speed $v$ and the right side from $\theta = \theta_0$ to $\theta = 0$, and show that the result is equivalent to $v = \sqrt{2gh}$, where $h$ is the original height of the bob above the bottom.

**93** ••• A rock climber is rappelling down the face of a cliff when his hold slips and he slides down over the rock face, supported only by the bungee cord he attached to the top of the cliff. The cliff face is in the form of a smooth quarter-cylinder with height (and radius) $H = 300$ m (see Figure 7-34). Treat the bungee cord as a spring with spring constant $k = 5$ N/m and unstretched length $L = 60$ m. The climber's mass is 85 kg. (a) Using a spreadsheet program, make a graph of the rock climber's potential energy as a function of $s$, his distance from the top of the cliff *measured along the curved surface*. Use values of $s$ between 60 m and 200 m. (b) If his fall began when he was a distance $s_i = 60$ m from the top of the cliff, and ends when he is a distance $s_f = 110$ m from the top, determine how much energy is dissipated by friction between the time he initially slipped and finally came to a stop.

**FIGURE 7-34** Problem 93

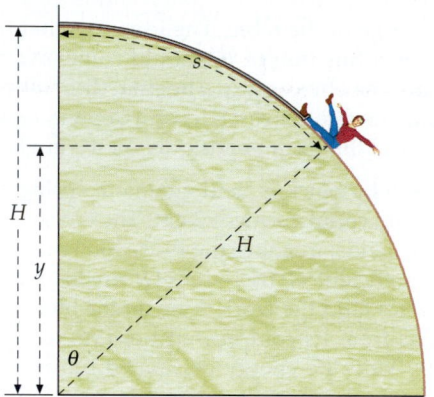

**94** ••• SSM A block of wood (mass $M$) is connected to two massless springs as shown in Figure 7-35. Each spring has unstretched length $L$ and spring constant $k$. (a) If the block is displaced a distance $x$, as shown, what is the change in the potential energy stored in the springs? (b) What is the magnitude of the force pulling the block back toward the equilibrium position? (c) Using a spreadsheet program or graphing calculator, make a graph of the potential energy $U$ as a function of $x$ for $0 \leq x \leq 0.2$ m. Assume $k = 1$ N/m, $L = 0.1$ m, and $M = 1$ kg. (d) If the block is displaced a distance $x = 0.1$ m and released, what is its speed as it passes the equilibrium point? Assume that the block is resting on a frictionless surface.

**FIGURE 7-35** Problem 94

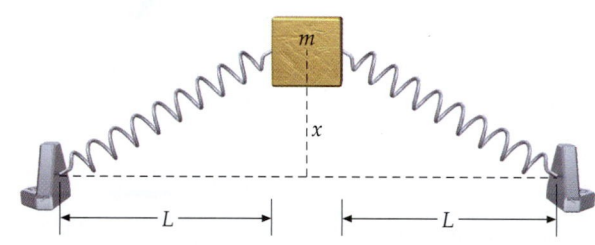

# Systems of Particles and Conservation of Linear Momentum

**If the golfer uses a more massive club, how will the distance traveled by the ball be affected? (See Example 8-12.)**

8-1    The Center of Mass

*8-2    Finding the Center of Mass by Integration

8-3    Motion of the Center of Mass

8-4    Conservation of Linear Momentum

8-5    Kinetic Energy of a System

8-6    Collisions

*8-7    The Center-of-Mass Reference Frame

*8-8    Systems With Continuously Varying Mass: Rocket Propulsion

We saw in Chapters 6 and 7 that the work–kinetic energy theorem for particles and the work–energy theorem for systems of particles are useful tools for analyzing a variety of situations and solving a wide variety of problems. But in order to analyze the motion of extended objects, and to understand collisions between objects like the golf club and ball in the photo above, we need to add two more theorems to our problem-solving toolbox. These are **Newton's second law for systems** (Section 8-3) and **the impulse–momentum theorem** (Section 8-6). The impulse–momentum theorem for systems introduces the conservation of momentum, one of the universal laws of physics.

The mass of a particle times its velocity is called the **momentum** of the particle. The momentum of a system of particles is the vector sum of the momenta of the individual particles in the system. When the net external force acting on a system remains zero, the system's momentum remains constant. The momentum of an isolated system is a conserved quantity, just like the energy of an isolated system.

➤ In this chapter, we will use conservation of momentum to analyze collisions between billiard balls, cars, and subatomic particles, and the decay of radioactive nuclei. We will apply Newton's laws to the motion of extended objects, such as cars, rockets, and people, by realizing that there is one point of a system, the *center of mass*, that moves as if all the mass of the system were concentrated at that point and all the external forces acting on the system were acting exclusively on that point.

## **8-1** The Center of Mass

The motion of any object or system of particles can be described in terms of the motion of the center of mass (which may be thought of as the bulk motion of the system) plus the motion of individual particles in the system relative to the center of mass. Let's first consider a simple system of two particles in one dimension. If two point particles with masses $m_1$ and $m_2$ have coordinates $x_1$ and $x_2$ on the $x$ axis, then the center-of-mass coordinate $x_{cm}$ is defined by

$$Mx_{cm} = m_1 x_1 + m_2 x_2 \qquad \text{8-1}$$

where $M = m_1 + m_2$ is the total mass of the system. In the case of just two particles, the center of mass lies at some point on the line between the particles; if the particles have equal masses, then the center of mass is midway between them (Figure 8-1).

If the two particles are of unequal mass, then the center of mass is closer to the more massive particle (Figure 8-2).

If we choose the origin and direction of the $x$ axis such that the position of $m_1$ is at the origin and that of $m_2$ is on the positive $x$ axis, then $x_1 = 0$ and $x_2 = d$, where $d$ is the distance between the particles (Figure 8-3) and the center of mass is given by

$$Mx_{cm} = m_1 x_1 + m_2 x_2 = m_1(0) + m_2 d$$

$$x_{cm} = \frac{m_2}{M} d = \frac{m_2}{m_1 + m_2} d \qquad \text{8-2}$$

**EXERCISE** A 4-kg mass is at the origin and a 2-kg mass is at $x = 6$ cm. Find $x_{cm}$. (*Answer* $x_{cm} = 2$ cm)

We can generalize from two particles in one dimension to a system of many particles in three dimensions. For $N$ particles,

$$Mx_{cm} = m_1 x_1 + m_2 x_2 + m_3 x_3 + \cdots + m_N x_N = \sum_i m_i x_i \qquad \text{8-3}$$

where again $M = \sum_i m_i$ is the total mass of the system. Similarly,

$$My_{cm} = \sum_i m_i y_i \qquad \text{and} \qquad Mz_{cm} = \sum_i m_i z_i \qquad \text{8-4}$$

In vector notation, $\vec{r}_i = x_i \hat{i} + y_i \hat{j} + z_i \hat{k}$ is the position vector of the $i$th particle. The position vector of the **center of mass**, $\vec{r}_{cm}$, is defined by

$$M\vec{r}_{cm} = m_1 \vec{r}_1 + m_2 \vec{r}_2 + \cdots = \sum_i m_i \vec{r}_i \qquad \text{8-5}$$

DEFINITION—CENTER OF MASS

where $\vec{r}_{cm} = x_{cm}\hat{i} + y_{cm}\hat{j} + z_{cm}\hat{k}.$

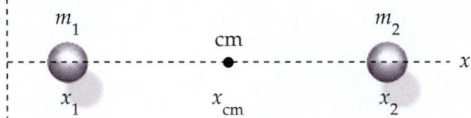

FIGURE 8-1

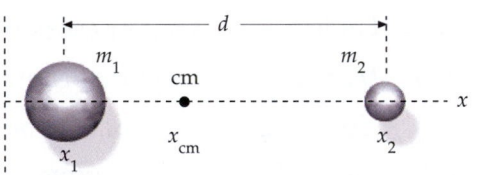

FIGURE 8-2

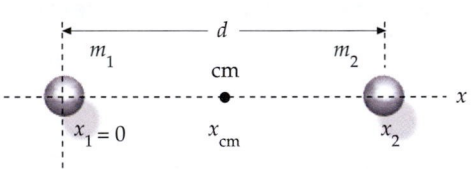

FIGURE 8-3

To find the center of mass of a continuous object we replace the sum in Equation (8-5) with an integral:

$$M\vec{r}_{cm} = \int \vec{r}\, dm \qquad\qquad 8\text{-}6$$

CENTER OF MASS, CONTINUOUS OBJECT

where $dm$ is an element of mass located at position $\vec{r}$, as shown in Figure 8-4. Examples involving Equation 8-6 are presented in Section 8-2. For highly symmetric objects, the center of mass is at the center of symmetry. For example, the center of mass of a uniform cylinder is located at its geometric center. Consider the following examples.

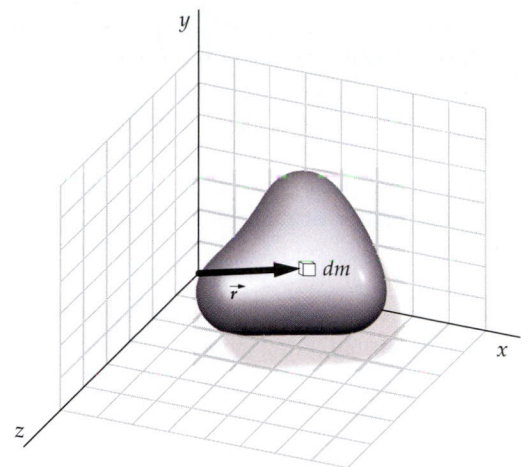

**FIGURE 8-4** Mass element $dm$ located at $\vec{r}$ for finding the center of mass by integration.

---

*THE CENTER OF MASS OF A WATER MOLECULE*      **EXAMPLE 8-1**

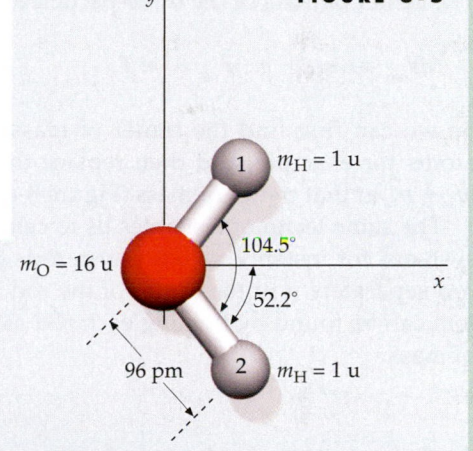

**FIGURE 8-5**

$m_H = 1\,u$

$104.5°$

$m_O = 16\,u$

$52.2°$

$96\,pm$      $m_H = 1\,u$

A water molecule consists of an oxygen atom and two hydrogen atoms. Oxygen has a mass of 16 unified mass units (u) and each hydrogen has a mass of 1 u. The hydrogen atoms are each at an average distance of 96 pm ($96 \times 10^{-12}$ m) from the oxygen atom, and are separated from one another by an angle of 104.5°. Find the center of mass of a water molecule.

**PICTURE THE PROBLEM** The calculation is simplified if we place the origin at the location of the oxygen atom, with the $x$ axis bisecting the angle between the hydrogen atoms (Figure 8-5). Then, given the symmetries of the molecule, the center of mass will be on the $x$ axis, and the line from the oxygen atom to each hydrogen atom will make an angle of 52.2°.

1. The location of the center of mass is given by its coordinates, $x_{cm}$ and $y_{cm}$:

$$x_{cm} = \frac{\Sigma\, m_i x_i}{M}, \qquad y_{cm} = \frac{\Sigma\, m_i y_i}{M}$$

2. Writing these out explicitly gives:

$$x_{cm} = \frac{m_{H1}x_{H1} + m_{H2}x_{H2} + m_O x_O}{m_{H1} + m_{H2} + m_O}$$

$$y_{cm} = \frac{m_{H1}y_{H1} + m_{H2}y_{H2} + m_O y_O}{m_{H1} + m_{H2} + m_O}$$

3. We have chosen the origin to be the location of the oxygen atom, so both the $x$ and $y$ coordinates of the oxygen atom are zero. The $x$ and $y$ coordinates of the hydrogen atoms are calculated from the 52.2° angle each hydrogen makes with the $x$ axis:

$$x_O = y_O = 0$$

$$x_{H1} = x_{H2} = 96\,pm \cos 52.2° = 59\,pm$$

$$y_{H1} = 96\,pm \sin 52.2° = 76\,pm$$

$$y_{H2} = -96\,pm \sin 52.2° = -76\,pm$$

4. Substituting the coordinate and mass values into step 2 gives $x_{cm}$:

$$x_{cm} = \frac{(1\,u)(59\,pm) + (1\,u)(59\,pm) + (16\,u)0}{1\,u + 1\,u + 16\,u}$$

$$= 6.6\,pm$$

$$y_{cm} = \frac{(1\,u)(76\,pm) + (1\,u)(-76\,pm) + (16\,u)0}{1\,u + 1\,u + 16\,u}$$

$$= 0$$

5. The center of mass is on the $x$ axis:

$$\vec{r}_{cm} = x_{cm}\,\hat{i} + y_{cm}\,\hat{j}$$
$$= 6.6\text{ pm }\hat{i} + 0\hat{j} = \boxed{6.6\text{ pm }\hat{i}}$$

**REMARKS** That $y_{cm} = 0$ can be seen from the symmetry of the mass distribution. Also, the center of mass is very close to the relatively massive oxygen atom, as expected.

We also can solve Example 8-1 by first finding the center of mass of just the two hydrogen atoms. For a system of three particles Equation 8-5 is

$$M\vec{r}_{cm} = m_1\vec{r}_1 + m_2\vec{r}_2 + m_3\vec{r}_3$$

The first two terms on the right side of this equation are related to the center of mass of the first two particles $\vec{r}'_{cm}$:

$$m_1\vec{r}_1 + m_2\vec{r}_2 = (m_1 + m_2)\vec{r}'_{cm}$$

The center of mass of the three-particle system can then be written

$$M\vec{r}_{cm} = (m_1 + m_2)\vec{r}'_{cm} + m_3\vec{r}_3$$

So we can first find the center of mass for two of the particles, the hydrogen atoms for example, and then replace them with a single particle of total mass $m_1 + m_2$ at that center of mass (Figure 8-6).

The same technique enables us to calculate centers of mass for more complex systems, for instance, two uniform rods (Figure 8-7). The center of mass of each rod separately is at the center of the rod. The center of mass of the two-rod system can be found by treating each rod as a point particle at its individual center of mass.

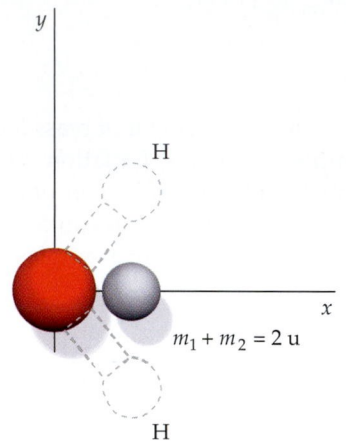

**FIGURE 8-6** Example 8-1 with the two H atoms replaced by a single particle of mass $m_1 + m_2 = 2$ u on the $x$ axis at the center of mass of the original atoms. The center of mass then falls between the oxygen atom at the origin and the calculated center of mass of the two hydrogen atoms.

**FIGURE 8-7**

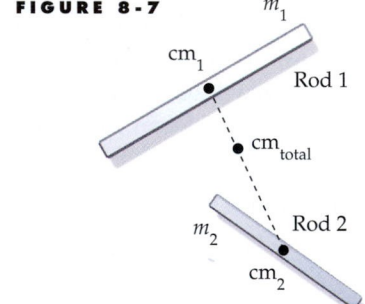

---

*The Center of Mass of a Plywood Sheet*      **EXAMPLE 8-2**

**Find the center of mass of the uniform sheet of plywood shown in Figure 8-8a.**

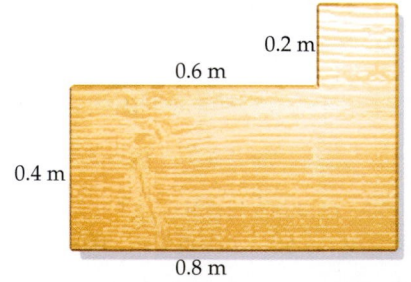

**FIGURE 8-8a**

**Try It Yourself**

**PICTURE THE PROBLEM** The sheet can be divided into two symmetrical parts (Figure 8-8b). The center of mass of each part is at its geometric center. Let $m_1$ be the mass of part 1 and $m_2$ be the mass of part 2. The total mass is $M = m_1 + m_2$. The masses are proportional to the areas.

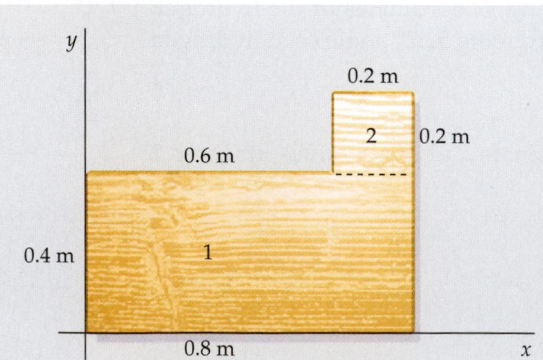

**FIGURE 8-8b**

**Cover the column to the right and try these on your own before looking at the answers.**

| Steps | Answers |
|---|---|
| 1. Write the $x$ and $y$ coordinates of the center of mass in terms of $m_1$ and $m_2$. | $Mx_{cm} = m_1 x_{1,cm} + m_2 x_{2,cm}$<br><br>$My_{cm} = m_1 y_{1,cm} + m_2 y_{2,cm}$ |
| 2. Divide through these equation by $M$ and then substitute area ratios for the mass ratios. | $x_{cm} = \dfrac{A_1}{A} x_{1,cm} + \dfrac{A_2}{A} x_{2,cm}$<br><br>$y_{cm} = \dfrac{A_1}{A} y_{1,cm} + \dfrac{A_2}{A} y_{2,cm}$ |
| 3. Calculate the areas and the ratios of the areas. | $A_1 = 0.32 \text{ m}^2, \quad A_2 = 0.04 \text{ m}^2, \quad A = 0.36 \text{ m}^2$<br><br>$\dfrac{A_1}{A} = \dfrac{8}{9} \qquad \dfrac{A_2}{A} = \dfrac{1}{9}$ |
| 4. Write the $x$ and $y$ coordinates of the center-of-mass coordinates for each part by inspection of the figure. | $x_{1,cm} = 0.4 \text{ m}, \qquad y_{1,cm} = 0.2 \text{ m}$<br><br>$x_{2,cm} = 0.7 \text{ m}, \qquad y_{2,cm} = 0.5 \text{ m}$ |
| 5. Substitute these results to calculate $x_{cm}$ and $y_{cm}$. | $x_{cm} = \boxed{0.433 \text{ m}}, \qquad y_{cm} = \boxed{0.233 \text{ m}}$ |

**REMARKS** 1. Note that the center of mass is very near the center of mass of part 1 because $m_1 = 8m_2$. 2. Placing the origin at the geometric center of part 1 of the sheet and drawing the $x$ axis through the center of part 2 would make the calculation of the center of mass somewhat simpler.

## Gravitational Potential Energy of a System

The gravitational potential energy of a system of particles in a uniform gravitational field is the same as if all the mass were concentrated at the center of mass. Let $h_i$ be the height of the $i$th particle in a system above some reference level. The gravitational potential energy of the system is

$$U = \sum_i m_i g h_i = g \sum_i m_i h_i$$

But, by definition of the center of mass, the height of the center of mass is given by

$$Mh_{cm} = \sum_i m_i h_i$$

so

$$U = Mgh_{cm} \qquad \qquad 8\text{-}7$$

We can use this result to locate the center of mass of an object experimentally. For example, two objects connected by a light rod will balance if the pivot is at the center of mass (Figure 8-9). If we pivot the system at any other point, the system will rotate until the potential energy is at a minimum, which occurs when the center of mass is at its lowest possible point directly below the pivot (Figure 8-10).

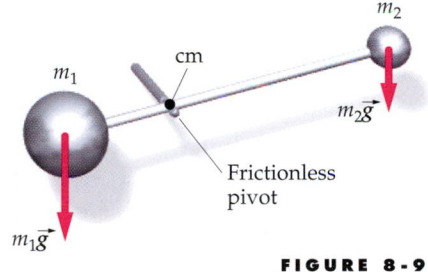

**FIGURE 8-9**

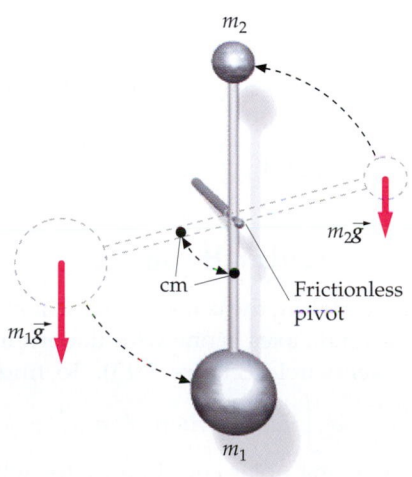

**FIGURE 8-10**

If we suspend any irregular object from a pivot, the object will hang so that its center of mass lies somewhere on the vertical line drawn directly downward from the pivot. Now suspend the object from another point and note where the vertical line now passes across the object. The center of mass will lie at the intersection of the two lines (Figure 8-11).

## *8-2 Finding the Center of Mass by Integration

In this section we find the center of mass by integration (Equation 8-6):

$$M\vec{r}_{cm} = \int \vec{r}\, dm$$

We will use the simple problem of finding the center of mass of a uniform thin rod to illustrate the technique for setting up the integration. While we can find the answer to this problem by symmetry, here we will use integration.

### Uniform Rod

We first choose a coordinate system. A good choice for a coordinate system is one with an $x$ axis through the length of the rod, with the origin at one end of the rod (Figure 8-12). Shown on the figure is a mass element $dm$ of length $dx$ a distance $x$ from the origin. Equation 8-6 thus gives

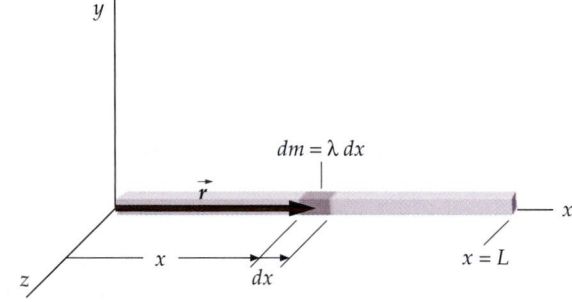

FIGURE 8-12

$$M\vec{r}_{cm} = \int \vec{r}\, dm = \int x\hat{i}\, dm$$

The mass is distributed on the $x$ axis along the interval $0 \le x \le L$. To sweep $dm$ along the mass distribution means the limits of the integral are 0 and $L$. (We sweep in the direction of increasing $x$.) The ratio $dm/dx$ is the mass per unit length $\lambda$, so $dm = \lambda\, dx$:

$$M\vec{r}_{cm} = \hat{i} \int x\, dm = \hat{i} \int_0^L x\lambda\, dx$$

Because the rod is uniform, $\lambda$ is constant and equal to $M/L$. Substituting for $\lambda$, we complete the calculation and obtain the expected result

$$\vec{r}_{cm} = \frac{1}{M}\hat{i}\,\lambda\,\frac{x^2}{2}\bigg|_0^L = \frac{1}{M}\hat{i}\,\frac{M}{L}\frac{L^2}{2} = \tfrac{1}{2}L\hat{i}$$

### Semicircular Hoop

In calculating the center of mass of a uniform semicircular hoop, a good choice of coordinate axes is one with the origin at the center and with the x axis bisecting the semicircle (Figure 8-13). To find the center of mass we use Equation 8-6 $\left(M\vec{r}_{cm} = \int \vec{r}\, dm\right)$, where $\vec{r} = x\hat{i} + y\hat{j}$. The semicircular mass distribution suggests using polar coordinates, for which $x = r\cos\theta$ and $y = r\sin\theta$. With these substitutions we have

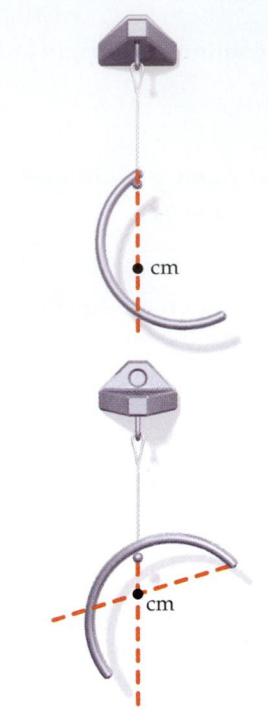

**FIGURE 8-11** The center of mass of an irregular object can be found by suspending it from two points.

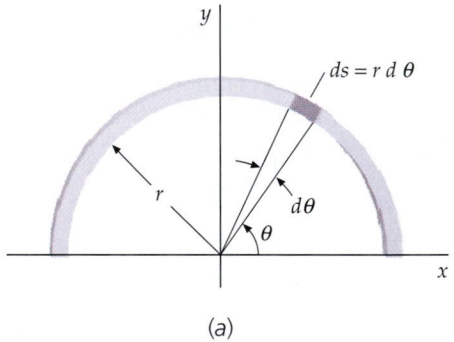

(a)

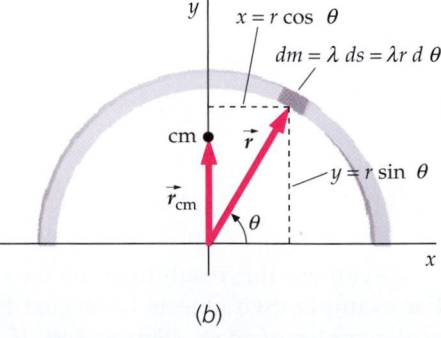

(b)

**FIGURE 8-13** Geometry for calculating the center of mass of a semicircular hoop by integration.

$$M\vec{r}_{cm} = \int (x\hat{i} + y\hat{j})\, dm = \int r(\cos\theta\,\hat{i} + \sin\theta\,\hat{j})\, dm$$

Next we express $dm$ in terms of $d\theta$. First, the mass element $dm$ has length $ds = r\, d\theta$, so

$$dm = \lambda\, ds = \lambda r\, d\theta$$

where $\lambda = dm/ds$ is the mass per unit length. Thus, we have

$$M\vec{r}_{cm} = \int r(\cos\theta\,\hat{i} + \sin\theta\,\hat{j})\,\lambda r\, d\theta$$

Evaluating this integral involves sweeping $dm$ along the semicircular mass distribution. This means $0 \le \theta \le \pi$. Sweeping in the direction of increasing $\theta$, the integration limits go from 0 to $\pi$. That is,

$$\vec{r}_{cm} = \frac{r^2}{M}\int_0^\pi (\cos\theta\,\hat{i} + \sin\theta\,\hat{j})\lambda\, d\theta$$

Because the hoop is uniform, we know that $\lambda = M/\pi r$, where $\pi r$ is the length of the semicircle. Substituting for $\lambda$ and rearranging gives

$$\vec{r}_{cm} = \frac{r}{\pi}\hat{i}\int_0^\pi \cos\theta\, d\theta + \frac{r}{\pi}\hat{j}\int_0^\pi \sin\theta\, d\theta = \frac{r}{\pi}\hat{i}\sin\theta\,\Big|_0^\pi - \frac{r}{\pi}\hat{j}\cos\theta\,\Big|_0^\pi = \frac{2r}{\pi}\hat{j}$$

The center of mass is on the $y$ axis a distance of $\dfrac{2}{\pi} r$ from the origin. Interestingly, it is outside of the material of the object.

## 8-3 Motion of the Center of Mass

Figure 8-14 is a multiflash photograph of a baton thrown into the air. Although the motion of the baton is complicated, the motion of the center of mass is simple. While the baton is in the air, the center of mass follows a parabolic path, the same path that would be followed by a point particle. We will show in general that the acceleration of the center of mass of a system of particles equals the net external force acting on the system divided by the total mass of the system. For the baton thrown into the air, the acceleration of the center of mass is $g$ downward; to find it, we first find the velocity by differentiating Equation 8-5 with respect to time:

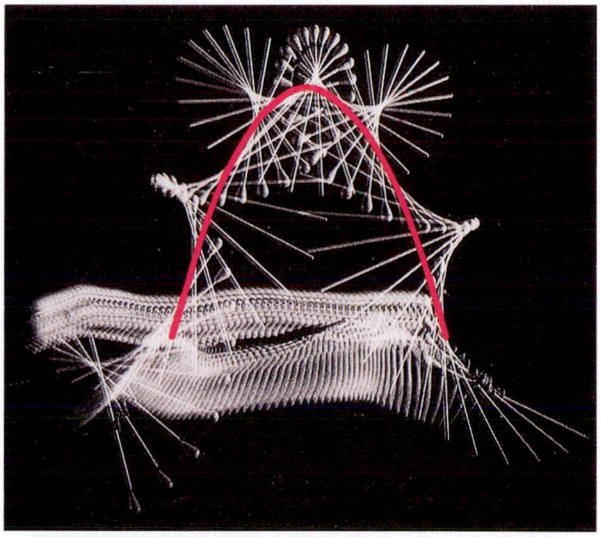

$$M\frac{d\vec{r}_{cm}}{dt} = m_1\frac{d\vec{r}_1}{dt} + m_2\frac{d\vec{r}_2}{dt} + \cdots = \sum_i m_i\frac{d\vec{r}_i}{dt}$$

**FIGURE 8-14**

Because the time derivative of the position is the velocity, this gives

$$M\vec{v}_{cm} = m_1\vec{v}_1 + m_2\vec{v}_2 + \cdots = \sum_i m_i\vec{v}_i \qquad\qquad 8\text{-}8$$

Differentiating again, we obtain the accelerations:

$$M\vec{a}_{cm} = m_1\vec{a}_1 + m_2\vec{a}_2 + \cdots = \sum_i m_i\vec{a}_i \qquad\qquad 8\text{-}9$$

However, in accord with Newton's second law, $m_i\vec{a}_i$ equals the sum of the forces acting on the $i$th particle, so

$$\sum_i m_i\vec{a}_i = \sum_i \vec{F}_i$$

where the term on the right is the sum of all the forces acting on each and every particle in the system. Some of these forces are *internal* forces (exerted on a particle in the system by some other particle in the system) and others are *external* forces (exerted on a particle in the system by a particle not in the system). Thus,

$$M\vec{a}_{cm} = \sum_i \vec{F}_{i,int} + \sum_i \vec{F}_{i,ext} \qquad 8\text{-}10$$

According to Newton's third law, forces come in action–reaction pairs. Thus, for each internal force acting on a particle in the system there is an equal and opposite internal force acting on some other particle in the system. When we sum all the internal forces, each action–reaction force pair sums to zero, so $\sum \vec{F}_{i,int} = 0$. Equation 8-10 then becomes

$$\vec{F}_{net,ext} = \sum_i \vec{F}_{i,ext} = M\vec{a}_{cm} \qquad 8\text{-}11$$

NEWTON'S SECOND LAW FOR A SYSTEM

That is, the net external force acting on the system equals the product of the total mass $M$ of the system and the acceleration of the center of mass $\vec{a}_{cm}$ of the system. Thus,

The center of mass of a system moves like a particle of mass $M = \sum m_i$ under the influence of the net external force acting on the system.

This theorem is important because it describes the motion of the center of mass for *any* system of particles: The center of mass moves exactly like a single point particle of mass $M$ acted on by only the external forces. The individual motion of any particle in the system is typically much more complex and is not described by Equation 8-11. The baton thrown into the air in Figure 8-14 is an example. The only external force acting is gravity, so the center of mass of the baton moves in a simple parabolic path, as would a point particle. (Equation 8-11 does not describe the rotational motion of one end of the baton about the center of mass.)

---

*An Exploding Projectile*  **EXAMPLE 8-3**

A projectile is fired into the air over level ground on a trajectory that would result in it landing 55 m away. However, at its highest point it explodes into two fragments of equal mass. Immediately following the explosion one fragment has a momentary speed of zero and then falls straight down to the ground. Where does the other fragment land? Neglect air resistance.

**PICTURE THE PROBLEM** Let the projectile be the system. Then the forces of the explosion are all internal forces. Since the only *external* force acting on the system is that due to gravity, the center of mass, which is midway between the two fragments, continues on its parabolic path as if there had been no explosion (Figure 8-15).

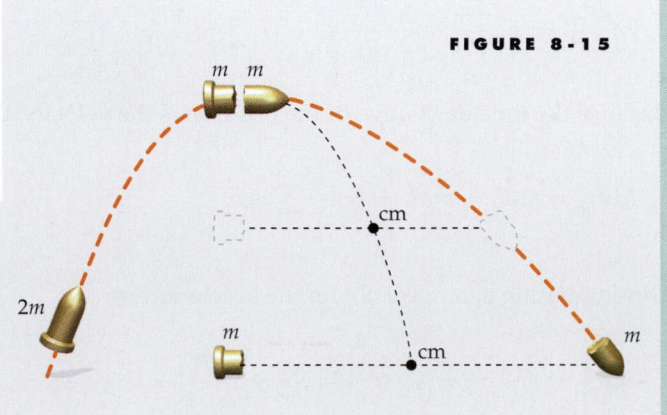

FIGURE 8-15

1. Let $x = 0$ be the initial position of the projectile. The landing positions $x_1$ and $x_2$ of the fragments are related to the final position of the center of mass by:

$$(2m)x_{cm} = mx_1 + mx_2$$

or

$$2x_{cm} = x_1 + x_2$$

2. At impact, $x_{cm} = R$ and $x_1 = 0.5R$, where $R = 55$ m is the range for the unexploded projectile. Solve for $x_2$:

$$x_2 = 2x_{cm} - x_1 = 2R - 0.5R = 1.5R$$

$$= 1.5(55 \text{ m}) = \boxed{82.5 \text{ m}}$$

**REMARKS** In Figure 8-16 height versus distance is plotted for exploding projectiles when fragment 1 has a horizontal velocity of half of the initial horizontal velocity. As in the original example in which fragment 1 falls straight down, the center of mass follows a normal parabolic trajectory.

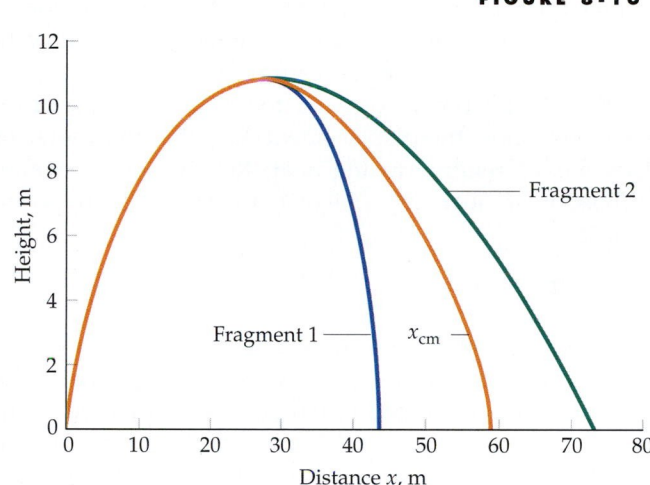

**FIGURE 8-16**

**EXERCISE** If the fragment that falls straight down has twice the mass of the other fragment, how far from the launch position does the lighter fragment land? (*Answer* 2R)

**REMARKS** If both fragments have the same vertical component of velocity after the explosion, they land at the same time. If just after the explosion the vertical component of the velocity of one fragment is less than that of the other, the fragment with the smaller vertical velocity component will hit the ground first. As soon as it does, the ground exerts a force on it and the net external force on the system is no longer just the gravitational force. From that moment on, our analysis is invalid.

**EXERCISE** A cylinder rests on a sheet of paper on a table (Figure 8-17). You pull the paper to the right, causing the cylinder to roll leftward *relative to the paper*. How does the cylinder's center of mass move relative to the table? (*Answer* It accelerates to the right, because the net external force acting on the cylinder is the frictional force to the right exerted on it by the paper. Try it. The cylinder may *appear* to move to the left, because relative to the paper it rolls leftward. However, relative to the table, which serves as an inertial reference frame, it moves to the right. If you mark the table with the original position of the cylinder, you will observe the center of mass move to the right *while the cylinder remains in contact with the paper*.)

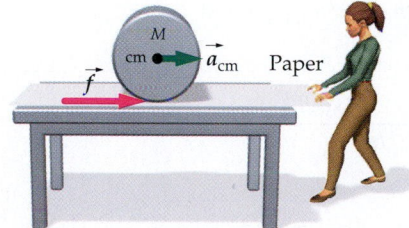

**FIGURE 8-17**

A special case of a system's center of mass in motion is the system with zero net external force acting on it. Here, $\vec{a}_{cm} = 0$, so the center of mass either remains at rest or moves with constant velocity. The internal forces and motion may be complex, but the behavior of the center of mass is simple. Further, if the net external force is not zero, but if a component of it in a given direction, say the $x$ direction, remains zero, then $a_{cmx} = 0$ and $v_{cmx}$ remains constant. An example of this is a projectile in the absence of air drag. The net external force on the projectile is the gravitational force. This force acts straight downward, so its component in any horizontal direction remains zero. It follows that the horizontal component of the velocity of the center of mass remains constant.

*CHANGING PLACES IN A ROWBOAT* **EXAMPLE 8-4**

Pete (mass 80 kg) and Dave (mass 120 kg) are in a rowboat (mass 60 kg) on a calm lake. Dave is near the bow of the boat, rowing, and Pete is at the stern, 2 m from the center. Dave gets tired and stops rowing. Pete offers to row, and after the boat comes to rest, they change places. How far does the boat move as Pete and Dave change places? (Neglect any horizontal force exerted by the water.)

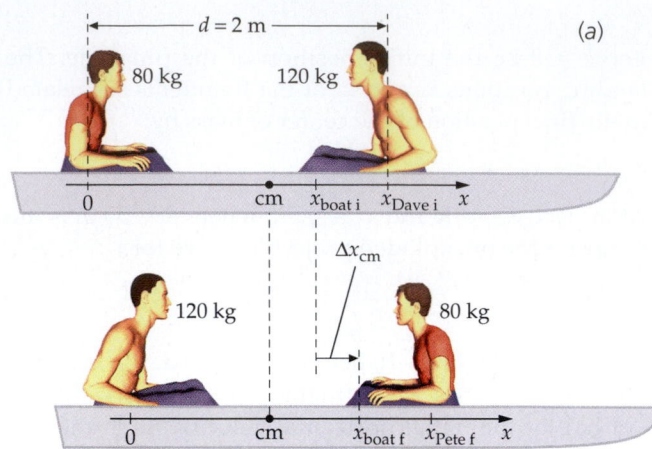

**PICTURE THE PROBLEM** Let the system be Dave, Pete, and the boat. There are no *external* forces in the horizontal direction, so the center of mass does not move horizontally relative to an inertial reference frame, such as the water. Relative to the boat, however, the center of mass does move. First find the displacement $\Delta x_{cm}$ that the center of mass moves relative to the boat. Relative to the water, the boat must move the same distance $\Delta x_{cm}$ in the opposite direction.

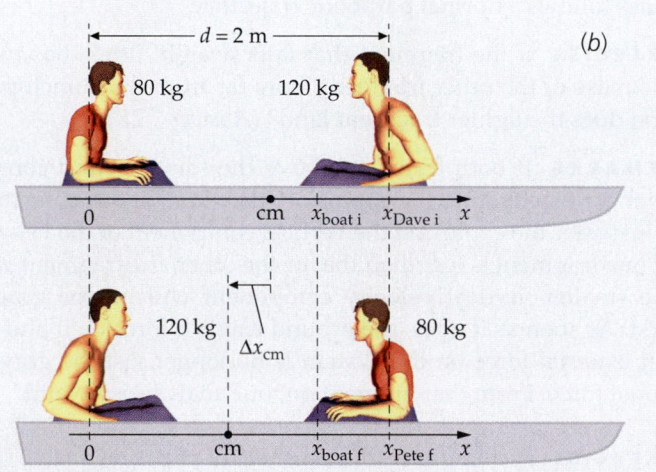

1. Make a sketch of the system in its initial and final configurations (Figure 8-18). Choose the origin at the back of the boat (Pete's initial position) and let $d = 2$ m so $x_{\text{Pete i}} = 0$ and $x_{\text{Dave i}} = d$. Let $x_{\text{boat}}$ be the position of the center of mass of the boat. When Pete and Dave switch places they merely exchange positions.

**FIGURE 8-18** (*a*) Pete and Dave changing places viewed from the reference frame of the water. (*b*) Pete and Dave changing places viewed from the reference frame of the boat.

2. Express the total mass $M$ times the initial value of the center of mass. Repeat this for the final value of the center of mass.

$$Mx_{\text{cm i}} = m_{\text{Pete}}x_{\text{Pete i}} + m_{\text{Dave}}x_{\text{Dave i}} + m_{\text{boat}}x_{\text{boat i}}$$
$$= 0 + m_{\text{Dave}}d + m_{\text{boat}}x_{\text{boat}}$$

$$Mx_{\text{cm f}} = m_{\text{Pete}}x_{\text{Pete f}} + m_{\text{Dave}}x_{\text{Dave f}} + m_{\text{boat}}x_{\text{boat f}}$$
$$= m_{\text{Pete}}d + 0 + m_{\text{boat}}x_{\text{boat}}$$

3. Find the displacement of the center of mass relative to the boat by subtracting the first step 2 result from the second:

$$Mx_{\text{cm f}} - Mx_{\text{cm i}} = (m_{\text{Pete}} - m_{\text{Dave}})d + 0$$

so

$$x_{\text{cm f}} - x_{\text{cm i}} = \frac{(m_{\text{Pete}} - m_{\text{Dave}})d}{M}$$

4. Because the center of mass does not move relative to the water, the boat must move an equal distance in the opposite direction:

$$-(x_{\text{cm f}} - x_{\text{cm i}}) = \frac{(m_{\text{Dave}} - m_{\text{Pete}})d}{M}$$
$$= \frac{(120 \text{ kg} - 80 \text{ kg})}{120 \text{ kg} + 80 \text{ kg} + 60 \text{ kg}}(2 \text{ m})$$
$$= \boxed{0.308 \text{ m}}$$

**REMARKS** The location of the center of mass of the boat, relative to the boat, subtracted out in step 3. The displacement of the boat is independent of this location.

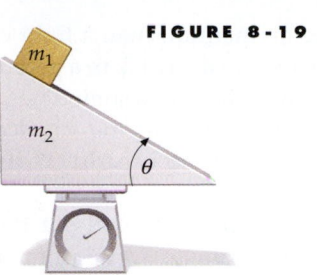

**FIGURE 8-19**

A SLIDING BLOCK                    **EXAMPLE  8 - 5**

A wedge of mass $m_2$ sits at rest on a scale as shown in Figure 8-19. A small block of mass $m_1$ slides down the frictionless incline of the wedge. Find the scale reading while the block slides.

**PICTURE THE PROBLEM** We choose the wedge plus block to be the system. Because the block accelerates down the wedge, the center of mass has acceleration components to the right and downward. The external forces on the system are the weights of the block and wedge, the force $F_x$ exerted by the scale on the wedge, and the normal force $F_n$ exerted upward by the scale. The scale reading is equal to the magnitude of $F_n$.

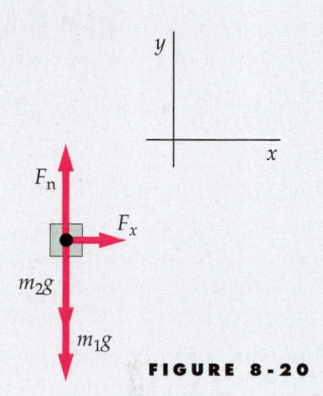

1. Draw a free-body diagram for the wedge–block system (Figure 8-20):

2. Write the vertical component of Newton's second law for the system and solve for $F_n$:

$$F_n - m_1g - m_2g = Ma_{cm,y} = (m_1 + m_2)a_{cm,y}$$
$$F_n = (m_1 + m_2)g + (m_1 + m_2)a_{cm,y}$$

3. Using Equation 8-9, express $a_{cm,y}$ in terms of the acceleration of the block $a_{1y}$:

$$Ma_{cm,y} = m_1a_{1y} + m_2a_{2y} = m_1a_{1y} + 0$$

$$a_{cm,y} = \frac{m_1}{m_1 + m_2}a_{1y}$$

**FIGURE 8-20**

4. From Example 4-7, a block sliding down a stationary incline has acceleration $g \sin \theta$ down the incline. Use trigonometry to find the $y$ component of this acceleration and use it to find $a_{cm,y}$:

$$a_{1y} = -a_1 \sin \theta = -g \sin^2 \theta$$

so

$$a_{cm,y} = \frac{m_1}{m_1 + m_2}a_{1y}$$

$$= -\frac{m_1}{m_1 + m_2}g \sin^2 \theta$$

5. Substitute for $a_{cm,y}$ in the step 2 result and solve for $F_n$:

$$F_n = (m_1 + m_2)g + (m_1 + m_2)a_{cm,y}$$
$$= (m_1 + m_2)g - m_1g \sin^2 \theta = [(1 - \sin^2 \theta)m_1 + m_2]g$$
$$= \boxed{(m_1 \cos^2 \theta + m_2)g}$$

**EXERCISE** What values of $F_n$ do you intuitively expect for $\theta = 0$ and for $\theta = 90°$? Show that $(m_1 \cos^2 \theta + m_2)g$ gives these expected values. (*Answer*   For $\theta = 0$ we expect $F_n = (m_1 + m_2)g$; for $\theta = 90°$ we expect $F_n = m_2g$.)

**EXERCISE** Find the force component $F_x$ exerted on the wedge by the scale. (*Answer*   $F_x = m_1g \sin \theta \cos \theta$.)

# 8-4  Conservation of Linear Momentum

A particle's **linear momentum** $\vec{p}$ is defined as the product of its mass and velocity:

$$\vec{p} = m\vec{v}$$                                    8-12

DEFINITION—MOMENTUM OF A PARTICLE

Linear momentum is a vector quantity that may be thought of as a measurement of the effort needed to bring a particle to rest. For example, a heavy truck has more momentum than a light car traveling at the same speed. It takes a greater force to stop the truck in a given time than it does to stop the car in the same time. (The quantity $m\vec{v}$ is sometimes referred to as the *linear momentum* of a particle to distinguish it from the *angular momentum*, which is discussed in Chapter 10. When there is no need to clarify that the momentum being referred to is the linear momentum, the adjective *linear* will be suppressed and we will just refer to the *momentum*. Throughout the remainder of this chapter momentum is used when referring to linear momentum.)

Newton's second law can be written in terms of the momentum of a particle. Differentiating Equation 8-12 we obtain

$$\frac{d\vec{p}}{dt} = \frac{d(m\vec{v})}{dt} = m\frac{d\vec{v}}{dt} = m\vec{a}$$

Then substituting the force $\vec{F}_{net}$ for $m\vec{a}$,

$$\vec{F}_{net} = \frac{d\vec{p}}{dt} \qquad\qquad 8\text{-}13$$

Thus, the net force acting on a particle equals the time rate of change of the particle's linear momentum. (Newton's original statement of his second law was in fact in this form.)

The total momentum $\vec{P}_{sys}$ of a system of particles is the sum of the momenta of the individual particles:

$$\vec{P}_{sys} = \sum_i m_i\vec{v}_i = \sum_i \vec{p}_i$$

According to Equation 8-8, $\Sigma m_i\vec{v}_i$ equals the total mass $M$ times the velocity of the center of mass:

$$\vec{P}_{sys} = \sum_i m_i\vec{v}_i = M\vec{v}_{cm} \qquad\qquad 8\text{-}14$$

TOTAL MOMENTUM OF A SYSTEM

Differentiating this equation, we obtain

$$\frac{d\vec{P}_{sys}}{dt} = M\frac{d\vec{v}_{cm}}{dt} = M\vec{a}_{cm}$$

But according to Newton's second law (Equation 8-11), $M\vec{a}_{cm}$ equals the net external force acting on the system. Thus,

$$\sum_i \vec{F}_{ext} = \vec{F}_{net,ext} = \frac{d\vec{P}_{sys}}{dt} \qquad\qquad 8\text{-}15$$

543 gm            209 gm

The two pucks are moving on an air cushion on a horizontal flat surface. (The hoses supplying the air are not shown.) The velocity of each puck changes in both magnitude and direction during the collision, but the velocity of the center of mass remains constant—unaffected by the internal forces of the collision.

When the net external force acting on a system of particles remains zero, the rate of change of the total momentum remains zero and the total momentum of the system remains constant:

$$\vec{P}_{sys} = \sum_i m_i\vec{v}_i = M\vec{v}_{cm} = \text{constant} \qquad (\vec{F}_{net,ext} = 0) \qquad\qquad 8\text{-}16$$

CONSERVATION OF MOMENTUM

This result is known as the **law of conservation of momentum:**

> If the net external force on a system remains zero, the total momentum of the system remains constant.

This law is one of the most important in physics. It is more widely applicable than the law of conservation of mechanical energy because internal forces exerted by one particle in a system on another are often not conservative. Thus, these internal forces can change the total mechanical energy of the system, though they have no effect on the system's total momentum. If the total momentum of a system is constant, then the velocity of the center of mass of the system is constant. The law of conservation of momentum is a vector relation so it is valid component by component. For example, if the $x$ component of the net external force on a system remains zero, then the $x$ component of the total momentum of the system remains constant.

---

*A SPACE REPAIR*                                          **EXAMPLE   8 - 6**

During repair of the Hubble Space Telescope, an astronaut replaces a solar panel whose frame is bent. Pushing the detached panel away into space, she is propelled in the opposite direction. The astronaut's mass is 60 kg and the panel's mass is 80 kg. The astronaut is at rest relative to her spaceship when she shoves away the panel, and she shoves it at 0.3 m/s relative to the spaceship. What is her subsequent velocity relative to the space ship? (During this operation the astronaut is tethered to the ship; for our calculation, assume that the tether remains slack.)

**PICTURE THE PROBLEM**  The velocity of the astronaut can be found from the velocity of the panel using conservation of momentum. Choose the direction of motion of the panel to be positive.

1. Apply conservation of momentum to find the velocity of the astronaut. Since the total momentum is initially zero, it remains zero:

$$p_p + p_a = m_p v_p + m_a v_a = 0$$

2. Solve for the astronaut's velocity:

$$v_a = -\frac{m_p}{m_a} v_p$$

$$= -\frac{80 \text{ kg}}{60 \text{ kg}}(0.3 \text{ m/s}) = \boxed{-0.4 \text{ m/s}}$$

**REMARKS**  Although momentum is conserved, the mechanical energy of this system increased because chemical energy of the astronaut was converted to kinetic energy.

**EXERCISE**  Find the final kinetic energy of the astronaut–panel system. (*Answer*   8.4 J)

---

*A RUNAWAY RAILROAD CAR*                                  **EXAMPLE   8 - 7**

A runaway 14,000-kg railroad car is rolling horizontally at 4 m/s toward a switchyard. As it passes by a grain elevator, 2000 kg of grain are suddenly dropped into the car. How long does it take the car to cover the 500-m distance from the elevator to the switchyard? Assume that the grain falls straight down and that slowing due to rolling friction or air drag is negligible.

**FIGURE 8-21**

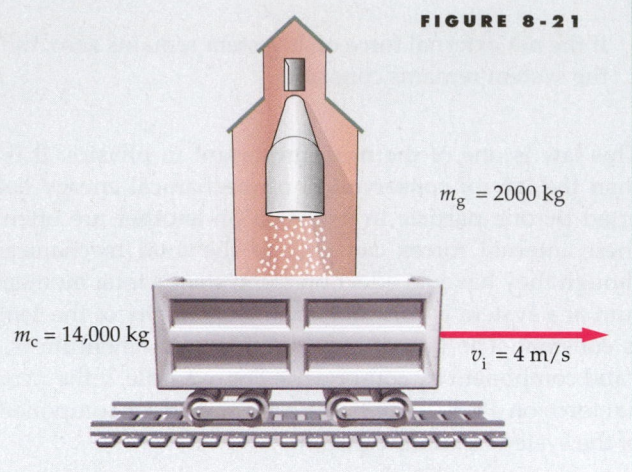

$m_g = 2000 \text{ kg}$

**PICTURE THE PROBLEM** We find the travel time that we seek from the distance traveled and the speed of the car. Consider the car and the grain as our system (Figure 8-21). There are no horizontal external forces acting on this system, so the horizontal component of the momentum of the system is conserved. The final speed of the grain-filled car is found from its final momentum, which equals the car's initial momentum. (The grain initially has no horizontal momentum.) Let $m_c$ and $m_g$ be the masses of the car and grain, respectively.

1. The time for the car to travel from the elevator to the yard is the distance to the yard $d$ divided by the car's speed $v_f$ following the grain dump:

$$\Delta t = \frac{d}{v_f}$$

$m_c = 14{,}000 \text{ kg}$

$v_i = 4 \text{ m/s}$

2. Apply conservation of momentum to relate the final velocity $v_f$ to the initial velocity $v_i$. Be careful, only the horizontal component of the system's momentum is conserved:

$$(m_c + m_g)v_f = m_c v_i + m_g(0)$$

3. Solve for $v_f$:

$$v_f = \frac{m_c v_i}{m_c + m_g}$$

4. Substitute the result for $v_f$ into step 1 and solve for the time:

$$\Delta t = \frac{d}{v_f} = \frac{(m_c + m_g)d}{m_c v_i}$$

$$= \frac{(14000 \text{ kg} + 2000 \text{ kg})(500 \text{ m})}{(14000 \text{ kg})(4 \text{ m/s})}$$

$$= \boxed{143 \text{ s}}$$

**REMARKS** Mechanical energy of the system is converted to thermal energy. Let $K_{gi}$ be the kinetic energy of the grain just as it hits the car. The initial mechanical energy is $K_{gi} + \frac{1}{2}m_c v_i^2 = K_{gi} + \frac{1}{2}(14{,}000 \text{ kg})(4 \text{ m/s})^2 = K_{gi} + 112 \text{ kJ}$. The final kinetic energy is $\frac{1}{2}(m_c + m_g)v_f^2$, where $v_f = m_c v_i/(m_c + m_g) = 14(4 \text{ m/s})/16 = 3.5 \text{ m/s}$, so the final kinetic energy is $\frac{1}{2}(16{,}000 \text{ kg})(3.5 \text{ m/s})^2 = 98 \text{ kJ}$. The final kinetic energy is less than the initial kinetic energy by 14 kJ.

**EXERCISE** Suppose that there is a small vertical chute in the bottom of the car so that the grain leaks out at 10 kg/s. Now how long does it take the car to cover the 500 m? (*Answer* 143 s. The grain leaking out does not impart any momentum to the rest of the system. If the ground were frictionless and flat, all of the grain initially in the car would arrive at the switchyard along with the car.)

---

*A SKATEBOARD WORKOUT* **EXAMPLE 8-8**

A 40-kg skateboarder on a 3-kg board is training with two 5-kg weights. Beginning from rest, she throws the weights horizontally, one at a time from her board. The speed of each weight is 7 m/s relative to her and the board after it is thrown. Assume the board rolls without friction. (*a*) How fast is she propelled in the opposite direction after throwing the first weight? (*b*) the second weight?

**PICTURE THE PROBLEM** Because no external forces act on the skateboarder–weights–board system with horizontal components, the horizontal component of its momentum is conserved. We need to find the velocity of the skateboarder after throwing each weight (Figure 8-22).

(a) 1. The mass $m$ of each weight is 5 kg and the mass $M$ of the skateboard and skateboarder is 43 kg. Choose the direction of the skateboarder's motion to be the positive direction. Then $v_{ws} = -7\,\text{m/s}$ is the velocity of the thrown weight relative to the skateboarder. Let $V_{sg1}$ and $v_{wg1}$ be the respective velocities of the skateboarder and the thrown weight relative to the ground. Apply conservation of momentum for the first throw.

$$P_{sys1} = P_{sys0}$$

$$(M + m)V_{sg1} + mv_{wg1} = 0$$

(The numbers in the subscripts stand for the time. Time 0 is before the first throw, time 1 is between the two throws, and time 2 is following the second throw.)

**FIGURE 8-22**

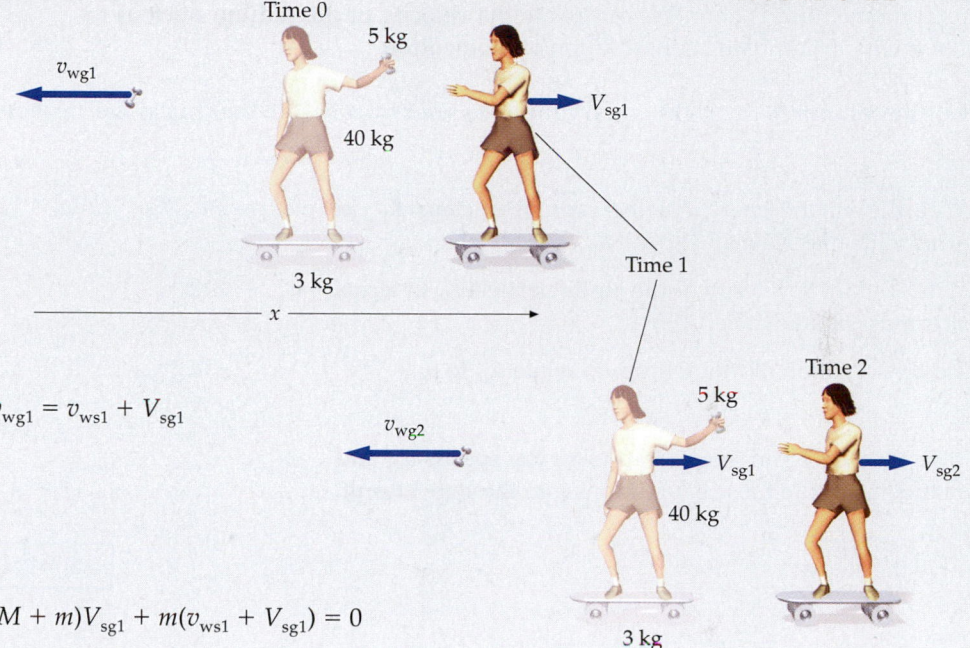

2. The velocity of the thrown weight relative to the ground equals the velocity of the weight relative to the skateboarder plus the velocity of the skateboarder relative to the ground:

$$v_{wg1} = v_{ws1} + V_{sg1}$$

3. Substitute for $v_{wg1}$ in the step 1 result and solve for $V_{sg1}$:

$$(M + m)V_{sg1} + m(v_{ws1} + V_{sg1}) = 0$$

so

$$V_{sg1} = -\frac{m}{M + 2m}v_{ws1} = -\frac{5\,\text{kg}}{43\,\text{kg} + 10\,\text{kg}}(-7\,\text{m/s})$$

$$= \boxed{0.66\,\text{m/s}}$$

(b) 1. Repeat step 1 of Part (a) for the second throw. Let $V_{sg2}$ and $v_{w'g2}$ be the respective velocities of the skateboarder and the second thrown weight relative to the ground:

$$P_{sys2} = P_{sys1}$$

$$MV_{sg2} + mv_{w'g2} = (M + m)V_{sg1}$$

2. Repeat step 2 of Part (a) for the second throw.

$$v_{w'g2} = v_{w's2} + V_{sg2}$$

3. Substitute for $v_{w'g2}$ in the Part (b) step 1 result and solve for $V_{sg2}$:

$$MV_{sg2} + m(v_{w's2} + V_{sg2}) = (M + m)V_{sg1}$$

so

$$V_{sg2} = \frac{(M + m)V_{sg1} - mv_{w's2}}{M + m} = V_{sg1} - \frac{m}{M + m}v_{w's2}$$

$$= 0.66\,\text{m/s} - \frac{5\,\text{kg}}{48\,\text{kg}}(-7\,\text{m/s}) = \boxed{1.39\,\text{m/s}}$$

**REMARKS** This example illustrates the principle of the rocket; a rocket moves forward by throwing its fuel out backward in the form of exhaust gases.

**EXERCISE** How fast is the skateboarder moving if, starting from rest, she throws both weights together, and the weights have speed 7 m/s relative to herself *after they are thrown*? (*Answer* 1.32 m/s)

**EXAMPLE 8-9** **Try It Yourself**

A thorium-227 nucleus at rest decays into a radium-223 nucleus (mass 223 u) by emitting an $\alpha$ particle (mass 4 u) (Figure 8-23). The kinetic energy of the $\alpha$ particle is measured to be 6.00 MeV. What is the kinetic energy of the recoiling radium nucleus?

Thorium-227  Radium-223

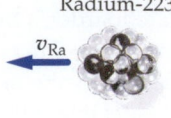 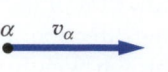

**FIGURE 8-23**

**PICTURE THE PROBLEM** Because the thorium nucleus before decay is at rest, its total momentum is zero. We can relate the velocity of the radium nucleus to that of the $\alpha$ particle using conservation of momentum.

**Cover the column to the right and try these on your own before looking at the answers.**

| Steps | Answers |
|---|---|
| 1. Write the kinetic energy of the radium nucleus $K_{Ra}$ in terms of its mass $m_{Ra}$ and speed $v_{Ra}$. | $K_{Ra} = \frac{1}{2}m_{Ra}v_{Ra}^2$ |
| 2. Write the kinetic energy of the alpha particle $K_\alpha$ in terms of its mass $m_\alpha$ and speed $v_\alpha$. | $K_\alpha = \frac{1}{2}m_\alpha v_\alpha^2$ |
| 3. Use conservation of momentum to relate $v_{Ra}$ to $v_\alpha$. | $m_\alpha v_\alpha = m_{Ra}v_{Ra}$ |
| 4. Solve the step 1 and step 2 results for the speeds $v_{Ra}$ and $v_\alpha$, and substitute these expressions into the step 3 result. | $m_\alpha\left(\dfrac{2K_\alpha}{m_\alpha}\right)^{1/2} = m_{Ra}\left(\dfrac{2K_{Ra}}{m_{Ra}}\right)^{1/2}$ |
| 5. Solve the step 4 result for $K_{Ra}$. | $K_{Ra} = \dfrac{m_\alpha}{m_{Ra}}K_\alpha = \boxed{0.108 \text{ MeV}}$ |

**REMARKS** In this process, rest energy of the thorium nucleus is converted into kinetic energy of the alpha particle plus radium nucleus. The mass of the thorium nucleus is greater than that of the alpha particle plus radium nucleus by 6.1 MeV/$c^2$.

# 8-5 Kinetic Energy of a System

If the net external force on the system remains zero, then the total momentum of a system of particles must remain constant; however, the total mechanical energy of the system can change. As we saw in the examples of the previous section, internal forces that cannot change the total momentum may be nonconservative and thus change the total mechanical energy of the system. There is an important theorem concerning the kinetic energy of a system of particles that allows us to treat the energy of complex systems more easily and gives us insight into energy changes within a system:

The kinetic energy of a system of particles can be written as the sum of two terms: (1) the kinetic energy associated with the motion of the center of mass, $\frac{1}{2}Mv_{cm}^2$, where $M$ is the total mass of the system; and (2) the kinetic energy associated with the motion of the particles of the system relative to the center of mass, $\Sigma \frac{1}{2}m_i u_i^2$, where $\vec{u}_i$ is the velocity of the $i$th particle relative to the center of mass.

THEOREM FOR THE KINETIC ENERGY OF A SYSTEM

The kinetic energy $K$ of a system of particles is the sum of the kinetic energies of the individual particles:

$$K = \sum_i K_i = \sum_i \tfrac{1}{2}m_i v_i^2 = \sum_i \tfrac{1}{2}m_i (\vec{v}_i \cdot \vec{v}_i)$$

The velocity of each particle can be written as the sum of the velocity of the center of mass $\vec{v}_{cm}$ and the velocity of the particle relative to the center of mass $\vec{u}_i$:

$$\vec{v}_i = \vec{v}_{cm} + \vec{u}_i \qquad\qquad 8\text{-}17$$

Then

$$K = \sum_i \tfrac{1}{2}m_i(\vec{v}_i \cdot \vec{v}_i) = \sum_i \tfrac{1}{2}m_i(\vec{v}_{cm} + \vec{u}_i) \cdot (\vec{v}_{cm} + \vec{u}_i)$$

$$= \sum_i \tfrac{1}{2}m_i(v_{cm}^2 + 2\vec{v}_{cm} \cdot \vec{u}_i + u_i^2)$$

$$= \sum_i \tfrac{1}{2}m_i v_{cm}^2 + \vec{v}_{cm} \cdot \sum_i m_i\vec{u}_i + \sum_i \tfrac{1}{2}m_i u_i^2$$

where in the middle term we have factored $\vec{v}_{cm}$ from the sum because it is the same for each term in the sum. That is, $\vec{v}_{cm}$ does not change from particle to particle. The quantity $\Sigma m_i\vec{u}_i$ is the total momentum of the system *relative to the center of mass*. This quantity, which equals $M\vec{u}_{cm}$ is necessarily zero. (Relative to the center of mass, the velocity of the center of mass $\vec{u}_{cm}$ is zero, so the total momentum $M\vec{u}_{cm}$ is also zero.) Thus,

$$K = \sum_i \tfrac{1}{2}m_i v_{cm}^2 + \sum_i \tfrac{1}{2}m_i u_i^2 = \tfrac{1}{2}Mv_{cm}^2 + K_{rel} \qquad\qquad 8\text{-}18$$

KINETIC ENERGY OF A SYSTEM OF PARTICLES

where $M$ is the total mass and $K_{rel}$ is the kinetic energy of the particles *relative to the center of mass*. If the net external force is zero, $\vec{v}_{cm}$ remains constant and the kinetic energy associated with bulk motion ($\tfrac{1}{2}Mv_{cm}^2$) does not change. Only the relative kinetic energy can change in an isolated system.

# 8-6 Collisions

In a **collision,** two objects approach and interact strongly for a very short time. During the brief time of collision, any external forces on the objects are assumed to be much weaker than the forces of interaction between the objects. Thus, during the collision the only important forces acting on the two-object system are the interaction forces, which are equal and opposite, so the total momentum of the system remains unchanged. Also, the collision time is so short that during the collision any displacements of the colliding objects can be neglected. Both prior to and following the collision, the interaction of the two objects is weak compared with their interaction during the collision. Examples of collisions are a cue ball hitting a billiard ball, a bat hitting a baseball, or a dart colliding with a dart board. **A comet swinging around the sun can be considered a collision, even though the two do not physically touch, and in the lab you may see a collision between two carts with magnetic bumpers that repel without touching.**

When the total kinetic energy of the two-object system is the same after the collision as before, the collision is called an **elastic collision.** Otherwise it is called an **inelastic collision.** An extreme case is the **perfectly inelastic collision,** in which all of the kinetic energy relative to the center of mass is converted to thermal or internal energy of the system, and the two objects stick together after the collision.

## Impulse and Average Force

Figure 8-24 shows the time variation of the magnitude of a typical force exerted by one object on another during a collision. During the collision time $\Delta t = t_f - t_i$ the force is large. For other times the force is negligibly small. The **impulse** $\vec{I}$ of the force is a vector defined as

$$\vec{I} = \int_{t_i}^{t_f} \vec{F}\, dt$$

8-19

DEFINITION—IMPULSE

The magnitude of the impulse of the force is the area under its $F$-versus-$t$ curve. Units of impulse are N·s. The net force $\vec{F}_{net}$ acting on a particle is related to the rate of change of momentum of the particle by Newton's second law: $\vec{F}_{net} = d\vec{p}/dt$, so the impulse of the net force equals the total change in momentum $\Delta\vec{p}$ during the time interval:

$$\vec{I}_{net} = \int_{t_i}^{t_f} \vec{F}_{net}\, dt = \int_{t_i}^{t_f} \frac{d\vec{p}}{dt}\, dt = \vec{p}_f - \vec{p}_i = \Delta\vec{p}$$

8-20

IMPULSE–MOMENTUM THEOREM FOR A PARTICLE

Also, the net impulse on a system due to external forces equals the change in the total momentum of the system:

$$\vec{I}_{net,ext} = \int_{t_i}^{t_f} \vec{F}_{net,ext}\, dt = \Delta\vec{P}_{sys}$$

8-21

IMPULSE–MOMENTUM THEOREM FOR A SYSTEM

The **average force** for the interval $\Delta t = t_f - t_i$ is defined as

$$\vec{F}_{av} = \frac{1}{\Delta t} \int_{t_i}^{t_f} \vec{F}\, dt = \frac{\vec{I}}{\Delta t}$$

8-22

DEFINITION—AVERAGE FORCE

The average force is the constant force that gives the same impulse as the actual force in the time interval $\Delta t$, as shown by the rectangle in Figure 8-24. The average net force can be calculated from the change in momentum if a collision time is known. This time can often be estimated using the displacement of one of the bodies during the collision.

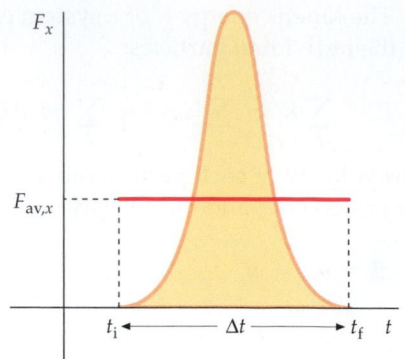

**FIGURE 8-24** Typical time variation of force during a collision. The area under the $F_x$-versus-$t$ curve is the $x$ component of the impulse, $I_x$. $F_{av,x}$ is the average force for time interval $\Delta t$. The rectangular area $F_{av,x}\,\Delta t$ is the same as the area under the $F_x$-versus-$t$ curve.

---

A KARATE COLLISION

**E X A M P L E    8 - 1 0**

With an expert karate blow, you shatter a concrete block. Consider your fist to have a mass 0.70 kg, to be moving 5.0 m/s as it strikes the block, and to stop within 6 mm of the point of contact. (*a*) What impulse does the block exert on your fist? (*b*) What is the approximate collision time and the average force the block exerts on your fist?

**PICTURE THE PROBLEM** The net impulse equals the change in momentum $\Delta \vec{p}$. We find $\Delta \vec{p}$ from the mass and velocity of the fist. The time of collision for Part (b) comes from the given displacement $\Delta y = -6$ mm and the average velocity $v_{av}$ during the collision, which we can estimate by assuming constant acceleration. Make a sketch of the fist and block. Include a vertical coordinate axis on the sketch (Figure 8-25).

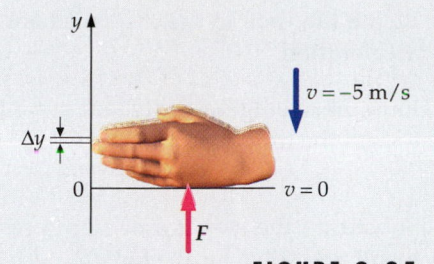

**FIGURE 8-25**

(a) 1. Set the impulse equal to the change in momentum:

$$\vec{I} = \Delta \vec{p} = \vec{p}_f - \vec{p}_i$$

2. The initial momentum is that of the fist just before it hits the block with velocity $\vec{v}$, and the final momentum is zero:

$$\vec{p}_i = m\vec{v} = (0.7 \text{ kg})(-5.0 \text{ m/s})\hat{j}$$
$$= -3.5 \text{ kg·m/s}\,\hat{j}$$
$$\vec{p}_f = 0$$

3. Find the impulse exerted by the block on the fist:

$$\vec{I} = \vec{p}_f - \vec{p}_i = 0 - (-3.5 \text{ kg} \cdot \text{m/s}\,\hat{j})$$
$$= \boxed{3.5 \text{ N·s}\,\hat{j}}$$

(b) 1. The collision time is the displacement divided by the average velocity:

$$\Delta t = \frac{\Delta y}{v_{av}}$$

2. We estimate the average speed by assuming constant acceleration, $v_{av} = \frac{1}{2}v$. Since we have chosen up to be positive, both $\Delta y$ and $v_{av}$ are negative. Calculate $\Delta t$:

$$\Delta t = \frac{\Delta y}{\frac{1}{2}v} = \frac{-0.006 \text{ m}}{-2.5 \text{ m/s}} = 0.0024 \text{ s}$$
$$= 2.4 \text{ ms}$$

3. The average force is the impulse divided by the collision time. It is upward, as expected:

$$\vec{F}_{av} = \frac{\vec{I}}{\Delta t} = \frac{3.5 \text{ N·s}\,\hat{j}}{0.0024 \text{ s}} = \boxed{1.46 \text{ kN}\,\hat{j}}$$

**REMARKS** Note that the average force is large—about 212 times the weight of the fist.

---

*A CRASH TEST*                    **EXAMPLE 8-11    Try It Yourself**

A car equipped with an 80-kg crash-test dummy (Figure 8-26) drives into a wall at 25 m/s (about 56 mi/h). Estimate the force that the seatbelt exerts on the dummy upon impact.

**PICTURE THE PROBLEM** Assume that the car and dummy travel about 1 m as the front end of the car crumples, and that the acceleration is constant during the crash. To find the force, calculate the impulse $I$, then divide it by the collision time $\Delta t$. Choose forward as the positive $x$ direction.

**FIGURE 8-26**

Cover the column on the right and try these on your own before looking at the answers.

**Steps**

**Answers**

1. Relate the average force to the impulse, and thus to the change in momentum.

$$\vec{F}_{av}\Delta t = \vec{I} = \Delta \vec{p}, \quad \text{so} \quad \vec{F}_{av} = \frac{\Delta \vec{p}}{\Delta t}$$

2. Find the change in the dummy's momentum.

$$\Delta \vec{p} = m\vec{v}_f - m\vec{v}_i = -2000 \text{ N·s}\,\hat{i}$$

3. Relate the time to the displacement, assuming constant acceleration.

$$\Delta t = \frac{\Delta x}{v_{av}}$$

4. Find the average velocity and use it and the step 3 result to find the time.

$$\vec{v}_{av} = \tfrac{1}{2}(\vec{v}_f + \vec{v}_i) = 12.5 \text{ m/s}\,\hat{\imath}, \quad \text{so}$$

$$\Delta t = 0.08 \text{ s}$$

5. Substitute the step 2 and step 4 results into the step 1 result and solve for the force.

$$\vec{F}_{av} = \boxed{-25 \text{ kN}\,\hat{\imath}}$$

**REMARKS** The magnitude of the average acceleration is $a_{av} = \Delta v/\Delta t = 313 \text{ m/s}^2$, or roughly $32g$. Such an acceleration means a net force about 32 times the weight of the dummy, clearly enough to cause serious injuries. An air bag increases the stopping distance somewhat, which helps to reduce the force and also allows the force to be distributed over a much larger area.

**REMARKS** In Figure 8-27, plot (a) shows the average force on the dummy as a function of the stopping distance. With no seat belt or air bag, you either fly though the windshield, or are stopped in a fraction of a meter by the dashboard or steering wheel. Plot (b) shows the force as a function of the initial velocity for three stopping distances: 2 m, 1.5 m, and 1 m.

**FIGURE 8-27**

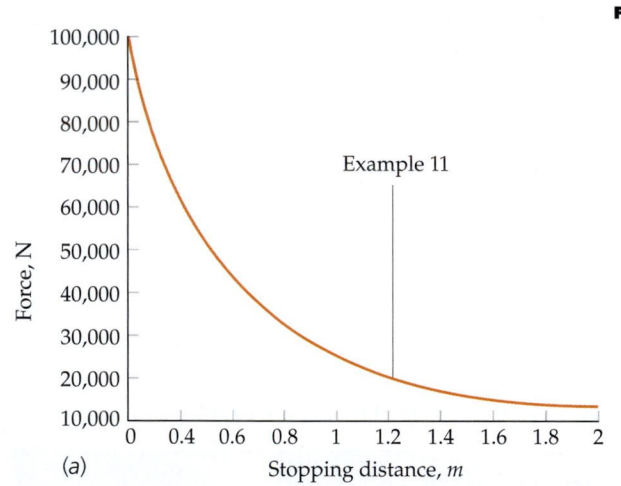

(a)

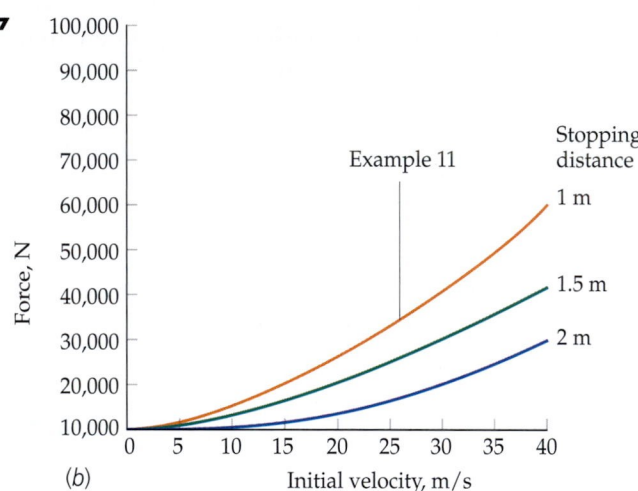

(b)

---

*A Golf Ball*

**EXAMPLE 8-12**

You strike a golf ball with a driving iron. What are reasonable estimates for the (a) impulse $I$, (b) collision time $\Delta t$, and (c) average force $F_{av}$? A typical golf ball has a mass $m = 45$ g and a radius $r = 2$ cm. For a typical drive, the range $R$ is roughly 192 m (210 yd) (Figure 8-28).

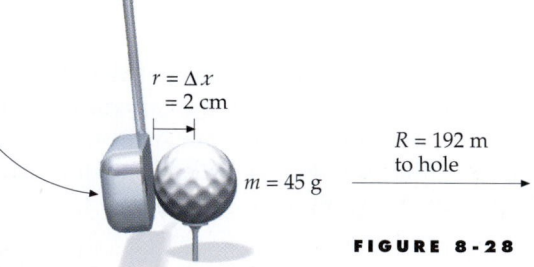

**FIGURE 8-28**

**PICTURE THE PROBLEM** Let $v_0$ denote the speed of the ball as it leaves the club face. The impulse on the ball equals its change in momentum, which is $mv_0$. We estimate $v_0$ from the range. We estimate the collision time from the distance traveled $\Delta x$ and the average speed $\frac{1}{2}v_0$, assuming constant acceleration. Taking $\Delta x = 2$ cm, the average force is then obtained from the impulse $I$ and collision time $\Delta t$.

(a) 1. Set the impulse equal to the change in momentum of the ball:

$$I = F_{av} \Delta t = \Delta p$$

2. The initial speed is related to the range $R$, which is given by Equation 2-23:

$$R = \frac{v_0^2}{g} \sin 2\theta_0$$

3. Take $\theta_0 = 13°$ and calculate the initial speed:

$$v_0 = \sqrt{\frac{Rg}{\sin 2\theta_0}}$$

$$= \sqrt{\frac{(192 \text{ m})(9.81 \text{ m/s}^2)}{\sin 26°}} = 65.5 \text{ m/s}$$

4. Use this value of $v_0$ to calculate the magnitude of the impulse:

$$I = \Delta p = m(v_0 - 0) = (0.045 \text{ kg})(65.5 \text{ m/s})$$

$$= 2.95 \text{ kg·m/s} = \boxed{2.95 \text{ N·s}}$$

(b) Calculate the collision time $\Delta t$ using $\Delta x = 2$ cm and $v_{av} = \frac{1}{2}(v_f - v_i)$:

$$\Delta t = \frac{\Delta x}{v_{av}} = \frac{\Delta x}{\frac{1}{2}v_0}$$

$$= \frac{0.02 \text{ m}}{\frac{1}{2}(65.5 \text{ m/s})} = \boxed{0.610 \times 10^{-3} \text{ s}}$$

(c) Use the calculated values of $I$ and $\Delta t$ to find the magnitude of the average force:

$$F_{av} = \frac{I}{\Delta t} = \frac{2.95 \text{ N·s}}{6.10 \times 10^{-4} \text{ s}} = \boxed{4.83 \text{ kN}}$$

**REMARKS** Again we see that forces exerted during a collision are very large. Here the force exerted on the golf ball by the club is roughly 10,000 times the weight of the ball, giving it an average acceleration of $10,000g$ for 0.61 ms. Also, the force of the air on the ball has been left out of our analysis. For an actual golf shot the effects of the air are not negligible.

## Collisions in One Dimension (Head-on Collisions)

Consider an object of mass $m_1$ with initial velocity $v_{1i}$ approaching a second object of mass $m_2$ that is moving in the same direction with initial velocity $v_{2i}$. If $v_{2i} < v_{1i}$, the objects collide. Let $v_{1f}$ and $v_{2f}$ be their final velocities after the collision. (These velocities can be positive or negative, depending on whether the objects are moving to the right or to the left.) Conservation of momentum gives one relation between the two unknown velocities $v_{1f}$ and $v_{2f}$:

$$m_1 v_{1f} + m_2 v_{2f} = m_1 v_{1i} + m_2 v_{2i} \qquad \text{8-23}$$

To determine $v_{1f}$ and $v_{2f}$, we must have a second relation. That second relation, which we will develop here, depends on the type of collision.

**Perfectly Inelastic Head-on Collisions** In perfectly inelastic collisions, the particles stick together after the collision. For example, the collision between a catcher's mitt and a baseball is perfectly inelastic if the catcher does not drop the ball. The second kind of relation between the velocities is that the final velocities are equal to each other and to the velocity of the center of mass:

$$v_{1f} = v_{2f} = v_{cm}$$

Substituting this result combined with conservation of momentum gives

$$(m_1 + m_2)v_{cm} = m_1 v_{1i} + m_2 v_{2i} \qquad \text{8-24}$$

**Perfectly inelastic collision of two cars.**

*A CATCH IN SPACE*                    **EXAMPLE 8-13**                    **FIGURE 8-29**

An astronaut of mass 60 kg is on a space walk to repair a communications satellite. Suddenly she needs to consult her physics book. You happen to have it with you, so you throw it to her with speed 4 m/s relative to your spacecraft. She is at rest relative to the spacecraft before catching the 3.0-kg book (Figure 8-29). Find (*a*) her velocity just after she catches the book, (*b*) the initial and final kinetic energies of the book–astronaut system, and (*c*) the impulse exerted by the book on the astronaut.

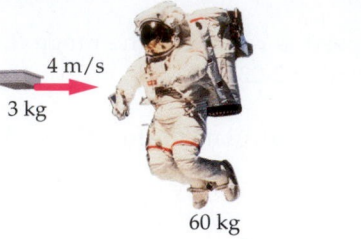

4 m/s

3 kg

60 kg          63 kg

**PICTURE THE PROBLEM** (*a*) The final velocity of the book and astronaut is the velocity of their center of mass. We find this using conservation of momentum, as expressed in Equation 8-24. The initial and final kinetic energies are calculated from the initial and final velocities. Because the book and astronaut move with the same final velocity, the collision is perfectly inelastic. (*b*) The kinetic energies of the book and astronaut are calculated directly from their masses and speeds. (*c*) The impulse exerted by the book on the astronaut equals the change in momentum of the astronaut.

(*a*) 1. Use conservation of momentum to relate the final velocity of the system $v_{cm}$ to the initial velocities:

$$m_b v_b + m_a v_a = (m_b + m_a)v_{cm}$$

2. Solve for $v_{cm}$:

$$v_{cm} = \frac{m_b v_b + m_a v_a}{m_b + m_a}$$

$$= \frac{(3.0 \text{ kg})(4 \text{ m/s}) + 0}{60 \text{ kg} + 3 \text{ kg}} = \boxed{0.19 \text{ m/s}}$$

(*b*) 1. The initial mechanical energy of the book–astronaut system is the initial kinetic energy of the book:

$$K_i = K_b = \tfrac{1}{2}m_b v_b^2$$

$$= \tfrac{1}{2}(3.0 \text{ kg})(4 \text{ m/s})^2 = \boxed{24 \text{ J}}$$

2. The final kinetic energy is the kinetic energy of the book and astronaut moving together at $v_{cm}$:

$$K_f = \tfrac{1}{2}(m_b + m_a)v_{cm}^2$$

$$= \tfrac{1}{2}(63 \text{ kg})(0.19 \text{ m/s})^2 = \boxed{1.14 \text{ J}}$$

(*c*) Set the impulse exerted on the astronaut equal to the change in momentum of the astronaut:

$$I = \Delta p_a = m_a \Delta v_a$$

$$= (60 \text{ kg})(0.19 \text{ m/s} - 0)$$

$$= 11.4 \text{ kg·m/s} = \boxed{11.4 \text{ N·s}}$$

**REMARKS** Most of the initial kinetic energy in this collision is lost by conversion to thermal energy. The impulse exerted by the book on the astronaut is equal and opposite to that exerted by the astronaut on the book, so the total change in momentum of the book–astronaut system is zero.

It is useful to express the kinetic energy $K$ of a particle in terms of its momentum $p$. For a mass $m$ moving with speed $v$, we have

$$K = \tfrac{1}{2}mv^2 = \frac{(mv)^2}{2m}$$

Since $p = mv$,

$$K = \frac{p^2}{2m}$$

8-25

We can apply this to a perfectly inelastic collision where one object is initially at rest. The momentum of the system is that of the incoming object:

$$P_{sys} = p_{1i} = m_1 v_{1i}$$

The initial kinetic energy is

$$K_i = \frac{P_{sys}^2}{2m_1} \qquad\qquad 8\text{-}26$$

After colliding, the objects move together as a single mass $m_1 + m_2$ with $v_{cm}$. Momentum is conserved, so the final momentum equals $P_{sys}$. The final kinetic energy is then

$$K_f = \frac{P_{sys}^2}{2(m_1 + m_2)} \qquad\qquad 8\text{-}27$$

Comparing Equations 8-26 and 8-27, we see that the final kinetic energy is less than the initial kinetic energy.

---

*A BALLISTIC PENDULUM*                    **E X A M P L E    8 - 1 4**

**In a feat of public marksmanship, you fire a bullet into a hanging wood block. The block, with bullet embedded, swings upward. Noting the height reached at the top of the swing, you immediately inform the crowd of the bullet's speed. How fast was the bullet traveling?**

**PICTURE THE PROBLEM** The initial speed of the bullet $v_{1i}$ is related to the postcollision speed of the bullet–block system $v_f$ by conservation of momentum. The speed $v_f$ is related to the height $h$ by conservation of mechanical energy (Figure 8-30). Let $m_1$ be the mass of the bullet and $m_2$ be the mass of the target.

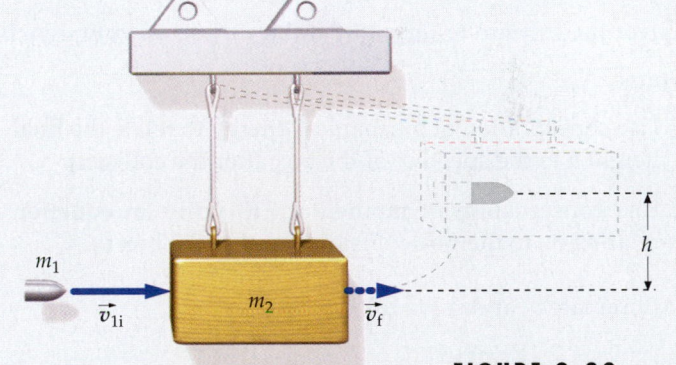

**FIGURE 8-30**

1. Using conservation of mechanical energy *after* the collision we relate the postcollision speed $v_f$ to the maximum height $h$:

$$\tfrac{1}{2}(m_1 + m_2)v_f^2 = (m_1 + m_2)gh$$

2. Using conservation of momentum *during* the collision we relate $v_{1i}$ to $v_f$:

$$m_1 v_{1i} = (m_1 + m_2)v_f$$

3. Solving the step 1 result for $v_f$ and then substituting for $v_f$ in the step 2 result, we can solve for $v_{1i}$:

$$v_f = \sqrt{2gh}$$

so

$$v_{1i} = \frac{m_1 + m_2}{m_1} v_f = \boxed{\frac{m_1 + m_2}{m_1}\sqrt{2gh}}$$

**REMARKS** We assumed that the time of the collision is so short that the displacement of the block during the collision is negligible. This meant the block had the postcollision speed while still at the lowest point in the arc.

Devices such as the one pictured are called *ballistic pendulums*.

**EXERCISE** Find the initial speed of the bullet if the mass of the bullet is 12 g, the mass of the block on the ballistic pendulum is 2 kg, and the final height is 10.4 cm. (*Answer* 240 m/s)

**EXERCISE** Can this example be solved by equating the initial kinetic energy of the bullet with the potential energy of the block–bullet composite at maximum height? That is, is mechanical energy conserved both during the inelastic collision and during the rise of the pendulum? (*Answer* No)

**EXERCISE** A 2000-kg car moving 25 m/s runs head-on into a 1500-kg car initially at rest. If the collision is perfectly inelastic, find (*a*) each car's speed after the collision, and (*b*) the ratio of the system's final kinetic energy to its initial kinetic energy. (*Answer* (*a*) 14.3 m/s, (*b*) 0.57)

---

COLLISION WITH AN EMPTY BOX | **EXAMPLE 8-15** **Try It Yourself**

You repeat your feat of Example 8-14, this time with an empty box as the target. The bullet strikes the box and passes through it completely. A laser ranging device indicates that the bullet emerged with half its initial velocity. Hearing this, you correctly report how high the target must have swung. How high did it swing?

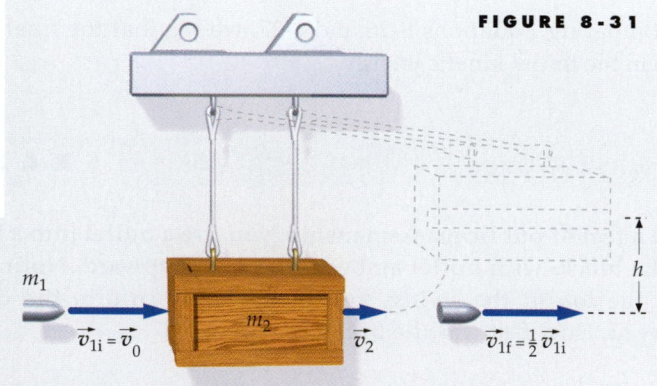

**FIGURE 8-31**

**PICTURE THE PROBLEM** The height $h$ is related to the box's speed $v_2$ after colliding by conservation of mechanical energy (Figure 8-31). This speed can be determined using conservation of momentum.

**Cover the column to the right and try these on your own before looking at the answers.**

| Steps | Answers |
|---|---|
| 1. Use conservation of mechanical energy to relate the final height $h$ to the speed $v_2$ of the box after the collision. | $m_2 gh = \frac{1}{2} m_2 v_2^2$ |
| 2. Use conservation of momentum to write an equation relating $v_{1i}$ to the postcollision speed of the box $v_2$. | $m_2 v_2 + m_1(\frac{1}{2}v_{1i}) = m_1 v_{1i}$ |
| 3. Eliminate $v_2$ and solve for $h$. | $h = \dfrac{m_1^2 v_{1i}^2}{8 m_2^2 g}$ |

---

**REMARKS** The collision of the bullet and the box is an inelastic collision, but not a perfectly inelastic collision because the two objects do not have the same velocity after the collision. Inelastic collisions also occur in microscopic systems. For example, when an electron collides with an atom, the atom is sometimes excited to a higher internal energy state. As a result, the total kinetic energy of the atom and the electron is less after the collision.

**Elastic Head-on Collisions** In some collisions, called **elastic collisions,** the initial and final kinetic energies are equal. Elastic collisions are an ideal that is sometimes approached but never realized in the macroscopic world. If a ball dropped onto a concrete platform bounces back to its original height, then the collision between the ball and the concrete would be elastic. That has never been observed. At the microscopic level, elastic collisions are common. For example, the collisions between air molecules are almost always elastic.

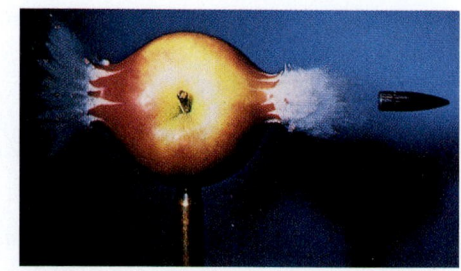

A bullet traveling 850 m/s collides inelastically with an apple, which moments later disintegrates completely. Exposure time is less than a millionth of a second.

For all elastic collisions we have:

$$\tfrac{1}{2}m_1v_{1f}^2 + \tfrac{1}{2}m_2v_{2f}^2 = \tfrac{1}{2}m_1v_{1i}^2 + \tfrac{1}{2}m_2v_{2i}^2 \qquad \text{8-28}$$

This equation, together with the head-on conservation of momentum equation (Equation 8-23), is sufficient to determine the final velocities of the two objects. However, the quadratic nature of Equation 8-28 often complicates the solution of an elastic collision problem. Such problems can be treated more easily if we express the velocity of the two particles relative to each other after the collision in terms of their relative velocity before the collision. Rearranging Equation 8-28 gives

$$m_2(v_{2f}^2 - v_{2i}^2) = m_1(v_{1i}^2 - v_{1f}^2)$$

or

$$m_2(v_{2f} - v_{2i})(v_{2f} + v_{2i}) = m_1(v_{1i} - v_{1f})(v_{1i} + v_{1f}) \qquad \text{8-29}$$

From conservation of momentum, we know that

$$m_1v_{1f} + m_2v_{2f} = m_1v_{1i} + m_2v_{2i}$$

or

$$m_2(v_{2f} - v_{2i}) = m_1(v_{1i} - v_{1f}) \qquad \text{8-30}$$

Then dividing Equation 8-29 by Equation 8-30, we get

$$v_{2f} + v_{2i} = v_{1i} + v_{1f}$$

Rearranging, we obtain

$$v_{2f} - v_{1f} = -(v_{2i} - v_{1i}) \qquad \text{8-31}$$

RELATIVE VELOCITIES IN AN ELASTIC COLLISION

If two objects are to collide, then $v_{2i} - v_{1i}$ must be negative (Figure 8-32), making their **speed of approach** $-(v_{2i} - v_{1i})$. After colliding, the objects' **speed of recession** is $v_{2f} - v_{1f}$. (Both of these terms give the speed of one object relative to the other.) Equation 8-31 states

In elastic collisions, the speed of recession equals the speed of approach.

Solving elastic-collision problems is usually easier using Equation 8-31 rather than Equation 8-28. But beware! Equation 8-31 is valid only if the initial and final kinetic energies are equal, so it applies *only* to elastic collisions.

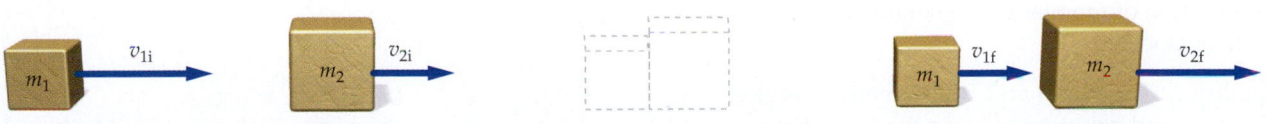

**FIGURE 8-32** Closing (approaching) and separating (receding) in a head-on elastic collision.

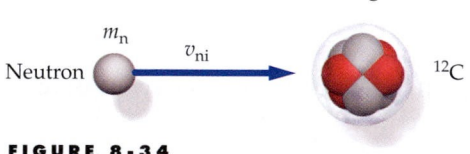

*ELASTIC COLLISION OF TWO BLOCKS*    **E X A M P L E    8 - 1 6**

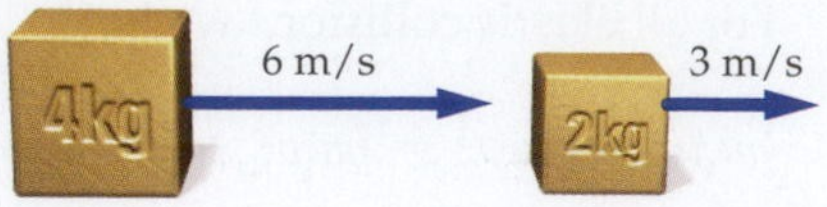

A 4-kg block moving right at 6 m/s collides elastically with a 2-kg block moving right at 3 m/s (Figure 8-33). Find their final velocities.

**FIGURE 8-33**

**PICTURE THE PROBLEM** Conservation of momentum and the equality of the initial and final kinetic energies (expressed as a reversal of relative velocities) give two equations for the two unknown final velocities. Let subscript 1 denote the 4-kg block, subscript 2 the 2-kg block.

1. Apply conservation of momentum and simplify to obtain an equation relating the two final velocities:

$$m_1 v_{1f} + m_2 v_{2f} = m_1 v_{1i} + m_2 v_{2i}$$

$$(4 \text{ kg})v_{1f} + (2 \text{ kg})v_{2f} = (4 \text{ kg})(6 \text{ m/s}) + (2 \text{ kg})(3 \text{ m/s})$$

so

$$2v_{1f} + v_{2f} = 15 \text{ m/s}$$

2. The equality of the initial and final kinetic energies provides a second equation relating the two final velocities. This is implemented by equating the speeds of recession and approach:

$$v_{2f} - v_{1f} = -(v_{2i} - v_{1i})$$
$$= -(3 \text{ m/s} - 6 \text{ m/s}) = 3 \text{ m/s}$$

3. Subtract the step 2 result from the step 1 result and solve for $v_{1f}$:

$$2v_{1f} + v_{1f} = 12 \text{ m/s}$$

so

$$v_{1f} = \boxed{4 \text{ m/s}}$$

4. Substitute into the step 2 result and solve for $v_{2f}$:

$$v_{2f} - 4 \text{ m/s} = 3 \text{ m/s}$$

so

$$v_{2f} = \boxed{7 \text{ m/s}}$$

**PLAUSIBILITY CHECK** As a check, we calculate the initial and final kinetic energies.

$$K_i = \tfrac{1}{2}(4 \text{ kg})(6 \text{ m/s})^2 + \tfrac{1}{2}(2 \text{ kg})(3 \text{ m/s})^2 = 72 \text{ J} + 9 \text{ J} = 81 \text{ J}.$$
$$K_f = \tfrac{1}{2}(4 \text{ kg})(4 \text{ m/s})^2 + \tfrac{1}{2}(2 \text{ kg})(7 \text{ m/s})^2 = 32 \text{ J} + 49 \text{ J} = 81 \text{ J} = K_i.$$

*ELASTIC COLLISION OF A NEUTRON AND A NUCLEUS*    **E X A M P L E    8 - 1 7**

A neutron of mass $m_n$ and speed $v_{ni}$ collides elastically with a carbon nucleus of mass $m_C$ initially at rest (Figure 8-34). (*a*) What are the final velocities of both particles? (*b*) What fraction *f* of its initial kinetic energy does the neutron lose?

**FIGURE 8-34**

**PICTURE THE PROBLEM** Conservation of momentum and conservation of energy allow us to find the final velocities. Since the initial kinetic energy of the carbon nucleus is zero, its final kinetic energy equals the energy lost by the neutron.

(*a*) 1. Use conservation of momentum to obtain one relation for the final velocities:

$$m_n v_{ni} = m_n v_{nf} + m_C v_{Cf}$$

2. The equality of the initial and final kinetic energies provides a second equation relating the two final velocities. This is implemented by equating the speeds of recession and approach:

$$v_{Cf} - v_{nf} = -(v_{Ci} - v_{ni}) = 0 + v_{ni}$$

so

$$v_{Cf} = v_{ni} + v_{nf}$$

3. To eliminate $v_{Cf}$, substitute the expression for $v_{Cf}$ from step 2 into the Step 1 result:

$$m_n v_{ni} = m_n v_{nf} + m_C(v_{ni} + v_{nf})$$

4. Solve for $v_{nf}$: (Note that $v_{nf}$ is negative. The neutron $m_n$ bounces back from the more massive carbon nucleus $m_C$.)

$$v_{nf} = -\frac{m_C - m_n}{m_n + m_C} v_{ni}$$

5. Substitute the step 4 result into the step 2 result and solve for $v_{Cf}$:

$$v_{Cf} = v_{ni} - \frac{m_C - m_n}{m_n + m_C} v_{ni} = \frac{2m_n}{m_n + m_C} v_{ni}$$

(b) 1. The collision is elastic, so the kinetic energy lost by the neutron is the final kinetic energy of the carbon nucleus:

$$f = \frac{-\Delta K_n}{K_{ni}} = \frac{K_{Cf}}{K_{ni}}$$

$$= \frac{\frac{1}{2}m_C v_{Cf}^2}{\frac{1}{2}m_n v_{ni}^2} = \frac{m_C}{m_n}\left(\frac{v_{Cf}}{v_{ni}}\right)^2$$

2. Solve the Part (a) step 5 result for the ratio of the velocities, substitute into the Part (b) step 1 result, and solve for the fractional energy loss of the neutron:

$$f = \frac{m_C}{m_n}\left(\frac{2m_n}{m_n + m_C}\right)^2 = \frac{4m_n m_C}{(m_n + m_C)^2}$$

**REMARKS**  An important application of energy transfer in elastic collisions is the slowing down of neutrons in a nuclear reactor. High-energy neutrons are emitted in the fission of a uranium nucleus. If these neutrons are to cause another uranium nucleus to fission, their energy must be reduced; that is, they must be slowed down or "moderated." One mechanism for slowing down neutrons is the elastic scattering of the neutrons with the nuclei in the reactor. The fractional energy loss $f = -\Delta K_n/K_n$ depends on the ratio of the mass of the moderator nucleus to that of the neutron, as shown in Figure 8-35. For uranium, $m_U \approx 235m_n$ and $f \approx 0.017 = 1.7$ percent. For carbon, $m_C \approx 12m_n$ and $f = 0.28 = 28$ percent; for deuterium, $m_D \approx 2m_n$ and $f \approx 0.89 = 89$ percent; and for Hydrogen, $m_H \approx 1m_n$ and $f \approx 1 = 100$ percent. A moderator such as graphite or water is added to a reactor to slow down the neutrons so that they can be captured by uranium nuclei.

**EXERCISE**  Show that for an elastic head-on collision between a moving object and a stationary object of equal mass, the fractional loss in kinetic energy is one and the final speed of the initially stationary object equals the initial speed of the initially moving object.

**EXERCISE**  A 2-kg box moving at 3 m/s makes an elastic collision with a stationary 4-kg box. (a) What is the original kinetic energy? (b) How much energy is transferred to the 4-kg box? (*Answer*  (a) 9 J, (b) 8 J)

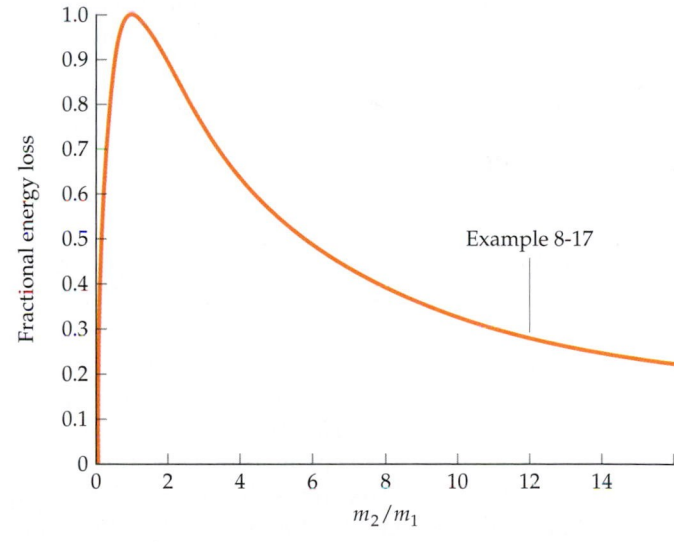

**FIGURE 8-35** Fractional energy loss as a function of the ratio of the two masses. The maximum energy loss occurs when $m_1 = m_2$.

The results of Example 8-17 for the final velocities of an incoming particle colliding with a second particle initially at rest are worth noting. The final velocity of the incoming particle $v_{1f}$ and that of the originally stationary particle $v_{2f}$ are related to the initial velocity of the incoming particle by

$$v_{1f} = \frac{m_1 - m_2}{m_1 + m_2} v_{1i} \qquad\qquad\qquad 8\text{-}32a$$

and

$$v_{2f} = \frac{2m_1}{m_1 + m_2} v_{1i} \qquad\qquad\qquad 8\text{-}32b$$

When a very massive object (say a bowling ball) collides with a light stationery object (say a Ping-Pong ball), the massive object is essentially unaffected. Before the collision, the relative velocity of approach is $v_{1i}$. If the massive object continues with a velocity that is essentially $v_{1i}$ after the collision, the velocity of the smaller object must be $2v_{1i}$ so that the speed of recession is equal to the speed of approach. This result also follows from Equations 8-32a and 8-32b if we take $m_2$ to be much smaller than $m_1$, in which case $v_{1f} \approx v_{1i}$ and $v_{2f} \approx 2v_{1i}$, as expected.

**\*The Coefficient of Restitution**   Most collisions lie somewhere between the extreme cases of elastic, in which the relative velocities are reversed, and perfectly inelastic, in which there is no relative velocity after the collision. The **coefficient of restitution** $e$, is a measure of the elasticity of a collision. It is defined as the ratio of the speed of recession to the speed of approach.

$$e = \frac{v_{rec}}{v_{app}} = -\frac{v_{2f} - v_{1f}}{v_{2i} - v_{1i}} \qquad \text{8-33}$$

DEFINITION—COEFFICIENT OF RESTITUTION

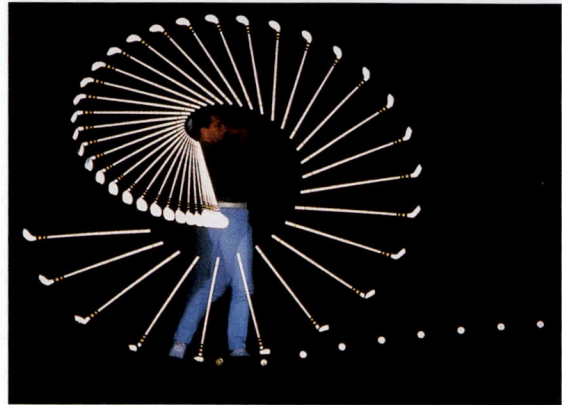

For an elastic collision, $e = 1$. For a perfectly inelastic collision, $e = 0$.

## Collisions in Three Dimensions

**Perfectly Inelastic Collisions in Three Dimensions**   For collisions in three dimensions, the total initial momentum is the sum of the initial momentum vectors of each object involved in the collision. Because the objects stick together and because momentum is conserved we have $m_{1i}\vec{v}_{1i} + m_2\vec{v}_{2i} = (m_1 + m_2)\vec{v}_f$. Because of this relation we know that the three velocity vectors, and thus the collision, are in the same plane. Also, from the definition of the center of mass we know that $\vec{v}_f = \vec{v}_{cm}$.

---

*A CAR–TRUCK COLLISION*                    **EXAMPLE  8-18**    **Put It in Context**

You are at the wheel of a 1200-kg car traveling east through an intersection when a 3000-kg truck traveling north through the intersection crashes into your car, as shown in Figure 8-36. Your car and the truck stick together after impact. The driver of the truck claims you were at fault because you were speeding. You look for evidence to disprove this claim. First, there are no skid marks, indicating that neither you nor the truck driver saw the accident coming and braked hard; second, there is a sign reading "Speed Limit 80 km/h" on the road you were driving on; third, the speedometer of the truck was smashed with the needle stuck at 50 km/h; and fourth, the wreck initially skidded from the impact zone at an angle of no less than 59° north of east. Does this evidence support or undermine the claim that you were speeding?

**FIGURE 8-36**

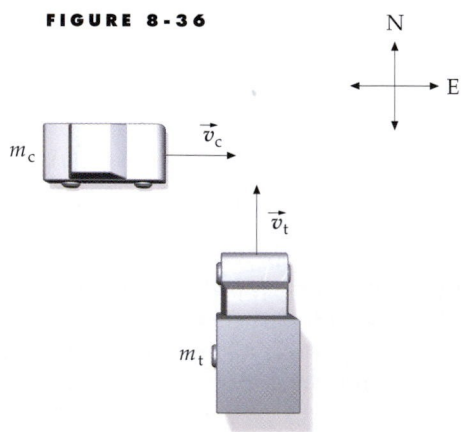

**PICTURE THE PROBLEM**   Choose your coordinate system so that initially the car is traveling in the $+x$ direction and the truck is traveling in the $+y$ direction (Figure 8-37). Then write the momentum of each object in vector form, and use conservation of momentum.

1. Write out the conservation of momentum equation in vector form in terms of masses and velocities:

$$m_c\vec{v}_c + m_t\vec{v}_t = (m_c + m_t)\vec{v}_f$$

**FIGURE 8-37**

2. Equate the $x$ component of the initial momentum to the $x$ component of the final momentum:

$$m_c v_c + 0 = (m_c + m_t)v_f \cos\theta$$

3. Equate the $y$ component of the initial momentum to the $y$ component of the final momentum:

$$0 + m_t v_t = (m_c + m_t)v_f \sin\theta$$

4. Eliminate $v_f$ by dividing the $y$ component equation by the $x$ component equation:

$$\frac{m_t v_t}{m_c v_c} = \frac{\sin\theta}{\cos\theta} = \tan\theta$$

so

$$v_c = \frac{m_t v_t}{m_c \tan\theta}$$

$$= \frac{(3000 \text{ kg})(50 \text{ km/h})}{(1200 \text{ kg}) \tan 59°}$$

$$= \boxed{75.1 \text{ km/h}}$$

5. Does this undermine the truck driver's claim that you were speeding?

> Because 75.1 km/h is less than the 80 km/h speed limit, the truck driver's claim is undermined by the careful application of physics.

**REMARKS** A court would probably require that these arguments be presented by an expert witness.

### *Elastic Collisions in Three Dimensions

Elastic collisions in three dimensions are more complicated than those we have already covered. Figure 8-38 shows an off-center collision between an object of mass $m_1$ moving with velocity $\vec{v}_{1i}$ parallel to the $x$ axis toward an object of mass $m_2$ that is initially at rest at the origin. The distance $b$ between the centers measured perpendicular to the direction of $\vec{v}_{1i}$ is called the **impact parameter**. After the collision, object 1 moves off with velocity $\vec{v}_{1i}$, making an angle $\theta_1$ with its initial velocity, and object 2 moves with velocity $\vec{v}_{2f}$, making an angle $\theta_2$ with $\vec{v}_{1f}$. Conservation of momentum gives

$$\vec{P}_{sys} = m_1\vec{v}_{1i} = m_1\vec{v}_{1f} + m_2\vec{v}_{2f}$$

We can see from this equation that the vector $\vec{v}_{2f}$ must lie in the plane formed by $\vec{v}_{1i}$ and $\vec{v}_{1f}$, which we will take to be the $xy$ plane. Assuming that we know the initial velocity $\vec{v}_{1i}$, we have four unknowns: the $x$ and $y$ components of both final velocities; or alternatively, the two final speeds and the two angles of deflection. The $x$ and $y$ components of the conservation-of-momentum equation give us two of the needed relations among these quantities. Conservation of mechanical energy gives a third relation. To find the four unknowns, we need another relation. The fourth relation depends on the impact parameter $b$ and on the type of interacting force exerted by the bodies on each other. In practice, the fourth relation is often found experimentally, by measuring the angle of deflection or the angle of recoil. Such a measurement can then give us information about the type of interacting force between the bodies.

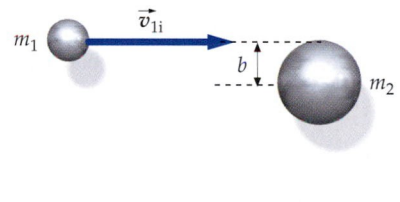

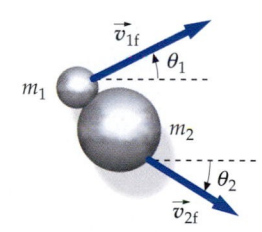

**FIGURE 8-38** Off-center collision. The final velocities depend on the impact parameter $b$ and on the type of force exerted by one object on the other.

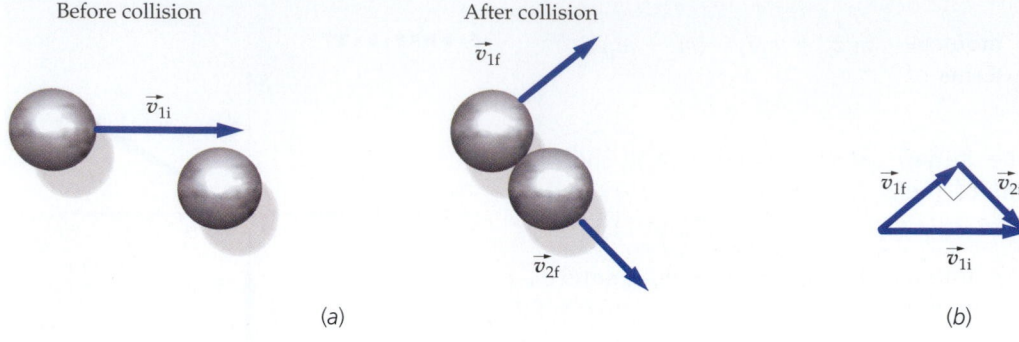

(a)                                                                                     (b)

**FIGURE 8-39** (*a*) Off-center elastic collision of two spheres of equal mass when one sphere is initially at rest. After the collision, the spheres move off at right angles to each other. (*b*) The velocity vectors for this collision form a right triangle.

Consider the interesting special case of the off-center elastic collision of two objects *of equal mass* when one is initially at rest (Figure 8-39*a*). If $\vec{v}_{1i}$ and $\vec{v}_{1f}$ are the initial and final velocities of object 1, and $\vec{v}_{2f}$ is the final velocity of object 2, conservation of momentum gives

$$m\vec{v}_{1i} = m\vec{v}_{1f} + m\vec{v}_{2f}$$

or

$$\vec{v}_{1i} = \vec{v}_{1f} + \vec{v}_{2f}$$

These vectors form the triangle shown in Figure 8-39*b*. Because energy is conserved in the collision,

$$\tfrac{1}{2}mv_{1i}^2 = \tfrac{1}{2}mv_{1f}^2 + \tfrac{1}{2}mv_{2f}^2$$

or

$$v_{1i}^2 = v_{1f}^2 + v_{2f}^2 \qquad\qquad 8\text{-}34$$

Equation 8-34 is the Pythagorean theorem for a right triangle formed by the vectors $\vec{v}_{1f}$, $\vec{v}_{2f}$, and $\vec{v}_{1i}$, with the hypotenuse of the triangle being $\vec{v}_{1i}$. So for this special case, the final velocity vectors $\vec{v}_{1f}$ and $\vec{v}_{2f}$ are perpendicular to each other, as shown in Figure 8-39*b*.

Multiflash photograph of an off-center elastic collision of two balls of equal mass. The dotted ball, entering from the left, strikes the striped ball, which is initially at rest. The final velocities of the two balls are perpendicular to each other.

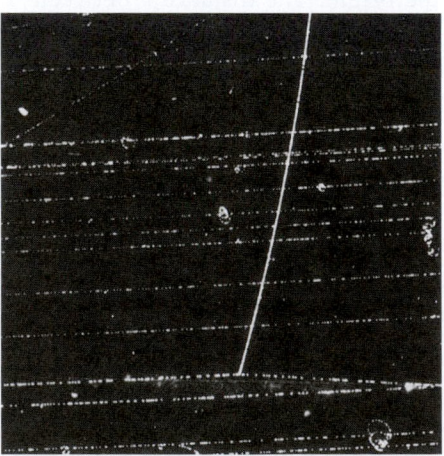

Proton–proton collision in a liquid-hydrogen bubble chamber. A proton entering from the left interacts with a stationary proton. The two then move off at right angles. The slight curvature of the tracks is due to a magnetic field.

# *8-7  The Center-of-Mass Reference Frame

If the net external force on a system remains zero, the velocity of the center of mass remains constant. It is often convenient to choose a coordinate system with the origin at the center of mass. Then, relative to the original coordinate system, this coordinate system moves with a constant velocity $\vec{v}_{cm}$. The frame of reference that moves with the center of mass is called the **center-of-mass reference frame.** If a particle has velocity $\vec{v}$ in the original reference frame, then its velocity relative to the center of mass is $\vec{u} = \vec{v} - \vec{v}_{cm}$. In the center-of-mass frame, the velocity of the center of mass is zero. Since the total momentum of a system equals the total mass times the velocity of the center of mass, the total momentum is also zero in the center-of-mass frame. Thus, the center-of-mass reference frame is also a **zero-momentum reference frame.**

The mathematics of collisions is greatly simplified when considered within the center-of-mass reference frame. The momenta of the two incoming objects are equal and opposite. After a perfectly inelastic collision, the objects remain at rest. A perfectly elastic collision in one dimension between two objects reverses the direction of each velocity vector without changing the magnitudes of these vectors (you will derive this in *Problem 8-101*).

Consider a simple two-particle system in a reference frame in which one particle of mass $m_1$ is moving with a velocity $\vec{v}_1$ and a second particle of mass $m_2$ is moving with a velocity $\vec{v}_2$ (Figure 8-40). In this frame, the velocity of the center of mass is

$$\vec{v}_{cm} = \frac{m_1\vec{v}_1 + m_2\vec{v}_2}{m_1 + m_2}$$

We can transform the velocities of the two particles to their velocities in the center-of-mass reference frame by subtracting $\vec{v}_{cm}$. The velocities of the particles in the center-of-mass frame are $\vec{u}_1$ and $\vec{u}_2$, given by

$$\vec{u}_1 = \vec{v}_1 - \vec{v}_{cm} \qquad\qquad 8\text{-}35a$$

and

$$\vec{u}_2 = \vec{v}_2 - \vec{v}_{cm} \qquad\qquad 8\text{-}35b$$

Since the total momentum is zero in the center-of-mass frame, the particles have equal and opposite momenta in this frame.

Original reference frame

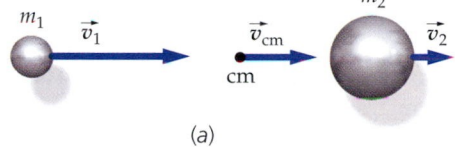

(a)

Center-of-mass reference frame

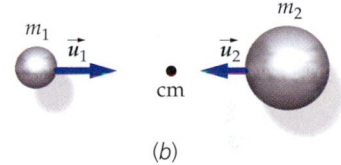

(b)

**FIGURE 8-40** (a) Two particles viewed from a frame in which the center of mass has a velocity $\vec{v}_{cm}$. (b) The same two particles viewed from a reference frame for which the center of mass is at rest.

---

*The Elastic Collision of Two Blocks*    **EXAMPLE 8-19**

**Find the final velocities for the elastic head-on collision in Example 8-16 (in which a 4-kg block moving right at 6 m/s collides elastically with a 2-kg block moving right at 3 m/s) by transforming their velocities to the center-of-mass reference frame.**

**PICTURE THE PROBLEM** We transform to the center-of-mass reference frame by first finding $v_{cm}$ and subtracting it from each velocity. We then solve the collision by reversing the velocities and transforming back to the original frame.

1. Calculate the velocity of the center of mass $v_{cm}$ (Figure 8-41):

$$v_{cm} = \frac{m_1 v_{1i} + m_2 v_{2i}}{m_1 + m_2}$$

$$= \frac{(4\ \text{kg})(6\ \text{m/s}) + (2\ \text{kg})(3\ \text{m/s})}{4\ \text{kg} + 2\ \text{kg}}$$

$$= 5\ \text{m/s}$$

**FIGURE 8-41**        Initial conditions

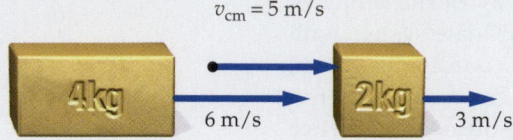

$v_{cm} = 5$ m/s

6 m/s      3 m/s

2. Transform the initial velocities to the center-of-mass reference frame by subtracting $v_{cm}$ from the initial velocities (Figure 8-42):

$$u_{1i} = v_{1i} - v_{cm}$$

$$= 6\ \text{m/s} - 5\ \text{m/s} = 1\ \text{m/s}$$

$$u_{2i} = v_{2i} - v_{cm}$$

$$= 3\ \text{m/s} - 5\ \text{m/s} = -2\ \text{m/s}$$

**FIGURE 8-42**      Transform to the center-of-mass frame by subtracting $v_{cm}$

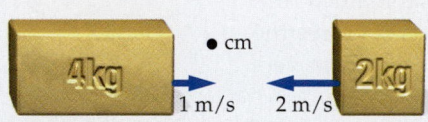

$v_{cm} = 0$

1 m/s    2 m/s

3. Solve the collision in the center-of-mass reference frame by reversing the velocity of each object (Figure 8-43):

$$u_{1f} = -u_{1i} = -1\ \text{m/s}$$

$$u_{2f} = -u_{2i} = +2\ \text{m/s}$$

**FIGURE 8-43**      Solve collision

 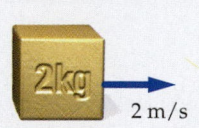

$v_{cm} = 0$

• cm

1 m/s          2 m/s

4. To find the final velocities in the original frame, add $v_{cm}$ to each final velocity (Figure 8-44).

$$v_{1f} = u_{1f} + v_{cm}$$

$$= -1\ \text{m/s} + 5\ \text{m/s} = \boxed{4\ \text{m/s}}$$

$$v_{2f} = u_{2f} + v_{cm}$$

$$= 2\ \text{m/s} + 5\ \text{m/s} = \boxed{7\ \text{m/s}}$$

**FIGURE 8-44**      Transform back to the original frame by adding $v_{cm}$

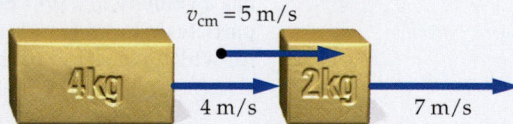

$v_{cm} = 5$ m/s

4 m/s      7 m/s

**REMARKS** This is the same result found in Example 8-16.

**EXERCISE** Show that the total momentum of the system both before the collision and after the collision is zero in the center-of-mass reference frame. (*Answer* before: $P_{\text{sys i}} = (4\ \text{kg})(1\ \text{m/s}) + (2\ \text{kg})(-2\ \text{m/s}) = 0$; after: $P_{\text{sys f}} = (4\ \text{kg})(-1\ \text{m/s}) + (2\ \text{kg})(2\ \text{m/s}) = 0$)

## *8-8 Systems With Continuously Varying Mass: Rocket Propulsion

A creative and important step in solving physics problems is specifying the system. In this section we explore situations in which the system has a continuously changing mass. One example of such a system is a rocket. For a rocket, we specify the system to be the rocket plus any unspent fuel in it. As the spent fuel (the exhaust) spews out the back, the mass of the system decreases.

Another example is a stream of effluence from a volcano on the Jovian moon Io impacting on an asteroid passing through the stream. We specify the system at time $t$ to be the asteroid, plus the volcanic effluence accumulated on the asteroid, plus the effluence that will impact the asteroid between time $t$ and time $t + \Delta t$. As the effluence continues to accumulate, the mass of this system increases. To solve such problems we will apply the impulse–momentum theorem to the system during the short time interval $\Delta t$, and then take the limit as $\Delta t \rightarrow 0$.

Consider the following derivation of Newton's second law for objects with continuously varying mass. A continuous stream of matter moving at velocity $\vec{u}$ is impacting an object of mass $M$ that is moving with velocity $\vec{v}$ (Figure 8-45). The mass of particles that impact the object during time $\Delta t$ is $\Delta M$, and these impacting particles stick to the object, increasing its mass by $\Delta M$ during time $\Delta t$. Also, during time $\Delta t$ the velocity $\vec{v}$ changes by $\Delta \vec{v}$, as shown. At time $t$ we define the system to be the object (including all matter that impacted the object prior to time $t$) plus all matter that will impact the object between time $t$ and time $t + \Delta t$. Applying the impulse–momentum theorem to this system gives

$$\vec{F}_{\text{net ext}}\Delta t = \Delta \vec{P} = [(M + \Delta M)(\vec{v} + \Delta \vec{v})] - [M\vec{v} + \Delta M \vec{u}]$$

where the first term (in square brackets) is the momentum at time $t + \Delta t$ and the second term in square brackets is the momentum at time $t$. Reorganizing terms gives

$$\vec{F}_{\text{net ext}}\Delta t = M \Delta \vec{v} + \Delta M(\vec{v} - \vec{u}) + \Delta M \Delta \vec{v} \qquad \text{8-36}$$

(where the net external force $\vec{F}_{\text{net ext}}$ would likely include a gravitational force plus one or more contact forces). Dividing through Equation 8-36 by $\Delta t$ gives

$$\vec{F}_{\text{net ext}} = M\frac{\Delta \vec{v}}{\Delta t} + \frac{\Delta M}{\Delta t}(\vec{v} - \vec{u}) + \frac{\Delta M}{\Delta t} \Delta \vec{v}$$

and taking the limit as $\Delta t \rightarrow 0$ (which also means as $\Delta M \rightarrow 0$ and as $\Delta \vec{v} \rightarrow 0$) gives

$$\vec{F}_{\text{net ext}} = M\frac{d\vec{v}}{dt} - \frac{dM}{dt}(\vec{u} - \vec{v}) + 0$$

Rearranging, we have

$$\vec{F}_{\text{net ext}} + \frac{dM}{dt}\vec{v}_{\text{rel}} = M\frac{d\vec{v}}{dt} \qquad \text{8-37}$$

NEWTON'S SECOND LAW—CONTINUOUSLY VARIABLE MASS

where $\vec{v}_{\text{rel}} = \vec{u} - \vec{v}$, the velocity of $dM$ relative to $M$. Note that except for the term $(dM/dt)\vec{v}_{\text{rel}}$, Equation 8-37 is just Newton's second law for a system with constant mass.

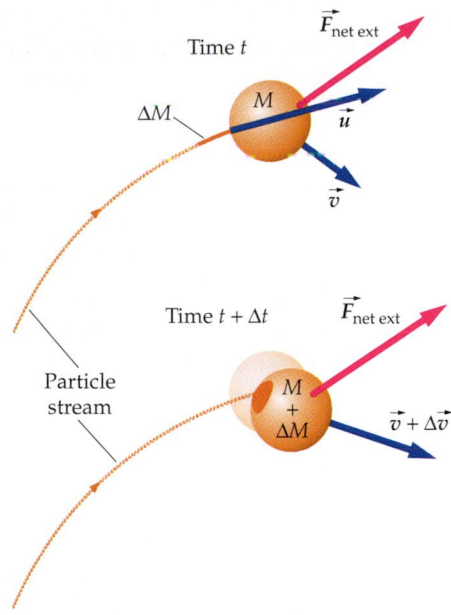

**FIGURE 8-45** Particles in a continuous stream and moving at velocity $\vec{u}$ undergo perfectly inelastic collisions with an object of mass $M$ moving at velocity $\vec{v}$.

---

*A FALLING ROPE*                    **EXAMPLE  8-20**

A uniform rope of mass $M$ and length $L$ is held with its lower end just touching the surface of a scale. The rope is released ands begins to fall. Find the force of the scale on the rope just as the midpoint of the rope first touches the scale.

**PICTURE THE PROBLEM** Apply Equation 8-37 to that portion of the rope on the scale. There are two external forces on that portion, the force of gravity and the normal force exerted by the scale. The impact velocities of the different points along the falling rope depend upon their initial heights above the scale.

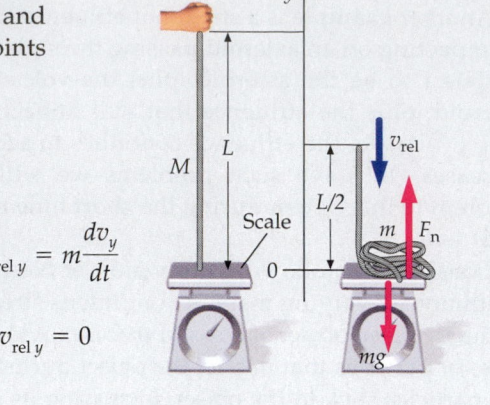

1. Draw a sketch of the situation (Figure 8-46). Include the initial configuration and the configuration at an arbitrary time later. Include a coordinate axis.

2. Express Equation 8-37 in component form. Let $m$ denote the mass of the system (that part of the rope on the scale). The velocity of the system remains zero, so the $dv_y/dt$ is zero:

$$F_{\text{net ext } y} + \frac{dm}{dt}v_{\text{rel } y} = m\frac{dv_y}{dt}$$

$$F_n - mg + \frac{dm}{dt}v_{\text{rel } y} = 0$$

3. Let $dm$ denote the mass of a short rope segment of length $d\ell$ that falls on the scale during time $dt$. Since the rope is uniform, the relation between $dm$ and $d\ell$ is:

$$\frac{dm}{d\ell} = \frac{M}{L}$$

**FIGURE 8-46** A very flexible rope of length $L$ and mass $M$ is released from rest and falls on the surface of a scale.

4. Solve for $dm/dt$ by multiplying both sides of the step 3 result by $d\ell/dt$:

$$\frac{dm}{dt} = \frac{M}{L}\frac{d\ell}{dt}$$

5. $d\ell/dt$ is the impact speed of the segment, so $v_{\text{rel } y} = -d\ell/dt$. ($v_{\text{rel } y}$ is negative because up is the positive $y$ direction and the rope is falling.) Substituting this into the step 4 result gives:

$$\frac{dm}{dt} = -\frac{M}{L}v_{\text{rel } y}$$

6. Substituting the step 5 result into the step 2 result and solving for $F_n$ gives:

$$F_n = mg + \frac{M}{L}v_{\text{rel } y}^2$$

7. Until it touches the scale, each point along the rope falls with constant acceleration $g$. Using Equation 2-23 with $\Delta y = -L/2$ gives:

$$v_{\text{rel } y}^2 = v_{\text{rel } y 0}^2 + 2a_y\Delta y$$
$$= 0 + 2(-g)(-L/2) = gL$$

8. Substituting the step 7 result into the step 6 result, with $m = M/2$, gives:

$$F_n = \frac{M}{2}g + \frac{M}{L}gL = \boxed{\frac{3}{2}Mg}$$

**REMARKS** As the midpoint of the rope strikes the surface, the weight of the system (the rope already on the scale) is $\frac{1}{2}Mg$, which is one third of the normal force. This means $F_{\text{net ext } y} = Mg$ (directed upward). This is the force required to stop the momentum of the falling rope striking the scale at that instant.

**EXERCISE** Find the normal force exerted by the scale on the rope (a) just before the upper end of the rope reaches the scale and (b) just after the upper end of the rope reaches the scale. (*Answer* (a) $3Mg$, (b) $Mg$)

Rocket propulsion is a striking example of the conservation of momentum in action. The mass of the rocket changes continuously as it burns fuel and expels exhaust gas. Consider a rocket moving straight up with velocity $\vec{v}$ relative to the earth, as shown in Figure 8-47. Assuming that the fuel is burned at a constant rate $R$, the rocket's mass at time $t$ is

$$M = M_0 - Rt \qquad\qquad 8\text{-}38$$

where $M_0$ is the initial mass of the rocket. The exhaust gas leaves the rocket with velocity $\vec{u}_{\text{ex}}$ relative to the rocket, and the rate at which the fuel is burned is the rate at which the mass $M$ decreases. We choose the rocket and the unspent fuel within it as the system. Neglecting air drag, the only external force on the system is that of gravity. With $\vec{F}_{\text{net ext}} = M\vec{g}$ and $dM/dt = -R$, Equation 8-37 becomes the **rocket equation:**

$$M\vec{g} - R\vec{u}_{ex} = M\frac{d\vec{v}}{dt} \qquad \text{8-39}$$

ROCKET EQUATION

The quantity $-R\vec{u}_{ex}$ is the force exerted on the rocket by the exhausting fuel. It is called the **thrust** $\vec{F}_{th}$:

$$\vec{F}_{th} = -R\vec{u}_{ex} = -\left|\frac{dM}{dt}\right|\vec{u}_{ex} \qquad \text{8-40}$$

DEFINITION—ROCKET THRUST

The rocket is moving straight up, so we choose upward as the positive $y$ direction and express Equation 8-39 as

$$-Mg + Ru_{ex} = M\frac{dv_y}{dt} \qquad \text{8-41}$$

ROCKET EQUATION (VERTICAL COMPONENT)

Dividing through by $M$, rearranging terms, and substituting from Equation 8-38 gives

$$\frac{dv_y}{dt} = \frac{Ru_{ex}}{M} - g = \frac{Ru_{ex}}{M_0 - Rt} - g \qquad \text{8-42}$$

ACCELERATION OF ROCKET (VERTICAL COMPONENT)

where $M_0$ is the initial value of $M$. Equation 8-42 is solved by integrating both sides with respect to time. For a rocket starting at rest at $t = 0$, the result is

$$v_y = u_{ex}\ln\left(\frac{M_0}{M_0 - Rt}\right) - gt \qquad \text{8-43}$$

VELOCITY OF ROCKET (VERTICAL COMPONENT)

**FIGURE 8-47**

assuming the acceleration of gravity to be constant.

---

*LIFTOFF*                     **EXAMPLE  8-21**    **Try It Yourself**

The Saturn V rocket used in the Apollo moon-landing program had an initial mass $M_0$ of $2.85 \times 10^6$ kg, a payload of 27%, a burn rate $R$ of $13.84 \times 10^3$ kg/s, and a thrust $F_{th}$ of $34 \times 10^6$ N. Find (*a*) the exhaust speed, (*b*) the burn time $t_b$, (*c*) the acceleration at liftoff, (*d*) the acceleration at burnout $t_b$, and (*e*) the final speed of the rocket.

**PICTURE THE PROBLEM** (*a*) The exhaust speed can be found from the thrust and burn rate. (*b*) To find the burn time, you need to find the total mass of fuel burned, which is the initial mass minus the payload. (*c*) The acceleration is found from Equation 8-42. (*d*) The final speed is given by Equation 8-43.

**Cover the column to the right and try these on your own before looking at the answers.**

| Steps | Answers |
|---|---|
| (a) Calculate $u_{ex}$ from the given thrust and burn rate. | $u_{ex} = \boxed{2.46 \text{ km/s}}$ |
| (b) 1. Calculate the final mass $M_f$ of the rocket. | $M_f = (0.27)M_0 = 7.70 \times 10^5 \text{ kg}$ |
| 2. Use your result to calculate the burn time $t_b$. | $t_b = \dfrac{M_0 - M_f}{R} = \boxed{150 \text{ s}}$ |
| (c) Calculate $dv_y/dt$ for $M = M_0$ and for $M = M_f$. | Initially, $\dfrac{dv_y}{dt} = 2.14 \text{ m/s}^2$; |
| | finally, $\dfrac{dv_y}{dt} = \boxed{34.3 \text{ m/s}^2}$ |
| (d) Calculate the final speed from Equation 8-43. | $v_{yf} = \boxed{1.75 \text{ km/s}}$ |

**REMARKS** The initial acceleration is small—only $0.21g$. At burnout (also called flameout), the rocket's acceleration has increased to $3.5g$. Immediately following burnout the acceleration is $-g$. The speed of the rocket at burnout, after two and a half minutes of burning, is roughly 6,300 km/h (3900 mi/h).

# SUMMARY

The conservation of momentum for an isolated system is a fundamental law of nature that has applications in all areas of physics.

| Topic | Relevant Equations and Remarks | |
|---|---|---|
| **1. Center of Mass** | | |
| System of particles | $M\vec{r}_{cm} = m_1\vec{r}_1 + m_2\vec{r}_2 + \cdots$ | 8-5 |
| | $M\vec{v}_{cm} = m_1\vec{v}_1 + m_2\vec{v}_2 + \cdots$ | 8-8 |
| | $M\vec{a}_{cm} = m_1\vec{a}_1 + m_2\vec{a}_2 + \cdots$ | 8-9 |
| Continuous objects | $M\vec{r}_{cm} = \displaystyle\int \vec{r}\, dm$ | 8-6 |
| Newton's second law for a system | $\vec{F}_{net\,ext} = \displaystyle\sum_i \vec{F}_{i\,ext} = M\vec{a}_{cm}$ | 8-11 |
| **2. Momentum** | | |
| Definition for a particle | $\vec{p} = m\vec{v}$ | 8-12 |
| Kinetic energy of a particle | $K = \dfrac{p^2}{2m}$ | 8-25 |
| **Momentum of a system** | $\vec{P}_{sys} = \displaystyle\sum_i m_i\vec{v}_i = M\vec{v}_{cm}$ | 8-14 |

| | | |
|---|---|---|
| Newton's second law for a system | $$\vec{F}_{\text{net ext}} = \frac{d\vec{P}_{\text{sys}}}{dt}$$ | 8-15 |
| Law of conservation of momentum | If the net external force acting on a system remains zero, the total momentum of the system is conserved. | |

### 3. Energy of a System

| | | |
|---|---|---|
| Kinetic energy | The kinetic energy associated with the motion of the particles of a system relative to its center of mass is $K_{\text{rel}} = \Sigma \frac{1}{2} m_i u_i^2$, where $u_i$ is the speed of the $i$th particle relative to the center of mass. | |
| | $$K = \tfrac{1}{2} M v_{\text{cm}}^2 + K_{\text{rel}}$$ | 8-18 |
| Gravitational potential energy | $$U = Mgh_{\text{cm}}$$ | 8-7 |

### 4. Collisions

| | | |
|---|---|---|
| Impulse | The impulse of a force is defined as the integral of the force over the time interval during which the force acts. | |
| | $$\vec{I} = \int_{t_i}^{t_f} \vec{F}\, dt$$ | 8-19 |
| Impulse–momentum theorem | $$\vec{I}_{\text{net}} = \int_{t_i}^{t_f} \vec{F}_{\text{net}}\, dt = \Delta\vec{p}$$ | 8-20 |
| Average force | $$\vec{F}_{\text{av}} = \frac{1}{\Delta t} \int_{t_i}^{t_f} \vec{F}\, dt = \frac{\vec{I}}{\Delta t}$$ | 8-22 |
| Elastic collisions | An elastic collision is one in which the sum of the kinetic energies of the two objects is the same before and after the collision. | |
| Relative speeds of approach and recession | For an elastic collision, the speed of separation of the two objects following an elastic collision equals their speed of approach prior to the collision. For a head-on collision, | |
| | $$v_{2f} - v_{1f} = -(v_{2i} - v_{1i})$$ | 8-31 |
| Perfectly inelastic collisions | Following a perfectly inelastic collision, the two objects stick together and move with the velocity of the center of mass. | |
| *Coefficient of restitution | The coefficient of restitution $e$ is a measure of the elasticity. It is the ratio of the separation speed to the closing speed: | |
| | $$e = -\frac{v_{2f} - v_{1f}}{(v_{2i} - v_{1i})}$$ | 8-33 |
| | For an elastic collision, $e = 1$; for a perfectly inelastic collision $e = 0$. | |

### *4. Continuously Variable Mass

| | | |
|---|---|---|
| Newton's second law | $$\vec{F}_{\text{net ext}} + \frac{dM}{dt}\vec{v}_{\text{rel}} = M\frac{d\vec{v}}{dt}$$ | 8-37 |
| Rocket equation | $$M\vec{g} - R\vec{u}_{\text{ex}} = M\frac{d\vec{v}}{dt}$$ | 8-39 |
| Thrust | $$\vec{F}_{\text{th}} = -R\vec{u}_{\text{ex}} = -\left|\frac{dM}{dt}\right|\vec{u}_{\text{ex}}$$ | 8-40 |

- • Single-concept, single-step, relatively easy
- •• Intermediate-level, may require synthesis of concepts
- ••• Challenging
- SSM Solution is in the *Student Solutions Manual*
- iSOLVE Problems available on iSOLVE online homework service
- iSOLVE ✓ These "Checkpoint" online homework service problems ask students additional questions about their confidence level, and how they arrived at their answer

In a few problems, you are given more data than you actually need; in a few other problems, you are required to supply data from your general knowledge, outside sources, or informed estimates.

**Take $g = 9.81$ N/kg $= 9.81$ m/s$^2$ and neglect friction in all problems unless otherwise stated.**

## Conceptual Problems

**1** • Give an example of a three-dimensional object that has no matter at its center of mass.

**2** • SSM A cannonball is dropped off a high tower while, simultaneously, an identical cannonball is launched directly upward into the air. The center of mass of the two cannonballs (*a*) stays in the same place, (*b*) initially rises, then falls, but begins to fall *before* the cannonball launched into the air starts falling, (*c*) initially rises, then falls, but begins to fall *at the same time* as the cannonball launched upward begins to fall, (*d*) initially rises, then falls, but begins to fall *after* the cannonball launched upward begins to fall.

**3** • Two pucks of masses $m_1$ and $m_2$ lie unconnected on a frictionless table. A horizontal force $F_1$ is exerted on $m_1$ only. What is the magnitude of the acceleration of the center of mass of the two-puck system? (*a*) $F_1/m_1$. (*b*) $F_1/(m_1 + m_2)$. (*c*) $F_1/m_2$. (*d*) $(m_1 + m_2)F_1/m_1m_2$.

**4** • The two pucks in Problem 3 are lying on a frictionless table and connected by a massless spring of force constant $k$. A horizontal force $F_1$ is again exerted only on $m_1$, along the spring and away from $m_2$. What is the magnitude of the acceleration of the center of mass? (*a*) $F_1/m_1$. (*b*) $F_1/(m_1 + m_2)$. (*c*) $(F_1 + kx)/(m_1 + m_2)$, where $x$ is the amount the spring is stretched. (*d*) $(m_1 + m_2)F_1/m_1m_2$.

**5** • SSM If two particles have equal kinetic energies, are the magnitudes of their momenta necessarily equal? Explain and give an example.

**6** • True or false:

(*a*) The momentum of a heavy object is greater than that of a light object moving at the same speed.

(*b*) The momentum of a system may be conserved even when mechanical energy is not.

(*c*) The velocity of the center of mass of a system equals the total momentum of the system divided by its total mass.

**7** • How is the recoil of a rifle related to momentum conservation?

**8** • SSM A child jumps from a small boat to a dock. Why does she have to jump with more energy than she would

need if she were jumping the same distance from one dock to another?

**9** •• SSM Much early research in rocket motion was done by Robert Goddard, physics professor at Clark College in Worcester, Massachusetts. A quotation from a 1920 editorial in the *New York Times* illustrates the public opinion of his work: "That Professor Goddard with his 'chair' at Clark College and the countenance of the Smithsonian Institution does not know the relation between action and reaction, and the need to have something better than a vacuum against which to react—to say that would be absurd. Of course, he only seems to lack the knowledge ladled out daily in high schools."[†] The belief that a rocket needs something to push against was a prevalent misconception before rockets in space were commonplace. Explain why that belief is wrong.

**10** • Two bowling balls are moving with the same velocity, but one just slides down the alley, whereas the other rolls down the alley. Which ball has more kinetic energy? Explain.

**11** • A philospher tells you, "All motion is impossible. Forces always come in equal but opposite action–reaction pairs. Therefore, all forces must cancel out." Answer his argument.

**12** • If two objects collide and one is initially at rest, is it still possible for both to be at rest after the collision? (Any external forces acting on the two-object system are negligibly small.) Is it possible for one to be at rest? Explain.

**13** • From the standpoint of physical laws, what is the problem with superhero comics where the hero flies around with no jets or hovers in midair while tossing huge objects like cars at the villains?

**14** •• SSM If only external forces can cause the center of mass of a system of particles to accelerate, how can a car move? We normally think of the car's engine as supplying the force needed to accelerate the car, but is this true? Where does the external force that accelerates the car come from?

---

[†] On page 43 of the July 17, 1969, edition of the *New York Times* "A Correction" to their editorial of 1920 was printed. This commentary, which was published three days before man's first walk on the moon, stated that "it is now definitely established that a rocket can function in a vacuum as well as in an atmosphere. *The Times* regrets the error."

**15** •• When we push on the brake pedal to slow down a car, a brake pad is pressed against the rotor so that the friction of the pad slows the wheel's rotation. However, the friction of the pad against the rotor can't be the force that slows the car down, because it is an internal force—both the rotor and the wheel are parts of the car, so any forces between them are purely internal to the system. What is the external force that slows down the car? Give a detailed explanation of how this force operates.

**16** • Explain why a safety net can save the life of a circus performer.

**17** • How might you estimate the collision time of a baseball and bat?

**18** • Why does a wine glass survive a fall onto a carpet but not onto a concrete floor?

**19** • True or false:

(a) In any perfectly inelastic collision, all the kinetic energy of the bodies is lost.
(b) In a head-on elastic collision, the relative speed of recession after the collision equals the relative speed of approach before the collision.
(c) Mechanical energy is conserved in an elastic collision.

**20** •• SSM Under what conditions can all the initial kinetic energy of an isolated system consisting of two colliding bodies be lost in a collision?

**21** •• Consider a perfectly inelastic collision of two objects of equal mass. (a) Is the loss of kinetic energy greater if the two objects have oppositely directed velocities of equal magnitude $v/2$ or if one of the two objects is initially at rest and the other has an initial velocity of $v$? (b) In which situation is the percentage loss in kinetic energy the greatest?

**22** •• SSM A double-barrelled pea shooter is shown in Figure 8-48. Air is blown from the left end of the straw, and identical peas A and B are positioned inside the straw as shown. If the shooter is held horizontally while the peas are shot off, which pea, A or B, will travel farther after leaving the straw? Why? *Hint: The answer has to do with the impulse–momentum theorem.*

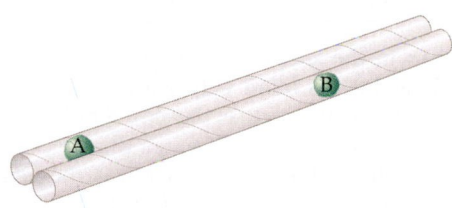

**FIGURE 8-48** Problem 22

**23** •• A particle of mass $m_1$ traveling with a speed $v$ makes a head-on elastic collision with a stationary mass $m_2$. In which scenario will the energy imparted to the particle of mass $m_2$ be greatest? (a) $m_2 \ll m_1$. (b) $m_2 = m_1$. (c) $m_2 \gg m_1$. (d) None of these.

### Optional Section

**24** • Describe a perfectly inelastic collision as viewed in the center-of-mass reference frame.

**25** • Nozzles for a garden hose are often made with a right-angle shape as shown in Figure 8-49. If you open the nozzle and spray water out, you will find that the nozzle presses against your hand with a pretty strong force—much stronger than if you used a nozzle not bent into a right angle. Why is this?

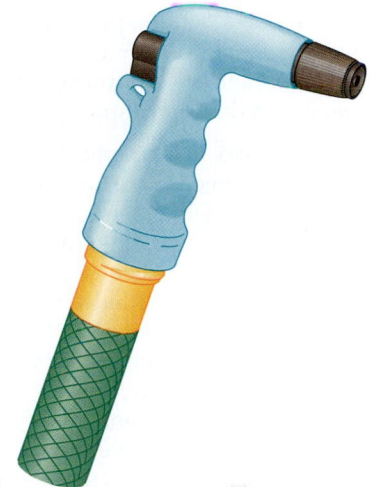

**FIGURE 8-49** Problem 25

**26** • Why can friction and the force of gravity usually be neglected in collision problems?

**27** • As a pendulum bob swings back and forth, is the momentum of the bob conserved? Explain.

**28** •• SSM A railroad car is passing by a grain elevator, which is dumping grain into it at a constant rate. (a) Does momentum conservation imply that the railroad car should be slowing down as it passes the grain elevator? Assume that the track is frictionless and perfectly level. (b) If the car is slowing down, this implies that there is some external force acting on the car to slow it down. Where does this force come from? (c) After passing the elevator, the railroad car springs a leak, and grain starts leaking out of a vertical hole in its floor at a constant rate. Should the car speed up as it loses mass?

**29** •• SSM To show that even really bright people can make mistakes, consider the following problem which was given to the freshman class at Caltech on an exam (paraphrased): *A sailboat is sitting in the water on a windless day. In order to make the boat move, a misguided sailor sets up a fan in the back of the boat to blow into the sails to make the boat move forward.* Explain why the boat won't move. The idea was that the net force of the wind pushing the sail forward would be counteracted by the force pushing the fan back (Newton's third law). However, as one of the students pointed out to his professor, the sailboat *could* in fact move forward. Why is that?

### Estimation and Approximation

**30** •• A 2000-kg car traveling at 90 km/h crashes into a concrete wall that does not give at all. (a) Estimate the time of collision, assuming that the center of the car travels halfway to the wall with constant deceleration. (Use any reasonable length for the car.) (b) Estimate the average force exerted by the wall on the car.

**31** •• In hand-pumped railcar races, a speed of 32 km/h has been achieved by teams of four. A car of mass 350 kg is moving at that speed toward a river when Carlos, the chief pumper, notices that the bridge ahead is out. All four people (of mass 75 kg each) jump simultaneously backward off the car with a velocity that has a horizontal component of 4 m/s relative to the car after jumping. The car proceeds off the bank and falls into the water a horizontal distance of 25.0 m off the bank. (a) Estimate the time of the fall of the railcar. (b) What happens to the team of pumpers?

**32** •• **SSM** A counterintuitive physics demonstration can be performed by firing a rifle bullet into a melon. (Don't try this at home!) When hit, nine times out of ten the melon will jump backward, toward the rifle, *opposite* to the direction in which the bullet was moving. (The tenth time, the melon simply explodes.) Doesn't this violate the laws of conservation of momentum? It doesn't, because we're not dealing simply with a two-body collision. Instead, a significant fraction of the energy of the bullet can be dumped into a jet of melon that is violently ejected out of the front of the melon. This jet can have a momentum greater than the momentum of the bullet, so that the rest of the melon must jump backward to conserve momentum. Let's make the following assumptions:

1. The mass of the melon is 2.50 kg.

2. The mass of the rifle bullet is 10.4 g and its velocity is 1800 ft/s.

3. 10 percent of the energy of the bullet is deposited as kinetic energy into a jet flying out of the front of the melon.

4. The mass of the matter in the jet is 0.14 kg.

5. All collisions occur in a straight line. What would be the speed of the melon's recoil? Compare this to a typical measured recoil speed of about 1.6 ft/s.

### Finding the Center of Mass

**33** • Three point masses of 2 kg each are located on the $x$ axis at the origin, at $x = 0.20$ m, and at $x = 0.50$ m. Find the center of mass of the system.

**34** • **SSM** Alley Oop's club-ax consists of a symmetrical 8-kg stone attached to the end of a uniform 2.5-kg stick that is 98 cm long. The dimensions of the club-ax are shown in Figure 8-50. How far is the center of mass from the handle end of the club-ax?

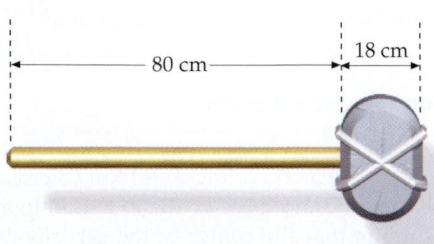

**FIGURE 8-50** Problem 34

**35** • **iSOLVE✓** Three balls A, B, and C, with masses of 3 kg, 1 kg, and 1 kg, respectively, are connected by massless rods. The balls are located as in Figure 8-51. What are the coordinates of the center of mass?

**FIGURE 8-51** Problems 35, 45

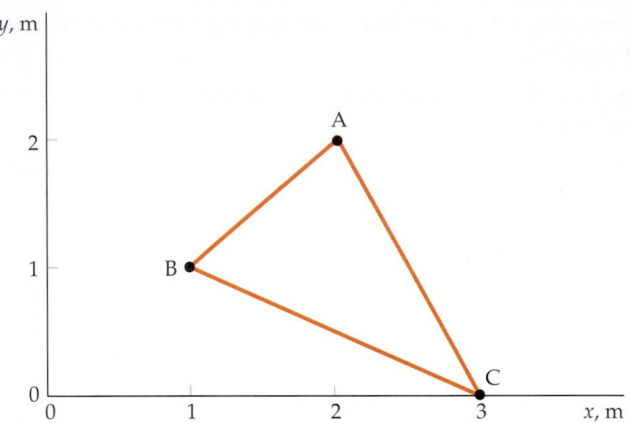

**36** • **iSOLVE** By symmetry, locate the center of mass of an equilateral triangle of side length $a$ with one vertex on the $y$ axis and the others at $(-a/2, 0)$ and $(+a/2, 0)$.

**37** •• **SSM** Find the center of mass of the uniform sheet of plywood in Figure 8-52. We shall consider this as two sheets, a square sheet of 3-m edge length and mass $m_1$ and a rectangular sheet 1 m × 2 m with a mass of $-m_2$. Let the coordinate origin be at the lower left-hand corner of the sheet.

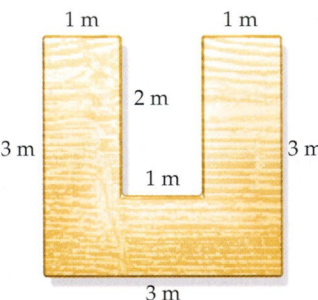

**FIGURE 8-52** Problem 37

**38** •• A can in the shape of a symmetrical cylinder with mass $M$ and height $H$ is filled with water. The initial mass of the water is $M$, the same mass as the can. A hole is punched in the bottom of the can, and the water drains out. (a) If the height of the water in the can is $x$, what is the height of the center of mass of the can + water? (b) What is the minimum height of the center of mass as the water drains out?

### *Finding the Center of Mass by Integration

**39** •• **SSM** Show that the center of mass of a uniform semicircular disk of radius $R$ is at a point $4R/(3\pi)$ from the center of the circle.

**40 •••** Find the center of mass of a homogeneous solid hemisphere of radius $R$ and mass $M$.

**41 •••** Find the center of mass of a thin homogeneous hemispherical shell.

**42 •••** A sheet of metal is cut in the shape of a parabola (Figure 8-53). The curved edge of the sheet is specified by the equation $y = ax^2$, and $y$ ranges from 0 to $b$. Find the center of mass in terms of $a$ and $b$. (You will need to find the area first.)

**FIGURE 8-53** Problem 42

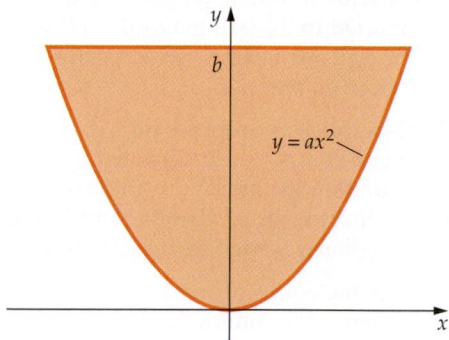

## Motion of the Center of Mass

**43 •** **SOLVE ✓** Two 3-kg particles have velocities $\vec{v}_1 = 2 \text{ m/s } \hat{i} + 3 \text{ m/s } \hat{j}$ and $\vec{v}_2 = 4 \text{ m/s } \hat{i} - 6 \text{ m/s } \hat{j}$. Find the velocity of the center of mass for the system.

**44 •** **SSM** **SOLVE ✓** A 1500-kg car is moving westward with a speed of 20 m/s, and a 3000-kg truck is traveling east with a speed of 16 m/s. Find the velocity of the center of mass of the system.

**45 •** **SOLVE** A force $\vec{F} = 12 \text{ N } \hat{i}$ is applied to the 3-kg ball in Figure 8-51 in Problem 35. What is the acceleration of the center of mass?

**46 ••** A block of mass $m$ is attached to a string and suspended inside a hollow box of mass $M$. The box rests on a scale that measures the system's weight. (a) If the string breaks, does the reading on the scale change? Explain your reasoning. (b) Assume that the string breaks and the mass $m$ falls with constant acceleration $g$. Find the acceleration of the center of mass of the box–block system, giving both direction and magnitude. (c) Using the result from (b), determine the reading on the scale while $m$ is in free-fall.

**47 ••** **SSM** A massless, vertical spring of force constant $k$ is attached at the bottom to a platform of mass $m_p$, and at the top to a massless cup, as in Figure 8-54. The platform rests on a scale. A ball of mass $m_b$ is dropped into the cup from a negligible height. What is the reading on the scale (a) when the spring is compressed an amount $d = m_b g/k$; (b) when the ball comes to rest momentarily with the spring compressed; (c) when the ball again comes to rest in its original position?

**FIGURE 8-54** Problems 47, 49, 136

**48 ••** **SSM** In the Atwood's machine in Figure 8-55, the string passes over a fixed cylinder of mass $m_c$. The cylinder does not rotate. Instead, the string slides on its frictionless surface. (a) Find the acceleration of the center of mass of the two-block-and-cylinder system. (b) Use Newton's second law for systems to find the force $F$ exerted by the support. (c) Find the tension $T$ in the string connecting the blocks and show that $F = m_c g + 2T$.

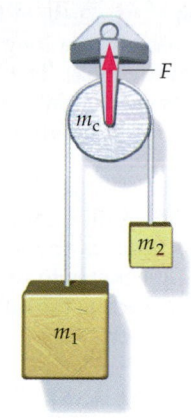

**FIGURE 8-55** Problem 48

**49 ••** **SOLVE** Repeat Problem 47(a) and 47(b) if the ball is dropped into the cup from a height $h$ above the cup.

## The Conservation of Momentum

**50 •** **SOLVE** A woman of mass 55 kg jumps off the bow of a 75-kg canoe that is initially at rest. If her velocity is 2.5 m/s to the right, what is the velocity of the canoe after she jumps?

**51 •** **SOLVE** A 5-kg object and a 10-kg object are connected by a massless compressed spring and rest on a frictionless table. After the spring is released, the object with the smaller mass has a velocity of 8 m/s to the left. What is the velocity of the object with the larger mass?

**52 •** **SSM** Figure 8-56 shows the behavior of a projectile just after it has broken up into three pieces. What was the speed of the projectile the instant before it broke up? (a) $v_3$. (b) $v_3/3$. (c) $v_3/4$. (d) $4v_3$. (e) $(v_1 + v_2 + v_3)/4$.

**FIGURE 8-56**
**Problem 52**

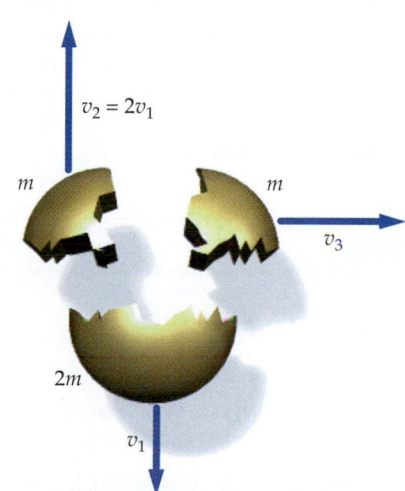

**53 •** A shell of mass $m$ and speed $v$ explodes into two identical fragments. If the shell was moving horizontally with respect to the earth, and one of the fragments is subsequently moving vertically with speed $v$, find the velocity $\vec{v}'$ of the other fragment.

**54** •• [SSM] A block and a gun are firmly fixed to opposite ends of a long glider mounted on a frictionless air track (Figure 8-57). The block and gun are a distance $L$ apart. The system is initially at rest. The gun is fired and the bullet leaves the muzzle with a velocity $v_b$ and impacts the block, becoming imbedded in it. The mass of the bullet is $m_b$ and the mass of the gun–glider–block system is $m_p$. (a) What is the velocity of the glider immediately after the bullet leaves the muzzle? (b) What is the velocity of the glider immediately after the bullet comes to rest in the block? (c) How far does the glider move while the bullet is in transit between the gun at rest and the block at rest?

**FIGURE 8-57** Problem 54

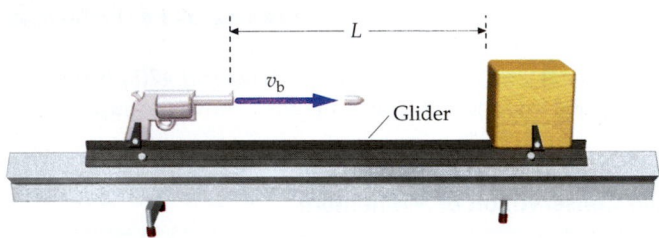

**55** •• A small object of mass $m$ slides down a wedge of mass $2m$ and exits smoothly onto a frictionless table. The wedge is initially at rest on the table. If the object is initially at rest at a height $h$ above the table, find the velocity of the wedge when the object leaves it.

**56** •• [SSM] Each glider on a frictionless air track (Figure 8-58) supports a strong magnet. These magnets attract each other. The mass of glider 1 is 0.10 kg and the mass of glider 2 is 0.20 kg. (Each mass includes the mass of the glider plus the supported magnet.) The origin is at the left end of the track, the center of glider 1 is at $x_1 = 0.10$ m and the center of glider 2 is at $x_2 = 1.60$ m. Glider 1 is 10 cm long, while glider 2 is 20 cm long. Each glider has its center of mass at its center. (a) When the two gliders are released from rest, they move toward each other and stick together. What is the position of the center of each glider when they first touch? (b) Will the two gliders continue to move after they stick together? Explain.

**FIGURE 8-58** Problem 56

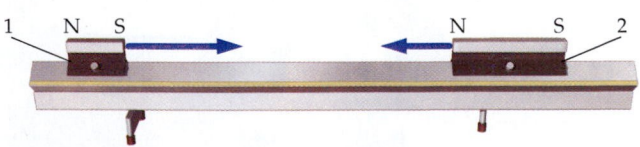

## Kinetic Energy of a System of Particles

**57** • [SSM] [iSOLVE] A 3-kg block is traveling to the right at 5 m/s, and a second 3-kg block is traveling to the left at 2 m/s. (a) Find the total kinetic energy of the two blocks. (b) Find the velocity of the center of mass of the two-block system. (c) Find the velocities of each block relative to the center of mass. (d) Find the kinetic energy of the motion of the blocks relative to the center of mass. (e) Show that your answer for part (a) is greater than your answer for part (d) by an amount equal to the kinetic energy associated with the motion of the center of mass.

**58** • Repeat Problem 57 with the second 3-kg block replaced by a 5-kg block moving to the right at 3 m/s.

## Impulse and Average Force

**59** • You kick a soccer ball of mass 0.43 kg. The ball leaves your foot with an initial speed of 25 m/s. (a) What is the impulse imparted to the ball by you? (b) If your foot is in contact with the ball for 0.008 s, what is the average force exerted by your foot on the ball?

**60** • A 0.3-kg brick is dropped from a height of 8 m. It hits the ground and comes to rest. (a) What is the impulse exerted by the ground on the brick? (b) If it takes 0.0013 s from the time the brick first touches the ground until it comes to rest, what is the average force exerted by the ground on the brick?

**61** • [SSM] A meteorite of mass 30.8 tonne (1 tonne = 1000 kg) is exhibited in the American Museum of Natural History in New York City. Suppose that the kinetic energy of the meteorite as it hit the ground was 617 MJ. Find the impulse $I$ experienced by the meteorite up to the time its kinetic energy was halved (which took about $t = 3.0$ s). Find also the average force $F$ exerted on the meteorite during this time interval.

**62** •• [iSOLVE] When a 0.15-kg baseball is hit, its velocity changes from +20 m/s to −20 m/s. (a) What is the magnitude of the impulse delivered by the bat to the ball? (b) If the baseball is in contact with the bat for 1.3 ms, what is the average force exerted by the bat on the ball?

**63** •• [SSM] [iSOLVE ✓] A 60-g handball moving with a speed of 5.0 m/s strikes the wall at an angle of 40° and then bounces off with the same speed at the same angle. It is in contact with the wall for 2 ms. What is the average force exerted by the ball on the wall?

**64** •• [iSOLVE] You throw a 150-g ball to a height of 40 m. (a) Use a reasonable value for the distance the ball moves while it is in your hand to calculate the average force exerted by your hand and the time the ball is in your hand while you throw it. (b) Is it reasonable to neglect the weight of the ball while it is being thrown?

**65** •• A handball of mass 60 g is thrown straight against a wall with a speed of 10 m/s. It rebounds with a speed of 8 m/s. (a) What impulse is delivered to the wall? (b) If the ball is in contact with the wall for 0.003 s, what average force is exerted on the wall by the ball? (c) The ball is caught by a player who brings it to rest. In the process, her hand moves back 0.5 m. What is the impulse received by the player? (d) What average force was exerted on the player by the ball?

**66** ••• [iSOLVE] The great limestone caverns were formed by dripping water. (a) If water droplets of 0.03 mL fall from a height of 5 m at a rate of 10 per minute, what is the average force exerted on the limestone floor by the droplets of water during a 1-min period? (b) Compare this force to the weight of a water droplet.

## Collisions in One Dimension

**67** • **SSM** A 2000-kg car traveling to the right at 30 m/s is chasing a second car of the same mass that is traveling to the right at 10 m/s. (a) If the two cars collide and stick together, what is their speed just after the collision? (b) What fraction of the initial kinetic energy of the cars is lost during this collision? Where does it go?

**68** • An 85-kg running back moving at 7 m/s makes a perfectly inelastic collision with a 105-kg linebacker who is initially at rest. What is the speed of the players just after their collision?

**69** • A 5.0-kg object with a speed of 4.0 m/s collides head-on with a 10-kg object moving toward it with a speed of 3.0 m/s. The 10-kg object stops dead after the collision. (a) What is the final speed of the 5-kg object? (b) Is the collision elastic?

**70** • A ball of mass $m$ moves with speed $v$ to the right toward a much heavier bat that is moving to the left with speed $v$. Find the speed of the ball after it makes an elastic collision with the bat.

**71** •• **SSM** **iSOLVE** A proton of mass $m$ undergoes a head-on elastic collision with a stationary carbon nucleus of mass $12m$. The speed of the proton is 300 m/s. (a) Find the velocity of the center of mass of the system. (b) Find the velocity of the proton after the collision.

**72** •• A 3-kg block moving at 4 m/s makes a head-on elastic collision with a stationary block of mass 2 kg. Use conservation of momentum and the fact that the relative speed of recession equals the relative speed of approach to find the velocity of each block after the collision. Check your answer by calculating the initial and final kinetic energies of each block.

**73** •• **iSOLVE** A block of mass $m_1 = 2$ kg slides along a frictionless table with a speed of 10 m/s. Directly in front of it, and moving in the same direction with a speed of 3 m/s, is a block of mass $m_2 = 5$ kg. A massless spring with spring constant $k = 1120$ N/m is attached to the second block as in Figure 8-59. (a) Before $m_1$ makes contact with the spring, what is the velocity of the center of mass of the system? (b) During the collision, the spring is compressed by a maximum amount $\Delta x$. What is the value of $\Delta x$? (c) The blocks will eventually separate again. What are the final velocities of the two blocks measured in the reference frame of the table?

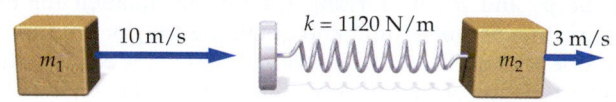

**FIGURE 8-59** Problem 73

**74** •• **SSM** **iSOLVE** A bullet of mass $m$ is fired vertically from below into a thin sheet of plywood of mass $M$ that is initially at rest, supported by a thin sheet of paper. The bullet blasts through the plywood, which rises to a height $H$ above the paper before falling back down. The bullet continues rising to a height $h$ above the paper. (a) Express the upward velocity of the bullet and the plywood immediately after the bullet exits the plywood in terms of $h$ and $H$. (b) Use conservation of momentum to express the speed of the bullet before it enters the sheet of plywood in terms of $m$, $h$, $M$, and $H$. (c) Obtain expressions for the mechanical energy of the system before and after the inelastic collision. (d) Express the energy dissipated in terms of $m$, $h$, $M$, and $H$.

**75** •• A proton of mass $m$ is moving with initial speed $v_0$ directly toward the center of an $\alpha$ particle of mass $4m$, which is initially at rest. Because both particles carry positive electrical charge, they repel each other. Find the speed $v'$ of the $\alpha$ particle (a) when the distance between the two particles is least and (b) later when the two particles are far apart.

**76** • An electron collides elastically with a hydrogen atom initially at rest. All motion occurs along a straight line. What fraction of the electron's initial kinetic energy is transferred to the atom? (Take the mass of the hydrogen atom to be 1840 times the mass of an electron.)

**77** •• **iSOLVE** ✓ A 16-g bullet is fired into the bob of a ballistic pendulum of mass 1.5 kg. When the bob is at its maximum height, the strings make an angle of 60° with the vertical. The length of the pendulum is 2.3 m. Find the speed of the bullet.

**78** •• **SSM** Show that in a one-dimensional elastic collision, if the mass and velocity of object 1 are $m_1$ and $v_{1i}$, and if the mass and velocity of object 2 are $m_2$ and $v_{2i}$, then the final velocities $v_{1f}$ and $v_{2f}$ are given by

$$v_{1f} = \frac{m_1 - m_2}{m_1 + m_2}v_{1i} + \frac{2m_2}{m_1 + m_2}v_{2i}$$

$$v_{2f} = \frac{2m_1}{m_1 + m_2}v_{1i} + \frac{m_2 - m_1}{m_1 + m_2}v_{2i}$$

**79** •• Use the results of Problem 78 to investigate one-dimensional elastic collisions in the following limits: (a) When the two masses are equal, show that the particles "swap" velocities: $v_{1f} = v_{2i}$ and $v_{2f} = v_{1i}$. (b) If $m_2 \gg m_1$, show that $v_{1f} \approx -v_{1i} + 2v_{2i}$ and $v_{2f} \approx v_{2i}$.

## Perfectly Inelastic Collisions and the Ballistic Pendulum

**80** •• A bullet of mass $m_1$ is fired with a speed $v$ into the bob of a ballistic pendulum of mass $m_2$. The bob is attached to a very light rod of length $L$ that is pivoted at the other end. The bullet is stopped in the bob. Find the minimum $v$ such that the bob will swing through a complete circle.

**81** •• **SSM** A bullet of mass $m_1$ is fired with a speed $v$ into the bob of a ballistic pendulum of mass $m_2$. Find the maximum height $h$ attained by the bob if the bullet passes through the bob and emerges with a speed $v/2$.

**82** • If we take a heavy wooden block normally used as a ballistic pendulum and put it on a flat table before firing a bullet into it, about how far will it slide before coming to a stop? Assume that the mass of the rifle bullet is 10.5 g, the mass of the wooden block is 10.5 kg, the bullet's velocity is 750 m/s, and the coefficient of kinetic friction of the block against the table is 0.22. Also assume that the bullet doesn't cause the block to rotate.

**83** •• A ball with mass $M = 0.425$ kg and speed $V = 1.3$ m/s rolls across a level table into an open box with mass $m = 0.327$ kg. The box with the ball inside it then slides across the table for a distance $x = 0.520$ m. What is the coefficient of kinetic friction of the table?

**84** •• **SSM** Tarzan is in the path of a pack of stampeding elephants when Jane swings in to the rescue on a rope vine, hauling him off to safety. The length of the vine is 25 m, and Jane starts her swing with the rope horizontal. If Jane's mass is 54 kg, and Tarzan's is 82 kg, to what height above the ground will the pair swing after she grabs him?

## Exploding Objects and Radioactive Decay

**85** •• **SOLVE** The beryllium isotope $^{8}$Be is unstable and decays into two $\alpha$ particles (helium nuclei of mass $m = 6.68 \times 10^{-27}$ kg) with the release of $1.5 \times 10^{-14}$ J of energy. Determine the velocities of the two $\alpha$ particles that arise from the decay of a $^{8}$Be nucleus at rest, assuming that all the energy appears as kinetic energy of the particles.

**86** •• The light isotope of lithium, $^{5}$Li, is unstable and breaks up spontaneously into a proton (hydrogen nucleus) and an $\alpha$ particle (helium nucleus). In this process, a total energy of $3.15 \times 10^{-13}$ J is released, appearing as the kinetic energy of the two decay products. Determine the velocities of the proton and $\alpha$ particle that arise from the decay of a $^{5}$Li nucleus at rest. (*Note*: The masses of the proton and alpha particle are $m_p = 1.67 \times 10^{-27}$ kg and $m_\alpha = 4m_p = 6.68 \times 10^{-27}$ kg.)

**87** ••• **SOLVE** A projectile of mass $m = 3$ kg is fired with an initial speed of 120 m/s at an angle of 30° with the horizontal. At the top of its trajectory, the projectile explodes into two fragments of masses 1 kg and 2 kg. The 2-kg fragment lands on the ground directly below the point of explosion 3.6 s after the explosion. (*a*) Determine the velocity of the 1-kg fragment immediately after the explosion. (*b*) Find the distance between the point of firing and the point at which the 1-kg fragment strikes the ground. (*c*) Determine the energy released in the explosion.

**88** ••• **SSM** The boron isotope $^{9}$B is unstable and disintegrates into a proton and two $\alpha$ particles. The total energy released as kinetic energy of the decay products is $4.4 \times 10^{-14}$ J. In one such event, with the $^{9}$B nucleus at rest prior to decay, the velocity of the proton is measured as $6.0 \times 10^{6}$ m/s. If the two $\alpha$ particles have equal energies, find the magnitude and the direction of their velocities with respect to that of the proton.

## *Coefficient of Restitution

**89** • The coefficient of restitution for steel on steel is measured by dropping a steel ball onto a steel plate that is rigidly attached to the earth. If the ball is dropped from a height of 3 m and rebounds to a height of 2.5 m, what is the coefficient of restitution?

**90** • **SSM** According to the official rules of racquetball, a ball acceptable for tournament play must bounce to a height of between 173 and 183 cm when dropped from a height of 254 cm at room temperature. What is the acceptable range of values for the coefficient of restitution for the racquetball–floor system?

**91** • **SOLVE** ✓ A ball bounces to 80 percent of its original height. (*a*) What fraction of its mechanical energy is lost each time it bounces? (*b*) What is the coefficient of restitution of the ball–floor system?

**92** •• **SOLVE** A 2-kg object moving at 6 m/s collides with a 4-kg object that is initially at rest. After the collision, the 2-kg object moves backward at 1 m/s. (*a*) Find the velocity of the 4-kg object after the collision. (*b*) Find the energy lost in the collision. (*c*) What is the coefficient of restitution for this collision?

**93** •• A 2-kg block moving to the right with speed 5 m/s collides with a 3-kg block that is moving in the same direction at 2 m/s, as in Figure 8-60. After the collision, the 3-kg block moves at 4.2 m/s. Find (*a*) the velocity of the 2-kg block after the collision and (*b*) the coefficient of restitution for the collision.

**FIGURE 8-60** Problem 93

## *Collisions in Three Dimensions

**94** •• **SSM** In Section 8-6 it was proven by geometrical means that when a particle elastically collides with another particle of equal mass that is initially at rest, the two separate at right angles. In this problem we will examine another way of proving this, one that illustrates the power of abstract vector notation. (*a*) Given $\vec{A} = \vec{B} + \vec{C}$, show that $A^2 = B^2 + C^2 + 2\vec{B} \cdot \vec{C}$. (*b*) Let the momentum of the initially moving particle be $\vec{P}$ and the momenta of the particles after the collision be $\vec{p}_1$ and $\vec{p}_2$. By writing the vector equation for the conservation of momentum and obtaining the dot product of each side with itself, and then comparing it to the equation for the conservation of energy, show that $\vec{p}_1 \cdot \vec{p}_2 = 0$.

**95** • **SOLVE** In a pool game, the cue ball, which has an initial speed of 5 m/s, makes an elastic collision with the eight ball, which is initially at rest. After the collision, the eight ball moves at an angle of 30° to the original direction of the cue ball. (*a*) Find the direction of motion of the cue ball after the collision. (*b*) Find the speed of each ball. Assume that the balls have equal mass.

**96** •• Object A with mass $m$ and velocity $v_0\hat{i}$ collides head-on with object B of mass $2m$ and velocity $\frac{1}{2}v_0\hat{j}$. Following the collision, object B has a velocity of $\frac{1}{4}v_0\hat{i}$. (a) Determine the velocity of object A after the collision. (b) Is the collision elastic? If not, express the change in the kinetic energy in terms of $m$ and $v_0$.

**97** •• **SSM** **iSOLVE** A puck of mass 5 kg moving at 2 m/s approaches an identical puck that is stationary on frictionless ice. After the collision, the first puck leaves with a speed $v_1$ at 30° to the original line of motion; the second puck leaves with speed $v_2$ at 60°, as in Figure 8-61. (a) Calculate $v_1$ and $v_2$. (b) Was the collision elastic?

**FIGURE 8-61** Problem 97

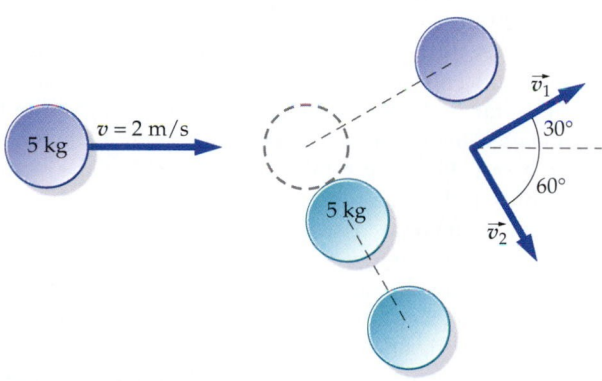

**98** •• Figure 8-62 shows the result of a collision between two objects of unequal mass. (a) Find the speed $v_2$ of the larger mass after the collision and the angle $\theta_2$. (b) Show that the collision is elastic.

**FIGURE 8-62** Problem 98

**99** •• **SSM** **iSOLVE** A ball moving at 10 m/s makes an off-center elastic collision with another ball of equal mass that is initially at rest. The incoming ball is deflected at an angle of 30° from its original direction of motion. Find the velocity of each ball after the collision.

**100** •• A particle has initial speed $v_0$. It collides with a second particle *with the same mass* that is at rest, and is deflected through an angle $\phi$. Its speed after the collision is $v$. The second particle recoils, and its velocity makes an angle $\theta$ with the initial direction of the first particle. (a) Show that

$$\tan \theta = \frac{v \sin \phi}{(v_0 - v \cos \phi)}$$

(b) Show that if the collision is elastic, $v = v_0 \cos \phi$.

### Center-of-Mass Frame

**101** •• In the center-of-mass reference frame a particle with mass $m_1$ and momentum $p_1$ makes an elastic head-on collision with a second particle of mass $m_2$ and momentum $p_2 = -p_1$. After the collision its momentum is $p_1'$. Write the total initial energy in terms of $m_1$, $m_2$, and $p_1$ and the total final energy in terms of $m_1$, $m_2$, and $p_1'$, and show that $p_1' = \pm p_1$. If $p_1' = -p_1$, the particle is merely turned around by the collision and leaves with the speed it had initially. What is the significance of the plus sign in your solution?

**102** •• **SSM** **iSOLVE** A 3-kg block is traveling in the negative $x$ direction at 5 m/s, and a 1-kg block is traveling in the positive $x$ direction at 3 m/s. (a) Find the velocity $v_{cm}$ of the center of mass. (b) Subtract $v_{cm}$ from the velocity of each block to find the velocity of each block in the center-of-mass reference frame. (c) After they make a head-on elastic collision, the velocity of each block is reversed (in this frame). Find the velocity of each block in the center-of-mass frame after the collision. (d) Transform back into the original frame by adding $v_{cm}$ to the velocity of each block. (e) Check your result by finding the initial and final kinetic energies of the blocks in the original frame and comparing them.

**103** •• Repeat Problem 102 with the second block having a mass of 5 kg and moving to the right at 3 m/s.

### *Systems With Continuously Varying Mass: Rocket Propulsion

**104** •• **iSOLVE** A rocket burns fuel at a rate of 200 kg/s and exhausts the gas at a relative speed of 6 km/s. Find the thrust of the rocket.

**105** •• A rocket has an initial mass of 30,000 kg, of which 80 percent is the fuel. It burns fuel at a rate of 200 kg/s and exhausts its gas at a relative speed of 1.8 km/s. Find (a) the thrust of the rocket, (b) the time until burnout, and (c) its speed at burnout assuming it moves straight upward near the surface of the earth where the gravitational field $g$ is constant.

**106** •• **SSM** The *specific impulse* of a rocket propellant is defined as $I_{sp} = F_{th}/(Rg)$, where $F_{th}$ is the thrust of the propellant, $g$ the magnitude of free-fall acceleration at the surface of the earth, and $R$ the rate at which the propellant is burned. The rate depends predominantly on the type and exact mixture of the propellant. (a) Show that the specific impulse has the dimension of time. (b) Show that $u_{ex} = gI_{sp}$, where $u_{ex}$ is the exhaust velocity of the propellant. (c) What is the specific impulse (in seconds) of the propellant used in the Saturn V rocket of Example 8-21?

**107 ••• SSM** The initial *thrust-to-weight ratio* $\tau_0$ of a rocket is $\tau_0 = F_{th}/(m_0\,g)$, where $F_{th}$ is the rocket's thrust and $m_0$ the initial mass of the rocket, including the propellant. (*a*) For a rocket launched straight up from the earth's surface, show that $\tau_0 = 1 + (a_0/g)$, where $a_0$ is the initial acceleration of the rocket. For manned rocket flight, $\tau_0$ cannot be made much larger than 4 for the comfort and safety of the astronauts. (The astronauts will feel that their weight as the rocket lifts off is equal to $\tau_0$ times their normal weight.) (*b*) Show that the final velocity of a rocket launched from the earth's surface, in terms of $\tau_0$ and $I_{sp}$ (see Problem 106) can be written as

$$v_f = gI_{sp}\left[\ln\left(\frac{m_0}{m_f}\right) - \frac{1}{\tau_0}\left(1 - \frac{m_f}{m_0}\right)\right]$$

where $m_f$ is the mass of the rocket (not including the spent propellant). (*c*) Using a spreadsheet program or graphing calculator, graph $v_f$ as a function of the mass ratio $m_0/m_f$ for $I_{sp} = 250$ s and $\tau_0 = 2$ for values of the mass ratio from 2 to 10. (Note that the mass ratio cannot be less than 1.) (*d*) To lift a rocket into orbit, a final velocity after burnout of $v_f = 7$ km/s is needed. Calculate the mass ratio required of a single stage rocket to do this, using the values of specific impulse and thrust ratio given in Part (*b*). For engineering reasons, it is difficult to make a rocket with a mass ratio much greater than 10. Can you see why multistage rockets are usually used to put payloads into orbit around the earth?

**108 ••** The height that a model rocket launched from earth can reach can be estimated by assuming that the burn time is short compared to the total flight time, so for most of the flight the rocket is in free-fall. (This estimate neglects the burn time in calculations of both time and displacement.) For a model rocket with specific impulse $I_{sp} = 100$ s, mass ratio $m_0/m_f = 1.2$, and initial thrust-to-weight ratio $\tau_0 = 5$ (these parameters are defined in Problems 106 and 107), estimate (*a*) the height the rocket can reach, and (*b*) the total flight time. (*c*) Justify the assumption used in the estimates by comparing the flight time from Part (*b*) to the time it takes for the fuel to be spent.

## General Problems

**109 •** A model-train car of mass 250 g traveling with a speed of 0.50 m/s links up with another car of mass 400 g that is initially at rest. What is the speed of the cars immediately after they have linked together? Find the initial and final kinetic energies.

**110 •** (*a*) Find the total kinetic energy of the two model-train cars of Problem 109 before they couple. (*b*) Find the initial velocities of the two cars relative to the center of mass of the system, and use them to calculate the initial center-of-mass kinetic energy of the system. (*c*) Find the kinetic energy of the center of mass. (*d*) Compare the sum of your answers for (*b*) and (*c*) with that for (*a*).

**111 • SSM** A 4-kg fish is swimming at 1.5 m/s to the right. He swallows a 1.2-kg fish swimming toward him at

3 m/s. Neglecting water resistance, what is the velocity of the larger fish immediately after his lunch?

**112 •** A 3-kg block moves at 6 m/s to the right while a 6-kg block moves at 3 m/s to the right. Find (*a*) the total kinetic energy of the two-block system, (*b*) the velocity of the center of mass, (*c*) the center-of-mass kinetic energy, and (*d*) the kinetic energy relative to the center of mass.

**113 •** A 1500-kg car traveling north at 70 km/h collides at an intersection with a 2000-kg car traveling west at 55 km/h. The two cars stick together. (*a*) What is the total momentum of the system before the collision? (*b*) Find the magnitude and direction of the velocity of the wreckage just after the collision.

**114 •• SSM** A 60-kg woman stands on the back of a 6-m-long, 120-kg raft that is floating at rest in still water with no friction. The raft is 0.5 m from a fixed pier, as shown in Figure 8-63. (*a*) The woman walks to the front of the raft and stops. How far is the raft from the pier now? (*b*) While the woman walks, she maintains a constant speed of 3 m/s relative to the raft. Find the total kinetic energy of the system (woman plus raft), and compare with the kinetic energy if the woman walked at 3 m/s on a raft tied to the pier. (*c*) Where does this energy come from, and where does it go when the woman stops at the front of the raft? (*d*) On land, the woman puts a lead shot 6 m. She stands at the back of the raft, aims forward, and puts the shot so that just after it leaves her hand, it has the same velocity relative to her as it did when she threw it from the ground. Approximately, where does her shot land?

**FIGURE 8-63** Problem 114

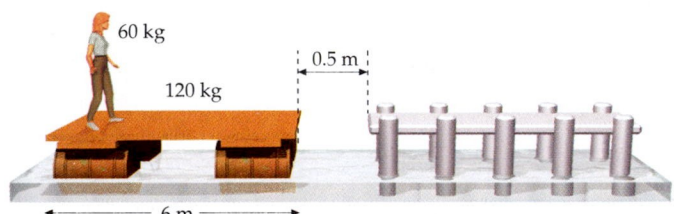

**115 ••** **iSOLVE** A 1-kg steel ball and a 2-m cord of negligible mass make up a simple pendulum that can pivot without friction about the point *O*, as in Figure 8-64. This pendulum is released from rest in a horizontal position and when the ball is at its lowest point it strikes a 1-kg block sitting at rest on a shelf. Assume that the collision is perfectly elastic and take the coefficient of friction between the block and shelf to be 0.1. (*a*) What is the velocity of the block just after impact? (*b*) How far does the block move before coming to rest?

**FIGURE 8-64** Problem 115

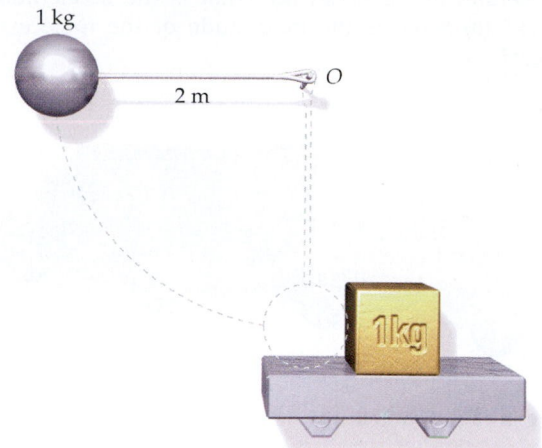

1 kg

2 m

O

1kg

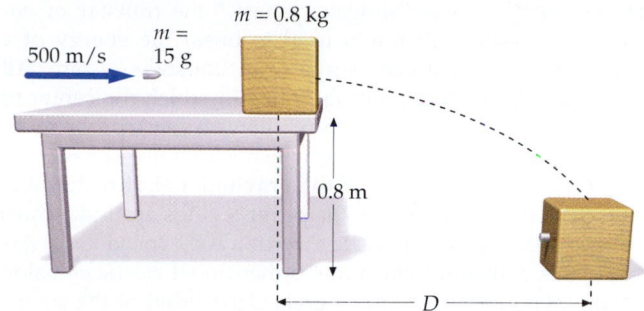

m = 0.8 kg

500 m/s

m = 15 g

0.8 m

D

**FIGURE 8-66** Problem 117

**116** •• **SSM** Figure 8-65 shows a World War I cannon mounted on a railcar so that it will project a shell at an angle of 30°. With the car initially at rest, the cannon fires a 200-kg projectile at 125 m/s. (All values are for the frame of reference of the track.) Now consider a system composed of a cannon, shell, and railcar, all on the frictionless track. (a) Will the total vector momentum of that system be the same (that is, "conserved") before and after the shell is fired? Explain your answer. (b) If the mass of the railcar plus cannon is 5000 kg, what will be the recoil velocity of the car along the track after the firing? (c) The shell is observed to rise to a maximum height of 180 m as it moves through its trajectory. At this point, its speed is 80 m/s. On the basis of this information, calculate the amount of thermal energy produced by air friction on the shell on its way from firing to this maximum height.

**FIGURE 8-65** Problem 116

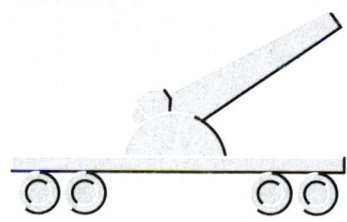

**117** •• A 15-g bullet traveling at 500 m/s strikes an 0.8-kg block of wood that is balanced on a table edge 0.8 m above the ground (Figure 8-66). If the bullet buries itself in the block, find the distance $D$ at which the block hits the floor.

**118** •• Two particles of masses $m$ and $4m$ are moving in a vacuum at right angles as in Figure 8-67. A force $\vec{F}$ acts on the particle of mass $m$ for a time $T$. As a result, the velocity of this particle becomes $4v$ in its original direction. A force equal to force $\vec{F}$ acts on the other particle for the same time $T$. Find the new velocity $\vec{v}\,'$ of the particle of mass $4m$.

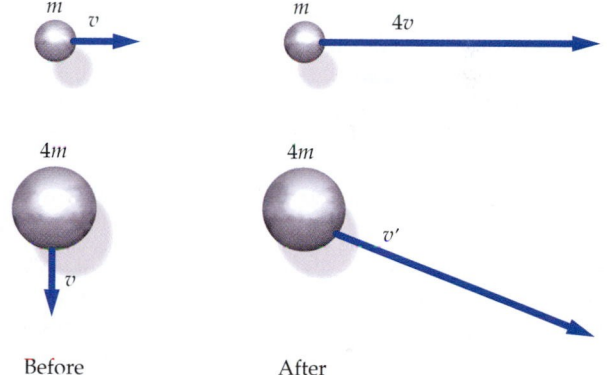

m

v

m

4v

4m

v

4m

v'

Before

After

**FIGURE 8-67** Problem 118

**119** •• One popular, if dangerous, classroom demonstration involves holding a baseball an inch or so directly above a basketball, holding the basketball a few feet above a hard floor, and dropping the two balls simultaneously. The two balls will collide just after the basketball bounces from the floor; the baseball will then rocket off into the ceiling tiles with a hard "thud" while the basketball will stop in midair. (The author of this problem once broke a light doing this.) (a) Assuming that the collision of the basketball with the floor is elastic, what is the relation between the velocities of the balls just before they collide? (b) Assuming the collision between the two balls is elastic, use the result of Part (a) and the conservation of momentum and energy to show that, if the basketball is three times as heavy as the baseball, the final velocity of the basketball will be zero. (This is approximately the true mass ratio, which is why the demonstration is so dramatic.) (c) If the speed of the top ball is $v$ just before the collision, what is its speed just after the collision?

**120** ••• (a) In Problem 119, if we held a third ball above the other two, and wanted both the lowest and middle balls to stop in mid-air, what should be the ratio of the mass of the top ball to the mass of the middle ball? (b) If the speed of the top ball is $v$ just before the collision, what is its speed just after the collision?

**121** •• [SSM] In the "slingshot effect," the transfer of energy in an elastic collision is used to boost the energy of a space probe so that it can escape from the solar system. All speeds are relative to an inertial frame in which the center of the sun remains at rest. Figure 8-68 shows a space probe moving at 10.4 km/s toward Saturn, which is moving at 9.6 km/s toward the probe. Because of the gravitational attraction between Saturn and the probe, the probe swings around Saturn and heads back in the opposite direction with speed $v_f$. (a) Assuming this collision to be a one-dimensional elastic collision with the mass of Saturn much greater than that of the probe, find $v_f$. (b) By what factor is the kinetic energy of the probe increased? Where does this energy come from?

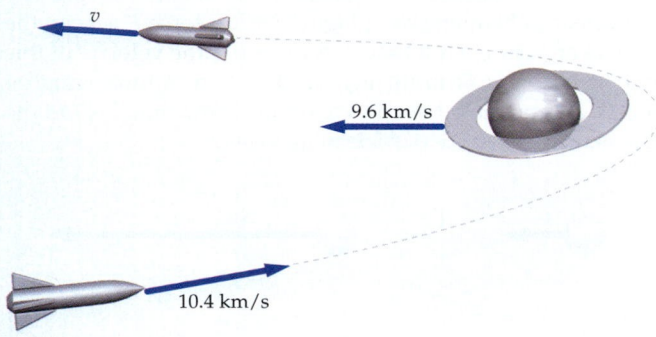

**FIGURE 8-68**  Problem 121

**122** •• [SSM] Imagine that a flashlight is floating in intergalactic space, far from any planet or star. It has batteries in it that are charged with a total energy $E = 1.5$ kJ. When the flashlight is turned on, it loses energy in the form of light; because of this, it also loses mass slightly, from $E = mc^2$. Because the lost "mass" is traveling away at the speed of light, the flashlight should start moving in the opposite direction. (a) If the flashlight loses energy $\Delta E$, argue that the change in momentum of the flashlight should be $P = \Delta E/c$. (b) If the mass of the flashlight is 1.5 kg, and it is turned on and left on until the batteries are discharged, what is its final velocity? Assume that the flashlight is 100 percent efficient in converting battery power into radiant power, and that it is initially at rest.

**123** • A block has a mass $M = 1$ kg. If we could convert 1 g of the mass of the block into a well-collimated beam of light shining in one direction, calculate the speed attained by the rest of the block. (See Problem 122 for a discussion of the momentum carried by a light beam.)

**124** •• You (mass 80 kg) and your friend (mass unknown) are in a rowboat (mass 60 kg) on a calm lake. You are at the center of the boat rowing and she is at the back, 2 m from the center. You get tired and stop rowing. She offers to row and after the boat comes to rest, you change places. You notice that after changing places the boat has moved 20 cm relative to a fixed log. What is your friend's mass?

**125** •• [i·SOLVE ✓] A small car of mass 800 kg is parked behind a small truck of mass 1550 kg on a level road (Figure 8-69). The brakes of both the car and the truck are off so that they are free to roll with negligible friction. A 50-kg woman sitting on the tailgate of the truck shoves the car away

by exerting a constant force on the car with her feet. The car accelerates at 1.2 m/s². (a) What is the acceleration of the truck? (b) What is the magnitude of the force exerted on the car?

**FIGURE 8-69**  Problem 125

**126** •• A 13-kg block is at rest on a level floor. A 400-g glob of putty is thrown at the block so that the putty travels horizontally, hits the block, and sticks to it. The block and putty slide 15 cm along the floor. If the coefficient of sliding friction is 0.4, what is the initial speed of the putty?

**127** •• [SSM] A careless driver rear-ends a car that is halted at a stop sign. Just before impact, the driver slams on his brakes, locking the wheels. The driver of the struck car also has his foot solidly on the brake pedal, locking his brakes. The mass of the struck car is 900 kg, and that of the initially moving vehicle is 1200 kg. On collision, the bumpers of the two cars mesh. Police determine from the skid marks that after the collision the two cars moved 0.76 m together. Tests revealed that the coefficient of sliding friction between the tires and pavement was 0.92. The driver of the moving car claims that he was traveling at less than 15 km/h as he approached the intersection. Is he telling the truth?

**128** •• A pendulum consists of a 0.4-kg bob attached to a string of length 1.6 m. A block of mass $m$ rests on a horizontal frictionless surface (Figure 8-70). The pendulum is released from rest at an angle of 53° with the vertical and the bob collides elastically with the block. Following the collision, the maximum angle of the pendulum with the vertical is 5.73°. Determine the mass $m$.

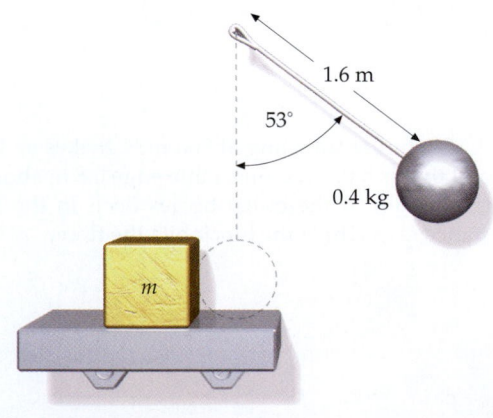

**FIGURE 8-70**  Problem 128

**129** •• Initially, the block whose mass $m = 1.0$ kg and the block whose mass is $M$ are both at rest on a frictionless inclined plane (Figure 8-71). The block of mass $M$ rests against a spring that has a spring constant of 11,000 N/m. The distance along the plane between the two blocks is 4.0 m. The block of mass $m$ is released, makes an elastic collision with the block of mass $M$, and rebounds a distance of 2.56 m back up the inclined plane. The block whose mass is $M$ comes to rest momentarily 4.0 cm from its initial position. Find $M$.

**FIGURE 8-71** Problem 129

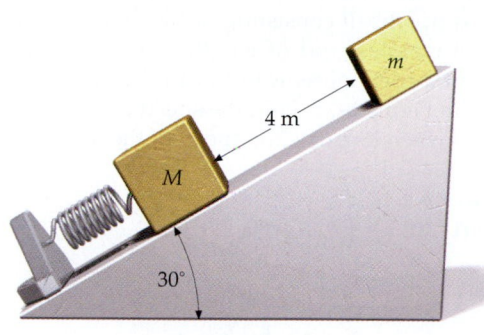

**130** •• $\boxed{\text{SSM}}$ A circular plate of radius $r$ has a circular hole of radius $r/2$ cut out of it (Figure 8-72). Find the center of mass of the plate. *Hint: The hole can be represented by two disks superimposed, one of mass m and the other of mass* $-$m.

**FIGURE 8-72**
**Problem 130**

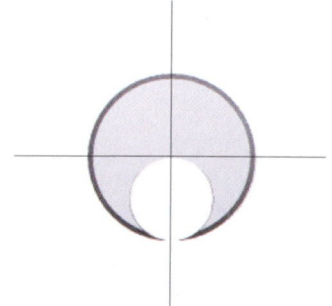

**131** •• Using the hint from Problem 130, find the center of mass of a solid sphere of radius $r$ that has a spherical cavity of radius $r/2$, as in Figure 8-73.

**FIGURE 8-73**
**Problem 131**

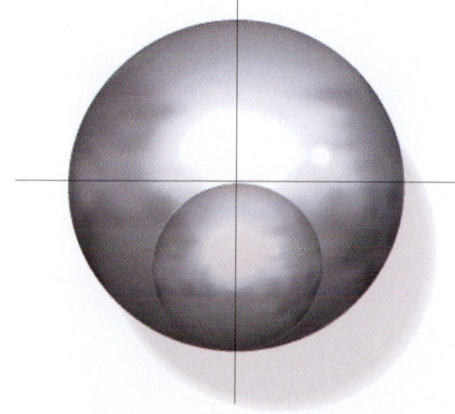

**132** •• $\boxed{\text{SSM}}$ A neutron of mass $m$ makes an elastic head-on collision with a stationary nucleus of mass $M$. (a) Show that the energy of the nucleus after the collision is $K_{\text{nucleus}} = [4mM/(m + M)^2]K_n$, where $K_n$ is the initial energy of the neutron. (b) Show that the fraction of energy lost by the neutron in this collision is

$$\frac{-\Delta K_n}{K_n} = \frac{4mM}{(m + M)^2} = \frac{4(m/M)}{(1 + m/M)^2}$$

**133** •• The mass of a carbon nucleus is approximately 12 times the mass of a neutron. (a) Use the results of Problem 132 to show that after $N$ head-on collisions of a neutron with carbon nuclei at rest, the energy of the neutron is approximately $0.716^N E_0$, where $E_0$ is its original energy. Neutrons emitted in the fission of a uranium nucleus have an energy of about 2 MeV. For such a neutron to cause the fission of another uranium nucleus in a reactor, its energy must be reduced to about 0.02 eV. (b) How many head-on collisions are needed to reduce the energy of a neutron from 2 MeV to 0.02 eV, assuming elastic head-on collisions with stationary carbon nuclei?

**134** •• On average, a neutron loses 63 percent of its energy in an elastic collision with a hydrogen atom and 11 percent of its energy in an elastic collision with a carbon atom. (The numbers are lower than the ones we have been using in earlier problems because most collisions are not head-on.) Calculate the number of collisions, on average, needed to reduce the energy of a neutron from 2 MeV to 0.02 eV (a desirable outcome for reasons explained in Problem 133) if the neutron collides with (a) hydrogen atoms and (b) carbon atoms.

**135** •• A rope of length $L$ and mass $M$ lies coiled on a table. Starting at $t = 0$, one end of the rope is lifted from the table with a force $F$ such that it moves with a constant velocity $v$. (a) Find the height of the center of mass of the rope as a function of time. (b) Differentiate your result in (a) twice to find the acceleration of the center of mass. (c) Assuming that the force exerted by the table equals the weight of the rope still there, find the force $F$ you exert on the top of the rope.

**136** •• Repeat Problem 47 if the cup has a mass $m_c$ and the ball collides with it inelastically.

**137** •• Two astronauts at rest face each other in space. One, with mass $m_1$, throws a ball of mass $m_b$ to the other, whose mass is $m_2$. She catches the ball and throws it back to the first astronaut. Following each throw the ball has a speed of $v$ relative to the thrower. How fast are they moving after each has made one throw and one catch?

**138** •• **SSM** The ratio of the mass of the earth to the mass of the moon is $M_e/m_m = 81.3$. The radius of the earth is about 6370 km, and the distance from the earth to the moon is about 384,000 km. (*a*) Locate the center of mass of the earth–moon system. Is it above or below the surface of the earth? (*b*) What external forces act on the earth–moon system? (*c*) In what direction is the acceleration of the center of mass of this system? (*d*) Assume that the center of mass of this system moves in a circular orbit around the sun. How far must the center of the earth move in the radial direction (toward or away from the sun) during the 14 days between the time the moon is farthest from the sun (full moon) and the time it is closest to the sun (new moon)?

**139** •• **iSOLVE** ✓ You want to enlarge a skating surface so you stand on the ice at one end and aim a hose horizontally to spray water on the schoolyard pavement. Water leaves the hose at 2.4 kg/s with a speed 30 m/s. If your mass is 75 kg, what is your recoil acceleration? (Neglect friction and the mass of the hose.)

**140** ••• **SSM** A stream of glass beads, each with a mass of 0.5 g, comes out of a horizontal tube at a rate of 100 per second (Figure 8-74). The beads fall a distance of 0.5 m to a balance pan and bounce back to their original height. How much mass must be placed in the other pan of the balance to keep the pointer at zero?

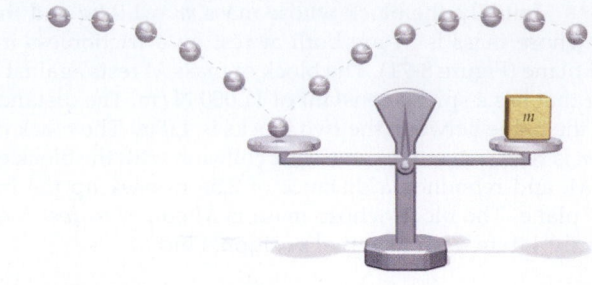

**FIGURE 8-74** Problem 140

**141** ••• A dumbbell consisting of two balls of mass *m* connected by a massless rod of length *L* rests on a frictionless floor against a frictionless wall until it begins to slide down the wall as in Figure 8-75. Find the speed of the bottom ball at the moment when it equals the speed of the top ball.

**FIGURE 8-75**
**Problem 141**

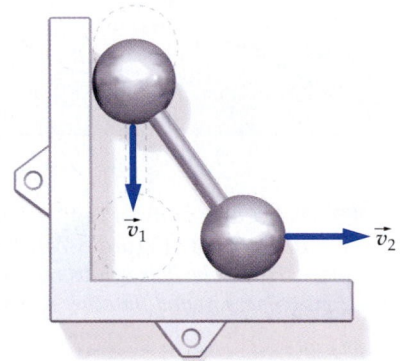

# Rotation

THESE BOWLING BALLS ROLL ON A HORIZONTAL BALL RETURN WITHOUT SLIPPING.

**?** **How might you design a ball return to stop the balls' rotation? (See Example 9-13.)**

9-1   Rotational Kinematics: Angular Velocity and Angular Acceleration

9-2   Rotational Kinetic Energy

9-3   Calculating the Moment of Inertia

9-4   Newton's Second Law for Rotation

9-5   Applications of Newton's Second Law for Rotation

9-6   Rolling Objects

In Chapters 4 and 5 we explored Newton's laws. In Chapters 6 and 7 we examined the conservation of energy, and in Chapter 8 we studied the conservation of momentum. In those chapters we discovered tools (laws and theorems) that are useful in analyzing new situations and in solving new problems. We will use those tools now as we explore rotational motion.

Rotational motion is all around us. The earth rotates about its axis. Wheels, gears, propellers, motors, the drive shaft in a car, a CD in its player, a pirouetting ice skater, all rotate.

➤ **In this chapter, we consider rotation about an axis that is fixed in space, as in a spinning top or a merry-go-round, or an axis that is moving parallel to itself, as in a rolling ball. (More general examples of rotational motion will be discussed in Chapter 10.)**

## 9-1 Rotational Kinematics: Angular Velocity and Angular Acceleration

Every point in a body rotating about a fixed axis moves in a circle whose center is on the axis and whose radius is the distance of that point from the axis of rotation. A line drawn from this axis to any point sweeps out the same angle in the same time. Imagine a disk spinning about a fixed axis perpendicular to the disk and through its center (Figure 9-1). Let $r_i$ be the distance from the center of the disk to the $i$th particle (Figure 9-2), and let $\theta_i$ be the angle measured counterclockwise from a fixed reference line in space to a radial line from the center to the particle. As the disk rotates through an angle $d\theta$, the particle moves through a circular arc of length $ds_i$, such that

$$ds_i = r_i|d\theta| \qquad\qquad 9\text{-}1$$

where $d\theta$ is measured in radians. The distances $ds_i$ and $r_i$ vary from particle to particle, but their ratio, called the **angular displacement** $d\theta$, is the same for all particles of the disk. For one complete revolution, the arc length $\Delta s_i$ is $2\pi r$ and the angular displacement $\Delta\theta$ is

$$\Delta\theta = \frac{2\pi r_i}{r_i} = 2\pi\,\text{rad} = 360° = 1\,\text{rev}$$

The time rate of change of the angle is the same for all particles of the disk, and is called the **angular velocity** $\omega$ of the disk:

$$\omega = \frac{d\theta}{dt} \qquad\qquad 9\text{-}2$$

DEFINITION—ANGULAR VELOCITY

where $\omega$ is a lowercase Greek omega. For counterclockwise rotation, $\theta$ increases, so $\omega$ is positive. For clockwise rotation, $\theta$ decreases, and $\omega$ is negative. The units of $\omega$ are radians per second. (In Chapter 10 we will see that, for rotation in general, the angular velocity is a vector quantity that points along the axis of rotation.) Since radians are dimensionless, the dimension of angular velocity is that of reciprocal time, $[T]^{-1}$. The magnitude of the angular velocity is called the **angular speed.** We often use revolutions per minute (rev/min or RPM) to describe rotation. To convert between revolutions, radians, and degrees, we use

$$1\,\text{rev} = 2\pi\,\text{rad} = 360°$$

**EXERCISE** A compact disk is rotating at 3000 rev/min. What is its angular speed in radians per second? (*Answer* 314 rad/s)

The time rate of change of angular velocity is called the **angular acceleration** $\alpha$:

$$\alpha = \frac{d\omega}{dt} = \frac{d^2\theta}{dt^2} \qquad\qquad 9\text{-}3$$

DEFINITION—ANGULAR ACCELERATION

The units of $\alpha$ are radians per second per second (rad/s$^2$). If $\omega$ is increasing, $\alpha$ is positive; if $\omega$ is decreasing, $\alpha$ is negative.

The angular displacement $\theta$, angular velocity $\omega$, and angular acceleration $\alpha$ are analogous to the linear displacement $x$, linear velocity $v$, and linear

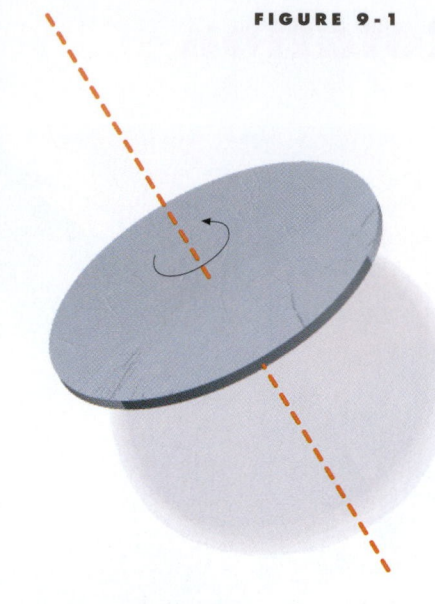

**FIGURE 9-1**

**FIGURE 9-2**

$d\theta$

$ds_i$ $P_i$

$r_i$ $\theta_i$

Reference line

acceleration $a$ in one-dimensional motion. If the angular acceleration $\alpha$ is constant, we can integrate Equation 9-3 to find $\omega$:

$$\omega = \omega_0 + \alpha t \qquad\qquad 9\text{-}4$$

where the constant of integration $\omega_0$ is the initial angular velocity. This is the rotational analog of $v = v_0 + at$. Integrating again, we obtain

$$\theta = \theta_0 + \omega_0 t + \tfrac{1}{2}\alpha t^2 \qquad\qquad 9\text{-}5$$

which is the rotational analog of $x = x_0 + v_0 t + \tfrac{1}{2}at^2$ with $\theta$ replacing $x$, $\omega$ replacing $v$, and $\alpha$ replacing $a$. Similarly, by eliminating $t$ from Equations 9-4 and 9-5, we get

$$\omega^2 = \omega_0^2 + 2\alpha(\theta - \theta_0) \qquad\qquad 9\text{-}6$$

which is the rotational analog of $v^2 = v_0^2 + 2a(x - x_0)$. The equations for constant angular acceleration have the same form as those for constant linear acceleration.

Star tracks in a time exposure of the night sky.

*A CD PLAYER*                                   **EXAMPLE   9 - 1**

**A compact disk rotates from rest to 500 rev/min in 5.5 s. (a) What is its angular acceleration, assuming it is constant? (b) How many revolutions does the disk make in 5.5 s? (c) How far does a point on the rim 6 cm from the center of the disk travel during the 5.5 s it takes to get to 500 rev/min?**

**PICTURE THE PROBLEM**  Part (*a*) is analogous to the linear problem of finding the acceleration given the time and the final velocity. To find $\alpha$ in rad/s² we first convert $\omega$ to rad/s. Part (*b*) is analogous to finding the distance traveled given the time and the final velocity.

(*a*) 1. The angular acceleration is related to the initial and final angular velocities:

$$\omega = \omega_0 + \alpha t = 0 + \alpha t$$

2. Solve for $\alpha$:

$$\alpha = \frac{\omega}{t}$$

$$= \frac{500 \text{ rev/min}}{5.5 \text{ s}} \times \frac{2\pi \text{ rad}}{1 \text{ rev}} \times \frac{1 \text{ min}}{60 \text{ s}}$$

$$= \boxed{9.52 \text{ rad/s}^2}$$

(*b*) 1. The angular displacement is related to the time by Equation 9-5:

$$\theta - \theta_0 = \omega_0 t + \tfrac{1}{2}\alpha t^2$$

$$= 0 + \tfrac{1}{2}(9.52 \text{ rad/s}^2)(5.5 \text{ s})^2$$

$$= 144 \text{ rad}$$

2. Convert radians to revolutions:

$$\theta - \theta_0 = 144 \text{ rad} \times \frac{1 \text{ rev}}{2\pi \text{ rad}} = \boxed{22.9 \text{ rev}}$$

(*c*) The distance traveled $\Delta s$ is $r$ times the angular displacement:

$$\Delta s = r\,\Delta\theta = (6 \text{ cm})(144 \text{ rad}) = \boxed{8.64 \text{ m}}$$

**PLAUSIBILITY CHECK** The average angular velocity in revolutions per minute is 250 rev/min. In 5.5 s, the compact disk rotates (250 rev/60 s)(5.5 s) = 22.9 rev.

**REMARKS** A compact disk is scanned by a laser that begins at the inner radius of about 2.4 cm and moves out to the edge at 6.0 cm. As the laser moves outward, the angular velocity of the disk decreases from 500 rev/min to 200 rev/min so that the linear (tangential) velocity of the disk at the point where the laser beam strikes remains constant.

**EXERCISE** (*a*) Convert 500 rev/min to rad/s. (*b*) Check the result of Part (*b*) in the example using $\omega^2 = \omega_0^2 + 2\alpha(\theta - \theta_0)$. (*Answer* (*a*) 500 rev/min = 52.4 rad/s)

The linear velocity $v_t$ of a particle on the disk is tangent to the circular path of the particle and has magnitude $ds_i/dt$. We can relate this "tangential" velocity to the angular velocity of the disk using Equations 9-1 and 9-2:

$$v_t = \frac{r_i\,d\theta}{dt}$$

so

$$v_t = r_i\omega \qquad\qquad 9\text{-}7$$

Similarly, the tangential acceleration of a particle on the disk is

$$a_t = \frac{dv_t}{dt} = r_i\frac{d\omega}{dt}$$

so

$$a_t = r\alpha \qquad\qquad 9\text{-}8$$

Each particle of the disk also has a centripetal acceleration, which points inward along the radial line and has the magnitude

$$a_c = \frac{v_t^2}{r_i} = \frac{(r_i\omega)^2}{r_i}$$

so

$$a_c = r_i\omega^2 \qquad\qquad 9\text{-}9$$

**EXERCISE** A point on the rim of a compact disk is 6.0 cm from the axis of rotation. Find the tangential speed $v_t$, tangential acceleration $a_t$, and centripetal acceleration $a_c$ of the point when the disk is rotating at a constant angular speed of 300 rev/min. (*Answer* $v_t$ = 188 cm/s, $a_t$ = 0, $a_c$ =5.92 × 10³ cm/s²)

**EXERCISE** Find the linear speed of a point on the CD in Example 9-1 at (*a*) *r* = 2.4 cm when the disk rotates at 500 rev/min, and (*b*) *r* = 6.0 cm when the disk rotates at 200 rev/min. (*Answer* (*a*) 126 cm/s (*b*) 126 cm/s)

# 9-2 Rotational Kinetic Energy

The kinetic energy of a rigid object rotating about a fixed axis is the sum of the kinetic energies of the individual particles that collectively constitute the object. The kinetic energy of the *i*th particle, with mass $m_i$, is

$$K_i = \tfrac{1}{2}m_i v_i^2$$

Summing over all the particles and using $v_i = r_i \omega$ gives

$$K = \sum_i (\tfrac{1}{2}m_i v_i^2) = \tfrac{1}{2}\sum_i (m_i r_i^2 \omega^2) = \tfrac{1}{2}\left(\sum_i m_i r_i^2\right)\omega^2$$

The sum in the term on the right is the object's **moment of inertia** *I* for the axis of rotation.

$$I = \sum_i m_i r_i^2 \qquad\qquad\qquad 9\text{-}10$$

MOMENT OF INERTIA DEFINED

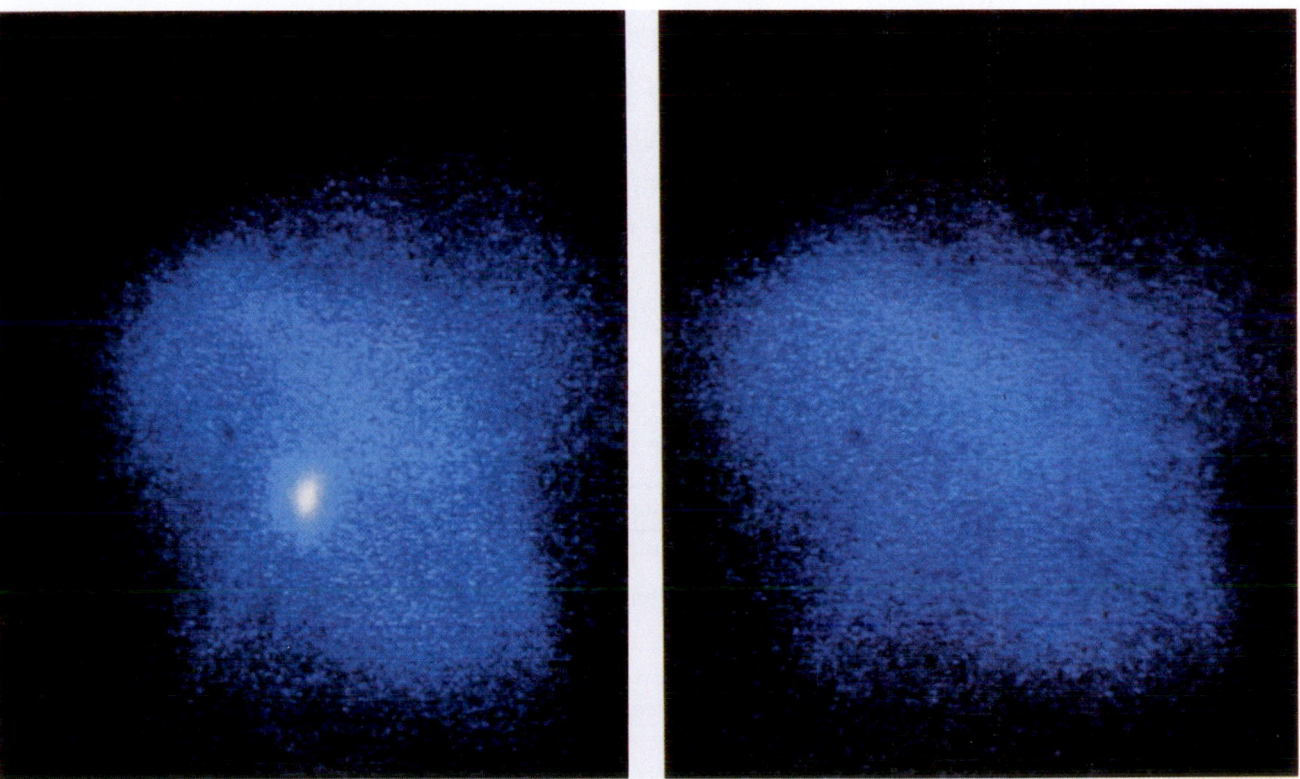

The Crab Pulsar is one of the fastest-rotating neutron stars known, but it is slowing down. It appears to blink on (left) and off (right) like the rotating lamp in a lighthouse, at the fast rate of about 30 times per second, but the period is increasing by about $10^{-5}$ s/y. The loss in rotational energy, which is equivalent to the power output of 100,000 suns, appears as light emitted by electrons accelerated in the magnetic field of the pulsar.

The kinetic energy is thus

$$K = \tfrac{1}{2}I\omega^2 \qquad\qquad 9\text{-}11$$

KINETIC ENERGY OF ROTATING OBJECT

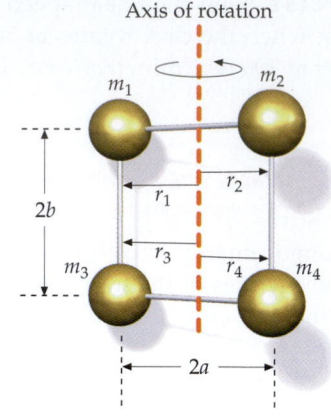

FIGURE 9-3

*A ROTATING SYSTEM OF PARTICLES*

**EXAMPLE 9-2**

An object consists of four point particles, each of mass $m$, that are connected by rigid massless rods to form a rectangle of sides $2a$ and $2b$ as shown in Figure 9-3. The system rotates with angular speed $\omega$ about an axis in the plane of the figure through the center as shown. (*a*) Find the kinetic energy of this object using Equations 9-10 and 9-11. (*b*) Check your result by individually calculating the kinetic energy of each particle and then taking their sum.

**PICTURE THE PROBLEM** Because we are given that the objects are particles, we use Equation 9-10 to calculate $I$ and then use Equation 9-11. In Equation 9-10, $r_i$ is the radial distance from the rotation axis to the particle of mass $m_i$.

(*a*) 1. Apply the definition of moment of inertia (Equation 9-10): 

$$I = \sum_i m_i r_i^2$$

$$= m_1 r_1^2 + m_2 r_2^2 + m_3 r_3^2 + m_4 r_4^2$$

2. The masses $m_i$ and the distances $r_i$ are given:

$$m_1 = m_2 = m_3 = m_4 = m$$

$$r_1 = r_2 = r_3 = r_4 = a$$

3. Substitution gives the moment of inertia:

$$I = ma^2 + ma^2 + ma^2 + ma^2 = 4ma^2$$

4. Using Equation 9-11, solve for the kinetic energy:

$$K = \tfrac{1}{2}I\omega^2 = \tfrac{1}{2}4ma^2\omega^2 = \boxed{2ma^2\omega^2}$$

(*b*) 1. To find the kinetic energy of the $i$th particle we must first find its speed:

$$K_i = \tfrac{1}{2}m_i v_i^2$$

2. The particles are all moving in circles of radius $a$. Find the speed of each particle:

$$v_i = r_i\omega = a\omega \qquad (i = 1, \ldots, 4)$$

3. Substitute into the Part (*b*) step 1 result:

$$K_i = \tfrac{1}{2}m_i v_i^2 = \tfrac{1}{2}ma^2\omega^2$$

4. Each particle has the same kinetic energy. Sum the kinetic energies to get the total:

$$K = \sum_{i=1}^{4} K_i = \tfrac{1}{2}m_1 v_1^2 + \tfrac{1}{2}m_2 v_2^2 + \tfrac{1}{2}m_3 v_3^2 + \tfrac{1}{2}m_4 v_4^2$$

$$= 4(\tfrac{1}{2}ma^2\omega^2) = 2ma^2\omega^2$$

5. Compare with the Part (*a*) result:

The two calculations give the same result.

**REMARKS** Notice that $I$ is independent of the length $b$, which has no effect on how far the masses are from the axis of rotation.

**EXERCISE** Find the moment of inertia of this system for rotation about an axis parallel to the first axis but passing through two of the particles, as shown in Figure 9-4. (*Answer* $I = 8ma^2$)

# 9-3 Calculating the Moment of Inertia

The moment of inertia about an axis is a measure of the inertial resistance of the object to changes in its rotational motion about the axis. It is the rotational analog of mass. The moment of inertia about an axis depends on the distribution of mass

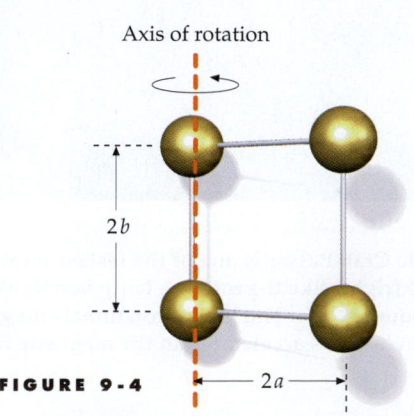

FIGURE 9-4

within the object relative to the axis. The farther an element of mass is from the axis, the greater its contribution to the moment of inertia about that axis. Thus, unlike the mass of an object, which is a property of the object itself, the moment of inertia depends on the location of the axis of rotation.

## Systems of Discrete Particles

For systems consisting of discrete particles, we can compute the moment of inertia about a given axis directly from Equation 9-10.

## Continuous Objects

To calculate the moment of inertia for continuous objects, we imagine the object to consist of a continuum of very small mass elements. Thus, the finite sum $\Sigma m_i r_i^2$ in Equation 9-10 becomes the integral

$$I = \int r^2 \, dm \qquad\qquad 9\text{-}12$$

where $r$ is the radial distance from the axis to mass element $dm$.

---

*MOMENT OF INERTIA OF A UNIFORM ROD*                    **EXAMPLE 9-3**

**Find the moment of inertia of a uniform rod of length $L$ and mass $M$ about an axis perpendicular to the rod and through one end. Assume that the rod has negligible thickness.**

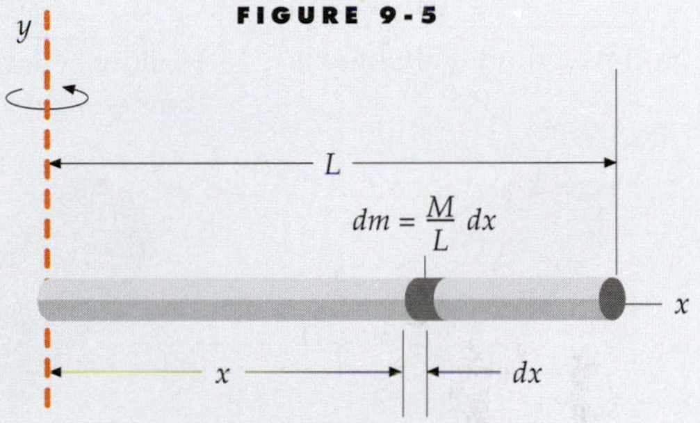

**FIGURE 9-5**

$$dm = \frac{M}{L} \, dx$$

**PICTURE THE PROBLEM** Let the rod lie along the $x$ axis with its end at the origin. To calculate $I$ about the $y$ axis, we choose a mass element $dm$ at a distance $x$ from the axis (Figure 9-5). Since the total mass $M$ is uniformly distributed along the length $L$, the mass per unit length (linear mass density) is $\lambda = M/L$.

1. The moment of inertia is given by the integral:

$$I = \int_0^L x^2 \, dm$$

2. To compute the integral, we first relate $dm$ to $dx$. Express $dm$ in terms of the mass density $\lambda$ and $dx$:

$$dm = \lambda \, dx = \frac{M}{L} \, dx$$

3. Substitute and perform the integration. We choose integration limits so that we sweep $dm$ through the mass distribution in the direction of increasing $x$:

$$I_y = \int x^2 \, dm = \int_0^L x^2 \frac{M}{L} \, dx = \frac{M}{L} \int_0^L x^2 \, dx$$

$$= \frac{M}{L} \frac{1}{3} x^3 \Big|_0^L = \frac{M}{L} \frac{L^3}{3} = \boxed{\frac{1}{3} ML^2}$$

**REMARKS** The moment of inertia about the $z$ axis is also $\frac{1}{3} ML^2$, and that about the $x$ axis is zero (assuming that all of the mass is right on the $x$ axis).

We can calculate $I$ for continuous objects of various shapes, again using Equation 9-12 (see Table 9-1).

### *Hoop About a Perpendicular Axis Through Its Center

Assume that a hoop has mass $M$ and radius $R$ (Figure 9-6). The axis of rotation is the axis of the hoop, which is perpendicular to the plane of the hoop. All the mass is at a distance $r = R$, and the moment of inertia is

$$I = \int r^2 \, dm = \int R^2 \, dm = R^2 \int dm = MR^2$$

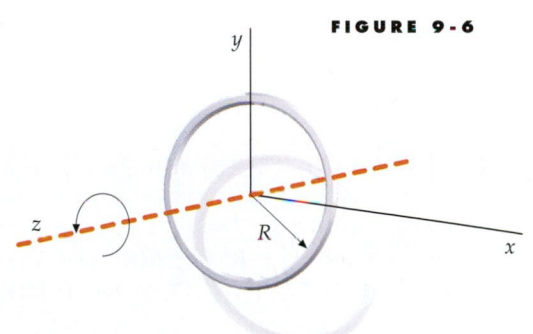

**FIGURE 9-6**

## TABLE 9-1

**Moments of Inertia of Uniform Bodies of Various Shapes**

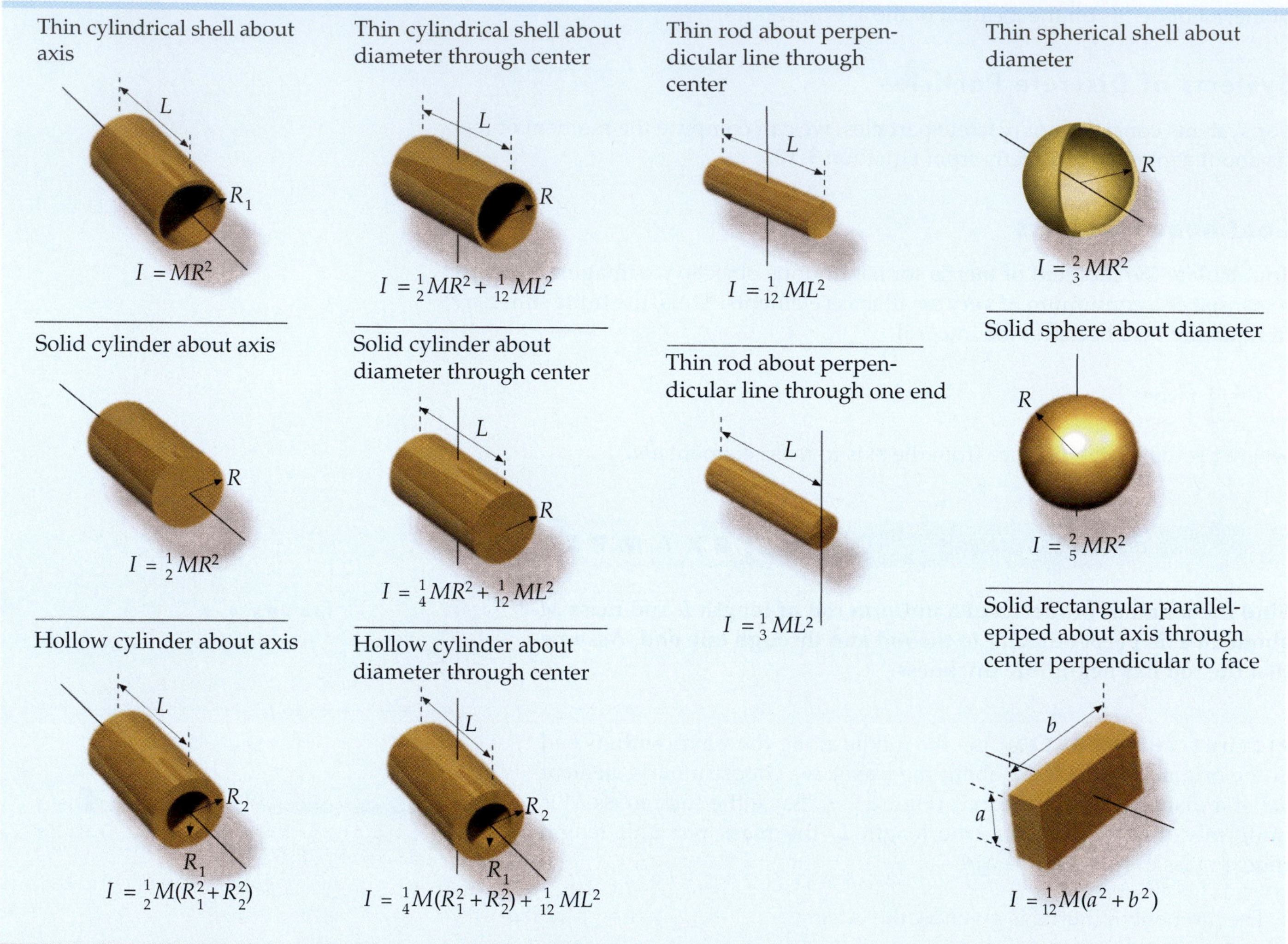

Thin cylindrical shell about axis
L
$R_1$
$I = MR^2$
Thin cylindrical shell about diameter through center
L
R
$I = \frac{1}{2}MR^2 + \frac{1}{12}ML^2$
Thin rod about perpendicular line through center
L
$I = \frac{1}{12}ML^2$
Thin spherical shell about diameter
R
$I = \frac{2}{3}MR^2$
Solid cylinder about axis
R
$I = \frac{1}{2}MR^2$
Solid cylinder about diameter through center
L
R
$I = \frac{1}{4}MR^2 + \frac{1}{12}ML^2$
Thin rod about perpendicular line through one end
L
$I = \frac{1}{3}ML^2$
Solid sphere about diameter
R
$I = \frac{2}{5}MR^2$
Hollow cylinder about axis
L
$R_2$
$R_1$
$I = \frac{1}{2}M(R_1^2 + R_2^2)$
Hollow cylinder about diameter through center
L
$R_2$
$R_1$
$I = \frac{1}{4}M(R_1^2 + R_2^2) + \frac{1}{12}ML^2$
Solid rectangular parallelepiped about axis through center perpendicular to face
b
a
$I = \frac{1}{12}M(a^2 + b^2)$

A disk is a cylinder whose length $L$ is negligible. By setting $L = 0$, the above formulas for cylinders hold for disks.

***Uniform Disk About a Perpendicular Axis Through Its Center** For the case of a uniform disk, we expect that $I$ will be smaller than $MR^2$ since the mass is uniformly distributed from $r = 0$ to $r = R$ rather than being concentrated at $r = R$ as it is in a hoop. In Figure 9-7, each mass element is a hoop of radius $r$ and thickness $dr$. The moment of inertia of any given mass element is $r^2\,dm$. Since the disk is uniform, mass per unit area $\sigma$ is constant. $\sigma = M/A$, where $A = \pi R^2$ is the area of the disk. Since the area of each mass element is $dA = 2\pi r\,dr$, the mass of each element is

$$dm = \sigma\,dA = \frac{M}{A}2\pi r\,dr$$

We thus have

$$I = \int r^2\,dm = \int_0^R r^2\sigma 2\pi r\,dr = 2\pi\sigma\int_0^R r^3\,dr$$

$$= \frac{2\pi M}{A}\frac{R^4}{4} = \frac{\pi M}{2\pi R^2}R^4 = \frac{1}{2}MR^2$$

**FIGURE 9-7**

y
dm
r
z
dr
R
x

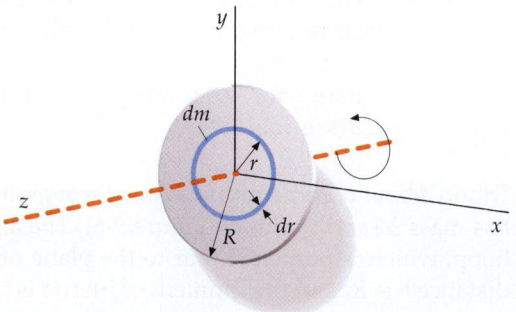

**\*Uniform Solid Cylinder About Its Axis**  We consider a cylinder to be a set of disks, each with mass $dm$ and moment of inertia $\frac{1}{2}dm\,R^2$ (Figure 9-8). The moment of inertia of the complete cylinder is then

$$I = \int \tfrac{1}{2}dm\,R^2 = \tfrac{1}{2}R^2 \int dm = \tfrac{1}{2}MR^2$$

where $M$ is the total mass of the cylinder.

## The Parallel-Axis Theorem

We can often simplify the calculation of moments of inertia for various bodies by using the **parallel-axis theorem,** which relates the moment of inertia about an axis through the center of mass of an object to the moment of inertia about a second, parallel axis (Figure 9-9). Let $I_{cm}$ be the moment of inertia about an axis through the center of mass of an object of total mass $M$, and let $I$ be that about a parallel axis a distance $h$ away. The parallel-axis theorem states that

$$I = I_{cm} + Mh^2 \qquad\qquad 9\text{-}13$$

PARALLEL-AXIS THEOREM

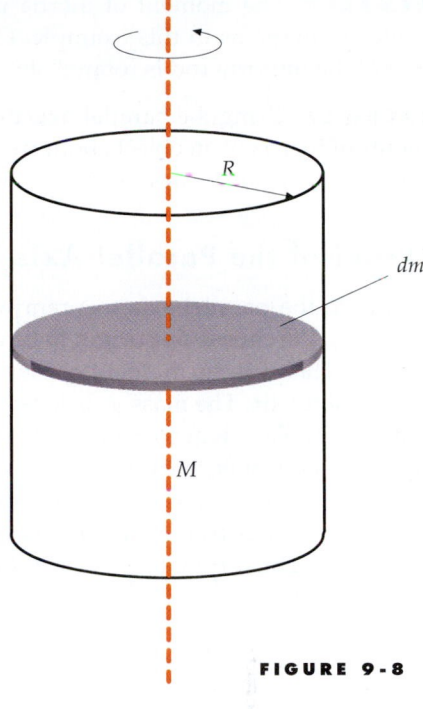

**FIGURE 9-8**

Example 9-2 and the exercise following it illustrate a special case of this theorem with $h = a$, $M = 4m$, and $I_{cm} = 4ma^2$. A proof of the parallel-axis theorem is given at the end of this section.

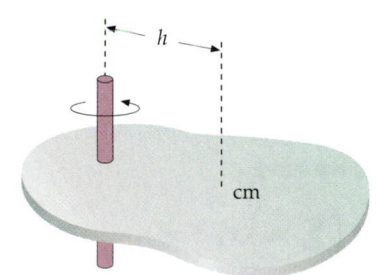

**FIGURE 9-9**  An object rotating about an axis parallel to an axis through the center of mass and a distance $h$ from it.

---

*MOMENT OF INERTIA ABOUT AN AXIS THROUGH THE CENTER OF MASS OF A UNIFORM ROD*  **EXAMPLE 9-4**  **Try It Yourself**

**Find the moment of inertia of a uniform rod about the $y'$ axis through the center of mass (Figure 9-10).**

**PICTURE THE PROBLEM**  Here you know that $I = \frac{1}{3}ML^2$ about one end and want to find $I_{cm}$. Use the parallel-axis theorem with $h = \frac{1}{2}L$.

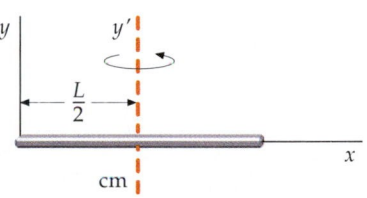

**FIGURE 9-10**

**Cover the column to the right and try these on your own before looking at the answers.**

**Steps**

**Answers**

1. Apply the parallel-axis theorem to write $I$ about the end in terms of $I_{cm}$.

$$I = I_{cm} + Mh^2$$

2. Substitute $I = \frac{1}{3}ML^2$ about the end and solve for $I_{cm}$.

$$I_{cm} = I - Mh^2 = \frac{1}{3}ML^2 - M\left(\frac{L}{2}\right)^2$$

$$= \boxed{\frac{1}{12}ML^2}$$

**REMARKS** The moment of inertia is least when an object is rotated about its center of mass, as in this example. Compare this result to that of Example 9-3, where the uniform rod is rotated about an axis through one end.

**EXERCISE** Using the parallel-axis theorem, show that when comparing the moments of inertia of an object about two parallel axes, the moment of inertia is least about the axis that is nearest to the center of mass.

## *Proof of the Parallel-Axis Theorem

To prove the parallel-axis theorem we start with an object and an axis $A$ (Figure 9-11). We choose our origin to be at the center of mass with the $z$ axis parallel with $A$. Then $I$ is the moment of inertia about $A$ and $I_{cm}$ is the moment of inertia about the $z$ axis. The mass $m_i$ is located at $(x_i, y_i, z_i)$, and the intersection of axis $A$ with the $xy$ plane is at coordinates $x_A, y_A$. The distances $r_i$ and $R_i$ are the perpendicular distances of $m_i$ from $A$ and the $z$ axis, respectively, and $h$ is the perpendicular distance between axes. It follows that $r_i$ is the distance from $(x_A, y_A, 0)$ to $(x_i, y_i, 0)$, $R_i$ is the distance from $(0, 0, 0)$ to $(x_i, y_i, 0)$, and $h$ is the distance from $(0, 0, 0)$ to $(x_A, y_A, 0)$. Thus, $r_i^2 = (x_i - x_A)^2 + (y_i - y_A)^2$, $R_i^2 = x_i^2 + y_i^2$, and $h^2 = x_A^2 + y_A^2$. Consequently,

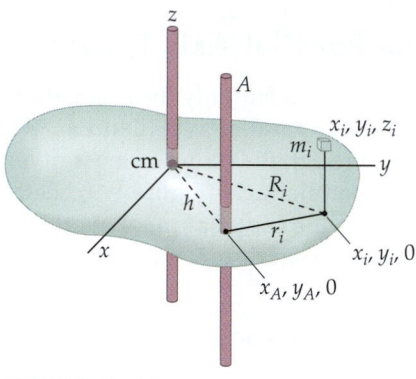

**FIGURE 9-11**

$$I_{cm} = \sum m_i R_i^2 = \sum m_i(x_i^2 + y_i^2)$$

and

$$I = \sum m_i r_i^2 = \sum m_i[(x_i - x_A)^2 + (y_i - y_A)^2]$$
$$= \sum m_i(x_i^2 + y_i^2) - \sum m_i 2x_i x_A - \sum m_i 2y_i y_A + \sum m_i(x_A^2 + y_A^2)$$

By factoring the common factors from these sums we have

$$I = \sum m_i(x_i^2 + y_i^2) - 2x_A \sum m_i x_i - 2y_A \sum m_i y_i + \left(\sum m_i\right)(x_A^2 + y_A^2)$$

The first term is $I_{cm}$. The second and third terms can be simplified using $\sum m_i x_i = Mx_{cm}$ and $\sum m_i y_i = My_{cm}$, and because $x_{cm} = y_{cm} = 0$, both the second and the third terms are equal to zero. The fourth term is $Mh^2$. Thus,

$$I = I_{cm} + Mh^2$$

which is the parallel-axis theorem.

---

*A FLYWHEEL–POWERED CAR*  **EXAMPLE 9-5** **Put It in Context**

You are driving an experimental hybrid vehicle that is designed for use in stop-and-go traffic. In a conventional car, each time you brake to a stop, the kinetic energy is dissipated as heat. In this hybrid vehicle, the braking mechanism transforms the translational kinetic energy of the vehicle's motion to the rotational kinetic energy of a massive flywheel. When the car returns to cruising speed this energy is transferred back into the translational energy of the car. The 100-kg flywheel is a hollow cylinder with an inner diameter $R_1$ of 25 cm, an outer diameter $R_2$ of 40 cm, and a maximum angular speed of 30,000 rev/min. On a dark and dreary night, the car runs out of gas 15 mi from home with the flywheel spinning at maximum speed. Is there sufficient energy stored in the flywheel for you and your nervous grandmother to make it home? (When driving at the minimum highway speed of 40 mi/h, air drag and rolling friction dissipate energy at 10 kW.)

**PICTURE THE PROBLEM** The kinetic energy is calculated directly from $K = \frac{1}{2}I\omega^2$.

1. The kinetic energy of rotation is:

$$K = \frac{1}{2}I\omega^2$$

2. Calculate the moment of inertia of the hollow cylinder:

$$I = \frac{1}{2}m(R_1^2 + R_2^2) = 11.1 \text{ kg·m}^2$$

3. Convert $\omega$ to rad/s:

$$\omega = 30{,}000 \text{ rev/min} = 3140 \text{ rad/s}$$

4. Substitute these values to find the kinetic energy:

$$K = \frac{1}{2}I\omega^2 = 54.8 \text{ MJ}$$

5. Energy is dissipated at 10 kW at a speed of 40 mi/h. To find the energy dissipated during the 15-mi trip, we first need to find the time required for the trip:

$$\Delta x = v\,\Delta t, \quad \text{so} \quad \Delta t = 1350 \text{ s}$$

6. The energy is dissipated at 10 kW for 1350 s. The total energy dissipated is:

13.5 MJ

7. Is there enough energy stored in the flywheel?

54.8 MJ are available and 13.5 MJ are dissipated.

$$\boxed{\text{Yes, there is more than enough energy stored.}}$$

**REMARKS** There are 130 MJ of energy in a gallon of gasoline. If the engine is 10% efficient, only 13 MJ/gal are available to move the car. We estimate the mileage for the car in this example to be about 15 mi/gal.

---

*THE PIVOTED ROD*                                             **EXAMPLE 9-6**

**A uniform thin rod of length *L* and mass *M*, pivoted at one end as shown in Figure 9-12, is held horizontal and then released from rest. Assuming the pivot to be frictionless, find (*a*) the angular velocity of the rod when it reaches its vertical position, and (*b*) the force exerted by the pivot at this time. (*c*) What initial angular velocity is needed for the rod to reach a vertical position at the top of its swing?**

**PICTURE THE PROBLEM** (*a*) As the rod swings down, its potential energy decreases and its kinetic energy increases. Since the pivot is frictionless, we use conservation of mechanical energy. The angular velocity of the rod is then found from its rotational kinetic energy. (*b*) To find the force of the pivot use Newton's second law for a system. (*c*) As in Part (*a*), use conservation of mechanical energy.

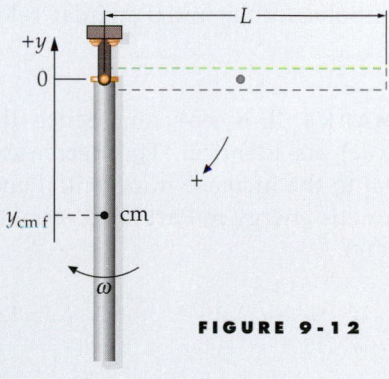

**FIGURE 9-12**

(*a*) 1. Make a diagram of the rod showing both its initial and its final configuration (Figure 9-12). Put a vertical coordinate axis on the figure with up as the positive direction and with its origin at the rotation axis.

2. Apply conservation of mechanical energy to relate the initial and final mechanical energies:

$$K_f + U_f = K_i + U_i$$

$$\frac{1}{2}I\omega_f^2 + Mgy_{cm\,f} = \frac{1}{2}I\omega_i^2 + Mgy_{cm\,i}$$

$$\frac{1}{2}I\omega_f^2 + Mg\left(-\frac{L}{2}\right) = 0 + 0$$

3. Solve for $\omega_f$:

$$\omega_f = \sqrt{\frac{MgL}{I}}$$

4. Obtain *I* from Table 9-1 and substitute into the step 3 result:

$$\omega_f = \sqrt{\frac{MgL}{\frac{1}{3}ML^2}} = \boxed{\sqrt{\frac{3g}{L}}}$$

(b) 1. Make a free-body diagram of the rod as it passes through the vertical position at the bottom of its swing (Figure 9-13).

2. Apply Newton's second law for a system to the rod. At the bottom of the swing the acceleration of the center of mass is in the centripetal (upward) direction:

$$\Sigma F_{ext\,y} = Ma_{cm}$$
$$F_p - Mg = Ma_{cm}$$

3. Relate the acceleration of the center of mass to the angular speed using $a_c = r\omega^2$. Substitute the Part (a) step 4 result for $\omega$ and solve for $a_{cm}$:

$$a_c = r\omega^2$$
$$a_{cm} = \frac{L}{2}\frac{3g}{L} = \frac{3}{2}g$$

4. Substitute into the Part (b) step 2 result and calculate $F_p$:

$$F_p = Mg + Ma_{cm} = Mg + M\tfrac{3}{2}g$$
$$= \boxed{\tfrac{5}{2}Mg}$$

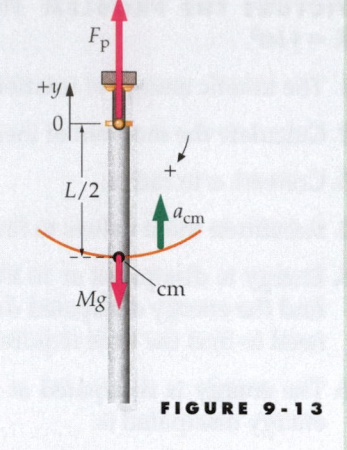

**FIGURE 9-13**

(c) 1. The initial angular velocity $\omega_i$ is related to the initial kinetic energy:

$$K_i = \tfrac{1}{2}I\omega_i^2$$

2. Make a diagram of the rod showing both its initial and its final configuration (Figure 9-14). Put a vertical coordinate axis on the figure with up as the positive direction and with its origin at the rotation axis.

3. Apply conservation of mechanical energy with $K_f = 0$ and $U_i = 0$ to relate the initial kinetic energy to the final position:

$$K_f + U_f = K_i + U_i$$
$$\tfrac{1}{2}I\omega_f^2 + Mgy_{cm\,f} = \tfrac{1}{2}I\omega_i^2 + Mgy_{cm\,i}$$
$$0 + Mg\frac{L}{2} = \frac{1}{2}I\omega_i^2 + 0$$

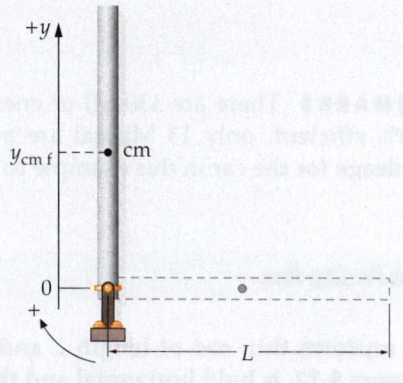

**FIGURE 9-14**

4. Solve for the initial angular velocity:

$$\omega_i = \sqrt{\frac{MgL}{I}} = \sqrt{\frac{MgL}{\tfrac{1}{3}ML^2}} = \boxed{\sqrt{\frac{3g}{L}}}$$

**REMARKS** It is no coincidence that the answers to Part (a) and Part (c) are identical. The decrease in potential energy in Part (a) is equal to the increase in potential energy in Part (c). Thus the increase in kinetic energy in Part (a) is equal to the decrease in kinetic energy in Part (c).

*Try It Yourself*

A WINCH AND A BUCKET                    **EXAMPLE 9-7**

A winch is at the top of a deep well. The drum of the winch has mass $m_w$ and radius $R$. Virtually all its mass is concentrated a distance $R$ from the axis. A cable wound around the drum suspends a bucket of water of mass $m_b$. The entire cable has mass $m_c$ and length $L$. Just when you have the bucket at the highest point your hand slips and the bucket falls back down the well, unwinding the winch cable as it does. How fast is the bucket moving after it has fallen a distance $d$, where $d$ is less than $L$?

**PICTURE THE PROBLEM** As the load falls, mechanical energy is conserved. Choose the initial potential energy to be zero. When the load has fallen a distance $d$, the center of mass of the hanging cable has dropped a distance $d/2$. Since the hanging part of the cable moves with speed $v$ and the cable does not stretch or become slack, the entire cable must move at speed $v$. We find $v$ from the conservation of mechanical energy.

**Cover the column to the right and try these on your own before looking at the answers.**

| Steps | Answers |
|---|---|

1. Make a diagram of the winch–cable–bucket system in both its initial and its final configuration, as shown in Figure 9-15. Include a $y$ axis with the origin at the height of the center of the winch.

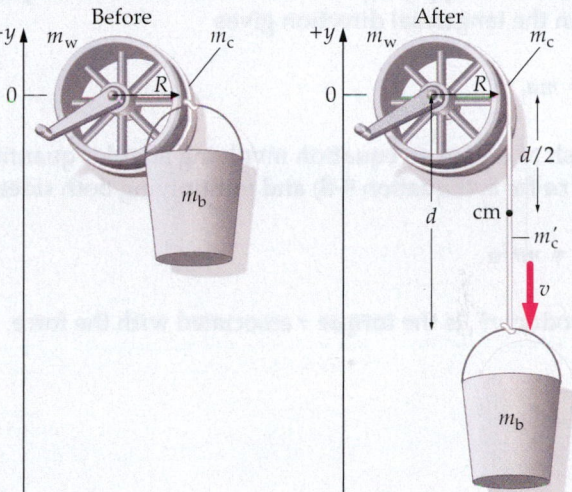

**FIGURE 9-15**

2. Apply conservation of mechanical energy. Choose the potential energy to be zero when the bucket of water is at the highest point.

$$U_f + K_f = U_i + K_i$$
$$= 0 + 0 = 0$$

3. Write an expression for the total potential energy when the bucket has fallen a distance $d$. Let $m_c'$ denote the mass of the hanging part of the cable.

$$U_f = U_{bf} + U_{cf} + U_{wf} = m_b g(-d) + m_c' g\left(-\frac{d}{2}\right) + 0$$
$$= -(m_b + \tfrac{1}{2} m_c') gd$$

4. Express the total kinetic energy when the bucket is falling with speed $v$. All the cable and the entire mass of the drum move with the same speed $v$ as the bucket.

$$K_f = K_{fc} + K_{fb} + K_{fw} = \tfrac{1}{2} m_c v^2 + \tfrac{1}{2} m_b v^2 + \tfrac{1}{2} m_w v^2$$
$$= \tfrac{1}{2}(m_c + m_b + m_w) v^2$$

5. Substitute into the conservation of mechanical energy equation (step 2) and solve for $v$.

$$-(m_b + \tfrac{1}{2} m_c') gd + \tfrac{1}{2}(m_c + m_b + m_w) v^2 = 0$$

so

$$v = \sqrt{\frac{(2m_b + m_c') gd}{m_c + m_b + m_w}}$$

6. Assume the cable is uniform and express $m_c'$ in terms of $m_c$, $d$, and $L$.

$$\frac{m_c'}{d} = \frac{m_c}{L} \quad \Rightarrow \quad m_c' = \frac{d}{L} m_c$$

7. Substitute the step 6 result into the step 5 result.

$$v = \sqrt{\frac{(2m_b L + m_c d) gd}{(m_c + m_b + m_w)L}}$$

**REMARKS** Because the entire mass of the drum is moving at the same speed $v$ we can express its kinetic energy as $\tfrac{1}{2} m_w v^2$. However, we can express it as $\tfrac{1}{2} I_w \omega^2$, where $I_w = m_w R^2$ and $\omega = v/R$. With these substitutions $K_w = \tfrac{1}{2} I_w \omega^2 = \tfrac{1}{2} m_w R^2 (v^2/R^2) = \tfrac{1}{2} m_w v^2$.

## 9-4 Newton's Second Law for Rotation

To set a top spinning, you twist it. In Figure 9-16, a disk is set spinning by the forces $\vec{F}_1$ and $\vec{F}_2$ exerted at the edges of the disk in the tangential direction. The directions of these forces are important. If the same forces are applied in the radial direction (Figure 9-17), the disk will not start to spin.

Figure 9-18 shows a particle of mass $m$ attached to one end of a massless rigid rod of length $r$. There is an axis perpendicular to the rod and passing through its other end, and the rod is free to rotate about this axis. Consequently, the particle is constrained to move in a circle of radius $r$. A single force $\vec{F}$ is applied to the particle as shown. Applying Newton's second law to the particle and taking components in the tangential direction gives

$$F_t = ma_t$$

We wish to obtain an equation involving angular quantities. Substituting $r\alpha$ for $a_t$ (Equation 9-8) and multiplying both sides by $r$ gives

$$rF_t = mr^2\alpha \qquad 9\text{-}14$$

The product $rF_t$ is the **torque** $\tau$ associated with the force. That is,

$$\tau = F_t r \qquad 9\text{-}15$$

TORQUE

Substituting into Equation 9-14 gives

$$\tau = mr^2\alpha \qquad 9\text{-}16$$

A rigid object that rotates about a fixed axis is just a collection of individual particles, each of which is constrained to move in a circular path with the same angular velocity $\omega$ and acceleration $\alpha$. Applying Equation 9-16 to the $i$th of these particles gives

$$\tau_{inet} = m_i r_i^2 \alpha$$

where $\tau_{inet}$ is the torque due to the net force on the $i$th particle. Summing both sides over all particles gives

$$\sum \tau_{inet} = \sum m_i r_i^2 \alpha = \left(\sum m_i r_i^2\right)\alpha = I\alpha \qquad 9\text{-}17$$

In Chapter 8 we saw that the net force acting on a system of particles is equal to the net *external* force acting on the system because the internal forces (those exerted by the particles within the system on one another) cancel in pairs. The treatment of internal torques exerted by the particles within a system on one another leads to a similar result, that is, the net torque acting on a system equals the net *external* torque acting on the system. (We discuss this further in Chapter 10.) We can thus write Equation 9-17 as

$$\tau_{net\ ext} = \sum \tau_{ext} = I\alpha \qquad 9\text{-}18$$

NEWTON'S SECOND LAW FOR ROTATION

This is the rotational analog of Newton's second law for linear motion, $\Sigma \vec{F} = m\vec{a}$.

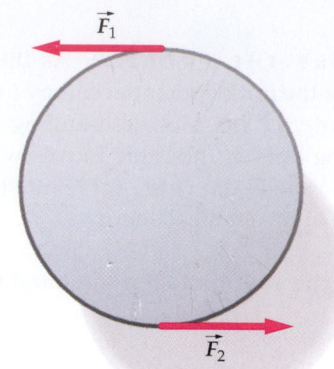

FIGURE 9-16

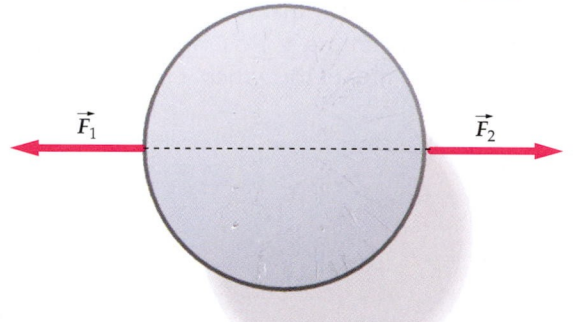

FIGURE 9-17

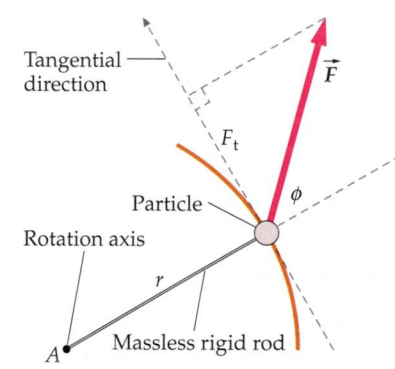

FIGURE 9-18

## Calculating Torques

Figure 9-19 shows a force $\vec{F}$ acting on an object constrained to rotate about a fixed axis $A$, not shown, which passes through $O$ and is perpendicular to the page. The positive tangential direction is shown at the point of application of the force, and $r$ is the radial distance of this point of application from $A$. The torque $\tau$ due to this force about axis $A$ is $\tau = F_t r$ (Equation 9-15). In principle, the expression $F_t r$ is all that is needed to calculate torques. However, in practice, calculations are often simpler if alternative expressions for torque are used. From the figure we can see that

$$F_t = F \sin \phi$$

where $\phi$ is the angle between the radial direction and the direction of the force. Thus, we can express the torque as $\tau = F_t r = (F \sin \phi)r$. The *line of action* of a force is the line parallel to the force and passing through the point of application of the force. From Figure 9-20 we can see that $r \sin \phi = \ell$, where the **lever arm** $\ell$ is the perpendicular distance between $A$ and the line of action. Consequently, the torque is also given by $\tau = F\ell$. Putting all three equivalent expressions for the torque in one place, we have

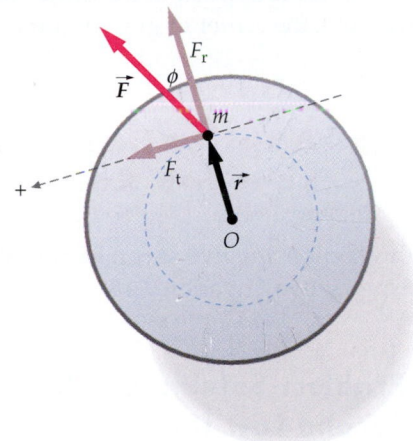

**FIGURE 9-19** The force $\vec{F}$ produces a torque $F_t r$ about the center.

$$\tau = F_t r = Fr \sin \phi = F\ell \qquad \text{9-19}$$

EQUIVALENT EXPRESSIONS FOR TORQUE

## Torque Due to Gravity

We can model an extended object as an assembly of microscopic point particles, and there is a microscopic gravitational force on each particle. Each of these microscopic gravitational forces exerts a microscopic torque about a given axis, and the net gravitational torque on the object is the sum of these microscopic torques. The net gravitational torque can be calculated by considering the entire weight (the sum of the microscopic gravitational forces) to act at a single point—**the center of gravity.** Consider an object (Figure 9-21) constrained to rotate about a horizontal axis $A$ coming out of the page. We choose the $z$ axis of our coordinate system to coincide with axis $A$, and choose the $x$ axis direction to be horizontal and the $y$ axis vertical as shown. The torque on a particle of mass $m_i$ due to gravity is $m_i g x_i$, where $x_i$ is the lever arm of the force $m_i \vec{g}$. The net gravitational torque on the object is the sum of the gravitational torques on the particles that make it up. That is, $\tau_{\text{grav}} = \Sigma m_i g x_i$. If $\vec{g}$ is the same throughout the object, then $g$ can be factored out of the sum. Factoring $g$ out of the sum gives $\tau_{\text{grav}} = (\Sigma m_i x_i)g$. You should recognize the sum in the parentheses as $M x_{\text{cm}}$ (see Equation 8-1). Substituting this for the sum gives

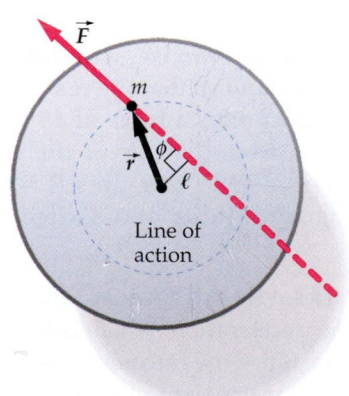

**FIGURE 9-20** The force $\vec{F}$ produces a torque $F\ell$ about the center.

$$\tau_{\text{grav}} = M g x_{\text{cm}} \qquad \text{9-20}$$

TORQUE DUE TO GRAVITY

**FIGURE 9-21**

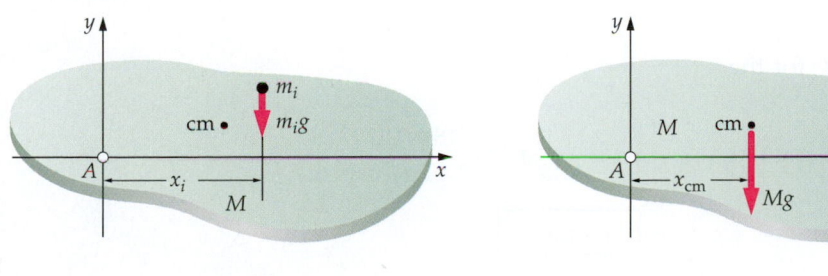

(a)                                                    (b)

The torque due to a uniform gravitational field is calculated as if the entire gravitational force is applied at the center of mass. For an object in a uniform gravitational field, the center of gravity coincides with the center of mass.

# 9-5 Applications of Newton's Second Law for Rotation

In this section we give several applications of Newton's second law for rotation as expressed in Equation 9-18.

## Problem-Solving Guidelines for Applying Newton's Second Law for Rotation

**Free-Body Diagram** Draw the free-body diagram with the object shown as a picture, not just a dot. Draw each force vector along the line of action of the force. On the diagram indicate the positive rotational direction (clockwise or counterclockwise).

---

*A STATIONARY BIKE*                                    **EXAMPLE 9-8**

To get some exercise without going anywhere, you set your bike on a stand so that the rear wheel is free to turn. As you pedal, the chain applies a force of 18 N to the sprocket at a distance of $r_s$ = 7 cm from the axle of the wheel. Consider the wheel to be a hoop ($I = MR^2$) of radius $R$ = 35 cm and mass 2.4 kg. What is the angular velocity of the wheel after 5 s?

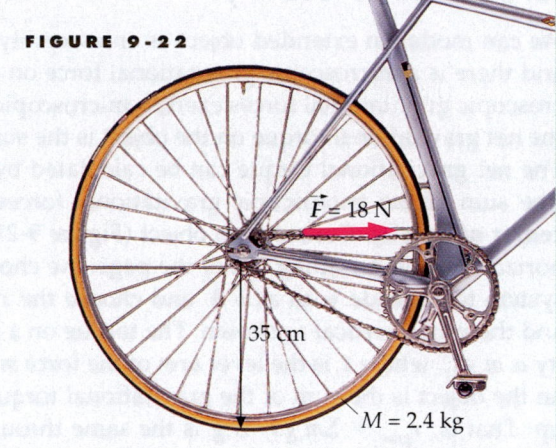

**FIGURE 9-22**

$\vec{F} = 18$ N

35 cm

$M = 2.4$ kg

**PICTURE THE PROBLEM** The angular velocity is found from the angular acceleration, which is found from Newton's second law for rotation. Since the forces are constant, the torques are also constant and the constant angular acceleration equations apply. Note that $\vec{F}$ acts in the direction of the chain, so the line of force is tangent to the sprocket and the lever arm is the radius $r_s$ of the sprocket (see Figure 9-22).

1. The angular velocity is related to the angular acceleration and the time:

$$\omega = \omega_0 + \alpha t = 0 + \alpha t$$

2. Apply Newton's second law for rotational motion to relate $\alpha$ to the net torque and the moment of inertia:

$$\Sigma \tau_{ext} = I\alpha$$

3. The only torque acting on the system is the applied force $F$ with lever arm $r_s$:

$$\tau_{ext} = Fr_s$$

4. Substitute this value for the torque and $I = MR^2$ for the moment of inertia:

$$\alpha = \frac{\Sigma \tau_{ext}}{I} = \frac{Fr_s}{MR^2}$$

5. Substitute into the step 1 result and solve for the angular velocity after 5 s:

$$\omega = \alpha t = \frac{Fr_s}{MR^2}t = \frac{(18 \text{ N})(0.07 \text{ m})}{(2.4 \text{ kg})(0.35 \text{ m})^2} 5 \text{ s}$$

$$= \boxed{21.4 \text{ rad/s}}$$

A uniform thin rod of length $L$ and mass $M$ is pivoted at one end. It is held horizontal and released. The pivot is frictionless. Find (a) the angular acceleration of the rod immediately following its release and (b) the force $F_A$ exerted on the rod by the pivot at this instant.

**PICTURE   THE   PROBLEM**  The angular acceleration is found from Newton's second law for rotation (Equation 9-18). The force $F_A$ is found from Newton's second law for a system (Equation 8-10). The tangential acceleration of the center of mass is related to the angular acceleration (Equation 9-5) and the centripetal acceleration of the center of mass is related to the angular speed (Equation 9-6).

**FIGURE 9-23**

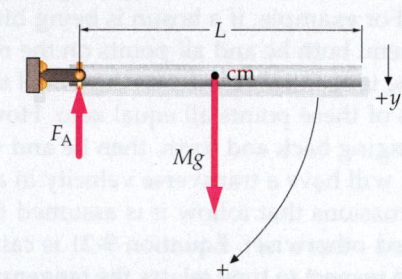

(a) 1. Sketch a free-body diagram of the rod (Figure 9-23).

2. Write Newton's second law for rotation:

$$\sum \tau_{ext} = I\alpha$$

3. Compute the torque due to gravity about the given axis. The rod is uniform so its center of mass is at its center, a distance $L/2$ from the axis:

$$\tau_{grav} = Mg\frac{L}{2}$$

4. Find the moment of inertia about the end of the rod from Table 9.1:

$$I = \tfrac{1}{3}ML^2$$

5. Substitute these values into the step 2 equation to compute $\alpha$:

$$\alpha = \frac{\tau_{grav}}{I} = \frac{Mg(L/2)}{(1/3)ML^2} = \boxed{\frac{3g}{2L}}$$

(b) 1. Write Newton's second law for a system for the rod:

$$\sum F_{exty} = Ma_{cmy}$$

$$Mg - F_A = Ma_{cmy}$$

2. Use the relation $a_c = r\omega^2$ to find $a_{cm\,c}$. Just following release, $\omega = 0$:

$$a_{cm\,c} = r_{cm}\omega^2 = \frac{L}{2}\omega^2 = 0$$

3. We now have two equations and three unknowns, $\alpha$, $a_{cmy}$, and $F_A$. Use the relation $a_t = r\alpha$ to obtain an equation relating $a_{cmy}$ to $\alpha$:

$$a_t = r\alpha$$

$$a_{cmy} = r_{cm}\alpha = \frac{L}{2}\alpha$$

4. Substitute both the Part (a) step 5 result and the Part (b) step 1 result into the Part (b) step 3 result and solve for $F_A$:

$$\frac{Mg - F_A}{M} = \frac{L}{2}\frac{3g}{2L}$$

so

$$F_A = \boxed{\tfrac{1}{4}Mg}$$

**REMARKS**  Just after the rod is released, the acceleration of the center of mass is directed straight down. Since the net external force and the acceleration must be in the same direction, it follows that $\vec{F}_A$ can not have a horizontal component at this moment.

**EXERCISE**  A small pebble of mass $m \ll M$ is placed on top of the rod at its center. Find (a) the acceleration of the pebble and (b) the force it exerts on the rod just after the rod is released. (*Answer*  (a) $a = 3g/4$ downward (b) $f = mg/4$ downward)

## Nonslip Conditions

There are many situations in which a string is wrapped around a rotating wheel or cylinder. The string must move with a tangential velocity $v_t$ that is equal to the tangential velocity of the rim of the wheel, provided the string remains taut and does not slip:

$$v_t = R\omega \qquad\qquad 9\text{-}21$$

NONSLIP CONDITION FOR $v$ AND $\omega$

For a point on a string, the velocity component tangent to the string is its *tangential* velocity, and the velocity perpendicular to the tangent is its *transverse* velocity. For example, if a bosun is being lifted straight upward by a rope and pulley system, both he and all points on the rope between him and the pulley have the same tangential velocity as the rim of the pulley wheel, but the transverse velocities of these points all equal zero. However, if, while being raised, the bosun is swinging back and forth, then he and the points along the swinging rope above him will have a transverse velocity in addition to their tangential velocity. In the discussions that follow it is assumed that there is no transverse motion (unless stated otherwise). Equation 9-21 is called a nonslip condition. Differentiating it with respect to time relates the tangential acceleration of the string to the angular acceleration of the rim.

$$a_t = R\alpha \qquad\qquad 9\text{-}22$$

$a$ AND $\alpha$ UNDER NONSLIP CONDITIONS

## Problem-Solving Guidelines for Applying Newton's Second Law for Rotation

**Tension**   Because of friction and inertia, if a string goes around a pulley wheel, the tension in it is greater on one side of the wheel than on the other. (This is necessary for the string to exert a torque on the wheel.) Use two different labels, $T_1$ and $T_2$, for these two tensions.

*TENSION IN A STRING*                    **EXAMPLE 9-10**

An object of mass $m$ is tied to a light string wound around a pulley wheel that has a moment of inertia $I$ and radius $R$. The wheel bearing is frictionless and the string does not slip on the rim. Find the tension in the string and the acceleration of the object.

**FIGURE 9-24**

**PICTURE THE PROBLEM**   In this system, the object descends with a downward acceleration $a$, while the wheel turns with an angular acceleration $\alpha$ (Figure 9-24). We apply Newton's second law for rotation to the wheel to determine $\alpha$, and Newton's second law to the object to obtain $a$. Relate $a_t$ and $\alpha$, using the nonslip condition.

1. Draw a free-body diagram of the pulley wheel, drawing each force vector with its tail at the point of application of the force. Put labels on the diagram and indicate the positive rotational direction as shown in Figure 9-25:

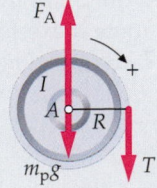

**FIGURE 9-25**

2. The only force that exerts a torque on the wheel is the tension $T$, which has lever arm $R$. Apply Newton's second law for rotational motion to relate $T$ and the angular acceleration $\alpha$:

$$\Sigma \tau_{\text{ext}} = I\alpha$$

$$TR = I\alpha$$

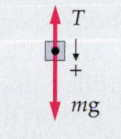

3. Draw a free-body diagram for the suspended object, and apply Newton's second law to relate $T$ to the tangential acceleration $a_t$ (Figure 9-26):

$$\Sigma F_{\text{exty}} = ma_y$$

$$mg - T = ma_t$$

**FIGURE 9-26**

4. We have two equations for three unknowns, $T$, $a_t$, and $\alpha$. A third equation is the relation between $a_t$ and $\alpha$ for the nonslip condition:

$$a_t = R\alpha$$

5. We now have three equations enabling us to determine $T$, $a_t$, and $\alpha$. To solve for $T$ use the step 2 result to obtain an expression for $\alpha$ and use the step 3 result to obtain an expression for $a_t$. Substitute these expressions into the step 4 result and solve for $T$.

$$\frac{mg - T}{m} = R\frac{TR}{I}$$

so

$$\boxed{T = \frac{mg}{1 + (mR^2/I)}}$$

6. Substitute this result for $T$ into the step 3 result and solve for $a_t$:

$$mg - \frac{mg}{1 + (mR^2/I)} = ma_t$$

so

$$\boxed{a_t = \frac{1}{1 + \dfrac{I}{mR^2}}\, g}$$

**❶ PLAUSIBILITY CHECK**  Let's check a couple of extreme limits. If $I = 0$, the object should fall freely, and the string should be slack; our results give $T = 0$, $a_t = g$. What happens if $I$ approaches $\infty$? For $I \gg mR^2$, our equations give $T \approx mg$ and $a_t \approx 0$.

---

*A PULLEY*                **EXAMPLE 9-11    Try It Yourself**

Two blocks are connected by a string that **passes over a pulley of radius $R$ and moment of inertia $I$. The block of mass $m_1$ slides on a frictionless, horizontal surface; the block of mass $m_2$ is suspended from the string** (Figure 9-27). **Find the acceleration $a$ of the blocks and the tensions $T_1$ and $T_2$ assuming the string does not slip on the pulley.**

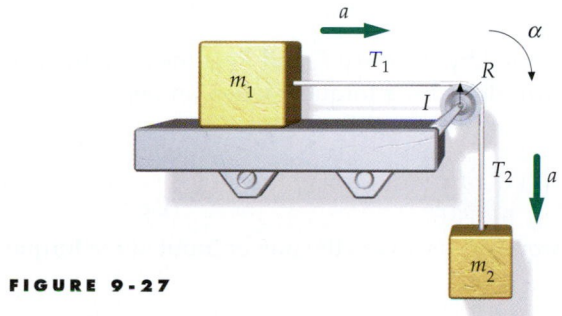

**FIGURE 9-27**

**PICTURE THE PROBLEM**  In this problem, the tensions $T_1$ and $T_2$ are not equal because the pulley has mass and because there is static friction between the string and the pulley (Figure 9-27). (Otherwise the pulley would not turn.) Note that $T_2$ exerts a clockwise torque and $T_1$ exerts a counterclockwise torque on the pulley. Use Newton's second law for each block and Newton's second law for rotational motion for the pulley. Relate $\alpha$ and $a$ using the nonslip condition.

**Cover the column to the right and try these on your own before looking at the answers.**

| Steps | Answers |
|---|---|

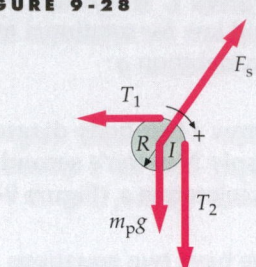

FIGURE 9-28

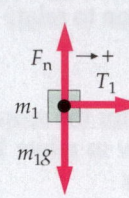

1. Draw a free-body diagram for each block and for the pulley, as shown in Figure 9-28. Note that the center of mass of the pulley does not accelerate, so the support must exert a force on the axle $F_s$ that balances the resultant of the gravitational force and the forces exerted by the string.

2. Apply Newton's second law to each block.

$T_1 = m_1 a; \quad m_2 g - T_2 = m_2 a$

3. Apply Newton's second law for rotation to the pulley wheel.

$T_2 R - T_1 R = I\alpha$

4. We have three equations and four unknowns. To get a fourth equation, use the nonslip condition to relate $a$ and $\alpha$.

$a = R\alpha$

5. Now we have four equations and four unknowns, so the rest is algebra. Do the algebra and obtain expressions for $a$, $T_1$, and $T_2$. (*Hint: To find $a$, obtain expressions for $T_1$ and $T_2$ from the step 2 results. Substitute these into the step 3 result to obtain an equation with unknowns $a$ and $\alpha$. Use the step 4 result to eliminate $\alpha$ and solve for $a$.*)

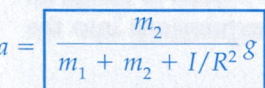

$$a = \frac{m_2}{m_1 + m_2 + I/R^2} g$$

$$T_1 = m_1 a = \frac{m_2}{m_1 + m_2 + I/R^2} m_1 g$$

$$T_2 = m_2(g - a) = \frac{(m_1 + I/R^2)}{m_1 + m_2 + I/R^2} m_2 g$$

**PLAUSIBILITY CHECK** If $I = 0$, $T_1 = T_2$, and the acceleration is $a = m_2 g/(m_1 + m_2)$, as expected. If $I$ is very large ($I/R^2 >> m_1 + m_2$), then $T_1 \approx 0$, $T_2 \approx m_2 g$, and $a \approx 0$.

## Power

When you spin an object you do work on it, increasing its kinetic energy. Consider a force $F_i$ acting on a rotating object. As the object rotates through an angle $d\theta$, the point of application of the force moves a distance $ds_i = r_i\, d\theta$, and the force does work

$$dW_i = F_{it}\, ds_i = F_{it} r_i\, d\theta = \tau_i\, d\theta$$

where $\tau_i$ is the torque exerted by the force $F_i$ and $F_{it}$ is the tangential component of $F_i$. In general, the work done by a torque $\tau$ when an object turns through a small angle $d\theta$ is

$$dW = \tau\, d\theta \qquad\qquad 9\text{-}23$$

The rate at which the torque does work is the power input of the torque:

$$P = \frac{dW}{dt} = \tau \frac{d\theta}{dt}$$

or

$$P = \tau\omega \qquad\qquad 9\text{-}24$$

POWER

Equations 9-23 and 9-24 are the rotational analogs of $dW = F_s\, ds$ and $P = F_s v_s$.

---

*TORQUE EXERTED BY AN AUTOMOBILE ENGINE*                    **EXAMPLE  9-12**    **Try It Yourself**

The maximum torque produced by the 8.0-L V10 engine of a 2002 Dodge Viper is 675 N·m of torque at 3700 rev/min. Find the power output of the engine operating at these maximum torque conditions.

**PICTURE THE PROBLEM** The power equals the product of the torque and angular velocity, which are given. You must express $\omega$ in radians per second to obtain the power in watts.

**Cover the column on the right and try these on your own before looking at the answers.**

| Steps | Answers |
|---|---|
| 1. Write the power in terms of $\tau$ and $\omega$. | $P = \tau\omega$ |
| 2. Convert rev/min to rad/s. | $\omega = 387$ rad/s |
| 3. Calculate the power. | $P = \boxed{262 \text{ kW}}$ |

**REMARKS** This power output is about 350 hp.

**EXERCISE** The maximum power produced by the Viper engine is 450 hp at 5200 rev/min. What is the torque when the engine is operating at maximum horsepower? (*Answer*   616 N·m)

There are many parallels between one-dimensional linear motion and rotational motion about a fixed axis. The similarities of the formulas can be seen in Table 9-2. The formulas are the same, but the symbols are different.

## TABLE 9-2

**Analogs in Rotational and Linear Motion**

| Rotational Motion | | Linear Motion | |
|---|---|---|---|
| Angular displacement | $\Delta\theta$ | Displacement | $\Delta x$ |
| Angular velocity | $\omega = \dfrac{d\theta}{dt}$ | Velocity | $v = \dfrac{dx}{dt}$ |
| Angular acceleration | $\alpha = \dfrac{d\omega}{dt} = \dfrac{d^2\theta}{dt^2}$ | Acceleration | $a = \dfrac{dv}{dt} = \dfrac{d^2x}{dt^2}$ |
| Constant angular acceleration equations | $\omega = \omega_0 + \alpha t$ | Constant acceleration equations | $v = v_0 + at$ |
| | $\Delta\theta = \omega_{av}\,\Delta t$ | | $\Delta x = v_{av}\,\Delta t$ |
| | $\omega_{av} = \frac{1}{2}(\omega_0 + \omega)$ | | $v_{av} = \frac{1}{2}(v_0 + v)$ |
| | $\theta = \theta_0 + \omega_0 t + \frac{1}{2}\alpha t^2$ | | $x = x_0 + v_0 t + \frac{1}{2}at^2$ |
| | $\omega^2 = \omega_0^2 + 2\alpha\,\Delta\theta$ | | $v^2 = v_0^2 + 2a\,\Delta x$ |
| Torque | $\tau$ | Force | $F$ |
| Moment of inertia | $I$ | Mass | $m$ |
| Work | $dW = \tau\,d\theta$ | Work | $dW = F_s\,ds$ |
| Kinetic energy | $K = \frac{1}{2}I\omega^2$ | Kinetic energy | $K = \frac{1}{2}mv^2$ |
| Power | $P = \tau\omega$ | Power | $P = Fv$ |
| Angular momentum[†] | $L = I\omega$ | Momentum | $p = mv$ |
| Newton's second law | $\tau_{net} = I\alpha = \dfrac{dL}{dt}$ | Newton's second law | $F_{net} = ma = \dfrac{dp}{dt}$ |

† Angular momentum is introduced in Chapter 10.

# 9-6 Rolling Objects

## Rolling Without Slipping

When a spool rolls without slipping down an incline (Figure 9-29), the points in contact with the incline are instantaneously at rest and the spool rotates about a rotation axis through the contact point. This can be observed because rapid motion causes blurring, so the part of the spool that is moving slowest is blurred the least. In Figure 9-30 a wheel of radius $R$ is rolling without slipping along a flat surface. Point $P$ on the wheel moves as shown with speed

$$v = r\omega \tag{9-25}$$

NONSLIP CONDITION FOR SPEED

FIGURE 9-29

where $r$ is the perpendicular distance from $P$ to the rotation axis. The center of mass of the wheel moves with speed

$$v_{cm} = R\omega \tag{9-26}$$

NONSLIP CONDITION FOR $v_{CM}$

Interestingly, for a point on the top of the wheel, $r = 2R$, so the top of the wheel is moving at twice the speed of the center of mass.

Differentiating each side of Equation (9-26) gives

$$a_{cm} = R\alpha \tag{9-27}$$

NONSLIP CONDITION FOR ACCELERATION

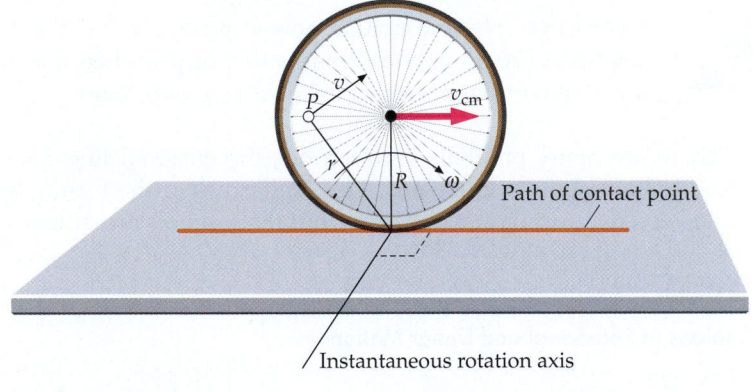

FIGURE 9-30

A falling yo-yo that is unwinding from a string, the top end of which is held fixed, follows the same nonslip conditions as the wheel.

A wheel of radius $R$ is rolling without slipping along a straight path. As the wheel rotates through angle $\phi$ (Figure 9-31), the point of contact between the wheel and the plane moves a distance $s$ that is related to $\phi$ by

$$s = R\phi \tag{9-28}$$

NONSLIP CONDITION FOR DISTANCE

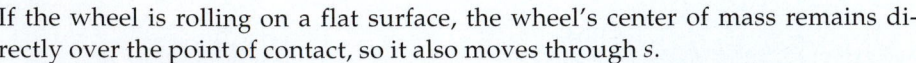

FIGURE 9-31

If the wheel is rolling on a flat surface, the wheel's center of mass remains directly over the point of contact, so it also moves through $s$.

We saw in Chapter 8 that the kinetic energy of a system can be written as the sum of the kinetic energy of motion of the center of mass plus the kinetic energy relative to the center of mass. For a rotation object, the relative kinetic energy is $\frac{1}{2}I_{cm}\omega^2$. Thus, the kinetic energy of a rotating object is

$$K = \frac{1}{2}Mv_{cm}^2 + \frac{1}{2}I_{cm}\omega^2 \tag{9-29}$$

KINETIC ENERGY OF A ROTATING OBJECT

*A BOWLING BALL*                              **EXAMPLE 9-13** **Try It Yourself**

A bowling ball of radius 11 cm and mass $M = 7.2$ kg is rolling without slipping on a horizontal ball return at 2 m/s. It then rolls without slipping up a hill to a height $h$ before momentarily stopping before rolling back down the hill. Find $h$.

**PICTURE THE PROBLEM** Mechanical energy is conserved. The initial kinetic energy, which is the translational kinetic energy of the center of mass, $\frac{1}{2}mv_{cm}^2$, plus the kinetic energy of rotation about the center of mass, $\frac{1}{2}I_{cm}\omega^2$, is converted to potential energy $mgh$. Because the sphere rolls without slipping, the initial linear and angular speeds are related by $v_{cm} = R\omega$. Make a labeled sketch showing the ball in both its initial and final positions (Figure 9-32).

**FIGURE 9-32**

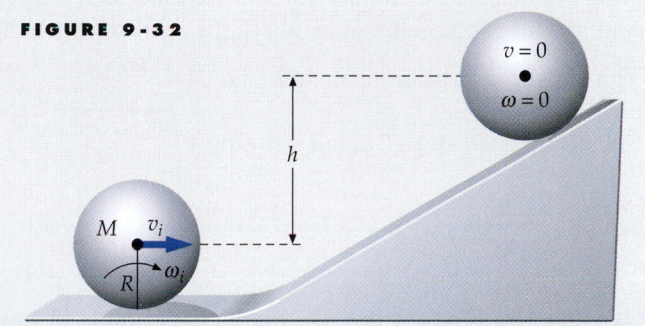

**Cover the column to the right and try these on your own before looking at the answers.**

| Steps | Answers |
|---|---|
| 1. Apply conservation of mechanical energy with $U_i = 0$ and $K_f = 0$. Write the total initial kinetic energy $K_i$ in terms of the speed $v_{cm}$ and the angular speed $\omega$. | $U_f + K_f = U_i + K_i$ <br><br> $Mgh + 0 = 0 + \frac{1}{2}Mv_i^2 + \frac{1}{2}I_{cm}\omega_i^2$ |
| 2. Substitute from $\omega = v_i/R$ and $I_{cm} = \frac{2}{5}MR^2$ and solve for $h$. | $Mgh = \frac{1}{2}Mv_i^2 + \frac{1}{2}\left(\frac{2}{5}MR^2\right)\frac{v_i^2}{R^2}$ <br><br> $= \frac{7}{10}Mv_i^2$ <br><br> so <br><br> $h = \dfrac{7v_i^2}{10g} = 0.285 \text{ m} = \boxed{28.5 \text{ cm}}$ |

**REMARKS** The height $h$ is independent of the mass or radius of the ball.

**EXERCISE** Find the initial kinetic energy of the ball. (*Answer* 20.2 J)

---

*PLAYING POOL*                              **EXAMPLE 9-14**

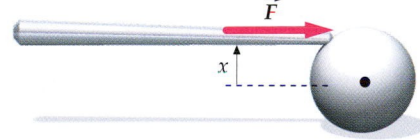

A cue stick hits a cue ball horizontally a distance $x$ above the center of the ball (Figure 9-33). Find the value of $x$ for which the cue ball will roll without slipping from the beginning. Express your answer in terms of the radius $R$ of the ball.

**FIGURE 9-33**

**PICTURE THE PROBLEM** The lines of action of the weight and normal forces pass through the center of mass and thus exert no torque about it. The frictional force is much smaller than the collision force of the stick and can be neglected. If the stick hits the ball at the level of the ball's center, the ball initially translates with no rotation. If the stick hits below the center, the ball initially has backspin. At a certain value of $x$, the ball has just the right forward spin and forward acceleration to satisfy the nonslip condition. The value of $x$ determines the torque to force ratio on the ball, and hence its angular acceleration $\alpha$ to linear acceleration ratio $a$. The linear acceleration $a$ is $F/m$, independent of $x$. For the ball to roll without slipping from the start, we find $\alpha$ and $a$, then set $a = R\alpha$ (nonslip condition) to find $x$.

1. Sketch a free-body diagram of the ball (Figure 9-34). We are assuming friction between the ball and the table is negligible, so do not include a frictional force:

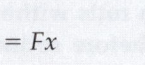

2. The torque about the axis through the center of the ball equals $F$ times $x$:

$$\tau = Fx$$

3. Apply Newton's second law for a system and Newton's second law for rotational motion about the center of the ball:

$$F = ma_{cm} \quad \text{and} \quad \tau = I_{cm}\alpha$$

4. The nonslip condition relates $a$ and $\alpha$:

$$a_{cm} = R\alpha$$

5. Substitute from steps 2 and 3 into step 4:

$$\frac{F}{m} = R\frac{Fx}{I_{cm}}$$

6. Find the moment of inertia from Table 9.1 and solve for $x$:

$$x = \frac{I_{cm}}{mR} = \frac{(2/5)mR^2}{mR} = \boxed{(2/5)R}$$

**FIGURE 9-34**

**REMARKS** Striking the ball at a point either higher or lower than $2R/5$ from the center will result in the ball rolling *and* slipping (skidding). Skidding is often desirable in the game of pool. Rolling and slipping is discussed in the next subsection.

When an object rolls down an incline, its center of mass is accelerated. The analysis of such a problem is simplified by an important theorem concerning the center of mass:

If the torques are computed from a reference frame moving with the center of mass, then Newton's second law for rotation holds even when the center of mass is accelerating, and the torques are computed in the center-of-mass reference frame. That is,

$$\tau_{net,cm} = I_{cm}\alpha \qquad 9\text{-}30$$

This is the same as Equation 9-18 except that here the torques and the moment of inertia are computed from a reference frame moving with the center of mass. (A derivation may be found at www.whfreeman.com/tipler.) When the center of mass is accelerating (a ball rolling down an incline, for example), the center-of-mass reference frame is a noninertial one, where we would not necessarily expect our equations for Newton's second law for rotation to be valid. Nevertheless, they are.

*ACCELERATION OF A BALL THAT IS ROLLING WITHOUT SLIPPING*   **EXAMPLE 9-15**

**A uniform solid ball of mass $m$ and radius $R$ rolls without slipping down a plane inclined at an angle $\phi$ above the horizontal. Find the acceleration of the center of mass and the frictional force.**

**PICTURE THE PROBLEM** From Newton's second law, the acceleration of the center of mass equals the net force divided by the mass. The forces acting are the weight $m\vec{g}$ downward, the normal force $\vec{F}_n$ that balances the normal component of the weight, and the force of friction $\vec{f}$ acting up the incline (Figure 9-35). As the object accelerates down the incline, its angular velocity must increase to maintain the nonslip condition. This angular acceleration requires a net external torque about the axis through the center of mass. We apply Newton's second law for rotation to find $\alpha$. The nonslip condition relates $\alpha$ and $a_{cm}$.

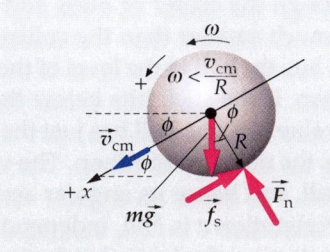

**FIGURE 9-35**

1. Apply Newton's second law for a system along the $x$ axis:

$$\Sigma F_x = ma_{cmx}$$

$$mg \sin \phi - f_s = ma_{cm}$$

2. Apply Newton's second law for rotational motion, for rotations about an axis parallel to the instantaneous rotation axis and passing through the center of mass:

$$\Sigma \tau_i = I_{cm}\alpha$$

$$f_s R + 0 + 0 = I_{cm}\alpha$$

3. Relate $a_{cm}$ and $\alpha$ using the nonslip condition:

$$a_{cm} = R\alpha$$

4. We now have three equations and three unknowns. Solve the step 1 result for $f_s$ and the step 3 result for $\alpha$, substitute for these quantities in the step 2 result and solve for $a_{cm}$:

$$f_s R = I_{cm}\alpha$$

$$(mg \sin \phi - ma_{cm})R = I_{cm}\frac{a_{cm}}{R}$$

so

$$a_{cm} = \frac{g \sin \phi}{1 + \dfrac{I_{cm}}{mR^2}}$$

5. Substitute the step 4 result into the step 1 result and solve for $f_s$:

$$f_s = mg \sin \phi - ma_{cm}$$

$$= mg \sin \phi - \frac{mg \sin \phi}{1 + \dfrac{I_{cm}}{mR^2}} = \frac{mg \sin \phi}{1 + \dfrac{mR^2}{I_{cm}}}$$

6. For a solid sphere, $I_{cm} = \frac{2}{5}mR^2$ (see Table 9.1):

$$a_{cm} = \frac{g \sin \phi}{1 + \frac{2}{5}} = \boxed{\frac{5}{7}g \sin \phi}$$

$$f_s = \frac{mg \sin \phi}{\frac{7}{2}} = \boxed{\frac{2}{7}mg \sin \phi}$$

**REMARKS** Because the ball rolls without slipping, the friction is static friction. Note that the result seems independent of the coefficient of static friction. However, we have assumed that the coefficient of static friction was large enough to prevent slipping.

The results of steps 5 and 6 in Example 9-15 apply to any round object with the center of mass at the geometric center that is rolling without slipping. For such objects, $I_{cm} = \beta mR^2$, where $\beta = \frac{2}{5}$ for a sphere, $\frac{1}{2}$ for a rolling solid cylinder, 1 for a thin cylindrical shell, and so forth. For such objects the step 5 and step 6 results can be expressed

$$f_s = \frac{mg \sin \phi}{1 + \beta^{-1}} \qquad\qquad 9\text{-}31$$

$$a_{cm} = \frac{g \sin \phi}{1 + \beta} \qquad\qquad 9\text{-}32$$

The linear acceleration of any object rolling down an incline is less than $g \sin \phi$ because of the frictional force directed up the incline. Note that these accelerations are independent of both the mass and the radius of the objects. That is, all solid spheres rolling without slipping down the same incline will have identical accelerations. However, if we release a sphere, a cylinder, and a hoop at the top of an incline, and if they all roll without slipping, the sphere will reach the bottom first because it has the greatest acceleration. The cylinder will be second and the hoop last (Figure 9-36). A block that slides without friction down the incline will arrive at the bottom ahead of all three rolling objects.

Because the friction is static, it does no work, and there is no dissipation of mechanical energy. We can therefore use the conservation of mechanical energy to find the speed of an object rolling without slipping down an incline. At the top of the incline, the total energy is the potential energy $mgh$. At the bottom, the total energy is kinetic energy. Conservation of mechanical energy therefore gives

**FIGURE 9-36** A sphere, a cylinder, and a hoop are released together from rest at the top of an incline. The sphere reaches the bottom first, followed by the cylinder and then the hoop.

$$\tfrac{1}{2} m v_{cm}^2 + \tfrac{1}{2} I_{cm} \omega^2 = mgh$$

We can use the nonslip condition to eliminate either $v_{cm}$ or $\omega$. Substituting $I_{cm} = \beta mR^2$ and $\omega = v_{cm}/R$, we obtain $\tfrac{1}{2} m v_{cm}^2 + \tfrac{1}{2}\beta mR^2 (v_{cm}^2/R^2) = mgh$. Solving for $v_{cm}^2$ gives

$$v_{cm}^2 = \frac{2gh}{1 + \beta} \qquad\qquad 9\text{-}33$$

For a cylinder, with $\beta = \tfrac{1}{2}$, we obtain $v_{cm} = \sqrt{\tfrac{4}{3} gh}$. Note that the speed is independent of the mass and radius of the cylinder, and is less than $\sqrt{2gh}$, the speed of an object sliding with no friction down the incline.

For an object rolling without slipping down an incline, the frictional force $f_s$ is less than or equal to its maximum value; that is, $f_s \le \mu_s F_n$, where $F_n = mg \cos \phi$. Substituting the expression from Equation 9-31 for the frictional force, we have

$$\frac{mg \sin \phi}{1 + \beta^{-1}} \le \mu_s mg \cos \phi$$

or

$$\tan \phi \le (1 + \beta^{-1})\mu_s \qquad\qquad 9\text{-}34$$

(For a uniform cylinder, $I_{cm} = \tfrac{1}{2} MR^2$, so $\beta = \tfrac{1}{2}$ and $\tan \phi \le 3\mu_s$.) If the tangent of the angle of incline is greater than $(1 + \beta^{-1})\mu_s$, the object will slip as it moves down the incline.

**EXERCISE** A uniform cylinder rolls down a plane inclined at $\theta = 50°$. What is the minimum value of the coefficient of static friction for which the cylinder will roll without slipping? (*Answer* 0.40)

**EXERCISE** For a hoop rolling down an incline, (*a*) what is the force of friction and (*b*) what is the maximum value of $\tan \theta$ for which the hoop will roll without slipping? (*Answer* (*a*) $f = \tfrac{1}{2}mg \sin \theta$ (*b*) $\tan \theta \le 2\mu_s$)

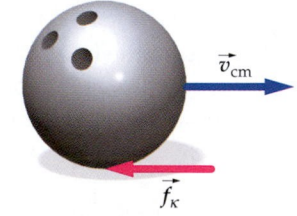

**FIGURE 9-37** A bowling ball moving with no initial rotation. The frictional force $\vec{f}_k$ exerted by the floor reduces the speed $v_{cm}$ and increases the angular speed $\omega$ until $v_{cm} = R\omega$.

## *Rolling With Slipping

When an object slides as it rolls, the nonslip condition $v_{cm} = R\omega$ does not hold. Suppose a bowling ball is thrown with no initial rotation. As the ball slides along the bowling lane, $v_{cm} > R\omega$. However, the frictional force both reduces its linear speed $v_{cm}$ (Figure 9-37) and increases its angular speed $\omega$ until the nonslip condition $v_{cm} = R\omega$ is reached, after which the ball rolls without slipping.

Another example of rolling with slipping is a ball with topspin, such as a cue ball struck at a point higher than $(2/5)R$ above the center (see Example 9-14) so that $v_{cm} < R\omega$. Then the frictional force both increases $v$ and decreases $\omega$ until the nonslip condition $v_{cm} = R\omega$ is reached (Figure 9-38).

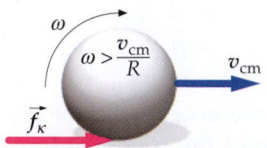

**FIGURE 9-38** Ball with excess topspin. The frictional force accelerates the ball in the direction of motion.

*A SLIDING BOWLING BALL*        **EXAMPLE 9-16**

A bowling ball of mass $M$ and radius $R$ is thrown so that the instant it touches the floor it is moving horizontally with speed $v_0 = 5$ m/s and is not rotating. The coefficient of kinetic friction between the ball and the floor is $\mu_k = 0.08$. Find (a) the time the ball slides and (b) the distance the ball slides before it rolls without slipping.

**PICTURE THE PROBLEM** We calculate $v_{cm}$ and $\omega$ as functions of time, set $v_{cm} = R\omega$, and solve for $t$. The linear and angular accelerations are found from $\Sigma F = ma$ and $\tau = I\alpha$. Let the direction of motion be positive. There is slipping and kinetic friction, so mechanical energy is dissipated. Therefore, conservation of mechanical energy cannot be used to solve this problem.

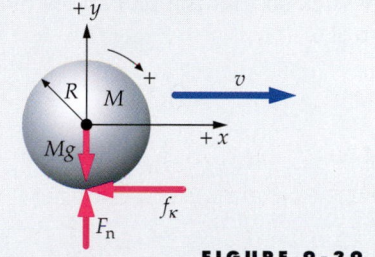

**FIGURE 9-39**

(a) 1. Sketch a free-body diagram of the ball (Figure 9-39).

2. The net force on the ball is the force of kinetic friction $f_k$, which acts in the negative $x$ direction. Apply Newton's second law:

$$\Sigma F_x = Ma_{cmx}$$

$$-f_k = Ma_{cmx}$$

3. The acceleration is in the negative $x$ direction and $a_{cmy} = 0$. Find $f_k$ by first finding $F_n$:

$$\Sigma F_y = Ma_{cmy} = 0 \;\Rightarrow\; F_n = Mg$$

so

$$f_k = \mu_k F_n = \mu_k Mg$$

4. Find the acceleration using the step 2 and step 3 results:

$$-\mu_k Mg = Ma_{cmx} \;\Rightarrow\; a_{cmx} = -\mu_k g$$

5. Relate the linear velocity to the acceleration and the time using a kinematic equation:

$$v_{cmx} = v_0 + a_{cmx}t = v_0 - \mu_k g t$$

6. Find $\alpha$ by applying Newton's second law for rotational motion to the ball. Compute the torques about the axis through the center of mass. Note that the free-body diagram has clockwise as positive:

$$\Sigma \tau = I_{cm}\alpha$$

$$\mu_k MgR + 0 + 0 = \tfrac{2}{5}MR^2\alpha$$

so

$$\alpha = \frac{5}{2}\frac{\mu_k g}{R}$$

7. Relate the angular velocity to the angular acceleration and the time using a kinematic equation:

$$\omega = \omega_0 + \alpha t = 0 + \alpha t = \frac{5}{2}\frac{\mu_k g}{R}t$$

8. Solve for the time $t$ at which $v_{cm} = R\omega$:

$$v_{cm} = R\omega$$

$$v_0 - \mu_k g t = \tfrac{5}{2}\mu_k g t$$

so

$$t = \frac{2v_0}{7\,\mu_k g} = \frac{2(5 \text{ m/s})}{7(0.08)(9.81 \text{ m/s}^2)} = \boxed{1.82 \text{ s}}$$

(b) The distance traveled in 1.82 s is:

$$\Delta x = v_0 t + \tfrac{1}{2}a_{cm}t^2$$

$$= v_0 \frac{2v_0}{7\,\mu_k g} + \frac{1}{2}(-\mu_k g)\left(\frac{2v_0}{7\,\mu_k g}\right)^2 = \frac{12}{49}\frac{v_0^2}{\mu_k g}$$

$$= \frac{12}{49}\frac{(5 \text{ m/s})^2}{(0.08)(9.81 \text{ m/s}^2)} = \boxed{7.80 \text{ m}}$$

**EXERCISE** Find the speed of the bowling ball when it begins to roll without slipping. (*Answer* $v_{cm} = \frac{5}{7}v_0$). This result is independent of the coefficient of kinetic friction. The rolling speed is $\frac{5}{7}v_0$ whether $\mu_k$ is large or small. The total mechanical energy lost is thus independent of $\mu_k$. (The time and the distance are sensitive to the value of $\mu_k$, however.)

**EXERCISE** Find the total kinetic energy of the ball after it begins to roll without slipping. (*Answer* $K = \frac{5}{14}mv_0^2$)

**REMARKS** In a well-maintained bowling alley, the lanes are lightly oiled and very slick so that the ball slides over a great distance, giving the bowler added control.

# SUMMARY

1. Angular displacement, angular velocity, and angular acceleration are fundamental defined quantities in rotational kinematics.

2. Torque and moment of inertia are important derived dynamic concepts. Torque is a measure of the effect of a force in causing an object to start or stop rotating. Moment of inertia is the measure of an object's inertial resistance to angular accelerations. The moment of inertia depends on the distribution of the mass relative to the rotation axis.

3. The parallel-axis theorem, which follows from the definition of the moment of inertia, often simplifies the calculation of $I$.

4. Newton's second law for rotation, $\Sigma \tau_{ext} = I\alpha$, is derived from Newton's second law and the definitions of $\tau$, $I$, and $\alpha$. It is an important relation for problems involving the rotation of a rigid object about a rotation axis of fixed direction.

| Topic | Relevant Equations and Remarks | |
|---|---|---|
| **1. Angular Velocity and Angular Acceleration** | | |
| Angular velocity | $\omega = \dfrac{d\theta}{dt}$  (Definition) | 9-2 |
| Angular acceleration | $\alpha = \dfrac{d\omega}{dt} = \dfrac{d^2\theta}{dt^2}$  (Definition) | 9-3 |
| Tangential velocity | $v_t = r\omega$ | 9-7 |
| Tangential acceleration | $a_t = r\alpha$ | 9-8 |
| Centripetal acceleration | $a_c = \dfrac{v^2}{r} = r\omega^2$ | 9-9 |

**2. Equations for Rotation With**

| | | |
|---|---|---|
| **Constant Angular Acceleration** | $\omega = \omega_0 + \alpha t$ | 9-4 |
| | $\theta = \theta_0 + \omega_0 t + \frac{1}{2}\alpha t^2$ | 9-5 |
| | $\omega^2 = \omega_0^2 + 2\alpha(\theta - \theta_0)$ | 9-6 |

**3. Moment of Inertia**

| | | |
|---|---|---|
| System of particles | $I = \Sigma\, m_i r_i^2$     (Definition) | 9-10 |
| Continuous object | $I = \int r^2\, dm$ | 9-12 |

| | | |
|---|---|---|
| Parallel-axis theorem | The moment of inertia about an axis a distance $h$ from a parallel axis through the center of mass is | |
| | $I = I_{cm} + Mh^2$ | 9-13 |
| | where $I_{cm}$ is the moment of inertia about the axis through the center of mass and $M$ is the total mass of the object. | |

**4. Energy**

| | | |
|---|---|---|
| Kinetic energy for rotation about a fixed axis | $K = \frac{1}{2}I\omega^2$ | 9-11 |
| Kinetic energy for rotating object | $K = \frac{1}{2}Mv_{cm}^2 + \frac{1}{2}I_{cm}\omega^2$ | 9-29 |
| Power | $P = \tau\omega$ | 9-24 |

**5. Torque**

| | | |
|---|---|---|
| | The torque due to a force equals the product of the tangential component of the force and the radial distance from the axis: | |
| | $\tau = F_t r = Fr \sin\phi = F\ell$ | 9-19 |

**6. Newton's Second Law for Rotation**

| | | |
|---|---|---|
| | $\tau_{net,ext} = \displaystyle\sum_i \tau_{i,ext} = I\alpha$ | 9-18 |
| | If torques about an axis through the center of mass are computed from a reference frame moving with the center of mass, Newton's second law for rotation holds for rotation about an axis through the center of mass, even if the reference frame is noninertial. | |

**7. Nonslip Conditions**

| | | |
|---|---|---|
| | When a string that is wrapped around a pulley or disk does not slip, the linear and angular quantities are related by | |
| | $v_t = R\omega$ | 9-21 |
| | $a_t = R\alpha$ | 9-22 |

**8. Rolling Objects**

| | | |
|---|---|---|
| Rolling without slipping | $v_{cm} = R\omega$ | 9-26 |
| *Rolling with slipping | When an object rolls and slips, $v_{cm} \neq R\omega$. Kinetic friction exerts a force that tends to change $v_{cm}$, and also exerts a torque that changes $\omega$ until $v_{cm} = R\omega$ and pure rolling sets in. | |

- • Single-concept, single-step, relatively easy
- •• Intermediate-level, may require synthesis of concepts
- ••• Challenging
- SSM Solution is in the *Student Solutions Manual*
- iSOLVE Problems available on iSOLVE online homework service
- iSOLVE✓ These "Checkpoint" online homework service problems ask students additional questions about their confidence level, and how they arrived at their answer

In a few problems, you are given more data than you actually need; in a few other problems, you are required to supply data from your general knowledge, outside sources, or informed estimates.

Take $g$ = 9.81 N/kg = 9.81 m/s² and neglect friction in all problems unless otherwise stated.

## Conceptual Problems

**1** • SSM Two points are on a disk turning at constant angular velocity, one point on the rim and the other halfway between the rim and the axis. Which point moves the greater distance in a given time? Which turns through the greater angle? Which has the greater speed? The greater angular velocity? The greater tangential acceleration? The greater angular acceleration? The greater centripetal acceleration?

**2** • True or false:

(a) Angular velocity and linear velocity have the same dimensions.

(b) All parts of a rotating wheel must have the same angular velocity.

(c) All parts of a rotating wheel must have the same angular acceleration.

**3** •• Starting from rest, a disk takes 10 revolutions to reach an angular velocity $\omega$ at constant angular acceleration. How many additional revolutions are required to reach an angular velocity of $2\omega$? (a) 10 rev. (b) 20 rev. (c) 30 rev. (d) 40 rev. (e) 50 rev.

**4** • SSM The dimension of torque is the same as that of (a) impulse, (b) energy, (c) momentum, (d) none of these.

**5** • The moment of inertia of an object of mass $M$ (a) is an intrinsic property of the object, (b) depends on the choice of axis of rotation, (c) is proportional to $M$ regardless of the choice of axis, (d) both (b) and (c) are correct.

**6** • SSM Can an object continue to rotate in the absence of torque?

**7** • Does an applied net torque always increase the angular speed of an object?

**8** • True or false:

(a) If the angular velocity of an object is zero at some instant, the net torque on the object must be zero at that instant.

(b) The moment of inertia of an object depends on the location of the axis of rotation.

(c) The moment of inertia of an object depends on the angular velocity of the object.

**9** • A disk is free to rotate about an axis. A tangential force applied a distance $d$ from the axis causes an angular acceleration $\alpha$. What angular acceleration is produced if the same force is applied a distance $2d$ from the axis? (a) $\alpha$. (b) $2\alpha$. (c) $\alpha/2$. (d) $4\alpha$. (e) $\alpha/4$.

**10** • SSM The moment of inertia of an object about an axis that does not pass through its center of mass is _____ the moment of inertia about a parallel axis through its center of mass. (a) always less than, (b) sometimes less than, (c) sometimes equal to, (d) always greater than.

**11** • A constant torque acts on a merry-go-round. The power input of the torque is (a) constant, (b) proportional to the angular speed of the merry-go-round, (c) zero, (d) none of these.

**12** • True or false: When an object rolls without slipping, friction does no work on the object.

**13** • Why do we put the knob on a door about as far away from the hinges as possible?

**14** • SSM A wheel of radius $R$ is rolling without slipping on a flat stationary surface. The velocity of the point on the rim that is in contact with the surface is (a) equal to $R\omega$ in the direction of motion of the center of mass, (b) equal to $R\omega$ opposite to the direction of motion of the center of mass, (c) zero, (d) equal to the velocity of the center of mass and in the same direction, (e) equal to the velocity of the center of mass but in the opposite direction.

**15** •• A solid uniform cylinder and a solid uniform sphere have equal masses. Both roll without slipping on a horizontal surface. If their kinetic energies are the same, then (a) the translational speed of the cylinder is greater than that of the sphere, (b) the translational speed of the cylinder is less than that of the sphere, (c) the translational speeds of the two objects are the same, (d) (a), (b), or (c) could be correct depending on the radii of the objects.

**16** • SSM Two identical-looking 1-m-long pipes enclose slugs of lead whose total mass is 10 kg (much larger than the mass of the pipe). In the first pipe the lead is concentrated at the center of the pipe, while in the second the lead is divided into two equal masses placed at opposite ends of the pipe. Without opening either pipe, how could you determine which is which?

**17** •• Starting from rest at the same time, a coin and a ring roll down an incline without slipping. Which of the following is true? (a) The ring reaches the bottom first. (b) The coin reaches the bottom first. (c) The coin and ring arrive at the bottom simultaneously. (d) The race to the bottom depends on their relative masses. (e) The race to the bottom depends on their relative diameters.

**18** •• For a hoop of mass $M$ and radius $R$ that is rolling without slipping, which is larger, its translational kinetic energy or its rotational kinetic energy? (a) Translational kinetic energy is larger. (b) Rotational kinetic energy is larger. (c) Both are the same size. (d) The answer depends on the radius. (e) The answer depends on the mass.

**19** •• For a disk of mass $M$ and radius $R$ that is rolling without slipping, which is larger, its translational kinetic energy or its rotational kinetic energy? (a) Translational kinetic energy is larger. (b) Rotational kinetic energy is larger. (c) Both are the same size. (d) The answer depends on the radius. (e) The answer depends on the mass.

**20** •• A perfectly rigid ball rolls without slipping along a perfectly rigid horizontal plane. Show that the frictional force acting on the ball must be zero. *Hint: Consider a possible direction for the action of the frictional force and what effects such a force would have on the velocity of the center of mass and on the angular velocity.*

**21** • True or false: When a sphere rolls and slips on a rough surface, mechanical energy is dissipated.

**22** • A cue ball is hit very near the top so that it starts to move with topspin. As it slides, the force of friction (a) increases $v_{cm}$, (b) decreases $v_{cm}$, (c) has no effect on $v_{cm}$.

### Estimation and Approximation

**23** •• A bicycle of mass 14 kg has 1.2-m-diameter wheels, each of mass 3 kg. The mass of the rider is 38 kg. Estimate the fraction of the total kinetic energy of bicycle and rider associated with rotation of the wheels.

**24** •• Why does toast falling off a table always land jelly-side down? The question may sound silly, but it has been a subject of serious scientific enquiry. The analysis is too complicated to reproduce here, but R. D. Edge and Darryl Steinert showed that a piece of toast, pushed gently over the edge of a table until it tilts off, typically falls off the table when it makes an angle of about 30° with the horizontal (see Figure 9-40) and has an angular velocity of $\omega = 0.956\sqrt{g/\ell}$, where $\ell$ is the length of one side of the piece of toast (assumed square). Assuming that it starts jelly-side up, what side will it land on if it falls from a table of height 0.5 m? How about a height of 1 m? Assume that the toast has a side length $\ell = 0.1$ m, and that it will land jelly-side down if the angle is between 180° and 270°. (If the toast rotates through an angle greater than 360°, we will need to reduce it to the range 0°–360°.) Ignore any forces due to air resistance. For readers interested in this problem and a host of others, we highly recommend Robert Erlich's wonderful book, *Why Toast Lands Jelly-Side Down: Zen and the Art of Physics Demonstrations.*[†]

---

[†] Robert Erlich, *Why Toast Lands Jelly-Side Down; Princeton, NJ: Princeton University Press (1997).

**FIGURE 9-40** Problem 24

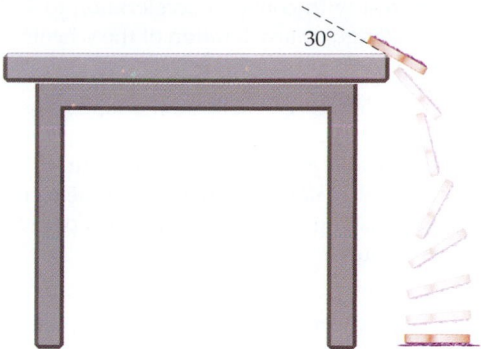

**25** •• **SSM** Consider the moment of inertia of an average adult man about an axis running vertically through the center of his body when he is standing straight up with arms flat at his sides and again when he is standing straight up holding his arms straight out. Estimate the ratio of his moment of inertia with his arms straight out to the moment of inertia with his arms flat against his sides.

### Angular Velocity and Angular Acceleration

**26** • A particle moves in a circle of radius 90 m with a constant speed of 25 m/s. (a) What is its angular velocity in radians per second about the center of the circle? (b) How many revolutions does it make in 30 s?

**27** • **ISOLVE** A wheel starts from rest with constant angular acceleration of 2.6 rad/s². After 6 s: (a) What is its angular velocity? (b) Through what angle has the wheel turned? (c) How many revolutions has it made? (d) What is the speed and acceleration of a point 0.3 m from the axis of rotation?

**28** • **SSM** When a turntable rotating at $33\frac{1}{3}$ rev/min is shut off, it comes to rest in 26 s. Assuming constant angular acceleration, find (a) the angular acceleration, (b) the average angular velocity of the turntable, and (c) the number of revolutions it makes before stopping.

**29** • A disk of radius 12 cm, initially at rest, begins rotating about its axis with a constant angular acceleration of 8 rad/s². At $t = 5$ s, what are (a) the angular velocity of the disk and (b) the tangential acceleration $a_t$ and the centripetal acceleration $a_c$ of a point on the edge of the disk?

**30** • **ISOLVE** A Ferris wheel of radius 12 m rotates once in 27 s. (a) What is its angular velocity in radians per second? (b) What is the linear speed of a passenger? What is the centripetal acceleration of a passenger?

**31** • A cyclist accelerates from rest. After 8 s, the wheels have made 3 revolutions. (a) What is the angular acceleration of the wheels? (b) What is the angular velocity of the wheels after 8 s?

**32** • What is the angular velocity of the earth in radians per second as it rotates about its axis?

**33** • A wheel rotates through 5.0 rad in 2.8 s as it is brought to rest with constant angular acceleration. The initial angular velocity of the wheel before braking began was (a) 0.6 rad/s, (b) 0.9 rad/s, (c) 1.8 rad/s, (d) 3.6 rad/s, (e) 7.2 rad/s.

**34** • A bicycle has wheels of 1.2-m diameter. The bicyclist accelerates from rest with constant acceleration to 24 km/h in 14.0 s. What is the angular acceleration of the wheels?

**35** •• **SSM** **iSOLVE ✓** The tape in a standard VHS video-tape cassette has a length $L = 246$ m; the tape plays for 2.0 h (Figure 9-41). As the tape starts, the full reel has an outer radius of about $R = 45$ mm and an inner radius of about $r = 12$ mm. At some point during the play, both reels have the same angular speed. Calculate this angular speed in radians per second and revolutions per minute.

**FIGURE 9-41** Problem 35

45 mm
12 mm

## Torque, Moment of Inertia, and Newton's Second Law for Rotation

**36** • **iSOLVE** A disk-shaped grindstone of mass 1.7 kg and radius 8 cm is spinning at 730 rev/min. After the power is shut off, a woman continues to sharpen her ax by holding it against the grindstone for 9 s until the grindstone stops rotating. (a) What is the angular acceleration of the grindstone? (b) What is the torque exerted by the ax on the grindstone? (Assume constant angular acceleration and a lack of other frictional torques.)

**37** • **SSM** A 2.5-kg cylinder of radius 11 cm, initially at rest, is free to rotate about the axis of the cylinder. A rope of negligible mass is wrapped around it and pulled with a force of 17 N. Find (a) the torque exerted by the rope, (b) the angular acceleration of the cylinder, and (c) the angular velocity of the cylinder at $t = 5$ s.

**38** •• **iSOLVE ✓** A wheel free to rotate about its axis that is not frictionless is initially at rest. A constant external torque of 50 N·m is applied to the wheel for 20 s, giving the wheel an angular velocity of 600 rev/min. The external torque is then removed, and the wheel comes to rest 120 s later. Find (a) the moment of inertia of the wheel and (b) the frictional torque, which is assumed to be constant.

**39** •• A pendulum consisting of a string of length $L$ attached to a bob of mass $m$ swings in a vertical plane. When the string is at an angle $\theta$ to the vertical, (a) what is the tangential component of acceleration of the bob? (b) What is the torque exerted about the pivot point? (c) Show that $\tau = I\alpha$ with $a_t = L\alpha$ gives the same tangential acceleration as found in Part (a).

**40** ••• **SSM** A uniform rod of mass $M$ and length $L$ is pivoted at one end and hangs as in Figure 9-42 so that it is free to rotate without friction about its pivot. It is struck by a horizontal force $F_0$ for a short time $\Delta t$ at a distance $x$ below the pivot as shown. (a) Show that the speed of the center of mass of the rod just after being struck is given by $v_0 = 3F_0 x \Delta t/2ML$. (b) Find the horizontal component of the force delivered by the pivot, and show that this force component is zero if $x = \frac{2}{3}L$. (Note: The point $x = \frac{2}{3}L$ is called the center of percussion of the rod.)

**FIGURE 9-42**
**Problem 40**

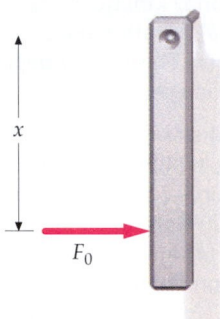

$x$

$F_0$

**41** ••• A uniform horizontal disk of mass $M$ and radius $R$ is rotating about the vertical axis through its center with an angular velocity $\omega$. When it is placed on a horizontal surface, the coefficient of kinetic friction between the disk and the surface is $\mu_k$. (a) Find the torque $d\tau$ exerted by the force of friction on a circular element of radius $r$ and width $dr$. (b) Find the total torque exerted by friction on the disk. (c) Find the time required for the disk to stop rotating.

## Calculating the Moment of Inertia

**42** • **iSOLVE** A tennis ball has a mass of 57 g and a diameter of 7 cm. Find the moment of inertia about its diameter. Assume that the ball is a thin spherical shell.

**43** • **SSM** **iSOLVE** Four particles at the corners of a square with side length $L = 2$ m are connected by massless rods (Figure 9-43). The masses of the particles are $m_1 = m_3 = 3$ kg and $m_2 = m_4 = 4$ kg. Find the moment of inertia of the system about the $z$ axis.

**FIGURE 9-43** Problems 43–45

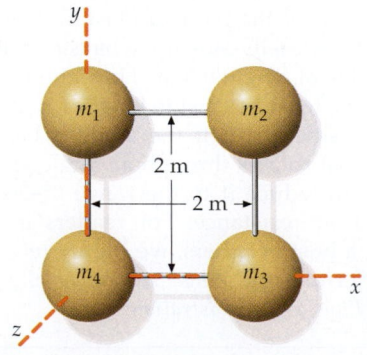

$y$

$m_1$    $m_2$

2 m

2 m

$m_4$    $m_3$

$z$    $x$

**44** • Use the parallel-axis theorem and your results for Problem 43 to find the moment of inertia of the four-particle system in Figure 9-43 about an axis that is perpendicular to the plane of the configuration and passes through the center of mass of the system. Check your result by direct computation.

**45** • For the four-particle system of Figure 9-43, (a) find the moment of inertia $I_x$ about the x axis, which passes through $m_3$ and $m_4$, and (b) find $I_y$ about the y axis, which passes through $m_1$ and $m_4$.

**46** • [SOLVE] Use the parallel-axis theorem to find the moment of inertia of a solid sphere of mass M and radius R about an axis that is tangent to the sphere (Figure 9-44).

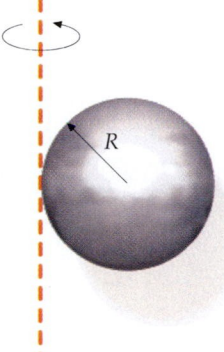

**FIGURE 9-44** Problem 46

**47** •• [SOLVE ✓] A 1.0-m-diameter wagon wheel consists of a thin rim having a mass of 8 kg and six spokes, each having a mass of 1.2 kg. Determine the moment of inertia of the wagon wheel for rotation about its axis.

**48** •• [SSM] [SOLVE] Two point masses $m_1$ and $m_2$ are separated by a massless rod of length L. (a) Write an expression for the moment of inertia about an axis perpendicular to the rod and passing through it at a distance x from mass $m_1$. (b) Calculate $dI/dx$ and show that I is at a minimum when the axis passes through the center of mass of the system.

**49** •• A uniform rectangular plate has mass m and sides of lengths a and b. (a) Show by integration that the moment of inertia of the plate about an axis that is perpendicular to the plate and passes through one corner is $m(a^2 + b^2)/3$. (b) What is the moment of inertia about an axis that is perpendicular to the plate and passes through its center of mass?

**50** •• [SSM] Tracey and Corey are doing intensive research on baton-twirling. Each is using "The Beast" as a model baton: two uniform spheres, each of mass 500 g and radius 5 cm, mounted at the ends of a 30-cm uniform rod of mass 60 g (Figure 9-45). They want to calculate the moment of inertia of The Beast about an axis perpendicular to the rod and passing through its center. Corey uses the approximation that the two spheres can be treated as point particles that are 20 cm from the axis of rotation, and that the mass of the rod is negligible. Tracey, however, makes her calculations without approximations. (a) Compare the two results. (b) If the spheres retained the same mass but were hollow, would the rotational inertia increase or decrease? Justify your choice with a sentence or two. It is not necessary to calculate the new value of I.

**FIGURE 9-45** Problem 50

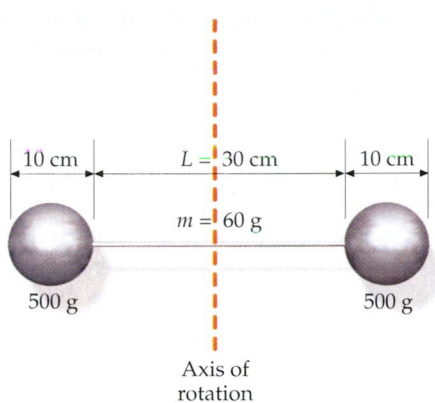

$$L = 30 \text{ cm}$$
10 cm ⟷ ⟷ 10 cm

$m = 60$ g

500 g          500 g

Axis of
rotation

**51** •• The methane molecule ($CH_4$) has four hydrogen atoms located at the vertices of a regular tetrahedron of side length 0.18 nm, with the carbon atom at the center of the tetrahedron (Figure 9-46). Find the moment of inertia of this molecule for rotation about an axis that passes through the centers of the carbon atom and one of the hydrogen atoms.

**FIGURE 9-46** Problem 51

**52** •• A hollow cylinder has mass m, an outside radius $R_2$, and an inside radius $R_1$. Show that its moment of inertia about its symmetry axis is given by $I = \frac{1}{2}m(R_2^2 + R_1^2)$. In Section 9-3 the moment of inertia was calculated for a solid cylinder by direct integration, first finding the moment of inertia of a disk. This calculation is the same as that one, except for the value of one of the integration limits and the expression for the area of a cross section.

**53** ••• Show that the moment of inertia of a spherical shell of radius R and mass m is $2mR^2/3$. This can be done by direct integration or, more easily, by finding the increase in the moment of inertia of a solid sphere when its radius changes. To do this, first show that the moment of inertia of a solid sphere of density $\rho$ is $I = (8/15)\pi\rho R^5$. Then compute the change $dI$ in I for a change $dR$, and use the fact that the mass of this shell is $dm = 4\pi R^2\rho\, dR$.

**54** ••• [SSM] The density of the earth is not quite uniform. It varies with the distance r from the center of the earth as $\rho = C[1.22 - (r/R)]$, where R is the radius of the earth and C is a constant. (a) Find C in terms of the total mass M and the radius R. (b) Find the moment of inertia of the earth. (See Problem 53.)

**55** ••• Use integration to determine the moment of inertia of a right circular homogeneous solid cone of height H, base radius R, and mass density $\rho$ about its symmetry axis.

**56** ••• Use integration to determine the moment of inertia of a thin uniform disk of mass $M$ and radius $R$ for rotation about a diameter. Check your answer by referring to Table 9-1.

**57** ••• A roadside ice-cream stand uses rotating solid cones to catch the eye of travelers. Each cone rotates about an axis perpendicular to its axis of symmetry and passing through its apex. The sizes of the cones vary, and the owner wonders if it would be more energy-efficient to use several smaller cones or a few big ones. To answer this, he must calculate the moment of inertia of a homogeneous right circular cone of height $H$, base radius $R$, and mass density $\rho$. What is the result?

## Rotational Kinetic Energy

**58** • The particles in Figure 9-47 are connected by a very light rod whose moment of inertia can be neglected. They rotate about the $y$ axis with angular velocity $\omega = 2$ rad/s. (a) Find the speed of each particle, and use it to calculate the kinetic energy of this system directly from $\Sigma \frac{1}{2} m_i v_i^2$. (b) Find the moment of inertia about the $y$ axis, and calculate the kinetic energy from $K = \frac{1}{2} I \omega^2$.

**FIGURE 9-47** Problem 58

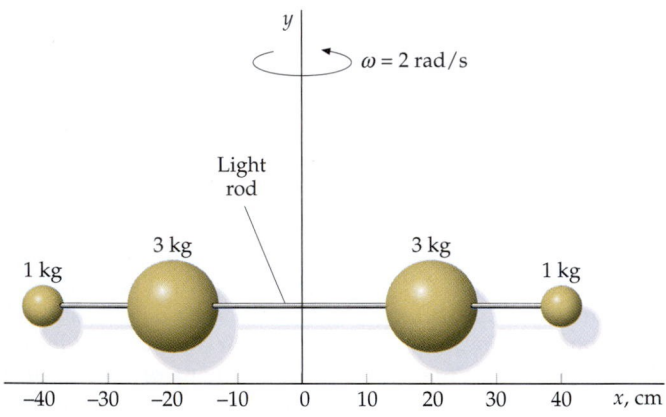

**59** • **SSM** **ISOLVE** A solid ball of mass 1.4 kg and diameter 15 cm is rotating about its diameter at 70 rev/min. (a) What is its kinetic energy? (b) If an additional 2 J of energy are supplied to the rotational energy, what is the new angular speed of the ball?

**60** • An engine develops 400 N·m of torque at 3700 rev/min. Find the power developed by the engine.

**61** •• Two point particles with masses $m_1$ and $m_2$ are connected by a massless rod of length $L$ to form a dumbbell that rotates about its center of mass with angular velocity $\omega$. Show that the ratio of kinetic energies of the particles is $K_1/K_2 = m_2/m_1$.

**62** •• **ISOLVE✓** Calculate the kinetic energy of rotation of the earth about its axis, and compare it with the kinetic energy of the orbital motion of the earth's center of mass about the sun. Assume the earth to be a homogeneous sphere of mass $6.0 \times 10^{24}$ kg and radius $6.4 \times 10^6$ m. The radius of the earth's orbit is $1.5 \times 10^{11}$ m.

**63** •• **SSM** **ISOLVE** A 2000-kg block is lifted at a constant speed of 8 cm/s by a steel cable that passes over a massless pulley to a motor-driven winch (Figure 9-48). The radius of the winch drum is 30 cm. (a) What force must be exerted by the cable? (b) What torque does the cable exert on the winch drum? (c) What is the angular velocity of the winch drum? (d) What power must be developed by the motor to drive the winch drum?

**FIGURE 9-48** Problem 63

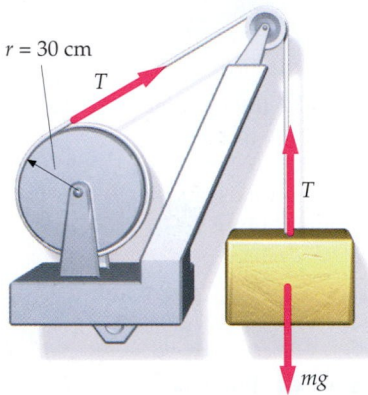

**64** •• A uniform disk of mass $M$ and radius $R$ can rotate freely about a horizontal axis through its center and perpendicular to the plane of the disk. A small particle of mass $m$ is attached to the rim of the disk at the top, directly above the pivot. The system is given a gentle nudge, and the disk begins to rotate. (a) What is the angular velocity of the disk when the particle is at its lowest point? (b) At this point, what force must be exerted by the disk on the particle to keep it stuck to the disk?

**65** •• A uniform ring 1.5 m in diameter is pivoted at one point on its perimeter so that it is free to rotate about a horizontal axis. Initially, the line joining the support point and the center is horizontal (Figure 9-49). (a) If the ring is released from rest, what is its maximum angular velocity? (b) What minimum initial angular velocity must it be given if it is to rotate a full 360°?

**FIGURE 9-49** Problem 65

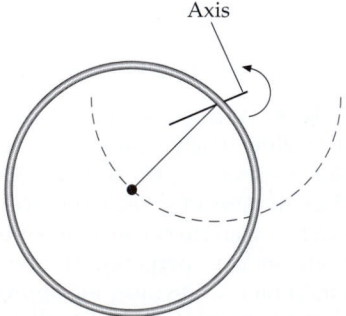

**66** •• **ISOLVE** You set out to design a car that uses the energy stored in a flywheel consisting of a uniform 100-kg cylinder of radius $R$. The flywheel must deliver an average of 2 MJ/km of mechanical energy, with a maximum angular velocity of 400 rev/s. Find the smallest value of $R$ for which the car can travel 300 km without the flywheel having to be recharged.

**67** •• A ladder that is 8.6 m long and has a mass of 60 kg is placed in a nearly vertical position against the wall of a building. You stand on a rung with your center of mass at the top of the ladder. Assume that your mass is 80 kg. As you lean back slightly, the ladder begins to rotate about its base away from the wall. Is it better to quickly step off the ladder and drop to the ground or to hold onto the ladder and step off just before the top end hits the ground?

## Pulleys, Yo-Yos, and Hanging Things

**68** •• **SSM** **iSOLVE** ✓ A 4-kg block resting on a friction-less horizontal ledge is attached to a string that passes over a pulley and is attached to a hanging 2-kg block (Figure 9-50). The pulley is a uniform disk of radius 8 cm and mass 0.6 kg. (a) Find the speed of the 2-kg block after it falls from rest a distance of 2.5 m. (b) What is the angular velocity of the pulley at this time?

**FIGURE 9-50**
**Problems 68–70**

**69** •• For the system in Problem 68, find the linear acceleration of each block and the tension in the string.

**70** •• Solve Problem 68 for the case in which the coefficient of friction between the ledge and the 4-kg block is 0.25.

**71** •• A 1200-kg car is being unloaded by a winch. At the moment shown in Figure 9-51, the gearbox shaft of the winch breaks and the car falls from rest. During the car's fall, there is no slipping between the (massless) rope, the pulley, and the winch drum. The moment of inertia of the winch drum is 320 kg·m² and that of the pulley is 4 kg·m². The radius of the winch drum is 0.80 m and that of the pulley is 0.30 m. Find the speed of the car as it hits the water.

**FIGURE 9-51**
**Problem 71**

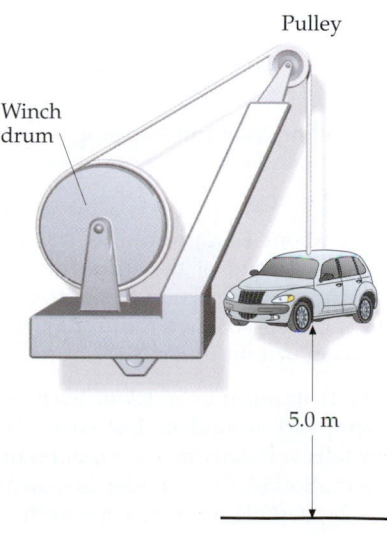

Pulley

Winch drum

5.0 m

**72** •• **SSM** The system in Figure 9-52 is released from rest. The 30-kg block is 2 m above the ledge. The pulley is a uniform disk with a radius of 10 cm and a mass of 5 kg. Find (a) the speed of the 30-kg block just before it hits the ledge, (b) the angular speed of the pulley at that time, (c) the tensions in the strings, and (d) the time it takes for the 30-kg block to reach the ledge. Assume that the string does not slip on the pulley.

**FIGURE 9-52**
**Problem 72**

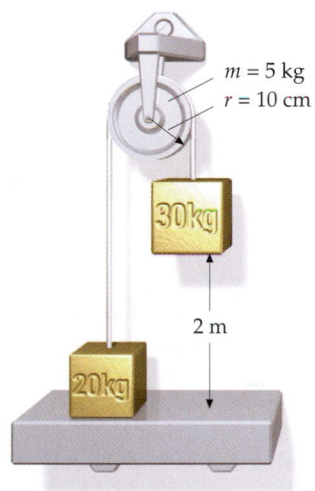

$m = 5$ kg
$r = 10$ cm

30kg

2 m

20kg

**73** •• A uniform sphere of mass $M$ and radius $R$ is free to rotate about a horizontal axis through its center. A string is wrapped around the sphere and is attached to an object of mass $m$ as shown in Figure 9-53. Find (a) the acceleration of the object and (b) the tension in the string.

**FIGURE 9-53** Problem 73

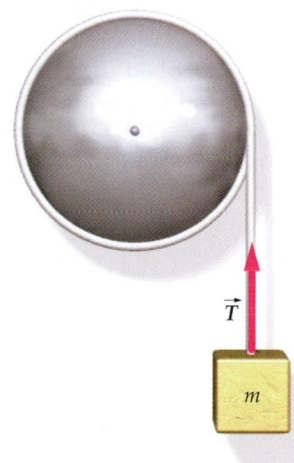

$\vec{T}$

$m$

**74** •• **iSOLVE** An Atwood's machine has two objects of masses $m_1 = 500$ g and $m_2 = 510$ g, connected by a string of negligible mass that passes over a pulley (Figure 9-54) with frictionless bearings. The pulley is a uniform disk with a mass of 50 g and a radius of 4 cm. The string does not slip on the pulley. (a) Find the acceleration of the objects. (b) What is the tension in the string supporting $m_1$? In the string supporting $m_2$? By how much do they differ? (c) What would your answers have been if you had neglected the mass of the pulley?

**FIGURE 9-54** Problem 74

The system is released from rest with $m_2$ at a height $h$ above the bottom of the incline. (a) What is the acceleration of the block? (b) What is the tension in the string? (c) What is the total energy of the cylinder–block–earth system when the block is at height $h$? (d) What is the total energy of the system when the block is at the bottom of the incline and has a speed $v$? (e) What is the speed $v$? (f) Evaluate your answers for the extreme cases of $\theta = 0°$, $\theta = 90°$, and $m_1 = 0$.

**FIGURE 9-57**
Problem 77

**75** •• **SSM** Two objects are attached to ropes that are attached to wheels on a common axle as shown in Figure 9-55. The two wheels are glued together so that they form a single object. The total moment of inertia of the object is 40 kg·m². The radii of the wheels are $R_1 = 1.2$ m and $R_2 = 0.4$ m. (a) If $m_1 = 24$ kg, find $m_2$ such that there is no angular acceleration of the wheels. (b) If 12 kg is gently added to the top of $m_1$, find the angular acceleration of the wheels and the tensions in the ropes.

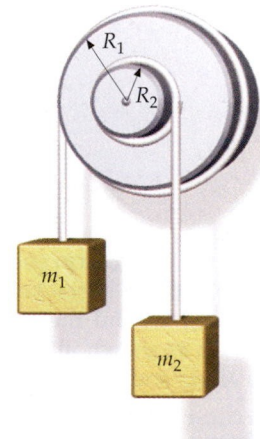

**FIGURE 9-55** Problem 75

**78** •• **SSM** A device for measuring the moment of inertia of an object is shown in Figure 9-58. A circular platform has a concentric drum of radius 10 cm about which a string is wound. The string passes over a frictionless pulley to a weight of mass $M$. The weight is released from rest, and the time required for it to drop a distance $D$ is measured. The system is then rewound, the object placed on the platform, and the system is again released from rest. The time required for the weight to drop the same distance $D$ then provides the data needed to calculate $I$. With $M = 2.5$ kg and $D = 1.8$ m, the time is 4.2 s. (a) Find the combined moment of inertia of the platform, drum, shaft, and pulley. (b) With the object placed on the platform, the time is 6.8 s for $D = 1.8$ m. Find $I$ of that object about the axis of the platform.

**FIGURE 9-58**
Problem 78

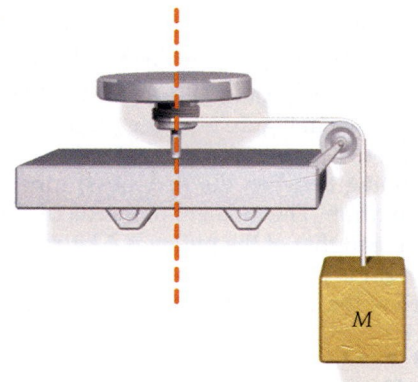

**76** •• The string wrapped around the cylinder in Figure 9-56 is held by a hand that is accelerated upward so that the center of mass of the cylinder does not move. Find (a) the tension in the string, (b) the angular acceleration of the cylinder, and (c) the acceleration of the hand.

**FIGURE 9-56**
Problems 76, 80

### Objects Rolling Without Slipping

**79** •• **iSOLVE** ✓ In 1993, a giant yo-yo of mass 400 kg measuring about 1.5 m in radius was dropped from a crane 57 m high. One end of the string was tied to the top of the crane, so the yo-yo unwound as it descended. Assuming that the axle of the yo-yo had a radius of $r = 0.1$ m, find the velocity of descent $v$ at the end of the fall.

**80** •• A uniform cylinder of mass $M$ and radius $R$ has a string wrapped around it. The string is held fixed, and the cylinder falls vertically as shown in Figure 9-56. (a) Show that the acceleration of the cylinder is downward with a magnitude $a = 2g/3$. (b) Find the tension in the string.

**77** •• A uniform cylinder of mass $m_1$ and radius $R$ is pivoted on frictionless bearings. A massless string wrapped around the cylinder is connected to a block of mass $m_2$ that is on a frictionless incline of angle $\theta$ as shown in Figure 9-57.

**81** •• A 0.1-kg yo-yo consists of two solid disks of radius 10 cm joined together by a massless rod of radius 1 cm and a string wrapped around the rod. One end of the string is held fixed and is under constant tension $T$ as the yo-yo is released. Find the acceleration of the yo-yo and the tension $T$.

**82** • **SSM** A homogeneous solid cylinder rolls without slipping on a horizontal surface. The total kinetic energy is $K$. The kinetic energy due to rotation about its center of mass is (a) $\frac{1}{2}K$, (b) $\frac{1}{3}K$, (c) $\frac{4}{7}K$, (d) none of these.

**83** • **iSOLVE** A homogeneous cylinder of radius 18 cm and mass 60 kg is rolling without slipping along a horizontal floor at 5 m/s. How much work was required to give it this motion?

**84** • Find the percentages of the total kinetic energy associated with rotation and translation for an object that is rolling without slipping if the object is (a) a uniform sphere, (b) a uniform cylinder, or (c) a hoop.

**85** • A hoop of radius 0.40 m and mass 0.6 kg is rolling without slipping at a speed of 15 m/s toward an incline of slope 30°. How far up the incline will the hoop roll, assuming that it rolls without slipping?

**86** •• **SSM** A uniform sphere rolls without slipping down an incline. What must be the angle of the incline if the linear acceleration of the center of mass of the sphere is $0.2g$?

**87** •• Repeat Problem 86 for a thin spherical shell.

**88** •• A basketball rolls without slipping down an incline of angle $\theta$. The coefficient of static friction is $\mu_s$. Find (a) the acceleration of the center of mass of the ball, (b) the frictional force acting on the ball, and (c) the maximum angle of the incline for which the ball will roll without slipping.

**89** •• Repeat Problem 88 for a solid cylinder of wood.

**90** •• **SSM** A hollow sphere and uniform sphere of the same mass $m$ and radius $R$ roll down an inclined plane from the same height $H$ without slipping (Figure 9-59). Each is moving horizontally as it leaves the ramp. When the spheres hit the ground, the range of the hollow sphere is $L$. Find the range $L'$ of the uniform sphere.

**FIGURE 9-59** Problem 90

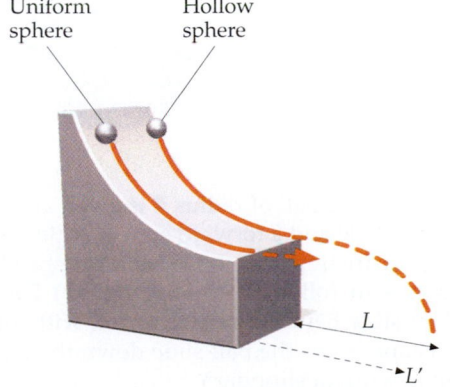

Uniform sphere    Hollow sphere

$L$
$L'$

**91** •• A uniform thin-walled cylinder and a uniform solid cylinder are rolling horizontally without slipping. The speed of the thin-walled cylinder is $v$. The cylinders encounter an incline that they climb without slipping. If the maximum height they reach is the same, find the initial speed $v'$ of the solid cylinder.

**92** •• A hollow, thin-walled cylinder and a solid sphere start from rest and roll without slipping down an inclined plane of length 3 m. The cylinder arrives at the bottom of the plane 2.4 s after the sphere. Determine the angle between the inclined plane and the horizontal.

**93** ••• A wheel has a thin 3.0-kg rim and four spokes, each of mass 1.2 kg. Find the kinetic energy of the wheel when it rolls at 6 m/s on a horizontal surface.

**94** •• **iSOLVE** ✓ Two uniform 20-kg disks of radius 30 cm are connected by a short rod of radius 2 cm and mass 1 kg. When the rod is placed on a plane inclined at 30°, so that the disks hang over the sides, the assembly rolls without slipping. Find (a) the linear acceleration of the system and (b) the angular acceleration of the system. (c) Find the kinetic energy of translation of the system after it has rolled 2 m down the incline, starting from rest. (d) Find the kinetic energy of rotation of the system at the same point.

**95** ••• A wheel of radius $R$ rolls without slipping at a speed $V$. The coordinates of the center of the wheel are $X$, $Y$. (a) Show that the $x$ and $y$ coordinates of point $P$ in Figure 9-60 are $X + r_0 \cos \theta$ and $R + r_0 \sin \theta$, respectively. (b) Show that the total velocity $\vec{v}$ of point $P$ has the components $v_x = V + (r_0 V \sin \theta)/R$ and $v_y = -(r_0 V \cos \theta)/R$. (c) Show that at the instant that $X = 0$, $\vec{v}$ and $\vec{r}$ are perpendicular to each other by calculating $\vec{v} \cdot \vec{r}$. (d) Show that $v = r\omega$, where $\omega = V/R$ is the angular velocity of the wheel. These results demonstrate that, in the case of rolling without slipping, the motion is the same as if the rolling object were instantaneously rotating about the point of contact with an angular speed $\omega = V/R$.

**FIGURE 9-60**
**Problem 95**

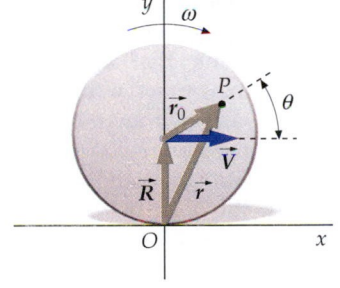

**96** ••• **SSM** A uniform cylinder of mass $M$ and radius $R$ is at rest on a block of mass $m$, which in turn rests on a horizontal, frictionless table (Figure 9-61). If a horizontal force $\vec{F}$ is applied to the block, it accelerates and the cylinder rolls without slipping. Find the acceleration of the block.

**FIGURE 9-61**
**Problems 96–98**

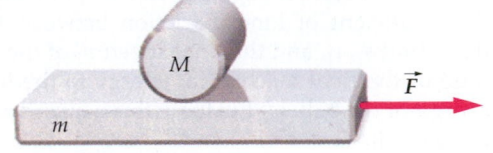

**97** ••• (a) Find the angular acceleration of the cylinder in Problem 96. Is the cylinder rotating clockwise or counterclockwise? (b) What is the cylinder's linear acceleration relative to the table? Let the direction of $\vec{F}$ be the positive direction. (c) What is the linear acceleration of the cylinder relative to the block?

**98** ••• If the force in Problem 96 acts over a distance $d$, find (a) the kinetic energy of the block and (b) the kinetic energy of the cylinder. (c) Show that the total kinetic energy is equal to the work done on the system.

**99** •• Two large gears are shown in Figure 9-62; each is free to rotate about a fixed axis running through its center. The radius of the first gear is $R_1 = 0.5$ m and the radius of the second gear is $R_2 = 1$ m. The moment of inertia of gear 1 is $I_1 = 1$ kg·m$^2$ and the moment of inertia of gear 2 is $I_2 = 16$ kg·m$^2$. The lever attached to gear 1 is 1 m long. (a) If a force of 2 N is applied to the end of the lever, what will the angular accelerations of gears 1 and 2 be? (b) What force must be applied tangentially to the edge of gear 2 to keep the gear system from rotating?

**FIGURE 9-62**
**Problem 99**

2 N

**100** •• SSM A marble of radius 1 cm rolls from rest from the top of a large sphere of radius 80 cm, which is held fixed. (a) Assuming that the marble rolls without slipping while it is in contact with the sphere (which is unrealistic), find the angle from the top of the sphere to the point where the marble breaks contact with the sphere. (b) Why is it unrealistic to assume that the marble rolls without slipping all the way down to the point where it breaks contact?

## Rolling With Slipping

**101** • A bowling ball of mass $M$ and radius $R$ is released so that at the instant it touches the floor it is moving horizontally with a speed $v_0$ and is not rotating. It slides for a time $t_1$ a distance $s_1$ before it begins to roll without slipping. (a) If $\mu_k$ is the coefficient of kinetic friction between the ball and the floor, find $s_1$, $t_1$, and the final speed $v_1$ of the ball. (b) Find the ratio of the final mechanical energy to the initial mechanical energy of the ball. (c) Evaluate these quantities for $v_0 = 8$ m/s and $\mu_k = 0.06$.

**102** •• SSM A cue ball of radius $r$ is initially at rest on a horizontal pool table (Figure 9-63). It is struck by a horizontal cue stick that delivers a force of magnitude $F_0$ for a very short time $\Delta t$. The stick strikes the ball at a point $h$ above the ball's point of contact with the table. Show that the ball's initial angular velocity $\omega_0$ is related to the initial linear velocity of its center of mass $v_0$ by $\omega_0 = (5/2)v_0(h - r)/r^2$.

**FIGURE 9-63**
**Problem 102**

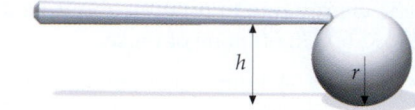

$h$    $r$

**103** •• A uniform solid sphere is set rotating about a horizontal axis with an angular speed $\omega_0$ and is placed on the floor. If the coefficient of sliding friction between the sphere and the floor is $\mu_k$, find the speed of the center of mass of the sphere when it begins to roll without slipping.

**104** •• A uniform solid ball resting on a horizontal surface has a mass of 20 g and a radius of 5 cm. A sharp force is applied to the ball in a horizontal direction 9 cm above the horizontal surface. The force increases linearly from 0 to a peak value of 40,000 N in $10^{-4}$ s and then decreases linearly to 0 in $10^{-4}$ s. (a) What is the velocity of the ball after impact? (b) What is the angular velocity of the ball after impact? (c) What is the velocity of the ball when it begins to roll without sliding? (d) For how long does the ball slide on the surface? Assume that $\mu_k = 0.5$.

**105** •• ISOLVE A 0.16-kg billiard ball of radius 3 cm is given a sharp blow by a cue stick. The applied force is horizontal and passes through the center of the ball. The initial velocity of the ball is 4 m/s. The coefficient of kinetic friction is 0.6. (a) For how many seconds does the ball slide before it begins to roll without slipping? (b) How far does it slide? (c) What is its velocity once it begins rolling without slipping?

**106** •• A billiard ball initially at rest is given a sharp blow by a cue stick. The force is horizontal and is applied at a distance $2R/3$ below the centerline, as shown in Figure 9-64. The initial speed of the ball is $v_0$ and the coefficient of kinetic friction is $\mu_k$. (a) What is the initial angular speed $\omega_0$? (b) What is the speed of the ball once it begins to roll without slipping? (c) What is the initial kinetic energy of the ball? (d) What is the frictional work done as it slides on the table?

**FIGURE 9-64**
**Problem 106**

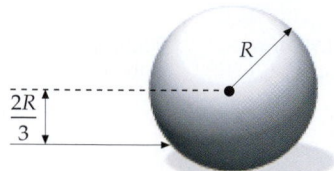

$R$

$\dfrac{2R}{3}$

**107** •• A bowling ball of radius $R$ is given an initial velocity $v_0$ down the lane and a forward spin $\omega_0 = 3v_0/R$. The coefficient of kinetic friction is $\mu_k$. (a) What is the speed of the ball when it begins to roll without slipping? (b) For how long does the ball slide before it begins to roll without slipping? (c) What distance does the ball slide down the lane before it begins rolling without slipping?

**108** •• **SSM** A solid cylinder of mass $M$ resting on its side on a horizontal surface is given a sharp blow by a cue stick. The applied force is horizontal and passes through the center of the cylinder so that the cylinder begins translating with initial velocity $v_0$. The coefficient of sliding friction between the cylinder and surface is $\mu_k$. (a) What is the translational velocity of the cylinder when it is rolling without slipping? (b) How far does the cylinder travel before it rolls without slipping? (c) What fraction of its initial mechanical energy is dissipated in friction?

**109** •• Consider a ball of radius $r$ and total mass $m$, with a nonuniform but radially symmetric mass distribution inside it, so that it can have an almost arbitrary moment of inertia $I$.

(a) Show that if this ball is projected across a floor with an initial velocity $v$, and is at first purely sliding across the floor (that is, its initial rotational velocity $\omega = 0$), its final velocity when it is rolling without slipping will be

$$v_f = \frac{1}{1 + I/mr^2} v,$$

independent of the coefficient of kinetic friction of the floor.

(b) Show that the total kinetic energy of the ball will be

$$K = \frac{1}{2} \frac{m}{(1 + I/mr^2)} v^2$$

(Note that $I/mr^2$ is independent of the mass and radius of the ball, but depends only on the distribution of mass inside the ball.)

## General Problems

**110** • **SSM** The moon rotates as it revolves around the earth so that we always see the same side. Use this fact to find the angular velocity in radians per second of the moon about its axis. (The period of revolution of the moon about the earth is 27.3 d.)

**111** • Find the moment of inertia of a hoop about an axis perpendicular to the plane of the hoop and through its edge.

**112** •• The radius of a park merry-go-round is 2.2 m. To start it rotating, you wrap a rope around it and pull with a force of 260 N for 12 s. During this time, the merry-go-round makes one complete rotation. (a) Find the angular acceleration of the merry-go-round. (b) What torque is exerted by the rope on the merry-go-round? (c) What is the moment of inertia of the merry-go-round?

**113** • A uniform stick of length 2 m is raised at an angle of 30° to the horizontal above a sheet of ice. The bottom end of the stick rests on the ice. When dropped from rest, the bottom of the stick remains in contact with the ice at all times. How far will the bottom of the stick have moved when the stick falls to the ice? Assume that the ice is frictionless.

**114** •• **ISOLVE ✓** A uniform disk of radius 0.12 m and mass 5 kg is pivoted so that it rotates freely about its central axis (Figure 9-65). A string wrapped around the disk is pulled with a force of 20 N. (a) What is the torque exerted on the disk? (b) What is the angular acceleration of the disk? (c) If the

disk starts from rest, what is its angular velocity after 5 s? (d) What is its kinetic energy after 5 s? (e) What is the total angle $\theta$ that the disk turns through in 5 s? (f) Show that the work done by the torque, $\tau \Delta \theta$, equals the kinetic energy.

**FIGURE 9-65** Problem 114

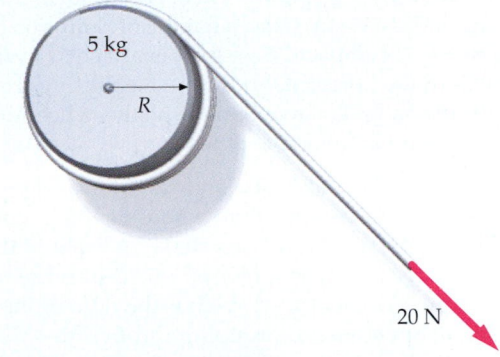

**115** •• A 0.25-kg rod of length 80 cm is suspended by a frictionless pivot at one end. It is held horizontal and released. Immediately after it is released, what is (a) the acceleration of the center of the rod and (b) the initial acceleration of a point on the end of the rod? (c) Find the linear velocity of the center of mass of the rod when it is vertical.

**116** •• A marble of mass $M$ and radius $R$ rolls without slipping down the track on the left from a height $h_1$ as shown in Figure 9-66. The marble then goes up the *frictionless* track on the right to a height $h_2$. Find $h_2$.

**FIGURE 9-66**
Problem 116

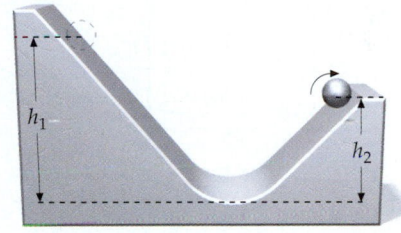

**117** •• **SSM** **ISOLVE** A uniform disk with a mass of 120 kg and a radius of 1.4 m rotates initially with an angular speed of 1100 rev/min. (a) A constant tangential force is applied at a radial distance of 0.6 m. How much work must this force do to stop the wheel? (b) If the wheel is brought to rest in 2.5 min, what torque does the force produce? What is the magnitude of the force? (c) How many revolutions does the wheel make in these 2.5 min?

**118** •• **ISOLVE** A park merry-go-round consists of a 240-kg circular wooden platform 4.00 m in diameter. Four children running alongside push tangentially along the platform's circumference until, starting from rest, the merry-go-round reaches a steady speed of one complete revolution every 2.8 s. (a) If each child exerts a force of 26 N, how far does each child run? (b) What is the angular acceleration of the merry-go-round? (c) How much work does each child do? (d) What is the kinetic energy of the merry-go-round?

**119** •• ██SOLVE██ A hoop of mass 1.5 kg and radius 65 cm has a string wrapped around its circumference and lies flat on a horizontal frictionless table. The string is pulled with a constant force of 5 N. (*a*) How far does the center of the hoop travel in 3 s? (*b*) What is the angular velocity of the hoop about its center of mass after 3 s?

**120** •• A vertical grinding wheel is a uniform disk of mass 60 kg and radius 45 cm. It has a handle of radius 65 cm of negligible mass. A compact 25-kg load is attached to the handle when it is in the horizontal position. Neglecting friction, find (*a*) the initial angular acceleration of the wheel and (*b*) the maximum angular velocity of the wheel.

**121** •• ██SSM██ Consider two uniform blocks of wood, identical in shape and composition, where one is larger than the other by a factor $S$ in all dimensions. (*a*) What is the ratio of the surface areas of the two blocks? (*b*) What is the ratio of the masses of the two blocks? (*c*) What is the ratio of the moments of inertia about some axis running through the block (in the same relative position and orientation in each)? These are examples of *scaling laws:* How do surface area, mass, and moment of inertia vary with the size of an object?

**122** •• In this problem, you are to derive the perpendicular-axis theorem for planar objects, which relates the moments of inertia about two perpendicular axes in the plane of Figure 9-67 to the moment of inertia about a third axis that is perpendicular to the plane of the figure. Consider the mass element $dm$ for the figure shown in the $xy$ plane. (*a*) Write an expression for the moment of inertia of the figure about the $z$ axis in terms of $dm$ and $r$. (*b*) Relate the distance $r$ of $dm$ to the distances $x$ and $y$ and show that $I_z = I_y + I_x$. (*c*) Apply your result to find the moment of inertia of a uniform disk of radius $R$ about a diameter of the disk.

**FIGURE 9-67**
**Problem 122**

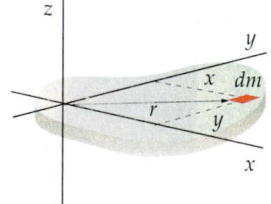

**123** •• ██SOLVE██ A uniform disk of radius $R$ and mass $M$ is pivoted about a horizontal axis parallel to its symmetry axis and passing through its edge so that it can swing freely in a vertical plane (Figure 9-68). It is released from rest with its center of mass at the same height as the pivot. (*a*) What is the angular velocity of the disk when its center of mass is directly below the pivot? (*b*) What force is exerted by the pivot at this time?

**FIGURE 9-68**
**Problem 123**

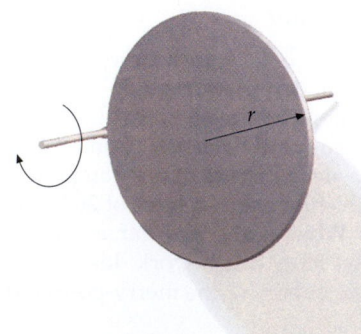

**124** •• The roof of the student dining hall at St. Mary's College is supported by high cross-braced wooden beams attached together in the shape of an upside-down L (see Figure 9-69). Each vertical beam is 12 ft high and 2 ft wide and the horizontal cross-member is 6 ft long. The mass of the vertical beam is 350 kg and the mass of the horizontal beam is 175 kg. When workers were building the hall, one of the structures started to fall over before it was anchored into place. (Luckily, the workers stopped it before it fell.) (*a*) If it started falling from an upright position, what was the initial angular acceleration of the structure? Assume that the bottom didn't slide across the floor and that it didn't fall "out of plane," that is, that the fall occurred in the vertical plane defined by the vertical and horizontal beams. (*b*) If a sparrow were sitting on the right end of the horizontal beam, what would the magnitude of its initial linear acceleration be when the structure started falling? (*c*) What would the horizontal component of the sparrow's initial linear acceleration be?

**FIGURE 9-69** Problem 124

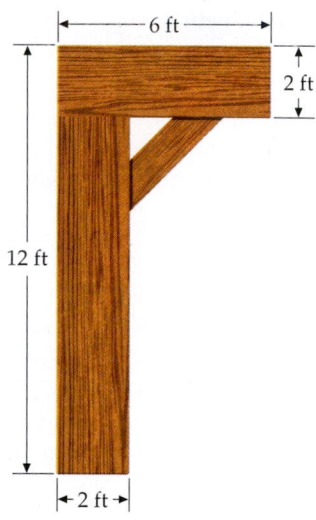

**125** •• A spool of mass $M$ rests on an inclined plane at a distance $D$ from the bottom. The ends of the spool have radius $R$, the center has radius $r$, and the moment of inertia of the spool about its axis is $I$. A long string of negligible mass is wound many times around the center of the spool. The other end of the string is fastened to a hook at the top of the inclined plane so that the string always pulls parallel to the slope as shown in Figure 9-70. (*a*) Suppose that initially the slope is so icy that there is *no* friction. How does the spool move as it slips down the slope? Use energy considerations to determine the speed of the center of mass of the spool when it reaches the bottom of the slope. Give your answer in terms of $M$, $I$, $r$, $R$, $g$, $D$, and $\theta$. (*b*) Now suppose that the ice is gone and that when the spool is set up in the same way, there is enough friction to keep it from slipping on the slope. What is the direction and magnitude of the friction force in this case?

**FIGURE 9-70** Problem 125

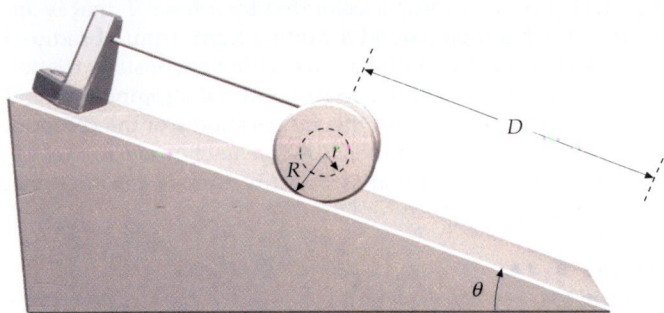

**126** •• ✓ A solid metal rod 1.5 m long is free to rotate without friction about a fixed horizontal axis perpendicular to the rod and passing through one end. The other end is held in a horizontal position. Small coins of mass $m$ are placed on the rod 25 cm, 50 cm, 75 cm, 1 m, 1.25 m, and 1.5 m from the bearing. If the free end is now released, calculate the initial force exerted on each coin by the rod. Assume that the mass of the coins may be neglected in comparison to the mass of the rod.

**127** •• **SSM** A popular classroom demonstration involves taking a meterstick and holding it horizontally at one end with a number of pennies spaced evenly along the meterstick. If the hand is relaxed so that the meterstick pivots about the hand under the influence of gravity, an interesting thing is seen: Pennies near the pivot point stay on the meterstick, while those farther away than a certain distance from the pivot are left behind by the falling meterstick. (This is often called the "faster than gravity" demonstration.) (a) What is the acceleration of the far end of the meterstick? (b) How far should a penny be from the end of the meterstick in order for it to be "left behind"?

**128** •• Figure 9-71 shows a hollow cylinder of length 1.8 m, mass 0.8 kg, and radius 0.2 m. The cylinder is free to rotate about a vertical axis that passes through its center and is perpendicular to the cylinder's axis. Inside the cylinder are two masses of 0.2 kg each, attached to springs of spring constant $k$ and unstretched lengths 0.4 m. The inside walls of the cylinder are frictionless. (a) Determine the value of the spring constant if the masses are located 0.8 m from the center of the cylinder when the cylinder rotates at 24 rad/s. (b) How much work was needed to bring the system from $\omega = 0$ to $\omega = 24$ rad/s?

**FIGURE 9-71** Problem 128

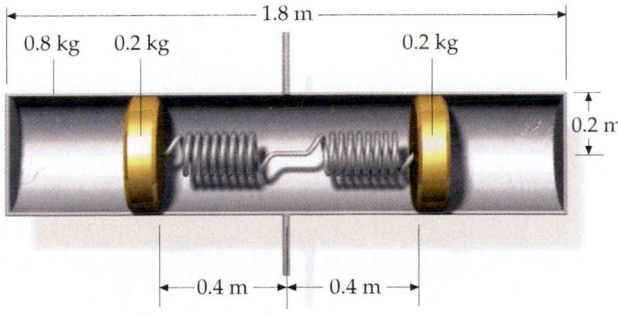

**129** •• Suppose that for the system described in Problem 128, the spring constants are each $k = 60$ N/m. The system starts from rest and slowly accelerates until the masses are 0.8 m from the center of the cylinder. How much work was done in the process?

**130** •• A string is wrapped around a uniform cylinder of radius $R$ and mass $M$ that rests on a horizontal frictionless surface. The string is pulled horizontally from the top with force $F$. (a) Show that the angular acceleration of the cylinder is twice that needed for rolling without slipping, so that the bottom point on the cylinder slides backward against the table. (b) Find the magnitude and direction of the frictional force between the table and cylinder needed for the cylinder to roll without slipping. What is the acceleration of the cylinder in this case?

**131** •• **SSM** Let's calculate the position $y$ of the falling load attached to the winch in Example 9-7 as a function of time by numerical integration. To do this, note that $v(y) = dy/dt$, or

$$t = \int_0^y \frac{1}{v(y')} dy' \approx \sum_{i=1}^N \frac{1}{v(y_i')} \Delta y'$$

where $\Delta y'$ is some (small) increment, $y_i' = i(\Delta y)$, and $y = y_N' = N(\Delta y)$. Hence, we can calculate $t$ as a function of $y$ by numerical summation. Make a graph of $y'$ versus $t$ for $t$ between 0 s and 2 s. Assume $M = 10$ kg, $R = 0.5$ m, $m = 5$ kg, $L = 10$ m, and $m_c = 3.5$ kg. Use $\Delta y' = 0.1$ m. Compare this to the position of the falling load if it were in free-fall.

**132** •• Figure 9-72 shows a solid cylinder of mass $M$ and radius $R$ to which a hollow cylinder of radius $r$ is attached. A string is wound about the hollow cylinder. The solid cylinder rests on a horizontal surface. The coefficient of static friction between the cylinder and surface is $\mu_s$. If a light tension is applied to the string in the vertical direction, the cylinder will roll to the left; if the tension is applied with the string horizontally to the right, the cylinder rolls to the right. Find the angle of the string with the horizontal that will allow the cylinder to remain stationary when a small tension is applied to the string.

**FIGURE 9-72**
**Problem 132**

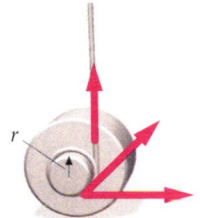

**133** •• **SSM** In problems dealing with a pulley with a nonzero moment of inertia, the magnitude of the tensions in the ropes hanging on either side of the pulley are not equal. The difference in the tension is due to the static frictional force between the rope and the pulley; however, the static frictional force cannot be made arbitrarily large. If you consider a massless rope wrapped partly around a cylinder through an angle $\Delta\theta$ (measured in radians), then you can show that if the tension on one side of the pulley is $T$, while the tension on the other side is $T'$ ($T' > T$), the maximum value of $T'$ in relation to $T$ that can be maintained without the rope slipping is $T'_{max} = Te^{\mu_s \Delta\theta}$, where $\mu_s$ is the coefficient of static friction. Consider the Atwood's machine in Figure 9-73: the pulley has a radius $r = 0.15$ m, the moment of inertia $I = 0.35$ kg·m², and the coefficient of static friction $\mu_s = 0.30$. (a) If the tension on one side of the pulley is 10 N, what is the maximum tension on the other side that will prevent the rope from slipping on the pulley? (b) If the mass of one of the hanging blocks is 1 kg, what is the maximum mass of the other block if, after the blocks are released, the pulley is to rotate without slipping? (c) What is the acceleration of the blocks in this case?

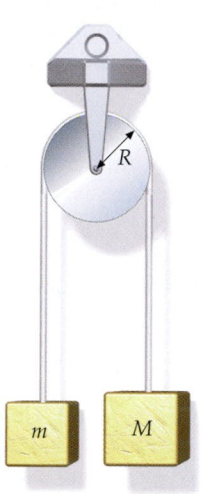

**FIGURE 9-73** Problem 133

**134** ••• A heavy, uniform cylinder has a mass $m$ and a radius $R$ (Figure 9-74). It is accelerated by a force $\vec{T}$ that is applied through a rope wound around a light drum of radius $r$ that is attached to the cylinder. The coefficient of static friction is sufficient for the cylinder to roll without slipping. (a) Find the frictional force. (b) Find the acceleration $a$ of the center of the cylinder. (c) Show that it is possible to choose $r$ so that $a$ is greater than $T/m$. (d) What is the direction of the frictional force in the circumstances of Part (c)?

**FIGURE 9-74**
**Problem 134**

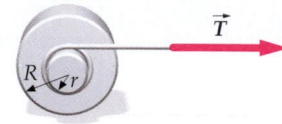

**135** ••• A uniform rod of length $L$ and mass $M$ is free to rotate about a horizontal axis through one end as shown in Figure 9-75. The rod is released from rest at $\theta = \theta_0$. Show that the force exerted by the axis on the rod is given by $F_{\parallel} = \frac{1}{2}Mg(5\cos\theta - 3\cos\theta_0)$ and $F_{\perp} = \frac{1}{4}Mg\sin\theta$, where $F_{\parallel}$ is the component of the force parallel with the rod and $F_{\perp}$ is the component of the force perpendicular to the rod.

**FIGURE 9-75**
**Problem 135**

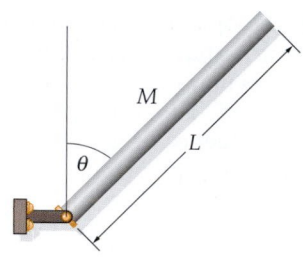

# Conservation of Angular Momentum

ANGULAR MOMENTUM IS THE
ROTATIONAL ANALOG OF LINEAR
MOMENTUM.

**?** **When the riders move closer
to the center, the merry-go-round
gains speed. Why does this
happen? (See Example 10-4.)**

10-1    The Vector Nature of Rotation

10-2    Torque and Angular Momentum

10-3    Conservation of Angular Momentum

*10-4   Quantization of Angular Momentum

➤ In this chapter, we extend our study of rotational motion to situations in
which the direction of the axis of rotation may change.

We begin with an examination of the vector properties of angular velocity
and torque and then introduce the concept of angular momentum, which is the
rotational analog of linear momentum. We then show that the net torque acting on a system equals the rate of change of its angular momentum, a result
that is equivalent to Newton's second law for rotational motion. Angular
momentum is therefore conserved in systems with zero net external torque.
Like conservation of linear momentum, conservation of angular momentum
is a fundamental law of nature, applying even in the atomic domain where
Newtonian mechanics fails.

## 10-1  The Vector Nature of Rotation

In Chapter 9, we indicated the direction of rotation about an axis with a fixed
direction by assigning plus and minus signs to indicate the direction of the angular velocity, just as we used them to indicate the direction of the velocity in one-
dimensional motion. But when the direction of the axis of rotation is *not* fixed in
space, plus and minus signs are not adequate to describe the direction of the angular velocity. This inadequacy is overcome by treating the angular velocity as a
vector $\vec{\omega}$ directed along the rotation axis. Consider, for example, the rotating disk
in Figure 10-1. We determine the direction of $\vec{\omega}$ by a convention known as the

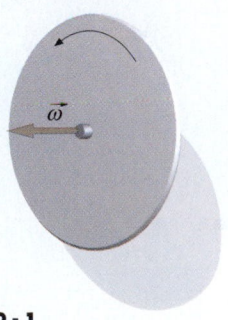

**FIGURE 10-1**

309

**right-hand rule,** which is illustrated in Figure 10-2. Thus, if the rotation is directed as shown in Figure 10-1, $\vec{\omega}$ is directed as shown; if the direction of the rotation is reversed, so is the direction of $\vec{\omega}$.

We apply similar considerations to the torque. Figure 10-3 shows a force $\vec{F}$ acting on a particle at some position $\vec{r}$ relative to the origin $O$. The torque $\vec{\tau}$ exerted by this force relative to the origin $O$ is defined as a vector that is perpendicular to the plane formed by $\vec{F}$ and $\vec{r}$ and has magnitude $Fr \sin \phi$, where $\phi$ is the angle between $\vec{F}$ and $\vec{r}$. (The angle between two vectors refers to the angle *between the directions* of the two vectors.) If $\vec{F}$ and $\vec{r}$ are in the $xy$ plane, as in Figure 10-3, the torque vector is along the $z$ axis. If $\vec{F}$ is applied to the rim of a disk of radius $r$, as shown in Figure 10-4, the torque vector has the magnitude $Fr$, and is along the axis of rotation as shown.

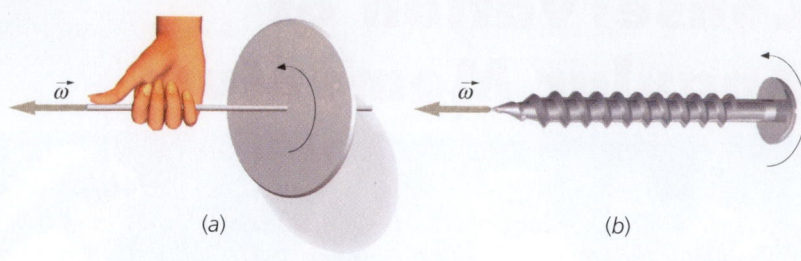

(a)     (b)

**FIGURE 10-2** (a) When the fingers of the right hand curl in the direction of rotation, the thumb points in the direction of $\vec{\omega}$. (b) Looked at another way, the direction of $\vec{\omega}$ is that of the advance of a rotating right-hand screw.

## The Cross Product

Torque is expressed mathematically as the **cross product** (or **vector product**) of $\vec{r}$ and $\vec{F}$:

$$\vec{\tau} = \vec{r} \times \vec{F}$$     10-1

The cross product of two vectors $\vec{A}$ and $\vec{B}$ is defined to be a vector $\vec{C} = \vec{A} \times \vec{B}$ whose magnitude equals the area of the parallelogram formed by the two vectors (Figure 10-5). The vector $\vec{C}$ is perpendicular to the plane containing $\vec{A}$ and $\vec{B}$ in the direction given by the right-hand rule, that is, as the fingers curl from the direction of $\vec{A}$ toward the direction of $\vec{B}$ (Figure 10-6). If $\phi$ is the angle between the two vectors and $\hat{n}$ is a unit vector that is perpendicular to each vector in the direction of $\vec{C}$, the cross product of $\vec{A}$ and $\vec{B}$ is

$$\vec{A} \times \vec{B} = AB \sin \phi \, \hat{n}$$     10-2

DEFINITION—CROSS PRODUCT

If $\vec{A}$ and $\vec{B}$ are parallel, $\vec{A} \times \vec{B}$ is zero. It follows from the definition of the cross product that

$$\vec{A} \times \vec{A} = 0$$     10-3

and

$$\vec{A} \times \vec{B} = -\vec{B} \times \vec{A}$$     10-4

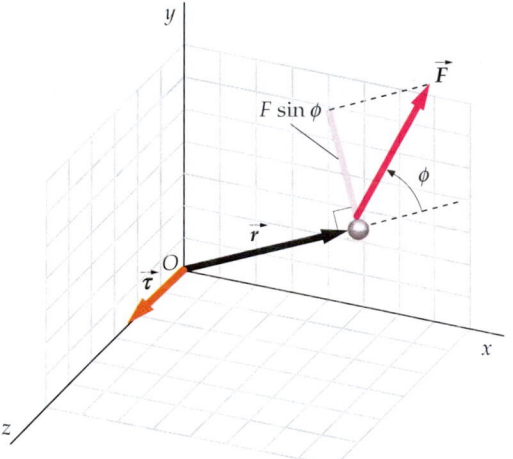

**FIGURE 10-3**

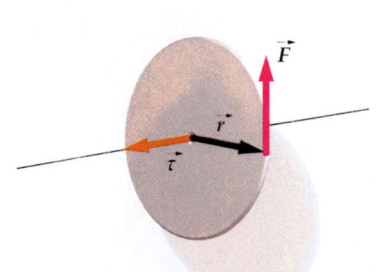

**FIGURE 10-4**

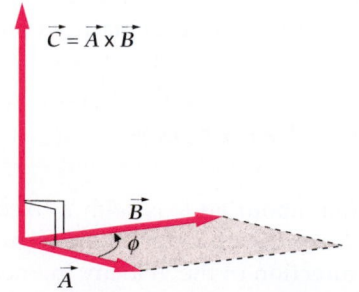

**FIGURE 10-5** The cross product $\vec{A} \times \vec{B}$ is a vector $\vec{C}$ that is perpendicular to both $\vec{A}$ and $\vec{B}$ and has a magnitude $AB \sin \phi$, which equals the area of the parallelogram shown.

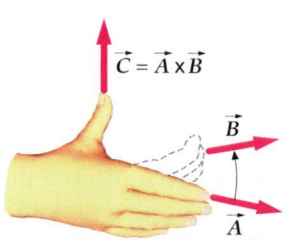

**FIGURE 10-6** The direction of $\vec{A} \times \vec{B}$ is given by the right-hand rule when the fingers are rotated from the direction of $\vec{A}$ toward $\vec{B}$ through the angle $\phi$.

Note that the order in which two vectors are multiplied in a cross product makes a difference. Below are some properties of the cross product of two vectors:

1. The cross product obeys a distributive law under addition:

$$\vec{A} \times (\vec{B} + \vec{C}) = (\vec{A} \times \vec{B}) + (\vec{A} \times \vec{C})$$  10-5

2. If $\vec{A}$ and $\vec{B}$ are functions of some variable such as $t$, the derivative of $\vec{A} \times \vec{B}$ follows the usual product rule for derivatives:

$$\frac{d}{dt}(\vec{A} \times \vec{B}) = \left(\vec{A} \times \frac{d\vec{B}}{dt}\right) + \left(\frac{d\vec{A}}{dt} \times \vec{B}\right)$$  10-6

3. The unit vectors $\hat{i}$, $\hat{j}$, and $\hat{k}$ (Figure 10-7), which are mutually perpendicular, have cross products given by

$$\hat{i} \times \hat{j} = \hat{k}, \qquad \hat{j} \times \hat{k} = \hat{i}, \qquad \text{and} \qquad \hat{k} \times \hat{i} = \hat{j}$$  10-7a

Furthermore,

$$\hat{i} \times \hat{i} = \hat{j} \times \hat{j} = \hat{k} \times \hat{k} = 0$$  10-7b

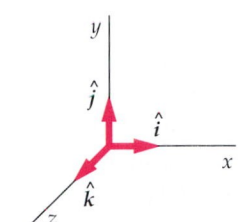

**FIGURE 10-7**

## 10-2 Torque and Angular Momentum

Figure 10-8 shows a particle of mass $m$ moving with a velocity $\vec{v}$ at a position $\vec{r}$ relative to the origin $O$. The linear momentum of the particle is $\vec{p} = m\vec{v}$. The **angular momentum** $\vec{L}$ of the particle relative to the origin $O$ is defined to be the cross product of $\vec{r}$ and $\vec{p}$:

$$\vec{L} = \vec{r} \times \vec{p}$$  10-8

ANGULAR MOMENTUM OF A PARTICLE DEFINED

If $\vec{r}$ and $\vec{p}$ are in the $xy$ plane, as in Figure 10-8, $\vec{L}$ is parallel with the $z$ axis and is given by $\vec{L} = \vec{r} \times \vec{p} = mvr \sin \phi \, \hat{k}$. Like torque, angular momentum is defined *relative to a point in space.*

Figure 10-9 shows a particle of mass $m$ attached to a circular disk of negligible mass in the $xy$ plane with its center at the origin. The disk is spinning about its axis with angular speed $\omega$. The speed $v$ of the particle and its angular speed are related by $v = r\omega$. The angular momentum of the particle relative to the center of the disk is

$$\vec{L} = \vec{r} \times \vec{p} = \vec{r} \times m\vec{v} = rmv \sin 90° \, \hat{k} = rmv\hat{k} = mr^2\omega\hat{k} = mr^2\vec{\omega}$$

**FIGURE 10-8**

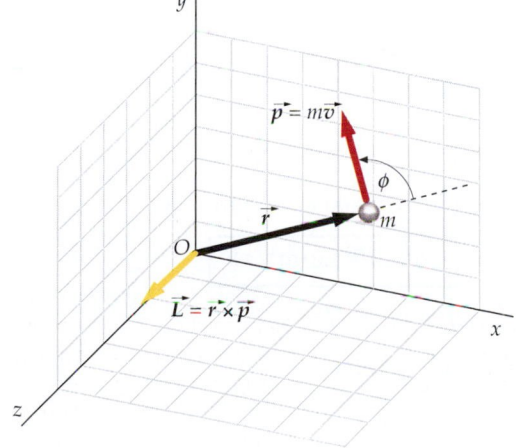

**FIGURE 10-9**

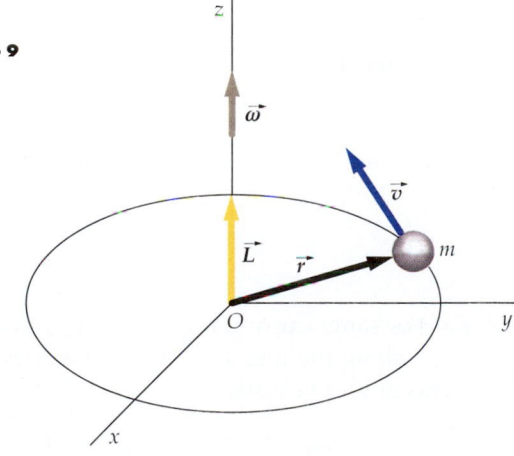

The angular momentum vector is in the same direction as the angular velocity vector.

Since $mr^2$ is the moment of inertia for a single particle about the $z$ axis, we have

$$\vec{L} = mr^2\vec{\omega} = I\vec{\omega}$$

This result does not hold for the angular momentum about a general point on the $z$ axis. Figure 10-10 shows the angular momentum vector $\vec{L}'$ for the same particle attached to the same disk but with $\vec{L}'$ computed about a point on the $z$ axis that is not at the center of the circle. In this case, the angular momentum is not parallel to the angular velocity vector $\vec{\omega}$, which is parallel with the $z$ axis.

In Figure 10-11, we attach a second particle of equal mass to the spinning disk. The angular momentum vectors $\vec{L}'_1$ and $\vec{L}'_2$ are shown relative to the same point $O'$. The total angular momentum $\vec{L}'_1 + \vec{L}'_2$ of the two-particle system is again parallel to the angular velocity vector $\vec{\omega}$. In this case, the axis of rotation, the $z$ axis, passes through the center of mass of the two-particle system, and the mass distribution is symmetric about this axis. Such an axis is called a **symmetry axis.** For any system of particles that rotates about a symmetry axis, the total angular momentum (which is the sum of the angular momenta of the individual particles) is parallel to the angular velocity and is given by

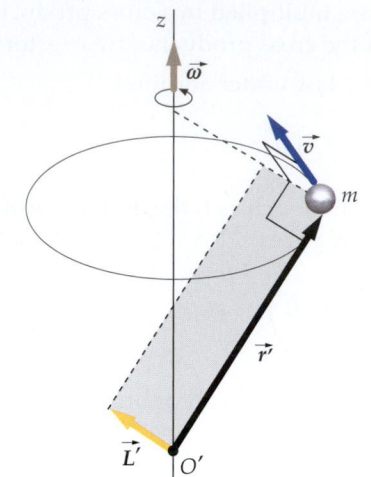

**FIGURE 10-10**

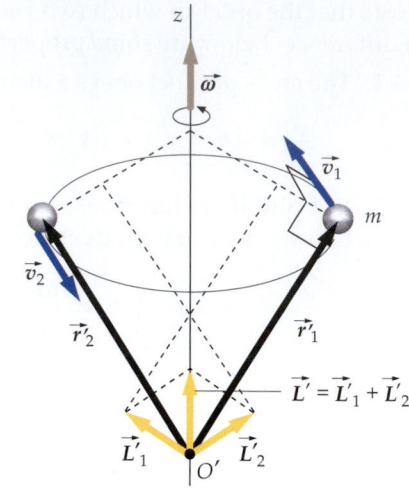

**FIGURE 10-11**

$$\vec{L} = I\vec{\omega} \qquad \text{10-9}$$

ANGULAR MOMENTUM OF A SYSTEM ROTATING ABOUT A SYMMETRY AXIS

---

*ANGULAR MOMENTUM ABOUT THE ORIGIN*          **EXAMPLE 10-1**

**Find the angular momentum about the origin for the following situations.** (*a*) **A car of mass 1200 kg moves in a circle of radius 20 m with a speed of 15 m/s. The circle is in the *xy* plane, centered at the origin. When viewed from a point on the positive *z* axis, the car moves counterclockwise.** (*b*) **The same car moves in the *xy* plane with velocity $\vec{v} = -(15 \text{ m/s})\hat{i}$ along the line $y = y_0 = 20$ m parallel to the *x* axis.** (*c*) **A disk in the *xy* plane of radius 20 m and mass 1200 kg rotates at 0.75 rad/s about its axis, which is also the *z* axis. When viewed from a point on the positive *z* axis, the disk rotates counterclockwise.**

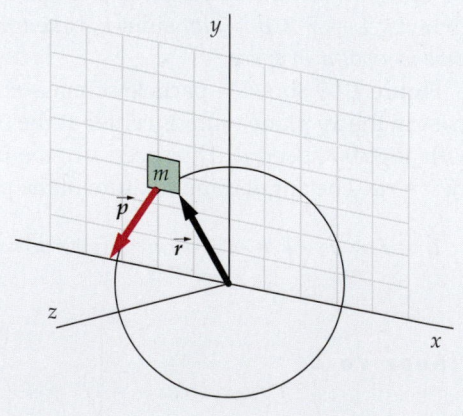

**PICTURE THE PROBLEM** For (*a*) and (*b*) use $\vec{L} = \vec{r} \times \vec{p}$. For (*c*) use $\vec{L} = I\vec{\omega}$. Draw a figure and apply the right-hand rule to find the direction of $\vec{L}$.

**FIGURE 10-12**

(*a*) $\vec{r}$ and $\vec{p}$ are perpendicular and $\vec{r} \times \vec{p}$ is along the $z$ axis (Figure 10-12):

$$\vec{L} = \vec{r} \times \vec{p} = rmv \sin 90° \,\hat{k}$$
$$= (20 \text{ m})(1200 \text{ kg})(15 \text{ m/s})\hat{k}$$
$$= \boxed{3.6 \times 10^5 \text{ kg·m}^2/\text{s}\,\hat{k}}$$

(*b*) 1. For the same car moving in the direction of decreasing $x$ along the line $y = 20$ m, we express $\vec{r}$ and $\vec{p}$ in terms of unit vectors:

$$\vec{r} = x\hat{i} + y\hat{j} = x\hat{i} + y_0\hat{j}$$
$$\vec{p} = m\vec{v} = -mv\hat{i}$$

2. Now compute $\vec{r} \times \vec{p}$ (Figure 10-13):

$$\vec{L} = \vec{r} \times \vec{p} = (x\hat{i} + y_0\hat{j}) \times (-mv\hat{i})$$

$$= -xmv(\hat{i} \times \hat{i}) - y_0\,mv(\hat{j} \times \hat{i})$$

$$= 0 - y_0mv(-\hat{k}) = y_0mv\hat{k}$$

$$= (20\text{ m})(1200\text{ kg})(15\text{ m/s})\hat{k}$$

$$= \boxed{3.6 \times 10^5 \text{ kg·m}^2/\text{s}\,\hat{k}}$$

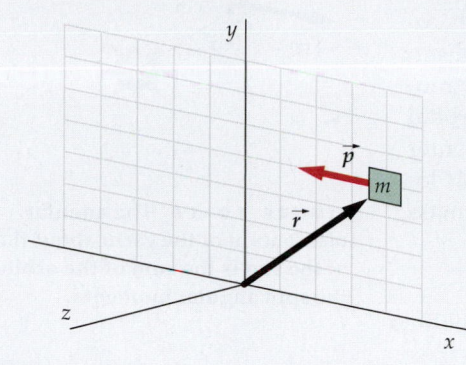

**FIGURE 10-13**

(c) Use $\vec{L} = I\vec{\omega}$
(Figure 10-14):

$$\vec{L} = I\vec{\omega} = I\omega\hat{k} = \tfrac{1}{2}mR^2\omega\,\hat{k}$$

$$= \tfrac{1}{2}(1200\text{ kg})(20\text{ m})^2\,(0.75\text{ rad/s})\hat{k}$$

$$= \boxed{1.8 \times 10^5 \text{ kg·m}^2/\text{s}\,\hat{k}}$$

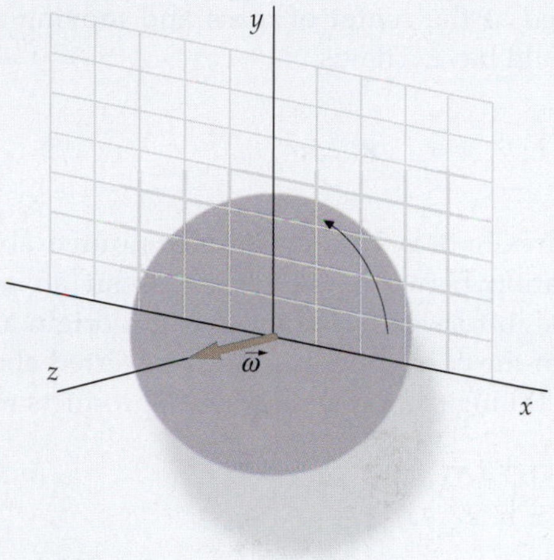

**FIGURE 10-14**

**REMARKS** The angular momentum of the car moving in a circle in (a) is the same as that of the car moving along a straight line in (b). In (c), the velocity of a point on the rim is $v = R\omega = (20\text{ m})(0.75\text{ rad/s}) = 15\text{ m/s}$, the same as the velocity of the car in Parts (a) and (b). The moment of inertia of a 1200-kg disk of radius 20 m is less than that of a 1200-kg car at 20 m from the axis because much of the mass of the disk is closer to the axis of rotation.

There are several additional results concerning torque and angular momentum for a system of particles. The first of these is

$$\vec{\tau}_{\text{net ext}} = \frac{d\vec{L}_{\text{sys}}}{dt}$$

10-10

The net external torque acting on a system equals the rate of change of the angular momentum of the system.

NEWTON'S SECOND LAW FOR ANGULAR MOTION

In Equation 10-10 the net external torque is the vector sum of the external torques acting on the system. Integrating both sides of this equation with respect to time gives

$$\Delta\vec{L}_{\text{sys}} = \int_{t_i}^{t_f} \vec{\tau}_{\text{net ext}}\, dt$$

10-11

ANGULAR IMPULSE–ANGULAR MOMENTUM EQUATION

It is often useful to split the total angular momentum of a system about an arbitrary point $O$ into orbital angular momentum and spin angular momentum:

$$\vec{L}_{sys} = \vec{L}_{orbit} + \vec{L}_{spin} \qquad \text{10-12}$$

The earth has spin angular momentum due to its spinning motion about its rotational axis and it has orbital angular momentum about the center of the sun due to its orbital motion around the sun (Figure 10-15). The total angular momentum of the earth relative to the sun is the vector sum of the spin and orbital angular momenta. $\vec{L}_{spin}$ is the angular momentum of a system about its center of mass and $\vec{L}_{orbit}$ is the angular momentum that a point particle of mass $M$ located at the center of mass and moving at the velocity of the center of mass would have. That is,

$$\vec{L}_{orbit} = \vec{r}_{cm} \times M\vec{v}_{cm} \qquad \text{10-13}$$

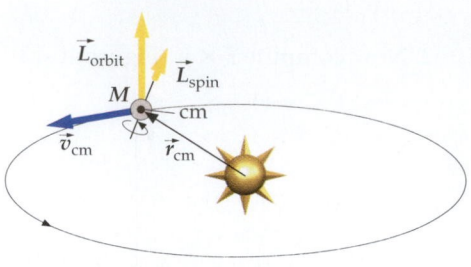

**FIGURE 10-15** The angular momentum of the earth about the center of the sun is the sum of the orbital and the spin angular momenta.

In Chapter 9 torques are computed about axes instead of about points. The relation between the torque about an axis and the torque about a point is straightforward. If point $O$ is the origin and if force $\vec{F}$ exerts torque $\vec{\tau}$ about $O$, then the corresponding torque exerted about the $z$ axis is $\tau_z$ (the $z$ component of $\vec{\tau}$.) Taking components of cross products requires some care. If $\vec{\tau} = \vec{r} \times \vec{F}$, then

$$\vec{\tau}_z = \vec{r}_{rad} \times \vec{F}_{xy} \qquad \text{10-14}$$

TORQUE ABOUT Z AXIS

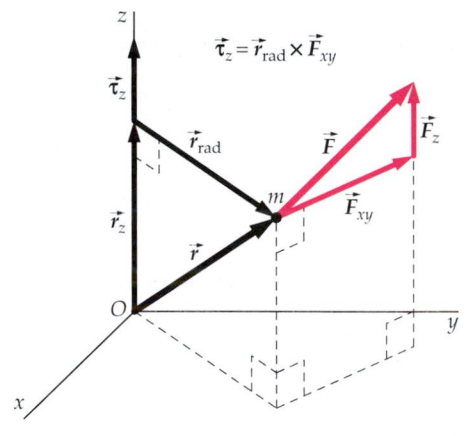

**FIGURE 10-16**

where $\vec{\tau}_z$, $\vec{r}_{rad}$, and $\vec{F}_{xy}$ (see Figure 10-16) are vector components of $\vec{\tau}$, $\vec{r}$, and $\vec{F}$. (The vector component in a given direction is the scalar component in that direction times the unit vector in that direction. For example, $\vec{\tau}_z = \tau_z \hat{k}$.) $\vec{r}_{rad}$ is the component of $\vec{r}$ directed radially away from the $z$ axis and $\vec{F}_{xy}$ is the component of $\vec{F}$ in the $xy$ plane (the plane perpendicular to the $z$ axis). The relation between angular momentum about an axis and angular momentum about a point is also straightforward. The angular momentum about an axis is

$$\vec{L}_z = \vec{r}_{rad} \times \vec{p}_{xy} \qquad \text{10-15}$$

ANGULAR MOMENTUM ABOUT Z AXIS

where $\vec{p}_{xy}$ is the projection of the linear momentum $\vec{p}$ in the $xy$ plane. Taking the $z$ vector components of both sides of Equation 10-10 gives

$$\vec{\tau}_{net\,ext,z} = \frac{d\vec{L}_{sys,z}}{dt} \qquad \text{10-16}$$

For a rigid object rotating about the $z$ axis, $\vec{L}_{sys,z} = I_z\vec{\omega}$, where $I_z$ is the moment of inertia about the $z$ axis. Substituting this into Equation 10-16 gives

$$\vec{\tau}_{net\,ext,z} = \frac{d\vec{L}_{sys,z}}{dt} = \frac{d}{dt}(I_z\vec{\omega}) = I_z\vec{\alpha} \qquad \text{10-17}$$

where the angular acceleration vector $\vec{\alpha}$ is defined by $\vec{\alpha} = d\vec{\omega}/dt$. (Equation 10-17 is the vector form of Equation 9-18.)

For a system of particles, the total angular momentum about the $z$ axis equals the sum of the angular momenta about the $z$ axis of the individual particles. Also, the total torque about the $z$ axis is the sum of the torques about the $z$ axis acting on the system.

THE ATWOOD'S MACHINE REVISITED

**EXAMPLE  10-2**

An Atwood's machine has two blocks with masses $m_1$ and $m_2$ ($m_1 > m_2$) connected by a string of negligible mass that passes over a pulley with frictionless bearings. The pulley is a uniform disk of mass $M$ and radius $R$. The string does not slip on the pulley. Apply Equation 10-16 to the system consisting of the two blocks, the string, and the pulley, to find the angular acceleration of the pulley and the linear acceleration of the blocks.

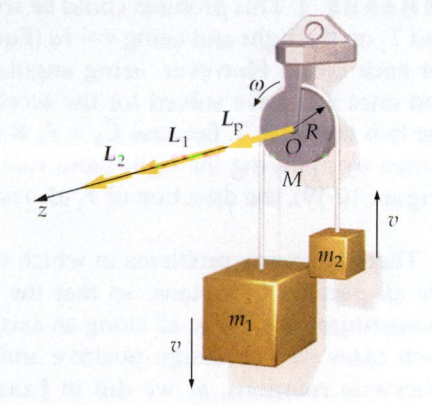

**PICTURE THE PROBLEM** Let the pulley and blocks be in the $xy$ plane with the $z$ axis out of the page through the center of the pulley at point $O$ as shown in Figure 10-17. We compute the torques and angular momenta about the $z$ axis and apply Newton's second law for angular motion (Equation 10-10). Since $m_1$ is greater than $m_2$, the disk will rotate counterclockwise corresponding to $\vec{\omega}$ out of the page in the positive $z$ direction. All the forces are in the $xy$ plane so all torques are along the $z$ axis. Also, all the velocities are in the $xy$ plane so all the angular momentum vectors are along the $z$ axis. Since the torque, angular velocity, and angular momentum vectors are all along the $z$ axis, we can treat this as a one-dimensional problem with positive assigned to counterclockwise motion and negative to clockwise motion. The acceleration $a$ of the blocks is related to the angular acceleration $\alpha$ of the pulley by the nonslip condition $a = R\alpha$.

**FIGURE 10-17**

1. Draw a free-body diagram of the system (Figure 10-18). The only thing touching the system is the pulley bearings. The external forces on the system are the normal force of the pulley bearings on the pulley and the gravity forces on the two blocks and the pulley:

2. Express Newton's second law for rotation, $z$ components only (Equation 10-16):

$$\sum \tau_{ext,z} = \frac{dL_z}{dt}$$

3. The total external torque about the $z$ axis is the sum of the torques exerted by the external forces. The lever arms $F_1$ and $F_2$ each equal $R$. (The lever arms of the other two forces are each zero.) $F_1 = m_1 g$ and $F_2 = m_2 g$:

$$\sum \tau_{ext,z} = \tau_n + \tau_p + \tau_1 + \tau_2$$

$$= 0 + 0 + m_1 gR - m_2 gR$$

**FIGURE 10-18**

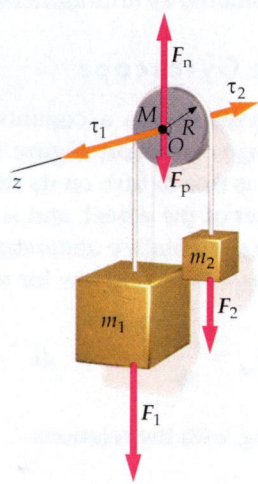

4. The total angular momentum about the $z$ axis equals the angular momentum of the pulley $\vec{L}_p$ plus the angular momenta of block 1, $\vec{L}_1$, and block 2, $\vec{L}_2$, each in the positive $z$ direction. The pulley has spin angular momentum but no orbital angular momentum, whereas each block has orbital angular momentum but no spin angular momentum.

$$L_z = L_p + L_1 + L_2$$

$$= I\omega + m_1 vR + m_2 vR$$

5. Substitute these results into Newton's second law for rotation in step 2:

$$\sum \tau_{ext,z} = \frac{dL_z}{dt}$$

$$m_1 gR - m_2 gR = \frac{d}{dt}(I\omega + m_1 vR + m_2 vR)$$

$$m_1 gR - m_2 gR = I\alpha + (m_1 + m_2)Ra$$

6. Relate $I$ to $M$ and $R$, and use the nonslip condition to relate $\alpha$ to $a$ and solve for both $a$ and $\alpha$:

$$m_1 gR - m_2 gR = \frac{1}{2}MR^2 \frac{a}{R} + (m_1 + m_2)Ra$$

so

$$\boxed{a = (m_1 - m_2)g/(\tfrac{1}{2}M + m_1 + m_2)}, \text{ and}$$

$$\boxed{\alpha = a/R = (m_1 - m_2)g/[(\tfrac{1}{2}M + m_1 + m_2)R]}$$

**REMARKS** 1. This problem could be solved by writing the tensions $T_1$ on the left and $T_2$ on the right and using $\tau = I\alpha$ (Equation 10-17) for the pulley and $\Sigma F_y = ma_y$ for each block. However, using angular momentum (Equation 10-16) is easier, and once you have solved for the acceleration, it is straightforward to solve for the two tensions. 2. Because $\vec{L}_2 = \vec{r}_2 \times m_2\vec{v}_2$ (Figure 10-19), the direction of $\vec{L}_2$ is gotten by applying the right-hand rule (Figure 10-6). And because $\vec{\tau}_2 = \vec{r}_2 \times \vec{F}_2$ (Figure 10-19), the direction of $\vec{\tau}_2$ also is gotten by applying the right-hand rule.

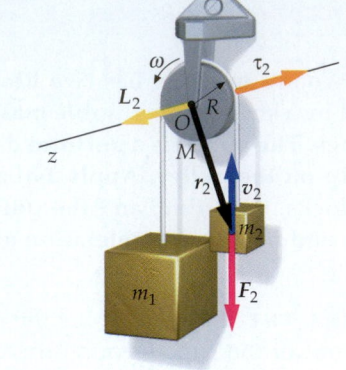

**FIGURE 10-19**

There are many problems in which the forces, position vectors, and velocities are all parallel to a plane, so that the torques, angular velocities, and angular momentum vectors are all along an axis of rotation that remains fixed in space. In such cases we can assign positive and negative values to counterclockwise or clockwise rotations, as we did in Example 10-2, and treat the case like a one-dimensional problem. However, there are other situations, such as the motion of a gyroscope, where torque, angular velocity, and angular momentum must be considered as multidimensional vectors.

## The Gyroscope

A gyroscope is a common example of motion in which the axis of rotation changes direction. Figure 10-20 shows a gyroscope consisting of a bicycle wheel that is free to turn on its axle. The axle is pivoted at a point a distance $D$ from the center of the wheel, and is free to rotate about the pivot in any direction. We can give a qualitative understanding of the complex motion of such a system by using Newton's second law for rotation,

$$\vec{\tau}_{net} = \frac{d\vec{L}}{dt} \quad \left( \text{or} \quad d\vec{L} = \vec{\tau}_{net}dt \right)$$

along with the relations

$$\vec{\tau}_{net} = \vec{r}_{cm} \times M\vec{g}$$

and

$$\vec{L} = I_s\vec{\omega}_s$$

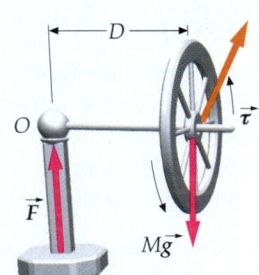

**FIGURE 10-20**

where $I_s$ and $\omega_s$ are the moment of inertia and angular velocity of the wheel about its spin axis. All we need to remember in order to describe the motion of a gyroscope is that the *change* in angular momentum of the wheel must be in the direction of the net torque acting on it.

Suppose the axle is held horizontally and then released. If the wheel isn't spinning, it simply falls, rotating about a horizontal axis through $O$ and perpendicular to $\vec{r}$. The torque vector is horizontal, into the page. For this case the initial angular momentum is zero, so the *change* in angular momentum equals the angular momentum itself, which, in this case, is also horizontal and into the page. However, if the wheel *is* spinning and has a large angular momentum along its axle, it does not fall when the axle is released. If it were to fall, the axle would point downward, resulting in a large component of angular momentum in the downward direction. But there is no torque in the downward direction. The torque is horizontal. What actually happens is that the axle moves horizontally (into the paper in the figure). The wheel must move this way so that the *change* in angular momentum is in the direction of the net torque. This is illustrated in Figure 10-21a, where we see a large angular momentum $\vec{L}$ along the axis of the wheel and a change in angular momentum $d\vec{L}$ in the direction of the torque $\vec{\tau}$. This motion, which is always surprising when first encountered, is called **precession.** We can calculate the angular velocity $\omega_p$ of precession. In a small time interval $dt$, the change in the angular momentum has a magnitude $dL$:

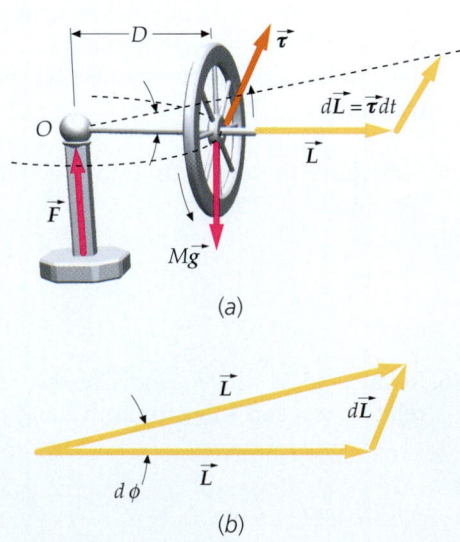

(a)

(b)

**FIGURE 10-21**

$$dL = \tau\, dt = MgD\, dt$$

where $MgD$ is the magnitude of the torque about the pivot point. From Figure 10-21b, the angle $d\phi$ through which the axle moves is

$$d\phi = \frac{dL}{L} = \frac{\tau\, dt}{L} = \frac{MgD\, dt}{L}$$

The angular velocity of the precession is thus

$$\omega_{\mathrm{p}} = \frac{d\phi}{dt} = \frac{MgD}{L} = \frac{MgD}{I_s\omega_s}$$
10-18

If the angular momentum due to the spin of the wheel is large, the precession can be very slow.

In the preceding analysis it is assumed that the spin angular momentum of the wheel is very large compared to the orbital angular momentum associated with the precessional motion.

If you prevent a spinning gyroscope from precessing using your hand, upon release it will initiate precessional motion with an up and down bouncing motion called nutation. This initial bouncing motion can be avoided by giving the gyroscope axis an initial angular velocity equal to that associated with its precessional motion.

# 10-3 Conservation of Angular Momentum

When the net external torque acting on a system remains zero, we have

$$\frac{d\vec{L}_{\mathrm{sys}}}{dt} = 0$$

or

$$\vec{L}_{\mathrm{sys}} = \text{constant}$$
10-19

Equation 10-19 is a statement of the **law of conservation of angular momentum.**

> If the net external torque acting on a system is zero, the total angular momentum of the system is constant.

CONSERVATION OF ANGULAR MOMENTUM

This is the rotational analog of the law of conservation of linear momentum. If a system is isolated from its surroundings, so that there are no external forces or torques acting on it, three quantities are conserved: energy, linear momentum, and angular momentum. The law of conservation of angular momentum is a fundamental law of nature. Even on the microscopic scale of atomic and nuclear physics, where Newtonian mechanics does not hold, the angular momentum of an isolated system is found to be constant over time.

Although conservation of angular momentum is an experimental law, independent of Newton's laws of motion, the fact that the internal torques of a system cancel is suggested by Newton's third law. Consider the two particles shown in Figure 10-22. Let $\vec{F}_{1,2}$ be the force exerted by particle 1 on particle 2 and $\vec{F}_{2,1}$ be

The Segway™ HT, a Dean Kamen invention, has been described as "the world's first self-balancing human transporter." When the person leans forward, the rotational motion of the device is detected by silicon "gyroscopes," which detect the rotational motion and then supply the internal control system with a signal. The control system interprets the signal and then activates motors that drive the wheels at just the right speed to maintain balance.

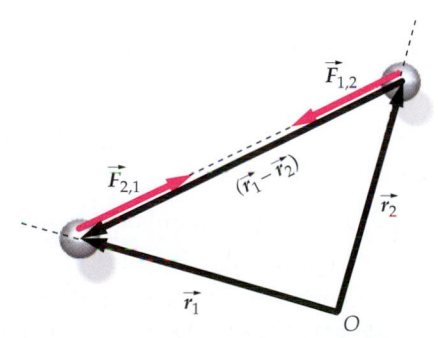

**FIGURE 10-22**

that exerted by particle 2 on particle 1. By Newton's third law, $\vec{F}_{2,1} = -\vec{F}_{1,2}$. The sum of the torques exerted by these forces about the origin $O$ is

$$\vec{\tau}_1 + \vec{\tau}_2 = \vec{r}_1 \times \vec{F}_{2,1} + \vec{r}_2 \times \vec{F}_{1,2} = \vec{r}_1 \times \vec{F}_{2,1} + \vec{r}_2 \times (-\vec{F}_{2,1}) = (\vec{r}_1 - \vec{r}_2) \times \vec{F}_{2,1}$$

The vector $\vec{r}_1 - \vec{r}_2$ is along the line joining the two particles. If $\vec{F}_{2,1}$ acts parallel to the line joining $m_1$ and $m_2$, $\vec{F}_{2,1}$ and $\vec{r}_1 - \vec{r}_2$ are either parallel or antiparallel and

$$(\vec{r}_1 - \vec{r}_2) \times \vec{F}_{2,1} = 0$$

If this is true for all the internal forces, the internal torques cancel in pairs.

There are many examples of the conservation of angular momentum in everyday life. Figures 10-23 and 10-24 illustrate angular momentum conservation in diving and ice skating.

**FIGURE 10-23** Multiflash photograph of a diver. The diver's center of mass moves along a parabolic path after he leaves the board. The angular momentum is provided by the initial external torque due to the force of the board, which does not pass through the diver's center of mass if he leans forward as he jumps. If the diver wanted to undergo one or more somersaults in the air, he would draw in his arms and legs, decreasing his moment of inertia to increase his angular velocity.

**FIGURE 10-24** A spinning skater. Because the torque exerted by the ice is small, the angular momentum of the skater is approximately constant. When she reduces her moment of inertia by drawing in her arms, her angular velocity increases.

---

*A Rotating Disk*          **EXAMPLE 10-3**

A disk is rotating with an initial angular velocity $\omega_i$ about a frictionless shaft through its symmetry axis as shown in Figure 10-25. Its moment of inertia about this axis is $I_1$. It drops onto another disk of moment of inertia $I_2$ that is initially at rest on the same shaft. Because of surface friction, the two disks eventually attain a common angular velocity $\omega_f$. Find $\omega_f$.

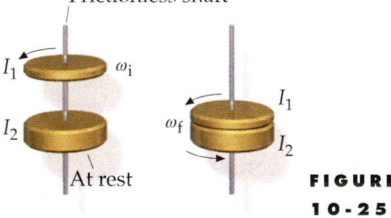

**FIGURE 10-25**

**PICTURE THE PROBLEM** We find the final angular velocity from the final angular momentum, which is equal to the initial angular momentum because there are no external torques acting on the two-disk system. Note that we do *not* use conservation of mechanical energy. The angular speed of the upper disk is reduced while that of the lower disk is increased by the forces of kinetic friction. Because kinetic friction dissipates mechanical energy, we expect that the total mechanical energy is decreased.

1. The final angular velocity is related to the initial angular velocity by conservation of angular momentum:

$$L_f = L_i$$
$$(I_1 + I_2)\omega_f = I_1\omega_i$$

2. Solve for the final angular velocity:

$$\boxed{\omega_f = \frac{I_1}{I_1 + I_2}\omega_i}$$

**PLAUSIBILITY CHECK** If $I_2 \ll I_1$, the collision should have little effect on disk 1. Our results agree, and give $\omega_f \rightarrow \omega_i$. If $I_2 \gg I_1$, then disk 1 should slow to a stop without causing disk 2 to rotate appreciably. Our results give $\omega_f \rightarrow 0$, as expected.

In the collision of the two disks in Example 10-3, mechanical energy is not conserved. We can see this by writing the energy in terms of the angular momentum. An object rotating with an angular velocity $\omega$ has kinetic energy

$$K = \frac{1}{2} I \omega^2 = \frac{(I\omega)^2}{2I}$$

Using $L = I\omega$ gives

$$K = \frac{L^2}{2I} \qquad \text{10-20}$$

(Compare this result with that for linear motion $K = p^2/2m$, Equation 8-25.) The initial kinetic energy in Example 10-3 is

$$K_i = \frac{L_i^2}{2I_1}$$

and the final kinetic energy is

$$K_f = \frac{L_f^2}{2(I_1 + I_2)}$$

Since $L_f = L_i$, the ratio of the final to the initial kinetic energy is

$$\frac{K_f}{K_i} = \frac{I_1}{I_1 + I_2}$$

which is less than one. This interaction of the disks is analogous to a one-dimensional perfectly inelastic collision of two objects.

The rotating plates in the transmission of a truck make inelastic collisions when engaged.

---

*MUD IN YOUR EYE*                    **EXAMPLE 10-4** **Put It in Context**

Benny, a high school physics student, has been the schoolyard bully for many years, so four of his fellow students decide to teach him a lesson using conservation of angular momentum. Here is their plan. The schoolyard has a small merry-go-round (Figure 10-26), a 3-m diameter turntable with a 130-kg·m$^2$ moment of inertia, and Benny loves to ride it. Initially, they need to get all five students to stand on the merry-go-round next to the rim while the merry-go-round is rotating at a modest 20 rev/min. When the signal is given, the four students will quickly move to the center of the merry-go-round leaving Benny near the rim. The merry-go-round will speed up, throwing Benny off and into the mud. (They plan to do this after a heavy rain.) Benny is very quick and very strong, so throwing him off will require that the centripetal acceleration of the rim be at least 4$g$. Will this plan work? (Assume that each student has a mass of 60 kg.)

**FIGURE 10-26**

**PICTURE THE PROBLEM** By moving to the center of the merry-go-round, the four students are decreasing the moment of inertia of the students–merry-go-round system. No external torques act on the system, so the angular momentum about the axis remains constant. The angular momentum is the moment of inertia times the angular velocity, so a decrease in the moment of inertia means an increase in the angular velocity. The angular velocity can be used to find the centripetal acceleration at the rim.

1. The centripetal acceleration depends on the angular speed $\omega$ and the radius $R$:

$$a_c = \omega^2 R$$

2. Angular momentum is conserved. For rotations about a fixed axis, $L = I\omega$:

$$L_f = L_i$$

$$I_f \omega_f = I_i \omega_i$$

3. The moment of inertia of the system is the sum of the moments of inertia of the students plus that of the merry-go-round. Each student has mass $m$:

$$I_i = 5 \times mR^2 + I_{mgr}$$
$$= 5(60\ \text{kg})(1.5\ \text{m})^2 + 130\ \text{kg} \cdot \text{m}^2$$
$$= 805\ \text{kg} \cdot \text{m}^2$$

4. To find the final moment of inertia, assume that the four students are 30 cm ($\approx$1 ft) from the center:

$$I_f = mR^2 + 4 \times mr^2 + I_{mgr}$$
$$= (60\ \text{kg})(1.5\ \text{m})^2 + 4(60\ \text{kg})(0.3\ \text{m})^2 + 130\ \text{kg} \cdot \text{m}^2$$
$$= 287\ \text{kg} \cdot \text{m}^2$$

5. Using conservation of angular momentum, solve for the final angular velocity:

$$\omega_f = \frac{I_i}{I_f}\omega_i = \frac{805\ \text{kg} \cdot \text{m}^2}{287\ \text{kg} \cdot \text{m}^2} 20\ \text{rev/min}$$
$$= 56.2\ \text{rev/min} = 5.88\ \text{rad/s}$$

6. Solve for the centripetal acceleration of the rim:

$$a_c = \omega^2 R = (5.88\ \text{rad/s})^2(1.5\ \text{m}) = 51.9\ \text{m/s}^2$$

7. Convert to $g$s:

$$a_c = 51.9\ \text{m/s}^2 \times \frac{1\ g}{9.81\ \text{m/s}^2} = 5.29g$$

8. Does Benny end up in the mud?

> Success! The acceleration is much greater than $4g$, so Benny flies off and lands in the mud.

**REMARKS** The linear speed of the rotating merry-go-round is greatest at the rim and decreases to zero at the center. At the rim the students are moving in a circle. As they walk toward the center they are stepping onto a part of the merry-go-round that is moving more slowly than they are, so they drag the merry-go-round in the direction of its motion, thus speeding it up. Also, the merry-go-round drags the students back, slowing their motion. The static frictional forces exerted by their feet exerts a net torque on the merry-go-round, increasing its angular momentum about the rotation axis. The merry-go-round exerts equal and opposite static frictional forces on the students' feet, so the torques associated with these forces decreases their angular momentum. The two torques are equal and opposite, as are the associated angular momentum changes. Thus the angular momentum of the students–merry-go-round system remains constant.

The moment of inertia of the students–merry-go-round system decreases as the students walk toward the center. Thus, the system's moment of inertia decreases while its angular momentum remains constant. As a result, we can see from Equation 10-20 that the kinetic energy of the students–merry-go-round system increases. The energy for this kinetic energy increase comes from the internal energy of the students.

ANOTHER RIDE ON THE MERRY-GO-ROUND    **EXAMPLE  10-5**  Try It Yourself

A 25-kg child in a playground runs with an initial speed of 2.5 m/s along a path *tangent* to the rim of a merry-go-round with a moment of inertia of 500 kg·m² that is initially at rest, and then jumps on (Figure 10-27). Find the final angular velocity of the child and the merry-go-round together.

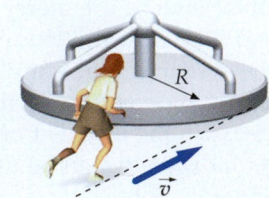

**FIGURE 10-27**

**PICTURE THE PROBLEM** Once the child's feet leave the ground, no external torques act on the child–merry-go-round system, hence the total angular momentum of the system about the rotation axis of the merry-go-round is conserved. The mass of the child is $m = 25$ kg, her initial speed is $v = 2.5$ m/s, the moment of inertia of the merry-go-round is $I = 500$ kg $\cdot$ m$^2$, and the radius of the merry-go-round is $R = 2.0$ m. The initial angular speed of the merry-go-round is $\omega_i = 0$.

**Cover the column at the right and solve it yourself before looking at the answers.**

| Steps | Answers |
|---|---|
| 1. Write an expression for the initial angular momentum of the child running about the axis of the merry-go-round. | $L_i = mvR$ |
| 2. Write an expression for the total final angular momentum of the child–merry-go-round system in terms of the final angular velocity $\omega_f$. | $L_f = (mR^2 + I_m)\omega_f$ |
| 3. Set your expressions in steps 1 and 2 equal and solve for $\omega_f$. | $\omega_f = \dfrac{mvR}{mR^2 + I_m} = \boxed{0.208 \text{ rad/s}}$ |

**EXERCISE** Calculate the initial and final kinetic energies of the child–merry-go-round system. (*Answer* $K_i = 78.2$ J, $K_f = 13.0$ J)

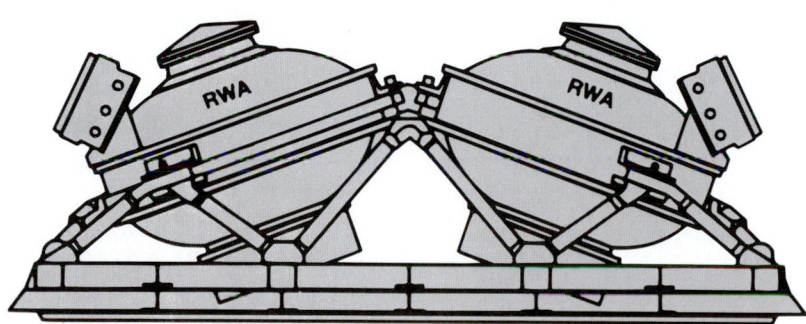

The Hubble Space Telescope is aimed by regulating the spin rates of 45-kg flywheels arranged off-axis from each other and spinning at up to 3000 rpm. Software-controlled changes in the spin rates create angular momentum that causes the satellite to slew into new positions. This aiming mechanism can achieve and hold a target to within 0.005 arcsec—equivalent to holding a flashlight beam in Los Angeles on a dime in San Francisco.

*PULLING THROUGH A HOLE*

**EXAMPLE 10-6**

A particle of mass $m$ moves with speed $v_0$ in a circle of radius $r_0$ on a frictionless tabletop. The particle is attached to a string that passes through a hole in the table, as shown in Figure 10-28. The string is slowly pulled downward so that the particle moves in a smaller circle of radius $r_f$. (*a*) Find the final velocity in terms of $r_0$, $v_0$, and $r_f$. (*b*) Find the tension when the particle is moving in a circle of radius $r$ in terms of $m$, $r$, and the angular momentum $\vec{L}$. (*c*) Calculate the work done on the particle by the tension $\vec{T}$ by integrating $\vec{T} \cdot d\vec{\ell}$. Express your answer in terms of $r$ and $L_0$.

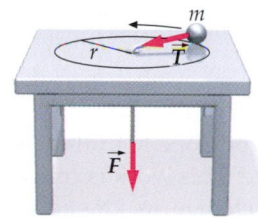

**FIGURE 10-28**

**PICTURE THE PROBLEM** The speed of the particle is related to its angular momentum. The net torque is equal to the rate of change of the angular momentum. Since the net force acting on the particle is the tension force $\vec{T}$ exerted by the string, which is always directed toward the hole, the torque about the axis through the hole is zero. Thus, the angular momentum about this axis remains constant.

An astronaut examines the flywheel of the Hubble Space Telescope.

(a) Conservation of angular momentum relates the final speed to the initial speed and the initial and final radii:

$$L_f = L_0$$

$$mv_f r_f = mv_0 r_0$$

so

$$v_f = \boxed{\frac{r_0}{r_f} v_0}$$

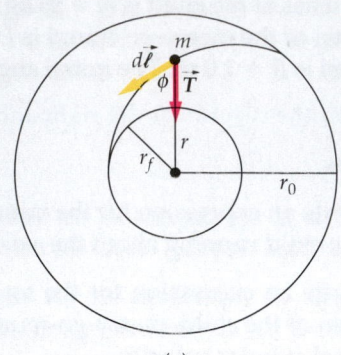

(a)

(b) 1. Apply Newton's second law to relate $T$ to $v$ and $r$:

$$T = m\frac{v^2}{r}$$

2. Obtain a relation between $L$, $r$, and $v$ using the definition of angular momentum:

$$\vec{L} = \vec{r} \times \vec{p}$$

$$L = rmv \sin 90° = mvr$$

3. Eliminate $v$ by solving the Part (b) step 2 result for $v$ and then substituting into the Part (a) result:

$$T = m\frac{v^2}{r} = \frac{m}{r}\left(\frac{L}{mr}\right)^2 = \boxed{\frac{L^2}{mr^3}}$$

(c) 1. Make a drawing of the particle as it moves closer to the hole (Figure 10-29). When the particle undergoes displacement $d\vec{\ell}$, its distance $r$ from the axis changes by $dr$. Since $r$ is decreasing, $dr$ is negative. Thus:

$$dr = -|dr|$$

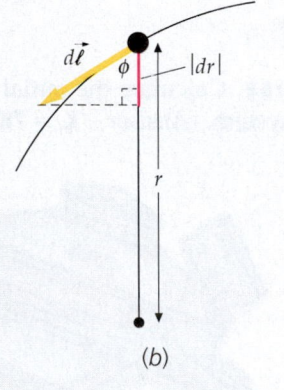

(b)

**FIGURE 10-29**

2. Using step 1, write $dW = \vec{T} \cdot d\vec{\ell}$ in terms of $T$ and $dr$, with $T = L^2/mr^3$ from Part (b):

$$dW = \vec{T} \cdot d\vec{\ell} = T\,d\ell \cos \phi$$

Since

$$|dr| = d\ell \cos \phi$$

$$dW = T|dr| = -T\,dr$$

3. Integrate from $r_0$ to $r_f$ after substituting for $T$ from the Part (b) step 3 result:

$$W = -\int_{r_0}^{r_f} T\,dr = -\int_{r_0}^{r_f} \frac{L^2}{mr^3} dr$$

$$= -\frac{L^2}{m}\int_{r_0}^{r_f} r^{-3}\,dr = -\frac{L^2}{m}\frac{r^{-2}}{-2}\Big|_{r_0}^{r_f}$$

$$= \boxed{\frac{L^2}{2m}\left(\frac{1}{r_f^2} - \frac{1}{r_0^2}\right)}$$

**PLAUSIBILITY CHECK** Note that work must be done to pull the string downward. Since $r_f$ is less than $r_0$, the work is positive. This work is converted into increased kinetic energy. We can calculate the change in kinetic energy of the particle directly. Using $K = L^2/2I$, with $L_0 = L_f = L$, and $I = mr^2$, the change in kinetic energy is $K_f - K_i = (L^2/2mr_f^2) - (L^2/2mr_0^2) = (L^2/2m)(r_f^{-2} - r_0^{-2})$, which is the same as the result found by direct integration.

**REMARKS** The increment of work $dW$ can also be obtained by expressing the increment of displacement $d\vec{\ell}$ as $d\vec{r}$, the change in the position vector $\vec{r}$. The dot product $\vec{T} \cdot d\vec{r}$ is then expanded via components giving $dW = \vec{T} \cdot d\vec{r} = T_r\,dr = -T\,dr$. In this expansion $T_r = -T$ is the radial component of $\vec{T}$ and $dr$ is the radial component of $d\vec{r}$.

**EXERCISE** At what radius $r_N$ is the tension $N$ times the tension at radius $r_0$? (*Answer* $r_N = r_0/\sqrt[3]{N}$)

In Figure 10-30, a puck on a frictionless plane is given an initial speed $v_0$. The puck is attached to a string that wraps around a vertical post. This situation looks similar to Example 10-6, but it is not the same. **There is no agent that can do work on the puck, nor is there any mechanism for energy dissipation.** Thus, mechanical energy must be conserved. Since $K = L^2/2I$ is constant and $I$ decreases as $r_0$ decreases, $L$ must also decrease. Note that the tension force does not act toward the axis of the post. The tension force on the puck produces a torque vector about the axis of the post in the downward direction, which reduces the angular momentum vector of the puck, which is in the upward direction.

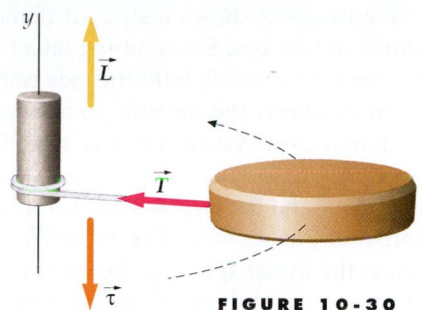

**FIGURE 10-30**

---

*THE BALLISTIC PENDULUM REVISITED*          **E X A M P L E   1 0 - 7**   **Try It Yourself**

A thin rod of mass $M$ and length $d$ is attached to a pivot at the top. A piece of clay of mass $m$ and speed $v$ hits the rod a distance $x$ from the pivot and sticks to it (Figure 10-31). Find the ratio of the system's kinetic energy just after the collision to the kinetic energy just before the collision.

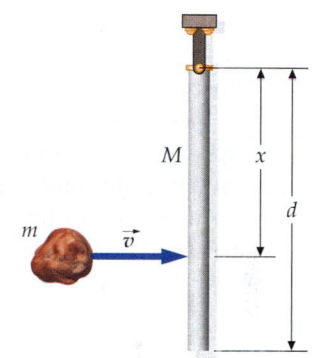

**PICTURE THE PROBLEM**   The collision is inelastic, so we do not expect mechanical energy to be conserved. During the collision, the pivot exerts a large force on the rod, so linear momentum is also not conserved. However, there are no external torques about the pivot point of the clay–rod system, so angular momentum about the pivot point is conserved. The kinetic energy after the inelastic collision can be written in terms of the angular momentum $L_{sys}$ and the moment of inertia $I'$ of the combined clay–rod system. Conservation of angular momentum allows you to relate $L_{sys}$ to the mass $m$ and velocity $v$ of the clay.

**FIGURE 10-31**

**Cover the column on the right and try these on your own before looking at the answers.**

| Steps | Answers |
|---|---|
| 1. Before the collision the kinetic energy of the system is that of the moving clay ball. | $K_i = \frac{1}{2}mv^2$ |
| 2. After the collision it is that of the swinging clay–rod object. Write the kinetic energy after the collision in terms of the angular momentum $L_{sys}$ and the moment of inertia $I'$ of the clay–rod object. | $K_f = \dfrac{L_{sys}^2}{2I'}$ |
| 3. During the collision angular momentum is conserved. Write the angular momentum $L_{sys}$ in terms of $m$, $v$, and $x$. | $L_{sys} = mvx$ |
| 4. Write $I'$ in terms of $m$, $x$, $M$, and $d$. | $I' = mx^2 + \frac{1}{3}Md^2$ |
| 5. Substitute these expressions for $L_f$ and $I'$ into your equation for $K_f$. | $K_f = \dfrac{L_{sys}^2}{2I'} = \dfrac{(mvx)^2}{2(mx^2 + \frac{1}{3}Md^2)}$ $= \dfrac{3}{2}\dfrac{m^2x^2v^2}{(3mx^2 + Md^2)}$ |
| 6. Divide the kinetic energy after the collision by the initial kinetic energy of the clay. | $\dfrac{K_f}{K_i} = \dfrac{\frac{3}{2}\dfrac{m^2x^2v^2}{(3mx^2 + Md^2)}}{\frac{1}{2}mv^2} = \boxed{\dfrac{1}{1 + \dfrac{Md^2}{3mx^2}}}$ |

**REMARKS**   This example is the rotational analog of the ballistic pendulum discussed in Example 8-14. In that example we used conservation of linear momentum to find the energy of the pendulum after the collision.

Figure 10-32 shows a student demonstrating conservation of angular momentum about an axis. She is sitting on a turntable, holding a bicycle wheel. We choose the z axis to coincide with the axis of the turntable. Because the turntable bearings are frictionless, the vertical component of the torque exerted on the turntable–student–wheel system remains zero. Consequently, the vertical component of the total angular momentum of the system remains constant. Initially, the turntable is stationary and the bicycle wheel is spinning rapidly about its axis, which is initially horizontal, as shown in Figure 10-32a. The system's initial angular momentum $\vec{L}_{\text{sys i}}$ is just the initial spin angular momentum of the wheel $\vec{L}_{\text{w,s i}}$, which is horizontal. This means that the vertical component of $\vec{L}_{\text{sys i}}$ is zero. If the student now (Figure 10-32b) tips the axis of the spinning wheel upward, the wheel's spin angular momentum vector $\vec{L}_{\text{w,s}}$ tips upward with it. As she tips the wheel upward the system starts rotating clockwise (viewed from above) about the turntable axis. The angular momentum vector associated with this clockwise rotation of the system about the turntable axis has a z component $L_{\text{t }z}$ directed downward. Since the vertical component of $\vec{L}_{\text{sys}}$ remains zero, we can conclude that $L_{\text{t }z}\hat{k}$ is equal and opposite to $L_{\text{w,s }z}\hat{k}$.

Unless you have seen this demonstration before, the rotation of the turntable is unexpected. Let's examine what causes it. The student must exert an upward torque on the spinning wheel in order to tip its angular momentum vector upward. (Due to the cross product, an upward torque requires horizontal forces.) The wheel exerts a downward torque (also horizontal forces) of equal magnitude back on the student. It is this downward torque on the student that causes the system to start rotating clockwise about the turntable axis.

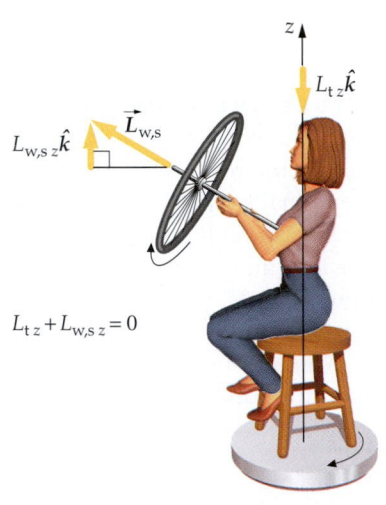

**EXERCISE** The bicycle wheel is spinning with its axis vertical and with its angular momentum vector upward when it is handed to the student on the turntable. In what direction will the turntable rotate when the student tips and then rotates the axis of the wheel toward the horizontal? (*Answer* Counterclockwise, if viewed from above)

### Proofs of Equations 10-10, 10-12, 10-13, 10-14, and 10-15

**Proof of Equation 10-10** We will now show that Newton's second law implies that the rate of change of the angular momentum of a particle equals the net torque acting on the particle. If more than one force acts on a particle, then the net torque relative to the origin $O$ is the sum of the torques due to each force:

$$\vec{\tau}_{\text{net}} = \vec{r} \times \vec{F}_1 + \vec{r} \times \vec{F}_2 + \cdots = \vec{r} \times \sum_i \vec{F}_i = \vec{r} \times \vec{F}_{\text{net}}$$

**FIGURE 10-32**

According to Newton's second law, the net force equals the rate of change of the linear momentum $d\vec{p}/dt$. Thus

$$\vec{\tau}_{\text{net}} = \vec{r} \times F_{\text{net}} = \vec{r} \times \frac{d\vec{p}}{dt} \qquad 10\text{-}21$$

We now compare this with the expression for the time rate of change of the angular momentum. The definition of the angular momentum of a particle (Equation 10-8) is

$$\vec{L} = \vec{r} \times \vec{p}$$

We can compute $d\vec{L}/dt$ using the product rule for derivatives:

$$\frac{d\vec{L}}{dt} = \frac{d}{dt}(\vec{r} \times \vec{p}) = \left(\frac{d\vec{r}}{dt} \times \vec{p}\right) + \left(\vec{r} \times \frac{d\vec{p}}{dt}\right)$$

The second term from the right of this equation is zero because $\vec{p} = m\vec{v}$ and $\vec{v} = d\vec{r}/dt$, so

$$\frac{d\vec{r}}{dt} \times \vec{p} = \vec{v} \times m\vec{v} = 0$$

Thus

$$\frac{d\vec{L}}{dt} = \vec{r} \times \frac{d\vec{p}}{dt}$$

Comparing with Equation 10-21 gives

$$\vec{\tau}_{net} = \frac{d\vec{L}}{dt} \qquad\qquad 10\text{-}22$$

The net torque acting on a system of particles is the sum of the individual torques. The generalization of Equation 10-22 to a system of particles is then

$$\sum_i \vec{\tau} = \sum_i \frac{d\vec{L}_i}{dt} = \frac{d}{dt} \sum_i \vec{L}_i = \frac{d\vec{L}_{sys}}{dt}$$

In this equation, the sum of the torques may include internal as well as external torques. The sum of the internal torques equals zero, so

$$\vec{\tau}_{net\,ext} = \frac{d\vec{L}_{sys}}{dt} \qquad\qquad 10\text{-}10$$

NEWTON'S SECOND LAW FOR ANGULAR MOTION

**\*Proofs of Equations 10-12 and 10-13**   We will now show that the angular momentum of a system of particles can be written as the sum of the orbital angular momentum and the spin angular momentum.

Figure 10-33 shows a system of particles. The angular momentum $\vec{L}_i$ of the $i$th particle about arbitrary point $O$ is given by

$$\vec{L}_i = \vec{r}_i \times \vec{p}_i = \vec{r}_i \times m_i\vec{v}_i \qquad\qquad 10\text{-}23$$

and the angular moment of the system about $O$ is

$$\vec{L} = \Sigma\vec{L}_i = \Sigma(\vec{r}_i \times m_i\vec{v}_i)$$

The angular momentum about the center of mass is given by

$$\vec{L}_{cm} = \Sigma(\vec{r}_i' \times m_i\vec{u}_i)$$

where $\vec{r}_i'$ and $\vec{u}_i$ are the position and velocity of the $i$th particle relative to the center of mass. It can be seen from the figure that

$$\vec{r}_i = \vec{r}_{cm} + \vec{r}_i'$$

Differentiating both sides gives

$$\vec{v}_i = \vec{v}_{cm} + \vec{u}_i$$

Substituting these into Equation 10-23, we have

$$\vec{L}_i = \vec{r}_i \times m_i\vec{v}_i = (\vec{r}_{cm} + \vec{r}_i') \times m_i(\vec{v}_{cm} + \vec{u}_i)$$

Expanding the right side, we obtain

$$\vec{L}_i = (\vec{r}_{cm} \times m_i\vec{v}_{cm}) + (\vec{r}_{cm} \times m_i\vec{u}_i) + (m_i\vec{r}_i' \times \vec{v}_{cm}) + (\vec{r}_i' \times m_i\vec{u}_i)$$

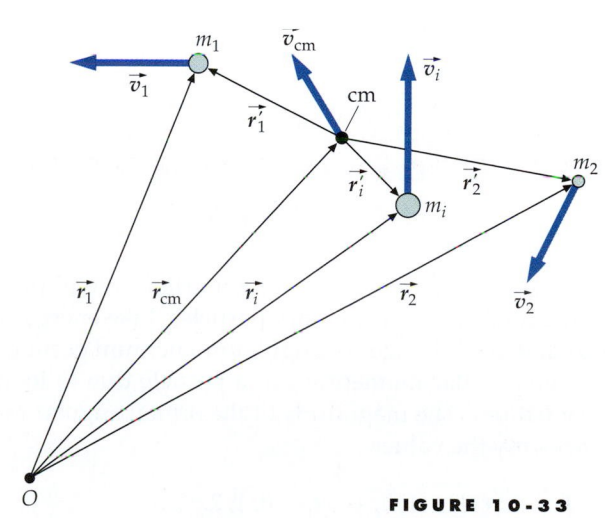

**FIGURE 10-33**

Summing both sides and factoring common terms out of the sums gives

$$\vec{L}_{sys} = \Sigma \vec{L}_i = \left( \vec{r}_{cm} \times \Sigma (m_i) \vec{v}_{cm} \right) + \left( \vec{r}_{cm} \times \Sigma (m_i \vec{u}_i) \right) + \left( \Sigma (m_i \vec{r}_i') \times \vec{v}_{cm} \right) + \left( \Sigma (\vec{r}_i' \times m_i \vec{u}_i) \right)$$

Because $\Sigma(m_i \vec{r}_i')$ and $\Sigma(m_i \vec{u}_i)$ are both zero, and because $\Sigma m_i = M$ and $\Sigma(\vec{r}_i' \times m_i \vec{u}_i) = \vec{L}_{cm}$, we have $\vec{L}_{sys} = \vec{r}_{cm} \times M\vec{v}_{cm} + \vec{L}_{cm}$, or

$$\vec{L}_{sys} = \vec{L}_{orbit} + \vec{L}_{spin} \qquad \text{10-12}$$

where $\vec{L}_{spin} = \vec{L}_{cm}$ and $\vec{L}_{orbit} = \vec{v}_{cm} \times M\vec{v}_{cm}$.     10-13

**\*Proofs of Equations 10-14 and 10-15** We will now take the $z$ components of the vectors for the torque and the angular momentum about a point to obtain the formulas for the torque and angular momentum about a fixed axis. The angular momentum of a particle about the origin is $\vec{L} = \vec{r} \times \vec{p}$, so finding the $z$ component of the angular momentum means finding the $z$ component of the product $\vec{r} \times \vec{p}$. To do this we express $\vec{r}$ and $\vec{p}$ as

$$\vec{r} = \vec{r}_{rad} + \vec{r}_z \qquad \text{and} \qquad \vec{p} = \vec{p}_{xy} + \vec{p}_z$$

where $\vec{r}_{rad}, \vec{r}_z, \vec{p}_{xy},$ and $\vec{p}_z$ are vector components (see Figure 10-34) of $\vec{r}$ and $\vec{p}$. Substituting $\vec{r}$ and $\vec{p}$ gives

$$\vec{L} = \vec{r} \times \vec{p} = (\vec{r}_{rad} + \vec{r}_z) \times (\vec{p}_{xy} + \vec{p}_z)$$

and expanding the right side, we have

$$\vec{L} = (\vec{r}_{rad} \times \vec{p}_{xy}) + (\vec{r}_{rad} \times \vec{p}_z) + (\vec{r}_z \times \vec{p}_{xy}) + (\vec{r}_z \times \vec{p}_z)$$

The cross product of any two vectors is perpendicular to both vectors, so the product $\vec{r}_{rad} \times \vec{p}_{xy}$ is parallel to the $z$ axis. In each of the other three products at least one of the two vectors is parallel to the $z$ axis, so the $z$ component of each of these cross products is zero. Therefore,

$$\vec{L}_z = \vec{r}_{rad} \times \vec{p}_{xy} \qquad \text{10-14}$$

ANGULAR MOMENTUM ABOUT $Z$ AXIS

The torque about the origin associated with a force acting on the particle is given by $\vec{\tau} = \vec{r} \times \vec{F}$ (Equation 10-1). Following the same procedure with the torque that we followed with the angular momentum gives

$$\vec{\tau}_z = \vec{r}_{rad} \times \vec{F}_{xy} \qquad \text{10-15}$$

TORQUE ABOUT $Z$ AXIS

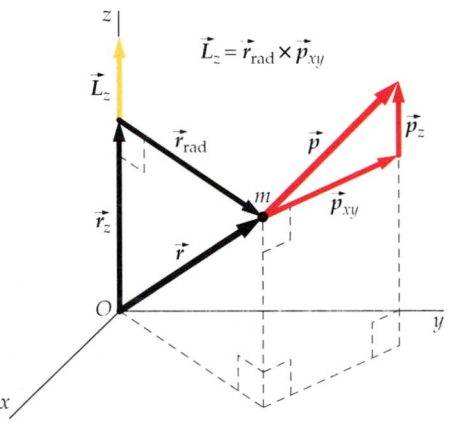

**FIGURE 10-34** The vector components $\vec{r}_{rad}, \vec{r}_z, \vec{p}_{xy},$ and $\vec{p}_z$ of $\vec{r}$ and $\vec{p}$ that are used for calculating the angular momentum about the $z$ axis $\vec{L}_z$.

# **\*10-4** Quantization of Angular Momentum

Angular momentum plays an important role in the description of atoms, molecules, nuclei, and elementary particles. Like energy, angular momentum is **quantized**, that is, changes in angular momentum occur only in discrete amounts.

The angular momentum of a particle due to its motion is its orbital angular momentum. The magnitude of the orbital angular momentum $L$ of a particle can have only the values

$$L = \sqrt{\ell(\ell + 1)}\hbar, \qquad \ell = 0, 1, 2, \ldots \qquad \text{10-24}$$

where $\hbar$ (read "h-bar") is the **fundamental unit of angular momentum,** which is related to Planck's constant $h$:

$$\hbar = \frac{h}{2\pi} = 1.05 \times 10^{-34} \, \text{J} \cdot \text{s} \qquad \text{10-25}$$

The component of orbital angular momentum along any line in space is also quantized and can have only the values $\pm m\hbar$, where $m$ is a nonnegative integer that is less than or equal to $\ell$. For example, if $\ell = 2$, $m$ can equal 2, 1, or 0.

Because the quantum of angular momentum $\hbar$ is so small, the quantization of angular momentum is not noticed in the macroscopic world. Consider a particle of mass $1 \, \text{g} = 10^{-3} \, \text{kg}$ moving in a circle of radius 1 cm with a period of 1 s. Its orbital angular momentum is

$$L = mvr = mr^2\omega = mr^2\frac{2\pi}{T} = (10^{-3}\,\text{kg})(10^{-2}\,\text{m})^2\frac{2\pi}{1\,\text{s}} = 6.28 \times 10^{-7} \, \text{J} \cdot \text{s}$$

If we divide by $\hbar$, we obtain

$$\frac{L}{\hbar} = \frac{6.28 \times 10^{-7} \, \text{J} \cdot \text{s}}{1.05 \times 10^{-34} \, \text{J} \cdot \text{s}} = 6 \times 10^{27}$$

Thus this typical macroscopic angular momentum contains $6 \times 10^{27}$ units of the fundamental unit of angular momentum. Even if we could measure $L$ to one part in a billion, we would never notice the quantization of this macroscopic angular momentum.

The quantization of orbital angular momentum leads to the quantization of rotational kinetic energy. Consider a molecule rotating about its center of mass with angular momentum $L$ (Figure 10-35). Let $I$ be its moment of inertia. Its kinetic energy is

$$K = \frac{L^2}{2I} \qquad \text{10-26}$$

But $L^2$ is quantized to the values $L^2 = \ell(\ell + 1)\hbar^2$ with $\ell = 0, 1, 2, \ldots$. Thus, the kinetic energy is quantized to the values $K_\ell$ given by

$$K_\ell = \frac{L^2}{2I} = \frac{\ell(\ell + 1)\hbar^2}{2I} = \ell(\ell + 1)E_{0r} \qquad \text{10-27}a$$

where

$$E_{0r} = \frac{\hbar^2}{2I} \qquad \text{10-27}b$$

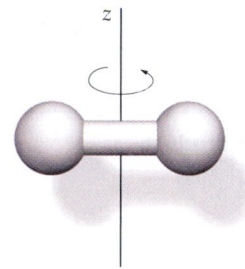

**FIGURE 10-35** Model of a rigid diatomic molecule rotating about the z axis.

Figure 10-36 shows an energy-level diagram for a rotating molecule with constant moment of inertia $I$. Note that, unlike the energy levels for a vibrating system (Section 7-4), the rotational energy levels are not equally spaced, and the lowest level is zero.

Stable matter contains just three kinds of particles: electrons, protons, and neutrons. In addition to its orbital angular momentum each of these particles also has an intrinsic angular momentum called its **spin.** The spin angular momentum of a particle, like its mass and electric charge, is a fundamental property of the particle that cannot be changed. The magnitude of the spin angular momentum vector for electrons and other fermions is $s = \sqrt{\frac{1}{2}(\frac{1}{2} + 1)}\hbar$ and the component along any line in space can have just two values: $+\frac{1}{2}\hbar$ and $-\frac{1}{2}\hbar$. Such particles are called "spin one-half" particles. Electrons, protons, neutrons, and other spin-one-half particles are called **fermions.** Other particles such as photons and $\alpha$ particles, called

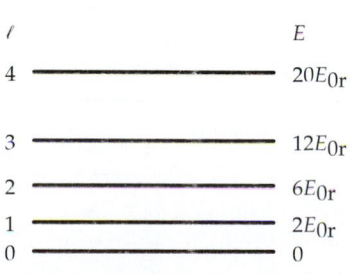

**FIGURE 10-36** Energy-level diagram for a rotating molecule.

**bosons**, have zero spin or integral spin. Spin is a quantum property of the particle that has nothing to do with the motion of the particle.

The picture of an electron as a spinning ball that orbits the nucleus in an atom (like the spinning earth orbiting the sun) is often a useful visualization. However, the angular momentum of a spinning ball can be increased or decreased, whereas the spin of the electron is a fixed property like its charge and mass that does not change. Furthermore, as far as we know, electrons are point particles that have no size.

# SUMMARY

1. Angular momentum is an important derived dynamic quantity in macroscopic physics. In microscopic physics, angular momentum is an intrinsic, fundamental property of elementary particles.
2. Conservation of angular momentum is a fundamental law of nature.
3. Quantization of angular momentum is a fundamental law of nature.

| Topic | Relevant Equations and Remarks | |
|---|---|---|
| **1. Vector Nature of Rotation** | When the direction of the axis of rotation is not fixed in space, plus and minus signs are inadequate to describe the direction of the angular velocity direction. | |
| Angular velocity $\vec{\omega}$ | The direction of the angular velocity $\vec{\omega}$ is along the axis of rotation in the sense given by the right-hand rule. | |
| Torque $\vec{\tau}$ | $\vec{\tau} = \vec{r} \times \vec{F}$ | 10-1 |
| **2. Vector Product** | $\vec{A} \times \vec{B} = AB \sin \phi \, \hat{n}$ | 10-2 |
| | where $\phi$ is the angle between the vectors and $\hat{n}$ is a unit vector perpendicular to the plane of $\vec{A}$ and $\vec{B}$ in the sense given by the right-hand rule as $\vec{A}$ is rotated into $\vec{B}$. | |
| Properties | $\vec{A} \times \vec{A} = 0$ | 10-3 |
| | $\vec{A} \times \vec{B} = -\vec{B} \times \vec{A}$ | 10-4 |
| | $\dfrac{d}{dt}(\vec{A} \times \vec{B}) = \left(\vec{A} \times \dfrac{d\vec{B}}{dt}\right) + \left(\dfrac{d\vec{A}}{dt} \times \vec{B}\right)$ | 10-6 |
| | $\hat{i} \times \hat{j} = \hat{k} \qquad \hat{j} \times \hat{k} = \hat{i} \qquad \hat{k} \times \hat{i} = \hat{j}$ | 10-7a |
| **3. Angular Momentum** | | |
| For a particle | $\vec{L} = \vec{r} \times \vec{p}$ | 10-8 |
| For a system rotating about a symmetry axis | $\vec{L} = I\vec{\omega}$ | 10-9 |
| For any system | The angular momentum about any point $O$ is the angular momentum about the center of mass (spin angular momentum) plus the angular momentum associated with center-of-mass motion about $O$ (orbital angular momentum).  $\vec{L} = \vec{L}_{\text{orbit}} + \vec{L}_{\text{spin}} = \left(\vec{r}_{\text{cm}} \times M\vec{v}_{\text{cm}}\right) + \left(\sum_i \vec{r}_i \times m_i \vec{u}_i\right)$ | 10-12 |
| Newton's second law for angular motion | $\vec{\tau}_{\text{net ext}} = \dfrac{d\vec{L}}{dt}$ | 10-10 |

| Conservation of angular momentum | If the net external torque is zero, the angular momentum of the system is conserved. (If the component of the net external torque in a given direction is zero, the component of the angular momentum of the system in that direction is conserved.) | |
|---|---|---|
| Kinetic energy of a rotating object | $K = \dfrac{L^2}{2I}$ | **10-20** |
| Quantization of angular momentum | The magnitude of the orbital angular momentum of a particle can have only the values $$L = \sqrt{\ell(\ell + 1)}\hbar, \qquad \ell = 0, 1, 2, \ldots$$ where | |
| Fundamental unit of angular momentum | $\hbar = \dfrac{h}{2\pi} = 1.05 \times 10^{-34}\text{ J·s}$ is the fundamental unit of angular momentum and $h$ is Planck's constant. | **10-25** |
| *Quantization of any component of orbital angular momentum | The component of orbital angular momentum along any line in space is also quantized and can have only the values $\pm m\,\hbar$, where $m$ is a nonnegative integer that is less than or equal to $\ell$. | |
| Spin | Electrons, protons, and neutrons have an intrinsic angular momentum called spin. The magnitude of the spin angular momentum vector for these particles is $$s = \sqrt{\tfrac{1}{2}(\tfrac{1}{2} + 1)}\,\hbar$$ and the component along any line in space can have just two values, $+\tfrac{1}{2}\hbar$ and $-\tfrac{1}{2}\hbar$. | |

# PROBLEMS

- • Single-concept, single-step, relatively easy
- •• Intermediate-level, may require synthesis of concepts
- ••• Challenging, for advanced students
- **SSM** Solution is in the *Student Solutions Manual*
- **iSOLVE** Problems available on iSOLVE online homework service
- **iSOLVE✓** These "Checkpoint" online homework service problems ask students additional questions about their confidence level, and how they arrived at their answer

In a few problems, you are given more data than you actually need; in a few other problems, you are required to supply data from your general knowledge, outside sources, or informed estimates.

Take $g = 9.81$ N/kg $= 9.81$ m/s$^2$ and neglect friction in all problems unless otherwise stated.

## Conceptual Problems

**1** • **SSM** True or false:

(a) If two vectors are parallel, their cross product must be zero.
(b) When a disk rotates about its symmetry axis, $\vec{\omega}$ is along the axis.
(c) The torque exerted by a force is always perpendicular to the force.

**2** • Consider two nonzero vectors $\vec{A}$ and $\vec{B}$. Their cross product has the greatest magnitude if $\vec{A}$ and $\vec{B}$ are (a) parallel, (b) equal in magnitude, (c) perpendicular, (d) antiparallel, (e) at an angle of 45° to each other.

**3** • What is the angle between a particle's linear momentum $\vec{p}$ and its angular momentum $\vec{L}$?

**4** • A particle of mass $m$ is moving with speed $v$ along a line that passes through point $P$. What is the angular momentum of the particle about point $P$? (a) $mv$. (b) Zero. (c) It changes sign as the particle passes through point $P$. (d) It depends on the distance of point $P$ from the origin of the coordinates.

**5** •• **SSM** A particle travels in a circular path and point $P$ is at the center of the circle. (a) If its linear momentum $\vec{p}$ is doubled, how is its angular momentum about $P$ affected? (b) If the radius of the circle is doubled but the speed is unchanged, how is the angular momentum of the particle about $P$ affected?

**6** •• A particle moves along a straight line at constant speed. How does its angular momentum about any fixed point vary over time?

**7** • True or false: If the net torque on a rotating system is zero, the angular velocity of the system cannot change.

**8** •• **SSM** Standing on a turntable that is initially not rotating, can you rotate yourself through 180°? Assume that no external torques act on the you–turntable system. *Hint: While you cannot change your mass easily, there are ways to change your moment of inertia.*

**9** • If the angular momentum of a system is constant, which of the following statements must be true?

(a) No torque acts on any part of the system.
(b) A constant torque acts on each part of the system.
(c) Zero net torque acts on each part of the system.
(d) A constant external torque acts on the system.
(e) Zero net external torque acts on the system.

**10** •• In Example 10-4, does the force exerted by the merry-go-round on Benny do work?

**11** •• Is it easier to crawl radially outward or radially inward on a rotating merry-go-round? Why?

**12** •• **SSM** A block sliding on a frictionless table is attached to a string that passes through a hole in the table. Initially, the block is sliding with speed $v_0$ in a circle of radius $r_0$. A student under the table pulls slowly on the string. What happens as the block spirals inward? Give supporting arguments for your choice. (a) Its energy and angular momentum are conserved. (b) Its angular momentum is conserved and its energy increases. (c) Its angular momentum is conserved and its energy decreases. (d) Its energy is conserved and its angular momentum increases. (e) Its energy is conserved and its angular momentum decreases.

**13** •• **SSM** How can you tell a hardboiled egg from an uncooked one without breaking it? One way is to lay the egg flat on a hard surface and try to spin it. A hardboiled egg will spin easily, while it takes a lot of effort to make an uncooked egg spin. However, once spinning, the uncooked egg will do something unusual: If you stop it with your finger, it may start spinning again. Explain the difference in the behavior of the two types of eggs.

**14** • True or false:

When a gyroscope is not spinning, $\vec{\tau} = \dfrac{d\vec{L}}{dt}$ does not hold.

**15** • **iSOLVE** An object of mass $M$ is rotating about a fixed axis with angular momentum $L$. Its moment of inertia about this axis is $I$. What is its kinetic energy? (a) $IL^2/2$. (b) $L^2/2I$. (c) $ML^2/2$. (d) $IL^2/2M$.

**16** • Explain why a helicopter with just one main rotor has a second smaller rotor mounted on a horizontal axis at the rear as in Figure 10-37. Describe the resultant motion of the helicopter if this rear rotor fails during flight.

**FIGURE 10-37** Problem 16

**17** •• The angular momentum vector for a spinning wheel lies along its axle and is pointed east. To make this vector point south, it is necessary to exert a force on the east end of the axle in which direction? (a) Up. (b) Down. (c) North. (d) South. (e) East.

**18** •• **SSM** You are walking north and with your left hand you are carrying a suitcase that contains a spinning gyroscope mounted on an axle attached to the front and back of the case. The angular velocity of the gyroscope points north. You now begin to turn to walk east. As a result, the front end of the suitcase will (a) resist your attempt to turn and will try to remain pointed north, (b) fight your attempt to turn and will pull to the west, (c) rise upward, (d) dip downward, (e) show no effect whatsoever.

**19** •• The angular momentum of the propeller of a small single-engine airplane points forward. The propeller rotates clockwise if viewed from behind. (a) As the plane takes off, the nose lifts and the airplane tends to veer to one side. To which side does it veer and why? (b) If the plane is flying horizontally and suddenly turns to the right, does the nose of the plane tend to move up or down? Why?

**20** •• A car is powered by the energy stored in a single flywheel with an angular momentum $\vec{L}$. Discuss the problems that would arise for various orientations of $\vec{L}$ and various maneuvers of the car. For example, what would happen if $\vec{L}$ points vertically upward and the car travels over a hilltop or through a valley? What would happen if $\vec{L}$ points forward or to one side and the car attempts to turn to the left or right? In each case that you examine, consider the direction of the torque exerted on the car by the road.

**21** •• **iSOLVE** You sit on a spinning piano stool with your arms folded. When you extend your arms out to the side, your kinetic energy (a) increases, (b) decreases, (c) remains the same.

**22** •• **SSM** In tetherball, a ball is attached to a string that is attached to a pole. When the ball is hit, the string wraps around the pole and the ball spirals inward. Neglecting air resistance, what happens as the ball swings around the pole? Give supporting arguments for your choice. (a) The mechanical energy and angular momentum of the ball are conserved. (b) The angular momentum of the ball is conserved, but the mechanical energy of the ball increases. (c) The angular momentum of the ball is conserved, and the mechanical energy of the ball decreases. (d) The mechanical energy of the ball is conserved and the angular momentum of the ball increases. (e) The mechanical energy of the ball is conserved and the angular momentum of the ball decreases.

**23** •• A uniform rod of mass $M$ and length $L$ lies on a horizontal frictionless table. A piece of putty of mass $m = M/4$ moves along a line perpendicular to the rod, strikes the rod near its end, and sticks to the rod. Describe qualitatively the subsequent motion of the rod and putty.

**24** • [SSM] True or false:

(a) The rate of change of a system's angular momentum is always parallel to the net external torque.

(b) If the net torque on a body is zero, the angular momentum must be zero.

## Estimation and Approximation

**25** •• [SSM] [iSOLVE]✓ An ice skater starts her pirouette with arms outstretched, rotating at 1.5 rev/s. Estimate her rotational speed (in revolutions per second) when she brings her arms flat against her body.

**26** •• [iSOLVE] The polar ice caps contain about $2.3 \times 10^{19}$ kg of ice. This mass contributes negligibly to the moment of inertia of the earth because it is located at the poles, close to the axis of rotation. Estimate the change in the length of the day that would be expected if the polar ice caps were to melt and the water were distributed uniformly over the surface of the earth. (The moment of inertia of a spherical shell of mass $m$ and radius $r$ is $2mr^2/3$.)

**27** • A 2-g particle moves at a constant speed of 3 mm/s around a circle of radius 4 mm. (a) Find the magnitude of the angular momentum of the particle. (b) If $L = \sqrt{\ell(\ell + 1)}\hbar$, where $\ell$ is an integer, find the value of $\ell(\ell + 1)$ and the approximate value of $\ell$. (c) Explain why the quantization of angular momentum is not noticed in macroscopic physics.

**28** •• [SSM] One problem in astrophysics in the 1960s was explaining pulsars—extremely regular astronomical sources of radio pulses whose periods ranged from seconds to milliseconds. At one point, these radio sources were given the acronym LGM, standing for "Little Green Men," a reference to the idea that they might be signals of extraterrestrial civilizations. The explanation given today is no less interesting: The sun, which is a fairly typical star, has a mass of $1.99 \times 10^{30}$ kg and a radius of $6.96 \times 10^8$ m. While it doesn't rotate uniformly, because it isn't a solid body, its average rate of rotation can be taken as about 1 rev/25 d. Stars somewhat larger than the sun can end their life in spectacular explosions—supernovae—leaving behind a collapsed remnant of the star called a neutron star. These neutron-star remnants have masses comparable to the original mass of the star, but radii of only a few kilometers! The high rotation rate is due to the conservation of angular momentum during the collapse. These stars emit beams of radio waves. Because of the rapid angular speed of the stars, the beam sweeps past the earth at regular intervals. To produce the observed radio-wave pulses, the star has to rotate at rates from about 1 rev/s to 1 rev/ms. (a) Using data from the textbook, estimate the rotation rate of the sun if it were to collapse into a neutron star of radius 10 km. Because the sun is not a uniform sphere of gas, its moment of inertia is given by the formula $I = 0.059MR^2$. Assume that the neutron star is spherical and has a uniform mass distribution. (b) Is the rotational kinetic energy of the sun greater or smaller after the collapse? By what factor does it change, and where does the energy go to or come from?

**29** •• The moment of inertia of the earth about its spin axis is approximately $8.03 \times 10^{37}$ kg·m². (a) Since the earth is nearly spherical, assume that the moment of inertia can be written as $I = CMR^2$, where $C$ is a dimensionless constant, $M = 5.98 \times 10^{24}$ kg is the mass of the earth, and $R = 6370$ km is its radius. Determine $C$. (b) If the earth's mass were distributed uniformly, $C$ would equal 2/5. From the value of $C$ calculated in Part (a), is the earth's mass density greater near the core or near the crust?

**30** •• [SSM] Estimate the angular velocity and angular momentum of the diver in Figure 10-23 (page 318) about his center of mass. Make any approximations that you think reasonable.

**31** •• Estimate the angular velocity of the diver in Figure 10-23 (page 318) if he curled himself into a ball in middive.

**32** •• [SSM] Estimate Timothy Goebel's initial takeoff speed, rotational velocity, and angular momentum when he performs a quadruple Lutz (Figure 10-38). Make any assumptions you think reasonable, but be prepared to justify them. Goebel's mass is about 60 kg and the height of the jump is about 0.6 m. Note that the angular velocity will change quite a bit during the jump, as he begins with arms outstretched and pulls them in. Your answer should be accurate to within a factor of 2 if you're careful.

**FIGURE 10-38**
**Problem 32**

## Vector Nature of Rotation

**33** • A force of magnitude $F$ is applied horizontally in the negative $x$ direction to the rim of a disk of radius $R$ as shown in Figure 10-39. Write $\vec{F}$ and $\vec{r}$ in terms of the unit vectors $\hat{i}$, $\hat{j}$, and $\hat{k}$, and compute the torque produced by the force about the origin at the center of the disk.

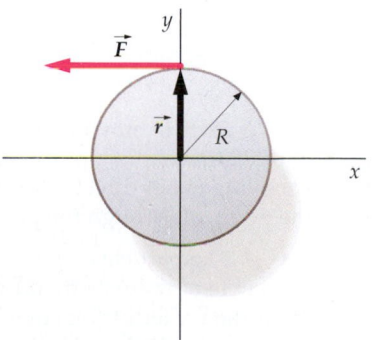

**FIGURE 10-39** Problem 33

**34** • Compute the torque about the origin for the force $\vec{F} = -mg\hat{j}$ acting on a particle at $\vec{r} = x\hat{i} + y\hat{j}$ and show that this torque is independent of the $y$ coordinate.

**35** • **SOLVE✔** Find $\vec{A} \times \vec{B}$ for (a) $\vec{A} = 4\hat{i}$ and $\vec{B} = 6\hat{i} + 6\hat{j}$, (b) $\vec{A} = 4\hat{i}$ and $\vec{B} = 6\hat{i} + 6\hat{k}$, and (c) $\vec{A} = 2\hat{i} + 3\hat{j}$ and $\vec{B} = 3\hat{i} + 2\hat{j}$.

**36** • **SSM** Under what conditions is the magnitude of $\vec{A} \times \vec{B}$ equal to $\vec{A} \cdot \vec{B}$?

**37** •• A particle moves in a circle that is centered at the origin. The particle has position $\vec{r}$ and angular velocity $\vec{\omega}$. (a) Show that its velocity is $\vec{v} = \vec{\omega} \times \vec{r}$. (b) Show that its centripetal acceleration is $\vec{a}_c = \vec{\omega} \times \vec{v} = \vec{\omega} \times (\vec{\omega} \times \vec{r})$.

**38** •• **SOLVE✔** If $\vec{A} = 4\hat{i}$, $B_z = 0$, $|\vec{B}| = 5$, and $\vec{A} \times \vec{B} = 12\hat{k}$, determine $\vec{B}$.

**39** • If $\vec{A} = 3\hat{j}$, $\vec{A} \times \vec{B} = 9\hat{i}$, and $\vec{A} \cdot \vec{B} = 12$, find $\vec{B}$.

**40** •• Use the rules for evaluating a determinant to show that if

$$\vec{A} = a_x\hat{i} + a_y\hat{j} + a_z\hat{k}$$

$$\vec{B} = b_x\hat{i} + b_y\hat{j} + b_z\hat{k}$$

$$\vec{C} = c_x\hat{i} + c_y\hat{j} + c_z\hat{k}$$

then

$$\vec{A} \cdot (\vec{B} \times \vec{C}) = \begin{vmatrix} a_x & a_y & a_z \\ b_x & b_y & b_z \\ c_x & c_y & c_z \end{vmatrix} = \vec{C} \cdot (\vec{A} \times \vec{B}) = \vec{B} \cdot (\vec{C} \times \vec{A})$$

**41** •• Given three noncoplanar vectors $\vec{A}$, $\vec{B}$, and $\vec{C}$, show that $\vec{A} \cdot (\vec{B} \times \vec{C})$ is the volume of the parallelopiped formed by the three vectors.

**42** •• **SSM** Using the cross product, prove the *law of sines* for the triangle shown in Figure 10-40: if $A$, $B$, and $C$ are the lengths of each side of the triangle, show that $A/\sin a = B/\sin b = C/\sin c$.

**FIGURE 10-40**
**Problem 42**

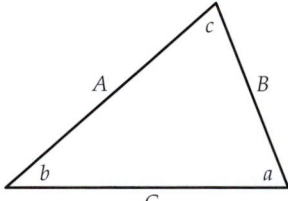

## Angular Momentum

**43** • A particle moving at constant velocity has zero angular momentum about a particular point. Show that the particle is moving either directly toward the point, directly away from the point, or through the point.

**44** • A 2-kg particle moves at a constant speed of 3.5 m/s around a circle of radius 4 m. (a) What is its angular momentum about the center of the circle and per-pendicular to the plane of the motion? (c) What is the angular speed of the particle?

**45** • **SOLVE✔** A 2-kg particle moves at constant speed of 4.5 m/s along a straight line. (a) What is the magnitude of its angular momentum about a point 6 m from the line? (b) Describe qualitatively how its angular speed about that point varies with time.

**46** •• **SSM** A particle is traveling with a constant velocity $\vec{v}$ along a line that is a distance $b$ from the origin $O$ (Figure 10-41). Let $dA$ be the area swept out by the position vector from $O$ to the particle in time $dt$. Show that $dA/dt$ is constant and is equal to $\frac{1}{2}L/m$, where $L$ is the angular momentum of the particle about the origin.

**FIGURE 10-41** Problem 46

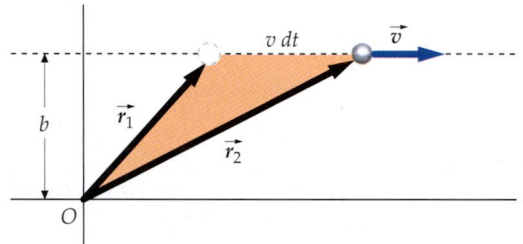

**47** •• **SOLVE** A 15-g coin of diameter 1.5 cm is spinning at 10 rev/s about a vertical diameter at a fixed point on a tabletop. (a) What is the angular momentum of the coin about its center of mass? (A coin is a cylinder of length $L$ and radius $R$, where $L$ is negligible compared to $R$. You can obtain its moment of inertia about an axis through a diameter in Table 9-1.) (b) What is its angular momentum about a point on the table 10 cm from the coin? If the coin spins about a vertical diameter at 10 rev/s while its center of mass travels in a straight line across the tabletop at 5 cm/s, (c) what is the angular momentum of the coin about a point on the line of motion of the center of mass? (d) What is the angular momentum of the coin about a point 10 cm to either side of the line of motion of the center of mass? (There are two answers to this question. Explain why and give both.)

**48** •• Two particles of masses $m_1$ and $m_2$ are located at $\vec{r}_1$ and $\vec{r}_2$ relative to some origin $O$ as in Figure 10-42. They exert equal and opposite forces on each other. Calculate the resultant torque exerted by these internal forces about the origin $O$ and show that it is zero if the forces $\vec{F}_1$ and $\vec{F}_2$ lie along the line joining the particles.

**FIGURE 10-42**
**Problem 48**

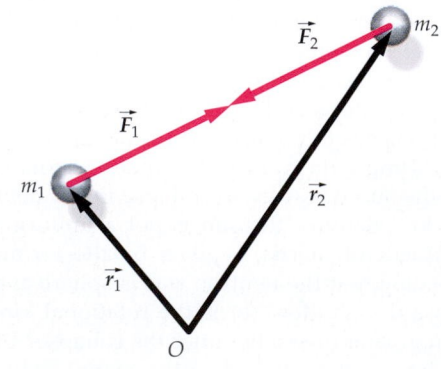

## Torque and Angular Momentum

**49** • **SOLVE** A 1.8-kg particle moves in a circle of radius 3.4 m. The magnitude of its angular momentum relative to the center of the circle depends on time according to $L = (4 \text{ N·m})t$. (a) Find the magnitude of the torque acting on the particle. (b) Find the angular speed of the particle as a function of time.

**50** •• **SOLVE** ✓ A uniform cylinder of mass 90 kg and radius 0.4 m is mounted so that it turns without friction on its fixed symmetry axis. It is rotated by a drive belt that wraps around its perimeter and exerts a constant torque. At time $t = 0$, its angular velocity is zero. At time $t = 25$ s, its angular velocity is 500 rev/min. (a) What is its angular momentum at $t = 25$ s? (b) At what rate is the angular momentum increasing? (c) What is the torque acting on the cylinder? (d) What is the magnitude of the frictional force acting on the rim of the cylinder?

**51** •• **SSM** In Figure 10-43, the incline is frictionless and the string passes through the center of mass of each block. The pulley has a moment of inertia $I$ and a radius $R$. (a) Find the net torque acting on the system (the two masses, string, and pulley) about the center of the pulley. (b) Write an expression for the total angular momentum of the system about the center of the pulley when the masses are moving with a speed $v$. (c) Find the acceleration of the masses from your results for Parts (a) and (b) by setting the net torque equal to the rate of change of the angular momentum of the system.

**FIGURE 10-43** Problem 51

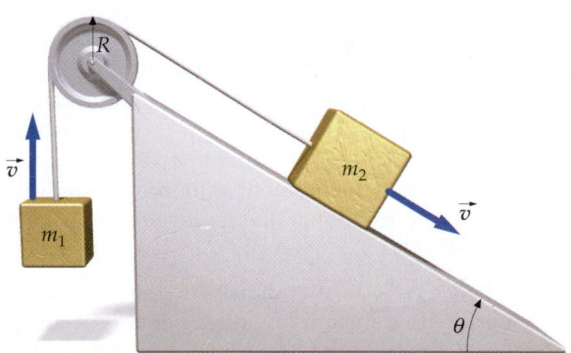

**52** •• **SOLVE** ✓ Figure 10-44 shows the rear view of a space capsule that is rotating about its longitudinal axis at 6 rev/min. The occupants want to stop this rotation. They have small jets mounted tangentially at a distance $R = 3$ m from the axis, as indicated, and can eject 10 g/s of gas from each jet with a nozzle velocity of 800 m/s. For how long must they turn on these jets to stop the rotation? The moment of inertia of the ship about its axis (assumed to be constant) is 4000 kg·m².

**FIGURE 10-44** Problem 52

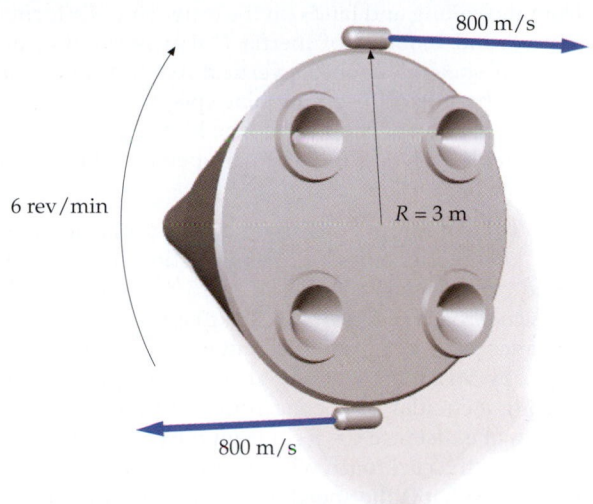

**53** •• A projectile (mass $M$) is launched at an angle $\theta$ with an initial speed $V$. Considering the torque and angular momentum about the launch point, explicitly show that $dL/dt = \tau$. Ignore the effects of air resistance. (The equations for projectile motion are found in Chapter 3.)

## Conservation of Angular Momentum

**54** • **SSM** **SOLVE** A planet moves in an elliptical orbit about the sun with the sun at one focus of the ellipse as in Figure 10-45. (a) What is the torque about the center of the sun due to the gravitational force of attraction of the sun for the planet? (b) At position $A$, the planet is a distance $r_1$ from the sun and is moving with a speed $v_1$ perpendicular to the line from the sun to the planet. At position $B$, it is at distance $r_2$ and is moving with speed $v_2$, again perpendicular to the line from the sun to the planet. What is the ratio of $v_1$ to $v_2$ in terms of $r_1$ and $r_2$?

**FIGURE 10-45** Problem 54

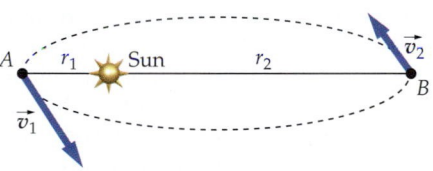

**55** •• **SOLVE** You stand on a frictionless platform that is rotating with an angular speed of 1.5 rev/s. Your arms are outstretched, and you hold a heavy weight in each hand. The moment of inertia of you, the extended weights, and the platform is 6 kg·m². When you pull the weights in toward your body, the moment of inertia decreases to 1.8 kg·m². (a) What is the resulting angular speed of the platform? (b) What is the change in kinetic energy of the system? (c) Where did this increase in energy come from?

**56** •• SSM ISOLVE✓ A small blob of putty of mass $m$ falls from the ceiling and lands on the outer rim of a turntable of radius $R$ and moment of inertia $I_0$ that is rotating freely with angular speed $\omega_i$ about its vertical fixed symmetry axis. (a) What is the postcollision angular speed of the turntable plus putty? (b) After several turns, the blob flies off the edge of the turntable. What is the angular speed of the turntable after the blob flies off?

**57** •• A Lazy Susan consists of a heavy plastic cylinder mounted on a frictionless bearing resting on a vertical shaft. The cylinder has a radius $R = 15$ cm and mass $M = 0.25$ kg. A cockroach (mass $m = 0.015$ kg) is on the Lazy Susan, at a distance of 8 cm from the center. Both the cockroach and the Lazy Susan are initially at rest. The cockroach then walks along a circular path concentric with the center of the Lazy Susan at a distance $r = 8$ cm from the center of the shaft. If the speed of the cockroach with respect to the Lazy Susan is $v = 0.01$ m/s, what is the speed of the cockroach with respect to the room?

**58** •• SSM ISOLVE Two disks of identical mass but different radii ($r$ and $2r$) are spinning on frictionless bearings at the same angular speed $\omega_0$ but in opposite directions (Figure 10-46). The two disks are brought slowly together. The resulting frictional force between the surfaces eventually brings them to a common angular velocity. What is the magnitude of that final angular velocity in terms of $\omega_0$?

**FIGURE 10-46** Problem 58

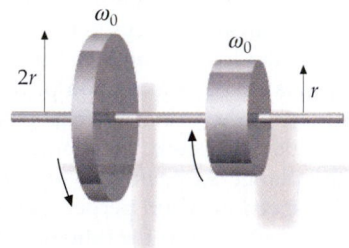

**59** •• ISOLVE✓ A block of mass $m$ sliding on a frictionless table is attached to a string that passes through a hole in the table. Initially, the block is sliding with speed $v_0$ in a circle of radius $r_0$. Find (a) the angular momentum of the block, (b) the kinetic energy of the block, and (c) the tension in the string. A student under the table now slowly pulls the string downward. How much work is required to reduce the radius of the circle from $r_0$ to $r_0/2$?

**60** •• SSM A 0.2-kg point mass moving on a frictionless horizontal surface is attached to a rubber band whose other end is fixed at point $P$. The rubber band exerts a force $F = bx$ toward $P$, where $x$ is the length of the rubber band and $b$ is an unknown constant. The mass moves along the dotted line in Figure 10-47. When it passes point $A$, its velocity is 4 m/s, directed as shown. The distance $AP$ is 0.6 m and $BP$ is 1.0 m. (a) Find the velocity of the mass at points $B$ and $C$. (b) Find $b$.

**FIGURE 10-47** Problem 60

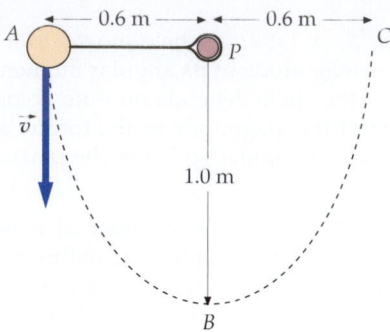

## *Quantization of Angular Momentum

**61** • SSM The $z$ component of the spin of an electron is $\frac{1}{2}\hbar$, but the magnitude of the spin vector is $\sqrt{0.75}\hbar$. What is the angle between the electron's spin angular momentum vector and the $z$ axis?

**62** •• Show that the energy difference between one rotational state and the next higher state is proportional to $\ell + 1$ (see Equation 10-27a).

**63** •• In the HBr molecule, the mass of the bromine nucleus is 80 times that of the hydrogen nucleus (a single proton); consequently, in calculating the rotational motion of the molecule, one may, to a good approximation, assume that the Br nucleus remains stationary as the H atom (mass $1.67 \times 10^{-27}$ kg) revolves around it. The separation between the H atom and bromine nucleus is 0.144 nm. Calculate (a) the moment of inertia of the HBr molecule about the bromine nucleus and (b) the rotational energies for $\ell = 1$, $\ell = 2$, and $\ell = 3$.

**64** •• ISOLVE The equilibrium separation between the nuclei of the nitrogen molecule is 0.11 nm. The mass of each nitrogen nucleus is 14 u, where u $= 1.66 \times 10^{-27}$ kg. We wish to calculate the energies of the three lowest angular momentum states of the nitrogen molecule. (a) Approximate the nitrogen molecule as a rigid dumbbell of two equal point masses and calculate the moment of inertia about its center of mass. (b) Find the rotational energy levels using the relation $E\ell = \ell(\ell + 1)\hbar^2/(2I)$.

**65** •• SSM How fast would a nitrogen molecule have to be moving for its translational kinetic energy to be equal to the rotational kinetic energy of its $l = 1$ quantum state?

## Collision Problems

**66** •• ISOLVE A 16.0-kg, 2.4-m-long rod is supported on a knife edge at its midpoint. A 3.2-kg ball of clay is dropped from rest from a height of 1.2 m and makes a perfectly inelastic collision with the rod 0.9 m from the point of support (Figure 10-48). Find the angular momentum of the rod and clay system about the point of support immediately after the inelastic collision.

**FIGURE 10-48** Problem 66

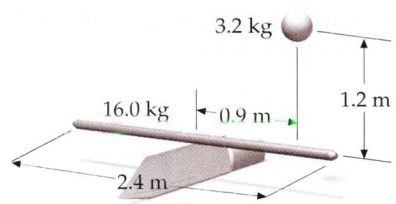

3.2 kg

16.0 kg — 0.9 m

1.2 m

2.4 m

**67** •• **SSM** Figure 10-49 shows a thin uniform bar of length $L$ and mass $M$ and a small blob of putty of mass $m$. The system is supported by a frictionless horizontal surface. The putty moves to the right with velocity $v$, strikes the bar at a distance $d$ from the center of the bar, and sticks to the bar at the point of contact. Obtain expressions for the velocity of the system's center of mass and for the angular velocity of the system about its center of mass.

**FIGURE 10-49**
**Problems 67, 68**

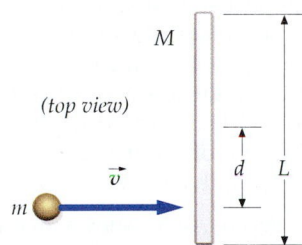

(top view)

$M$

$\vec{v}$

$m$

$d$ $L$

**68** •• In Problem 67, replace the blob of putty with a small hard sphere of negligible size that collides elastically with the bar. Find $d$ so that the sphere is at rest after the collision.

**69** •• **iSOLVE** Figure 10-50 shows a uniform rod of length $d$ and mass $M$ pivoted at the top. The rod, which is initially at rest, is struck by a particle of mass $m$ at a point $x = 0.8d$ below the pivot. Assume that the particle sticks to the rod. What must be the speed $v$ of the particle so that the maximum angle between the rod and the vertical is 90°?

**FIGURE 10-50** Problems 69, 70

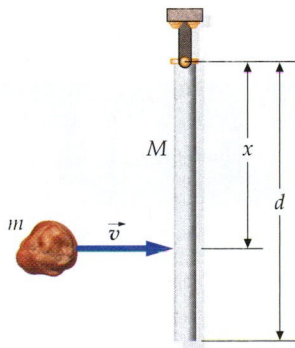

$M$ $x$

$d$

$m$ $\vec{v}$

**70** •• If, for the system of Problem 69, $d = 1.2$ m, $M = 0.8$ kg, $m = 0.3$ kg, and the maximum angle between the rod and the vertical is 60°, find the speed of the particle before impact.

**71** •• A uniform rod is resting on a frictionless table when it is suddenly struck near one end by a horizontal blow in a direction perpendicular to the rod. The mass of the rod is $M$ and the impulse applied by the blow is $K$. Immediately after the rod is struck, (a) what is the velocity of the center of mass of the rod, (b) what is the velocity of the end that is struck, (c) what is the velocity of the other end of the rod, and (d) is there a point on the rod that is motionless?

**72** •• A projectile of mass $m_p$ is traveling at a constant velocity $\vec{v}_0$ toward a stationary disk of mass $M$ and radius $R$ that is free to rotate about a pivot through its axis $O$ (Figure 10-51). Before impact, the projectile is traveling along a line displaced a distance $b$ below the axis. The projectile strikes the disk and sticks to point $B$. Treat the projectile as a point mass. (a) Before impact, what is the total angular momentum $L_0$ of the projectile and disk about the $O$ axis? (b) What is the angular speed $\omega$ of the disk and projectile system just after the impact? (c) What is the kinetic energy of the disk and projectile system after impact? (d) How much mechanical energy is lost in this collision?

**FIGURE 10-51** Problem 72

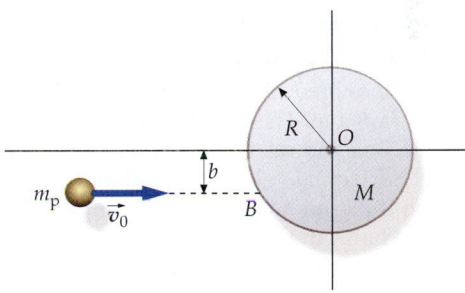

$R$ $O$

$b$

$m_p$ $\vec{v}_0$ $B$ $M$

**73** •• **SSM** A uniform rod of length $L_1$ and mass $M = 0.75$ kg is supported by a hinge at one end and is free to rotate in the vertical plane (Figure 10-52). The rod is released from rest in the position shown. A particle of mass $m = 0.5$ kg is supported by a thin string of length $L_2$ from the hinge. The particle sticks to the rod on contact. What should be the ratio $L_2/L_1$ so that $\theta_{max} = 60°$ after the collision?

**FIGURE 10-52**
**Problems 73–76**

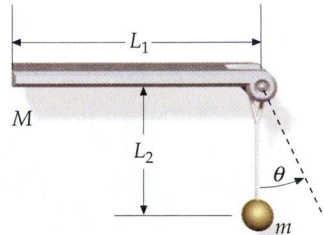

$L_1$

$M$

$L_2$

$\theta$

$m$

**74** •• **iSOLVE** A uniform rod of length $L_1 = 1.2$ m and mass $M = 2.0$ kg is supported by a hinge at one end and is free to rotate in the vertical plane (Figure 10-52). The rod is released from rest in the position shown. A particle of mass $m$ is supported by a thin string of length $L_2 = 0.8$ m from the hinge. The particle sticks to the rod on contact, and after the collision the rod continues to rotate until $\theta = \theta_{max} = 37°$. (a) Find $m$. (b) How much energy is dissipated during the collision?

**75** •• A uniform rod of length $L_1 = 1.2$ m and mass $M = 2.0$ kg is supported by a hinge at one end and is free to rotate in the vertical plane (Figure 10-52). The rod is given a sharp downward blow by a hammer, giving the rod an initial angular velocity. A particle of mass $m = 0.4$ kg is supported by a thin string of length $L_2 = 0.8$ m from the hinge. The particle sticks to the rod on contact, and after the collision the rod continues to rotate through a complete revolution. What is the minimum angular velocity given to the rod by the blow? How much energy is dissipated during the collision?

**76** ••• Repeat Problem 74, assuming that the collision between the rod and the particle is elastic.

**77** •• **SOLVE** A bicycle wheel of radius 28 cm is mounted at the middle of an axle 50 cm long. The tire and rim weigh 30 N. The wheel is spun at 12 rev/s, and the axle is then placed in a horizontal position with one end resting on a pivot. (a) What is the angular momentum due to the spinning of the wheel? (Treat the wheel as a hoop.) (b) What is the angular velocity of precession? (c) How long does it take for the axle to swing through 360° around the pivot? (d) What is the angular momentum associated with the motion of the center of mass, that is, due to the precession? In what direction is this angular momentum?

**78** •• **SSM** **SOLVE** A uniform disk of mass 2.5 kg and radius 6.4 cm is mounted in the center of a 10-cm-long axle and spun at 700 rev/min. The axle is then placed in a horizontal position with one end resting on a pivot. The other end is given an initial horizontal velocity such that the precession is smooth with no nutation. (a) What is the angular velocity of precession? (b) What is the speed of the center of mass during the precession? (c) What are the magnitude and direction of the acceleration of the center of mass? (d) What are the vertical and horizontal components of the force exerted by the pivot?

## General Problems

**79** • **SOLVE** ✓ A particle of mass 3 kg moves with velocity $\vec{v} = 3$ m/s$\hat{i}$ along the line $z = 0$, $y = 5.3$ m. (a) Find the angular momentum $\vec{L}$ relative to the origin when the particle is at $x = 12$ m, $y = 5.3$ m. (b) A force $\vec{F} = -3$ N$\hat{i}$ is applied to the particle. Find the torque relative to the origin due to this force.

**80** • **SOLVE** The position vector of a particle of mass 3 kg is given by $\vec{r} = 4\hat{i} + 3t^2\hat{j}$, where $\vec{r}$ is in meters and $t$ is in seconds. Determine the angular momentum and torque about the origin acting on the particle.

**81** •• **SOLVE** Two ice skaters hold hands and rotate, making one revolution in 2.5 s. Their masses are 55 kg and 85 kg, and they are separated by 1.7 m. Find (a) the angular momentum of the system about their center of mass and (b) the total kinetic energy of the system.

**82** •• **SSM** A 2-kg ball attached to a string of length 1.5 m moves counterclockwise (as viewed from above) in a horizontal circle (Figure 10-53). The string makes an angle $\theta = 30°$ with the vertical. (a) Show that the angular momentum of the ball about the point of support $P$ has a horizontal component

away from the center of the circle as well as a vertical component, and find these components. (b) Find the magnitude of $d\vec{L}/dt$ and show that it equals the magnitude of the torque exerted by gravity about the point of support.

**FIGURE 10-53**
Problem 82

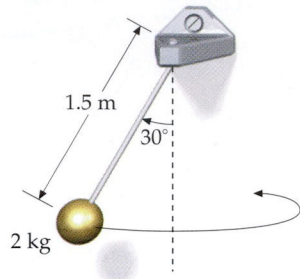

**83** •• An object of mass $m$ on a horizontal, frictionless surface is attached to a string that wraps around a vertical cylindrical post so that when the object is set into motion, it follows a path that spirals inward. (a) Is the angular momentum of the object about the point on the axis of the post in the plane of the spiral conserved? (b) Is the energy of the object conserved? (c) If the speed of the object is $v_0$ when the unwrapped length of the string is $r$, what is its speed when the unwrapped length has shortened to $r/2$?

**84** •• Figure 10-54 shows a hollow cylindrical tube of mass $M$, length $L$, and moment of inertia $ML^2/10$. Inside the cylinder are two disks of mass $m$, separated by a distance $\ell$ and tied to a central post by a thin string. The system can rotate about a vertical axis through the center of the cylinder. With the system rotating at $\omega$, the strings holding the disks suddenly break. When the disks reach the ends of the cylinder, they stick. Obtain expressions for the final angular velocity and the initial and final energies of the system. Assume that the inside walls of the cylinder are frictionless.

**FIGURE 10-54** Problems 84–88

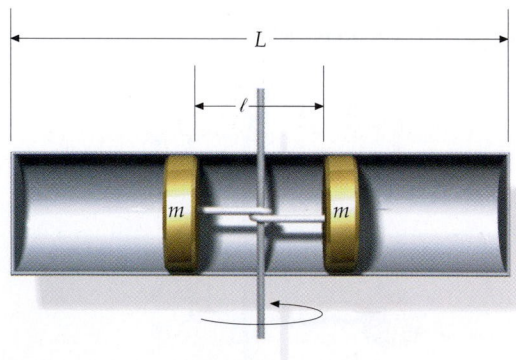

**85** •• Repeat Problem 84, this time allowing for friction between the disks and the walls of the cylinder. However, the coefficient of friction is not great enough to prevent the disks from reaching the ends of the cylinder. Can the final energy of the system be determined without knowing the coefficient of kinetic friction?

**86** •• **SOLVE✓** Suppose that in Figure 10-54, $\ell = 0.6$ m, $L = 2.0$ m, $M = 0.8$ kg, and $m = 0.4$ kg. The system rotates at $\omega$ such that the tension in the string is 108 N just before it breaks. Determine the initial and final angular velocities and initial and final energies of the system. Assume that the inside walls of the cylinder are frictionless.

**87** •• For Problem 84, determine the radial velocity of each disk just before it reaches the end of the cylinder.

**88** •• Given the numerical values of Problem 86, suppose that the coefficient of friction between the disks and the walls of the cylinder is such that the disks cease sliding 0.2 m from the ends of the cylinder. Determine the initial and final angular velocities of the system and the energy dissipated in friction.

**89** •• **SSM** Kepler's second law states: *The radius vector from the sun to a planet sweeps out equal areas in equal times.* Show that this law follows directly from the law of conservation of angular momentum and the fact that the force of gravitational attraction between a planet and the sun acts along the line joining the two celestial objects.

**90** •• Figure 10-55 shows a hollow cylindrical shell of length 1.8 m, mass 0.8 kg, and radius 0.2 m that is free to rotate about a vertical axis through its center and perpendicular to the cylinder's axis. Inside the cylinder are two thin disks of 0.2 kg each, attached to springs of spring constant $k$ and unstretched lengths 0.4 m. The system is brought to a rotational speed of 8 rad/s with the springs clamped so they do not stretch. The springs are then suddenly unclamped. When the disks have stopped their radial motion due to friction between the disks and the wall, they come to rest 0.6 m from the central axis. What is the angular velocity of the cylinder when the disks have stopped their radial motion? How much energy was dissipated in friction between the disks and cylinder wall?

**FIGURE 10-55** Problem 90

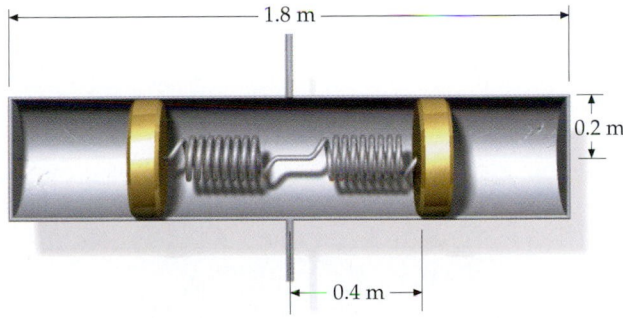

**91** •• You are given a heavy metal disk (like a coin, but larger; Figure 10-56). (Objects like this called *Euler disks* are sold through Tangent Toys.) Placing it on a turntable, you spin the disk about a vertical axis through a diameter of the disk and the center of the turntable. As you do this, you hold the turntable still with your other hand, letting it go immediately after you spin the disk. The mass of the disk is 0.500 kg and it has a radius of 0.125 m and negligible thickness. The turntable is a uniform solid cylinder with a radius of 0.250 m and a mass of 0.735 kg and rotates on a frictionless bearing.

The disk has an initial angular speed of 30 rev/min. (a) If the disk spins down to rest on the turntable with its symmetry axis coinciding with the turntable's, what is the final angular speed of the turntable? (b) What will be the final angular speed if the disk's axis is 0.10 m from the axis of the turntable when it comes to rest?

**FIGURE 10-56** Problem 91

**92** •• (a) Assuming the earth to be a homogeneous sphere of radius $r$ and mass $m$, show that the period $T$ of the earth's rotation about its axis is related to its radius by $T = br^2$, where $b = (4/5)\,\pi m/L$, and where $L$ is the spin angular momentum of the earth. (b) Suppose that the radius $r$ changes by a very small amount $\Delta r$ due to some internal cause such as thermal expansion. Show that the fractional change in the period $\Delta T$ is given approximately by $\Delta T/T = 2\Delta r/r$. *Hint: Use the differentials dr and dT to approximate the changes in these quantities.* (c) By how many kilometers would the earth need to expand for the period to change by 0.25 d/y so that leap years would no longer be necessary?

**93** •• **SSM** The Precession of the Equinoxes refers to the fact that the direction of the earth's spin axis does not stay fixed in the sky, but moves in a circle of radius 23° with a period of about 26,000 y. This is why our pole star, Polaris, will not remain the pole star forever. The reason for this is that the earth is a giant gyroscope, with the torque on the earth provided by the gravitational forces of the sun and moon. Calculate an approximate value for this torque, given that the period of rotation of the earth is 1 d and its moment of inertia is $8.03 \times 10^{37}$ kg·m².

**94** ••• **SOLVE** Figure 10-57 shows a cylindrical shell of mass $M = 1.2$ kg and length $L = 1.6$ m that is free to rotate about a vertical axis through its center. Inside the cylinder are two disks, each of mass 0.4 kg, that are tied to a central post by a thin string and separated by a distance $\ell = 0.8$ m. The string breaks if the tension exceeds 100 N. Starting from rest, a torque is applied to the system until the string breaks. Assuming that the disks are point masses and the radius of the cylindrical shell is negligible, find the amount of work done up to that instant. Suppose that, at that instant, the applied torque is removed and the walls of the cylindrical shell are frictionless. Obtain an expression for the angular velocity of the system as a function of $x$ for $x < L/2$, where $x$ is the distance between each mass and the central post.

**FIGURE 10-57** Problems 94–96

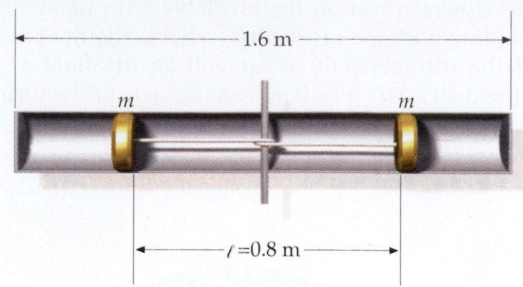

95 ••• For the system of Problem 94, find the angular velocity of the system just before and just after the point masses pass the ends of the cylindrical shell.

96 ••• Repeat Problem 94 with the radius of the cylindrical shell equal to 0.4 m and the masses treated as thin disks rather than point masses.

97 ••• SSM Figure 10-58 shows a pulley in the form of a uniform disk with a heavy rope hanging over it. The circumference of the pulley is 1.2 m and its mass is 2.2 kg. The rope is 8.0 m long and its mass is 4.8 kg. At the instant shown in the figure, the system is at rest and the difference in height of the two ends of the rope is 0.6 m. (a) What is the angular velocity of the pulley when the difference in height between the two ends of the rope is 7.2 m? (b) Obtain an expression for the angular momentum of the system as a function of time while neither end of the rope is above the center of the pulley. There is no slippage between rope and pulley wheel.

**FIGURE 10-58**
**Problem 97**

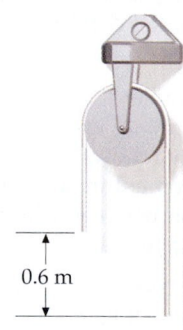

0.6 m

# Special Relativity

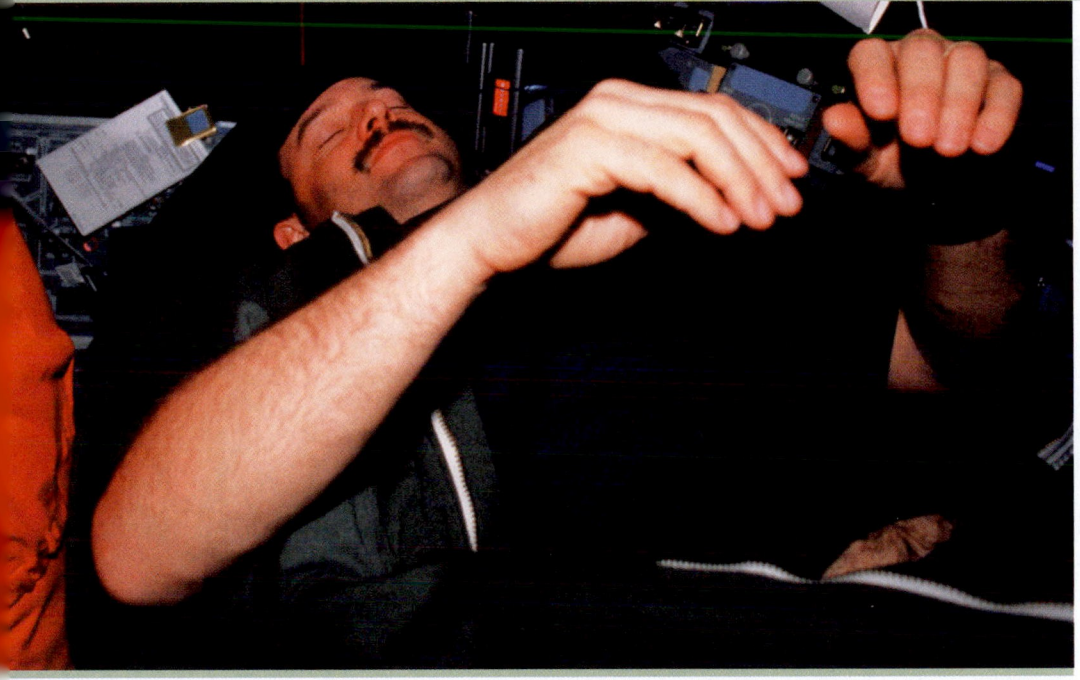

STS109-335-020 (1-12 MARCH 2002)—
ASTRONAUT SCOTT D. ALTMAN,
STS-109 MISSION COMMANDER,
SLEEPS ON THE FLIGHT DECK OF THE
SPACE SHUTTLE COLUMBIA. THE
ORBITAL SPEED OF THE SHUTTLE IS
ABOUT 7.7 KM/S (5 MI/S). THIS IS A
SMALL FRACTION OF THE SPEED OF
LIGHT, WHICH IS $3 \times 10^5$ KM/S.

**?** If an astronaut on a
spaceship that is traveling at
$0.6c$ relative to the earth takes
a one-hour-long nap, does that
nap take one hour according to
observers on earth? (See
Example R-1.)

R-1    The Principle of Relativity and the Constancy of the Speed of Light

R-2    Moving Sticks

R-3    Moving Clocks

R-4    Moving Sticks Again

R-5    Distant Clocks and Simultaneity

R-6    Applying the Rules

R-7    Relativistic Momentum, Mass, and Energy

The theory of relativity consists of two rather different theories, the special theory and the general theory. The special theory, developed by Albert Einstein and others in 1905, concerns the comparison of measurements made in different inertial reference frames moving with constant velocity relative to one another. Its consequences, which can be derived with a minimum of mathematics, are applicable in a wide variety of situations encountered in physics and engineering. On the other hand, the general theory, also developed by Einstein and others around 1916, is concerned with accelerated reference frames and gravity. A thorough understanding of the general theory requires sophisticated mathematics, and the applications of this theory are chiefly in the area of gravitation.

➤ In this chapter we concentrate on the special theory of relativity (often referred to as *special relativity*). In the early 1900s the special theory of relativity was accepted with enthusiasm by some, but by many it was either reluctantly accepted or dismissed as folly. Today it is not only widely accepted, but embraced as a window into the workings of nature. We will see how this theory challenges our everyday experience of time and distance, as we describe the slowing down of moving clocks, the shortening of moving sticks, the relativity of simultaneity for events that occur in different locations, and the relativity of momentum and energy relation.

# R-1 The Principle of Relativity and the Constancy of the Speed of Light

The principle of relativity can be stated as follows:

> It is impossible to devise an experiment that determines whether you are at rest or moving uniformly.

POSTULATE I, THE PRINCIPLE OF RELATIVITY

Moving uniformly means moving at constant velocity relative to an inertial reference frame. For example, suppose that you are in your seat on board a high-speed airplane moving uniformly relative to the surface of the earth. If you drop your fork, it will fall to the floor in exactly the same way that it would if the plane were parked on a runway. When the airplane is in flight, you can consider yourself and the airplane to be at rest and the surface of the earth below you to be moving. There is nothing to distinguish whether you and the plane are moving and the surface of the earth is at rest, or vice versa.

Any reference frame in which a particle with no forces acting on it moves with constant velocity is, by definition, an inertial reference frame.[†] The surface of the earth is, to a good approximation, an inertial reference frame. The airplane is also an inertial reference frame as long as it moves with constant velocity relative to the surface of the earth. As long as you remain seated or standing still on the airplane you can consider yourself and the airplane to be at rest and the surface of the earth to be moving, or you can consider the surface of the earth to be at rest and yourself and the airplane to be moving.

In the nineteenth century the existence of a preferred frame of reference that could be considered to be at rest was widely accepted. This was thought to be the reference frame of the *ether,* the medium filling all of space through which light was thought to propagate. (It was then accepted that light waves needed a medium to propagate through, just as it is now accepted that sound waves need air or some other material medium through which to propagate.) The ether was considered to be the preferred "at rest" reference frame.

A carefully devised series of measurements to measure the orbital speed of the earth relative to the ether were carried out in 1887 by Albert Michelson and Edward Morley. These measurements were considered challenging because the orbital speed of the earth is less than 1/10,000 the speed of light in vacuum. Much to the surprise of nearly everyone, the observations always found the speed of the earth relative to the ether to be zero. It was Albert Einstein who came up with a theory that was consistent with these observations. His explanation was that light is capable of traveling through empty space and that the ether was an unnecessary construct that did not exist. Einstein also postulated:

> The speed of light is independent of the speed of the light source.

POSTULATE II

The *speed of light* refers to the speed at which light travels through the vacuum of empty space.

A consequence of Postulate II and the principle of relativity is that all inertial observers measure the same value for the speed of light. (An inertial observer

---

† Further discussion on reference frames can be found in Chapters 2 and 4.

is one that remains at rest in an inertial reference frame.) To establish that all inertial observers measure the same value for the speed of light, we consider inertial observers A and B, where observer A is moving relative to observer B. The principle of relativity states that it is impossible to devise an experiment that determines whether an inertial observer is at rest or moving uniformly. If observer A measures a different value for the speed of light than observer B, then observers A and B could not both consider themselves to be at rest—a result in direct contradiction with the principle of relativity. Thus, a consequence of both the principle of relativity and Postulate II (that the speed of light is independent of the speed of the source) leads to the **constancy of the speed of light:**

> The speed of light $c$ is the same in any inertial reference frame.

THE CONSTANCY OF THE SPEED OF LIGHT

That is, anything (not just light) that travels at speed $c$ relative to one inertial reference frame travels at the same speed $c$ relative to any inertial reference frame.

Suppose you are in your backyard here on earth and Bob is on a spaceship moving away from you at half the speed of light ($\frac{1}{2}c$). You point a flashlight in Bob's direction and turn it on. The light leaves the flashlight, traveling at speed $c$ (relative to the flashlight), and passes by your neighbor Keisha who is standing on the roof of her house next door. Keisha measures the speed of the light going by and finds it to be traveling at speed $c$. A few minutes later the light travels past Bob and his spaceship. Like Keisha, Bob measures the speed of the light going by him and also finds it to be traveling at speed $c$. This surprises Bob because he expected the light to be traveling past him at speed $\frac{1}{2}c$ rather than at speed $c$—after all, Bob is moving at speed $\frac{1}{2}c$ relative to the source of the light (the flashlight in your backyard). Like many people, Bob finds the constancy of the speed of light to be counterintuitive. This leaves him with a dilemma. Should he trust his measuring instruments or trust his intuition? It turns out that it is Bob's intuition that needs adjusting, not his instruments. Bob must change his concepts of both space and time.

Suppose that instead of pointing a flashlight you point a high-speed particle beam in his direction, where by "high speed" we mean a speed very close to the speed of light $c$. (A particle such as an electron or proton cannot travel at the speed of light, but it can travel at speeds extremely close to the speed of light.) If Keisha measures the particles going by her to be traveling at $0.9999c$ (relative to her), then how fast will Bob measure the particles going by him? Bob's intuition tells him that, because he is moving away from the source of the particles, they will be traveling past him at the slower speed of $0.4999c$, but that is not the case. When Bob measures the speed of the particles (relative to him) he finds it to be extremely close to $0.9999c$. (The actual value is $0.9997c$.)[†]

We tend to think of distances between cities as fixed. However, this too is not the case. According to a certain road map the distance between Baltimore and Philadelphia is 160 km. However, if you travel from Baltimore to Philadelphia at a significant fraction of the speed of light, the distance between the two cities will be much shorter than it is if you travel at 100 km/h (62 mi/h). For someone driving at 100 km/h, the distance between Baltimore and Philadelphia is very close to 160 km. However, for someone traveling at a speed of $0.866c$ (relative to the earth's surface), the distance is only 80 km, and for someone traveling at $0.9999c$, the distance is only 2.2 km.

The fastest speed that a human being has ever traveled relative to the earth (which occurred during the Apollo missions to the moon) is only about 10 km/s = $3.3 \times 10^{-5}c$. This speed is so slow compared to the speed of light, that for someone traveling from Baltimore to Philadelphia at that speed, the distance between those cities would be shorter by less than the diameter of a human hair. The logic explaining how this is determined is presented in the next three sections.

---

† Relative velocity in special relativity is covered in Chapter 37.

# R-2 Moving Sticks

We wish to show that if a stick moves perpendicular to its length, its length does not change. We do this by showing that any increase or decrease in length contradicts the principle of relativity. Showing that a stick does not change its length may seem mundane. However, we show it because an immediate consequence is that moving clocks run slow.

Suppose that we have two identical metersticks, stick A and stick B. We verify that the sticks are identical by placing them side by side in a reference frame in which they are both at rest and examining them visually. We then give stick B to Bob before he takes off on another trip in his spaceship. On this trip Bob makes sure to hold the stick at right angles to the velocity of the spaceship relative to the earth. Is stick B now shorter than stick A, which remains back on the earth with us?

To answer this question we conduct a thought experiment. We attach felt-tipped marking pens to stick A, one at the 20-cm mark, the other at the 80-cm mark. Then Bob and his spaceship execute a flyby during which Bob holds stick B out a porthole, keeping it at right angles to the ship's velocity. During the flyby we hold up our stick (stick A), keeping it parallel with stick B. As the sticks pass by each other, two marks are drawn on stick B (Figure R-1). Bob then returns to earth with stick B and the two sticks are again placed side by side (Figure R-2), and the distance between the two marks on stick B and the distance between the two marking pens on stick A are compared. Let us assume that a stick moving perpendicular to its length is shorter than is an identical stationary stick. Then the distance between the two pens will be less than the distance between the two marks (Figure R-2)—clear evidence that during the flyby the moving stick (stick B) was shorter than the stationary stick. However, according to the principle of relativity it is equally valid to think of stick B as stationary and stick A as moving during the flyby. From this perspective, the same evidence (Figure R-2) demonstrates that the moving stick—now stick A—is longer than the stationary stick. Thus our assumption—that a stick moving perpendicular to its length is shorter than an identical stationary stick—leads to a contradiction and must be rejected. The assumption that a stick moving perpendicular to its length is longer than is an identical stationary stick also leads to a contradiction, as can be shown using an analogous argument. Thus we conclude:

A stick moving perpendicular to its length has the same length as an identical stationary stick.

This rule is established without any consideration of the material from which the two sticks are made. Thus, the rule does not reflect a property of sticks. Instead it reflects a property of space.

The frame of reference in which the stick is at rest is called its **proper frame** or **rest frame**, and the length of a stick in its proper reference frame is called its **proper length** or **rest length**.

# R-3 Moving Clocks

Clocks are used to measure time. In this section we will show that clocks moving at high speeds run slow, so if a high-speed spaceship travels by us, we would observe that all the clocks on the ship run slower than our clocks. However, the people on the ship are free to consider themselves to be at rest and us to be moving, and they would observe our clocks to run slow compared to their clocks. Let's examine how these observations are consistent with the constancy of the speed of light and the principle of relativity.

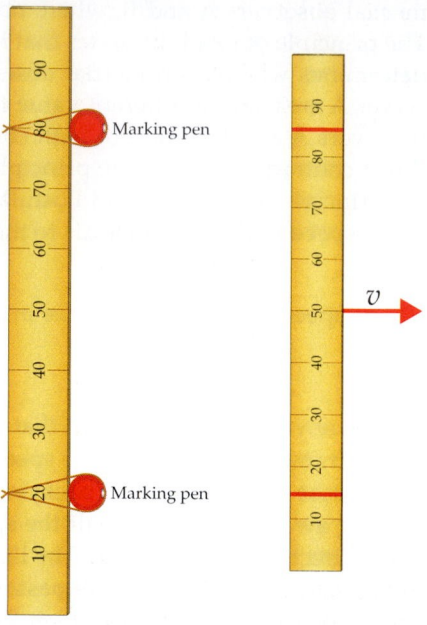

**FIGURE R-1** During the flyby marks like this would be made on stick B by marking pens attached to stick A if the moving stick was shortened.

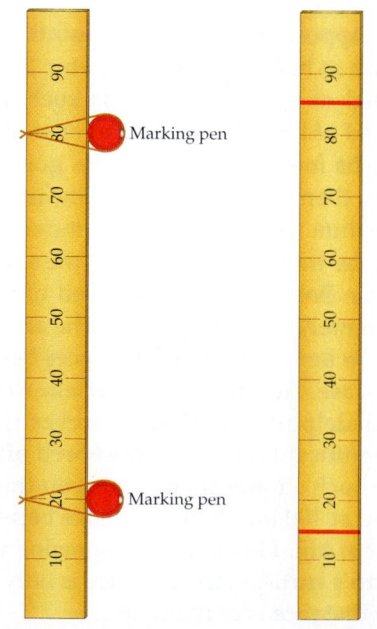

**FIGURE R-2** If the distance between marks is greater than the distance between marker pens, this would demonstrate that stick B was shorter than stick A when the marks were made.

We construct a clock, called a *light clock*, using a stick of proper length $L_0$ and two mirrors (Figure R-3). The two mirrors face each other, and a light pulse bounces back and forth between them. Each time the light pulse strikes one of the mirrors, say the lower mirror, the clock is said to tick. Between successive ticks the light pulse travels a distance $2L_0$ in the proper reference frame of the clock. Thus the time between ticks $T_0$ is related to $L_0$ by

$$2L_0 = cT_0 \qquad\qquad \text{R-1}$$

Next we consider the time between ticks $T$ of the same light clock, but this time we observe it from a reference frame in which the clock is moving with speed $v$ perpendicular to the stick (Figure R-4). In this reference frame the clock moves a distance $vT$ between ticks and the light pulse moves a distance $cT$ between ticks. The distance the pulse moves in traveling from the bottom mirror to the top mirror is $\sqrt{L_0^2 + (\frac{1}{2}vT)^2}$. The light pulse travels the same distance in traveling from the top mirror to the bottom mirror. Thus,

$$2\sqrt{L_0^2 + (\tfrac{1}{2}vT)^2} = cT \qquad\qquad \text{R-2}$$

(Note that we have used the same symbol $c$ for the speed of light in Equations R-1 and R-2.) Solving Equation R-1 for $L_0$ and substituting into Equation R-2 gives

$$\sqrt{(\tfrac{1}{2}cT_0)^2 + (\tfrac{1}{2}vT)^2} = \tfrac{1}{2}cT$$

Solving for $T$ gives

$$T = \frac{T_0}{\sqrt{1 - (v^2/c^2)}} \qquad\qquad \text{R-3}$$

TIME DILATION

According to Equation R-3, the time between ticks in the reference frame in which the clock moves at speed $v$ is greater than the time between ticks in the proper reference frame of the clock.

This raises the question, do other clocks run slow when they move with speed $v$ according to Equation R-3 or is Equation R-3 valid only for light clocks? To answer this question we attach a conventional clock (with a conventional clock mechanism) to the lower mirror of the light clock (Figure R-5). The conventional clock has no minute or hour hands. Instead of a second hand it has an opaque disk with a narrow slot to indicate the time. The clock's face contains 60 equal-spaced marks (called tick marks) around its perimeter—one for each second. The clock ticks each time the slot passes over one of the tick marks. We adjust the length $L_0$ of the light-clock stick so that the time between ticks of both clocks is the same in the proper reference frame of the clocks. Next, we synchronize the clocks so each tick of the light clock occurs simultaneously with a tick of the conventional clock.

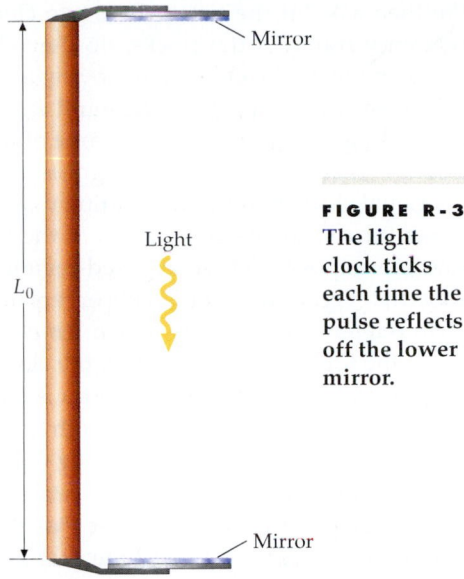

**FIGURE R-3**
**The light clock ticks each time the pulse reflects off the lower mirror.**

Mirror

Light

$L_0$

Mirror

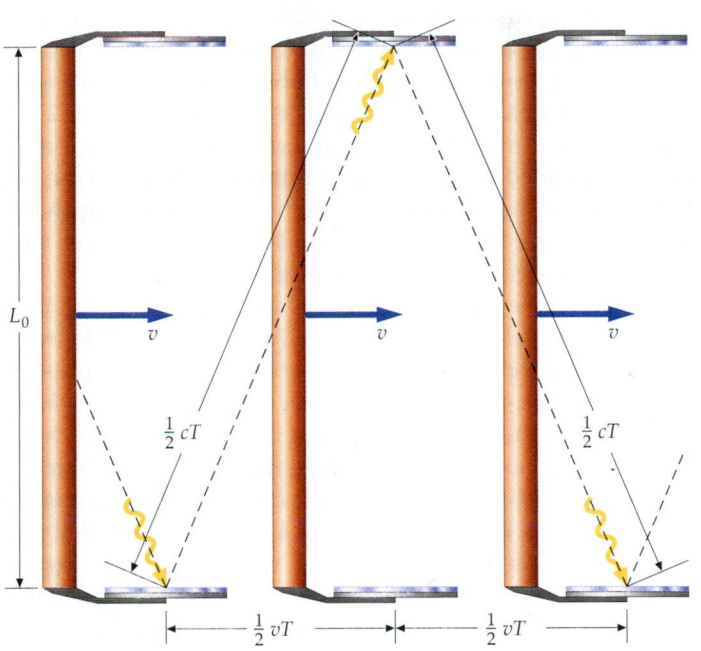

**FIGURE R-4**

$L_0$

$v$   $v$   $v$

$\tfrac{1}{2}cT$     $\tfrac{1}{2}cT$

$\tfrac{1}{2}vT$     $\tfrac{1}{2}vT$

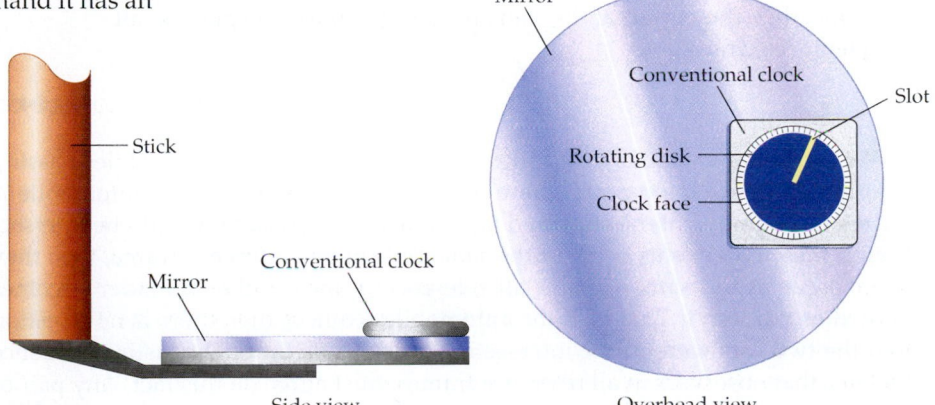

**FIGURE R-5**

Stick

Mirror

Conventional clock

Side view

Mirror

Conventional clock

Rotating disk

Clock face

Slot

Overhead view

We then ask, "If the ticks of the two clocks occur simultaneously in the proper reference frame of the clocks, do they also occur simultaneously in a reference frame in which the clocks are moving at speed $v$?"

The answer is yes. To understand why this is so, consider the following experiment. In the proper reference frame of the clocks, the time between ticks of both clocks is exactly one second. A light-sensitive film is placed on the face of the conventional clock, behind the rotating disk. Each time the light pulse reflects off the lower mirror, the narrow region of the light-sensitive paper directly behind the slot gets exposed. These exposed regions will be aligned with the tick marks as shown in Figure R-6, and all observers must agree with this permanent record.

In reference frame A in which the clocks are both moving, the light pulse exposes the film behind the slot on the clock face each time the pulse reflects off the lower mirror. Because the light clock is moving, the time between these reflections is greater than 1 s, in accordance with Equation R-3. When an observer of reference frame A sees that the lines appearing where the film was exposed are aligned with the tick marks, she realizes that in her reference frame the conventional clock runs slow in exactly the same manner that the light clock runs slow—in accord with Equation R-3. Thus we conclude that all moving clocks run slow in exactly the same manner that a moving light clock runs slow. Because this is the case, we conclude that it is time itself that runs slow, a phenomenon known as **time dilation.**

Something that occurs at a specific instant in time and at a specific location in space is called a **spacetime event,** or just an **event.** Each reflection of the light pulse off the lower mirror of the light clock is a spacetime event. If we call one of these reflections event 1, and the next reflection event 2, then the time between events 1 and 2 in a frame of reference in which the two events occur at the same location is called the **proper time interval** $T_0$ between the two events. Let $T$ be the time between the same two events in a reference frame in which they occur at different locations. Equation R-3 relates the time $T$ between two events to the proper time $T_0$ between the same two events.

Each time the light pulse reflects off the lower mirror, the slot (second hand) of the conventional clock passes directly over a tick mark. In the proper frame of the two clocks, these two events—the arrival of the light pulse and the passing of the slot over a tick mark—occur at the same time *and* at the same place. Any two events that occur both at the same time and at the same place in one reference frame will occur both at the same time and at the same place in all reference frames. This is because such events can have lasting consequences—like producing lines on the light-sensitive film aligned with the tick marks on the clock face. We cannot have the marks aligned with the ticks marks in one reference frame and not aligned with the tick marks in another reference frame. After all, there is only one clock face and one set of marks. This conclusion can be generalized into a principle, called the **principle of invariance of coincidences:**

If two events occur at the same time and at the same place in one reference frame, then they occur at the same time and at the same place in all reference frames.

PRINCIPLE OF INVARIANCE OF COINCIDENCES

We can better visualize this principle by considering two automobiles passing through an intersection at the same time. The two events are: (1) automobile A passes through the intersection and (2) automobile B passes through the intersection. If these two events occur at the same time in one reference frame, then they must occur at the same time in all reference frames. Either a fender becomes dented or it doesn't. That is, if the automobiles collide, then there is no question that the two cars were in the intersection at the same time. The lasting evidence dictates that observers in all reference frames must agree on this fact. Any pair of events that occur at the same time *and* at the same location are referred to as a **spacetime coincidence.**

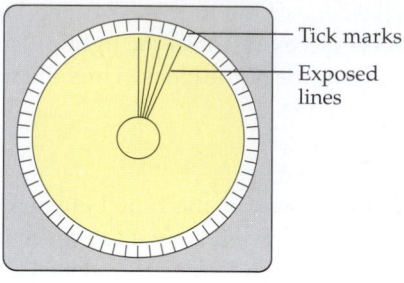

Tick marks
Exposed lines

Conventional clock

**FIGURE R-6**

*THE NAPPING ASTRONAUTS*                    **EXAMPLE  R-1**   **Put It in Context**

Astronauts in a spaceship traveling at $v = 0.6c$ relative to the earth sign off from space control, saying that they are going to nap for 1 h and then will call back. How long does their nap last according to observers on the earth?

**PICTURE THE PROBLEM**  Clock S on the ship reads $t_0$ when the nap begins (a spacetime coincidence) and reads $t_0 + 1$ h when the nap ends (also a spacetime co-incidence). Observers on the ship agree that, because clock S is stationary it does not run slow, so the nap lasted 1 h. In the reference frame of the ship the two events (the beginning of the nap and the end of the nap) occur at the same location, so the time interval between the events is the proper time interval between them. Ob-servers on the earth agree that clock S reads $t_0$ when the nap begins and it reads $t_0 + 1$ h when the nap ends. However, the observers on the earth agree that because clock S is moving at speed $v$, it is running slow so the nap lasted more than 1 h. In the reference frame of the earth the ship is moving so the nap begins and ends at different locations. Therefore, in the reference frame of the earth the time interval between the events is not the proper time interval between the events.

1. Event 1 is the beginning of the nap and event 2 is the end of the nap. Clock S on the ship advances 1 h between these events. Determine the proper time interval $T_0$ between these events:

$$T_0 = 1\,h$$

2. Find the time interval $T$ between events 1 and 2 for observers on earth:

$$T = \frac{T_0}{\sqrt{1 - (v^2/c^2)}} = 1\,h/\sqrt{1 - \frac{(0.6c)^2}{c^2}}$$

$$= \frac{1\,h}{\sqrt{1 - 0.36}} = \frac{1\,h}{\sqrt{0.64}} = \frac{1\,h}{0.8} = \boxed{1.25\,h}$$

**REMARKS**  Clock S is an unnecessary construct as the astronauts themselves serve as clocks. What it is necessary to realize is that the proper time between the beginning and the end of the nap is 1 h, so the time $T$ between the same events in a reference frame where the clocks (astronauts) are moving with speed $v$ is given by Equation R-3.

**EXERCISE**  A pion[†] has a mean proper lifetime of 26 ns (1 ns = $1 \times 10^{-9}$ s) (mea-sured when the pion is at rest). What is the mean lifetime if measured when the pion is moving at 0.995c? (*Answer*   260 ns)

**EXERCISE**  A pion has a proper half life of 18 ns (measured when the pion is at rest). A beam of pions moving at 0.995c passes point P. How far from P do the pi-ons travel before only half of the pions in the beam remain? (*Answer*   38 m)

## R-4  Moving Sticks Again

In Section R-2 the length of a stick moving perpendicular to its length and the length of an identical stationary stick are compared and found to be equal. However, the technique used for this com-parison works only if the velocity of the moving stick is perpen-dicular to its length. Here we apply a different technique to compare the length of a stick at rest to its length when it is moving parallel to its velocity.

A light clock is shown in its proper frame in Figure R-7. This clock ticks each time the light pulse reflects off the mirror on the left. In its proper reference frame the length of the clock is $L_0$ and the time between ticks is $T_0 = 2L_0/c$ (Equation R-1). To find the

**FIGURE R-7**

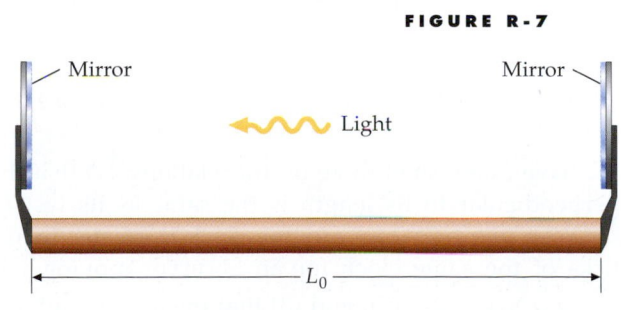

---

[†] A pion (short for pi meson) is a subatomic particle.

length of the clock in a reference frame in which it is moving to the right at speed $v$ we consider three sequential events:

Event 0   Light pulse reflects off the mirror at the left end.

Event 1   Light pulse reflects off the mirror at the right end.

Event 2   Light pulse reflects off at the mirror at the left end.

In Figure R-8 the clock is shown at the time of each of these events in a reference frame in which the clocks move to the right with speed $v$. (The clocks are drawn lower down the page at later times to avoid visual overlap.) The times of occurrence for events 0, 1, and 2 in this reference frame are $t'_0$, $t'_1$, and $t'_2$, respectively. In the time between events 0 and 1 the clock moves a distance $v(t'_1 - t'_0)$ and the light pulse travels a distance $c(t'_1 - t'_0)$. Thus,

$$c(t'_1 - t'_0) = L + v(t'_1 - t'_0) \qquad \text{R-4}$$

In the time between events 1 and 2 the clock moves a distance $v(t'_2 - t'_1)$ and the light pulse travels a distance $c(t'_2 - t'_1)$, so

$$c(t'_2 - t'_1) = L - v(t'_2 - t'_1) \qquad \text{R-5}$$

Eliminating $t'_1$ by solving Equation R-4 for $t'_1$, substituting the result into Equation R-5, and then solving for $t'_2 - t'_0$ gives

$$t'_2 - t'_0 = \frac{2L/c}{1 - (v^2/c^2)} \qquad \text{R-6}$$

The time interval $t'_2 - t'_0$ is related to the proper time interval $t_2 - t_0$ between events 0 and 2 (Equation R-3) by

$$t'_2 - t'_0 = \frac{t_2 - t_0}{\sqrt{1 - (v^2/c^2)}} \qquad \text{R-7}$$

where $t_2 - t_0 = 2L_0/c$ (Equation R-1). Substituting $2L_0/c$ for $t_2 - t_0$ gives

$$t'_2 - t'_0 = \frac{2L_0/c}{\sqrt{1 - (v^2/c^2)}} \qquad \text{R-8}$$

Equating the right sides of Equations R-6 and R-8, and then solving for $L$ gives

$$L = L_0\sqrt{1 - (v^2/c^2)} \qquad \text{R-9}$$

LENGTH CONTRACTION

Establishing this result did not involve any properties of the stick. Thus Equation R-9 reflects the nature of space and time, and not the nature of the sticks.

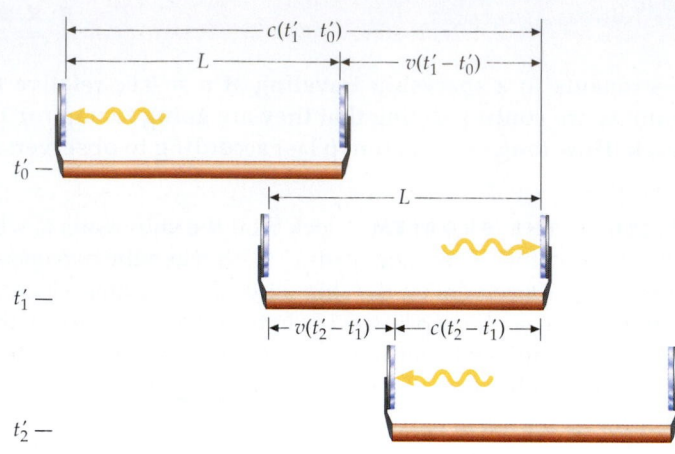

**FIGURE R-8**  A light clock moving to the right at speed $v$ is shown at 3 instants of time.

# R-5  Distant Clocks and Simultaneity

We have established three useful relations: (1) that the length of a stick moving perpendicular to its length is the same as its rest length; (2) that the time $T$ between two ticks of a moving clock is greater than the time between the two ticks of the same clock for an observer moving with the clock according to $T = T_0/\sqrt{1 - (v^2/c^2)}$; and (3) that the length $L$ of a stick moving parallel to its length is less than its rest length $L_0$ according to $L = L_0\sqrt{1 - (v^2/c^2)}$. But in order to analyze events from the perspective of observers in reference frames moving at different velocities, we need one more relation, one that concerns the readings on clocks at different locations.

Clocks A and B (Figure R-9a) are at rest relative to each other, and in their rest frame the clocks are separated by a distance $L_0$. To synchronize these clocks there is a flash lamp on clock A and a light sensitive film on the face of clock B. The alarm on clock A is set to energize the flash lamp when the second hand on clock A passes zero. Like the conventional clock in Section R-3, the second hand on clock B is a rotating opaque disk with a slot to indicate the time. Behind the disk is a light-sensitive film. When the light from the flash reaches clock B, the film is illuminated on the narrow region behind the slot. This provides a lasting record of the reading on clock B when the light from the flash lamp reaches it. Let this reading be $t_1$. In the rest frame of the clocks, the time for the light to travel at speed $c$ from clock A to clock B is $L_0/c$, so when the light arrives at clock B, clock A reads $L_0/c$ and clock B reads $t_1$. To synchronize the two clocks we turn clock B back by $\Delta t = t_1 - L_0/c$.

With the two clocks synchronized in their rest frame (frame 1), we then determine whether they are also synchronized in a reference frame (frame 2) in which they are moving at speed $v$ parallel to the line joining them, shown in Figure R-9b. We reset the alarm to energize the flash lamp when clock A next reads zero. These two events—clock A reads zero and the lamp flashes—are a spacetime coincidence, so we know they occur simultaneously in all reference frames. Also, the light reaching clock B and clock B reading $L_0/c$, are a spacetime coincidence, so we know they occur simultaneously in all reference frames.

In frame 2, the distance $L$ between the clocks is given by

$$L = L_0\sqrt{1 - (v^2/c^2)}$$

and clock B is moving toward the flash lamp. In this frame the light traveling from clock A to clock B travels a distance $L - vt$, where $t$ is the time required for the light to travel the distance. The time $t$, the distance $L$, and the speed $v$ are related by

$$ct = L - vt$$

Solving for the time gives $t = L/(c + v)$.

Moving clocks run slow, so during time $t$ the readings on both clocks advance not by $L/(c + v)$ but by

$$\frac{L}{c + v}\sqrt{1 - (v^2/c^2)} = \frac{L_0}{c + v}\left(1 - \frac{v^2}{c^2}\right) = \frac{L_0}{(c + v)}\frac{(c + v)(c - v)}{c^2} = \frac{L_0}{c} - \frac{vL_0}{c^2}$$

Thus, when the light arrives at clock B, clock B reads $L_0/c$ and clock A reads $L_0/c - vL_0/c^2$. Therefore, in frame 2 clock B is ahead of clock A by $vL_0/c^2$:

> If two clocks are synchronized in their rest frame, then in a frame where they move with speed $v$ parallel to the line joining them the clock in the rear is ahead of the clock in the front by $vL_0/c^2$.

THE RELATIVITY OF SIMULTANEITY

In this case $L_0$ is the distance between the clocks in their rest frame. It is also true that if two clocks are synchronized in their rest frame, they are also synchronized in a frame in which they move with speed $v$ perpendicular to the line joining them. This follows from the symmetry of the situation. (For one thing, there is no way to state a rule specifying which of the two clocks is ahead.)

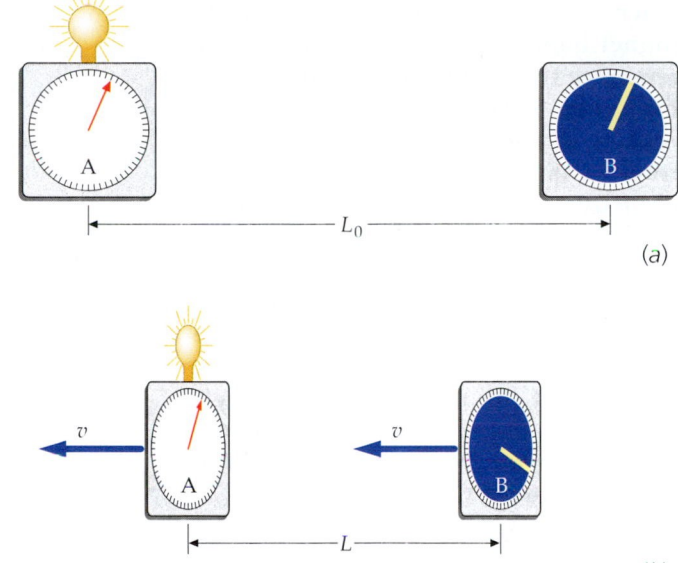

**FIGURE R-9** (*a*) The clocks are synchronized in the reference frame in which they are at rest. (*b*) Are the clocks synchronized in the reference frame in which they are moving with speed $v$ parallel to the line joining them?

# R-6 Applying the Rules

**FIGURE R-10**

A high-speed train is about to enter a tunnel through a mountain. The tunnel has a proper length of 1.2 km. The length of the train in the reference frame of the mountain is also 1.2 km, and the proper length of the train is 2.0 km. Clock A is attached to the mountain at the entrance to the tunnel, and clock B is attached to the mountain at the exit to the tunnel. In the reference frame of the mountain, when the front of the train enters the tunnel both clocks read zero. (*a*) In the reference frame of the mountain, what is the speed of the train and what is the reading of both clocks at the instant the front of the train exits the tunnel (Figure R-10*a*)? (*b*) In the reference frame of the train, what is the length of the tunnel, what is the reading of both clocks at the instant the front of the train enters the tunnel (Figure R-10*b*), and what is the reading of both clocks at the instant the front of the train exits the tunnel? (*c*) For a passenger on the train, how long does it take for the front of the train to pass through the tunnel?

(a)

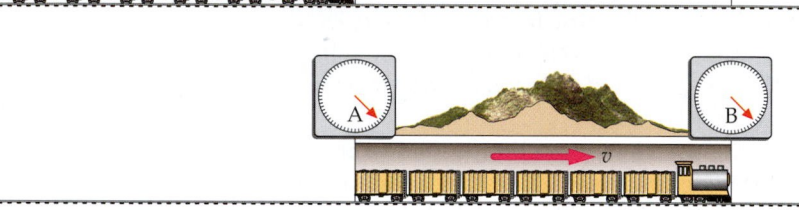

(b)

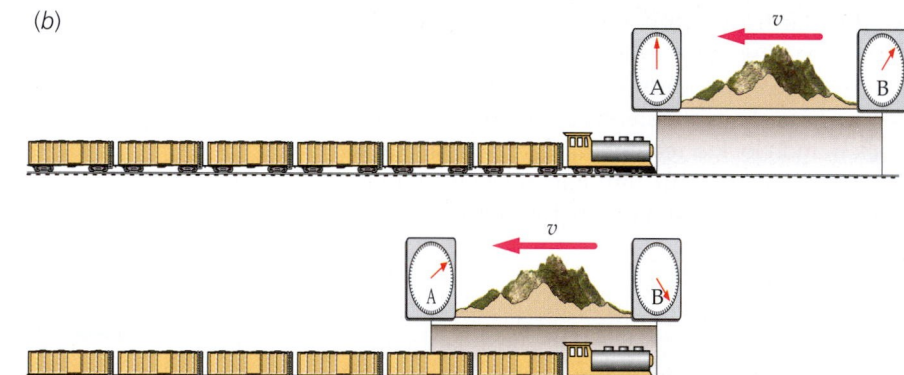

**PICTURE THE PROBLEM** The speed of the train and the length of the train are related by the length-contraction formula. Some of the clock readings in the two reference frames can be equated because they are event pairs that form spacetime coincidences. Other clock readings can be related by the relativity–of–simultaneity relation.

(*a*) 1. Using the length-contraction formula, solve for the speed of the train:

$$L = L_0\sqrt{1 - (v^2/c^2)}$$

$$1.2 \text{ km} = 2.0 \text{ km}\sqrt{1 - (v^2/c^2)}$$

so

$$v = 0.8c = 0.8(3.00 \times 10^8 \text{ m/s}) = \boxed{2.4 \times 10^8 \text{ m/s}}$$

2. The length of the tunnel equals its proper length and because the clocks are not moving they do not run slow. The reading on both clocks is the time $t$ that it takes for the front of the train to travel the length of the tunnel:

$$L_{\text{tunnel},0} = vt$$

so

$$t = \frac{L_{\text{tunnel},0}}{v} = \frac{1.2 \times 10^3 \text{ m}}{2.4 \times 10^8 \text{ m/s}} = 5 \times 10^{-6} \text{ s} = 5 \text{ } \mu s$$

3. The clocks are synchronized, so when the front of the train exits the tunnel both clocks read 5 $\mu$s:

Clock A reading = Clock B reading = $\boxed{5 \text{ } \mu s}$

(*b*) 1. In this frame the mountain is moving at 0.8*c*. Using the length-contraction formula, solve for the length of the tunnel:

$$L_{\text{tunnel}} = L_{\text{tunnel},0}\sqrt{1 - (v^2/c^2)} = 1.2 \text{ km}\sqrt{1 - \frac{(0.8c)^2}{c^2}}$$

$$= 1.2 \text{ km}\sqrt{1 - 0.8^2} = \boxed{0.72 \text{ km} = 720 \text{ m}}$$

2. The front of the train entering the tunnel and a zero reading on clock A are a spacetime coincidence:

Clock A reads zero

3. The two clocks are moving toward the train with clock B in the rear, so clock B is ahead of clock A by $vL_0/c^2$. When the train enters the tunnel clock A reads zero, so clock B reads $vL_0/c^2$:

$$\frac{vL_{\text{tunnel},0}}{c^2} = \frac{0.8cL_{\text{tunnel},0}}{c^2} = \frac{0.8L_{\text{tunnel},0}}{c}$$

$$= \frac{0.8(1.2 \times 10^3 \text{ m})}{3.0 \times 10^8 \text{ m/s}} = 3.2 \ \mu\text{s}$$

Clock B reading = $\boxed{3.2 \ \mu\text{s}}$

4. The front of the train exiting the tunnel and a 5-$\mu$s reading of clock B are a spacetime coincidence:

Clock B reading = $\boxed{5 \ \mu\text{s}}$

5. Clock B is in the rear, so clock A lags behind clock B by $vL_0/c^2$.

Clock A reading = Clock B reading $- \dfrac{vL_{\text{tunnel},0}}{c^2}$

$$= 5 \ \mu\text{s} - 3.2 \ \mu\text{s} = \boxed{2.6 \ \mu\text{s}}$$

(c) For an observer in the reference frame of the train, the mountain is traveling at 0.8c and the tunnel is 720-m long:

$L_{\text{tunnel}} = vt$

so

$$t = \frac{L_{\text{tunnel}}}{v} = \frac{L_{\text{tunnel}}}{0.8c} = \frac{720 \text{ m}}{2.4 \times 10^8 \text{ m/s}} = \boxed{3 \ \mu\text{s}}$$

**REMARKS** In the frame of reference of the mountain, the length of the train and the length of the tunnel are both 1.2 km. In the frame of reference of the train, the length of the train is 2.0 km and the length of the tunnel is 720 m. In the frame of reference of the mountain, when both clocks read 5 $\mu$s the entire train is in the tunnel. In the reference frame of the train, the train is longer than the tunnel so at no time is the entire train within the tunnel.

**EXERCISE** Event 1 is the front of the train entering the tunnel, and event 2 is the front of the train exiting the tunnel. (a) In which reference frame do these two events occur at the same location? (b) What is the proper time interval between events 1 and 2? (*Answer* (a) The reference frame of the train, because both events occur at the front end of the train. (b) 3 $\mu$s)

It is often convenient to measure large distances in light-years, where a light-year is the distance traveled when traveling at the speed of light for a time of one year. That is,

1 light-year = 1 $c\cdot$y

**EXPLORING**

*PDF files of two worked-out examples on the twin paradox are available at* www.whfreeman.com/tipler5e.

where 1 $c\cdot$y = $c$ (1 y). This notation is particularly convenient when distances are divided by speeds. For example, the time $T$ for a particle traveling at $v = 0.1c$ to travel a distance of $L = 25$ light-years is

$$T = \frac{L}{v} = \frac{25 \ c\cdot\text{y}}{0.1c} = 250 \text{ y}$$

where the $c$'s cancel.

**EXERCISE** In the reference frame of the earth, it takes 8 minutes for light to travel from the sun to the earth, so the distance between the sun and the earth is 8 $c\cdot$min. How many minutes does it take a particle from the sun to reach the earth if the particle travels at 0.1c? (*Answer* $(8 \ c\cdot\text{min})/0.1c = (8 \text{ min})/0.1 = 80 \text{ min}$)

# R-7 Relativistic Momentum, Mass, and Energy

## Momentum and Mass

In special relativity both momentum and energy are conserved, just as they are in classical physics. However, there are differences that have to be accounted for. The momentum of a particle moving with velocity $v$ is given by

$$p = \frac{mv}{\sqrt{1 - (v^2/c^2)}} \qquad \text{R-10}$$

RELATIVISTIC MOMENTUM

where $m$ is the mass of the particle. Equation R-10 is sometimes written $p = m_r v$, where $m_r$ is called the relativistic mass: $m_r = m/\sqrt{1 - (v^2/c^2)}$. In the rest frame of the particle, $v = 0$ and the relativistic mass equals $m$. (The mass $m$ is sometimes called the rest mass to differentiate it from the relativistic mass.) Relativistic momentum and mass are discussed further in Chapter 39.

## Energy

In relativistic mechanics, as in classical mechanics, the net force on a particle is equal to the time rate of change of the momentum of the particle. Considering one-dimensional motion only, we have

$$F_{net} = \frac{dp}{dt} \qquad \text{R-11}$$

We wish to find an expression for the kinetic energy. To do this we will multiply both sides of Equation R-11 by the displacement $ds$. This gives

$$F_{net}\,ds = \frac{dp}{dt}ds \qquad \text{R-12}$$

where we identify the term on the left as the work and the term on the right as the change in kinetic energy $dK$. Substituting $v\,dt$ for $ds$ in the term on the right we obtain

$$dK = \frac{dp}{dt}v\,dt = v\,dp$$

Integrating both sides gives

$$K = \int v\,dp \qquad \text{R-13}$$

This integral is evaluated in Chapter 37. The result is

$$K = \frac{mc^2}{\sqrt{1 - (v^2/c^2)}} - mc^2 \qquad \text{R-14}$$

Defining $\dfrac{mc^2}{\sqrt{1 - (v^2/c^2)}}$ as the **total relativistic energy** $E$ gives

$$E = K + mc^2 = \frac{mc^2}{\sqrt{1 - (v^2/c^2)}} \qquad \text{R-15}$$

where $m_0c^2$, called the **rest energy** $E_0$, is energy the particle has when it is at rest.

By multiplying both sides of Equation R-10 by $c$ and then dividing the resulting equation by Equation R-15 we obtain

$$\frac{v}{c} = \frac{pc}{E}$$ R-16

which can be useful when trying to solve for the speed $v$. Eliminating $v$ from Equations R-10 and R-16, and solving for $E^2$ (see Problem R-41) gives

$$E^2 = p^2c^2 + m^2c^4$$ R-17

The relation between mass and energy is discussed in Section 3 of Chapter 7.

# SUMMARY

| Topic | Relevant Equations and Remarks |
|---|---|
| **1. Postulates of Special Relativity** | |
| Postulate I: Principle of relativity | It is impossible to devise an experiment that determines whether you are at rest or moving uniformly, where moving uniformly means moving at constant velocity relative to an inertial reference frame. |
| Postulate II | The speed of light is independent of the speed of the source. |
| Constancy of the speed of light | It follows that the speed of light is the same in any inertial reference frame. |
| **2. Moving Sticks** | The length of a stick moving perpendicular to its length is equal to its proper length. |
| | The length of a stick moving with speed $v$ parallel to its length is shorter than its proper length $L_0$ according to $$L = L_0\sqrt{1 - (v^2/c^2)} \qquad \text{R-9}$$ |
| **3. Moving Clocks** | |
| Time dilation | The time between ticks of a clock moving with speed $v$ is longer than the proper time between ticks of the same clock by $$T = \frac{T_0}{\sqrt{1 - (v^2/c^2)}} \qquad \text{R-3}$$ |
| Relativity of simultaneity | If two clocks are synchronized in their rest frame, in a frame where they move with speed $v$ parallel to the line joining them the clock in the rear is ahead of the clock in front by $vL_0/c^2$, where $L_0$ is the distance between them in their rest frame. |
| | If two clocks are synchronized in their rest frame they are also synchronized in a frame where they move with speed $v$ perpendicular to the line joining them. |
| **4. Spacetime Coincidence** | If two events occur both at the same time and at the same place in one reference frame, they occur both at the same time and at the same place in any reference frame. |
| **5. Momentum, Mass, and Energy** | |
| Momentum | The momentum of a particle is given by $$p = \frac{mv}{\sqrt{1 - (v^2/c^2)}} \qquad \text{R-10}$$ |

Mass and energy

The total relativistic energy of a particle equals its rest energy plus its kinetic energy.

$$E = K + mc^2 = \frac{mc^2}{\sqrt{1 - (v^2/c^2)}}$$

R-15

where $mc^2$ is the rest energy $E_0$.

Momentum and Energy

$$\frac{v}{c} = \frac{pc}{E} \quad \text{and} \quad E^2 = p^2c^2 + m^2c^4$$

R-16, R-17

# PROBLEMS

- Single-concept, single-step, relatively easy
- •• Intermediate-level, may require synthesis of concepts
- ••• Challenging
- SSM Solution is in the *Student Solutions Manual*
- iSOLVE Problems available on iSOLVE online homework service
- iSOLVE✓ These "Checkpoint" online homework service problems ask students additional questions about their confidence level, and how they arrived at their answer

In a few problems, you are given more data than you actually need; in a few other problems, you are required to supply data from your general knowledge, outside sources, or informed estimates.

## Conceptual Problems

**1** • You are standing on a corner and a friend is driving past in an automobile. Each of you is wearing a wrist watch. Both of you note the times when the car passes two different intersections and determine from your watch readings the time that elapses between the two events. Which of you has determined the proper time interval?

**2** • SSM If event A occurs before event B in some frame, might it be possible for there to be a reference frame in which event B occurs before event A?

**3** • Two events are simultaneous in a frame in which they also occur at the same point in space. Are they simultaneous in all other reference frames?

**4** •• Two inertial observers are in relative motion. In what circumstances can they agree on the simultaneity of two different events?

**5** • The approximate total energy of a particle of mass $m = m_0$ moving at speed $v \ll c$ is (a) $mc^2 + \frac{1}{2}mv^2$, (b) $mv^2$, (c) $cmv$, (d) $\frac{1}{2}mc^2$, (e) $cmv$.

**6** • SSM True or false:

(a) The speed of light is the same in all reference frames.
(b) Proper time is the shortest time interval between two events.
(c) Absolute motion can be determined by means of length contraction.
(d) The light-year is a unit of distance.
(e) For two events to form a spacetime coincidence they must occur at the same place.
(f) If two events are not simultaneous in one frame, they cannot be simultaneous in any other frame.

## Estimation and Approximation

**7** •• The satellites used for the Global Positioning System (GPS) have extremely accurate clocks on board. In fact, the positioning system is so accurate that the relativistic time dilation of the clocks with respect to earth-bound observers must be taken into account. If the GPS satellite orbits have an average radius of about 26,000 km, estimate the time-dilation factor of the satellite clocks with respect to an earth-bound observer.

**8** •• It is said that exercising by riding a bike will extend your life. From the standpoint of relativity, this is certainly true. Estimate by what factor your life will be extended through relativistic time dilation with respect to a stationary observer if you ride a bike as part of your regular exercise program. Make any assumptions that you feel are reasonable, but be certain to state them in your answer.

**9** •• SSM iSOLVE✓ In 1975, an airplane carrying an atomic clock flew back and forth for 15 h at an average speed of 140 m/s as part of a time-dilation experiment. The time on the clock was compared to the time on an atomic clock kept on the ground. How much time did the airborne clock "lose" with respect to the clock on the ground?

## Length Contraction and Time Dilation

**10** • iSOLVE✓ The proper mean lifetime of subnuclear particles called pions is $2.6 \times 10^{-8}$ s. A beam of pions has a speed of $0.85c$ relative to a laboratory. (a) What would be their mean lifetime as measured in the laboratory? (b) How far would they travel, on average, before they decay? (c) What would be your answer to Part (b) if you neglect time dilation?

**11** • **iSOLVE** (*a*) In the reference frame of the pion in Problem 10, how far does the laboratory travel in a typical lifetime of $2.6 \times 10^{-8}$ s? (*b*) What is this distance in the laboratory's frame?

**12** • **SSM** **iSOLVE** ✓ The proper mean lifetime of a subnuclear particle called a muon is 2 $\mu$s. Muons in a beam are traveling at $0.999c$ relative to a laboratory. (*a*) What is their mean lifetime as measured in the laboratory? (*b*) How far do they travel, on average, before they decay?

**13** • **iSOLVE** (*a*) In the reference frame of the muon in Problem 12, how far does the laboratory travel in a typical lifetime of 2 $\mu$s? (*b*) What is this distance in the laboratory's frame?

**14** • You have been posted to a remote region of space to monitor traffic. Toward the end of a quiet shift, a spacecraft goes by and you measure its length using a laser device, which reports a length of 85 m. You flip open your handy reference catalogue and identify the craft as a CCCNX-22, which has a proper length of 100 m. When you phone in your report, what speed should you give for this spacecraft?

**15** • **SSM** A spaceship travels from earth to a star 95 light-years away at a speed of $2.2 \times 10^8$ m/s. How long does it take to get there (*a*) as measured on the earth and (*b*) as measured by a passenger on the spaceship?

**16** • The average lifetime of a beam of subnuclear particles called pions traveling at high speed is measured to be $7.5 \times 10^{-8}$ s. Their average lifetime when measured at rest is $2.6 \times 10^{-8}$ s. How fast is the pion beam traveling?

**17** • A meterstick moves with speed $0.8c$ relative to you in the direction parallel to the stick. (*a*) Find the length of the stick as measured by you. (*b*) How long does it take for the stick to pass you?

**18** • The half life of charged subnuclear particles called pions in their rest frame is $1.8 \times 10^{-8}$ s (that is, in the rest frame of the pions, if there are $N$ pions at time $t = 0$, there will be only $N/2$ pions at time $t = 1.8 \times 10^{-8}$ s). Pions are produced in an accelerator and emerge with a speed of $0.998c$. How far do these particles travel in the accelerator laboratory before half of them have decayed?

**19** •• **iSOLVE** Your friend, who is the same age as you, travels to the star Alpha Centauri, which is 4 light-years away, and returns immediately. He claims that the entire trip took just 6 y. How fast did he travel?

**20** •• **SSM** Two spaceships pass each other traveling in opposite directions. A passenger in ship A, who happens to know that her ship is 100 m long, notes that ship B is moving with a speed of $0.92c$ relative to A and that the length of B is 36 m. What are the lengths of the two spaceships as measured by a passenger in ship B?

**21** •• Supersonic jets achieve maximum speeds of about $3 \times 10^{-6}c$. (*a*) By what percentage would a jet traveling at this speed contract in length? (*b*) During a time of 1 y = $3.15 \times 10^7$ s on your clock, how much time would elapse on the pilot's clock? How many minutes are lost by the pilot's clock in 1 y of your time?

## The Relativity of Simultaneity

*Problems 22 through 26 refer to the following situation: Mary is a worker on a large space platform. She places a clock A at point A and clock B at point B, which is 100 light-minutes from point A (Figure R-12). She also places a flashbulb at a point midway between points A and B. Jamal, a worker on a different platform, is standing next to clock C. Each clock immediately starts at zero when the flash reaches it. Mary's platform moves at speed of 0.6c relative to Jamal's. As Mary's platform passes by, clock B, then the flashbulb, and then clock A pass directly over clock C—just missing it as they go by. As the flashbulb passes next to clock C, it flashes and clock C immediately starts at zero.*

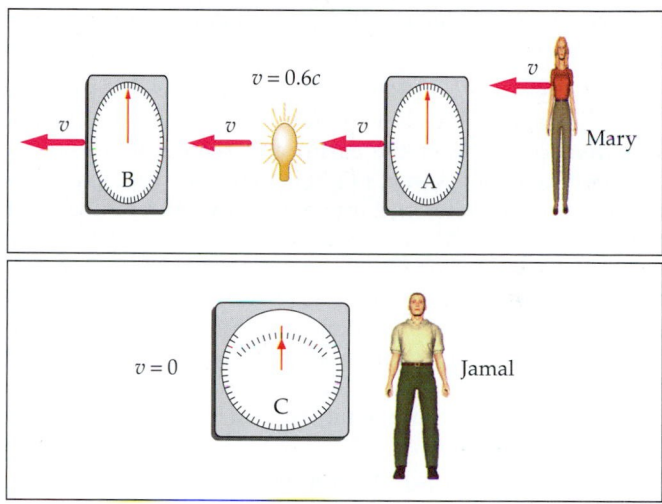

**FIGURE R-12**

**22** • **SSM** According to Jamal, (*a*) what is the distance between the flashbulb and clock A, (*b*) how far does the flash travel to reach clock A, and (*c*) how far does clock A travel while the flash is traveling from the flashbulb to it?

**23** •• According to Jamal, how long does it take the flash to travel to clock A, and what does clock C read as the flash reaches clock A.

**24** •• Show that clock C reads 100 min as the light flash reaches clock B, which is traveling away from clock C with speed 0.6c.

**25** •• According to Jamal, the reading on clock C advances from 25 min to 100 min between the reception of the flashes by clocks A and B in Problems 23 and 24. Again according to Jamal, how much will the reading on clock A advance during this 75-min interval?

**26** •• **SSM** The advance of clock A calculated in Problem 25 is the amount that clock A leads clock B according to Jamal. Compare this result with $vL_0/c^2$, where $v = 0.6c$.

**27** •• In inertial reference frame $S$, event B occurs 2 $\mu$s after event A, which occurs 1.5 km from event A. How fast must an observer be moving along the line joining the two events so that the two events occur simultaneously? For an observer traveling fast enough is it possible for event B to precede event A?

**28** •• A large flat space platform has an $x$ axis painted on it. A firecracker explodes on the $x$ axis at $x_1 = 480$ m, and a second firecracker explodes on the $x$ axis 5 $\mu$s later at $x_2 = 1200$ m. In the reference frame of a train traveling alongside the $x$ axis at speed $v$ relative to the platform, these two explosions occur at the same place in space. What is the separation in time between the two explosions in the reference frame of the train?

**29** • **iSOLVE** Herb and Randy are twin jazz musicians who perform as a trombone–saxophone duo. At the age of twenty, however, Randy got an irresistible offer to join a road trip to perform on a star 15 light-years away. To celebrate his good fortune, he bought a new vehicle for the trip—a deluxe space-coupe that travels at 0.999c. Each of the twins promises to practice diligently, so they can reunite afterward. However, Randy's gig goes so well that he stays for a full 10 y before returning to Herb. After their reunion, (a) how many years of practice will Randy have had; (b) how many years of practice will Herb have had?

**30** •• **SSM** Al and Bert are twins. Al travels at 0.6c to Alpha Centauri (which is 4 c·y from the earth as measured in the reference frame of the earth) and returns immediately. Each twin sends the other a light signal every 0.01 y as measured in his own reference frame. (a) At what rate does Bert receive signals as Al is moving away from him? (b) How many signals does Bert receive at this rate? (c) How many total signals are received by Bert before Al has returned? (d) At what rate does Al receive signals as Bert is receding from him? (e) How many signals does Al receive at this rate? (f) How many total signals are received by Al? (g) Which twin is younger at the end of the trip and by how many years?

## Relativistic Energy and Momentum

**31** • **iSOLVE** Find the ratio of the total energy to the rest energy of a particle of rest mass $m_0$ moving with speed (a) 0.1c, (b) 0.5c, (c) 0.8c, and (d) 0.99c.

**32** • **iSOLVE** ✓ A proton (rest energy 938 MeV) has a total energy of 1400 MeV. (a) What is its speed? (b) What is its momentum?

**33** • **SSM** How much energy would be required to accelerate a particle of mass $m$ from rest to (a) 0.5c, (b) 0.9c, and (c) 0.99c? Express your answers as multiples of the rest energy.

**34** • If the kinetic energy of a particle equals its rest energy, what error is made by using $p = mv$ for its momentum?

**35** • What is the total energy of a proton whose momentum is $3mc$?

**36** •• **SSM** Using a spreadsheet program or graphing calculator, make a graph of the kinetic energy of a particle with mass $m = 100$ MeV/$c^2$ for speeds between 0 and c. On the same graph, plot $\frac{1}{2}mv^2$ by way of comparison. Using the graph, estimate at about what velocity this formula is no longer a good approximation to the kinetic energy. As a suggestion, plot the energy in units of MeV and the velocity in the dimensionless form $v/c$.

**37** •• Derive the equation $E^2 = p^2c^2 + m^2c^4$ (Equation R-17) by eliminating $v$ from Equations R-10 and R-16.

**38** •• Use the binomial expansion and Equation R-17 to show that when $pc \ll mc^2$, the total energy is given approximately by $E \approx mc^2 + p^2/(2m)$

**39** •• (a) Show that the speed $v$ of a particle of mass $m$ and total energy $E$ is given by

$$\frac{v}{c} = \left[1 - \frac{(mc^2)^2}{E^2}\right]^{1/2}$$

and that when $E$ is much greater than $mc^2$, this can be approximated by

$$\frac{v}{c} \approx 1 - \frac{(mc^2)^2}{2E^2}.$$

Find the speed of an electron with kinetic energy of (b) 0.51 MeV and (c) 10 MeV.

**40** •• **SSM** The rest energy of a proton is about 938 MeV. If its kinetic energy is also 938 MeV, find (a) its momentum and (b) its speed.

**41** •• What percentage error is made in using $\frac{1}{2}m_0v^2$ for the kinetic energy of a particle if its speed is (a) 0.1c and (b) 0.9c?

## General Problems

**42** • A spaceship departs from Earth for the star Alpha Centauri, which is 4 light-years away in the reference frame of Earth. The spaceship travels at 0.75c. How long does it take to get there (a) as measured on Earth and (b) as measured by a passenger on the spaceship?

**43** • The total energy of a particle is three times its rest energy. (a) Find $v/c$ for the particle. (b) Show that its momentum is given by $p = \sqrt{8}mc$.

**44** • **iSOLVE** A subnuclear particle called a muon has a mean lifetime of 2 μs when stationary. If you measure the mean lifetime of the muons coming out of a nuclear reactor port to be 46 μs, how fast are they moving?

**45** •• **SSM** The rest mass of the neutrino (the "ghost particle" of physics) is known to have a small but as yet unmeasured value. Both high-energy neutrinos and light are produced by a supernova explosion, so it is possible to estimate the neutrino's mass by timing the relative arrival of light versus neutrinos from a supernova. If a supernova explodes 100,000 light-years from the earth, calculate the rest mass necessary for a neutrino with total energy of 100 MeV to trail the light by (a) 1 min, (b) 1 s, and (c) 0.01 s.

**46** • **iSOLVE** ✓ Relative to you, how fast must a meter-stick travel in the direction parallel to itself so that its length as measured by you is 50 cm?

**47** ••• **SSM** Keisha and Ernie are trying to fit a 15-ft-long ladder into a 10-ft-long shed with doors at each end. Recalling her physics lessons, Keisha suggests to Ernie that they open the front door to the shed and have Ernie run toward it with the ladder at a speed such that the length contraction of the ladder shortens it enough so that it fits in the shed. As soon as the back end of the ladder passes through the door, Keisha will slam it shut. (a) What is the minimum speed at which Ernie must run to fit the ladder into the shed? Express it as a fraction of the speed of light. (b) As Ernie runs toward the shed at a speed of 0.866c, he realizes that in the reference frame of the ladder and himself, it is the shed which is shorter, not the ladder. How long is the shed in the rest frame of the ladder? (c) In the reference frame of the ladder is there any instant that both ends of the ladder are simultaneously inside the shed? Examine this from the point of view of relativistic simultaneity.

# Gravity

ON A CLEAR AUTUMN NIGHT YOU ARE ON THE ROOF OF YOUR APARTMENT BUILDING IN NEW YORK CITY, TALKING ON YOUR CELL PHONE TO YOUR FRIEND IN KANSAS CITY, WHEN YOU SEE THE INTERNATIONAL SPACE STATION (ISS) PASS DIRECTLY OVERHEAD.

**?** **How might you use your understanding of gravity to determine when the ISS will next pass directly over Kansas City? (See Example 11-3.)**

11-1    Kepler's Laws

11-2    Newton's Law of Gravity

11-3    Gravitational Potential Energy

11-4    The Gravitational Field $\vec{g}$

*11-5    Finding the Gravitational Field of a Spherical Shell by Integration

G ravity is the weakest of the four basic forces. It is negligible in the interactions of elementary particles and thus plays no role in the behavior of molecules, atoms, and nuclei. The gravitational attraction between objects of the size we ordinarily encounter, for example, the gravitational force of attraction exerted by a building on a car, is too small to be readily noticed. Yet when we consider objects of astronomical size, such as moons, planets, and stars, gravity is of primary importance. The gravitational force exerted by the earth on us and on the objects around us is a fundamental part of our experience. It is gravity that binds us to the earth and keeps the earth and the other planets on course within the solar system. The gravitational force plays an important role in the life history of stars and in the behavior of galaxies. On the largest of all scales, it is gravity that controls the evolution of the universe.

At the time of Newton many believed that nature followed different rules in the larger universe than here on earth. Newton's law of universal gravity, along with his three laws of motion, revealed that nature follows the same rules everywhere, and this revelation has had a profound effect on our view of the universe. ➤ **In this chapter we use the tools of conservation of angular momentum, conservation of energy, Newton's laws of motion, and Newton's law of gravity to predict the motion of the planets and other celestial bodies, including those we have put there.**

# 11-1 Kepler's Laws

The nighttime sky with its myriad stars and shining planets has always fascinated us on earth. Toward the end of the sixteenth century, the astronomer Tycho Brahe studied the motions of the planets and made observations that were considerably more accurate than those previously available. Using Brahe's data, Johannes Kepler discovered that the paths of the planets about the sun are ellipses (Figure 11-1). He also showed that each planet moves faster when its orbit brings it closer to the sun and slower when its orbit takes it farther away. Finally, Kepler developed a precise mathematical relation between the period of a planet and its average distance from the sun (see Table 11.1). He stated these results in three empirical laws of planetary motion. Ultimately, these laws provided the basis for Newton's discovery of the law of gravity. Kepler's three laws follow.

A mechanical model of the solar system, called an orrery, in the collection of Historical Scientific Instruments at Harvard University.

> **Law 1.** All planets move in elliptical orbits with the sun at one focus.

An ellipse is the locus of points for which the sum of the distances from two foci $F$ is constant, as shown in Figure 11-2. Figure 11-3 shows a planet following an elliptical path with the sun at one focus. The earth's orbit is nearly circular, with the distance to the sun at perihelion (closest point) being $1.48 \times 10^{11}$ m and at aphelion (farthest point) being $1.52 \times 10^{11}$ m. The semimajor axis equals the average of these two distances, which is $1.50 \times 10^{11}$ m (93 million miles) for the earth's orbit. The mean earth–sun distance defines the astronomical unit (AU):

$$1 \text{ AU} = 1.50 \times 10^{11} \text{ m} = 93.0 \times 10^{6} \text{ mi} \qquad \text{11-1}$$

The AU is used frequently in problems dealing with the solar system.

## TABLE 11-1

**Mean Orbital Radii and Orbital Periods for the Planets**

| Planet | Mean Radius $r$ ($\times 10^{10}$ m) | Period $T$ (y) |
|---|---|---|
| Mercury | 5.79 | 0.241 |
| Venus | 10.8 | 0.615 |
| Earth | 15.0 | 1.00 |
| Mars | 22.8 | 1.88 |
| Jupiter | 77.8 | 11.9 |
| Saturn | 143 | 29.5 |
| Uranus | 287 | 84 |
| Neptune | 450 | 165 |
| Pluto | 590 | 248 |

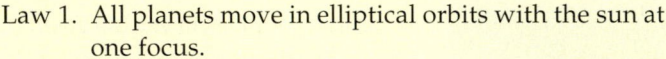

**FIGURE 11-1** Orbits of the planets around the sun.

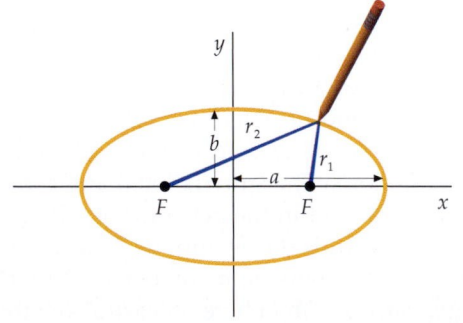

**FIGURE 11-2** An ellipse is the locus of points for which $r_1 + r_2 = $ constant. The distance $a$ is called the *semimajor* axis, and $b$ is the *semiminor* axis. You can draw an ellipse with a piece of string by fixing each end at a focus $F$ and using it to guide the pencil. Circles are special cases in which the two foci coincide.

Law 2. A line joining any planet to the sun sweeps out equal areas in equal times.

Figure 11-4 illustrates Kepler's second law, the law of equal areas. A planet moves faster when it is closer to the sun than when it is farther away, so that the area swept out by the radius vector in a given time interval is the same throughout the orbit. The law of equal areas is a consequence of the conservation of angular momentum, as we will see in the next section.

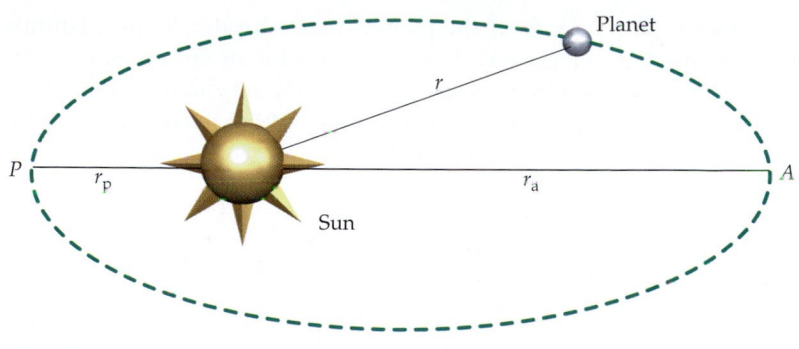

**FIGURE 11-3** The elliptical path of a planet with the sun at one focus. Point $P$, where the planet is closest to the sun, is called the perihelion, and point $A$, where it is farthest, is called the aphelion. The average distance between the planet and the sun, defined as $(r_p + r_a)/2$, is equal to the semimajor axis.

Law 3. The square of the period of any planet is proportional to the cube of the semimajor axis of its orbit.

Kepler's third law relates the period of any planet to its mean distance from the sun, which equals the semimajor axis of its elliptical path. In algebraic form, if $r$ is the mean distance between a planet and the sun and $T$ is the planet's period of revolution, Kepler's third law states that

$$T^2 = Cr^3 \qquad \text{11-2}$$

where the constant $C$ has the same value for all the planets. This law is a consequence of the fact that the force exerted by the sun on a planet varies inversely with the square of the distance from the sun to the planet. We will demonstrate this in Section 11-2 for the special case of a circular orbit.

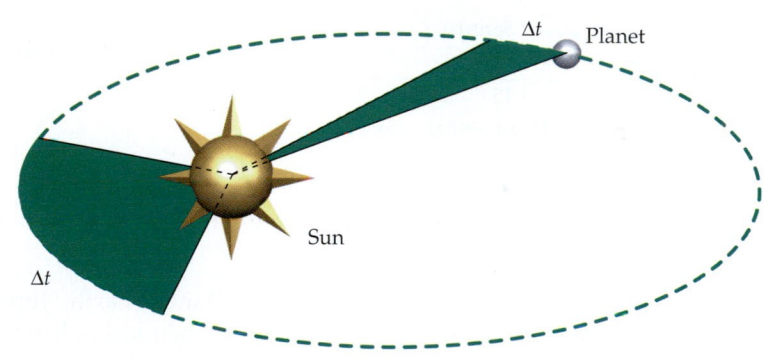

**FIGURE 11-4** When a planet is close to the sun, it moves faster than when it is farther away. The areas swept out by the radius vector in a given time interval are equal.

*JUPITER'S ORBIT*

**E X A M P L E  1 1 - 1**

**The mean distance from the sun to Jupiter is 5.20 AU. What is the period of Jupiter's orbit around the sun?**

**PICTURE THE PROBLEM** We use Kepler's third law to relate the period of Jupiter to its mean orbital radius. The constant $C$ can be obtained from the known mean distance and period of the earth. Let $T_E = 1$ y and $r_E = 1$ AU be the period and mean distance for the earth, and let $T_J$ and $r_J = 5.20$ AU be the period and mean distance for Jupiter.

1. Kepler's third law relates Jupiter's period $T_J$ and mean distance $r_J$:  $T_J^2 = Cr_J^3$

2. Apply Kepler's third law to the earth to obtain a second equation relating the same constant $C$ to $T_E$ and $r_E$:  $T_E^2 = Cr_E^3$

3. Divide the two equations, eliminating $C$, and solve for $T_J$:  $\dfrac{T_J^2}{T_E^2} = \dfrac{r_J^3}{r_E^3}$

so

$$T_J = T_E\left(\frac{r_J}{r_E}\right)^{3/2} = (1\text{ y})\left(\frac{5.20\text{ AU}}{1\text{ AU}}\right)^{3/2}$$

$$= \boxed{11.9\text{ y}}$$

**REMARKS** The periods of the planets Earth, Jupiter, Saturn, Uranus, and Neptune are plotted in Figure 11-5 as functions of their mean distances from the sun. In (a), periods are plotted versus mean distances from the sun. In (b), the squares of the periods are plotted versus the cubes of the mean distances from the sun. Here the points fall on a straight line.

**FIGURE 11-5**

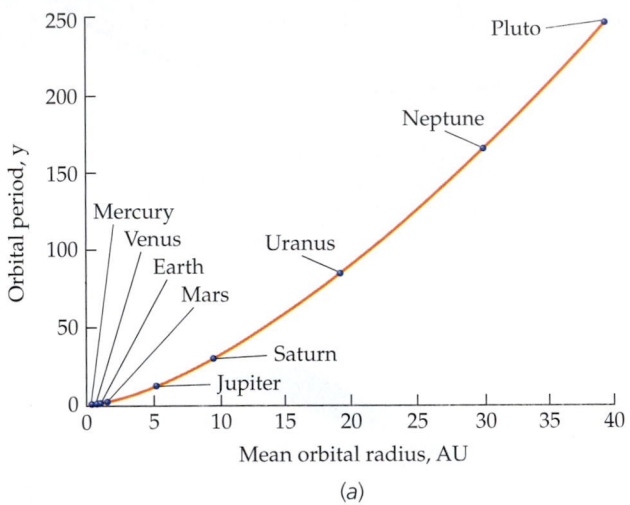

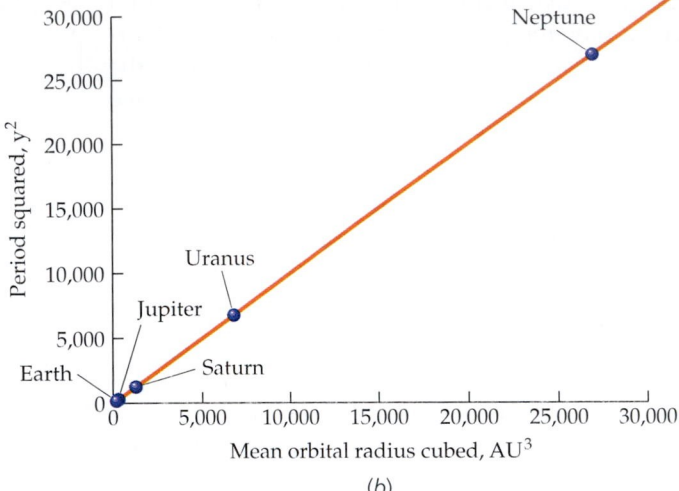

(a)

(b)

**EXERCISE** The period of Neptune is 164.8 y. What is its mean distance from the sun? (*Answer* 30.1 AU)

**EXERCISE** If the logs of the periods of the planets Earth, Jupiter, Saturn, Uranus, and Neptune are plotted versus the logs of their mean distances from the sun, the points fall on a curve. What is the shape of this curve? (*Answer* A straight line)

# 11-2 Newton's Law of Gravity

Although Kepler's laws were an important first step in understanding the motion of planets, they were nothing more than empirical rules obtained from the astronomical observations of Brahe. It remained for Newton to take the next giant step by attributing the acceleration of a planet in its orbit to a specific force exerted on it by the sun. Newton proved that a force that varies inversely with the square of the distance between the sun and a planet results in an elliptical orbit, as observed by Kepler. He then made the bold assumption that this force acts between any two objects in the universe. Before Newton, it was not even generally accepted that the laws of physics observed on earth were applicable to the heavenly bodies. **Newton's law of gravity** postulates that there is a force of attraction between each pair of point particles that is proportional to the product of the masses of the particles and inversely proportional to the square of the distance separating them. Let $m_1$ and $m_2$ be the masses of point particles 1 and 2 (at positions $\vec{r}_1$ and $\vec{r}_2$, respectively) and $\vec{r}_{1,2}$ be the vector pointing from particle 1 to particle 2 (Figure 11-6a).

The force $\vec{F}_{1,2}$ exerted by particle 1 on particle 2 is then

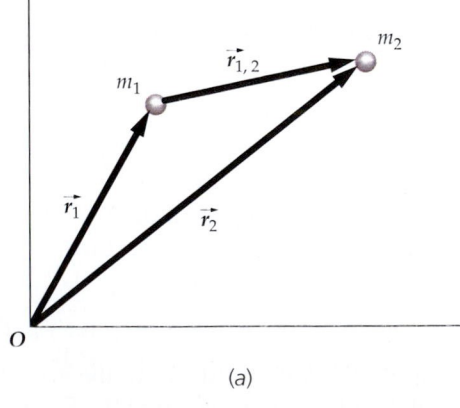

**FIGURE 11-6** (a) Particles at $\vec{r}_1$ and $\vec{r}_2$. (b) The particles exert equal and opposite forces on each other.

$$\vec{F}_{1,2} = -\frac{Gm_1m_2}{r_{1,2}^2}\hat{r}_{1,2}$$

11-3

NEWTON'S LAW OF GRAVITY

where $\hat{r}_{1,2} = \vec{r}_{1,2}/r_{1,2}$ is a unit vector pointing from 1 to 2 and $G$ is the **universal gravitational constant,** which has the value

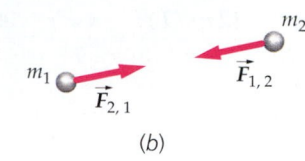

$$G = 6.67 \times 10^{-11}\,\text{N}\cdot\text{m}^2/\text{kg}^2 \qquad \text{11-4}$$

The force $\vec{F}_{2,1}$ exerted by 2 on 1 is the negative of $\vec{F}_{1,2}$, according to Newton's third law (Figure 11-6b). The magnitude of the gravitational force exerted by a point particle of mass $m_1$ on another point particle of mass $m_2$ a distance $r$ away is thus given by

**FIGURE 11-6** (b) The particles exert equal and opposite forces on each other.

$$F = \frac{Gm_1m_2}{r^2} \qquad \text{11-5}$$

Newton published his theory of gravitation in 1686, but it was not until a century later that an accurate experimental determination of $G$ was made by Cavendish, as will be discussed in Section 11-4.

We can use the known value of $G$ to compute the gravitational attraction between two ordinary objects.

**EXERCISE** Find the gravitational force that attracts a 65-kg man to a 50-kg woman when they are 0.5 m apart. Model them as point particles. (*Answer* $8.67 \times 10^{-7}$ N)

This exercise demonstrates that the gravitational force exerted by an object of ordinary size on another such object is so small as to be unnoticeable. For comparison, a mosquito weighs about $1 \times 10^{-7}$ N. The weight of a 50-kg person is 491 N, about half a billion times the force of attraction calculated in the exercise! Gravitational attraction is easily noticed only when at least one of the objects is astronomically massive. The gravitational attraction between the girl and the earth for example, is readily apparent.

To check the validity of the inverse-square nature of the gravitational force, Newton compared the acceleration of the moon in its orbit with the free-fall acceleration of objects near the surface of the earth (such as the legendary apple). He assumed that the gravitational attraction due to the earth causes both accelerations. He first assumed that the earth and moon could be treated as point particles with their total masses concentrated at their centers. The force on a particle of mass $m$ a distance $r$ from the center of the earth is

$$F = \frac{GM_Em}{r^2} \qquad \text{11-6}$$

If this is the only force acting on the particle, then its acceleration is

$$a = \frac{F}{m} = \frac{GM_E}{r^2} \qquad \text{11-7}$$

For objects near the surface of the earth, $r = R_E$ and the free-fall acceleration is $g$:

$$g = \frac{GM_E}{R_E^2} \qquad \text{11-8}$$

The distance to the moon is about 60 times the radius of the earth ($r = 60R_E$). Substituting this into Equation 11-7 gives $a = g/60^2$, so the acceleration of the moon in its near-circular orbit is the free-fall acceleration $g$ near the surface of the earth divided by $60^2$. That is, the acceleration of the moon $a_m$ should be $(9.81\ \text{m/s}^2)/60^2$. The moon's acceleration can be calculated from its known distance from the center of the earth, $r = 3.84 \times 10^8$ m, and its period $T = 27.3\ \text{d} = 2.36 \times 10^6$ s:

$$a_\mathrm{m} = \frac{v^2}{r} = \frac{(2\pi r/T)^2}{r} = \frac{4\pi^2 r}{T^2} = \frac{4\pi^2(3.84 \times 10^8 \text{ m})}{(2.36 \times 10^6 \text{ s})^2} = 2.72 \times 10^{-3} \text{ m/s}^2$$

Then

$$\frac{g}{a_\mathrm{m}} = \frac{9.81 \text{ m/s}^2}{2.72 \times 10^{-3} \text{ m/s}^2} = 3604 \approx 60^2$$

In Newton's words, "I thereby compared the force requisite to keep the Moon in her orb with the force of gravity at the surface of the Earth, and found them answer pretty nearly."

The assumption that the earth and moon can be treated as point particles in the calculation of the force on the moon is reasonable because the earth-to-moon distance is large compared with the radius of either the earth or the moon, but such an assumption is certainly questionable when applied to an object near the earth's surface. After considerable effort, Newton was able to prove that the force exerted by any object with a spherically symmetric mass distribution on a point mass either on or outside its surface is the same as if all the mass of the object were concentrated at its center. The proof involves integral calculus, which Newton developed to solve this problem.

**Earth as seen from Apollo 11 orbiting the moon on July 16, 1969.**

Because $g = 9.81$ m/s$^2$ is readily measured and the radius of the earth is known, Equation 11-8 can be used to determine the value of the product $GM_\mathrm{E}$. Newton estimated the value of $G$ from an approximation of the mass of the earth. When Cavendish determined $G$ some 100 years later by measuring the force between small spheres of known mass and separation, he called his experiment "weighing the earth." Knowing the value of $G$ meant that the mass of the sun and the mass of any planet with a satellite could be determined. The method for doing this is described in Section 11-4.

---

*FALLING TO EARTH* **EXAMPLE 11-2**

**What is the free-fall acceleration of an object at the altitude of the space shuttle's orbit, about 400 km above the earth's surface?**

**PICTURE THE PROBLEM** The force is given by Equation 11-6 with $r = R_\mathrm{E} + 400$ km.

1. The acceleration is given by $a = F/m$, where $F$ is given by Newton's law of gravity:

$$a = \frac{F}{m} = \frac{GmM_\mathrm{E}/r^2}{m} = \frac{GM_\mathrm{E}}{r^2}$$

2. The distance $r$ is related to the radius of the earth $R_\mathrm{E}$ and the altitude $h$:

$$r = R_\mathrm{E} + h = 6370 \text{ km} + 400 \text{ km}$$
$$= 6770 \text{ km}$$

3. The acceleration is then:

$$a = \frac{GM_\mathrm{E}}{r^2}$$

$$= \frac{(6.67 \times 10^{-11} \text{ N·m}^2/\text{kg}^2)(5.98 \times 10^{24} \text{ kg})}{(6.77 \times 10^6 \text{ m})^2}$$

$$= \boxed{8.70 \text{ m/s}^2}$$

**REMARKS** This is also the acceleration of the "weightless" shuttle astronauts as they accelerate in their circular orbit.

The calculation in Example 11-2 can be simplified by using Equation 11-8 to eliminate $GM_\mathrm{E}$ from Equation 11-7. Then the acceleration at a distance $r$ is

$$a = \frac{F}{m} = \frac{GM_E}{r^2} = g\frac{R_E^2}{r^2}$$  11-9

**EXERCISE** At what distance $h$ above the surface of the earth is the acceleration of gravity half its value at sea level? (*Answer* 2640 km)

## Measurement of G

The universal gravitational constant $G$ was first measured in 1798 by Henry Cavendish, who used the apparatus shown in Figure 11-7. Cavendish's measurement of $G$ has been repeated by other experimenters with various improvements and refinements. All measurements of $G$ are difficult because of the extreme weakness of the gravitational attraction. Consequently, the value of $G$ is known today only to about 1 part in 10,000. Although $G$ was one of the first physical constants ever measured, it remains one of the least accurately known.

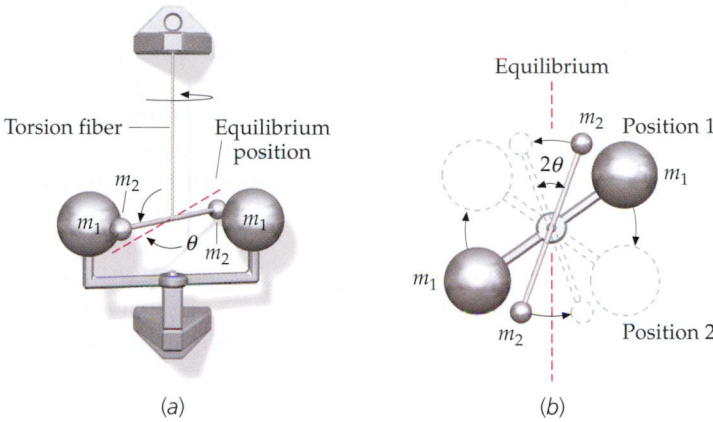

(a)

(b)

## Gravitational and Inertial Mass

The property of an object responsible for the gravitational force it exerts on another object is its *gravitational* mass. On the other hand, the property of an object that measures its resistance to acceleration is its *inertial* mass. We have used the same symbol $m$ for these two properties because, experimentally, they are proportional. For convenience, units are judiciously defined to make the proportionality constant one. The fact that the gravitational force exerted on an object is proportional to its inertial mass is a characteristic unique to the force of gravity. One consequence is that all objects near the surface of the earth fall with the same acceleration if air resistance is neglected. The well-known story of Galileo dropping objects from the Leaning Tower of Pisa to demonstrate that the free-fall acceleration is the same for objects with different inertial masses is just one example of the excitement this discovery aroused in the sixteenth century.

We could easily imagine that the gravitational and inertial masses of an object were not the same. Suppose we write $m_G$ for the gravitational mass and $m$ for the inertial mass. The force exerted by the earth on an object near its surface would then be

$$F = \frac{GM_E m_G}{R_E^2}$$  11-10

where $M_E$ is the gravitational mass of the earth. The free-fall acceleration of the object near the earth's surface would then be

$$a = \frac{F}{m} = \left(\frac{GM_E}{R_E^2}\right)\frac{m_G}{m}$$  11-11

If gravity were just another property of matter, like color or hardness, it might be

**FIGURE 11-7** (a) Two small spheres, each of mass $m_2$, are at the ends of a light rod that is suspended by a fine fiber. Careful measurements determine the torque required to turn the fiber through a given angle. Two large spheres, each of mass $m_1$, are then placed near the small spheres. Because of the gravitational attraction of the large spheres of mass $m_1$ for the small spheres, the fiber is turned through a very small angle $\theta$ from its equilibrium position. (b) The apparatus as seen from above. After the apparatus comes to rest, the positions of the large spheres are reversed, as shown by the dashed lines, so that they are at the same distance from the equilibrium position of the balance but on the other side. If the apparatus is again allowed to come to rest, the fiber will turn through angle $2\theta$ in response to the reversal of the torque. Once the torsion constant has been determined, the forces between the masses $m_1$ and $m_2$ can be determined from the measurement of this angle. Since the masses and their separations are known, $G$ can be calculated. Cavendish obtained a value for $G$ within about 1 percent of the presently accepted value given by Equation 11-4.

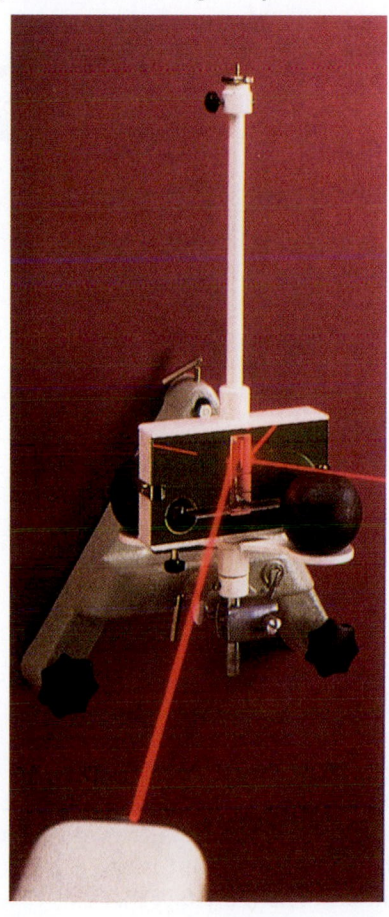

A gravitational torsion balance used in student labs for the measurement of $G$. A tiny angular deflection of the balance results in a large angular deflection of the laser beam that reflects from a mirror on the balance.

reasonable to expect that the ratio $m_G/m$ would depend on such things as the chemical composition of the object or its temperature. The free-fall acceleration would then be different for different objects. **The experimental evidence, however, is that a is the same for all objects.** Thus we need not maintain the distinction between $m_G$ and $m$ and can set $m_G = m$. We must keep in mind, however, that the equivalence of gravitational and inertial mass is an empirical law that is limited by the accuracy of experiment. Experiments testing this equivalence were carried out by Simon Stevin in the 1580s. Galileo publicized this law widely, and his contemporaries made considerable improvements in the experimental accuracy with which the law was established.

The most precise early comparisons of gravitational and inertial mass were made by Newton. Through experiments using simple pendulums rather than falling bodies, Newton was able to establish the equivalence between gravitational and inertial mass to an accuracy of about 1 part in 1000. Experiments comparing gravitational and inertial mass have improved steadily over the years. Their equivalence is now established to about 1 part in $10^{12}$. Thus, the equivalence of gravitational and inertial mass is one of the best established of all physical laws. It is the basis for the principle of equivalence, which is the foundation of Einstein's general theory of relativity.

## Derivation of Kepler's Laws

Newton showed that when an object such as a planet or comet moves around a $1/r^2$ force center such as the sun, the object's path is a conic section (an ellipse, a parabola, or a hyperbola). The parabolic and hyperbolic paths apply to objects that make one pass by the sun and never return. Such orbits are not closed. The only closed orbits in an inverse-square force field are ellipses. Thus, Kepler's first law is a direct consequence of Newton's law of gravity. Kepler's second law, the law of equal areas, follows from the fact that the force exerted by the sun on a planet is directed toward the sun. Such a force is called a **central force.** Figure 11-8a shows a planet moving in an elliptical orbit around the sun. In time $dt$, the planet moves a distance $v\,dt$ and the radius vector $\vec{r}$ sweeps out the area shaded in the figure. This is half the area of the parallelogram formed by the vectors $\vec{r}$ and $\vec{v}\,dt$, which is $|\vec{r} \times \vec{v}\,dt|$. Thus the area $dA$ swept out by the radius vector $\vec{r}$ in time $dt$ is

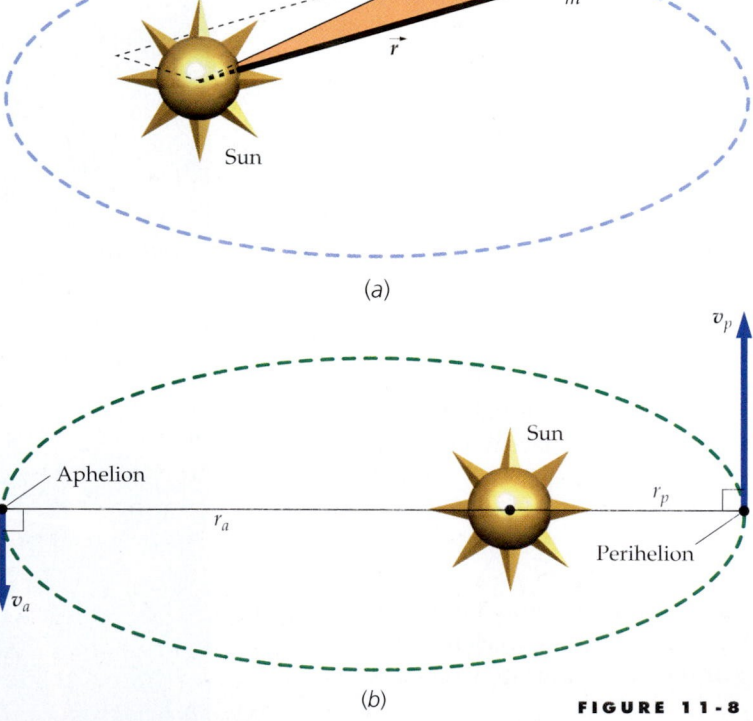

(a)

$$dA = \frac{1}{2}|\vec{r} \times \vec{v}\,dt| = \frac{1}{2m}|\vec{r} \times m\vec{v}|\,dt$$

or

$$\frac{dA}{dt} = \frac{1}{2m}L \qquad\qquad 11\text{-}12$$

(b)

**FIGURE 11-8**

where $L = |\vec{r} \times m\vec{v}|$ is the magnitude of the orbital angular momentum of the planet about the sun. The area $dA$ swept out in a given time interval $dt$ is therefore proportional to the magnitude of the angular momentum $L$. Since the force on a planet is along the line from the planet to the sun, it has no torque about the sun. Thus the magnitude of the angular momentum of the planet is conserved; that is, $L$ is constant. Therefore, the rate at which area is swept out is the same for all parts of the orbit, which is Kepler's second law. Also, the fact that $L$ is constant means that $rv \sin \phi$ is constant. At aphelion and perihelion $\phi = 90°$ (Figure 11-8*b*), so $r_a v_a = r_p v_p$.

We will now show that Newton's law of gravity implies Kepler's third law for the special case of a circular orbit. Consider a planet moving with speed $v$ in a circular orbit of radius $r$ about the sun. The gravitational force of attraction between the sun and the planet provides the centripetal acceleration $v^2/r$. Newton's second law gives

$$F = M_p a$$

$$\frac{GM_s M_p}{r^2} = M_p \frac{v^2}{r} \qquad \text{11-13}$$

where $M_s$ is the mass of the sun and $M_p$ is that of the planet. Solving for $v^2$,

$$v^2 = \frac{GM_s}{r} \qquad \text{11-14}$$

Because the planet moves a distance $2\pi r$ in time $T$, its speed is related to the period by

$$v = \frac{2\pi r}{T} \qquad \text{11-15}$$

Substituting this expression for $v$ in Equation 11-14, we obtain

$$v^2 = \frac{4\pi^2 r^2}{T^2} = \frac{GM_s}{r}$$

or

$$T^2 = \frac{4\pi^2}{GM_s} r^3 \qquad \text{11-16}$$

KEPLER'S THIRD LAW

Equation 11-16 is Kepler's third law, which is the same as Equation 11-2 with $C = 4\pi^2/GM_s$. Equation 11-16 also applies to the orbits of the satellites of any planet if we replace the mass of the sun $M_s$ with the mass of the planet.

---

*THE ORBITING SPACE STATION*      **EXAMPLE 11-3**    **Put It in Context**

**The International Space Station travels in a roughly circular orbit around the earth. If its altitude is 385 km above the earth's surface, how long do you have to wait between sightings? (Assume that air resistance can be neglected.)**

**PICTURE THE PROBLEM** The sightings occur only at night and then only if the space station is above the horizon at your location. Thus, the minimum time between sightings is approximately equal to the orbital period. To find the orbital period we use Kepler's third law with $M_s$ in Equation 11-16 replaced by the mass of the earth $M_E$. The numerical calculation is simplified somewhat by using $GM_E = R_E^2 g$ from Equation 11-8.

1. Apply Kepler's third law to the space station:      $T^2 = \frac{4\pi^2}{GM_E} r^3$

2. At an altitude $h = 385$ km, $r = R_E + h = 6760$ km. Substitute $r = R_E + h$ and solve for the period:

$$T^2 = \frac{4\pi^2}{GM_E}(R_E + h)^3$$

3. Use $GM_E = R_E^2 g$ to write $T$ in terms of $g$:

$$T^2 = \frac{4\pi^2}{R_E^2 g}(R_E + h)^3$$

so

$$T = \frac{2\pi}{R_E\sqrt{g}}(R_E + h)^{3/2}$$

$$= \frac{2\pi(6.755 \times 10^6 \text{ m})^{3/2}}{(6.37 \times 10^6 \text{ m})(9.81 \text{ m/s}^2)^{1/2}}$$

$$= 5529 \text{ s} = \boxed{92.1 \text{ min}}$$

**EXERCISE** How many degrees does the earth rotate in a single orbital period of the ISS? (*Answer* The earth rotates $360°$ in 24 h = 1440 min. In 92.1 min it rotates $23.0°$.)

**REMARKS** The near-circular orbit of the ISS, which is inclined $\approx 52°$ to the equatorial plane, does not rotate with the earth. If the ISS is directly over your home at time $t$, 92.1 min later it will be directly over a location $23.0°$ due west of your home. For example, Kansas City is $23°$ due west of New York City. If the ISS passes over your home in New York City at midnight Eastern Time, you could tell your friend in Kansas City that it will pass over Kansas City at 12:32 A.M. Central Time (1:32 A.M. Eastern Time).

**EXERCISE** Find the radius of a circular orbit of a satellite that orbits the earth with a period of 1 d. (*Answer* $r = 6.63R_E = 4.22 \times 10^7$ m = 26,200 mi. If such a satellite is in orbit over the equator and moves in the same direction as the rotation of the earth, it appears stationary relative to the earth. Many satellites are "parked" in such an orbit, called a geosynchronous orbit.)

Because $G$ is known, by using Equation 11-16 we can determine the mass of an astronomical object by measuring the period $T$ and the mean orbital radius $r$ of a satellite orbiting it. In establishing Equation 11-16 the mass of the satellite was assumed negligible compared to the mass of the central object. This means that the central object remains stationary as the satellite revolves around it. In fact, the central object and satellite both revolve around a common point, their center of mass. If the mass of the satellite is not assumed negligible, the result is

$$T^2 = \frac{4\pi^2}{G(M_1 + M_2)}r^3 \qquad\qquad 11\text{-}17$$

where $r$ is the center-to-center separation of the objects. (For the more general elliptical orbits, the math is more challenging but the result is the same.) If the mass of the satellite is not negligible, as is the case with most binary star systems, then only the sum of the masses is determined, as revealed by Equation 11-17. The moon, along with planets Mercury and Venus, have no natural satellites, so their masses were not well known until the 1960s when artificial satellites were first placed in orbit around them.

**EXERCISE** The Martian moon Phobos has a period of 460 min and a mean orbital radius of 9400 km. What is the mass of Mars? (*Answer* $6.45 \times 10^{23}$ kg = $0.108 M_E$)

EXPLORING

*What is the difference between gravitational mass and inertial mass? Find out this, and more,*
*at www.whfreeman.com/tipler5e.*

# 11-3    Gravitational Potential Energy

Near the surface of the earth, the gravitational force exerted by the earth on
an object is essentially uniform because the distance to the center of the earth, $r =
R_E + h$, is always approximately $R_E$ for $h \ll R_E$. The potential energy of an object
near the earth's surface is $mgh = mg(r - R_E)$, where we have chosen $U = 0$ at the
earth's surface, $r = R_E$. When we are far from the surface of the earth, we must
take into account the fact that the gravitational force exerted by the earth is
not uniform but decreases as $1/r^2$. The general definition of potential energy
(Equation 6-19$b$) is

$$dU = -\vec{F} \cdot d\vec{s}$$

where $\vec{F}$ is the conservative force on a particle and $d\vec{s}$ is a general displacement
of the particle. For the radial gravitational force $\vec{F}$ given by Equation 11-6 we have

$$dU = -\vec{F} \cdot d\vec{s} = -F\hat{r} \cdot d\vec{s} = -F_r \, dr = -\left( -\frac{GM_Em}{r^2} \right) dr = \frac{GM_Em}{r^2} \, dr \qquad \text{11-18}$$

Integrating both sides of this equation we obtain

$$U = -\frac{GM_Em}{r} + U_0 \qquad \text{11-19}$$

where $U_0$ is a constant of integration. Since only changes in potential energy are
important, we can choose the potential energy to be zero at any position. The
earth's surface is a good choice for many everyday problems, but it is not always
a convenient choice. For example, when considering the potential energy associated with a planet and the sun, there is no reason to want the potential energy to
be zero at the surface of the sun. In fact, it is nearly always more convenient to
choose the gravitational potential energy of a two-object system to be zero when
the separation of the objects is infinite. Thus, $U_0 = 0$ is often a convenient choice.
Then

$$U(r) = -\frac{GMm}{r} \qquad \text{11-20}$$

GRAVITATIONAL POTENTIAL ENERGY WITH $U = 0$ AT
INFINITE SEPARATION

where $U = 0$ if $r = \infty$. Figure 11-9 is a plot of $U(r)$ versus $r$
for this choice of $U = 0$ at $r = \infty$ for an object of mass $m$
and the earth of mass $M_E$. This function begins at
the negative value $U = -GM_Em/R_E = -mgR_E$ at the
earth's surface and increases as $r$ increases, approaching
zero at infinite $r$. The slope of this curve at $r = R_E$ is

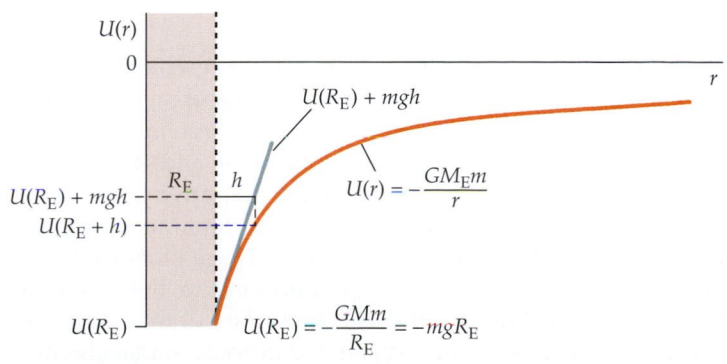

**FIGURE 11-9**

$GM_E m/R_E^2 = mg$, so the equation of the tangent line, drawn in blue, is $f(h) = U(R_E) + mgh$, where $h = r - R_E$ is the distance above the earth's surface. From the figure you can see that for small $h$, $U(R_E) + mgh \approx U(r)$.

## Escape Speed

During the last half-century, the idea of escaping from the earth's gravity has changed from fantasy to reality. Space probes have been sent out to the far reaches of the solar system. Some of these probes orbit the sun, while others leave the solar system and drift on into outer space. We will see that there is a minimum initial speed, called the **escape speed,** that is required for an object in free-fall to escape from the earth.

If we project an object upward from the earth with some initial kinetic energy, the kinetic energy decreases and the potential energy increases as the object rises. But the maximum increase in potential energy is $GM_E m/R_E$. Therefore, this amount is the most that the kinetic energy can decrease. If the initial kinetic energy is greater than $GM_E m/R_E$, then the total energy $E$ will be greater than zero ($E_2$ in Figure 11-10), and the object will still have some kinetic energy when $r$ is very large (or even when $r$ approaches infinity). Thus, if the initial kinetic energy is greater than $GM_E m/R_E$, the object will escape from the earth's gravity. Since the potential energy at the earth's surface is $-GM_E m/R_E$, the total energy $E = K + U$ must be greater than or equal to zero in order for the object to escape. The speed near the earth's surface corresponding to zero total energy is called the escape speed $v_e$. It is found from

$$K_f + U_f = K_i + U_i$$

$$0 + 0 = \frac{1}{2} mv_e^2 - \frac{GM_E m}{R_E}$$

so

$$v_e = \sqrt{\frac{2GM_E}{R_E}} = \sqrt{2gR_E} \qquad \text{11-21}$$

ESCAPE SPEED

Using $g = 9.81$ m/s$^2$ and $R_E = 6.37 \times 10^6$ m, we obtain

$$v_e = \sqrt{2(9.81 \text{ m/s}^2)(6.37 \times 10^6 \text{ m})} = 11.2 \text{ km/s}$$

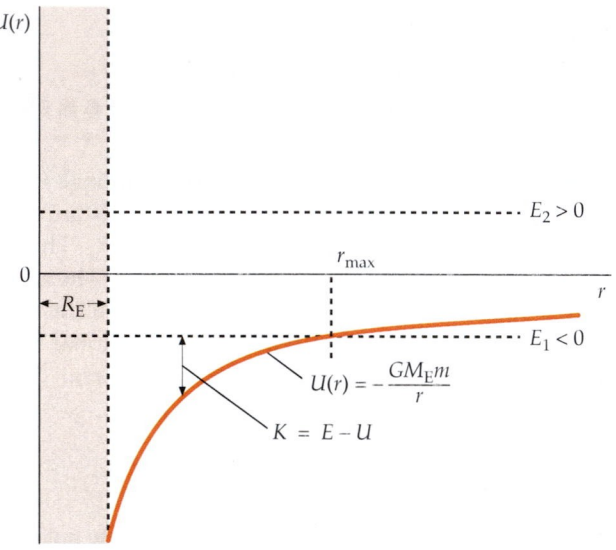

**FIGURE 11-10** The kinetic energy of an object at a distance $r$ from the center of the earth is $E - U(r)$. When the total energy is less than zero ($E_1$ in the figure), the kinetic energy is zero at $r = r_{max}$ and the object is bound to the earth. When the total energy is greater than zero ($E_2$ in the figure), the object can escape the earth.

This is about 7 mi/s or 25,000 mi/h. An object with this speed will escape the earth's gravity but it will not escape the solar system because we have neglected the gravitational attraction of the sun and other planets (see Problems 46 and 48).

The escape speed for a planet or moon relative to the thermal speeds of gas molecules determines the kind of atmosphere a planet or moon can have. The average kinetic energy of gas molecules, $(\frac{1}{2} mv^2)_{av}$, is proportional to the absolute temperature $T$ (Chapter 18). Near the surface of the earth, the speeds of nearly all of the oxygen and nitrogen molecules are much lower than the escape speed, so these gases are retained in our atmosphere. For the lighter molecules hydrogen and helium, however, a significant fraction have speeds greater than the escape speed. Hydrogen and helium gases are therefore not found in our atmosphere. The escape speed at the surface of the moon is 2.3 km/s, which can be calculated from Equation 11-21 with the mass and radius of the moon replacing $M_E$ and $R_E$. This is considerably smaller than the escape speed for earth and, in fact, is too small for any atmosphere to exist.

**EXERCISE** Find the escape speed at the surface of Mercury, which has a mass $M = 3.31 \times 10^{23}$ kg and a radius $R = 2440$ km. (*Answer* $v_e = \sqrt{2GM/R} = 4.25$ km/s)

## Classification of Orbits by Energy

In Figure 11-10, two possible values for the total energy $E$ are indicated on a graph of $U$ versus $r$: $E_1$, which is negative, and $E_2$, which is positive. A negative total energy simply means that the kinetic energy at the earth's surface is less than $+GM_Em/R_E$, so that $K + U$ is never greater than zero. From this figure we see that, if the total energy is negative, the total-energy line intersects the potential-energy curve at some maximum separation $r_{max}$ and the system is bound. On the other hand, if the total energy is zero or positive, there is no such intersection and the system is unbound. The criteria for a bound or unbound system are simply stated:

>    If $E < 0$, the system is bound.

>    If $E \geq 0$, the system is unbound.

When $E$ is negative, its absolute value $|E|$ is called the binding energy. The binding energy is the energy that must be added to the system to bring the total energy up to zero.

The potential energy of an object such as a planet or comet of mass $m$ at a distance $r$ from the sun is

$$U(r) = -\frac{GM_s m}{r}$$                                          11-22

where $M_s$ is the mass of the sun. The kinetic energy of the object is $\frac{1}{2}mv^2$. If the total energy, kinetic plus potential, is less than zero, then the orbit will be an ellipse (possibly a circle), and the object will be bound to the sun. If, instead, the total energy is positive, then the orbit will be a hyperbola, and the object will make one swing around the sun and leave the solar system, never to return. If the total energy is exactly zero, the orbit will be a parabola, and again the object will make one pass and then escape. To summarize, when the total energy is zero or positive the object is not bound to the sun, but will escape. Curiously, there have not been any measurements of the energy $E$ of a comet or an asteroid that are definitely nonnegative. Thus, all observed objects of this type are bound to the solar system.

*HEIGHT OF A PROJECTILE*                    **EXAMPLE    1 1 - 4**

**A projectile is fired straight up from the surface of the earth with an initial speed $v_i = 8$ km/s. Find the maximum height it reaches, neglecting air resistance.**

**PICTURE THE PROBLEM** The maximum height is found using energy conservation. We take the surface of the earth as the initial point, with $U_i = -GM_Em/R_E$ and $K_i = \frac{1}{2}mv_i^2$. At the greatest height, $K_f = 0$.

1. Apply conservation of mechanical energy:

$$K_f + U_f = K_i + U_i$$

$$0 - \frac{GM_Em}{r_f} = \frac{1}{2}mv_i^2 - \frac{GM_Em}{R_E}$$

2. Multiply through by $-1/(GM_Em)$, use $GM_E = gR_E^2$, and solve for $r_f$:

$$\frac{1}{r_f} = -\frac{v_i^2}{2GM_E} + \frac{1}{R_E} = -\frac{v_i^2}{2gR_E^2} + \frac{1}{R_E}$$

so

$$r_f = 1/\left(\frac{1}{R_E} - \frac{v_i^2}{2gR_E^2}\right) = R_E/\left(1 - \frac{v_i^2}{2gR_E}\right)$$

3. Substitute numerical values to find $r$ and $h = r - R_E$:

$$r_f = R_E/\left(1 - \frac{(8000 \text{ m/s})^2}{2(9.81 \text{ m/s}^2)(6.37 \times 10^6 \text{ m})}\right)$$

$$= 2.05R_E$$

$$h = r - R_E = \boxed{1.05R_E = 6.69 \times 10^6 \text{ m}}$$

---

*SPEED OF A PROJECTILE*    **EXAMPLE 11-5**    **Try It Yourself**

**A projectile is fired straight up from the surface of the earth with an initial speed $v_i = 15$ km/s. Find the speed of the projectile when it is very far from the earth, neglecting air resistance.**

**PICTURE THE PROBLEM** Very far from earth means $r \gg R_E$. The initial speed is greater than the escape speed of 11 km/s, so the total energy of the projectile is positive and the projectile will escape the earth with some final kinetic energy. Use conservation of mechanical energy to find this kinetic energy and then solve for the final speed.

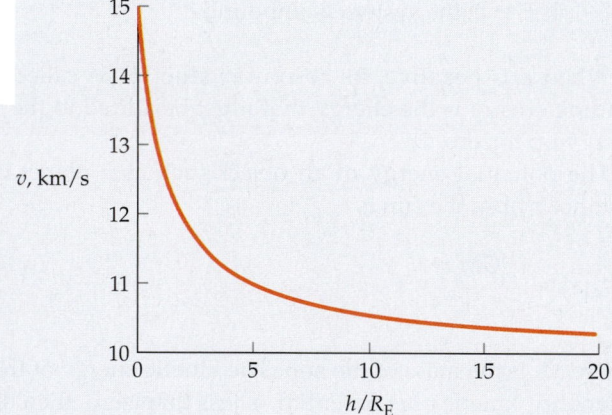

$v$, km/s

$h/R_E$

**FIGURE 11-11**

**Cover the column to the right and try these on your own before looking at the answers.**

**Steps**

**Answers**

1. Apply conservation of mechanical energy, noting that $r_f = \infty$, so $U_f = 0$.

$$K_f + U_f = K_i + U_i$$

$$\frac{1}{2}mv_f^2 + 0 = \frac{1}{2}mv_i^2 - \frac{GM_Em}{R_E}$$

2. Solve for $v_f^2$ using $GM_E/R_E^2 = g$ to simplify.

$$v_f^2 = v_i^2 - \frac{2GM_E}{R_E} = v_i^2 - \frac{2gR_E^2}{R_E}$$

$$= v_i^2 - 2gR_E$$

3. Substitute known values for $g$ and $R_E$ to calculate $v_f$.

$$v_f = \sqrt{(15 \times 10^3 \text{ m/s})^2 - 2(9.81 \text{ m/s}^2)(6.37 \times 10^6 \text{ m})}$$

$$= 10^4 \text{ m/s} = \boxed{10 \text{ km/s}}$$

**REMARKS** In Figure 11-11, the speed of the projectile in kilometers per second is plotted versus $h/R_E$, where $h$ is the height above the earth's surface. At very large values of $h/R_E$, the speed approaches the horizontal line $v = 10$ km/s.

**EXAMPLE 11-6** Try It Yourself

**Show that the total energy of a satellite in a circular orbit is half its potential energy.**

**PICTURE THE PROBLEM** The total energy of a satellite is the sum of its potential and kinetic energies, $E = U + K$. Newton's second law will allow us to find the speed $v$ of the satellite in terms of its orbital radius $r$. The kinetic energy depends on the speed, so we can find the kinetic energy in terms of $r$. Assume the mass of the earth is much greater than that of the satellite.

**Cover the column to the right and try these on your own before looking at the answers.**

| Steps | Answers |
|---|---|
| 1. Write the total energy equal to the sum of the potential energy and the kinetic energy. | $E = K + U$ $= \dfrac{1}{2}mv^2 - \dfrac{GM_{E}m}{r}$ |
| 2. Apply Newton's second law to the satellite and solve for the square of the speed. | $F = ma$ $\dfrac{GM_{E}m}{r^2} = m\dfrac{v^2}{r}$ so $v^2 = \dfrac{GM_{E}}{r}$ |
| 3. Substitute into the step 1 result and simplify. | $E = \dfrac{1}{2}m\dfrac{GM_{E}}{r} - \dfrac{GM_{E}m}{r}$ $= \dfrac{GM_{E}m - 2GM_{E}m}{2r} = -\dfrac{GM_{E}m}{2r}$ |
| 4. Compare the step 3 result with $U$ in step 1. | $E = -\dfrac{GM_{E}m}{2r} = \dfrac{1}{2}\left(-\dfrac{GM_{E}m}{r}\right) = \boxed{\dfrac{1}{2}U}$ |

**EXERCISE** A satellite of mass 450 kg orbits the earth in a circular orbit at 6830 km above the earth's surface. Find (*a*) the potential energy, (*b*) the kinetic energy, and (*c*) the total energy of the satellite. (*Answer* Note that $r = R_{E} + h = 13{,}200$ km (*a*) $U = -13.6 \times 10^9$ J (*b*) $K = 6.80 \times 10^9$ J (*c*) $E = -6.80 \times 10^9$ J)

# 11-4 The Gravitational Field $\vec{g}$

The gravitational force exerted by a point particle of mass $m_1$ on a second point particle of mass $m_2$ a distance $r_{1,2}$ away is given by

$$\vec{F}_{1,2} = -\frac{Gm_1m_2}{r_{1,2}^2}\hat{r}_{1,2}$$

where $\hat{r}_{1,2} = \vec{r}_{1,2}/r_{1,2}$ is a unit vector pointing from particle 1 to particle 2. The gravitational field at point $P$ is determined by placing a point particle of mass $m$ at $P$ and calculating the gravitational force $\vec{F}$ on it due to all other particles. The gravitational force $\vec{F}$ divided by the mass $m$ is the **gravitational field** $\vec{g}$ at $P$:

$$\vec{g} = \frac{\vec{F}}{m}$$  11-23

DEFINITION—GRAVITATIONAL FIELD

The point $P$ is called a **field point**. The gravitational field at a field point due to the masses of a collection of point particles is the vector sum of the fields due to the individual masses at that point:

$$\vec{g} = \sum_i \vec{g}_i$$  11-24a

The points where these point particles are located are called **source points**. To find the gravitational field at a field point due to a continuous object, we find the field $d\vec{g}$ due to a small element of volume with mass $dm$ and integrate over the entire mass distribution of the object (the entire set of source points).

$$\vec{g} = \int d\vec{g}$$  11-24b

The gravitational field of the earth at a distance $r \geq R_E$ points toward the earth and has the magnitude $g(r)$ given by

$$g(r) = \frac{F}{m} = \frac{GM_E}{r^2}$$  11-25

GRAVITATIONAL FIELD OF THE EARTH

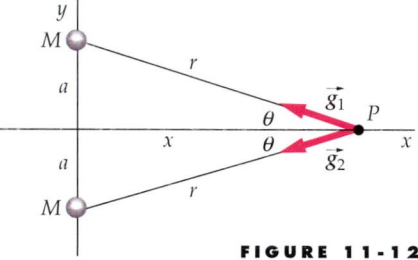

FIGURE 11-12

GRAVITATIONAL FIELD OF TWO POINT PARTICLES          **E X A M P L E   1 1 - 7**

Two point particles, each of mass $M$, are fixed on the $y$ axis at $y = +a$ and $y = -a$ (see Figure 11-12). Find the gravitational field at all points on the $x$ axis as a function of $x$.

**PICTURE THE PROBLEM** Two particles of mass $M$ each produce a gravitational field at point $P$ located at $x\hat{i}$. The distance between $P$ and either particle is $r = \sqrt{x^2 + a^2}$. The resultant field $\vec{g}$ is the vector sum of the fields $\vec{g}_1$ and $\vec{g}_2$ due to each particle.

1. Calculate the magnitude of either $\vec{g}_1$ or $\vec{g}_2$:

$$g_1 = \frac{GM}{r^2}$$

2. The $y$ component of the resultant field is zero. The $x$ component is the sum of $g_{1x}$ and $g_{2x}$:

$$g_x = g_{1x} + g_{2x} = 2g_{1x} = 2g_1 \cos\theta$$

3. Express $\cos\theta$ in terms of $x$ and $r$ from the figure:

$$\cos\theta = \frac{x}{r}$$

4. Combining the last two results yields $\vec{g}$. To express $\vec{g}$ as a function of $x$, substitute $(x^2 + a^2)^{1/2}$ for $r$:

$$\vec{g} = g_x\hat{i} = -2\frac{GM}{r^2}\frac{x}{r}\hat{i} = -\frac{2GMx}{r^3}\hat{i}$$

$$= -\frac{2GMx}{(x^2 + a^2)^{3/2}}\hat{i}$$

**PLAUSIBILITY CHECK** For $x < 0$, $\vec{g}$ is in the positive $x$ direction and for $x > 0$, $\vec{g}$ is in the negative direction, as expected. If $x = 0$, we find that $\vec{g} = 0$; the fields due to $m_1$ and $m_2$ are equal and opposite at $x = 0$, and hence they cancel. For $x \gg a$, $\vec{g} \approx -(2GM/x^2)\hat{i}$. The field is the same as if a single mass of $2M$ were at the origin.

GRAVITATIONAL FIELD OF A UNIFORM ROD          **EXAMPLE   11-8**

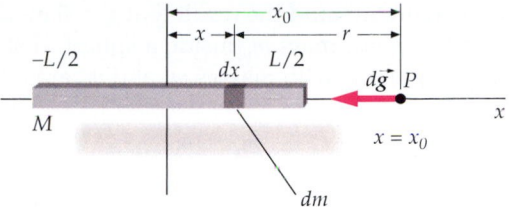

A uniform rod of mass $M$ and length $L$ is centered on the origin and lies along the $x$ axis (Figure 11-13). Find the gravitational field due to the rod at all points on the $x$ axis for $x > L/2$.

**FIGURE 11-13**

**PICTURE THE PROBLEM** Choose a mass element $dm$ of length $dx$ at $x\hat{i}$, and choose a field point $P$ on the $x$ axis at $x = x_0$ in the region $x > L/2$. Each mass element produces a gravitational field at $P$ that points in the negative $x$ direction. We can calculate the total field by integrating the $x$ component of the field produced by $dm$ from $x = -L/2$ to $x = +L/2$.

1. Find the $x$ component of the field at $P$ due to the element $dm$:

$$dg_x = -\frac{G\,dm}{r^2}$$

2. The mass $dm$ is proportional to the size of the element $dx$:

$$dm = \frac{M}{L}\,dx$$

3. Write the distance $r$ between $dm$ and point $P$ in terms of $x$ and $x_0$:

$$r = x_0 - x$$

4. Substitute these results to express $dg$ in terms of $x$:

$$dg_x = -\frac{G\,dm}{r^2} = -\frac{G(M/L)dx}{(x_0 - x)^2}$$

5. Integrate to find the $x$ component of the total field:

$$g_x = \int dg_x = -\frac{GM}{L}\int_{-L/2}^{L/2}\frac{dx}{(x_0 - x)^2} = -\frac{GM}{L}\left[\frac{1}{x_0 - x}\right]_{-L/2}^{L/2}$$

$$= -\frac{GM}{L}\left(\frac{1}{x_0 - L/2} - \frac{1}{x_0 + L/2}\right) = -\frac{GM}{x_0^2 - (L/2)^2}$$

6. Express the resultant field as a vector:

$$\vec{g} = g_x\hat{i} = -\frac{GM}{x_0^2 - (L/2)^2}\hat{i}$$

7. $x_0$ is an arbitrary point on the $x$ axis in the region $x > L/2$, so we can replace it with $x$:

$$\boxed{\vec{g} = -\frac{GM}{x^2 - (L/2)^2}\hat{i}}$$

**PLAUSIBILITY CHECK** For $x \gg L/2$, the field approaches that of a point particle of mass $M$, $\vec{g} = -(GM/x^2)\hat{i}$.

## $\vec{g}$ of a Spherical Shell and of a Solid Sphere

One of Newton's motivations for developing calculus was to prove that the gravitational field outside a solid sphere is the same as if all the mass of the sphere were concentrated at its center. We will prove this in the next section. Here we merely discuss the results of this proof. We first consider a uniform spherical shell of mass $M$ and radius $R$ (Figure 11-14). We will show that the gravitational field due to the shell at a distance $r$ from the center of the shell is given by

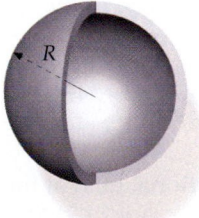

**FIGURE 11-14** A uniform spherical shell of mass $M$ and radius $R$.

$$\vec{g} = -\frac{GM}{r^2}\hat{r} \quad \text{for} \quad r > R \qquad\qquad 11\text{-}26a$$

$$\vec{g} = 0 \quad \text{for} \quad r < R \qquad\qquad 11\text{-}26b$$

GRAVITATIONAL FIELD OF A SPHERICAL SHELL

We can understand the result that $\vec{g} = 0$ inside the shell from Figure 11-15, which shows a point mass $m_0$ inside a spherical shell. In this figure, the masses of the shell segments with masses $m_1$ and $m_2$ are related by

$$\frac{m_1}{m_2} = \frac{r_1^2}{r_2^2} \qquad \text{or} \qquad \frac{m_1}{r_1^2} = \frac{m_2}{r_2^2}$$

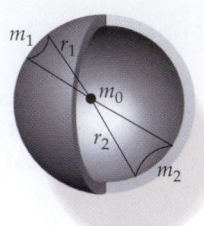

Since the gravitational force falls off inversely as the square of the distance, the force due to the smaller mass on the left is exactly balanced by that due to the more distant, larger mass on the right.

The gravitational field outside a solid sphere is a simple extension of Equation 11-26a. We merely consider the solid sphere to consist of a continuous set of spherical shells. Since the field due to each shell is the same as if its mass were concentrated at the center of the shell, the field due to the entire sphere is the same as if the entire mass of the sphere were concentrated at its center:

**FIGURE 11-15** A point mass $m_0$ inside a uniform spherical shell feels no net force.

$$g_r = -\frac{GM}{r^2} \qquad \text{for} \quad r > R \qquad\qquad 11\text{-}27$$

This result holds whether or not the sphere has a constant density, as long as the density depends only on $r$ so that spherical symmetry is maintained.

$M$ = total mass

## $\vec{g}$ Inside a Solid Sphere

We now use Equations 11-26a and 11-26b to find the gravitational field inside a solid sphere of constant density at a point a distance $r$ from the center, where $r$ is less than the radius $R$ of the sphere. This would apply, for example, to finding the weight of an object at the bottom of a deep mine shaft. As we have seen, the field inside a spherical shell is zero. Thus, in Figure 11-16 the mass of that part of the sphere outside $r$ exerts no force at or inside $r$. Therefore, only the mass $M'$ within the radius $r$ contributes to the gravitational field at $r$. This mass produces a field equal to that of a point mass $M'$ at the center of the sphere. The fraction of the total mass of the sphere within $r$ is equal to the ratio of the volume of a sphere of radius $r$ to that of a sphere of radius $R$. Thus, for a uniform mass distribution, if $M$ is the total mass of the sphere, $M'$ is given by

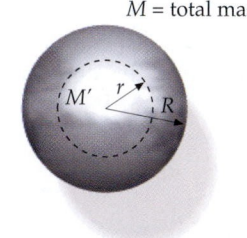

**FIGURE 11-16** A uniform solid sphere of radius $R$ and mass $M$. Only the mass $M'$, which is inside the sphere of radius $r$, contributes to the gravitational field at the distance $r$.

$$M' = M\frac{\frac{4}{3}\pi r^3}{\frac{4}{3}\pi R^3} = M\frac{r^3}{R^3} \qquad\qquad 11\text{-}28$$

The gravitational field at the distance $r$ is thus

$$g_r = -\frac{GM'}{r^2} = -\frac{GM}{r^2}\frac{r^3}{R^3}$$

or

$$g_r = -\frac{GM}{R^3}r \qquad \text{for} \quad r < R \qquad\qquad 11\text{-}29$$

The magnitude of the field is zero at the center and increases with distance $r$ inside the sphere. Figure 11-17 shows a plot of the field $g_r$ as a function of $r$ for a solid sphere of uniform mass density.

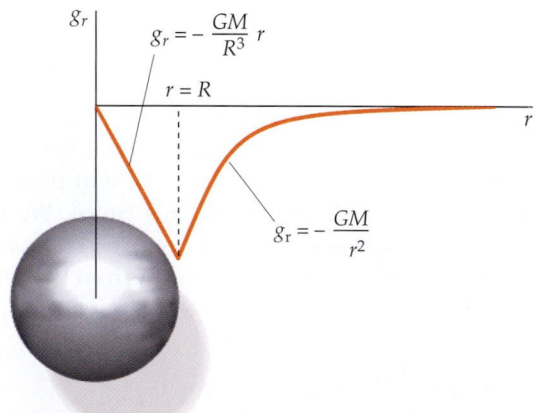

**FIGURE 11-17** A plot of $g_r$ versus $r$ for a uniform solid sphere of mass $M$. The magnitude of the field increases linearly with $r$ inside the sphere and decreases as $1/r^2$ outside the sphere.

A HOLLOW PLANET                    E X A M P L E    1 1 - 9

A planet with a hollow core consists of a uniform, thick spherical shell with mass $M$, outer radius $R$, and inner radius $R/2$. (a) What amount of mass is closer than $\frac{3}{4}R$ to the center of the planet? (b) What is the gravitational field a distance $\frac{3}{4}R$ from the center?

**PICTURE THE PROBLEM**  The mass of that part of the thick shell that is closer to the center than $\frac{3}{4}R$ is the density times the volume of the thick shell with outer radius $\frac{3}{4}R$ and inner radius $\frac{1}{2}R$. First find the density and the volume, then find the mass. The gravitational field at $r = \frac{3}{4}R$ is due only to the mass closer than this to the center.

(a) 1. The mass $M'$ (the mass of the thick shell with outer radius $\frac{3}{4}R$ and inner radius $\frac{1}{2}R$) is the density $\rho$ times the volume $V'$:

$$M' = \rho V'$$

2. The density is the mass $M$ divided by the volume $V$:

$$\rho = \frac{M}{V} = M \bigg/ \left[ \frac{4}{3}\pi R^3 - \frac{4}{3}\pi \left(\frac{R}{2}\right)^3 \right] = \frac{M}{7\pi R^3 / 6}$$

3. Find the volume $V'$ of the thick shell with outer radius $\frac{3}{4}R$ and inner radius $\frac{1}{2}R$:

$$V' = \frac{4}{3}\pi \left(\frac{3R}{4}\right)^3 - \frac{4}{3}\pi \left(\frac{R}{2}\right)^3 = \frac{19}{48}\pi R^3$$

4. Find the mass $M'$:

$$M' = \rho V' = \frac{6}{7}\frac{M}{\pi R^3}\frac{19}{48}\pi R^3 = \boxed{\frac{19}{56}M}$$

(b) The gravitational field at $r = \frac{3}{4}R$ is due only to the mass $M'$:

$$\vec{g} = -\frac{GM'}{r^2}\hat{r} = -\frac{G\frac{19}{56}M}{(\frac{3}{4}R)^2}\hat{r}$$

$$= \boxed{-\frac{38}{63}\frac{GM}{R^2}\hat{r}}$$

**PLAUSIBILITY CHECK**  We expect less than half the mass $M$ to be in the region $\frac{1}{2}R < r < \frac{3}{4}R$, and that is the case here, 19/56 is less than half.

**REMARKS**  Note that the volume $V$ in the denominator of Part (a) step 2 is $7\pi R^3 16$. That is more than half the volume of a sphere of radius $R$. We expect the volume of the region $\frac{1}{2}R < r < R$ to be more than half the volume of a sphere of radius $R$. Also, note that the volume $V'$ (Part (a) step 3) and mass $M'$ (Part (a) step 4) are the same fraction of $V$ and $M$. This is as expected because the shell is uniform.

RADIALLY DEPENDENT DENSITY          E X A M P L E    1 1 - 1 0    Try It Yourself

A solid sphere of radius $R$ and mass $M$ is spherically symmetric but not uniform. Its density $\rho$, defined as its mass per unit volume, is proportional to the distance from the center $r$ for $r \le R$. That is, $\rho = Cr$ for $r \le R$, where $C$ is a constant. (a) Find $C$. (b) Find $g_r$ for $r \ge R$. (c) Find $g_r$ at $r = R/2$.

**PICTURE THE PROBLEM**  (a) You can find $C$ by integrating the density over the volume of the sphere and setting the result equal to $M$. For a volume element, take a spherical shell of radius $r$ and thickness $dr$ (Figure 11-18). Its volume is $4\pi r^2\, dr$ and its mass is $dm = \rho\, dV = Cr(4\pi r^2\, dr)$. (b) The field at $r \ge R$ is the same as if the total mass $M$ were at the center of the sphere. (c) The field at $r = R/2$ is

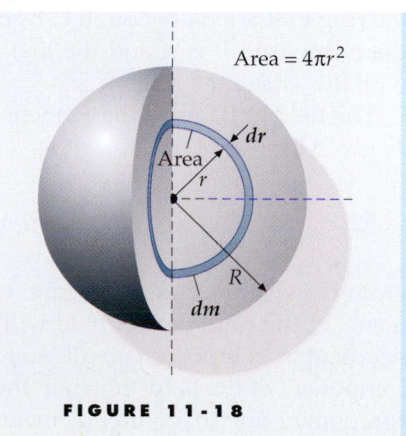

Area $= 4\pi r^2$

$dr$

Area

$r$

$R$

$dm$

**FIGURE 11-18**

the same as if mass $M'$ were at the center of the sphere, where $M'$ is the amount of mass within the sphere of radius $R/2$. The mass between $r = R/2$ and $r = R$ produces zero field at $r = R/2$.

**Cover the column to the right and try these on your own before looking at the answers.**

**Steps**

**Answers**

(a) 1. Integrate $dm$ from $r = 0$ to $r = R$.

$$M = \int dM = \int \rho\, dV = \int_0^R Cr(4\pi r^2 dr) = C\pi R^4$$

2. Solve for $C$ in terms of the given quantities $M$ and $R$.

$$C = \boxed{\dfrac{M}{\pi R^4}}$$

(b) Write an expression for the field outside the sphere in terms of the mass $M$, the distance $r$ from the center, and the unit vector in the $\hat{r}$ direction of increasing $r$.

$$\vec{g} = \boxed{-\dfrac{GM}{r^2}\hat{r}} \quad (r > R)$$

(c) 1. Compute the mass $M'$ that is within the radius $R/2$ by integrating $dm = \rho\, dV$ from $r = 0$ to $r = R/2$ and use the value of $C$ found in Part (a) step 2.

$$M' = \dfrac{M}{16}$$

2. Write an expression for the field at $r = R/2$ in terms of $M$ and $R$.

$$\vec{g} = -\dfrac{GM'}{r^2}\hat{r} = \boxed{-\dfrac{GM}{4R^2}\hat{r}} \quad \text{at } r = \dfrac{R}{2}$$

**PLAUSIBILITY CHECK** For a uniform sphere, Equation 11-29 gives the field at $r = R/2$ as $g_r = -GM/(2R^2)$, twice as large as our Part (c) result. This is as expected since a uniform sphere has a larger fraction of its total mass in the region $0 < r < R/2$ than does the sphere in Equation 11-29.

**REMARKS** Note that the units for $C$ are kg/m$^4$, so the units for $\rho$ are kg/m$^3$, which is mass per volume.

## *11-5 Finding the Gravitational Field of a Spherical Shell by Integration

We will derive the equation for the gravitational field of a spherical shell in two steps. First, we find the gravitational field on the axis of a ring of uniform mass. We then apply our result to a spherical shell, which we can consider to be a set of coaxial rings.

Figure 11-19 shows a ring of total mass $m$ and radius $a$ and a field point $P$ on the axis of the ring a distance $x$ from its center. We choose a mass element $dm$ on the ring that is small enough to be considered a point particle. The distance from the element to $P$ is $s$, and the line joining the element and $P$ makes an angle $\alpha$ with the axis of the ring.

The field at $P$ due to the element $dm$ is toward the element and has magnitude $dg$ given by

$$dg = \frac{G\, dm}{s^2}$$

From the symmetry of the figure, we can see that when we sum over all the elements of the ring, the net field will be along the axis of the ring; that is, the perpendicular components will sum to zero. For example, the perpendicular component of the field shown in the figure will be canceled by the perpendicular component due to another element of the ring directly opposite the one shown.

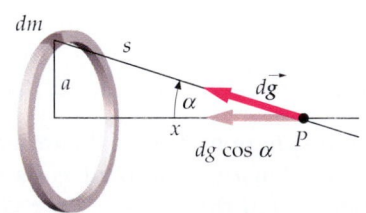

**FIGURE 11-19** The gravitational field at a point $P$ a distance $x$ from a uniform ring. The field due to the element $dm$ points toward the element. The total field due to the ring is along the axis of the ring.

The net field will therefore be in the negative $x$ direction. The $x$ component of the field due to the element $dm$ is

$$dg_x = -dg \cos \alpha = -\frac{G\, dm}{s^2} \cos \alpha$$

We obtain the total field by summing over all the elements of the ring:

$$g_x = -\int \frac{G \cos \alpha}{s^2} dm$$

Since $s$ and $\alpha$ are the same for all points on the ring, they are constants as far as the integration is concerned. Thus,

$$g_x = -\frac{G \cos \alpha}{s^2} \int dm = -\frac{Gm}{s^2} \cos \alpha \qquad \text{11-30}$$

where $m = \int dm$ is the total mass of the ring.

We now use this result to calculate the gravitational field of a spherical shell of mass $M$ and radius $R$ at a point a distance $r$ from the center of the shell. We first consider the case in which the field point $P$ is outside the shell, as in Figure 11-20. By symmetry, the field must be radial. We choose for our element of mass the strip shown, which can be considered to be a ring of mass $dM$. The field due to this strip is given by Equation 11-30 with $m$ replaced by $dM$.

$$dg_r = -\frac{G\, dM}{s^2} \cos \alpha \qquad \text{11-31}$$

The mass $dM$ is proportional to the area of the strip $dA$, which equals the circumference times the width. The radius of the strip is $R \sin \theta$, so the circumference is $2\pi R \sin \theta$. The width is $R\, d\theta$. If $M$ is the total mass of the shell and $A = 4\pi R^2$ is its total area, the mass of the strip of area $dA$ is

$$dM = \frac{M}{A} dA = \frac{M}{4\pi R^2} 2\pi R^2 \sin \theta\, d\theta = \frac{1}{2} M \sin \theta\, d\theta \qquad \text{11-32}$$

Substituting this result into Equation 11-31 gives

$$dg_r = -\frac{G\, dM}{s^2} \cos \alpha = -\frac{GM \sin \theta\, d\theta}{2s^2} \cos \alpha \qquad \text{11-33}$$

Before integrating over the entire shell, we must eliminate two of the three related variables $s$, $\theta$, and $\alpha$. It turns out to be easiest to write everything in terms of $s$. By the law of cosines, we have

$$s^2 = r^2 + R^2 - 2rR \cos \theta$$

Differentiating gives

$$2s\, ds = +2rR \sin \theta\, d\theta$$

or

$$\sin \theta\, d\theta = \frac{s\, ds}{rR}$$

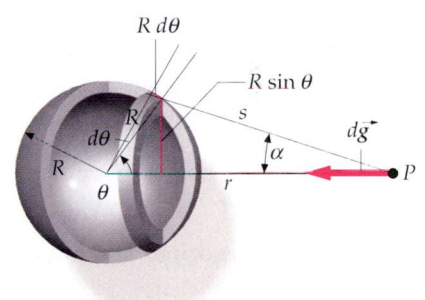

**FIGURE 11-20** A uniform thin spherical shell of radius $R$ and total mass $M$. The strip shown can be considered to be a ring of width $R\, d\theta$ and circumference $2\pi R \sin \theta$.

An expression for $\cos \alpha$ can be obtained by again applying the law of cosines to the same triangle. We have

$$R^2 = s^2 + r^2 - 2sr \cos \alpha$$

or

$$\cos \alpha = \frac{s^2 + r^2 - R^2}{2sr}$$

Substituting these results into Equation 11-33 gives

$$dg_r = -\frac{GM \sin \theta \, d\theta}{2s^2} \cos \alpha = -\frac{GM}{2s^2} \left(\frac{s \, ds}{rR}\right) \frac{s^2 + r^2 - R^2}{2sr}$$

$$= -\frac{GM \, ds}{4s^2 r^2 R}(s^2 + r^2 - R^2) = -\frac{GM}{4r^2 R}\left(1 + \frac{r^2 - R^2}{s^2}\right) ds \qquad \text{11-34}$$

To find the field at $P$ we integrate this over the entire shell. The integration limits for this step depend on whether the field point $P$ lies outside the shell or inside it. For $P$ outside the shell $s$ varies from $r - R$ ($\theta = 0$) to $s = r + R$ ($\theta = 180°$), so the field due to the entire shell is found by integrating from $s = r - R$ to $s = r + R$.

$$g_r = -\frac{GM}{4r^2 R} \int_{r-R}^{r+R} \left(1 + \frac{(r-R)(r+R)}{s^2}\right) ds = -\frac{GM}{4r^2 R}\left[s - \frac{(r-R)(r+R)}{s}\right]_{r-R}^{r+R}$$

Substitution of the upper and lower limits yields $4R$ for the quantity in brackets. Thus,

$$g_r = -\frac{GM}{r^2} \qquad \text{for} \quad r > R$$

which is the same result as Equation 11-26$a$.

If the field point $P$ is inside the shell (Figure 11-21), the calculation is identical except that $s$ now varies from $R - r$ to $R + r$. Thus,

$$g_r = -\frac{GM}{4r^2 R}\left[s - \frac{(r-R)(r+R)}{s}\right]_{R-r}^{R+r}$$

Substitution of these upper and lower limits yields 0. Therefore,

$$g_r = 0 \qquad \text{for} \quad r < R$$

which is the same as Equation 11-26$b$.

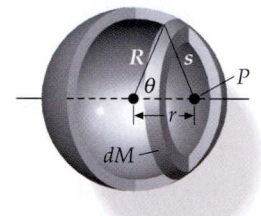

**FIGURE 11-21**

## EXPLORING

*Are tidal forces just as strong as gravitational forces? Find out this, and more, at* www.whfreeman.com/tipler5e.

# SUMMARY

1. Kepler's laws are *empirical* observations. They can also be derived from Newton's laws.

2. Newton's law of gravity is a *fundamental law* of physics.

3. The gravitational potential energy of a two-particle system relative to $U = 0$ at infinite separation is given by $U = -Gm_1m_2/r$. If the system is bound, its total energy is negative.

4. The gravitational field is a *fundamental physical concept* that describes the condition in space set up by a mass distribution.

| Topic | Relevant Equations and Remarks |
|---|---|
| **1. Kepler's Three Laws** | Law 1. All planets move in elliptical orbits with the sun at one focus. |
| | Law 2. A line joining any planet to the sun sweeps out equal areas in equal times. |
| | Law 3. The square of the period of any planet is proportional to the cube of the planet's mean distance from the sun: |

$$T^2 = Cr^3 \qquad \text{11-2}$$

where $C$ has almost the same value for all planets; from Newton's law of gravity, $C$ can be shown to be $4\pi^2/[G(M_s + M_p)]$. If $M_s \gg M_p$, this can be expressed as

$$T^2 = \frac{4\pi^2}{GM_s}r^3 \qquad \text{11-16}$$

Kepler's laws can be derived from Newton's law of gravity. The first and third laws follow from the fact that the force exerted by the sun on the planets varies inversely as the square of the separation distance. The second law follows from the fact that the force exerted by the sun on a planet is along the line joining them, so the orbital angular momentum of the planet is conserved. Kepler's laws also hold for any object orbiting another in an inverse-square field, such as a satellite orbiting a planet.

| | |
|---|---|
| **2. Newton's Law of Gravity** | Every point particle exerts on every other point particle an attractive force that is proportional to the masses of the two particles and inversely proportional to the square of the distance separating them. |

$$\vec{F}_{1,2} = -\frac{Gm_1m_2}{r_{1,2}^2}\hat{r}_{1,2} \qquad \text{11-3}$$

| | |
|---|---|
| Universal gravitational constant | $G = 6.67 \times 10^{-11} \text{ N·m}^2/\text{kg}^2 \qquad$ 11-4 |

| | |
|---|---|
| **3. Gravitational Potential Energy** | The gravitational potential energy $U$ for a system consisting of a particle of mass $m$ outside a spherically symmetric object of mass $M$ and at a distance $r$ from its center is |

$$U(r) = -\frac{GMm}{r} \qquad \text{11-20}$$

This potential-energy function approaches zero as $r$ approaches infinity.

| | |
|---|---|
| **4. Mechanical Energy** | The mechanical energy $E$ for a system consisting of a particle of mass $m$ outside a spherically symmetric object of mass $M$ and at a distance $r$ from its center is |

$$E = \frac{1}{2}mv^2 - \frac{GMm}{r}$$

| | | |
|---|---|---|
| Escape speed | For a given value of $r$, the speed of the particle for which $E = 0$ is called the escape speed $v_e$. That is, if $v = v_e$, then $E = 0$. | |

| | | |
|---|---|---|
| 5. **Classification of Orbits** | If $E < 0$, the system is bound and the orbit is an ellipse (or circle, which is a type of ellipse). <br> If $E \geq 0$, the system is unbound and the orbit is a hyperbola (or a parabola for $E = 0$). | |

6. **Gravitational Field**

| | | |
|---|---|---|
| Definition | $$\vec{g} = \frac{\vec{F}}{m}$$ | 11-23 |
| Due to the earth | $$\vec{g}(r) = \frac{\vec{F}}{m} = -\frac{GM_E}{r^2}\hat{r} \quad (r \geq R_E)$$ | 11-29 |
| Due to thin spherical shell | Outside the shell the gravitational field is the same as if all the mass of the shell were concentrated at the center. The field inside the shell is zero. <br><br> $$\vec{g} = -\frac{GM}{r^2}\hat{r} \quad \text{for} \quad r > R$$ <br><br> $$\vec{g} = 0 \quad \text{for} \quad r < R$$ | 11-26$a$ <br><br> 11-26$b$ |

# PROBLEMS

- • Single-concept, single-step, relatively easy
- •• Intermediate-level, may require synthesis of concepts
- ••• Challenging
- SSM Solution is in the *Student Solutions Manual*
- iSOLVE Problems available on iSOLVE online homework service
- iSOLVE ✓ These "Checkpoint" online homework service problems ask students additional questions about their confidence level, and how they arrived at their answer

In a few problems, you are given more data than you actually need; in a few other problems, you are required to supply data from your general knowledge, outside sources, or informed estimates.

**Take $g = 9.81$ N/kg $= 9.81$ m/s$^2$ and neglect friction in all problems unless otherwise stated.**

## Conceptual Problems

**1** • SSM True or false:

(a) Kepler's law of equal areas implies that gravity varies inversely with the square of the distance.

(b) The planet closest to the sun, on the average, has the shortest orbital period.

**2** • If the mass of a satellite is doubled, the radius of its orbit can remain constant if the speed of the satellite (a) increases by a factor of 8, (b) increases by a factor of 2, (c) does not change, (d) is reduced by a factor of 8, (e) is reduced by a factor of 2.

**3** •• SSM At the surface of the moon, the acceleration due to the gravity of the moon is $a$. At a distance from the center of the moon equal to four times the radius of the moon, the acceleration due to the gravity of the moon is (a) $16a$, (b) $a/4$, (c) $a/3$, (d) $a/16$, (e) none of the answers given.

**4** • Why is $G$ so difficult to measure?

**5** • If the mass of a planet is doubled with no increase in its radius, the escape speed for that planet will be (a) increased by a factor of 1.4, (b) increased by a factor of 2, (c) unchanged, (d) reduced by a factor of 1.4, (e) reduced by a factor of 2.

**6** •• A newly discovered comet-like object is observed to make a pass around the sun. How can we tell if the object will return many years later, or if it will never return?

**7** •• Explain why the gravitational field is directly proportional to $r$ rather than inversely proportional to $r$ inside a solid sphere of uniform mass.

**8** • SSM If $K$ is the kinetic energy of Mercury in its orbit around the sun, and $U$ is the potential energy of the Mercury–sun system, what is the relationship between $K$ and $U$?

**9** •• A student whose weight on the surface of the earth is $w$ is lifted to a height that is two earth radii above the surface. Her weight there is (a) $w/2$, (b) $w/4$, (c) $w/3$, (d) $w/9$.

**10** •• In the novel "First Men in the Moon" by H.G. Wells, the trip was accomplished in a slightly different manner than by Jules Verne's heroes. The novel's hero, Professor Cavour, invented a material, Cavourite, that shielded objects from the gravitational force. Simply by surrounding an object with Cavourite, the gravitational pull on it by its surroundings was nullified. The adventurers in this book built a ship whose bottom was covered by Cavourite, but whose top was unshielded. Then they pointed the top of the ship at the moon to let its attraction pull them up. As far as real physics goes, Cavourite is an impossibility, because it would allow the violation of the law of conservation of energy. To show this, invent a perpetual motion machine using Cavourite.

## Estimation and Approximation

**11** • Estimate the mass of the galaxy if the sun orbits the center of the galaxy with a period of 250 million years at a mean distance of 30,000 light-years. Express the mass in terms of multiples of the solar mass $M_s$. (Neglect the mass farther from the center than the sun, and assume that the mass closer to the center than the sun exerts the same force on the sun as would a point particle of the same mass.)

**12** ••• **SSM** One of the great discoveries in astronomy in recent years is the detection of planets outside the solar system. Since 1996, 100 planets have been detected orbiting stars other than the sun. While the planets themselves cannot be seen directly, telescopes can detect the small periodic motion of the star as the star and planet orbit around their common center of mass. (This is measured using the *Doppler effect*, which will be discussed in Chapter 15.) Both the period of this motion and the variation in the velocity of the star over the course of time can be determined observationally. The mass of the star is found from its observed luminance and from the theory of stellar structure. Iota Draconis is the 8th brightest star in the constellation Draco. Observations show that there is a planet orbiting this star with an orbital period of 1.50 y. The mass of Iota Draconis is $1.05M_{sun}$. (*a*) What is the size (in AU) of the semimajor axis of this planet's orbit? (*b*) The radial velocity of the star is observed to vary by 592 m/s. Use conservation of momentum to find the mass of the planet. Assume the orbit is circular, we are observing the orbit edge-on, and no other planets orbit Iota Draconis. Express the mass as a multiple of the mass of Jupiter.

**13** ••• One of the biggest unresolved problems in the theory of the formation of the solar system is that, while the mass of the sun is 99.9 percent of the total mass of the solar system, it carries only about 2 percent of the total angular momentum. The most widely accepted theory of solar system formation has as its central hypothesis the collapse of a cloud of dust and gas under the force of gravity, with most of the mass forming the sun. However, because the net angular momentum of this cloud is conserved, a simple theory would indicate that the sun should be rotating much more rapidly than it currently is. In this problem, you will show why it is important that most of the angular momentum was somehow transferred to the planets. (*a*) The sun is a cloud of gas held together by the force of gravity. If the sun were rotating too rapidly, gravity couldn't hold it together. Using the known mass of the sun ($1.99 \times 10^{30}$ kg) and its radius ($6.96 \times 10^8$ m),

estimate the maximum angular velocity that the sun can have if it is to stay intact. What is the period of rotation corresponding to this rotation rate? (*b*) Calculate the orbital angular momentum of Jupiter and of Saturn from their masses (318 and 95.1 Earth masses, respectively), mean distances from the sun (778 and 1430 million km, respectively), and orbital periods (11.9 and 29.5 y, respectively). Compare them to the experimentally measured value of the sun's angular momentum of $1.91 \times 10^{41}$ kg m²/s. (*c*) If we were to somehow transfer all of Jupiter's and Saturn's angular momentum to the sun, what would be the sun's new rotational period? Because the sun is not a uniform sphere of gas, its moment of inertia is given by the formula $I = 0.059MR^2$. Compare this to the maximum rotation rate of Part(*a*).

## Kepler's Laws

**14** • **SSM** The new comet Alex-Casey is discovered on a very elliptical orbit with a period of 127.4 y. If the closest approach of Alex-Casey to the sun is 0.1 AU, what is its greatest distance from the sun?

**15** • The radius of the earth's orbit is $1.496 \times 10^{11}$ m and that of Uranus is $2.87 \times 10^{12}$ m. What is the period of Uranus?

**16** • **iSOLVE** ✔ The asteroid Hektor, discovered in 1907, is in a nearly circular orbit of radius 5.16 AU about the sun. Determine the period of this asteroid.

**17** •• The asteroid Icarus, discovered in 1949, was so named because its highly eccentric elliptical orbit brings it close to the sun at perihelion. The eccentricity $e$ of an ellipse is defined by the relation $r_p = a(1 - e)$, where $r_p$ is the perihelion distance and $a$ is the semimajor axis. Icarus has an eccentricity of 0.83 and a period of 1.1 y. (*a*) Determine the semimajor axis of the orbit of Icarus. (*b*) Find the perihelion and aphelion distances of the orbit of Icarus.

**18** •• There has been much discussion of a manned mission to Mars and the problems that it would entail due to the extremely long time spent in space by the astronauts. To examine this issue in a simple way, consider one possible trajectory for the spacecraft: the "Hohmann transfer orbit." This orbit consists of an elliptical orbit tangent to the orbit of Earth at its point nearest the sun and tangent to the orbit of Mars at the point farthest from the sun. Given that Mars has a mean distance from the sun of 1.52 times the mean sun–Earth distance, calculate the total time spent by the astronauts in transit.

**19** •• **SSM** Kepler determined distances in the solar system from his data. For example, he found the relative distance from the sun to Venus (as compared to the distance from the sun to Earth) as follows. Because Venus's orbit is closer to the sun than is Earth's, Venus is a morning or evening star— its position in the sky is never very far from the sun (see Figure 11-22). If we consider the orbit of Venus as a perfect circle, then consider the relative orientation of Venus, Earth, and the sun at maximum extension—when Venus is farthest from the sun in the sky. (*a*) Under this condition, show that angle *b* in Figure 11-22 is 90°. (*b*) If the maximum elongation angle between Venus and the sun is 47°, what is the distance between Venus and the sun in AU?

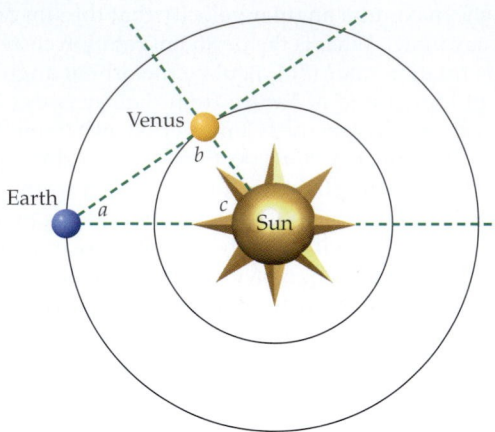

**FIGURE 11-22** Problem 19

**20** •• At apogee the moon is 406,395 km from the earth and at perigee it is 357,643 km. What is the velocity of the moon at perigee and apogee? Its orbital period is 27.3 d.

## Newton's Law of Gravity

**21** •• **SSM** Jupiter's satellite Europa orbits Jupiter with a period of 3.55 d at an average distance of $6.71 \times 10^8$ m. (a) Assuming that the orbit is circular, determine the mass of Jupiter from the data given. (b) Another satellite of Jupiter, Callisto, orbits at an average distance of $18.8 \times 10^8$ m with an orbital period of 16.7 d. Determine the acceleration of Callisto and Europa from the data given, and show that this data is consistent with an inverse square law for gravity [Note: Do NOT use the value of $G$ anywhere in Part (b)].

**22** • **SSM** Some people think that shuttle astronauts are "weightless" because they are "beyond the pull of Earth's gravity." In fact, this is completely untrue. (a) What is the magnitude of the acceleration of gravity for shuttle astronauts? A shuttle orbit is about 400 km above the ground. (b) Given the answer in Part (a), why are shuttle astronauts "weightless"?

**23** • **SOLVE** The mass of Saturn is $5.69 \times 10^{26}$ kg. (a) Find the period of its moon Mimas, whose mean orbital radius is $1.86 \times 10^8$ m. (b) Find the mean orbital radius of its moon Titan, whose period is $1.38 \times 10^6$ s.

**24** • **SOLVE** Calculate the mass of the earth from the period of the moon, $T = 27.3$ d; its mean orbital radius, $r_m = 3.84 \times 10^8$ m; and the known value of $G$.

**25** • **SOLVE** Use the period of the earth (1 y), its mean orbital radius ($1.496 \times 10^{11}$ m), and the value of $G$ to calculate the mass of the sun.

**26** • **SSM** **SOLVE** An object is dropped from a height of $6.37 \times 10^6$ m above the surface of the earth. What is its initial acceleration?

**27** • Suppose you leave the solar system and arrive at a planet that has the same mass-to-volume ratio as the earth but has 10 times the earth's radius. What would you weigh on this planet compared with what you weigh on the earth?

**28** • Suppose that the earth retained its present mass but was somehow compressed to half its present radius. What

would be the value of $g$ at the surface of this new, compact planet?

**29** • A planet orbits a massive sun. When the planet is at perihelion, it has a speed of $5 \times 10^4$ m/s and is $1.0 \times 10^{15}$ m from the sun. The orbital radius increases to $2.2 \times 10^{15}$ m at aphelion. What is the planet's speed at aphelion?

**30** • What is the acceleration of gravity at the surface of a neutron star whose mass is 1.60 times the mass of the sun and whose radius is 10.5 km?

**31** •• **SSM** The speed of an asteroid is 20 km/s at perihelion and 14 km/s at aphelion. Determine the ratio of the aphelion to perihelion distances.

**32** •• A satellite with a mass of 300 kg moves in a circular orbit $5 \times 10^7$ m above the earth's surface. (a) What is the gravitational force on the satellite? (b) What is the speed of the satellite? (c) What is the period of the satellite?

**33** •• **SSM** **SOLVE** A superconducting gravity meter can measure changes in gravity of the order $\Delta g/g = 10^{-11}$. (a) You are hiding behind a tree holding the meter, and your 80-kg friend approaches the tree from the other side. How close to you can your friend get before the meter detects a change in $g$ due to his presence? (b) You are in a hot air balloon and are using the meter to determine the rate of ascent (presumed constant). What is the smallest change in altitude that results in a detectable change in the gravitational field of the earth?

**34** •• Suppose that the attractive interaction between a star of mass $M$ and a planet of mass $m \ll M$ were of the form $F = KMm/r$, where $K$ is the gravitational constant. What would be the relation between the radius of the planet's circular orbit and its period?

**35** •• **SSM** The mass of the earth is $5.97 \times 10^{24}$ kg and its radius is 6370 km. The radius of the moon is 1738 km. The acceleration of gravity at the surface of the moon is $1.62$ m/s². What is the ratio of the average density of the moon to that of the earth?

## Measurement of G

**36** • **SOLVE** The large and small spheres in a Cavendish balance (Figure 11-7) have masses $m_1 = 10$ kg and $m_2 = 0.010$ kg, respectively, and the center-to-center separation of the two small spheres is 20 cm. The center-to-center separation of each large sphere with the small sphere adjacent to it is 6 cm. (a) What is the force by which each large sphere attracts the small sphere adjacent to it? (b) What torque must be exerted by the suspension to balance the torque due to these forces?

## Gravitational and Inertial Mass

**37** • **SOLVE** A standard object defined as having a mass of exactly 1 kg is given an acceleration of $2.6587$ m/s² when a certain force is applied to it. A second object of unknown mass acquires an acceleration of $1.1705$ m/s² when the same force is applied to it. (a) What is the mass of the second object? (b) Is the mass that you determined in Part (a) gravitational or inertial mass?

**38** • **ISOLVE✓** The weight of a standard object defined as having a mass of exactly 1 kg is measured to be 9.81 N. In the same laboratory, a second object weighs 56.6 N. (a) What is the mass of the second object? (b) Is the mass you determined in Part (a) gravitational or inertial mass?

**39** • **SSM** The principle of equivalence states that the free-fall acceleration of any object in a gravitational field is independent of the mass of the object. This can be seen from the form of the law of universal gravitation—but how well does it hold experimentally? The Roll-Krotkov-Dicke experiment performed in the 1960s indicates that the free-fall acceleration is independent of mass to at least 1 part in $10^{12}$. Suppose two objects are simultaneously released from rest in a uniform gravitational field. Also, suppose one of the objects falls with a constant acceleration of exactly $9.8 \text{m/s}^2$ while the other falls with a constant acceleration that is greater than 9.8 m/s² by one part in $10^{12}$. How far will the first object have fallen when the second object has fallen 1 mm further than it has? Note that this estimate is an upper bound on the difference in the accelerations; most physicists believe that there is no difference whatsoever.

## Gravitational Potential Energy

**40** • (a) Taking the potential energy to be zero at infinite separation, find the potential energy of a 100-kg object at the surface of the earth. (Use $6.37 \times 10^6$ m for the earth's radius.) (b) Find the potential energy of the same object at a height above the earth's surface equal to the earth's radius. (c) Find the escape speed for a body projected from this height.

**41** • A point particle of mass $m_0$ is initially at the surface of a large sphere of mass $M$ and radius $R$. How much work is needed to remove it to a very large distance away from the large sphere?

**42** • Suppose that in space there is a duplicate earth, except that it has no atmosphere, is not rotating, and is not in motion around any sun. What initial velocity must a spacecraft on this planet's surface have to travel vertically upward a distance equal to one earth radius above the surface of the planet?

**43** •• **SSM** **ISOLVE✓** An object is dropped from rest from a height of $4 \times 10^6$ m above the surface of the earth. If there is no air resistance, what is its speed when it strikes the earth?

**44** •• **ISOLVE** An object is projected upward from the surface of the earth with an initial speed of 4 km/s. Find the maximum height it reaches.

**45** •• **ISOLVE** A uniform thin spherical shell has a radius $R$ and a mass $M$. (a) Write expressions for the force exerted by the shell on a point mass $m_0$ when $m_0$ is outside the shell and when it is inside the shell. (b) What is the potential-energy function $U(r)$ for this system when the mass $m_0$ is at a distance $r$ $(r \geq R)$ if $U = 0$ at $r = \infty$? Evaluate this function at $r = R$. (c) Using the general relation for $dU = -\vec{F} \cdot d\vec{r} = -F_r \, dr$, show that $U$ is constant everywhere inside the shell. (d) Using the fact that $U$ is continuous everywhere, including at $r = R$, find the value of the constant $U$ inside the shell. (e) Sketch $U(r)$ versus $r$ for all possible values of $r$.

**46** • **ISOLVE** The planet Saturn has a mass 95.2 times that of the earth and a radius 9.47 times that of the earth. Find the escape speed for objects on the surface of Saturn.

**47** • **ISOLVE✓** Find the escape speed for a rocket leaving the moon. The acceleration of gravity on the moon is 0.166 times that on earth and the moon's radius is $0.273R_E$.

**48** • **SSM** The science fiction writer Robert Heinlein once said, "If you can get into orbit, then you're halfway to anywhere." Justify this statement by comparing the kinetic energy needed to place a satellite into low earth orbit ($h = 400$ km) to that needed to set it completely free from the bonds of earth's gravity.

**49** •• A particle is projected from the surface of the earth with a speed twice the escape speed. When it is very far from the earth, what is its speed?

**50** •• What initial speed should a particle be given if it is to have a final speed when it is very far from the earth that is equal to its escape speed?

**51** •• **ISOLVE** (a) Calculate the energy in joules necessary to launch an object whose mass is 1 kg from the earth at escape speed. (b) Convert this energy to kilowatt-hours. (c) If energy can be obtained at \$0.10/kW·h, what is the minimum cost of giving an 80-kg astronaut enough energy to escape the earth's gravitational field?

**52** •• An object is projected vertically from the surface of the earth at less than the escape speed. Show that the maximum height reached by the object is $H = R_E H'/(R_E - H')$, where $H'$ is the height that it would reach if the gravitational field were constant.

## Orbits

**53** •• **ISOLVE✓** A spacecraft of 100-kg mass is in a circular orbit about the earth at a height $h = 2R_E$. (a) What is the period of the spacecraft's orbit about the earth? (b) What is the spacecraft's kinetic energy? (c) Express the angular momentum $L$ of the spacecraft about the center of the earth in terms of its kinetic energy $K$ and find its numerical value.

**54** • **SSM** We normally say that the moon orbits the earth, but this is not quite true: instead, both the moon and the earth orbit their common center of mass, which is not at the center of the earth. (a) The mass of the earth is $5.98 \times 10^{24}$ kg and the mass of the moon is $7.36 \times 10^{22}$ kg. The mean distance from the center of the earth to the center of the moon is $3.82 \times 10^8$ m. How far above the surface of the earth is the center of mass of the earth–moon system? (b) Estimate the mean "orbital speed" of the earth as it orbits the center of mass of the earth–moon system, using the moon's mass, period, and average distance from the earth. Ignore any external forces acting on the system. The orbital period of the moon is 27.3 d.

**55** •• **ISOLVE✓** Many satellites orbit the earth about 1000 km above the earth's surface. Geosynchronous satellites orbit at a distance of $4.22 \times 10^7$ m from the center of the earth. How much more energy is required to launch a 500-kg satellite into a geosynchronous orbit than into an orbit 1000 km above the surface of the earth?

**56** •• Using the period of the moon's orbit around the earth (27.3 d), its average distance from the center of the earth ($3.82 \times 10^8$ m), the length of a year (365.25 d), and the average distance from the earth to the sun ($1.50 \times 10^{11}$ m), establish the ratio of the mass of the sun to the mass of the earth. Compare this to the measured ratio of $3.33 \times 10^5$.

## The Gravitational Field

**57** • **iSOLVE** A 3-kg mass experiences a gravitational force of 12 N $\hat{i}$ at some point $P$. What is the gravitational field at that point?

**58** • **SSM** **iSOLVE** The gravitational field at some point is given by $\vec{g} = 2.5 \times 10^{-6}$ N/kg $\hat{j}$. What is the gravitational force on a mass of 4 g at that point?

**59** •• A point mass $m$ is on the $x$ axis at $x = L$ and an identical point mass is on the $y$ axis at $y = L$. (a) Find the gravitational field at the origin. (b) What is the magnitude of this field?

**60** •• Five objects of mass $M$ are equally spaced on the arc of a semicircle of radius $R$ as in Figure 11-23. An object of mass $m$ is located at the center of curvature of the arc. (a) If $M$ is 3 kg, $m$ is 2 kg, and $R$ is 10 cm, what is the force on $m$ due to the five objects? (b) If the object whose mass is $m$ is removed, what is the gravitational field at the center of curvature of the arc?

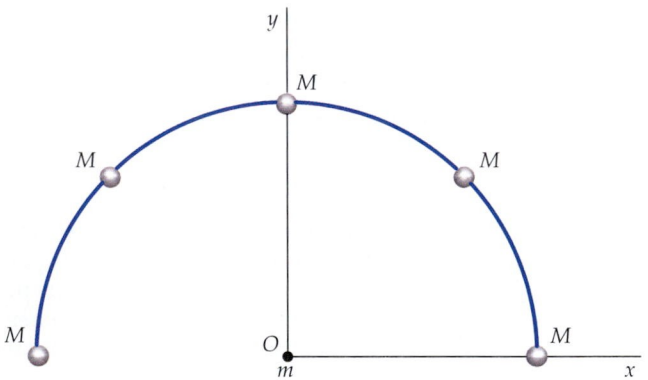

**FIGURE 11-23** Problem 60

**61** •• A point particle of mass $m_1 = 2$ kg is at the origin and a second point particle of mass $m_2 = 4$ kg is on the $x$ axis at $x = 6$ m. Find the gravitational field at (a) $x = 2$ m and (b) $x = 12$ m. (c) Find the point on the $x$ axis for which $g = 0$.

**62** •• Show that the maximum value of $|g_x|$ for the field of Example 11-7 occurs at the points $x = \pm a/\sqrt{2}$.

**63** •• A nonuniform rod of length $L$ lies on the $x$ axis with one end at the origin and the other at $x = L$. Its mass per unit length $\lambda$ varies as $\lambda = Cx$, where $C$ is a constant. (Thus an element of the rod has mass $dm = \lambda \, dx$.) (a) What is the total mass of the rod? (b) Find the gravitational field due to the rod on the $x$ axis at a point $x_0$, where $x_0 > L$.

**64** ••• A uniform rod of mass $M$ and length $L$ lies along the $x$ axis with its center at the origin. Consider an element of length $dx$ at point $x$, where $-\frac{1}{2}L < x < \frac{1}{2}L$. (a) Show that this element produces a gravitational field at a point $x_0$ on the $x$ axis ($x_0 > \frac{1}{2}L$) given by

$$dg_x = -\frac{GM}{L(x_0 - x)^2} \, dx$$

(b) Integrate this result over the length of the rod to find the total gravitational field at the point $x_0$ due to the rod. (c) What is the force on an object of mass $m_0$ at $x_0$? (d) Show that for $x_0 \gg L$, the field is approximately equal to that of a point particle of mass $M$ at $x = 0$.

## $\vec{g}$ due to Spherical Objects

**65** • **iSOLVE** A thin spherical shell has a radius of 2 m and a mass of 300 kg. What is the gravitational field at the following distances from the center of the shell: (a) 0.5 m, (b) 1.9 m, (c) 2.5 m?

**66** • **iSOLVE** A thin spherical shell has a radius of 2 m and a mass of 300 kg, and its center is located at the origin of a coordinate system. Another thin spherical shell with a radius of 1 m and mass 150 kg is inside the larger shell with its center at 0.6 m on the $x$ axis. What is the gravitational force of attraction between the two shells?

**67** • **SSM** Two solid spheres, $S_1$ and $S_2$, have equal radii $R$ and equal masses $M$. The density of sphere $S_1$ is constant, whereas that of sphere $S_2$ depends on the radial distance according to $\rho(r) = C/r$. If the acceleration of gravity at the surface of sphere $S_1$ is $g_1$, what is the acceleration of gravity at the surface of sphere $S_2$?

**68** •• Two homogeneous solid spheres, $S_1$ and $S_2$, have equal masses but different radii, $R_1$ and $R_2$. If the acceleration of gravity on the surface of sphere $S_1$ is $g_1$, what is the acceleration of gravity on the surface of sphere $S_2$?

**69** •• Two concentric uniform thin spherical shells have masses $M_1$ and $M_2$ and radii $a$ and $2a$ as in Figure 11-24. What is the magnitude of the gravitational force on a point particle of mass $m$ located (a) a distance $3a$ from the center of the shells? (b) a distance $1.9a$ from the center of the shells? (c) a distance $0.9a$ from the center of the shells?

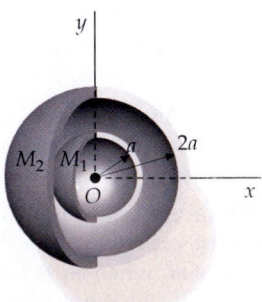

**FIGURE 11-24** Problems 69, 70

**70** •• The inner spherical shell in Problem 69 is shifted so that its center is now on the $x$ axis at $x = 0.8a$. What is the magnitude of the gravitational force on a particle of point mass $m$ located on the $x$ axis at (a) $x = 3a$, (b) $x = 1.9a$, (c) $x = 0.9a$?

## $\vec{g}$ Inside Solid Spheres

**71** •• **SSM** Is your "weight" (as measured on a spring scale) at the bottom of a deep mine shaft greater than or less than your weight at the surface of the earth? Model the earth as a homogeneous sphere. Consider the effects in (*a*) and (*b*).
(*a*) Show that the force of gravity on you from a uniform-density, perfectly spherical planet is proportional to your distance from the center of the planet.
(*b*) Show that your effective "weight" increases linearly due to the effects of rotation as you approach the center. (Consider a mine shaft located on the equator.)
(*c*) Which effect is more important for the earth? Use $M = 5.98 \times 10^{24}$ kg, $R = 6370$ km, and $T = 24$ h.

**72** •• Suppose the earth were a nonrotating uniform sphere. If there were an elevator shaft going 15,000 m into the earth, what would be the decrease in the weight of a student who weighs 800 N at the surface of the earth as he is lowered from the surface to the bottom of the shaft?

**73** •• A solid sphere of radius $R$ has its center at the origin. It has a uniform mass density $\rho_0$, except that there is a spherical cavity in it of radius $r = \frac{1}{2}R$ centered at $x = \frac{1}{2}R$ as in Figure 11-25. Find the gravitational field at points on the $x$ axis for $|x| > R$. (*Hint: The cavity may be thought of as a sphere of mass $m = (4/3)\pi r^3 \rho_0$ plus a sphere of mass $-m$.*)

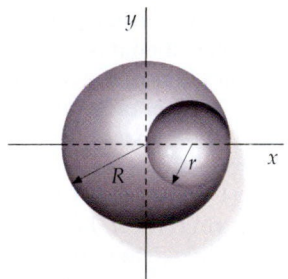

**FIGURE 11-25** Problems 73, 74

**74** ••• For the sphere with the cavity in Problem 73, show that the gravitational field inside the cavity is uniform and find its magnitude and direction.

**75** ••• A straight, smooth tunnel is dug through a spherical planet whose mass density $\rho_0$ is constant. The tunnel passes through the center of the planet and is perpendicular to the planet's axis of rotation, which is fixed in space. The planet rotates with a constant angular velocity $\omega$ so that objects in the tunnel have no apparent weight. Find $\omega$.

**76** ••• **ISOLVE** The density of a sphere is given by $\rho(r) = C/r$. The sphere has a radius of 5 m and a mass of 1011 kg. (*a*) Determine the constant $C$. (*b*) Obtain expressions for the gravitational field for (1) $r > 5$ m and (2) $r < 5$ m.

**77** ••• **SSM** **ISOLVE** A small diameter hole is drilled into the sphere of Problem 76 toward the center of the sphere to a depth of 2 m below the sphere's surface. A small mass is dropped from the surface into the hole. Determine the speed of the small mass as it strikes the bottom of the hole.

**78** ••• **ISOLVE** ✓ The crust of the earth is about 40 km thick and has a density of about 3000 kg/m³. A spherical de-

posit of heavy metals with a density of 8000 kg/m³ and radius of 1000 m is centered 2000 m below the surface. Find $\Delta g/g$ at the surface directly above this deposit, where $\Delta g$ is the increase in the gravitational field due to the deposit.

**79** ••• **SSM** Two identical spherical cavities are made in a lead sphere of radius $R$. The cavities have a radius $R/2$. They touch the outside surface of the sphere and its center as in Figure 11-26. The mass of a solid uniform lead sphere is $M$. (*a*) Find the force of attraction of a point particle of mass $m$ at the position shown in the figure to the cavitated lead sphere. (*b*) What is the attractive force on the point particle if it is located on the $x$ axis at $x = R$?

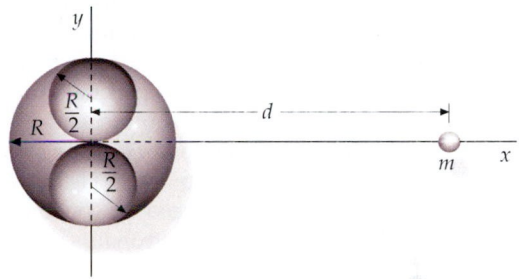

**FIGURE 11-26** Problem 79

**80** •• A globular cluster is a roughly spherical collection of up to millions of stars bound together by the force of gravity. Astronomers can measure the velocities of stars in the cluster to study its composition and to get an idea of the mass distribution within the cluster. Assuming that all of the stars have about the same mass and are distributed uniformly within the cluster, show that the mean speed of a star in a circular orbit around the center of the cluster should increase linearly with its distance from the center.

## General Problems

**81** • **SSM** The mean distance of Pluto from the sun is 39.5 AU. Find the period of Pluto.

**82** • Calculate the mass of the earth using the known values of $G$, $g$, and $R_E$.

**83** •• The force exerted by the earth on a particle of mass $m$ at a distance $r$ ($r > R_E$) from the center of the earth has the magnitude $mgR_E^2/r^2$, where $g = GM_E/R_E^2$. (*a*) Calculate the work you must do to move the particle from distance $r_1$ to distance $r_2$. (*b*) Show that when $r_1 = R_E$ and $r_2 = R_E + h$, the result can be written $W = mgR_E^2[(1/R_E) - 1/(R_E + h)]$. (*c*) Show that when $h \ll R_E$, the work is given approximately by $W = mgh$.

**84** •• A uniform sphere of radius 100 m and density 2000 kg/m³ is in free space far from other massive objects. (*a*) Find the gravitational field outside of the sphere as a function of $r$. (*b*) Find the gravitational field inside the sphere as a function of $r$.

**85** •• Jupiter has a mass 320 times that of Earth and a volume 1320 times that of Earth. A "day" on Jupiter is 9 h 50 min long. Find the height $h$ above Jupiter at which a satellite must be orbiting to have a period equal to one Jovian day.

**86** •• The average density of the moon is $\rho = 3340 \text{ kg/m}^3$. Find the minimum possible period $T$ of a spacecraft orbiting the moon.

**87** •• A satellite is circling around the moon (radius 1700 km) close to the surface at a speed $v$. A projectile is launched from the moon vertically up at the same initial speed $v$. How high will it rise?

**88** •• **SSM** In the novel "A Voyage to the Moon" by Jules Verne, astronauts were launched by a giant cannon from the earth to the moon. (a) If we estimate the velocity needed to reach the moon as the earth's escape velocity, and the length of the cannon as 900 ft (as stated in the book), what is the probability that the astronauts would have survived "liftoff"? (b) In the book, the astronauts in their ship feel the effects of the earth's gravity until they reach the balance point where the moon's gravitational pull on the ship equals that of the earth. At this point, everything flips around and what was the ceiling in the ship becomes the floor. How far away from the center of the earth is this balance point? (c) Is this description of what the astronauts experienced reasonable? Is this what actually happened to the Apollo astronauts on their visit to the moon?

**89** •• In a binary star system, two stars follow circular orbits about their common center of mass. If the stars have masses $m_1$ and $m_2$ and are separated by a distance $r$, show that the period of rotation is related to $r$ by $T^2 = 4\pi r^3/[G(m_1 + m_2)]$.

**90** •• Two particles of masses $m_1$ and $m_2$ are released from rest at a large separation distance. Find their speeds $v_1$ and $v_2$ when their separation distance is $r$. The initial separation distance is given as large, but large is a relative term. Relative to what distance is it large?

**91** •• **SSM** Four identical planets are arranged in a square as shown in Figure 11-27. If the mass of each planet is $M$ and the edge length of the square is $a$, what must be their speed if they are to orbit their common center under the influence of their mutual attraction?

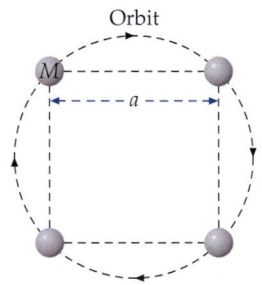

**FIGURE 11-27** Problem 91

**92** •• A hole is drilled from the surface of the earth to its center as in Figure 11-28. Ignore the earth's rotation and air resistance, and model the earth as a uniform sphere. (a) How much work is required to lift a particle of mass $m$ from the center of the earth to the earth's surface? (b) If the particle is dropped from rest at the surface of the earth, what is its speed when it reaches the center of the earth? (c) What is the escape speed for a particle projected from the center of the earth? Express your answers in terms of $m$, $g$, and $R_E$.

**FIGURE 11-28** Problem 92

**93** •• A thick spherical shell of mass $M$ and uniform density has an inner radius $R_1$ and an outer radius $R_2$. Find the gravitational field $g_r$ as a function of $r$ for $0 < r < \infty$. Sketch a graph of $g_r$ versus $r$.

**94** •• (a) Sketch a plot of the gravitational field $g_x$ versus $x$ due to a uniform ring of mass $M$ and radius $R$ whose axis is the $x$ axis. (b) At what points is the magnitude of $g_x$ a maximum?

**95** ••• Find the magnitude of the gravitational field a distance $r$ from an infinitely long wire whose mass per unit length is $\lambda$.

**96** ••• One big question in planetary science is whether each of the rings of Saturn is solid or composed of many smaller satellites. There is a simple observation that can now be made to resolve this issue. Measure the velocity of the inner and outer portion of the ring: if the inner portion of the ring moves more slowly than the outer portion, then the ring is solid; if the opposite is true, then it is composed of many separate chunks. (a) If the thickness of the ring is $r$, the average distance of the ring from the center of Saturn is $R$, and the average velocity of the ring is $v$, show that $v_{out} - v_{in} \approx rv/R$ if the ring is solid. Here, $v_{out}$ is the speed of the outermost portion of the ring, $v_{in}$ is the speed of the innermost portion, and $v$ is the average velocity of the ring. (b) If, however, the ring is composed of many small chunks, show that $v_{out} - v_{in} \approx -\frac{1}{2}rv/R$. (Assume that $r \ll R$.)

**97** ••• In this problem you are to find the gravitational potential energy of the rod in Example 11-8 and a point mass $m_0$ that is on the $x$ axis at $x_0$. (a) Show that the potential energy shared by an element of the rod of mass $dm$ (shown in Figure 11-13) and a point particle of mass $m_0$ located on the $x$ axis at $x_0 \geq \frac{1}{2}L$ is given by

$$dU = -\frac{Gm_0 \, dm}{x_0 - x} = \frac{GMm_0}{L(x_0 - x)} \, dx$$

where $U = 0$ at $x_0 = \infty$. (b) Integrate your result for Part (a) over the length of the rod to find the total potential energy for the system. Write your result as a general function $U(x)$ by setting $x_0$ equal to a general point $x$. (c) Compute the force on $m_0$ at a general point $x$ from $F_x = -dU/dx$ and compare your result with $m_0 g$, where $g$ is the field at $x_0$ calculated in Example 11-8.

**98** ••• **SSM** A uniform sphere of mass $M$ is located near a thin, uniform rod of mass $m$ and length $L$ as in Figure 11-29. Find the gravitational force of attraction exerted by the sphere on the rod.

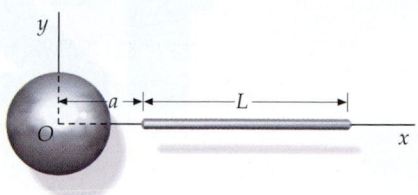

**FIGURE 11-29** Problem 98

**99** ••• A uniform rod of mass $M = 20$ kg and length $L = 5$ m is bent into a semicircle. What is the gravitational force exerted by the rod on a point mass $m = 0.1$ kg located at the center of curvature of the circular arc?

**100** ••• **SSM** Both the sun and the moon exert gravitational forces on the oceans of the earth, causing tides. (a) Show that the ratio of the force exerted on a point particle on earth by the sun to that exerted by the moon is $M_s r_m^2/M_m r_s^2$, where $M_s$ and $M_m$ are the masses of the sun and moon and $r_s$ and $r_m$ are the distances from the earth to the sun and to the moon. Evaluate this ratio. (b) Even though the sun exerts a much greater force on the oceans than does the moon, the moon has a greater ef-

fect on the tides because it is the difference in the force from one side of the earth to the other that is important. Differentiate the expression $F = Gm_1m_2/r^2$ to calculate the change in $F$ due to a small change in $r$. Show that $dF/F = (-2\ dr)/r$. (c) During one full day, the rotation of the earth can cause the distance from the sun or moon to an ocean to change by, at most, the diameter of the earth. Show that for a small change in distance, the change in the force exerted by the sun is related to the change in the force exerted by the moon by $\Delta F_s/\Delta F_m \approx (M_s r_m^3)/(M_m r_s^3)$. Calculate this ratio.

**101** •• United Federation Spaceship *Excelsior* is dropping two robot probes to the surface of a neutron star for exploration. The mass of the star is the same as that of the sun, but the star's diameter is only 10 km. The robot probes are linked together by a 1-m-long steel cord, and are dropped vertically (that is, one always above the other). (a) Explain why there seems to be a "force" trying to pull the robots apart. (See Problem 99 for a hint.) (b) How close will the robots be to the surface of the star before the cord breaks? Assume that the cord has a breaking tension of 25,000 N and that the robots each have a mass of 1 kg.

# Static Equilibrium and Elasticity

THE PLANK IN THIS PHOTO IS IN STATIC EQUILIBRIUM. THE OVERHANG IS ONE-SIXTH THE LENGTH OF THE PLANK, AND THE PLANK REMAINS BALANCED WHEN A STUDENT STANDS WITH HIS CENTER OF MASS OVER THE END OF THE PLANK.

**?** **What is the maximum ratio of the mass of the student to the mass of the plank if the plank remains balanced when the student stands with his center of mass over the very end of the plank? (See Example 12-1.)**

12-1   Conditions for Equilibrium

12-2   The Center of Gravity

12-3   Some Examples of Static Equilibrium

12-4   Couples

12-5   Static Equilibrium in an Accelerated Frame

12-6   Stability of Rotational Equilibrium

12-7   Indeterminate Problems

12-8   Stress and Strain

I f an object is stationary and remains stationary, it is said to be in static equilibrium. Being able to determine the forces acting on an object in static equilibrium has many important applications. For example, the forces exerted by the cables of a suspension bridge must be known so that the cables can be designed to be strong enough to support the bridge. Similarly, cranes must be designed so that they do not topple over when lifting a weight.

The forces exerted by the cables and beams in a structure are called elastic forces. They are the result of slight deformations—the stretching or compression of solid objects under stress from bearing loads.

➤ **In this chapter, we study the equilibrium of rigid bodies and then briefly consider the deformations and elastic forces that arise when real solids are under stress.**

## 12-1  Conditions for Equilibrium

A necessary condition for a particle at rest to remain at rest is that the net force acting on it remain zero. Similarly, a necessary condition for the center of mass of a rigid object to remain at rest is that the net force acting on the object remain zero. A rigid object can rotate, even when its center of mass is at rest, but then the object is not in static equilibrium. Therefore, a second necessary condition for a rigid object to remain in static equilibrium is that the net torque acting on it about *any* point must remain zero. This gives us the option to choose the point $P$ when calculating torques, which greatly simplifies the solution of most static problems.

The two necessary conditions for a rigid body to be in static equilibrium are as follows:

> 1. The net external force acting on the body must remain zero:
>
> $$\sum \vec{F} = 0 \qquad\qquad 12\text{-}1$$
>
> 2. The net external torque about *any* point must remain zero:
>
> $$\sum \vec{\tau} = 0 \qquad\qquad 12\text{-}2$$

CONDITIONS FOR EQUILIBRIUM

## 12-2  The Center of Gravity

Figure 12-1a shows a rigid object in static equilibrium and a point $O$. We consider the object to be composed of many small particles. The weight of the $i$th small particle is $\vec{w}_i$, and the total weight of the object is $\vec{W} = \Sigma \vec{w}_i$. If $\vec{r}_i$ is the position vector of the $i$th particle relative to $O$, then $\vec{\tau}_i = \vec{r}_i \times \vec{w}_i$, where $\vec{\tau}_i$ the torque due to $\vec{w}_i$ about $O$. The net gravitational torque about $O$ is then $\vec{\tau}_{net} = \Sigma (\vec{r}_i \times \vec{w}_i)$. Conveniently, the net torque due to gravity about any point can be calculated as if the entire weight $\vec{W}$ were applied at a single point, the **center of gravity** (see Figure 12-1b). That is,

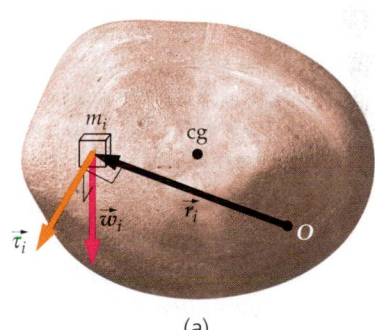
(a)

$$\vec{\tau}_{net} = \vec{r}_{cg} \times \vec{W} \qquad\qquad 12\text{-}3$$

CENTER OF GRAVITY DEFINED

where $\vec{r}_{cg}$ is the position vector of the center of gravity relative to $O$.

If the gravitational field $\vec{g}$ is uniform over the object (as is nearly always the case for objects of less than astronomical size), we can write $\vec{w}_i = m_i \vec{g}$. Summing both sides of this gives $\vec{W} = M\vec{g}$, where $M = \Sigma m_i$ is the mass of the object. The net torque is the sum of the individual torques. That is,

$$\vec{\tau}_{net} = \sum_i (\vec{r}_i \times \vec{w}_i) = \sum_i (\vec{r}_i \times m_i \vec{g}) = \sum_i (m_i \vec{r}_i \times \vec{g})$$

Factoring $\vec{g}$ from the term on the right gives

$$\vec{\tau}_{net} = \left( \sum_i m_i \vec{r}_i \right) \times \vec{g}$$

and substituting $M\vec{r}_{cm}$ for $\Sigma m_i \vec{r}_i$ using the definition of center of mass ($M\vec{r}_{cm} = \Sigma m_i \vec{r}_i$), we obtain

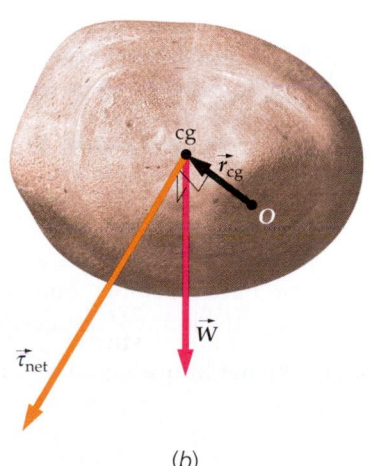
(b)

**FIGURE 12-1**

$$\vec{\tau}_{net} = M\vec{r}_{cm} \times \vec{g} = \vec{r}_{cm} \times M\vec{g} = \vec{r}_{cm} \times \vec{W} \qquad\qquad 12\text{-}4$$

Equations 12-3 and 12-4 are valid for any choice of the point $O$ only if $\vec{r}_{cg} = \vec{r}_{cm}$. That is, the center of gravity and the center of mass coincide if the object is in a uniform gravitational field.

If $O$ is directly above the center of gravity, then $\vec{r}_{cg}$ and $\vec{W}$ are both in the same direction (downward), so $\vec{\tau}_{net} = \vec{r}_{cg} \times \vec{W} = 0$. For example, when a mobile is suspended with its center of gravity directly below its suspension point, the net torque on the mobile about the suspension point is zero, so it is in static equilibrium.

## 12-3 Some Examples of Static Equilibrium

For most examples and problems in this chapter, all the forces are perpendicular to the $z$ axis. For such problems it is best to calculate torques about an axis parallel to the $z$ axis. Out of the page is frequently chosen as the positive $z$ direction. This is equivalent to choosing counterclockwise as positive and clockwise as negative.

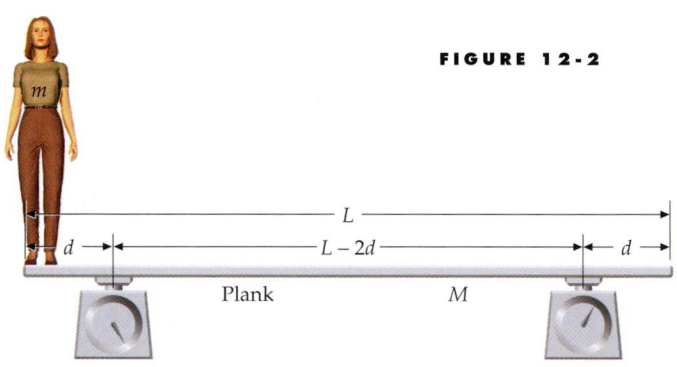

FIGURE 12-2

*WALKING THE PLANK*        **EXAMPLE 12-1**

A uniform plank of length $L = 3$ m and mass $M = 35$ kg is supported by scales a distance $d = 0.5$ m from the ends of the board, as shown in Figure 12-2. (*a*) Find the reading on the scales when Mary, whose mass $m = 45$ kg, stands on the left end of the plank. (*b*) Sergio climbs onto the plank and walks toward Mary, who jumps to the floor when the plank starts to tip. Sergio keeps walking all the way to the left end of the plank, and when he gets there the scale supporting the right end of the plank reads zero. What is Sergio's mass?

**PICTURE THE PROBLEM** The readings on the scales are the magnitudes of the forces they exert on the boards. To find these magnitudes we apply the two conditions for equilibrium.

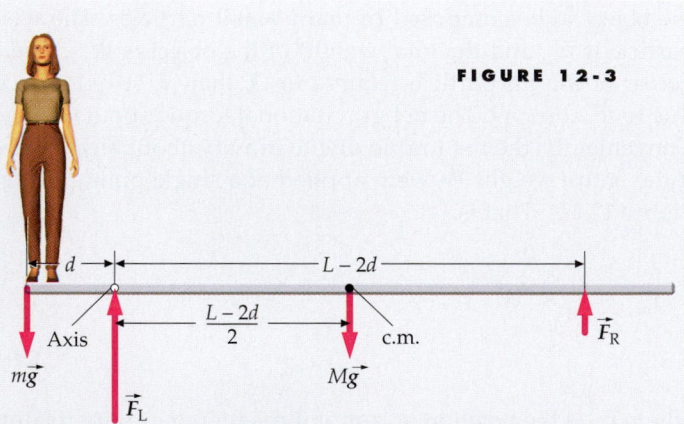

FIGURE 12-3

(*a*) 1. Draw a free-body diagram of the system consisting of Mary and the plank (Figure 12-3). Forces $\vec{F}_L$ and $\vec{F}_R$ are the forces exerted by the left and right scales:

2. Set the net force equal to zero, taking upward as positive:

$$\sum F_y = 0$$

$$F_L + F_R - Mg - mg = 0$$

3. Calculate the net torque about the axis directed out of the page (making counterclockwise positive) and through the point of application of $\vec{F}_L$:

$$\sum \tau = F_R(L - 2d) - Mg\frac{L - 2d}{2} + mgd$$

4. Set the net torque equal to zero and solve for $F_R$:

$$0 = F_R(L - 2d) - Mg\frac{L - 2d}{2} + mgd$$

so

$$F_R = \left(\frac{1}{2}M - \frac{d}{L - 2d}m\right)g$$

5. Substitute this result for $F_R$ into step 2 and solve for $F_L$:

$$F_L = Mg + mg - F_R = \left(\frac{1}{2}M + \frac{L-d}{L-2d}m\right)g$$

6. Substitute numerical values to obtain numerical values for the forces:

$$F_R = \left(\frac{1}{2}(35 \text{ kg}) - \frac{0.5 \text{ m}}{2 \text{ m}}45 \text{ kg}\right)(9.81 \text{ N/kg})$$

$$= \boxed{61.3 \text{ N}}$$

$$F_L = \left(\frac{1}{2}(35 \text{ kg}) + \frac{2.5 \text{ m}}{2 \text{ m}}45 \text{ kg}\right)(9.81 \text{ N/kg})$$

$$= \boxed{723.5 \text{ N}}$$

(b) Using the Part (a) step 4 result, set $F_R = 0$ and solve for $m$:

$$0 = \left(\frac{1}{2}M - \frac{d}{L-2d}m\right)g$$

so

$$m = \frac{L-2d}{2d}M = \frac{2 \text{ m}}{1 \text{ m}}(35 \text{ kg}) = \boxed{70 \text{ kg}}$$

**PLAUSIBILITY CHECK**  The sum of the two forces in the Part (a) step 6 results should equal Mary's weight plus the weight of the plank. The total weight is $(M + m)g = (35 \text{ kg} + 45 \text{ kg})(9.81 \text{ N/kg}) = 785 \text{ N}$. Also, $F_L + F_R = 723.5 \text{ N} + 61.3 \text{ N} = 785 \text{ N}$. Also, $F_R < F_L$ as one would expect.

**REMARKS**  Sergio is 0.5 m from the axis and the center of mass of the plank is 1 m from the axis when the system is balanced with $F_R = 0$. Thus, Sergio's mass is twice the mass of the plank.

Example 12-1 can be solved using an axis through the center of the plank, but in this case both $F_L$ and $F_R$ appear in the torque equation, hence the algebra is a bit more complex. In general, a statics problem can be simplified by computing the torques about an axis through the line of action of one of the unknown forces, as when we chose the axis through the point of application of force $F_L$ in the example.

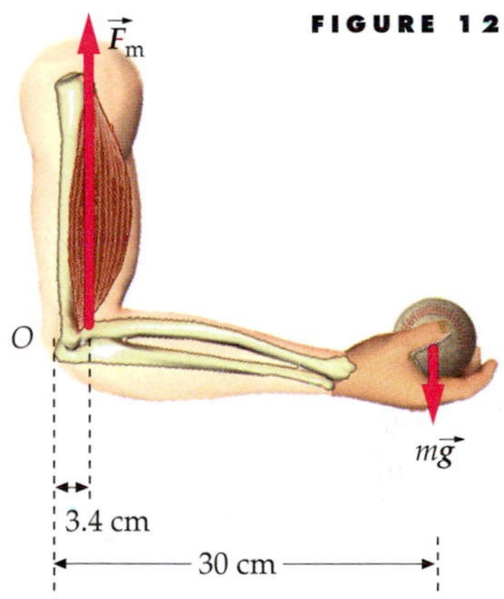

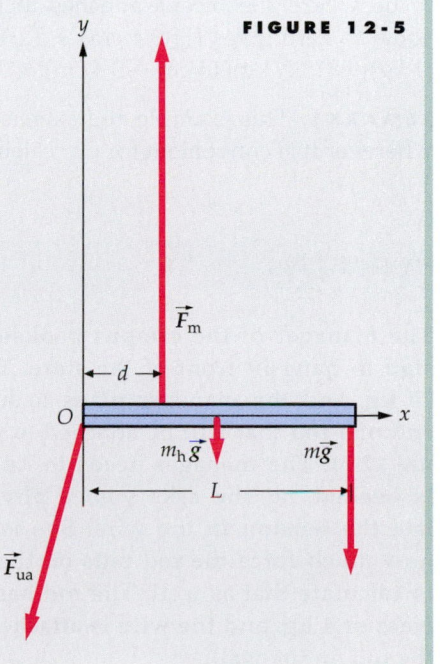

**FIGURE 12-4**

3.4 cm

30 cm

> Obtain a simple solution by choosing an axis through the point of application of the force you have the least information about to calculate the torques.

PROBLEM SOLVING GUIDELINE

*FORCE ON AN ELBOW*    **EXAMPLE  12-2**

You hold a 6-kg weight in your hand with your forearm making a 90° angle with your upper arm, as shown in Figure 12-4. Your biceps muscle exerts an upward force $\vec{F}_m$ that acts 3.4 cm from the pivot point $O$ at the elbow joint. Model the forearm and hand as a 30-cm-long uniform rod with a mass of 1 kg. (a) Find the magnitude of $\vec{F}_m$ if the distance from the weight to the pivot point (elbow joint) is 30 cm and (b) find the magnitude and direction of the force exerted on the elbow joint by the upper arm.

**FIGURE 12-5**

**PICTURE THE PROBLEM**  To find the two forces, apply the two conditions for static equilibrium ($\Sigma F = 0$ and $\Sigma\tau = 0$) to the forearm.

(a) 1. Draw a free-body diagram of the forearm (Figure 12-5). Model the forearm as a horizontal rod.

2. The force we know least about is the force of the upper arm on the elbow joint $\vec{F}_{ua}$ (we know neither its magnitude nor its direction). Apply $\Sigma \tau = 0$ about an axis directed out of the page and through the point of application of $\vec{F}_{ua}$:

$$F_{ua}(0) - m_h g \frac{L}{2} + F_m d - mgL = 0$$

so

$$F_m = \left(\frac{1}{2}m_h + m\right)g\frac{L}{d}$$

$$= \left(\frac{1}{2}(1\text{ kg}) + 6\text{ kg}\right)(9.81\text{ N/kg})\frac{30\text{ cm}}{3.4\text{ cm}}$$

$$= \boxed{563\text{ N}}$$

(b) Apply $\Sigma F_x = 0$ and $\Sigma F_y = 0$ to obtain $\vec{F}_{ua}$:

$$F_{ua,x} + 0 + 0 + 0 = 0$$

and

$$F_{ua,y} + F_m - m_h g - mg = 0$$

so

$$F_{ua,x} = 0$$

and

$$F_{ua,y} = (m + m_h)g - F_m$$

$$= (7\text{ kg})(9.81\text{ N/kg}) - 563\text{ N}$$

$$= -494\text{ N}$$

Therefore

$$\vec{F}_{ua} = \boxed{494\text{ N, down}}$$

**REMARKS** The force that must be exerted by the muscle is 9.6 times the weight of the object! In addition, as the muscle pulls upward, the upper arm must push downward to keep the forearm in equilibrium. The force exerted by the upper arm is 8.4 times greater than the object's weight.

**EXERCISE** Show that $F_{ua}$ can be found in one step by choosing the pivot point to be where the biceps attaches to the forearm. (*Answer* Setting net torque equal to zero gives $F_{ua}(3.4\text{ cm}) + F_m(0) - (6\text{ kg})(9.81\text{ N/kg})(30\text{ cm} - 3.4\text{ cm}) - (1\text{ kg})(9.81\text{ N/kg})(15\text{ cm} - 3.4\text{ cm}) = 0$. This yields $F_{ua} = 494\text{ N}$.)

**REMARKS** This example and exercise show that we can choose the pivot point wherever it is convenient for our calculation.

---

*HANGING A SIGN*                 **EXAMPLE 12-3**    **Try It Yourself**

The manager of the campus bookstore has ordered a new sign to hang in front of the store. The sign has a mass of 20 kg, and the manager plans to hang the sign from the end of a rod that will be attached to the wall by a wire (Figure 12-6). The manager needs to know how strong a wire is needed, so she asks you, a physics student, to calculate the tension in the wire. She is also concerned about how much force the rod puts on the wall, so she asks you to calculate that as well. The rod has a length of 2 m and a mass of 4 kg, and the wire is attached to a point 1 m above the rod on the wall.

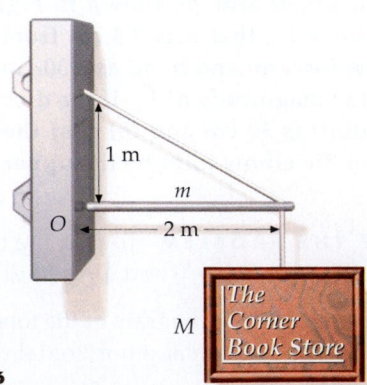

**FIGURE 12-6**

**PICTURE THE PROBLEM** There are three conditions for the rod to be in equilibrium: $\Sigma F_x = 0$, $\Sigma F_y = 0$, and $\Sigma \tau = 0$, and we have three unknowns: $T$ and the components $F_x$ and $F_y$ of the force exerted by the wall on the rod. The force exerted by the rod on the wall is equal but opposite to the force exerted by the wall on the rod.

**Cover the column to the right and try these on your own before looking at the answers.**

| Steps | Answers |
|---|---|
| 1. Draw a free-body diagram for the rod (Figure 12-7). | |

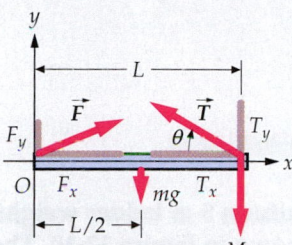

2. Set $\Sigma \tau = 0$ about point $O$.

$$TL \sin \theta - MgL - mg\frac{L}{2} = 0$$

3. Use trigonometry to solve for $\theta$.

$$\theta = \tan^{-1}\tfrac{1}{2} = 26.6°$$

4. Solve the step 2 result for $T$.

$$T = \boxed{483 \text{ N}}$$

5. Set $\Sigma F_x = 0$ and $\Sigma F_y = 0$ and, using your values for $T$ and $\theta$, solve for $F_x$ and $F_y$.

$$F_x + T_x = 0$$

$$F_y + T_y - Mg - mg = 0$$

so

$$F_x = 432 \text{ N}, \qquad F_y = 19.2 \text{ N}$$

**FIGURE 12-7**

6. Solve for the force $\vec{F}'$ exerted by the rod on the wall. The force exerted by the rod on the wall and that by the wall on the rod constitute an action–reaction pair.

$$\vec{F}' = -\vec{F} = \boxed{-432 \text{ N}\hat{\imath} - 19.2 \text{ N}\hat{\jmath}}$$

---

*RAISING A WHEEL*         **EXAMPLE 12-4**   **Try It Yourself**

A wheel of mass $M$ and radius $R$ (Figure 12-8) rests on a horizontal surface against a step of height $h$ ($h < R$). The wheel is to be raised over the step by a horizontal force $\vec{F}$ applied to the axle of the wheel as shown. Find the minimum force $F_{min}$ necessary to raise the wheel over the step.

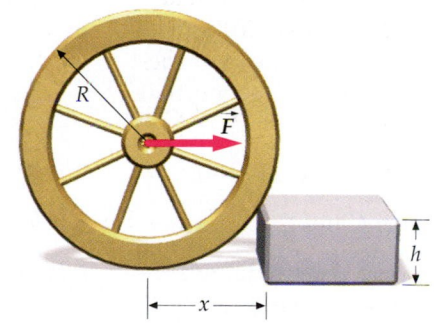

**PICTURE THE PROBLEM** If the magnitude of $F$ is less than $F_{min}$, the surface at the bottom of the wheel exerts an upward normal force on the wheel. If $F$ is increased, this normal force decreases. Apply the conditions for static equilibrium to find the value of $F$ that will hold the wheel in place when the normal force is zero.

**FIGURE 12-8**

**Cover the column to the right and try these on your own before looking at the answers.**

| Steps | Answers |
|---|---|
| 1. Draw a free-body diagram of the wheel (Figure 12-9). | |

2. Apply $\Sigma \tau = 0$ to the wheel. Both the direction and the magnitude of $\vec{F}'$ are unknown, so follow the guidelines and calculate torques about an axis through its point of application. Obtain expressions for the lever arms from the free-body diagram and solve for $F_{min}$.

$$F_{min}(R - h) - Mgx = 0$$

so

$$F_{min} = \frac{Mgx}{R - h}$$

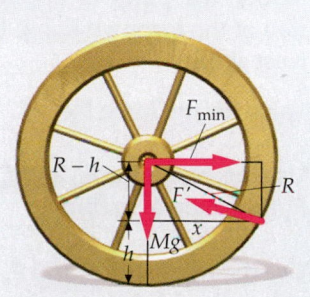

**FIGURE 12-9**

3. Use the Pythagorean theorem to express $x$ in terms of $h$ and $R$.

$$x = \sqrt{h(2R - h)}$$

4. Set the magnitudes of the torques equal to each other and solve for $F$.

$$F_{min} = \left| \frac{Mg\sqrt{h(2R - h)}}{R - h} \right|$$

**REMARKS** Applying $\Sigma\tau = 0$ about the axis through the center of the wheel shows that $\vec{F}'$ is directed toward the wheel's center; otherwise there would be a nonzero net torque.

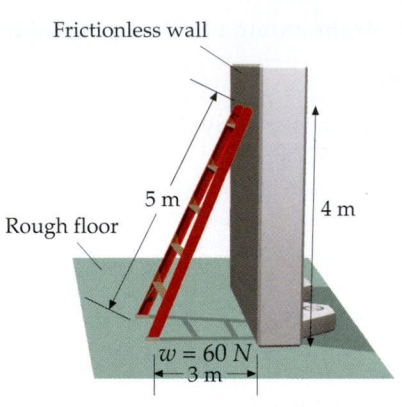

Frictionless wall

Rough floor

5 m   4 m

$w = 60$ N
3 m

**FIGURE 12-10**

*A LEANING LADDER*                    **E X A M P L E    1 2 - 5**

**A uniform 5-m ladder weighing 60 N leans against a frictionless vertical wall, as shown in Figure 12-10. The foot of the ladder is 3 m from the wall. What is the minimum coefficient of static friction necessary between the ladder and the floor if the ladder is not to slip?**

**PICTURE THE PROBLEM** There are three conditions for the ladder to be in equilibrium: $\Sigma F_x = 0$, $\Sigma F_y = 0$, and $\Sigma\tau = 0$. Apply these along with $f_s \leq \mu_s F_n$ to solve for $\mu_s$.

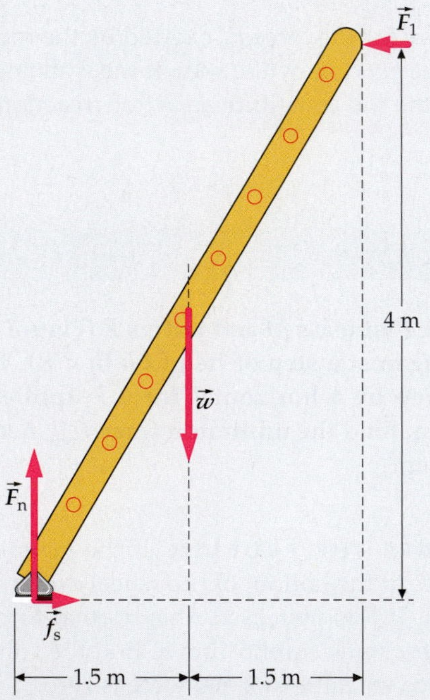

1. Draw a free-body diagram of the ladder as shown in Figure 12-11. The forces acting on the ladder are the force due to gravity $\vec{w}$, the force $\vec{F}_1$ exerted by the wall (since the wall is frictionless, it exerts only a normal force), and the force exerted by the floor, which consists of a normal component $F_n$ and a static frictional component $f_s$. The three conditions for static equilibrium determine $F_1, f_s$, and $F_n$.

2. The minimum coefficient of static friction relates the frictional force $f_s$ and normal force $F_n$:

$$\mu_s \geq \frac{f_s}{F_n} \quad \text{so} \quad \mu_{s,min} = \frac{f_s}{F_n}$$

3. Set $\Sigma F_x = 0$ and $\Sigma F_y = 0$:

$$f_s - F_1 = 0 \quad \text{and} \quad F_n - w = 0$$

4. Solve for $f_s$ and $F_n$:

$$f_s = F_1 \quad \text{and} \quad F_n = w = 60 \text{ N}$$

5. Following the guideline, we set $\Sigma\tau = 0$ about an axis directed out of the page and through the foot of the ladder, the point of application of the force we know the least about:

$$F_1(4 \text{ m}) - w(1.5 \text{ m}) = 0$$

$\vec{F}_1$

4 m

$\vec{w}$

$\vec{F}_n$

$\vec{f}_s$

1.5 m   1.5 m

**FIGURE 12-11**

6. Solve for the force $F_1$:

$$F_1 = \frac{w(15 \text{ m})}{4 \text{ m}} = \frac{(60 \text{ N})(1.5 \text{ m})}{4 \text{ m}} = 22.5 \text{ N}$$

7. Use this result for $F_1$, and $f_s = F_1$ from step 4, to find $f_s$:

$$f_s = F_1 = 22.5 \text{ N}$$

8. Use the results for $f_s$ and $F_n$ to obtain the minimum value of $\mu_s$ from step 2:

$$\mu_{s,min} = \frac{f_s}{F_n} = \frac{22.5 \text{ N}}{60 \text{ N}} = \boxed{0.375}$$

**REMARKS** There is another way to look at this problem. In the free-body diagram for the ladder shown in Figure 12-12, the lines of action of the weight $\vec{w}$ and the force $\vec{F}_1$ exerted by the wall intersect at point $P$. The line of action of the resultant force exerted by the ground, $\vec{f}_s + \vec{F}_n$, must also go through point $P$ or there would be an unbalanced torque about this point. The tangent of $\theta'$ equals 4 m/1.5 m = 2.67 = $F_n/f_s$. Whenever an object is in static equilibrium under the influence of three nonparallel forces, the lines of action of the forces must intersect at one point.

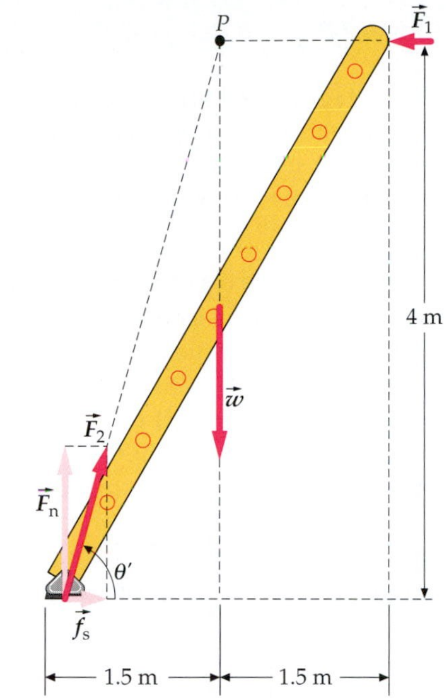

## 12-4 Couples

The forces $\vec{F}_n$ and $\vec{w}$ in Figure 12-11 of Example 12-5 are equal and opposite. Such a pair of forces, called a couple, tends to produce an angular acceleration, but its net force is zero. The forces $\vec{f}_s$ and $\vec{F}_1$ in those figures also constitute a couple. Figure 12-13 shows a couple consisting of forces $\vec{F}_1$ and $\vec{F}_2$ a distance $D$ apart. The torque produced by this couple about an arbitrary point $O$ is

$$\vec{\tau} = \vec{r}_1 \times \vec{F}_1 + \vec{r}_2 \times \vec{F}_2 = \vec{r}_1 \times \vec{F}_1 + \vec{r}_2 \times (-\vec{F}_1) = (\vec{r}_1 - \vec{r}_2) \times \vec{F}_1 \qquad 12\text{-}5$$

This result does not depend on the choice of the point $O$. The magnitude of this torque is

$$\tau = FD \qquad 12\text{-}6$$

where $F$ is the magnitude of either force and $D$ is the distance between their lines of action.

> The torque produced by a couple is the same about all points in space.

**EXERCISE** Show that the magnitude of the torque exerted by the couple in Figure 12-13 is given by $FD$, where $D$ is the distance between the lines of action of the two forces and $F$ is the magnitude of either force. (*Answer* The angle $\beta$ between $\vec{r}_1 - \vec{r}_2$ and $\vec{F}_1$ is $90° - \phi$, so $|\vec{\tau}| = |(\vec{r}_1 - \vec{r}_2) \times \vec{F}_1| = |\vec{r}_1 - \vec{r}_2| F \sin(90° - \phi) = F|\vec{r}_1 - \vec{r}_2| \cos \phi = FD$.)

**FIGURE 12-12**

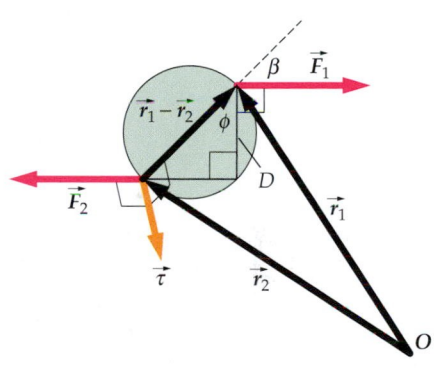

**FIGURE 12-13**

## 12-5 Static Equilibrium in an Accelerated Frame

By an accelerated frame we mean a reference frame that is accelerating relative to an inertial frame of reference. The net force on an object that remains at rest relative to an accelerated reference frame is not equal to zero. An object at rest relative to the accelerated frame has the same acceleration as the frame. The two conditions for an object to be in static equilibrium in an accelerated reference frame are

1. $\Sigma \vec{F} = m\vec{a}_{cm}$

   where $\vec{a}_{cm}$ is the acceleration of the center of mass, which is the acceleration of the reference frame.

2. $\Sigma \vec{\tau}_{cm} = 0$

   The sum of the torques about the center of mass must be zero.

The second condition follows from the fact that Newton's second law for rotation, $\Sigma \vec{\tau}_{cm} = I_{cm}\vec{\alpha}$, holds for torques about the center of mass whether or not the center of mass is accelerating.[†]

---

† See the discussion surrounding Equation 9-30.

MOVING A BOX                    **EXAMPLE 1 2 - 6**

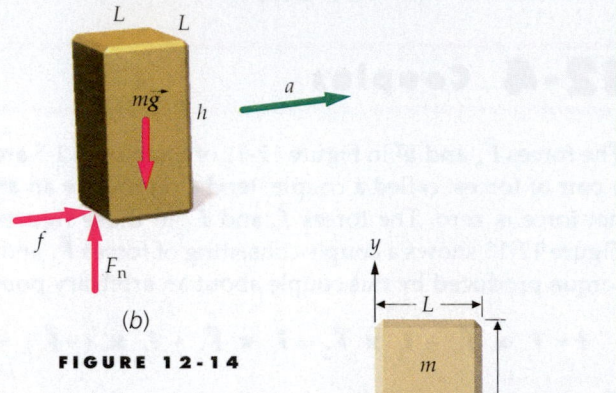

(a)

A truck (Figure 12-14a) carries a uniform box of mass $m$, height $h$, and square cross section of edge-length $L$. What is the greatest acceleration the truck can have without the box tipping over? Assume that the box tips before it slides.

**PICTURE THE PROBLEM** The acceleration of the box is due to the frictional force, as shown in Figure 12-14b. This force exerts a counterclockwise torque about the center of mass of the box. The only other force that exerts a torque about the center of mass of the box is the normal force. If the box is not accelerating, the normal force is distributed uniformly across the bottom of the box. If the acceleration is small, this distribution shifts and the effective normal force moves to the left to provide a balancing torque about the center of mass. The greatest balancing torque this force can exert is when the effective normal force is at the edge of the box, as shown.

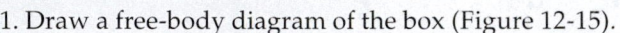

(b)

**FIGURE 12-14**

1. Draw a free-body diagram of the box (Figure 12-15).

2. Apply $\Sigma F_y = ma_{cmy}$ to the box and then solve for the normal force:
$$F_n - mg = 0 \quad \text{so} \quad F_n = mg$$

3. Apply $\Sigma F_x = ma_{cmx}$ to the box:
$$f_s = ma$$

4. Apply $\Sigma \tau_{cm} = 0$:
$$f_s \frac{h}{2} - F_n d = 0$$

5. Substitute for $d, f_s,$ and $F_n$ and solve for $a$:
$$ma\frac{h}{2} - mg\frac{L}{2} = 0 \quad \text{so} \quad \boxed{a = \frac{L}{h}g}$$

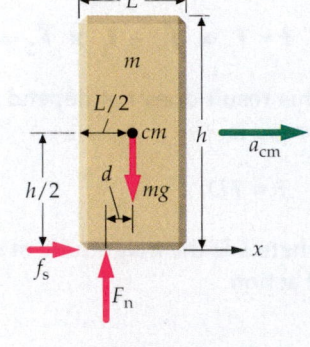

**FIGURE 12-15**

**REMARKS** The maximum acceleration is proportional to $L/h$. This maximum acceleration is small for a tall, narrow box ($L/h$ small) and large for a short, wide box ($L/h$ large). Thus, a short, wide box is more stable.

# **12-6** Stability of Rotational Equilibrium

There are three categories of rotational equilibrium for an object: stable, unstable, or neutral. **Stable rotational equilibrium** occurs when the torques that arise from a small angular displacement of the object urge the object back toward its equilibrium position. Stable equilibrium is illustrated in Figure 12-16a. When the

**FIGURE 12-16** If a slight rotation raises the center of gravity, as in (a), the equilibrium is stable. If a slight rotation lowers the center of gravity, as in (b), the equilibrium is unstable. If a slight rotation neither raises nor lowers the center of gravity, as in (c), the equilibrium is neutral.

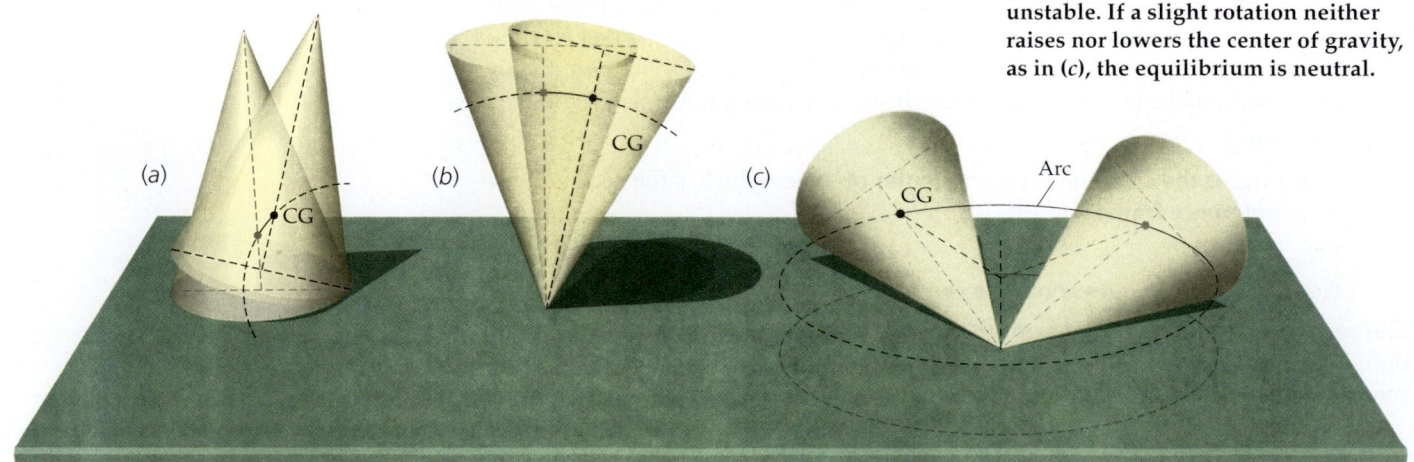

cone is tipped slightly as shown, the resulting torque about the pivot point tends to restore the cone to its original position. Note that this slight tipping lifts the center of gravity, increasing the potential energy of the cone.

**Unstable rotational equilibrium,** illustrated in Figure 12-16*b*, occurs when the torques that arise from a small angular displacement of the object urge the object away from its equilibrium position. A slight tipping of the cone causes it to fall over because the torque due to its weight tends to rotate it away from its original position. Here the rotation lowers the center of gravity and decreases the potential energy of the cone.

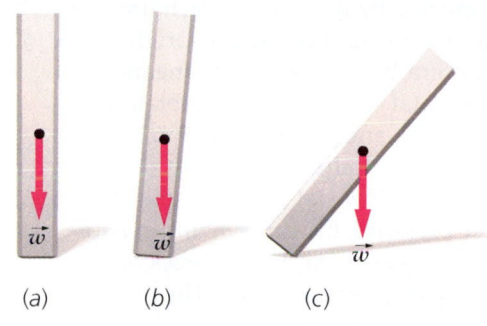

(a)   (b)   (c)

**FIGURE 12-17** Stability of equilibrium is relative. If the rod in (*a*) is rotated slightly, as in (*b*), it returns to its original equilibrium position as long as the center of gravity lies over the base of support. (*c*) If the rotation is too great, the center of gravity is no longer over the base of support, and the rod falls over.

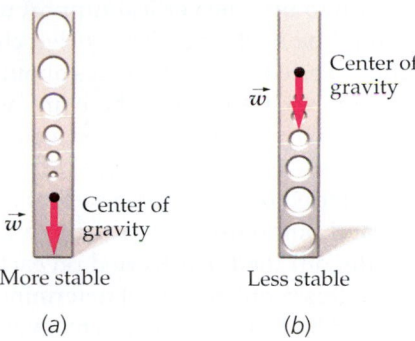

More stable   Less stable
(a)   (b)

**FIGURE 12-18** When a nonuniform rod rests on its heavy end with its center of gravity low, as in (*a*), the equilibrium is more stable than when its center of gravity is high, as in (*b*).

The cone resting on a horizontal surface in Figure 12-16*c* illustrates **neutral rotational equilibrium.** If the cone is rolled slightly, there is no torque or force that urges it either back toward or away from its original position. As the cone rotates, the height of the center of gravity remains unchanged, so the potential energy does not change.

In summary, if a system is disturbed slightly from its equilibrium position, the equilibrium is stable if the system returns to its original position, unstable if it moves farther away, and neutral if there are no torques or forces tending to rotate it in either direction.

Because "disturbed slightly" is a relative term, stability is also relative. One example of equilibrium may be more or less stable than another. A rod is balanced on one end, as in Figure 12-17*a*. Here, if the disturbance is very small (Figure 12-17*b*), the rod will move back toward its original position, but if the disturbance is great enough so that the center of gravity no longer lies over the base of support (Figure 12-17*c*), the rod will fall.

We can improve the stability of a system by either lowering the center of gravity or widening the base of support. Figure 12-18 shows a nonuniform rod that is loaded so that its center of gravity is near one end. If it stands on its heavy end so that the center of gravity is low (Figure 12-18*a*), it is much more stable than if it stands on the other end so that the center of gravity is high (Figure 12-18*b*).

In Figure 12-19 the system is stable for any angular displacement because the resulting torque always rotates the system back toward its equilibrium position.

Standing or walking upright is difficult for a human because the center of gravity is high, and must be kept over a relatively small base of support, the feet. Human infants take about a year to learn to walk. A four-footed creature has a much easier time because its base of support is larger and its center of gravity is lower. Newborn kittens can walk almost immediately.

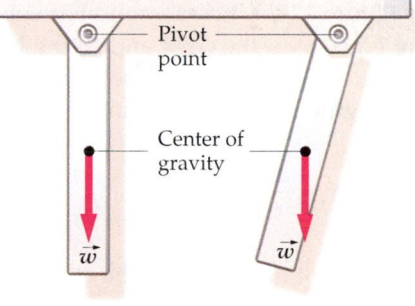

Pivot point

Center of gravity

**FIGURE 12-19**

# 12-7 Indeterminate Problems

When objects are not rigid, but deformable, we need more information to determine the forces required for equilibrium. Consider a car resting on a horizontal surface. Suppose there is a very heavy object on one side of the trunk.

We wish to find the vertical support force exerted by the road on each tire. Let the road be in the $xy$ plane. If we choose one of the tires as our origin, the torque exerted by all the forces about that point has $x$ and $y$ components, but no $z$ component because there are no horizontal forces. We thus obtain two equations by setting the net torque equal to zero, and a third equation by setting the net vertical force equal to zero. We need another equation to find the force exerted by the road on each of the four tires. If we let air out of one of the tires and pump up another tire to a greater pressure, the car remains in equilibrium, but the force exerted on each tire changes. Clearly, the forces on the tires in this problem are not determined by the information given. The tires are not rigid bodies. To some extent, every object is deformable.

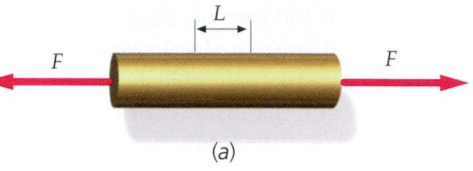

(a)

# 12-8 Stress and Strain

If a solid object is subjected to forces that tend to stretch, shear, or compress the object, its shape changes. If the object returns to its original shape when the forces are removed, it is said to be **elastic.** Most objects are elastic for forces up to a certain maximum, called the **elastic limit.** If the forces exceed the elastic limit, the object does not return to its original shape but is permanently deformed.

Figure 12-20 shows a solid bar subjected to a stretching or **tensile force $F$** acting equally to the right and to the left. The bar is in equilibrium, but the forces acting on it tend to increase its length. The fractional change in the length $\Delta L/L$ of a segment of the bar is called the **strain:**

$$\text{Strain} = \frac{\Delta L}{L} \qquad 12\text{-}7$$

The ratio of the force $F$ to the cross-sectional area $A$ is called the **tensile stress:**

$$\text{Stress} = \frac{F}{A} \qquad 12\text{-}8$$

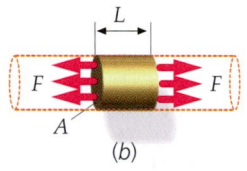

(b)

**FIGURE 12-20** (a) A solid bar subjected to stretching forces of magnitude $F$ acting on each end. (b) A small section of the bar of length $L$. The elements of the bar to the left and right of this section exert forces on the section. If the section is not too close to the end, these forces are distributed equally over the cross-sectional area. The force per unit area is the stress.

Figure 12-21 shows a graph of stress versus strain for a typical solid bar. The graph is linear until point A. Up to this point, known as the proportional limit, the strain is proportional to the stress. The result that strain varies linearly with stress is known as Hooke's law. Point B in Figure 12-21 is the elastic limit of the material. If the bar is stretched beyond this point, it is permanently deformed. If an even greater stress is applied, the material eventually breaks, shown happening at point C. The ratio of stress to strain in the linear region of the graph is a constant called **Young's modulus $Y$:**

$$Y = \frac{\text{stress}}{\text{strain}} = \frac{F/A}{\Delta L/L} \qquad 12\text{-}9$$

YOUNG'S MODULUS DEFINED

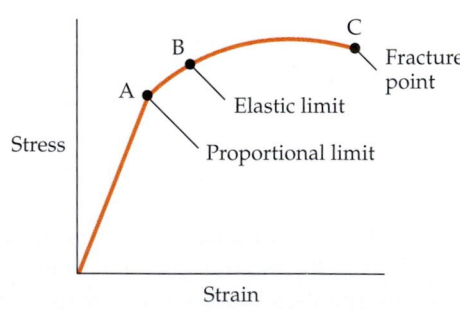

**FIGURE 12-21** A graph of stress versus strain. Up to point A, the strain is proportional to the stress. Beyond the elastic limit at point B, the bar will not return to its original length when the stress is removed. At point C, the bar fractures.

The units of Young's modulus are newtons per square meter (or pounds per square inch). Approximate values of Young's modulus for various materials are listed in Table 12-1.

**EXERCISE** Suppose that the biceps muscle of your right arm has a maximum cross-sectional area of $12 \text{ cm}^2 = 1.2 \times 10^{-3} \text{ m}^2$. What is the stress in the muscle if it exerts a force of 300 N? (*Answer* Stress $= F/A = 2.5 \times 10^5 \text{ N/m}^2$. The maximum stress that can be exerted is approximately the same for all human muscles. Greater forces can be exerted by muscles with greater cross-sectional areas.)

## TABLE 12-1

**Young's Modulus $Y$ and Strengths of Various Materials[†]**

| Material | $Y$, GN/m²[‡] | Tensile strength, MN/m² | Compressive strength, MN/m² |
|---|---|---|---|
| Aluminum | 70 | 90 | |
| Bone | | | |
|    Tensile | 16 | 200 | |
|    Compressive | 9 | | 270 |
| Brass | 90 | 370 | |
| Concrete | 23 | 2 | 17 |
| Copper | 110 | 230 | |
| Iron (wrought) | 190 | 390 | |
| Lead | 16 | 12 | |
| Steel | 200 | 520 | 520 |

† These values are representative. Actual values for particular samples may differ.
‡ 1 GN = $10^3$ MN = $1 \times 10^9$ N.

If a bar is subjected to forces that tend to compress it rather than stretch it, the stress is called **compressive stress.** For many materials, Young's modulus for compressive stress is the same as that for tensile stress. Note that $\Delta L$ in Equation 12-7 is then taken to be the *decrease* in the length of the bar. If the tensile or compressive stress is too great, the bar breaks. The stress at which breakage occurs is called the **tensile strength,** or in the case of compression, the **compressive strength.** Approximate values of the tensile and compressive strengths of various materials are listed in Table 12-1. Note from the table that the compressive strength of bone is greater than the tensile strength. Also note that, for bone, Young's modulus is significantly larger for tensile stress than for compressive stress. These differences have biological significance, because the major job of bone is to resist the compressive load exerted by contracting muscles.

*ELEVATOR SAFETY*    **EXAMPLE  12-7**    **Put It in Context**

**While working with an engineering company during the summer, you are assigned to check the safety of an elevator system in a new office building. The elevator has a maximum load of 1000 kg including its own mass, and is supported by a steel cable 3.0 cm in diameter and 300 m long at full extension. There will be safety concerns if the steel stretches more than 3.0 cm. Your job is to determine whether or not the elevator is safe as planned, given a maximum acceleration of the system of 1.5 m/s².**

**PICTURE THE PROBLEM**  $L$ is the unstretched length of cable, $F$ is the force acting on it, and $A$ is its cross-sectional area. The stretch in the cable $\Delta L$ is related to Young's modulus by $Y = (F/A)/(\Delta L/L)$. From Table 12-1 we find the numerical value of Young's modulus for steel, $Y = 2.0 \times 10^{11}$ N/m².

1. The amount the cable is stretched, $\Delta L$, is found from Young's modulus:

$$Y = \frac{F/A}{\Delta L/L} \quad \text{so} \quad \Delta L = \frac{FL}{AY}$$

2. To find the force acting on the cable we apply Newton's second law to the elevator. There are two forces on the elevator, the force $F$ of the cable and the weight:

$$F - mg = ma_y$$

so

$$F_{max} = m(g + a_{y,max})$$

$$= (1000 \text{ kg})(9.81 \text{ N/kg} + 1.5 \text{ N/kg})$$

$$= 1.13 \times 10^4 \text{ N}$$

3. Substitute into the step 1 result and obtain the maximum amount of stretch:

$$\Delta L = \frac{F_{max}L}{AY} = \frac{F_{max}L}{\pi r^2 Y}$$

$$= \frac{(1.13 \times 10^4 \text{ N})(300 \text{ m})}{\pi(0.015 \text{ m})^2(2.0 \times 10^{11} \text{ N/m}^2)}$$

$$= 2.40 \text{ cm}$$

4. Report your results to your boss:

> According to my calculations, the most the cable will stretch is 2.4 cm, only 20% less than the 3.0-cm limit. However, in reading the footnote to the table, I note that the values given for Young's modulus are representative values, and that actual values vary from sample to sample. I recommend that you consult a engineer and get a professional evaluation.

**EXERCISE** A wire 1.5 m long has a cross-sectional area of 2.4 mm². It is hung vertically and stretches 0.32 mm when a 10-kg block is attached to it. Find (a) the stress, (b) the strain, and (c) Young's modulus for the wire. (*Answer* (a) $4.09 \times 10^7 \text{ N/m}^2$ (b) $2.13 \times 10^{-4}$ (c) $192 \text{ GN/m}^2$)

In Figure 12-22, a force $F_s$ is applied tangentially to the top of a book. Such a force is called a **shear force**. The ratio of the shear force $F_s$ to the area $A$ is called the **shear stress**:

$$\text{Shear stress} = \frac{F_s}{A} \qquad \text{12-10}$$

A shear stress tends to deform an object, as shown in Figure 12-22. The ratio $\Delta X/L$ is called the **shear strain**:

$$\text{Shear strain} = \frac{\Delta X}{L} = \tan\theta \qquad \text{12-11}$$

where $\theta$ is the shear angle shown in the figure. The ratio of the shear stress to the shear strain is called the **shear modulus** $M_s$:

$$M_s = \frac{\text{shear stress}}{\text{shear strain}} = \frac{F_s/A}{\Delta X/L} = \frac{F_s/A}{\tan\theta} \qquad \text{12-12}$$

DEFINITION — SHEAR MODULUS

The shear modulus is also known as the **torsion modulus**. The torsion modulus is approximately constant for small stresses, which implies that the shear strain varies linearly with the shear stress. This observation is known as Hooke's law for torsional stress. In a torsion balance, such as that used in Cavendish's apparatus for measuring the universal gravitational constant $G$, the torque (which is related to the stress) is proportional to the angle of twist (which equals the strain for small angles). Approximate values of the shear modulus for various materials are listed in Table 12-2.

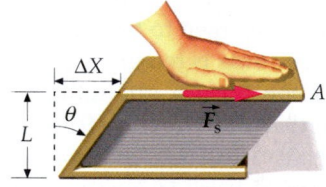

**FIGURE 12-22** The application of the horizontal force $\vec{F}_s$ to the book causes a shear stress defined as the force per unit area. The ratio $\Delta X/L = \tan\theta$ is the shear strain.

## TABLE 12-2

**Approximate Values of the Shear Modulus $M_s$ of Various Materials**

| Material | $M_s$, GN/m² |
| --- | --- |
| Aluminum | 30 |
| Brass | 36 |
| Copper | 42 |
| Iron | 70 |
| Lead | 5.6 |
| Steel | 84 |
| Tungsten | 150 |

| Topic | Relevant Equations and Remarks |
|---|---|
| **1. Equilibrium of a Rigid Object** | |
| Conditions | 1. The net external force acting on the object must be zero: $$\Sigma \vec{F} = 0 \qquad \text{12-1}$$ 2. The net external torque about any point must be zero: $$\Sigma \vec{\tau} = 0 \qquad \text{12-2}$$ If all the forces acting on a rigid object are perpendicular to an axis, then the sum of the torques about that axis equals zero. |
| Stability | The equilibrium of an object can be classified as stable, unstable, or neutral. An object resting on some surface will be in equilibrium if its center of gravity lies over its base of support. Stability can be improved by lowering the center of gravity or by increasing the size of the base. |
| **2. Center of Gravity** | The force of gravity exerted on the various parts of an object can be replaced by a single force, the total weight of the object $\vec{W}$, acting at the center of gravity. $$\vec{\tau}_{net} = \sum_i (\vec{r}_i \times \vec{w}_i) = \vec{r}_{cg} \times \vec{W} \qquad \text{12-3}$$ For an object in a uniform gravitational field, the center of gravity coincides with the center of mass. |
| **3. Couples** | A pair of equal and opposite forces constitutes a couple. The torque produced by a couple is the same about any point in space. $$\vec{\tau} = (\vec{r}_1 - \vec{r}_2) \times \vec{F}_1, \quad \text{so} \quad \tau = FD \qquad \text{12-5, 12-6}$$ where $D$ is the distance between the lines of action of the forces. |
| **5. Accelerated Reference Frame** | The conditions for static equilibrium in an accelerated reference frame are 1. $\Sigma \vec{F} = m\vec{a}_{cm}$ where $\vec{a}_{cm}$ is the acceleration of the center of mass, which is the acceleration of the reference frame. 2. $\Sigma \vec{\tau}_{cm} = 0$ The sum of the torques about the center of mass must be zero. |
| **6. Stress and Strain** | |
| Young's modulus | $$Y = \frac{\text{stress}}{\text{strain}} = \frac{F/A}{\Delta L/L} \qquad \text{12-9}$$ |
| Shear modulus | $$M_s = \frac{\text{shear stress}}{\text{shear strain}} = \frac{F_s/A}{\Delta X/L} = \frac{F_s/A}{\tan \theta} \qquad \text{12-12}$$ |

# PROBLEMS

- Single-concept, single-step, relatively easy
- •• Intermediate-level, may require synthesis of concepts
- ••• Challenging
- **SSM** Solution is in the *Student Solutions Manual*
- **iSOLVE** Problems available on iSOLVE online homework service
- **iSOLVE✓** These "Checkpoint" online homework service problems ask students additional questions about their confidence level, and how they arrived at their answer

In a few problems, you are given more data than you actually need; in a few other problems, you are required to supply data from your general knowledge, outside sources, or informed estimates.

Take $g = 9.81$ N/kg $= 9.81$ m/s$^2$ and neglect friction in all problems unless otherwise stated.

## Conceptual Problems

**1** • **SSM** True or false:

(a) $\sum_i \vec{F}_i = 0$ is sufficient for static equilibrium to exist.

(b) $\sum_i \vec{F}_i = 0$ is necessary for static equilibrium to exist.

(c) In static equilibrium, the net torque about any point is zero.

(d) An object is in equilibrium only when there are no forces acting on it.

**2** • True or false: The center of gravity is always at the geometric center of a body.

**3** • Must there be any material at the center of gravity of an object?

**4** • **SSM** If the gravitational field $\vec{g}$ is not constant over an object, is it the center of mass or the center of gravity that is the pivot point when the object is balanced?

**5** •• To find the center of mass of an irregular plane figure, the following method can be used: hang the figure from one corner, by a string and extend the line that the string makes downward across the figure. Repeat, hanging it from another corner. The intersection of the two lines will be at the center of mass. Explain why this technique works.

**6** • **SSM** Is it possible to climb a ladder placed against a wall if the ground is frictionless but the wall is not? Explain.

**7** • An aluminum wire and a steel wire of the same length $L$ and diameter $D$ are joined to form a wire of length $2L$. The wire is fastened to the ceiling and a weight $W$ is attached to the other end. Neglecting the mass of the wires, which of the following statements is true? (a) The aluminum portion will stretch by the same amount as the steel portion. (b) The tensions in the aluminum portion and the steel portion are equal. (c) The tension in the aluminum portion is greater than that in the steel portion. (d) None of these statements is true.

**8** • If the net torque about some point is zero, must it be zero about any other point? Explain.

**9** • **SSM** The horizontal bar in Figure 12-23 will remain horizontal if (a) $L_1 = L_2$ and $R_1 = R_2$, (b) $L_1 = L_2$ and $M_1 = M_2$, (c) $R_1 = R_2$ and $M_1 = M_2$, (d) $L_1 M_1 = L_2 M_2$, (e) $R_1 L_1 = R_2 L_2$.

**FIGURE 12-23** Problem 9

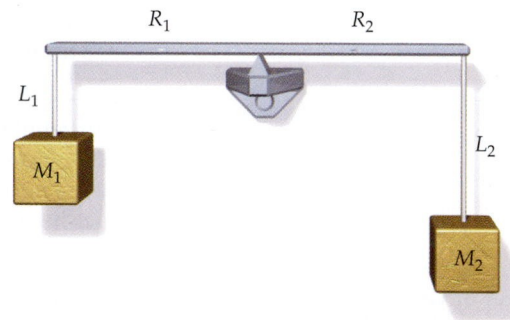

**10** •• Sit in a chair with your back straight. Now try to stand up without leaning forward. Explain why you cannot do it.

**11** •• **SSM** The great engineering feats of ancient times (Roman arch bridges, the great cathedrals, and the pyramids, to name a few) all have two things in common: they are made of stone, and they are compressive structures—that is, they are built so that all strains in the structure are compressive rather than tensile in nature. Look up the tensile and compressive strengths of stone and cement to explain why this is true.

## Estimation and Approximation

**12** •• A large crate weighing 4500 N rests on four 12-cm-high blocks on a horizontal surface (Figure 12-24). The crate is 2 m long, 1.2 m high, and 1.2 m deep. You are asked to lift one end of the crate using a long steel pry bar. The fulcrum on the pry bar is 10 cm from the end that lifts the crate. Estimate the length of the bar you will need to lift the end of the crate.

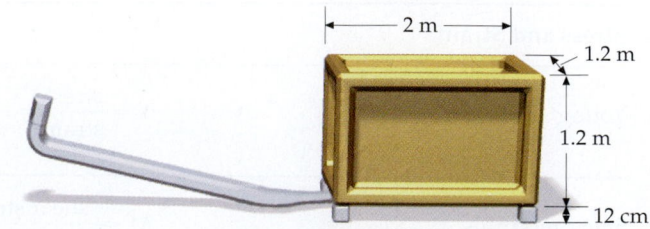

**FIGURE 12-24** Problem 12

**13** •• SSM Consider an atomic model for Young's modulus: assume that we have a large number of atoms arranged in a cubic array separated by distance $a$. Imagine that each atom is attached to its 6 nearest neighbors by little springs with spring constant $k$. (Atoms are not really attached by springs, but the forces between them act enough like springs to make this a good model.) (a) Show that this material, if stretched, will have a Young's modulus $Y = k/a$. (b) From Table 12-1, and assuming that $a \approx 1$ nm, estimate a typical value for the "atomic spring constant" $k$ in a metal.

## Conditions for Equilibrium

**14** • A seesaw consists of a 4-m board pivoted at the center. A 28-kg child sits on one end of the board. Where should a 40-kg child sit to balance the seesaw?

**15** • iSOLVE✓ In Figure 12-25, Misako is about to do a push-up. Her center of gravity lies directly above point $P$ on the floor, which is 0.9 m from her feet and 0.6 m from her hands. If her mass is 54 kg, what is the force exerted by the floor on her hands?

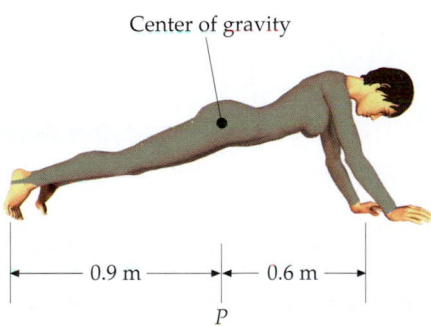

Center of gravity

|← 0.9 m →|← 0.6 m →|

$P$

FIGURE 12-25 Problem 15

**16** • SSM iSOLVE✓ Misako wants to measure the strength of her biceps muscle by exerting a force on a test strap as shown in Figure 12-26. The strap is 28 cm from the pivot point at the elbow, and her biceps muscle is attached at a point 5 cm from the pivot point. If the scale reads 18 N when she exerts her maximum force, what force is exerted by the biceps muscle?

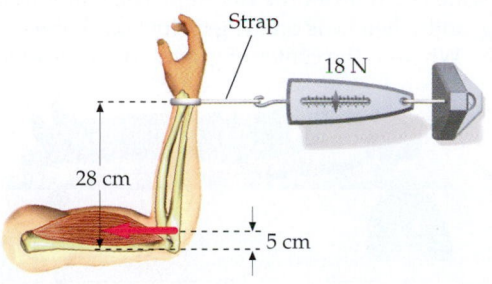

Strap

18 N

28 cm

5 cm

FIGURE 12-26 Problem 16

**17** • A crutch is pressed against the sidewalk with a force $\vec{F}_c$ along its own direction as in Figure 12-27. This force is balanced by the normal force $\vec{F}_n$ and a frictional force $\vec{f}_s$. (a) Show that when the force of friction is at its maximum value, the coefficient of friction is related to the angle $\theta$ by $\mu_s = \tan \theta$. (b) Explain how this result applies to the forces on your foot when you are not using a crutch. (c) Why is it advantageous to take short steps when walking on ice?

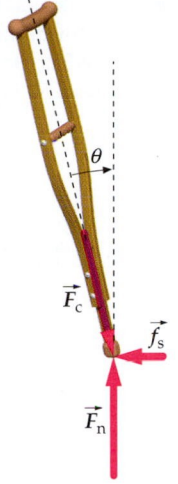

$\theta$

$\vec{F}_c$

$\vec{f}_s$

$\vec{F}_n$

FIGURE 12-27 Problem 17

## The Center of Gravity

**18** • iSOLVE An automobile has 58 percent of its weight on the front wheels. The front and back wheels are separated by 2 m. Where is the center of gravity located with respect to the front wheels?

**19** • SSM Each of the objects shown in Figure 12-28 is suspended from the ceiling by a thread attached to the point marked × on the object. Describe the orientation of each suspended object with a diagram.

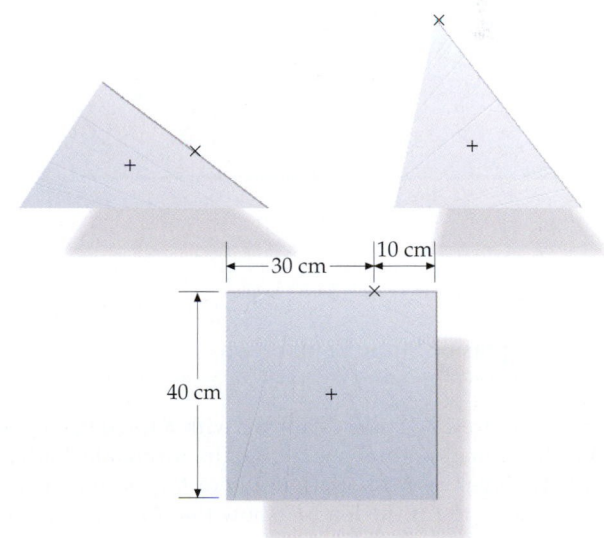

|← 30 cm →|10 cm|

40 cm

FIGURE 12-28 Problem 19

**20** •• **ISOLVE✓** A square plate is produced by welding together four smaller square plates, each of side $a$ as shown in Figure 12-29. Plate 1 weighs 40 N; plate 2, 60 N; plate 3, 30 N; and plate 4, 50 N. Find the center of gravity $(x_{cg}, y_{cg})$.

**FIGURE 12-29** Problem 20

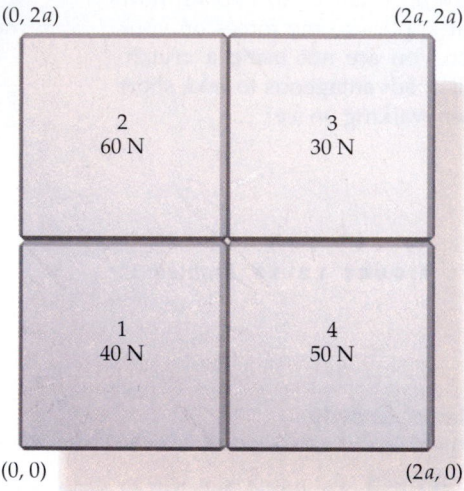

(0, 2a)          (2a, 2a)

|  |  |
|---|---|
| **2**<br>60 N | **3**<br>30 N |
| **1**<br>40 N | **4**<br>50 N |

(0, 0)          (2a, 0)

**21** •• **ISOLVE** A uniform rectangular plate has a circular section of radius $R$ cut out as shown in Figure 12-30. Find the center of gravity of the system. *Hint: Do not integrate. Use superposition of a rectangular plate minus a circular plate.*

**FIGURE 12-30** Problem 21

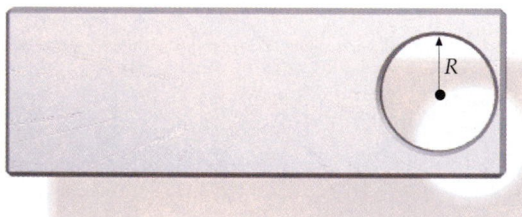

## Some Examples of Static Equilibrium

**22** • Figure 12-31 shows a lever with a force $f$ being applied to lift a load $F$. (*a*) If we define the mechanical advantage of the lever as $M = F/f_{min}$, where $f_{min}$ is the smallest force necessary to lift the load $F$, show that $M = x/X$, where $x$ is the moment arm (distance to the pivot) for the applied force and $X$ is the moment arm for the load. (*b*) To lift a heavy load, the moment arm for the applied force is usually larger than the load, so that the lifting force is smaller than the applied force. However, sometimes the moment arm for the applied force is much smaller than that for the load, so that the applied force must be much larger than the load (the arrangements of muscles in the forearm, as shown in Figure 12-4, is a good example). Why is this arrangement useful under some circumstances?

**FIGURE 12-31** Problem 22

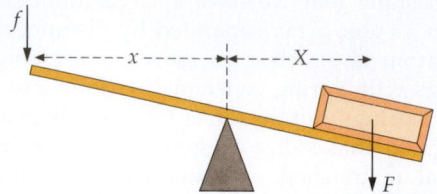

**23** • **ISOLVE** Figure 12-32 shows a 25-foot sailboat. The mast is a uniform pole of 120 kg and is supported on the deck and held fore and aft by wires as shown. The tension in the forestay (wire leading to the bow) is 1000 N. Determine the tension in the backstay and the force that the deck exerts on the mast. Is there a tendency for the mast to slide forward or aft? If so, where should a block be placed to prevent the mast from moving?

**FIGURE 12-32** Problem 23

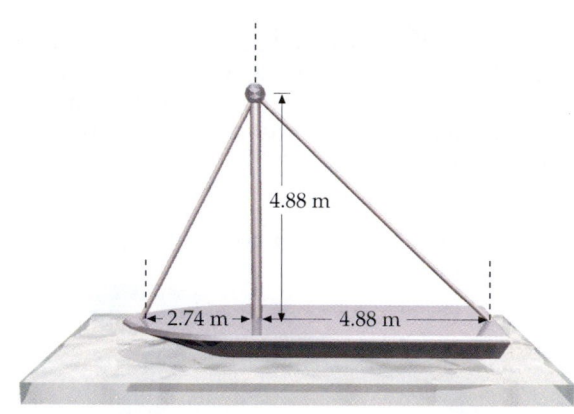

4.88 m

← 2.74 m → ← 4.88 m →

**24** •• **ISOLVE** A 10-m beam of mass 300 kg extends over a ledge as in Figure 12-33. The beam is not attached, but simply rests on the surface. A 60-kg student intends to position the beam so that he can walk to the end of it. How far from the edge of the ledge can the beam extend?

**FIGURE 12-33**
**Problem 24**

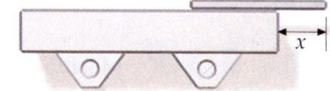

$x$

**25** •• **SSM** A gravity board for locating the center of gravity of a person consists of a horizontal board supported by a fulcrum at one end and by a scale at the other end. A physics student lies horizontally on the board with the top of his head above the fulcrum point as shown in Figure 12-34. The scale is 2 m from the fulcrum. The student has a mass of 70 kg, and when he is on the gravity board, the scale advances 250 N. Where is the center of gravity of the student?

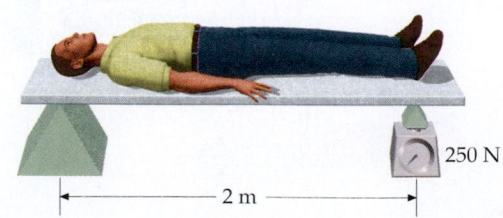

250 N

← 2 m →

**FIGURE 12-34** Problem 25

**26** •• **ISOLVE** ✔ A 3-m board of mass 5 kg is hinged at one end. A force $\vec{F}$ is applied vertically at the other end to lift a 60-kg block, which rests on the board 80 cm from the hinge, as shown in Figure 12-35. (*a*) Find the magnitude of the force needed to hold the board stationary at $\theta = 30°$. (*b*) Find the force exerted by the hinge at this angle. (*c*) Find the magnitude of the force $\vec{F}$ and the force exerted by the hinge if $\vec{F}$ is exerted perpendicular to the board when $\theta = 30°$.

**FIGURE 12-35**
**Problem 26**

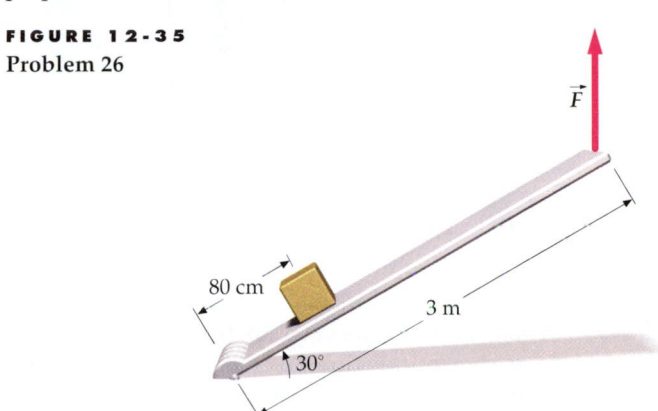

**27** •• **SSM** **ISOLVE** A cylinder of weight $W$ is supported by a frictionless trough formed by a plane inclined at 30° to the horizontal on the left and one inclined at 60° on the right as shown in Figure 12-36. Find the force exerted by each plane on the cylinder.

**FIGURE 12-36**
**Problem 27**

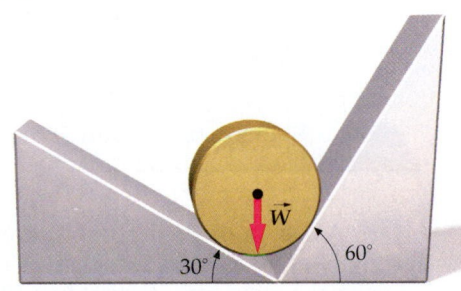

**28** •• **ISOLVE** An 80-N weight is supported by a cable attached to a strut hinged at point $A$ as in Figure 12-37. The strut is supported by a second cable under tension $T_2$. The mass of the strut is negligible. (*a*) What are the three nonnegligible forces acting on the strut? (*b*) Show that the vertical component of the tension $T_2$ must equal 80 N. (*c*) Find the force exerted on the strut by the hinge.

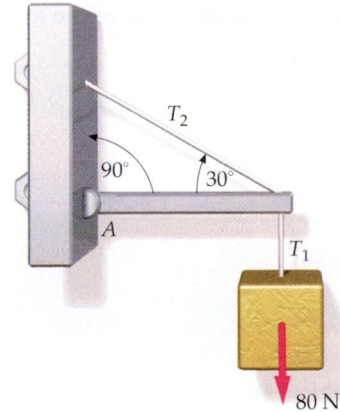

**FIGURE 12-37** Problem 28

**29** •• A horizontal board 8.0 m long is used by pirates to make their victims walk the plank. A pirate of mass 105 kg stands on the shipboard end of the plank to prevent it from tipping. Find the maximum distance the plank can overhang for a 63-kg victim to be able to walk to the end if (*a*) the mass of the plank is negligible and (*b*) the mass of the plank is 25 kg.

**30** •• A uniform 18-kg door that is 2.0 m high by 0.8 m wide is hung from two hinges that are 20 cm from the top and 20 cm from the bottom. If each hinge supports half the weight of the door, find the magnitude and direction of the horizontal components of the forces exerted by the two hinges on the door.

**31** •• Find the force exerted by the edge of the block on the wheel in Example 12-4, just as the wheel lifts off the surface.

**32** •• **SSM** **ISOLVE** The diving board shown in Figure 12-38 has a mass of 30 kg. Find the force on the supports when a 70-kg diver stands at the end of the diving board. Give the direction of each support force as a tension or a compression.

**FIGURE 12-38** Problem 32

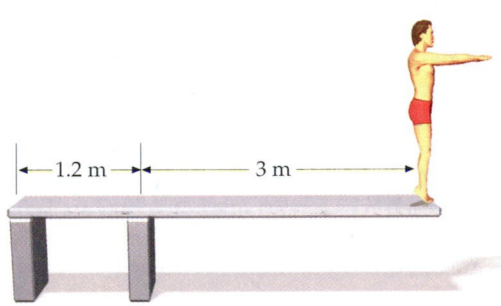

**33** •• Find the force exerted on the strut by the hinge at $A$ for the arrangement in Figure 12-39 if (*a*) the strut is weightless and (*b*) the strut weighs 20 N.

**FIGURE 12-39**
**Problem 33**

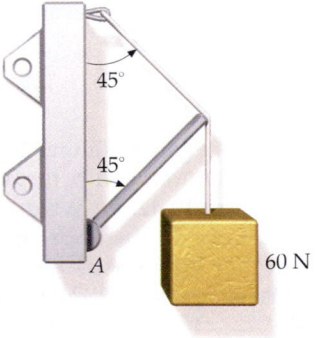

**34** •• Julie has been hired to help paint the trim of a building, but she is not convinced of the safety of the apparatus. A 5.0-m plank is suspended horizontally from the top of the building by ropes attached at each end. Julie knows from previous experience that the ropes being used will break if the tension exceeds 1 kN. Her 80-kg boss dismisses Julie's worries and begins painting while standing 1 m from the end of the plank. If Julie's mass is 60 kg and the plank has a mass of 20 kg, then over what range of positions can Julie stand to join her boss without causing the ropes to break?

**35** •• A cylinder of mass $M$ and radius $R$ rolls against a step of height $h$ as shown in Figure 12-40. When a horizontal force $\vec{F}$ is applied to the top of the cylinder, the cylinder remains at rest. (*a*) What is the normal force exerted by the floor on the cylinder? (*b*) What is the horizontal force exerted by the edge of the step on the cylinder? (*c*) What is the vertical component of the force exerted by the edge of the step on the cylinder?

**FIGURE 12-40**
**Problems 35, 36**

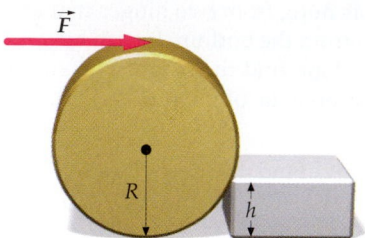

**36** •• For the cylinder in Problem 35, find the minimum horizontal force $\vec{F}$ that will roll the cylinder over the step if the cylinder does not slide on the edge.

**37** •• SSM Figure 12-41 shows a hand holding an epee, a weapon used in the sport of fencing. The center of mass of the epee, indicated in Figure 12-41, is 24 cm from the pommel; its total mass is 0.700 kg and its length is 110 cm. (*a*) Apply one of the conditions for static equilibrium to find the (total) force exerted by the hand on the epee. (*b*) Apply the other condition for static equilibrium to find the torque exerted by the hand on the epee. (*c*) Model the forces exerted by the hand as two oppositely directed forces whose lines of action are separated by the width of the fencer's hand ($\approx$10 cm). What are the magnitudes and directions of these two forces?

**FIGURE 12-41** Problem 37

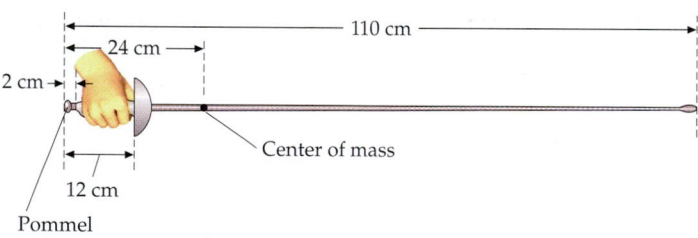

**38** •• A large gate weighing 200 N is supported by hinges at the top and bottom and is further supported by a wire as shown in Figure 12-42. (*a*) What must be the tension in the wire for the force on the upper hinge to have no horizontal component? (*b*) What is the horizontal force on the lower hinge? (*c*) What are the vertical forces on the hinges?

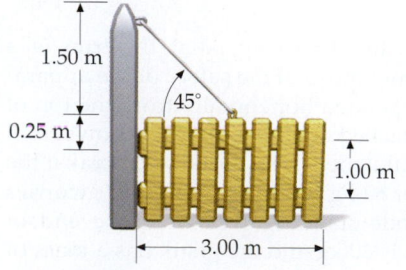

**FIGURE 12-42** Problem 38

**39** ••• SSM A uniform log with a mass of 100 kg, a length of 4 m, and a radius of 12 cm is held in an inclined position, as shown in Figure 12-43. The coefficient of static friction between the log and the horizontal surface is 0.6. The log is on the verge of slipping to the right. Find the tension in the support wire and the angle the wire makes with the vertical wall.

**FIGURE 12-43** Problem 39

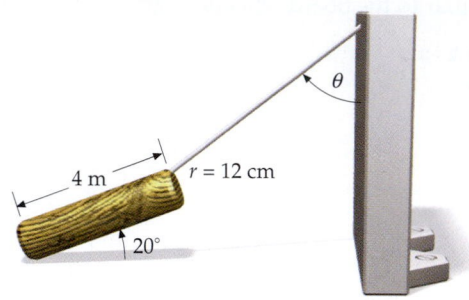

**40** ••• A tall, uniform, rectangular block sits on an inclined plane as shown in Figure 12-44. A cord is attached to the top of the block to prevent it from falling down the incline. What is the maximum angle $\theta$ for which the block will not slide on the incline? Let $b/a$ be 4 and $\mu_s = 0.8$.

**FIGURE 12-44** Problem 40

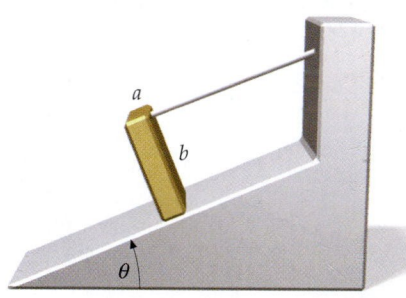

**41** •• SSM A boat is moored at the end of a dock in a rapidly flowing river by a chain 5 m long, as shown in Figure 12-45. To give the chain some flexibility, a 100-N weight is attached in the center in the chain, to allow for variations in the force pulling the boat away from the dock. (*a*) If the drag force on the boat is 50 N, what is the tension in the chain? (*b*) How far will the chain sag? Ignore the weight of the chain itself. (*c*) How far is the boat from the dock? (*d*) If the maximum tension that the chain can support is 500 N, what is the maximum value of the force that the river can exert on the boat?

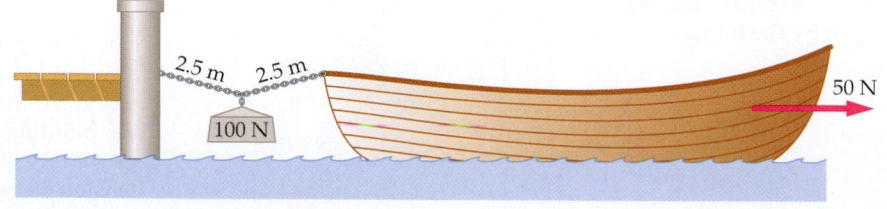

**FIGURE 12-45** Problem 41

**42** •• **iSOLVE** A thin rod of length 10 m and mass 20 kg is supported at a 30° incline. One support is 2 m and the other is 6 m from the lower end of the rod. Friction prevents the rod from sliding off the supports. Find the normal force exerted on the rod by each support.

**43** • Two 80-N forces are applied to opposite corners of a rectangular plate as shown in Figure 12-46. Find the torque produced by this couple.

**FIGURE 12-46** Problems 43, 45

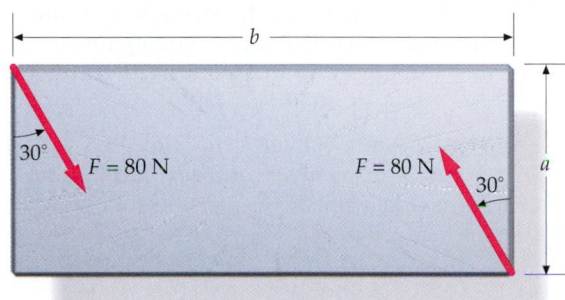

**44** •• **SSM** A uniform cube of side $a$ and mass $M$ rests on a horizontal surface. A horizontal force $\vec{F}$ is applied to the top of the cube as in Figure 12-47. This force is not sufficient to move or tip the cube. (*a*) Show that the force of static friction exerted by the surface and the applied force constitute a couple, and find the torque exerted by the couple. (*b*) This couple is balanced by the couple consisting of the normal force exerted by the surface and the weight of the cube. Use this fact to find the effective point of application of the normal force when $F = Mg/3$. (*c*) What is the greatest magnitude of $\vec{F}$ for which the cube will not tip?

**FIGURE 12-47**
**Problem 44**

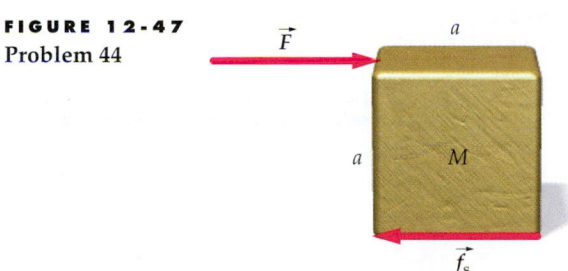

**45** •• Resolve each force in Problem 43 into its horizontal and vertical components, producing two couples. The algebraic sum of the two component couples equals the resultant couple. Use this result to find the perpendicular distance between the lines of action of the two forces.

**46** •• **SSM** A section of a cathedral wall is shown in Figure 12-48. The arch attached to the wall exerts a force of $2 \times 10^5$ N directed at an angle of 30° below the horizontal at a point 10 m above the ground and the mass of the wall itself is 30,000 kg. The coefficient of static friction between the wall and the ground is $\mu_k = 0.8$ and the base of the wall is 1.25 m long. (*a*) Calculate the effective normal force, the frictional force, and the point at which the effective normal force acts on the wall. (This is called the *thrust point* by architects and civil

engineers.) (*b*) If the thrust point ever moves outside the base of the wall, the wall will overturn. Apart from aesthetics, placing a heavy statue on top of the wall has good engineering practicality: it will move the thrust point toward the center of the wall. Explain why.

**FIGURE 12-48**
**Problems 46, 47**

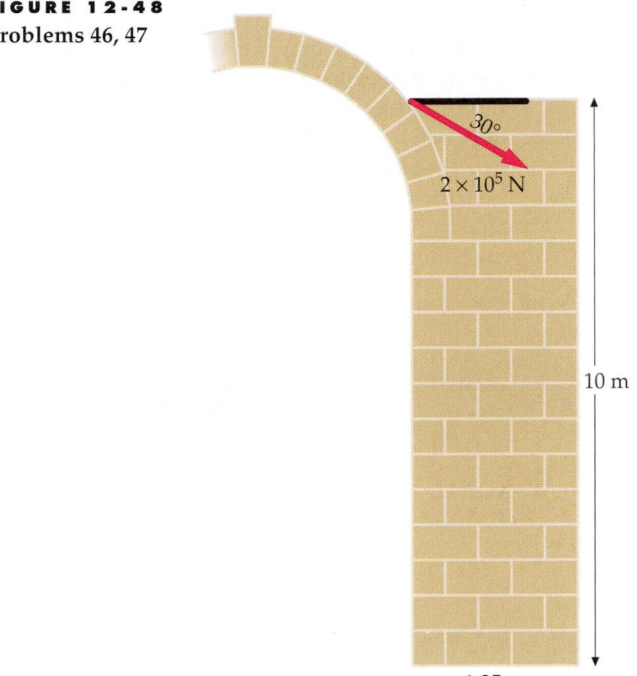

**47** •• In civil engineering, the *thrust line* can be found by determining the thrust point at *any* height inside a wall or similar structure. Determine the thrust line for the wall in Problem 46 and graph it using a spreadsheet program or graphing calculator. (Despite its name, the thrust line is really a curve.)

## Ladder Problems

**48** •• **SSM** **iSOLVE ✓** Romeo takes a uniform 10-m ladder and leans it against the smooth (frictionless) wall of the Capulet residence. The ladder's mass is 22.0 kg and the bottom rests on the ground 2.8 m from the wall. When Romeo, whose mass is 70 kg, gets 90 percent of the way to the top, the ladder begins to slip. What is the coefficient of static friction between the ground and the ladder?

**49** •• **SSM** A massless ladder of length $L$ leans against a smooth wall making an angle of $\theta$ with the horizontal floor. The coefficient of friction between the ladder and the floor is $\mu_s$. A man of mass $M$ climbs the ladder. What height $h$ can he reach before the ladder slips?

**50** •• A uniform ladder of length $L$ and mass $m$ leans against a frictionless vertical wall with its lower end on the ground. It makes an angle of 60° with the horizontal ground. The coefficient of static friction between the ladder and the ground is 0.45. If your mass is four times that of the ladder, how far up the ladder can you climb before it begins to slip?

**51** •• A ladder making an angle $\theta$ with the horizontal, of mass $m$ and length $L$, leans against a frictionless vertical wall. The center of mass is at a height $h$ from the floor. A force $F$ pulls horizontally against the ladder at the midpoint. Find the minimum coefficient of static friction $\mu_s$ for which the top end of the ladder will separate from the wall while the lower end does not slip.

**52** •• **iSOLVE✓** A 900-N man sits on top of a stepladder of negligible weight that rests on a frictionless floor as in Figure 12-49. There is a cross brace halfway up the ladder. The angle at the apex is $\theta = 30°$. (a) What is the force exerted by the floor on each leg of the ladder? (b) Find the tension in the cross brace. (c) If the cross brace is moved down toward the bottom of the ladder (maintaining the same angle $\theta$), will its tension be greater or less?

**FIGURE 12-49**
**Problem 52**

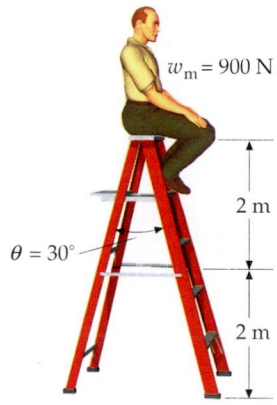

$w_m = 900\ N$

2 m

$\theta = 30°$

2 m

**53** •• **SSM** A uniform ladder rests against a frictionless vertical wall. The coefficient of static friction between the ladder and the floor is 0.3. What is the smallest angle at which the ladder will remain stationary?

## Stress and Strain

**54** • **SSM** A 50-kg ball is suspended from a steel wire of length 5 m and radius 2 mm. By how much does the wire stretch?

**55** • **iSOLVE✓** Copper has a breaking stress of about $3 \times 10^8\ N/m^2$. (a) What is the maximum load that can be hung from a copper wire of diameter 0.42 mm? (b) If half this maximum load is hung from the copper wire, by what percentage of its length will it stretch?

**56** • **iSOLVE** A 4-kg mass is supported by a steel wire of diameter 0.6 mm and length 1.2 m. How much will the wire stretch under this load?

**57** • **SSM** **iSOLVE** As a runner's foot pushes off on the ground, the shearing force acting on an 8-mm-thick sole is shown in Figure 12-50. If the force of 25 N is distributed over an area of 15 cm², find the angle of shear $\theta$, given that the shear modulus of the sole is $1.9 \times 10^5\ N/m^2$.

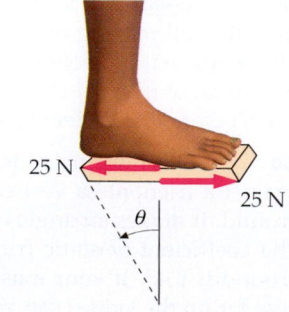

25 N

25 N

$\theta$

**FIGURE 12-50** Problem 57

**58** •• **iSOLVE** A steel wire of length 1.5 m and diameter 1 mm is joined to an aluminum wire of identical dimensions to make a composite wire of length 3.0 m. What is the length of the composite wire if it is used to support a mass of 5 kg?

**59** •• A force $F$ is applied to a long wire of length $L$ and cross-sectional area $A$. Show that if the wire is considered to be a spring, the force constant $k$ is given by $k = AY/L$ and the energy stored in the wire is $U = \frac{1}{2}F\Delta L$, where $Y$ is Young's modulus and $\Delta L$ is the amount the wire has stretched.

**60** •• **iSOLVE** The steel E string of a violin is under a tension of 53 N. The diameter of the string is 0.20 mm and its length under tension is 35.0 cm. Find (a) the unstretched length of this string and (b) the work needed to stretch the string.

**61** •• **SSM** When a rubber strip with a cross section of 3 mm × 1.5 mm is suspended vertically and various masses are attached to it, a student obtains the following data for length versus load:

| Load, kg | 0 | 0.1 | 0.2 | 0.3 | 0.4 | 0.5 |
|---|---|---|---|---|---|---|
| Length, cm | 5.0 | 5.6 | 6.2 | 6.9 | 7.8 | 10.0 |

(a) Find Young's modulus for the rubber strip for small loads.
(b) Find the energy stored in the strip when the load is 0.15 kg. (See Problem 59.)

**62** •• A large mirror is hung from a nail as shown in Figure 12-51. The supporting steel wire has a diameter of 0.2 mm and an unstretched length of 1.7 m. The distance between the points of support at the top of the mirror's frame is 1.5 m. The mass of the mirror is 2.4 kg. What is the distance between the nail and the top of the frame when the mirror is hung?

**FIGURE 12-51**
**Problems 62, 86**

**63** •• **iSOLVE✓** Two masses, $M_1$ and $M_2$, are supported by wires that have equal lengths when unstretched. The wire supporting $M_1$ is an aluminum wire 0.7 mm in diameter, and the one supporting $M_2$ is a steel wire 0.5 mm in diameter. What is the ratio $M_1/M_2$ if the two wires stretch by the same amount?

**64** •• A 0.5-kg ball is attached to an aluminum wire having a diameter of 1.6 mm and an unstretched length of 0.7 m. The other end of the wire is fixed to a post. The ball rotates about the post in a horizontal plane at a rotational speed such that the angle between the wire and the horizontal is 5.0°. Find the tension in the wire and its length.

**65** •• **SSM** An elevator cable is to be made of a new type of composite developed by Acme Laboratories. In the lab, a sample of the cable that is 2 m long and has a cross-sectional area of 0.2 mm² fails under a load of 1000 N. The cable in the elevator will be 20 m long and have a cross-sectional area of 1.2 mm². It will need to support a load of 20,000 N safely. Will it?

**66** ••• **SSM** When a material is stretched in one direction, if its density remains constant, then (because its total volume remains constant), its length must decrease in one or both of the other directions. Take a rectangular block of length $x$, width $y$, and depth $z$, and pull on it so that its new length $x'' = x + \Delta x$. If $\Delta x \ll x$ and $\Delta y/y = \Delta z/z$, show that $\Delta y/y = -\frac{1}{2} \Delta x/x$.

**67** •• If you are given a wire with a circular cross-section of radius $r$ and a length $L$, show that $\Delta r/r = -\frac{1}{2} \Delta L/L$, assuming that $\Delta L \ll L$. (See Problem 66.)

**68** ••• **SSM** For most materials listed in Table 12-1, the tensile strength is two to three orders of magnitude lower than Young's modulus. Consequently, most of these materials will break before their strain exceeds 1 percent. Of man-made materials, nylon has about the greatest extensibility—it can take strains of about 0.2 before breaking. But spider silk beats anything man-made. Certain forms of spider silk can take strains on the order of 10 before breaking! (a) If such a thread has a circular cross-section of radius $r_0$ and unstretched length $L_0$, find its new radius $r$ when stretched to a length $L = 10L_0$. (b) If the Young's modulus of the spider thread is $Y$, calculate the tension needed to break the thread in terms of $Y$ and $r_0$.

## General Problems

**69** • A 90-N board 12 m long rests on two supports, each 1 m from the end of the board. A 360-N block is placed on the board 3 m from one end as shown in Figure 12-52. Find the force exerted by each support on the board.

**FIGURE 12-52**
**Problem 69**

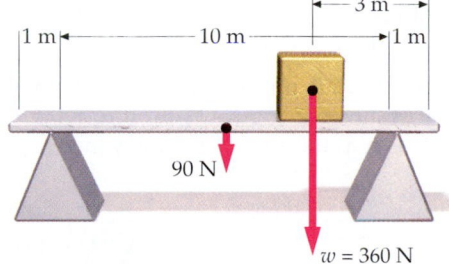

**70** • The height of the center of gravity of a man standing erect is determined by weighing the man as he lies on a board of negligible weight supported by two scales as shown in Figure 12-53. If the man's height is 188 cm and the left scale reads 445 N while the right scale reads 400 N, where is his center of gravity relative to his feet? Would the reading on the scales be different if he were holding his head up—keeping it slightly above the board?

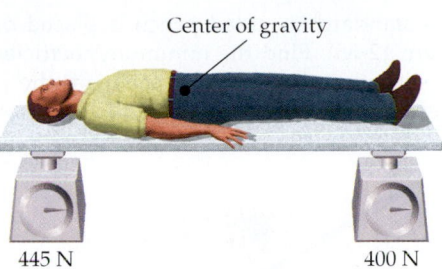

**FIGURE 12-53** Problem 70

**71** • **SSM** **iSOLVE** ✓ Figure 12-54 shows a mobile consisting of four weights hanging on three rods of negligible mass. Find the value of each of the unknown weights if the mobile is to balance. *Hint: Find the weight $w_1$ first.*

**FIGURE 12-54** Problem 71

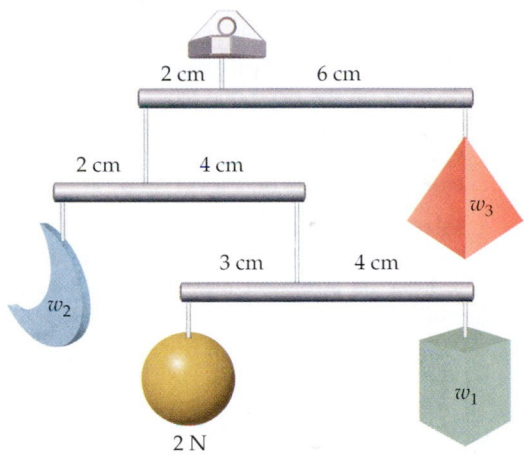

**72** • A rope and pulley system, called a block and tackle, is used to support an object of mass $m$, as shown in Figure 12-55. The object is being raised at constant speed. When a segment of rope of length $L$ passes over the uppermost pulley wheel, the height of the lower pulley is increased by $h$. (a) What is the ratio $L/h$? (b) Assume that the block and tackle are massless and frictionless. Show that $FL = mgh$ by applying the work–energy principle to the block–tackle object.

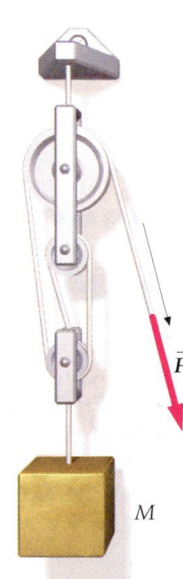

**FIGURE 12-55** Problem 72

**73** •• A plate of mass $M$ in the shape of an equilateral triangle is suspended from one corner and a mass $m$ is suspended from another of its corners. If the base of the triangle makes an angle of 6.0° with the horizontal, what is the ratio $m/M$?

**74** •• A standard six-sided pencil is placed on a pad of paper (Figure 12-56). Find the minimum coefficient of static friction $\mu_s$ such that, if the pad is inclined, the pencil rolls down rather than slides.

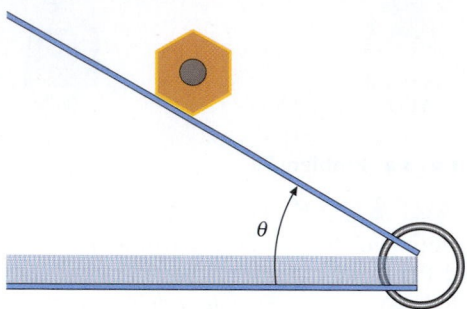

**FIGURE 12-56** Problem 74

**75** •• SSM A uniform box of mass 8 kg that is twice as tall as it is wide rests on the floor of a truck. What is the maximum coefficient of static friction between the box and floor so that the box will slide toward the rear of the truck rather than tip when the truck accelerates on a level road?

**76** •• A balance scale has unequal arms. A 1.5-kg block appears to have a mass of 1.95 kg on the left pan of the scale (Figure 12-57). Find its apparent mass if the 1.5-kg block is placed on the right pan.

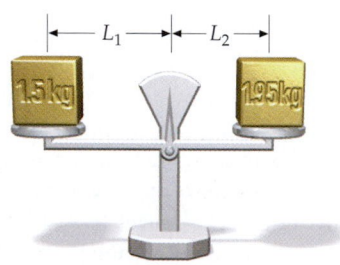

**FIGURE 12-57** Problem 76

**77** •• SSM A cube of mass $M$ leans against a frictionless wall making an angle of $\theta$ with the floor as shown in Figure 12-58. Find the minimum coefficient of static friction $\mu_s$ between the cube and the floor that allows the cube to stay at rest.

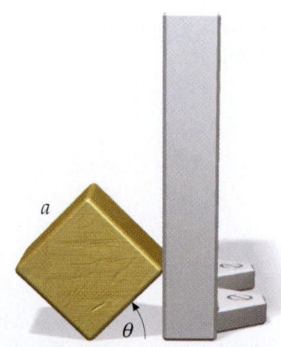

**FIGURE 12-58** Problem 77

**78** •• Figure 12-59 shows a steel meter stick hinged to a vertical wall and supported by a thin wire. The wire and meter stick make angles of 45° with the vertical. The mass of the meter stick is 5.0 kg. When a 10.0-kg block is suspended from the midpoint of the meter stick, the tension $T$ in the supporting wire is 52 N. If the wire will break should the tension exceed 75 N, what is the maximum distance along the meter stick at which the block can be suspended?

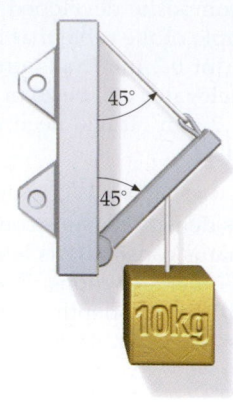

**FIGURE 12-59** Problem 78

**79** •• iSOLVE✓ Figure 12-60 shows a 20-kg ladder leaning against a frictionless wall and resting on a frictionless horizontal surface. To keep the ladder from slipping, the bottom of the ladder is tied to the wall with a thin wire; the tension in the wire is 29.4 N. The wire will break if the tension exceeds 200 N. (a) If an 80-kg person climbs halfway up the ladder, what force will be exerted by the ladder against the wall? (b) How far up can an 80-kg person climb this ladder?

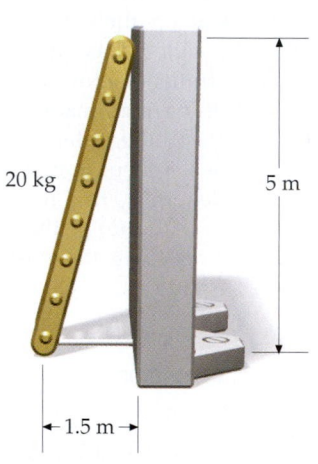

**FIGURE 12-60** Problem 79

**80** •• SSM A uniform cube can be moved along a horizontal plane either by pushing the cube so that it slips or by turning it over ("rolling"). What coefficient of kinetic friction $\mu_k$ between the cube and the floor makes both ways equal in terms of the work needed?

**81** •• A tall, uniform, rectangular block sits on an inclined plane as shown in Figure 12-61. If $\mu_s = 0.4$, does the block slide or fall over as the angle $\theta$ is slowly increased?

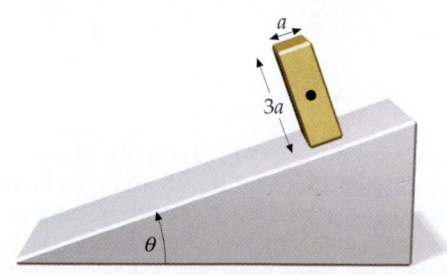

**FIGURE 12-61** Problem 81

**82** •• A 360-kg object is supported on a wire attached to a 15-m-long steel bar that is pivoted at a vertical wall and supported by a cable as shown in Figure 12-62. The mass of the bar is 85 kg. (*a*) With the cable attached to the bar 5.0 m from the lower end as shown, find the tension in the cable and the force exerted by the wall on the steel bar. (*b*) Repeat if a somewhat longer cable is attached to the steel bar 5.0 m from its upper end, maintaining the same angle between the bar and the wall.

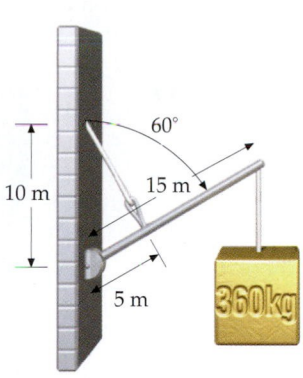

**FIGURE 12-62** Problem 82

**83** •• Repeat Problem 75 if the truck accelerates up a hill that makes an angle of 9.0° with the horizontal.

**84** •• **SSM** A thin rod 60 cm long is balanced 20 cm from one end when an object whose mass is (2*m* + 2 grams) is at the end nearest the pivot and an object of mass *m* is at the opposite end (Figure 12-63*a*). Balance is again achieved if the object whose mass is (2*m* + 2 grams) is replaced by the object of mass *m* and no object is placed at the other end (Figure 12-63*b*). Determine the mass of the rod.

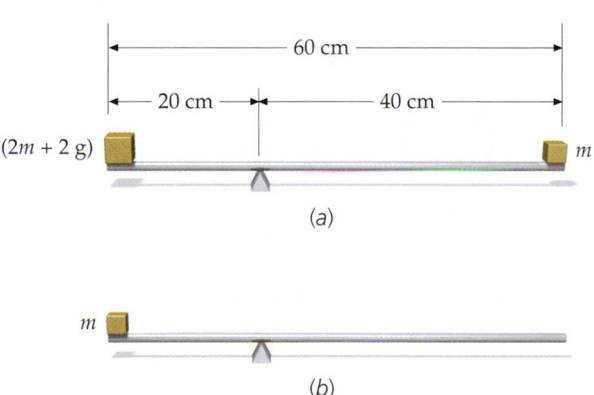

**FIGURE 12-63** Problem 84

**85** •• **SSM** If you balance a meter stick across two fingers (one from each hand) and slowly bring your hands together, the fingers will always meet in the middle of the stick, no matter where they are initially placed. (*a*) Explain why this is true. (*b*) As you move your fingers together, first one will move, and then the other; both will not move at the same time. Explain quantitatively why this is true; assume that the coefficient of static friction between the meter stick and your fingers is $\mu_s$ and the coefficient of kinetic friction is $\mu_k$.

**86** •• When a picture is hung on a smooth vertical wall using a wire and a nail, as in Figure 12-51, the picture almost always tips slightly forward, that is, the plane of the picture makes a small angle with the vertical. (*a*) Explain why pictures supported in this manner generally do not hang flush against the wall. (*b*) A framed picture 1.5 m wide and 1.2 m high with a mass of 8.0 kg is hung as in Figure 12-51 using a wire 1.7 m long. The ends of the wire are fastened to the sides of the frame at the rear, 0.4 m below the top. When the picture is hung, the angle between the plane of the frame and the wall is 5.0°. Determine the force that the wall exerts on the bottom of the frame.

**87** •• **SSM** If a train travels around an unbanked bend in the rail bed too fast, the freight cars will tip over. Assume that the cargo portion of each freight car is a regular parallelepiped of uniform density and $1.5 \times 10^4$ kg mass, 10 m long, 3.0 m high, and 2.20 m wide, and that its base is 0.65 m above the rails. The car axles are 7.6 m apart, 1.2 m from each end of the boxcar. The separation between the rails is 1.55 m. Find the maximum safe speed of the train if the radius of curvature of the bend is (*a*) 150 m and (*b*) 240 m. (Neglect the mass of the flatcar.)

**88** •• For balance, a tightrope walker uses a thin rod 8 m long and bowed in a circular arc shape. At each end of the rod is a lead mass of 20 kg. The tightrope walker, whose mass is 58 kg and whose center of gravity is 0.90 m above the rope, holds the rod tightly at its center 0.65 m above the rope. What should be the radius of curvature of the arc of the rod so that the tightrope walker will be in neutral equilibrium as he slowly makes his way across the rope? Neglect the mass of the rod.

**89** ••• **SSM** We have a large number of identical uniform bricks, each of length *L*. If we stack one on top of another lengthwise (see Figure 12-64), the maximum offset that will allow the top brick to rest on the bottom brick is *L*/2. (*a*) Show that if we place this two-brick stack on top of a third brick, the maximum offset of the second brick on the third brick is *L*/3. (*b*) Show that, in general, if we have a stack of *N* bricks, the maximum offset of the (*n* − 1)th brick (counting down from the top) on the *n*th brick is *L*/*n*, where *n* ≤ *N*. (*c*) Write a spreadsheet program to calculate total offset (the sum of the individual offsets) for a stack of *N* bricks, and calculate this for *L* = 1 m and *N* = 5, 10, and 100. (*d*) Does the sum of the individual offsets approach a finite limit as *N* → ∞? If so, what is that limit?

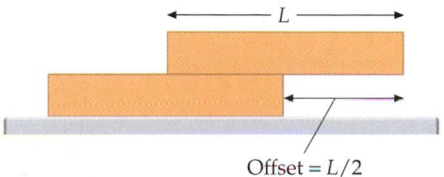

**FIGURE 12-64** Problem 89

**90 •••** A uniform sphere of radius $R$ and mass $M$ is held at rest on an inclined plane of angle $\theta$ by a horizontal string, as shown in Figure 12-65. Let $R = 20$ cm, $M = 3$ kg, and $\theta = 30°$. (*a*) Find the tension in the string. (*b*) What is the normal force exerted on the sphere by the inclined plane? (*c*) What is the frictional force acting on the sphere?

**FIGURE 12-65**
Problem 90

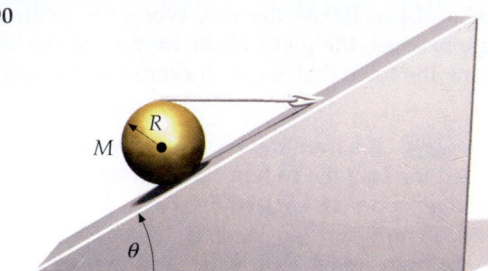

**91 •••** The legs of a tripod make equal angles of 90° with each other at the apex, where they join together. A 100-kg block hangs from the apex. What are the compressional forces in the three legs?

**92 ••** Figure 12-66 shows a 20-cm-long uniform beam resting on a cylinder of 4-cm radius. The mass of the beam is 5.0 kg and that of the cylinder is 8.0 kg. The coefficient of static friction between beam and cylinder is zero, whereas the coefficient of static friction between the cylinder and the floor, and between the beam and the floor, are each greater than zero. Are there any values for these coefficients of static friction such that the system is in static equilibrium? If so, what are these values? If not, explain.

**FIGURE 12-66**
Problem 92

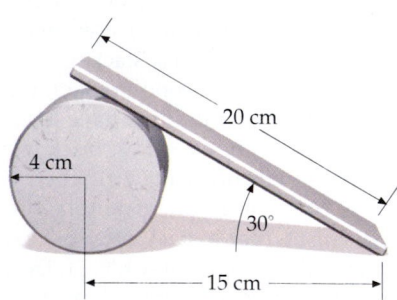

**93 •••** Two solid smooth (frictionless) spheres of radius $r$ are placed inside a cylinder of radius $R$ as in Figure 12-67. The mass of each sphere is $m$. Find the force exerted by the bottom of the cylinder on the bottom sphere, the force exerted by the wall of the cylinder on each sphere, and the force exerted by one sphere on the other.

**FIGURE 12-67**
Problem 93

**94 •••** **SSM** A solid cube of side length $a$ balanced atop a cylinder of diameter $d$ is in unstable equilibrium if $d \ll a$ and is in stable equilibrium if $d \gg a$ (Figure 12-68). Determine the minimum value of $d/a$ for which the cube is in stable equilibrium.

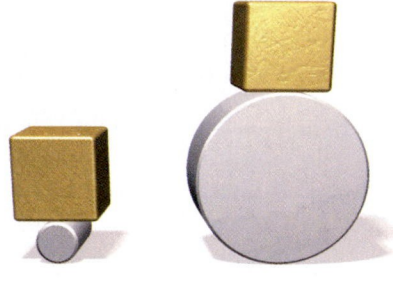

**FIGURE 12-68** Problem 94

# Fluids

THE HOOVER DAM IS 660 FEET THICK
AT THE BOTTOM, BUT ONLY 45 FEET
THICK AT THE TOP OF THE SPAN.

**?** **Dams are designed to be
thicker at the bottom than at the
top. Why is this so? (See
Example 13.2.)**

13-1    Density

13-2    Pressure in a Fluid

13-3    Buoyancy and Archimedes' Principle

13-4    Fluids in Motion

Fluids include both liquids and gases. Liquids flow under gravity until they occupy the lowest possible regions of their containers. Gases expand to fill their containers. Fluids permeate our environment as well as our bodies. To understand fluid behavior is to understand much about ourselves and our interactions with the world around us.

➤ **We begin this chapter by studying fluids at rest, followed by the study of steady-state flow with an emphasis on laminar flow.**

In a gas, the average distance between two molecules is large compared with the size of a molecule. The molecules have little influence on one another except during their brief collisions. In a liquid or solid, the molecules are close together and exert forces on one another that are comparable to the forces that bind atoms into molecules. Molecules in a liquid form temporary short-range bonds that are continually broken and reformed due to the thermal kinetic energy of the molecules. These bonds hold the liquid together; if the bonds were not present, the liquid would immediately evaporate and the molecules would escape as a vapor. The strength of the bonds in a liquid depends on the type of molecule that makes up the liquid. For example, the bonds between helium molecules are very weak and, for this reason, helium does not liquefy at atmospheric pressure unless the temperature is 4.2 K or lower.

# 13-1 Density

An important property of a substance is the ratio of its mass to its volume, which is called its **density**:

$$\text{Density} = \frac{\text{mass}}{\text{volume}}$$

The Greek letter $\rho$ (rho) is usually used to denote density:

$$\rho = \frac{m}{V} \qquad \text{13-1}$$

DEFINITION—DENSITY

Because the gram was originally defined as the mass of one cubic centimeter of water, the density of water in cgs units is $1 \text{ g/cm}^3$. Converting to SI units, we obtain for the density of water

$$\rho_w = \frac{1 \text{ g}}{\text{cm}^3} \times \frac{1 \text{ kg}}{10^3 \text{ g}} \times \left(\frac{100 \text{ cm}}{1 \text{ m}}\right)^3 = 10^3 \text{ kg/m}^3 \qquad \text{13-2}$$

Precise measurements of density must take temperature into account, since the densities of most materials, including water, vary with temperature. Equation 13-2 gives the maximum value for the density of water, which occurs at 4°C. Table 13-1 lists the densities of some common materials.

A convenient unit of volume for fluids is the **liter** (L):

$$1 \text{ L} = 10^3 \text{ cm}^3 = 10^{-3} \text{ m}^3$$

In terms of this unit, the density of water at 4°C is $1.00 \text{ kg/L} = 1.00 \text{ g/mL}$. When an object's density is greater than that of water, it sinks in water. When its density is less, it floats. The ratio of the density of a substance to that of water is called its **specific gravity.** For example, the specific gravity of aluminum is 2.7, meaning that a volume of aluminum has 2.7 times the mass of an equal volume of water. The specific gravities of objects that sink in water range from 1 to about 22.5 (for the densest element, osmium).

Most solids and liquids expand only slightly when heated, and contract slightly when subjected to an increase in external pressure. Since these changes in volume are relatively small, we often treat the densities of solids and liquids as approximately independent of temperature and pressure. The density of a gas, on the other hand, depends strongly on the pressure and temperature, so these variables must be specified when reporting the densities of gases. By convention, the standard conditions for the measurement of physical properties are atmospheric pressure at sea level and a temperature of 0°C. The densities for the substances listed in Table 13-1 are for these conditions. Note that the densities of liquids and solids are considerably greater than those of gases. For example, the density of water is about 800 times that of air under standard conditions.

## TABLE 13-1

**Densities of Selected Substances**

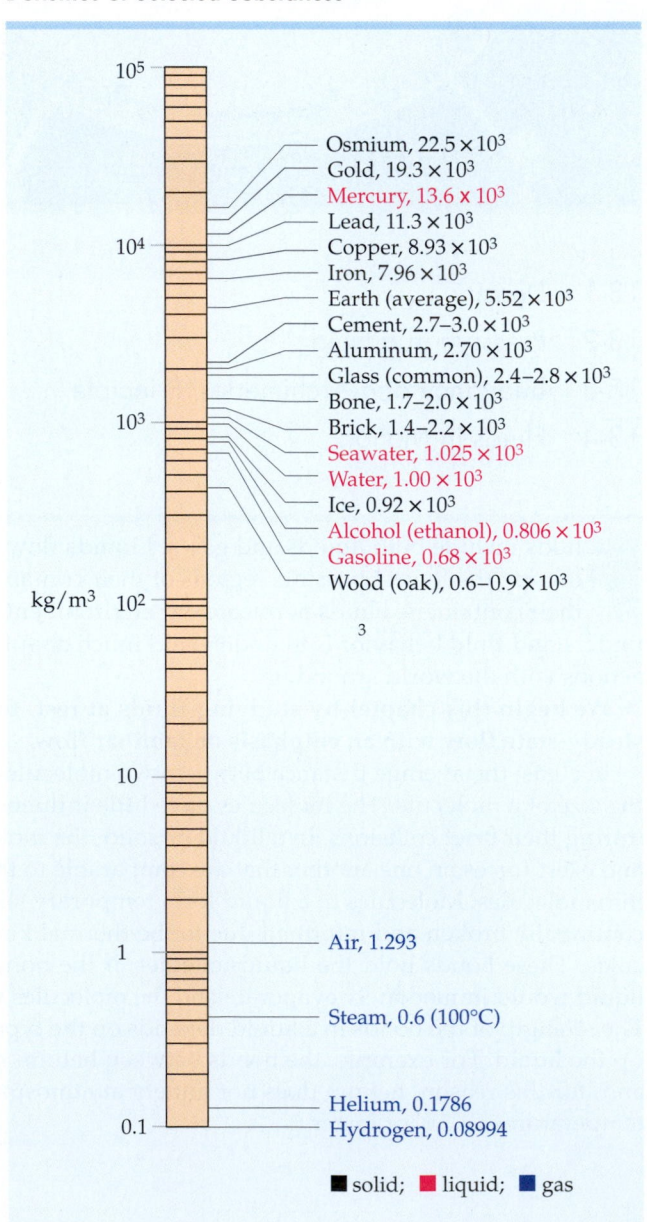

Osmium, $22.5 \times 10^3$
Gold, $19.3 \times 10^3$
Mercury, $13.6 \times 10^3$
Lead, $11.3 \times 10^3$
Copper, $8.93 \times 10^3$
Iron, $7.96 \times 10^3$
Earth (average), $5.52 \times 10^3$
Cement, $2.7–3.0 \times 10^3$
Aluminum, $2.70 \times 10^3$
Glass (common), $2.4–2.8 \times 10^3$
Bone, $1.7–2.0 \times 10^3$
Brick, $1.4–2.2 \times 10^3$
Seawater, $1.025 \times 10^3$
Water, $1.00 \times 10^3$
Ice, $0.92 \times 10^3$
Alcohol (ethanol), $0.806 \times 10^3$
Gasoline, $0.68 \times 10^3$
Wood (oak), $0.6–0.9 \times 10^3$

Air, 1.293

Steam, 0.6 (100°C)

Helium, 0.1786
Hydrogen, 0.08994

■ solid; ■ liquid; ■ gas

**EXAMPLE 13-1**

**A 200-mL flask is filled with water at 4°C. When the flask is heated to 80°C, 6 g of water spill out. What is the density of water at 80°C? (Assume that the expansion of the flask is negligible.)**

**PICTURE THE PROBLEM** The density of water at 80°C is $\rho' = m'/V$, where $V = 200$ mL $= 200$ cm$^3$ is the volume of the flask, and $m'$ is the mass remaining in the flask after 6 g spill out. We find $m'$ by first finding the mass of water originally in the flask.

1. Calculate the original mass $m$ of water in the flask at 4°C using $\rho = 1.00$ g/cm:

$$m = \rho V = (1.00 \text{ g/cm}^3)(200 \text{ cm}^3) = 200 \text{ g}$$

2. Calculate the mass of water remaining $m'$ after 6 g spill out:

$$m' = m - 6 \text{ g} = 200 \text{ g} - 6 \text{ g} = 194 \text{ g}$$

3. Use this value of $m'$ to find the density of water at 80°:

$$\rho' = \frac{m'}{V} = \frac{194 \text{ g}}{200 \text{ cm}^3} = \boxed{0.97 \text{ g/cm}^3}$$

**EXERCISE** A solid metal cube 8 cm on a side has a mass of 4.08 kg. (*a*) What is the density of the cube? (*b*) If the cube is made from a single element listed in Table 13-1, what is the element? (*Answer* (*a*) 7.97 kg/L (*b*) iron)

**EXERCISE** A gold brick is 5 cm $\times$ 10 cm $\times$ 20 cm. What is its mass? (*Answer* 19.3 kg)

# 13-2 Pressure in a Fluid

When a body is submerged in a fluid such as water, the fluid exerts a force perpendicular to the surface of the body at each point on the surface. This force per unit area is called the **pressure** $P$ of the fluid:

$$P = \frac{F}{A} \qquad\qquad 13\text{-}3$$

DEFINITION — PRESSURE

The SI unit of pressure is the newton per square meter (N/m$^2$), which is called the **pascal** (Pa):

$$1 \text{ Pa} = 1 \text{ N/m}^2 \qquad\qquad 13\text{-}4$$

In the U.S. customary system, pressure is usually given in pounds per square inch (lb/in.$^2$). Another common unit of pressure is the atmosphere (atm), which equals approximately the air pressure at sea level. One atmosphere is now defined to be 101.325 kilopascals, which is approximately 14.70 lb/in.$^2$:

$$1 \text{ atm} = 101.325 \text{ kPa} \approx 14.70 \text{ lb/in.}^2 \qquad\qquad 13\text{-}5$$

Other units of pressure in common use will be discussed later in this chapter.

The pressure due to a fluid pressing in on an object tends to compress the object. The ratio of the increase in pressure $\Delta P$ to the fractional decrease in volume $(-\Delta V/V)$ is called the **bulk modulus**:

$$B = -\frac{\Delta P}{\Delta V/V} \qquad\qquad 13\text{-}6$$

DEFINITION — BULK MODULUS

(The minus sign in Equation 13-6 is introduced to make $B$ positive, since all materials decrease in volume when subjected to an increase in external pressure.)

The more difficult it is to compress a material, the smaller is the fractional volume decrease $-\Delta V/V$ for a given pressure increase $\Delta P$, and hence the greater the bulk modulus. The **compressibility** is the reciprocal of the bulk modulus. (The easier it is to compress a material, the larger the compressibility.) Liquids, gases, and solids all have a bulk modulus. Since liquids and solids are relatively incompressible, they have large values of $B$, and these values are relatively independent of temperature and pressure. Gases, on the other hand, are easily compressed, and their values for $B$ depend strongly on pressure and temperature. Table 13-2 charts values for the bulk modulus of various materials.

As any scuba diver knows, the pressure in a lake or ocean increases with depth. Similarly, the pressure of the atmosphere decreases with altitude. For a liquid such as water, whose density is approximately constant throughout, the pressure increases linearly with depth. We can see this by considering a column of liquid of cross-sectional area $A$, as shown in Figure 13-1. To support the weight of the liquid in the column of height $\Delta h$, the pressure at the bottom of the column must be greater than the pressure at the top by the weight of the column. The weight of the liquid in the column is

$$w = mg = (\rho V)g = \rho A \, \Delta h g$$

If $P_0$ is the pressure at the top and $P$ is the pressure at the bottom, the net upward force exerted by this pressure difference is $PA - P_0 A$. Setting this net upward force equal to the weight of the column, we obtain

$$PA - P_0 A = \rho A \Delta h g$$

or

$$P = P_0 + \rho g \Delta h \qquad (\rho \text{ constant}) \qquad \text{13-7}$$

**EXERCISE** How far below the surface of a lake is the pressure equal to 2 atm? (The pressure at the surface is 1 atm.) (*Answer* With $P_0 = 1$ atm $= 101$ kPa, $P = 2$ atm, $\rho = 1000$ kg/m³, and $g = 9.81$ N/kg, we have $\Delta h = \Delta P/\rho g = 10.3$ m. The pressure at a depth of 10.3 m is twice that at the surface.)

**TABLE 13-2**

**Approximate Values for the Bulk Modulus $B$ of Various Materials**

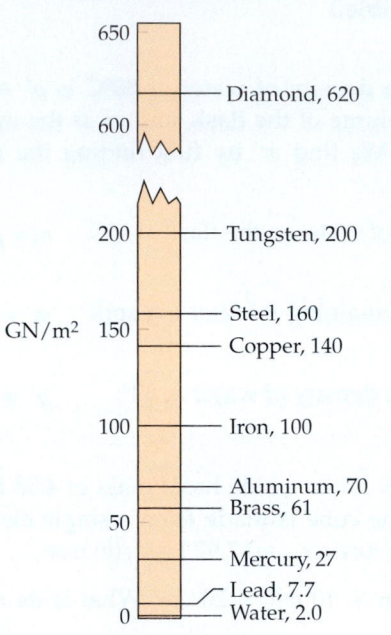

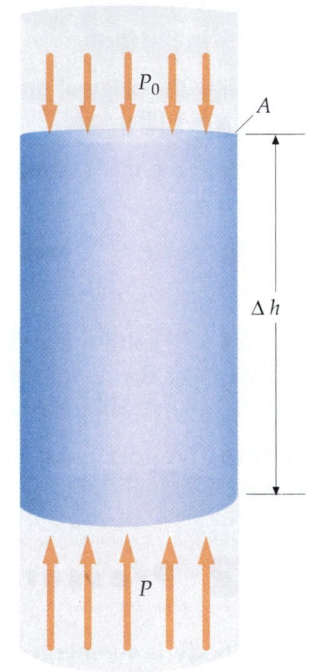

**FIGURE 13-1** Column of water of height $\Delta h$ and cross-sectional area $A$. The pressure $P$ at the bottom must be greater than the pressure $P_0$ at the top to balance the weight of the water.

---

*FORCE ON A DAM*                **EXAMPLE 13-2**

A rectangular dam 30 m wide supports a body of water to a depth of 25 m. Find the total horizontal force on the dam.

**PICTURE THE PROBLEM** Because the pressure varies with depth, we cannot merely multiply the pressure times the area of the dam to find the force exerted by the water. Instead we can consider the force exerted on a strip of width $L = 30$ m, height $dh$, and area $dA = Ldh$ at a depth $h$ (Figure 13-2), and then integrate from $h = 0$ to $h = H = 25$ m. The water pressure at depth $h$ is $P_{atm} + \rho g h$. We can omit the atmospheric pressure because it is exerted on each side of the wall.

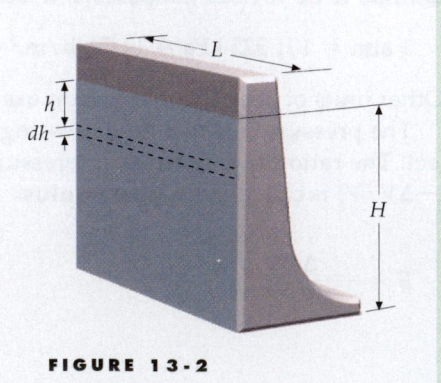

**FIGURE 13-2**

1. Express the force $dF$ on the element of width $L$ and height $dh$ in terms of the net pressure $\rho hg$

$$dF = P\,dA = \rho ghL\,dh$$

2. Integrate from $h = 0$ to $h = H$:

$$F = \int_{h=0}^{h=H} dF = \int_{0}^{H} \rho ghL\,dh = \rho gL\,\frac{h^2}{2}\bigg|_{0}^{H} = \frac{1}{2}\rho gLH^2$$

3. Substitute the given values to find the numerical result:

$$F = \tfrac{1}{2}\rho gLH^2 = \tfrac{1}{2}(1000\ \text{kg/m}^3)(9.81\ \text{N/kg})(30\ \text{m})(25\ \text{m})^2$$

$$\boxed{= 9.20 \times 10^7\ \text{N}}$$

**REMARKS** Dams typically are thicker at the bottom than at the top because the pressure on the dam increases with the depth of the water. Because the pressure on the dam is *proportional* to the depth of the water, the average pressure is $P_{\text{atm}} + \rho g H/2$ (the pressure at depth $h = H/2$).

The result that the pressure increases linearly with depth holds for a liquid in any container, independent of the shape of the container. Furthermore, the pressure is the same at all points at the same depth. We can see this by comparing the pressure at point 1 in Figure 13-3*a* with the pressure at point 2, which is inside an underwater cave. First we compare the pressure at points 1 and 3, a point directly below 1 at the same depth as 2, as shown in Figure 13-3*b*. Consider the vertical forces on the vertical column of water of height $\Delta h$ and cross-sectional area $A$ joining points 1 and 3. The upward force on the column, $P_3A$, balances the two downward forces $P_1A$ and $mg$, where $m = \rho A\Delta h$ is the mass of the water in the column ($A\Delta h$ is the volume of the column). That is, $P_3A = P_1A + \rho A\Delta hg$. Dividing through by $A$ gives

$$P_3 = P_1 + \rho g\Delta h$$

Next consider the forces on the horizontal cylinder of water, also of cross-sectional area $A$, connecting points 2 and 3 (Figure 13-3*c*). There are two forces with components along the

**FIGURE 13-3**

(a)

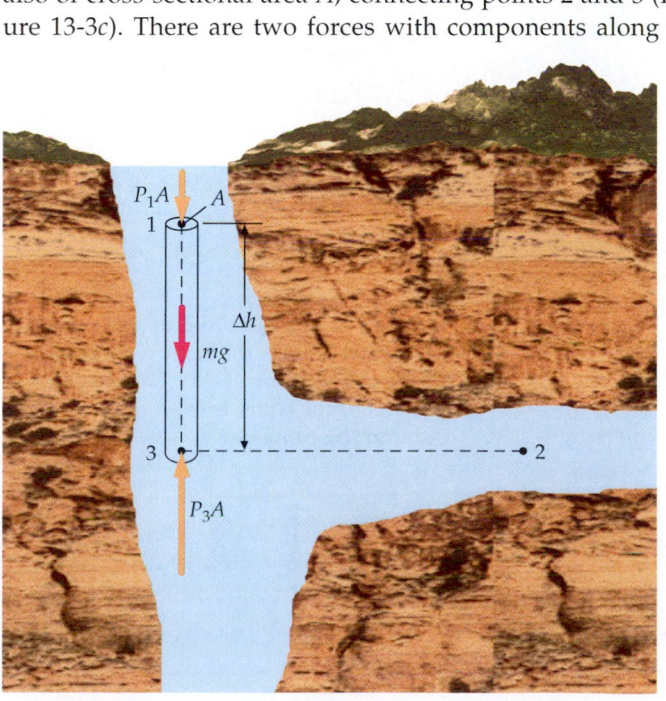

(b)

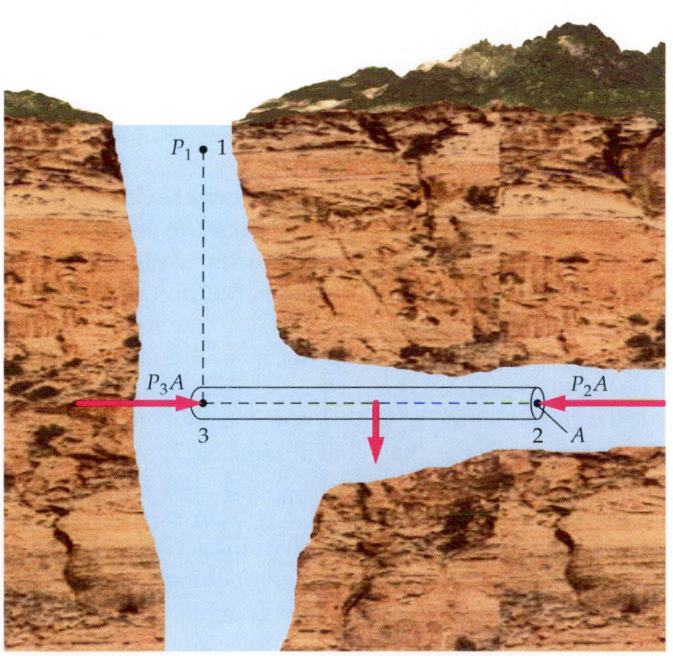

(c)

cylinder's axis, $P_3A$ and $P_2A$. The fact that these forces balance each other means that $P_3 = P_2$. It follows that

$$P_2 = P_1 + \rho g \Delta h$$

If we increase the pressure in a container of water by pressing down on the top surface with a piston, the increase in pressure is the same throughout the liquid. This is known as **Pascal's principle,** named after Blaise Pascal (1623–1662):

> A pressure change applied to an enclosed liquid is transmitted undiminished to every point in the liquid and to the walls of the container.

PASCAL'S PRINCIPLE

A common application of Pascal's principle is the hydraulic lift shown in Figure 13-4.

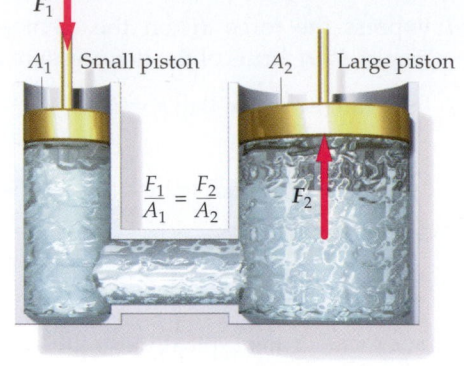

**FIGURE 13-4** Hydraulic lift. A small force $F_1$ on the small piston produces a change in pressure $F_1/A_1$ that is transmitted by the liquid to the large piston. Since the pressures at the small and large pistons are the same, the forces are related by $F_2/A_2 = F_1/A_1$. Since the area of the large piston is much greater than that of the small piston, the force on the large piston $F_2 = (A_2/A_1)F_1$ is much greater than $F_1$.

---

*A Hydraulic Lift*                                    **EXAMPLE 13-3**

**The large piston in a hydraulic lift has a radius of 20 cm. What force must be applied to the small piston of radius 2 cm to raise a car of mass 1500 kg?**

**PICTURE THE PROBLEM** The pressure $P$ times the area $A_2$ of the large piston must equal the weight $mg$ of the car. The force $F_1$ that must be exerted on the small piston is this pressure times the area $A_1$. (See Figure 13-4.)

1. The force $F_1$ is the pressure $P$ times the area $A_1$:

$$F_1 = PA_1$$

2. The pressure $P$ times the area $A_2$ equals the weight of the car:

$$PA_2 = mg \quad \text{so} \quad P = \frac{mg}{A_2}$$

3. Substitute this result for $P$ into the step 1 result and calculate $F_1$:

$$F_1 = PA_1$$

$$= \frac{mg}{A_2}A_1 = mg\frac{A_1}{A_2} = mg\frac{\pi r_1^2}{\pi r_2^2}$$

$$= (1500 \text{ kg})(9.81 \text{ N/kg})\left(\frac{2 \text{ cm}}{20 \text{ cm}}\right)^2$$

$$= \boxed{147 \text{ N}}$$

---

Figure 13-5 shows water in a container with sections of different shapes. At first glance, it might seem that the pressure at the bottom of section 3 of the container would be greatest and that water would therefore be forced to a greater height in section 2, which is smaller. But that does not happen, a result known as the **hydrostatic paradox. The pressure depends only on the depth of the water, not on the shape of the container, so at the same depth the pressure is the same in all parts of the container, a finding that can be shown experimentally.** Although the water in section 4 of the container weighs more than that in section 2, the portion of the water in section 4 that is not above the opening at the bottom is supported by the horizontal shelf of the section. In fact, the water above the opening at the bottom of section 5 weighs less than the water above an opening of the same size at the bottom of section 1. However, the

**FIGURE 13-5** The hydrostatic paradox. The water level is the same regardless of the shape of the vessel. The shaded portion of the water is supported by the sides of the container.

1          2          3          4          5

horizontal shelf of section 5 exerts a downward force on the water—exactly compensating for the shortfall of weight.

We can use the fact that the pressure difference is proportional to the depth of a liquid to measure unknown pressures. Figure 13-6 shows a simple pressure gauge, the open-tube manometer. The top of the tube is open to the atmosphere at pressure $P_{at}$. The other end of the tube is at pressure $P$, which is to be measured. The difference $P - P_{at}$, called the **gauge pressure** $P_{gauge}$, is equal to $\rho g h$, where $\rho$ is the density of the liquid in the tube. The pressure you measure in your automobile tire is gauge pressure. When the tire is entirely flat, the gauge pressure is zero, and the absolute pressure in the tire is atmospheric pressure. The absolute pressure $P$ is obtained from the gauge pressure by adding atmospheric pressure to it:

$$P = P_{gauge} + P_{at} \qquad \text{13-8}$$

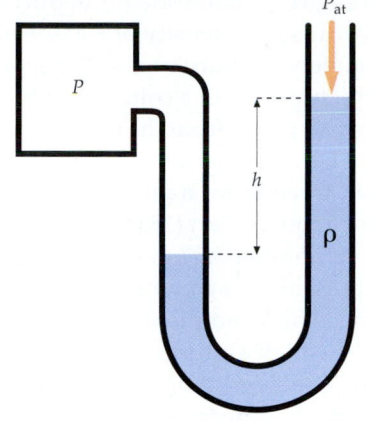

**FIGURE 13-6** Open-tube manometer for measuring an unknown pressure $P$. The difference $P - P_{at}$ equals $\rho g h$.

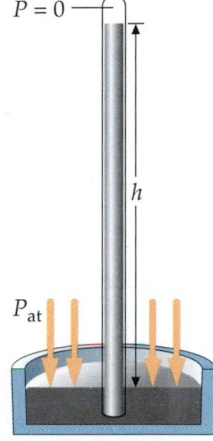

**FIGURE 13-7**

Figure 13-7 shows a mercury barometer, which is used to measure atmospheric pressure. The top end of the tube has been closed off and evacuated so that the pressure there is zero. The other end is submerged in a pool of mercury that is open to the atmosphere at pressure $P_{at}$. The pressure $P_{at}$ is $\rho g h$, where $\rho$ is the density of mercury.

**EXERCISE** At 0°C, the density of mercury is $13.595 \times 10^3$ kg/m³. What is the height of the mercury column in a barometer if the pressure is 1 atm = 101.325 kPa? (*Answer* $h = P/\rho g = 0.760$ m $= 760$ mm)

In practice, pressure is often measured in millimeters of mercury, a unit called the **torr,** after the Italian physicist Evangelista Torricelli, or in inches of mercury (written inHg). The various units of pressure are related as follows:

$$1 \text{ atm} = 760 \text{ mmHg} = 760 \text{ torr} = 29.9 \text{ in.Hg}$$
$$= 101.325 \text{ kPa} = 14.7 \text{ lb/in.}^2 \qquad \text{13-9}$$

Other units commonly used on weather maps are the **bar** and the **millibar,** which are defined as

$$1 \text{ bar} = 10^3 \text{ millibars} = 100 \text{ kPa} \qquad \text{13-10}$$

A pressure of 1 atm is about 1.3% greater than a pressure of 1 bar.

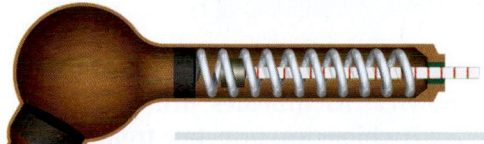

Tire-pressure gauge. The piston pushes the rod to the right until the force of the spring plus the force due to atmospheric pressure balances the force due to the air pressure in the tire.

---

*BLOOD PRESSURE IN THE AORTA* **EXAMPLE 13-4**

**The average gauge pressure in the human aorta is about 100 mmHg. Convert this average blood pressure to pascals.**

We use the conversion factors that can be obtained from Equation 13-9:

$$P = 100 \text{ mmHg} \times \frac{101.325 \text{ kPa}}{760 \text{ mmHg}}$$

$$= \boxed{13.3 \text{ kPa}}$$

---

**EXERCISE** Convert a pressure of 45 kPa to (*a*) millimeters of mercury and (*b*) atmospheres. (*Answer* (*a*) 338 mmHg (*b*) 0.444 atm)

The relation between pressure and altitude (or depth) is more complicated for a gas than for a liquid because the density of a gas is not constant like that of a liquid, but is approximately proportional to the pressure. As you go up from the surface of the earth, pressure in a column of air decreases, just as the pressure would decrease as you go up from the bottom in a water column. But the decrease in air pressure is not linear with distance. Instead, the air pressure decreases by a constant fraction for a given increase in height, as shown in Figure 13-8. At a height of about 5.5 km (18,000 ft), the air pressure is half its value at sea level. If we go up another 5.5 km to an altitude of 11 km (a typical altitude for airliners), the pressure is again halved so that it is one-fourth its value at sea level, and so on. This example of an *exponential decrease* is called the law of atmospheres. At the high altitudes at which commercial jets fly, the cabins must be pressurized. The density of air is approximately proportional to the pressure, so the density of air decreases with altitude. There is less oxygen available on a mountain top than at normal elevations. This makes exercising in the Rockies difficult, and climbing in the Himalayas dangerous.

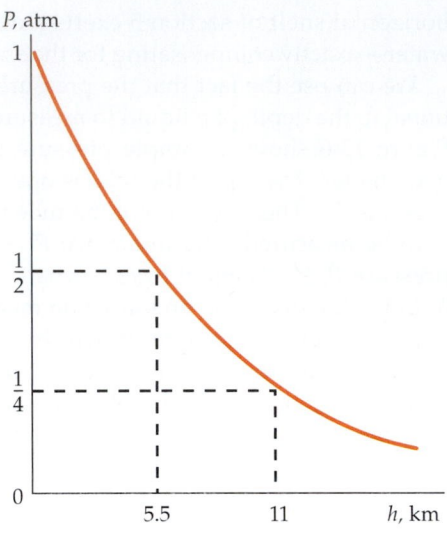

**FIGURE 13-8** Variation in pressure with height above the earth's surface. For each 5.5-km increase in height, the pressure decreases by half.

## 13-3 Buoyancy and Archimedes' Principle

If a dense object submerged in water is weighed by suspending it from a spring scale (Figure 13-9a), the apparent weight of the object when submerged (the reading on the scale) is less than the weight of the object. This is so because the water exerts an upward force that partially balances the force of gravity. This upward force is even more evident when we submerge a piece of cork. When completely submerged, the cork experiences an upward force from the water pressure that is greater than the force of gravity, so when released it accelerates up toward the surface. The force exerted by a fluid on a body wholly or partially submerged in it is called the **buoyant force.**[†] It is equal to the weight of the fluid displaced by the body.

> A body wholly or partially submerged in a fluid is buoyed up by a force equal to the weight of the displaced fluid.
>
> ARCHIMEDES' PRINCIPLE

This result is known as **Archimedes' principle.**

We can derive Archimedes' principle from Newton's laws by considering the forces acting on a portion of a fluid and noting that in static equilibrium the net force must be zero. Figure 13-9b shows the vertical forces acting on an object being weighed while submerged. These are the force of gravity $\vec{w}$ acting down, the force of the spring scale $\vec{F}_s$ acting up, a force $\vec{F}_1$ acting down because of the fluid pressure on the top surface of the object, and a force $\vec{F}_2$ acting up because of the fluid pressing on the bottom surface of the object. Since the spring scale reads a force less than the weight, the force $\vec{F}_2$ must be greater in magnitude than the force $\vec{F}_1$. The difference in magnitude of these two forces is the buoyant force $\vec{B} = \vec{F}_2 - \vec{F}_1$ (Figure 13-9c). The buoyant force occurs because the pressure of the fluid at the bottom of the object is greater than that at the top.

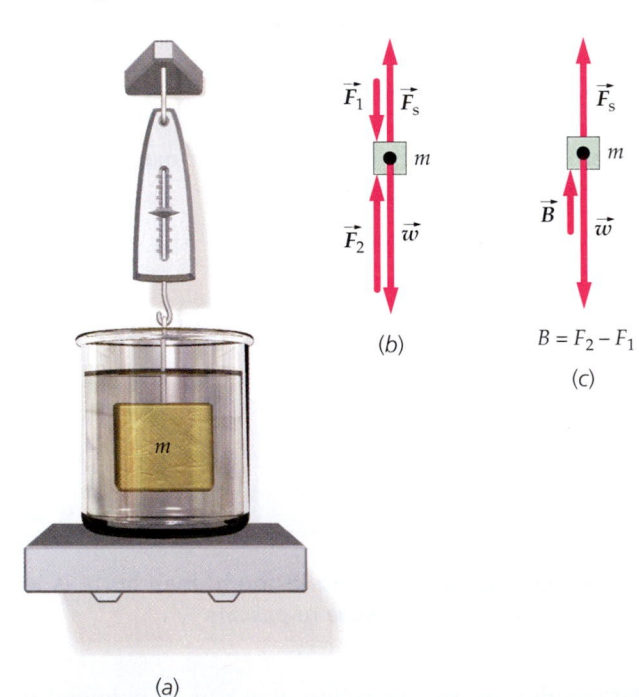

(a)

(b)

(c)

$B = F_2 - F_1$

**FIGURE 13-9** (a) Weighing an object submerged in a fluid. (b) Free-body diagram showing the weight $\vec{w}$, the force $\vec{F}_s$ of the spring, and the forces $\vec{F}_1$ and $\vec{F}_2$ exerted by the surrounding field. (c) The buoyant force $B = F_2 - F_1$ is the net force exerted on the object by the fluid.

---

[†] The definition of buoyant force is further refined later in this section.

In Figure 13-10, the spring scale has been eliminated and the submerged object has been replaced by an equal volume of fluid (outlined by the dashed lines). The buoyant force $\vec{B} = \vec{F}_2 - \vec{F}_1$ acting on this volume of fluid is the same as the buoyant force that acted on our original object since the fluid surrounding the space is the same. Because this volume of fluid is in equilibrium, the net force acting on it must be zero. The upward buoyant force thus equals the downward weight of this volume of the fluid:

$$B = w_f \qquad \qquad 13\text{-}11$$

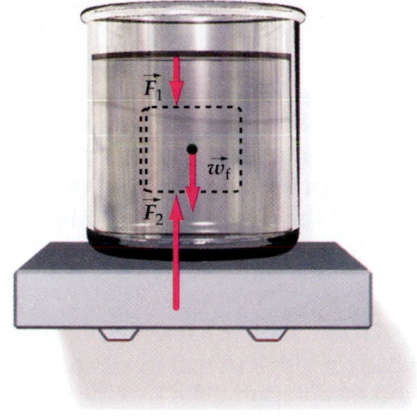

**FIGURE 13-10** Figure 13-9 with the submerged body replaced by an equal volume of fluid. The forces $\vec{F}_1$ and $\vec{F}_2$, due to the pressure of the fluid are the same as in Figure 13-9. The magnitude of the buoyant force is thus equal to the weight $w_f$ of the displaced fluid.

Note that this result does not depend on the shape of the submerged object. If we consider any irregularly shaped portion of a static fluid, there must be a buoyant force acting on it by the surrounding fluid that exactly supports its weight. Thus, we have derived Archimedes' principle.

Archimedes (287–212 B.C.) had been given the task of determining whether a crown made for King Hieron II was of pure gold or had been adulterated with some cheaper metal such as silver, and to do this without destroying the crown. For Archimedes, the problem was to determine if the density of the irregularly shaped crown was the same as the density of gold. As the story goes, he came upon the solution while sinking himself into a bathtub and immediately rushed naked through the streets of Syracuse shouting "Eureka!" ("I have found it!"). This flash of insight preceded Newton's laws, which we used to derive Archimedes' principle, by some 1900 years. What Archimedes found was a simple and accurate way to compare the density of the crown with the density of gold, using a pan balance. He placed the balance in a large basin, and placed the crown on one pan and an equal mass of pure gold on the other pan. He then added water to the basin, submerging the crown and the pure gold. The balance tilted, with the crown rising—indicating that the buoyant force on the crown was greater than that on the pure gold because the volume of water displaced by the crown was greater than that displaced by the pure gold. The crown was less dense than the pure gold.

The specific gravity of an object equals the weight of the object divided by the weight of an equal volume of water. According to Archimedes' principle, the weight of an equal volume of water equals the buoyant force on the object when it is submerged in water. Therefore, the specific gravity is equal to the weight of the object divided by the buoyant force on it when it is submerged in water:

$$\text{Specific gravity} = \frac{\text{weight}}{\text{buoyant force when submerged in water}} = \frac{w}{B_{\text{water}}} \qquad 13\text{-}12$$

The apparent weight $w_{\text{app}}$ of an object submerged in a fluid is the difference between its weight $w$ and the buoyant force $B$:

$$w_{\text{app}} = w - B \qquad \qquad 13\text{-}13$$

Hot-air balloons rising in the night sky over Albuquerque during a balloon festival.

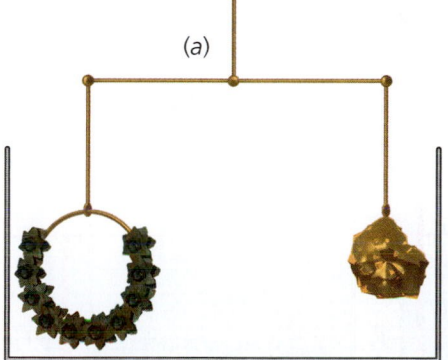 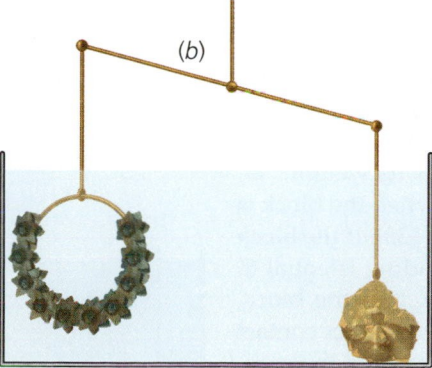

(a)     (b)

(a) The crown and the gold nugget have equal weight. (b) The wreath displaces more water than the gold nugget.

Your friend is concerned about a gold ring she bought on a recent trip. The ring was expensive, and she would like to know whether it is really made of gold or of something else. You decide to help her, using your knowledge of physics. You weigh the ring and find that it has a weight of 0.158 N. Using a string, you suspend the ring from the scale and, with the ring submerged in water, weigh it again to find a new reading of 0.150 N. Is the ring pure gold?

**PICTURE THE PROBLEM** If the ring is pure gold, its specific gravity (its density relative to that of water) is 19.3 (see Table 13-1). Using Equation 13-12 as a guide, determine the specific gravity of the ring.

1. Equation 13-12 relates the specific gravity of the ring to the ratio of its weight $w$ to the buoyant force $B$ when submerged in water:

$$\text{Specific gravity} = \frac{w}{B_{\text{water}}} = \frac{w}{B}$$

2. $B$ equals the weight minus the apparent weight $w_{\text{app}}$ when submerged:

$$B = w - w_{\text{app}}$$

3. Combine steps 1 and 2 and solve for the specific gravity:

$$\text{Specific gravity} = \frac{w}{B} = \frac{w}{w - w_{\text{app}}}$$

$$= \frac{0.158\ \text{N}}{0.158\ \text{N} - 0.150\ \text{N}} = 19.3$$

4. Compare the specific gravity of the ring with the specific gravity of gold, which is 19.3:

$$19.3 \sim 19.3$$

$$\boxed{\text{The ring is pure gold.}}$$

**EXERCISE** A block of an unknown material weighs 3 N and has an apparent weight of 1.89 N when submerged in water. What is the material? (*Answer* The specific gravity of the material is 2.70, which is the specific gravity of aluminum. The material is aluminum.)

**EXERCISE** An aluminum block has an apparent weight of 3 N when surrounded by air. What is the weight of the block? (*Answer* $w_{\text{app}} = w - B$ where $B = \rho_{\text{air}} Vg$ and $w = \rho_{\text{alum}} Vg$. From Table 13-1 we obtain $\rho_{\text{air}} = 1.293\ \text{kg/m}^3$ and $\rho_{\text{alum}} = 2.70 \times 10^3\ \text{kg/m}^3$. We have three equations and three unknowns: $w$, $B$, and $V$. Thus, $w = 3.0014\ \text{N}$, which is only 0.048 percent greater than the apparent weight in air. Clearly, buoyant forces on solids and liquids due to air can usually be ignored.)

**EXERCISE** A piece of lead (specific gravity = 11.3) weighs 80 N in air. What does it weigh when submerged in water? (*Answer* 72.9 N)

**FIGURE 13-11**

The density of the block shown in Figure 13-11 is greater than the density of the surrounding fluid, and both the block and the scale pan are completely submerged in the fluid. The gravitation force on the block is its weight, $\vec{w}$, and the scale is adjusted, so it reads zero when the block is not being supported by the pan (Figure 11-13b). If the block is on the pan (Figure 13-11a), the scale reading is equal to the magnitude of the apparent weight $w_{\text{app}}$ of the block. When the block is on the pan, the fluid is in direct contact with the entire surface of the block—except for those regions of the bottom surface of the block that are in direct contact with the pan. We assume the pan is not perfectly flat, but instead has some high and some low regions, and that the pan is in direct contact with the bottom surface of

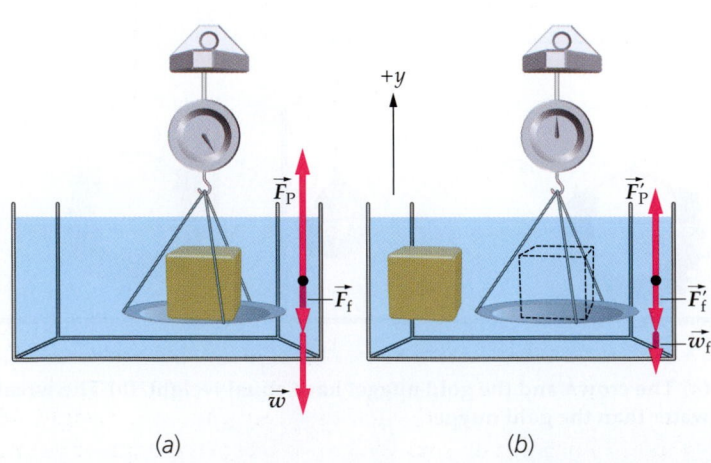

(a)    (b)

the block only at the high regions of the pan. (At the low regions of the pan, there is fluid in direct contact with bottom surface of the block.) We now analyze this situation to determine if the scale reading depends on the area of contact of the bottom surface of the block with the pan.

While resting on the pan, the net force exerted by the fluid on the block $\vec{F}_f$ is a combination of the downward force of the fluid on the top surface of the block and the upward force of the fluid on those regions of the bottom surface of the block that are in direct contact with the fluid. (We have drawn $\vec{F}_f$ downward. However, if the fluid were in direct contact with a large enough area of the bottom surface of the block, this force would be upward.) The two other vertical forces acting on the block are the gravitational force $\vec{w}$ and the upward force $\vec{F}_P$ exerted on the block by the pan at the regions of direct contact. In Figure 13-11b, the block has been moved off the pan, and in its place is a sample of fluid of identical size and shape (outlined by the dashed lines). The same regions of the surface of the pan are in direct contact with the bottom surface of this sample of fluid as if it were in direct contact with the block before it was moved. The forces acting on the fluid sample are the forces acting on it by the surrounding fluid $\vec{F}'_f$, the upward force on it by the pan $\vec{F}'_P$, and the gravitational force $\vec{w}_f$. The forces $\vec{F}_f$ and $\vec{F}'_f$ are equal because, at every point where the sample and the surrounding fluid are in direct contact, the pressure of the surrounding fluid is the same as it was at the same point before the block was moved off the pan.

When the submerged block rests on the pan, the block is in equilibrium, so $F_{f,y} + F_{P,y} + w_y = 0$, or

$$F_{f,y} + F_P - w = 0$$

and when the block is moved off the pan, the fluid sample in its place is in equilibrium, so

$$F'_{f,y} + F'_P - w_f = 0$$

Subtracting these equations, exploiting that $F'_{f,y} = F_{f,y}$, and rearranging the result gives

$$F_P - F'_P = w - w_f$$

where $F_P - F'_P$ is the change in the scale reading when the block is moved off the pan, so $F_P - F'_P$ is the magnitude of the apparent weight of the submerged block. Thus,

$$w_{app} = w - w_f$$

It is common parlance to refer to $w - w_{app}$ as the buoyant force $B$. Rearranging gives

$$B = w - w_{app} = w_f$$

This is the same expression for the buoyant force as is in Equation 13-13, which was established with the fluid in direct contact with 100 percent of surface of the submerged object.

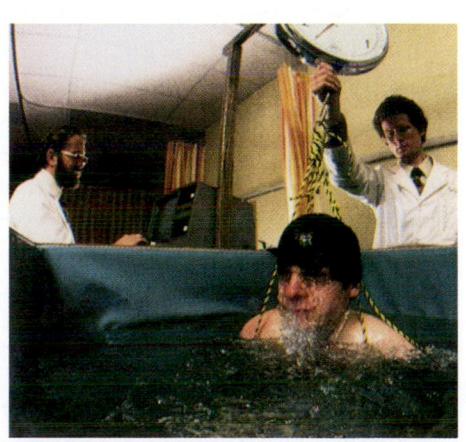

To determine the percentage of fat in this man's body, his density is measured by weighing him while he is submerged under water.

---

*MEASURING THE FAT*                                   **EXAMPLE    13-6**

**The percentage of body fat can be estimated by measuring the density of the body. Fat is less dense than muscle or bone. The density of fat $\approx 0.9 \times 10^3$ kg/m³ and the density of lean tissue (everything except fat) $\approx 1.1 \times 10^3$ kg/m³. Measuring the density of the body involves measuring the apparent weight while the body is submerged in water with the air completely exhaled from the lungs. (In practice, the amount of air remaining in the lungs is estimated and corrected for.) Suppose that your apparent weight when submerged in water is 5 percent of your weight. What percentage of your body mass is fat?**

**PICTURE THE PROBLEM** For the person, the total volume equals the volume of the fat plus the volume of the lean, and the total mass equals the mass of the fat plus the mass of the lean. The volume and density are related to the mass by $m = \rho V$. The fraction of fat equals the mass of the fat divided by the total mass and the fraction of lean equals the mass of the lean divided by the total mass. Also, the fraction of fat plus the fraction of lean equals 1.

1. Using Equations 13-2 and 13-3, find the ratio of your body's density to the density of water:

$$\frac{\rho}{\rho_{\text{water}}} = \frac{w}{w - w_{\text{app}}} = \frac{w}{w - 0.05w} = 1.05$$

2. Your total body volume equals the volume of fat plus the volume of lean tissue:

$$V_{\text{tot}} = V_{\text{fat}} + V_{\text{lean}}$$

3. Because density equals mass times volume, volume equals mass divided by density. Substitute the corresponding mass-to-density ratio for each volume in the step 2 result:

$$\frac{m_{\text{tot}}}{\rho} = \frac{m_{\text{fat}}}{\rho_{\text{fat}}} + \frac{m_{\text{lean}}}{\rho_{\text{lean}}}$$

4. The ratio $m_{\text{fat}}/m_{\text{tot}}$ is the fraction of fat $f_{\text{fat}}$, and $m_{\text{lean}}/m_{\text{tot}}$ is the fraction of lean $f_{\text{lean}}$. Substitute for $m_{\text{fat}}$ and $m_{\text{lean}}$ in the step 3 result:

$$\frac{m_{\text{tot}}}{\rho} = \frac{f_{\text{fat}}m_{\text{tot}}}{\rho_{\text{fat}}} + \frac{f_{\text{lean}}m_{\text{tot}}}{\rho_{\text{lean}}}$$

5. The fraction of fat plus the fraction of lean tissue equals 1:

$$f_{\text{fat}} + f_{\text{lean}} = 1$$

6. Divide both sides of the step 4 result by $m_{\text{tot}}$ and substitute $1 - f_{\text{fat}}$ for $f_{\text{lean}}$:

$$\frac{1}{\rho} = \frac{f_{\text{fat}}}{\rho_{\text{fat}}} + \frac{(1 - f_{\text{fat}})}{\rho_{\text{lean}}}$$

7. Solve the step 6 result for $f_{\text{fat}}$:

$$f_{\text{fat}} = \frac{1 - (\rho_{\text{lean}}/\rho)}{1 - (\rho_{\text{lean}}/\rho_{\text{fat}})}$$

8. Using the step 1 result, substitute for $\rho$ in the step 7 result and solve for $f_{\text{fat}}$:

$$f_{\text{fat}} = \frac{1 - (\rho_{\text{lean}}/1.05\rho_{\text{water}})}{1 - (\rho_{\text{lean}}/\rho_{\text{fat}})} = \frac{1 - (1.1/1.05)}{1 - (1.1/0.9)} = 0.21$$

9. Convert to a percentage:

$$100\% \times f_{\text{fat}} = \boxed{21\%}$$

**EXERCISE** If Ed's apparent weight when submerged is zero, what is his body-fat percentage? (*Answer* 45 percent)

---

**FIGURE 13-12**

*FLOATING ON A RAFT*              **EXAMPLE 13-7**

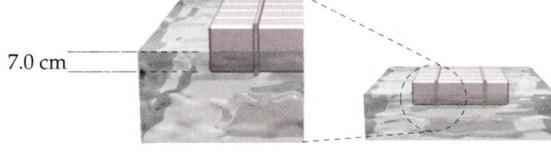

A raft of area $A$, thickness $h$, and mass $M = 600$ kg floats in calm water with 7 cm submerged (Figure 13-12). When Bob stands on the raft, 8.4 cm are submerged. What is Bob's mass $m$?

7.0 cm

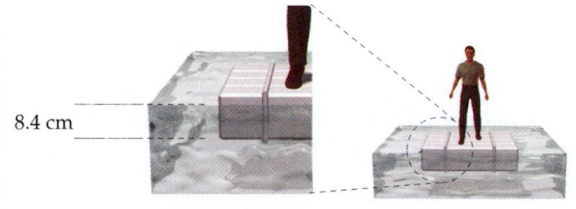

8.4 cm

**PICTURE THE PROBLEM** Let $A$ be the area of the raft. The weight of the displaced fluid is then $\rho_{\text{w}}Ad_1 g$ with just the raft and $\rho_{\text{w}}Ad_2 g$ with Bob on the raft, where $\rho_{\text{w}}$ is the density of the water, $d_1 = 7$ cm, and $d_2 = 8.4$ cm. If we set the weight of displaced fluid equal to the weight of the floating objects in each case, we can eliminate $A$ and $\rho_{\text{w}}$ and solve for $m$.

1. Set the buoyant force with $d_1 = 7$ cm submerged equal to the weight of the raft, and with $d_2 = 8.4$ cm submerged equal to the weight of the raft plus Bob:

$$\rho_{\text{w}}Ad_1 g = Mg$$
$$\rho_{\text{w}}Ad_2 g = (M + m)g$$

2. Divide these two equations to eliminate the unknowns, $A$ and $\rho_{\text{w}}$:

$$\frac{d_2}{d_1} = \frac{M + m}{M}$$

3. Solve for $m$:

$$m = \left(\frac{d_2}{d_1} - 1\right)M = \left(\frac{8.4 \text{ cm}}{7.0 \text{ cm}} - 1\right)(600 \text{ kg}) = \boxed{120 \text{ kg}}$$

**EXAMPLE 13-8**

**A cork has a density of 200 kg/m³. Find the fraction of the volume of the cork that is submerged when the cork floats in water.**

**PICTURE THE PROBLEM** Let $V$ be the volume of the cork and $V_{sub}$ be the volume that is submerged when the cork floats. The weight of the cork is $\rho_c V g$ and the buoyant force due to the water is $\rho_w V_{sub} g$.

1. Since the cork is in equilibrium, the buoyant force equals its weight:

$$w = B$$
$$\rho_c V g = \rho_w V_{sub} g$$

2. Solve for $V_{sub}/V$:

$$\frac{V_{sub}}{V} = \frac{\rho_c}{\rho_w} = \frac{200 \text{ kg/m}^3}{1000 \text{ kg/m}^3} = \boxed{\frac{1}{5}}$$

MASTER the CONCEPT WEB

**REMARKS** We see that only one-fifth of the cork is submerged. This result is independent of the shape of the cork.

If we replace $\rho_w$ in the calculation above with $\rho_f$, the density of the fluid, we can determine the submerged fraction of an object floating in any fluid. From Example 13-8, the fraction of a floating object that is submerged equals the ratio of its density to the density of the fluid.

$$\frac{V_{sub}}{V} = \frac{\rho}{\rho_f}$$                    13-14

Because the density of ice is 920 kg/m³ and that of sea water is 1025 kg/m³, the fraction of an iceberg that is submerged in sea water is

$$\frac{V_{sub}}{V} = \frac{\rho}{\rho_f} = \frac{920 \text{kg/m}^3}{1025 \text{kg/m}^3} = 0.898$$

The great danger that icebergs pose to ships springs directly from the fact that only about 10 percent of an iceberg is visible above the water, and the visible portion gives little hint of where the submerged portion may extend, posing extreme danger to passing vessels.

Smoke from a burning cigarette. At first the smoke rises in a regular stream, but the simple streamlined flow quickly becomes turbulent and the smoke begins to swirl irregularly.

# 13-4 Fluids in Motion

The behavior of a fluid in motion can be complex. Consider, for example, the rise of smoke from a burning cigarette. At first the smoke rises in a regular stream of warm gas, but the simple streamlined flow quickly becomes turbulent and the smoke begins to swirl irregularly. Turbulent flow is very difficult to describe, even qualitatively. We will therefore consider only nonturbulent, steady-state flow of an "ideal" fluid, one that is nonviscous; that is, one that flows with no dissipation of mechanical energy. We also assume that the fluid is incompressible, which is an excellent approximation for liquids in most situations. In an incompressible fluid, the density is constant throughout the fluid.

Figure 13-13 shows a tube full of liquid. The tube contains a tapered section with decreasing cross-sectional area. The fluid is flowing from left to right, and the shaded portion on the left depicts the volume $\Delta V$ of fluid that flows past point 1 during time $\Delta t$. If the speed of the fluid at this point is $v_1$ and the cross-sectional area of the tube is $A_1$, then the volume flowing past point 1 in time $\Delta t$ is

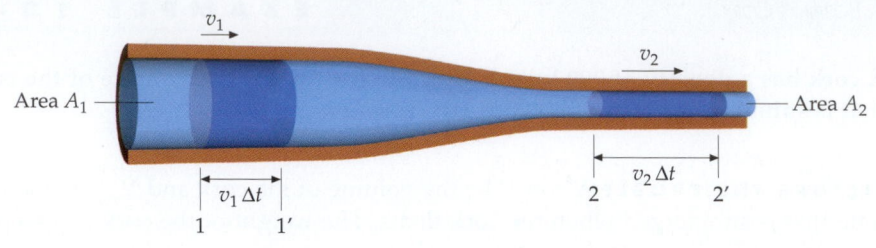

**FIGURE 13-13**

$$\Delta V = A_1 v_1 \, \Delta t$$

Since we assume that the fluid is incompressible, an equal volume of fluid must flow past any point along the tube during the same time $\Delta t$. The volume that flows past point 2 is depicted by the shaded portion on the right. If the speed of the fluid at this point is $v_2$ and the cross-sectional area is $A_2$, then the volume is $\Delta V = A_2 v_2 \, \Delta t$. Since these volumes must be equal, we have

$$A_1 v_1 \, \Delta t = A_2 v_2 \, \Delta t \quad \text{so} \quad A_1 v_1 = A_2 v_2 \qquad \qquad \text{13-15}$$

The quantity $Av$ is called the **volume flow rate** $I_v$. The dimensions of $I_v$ are volume divided by time. In the flow of an incompressible fluid, the volume flow rate is the same at any point in the fluid:

$$I_v = Av = \text{constant} \qquad \qquad \text{13-16}$$

CONTINUITY EQUATION FOR INCOMPRESSIBLE FLUID

Equation 13-16 is called the **continuity equation** for an incompressible fluid.

**EXERCISE** Blood flows in an aorta of radius 1.0 cm at 30 cm/s. What is the volume flow rate? (*Answer*  $I_V = vA = 9.42 \times 10^{-5} \, \text{m}^3/\text{s}$. It is customary to give the pumping rate of the heart in liters per minute. Using $1 \, \text{m}^3 = 1000 \, \text{L}$, we have $I_v = 5.65 \, \text{L/min}$.)

**EXERCISE** Blood flows from a large artery of radius 0.3 cm, where its speed is 10 cm/s, into a region where the radius has been reduced to 0.2 cm because of thickening of the arterial walls (arteriosclerosis). What is the speed of the blood in the narrower region? (*Answer*  If $v_1$ and $v_2$ are the initial and final speeds and $A_1$ and $A_2$ are the initial and final areas, Equation 13-16 gives

$$v_2 = \frac{A_1}{A_2} v_1 = \frac{\pi (0.3 \, \text{cm})^2}{\pi (0.2 \, \text{cm})^2} (10 \, \text{cm/s}) = 22.5 \, \text{cm/s} \Big)$$

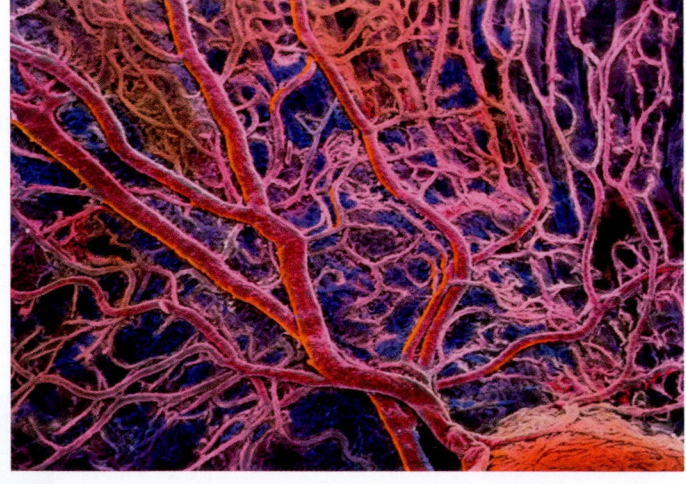

## Bernoulli's Equation

When a small sample of a fluid flows into a narrow region of a tube, it gains speed because the pressure behind it that pushes it forward is greater than the pressure in front of it opposing its motion. Bernoulli's equation relates the pressure, elevation, and speed of an *incompressible* fluid in *steady* flow. It follows from Newton's laws and is most easily derived by applying the work–energy theorem to a segment of the fluid.

Consider a fluid flowing in a tube that varies in elevation as well as in cross-sectional area, as shown in Figure 13-14. We apply the work–energy theorem to a sample of fluid that initially is contained between points 1 and 2 in Figure 13-14a.

During time $\Delta t$ this sample moves along the tube to the region between points 1′ and 2′ as shown in Figure 13-14b. Let $\Delta V$ be the volume of fluid passing point 1′ during time $\Delta t$. The same volume passes point 2 during the same time. Also, let $\Delta m = \rho \Delta V$ be the mass of the fluid with volume $\Delta V$. The net effect on the sample during time $\Delta t$ is that a mass $\Delta m$ initially at height $h_1$ moving with speed $v_1$ is "transferred" to height $h_2$ with speed $v_2$. The change in the potential energy of the sample is thus

$$\Delta U = (\Delta m)gh_2 - (\Delta m)gh_1 = \rho \Delta V g(h_2 - h_1)$$

and the change in its kinetic energy is

$$\Delta K = \tfrac{1}{2}(\Delta m)v_2^2 - \tfrac{1}{2}(\Delta m)v_1^2 = \tfrac{1}{2}\rho \Delta V(v_2^2 - v_1^2)$$

The fluid behind the sample (to the sample's left in Figure 13-15) pushes on the sample with a force of magnitude $F_1 = P_1 A_1$, where $P_1$ is the pressure at point 1. This force does work:

$$W_1 = F_1 \Delta x_1 = P_1 A_1 \Delta x_1 = P_1 \Delta V$$

At the same time, the fluid in front of the sample (to the sample's right) exerts a force $F_2 = P_2 A_2$ opposing the sample's motion (pushing it to the left). This force does negative work:

$$W_2 = -F_2 \Delta x_2 = -P_2 A_2 \Delta x_2 = -P_2 \Delta V$$

The total work done by these forces is

$$W_{\text{total}} = P_1 \Delta V - P_2 \Delta V = (P_1 - P_2)\Delta V$$

The work–energy theorem gives

$$W_{\text{total}} = \Delta U + \Delta K$$

so

$$(P_1 - P_2)\Delta V = \rho \Delta V g(h_2 - h_1) + \tfrac{1}{2}\rho \Delta V(v_2^2 - v_1^2)$$

If we divide both sides by $\Delta V$, we obtain

$$P_1 - P_2 = \rho g h_2 - \rho g h_1 + \tfrac{1}{2}\rho v_2^2 - \tfrac{1}{2}\rho v_1^2$$

When we collect all the quantities having a subscript 1 on one side and those having a subscript 2 on the other, this equation becomes

$$P_1 + \rho g h_1 + \tfrac{1}{2}\rho v_1^2 = P_2 + \rho g h_2 + \tfrac{1}{2}\rho v_2^2 \qquad 13\text{-}17a$$

This result can be restated as

$$P + \rho g h + \tfrac{1}{2}\rho v^2 = \text{constant} \qquad 13\text{-}17b$$

BERNOULLI'S EQUATION

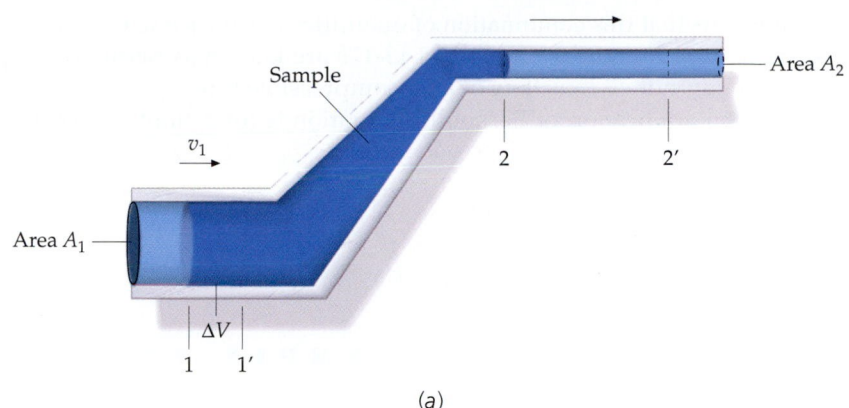

(a)

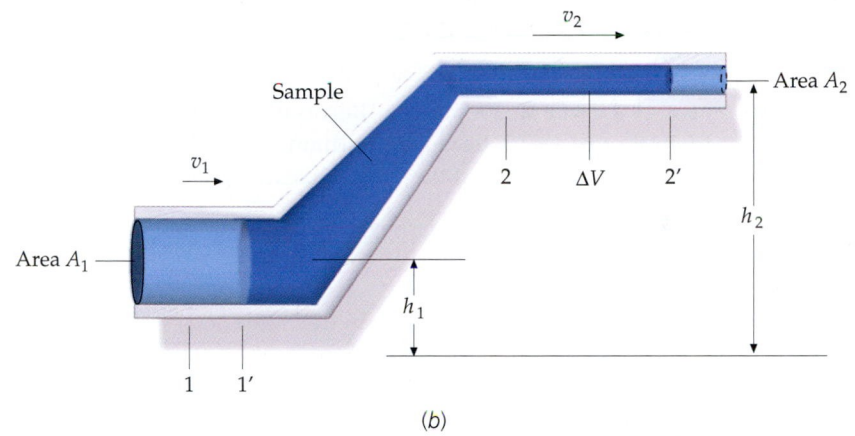

(b)

**FIGURE 13-14**

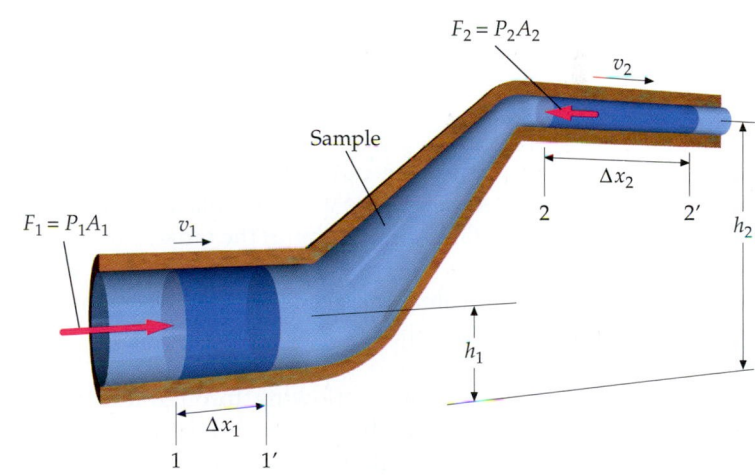

**FIGURE 13-15** Fluid moving in a pipe that varies in both height and cross-sectional area. The total work done by the forces $F_1 = P_1 A_1$ and $F_2 = P_2 A_2$ has the effect of raising the shaded portion of the fluid from height $h_1$ to $h_2$ and changing its speed from $v_1$ to $v_2$.

which means that this combination of quantities has the same value at any point along the tube. Equations 13-17a and 13-17b are known as **Bernoulli's equation** for the steady, nonviscous flow of an incompressible fluid.

A special application of Bernoulli's equation is for a fluid at rest. Then $v_1 = v_2 = 0$, and we obtain

$$P_1 - P_2 = \rho g h_2 - \rho g h_1 = \rho g \Delta h$$

This is the same as Equation 13-7.

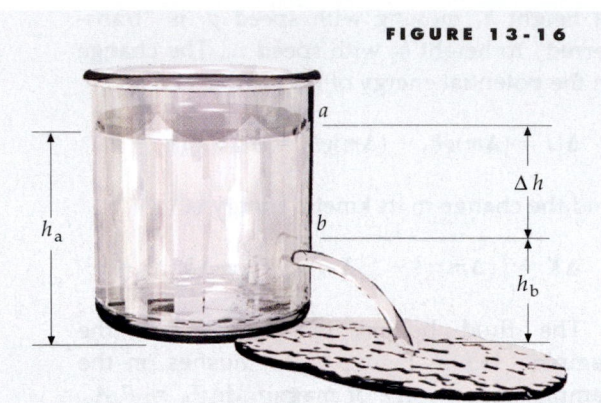

**FIGURE 13-16**

---

*A Leaking Tank* **EXAMPLE 13-9**

**A large tank of water, open at the top, has a small hole through its side a distance $h$ below the surface of the water. Find the speed of the water as it flows out the hole.**

**PICTURE THE PROBLEM** We apply Bernoulli's equation to points $a$ and $b$ in Figure 13-16. Because the diameter of the hole is much smaller than the diameter of the tank, we can neglect the velocity of the water at the top (point $a$).

1. Bernoulli's equation with $v_a = 0$ gives:

$$P_a + \rho g h_a + 0 = P_b + \rho g h_b + \tfrac{1}{2}\rho v_b^2$$

2. The pressure at point $a$ and at point $b$ is the same, $P_{at}$, since both points are open to the atmosphere:

$$P_a = P_{at} \quad \text{and} \quad P_b = P_{at}$$

so

$$P_{at} + \rho g h_a + 0 = P_{at} + \rho g h_b + \tfrac{1}{2}\rho v_b^2$$

3. Solve the step 2 result for the speed $v_b$ of the water flowing from the hole:

$$v_b^2 = 2g(h_a - h_b) = 2g\Delta h$$

so

$$\boxed{v_b = \sqrt{2g\Delta h}}$$

**EXERCISE** If the water flowing out of the hole is directed vertically upward, how high does it rise? (*Answer* The water shoots upward a distance $h$; that is, to the same level as the surface of the water in the tank.)

---

In Example 13-9, the water emerges from the hole with a speed equal to the speed it would have if it dropped in free-fall a distance $h$. This finding is known as *Torricelli's law*.

In Figure 13-17, water is shown flowing through a horizontal pipe that has a constricted section. Because both sections of the pipe are at the same elevation, $h_1 = h_2$ in Equation 13-17a. Then Bernoulli's equation becomes

$$P + \tfrac{1}{2}\rho v^2 = \text{constant} \qquad \qquad 13\text{-}18$$

When the fluid moves into the constriction, the area $A$ gets smaller, so the speed $v$ must get larger since $Av$ remains constant. But because $P + \tfrac{1}{2}\rho v^2$ is constant, when the speed gets larger the pressure must get smaller. Thus, the pressure in the constriction is reduced.

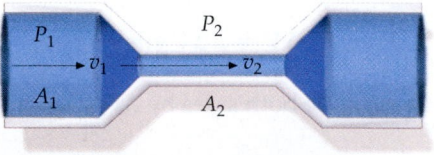

**FIGURE 13-17** Constriction in a pipe carrying a moving fluid. The pressure is lower in the narrow section of the pipe where the fluid is moving faster.

> When the speed of a fluid increases, the pressure drops.

VENTURI EFFECT

This result is often referred to as the **Venturi effect.** Equation 13-18 is an important result that applies to many situations in which we can ignore changes in height.

In Figure 13-18, lines, called **streamlines,** are drawn to pictorially represent the flow of the fluid. The direction of the lines denotes the direction of flow and the spacing between lines represents the speed of the flow. The smaller the spacing, the greater the speed. For horizontal flow, where the speed increases, the pressure decreases, so a decrease in streamline spacing is accompanied by a decrease in pressure.

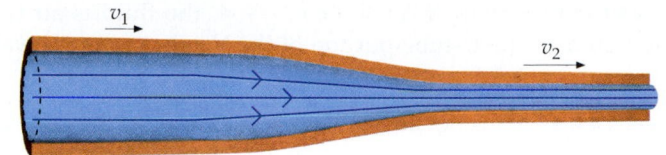

**FIGURE 13-18**

---

*A VENTURI METER*                **EXAMPLE  13-10**  **Try It Yourself**

A *Venturi meter,* which is used to measure the flow rate of a fluid, is shown in Figure 13-19. The fluid of density $\rho_F$ passes through a pipe of cross-sectional area $A_1$ that has a constriction of cross-sectional area $A_2$. The two parts of the pipe are connected with a U-tube manometer partially filled with a liquid of density $\rho_L$. Since the velocity of flow is greater in the constricted region, the pressure in that section is less than in the other portion of the pipe. The pressure difference is measured by the difference in the levels of the liquid in the U-tube, $\Delta h$. Express the velocity $v_1$ in terms of the measured height $\Delta h$ and the known quantities $\rho_F$, $\rho_L$, and $r = A_1/A_2$.

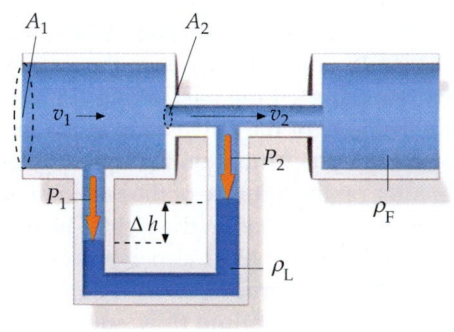

**FIGURE 13-19** A Venturi meter.

**PICTURE THE PROBLEM** The pressures $P_1$ and $P_2$ in the two regions are related to the speeds $v_1$ and $v_2$ by Bernoulli's equation. The pressure difference is related to the height $h$ by $P_1 - P_2 = \rho_L gh$. You can express $v_2$ in terms of $v_1$ and the areas $A_1$ and $A_2$ by the continuity equation.

**Cover the column to the right and try these on your own before looking at the answers.**

| Steps | Answers |
|---|---|
| 1. Write Bernoulli's equation for constant elevation for the two regions. | $P_1 + \frac{1}{2}\rho_F v_1^2 = P_2 + \frac{1}{2}\rho_F v_2^2$ |
| 2. Write the continuity equation for the two regions, and solve for $v_2$ in terms of $v_1$ and the areas $A_1$ and $A_2$. | $v_2 A_2 = v_1 A_1$ <br> so <br> $v_2 = \dfrac{A_1}{A_2} v_1 = rv_1$ |
| 3. Substitute your result for $v_2$ into the equation in step 1 and obtain an equation for $P_1 - P_2$. | $P_1 - P_2 = \frac{1}{2}\rho_F(v_2^2 - v_1^2)$ <br> $= \frac{1}{2}\rho_F(r^2 - 1)v_1^2$ |
| 4. Write $P_1 - P_2$ in terms of the difference in height $\Delta h$ of the liquid in the arms of the U-tube. This pressure difference equals the pressure drop in a column of the liquid of height $\Delta h$ and that in a column of the fluid of the same height. | $P_1 - P_2 = \rho_L g\Delta h - \rho_F g\Delta h$ <br> $= (\rho_L - \rho_F)g\Delta h$ |
| 5. Equate the two expressions for $P_1 - P_2$ and solve for $v_1$ in terms of $\Delta h$. | $\frac{1}{2}\rho_F(r^2 - 1)v_1^2 = (\rho_L - \rho_F)g\Delta h$ <br> so <br> $v_1 = \sqrt{\dfrac{2(\rho_L - \rho_F)g\Delta h}{\rho_F(r^2 - 1)}}$ |

**EXERCISE** Find $v_1$ if $\Delta h = 3$ cm, $r = 4$, the fluid is air ($\rho_F = 1.29$ kg/m³), and the liquid in the U-tube portion of the Venturi meter is water ($\rho_w = 10^3$ kg/m³). (*Answer* $v_1 = 5.51$ m/s)

**REMARKS** Air is not an incompressible fluid, so the calculation in the Exercise is not as accurate as the calculation in the Example 13-9. Strictly speaking, Bernoulli's equation and the continuity equation hold only for incompressible fluids.

The Venturi effect can be used to give a qualitative understanding of the lift of an airplane wing and the path of a pitcher's curveball. An airplane wing is designed so that air moves faster over the top of the wing than it does under the wing, thus making the air pressure less on top than underneath. This difference in pressure results in a net force upward on the wing. Figure 13-20a shows a top view of the motion of a curveball. As the ball spins, it tends to drag air around with it. Figure 13-20b is drawn from the point of view of a stationary (but spinning) ball with the air rushing past it. The air movement caused by the spinning ball adds to the velocity of the air rushing by on the left side of the ball (the top in the figure) and subtracts from it on the right (the bottom in the figure). Thus, the air speed is greater on the left side of the ball than on the right, causing the pressure on the left to be less than that on the right. The ball therefore curves to the left. The atomizer shown in Figure 13-21 also works on the principle of the Venturi effect.

Although Bernoulli's equation is very useful for qualitative descriptions of many features of fluid flow, such descriptions are often grossly inaccurate when compared with the quantitative results of experiments. Prominent reasons for the discrepancies are that gases like air are hardly incompressible, and liquids like water have viscosity, which invalidates the assumption of the conservation of mechanical energy. In addition, it is often difficult to maintain steady-state, streamlined flow without turbulence, and the introduction of turbulence can greatly affect the results.

Airflow above and below the wing of this Indy race car creates greater pressure above the wing, increasing the effective weight of the car for better control at high speeds. An airplane wing is designed so that the flow creates greater pressure below the wing to lift the plane.

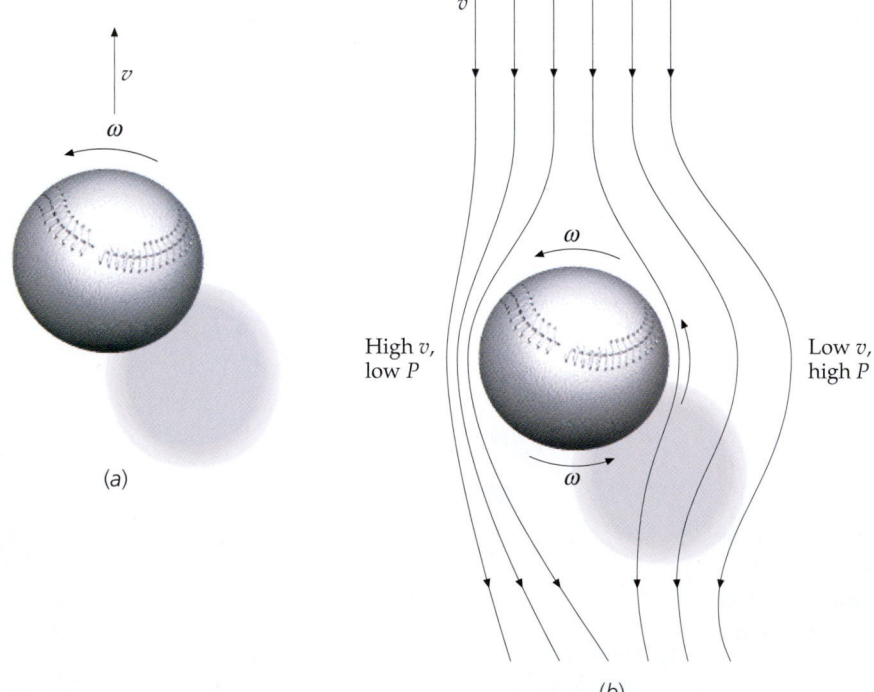

(a)

(b)

**FIGURE 13-20** (a) Top view of a baseball thrown with a counterclockwise spin $\omega$, like an overhead view of a curveball thrown by a right-handed pitcher. (b) In the frame moving with the ball, the ball is stationary (but spinning) and the air rushes past it. Because of its rough cover, the spinning ball drags the air around with it, making the air speed higher on the left side and lower on the right. The pressure is therefore lower on the left side, so the ball curves to the left.

**FIGURE 13-21** When the bulb of an atomizer is squeezed, the air is forced through the constriction in the horizontal tube, which reduces the pressure there below atmospheric pressure. Because of the resulting pressure difference, the liquid in the jar is pumped up through the vertical tube, enters the air stream, and emerges from the nozzle. A similar effect occurs in the carburetor of a gasoline lawnmower engine.

## *Viscous Flow

According to Bernoulli's equation, when a fluid flows steadily through a long, narrow, horizontal pipe of constant cross section, the pressure along the pipe will be constant. In practice, however, we observe a pressure drop as we move along the direction of the flow. Put another way, a pressure difference is required to push a fluid through a horizontal pipe. This pressure difference is needed because of the drag force that is exerted by the pipe on the layer of fluid in contact with the pipe, and because of the drag force exerted by each layer of the fluid on an adjacent layer that is moving with a slightly different velocity. These drag forces are called **viscous forces**. As a result of viscous forces, the velocity of the fluid is not constant across the diameter of the pipe. Instead, it is greatest near the center of the pipe and approaches zero where the fluid is in contact with the walls of the pipe (Figure 13-22). Let $P_1$ be the pressure at point 1 and $P_2$ be that at point 2, a distance $L$ downstream from point 1. The pressure drop $\Delta P = P_1 - P_2$ is proportional to the volume flow rate:

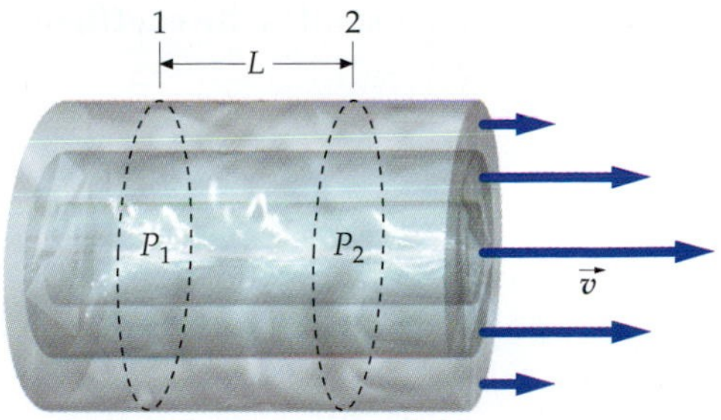

$$\Delta P = P_1 - P_2 = I_V R \qquad\qquad 13\text{-}19$$

where $I_V = vA$ is the volume flow rate and the proportionality constant $R$ is the resistance to flow, which depends on the length of the pipe $L$, the radius $r$, and the viscosity of the fluid.

**FIGURE 13-22** When a viscous fluid flows through a pipe, the speed is greatest at the center of the pipe. At the walls of the pipe, the fluid flow rate approaches zero.

---

*RESISTANCE TO BLOOD FLOW*       **EXAMPLE 13-11**

**Blood flows from the aorta through the major arteries, the small arteries, the capillaries, and the veins until it reaches the right atrium. In the course of that flow, the (gauge) pressure drops from about 100 torr to zero. If the volume flow rate is 0.8 L/s, find the total resistance of the circulatory system.**

**PICTURE THE PROBLEM** The resistance is related to the pressure drop and volume flow rate by Equation 13-19. We can use Equation 13-9 to convert from torr to kPa.

Write the resistance in terms of the pressure drop and volume flow rate, and convert all terms to SI units:

$$R = \frac{\Delta P}{I_v}$$

$$= \frac{100 \text{ torr}}{0.8 \text{ L/s}} \times \frac{101 \text{ kPa}}{760 \text{ torr}} \times \frac{1 \text{ L}}{10^3 \text{ cm}^3} \times \frac{1 \text{ cm}^3}{10^{-6} \text{ m}^3}$$

$$= \boxed{16.6 \text{ kPa·s/m}^3}$$

**REMARKS** We could have used $1 \text{ Pa} = 1 \text{ N/m}^2$ to write the result as $16.6 \text{ kN·s/m}^5$.

To define the coefficient of viscosity of a fluid, we consider a fluid that is confined between two parallel plates, each of area $A$, separated by a distance $z$ as shown in Figure 13-23. The upper plate is pulled at a constant speed $v$ by a force $\vec{F}$ while the bottom plate is held at rest. A force is needed to pull the upper plate because the fluid next to the plate exerts a viscous drag force opposing its motion. The speed of the fluid between the plates approaches $v$ near the upper plate and zero near the lower plate, and it varies linearly with separation between the plates. The force $\vec{F}$ is found to be directly

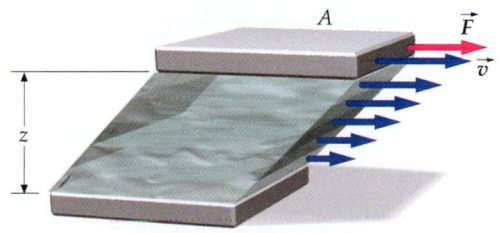

**FIGURE 13-23** Two plates of equal area with a viscous fluid between them. When the upper plate is moved relative to the lower one, each layer of fluid exerts a drag force on adjacent layers. The force needed to pull the upper plate is directly proportional to $v$ and the area $A$ and inversely proportional to $z$, the separation between the plates.

proportional to $v$ and $A$ and inversely proportional to the plate separation $z$. The proportionality constant is the **coefficient of viscosity** $\eta$:

$$F = \eta \frac{vA}{z}$$ 13-20

The SI unit of viscosity is the N·s/m² = Pa·s. An older cgs unit still in common use is the **poise,** named after the French physicist Jean Poiseuille. These units are related by

$$1 \text{ Pa·s} = 10 \text{ poise}$$ 13-21

Table 13-3 lists the coefficients of viscosity for several fluids at various temperatures. Typically, the viscosity of a liquid increases as the temperature decreases. Thus, in cold climates, a less viscous grade of oil is used to lubricate automobile engines in the winter than in summer.

**Poiseuille's Law**   The resistance to flow $R$ in Equation 13-19 for steady flow through a circular tube of radius $r$ can be shown to be

$$R = \frac{8\eta L}{\pi r^4}$$ 13-22

Equations 13-19 and 13-22 can be combined to give the pressure drop over a length $L$ of a circular tube of radius $r$:

$$\Delta P = \frac{8\eta L}{\pi r^4} I_V$$ 13-23

POISEUILLE'S LAW

Equation 13-23 is known as **Poiseuille's law.** Note the inverse $r^4$ dependence of the pressure drop. If the radius of the tube is halved, the pressure drop for a given volume flow rate is increased by a factor of 16; or a pressure 16 times as

## TABLE 13-3

**Coefficients of Viscosity for Various Fluids**

| Fluid | $t$, °C | $\eta$, mPa·s |
|---|---|---|
| Water | 0 | 1.8 |
|  | 20 | 1.00 |
|  | 60 | 0.65 |
| Blood (whole) | 37 | 4.0 |
| Engine oil (SAE 10W) | 30 | 200 |
| Glycerin | 0 | 10,000 |
|  | 20 | 1,410 |
|  | 60 | 81 |
| Air | 20 | 0.018 |

great is needed to pump the fluid through the tube at the original volume flow rate. Thus, for example, if the diameter of a person's blood vessels or arteries is reduced for some reason, either the volume flow rate of the blood is greatly reduced, or the blood pressure must escalate to maintain the volume flow rate. For water flowing through a long garden hose, the pressure drop is pretty much fixed. It equals the difference in pressure between that at the water source and atmospheric pressure at the open end. The volume flow rate is then proportional to the fourth power of the radius. Thus, if the radius is halved, the volume flow rate drops by a factor of 16.

Poiseuille's law applies only to the laminar (nonturbulent) flow of a fluid of constant viscosity. In some fluids, viscosity changes with velocity, violating Poiseuille's law. Blood, for example, is a complex fluid consisting of solid particles of various shapes suspended in a liquid. Red blood cells are disk-shaped objects that are randomly oriented at low velocities but at high velocities tend to become oriented to facilitate the flow. Thus, the viscosity of blood decreases as the flow velocity increases, so Poiseuille's law cannot be strictly applied. Nevertheless, Poiseuille's law is a good approximation that is very useful for obtaining a qualitative understanding of blood flow.

In Chapter 25 the flow of electrical current $I$ through metal wires is studied. One of the basic relations in that chapter is Ohm's law, $\Delta V = IR$, where $\Delta V$ is the potential difference and $R$ is the electrical resistance of the wire. As we shall see, Ohm's law is analagous to Poiseuille's law $\Delta P = I_V R$.

**Turbulence—Reynolds Number**   When the flow velocity of a fluid becomes sufficiently great, laminar flow breaks down and turbulence sets in. The critical velocity above which the flow through a tube is turbulent depends on the density and viscosity of the fluid and on the radius of the tube. The flow of a fluid can be characterized by a dimensionless number called the **Reynolds number** $N_R$, which is defined by

$$N_R = \frac{2r\rho v}{\eta}$$

13-24

where $v$ is the average velocity of the fluid. Experiments have shown that the flow will be laminar if the Reynolds number is less than about 2000 and turbulent if it is greater than 3000. Between these values, the flow is unstable and may change from one type to the other.

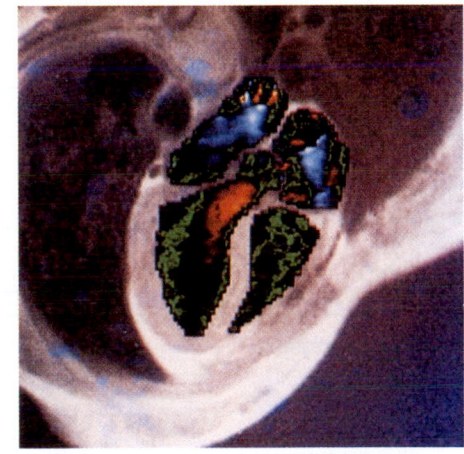

False-color view of turbulence of blood flowing into and out of the heart as seen by magnetic resonance imaging (MRI). Systolic ejection from the left ventricle into the aorta is seen in red, and diastolic filling of the ventricles in blue.

---

*BLOOD FLOW IN THE AORTA*                **EXAMPLE   13-12**

Calculate the Reynolds number for blood flowing at 30 cm/s through an aorta of radius 1.0 cm. Assume that blood has a viscosity of 4 mPa·s and a density of 1060 kg/m³.

**PICTURE THE PROBLEM**   Because $N_R$ is dimensionless, we can use any set of units as long as we are consistent.

Write Equation 13-24 for the Reynolds number, expressing each quantity in SI units:

$$N_R = \frac{2r\rho v}{\eta}$$

$$= \frac{2(0.01 \text{ m})(1060 \text{ kg/m}^3)(0.3 \text{ m/s})}{4 \times 10^{-3} \text{ Pa·s}} = \boxed{1590}$$

**REMARKS**   Since the Reynolds number is less than 2000, this flow will be laminar rather than turbulent.

1. Density, specific gravity, and pressure are defined quantities that are important in fluid statics and dynamics.
2. Pascal's principle and Archimedes' principle are derived from Newton's laws.
3. Bernoulli's equation is derived from the conservation of mechanical energy.
*4. The Venturi effect is a special case of Bernoulli's equation.
*5. Poiseuille's law accounts for pressure drops due to viscosity, Reynolds number is used to predict whether flow is laminar or turbulent.

| Topic | Relevant Equations and Remarks |
|---|---|
| **1. Density** | The density of a substance is the ratio of its mass to its volume: |
| | $$\rho = \frac{m}{V} \qquad \text{13-1}$$ |
| | The densities of most solids and liquids are approximately independent of temperature and pressure, whereas those of gases depend strongly on these quantities. |
| **2. Specific Gravity** | The specific gravity of a substance is the ratio of its density to that of water. An object sinks or floats in a given fluid depending on whether its density is greater than or less than that of the fluid. |
| **3. Pressure** | $$P = \frac{F}{A} \qquad \text{13-3}$$ |
| Units | $1\,\text{Pa} = 1\,\text{N/m}^2 \qquad \text{13-4}$ |
| | $1\,\text{atm} = 760\,\text{mmHg} = 760\,\text{torr} = 29.9\,\text{in.Hg}$ |
| | $\phantom{1\,\text{atm}} = 101.325\,\text{kPa} = 14.7\,\text{lb/in.}^2 \qquad \text{13-9}$ |
| | $1\,\text{bar} = 10^3\,\text{millibars} = 100\,\text{kPa} \qquad \text{13-10}$ |
| Gauge pressure | Gauge pressure is the difference between the absolute pressure and atmospheric pressure: |
| | $$P = P_{\text{gauge}} + P_{\text{at}} \qquad \text{13-8}$$ |
| In a liquid | $$P = P_0 + \rho g \Delta h \quad (\rho\ \text{constant}) \qquad \text{13-7}$$ |
| In a gas | In a gas such as air, pressure decreases exponentially with altitude. |
| Bulk modulus | $$B = -\frac{\Delta P}{\Delta V / V} \qquad \text{13-6}$$ |
| **4. Pascal's Principle** | Pressure changes applied to an enclosed liquid are transmitted undiminished to every point in the fluid and to the walls of the container. |
| **5. Archimedes' Principle** | A body wholly or partially submerged in a fluid is buoyed up by a force equal to the weight of the displaced fluid. |
| **\*6. Fluid Flow** | |
| Volume flow rate | $$I_{\text{v}} = Av$$ |

| | | |
|---|---|---|
| Continuity equation for an incompressible fluid | $I_v = Av = \text{constant}$ | 13-16 |
| Bernoulli's equation | $P + \rho gh + \frac{1}{2}\rho v^2 = \text{constant}$ | 13-17b |
| Venturi effect | When the speed of a fluid increases, the pressure drops. | |
| Resistance to fluid flow | $\Delta P_2 = I_V R$ | 13-19 |
| Coefficient of viscosity | $F = \eta \dfrac{vA}{z}$ | 13-20 |
| Poiseuille's law for viscous flow | $\Delta P = RI_V = \dfrac{8\eta L}{\pi r^4} I_V$ | 13-23 |
| Laminar flow, turbulent flow, and the Reynolds number | The flow will be laminar if the Reynolds number $N_R$ is less than about 2000 and turbulent if it is greater than 3000, where $N_R$ is given by $$N_R = \frac{2r\rho v}{\eta}$$ | 13-24 |

# PROBLEMS

- • Single-concept, single-step, relatively easy
- •• Intermediate-level, may require synthesis of concepts
- ••• Challenging
- **SSM** Solution is in the *Student Solutions Manual*
- **iSOLVE** Problems available on iSOLVE online homework service
- **iSOLVE** ✓ These "Checkpoint" online homework service problems ask students additional questions about their confidence level, and how they arrived at their answer

In a few problems, you are given more data than you actually need; in a few other problems, you are required to supply data from your general knowledge, outside sources, or informed estimates.

## Conceptual Problems

**1** • If the gauge pressure is doubled, the absolute pressure will be (*a*) halved, (*b*) doubled, (*c*) unchanged, (*d*) squared, (*e*) not enough information is given to determine the effect.

**2** • **SSM** Does Archimedes' principle hold in a satellite orbiting the earth in a circular orbit? Explain.

**3** •• A rock of mass $M$ with a density three times that of water is suspended by a thread. Holding the free end of the thread in your hand, you lower the rock into an aquarium tank with water almost up to the rim. The tank is resting on the platform of a scale. When the rock is just above the bottom of the tank the thread breaks. From the moment the thread breaks to the time the rock is resting on the bottom of the tank the reading on the scale increases by (*a*) $2Mg$, (*b*) $Mg$, (*c*) $\frac{2}{3}Mg$, (*d*) $\frac{1}{3}Mg$, (*e*) zero.

**4** •• A rock is thrown into a swimming pool filled with water of uniform temperature. Which of the following statements is true?

(*a*) The buoyant force on the rock is zero as it sinks.
(*b*) The buoyant force on the rock increases as it sinks.
(*c*) The buoyant force on the rock decreases as it sinks.
(*d*) The buoyant force on the rock is constant as it sinks.
(*e*) The buoyant force on the rock as it sinks is nonzero at first but becomes zero once the terminal velocity is reached.

**5** •• A fishbowl rests on a scale. The fish suddenly swims upward to get food. What happens to the scale reading?

**6** •• **SSM** Two objects are balanced as in Figure 13-24. The objects have identical volumes but different masses. Will the equilibrium be disturbed if the entire system is completely immersed in water? Explain.

**FIGURE 13-24** Problem 6

**7** •• A 200-g block of lead and a 200-g block of copper are completely under water. Each is suspended by a thread just above the bottom of an aquarium filled with water. Which of the following is true?

(a) The buoyant force is greater on the lead than on the copper.
(b) The buoyant force is greater on the copper than on the lead.
(c) The buoyant force is the same on both blocks.
(d) More information is needed to choose the correct answer.

**8** •• A 20-cm³ block of lead and a 20-cm³ block of copper are completely under water. Each is suspended by a thread just above the bottom of an aquarium filled with water. Which of the following is true?

(a) The buoyant force is greater on the lead than on the copper.
(b) The buoyant force is greater on the copper than on the lead.
(c) The buoyant force is the same on both blocks.
(d) More information is needed to choose the correct answer.

**9** • In a department store, a beach ball is supported by the airstream from a hose connected to the exhaust of a vacuum cleaner. Does the air blow under or over the ball to support it? Why?

**10** • A horizontal pipe narrows from a diameter of 10 cm to 5 cm. For a nonviscous liquid flowing without turbulance from the larger diameter to the smaller, (a) the velocity and pressure both increase, (b) the velocity increases and the pressure decreases, (c) the velocity decreases and the pressure increases, (d) the velocity and pressure both decrease, (e) either the velocity or pressure changes but not both.

**11** • **SSM** True or false: The buoyant force on a submerged object depends on the shape of the object.

**12** • Figure 13-25 shows a Cartesian diver. The diver consists of a small tube, open at the bottom, with an air bubble at the top, inside a closed plastic soda bottle that is partly filled with water. Normally, the diver floats, but sinks when the bottle is squeezed hard. Explain why this happens.

**FIGURE 13-25**
**Problem 12**

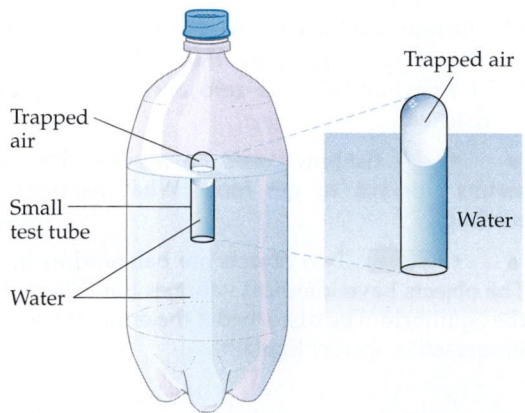

Trapped air

Trapped air

Small test tube

Water

Water

**13** • A glass of water has ice cubes floating in it. What happens to the water level when the ice melts?

**14** • Why do you float higher out of the water in salt water than in fresh water?

**15** •• A certain object has a density just slightly less than that of water so that it floats almost completely submerged. However, the object is more compressible than water. What happens if the floating object is given a slight push to submerge it?

**16** •• In Example 13-10, the fluid is accelerated to a greater speed as it enters the narrow part of the pipe. Identify the forces that act on the fluid to produce this acceleration.

**17** •• A glass of water is accelerating to the right along a horizontal surface. What is the origin of the force that produces the acceleration on a small element of water in the middle of the glass? Explain with a picture.

**18** •• **SSM** You are sitting in a boat floating on a very small pond. You take the anchor out of the boat and drop it into the water. What happens to the water level in the pond?

**19** •• Figure 13-26 is a diagram of a prairie dog tunnel. The geometry of the two holes and their positioning ensures that the wind blowing above hole number 2 will always have a lower speed than that above hole number 1. Explain how Bernoulli's principle keeps the tunnel ventilated, and indicate in which direction air will flow through the tunnel.

**FIGURE 13-26**
**Problem 19**

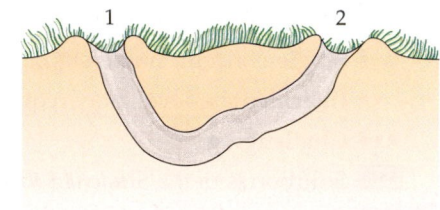

**20** • **SSM** Three bottles of different shapes are filled to the same level with water, as shown in Figure 13-27. The area of the bottom of each bottle is the same. The hydrostatic pressure is the same on the bottom of each of the bottles, but the total force must be different, as each bottle has a different amount of liquid in it. Explain this apparent paradox.

**FIGURE 13-27**
**Problem 20**

### Density

**21** • Find the mass of a copper cylinder that is 6 cm long with a radius of 2 cm.

**22** • Find the mass of a lead sphere of radius 2 cm.

**23** • **iSOLVE** Find the mass of air in a room 4 m × 5 m × 4 m.

**24** •• [SSM] A 60-mL flask is filled with mercury at 0°C (Figure 13-28). When the temperature rises to 80°C, 1.47 g of mercury spill out of the flask. Assuming that the volume of the flask is constant, find the density of mercury at 80°C if its density at 0°C is 13,645 kg/m³.

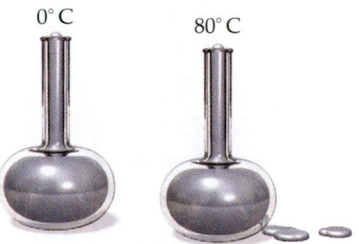

**FIGURE 13-28** Problem 24

## Pressure

**25** • [iSOLVE] Barometer readings are commonly given in inches of mercury. Find the pressure in inches of mercury equal to 101 kPa.

**26** • The pressure on the surface of a lake is atmospheric pressure $P_{at}$ = 101 kPa. (a) At what depth is the pressure twice atmospheric pressure? (b) If the pressure at the top of a deep pool of mercury is $P_{at}$, at what depth is the pressure $2P_{at}$?

**27** • [SSM] [iSOLVE] A hydraulic lift is used to raise an automobile of mass 1500 kg. The radius of the shaft of the lift is 8 cm and that of the piston is 1 cm. How much force must be applied to the piston to raise the automobile?

**28** •• When a woman in high-heeled shoes takes a step, she momentarily places her entire weight on one heel of her shoe. If her mass is 56 kg and if the area of the heel is 1 cm², what is the pressure exerted on the floor by her heel?

**29** • [SSM] What pressure increase is required to compress the volume of 1 kg of water from 1.00 L to 0.99 L?

**30** • [iSOLVE] A 1500-kg car rests on four tires, each of which is inflated to a gauge pressure of 200 kPa. What is the area of contact of each tire with the road, if the four tires support the weight equally?

**31** •• In the seventeenth century, Blaise Pascal performed the experiment shown in Figure 13-29. A wine barrel filled with water was coupled to a long tube. Water was added to the tube until the barrel burst. (a) If the radius of the lid was 20 cm and the height of the water in the tube was 12 m, calculate the force exerted on the lid. (b) If the tube had an inner radius of 3 mm, what mass of water in the tube caused the pressure that burst the barrel?

**FIGURE 13-29** Problem 31

**32** •• [iSOLVE] Blood plasma flows from a bag through a tube into a patient's vein, where the blood pressure is 12 mmHg. The specific gravity of blood plasma at 37°C is 1.03. What is the minimum elevation the bag must have so the plasma flows into the vein?

**33** •• Many people have imagined that if they were to float the top of a flexible snorkel tube out of the water, they would be able to breathe through it while walking underwater (Figure 13-30). However, they generally do not take into account just how much water pressure opposes the expansion of the chest and the inflation of the lungs. Suppose you can just breathe while lying on the floor with a 400-N (90-lb) weight on your chest. How far below the surface of the water could your chest be for you still to be able to breathe, assuming your chest has a frontal area of 0.09 m²?

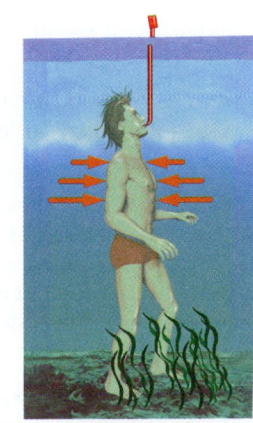

**FIGURE 13-30** Problem 33

**34** •• In Example 13-3, a force of 147 N is applied to a small piston to lift a car that weighs 14,700 N. Demonstrate that this does not violate the law of conservation of mechanical energy by showing that, when the car is lifted some distance h, the work done by the 147-N force acting on the small piston equals the work done by the large piston on the car.

**35** • [iSOLVE] A hollow cube with edge length a is half-filled with water of density ρ. Find the force exerted on a side of the cube by the water. (The edges of the cube are either horizontal or vertical.)

**36** ••• [SSM] The volume of a cone of height h and base radius r is $V = \pi r^2 h/3$. A conical vessel of height 25 cm resting on its base of radius 15 cm is filled with water. (a) Find the volume and weight of the water in the vessel. (b) Find the force exerted by the water on the base of the vessel. Explain how this force can be greater than the weight of the water.

## Buoyancy

**37** • [SSM] [iSOLVE] A 500-g piece of copper (specific gravity 9.0) is suspended from a spring scale and is submerged in water (Figure 13-31). What force does the spring scale read?

**FIGURE 13-31** Problem 37

**38** • **iSOLVE** When a 60-N stone is attached to a spring scale and is submerged in water, the spring scale reads 40 N. What is the density of the stone?

**39** • **iSOLVE** A block of an unknown material weighs 5 N in air and 4.55 N when submerged in water. (*a*) What is the density of the material? (*b*) Of what material is the block made?

**40** • A solid piece of metal weighs 90 N in air and 56.6 N when submerged in water. Determine the density of this metal.

**41** •• **iSOLVE✓** A homogeneous solid object floats on water with 80 percent of its volume below the surface. The same object when placed in a second liquid floats on that liquid with 72 percent of its volume below the surface. Determine the density of the object and the specific gravity of the liquid.

**42** •• **SSM** A 5-kg iron block is suspended from a spring scale and is submerged in a fluid of unknown density. The spring scale reads 6.16 N. What is the density of the fluid?

**43** •• A large piece of cork weighs 0.285 N in air. When held submerged underwater by a spring scale as shown in Figure 13-32, the spring scale reads 0.855 N. Find the density of the cork.

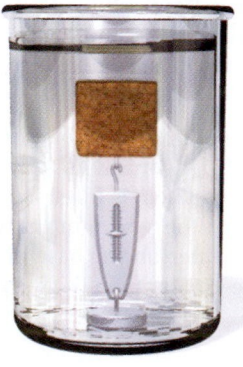

**FIGURE 13-32** Problem 43

**44** •• **iSOLVE✓** A helium balloon lifts a basket and cargo of total weight 2000 N under standard conditions, at which the density of air is 1.29 kg/m³ and the density of helium is 0.178 kg/m³. What is the minimum volume of the balloon?

**45** •• **SSM** An object has neutral buoyancy when its density equals that of the liquid in which it is submerged, which means that it neither floats nor sinks. If the average density of an 85-kg diver is 0.96 kg/L, what mass of lead should be added to give him neutral buoyancy?

**46** •• **iSOLVE✓** A beaker of mass 1 kg containing 2 kg of water rests on a scale. A 2-kg block of aluminum (density 2.70 × 10³ kg/m³) suspended from a spring scale is submerged in the water as in Figure 13-33. Find the readings of both scales.

**FIGURE 13-33** Problem 46

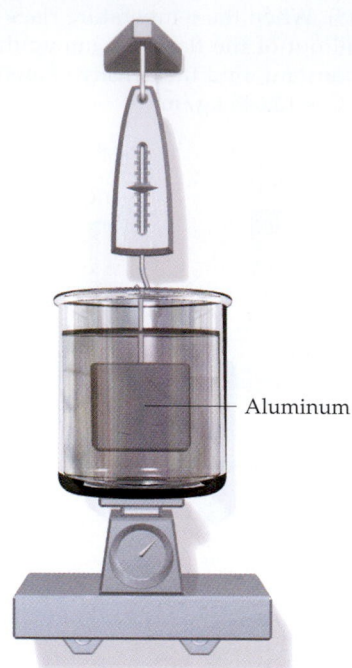

Aluminum

**47** ••• A ship sails from seawater (specific gravity 1.025) into freshwater and therefore sinks slightly. When its load of 600,000 kg is removed, it returns to its original level. Assuming that the sides of the ship are vertical at the water line, find the mass of the ship before it was unloaded.

**48** ••• **SSM** The hydrometer shown in Figure 13-34 is a device for measuring the specific gravity of liquids. The bulb contains lead shot, and the specific gravity can be read directly from the liquid level on the stem after the hydrometer has been calibrated. The volume of the bulb is 20 mL, the stem is 15 cm long and has a diameter of 5.00 mm, and the mass of the glass is 6.0 g. (*a*) What mass of lead shot must be added so that the lowest specific gravity of liquid that can be measured is 0.9? (*b*) What is the maximum specific gravity of liquid that can be measured?

**FIGURE 13-34** Problems 48, 98

**49** • When cracks form at the base of a dam, the water seeping in the cracks exerts a buoyant force that tends to lift the dam. This can topple the dam. Estimate the buoyant force exerted on a 2-m thick by 5-m long dam wall by water seeping into cracks at its base. The water level is 5 m above the cracks.

**50** •• A large helium weather balloon is spherical in shape, with a radius of 2.5 m and a total mass of 15 kg (balloon plus helium plus equipment). (*a*) What is the initial upward acceleration of the balloon when it is released from sea level? (*b*) If the drag force on the balloon is given by $F_D = \frac{1}{2} \pi r^2 \rho v^2$, where $r$ is the balloon radius, $\rho$ is the density of air, and $v$ the balloon's ascension speed, calculate the terminal velocity of the ascending balloon. (*c*) Roughly how long will it take for the balloon to ascend to a height of 10 km?

## Continuity and Bernoulli's Equation

**51** •• SSM Water exits a circular tap moving straight down with a flow rate of 10.5 cm³/s. (*a*) If the diameter of the tap is 1.2 cm, what is the speed of the water? (*b*) As the fluid falls from the tap, the stream of water narrows. Find the new diameter of the stream at a point 7.5 cm below the tap. Assume that the stream still has a circular cross section and neglect any drag forces acting on the water. (*c*) If turbulent flows are characterized by Reynolds numbers above 2300 or so, how far does the water have to fall before it becomes turbulent? Does this match everyday experience?

**52** • ISOLVE✓ Water flows at 0.65 m/s through a 3-cm-diameter hose. At the end of the hose is a 0.30-cm-diameter nozzle. (*a*) At what speed does the water pass through the nozzle? (*b*) If the pump at one end of the hose and the nozzle at the other end are at the same height, and if the pressure at the nozzle is 1 atm, what is the pressure at the pump? Assume laminar nonviscous flow.

**53** • ISOLVE✓ Water is flowing at 3 m/s in a horizontal pipe under a pressure of 200 kPa. The pipe narrows to half its original diameter. (*a*) What is the speed of flow in the narrow section? (*b*) What is the pressure in the narrow section? (*c*) How do the volume flow rates in the two sections compare? Assume laminar nonviscous flow.

**54** • ISOLVE The pressure in a section of horizontal pipe with a diameter of 2 cm is 142 kPa. Water flows through the pipe at 2.80 L/s. If the pressure at a certain point is to be reduced to 101 kPa by constricting a section of the pipe, what should the diameter of the constricted section be? Assume laminar nonviscous flow.

**55** •• SSM Blood flows in an aorta of radius 9 mm at 30 cm/s. (*a*) Calculate the volume flow rate in liters per minute. (*b*) Although the cross-sectional area of a capillary is much smaller than that of the aorta, there are many capillaries, so their total cross-sectional area is much larger. If all the blood from the aorta flows into the capillaries and the speed of flow through the capillaries is 1.0 mm/s, calculate the total cross-sectional area of the capillaries.

**56** •• ISOLVE✓ A large tank of water is tapped a distance $h$ below the water surface by a small pipe as in Figure 13-35. Find the distance $x$ reached by the water flowing out of the pipe. Assume laminar nonviscous flow.

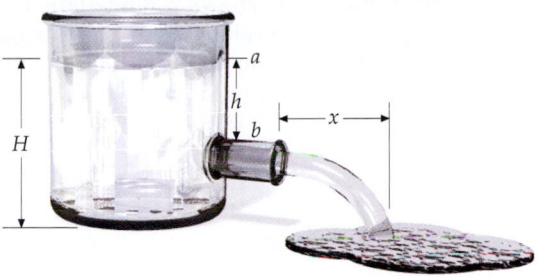

**FIGURE 13-35** Problems 56, 61

**57** •• ISOLVE The \$8-billion, 800-mile long Alaskan Pipeline has a capacity of 240,000 m³ of oil per day. Along most of the pipeline the radius is 60 cm. Find the pressure $P'$ at a point where the pipe has a 30-cm radius. Take the pressure in the 60-cm-diameter sections to be $P = 180$ kPa and the density of oil to be 800 kg/m³. Assume laminar nonviscous flow.

**58** •• SSM Water flows through a Venturi meter like that in Example 13-10 with a pipe diameter of 9.5 cm and a constriction diameter of 5.6 cm. The U-tube manometer is partially filled with mercury. Find the volume flow rate of the water if the difference in the mercury level in the U-tube is 2.40 cm.

**59** •• A firefighter holds a hose with a bend in it as in Figure 13-36. Water is expelled from the hose in a stream of radius 1.5 cm at a speed of 30 m/s. (*a*) What mass of water emerges from the hose in 1 s? (*b*) What is the horizontal momentum of this water? (*c*) Before reaching the bend, the water has momentum upward, whereas afterward, its momentum is horizontal. Draw a vector diagram of the initial and final momentum vectors, and find the change in the momentum of the water at the bend during 1 s. From this, find the force exerted on the water by the hose.

**FIGURE 13-36** Problem 59

**60** •• ISOLVE✓ A fountain designed to spray a column of water 12 m into the air has a 1-cm-diameter nozzle at ground level. The water pump is 3 m below the ground. The pipe to the nozzle has a diameter of 2 cm. Find the necessary pump pressure. Assume laminar nonviscous flow.

**61** ••• In Figure 13-35, $H$ is the depth of the liquid and $h$ is the distance of the opening below the surface of the liquid. (*a*) Find the distance $x$ at which the water strikes the ground as a function of $h$ and $H$. (*b*) Show that, for a given value of $H$, there are two values of $h$ (whose average value is $\frac{1}{2}H$) that give the same distance $x$. (*c*) Show that $x$ is a maximum when $h = \frac{1}{2}H$. What is the value of this maximum distance $x$? Assume laminar nonviscous flow.

**62** •• [SSM] Figure 13-37 shows a Pitot-static tube, a device used for measuring the velocity of a gas. The inner pipe faces the incoming fluid, while the ring of holes in the outer tube are parallel to the gas flow. Show that the speed of the gas is given by $v^2 = 2gh(\rho - \rho_g)/\rho_g$, where $\rho$ is the density of the liquid used in the manometer and $\rho_g$ is the density of the gas.

**FIGURE 13-37** Problem 62

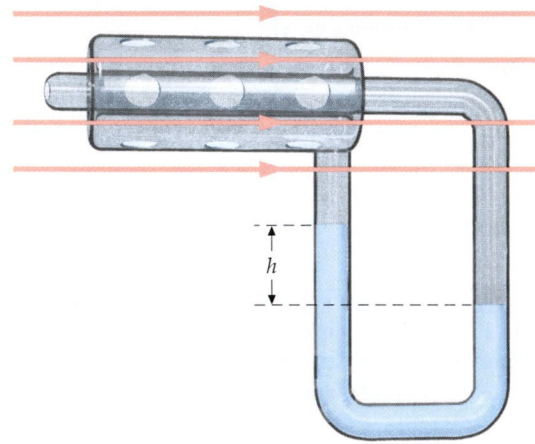

**63** •• A siphon is a device for transferring a liquid from container to container. The tube shown in Figure 13-38 must be filled to start the siphon, but once this has been done, fluid will flow through the tube until the liquid surfaces in the containers are at the same level. (a) Using Bernoulli's equation, show that the velocity of water in the tube is $v = \sqrt{2gd}$ (b) What is the pressure at the highest part of the tube? Assume laminar nonviscous flow.

**FIGURE 13-38** Problem 63

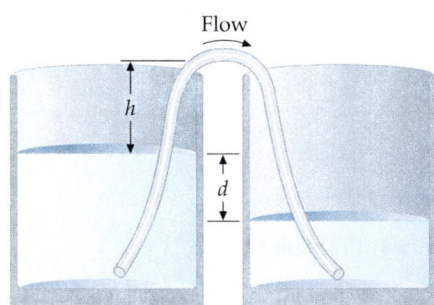

Flow

## *Viscous Flow

**64** • [ISOLVE] A horizontal tube with an inside diameter of 1.2 mm and a length of 25 cm has water flowing through it at 0.30 mL/s. Find the pressure difference required to drive this flow if the viscosity of water is 1.00 mPa·s. Assume laminar flow.

**65** • Find the diameter of a tube that would give double the flow rate for the pressure difference in Problem 64.

**66** • [SSM] Blood takes about 1.0 s to pass through a 1-mm-long capillary of the human circulatory system. If the diameter of the capillary is 7 $\mu$m and the pressure drop is 2.60 kPa, find the viscosity of blood. Assume laminar flow.

**67** • [SSM] At Reynolds numbers of about $3 \times 10^5$ there is an abrupt transition, where the drag on a sphere abruptly decreases. Estimate the velocity at which this drag crisis occurs for a baseball, and comment on whether or not it should play a role in the physics of the game.

**68** ••• Stokes' law states that the drag force on a sphere at very low Reynolds number is given by $F_D = 6\pi\eta av$, where $\eta$ is the viscosity of the surrounding fluid and $a$ is the radius of the sphere. Using this, find the terminal velocity of ascent for a spherical carbon dioxide bubble of 1-mm diameter rising in a glass of soda (density $1.1 \times 10^3$ kg/m³). How long should it take for the bubble to rise the height of a "typical" soda glass? Is this consistent with your experience?

## General Problems

**69** •• [SSM] Very roughly speaking, the mass of a person should increase as the cube of his or her height—that is, $M = C\rho h^3$, where $M$ is the mass, $h$ is the height, $\rho$ is body density, and $C$ is a person's "coefficient of roundness." Estimate $C$ for an adult male and female, using "typical" values for height and weight. Assume $\rho = 1000$ kg/m³.

**70** • Using that weight is proportional to height cubed (see problem 69) what should be the difference in weight for two men, one of whom is 5′ 9″ tall and the other 6′ 0″?

**71** • The top of a card table is 80 cm × 80 cm. What is the force exerted on it by the atmosphere? Why doesn't the table collapse?

**72** • [ISOLVE] ✓ A 4.0-g Ping-Pong ball is attached by a thread to the bottom of a beaker. When the beaker is filled with water so that the ball is totally submerged, the tension in the thread is $2.8 \times 10^{-2}$ N. Determine the diameter of the ball.

**73** • [ISOLVE] Seawater has a bulk modulus of $2.3 \times 10^9$ N/m². Find the density of seawater at a depth where the pressure is 800 atm if the density at the surface is 1025 kg/m³.

**74** • A solid cubical block of edge length 0.6 m is suspended from a spring balance. When the block is submerged in water, the spring balance reads 80 percent of the reading when the block is in air. Determine the density of the block.

**75** • [SSM] When submerged in water, a block of copper has an apparent weight of 56 N. What fraction of this copper block will be submerged if it is floated on a pool of mercury?

**76** • [ISOLVE] A 4.5-kg block of material floats on ethanol with 10 percent of its volume above the liquid surface. What fraction of the volume of this block will be submerged if it is floated on water?

**77** • What is the buoyant force on your body when you are floating (a) in a freshwater lake (specific gravity = 1.00) and (b) in the ocean (specific gravity = 1.03)?

**78** • Suppose that when you are floating in fresh water, 96 percent of your body volume is submerged. What is the volume of water your body displaces when it is fully submerged?

**79** •• A block of wood of 1.5-kg mass floats on water with 68 percent of its volume submerged. A lead block is placed on the wood and the wood is then fully submerged with the lead entirely out of the water. Find the mass of the lead block.

**80** •• **SSM** **ISOLVE** ✓ A Styrofoam cube, 25 cm on an edge, is placed on one pan of a balance. The balance is in equilibrium when a 20-g mass of brass is placed on the opposite pan of the balance. Find the mass of the Styrofoam cube.

**81** •• **ISOLVE** A spherical shell of copper with an outer diameter of 12 cm floats on water with half its volume above the water's surface. Determine the inner diameter of the shell.

**82** •• A beaker filled with water is balanced on the left pan of a balance. A cube 4 cm on an edge is attached to a string and lowered into the water so that it is completely submerged. The cube is not touching the bottom of the beaker. A weight of mass $m$ is added to the right pan to restore equilibrium. What is $m$?

**83** •• **SSM** Crude oil has a viscosity of about 0.8 Pa·s at normal temperature. A 50-km pipeline is to be constructed from an oil field to a tanker terminal. The pipeline is to deliver oil at the terminal at a rate of 500 L/s and the flow through the pipeline is to be laminar to minimize the pressure needed to push the fluid through the pipeline. Assuming that the density of crude oil is 700 kg/m$^3$, estimate the diameter of the pipeline that should be used.

**84** •• **ISOLVE** Water flows through the pipe in Figure 13-39 and exits to the atmosphere at C. The diameter of the pipe is 2.0 cm at A, 1.0 cm at B, and 0.8 cm at C. The gauge pressure in the pipe at A is 1.22 atm and the flow rate is 0.8 L/s. The vertical pipes are open to the air. Find the level of the liquid–air interfaces in the two vertical pipes. Assume laminar nonviscous flow.

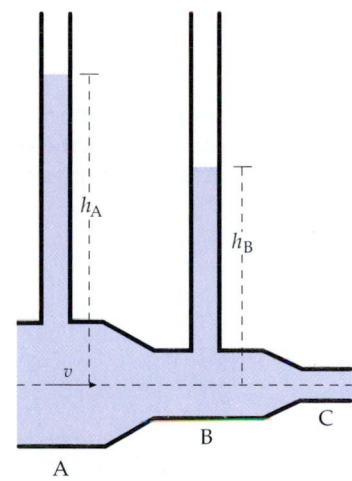

**FIGURE 13-39** Problems 84, 85

**85** •• Repeat Problem 84 with the flow rate reduced to 0.6 L/s and the size of the opening at C reduced so that the pressure in the pipe at A remains unchanged.

**86** •• **SSM** **ISOLVE** Figure 13-40 is a sketch of an *aspirator*, a simple device that can be used to achieve a partial vacuum in a reservoir connected to the vertical tube at B. An aspirator attached to the end of a garden hose may be used to deliver soap or fertilizer from the reservoir. Suppose that the diameter at A is 2.0 cm and at C, where the water exits to the atmosphere, it is 1.0 cm. If the flow rate is 0.5 L/s and

the gauge pressure at A is 0.187 atm, what diameter of the constriction at B will achieve a pressure of 0.1 atm in the container? Assume laminar nonviscous flow.

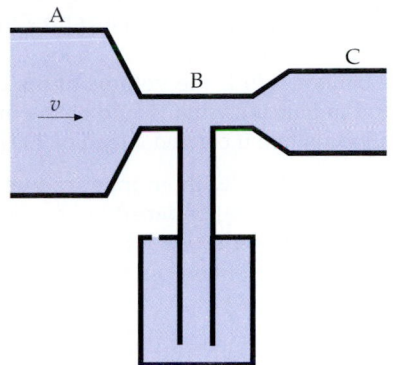

**FIGURE 13-40** Problem 86

**87** •• A cylindrical buoy at the entrance of a harbor has a diameter of 0.9 m and a height of 2.6 m. The mass of the buoy is 600 kg. It is attached to the bottom of the sea with a nylon cable of negligible mass. The specific gravity of the seawater is 1.025. (*a*) How much of the buoy is visible when the cable is slack? (*b*) If a tsunami completely submerges the buoy, what is the tension in the taut cable? (*c*) If the cable breaks, what is the initial upward acceleration of the buoy?

**88** •• **ISOLVE** Two connected vessels contain a liquid of density $\rho_0$ (Figure 13-41). The cross-sectional areas of the vessels are $A$ and $3A$. Find the change in elevation of the liquid level if an object of mass $m$ and density $\rho' = 0.8\rho_0$ is placed into one of the vessels.

**FIGURE 13-41** Problem 88

**89** •• If an oil-filled manometer ($\rho = 900$ kg/m$^3$) can be read to $\pm 0.05$ mm, what is the smallest pressure change that can be detected?

**90** •• **ISOLVE** ✓ A U-tube is filled with water until the liquid level is 28 cm above the bottom of the tube. An oil of specific gravity 0.78 is now poured into one arm of the U-tube until the level of the water in the other arm of the tube is 34 cm above the bottom of the tube. Find the levels of the oil–water and oil–air interfaces in the other arm of the tube.

**91** •• A U-tube contains liquid of unknown specific gravity. An oil of density 800 kg/m$^3$ is poured into one arm of the tube until the oil column is 12 cm high. The oil–air interface is then 5.0 cm above the liquid level in the other arm of the U-tube. Find the specific gravity of the liquid.

**92** •• A lead block is suspended from the underside of a 0.5-kg block of wood of specific gravity 0.7. If the upper surface of the wood is just level with the surface of the water, what is the mass of the lead block?

**93** •• [SSM] A helium balloon can just lift a load of 750 N. The skin of the balloon has a mass of 1.5 kg. (a) What is the volume of the balloon? (b) If the volume of the balloon were twice that found in Part (a), what would be the initial acceleration of the balloon when it carried a load of 900 N?

**94** •• A hollow sphere with an inner radius $R$ and outer radius $2R$ is made of material of density $\rho_0$ and is floating in a liquid of density $2\rho_0$. The interior is now filled with material of density $\rho'$ so that the sphere just floats completely submerged. Find $\rho'$.

**95** •• [SSM] As mentioned in the discussion of *the law of atmospheres*, the fractional decrease in atmospheric pressure is proportional to the change in altitude. Expressed as a differential equation we have $dP/P = -C/dh$, where $C$ is a constant. (a) Show that $P(h) = P_0e^{-Ch}$ is a solution of the differential equation. (b) Show that if $\Delta h \ll h_0$ where $h_0 = 1/C$, then $P(h + \Delta h) \approx P(h)(1 - \Delta h/h_0)$. (c) Given that the pressure at $h = 5.5$ km is half that at sea level, find the constant $C$.

**96** •• A submarine has a total mass of $2.4 \times 10^6$ kg, including crew and equipment. The vessel consists of two parts, the pressure hull, which has a volume of $2 \times 10^3$ m$^3$, and the ballast tanks, which have a volume of $4 \times 10^2$ m$^3$. When the sub cruises on the surface, the ballast tanks are filled with air; when it is cruising below the surface, seawater is admitted into the tanks. (a) What fraction of the submarine's volume is above the water surface when the tanks are filled with air? (b) How much water must be admitted into the tanks to give the submarine neutral buoyancy? Neglect the mass of air in the tanks and use 1.025 as the specific gravity of seawater.

**97** •• A marine salvage crew raises a crate that measures 1.4 m × 0.75 m × 0.5 m. The average density of the empty crate is the same as seawater, $1.025 \times 10^3$ kg/m$^3$, and its mass

when empty is 32 kg. The crate contains gold bullion that fills 36 percent of its volume; the remaining volume is filled with seawater. (a) What is the tension in the cable that raises the crate and bullion while the crate is below the surface of the sea? (b) What is the tension in the cable while the crate is lifted to the deck of the ship if (1) none of the seawater leaks out of the crate, and (2) the crate is lifted so slowly that all of the seawater leaks out of the crate?

**98** ••• When the hydrometer of Problem 48 (Figure 13-34) is placed in a liquid whose specific gravity is greater than some minimum value, the device floats with part of the glass tube above the liquid level. Consider a hydrometer that has a spherical bulb 2.4 cm in diameter. The glass tube attached to the bulb is 20 cm long and has a diameter of 7.5 mm. The mass of the hydrometer before lead pellets are dropped into the bulb and the tube is sealed is 7.28 g. (a) What mass of lead should be placed in the bulb so that the hydrometer just floats in a liquid of specific gravity 0.78? (b) If the hydrometer is now placed in water, what is the length of the tube that shows above the surface of the water? (c) The hydrometer is placed in a liquid of unknown specific gravity; the length of the tube above the surface of the liquid is 12.2 cm. Determine the specific gravity of the liquid.

**99** ••• A large root beer keg of height $H$ and cross-sectional area $A_1$ is filled with root beer. The top is open to atmospheric pressure. At the bottom is a spigot opening of area $A_2$, which is much smaller than $A_1$. (a) Show that when the height of the root beer is $h$, the speed of the root beer leaving the spigot is approximately $\sqrt{2gh}$. (b) Show that for the approximation $A_2 \ll A_1$, the rate of change of the height $h$ of the root beer is given by

$$\frac{dh}{dt} = -\frac{A_2}{A_1}(2gh)^{1/2}$$

(c) Find $h$ as a function of time if $h = H$ at $t = 0$. (d) Find the total time needed to drain the keg if $H = 2$ m, $A_1 = 0.8$ m$^2$, and $A_2 = 10^{-4}A_1$. Assume laminar nonviscous flow.

CHAPTER

# 14

# Oscillations

THIS BOAT IS RISING AND FALLING ON THE SWELLS OF THE SEA. ITS MOTION IS AN EXAMPLE OF OSCILLATORY MOTION. THE MAXIMUM CHANGE IN THE VERTICAL POSITION OF THE BOAT CAN BE READILY MEASURED, AS CAN THE TIME FOR THE BOAT TO COMPLETE ONE CYCLE OF THIS UP AND DOWN MOTION.

**?** How can the vertical position of the boat be expressed as a function of time? (See Example 14-1.)

14-1    Simple Harmonic Motion

14-2    Energy in Simple Harmonic Motion

14-3    Some Oscillating Systems

14-4    Damped Oscillations

14-5    Driven Oscillations and Resonance

Oscillation occurs when a system is disturbed from a position of stable equilibrium. There are many familiar examples: boats bob up and down, clock pendulums swing back and forth, and the strings and reeds of musical instruments vibrate. Other, less familiar examples are the oscillations of air molecules in a sound wave and the oscillations of electric currents in radios and television sets.

➤ **In this chapter, we deal mostly with simple harmonic motion—the most basic type of oscillatory motion. Applying the kinematics and dynamics of simple harmonic motion provides the analysis of the oscillatory motion of a variety of interesting systems. In some situations dissipative forces dampen the oscillatory motion, but in other situations driving forces sustain the motion by compensating for the damping.**

# 14-1 Simple Harmonic Motion

A common, very important, and very basic kind of oscillatory motion is **simple harmonic motion** such as the motion of an object attached to a spring (Figure 14-1). In equilibrium, the spring exerts no force on the object. When the object is displaced an amount $x$ from its equilibrium position, the spring exerts a force $-kx$, as given by Hooke's law[†]:

$$F_x = -kx \qquad 14\text{-}1$$

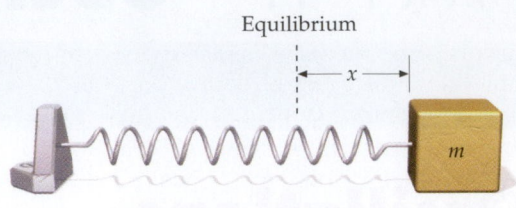

**FIGURE 14-1** An object and spring on a frictionless surface. The displacement $x$, measured from the equilibrium position, is positive if the spring is stretched and negative if the spring is compressed.

where $k$ is the force constant of the spring, a measure of the spring's stiffness. The minus sign indicates that the force is a restoring force; that is, it is opposite to the direction of the displacement from the equilibrium position. Combining Equation 14-1 with Newton's second law ($F_x = ma_x$), we have

$$-kx = ma_x$$

or

$$a_x = -\frac{k}{m}x \qquad \left(\text{or} \quad \frac{d^2x}{dt^2} = -\frac{k}{m}x\right) \qquad 14\text{-}2$$

The acceleration is proportional to the displacement and is oppositely directed. This is the defining characteristic of simple harmonic motion and can be used to identify systems that will exhibit it:

> Whenever the acceleration of an object is proportional to its displacement and is oppositely directed, the object will move with simple harmonic motion.

CONDITIONS FOR SIMPLE HARMONIC MOTION IN TERMS OF ACCELERATION

Because the acceleration is proportional to the net force, whenever the net force on an object is proportional to its displacement and is oppositely directed, the object will move with simple harmonic motion.

The time it takes for a displaced object to execute a complete cycle of oscillatory motion—from one extreme to the other extreme and back—is called the **period** $T$. The reciprocal of the period is the **frequency** $f$, which is the number of cycles per second:

$$f = \frac{1}{T} \qquad 14\text{-}3$$

The unit of frequency is the cycle per second (cy/s), which is called a **hertz** (Hz). For example, if the time for one complete cycle of oscillation is 0.25 s, the frequency is 4 Hz.

Figure 14-2 shows how we can experimentally obtain $x$ versus $t$ for a mass on a spring. The general equation for such a curve is

$$x = A \cos(\omega t + \delta) \qquad 14\text{-}4$$

POSITION IN SIMPLE HARMONIC MOTION

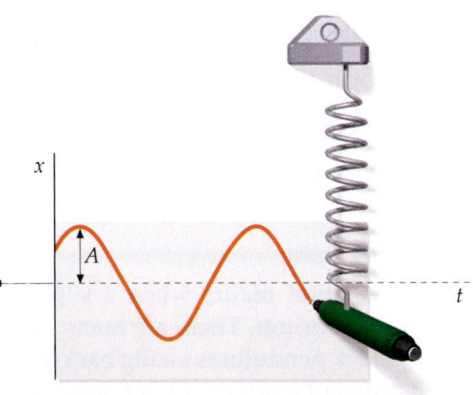

**FIGURE 14-2** A marking pen is attached to a mass on a spring, and the paper is pulled to the left. As the paper moves with constant speed, the pen traces out the displacement $x$ as a function of time $t$. (Here we have chosen $x$ to be positive when the spring is compressed.)

---

† Hooke's law is introduced in Chapter 4, Section 4.

where $A$, $\omega$, and $\delta$ are constants. The maximum displacement $x_{max}$ from equilibrium is called the **amplitude** $A$. The argument of the cosine function, $\omega t + \delta$, is called the **phase** of the motion, and the constant $\delta$ is called the **phase constant,** which is the phase at $t = 0$. (Note that $\cos(\omega t + \delta) = \sin(\omega t + \delta + \pi/2)$; thus, whether the equation is expressed as a cosine function or a sine function simply depends on the phase of the oscillation at the moment we designate to be $t = 0$.) If we have just one oscillating system, we can always choose $t = 0$ at which $\delta = 0$. If we have two systems oscillating with the same amplitude and frequency but different phase, we can choose $\delta = 0$ for one of them. The equations for the two systems are then

$$x_1 = A\cos(\omega t)$$

and

$$x_2 = A\cos(\omega t + \delta)$$

If the phase difference $\delta$ is 0 or an integer times $2\pi$, then $x_2 = x_1$ and the systems are said to be in phase. If the phase difference $\delta$ is $\pi$ or an odd integer times $\pi$, then $x_2 = -x_1$, and the systems are said to be 180° out of phase.

The swaying of the Citicorp Building in New York City during high winds is reduced by this tuned-mass damper mounted on an upper floor. It consists of a 400-ton sliding block connected to the building by a spring. The spring constant is chosen so that the natural frequency of the spring–block system is the same as the natural sway frequency of the building. Set into motion by winds, the building and damper oscillate 180° out of phase with each other, thereby significantly reducing the swaying.

We can show that Equation 14-4 is a solution of Equation 14-2 by differentiating $x$ twice with respect to time. The first derivative of $x$ gives the velocity $v$:

$$v = \frac{dx}{dt} = -\omega A\sin(\omega t + \delta) \qquad 14\text{-}5$$

VELOCITY IN SIMPLE HARMONIC MOTION

Differentiating velocity with respect to time gives the acceleration:

$$a = \frac{dv}{dt} = \frac{d^2x}{dt^2} = -\omega^2 A\cos(\omega t + \delta) \qquad 14\text{-}6$$

Substituting $x$ for $A \cos(\omega t + \delta)$ (see Equation 14-4) gives

$$a = -\omega^2 x \qquad \text{14-7}$$

ACCELERATION IN SIMPLE HARMONIC MOTION

Comparing $a = -\omega^2 x$ (Equation 14-7) with $a = -(k/m)x$ (Equation 14-2), we see that $x = A \cos(\omega t + \delta)$ is a solution of Equation 14-2 (which can be expressed $d^2x/dt^2 = -(k/m)x$) if

$$\omega = \sqrt{\frac{k}{m}} \qquad \text{14-8}$$

The amplitude $A$ and the phase constant $\delta$ can be determined from the initial position $x_0$ and the initial velocity $v_0$ of the system. Setting $t = 0$ in $x = A \cos(\omega t + \delta)$ gives

$$x_0 = A \cos \delta \qquad \text{14-9}$$

Similarly, setting $t = 0$ in $v = dx/dt = -A\omega \sin(\omega t + \delta)$ gives

$$v_0 = -A\omega \sin \delta \qquad \text{14-10}$$

These equations can be solved for $A$ and $\delta$ in terms of $x_0$ and $v_0$.

The period $T$ is the shortest time satisfying the relation

$$x(t) = x(t + T)$$

for all $t$. Substituting into this relation using Equation 14-4 gives

$$A \cos(\omega t + \delta) = A \cos[\omega(t + T) + \delta]$$
$$= A \cos(\omega t + \delta + \omega T)$$

The cosine (and sine) function repeats in value when the phase increases by $2\pi$, so

$$\omega T = 2\pi \qquad \left( \text{or} \quad \omega = \frac{2\pi}{T} \right)$$

The constant $\omega$ is called the **angular frequency**. It has units of radians per second and dimensions of inverse time, the same as angular speed, which is also designated by $\omega$. Substituting $2\pi/T$ for $\omega$ in Equation 14-4 gives

$$x = A \cos\left( 2\pi \frac{t}{T} + \delta \right)$$

We can see by inspection that each time $t$ increases by $T$, the phase increases by $2\pi$ and one cycle of the motion is completed.

The frequency is the reciprocal of the period:

$$f = \frac{1}{T} = \frac{\omega}{2\pi} \qquad \text{14-11}$$

DEFINITION—FREQUENCY, PERIOD, AND ANGULAR FREQUENCY

Because $\omega = \sqrt{k/m}$, the frequency and period of an object on a spring are related to the force constant $k$ and the mass $m$ by

$$f = \frac{1}{T} = \frac{1}{2\pi}\sqrt{\frac{k}{m}}$$   14-12

FREQUENCY AND PERIOD FOR AN OBJECT ON A SPRING

The frequency increases with increasing $k$ (spring stiffness) and decreases with increasing mass.

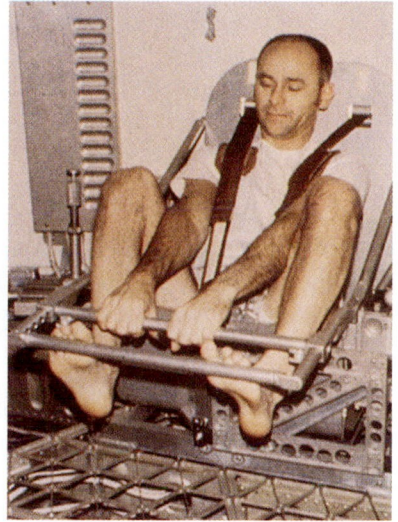

Astronaut Alan L. Bean measures his body mass during the second Skylab mission by sitting in a seat attached to a spring and oscillating back and forth. The total mass of the astronaut plus the seat is related to his frequency of vibration by Equation 14-12.

---

*RIDING THE WAVES*                    **E X A M P L E   1 4 - 1**

You are on a boat, which is bobbing up and down. The boat's vertical displacement $y$ is given by

$$y = (1.2 \text{ m}) \cos\left(\frac{1}{2\text{ s}}t + \frac{\pi}{6}\right)$$

(*a*) Find the amplitude, angular frequency, phase constant, frequency, and period of the motion. (*b*) Where is the boat at $t = 1$ s? (*c*) Find the velocity and acceleration as functions of time $t$. (*d*) Find the initial values of the position, velocity, and acceleration of the boat.

**PICTURE THE PROBLEM** We find the quantities asked for in (*a*) by comparing the equation of motion

$$y = (1.2 \text{ m}) \cos\left(\frac{1}{2\text{ s}}t + \frac{\pi}{6}\right)$$

with the standard equation for simple harmonic motion, Equation 14-4. The velocity and acceleration are found by differentiating $y(t)$.

(*a*) 1. Compare this equation with Equation 14-4, $y = A\cos(\omega t + \delta)$, to get $A$, $\omega$, and $\delta$:

$$y = (1.2 \text{ m}) \cos\left(\frac{1}{2\text{ s}}t + \frac{\pi}{6}\right)$$

$$A = \boxed{1.2 \text{ m}} \qquad \omega = \boxed{1/2 \text{ rad/s}} \qquad \delta = \boxed{\pi/6 \text{ rad}}$$

2. The frequency and period are found from $\omega$:

$$f = \frac{\omega}{2\pi} = \boxed{0.0796 \text{ Hz}} \qquad T = \frac{1}{f} = \boxed{12.6 \text{ s}}$$

(*b*) Set $t = 1$ s to find the boat's position:

$$y = (1.2 \text{ m}) \cos\left[\frac{1}{2\text{ s}}(1 \text{ s}) + \frac{\pi}{6}\right] = \boxed{0.624 \text{ m}}$$

(*c*) The velocity and acceleration are obtained from the position by differentiation with respect to time:

$$v_y = \frac{dy}{dt} = \frac{d}{dt}[A\cos(\omega t + \delta)]$$

$$= -\omega A \sin(\omega t + \delta)$$

$$= -\frac{1}{2\text{ s}}(1.2 \text{ m})\sin\left(\frac{1}{2\text{ s}}t + \frac{\pi}{6}\right)$$

$$= \boxed{-(0.6 \text{ m/s})\sin\left(\frac{1}{2\text{ s}}t + \frac{\pi}{6}\right)}$$

$$a_y = \frac{dv_y}{dt} = \frac{d}{dt}[-\omega A \sin(\omega t + \delta)]$$

$$= -\omega^2 A \cos(\omega t + \delta)$$

$$= -\left(\frac{1}{2\,\text{s}}\right)^2 (1.2\,\text{m}) \cos\left(\frac{1}{2\,\text{s}}t + \frac{\pi}{6}\right)$$

$$= \boxed{-(0.3\,\text{m/s}^2)\cos\left(\frac{1}{2\,\text{s}}t + \frac{\pi}{6}\right)}$$

(d) Set $t = 0$ to find $y_0$, $v_{y0}$, and $a_{y0}$:

$$y_0 = (1.2\,\text{m})\cos\frac{\pi}{6} = \boxed{1.04\,\text{m}}$$

$$v_{y0} = -(0.6\,\text{m/s})\sin\frac{\pi}{6} = \boxed{-0.300\,\text{m/s}}$$

$$a_{y0} = -(0.3\,\text{m/s}^2)\cos\frac{\pi}{6} = \boxed{-0.260\,\text{m/s}^2}$$

**EXERCISE** A 0.8-kg object is attached to a spring of force constant $k = 400\,\text{N/m}$. Find the frequency and period of motion of the object when it is displaced from equilibrium. (*Answer* $f = 3.56\,\text{Hz}$, $T = 0.281\,\text{s}$)

Figure 14-3 shows two identical masses attached to identical springs and resting on a frictionless surface. One spring is stretched 10 cm and the other 5 cm. If they are released at the same time, which object reaches the equilibrium position first?

According to Equation 14-12, the period depends only on $k$ and $m$ and not on the amplitude. Since $k$ and $m$ are the same for both systems, the periods are the same. Thus, the objects reach the equilibrium position at the same time. The second object has twice as far to go to reach equilibrium, but it will also have twice the speed at any given instant. Figure 14-4 shows a sketch of the position functions for the two objects. This illustrates an important general property of simple harmonic motion:

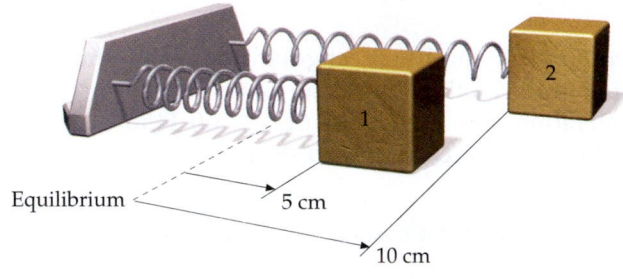

**FIGURE 14-3** Two identical mass–spring systems.

The frequency and period of simple harmonic motion are independent of the amplitude.

The fact that the frequency in simple harmonic motion is independent of the amplitude has important consequences in many fields. In music, for example, it means that when a note is struck on the piano, the pitch (which corresponds to the frequency) does not depend on how loudly the note is played (which corresponds to the amplitude).[†] If changes in amplitude had a large effect on the frequency, then musical instruments would be unplayable.

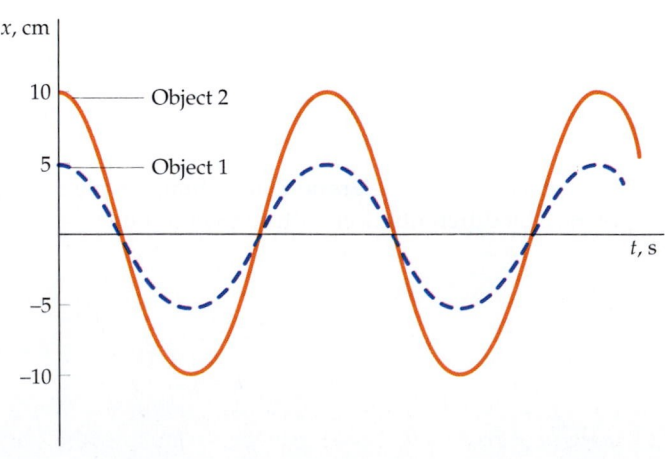

**FIGURE 14-4** Plots of $x$ versus $t$ for the systems in Figure 14-3. Both reach their equilibrium positions at the same time.

---

† For many musical instruments, there is a slight dependence of frequency on amplitude. The vibration of an oboe reed, for example, is not exactly simple harmonic, thus its pitch depends slightly on how hard it is blown. This effect can be corrected for by a skilled musician.

*AN OSCILLATING OBJECT*                                    **EXAMPLE 14-2**

An object oscillates with angular frequency $\omega = 8.0$ rad/s. At $t = 0$, the object is at $x = 4$ cm with an initial velocity $v = -25$ cm/s. (*a*) Find the amplitude and phase constant for the motion. (*b*) Write $x$ as a function of time.

**PICTURE THE PROBLEM** The initial position and velocity give us two equations from which to determine the amplitude $A$ and the phase constant $\delta$.

(*a*) 1. The initial position and velocity are related to the amplitude and phase constant. The position is given by Equation 14-4. The velocity is found by taking the derivative with respect to time:

$$x = A \cos(\omega t + \delta)$$

and

$$v = \frac{dx}{dt} = -\omega A \sin(\omega t + \delta)$$

2. At $t = 0$ the position and velocity are:

$$x_0 = A \cos \delta \quad \text{and} \quad v_0 = -\omega A \sin \delta$$

3. Divide these equations to eliminate $A$:

$$\frac{v_0}{x_0} = \frac{-\omega A \sin \delta}{A \cos \delta} = -\omega \tan \delta$$

4. Substituting numerical values yields $\delta$:

$$\tan \delta = -\frac{v_0}{\omega x_0}$$

so

$$\delta = \tan^{-1}\left(-\frac{v_0}{\omega x_0}\right)$$

$$= \tan^{-1}\left[-\frac{-25 \text{ cm/s}}{(8.0 \text{ rad/s})(4 \text{ cm})}\right]$$

$$= \boxed{0.663 \text{ rad}}$$

5. The amplitude can be found using either the $x_0$ or $v_0$ equation. Here we use $x_0$:

$$A = \frac{x_0}{\cos \delta} = \frac{4 \text{ cm}}{\cos 0.663} = \boxed{5.08 \text{ cm}}$$

(*b*) Comparing with Equation 14-4 yields $x$:

$$x = \boxed{(5.08 \text{ cm}) \cos[(8.0 \text{ s}^{-1})t + 0.663]}$$

When the phase constant is $\delta = 0$, Equations 14-4, 14-5, and 14-6 then become

$$x = A \cos \omega t \qquad\qquad\qquad 14\text{-}13a$$
$$v = -\omega A \sin \omega t \qquad\qquad 14\text{-}13b$$

and

$$a = -\omega^2 A \cos \omega t \qquad\qquad 14\text{-}13c$$

These functions are plotted in Figure 14-5.

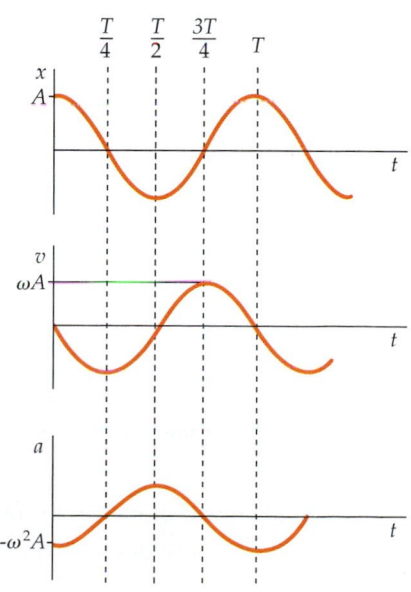

**FIGURE 14-5** Plots of $x$, $v$, and $a$ as functions of time $t$ for $\delta = 0$. At $t = 0$, the displacement is maximum, the velocity is zero, and the acceleration is negative and equal to $-\omega^2 A$. The velocity becomes negative as the object moves back toward its equilibrium position. After one quarter-period ($t = T/4$), the object is at equilibrium, $x = 0$, $a = 0$, and the speed has its maximum value of $\omega A$. At $t = T/2$, the displacement is $-A$, the velocity is again zero, and the acceleration is $+\omega^2 A$. At $t = 3T/4$, $x = 0$, $a = 0$, and $v = +\omega A$.

*A BLOCK ON A SPRING* **EXAMPLE 14-3** **Try It Yourself**

A 2-kg block is attached to a spring as in Figure 14-1. The force constant of the spring is $k = 196$ N/m. The block is held a distance 5 cm from the equilibrium position and is released at $t = 0$. (a) Find the angular frequency $\omega$, the frequency $f$, and the period $T$. (b) Write $x$ as a function of time.

**Cover the column to the right and try these on your own before looking at the answers.**

**Steps** | **Answers**

(a) 1. Calculate $\omega$ from $\omega = \sqrt{k/m}$.

$\omega = \boxed{9.90 \text{ rad/s}}$

2. Use your result to find $f$ and $T$.

$f = \boxed{1.58 \text{ Hz}}$ $T = \boxed{0.635 \text{ s}}$

3. Find $A$ and $\delta$ from the initial conditions.

$A = 5$ cm $\quad \delta = 0$

(b) Write $x(t)$ using your results for $A$, $\omega$, and $\delta$.

$x = \boxed{(5 \text{ cm}) \cos[(9.90 \text{ s}^{-1})t]}$

---

*SPEED AND ACCELERATION OF AN OBJECT ON A SPRING* **EXAMPLE 14-4**

Consider an object on a spring whose position is given by $x = (5$ cm$)$ cos $(9.90$ s$^{-1}$ $t)$. (a) What is the maximum speed of the object? (b) When does this maximum speed first occur? (c) What is the maximum of the acceleration of the object? (d) When does maximum acceleration first occur after $t = 0$?

**PICTURE THE PROBLEM** Because the object is released from rest, $\delta = 0$, and the position, velocity, and acceleration are given by Equations 14-13a, b, and c.

(a) 1. Equation 14-13a, with $\delta = 0$, gives the position. We get the velocity by taking the derivative with respect to time:

$x = A \cos \omega t$

so

$v = \dfrac{dx}{dt} = -\omega A \sin \omega t$

2. Maximum speed occurs when $|\sin \omega t| = 1$:

$|v| = \omega A |\sin \omega t|$

so

$|v|_{\max} = \omega A = (9.90 \text{ rad/s})(5 \text{ cm})$

$= \boxed{49.5 \text{ cm/s}}$

(b) 1. $|\sin \omega t| = 1$ first occurs when $\omega t = \pi/2$:

$|\sin \omega t| = 1 \Rightarrow \omega t = \dfrac{\pi}{2}, \dfrac{3\pi}{2}, \dfrac{5\pi}{2}, \cdots$

2. Solve for $t$ when $\omega t = \pi/2$:

$t = \dfrac{\pi}{2\omega} = \dfrac{\pi}{2(9.90 \text{ s}^{-1})} = \boxed{0.159 \text{ s}}$

(c) 1. We find the acceleration by taking the derivative of the velocity, obtained in step 1 of Part (a):

$a = \dfrac{dv}{dt} = -\omega^2 A \cos \omega t$

2. Maximum acceleration corresponds to $\cos \omega t = -1$.

$$a_{max} = \omega^2 A = (9.90 \text{ rad/s})^2 (5 \text{ cm})$$

$$= \boxed{490 \text{ cm/s}^2 \approx \tfrac{1}{2}g}$$

(d) The maximum acceleration occurs when $|\cos \omega t| = 1$, which is when $\omega t = 0, \pi, 2\pi, \ldots$ :

$\omega t = \pi$

so

$$t = \frac{\pi}{\omega} = \frac{\pi}{9.90 \text{ s}^{-1}} = \boxed{0.317 \text{ s}}$$

**REMARKS** The maximum speed first occurs after one quarter-period

$$t = \frac{\pi}{2\omega} = \frac{\pi}{2(2\pi/T)} = \frac{1}{4}T$$

The maxima of the magnitude of the acceleration occur when $\omega t = 0, \pi, 2\pi, \ldots$. These correspond to $t = 0, \tfrac{1}{2}T, \tfrac{2}{2}T, \tfrac{3}{2}T, \ldots$.

## Simple Harmonic Motion and Circular Motion

There is a relation between simple harmonic motion and circular motion with constant speed. Imagine a particle moving with constant speed $v$ in a circle of radius $A$ (Figure 14-6a). Its angular displacement relative to the x axis is

$$\theta = \omega t + \delta \qquad\qquad 14\text{-}14$$

where $\delta$ is the angular displacement at time $t = 0$ and $\omega = v/A$ is the angular speed of the particle. The x component of the particle's position (Figure 14-6b) is

$$x = A \cos \theta = A \cos(\omega t + \delta)$$

which is the same as Equation 14-4 for simple harmonic motion.

> When a particle moves with constant speed in a circle, its projection onto a diameter of the circle moves with simple harmonic motion (Figure 14-6).

The speed of a particle moving in a circle is $r\omega$, where $r$ is the radius. For the particle in Figure 14-6b, $r = A$, so its speed is $A\omega$. The projection of the velocity vector onto the x axis gives $v_x = -v \sin \theta$. Substituting for $v$ and $\theta$ gives

$$v_x = -\omega A \sin \theta = -\omega A \sin(\omega t + \delta)$$

which is the same as Equation 14-5 for simple harmonic motion.

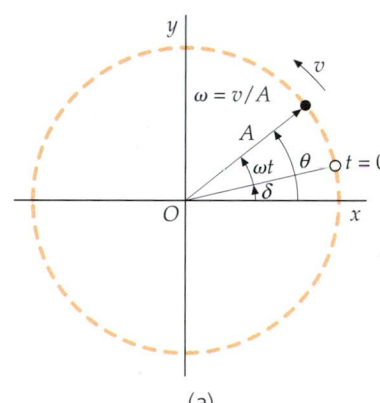

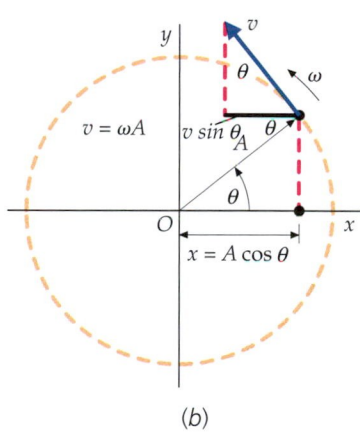

**FIGURE 14-6** A particle moves in a circular path with constant speed. (a) Its x component of position describes simple harmonic motion, and (b) its x component of velocity describes the velocity of the simple harmonic motion.

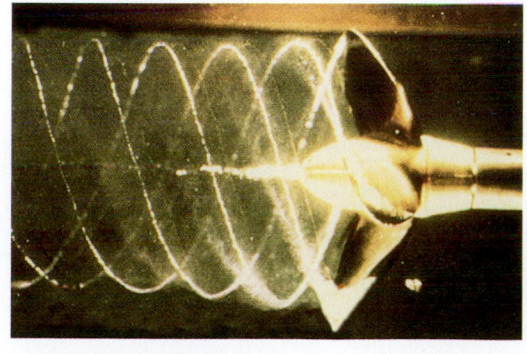

Bubbles foaming off the edge of a rotating propeller that is moving through water produce a sinusoidal pattern.

# 14-2 Energy in Simple Harmonic Motion

When an object on a spring undergoes simple harmonic motion, the system's potential energy and kinetic energy vary with time. Their sum, the total mechanical energy $E = K + U$, is constant. Consider an object a distance $x$ from equilibrium, acted on by a restoring force $-kx$. The system's potential energy is

$$U = \tfrac{1}{2}kx^2$$

This is Equation 6-21. For simple harmonic motion, $x = A\cos(\omega t + \delta)$. Substituting gives

$$U = \tfrac{1}{2}kA^2 \cos^2(\omega t + \delta) \qquad\qquad 14\text{-}15$$

POTENTIAL ENERGY IN SIMPLE HARMONIC MOTION

The kinetic energy of the system is

$$K = \tfrac{1}{2}mv^2$$

where $m$ is the object's mass and $v$ is its speed. For simple harmonic motion, $v_x = -\omega A \sin(\omega t + \delta)$. Substituting gives

$$K = \tfrac{1}{2}m\omega^2 A^2 \sin^2(\omega t + \delta)$$

Then using $\omega^2 = k/m$,

$$K = \tfrac{1}{2}kA^2 \sin^2(\omega t + \delta) \qquad\qquad 14\text{-}16$$

KINETIC ENERGY IN SIMPLE HARMONIC MOTION

The total mechanical energy is the sum of the potential and kinetic energies:

$$E_{\text{total}} = U + K = \tfrac{1}{2}kA^2 \cos^2(\omega t + \delta) + \tfrac{1}{2}kA \sin^2(\omega t + \delta)$$
$$= \tfrac{1}{2}kA^2\left[\cos^2(\omega t + \delta) + \sin^2(\omega t + \delta)\right]$$

Since $\sin^2(\omega t + \delta) + \cos^2(\omega t + \delta) = 1$,

$$E_{\text{total}} = \tfrac{1}{2}kA^2 \qquad\qquad 14\text{-}17$$

TOTAL MECHANICAL ENERGY IN SIMPLE HARMONIC MOTION

This equation reveals an important general property of simple harmonic motion:

The total mechanical energy in simple harmonic motion is proportional to the square of the amplitude.

For an object at its maximum displacement, the total energy is all potential energy. As the object moves toward its equilibrium position, the kinetic energy of the system increases and its potential energy decreases. As it moves through its equilibrium position, the kinetic energy of the object is maximum, the potential energy of the system is zero, and the total energy is kinetic.

As the object moves past the equilibrium point, its kinetic energy begins to decrease, and the potential energy of the system increases until the object again stops momentarily at its maximum displacement (now in the other direction). At all times, the sum of the potential and kinetic energies is constant. Figures 14-7b and c show plots of U and K versus time. These curves have the same shape except that one is zero when the other is maximum. Their average values over one or more cycles are equal, and because $U + K = E$, their average values are given by

$$U_{av} = K_{av} = \tfrac{1}{2}E \qquad\qquad 14\text{-}18$$

In Figure 14-8, the potential energy U is graphed as a function of x. The total energy $E_{total}$ is constant and is therefore plotted as a horizontal line. This line intersects the potential-energy curve at $x = A$ and $x = -A$. These are the points at which oscillating objects reverse direction and head back toward the equilibrium position, and are called the **turning points**. Because $U \leq E_{total}$, the motion is restricted to $A \leq x \leq +A$.

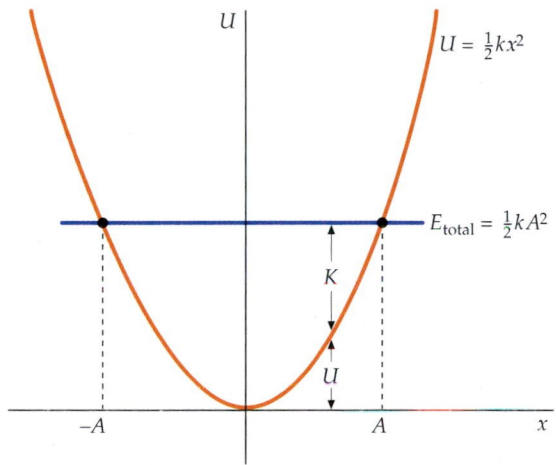

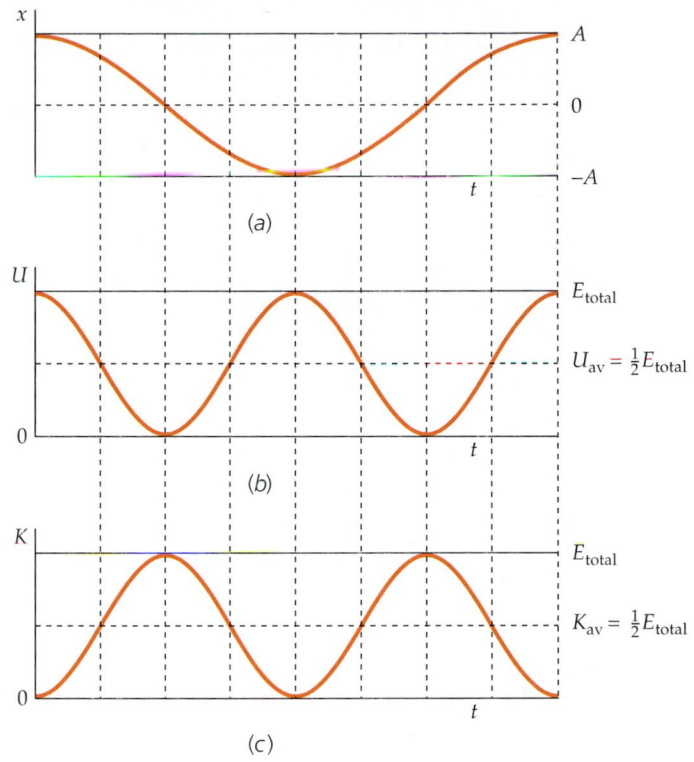

**FIGURE 14-7** Plots of x, U, and K versus t.

**FIGURE 14-8** The potential-energy function $U = \tfrac{1}{2}kx^2$ for an object of mechanical mass m on a (massless) spring of force constant k. The horizontal blue line represents the total mechanical energy $E_{total}$ for an amplitude of A. The kinetic energy K is represented by the vertical distance $K = E_{total} - U$. Since $E_{total} \geq U$, the motion is restricted to $-A \leq x \leq +A$.

---

*ENERGY AND SPEED OF AN OSCILLATING OBJECT*     **EXAMPLE 14-5**

A 3-kg object attached to a spring oscillates with an amplitude of 4 cm and a period of 2 s. (a) What is the total energy? (b) What is the maximum speed of the object? (c) At what position $x_1$ is the speed equal to half its maximum value?

**PICTURE THE PROBLEM** (a) The total energy can be found from the amplitude and force constant, and the force constant can be found from the mass and period. (b) The maximum speed occurs when the kinetic energy equals the total energy. (c) We can relate the position to the speed by using conservation of energy.

(a) 1. Write the total energy E in terms of the force constant k and amplitude A:

$$E = \tfrac{1}{2}kA^2$$

2. The force constant is related to the period and mass:

$$k = m\omega^2 = m\left(\frac{2\pi}{T}\right)^2$$

3. Substitute the given values to find E:

$$E = \frac{1}{2}kA^2 = \frac{1}{2}m\left(\frac{2\pi}{T}\right)^2 A^2$$

$$= \frac{1}{2}(3\text{ kg})\left(\frac{2\pi}{2\text{ s}}\right)^2 (0.04\text{ m})^2 = \boxed{2.37 \times 10^{-2}\text{ J}}$$

(b) To find $v_{max}$, set the kinetic energy equal to the total energy and solve for $v$:

$$\tfrac{1}{2}mv_{max}^2 = E$$

so

$$v_{max} = \sqrt{\frac{2E}{m}} = \sqrt{\frac{2(2.37 \times 10^{-2}\,\text{J})}{3\,\text{kg}}}$$

$$= \boxed{0.126\,\text{m/s}}$$

(c) 1. Conservation of energy relates the position $x$ to the speed $v$:

$$E = \tfrac{1}{2}mv^2 + \tfrac{1}{2}kx^2$$

2. Substitute $v = \tfrac{1}{2}v_{max}$ and solve for $x_1$. It is convenient to find $x$ in terms of $E$ and then write $E = \tfrac{1}{2}kA^2$ to obtain an expression for $x$ in terms of $A$:

$$E = \tfrac{1}{2}m\left(\tfrac{1}{2}v_{max}\right)^2 + \tfrac{1}{2}kx_1^2$$

$$= \tfrac{1}{4}\left(\tfrac{1}{2}mv_{max}^2\right) + \tfrac{1}{2}kx_1^2 = \tfrac{1}{4}E + \tfrac{1}{2}kx_1^2$$

so

$$\tfrac{1}{2}kx_1^2 = E - \tfrac{1}{4}E = \tfrac{3}{4}E$$

and

$$x_1 = \sqrt{\frac{3E}{2k}} = \sqrt{\frac{3}{2k}\left(\frac{1}{2}kA^2\right)} = \frac{\sqrt{3}}{2}A$$

$$= \frac{\sqrt{3}}{2}(4\,\text{cm}) = \boxed{3.46\,\text{cm}}$$

**EXERCISE** Calculate $\omega$ for this example and find $v_{max}$ from $v_{max} = \omega A$. (*Answer* $\omega = 3.14$ rad/s, $v_{max} = 0.126$ m/s)

**EXERCISE** An object of mass 2 kg is attached to a spring of force constant 40 N/m. The object is moving at 25 cm/s when it is at its equilibrium position. (a) What is the total energy of the object? (b) What is the amplitude of the motion? (*Answer* (a) $E_{total} = \tfrac{1}{2}mv_{max}^2 = 0.0625$ J (b) $A = \sqrt{2E_{total}/k} = 5.59$ cm)

## *General Motion Near Equilibrium

Simple harmonic motion typically occurs when a particle is displaced slightly from a position of stable equilibrium. Figure 14-9 is a graph of the potential energy $U$ versus $x$ for a force that has a position of stable equilibrium and a position of unstable equilibrium. As discussed in Chapter 6, the maximum at $x_2$ on Figure 14-9 corresponds to unstable equilibrium, whereas the minimum at $x_1$ corresponds to stable equilibrium. Many smooth curves with a minimum as in Figure 14-9 can be approximated near the minimum by a parabola. The dashed curve in this figure is a parabolic curve that approximately fits $U$ near the stable equilibrium point. The general equation for a parabola that has a minimum at point $x_1$ can be written

$$U = A + B(x - x_1)^2 \qquad \text{14-19}$$

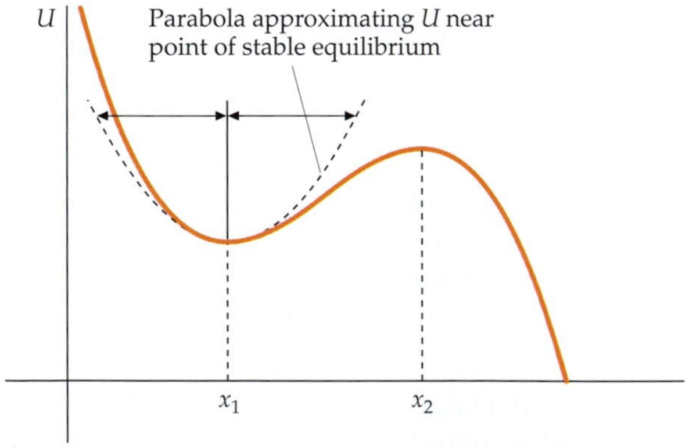

where $A$ and $B$ are constants. The constant $A$ is the value of $U$ at the equilibrium position $x = x_1$. The force is related to the potential energy curve by $F_x = -dU/dx$. Then

$$F_x = -\frac{dU}{dx} = -2B(x - x_1)$$

**FIGURE 14-9** Plot of $U$ versus $x$ for a force that has a position of stable equilibrium ($x_1$) and a position of unstable equilibrium ($x_2$).

If we set $2B = k$, this equation reduces to

$$F_x = -\frac{dU}{dx} = -k(x - x_1)$$  14-20

According to Equation 14-20, the force is proportional to the displacement and oppositely directed, so the motion will be simple harmonic. Figure 14-9 shows a graph of this system's potential energy function $U(x)$, which has a position of stable equilibrium at $x = x_1$. Figure 14-10 shows a potential energy function that has a position of stable equilibrium at $x = 0$. The system for this function is a small particle of mass $m$ oscillating back and forth at the bottom of a frictionless spherical bowl.

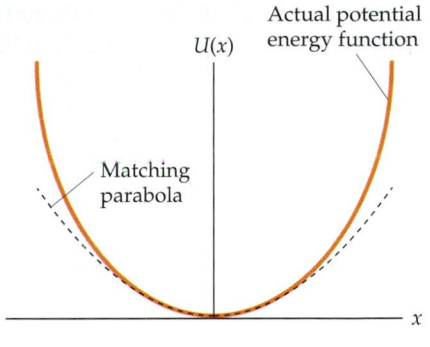

**FIGURE 14-10** Plot of $U$ versus $x$ for a small particle oscillating back and forth at the bottom of a spherical bowl.

## 14-3 Some Oscillating Systems

### Object on a Vertical Spring

When an object hangs from a vertical spring, there is a downward force $mg$ in addition to the force of the spring (Figure 14-11). If we choose downward as the positive $y$ direction, then the spring's force on the object is $-ky$, where $y$ is the extension of the spring. The net force on the object is then

$$\sum F_y = -ky + mg$$  14-21

We can simplify this equation by changing to a new variable $y' = y - y_0$, where $y_0 = mg/k$ is the amount the spring is stretched when the object is in equilibrium. Substituting $y = y' + y_0$ gives

$$\sum F_y = -k(y' + y_0) + mg$$

But $ky_0 = mg$, so

$$\sum F_y = -ky'$$  14-22

Newton's second law ($\sum F_y = ma_y$) gives

$$-ky' = m\frac{d^2y}{dt^2}$$

However, $y = y' + y_0$, where $y_0 = mg/k$ is a constant. Thus $d^2y/dt^2 = d^2y'/dt^2$, so

$$-ky' = m\frac{d^2y'}{dt^2}$$

Rearranging gives

$$\frac{d^2y'}{dt^2} = -\frac{k}{m}y'$$

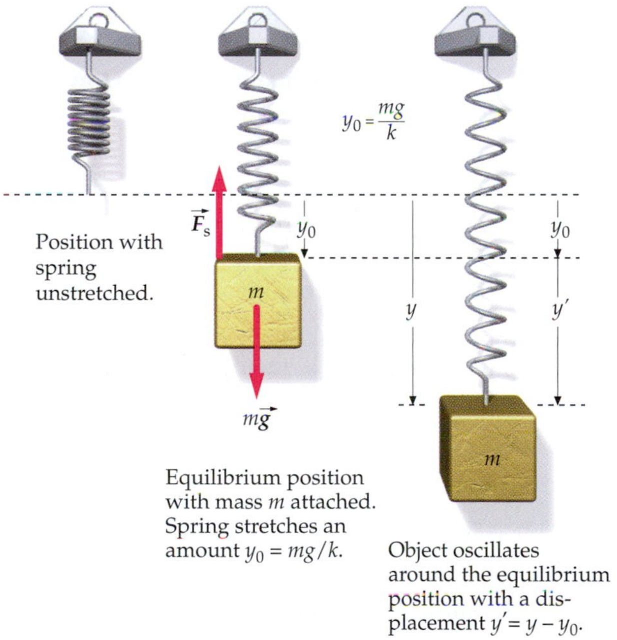

**FIGURE 14-11** The problem of a mass on a vertical spring is simplified if the displacement ($y'$) is measured from the equilibrium position of the spring with the mass attached.

Position with spring unstretched.

Equilibrium position with mass $m$ attached. Spring stretches an amount $y_0 = mg/k$.

Object oscillates around the equilibrium position with a displacement $y' = y - y_0$.

$y_0 = \frac{mg}{k}$

which is the same as Equation 14-2 with $y'$ replacing $x$. It has the now familiar solution

$$y' = A\cos(\omega t + \delta)$$

where $\omega = \sqrt{k/m}$.

Thus the effect of the gravitational force $mg$ is merely to shift the equilibrium position from $y = 0$ to $y' = 0$. When the object is displaced from this equilibrium position by the amount $y'$, the net force is $-ky'$. The object oscillates about this equilibrium position with an angular frequency $\omega = \sqrt{k/m}$, the same angular frequency as that for an object on a horizontal spring.

A force is conservative if the work done by it is independent of path. Both the force of the spring and the force of gravity are conservative, and the sum of these forces (Equations 14-21 and 14-22) also is conservative. The potential energy function $U$ associated with the sum of these forces is the negative of the work done plus an arbitrary integration constant. That is,

$$U = -\int -ky' \, dy' = \tfrac{1}{2}ky'^2 + U_0$$

where the integration constant $U_0$ is the value of $U$ at the equilibrium position ($y' = 0$). Thus,

$$U = \tfrac{1}{2}ky'^2 + U_0 \qquad\qquad 14\text{-}23$$

---

*PAPER SPRINGS*

**EXAMPLE 14-6**  **Put It in Context**

**You are showing your nieces how to make paper party decorations using paper springs. One niece makes a paper spring and, with a single sheet of colored paper suspended from it, the spring is stretched 8 cm. You want the decorations to bounce at approximately 1 cy/s. How many sheets of colored paper should be used for the decoration on that spring if it is to bounce at 1 cy/s?**

**PICTURE THE PROBLEM** The frequency depends on the ratio of the spring constant to the suspended mass (Equation 14-12), and you do not know either the spring constant or the mass. However, Hooke's law (Equation 14-1) can be used to find the required ratio from the information given.

1. Write the frequency in terms of the force constant $k$ and the mass $M$ (Equation 14-12), where $M$ is the mass of $N$ sheets. We need to find $N$:

$$f = \frac{\omega}{2\pi} = \frac{1}{2\pi}\sqrt{\frac{k}{M}}$$

2. The spring stretches a distance of $y_0 = 8$ cm when a single sheet of mass $m$ is suspended:

$$ky_0 = mg \quad\text{so}\quad \frac{k}{m} = \frac{g}{y_0}$$

3. The mass of $N$ sheets equals $N$ times the mass of a single sheet:

$$M = Nm$$

4. Using the steps 2 and 3 results, solve for $k/M$:

$$\frac{k}{M} = \frac{k}{Nm} = \frac{1}{N}\frac{g}{y_0}$$

5. Substitute the step 4 result into the step 1 result and solve for $N$:

$$f = \frac{1}{2\pi}\sqrt{\frac{k}{M}} = \frac{1}{2\pi}\sqrt{\frac{1}{N}\frac{g}{y_0}}$$

so

$$N = \frac{g}{(2\pi f)^2 y_0} = \frac{9.81 \text{ m/s}^2}{4\pi^2 (1\text{ Hz})^2 (0.08\text{ m})} = 3.11$$

$\boxed{\text{Three sheets are needed.}}$

**REMARKS** Note that in this example we didn't need to use the value of $m$ or $k$ because the frequency depends on the ratio $k/m$, which equals $g/y_0$. Also, the units work out since 1 Hz = 1 cy/s, and a cycle is a dimensionless unit.

**EXERCISE** How much is the paper spring stretched when a decoration made from three sheets of paper is suspended from it and the paper is in equilibrium? (*Answer* 24 cm)

---

*A Bead on a Block* **EXAMPLE 14-7**

A block attached to a spring oscillates vertically with a frequency of 4 Hz and an amplitude of 7 cm. A tiny bead is placed on top of the oscillating block just as it reaches its lowest point. Assume that the bead's mass is so small that its effect on the motion of the block is negligible. At what distance from the block's equilibrium position does the bead lose contact with the block?

**PICTURE THE PROBLEM** The forces on the bead are its weight $mg$ downward and the upward normal force exerted by the block. The magnitude of this normal force changes as the acceleration changes. As the block moves upward *from equilibrium*, its acceleration and that of the bead is *downward* and increasing in magnitude. When the acceleration reaches $g$ downward, the normal force will be zero. If the block's downward acceleration becomes even slightly larger, the bead will leave the block.

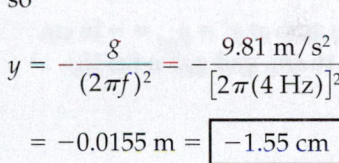

1. Draw a sketch of the system (Figure 14-12). Include a $y$ coordinate axis with its origin at the equilibrium position and with down as the positive direction:

2. We are looking for the value of $y$ when the acceleration is $g$ downward. Use Equation 14-7:

$$a_y = -\omega^2 y$$
$$g = -\omega^2 y$$

3. Substitute $2\pi f$ for $\omega$ and solve for $y$:

$$g = -(2\pi f)^2 y$$

so

$$y = -\frac{g}{(2\pi f)^2} = -\frac{9.81 \text{ m/s}^2}{[2\pi(4 \text{ Hz})]^2}$$

$$= -0.0155 \text{ m} = \boxed{-1.55 \text{ cm}}$$

**FIGURE 14-12**

**PLAUSIBILITY CHECK** The bead leaves the block when $y$ is negative, which is when the bead is above the equilibrium position, as expected.

---

*Potential Energy of the Spring–Earth System* **EXAMPLE 14-8**

A 3-kg object stretches a spring 16 cm when it hangs vertically in equilibrium. The spring is then stretched an additional 5 cm and the object is released. Let $U$ be the total potential energy of the spring-object-planet system. When the mass is at its maximum displacement from equilibrium find $U$ (*a*) with $U = 0$ at the equilibrium position and (*b*) with $U = 0$ when the spring is unstretched.

**PICTURE THE PROBLEM** (*a*) With $U = 0$ at the equilibrium position the total potential energy $U$ is $\frac{1}{2}ky'^2$, where $y'$ is the displacement from the equilibrium position. (*b*) With $U = 0$ when the spring is unstretched the total potential energy is the potential energy of the spring plus the gravitational potential energy.

(a) 1. Make three sketches of the system, one with the spring unstretched, one with it stretched 16 cm, and a third with it stretched 21 cm (Figure 14-13).

2. Let the positive $y'$ direction be downward and let $y' = 0$ at the equilibrium position. The total potential energy function (Equation 14-23) is:

$$U = \tfrac{1}{2}ky'^2$$

3. To determine $U$ we first need to find the spring constant $k$. At equilibrium the upward force of the spring equals the downward force of gravity. Use this to calculate the value of $k$:

$$ky_0 = mg$$

so

$$k = \frac{mg}{y_0} = \frac{(3\text{ kg})(9.81\text{ m/s}^2)}{0.16\text{ m}} = 184\text{ N/m}$$

4. Substituting for $k$ in the step 1 result and solving for $U$ gives:

$$U = \tfrac{1}{2}ky'^2 = \tfrac{1}{2}(184\text{ N/m})(0.05\text{ m})^2 = \boxed{0.230\text{ J}}$$

(b) 1. The total potential energy function is given by Equation 14-23:

$$U = \tfrac{1}{2}ky'^2 + U_0$$

2. The potential energy equals zero at $y' = y'_{\text{ref}} = -16$ cm. Set $U = 0$, set $y' = y'_{\text{ref}} = -16$ cm, and solve for $U_0$:

$$0 = \tfrac{1}{2}ky'^2_{\text{ref}} + U_0$$

so

$$U_0 = -\tfrac{1}{2}ky'^2_{\text{ref}} = -\tfrac{1}{2}(184\text{ N/m})(-0.16\text{ m})^2$$

$$= -2.35\text{ J}$$

3. Substituting for $U_0$ in the Part (b) step 1 result gives:

$$U = \tfrac{1}{2}ky'^2 + U_0$$

$$= \tfrac{1}{2}(184\text{ N/m})(0.05\text{ m})^2 - 2.35\text{ J} = \boxed{-2.12\text{ J}}$$

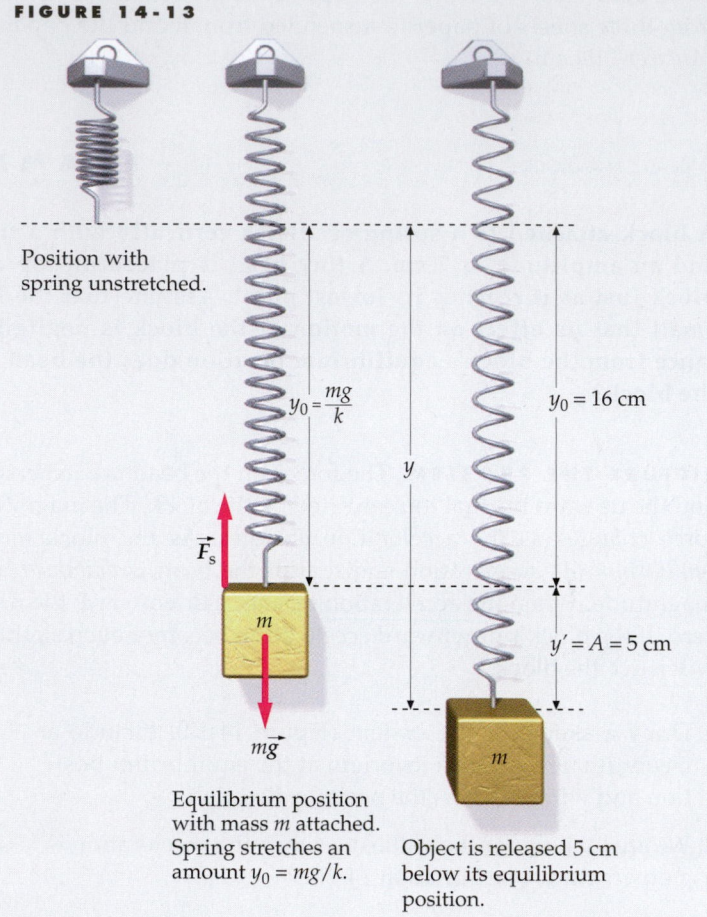

**FIGURE 14-13**

Position with spring unstretched.

$y_0 = \dfrac{mg}{k}$

$\vec{F}_s$

$m$

$m\vec{g}$

Equilibrium position with mass $m$ attached. Spring stretches an amount $y_0 = mg/k$.

$y_0 = 16$ cm

$y$

$y' = A = 5$ cm

$m$

Object is released 5 cm below its equilibrium position.

**PLAUSIBILITY CHECK** The potential energy calculated in Part (b) should equal the sum of the spring's potential energy $U_s$ at $y = 21$ cm plus the gravitational potential energy $U_g$ at $y = 21$ cm, where each of these potential energies is zero if the spring is unstretched, the positive $y$ direction is downward, and $y = 0$ if the spring is unstretched. $U_s = \tfrac{1}{2}ky^2 = \tfrac{1}{2}(184\text{ N/m})(0.21\text{ m})^2 = 4.06$ J and $U_g = mg(-y) = (3\text{ kg})(9.81\text{ N/kg})(-0.21\text{ m}) = -6.18$ J. Adding these gives $4.06\text{ J} - 6.18\text{ J} = -2.12$ J, which agrees with the Part (b) result.

MASTER the CONCEPT WEB

## The Simple Pendulum

A simple pendulum consists of a string of length $L$ and a bob of mass $m$. When the bob is released from an initial angle $\phi_0$ with the vertical, it swings back and forth with some period $T$.

**EXERCISE IN DIMENSIONAL ANALYSIS** We might expect the period of a simple pendulum to depend on the mass $m$ of a pendulum bob, the length $L$ of the pendulum, the acceleration due to gravity $g$, and the initial angle $\phi_0$. Find a simple combination of some or all of these quantities that gives the correct dimensions for the period. (*Answer* $\sqrt{L/g}$)

**REMARKS** The units of length, mass, and $g$, are m, kg, and m/s$^2$, respectively. If we divide $L$ by $g$, the meters cancel and we are left with seconds squared, suggesting the form $\sqrt{L/g}$. If the formula for the period contains the mass, then the unit kg must be canceled by some other quantity. But there is no combination of $L$ and $g$ that can cancel mass units. So the period cannot depend on the mass of the bob. Since the initial angle $\phi_0$ is dimensionless, we cannot tell whether or not it is a factor in the period. We will see below that for small $\phi_0$, the period is given by $T = 2\pi\sqrt{L/g}$.

The forces on the bob are its weight $m\vec{g}$ and the string tension $\vec{T}$ (Figure 14-14). At an angle $\phi$ with the vertical, the weight has components $mg \cos \phi$ along the string and $mg \sin \phi$ tangential to the circular arc in the direction of decreasing $\phi$. Using tangential components, Newton's second law ($\Sigma F_t = m\,a_t$) gives

$$-mg \sin \phi = m\frac{d^2s}{dt^2} \qquad\qquad \text{14-24}$$

where the arc length $s$ is related to the angle $\phi$ by $s = L\phi$. Repeatedly differentiating both sides of $s = L\phi$ gives

$$\frac{d^2s}{dt^2} = L\frac{d^2\phi}{dt^2}$$

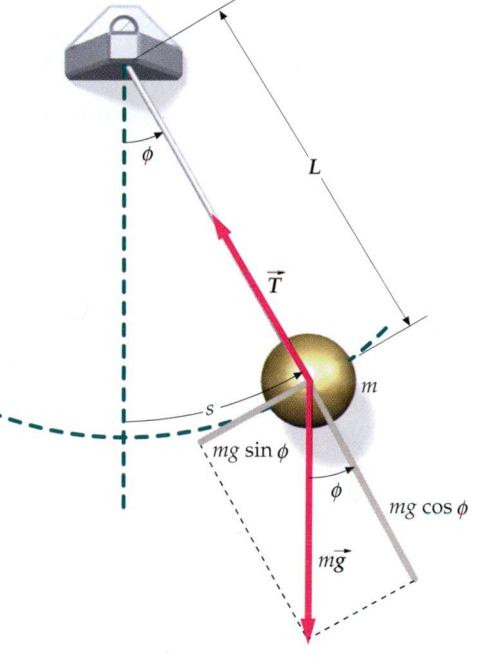

**FIGURE 14-14** Forces on a pendulum bob.

Substituting $L\,d^2\phi/dt^2$ into Equation 14-24 for $d^2s/dt^2$ and rearranging gives

$$\frac{d^2\phi}{dt^2} = -\frac{g}{L} \sin \phi \qquad\qquad \text{14-25}$$

Note that the mass $m$ does not appear in Equation 14-25—the motion of a pendulum does not depend on its mass. For small $\phi$, $\sin \phi \approx \phi$, and

$$\frac{d^2\phi}{dt^2} \approx -\frac{g}{L}\phi \qquad\qquad \text{14-26}$$

Equation 14-26 is of the same form as Equation 14-2 for an object on a spring. Thus, the motion of a pendulum approximates simple harmonic motion for small angular displacements.

Equation 14-26 can be written

$$\frac{d^2\phi}{dt^2} = -\omega^2\phi, \qquad \text{where} \quad \omega^2 = \frac{g}{L} \qquad\qquad \text{14-27}$$

and $\omega$ is the angular frequency—not the angular speed—of the motion of the pendulum. The period of the motion is thus

$$T = \frac{2\pi}{\omega} = 2\pi\sqrt{\frac{L}{g}} \qquad\qquad \text{14-28}$$

PERIOD OF A SIMPLE PENDULUM

The solution of Equation 14-27 is

$$\phi = \phi_0 \cos(\omega t + \delta)$$

where $\phi_0$ is the maximum angular displacement.

According to Equation 14-28, the greater the length of a pendulum, the greater the period, which is consistent with experimental observation. The period, and therefore the frequency, are independent of the amplitude of oscillation (as long as the amplitude is small), a general feature of simple harmonic motion.

**EXERCISE** Find the period of a simple pendulum of length 1 m. (*Answer* 2.01 s)

The acceleration due to gravity can be measured using a simple pendulum. We need only measure the length $L$ and period $T$ of the pendulum, and using Equation 14-28, solve for $g$. (When finding $T$, we usually measure the time for $n$ oscillations and then divide by $n$, which minimizes measurement error.)

**Pendulum in an Accelerated Reference Frame** Figure 14-15*a* shows a simple pendulum suspended from the ceiling of a boxcar that has acceleration $\vec{a}_0$ relative to the ground, to the right, and $\vec{a}$ is the acceleration of the bob relative to the ground. Applying Newton's second law to the bob gives

$$\sum \vec{F} = \vec{T} + m\vec{g} = m\vec{a} \qquad \text{14-29}$$

If the bob remains at rest relative to the boxcar, then $\vec{a} = \vec{a}_0$ and

$$\sum F_x = T \sin \theta_0 = ma_0$$

$$\sum F_y = T \cos \theta_0 - mg = 0$$

where $\theta_0$ is the equilibrium angle. $\theta_0$ is thus given by $\tan \theta_0 = a_0/g$. If the bob is moving relative to the boxcar, then $\vec{a}' = \vec{a} - \vec{a}_0$, where $\vec{a}'$ is the acceleration of the bob relative to the boxcar. Substituting for $\vec{a}$ in Equation 14-29 gives

$$\sum \vec{F} = \vec{T} + m\vec{g} = m(\vec{a}' + \vec{a}_0)$$

Subtracting $m\vec{a}_0$ from both sides of this equation and rearranging terms gives

$$\vec{T} + m\vec{g}' = m\vec{a}'$$

where $\vec{g}' = \vec{g} - \vec{a}_0$. Thus by replacing $\vec{g}$ by $\vec{g}'$ and $\vec{a}$ by $\vec{a}'$ in Equation 14-29 we can solve for the motion of the bob relative to the boxcar. The vectors $\vec{T}$ and $m\vec{g}'$ are shown in Figure 14-15*b*. If the string breaks so that $\vec{T} = 0$, then our equation gives $\vec{a}' = \vec{g}'$, which means that $\vec{g}'$ is the free-fall acceleration in the reference frame of the boxcar. If the bob is displaced slightly from equilibrium, it will oscillate with a period $T$ given by Equation 14-28 with $g$ replaced by $g'$.

**EXERCISE** A simple pendulum of length 1 m is in a boxcar that is accelerating horizontally with acceleration $a_0 = 3$ m/s$^2$. Find $g'$ and the period $T$. (*Answer* $g' = 10.26$ m/s$^2$, $T = 1.96$ s)

**Large-Amplitude Oscillations** When the amplitude of a pendulum's oscillation becomes large, its motion continues to be periodic, but it is no longer simple harmonic. A slight dependence on the amplitude

All mechanical clocks keep time because the period of the oscillating part of the mechanism remains constant. The period of any pendulum changes with changes in amplitude. However, the driving mechanism of a pendulum clock maintains the amplitude at a constant value.

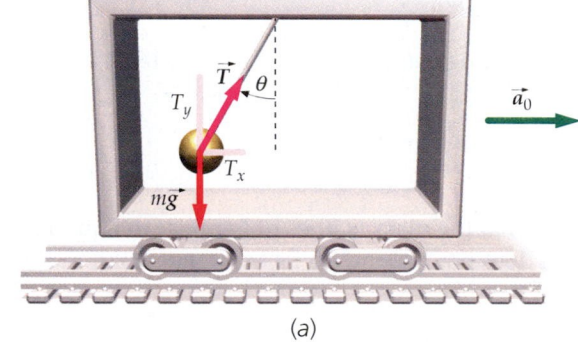

(a)

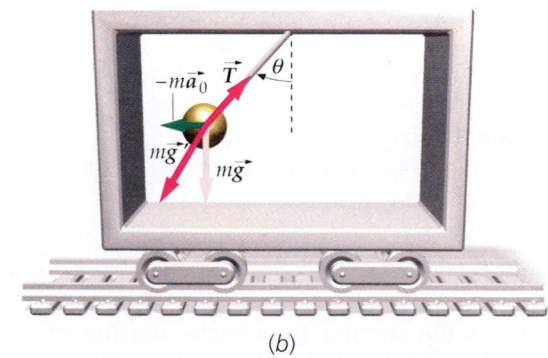

(b)

**FIGURE 14-15** (*a*) Simple pendulum in apparent equilibrium in an accelerating boxcar. Forces are those as seen from a separate stationary frame. (*b*) Forces on the bob as seen in the accelerated frame. Adding the pseudoforce $-m\vec{a}_0$ is equivalent to replacing $\vec{g}$ by $\vec{g}'$.

**FIGURE 14-16** Note that the ordinate values range from 1 to 1.06. Over a range of $\phi$ from 0 to 0.8 rad (46°), the period varies by about 5 percent.

must be accounted for when determining the period. For a general angular amplitude $\phi_0$, the period can be shown to be

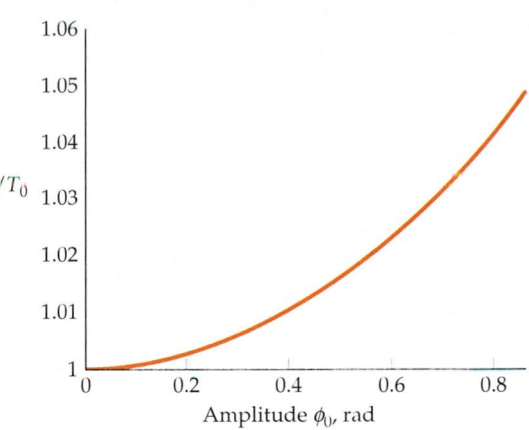

$$T = T_0\left[1 + \frac{1}{2^2}\sin^2\frac{1}{2}\phi_0 + \frac{1}{2^2}\left(\frac{3}{4}\right)^2\sin^4\frac{1}{2}\phi_0 + \cdots\right]$$    14-30

PERIOD FOR LARGE-AMPLITUDE OSCILLATIONS

where $T_0 = 2\pi\sqrt{L/g}$ is the period for very small amplitudes. Figure 14-16 shows $T/T_0$ as a function of amplitude $\phi_0$.

---

*A PENDULUM CLOCK*                    **EXAMPLE 14-9**    **Try It Yourself**

A simple pendulum clock is calibrated to keep accurate time at an angular amplitude of $\phi_0 = 10°$. When the amplitude has decreased to the point where it is very small, does the clock gain or lose time? How much time will the clock gain or lose in one day if the amplitude remains very small.

**Cover the column to the right and try these on your own before looking at the answers.**

**Steps**

1. Answer the first question by finding out if the period increases or decreases.

2. Use Equation 14-30 to find the percentage change $[(T - T_0)/T] \times 100\%$ for $\phi = 10°$. Use only the first correction term.

3. Find the number of minutes in a day.

4. Combine the steps 2 and 3 to find the change in the number of minutes in a day.

**Answers**

$T$ decreases as $\phi$ decreases, so the clock gains time.

0.190%

There are 1440 minutes in a day.

The gain is 2.73 min/d

**REMARKS** To avoid this gain, pendulum-clock mechanisms are designed to keep the amplitude fairly constant.

## *The Physical Pendulum

A rigid object free to rotate about a horizontal axis that is not through its center of mass will oscillate when displaced from equilibrium. Such a system is called a **physical pendulum.** Consider a plane figure with a rotation axis a distance $D$ from the figure's center of mass and displaced from equilibrium by the angle $\phi$ (Figure 14-17). The torque about the axis has a magnitude $MgD\sin\phi$ and tends to decrease $|\phi|$. Newton's second law applied to rotation is

$$\tau = I\alpha$$

where $\alpha$ is the angular acceleration and $I$ is the moment of inertia about the axis. Substituting $-MgD\sin\phi$ for the net torque and $d^2\phi/dt^2$ for $\alpha$, we have

$$-MgD\sin\phi = I\frac{d^2\phi}{dt^2}$$

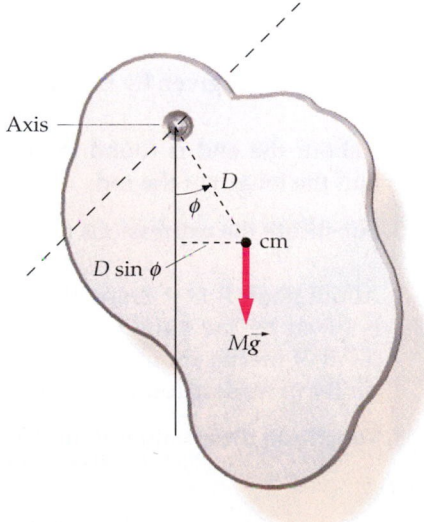

**FIGURE 14-17** A physical pendulum.

or

$$\frac{d^2\phi}{dt^2} = -\frac{MgD}{I}\sin\phi \qquad\qquad 14\text{-}31$$

As with the simple pendulum, the motion is approximately simple harmonic if the angular displacements are small, so $\sin\phi \approx \phi$. In this case, we have

$$\frac{d^2\phi}{dt^2} \approx -\frac{MgD}{I}\phi = -\omega^2\phi \qquad\qquad 14\text{-}32$$

where $\omega = \sqrt{MgD/I}$ is the angular frequency—not the angular speed—of the motion. The period is therefore

$$T = \frac{2\pi}{\omega} = 2\pi\sqrt{\frac{I}{MgD}} \qquad\qquad 14\text{-}33$$

PERIOD OF A PHYSICAL PENDULUM

For large amplitudes, the period is given by Equation 14-30, with $T_0$ given by Equation 14-33. For a simple pendulum of length $L$, the moment of inertia is $I = ML^2$ and $D = L$. Then Equation 14-33 gives $T = 2\pi\sqrt{ML^2/(MgL)} = 2\pi\sqrt{L/g}$, the same as Equation 14-28.

---

*A ROTATING ROD*                                    **EXAMPLE 14-10**

**A uniform rod of mass $M$ and length $L$ is free to rotate about a horizontal axis perpendicular to the rod and through one end. (a) Find the period of oscillation for small angular displacements. (b) Find the period of oscillation if the rotation axis is a distance $x$ from the center of mass.**

**PICTURE THE PROBLEM** (a) The period is given by Equation 14-33. The center of mass is at the center of the rod, so the distance from the center of mass to the rotation axis is half the length of the rod (Figure 14-18a). The moment of inertia of a uniform rod can be found in Table 9-1. (b) For rotations around an axis through point $P$ (Figure 14-18b), the moment of inertia can be found from the parallel-axis theorem $I = I_{cm} + MD^2$ (Equation 9-44), where $I_{cm}$ can be found in Table 9-1.

(a) 1. The period is given by Equation 14-33:

$$T = 2\pi\sqrt{\frac{I}{MgD}}$$

2. $I$ about the end is found in Table 9-1 and $D$ is half the length of the rod:

$$I = \tfrac{1}{3}ML^2; \qquad D = \tfrac{1}{2}L$$

3. Substitute the expressions for $I$ and $D$ to find $T$:

$$T = 2\pi\sqrt{\frac{\tfrac{1}{3}ML^2}{Mg(\tfrac{1}{2}L)}} = \boxed{2\pi\sqrt{\frac{2L}{3g}}}$$

(b) 1. About point $P$, $D = x$, and the moment of inertia is given by the parallel-axis theorem. The moment of inertia about a parallel axis through the center of mass is found in Table 9-1:

$$D = x$$

$$I = I_{cm} + MD^2 = \tfrac{1}{12}ML^2 + Mx^2$$

2. Substitute these values to find $T$:

$$T = 2\pi\sqrt{\frac{I}{MgD}} = 2\pi\sqrt{\frac{(\tfrac{1}{12}ML^2 + Mx^2)}{Mgx}}$$

$$= \boxed{2\pi\sqrt{\frac{(\tfrac{1}{12}L^2 + x^2)}{gx}}}$$

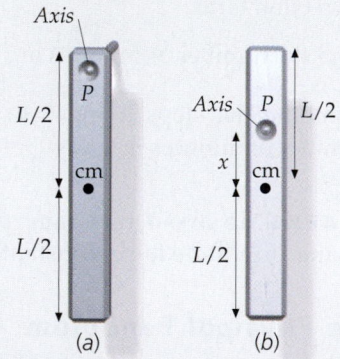

**FIGURE 14-18** Plot of the period versus the distance from the pivot to the center of mass. For $x > 0.5$ m the pivot is beyond the end of the rod.

🅞 **PLAUSIBILITY CHECK** As $x \to 0$, $T \to \infty$ as expected. (If the rotation axis of the rod passes through its center of mass, we do not expect gravity to exert a restoring torque.) Also, if $x = L/2$, we get the same result as found in Part (*a*), and if $x \gg L$, the expression for the period approaches $T = 2\pi\sqrt{x/g}$, which is the expression for the period of a simple pendulum of length $x$ (Equation 14-28).

**EXERCISE** What is the period of oscillation for small angular displacements of a 1-m-long uniform rod about an axis through one end? (*Answer* $T = 1.64$ s) Note that this is a smaller period than for a simple pendulum of length $L = 1$ m. The period of the simple pendulum is greater because the ratio of its moment of inertia to the restoring torque is greater.

**EXERCISE** Show that when $x = L/6$, the period is the same as when $x = L/2$.

**REMARKS** The period $T$ versus distance $x$ from the center of mass for a rod of length 1 m is shown in Figure 14-19.

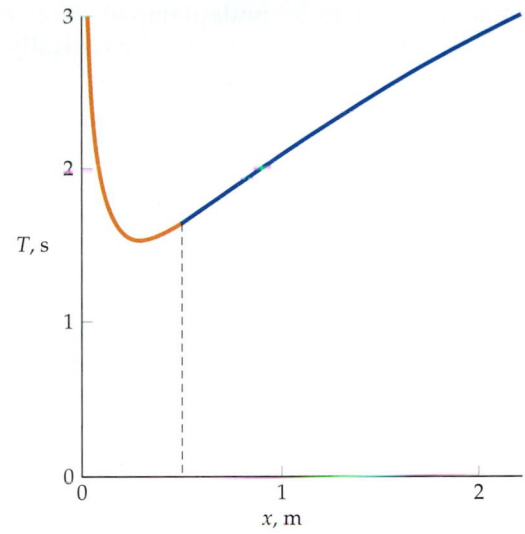

**FIGURE 14-19**

THE ROTATING ROD REVISITED        **EXAMPLE 14-11**    **Try It Yourself**

**Find the value of $x$ in Example 14-10 for which the period is a minimum.**

**PICTURE THE PROBLEM** At the value of $x$ for which $T$ is a minimum, $dT/dx = 0$.

**Cover the column to the right and try these on your own before looking at the answers.**

**Steps**

1. The period, given by the Example 14-10 Part (*b*) result, equals $T = 2\pi\sqrt{Z/g}$, where $Z = (\frac{1}{12}L^2 + x^2)/x$. Find the period both as $x$ approaches zero and as $x$ approaches infinity.

2. The period goes to infinity as $x$ approaches zero and as $x$ approaches infinity. Somewhere in the range $0 < x < \infty$ the period is a minimum. To find the minimum, evaluate $dT/dx$, set it equal to zero, and solve for $x$.

**Answers**

As $x \to 0$, $Z \to \infty$, and $T \to \infty$.

As $x \to \infty$, $Z \to \infty$, and $T \to \infty$.

$$\frac{dT}{dx} = \frac{dT}{dZ}\frac{dZ}{dx} = \frac{\pi}{\sqrt{g}}Z^{-1/2}\frac{dZ}{dx}$$

$Z > 0$ throughout the range $0 < x < \infty$, so $\dfrac{dT}{dx} = 0 \Rightarrow \dfrac{dZ}{dx} = 0.$

$$\frac{dZ}{dx} = 0 \Rightarrow x = \boxed{\frac{L}{\sqrt{12}} = 0.289L}$$

# 14-4 Damped Oscillations

Left to itself, a spring or a pendulum eventually stops oscillating because the mechanical energy is dissipated by frictional forces. Such motion is said to be **damped.** If the damping is large enough, as, for example, a pendulum submerged in molasses, the oscillator fails to complete even one cycle of oscillation. Instead it just moves toward the equilibrium position with a speed that approaches zero as the object approaches the equilibrium position. This type of motion is referred to as **overdamped.** If the damping is small enough that the system oscillates with an amplitude that decreases slowly with time—like a child on a playground swing when Mom stops providing a push each cycle—the motion is said to be **underdamped.** Motion with the minimum damping for nonoscillatory motion is said to be **critically damped.** (With any less damping, the motion would be underdamped.)

**Underdamped Motion**   The damping force exerted on an oscillator such as the one shown in Figure 14-20*a* can be represented by the empirical expression

$$\vec{F}_d = -b\vec{v}$$

(a)

(b)

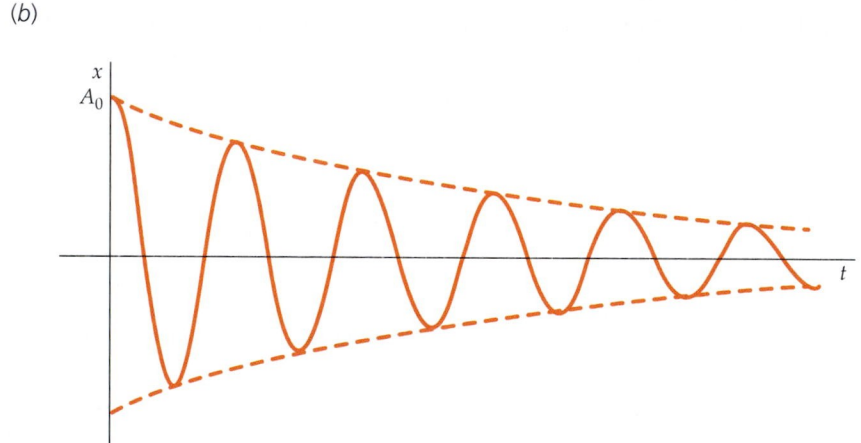

**FIGURE 14-20** (*a*) A damped oscillator. The motion is damped by the plunger immersed in the liquid. (*b*) Damped oscillation curve.

where *b* is a constant. Such a system is said to be linearly damped. The discussion here is for linearly damped motion. Because the damping force is opposite to the direction of motion, it does negative work and causes the mechanical energy of the system to decrease. This energy is proportional to the square of the amplitude (Equation 14-17), and the square of the amplitude decreases exponentially with increasing time. That is,

$$A^2 = A_0^2 e^{-t/\tau} \qquad\qquad 14\text{-}34$$

DEFINITION—TIME CONSTANT

where *A* is the amplitude, $A_0$ is the amplitude at $t = 0$, and $\tau$ is the **decay time** or **time constant.** The time constant is the time for the energy to decrease by a factor of *e*.

   The motion of a damped system can be obtained from Newton's second law. For an object of mass *m* on a spring of force constant *k*, the net force is $-kx - b(dx/dt)$. Setting the net force equal to the mass times the acceleration $d^2x/dt^2$, we obtain

$$-kx - b\frac{dx}{dt} = m\frac{d^2x}{dt^2} \qquad\qquad 14\text{-}35$$

DIFFERENTIAL EQUATION FOR A DAMPED OSCILLATOR

The exact solution of this equation can be found using standard methods for solving differential equations. The solution for the underdamped case is

$$x = A_0 e^{-(b/2m)t} \cos(\omega' t + \delta)$$  14-36

where $A_0$ is the initial amplitude. The frequency $\omega'$ is given by

$$\omega' = \omega_0 \sqrt{1 - \left(\frac{b}{2m\omega_0}\right)^2}$$  14-37

where $\omega_0$ is the frequency with no damping ($\omega_0 = \sqrt{k/m}$ for a mass on a spring). For weak damping, $b/(2m\omega_0) \ll 1$ and $\omega'$ is nearly equal to $\omega_0$. The dashed curves in Figure 14-20b correspond to $x = A$ and $x = -A$, where $A$ is given by

$$A = A_0 e^{-(b/2m)t}$$  14-38

By squaring both sides of this equation and comparing the results with Equation 14-34 we have

$$\tau = \frac{m}{b}$$  14-39

If the damping constant $b$ is gradually increased, the angular frequency $\omega'$ decreases until it becomes zero at the critical value

$$b_c = 2m\omega_0$$  14-40

When $b$ is greater than or equal to $b_c$, the system does not oscillate. If $b > b_c$, the system is overdamped. The smaller $b$ is, the more rapidly the object returns to equilibrium. If $b = b_c$, the system is said to be critically damped and the object returns to equilibrium (without oscillation) most rapidly. Figure 14-21 shows plots of the displacement versus time for a critically damped and an overdamped oscillator. We often use critical damping when we want a system to avoid oscillations and yet return to equilibrium quickly. For example, shock absorbers are used to damp the oscillations of an automobile on its springs. You can test the damping of a car's shock absorbers by pushing down on one fender of the car and then releasing it. If the car returns to equilibrium with no oscillation, then the system is critically damped or overdamped. (You will usually observe one or two oscillations for an unoccupied vehicle, indicating that the damping constant is just under the critical value.)

Because the energy of an oscillator is proportional to the square of its amplitude, the energy of an underdamped oscillator (averaged over a cycle) also decreases exponentially with time:

$$E = \tfrac{1}{2}m\omega^2 A^2 = \tfrac{1}{2}m\omega^2(A_0 e^{-(b/2m)t})^2 = \tfrac{1}{2}m\omega^2 A_0^2 e^{-(b/m)t} = E_0 e^{-t/\tau}$$  14-41

where $E_0 = \tfrac{1}{2}m\omega^2 A_0^2$ and

$$\tau = \frac{m}{b}$$  14-42

A damped oscillator is often described by its $Q$ factor (for quality factor),

$$Q = \omega_0 \tau$$  14-43

DEFINITION—$Q$ FACTOR

Shock absorbers (yellow cylinders) are used to damp the oscillations of this truck.

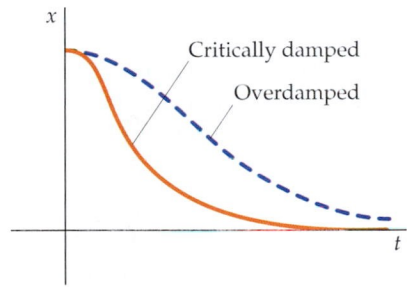

**FIGURE 14-21** Plots of displacement versus time for a critically damped and an overdamped oscillator.

Weights are placed in automobile wheels when the wheels are "balanced." The purpose of balancing the wheels is to prevent vibrations that will drive oscillations of the wheel assembly.

The $Q$ factor is dimensionless. (Since $\omega_0$ has dimensions of reciprocal time, $\omega_0\tau$ is without dimension.) We can relate $Q$ to the fractional energy loss per cycle. Differentiating Equation 14-41 gives

$$dE = -(1/\tau)E_0 e^{-t/\tau}\, dt = -(1/\tau)E\, dt$$

If the damping is weak so that the energy loss per cycle is small, we can replace $dE$ by $\Delta E$ and $dt$ by the period $T$. Then $|\Delta E|/E$ in one cycle (one period) is given by

$$\left(\frac{|\Delta E|}{E}\right)_{\text{cycle}} = \frac{T}{\tau} = \frac{2\pi}{\omega_0\tau} = \frac{2\pi}{Q} \qquad\qquad \text{14-44}$$

by

$$Q = \frac{2\pi}{(|\Delta E|/E)_{\text{cycle}}}, \qquad \frac{|\Delta E|}{E} \ll 1 \qquad\qquad \text{14-45}$$

PHYSICAL INTERPRETATION OF $Q$ FOR WEAK DAMPING

$Q$ is thus inversely proportional to the fractional energy loss per cycle.

MAKING MUSIC                    E X A M P L E   1 4 - 1 2

When middle C on a piano (frequency 262 Hz) is struck, it loses half its energy after 4 s. (a) What is the decay time $\tau$? (b) What is the $Q$ factor for this piano wire? (c) What is the fractional energy loss per cycle?

PICTURE THE PROBLEM (a) We use $E = E_0 e^{-t/\tau}$ and set $E$ equal to $\frac{1}{2}E_0$. (b) The $Q$ value can then be found from the decay time and the frequency.

(a) 1. Set the energy at time $t = 4$ s equal to half the original energy:

$$E = E_0 e^{-t/\tau} \quad\text{so}\quad \tfrac{1}{2}E_0 = E_0 e^{-4\,\text{s}/\tau}$$

$$\tfrac{1}{2} = e^{-4\,\text{s}/\tau}$$

2. Solve for the time $\tau$ by taking the natural log of both sides:

$$\ln\frac{1}{2} = -\frac{4\,\text{s}}{\tau}$$

so

$$\tau = \frac{4\,\text{s}}{\ln 2} = \boxed{5.77\,\text{s}}$$

(b) Calculate $Q$ from $\tau$ and $\omega_0$:

$$Q = \omega_0\tau = 2\pi f\tau$$

$$= 2\pi(262\,\text{Hz})(5.77\,\text{s}) = \boxed{9.50 \times 10^3}$$

(c) The fractional energy loss in a cycle is given by Equation 14-44 and the frequency $f = 1/T$:

$$\left(\frac{|\Delta E|}{E}\right)_{\text{cycle}} = \frac{T}{\tau} = \frac{1}{f\tau} = \frac{1}{(262\,\text{Hz})(5.77\,\text{s})} = \boxed{6.61 \times 10^{-4}}$$

PLAUSIBILITY CHECK $Q$ can also be calculated from $Q = 2\pi/(\Delta E/E)_{\text{cycle}} = 2\pi/(6.61 \times 10^{-4}) = 9.50 \times 10^3$. Note that the fractional energy loss after 4 s is not just the number of cycles ($4 \times 262$) times the fractional energy loss per cycle, because the energy decrease is exponential, not constant.

**REMARKS** Figure 14-22 shows the relative amplitude $A/A_0$ versus time and the relative energy $E/E_0$ versus time for the oscillation of a piano string after middle C is struck. After 4 s, the amplitude has decreased to about 0.7 times its initial value, and the energy, which is proportional to the amplitude squared, drops to about half its initial value.

Note that $Q$ is quite large. You can estimate $\tau$ and $Q$ of various oscillating systems. Tap a crystal wine glass and see how long it rings. The longer it rings, the greater the value of $\tau$ and $Q$ and the lower the damping. Glass beakers from the laboratory may also have a high $Q$. Try tapping a plastic cup. How does the damping compare to that of the beaker?

In terms of $Q$, the exact frequency of an underdamped oscillator is

$$\omega' = \omega_0\sqrt{1 - \left(\frac{b}{2m\omega_0}\right)^2} = \omega_0\sqrt{1 - \frac{1}{4Q^2}} \qquad 14\text{-}46$$

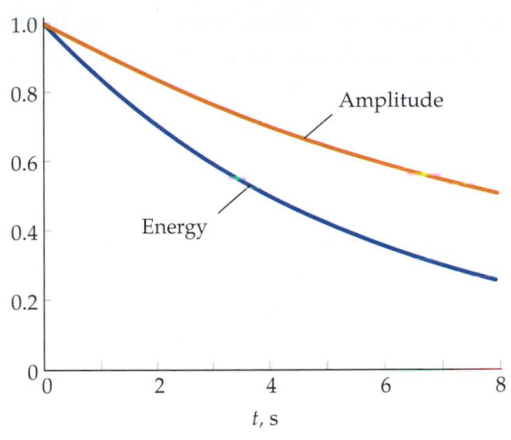

**FIGURE 14-22** Plots of $A/A_0$ and $E/E_0$ for a struck piano string.

Because $b$ is quite small (and $Q$ is quite large) for a weakly damped oscillator (Example 14-12), we see that $\omega'$ is nearly equal to $\omega_0$.

We can understand much of the behavior of a weakly damped oscillator by considering its energy. The power dissipated by the damping force equals the instantaneous rate of change of the total mechanical energy

$$P = \frac{dE}{dt} = \vec{F}_d \cdot \vec{v} = -b\vec{v} \cdot \vec{v} = -bv^2 \qquad 14\text{-}47$$

For a weakly damped oscillator, the total mechanical energy decreases slowly with time. The average kinetic energy per cycle equals half the total energy

$$\left(\frac{1}{2}mv^2\right)_{av} = \frac{1}{2}E \qquad \text{or} \qquad (v^2)_{av} = \frac{E}{m}$$

If we substitute $(v^2)_{av} = E/m$ for $v^2$ in Equation 14-47, we have

$$\frac{dE}{dt} = -bv^2 \approx -b(v^2)_{av} = -\frac{b}{m}E \qquad 14\text{-}48$$

Rearranging Equation 14-48 gives

$$\frac{dE}{E} = -\frac{b}{m}dt$$

which upon integration gives

$$E = E_0 e^{-(b/m)t} = E_0 e^{-t/\tau}$$

which is Equation 14-41.

# 14-5 Driven Oscillations and Resonance

To keep a damped system going, mechanical energy must be put into the system. When this is done, the oscillator is said to be driven or forced. When you keep a swing going by "pumping," that is, by moving your body and legs, you are driving an oscillator. If you put mechanical energy into the system faster than it is dissipated, the mechanical energy increases with time, and the amplitude increases. If you put mechanical energy in at the same rate it is being dissipated,

By pumping the swing, she is transferring her internal energy into the mechanical energy of the oscillator.

the amplitude remains constant over time. The motion of the oscillator is then said to be in steady state.

Figure 14-23 shows a system consisting of an object on a spring that is being driven by moving the point of support up and down with simple harmonic motion of frequency $\omega$. At first the motion is complicated, but eventually steady-state motion is reached in which the system oscillates with the same frequency as that of the driver and with a constant amplitude and, therefore, at constant energy. In the steady state, the energy put into the system per cycle by the driving force equals the energy dissipated per cycle due to the damping.

The amplitude, and therefore the energy, of a system in the steady state depends not only on the amplitude of the driving force, but also on its frequency. The **natural frequency** of an oscillator, $\omega_0$, is its frequency when no driving or damping forces are present. (In the case of a spring, for example, $\omega_0 = \sqrt{k/m}$.) If the driving frequency is approximately equal to the natural frequency of the system, the system will oscillate with a relatively large amplitude. For example, if the support in Figure 14-23 oscillates at a frequency close to the natural frequency of the mass–spring system, the mass will oscillate with a much greater amplitude than it would if the support oscillates at higher or lower frequencies. This phenomenon is called **resonance**. When the driving frequency equals the natural frequency of the oscillator, the energy per cycle transferred to the oscillator is maximum. The natural frequency of the system is thus called the **resonance frequency**. (Mathematically, the angular frequency $\omega$ is more convenient to use than the frequency $f = \omega/(2\pi)$. Because $\omega$ and $f$ are proportional, most statements concerning angular frequency also hold for frequency. In verbal descriptions, we usually omit the word angular when the omission will not cause confusion.) Figure 14-24 shows plots of the average power delivered to an oscillator as a function of the driving frequency for two different values of damping. These curves are called **resonance curves**. When the damping is weak (large Q), the width of the peak of the resonance curve is correspondingly narrow, and we speak of the resonance as being sharp. For strong damping, the resonance curve is broad. The width of each resonance curve $\Delta\omega$, indicated in the figure, is the width at half the maximum height. For weak damping, the ratio of the width of the resonance to the resonant frequency can be shown to equal the reciprocal of the Q factor (see Problem 126):

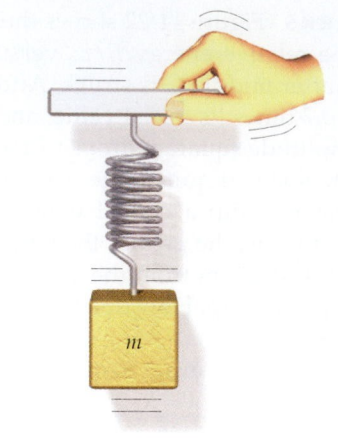

**FIGURE 14-23** An object on a vertical spring can be driven by moving the support up and down.

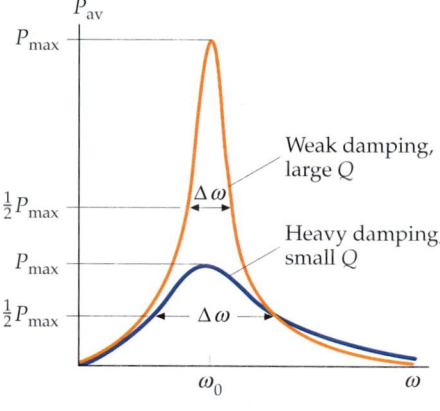

$$\frac{\Delta\omega}{\omega_0} = \frac{1}{Q} \qquad \text{14-49}$$

RESONANCE WIDTH FOR WEAK DAMPING

**FIGURE 14-24** Resonance for an oscillator. The width $\Delta\omega$ of the resonance peak for a high-Q oscillator is small compared to the natural frequency of $\omega_0$.

Thus, the Q factor is a direct measure of the sharpness of resonance.

You can do a simple experiment to demonstrate resonance. Hold a meterstick at one end between two fingers so that it acts like a pendulum. (If a meterstick is not available, use whatever is convenient. A golf club works fine.) Release the stick from some initial angular displacement and observe the natural frequency of its motion. Then move your hand back and forth horizontally, driving it at its natural frequency. Even if the amplitude of the motion of your hand is small, the stick will oscillate with a substantial amplitude. Now move your hand back and forth at a frequency two or three times the natural frequency and note the decrease in amplitude of the oscillating stick.

There are many familiar examples of resonance. When you sit on a swing, you learn intuitively to pump with the same frequency as the natural frequency of the swing. Many machines vibrate because they have rotating parts that are not in perfect balance. (Observe a washing machine in the spin cycle for an example.) If such a machine is attached to a structure that can vibrate, the structure becomes a driven oscillatory system that is set in motion by the machine. Engineers pay great attention to balancing the rotary parts of such machines, damping their vibrations, and isolating them from building supports.

A glass with low damping can be broken by an intense sound wave at a frequency equal to or very nearly equal to the natural frequency of vibration of the glass. This is often done in physics demonstrations using an audio oscillator and an amplifier.

## *Mathematical Treatment of Resonance

We can treat a driven oscillator mathematically by assuming that, in addition to the restoring force and a damping force, the oscillator is subject to an external driving force that varies harmonically with time:

$$F_{ext} = F_0 \cos \omega t \qquad \text{14-50}$$

where $F_0$ and $\omega$ are the amplitude and angular frequency of the driving force. This frequency is generally not related to the natural angular frequency of the system $\omega_0$.

Newton's second law applied to an object of mass $m$ attached to a spring of force constant $k$ and subject to a damping force $-bv_x$ and an external force $F_0 \cos \omega t$ gives

$$\sum F_x = ma_x$$

$$-kx - bv_x + F_0 \cos \omega t = m\frac{d^2x}{dt^2}$$

where we have used $a_x = d^2x/dt^2$. Substituting $m\omega_0^2$ for $k$ (Equation 14-8) and rearranging gives

$$m\frac{d^2x}{dt^2} + b\frac{dx}{dt} + m\omega_0^2x = F_0 \cos \omega t \qquad \text{14-51}$$

DIFFERENTIAL EQUATION FOR A DRIVEN OSCILLATOR

We will discuss the general solution of Equation 14-51 qualitatively. It consists of two parts, the **transient solution** and the **steady-state solution**. The transient part of the solution is identical to that for a damped oscillator given in Equation 14-36. The constants in this part of the solution depend on the initial conditions. Over time, this part of the solution becomes negligible because of the exponential decrease of the amplitude. We are then left with the steady-state solution, which can be written as

$$x = A \cos(\omega t - \delta) \qquad \text{14-52}$$

POSITION FOR A DRIVEN OSCILLATOR

where the angular frequency $\omega$ is the same as that of the driving force. The amplitude $A$ is given by

268 Hz ($Q = 52$)

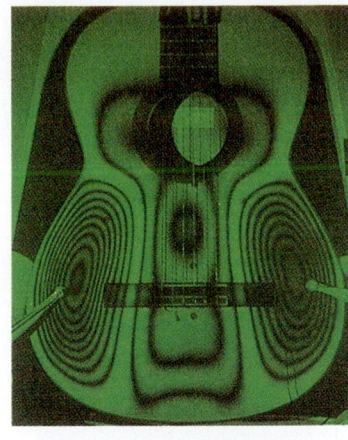

553 Hz ($Q = 66$)

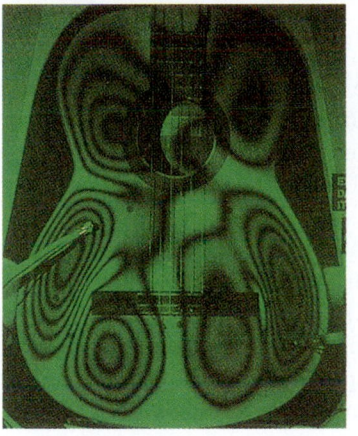

672 Hz ($Q = 61$)

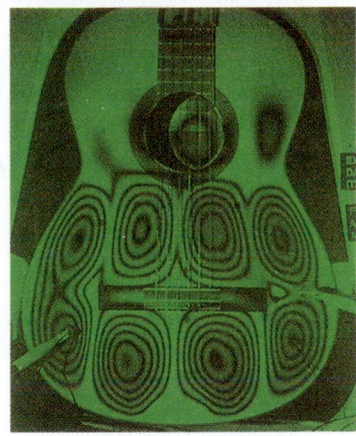

1010 Hz ($Q = 80$)

Extended objects have more than one resonance frequency. When plucked, a guitar string transmits its energy to the body of the guitar. The body's oscillations, coupled to those of the air mass it encloses, produce the resonance patterns shown.

$$A = \frac{F_0}{\sqrt{m^2(\omega_0^2 - \omega^2)^2 + b^2\omega^2}} \qquad \text{14-53}$$

AMPLITUDE FOR A DRIVEN OSCILLATOR

and the phase constant $\delta$ is given by

$$\tan \delta = \frac{b\omega}{m(\omega_0^2 - \omega^2)} \qquad \text{14-54}$$

PHASE CONSTANT FOR A DRIVEN OSCILLATOR

Comparing Equations 14-50 and 14-52, we can see that the displacement and the driving force oscillate with the same frequency, but they differ in phase by $\delta$. When the driving frequency $\omega$ is much less than the natural frequency $\omega_0$, $\delta \approx 0$, as can be seen from Equation 14-54. At resonance, $\omega = \omega_0$, $\delta = \pi/2$, and when $\omega$ is much greater than $\omega_0$, $\delta \approx \pi$. At the beginning of this chapter the displacement of a particle undergoing simple harmonic motion is written $x = A \cos(\omega t + \delta)$ (Equation 14-4). This equation is identical with Equation 14-52 except for the sign preceding the phase constant $\delta$. The phase of a driven oscillator always lags behind the phase of the driving force. The negative sign in Equation 14-52 ensures that $\delta$ is always positive (rather than always negative).

In your simple experiment to drive a meterstick by moving your hand back and forth, you should note that at resonance the oscillation of your hand is neither in phase nor 180° out of phase with the oscillation of the stick. If you move your hand back and forth at a frequency several times the natural frequency of the stick, the stick's steady state motion will be almost 180° out of phase with your hand.

The velocity of the object in the steady state is obtained by differentiating $x$ with respect to $t$:

$$v = \frac{dx}{dt} = -\omega A \sin(\omega t - \delta)$$

At resonance, $\delta = \pi/2$, and the velocity is in phase with the driving force:

$$v = -\omega A \sin\left(\omega t - \frac{\pi}{2}\right) = +\omega A \cos \omega t$$

Thus, at resonance the object is always moving in the direction of the driving force, as would be expected for maximum power input. The velocity amplitude $\omega A$ is maximum at $\omega = \omega_0$.

 **EXPLORING**

*Is there a numerical solution to the equations for damped and driven oscillators? Find out this, and more, at* www.whfreeman.com/tipler5e.

---

AN OBJECT ON A SPRING          **EXAMPLE 14-13**   **Try It Yourself**

An object of mass 1.5 kg on a spring of force constant 600 N/m loses 3 percent of its energy in each cycle. The same system is driven by a sinusoidal force with a maximum value of $F_0 = 0.5$ N. (*a*) What is $Q$ for this system? (*b*) What is the resonance (angular) frequency? (*c*) If the driving frequency is slowly varied through resonance, what is the width $\Delta\omega$ of the resonance? (*d*) What is the amplitude at resonance? (*e*) What is the amplitude if the driving frequency is $\omega = 19$ rad/s?

**PICTURE THE PROBLEM** The energy loss per cycle is only 3 percent, so the damping is weak. We can find $Q$ from $Q = 2\pi/(\Delta E/E)_{cycle}$ (Equation 14-45) and then use this result and $\Delta\omega/\omega_0 = 1/Q$ (Equation 14-49) to find the width $\Delta\omega$ of the resonance. The resonance frequency is the natural frequency. The amplitude both at resonance and off resonance can be found from Equation 14-53, with the damping constant calculated from $Q$ using the definition of $Q$ (Equation 14-43) $Q = \omega_0\tau = \omega_0 m/b$.

**Cover the column to the right and try these on your own before looking at the answers.**

| Steps | Answers |
|---|---|
| (a) The damping is weak. Relate $Q$ to the fractional energy loss using Equation 14-45 and solve for $Q$. | $Q \approx \dfrac{2\pi}{(\lvert\Delta E\rvert/E)_{cycle}} = \dfrac{2\pi}{0.03} = \boxed{209}$ |
| (b) Relate the resonance frequency to the natural frequency of the system. | $\omega_0 = \sqrt{\dfrac{k}{m}} = \boxed{20 \text{ rad/s}}$ |
| (c) Relate the width of the resonance $\Delta\omega$ to $Q$. | $\Delta\omega = \dfrac{\omega_0}{Q} = \boxed{0.0957 \text{ rad/s}}$ |
| (d) 1. Write an expression for the amplitude $A$ for any driving frequency $\omega$. | $A(\omega) = \dfrac{F_0}{\sqrt{m^2(\omega_0^2 - \omega^2)^2 + b^2\omega^2}}$ |
| 2. Substitute $\omega = \omega_0$ to calculate $A$ at resonance. | $A(\omega_0) = \dfrac{F_0}{b\omega_0}$ |
| 3. Use Equation 14-43 to relate the damping constant $b$ to $Q$. | $b = \dfrac{m\omega_0}{Q} = 0.144 \text{ kg/s}$ |
| 4. Use the results of the previous two steps to calculate the amplitude at resonance. | $A(\omega_0) = \dfrac{F_0}{b\omega_0} = \boxed{17.4 \text{ cm}}$ |
| (e) Calculate the amplitude for $\omega = 19$ rad/s. (We can omit the units to simplify the equation. Since all quantities are in SI units, $A$ will be in meters.) | $A(19 \text{ s}^{-1}) = \dfrac{0.5}{\sqrt{1.5^2(20^2 - 19^2)^2 + 0.144^2(19)^2}}$ $= \boxed{0.854 \text{ cm}}$ |

**REMARKS** At just 1 rad/s off resonance, the amplitude drops by a factor of 20. This is not surprising, because the width $\Delta\omega$ of the resonance is only 0.0957 rad/s. Note that off resonance the term $b^2\omega^2$ is negligible compared with the other term in the denominator of the expression for $A$. When $\omega - \omega_0$ is more than several times the half width $\Delta\omega$, as it was in this example, we can neglect the $b^2\omega^2$ term and calculate $A$ from $A \approx F_0/[m(\omega_0^2 - \omega^2)]$. Figure 14-25 shows the amplitude versus driving frequency $\omega$. Note that the horizontal scale is over a small range of $\omega$.

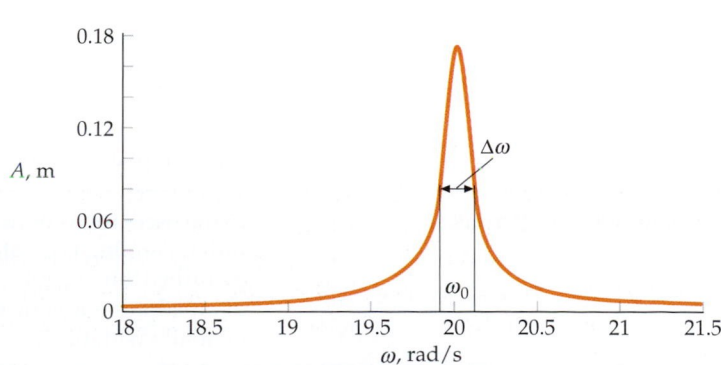

**FIGURE 14-25**

1. Simple harmonic motion occurs whenever the restoring force is proportional to the displacement from equilibrium. It has wide application in the study of oscillations, waves, electrical circuits, and molecular dynamics.

2. Resonance is an important phenomenon in many areas of physics. It occurs when the frequency of the driving force is close to the natural frequency of the oscillating system.

| Topic | Relevant Equations and Remarks | |
|---|---|---|
| **1. Simple Harmonic Motion** | In simple harmonic motion, the net force and acceleration are both proportional to the displacement and oppositely directed. | |
| Position function | $x = A \cos(\omega t + \delta)$ | 14-4 |
| Velocity | $v = -\omega A \sin(\omega t + \delta)$ | 14-5 |
| Acceleration | $a_x = -\omega^2 A \cos(\omega t + \delta)$ | 14-6 |
| | $a_x = -\omega^2 x$ | 14-7 |
| Angular frequency | $\omega = 2\pi f = \dfrac{2\pi}{T}$ | 14-11 |
| Total energy | $E_{\text{total}} = K + U = \frac{1}{2} kA^2$ | 14-17 |
| Average kinetic or potential energy | $K_{\text{av}} = U_{\text{av}} = \frac{1}{2} E_{\text{total}}$ | 14-18 |
| Circular motion | If a particle moves in a circle with constant speed, the projection of the particle onto a diameter of the circle moves in simple harmonic motion. | |
| General motion near equilibrium | If an object is given a small displacement from a position of stable equilibrium, it typically oscillates about this position with simple harmonic motion. | |
| **2. Angular Frequencies for Various Systems** | | |
| Mass on a spring | $\omega = \sqrt{\dfrac{k}{m}}$ | 14-8 |
| Simple pendulum | $\omega = \sqrt{\dfrac{g}{L}}$ | 14-27 |
| *Physical pendulum | $\omega = \sqrt{\dfrac{MgD}{I}}$ | 14-32 |
| | where $D$ is the distance from the center of mass to the rotation axis and $I$ is the moment of inertia about the rotation axis. | |
| **3. Damped Oscillations** | In the oscillations of real systems, the motion is damped because of dissipative forces. If the damping is greater than some critical value, the system does not oscillate when disturbed but merely returns to its equilibrium position. The motion of a weakly damped system is nearly simple harmonic with an amplitude that decreases exponentially with time. | |

| | | |
|---|---|---|
| Frequency | $\omega' = \omega_0\sqrt{1 - \dfrac{1}{4Q^2}}$ | 14-46 |
| Amplitude | $A = A_0 e^{-(b/2m)t}$ | 14-38 |
| Energy | $E = E_0 e^{-t/\tau}$ | 14-41 |
| Decay time | $\tau = \dfrac{m}{b}$ | 14-42 |
| Q factor defined | $Q = \omega_0 \tau$ | 14-43 |
| Q factor for weak damping | $Q = \dfrac{2\pi}{(|\Delta E|/E)_{cycle}}, \quad \dfrac{|\Delta E|}{E} \ll 1$ | 14-45 |

**4. Driven Oscillations**

When an underdamped ($b < b_c$) system is driven by an external sinusoidal force $F_{ext} = F_0 \cos \omega t$, the system oscillates with a frequency $\omega$ equal to the driving frequency and an amplitude $A$ that depends on the driving frequency.

| | | |
|---|---|---|
| Resonance frequency | $\omega = \omega_0$ | |
| Resonance width for weak damping | $\dfrac{\Delta \omega}{\omega_0} = \dfrac{1}{Q}$ | 14-49 |
| *Position function | $x = A \cos(\omega t - \delta)$ | 14-52 |
| *Amplitude | $A = \dfrac{F_0}{\sqrt{m^2(\omega_0^2 - \omega^2)^2 + b^2\omega^2}}$ | 14-53 |
| *Phase constant | $\tan \delta = \dfrac{b\omega}{m(\omega_0^2 - \omega^2)}$ | 14-54 |

# PROBLEMS

- • Single-concept, single-step, relatively easy
- •• Intermediate-level, may require synthesis of concepts
- ••• Challenging
- **SSM** Solution is in the *Student Solutions Manual*
- **iSOLVE** Problems available on iSOLVE online homework service
- **iSOLVE ✓** These "Checkpoint" online homework service problems ask students additional questions about their confidence level, and how they arrived at their answer

In a few problems, you are given more data than you actually need; in a few other problems, you are required to supply data from your general knowledge, outside sources, or informed estimates.

## Conceptual Problems

**1** • What is the magnitude of the acceleration of an oscillator of amplitude $A$ and frequency $f$ when its speed is maximum? When its displacement from equilibrium is maximum?

**2** • Are the acceleration and the displacement (from equilibrium) of a simple harmonic oscillator ever in the same direction? The acceleration and the velocity? The velocity and the displacement? Explain.

**3** • True or false:

(a) For a simple harmonic oscillator, the period is proportional to the square of the amplitude.

(b) For a simple harmonic oscillator, the frequency does not depend on the amplitude.

(c) If the acceleration of a particle undergoing 1-dimensional motion is proportional to the displacement from equilibrium and oppositely directed, the motion is simple harmonic.

**4** • SSM If the amplitude of a simple harmonic oscillator is tripled, by what factor is the energy changed?

**5** •• An object attached to a spring has simple harmonic motion with an amplitude of 4.0 cm. When the object is 2.0 cm from the equilibrium position, what fraction of its total energy is potential energy? (a) One-quarter. (b) One-third. (c) One-half. (d) Two-thirds. (e) Three-quarters.

**6** • True or false:

(a) For a given object on a given spring, the period is the same whether the spring is vertical or horizontal.
(b) For a given object oscillating with amplitude A on a given spring, the maximum speed is the same whether the spring is vertical or horizontal.

**7** • True or false: The motion of a simple pendulum is simple harmonic for any initial angular displacement.

**8** • True or false: The motion of a simple pendulum is periodic for any initial angular displacement.

**9** •• SSM Two identical carts on a frictionless air track are attached by a spring. One is suddenly struck a blow that sends it moving away from the other cart. The motion of the carts is seen to be very jerky—first one cart moves, then stops as the other cart moves and stops in its turn. Explain the motion in a qualitative way.

**10** •• SSM The length of the string or wire supporting a pendulum bob increases slightly when its temperature is raised. How would this affect a clock operated by a simple pendulum?

**11** • True or false: The mechanical energy of a damped, undriven oscillator decreases exponentially with time.

**12** • True or false:

(a) Resonance occurs when the driving frequency equals the natural frequency.
(b) If the Q value is high, the resonance is sharp.

**13** • Give some examples of common systems that can be considered to be driven oscillators.

**14** • A crystal wineglass shattered by an intense sound is an example of (a) resonance, (b) critical damping, (c) an exponential decrease in energy, (d) overdamping.

**15** • SSM The effect of the mass of a spring on the motion of an object attached to it is usually neglected. Describe qualitatively its effect when it is not neglected.

**16** •• A lamp hanging from the ceiling of the club car in a train oscillates with period $T_0$ when the train is at rest. The period will be (match left and right columns)

1. greater than $T_0$ when    A. the train moves horizontally with constant velocity.

2. less than $T_0$ when    B. the train rounds a curve of radius R with speed v.

3. equal to $T_0$ when    C. the train climbs a hill of inclination $\theta$ at constant speed.

D. the train goes over the crest of a hill of radius of curvature R with constant speed.

**17** •• Two mass–spring systems oscillate at frequencies $f_A$ and $f_B$. If $f_A = 2f_B$ and the spring constants of the two springs are equal, it follows that the masses are related by (a) $M_A = 4M_B$, (b) $M_A = M_B/\sqrt{2}$, (c) $M_A = M_B/2$, (d) $M_A = M_B/4$.

**18** •• Two mass–spring systems A and B oscillate so that their energies are equal. If $M_A = 2M_B$, which formula relates the amplitudes of oscillation? (a) $A_A = A_B/4$. (b) $A_A = A_B/\sqrt{2}$. (c) $A_A = A_B$. (d) Not enough information is given to determine the ratio of the amplitudes.

**19** •• Two mass–spring systems A and B oscillate so that their energies are equal. If $k_A = 2k_B$, then which formula relates the amplitudes of oscillation? (a) $A_A = A_B/4$. (b) $A_A = A_B/\sqrt{2}$. (c) $A_A = A_B$. (d) Not enough information is given to determine the ratio of the amplitudes.

**20** •• Pendulum A has a bob of mass $M_A$ and a length $L_A$; pendulum B has a bob of mass $M_B$ and a length $L_B$. If the period of A is twice that of B, then (a) $L_A = 2L_B$ and $M_A = 2M_B$, (b) $L_A = 4L_B$ and $M_A = M_B$, (c) $L_A = 4L_B$ whatever the ratio $M_A/M_B$, (d) $L_A = \sqrt{2}\,L_B$ whatever the ratio $M_A/M_B$.

### Estimation and Approximation

**21** •• For a child on a swing, the amplitude drops by a factor of $1/e$ in about eight periods if no mechanical energy is fed in. Estimate the Q factor for this system.

**22** •• SSM (a) Estimate the natural period of oscillation for swinging your arms as you walk, with your hands empty. (b) Now estimate it when carrying a heavy briefcase. Look around at other people as they walk by—do these two estimates seem on target?

### Simple Harmonic Motion

**23** • ISOLVE The position of a particle is given by $x = (7\ \text{cm})\cos 6\pi t$, where t is in seconds. What are (a) the frequency, (b) the period, and (c) the amplitude of the particle's motion? (d) What is the first time after $t = 0$ that the particle is at its equilibrium position? In what direction is it moving at that time?

**24** • What is the phase constant $\delta$ in Equation 14-4 if the position of the oscillating particle at time $t = 0$ is (a) 0, (b) $-A$, (c) A, (d) A/2?

**25** • SSM A particle of mass m begins at rest from $x = +25$ cm and oscillates about its equilibrium position at $x = 0$ with a period of 1.5 s. Write equations for (a) the position x as a function of t, (b) the velocity v as a function of t, and (c) the acceleration a as a function of t.

**26** • Find (a) the maximum speed and (b) the maximum acceleration of the particle in Problem 23. (c) What is the first time that the particle is at $x = 0$ and moving to the right?

**27** •• Work Problem 25 with the particle initially at $x = 25$ cm and moving with velocity $v_0 = +50$ cm/s.

**28** •• The period of an oscillating particle is 8 s and its amplitude is 12 cm. At $t = 0$ it is at its equilibrium position. Find the distance traveled during the intervals (a) $t = 0$ to

$t = 2$ s, (b) $t = 2$ s to $t = 4$ s, (c) $t = 0$ to $t = 1$ s, and (d) $t = 1$ s to $t = 2$ s.

**29** •• The period of an oscillating particle is 8 s. At $t = 0$, the particle is at rest at $x = A = 10$ cm. (a) Sketch $x$ as a function of $t$. (b) Find the distance traveled in the first, second, third, and fourth second after $t = 0$.

**30** •• **SSM** **iSOLVE** Military specifications often call for electronic devices to be able to withstand accelerations of $10g = 98.1$ m/s². To make sure that their products meet this specification, manufacturers test them using a shaking table that can vibrate a device at various specified frequencies and amplitudes. If a device is given a vibration of amplitude 1.5 cm, what should its frequency be in order to test for compliance with the $10g$ military specification?

**31** •• **iSOLVE** ✓ The position of a particle is given by $x = 2.5 \cos \pi t$, where $x$ is in meters and $t$ is in seconds. (a) Find the maximum speed and maximum acceleration of the particle. (b) Find the speed and acceleration of the particle when $x = 1.5$ m.

**32** •• **SSM** (a) Show that $A_0 \cos(\omega t + \delta)$ can be written as $A_s \sin(\omega t) + A_c \cos(\omega t)$, and determine $A_s$ and $A_c$ in terms of $A_0$ and $\delta$. (b) Relate $A_c$ and $A_s$ to the initial position and velocity of a particle undergoing simple harmonic motion.

## Simple Harmonic Motion and Circular Motion

**33** • **iSOLVE** A particle moves in a circle of radius 40 cm with a constant speed of 80 cm/s. Find (a) the frequency of the motion and (b) the period of the motion. (c) Write an equation for the $x$ component of the position of the particle as a function of time $t$, assuming that the particle is on the positive $x$ axis at time $t = 0$.

**34** • **SSM** A particle moves in a circle of radius 15 cm, making 1 revolution every 3 s. (a) What is the speed of the particle? (b) What is its angular velocity $\omega$? (c) Write an equation for the $x$ component of the position of the particle as a function of time $t$, assuming that the particle is on the positive $x$ axis at time $t = 0$.

## Energy in Simple Harmonic Motion

**35** • A 2.4-kg object is attached to a horizontal spring of force constant $k = 4.5$ kN/m. The spring is stretched 10 cm from equilibrium and released. Find its total energy.

**36** • Find the total energy of a 3-kg object oscillating on a horizontal spring with an amplitude of 10 cm and a frequency of 2.4 Hz.

**37** • A 1.5-kg object oscillates with simple harmonic motion on a spring of force constant $k = 500$ N/m. Its maximum speed is 70 cm/s. (a) What is the total mechanical energy? (b) What is the amplitude of the oscillation?

**38** • A 3-kg object oscillating on a spring of force constant 2 kN/m has a total energy of 0.9 J. (a) What is the amplitude of the motion? (b) What is the maximum speed?

**39** • An object oscillates on a spring with an amplitude of 4.5 cm. Its total energy is 1.4 J. What is the force constant of the spring?

**40** •• **SSM** **iSOLVE** ✓ A 3-kg object oscillates on a spring with an amplitude of 8 cm. Its maximum acceleration is 3.50 m/s². Find the total energy.

## Springs

**41** • A 2.4-kg object is attached to a horizontal spring of force constant $k = 4.5$ kN/m. The spring is stretched 10 cm from equilibrium and released. What are (a) the frequency of the motion, (b) the period, (c) the amplitude, (d) the maximum speed, and (e) the maximum acceleration? (f) When does the object first reach its equilibrium position? What is its acceleration at this time?

**42** • Answer the questions in Problem 41 for a 5-kg object attached to a spring of force constant $k = 700$ N/m when the spring is initially stretched 8 cm from equilibrium.

**43** • A 3-kg object attached to a horizontal spring oscillates with an amplitude $A = 10$ cm and a frequency $f = 2.4$ Hz. (a) What is the force constant of the spring? (b) What is the period of the motion? (c) What is the maximum speed of the object? (d) What is the maximum acceleration of the object?

**44** • **SSM** **iSOLVE** ✓ An 85-kg person steps into a car of mass 2400 kg, causing it to sink 2.35 cm on its springs. Assuming no damping, with what frequency will the car and passenger vibrate on the springs?

**45** • A 4.5-kg object oscillates on a horizontal spring with an amplitude of 3.8 cm. Its maximum acceleration is 26 m/s². Find (a) the force constant $k$, (b) the frequency, and (c) the period of the motion.

**46** • An object oscillates with an amplitude of 5.8 cm on a horizontal spring of force constant 1.8 kN/m. Its maximum speed is 2.20 m/s. Find (a) the mass of the object, (b) the frequency of the motion, and (c) the period of the motion.

**47** •• **iSOLVE** ✓ A 0.4-kg block attached to a spring of force constant 12 N/m oscillates with an amplitude of 8 cm. Find (a) the maximum speed of the block, (b) the speed and acceleration of the block when it is at $x = 4$ cm from the equilibrium position, and (c) the time it takes the block to move from $x = 0$ to $x = 4$ cm.

**48** •• **SSM** An object of mass $m$ is supported by a vertical spring of force constant 1800 N/m. When pulled down 2.5 cm from equilibrium and released from rest, the object oscillates at 5.5 Hz. (a) Find $m$. (b) Find the amount the spring is stretched from its natural length when the object is in equilibrium. (c) Write expressions for the displacement $x$, the velocity $v$, and the acceleration $a$ as functions of time $t$.

**49** •• **iSOLVE** ✓ An object of unknown mass is hung on the end of an unstretched spring and is released from rest. If the object falls 3.42 cm before first coming to rest, find the period of the motion.

**50** •• A spring of force constant $k = 250$ N/m is suspended from a rigid support. An object of mass 1 kg is attached to the unstretched spring and the object is released from rest. (a) How far below the starting point is the equilibrium position for the object? (b) How far down does the object move before it starts up again? (c) What is the period of oscillation? (d) What is the speed of the object when it first reaches its equilibrium position? (*e*) When does it first reach its equilibrium position?

**51** •• **iSOLVE** The St. Louis Arch has a height of 192 m. Suppose that a stunt woman of mass 60 kg jumps off the top of the arch with an elastic band attached to her feet. She reaches the ground at zero speed. Find her kinetic energy *K* after 2.00 s of the flight. (Assume that the elastic band obeys Hooke's law, and neglect its length when relaxed.)

**52** •• **SSM** A suitcase of mass 20 kg is hung from two bungie cords, as shown in Figure 14-26. Each cord is stretched 5 cm when the suitcase is in equilibrium. If the suitcase is pulled down a little and released, what will be its oscillation frequency?

**FIGURE 14-26** Problem 52

**53** •• A 0.12-kg block is suspended from a spring. When a small stone of mass 30 g is placed on the block, the spring stretches an additional 5 cm. With the stone on the block, the spring oscillates with an amplitude of 12 cm. (*a*) What is the frequency of the motion? (*b*) How long does the block take to travel from its lowest point to its highest point? (*c*) What is the net force on the stone when it is at the point of maximum upward displacement?

**54** •• In Problem 53, find the maximum amplitude of oscillation at which the stone will remain in contact with the block.

**55** •• An object of mass 2.0 kg is attached to the top of a vertical spring that is anchored to the floor. The uncompressed length of the spring is 8.0 cm and the length of the spring when the object is in equilibrium is 5.0 cm. When the object is resting at its equilibrium position, it is given a downward impulse with a hammer so that its initial speed is 0.3 m/s. (*a*) To what maximum height above the floor does the object eventually rise? (*b*) How long does it take for the object to reach its maximum height for the first time? (*c*) Does the spring ever become uncompressed? What minimum initial velocity must be given to the object for the spring to be uncompressed at some time?

**56** •• **SSM** A winch cable has a cross-sectional area of 1.5 cm² and a length of 2.5 m. Young's modulus for the cable is 150 GN/m². A 950-kg engine block is hung from the end of the cable. (*a*) By what length does the cable stretch? (*b*) Treating the cable as a simple spring, what is the oscillation frequency of the engine block at the end of the cable?

## Energy of an Object on a Vertical Spring

**57** •• A 2.5-kg object hanging from a vertical spring of force constant 600 N/m oscillates with an amplitude of 3 cm. When the object is at its maximum downward displacement, find (*a*) the total energy of the system, (*b*) the gravitational potential energy, and (*c*) the potential energy in the spring. (*d*) What is the maximum kinetic energy of the object? (Choose $U = 0$ when the object is in equilibrium.)

**58** •• **iSOLVE** A 1.5-kg object that stretches a spring 2.8 cm from its natural length when hanging at rest oscillates with an amplitude of 2.2 cm. (*a*) Find the total energy of the system. (*b*) Find the gravitational potential energy at maximum downward displacement. (*c*) Find the potential energy in the spring at maximum downward displacement. (*d*) What is the maximum kinetic energy of the object?

**59** •• **SSM** A 1.2-kg object hanging from a spring of force constant 300 N/m oscillates with a maximum speed of 30 cm/s. (*a*) What is its maximum displacement? When the object is at its maximum displacement, find (*b*) the total energy of the system, (*c*) the gravitational potential energy, and (*d*) the potential energy in the spring.

## Simple Pendulums

**60** • Find the length of a simple pendulum if the period is 5 s at a location where $g = 9.81$ m/s².

**61** • What would be the period of the pendulum in Problem 60 if the pendulum were on the moon, where the acceleration due to gravity is one-sixth that on earth?

**62** • **iSOLVE** ✓ If the period of a pendulum 70 cm long is 1.68 s, what is the value of $g$ at the location of the pendulum?

**63** • **SSM** **iSOLVE** A pendulum set up in the stairwell of a 10-story building consists of a heavy weight suspended on a 34.0-m wire. If $g = 9.81$ m/s², what is the period of oscillation?

**64** •• Show that the total energy of a simple pendulum undergoing oscillations of small amplitude $\phi_0$ is approximately $E \approx \frac{1}{2}mgL\phi_0^2$. Hint: Use the approximation $\cos \phi \approx 1 - \frac{1}{2}\phi^2$ for small $\phi$.

**65** •• A simple pendulum of length *L* is attached to a massive cart that slides without friction down a plane inclined at angle $\theta$ with the horizontal as shown in Figure 14-27. Find the period of oscillation of the pendulum on the sliding cart.

**FIGURE 14-27** Problem 65

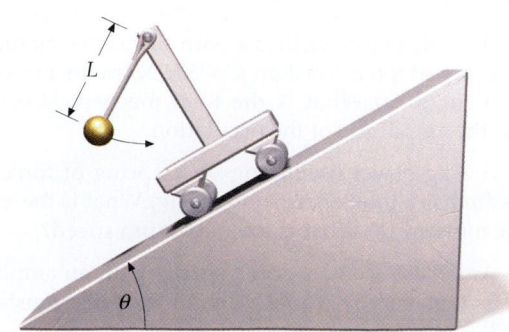

**66** •• A simple pendulum of length $L$ is released from rest from an angle $\phi_0$. (a) Assuming that the pendulum undergoes simple harmonic motion, find its speed as it passes through $\phi = 0$. (b) Using the conservation of energy, find this speed exactly. (c) Show that your results for (a) and (b) are the same when $\phi_0$ is small. (d) Find the difference in your results for $\phi_0 = 0.20$ rad and $L = 1$ m.

## *Physical Pendulums

**67** • **iSOLVE** ✔ A thin disk of mass 5 kg and radius 20 cm is suspended by a horizontal axis perpendicular to the disk through its rim. The disk is displaced slightly from equilibrium and released. Find the period of the subsequent simple harmonic motion.

**68** • A circular hoop of radius 50 cm is hung on a narrow horizontal rod and allowed to swing in the plane of the hoop. What is the period of its oscillation, assuming that the amplitude is small?

**69** • A 3-kg plane figure is suspended at a point 10 cm from its center of mass. When it is oscillating with small amplitude, the period of oscillation is 2.6 s. Find the moment of inertia $I$ about an axis perpendicular to the plane of the figure through the pivot point.

**70** •• The pendulum bob of a large town-hall clock has a length of 4 m. (a) What is its period of oscillation? Treat it as a simple pendulum with small amplitude oscillations. (b) To regulate the period of the pendulum there is a tray attached to its shaft, halfway up. The tray holds a stack of coins. To change the period by a little bit, coins are added or removed from the tray. Explain in detail why this works. Will adding coins increase or decrease the period of the pendulum?

**71** •• Figure 14-28 shows a dumbbell with two equal masses, to be considered as point masses attached to a thin massless rod of length $L$. (a) Show that the period of this pendulum is a minimum when the pivot point $P$ is at one of the masses. (b) Find the period of this physical pendulum if the distance between $P$ and the upper mass is $L/4$.

**FIGURE 14-28** Problem 71

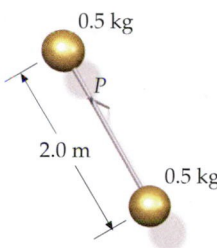

0.5 kg

$P$

2.0 m

0.5 kg

**72** •• Suppose the rod in Problem 71 has a mass of $2m$ (Figure 14-29). Determine the distance between the upper mass and the pivot point $P$ when the period of this physical pendulum is a minimum.

**FIGURE 14-29** Problem 72

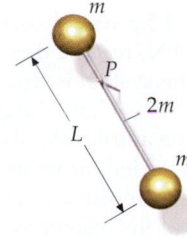

$m$

$P$

$2m$

$L$

$m$

**73** •• **SSM** You are given a meterstick and asked to drill a narrow hole in it so that, when the stick is pivoted about the hole, the period of the pendulum will be a minimum. Where should you drill the hole?

**74** •• **iSOLVE** Figure 14-30 shows a uniform disk of radius $R = 0.8$ m and 6-kg mass with a small hole a distance $d$ from the disk's center that can serve as a pivot point. (a) What should be the distance $d$ so that the period of this physical pendulum is 2.5 s? (b) What should be the distance $d$ so that this physical pendulum will have the shortest possible period? What is this period?

**FIGURE 14-30** Problem 74

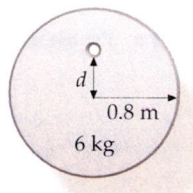

$d$

0.8 m

6 kg

**75** ••• A plane object has a moment of inertia $I$ about its center of mass. When pivoted at point $P_1$, as shown in Figure 14-31, it oscillates about the pivot with a period $T$. There is a second point $P_2$ on the opposite side of the center of mass about which the object can be pivoted so that the period of oscillation is also $T$. Show that $h_1 + h_2 = gT^2/(4\pi^2)$.

**FIGURE 14-31** Problem 75

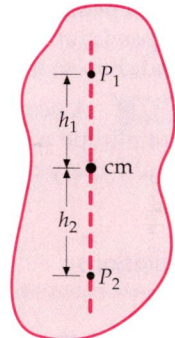

$P_1$

$h_1$

cm

$h_2$

$P_2$

**76** ••• A physical pendulum consists of a spherical bob of radius $r$ and mass $m$ suspended from a string (Figure 14-32). The distance from the center of the sphere to the point of support is $L$. When $r$ is much less than $L$, such a pendulum is often treated as a simple pendulum of length $L$. (a) Show that the period for small oscillations is given by

**FIGURE 14-32** Problem 76

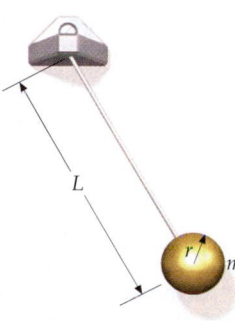

$L$

$r$

$m$

$$T = T_0\sqrt{1 + \frac{2r^2}{5L^2}}$$

where $T_0 = 2\pi\sqrt{L/g}$ is the period of a simple pendulum of length $L$. (b) Show that when $r$ is much smaller than $L$, the period is approximately $T \approx T_0(1 + r^2/5L^2)$. (c) If $L = 1$ m and

$r = 2$ cm, find the error when the approximation $T = T_0$ is used for this pendulum. How large must be the radius of the bob for the error to be 1 percent?

**77 •••** Figure 14-33 shows the pendulum of a clock. The uniform rod of length $L = 2.0$ m has a mass $m = 0.8$ kg. Attached to the rod is a uniform disk of mass $M = 1.2$ kg and radius 0.15 m. The clock is constructed to keep perfect time if the period of the pendulum is exactly 3.50 s. (a) What should be the distance $d$ so that the period of this pendulum is 2.50 s? (b) Suppose that the pendulum clock loses 5.0 min/d. How far and in what direction should the disk be moved to ensure that the clock will keep perfect time?

**FIGURE 14-33** Problem 77

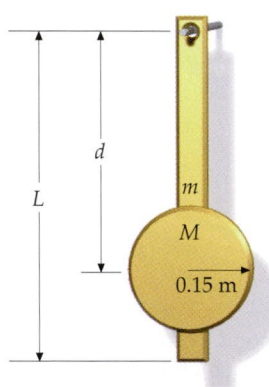

**78 ••** **SSM** A pendulum clock loses 48 s/d when the amplitude of the pendulum is 8.4°. What should be the amplitude of the pendulum so that the clock keeps perfect time?

**79 ••** **iSOLVE** ✓ A pendulum clock that has run down to a very small amplitude gains 5 min each day. What angular amplitude should the pendulum have to keep the correct time?

## Damped Oscillations

**80 •** Show that the damping constant $b$ has units of kg/s.

**81 •** An oscillator has a $Q$ factor of 200. By what percentage does its energy decrease during one period?

**82 •** **iSOLVE** A 2-kg object oscillates with an initial amplitude of 3 cm on a spring of force constant $k = 400$ N/m. Find (a) the period and (b) the total initial energy. (c) If the energy decreases by 1 percent per period, find the damping constant $b$ and the $Q$ factor.

**83 ••** Show that the ratio of the amplitudes for two successive oscillations is constant for a damped oscillator.

**84 ••** An oscillator has a period of 3 s. Its amplitude decreases by 5 percent during each cycle. (a) By how much does its energy decrease during each cycle? (b) What is the time constant $\tau$? (c) What is the $Q$ factor?

**85 ••** An oscillator has a $Q$ factor of 20. (a) By what fraction does the energy decrease during each cycle? (b) Use Equation 14-37 to find the percentage difference between $\omega'$ and $\omega_0$. Hint: Use the approximation $(1 + x)^{\frac{1}{2}} \approx 1 + \frac{1}{2}x$ for small x.

**86 ••** **iSOLVE** A damped mass–spring system oscillates at 200 Hz. The time constant of the system is 2.0 s. At $t = 0$ the amplitude of oscillation is 6.0 cm and the energy of the oscillating system is 60 J. (a) What are the amplitudes of oscillation at $t = 2.0$ s and $t = 4.0$ s? (b) How much energy is dissipated in the first 2-s interval and in the second 2-s interval?

**87 ••** **SSM** **iSOLVE** It has been stated that the vibrating earth has a resonance period of 54 min and a $Q$ factor of about 400 and that after a large earthquake, the earth "rings" (continues to vibrate) for about 2 months. (a) Find the percentage of the energy of vibration lost to damping forces during each cycle. (b) Show that after $n$ periods the energy is $E_n = (0.984)^n E_0$, where $E_0$ is the original energy. (c) If the original energy of vibration of an earthquake is $E_0$, what is the energy after 2 d?

**88 ••** A compact pendulum used in a physics experiment has a mass of 15 g; the length of the pendulum is 75 cm. To start the bob oscillating, a physics student puts a fan next to it that blows a horizontal stream of air on the bob. With the fan on, the bob is in equilibrium when the pendulum is displaced by an angle of 5° from the vertical. The speed of the air from the fan is 7 m/s. The fan is then turned off, and the pendulum is allowed to oscillate. (a) If we assume that the drag force due to the air is of the form $-bv$, what is the decay time constant $\tau$ for the oscillations of the pendulum? (b) How long will it take for the amplitude of oscillation of the pendulum to reach 1°?

## Driven Oscillations and Resonance

**89 •** Find the resonance frequency for each of the three systems shown in Figure 14-34.

**FIGURE 14-34** Problem 89

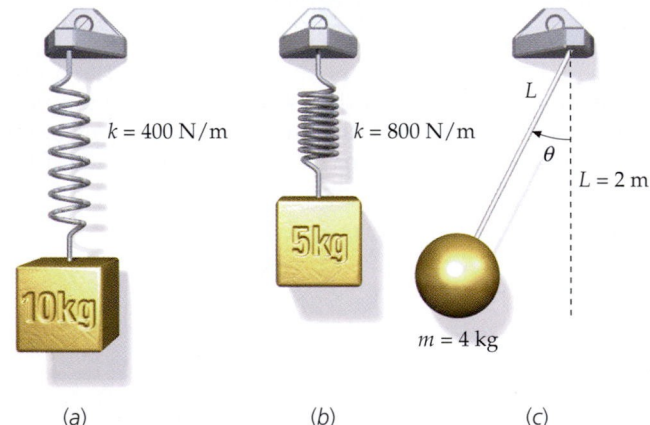

(a)  (b)  (c)

**90 •** A damped oscillator loses 2 percent of its energy during each cycle. (a) What is its $Q$ factor? (b) If its resonance frequency is 300 Hz, what is the width of the resonance curve $\Delta\omega$ when the oscillator is driven?

**91 ••** A 2-kg object oscillates on a spring of force constant $k = 400$ N/m. The damping constant has a value of $b = 2.00$ kg/s. The system is driven by a sinusoidal force of maximum value 10 N and angular frequency $\omega = 10$ rad/s. (a) What is the amplitude of the oscillations? (b) If the driving frequency is varied, at what frequency will resonance occur? (c) What is the amplitude of oscillation at resonance? (d) What is the width of the resonance curve $\Delta\omega$?

**92** •• A damped oscillator loses 3.5 percent of its energy during each cycle. (a) How many cycles elapse before half of its original energy is dissipated? (b) What is its $Q$ factor? (c) If the natural frequency is 100 Hz, what is the width of the resonance curve when the oscillator is driven?

## Collisions

**93** ••• Figure 14-35 shows a vibrating mass–spring system supported on a frictionless surface and a second, equal mass that is moving toward the vibrating mass with velocity $v$. The motion of the vibrating mass is given by $x(t) = (0.1 \text{ m}) \cos(40 \text{ s}^{-1} t)$, where $x$ is the displacement of the mass from its equilibrium position. The two masses collide elastically just as the vibrating mass passes through its equilibrium position traveling to the right. (a) What should be the velocity $v$ of the second mass so that the mass–spring system is at rest following the elastic collision? (b) What is the velocity of the second mass after the elastic collision?

**FIGURE 14-35** Problem 93

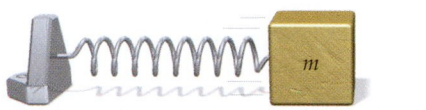 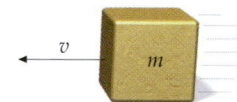

**94** ••• Following the elastic collision in Problem 93, the kinetic energy of the recoiling mass is 8.0 J. Find the masses $m$ and the spring constant $k$.

**95** ••• An object of mass 2 kg resting on a frictionless horizontal surface is attached to a spring of force constant 600 N/m. A second object of mass 1 kg slides along the surface toward the first object at 6 m/s. (a) Find the amplitude of oscillation if the objects make a perfectly inelastic collision and remain together on the spring. What is the period of oscillation? (b) Find the amplitude and period of oscillation if the collision is elastic. (c) For each type of collision, write an expression for the position $x$ as a function of time $t$ for the object attached to the spring, assuming that the collision occurs at time $t = 0$.

## General Problems

**96** • A particle has a displacement $x = 0.4 \cos(3t + \pi/4)$, where $x$ is in meters and $t$ is in seconds. (a) Find the frequency $f$ and period $T$ of the motion. (b) Where is the particle at $t = 0$? (c) Where is the particle at $t = 0.5$ s?

**97** • (a) Find an expression for the velocity of the particle whose position is given in Problem 96. (b) What is the velocity at time $t = 0$? (c) What is the maximum velocity? (d) At what time after $t = 0$ does this maximum velocity first occur?

**98** • An object on a horizontal spring oscillates with a period of 4.5 s. If the object is suspended from the spring vertically, by how much is the spring stretched from its natural length when the object is in equilibrium?

**99** •• **SSM** A small particle of mass $m$ slides without friction in a spherical bowl of radius $r$. (a) Show that the motion

of the particle is the same as if it were attached to a string of length $r$. (b) Figure 14-36 shows a particle of mass $m_1$ that is displaced a small distance $s_1$ from the bottom of the bowl, where $s_1$ is much smaller than $r$. A second particle of mass $m_2$ is displaced in the opposite direction a distance $s_2 = 3s_1$, where $s_2$ is also much smaller than $r$. If the particles are released at the same time, where do they meet? Explain.

**FIGURE 14-36** Problems 99, 100

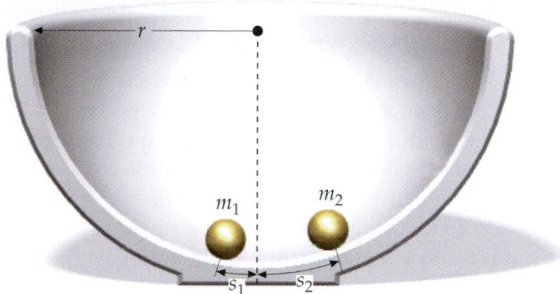

**100** •• Now consider a very small uniform ball of mass $m$ and radius $R$ rolling without slipping near the bottom of the bowl in Figure 14-36. (a) Write an expression for the total energy of the ball in terms of its velocity and the distance (assumed small) from the center of the bowl. (b) By comparing this expression with that for the total energy of a frictionless ball of mass $m$ sliding down the side of the bowl, determine the oscillation frequency of the ball about the center of the bowl.

**101** •• **iSOLVE** ✔ As your jet plane speeds down the runway on takeoff, you measure its acceleration by suspending your yo-yo as a simple pendulum and noting that when the bob (mass 40 g) is at rest relative to you, the string (length 70 cm) makes an angle of 22° with the vertical. Find the period $T$ for small oscillations of this pendulum.

**102** •• If a wire is twisted, there will be a restoring torque $\tau = -\kappa\theta$, where $\kappa$ is a torsional spring constant and $\theta$ is the total twist angle. A torsion balance consists of an object with moment of inertia $I$ hung at the end of a wire. If the object is given a twist, show that the frequency of small torsional oscillations is $\omega = \sqrt{\kappa/I}$.

**103** •• A simple torsion balance (see Problem 102) used in a variety of physics experiments is shown in Figure 14-37. There is a cross-arm of negligible mass at the end of the wire with identical particles attached at each end. If each particle has a mass of 50 g, the length of the cross-arm is 5.0 cm, and the oscillation period of the balance is 80 s, what is the wire's torsion constant $\kappa$?

**FIGURE 14-37** Problem 103

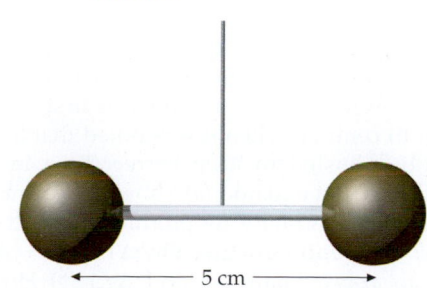

5 cm

**104** •• SSM ISOLVE A wooden cube with edge length $a$ and mass $m$ floats in water with one of its faces parallel to the water surface. The density of the water is $\rho$. Find the period of oscillation in the vertical direction if the cube is pushed down slightly.

**105** •• A clock with a pendulum keeps perfect time on the earth's surface. In which case will the error be greater: if the clock is placed in a mine of depth $h$ or if the clock is elevated to a height $h$? Assume that $h \ll R_E$.

**106** •• ISOLVE Figure 14-38 shows a pendulum of length $L$ with a bob of mass $M$. The bob is attached to a spring of spring constant $k$ as shown. When the bob is directly below the pendulum support, the spring is at its equilibrium length. (a) Derive an expression for the period of this oscillating system for small amplitude vibrations. (b) Suppose that $M = 1$ kg and $L$ is such that in the absence of the spring the period is 2.0 s. What is the spring constant $k$ if the period of the oscillating system is 1.0 s?

**FIGURE 14-38**
**Problem 106**

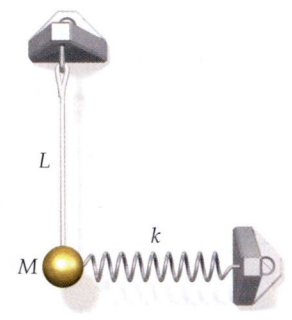

**107** •• An object of mass $m_1$ sliding on a frictionless horizontal surface is attached to a spring of force constant $k$ and oscillates with an amplitude $A$. When the spring is at its greatest extension and the mass is instantaneously at rest, a second object of mass $m_2$ is placed on top of it. (a) What is the smallest value for the coefficient of static friction $\mu_s$ such that the second object does not slip on the first? (b) Explain how the total energy $E$, the amplitude $A$, the angular frequency $\omega$, and the period $T$ of the system are changed by placing $m_2$ on $m_1$, assuming that the friction is great enough so that there is no slippage.

**108** •• A box with a mass of 100 kg hangs from the ceiling of a room by a spring with a spring constant of 500 N/m. The uncompressed length of the spring is 0.5 m. (a) Find the equilibrium position of the box. (b) An identical spring is stretched and attached to the ceiling and box in parallel with the first spring. Find the frequency of the oscillations when the box is released. (c) What is the new equilibrium position of the box once it comes to rest?

**109** •• The acceleration due to gravity $g$ varies with geographical location because of the earth's rotation and because the earth is not exactly spherical. This was first discovered in the seventeenth century, when it was noted that a pendulum clock carefully adjusted to keep correct time in Paris lost about 90 s/d near the equator. (a) Show that a small change in the acceleration of gravity $\Delta g$ produces a small change in the period $\Delta T$ of a pendulum given by $\Delta T/T \approx -\frac{1}{2}\Delta g/g$. (Use differentials to approximate $\Delta T$ and $\Delta g$.) (b) How great a

change in $g$ is needed to account for a change in the period of 90 s/d?

**110** •• Figure 14-39 shows two 0.6-kg blocks glued to each other and connected to a spring of spring constant $k = 240$ N/m. The blocks, which rest on a frictionless horizontal surface, are displaced 0.6 m from their equilibrium position and released. Before they are released, a few drops of solvent are deposited on the glue. (a) Find the frequency of vibration and total energy of the vibrating system before the glue has dissolved. (b) Find the frequency, amplitude of vibration, and energy of the vibrating system if the glue dissolves when the spring is (1) at maximum compression and (2) at maximum extension.

**FIGURE 14-39** Problem 110

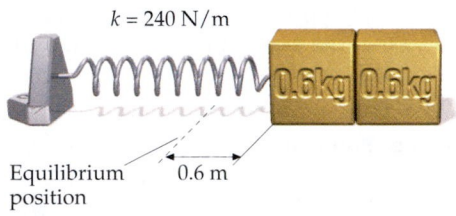

Equilibrium position    0.6 m

**111** •• Show that for the situations in Figure 14-40(a) and (b), the object oscillates with a frequency $f = (1/2\pi)\sqrt{k_{eff}/m}$, where $k_{eff}$ is given by (a) $k_{eff} = k_1 + k_2$ and (b) $1/k_{eff} = 1/k_1 + 1/k_2$. Hint: Find the net force F on the object for a small displacement $x$ and write $F = -k_{eff}x$. Note that in (b) the springs stretch by different amounts, the sum of which is $x$.

**FIGURE 14-40** Problem 111

(a)

(b)

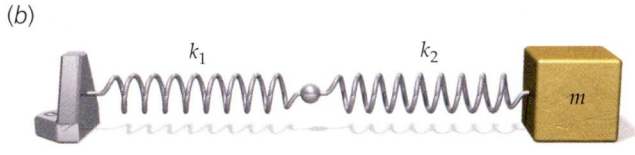

**112** •• SSM A small block of mass $m_1$ rests on a piston that is vibrating vertically with simple harmonic motion given by $y = A \sin \omega t$. (a) Show that the block will leave the piston if $\omega^2 A > g$. (b) If $\omega^2 A = 3g$ and $A = 15$ cm, at what time will the block leave the piston?

**113** •• ISOLVE The plunger of a pinball machine has mass $m_p$ and is attached to a spring of force constant $k$ (Figure 14-41). The spring is compressed a distance $x_0$ from its equilibrium position $x = 0$ and released. A ball of mass $m_b$ is next to the plunger. (a) Where does the ball leave the plunger? (b) What is the speed $v_s$ of the ball when it separates? (c) At what distance $x_f$ does the plunger come to rest momentarily? (Assume that the surface is horizontal and frictionless so that the ball slides rather than rolls.)

**FIGURE 14-41** Problem 113

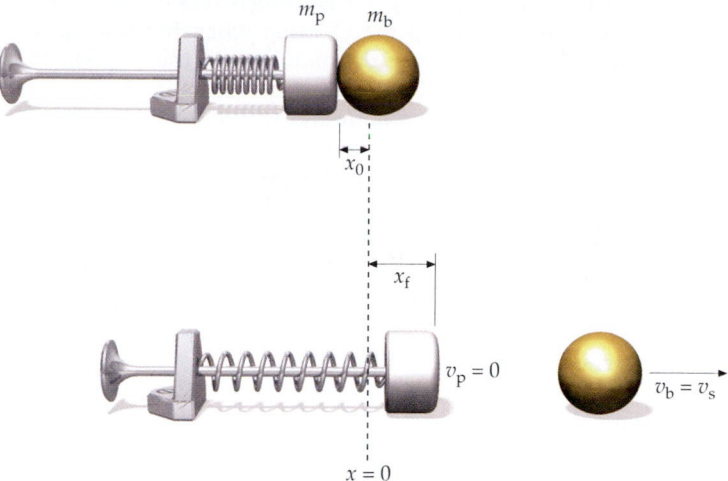

**114** •• A level platform vibrates horizontally with simple harmonic motion with a period of 0.8 s. (*a*) A box on the platform starts to slide when the amplitude of vibration reaches 40 cm; what is the coefficient of static friction between the box and the platform? (*b*) If the coefficient of friction between box and platform were 0.40, what would be the maximum amplitude of vibration before the box would slip?

**115** ••• The potential energy of a particle of mass $m$ as a function of position is given by $U(x) = U_0(\alpha + 1/\alpha)$, where $\alpha = x/a$ and $a$ is a constant. (*a*) Plot $U(x)$ versus $x$ for $0.1a < x < 3a$. (*b*) Find the value of $x = x_0$ at stable equilibrium. (*c*) Write the potential energy $U(x)$ for $x = x_0 + \varepsilon$, where $\varepsilon$ is a small displacement from the equilibrium position $x_0$. (*d*) Approximate the $1/x$ term using the binomial expansion

$$(1 + r)^n = 1 + nr + \frac{n(n-1)}{(2)(1)}r^2 + \frac{n(n-1)(n-2)}{(3)(2)(1)}r^3 + \cdots,$$

with $r = \varepsilon/x_0 << 1$ and discarding all terms of power greater than $r^2$. (*e*) Compare your result with the potential for a simple harmonic oscillator. Show that the mass will undergo simple harmonic motion for small displacements from equilibrium and determine the frequency of this motion.

**116** ••• A solid cylindrical drum of mass 6.0 kg and diameter 0.06 m rolls without slipping on a horizontal surface (Figure 14-42). The axle of the drum is attached to a spring of spring constant $k = 4000$ N/m as shown. (*a*) Determine the frequency of oscillation of this system for small displacements from equilibrium. (*b*) What is the minimum value of the coefficient of static friction so that the drum will not slip when the vibrational energy is 5.0 J?

**FIGURE 14-42** Problem 116

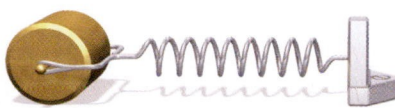

**117** ••• **SSM** If we attach two blocks of masses $m_1$ and $m_2$, to either end of a spring of spring constant $k$ and set them into oscillation, show that the oscillation frequency $\omega = (k/\mu)^{1/2}$, where $\mu = m_1m_2/(m_1 + m_2)$ is the reduced mass of the system.

**118** •• One of the vibrational modes of the HCl molecule has a frequency of $8.969 \times 10^{13}$ s$^{-1}$. Using the relation stated in Problem 117, find the "spring constant" for the HCl molecule.

**119** •• In Problem 118, if we were to replace the hydrogen atom in HCl by a deuterium atom, what would be the new vibration frequency of the molecule? Deuterium consists of 1 proton and 1 neutron.

**120** ••• A block of mass $m$ on a horizontal table is attached to a spring of force constant $k$ as shown in Figure 14-43. The coefficient of kinetic friction between the block and the table is $\mu_k$. The spring is unstretched if the block is at the origin ($x = 0$). The block is released with the spring stretched a distance $A$, where $kA > \mu_k mg$. (*a*) Apply Newton's second law to the block to obtain an equation for its acceleration $d^2x/dt^2$ for the first half-cycle, during which the block is moving to the left. Show that the resulting equation can be written $d^2x'/dt^2 = -\omega^2x'$, where $\omega = \sqrt{k/m}$ and $x' = x - x_0$, with $x_0 = \mu_k mg/k = \mu_k g/\omega^2$. (*b*) Repeat Part (*a*) for the second half-cycle as the block moves to the right, and show that $d^2x''/dt^2 = -\omega^2x''$, where $x'' = x + x_0$ and $x_0$ has the same value. (*c*) Use a spreadsheet program to graph the first 5 half-cycles for $A = 10x_0$. Describe the motion, if any, after the fifth half-cycle.

**FIGURE 14-43**
**Problem 120**

**121** ••• Figure 14-44 shows a uniform solid half-cylinder of mass $M$ and radius $R$ resting on a horizontal surface. If one side of this cylinder is pushed down slightly and then released, the object will oscillate about its equilibrium position. Determine the period of this oscillation.

**FIGURE 14-44**
**Problem 121**

**122** ••• **SSM** A straight tunnel is dug through the earth as shown in Figure 14-45. Assume that the walls of the tunnel are frictionless. (*a*) The gravitational force exerted by the earth on a particle of mass $m$ at a distance $r$ from the center of the earth when $r < R_E$ is $F_r = -(GmM_E/R_E^3)r$, where $M_E$ is the mass of the earth and $R_E$ is its radius. Show that the net force on a particle of mass $m$ at a distance $x$ from the middle of the tunnel is given by $F_x = -(GmM_E/R_E^3)x$, and that the motion of the particle is therefore simple harmonic motion. (*b*) Show that the period of the motion is given by $T = 2\pi\sqrt{R_E/g}$ and find its value in minutes. (This is the same period as that of a satellite orbiting near the surface of the earth and is independent of the length of the tunnel.)

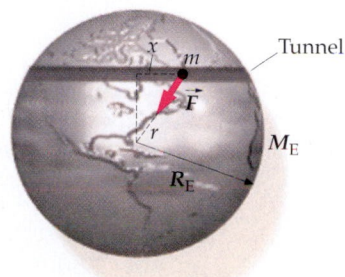

**FIGURE 14-45**
**Problem 122**

**123 •••** A damped oscillator has a frequency $\omega'$ that is 10 percent less than its undamped frequency. (*a*) By what factor is the amplitude of the oscillator decreased during each oscillation? (*b*) By what factor is its energy reduced during each oscillation?

**124 •••** Show by direct substitution that Equation 14-52 is a solution of Equation 14-51.

**125 •••** **SSM** In this problem, you will derive the expression for the average power delivered by a driving force to a driven oscillator (Figure 14-24, page 450).

(*a*) Show that the instantaneous power input of the driving force is given by

$$P = Fv = -A\omega F_0 \cos \omega t \sin(\omega t - \delta).$$

(*b*) Use the trigonometric identity $\sin(\theta_1 - \theta_2) = \sin \theta_1 \cos \theta_2 - \cos \theta_1 \sin \theta_2$ to show that the equation in (*a*) can be written

$$P = A\omega F_0 \sin \delta \cos^2 \omega t - A\omega F_0 \cos \delta \cos \omega t \sin \omega t.$$

(*c*) Show that the average value of the second term in your result for (*b*) over one or more periods is zero and that therefore

$$P_{av} = \tfrac{1}{2}A\omega F_0 \sin \delta.$$

(*d*) From Equation 14-54 for $\tan \delta$, construct a right triangle in which the side opposite the angle $\delta$ is $b\omega$ and the side adjacent is $m(\omega_0^2 - \omega^2)$, and use this triangle to show that

$$\sin \delta = \frac{b\omega}{\sqrt{m^2(\omega_0^2 - \omega^2)^2 + b^2\omega^2}} = \frac{b\omega A}{F_0}$$

(*e*) Use your result for (*d*) to eliminate $\omega A$ from your result for (*c*) so that the average power input can be written

$$P_{av} = \frac{1}{2}\frac{F_0^2}{b} \sin^2 \delta = \frac{1}{2}\left[\frac{b\omega^2 F_0^2}{m^2(\omega_0^2 - \omega^2)^2 + b^2\omega^2}\right] \qquad 14\text{-}55$$

**126 •••** In this problem, you are to use the result of Problem 125 to derive Equation 14-49, which relates the width of the resonance curve to the $Q$ value when the resonance is sharp. At resonance, the denominator of the fraction in brackets in Equation 14-55 is $b^2\omega_0^2$ and $P_{av}$ has its maximum value. (Equation 14-55 can be found in Problem 125.) For a sharp resonance, the variation in $\omega$ in the numerator in Equation 14-55 can be neglected. Then the power input will be half its maximum value at the values of $\omega$, for which the denominator is $2b^2\omega_0^2$.

(*a*) Show that $\omega$ then satisfies $m^2(\omega - \omega_0)^2(\omega + \omega_0)^2 \approx b^2\omega_0^2$.
(*b*) Using the approximation $\omega + \omega_0 \approx 2\omega_0$, show that $\omega - \omega_0 \approx \pm b/2m$.
(*c*) Express $b$ in terms of $Q$.
(*d*) Combine the results of (*b*) and (*c*) to show that there are two values of $\omega$ for which the power input is half that at resonance and that they are given by

$$\omega_1 = \omega_0 - \frac{\omega_0}{2Q} \quad \text{and} \quad \omega_2 = \omega_0 + \frac{\omega_0}{2Q}$$

Therefore, $\omega_2 - \omega_1 = \Delta\omega = \omega_0/Q$, which is equivalent to Equation 14-49.

**127 •••** The Morse potential, which is often used to model interatomic forces, can be written in the form $U(r) = D(1 - e^{-\beta(r-r_0)})^2$, where $r$ is the distance between the two atomic nuclei. (*a*) Using a spreadsheet program or graphing calculator, make a graph of the Morse potential using $D = 5$ eV, $\beta = 0.2$ nm$^{-1}$, and $r_0 = 0.75$ nm. (*b*) Determine the equilibrium separation and "spring constant" for small displacements from equilibrium for the Morse potential. (*c*) Determine a formula for the oscillation frequency for a homonuclear diatomic molecule (that is, two of the same atoms), where the atoms have mass $m$.

# Traveling Waves

THIS POLICE RADAR UNIT SENDS OUT ELECTROMAGNETIC WAVES THAT TRAVEL AT THE SPEED OF LIGHT AND REFLECT FROM THE MOVING CAR.

**?** **How does the police officer determine the speed of the car? (See Example 15-12.)**

15-1    Simple Wave Motion

15-2    Periodic Waves

15-3    Waves in Three Dimensions

15-4    Waves Encountering Barriers

15-5    The Doppler Effect

**W**aves transport energy and momentum through space without transporting matter. As a water wave moves across a pond, for example, the molecules of water oscillate up and down, but do not cross the pond with the wave. Energy and momentum are transported by the wave, but matter is not. A rowboat will bob up and down on the waves but will not be moved by them across the pond. Water waves, waves on a stretched guitar string, and sound waves all involve oscillation.

➤ **In this chapter we continue the study of oscillatory motion that we began in Chapter 14 by examining periodic waves, particularly harmonic waves. We will see that mechanical waves occur when there is a disturbance in a medium, such as air or water, while electromagnetic waves exist without a material medium.**

## 15-1 Simple Wave Motion

### Transverse and Longitudinal Waves

A mechanical wave is caused by a disturbance in a medium. For example, when a taut string is plucked, the pulses produced travel down the string as waves. The disturbance in this case is the change in shape of the string from its equilibrium shape. Its propagation arises from the interaction of each string segment with the adjacent segments. The segments of the string (the medium) move in the direction perpendicular to the string as the pulses propagate up and down the string. Waves such as these, in which the disturbance is perpendicular to the direction of propagation, are called **transverse** (Figure 15-1). Waves in which the disturbance is parallel to the direction of propagation are called **longitudinal** (Figure 15-2). Sound waves are examples of longitudinal waves—the molecules of a gas, liquid, or solid, through which sound travels oscillate (move back and forth) along the line of propagation, alternately compressing and rarefying (expanding) the medium.

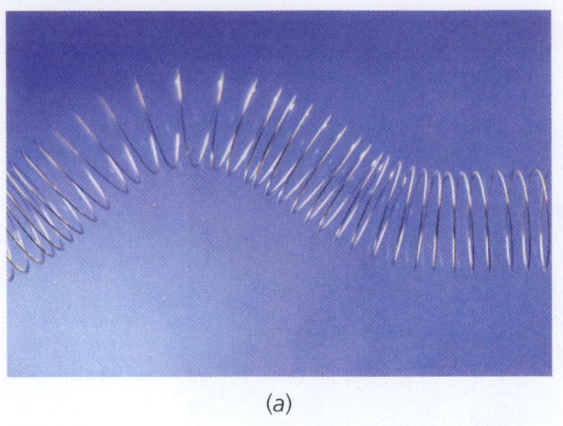

(a)

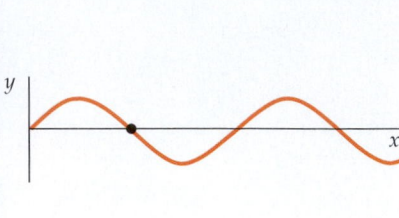

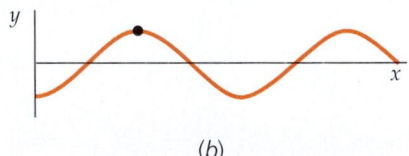

(b)

**FIGURE 15-1** (*a*) Transverse wave pulse on a spring. The disturbance is perpendicular to the direction of the motion of the wave. (*b*) Three successive drawings of a transverse wave on a string traveling to the right. An element of the string moves up and down.

## Wave Pulses

Figure 15-3*a* shows a pulse on a string at time $t = 0$. The shape of the string at this time can be represented by some function $y = f(x)$. At some later time (Figure 15-3*b*), the pulse is farther down the string. In a new coordinate system with origin $O'$ that moves with the speed of the pulse, the pulse is stationary. The string is described in this frame by $f(x')$ for all times. The $x$ coordinates of the two reference frames are related by

$$x' = x - vt$$

so $f(x') = f(x - vt)$.

Thus, the shape of the string in the original frame is

$$y = f(x - vt), \quad \text{wave moving in the positive } x \text{ direction} \qquad 15\text{-}1$$

The same line of reasoning for a pulse moving to the left leads to

$$y = f(x + vt), \quad \text{wave moving in the negative } x \text{ direction} \qquad 15\text{-}2$$

In both expressions, $v$ is the speed of propagation of the wave. The function $y = f(x - vt)$ is called a **wave function**. For waves on a string, the wave function represents the transverse displacement of the string. For sound waves in air, the wave function can be the longitudinal displacement of the air molecules, or the pressure of the air. These wave functions are solutions of a differential equation called the wave equation, which can be derived from Newton's laws.

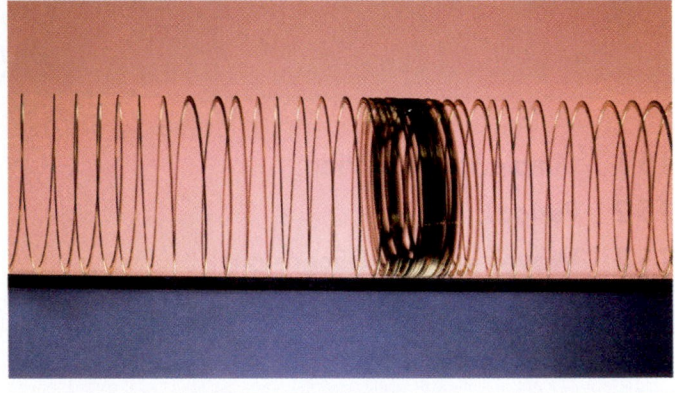

**FIGURE 15-2** Longitudinal wave pulse on a spring. The disturbance is in the direction of the motion of the wave.

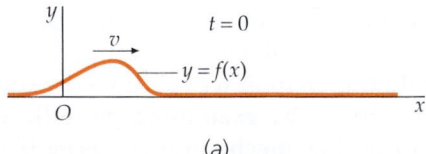

(a)

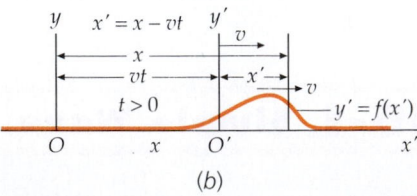

(b)

**FIGURE 15-3**

## Speed of Waves

A general property of waves is that their speed relative to the medium depends on the properties of the medium but is independent of the motion of the source of the waves. For example, the speed of a sound from a car horn depends only on the properties of air and not on the motion of the car.

For wave pulses on a rope, we can easily demonstrate that the greater the tension, the faster the propagation of the waves. Furthermore, waves propagate faster in a light rope than in a heavy rope under the same tension. If $F_T$ is the tension (we use $F_T$ for tension because we use $T$ for the period) and $\mu$ is the linear mass density (mass per unit length), then the wave speed is

$$v = \sqrt{\frac{F_T}{\mu}} \qquad\qquad 15\text{-}3$$

SPEED OF WAVES ON A STRING

---

*INCHY RUNS FOR HIS LIFE*  **EXAMPLE  15-1**  **Put It in Context**

Inchy, an inchworm, is inching along a cotton clothesline. The 25-m-long clothesline has a mass of 0.25 kg and is kept taut by a hanging object of mass 10 kg as shown in Figure 15-4. Vivian is hanging up her swimsuit 5 m from one end when she sees Inchy 2.5 cm from the opposite end. She plucks the line sending a terrifying 3-cm-high pulse toward Inchy. If Inchy crawls at 1 in./s, will he get to the end of the clothesline before the pulse reaches him?

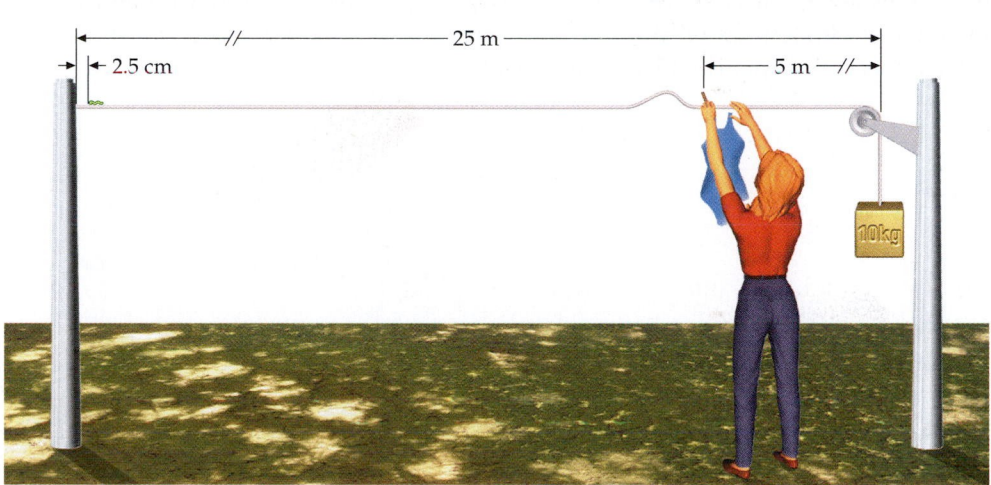

**FIGURE 15-4**

**PICTURE THE PROBLEM** We need to know how fast the wave travels. To find the wave speed we use the formula $v = \sqrt{F_T/\mu}$.

1. The speed of the pulse is related to the tension $F_T$ and mass density $\mu$:

$$v = \sqrt{\frac{F_T}{\mu}}$$

2. Express the mass density and tension in terms of the given parameters:

$$\mu = \frac{m_s}{L} \qquad \text{and} \qquad F_T = mg$$

3. Substitute these values to calculate the speed:

$$v = \sqrt{\frac{F_T}{\mu}} = \sqrt{\frac{mgL}{m_s}} = \sqrt{\frac{(10 \text{ kg})(9.81 \text{ m/s}^2)(25 \text{ m})}{0.25 \text{ kg}}}$$

$$= 99.0 \text{ m/s}$$

4. Use this speed to find the time for the pulse to travel the 20 m to the far end.

$$\Delta t = \frac{\Delta x}{v} = \frac{20 \text{ m}}{99.0 \text{ s}} = 0.202 \text{ s}$$

5. Find the time it takes Inchy to travel the 2.5 cm to the end traveling at 1 in./s:

$$\Delta t' = \frac{\Delta x'}{v'} = \frac{2.5 \text{ cm}}{1 \text{ in./s}} \times \frac{1 \text{ in.}}{2.54 \text{ cm}} = 0.984 \text{ s}$$

Inchy does not beat the pulse.

**EXERCISE** If the 10-kg mass is replaced with a 20-kg mass, what is the speed of waves on the clothesline? (*Answer* 140 m/s)

**EXERCISE** Show that the units of $\sqrt{F_T/\mu}$ are m/s when $F_T$ is in newtons and $\mu$ is in kg/m.

For sound waves in a fluid such as air or water, the speed $v$ is given by

$$v = \sqrt{\frac{B}{\rho}}$$
15-4

where $\rho$ is the equilibrium density of the medium and $B$ is the bulk modulus[†] (Equation 13-6). Comparing Equations 15-3 and 15-4, we can see that, in general, the speed of waves depends on an elastic property of the medium (the tension for string waves and the bulk modulus for sound waves) and on an inertial property of the medium (the linear mass density or the volume mass density).

For sound waves in a gas such as air, the bulk modulus[‡] is proportional to the pressure, which in turn is proportional to the density $\rho$ and to the absolute temperature $T$ of the gas. The ratio $B/\rho$ is thus independent of density and is merely proportional to the absolute temperature $T$. In Chapter 17 we will show that, in this case, Equation 15-4 is equivalent to

$$v = \sqrt{\frac{\gamma RT}{M}}$$
15-5

SPEED OF SOUND IN A GAS

In this equation $T$ is the absolute temperature measured in kelvins (K), which is related to the Celsius temperature $t_C$ by

$$T = t_C + 273$$
15-6

The constant $\gamma$ depends on the kind of gas. For diatomic molecules, such as $O_2$ and $N_2$, $\gamma$ has the value 1.4, and since $O_2$ and $N_2$ comprise 98 percent of the atmosphere, that is the value for air. (For monatomic molecules such as He, $\gamma$ has the value 1.67.) The constant $R$ is the universal gas constant

$$R = 8.314 \text{ J}/(\text{mol·K})$$
15-7

and $M$ is the molar mass of the gas (that is, the mass of 1 mol of the gas), which for air is

$$M = 29 \times 10^{-3} \text{ kg/mol}$$

---

† The bulk modulus is the negative ratio of the pressure change to the fractional change in volume (Chapter 13):

$$B = -\frac{\Delta P}{\Delta V/V}$$

‡ The **isothermal bulk modulus**, which describes changes that occur at constant temperature, differs from the **adiabatic bulk modulus**, which describes changes with no heat transfer. For sound waves at audible frequencies the changes occur too rapidly for appreciable heat flow, so the appropriate bulk modulus is the adiabatic bulk modulus.

---

*SPEED OF SOUND IN AIR*        **EXAMPLE 15-2** **Try It Yourself**

Calculate the speed of sound in air at (*a*) 0°C and (*b*) 20°C.

Cover the column to the right and try these on your own before looking at the answers.

| Steps | Answers |
|---|---|
| (*a*) 1. Write Equation 15-5. | $v_a = \sqrt{\dfrac{\gamma R T_a}{M}}$ |
| 2. Enter the given values into the equation and solve for the speed. (Be sure to convert the temperature to kelvins.) | $v_a = \boxed{331 \text{ m/s}}$ |
| (*b*) 1. From Equation 15-5 we can see that $v$ is proportional to $\sqrt{T}$. Use this to express the ratio of the speed at 293 K to the speed at 273 K. | $\dfrac{v_b}{v_a} = \sqrt{\dfrac{T_b}{T_a}}$ |
| 2. Calculate $v$ at 293 K. | $v_b = \boxed{343 \text{ m/s}}$ |

**REMARKS** We see from this example that the speed of sound in air is about 340 m/s at normal temperatures.

**EXERCISE** For helium, $M = 4 \times 10^{-3}$ kg/mol and $\gamma = 1.67$. What is the speed of sound waves in helium at 20°C? (*Answer* 1.01 km/s)

**Derivation of $v$ for Waves on a String** Equation 15-3 can be obtained from Newton's laws. Consider a pulse traveling along a string with a speed $v$ to the right (Figure 15-5*a*). If the amplitude of the pulse is small compared to the length of the string, then the tension $F_T$ will be approximately constant along the string. In a reference frame moving with speed $v$ to the right, the pulse is stationary and the string moves with a speed $v$ to the left. Figure 15-5*b* shows a small segment of the string of length $\Delta s$ at the top of the pulse. The segment forms part of a circular arc of radius $R$. Instantaneously it is moving with speed $v$ in a circular path, so it has an acceleration $v^2/R$ in the centripetal direction. The magnitudes of the forces acting on the segment are the tension $F_T$ at each end. The horizontal components of these forces are equal and opposite and thus cancel. The vertical components of these forces point radially inward toward the center of the circular arc for sufficiently small $\Delta s$. These radial forces provide the centripetal acceleration.

Let the angle subtended by the string be $\theta$. The centripetal component of the net force on the segment is

$$\sum F_c = 2F_T \sin \tfrac{1}{2}\theta \approx 2F_T (\tfrac{1}{2}\theta) = F_T \theta$$

where we have used the small angle approximation $\sin \tfrac{1}{2}\theta \approx \tfrac{1}{2}\theta$. If $\mu$ is the mass per unit length of the string, the mass of a segment of length $\Delta s$ is $m = \mu \, \Delta s$. The angle $\theta$ is related to $\Delta s$ by

$$\theta = \frac{\Delta s}{R}$$

The mass of the element is thus

$$m = \mu \, \Delta s = \mu R \theta$$

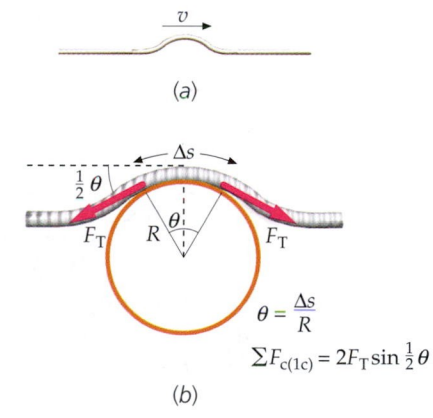

$$\theta = \frac{\Delta s}{R}$$

$$\sum F_{c(1c)} = 2F_T \sin \tfrac{1}{2}\theta$$

**FIGURE 15-5** (*a*) Wave pulse moving with a speed $v$ along a string. (*b*) In a frame in which the wave pulse of (*a*) is at rest, the string is moving with a speed $v$ to the left. A small segment of the string of length $\Delta s$ is moving in a circular arc of radius $R$. The centripetal acceleration of the segment is provided by the radial components of the tension.

Newton's second law ($\Sigma F_c = ma_c$) gives

$$F_T\theta = \mu R\theta \frac{v^2}{R}$$

Solving for $v$, we obtain $v = \sqrt{F_T/\mu}$.

In the original frame, the string is fixed, and the pulse moves with speed $v = \sqrt{F_T/\mu}$, which is Equation 15-3. Since $v$ is independent of $R$ and $\theta$, this result holds for the peak of any pulse. In the following discussion we will show that this result is true not only for the peak but for other parts of the pulse.

## *The Wave Equation

We can apply Newton's laws to a segment of the string to derive a differential equation known as the wave equation, which relates the spatial derivatives of $y(x,t)$ to its time derivatives. Figure 15-6 shows one segment of a string. We consider only small angles $\theta_1$ and $\theta_2$. Then the length of the segment is approximately $\Delta x$ and its mass is $m = \mu\,\Delta x$, where $\mu$ is the string's mass per unit length. First we will show that, for small vertical displacements, the net horizontal force on a segment is zero and the tension is uniform and constant. The net force in the horizontal direction is zero. That is,

$$\sum F_x = F_{T2}\cos\theta_2 - F_{T1}\cos\theta_1 = 0$$

where $\theta_2$ and $\theta_1$ are the angles shown and $F_T$ is the tension in the string. Since the angles are assumed to be small, we may approximate $\cos\theta$ by 1 for each angle. Then the net horizontal force on the segment can be written

$$\sum F_x = F_{T2} - F_{T1} = 0$$

Thus

$$F_{T2} = F_{T1} = F_T$$

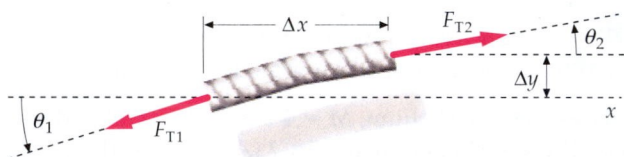

**FIGURE 15-6** Segment of a stretched string used for the derivation of the wave equation. The net vertical force on the segment is $F_{T2}\sin\theta_2 - F_{T1}\sin\theta_1$, where $F$ is the tension in the string. The wave equation is derived by applying Newton's second law to the segment.

The segment moves vertically, and the net force in this direction is

$$\sum F_y = F_T\sin\theta_2 - F_T\sin\theta_1$$

Since the angles are assumed to be small, we may approximate $\sin\theta$ by $\tan\theta$ for each angle. Then the net vertical force on the string segment can be written

$$\sum F_y = F_T(\sin\theta_2 - \sin\theta_1) \approx F_T(\tan\theta_2 - \tan\theta_1)$$

The tangent of the angle made by the string with the horizontal is the slope of the curve formed by the string. The slope $S$ is the first derivative of $y(x,t)$ with respect to $x$ for constant $t$. A derivative of a function of two variables with respect to one of the variables with the other held constant is called a **partial derivative.** The partial derivative of $y$ with respect to $x$ is written $\partial y/\partial x$. Thus we have

$$S = \tan\theta = \frac{\partial y}{\partial x}$$

Then

$$\sum F_y = F_T(S_2 - S_1) = F_T\,\Delta S$$

where $S_1$ and $S_2$ are the slopes of either end of the string segment and $\Delta S$ is the change in the slope. Setting this net force equal to the mass $\mu \, \Delta x$ times the acceleration $\partial^2 y / \partial t^2$ gives

$$F_T \, \Delta S = \mu \, \Delta x \, \frac{\partial^2 y}{\partial t^2}$$

or

$$F_T \frac{\Delta S}{\Delta x} = \mu \frac{\partial^2 y}{\partial t^2} \qquad \qquad 15\text{-}8$$

In the limit $\Delta x \rightarrow 0$, we have

$$\lim_{\Delta x \rightarrow 0} \frac{\Delta S}{\Delta x} = \frac{\partial S}{\partial x} = \frac{\partial}{\partial x} \frac{\partial y}{\partial x} = \frac{\partial^2 y}{\partial x^2}$$

Thus Equation 15-8 becomes

$$\frac{\partial^2 y}{\partial x^2} = \frac{\mu}{F_T} \frac{\partial^2 y}{\partial t^2} \qquad \qquad 15\text{-}9a$$

Equation 15-9a is the **wave equation** for a stretched string.

We now show that the wave equation is satisfied by any function $x - vt$. Let $\alpha = x - vt$ and consider any wave function

$$y = y(x - vt) = y(\alpha)$$

We will use $y'$ for the derivative of $y$ with respect to $\alpha$. Then, by the chain rule for derivatives,

$$\frac{\partial y}{\partial x} = \frac{\partial y}{\partial \alpha} \frac{\partial \alpha}{\partial x} = y' \frac{\partial \alpha}{\partial x}$$

and

$$\frac{\partial y}{\partial t} = \frac{\partial y}{\partial \alpha} \frac{\partial \alpha}{\partial t} = y' \frac{\partial \alpha}{\partial t}$$

Since

$$\frac{\partial \alpha}{\partial x} = \frac{\partial (x - vt)}{\partial x} = 1 \qquad \text{and} \qquad \frac{\partial \alpha}{\partial t} = \frac{\partial (x - vt)}{\partial t} = -v$$

we have

$$\frac{\partial y}{\partial x} = y' \qquad \text{and} \qquad \frac{\partial y}{\partial t} = -vy'$$

Taking the second derivatives, we obtain

$$\frac{\partial^2 y}{\partial x^2} = y''$$

and

$$\frac{\partial^2 y}{\partial t^2} = -v\frac{\partial y'}{\partial t} = -v\frac{\partial y'}{\partial \alpha}\frac{\partial \alpha}{\partial t} = +v^2\, y''$$

Thus

$$\frac{\partial^2 y}{\partial x^2} = \frac{1}{v^2}\frac{\partial^2 y}{\partial t^2}$$

15-9b

WAVE EQUATION

The same result can be obtained for any function of $x + vt$. Comparing Equations 15-9a and 15-9b, we see that the speed of propagation of the wave is $v = \sqrt{F_T/\mu}$, which is Equation 15-3.

---

*HARMONIC WAVE FUNCTION*                    **EXAMPLE 15-3**

In the following section harmonic waves are defined by the wave function $y(x,t) = A\sin(kx - \omega t)$, where $v = \omega/k$. Show by explicitly calculating the derivatives that this wave function satisfies Equation 15-9b.

1. Calculate the second derivative of $y$ with respect to $x$:

$$\frac{\partial y}{\partial x} = \frac{\partial}{\partial x}[A\sin(kx - \omega t)] = A\cos(kx - \omega t)\frac{\partial(kx - \omega t)}{\partial x}$$

$$= kA\cos(kx - \omega t)$$

$$\frac{\partial^2 y}{\partial x^2} = \frac{\partial}{\partial x}\frac{\partial y}{\partial x} = \frac{\partial}{\partial x}kA\cos(kx - \omega t)$$

$$= -kA\sin(kx - \omega t)\frac{\partial(kx - \omega t)}{\partial x}$$

$$= -k^2 A\sin(kx - \omega t)$$

2. Similarly, the two partial derivatives with respect to $t$ are:

$$\frac{\partial y}{\partial t} = \frac{\partial}{\partial t}[A\sin(kx - \omega t)] = A\cos(kx - \omega t)\frac{\partial(kx - \omega t)}{\partial t}$$

$$= -\omega A\cos(kx - \omega t)$$

$$\frac{\partial^2 y}{\partial t^2} = \omega A\sin(kx - \omega t)\frac{\partial(kx - \omega t)}{\partial t} = -\omega^2 A\sin(kx - \omega t)$$

3. Substituting these results in Equation 15-9b gives:

$$-k^2 A\sin(kx - \omega t) = \frac{1}{v^2}[-\omega^2 A\sin(kx - \omega t)]$$

or

$$A\sin(kx - \omega t) = \frac{\omega^2/k^2}{v^2}A\sin(kx - \omega t)$$

4. Substituting for $k$ using $k = \omega/v$ gives:

$$A\sin(kx - \omega t) = \frac{v^2}{v^2}A\sin(kx - \omega t) = \boxed{A\sin(kx - \omega t)}$$

**REMARKS** We have shown that the function $y = A\sin(kx - \omega t)$ is a solution to the wave equation provided $v = \omega/k$.

**EXERCISE** Show that any function $y(x + vt)$ satisfies Equation 15-9b.

A wave equation for sound waves can also be derived using Newton's laws. In one dimension, this equation is

$$\frac{\partial^2 s}{\partial x^2} = \frac{1}{v_s^2} \frac{\partial^2 s}{\partial t^2}$$

where $s$ is the displacement of the medium in the $x$ direction and $v_s$ is the speed of sound.

## 15-2 Periodic Waves

If one end of a long taut string is shaken back and forth in periodic motion, then a **periodic wave** is generated. If a periodic wave is traveling along a taut string or any other medium, each point along the medium oscillates with the same period.

### Harmonic Waves

Harmonic waves are the most basic type of periodic waves. All waves, whether they are periodic or not, can be modeled as a superposition of harmonic waves. Consequently, an understanding of harmonic wave motion can be generalized to form an understanding of any type of wave motion. If a **harmonic wave** is traveling through a medium, each point of the medium oscillates in simple harmonic motion.

If one end of a string is attached to a vibrating tuning fork that is moving up and down with simple harmonic motion, a sinusoidal wave train propagates along the string. This wave train is a harmonic wave. As shown in Figure 15-7, the shape of the string is that of a sine function. The minimum distance after which the wave repeats (the distance between crests, for example) in this figure is called the **wavelength** $\lambda$.

As the wave propagates along the string, each point on the string moves up and down—perpendicular to the direction of propagation—in simple harmonic motion with the frequency $f$ of the tuning fork. During one period $T = 1/f$, the wave moves a distance of one wavelength, so its speed is given by

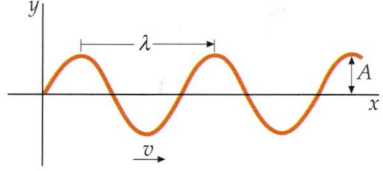

**FIGURE 15-7** Harmonic wave at some instant in time. $A$ is amplitude and $\lambda$ is the wavelength. For a wave on a string, this figure can be obtained by taking a photographic snapshot of the string.

$$v = \frac{\lambda}{T} = f\lambda \qquad\qquad 15\text{-}10$$

Since this relation arises only from the definitions of wavelength and frequency, it applies to all harmonic waves.

The sine function that describes the displacements in Figure 15-7 is

$$y(x) = A \sin\left(2\pi\frac{x}{\lambda} + \delta\right)$$

where $A$ is the amplitude, $\lambda$ is the wavelength, and $\delta$ is a phase constant that depends on the choice of the origin $x = 0$. This equation is expressed more simply as

$$y(x) = A \sin(kx + \delta) \qquad\qquad 15\text{-}11$$

where $k$, called the **wave number,** is given by

$$k = \frac{2\pi}{\lambda} \qquad\qquad 15\text{-}12$$

Note that $k$ has dimensions of m$^{-1}$. (Because the angle must be in radians, we sometimes write the units of $k$ as rad/m.) When dealing with a single harmonic wave we usually choose the location of the origin so that $\delta = 0$.

For a wave traveling in the direction of increasing $x$ with speed $v$, replace $x$ in Equation 15-11 with $x - vt$ (see "Wave Pulses" in Section 15-1). With $\delta$ chosen to be zero, this gives

$$y(x,t) = A \sin k(x - vt) = A \sin(kx - kvt)$$

or

$$y(x,t) = A \sin(kx - \omega t) \qquad 15\text{-}13$$

HARMONIC WAVE FUNCTION

where

$$\omega = kv \qquad 15\text{-}14$$

is the angular frequency, and the argument of the sine function, $(kx - wt)$, is called the **phase.** The angular frequency is related to the frequency $f$ and period $T$ by

$$\omega = 2\pi f = \frac{2\pi}{T} \qquad 15\text{-}15$$

Substituting $\omega = 2\pi f$ into Equation 15-14 and using $k = 2\pi/\lambda$, we obtain

$$2\pi f = kv = \frac{2\pi}{\lambda} v$$

or $v = f\lambda$, which is Equation 15-10.

If a harmonic wave traveling along a string is described by $y(x,t) = A \sin(kx - \omega t)$, the velocity of a point on the string at a fixed value of $x$ is

$$v_y = \frac{\partial y}{\partial t} = \frac{\partial}{\partial t}\left[A \sin(kx - \omega t)\right] = -\omega A \cos(kx - \omega t) \qquad 15\text{-}16$$

TRANSVERSE VELOCITY

The acceleration of this point is given by $\partial^2 y/\partial t^2$.

---

*A HARMONIC WAVE ON A STRING*       **E X A M P L E   1 5 - 4**

The wave function for a harmonic wave on a string is $y(x,t) = (0.03 \text{ m}) \times \sin[(2.2 \text{ m}^{-1})x - (3.5 \text{ s}^{-1})t]$. (*a*) In what direction does this wave travel and what is its speed? (*b*) Find the wavelength, frequency, and period of this wave. (*c*) What is the maximum displacement of any string segment? (*d*) What is the maximum speed of any short string segment?

**PICTURE THE PROBLEM** (*a*) To find the direction of travel, express $y(x,t)$ as either a function of $(x - vt)$ or as a function of $(x + vt)$ and use Equations 15-1 and 15-2. To find the wave speed use $\omega = kv$ (Equation 15-14). (*b*) The wavelength, frequency, and period can be found from the wave number $k$ and the angular frequency $\omega$. (*c*) The maximum displacement of a point on the string is the amplitude $A$. (*d*) The velocity of any short string segment is $\partial y/\partial t$.

(a) 1. The given wave function is of the form $y(x,t) = A \sin(kx - \omega t)$. Using $\omega = kv$ (Equation 15-14), write the wave function as a function of $x - vt$. Then use Equations 15-1 and 15-2 to find the direction of travel:

$y(x,t) = A \sin(kx - \omega t)$ and $\omega = kv$

so

$y(x,t) = A \sin(kx - kvt) = A \sin[k(x - vt)]$

The wave travels in the $\boxed{+x \text{ direction}}$

2. Since the form is $y = A \sin(kx - \omega t)$, we know $A$ as well as both $\omega$ and $k$. Use these to calculate the speed:

$v = \dfrac{\lambda}{T} = \dfrac{\lambda}{2\pi}\dfrac{2\pi}{T} = \dfrac{\omega}{k} = \dfrac{3.5 \text{ s}^{-1}}{2.2 \text{ m}^{-1}} = \boxed{1.59 \text{ m/s}}$

(b) The wavelength $\lambda$ is related to the wave number $k$, and the period $T$ and frequency $f$ are related to $\omega$:

$\lambda = \dfrac{2\pi}{k} = \dfrac{2\pi}{2.2 \text{ m}^{-1}} = \boxed{2.86 \text{ m}}$

$T = \dfrac{2\pi}{\omega} = \dfrac{2\pi}{3.5 \text{ s}^{-1}} = \boxed{1.80 \text{ s}}$

$f = \dfrac{1}{T} = \dfrac{1}{1.80 \text{ s}} = \boxed{0.557 \text{ Hz}}$

(c) The maximum displacement of a string segment is the amplitude $A$:

$A = \boxed{0.03 \text{ m}}$

(d) 1. Compute $\partial y / \partial t$ to find the velocity of a point on the string:

$v_y = \dfrac{\partial y}{\partial t} = (0.03 \text{ m})\dfrac{\partial\left[\sin(2.2 \text{ m}^{-1} x - 3.5 \text{ s}^{-1} t)\right]}{\partial t}$

$= (0.03 \text{ m})(-3.5 \text{ s}^{-1}) \cos(2.2 \text{ m}^{-1} x - 3.5 \text{ s}^{-1} t)$

$= -(0.105 \text{ m/s}) \cos(2.2 \text{ m}^{-1} x - 3.5 \text{ s}^{-1} t)$

2. The maximum transverse speed occurs when the cosine function has the value of $\pm 1$:

$v_{y,\text{max}} = \boxed{0.105 \text{ m/s}}$

**REMARKS**  We have included the units explicitly to show how they work out. Often we will omit the units for simplicity.

### Energy Transfer via Waves on a String

Consider again a string attached to a tuning fork. As the fork vibrates, it transfers energy to the segment of the string attached to it. For example, as the fork moves through its equilibrium position, it stretches the adjacent string segment slightly, increasing its potential energy, and the fork imparts a transverse speed to it, increasing its kinetic energy. As a wave moves along the string, energy is transferred to the other segments of the string.

Power is the rate of energy transfer. We can calculate the power by considering work done by the force that one segment of the string exerts on a neighboring segment. The rate of work done by this force is the power. Figure 15-8 shows a harmonic wave moving to the right along a string segment. The tension force $\vec{F}_T$ on the left end of the segment is directed tangent to the string as shown. To calculate the power transferred by this force we use the formula $P = \vec{F}_T \cdot \vec{v}_t$ (Equation 6-16), where $F_T$ is the tension and $\vec{v}_t$, the transverse velocity, is the velocity of the end of the segment. To obtain an expression for the power we first express the vectors in component form. That is, $\vec{F}_T = F_{Tx}\hat{i} + F_{Ty}\hat{j}$ and $\vec{v}_t = v_y\hat{j}$, so $P = F_{Ty}v_y$. We obtain $v_y$ from Equation 15-16. From the figure we see that $F_{Ty} = -F_T \sin\theta \approx -F_T \tan\theta$, where we have used the small angle approximation $\sin\theta \approx \tan\theta$. Since $\tan\theta$ is the slope of the string, we have $\tan\theta = \partial y/\partial x$. Thus

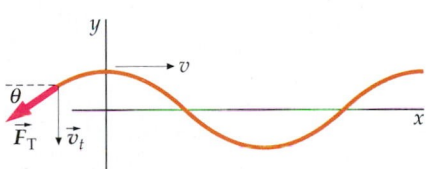

**FIGURE 15-8**

$P = F_{Ty}v_y \approx -F_T \tan\theta v_y = -F_T \dfrac{\partial y}{\partial x}\dfrac{\partial y}{\partial t}$

$= -F_T[kA\cos(kx - \omega t)][-A\omega\cos(kx - \omega t)]$

Using Equations 15-3 and 15-14, we substitute for $F_T$ and the leading $k$ to obtain

$$P = \mu v \omega^2 A^2 \cos^2(kx - \omega t) \qquad\qquad 15\text{-}17$$

where $v$ is the wave speed. The average power is

$$P_{av} = \tfrac{1}{2}\mu v \omega^2 A^2 \qquad\qquad 15\text{-}18$$

**FIGURE 15-9**

since the average value of $\cos^2(kx - \omega t)$ is $\tfrac{1}{2}$ if the average is taken over an entire period of the motion and $x$ remains constant.

The energy travels along a string at the wave speed $v$, so the average energy $(\Delta E)_{av}$ flowing past point $P_1$ during time $\Delta t$ (Figures 15-9a and 15-9b) is

$$(\Delta E)_{av} = P_{av}\,\Delta t = \tfrac{1}{2}\mu v \omega^2 A^2\,\Delta t$$

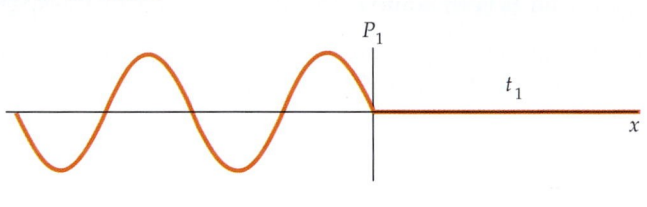

(a)

This energy is distributed over a length $\Delta x = v\,\Delta t$ so the average energy in length $\Delta x$ is

$$(\Delta E)_{av} = \tfrac{1}{2}\mu \omega^2 A^2\,\Delta x \qquad\qquad 15\text{-}19$$

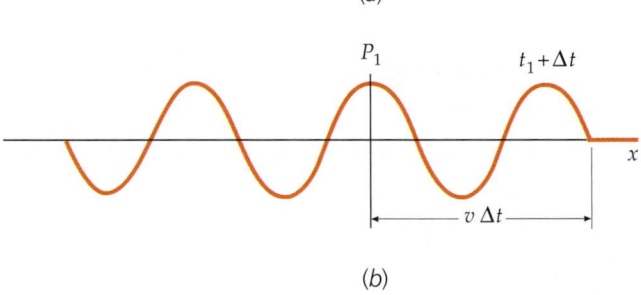

(b)

Note that both the average energy and the power transmitted are proportional to the square of the amplitude of the wave.

---

*AVERAGE TOTAL ENERGY OF A WAVE ON A STRING*  **E X A M P L E   1 5 - 5**

**A harmonic wave of wavelength 25 cm and amplitude 1.2 cm moves along a 15-m-long segment of a 60-m-long string that has a mass of 320 g and a tension of 12 N. (a) What is the speed and angular frequency of the wave? (b) What is the average total energy of the wave?**

**PICTURE THE PROBLEM** The speed of the waves is $v = \sqrt{F_T/\mu}$, where $F_T$ is given and $\mu = m/L$. We find $\omega$ from $\omega = 2\pi f$, where $f = v/\lambda$. The energy is found using Equation 15-19.

(a) 1. The speed is related to the tension and mass density:

$$v = \sqrt{\frac{F_T}{\mu}}$$

2. Calculate the linear mass density:

$$\mu = \frac{m}{L}$$

so

$$v = \sqrt{\frac{F_T}{\mu}} = \sqrt{\frac{F_T L}{m}} = \sqrt{\frac{(12\text{ N})(60\text{ m})}{(0.32\text{ kg})}} = \boxed{47.4\text{ m/s}}$$

3. The angular frequency is found from the frequency, which is found from the speed and wavelength:

$$\omega = 2\pi f = 2\pi\frac{v}{\lambda} = 2\pi\frac{47.4\text{ m/s}}{0.25\text{ m}} = \boxed{1190\text{ rad/s}}$$

(b) The average total energy of waves on the string is given by Equation 15-19 with $\mu\,\Delta x = m = 80$ g:

$$(\Delta E)_{av} = \frac{1}{2}\mu \omega^2 A^2 \Delta x = \frac{1}{2}\frac{m}{L}\omega^2 A^2 \Delta x$$

$$= \frac{1}{2}\frac{0.32\text{ kg}}{60\text{m}}(1190\text{ s}^{-1})^2\,(0.012\text{ m})^2\,(15\text{ m})$$

$$= \boxed{8.19\text{ J}}$$

**EXERCISE** Calculate the rate at which energy is transmitted along the string.
(*Answer* 25.9 W)

## Harmonic Sound Waves

Harmonic sound waves can be generated by a tuning fork or loud-speaker that is vibrating with simple harmonic motion. The vibrating source causes the air molecules next to it to oscillate with simple harmonic motion about their equilibrium positions. These molecules collide with neighboring molecules, causing them to oscillate, thereby propagating the sound wave. Equation 15-13 describes a harmonic sound wave if the wave function $y(x,t)$ is replaced by $s(x,t)$, the displacement of the molecules from equilibrium:

$$s(x, t) = s_0 \sin(kx - \omega t) \qquad 15\text{-}20$$

These displacements are along the direction of the motion of the wave, and lead to variations in the density and pressure of the air. Figure 15-10 shows the displacement of air molecules and the density changes caused by a sound wave at some fixed time. Because the pressure in a gas is proportional to its density, the change in pressure is maximum where the change in density is maximum. We see from this figure that the pressure or density wave is 90° out of phase with the displacement wave. (In the arguments of sine or cosine functions we will always express phase angles in radians. However, in verbal descriptions we usually say that "two waves are 90° out of phase" rather than "two waves are out of phase by $\pi/2$ rad.) When the displacement is zero, the pressure and density changes are either maximum or minimum. When the displacement is a maximum or minimum, the pressure and density changes are zero. A displacement wave given by Equation 15-20 thus implies a pressure wave given by

$$p = p_0 \sin\left(kx - \omega t - \frac{\pi}{2}\right) \qquad 15\text{-}21$$

where $p$ stands for the *change* in pressure from the equilibrium pressure and $p_0$, the maximum value of this change, is called the pressure amplitude. It can be shown that the pressure amplitude $p_0$ is related to the displacement amplitude $s_0$ by

$$p_0 = \rho \omega v s_0 \qquad 15\text{-}22$$

where $v$ is the speed of propagation and $\rho$ is the equilibrium density of the gas. Thus, as a sound wave moves in time, the displacement of air molecules, the pressure, and the density all vary sinusoidally with the frequency of the vibrating source.

**EXERCISE** We can hear sound of frequencies from about 20 Hz to about 20,000 Hz (although many people have rather limited hearing above 15,000 Hz). If the speed of sound in air is 340 m/s, what are the wavelengths that correspond to these extreme frequencies? (*Answer* $\lambda = 17$ m at 20 Hz, 1.7 cm at 20,000 Hz)

**Energy of Sound Waves** The average energy of a harmonic sound wave in a volume element $\Delta V$ is given by Equation 15-19 with $A$ replaced by $s_0$ and $\mu \, \Delta x$ replaced by $\rho \, \Delta V$, where $\rho$ is the average density of the medium.

$$(\Delta E)_{\text{av}} = \tfrac{1}{2} \rho \omega^2 s_0^2 \, \Delta V \qquad 15\text{-}23$$

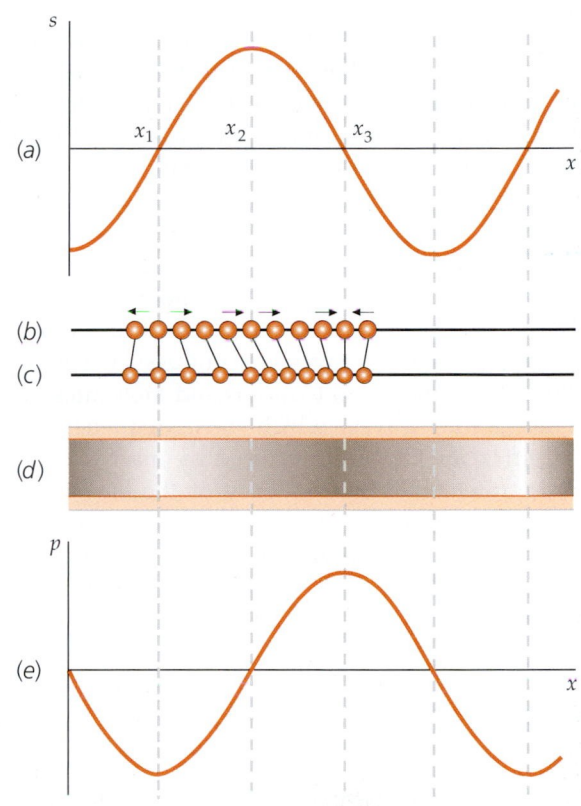

**FIGURE 15-10** (*a*) Displacement from equilibrium of air molecules in a harmonic sound wave versus position at some instant. Points $x_1$ and $x_3$ are points of zero displacement. (*b*) Some representative molecules equally spaced at their equilibrium positions 1/4 cycle earlier. The arrows indicate the directions of their velocities at that instant. (*c*) Molecules near points $x_1$, $x_2$, and $x_3$ after the sound wave arrives. Just to the left of $x_1$, the displacement is negative, indicating that the gas molecules are displaced to the left, away from point $x_1$, at this time. Just to the right of $x_1$, the displacement is positive, indicating that the molecules are displaced to the right, which is again away from point $x_1$. So at point $x_1$, the density is a minimum because the gas molecules on both sides are displaced away from that point. At point $x_3$, the density is a maximum because the molecules on both sides of that point are displaced toward point $x_3$. At point $x_2$, the density does not change because the gas molecules on both sides of that point have equal displacements in the same direction. (*d*) Density of the air at this instant. The density is maximum at $x_3$ and minimum at $x_1$, which are both points of zero displacement. It is equal to the equilibrium value at point $x_2$, which is a maximum in displacement. (*e*) Pressure change, which is proportional to the density change, versus position. The pressure change and displacement (position change) are 90° out of phase.

The energy per unit volume is the average **energy density** $\eta_{av}$:

$$\eta_{av} = \frac{\Delta E_{av}}{\Delta V} = \frac{1}{2}\rho\omega^2 s_0^2 \qquad\qquad 15\text{-}24$$

## Electromagnetic Waves

Electromagnetic waves include light, radio waves, X rays, gamma rays, and microwaves, among others. The various types of electromagnetic waves differ only in wavelength and frequency. Unlike mechanical waves, electromagnetic waves do not require a medium for propagation. They travel through a vacuum with speed $c$, which is a universal constant, $c \approx 3 \times 10^8$ m/s. The wave function for electromagnetic waves is an electric field associated with the wave, $\vec{E}(x,t)$. (Electric fields are discussed in Chapter 21. A wave equation, similar to those for string waves and sound waves, is derived from the laws of electricity and magnetism in Chapter 30.) The electric field is perpendicular to the direction of propagation, so electromagnetic waves are transverse waves.

Electromagnetic waves are produced when free electric charges accelerate or when electrons bound to atoms and molecules make transitions to lower energy states. Radio waves, which have frequencies of about 1 MHz for AM and 100 MHz for FM, are produced by macroscopic electric currents oscillating in radio antennas. The frequency of the emitted waves equals the frequency of oscillation of the charges. Light waves, which have frequencies of the order of $10^{14}$ Hz, are generally produced by atomic or molecular transitions involving bound electrons. The spectrum of electromagnetic waves is discussed in Chapter 31.

## 15-3 Waves in Three Dimensions

Figure 15-11 shows two-dimensional circular waves on the surface of water in a ripple tank. These waves are generated by a point source moving up and down with simple harmonic motion. The wavelength is the distance between successive wave crests, which in this case are concentric circles. These circles are called **wavefronts.** For a point source of sound, the waves move out in three dimensions, and the wavefronts are concentric spherical surfaces.

The motion of any set of wavefronts can be indicated by **rays,** which are directed lines perpendicular to the wavefronts (Figure 15-12). For circular or spherical waves, the rays are radial lines.

**FIGURE 15-11**
Circular wavefronts diverging from a point source in a ripple tank.

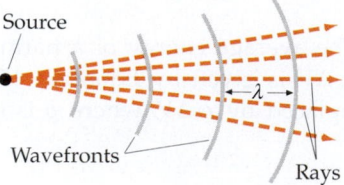

Source

Wavefronts

Rays

$\lambda$

**FIGURE 15-12** The motion of wavefronts can be represented by rays drawn perpendicular to the wavefronts. For a point source, the rays are radial lines diverging from the source.

In a homogeneous medium, such as air at constant density, a wave travels in straight lines in the direction of the rays, much like a beam of particles. At a great distance from a point source, a small part of the wavefront can be approximated by a plane, and the rays are approximately parallel lines; such a wave is called a **plane wave** (Figure 15-13). The two-dimensional analog of a plane wave is a line wave, which is a small part of a circular wavefront at a great distance from the source. Such waves can also be produced in a ripple tank by a line source, as in Figure 15-14.

## Wave Intensity

If a point source emits waves uniformly in all directions, then the energy at a distance $r$ from the source is distributed uniformly on a spherical surface of radius $r$ and area $A = 4\pi r^2$. If $P$ is the power emitted by the source, then the power per unit area at a distance $r$ from the source is $P/(4\pi r^2)$. The average power per unit area that is incident perpendicular to the direction of propagation is called the **intensity:**

**FIGURE 15-13** Plane waves. At great distances from a point source, the wavefronts are approximately parallel planes, and the rays are approximately parallel lines perpendicular to the wavefronts.

$$I = \frac{P_{av}}{A} \qquad 15\text{-}25$$

INTENSITY DEFINED

The units of intensity are watts per square meter. At a distance $r$ from a point source, the intensity is

$$I = \frac{P_{av}}{4\pi r^2} \qquad 15\text{-}26$$

INTENSITY DUE TO A POINT SOURCE

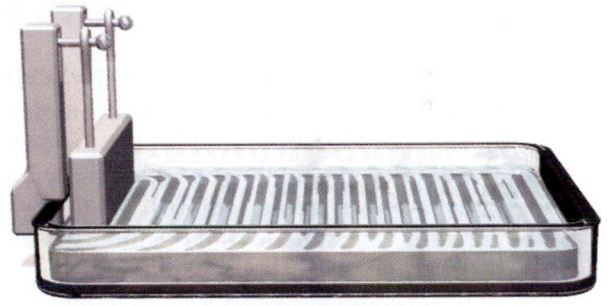

The intensity of a three-dimensional wave varies inversely with the square of the distance from a point source.

There is a simple relation between the intensity of a wave and the energy density in the medium through which it propagates. Figure 15-15 shows a spherical wave that has just reached the radius $r_1$. The volume inside the radius $r_1$ contains energy because the particles in that region are oscillating with simple harmonic motion. The region outside $r_1$ contains no energy because the wave has not yet reached it. After a short time $\Delta t$, the wave moves out a short distance $\Delta r = v\,\Delta t$ past $r_1$. The average energy in the spherical shell of surface area $A$, thickness $v\,\Delta t$, and volume $\Delta V = A\,\Delta r = Av\,\Delta t$ is

**FIGURE 15-14** A two-dimensional analog of a plane wave can be generated in a ripple tank by a flat board that oscillates up and down in the water to produce the wavefronts, which are straight lines.

$$(\Delta E)_{av} = \eta_{av}\,\Delta V = \eta_{av}\,Av\,\Delta t$$

The rate of transfer of energy is the power passing into the shell. The average incident power is

$$P_{av} = \frac{(\Delta E)_{av}}{\Delta t} = \eta_{av}Av$$

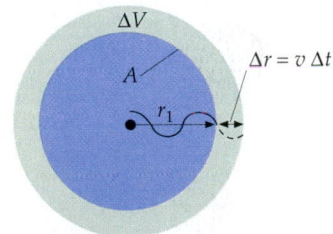

Volume of shell $= \Delta V = A\,\Delta r = Av\,\Delta t$

**FIGURE 15-15**

and the intensity of the wave is

$$I = \frac{P_{av}}{A} = \eta_{av}v \qquad 15\text{-}27$$

Thus the intensity equals the product of the wave speed $v$ and the average energy density $\eta_{av}$. Substituting $\eta_{av} = \frac{1}{2}\rho\omega^2 s_0^2$ from Equation 15-24 for the energy density in a sound wave, we obtain

$$I = \eta_{av}v = \frac{1}{2}\rho\omega^2 s_0^2 v = \frac{1}{2}\frac{p_0^2}{\rho v} \qquad\qquad 15\text{-}28$$

where we have used $s_0 = p_0/(\rho\omega v)$ from Equation 15-22. This result—that the intensity of a sound wave is proportional to the square of the amplitude—is a general property of harmonic waves.

The human ear can accommodate a large range of sound-wave intensities, from about $10^{-12}$ W/m$^2$ (which is usually taken to be the threshold of hearing) to about 1 W/m$^2$ (an intensity great enough to stimulate pain in most people). The pressure variations that correspond to these extreme intensities are about $3 \times 10^{-5}$ Pa for the hearing threshold and 30 Pa for the pain threshold. (Recall that a pascal is a newton per square meter.) These very small pressure variations add to or subtract from the normal atmospheric pressure of about 101 kPa.

Sound waves from a telephone handset spreading out in the air. The waves have been made visible by sweeping out the space in front of the handset with a light source whose brightness is controlled by a microphone.

---

*A LOUDSPEAKER*

## EXAMPLE 15-6

A loudspeaker diaphragm 30 cm in diameter is vibrating at 1 kHz with an amplitude of 0.020 mm. Assuming that the air molecules in the vicinity have the same amplitude of vibration, find (*a*) the pressure amplitude immediately in front of the diaphragm, (*b*) the sound intensity in front of the diaphragm, and (*c*) the acoustic power being radiated. (*d*) If the sound is radiated uniformly into the forward hemisphere, find the intensity at 5 m from the loudspeaker.

**PICTURE THE PROBLEM** (*a*) and (*b*) The pressure amplitude is calculated directly from $p_0 = \rho\omega v s_0$ (Equation 15-22), and the intensity from $I = \frac{1}{2}\rho\omega^2 s_0^2 v$ (Equation 15-28). (*c*) The power radiated is the intensity times the area of the diaphragm. (*d*) The area of a hemisphere of radius $r$ is $2\pi r^2$. We can use Equation 15-25 with $A = 2\pi r^2$.

(*a*) Equation 15-22 relates the pressure amplitude to the displacement amplitude, frequency, wave velocity, and air density:

$$p_0 = \rho\omega v s_0$$
$$= (1.29 \text{ kg/m}^3)2\pi(10^3 \text{ Hz})(340 \text{ m/s})(2 \times 10^{-5} \text{ m})$$
$$= \boxed{55.1 \text{ N/m}^2}$$

(*b*) Equation 15-28 relates the intensity to these same known quantities:

$$I = \frac{1}{2}\rho\omega^2 s_0^2 v$$
$$= \frac{1}{2}(1.29 \text{ kg/m}^3)[2\pi(10^3 \text{ Hz})]^2 (2 \times 10^{-5} \text{ m})^2 (340 \text{ m/s})$$
$$= \boxed{3.46 \text{ W/m}^2}$$

(*c*) The power is the intensity times the area of the diaphragm:

$$P = IA = (3.46 \text{ W/m}^2)\pi(0.15 \text{ m})^2 = \boxed{0.245 \text{ W}}$$

(*d*) Calculate the intensity at $r = 5$ m, assuming uniform radiation into the forward hemisphere:

$$I = \frac{P_{av}}{A} = \frac{0.245 \text{ W}}{2\pi(5 \text{ m})^2} = \boxed{1.56 \times 10^{-3} \text{ W/m}^2}$$

---

**REMARKS** The assumption of uniform radiation in the forward hemisphere is not a very good one because the wavelength in this case [$\lambda = v/f = (340 \text{ m/s})/(1000\text{s}^{-1}) = 34$ cm] is not large compared with the speaker diameter. There is also some radiation in the backward direction, as can be observed if you stand behind a loudspeaker.

Loudspeakers at a rock concert may put out more than 100 times as much power as the speaker in this example.

**\*Intensity Level and Loudness** Our perception of loudness is not proportional to the intensity but varies logarithmically. We therefore use a logarithmic scale to describe the **intensity level** $\beta$ of a sound wave, which is measured in **decibels** (dB) and defined by

$$\beta = 10 \log \frac{I}{I_0}$$ 15-29

DEFINITION—INTENSITY LEVEL IN DB

Here $I$ is the intensity of the sound and $I_0$ is a reference level, which usually is taken to be the threshold of hearing:

$$I_0 = 10^{-12} \text{ W/m}^2$$ 15-30

On this scale, the threshold of hearing is $\beta = 10 \log(I/I_0) = 0$ dB and the pain threshold ($I = 1$ W/m$^2$) is $\beta = 10 \log(1/10^{-12}) = 10 \log 10^{12} = 120$ dB. Thus, the range of sound intensities from $10^{-12}$ W/m$^2$ to 1 W/m$^2$ corresponds to intensity levels from 0 dB to 120 dB. Table 15-1 lists the intensity levels of some common sounds.

---

*SOUNDPROOFING*                                    **EXAMPLE 15-7**

**A sound absorber attenuates the sound level by 30 dB. By what factor is the intensity decreased?**

From Table 15-1, we can see that for every 10-dB drop in the intensity level, the intensity decreases by a factor of 10. Thus, if the sound level drops 30 db then the intensity drops by a factor of $10^3 = \boxed{1000}$.

---

## TABLE 15-1

**Intensity and Intensity Level of Some Common Sounds ($I_0 = 10^{-12}$ W/m$^2$)**

| Source | $I/I_0$ | dB | Description |
|---|---|---|---|
| | $10^0$ | 0 | Hearing threshold |
| Normal breathing | $10^1$ | 10 | Barely audible |
| Rustling leaves | $10^2$ | 20 | |
| Soft whisper (at 5 m) | $10^3$ | 30 | Very quiet |
| Library | $10^4$ | 40 | |
| Quiet office | $10^5$ | 50 | Quiet |
| Normal conversation (at 1 m) | $10^6$ | 60 | |
| Busy traffic | $10^7$ | 70 | |
| Noisy office with machines; average factory | $10^8$ | 80 | |
| Heavy truck (at 15 m); Niagara Falls | $10^9$ | 90 | Constant exposure endangers hearing |
| Old subway train | $10^{10}$ | 100 | |
| Construction noise (at 3 m) | $10^{11}$ | 110 | |
| Rock concert with amplifiers (at 2 m); jet takeoff (at 60 m) | $10^{12}$ | 120 | Pain threshold |
| Pneumatic riveter; machine gun | $10^{13}$ | 130 | |
| Jet takeoff (nearby) | $10^{15}$ | 150 | |
| Large rocket engine (nearby) | $10^{18}$ | 180 | |

The sensation of loudness depends on the frequency as well as the intensity of a sound. Figure 15-16 is a plot of intensity level versus frequency for sounds of equal loudness to the human ear. (In this figure, the frequency is plotted on a logarithmic scale to display the wide range of frequencies from 20 Hz to 10 kHz.) We note from this figure that the human ear is most sensitive at about 4 kHz for all intensity levels.

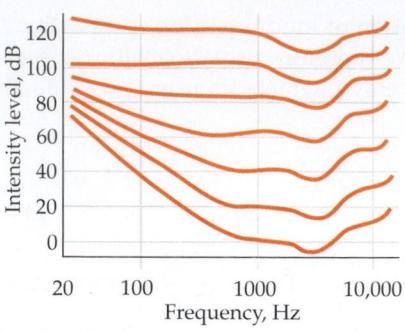

**FIGURE 15-16** Intensity level versus frequency for sounds perceived to be of equal loudness. The lowest curve is below the threshold for hearing of all but about 1 percent of the population. The second lowest curve is approximately the hearing threshold for about 50 percent of the population.

---

*BARKING DOGS* **EXAMPLE 15-8**

A barking dog delivers about 1 mW of power. (a) If this power is uniformly distributed in all directions, what is the sound intensity level at a distance of 5 m? (b) What would be the intensity level of two dogs barking at the same time if each delivered 1 mW of power?

**PICTURE THE PROBLEM** The intensity level is found from the intensity, which is found from $I = P/(4\pi r^2)$. For two dogs, the intensities are added.

(a) 1. The intensity level is related to the intensity:

$$\beta = 10 \log \frac{I}{I_0}$$

2. Calculate the intensity at $r = 5$ m:

$$I_1 = \frac{P_1}{4\pi r^2} = \frac{10^{-3}\ \text{W}}{4\pi (5\ \text{m})^2} = 3.18 \times 10^{-6}\ \text{W/m}^2$$

3. Use your result to find the intensity level at 5 m:

$$\beta_1 = 10 \log \frac{I_1}{I_0} = 10 \log \frac{3.18 \times 10^{-6}}{10^{-12}} = \boxed{65.0\ \text{dB}}$$

(b) If $I_1$ is the intensity for one dog, the intensity for two dogs is $I_2 = 2I_1$:

$$\beta_2 = 10 \log \frac{I_2}{I_0} = 10 \log \frac{2I_1}{I_0} = 10 \log 2 + 10 \log \frac{I_1}{I_0}$$

$$= 10 \log 2 + \beta_1 = 3.01 + 65.0 = \boxed{68.0\ \text{dB}}$$

**REMARKS** We can see from this example that whenever the intensity is doubled, the intensity level increases by 3 dB.

# 15-4 Waves Encountering Barriers

## Reflection and Refraction

When a wave is incident on a boundary that separates two regions of differing wave speed, part of the wave is reflected and part is transmitted. Figure 15-17a shows a pulse on a light string that is attached to a heavier string. In this case, the pulse reflected at the boundary is inverted. If the second string is lighter than the first (Figure 15-17b), then the reflected pulse is not inverted. In either case,

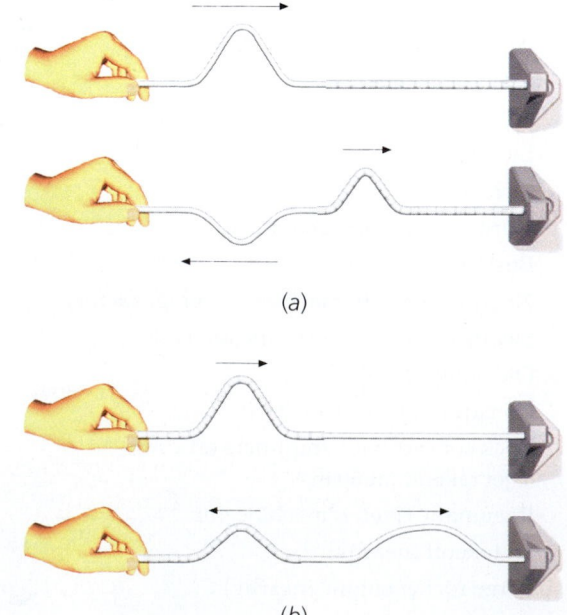

(a)

(b)

**FIGURE 15-17** (a) A wave pulse traveling on a light string attached to a heavier string in which the wave speed is smaller. The reflected pulse is inverted, whereas the transmitted pulse is not. (b) A wave pulse traveling on a heavy string attached to a light string in which the wave speed is greater. In this case, the reflected pulse is not inverted.

the transmitted pulse is not inverted. If the string is tied to a fixed point, then the pulse is reflected and inverted. If it is tied to a string of negligible mass, then the pulse is reflected, but not inverted.

*TWO SOLDERED WIRES*                    **EXAMPLE 15-9**

**Two wires of different linear mass densities are soldered together end to end and then stretched under a tension $F_T$ (the tension is the same in both wires). The wave speed in the first wire is twice that in the second. If a harmonic wave traveling in the first wire is incident on the junction of the wires, the amplitude of the reflected wave is half the amplitude of the transmitted wave. (a) If the amplitude of the incident wave is $A$, what are the amplitudes of the reflected and transmitted waves? (b) What fraction of the incident power is reflected at the junction and what fraction is transmitted?**

**PICTURE THE PROBLEM** By conservation of energy, the power incident on the junction equals the power reflected plus the power transmitted. Each power is expressed in Equation 15-18 as a function of the density $\mu$, amplitude $A$, frequency $\omega$, and wave speed $v$ (Figure 15-18). The angular frequencies of all the waves are equal. Since the reflected wave and incident wave are in the same medium, they have the same wave speed $v_1$. We are given that the speed in the second wire is $v_2 = \frac{1}{2} v_1$.

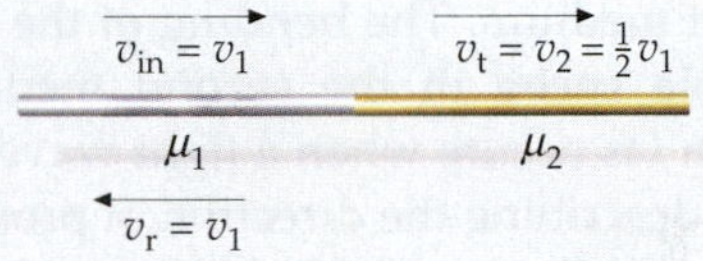

$v_{in} = v_1$ $\qquad$ $v_t = v_2 = \frac{1}{2} v_1$

$\mu_1$ $\qquad$ $\mu_2$

$v_r = v_1$

**FIGURE 15-18**

(a) 1. By conservation of energy, the incident power equals the transmitted power plus the reflected power:

$$P_{in} = P_t + P_r$$

2. Write Equation 15-18:

$$P_{av} = \tfrac{1}{2} \mu v \omega^2 A^2$$

3. Substitute into the step 1 result and simplify: The angular frequency is the same for all three waves.

$$\tfrac{1}{2} \mu_1 \omega^2 A_{in}^2 v_1 = \tfrac{1}{2} \mu_2 \omega^2 A_t^2 v_2 + \tfrac{1}{2} \mu_1 \omega^2 A_r^2 v_1$$

$$\mu_1 A_{in}^2 v_1 = \mu_2 A_t^2 v_2 + \mu_1 A_r^2 v_1$$

4. Using the relation $v = \sqrt{F_T/\mu}$ (Equation 15-3), substitute for $\mu_1$ and $\mu_2$ and simplify. $F_T$ is the same on either side of the junction:

$$\frac{F_T}{v_1^2} A_{in}^2 v_1 = \frac{F_T}{v_2^2} A_t^2 v_2 + \frac{F_T}{v_1^2} A_r^2 v_1$$

$$\frac{A_{in}^2}{v_1} = \frac{A_t^2}{v_2} + \frac{A_r^2}{v_1}$$

5. Using the given relations $v_2 = \frac{1}{2} v_1$ and $A_r = \frac{1}{2} A_t$, substitute and solve for the amplitudes:

$$\frac{A_{in}^2}{v_1} = \frac{A_t^2}{\frac{1}{2} v_1} + \frac{(\frac{1}{2} A_t)^2}{v_1} = \frac{9}{4} \frac{A_t^2}{v_1}$$

so

$$A_t = \boxed{\tfrac{2}{3} A_{in}} \quad \text{and} \quad A_r = \boxed{\tfrac{1}{3} A_{in}}$$

(b) 1. In Part (a) steps 1–4 it was shown that the power is proportional to $A^2/v$. Express each of the three powers, using $b$ as the proportionality constant:

$$P_{in} = b \frac{A_{in}^2}{v_1} \qquad P_t = b \frac{A_t^2}{v_2} \qquad P_r = b \frac{A_r^2}{v_1}$$

2. Using the Part (a) step 5 results, eliminate $v_2$, $A_t$, and $A_r$ from the expressions for $P_t$ and $P_r$:

$$P_t = b \frac{(\tfrac{2}{3} A_{in})^2}{\frac{1}{2} v_1} = \frac{8}{9} b \frac{A_{in}^2}{v_1} = \boxed{\frac{8}{9} P_{in}}$$

$$P_r = b \frac{(\tfrac{1}{3} A_{in})^2}{v_1} = \frac{1}{9} b \frac{A_{in}^2}{v_1} = \boxed{\frac{1}{9} P_{in}}$$

**REMARKS** The reflected wave is inverted relative to the incident wave, so it is 180° out of phase with it. When the displacement of the wire just to the left of the junction would be $y_1$ due only to the incident wave, it would be $-(y_1/3)$ due only to the reflected wave. These add (according to the principle of superposition to be studied in the next chapter), giving a resultant displacement of $2y_1/3$, which is also the displacement that occurs just to the right of the junction due to the transmitted wave. It can be shown that, given the ratio of the wave speeds, the amplitudes of the transmitted and reflected waves can be determined by requiring that the displacement and the slope be continuous at the junction.

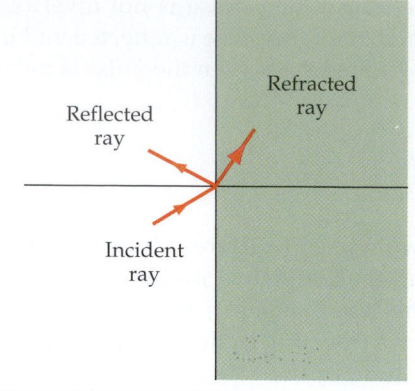

**FIGURE 15-19** A wave striking a boundary surface between two media in which the wave speed differs. Part of the wave is reflected and part is transmitted. The change in direction of the transmitted (refracted) ray is called refraction.

In three dimensions, a boundary between two regions of differing wave speed is a surface. Figure 15-19 shows a ray incident on such a boundary surface. This example could be a sound wave in air striking a solid or liquid surface. The reflected ray makes an angle with the normal to the surface equal to that of the incident ray, as shown.

The transmitted ray is bent toward or away from the normal—depending on whether the wave speed in the second medium is less or greater than that in the incident medium. The bending of the transmitted ray is called **refraction.** When the wave speed in the second medium is greater than that in the incident medium (as occurs when a light wave in glass or water is refracted into the air), the ray describing the direction of propagation is bent away from the normal, as shown in Figure 15-20. As the angle of incidence is increased, the angle of refraction increases, until a critical angle of incidence is reached for which the angle of refraction is 90°. For incident angles greater than the critical angle, there is no refracted ray, a phenomenon known as **total internal reflection.**

The amount of energy reflected from a surface depends on the surface. Flat walls, floors, and ceilings make good reflectors for sound waves, whereas porous and less rigid materials, such as cloth in draperies and furniture coverings, absorb much of the incident sound. The reflection of sound waves plays an important role in the design of a lecture hall, a library, or a music auditorium. If a lecture hall has many flat reflecting surfaces, speech is difficult to understand because of the many echoes that arrive at different times at the listener's ear. Absorbent material is often placed on the walls and ceiling to reduce such reflections. In a concert hall, a reflecting shell is placed behind the orchestra, and reflecting panels are hung from the ceiling to reflect and direct the sound back toward the listeners.

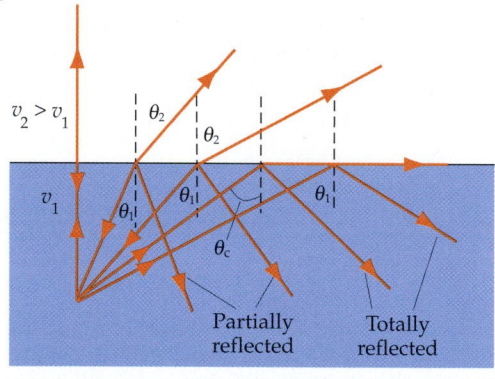

**FIGURE 15-20** Light from a source in the water is bent away from the normal when it enters the air. For angles of incidence above a critical angle, there is no transmitted ray, a condition known as total internal reflection.

## Diffraction

If a wavefront is partially blocked by an obstacle, the unblocked part of the wavefront bends behind the obstacle. This bending of the wavefronts is called **diffraction.** Almost all of the diffraction occurs for that part of the wavefront that passes within a few wavelengths of the edge of the obstacle. For the parts of the wavefront that pass farther than a few wavelengths from the edge, diffraction is negligible and the wave propagates in straight lines in the direction of the incident rays. When wavefronts encounter a barrier with an aperture (hole) only a few wavelengths across, the part of the wavefronts passing through the aperture all pass within a few wavelengths of an edge. Thus, flat

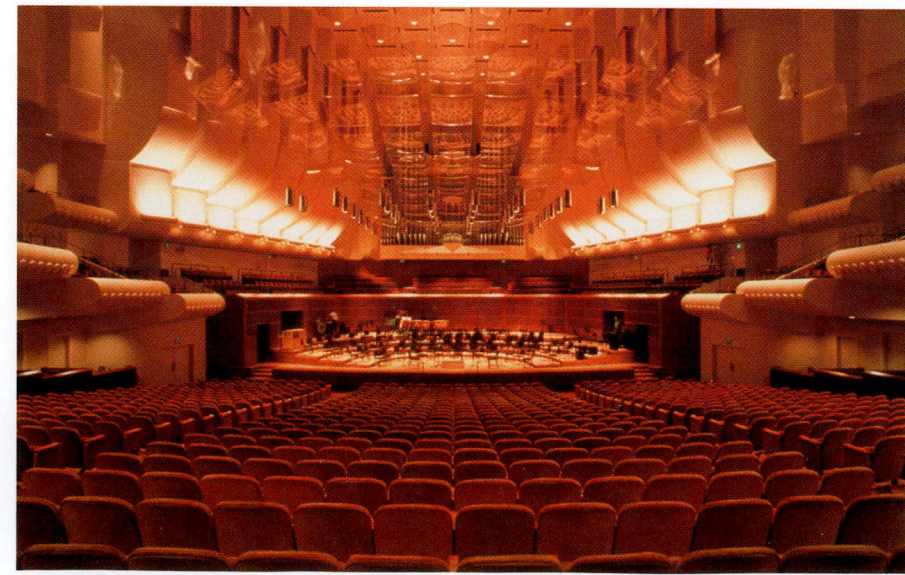

wavefronts bend and spread out and become spherical or circular (Figure 15-21). In contrast, for a beam of *particles* falling upon a barrier with an aperture, the part of the beam passing through the aperture does so with no change in the direction of the particles (Figure 15-22). Diffraction is one of the key characteristics that distinguishes waves from particles. We will discuss how diffraction arises when we study the interference and diffraction of light in Chapter 35.

Though waves passing through an aperture always bend, or diffract, to some extent, the amount of diffraction depends on whether the wavelength is small or large relative to the size of the aperture. If the wavelength is large relative to the aperture, as in Figure 15-21, the diffraction effects are large, and the waves spread out as they pass through the aperture—as if the waves were originating from a point source. On the other hand, if the wavelength is small relative to the aperture, the effect of diffraction is small, as shown in Figure 15-23. Near the edges of the aperture the wavefronts are distorted and the waves appear to bend slightly. For the most part, however, the wavefronts are not affected and the waves propagate in straight lines, much like a beam of particles. The approximation that waves propagate in straight lines in the direction of the rays with no diffraction is known as the **ray approximation**. Wavefronts are distorted *near* the edges of any obstacle blocking part of the wavefronts. By *near* we mean within a few wavelengths of the edges.

Because the wavelengths of audible sound (which range from a few centimeters to several meters) are generally large compared with apertures and obstacles (doors or windows, and people, for example), diffraction of sound waves is a phenomenon that is often observed. On the other hand, the wavelengths of visible light ($4 \times 10^{-7}$ to $7 \times 10^{-7}$ m) are so small compared with the size of ordinary objects and apertures that the diffraction of light is not easily noticed; light appears to travel in straight lines. Nevertheless, the diffraction of light is an important phenomena, one we will study in detail in Chapter 35.

Diffraction places a limitation on how accurately small objects can be located by reflecting waves off them and on how well details of the objects can be resolved. Waves are not reflected appreciably from objects smaller than the wavelength, so detail cannot be observed on a scale smaller than the wavelength used. If waves of wavelength $\lambda$ are used to locate an object, then its position can be known only to within an uncertainty of one wavelength.

**FIGURE 15-21** Plane waves in a ripple tank meeting a barrier with an opening that is only a few wavelengths wide. Beyond the barrier are circular waves that are concentric about the opening, just as if there were a point source at the opening.

**FIGURE 15-23** Plane waves in a ripple tank meeting a barrier with an opening width that is large compared to $\lambda$. The wave continues in the forward direction, with only a small amount of spreading into the regions to either side of the opening.

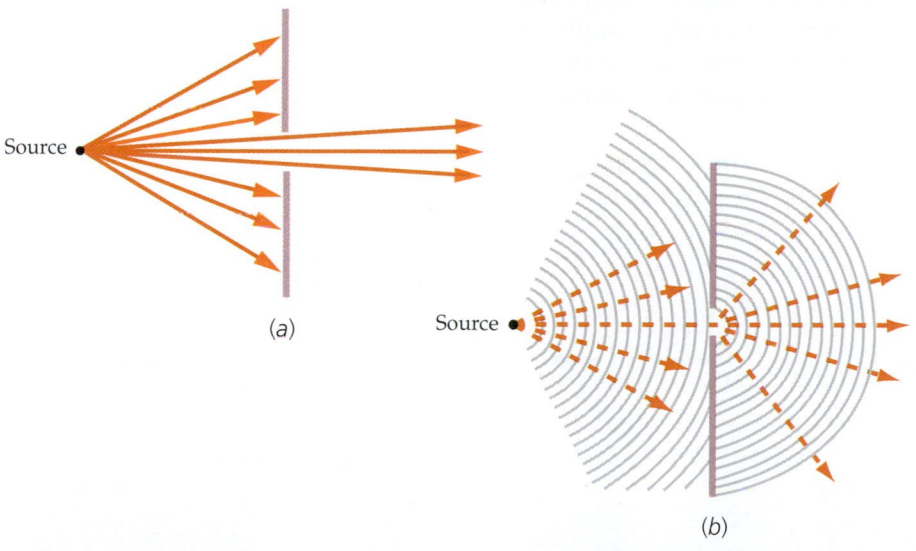

**FIGURE 15-22** Comparison of particles and waves passing through a narrow opening in a barrier. (*a*) Transmitted particles are confined to a narrow-angle beam. (*b*) Transmitted waves spread out (radiate widely) from the aperture, which acts like a point source of circular waves.

Sound waves with frequencies above 20,000 Hz are called **ultrasonic waves.** Because of their very small wavelengths, narrow beams of ultrasonic waves can be sent out and reflected from small objects. Bats can emit and detect frequencies up to about 120,000 Hz, corresponding to a wavelength of 2.8 mm, which they use to locate small prey such as moths. Echolocation systems, called sonar (from *sound* and *na*vigation *r*anging), are used to detect the outlines of submerged objects with sound waves. The frequency used by commercially available fish finders ranges from about 25 to 200 kHz, and porpoises produce echolocation clicks in the same frequency range. In medicine, ultrasonic waves are used for diagnostic purposes. Ultrasonic waves are passed through the human body and information about the frequency and intensity of the transmitted and reflected waves is processed to construct a three-dimensional picture of the body's interior, called a sonogram.

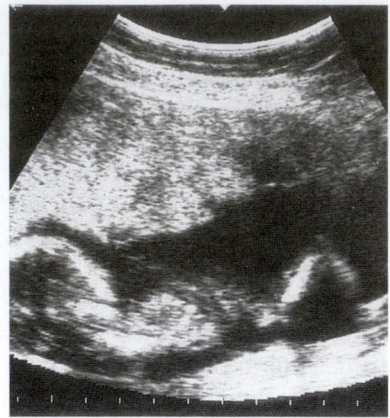

# 15-5 The Doppler Effect

If a wave source and a receiver are moving relative to each other, the received frequency is not the same as the frequency of the source. If they are moving closer together, the received frequency is greater than the source frequency; and if they are moving farther apart, the received frequency is less than the source frequency. This is called the **Doppler effect.** A familiar example is the drop in pitch of the sound of the horn of an approaching car as the car passes by—and then recedes.

In the following discussion, all motions are relative to the medium. Consider the source moving with speed $u_s$, shown in Figure 15-24a and b, and a stationary receiver. The source has frequency $f_s$ (and period $T_s = 1/f_s$). The received frequency $f_r$, the number of wave crests passing the receiver per unit time, is

$$f_r = \frac{v}{\lambda} \qquad \text{(stationary receiver)} \qquad \qquad 15\text{-}31$$

where $v$ is the wave speed and $\lambda$ is the wavelength (the distance between successive crests). To find $f_r$ we first need to find $\lambda$. Consider event 1—a wave crest leaves the source—and event two—the next wave crest leaves the source—as shown in Figure 15-24c. The time between these two events is $T_s$, and between these events the crest leaving the source first travels a distance $vT_s$ while the source itself travels a distance $u_sT_s$. Consequently, at the time of the second event, the distance between the source and the crest leaving first equals the wavelength $\lambda$.

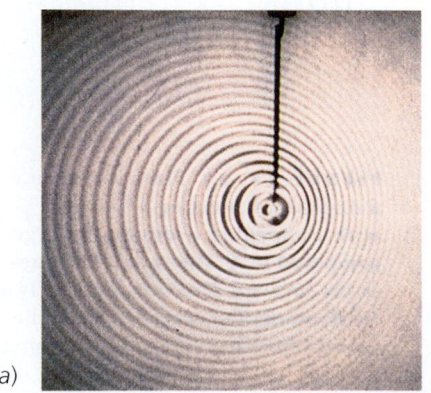

(a)

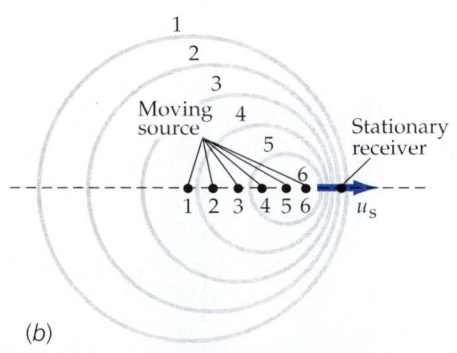

(b)

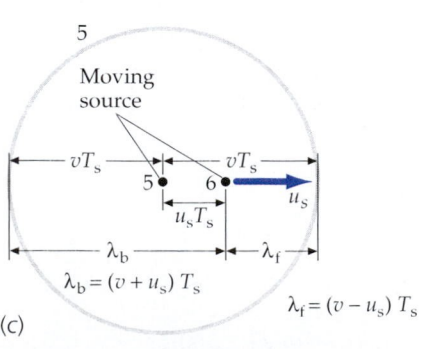

(c)

**FIGURE 15-24** (a) Waves in a ripple tank produced by a point source moving to the right. The wavefronts are closer together in front of the source and farther apart behind the source. (b) Successive wavefronts emitted by a point source moving with speed $u_s$ to the right. The numbers of the wavefronts correspond to the positions of the source when the wave was emitted. (c) The source vibrates one cycle in time $T_s$. During time $T_s$ the source moves a distance $u_sT_s$ and the 5th wavefront travels a distance $vT_s$. In front of the source the wavelength $\lambda_f = (v - u_s)T_s$, while behind the source $\lambda_b = (v + u_s)T_s$.

Behind the source $\lambda = \lambda_b = (v + u_s)T_s$, and in front of the source $\lambda = \lambda_f = (v - u_s)T_s$, provided $u_s < v$. (If $u_s \geq v$, no wavefronts reach the region ahead of the source.) We can express both these wavelengths as

$$\lambda = (v \pm u_s)T_s = \frac{v \pm u_s}{f_s} \qquad\qquad 15\text{-}32$$

where we have substituted $1/f_s$ for $T_s$. In front of the source the wavelength is shortest, so the minus sign applies. Behind the source the plus sign applies. Substituting for $\lambda$ in Equation 15-31 gives

$$f_r = \frac{v}{\lambda} = \frac{v}{v \pm u_s}f \qquad \text{(stationary receiver)} \qquad\qquad 15\text{-}33$$

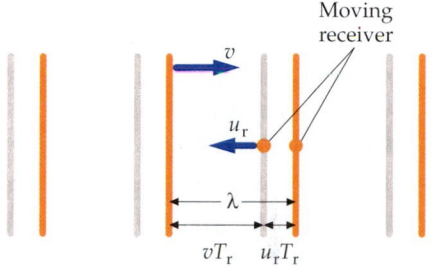

When the receiver moves relative to the medium, the received frequency is different simply because the receiver moves past more or fewer wave crests in a given time. For a receiver moving with speed $u_r$, let $T_r$ denote the time between arrivals of successive crests. Then, during the time between the arrivals of two successive crests, each crest will have traveled a distance $vT_r$, and during the same time the receiver will have traveled a distance $u_rT_r$. If the receiver moves in the direction opposite to that of the wave (Figure 15-25), then during time $T_r$ the distance each crest moves plus the distance the receiver moves equals the wavelength. That is, $vT_r + u_rT_r = \lambda$, or $T_r = \lambda/(v + u_r)$. [If the receiver moves in the same direction as the wave, then $vT_r - \lambda = u_rT_r$, so $T_r = \lambda/(v - u_r)$]. Since $f_r = 1/T_r$ we have

**FIGURE 15-25**

$$f_r = \frac{1}{T_r} = \frac{v \pm u_r}{\lambda} \qquad\qquad 15\text{-}34$$

where, if the receiver moves in the same direction as the wave, the received frequency is lower, so we choose the negative sign. If the receiver moves in the direction opposite to that of the wave, the frequency is higher, so we choose the positive sign. Substituting for $\lambda$ from Equation 15-32 we obtain

$$f_r = \frac{v \pm u_r}{v \pm u_s}f_s \qquad\qquad 15\text{-}35a$$

The correct choices for the plus or minus signs are most easily determined by remembering that the frequency tends to increase both when the source moves toward the receiver and when the receiver moves toward the source. For example, if the receiver is moving toward the source the plus sign is selected in the numerator, which tends to increase the received frequency, and if the source is moving away from the receiver the plus sign is selected in the denominator, which tends to decrease the received frequency. Equation 15-35a appears more symmetric, and thus is easier to remember, if expressed in the form

$$\frac{f_r}{v \pm u_r} = \frac{f_s}{v \pm u_s} \qquad\qquad 15\text{-}35b$$

It can be shown (see Problem 89) that if both $u_s$ and $u_r$ are much smaller than the wave speed $v$, then the shift in frequency $\Delta f = f_r - f_s$ is given approximately by

$$\frac{\Delta f}{f_s} \approx \pm\frac{u}{v} \qquad (u \ll v) \qquad\qquad 15\text{-}36$$

where $u = u_s \pm u_r$ is the speed of the source relative to the receiver.

Equations 15-31 through 15-36 are valid *only* in the reference frame of the medium. In a reference frame in which the medium is moving (for example, the reference frame of the ground if air is the medium and there is a wind blowing), the wave speed $v$ is replaced by $v' = v \pm u_w$, where $u_w$ is the speed of the wind relative to the ground.

---

*SOUNDING THE HORN*  **EXAMPLE 15-10**

The frequency of a car horn is 400 Hz. If the horn is honked as the car moves with a speed $u_s = 34$ m/s (about 122 km/h) through still air toward a stationary receiver, find (a) the wavelength of the sound passing the receiver and (b) the frequency received. Take the speed of sound in air to be 340 m/s. (c) Find the wavelength of the sound passing the receiver and find the frequency received if the car is stationary as the horn is honked and a receiver moves with a speed $u_r = 34$ m/s toward the car.

**PICTURE THE PROBLEM** (a) The waves in front of the source are compressed, so we use the minus sign in Equation 15-32. (b) We calculate the received frequency from Equation 15-35a. (c) For a moving receiver we use the same equations as in Parts (a) and (b).

(a) Using Equation 15-32, calculate the wavelength in front of the car. In front of the source the wavelength is shorter, so choose the sign accordingly:

$$\lambda = \frac{v - u_s}{f_s} = \frac{340 \text{ m/s} - 34 \text{ m/s}}{400 \text{ Hz}} = \boxed{0.765 \text{ m}}$$

(b) Using Equation 15-35a, solve for the received frequency:

$$f_r = \frac{v \pm u_r}{v \pm u_s} f_s = \frac{v + 0}{v - u_s} f_s$$

$$= \left(\frac{340}{340 - 34}\right)(400 \text{ Hz}) = \boxed{444 \text{ Hz}}$$

(c) 1. Using Equation 15-32, calculate the wavelength in the vicinity of the receiver:

$$\lambda = \frac{v \pm u_s}{f_s} = \frac{340 \text{ m/s} \pm 0}{400 \text{ Hz}} = \boxed{0.850 \text{ m}}$$

2. The received frequency is given by Equation 15-35a. The source is approaching the receiver so the frequency is shifted upward. Choose the sign accordingly:

$$f_r = \frac{v \pm u_r}{v \pm u_s} f_s = \frac{v + u_r}{v \pm 0} f_s = \left(1 + \frac{u_r}{v}\right) f_s$$

$$= \left(1 + \frac{34}{340}\right)(400 \text{ Hz}) = \boxed{440 \text{ Hz}}$$

**REMARKS** The frequency $f_r$ can also be obtained using Equation 15-34.

**EXERCISE** As a train moving at 90 km/h is approaching a stationary listener, it blows its horn, which has a frequency of 630 Hz. There is no wind. (a) What is the wavelength of the sound waves in front of the train? (b) What frequency is heard by the listener? (Use 340 m/s for the speed of sound.) (*Answer* (a) $\lambda = 0.5$ m (b) $f_r = 680$ Hz)

---

*ANOTHER CAR HORN*  **EXAMPLE 15-11**  **Try It Yourself**

The ratio of the frequency of a note to the frequency of the semitone above it on the diatonic scale is about 15:16. How fast is a car going if its horn drops a semitone as it passes you? There is no wind and you are standing next to the road. Take the speed of sound in air to be 340 m/s.

**PICTURE THE PROBLEM** Let $u_s$ be the speed of the car and $f_s$ be the frequency of the horn. The frequency received as the car approaches, $f_r$, is greater than $f_s$ and the frequency received as the car recedes, $f_r'$, is less than $f_s$. Set the ratio $f_r'/f_r = 15/16$ and solve for $u_s$.

**Cover the column to the right and try these on your own before looking at the answers.**

| Steps | Answers |
|---|---|

1. Write the frequency received as the car approaches in terms of $f_s$.

$$f_r = \frac{v \pm u_r}{v \pm u_s} f_s = \frac{v}{v - u_s} f_s$$

2. Write the frequency received as the car recedes in terms of $f_s$.

$$f'_r = \frac{v \pm u_r}{v \pm u_s} f_s = \frac{v}{v + u_s} f_s$$

3. Set the ratio $f'_r / f_r$ equal to 15/16.

$$\frac{f'_r}{f_r} = \frac{v - u_s}{v + u_s} = \frac{15}{16}$$

4. Solve for $u_s$.

$$u_s = 0.0323v = \boxed{39.5 \text{ km/h} = 24.5 \text{ mi/h}}$$

**REMARKS**  The wavelength of the sound behind the car is longer than the wavelength of the sound in front of the car. Also, 1 m/s = 3.6 km/h.

Another familiar example of the Doppler effect is the radar used by police to measure the speed of a car. Electromagnetic waves emitted by the radar transmitter strike the moving car. The car acts as both a moving receiver and a moving source as the waves reflect off it back to the radar receiver. Since electromagnetic waves travel at the speed of light, $v = c = 3 \times 10^8$ m/s, the condition $u \ll v$ is certainly met and Equation 15-36 can be used to calculate the Doppler shift.

---

*POLICE RADAR*                    **EXAMPLE  15-12**    **Try It Yourself**

The radar unit in a police car sends out electromagnetic waves that travel at the speed of light $c$. The electric current in the antenna of the radar unit oscillates at frequency $f_s$. The waves reflect from a speeding car moving away from the police car at speed $u$ relative to the police car. There is a frequency difference of $\Delta f$ between $f_s$ and $f'_r$, the frequency received at the police car. Find $u$ in terms of $f_s$ and $\Delta f$.

**PICTURE THE PROBLEM**  The radar wave strikes the speeding car at frequency $f_r$. This frequency is less than $f_s$ because the car is moving away from the source. The frequency shift is given by Equation 15-36. The car then acts as a moving source emitting waves of frequency $f_r$. The police unit detects waves of frequency $f'_r < f_r$ because the source (the speeding car) is moving away from the police car. The frequency difference is $f'_r - f_s$.

**Cover the column to the right and try these on your own before looking at the answers.**

| Steps | Answers |
|---|---|

1. The radar unit must be able to determine the speed based only on what it transmits and what it detects.

   The radar unit must determine $u$ in terms of $f_s$ and $f'_r$. Because of the way Equation 15-36 is written, we will solve for $u$ in terms of $f_s$ and $\Delta f = f'_r - f_s$.

2. The frequency difference $\Delta f$ is the frequency difference $\Delta f_1 = f_r - f_s$ plus the frequency difference $\Delta f_2 = f'_r - f_r$.

$$\Delta f = \Delta f_1 + \Delta f_2$$

3. Using Equation 15-36, substitute for the frequency differences in step 2.

$$\Delta f = -\frac{u}{c} f_s - \frac{u}{c} f_r = -\frac{u}{c}(f_s + f_r)$$

4. Again using Equation 15-36, solve for $f_r$ in terms of $f_s$.

$$\frac{\Delta f_1}{f_s} = -\frac{u}{c} \quad \text{so} \quad f_r = \left(1 - \frac{u}{c}\right)f_s$$

5. Substitute your step 4 result into your step 3 result and simplify.

$$\Delta f = -\frac{u}{c}\left(2 - \frac{u}{c}\right)f_s$$

6. $u/c$ is negligible compared to 2. Use this to simplify the step 5 result and solve for $u$ in terms of $\Delta f$ and $f_s$.

$$\Delta f \approx -2f_s\frac{u}{c} \quad \text{so} \quad u = -\frac{\Delta f}{2f_s}c = \boxed{\frac{|\Delta f|}{2f_s}c}$$

**REMARKS** The difference in frequency between two waves of nearly equal frequency is easy to detect because the two waves interfere to produce a wave whose amplitude oscillates with frequency $|\Delta f|$, which is called the beat frequency. Interference and beats are discussed in Chapter 16.

**EXERCISE** Calculate $\Delta f$ if $f_s = 1.5 \times 10^9$ Hz, $c = 3 \times 10^8$ m/s, and $u = 50$ m/s (*Answer* $\Delta f = 500$ Hz)

**The Doppler Shift and Relativity** We see from Example 15-10 (and Equations 15-33, 15-34, and 15-35) that the magnitude of the Doppler shift in frequency depends on whether it is the source or the receiver that is moving relative to the medium. For sound, these two situations are physically different. For example, if you move relative to still air, you feel air rushing past you. In your reference frame, there is a wind. For sound waves in air, therefore, we can tell whether the source or receiver is moving by noting if there is a wind in the reference frame of the source or the receiver. However, light and other electromagnetic waves propagate through empty space in which there is no medium. There is no "wind" to tell us whether the source or receiver is moving. According to Einstein's theory of relativity, absolute motion cannot be detected, and all observers measure the same speed $c$ for light, independent of their motion relative to the source. Thus Equation 15-35 cannot be correct for the Doppler shift for light. Two modifications must be made in calculating the relativistic Doppler effect for light. First, the speed of waves passing a receiver is $c$ independent of the motion of the receiver. Second, the time interval between the emission of successive wave crests, which is $T_s = 1/f_s$ in the reference frame of the source, is different in the reference frame of the receiver when the two reference frames are in relative motion because of relativistic time dilation and length contraction (Equations R-9 and R-3). (We will discuss the relativistic Doppler effect in Chapter 39.) The result is that the frequency received depends only on the relative speed of approach (or recession) $u$, and is related to the frequency emitted by

$$f_r = \sqrt{\frac{c \pm u}{c \mp u}}\,f_s \qquad\qquad 15\text{-}37$$

Choose the signs that give an up-shift in frequency when the source and receiver are approaching, and vice versa. (The upper signs are used if the source and receiver are approaching, and the lower signs are used if they are separating.) Again, when $u \ll c$, $\Delta f/f_s \approx \pm u/c$, as given by Equation 15-36.

## Shock Waves

In our derivations of the Doppler-shift expressions, we assumed that the speed $u$ of the source was less than the wave speed $v$. If a source moves with speed greater than the wave speed, then there will be no waves in front of the source. Instead, the waves pile up behind the source to form a shock wave. In the case of sound waves, this shock wave is heard as a sonic boom when it arrives at the receiver.

(a)

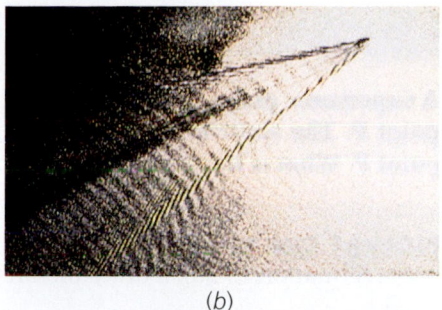

(b)

(c)

Figure 15-26 shows a source originally at point $P_1$ moving to the right with speed $u$. After some time $t$, the wave emitted from point $P_1$ has traveled a distance $vt$. The source has traveled a distance $ut$ and will be at point $P_2$. The line from this new position of the source to the wavefront emitted when the source was at $P_1$ makes an angle $\theta$ with the path of the source, given by

$$\sin \theta = \frac{vt}{ut} = \frac{v}{u} \qquad\qquad 15\text{-}38$$

Thus the shock wave is confined to a cone that narrows as $u$ increases. The ratio of the source speed $u$ to the wave speed $v$ is called the Mach number:

$$\text{Mach number} = \frac{u}{v} \qquad\qquad 15\text{-}39$$

(a) Shock waves from a supersonic airplane. (b) Bow waves from a boat. (c) Shock waves produced by a bullet traversing a helium balloon.

Equation 15-38 also applies to the electromagnetic radiation called Cerenkov radiation, which is given off when a charged particle moves in a medium with speed $u$ that is greater than the speed of light $v$ in that medium. (According to the special theory of relativity, it is impossible for a particle to move faster than $c$, the speed of light in vacuum. In a medium such as glass however, electrons and other particles can move faster than the speed of light in that medium.) The blue glow surrounding the fuel elements of a nuclear reactor is an example of Cerenkov radiation.

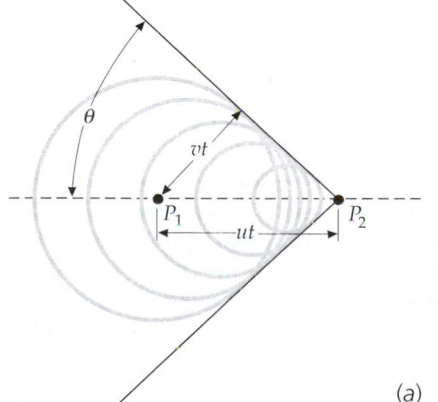

(a)

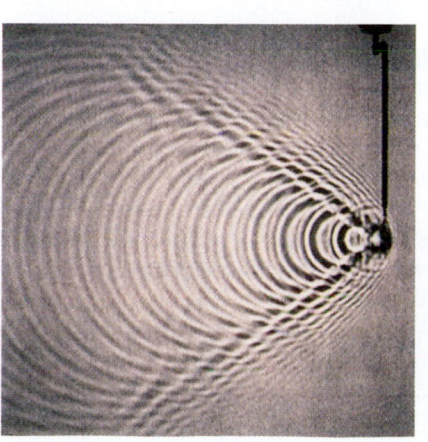

(b)

**FIGURE 15-26** (a) Source moving with a speed $u$ that is greater than the wave speed $v$. The envelope of the wavefronts forms a cone with the source at the apex. (b) Waves in a ripple tank produced by a source moving with a speed $u > v$.

A SONIC BOOM                    **EXAMPLE 15-13**  Try It Yourself

A supersonic plane flying due east at an altitude of 15 km passes directly over point $P$. The sonic boom is heard at point $P$ when the plane is 22 km east of point $P$. What is the speed of the supersonic plane?

**PICTURE THE PROBLEM** The speed of the plane is related the sine of the Mach angle (Equation 15-38). Draw a picture so the sine of the Mach angle can be calculated.

**Cover the column to the right and try these on your own before looking at the answers.**

**Steps**

1. Sketch the position of the plane (Figure 15-27) both at the instant the sonic boom is heard at point $P$ and at the instant that sound was produced. Label the distance the sound travels $v\Delta t$, and the distance the plane travels $u\Delta t$.

**Answers**                                    **FIGURE 15-27**

2. From your sketch and Equation 15-38, calculate $u$.

$$\tan \theta = \frac{15 \text{ km}}{22 \text{ km}} \quad \text{so} \quad \theta = 34.3°$$

$$\sin \theta = \frac{v\Delta t}{u\Delta t} = \frac{v}{u} \quad \text{so}$$

$$u = \frac{v}{\sin \theta} = \boxed{604 \text{ m/s}}$$

# SUMMARY

1. In wave motion, energy and momentum are transported from one point in space to another without the transport of matter.

2. The relation $v = f\lambda$ holds for all harmonic waves.

| Topic | Relevant Equations and Remarks |
|---|---|
| 1. **Transverse and Longitudinal Waves** | In transverse waves, such as waves on a string, the disturbance is perpendicular to the direction of propagation. In longitudinal waves, such as sound waves, the disturbance is along the direction of propagation. |
| 2. **Speed of Waves** | The wave speed $v$ is independent of the motion of the wave source. The speed of a wave relative to the medium depends on the density and elastic properties of the medium. |
| Waves on a string | $v = \sqrt{F_T/\mu}$      15-3 |
| Sound waves | $v = \sqrt{B/\rho}$      15-4 |

| | | |
|---|---|---|
| Sound waves in a gas | $v = \sqrt{\gamma RT/M}$ | 15-5 |
| | where $T$ is the absolute temperature, | |
| | $T = t_C + 273$ | 15-6 |
| | $R$ is the universal gas constant, | |
| | $R = 8.314 \, \text{J/mol·K}$ | 15-7 |
| | $M$ is the molar mass of the gas, which for air is $29 \times 10^{-3} \, \text{kg/mol}$, and $\gamma$ is a constant that depends on the kind of gas. For a diatomic gas such as air, $\gamma = 1.4$. For a monatomic gas such as helium, $\gamma = 1.67$. | |
| Electromagnetic waves | The speed of electromagnetic waves in vacuum is a universal constant | |
| | $c = 3 \times 10^8 \, \text{m/s}$ | |

**\*3. Wave Equation**

$$\frac{\partial^2 y}{\partial x^2} = \frac{1}{v^2}\frac{\partial^2 y}{\partial t^2}$$     15-9*b*

**4. Harmonic Waves**

| | | |
|---|---|---|
| Wave function | $y(x,t) = A \sin(kx \pm \omega t)$ | 15-13 |
| | where $A$ is the amplitude, $k$ is the wave number, and $\omega$ is the angular frequency. Use $-$ for a wave traveling in the positive $x$ direction, and $+$ for a wave traveling in the negative $x$ direction. | |
| Wave number | $k = \dfrac{2\pi}{\lambda}$ | 15-12 |
| Angular frequency | $\omega = 2\pi f = \dfrac{2\pi}{T}$ | 15-15 |
| Speed | $v = f\lambda = \omega/k$ | 15-10, 15-14 |
| Energy | The energy in a harmonic wave is proportional to the square of the amplitude. | |
| Power for harmonic waves on a string | $p_{av} = \frac{1}{2}\mu v \omega^2 A^2$ | 15-18 |

**5. Harmonic Sound Waves**

Sound waves can be considered to be either displacement waves or pressure waves. The human ear is sensitive to sound waves of frequencies from about 20 Hz to 20 kHz. In a harmonic sound wave, the pressure and displacement are 90° out of phase.

| | | |
|---|---|---|
| Amplitudes | The pressure and displacement amplitudes are related by | |
| | $p_0 = \rho \omega v s_0$ | 15-22 |
| | where $\rho$ is the density of the medium. | |
| Energy density | $\eta_{av} = \dfrac{(\Delta E)_{av}}{\Delta V} = \dfrac{1}{2}\rho \omega^2 s_0^2$ | 15-24 |

**6. Intensity**

The intensity of a wave is the average power per unit area.

$$I = \frac{P_{av}}{A}$$     15-25

| Average energy density $\eta_{av}$ of a sound wave | $I = \eta_{av}\,v = \dfrac{1}{2}\rho\omega^2 s_0^2 v = \dfrac{1}{2}\dfrac{p_0^2}{\rho v}$ | 15-28 |
|---|---|---|

| *Intensity level $\beta$ in dB | Sound intensity levels are measured on a logarithmic scale. | |
| | $\beta = 10\log\dfrac{I}{I_0}$ | 15-29 |
| | where $I_0 = 10^{-12}\ \text{W/m}^2$ is approximately the threshold of hearing. | |

| **7. Reflection and Refraction** | When a wave is incident on a boundary surface that separates two regions of differing wave speed, part of the wave is reflected and part is transmitted. | |
|---|---|---|

| **8. Diffraction** | If a wavefront is partially blocked by an obstacle, the unblocked part of the wavefront diffracts (bends) into the region behind the obstacle. | |

| Ray approximation | If a wavefront is partially blocked by an obstacle, almost all of the diffraction occurs for that part of the wavefront that passes within a few wavelengths of the edge. For those parts of the wavefront that pass farther from the edge than a few wavelengths, diffraction is negligible and the wave propagates in straight lines in the direction of the incident rays. | |

| **9. Doppler Effect** | When a sound source and receiver are in relative motion, the received frequency $f_r$ is higher than the frequency of the source $f_s$ if their separation is decreasing, and lower if their separation is increasing. | |
|---|---|---|

| Moving source | $\lambda = \dfrac{v \pm u_s}{f_s}$ | 15-32† |

| Moving receiver | $f_r = \dfrac{v \pm u_r}{\lambda}$ | 15-34† |

| Either source or receiver moving | $f_r = \dfrac{v \pm u_r}{v \pm u_s}f_s \qquad \text{or} \qquad \dfrac{f_r}{v \pm u_r} = \dfrac{f_s}{v \pm u_s}$ | 15-35† |
| | Choose the signs that give an up-shift in frequency for an approaching source or receiver, and vice versa. | |

| Small speeds of source or receiver | $\dfrac{\Delta f}{f_s} \approx \pm\dfrac{u}{v} \qquad (u \ll v)$ | 15-36† |

| Relativistic Doppler shift | $f_r = \sqrt{\dfrac{c \pm u}{c \mp u}}\,f_s$ | 15-37 |
| | Choose the signs that give an up-shift in frequency for an approaching source or receiver, and vice versa. | |

| **10. Shock Waves** | When the source speed is greater than the wave speed, the waves behind the source are confined to a cone of angle $\theta$ given by | |
|---|---|---|

| Mach angle | $\sin\theta = \dfrac{v}{u}$ | 15-38 |

| Mach number | $\text{Mach number} = \dfrac{u}{v}$ | 15-39 |

† Equations 15-32 through 15-36 are valid *only* in the reference frame of the medium. If the medium is moving, the wave speed $v$ is replaced by $v' = v \pm u_m$, where $u_m$ is the speed of the medium.

# PROBLEMS

- Single-concept, single-step, relatively easy
- •• Intermediate-level, may require synthesis of concepts
- ••• Challenging
- SSM Solution is in the *Student Solutions Manual*
- iSOLVE Problems available on iSOLVE online homework service
- iSOLVE ✓ These "Checkpoint" online homework service problems ask students additional questions about their confidence level, and how they arrived at their answer

In a few problems, you are given more data than you actually need; in a few other problems, you are required to supply data from your general knowledge, outside sources, or informed estimates.

**Use $v = 340$ m/s for the speed of sound in air unless otherwise indicated.**

## Conceptual Problems

**1** • SSM A rope hangs vertically from the ceiling. Do waves on the rope move faster, slower, or at the same speed as they move from bottom to top? Explain.

**2** • A traveling wave passes a point of observation. At this point, the time between successive crests is 0.2 s. Which of the following is true? (*a*) The wavelength is 5 m. (*b*) The frequency is 5 Hz. (*c*) The velocity of propagation is 5 m/s. (*d*) The wavelength is 0.2 m. (*e*) There is not enough information to justify any of these statements.

**3** • True or false: The energy in a wave is proportional to the square of the amplitude of the wave.

**4** • A rope hangs vertically. You shake the bottom back and forth, creating a sinusoidal wave train. Is the wavelength near the top the same as, less than, or greater than the wavelength near the bottom?

**5** • SSM The crack of a bullwhip is caused by the speed of the tip breaking the sound barrier. Explain how the tapered shape of the whip helps the tip move much faster than the hand holding the whip.

**6** • True or false: A 60-dB sound has twice the intensity of a 30-dB sound.

**7** • If the source and receiver are at rest relative to each other but the wave medium is moving relative to them, will there be any Doppler shift in frequency?

**8** • The frequency of a car horn is $f_0$. What frequency is observed if both the car and the observer are at rest, but a wind blows toward the observer? (*a*) $f_0$. (*b*) Greater than $f_0$. (*c*) Less than $f_0$. (*d*) It could be either greater or less than $f_0$. (*e*) It could be $f_0$ or greater than $f_0$, depending on how wind speed compares to speed of sound.

**9** •• SSM Stars often occur in pairs revolving around their common center of mass. If one of the stars is a black hole, it is invisible. Explain how the existence of such a black hole might be inferred from the light observed from the other, visible star.

**10** • When a guitar string is plucked, is the wavelength of the wave it produces in air the same as the wavelength of the wave on the string?

**11** • True or false:
(*a*) Wave pulses on strings are transverse waves.
(*b*) Sound waves in air are transverse waves of compression and rarefaction.
(*c*) The speed of sound in air at 20°C is twice that at 5°C.

**12** • Sound travels at 340 m/s in air and 1500 m/s in water. A sound of 256 Hz is made under water. In the air, the frequency will be (*a*) the same, but the wavelength will be shorter, (*b*) higher, but the wavelength will stay the same, (*c*) lower, but the wavelength will be longer, (*d*) lower, and the wavelength will be shorter, (*e*) the same, and the wavelength too will stay the same.

**13** • SSM While out on patrol, the battleship *Rodger Young* hits a mine and begins to burn, ultimately exploding. Sailor Abel jumps into the water and begins swimming away from the doomed ship, while Sailor Baker gets into a life raft. Comparing their experiences later, Abel tells Baker, "I was swimming underwater, and heard a big explosion from the ship. When I surfaced, I heard a second explosion. What do you think it could be?" Baker says, "I think it was your imagination—I only heard one explosion." Explain why Baker only heard one explosion, while Abel heard two.

**14** •• Figure 15-28 shows a wave pulse at time $t = 0$ moving to the right. At this particular time, which segments of the string are moving up? Which are moving down? Is there any segment of the string at the pulse that is instantaneously at rest? Answer these questions by sketching the pulse at a slightly later time and a slightly earlier time to see how the segments of the string are moving.

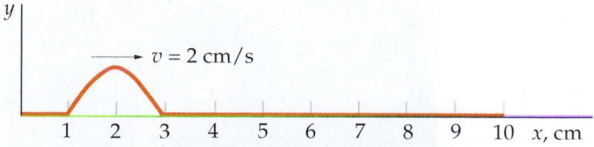

**FIGURE 15-28** Problems 14, 15

**15** •• Make a sketch of the velocity of each string segment versus position for the pulse shown in Figure 15-28.

**16** •• In a classic physics experiment, a bell is placed in a sealed jar and rung while the air is slowly removed. After a while, the bell becomes inaudible. This is commonly cited as proof that sound waves can't travel through a vacuum, but in fact the sound becomes inaudible well before the jar is completely evacuated. Can you give another reason why the sound from the bell can't be heard?

**17** •• **SSM** The explosion of a depth charge beneath the surface of the water is recorded by a helicopter hovering above its surface, as shown in Figure 15-29. Along which path, A, B, or C, will the sound wave take the least time to reach the helicopter?

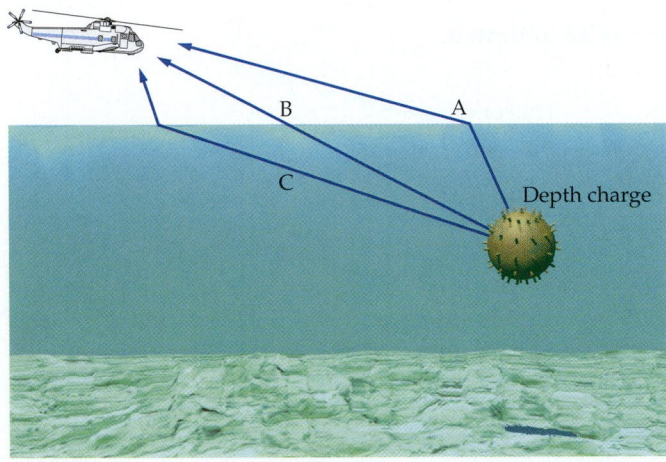

**FIGURE 15-29** Problem 17

### Estimation and Approximation

**18** •• Normal human speech has a sound intensity level of about 65 dB at a distance of 1 m. Estimate the power of human speech.

**19** •• A man drops a stone from a high bridge and hears it strike the water below exactly 4 s later. (*a*) Estimate the distance to the water based on the assumption that the travel time for the sound to reach the man is negligible. (*b*) Improve your estimate by using your result from Part (*a*) for the distance to the water to estimate the time it takes for sound to travel this distance and then calculate the distance the rock falls in 4 s minus this time. (*c*) Calculate the exact distance and compare your result with your previous estimates.

**20** •• **SSM** Estimate the speed of the bullet as it passes through the helium balloon in Figure 15-30 from the angle of its shock cone.

**FIGURE 15-30** Problem 20

**21** •• The new student townhouses at a local college are in the form of a semicircle half-enclosing the track field. To estimate the speed of sound in air, an ambitious physics student stood at the center of the semicircle and clapped his hands rhythmically at a frequency at which he couldn't hear the echo of the clap, because it reached him at the same time as his next clap. This frequency was about 2.5 claps/s. Once he established this frequency, he paced off the distance to the townhouses, which was 30 double paces. Assuming that the distance of a double-pace stride is the same as his height (5 ft 11 in), estimate the speed of sound in air using this data. How far off is this from the commonly accepted value?

### Speed of Waves

**22** • **iSOLVE** ✓ (*a*) The bulk modulus for water is $2.0 \times 10^9$ N/m². Use it to find the speed of sound in water. (*b*) The speed of sound in mercury is 1410 m/s. What is the bulk modulus for mercury ($\rho = 13.6 \times 10^3$ kg/m³)?

**23** • **SSM** **iSOLVE** ✓ Calculate the speed of sound waves in hydrogen gas at $T = 300$ K. (Take $M = 2$ g/mol and $\gamma = 1.4$.)

**24** • A steel wire 7 m long has a mass of 100 g. It is under a tension of 900 N. What is the speed of a transverse wave pulse on this wire?

**25** • Transverse waves travel at 150 m/s on a wire of length 80 cm that is under a tension of 550 N. What is the mass of the wire?

**26** • **SSM** A wave pulse propagates along a wire in the positive $x$ direction at 20 m/s. What will be the pulse velocity if we (*a*) double the length of the wire but keep the tension and mass per unit length constant? (*b*) double the tension while holding the length and mass per unit length constant? (*c*) double the mass per unit length while holding the other variables constant?

**27** • **iSOLVE** A steel piano wire is 0.7 m long and has a mass of 5 g. It is stretched with a tension of 500 N. (*a*) What is the speed of transverse waves on the wire? (*b*) To reduce the wave speed by a factor of 2 without changing the tension, what mass of copper wire would have to be wrapped around the steel wire?

**28** •• A common method for estimating the distance to a lightning bolt is to begin counting when the flash is observed and continue until the thunder clap is heard. The number of seconds counted is then divided by 3 to get the distance in kilometers. (*a*) What is the velocity of sound in kilometers per second? (*b*) How accurate is this procedure? (*c*) Is a correction for the time it takes for the light to reach you important? (The speed of light is $3 \times 10^8$ m/s.)

**29** •• **SSM** (*a*) Compute the derivative of the speed of a wave on a string with respect to the tension $dv/dF$, and show that the differentials $dv$ and $dF$ obey $dv/v = \frac{1}{2}dF/F$. (*b*) A wave moves with a speed of 300 m/s on a wire that is under a tension of 500 N. Using $dF$ to approximate a change in tension, determine how much the tension must be changed to increase the speed to 312 m/s.

**30** •• (a) Compute the derivative of the velocity of sound in air with respect to the absolute temperature, and show that the differentials $dv$ and $dT$ obey $dv/v = \frac{1}{2}dT/T$. (b) Use this result to compute the percentage change in the velocity of sound when the temperature changes from 0 to 27°C. (c) If the speed of sound is 331 m/s at 0°C, what is it (approximately) at 27°C? How does this approximation compare with the result of an exact calculation?

**31** ••• In this problem, you will derive a convenient formula for the speed of sound in air at temperature $t$ in Celsius degrees. Begin by writing the temperature as $T = T_0 + \Delta T$, where $T_0 = 273$ K corresponds to 0°C and $\Delta T = t$, the Celsius temperature. The speed of sound is a function of $T$, $v(T)$. To a first-order approximation, you can write $v(T) \approx v(T_0) + (dv/dT)_{T_0} \Delta T$, where $(dv/dT)_{T_0}$ is the derivative evaluated at $T = T_0$. Compute this derivative, and show that the result leads to $v = (331 \text{ m/s})\left(1 + \dfrac{t}{2T_0}\right) = (331 + 0.606t)$ m/s.

**32** •• Various stories of psychic phenomena can often be explained by considering physical phenomena. For example, there is the often repeated story of a man being woken from a deep sleep for no reason, getting out of bed and walking to the window just in time to hear the sound of the explosion of a munitions plant across town. The story is often cited to give credence to the idea of clairvoyance, but can be explained instead by assuming that the man was woken by the tremor of the sound wave traveling through the earth, and then walked to the window in time to hear the sound wave traveling through the air. If he took 3 s to move from his bed to the window, and the average speed of sound through solid rock is 3000 m/s, how far was his house from the munitions plant?

**33** ••• While studying physics in her dorm room, a student is listening to a live radio broadcast of a baseball game. She is 1.6 km due south of the baseball field. Over her radio, the student hears a noise generated by the electromagnetic pulse of a lightning bolt. Two seconds later, she hears over the radio the thunder picked up by the microphone at the baseball field. Four seconds after she hears the noise of the electromagnetic pulse over the radio, thunder rattles her windows. Where, relative to the ballpark, did the lightning bolt occur?

**34** ••• SSM Weather station Beta is located 0.75 mi due east of weather station Alpha. Observers at the two stations see a lightning strike to the north of the stations; observers at station Alpha hear the thunder 3.4 s after seeing the strike, while observers at Beta hear it 2.5 s after seeing the strike. Locate the coordinates of the lightning strike relative to the position of station Alpha.

**35** ••• A coiled spring, such as a Slinky, is stretched to a length $L$. It has a force constant $k$ and a mass $m$. (a) Show that the velocity of longitudinal compression waves along the spring is given by $v = L\sqrt{k/m}$. (b) Show that this is also the velocity of transverse waves along the spring if the natural length of the spring is much less than $L$.

### The Wave Equation

**36** • Show explicitly that the following functions satisfy the wave equation: (a) $y(x,t) = k(x + vt)^3$; (b) $y(x, t) = Ae^{ik(x - vt)}$

where $A$ and $k$ are constants and $i = \sqrt{-1}$; and (c) $y(x,t) = \ln k(x - vt)$.

**37** • SSM Show that the function $y = A \sin kx \cos \omega t$ satisfies the wave equation.

### Harmonic Waves on a String

**38** • SOLVE One end of a string 6 m long is moved up and down with simple harmonic motion at a frequency of 60 Hz. The waves reach the other end of the string in 0.5 s. Find the wavelength of the waves on the string.

**39** • Equation 15-13 expresses the displacement of a harmonic wave as a function of $x$ and $t$ in terms of the wave parameters $k$ and $\omega$. Write the equivalent expressions that contain the following pairs of parameters instead of $k$ and $\omega$: (a) $k$ and $v$, (b) $\lambda$ and $f$, (c) $\lambda$ and $T$, (d) $\lambda$ and $v$, and (e) $f$ and $v$.

**40** • SSM Equation 15-10 applies to all types of periodic waves, including electromagnetic waves such as light waves and microwaves, which travel at $3 \times 10^8$ m/s in a vacuum. (a) The range of wavelengths of light to which the eye is sensitive is about $4 \times 10^{-7}$ to $7 \times 10^{-7}$ m. What are the frequencies that correspond to these wavelengths? (b) Find the frequency of a microwave that has a wavelength of 3 cm.

**41** • SOLVE A harmonic wave on a string with a mass per unit length of 0.05 kg/m and a tension of 80 N has an amplitude of 5 cm. Each section of the string moves with simple harmonic motion at a frequency of 10 Hz. Find the power propagated along the string.

**42** • SOLVE✓ A rope 2 m long has a mass of 0.1 kg. The tension is 60 N. A power source at one end sends a harmonic wave with an amplitude of 1 cm down the rope. The wave is extracted at the other end without any reflection. What is the frequency of the power source if the power transmitted is 100 W?

**43** •• The wave function for a harmonic wave on a string is $y(x, t) = (0.001 \text{ m}) \sin(62.8 \text{ m}^{-1} x + 314 \text{ s}^{-1} t)$. (a) In what direction does this wave travel, and what is its speed? (b) Find the wavelength, frequency, and period of this wave. (c) What is the maximum speed of any string segment?

**44** •• A harmonic wave with a frequency of 80 Hz and an amplitude of 0.025 m travels along a string to the right with a speed of 12 m/s. (a) Write a suitable wave function for this wave. (b) Find the maximum speed of a point on the string. (c) Find the maximum acceleration of a point on the string.

**45** •• SOLVE Waves of frequency 200 Hz and amplitude 1.2 cm move along a 20-m string that has a mass of 0.06 kg and a tension of 50 N. (a) What is the average total energy of the waves on the string? (b) Find the power transmitted past a given point on the string.

**46** •• SSM In a real string, a wave loses some energy as it travels down the string. Such a situation can be described by a wave function whose amplitude $A(x)$ depends on $x$: $y = A(x) \sin (kx - \omega t) = (A_0 e^{-bx}) \sin (kx - \omega t)$ (a) What is the power transported by the wave at the origin? (b) What is the power transported by the wave at point $x$, where $x > 0$?

**47** •• Power is to be transmitted along a stretched wire by means of transverse harmonic waves. The wave speed is 10 m/s and the linear mass density of the wire is 0.01 kg/m. The power source oscillates with an amplitude of 0.50 mm. (a) What average power is transmitted along the wire if the frequency is 400 Hz? (b) The power transmitted can be increased by increasing the tension in the wire, the frequency of the source, or the amplitude of the waves. By how much would each of these quantities have to increase to cause an increase in power by a factor of 100 if it is the only quantity changed? (c) Which of the quantities would probably be the easiest to vary?

**48** ••• **SSM** Two very long strings are tied together at the point $x = 0$. In the region $x < 0$, the wave speed is $v_1$, while in the region $x > 0$, the speed is $v_2$. A sinusoidal wave is incident from the left ($x < 0$); part of the wave is reflected and part is transmitted. For $x < 0$, the displacement of the wave is describable by $y(x,t) = A \sin(k_1 x - \omega t) + B \sin(k_1 x + \omega t)$, while for $x > 0$, $y(x,t) = C \sin(k_2 x - \omega t)$, where $\omega/k_1 = v_1$ and $\omega/k_2 = v_2$. (a) If we assume that both the wave function $y$ and its first spatial derivative $\partial y/\partial x$ must be continuous at $x = 0$, show that $C/A = 2/(1 + v_1/v_2)$, and that $B/A = (1 - v_1/v_2)/(1 + v_1/v_2)$. (b) Show that $B^2 + (v_1/v_2)C^2 = A^2$.

## Harmonic Sound Waves

**49** • **SSM** A sound wave in air produces a pressure variation given by

$$p(x,t) = 0.75 \cos \frac{\pi}{2} (x - 340t)$$

where $p$ is in pascals, $x$ is in meters, and $t$ is in seconds. Find (a) the pressure amplitude of the sound wave, (b) the wavelength, (c) the frequency, and (d) the speed.

**50** • **iSOLVE** (a) Middle C on the musical scale has a frequency of 262 Hz. What is the wavelength of this note in air? (b) The frequency of the C an octave above middle C is twice that of middle C. What is the wavelength of this note in air?

**51** • (a) What is the displacement amplitude for a sound wave having a frequency of 100 Hz and a pressure amplitude of $10^{-4}$ atm? (b) The displacement amplitude of a sound wave of frequency 300 Hz is $10^{-7}$ m. What is the pressure amplitude of this wave?

**52** • **iSOLVE**✓ (a) Find the displacement amplitude of a sound wave of frequency 500 Hz at the pain-threshold pressure amplitude of 29 Pa. (b) Find the displacement amplitude of a sound wave with the same pressure amplitude but a frequency of 1 kHz.

**53** • A typical loud sound wave with a frequency of 1 kHz has a pressure amplitude of about $10^{-4}$ atm. (a) At $t = 0$, the pressure is a maximum at some point $x_1$. What is the displacement at that point at $t = 0$? (b) What is the maximum value of the displacement at any time and place? (Take the density of air to be 1.29 kg/m³.)

**54** • **SSM** An octave represents a change in frequency by a factor of two. Over how many octaves can a typical person hear?

## Waves in Three Dimensions: Intensity

**55** • A piston at one end of a long tube filled with air at room temperature and normal pressure oscillates with a frequency of 500 Hz and an amplitude of 0.1 mm. The area of the piston is 100 cm². (a) What is the pressure amplitude of the sound waves generated in the tube? (b) What is the intensity of the waves? (c) What average power is required to keep the piston oscillating (neglecting friction)?

**56** • A spherical source radiates sound uniformly in all directions. At a distance of 10 m, the sound intensity level is $10^{-4}$ W/m². (a) At what distance from the source is the intensity $10^{-6}$ W/m²? (b) What power is radiated by this source?

**57** • **SSM** **iSOLVE** A loudspeaker at a rock concert generates $10^{-2}$ W/m² at 20 m at a frequency of 1 kHz. Assume that the speaker spreads its energy uniformly in three dimensions. (a) What is the total acoustic power output of the speaker? (b) At what distance will the intensity be at the pain threshold of 1 W/m²? (c) What is the intensity at 30 m?

**58** •• When a pin of mass 0.1 g is dropped from a height of 1 m, 0.05 percent of its energy is converted into a sound pulse with a duration of 0.1 s. (a) Estimate the range at which the dropped pin can be heard if the minimum audible intensity is $10^{-11}$ W/m². (b) Your result in (a) is much too large in practice because of background noise. If you assume that the intensity must be at least $10^{-8}$ W/m² for the sound to be heard, estimate the range at which the dropped pin can be heard. (In both parts, assume that the intensity is $P/4\pi r^2$.)

## *Intensity Level

**59** • What is the intensity level in decibels of a sound wave of intensity (a) $10^{-10}$ W/m² and (b) $10^{-2}$ W/m²?

**60** • **iSOLVE** Find the intensity of a sound wave if (a) $\beta = 10$ dB and (b) $\beta = 3$ dB. (c) Find the pressure amplitudes of sound waves in air for each of these intensities.

**61** • **SSM** The sound level of a dog's bark is 50 dB. The intensity of a rock concert is 10,000 times that of the dog's bark. What is the sound level of the rock concert?

**62** • **iSOLVE** Two sounds differ by 30 dB. The intensity of the louder sound is $I_L$ and that of the softer sound is $I_s$. The value of the ratio $I_L/I_s$ is (a) 1000, (b) 30, (c) 9, (d) 100, (e) 300.

**63** • Show that if the intensity is doubled, the intensity level increases by 3.0 dB.

**64** • **SSM** What fraction of the acoustic power of a noise would have to be eliminated to lower its sound intensity level from 90 to 70 dB?

**65** •• **iSOLVE** A spherical source radiates sound uniformly in all directions. At a distance of 10 m, the sound intensity level is 80 dB. (a) At what distance from the source is the intensity level 60 dB? (b) What power is radiated by this source?

**66** •• A spherical source of intensity $I_0$ radiates sound uniformly in all directions. Its intensity level is $\beta_1$ at a distance $r_1$ and $\beta_2$ at a distance $r_2$. Find $\beta_2/\beta_1$.

**67** •• **SOLVE ✓** A loudspeaker at a rock concert generates $10^{-2}$ W/m² at 20 m at a frequency of 1 kHz. Assume that the speaker spreads its energy uniformly in all directions. (a) What is the intensity level at 20 m? (b) What is the total acoustic power output of the speaker? (c) At what distance will the intensity level be at the pain threshold of 120 dB? (d) What is the intensity level at 30 m?

**68** •• An article on noise pollution claims that sound intensity levels in large cities have been increasing by about 1 dB annually. (a) To what percentage increase in intensity does this correspond? Does this increase seem plausible? (b) In about how many years will the intensity of sound double if it increases at 1 dB annually?

**69** •• Three noise sources produce intensity levels of 70, 73, and 80 dB when acting separately. When the sources act together, their intensities add. (a) Find the sound intensity level in decibels when the three sources act at the same time. (b) Discuss the effectiveness of eliminating the two least intense sources in reducing the intensity level of the noise.

**70** •• **SSM** If you double the distance between a source of sound and a receiver, the intensity at the receiver drops by approximately (a) 2 dB, (b) 3 dB, (c) 6 dB, (d) Amount cannot be determined from the information given.

**71** ••• **SOLVE** Everyone at a party is talking equally loudly. If only one person were talking, the sound level would be 72 dB. Find the sound level when all 38 people are talking.

**72** ••• **SSM** When a violinist pulls the bow across a string, the force with which the bow is pulled is fairly small, about 0.6 N. Suppose the bow travels across the A string, which vibrates at 440 Hz, at 0.5 m/s. A listener 35 m from the performer hears a sound of 60 dB intensity. With what efficiency is the mechanical energy of bowing converted to sound energy? (Assume that the sound radiates uniformly in all directions.)

**73** ••• The noise level in an empty examination hall is 40 dB. When 100 students are writing an exam, the sounds of heavy breathing and pens traveling rapidly over paper cause the noise level to rise to 60 dB (not counting the occasional groans). Assuming that each student contributes an equal amount of noise power, find the noise level to the nearest decibel when 50 students have left.

## The Doppler Effect

In Problems 74 through 79, a source emits sounds of frequency 200 Hz that travel through still air at 340 m/s.

**74** • The sound source described moves with a speed of 80 m/s relative to still air toward a stationary listener. (a) Find the wavelength of the sound in the region between the source and the listener. (b) Find the frequency heard by the listener.

**75** • Consider the situation in Problem 74 from the reference frame in which the source is at rest. In this frame, the listener moves toward the source with a speed of 80 m/s, and there is a wind blowing at 80 m/s from the listener to the source. (a) What is the speed at which the sound travels from the source to the listener in this frame? (b) Find the wavelength of the sound in the region between the source and the listener. (c) Find the frequency heard by the listener.

**76** • The source moves away from the stationary listener at 80 m/s. (a) Find the wavelength of the sound waves in the region between the source and the listener. (b) Find the frequency heard by the listener.

**77** • **SOLVE ✓** The listener moves at 80 m/s relative to still air toward the stationary source. (a) What is the wavelength of the sound in the region between the source and the listener? (b) What is the frequency heard by the listener?

**78** • Consider the situation in Problem 77 in a reference frame in which the listener is at rest. (a) What is the wind velocity in this frame? (b) What is the speed of the sound as it travels from the source to the listener in this frame, that is, relative to the listener? (c) Find the wavelength of the sound in the region between the source and the listener in this frame. (d) Find the frequency heard by the listener.

**79** • The listener moves at 80 m/s relative to the still air away from the stationary source. Find the frequency heard by the listener.

**80** • A jet is traveling at Mach 2.5 at an altitude of 5000 m. (a) What is the angle that the shock wave makes with the track of the jet? (Assume that the speed of sound at this altitude is still 340 m/s.) (b) Where is the jet when a person on the ground hears the shock wave?

**81** • If you are running at top speed toward a source of sound at 1000 Hz, estimate the frequency of the sound that you hear. Suppose that you can recognize a change in frequency of 3 percent. Can you use your sense of pitch to estimate your running speed?

**82** •• **SOLVE** A radar device emits microwaves with a frequency of 2.00 GHz. When the waves are reflected from a car moving directly away from the emitter, a frequency difference of 293 Hz is detected. Find the speed of the car.

**83** •• **SSM** The Doppler effect is routinely used to measure the speed of winds in storm systems. A weather station uses a Doppler radar system of frequency $f$ = 625 MHz to bounce a radar pulse off of the raindrops in a swirling thunderstorm system 50 km away; the reflected radar pulse is found to be up-shifted in frequency by 325 Hz. Assuming the wind is headed directly toward the radar antenna, how fast are the winds in the storm system moving? (The radar system can only measure the radial component of the velocity.)

**84** •• **SOLVE** A stationary destroyer is equipped with sonar that sends out pulses of sound at 40 MHz. Reflected pulses are received from a submarine directly below with a time delay of 80 ms at a frequency of 39.958 MHz. If the speed of sound in seawater is 1.54 km/s, find (a) the depth of the submarine and (b) its vertical speed.

**85** •• A police radar unit transmits microwaves of frequency $3 \times 10^{10}$ Hz. The speed of these waves in air is $3.0 \times 10^8$ m/s. Suppose a car is receding from the stationary police car at a speed of 140 km/h. What is the frequency difference between the transmitted signal and the signal received from the receding car?

**86** •• Suppose the police car of Problem 85 is moving in the same direction as the other vehicle at a speed of 60 km/h. What then is the difference in frequency between the emitted and the reflected signals?

**87** •• **SOLVE** At time $t = 0$, a supersonic plane is directly over point $P$, flying due west at an altitude of 12 km and a speed of Mach 1.6. Where is the plane when the sonic boom is heard?

**88** •• A small radio of mass 0.10 kg is attached to one end of an air track by a spring. The radio emits a sound of 800 Hz. A listener at the other end of the air track hears a sound whose frequency varies between 797 and 803 Hz. (a) Determine the energy of the vibrating mass–spring system. (b) If the spring constant is 200 N/m, what is the amplitude of vibration of the mass and what is the period of the oscillating system?

**89** •• A sound source of frequency $f_0$ moves with speed $u_s$ relative to still air toward a receiver who is moving with speed $u_r$ relative to still air away from the source. (a) Write an expression for the received frequency $f'$. (b) Use the result that $(1 - x)^{-1} \approx 1 + x$ to show that if both $u_s$ and $u_r$ are small compared to $v$, then the received frequency is approximately

$$f' \approx \left(1 + \frac{u_s - u_r}{v}\right)f_0 = \left(1 + \frac{u_{rel}}{v}\right)f_0$$

where $u_{rel} = u_s - u_r$ is the relative velocity of approach of the source and receiver.

**90** •• Two students with vibrating 440-Hz tuning forks walk away from each other with equal speeds. How fast must they walk so that they each hear a frequency of 438 Hz from the other fork?

**91** •• A physics student walks down a long hall carrying a vibrating 512-Hz tuning fork. The end of the hall is closed so that sound reflects from it. The student hears a sound of 516 Hz from the wall. How fast is the student walking?

**92** •• **SSM** A small speaker radiating sound at 1000 Hz is tied to one end of an 0.8-m-long rod that is free to rotate about its other end. The rod rotates in the horizontal plane at 4.0 rad/s. Derive an expression for the frequency heard by a stationary observer far from the rotating speaker.

**93** •• A balloon carried along by a 36-km/h wind emits a sound of 800 Hz as it approaches a tall building. (a) What is the frequency of the sound heard by an observer at the window of this building? (b) What is the frequency of the reflected sound heard by a person riding in the balloon?

**94** •• **SOLVE** A car is approaching a reflecting wall. A stationary observer behind the car hears a sound of frequency 745 Hz from the car horn and a sound of frequency 863 Hz from the wall. (a) How fast is the car traveling? (b) What is the frequency of the car horn? (c) What frequency does the car driver hear reflected from the wall?

**95** •• The driver of a car traveling at 100 km/h toward a vertical cliff briefly sounds the horn. Exactly 1 s later she hears the echo and notes that its frequency is 840 Hz. How far from the cliff was the car when the driver sounded the horn and what is the frequency of the horn?

**96** •• You are on a transatlantic flight traveling due west at 800 km/h. A Concorde flying at Mach 1.6 and 3 km to the north of your plane is also on an east-to-west course. What is the distance between the two planes when you hear the sonic boom from the Concorde?

**97** •• **SSM** The Hubble space telescope has been used to determine the existence of planets orbiting around distant stars. The planet orbiting the star will cause the star to "wobble" with the same period as the planet's orbit; because of this, light from the star will be Doppler-shifted up and down periodically. Estimate the maximum and minimum wavelengths of light of nominal wavelength 500nm emitted by the sun that is Doppler-shifted by the motion of the sun due to the planet Jupiter.

**98** ••• **SOLVE** A physics student drops a vibrating 440-Hz tuning fork down the elevator shaft of a tall building. When the student hears a frequency of 400 Hz, how far has the tuning fork fallen?

**99** •• The SuperKamiokande neutrino detector in Japan is a water tank the size of a 14-story building. It detects neutrinos, the "ghost particles" of physics, by the shock wave produced when a neutrino imparts most of its energy to an electron, which then goes flying off at near light-speed through the water. If the maximum angle of the Cerenkov shock-wave cone is 48.75°, what is the speed of light in water?

## General Problems

**100** • At time $t = 0$, the shape of a wave pulse on a string is given by the function

$$y(x,0) = \frac{0.12 \text{ m}^3}{(2.00 \text{ m})^2 + x^2}$$

where $x$ is in meters. (a) Sketch $y(x, 0)$ versus $x$. Give the wave function $y(x,t)$ at a general time $t$ if (b) the pulse is moving in the positive $x$ direction with a speed of 10 m/s and (c) the pulse is moving in the negative $x$ direction with a speed of 10 m/s.

**101** • **SOLVE** A wave with a frequency of 1200 Hz propagates along a wire that is under a tension of 800 N. Its wavelength is 24 cm. What will be the wavelength if the tension is decreased to 600 N and the frequency is kept constant?

**102** • **SOLVE** In a common lecture demonstration of wave pulses, a length of rubber tubing is tied at one end to a fixed post and is passed over a pulley to a weight hanging at the other end. Suppose that the distance from the fixed support to the pulley is 10 m, the mass of this length of tubing is 0.7 kg, and the suspended weight is 110 N. If the tubing is given a transverse blow at one end, how long will it take the resulting pulse to reach the other end?

**103** • A boat traveling at 10 m/s on a still lake makes a bow wave at an angle of 20° with its direction of motion. What is the speed of the bow wave?

**104** • If a wavelength is much larger than the diameter of a loudspeaker, the speaker radiates in all directions, much like a point source. On the other hand, if the wavelength is much smaller than the loudspeaker diameter, the sound travels in a beam in the forward direction—and does not spread out. Find the frequency of a sound wave that has a wavelength (a) 10 times the diameter of a 30-cm speaker and (b) one-tenth the diameter of a 30-cm speaker. (c) Repeat this problem for a 6-cm speaker.

**105** • A whistle of frequency 500 Hz moves in a circle of radius 1 m at 3 rev/s. What are the maximum and minimum frequencies heard by a stationary listener in the plane of the circle and 5 m away from its center?

**106** • Ocean waves move toward the beach with a speed of 8.9 m/s and a crest-to-crest separation of 15.0 m. You are in a small boat anchored off shore. (*a*) What is the frequency of the ocean waves? (*b*) You now lift anchor and head out to sea at a speed of 15 m/s. What frequency of the waves do you observe?

**107** •• A 12.0-m wire of mass 85 g is stretched under a tension of 180 N. A pulse is generated at the left end of the wire, and 25 ms later a second pulse is generated at the right end of the wire. Where do the pulses first meet?

**108** •• SSM ISOLVE✓ Find the speed of a car the tone of whose horn will drop by 10 percent as it passes you.

**109** •• A loudspeaker driver 20 cm in diameter is vibrating at 800 Hz with an amplitude of 0.025 mm. Assuming that the air molecules in the vicinity have the same amplitude of vibration, find (*a*) the pressure amplitude immediately in front of the driver, (*b*) the sound intensity, and (*c*) the acoustic power being radiated.

**110** •• ISOLVE A plane, harmonic, acoustical wave that oscillates in air with an amplitude of 1 $\mu$m has an intensity of 10 mW/m². What is the frequency of the sound wave?

**111** •• Water flows at 7 m/s in a pipe of radius 5 cm. A plate having an area equal to the cross-sectional area of the pipe is suddenly inserted to stop the flow. Find the force exerted on the plate. Take the speed of sound in water to be 1.4 km/s. *Hint: When the plate is inserted, a pressure wave propagates through the water at the speed of sound $v_s$. The mass of water brought to a stop in time $\Delta t$ is the water in a length of tube equal to $v_s \Delta t$.*

**112** •• A high-speed flash photography setup to capture a picture of a bullet exploding a soap bubble is shown in Figure 15-31. The shock wave from the bullet is to be detected by a microphone that will trigger the flash. The microphone is placed on a track that is parallel to the path of the bullet, at the same height as the bullet. The track is used to adjust the position of the microphone. If the bullet is traveling at 1.25 times the speed of sound, and the distance between the lab bench and the track is 0.35 m, how far back from the soap bubble must the microphone be set to trigger the flash? (Assume that the flash itself is instantaneous once the microphone is triggered.)

**113** •• A column of precision marchers keeps in step by listening to the band positioned at the head of the column. The beat of the music is for 100 paces/min. A television camera shows that only the marchers at the front and the rear of the column are actually in step. The marchers in the middle section are striding forward with the left foot when

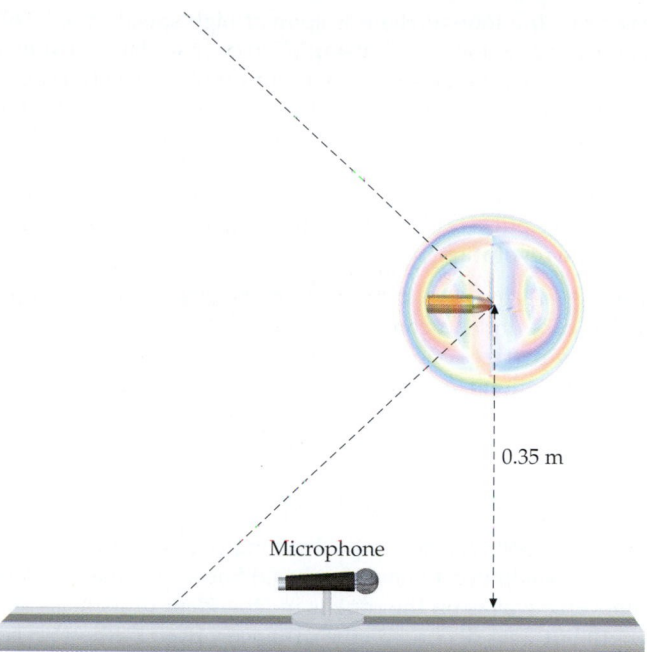

0.35 m

Microphone

**FIGURE 15-31** Problem 112

those at the front and rear are striding forward with the right foot. The marchers are so well trained, however, that they are all certain that they are in proper step with the music. Explain the source of the problem, and calculate the length of the column.

**114** •• A bat flying toward an obstacle at 12 m/s emits brief, high-frequency sound pulses at a repetition frequency of 80 Hz. What is the interval between the arrival times of the echo pulses heard by the bat?

**115** •• SSM Laser ranging to the moon is done routinely to accurately determine the earth–moon distance. However, to determine the distance accurately, corrections must be made for the speed of light in the earth's atmosphere, which is 99.997 percent of the speed of light in vacuum. Assuming that the earth's atmosphere is 8 km high, estimate the length of the correction.

**116** •• A tuning fork attached to a stretched wire generates transverse waves. The vibration of the fork is perpendicular to the wire. Its frequency is 400 Hz and the amplitude of its oscillation is 0.50 mm. The wire has a linear mass density of 0.01 kg/m and is under a tension of 1 kN. Assume that there are no reflected waves. (*a*) Find the period and frequency of waves on the wire. (*b*) What is the speed of the waves? (*c*) What are the wavelength and wave number? (*d*) Write a suitable wave function for the waves on the wire. (*e*) Calculate the maximum speed and acceleration of a point on the wire. (*f*) At what average rate must energy be supplied to the fork to keep it oscillating at a steady amplitude?

**117 •••** If a loop of chain is spun at high speed, it will roll like a hoop without collapsing. Consider a chain of linear mass density $\mu$ that is rolling without slipping at a high speed $v_0$. (a) Show that the tension in the chain is $F = \mu v_0^2$. (b) If the chain rolls over a small bump, a transverse wave pulse will be generated in the chain. At what speed will it travel along the chain? (c) How far around the loop (in degrees) will a transverse wave pulse travel in the time the hoop rolls through one complete revolution?

**118 •••** A long rope with a mass per unit length of 0.1 kg/m is under a constant tension of 10 N. A motor at the point $x = 0$ drives one end of the rope with harmonic motion at 5 oscillations per second and an amplitude of 4 cm. (a) What is the wave speed? (b) What is the wavelength? (c) What is the maximum transverse linear momentum of a 1-mm segment of the rope? (d) What is the maximum net force on a 1-mm segment of the rope?

**119 •••** **SSM** A heavy rope 3 m long is attached to the ceiling and is allowed to hang freely. (a) Show that the speed of transverse waves on the rope is independent of its mass and length but does depend on the distance $y$ from the bottom according to the formula $v = \sqrt{gy}$. (b) If the bottom end of the rope is given a sudden sideways displacement, how long does it take the resulting wave pulse to go to the ceiling, reflect, and return to the bottom of the rope?

**120 •••** In this problem you will derive an expression for the potential energy of a segment of a string carrying a traveling wave (Figure 15-32). The potential energy of a segment equals the work done by the tension in stretching the string, which is $\Delta U = F(\Delta \ell - \Delta x)$, where $F$ is the tension, $\Delta \ell$ is the length of the stretched segment, and $\Delta x$ is its original length. From the figure we see that

$$\Delta \ell = \sqrt{(\Delta x)^2 + (\Delta y)^2} = \Delta x [1 + (\Delta y/\Delta x)^2]^{1/2}$$

(a) Use the binomial expansion to show that $\Delta \ell - \Delta x \approx \frac{1}{2}(\Delta y/\Delta x)^2 \Delta x$, and therefore $\Delta U \approx \frac{1}{2}F(\Delta y/\Delta x)^2 \Delta x$. (b) Compute $dy/dx$ from the wave function in Equation 15-13 and show that $\Delta U \approx \frac{1}{2}Fk^2 \cos^2(kx - \omega t) \Delta x$.

**FIGURE 15-32**
**Problem 120**

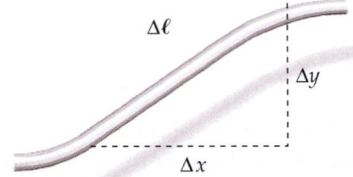

# Superposition and Standing Waves

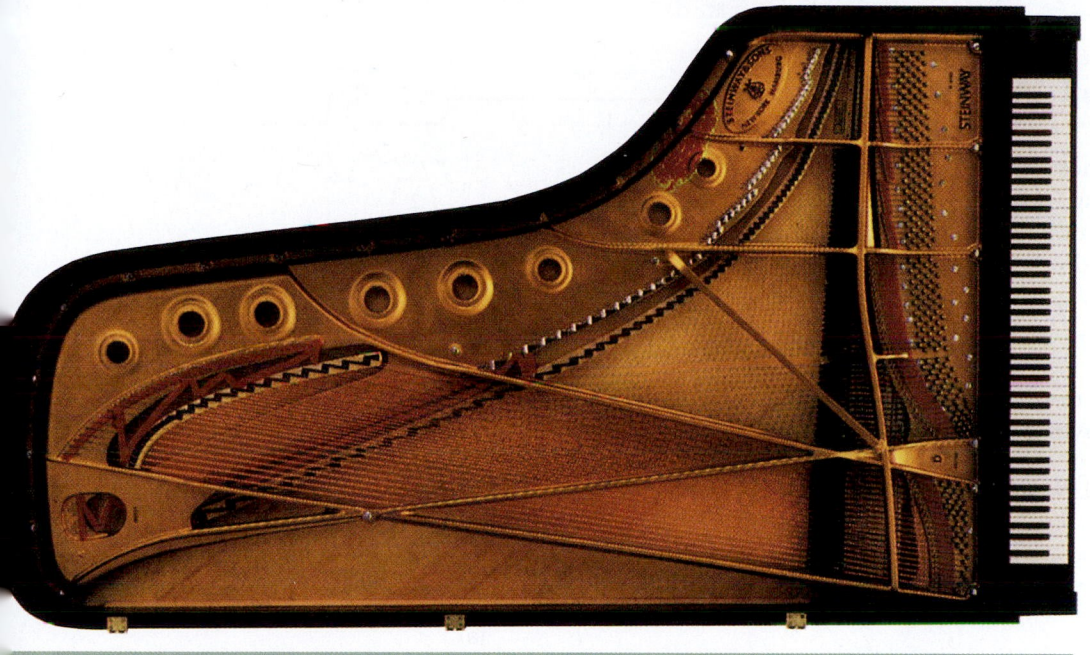

THE STRINGS IN THIS STEINWAY
GRAND PIANO VIBRATE WHEN
STRUCK BY THE HAMMERS, WHICH
ARE CONTROLLED BY THE KEYS.
THE LONGER STRINGS VIBRATE AT
LOWER FREQUENCIES THAN THE
SHORTER STRINGS.

**?** **What other factors come
into play when tuning a piano?
(See Example 16-6.)**

16-1   Superposition of Waves

16-2   Standing Waves

*16-3   The Superposition of Standing Waves

*16-4   Harmonic Analysis and Synthesis

*16-5   Wave Packets and Dispersion

**W**hen two or more waves overlap in space, their individual disturbances (represented mathematically by their wave functions) superimpose and add algebraically, creating a resultant wave. This property of waves is called the principle of superposition. Under certain circumstances the superposition of harmonic waves of the same frequency produces sustained wave patterns in space. This phenomenon is called interference. Interference and diffraction are what distinguish wave motion from particle motion. Thomas Young's observation in 1801 of interference of light led to the understanding that light propagates via wave motion, not via particle motion as had been proposed by Newton. The observation of interference of electron waves by C. J. Davisson and L. H. Germer in 1927 led to our understanding of the wave nature of electrons and other material objects. These ideas are integral to understanding quantum physics, which is presented in Chapter 34.

➤ **In this chapter, we begin with the superposition of wave pulses on a string and then consider the superposition and interference of harmonic waves. We examine the phenomenon of beats, which result from the interference of two waves of slightly different frequencies, and study standing waves, which occur when harmonic waves are confined in space. We also consider the analysis of complex musical tones in terms of their component harmonic waves, and** the inverse problem of the synthesis of harmonic waves to produce complex tones. We conclude with a qualitative discussion of the extension of harmonic analysis to aperiodic waves such as wave pulses.

# 16-1 Superposition of Waves

Figure 16-1a shows small wave pulses moving in opposite directions on a string. The shape of the string when they meet can be found by adding the displacements produced by each pulse separately. The **principle of superposition** is a property of wave motion which states:

> When two or more waves overlap, the resultant wave is the algebraic sum of the individual waves.

PRINCIPLE OF SUPERPOSITION

Mathematically, when there are two pulses on the string, the total wave function is the algebraic sum of the individual wave functions.

In the special case of two pulses that are identical except that one is inverted relative to the other, as in Figure 16-1b, there is an instant when the pulses exactly overlap and add to zero. At this instant the string is horizontal, but it is not stationary. At the right edge of the overlap region the string is moving upward and at the left edge it is moving downward. A short time later the individual pulses emerge, each continuing in its original direction.

Superposition is a characteristic and unique property of wave motion. There is no analogous situation in Newtonian particle motion; that is, two Newtonian particles never overlap or add together in this way.

## *Superposition and the Wave Equation

The principle of superposition follows from the fact that the wave equation (Equation 15-9) is linear for small transverse displacements. That is, the function $y(x,t)$ and its derivatives occur only to the first power. The defining property of a linear equation is that if $y_1$ and $y_2$ are two solutions of the equation, then the linear combination

$$y_3 = C_1y_1 + C_2y_2 \qquad 16\text{-}1$$

is also a solution, where $C_1$ and $C_2$ are any constants. The linearity of the wave equation can be shown by the direct substitution of $y_3$ into the equation. The result is the mathematical statement of the principle of superposition. If any two waves satisfy a wave equation, then their algebraic sum also satisfies the same wave equation.

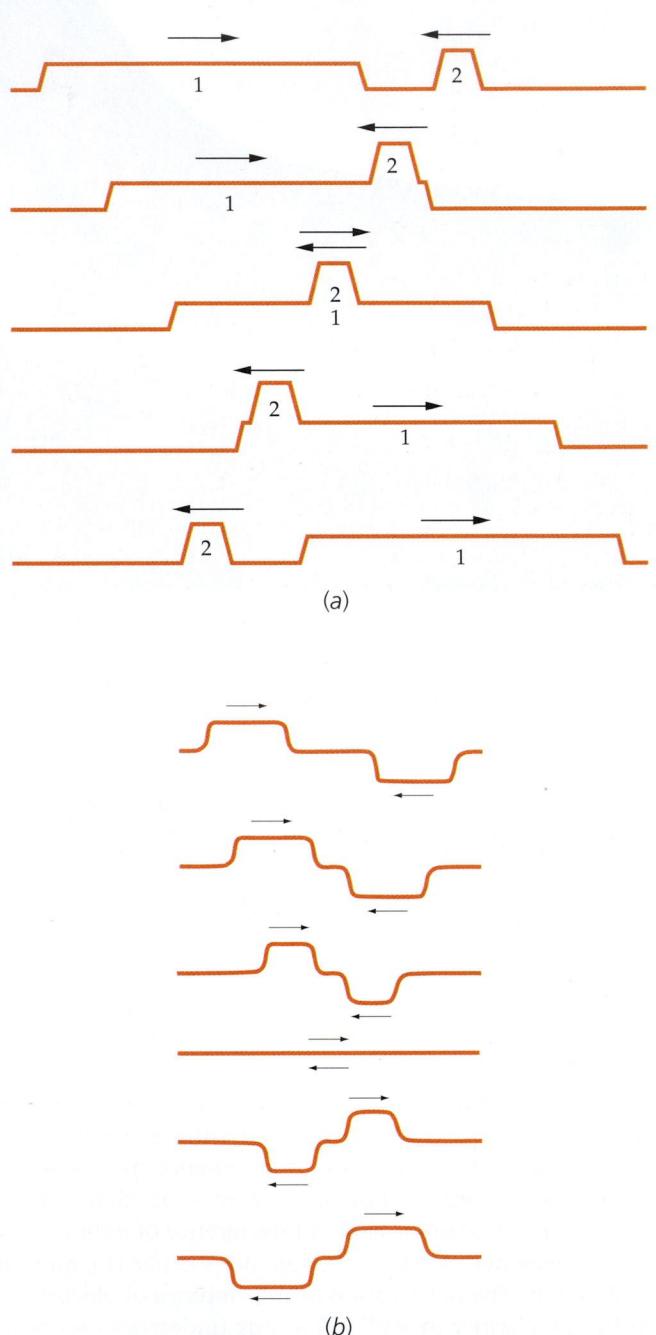

(a)

(b)

**FIGURE 16-1** Wave pulses moving in opposite directions on a string. The shape of the string when the pulses meet is found by adding the displacements of each separate pulse. (a) Superposition of pulses having displacements in the same direction. (b) Superposition of pulses having equal but opposite displacements. Here the algebraic addition of the displacement amounts to the subtraction of the magnitudes.

*SUPERPOSITION AND THE WAVE EQUATION* **EXAMPLE 16-1**

Show that if functions $y_1$ and $y_2$ both satisfy wave equation

$$\frac{\partial^2 y}{\partial x^2} = \frac{1}{v^2}\frac{\partial^2 y}{\partial t^2} \qquad \text{(Equation 15-9}b)$$

then the function $y_3$ given by Equation 16-1 also satisfies it.

**PICTURE THE PROBLEM** Substitute $y_3$ into the wave equation, assume that $y_1$ and $y_2$ each satisfy the wave equation, and show that, as a consequence, the linear combination $C_1 y_1 + C_2 y_2$ satisfies the wave equation.

1. Substitute the expression for $y_3$ in Equation 16-1 into the left side of the wave equation, then break it into separate terms for $y_1$ and $y_2$:

$$\frac{\partial^2 y_3}{\partial x^2} = \frac{\partial^2}{\partial x^2}(C_1 y_1 + C_2 y_2) = C_1\frac{\partial^2 y_1}{\partial x^2} + C_2\frac{\partial^2 y_2}{\partial x^2}$$

2. Both $y_1$ and $y_2$ satisfy the wave function. Write the wave equation for both $y_1$ and $y_2$:

$$\frac{\partial^2 y_1}{\partial x^2} = \frac{1}{v^2}\frac{\partial^2 y_1}{\partial t^2} \quad\text{and}\quad \frac{\partial^2 y_2}{\partial x^2} = \frac{1}{v^2}\frac{\partial^2 y_2}{\partial t^2}$$

3. Substitute the step 2 results into the step 1 result and factor out any common terms:

$$\frac{\partial^2 y_3}{\partial x^2} = C_1\frac{1}{v^2}\frac{\partial^2 y_1}{\partial t^2} + C_2\frac{1}{v^2}\frac{\partial^2 y_2}{\partial t^2} = \frac{1}{v^2}\left(C_1\frac{\partial^2 y_1}{\partial t^2} + C_2\frac{\partial^2 y_2}{\partial t^2}\right)$$

4. Move the constants inside the arguments of the derivatives and express the sum of the derivatives as the derivative of the sum:

$$\frac{\partial^2 y_3}{\partial x^2} = \frac{1}{v^2}\left(\frac{\partial^2 C_1 y_1}{\partial t^2} + \frac{\partial^2 C_2 y_2}{\partial t^2}\right) = \frac{1}{v^2}\frac{\partial^2}{\partial t^2}(C_1 y_1 + C_2 y_2)$$

5. The argument of the derivative in step 4 is $y_3$:

$$\therefore \boxed{\frac{\partial^2 y_3}{\partial x^2} = \frac{1}{v^2}\frac{\partial^2 y_3}{\partial t^2}}$$

## Interference of Harmonic Waves

The result of the superposition of two harmonic waves of the same frequency depends on the phase difference $\delta$ between the waves. Let $y_1(x,t)$ be the wave function for a harmonic wave traveling to the right with amplitude $A$, angular frequency $\omega$, and wave number $k$:

$$y_1 = A\sin(kx - \omega t) \qquad\qquad 16\text{-}2$$

For this wave function, we have chosen the phase constant to be zero.[†] If we have another harmonic wave also traveling to the right with the same amplitude, frequency, and wave number, then the general equation for its wave function can be written

$$y_2 = A\sin(kx - \omega t + \delta) \qquad\qquad 16\text{-}3$$

where $\delta$ is the phase constant. The two waves described by Equations 16-2 and 16-3 differ in phase by $\delta$. Figure 16-2 shows a plot of the two wave functions versus position for a fixed time. The resultant wave is the sum

$$y_1 + y_2 = A\sin(kx - \omega t) + A\sin(kx - \omega t + \delta) \qquad 16\text{-}4$$

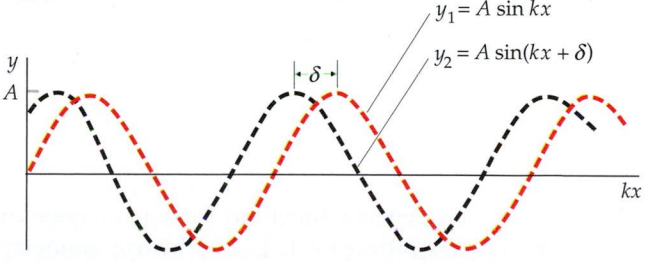

$y_1 = A\sin kx$

$y_2 = A\sin(kx + \delta)$

**FIGURE 16-2** Displacement versus position for two harmonic waves having the same amplitude, frequency, and wavelength, but differing in phase by $\delta$.

---

[†] This choice is convenient but not mandatory. If, for example, we chose $t = 0$ when the displacement was maximum at $x = 0$, we would write $y_1 = A\cos(kx - \omega t) = A\sin(kx - \omega t - \pi/2)$.

We can simplify Equation 16-4 by using the trigonometric identity

$$\sin\theta_1 + \sin\theta_2 = 2\cos\tfrac{1}{2}(\theta_1 - \theta_2)\sin\tfrac{1}{2}(\theta_1 + \theta_2) \qquad 16\text{-}5$$

For this case, $\theta_1 = kx - \omega t$ and $\theta_2 = kx - \omega t + \delta$, so that

$$\tfrac{1}{2}(\theta_1 - \theta_2) = -\tfrac{1}{2}\delta$$

and

$$\tfrac{1}{2}(\theta_1 + \theta_2) = kx - \omega t + \tfrac{1}{2}\delta$$

Thus, Equation 16-4 becomes

$$y_1 + y_2 = \left[2A\cos\tfrac{1}{2}\delta\right]\sin(kx - \omega t + \tfrac{1}{2}\delta) \qquad 16\text{-}6$$

SUPERPOSITION OF TWO WAVES OF THE SAME AMPLITUDE AND FREQUENCY

where we have used $\cos(-\tfrac{1}{2}\delta) = \cos\tfrac{1}{2}\delta$. We see that the result of the superposition of two harmonic waves of equal wave number and frequency is a harmonic wave having the same wave number and frequency. The resultant wave has amplitude $2A\cos\tfrac{1}{2}\delta$ and a phase equal to half the difference between the phases of the original waves. The phenomenon of two or more waves of the same, or almost the same, frequency superposing to produce an observable pattern in the intensity is called **interference.** In this example, the intensity, which is proportional to the square of the amplitude, is uniform. If the two waves are in phase, then $\delta = 0$, $\cos 0 = 1$, and the amplitude of the resultant wave is $2A$. The interference of two waves in phase is called **constructive interference** (Figure 16-3). If the two waves are 180° out of phase, then $\delta = \pi$, $\cos(\pi/2) = 0$, and the amplitude of the resultant wave is zero. The interference of two waves 180° out of phase is called **destructive interference** (Figure 16-4).

**EXERCISE** Two waves with the same frequency, wavelength, and amplitude are traveling in the same direction. (*a*) If they differ in phase by $\pi/2$ and each has an amplitude of 4.0 cm, what is the amplitude of the resultant wave? (*b*) For what phase difference $\delta$ will the resultant amplitude be equal to 4.0 cm? (*Answer*   (*a*) 5.66 cm (*b*) 120° or 240°)

**Beats**   The interference of two sound waves with slightly different frequencies produces the interesting phenomenon known as **beats.** Consider two sound waves that have angular frequencies of $\omega_1$ and $\omega_2$ and the same pressure amplitude $p_0$. What do we hear? At a fixed point, the spatial dependence of the wave merely contributes a phase constant, so we can neglect it. The pressure at the ear due to either wave acting alone will be a simple harmonic function of the type

$$p_1 = p_0 \sin\omega_1 t$$

and

$$p_2 = p_0 \sin\omega_2 t$$

where we have chosen sine functions for convenience and have assumed that the waves are in phase at time $t = 0$. Using the trigonometry identity

$$\sin\theta_1 + \sin\theta_2 = 2\cos\tfrac{1}{2}(\theta_1 - \theta_2)\sin\tfrac{1}{2}(\theta_1 + \theta_2)$$

for the sum of two sine functions, we obtain for the resultant wave

$$p = p_0\sin\omega_1 t + p_0\sin\omega_2 t = 2p_0\cos\tfrac{1}{2}(\omega_1 - \omega_2)t\,\sin\tfrac{1}{2}(\omega_1 + \omega_2)t$$

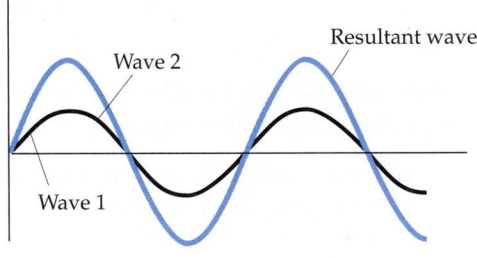

**FIGURE 16-3** Constructive interference. When two waves are in phase, the amplitude of the resultant wave is the sum of the amplitudes of the individual waves.

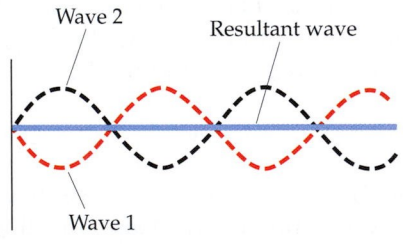

**FIGURE 16-4** Destructive interference. When two waves have a phase difference of $\pi$, the amplitude of the resultant wave is the difference between the amplitudes of the individual waves. If the original waves have equal amplitudes, they cancel completely.

If we write $\omega_{av} = \frac{1}{2}(\omega_1 + \omega_2)$ for the average angular frequency and $\Delta\omega = \omega_1 - \omega_2$ for the difference in angular frequencies, the resultant wave function is

$$p = 2p_0 \cos(\tfrac{1}{2}\Delta\omega t) \sin \omega_{av}t =$$
$$2p_0 \cos(2\pi\tfrac{1}{2}\Delta f t)\sin 2\pi f_{av}t \qquad 16\text{-}7$$

where $\Delta f = \Delta\omega/(2\pi)$ and $f_{av} = \omega_{av}/(2\pi)$.

Figure 16-5 shows a plot of pressure variations as a function of time. The waves are originally in phase and add constructively at time $t = 0$. Because their frequencies differ, the waves gradually become out of phase, and at time $t_1$ they are 180° out of phase and interfere destructively.[†] An equal time interval later (time $t_2$ in the figure), the two waves are again in phase and interfere constructively. The greater the difference in the frequencies of the two waves, the more rapidly they oscillate in and out of phase.

The tone we hear has a frequency of $f_{av} = \frac{1}{2}(f_1 + f_2)$ and amplitude $2p_0 \cos(2\pi\frac{1}{2}\Delta f t)$. (For some values of $t$ the amplitude is negative. Since $-\cos\theta = \cos(\theta + \pi)$, a change in the sign of the amplitude is equivalent to a 180° phase change.) The amplitude oscillates with the frequency $\frac{1}{2}\Delta f$. Since the sound intensity is proportional to the square of the amplitude, the sound is loud whenever the amplitude function is either a maximum or a minimum. Thus the frequency of this variation in intensity, called the **beat frequency**, is twice $\frac{1}{2}\Delta f$:

$$f_{beat} = \Delta f \qquad 16\text{-}8$$

BEAT FREQUENCY

The beat frequency equals the difference in the individual frequencies of the two waves. If we simultaneously strike two tuning forks having the frequencies 241 Hz and 243 Hz, we will hear a pulsating tone at the average frequency of 242 Hz that has a maximum intensity 2 times per second; that is, the beat frequency is 2 Hz. The ear can detect up to about 15 to 20 beats per second. Above this frequency, the fluctuations in loudness are too rapid to be distinguished.

The phenomenon of beats is often used to compare an unknown frequency with a known frequency, as when a tuning fork is used to tune a piano string. Pianos are tuned by simultaneously ringing the tuning fork and striking a key, while at the same time adjusting the tension of the piano string until the beats are far apart, indicating that the difference in frequency of the two sound generators is very small.

*FIGURE 16-5* Beats. (a) Two waves of different but nearly equal frequencies that are in phase at $t_0 = 0$ are 180° out of phase at some time later $t_1$. At a still later time, $t_2$, they are back in phase. (b) The resultant of the two waves shown in (a). The frequency of the resultant wave is about the same as those of the original waves, but the amplitude is modulated as indicated by the dashed envelope. The amplitude is maximum at times $t_0$ and $t_2$ and zero at times $t_1$ and $t_3$.

---

*TUNING A GUITAR*                    **EXAMPLE   16-2**

When a 440-Hz (concert A) tuning fork is struck simultaneously with the playing of the A string of a slightly out-of-tune guitar, 3 beats per second are heard. The guitar string is tightened a little to increase its frequency; the beat frequency increases to 6 beats per second. As the guitar string is slowly tightened, you hear the beat frequency slowly increase. What was the frequency of the guitar string before it is tightened?

**PICTURE THE PROBLEM** Because 3 beats per second were heard initially, the original frequency of the guitar string was either 443 Hz or 437 Hz. Had it been 437 Hz, slowly increasing the string's frequency by tightening it would decrease the beat frequency.

Because the beat frequency increases as the tension increases, from 3 to 6 beats per second, the original frequency must have been 443 Hz.

$$f = f_A + f_{beat} = 440 \text{ Hz} + 3 \text{ Hz} = \boxed{443 \text{ Hz}}$$

---

[†] Complete cancellation occurs only when the pressure amplitudes of the two waves are equal.

**Phase Difference due to Path Difference** A common cause of a phase difference between two waves is different path lengths between the sources of the waves and the point of interference. Suppose that two sources oscillate in phase (for example, positive crests leave the sources at the same time) and emit harmonic waves of the same frequency and wavelength. Now consider a point in space for which the path lengths to the two sources differ. If the path difference is one wavelength, as is the case in Figure 16-6a, or an integral number of wavelengths, the interference is constructive. If the path difference is one half of a wavelength or an odd number of half wavelengths, as in Figure 16-6b, the maximum of one wave falls at the minimum of the other and the interference is destructive.

The wave functions for waves from two sources oscillating in phase can be written

$$p_1 = p_0 \sin(kx_1 - \omega t)$$

and

$$p_2 = p_0 \sin(kx_2 - \omega t)$$

The phase difference for these two wave functions is

$$\delta = (kx_2 - \omega t) - (kx_1 - \omega t) = k(x_2 - x_1) = k\Delta x$$

Using $k = 2\pi/\lambda$, we have

$$\delta = k\Delta x = 2\pi\frac{\Delta x}{\lambda} \qquad \text{16-9}$$

PHASE DIFFERENCE DUE TO PATH DIFFERENCE

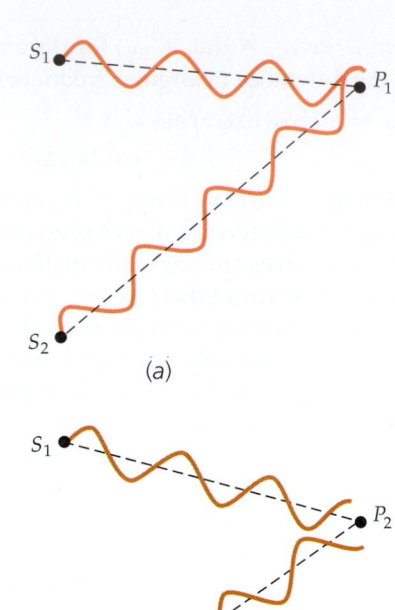

**FIGURE 16-6** Waves from two sources $S_1$ and $S_2$ that are in phase when they meet at a point $P_1$. (a) When the path difference is one wavelength $\lambda$, the waves are in phase at $P_1$ and interfere constructively. (b) When the path difference is $\frac{1}{2}\lambda$, the waves at $P_2$ are out of phase by 180° and therefore interfere destructively. If the waves are of equal amplitude at $P_2$, they will cancel completely at this point.

---

*A Resultant Sound Wave*   **EXAMPLE 16-3**

**Two sound sources oscillate in phase. At a point 5.00 m from one source and 5.17 m from the other, the amplitude of the sound from each source separately is $p_0$. Find the amplitude of the resultant wave if the frequency of the sound waves is (a) 1000 Hz, (b) 2000 Hz, and (c) 500 Hz. (Use 340 m/s for the speed of sound.)**

**PICTURE THE PROBLEM** The amplitude of the resultant wave due to superposition of two waves differing in phase by $\delta$ is given by $A = 2p_0 \cos\frac{1}{2}\delta$ (Equation 16-6), where $p_0$ is the amplitude of either wave and $\delta = 2\pi \Delta x/\lambda$ is the phase difference. We are given the path difference, $\Delta x = 5.17$ m $- 5$ m $= 0.17$ m, so all that is needed is the wavelength $\lambda$.

(a) 1. The wavelength equals the speed divided by the frequency. Calculate $\lambda$ for $f = 1000$ Hz:

$$\lambda = \frac{v}{f} = \frac{340 \text{ m/s}}{1000 \text{ Hz}} = 0.34 \text{ m}$$

2. For $\lambda = 0.34$ m, the given path difference ($\Delta x = 0.17$ m) is $\frac{1}{2}\lambda$, so we expect destructive interference. Use this value of $\lambda$ to calculate the phase difference $\delta$ and use $\delta$ to calculate the amplitude $A$:

$$\delta = 2\pi\frac{\Delta x}{\lambda} = 2\pi\frac{0.17 \text{ m}}{0.34 \text{ m}} = \pi$$

so

$$A = 2p_0 \cos\frac{1}{2}\delta = 2p_0 \cos\frac{\pi}{2} = \boxed{0}$$

(b) 1. Calculate $\lambda$ for $f = 2000$ Hz:

$$\lambda = \frac{v}{f} = \frac{340 \text{ m/s}}{2000 \text{ Hz}} = 0.17 \text{ m}$$

2. For $\lambda = 0.17$ m, the path difference equals $\lambda$, so we expect constructive interference. Calculate the phase difference and amplitude:

$$\delta = 2\pi\frac{\Delta x}{\lambda} = 2\pi\frac{0.17 \text{ m}}{0.17 \text{ m}} = 2\pi$$

so

$$A = 2p_0 \cos\tfrac{1}{2}\delta = 2p_0 \cos\pi = \boxed{-2p_0}$$

(c) 1. Calculate $\lambda$ for $f = 500$ Hz:

$$\lambda = \frac{v}{f} = \frac{340 \text{ m/s}}{500 \text{ Hz}} = 0.68 \text{ m}$$

2. Calculate the phase difference and amplitude:

$$\delta = 2\pi\frac{\Delta x}{\lambda} = 2\pi\frac{0.17 \text{ m}}{0.68 \text{ m}} = \frac{\pi}{2}$$

so

$$A = 2p_0 \cos\frac{1}{2}\delta = 2p_0 \cos\frac{\pi}{4} = \boxed{\sqrt{2}p_0}$$

**REMARKS**  In part (b), $A$ is found to be negative. Equation 16-6 can be written $y_1 + y_2 = A' \sin(kx - \omega t + \frac{\delta}{2})$, which can also be written $y_1 + y_2 = -A' \sin(kx - \omega t + \frac{\delta}{2} + \pi)$. A phase shift of $\pi = 180°$ is equivalent to multiplying by $-1$.

---

*SOUND INTENSITY OF TWO LOUDSPEAKERS*                **EXAMPLE  16-4**

**Two identical loudspeakers face each other at a distance of 180 cm and are driven by a common audio oscillator at 680 Hz. Locate the points between the speakers along a line joining them for which the sound intensity is (a) maximum and (b) minimum. (Neglect the variation in intensity from either speaker with distance, and use 340 m/s for the speed of sound.)**

**FIGURE  16-7**

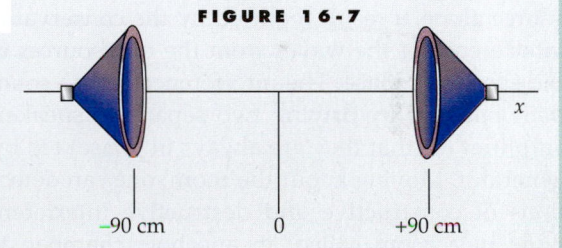

**PICTURE  THE  PROBLEM**  We choose the origin to be at the midpoint between the speakers (Figure 16-7). Since this point is equidistant from the speakers, it is a point of maximum intensity. When we move a distance $x$ toward one of the speakers, the path difference is $2x$. The intensity will be maximum when $2x = 0, \lambda, 2\lambda, 3\lambda, \ldots$, and minimum when $2x = (1/2)\lambda, (3/2)\lambda, (5/2)\lambda, \ldots$.

(a) 1. The intensity will be maximum when $2x$ equals an integral number of wavelengths:

$$2x = 0, \pm\lambda, \pm 2\lambda, \pm 3\lambda, \ldots$$

2. Calculate the wavelength:

$$\lambda = \frac{v}{f} = \frac{340 \text{ m/s}}{680 \text{ Hz}} = 0.5 \text{ m} = 50 \text{ cm}$$

3. Solve for $x$ using the calculated wavelength:

$$x = 0, \pm\tfrac{1}{2}\lambda, \pm\lambda, \pm\tfrac{3}{2}\lambda, \ldots$$

$$= \boxed{0, \pm 25 \text{ cm}, \pm 50 \text{ cm}, \pm 75 \text{ cm}}$$

(b) 1. The intensity will be minimum when $2x$ equals an odd number of half wavelengths:

$$2x = \pm\tfrac{1}{2}\lambda, \pm\tfrac{3}{2}\lambda, \pm\tfrac{5}{2}\lambda, \ldots$$

2. Solve for $x$ using the calculated wavelength:

$$x = \pm\tfrac{1}{4}\lambda, \pm\tfrac{3}{4}\lambda, \pm\tfrac{5}{4}\lambda, \ldots$$

$$= \boxed{\pm 12.5 \text{ cm}, \pm 37.5 \text{ cm}, \pm 62.5 \text{ cm}, \pm 87.5 \text{ cm}}$$

**REMARKS**  The maxima and minima will be relative maxima and relative minima because the amplitude from the near speaker will be slightly greater than that from the far speaker. Only seven terms were used for the maxima and only eight terms for the minima because any additional terms would be at a distance beyond one speaker.

Figure 16-8a shows the wave pattern produced by two point sources in a ripple tank that are oscillating in phase. Each source produces waves with circular wavefronts. The circular wavefronts shown all have the same phase and are separated by one wavelength. We can construct a similar pattern with a compass by drawing circular arcs representing the wave crests from each source at some particular time (Figure 16-8b). Where the crests from each source overlap, the waves interfere constructively. At these points, the path lengths from the two sources are either equal or they differ by an integral number of wavelengths. The dashed lines indicate the points that are equidistant from the sources or whose path differences are one wavelength, two wavelengths, or three wavelengths. At each point along any of these lines the interference is constructive, so these are lines of interference maxima. Between the lines of interference maxima are lines of interference minima. On a line of interference minima, the path length from any point on the line to each of the two sources differs by an odd number of half wavelengths. Throughout the region where the two waves are superposed, the amplitude of the resultant wave is given by $A = 2p_0 \cos \frac{1}{2}\delta$, where $p_0$ is the amplitude of each wave separately and $\delta$ is related to the path difference $\Delta r$ by $\delta = 2\pi \Delta r/\lambda$ (Equation 16-9).

Figure 16-9 shows the intensity $I$ of the resultant wave from two sources as a function of path difference $\Delta x$. At points where the interference is constructive, the amplitude of the resultant wave is twice that of either wave alone, and since the intensity is proportional to the square of the amplitude, the intensity is $4I_0$, where $I_0$ is the intensity due to either source alone. At points of destructive interference, the intensity is zero. The average intensity, shown by the dashed line in the figure, is twice the intensity due to either source alone, a result required by the conservation of energy. The interference of the waves from the two sources thus redistributes the energy in space. The interference of two sound sources can be demonstrated by driving two separated speakers with the same amplifier (so that they are always in phase) fed by an audio-signal generator. Moving about the room, one can detect by ear the positions of constructive and destructive interference.[†] This is best done in a room called an anechoic chamber, where reflections (echoes) off the walls of the room are minimized.

**Coherence**   Two sources need not be in phase to produce an interference pattern. Consider two sources that are 180° out of phase. (Two speakers that are in phase can be made to be out of phase by 180° merely by switching the leads to one of the speakers.) The intensity pattern is the same as that in Figure 16-9 except that the maxima and minima are interchanged. At points for which the distance differs by an integral number of wavelengths, the interference is destructive because the waves are 180° out of phase. At points where the path difference is an odd number of half wavelengths, the waves are now in phase because the 180° phase difference of the sources is offset by the 180° phase difference due to the path difference.

Similar interference patterns will be produced by any two sources whose phase difference remains constant. Two

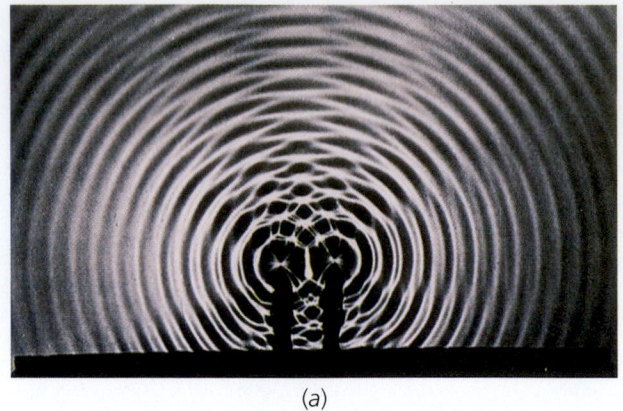

(a)

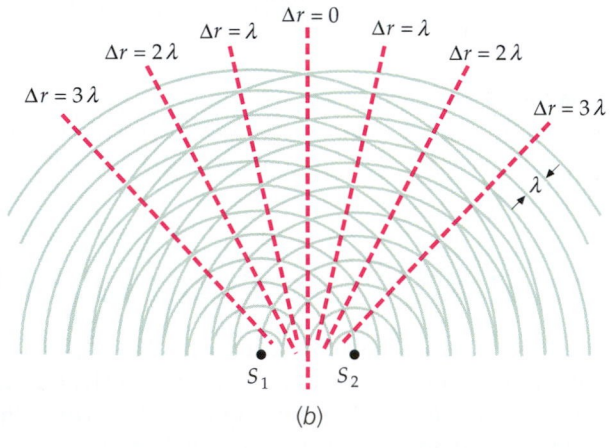

(b)

**FIGURE 16-8** (a) Water waves in a ripple tank produced by two nearby sources oscillating in phase. (b) Drawing of wave crests for the sources in (a). The dashed lines indicate points for which the path difference is an integral number of wavelengths.

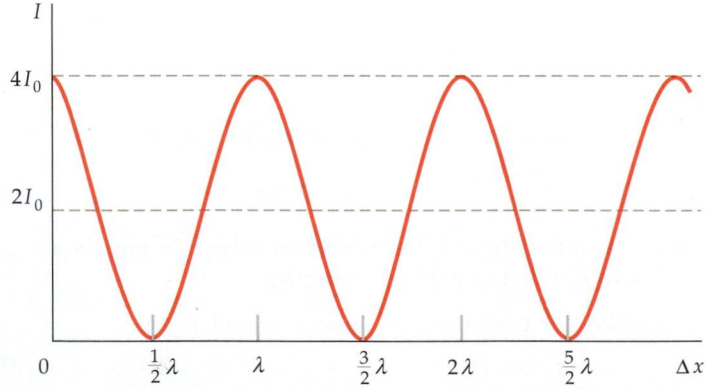

**FIGURE 16-9** Intensity versus path difference for two sources that are in phase. $I_0$ is the intensity due to each source individually.

---

† In this demonstration, the sound intensity will be not quite zero at the points of destructive interference because of sound reflections from the walls or objects in the room.

sources that remain in phase or maintain a constant phase difference are said to be **coherent.** Coherent sources of water waves in a ripple tank are easy to produce by driving both sources with the same motor. Coherent sound sources are obtained by driving two speakers with the same signal source and amplifier.

Wave sources whose difference in phase is not constant but varies randomly are said to be **incoherent sources.** There are many examples of incoherent sources, such as two speakers driven by different amplifiers or two violins played by different violinists. For incoherent sources, the interference at a particular point varies rapidly back and forth from constructive to destructive, and no interference pattern is maintained long enough to be observed. The resultant intensity of waves from two or more incoherent sources is simply the sum of the intensities due to the individual sources.

# 16-2 Standing Waves

If waves are confined in space, like the waves on a piano string, sound waves in an organ pipe, or light waves in a laser, reflections at both ends cause the waves to travel in both directions. These superposing waves interfere in accordance with the principle of superposition. For a given string or pipe, there are certain frequencies for which superposition results in a stationary vibration pattern called a **standing wave.** Standing waves have important applications in musical instruments and in quantum theory.

## Standing Waves on Strings

**String Fixed at Both Ends**  If we fix both ends of a string and move a portion of the string up and down with simple harmonic motion of small amplitude, we find that at certain frequencies, standing-wave patterns such as those shown in Figure 16-10 are produced. The frequencies that produce these patterns are called the **resonance frequencies** of the string system. Each such frequency, with its accompanying wave function, is called a **mode of vibration.** The lowest resonance frequency is called the **fundamental** frequency $f_1$. It produces the standing-wave pattern shown in Figure 16-10a, which is called the **fundamental mode** of vibration or the **first harmonic.** The second lowest frequency $f_2$ produces the pattern shown in Figure 16-10b. This mode of vibration has a frequency twice that of the fundamental frequency and is called the second harmonic. The third lowest frequency $f_3$ is three times the fundamental frequency, and it produces the third harmonic pattern shown in Figure 16-10c. The set of all resonant frequencies is called the **resonant frequency spectrum** of the string.

❶  Not all resonant frequencies are called harmonics. **Only if each frequency of a resonant frequency spectrum is an integral multiple of the fundamental (lowest) frequency are the frequencies referred to as harmonics.** Many systems that support standing waves have resonant frequency spectra in which the resonant frequencies are not integral multiples of the lowest frequency. In all resonant frequency spectra the lowest resonant frequency is called the fundamental frequency (or just the fundamental), the next lowest resonant frequency is called the first **overtone,** the next lowest the second overtone, and so forth. This terminology has its

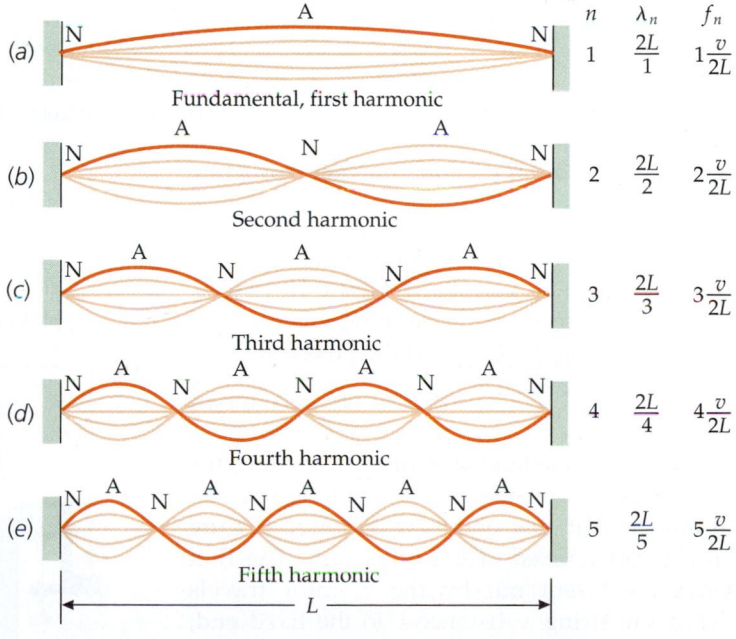

**FIGURE 16-10** Standing waves on a string that is fixed at both ends. Points labeled A are antinodes and those labeled N are nodes. In general, the $n$th harmonic has $n$ antinodes.

roots in music. Only if each resonant frequency is an integral multiple of the fundamental frequency are they referred to as harmonics.

We note from Figure 16-10 that for each harmonic there are certain points on the string (the midpoint in Figure 16-10b, for example) that do not move. Such points are called **nodes.** Midway between each pair of nodes is a point of maximum amplitude of vibration called an **antinode.** Both fixed ends of the string are, of course, nodes. (If one end is attached to a tuning fork or other vibrator rather than being fixed, it will still be approximately a node because the amplitude of the vibration at that end is so much smaller than the amplitude at the antinodes.) We note that the first harmonic has one antinode, the second harmonic has two antinodes, and so on.

We can relate the resonance frequencies to the wave speed in the string and the length of the string. The distance from a node to the nearest antinode is one-fourth of the wavelength. Therefore, the length of the string $L$ equals one-half the wavelength in the fundamental mode of vibration (Figure 16-11) and, as Figure 16-10 reveals, $L$ equals two half-wavelengths for the second harmonic, three half-wavelengths for the third harmonic, and so forth. In general, if $\lambda_n$ is the wavelength of the $n$th harmonic, we have

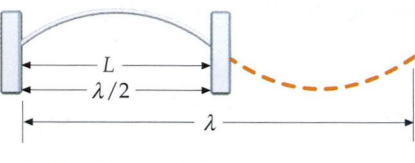

**FIGURE 16-11**

$$L = n\frac{\lambda_n}{2}, \qquad n = 1, 2, 3, \ldots \qquad \text{16-10}$$

STANDING-WAVE CONDITION, BOTH ENDS FIXED

This result is known as the **standing-wave condition.** We can find the frequency of the $n$th harmonic from the fact that the wave speed $v$ equals the frequency $f_n$ times the wavelength. Thus,

$$f_n = \frac{v}{\lambda_n} = \frac{v}{2L/n} \qquad n = 1, 2, 3, \ldots$$

or

$$f_n = n\frac{v}{2L} = nf_1 \qquad n = 1, 2, 3, \ldots \qquad \text{16-11}$$

RESONANCE FREQUENCIES, BOTH ENDS FIXED

where $f_1 = v/(2L)$ is the fundamental frequency.

You shouldn't bother to memorize Equation 16-11. Just sketch Figure 16-10 to remind yourself of the standing-wave condition, $\lambda_n = 2L/n$, and then use $v = f_n\lambda_n$.

PROBLEM-SOLVING GUIDELINE

We can understand standing waves in terms of resonance. Consider a string of length $L$ that is attached at one end to a vibrator (Figure 16-12) and is fixed at the other end. The first wave crest sent out by the vibrator travels down the string a distance $L$ to the fixed end, where it is reflected and inverted. It then travels back a distance $L$ and is again reflected and inverted at the vibrator. The total time for the round trip is $2L/v$. If this time equals the period

**FIGURE 16-12**

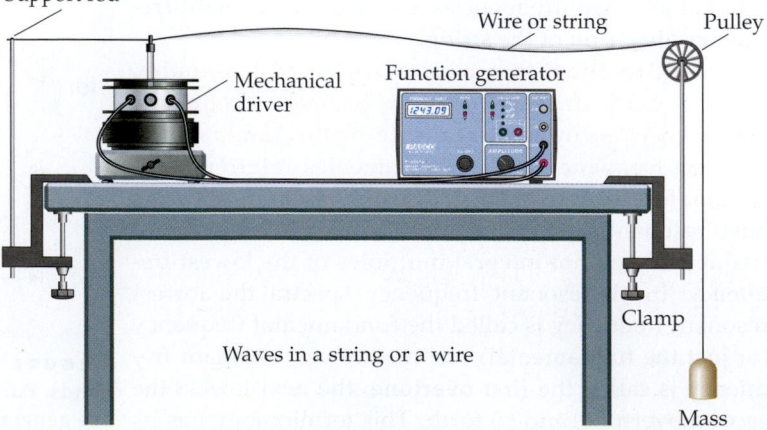

Waves in a string or a wire

of the vibrator, the twice-reflected wave crest exactly overlaps the second wave crest produced by the vibrator, and the two crests interfere constructively, producing a crest with twice the original amplitude. The combined wave crest travels down the string and back and is added to by the third crest produced by the vibrator, increasing the amplitude three-fold, and so on. Thus, the vibrator is in resonance with the string. The wavelength is equal to $2L$ and the frequency is equal to $v/(2L)$.

Resonance also occurs at other vibrator frequencies. The vibrator is in resonance with the string if the time it takes for the first wave crest to travel the distance $2L$ is any integer $n$ times the period $T_n$ of the vibrator. That is, if $2L/v = nT_n$, where $2L/v$ is the round trip time for a wave crest. Thus,

$$f_n = \frac{1}{T_n} = n\frac{v}{2L}, \qquad n = 1, 2, 3, \ldots$$

is the condition for resonance. This is the same result we found by fitting an integral number of half-wavelengths into the distance $L$. Various damping effects, such as the loss of energy during reflection and the imperfect flexibility of the string, put a limit on the maximum amplitude that can be reached.

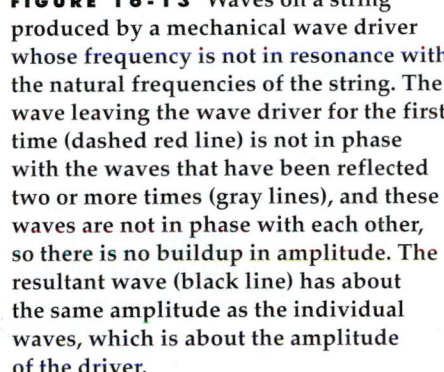

The resonance frequencies given by Equation 16-11 are also called the **natural frequencies** of the string. When the frequency of the vibrator is not one of the natural frequencies of the vibrating string, standing waves are not produced. After the first wave travels the distance $2L$ and is reflected from the fork, it differs in phase from the wave being generated at the vibrator (Figure 16-13). When this resultant wave has traveled the distance $2L$ and is again reflected at the vibrator, it will differ in phase from the next wave generated. In some cases, the new resultant wave will have an amplitude greater than that of the previous wave, in other cases the new amplitude will be less. On the average, the amplitude will not increase but will remain on the order of the amplitude of the first wave generated, which is the amplitude of the vibrator. This amplitude is very small compared with the amplitudes attained at resonance frequencies.

The resonance of standing waves is analogous to the resonance of a simple harmonic oscillator with a harmonic driving force. However, a vibrating string has not just one natural frequency but a sequence of natural frequencies that are integral multiples of the fundamental frequency. This sequence is called a **harmonic series.**

**FIGURE 16-13** Waves on a string produced by a mechanical wave driver whose frequency is not in resonance with the natural frequencies of the string. The wave leaving the wave driver for the first time (dashed red line) is not in phase with the waves that have been reflected two or more times (gray lines), and these waves are not in phase with each other, so there is no buildup in amplitude. The resultant wave (black line) has about the same amplitude as the individual waves, which is about the amplitude of the driver.

Turbulent winds set up standing waves in the Tacoma Narrows suspension bridge, leading to its collapse on November 7, 1940, just four months after it had been opened for traffic.

*GIVE ME AN A*                                    **EXAMPLE   16-5**

A string is stretched between two fixed supports 0.7 m apart and the tension is adjusted until the fundamental frequency of the string is concert A, 440 Hz. What is the speed of transverse waves on the string?

**PICTURE THE PROBLEM** The wave speed equals the frequency times the wavelength. For a string fixed at both ends, in the fundamental mode there is a single antinode in the middle of the string. Thus the length of the string equals one half-wavelength.

1. The wave speed is related to the frequency and wave-length: $\quad v = f_1\lambda_1$

2. For the fundamental, the length of the string is one half-wavelength: $\quad L = \lambda_1/2$

3. Use this wavelength and the given frequency to find the speed: $\quad v = f_1\lambda_1 = f_1 2L = (440 \text{ Hz}) \times 2(0.7 \text{ m})$

$$= \boxed{616 \text{ m/s}}$$

 **EXERCISE** The speed of transverse waves on a stretched string is 200 m/s. If the string is 5 m long, find the frequencies of the fundamental and the second and third harmonics. (*Answer* $f_1 = 20$ Hz, $f_2 = 40$ Hz, $f_3 = 60$ Hz)

---

*TESTING PIANO WIRE*                              **EXAMPLE   16-6**

You have a summer job at a music shop, helping the owner build instruments. He asks you to test a new wire for possible use in pianos. He tells you that the 3-m-long wire has a linear mass density of 0.0025 kg/m, and he has found two adjacent resonant frequencies at 252 Hz and at 336 Hz. He wants you to determine the fundamental frequency of the wire and determine whether or not the wire is a good choice for piano strings. You know that safety issues start to arise if the tension in the wire gets above 700 N.

**PICTURE THE PROBLEM** The tension $F_T$ is found from $v = \sqrt{F_T/\mu}$, where the speed $v$ can be found from $v = f\lambda$ using any harmonic. The wavelength for the fundamental is twice the length of the wire. To find the fundamental frequency let 252 Hz be the $n$th harmonic. Then $f_n = nf_1$ and $f_{n+1} = (n + 1)f_1$, where $f_{n+1} = 336$ Hz. We can solve these two equations for $f_1$.

1. The tension is related to the wave speed: $\quad v = \sqrt{F_T/\mu} \quad$ or $\quad F_T = \mu v^2$

2. Use the fundamental $f_1$, with $\lambda_1 = 2L$, to obtain the speed: $\quad v = f_1\lambda_1 = f_1(2L)$

3. Combine the two previous results to find the tension: $\quad F_T = \mu v^2 = \mu f_1^2(2L)^2$

4. The consecutive harmonics $f_n$ and $f_{n+1}$ are related to the fundamental frequency $f_1$: $\quad nf_1 = 252 \text{ Hz}$

$\quad (n + 1)f_1 = 336 \text{ Hz}$

5. Dividing these equations eliminates $f_1$ and allows us to determine $n$: $\quad \dfrac{n}{n+1} = \dfrac{252 \text{ Hz}}{336 \text{ Hz}} = 0.75 = \dfrac{3}{4}$

$\quad 4n = 3n + 3, \quad$ so $\quad n = 3$

6. Solve for $f_1$:

$$f_n = nf_1 \quad \text{so} \quad f_1 = \frac{f_n}{n} = \frac{f_3}{3} = \frac{252 \text{ Hz}}{3} = 84 \text{ Hz}$$

7. Using the step 3 result, solve for $F_T$:

$$F_T = \mu f_1^2 (2L)^2 = (0.0025 \text{ kg/m})(84 \text{ Hz})^2 (6 \text{ m})^2$$

$$= 635 \text{ N}$$

8. Is the tension safe?

> The tension is less than 700 N. It seems it is safe to use as long as the tension is not increased significantly

### String Fixed at One End, Free at the Other

Figure 16-14 shows a string that has one end fixed and one end attached to a massless ring that is free to slide up and down on a friction-free pole. The vertical motion of the end of the string attached to the ring is unconstrained, so it is said to be a free end. The ring is massless, that is, a finite vertical force on it by the string would give the ring an infinite vertical acceleration. This acceleration will remain finite if the slope of the string at its free end remains horizontal. This means that the free end of the string is an antinode. In the fundamental mode of vibration for a string fixed at one end only, there is a node at the fixed end and an antinode at the free end, so $L = \lambda_1/4$ (Figure 16-15). (The distance from a node to an adjacent antinode is equal to one-quarter wavelength.)

In each mode of vibration shown in Figure 16-16 there are an odd number of quarter-wavelengths in the length L. That is, $L = n\lambda_n/4$, where $n = 1, 3, 5, \dots$. The standing-wave condition can thus be written

**FIGURE 16-14** An approximation of a string fixed at one end and free at the other end can be produced by connecting the "free" end of the string to a ring that is free to move on a post. The end attached to the mechanical wave driver is approximately fixed because the amplitude of the driver is very small.

$$L = n\frac{\lambda_n}{4}, \qquad n = 1, 3, 5, \dots \qquad 16\text{-}12$$

STANDING-WAVE CONDITION, ONE END FREE

**FIGURE 16-15**

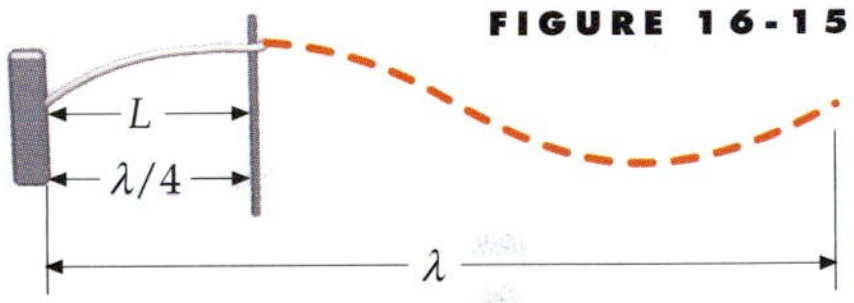

so $\lambda_n = 4L/n$. The resonance frequencies are therefore given by

$$f_n = \frac{v}{\lambda_n} = n\frac{v}{4L} = nf_1, \qquad n = 1, 3, 5, \dots \qquad 16\text{-}13$$

RESONANCE FREQUENCIES, ONE END FREE

where

$$f_1 = \frac{v}{4L} \qquad 16\text{-}14$$

is the fundamental frequency. The natural frequencies of this system occur in the ratios 1:3:5:7:..., which means that the even harmonics are missing.

> Again, an easy way to remember the resonance frequencies given by Equation 16-13 is to sketch Figure 16-16 to remind yourself of the standing-wave condition and use $v = f_n\lambda_n$

PROBLEM-SOLVING GUIDELINE

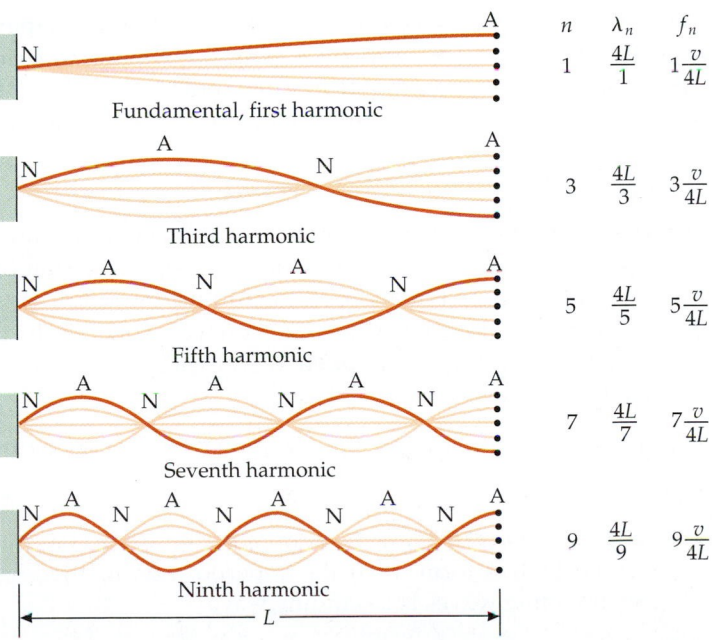

**FIGURE 16-16** Standing waves on a string fixed at only one end. The free end is an antinode.

**Wave Functions for Standing Waves** If a string vibrates in its $n$th mode, each point on the string moves with simple harmonic motion. Its displacement $y_n(x,t)$ is given by

$$y_n(x,t) = A_n(x) \cos(\omega_n t + \delta_n)$$

where $\omega_n$ is the angular frequency, $\delta_n$ is the phase constant, which depends on the initial conditions, and $A_n(x)$ is the amplitude, which depends on the position $x$ of the segment. The function $A_n(x)$ is the shape of the string when $\cos(\omega_n t + \delta_n) = 1$ (the instant that the vibration has its maximum displacement). The amplitude of a string vibrating in its $n$th mode is described by

$$A_n(x) = A_n \sin k_n x \qquad\qquad 16\text{-}15$$

where $k_n = 2\pi/\lambda_n$ is the wave number. The wave function for a standing wave in the $n$th harmonic can thus be written

$$y_n(x,t) = A_n \sin(k_n x)\cos(\omega_n t + \delta_n) \qquad\qquad 16\text{-}16$$

It is useful to remember the two conditions necessary for standing-wave motion, which are as follows:

> 1. Each point on the string either remains at rest or oscillates in simple harmonic motion. (Those points remaining at rest are at nodes.)
>
> 2. The motions of any two points on the string not at nodes oscillate either in phase or 180° out of phase.

NECESSARY CONDITIONS FOR A STANDING WAVE MOTION ON A LENGTH OF STRING

*STANDING WAVES*  **EXAMPLE 16-7** **Try It Yourself**

(*a*) The wave functions for two waves that have equal amplitude, frequency, and wavelength, but that travel in opposite directions, are given by $y_1 = y_0 \sin(kx - \omega t)$ and $y_2 = y_0 \sin(kx + \omega t)$. Show that the superposition of these two waves is a standing wave. (*b*) A standing wave on a string that is fixed at both ends is given by $y(x,t) = (0.024 \text{ m}) \sin(52.3 \text{ m}^{-1} x) \cos(480 \text{ s}^{-1} t)$. Find the speed of waves on the string and find the distance between adjacent nodes for the standing waves.

**PICTURE THE PROBLEM** To show that the superposition of the two given waves is a standing wave is to show that the algebraic sum of $y_1$ and $y_2$ can be written in the form of Equation 16-16. To find the wave speed and the wavelength we compare the given wave function with Equation 16-16 and identify the wave number and angular frequency. Knowing these, we can determine the wavelength and wave speed.

**Cover the column to the right and try these on your own before looking at the answers.**

| Steps | Answers |
|---|---|
| (*a*) 1. Write Equation 16-16. If the sum of $y_1$ and $y_2$ can be written in this form, then the superposition of the two traveling waves is a standing wave. | $y(x,t) = A \sin kx \cos \omega t$ |
| 2. Add the two wave functions and use the trigonometric identity $\sin \theta_1 + \sin \theta_2 = 2 \sin \tfrac{1}{2}(\theta_1 + \theta_2) \cos \tfrac{1}{2}(\theta_1 - \theta_2)$. | $y = y_0 \sin(kx - \omega t) + y_0 \sin(kx + \omega t)$ <br> $= 2y_0 \sin kx \cos \omega t$ |

Note: This is of the form given by Equation 16-16 with $A = 2y_0$, so the superposition is a standing wave.

(b) 1. Identify the wave number and the angular frequency.  $k = \boxed{52.3 \text{ m}^{-1}}$ ,   $\omega = \boxed{480 \text{ s}^{-1}}$

   2. Calculate the speed from $v = \omega/k$.  $v = \boxed{9.18 \text{ m/s}}$

   3. Find the wavelength $\lambda = 2\pi/k$, and use it to find the distance between nodes.  $\dfrac{\lambda}{2} = \boxed{6.01 \text{ cm}}$

## Standing Sound Waves

An organ pipe is a familiar example of the use of standing waves in air columns. In the flue-type organ pipe, a stream of air is directed against the sharp edge of an opening (point $A$ in Figure 16-17). The complicated swirling motion of the air near the edge sets up vibrations in the air column. The resonance frequencies of the pipe depend on the length of the pipe and on whether the top is stopped (closed) or open.

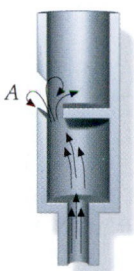

**FIGURE 16-17** Flue-type organ pipe. Air is blown against the edge, causing a swirling motion of the air near point $A$ that excites standing waves in the pipe. There is a pressure node near point $A$, which is open to the atmosphere.

In an open organ pipe, the pressure remains at one atmosphere near each open end. Since the pressure just beyond the ends does not vary, we say that there is a pressure node near each end. If the sound wave in the tube is a one-dimensional wave, which is largely correct if the tube diameter is much smaller than the wavelength, then the pressure node is at the opening of the tube. In practice, however, the pressure nodes lie slightly beyond the ends of the tube. The effective length of the pipe is $L_{\text{eff}} = L + \Delta L$ where $\Delta L$ is the end correction, which is of the order of the tube diameter. The standing-wave condition for this system is the same as that for a string fixed at both ends, where $L$ is replaced by $L_{\text{eff}}$ (the effective length of the tube), and all the same equations apply.

In a stopped organ pipe (open at one end, closed at the other), there is a pressure node near the opening (point $A$ in Figure 16-17 and a pressure antinode at the closed end. The standing-wave condition for this system is the same as that for a string with one end fixed and one end free. The effective length of the tube is equal to an odd integer times $\lambda/4$. That is, the wavelength of the fundamental mode is 4 times the effective length of the tube, and only the odd harmonics are present.

As we saw in Chapter 15, a sound wave can be thought of as either a pressure wave or a displacement wave. The pressure and displacement variations in a sound wave are 90° out of phase. Thus, in a standing sound wave, the pressure nodes are displacement antinodes and vice versa. Near the open end of an organ pipe there is a pressure node and a displacement antinode, whereas at a stopped end there is a pressure antinode and a displacement node.

---

*STANDING SOUND WAVES IN AN AIR COLUMN I*       **E X A M P L E   1 6 - 8**   **Try It Yourself**

**If the speed of sound is 340 m/s, what are the allowed frequencies and wavelengths for standing sound waves in an unstopped (open at both ends) organ pipe whose effective length is 1 m?**

**PICTURE THE PROBLEM** There is a displacement antinode (and a pressure node) at each end. Therefore, the effective length of the pipe is equal to an integral number of half-wavelengths.

**Cover the column to the right and try these on your own before looking at the answers.**

**Steps**

1. Calculate the fundamental wavelength from $\lambda_1 = 2L_{eff}$.

2. Use your value of $\lambda_1$ to calculate the fundamental frequency $f_1$.

3. Write expressions for the frequencies $f_n$ and wavelengths $\lambda_n$ of the other harmonics in terms of $n$.

**Answers**

$\lambda_1 = 2L_{eff} = 2\ m$

$f_1 = \dfrac{v}{\lambda_1} = 170\ Hz$

$\boxed{f_n = nf_1 = n(170\ Hz), \qquad n = 1, 2, 3, \ldots}$

$\boxed{\lambda_n = \dfrac{2L}{n} = \dfrac{2\ m}{n}, \qquad n = 1, 2, 3, \ldots}$

---

*STANDING SOUND WAVES IN AN AIR COLUMN II*  **EXAMPLE 16-9**

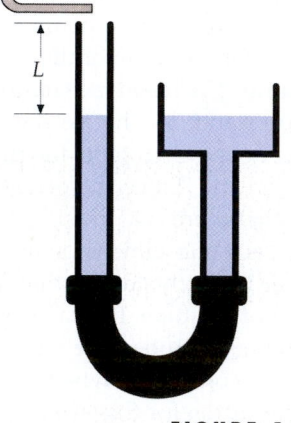

When a tuning fork of frequency 500 Hz is held above a tube that is partly filled with water as in Figure 16-18, resonances are found when the water level is at distances $L$ = 16.0, 50.5, 85.0, and 119.5 cm from the top of the tube. (*a*) What is the speed of sound in air? (*b*) How far from the open end of the tube is the displacement antinode?

**PICTURE THE PROBLEM** Sound waves of frequency 500 Hz are excited in the air column whose length $L$ can be adjusted (by adjusting the water level). The air column is stopped at one end, open at the other. Thus at resonance, the number of quarter-wavelengths in the effective length $L_{eff}$ of the tube is equal to an odd integer (Figure 16-19). There is a displacement node at the surface of the water and a displacement antinode a short distance above the open end of the tube. Since the frequency is fixed, so is the wavelength. The speed is then found from $v = f\lambda$, where $f$ is 500 Hz.

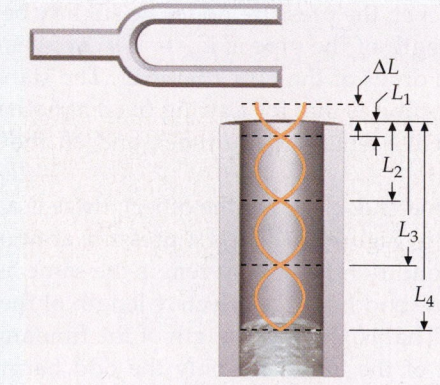

**FIGURE 16-18**

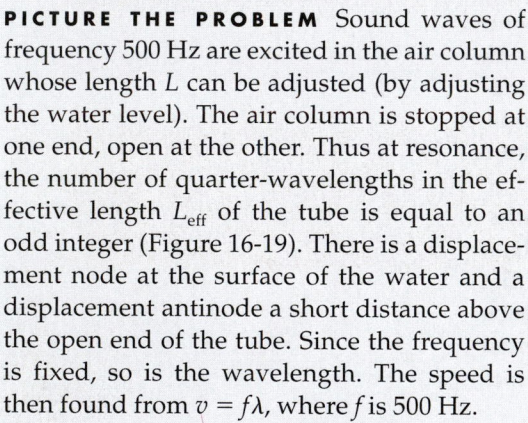

**FIGURE 16-19**

(*a*) 1. The speed of sound in air is related to the frequency and wavelength:

$v = f\lambda$

2. The wavelength is twice the distance between successive water levels at which resonance occurs:

$\lambda = 2(L_{n+1} - L_n), n = 1, 2, 3, 4$

3. The distance between successive levels is found from the data given in the problem:

$L_{n+1} - L_n = L_4 - L_3 = 119.5\ cm - 85\ cm = 34.5\ cm$

so

$\lambda = 2(34.5\ cm) = 69\ cm = 0.69\ m$

4. Substitute the values of $f$ and $\lambda$ to determine $v$:

$v = f\lambda = (500\ Hz)(0.69\ m) = \boxed{345\ m/s}$

(*b*) There will be a displacement antinode one quarter-wavelength above the displacement node at the surface of the water. Thus, the distance from the highest water level supporting resonance and the displacement antinode above the opening of the tube is one-quarter wavelength. $\frac{1}{4}\lambda = L_1 + \Delta L$:

$\Delta L = \frac{1}{4}\lambda - L_1 = \frac{1}{4}(69.0\ cm) - (16.0\ cm)$

$= \boxed{1.25\ cm}$

Most musical wind instruments are much more complicated than simple cylindrical tubes. The conical tube, which is the basis for the oboe, bassoon, English horn, and saxophone, has a complete harmonic series with its fundamental wavelength equal to twice the length of the cone. Brass instruments are combinations of cones and cylinders. The analysis of these instruments is extremely complex. The fact that they have nearly harmonic series is a triumph of educated trial and error rather than mathematical calculation.

**Holographic interferograms showing standing waves in a handbell. The "bull's eyes" locate the antinodes.**

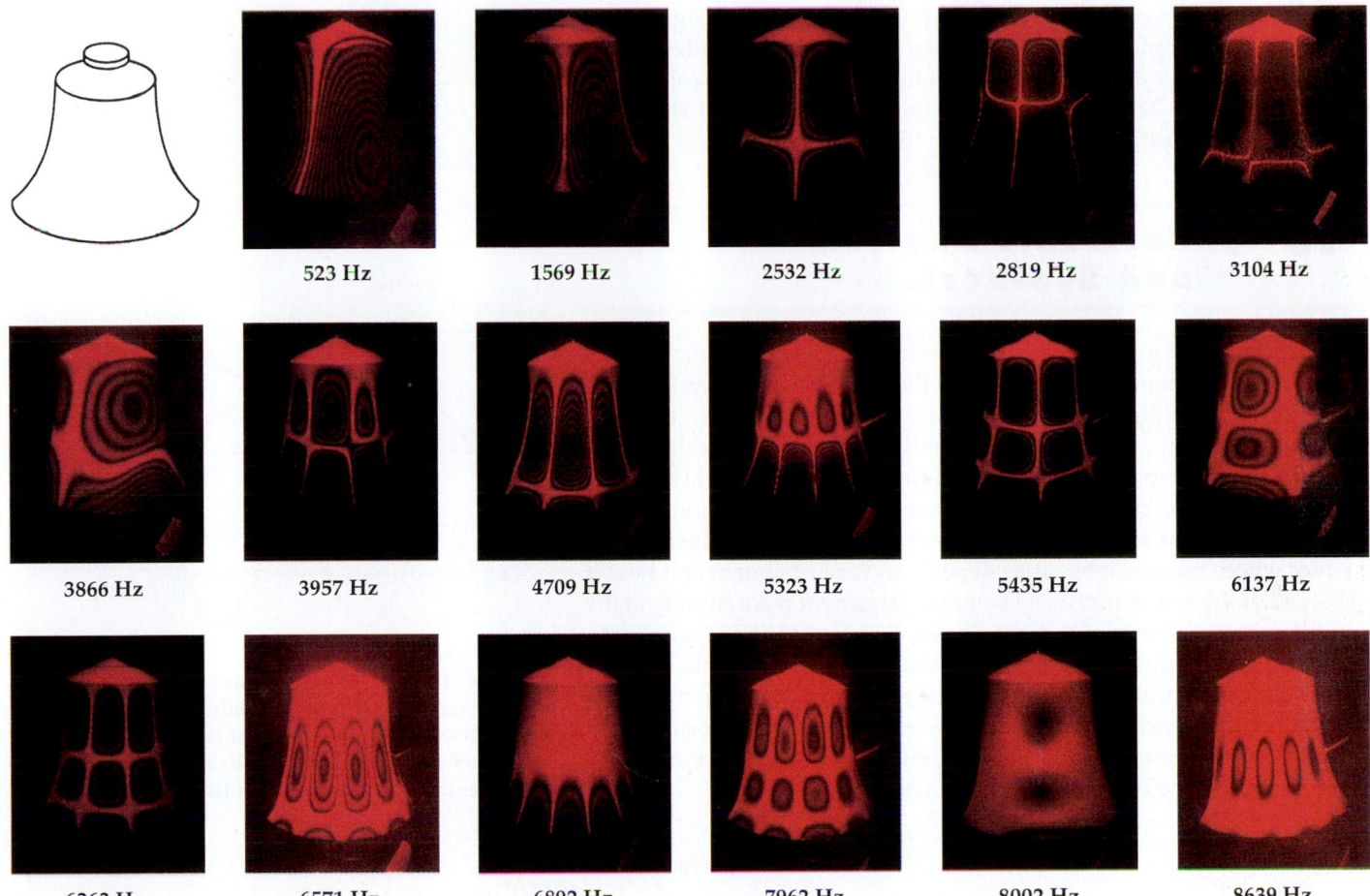

| | | | | |
|---|---|---|---|---|
| 523 Hz | 1569 Hz | 2532 Hz | 2819 Hz | 3104 Hz |
| 3866 Hz | 3957 Hz | 4709 Hz | 5323 Hz | 5435 Hz | 6137 Hz |
| 6263 Hz | 6571 Hz | 6892 Hz | 7962 Hz | 8002 Hz | 8639 Hz |

## *16-3 The Superposition of Standing Waves

As we saw in the preceding section, there is a set of natural resonance frequencies that produce standing waves for sound waves in air columns or vibrating strings that are fixed at one or both ends. For example, for a string fixed at both ends, the frequency of the fundamental mode of vibration is $f_1 = v/(2L)$, where $L$ is the length of the string and $v$ is the wave speed and the wave function is Equation 16-16:

$$y_1(x,t) = A_1 \sin k_1 x \cos(\omega_1 t + \delta_1)$$

In general, a vibrating system does not vibrate in a single harmonic mode. Instead, the motion consists of a mixture of the allowed harmonics. The wave function is a linear combination of the harmonic wave functions:

$$y(x,t) = \sum_n A_n \sin(k_n x) \cos(\omega_n t + \delta_n) \qquad \text{16-17}$$

where $k_n = 2\pi/\lambda_n$, $\omega_n = 2\pi f_n$, and $A_n$ and $\delta_n$ are constants. The constants $A_n$ and $\delta_n$ depend on the initial position and velocity of the string. If a harp string, for example, is plucked at the center and released, as in Figure 16-20, the initial shape of the string is symmetric about the point $x = \frac{1}{2}L$ and the initial velocity is zero throughout the length of the string. The motion of the string after it has been released will remain symmetric about $x = \frac{1}{2}L$. Only the odd harmonics, which are also symmetric about $x = \frac{1}{2}L$, will be excited. The even harmonics, which are antisymmetric about $x = \frac{1}{2}L$, are not excited; that is, the constant $A_n$ is zero for all even $n$. The shapes of the first four harmonics are shown in Figure 16-21. Most of the energy of the plucked string is associated with the fundamental, but small amounts of energy are associated with the third, fifth, and other odd harmonic modes. Figure 16-22 shows an approximation to the initial shape of the string using the superposition of only the first three odd harmonics.

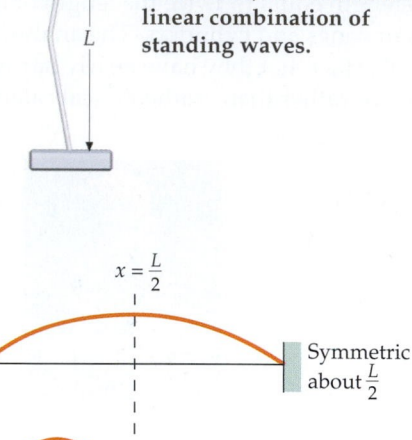

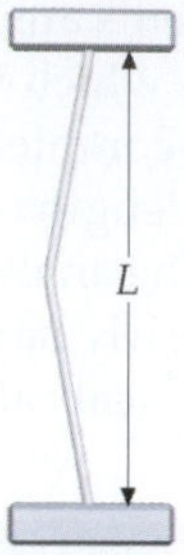

**FIGURE 16-20** A string plucked at the center. When it is released, its vibration is a linear combination of standing waves.

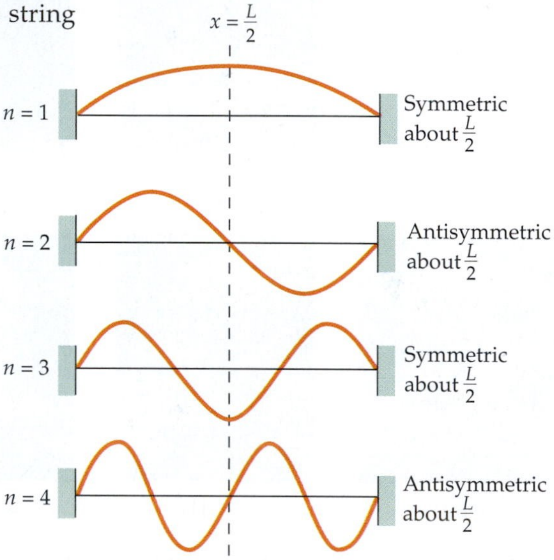

# *16-4 Harmonic Analysis and Synthesis

When a clarinet and an oboe play the same note, say, concert A, they sound quite different. Both notes have the same **pitch**, a physiological sensation of the highness or lowness of the note that is strongly correlated with frequency. However, the notes differ in what is called **tone quality**. The principal reason for the difference in tone quality is that, although both the clarinet and oboe are producing vibrations at the same fundamental frequency, each instrument is also producing harmonics whose relative intensities depend on the instrument and how it is played. If the sound produced by each instrument were entirely at the fundamental frequency of the instrument, they would sound identical.

Figure 16-23 shows plots of the pressure variations versus time for the sound from a tuning fork, a clarinet, and an oboe, each playing the same note. These patterns are called **waveforms**. The waveform for the sound from the tuning fork is nearly a pure sine wave, but those from the clarinet and the oboe are clearly more complex.

**FIGURE 16-21** The first four harmonics for a string fixed at both ends. The odd harmonics are symmetrical about the center of the string, whereas the even harmonics are not. When a string is plucked at the center, it vibrates only in its odd harmonics.

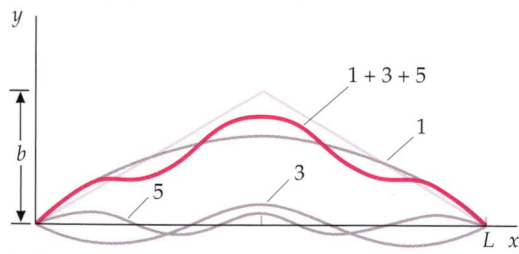

**FIGURE 16-22** Approximating the shape of a string plucked at the center, as in Figure 16-20, using harmonics. The red line is an approximation of the original shape of the string based on the first three odd harmonics. The height of the string is exaggerated in this drawing to show the relative amplitudes of the harmonics. Most of the energy is associated with the fundamental, but there is some energy in the third, fifth, and other odd harmonics.

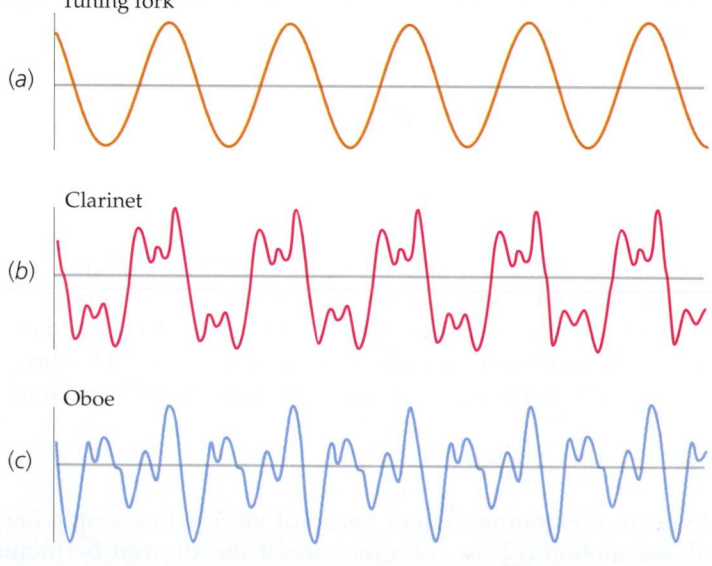

**FIGURE 16-23** Waveforms of (a) a tuning fork, (b) a clarinet, and (c) an oboe, each at a fundamental frequency of 440 Hz and at approximately the same intensity.

Waveforms can be analyzed in terms of the harmonics that comprise them by means of **harmonic analysis.** (Harmonic analysis is also called **Fourier analysis** after the French mathematician J.B.J. Fourier, who developed the techniques for analyzing periodic functions.) Figure 16-24 shows a plot of the relative intensities of the harmonics of the waveforms in Figure 16-23. The waveform of the sound from the tuning fork contains only the fundamental frequency. That for the sound from the clarinet contains the fundamental, large amounts of the third, fifth, and seventh harmonics, and lesser amounts of the second, fourth, and sixth harmonics. For the sound from the oboe, there is more intensity in the second and third harmonics than in the fundamental.

The inverse of harmonic analysis is **harmonic synthesis,** which is the construction of a periodic wave from harmonic components. Figure 16-25a shows the first three odd harmonics used to synthesize a square wave and Figure 16-25b shows the square wave that results from the sum of the three harmonics. The more harmonics used in a synthesis, the closer the approximation will be to the actual waveform (the gray line in the Figure). The relative amplitudes of the harmonics needed to synthesize the square wave are shown in Figure 16-26.

## *16-5 Wave Packets and Dispersion

The waveforms discussed in Section 16.4 are periodic in time. Pulses, which are aperiodic, can also be represented by a group of harmonic waves of different frequencies. However, the synthesis of a pulse requires a continuous

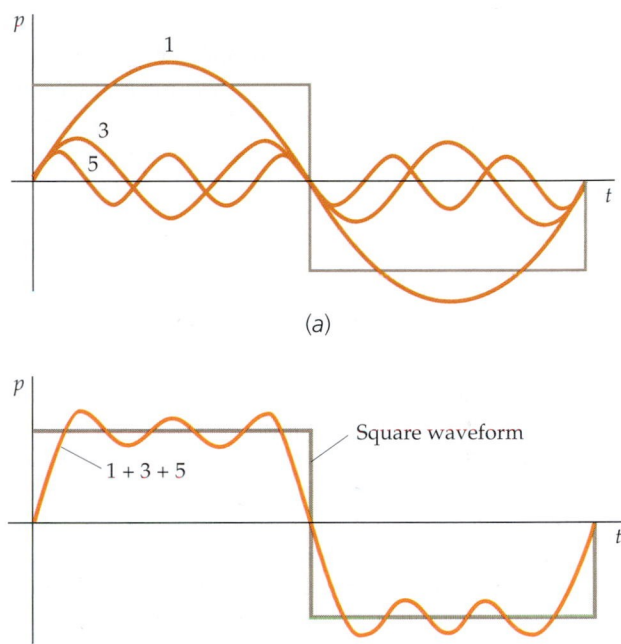

(a)

(b)

**FIGURE 16-25** (*a*) The first three odd harmonics of a single sine wave, used to synthesize a square wave. (*b*) The approximation of a square wave that results from summing the first three odd harmonics in (*a*).

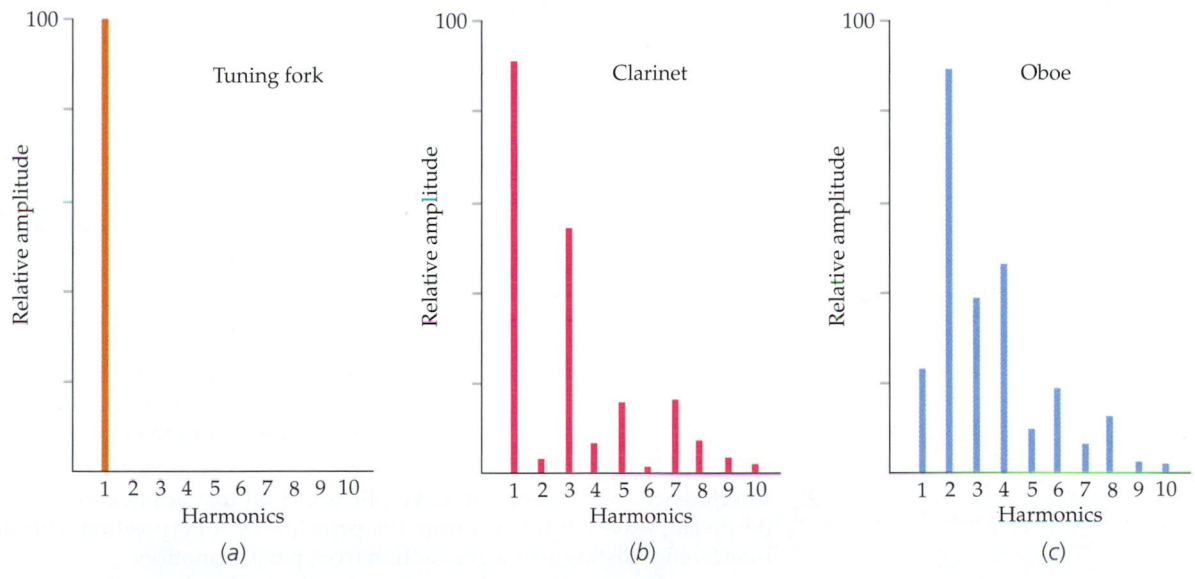

**FIGURE 16-24** Relative intensities of the harmonics in the waveforms shown in Figure 16-23 for (*a*) the tuning fork, (*b*) the clarinet, and (*c*) the oboe.

distribution of frequencies rather than a discrete set of harmonics as in Figure 16-26. Such a group is called a **wave packet.** The characteristic feature of a wave pulse is that it has a beginning and an end, whereas a harmonic wave repeats over and over. If the duration $\Delta t$ of the pulse is very short, the range of frequencies $\Delta \omega$ needed to describe the pulse is very large. The general relation between $\Delta t$ and $\Delta \omega$ is

$$\Delta \omega\, \Delta t \sim 1 \qquad\qquad 16\text{-}18$$

where the tilde $\sim$ means "of the order of."

The exact value of this product depends on just how the quantities $\Delta \omega$ and $\Delta t$ are defined. For any reasonable definitions, $\Delta \omega$ and $1/\Delta t$ have the same order of magnitude. A wave pulse produced by a source of short duration $\Delta t$, like the crack of a bat on a ball, has a narrow width in space $\Delta x = v\, \Delta t$, where $v$ is the wave speed. Each harmonic wave of frequency $\omega$ has a wave number $k = \omega/v$. A range of frequencies $\Delta \omega$ implies a range of wave numbers $\Delta k = \Delta \omega / v$. Substituting $v\, \Delta k$ for $\Delta \omega$ in Equation 16-18 gives $v\, \Delta k\, \Delta t \sim 1$ or

$$\Delta k\, \Delta x \sim 1 \qquad\qquad 16\text{-}19$$

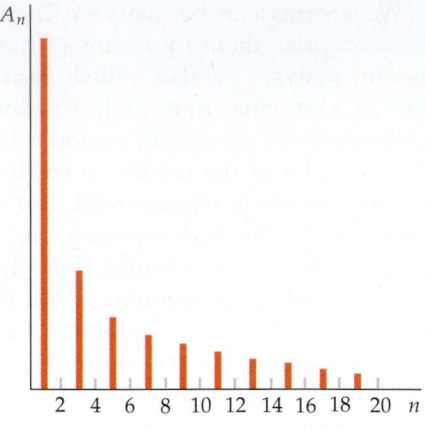

**FIGURE 16-26** Relative amplitudes $A_n$ of the first ten harmonics needed to synthesize a square wave. The more harmonics that are used, the closer the approximation is to the square wave.

If a wave packet is to maintain its shape as it travels, all of the component harmonic waves that make up the packet must travel with the same speed. This occurs if the speed of the component waves in a given medium is independent of frequency or wavelength. Such a medium is called a **nondispersive medium.** Air is, to an excellent approximation, a nondispersive medium for sound waves, but solids and liquids generally are not. (Probably the most familiar example of dispersion is the formation of a rainbow, due to the fact that the velocity of light waves in water depends slightly on the frequency of the light, so the different colors, corresponding to different frequencies, have slightly different angles of refraction.)

When the wave speed in a dispersive medium depends only slightly on the frequency (or wavelength), a wave packet changes shape very slowly as it travels, and it covers a considerable distance as a recognizable entity. But the speed of the packet, called the **group velocity,** is not the same as the (average) speed of the individual component harmonic waves, called the **phase velocity.** (By the speed of an individual harmonic wave we mean the speed of its wavefronts. Because wavefronts are lines or surfaces of constant phase, their speed is called the phase velocity of the wave.)

## SUMMARY

1. The principle of superposition, which holds for all electromagnetic waves in empty space, for waves on a flexible taut string in the small angle approximation, and for sound waves of small amplitude, follows from the linearity of the corresponding wave equations.

2. Interference is an important wave phenomenon that applies to all coherent superposing waves. It follows from the principle of superposition. Diffraction and interference distinguish wave motion from particle motion.

3. The standing-wave conditions can be recalled by sketching a string or tube and drawing waves that have nodes at a fixed or stopped end, and antinodes at a free or open end.

| Topic | Relevant Equations and Remarks |
|---|---|
| **1. Superposition and Interference** | The superposition of two harmonic waves of equal amplitude, wave number, and frequency but phase difference $\delta$ results in a harmonic wave of the same wave number and frequency, but differing in phase and amplitude from each of the two waves |

$$y = y_1 + y_2 = y_0 \sin(kx - \omega t) + y_0 \sin(kx - \omega t + \delta)$$
$$= \left[2y_0 \cos \tfrac{1}{2}\delta\right] \sin(kx - \omega t + \tfrac{1}{2}\delta) \qquad \text{16-6}$$

| Topic | Relevant Equations and Remarks |
|---|---|
| Constructive interference | If waves are in phase or differ in phase by an integer times $2\pi$, then the amplitudes of the waves add and the interference is constructive. |
| Destructive interference | If waves differ in phase by $\pi$ or by an odd integer times $\pi$, then the amplitudes subtract and the interference is destructive. |
| Beats | Beats are the result of the interference of two waves of slightly different frequencies. The beat frequency equals the difference in the frequencies of the two waves: |

$$f_{\text{beat}} = \Delta f \qquad \text{16-8}$$

| Topic | Relevant Equations and Remarks |
|---|---|
| Phase difference $\delta$ due to path difference $\Delta x$ | |

$$\delta = k\,\Delta x = 2\pi \frac{\Delta x}{\lambda} \qquad \text{16-9}$$

| Topic | Relevant Equations and Remarks |
|---|---|
| **2. Standing Waves** | Standing waves occur for certain frequencies and wavelengths when waves are confined in space. They occur only if each point of the system oscillates in simple harmonic motion and any two moving points move either in phase or 180° out of phase. |
| Wavelength | The distance between a node and an adjacent antinode is a quarter-wavelength. |
| String fixed at both ends | For a string fixed at both ends, there is a node at each end so that an integral number of half-wavelengths must fit into the length of the string. The standing-wave condition in this case is |

$$L = n\frac{\lambda_n}{2}, \qquad n = 1, 2, 3, \ldots \qquad \text{16-10}$$

| Topic | Relevant Equations and Remarks |
|---|---|
| Standing wave function for a string fixed at both ends | The allowed waves form a harmonic series, with the frequencies given by |

$$f_n = \frac{v}{\lambda_n} = n\frac{v}{2L} = nf_1, \qquad n = 1, 2, 3, \ldots \qquad \text{16-11}$$

where $f_1 = v/2L$ is the lowest frequency, called the fundamental.

| Topic | Relevant Equations and Remarks |
|---|---|
| Organ pipe open at both ends | Standing sound waves in the air in a pipe that is open at both ends have a pressure node (and a displacement antinode) near each end so that the standing wave condition is the same as for a string fixed at both ends. |
| String fixed at one end and free at the other | For a string with one end fixed and one end free, there is a node at the fixed end and an antinode at the free end, so that an integral number of quarter-wavelengths must fit into the length of the string. The standing-wave condition in this case is |

$$L = n\frac{\lambda_n}{4}, \qquad n = 1, 3, 5, \ldots \qquad \text{16-12}$$

Only the odd harmonics are present. Their frequencies are given by

$$f_n = \frac{v}{\lambda_n} = n\frac{v}{4L} = nf_1, \qquad n = 1, 3, 5, \ldots \qquad \text{16-13}$$

where $f_1 = v/4L$.

| Topic | Relevant Equations and Remarks |
|---|---|
| Organ pipe open at one end and stopped at the other | Standing sound waves in a pipe that is open at one end and stopped at the other end have a displacement antinode at the open end and a displacement node at the stopped end. The standing wave condition is the same as for a string fixed at one end. |

| | | |
|---|---|---|
| Wave Functions for Standing Waves | $y_n(x,t) = A_n \sin(k_n x) \cos(\omega_n t + \delta_n)$ | **16-16** |

where $k_n = 2\pi/\lambda_n$ and $\omega_n = 2\pi f_n$.

The necessary conditions for standing waves on a string are

1. Each point on the string either remains at rest or oscillates with simple harmonic motion. (Those points remaining at rest are nodes.)
2. The motions of any two points on the string that are not nodes oscillate either in phase or 180° out of phase.

---

**\*3. Superposition of Standing Waves**
A vibrating system typically does not vibrate in a single harmonic mode but in a superposition of the allowed harmonic modes.

---

**\*4. Harmonic Analysis and Synthesis**
Sounds of different tone quality contain different mixtures of harmonics. The analysis of a particular tone in terms of its harmonic content is called harmonic analysis. Harmonic synthesis is the construction of a tone by the addition of harmonics.

---

**\*5. Wave Packets**
A wave pulse can be represented by a continuous distribution of harmonic waves. The range of frequencies $\Delta\omega$ is related to the width in time $\Delta t$, and the range of wave numbers $\Delta k$ is related to the width in space $\Delta x$.

| | | |
|---|---|---|
| Frequency and time ranges | $\Delta\omega \, \Delta t \sim 1$ | **16-18** |
| Wave number and space ranges | $\Delta k \, \Delta x \sim 1$ | **16-19** |

---

**\*6. Dispersion**
In a nondispersive medium, the phase velocity is independent of frequency, and a pulse (wave packet) travels without change in shape. In a dispersive medium, the phase velocity varies with frequency, and the pulse changes shape as it moves. The pulse moves with a velocity called the group velocity of the packet.

---

# PROBLEMS

- • Single-concept, single-step, relatively easy
- •• Intermediate-level, may require synthesis of concepts
- ••• Challenging
- **SSM** Solution is in the *Student Solutions Manual*
-  Problems available on iSOLVE online homework service
-  These "Checkpoint" online homework service problems ask students additional questions about their confidence level, and how they arrived at their answer

In a few problems, you are given more data than you actually need; in a few other problems, you are required to supply data from your general knowledge, outside sources, or informed estimates.

## Conceptual Problems

**1** •• **SSM** Two rectangular wave pulses are traveling in opposite directions along a string. At $t = 0$, the two pulses are as shown in Figure 16-27. Sketch the wave functions for $t = 1$, 2, and 3 s.

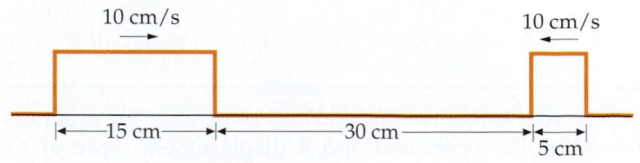

**FIGURE 16-27** Problems 1, 2

**2** •• Repeat Problem 1 for the case in which the pulse on the right is inverted.

**3** • Beats are produced by the superposition of two harmonic waves only if (*a*) their amplitudes and frequencies are equal, (*b*) their amplitudes are the same but their frequencies differ slightly, (*c*) their frequencies differ slightly even if their amplitudes are not equal, (*d*) their frequencies are equal but their amplitudes differ slightly.

**4** • True or false:

(*a*) The frequency of the third harmonic is three times that of the first harmonic.
(*b*) The frequency of the fifth harmonic is five times that of the fundamental.
(*c*) In a pipe that is open at one end and closed at the other, the even harmonics are not excited.

**5** •• Standing waves result from the superposition of two waves of (a) the same amplitude, frequency, and direction of propagation, (b) the same amplitude and frequency and opposite directions of propagation, (c) the same amplitude, slightly different frequency, and the same direction of propagation, (d) the same amplitude, slightly different frequency, and opposite directions of propagation.

**6** • **SSM** The resonant frequencies of a violin string are all integer multiples of the fundamental frequency, while the resonant frequencies of a circular drumhead are irregularly spaced. Given this information, explain the difference in the sounds of a violin and a drum.

**7** • An organ pipe open at both ends has a fundamental frequency of 400 Hz. If one end of this pipe is now closed, the fundamental frequency will be (a) 200 Hz, (b) 400 Hz, (c) 546 Hz, (d) 800 Hz.

**8** •• A string fixed at both ends resonates at a fundamental frequency of 180 Hz. Which of the following will reduce the fundamental frequency to 90 Hz? (a) Double the tension and double the length. (b) Halve the tension and keep the length fixed. (c) Keep the tension fixed and double the length. (d) Keep the tension fixed and halve the length.

**9** •• How do the resonance frequencies of an organ pipe change when the air temperature increases?

**10** • **SSM** When two waves moving in opposite directions superimpose as in Figure 16-1, does either impede the progress of the other?

**11** • When a guitar string is plucked, is the wavelength of the wave it produces in air the same as the wavelength of the wave on the string?

**12** • When two waves interfere constructively or destructively, is there any gain or loss in energy? Explain.

**13** • A musical instrument consists of drinking glasses partially filled with water that are struck with a small mallet. Explain how this works.

**14** •• During an organ recital, the air compressor that drives the organ pipes suddenly fails. An enterprising physics student in the audience comes to the rescue by connecting a tank of pure nitrogen gas under high pressure to the output of the compressor. What effect, if any, will this change have on the operation of the organ? What if the tank contained helium?

**15** •• **SSM** When the tension on a piano wire is increased, which of the following occurs? (a) Its wavelength decreases. (b) Its wavelength remains the same while its frequency increases. (c) Its wavelength and frequency increase. (d) None of the above occur.

**16** •• The following instructions are given for connecting stereo speakers to an amplifier so that they are in phase: "After both speakers are connected, play a monophonic record or program with the bass control turned up and the treble control turned down. While listening to the speakers, turn the balance control so that first one speaker is heard separately, then the two together, and then the other separately. If the bass is stronger when both speakers play together, they are connected properly. If the bass is weaker when both play together than when each plays separately, reverse the connections on one speaker." Explain why this method works. In particular, explain why a stereo source is not used and why only the bass is compared.

**17** •• The constant $\gamma$ for helium (and all monatomic gases) is 1.67. If a man inhales helium and then speaks, he sounds like Alvin of the Chipmunks. Why?

**18** •• **SSM** Figure 16-28 is a photograph of two pieces of very finely woven silk placed one on top of the other. Where the pieces overlap, a series of light and dark lines are seen. This moiré pattern can also be seen when a scanner is used to copy photos from a book or newspaper. What causes the moiré pattern, and how is it similar to the phenomenon of interference?

**FIGURE 16-28** Problem 18

### Estimation and Approximation

**19** •• About how accurately can you tune a piano string to a tuning-fork frequency?

**20** • **SSM** The shortest pipes used in organs are about 7.5 cm long. (a) What is the fundamental frequency of a pipe this long that is open at both ends? (b) For such a pipe, what is the highest harmonic that is within the audible range? (The normal range of hearing is about 20 to 20,000 Hz.)

**21** •• On a windy day, a drain pipe will sometimes resonate. Estimate the resonance frequency of a drain pipe on a single-story house. How much might this frequency change from winter to summer in your region?

### Superposition and Interference

**22** • Two waves traveling on a string in the same direction both have a frequency of 100 Hz, a wavelength of 2 cm, and an amplitude of 0.02 m. What is the amplitude of the resultant wave if the original waves differ in phase by (a) $\pi/6$ and (b) $\pi/3$?

**23** • Two waves having the same frequency, wavelength, and amplitude are traveling in the same direction. If they differ in phase by $\pi/2$ and each has an amplitude of 0.05 m, what is the amplitude of the resultant wave?

**24** • [SSM] [ISOLVE] Two sound sources oscillate in phase with the same amplitude $A$. They are separated in space by $\lambda/3$. What is the amplitude of the resultant wave formed from the two sources at a point that is on the line that passes through the sources but is not between the sources?

**25** • Two sound sources oscillate in phase with a frequency of 100 Hz. At a point 5.00 m from one source and 5.85 m from the other, the amplitude of the sound from each source separately is $A$. (a) What is the phase difference in the sound waves from the two sources at that point? (b) What is the amplitude of the resultant wave at that point?

**26** • [SSM] With a compass, draw circular arcs representing wave crests originating from each of two point sources a distance $d = 6$ cm apart for $\lambda = 1$ cm. Connect the intersections corresponding to points of constant path difference and label the path difference for each line. (See Figure 16-8.)

**27** • Two speakers separated by some distance emit sound waves of the same frequency. At some point $P$, the intensity due to each speaker separately is $I_0$. The path distance from $P$ to one of the speakers is $\frac{1}{2}\lambda$ greater than that from $P$ to the other speaker. What is the intensity at $P$ if (a) the speakers are coherent and in phase, (b) the speakers are incoherent, and (c) the speakers are coherent but have a phase difference of $\pi$ rad?

**28** • Answer the questions of Problem 27 for a point $P'$ for which the distance to the far speaker is $1\lambda$ greater than the distance to the near speaker. Assume that the intensity at point $P'$ due to each speaker separately is again $I_0$.

**29** • Two speakers separated by some distance emit sound waves of the same frequency, but the speakers are out of phase by 90°. Let $r_1$ be the distance from some point to speaker 1 and $r_2$ be the distance from that point to speaker 2. Find the smallest value of $r_2 - r_1$ at which the sound at that point will be (a) maximum and (b) minimum. (Express your answers in terms of the wavelength.)

**30** •• [SSM] Show that, if the separation between two sound sources radiating coherently in phase is less than half a wavelength, complete destructive interference will not be observed in any direction.

**31** •• [ISOLVE] A transverse wave of frequency 40 Hz propagates down a string. Two points 5 cm apart are out of phase by $\pi/6$. (a) What is the wavelength of the wave? (b) At a given point, what is the phase difference between two displacements for times 5 ms apart? (c) What is the wave velocity?

**32** •• It is thought that the brain determines the direction of the source of a sound by sensing the phase difference between the sound waves striking the eardrums. A distant source emits sound of frequency 680 Hz. When you are directly facing a sound source there should be no phase difference. Estimate the phase difference between the sounds received by each ear as you turn from facing directly toward the source through 90°.

**33** •• [ISOLVE✓] Sound source A is located at $x = 0$, $y = 0$, and sound source B is placed at $x = 0$, $y = 2.4$ m. The two sources radiate coherently in phase. An observer at $x = 40$ m, $y = 0$ notes that as she takes a few steps in either the positive

or negative $y$ direction away from $y = 0$, the sound intensity diminishes. What are the lowest and the next higher frequencies of the sources that can account for that observation?

**34** •• Suppose that the observer in Problem 33 finds herself at a point of minimum intensity at $x = 40$ m, $y = 0$. What are then the lowest and next higher frequencies of the sources consistent with this observation?

**35** •• [SSM] Two harmonic water waves of equal amplitudes but different frequencies, wave vectors, *and velocities* are superposed on each other. The total displacement of the wave can be written as $y(x,t) = A[\cos(k_1x - \omega_1t) + \cos(k_2x - \omega_2t)]$, where $\omega_1k_1 = v_1$ (the speed of the first wave) and $\omega_2/k_2 = v_2$ (the speed of the second wave). (a) Show that $y(x,t)$ can be written in the form $y(x,t) = 2A\cos[(\Delta k/2)x - (\Delta\omega/2)t]\cos(k_{av}x - \omega_{av}t)$ where $\omega_{av} = (\omega_1 + \omega_2)/2$. $k_{av} = (k_1 + k_2)/2$, $\Delta\omega = \omega_1 - \omega_2$, and $\Delta k = k_1 - k_2$. The factor $2A\cos[(\Delta k/2)x - (\Delta\omega/2)t]$ is called the *envelope* of the wave. (b) Using a spreadsheet program or graphing calculator, make a graph of $y(x,t)$ for $A = 1$, $\omega_1 = 1$ rad/s, $k_1 = 1$ m$^{-1}$, $\omega_2 = 0.9$ rad/s, and $k_2 = 0.8$ m$^{-1}$ at $t = 0$ s, 0.5 s, and 1 s for $x$ between 0 m and 50 m. (c) What is the speed at which the envelope moves?

**36** •• Two point sources that are in phase are separated by a distance $d$. An interference pattern is detected along a line parallel to the line through the sources and a large distance $D$ from the sources, as shown in Figure 16-29. (a) Show that the path difference from the two sources to some point on the line at a small angle $\theta$ is given approximately by $\Delta s = d\sin\theta$. Hint: Assume that $D \gg d$, so the lines from the sources to $P$ are approximately parallel. (b) Show that the distance $y_m$ from the central maximum point to the $m$th interference maximum is given approximately by $y_m = m(D\lambda/d)$.

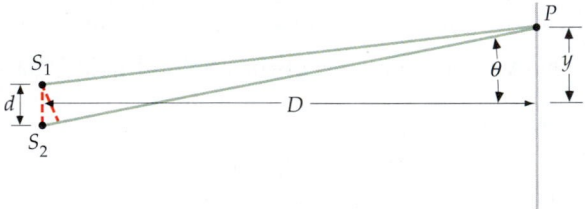

**FIGURE 16-29** Problems 36–40

**37** •• [ISOLVE✓] Two sound sources radiating in phase at a frequency of 480 Hz interfere such that maxima are heard at angles of 0° and 23° from a line perpendicular to that joining the two sources. Find the separation between the two sources and any other angles at which a maximum intensity will be heard. (Use the result of Problem 36.)

**38** ••• [SSM] Two loudspeakers are driven in phase by an audio amplifier at a frequency of 600 Hz. The speakers are on the $y$ axis, one at $y = +1.00$ m and the other at $y = -1.00$ m. A listener begins at $y = 0$ a very large distance $D$ away and walks along a line parallel to the $y$ axis. (See Problem 36.) (a) At what angle $\theta$ will she first hear a minimum in the sound intensity? (b) At what angle will she first hear a maximum (after $\theta = 0$)? (c) How many maxima can she possibly hear if she keeps walking in the same direction?

**39** ••• **iSOLVE** Two sound sources driven in phase by the same amplifier are 2 m apart on the $y$ axis. At a point a very large distance from the $y$ axis, constructive interference is first heard at an angle $\theta_1 = 0.140$ rad with the $x$ axis and is next heard at $\theta_2 = 0.283$ rad (see Figure 16-29). (a) What is the wavelength of the sound waves from the sources? (b) What is the frequency of the sources? (c) At what other angles is constructive interference heard? (d) What is the smallest angle for which the sound waves cancel?

**40** ••• The two sound sources from Problem 39 are now driven 90° out-of-phase, but at the same frequency as in Problem 39. At what angles are constructive and destructive interference heard?

**41** ••• A radio telescope consists of two antennas separated by a distance of 200 m. Both antennas are tuned to a particular frequency, such as 20 MHz. The signals from each antenna are fed into a common amplifier, but one signal first passes through a phase adjuster that delays its phase by a chosen amount so that the telescope can "look" in different directions. When the phase delay is zero, plane radio waves that are incident vertically on the antennas produce signals that add constructively at the amplifier. What should be the phase delay so that signals coming from an angle $\theta = 10°$ with the vertical (in the plane formed by the vertical and the line joining the antennas) will add constructively at the amplifier?

## Beats

**42** • **iSOLVE** When two tuning forks are struck simultaneously, 4 beats per second are heard. The frequency of one fork is 500 Hz. (a) What are the possible values for the frequency of the other fork? (b) A piece of wax is placed on the 500-Hz fork to lower its frequency slightly. Explain how the measurement of the new beat frequency can be used to determine which of your answers to Part (a) is the correct frequency of the second fork.

**43** •• **SSM** Two ambulances move toward each other on a straight road. Each travels at a speed of 50 mph. The siren on each ambulance produces a sound wave of frequency 500 Hz. (a) The driver of each ambulance hears the other's siren and hears a beat note between the frequency of his own siren and the siren of the other ambulance. What is the frequency of the beat note? (b) A passerby stands midway between the two ambulances. What is the frequency of the beat note between the two sirens that he hears?

## Standing Waves

**44** • **SSM** **iSOLVE✓** A string fixed at both ends is 3 m long. It resonates in its second harmonic at a frequency of 60 Hz. What is the speed of transverse waves on the string?

**45** • A string 3 m long and fixed at both ends is vibrating in its third harmonic. The maximum displacement of any point on the string is 4 mm. The speed of transverse waves on this string is 50 m/s. (a) What are the wavelength and frequency of this wave? (b) Write the wave function for this wave.

**46** • **iSOLVE✓** Calculate the fundamental frequency for a 10-m organ pipe that is (a) open at both ends and (b) closed at one end.

**47** • A steel wire having a mass of 5 g and a length of 1.4 m is fixed at both ends and has a tension of 968 N. (a) Find the speed of transverse waves on the wire. (b) Find the wavelength and frequency of the fundamental. (c) Find the frequencies of the second and third harmonics.

**48** • A rope 4 m long is fixed at one end; the other end is attached to a light string so that it is free to move. The speed of waves on the rope is 20 m/s. Find the frequency of (a) the fundamental, (b) the second harmonic, and (c) the third harmonic.

**49** • A piano wire without windings has a fundamental frequency of 200 Hz. When it is wound with wire, its linear mass density is doubled. What is its new fundamental frequency, assuming that the tension is unchanged?

**50** • **SSM** What is the greatest length that an organ pipe can have in order to have its fundamental note in the audible range (20 to 20,000 Hz) if (a) the pipe is closed at one end and (b) it is open at both ends?

**51** •• **iSOLVE✓** The wave function $y(x,t)$ for a certain standing wave on a string fixed at both ends is given by $y(x,t) = 4.2 \sin 0.20x \cos 300t$, where $y$ and $x$ are in centimeters and $t$ is in seconds. (a) What are the wavelength and frequency of this wave? (b) What is the speed of transverse waves on this string? (c) If the string is vibrating in its fourth harmonic, how long is it?

**52** •• The wave function $y(x,t)$ for a certain standing wave on a string fixed at both ends is given by $y(x,t) = (0.05 \text{ m}) \sin 2.5 \text{ m}^{-1} x \cos 500 \text{ s}^{-1} t$. (a) What are the speed and amplitude of the two traveling waves that result in this standing wave? (b) What is the distance between successive nodes on the string? (c) What is the shortest possible length of the string?

**53** •• A 2.51-m-long string has the wave function given in Problem 52. (a) Sketch the position of the string at the times $t = 0, t = T/4, t = T/2,$ and $t = 3T/4$, where $T = 1/f$ is the period of the vibration. (b) Find $T$ in seconds. (c) At a time $t$ when the string is horizontal, that is, $y(x) = 0$ for all $x$, what has become of the energy in the wave?

**54** •• **SSM** **iSOLVE** Three successive resonance frequencies for a certain string are 75, 125, and 175 Hz. (a) Find the ratios of each pair of successive resonance frequencies. (b) How can you tell that these frequencies are for a string fixed at one end only rather than for a string fixed at both ends? (c) What is the fundamental frequency? (d) Which harmonics are these resonance frequencies? (e) If the speed of transverse waves on this string is 400 m/s, find the length of the string.

**55** •• The space above the water in a tube like that shown in Example 16-9 is 120 cm long. Near the open end there is a loudspeaker that is driven by an audio oscillator whose frequency can be varied from 10 to 5000 Hz. (a) What is the lowest frequency of the oscillator that will produce resonance within the tube? (b) What is the highest frequency that will produce resonance? (c) How many different frequencies of the oscillator will produce resonance? (Neglect the end correction.)

**56** •• A 460-Hz tuning fork causes resonance in the tube in Example 16-9 when the top of the tube is 18.3 and 55.8 cm above the water surface. (a) Find the speed of sound in air. (b) What is the end correction to adjust for the fact that the antinode does not occur exactly at the end of the open tube?

**57** •• **SSM** **ISOLVE✓** At 16°C, the fundamental frequency of an organ pipe is 440.0 Hz. What will be the fundamental frequency of the pipe if the temperature increases to 32°C? Would it be better to construct the pipe with a material that expands substantially as the temperature increases or should the pipe be made of material that maintains the same length at all normal temperatures?

**58** •• The end correction for a circular pipe is approximately $\Delta L = 0.3186D$, where $D$ is the pipe diameter. Find the length of a pipe open at both ends that will produce a middle C (256 Hz) as its fundamental mode for pipes of diameter $D = 1$ cm, 10 cm, and 30 cm.

**59** •• **ISOLVE** A violin string of length 40 cm and mass 1.2 g has a frequency of 500 Hz when it is vibrating in its fundamental mode. (a) What is the wavelength of the standing wave on the string? (b) What is the tension in the string? (c) Where should you place your finger to increase the frequency to 650 Hz?

**60** •• The G string on a violin is 30 cm long. When played without fingering, it vibrates at a frequency of 196 Hz. The next higher notes on the C-major scale are A (220 Hz), B (247 Hz), C (262 Hz), and D (294 Hz). How far from the end of the string must a finger be placed to play each of these notes?

**61** •• A string with a mass density of $4 \times 10^{-3}$ kg/m is under a tension of 360 N and is fixed at both ends. One of its resonance frequencies is 375 Hz. The next higher resonance frequency is 450 Hz. (a) What is the fundamental frequency of this string? (b) Which harmonics are the ones given? (c) What is the length of the string?

**62** •• **ISOLVE✓** A string fastened at both ends has successive resonances with wavelengths of 0.54 m for the $n$th harmonic and 0.48 m for the $(n + 1)$th harmonic. (a) Which harmonics are these? (b) What is the length of the string?

**63** •• The strings of a violin are tuned to the tones G, D, A, and E, which are separated by a fifth from one another. That is, $f(D) = 1.5f(G)$, $f(A) = 1.5f(D) = 440$ Hz, and $f(E) = 1.5f(A)$. The distance between the two fixed points, the bridges at the scroll and over the body of the instrument, is 30 cm. The tension on the E string is 90 N. (a) What is the mass per meter of the E string? (b) To prevent distortion of the instrument over time, it is important that the tension on all strings be the same. Find the masses per meter of the other strings.

**64** •• An ambulance is driving at 50 mph towards the brick wall of the hospital, which reflects the sound of the siren back toward the ambulance. When the ambulance is stationary, the siren's frequency is 500 Hz. (a) What is the spatial period of the standing wave caused by the sound of the siren and its reflection? (b) A doctor standing between the ambulance and the wall will hear the siren grow alternately louder and softer as the ambulance drives toward her. Why is this?

**65** •• To tune a violin, the violinist first tunes the A string to the correct pitch of 440 Hz and then bows two adjoining strings simultaneously and listens for a beat pattern. While bowing the A and E strings, the violinist hears a beat frequency of 3 Hz and notes that the beat frequency increases as the tension on the E string is increased. (The E string is to be tuned to 660 Hz.) (a) Why is a beat produced by these two strings bowed simultaneously? (b) What is the frequency of the E string vibration when the beat frequency is 3 Hz? (c) If the tension on the E string is 80.0 N when the beat frequency is 3 Hz, what tension corresponds to perfect tuning of that string?

**66** •• Suppose that you carry a small oscillator and speaker as you walk very slowly down a long hall. The speaker emits a sound of frequency 680 Hz, which is reflected from the walls at each end of the hall. As you walk along, you note that the sound intensity that you hear passes through successive maxima and minima. What distance must you walk to pass from one maximum to the next?

**67** •• **SSM** Show that the standing wave function $A' \sin kx \cos(\omega t + \delta)$ can be written as the sum of two harmonic wave functions—one for a wave traveling in the positive $x$ direction and the other for a wave of the same amplitude traveling in the negative $x$ direction. The traveling waves each have the same wave number and angular frequency as does the standing wave.

**68** •• A 2-m string is fixed at one end and is vibrating in its third harmonic with amplitude 3 cm and frequency 100 Hz. (a) Write the wave function for this vibration. (b) Write an expression for the kinetic energy of a segment of the string of length $dx$ at a point $x$ at some time $t$. At what time is this kinetic energy maximum? What is the shape of the string at this time? (c) Find the maximum kinetic energy of the string by integrating your expression for Part (b) over the total length of the string.

**69** •• **SSM** A commonly used physics experiment that examines resonances of transverse waves on a string is shown in Figure 16-30. A weight is attached to the end of a string draped over a pulley; the other end of the string is attached to a mechanical oscillator that moves the string up and down at a set frequency $f$. The length $L$ between the oscillator and the pulley is fixed. For certain values of the weight the string resonates. If $L = 1$ m, $f = 80$ Hz, and the mass density of the string is $\mu = 0.75$ g/m, what weights are needed for each of the first three modes (standing waves) of the string?

**FIGURE 16-30** Problem 69

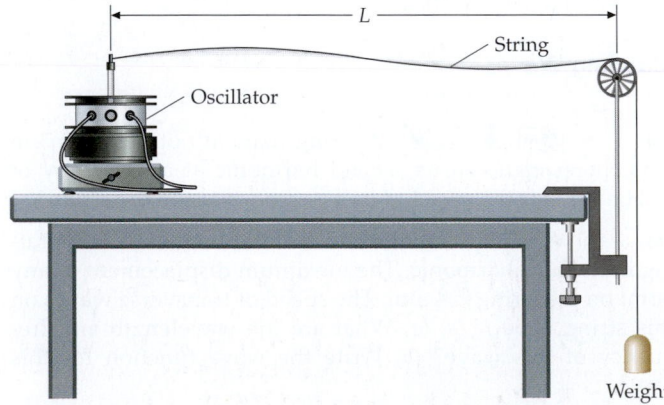

## *Wave Packets

**70** • Information used by computers is transmitted along a cable in the form of short electric pulses at the rate of $10^7$ pulses per second. (*a*) What is the maximum duration of each pulse if no two pulses overlap? (*b*) What is the range of frequencies to which the receiving equipment must respond?

**71** • **SSM** A tuning fork of frequency $f_0$ begins vibrating at time $t = 0$ and is stopped after a time interval $\Delta t$. The waveform of the sound at some later time is shown as a function of $x$. Let $N$ be the (approximate) number of cycles in this waveform. (*a*) How are $N, f_0,$ and $\Delta t$ related? (*b*) If $\Delta x$ is the length in space of this wave packet, what is the wavelength in terms of $\Delta x$ and $N$? (*c*) What is the wave number $k$ in terms of $N$ and $\Delta x$? (*d*) The number $N$ is uncertain by about $\pm 1$ cycle. Use Figure 16-31 to explain why. (*e*) Show that the uncertainty in the wave number due to the uncertainty in $N$ is $2\pi/\Delta x$.

**FIGURE 16-31** Problem 71

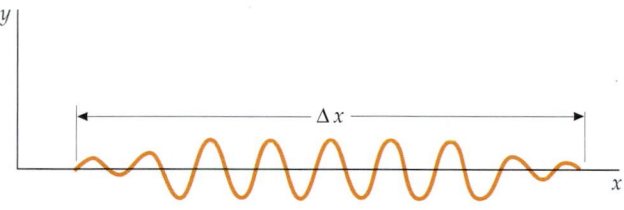

## General Problems

**72** • Middle C on the equal-temperament scale used by modern instrument makers has a frequency of 261.63 Hz. If a 7-g piano wire that is 80 cm long is to be tuned so that 261.63 is its fundamental frequency, what should be the tension in the wire?

**73** • **iSOLVE** The ear canal, which is about 2.5 cm long, roughly approximates a pipe that is open at one end and closed at the other. (*a*) What are the resonance frequencies of the ear canal? (*b*) Describe the possible effect of the resonance modes of the ear canal on the threshold of hearing.

**74** • A 4-m-long, 160-g rope is fixed at one end and is tied to a light string at the other end. Its tension is 400 N. (*a*) What are the wavelengths of the fundamental and the next two harmonics? (*b*) What are the frequencies of these standing waves?

**75** •• Two waves from two coherent sources have the same wavelength $\lambda$, frequency $\omega$, and amplitude $A$. What is the path difference if the resultant wave at some point has amplitude $A$?

**76** •• **iSOLVE** A 35-m string has a linear mass density of 0.0085 kg/m and is under a tension of 18 N. Find the frequencies of the lowest four harmonics if (*a*) the string is fixed at both ends and (*b*) the string is fixed at one end and attached to a long, thin, massless thread at the other end.

**77** •• **iSOLVE✓** You find an abandoned mine shaft and decide to measure its depth. Using an audio oscillator of

variable frequency, you note that you can produce successive resonances at frequencies of 63.58 and 89.25 Hz. What is the depth of the shaft?

**78** •• A string 5 m long that is fixed at one end only is vibrating in its fifth harmonic with a frequency of 400 Hz. The maximum displacement of any segment of the string is 3 cm. (*a*) What is the wavelength of this wave? (*b*) What is the wave number $k$? (*c*) What is the angular frequency? (*d*) Write the wave function for this standing wave.

**79** •• The wave function for a standing wave on a string is described by $y(x,t) = 0.02 \sin 4\pi x \cos 60\pi t$, where $y$ and $x$ are in meters and $t$ is in seconds. Determine the maximum displacement and maximum speed of a point on the string at (*a*) $x = 0.10$ m, (*b*) $x = 0.25$ m, (*c*) $x = 0.30$ m, and (*d*) $x = 0.50$ m.

**80** •• A 2.5-m-long wire with a mass of 0.10 kg is fixed at both ends and is under tension of 30 N. When the $n$th harmonic is excited, there is a node 0.50 m from one end. (*a*) What is $n$? (*b*) What are the frequencies of the first three allowed modes of vibration?

**81** •• **SSM** In an early method used to determine the speed of sound in gases, powder was spread along the bottom of a horizontal, cylinderical glass tube. One end of the tube was closed by a piston that oscillated at a known frequency $f$. The other end was closed by a movable piston whose position was adjusted until resonance occurred. At resonance, the powder collected in equally spaced piles along the bottom of the tube. (*a*) Explain why the powder collects in this way. (*b*) Derive a formula that gives the speed of sound in the gas in terms of $f$ and the distance between the piles of powder. (*c*) Give suitable values for the frequency $f$ and the distance between the piles of powder. (*d*) Give suitable values for the frequency $f$ and the length $L$ of the tube for which the speed of sound could be measured in either air or helium.

**82** •• In a lecture demonstration of standing waves, a string is attached to a tuning fork that vibrates at 60 Hz and sets up transverse waves of that frequency on the string. The other end of the string passes over a pulley, and the tension is varied by attaching weights to that end. The string has approximate nodes at the tuning fork and at the pulley. (*a*) If the string has a linear mass density of 8 g/m and is 2.5 m long (from the tuning fork to the pulley), what must be the tension for the string to vibrate in its fundamental mode? (*b*) Find the tension necessary for the string to vibrate in its second, third, and fourth harmonic.

**83** •• **iSOLVE✓** Three successive resonance frequencies in an organ pipe are 1310, 1834, and 2358 Hz. (*a*) Is the pipe closed at one end or open at both ends? (*b*) What is the fundamental frequency? (*c*) What is the length of the pipe?

**84** •• **iSOLVE✓** A wire of mass 1 g and length 50 cm is stretched with a tension of 440 N. It is then placed near the open end of the tube in Example 16-9 and stroked with a violin bow so that it oscillates at its fundamental frequency. The water level in the tube is then lowered until a resonance is obtained, which occurs at 18 cm below the top of the tube. Use the data given to determine the speed of sound in air. Why is this method not very accurate?

**85** •• A standing wave on a rope is represented by the wave function $y(x,t) = 0.02 \sin \frac{1}{2}\pi x \cos 40\pi t$, where $x$ and $y$ are in meters and $t$ is in seconds. (a) Write wave functions for two traveling waves that, when superimposed, will produce the resultant standing-wave pattern. (b) What is the distance between the nodes of the standing wave? (c) What is the velocity of a segment of the rope at $x = 1$ m? (d) What is the acceleration of a segment of the rope at $x = 1$ m?

**86** •• **iSOLVE** Two identical speakers emit sound waves of frequency 680 Hz uniformly in all directions. The total audio output of each speaker is 1 mW. A point $P$ is 2.00 m from one speaker and 3.00 m from the other. (a) Find the intensities $I_1$ and $I_2$ from each speaker separately at point $P$. (b) If the speakers are driven coherently and are in phase, what is the intensity at point $P$? (c) If they are driven coherently but are 180° out of phase, what is the intensity at point $P$? (d) If the speakers are incoherent, what is the intensity at point $P$?

**87** •• Three waves with the same frequency, wavelength, and amplitude are traveling in the same direction. The three waves are given by $y_1(x,t) = 0.05 \sin\left(kx - \omega t - \frac{\pi}{3}\right)$, $y_2(x,t) = 0.05 \sin(kx - \omega t)$, and $y_3(x,t) = 0.05 \sin\left(kx - \omega t + \frac{\pi}{3}\right)$. Find the resultant wave.

**88** •• A plane wave has the form $f(x, y, t) = A \cos(k_x x + k_y y - \omega t)$. Show that the direction in which the wave is traveling makes an angle $\theta = \tan^{-1}(k_y/k_x)$ with the positive $x$ direction and that the wave speed is $v = \omega/\sqrt{k_x^2 + k_y^2}$.

**89** •• **SSM** The speed of sound is proportional to the square root of the absolute temperature $T$ (Equation 15-5). (a) Show that if the temperature changes by a small amount $\Delta T$, the fundamental frequency of an organ pipe changes by approximately $\Delta f$, where $\Delta f/f = \frac{1}{2}\Delta T/T$. (b) Suppose that an organ pipe that is closed at one end has a fundamental frequency of 200 Hz when the temperature is 20°C. What will be its fundamental frequency when the temperature is 30°C? (Ignore any change in the length of the pipe due to thermal expansion.)

**90** •• Two traveling wave pulses on a string are represented by the wave functions

$$y_1(x,t) = \frac{0.02}{2 + (x - 2t)^2}$$

and

$$y_2(x,t) = \frac{-0.02}{2 + (x + 2t)^2}$$

where $x$ is in meters and $t$ is in seconds. (a) Using a spreadsheet program or graphing calculator, make a graph of each wave function separately as a function of $x$ at $t = 0$ and describe the behavior of each as time increases. (b) Find the resultant wave function at $t = 0$. (c) Find the resultant wave function at $t = 1$ s. (d) Graph the resultant wave function at $t = 1$ s.

**91** •• If you put your ear and your hand near the end of a long, open-ended tube and snap your fingers, you will hear a sound similar to that of a guitar string being plucked. (Tubes of about 1-m length are best.) (a) Explain what causes this sound. (b) What effective tube length do you need to make a sound like that of a guitar string with a pitch of A above middle C (440 Hz)?

**92** •• The kinetic energy of a segment of length $\Delta x$ and mass $\Delta m$ of a vibrating string is given by $\Delta K = \frac{1}{2}\Delta m(\partial y/\partial t)^2 = \frac{1}{2}\mu(\partial y/\partial t)^2 \Delta x$, where $\mu = \Delta m/\Delta x$. (a) Find the total kinetic energy of the $n$th mode of vibration of a string of length $L$ fixed at both ends. (b) Give the maximum kinetic energy of the string. (c) What is the wave function when the kinetic energy has its maximum value? (d) Show that the maximum kinetic energy in the $n$th mode is proportional to $n^2 A_n^2$.

**93** •• (a) Show that when the tension in a string fixed at both ends is changed by a small amount $dF$, the frequency of the fundamental is changed by approximately $df$, where $df/f = \frac{1}{2}dF/F$. Does this result apply to all harmonics? (b) Use this result to find the percentage change in the tension needed to increase the frequency of the fundamental of a piano wire from 260 to 262 Hz.

**94** •• **SSM** Two sources of harmonic waves on the $x$ axis have a phase difference that is proportional to time: $\delta_s = Ct$, where $C$ is a constant. The amplitude of the wave from each source at some point $P$ on the $x$ axis is $A_0$. (a) Write the wave functions for each of the two waves at point $P$, assuming this point to be a distance $x_1$ from one source and $x_1 + \Delta x$ from the other. (b) Find the resultant wave function and show that its amplitude is $2A_0 \cos[\frac{1}{2}(\delta + \delta_0)]$, where $\delta$ is the phase difference at $P$ due to the path difference. (c) Using a spreadsheet program or graphing calculator, graph the intensity at point $P$ versus time for a zero path difference. (Let $I_0$ be the intensity due to each wave separately.) What is the time average of the intensity? (d) Make the same graph for the intensity at a point for which the path difference is $\lambda/2$.

**95** ••• The wave functions of two standing waves on a string of length $L$ are $y_1(x,t) = A_1 \cos \omega_1 t \sin k_1 x$ and $y_2(x,t) = A_2 \cos \omega_2 t \sin k_2 x$, where $k_n = n\pi/L$ and $\omega_n = n\omega_1$. The wave function of the resultant wave is $y_r(x,t) = y_1(x,t) + y_2(x,t)$. (a) Find the velocity of a segment $dx$ of the string. (b) Find the kinetic energy of this segment. (c) By integration, find the total kinetic energy of the resultant wave. Notice the disappearance of the cross terms so that the total kinetic energy is proportional to $(n_1 A_1)^2 + (n_2 A_2)^2$.

**96** ••• A 2-m wire fixed at both ends is vibrating in its fundamental mode. The tension in the wire is 40 N and the mass of the wire is 0.1 kg. At the midpoint of the wire, the amplitude is 2 cm. (a) Find the maximum kinetic energy of the wire. (b) At the instant that the transverse displacement is given by $(0.02 \text{ m}) \sin(\pi x/2)$, what is the kinetic energy of the wire? (c) At what position on the wire does the kinetic energy per unit length have its largest value? (d) Where does the potential energy per unit length have its maximum value?

**97** ••• In principle, a wave with almost any arbitrary shape can be expressed as a sum of harmonic waves of different frequencies. (a) Consider the function defined by

$$f(x) = \frac{4}{\pi}\left(\frac{\cos x}{1} - \frac{\cos 3x}{3} + \frac{\cos 5x}{5} - \cdots\right)$$

$$= \frac{4}{\pi}\sum_{n=0}^{\infty}(-1)^n \frac{\cos[(2n+1)x]}{2n+1}$$

Write a spreadsheet program to calculate this series using a finite number of terms, and make three graphs of the function in the range $x = 0$ to $x = 4\pi$. For the first graph approximate the sum from $n = 0$ to $n = \infty$ with the first term of the sum. For the second and third graphs use only the first five term and the first ten terms, respectively. This function is sometimes called the *square wave* (or *θ function*). (*b*) What is the relation between this function and Liebnitz' series for $\pi$,

$$\frac{\pi}{4} = 1 - \frac{1}{3} + \frac{1}{5} - \frac{1}{7} + \cdots ?$$

**98** ••• Write a spreadsheet program to calculate and graph the function

$$f(x) = \frac{4}{\pi}\left(\sin x - \frac{\sin 3x}{9} + \frac{\sin 5x}{25} - \cdots\right)$$

$$= \frac{4}{\pi}\sum_n \frac{(-1)^n \sin(2n + 1)x}{(2n + 1)^2}$$

What kind of wave is this?

**99** ••• If you clap your hands at the end of a long, cylindrical tube, the echo you hear back will not sound like the handclap; instead, you will hear what sounds like a whistle,

initially at a very high frequency, but descending rapidly down to almost nothing. This "culvert whistler" can be explained by thinking of the sound from the clap as a single compression radiating outward from the hands. The echoes of the handclap arriving at your ear have traveled along different paths through the tube, as shown in Figure 16-32. The first echo to arrive travels straight down and straight back along the tube, while the second echo reflects once off of the center of the tube going out, and again going back, the third echo reflects twice at points 1/4 and 3/4 of the distance, etc. The tone of the sound you hear reflects the frequency at which these echoes reach your ears. (*a*) Show that the time delay between the $n_{th}$ echo and the $n+1_{th}$ is

$$\Delta t_n = \frac{2}{v}\left(\sqrt{(2n)^2 r^2 + L^2} - \sqrt{[2(n - 1)]^2 r^2 + L^2}\right),$$

where $v$ is the speed of sound, $L$ is the length of the tube and $r$ is its radius. (*b*) Using a spreadsheet program or graphing calculator, graph $\Delta t_n$ versus $n$ for $L = 90$ m, $r = 1$ m. (These are the approximate length and diameter of the long tube in the San Francisco Exploratorium.) Go to at least $n = 100$. (*c*) From your graph, explain why the frequency decreases over time. What are the highest and lowest frequencies you will hear in the whistler?

**FIGURE 16-32** Problem 99

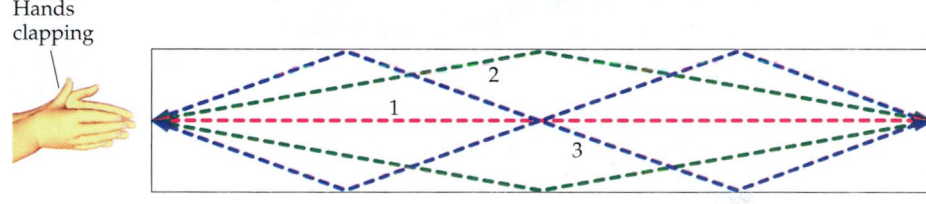

Hands clapping

# Temperature and the Kinetic Theory of Gases

THESE HELIUM BALLOONS ARE WELL INFLATED ON A WARM SUMMER DAY.

**?** **What might happen to them if they are taken indoors to an air-conditioned room? (See Example 17-6.)**

17-1  Thermal Equilibrium and Temperature

17-2  The Celsius and Fahrenheit Temperature Scales

17-3  Gas Thermometers and the Absolute Temperature Scale

17-4  The Ideal-Gas Law

17-5  The Kinetic Theory of Gases

**T**emperature is familiar to us as the measure of the hotness or coldness of objects or of our surroundings.

➤ **In this chapter, we will show that a consistent temperature scale can be defined in terms of the properties of gases at low densities, and that temperature is a measure of the average internal molecular kinetic energy of an object.**

## 17-1 Thermal Equilibrium and Temperature

Our sense of touch can usually tell us if an object is hot or cold. Early in childhood we learn that to make a cold object warmer, we place it in contact with a hot object. To make a hot object cooler, we place it in contact with a cold object.

When an object is heated or cooled, some of its physical properties change. Most solids and liquids expand when they are heated. A gas, if its pressure is kept constant, will also expand when it is heated, or, if its volume is kept constant, its pressure will rise. If an electrical conductor is heated, its electrical resistance changes. (This is discussed in Chapter 25.) A physical property that changes with temperature is called a **thermometric property**. A change in a thermometric property indicates a change in the temperature of the object.

Suppose that we place a warm copper bar in close contact with a cold iron bar so that the copper bar cools and the iron bar warms. We say that the two bars are in **thermal contact**. The copper bar contracts slightly as it cools, and the iron bar expands slightly as it warms. Eventually this process stops and the lengths of the bars remain constant. The two bars are then in **thermal equilibrium** with each other.

Suppose instead that we place the warm copper bar in a cool running stream. The bar cools until it stops contracting, at the point at which the bar and the water are in thermal equilibrium. Next we place a cold iron bar in the stream on the side opposite the copper bar. The iron bar will warm until it and the water are also in thermal equilibrium. If we remove the bars and place them in thermal contact with each other, we find that their lengths do not change. They are in thermal equilibrium with each other. Though it is common sense, there is no logical way to deduce this fact, which is called the **zeroth law of thermodynamics** (Figure 17-1):

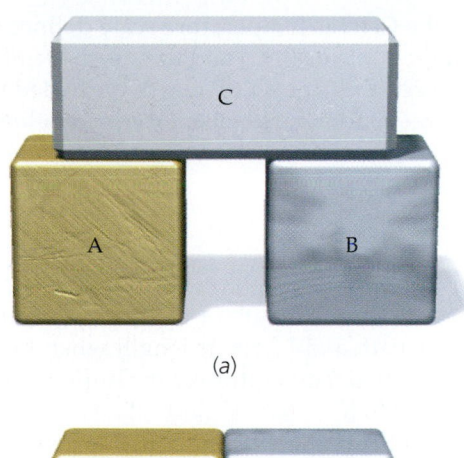

(a)

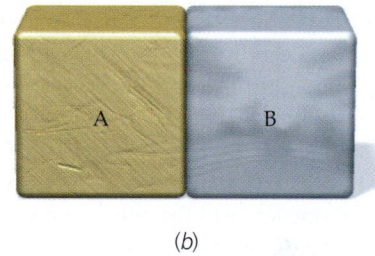

(b)

> If two objects are in thermal equilibrium with a third, then they are in thermal equilibrium with each other.

ZEROTH LAW OF THERMODYNAMICS

Two objects are defined to have the same *temperature* if they are in thermal equilibrium with each other. The zeroth law, as we will see, enables us to define a temperature scale.

**FIGURE 17-1** The zeroth law of thermodynamics. (*a*) Systems A and B are in thermal contact with system C but not with each other. When A and B are each in thermal equilibrium with C, they are in thermal equilibrium with each other, which can be checked by placing them in contact with each other as in (*b*).

## 17-2 The Celsius and Fahrenheit Temperature Scales

Any thermometric property can be used to establish a temperature scale. The common mercury thermometer consists of a glass bulb and tube containing a fixed amount of mercury.[†] When this thermometer is put in contact with a warmer object, the mercury expands, increasing the length of the mercury column (the glass expands too, but by a negligible amount). We can create a scale along the glass tube as follows. First the thermometer is placed in ice and water in equilibrium[‡] at a pressure of 1 atm. When the thermometer is in thermal equilibrium with the ice water, the position of the mercury column is marked on the glass tube. This is the **ice-point temperature** (also called the **normal freezing point** of water). Next, the thermometer is placed in boiling water at a pressure of 1 atm. When the thermometer is in thermal equilibrium with the boiling water, the new position of the mercury column is marked. This is the **steam-point temperature** (also called the **normal boiling point** of water).

---

[†] Because mercury is highly toxic, mercury thermometers are no longer sold in the United States. Today, alcohol is commonly used in thermometers.

[‡] Water and ice in equilibrium provide a constant-temperature bath. When ice is placed in warm water, the water cools as some of the ice melts. Eventually, thermal equilibrium is reached and no more ice melts. If the system is heated slightly, some more of the ice melts, but the temperature does not change as long as some ice remains.

The **Celsius temperature scale** defines the ice-point temperature as zero degrees Celsius (0°C) and the steam-point temperature as 100°C. The space between the 0° and 100° marks is divided into 100 equal intervals (degrees). Degree markings are also extended below and above these points. If $L_t$ is the length of the mercury column, the Celsius temperature $t_C$ is given by

$$t_C = \frac{L_t - L_0}{L_{100} - L_0} \times 100° \qquad 17\text{-}1$$

where $L_0$ is the length of the mercury column when the thermometer is in an ice bath and $L_{100}$ is its length when the thermometer is in a steam bath. The normal temperature of the human body measured on the Celsius scale is about 37°C.

The **Fahrenheit temperature scale** (which is used in the United States) defines the ice-point temperature as 32°F and the steam-point temperature as 212°F.[†] To convert temperatures between Fahrenheit and Celsius scales, we note there are 100 Celsius degrees and 180 Fahrenheit degrees between the ice and steam points. A temperature change of one Celsius degree therefore equals a change of 1.8 = 9/5 Fahrenheit degrees. To convert a temperature from one scale to the other, we must also take into account the fact that the zero temperatures of the two scales are not the same. The general relation between a Fahrenheit temperature $t_F$ and Celsius temperature $t_C$ is

$$t_C = \tfrac{5}{9}(t_F - 32°) \qquad 17\text{-}2$$

FAHRENHEIT–CELSIUS CONVERSION

**FIGURE 17-2** A bimetallic strip. When heated or cooled, the two metals expand or contract by different amounts, causing the strip to bend.

*CONVERTING FAHRENHEIT AND CELSIUS TEMPERATURES*     **EXAMPLE 17-1**

(*a*) Find the temperature on the Celsius scale equivalent to 41°F. (*b*) Find the temperature on the Fahrenheit scale equivalent to 37.0°C.

(*a*) Apply Equation 17-2 with $t_F = 41°$F:

$$t_C = \tfrac{5}{9}(t_F - 32°) = \tfrac{5}{9}(41° - 32°) = \tfrac{5}{9}(9°) = \boxed{5°C}$$

(*b*) 1. Solve Equation 17-2 for $t_F$ in terms of $t_C$:

$$t_F = \tfrac{9}{5}t_C + 32°$$

2. Substitute $t_C = 37°$C:

$$t_F = \tfrac{9}{5}(37.0°) + 32° = 66.6° + 32° = \boxed{98.6°F}$$

**EXERCISE** (*a*) Find the Celsius temperature equivalent to 68°F. (*b*) Find the Fahrenheit temperature equivalent to −40°C. (*Answer* (*a*) 20°C (*b*) −40°F)

Other thermometric properties can be used to set up thermometers and construct temperature scales. Figure 17-2 shows a bimetallic strip consisting of two different metals bonded together. When the strip is heated or cooled, it bends to accommodate the difference in the thermal expansion of the two metals. Figure 17-3 shows a thermometer consisting of a bimetallic coil with a pointer attached to indicate the temperature. When the thermometer is heated, the coil bends and the pointer moves. Like mercury thermometers, it is calibrated by dividing the interval between the ice point and the steam point into 100 Celsius degrees (or 180 Fahrenheit degrees).

---

† When the German physicist Daniel Fahrenheit devised his temperature scale, he wanted all measurable temperatures to be positive. Originally, he chose 0°F for the coldest temperature he could obtain with a mixture of ice and salt water and 96°F (a convenient number with many factors for subdivision) for the temperature of the human body. He then modified his scale slightly to make the ice-point and steam-point temperatures whole numbers. This resulted in the average temperature of the human body being between 98° and 99°F.

**FIGURE 17-3** (*a*) A thermometer using a bimetallic strip in the form of a coil. (The red pointer is attached to one end of the coil.) When the temperature of the coil increases, the needle rotates clockwise because the outer metal expands more than the inner metal. (*b*) A home thermostat controls the central air conditioner. When the air gets warmer, the coil expands, the glass bulb mounted on it tilts, and mercury in the tube slides to close an electrical switch, turning on the air conditioning. A slide lever (at the lower right), used to rotate the coil mount, is used to set the desired temperature. The circuit will be broken when the cooler air causes the bimetallic coil to contract.

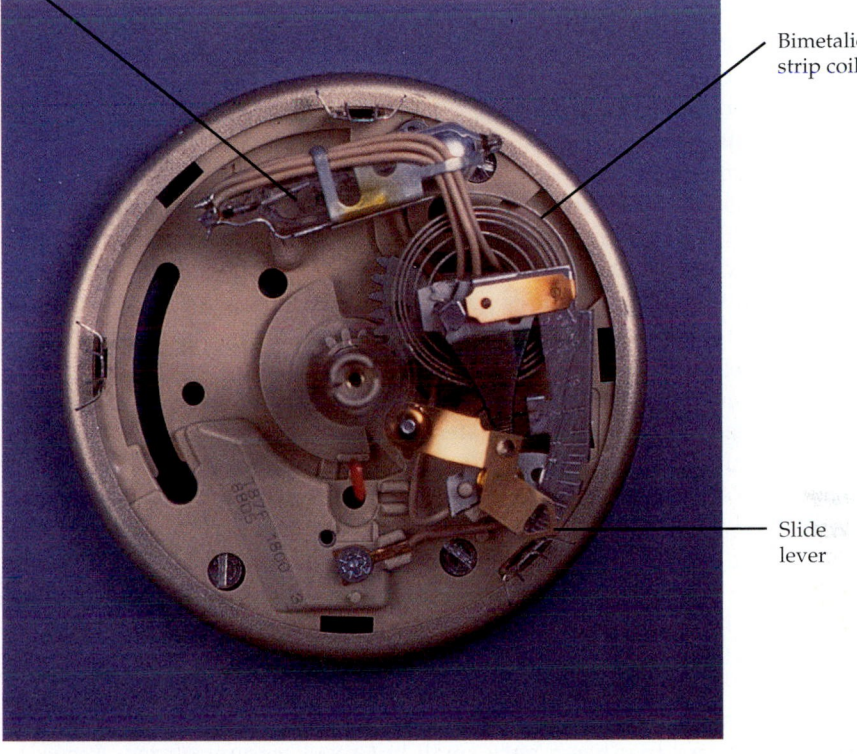

Glass Bulb
Mercury switch

Bimetalic strip coil

Slide lever

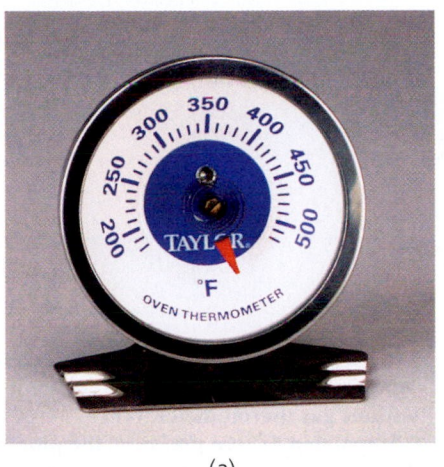

(a)

(b)

# 17-3 Gas Thermometers and the Absolute Temperature Scale

When different types of thermometers are calibrated in ice water and steam, they agree (by definition) at 0°C and 100°C, but they give slightly different readings at points in between. Discrepancies increase markedly above the steam point and below the ice point. However, in one group of thermometers, gas thermometers, measured temperatures agree closely with each other even far from the calibration points. In a **constant-volume gas thermometer,** the gas volume is kept constant, and change in gas pressure is used to indicate a change in temperature (Figure 17-4). An ice-point pressure $P_0$ and steam-point pressure $P_{100}$ are determined by placing the thermometer in ice–water and water–steam baths, and the interval between is divided into 100 equal degrees (for the Celsius scale). If the pressure is $P_t$ in a bath whose temperature is to be determined, that temperature in degrees Celsius is defined to be

$$t_C = \frac{P_t - P_0}{P_{100} - P_0} \times 100° \qquad 17\text{-}3$$

Suppose we measure a specific temperature, say the boiling point of sulfur at 1 atm pressure, using four constant-volume gas thermometers, each containing one of four gasses—air, hydrogen, nitrogen, and oxygen. The thermometers are calibrated, meaning values for $P_{100}$ and $P_0$ are determined for each. Each thermometer is then immersed in boiling sulfur, and when it is in thermal equilibrium with the sulfur, the pressure in the thermometer is measured. Next, the

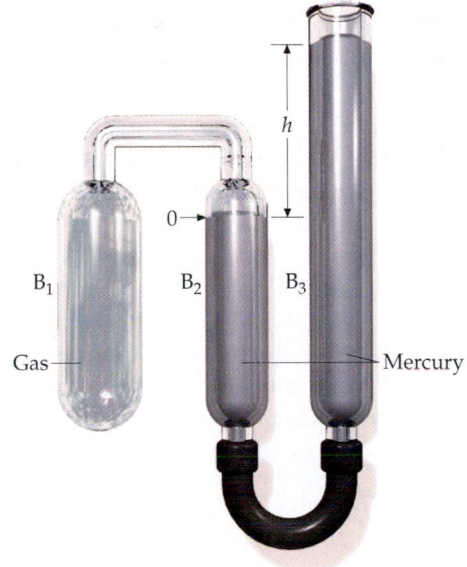

$B_1$  $B_2$  $B_3$

Gas

Mercury

$h$

0

**FIGURE 17-4** A constant-volume gas thermometer. The volume is kept constant by raising or lowering tube $B_3$ so that the mercury in tube $B_2$ remains at the zero mark. The temperature is chosen to be proportional to the pressure of the gas in tube $B_1$, which is indicated by the height $h$ of the mercury column in tube $B_3$.

temperature is calculated using Equation 17-3. Will this process give the same result for each of the four thermometers? Surprisingly perhaps, the answer is yes. All four thermometers measure the same temperature so long as the density of the gas in each is sufficiently low.

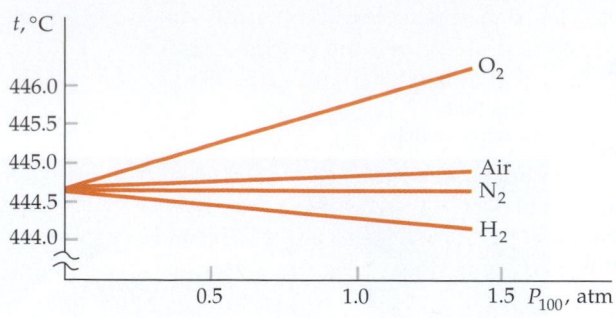

One measure of the density of the gas in the thermometer is its pressure at the steam point, $P_{100}$. If we vary the amount of gas in a constant-volume gas thermometer, by either adding or removing gas, we change both $P_{100}$ and $P_0$. As a result, each time the amount of gas is varied, the thermometer must be recalibrated. Figure 17-5 shows the results of measurements of the boiling point of sulfur using four constant-volume gas thermometers, each filled with air, hydrogen, nitrogen, or oxygen. For each thermometer the measured temperature is plotted as a function of the steam-point pressure $P_{100}$ of the thermometer. As the amount of a gas is reduced, its density and the steam-point pressure both decrease. We see that agreement among the thermometers is very close at low gas densities (low $P_{100}$). In the limit as gas density goes to zero, all gas thermometers give the same value for the temperature of boiling sulfur. This low-density temperature measurement is independent of the properties of any particular gas. Of course, there is nothing special about the boiling point of sulfur. Constant-volume gas thermometers at low densities are in agreement at any temperature. Thus, low-density gas thermometers can be used to define temperature.

Now consider a series of temperature measurements with a constant-volume gas thermometer that has a very small but fixed amount of gas. According to Equation 17-3, the pressure in the thermometer $P_t$ varies linearly with the measured temperature $t_C$. Figure 17-6 shows a plot of pressure versus measured temperature in a constant-volume gas thermometer. When we extrapolate this straight line to zero pressure, the temperature approaches $-273.15°C$. This limit is the same no matter what kind of gas is used.

A reference state that is much more precisely reproducible than either the ice or steam points is the **triple point of water**—the unique temperature and pressure at which water, water vapor, and ice coexist in equilibrium (see Figure 17-7). This equilibrium state occurs at 4.58 mmHg and 0.01°C. The **ideal-gas temperature scale** is defined so that the temperature of the triple point is 273.16 kelvins (K). (The kelvin is a degree unit that is the same size as the Celsius degree.) The temperature $T$ of any other state is defined to be proportional to the pressure in a constant-volume gas thermometer:

$$T = \frac{273.16\ \text{K}}{P_3} P \qquad\qquad 17\text{-}4$$

IDEAL-GAS TEMPERATURE SCALE

where $P$ is the observed pressure of the gas in the thermometer and $P_3$ is the pressure when the thermometer is immersed in a water–ice–vapor bath at its triple point. The value of $P_3$ depends

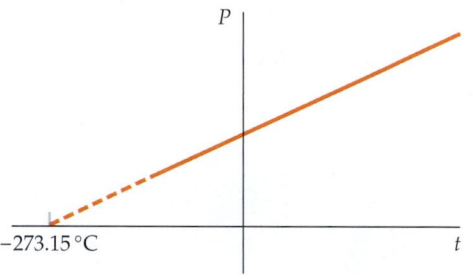

**FIGURE 17-5** Temperature of the boiling point of sulfur measured with constant-volume gas thermometers filled with various gases. Increasing or decreasing the amount of gas in the thermometer varies the pressure $P_{100}$ at the steam point of water. As the amount of gas is reduced, the temperatures measured by all the thermometers approach the value 444.60°C.

**FIGURE 17-6** Plot of pressure versus temperature as measured by a constant-volume gas thermometer. When extrapolated to zero pressure, the plot intersects the temperature axis at the value −273.15°C.

**FIGURE 17-7** H$_2$O at its triple point. The spherical flask contains water, ice, and water vapor in equilibrium.

on the amount of gas in the thermometer. The ideal-gas temperature scale, defined by Equation 17-4, has the advantage that any measured temperature does not depend on the properties of the particular gas that is used, but depends only on the general properties of gases.

The lowest temperature that can be measured with a gas thermometer is about 1 K, and requires helium for the gas. Below this temperature helium liquefies; all other gases liquefy at higher temperatures (see Table 17-1). In Chapter 19 we will see that the second law of thermodynamics can be used to define the **absolute temperature scale** independent of the properties of any substance, and with no limitations on the range of temperatures that can be measured. Temperatures as low as a millionth of a kelvin have been measured. The absolute scale so defined is identical to that defined by Equation 17-4 for the range of temperatures for which gas thermometers can be used. The symbol $T$ is used when referring to absolute temperature.

Because the Celsius degree and the kelvin are the same size, temperature *differences* are the same on both the Celsius and the absolute temperature scales (also called the **Kelvin scale**). That is, a temperature *change* of 1 K is identical to a temperature *change* of $1C°$.[†] The two scales differ only in the choice of zero temperature. To convert from degrees Celsius to kelvins, we merely add 273.15:[‡]

$$T = t_C + 273.15 \text{ K} \qquad\qquad 17\text{-}5$$

CELSIUS–ABSOLUTE CONVERSION

Although the Celsius and Fahrenheit scales are convenient for everyday use, the absolute scale is much more convenient for scientific purposes, partly because many formulas are more simply expressed in it, and partly because the absolute temperature can be given a more fundamental interpretation.

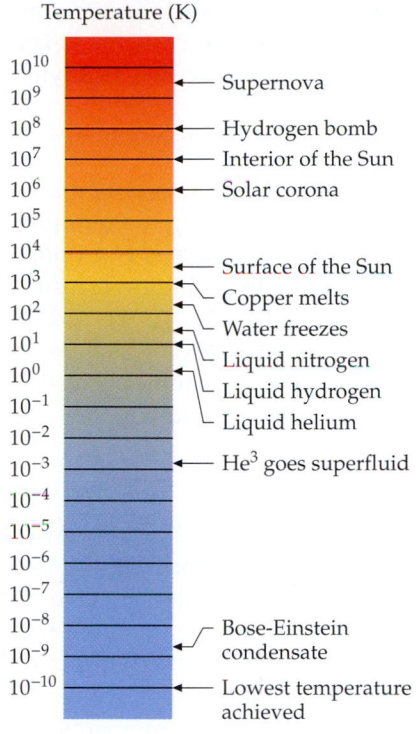

Temperature (K)

**TABLE 17-1** **The temperatures of various places and phenomena.**

---

*CONVERTING FROM KELVIN TO FAHRENHEIT*          **EXAMPLE   17-2**

**What is the Kelvin temperature corresponding to 70°F?**

**PICTURE THE PROBLEM**  First convert to degrees Celsius, then to kelvins.

1. Convert to degrees Celsius:
$$t_C = \tfrac{5}{9}(70° - 32°) = 21.1°C$$

2. To find the Kelvin temperature we add 273:
$$T = t_C + 273 = 21.1 + 273 = \boxed{294 \text{ K}}$$

**EXERCISE**  The "high-temperature" superconductor $YBa_2Cu_3O_7$ becomes superconducting when the temperature is lowered to 92 K. Find the superconducting threshold temperature in degrees Fahrenheit. (*Answer:*  $-294°F$)

## 17-4  The Ideal-Gas Law

The properties of gases at low densities allow the definition of the ideal-gas temperature scale. If we compress such a gas while keeping its temperature constant, the pressure increases. Similarly, if a gas expands at constant temperature, its pressure decreases. To a good approximation, the product of the pressure and

---

[†] We write $1C°$ to indicate a *temperature change* of one Celsius degree, in contrast to $1°C$, which means a temperature of one degree Celsius.

[‡] For most purposes, we can round off the temperature of absolute zero to $-273°C$.

volume of a low-density gas is constant at a constant temperature. This result was discovered experimentally by Robert Boyle (1627–1691), and is known as **Boyle's law:**

$$PV = \text{constant (constant temperature)}$$

A more general law exists that reproduces Boyle's law as a special case. According to Equation 17-4, the absolute temperature of a low-density gas is proportional to its pressure at constant volume. In addition—a result discovered experimentally by Jacques Charles (1746–1823) and Joseph Gay-Lussac (1778–1850)—the absolute temperature of a low-density gas is proportional to its volume at constant pressure. We can combine these two results by stating

$$PV = CT \qquad\qquad 17\text{-}6$$

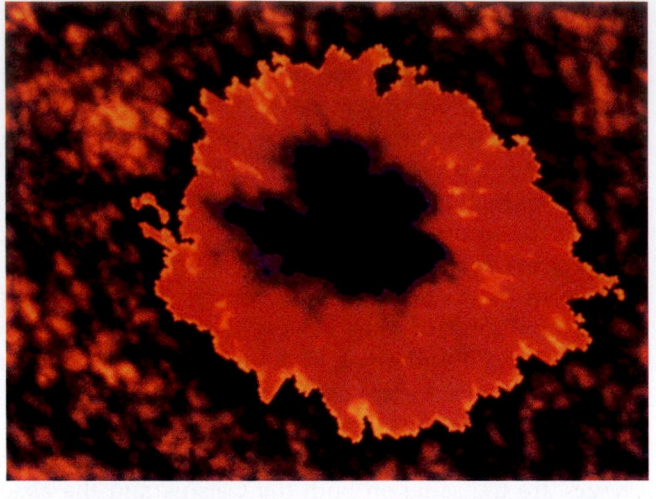

Sunspots appear on the surface of the sun when streams of gases slowly erupt from deep within the star. The solar "flower" is 10,000 miles in diameter. The temperature variation, indicated by computer-enhanced color changes, is not fully understood. The central portion of the sunspot is cooler than the outer regions as indicated by the dark area. The temperature at the sun's core is of the order of $10^7$ K, whereas at the surface the temperature is only about 6000 K.

where $C$ is a constant of proportionality. We can see that this constant is proportional to the amount of gas by considering the following. Suppose that we have two containers with identical volumes, each holding the same amount of the same kind of gas at the same temperature and pressure. If we consider the two containers as one system, we have twice the amount of gas at twice the volume, but at the same temperature and pressure. We have thus doubled the quantity $PV/T = C$ by doubling the amount of gas. We can therefore write $C$ as a constant $k$ times the number of molecules in the gas $N$:

$$C = kN$$

Equation 17-6 then becomes

$$PV = NkT \qquad\qquad 17\text{-}7$$

The constant $k$ is called **Boltzmann's constant.** It is found experimentally to have the same value for any kind of gas:

$$k = 1.381 \times 10^{-23} \, \text{J/K} = 8.617 \times 10^{-5} \, \text{eV/K} \qquad\qquad 17\text{-}8$$

An amount of gas is often expressed in moles. A **mole** (mol) of any substance is the amount of that substance that contains Avogadro's number $N_A$ of atoms or molecules, defined as the number of carbon atoms in 12 g of $^{12}$C:

$$N_A = 6.022 \times 10^{23} \qquad\qquad 17\text{-}9$$

AVOGADRO'S NUMBER

If we have $n$ moles of a substance, then the number of molecules is

$$N = nN_A \qquad\qquad 17\text{-}10$$

Equation 17-7 is then

$$PV = nN_A kT = nRT \qquad\qquad 17\text{-}11$$

where $R = N_A k$ is called the **universal gas constant.** Its value, which is the same for all gases, is

$$R = N_A k = 8.314 \, \text{J/(mol·K)} = 0.08206 \, \text{L·atm/(mol·K)} \qquad\qquad 17\text{-}12$$

Figure 17-8 shows plots of $PV/(nT)$ versus the pressure $P$ for several gases. For all gases, $PV/(nT)$ is nearly constant over a large range of pressures. Even oxygen, which varies the most in this graph, changes by only about 1 percent between 0 and 5 atm. An **ideal gas** is defined as one for which $PV/(nT)$ is constant for all pressures. The pressure, volume, and temperature of an ideal gas are related by

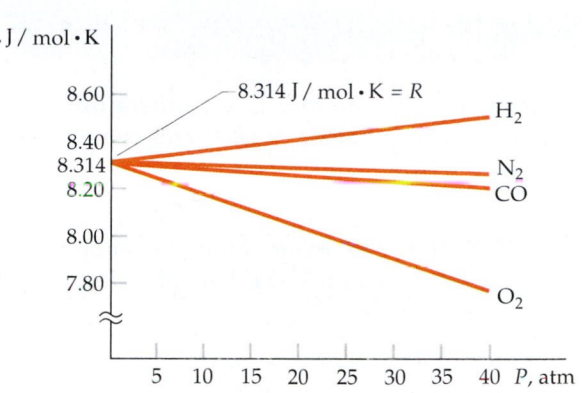

$$PV = nRT \qquad\qquad 17\text{-}13$$

IDEAL-GAS LAW

Equation 17-13, which relates the variables $P$, $V$, and $T$, is known as the ideal-gas law, and is an example of an **equation of state**. It describes the properties of real gases with low densities (and therefore low pressures). At higher densities, corrections must be made to this equation. In Chapter 20 we discuss another equation of state, the van der Waals equation, which includes such corrections. For any gas at any density, there is an equation of state relating $P$, $V$, and $T$ for a given amount of gas. Thus the state of a given amount of gas is determined by any two of the three **state variables** $P$, $V$, and $T$.

**FIGURE 17-8** Plot of $PV/nT$ versus $P$ for real gases. In these plots, varying the amount of gas varies the pressure. The ratio $PV/nT$ approaches the same value, 8.314 J/(mol·K), for all gases as we reduce the density, and thereby the pressure, of the gas. This value is the universal gas constant $R$.

---

*VOLUME OF AN IDEAL GAS*    **EXAMPLE 17-3**

**What volume is occupied by 1 mol of an ideal gas at a temperature of 0°C and a pressure of 1 atm?**

We can find the volume using the ideal-gas law, with $T = 273$ K:

$$V = \frac{nRT}{P}$$

$$= \frac{(1 \text{ mol})(0.0821 \text{ L·atm/[mol·K]})(273 \text{ K})}{1 \text{ atm}}$$

$$= \boxed{22.4 \text{ L}}$$

**REMARKS** Note that by writing $R$ in L·atm/(mol·K), we could write $P$ in atmospheres to get $V$ in liters.

**EXERCISE** Find (a) the number of moles $n$ and (b) the number of molecules $N$ in 1 cm³ of a gas at 0°C and 1 atm. (*Answer* (a) $n = 4.46 \times 10^{-5}$ mol (b) $N = 2.68 \times 10^{19}$ molecules)

The temperature of 0°C = 273 K and the pressure of 1 atm are often referred to as **standard conditions**. We see from Example 17-3 that under standard conditions, 1 mol of an ideal gas occupies a volume of 22.4 L.

Figure 17-9 shows plots of $P$ versus $V$ at several constant temperatures $T$. These curves are called **isotherms**. The isotherms for an ideal gas are hyperbolas. For a fixed amount of gas, we can see from Equation 17-13 that the quantity $PV/T$ is constant. Using the subscripts 1 for the initial values and 2 for the final values, we have

$$\frac{P_2 V_2}{T_2} = \frac{P_1 V_1}{T_1} \qquad\qquad 17\text{-}14$$

IDEAL-GAS LAW FOR FIXED AMOUNT OF GAS

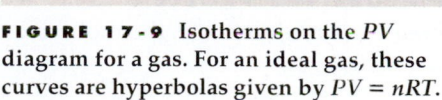

**FIGURE 17-9** Isotherms on the $PV$ diagram for a gas. For an ideal gas, these curves are hyperbolas given by $PV = nRT$.

*HEATING AND COMPRESSING A GAS*   **EXAMPLE 17-4**

A gas has a volume of 2 L, a temperature of 30°C, and a pressure of 1 atm. When the gas is heated to 60°C and compressed to a volume of 1.5 L, what is its new pressure?

**PICTURE THE PROBLEM** Since the amount of gas is fixed, the pressure can be found using Equation 17-14. Let subscripts 1 and 2 refer to the initial and final states, respectively.

1. Express the pressure $P_2$ in terms of $P_1$ and the initial and final volumes and temperatures:

$$\frac{P_1 V_1}{T_1} = \frac{P_2 V_2}{T_2}, \qquad P_2 = \frac{T_2 V_1}{T_1 V_2} P_1$$

2. Calculate the initial and final absolute temperatures:

$$T_1 = 273 + 30 = 303 \text{ K}$$
$$T_2 = 273 + 60 = 333 \text{ K}$$

3. Substitute numerical values in step 1 to find $P_2$:

$$P_2 = \frac{(333 \text{ K})(2 \text{ L})}{(303 \text{ K})(1.5 \text{ L})}(1 \text{ atm}) = \boxed{1.47 \text{ atm}}$$

**EXERCISE** How many moles of gas are in the system described in this example? (*Answer* $n = 0.0804$ mol)

The mass per mole of a substance is called its **molar mass** $M$. (The terms *molecular weight or molecular mass* are also sometimes used.) The molar mass of $^{12}C$ is, by definition, 12 g/mol or $12 \times 10^{-3}$ kg/mol. Molar masses of the elements are given in the periodic table in Appendix E. The molar mass of a molecule such as $CO_2$ is the sum of the molar masses of the elements in the molecule. Because the molar mass of oxygen is 16 g/mol (actually 15.999 g/mol), the molar mass of $O_2$ is 32 g/mol and that of $CO_2$ is $12 + 32 = 44$ g/mol.

The mass of $n$ moles of a gas is given by

$$m = nM$$

and the density $\rho$ of an ideal gas is

$$\rho = \frac{m}{V} = \frac{nM}{V}$$

Using $n/V = P/RT$ from Equation 17-13, we have

$$\rho = \frac{M}{RT} P \qquad\qquad 17\text{-}15$$

At a given temperature, the density of an ideal gas is proportional to its pressure.

*THE MASS OF A HYDROGEN ATOM*   **EXAMPLE 17-5**

The molar mass of hydrogen is 1.008 g/mol. What is the mass of one hydrogen atom?

**PICTURE THE PROBLEM** Let $m$ be the mass of a hydrogen atom. Since there are $N_A$ atoms in a mole, the molar mass $M$ is given by $M = m N_A$. We can use this to solve for $m$.

The mass of a hydrogen atom is the molar mass divided by Avogadro's number:

$$m = \frac{M}{N_A} = \frac{1.008 \text{ g/mol}}{6.022 \times 10^{23} \text{ atoms/mol}}$$

$$= \boxed{1.67 \times 10^{-24} \text{ g/atom}}$$

**REMARKS** Note that Avogadro's number is essentially the reciprocal of the mass of the hydrogen atom measured in grams.

---

*EXPANDING A GAS AT CONSTANT TEMPERATURE*　　　**EXAMPLE 17-6** **Try It Yourself**

A 100-g sample of $CO_2$ occupies a volume of 55 L at a pressure of 1 atm. (*a*) What is the temperature? (*b*) If the volume is increased to 80 L and the temperature is kept constant, what is the new pressure?

**PICTURE THE PROBLEM** Both questions can be answered using the ideal-gas law (Equation 17-13) if we first find the number of moles, $n$.

**Cover the column to the right and try these on your own before looking at the answers.**

| Steps | Answers |
|---|---|
| (*a*) 1. The number of moles $n$ is calculated from the mass of the sample $m$ and the molar mass $M$ of $CO_2$: The molar mass, from information in Appendix C, is 44 g/mol. | $n = \dfrac{m}{M} = 2.27 \text{ mol}$ |
| 2. Find the temperature $T$ from the ideal-gas law. | $T = \dfrac{PV}{nR} = \boxed{295 \text{ K}}$ |
| (*b*) Use $PV = $ constant to find the new pressure for $V = 80$ L. | $P_2 = \boxed{0.688 \text{ atm}}$ |

**EXERCISE** If the temperature is decreased at constant pressure, what happens to the volume? (*Answer* It decreases)

---

## 17-5 The Kinetic Theory of Gases

The description of the behavior of a gas in terms of the macroscopic state variables $P$, $V$, and $T$ can be related to simple averages of microscopic quantities such as the mass and speed of the molecules in the gas. The resulting theory is called **the kinetic theory of gases.**

From the point of view of kinetic theory, a gas consists of a large number of molecules making elastic collisions with each other and with the walls of a container. In the absence of external forces (we may neglect gravity), there is no preferred position for a molecule in the container,[†] and no preferred direction for its velocity vector. The molecules are separated, on the average, by distances that are large compared with their diameters, and they exert no forces on each other except when they collide. (This final assumption is equivalent to assuming a very low gas density, which, as we saw in the last section, is the same as assuming that the gas is an ideal gas. Because momentum is conserved, the collisions the

---

† Because of gravity, the density of molecules at the bottom of the container is slightly greater than at the top. As discussed in Chapter 13, the density of air decreases by half at a height of about 5.5 km, so the variation over a normal sized container is negligible.

molecules make with each other have no effect on the total momentum in any direction—thus such collisions can be neglected.)

## Calculating the Pressure Exerted by a Gas

The pressure that a gas exerts on its container is due to collisions between gas molecules and the container walls. This pressure is a force per unit area and, by Newton's second law, this force is the rate of change of momentum of the gas molecules colliding with the wall.

Consider a rectangular container of volume $V$ containing $N$ gas molecules, each of mass $m$ moving with a speed $v$. Let us calculate the force exerted by these molecules on the right-hand wall, which is perpendicular to the $x$ axis and has area $A$. The molecules hitting this wall in a time interval $\Delta t$ are those that are within distance $v_x \Delta t$ of the wall (Figure 17-10) and are moving to the right. Thus, the number of molecules hitting the wall during time $\Delta t$ is the number per unit volume $N/V$ times the volume $v_x \Delta t A$ times $\frac{1}{2}$ because, on average, only half the molecules are moving to the right. That is,

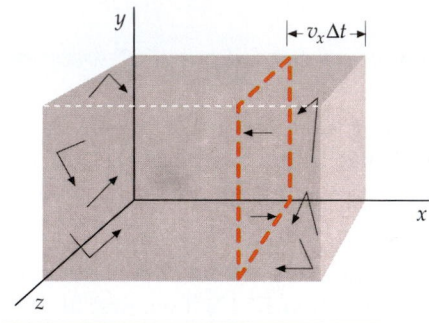

**FIGURE 17-10** Gas molecules in a rectangular container. In a time interval $\Delta t$, the molecules closer to the right wall than the distance $v_x \Delta t$ will hit the right wall if they are moving to the right.

$$\text{Molecules that hit the wall} = \frac{1}{2}\frac{N}{V}v_x\Delta t\, A$$

The $x$ component of momentum of a molecule is $+mv_x$ before it hits the wall, and $-mv_x$ after an elastic collision with the wall. The change in momentum has the magnitude $2mv_x$. The magnitude of the total change in momentum $|\Delta \vec{p}|$ of all molecules during a time interval $\Delta t$ is $2mv_x$ times the number of molecules that hit the wall during this interval:

$$|\Delta \vec{p}| = (2mv_x) \times \left(\frac{1}{2}\frac{N}{V}v_x\Delta t\, A\right) = \frac{N}{V}mv_x^2 A\,\Delta t \qquad 17\text{-}16$$

The magnitude of the force exerted by the wall on the molecules and by the molecules on the wall is $|\Delta \vec{p}|/\Delta t$. The pressure is the magnitude of this force divided by the area $A$:

$$P = \frac{F}{A} = \frac{1}{A}\frac{|\Delta \vec{p}|}{\Delta t} = \frac{N}{V}mv_x^2$$

or

$$PV = Nmv_x^2 \qquad 17\text{-}17$$

To allow for the fact that all the molecules in a container do not have the same speed, we merely replace $v_x^2$ with the average value $(v_x^2)_{av}$. Then, writing Equation 17-17 in terms of the kinetic energy $\frac{1}{2}mv_x^2$ associated with motion along the $x$ axis, we have

$$PV = 2N(\tfrac{1}{2}mv_x^2)_{av} \qquad 17\text{-}18$$

## The Molecular Interpretation of Temperature

Comparing Equation 17-18 with Equation 17-7, which was obtained experimentally for any gas at very low densities, we can see that

$$PV = NkT = 2N(\tfrac{1}{2}mv_x^2)_{av}$$

or

$$\left(\tfrac{1}{2}mv_x^2\right)_{\text{av}} = \tfrac{1}{2}kT \qquad\qquad 17\text{-}19$$

THE AVERAGE ENERGY ASSOCIATED WITH MOTION IN THE $x$ DIRECTION

Thus, the average kinetic energy associated with motion along the $x$ axis is $\tfrac{1}{2}kT$. But there is nothing special about the $x$ direction. On the average,

$$\left(v_x^2\right)_{\text{av}} = \left(v_y^2\right)_{\text{av}} = \left(v_z^2\right)_{\text{av}} \qquad\qquad 17\text{-}20$$

and

$$\left(v^2\right)_{\text{av}} = \left(v_x^2\right)_{\text{av}} + \left(v_y^2\right)_{\text{av}} + \left(v_z^2\right)_{\text{av}} = 3\left(v_x^2\right)_{\text{av}}$$

Writing $\left(v_x^2\right)_{\text{av}} = \tfrac{1}{3}\left(v^2\right)_{\text{av}}$ and $K_{\text{av}}$ for the average translational kinetic energy of the molecules,[†] Equation 17-19 becomes

$$K_{\text{av}} = \left(\tfrac{1}{2}mv^2\right)_{\text{av}} = \tfrac{3}{2}kT \qquad\qquad 17\text{-}21$$

AVERAGE KINETIC ENERGY OF A MOLECULE

The absolute temperature is thus a measure of the average translational kinetic energy of the molecules. The total translational kinetic energy of $n$ moles of a gas containing $N$ molecules is

$$K = N\left(\tfrac{1}{2}mv^2\right)_{\text{av}} = \tfrac{3}{2}NkT = \tfrac{3}{2}nRT \qquad\qquad 17\text{-}22$$

KINETIC ENERGY OF TRANSLATION FOR $n$ MOLES OF A GAS

where we've used $Nk = nN_Ak = nR$. Thus, the translational kinetic energy is $\tfrac{3}{2}kT$ per molecule and $\tfrac{3}{2}RT$ per mole.

We can use these results to estimate the order of magnitude of the speeds of the molecules in a gas. The average value of $v^2$ is, by Equation 17-21,

$$\left(v^2\right)_{\text{av}} = \frac{3kT}{m} = \frac{3N_AkT}{N_Am} = \frac{3RT}{M}$$

where $M = N_Am$ is the molar mass. The square root of $\left(v^2\right)_{\text{av}}$ is referred to as the **root mean square** (rms) speed:

$$v_{\text{rms}} = \sqrt{\left(v^2\right)_{\text{av}}} = \sqrt{\frac{3kT}{m}} = \sqrt{\frac{3RT}{M}} \qquad\qquad 17\text{-}23$$

Note that Equation 17-23 is similar to Equation 15-5 for the speed of sound in a gas:

$$v_{\text{sound}} = \sqrt{\frac{\gamma RT}{M}} \qquad\qquad 17\text{-}24$$

where $\gamma = 1.4$ for air. This is not surprising since a sound wave in air is a pressure disturbance propagated by collisions between air molecules.

---

† We include the word *translational* because the molecules may also have rotational or vibrational kinetic energy. Only the translational kinetic energy is relevant to the calculation of the pressure exerted by a gas on the walls of its container.

---

Oxygen gas ($O_2$) has a molar mass of about 32 g/mol and hydrogen gas ($H_2$) has a molar mass of about 2 g/mol. Calculate (*a*) the rms speed of an oxygen molecule when the temperature is 300 K and (*b*) the rms speed of a hydrogen molecule at the same temperature.

**PICTURE THE PROBLEM** (*a*) We find $v_{rms}$ using Equation 17-23. For the units to work out right, we use $R = 8.31$ J/(mol·K), and we express the molecular mass of $O_2$ in kg/mol. (*b*) Since $v_{rms}$ is proportional to $1/\sqrt{M}$, and the molar mass of hydrogen is one-sixteenth that of oxygen, the rms speed of hydrogen is 4 times that of oxygen.

(*a*) Substitute the given values into Equation 17-23:

$$v_{rms} = \sqrt{\frac{3RT}{M}} = \sqrt{\frac{3(8.31 \text{ J/[mol·K]})(300 \text{ K})}{32 \times 10^{-3} \text{ kg/mol}}}$$

$$= \boxed{483 \text{ m/s}}$$

(*b*) Use $v_{rms} \propto 1/\sqrt{M}$ to calculate $v_{rms}$ for hydrogen:

$$\frac{v_{rms}(H_2)}{v_{rms}(O_2)} = \frac{\sqrt{M_{O_2}}}{\sqrt{M_{H_2}}}$$

so

$$v_{rms}(H_2) = \sqrt{\frac{M_{O_2}}{M_{H_2}}} \, v_{rms}(O_2) = \sqrt{\frac{32 \text{ g/mol}}{2 \text{ g/mol}}} \, (483 \text{ m/s})$$

$$= \boxed{1930 \text{ m/s}}$$

**REMARKS** The rms speed of oxygen molecules is 483 m/s = 1080 mi/h, about 1.4 times the speed of sound in air, which at 300 K is about 347 m/s.

**EXERCISE** Find the rms speed of a nitrogen molecule ($M = 28$ g/mol) at 300 K. (*Answer* 516 m/s)

## The Equipartition Theorem

We have seen that the average kinetic energy associated with translational motion in any direction is $\frac{1}{2}kT$ per molecule (Equation 17-21) (or, equivalently, $\frac{1}{2}RT$ per mole), where $k$ is Boltzmann's constant. If the energy of a molecule associated with its motion in one direction is momentarily increased, say, by a collision between the molecule and a moving piston during a compression, collisions between that molecule and other molecules will quickly redistribute the added energy. When the gas is again in equilibrium, energy will be equally partitioned among the translational kinetic energies associated with motion in the $x$, $y$, and $z$ directions. This sharing of the energy equally among the three terms in the translational kinetic energy is a special case of the **equipartition theorem,** a result that follows from classical statistical mechanics. Each component of position and momentum (including angular position and angular momentum) that appears as a squared term in the expression for the energy of the system is called a **degree of freedom.** Typical degrees of freedom are associated with the kinetic energy of translation, rotation, and vibration, and with the potential energy of vibration. The equipartition theorem states that:

> When a substance is in equilibrium, there is an average energy of $\frac{1}{2}kT$ per molecule or $\frac{1}{2}RT$ per mole associated with each degree of freedom.

EQUIPARTITION THEOREM

In Chapter 18 we will use the equipartition theorem to relate the measured heat capacities of gases to their molecular structure.

## Mean Free Path

The average speed of molecules in a gas at normal pressures is several hundred meters per second, yet if somebody across the room from you opens a perfume bottle, you don't detect the odor for several minutes. The reason for the time delay is that the perfume molecules do not travel directly toward you, but instead travel a zigzag path due to collisions with the air molecules. The average distance $\lambda$ traveled by a molecule between collisions is called its **mean free path.** (The reason you smell the perfume at all is due to air currents (convection). The time for a perfume molecule to diffuse across a room is of the order of weeks.)

The mean free path of a gas molecule is related to its size, to the size of the surrounding gas molecules, and to the density of the gas. Consider one gas molecule of radius $r_1$ moving with speed $v$ through a region of stationary molecules (Figure 17-11). The moving molecule will collide with another molecule of radius $r_2$ if the centers of the two molecules come within a distance $d = r_1 + r_2$ from each other. (If all the molecules are the same type, then $d$ is the molecular diameter.) As the molecule moves, it will collide with any molecule whose center is in a circle of radius $d$ (Figure 17-12). In some time $t$, the molecule moves a distance $vt$ and collides with every molecule in the cylindrical volume $\pi d^2 vt$. The number of molecules in this volume is $n_V \pi d^2 vt$, where $n_V = N/V$ is the number of molecules per unit volume. (After each collision, the direction of the molecule changes, so the path actually zigs and zags.) The total path length divided by the number of collisions is the mean free path:

$$\lambda = \frac{vt}{n_V \pi d^2 vt} = \frac{1}{n_V \pi d^2}$$

This calculation of the mean free path assumes that all but one of the gas molecules are stationary, which is not a realistic situation. When the motion of all the molecules is taken into account, the correct expression for the mean free path is given by

$$\lambda = \frac{1}{\sqrt{2}\, n_V \pi d^2} \qquad \text{17-25}$$

The average time between collisions is called the collision time $\tau$. The reciprocal of the collision time, $1/\tau$, is equal to the average number of collisions per second, or the collision frequency. If $v_{av}$ is the average speed, then the average distance traveled between collisions is

$$\lambda = v_{av}\tau \qquad \text{17-26}$$

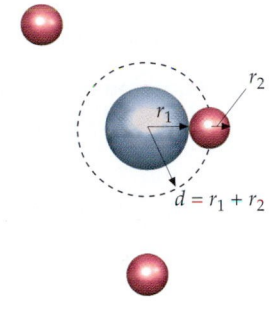

**FIGURE 17-11** Model of a molecule (center sphere) moving in a gas. The molecule of radius $r_1$ will collide with any molecule of radius $r_2$ if their centers are a distance $d = r_1 + r_2$ apart, which is any molecule whose center is on a sphere of radius $d = r_1 + r_2$ centered about the molecule.

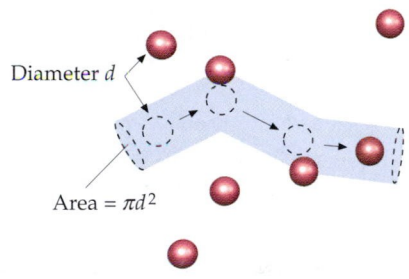

**FIGURE 17-12** Model of a molecule moving with speed $v$ in a gas of similar molecules. The motion is shown during time $t$. The molecule of diameter $d$ will collide with any similar molecule whose center is in a cylinder of volume $\pi d^2 vt$. In this picture, all collisions are assumed to be elastic and all the molecules but one are assumed to be at rest.

---

*MEAN FREE PATH OF A CO MOLECULE IN AIR* **EXAMPLE 17-8** **Put It in Context**

**The local poison control center wants to know more about carbon monoxide and how it spreads through a room. You are asked (a) to calculate the mean free path of a carbon monoxide molecule and (b) to estimate the mean time between collisions. The molecular mass of carbon monoxide is 28 g/mol. Assume that the CO molecule is traveling in air at 300K and 1 atm, and that the diameters of a CO molecule and air molecules are approximately $3.75 \times 10^{-10}$ m.**

**PICTURE THE PROBLEM** (a) Since $d$ is given, we can find $\lambda$ from $\lambda = 1/(\sqrt{2}\,n_V\pi d^2)$ using the ideal gas law to find $n_V = N/V$ (b) We can estimate the collision time by using $v_{rms}$ for the average speed.

(a) 1. Write $\lambda$ in terms of the number density $n_V$ and the molecular diameter $d$:

$$\lambda = \frac{1}{\sqrt{2}\,n_V\pi d^2}$$

2. Use the equation $PV = NkT$ to calculate $n_V = N/V$:

$$n_V = \frac{N}{V} = \frac{P}{kT} = \frac{101.3 \times 10^3\ \text{Pa}}{(1.38 \times 10^{-23}\ \text{J/K})(300\ \text{K})}$$

$$= 2.45 \times 10^{25}\ \text{molecules/m}^3$$

3. Substitute this value of $n_V$ and the given value of $d$ to calculate $\lambda$:

$$\lambda = \frac{1}{\sqrt{2}\,n_V\pi d^2}$$

$$= \frac{1}{\sqrt{2}\,(2.45 \times 10^{25}/\text{m}^3)\,\pi(3.75 \times 10^{-10}\ \text{m}^2)^2}$$

$$= \boxed{6.53 \times 10^{-8}\ \text{m}}$$

(b) 1. Write $\tau$ in terms of the mean free path $\lambda$:

$$\tau = \frac{\lambda}{v_{av}}$$

2. Estimate $v_{av}$ by calculating $v_{rms}$:

$$v_{rms} = \sqrt{\frac{3RT}{M}} = \sqrt{\frac{3(8.31\ \text{J/[mol}\cdot\text{K])}(300\ \text{K})}{28 \times 10^{-3}\ \text{kg/mol}}}$$

$$= 517\ \text{m/s}$$

3. Use $v_{av} \approx v_{rms}$ to estimate $\tau$:

$$\tau = \frac{\lambda}{v_{av}} = \frac{6.53 \times 10^{-8}\ \text{m}}{517\ \text{m/s}} = \boxed{1.26 \times 10^{-10}\ \text{s}}$$

**REMARKS** Note that we put atmospheric pressure in pascals to get the proper units for $\lambda$. The mean free path is about 200 times the diameter of the molecule, and the collision frequency is about $1/\tau \approx 8 \times 10^9$ collisions per second.

## *The Distribution of Molecular Speeds

We would not expect all of the molecules in a gas to have the same velocity. The calculation of the pressure of a gas allows us to calculate the square of the average speed and therefore the average energy of molecules in a gas, but it does not yield any details about the *distribution* of molecular velocities. Before we consider this problem, we will discuss the idea of distribution functions in general with some elementary examples from common experience.

**Distribution Functions** Suppose that a teacher gave a 25-point quiz to a large number $N$ of students. To describe the results, the teacher might give the average score, but this would not be a complete description. If all the students received a score of 12.5, for example, that would be quite different from half the students receiving 25 and the other half zero, but the average score would be the same in both cases. A complete description of the results would be to give the number of students $n_i$ who received a score $s_i$ for all the scores received. Alternatively, one could give the fraction of the students $f_i = n_i/N$ who received the score $s_i$. Both $n_i$ and $f_i$, which are functions of the variable $s$, are called **distribution functions.** The fractional distribution is somewhat more convenient to use. The probability that one of the $N$ students selected at random received the score $s_i$ equals the total number of students who received that score $n_i$ divided by $N$, that is, the probability equals $f_i$. Note that

$$\sum_i f_i = \sum_i \frac{n_i}{N} = \frac{1}{N}\sum_i n_i$$

and since $\Sigma n_i = N$,

$$\sum_i f_i = 1 \qquad\qquad 17\text{-}27$$

Equation 17-27 is called the **normalization condition** for fractional distributions.

To find the average score, we add all the scores and divide by $N$. Since each score $s_i$ was obtained by $n_i = N f_i$ students, this is equivalent to

$$s_{av} = \frac{1}{N} \sum_i n_i s_i = \sum_i s_i f_i \qquad\qquad 17\text{-}28$$

Similarly, the average of any function $g(s)$ is defined by

$$g(s)_{av} = \frac{1}{N} \sum_i g(s_i) n_i = \sum_i g(s_i) f_i \qquad\qquad 17\text{-}29$$

In particular, the square of the average score of the square of the scores is

$$(s^2)_{av} = \frac{1}{N} \sum_i s_i^2 n_i = \sum_i s_i^2 f_i \qquad\qquad 17\text{-}30$$

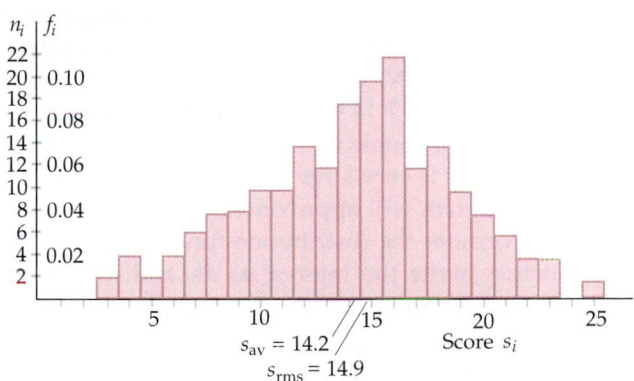

**FIGURE 17-13** Grade distribution for a 25-point quiz given to 200 students. $n_i$ is the number of students receiving grade $s_i$ and $f_i = n_i/N$ is the fraction of students receiving grade $s_i$.

The square root of $(s^2)_{av}$ is called the **root mean square score** or rms score. A possible distribution function is shown in Figure 17-13. For this distribution, the most probable score (that obtained by the most students) is 16, the average score is 14.2, and the rms score is 14.9.

---

*MAKING THE GRADE*                                          **EXAMPLE   17-9**

**Fifteen students took a 25-point quiz. Their scores were 25, 22, 22, 20, 20, 20, 18, 18, 18, 18, 18, 15, 15, 15, and 10. Find the average score and the rms score.**

**PICTURE THE PROBLEM**   The distribution function for this problem is $n_{25} = 1$, $n_{22} = 2$, $n_{20} = 3$, $n_{18} = 5$, $n_{15} = 3$, and $n_{10} = 1$. To find the average score, we use Equation 17-28. To find the rms score, we use Equation 17-30 and then take the square root.

1. By definition, $s_{av}$ is:

$$s_{av} = \frac{1}{N} \sum_i n_i s_i$$

$$= \frac{1}{15} \big[ 1(25) + 2(22) + 3(20) + 5(18) + 3(15) + 1(10) \big]$$

$$= \frac{1}{15}(274) = 18.3$$

2. To calculate $s_{rms}$, first find the average of $s^2$:

$$(s^2)_{av} = \frac{1}{N} \sum_i n_i s_i^2$$

$$= \frac{1}{15} \big[ 1(25)^2 + 2(22)^2 + 3(20)^2 + 5(18)^2 + 3(15)^2 + 1(10)^2 \big]$$

$$= \frac{1}{15}(5188) = 346$$

3. Take the square root of $(s^2)_{av}$:

$$s_{rms} = \sqrt{(s^2)_{av}} = \boxed{18.6}$$

Now consider the case of a continuous distribution, for example, the distribution of heights in a population. For any finite number $N$, the number of people who are *exactly* 2 m tall is zero. If we assume that height can be determined to any desired accuracy, there are an infinite number of possible heights, so the probability is zero that anybody has any one particular (exact) height. Therefore, we divide the heights into intervals $\Delta h$ (for example, $\Delta h$ might be 1 cm or 0.5 cm) and ask what fraction of people has heights that fall in any particular interval. For very large $N$, this number is proportional to the size of the interval, provided the interval is sufficiently small. We define the distribution function $f(h)$ as the fraction of the number of people with heights in the interval between $h$ and $h + \Delta h$. Then for $N$ people, $Nf(h)\,\Delta h$ is the number of people whose height is between $h$ and $h + \Delta h$. Figure 17-14 shows a possible height distribution.

The fraction of people with heights in a given interval $\Delta h$ is the area $f(h)\,\Delta h$. If $N$ is very large, we can choose $\Delta h$ to be very small, and the histogram will approximate a continuous curve. We can therefore consider the distribution function $f(h)$ to be a continuous function, write the interval as $dh$, and replace the sums in Equations 17-27 through 17-30 by integrals:

$$\int f(h)\,dh = 1$$

$$h_{av} = \int hf(h)\,dh \qquad \text{17-32}$$

$$[g(h)]_{av} = \int g(h)\,f(h)\,dh \qquad \text{17-33}$$

where $g(h)$ is an arbitrary function of $h$. Thus,

$$(h^2)_{av} = \int h^2 f(h)\,dh \qquad \text{17-34}$$

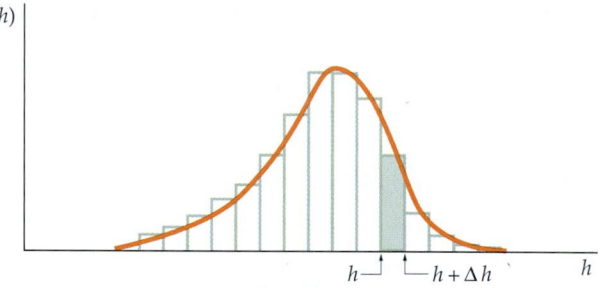

**FIGURE 17-14** A possible height distribution function. The fraction of the number of heights between $h$ and $h + \Delta h$ equals the shaded area $f(h)\,\Delta h$. The histogram can be approximated by a continuous curve as shown.

The probability of a person selected at random having a height between $h$ and $h + dh$ is $f(h)dh$. A useful quantity characterizing a distribution is the **standard deviation** $\sigma$ defined by

$$\sigma^2 = [(x - x_{av})^2]_{av} \qquad \text{17-35a}$$

Expanding the square on the right, we obtain

$$\sigma^2 = [x^2 - 2xx_{av} + x_{av}^2]_{av} = (x^2)_{av} - 2x_{av}x_{av} + x_{av}^2$$

or

$$\sigma^2 = (x^2)_{av} - x_{av}^2 \qquad \text{17-35b}$$

The standard deviation measures the spread of the values about the average value. For most distributions there will be few values that differ from $x_{av}$ by more than a few multiples of $\sigma$. For the familiar bell-shaped distribution (called a normal distribution), about 68 percent of the values are expected to fall within $x_{av} \pm \sigma$.

In Example 17-7, we found that the rms value was greater than the average value. This is a general feature for any distribution (unless all the values are identical, in which case $x_{rms} = x_{av}$). We can see this from Equation 17-35b by noting that $x_{rms}^2 = (x^2)_{av}$. Then $\sigma^2 = (x^2)_{av} - x_{av}^2 = x_{rms}^2 - x_{av}^2$. Since $\sigma^2$ and $x_{rms}$ are always positive, $x_{rms}$ must always be greater than $|x_{av}|$.

For the familiar bell-shaped distribution (called a normal distribution), 68 percent of the values fall within $x_{av} \pm \sigma$, 95 percent fall within $x_{av} \pm 2\sigma$, and 99.7 percent fall within $x_{av} \pm 3\sigma$. This is known as the 68, 95, 99.7 rule.

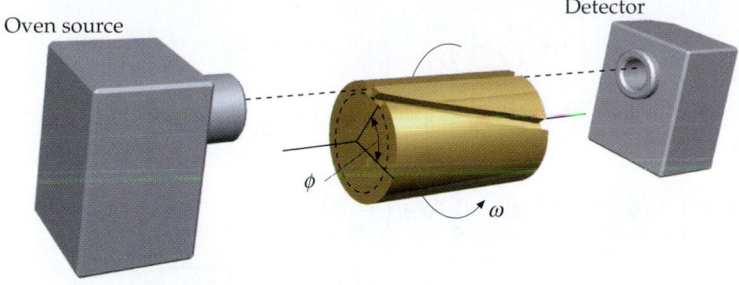

Oven source

Detector

**FIGURE 17-15** Schematic diagram of the apparatus for determining the speed distribution of the molecules of a gas. A substance is vaporized in an oven and the vapor molecules are allowed to escape through a hole in the oven wall into a vacuum chamber. The molecules are collimated into a narrow beam by a series of slits (not shown). The beam is aimed at a detector that counts the number of molecules that are incident on it in a given period of time. A rotating cylinder stops most of the beam. Small slits in the cylinder (only one of which is depicted here) allow the passage of molecules that have a narrow range of speeds that is determined by the angular velocity of rotation of the cylinder. Varying the angular velocity of the cylinder and counting the number of molecules that reach the detector for each angular velocity give a measure of the number of molecules in each range of speeds.

**The Maxwell–Boltzmann Distribution** The distribution of the molecular speeds of a gas can be measured directly using the apparatus illustrated in Figure 17-15. In Figure 17-16, these speeds are shown for two different temperatures. The quantity $f(v)$ in Figure 17-16 is called the **Maxwell–Boltzmann speed distribution function**. In a gas of $N$ molecules, the number with speeds in the range between $v$ and $v + dv$ is $dN$, given by

$$dN = N f(v)\, dv \qquad\qquad 17\text{-}36$$

The fraction $dN/N = f(v)\, dv$ in a particular range $dv$ is illustrated by the shaded region in the figure. The Maxwell–Boltzmann speed distribution function can be derived using statistical mechanics. The result is

$$f(v) = \frac{4}{\sqrt{\pi}}\left(\frac{m}{2kT}\right)^{3/2} v^2 e^{-mv^2/(2kT)} \qquad\qquad 17\text{-}37$$

MAXWELL–BOLTZMANN SPEED DISTRIBUTION FUNCTION

The most probable speed $v_{\max}$ is that speed for which $f(v)$ is maximum. It is left as a problem to show that

$$v_{\max} = \sqrt{\frac{2kT}{m}} = \sqrt{\frac{2RT}{M}} \qquad\qquad 17\text{-}38$$

Comparing Equation 17-38 with Equation 17-23, we see that the most probable speed is slightly less than the rms speed.

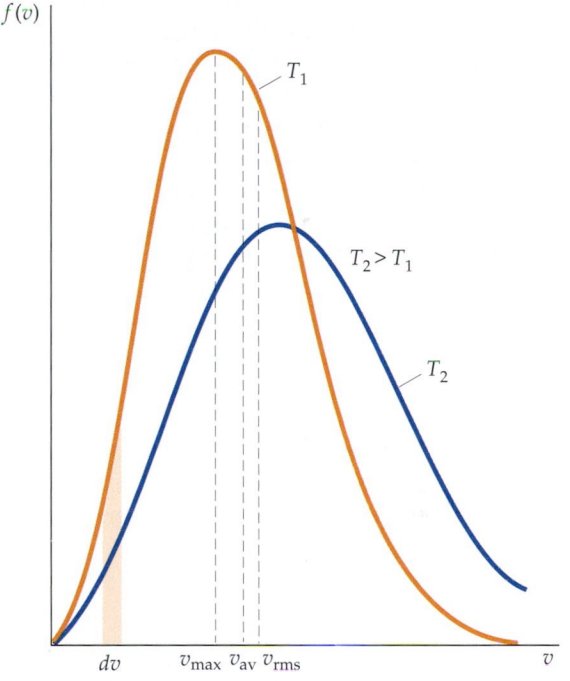

**FIGURE 17-16** Distributions of molecular speeds in a gas at two temperatures, $T_1$ and $T_2 > T_1$. The shaded area $f(v)\, dv$ equals the fraction of the number of molecules having a particular speed in a narrow range of speeds $dv$. The mean speed $v_{\mathrm{av}}$ and the rms speed $v_{\mathrm{rms}}$ are both slightly greater than the most probable speed $v_{\max}$.

---

*USING THE MAXWELL–BOLTZMANN DISTRIBUTION*    **EXAMPLE 17-10**

Calculate the average value of $v^2$ for the molecules in a gas using the Maxwell–Boltzmann distribution function.

**PICTURE THE PROBLEM** The average value of $v^2$ is calculated from Equation 17-34 with $v$ replacing $h$ and $f(v)$ given by Equation 17-37.

1. By definition, $(v^2)_{\mathrm{av}}$ is:
$$(v^2)_{\mathrm{av}} = \int_0^\infty v^2 f(v)\, dv$$

2. Use Equation 17-37 for $f(v)$:

$$(v^2)_{av} = \int_0^\infty v^2 \frac{4}{\sqrt{\pi}} \left(\frac{m}{2kT}\right)^{3/2} v^2 e^{-mv^2/(2kT)}\, dv$$

$$= \frac{4}{\sqrt{\pi}} \left(\frac{m}{2kT}\right)^{3/2} \int_0^\infty v^4 e^{-mv^2/(2kT)}\, dv$$

3. The integral in step 2 can be found in standard integral tables:

$$\int_0^\infty v^4 e^{-mv^2/(2\,kT)}\, dv = \frac{3}{8} \sqrt{\pi} \left(\frac{2kT}{m}\right)^{5/2}$$

4. Use this result to calculate $(v^2)_{av}$:

$$(v^2)_{av} = \frac{4}{\sqrt{\pi}} \left(\frac{m}{2kT}\right)^{3/2} \frac{3}{8} \sqrt{\pi} \left(\frac{2kT}{m}\right)^{5/2} = \boxed{\frac{3kT}{m}}$$

■ **REMARKS** Note that our result agrees with $v_{rms} = \sqrt{3kT/m}$ from Equation 17-23.

In Example 17-6 we found that the rms speed of hydrogen molecules is about 1.93 km/s. This is about one-sixth of the escape speed at the surface of the earth, which we found to be 11.2 km/s in Section 11-3. So why is there no free hydrogen in the earth's atmosphere? As we can see from Figure 17-16, a considerable fraction of the molecules of a gas in equilibrium have speeds greater than the rms speed. When the rms speed of the molecules of a particular gas is as great as 15 to 20 percent of the escape speed for a planet, enough of the molecules have speeds greater than the escape speed so that the gas does not remain in the atmosphere of that planet very long before escaping. Thus, there is virtually no hydrogen gas in the earth's atmosphere. The rms speed of oxygen molecules, on the other hand, is about one-fourth that of hydrogen molecules, which makes it only about 4 percent of the escape speed at the surface of the earth. Therefore, only a negligible fraction of the oxygen molecules have speeds greater than the escape speed, and oxygen remains in the earth's atmosphere.

**The Energy Distribution** The Maxwell–Boltzmann speed distribution as given by Equation 17-37 can also be written as an energy distribution. We write the number of molecules with energy $E$ in the range between $E$ and $E + dE$ as

$$dN = NF(E)\, dE$$

where $F(E)$ is the energy distribution function. This will be the same number as given by Equation 17-37 if the energy $E$ is related to the speed $v$ by $E = \frac{1}{2}mv^2$. Then

$$dE = mv\, dv$$

and

$$Nf(v)\, dv = NF(E)\, dE$$

We can write

$$f(v)\, dv = Cv^2 e^{-mv^2/(2kT)}\, dv = Cve^{-E/(kT)}\, v\, dv = C\left(\frac{2E}{m}\right)^{1/2} e^{-E/(kT)} \frac{dE}{m}$$

where $C = (4/\sqrt{\pi})[m/(2kT)]^{3/2}$ (from Equation 17-37). The energy distribution function $F(E)$ is thus given by

$$F(E) = \frac{4}{\sqrt{\pi}} \left(\frac{m}{2kT}\right)^{3/2} \left(\frac{2}{m}\right)^{1/2} \frac{1}{m} E^{1/2} e^{-E/(kT)}$$

Simplifying, we obtain the **Maxwell–Boltzmann energy distribution function:**

$$F(E) = \frac{2}{\sqrt{\pi}}\left(\frac{1}{kT}\right)^{3/2} E^{1/2}e^{-E/(kT)} \qquad 17\text{-}39$$

Maxwell–Boltzmann energy distribution function

In the language of statistical mechanics, the energy distribution is considered to be the product of two factors: one, called the **density of states,** is proportional to $E^{1/2}$; the other is the probability of a state being occupied, which is $e^{-E/(kT)}$ and is called the **Boltzmann factor.**

# SUMMARY

| Topic | Relevant Equations and Remarks |
|---|---|
| 1. **Celsius and Fahrenheit Scales** | On the Celsius scale, the ice point is defined to be 0°C and the steam point is 100°C. On the Fahrenheit scale, the ice point is 32°F and the steam point is 212°F. Temperatures on the Fahrenheit and Celsius scales are related by<br><br>$$t_C = \tfrac{5}{9}(t_F - 32°) \qquad 17\text{-}2$$ |
| 2. **Gas Thermometers** | Gas thermometers have the property that they all agree with each other in the measurement of any temperature as long as the density of the gas is very low. The ideal-gas temperature $T$ is defined by<br><br>$$T = \frac{273.16\ \text{K}}{P_3} P \qquad 17\text{-}4$$<br><br>where $P$ is the observed pressure of the gas in the thermometer and $P_3$ is the pressure when the thermometer is immersed in a water–ice–vapor bath at its triple point. |
| 3. **Kelvin Temperature Scale** | The absolute temperature or temperature in kelvins is related to the Celsius temperature by<br><br>$$T = t_C + 273.15\ \text{K} \qquad 17\text{-}5$$ |
| 4. **Ideal Gas** | At low densities, all gases obey the ideal-gas law. |
|    Equation of state | $$PV = nRT \qquad 17\text{-}13$$ |
|    Universal gas constant | $$R = N_A k = 8.314\ \text{J/(mol·K)}$$ $$= 0.08206\ \text{L·atm/(mol·K)} \qquad 17\text{-}12$$ |
|    Boltzmann's constant | $$k = 1.381 \times 10^{-23}\ \text{J/K} = 8.617 \times 10^{-5}\ \text{eV/K} \qquad 17\text{-}8$$ |
|    Avogadro's number | $$N_A = 6.022 \times 10^{23} \qquad 17\text{-}9$$ |
|    Equation for a fixed amount of gas | A form of the ideal-gas law that is useful for solving problems involving a fixed amount of gas is<br><br>$$\frac{P_2 V_2}{T_2} = \frac{P_1 V_1}{T_1} \qquad 17\text{-}14$$ |

**5.  Kinetic Theory of Gases**

| | |
|---|---|
| Molecular interpretation of temperature | The absolute temperature $T$ is a measure of the average molecular translational kinetic energy. |
| Equipartition theorem | When a system is in equilibrium, there is an average energy of $\frac{1}{2}kT$ per molecule (or $\frac{1}{2}RT$ per mole) associated with each degree of freedom. |
| Average kinetic energy | For an ideal gas, the average translational kinetic energy of the molecules is |

$$K_{av} = (\tfrac{1}{2}mv^2)_{av} = \tfrac{3}{2}kT \tag{17-21}$$

| | |
|---|---|
| Total kinetic energy | The total translational kinetic energy of $n$ moles of a gas containing $N$ molecules is given by |

$$K = N(\tfrac{1}{2}mv^2)_{av} = \tfrac{3}{2}NkT = \tfrac{3}{2}nRT \tag{17-22}$$

| | |
|---|---|
| rms speed of molecules | The rms speed of a molecule of a gas is related to the absolute temperature by |

$$v_{rms} = \sqrt{(v^2)_{av}} = \sqrt{\frac{3kT}{m}} = \sqrt{\frac{3RT}{M}} \tag{17-23}$$

where $m$ is the mass of the molecule and $M$ is the molar mass.

| | |
|---|---|
| Mean free path | The mean free path of a molecule is related to its diameter $d$ and the number of molecules per unit volume $n_V$ by |

$$\lambda = \frac{1}{\sqrt{2}n_V\pi d^2} \tag{17-25}$$

**\*6. Maxwell–Boltzmann Distribution**
$$f(v) = \frac{4}{\sqrt{\pi}}\left(\frac{m}{2kT}\right)^{3/2} v^2 e^{-mv^2/(2kT)} \tag{17-37}$$

| | |
|---|---|
| Energy distribution | $F(E) = \dfrac{2}{\sqrt{\pi}}\left(\dfrac{1}{kT}\right)^{3/2} E^{1/2}e^{-E/(kT)}$ |

$$\tag{17-39}$$

# PROBLEMS

- • Single-concept, single-step, relatively easy
- •• Intermediate-level, may require synthesis of concepts
- ••• Challenging
- **SSM** Solution is in the *Student Solutions Manual*
- **iSOLVE** Problems available on iSOLVE online homework service
- **iSOLVE ✔** These "Checkpoint" online homework service problems ask students additional questions about their confidence level, and how they arrived at their answer

In a few problems, you are given more data than you actually need; in a few other problems, you are required to supply data from your general knowledge, outside sources, or informed estimates.

## Conceptual Problems

**1** • **SSM** True or false:

(a) Two objects in thermal equilibrium with each other must be in thermal equilibrium with a third object.

(b) The Fahrenheit and Celsius temperature scales differ only in the choice of the zero temperature.

(c) The kelvin is the same size as the Celsius degree.

(d) All thermometers give the same result when measuring the temperature of a particular system.

**2** • How can you determine if two bodies are in thermal equilibrium with each other if it is impossible to put them into thermal contact with each other?

**3** • "One day I woke up and it was 20°F in my bedroom," said Mert to his old friend Mort. "That's nothing," replied Mort. "My room was once −5°C." Which room was colder?

**4** •• Two identical vessels contain different ideal gases at the same pressure and temperature. It follows that (*a*) the number of gas molecules is the same in both vessels, (*b*) the total mass of gas is the same in both vessels, (*c*) the average speed of the gas molecules is the same in both vessels, (*d*) none of these answers is correct.

**5** •• Figure 17-17 shows a plot of volume versus temperature for a process that takes an ideal gas from point A to point B. What happens to the pressure of the gas?

**FIGURE 17-17**
**Problem 5**

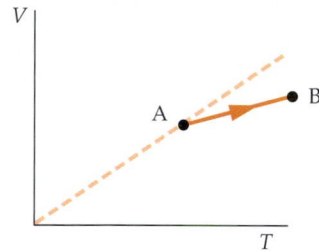

**6** •• **SSM** Figure 17-18 shows a plot of pressure versus temperature for a process that takes an ideal gas from point A to point B. What happens to the volume of the gas?

**FIGURE 17-18**
**Problem 6**

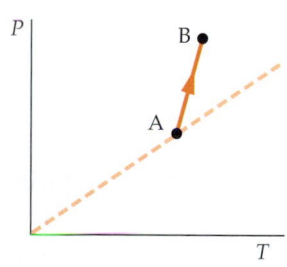

**7** • True or false: The absolute temperature of a gas is a measure of the average translational kinetic energy of the gas molecules.

**8** • By what factor must the absolute temperature of a gas be increased to double the rms speed of its molecules?

**9** • How does the average translational kinetic energy of a molecule of a gas change if the pressure is doubled while the volume is kept constant? If the volume is doubled while the pressure is kept constant?

**10** •• A vessel holds an equal number of moles of helium and methane, $CH_4$. The ratio of the rms speeds of the helium atoms to the $CH_4$ molecules is (*a*) 1, (*b*) 2, (*c*) 4, (*d*) 16.

**11** • True or false: If the pressure of a gas increases, the temperature must increase.

**12** • Why might the Celsius and Fahrenheit scales be more convenient than the absolute scale for ordinary, non-scientific purposes?

**13** • **SSM** The temperature of the interior of the sun is said to be about $10^7$ degrees. Do you think that this is degrees Celsius or kelvins, or does it matter?

**14** • **ISOLVE** If the temperature of an ideal gas is doubled while maintaining constant pressure, the average speed of the molecules (*a*) remains constant, (*b*) increases by a factor of 4, (*c*) increases by a factor of 2, (*d*) increases by a factor of $\sqrt{2}$.

**15** • **ISOLVE** If both temperature and volume of an ideal gas are halved, the pressure (*a*) diminishes by a factor of 2, (*b*) remains constant, (*c*) increases by a factor of 2, (*d*) diminishes by a factor of $\sqrt{2}$.

**16** • The average translational kinetic energy of the molecules of an ideal gas depends on (*a*) the number of moles of the gas and its temperature, (*b*) the pressure of the gas and its temperature, (*c*) the pressure of the gas only, (*d*) the temperature of the gas only.

**17** • If a vessel contains equal amounts, by weight, of helium and argon, which of the following are true?

(*a*) The pressure exerted by the two gases on the walls of the container is the same.
(*b*) The average speed of a helium atom is the same as that of an argon atom.
(*c*) The number of helium atoms and argon atoms in the vessel are equal.
(*d*) None of these statements is correct.

**18** •• Two rooms, A and B, have equal volumes and are connected by an open door. Room A, which is air-conditioned, is 5C° cooler than room B. Which room has more air in it?

**19** • Two different gases are at the same temperature. What can you say about the rms speeds of the gas molecules? What can you say about the average kinetic energies of the molecules?

**20** •• Explain in terms of molecular motion why the pressure on the walls of a container increases when a gas is heated at constant volume.

**21** •• **SSM** Explain in terms of molecular motion why the pressure on the walls of a container increases when the volume of a gas is reduced at constant temperature.

**22** •• Oxygen has a molar mass of 32 g/mol and nitrogen has a molar mass of 28 g/mol. The oxygen and nitrogen molecules in a room have

(*a*) equal average kinetic energies, but the oxygen molecules are faster.
(*b*) equal average kinetic energies, but the oxygen molecules are slower.
(*c*) equal average kinetic energies and speeds.
(*d*) equal average speeds, but the oxygen molecules have a higher average kinetic energy.
(*e*) equal average speeds, but the oxygen molecules have a lower average kinetic energy.
(*f*) None of these answers is correct.

**23** •• As any low-temperature physicist knows, liquid nitrogen is relatively cheap, while liquid helium is very expensive. One reason for this is that while nitrogen is the most common constituent of the atmosphere, helium is only found sealed in underground pockets of rock. Use ideas from this chapter to explain why this is true.

## Estimation and Approximation

**24** •• [SSM] A stoppered test tube that has a volume of 10 ml has 1 ml of water at its bottom and is at a temperature of 100°C and initially at a pressure of 1 atm ($1.01 \times 10^5$ N/m²). Estimate the pressure inside the test tube when the water is completely boiled away.

**25** ••• In Chapter 11, we found that the escape speed at the surface of a planet of radius $R$ is $v_e = \sqrt{2gR}$, where $g$ is the acceleration due to gravity at the surface of the planet. If the rms speed of a gas is greater than about 15 to 20 percent of the escape velocity of a planet, virtually all of the molecules of that gas will escape the atmosphere of the planet.

(a) At what temperature is $v_{rms}$ for $O_2$ equal to 15 percent of the escape speed for the earth?
(b) At what temperature is $v_{rms}$ for $H_2$ equal to 15 percent of the escape speed for the earth?
(c) Temperatures in the upper atmosphere reach 1000 K. How does this account for the low abundance of hydrogen in the earth's atmosphere?
(d) Compute the temperatures for which the rms speeds of $O_2$ and $H_2$ are equal to 15 percent of the escape velocity at the surface of the moon, where $g$ is about one-sixth of its value on earth and $R = 1738$ km. How does this account for the absence of an atmosphere on the moon?

**26** •• The escape velocity on Mars is 5.0 km/s and the surface temperature is typically 0°C. Calculate the rms speeds for (a) $H_2$, (b) $O_2$, and (c) $CO_2$ at this temperature. (d) Based on the criterion given in the chapter, are $H_2$, $O_2$, and $CO_2$ likely to be found in the atmosphere of Mars?

**27** •• [SSM] Repeat Problem 26 for Jupiter, whose escape velocity is 60 km/s and whose temperature is typically −150°C.

## Temperature Scales

**28** • A certain ski wax is rated for use between −12 and −7°C. What is this temperature range on the Fahrenheit scale?

**29** • [SOLVE] The melting point of gold (Au) is 1945.4°F. Express this temperature in degrees Celsius.

**30** • [SSM] [SOLVE] What is the Celsius temperature corresponding to the normal temperature of the human body, 98.6°F?

**31** • [SOLVE✓] The length of the column of mercury in a thermometer is 4.0 cm when the thermometer is immersed in ice water and 24.0 cm when the thermometer is immersed in boiling water. (a) What should be the length at room temperature, 22.0°C? (b) If the mercury column is 25.4 cm long when the thermometer is immersed in a chemical solution, what is the temperature of the solution?

**32** • The temperature of the interior of the sun is about $10^7$ K. What is this temperature on (a) the Celsius scale and (b) the Fahrenheit scale?

**33** • The boiling point of nitrogen $N_2$ is 77.35 K. Express this temperature in degrees Fahrenheit.

**34** • [SOLVE] The pressure of a constant-volume gas thermometer is 0.400 atm at the ice point and 0.546 atm at the steam point. (a) When the pressure is 0.100 atm, what is the temperature? (b) What is the pressure at 444.6°C, the boiling point of sulfur?

**35** • [SSM] [SOLVE] A constant-volume gas thermometer reads 50 torr at the triple point of water. (a) What will be the pressure when the thermometer measures a temperature of 300 K? (b) What ideal-gas temperature corresponds to a pressure of 678 torr?

**36** • A constant-volume gas thermometer has a pressure of 30 torr when it reads a temperature of 373 K. (a) What is its triple-point pressure $P_3$? (b) What temperature corresponds to a pressure of 0.175 torr?

**37** • At what temperature do the Fahrenheit and Celsius temperature scales give the same reading?

**38** • [SOLVE] Sodium melts at 371 K. What is the melting point of sodium on the Celsius and Fahrenheit temperature scales?

**39** • The boiling point of oxygen at 1 atm is 90.2 K. What is the boiling point of oxygen on the Celsius and Fahrenheit scales?

**40** •• On the Réaumur temperature scale, the melting point of ice is 0°R and the boiling point of water is 80°R. Derive expressions for converting temperatures on the Réaumur scale to the Celsius and Fahrenheit scales.

**41** ••• [SSM] A thermistor is a solid-state device whose resistance varies greatly with temperature. Its temperature dependence is given approximately by $R = R_0 e^{B/T}$, where $R$ is

Jupiter as seen from about twelve million miles. Because the escape speed at the surface of Jupiter is about 600 km/s, Jupiter easily retains hydrogen in its atmosphere.

in ohms ($\Omega$), $T$ is in kelvins, and $R_0$ and $B$ are constants that can be determined by measuring $R$ at calibration points such as the ice point and the steam point. (a) If $R = 7360 \, \Omega$ at the ice point and $153 \, \Omega$ at the steam point, find $R_0$ and $B$. (b) What is the resistance of the thermistor at $t = 98.6°F$? (c) What is the rate of change of the resistance with temperature ($dR/dT$) at the ice point and the steam point? (d) At which temperature is the thermistor most sensitive?

## The Ideal-Gas Law

**42** • **iSOLVE✓** A gas is kept at constant pressure. If its temperature is changed from 50 to 100°C, by what factor does the volume change?

**43** • **iSOLVE✓** A 10-L vessel contains gas at a temperature of 0°C and a pressure of 4 atm. How many moles of gas are in the vessel? How many molecules?

**44** •• **iSOLVE✓** A pressure as low as $1 \times 10^{-8}$ torr can be achieved using an oil diffusion pump. How many molecules are there in 1 cm$^3$ of a gas at this pressure if its temperature is 300 K?

**45** •• **SSM** You copy the following paragraph from a Martian physics textbook: "1 *snorf* of an ideal gas occupies a volume of 1.35 *zaks*. At a temperature of 22 *glips*, the gas has a pressure of 12.5 *klads*. At a temperature of $-10$ *glips*, the same gas now has a pressure of 8.7 *klads*." Determine the temperature of absolute zero in *glips*.

**46** •• **iSOLVE** A motorist inflates the tires of her car to a gauge pressure of 180 kPa on a day when the temperature is $-8.0°C$. When she arrives at her destination, the tire pressure has increased to 245 kPa. What is the temperature of the tires if we assume that (a) the tires do not expand or (b) that the tires expand by 7 percent?

**47** •• A room is 6 m by 5 m by 3 m. (a) If the air pressure in the room is 1 atm and the temperature is 300 K, find the number of moles of air in the room. (b) If the temperature rises by 5 K and the pressure remains constant, how many moles of air leave the room?

**48** •• **SSM** **iSOLVE** The boiling point of helium at 1 atm is 4.2 K. What is the volume occupied by helium gas due to evaporation of 10 g of liquid helium at 1 atm pressure and a temperature of (a) 4.2 K and (b) 293 K?

**49** •• A container with a volume of 6.0 L holds 10 g of liquid helium. As the container warms to room temperature, what is the pressure exerted by the gas on its walls?

**50** •• **SSM** **iSOLVE** An automobile tire is filled to a gauge pressure of 200 kPa when its temperature is 20°C. (Gauge pressure is the difference between the actual pressure and atmospheric pressure.) After the car has been driven at high speeds, the tire temperature increases to 50°C. (a) Assuming that the volume of the tire does not change and that air behaves as an ideal gas, find the gauge pressure of the air in the tire. (b) Calculate the gauge pressure if the volume of the tire expands by 10 percent.

**51** •• Calculate the mass density of air at a temperature of 24°C and a pressure of 1 atm ($1.01 \times 10^5$ N/m$^2$) using the ideal-gas law. Air is roughly 74 percent N$_2$ and 26 percent O$_2$.

**52** •• **iSOLVE✓** A scuba diver is 40 m below the surface of a lake, where the temperature is 5°C. He releases an air bubble with a volume of 15 cm$^3$. The bubble rises to the surface, where the temperature is 25°C. What is the volume of the bubble right before it breaks the surface? *Hint: Remember that the pressure also changes.*

**53** •• A hot-air balloon has a volume of 1.5 m$^3$ and is open at the bottom. If the air inside the balloon is at a temperature of 75°C, while the temperature of the air outside the balloon is 24°C, at a pressure of about 1 atm, what is the net force on the balloon and its contents? (Neglect the weight of the balloon itself.)

**54** ••• A helium balloon is used to lift a load of 110 N. The weight of the balloon's skin is 50 N and the volume of the balloon when fully inflated is 32 m$^3$. The temperature of the air is 0°C and the atmospheric pressure is 1 atm. The balloon is inflated with sufficient helium gas that the net force on the balloon and its load is 30 N. Neglect changes of temperature with altitude. (a) How many moles of helium gas are contained in the balloon? (b) At what altitude will the balloon be fully inflated? (c) Does the balloon ever reach the altitude at which it is fully inflated? (d) If the answer to (c) is affirmative, what is the maximum altitude attained by the balloon?

## Kinetic Theory of Gases

**55** • **SSM** (a) Find $v_{rms}$ for an argon atom if 1 mol of the gas is confined to a 1-L container at a pressure of 10 atm. (For argon, $M = 40 \times 10^{-3}$ kg/mol.) (b) Compare this with $v_{rms}$ for a helium atom under the same conditions. (For helium, $M = 4 \times 10^{-3}$ kg/mol.)

**56** • **iSOLVE✓** Find the total translational kinetic energy of 1 L of oxygen gas held at a temperature of 0°C and a pressure of 1 atm.

**57** • **iSOLVE** Find the rms speed and the average kinetic energy of a hydrogen atom at a temperature of $10^7$ K. (At this temperature, which is of the order of the temperature in the interior of a star, the hydrogen is ionized and consists of a single proton.)

**58** • **SSM** In one model of a solid, the material is assumed to consist of a regular array of atoms in which each atom has a fixed equilibrium position and is connected by springs to its neighbors. Each atom can vibrate in the $x$, $y$, and $z$ directions. The total energy of an atom in this model is

$$E = \tfrac{1}{2}mv_x^2 + \tfrac{1}{2}mv_y^2 + \tfrac{1}{2}mv_z^2 + \tfrac{1}{2}kx^2 + \tfrac{1}{2}ky^2 + kz^2$$

What is the average energy of an atom in the solid when the temperature is $T$? What is the total energy of 1 mol of such a solid?

**59** • Show that the mean free path for a molecule in an ideal gas at temperature $T$ and pressure $P$ is given by

$$\lambda = \frac{kT}{\sqrt{2}P\pi d^2}$$

**60** •• **iSOLVE✓** A pressure as low as $P = 7 \times 10^{-11}$ Pa has been obtained. Suppose that a chamber contains helium at this pressure and at room temperature (300 K). Estimate the mean free path $\lambda$ and the collision time $\tau$ for helium in the chamber. Take the diameter of a helium molecule to be $10^{-10}$ m.

**61** •• **SSM** Oxygen ($O_2$) is confined to a cubic container 15 cm on a side at a temperature of 300 K. Compare the average kinetic energy of a molecule of the gas to the change in its gravitational potential energy if it falls from the top of the container to the bottom.

### *The Distribution of Molecular Speeds

**62** •• Show that $f(v)$ given by Equation 17-37 is maximum when $v = \sqrt{2kT/m}$. Hint: Set $df/dv = 0$ and solve for $v$.

**63** •• **SSM** $f(v)$ is defined in Equation 17-37. Because $f(v)\,dv$ gives the fraction of molecules that have speeds in the range $dv$, the integral of $f(v)\,dv$ over all the possible ranges of speeds must equal 1. Given the integral

$$\int_0^\infty v^2 e^{-av^2}\,dv = \frac{\sqrt{\pi}}{4}a^{-3/2}$$

show that $\int_0^\infty f(v)\,dv = 1$, where $f(v)$ is given by Equation 17-37.

**64** •• Given the integral

$$\int_0^\infty v^3 e^{-av^2}\,dv = \frac{a^{-2}}{2}$$

calculate the average speed $v_{av}$ of molecules in a gas using the Maxwell–Boltzmann distribution function.

**65** •• **SSM** Current experiments in atomic trapping and cooling can create low-density gases of rubidium and other atoms with temperatures in the nanokelvin ($10^{-9}$ K) range. These atoms are trapped and cooled using magnetic fields and lasers in ultrahigh vacuum chambers. One method that is used to measure the temperature of a trapped gas is to turn the trap off and measure the time it takes for molecules of the gas to fall a given distance! Consider a gas of rubidium atoms at a temperature of 120 nK. Calculate how long it would take an atom traveling at the rms speed of the gas to fall a distance of 10 cm if (*a*) it were initially moving directly downward and (*b*) if it were initially moving directly upward. Assume that the atom doesn't collide with any others along its trajectory.

### General Problems

**66** • At what temperature will the rms speed of an $H_2$ molecule equal 331 m/s?

**67** •• (*a*) If 1 mol of a gas in a container occupies a volume of 10 L at a pressure of 1 atm, what is the temperature of the gas in kelvins? (*b*) The container is fitted with a piston so that the volume can change. When the gas is heated at constant pressure, it expands to a volume of 20 L. What is the temperature of the gas in kelvins? (*c*) The volume is fixed at 20 L, and the gas is heated at constant volume until its temperature is 350 K. What is the pressure of the gas?

**68** •• **iSOLVE** ✓ A cubic metal box with sides of 20 cm contains air at a pressure of 1 atm and a temperature of 300 K. The box is sealed so that the volume is constant and it is heated to a temperature of 400 K. Find the net force on each wall of the box.

**69** •• **SSM** Water, $H_2O$, can be converted into $H_2$ and $O_2$ gases by electrolysis. How many moles of these gases result from the electrolysis of 2 L of water?

**70** •• A massless cylinder 40 cm long rests on a horizontal frictionless table. The cylinder is divided into two equal sections by a membrane. One section contains nitrogen and the other contains oxygen. The pressure of the nitrogen is twice that of the oxygen. How far will the cylinder move if the membrane is removed?

**71** •• A cylinder contains a mixture of nitrogen gas ($N_2$) and hydrogen gas ($H_2$). At a temperature $T_1$ the nitrogen is completely dissociated but the hydrogen does not dissociate at all, and the pressure is $P_1$. If the temperature is doubled to $T_2 = 2T_1$, the pressure is tripled due to complete dissociation of hydrogen. If the mass of hydrogen is $m_H$, find the mass of nitrogen $m_N$.

**72** •• **SSM** Three insulated vessels of equal volume $V$ are connected by thin tubes that can transfer gas but do not transfer heat. Initially all vessels are filled with the same type of gas at a temperature $T_0$ and pressure $P_0$. Then the temperature in the first vessel is doubled and the temperature in the second vessel is tripled. The temperature in the third vessel remains unchanged. Find the final pressure $P'$ in the system in terms of the initial pressure $P_0$.

**73** •• A constant-volume gas thermometer with a triple-point pressure $P_3 = 500$ torr is used to measure the boiling point of some substance. When the thermometer is placed in thermal contact with the boiling substance, its pressure is 734 torr. Some of the gas in the thermometer is then allowed to escape so that its triple-point pressure is 200 torr. When it is again placed in thermal contact with the boiling substance, its pressure is 293.4 torr. Again, some of the gas is removed from the thermometer so that its triple-point pressure is 100 torr. When the thermometer is placed in thermal contact with the boiling substance once again, its pressure is 146.65 torr. Find the ideal-gas temperature of the boiling substance.

**74** •• **SSM** The mean free path for $O_2$ molecules at a temperature of 300 K at 1 atm pressure ($p = 1.01 \times 10^5$ Pa) is $\lambda = 7.1 \times 10^{-8}$ m. Use this data to estimate the size of an $O_2$ molecule.

**75** •• An experimental balloon contains hydrogen gas ($H_2$) at a temperature of 300 K and a pressure of 1 atm ($1.01 \times 10^5$ N/m$^2$). (*a*) Calculate the mean-free path of a hydrogen molecule. Assume that a $H_2$ molecule is effectively spherical, with a mean diameter of $1.6 \times 10^{-10}$ m. (*b*) Calculate the available volume per molecule ($V/N$), and find the average distance between each molecule and its nearest neighboring molecule (approximately the cube root of the available volume). Which is larger, the mean free path or the average nearest-neighbor distance between molecules?

**76** ••• A cylinder 2.4 m tall is filled with 0.1 mol of an ideal gas at standard temperature and pressure (Figure 17-19). The top of the cylinder is then sealed with a piston whose mass is 1.4 kg and the piston is allowed to drop until it is in equilibrium. (*a*) Find the height of the piston, assuming that the temperature of the gas does not change as it is compressed. (*b*) Suppose that the piston is pushed down below its equilibrium position by a small amount and then released. Assuming that the temperature of the gas remains constant, find the frequency of vibration of the piston.

— 1.4 kg

**FIGURE 17-19**
Problem 76

**77** ••• SSM The table below gives values of

$$\frac{4}{\sqrt{\pi}} \int_0^x z^2 e^{-z^2}\, dz$$

for different values of $x$. Use the table to answer the following questions: (*a*) For $O_2$ gas at 273 K, what fraction of molecules have speeds less than 400 m/s? (*b*) For the same gas, what percentage of molecules have speeds between 190 m/s and 565 m/s?

| $x$ | $\frac{4}{\sqrt{\pi}} \int_0^x z^2 e^{-z^2}\, dz$ | $x$ | $\frac{4}{\sqrt{\pi}} \int_0^x z^2 e^{-z^2}\, dz$ |
|---|---|---|---|
| 0.1 | $7.48 \times 10^{-4}$ | 0.7 | 0.194 |
| 0.2 | $5.88 \times 10^{-3}$ | 0.8 | 0.266 |
| 0.3 | 0.019 | 0.9 | 0.345 |
| 0.4 | 0.044 | 1.0 | 0.438 |
| 0.5 | 0.081 | 1.5 | 0.788 |
| 0.6 | 0.132 | 2.0 | 0.954 |

# Heat and the First Law of Thermodynamics

THE WARM LEMONADE IN THIS PITCHER IS COOLED BY ADDING ICE. HEAT IS TRANSFERRED FROM THE LEMONADE TO THE ICE BECAUSE OF A DIFFERENCE IN TEMPERATURE.

 **How much ice should you add to a cup of lemonade to reduce the temperature of the lemonade from 20°C to 0°C? (See Example 18-4.)**

18-1    Heat Capacity and Specific Heat

18-2    Change of Phase and Latent Heat

18-3    Joule's Experiment and the First Law of Thermodynamics

18-4    The Internal Energy of an Ideal Gas

18-5    Work and the *PV* Diagram for a Gas

18-6    Heat Capacities of Gases

18-7    Heat Capacities of Solids

18-8    Failure of the Equipartition Theorem

18-9    The Quasi-Static Adiabatic Compression of a Gas

**H**eat is energy that is being transferred from one system to another because of a difference in temperature. In the seventeenth century, Galileo, Newton, and other scientists generally supported the theory of the ancient Greek atomists who considered thermal energy to be a manifestation of molecular motion. In the next century, methods were developed for making quantitative measurements of the amount of heat that leaves or enters an object, and it was found that if objects are in thermal contact, the amount of heat that leaves one object equals the amount that enters the other. This discovery led to the caloric theory of heat as a conserved material substance. In this theory, an invisible fluid called "caloric" flowed out of one object and into another and this "caloric" could be neither created nor destroyed.

The caloric theory reigned until the nineteenth century, when it was found that friction between objects could generate an unlimited amount of thermal energy, deposing of the idea that caloric was a substance present in a fixed amount.

The modern theory of heat did not emerge until the 1840s, when James Joule (1818–1889) demonstrated that the increase or decrease of a given amount of thermal energy was always accompanied by the decrease or increase of an equivalent quantity of mechanical energy. Thermal energy, therefore, is not itself conserved. Instead, thermal energy is a form of internal energy, and it is energy that is conserved. ➤ **In this chapter, we define heat capacity, and examine how heating a system can cause either a change in its temperature or a change in its phase. We then examine the relationship between heat conduction, work, and internal energy of a system and express the law of conservation of energy for the thermal systems as the first law of thermodynamics. Finally, we shall see how the heat capacity of a system is related to its molecular structure.**

## 18-1 Heat Capacity and Specific Heat

When energy is transferred to a substance by heating it, the temperature of the substance usually rises.[†] The amount of heat energy $Q$ needed to raise the temperature of a substance is proportional to the temperature change and to the mass of the substance:

$$Q = C\Delta T = mc\Delta T \qquad \text{18-1}$$

where $C$ is the **heat capacity,** which is defined as the amount of energy transferred via heating necessary to raise the temperature of a substance by one degree. The **specific heat** $c$ is the heat capacity per unit mass:

$$c = \frac{C}{m} \qquad \text{18-2}$$

Steel ingots in a twin-tube tunnel furnace. The three 53-cm diameter carbon steel ingots seen here have been heated for about 7 hours to approximately 1340°C. Each 3200-kg ingot sits on a furnace car that transports it through the 81-m furnace, which is divided into twelve separate heating zones so that the temperature of the ingot is increased gradually to prevent cracking. The ingots, glowing a yellow-whitish color, exit the furnace to be milled into large, heavy-walled pipes.

The historical unit of heat energy, the **calorie,** was originally defined to be the amount of heat needed to raise the temperature of one gram of water one Celsius degree.[‡] Because we now recognize that heat is a measure of energy transfer, we can define the calorie in terms of the SI unit of energy, the joule:

$$1 \text{ cal} = 4.184 \text{ J} \qquad \text{18-3}$$

The U.S. customary unit of heat is the **Btu** (for British thermal unit), which was originally defined to be the amount of energy needed to raise the temperature of 1 pound of water by 1°F. The Btu is related to the calorie and to the joule by

$$1 \text{ Btu} = 252 \text{ cal} = 1.054 \text{ kJ} \qquad \text{18-4}$$

The original definition of the calorie implies that the specific heat of water is[§]

$$\begin{aligned} c_{\text{water}} &= 1 \text{ cal}/(\text{g·C°}) = 1 \text{ kcal}/(\text{kg·C°}) \\ &= 1 \text{ kcal}/(\text{kg·K}) = 4.184 \text{ kJ}/(\text{kg·K}) \end{aligned} \qquad \text{18-5}a$$

Similarly, from the definition of the Btu, the specific heat of water in U.S. customary units is

$$c_{\text{water}} = 1 \text{ Btu}/(\text{lb·F°}) \qquad \text{18-5}b$$

---

† An exception occurs during a change in phase, as when water freezes or evaporates. Changes of phase are discussed in Section 18-2.

‡ The kilocalorie is then the amount of heat energy needed to raise the temperature of 1 kg of water by 1°C. The "calorie" used in measuring the energy equivalent of foods is actually the kilocalorie.

§ Careful measurement shows that the specific heat of water varies by about 1 percent over the temperature range from 0 to 100°C. We will usually neglect this small variation.

The heat capacity per mole is called the **molar specific heat** $c'$,

$$c' = \frac{C}{n}$$

where $n$ is the number of moles. Since $C = mc$, the molar specific heat $c'$ and specific heat $c$ are related by

$$c' = \frac{C}{n} = \frac{mc}{n} = Mc \qquad \text{18-6}$$

where $M = m/n$ is the molar mass. Table 18-1 lists the specific heats and molar specific heats of some solids and liquids. Note that the molar heats of all the metals are about the same. We will discuss the significance of this in Section 18-7.

---

*RAISING THE TEMPERATURE*                                 **EXAMPLE 18-1**

**How much heat is needed to raise the temperature of 3 kg of copper by 20 C°?**

**PICTURE THE PROBLEM** The amount of heat needed to raise the temperature of the substance (copper) is proportional to the temperature change (20 C°) and to the mass (3 kg) of the substance.

The required heat is given by Equation 18-1 with $\qquad Q = mc\,\Delta T = (3\text{ kg})(0.386\text{ kJ/kg·K})(20\text{ K})$
$c = 0.386$ kJ/kg·K from Table 18-1:

$$= \boxed{23.2\text{ kJ}}$$

**REMARKS** Note that we use $\Delta T = 20$ C° $= 20$ K. Alternatively, we could express the specific heat as 0.386 kJ/kg·C° and write the temperature change as 20 C°.

**EXERCISE** A 2-kg aluminum block is originally at 10°C. If 36 kJ of heat energy are added to the block, what is its final temperature? (*Answer* 30°C)

## TABLE 18-1

**Specific Heats and Molar Specific Heats of Some Solids and Liquids**

| Substance | c, kJ/kg·K | c, kcal/kg·K or Btu/lb·F° | c', J/mol·K |
|---|---|---|---|
| Aluminum | 0.900 | 0.215 | 24.3 |
| Bismuth | 0.123 | 0.0294 | 25.7 |
| Copper | 0.386 | 0.0923 | 24.5 |
| Glass | 0.840 | 0.20 | — |
| Gold | 0.126 | 0.0301 | 25.6 |
| Ice (−10°C) | 2.05 | 0.49 | 36.9 |
| Lead | 0.128 | 0.0305 | 26.4 |
| Silver | 0.233 | 0.0558 | 24.9 |
| Tungsten | 0.134 | 0.0321 | 24.8 |
| Zinc | 0.387 | 0.0925 | 25.2 |
| Alcohol (ethyl) | 2.4 | 0.58 | 111 |
| Mercury | 0.140 | 0.033 | 28.3 |
| Water | 4.18 | 1.00 | 75.2 |

We see from Table 18-1 that the specific heat of water is considerably larger than that of other substances. Thus, water is an excellent material for storing thermal energy, as in a solar heating system. It is also an excellent coolant, as in a car engine.

## Calorimetry

To measure the specific heat of an object we can first heat it to some known temperature, say the boiling point of water, then transfer it to a water bath of known mass and initial temperature, and, finally, measure the final equilibrium temperature of the object (and the bath). If the system is isolated from its surroundings (by insulating the container, for example), then the heat leaving the object will equal the heat entering the water and its container. This procedure is called **calorimetry,** and the insulated water container is called a **calorimeter.**

Let $m$ be the mass of the object, let $c$ be its specific heat, and let $T_{io}$ be its initial temperature. If $T_f$ is the final temperature of the object in its water bath, the heat leaving the object is

$$Q_{out} = mc(T_{io} - T_f)$$

Similarly, if $T_{iw}$ is the initial temperature of the water and container, and $T_f$ is their final equilibrium temperature, then the heat absorbed by the water and container is

$$Q_{in} = m_w c_w(T_f - T_{iw}) + m_c c_c(T_f - T_{iw})$$

where $m_w$ and $c_w = 4.18$ kJ/kg·K are the mass and specific heat of the water, and $m_c$ and $c_c$ are the mass and specific heat of the container. (Note that we have chosen the temperature differences so that the heat in and heat out are both positive quantities.) Setting these amounts of heat equal yields the specific heat $c$ of the object:

$$Q_{out} = Q_{in} \qquad\qquad 18\text{-}7$$
$$mc\,(T_{io} - T_f) = m_w c_w(T_f - T_{iw}) + m_c c_c(T_f - T_{iw})$$

Because only temperature differences occur in Equation 18-7, and because the kelvin and Celsius degree are the same size, it doesn't matter whether we use kelvins or Celsius degrees.

Large bodies of water, such as lakes or oceans, tend to moderate fluctuations of the air temperature nearby because the bodies of water can absorb or release large quantities of thermal energy while undergoing only very small changes in temperature.

---

*MEASURING SPECIFIC HEAT*                              **EXAMPLE 18-2**

**To measure the specific heat of lead, you heat 600 g of lead shot to 100°C and place it in an aluminum calorimeter of mass 200 g that contains 500 g of water initially at 17.3°C. If the final temperature of the mixture is 20.0°C, what is the specific heat of lead? [The specific heat of the aluminum container is 0.900 kJ/kg·K.]**

**PICTURE THE PROBLEM** We set the heat leaving the lead equal to the heat entering the water and container and solve for the specific heat of lead $c_{Pb}$.

1. Write the heat leaving the lead in terms of its specific heat:

$$Q_{Pb} = m_{Pb}c_{Pb}|\Delta T_{Pb}|$$

2. Find the heat absorbed by the water:

$$Q_w = m_w c_w \Delta T_w$$

3. Find the heat absorbed by the container:

$$Q_c = m_c c_c \Delta T_c$$

4. Set the heat out equal to the heat in:

$$Q_{Pb} = Q_w + Q_c$$

$$m_{Pb}c_{Pb}|\Delta T_{Pb}| = m_w c_w \Delta T_w + m_c c_c \Delta T_c$$

where

$$\Delta T_c = \Delta T_w = 2.7 \text{ K and } |\Delta T_{Pb}| = 80 \text{ K}$$

5. Solve for $c_{Pb}$:

$$c_{Pb} = \frac{(m_w c_w + m_c c_c)\Delta T_w}{m_{Pb}|\Delta T_{Pb}|}$$

$$= \frac{[(0.5 \text{ kg})(4.18 \text{ kJ/kg·K}) + (0.2 \text{ kg})(0.9 \text{ kJ/kg·K})](2.7 \text{ K})}{(0.6 \text{ kg})(80 \text{ K})}$$

$$= \boxed{0.128 \text{ kJ/kg·K}}$$

**REMARKS** Note that the specific heat of lead is considerably less than that of water.

# 18-2 Change of Phase and Latent Heat

If heat is added to ice at 0°C, the temperature of the ice does not change. Instead, the ice melts. Melting is an example of a **phase change.** Common types of phase changes include fusion (liquid to solid), melting (solid to liquid), vaporization (liquid to vapor or gas), condensation (gas or vapor to liquid), and sublimation (solid directly to vapor, as when solid carbon dioxide [dry ice] changes to vapor). There are other types of phase changes as well, such as the change of a solid from one crystalline form to another. For example, carbon under intense pressure becomes a diamond.

Molecular theory can help us to understand why temperature remains constant during a phase change. The molecules in a liquid are close together and exert attractive forces on each other, whereas molecules in a gas are far apart. Because of this molecular attraction, it takes energy to remove molecules from a liquid to form a gas. Consider a pot of water sitting over a flame on the stove. At first, as the water is heated, the motion of its molecules increases and the temperature rises. When the temperature reaches the boiling point, the molecules can no longer increase their kinetic energy and remain in the liquid. As the liquid water vaporizes, the added heat energy is used to overcome the attractive forces between the water molecules as they spread farther apart in the gas phase. The added energy thus increases the potential energy of the molecules rather than their kinetic energy. Because temperature is a measure of the average translational *kinetic* energy of molecules, the temperature doesn't change.

For a pure substance, a change in phase at a given pressure occurs only at a particular temperature. For example, pure water at a pressure of 1 atm changes from solid to liquid at 0°C (the normal melting point of water) and from liquid to gas at 100°C (the normal boiling point of water).

The heat energy required to melt a substance of mass $m$ with no change in its temperature is proportional to the mass of the substance:

$$Q_f = mL_f \hspace{4em} \text{18-8}$$

Although melting indicates that the ice has experienced a change in phase, the temperature of the ice does not change.

where $L_f$ is called the **latent heat of fusion** of the substance. At a pressure of 1 atm, the latent heat of fusion for water is 333.5 kJ/kg = 79.7 kcal/kg. If the phase change is from liquid to gas, the heat required is

$$Q_v = mL_v \qquad\qquad 18\text{-}9$$

where $L_v$ is the **latent heat of vaporization.** For water at a pressure of 1 atm, the latent heat of vaporization is 2.26 MJ/kg = 540 kcal/kg. Table 18-2 gives the normal melting and boiling points, and the latent heats of fusion and vaporization at 1 atm, for various substances.

## TABLE 18-2

Normal Melting Point (MP), Latent Heat of Fusion ($L_f$), Normal Boiling Point (BP), and Latent Heat of Vaporization ($L_v$) for Various Substances at 1 atm

| Substance | MP, K | $L_f$, kJ/kg | BP, K | $L_v$, kJ/kg |
|---|---|---|---|---|
| Alcohol, ethyl | 159 | 109 | 351 | 879 |
| Bromine | 266 | 67.4 | 332 | 369 |
| Carbon dioxide | — | — | 194.6[†] | 573[†] |
| Copper | 1356 | 205 | 2839 | 4726 |
| Gold | 1336 | 62.8 | 3081 | 1701 |
| Helium | — | — | 4.2 | 21 |
| Lead | 600 | 24.7 | 2023 | 858 |
| Mercury | 234 | 11.3 | 630 | 296 |
| Nitrogen | 63 | 25.7 | 77.35 | 199 |
| Oxygen | 54.4 | 13.8 | 90.2 | 213 |
| Silver | 1234 | 105 | 2436 | 2323 |
| Sulfur | 388 | 38.5 | 717.75 | 287 |
| Water | 273.15 | 333.5 | 373.15 | 2257 |
| Zinc | 692 | 102 | 1184 | 1768 |

† These values are for sublimation. Carbon dioxide does not have a liquid state at 1 atm.

---

*CHANGING ICE INTO STEAM*     **E X A M P L E   1 8 - 3**   **Try It Yourself**

How much heat is needed to change 1.5 kg of ice at −20 C° and 1 atm into steam?

**PICTURE THE PROBLEM** The heat required to change the ice into steam consists of four parts: $Q_1$, the heat needed to warm the ice from −20°C to 0°C; $Q_2$, the heat needed to melt the ice; $Q_3$, the heat needed to warm the water from 0°C to 100°C; and $Q_4$, the heat needed to vaporize the water. In calculating $Q_1$ and $Q_3$, we will assume that the specific heats are constant, with the values 2.05 kJ/kg·K for ice and 4.18 kJ/kg·K for water.

**Cover the column to the right and try these on your own before looking at the answers.**

| Steps | Answers |
|---|---|
| 1. Use $Q_1 = mc\,\Delta T$ to find the heat needed to warm the ice to 0°C. | $Q_1 = 61.5$ kJ $= 0.0615$ MJ |

2. Use $L_f$ from Table 18-2 to find the heat $Q_2$ needed to melt the ice.

$Q_2 = 500 \text{ kJ} = 0.500 \text{ MJ}$

3. Find the heat $Q_3$ needed to warm the water from 0°C to 100°C.

$Q_3 = 627 \text{ kJ} = 0.627 \text{ MJ}$

4. Use $L_v$ from Table 18-2 to find the heat $Q_4$ needed to vaporize the water.

$Q_4 = 3.39 \text{ MJ}$

5. Sum your results to find the total heat $Q$.

$Q = Q_1 + Q_2 + Q_3 + Q_4 = \boxed{4.58 \text{ MJ}}$

**REMARKS** Notice that most of the heat was needed to vaporize the water, and that the amount needed to melt the ice was almost as much as that needed to raise the temperature of the water by 100 C°. A graph of temperature versus time for the case in which the heat is added at a constant rate of 1 kJ/s is shown in Figure 18-1. Note that it takes considerably longer to vaporize the water than it does to melt the ice or to raise the temperature of the water. When all of the water has vaporized, the temperature again rises as heat is added.

**EXERCISE** An 830-g piece of lead is heated to its melting point of 600 K. How much additional heat energy must be added to melt the lead? (*Answer* 20.5 kJ)

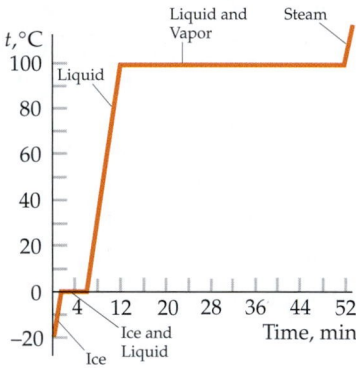

**FIGURE 18-1**

A COOL DRINK — **EXAMPLE 18-4 Put It in Context**

A 2-liter pitcher of lemonade has been sitting on the picnic table in the sun all day at 33°C. You pour 0.24 kg into a Styrofoam cup and add 2 ice cubes (each 0.025 kg at 0°C). (*a*) Assuming no heat lost to the surroundings, what is the final temperature of the lemonade? (*b*) What is the final temperature if you add 6 ice cubes?

**PICTURE THE PROBLEM** We set the heat lost by the lemonade equal to the heat gained by the ice cubes. Let $T_f$ be the final temperature of the lemonade and water. We assume that lemonade has the same specific heat as water.

(*a*) 1. Write the heat lost by the lemonade in terms of the final temperature $T_f$:

$Q_{out} = m_L c |\Delta T| = m_L c (T_{Li} - T_f)$

2. Write the heat gained by the ice cubes and resulting water in terms of the final temperature:

$Q_{in} = m_{ice} L_f + m_{ice} c \Delta T_w = m_{ice} L_f + m_{ice} c T_f$

3. Set the heat lost equal to the heat gained and solve for $T_f$:

$$Q_{out} = Q_{in}$$

$$m_L c (T_{Li} - T_f) = m_{ice} L_f + m_{ice} c T_f$$

so

$$T_f = \frac{m_L c T_{Li} - m_{ice} L_f}{(m_L + m_{ice}) c}$$

$$= \frac{(0.24 \text{ kg})(4.18 \text{ kJ}/(\text{kg} \cdot \text{C}°))(33°C) - (0.05 \text{ kg})(333.5 \text{ kJ}/\text{kg})}{(0.29 \text{ kg})(4.18 \text{ kJ}/(\text{kg} \cdot \text{C}°))}$$

$$= \boxed{13.6°C}$$

(b) 1. For 6 ice cubes, $m_{ice} = 0.15$ kg. Find the final temperature as in step 3 of Part (a):

$$T_f = \frac{m_L c T_{Li} - m_{ice} L_f}{(m_L + m_{ice})c}$$

$$= \frac{(0.24 \text{ kg})(4.18 \text{ kJ/kg·C°})(33°C) - (0.15 \text{ kg})(333.5 \text{ kJ/kg})}{(0.39 \text{ kg})(4.18 \text{ kJ/kg·C°})}$$

$$= -10.4°C$$

2. This cannot be correct! No amount of ice at 0°C can lower the temperature of warm lemonade to below 0°C. Our calculation is wrong because our assumption in step 2 of Part (a) that all of the ice melts was wrong. Instead, the heat given off by the lemonade as it cools from 32°C to 0°C is not enough to melt all of the ice. The final temperature is thus:

$$T_f = \boxed{0°C}$$

**MASTER the CONCEPT WEB** ❶ **PLAUSIBILITY CHECK** Let's calculate how much ice is melted. For the lemonade to cool from 33°C to 0°C, it must give off heat in the amount $Q_{out} = (0.24 \text{ kg})(4.18 \text{ kJ/kg·C°})(33°C) = 33.1$ kJ. The mass of ice that this amount of heat will melt is $m_{ice} = Q_{in}/L_f = 33.1 \text{ kJ}/(333.5 \text{ kJ/kg}) = 0.10$ kg. This is the mass of only 4 ice cubes. Adding more than 4 ice cubes does not lower the temperature below 0°C. It merely increases the amount of ice in the ice-lemonade mixture. In problems like this one, we should first find out how much ice must be melted to reduce the temperature of the liquid to 0°C. If less than that amount is added, we can proceed as in Part (a). If more ice is added, the final temperature is 0°C.

# 18-3 Joule's Experiment and the First Law of Thermodynamics

We can raise the temperature of a system by adding heat, but we can also raise its temperature by doing work on it.

Figure 18-2 is a diagram of the apparatus Joule used in a famous experiment in which he determined the amount of work needed to raise the temperature of 1 g of water by 1 C°. Here the system is a thermally insulated container of 1 g of water. Joule's apparatus converts the potential energy of falling weights into work done on the water by an attached paddle, as shown in the figure. Joule found that he could raise the temperature of his water sample by 1 F° by dropping 772 pounds of attached weights a distance of one foot. Converting to modern units, Joule found that it takes about 4.184 J (the energy units adopted by the scientific community in 1948) to raise the temperature of 1 g of water by 1 C°. The result that 4.184 J of mechanical energy is equivalent to 1 cal of heat energy is known as the **mechanical equivalence of heat.**

There are other ways of doing work on this system. For example, we could drop the insulated container of water from some height $h$, letting the system make an inelastic collision with the ground, or we could do mechanical work to generate electricity and then use the electricity to heat the water (Figure 18-3). In all such experiments, the same amount of work is required to produce a given temperature change. By the conservation of energy, the work done equals the increase in the internal energy of the system.

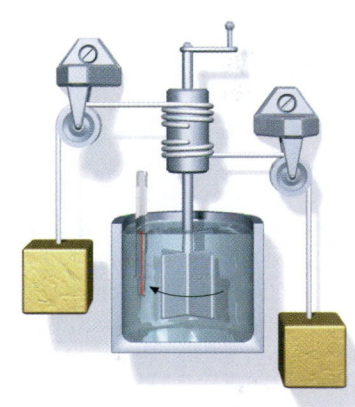

**FIGURE 18-2** Schematic diagram for Joule's experiment. Insulating walls to prevent heat transfer enclose water. As the weights fall at constant speed, they turn a paddle wheel, which does work on the water. If friction is negligible, the work done by the paddle wheel on the water equals the loss of mechanical energy of the weights, which is determined by calculating the loss in the potential energy of the weights.

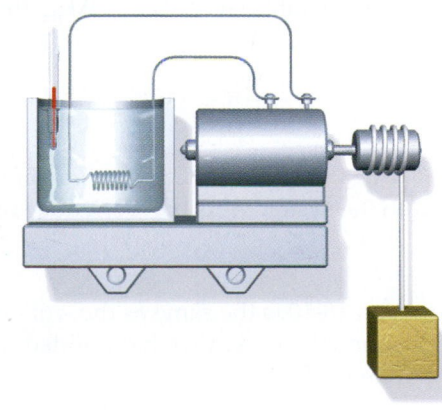

**FIGURE 18-3** Another method of doing work on a thermally insulated container of water. Electrical work is done on the system by the generator, which is driven by the falling weight.

---

**EXAMPLE 18-5**

You drop a thermally insulated container of water from a height $h$ to the ground. If the collision is perfectly inelastic and all of the mechanical energy lost goes into the internal energy of the water, what must $h$ be for the temperature of the water to increase by 1 C°?

**PICTURE THE PROBLEM** The kinetic energy of the water just before it hits the ground equals its original potential energy $mgh$. During the collision, this energy is converted into thermal energy $Q$, which in turn causes a rise in temperature given by $Q = mc\Delta T$.

1. Set the potential energy equal to the thermal energy:

$$mgh = mc\Delta T$$

2. Solve for the height $h$:

$$h = \frac{c\Delta T}{g}$$

3. Substitute $c = 4.18 \text{ kJ/kg·K}$ and $\Delta T = 1\text{ C}° = 1 \text{ K}$:

$$h = \frac{(4.18 \text{ kJ/kg·K})(1 \text{ K})}{9.81 \text{ N/kg}} = 0.426 \text{ km}$$

$$= \boxed{426 \text{ m}}$$

**REMARKS** Note that $h$ is independent of the mass of the water. It is also rather large, which illustrates one of the difficulties with Joule's experiment—a large amount of work must be done to produce a measurable change in the temperature of the water.

Suppose that we perform Joule's experiment but replace the insulating walls of the container with conducting walls. We find that the work needed to produce a given change in the temperature of the system depends on how much heat is added to or subtracted from the system by conduction through the walls. However, if we sum the work done on the system and the net heat added to the system, the result is always the same for a given temperature change. That is, the sum of the heat transfer *into* the system and the work done *on* the system equals the change in the internal energy of the system. This is the **first law of thermodynamics,** which is simply a statement of the conservation of energy.

Let $W_{on}$ stand for the work done by the surroundings *on* the system, and let $W_{by}$ stand for the work done *by* the system on its surroundings. For example, suppose our system is a gas confined to a cylinder by a piston. If the piston compresses the gas, the surroundings do positive work on the gas, and $W_{on}$ is positive. (However, if the gas expands against the piston, the surroundings do negative work on the gas, and $W_{on}$ is negative.) Also, let $Q_{in}$ stand for the heat transferred into the system. If heat is transferred into the system then $Q_{in}$ is positive; if heat is transferred out of the system then $Q_{in}$ is negative (Figure 18-4). Using these conventions, and denoting the internal energy by $\Delta E_{int}$, the first law of thermodynamics is written

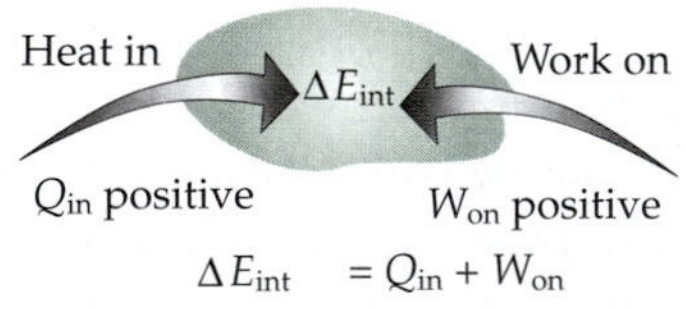

Heat in      Work on

$\Delta E_{int}$

$Q_{in}$ positive      $W_{on}$ positive

$\Delta E_{int} = Q_{in} + W_{on}$

**FIGURE 18-4** Sign convention for the first law of thermodynamics.

$$\Delta E_{int} = Q_{in} + W_{on} \qquad\qquad 18\text{-}10$$

The change in the internal energy of the system equals the heat transferred into the system plus the work done on the system.

FIRST LAW OF THERMODYNAMICS

Equation 18-10 is the same as the work energy theorem $W_{ext} = \Delta E_{sys}$ of Chapter 7 (Equation 7-9), except we have added the heat term $Q_{in}$ and called the energy of the system $\Delta E_{int}$.

*STIRRING THE WATER*    **EXAMPLE    18-6**

**You do 25 kJ of work on a system consisting of 3 kg of water by stirring it with a paddle wheel. During this time, 15 kcal of heat leaves the system due to poor insulation. What is the change in the internal energy of the system?**

**PICTURE THE PROBLEM**    We express all energies in joules and apply the first law of thermodynamics.

1. $\Delta E_{int}$ is found from the first law of thermodynamics:

$$\Delta E_{int} = Q_{in} + W_{on}$$

2. Heat is *removed* from the system, thus the heat *added* is negative:

$$Q_{in} = -15 \text{ kcal} = -(15 \text{ kcal})\left(\frac{4.18 \text{ kJ}}{1 \text{ kcal}}\right)$$

$$= -62.7 \text{ kJ}$$

3. The work done *on* the system is positive, thus:

$$W_{on} = 25 \text{ kJ}$$

4. Substitute these quantities and solve for $\Delta E_{int}$:

$$\Delta E_{int} = Q_{in} + W_{on} = (-62.7 \text{ kJ}) + (25 \text{ kJ})$$

$$= \boxed{-37.7 \text{ kJ}}$$

**REMARKS**    The internal energy decreases because the heat loss exceeds the work gain.

It is important to understand that the internal energy $E_{int}$ is a function of the state of the system, just as $P$, $V$, and $T$ are functions of the state of the system. Consider a gas in some initial state $(P_i, V_i)$. The temperature $T_i$ is determined by the equation of state. For example, if the gas is ideal, the $T_i = P_i V_i/(nR)$. The internal energy $E_{int,i}$ also depends only on the state of the gas, which is determined by any two state variables such as $P$ and $V$, $P$ and $T$, or $V$ and $T$. If we compress the gas or let it expand, add to or remove heat from it, do work on it or let it do work, the gas will move through a sequence of states; that is, it will have different values of the state functions $P$, $V$, $T$, and $E_{int}$. If the gas is then returned to its original state $(P_i, V_i)$, the temperature $T$ and the internal energy $E_{int}$ must equal their original values.

On the other hand, the net heat input $Q$ and the work $W$ done by the gas are not functions of the state of the system. There are no functions $Q$ or $W$ associated with any particular state of the gas. We could take the gas through a sequence of states beginning and ending at state $(P_i, V_i)$ during which the gas did positive work and absorbed an equal amount of heat. Or we could take it though a different sequence of states such that work was done on the gas and heat was removed from the gas. **It is correct to say that a system has a large amount of internal energy, but it is not correct to say that a system has a large amount of heat or a large amount of work.** Heat is not something that is contained in a system. Rather, it is a measure of the energy that flows from one system to another because of a difference in temperature. Work is a measure of the energy that flows from one system to another because the point of contact of a force exerted by one system on the other undergoes a displacement not perpendicular to the force.

For very small amounts of heat added, work done, or changes in internal energy, it is customary to write Equation 18-10 as

$$dE_{int} = dQ_{in} + dW_{on} \qquad\qquad 18\text{-}11$$

In this equation, $dE_{int}$ is the differential of the internal-energy function. However, neither $dQ_{in}$ nor $dW_{on}$ is a differential of any function. Instead, $dQ_{in}$ merely represents a small amount of heat added to the system, and $dW_{on}$ represents a small amount of work done on the system.

# 18-4 The Internal Energy of an Ideal Gas

The translational kinetic energy $K$ of the molecules in an *ideal* gas is related to the absolute temperature $T$ by Equation 17-22 in Chapter 17:

$$K = \tfrac{3}{2}nRT$$

where $n$ is the number of moles of gas and $R$ is the universal gas constant. If the internal energy of a gas is just this translational kinetic energy, then $E_{int} = K$, and

$$E_{int} = \tfrac{3}{2}nRT \tag{18-12}$$

Then the internal energy will depend only on the temperature of the gas, and not on its volume or pressure. If the molecules have other types of energy in addition to translational kinetic energy, such as energy of rotation, the internal energy will be greater than that given by Equation 18-12. But according to the equipartition theorem (Chapter 17, Section 5) the average energy associated with any degree of freedom will be $\tfrac{1}{2}RT$ per mole ($\tfrac{1}{2}kT$ per molecule), so again, the internal energy will depend only on the temperature and not on the volume or pressure.

We can imagine that the internal energy of a *real* gas might include other kinds of energy, which depend on the pressure and volume of the gas. Suppose, for example, that nearby gas molecules exert attractive forces on each other. Work is then required to increase the separation of the molecules. Then, if the average distance between the molecules is increased, the potential energy associated with the molecular attraction will increase. The internal energy of the gas will then depend on the volume of the gas as well as on its temperature.

Joule, using an apparatus like the one shown in Figure 18-5, performed a simple but interesting experiment to determine whether or not the internal energy of a gas depends on its volume. Initially, the compartment on the left in Figure 18-5 contains a gas and the compartment on the right has been evacuated. A stopcock that is initially closed connects the two compartments. The whole system is thermally insulated from its surroundings by rigid walls so that no heat can go into or out of the system *and* no work can be done. When the stopcock is opened, the gas rushes into the evacuated chamber. This process is called a **free expansion.** Eventually, the gas reaches thermal equilibrium with itself. Since no work has been done on the gas and no heat has been transferred to it, the final internal energy of the gas must equal its initial internal energy. If the gas molecules exert attractive forces on one another, the potential energy associated with these forces will increase as the volume increases. Since energy is conserved, the kinetic energy of translation must therefore decrease, which will result in a decrease in the temperature of the gas.

When Joule did this experiment, he found the final temperature to be equal to the initial temperature. Subsequent experiments verified this result for low gas densities. This implies that for a gas at low density—that is, for an ideal gas—the temperature depends only on the internal energy, or as we usually think of it, the internal energy depends only on the temperature. However, if the experiment is done with a large amount of gas initially in the left compartment so that the density is high, then the temperature after expansion is slightly lower than the temperature before the expansion. This indicates that there is a small attraction between the gas molecules of a real gas.

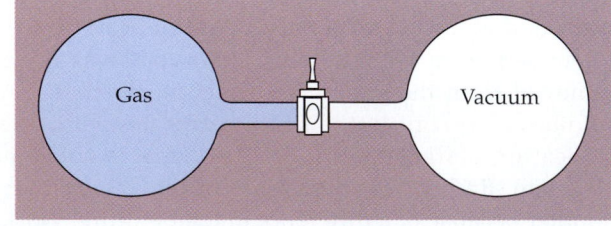

**FIGURE 18-5** Free expansion of a gas. When the stopcock on the gas is opened, the gas expands rapidly into the evacuated chamber. Because no work is done on the gas and the whole system is thermally insulated, the initial and final internal energies of the gas are equal.

# 18-5   Work and the *PV* Diagram for a Gas

In many types of engines, a gas expanding against a moveable piston does work. For example, in a steam engine, water is heated in a boiler to produce steam. The steam then does work as it expands and drives a piston. In an automobile engine, a mixture of gasoline vapor and air is ignited, causing it to explode. The resulting high temperatures and pressures cause the gas to expand rapidly, driving a piston and doing work. In this section, we will see how we can mathematically describe the work done by an expanding gas.

## Quasi-Static Processes

Figure 18-6 shows an ideal gas confined in a container with a tightly fitting piston that we assume to be frictionless. If the piston moves, the volume of the gas changes. The temperature or pressure or both must also change since these three variables are related by the equation of state $PV = nRT$. If we suddenly push in the piston to compress the gas, the pressure will initially be greater near the piston than far from it. Eventually the gas will settle down to a new equilibrium pressure and temperature. We cannot determine such macroscopic variables as $T$, $P$, or $E_{int}$ for the entire gas system until equilibrium is restored in the gas. However, if we move the piston slowly in small steps and allow equilibrium to be reestablished after each step, we can compress or expand the gas in such a way that the gas is never far from an equilibrium state. In this kind of process, called a **quasi-static process,** the gas moves through a series of equilibrium states. In practice, it is possible to approximate quasi-static processes fairly well.

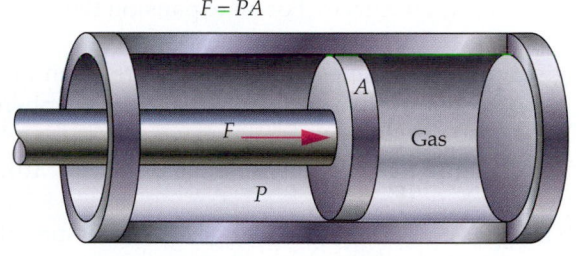

**FIGURE 18-6** Gas confined in a thermally insulated cylinder with a movable piston. If the piston moves a distance $dx$, the volume of the gas changes by $dV = A\,dx$. The work done by the gas is $PA\,dx = P\,dV$, where $P$ is the pressure.

Let us begin with a gas at a high pressure, and let it expand quasi-statically. The magnitude of the force $F$ exerted by the gas on the piston is $PA$, where $A$ is the area of the piston and $P$ is the gas pressure. As the piston moves a small distance $dx$, the work done *by* the gas on the piston is

$$dW_{by\ gas} = F_x\,dx = PA\,dx = P\,dV \qquad 18\text{-}13$$

where $dV = A\,dx$ is the increase in the volume of the gas. During the expansion the piston exerts a force of magnitude $PA$ on the gas, but opposite in direction to the force of the gas on the piston. Thus, work done by the piston *on* the gas is just the negative of the work done *by* the gas

$$dW_{on\ gas} = -dW_{by\ gas} = -P\,dV \qquad 18\text{-}14$$

Note that for an expansion, $dV$ is positive, the gas does work on the piston, so $dW_{on\ gas}$ is negative, and for a compression, $dV$ is negative, work is done on the gas, so $dW_{on\ gas}$ is positive.

The work done on the gas during an expansion or compression from a volume of $V_i$ to a volume of $V_f$ is

$$W_{on\ gas} = -\int_{V_i}^{V_f} P\,dV \qquad 18\text{-}15$$

WORK DONE ON A GAS

To calculate this work we need to know how the pressure varies during the expansion or compression. The various possibilities can be illustrated most easily using a *PV* diagram.

## PV Diagrams

We can represent the states of a gas on a diagram of $P$ versus $V$. Because by specifying both $P$ and $V$ we specify the state of the gas, each point on the $PV$ diagram indicates a particular state of the gas. Figure 18-7 shows a $PV$ diagram with a directed horizontal line representing a series of states that all have the same value of $P$. This line represents a *compression* at constant pressure. Such a process is called an **isobaric compression.** For a volume change of $\Delta V$ ($\Delta V$ is negative for a compression), we have

$$W_{\text{on}} = -\int_{V_i}^{V_f} P\,dV = -P\int_{V_i}^{V_f} dV = -P\Delta V$$

which is equal to the shaded area under the curve (directed line) in the figure. In general, for a compression the work done on the gas is equal to the area under the $P$-versus-$V$ curve. (For an expansion the work done on the gas is equal to the negative of the area under the $P$-versus-$V$ curve.) Because pressures are often given in atmospheres and volumes are often given in liters, it is convenient to have a conversion factor between liter-atmospheres and joules:

$$1\ \text{L·atm} = (10^{-3}\ \text{m}^3)(101.3 \times 10^3\ \text{N/m}^2) = 101.3\ \text{J} \qquad 18\text{-}16$$

**EXERCISE** If 5 L of an ideal gas at a pressure of 2 atm is cooled so that it contracts at constant pressure until its volume is 3 L, what is the work done on the gas? (*Answer* 405.2 J)

Figure 18-8 shows three different possible paths on a $PV$ diagram for a gas that is initially in state $(P_i, V_i)$ and is finally in state $(P_f, V_f)$. We assume that the gas is ideal and have chosen the original and final states to have the same temperature so that $P_i V_i = P_f V_f = nRT$. Since the internal energy depends only on the temperature, the initial and final internal energies are the same also.

In Figure 18-8a, the gas is heated at constant volume until its pressure is $P_f$, after which it is cooled at constant pressure until its volume is $V_f$. The work done on the gas along the constant-volume (vertical) part of path A is zero; along the constant pressure (horizontal) part of the path A, it is $P_f |V_f - V_i| = -P_f (V_f - V_i)$.

In Figure 18-8b, the gas is first cooled at constant pressure until its volume is $V_f$, after which it is heated at constant volume until its pressure is $P_f$. The work done on the gas along this path is $P_i |V_f - V_i| = -P_i(V_f - V_i)$, which is much less than that done along the path shown in Figure 18-8a as can be seen by comparing the shaded regions in Figure 18-8a and Figure 18-8b.

In Figure 18-8c, path C represents an **isothermal** compression, meaning that the temperature remains constant. (Keeping the temperature constant during the compression requires that heat be transferred out of the gas during the

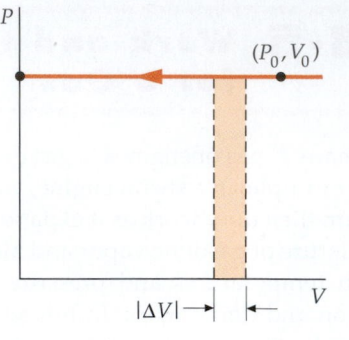

**FIGURE 18-7** Each point on a $PV$ diagram, such as $(P_0, V_0)$, represents a particular state of the gas. The horizontal line represents states with a constant pressure $P_0$. The shaded area, $P_0 |\Delta V|$, represents the work done on the gas as it is compressed an amount $|\Delta V|$.

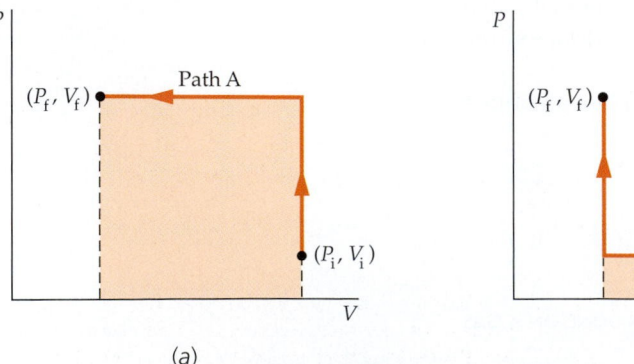

(a)

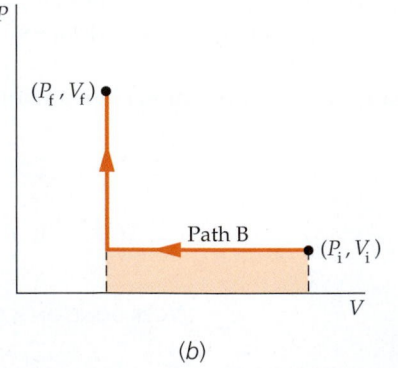

(b)

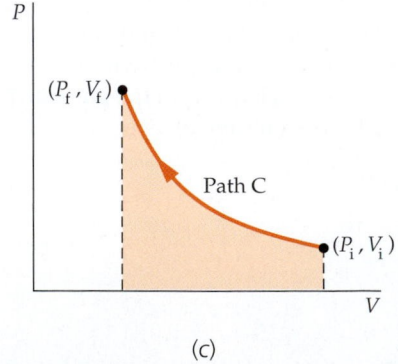

(c)

**FIGURE 18-8** Three paths on $PV$ diagrams connecting an initial state $(P_i, V_i)$ and a final state $(P_f, V_f)$. The corresponding shaded area indicates the work done on the gas along each path.

compression.) We can calculate the work done on the gas along path C by using $P = nRT/V$. Hence, the work done on the gas as it is compressed from $V_i$ to $V_f$ is

$$W_{on} = -\int_{V_i}^{V_f} P \, dV = -\int_{V_i}^{V_f} \frac{nRT}{V} \, dV$$

Since $T$ is constant for an isothermal process, we can remove it from the integral. We then have

$$W_{isothermal} = -nRT \int_{V_i}^{V_f} \frac{dV}{V} = -nRT \ln\frac{V_f}{V_i} = nRT \ln\frac{V_i}{V_f} \qquad \text{18-17}$$

WORK DONE ON GAS DURING ISOTHERMAL COMPRESSION

We see that the amount of work done on the gas is different for each process illustrated. Since for these states $E_{int,f} = E_{int,i}$, the net amount of heat added must also be different for each of the processes. This discussion illustrates the fact that both the work done and the heat added depend only on how a system moves from one state to another, but the change in the internal energy of the system does not.

---

**WORK DONE ON AN IDEAL GAS**                     **EXAMPLE 18-7**

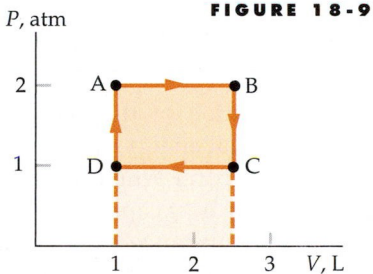

**FIGURE 18-9**

An ideal gas undergoes a cyclic process from point *A* to point *B* to point *C* to point *D* and back to point *A* as shown in Figure 18-9. The gas begins at a volume of 1 L and a pressure of 2 atm and expands at constant pressure until the volume is 2.5 L, after which it is cooled at constant volume until its pressure is 1 atm. It is then compressed at constant pressure until its volume is again 1 L, after which it is heated at constant volume until it is back in its original state. Find the total work done on the gas and the total heat added to it during the cycle.

**PICTURE THE PROBLEM**  We calculate the work done during each step. Since $\Delta E_{int} = 0$ for any complete cycle, the first law of thermodynamics implies that the total heat added to the gas plus the total work done on the gas equals zero.

1. From *A* to *B* the process is an isobaric (constant pressure) expansion, so the work done on the gas is negative. The work done on the gas equals the negative of the shaded area under the *AB* curve, shown in Figure 18-10*a*:

$$W_{AB} = -P\Delta V = -P(V_B - V_A)$$
$$= -(2 \text{ atm})(2.5 \text{ L} - 1 \text{ L}) = -3 \text{ L·atm}$$

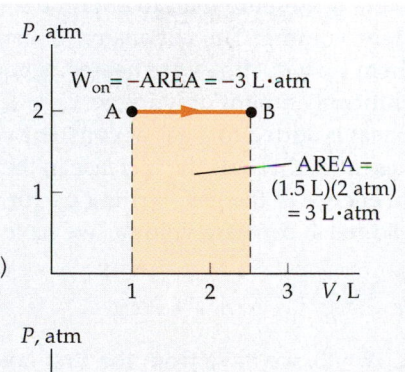

2. Convert the units to joules:

$$W_{AB} = -3 \text{ L·atm} \times \frac{101.3 \text{ J}}{1 \text{ L·atm}} = -304 \text{ J}$$

3. From *B* to *C* (Figure 18-9) the gas cools at constant volume so the work done is zero:

$$W_{BC} = 0$$

4. As the gas undergoes an isobaric compression from *C* to *D*, the work done on it is positive. This work equals the area under the *CD* curve, shown in Figure 18-10*b*:

$$W_{CD} = -P\Delta V = -P(V_D - V_C)$$
$$= -(1 \text{ atm})(1 \text{ L} - 2.5 \text{ L})$$
$$= 1.5 \text{ L·atm} = \boxed{152 \text{ J}}$$

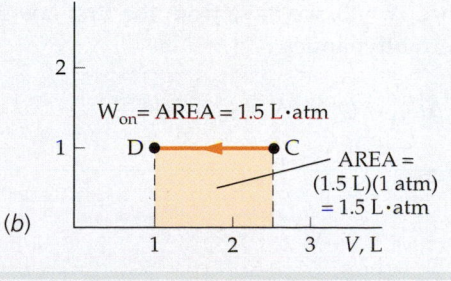

**FIGURE 18-10**  (*a*) The work done on the gas during the expansion from *A* to *B* is equal to the negative of the area under the curve. (*b*) The work done on the gas during the compression from *C* to *D* is equal to the area under the curve.

5. As the gas is heated back to its original state $A$, the volume is again constant (Figure 18-9), so no work is done:

$$W_{DA} = 0$$

6. The total work done by the gas is the sum of the work done along each step:

$$W_{total} = W_{AB} + W_{BC} + W_{CD} + W_{DA}$$
$$= (-304\ J) + 0 + 152\ J + 0 = \boxed{-152\ J}$$

7. Because the gas is back in its original state, the total change in internal energy is zero:

$$\Delta E_{int} = 0$$

8. The heat added is found from the first law:

$$\Delta E_{int} = Q_{in} + W_{on}$$

so

$$Q_{in} = \Delta E_{int} - W_{on} = 0 - (-152\ J) = \boxed{152\ J}$$

**REMARKS** The work done by the gas equals the negative of the work done on the gas, so the total work done by the gas during the cycle is $+152$ J. During the cycle the gas extracts 152 J of heat from its surroundings and does 152 J of work on its surroundings. This leaves the gas in its initial state. The total work done by the gas equals the area enclosed by the cycle in Figure 18-9. Such cyclic processes have important applications for heat engines, as we will see in Chapter 20.

# 18-6 Heat Capacities of Gases

The determination of the heat capacity of a substance provides information about its internal energy, which is related to its molecular structure. For all substances that expand when heated, the heat capacity at constant pressure $C_p$ is greater than the heat capacity at constant volume $C_v$. If heat is added at constant pressure, the substance expands and does positive work on its surroundings (Figure 18-11). Therefore, it takes more heat to obtain a given temperature change at constant pressure than to obtain the same temperature change when heated at constant volume. The expansion is usually negligible for solids and liquids, so for them $C_p \approx C_v$. But a gas heated at constant pressure readily expands and does a significant amount of work, so $C_p - C_v$ is not negligible.

If heat is added to a gas at constant volume, no work is done (Figure 18-12), so the heat added equals the increase in the internal energy of the gas. Writing $Q_v$ for the heat added at constant volume, we have

$$Q_v = C_v \Delta T$$

Since $W = 0$, we have from the first law of thermodynamics

$$\Delta E_{int} = Q_v + W = Q_v$$

Thus,

$$\Delta E_{int} = C_v \Delta T$$

Taking the limit as $\Delta T$ approaches zero, we obtain

$$dE_{int} = C_v\, dT \qquad\qquad 18\text{-}18a$$

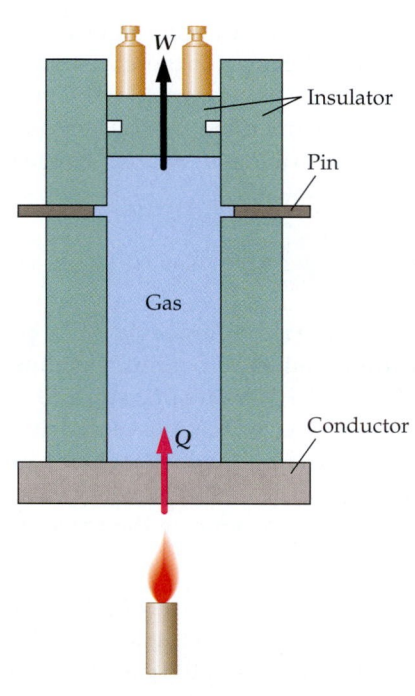

**FIGURE 18-11** Heat is added and the pressure remains constant. The gas expands, thus doing positive work on the piston.

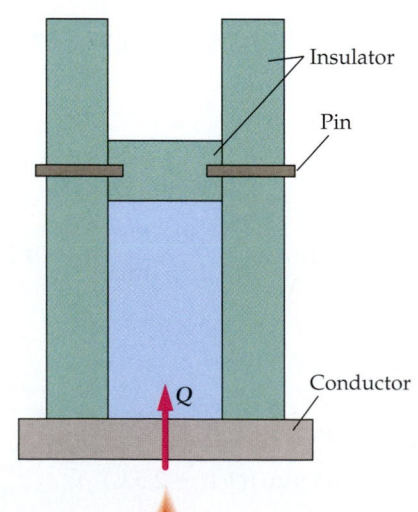

**FIGURE 18-12** Heat is added at constant volume, so no work is done and all the heat goes into the internal energy of the gas. The piston is held in place by pins.

and

$$C_v = \frac{dE_{int}}{dT} \qquad\qquad 18\text{-}18b$$

The heat capacity at constant volume is the rate of change of the internal energy with temperature. Since $E_{int}$ and $T$ are both state functions, Equations 18-18a and 18-18b hold for any process.

Now let's calculate the difference $C_p - C_v$ for an ideal gas. From the definition of $C_p$, the heat added at constant pressure is

$$Q_p = C_p \Delta T$$

From the first law of thermodynamics,

$$\Delta E_{int} = Q_p + W_{on} = Q_p - P\Delta V$$

Then

$$\Delta E_{int} = C_p \Delta T - P\Delta V \quad \text{or} \quad C_p \Delta T = \Delta E_{int} + P\Delta V$$

For infinitesimal changes, this becomes

$$C_p\, dT = dE_{int} + P\, dV$$

Using Equation 18-18a for $dE_{int}$, we obtain

$$C_p\, dT = C_v\, dT + P dV \qquad\qquad 18\text{-}19$$

The pressure, volume, and temperature for an ideal gas are related by

$$PV = nRT$$

Taking the differentials of both sides of the ideal-gas law, we obtain

$$P\, dV + V\, dP = nR\, dT$$

For a constant-pressure process $dP = 0$, so

$$P\, dV = nR\, dT$$

Substituting this into Equation 18-19 gives

$$C_p\, dT = C_v\, dT + nR\, dT = (C_v + nR)\, dT$$

Therefore,

$$C_p = C_v + nR \qquad\qquad 18\text{-}20$$

which shows that, for an ideal gas, the heat capacity at constant pressure is greater than the heat capacity at constant volume by the amount $nR$.

Table 18-3 lists measured molar heat capacities $c_p'$ and $c_v'$ for several gases. Note from this table that the ideal gas prediction, $c_p' - c_v' = R$, holds quite well for all gases. The table also shows that $c_v'$ is approximately 1.5R for all monatomic gases, 2.5R for all diatomic gases, and greater than 2.5R for gases consisting of more complex molecules. We can understand these results by considering the molecular model of a gas (Chapter 17.) The total translational kinetic energy of $n$ moles of a gas is $K = \frac{3}{2}nRT$ (Equation 17-22). Thus, if the internal energy of a gas consists of translational kinetic energy only, we have

$$E_{int} = \frac{3}{2}nRT \qquad \qquad 18\text{-}21$$

The heat capacities are then

$$C_v = \frac{dE_{int}}{dT} = \frac{3}{2}nR \qquad \qquad 18\text{-}22$$

$C_v$ FOR AN IDEAL MONATOMIC GAS

and

$$C_p = C_v + nR = \frac{5}{2}nR \qquad \qquad 18\text{-}23$$

$C_p$ FOR AN IDEAL MONATOMIC GAS

The results in Table 18-3 agree well with these predictions for monatomic gases, but for other gases, the heat capacities are greater than those predicted by Equations 18-22 and 18-23. The internal energy for a gas consisting of diatomic or more complicated molecules is evidently greater than $\frac{3}{2}nRT$. The reason is that such molecules can have other types of energy, such as rotational or vibrational energy, in addition to translational kinetic energy.

## TABLE 18-3

**Molar Heat Capacities in J/mol·K of Various Gases at 25°C**

| Gas | $c_p'$ | $c_v'$ | $c_v'/R$ | $c_p' - c_v'$ | $(c_p' - c_v')/R$ |
|---|---|---|---|---|---|
| *Monatomic* | | | | | |
| He | 20.79 | 12.52 | 1.51 | 8.27 | 0.99 |
| Ne | 20.79 | 12.68 | 1.52 | 8.11 | 0.98 |
| Ar | 20.79 | 12.45 | 1.50 | 8.34 | 1.00 |
| Kr | 20.79 | 12.45 | 1.50 | 8.34 | 1.00 |
| Xe | 20.79 | 12.52 | 1.51 | 8.27 | 0.99 |
| *Diatomic* | | | | | |
| $N_2$ | 29.12 | 20.80 | 2.50 | 8.32 | 1.00 |
| $H_2$ | 28.82 | 20.44 | 2.46 | 8.38 | 1.01 |
| $O_2$ | 29.37 | 20.98 | 2.52 | 8.39 | 1.01 |
| CO | 29.04 | 20.74 | 2.49 | 8.30 | 1.00 |
| *Polyatomic* | | | | | |
| $CO_2$ | 36.62 | 28.17 | 3.39 | 8.45 | 1.02 |
| $N_2O$ | 36.90 | 28.39 | 3.41 | 8.51 | 1.02 |
| $H_2S$ | 36.12 | 27.36 | 3.29 | 8.76 | 1.05 |

HEATING, COOLING, AND COMPRESSING AN IDEAL GAS  **EXAMPLE 18-8** Try It Yourself

A system consisting of 0.32 mol of a monatomic ideal gas with $c_v' = \frac{3}{2}RT$ occupies a volume of 2.2 L at a pressure of 2.4 atm, as represented by point $A$ in Figure 18-13. The system is carried through a cycle consisting of three processes:

1. The gas is heated at constant pressure until its volume is 4.4 L at point $B$.

2. The gas is cooled at constant volume until the pressure decreases to 1.2 atm (point $C$).

3. The gas undergoes an isothermal compression back to point $A$.

(a) What is the temperature at points $A$, $B$, and $C$? (b) Find $W$, $Q$, and $\Delta E_{int}$ for each process and for the entire cycle.

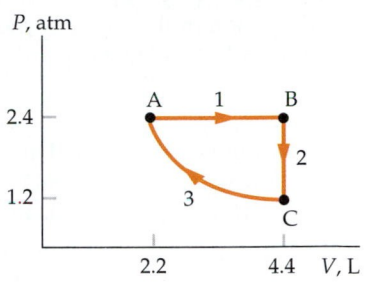

FIGURE 18-13

**PICTURE THE PROBLEM** You can find the temperatures at all points from the ideal-gas law. You can find the work for each process by finding the area under the curve, and the heat exchanged from the given heat capacity and the initial and final temperatures for each process. In process 3, $T$ is constant, so $\Delta E_{int} = 0$ and the heat input plus the work done on the gas equals zero.

**Cover the column to the right and try these on your own before looking at the answers.**

**Steps**

(a) Find the temperatures at points $A$, $B$, and $C$ using the ideal-gas law.

(b) 1. For process 1, use $W_1 = -P_c \Delta V$ to calculate the work, and $C_p = \frac{5}{2}nR$ to calculate the heat $Q_1$. Then use $W_1$ and $Q_1$ to calculate $\Delta E_{int,1}$.

2. For process 2, use $C_v = \frac{3}{2}nR$ and $T_C - T_B$ from step 1 to find $Q_2$. Then, since $W_2 = 0$, $\Delta E_{int,2} = Q_2$.

3. Calculate $W_3$ from $W = -nRT \ln(V_A/V_C)$ in the isothermal compression. Then, since $\Delta E_{int,3} = 0$, $Q_3 = -W_3$.

4. Find the total work $W$, the total heat $Q$, and the total change $\Delta E_{int}$ by summing the quantities found in steps 2, 3, and 4.

**Answers**

$T_A = T_C = \boxed{201 \text{ K}}$, $T_B = \boxed{402 \text{ K}}$

$W_1 = -5.28 \text{ L·atm} = \boxed{-535 \text{ J}}$, $Q_1 = \boxed{1337 \text{ J}}$

$\Delta E_{int,1} = Q_1 + W_1 = \boxed{802 \text{ J}}$

$W_2 = \boxed{0}$, $Q_2 = \boxed{-802 \text{ J}}$, $\Delta E_{int,2} = \boxed{-802 \text{ J}}$

$W_3 = \boxed{371 \text{ J}}$, $Q_3 = \boxed{-371 \text{ J}}$, $\Delta E_{int,3} = \boxed{0}$

$W_{total} = W_1 + W_2 + W_3$

$\quad = (-535 \text{ J}) + 0 + 371 \text{ J} = \boxed{-164 \text{ J}}$

$Q_{total} = Q_1 + Q_2 + Q_3$

$\quad = 1337 \text{ J} + (-802 \text{ J}) + (-371 \text{ J})$

$\quad = \boxed{164 \text{ J}}$

$\Delta E_{int,total} = \Delta E_{int,1} + \Delta E_{int,2} + \Delta E_{int,3}$

$\quad = 802 \text{ J} + (-802 \text{ J}) + 0 = \boxed{0}$

**REMARKS** The total change in internal energy is zero, as it must be for a cyclic process. The total work done on the gas plus the total heat absorbed by the gas equals zero. The total work done on the gas equals the area under the $CA$ curve minus the area under the $AB$ curve, which equals the negative of the area enclosed by the three curves in Figure 18-13.

## Heat Capacities and the Equipartition Theorem

According to the equipartition theorem stated in Section 5 of Chapter 17, the internal energy of $n$ moles of a gas should equal $\frac{1}{2}nRT$ for each degree of freedom of the gas molecule. The heat capacity at constant volume of a gas should then be $\frac{1}{2}nR$ times the number of degrees of freedom of the molecule. From Table 18-2, nitrogen, oxygen, hydrogen, and carbon monoxide all have molar heat capacities at constant volume of about $\frac{5}{2}R$. Thus, the molecules in each of these gases have five degrees of freedom. About 1880, Rudolf Clausius speculated that these gases must consist of diatomic molecules that can rotate about two axes, giving them two additional degrees of freedom (Figure 18-14). The two degrees of freedom besides the three for translation are now known to be associated with their rotation about each of the two axes, $x'$ and $y'$, perpendicular to the line joining the atoms. The kinetic energy of a diatomic molecule is therefore

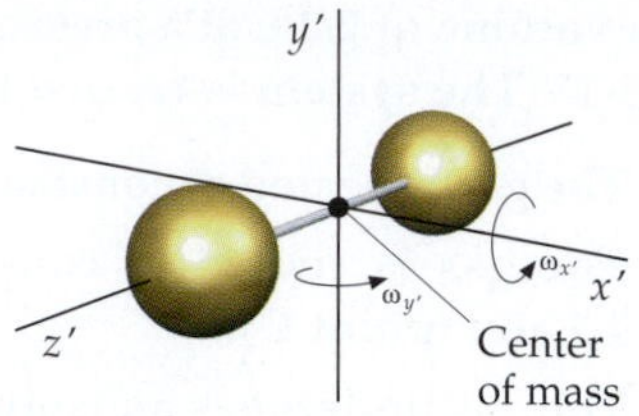

**FIGURE 18-14** Rigid-dumbbell model of a diatomic molecule.

$$K = \tfrac{1}{2}mv_x^2 + \tfrac{1}{2}mv_y^2 + \tfrac{1}{2}mv_z^2 + \tfrac{1}{2}I_{x'}\omega_{x'}^2 + \tfrac{1}{2}I_{y'}\omega_{y'}^2$$

The total internal energy of $n$ moles of such a gas is then

$$E_{\text{int}} = 5 \times \tfrac{1}{2}nRT = \tfrac{5}{2}nRT \qquad\qquad 18\text{-}24$$

and the heat capacity at constant volume is

$$C_v = \tfrac{5}{2}nR \qquad\qquad 18\text{-}25$$

Apparently, diatomic gases do not rotate about the line joining the two atoms—if they did, there would be six degrees of freedom and $C_v$ would be $\frac{6}{2}nR = 3nR$, which is contrary to experimental results. Furthermore, monatomic gases apparently do not rotate at all. We will see in Section 18-8 that these puzzling facts are easily explained when we take into account the quantization of energy.

---

*HEATING A DIATOMIC IDEAL GAS*                     **EXAMPLE 18-9**

**Two moles of oxygen gas are heated from a temperature of 20°C and a pressure of 1 atm to a temperature of 100°C. Assume that oxygen is an ideal gas. (*a*) How much heat must be supplied if the volume is kept constant during the heating? (*b*) How much heat must be supplied if the pressure is kept constant? (*c*) How much work does the gas do in part (*b*)?**

**PICTURE THE PROBLEM** The heat needed for constant-volume heating is $Q_v = C_v\,\Delta T$, where $C_v = \frac{5}{2}nR$ since oxygen is a diatomic gas. For constant-pressure heating, $Q_p = C_p\,\Delta T$, where $C_p = C_v + nR$. Finally, the amount of work done by the gas equals the negative of the work done on the gas, which can be found from $\Delta E_{\text{int}} = Q + W_{\text{on}}$. Alternatively, $W_{\text{by}} = P\Delta V$.

(*a*) 1. Write the heat needed for constant volume in terms of $C_v$ and $\Delta T$:

$$Q_v = C_v\Delta T$$

2. Calculate the heat for $\Delta T = 80\,\text{C}° = 80\,\text{K}$:

$$Q_V = C_V\,\Delta T = \tfrac{5}{2}nR\Delta T$$

$$= \frac{5}{2}(2\ \text{mol})\left(8.314\ \frac{\text{J}}{\text{mol·K}}\right)(80\ \text{K})$$

$$= \boxed{3.33\ \text{kJ}}$$

(b) 1. Write the heat needed for constant pressure in terms of $C_p$ and $\Delta T$:

$Q_p = C_p \Delta T$

2. Calculate the heat capacity at constant pressure:

$C_p = C_v + nR = \frac{5}{2}nR + nR = \frac{7}{2}nR$

3. Calculate the heat added at constant pressure for $\Delta T = 80$ K:

$Q_p = C_p \Delta T$

$$= \frac{7}{2}(2 \text{ mol})\left(8.314 \frac{J}{\text{mol} \cdot K}\right)(80 \text{ K})$$

$$= \boxed{4.66 \text{ kJ}}$$

(c) 1. The work $W_{on}$ can be found from the first law of thermodynamics:

$\Delta E_{int} = Q_{in} + W_{on}$, so $W_{on} = \Delta E_{int} - Q_{in}$

2. The internal energy change equals the heat added at constant volume which was calculated in Part (a):

$\Delta E_{int} = Q_v = C_v \Delta T = \frac{5}{2}nR\Delta T$

and

$Q_p = C_p \Delta T = \frac{7}{2}nR\Delta T$

so

$W_{on} = \Delta E_{int} - Q_p = \frac{5}{2}nR\Delta T - \frac{7}{2}nR\Delta T = -nR\Delta T$

$$= -(2 \text{ mol})\left(8.314 \frac{J}{\text{mol} \cdot K}\right)(80 \text{ K})$$

$$= -1.33 \text{ kJ}$$

3. The work done by the gas at constant pressure is then:

$W_{by} = -W_{on} = \boxed{1.33 \text{ kJ}}$

**REMARKS** Note that the change in internal energy is independent of the process. It depends only on the initial and final states.

**EXERCISE** Find the initial and final volumes of this gas from the ideal-gas law, and use them to calculate the work done by the gas if the heat is added at constant pressure from $W_{by} = P\Delta V$. (*Answer* $V_i = 48.0$ L, $V_f = 61.1$ L, $W = 13.1$ L·atm $= 1.33$ kJ)

# 18-7 Heat Capacities of Solids

In Section 18-1, we noted that all of the metals listed in Table 18-1 have approximately equal molar specific heats. Experimentally, most solids have molar heat capacities approximately equal to $3R$:

$c' = 3R = 24.9 \text{ J/mol·K}$         18-26

This result is known as the **Dulong–Petit law.** We can understand this law by applying the equipartition theorem to the simple model for a solid shown in Figure 18-15. According to this model, a solid consists of a regular array of atoms in which each of the atoms has a fixed equilibrium position and is connected by springs to its neighbors. Each atom can vibrate in the $x$, $y$, and $z$ directions. The total energy of an atom in a solid is

$E = \frac{1}{2}mv_x^2 + \frac{1}{2}mv_y^2 + \frac{1}{2}mv_z^2 + \frac{1}{2}k_{eff}x^2 + \frac{1}{2}k_{eff}y^2 + \frac{1}{2}k_{eff}z^2$

where $k_{eff}$ is the effective force constant of the hypothetical springs. Each atom thus has six degrees of freedom. The equipartition theorem states that a substance

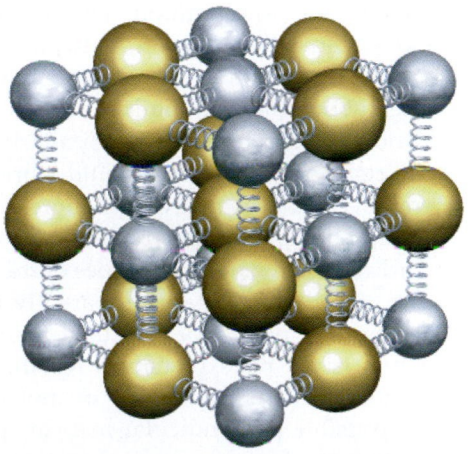

**FIGURE 18-15** Model of a solid in which the atoms are connected to each other by springs. The internal energy of the molecule consists of the kinetic and potential energies of vibration.

in equilibrium has an average energy of $\frac{1}{2}RT$ per mole for each degree of freedom. Thus, the internal energy of a mole of a solid is

$$E_{\text{int,m}} = 6 \times \tfrac{1}{2}RT = 3RT \qquad\qquad 18\text{-}27$$

which means that $c'$ is equal to $3R$.

---

*USING THE DULONG–PETIT LAW*                      **E X A M P L E   1 8 - 1 0**

**The molar mass of copper is 63.5 g/mol. Use the Dulong–Petit law to calculate the specific heat of copper.**

**PICTURE THE PROBLEM** The Dulong–Petit law gives the molar specific heat of a solid, $c'$. The specific heat is then $c = c'/M$ (Equation 18-6), where $M$ is the molar mass.

1. The Dulong–Petit law gives $c'$ in terms of $R$:

$$c' = 3R$$

2. Using $M = 63.5$ g/mol for copper, the specific heat is:

$$c = \frac{c'}{M} = \frac{3R}{M} = \frac{3(8.314 \text{ J/mol·K})}{63.5 \text{ g/mol}}$$

$$= 0.392 \text{ J/(g·K)} = \boxed{0.392 \text{ kJ/kg·K}}$$

**REMARKS** This differs from the measured value of 0.386 kJ/kg·K given in Table 18-1 by less than 2 percent.

**EXERCISE** The specific heat of a certain metal is measured to be 1.02 kJ/kg·K. (a) Calculate the molar mass of this metal, assuming that the metal obeys the Dulong–Petit law. (b) What is the metal? (*Answer* (a) $M = 24.4$ g/mol. (b) The metal must be magnesium, which has a molar mass of 24.31 g/mol)

---

# 18-8 Failure of the Equipartition Theorem

Although the equipartition theorem had spectacular successes in explaining the heat capacities of gases and solids, it had equally spectacular failures. For example, if a diatomic gas molecule like the one in Figure 18-14 rotates about the line joining the atoms, there should be an additional degree of freedom. Similarly, if a diatomic molecule is not rigid, the two atoms should vibrate along the line joining them. We would then have two more degrees of freedom corresponding to kinetic and potential energies of vibration. But according to the measured values of the molar heat capacities in Table 18-3, diatomic gases apparently do not rotate about the line joining them, nor do they vibrate. The equipartition theorem gives no explanation for this, nor for the fact that monatomic atoms apparently do not rotate about any of the three possible perpendicular axes in space. Furthermore, heat capacities are found to depend on temperature, contrary to the predictions of the equipartition theorem. The most spectacular case of the temperature dependence of heat capacity is that of $H_2$, as shown in Figure 18-16. At temperatures below 70 K, $c_v'$ for $H_2$ is $\frac{3}{2}R$, which is the same as that for a gas of molecules that translate, but do not rotate or vibrate. At temperatures between

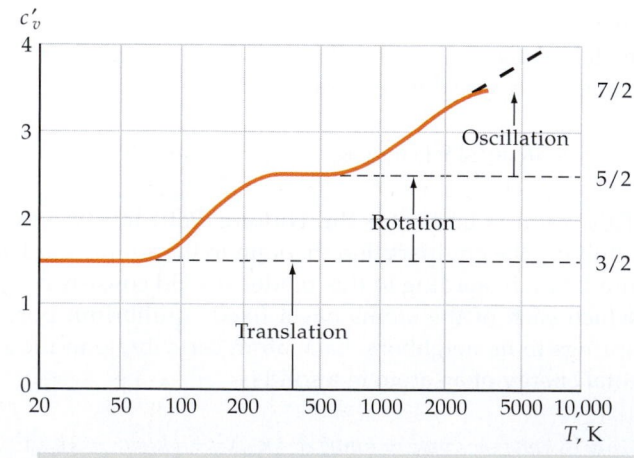

**FIGURE 18-16** Temperature dependence of the molar heat capacity of $H_2$. (The curve is qualitative in those regions where $c_v'$ is changing.)

250 K and 700 K, $c_v' = \frac{5}{2}R$, which is that for molecules that translate and rotate but do not vibrate. And at temperatures above 700 K, the $H_2$ molecules begin to vibrate. However, the molecules dissociate before $c_v'$ reaches $\frac{7}{2}R$. Finally, the equipartition theorem predicts a constant value of $3R$ for the heat capacity of solids. While this result holds for many, although not all, solids at high temperatures, it does not hold at very low temperatures.

The equipartition theorem fails because the energy is **quantized.** That is, a molecule can have only certain values of internal energy, as illustrated schematically by the energy-level diagram in Figure 18-17. The molecule can gain or lose energy only if the gain or loss takes it to another allowed level. For example, the energy that can be transferred between colliding gas molecules is of the order of $kT$, the typical thermal energy of a molecule. The validity of the equipartition theorem depends on the relative size of $kT$ and the spacing of the allowed energy levels.

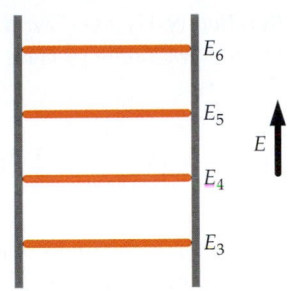

**FIGURE 18-17** Energy-level diagram. A system can have only certain discrete energies.

> If the spacing of the allowed energy levels is large compared with $kT$, energy cannot be transferred by collisions and the classical equipartition theorem is not valid. If the spacing of the levels is much smaller than $kT$, energy quantization will not be noticed and the equipartition theorem will hold.

CONDITIONS FOR THE VALIDITY OF THE EQUIPARTITION THEOREM

Consider the rotation of a molecule. The energy of rotation is

$$E = \frac{1}{2}I\omega^2 = \frac{(I\omega)^2}{2I} = \frac{L^2}{2I} \qquad\qquad 18\text{-}28$$

where $I$ is the moment of inertia of the molecule, $\omega$ is its angular velocity, and $L = I\omega$ is its angular momentum. In Section 10-5, we mentioned that angular momentum is quantized, and its magnitude is restricted to

$$L = \sqrt{\ell(\ell+1)}\,\hbar \qquad \ell = 0, 1, 2, \ldots \qquad\qquad 18\text{-}29$$

where $\hbar = h/(2\pi)$, and $h$ is Planck's constant. The energy of a rotating molecule is therefore quantized to the values

$$E = \frac{L^2}{2I} = \frac{\ell(\ell+1)\hbar^2}{2I} = \ell(\ell+1)\,E_{0r} \qquad\qquad 18\text{-}30$$

where

$$E_{0r} = \frac{\hbar^2}{2I} \qquad\qquad 18\text{-}31$$

is characteristic of the energy gap between levels. If this energy is much less than $kT$, we expect classical physics and the equipartition theorem to hold. Let us define a critical temperature $T_c$ by

$$kT_c = E_{0r} = \frac{\hbar^2}{2I} \qquad\qquad 18\text{-}32$$

If $T$ is much greater than this critical temperature, then $kT$ will be much greater than the spacing of the energy levels, which is of the order of $kT_c$, and we expect classical physics and the equipartition theorem to be valid. If $T$ is less than or of the order of $T_c$, then $kT$ will not be much greater than the energy-level spacing, and we expect classical physics and the equipartition theorem to break down. Let's estimate $T_c$ for some cases of interest.

1. *Rotation of $H_2$ about an axis perpendicular to the line joining the H atoms and through the center of mass (Figure 18-18):* The moment of inertia of $H_2$ about the axis is

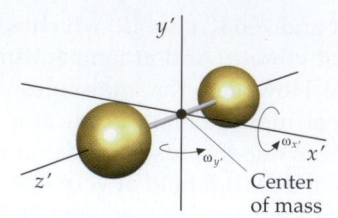

FIGURE 18-18 Rigid-dumbbell model of a diatomic molecule.

$$I_H = 2M_H \left(\frac{r_s}{2}\right)^2 = \frac{1}{2}M_H r_s^2$$

where $M_H$ is the mass of an H atom and $r_s$ is the separation distance. For hydrogen, $M_H = 1.67 \times 10^{-27}$ kg, and $r_s \approx 8 \times 10^{-11}$ m. The critical temperature is then

$$T_c = \frac{\hbar^2}{2kI} = \frac{\hbar^2}{kM_H r_s^2}$$

$$= \frac{(1.05 \times 10^{-34} \text{ J} \cdot \text{s})^2}{(1.38 \times 10^{-23} \text{ J/K})(1.67 \times 10^{-27} \text{ kg})(8 \times 10^{-11} \text{ m})^2} \approx 75 \text{ K}$$

As we see from Figure 18-16, this is approximately the temperature below which the rotational energy does not contribute to the heat capacity.

2. $O_2$: Because the mass of $O_2$ is about 16 times that of $H_2$, and the separation is about the same, the critical temperature for $O_2$ should be about $(75/16) \approx$ 4.6 K. For all temperatures for which $O_2$ exists as a gas, $T \gg T_c$, so $kT$ is much greater than the energy level spacing. Consequently, we expect the equipartition theorem of classical physics to apply.

3. *Rotation of a monatomic gas:* Consider the He atom, which consists of the He nucleus and two electrons. The mass of an electron is about 8000 times smaller than the mass of the He nucleus, but the radius of the nucleus is about 100,000 times smaller than the distance to the electron. Therefore, the moment of inertia of the He atom is almost entirely due to its two electrons. The distance from the He nucleus to its electrons is about half the separation distance of the H atoms in $H_2$, and the electron mass is about 2000 times smaller than that of the H nucleus. Thus, using $m_e = M_H/2000$ and $r = r_s/2$, we find the moment of inertia of the two electrons in He to be roughly

$$I_{He} = 2m_e r^2 \approx 2\frac{M_H}{2000}\left(\frac{r_s}{2}\right)^2 = \frac{I_H}{2000}$$

The critical temperature for He is thus about 2000 times that of $H_2$ or about 150,000 K. This is much higher than the dissociation temperature (the temperature at which electrons are stripped from their nuclei) for helium. So, the gap between allowed energy levels is always much greater than $kT$ and the He molecules cannot be induced to rotate by collisions occurring in the gas. Other monatomic gases have slightly greater moments of inertia because they have more electrons, but their critical temperatures are still tens of thousands of kelvins. Therefore, their molecules also cannot be induced to rotate by collisions occurring in the gas.

4. *Rotation of a diatomic gas about the axis joining the atoms:* We see from our discussion of monatomic gases that the moment of inertia for a diatomic gas molecule about this axis will also be due mainly to the electrons and will be of the same order of magnitude as for a monatomic gas. Again, the critical temperature, $T_c$, calculated in order for this rotation to occur due to collisions between molecules in the gas, exceeds the gas's dissociation temperature, making rotation under those circumstances impossible.

It is interesting to note that the successes of the equipartition theorem in explaining the measured heat capacities of gases and solids led to the first real understanding of molecular structure in the nineteenth century, whereas its failures played an important role in the development of quantum mechanics in the twentieth century.

---

*ROTATIONAL ENERGY OF THE HYDROGEN ATOM* **EXAMPLE 18-11**

(a) **Estimate the lowest (nonzero) rotational energy for the hydrogen atom and compare it to** $kT$ **at room temperature,** $T = 300$ K. (b) **Calculate the critical temperature** $T_c$.

**PICTURE THE PROBLEM** From Equation 18-30, the lowest rotational energy is for $\ell = 1$. We use Equation 18-30 to determine the energy in terms of the moment of inertia. We can neglect the moment of inertia of the nucleus because its radius is 100,000 times smaller than the radius of the atom. Therefore, the moment of inertia for the atom is essentially the moment of inertia of the electron about the nucleus. Then $I = m_e r^2$, where $r \approx 5 \times 10^{-11}$ m is the distance from the nucleus to the electron.

(a) 1. The lowest energy greater than zero occurs for $\ell = 1$:
$$E_\ell = \frac{\ell(\ell + 1)\hbar^2}{2I}, \qquad \ell = 0, 1, 2, \ldots$$

so

$$E_1 = \frac{1(1 + 1)\hbar^2}{2m_e r^2} = \frac{\hbar^2}{m_e r^2}$$

2. The numerical values are:
$$\hbar = 1.05 \times 10^{-34} \text{ J·s}$$
$$m_e = 9.11 \times 10^{-31} \text{ kg}$$
$$r = 5 \times 10^{-11} \text{ m}$$

3. Substitute the numerical values:
$$E_1 = \frac{\hbar^2}{m_e r^2} = \frac{(1.05 \times 10^{-34} \text{ J·s})^2}{(9.11 \times 10^{-31} \text{ kg})(5 \times 10^{-11} \text{ m})^2}$$

$$= \boxed{4.8 \times 10^{-18} \text{ J}}$$

4. The value of $kT$ at $T = 300$ K is:
$$kT = (1.38 \times 10^{-23} \text{ J/K})(300 \text{ K}) = 4.1 \times 10^{-21} \text{ J}$$

5. Compare $E_1$ and $kT$:
$$\frac{E_1}{kT} = \frac{4.8 \times 10^{-18} \text{ J}}{4.1 \times 10^{-21} \text{ J}} \approx 10^3$$

$\boxed{E_1 \text{ is about three orders of magnitude larger than } kT.}$

(b) Set $kT_c = E_1$ and solve for $T_c$:
$$kT_c = E_1$$

$$T_c = \frac{E_1}{k} = \frac{4.8 \times 10^{-18} \text{ J}}{1.38 \times 10^{-23} \text{ J/K}} = \boxed{3.48 \times 10^5 \text{ K}}$$

**REMARKS** The critical temperature of a hydrogen atom is so high that the atom would be ionized well before the critical temperature could be reached.

---

# 18-9 The Quasi-Static Adiabatic Compression of a Gas

A process in which no heat flows into or out of a system is called an **adiabatic process.** Such a process can occur when the system is extremely well insulated, or when the process happens very quickly. Consider the quasi-static adiabatic compression of a gas in which the gas in a thermally insulated container is slowly compressed by a piston, which is thereby doing work on the gas. Because no heat enters or leaves the gas, the work done on the gas equals the increase in the internal energy of the gas, and the temperature of the gas increases. The curve representing this process on a $PV$ diagram is shown in Figure 18-19.

We can find the equation for the adiabatic curve for an ideal gas by using the equation of state ($PV = nRT$) and the first law of thermodynamics ($dE_{int} = dQ_m + dW_{on}$). We have

$$C_v dT = 0 + (-PdV) \qquad 18\text{-}33$$

where we have used $dE_{int} = C_v dT$ (Equation 18-18a), $dQ_m = 0$ (the process is adiabatic), and $dW_{on} = -PdV$ (Equation 18-15). Then, substituting for $P$ using $P = nRT/V$, we obtain

$$C_v dT = -nRT \frac{dV}{V}$$

Rearranging,

$$\frac{dT}{T} + \frac{nR}{C_v} \frac{dV}{V} = 0$$

Integration gives

$$\ln T + \frac{nR}{C_v} \ln V = \text{constant}$$

Simplifying,

$$\ln T + \frac{nR}{C_v} \ln V = \ln T + \ln V^{nR/C_v} = \ln TV^{nR/C_v} = \text{constant}$$

Thus,

$$TV^{nR/C_v} = \text{constant} \qquad 18\text{-}34$$

where the constants in the two preceding equations are not the same. Equation 18-34 can be rewritten by noting that $C_p - C_v = nR$, so

$$\frac{nR}{C_v} = \frac{C_p - C_v}{C_v} = \frac{C_p}{C_v} - 1 = \gamma - 1 \qquad 18\text{-}35$$

where $\gamma$ is the ratio of the heat capacities:

$$\gamma = \frac{C_p}{C_v} \qquad 18\text{-}36$$

Therefore,

$$TV^{\gamma-1} = \text{constant} \qquad 18\text{-}37$$

We can eliminate $T$ from Equation 18-37 using $PV = nRT$. We then have

$$\frac{PV}{nR} V^{\gamma-1} = \text{constant}$$

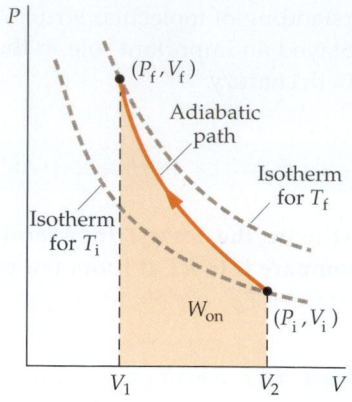

**FIGURE 18-19** Quasi-static adiabatic compression of an ideal gas. The dashed lines are the isotherms for the initial and final temperatures. The curve connecting the initial and final states of the adiabatic compression is steeper than the isotherms because the temperature increases during the compression.

Clouds form if rising moist air cools due to adiabatic expansion of the air. Cooling causes water vapor to condense into liquid droplets.

or

$$PV^\gamma = \text{constant} \qquad\qquad 18\text{-}38$$

QUASI-STATIC ADIABATIC PROCESS

Equation 18-38 relates $P$ and $V$ for adiabatic expansions and compressions.

**EXERCISE** Show that for quasi-static adiabatic process $T^\gamma/P^{\gamma-1} = \text{constant}$.

The work done on the gas in an adiabatic compression can be calculated from the first law of thermodynamics:

$$dE_{int} = dQ_{in} + dW_{on} \quad \text{or} \quad dW_{on} = dE_{int} - dQ_{in}$$

Since $dE_{int} = C_v\, dT$ and $dQ_{in} = 0$, we have

$$dW_{on} = C_v\, dT$$

Then

$$W_{adiabatic} = \int dW_{on} = \int C_v dT = C_v \Delta T \qquad\qquad 18\text{-}39$$

ADIABATIC WORK ON GAS

where we have assumed that $C_v$ is constant.[†] We note that the work done on the gas depends only on the change in the temperature of the gas. In an adiabatic compression, work is done on the gas, and its internal energy and temperature increase. In a quasi-static adiabatic *expansion,* work is done *by* the gas, and the internal energy and temperature decrease.

We can use the ideal-gas law to write Equation 18-39 in terms of the initial and final values of the pressure and volume. If $T_i$ is the initial temperature and $T_f$ is the final temperature, we have for the work done

$$W_{adiabatic} = C_v \Delta T = C_v(T_f - T_i)$$

Using $PV = nRT$, we obtain

$$W_{adiabatic} = C_v\left(\frac{P_f V_f}{nR} - \frac{P_i V_i}{nR}\right) = \frac{C_v}{nR}(P_f V_f - P_i V_i)$$

Using Equation 18-35 to simplify this expression, we have

$$W_{adiabatic} = \frac{P_f V_f - P_i V_i}{\gamma - 1} \qquad\qquad 18\text{-}40$$

ADIABATIC WORK ON GAS

---

QUASI-STATIC ADIABATIC COMPRESSION OF AIR          **EXAMPLE 18-12**

A quantity of air is compressed adiabatically and quasi-statically from an initial pressure of 1 atm and a volume of 4 L at temperature 20°C to half its original volume. Find (*a*) the final pressure, (*b*) the final temperature, and (*c*) the work done on the gas.

---

[†] For an ideal gas, $E_{int}$ is proportional to the absolute temperature, and therefore $C_v = dE_{int}/dT$ is a constant.

**PICTURE THE PROBLEM** Because the process is both quasi-static and adiabatic, we know that $PV^\gamma = $ constant, and $TV^{\gamma-1} = $ constant. These relations yield the final pressure and final temperature, respectively. Find $\gamma$ using Equations 18-36, 18-20, and 18-25. The work done is found from $W_{adiabatic} = C_v \Delta T$ (Equation 18-39). For a diatomic gas, $C_v = \frac{5}{2} nR$. Let subscript 1 refer to initial values, and subscript 2 to final values. Then $P_1 = 1$ atm, $V_1 = 4$ L, $V_2 = 2$ L, $T_1 = 20°C = 293$ K.

(a) 1. Write $PV^\gamma = $ constant in terms of initial and final values:

$$P_1 V_1^\gamma = P_2 V_2^\gamma$$

2. Find $\gamma$ for a diatomic gas using Equations 18-36, 18-20, and 18-25:

$$\gamma = \frac{C_P}{C_v} = \frac{C_v + nR}{C_v} = 1 + \frac{nR}{C_v} = 1 + \frac{nR}{\frac{5}{2}nR}$$

3. Solve for $P_2$:

$$P_2 = P_1 \left(\frac{V_1}{V_2}\right)^\gamma = (1 \text{ atm})\left(\frac{4 \text{ L}}{2 \text{ L}}\right)^{1.4} = \boxed{2.64 \text{ atm}}$$

(b) 1. Write $TV^{\gamma-1} = $ constant in terms of initial and final values:

$$T_1 V_1^{\gamma-1} = T_2 V_2^{\gamma-1}$$

2. Solve for $T_2$:

$$T_2 = T_1 \left(\frac{V_1}{V_2}\right)^{\gamma-1} = (293 \text{ K})\left(\frac{4 \text{ L}}{2 \text{ L}}\right)^{0.4}$$

$$= 387 \text{ K} = \boxed{114°C}$$

(c) 1. Equation 18-39 gives the work done:

$$W_{adiabatic} = C_v \Delta T = \frac{5}{2} nR\Delta T$$

2. Using the ideal-gas law for the initial conditions, express $nR$ in terms of $P_1$, $V_1$, and $T_1$:

$$W_{adiabatic} = \frac{5}{2} nR\Delta T = \frac{5}{2} \frac{P_1 V_1}{T_1}(T_2 - T_1)$$

$$= \frac{5}{2} \frac{(1 \text{ atm})(4 \text{ L})}{293 \text{ K}}(387 \text{ K} - 293 \text{ K})$$

$$= 3.20 \text{ L·atm} = \boxed{324 \text{ J}}$$

**REMARKS** The work can also be calculated using Equation 18-40, but using $W_{adiabatic} = C_v \Delta T$ is preferable because it is more directly connected to a principle, the first law of thermodynamics, and thus is easier to recall.

## Speed of Sound Waves

We can use Equation 18-38 to calculate the adiabatic bulk modulus of an ideal gas, which is related to the speed of sound waves in air. We first compute the differential of both sides of $PV^\gamma = $ constant (Equation 18-38):

$$Pd(V^\gamma) + V^\gamma dP = 0$$

or

$$\gamma PV^{\gamma-1} dV + V^\gamma dP = 0$$

Then

$$dP = -\frac{\gamma P dV}{V}$$

Referring to Equation 13-6, the adiabatic bulk modulus[†] is then:

$$B_{adiabatic} = -\frac{dP}{dV/V} = \gamma P \qquad \text{18-41}$$

[†] The bulk modulus, discussed in Chapter 13, is the negative ratio of the pressure change to the fractional change in volume, $B = -\Delta P/(\Delta V/V)$. The isothermal bulk modulus, which describes changes that occur at constant temperature, differs from the adiabatic bulk modulus, which describes changes with no heat transfer. For sound waves at audible frequencies, the pressure changes occur too rapidly for appreciable heat flow, so the appropriate bulk modulus is the adiabatic bulk modulus.

The speed of sound (Equation 15-4) is given by

$$v = \sqrt{\frac{B_{\text{adiabatic}}}{\rho}}$$

where the mass density $\rho$ is related to the number of moles $n$ and the molecular mass $M$ by $\rho = m/V = nM/V$. Using the ideal-gas law, $PV = nRT$, we can eliminate $V$ from the density

$$\rho = \frac{nM}{V} = \frac{nM}{nRT/P} = \frac{MP}{RT}$$

Using this result and $\gamma P$ for $B_{\text{adiabatic}}$, we obtain

$$v = \sqrt{\frac{B_{\text{adiabatic}}}{\rho}} = \sqrt{\frac{\gamma P}{MP/(RT)}} = \sqrt{\frac{\gamma RT}{M}}$$

which is Equation 15-5, the speed of sound in a gas.

# SUMMARY

1. The first law of thermodynamics, which is a statement of the conservation of energy, is a fundamental law of physics.
2. The equipartition theorem is a fundamental law of classical physics. It breaks down if the typical thermal energy $kT$ is small compared to the spacing of quantized energy levels.

| Topic | Relevant Equations and Remarks | |
|---|---|---|
| **1. Heat** | Heat is energy that is transferred from one object to another because of a temperature difference. | |
| Calorie | The calorie, originally defined as the heat necessary to raise the temperature of 1 g of water by 1°C, is now defined to be 4.184 joules. | |
| **2. Heat Capacity** | Heat capacity is the amount of heat needed to raise the temperature of a substance by one degree. | |
| | $$C = \frac{Q}{\Delta T}$$ | 18-1 |
| At constant volume | $$C_v = \frac{Q_v}{\Delta T}$$ | |
| At constant pressure | $$C_p = \frac{Q_p}{\Delta T}$$ | |
| Specific heat (heat capacity per unit mass) | $$c = \frac{C}{m}$$ | 18-2 |
| Molar specific heat (heat capacity per mole) | $$c' = \frac{C}{n}$$ | 18-6 |

| | | |
|---|---|---|
| Heat capacity–internal energy relation | $C_v = \dfrac{dE_{int}}{dT}$ | 18-18a |
| Ideal gas | $C_p = C_v + nR$ | 18-20 |
| Monatomic ideal gas | $C_v = \frac{3}{2}nR$ | 18-22 |
| Diatomic ideal gas | $C_v = \frac{5}{2}nR$ | 18-25 |

**3. Fusion and Vaporization**

Both melting and vaporization occur at a constant temperature.

| | | |
|---|---|---|
| Latent heat of fusion | The heat needed to melt a substance is the product of the mass of the substance and its latent heat of fusion $L_f$: $$Q_f = mL_f$$ | 18-8 |
| $L_f$ of water | $L_f = 333.5 \text{ kJ/kg}$ | |
| Latent heat of vaporization | The heat needed to vaporize a liquid is the product of the mass of the liquid and its latent heat of vaporization $L_v$: $$Q_v = mL_v$$ | 18-9 |
| $L_v$ of water | $L_v = 2257 \text{ kJ/kg}$ | |

**4. First Law of Thermodynamics**

The change in the internal energy of a system equals the heat transferred into the system plus the work done on the system:

$$\Delta E_{int} = Q_{in} + W_{on} \qquad \text{18-10}$$

**5. Internal Energy $E_{int}$**

The internal energy of a system is a property of the state of the system, as are the pressure, volume, and temperature. Heat and work are not properties of state.

| | | |
|---|---|---|
| Ideal gas | $E_{int}$ depends only on the temperature $T$. | |
| Monatomic ideal gas | $E_{int} = \frac{3}{2}nRT$ | 18-12 |
| Internal energy related to heat capacity | $dE_{int} = C_v\,dT$ | 18-18b |

**6. Quasi-Static Process**

A quasi-static process is one that occurs slowly so that the system moves through a series of equilibrium states.

| | | |
|---|---|---|
| Isobaric | $P = \text{constant}$ | |
| Isothermal | $T = \text{constant}$ | |
| Adiabatic | $Q = 0$ | |
| Adiabatic, ideal gas | $TV^{\gamma-1} = \text{constant}$ | 18-37 |
| | or | |
| | $PV^{\gamma} = \text{constant}$ | 18-38 |
| | where | |
| | $\gamma = \dfrac{C_p}{C_v}$ | 18-36 |

7. **Work Done on a Gas**

$$W_{on} = -\int_{V_i}^{V_f} P\,dV = C_v\Delta T - Q_{in}$$

18-10, 18-15, and 18-18

Constant volume

$$W_{on} = -\int_{V_i}^{V_f} P\,dV = 0 \qquad V_f = V_i$$

Isobaric

$$W_{on} = -\int_{V_i}^{V_f} P\,dV = -P\int_{V_i}^{V_f} dV = -P\Delta V$$

Isothermal

$$W_{isothermal} = -\int_{V_i}^{V_f} P\,dV = -nRT\int_{V_i}^{V_f} \frac{dV}{V} = nRT\ln\frac{V_i}{V_f}$$

18-17

Adiabatic

$$W_{adiabatic} = C_v\Delta T$$

18-39

8. **Equipartition Theorem**

The equipartition theorem states that if a system is in equilibrium, there is an average energy of $\frac{1}{2}kT$ per molecule or $\frac{1}{2}RT$ per mole associated with each degree of freedom.

Failure of the equipartition theorem

The equipartition theorem fails if the thermal energy ($\sim kT$) that can be transferred in collisions is smaller than the energy gap $\Delta E$ between quantized energy levels. For example, monatomic gas molecules cannot rotate because the first nonzero energy permitted is much greater than $kT$.

9. **Dulong–Petit Law**

The molar specific heat of most solids is $3R$. This is predicted by the equipartition theorem, assuming a solid atom has six degrees of freedom.

# PROBLEMS

- Single-concept, single-step, relatively easy
- •• Intermediate-level, may require synthesis of concepts
- ••• Challenging
- **SSM** Solution is in the *Student Solutions Manual*
- **iSOLVE** Problems available on iSOLVE online homework service
- **iSOLVE✓** These "Checkpoint" online homework service problems ask students additional questions about their confidence level, and how they arrived at their answer

In a few problems, you are given more data than you actually need; in a few other problems, you are required to supply data from your general knowledge, outside sources, or informed estimates.

**Use $v = 340$ m/s for the speed of sound in air unless otherwise indicated.**

## Conceptual Problems

**1** • Body A has twice the mass and twice the specific heat of body B. If they are supplied with equal amounts of heat, how do the subsequent changes in their temperatures compare?

**2** • **SSM** The temperature change of two blocks of masses $M_A$ and $M_B$ is the same when they absorb equal amounts of heat. It follows that the specific heats are related by (a) $c_A = (M_A/M_B)c_B$, (b) $c_A = (M_B/M_A)c_B$, (c) $c_A = c_B$, (d) none of the above.

**3** • The specific heat of aluminum is more than twice that of copper. Identical masses of copper and aluminum, both at 20°C, are dropped into a calorimeter containing water at 40°C. When thermal equilibrium is reached, (a) the aluminum is at a higher temperature than the copper, (b) the aluminum has absorbed less energy than the copper, (c) the aluminum has absorbed more energy than the copper, (d) both (a) and (c) are correct statements.

**4** • Joule's experiment establishing the mechanical equivalence of heat involved the conversion of mechanical energy into internal energy. Give some examples of the internal energy of a system being converted into mechanical energy.

**5** • SSM Can a system absorb heat with no change in its internal energy?

**6** • In the equation $\Delta E_{int} = Q + W$ (the formal statement of the first law of thermodynamics), the quantities $Q$ and $W$ represent (a) the heat supplied to the system and the work done by the system, (b) the heat supplied to the system and the work done on the system, (c) the heat released by the system and the work done by the system, (d) the heat released by the system and the work done on the system.

**7** • A real gas cools during a free expansion, though an ideal gas does not. Explain.

**8** • An ideal gas at one atmosphere pressure and 300 K is confined to half of an insulated container by a thin partition. The partition is then removed and equilibrium is established. At that point of equilibrium, which of the following is correct? (a) The pressure is half an atmosphere and the temperature is 150 K, (b) the pressure is one atmosphere and the temperature is 150 K, (c) the pressure is half an atmosphere and the temperature is 300 K, (d) none of the above.

**9** • A certain gas consists of ions that repel each other. The gas undergoes a free expansion with no heat exchange and no work done. How does the temperature of the gas change? Why?

**10** •• SSM Two gas-filled rubber balloons of (initially) equal volume are at the bottom of a dark, cold lake. The top of the lake is warmer than the bottom. One balloon rises rapidly and expands adiabatically as it rises. The other balloon rises more slowly and expands isothermally. Which balloon is larger when it reaches the surface of the lake?

**11** • A gas changes its state reversibly from A to C (Figure 18-20). The work done by the gas is (a) greatest for path A→B→C, (b) least for path A→C, (c) greatest for path A→D→C, (d) the same for all three paths.

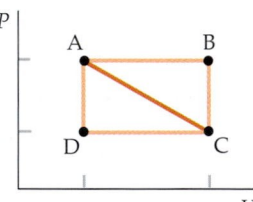

**FIGURE 18-20** Problem 11

**12** • When an ideal gas is subjected to an adiabatic process, (a) no work is done by the system, (b) no heat is supplied to the system, (c) the internal energy remains constant, (d) the heat supplied to the system equals the work done by the system.

**13** • True or false:

(a) The heat capacity of a body is the amount of heat it can store at a given temperature.
(b) When a system goes from state 1 to state 2, the amount of heat added to the system is the same for all processes.
(c) When a system goes from state 1 to state 2, the work done on the system is the same for all processes.
(d) When a system goes from state 1 to state 2, the change in the internal energy of the system is the same for all processes.
(e) The internal energy of a given amount of an ideal gas depends only on its absolute temperature.
(f) A quasi-static process is one in which the system is never far from being in equilibrium.
(g) For any material that expands when heated, $C_p$ is greater than $C_v$.

**14** • SSM If a system's volume remains constant while undergoing changes in temperature and pressure, then (a) the internal energy of the system is unchanged, (b) the system does no work, (c) the system absorbs no heat, (d) the change in internal energy equals the heat absorbed by the system.

**15** • When an ideal gas is subjected to an isothermal process, (a) no work is done by the system, (b) no heat is supplied to the system, (c) the heat supplied to the system equals the change in internal energy, (d) the heat supplied to the system equals the work done by the system.

**16** •• The 1-L fuel tank of a gas grill contains 600 g of propane ($C_3H_8$) at a pressure of 2 MPa. What can you say about the phase state of the propane?

**17** •• An ideal gas undergoes a process during which $P\sqrt{V}$ = constant and the volume of the gas decreases. What happens to the temperature?

**18** •• SSM Which would you expect to have a higher heat capacity per *unit mass*, lead or copper? Why? (Don't look up the heat capacities to answer this question.)

**19** •• Calculating the heat capacity of a liquid is very difficult, because of the strong intermolecular interactions and the random positions of the molecules in a liquid. However, simply based on counting degrees of freedom, would you expect a monatomic liquid to have a higher or lower heat capacity than the solid phase of the same substance it melted from? Assume that the melting temperature is high enough that you don't have to take quantum effects into consideration when calculating the heat capacity of the solid.

## Estimation and Approximation

**20** •• SSM A simple demonstration to show that heat is a form of energy is to take a bag of lead shot and drop it repeatedly onto a very rigid surface from a small height. The bag's temperature will increase, allowing an estimate of the heat capacity of lead. (a) Estimate the temperature increase of a bag filled with 1 kg of lead shot dropped 50 times from a height of 1 m. (b) In principle, the change in temperature is independent of the mass of the shot in the bag; in practice, it's better to use a larger mass than a smaller one. Why might this be true?

**21** •• A "typical" microwave oven has a power consumption of about 1200 W. Estimate how long it should take to boil a cup of water in the microwave assuming that 50% of the power consumption goes into heating the water. Does this estimate mesh with everyday experience?

**22** • A demonstration of the heating of a gas under adiabatic compression involves putting a few scraps of paper into a large glass test tube, which is then sealed off with a piston. If the piston compresses the trapped air very rapidly, the paper will catch fire. Assuming that the burning point of paper is 451°F, estimate the factor by which the volume of the air trapped by the piston must be reduced for this demonstration to work.

**23 ••** There is a small change in the volume of a liquid on heating at constant pressure. Use the following data for water to estimate the contribution this makes to the heat capacity of water between 4°C and 100°C:

Density of water at 4°C and 1 atm pressure: 1.000 g/cm³

Density of liquid water at 100°C and 1 atm pressure: 0.9584 g/cm³

**24 ••** **SSM** A certain molecule has vibrational energy levels that are equally spaced by 0.15 eV. Find the critical temperature $T_c$ so for $T \gg T_c$ you would expect the equipartition theorem to hold and for $T \ll T_c$ you would expect the equipartition theorem to fail.

## Heat Capacity; Specific Heat; Latent Heat

**25 •** **SSM** A "typical" adult male consumes about 2500 kcal of food in a day. (a) How many joules is this? (b) If this consumed energy is dissipated over the course of 24 hours, what is his average output power in watts?

**26 •** **iSOLVE** A solar home contains $10^5$ kg of concrete (specific heat = 1.00 kJ/kg·K). How much heat is given off by the concrete when it cools from 25 to 20°C?

**27 •** **iSOLVE** How many calories must be supplied to 60 g of ice at −10°C to melt it and raise the temperature of the water to 40°C?

**28 ••** **iSOLVE** How much heat must be removed when 100 g of steam at 150°C is cooled and frozen into 100 g of ice at 0°C? (Take the specific heat of steam to be 2.01 kJ/kg·K.)

**29 ••** **iSOLVE** A 50-g piece of aluminum at 20°C is cooled to −196°C by placing it in a large container of liquid nitrogen at that temperature. How much nitrogen is vaporized? (Assume that the specific heat of aluminum is constant and is equal to 0.90 J/kg·K.)

**30 ••** **iSOLVE** If 500 g of molten lead at 327°C is poured into a cavity in a large block of ice at 0°C, how much of the ice melts?

**31 ••** **SSM** **iSOLVE** A 30-g lead bullet initially at 20°C comes to rest in the block of a ballistic pendulum. Assume that half the initial kinetic energy of the bullet is converted into thermal energy within the bullet. If the speed of the bullet was 420 m/s, what is the temperature of the bullet immediately after coming to rest in the block?

**32 ••** **iSOLVE** A 1400-kg car traveling at 80 km/h is brought to rest by applying the brakes. If the specific heat of steel is 0.11 cal/g·K, what total mass of steel must be contained in the steel brake drums if the temperature of the brake drums is not to rise by more than 120 C°?

## Calorimetry

**33 •** **iSOLVE** A 200-g piece of lead is heated to 90°C and is then dropped into a calorimeter containing 500 g of water that is initially at 20°C. Neglecting the heat capacity of the container, find the final temperature of the lead and water.

**34 •** **SSM** **iSOLVE** The specific heat of a certain metal can be determined by measuring the temperature change that

occurs when a piece of the metal is heated and then placed in an insulated container made of the same material and containing water. Suppose a piece of metal has a mass of 100 g and is initially at 100°C. The container has a mass of 200 g and contains 500 g of water at an initial temperature of 20.0°C. The final temperature is 21.4°C. What is the specific heat of the metal?

**35 ••** In the 2002 Tour de France, champion bicyclist Lance Armstrong expended an average power of about 400 W, 5 hours a day for 20 days. What quantity of water, initially at 24°C, could be brought to the boiling point by harnessing all of that energy?

**36 ••** A 25-g glass tumbler contains 200 mL of water at 24°C. If two 15-g ice cubes, each at a temperature of −3°C, are dropped into the tumbler, what is the final temperature of the drink? Neglect thermal conduction between the tumbler and the room.

**37 ••** A 200-g piece of ice at 0°C is placed in 500 g of water at 20°C. The system is in a container of negligible heat capacity and is insulated from its surroundings. (a) What is the final equilibrium temperature of the system? (b) How much of the ice melts?

**38 ••** **iSOLVE✓** A 3.5-kg block of copper at a temperature of 80°C is dropped into a bucket containing a mixture of ice and water whose total mass is 1.2 kg. When thermal equilibrium is reached, the temperature of the water is 8°C. How much ice was in the bucket before the copper block was placed in it? (Neglect the heat capacity of the bucket.)

**39 ••** **iSOLVE✓** A well-insulated bucket contains 150 g of ice at 0°C. (a) If 20 g of steam at 100°C is injected into the bucket, what is the final equilibrium temperature of the system? (b) Is any ice left afterward?

**40 ••** **iSOLVE✓** A calorimeter of negligible mass contains 1 kg of water at 303 K and 50 g of ice at 273 K. Find the final temperature $T$. Solve the same problem if the mass of ice is 500 g.

**41 ••** **SSM** A 200-g aluminum calorimeter contains 500 g of water at 20°C. A 100-g piece of ice cooled to −20°C is placed in the calorimeter. (a) Find the final temperature of the system, assuming no heat loss. (Assume that the specific heat of ice is 2.0 kJ/kg·K.) (b) A second 200-g piece of ice at −20°C is added. How much ice remains in the system after it reaches equilibrium? (c) Would you give a different answer for (b) if both pieces of ice were added at the same time?

**42 ••** **iSOLVE** The specific heat of a 100-g block of material is to be determined. The block is placed in a 25-g copper calorimeter that also holds 60 g of water. The system is initially at 20°C. Then 120 mL of water at 80°C are added to the calorimeter vessel. When thermal equilibrium is attained, the temperature of the water is 54°C. Determine the specific heat of the block.

**43 ••** A 100-g piece of copper is heated in a furnace to a temperature $t$. The copper is then inserted into a 150-g copper calorimeter containing 200 g of water. The initial temperature of the water and calorimeter is 16°C, and the final temperature after equilibrium is established is 38°C. When the calorimeter and its contents are weighed, 1.2 g of water are found to have evaporated. What was the temperature $t$?

**44** •• A 200-g aluminum calorimeter contains 500 g of water at 20°C. Aluminum shot with a mass 300 g is heated to 100°C and is then placed in the calorimeter. (*a*) Using the value of the specific heat of aluminum given in Table 18-1, find the final temperature of the system, assuming that no heat is lost to the surroundings. (*b*) The error due to heat transfer between the system and its surroundings can be minimized if the initial temperature of the water and calorimeter is chosen to be below room temperature, where $\Delta t_w$ is the temperature change of the calorimeter and water during the measurement. Then the final temperature is $\frac{1}{2}\Delta t_w$ above room temperature. What should the initial temperature of the water and container be if the room temperature is 20°C?

## First Law of Thermodynamics

**45** • **ISOLVE** A diatomic gas does 300 J of work and also absorbs 600 cal of heat. What is the change in internal energy of the gas?

**46** • **SSM** **ISOLVE** If 400 kcal is added to a gas that expands and does 800 kJ of work, what is the change in the internal energy of the gas?

**47** • **ISOLVE** A lead bullet moving at 200 m/s is stopped in a block of wood. Assuming that all of the energy change goes into heating the bullet, find the final temperature of the bullet if its initial temperature is 20°C.

**48** • (*a*) At Niagara Falls, the water drops 50 m. If the change in potential energy goes into the internal energy of the water, compute the increase in its temperature. (*b*) Do the same for Yosemite Falls, where the water drops 740 m. (These temperature rises are not observed because the water cools by evaporation as it falls.)

**49** • When 20 cal of heat are absorbed by a gas, the system performs 30 J of work. What is the change in the internal energy of the gas?

**50** •• **ISOLVE** A lead bullet initially at 30°C just melts upon striking a target. Assuming that all of the initial kinetic energy of the bullet goes into the internal energy of the bullet to raise its temperature and melt it, calculate the speed of the bullet upon impact.

**51** •• **SSM** On a cold day you can warm your hands by rubbing them together. (*a*) Assume that the coefficient of friction between your hands is 0.5, that the normal force between your hands is 35 N, and that you rub them together at an average speed of 35 cm/s. What is the rate at which heat is generated? (*b*) Assume further that the mass of each of your hands is approximately 350 g, that the specific heat of your hands is about 4 kJ/kg·K, and that all the heat generated goes into raising the temperature of your hands. How long must you rub your hands together to produce a 5 C° increase in their temperature?

## Work and the PV Diagram for a Gas

*In Problems 52 through 55, the initial state of 1 mol of an ideal gas is $P_1 = 3$ atm, $V_1 = 1$ L, and $E_{int,1} = 456$ J, and its final state is $P_2 = 2$ atm, $V_2 = 3$ L, and $E_{int,2} = 912$ J.*

**52** • **ISOLVE** The gas is allowed to expand at constant pressure to a volume of 3 L. It is then cooled at constant

volume until its pressure is 2 atm. (*a*) Show this process on a *PV* diagram, and calculate the work done by the gas. (*b*) Find the heat added during this process.

**53** • **ISOLVE** The gas is first cooled at constant volume until its pressure is 2 atm. It is then allowed to expand at constant pressure until its volume is 3 L. (*a*) Show this process on a *PV* diagram, and calculate the work done by the gas. (*b*) Find the heat added during this process.

**54** •• **SSM** The gas is allowed to expand isothermally until its volume is 3 L and its pressure is 1 atm. It is then heated at constant volume until its pressure is 2 atm. (*a*) Show this process on a *PV* diagram, and calculate the work done by the gas. (*b*) Find the heat added during this process.

**55** •• The gas is heated and is allowed to expand such that it follows a straight-line path on a *PV* diagram from its initial state to its final state. (*a*) Show this process on a *PV* diagram, and calculate the work done by the gas. (*b*) Find the heat added during this process.

**56** •• **ISOLVE** One mole of the ideal gas is initially in the state $P_0 = 1$ atm, $V_0 = 25$ L. As the gas is slowly heated, the plot of its state on a *PV* diagram moves in a straight line to the state $P = 3$ atm, $V = 75$ L. Find the work done by the gas.

**57** •• One mole of the ideal gas is heated while its volume changes, so that $T = AP^2$, where $A$ is a constant. The temperature changes from $T_0$ to $4T_0$. Find the work done by the gas.

**58** • **SSM** An *isobaric* expansion is one carried out at constant pressure. Draw several isobars for an ideal gas on a diagram showing volume as a function of temperature.

**59** •• **ISOLVE** An ideal gas initially at 20°C and 200 kPa has a volume of 4 L. It undergoes a quasi-static, isothermal expansion until its pressure is reduced to 100 kPa. Find (*a*) the work done by the gas, and (*b*) the heat added to the gas during the expansion.

## Heat Capacities of Gases and the Equipartition Theorem

**60** • The heat capacity at constant volume of a certain amount of a monatomic gas is 49.8 J/K. (*a*) Find the number of moles of the gas. (*b*) What is the internal energy of the gas at $T = 300$ K? (*c*) What is the heat capacity of the gas at constant pressure?

**61** • The Dulong–Petit law was originally used to determine the molecular mass of a substance from its measured heat capacity. The specific heat of a certain solid is measured to be 0.447 kJ/kg·K. (*a*) Find the molecular mass of the substance. (*b*) What element is this?

**62** •• **SSM** (*a*) Calculate the specific heats per unit mass of air at constant volume and constant pressure. Assume a temperature of 300 K and a pressure of $10^5$ N/m². Assume that air is composed of 74% $N_2$ (molecular weight 28 g/mole) molecules and 26% $O_2$ molecules (molecular weight 32 g/mole) and that both components are ideal gases. (*b*) Compare your answer to the value listed in the *Handbook of Chemistry and Physics* for the heat capacity at constant pressure of 1.032 J/g·K.

**63** •• One mole of an ideal diatomic gas is heated at constant volume from 300 to 600 K. (a) Find the increase in internal energy, the work done, and the heat added. (b) Find the same quantities if this gas is heated from 300 to 600 K at constant pressure. Use the first law of thermodynamics and your results for (a) to calculate the work done. (c) Calculate the work done in (b) directly from $dW = P\,dV$.

**64** •• **ISOLVE✓** A diatomic gas (molar mass $M$) is confined to a closed container of volume $V$ at a pressure $P_0$. What amount of heat $Q$ should be transferred to the gas in order to triple the pressure? (Express your answer in terms of $P_0$ and $V$.)

**65** •• One mole of air ($c_v = 5R/2$) is confined at atmospheric pressure in a cylinder with a piston at 0°C. The initial volume, occupied by gas, is $V$. Find the volume of gas $V'$ after the equivalent of 13,200 J of heat is transferred to it.

**66** •• The heat capacity of a certain amount of a particular gas at constant pressure is greater than that at constant volume by 29.1 J/K. (a) How many moles of the gas are there? (b) If the gas is monatomic, what are $C_v$ and $C_p$? (c) If the gas consists of diatomic molecules that rotate but do not vibrate, what are $C_v$ and $C_p$?

**67** •• **SSM** Carbon dioxide ($CO_2$) at 1 atm of pressure and a temperature of $-78.5$°C sublimates directly from a solid to a gaseous state, without going through a liquid phase. What is the change in the heat capacity (at constant pressure) per mole of $CO_2$ when it undergoes sublimation? Assume that the gas molecules can rotate but do not vibrate. Is the change in the heat capacity positive or negative on sublimation? The $CO_2$ molecule is pictured in Figure 18-21.

**FIGURE 18-21** Problem 67

O    C    O

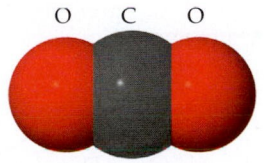

**68** •• One mole of a monatomic ideal gas is initially at 273 K and 1 atm. (a) What is its initial internal energy? (b) Find its final internal energy and the work done by the gas when 500 J of heat are added at constant pressure. (c) Find the same quantities when 500 J of heat are added at constant volume.

**69** •• List all of the degrees of freedom possible for a water molecule and estimate the heat capacity of water at a temperature very far above its boiling point. (Ignore the fact the molecule might dissociate at high temperatures.) Think carefully about all of the different ways in which a water molecule can vibrate.

## Quasi-Static Adiabatic Expansion of a Gas

**70** •• One mole of an ideal gas ($\gamma = \frac{5}{3}$) expands adiabatically and quasi-statically from a pressure of 10 atm and a temperature of 0°C to a pressure of 2 atm. Find (a) the initial and final volumes, (b) the final temperature, and (c) the work done by the gas.

**71** • An ideal gas at a temperature of 20°C is compressed quasi-statically and adiabatically to half its original volume. Find its final temperature if (a) $C_v = \frac{3}{2}nR$ and (b) $C_v = \frac{5}{2}nR$.

**72** • **ISOLVE✓** Two moles of neon gas initially at 20°C and a pressure of 1 atm are compressed adiabatically to one-fourth of their initial volume. Determine the temperature and pressure following compression.

**73** •• **SSM** Half a mole of an ideal monatomic gas at a pressure of 400 kPa and a temperature of 300 K expands until the pressure has diminished to 160 kPa. Find the final temperature and volume, the work done, and the heat absorbed by the gas if the expansion is (a) isothermal and (b) adiabatic.

**74** •• Repeat Problem 73 for a diatomic gas.

**75** •• One-half mole of helium is expanded adiabatically and quasi-statically from an initial pressure of 5 atm and temperature of 500 K to a final pressure of 1 atm. Find (a) the final temperature, (b) the final volume, (c) the work done by the gas, and (d) the change in the internal energy of the gas.

**76** ••• **SSM** A hand pump is used to inflate a bicycle tire to a gauge pressure of 482 kPa (about 70 lb/in.²). How much work must be done if each stroke of the pump is an adiabatic process? Atmospheric pressure is 1 atm, the air temperature is initially 20°C, and the volume of the air in the tire remains constant at 1 L.

**77** ••• An ideal gas at initial volume $V_1$ and pressure $P_1$ expands quasi-statically and adiabatically to volume $V_2$ and pressure $P_2$. Calculate the work done by the gas directly by integrating $P\,dV$, and show that your result is the same as that given by Equation 18-39.

## Cyclic Processes

**78** •• One mole of $N_2$ ($C_v = \frac{5}{2}nR$) gas is originally at room temperature (20°C) and a pressure of 5 atm. It is allowed to expand adiabatically and quasi-statically until its pressure equals the room pressure of 1 atm. It is then heated at constant pressure until its temperature is again 20°C. During this heating, the gas expands. After it reaches room temperature, it is heated at constant volume until its pressure is 5 atm. It is then compressed at constant pressure until it is back to its original state. (a) Construct an accurate $PV$ diagram showing each process in the cycle. (b) From your graph, determine the work done by the gas during the complete cycle. (c) How much heat is added or subtracted from the gas during the complete cycle? (d) Check your graphical determination of the work done by the gas in (b) by calculating the work done during each part of the cycle.

**79** •• **SSM** One mole of an ideal diatomic gas is allowed to expand along the straight line from 1 to 2 in the $PV$ diagram (Figure 18-22). It is then compressed back isothermally from 2 to 1. Calculate the total work done on the gas during this cycle.

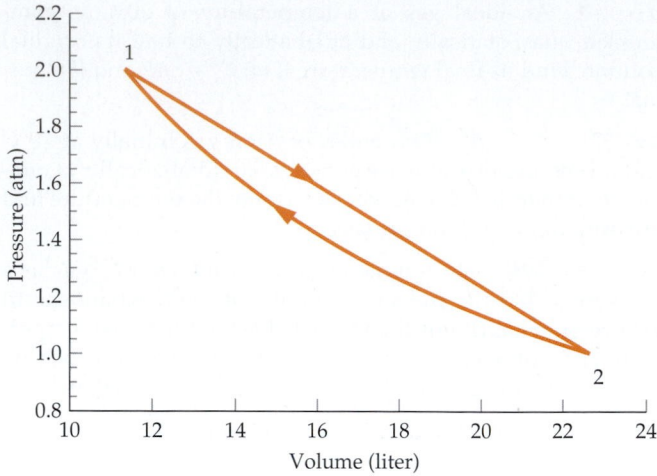

**FIGURE 18-22** Problem 79

**80** •• Two moles of an ideal monatomic gas have an initial pressure $P_1 = 2$ atm and an initial volume $V_1 = 2$ L. The gas is taken through the following quasi-static cycle: It is expanded isothermally until it has a volume $V_2 = 4$ L. It is then heated at constant volume until it has a pressure $P_3 = 2$ atm. It is then cooled at constant pressure until it is back to its initial state. (a) Show this cycle on a $PV$ diagram. (b) Calculate the heat added and the work done by the gas during each part of the cycle. (c) Find the temperatures $T_1$, $T_2$, and $T_3$.

**81** ••• At point D in Figure 18-23 the pressure and temperature of 2 mol of an ideal monatomic gas are 2 atm and 360 K. The volume of the gas at point B on the $PV$ diagram is three times that at point D and its pressure is twice that at point C. Paths AB and CD represent isothermal processes. The gas is carried through a complete cycle along the path DABCD. Determine the total amount of work done by the gas and the heat supplied to the gas along each portion of the cycle.

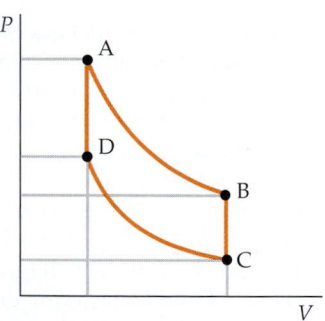

**FIGURE 18-23** Problems 81 and 82

**82** ••• **SSM** Repeat Problem 81 with a diatomic gas.

**83** ••• An ideal gas of $n$ mol is initially at pressure $P_1$, volume $V_1$, and temperature $T_h$. It expands isothermally until its pressure and volume are $P_2$ and $V_2$. It then expands adiabatically until its temperature is $T_c$ and its pressure and volume are $P_3$ and $V_3$. It is then compressed isothermally until it is at a pressure $P_4$ and a volume $V_4$, which is related to its initial volume $V_1$ by $T_cV_4^{\gamma-1} = T_hV_1^{\gamma-1}$. The gas is then compressed adiabatically until it is back in its original state. (a) Assuming that each process is quasi-static, plot this cycle on a $PV$ diagram. (This cycle is known as the Carnot cycle for an ideal gas.)

(b) Show that the heat $Q_h$ absorbed during the isothermal expansion at $T_h$ is $Q_h = nRT_h \ln(V_2/V_1)$. (c) Show that the heat $Q_c$ given off by the gas during the isothermal compression at $T_c$ is $Q_c = nRT_c \ln (V_3/V_4)$. (d) Using the result that $TV^{\gamma-1}$ is constant for an adiabatic expansion, show that $V_2/V_1 = V_3/V_4$. (e) The efficiency of a Carnot cycle is defined to be the net work done divided by the heat absorbed $Q_h$. Using the first law of thermodynamics, show that the efficiency is $1 - Q_c/Q_h$. (f) Using your results from the previous parts of this problem, show that $Q_c/Q_h = T_c/T_h$.

## General Problems

**84** • **iSOLVE** ✓ The volume of three moles of a monatomic gas is increased from 50 L to 200 L at constant pressure. The initial temperature of the gas is 300 K. How much heat must be supplied to the gas?

**85** • In the process of compressing $n$ moles of an ideal diatomic gas to one-fifth of its initial volume, 180 kJ of work is done on the gas. If this is accomplished isothermally at room temperature (293 K), how many calories of heat are removed from the gas?

**86** • **SSM** What is the number of moles $n$ of the gas in Problem 85?

**87** • The $PV$ diagram in Figure 18-24 represents 3 mol of an ideal monatomic gas. The gas is initially at point A. The paths AD and BC represent isothermal changes. If the system is brought to point C along the path AEC, find (a) the initial and final temperatures, (b) the work done by the gas, and (c) the heat absorbed by the gas.

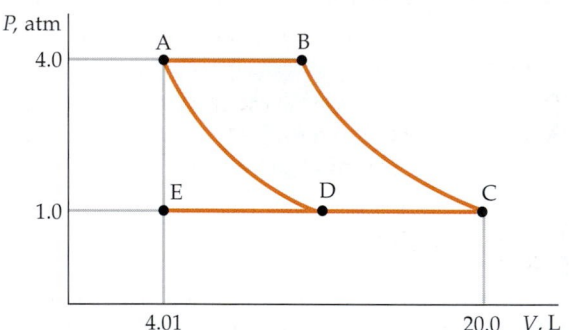

**FIGURE 18-24** Problems 87-90

**88** •• Repeat Problem 87 with the gas following path ABC.

**89** •• **SSM** Repeat Problem 87 with the gas following path ADC.

**90** •• Suppose that the paths AD and BC represent adiabatic processes. What then are the work done by the gas and the heat absorbed by the gas in following the path ABC?

**91** •• At very low temperatures, the specific heat of a metal is given by $c = aT + bT^3$. For the metal copper, $a = 0.0108$ J/kg·K² and $b = 7.62 \times 10^{-4}$ J/kg·K⁴. (a) What is the specific heat of copper at 4 K? (b) How much heat is required to heat copper from 1 to 3 K?

**92** •• Two moles of a diatomic ideal gas are compressed isothermally from 18 L to 8 L. In the process, 170 calories escape from the system. Determine the amount of work done by the gas, the change in internal energy, and the initial and final temperatures of the gas.

**93** •• Suppose the two moles of a diatomic ideal gas in Problem 92 are compressed from 18 L to 8 L adiabatically. The work done on the gas is 820 J. Find the initial temperature and the initial and final pressures.

**94** •• How much work must be done to 30 grams of CO at standard temperature and pressure to compress it to one-fifth of its initial volume if the process is (a) isothermal; (b) adiabatic?

**95** •• Repeat Problem 94 if the gas is $CO_2$.

**96** •• Repeat Problem 94 if the gas is argon.

**97** •• **iSOLVE** A thermally insulated system consists of 1 mol of a diatomic ideal gas at 100 K and 2 mol of a solid at 200 K that are separated by a rigid insulating wall. Find the equilibrium temperature of the system after the insulating wall is removed, assuming that the solid obeys the Dulong–Petit law.

**98** •• **SSM** When an ideal gas undergoes a temperature change at constant volume, its energy changes by $\Delta E_{int} = C_v \Delta T$. (a) Explain why this result holds for an ideal gas for any temperature change independent of the process. (b) Show explicitly that this result holds for the expansion of an ideal gas at constant pressure by first calculating the work done and showing that it can be written as $W = nR\Delta T$, and then by using $\Delta E_{int} = Q - W$, where $Q = C_p\Delta T$.

**99** •• One mole of an ideal monatomic gas is heated at constant volume from 300 to 600 K. (a) Find the heat added, the work done by the gas, and the change in its internal energy. (b) Find these same quantities if the gas is heated from 300 to 600 K at constant pressure.

**100** •• **SSM** Heat in the amount of 500 J is supplied to 2 mol of an ideal diatomic gas. (a) Find the change in temperature if the pressure is kept constant. (b) Find the work done by the gas. (c) Find the ratio of the final volume of the gas to the initial volume if the initial temperature is 20°C.

**101** •• An insulated cylinder is fitted with a movable piston to maintain constant pressure. The cylinder initially contains 100 g of ice at −10°C. Heat is supplied to the contents at a constant rate by a 100-W heater. Make a graph showing the temperature of the cylinder contents as a function of time starting at $t = 0$, when the temperature is −10°C, and ending when the temperature is 110°C. (Use $c = 2.0$ kJ/kg·K for the average specific heat of ice from −10 to 0°C and of steam from 100 to 110°C.)

**102** •• **SSM** Two moles of a diatomic ideal gas expand adiabatically. The initial temperature of the gas is 300 K. The work done by the gas during the expansion is 3.5 kJ. What is the final temperature of the gas?

**103** •• One mole of monatomic gas, initially at temperature $T$, undergoes a process in which its temperature is quadrupled and its volume is halved. Find the amount of heat $Q$ transferred to the gas. It is known that in this process the pressure was never less than the initial pressure, and the work done on the gas was the minimum possible.

**104** •• If a small amount of a substance is dissolved into a liquid, the liquid pressure will rise slightly. For a dilute solution, the change in pressure follows the ideal-gas law: $PV = NkT$, where $N$ is the number of solute molecules dissolved in the liquid, $V$ is the liquid volume, and $P$ is the increase in the liquid pressure. Calculate the increase in pressure when 20 g of table salt (NaCl) are dissolved in 1 L of water at a temperature of 24°C.

**105** •• A vertical heat-insulated cylinder is divided into two parts by a movable piston of mass $m$. Initially the piston is held at rest. The top part is evacuated and the bottom part is filled with 1 mole of diatomic ideal gas at temperature 300 K. After the piston is released and the system comes to equilibrium, the volume, occupied by gas, is halved. Find the final temperature of the gas.

**106** •• According to the Einstein model of a crystalline solid, the internal energy per mole is given by

$$U = \frac{3N_A kT_E}{e^{T_E/T} - 1}$$

where $T_E$ is a characteristic temperature called the Einstein temperature, and $T$ is the temperature of the solid in kelvins. Calculate the molar internal energy of diamond ($T_E = 1060$ K) at 300 K and 600 K, and thereby the increase in internal energy as diamond is heated from 300 K to 600 K.

**107** ••• **SSM** In an isothermal expansion, an ideal gas at an initial pressure $P_0$ expands until its volume is twice its initial volume. (a) Find its pressure after the expansion. (b) The gas is then compressed adiabatically and quasi-statically back to its original volume, at which point its pressure is $1.32P_0$. Is the gas monatomic, diatomic, or polyatomic? (c) How does the translational kinetic energy of the gas change in these processes?

*Note: Problems 108 and 109 involve nonquasi-static processes. Nevertheless, assuming that the gases participating in these processes approximate ideal gases, one can calculate the state functions of the end products of the reactions using the first law of thermodynamics and the ideal-gas law. For $T > 2000$ K, vibration of the atoms contributes to $C_p$ of $H_2O$ and $CO_2$ so that $C_p$ of these gases is $7.5R$ at high temperatures. Also, assume the gases do not dissociate.*

**108** ••• The combustion of benzene is represented by the chemical reaction $2(C_6H_6) + 15(O_2) \rightarrow 12(CO_2) + 6(H_2O)$. The amount of energy released in the combustion of 2 mol of benzene is 1516 kcal. One mol of benzene and 7.5 mol of oxygen at 300 K are confined in an insulated enclosure at a pressure of 1 atm. (a) Find the temperature and volume following combustion if the pressure is maintained at 1 atm. (b) If, following combustion, the thermal insulation about the container is removed and the system is cooled to 300 K, what is the final pressure?

**109** ••• **SSM** Carbon monoxide and oxygen combine to form carbon dioxide with an energy release of 280 kJ/mol of CO according to the reaction $2(CO) + O_2 \rightarrow 2(CO_2)$. Two mol of CO and one mol of $O_2$ at 300 K are confined in an 80-L container; the combustion reaction is initiated with a spark. (a) What is the pressure in the container prior to the reaction? (b) If the reaction proceeds adiabatically, what are the final temperature and pressure? (c) If the resulting $CO_2$ gas is cooled to 0°C, what is the pressure in the container?

**110 •••** Use the expression given in Problem 106 for the internal energy per mole of a solid according to the Einstein model to show that the molar heat capacity at constant volume is given by

$$c_v' = 3R\left(\frac{T_E}{T}\right)^2 \frac{e^{T_E/T}}{(e^{T_E/T} - 1)^2}$$

**111 •••** (a) Use the results of Problem 110 to show that the Dulong–Petit law, $c_v' \approx 3R$ holds for the Einstein model when $T > T_E$. (b) For diamond, $T_E$ is approximately 1060 K. Numerically integrate $dE_{int} = c_v' \, dT$ to find the increase in the internal energy if 1 mol of diamond is heated from 300 to 600 K. Compare your result to that obtained in Problem 106.

**112 •••** If a hole is punctured in a tire, the gas inside will gradually leak out of it. Let's assume the following: the area of the hole is $A$; the tire volume is $V$; and the time, $\tau$, it takes for most of the air to leak out of the tire can be expressed in terms of the ratio $A/V$, the temperature $T$, the Boltzmann constant $k$, and the mass of the gas molecules inside the tire, $m$. (a) Under these assumptions, use dimensional analysis to find an estimate for $\tau$. (b) Use the result of Part (a) to estimate the time it takes for a car tire with a nail hole punched in it to go flat.

# The Second Law of Thermodynamics

THIS OLD-FASHIONED TRAIN ENGINE PRODUCES STEAM WHICH DOES WORK ON A PISTON THAT CAUSES THE TRAIN'S WHEELS TO MOVE. THE STEAM ENGINE'S EFFICIENCY IS LIMITED BY THE SECOND LAW OF THERMODYNAMICS.

**?** **What is the maximum possible efficiency of this engine? (See Example 19-4.)**

19-1    Heat Engines and the Second Law of Thermodynamics

19-2    Refrigerators and the Second Law of Thermodynamics

19-3    Equivalence of the Heat-Engine and Refrigerator Statements

19-4    The Carnot Engine

*19-5   Heat Pumps

19-6    Irreversibility and Disorder

19-7    Entropy

19-8    Entropy and the Availability of Energy

19-9    Entropy and Probability

Solar energy is directed toward the solar oven at the center by this circular array of reflectors at Barstow, California.

**W**e are often asked to conserve energy. But according to the first law of thermodynamics, energy is always conserved. What then does it mean to conserve energy if the total amount of energy in the universe does not change regardless of what we do? The first law of thermodynamics does not tell the whole story. Energy is always conserved, but some forms of energy are more useful than others. The possibility or impossibility of putting energy to *use* is the subject of the **second law of thermodynamics.** For example, it is easy to convert work into thermal energy, but it is impossible to remove energy as heat from a single reservoir and convert it entirely into work with no other changes. This experimental fact is one statement of the second law of thermodynamics.

No system can take energy as heat from a single reservoir and convert it entirely into work without additional net changes in the system or its surroundings.

SECOND LAW OF THERMODYNAMICS: KELVIN STATEMENT

➤ **In this chapter, we will encounter several other formulations of this law.**

A common example of the conversion of work into heat is movement with friction. For example, suppose you spend two minutes pushing a block this way and that way along a tabletop in a closed path, leaving the block in its initial position. Also, suppose that the block-table system is initially in thermal equilibrium with its surroundings. The work you do on the system is converted into internal energy of the system, and as a result the block-table system becomes warmer. Consequently, the system is no longer in thermal equilibrium with its surroundings. However, the system will transfer energy as heat to its surroundings until it returns to thermal equilibrium with those surroundings. Because the final and initial states of the system are the same, the first law of thermodynamics dictates that the energy transferred to the environment as heat equals the work done by you on the system. The reverse process never occurs—a block and table that are warm will never spontaneously cool by converting their internal energy into work that causes the block to push your hand around the table! Yet such an amazing occurrence would not violate the first law of thermodynamics or any other physical laws we have encountered so far. It does, however, violate the second law of thermodynamics. Thus, there is a lack of symmetry in the roles played by heat and work that is not evident from the first law. This lack of symmetry is related to the fact that some processes are *irreversible*.

There are many other irreversible processes, seemingly quite different from one another, but all related to the second law. For example, heat conduction is an irreversible process. If we place a hot body in contact with a cold body, heat will flow from the hot body to the cold body until they are at the same temperature. However, the reverse does not occur. Two bodies in contact at the same temperature remain at the same temperature; heat does not flow from one to the other leaving one colder and the other warmer. This experimental fact gives us a second statement of the second law of thermodynamics.

A process whose only net result is to transfer energy as heat from a cooler object to a hotter one is impossible.

SECOND LAW OF THERMODYNAMICS: CLAUSIUS STATEMENT

We will show in this chapter that the Kelvin and Clausius statements of the second law are equivalent.

## 19-1 Heat Engines and the Second Law of Thermodynamics

The study of the efficiency of heat engines gave rise to the first clear statements of the second law. A **heat engine** is a cyclic device whose purpose is to convert as much heat input into work as possible. Heat engines contain a **working substance** (water in a steam engine, air and gasoline vapor in an internal-combustion engine) that absorbs a quantity of heat $Q_h$ from a high temperature reservoir, does work $W$ on its surroundings, and gives off heat $Q_c$ as it returns to its initial state, where

$Q_h$, $W$, and $Q_c$ represent magnitudes and are positive.

The earliest heat engines were steam engines, invented in the eighteenth century for pumping water from coal mines. Today steam engines are used to generate electricity. In a typical steam engine, liquid water is heated under several hundred atmospheres of pressure until it vaporizes at about 500°C (Figure 19-1). This steam expands against a piston (or turbine blades), doing work, then exits at a much lower temperature and is cooled further in the condenser where heat is transferred from it, causing it to condense. The water is then pumped back into the boiler and heated again.

Steam
Water
Heat
Work
Engine
Condenser
Heat

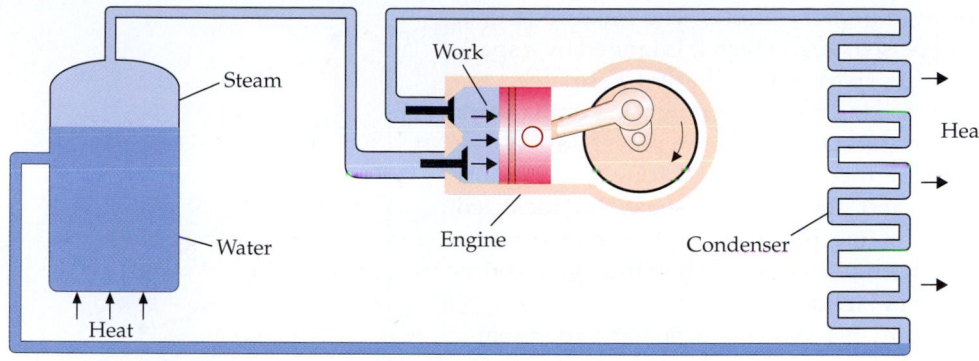

**FIGURE 19-1** Schematic drawing of a steam engine. High-pressure steam does work on the piston.

Figure 19-2 is a schematic diagram of the heat engine used in many automobiles —the internal-combustion engine. With the exhaust valve closed, a mixture of gasoline vapor and air enters the combustion chamber as the piston moves down

Intake valve open
Gas vapor and air mixture
Intake valve
Exhaust valve
A mixture of gasoline vapor and air enters the combustion chamber as the piston moves down.
Intake stroke (1)
Spark plug
Both valves closed
Cylinder
Piston
Connecting rod
Crankshaft
The piston then moves up, compressing the gas for ignition.
Compression stroke (2)
Both valves closed
When the gas ignites, it expands.
Ignition (3)
Both valves closed
The expanding gas moves the piston down, a stage called the power stroke.
Power stroke (4)
Exhaust valve open
To exhaust pipe
The piston moves up again to exhaust the burned gases.
Exhaust stroke (5)

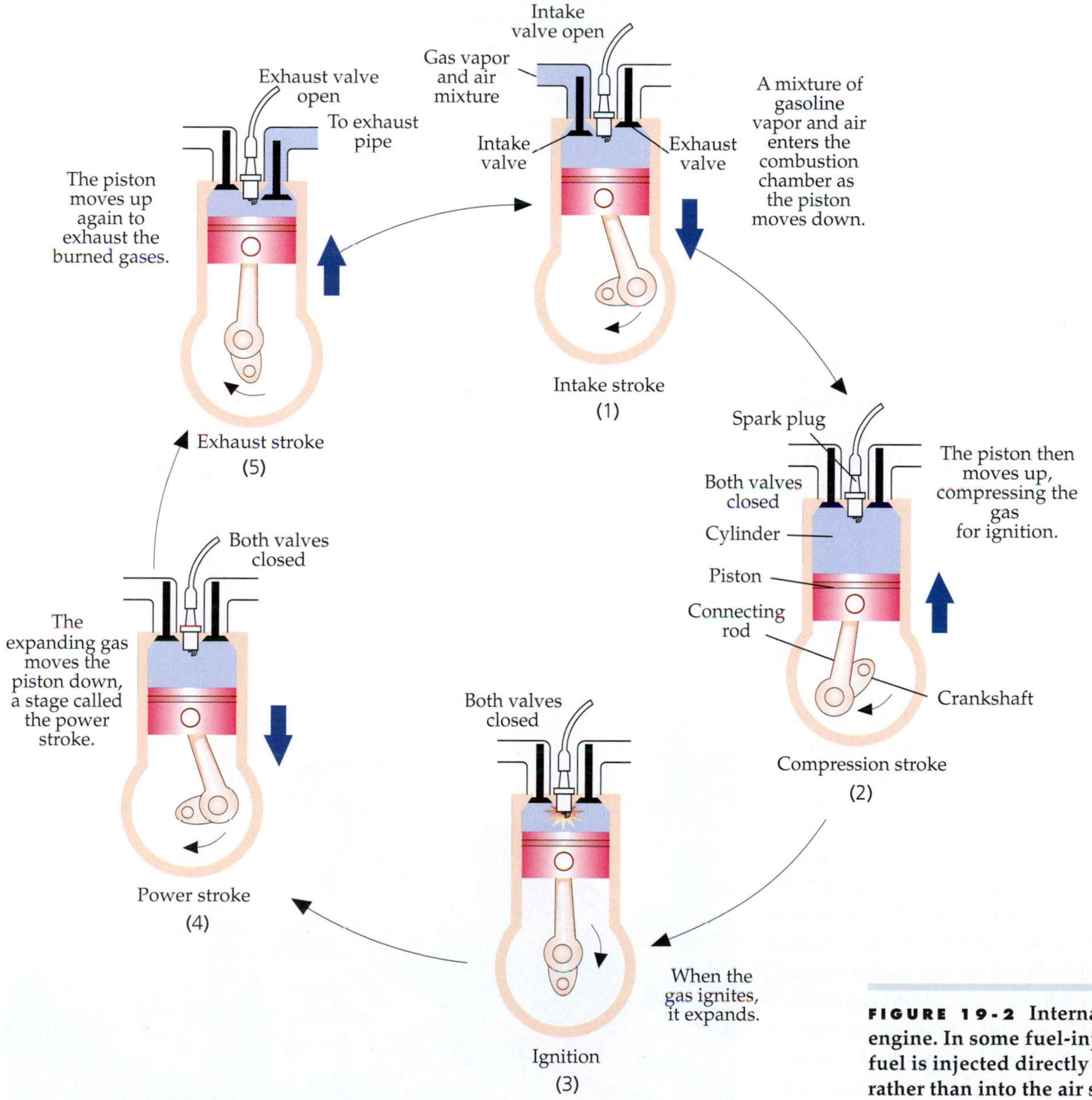

**FIGURE 19-2** Internal-combustion engine. In some fuel-injected engines, the fuel is injected directly into the cylinder rather than into the air stream.

during the intake stroke. The mixture is then compressed, after which it is ignited by a spark from the spark plug. The hot gases then expand, driving the piston down and doing work on it in the stage called the power stroke. The gases are then exhausted through the exhaust valve, and the cycle repeats. An idealized model of the processes in the internal combustion engine is called the **Otto cycle** and is shown in Figure 19-3.

Figure 19-4 shows a schematic representation of a basic heat engine. The heat input is represented as coming from a **hot heat reservoir** at temperature $T_h$, and the exhaust goes into a **cold heat reservoir** at a lower temperature $T_c$. A hot or cold heat reservoir is an idealized body or system that has a very large heat capacity so that it can absorb or give off energy as heat with no noticeable change in its temperature. In practice, burning fossil fuel often acts as the high-temperature reservoir, and the surrounding atmosphere or a lake often acts as the low-temperature reservoir. Applying the first law of thermodynamics ($\Delta E_{int} = Q_{in} + W_{on}$) to the heat engine gives

$$W = Q_h - Q_c \qquad\qquad 19\text{-}1$$

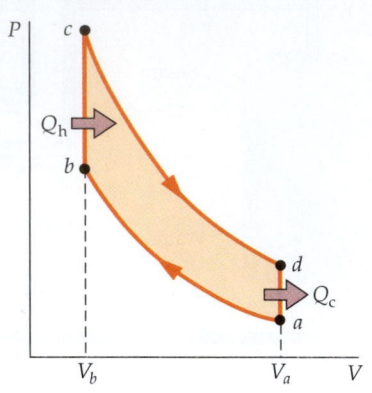

**FIGURE 19-3** Otto cycle, representing the internal-combustion engine. The fuel-air mixture enters at *a* and is adiabatically compressed to *b*. It is then heated (by ignition from the spark plug) at constant volume to *c*. The power stroke is represented by the adiabatic expansion from *c* to *d*. The cooling at constant volume from *d* to *a* represents the exhausting of the combustion products and the intake of a fresh fuel-air mixture.

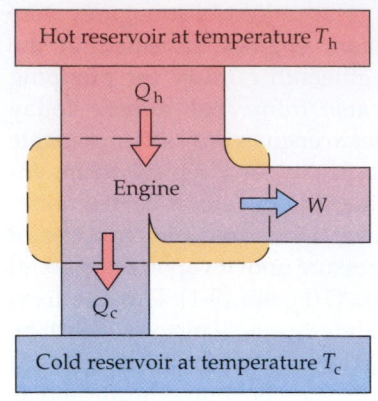

**FIGURE 19-4** Schematic representation of a heat engine. The engine removes heat energy $Q_h$ from a hot reservoir at a temperature $T_h$, does work $W$, and gives off heat $Q_c$ to a cold reservoir at a temperature $T_c$.

where $W$ is the work done *by* the engine during one complete cycle, $Q_h - Q_c$ is the total energy transferred to the engine as heat during one cycle, and $\Delta E_{int}$ is the change in internal energy of the engine (including the working substance) during one cycle. Since the initial and final states of the engine for a complete cycle are the same, the initial and final internal energies of the engine are equal. Thus, $\Delta E_{int} = 0$.

The **efficiency** $\varepsilon$ of a heat engine is defined as the ratio of the work done by the engine to the heat absorbed from the high temperature reservoir:

$$\varepsilon = \frac{W}{Q_h} = \frac{Q_h - Q_c}{Q_h} = 1 - \frac{Q_c}{Q_h} \qquad\qquad 19\text{-}2$$

DEFINITION—EFFICIENCY OF A HEAT ENGINE

The heat $Q_h$ is usually produced by burning some fuel like coal or oil that must be paid for, so it is desirable to get the most efficient use of the fuel as possible. The best steam engines operate near 40 percent efficiency; the best internal-combustion engines operate near 50 percent efficiency. At 100 percent efficiency ($\varepsilon = 1$), all the thermal energy absorbed from the hot reservoir would be converted into work and no thermal energy would be given off to the cold reservoir. However, *it is impossible to make a heat engine with an efficiency of 100 percent.* This experimental result is the **heat-engine statement of the second law of thermodynamics.** It is another way of expressing the Kelvin statement given earlier:

An exhaust manifold feeds the header pipes seen on this top-fuel dragster in order to carry heat away from the engine to reduce its temperature.

> It is impossible for a heat engine working in a cycle to produce *only the effect* of extracting heat from a single reservoir and performing an equivalent amount of work.

SECOND LAW OF THERMODYNAMICS: HEAT-ENGINE STATEMENT

**❶ The word *cycle* in this statement is important because it *is* possible to convert heat completely into work in a noncyclic process.** An ideal gas undergoing an isothermal expansion does just this. But after the expansion, the gas is not in its original state. To bring the gas back to its original state, work must be done on the gas, and some heat will be exhausted.

The second law tells us that to do work with energy extracted from a heat reservoir, we must have a colder reservoir available to receive part of the energy as exhaust. If this were not true, we could design a ship with a heat engine that was powered by simply extracting energy as heat from the ocean. Unfortunately, the lack of a colder reservoir for exhaust makes this enormous reservoir of energy unavailable for such use. (It is theoretically possible to run a heat engine between the warmer surface water of the ocean and the colder water at greater depths, but no practical scheme for using this temperature difference has yet emerged.) In order to convert completely disordered thermal energy at a single temperature into the completely ordered energy associated with work (with no other changes in the source or object), a separate cold reservoir must be used.

*EFFICIENCY OF A HEAT ENGINE*                 **EXAMPLE 19-1**

**During each cycle a heat engine absorbs 200 J of heat from a hot reservoir, does work, and exhausts 160 J to a cold reservoir. What is the efficiency of the engine?**

**PICTURE THE PROBLEM** We use the definition of the efficiency of a heat engine (Equation 19-2).

1. The efficiency is the work done divided by the heat absorbed:

$$\varepsilon = \frac{W}{Q_h}$$

2. The heat absorbed is given:

$$Q_h = 200 \text{ J}$$

3. The work is found from the first law:

$$W = Q_h - Q_c = 200 \text{ J} - 160 \text{ J} = 40 \text{ J}$$

4. Substitute the values of $Q_h$ and $W$ to calculate the efficiency:

$$\varepsilon = \frac{W}{Q_h} = \frac{40 \text{ J}}{200 \text{ J}} = 0.20 = \boxed{20\%}$$

**EXERCISE** A heat engine has an efficiency of 35%. (*a*) How much work does it perform in a cycle if it extracts 150 J of energy as heat from a hot reservoir per cycle? (*b*) How much energy as heat is exhausted to the cold reservoir per cycle? (*Answer* (*a*) 52.5 J (*b*) 97.5 J)

*EFFICIENCY OF AN IDEAL INTERNAL COMBUSTION ENGINE—*   **EXAMPLE 19-2**   **Try It Yourself**
*THE OTTO CYCLE*

(*a*) **Find the efficiency of the Otto cycle shown in Figure 19-3.** (*b*) **Express your answer in terms of the ratio of the volumes** $r = V_a/V_b = V_d/V_c$.

**PICTURE THE PROBLEM** (*a*) To find $\varepsilon$, you need to find $Q_h$ and $Q_c$. Heat transfer occurs only during the two constant-volume processes, *b* to *c* and *d* to *a*. You can thus find $Q_h$ and $Q_c$ and therefore $\varepsilon$ in terms of the temperatures $T_a$, $T_b$, $T_c$, and $T_d$. (*b*) The temperatures can be related to the volumes using $TV^{\gamma-1} =$ constant for adiabatic processes.

Cover the column to the right and try these on your own before looking at the answers.

**Steps**

**Answers**

(a) 1. Write the efficiency in terms of $Q_h$ and $Q_c$.

$$\varepsilon = 1 - \frac{Q_{cold}}{Q_{hot}} = 1 - \frac{Q_c}{Q_h}$$

2. The heat out occurs at constant volume from $d$ to $a$. Write $Q_c$ in terms of $C_v$ and the temperatures $T_a$ and $T_d$.

$$Q_c = |Q_{d\to a}| = C_v|T_a - T_d| = C_v(T_d - T_a)$$

3. The heat in occurs at constant volume from $b$ to $c$. Write $Q_h$ in terms of $C_v$ and the temperatures $T_c$ and $T_b$.

$$Q_h = Q_{b\to c} = C_v(T_c - T_b)$$

4. Substitute these values of $Q_c$ and $Q_h$ to find the efficiency in terms of the temperatures $T_a$, $T_b$, $T_c$, and $T_d$.

$$\varepsilon = \boxed{1 - \frac{T_d - T_a}{T_c - T_b}}$$

(b) 1. Relate $T_c$ to $T_d$ using $TV^{\gamma-1}$ = constant, and $V_a/V_c = r$.

$$T_c V_c^{\gamma-1} = T_d V_d^{\gamma-1}$$

$$T_c = T_d \frac{V_d^{\gamma-1}}{V_c^{\gamma-1}} = T_d r^{\gamma-1}$$

2. Relate $T_b$ to $T_a$ as in step 1.

$$T_b = T_a r^{\gamma-1}$$

3. Use these relations to eliminate $T_c$ and $T_b$ from $\varepsilon$ in Part (a) so that $\varepsilon$ is expressed in terms of $r$.

$$\varepsilon = 1 - \frac{T_d - T_a}{T_d r^{\gamma-1} - T_a r^{\gamma-1}} = \boxed{1 - \frac{1}{r^{\gamma-1}}}$$

**REMARKS** The ratio $r$ (volume before compression/volume after compression) is called the compression ratio.

# 19-2 Refrigerators and the Second Law of Thermodynamics

A **refrigerator** is essentially a heat engine run backwards (Figure 19-5a.). The refrigerator's engine extracts thermal energy from the interior of the refrigerator (cold reservoir) and transfers it to the surroundings (hot reservoir) (Figure 19-5b). Experience shows that such a transfer always requires work—a result known as the **refrigerator statement of the second law of thermodynamics,** which is another way of expressing the Clausius statement:

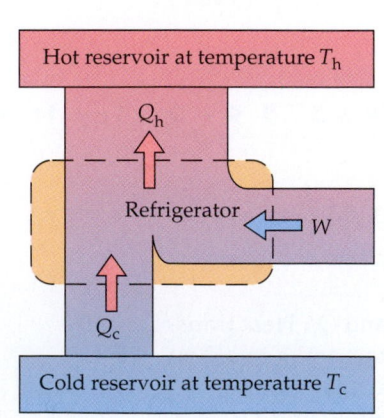

**FIGURE 19-5**
(a) Schematic representation of a refrigerator. Work $W$ is done on the refrigerator and it removes heat energy $Q_c$ from a cold reservoir and gives off heat $Q_h$.
(b) An actual refrigerator.

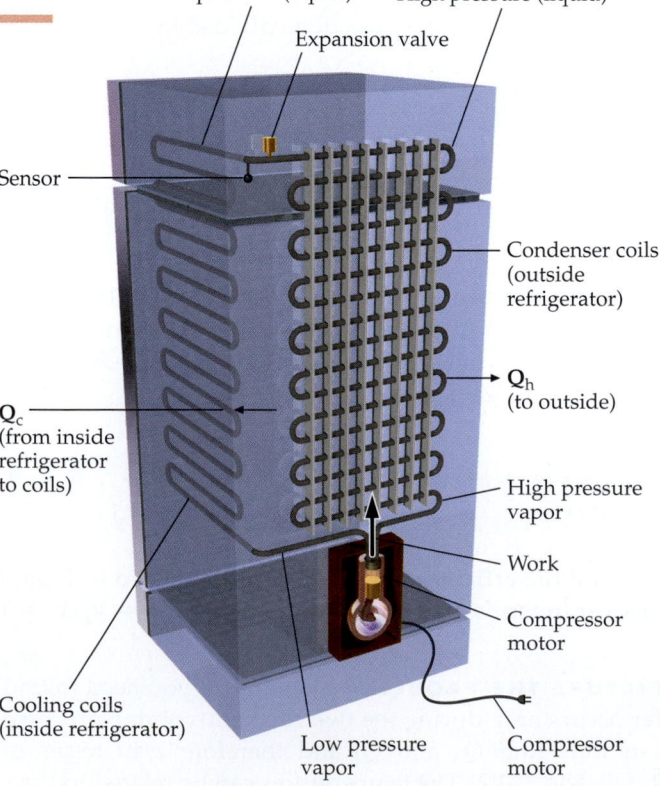

It is impossible for a refrigerator working in a cycle to produce *only the effect* of extracting heat from a cold object and reject the same amount of heat to a hot object.

SECOND LAW OF THERMODYNAMICS: REFRIGERATOR STATEMENT

Were the above statement not true, we could cool our homes in the summer with refrigerators that pumped thermal energy to the outside without using any electricity or any other energy.

A measure of a refrigerator's performance is the ratio $Q_c/W$ of the heat removed from the low temperature reservoir to the work done on the refrigerator. (This work equals the electrical energy that comes from the wall outlet.) The ratio $Q_c/W$ is called the **coefficient of performance** (COP):

$$\text{COP} = \frac{Q_c}{W} \qquad\qquad 19\text{-}3$$

DEFINITION—COEFFICIENT OF PERFORMANCE (REFRIGERATOR)

The greater the COP, the better the refrigerator. Typical refrigerators have coefficients of performance of about 5 or 6. In terms of this ratio, the refrigerator statement of the second law says that the COP of a refrigerator cannot be infinite.

---

*MAKING ICE CUBES*         **EXAMPLE 19-3**    **Put It in Context**

**You have one hour before guests start arriving for your party when you suddenly realize that you forgot to buy ice for drinks. You quickly put one liter of water at 10°C into your ice cube trays and pop them into the freezer. Will you have ice in time for your guests? The label on your refrigerator states that the appliance has a coefficient of performance of 5.5 and a power rating of 550 W. You estimate that only 10 percent of the power goes to freezing the ice.**

**PICTURE THE PROBLEM** Work equals power times time. We are given the power, so we need to find the work to determine the time. The work is related to $Q_c$ by Equation 19-3. To find $Q_c$ we calculate how much heat must be extracted from the water.

1. The time needed is related to the power available and the work required:

$$P = W/t$$
$$t = W/P$$

2. The work is related to the coefficient of performance and the heat extracted:

$$W = \frac{Q_c}{\text{COP}}$$

3. The heat $Q_c$ removed from inside of the refrigerator equals the heat $Q_{cool}$ to be removed from the water to cool it plus the heat $Q_{freeze}$ to be removed from the water to freeze it:

$$Q_c = Q_{cool} + Q_{freeze}$$

4. The heat needed to cool 1 L of water (mass 1 kg) by 10°C is:

$$Q_{cool} = mc\Delta T$$
$$= (1\text{ kg})(4.18\text{ kJ}/(\text{kg·K}))(10\text{ K})$$
$$= 41.8\text{ kJ}$$

5. The heat needed to freeze 1 L of water into ice cubes is:

$$Q_{freeze} = mL_f = (1\text{ kg})(333.5\text{ kJ}/\text{kg}) = 333.5\text{ kJ}$$

6. Add these heats to obtain $Q_c$:

$$Q_c = 41.8 \text{ kJ} + 333.5 \text{ kJ} = 375.3 \text{ kJ} \approx 375 \text{ kJ}$$

7. Substitute $Q_c$ into step 2 to find the work $W$:

$$W = \frac{Q_c}{\text{COP}} = \frac{375 \text{ kJ}}{5.5} = 68.2 \text{ kJ}$$

8. Use this value of $W$ and 55 W for the available power to find the time $t$:

$$t = \frac{W}{P} = \frac{68.2 \text{ kJ}}{55 \text{ J/s}} = 1.24 \text{ ks} \times \frac{1 \text{ min}}{60 \text{ s}} = \boxed{20.7 \text{ min}}$$

**REMARKS** Relax! You'll have ice in well under an hour.

**EXERCISE** A refrigerator has a coefficient of performance of 4.0. How much heat is exhausted to the hot reservoir if 200 kJ of heat are removed from the cold reservoir? (*Answer* 250 kJ)

# 19-3 Equivalence of the Heat-Engine and Refrigerator Statements

The heat-engine and refrigerator statements (or the Kelvin and Clausius statements, respectively) of the second law of thermodynamics seem quite different, but they are actually equivalent. We can prove this by showing that if either statement is assumed to be false, then the other must also be false. We'll use a numerical example to show that if the heat-engine statement is false, then the refrigerator statement is false.

Figure 19-6a shows an ordinary refrigerator that uses 50 J of work to remove 100 J of energy as heat from a cold reservoir and rejects 150 J of energy as heat to a hot reservoir. Suppose the heat-engine statement of the second law were not true. Then a "perfect" heat engine could remove energy from the hot reservoir and convert it completely into work with 100 percent efficiency. We could use this perfect heat engine to remove 50 J of energy from the hot reservoir and do 50 J of work (Figure 19-6b) on the ordinary refrigerator. Then, the combination of the perfect heat engine and the ordinary refrigerator would be a perfect refrigerator, transferring 100 J of energy as heat from the cold reservoir to the hot reservoir without requiring any work, as illustrated in Figure 19-6c. This violates the refrigerator statement of the second law. Thus, if the heat-engine statement is false, the refrigerator statement is also false. Similarly, if a perfect refrigerator existed, it could be used in conjunction with an ordinary heat engine to construct a perfect heat engine. Thus, if the refrigerator statement is false, the heat-engine statement is also false. It then follows that if one statement is true, the other is also true. Therefore, the heat engine statement and the refrigerator statement are equivalent.

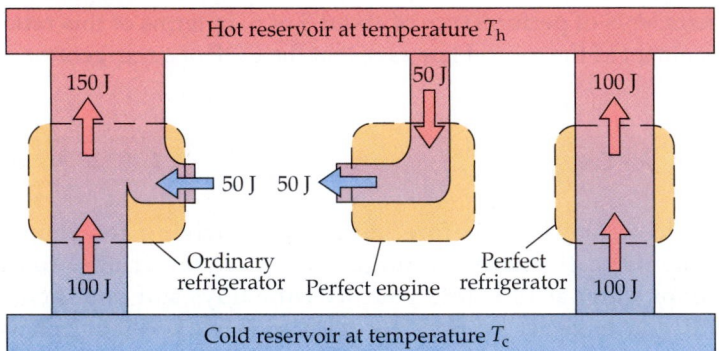

(a)
An ordinary refrigerator removes 100 J from a cold reservoir, requiring the input of 50 J of work.

(b)
A perfect heat engine violates the heat engine statement of the second law by removing 50 J from the hot reservoir and converting it completely into work.

(c)
Putting the two together makes a perfect refrigerator that violates the refrigerator statement of the second law by transferring 100 J from the cold reservoir to the hot reservoir with no other effect.

**FIGURE 19-6** Demonstration of the equivalence of the heat-engine and refrigerator statements of the second law of thermodynamics.

# 19-4 The Carnot Engine

According to the second law of thermodynamics, it is impossible for a heat engine working between two heat reservoirs to be 100% efficient. What, then, is the maximum possible efficiency for such an engine? A young French engineer,

Sadi Carnot answered this question in 1824, before either the first or the second law of thermodynamics had been established. Carnot found that a *reversible engine* is the most efficient engine that can operate between any two given reservoirs. This result is known as the Carnot theorem:

No engine working between two given heat reservoirs can be more efficient than a reversible engine working between those two reservoirs.

CARNOT THEOREM

A reversible engine working in a cycle between two heat reservoirs is called a **Carnot engine**, and its cycle is called a **Carnot cycle.** Figure 19-7 illustrates the Carnot theorem with a numerical example.

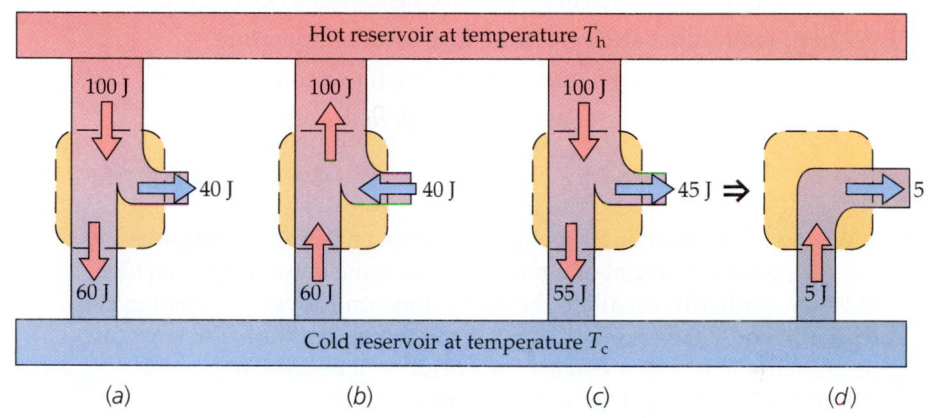

If no engine can have a greater efficiency than a Carnot engine, it follows that all Carnot engines working between the same two reservoirs have the same efficiency. This efficiency, called the **Carnot efficiency,** must be independent of the working substance of the engine and thus can depend only on the temperatures of the reservoirs.

**FIGURE 19-7** Illustration of the Carnot theorem. (*a*) A reversible heat engine with 40 percent efficiency removes 100 J from a hot reservoir, does 40 J work, and exhausts 60 J to the cold reservoir. (*b*) If the same engine runs backwards as a refrigerator, 40 J of work are done to remove 60 J from the cold reservoir and exhaust 100 J to the hot reservoir. (*c*) An assumed heat engine working between the same two reservoirs with an efficiency of 45 percent which is greater than that of the reversible engine in (*a*). (*d*) The net effect of running the engine in (*c*) in conjunction with the refrigerator in (*b*) is the same as that of a perfect heat engine that removes 5 J from the cold reservoir and converts it completely into work with no other effect, violating the second law of thermodynamics. Thus, the reversible engine in (*a*) is the most efficient engine that can operate between these two reservoirs.

Let us look at what makes a process reversible or irreversible. According to the second law, heat flows from hot objects to cold objects and never the other way around. Thus, the conduction of heat from a hot object to a cold one is *not* reversible. Also, friction can transform work into heat, but friction can never transform heat into work. The conversion of work into heat via friction is *not* reversible. Friction and other dissipative forces irreversibly transform mechanical energy into thermal energy. A third type of irreversibility occurs when a system passes through nonequilibrium states, such as when there is turbulence in a gas or when a gas explodes. For a process to be reversible, we must be able to move the system back through the same equilibrium states in the reverse order.

From these considerations and our statements of the second law of thermodynamics, we can list some conditions that are necessary for a process to be reversible:

1. No mechanical energy is transformed into thermal energy by friction, viscous forces, or other dissipative forces.

2. Energy transfer as heat can only occur between objects at the same temperature (or infinitesimally near the same temperature).

3. The process must be quasi-static so that the system is always in an equilibrium state (or infinitesimally near an equilibrium state).

CONDITIONS FOR REVERSIBILITY

Any process that violates any of the above conditions is irreversible. Most processes in nature are irreversible. To have a reversible process, great care must be taken to eliminate frictional and other dissipative forces and to make the process quasi-static. Because this can never be completely accomplished, a reversible process is an idealization similar to the idealization of motion without friction in mechanics problems. Reversibility can, nevertheless, be closely approximated in practice.

We can now understand the features of a Carnot cycle, which is a reversible cycle between two reservoirs only. Because all heat transfer must be done isothermally in order for the process to be reversible, the heat absorbed from the hot reservoir must be absorbed isothermally. The next step is a quasi-static adiabatic expansion to the lower temperature of the cold reservoir. Next, heat is given off isothermally to the cold reservoir. Finally, there is a quasi-static, adiabatic compression to the higher temperature of the hot reservoir. The Carnot cycle thus consists of four reversible steps:

1. A quasi-static isothermal absorption of heat from a hot reservoir
2. A quasi-static adiabatic expansion to a lower temperature
3. A quasi-static isothermal exhaustion of heat to a cold reservoir
4. A quasi-static adiabatic compression back to the original state

STEPS IN A CARNOT CYCLE

One way to calculate the efficiency of a Carnot engine is to choose as the working substance a material of which we have some knowledge—an ideal gas, and then explicitly calculate the work done on it over a Carnot cycle (Figures 19-8a and 8b). Since all Carnot cycles have the same efficiency independent of the working substance, our result will be valid in general.

The efficiency of the Carnot cycle (Equation 19-2) is

$$\varepsilon = 1 - \frac{Q_c}{Q_h}$$

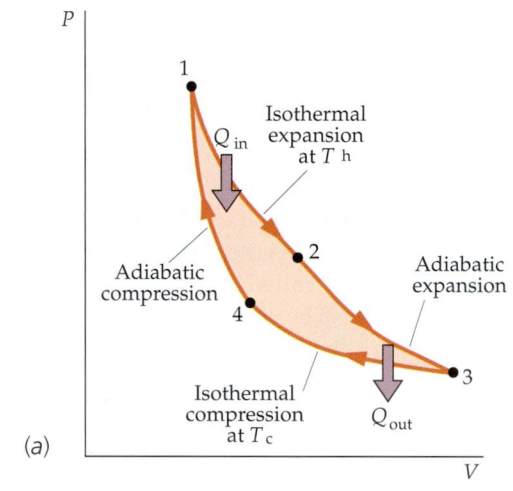

(a)

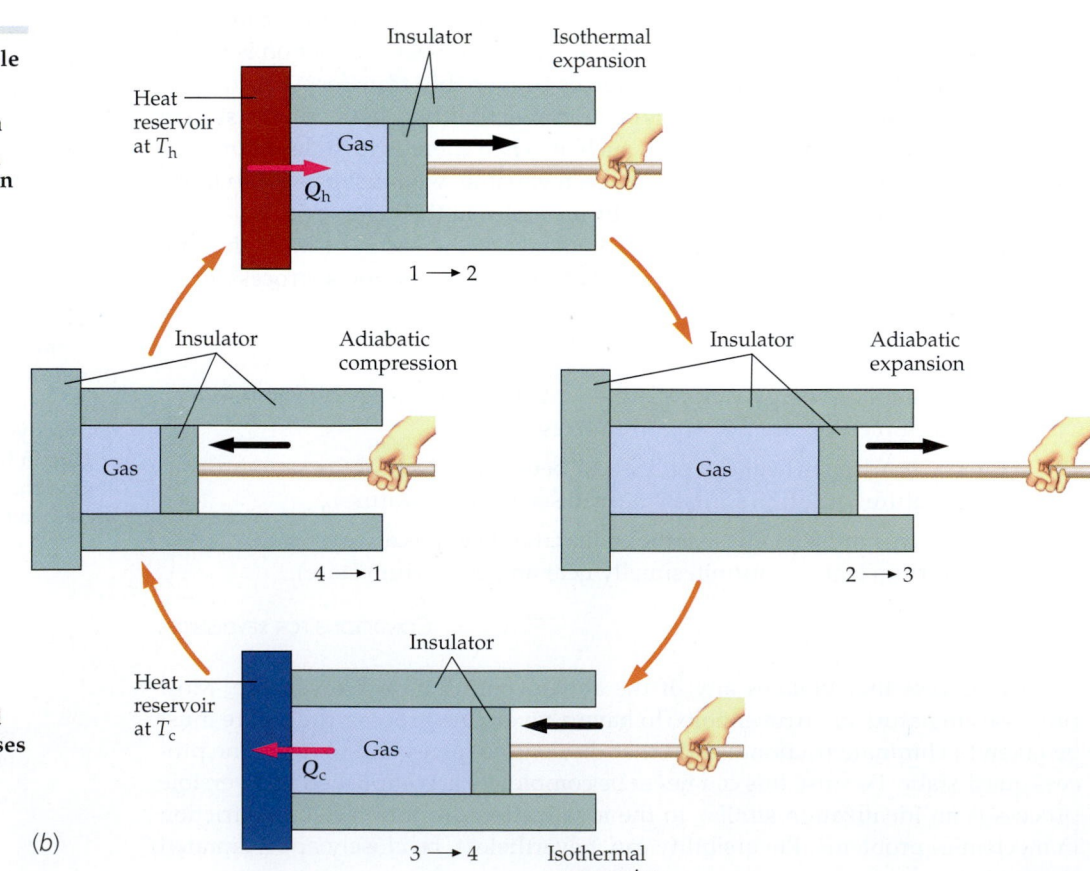

**FIGURE 19-8** (a) Carnot cycle for an ideal gas:

*Step 1:* Heat is absorbed from a hot reservoir at temperature $T_h$ during an isothermal expansion from state 1 to state 2.

*Step 2:* The gas expands adiabatically from state 2 to state 3, reducing its temperature to $T_c$.

*Step 3:* The gas gives off heat to the cold reservoir as it is compressed isothermally at $T_c$ from state 3 to state 4.

*Step 4:* The gas is compressed adiabatically until its temperature is again $T_h$.

(b) Work is done on the gas or by the gas during each step. The net work done during the cycle is represented by the shaded area. All processes are reversible. All steps are quasi-static.

(b)

The heat $Q_h$ is absorbed during the isothermal expansion from state 1 to state 2. The first law of thermodynamics is $\Delta E_{int} = Q_{in} + W_{on}$. For an isothermal expansion of an ideal gas $\Delta E_{int} = 0$. Applying the first law to the isothermal expansion from state 1 to state 2 we have $Q_h = Q_{in}$, so $Q_h$ equals the work done by the gas.

$$Q_h = W_{by\ gas} = \int_{V_1}^{V_2} P\,dV = \int_{V_1}^{V_2} \frac{nRT_h}{V}\,dV = nRT_h \int_{V_1}^{V_2} \frac{dV}{V} = nRT_h \ln \frac{V_2}{V_1}$$

Similarly, the heat given off to the cold reservoir equals the work done on the gas during the isothermal compression at temperature $T_c$ from state 3 to state 4. This work has the same magnitude as that done by the gas if it expands from state 4 to state 3. The heat rejected is thus

$$Q_c = W_{on\ gas} = nRT_c \ln \frac{V_3}{V_4}$$

The ratio of these heats is

$$\frac{Q_c}{Q_h} = \frac{T_c \ln \dfrac{V_3}{V_4}}{T_h \ln \dfrac{V_2}{V_1}} \qquad\qquad 19\text{-}4$$

We can relate the ratios $V_2/V_1$ and $V_3/V_4$ using Equation 18-37 for a quasi-static adiabatic expansion. For the expansion from state 2 to state 3, we have

$$T_h V_2^{\gamma-1} = T_c V_3^{\gamma-1}$$

Similarly, for the adiabatic compression from state 4 to state 1, we have

$$T_h V_1^{\gamma-1} = T_c V_4^{\gamma-1}$$

Dividing these two equations, we obtain

$$\left(\frac{V_2}{V_1}\right)^{\gamma-1} = \left(\frac{V_3}{V_4}\right)^{\gamma-1} \Rightarrow \frac{V_2}{V_1} = \frac{V_3}{V_4}$$

Coal-fueled electric generating plant at Four Corners, New Mexico.

Power plant at Wairakei, New Zealand, that converts geothermal energy into electricity.

Solar energy is focused and collected individually to produce electricity by these heliostats that are being tested at Sandia National Laboratory.

Control rods are inserted into this nuclear reactor at Tihange, Belgium.

Thus, Equation 19-4 gives

$$\frac{Q_c}{Q_h} = \frac{T_c \ln \dfrac{V_2}{V_1}}{T_h \ln \dfrac{V_2}{V_1}} = \frac{T_c}{T_h} \qquad \text{19-5}$$

The Carnot efficiency $\varepsilon_c$ is thus

$$\varepsilon_C = 1 - \frac{T_c}{T_h} \qquad \text{19-6}$$

CARNOT EFFICIENCY

Equation 19-6 demonstrates that because the Carnot efficiency must be independent of the working substance of any particular engine, it depends only on the temperatures of the two reservoirs.

An experimental wind-powered electric generator at Sandia National Laboratory. The propeller is designed for optimum transfer of wind energy to mechanical energy.

*EFFICIENCY OF A STEAM ENGINE* **EXAMPLE 19-4**

A steam engine works between a hot reservoir at 100°C (373 K) and a cold reservoir at 0°C (273 K). (*a*) What is the maximum possible efficiency of this engine? (*b*) If the engine is run backwards as a refrigerator, what is its maximum coefficient of performance?

**PICTURE THE PROBLEM** The maximum efficiency is the Carnot efficiency given by Equation 19-6. To find the maximum COP, we use the definition of efficiency ($\varepsilon = W/Q_h$), the definition of COP (COP = $Q_c/W$), and Equation 19-5.

(a) The maximum efficiency is the Carnot efficiency:

$$\varepsilon_{max} = \varepsilon_C = 1 - \frac{T_c}{T_h} = 1 - \frac{273 \text{ K}}{373 \text{ K}}$$

$$= 0.268 = \boxed{26.8\%}$$

(b) 1. Write the expression for the COP if the engine is run in reverse for a single cycle:

$$COP = \frac{Q_c}{W}$$

2. Write the expression for the efficiency if the engine is run forward for a single cycle. (Since for maximum possible performance the engine is reversible, the values for $Q_h$, $Q_c$, and $W$ are the same whether the engine is run backward or forward.):

$$\varepsilon = \frac{W}{Q_h}$$

3. Solve the step 2 result for the work and substitute it into the step 1 result:

$$COP = \frac{Q_c}{W} = \frac{Q_c}{\varepsilon Q_h}$$

4. Using Equation 19-5 and the Part (a) result, solve for the COP:

$$COP = \frac{Q_c}{\varepsilon Q_h} = \frac{T_c}{\varepsilon T_h} = \frac{273 \text{ K}}{0.268(373 \text{ K})}$$

$$COP = \boxed{2.73}$$

**REMARKS** Even though this maximum efficiency seems to be quite low, it is the greatest efficiency possible for any engine working between these temperatures. Real engines will have lower efficiencies because of friction, heat conduction, and other irreversible processes. Real refrigerators will have a lower coefficient of performance. It can be shown that the coefficient of performance of a Carnot refrigerator is $T_c/\Delta T$.

The Carnot efficiency gives us an upper limit on possible efficiencies, and is therefore useful to know. For example, we calculated in Example 19-4 that the Carnot efficiency is 26.8 percent. This means that, no matter how much we reduce friction and other irreversible losses, the best efficiency obtained between reservoirs at 373 K and 273 K is 26.8 percent. We would know, then, that an engine working between those two temperatures with an efficiency of 25 percent is a very good engine!

*WORK LOST BY AN ENGINE*                    **EXAMPLE 19-5**

An engine removes 200 J from a hot reservoir at 373 K, does 48 J of work, and exhausts 152 J to a cold reservoir at 273 K. How much work is "lost" per cycle due to irreversible processes in this engine?

**PICTURE THE PROBLEM** The difference between maximum amount of work that could be done using a Carnot engine and 48 J is the work lost.

1. The work lost is the maximum amount of work that could be done minus the work actually done:

$$W_{lost} = W_{max} - W$$

2. The maximum amount of work that could be done is the work done using a Carnot engine:

$$W_{max} = \varepsilon_C Q_h$$

3. The work lost is then:

$$W_{lost} = \varepsilon_C Q_h - W$$

4. The Carnot efficiency can be expressed in terms of the temperatures:

$$\varepsilon_C = 1 - \frac{T_c}{T_h}$$

5. Substituting for $\varepsilon_C$ gives:

$$W_{lost} = \left(1 - \frac{T_c}{T_h}\right)Q_h - W$$

$$= \left(1 - \frac{273\ K}{373\ K}\right)(200\ J) - 48\ J$$

$$= \boxed{5.6\ J}$$

**REMARKS** The 5.6 J of energy in the answer is not "lost" to the universe—total energy is conserved. That 5.6 J of energy exhausted into the cold reservoir by the non-ideal engine of the problem is only lost in that it would have been converted into useful work if an ideal (reversible) engine had been used.

---

*WORK LOST BETWEEN HEAT RESERVOIRS*                   **EXAMPLE 19-6**

**If 200 J of heat are conducted from a heat reservoir at 373 K to one at 273 K, how much work capability is "lost" in this process?**

We saw in the previous example that a Carnot engine working between these two reservoirs could do 53.6 J of work if it extracted 200 J from the 373-K reservoir and exhausted to a 273-K reservoir. Thus, if 200 J is conducted directly from the hot reservoir to the cold reservoir without any work being done, 53.6 J of this energy has been "lost" in the sense that it could have been converted into useful work.

**EXERCISE** A Carnot engine works between heat reservoirs at 500 K and 300 K. (*a*) What is its efficiency? (*b*) If it removes 200 kJ of heat from the hot reservoir, how much work does it do? (*Answer* (*a*) 40% (*b*) 80 kJ)

**EXERCISE** A real engine works between heat reservoirs at 500 K and 300 K. It removes 500 kJ of heat from the hot reservoir and does 150 kJ of work during each cycle. What is its efficiency? (*Answer* 30%)

## The Thermodynamic or Absolute Temperature Scale

In Chapter 17, the ideal-gas temperature scale was defined in terms of the properties of gases at low densities. Because the Carnot efficiency depends only on the temperatures of the two heat reservoirs, it can be used to define the ratio of the temperatures of the reservoirs independent of the properties of any substance. We *define* the ratio of the thermodynamic temperatures of the hot and cold reservoirs to be

$$\frac{T_c}{T_h} = \frac{Q_c}{Q_h} \qquad \qquad 19\text{-}7$$

DEFINITION OF THERMODYNAMIC TEMPERATURE

where $Q_h$ is the energy removed from the hot reservoir and $Q_c$ is the energy exhausted to the cold reservoir by a Carnot engine working between the two reservoirs. Thus, to find the ratio of two reservoir temperatures, we set up a reversible engine operating between them and measure the energy transferred as heat to or from each reservoir during one cycle. The **thermodynamic temperature** is completely specified by Equation 19-7 *and* the choice of one fixed point. If the fixed point is defined to be 273.16 K for the triple point of water, then the thermodynamic temperature scale matches the ideal-gas temperature scale for the range of temperatures over which a gas thermometer can be used. Any temperature that reads zero at absolute zero is called an *absolute temperature scale*.

## *19-5  Heat Pumps

A **heat pump** is a refrigerator with a different objective. Typically, the objective of a refrigerator is to cool an object or region of interest. The objective of a heat pump, however, is to heat an object or region of interest. For example, if you use a heat pump to heat your house you transfer heat from the cold air outside the house to the warmer air inside it. Your objective is to heat the region inside your house. If work $W$ is done on a heat pump to remove heat $Q_c$ from the cold reservoir and reject heat $Q_h$ to the hot reservoir, the coefficient of performance for a heat pump is defined as

$$COP_{HP} = \frac{Q_h}{W}$$

19-8

DEFINITION—COEFFICIENT OF PERFORMANCE (HEAT PUMP)

This coefficient of performance differs from that for the refrigerator, which is $Q_c/W$ (Equation 19-3). Using $W = Q_h - Q_c$, this can be written

$$COP_{HP} = \frac{Q_h}{Q_h - Q_c} = \frac{1}{1 - \dfrac{Q_c}{Q_h}}$$

19-9

The maximum coefficient of performance is obtained using a Carnot heat pump. Then $Q_c$ and $Q_h$ are related by Equation 19-5. Substituting $Q_c/Q_h = T_c/T_h$ into Equation 19-9, we obtain for the maximum coefficient of performance

$$COP_{HP\,max} = \frac{1}{1 - \dfrac{T_c}{T_h}} = \frac{T_h}{T_h - T_c} = \frac{T_h}{\Delta T}$$

19-10

where $\Delta T$ is the difference in temperature between the hot and cold reservoirs. Real heat pumps have coefficients of performance less than the $COP_{HP\,max}$ because of friction, heat conduction, and other irreversible processes.

The two coefficients are related. Using $Q_h = Q_c + W$, we can relate Equations 19-3 and 19-10:

$$COP_{HP} = \frac{Q_h}{W} = \frac{Q_c + W}{W} = 1 + \frac{Q_c}{W} = 1 + COP$$

19-11

---

AN *IDEAL HEAT PUMP*                    **EXAMPLE  19-7**  **Try It Yourself**

An ideal heat pump is used to pump heat from the outside air at −5°C to the hot-air supply for the heating fan in a house, which is at 40°C. How much work is required to pump 1 kJ of heat into the house?

**PICTURE THE PROBLEM** Use Equation 19-11 with $COP_{HP\,max}$ calculated from Equation 19-10 for $T_c = -5°C = 268$ K and $\Delta T = 45$ K.

| Steps | Answers |
|---|---|
| 1. Calculate the work from Equation 19-8: | $W = \dfrac{Q_h}{COP_{HP}}$ |
| 2. Calculate the $COP_{HP}$ from Equation 19-10: | $COP_{HP} = COP_{HP\,max} = \dfrac{T_h}{\Delta T}$ |
| 3. Solve for the work: | $W = \dfrac{Q_h}{COP_{HP}} = Q_h \dfrac{\Delta T}{T_h} = (1\ kJ)\dfrac{45\ K}{313\ K}$ |
| | $W = \boxed{0.144\ kJ}$ |

**REMARKS** The $COP_{HP\,max} = T_h/\Delta T = 6.96$. That is, the energy transferred inside the house as heat is 6.96 times larger than the work done. (Only 0.144 kJ of work is needed to pump 1 kJ of heat into the hot-air supply in the house.)

## 19-6 Irreversibility and Disorder

There are many irreversible processes that cannot be described by the heat-engine or refrigerator statements of the second law, such as a glass falling to the floor and breaking or a balloon popping. However, all irreversible processes have one thing in common—the system plus its surroundings moves toward a less ordered state.

Suppose a box containing a gas of mass $M$ at a temperature $T$ is moving along a frictionless table with a velocity $v_{cm}$ (Figure 19-9a). The total kinetic energy of the gas has two components: that associated with the movement of the center of mass $\frac{1}{2}Mv_{cm}^2$, and the energy of the motion of its molecules relative to its center of mass. The center of mass energy $\frac{1}{2}Mv_{cm}^2$ is ordered mechanical energy that could be converted entirely into work. (For example, if a weight were attached to the moving box by a string passing over a pulley, this energy could be used to lift the weight.) The relative energy is the internal thermal energy of the gas, which is related to its temperature $T$. It is random, non-ordered energy that cannot be converted entirely into work.

Now, suppose that the box hits a fixed wall and stops (Figure 19-9b). This inelastic collision is clearly an irreversible process. The ordered mechanical energy of the gas is converted into random internal energy and the temperature of the gas rises. The gas still has the same total energy, but now all of that energy is associated with the random motion of the gas molecules about the center of mass of the gas, which is now at rest. Thus, the gas has become less ordered (more disordered), and has lost some of its ability to do work.

## 19-7 Entropy

There is a thermodynamic function called **entropy** $S$ that is a measure of the disorder of a system. Entropy $S$, like pressure $P$, volume $V$, temperature $T$, and

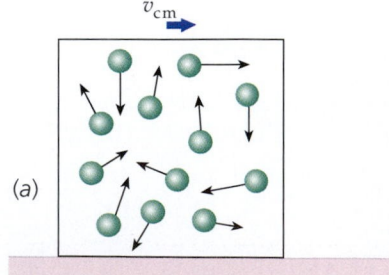
(a)

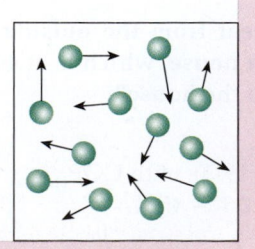
(b)

**FIGURE 19-9**

internal energy $U$, is a function of the state of a system. As with potential energy, it is the *change* in entropy that is important. The change in entropy $dS$ of a system as it goes from one state to another is defined as

$$dS = \frac{dQ_{rev}}{T}$$  19-12

DEFINITION—ENTROPY CHANGE

where $dQ_{rev}$ is the energy that must be transferred to the system as heat in a *reversible* process that brings the system from the initial state to the final state. If $dQ_{rev}$ is negative, then the entropy change of the system is negative.

The term $dQ_{rev}$ does not mean that a reversible heat transfer must take place in order for the entropy of a system to change. Indeed, there are many situations in which the entropy of a system changes when there is no transfer of heat whatsoever, for example, the box of gas colliding with the wall in Figure 19-9. Equation 19-12 simply gives us a method for *calculating* the entropy difference between two states of a system. Because entropy is a state function, the change in entropy when the system moves from one state to another depends only on the system's initial and final states, not on the process by which the change occurs.

## Entropy of an Ideal Gas

We can illustrate that $dQ_{rev}/T$ is in fact the differential of a state function for an ideal gas (even though $dQ_{rev}$ is not). Consider an arbitrary reversible quasi-static process in which a system consisting of an ideal gas absorbs an amount of heat $dQ_{rev}$. According to the first law, $dQ_{rev}$ is related to the change in the internal energy $dE_{int}$ of the gas and the work done on the gas ($dW_{on} = -PdV$) by

$$dE_{int} = dQ_{rev} + dW_{on} = dQ_{rev} - PdV$$

For an ideal gas, we can write $dE_{int}$ in terms of the heat capacity, $dE_{int} = C_v dT$, and we can substitute $nRT/V$ for $P$ from the equation of state. Then

$$C_v dT = dQ_{rev} - nRT \frac{dV}{V}$$  19-13

Equation 19-13 cannot be integrated unless we know how $T$ depends on $V$. This is just another way of saying that $dQ_{rev}$ is not a differential of a state function $Q_{rev}$. But if we divide each term by $T$, we obtain

$$C_v \frac{dT}{T} = \frac{dQ_{rev}}{T} - nR \frac{dV}{V}$$  19-14

Since $C_v$ depends only on $T$, the term on the left can be integrated as can the second term on the right.[†] Thus, $dQ_{rev}/T$ is the differential of a function, the entropy function $S$.

$$dS = \frac{dQ_{rev}}{T} = \left(C_v \times \frac{dT}{T}\right) + nR \frac{dV}{V}$$  19-15

For simplicity, we will assume that $C_v$ is constant. Integrating Equation 19-15, we obtain

$$\Delta S = \int \frac{dQ}{T} = C_v \ln \frac{T_2}{T_1} + nR \ln \frac{V_2}{V_1}$$  19-16

Equation 19-16 gives the entropy change of an ideal gas that undergoes a reversible expansion from an initial state of volume $V_1$ and temperature $T_1$ to a final state of volume $V_2$ and temperature $T_2$.

---

† Mathematically, the factor $1/T$ is called an integrating factor for Equation 19-13.

## Entropy Changes for Various Processes

**ΔS for an Isothermal Expansion of an Ideal Gas**   If an ideal gas undergoes an isothermal expansion, then $T_2 = T_1$ and its entropy change is

$$\Delta S = \int \frac{dQ}{T} = nR \ln \frac{V_2}{V_1} \qquad\qquad 19\text{-}17$$

The entropy change of the gas is positive because $V_2$ is greater than $V_1$. In this process, an amount of energy $Q$ is transferred as heat from the reservoir to the gas. This heat equals the work done by the gas:

$$Q = W_{by} = \int_{V_1}^{V_2} P\,dV = nRT \int_{V_1}^{V_2} \frac{dV}{V} = nRT \ln \frac{V_2}{V_1} \qquad\qquad 19\text{-}18$$

The entropy change of the gas is $+Q/T$. Because the same amount of heat leaves the reservoir at temperature $T$, the entropy change of the reservoir is $-Q/T$. The net entropy change of the gas plus the reservoir is zero. We will refer to the system under consideration plus its surroundings as the "universe." This example illustrates a general result:

> In a reversible process, the entropy change of the universe is zero.

**ΔS for a Free Expansion of an Ideal Gas**   In the free expansion of a gas discussed in Section 18-4, a gas is initially confined in one compartment of a container, which is connected by a stopcock to another compartment that is evacuated. The whole system has rigid walls and is thermally insulated from its surroundings so that no heat can flow in or out, and no work can be done on (or by) the system (Figure 19-10). When the stopcock is opened, the gas rushes into the evacuated chamber. Eventually, the gas reaches thermal equilibrium with itself. Since there is no work done and no heat transferred, the final internal energy of the gas must equal its initial internal energy. If we assume that the gas is ideal, the final temperature $T$ equals the initial temperature.

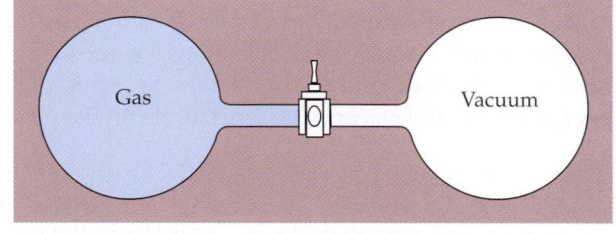

**FIGURE 19-10** Free expansion of a gas. When the stopcock is opened, the gas expands rapidly into the evacuated chamber. Since no work is done on the gas and the whole system is thermally insulated, the initial and final internal energies of the gas are equal.

We might think that there is no entropy change of the gas because there is no heat transfer. But this process is not reversible, so we cannot use $\int dQ/T$ to find the change in entropy of the gas. However, the initial and final states of the gas in the free expansion are the same as those of the gas in the isothermal expansion just discussed. *Because the change in the entropy of a system for any process depends only on the initial and final states of the system, the entropy change of the gas for the free expansion is the same as that for the isothermal expansion.* If $V_1$ is the initial volume of the gas and $V_2$ is its final volume, the entropy change of the gas is given by Equation 19-17, or

$$\Delta S_{gas} = nR \ln \frac{V_2}{V_1}$$

In this case, there is no change in the surroundings, so the entropy change of the gas is also the entropy change of the universe:

$$\Delta S_u = nR \ln \frac{V_2}{V_1} \qquad\qquad 19\text{-}19$$

Note that because $V_2$ is greater than $V_1$, the change in entropy of the universe for this irreversible process is positive; that is, the entropy of the universe increases. This is also a general result:

In an irreversible process, the entropy of the universe increases.

If the final volume in the free expansion were less than the initial volume, then the entropy of the universe would decrease—but this does not happen. A gas does not freely contract by itself into a smaller volume. This leads us to yet another statement of the second law of thermodynamics:

For any process, the entropy of the universe never decreases.

*FREE EXPANSION OF AN IDEAL GAS*                **EXAMPLE 19-8**

**Find the entropy change for the free expansion of 0.75 mol of an ideal gas from $V_1 = 1.5$ L to $V_2 = 3$ L.**

**PICTURE THE PROBLEM** For a free expansion of an ideal gas the initial and final temperatures are the same. Thus, the entropy change $\Delta S$ for a free expansion from $V_1$ to $V_2$ is the same as $\Delta S$ for an isothermal process from $V_1$ to $V_2$. For the isothermal process $\Delta E_{int} = 0$, so $Q = W_{by}$. First we calculate $Q$, then we set $\Delta S = Q/T$.

1. The entropy change is the same as for an isothermal expansion from $V_1$ to $V_2$:

$$\Delta S = \Delta S_{isothermal} = \frac{Q}{T}$$

2. The heat $Q$ that would enter the gas during an isothermal expansion at temperature $T$ equals the work done by the gas during the expansion:

$$Q = W_{by} = nRT \ln \frac{V_2}{V_1}$$

3. Substitute this value of $Q$ to calculate $\Delta S$:

$$\Delta S = \frac{Q}{T} = nR \ln \frac{V_2}{V_1}$$

$$\Delta S = (0.75 \text{ mol})(8.31 \text{ J/mol·K}) \ln 2$$

$$\Delta S = \boxed{4.32 \text{ J/K}}$$

**$\Delta S$ for Constant-Pressure Processes** If a substance is heated from temperature $T_1$ to temperature $T_2$ at constant pressure, the heat absorbed $dQ$ is related to its temperature change $dT$ by

$$dQ = C_p \, dT$$

We can approximate reversible heat conduction if we have a large number of heat reservoirs with temperatures ranging from just slightly greater than $T_1$ to $T_2$ in very small steps. We could place the substance, with initial temperature $T_1$, in contact with the first reservoir at a temperature just slightly greater than $T_1$ and let the substance absorb a small amount of heat. Because the heat transfer is approximately isothermal, the process will be approximately reversible. We then place the substance in contact with the next reservoir at a slightly higher temperature, and so on, until the final temperature $T_2$ is reached. If heat $dQ$ is absorbed reversibly, the entropy change of the substance is

$$dS = \frac{dQ}{T} = C_p \frac{dT}{T}$$

Integrating from $T_1$ to $T_2$, we obtain the total entropy change of the substance:

$$\Delta S = C_p \int_{T_1}^{T_2} \frac{dT}{T} = C_p \ln \frac{T_2}{T_1} \qquad\qquad 19\text{-}20$$

This result gives the entropy change of a substance that is heated from $T_1$ to $T_2$ by any process, reversible or irreversible, as long as the final pressure equals the initial pressure. It also gives the entropy change of a substance that is cooled. In the case of cooling, $T_2$ is less than $T_1$, and $\ln(T_2/T_1)$ is negative, giving a negative entropy change.

**EXERCISE** Find the change in entropy of 1 kg of water that is heated at constant pressure from 0°C to 100°C. (*Answer* $\Delta S = 1.31$ kJ/K)

---

*ENTROPY CHANGES DURING HEAT TRANSFER*          **EXAMPLE 19-9**   **Try It Yourself**

Suppose 1 kg of water at temperature $T_1 = 30°C$ is added to 2 kg of water at $T_2 = 90°C$ in a calorimeter of negligible heat capacity at a constant pressure of 1 atm. (*a*) Find the change in entropy of the system. (*b*) Find the change in entropy of the universe.

**PICTURE THE PROBLEM** When the two amounts of water are combined, they eventually come to a final equilibrium temperature, $T_f$, that can be found by setting the heat lost equal to the heat gained. To calculate the entropy change of each mass of water, we consider a reversible isobaric heating of the 1-kg mass of water from 30°C to $T_f$ and a reversible isobaric cooling of the 2-kg mass from 90°C to $T_f$ using Equation 19-18. The entropy change of the system is the sum of the entropy changes of each part. The entropy change of the universe is the entropy change of the system plus the entropy change of its surroundings. To find the entropy change of the surroundings, assume no heat leaves the calorimeter during the time it takes the water to reach its final temperature.

**Cover the column to the right and try these on your own before looking at the answers.**

| Steps | Answers |
|---|---|
| (*a*) 1. Calculate $T_f$ by setting the heat lost equal to the heat gained: | $T_f = 70°C = 343$ K |
| 2. Use your result for $T_f$ and the data given to calculate $\Delta S_1$ and $\Delta S_2$: | $\Delta S_1 = 0.519$ kJ/K<br>$\Delta S_2 = -0.474$ kJ/K |
| 3. Add $\Delta S_1$ and $\Delta S_2$ to find the total entropy change of the system: | $\Delta S_{\text{system}} = \boxed{+0.0453 \text{ kJ/K}}$ |
| (*b*) 1. Assuming no heat leaves the calorimeter, find the entropy change of the surroundings: | $\Delta S_{\text{surroundings}} = 0$ |
| 2. Add $\Delta S_{\text{system}}$ and $\Delta S_{\text{surroundings}}$ to find the entropy change of the universe: | $\Delta S_u = \boxed{+0.0453 \text{ kJ/K}}$ |

**REMARKS** Note that we had to convert the temperatures to the absolute scale to calculate the entropy changes. The entropy change of the universe is positive, as expected.

**$\Delta S$ for an Inelastic Collision** Because mechanical energy is converted into thermal energy in an inelastic collision, such a process is clearly irreversible. The entropy of the universe must therefore increase. Consider a block of mass $m$

falling from a height $h$ and making an inelastic collision with the ground. Let the block, ground, and atmosphere all be at a temperature $T$, which is not significantly changed by the process. If we consider the block, ground, and atmosphere as our isolated system, there is no heat conducted into or out of the system. The state of the system has been changed because its internal energy has been increased by an amount $mgh$. This change is the same as if we added heat $Q = mgh$ to the system at constant temperature $T$. To calculate the change in entropy of the system, we thus consider a reversible process in which heat $Q_{rev} = mgh$ is added at a constant temperature $T$. According to Equation 19-12, the change in entropy is then

$$\Delta S = \frac{Q_{rev}}{T} = \frac{mgh}{T}$$

This positive entropy change is also the entropy change of the universe.

**$\Delta S$ for Heat Conduction from One Reservoir to Another**   Heat conduction is also an irreversible process, and so we expect the entropy of the universe to increase when this occurs. Consider the simple case of heat $Q$ conducted from a hot reservoir at a temperature $T_h$ to a cold reservoir at a temperature $T_c$. The state of a heat reservoir is determined by its temperature and its internal energy only. The change in entropy of a heat reservoir due to a heat exchange is the same whether the heat exchange is reversible or not. If heat $Q$ is put into a reservoir at temperature $T$, then the entropy of the reservoir increases by $Q/T$. If the heat is removed, then the entropy of the reservoir decreases by $-Q/T$. In the case of heat conduction, the hot reservoir loses heat, so its entropy change is

$$\Delta S_h = -\frac{Q}{T_h}$$

The cold reservoir absorbs heat, so its entropy change is

$$\Delta S_c = +\frac{Q}{T_c}$$

The net entropy change of the universe is

$$\Delta S_u = \Delta S_c + \Delta S_h = \frac{Q}{T_c} - \frac{Q}{T_h} \qquad\qquad 19\text{-}21$$

Note that, because heat flows from a hot reservoir to a cold reservoir, the change in entropy of the universe is positive.

**$\Delta S$ for a Carnot Cycle**   Because a Carnot cycle is by definition reversible, the entropy change of the universe after a cycle must be zero. We demonstrate this by showing that the entropy change of the reservoirs in a Carnot engine is zero. (Since a Carnot engine works in a cycle, the entropy change of the engine itself is zero, so the entropy change of the universe is just the sum of the entropy changes of the reservoirs.) The entropy change of the hot reservoir is $\Delta S_h = -\frac{Q_h}{T_h}$ and the entropy change of the cold reservoir is $\Delta S_c = +\frac{Q_c}{T_c}$. These heats are related to the temperatures by the definition of thermodynamic temperature (Equation 19-7)

$$\frac{T_c}{T_h} = \frac{Q_c}{Q_h} \left( \text{or} \quad \frac{Q_h}{T_h} = \frac{Q_c}{T_c} \right)$$

The entropy change of the universe is thus

$$\Delta S_u = \Delta S_h + \Delta S_c = -\frac{Q_h}{T_h} + \frac{Q_c}{T_c} = -\frac{Q_h}{T_h} + \frac{Q_h}{T_h} = 0$$

The entropy change of the universe is zero as expected.

Notice that we have ignored any entropy change associated with the energy transferred via work from the Carnot engine to its surroundings. If this work is used to raise a weight, or some other ordered process, then there is no entropy change. However, if this work is used to push a block across a table top where friction is involved, then there is an additional entropy increase associated with this work.

---

*ENTROPY CHANGES IN A CARNOT CYCLE*                    **E X A M P L E   1 9 - 1 0**

**During each cycle, a Carnot engine removes 100 J of energy from a reservoir at 400 K, does work, and exhausts heat to a reservoir at 300 K. Compute the entropy change of each reservoir for each cycle, and show explicitly that the entropy change of the universe is zero for this reversible process.**

**PICTURE THE PROBLEM** Since the engine works in a cycle, its entropy change is zero. We therefore compute the entropy change of each reservoir and add them to obtain the entropy change of the universe.

1. The entropy change of the universe equals the sum of the entropy changes of the reservoirs:

$$\Delta S_u = \Delta S_{400} + \Delta S_{300}$$

2. Calculate the entropy change of the hot reservoir:

$$\Delta S_{400} = -\frac{Q_h}{T_h} = -\frac{100\,\text{J}}{400\,\text{K}} = \boxed{-0.250\,\text{J/K}}$$

3. The entropy change of the cold reservoir is $Q_c$ divided by $T_c$, where $Q_c = Q_h - W$:

$$\Delta S_{300} = \frac{Q_c}{T_c} = \frac{Q_h - W}{T_c}$$

4. We use $W = \varepsilon_C Q_h$ (Equation 19-2) to relate $W$ to $Q_h$. The efficiency is the Carnot efficiency (Equation 19-6):

$$W = \varepsilon Q_h, \text{ where } \varepsilon = \varepsilon_C = 1 - \frac{T_c}{T_h}$$

so

$$W = \left(1 - \frac{T_c}{T_h}\right) Q_h$$

5. Calculate the entropy change of the cold reservoir:

$$\Delta S_{300} = \frac{Q_h - W}{T_c} = \frac{Q_h - Q_h\left(1 - \dfrac{T_c}{T_h}\right)}{T_c} = \frac{Q_h}{T_h}$$

$$= \frac{100\,\text{J}}{400\,\text{K}} = \boxed{0.250\,\text{J/K}}$$

6. Substitute these results into step 1 to find the entropy change of the universe:

$$\Delta S_u = \Delta S_{400} + \Delta S_{300}$$

$$\Delta S_u = -0.250\,\text{J/K} + 0.250\,\text{J/K} = \boxed{0}$$

**REMARKS** Suppose that an ordinary, nonreversible engine removed 100 J from the hot reservoir. Because its efficiency must be less than that of a Carnot engine, it would do less work and exhaust more heat to the cold reservoir. Then the entropy increase of the cold reservoir would be greater than the entropy decrease of the hot reservoir, and the entropy change of the universe would be positive.

**EXAMPLE 19-11**

Because entropy is a state function, thermodynamic processes can be represented as *ST*, *SV*, or *SP* diagrams instead of the *PV* diagrams we have used so far. Make a sketch of the Carnot cycle on an *ST* plot.

**PICTURE THE PROBLEM** The Carnot cycle consists of a reversible isothermal expansion followed by a reversible adiabatic expansion, then a reversible isothermal compression followed by a reversible adiabatic compression. During the isothermal processes, heat is absorbed or expelled at constant temperature, so $S$ increases or decreases at constant $T$. During the adiabatic processes, the temperature changes, but since $\Delta Q_{\text{rev}} = 0$, $S$ is constant.

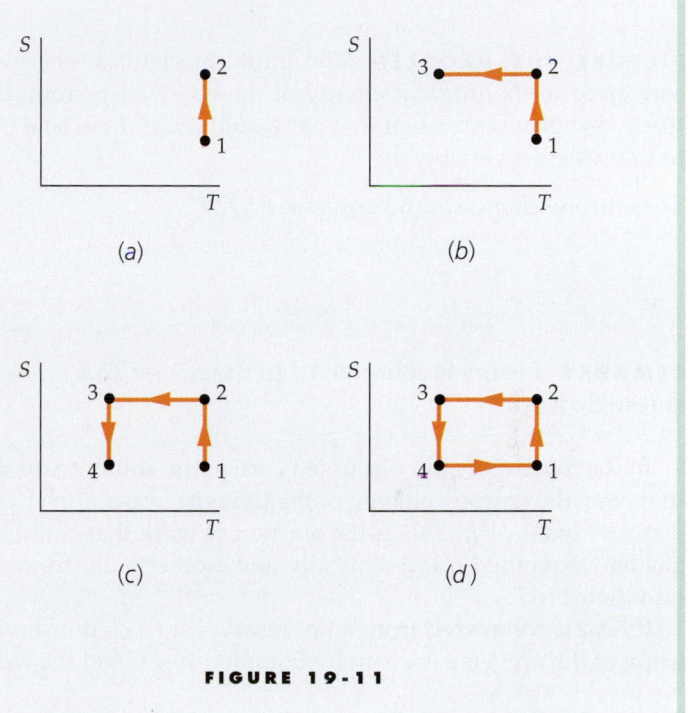

(a)

(b)

(c)

(d)

**FIGURE 19-11**

1. During the isothermal expansion (1 to 2 in Figure 19-11*a*), heat is absorbed reversibly so, $S$ increases at constant $T$:

2. During the reversible adiabatic expansion (2 to 3 in Figure 19-11*b*), the temperature decreases while $S$ is constant:

3. During the isothermal compression (3 to 4 in Figure 19-11*c*) heat is rejected reversibly, so $S$ decreases at constant $T$:

4. During the reversible adiabatic compression (4 to 1 in Figure 19-11*d*) the temperature increases while $S$ is constant:

■ **REMARKS** The Carnot cycle is a rectangle if plotted on an $S$ versus $T$ diagram.

# 19-8 Entropy and the Availability of Energy

If an irreversible process occurs, energy is conserved, but some of the energy becomes unavailable to do work and is "wasted." Consider a block falling to the ground. The entropy change of the universe for this process is $mgh/T$. When the block was at a height $h$, its potential energy $mgh$ could have been used to do useful work. But after the inelastic collision of the block with the ground, this energy is no longer available because it has become the disordered internal energy of the block and its surroundings. The energy that has become unavailable (wasted) is equal to $mgh = T\Delta S_{\text{u}}$. This is a general result:

In an irreversible process, energy equal to $T\Delta S_{\text{u}}$ becomes unavailable to do work, where $T$ is the temperature of the coldest available reservoir.

For simplicity, we will call the energy that becomes unavailable to do work the "work lost":

$$W_{\text{lost}} = T\Delta S_{\text{u}} \qquad\qquad 19\text{-}22$$

---

A SLIDING BOX REVISITED          **E X A M P L E    1 9 - 1 2**

Suppose that the box shown in Figure 19-9a and b has a mass of 2.4 kg and slides with a speed of $v = 3$ m/s before crashing into a fixed wall and stopping. The temperature $T$ of the box, table, and surroundings is 293 K and does not change appreciably as the box comes to rest. Find the entropy change of the universe.

**PICTURE THE PROBLEM** The initial mechanical energy of the box $\frac{1}{2}Mv^2$ is converted to the internal energy of the box-wall-surroundings system. The entropy change is equivalent to what would occur if the heat $Q = \frac{1}{2}Mv^2$ were added to the system reversibly.

The entropy change of the universe is $Q/T$:

$$\Delta S_u = \frac{Q}{T} = \frac{\frac{1}{2}Mv^2}{T} = \frac{\frac{1}{2}(2.4\,\text{kg})(3\,\text{m/s})^2}{293\,\text{K}}$$

$$\Delta S_u = \boxed{0.0369\ \text{J/K}}$$

**REMARKS** Energy is conserved, but the energy $T\Delta S_u = \frac{1}{2}Mv^2$ is no longer available to do work.

In the free expansion discussed earlier, the ability to do work was also lost. In that case, the entropy change of the universe was $nR \ln (V_2/V_1)$, so the work lost was $nRT \ln (V_2/V_1)$. This is the amount of work that could have been done if the gas had expanded quasi-statically and isothermally from $V_1$ to $V_2$, as given by Equation 19-17.

If heat is conducted from a hot reservoir to a cold reservoir, the change in entropy of the universe is given by Equation 19-21, and the work lost is

$$W_{\text{lost}} = T_c\,\Delta S_u = T_c\left(\frac{Q}{T_c} - \frac{Q}{T_h}\right) = Q\left(1 - \frac{T_c}{T_h}\right)$$

We can see that this is just the work that could have been done by a Carnot engine running between these reservoirs, removing heat $Q$ from the hot reservoir and doing work $W = \varepsilon_C Q$, where $\varepsilon_C = 1 - T_c/T_h$.

## 19-9 Entropy and Probability

Entropy, which is a measure of the disorder of a system, is related to probability. Essentially, a state of high order has a low probability, whereas a state of low order has a high probability. Thus, in an irreversible process, the universe moves from a state of low probability to one of high probability.

Let us consider a free expansion in which a gas expands from an initial volume $V_1$ to a final volume $V_2 = 2V_1$. The entropy change of the universe for this process is given by Equation 19-19:

$$\Delta S = nR \ln\frac{V_2}{V_1} = nR \ln 2$$

Why is this process irreversible? Why can't the gas spontaneously compress back into its original volume? Such a compression would not violate the first law of thermodynamics, as there is no energy change involved. The reason that the gas does not compress to its original volume is merely that such a compression is extremely *improbable*. To see this, let's assume that the gas consists of only

10 molecules and that, initially, these molecules occupy the entire volume of their container. Then the chance that any one particular molecule will be in the left half of the container at any given time is $\frac{1}{2}$. The chance that any two particular molecules will both be in the left half is $\frac{1}{2} \times \frac{1}{2} = \frac{1}{4}$. (This is the same as the chance that a coin flipped twice will come up heads both times.) The chance that three particular molecules will be in the left half is $\frac{1}{2} \times \frac{1}{2} \times \frac{1}{2} = (\frac{1}{2})^3 = \frac{1}{8}$. The chance that all 10 molecules will be in the left half is $(\frac{1}{2})^{10} = \frac{1}{1024}$. That is, there is 1 chance in 1024 that all 10 molecules will be in the left half of the container at any given time.

Though the probability of all 10 molecules being on one side of the container is small, we would not be completely surprised to see it occur. If we look at the gas once each second, we could expect to see it happen once in every 1024 sec, or about once every 17 min. If we started with the 10 molecules randomly distributed and then found them all in the left half of the original volume, the entropy of the universe would have *decreased* by $nR \ln 2$. However, this decrease is extremely small, since the number of moles $n$ corresponding to 10 molecules is only about $10^{-23}$. Still, it would violate the entropy statement of the second law of thermodynamics, which says that for any process, the entropy of the universe never decreases. Therefore, if we wish to apply the second law of thermodynamics to microscopic systems such as a small number of molecules, we should consider the second law to be a statement of *probability*.

We can relate the probability of a gas spontaneously compressing itself into a smaller volume to the change in its entropy. If the original volume is $V_1$, the probability $p$ of finding $N$ molecules in a smaller volume $V_2$ is

$$ p = \left( \frac{V_2}{V_1} \right)^N $$

Taking the natural logarithm of both sides of this equation, we obtain

$$ \ln p = N \ln \frac{V_2}{V_1} = nN_A \ln \frac{V_2}{V_1} \qquad\qquad 19\text{-}23 $$

where $n$ is the number of moles and $N_A$ is Avogadro's number. The entropy change of the gas is

$$ \Delta S = nR \ln \frac{V_2}{V_1} \qquad\qquad 19\text{-}24 $$

Comparing Equations 19-23 and 19-24, we see that

$$ \Delta S = \frac{R}{N_A} \ln p = k \ln p \qquad\qquad 19\text{-}25 $$

where $k$ is Boltzmann's constant.

It may be disturbing to learn that irreversible processes, such as the spontaneous compression of a gas or the spontaneous conduction of heat from a cold body to a hot body, are not impossible—they are just improbable. As we have just seen, there is a reasonable chance that an irreversible process will occur in a system consisting of a very small number of molecules; however, *thermodynamics itself is applicable only to macroscopic systems*, that is, to systems that have a very large number of molecules. Consider trying to measure the pressure of a gas consisting of only 10 molecules. The pressure would vary wildly depending on whether no molecule, 2 molecules, or 10 molecules were colliding with the wall of the container at the time of measurement. The macroscopic variables of pressure and temperature are not applicable to a microscopic system with only 10 molecules.

As we increase the number of molecules in a system, the chance of an irreversible process occurring decreases dramatically. For example, if we have 50 molecules in a container, the chance that they will all be in the left half of the container is $(\frac{1}{2})^{50} \approx 10^{-15}$. Thus, if we look at the gas once each second, we could expect to see all 50 molecules in the left half of the volume about once in every $10^{15}$ seconds or once in every 36 million years! For 1 mole ($6 \times 10^{23}$ molecules), the chance that all will wind up in half of the volume is vanishingly small, essentially zero. For macroscopic systems, then, the probability of a process resulting in a decrease in the entropy of the universe is so extremely small that the distinction between improbable and impossible becomes blurred.

# SUMMARY

The second law of thermodynamics is a fundamental law of nature.

| Topic | Relevant Equations and Remarks |
|---|---|
| **1. Efficiency of a Heat Engine** | If the engine removes $Q_h$ from a hot reservoir, does work $W$, and exhausts heat $Q_c$ to a cold reservoir, its efficiency is $$\varepsilon = \frac{W}{Q_h} = \frac{Q_h - Q_c}{Q_h} = 1 - \frac{Q_c}{Q_h} \qquad \text{19-2}$$ |
| **2. Coefficient of Performance of a Refrigerator** | $$\mathrm{COP} = \frac{Q_c}{W} \qquad \text{19-3}$$ |
| **3. Coefficient of Performance of a Heat Pump** | $$\mathrm{COP}_{\mathrm{HP}} = \frac{Q_h}{W} \qquad \text{19-8}$$ |
| **4. Equivalent Statements of the Second Law of Thermodynamics** | |
| The Kelvin statement | No system can take energy as heat from a single reservoir and convert it entirely into work without additional net changes in the system or its surroundings. |
| The heat-engine statement | It is impossible for a heat engine working in a cycle to produce *only the effect* of extracting heat from a single reservoir and performing an equivalent amount of work. |
| The Clausius statement | A process whose only net result is to transfer energy as heat from a cooler object to a hotter one is impossible. |
| The refrigerator statement | It is impossible for a refrigerator working in a cycle to produce *only the effect* of extracting heat from a cold object and rejecting the same amount of heat to a hot object. |
| The entropy statement | The entropy of the universe (system plus surroundings) can never decrease. |
| **5. Conditions for a Reversible Process** | 1. No mechanical energy is transformed into thermal energy by friction, viscous forces, or other dissipative forces. <br> 2. Energy transfer as heat can only occur between objects at the same temperature (or infinitesimally near the same temperature). <br> 3. The process must be quasi-static so that the system is always in an equilibrium state (or infinitesimally near an equilibrium state). |

| 6. **Carnot Engine** | A Carnot engine is a reversible engine that works between two reservoirs. It uses a Carnot cycle, which consists of |
|---|---|
| Carnot cycle | 1. A quasi-static isothermal absorption of heat at temperature $T_h$ <br> 2. A quasi-static adiabatic expansion <br> 3. A quasi-static isothermal exhaustion of heat at temperature $T_c$ <br> 4. A quasi-static adiabatic compression back to the original state |
| Carnot efficiency | $$\varepsilon_C = 1 - \frac{Q_c}{Q_h} = 1 - \frac{T_c}{T_h} \qquad \textbf{19-6}$$ |

| 7. **Thermodynamic Temperature** | The ratio of the thermodynamic temperatures of two reservoirs is defined to be the ratio of the heat exhausted to the heat intake of a Carnot engine running between the reservoirs. |
|---|---|
| | $$\frac{T_c}{T_h} = \frac{Q_c}{Q_h} \qquad \textbf{19-7}$$ |

| 8. **Entropy** | Entropy is a measure of the disorder of a system. The difference in entropy between two nearby states is given by |
|---|---|
| | $$dS = \frac{dQ_{rev}}{T} \qquad \textbf{19-12}$$ <br> where $dQ_{rev}$ is the heat added in a reversible process connecting the states. The entropy change of a system can be positive or negative. |
| Entropy and loss of work capability | During an irreversible process, the entropy of the universe $S_u$ increases and an amount of energy <br><br> $$W_{lost} = T\Delta S_u \qquad \textbf{19-22}$$ <br><br> becomes unavailable for doing work. |
| Entropy and probability | Entropy is related to probability. A highly ordered system is one of low probability and low entropy. An isolated system moves towards a state of high probability, low order, and high entropy. |

# PROBLEMS

- • Single-concept, single-step, relatively easy
- •• Intermediate-level, may require synthesis of concepts
- ••• Challenging
- SSM Solution is in the *Student Solutions Manual*
- iSOLVE Problems available on iSOLVE online homework service
- iSOLVE✓ These "Checkpoint" online homework service problems ask students additional questions about their confidence level and how they arrived at their answer

In a few problems, you are given more data than you actually need; in a few other problems, you are required to supply data from your general knowledge, outside sources, or informed estimates.

## Conceptual Problems

**1** • How does kinetic friction in an engine affect its efficiency?

**2** • SSM Explain why you can't just open your refrigerator to cool your kitchen on a hot day. Why is it that turning on a room air conditioner will cool down the room but opening a refrigerator door will not?

**3** • Why do power-plant designers try to increase the temperature of the steam fed to engines as much as possible?

**4** •• On a humid day, water vapor condenses on a cold surface. During condensation, the entropy of the water (a) increases, (b) remains constant, (c) decreases, (d) may decrease or remain unchanged.

**5** • SSM In a reversible adiabatic process, (a) the internal energy of the system remains constant, (b) no work is done by the system, (c) the entropy of the system remains constant, (d) the temperature of the system remains constant.

**6** •• True or false:

(a) Work can never be converted completely into heat.
(b) Heat can never be converted completely into work.
(c) All heat engines have the same efficiency.
(d) It is impossible to transfer a given quantity of heat from a cold reservoir to a hot reservoir.
(e) The coefficient of performance of a refrigerator cannot be greater than 1.
(f) All Carnot engines are reversible.
(g) The entropy of a system can never decrease.
(h) The entropy of the universe can never decrease.

**7** •• An ideal gas is taken reversibly from an initial state $P_i$, $V_i$, $T_i$ to the final state $P_f$, $V_f$, $T_f$. Two possible paths are (A) an isothermal expansion followed by an adiabatic compression, and (B) an adiabatic compression followed by an isothermal expansion. For these two paths, (a) $\Delta E_{int\,A} > \Delta E_{int\,B}$, (b) $\Delta S_A > \Delta S_B$, (c) $\Delta S_A < \Delta S_B$, (d) none of the above is correct.

**8** •• SSM Figure 19-12 shows a thermodynamic cycle on an $ST$ diagram. Identify this cycle and sketch it on a $PV$ diagram.

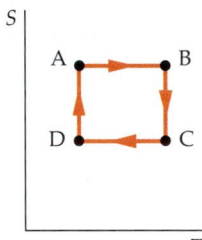

**FIGURE 19-12**
**Problems 8 and 68**

**9** •• Figure 19-13 shows a thermodynamic cycle on an $SV$ diagram. Identify the type of engine represented by this diagram.

**FIGURE 19-13**
**Problem 9**

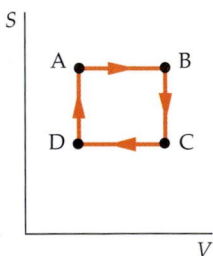

**10** •• Sketch an $ST$ diagram of the Otto cycle.

**11** •• Sketch an $SV$ diagram of the Carnot cycle.

**12** •• Sketch an $SV$ diagram of the Otto cycle.

**13** •• Figure 19-14 shows a thermodynamic cycle on an $SP$ diagram. Make a sketch of this cycle on a $PV$ diagram.

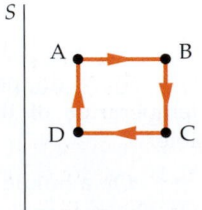

**FIGURE 19-14** Problem 13

**14** • SSM Which has a greater effect on increasing the efficiency of a Carnot engine, a 5-K increase in the temperature of the hot reservoir or a 5-K decrease in the temperature of the cold reservoir?

## Estimation and Approximation

**15** •• Estimate the maximum efficiency of an automobile engine with a compression ratio of 8:1. Assume the Otto cycle and assume $\gamma = 1.4$.

**16** •• SSM (a) Estimate the highest COP possible for a "typical" household refrigerator. (b) If the refrigerator draws 600 W of electrical power, estimate the rate at which heat is being drawn from the refrigerator compartment.

**17** •• The temperature of the sun is about 5400 K, the Earth's average temperature is about 290 K, and the solar constant (the intensity of sunlight reaching the Earth's orbit) is about 1.3 kW/m². (a) Calculate the total power of sunlight hitting the Earth. (b) Calculate the net rate at which the Earth's entropy is increasing due to the flow of solar radiation. (c) Calculate the net rate at which the sun's entropy is decreasing just due to the outflow of solar radiation *hitting the Earth*.

**18** •• (a) Using the information given in Problem 17 and the known distance from the Earth to the sun ($1.5 \times 10^{11}$ m), calculate the total power that the sun radiates into space. (b) There are about $10^{11}$ stars like the sun in the Milky Way galaxy, and $10^{11}$ galaxies in the universe. Use this information to estimate the rate at which the entropy of the universe is increasing, assuming that the average temperature of the universe is 2.73 K.

**19** •• A typical human body produces about 100 W of heat. Estimate the increase in entropy of the universe produced by a single human body over the course of a spring day where the temperature is 70°F during the day and 55°F at night.

**20** ••• SSM How long, on average, should we have to wait until all of the air molecules in a room rush to one half of the room? (As a friend of mine put it, "Don't hold your breath....") Assume that the air molecules are contained in a 1 m × 1 m × 1 m box and that they reshuffle their positions 100 times per second. Calculate the average time it should take for all the molecules to occupy only one half of the box if there are (a) 10 molecules, (b) 100 molecules, (c) 1000 molecules, and (d) 1 mole of molecules in the box. (e) The highest vacuums that have been created to date have pressures of about $10^{-12}$ torr. If a typical vacuum chamber has a capacity of about 1 liter, how long will a physicist have to wait before all of the gas molecules in the vacuum chamber occupy only one half of it? Compare that to the expected lifetime of the universe, which is about $10^{10}$ years.

## Heat Engines and Refrigerators

**21** • iSOLVE An engine with 20% efficiency does 100 J of work in each cycle. (a) How much heat is absorbed in each cycle? (b) How much heat is rejected in each cycle?

**22** • iSOLVE An engine absorbs 400 J of heat and does 120 J of work in each cycle. (a) What is its efficiency? (b) How much heat is rejected in each cycle?

**23** • ✦SOLVE✓ An engine absorbs 100 J and rejects 60 J in each cycle. (a) What is its efficiency? (b) If each cycle takes 0.5 s, find the power output of this engine in watts.

**24** • SSM ✦SOLVE A refrigerator absorbs 5 kJ of energy from a cold reservoir and rejects 8 kJ to a hot reservoir. (a) Find the coefficient of performance of the refrigerator. (b) The refrigerator is reversible and is run backward as a heat engine between the same two reservoirs. What is its efficiency?

**25** •• An engine operates with 1 mol of an ideal gas, for which $C_v = \frac{3}{2}R$ and $C_p = \frac{5}{2}R$, as its working substance. The cycle begins at $P_1 = 1$ atm and $V_1 = 24.6$ L. The gas is heated at constant volume to $P_2 = 2$ atm. It then expands at constant pressure until $V_2 = 49.2$ L. During these two steps, heat is absorbed by the gas. The gas is then cooled at constant volume until its pressure is again 1 atm. It is then compressed at constant pressure to its original state. During the last two steps, heat is rejected by the gas. All the steps are quasi-static and reversible. (a) Show this cycle on a PV diagram. Find the work done, the heat added, and the change in the internal energy of the gas for each step of the cycle. (b) Find the efficiency of the cycle.

**26** •• An engine using 1 mol of a diatomic ideal gas performs a cycle consisting of three steps: (1) an adiabatic expansion from an initial pressure of 2.64 atm and an initial volume of 10 L to a pressure of 1 atm and a volume of 20 L, (2) a compression at constant pressure to its original volume of 10 L, and (3) heating at constant volume to its original pressure of 2.64 atm. Find the efficiency of this cycle.

**27** •• ✦SOLVE An engine using 1 mol of an ideal gas initially at $V_1 = 24.6$ L and $T = 400$ K performs a cycle consisting of four steps: (1) an isothermal expansion at $T = 400$ K to twice its initial volume, (2) cooling at constant volume to $T = 300$ K, (3) an isothermal compression to its original volume, and (4) heating at constant volume to its original temperature of 400 K. Assume that $C_v = 21$ J/K. Sketch the cycle on a PV diagram and find its efficiency.

**28** •• SSM One mole of an ideal monatomic gas at an initial volume $V_1 = 25$ L follows the cycle shown in Figure 19-15. All the processes are quasi-static. Find (a) the temperature of each state of the cycle, (b) the heat flow for each part of the cycle, and (c) the efficiency of the cycle.

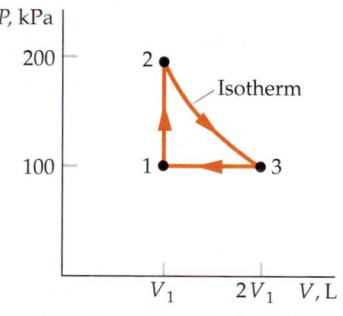

**FIGURE 19-15** Problem 28

**29** •• An ideal gas ($\gamma = 1.4$) follows the cycle shown in Figure 19-16. The temperature of state 1 is 200 K. Find (a) the temperatures of the other three states of the cycle and (b) the efficiency of the cycle.

**FIGURE 19-16** Problem 29

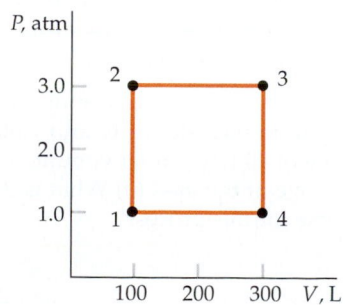

**30** ••• The *diesel cycle* shown in Figure 19-17 approximates the behavior of a diesel engine. Process ab is an adiabatic compression, process bc is an expansion at constant pressure, process cd is an adiabatic expansion, and process da is cooling at constant volume. Find the efficiency of this cycle in terms of the volumes $V_a$, $V_b$, $V_c$, and $V_d$.

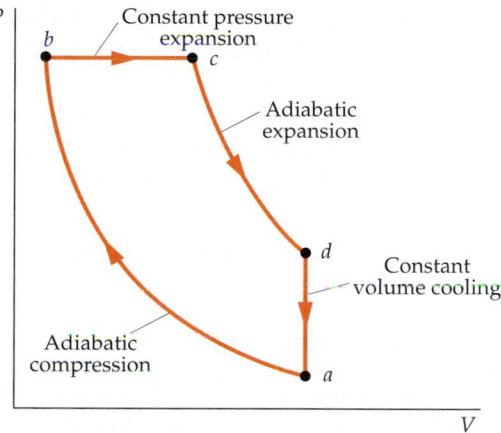

**FIGURE 19-17** Diesel cycle for Problem 30

**31** •• SSM "As far as we know, Nature has never evolved a heat engine"—Steven Vogel, *Life's Devices*, Princeton University Press (1988). (a) Calculate the efficiency of a heat engine operating between body temperature (98.6°F) and a typical outdoor temperature (70°F), and compare this to the human body's efficiency for converting chemical energy into work (approximately 20%). Does this contradict the Second Law of Thermodynamics? (b) From the result of Part (a), and a general knowledge of the conditions under which most warm-blooded animal life exists, explain why no warm-blooded animals have evolved heat engines to supply their internal energy.

**32** ••• The Clausius equation of state is $P(V - bn) = nRT$, where b is a constant. Show that the efficiency of a Carnot cycle is the same for a gas that obeys this equation of state as it is for one that obeys the ideal-gas equation of state, $PV = nRT$.

## Second Law of Thermodynamics

**33** •• A refrigerator takes in 500 J of heat from a cold reservoir and gives off 800 J to a hot reservoir. Assume that the heat-engine statement of the second law of thermodynamics is false, and show how a perfect engine working with this refrigerator can violate the refrigerator statement of the second law.

**34** •• SSM If two adiabatic curves intersect on a PV diagram, a cycle could be completed by an isothermal path between the two adiabatic curves shown in Figure 19-18. Show that such a cycle could violate the second law of thermodynamics.

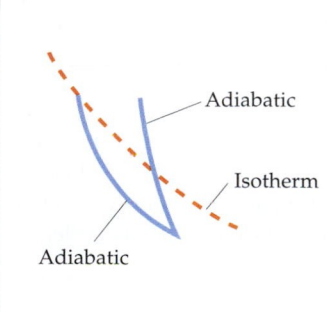

**FIGURE 19-18** Problem 34

## Carnot Engines

**35** • 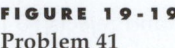 ✓ A Carnot engine works between two heat reservoirs at temperatures $T_h = 300$ K and $T_c = 200$ K. (*a*) What is its efficiency? (*b*) If it absorbs 100 J from the hot reservoir during each cycle, how much work does it do? (*c*) How much heat does it give off during each cycle? (*d*) What is the COP of this engine when it works as a refrigerator between the same two reservoirs?

**36** • **iSOLVE** An engine removes 250 J from a reservoir at 300 K and exhausts 200 J to a reservoir at 200 K. (*a*) What is its efficiency? (*b*) How much more work could be done if the engine were reversible?

**37** •• A reversible engine working between two reservoirs at temperatures $T_h$ and $T_c$ has an efficiency of 30%. Working as a heat engine, it gives off 140 J of heat to the cold reservoir. A second engine working between the same two reservoirs also gives off 140 J to the cold reservoir. Show that if the second engine has an efficiency greater than 30%, the two engines working together would violate the heat-engine statement of the second law.

**38** •• A reversible engine working between two reservoirs at temperatures $T_h$ and $T_c$ has an efficiency of 20%. Working as a heat engine, it does 100 J of work in each cycle. A second engine working between the same two reservoirs also does 100 J of work in each cycle. Show that if the efficiency of the second engine is greater than 20%, the two engines working together would violate the refrigerator statement of the second law.

**39** •• **SSM** A Carnot engine works between two heat reservoirs as a refrigerator. It does 50 J of work to remove 100 J from the cold reservoir and gives off 150 J to the hot reservoir during each cycle. Its coefficient of performance COP = $Q_c/W$ = (100 J)/(50 J) = 2. (*a*) What is the efficiency of the Carnot engine when it works as a heat engine between the same two reservoirs? (*b*) Show that no other engine working as a refrigerator between the same two reservoirs can have a COP greater than 2.

**40** •• **iSOLVE** A Carnot engine works between two heat reservoirs at temperatures $T_h = 300$ K and $T_c = 77$ K. (*a*) What is its efficiency? (*b*) If it absorbs 100 J from the hot reservoir during each cycle, how much work does it do? (*c*) How much heat does it give off in each cycle? (*d*) What is the coefficient of performance of this engine when it works as a refrigerator between these two reservoirs?

**41** •• In the cycle shown in Figure 19-19, 1 mol of an ideal gas ($\gamma = 1.4$) is initially at a pressure of 1 atm and a temperature of 0°C. The gas is heated at constant volume to $T_2 = 150$°C and is then expanded adiabatically until its pressure is again 1 atm. It is then compressed at constant pressure back to its original state. Find (*a*) the temperature $T_3$ after the adiabatic expansion, (*b*) the heat entering or leaving the system during each process, (*c*) the efficiency of this cycle, and (*d*) the efficiency of a Carnot cycle operating between the temperature extremes of this cycle.

**FIGURE 19-19**
**Problem 41**

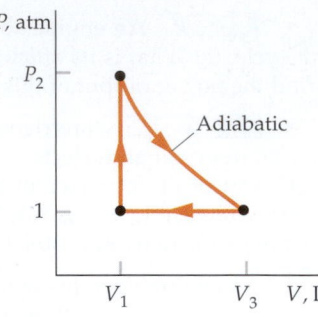

**42** •• **iSOLVE** A steam engine takes in superheated steam at 270°C and discharges condensed steam from its cylinder at 50°C. Its efficiency is 30%. (*a*) How does this efficiency compare with the maximum possible efficiency for these temperatures? (*b*) If the useful power output of the engine is 200 kW, how much heat does the engine discharge to its surroundings in 1 h?

## *Heat Pumps

**43** • **SSM** **iSOLVE** ✓ A heat pump delivers 20 kW to heat a house. The outside temperature is −10°C and the inside temperature of the hot-air supply for the heating fan is 40°C. (*a*) What is the coefficient of performance of a Carnot heat pump operating between these temperatures? (*b*) What must be the minimum power of the engine needed to run the heat pump? (*c*) If the COP of the heat pump is 60% of the efficiency of an ideal pump, what must be the minimum power of the engine?

**44** • **iSOLVE** ✓ A refrigerator is rated at 370 W. (*a*) What is the maximum amount of heat it can remove in 1 min if the inside temperature of the refrigerator is 0°C and it exhausts into a room at 20°C? (*b*) If the COP of the refrigerator is 70% of that of an ideal pump, how much heat can it remove in 1 min?

**45** • Rework Problem 44 for a room temperature of 35°C.

## Entropy Changes

**46** • What is the change in entropy of 1 mol of water at 0°C that freezes?

**47** •• Consider the freezing of 50 g of water by placing it in the freezer compartment of a refrigerator. Assume the walls of the freezer are maintained at −10°C. The water, initially liquid at 0°C, is frozen into ice and cooled to −10°C. Show that even though the entropy of the ice decreases, the net entropy of the universe increases.

**48** • **iSOLVE** Two moles of an ideal gas at $T = 400$ K expand quasi-statically and isothermally from an initial volume of 40 L to a final volume of 80 L. (*a*) What is the entropy change of the gas? (*b*) What is the entropy change of the universe for this process?

**49** • The gas in Problem 48 is taken from the same initial state ($T = 400$ K, $V_1 = 40$ L) to the same final state ($T = 400$ K, $V_2 = 80$ L) by a process that is not quasi-static. (*a*) What is the entropy change of the gas? (*b*) What can be said about the entropy change of the universe?

**50** • What is the change in entropy of 1.0 kg of water when it changes to steam at 100°C and a pressure of 1 atm?

**51** • [ISOLVE✔] What is the change in entropy of 1.0 kg of ice when it changes to water at 0°C and a pressure of 1 atm?

**52** •• A system absorbs 200 J of heat reversibly from a reservoir at 300 K and gives off 100 J reversibly to a reservoir at 200 K as it moves from state A to state B. During this process, the system does 50 J of work. (*a*) What is the change in the internal energy of the system? (*b*) What is the change in entropy of the system? (*c*) What is the change in entropy of the universe? (*d*) If the system goes from state A to state B by a nonreversible process, how would your answers for Parts (*a*), (*b*), and (*c*) differ?

**53** •• [SSM] [ISOLVE] A system absorbs 300 J from a reservoir at 300 K and 200 J from a reservoir at 400 K. It then returns to its original state, doing 100 J of work and rejecting 400 J of heat to a reservoir at a temperature $T$. (*a*) What is the entropy change of the system for the complete cycle? (*b*) If the cycle is reversible, what is the temperature $T$?

**54** •• [ISOLVE] Two moles of an ideal gas originally at $T = 400$ K and $V = 40$ L undergo a free expansion to twice their initial volume. What is (*a*) the entropy change of the gas and (*b*) the entropy change of the universe?

**55** •• A 200-kg block of ice at 0°C is placed in a large lake. The temperature of the lake is just slightly higher than 0°C, and the ice melts. (*a*) What is the entropy change of the ice? (*b*) What is the entropy change of the lake? (*c*) What is the entropy change of the universe (the ice plus the lake)?

**56** •• A 100-g piece of ice at 0°C is placed in an insulated container with 100 g of water at 100°C. (*a*) When thermal equilibrium is established, what is the final temperature of the water (Ignore the heat capacity of the container.) (*b*) Find the entropy change of the universe for this process.

**57** •• [SSM] A 1-kg block of copper at 100°C is placed in a calorimeter of negligible heat capacity containing 4 L of water at 0°C. Find the entropy change of (*a*) the copper block, (*b*) the water, and (*c*) the universe.

**58** •• If a 2-kg piece of lead at 100°C is dropped into a lake at 10°C, find the entropy change of the universe.

**59** •• [ISOLVE] A 1500-kg car traveling at 100 km/h crashes into a concrete wall. If the temperature of the air is 20°C, calculate the entropy change of the universe.

**60** •• [SSM] A box is divided into two identical halves by an impermeable partition through its middle. On one side is 1 mole of ideal gas A; on the other, 1 mole of ideal gas B (which is different from A). (*a*) Calculate the change in entropy when the partition is lifted, and the two gases mix together. (*b*) If we repeat the process with the same type of gas in each side, should the entropy change when the partition is lifted? Explain. (Think carefully about this question!)

## Entropy and Work Lost

**61** •• [SSM] [ISOLVE] If 500 J of heat is conducted from a reservoir at 400 K to one at 300 K, (*a*) what is the change in entropy of the universe, and (*b*) how much of the 500 J of heat conducted could have been converted into work using a cold reservoir at 300 K?

**62** •• One mole of an ideal gas undergoes a free expansion from $V_1 = 12.3$ L and $T_1 = 300$ K to $V_2 = 24.6$ L and $T_2 = 300$ K. It is then compressed isothermally and quasi-statically back to its original state. (*a*) What is the entropy change of the universe for the complete cycle? (*b*) How much work is wasted in this cycle? (*c*) Show that the work wasted is $T\Delta S_u$.

## General Problems

**63** • [ISOLVE✔] An engine with an output of 200 W has an efficiency of 30%. It works at 10 cycles/s. (*a*) How much work is done in each cycle? (*b*) How much heat is absorbed and how much is given off in each cycle?

**64** • [ISOLVE✔] In each cycle, an engine removes 150 J from a reservoir at 100°C and gives off 125 J to a reservoir at 20°C. (*a*) What is the efficiency of this engine? (*b*) What is the ratio of its efficiency to that of a Carnot engine working between the same reservoirs? (This ratio is called the *second law efficiency*.)

**65** • An engine removes 200 kJ of heat from a hot reservoir at 500 K in each cycle and exhausts heat to a cold reservoir at 200 K. Its efficiency is 85% of a Carnot engine working between the same reservoirs. (*a*) What is the efficiency of this engine? (*b*) How much work is done in each cycle? (*c*) How much heat is exhausted in each cycle?

**66** •• (*a*) Calvin Cliffs Nuclear Power Plant, located on the Hobbes River, generates 1 GW of power. In this plant liquid sodium circulates between the reactor core and a heat exchanger located in the superheated steam that drives the turbine. Heat is transferred into the liquid sodium in the core, and out of the liquid sodium (and into the superheated steam) in the heat exchanger. The temperature of the superheated steam is 500 K. Waste heat is dumped into the river, which flows by at a temperature of 25°C. (*a*) What is the highest efficiency that this plant can have? (*b*) How much waste heat is dumped into the river every second? (*c*) How much heat must be generated to supply 1 GW of power? (*d*) Assume that new, tough environmental laws have been passed (to preserve the unique wildlife of the river). Because of this, the plant is not allowed to heat the river by more than 0.5°C. What is the minimum flow rate that the Hobbes river must have (in L/sec)?

**67** • [ISOLVE] To maintain the temperature inside a house at 20°C, the power consumption of the electric baseboard heaters is 30 kW on a day when the outside temperature is −7°C. At what rate does this house contribute to the increase in the entropy of the universe?

**68** •• The system represented in Figure 19-12 (Problem 8) is 1 mol of an ideal monatomic gas. The temperatures at points A and B are 300 and 750 K, respectively. What is the thermodynamic efficiency of the cyclic process ABCDA?

**69** •• (a) Which process is more wasteful: (1) a block moving with 500 J of kinetic energy being slowed to rest by friction when the temperature of the atmosphere is 300 K, or (2) 1 kJ of heat being conducted from a reservoir at 400 K to one at 300 K? *Hint: How much of the 1 kJ of heat could be converted into work in an ideal situation?* (b) What is the change in entropy of the universe for each process?

**70** •• Helium gas ($\gamma = 1.67$) is initially at a pressure of 16 atm, a volume of 1 L, and a temperature of 600 K. It is expanded isothermally until its volume is 4 L and is then compressed at constant pressure until its volume and temperature are such that an adiabatic compression will return the gas to its original state. (a) Sketch this cycle on a *PV* diagram. (b) Find the volume and temperature after the isobaric compression. (c) Find the work done during each cycle. (d) Find the efficiency of the cycle.

**71** •• **SSM** A heat engine that does the work of blowing up a balloon at a pressure of 1 atm extracts 4 kJ from a hot reservoir at 120°C. The volume of the balloon increases by 4 L, and heat is exhausted to a cold reservoir at a temperature $T_c$. If the efficiency of the heat engine is 50% of the efficiency of a Carnot engine working between the same reservoirs, find the temperature $T_c$.

**72** •• Show that the COP of a Carnot refrigerator is related to the efficiency of a Carnot engine by $COP = T_c/(\varepsilon_C T_h)$.

**73** •• **iSOLVE** ✔ A freezer has a temperature $T_c = -23°C$. The air in the kitchen has a temperature $T_h = 27°C$. Since the heat insulation is not perfect, some heat flows into the freezer at a rate of 50 W. Find the power of the motor that is needed to maintain the temperature in the freezer.

**74** •• Two moles of a diatomic gas are taken through the cycle ABCA as shown on the *PV* diagram in Figure 19-20. At A the pressure and temperature are 5 atm and 600 K. The volume at B is twice that at A. The segment BC is an adiabatic expansion and the segment CA is an isothermal compression. (a) What is the volume of the gas at A? (b) What are the volume and temperature of the gas at B? (c) What is the temperature of the gas at C? (d) What is the volume of the gas at C? (e) How much work is done by the gas in each of the three segments of the cycle? (f) How much heat is absorbed by the gas in each segment of the cycle? (g) What is the thermodynamic efficiency of this cycle?

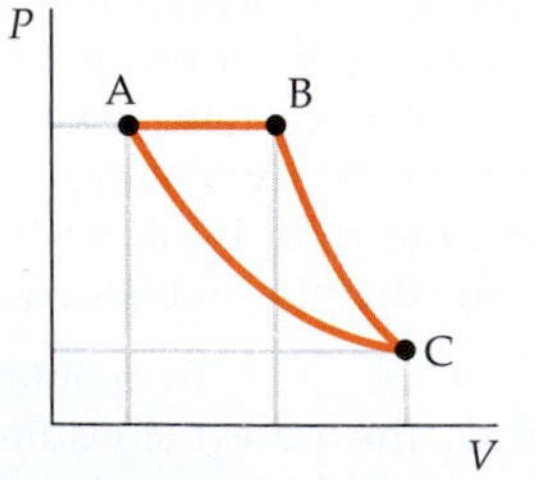

**FIGURE 19-20**
Problems 74 and 76

**75** •• Two moles of a diatomic gas are carried through the cycle ABCDA shown in the *PV* diagram in Figure 19-21. The segment AB represents an isothermal expansion, the

segment BC an adiabatic expansion. The pressure and temperature at A are 5 atm and 600 K. The volume at B is twice that at A. The pressure at D is 1 atm. (a) What is the pressure at B? (b) What is the temperature at C? (c) Find the work done by the gas in one cycle and the thermodynamic efficiency of this cycle.

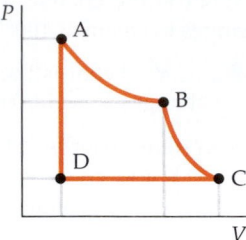

**FIGURE 19-21**
Problems 75 and 77

**76** •• Repeat Problem 74 for a monatomic gas.

**77** •• Repeat Problem 75 for a monatomic gas.

**78** •• Compare the efficiency of the Otto engine and the Carnot engine operating between the same maximum and minimum temperatures.

**79** ••• **SSM** Using the equation for the entropy change of an ideal gas when the volume and temperature change and $TV^{\gamma-1}$ is a constant, show explicitly that the entropy change is zero for a quasi-static adiabatic expansion from state ($V_1, T_1$) to state ($V_2, T_2$).

**80** ••• (a) Show that if the refrigerator statement of the second law of thermodynamics were not true, then the entropy of the universe could decrease. (b) Show that if the heat-engine statement of the second law were not true, then the entropy of the universe could decrease. (c) An alternative statement of the second law is that the entropy of the universe cannot decrease. Have you just proved that this statement is equivalent to the refrigerator and heat-engine statements?

**81** ••• Suppose that two heat engines are connected in series, such that the heat exhaust of the first engine is used as the heat input of the second engine as shown in Figure 19-22. The efficiencies of the engines are $\varepsilon_1$ and $\varepsilon_2$, respectively. Show that the net efficiency of the combination is given by

$$\varepsilon_{net} = \varepsilon_1 + (1 - \varepsilon_1)\varepsilon_2.$$

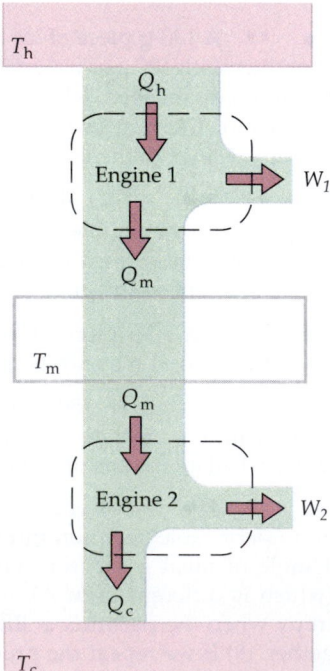

**FIGURE 19-22**
Problems 81 and 82

**82** ••• **SSM** Suppose that each engine in Figure 19-22 is an ideal reversible heat engine. Engine 1 operates between temperatures $T_h$ and $T_m$ and Engine 2 operates between $T_m$ and $T_c$, where $T_h > T_m > T_c$. Show that

$$\varepsilon_{net} = 1 - \frac{T_c}{T_h}$$

This means that two reversible heat engines in series are equivalent to one reversible heat engine operating between the hottest and coldest reservoirs.

**83** ••• Bertrand Russell once said that if a million monkeys were given a million typewriters and typed away at random for a million years, they would produce all of Shakespeare's works. Let's limit ourselves to the following fragment of Shakespeare (*Julius Caesar* III:ii):

*Friends, Romans, countrymen! Lend me your ears.*
*I come to bury Caesar, not to praise him.*
*The evil that men do lives on after them,*
*The good is oft interred with the bones.*
*So let it be with Caesar.*
*The noble Brutus hath told you that Caesar was ambitious,*
*And, if so, it were a grievous fault,*
*And grievously hath Caesar answered it . . .*

Even with this small fragment, it will take a lot longer than a million years! By what factor (roughly speaking) was Russell in error? Make any reasonable assumptions you want. (You may even assume that the monkeys are immortal.)

# Thermal Properties and Processes

EXPANSION JOINTS LIKE THIS ONE ARE INCLUDED ON STEEL BRIDGES TO ALLOW THE BRIDGE TO EXPAND AND CONTRACT WITH CHANGES IN TEMPERATURE WITHOUT PUTTING STRESS ON THE STEEL STRUCTURE.

**?** **What might eventually happen to this bridge if it did not have expansion joints? (See Example 20-1.)**

20-1    Thermal Expansion

20-2    The van der Waals Equation and Liquid–Vapor Isotherms

20-3    Phase Diagrams

20-4    The Transfer of Thermal Energy

**W**hen an object absorbs thermal energy, various changes may occur in the physical properties of the object. For example, its temperature may rise, accompanied by an expansion or contraction of the object, or the object may liquefy or vaporize, during which its temperature remains constant.

➤ **In this chapter, we examine some of the thermal properties of matter and some important processes involving thermal energy.**

## 20-1 Thermal Expansion

When the temperature of an object increases, the object typically expands. (Consider that on concrete highways, expansion joints appear every 10 to 15 m, allowing the road to expand without cracking.) Suppose that we have a long rod of length $L$ at a temperature $T$. When the temperature changes by $\Delta T$, the fractional change in length $\Delta L$ is proportional to $\Delta T$:

$$\frac{\Delta L}{L} = \alpha \Delta T \qquad\qquad 20\text{-}1$$

where $\alpha$, called the **coefficient of linear expansion,** is the ratio of the fractional change in length to the change in temperature:

$$\alpha = \frac{\Delta L/L}{\Delta T} \qquad\qquad 20\text{-}2$$

The units for the coefficient of linear expansion are reciprocal Celsius degrees (1/°C), which are the same as reciprocal kelvins (1/K). The value of $\alpha$ for a solid or liquid doesn't vary much with pressure, but it may vary significantly with temperature. Equation 20-2 gives the average value over the temperature interval $\Delta T$. The coefficient of linear expansion at a particular temperature $T$ is found by taking the limit as $\Delta T$ approaches zero:

$$\alpha = \lim_{\Delta T \to 0} \frac{\Delta L/L}{\Delta T} = \frac{1}{L}\frac{dL}{dT} \qquad\qquad 20\text{-}3$$

The accuracy obtained by using the average value of $\alpha$ over a wide temperature range is sufficient for most purposes.

The **coefficient of volume expansion** $\beta$ is similarly defined as the ratio of the fractional change in volume to the change in temperature (at constant pressure):

$$\beta = \lim_{\Delta T \to 0} \frac{\Delta V/V}{\Delta T} = \frac{1}{V}\frac{dV}{dT} \qquad\qquad 20\text{-}4$$

Like $\alpha$, $\beta$ does not usually vary with pressure for solids and liquids, but may vary with temperature. Average values for $\alpha$ and $\beta$ for various substances are given in Table 20-1.

For a given material, $\beta = 3\alpha$. We can show this by considering a box of dimensions $L_1$, $L_2$, and $L_3$. Its volume at a temperature $T$ is

$$V = L_1 L_2 L_3$$

The rate of change of the volume with respect to temperature is

$$\frac{dV}{dT} = L_1 L_2 \frac{dL_3}{dT} + L_1 L_3 \frac{dL_2}{dT} + L_2 L_3 \frac{dL_1}{dT}$$

Dividing each side of the equation by the volume, we obtain

$$\beta = \frac{1}{V}\frac{dV}{dT} = \frac{1}{L_3}\frac{dL_3}{dT} + \frac{1}{L_2}\frac{dL_2}{dT} + \frac{1}{L_1}\frac{dL_1}{dT}$$

We can see that each term on the right side of the above equation equals $\alpha$, and so we have

$$\beta = 3\alpha \qquad\qquad 20\text{-}5$$

Similarly, the coefficient of area expansion is twice that of linear expansion.

The increase in size of any part of an object for a given temperature change is proportional to the original size of that part of the body. For example, if we increase the temperature of a steel ruler, the effect will be similar to that of a (very slight) photographic enlargement. That is, the dimensions of the ruler itself will be larger, as will the distance between the equally spaced lines. **If the ruler has a 1-cm-diameter hole in it, say between the 3-cm and 4-cm lines, the hole will get larger, just as the distance between the 3-cm and 4-cm lines does.**

## TABLE 20-1

**Approximate Values of the Coefficients of Thermal Expansion for Various Substances**

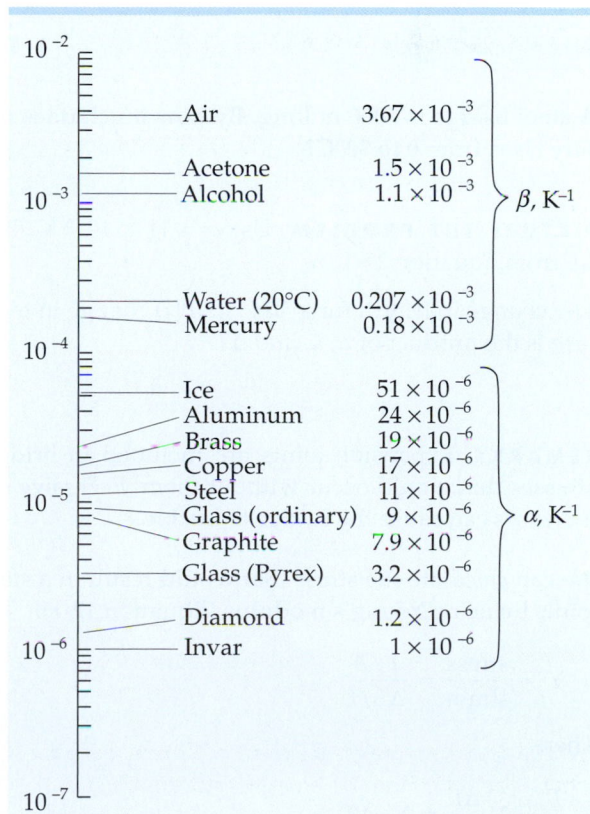

| | | |
|---|---|---|
| Air | $3.67 \times 10^{-3}$ | |
| Acetone | $1.5 \times 10^{-3}$ | |
| Alcohol | $1.1 \times 10^{-3}$ | $\beta$, K$^{-1}$ |
| Water (20°C) | $0.207 \times 10^{-3}$ | |
| Mercury | $0.18 \times 10^{-3}$ | |
| Ice | $51 \times 10^{-6}$ | |
| Aluminum | $24 \times 10^{-6}$ | |
| Brass | $19 \times 10^{-6}$ | |
| Copper | $17 \times 10^{-6}$ | |
| Steel | $11 \times 10^{-6}$ | |
| Glass (ordinary) | $9 \times 10^{-6}$ | $\alpha$, K$^{-1}$ |
| Graphite | $7.9 \times 10^{-6}$ | |
| Glass (Pyrex) | $3.2 \times 10^{-6}$ | |
| Diamond | $1.2 \times 10^{-6}$ | |
| Invar | $1 \times 10^{-6}$ | |

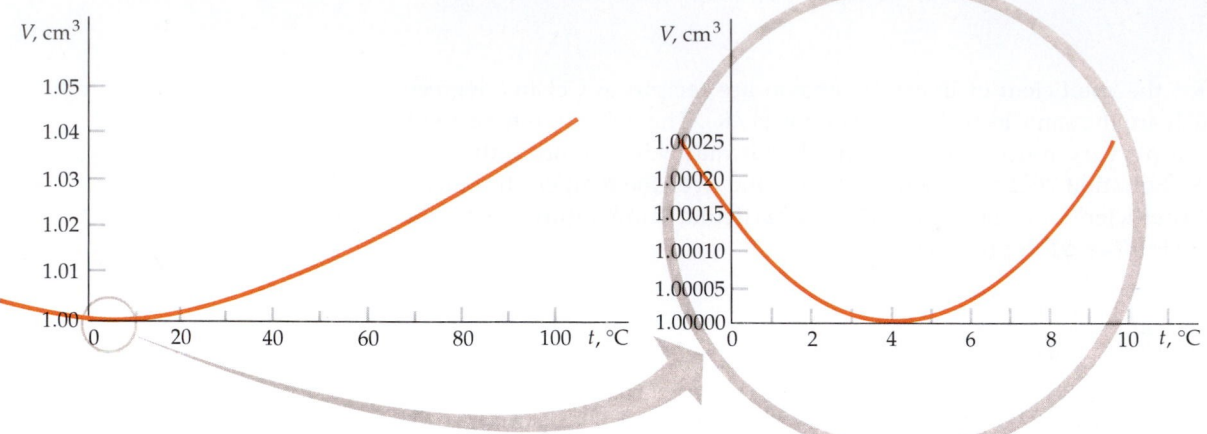

**FIGURE 20-1** Volume of 1 g of water at atmospheric pressure versus temperature. The minimum volume, which corresponds to the maximum density, occurs at 4°C. At temperatures below 0°C, the curve shown is for supercooled water. (Supercooled water is water that is cooled below the normal freezing point without solidifying.)

Most materials expand when heated and contract when cooled. Water, however, presents an important exception. Figure 20-1 shows the volume occupied by 1 g of water as a function of temperature. The minimum volume, and therefore the maximum density, is at 4°C. Thus, when water below 4°C is cooled, it expands rather than contracts and vice versa. This property of water has important consequences for the ecology of lakes. At temperatures above 4°C, the water in a lake becomes denser as it cools and therefore sinks to the bottom. But as the water cools below 4°C, it becomes less dense and rises to the surface. This is the reason that ice forms first on the surface of a lake. Since ice is less dense than liquid water, it remains at the surface and acts as a thermal insulator for the water below. If water behaved like most substances and contracted when it froze, then ice would sink and expose more water at the surface that would then freeze. Lakes would fill with ice from the bottom up and would be much more likely to freeze completely in the winter, killing fish and other aquatic life.

---

*AN EXPANDING BRIDGE*                       **EXAMPLE 20-1**

**A steel bridge is 1000 m long. By how much does it expand when the temperature rises from 0 to 30°C?**

**PICTURE THE PROBLEM** Use $\alpha = 11 \times 10^{-6}$ K$^{-1}$ from Table 20-1 and calculate $\Delta L$ from Equation 20-1.

The change in length for a 30 C° (30 K) change in temperature is the product of $\alpha$, $L$, and $\Delta T$:

$$\Delta L = \alpha L \Delta T = (11 \times 10^{-6} \text{ K}^{-1})(1000 \text{ m})(30 \text{ K})$$

$$= \boxed{0.33 \text{ m} = 33 \text{ cm}}$$

**REMARKS** Expansion joints are included in bridges to relieve the enormous stresses that would occur without them. Excessive stress caused by temperature increases can cause the bridge to buckle.

We can calculate the stress that would result in a steel bridge without expansion joints by using Young's modulus (Equation 12-1):

$$Y = \frac{\text{stress}}{\text{strain}} = \frac{F/A}{\Delta L/L}$$

Then

$$\frac{F}{A} = Y \frac{\Delta L}{L} = Y \alpha \Delta T$$

For $\Delta T = 30$ K, $\Delta L/L = 0.33$ m/1000 m as found in Example 20-1. Then using $Y = 2 \times 10^{11}$ N/m² (from Table 12-1),

$$\frac{F}{A} = Y\frac{\Delta L}{L} = (2 \times 10^{11}\,\text{N/m}^2)\,\frac{0.33\,\text{m}}{1000\,\text{m}} = 6.6 \times 10^7\,\text{N/m}^2$$

This stress is about one-third of the breaking stress for steel under compression. A compression stress of this magnitude would cause a steel bridge to buckle and become permanently deformed.

---

*A COMPLETELY FILLED GLASS*                     **EXAMPLE 20-2**

**While working in the laboratory, you fill a 1-L glass flask to the brim with water at 10°C. You heat the flask, raising the temperature of the water and flask to 30°C. How much water spills out of the flask?**

**PICTURE THE PROBLEM** The glass flask and the water both expand when heated, but the water expands more, so some spills out. We calculate the amount spilled by finding the changes in volume for $\Delta T = 20$ K using $\Delta V_a = \beta V\Delta T$ with $\beta = 1.1 \times 10^{-3}$ K$^{-1}$ for water (from Table 20-1), and $\Delta V_g = \beta V\Delta T = 3\alpha V\Delta T$ with $\alpha = 9 \times 10^{-6}$ K$^{-1}$ for glass. The difference in these volume changes equals the volume spilled.

1. The volume of water spilled $\Delta V_s$ is the difference in the changes in volume of the water and glass:

$$\Delta V_s = \Delta V_a - \Delta V_g$$

2. Find the increase in the volume of the water:

$$\Delta V_a = \beta_a V\Delta T$$

3. Find the increase in the volume of the glass flask:

$$\Delta V_g = \beta_g V\Delta T = 3\alpha_g V\Delta T$$

4. Subtract to find the amount of water spilled:

$$\Delta V_s = \Delta V_a - \Delta V_g = \beta_a V\Delta T - \beta_g V\Delta T$$
$$= (\beta_a - \beta_g)\,V\Delta T = (\beta_a - 3\alpha_g)\,V\Delta T$$
$$= [0.207 \times 10^{-3}\,\text{K}^{-1} - 3(9 \times 10^{-6}\,\text{K}^{-1})](1\,\text{L})(20\,\text{K})$$
$$= 3.6 \times 10^{-3}\,\text{L} = \boxed{3.6\,\text{mL}}$$

---

*BREAKING COPPER*                               **EXAMPLE 20-3**

**A copper bar is heated to 300°C. Then it is clamped rigidly between two fixed points so that it can neither expand nor contract. If the breaking stress of copper is 230 MN/m², at what temperature will the bar break as it cools?**

**PICTURE THE PROBLEM** As the bar cools, the change in length $\Delta L$ that *would* occur if the bar contracted is offset by an equal stretching due to tensile stress in the bar. The stress $F/A$ is related to the stretching $\Delta L$ by $Y = (F/A)/(\Delta L/L)$, where Young's modulus for copper is $Y = 110$ GN/m² (from Table 12-1). The maximum allowable stretching occurs when $F/A$ equals 230 MN/m². Thus, we find the temperature change that would produce this maximum contraction.

1. Calculate the change in length $\Delta L_1$ that would occur if the bar were unclamped and cooled by $\Delta T$:

$$\Delta L_1 = \alpha L\Delta T$$

2. A tensile stress $F/A$ stretches the bar by $\Delta L_2$, where $L_1 + \Delta L_2 = 0$:

$$Y = \frac{F/A}{\Delta L_2/L}, \quad \text{so} \quad \Delta L_2 = L\frac{F/A}{Y}$$

3. Substitute the step 1 and step 2 results into $\Delta L_1 + \Delta L_2 = 0$ and solve for $\Delta T$ with the stress equal to the breaking value:

$$\Delta L_1 + \Delta L_2 = 0$$

$$\alpha L \Delta T + L\frac{F/A}{Y} = 0$$

so

$$\Delta T = -\frac{F/A}{\alpha Y}$$

$$= -\frac{230 \times 10^6 \ \text{N/m}^2}{(17 \times 10^{-6} \ \text{K}^{-1})(110 \times 10^9 \ \text{N/m}^2)}$$

$$= -123 \ \text{K} = -123 \ \text{C}°$$

4. Add this result to the original temperature to find the final temperature at which the bar breaks:

$$T_f = T_1 + \Delta T = 300°\text{C} - 123 \ \text{C}° = \boxed{177°\text{C}}$$

# 20-2 The van der Waals Equation and Liquid–Vapor Isotherms

At ordinary pressures most gases behave like an ideal gas. However, this ideal behavior breaks down when the pressure is high enough or the temperature is low enough such that the density of the gas is high and the molecules are, on average, closer together. An equation of state called the **van der Waals equation** describes the behavior of many real gases over a wide range of pressures more accurately than does the ideal-gas equation of state ($PV = nRT$). The van der Waals equation for $n$ moles of gas is

$$\left(P + \frac{an^2}{V^2}\right)(V - bn) = nRT \qquad \text{20-6}$$

The constant $b$ in this equation arises because the gas molecules are not point particles but objects that have a finite size; therefore, the volume available to each molecule is reduced. The magnitude of $b$ is the volume of one mole of gas molecules. The term $an^2/V^2$ arises from the attraction of the gas molecules to each other. As a molecule approaches the wall of the container, it is pulled back by the molecules surrounding it with a force that is proportional to the density of those molecules $n/V$. Because the number of molecules that hit the wall in a given time is also proportional to the density of the molecules, the decrease in pressure due to the attraction of the molecules is proportional to the square of the density and therefore to $n^2/V^2$. The constant $a$ depends on the gas and is small for inert gases, which have very weak chemical interactions. The terms $bn$ and $an^2/V^2$ are both negligible when the volume $V$ is large, so at low densities the van der Waals equation approaches the ideal-gas law. At high densities the van der Waals equation provides a much better description of the behavior of real gases than does the ideal-gas law.

Figure 20-2 shows $PV$ isothermal curves for a substance at various temperatures. Except for the region where the liquid and vapor coexist, these curves are described quite accurately by the van der Waals equation and can be used to determine the constants $a$ and $b$. For example, the values of these constants that give the best fit to the experimental curves for nitrogen are $a = 0.14 \ \text{Pa·m}^6/\text{mol}^2$ and $b = 39.1 \ \text{mL/mol}$. This volume of 39.1 mL per mole is about 0.2 percent of the

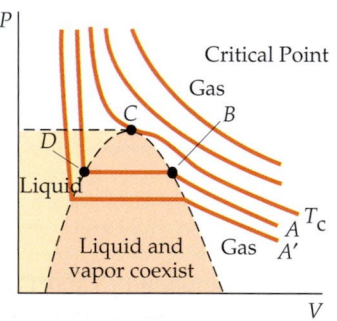

**FIGURE 20-2** Isotherms on the $PV$ diagram for a substance. For temperatures above the critical temperature $T_c$, the substance remains a gas at all pressures. Except for the region where the liquid and vapor coexist, these curves are described quite well by the van der Waals equation. The pressure for the horizontal portions of the curves in the shaded region is the vapor pressure, which is the pressure at which the vapor and liquid are in equilibrium. To the left of the shaded region for temperatures below the critical temperature, the substance is a liquid and is nearly incompressible.

volume of 22.4 L occupied by 1 mol of nitrogen under standard conditions. Since the molar mass of nitrogen is 28 g/mol, if 1 mol of nitrogen molecules were packed into a volume of 39.1 mL, then the density would be

$$\rho = \frac{M}{V} = \frac{28 \text{ g}}{39.1 \text{ mL}} = 0.72 \text{ g/mL} = 0.72 \text{ kg/L}$$

which is almost the same as the density of liquid nitrogen, 0.80 kg/L.

The value of the constant $b$ can be used to estimate the size of a molecule. Since 1 mol ($N_A$ molecules) of nitrogen has a volume of 39.1 cm³, the volume of one nitrogen molecule is

$$V = \frac{b}{N_A} = \frac{39.1 \text{ cm}^3/\text{mol}}{6.02 \times 10^{23} \text{ molecules/mol}} = 6.50 \times 10^{-23} \text{ cm}^3/\text{molecule}$$

If we assume that each molecule occupies a cube of side $d$, we obtain

$$d^3 = 6.50 \times 10^{-23} \text{ cm}^3$$

or

$$d = 4.0 \times 10^{-8} \text{ cm} = 0.4 \text{ nm}$$

which is a plausible estimate for the "diameter" of a nitrogen molecule.

At temperatures below $T_c$, the van der Waals equation describes those portions of the isotherms outside the shaded region in Figure 20-2 but not those portions inside the shaded region. Suppose we have a gas at a temperature below $T_c$ that initially has a low pressure and a large volume. We begin to compress the gas while holding the temperature constant (isotherm $A$ in the figure). At first the pressure rises, but when we reach point $B$ on the dashed curve, the pressure ceases to rise and the gas begins to liquefy at constant pressure. Along the horizontal line $BD$ in the figure, the gas and liquid are in equilibrium. As we continue to compress the gas, more and more gas liquefies until point $D$ on the dashed curve, at which point we have only liquid. Then, if we try to compress the substance further, the pressure rises sharply because a liquid is nearly incompressible.

Now consider injecting a liquid such as water into a sealed evacuated container. As some of the water evaporates, water-vapor molecules fill the previously empty space in the container. Some of these molecules will hit the liquid surface and rejoin the liquid water in a process called condensation. Initially, the rate of evaporation will be greater than the rate of condensation, but eventually equilibrium will be reached. The pressure at which a liquid is in equilibrium with its own vapor is called the **vapor pressure.** If we now heat the container slightly, the liquid boils, more liquid evaporates, and a new equilibrium is established at a higher vapor pressure. Vapor pressure thus depends on the temperature. We can see this from Figure 20-2. If we had started compressing the gas at a lower temperature, as with isotherm $A'$ in Figure 20-2, the vapor pressure would be lower, as is indicated by the horizontal constant-pressure line for $A'$ at a lower value of pressure. The temperature for which the vapor pressure for a substance equals 1 atm is the **normal boiling point** of that substance. For example, the temperature at which the vapor pressure of water is 1 atm is 373 K (= 100°C), so this temperature is the normal boiling point of water. At high altitudes, such

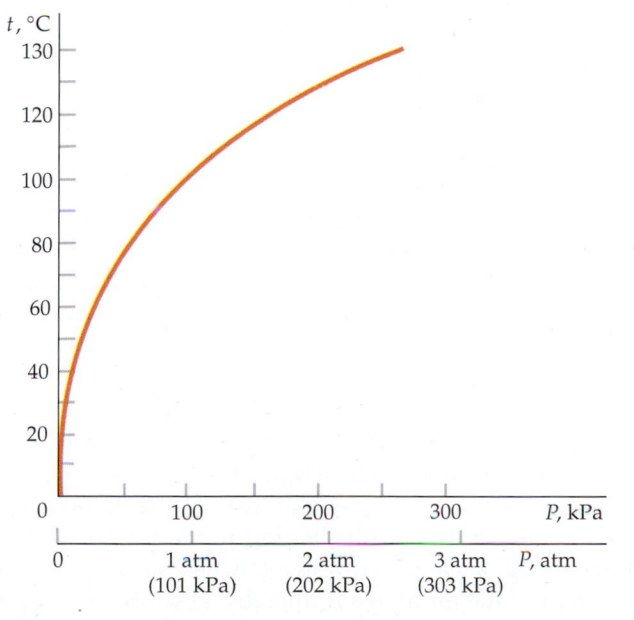

**FIGURE 20-3** Boiling point of water versus temperature.

as on the top of a mountain, the pressure is less than 1 atm, therefore, water boils at a temperature lower than 373 K. Figure 20-3 gives the vapor pressures of water at various temperatures.

At temperatures greater than the critical temperature $T_c$, a gas will not liquefy at any pressure. The critical temperature for water vapor is 647 K (= 374°C). The point at which the critical isotherm intersects the dashed curve in Figure 20-2 (point C) is called the **critical point.**

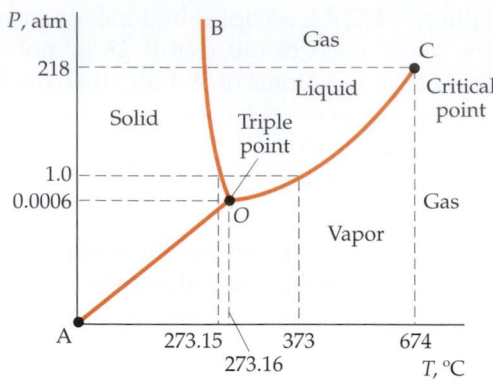

# 20-3 Phase Diagrams

Figure 20-4 is a plot of pressure versus temperature at a constant volume for water. Such a plot is called a **phase diagram.** The portion of the diagram between points $O$ and $C$ shows vapor pressure versus temperature. As we continue to heat the container, the density of the liquid decreases and the density of the vapor increases. At point $C$ on the diagram, these densities are equal. Point $C$ is called the **critical point.** At this point and above it, there is no distinction between the liquid and the gas. Critical-point temperatures $T_c$ for various substances are listed in Table 20-2. At temperatures greater than the critical temperature a gas will not liquefy at any pressure.

If we now cool our container, some of the vapor condenses into a liquid as we move back down the curve $OC$ until the substance reaches point $O$ in Figure 20-4. At this point, the liquid begins to solidify. Point $O$ is the **triple point,** that one point at which the vapor, liquid, and solid phases of a substance can coexist in equilibrium. Every substance has a unique triple point at a specific temperature and pressure. The triple-point temperature for water is 273.16 K (= 0.01°C) and the triple-point pressure is 4.58 mmHg.

At temperatures and pressures below the triple point, the liquid cannot exist. The curve $OA$ in the phase diagram of Figure 20-4 is the locus of pressures and temperatures for which the solid and vapor coexist in equilibrium. The direct change from a solid to a vapor is called **sublimation.** You can observe sublimation by putting a few loose ice cubes in the freezer compartment of a no-frost (self-defrosting) refrigerator. Over time, the ice cubes will shrink and eventually disappear due to sublimation. This happens because the atmospheric pressure is well above the triple-point pressure of water, and therefore, equilibrium is never established between the ice and water vapor. The triple-point temperature and pressure of carbon dioxide ($CO_2$) are 216.55 K and 3880 mmHg, which means that liquid $CO_2$ can only exist at pressures above 3880 mmHG (= 5.1 atm). Thus, at ordinary atmospheric pressures, liquid carbon dioxide cannot exist at any temperature. When solid carbon dioxide "melts," it sublimates directly into gaseous $CO_2$ without going through the liquid phase, hence the name "dry ice."

The curve $OB$ in Figure 20-4 is the melting curve separating the liquid and solid phases. For a substance like water for which the melting temperature decreases as the pressure increases, curve $OB$ slopes upward to the left from the triple point, as in this figure. For most other substances, the melting temperature increases as the pressure increases. For such a substance, curve $OB$ slopes upward to the right from the triple point.

For a molecule to escape (evaporate) from a substance in the liquid state, energy is required to break the molecular bonds at the liquid's surface. Vaporization cools the liquid left behind. If water is brought to a boil over heat, this cooling effect keeps the temperature of the liquid constant at the boiling point. This is the reason that the boiling point of a substance can be used to calibrate thermometers. However, water can also be made to boil without adding heat by evacuating the air above it, thereby lowering the applied pressure. The energy needed for vaporization is then taken from the water left behind. As a result, the water will cool down, even to the point that ice forms on top of the boiling water!

**FIGURE 20-4** Phase diagram for water. The pressure and temperature scales are not linear but are compressed to show the interesting points. Curve $OC$ is the curve of vapor pressure versus temperature. Curve $OB$ is the melting curve, and curve $OA$ is the sublimation curve.

## TABLE 20-2

**Critical Temperatures $T_c$ for Various Substances**

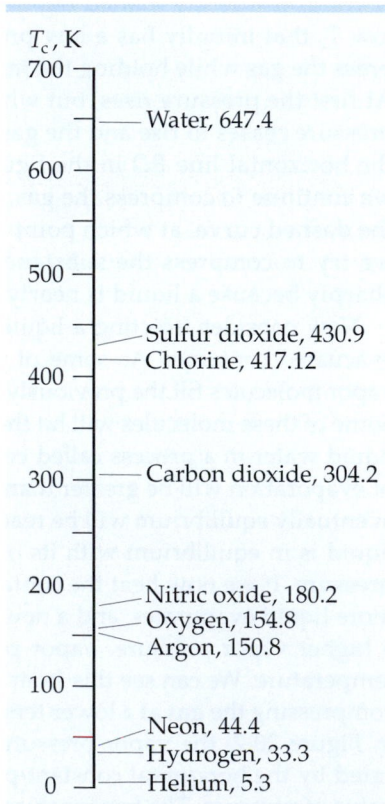

# 20-4 The Transfer of Thermal Energy

Thermal energy is transferred from one place to another by three processes: conduction, convection, and radiation.

In **conduction,** energy is transferred as heat by interactions among atoms or molecules, although there is no transport of the atoms or molecules themselves. For example, if one end of a solid bar is heated, the atoms in the heated end vibrate with greater energy than do those at the cooler end. The interaction of the more energetic atoms with their neighbors causes this energy to be transported along the bar.[†]

In **convection,** energy is transported as heat by direct mass transport. For example, warm air in a region of a room expands, its density decreases, and the buoyant force on it due to the surrounding air causes it to rise. Energy is thus transported upward along with the mass of warm air.

In **radiation,** energy is transported as heat through space in the form of electromagnetic waves that move at the speed of light. Thermal radiation, light waves, radio waves, television waves, and X rays are all forms of electromagnetic radiation that differ from one another in their wavelengths and frequencies.

In all mechanisms of heat transfer, the rate of cooling of a body is approximately proportional to the temperature difference between the body and its surroundings. This result is known as **Newton's law of cooling.**

In many real situations, all three mechanisms for heat transfer occur simultaneously, though one may be more dominant than the others. For example, an ordinary space heater uses both radiation and convection. If the heating element is quartz, then the main mechanism of heat transference is radiation. If the heating element is metal (which does not radiate as efficiently as quartz), then convection is the main mechanism by which heat is transmitted, with the heated air rising to be replaced by cooler air. Fans are often included in heaters to speed the convection process.

## Conduction

Figure 20-5a shows an insulated uniform solid bar of cross-sectional area $A$. If we keep one end of the bar at a high temperature and the other end at a low temperature, energy is conducted down the bar from the hot end to the cold end. In the steady state, the temperature varies linearly from the hot end to the cold end. The rate of change of the temperature along the bar $dT/dx$ is called the **temperature gradient.**

Let $\Delta T$ be the temperature difference across a small segment of length $\Delta x$ (Figure 20-5b). If $\Delta Q$ is the amount of heat conducted through the segment in some time $\Delta t$, then the rate of conduction of heat $\Delta Q/\Delta t$ is called the thermal current $I$. Experimentally, it is found that the thermal current is proportional to the temperature gradient and to the cross-sectional area $A$:

$$I = \frac{\Delta Q}{\Delta t} = kA\frac{\Delta T}{\Delta x} \qquad\qquad 20\text{-}7$$

DEFINITION — THERMAL CURRENT

The proportionality constant $k$ called the *thermal conductivity,* depends on the composition of the bar.[‡] In SI units, thermal current is expressed in watts, and the thermal conductivity has units of W/(m·K).[§] In practical calculations in the

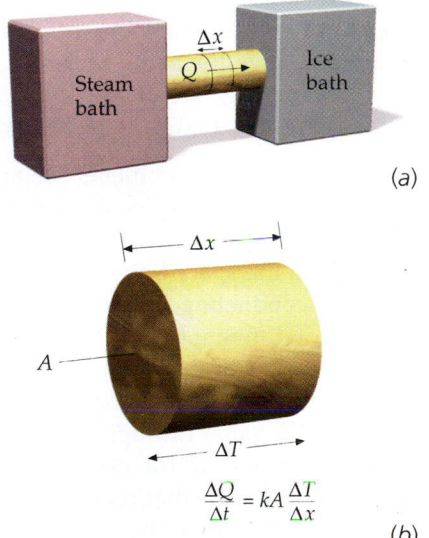

(a)

$$\frac{\Delta Q}{\Delta t} = kA\frac{\Delta T}{\Delta x}$$

(b)

**FIGURE 20-5** (*a*) An insulated conducting bar with its ends at two different temperatures. (*b*) A segment of the bar of length $\Delta x$. The rate at which thermal energy is conducted across the segment is proportional to the cross-sectional area of the bar and the temperature difference across the segment, and it is inversely proportional to the length of the segment.

---

† If the solid is a metal, the transport of thermal energy is helped by free electrons, which move throughout the metal.
‡ Don't confuse the thermal conductivity with Boltzmann's constant, which is also designated by $k$.
§ In some tables, the energy may be given in calories or kilocalories and the thickness in centimeters.

United States, the thermal current is usually expressed in Btu per hour, the area in square feet, the length (or thickness) in inches, and the temperature in degrees Fahrenheit. The thermal conductivity is then given in Btu·in./(h·ft²·F°). Table 20-3 gives the thermal conductivities of various materials.

If we solve Equation 20-7 for the temperature difference, we obtain

$$\Delta T = I\frac{\Delta x}{kA} \qquad \text{20-8}$$

or

$$\Delta T = IR \qquad \text{20-9}$$

TEMPERATURE CHANGE VERSUS CURRENT

where $\Delta x/(kA)$ is the **thermal resistance** $R$:

$$R = \frac{\Delta x}{kA} \qquad \text{20-10}$$

DEFINITION—THERMAL RESISTANCE

**EXERCISE** Calculate the thermal resistance of an aluminum slab of cross-sectional area 15 cm² and thickness 2 cm. (*Answer* 0.0563 K/W = 56.3 mK/W)

**EXERCISE** What thickness of silver would be required to give the same thermal resistance as a 1-cm thickness of air of the same area? (*Answer* $\Delta x = (1 \text{ cm})(429)/(0.026) = 16{,}500 \text{ cm} = 165 \text{ m}$)

## TABLE 20-3

**Thermal Conductivities $k$ for Various Materials**

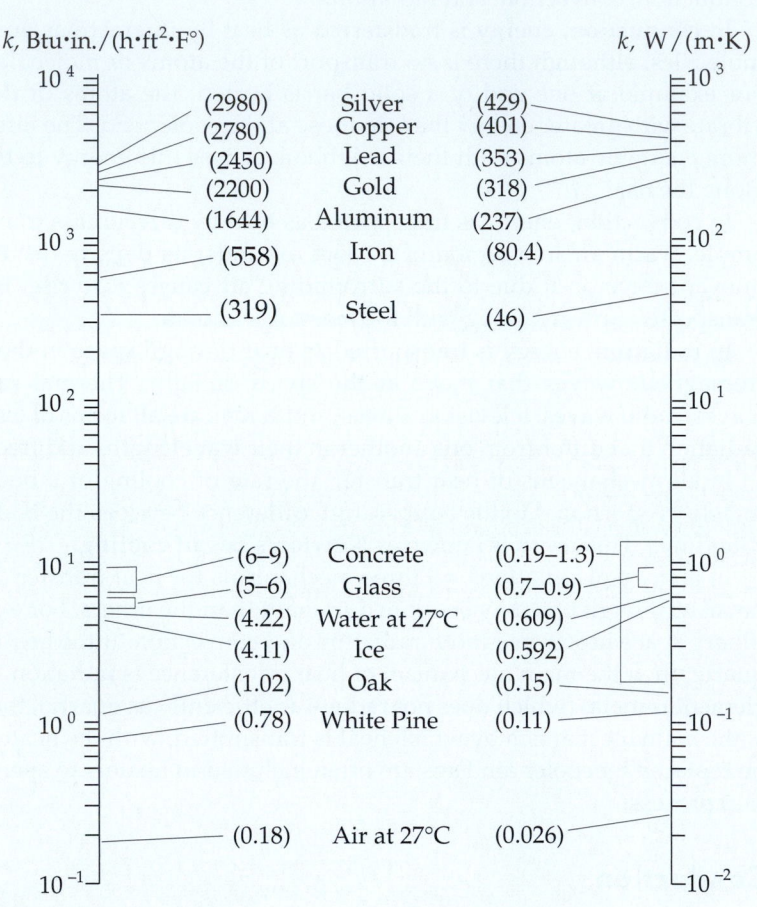

| $k$, Btu·in./(h·ft²·F°) | | $k$, W/(m·K) |
|---|---|---|
| (2980) | Silver | (429) |
| (2780) | Copper | (401) |
| (2450) | Lead | (353) |
| (2200) | Gold | (318) |
| (1644) | Aluminum | (237) |
| (558) | Iron | (80.4) |
| (319) | Steel | (46) |
| (6–9) | Concrete | (0.19–1.3) |
| (5–6) | Glass | (0.7–0.9) |
| (4.22) | Water at 27°C | (0.609) |
| (4.11) | Ice | (0.592) |
| (1.02) | Oak | (0.15) |
| (0.78) | White Pine | (0.11) |
| (0.18) | Air at 27°C | (0.026) |

In many practical problems, we are interested in the flow of heat through two or more conductors (or insulators) in series. For example, we may want to know the effect of adding insulating material of a certain thickness and thermal conductivity to the space between two layers of wallboard. Figure 20-6 shows two thermally conducting slabs of the same cross-sectional area but of different materials and of different thickness. Let $T_1$ be the temperature on the warm side, $T_2$ be the temperature at the interface between the slabs, and $T_3$ be the temperature on the cool side. Under the conditions of steady-state heat flow, the thermal current $I$ must be the same through both slabs. This follows from energy conservation; for steady-state flow, the rate at which energy enters any region must equal the rate at which it exits that region.

If $R_1$ and $R_2$ are the thermal resistances of the two slabs, we have from Equation 20-9 for each slab

$$T_1 - T_2 = IR_1$$

and

$$T_2 - T_3 = IR_2$$

Adding these equations gives

$$\Delta T = T_1 - T_3 = I(R_1 + R_2) = IR_{eq}$$

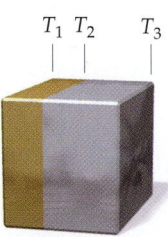

**FIGURE 20-6** Two thermally conducting slabs of different materials in series. The equivalent thermal resistance of the slabs in series is the sum of their individual thermal resistances. The thermal current is the same through both slabs.

or

$$I = \frac{\Delta T}{R_{eq}}$$
20-11

where $R_{eq}$ is the **equivalent resistance.** Thus, for thermal resistances in series, the equivalent resistance is the sum of the individual resistances:

$$R_{eq} = R_1 + R_2 + \cdots$$
20-12

THERMAL RESISTANCES IN SERIES

This result can be applied to any number of resistances in series. In Chapter 25, we will find that the same formula applies to electrical resistances in series.

To calculate the amount of heat leaving a room by conduction in a given time, we need to know how much heat leaves through the walls, the windows, the floor, and the ceiling. For this type of problem, in which there are several paths for heat flow, the resistances are said to be in parallel. The temperature difference is the same for each path, but the thermal current is different. The total thermal current is the sum of the thermal currents through each of the parallel paths:

This thermogram of a house shows the heat energy being radiated to its surroundings.

$$I_{total} = I_1 + I_2 + \cdots = \frac{\Delta T}{R_1} + \frac{\Delta T}{R_2} + \cdots = \Delta T\left(\frac{1}{R_1} + \frac{1}{R_2} + \cdots\right)$$

or

$$I_{total} = \frac{\Delta T}{R_{eq}}$$
20-13

where the equivalent thermal resistance is given by

$$\frac{1}{R_{eq}} = \frac{1}{R_1} + \frac{1}{R_2} + \cdots$$
20-14

THERMAL RESISTANCES IN PARALLEL

We will encounter this equation again in Chapter 25 when we study electric conduction through parallel resistances. Note that for both resistors in series (Equation 20-11) and resistors in parallel (Equation 20-13) $I$ is proportional to $\Delta T$, which is in agreement with Newton's law of cooling.

**Try It Yourself**

THERMAL CURRENT BETWEEN TWO METAL BARS **EXAMPLE 20-4**

Two insulated metal bars, each of length 5 cm and rectangular cross section with sides 2 cm and 3 cm, are wedged between two walls, one held at 100°C and the other at 0°C. (Figure 20-7). The bars are lead and silver. Find (a) the total thermal current through the two-bar combination, and (b) the temperature at the interface.

**FIGURE 20-7** Two thermally conducting slabs of different materials in parallel.

**PICTURE THE PROBLEM** The bars are thermal resistors connected in series. (a) You can find the total thermal current from $I = R_{eq}/\Delta T$, where the equivalent resistance $R_{eq}$ is the sum of the individual resistances. Using Equation 20-10 and the thermal conductivities given in Table 20-3, the individual resistances can be determined. (b) You can find the temperature at the interface by applying $I = R_1/\Delta T$ to the lead bar only, and solving for $\Delta T$ in terms of the value for $I$ found in Part (a).

**Cover the column to the right and try these on your own before looking at the answers.**

| Steps | Answers |
|---|---|
| (a) 1. Write the equivalent thermal resistance in terms of the thermal resistances of the two bars. | $R_{eq} = R_{Pb} + R_{Ag}$ |
| 2. Using Equation 20-10, write each resistance in terms of the individual thermal conductivities and geometric parameters: | $R_{Pb} = \dfrac{\Delta x_{Pb}}{k_{Pb}A_{Pb}}, R_{Ag} = \dfrac{\Delta x_{Ag}}{k_{Ag}A_{Ag}}$ |
| 3. Use Equation 20-13 to find the thermal current. | $I = \Delta T/R_{eq} = \boxed{232 \text{ W}}$ |
| (b) 1. Calculate the temperature difference across the lead bar using the current and thermal resistance found in Part (a). | $\Delta T_{Pb} = IR_{Pb} = 54.9 \text{ K} = 54.9°C$ |
| 2. Use your result from the previous step to find the temperature at the interface. | $T_{if} = 100°C - \Delta T_{Pb} = \boxed{45.1°C}$ |
| 3. Check your result by finding the temperature difference across the silver bar. | $\Delta T_{Ag} = IR_{Ag} = 45.1°C$ |

---

*THE METAL BARS REARRANGED*      **EXAMPLE 20-5**

The metal bars in Example 20-4 are rearranged as shown in Figure 20-8. Find (a) the thermal current in each bar, (b) the total thermal current, and (c) the equivalent thermal resistance of the two-bar system.

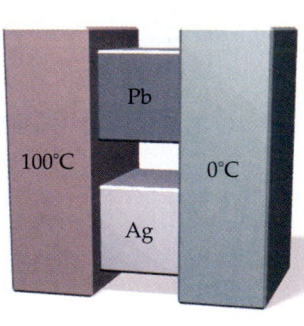

**FIGURE 20-8**

**PICTURE THE PROBLEM** The current in each bar is found from $I = \Delta T/R$, where $R$ is the thermal resistance of the bar (found in Example 20-4). The total current is the sum of the currents. The equivalent resistance can be found from Equation 20-14 or from $I_{total} = \Delta T/R_{eq}$.

(a) Calculate the thermal current for each bar:

$$I_{Pb} = \frac{\Delta T}{R_{Pb}} = \frac{100 \text{ K}}{0.236 \text{ K/W}} = \boxed{424 \text{ W}}$$

$$I_{Ag} = \frac{\Delta T}{R_{Ag}} = \frac{100 \text{ K}}{0.194 \text{ K/W}} = \boxed{515 \text{ W}}$$

(b) Add your results to find the total thermal current:

$$I_{total} = I_{Pb} + I_{Ag}$$

$$= 424 \text{ W} + 515 \text{ W} = \boxed{938 \text{ W}}$$

(c) 1. Use Equation 20-14 to calculate the equivalent resistance of the two bars in parallel:

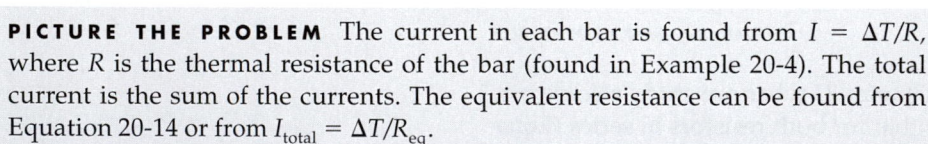

$$\frac{1}{R_{eq}} = \frac{1}{R_{Pb}} + \frac{1}{R_{Ag}} = \frac{1}{0.236 \text{ W}} + \frac{1}{0.194 \text{ W}}, \quad \text{so}$$

$$R_{eq} = \boxed{0.107 \text{ K/W}}$$

2. Check your result using, $I_{total} = \Delta T / R_{eq}$:

$$I_{total} = \frac{\Delta T}{R_{eq}};$$

$$R_{eq} = \frac{\Delta T}{I_{total}} = \frac{100 \text{ K}}{938 \text{ W}} = 0.107 \text{ K/W}$$

**REMARKS** Note that the equivalent resistance is less than either of the individual resistances. This is always the case for parallel resistors.

In the building industry, the thermal resistance of a square foot of cross-sectional area of a material is called its **R factor** $R_f$. Consider a 32 ft² sheet of insulating material with thickness $\Delta x$ and R factor $R_f$ of 7.2. That is, each square foot (Figure 20-9) has a thermal resistance of 7.2 F°/(Btu/h). The 32 square feet are in parallel, so the net resistance $R_{net}$ is calculated using Equation 20-14 giving

$$\frac{1}{R_{net}} = \frac{1}{R_f} + \frac{1}{R_f} + \cdots = \frac{32}{R_f}, \quad \text{so} \quad R_{total} = \frac{R_f}{32}$$

Thus, the total thermal resistance $R$ in F°/(Btu/h) equals the $R$ factor divided by the area $A$ in square feet. That is

$$R_{net} = \frac{R_f}{A}$$

**FIGURE 20-9** For a 1-in. thickness of this material, the $R_f = 7.2$.

Since the total resistance $R_{total}$ is related to the conductivity by $R_{net} = \Delta x/(kA)$ (Equation 20-10), we can express the $R$ factor by

$$R_f = R_{net}A = \frac{\Delta x}{k} \qquad \text{20-15}$$

DEFINITION—R FACTOR

where $\Delta x$ is the thickness in inches and $k$ is the conductivity in Btu·in./(h·ft²·F°). Table 20-4 lists $R$ factors for several materials. In terms of the $R$ factor, Equation 20-9 for the thermal current is

$$\Delta T = IR_{net} = \frac{I}{A}R_f \qquad \text{20-16}$$

For slabs of insulating material of the same area in series, $R_f$ is replaced by the equivalent $R$ factor $R_{f,eq}$

$$R_{f,eq} = R_{f1} + R_{f2} + \ldots$$

For parallel slabs, we calculate the thermal current through each slab and add all these currents together in order to obtain the total current.

## TABLE 20-4

**R Factors $\Delta x/k$ for Various Building Materials**

| Material | Thickness, in. | $R_f$, h·ft²·F°/Btu |
|---|---|---|
| Building board | | |
| Gypsum or plasterboard | 0.375 | 0.32 |
| Plywood (Douglas fir) | 0.5 | 0.62 |
| Plywood or wood panels | 0.75 | 0.93 |
| Particle board, medium density | 1.0 | 1.06 |
| Finish flooring materials | | |
| Carpet and fibrous pad | 1.0 | 2.08 |
| Tile | | 0.5 |
| Wood, hardwood finish | 0.75 | 0.68 |
| Roof insulation | 1.0 | 2.8 |
| Roofing | | |
| Asphalt roll roofing | | 0.15 |
| Asphalt shingles | | 0.44 |
| Windows | | |
| Single-pane | | 0.9 |
| Double pane | | 1.8 |

**EXAMPLE 20-6** **Put It in Context**

You are helping your friend's family put new asphalt shingles on the roof of their winter cabin. The 60 ft × 20 ft roof is made of 1-in. pine board covered with asphalt shingles. There is room for 2 in. of roof insulation, and your friend's family is wondering how much of a difference it would make to their energy bill if they were to install the two inches of insulation. Knowing that you are studying physics, they ask for your opinion.

**PICTURE THE PROBLEM** To assess the situation, you first calculate the $R$ factor for each layer of the roof. Since the layers are in series, the equivalent $R$ factor is just the sum of the individual $R$ factors. The aim is to calculate the equivalent $R$ factor of the roof with and without the insulation. The $R$ factors for asphalt shingles and for roof insulation are found in Table 20-4. The $R$ factor for the pine board is calculated from its thermal conductivity, which is found in Table 20-3. Note that when you shingle a roof you have to overlap the shingles, so there are two layers of asphalt shingling on the roof.

1. The equivalent $R$ factor is the sum of the individual $R$ factors:

$$R_{f,eq} = R_{f,pine} + R_{f,asph} + R_{f,insul}$$

2. The $R$ factor for the double layer of shingles is twice the $R$ factor for one layer:

$$R_{f,asph} = 2(0.44 \text{ h·ft}^2\text{·F°/Btu})$$
$$= 0.88 \text{ h·ft}^2\text{·F°/Btu}$$

3. The $R$ factor for 2 in. of roof insulation is twice that for 1 in.:

$$R_{f,insul} = 2(2.8 \text{ h·ft}^2\text{·F°/Btu})$$
$$= 5.6 \text{ h·ft}^2\text{·F°/Btu}$$

4. The $R$ factor for 1-in.-thick pine is obtained from the conductivity:

$$R_{f,p} = \frac{\Delta x_p}{k_p} = \frac{1 \text{ in.}}{0.78 \text{ Btu·in.}/(\text{h·ft}^2\text{·F°})}$$
$$= 1.28 \text{ h·ft}^2\text{·F°/Btu}$$

5. The equivalent $R$ factor without the insulation is:

$$R'_{f,eq} = R_{f,pine} + R_{f,asph}$$
$$= 1.28 \text{ h·ft}^2\text{·F°/Btu} + 0.88 \text{ h·ft}^2\text{·F°/Btu}$$
$$= 2.16 \text{ h·ft}^2\text{·F°/Btu}$$

6. The equivalent $R$ factor with insulation is:

$$R_{f,eq} = R_{f,pine} + R_{f,asph} + R_{f,insul} = R'_{f,eq} + R_{f,insul}$$
$$= 2.16 \text{ h·ft}^2\text{·F°/Btu} + 5.6 \text{ h·ft}^2\text{·F°/Btu}$$
$$= 7.76 \text{ h·ft}^2\text{·F°/Btu}$$

7. One comparison of the two equivalent $R$ factors is their ratio:

$$\frac{R'_{f,eq}}{R_{f,eq}} = \frac{2.16}{7.76} = 0.28$$

8. By adding the insulation the heat loss rate per square foot is reduced by 78%. Is it 78% of a large heat loss or a small heat loss? Using Equation 20-16 we calculate the thermal current $I$ through the entire roof.

$$\Delta T = IR_{net} = \frac{I}{A} R_f$$

$$I' = \frac{A}{R'_{f,eq}} \Delta T = \frac{(60 \text{ ft})(20 \text{ ft})}{2.16 \text{ h·ft}^2\text{·F°/Btu}} \Delta T$$
$$= \left[556 \text{ (Btu/h)/F°}\right] \Delta T$$

9. To complete the calculation we estimate that the temperature inside the cabin is maintained at 70°F and the temperature outside the cabin during the winter is typically 40°F colder.

$$I' = \left[556 \text{ (Btu/h)/F°}\right] \Delta T$$
$$= \left[556 \text{ (Btu/h)/F°}\right](40°\text{F}) = 22{,}200 \text{ Btu/h}$$

and

$$I = 0.28I' = 0.28(22{,}200 \text{ Btu/h}) = 6200 \text{ Btu/h}$$

so the reduction due to the insulation is

$$I - I' = 22{,}200 \text{ Btu/h} - 6200 \text{ Btu/h}$$

$$= 16{,}000 \text{ Btu/h}$$

10. Estimate the savings that would result from adding the 2 in. of insulation.

See the following Remarks for an estimate of the cost.

**REMARKS** Installing 2 in. of roof insulation reduces the heat loss through the roof by 22,200 Btu/h. The cabin is heated with propane, and the energy content of propane is about 92,000 Btu/gal. Insulating the roof reduces consumption by approximately 6 gal of propane every 24 h of use. Propane costs about $1.40/gal, so this amounts to a savings of approximately $8.40 per day, or $252 per month. Your friend's family is impressed by the potential savings (and by the benefits of your physics knowledge). They decide to install the 2 in. of roofing insulation.

**EXERCISE** How much additional savings can be had by adding even more insulation to the roof? (*Answer* The maximum additional savings is 6200 Btu/h which would save $68 per month.)

**REMARKS** These cost estimates do not include the cost of purchasing and installing the insulation.

The thermal conductivity of air is very small compared with that of solid materials, which makes air a very good insulator. However, when there is a large air gap—say, between a storm window and the inside window—the insulating efficiency of air is greatly reduced because of convection. Whenever there is a temperature difference between different parts of the air space, convection currents act quickly to equalize the temperature, so the effective conductivity is greatly increased. For storm windows, air gaps of about 1 to 2 cm are optimal. Wider air gaps actually reduce the thermal resistance of a double-pane window due to convection.

The insulating properties of air are most effectively used when the air is trapped in small pockets that prevent convection from taking place. This is the principle underlying the excellent insulating properties of both goose down and Styrofoam.

If you touch the inside surface of a glass window when it is cold outside, you will observe that the surface is considerably colder than the inside air. The thermal resistance of windows is mainly due to thin films of insulating air that adhere to either side of the glass surface. The thickness of the glass has little effect on the overall thermal resistance. The air film on each side typically adds an $R$ factor of about 0.45 per side. Thus, the $R$ factor of a window with $N$ separated glass layers is approximately $0.9N$ because of the two sides of each layer. Under windy conditions, the outside air film may be greatly decreased, leading to a smaller $R$ factor for the window.

## Convection

Convection is the transport of energy as heat by the transport of the material medium itself. This thermal property is responsible for the great ocean currents as well as the global circulation of the atmosphere. In the simplest case, convection arises when a fluid (gas or liquid) is heated from below. The warm fluid then expands and rises as the cooler fluid sinks. The mathematical description of convection is very complex because the flow depends on the temperature difference in different parts of the fluid, and this temperature difference is affected by the flow itself.

The heat transferred from an object to its surroundings by convection is approximately proportional to the area of the object and to the difference in temperature between the object and the surrounding fluid. It is possible to write an equation for the energy transported as heat by convection and to define a coefficient of convection, but the analyses of practical problems involving convection is quite complex and will not be discussed here.

## Radiation

All objects emit and absorb electromagnetic radiation. When an object is in thermal equilibrium with its surroundings, it emits heat and absorbs heat at the same rate. The rate at which an object radiates energy is proportional to both the area of the object and to the fourth power of its absolute temperature. This result, found empirically by Josef Stefan in 1879 and derived theoretically by Ludwig Boltzmann about five years later, is called the **Stefan-Boltzmann law:**

$$P_r = e\sigma A T^4 \qquad\qquad 20\text{-}17$$

STEFAN–BOLTZMANN LAW

where $P_r$ is the power radiated, $A$ is the area, $\sigma$ is a universal constant called Stefan's constant, which has the value

$$\sigma = 5.6703 \times 10^{-8}\,\text{W}/(\text{m}^2{\cdot}\text{K}^4) \qquad\qquad 20\text{-}18$$

and $e$ is the **emissivity** of the object, a fractional quantity between 0 and 1 that is dependent upon the composition of the surface of the object.

When electromagnetic radiation falls on an opaque object, part of the radiation is reflected and part is absorbed. Light-colored objects reflect most visible radiation, whereas dark objects absorb most of it. The rate at which an object absorbs radiation is given by

$$P_a = e\sigma A T_0^4 \qquad\qquad 20\text{-}19$$

where $T_0$ is the temperature of the source of the radiation.

If an object emits more radiation than it absorbs, then it cools, while the object's surroundings absorb radiation from the object and become warmer. If the object absorbs more radiation than it emits, then the object warms and its surroundings cool. The net power radiated by an object at temperature $T$ in an environment at temperature $T_0$ is

$$P_{net} = e\sigma A(T^4 - T_0^4) \qquad\qquad 20\text{-}20$$

When an object is in thermal equilibrium with its surroundings, $T = T_0$, and the object emits and absorbs radiation at the same rate.

An object that absorbs all the radiation incident upon it has an emissivity equal to 1, and it is called a **blackbody.** A blackbody is also an ideal radiator. The concept of a blackbody is important because the characteristics of the radiation emitted by such an ideal object can be calculated theoretically. Materials such as black velvet come close to being ideal blackbodies. The best practical approximation of an ideal blackbody is a small hole leading into a cavity, such as a keyhole in a closet door (Figure 20-10). Radiation incident on the hole has little chance of being reflected out the hole before the walls of the cavity absorb it. Thus, the radiation emitted out of the hole is characteristic of the temperature of the walls of the cavity.

The radiation emitted by an object at temperatures below approximately 600°C is not visible. Most radiation emissions are concentrated at wavelengths much longer than those of visible light.[†] As an object is heated, the rate of energy

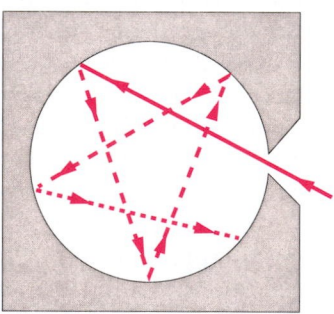

**FIGURE 20-10** A hole in a cavity approximates an ideal blackbody. Radiation entering the cavity has little chance of leaving the cavity before it is completely absorbed. The radiation emitted through the hole (not shown) is therefore characteristic of the temperature of the walls of the cavity.

---

† When we study light, we will see that visible light is electromagnetic radiation with wavelengths between about 400 and 700 nm.

emission increases, and the energy radiated extends to higher frequencies (and shorter wavelengths). Between about 600 and 700°C, enough of the radiated energy is in the visible spectrum for the object to glow a dull red. At higher temperatures, it may become bright red or even "white hot." Figure 20-11 shows the power radiated by a blackbody as a function of wavelength for several different temperatures. The wavelength at which the power is a maximum varies inversely with the temperature, a result known as Wien's displacement law:

$$\lambda_{max} = \frac{2.898 \text{ mm·K}}{T} \qquad \qquad 20\text{-}21$$

WIEN'S DISPLACEMENT LAW

This law is used to determine the surface temperatures of stars by analyzing their radiation. It can also be used to map out the variation in temperature over different regions of the surface of an object. Such a map is called a thermograph. Thermographs can be used to detect cancer because cancerous tissue results in increased circulation which produces a slight increase in skin temperature.

The spectral-distribution curves shown in Figure 20-11 played an important role in the history of physics. It was the discrepancy between theoretical calculations (using classical thermodynamics) of what the blackbody spectral distribution should be and the actual experimental measurements of spectral distributions that led to Max Planck's first ideas about the quantization of energy in 1900.

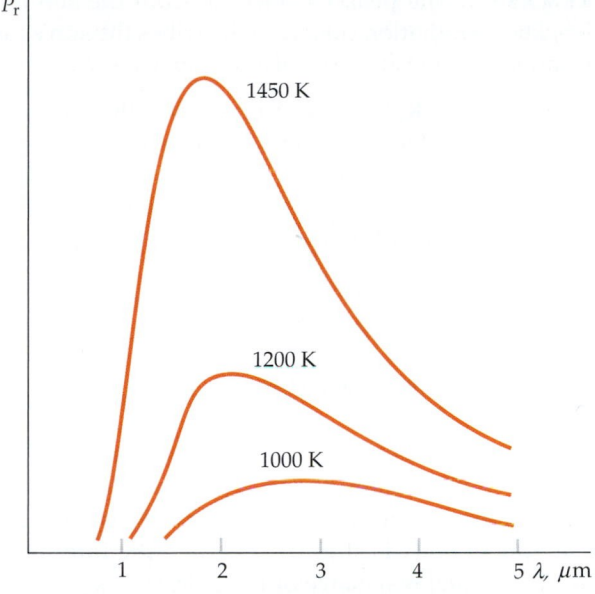

**FIGURE 20-11** Radiated power versus wavelength for radiation emitted by a blackbody. The wavelength of the maximum power varies inversely with the absolute temperature of the blackbody.

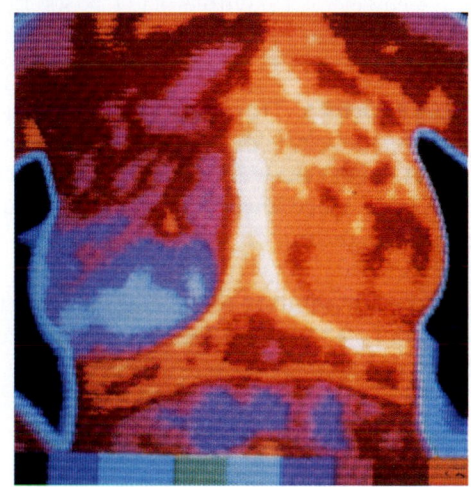

A thermograph was used to detect this cancerous tumor.

---

RADIATION FROM THE SUN

**EXAMPLE 20-7**

(a) The radiation emitted by the surface of the sun emits maximum power at a wavelength of about 500 nm. Assuming the sun to be a blackbody emitter, what is its surface temperature? (b) Calculate $\lambda_{max}$ for a blackbody at room temperature, $T = 300$ K.

(a) We can find $T$ given $\lambda_{max}$ using Wien's displacement law:

$$\lambda_{max} = \frac{2.898 \text{ mm·K}}{T}$$

so

$$T = \frac{2.898 \text{ mm·K}}{\lambda_{max}} = \frac{2.898 \text{ mm·K}}{500 \text{ nm}}$$

$$= \boxed{5800 \text{ K}}$$

(b) We can find $\lambda_{max}$ from Wien's displacement law for $T = 300$ K:

$$\lambda_{max} = \frac{2.898 \text{ mm·K}}{300 \text{ K}} = 9.66 \times 10^{-3} \text{ mm}$$

$$= \boxed{9.66 \text{ } \mu\text{m}}$$

**REMARKS** The peak wavelength from the sun is in the visible spectrum. The blackbody radiation spectrum describes the sun's radiation fairly well, so the sun is indeed a good example of a blackbody.

For $T = 300$ K, the spectrum peaks in the infrared at wavelengths much longer than the wavelengths visible to the eye. Surfaces that are not black to our eyes may act as blackbodies for infrared radiation and absorption. For example, it has been found experimentally that the skin of human beings of all races is black to infrared radiation; hence, the emissivity of skin is 1.00 for its own radiation process.

---

RADIATION FROM THE HUMAN BODY          **EXAMPLE 20-8** Try It Yourself

Calculate the net rate of heat loss in radiated energy for a naked person in a room at 20°C, assuming the person to be a blackbody with a surface area of 1.4 m² and a surface temperature of 33°C (= 306 K). (The surface temperature of the human body is slightly less than the internal temperature of 37°C because of the thermal resistance of the skin.)

Cover the column to the right and try these on your own before looking at the answers.

| Steps | Answer |
|---|---|
| Use $P_{net} = e\sigma A(T^4 - T_0^4)$ with $e = 1$, $T = 306$ K, and $T_0 = 293$ K. | $P_{net} = 111$ W |

**REMARKS** This large energy loss is approximately equal to the basal metabolic rate of about 120 W. We protect ourselves from this great loss of energy by wearing clothing, which, because of its low thermal conductivity, has a much lower outside temperature and therefore a much lower rate of thermal radiation.

When the temperature of an object $T$ is not too different from the surrounding temperature $T_0$, a radiating object obeys Newton's law of cooling. We can see this by writing Equation 20-20 as

$$P_{net} = e\sigma A(T^4 - T_0^4) = e\sigma A(T^2 + T_0^2)(T^2 - T_0^2)$$

$$= e\sigma A(T^2 + T_0^2)(T + T_0)(T - T_0)$$

When $T - T_0$ is small, we can replace $T$ by $T_0$ in the sums with little change in the result. Then

$$P_{net} = e\sigma A(T^4 - T_0^4) \approx e\sigma A(T_0^2 + T_0^2)(T_0 + T_0)(T - T_0) = 4e\sigma AT_0^3 \, \Delta T$$

The net power radiated is approximately proportional to the temperature difference, in agreement with Newton's law of cooling. This result can also be obtained by using the differential approximation.

$$\Delta P_r \approx \frac{dP_r}{dT}\bigg|_{T=T_0} (T - T_0)$$

where $P_r = e\sigma A(T^4 - T_0^4)$. For a small temperature difference $T - T_0$ we have

$$\Delta P_r \approx e\sigma A \, 4T^3\big|_{T=T_0} (T - T_0) = 4e\sigma AT_0^3 \, \Delta T$$

| Topic | Relevant Equations and Remarks | |
|---|---|---|
| **1. Thermal Expansion** | | |
| Coefficient of linear expansion | $$\alpha = \frac{\Delta L/L}{\Delta T}$$ | 20-2 |
| Coefficient of volume expansion | $$\beta = \frac{\Delta V/V}{\Delta T} = 3\alpha$$ | 20-4, 20-5 |
| **2. The van der Waals Equation of State** | The van der Waals equation of state describes the behavior of real gases over a wide range of temperatures and pressures, taking into account the space occupied by the gas molecules themselves and the attraction of the molecules to one another. | |
| | $$\left(P + \frac{an^2}{V^2}\right)(V - bn) = nRT$$ | 20-6 |
| **3. Vapor Pressure** | Vapor pressure is the pressure at which the liquid and gas phases of a substance are in equilibrium at a given temperature. The liquid boils at that temperature for which the external pressure equals the vapor pressure. | |
| **4. The Triple Point** | The triple point is the unique temperature and pressure at which the gas, liquid, and solid phases of a substance can coexist in equilibrium. At temperatures and pressures below the triple point, the liquid phase of a substance cannot exist. | |
| **5. Heat Transfer** | The three mechanisms by which thermal energy is transferred are radiation, conduction, and convection. | |
| Newton's law of cooling | For all mechanisms of heat transfer, if the temperature difference between the body and its surroundings is small, the rate of cooling of a body is approximately proportional to the temperature difference. | |
| **6. Heat Conduction** | | |
| Current | The rate of conduction of thermal energy is given by | |
| | $$I = \frac{\Delta Q}{\Delta t} = kA\frac{\Delta T}{\Delta x}$$ | 20-7 |
| | where $I$ is the thermal current, $k$ is the coefficient of thermal conductivity, and $\Delta T/\Delta x$ is the temperature gradient. | |
| Thermal resistance | $$\Delta T = IR$$ | 20-9 |
| | where $R$ is the thermal resistance: | |
| | $$R = \frac{\Delta x}{kA}$$ | 20-10 |
| Equivalent resistance: | | |
| series | $$R_{eq} = R_1 + R_2 + \ldots$$ | 20-12 |
| parallel | $$\frac{1}{R_{eq}} = \frac{1}{R_1} + \frac{1}{R_2} + \ldots$$ | 20-14 |

| | |
|---|---|
| R factor | The $R$ factor is the thermal resistance in in.·ft²·°F/(Btu/h) for a square foot of a slab of material |

$$R_f = R_{net}A = \frac{\Delta x}{k}$$ 20-15

## 7. Thermal Radiation

| | | |
|---|---|---|
| Rate of power radiated | $P_r = e\sigma AT^4$ | 20-17 |

where $\sigma = 5.6703 \times 10^{-8}$ W/m²·K⁴ is Stefan's constant and $e$ is the emissivity, which varies between 0 and 1 (depending on the composition of the surface of the object). Materials that are good heat absorbers are also good heat radiators.

| | | |
|---|---|---|
| Net power radiated by an object at $T$ to its environment at $T_0$ | $P_{net} = e\sigma A(T^4 - T_0^4)$ | 20-20 |

| | |
|---|---|
| Blackbody | A blackbody has an emissivity of 1. It is a perfect radiator, and it absorbs all radiation incident upon it. |

| | |
|---|---|
| Wein's law | The power spectrum of electromagnetic energy radiated by a blackbody has a maximum at a wavelength $\lambda_{max}$, which varies inversely with the absolute temperature of the body: |

$$\lambda_{max} = \frac{2.898 \text{ mm·K}}{T}$$ 20-21

# PROBLEMS

- • Single-concept, single-step, relatively easy
- •• Intermediate-level, may require synthesis of concepts
- ••• Challenging
- **SSM** Solution is in the *Student Solutions Manual*
- **iSOLVE** Problems available on iSOLVE online homework service
- **iSOLVE✓** These "Checkpoint" online homework service problems ask students additional questions about their confidence level, and how they arrived at their answer

In a few problems, you are given more data than you actually need; in a few other problems, you are required to supply data from your general knowledge, outside sources, or informed estimates.

## Conceptual Problems

**1** • **SSM** Why does the mercury level first decrease slightly when a thermometer is placed in warm water?

**2** • A large sheet of metal has a hole cut in the middle of it. When the sheet is heated, the area of the hole will (a) not change, (b) always increase, (c) always decrease, (d) increase if the hole is not in the exact center of the sheet, (e) decrease only if the hole is in the exact center of the sheet.

**3** • Mountaineers say that you cannot hard boil an egg on the top of Mount Rainier. This is true because (a) the air is too cold to boil water, (b) the air pressure is too low for stoves to burn, (c) boiling water is not hot enough to hard boil the egg, (d) the oxygen content of the air is too low, (e) eggs always break in their backpacks.

**4** • Which gases in Table 20-2 cannot be liquefied by applying pressure at 20°C?

**5** •• **SSM** The phase diagram in Figure 20-12 can be interpreted to yield information on how the boiling and melting points of water change with altitude. (a) Explain how this information can be obtained. (b) How might this information affect cooking procedures in the mountains?

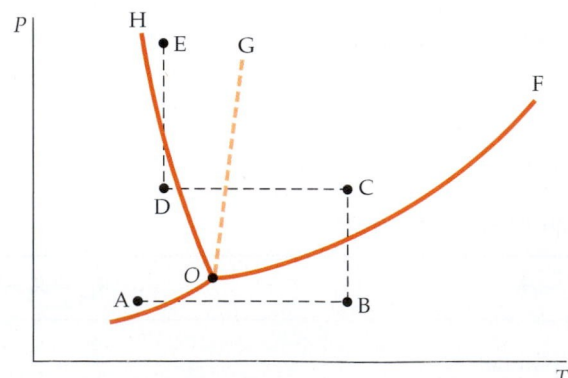

**FIGURE 20-12** Problem 5

**6** • If the absolute temperature of an object is tripled, the rate at which it radiates thermal energy (a) triples, (b) increases by a factor of 9, (c) increases by a factor of 27, (d) increases by a factor of 81, (e) depends on whether the absolute temperature is above or below zero.

**7** • [SSM] In a cool room, a metal or marble table top feels much colder to the touch than does a wood surface even though they are at the same temperature. Why?

**8** • True or false:

(a) During a phase change, the temperature of a substance remains constant.
(b) The rate of conduction of thermal energy is proportional to the temperature gradient.
(c) The rate at which an object radiates energy is proportional to the square of its absolute temperature.
(d) All materials expand when they are heated.
(e) The vapor pressure of a liquid depends on the temperature.

**9** • The earth loses heat by (a) conduction, (b) convection, (c) radiation, (d) all of the above.

**10** • Which heat-transfer mechanisms are the most important in the warming effect of a fire in a fireplace?

**11** • Which heat-transfer mechanism is important in the transfer of energy from the sun to the earth?

**12** •• Explain why turning down the temperature of a house at night in winter can save money on heating costs. Why doesn't the cost of the fuel consumed to heat the house back up in the morning equal the savings realized by cooling it down?

**13** •• Two cylinders made of materials A and B have the same lengths; their diameters are related by $d_A = 2d_B$. When the same temperature difference is maintained between the ends of the cylinders they conduct heat at the same rate. Their thermal conductivities are related by (a) $k_A = k_B/4$, (b) $k_A = k_B/2$, (c) $k_A = k_B$, (d) $k_A = 2k_B$, (e) $k_A = 4k_B$.

**14** • Infrared light is sometimes referred to as "heat waves." Explain why infrared light has received this label, and why the label is inaccurate.

**15** • [SSM] In artistic nomenclature, blue is often referred to as a "cool" color, while red is referred to as a "warm" color. In physics, however, red is considered a "cooler" color than blue. Explain why.

### Estimation and Approximation

**16** ••• Liquid helium is stored in containers fitted with 7-cm-thick "superinsulation" consisting of numerous layers of very thin aluminized Mylar sheets. The rate of evaporation of liquid in a 200-L container is about 0.7 L per day. Assume the container is spherical and that the external temperature is 20°C. The specific gravity of liquid helium is 0.125 and the latent heat of vaporization is 21 kJ/kg. Estimate the thermal conductivity of superinsulation.

**17** •• Estimate the thermal conductivity of the skin, given that the body of an "average" adult male has about 1.8 m² of skin area, and produces about 130 W of heat when resting. Use an internal temperature of 37°C (98.6°F) and an external skin temperature of 33°C. Assume that the skin has an average thickness of about 1 mm.

**18** •• [SSM] Estimate the effective emissivity of the earth, given the following information: the solar constant (the intensity of light incident on the earth from the sun) is 1370 W/m², 70 percent of this light is absorbed by the earth, and the earth's average temperature is 288 K. (Assume that the effective area that is absorbing the light is $\pi R^2$, where $R$ is the earth's radius, while the blackbody-emission area is $4\pi R^2$.)

**19** •• Black holes in orbit around a normal star are detected from earth due to the frictional heating of infalling gas into the black hole, which can reach temperatures greater than $10^6$ K. Assuming that the infalling gas can be modeled as a blackbody radiator, estimate $\lambda_{max}$ for use in an astronomical detection of a black hole. (Remark: This is in the x-ray region of the electromagnetic spectrum.)

### Thermal Expansion

**20** • [iSOLVE] A steel ruler has a length of 30 cm at 20°C. What is its length at 100°C?

**21** •• (a) Define a coefficient of area expansion. (b) Calculate it for a square and a circle, and show that it is two times the coefficient of linear expansion.

**22** •• [iSOLVE] The density of aluminum is $2.70 \times 10^3$ kg/m³ at 0°C. What is the density of aluminum at 200°C?

**23** •• [iSOLVE] A copper collar is to fit tightly about a steel shaft that has a diameter of 6.0000 cm at 20°C. The inside diameter of the copper collar at that temperature is 5.9800 cm. To what temperature must the copper collar be raised so that it will just slip on the steel shaft, assuming the steel shaft remains at 20°C?

**24** •• [SSM] [iSOLVE] Repeat Problem 23 when the temperature of both the steel shaft and copper collar are raised simultaneously.

**25** •• A container is filled to the brim with 1.4 L of mercury at 20°C. When the temperature of container and mercury is raised to 60°C, 7.5 mL of mercury spill over the brim of the container. Determine the linear expansion coefficient of the container.

**26** •• [iSOLVE]✔ A hole is drilled in an aluminum plate with a steel drill bit whose diameter at 20°C is 6.245 cm. In the process of drilling, the temperature of the drill bit and of the aluminum plate rise to 168°C. What is the diameter of the hole in the aluminum plate when it has cooled to room temperature?

**27** •• [SSM] A rookie crew was left to put in the final 1 km of rail for a stretch of railroad track. When they finished, the temperature was 20°C, and they headed to town for some refreshments. After an hour or two, one of the old-timers noticed that the temperature had gone up to 25°C, so he said, "I hope you left some gaps to allow for expansion." From the look on their faces, he knew that they had not, and they all rushed back to the work site. The rail had buckled into an isosceles triangle. How high was the buckle?

**28** •• [iSOLVE] A car has a 60-L steel gas tank filled to the top with gasoline when the temperature is 10°C. The coefficient of volume expansion of gasoline is $\beta = 0.900 \times 10^{-3}$ K⁻¹. Taking the expansion of the steel tank into account, how much gasoline spills out of the tank when the car is parked in the sun and its temperature rises to 25°C?

**29 ••** A thermometer has an ordinary glass bulb and thin glass tube filled with 1 mL of mercury. A temperature change of 1°C changes the level of mercury in the thin tube by 3.0 mm. Find the inside diameter of the thin glass tube.

**30 •• ISOLVE✓** A mercury thermometer consists of a 0.4-mm capillary tube connected to a glass bulb. The mercury level rises 7.5 cm as the temperature of the thermometer increases from 35°C to 43°C. Find the volume of the thermometer bulb.

**31 •••** A grandfather's clock is calibrated at a temperature of 20°C. (a) On a hot day, when the temperature is 30°C, does the clock run fast or slow? (b) How much does it gain or lose in a 24-h period? Assume that the pendulum is a thin brass rod of negligible mass with a heavy bob attached to the end.

**32 ••• ISOLVE✓** A steel tube has an outside diameter of 3.000 cm at room temperature (20°C). A brass tube has an inside diameter of 2.997 cm at the same temperature. To what temperature must the ends of the tubes be heated if the steel tube is to be inserted into the brass tube?

**33 ••• SSM** What is the tensile stress in the copper collar of Problem 23 when its temperature returns to 20°C?

## The van der Waals Equation, Liquid–Vapor Isotherms, and Phase Diagrams

**34 • ISOLVE✓** (a) Calculate the volume of 1 mol of steam at 100°C and a pressure of 1 atm, assuming that it is an ideal gas. (b) Find the temperature at which the steam will occupy the volume found in Part (a) if it obeys the van der Waals equation with $a = 0.55$ Pa·m$^6$/mol$^2$ and $b = 30$ cm$^3$/mol.

**35 ••** From Figure 20-3, find (a) the temperature at which water boils on a mountain where the atmospheric pressure is 70 kPa, (b) the temperature at which water will boil in a container in which the pressure has been reduced to 0.5 atm, and (c) the pressure at which water will boil at 115°C.

**36 •• SSM** The van der Waals constants for helium are $a = 0.03412$ L$^2$·atm/mol$^2$ and $b = 0.0237$ L/mol. Use these data to find the volume in cubic centimeters occupied by one helium atom and to estimate the radius of the atom.

**37 •••** (a) For a van der Waals gas, show that the critical temperature is $8a/27Rb$ and the critical pressure is $a/27b^2$. (b) Rewrite the van der Waals equation of state in terms of the reduced variable $V_r = V/V_c$, $P_r = P/P_c$, and $T_r = T/T_c$.

## Heat Conduction

**38 • ISOLVE** A copper bar 2 m long has a circular cross section of radius 1 cm. One end is kept at 100°C and the other end is kept at 0°C. The surface of the bar is insulated so that there is negligible heat loss through it. Find (a) the thermal resistance of the bar, (b) the thermal current I, (c) the temperature gradient $\Delta T/\Delta x$, and (d) the temperature of the bar 25 cm from the hot end.

**39 • ISOLVE✓** A 20 × 30-ft slab of insulation has an R factor of 11. How much heat (in Btu/h) is conducted through the slab if the temperature on one side is 68°F and on the other side it is 30°F?

**40 •• ISOLVE✓** Two metal cubes with 3-cm edges, one copper (Cu) and one aluminum (Al), are arranged as shown in Figure 20-13. Find (a) the thermal resistance of each cube, (b) the thermal resistance of the two-cube system, (c) the thermal current I, and (d) the temperature at the interface of the two cubes.

**FIGURE 20-13**
**Problem 40**

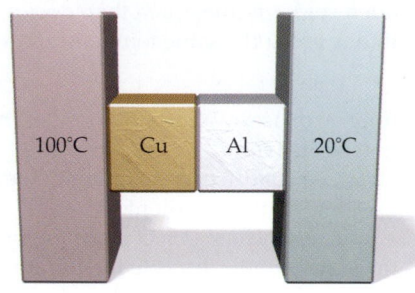

**41 ••** The cubes in Problem 40 are rearranged in parallel as shown in Figure 20-14. Find (a) the thermal current carried by each cube from one side to the other, (b) the total thermal current, and (c) the equivalent thermal resistance of the two-cube system.

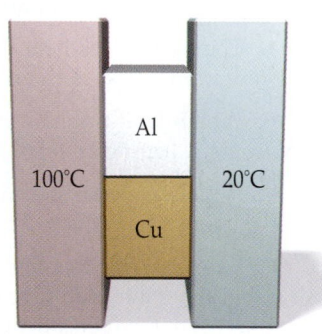

**FIGURE 20-14** Problem 41

**42 •• SSM** The cost of air conditioning a house is proportional to the rate at which heat flows from the house and is inversely proportional to the coefficient of performance (COP) of the air conditioner. We denote the temperature difference between the house and its surroundings as $\Delta T$. Assuming that the rate at which heat flows from a house is proportional to $\Delta T$ and that the air conditioner is operating ideally, show that the cost of air conditioning is proportional to $(\Delta T)^2$.

**43 •••** A spherical shell of thermal conductivity $k$ has inside radius $r_1$ and outside radius $r_2$ (Figure 20-15). The inside of the shell is held at a temperature $T_1$, and the outside at temperature $T_2$. In this problem, you are to show that the thermal current through the shell is given by

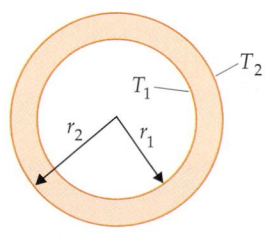

**FIGURE 20-15** Problem 43

$$I = \frac{4\pi k r_1 r_2}{r_2 - r_1}(T_2 - T_1) \qquad\qquad 20\text{-}22$$

Consider a spherical element of the shell of radius $r$ and thickness $dr$. (a) Why must the thermal current through each such element be the same? (b) Write the thermal current I through such a shell element in terms of the area $A = 4\pi r^2$, the thickness $dr$, and the temperature difference $dT$ across the element. (c) Solve for $dT$ in terms of $dr$ and integrate from $r = r_1$ to $r = r_2$. (d) Show that when $r_1$ and $r_2$ are much larger than $r_2 - r_1$, Equation 20-22 (shown above) is the same as Equation 20-7.

**44** •• SSM SOLVE For a boiler at a power station, heat must be transferred to boiling water at the rate of 3 GW. The boiling water passes through copper pipes having a wall thickness of 4.0 mm and a surface area of 0.12 m² per meter length of pipe. Find the total length of pipe (actually there are many pipes in parallel) that must pass through the furnace if the steam temperature is 225°C and the external temperature of the pipes is 600°C.

**45** ••• A steam pipe of length L is insulated with a layer of material of thermal conductivity k. Find the rate of heat transfer if the temperature outside the insulation is $T_1$, the temperature inside is $T_2$, the outside radius of the insulation is $r_1$, and the inside radius is $r_2$.

## Radiation

**46** • SSM SOLVE✓ Calculate $\lambda_{max}$ for a human blackbody radiator, assuming the surface temperature of the skin to be 33°C.

**47** • SOLVE✓ The heating wires of a 1-kW electric heater are red hot at a temperature of 900°C. Assuming that 100% of the heat output is due to radiation and that the wires act as blackbody radiators, what is the effective area of the radiating surface? (Assume a room temperature of 20°C.)

**48** •• SOLVE✓ A blackened, solid copper sphere of radius 4.0 cm hangs in a vacuum in an enclosure whose walls have a temperature of 20°C. If the sphere is initially at 0°C, find the rate at which its temperature changes, assuming that heat is transferred by radiation only.

**49** •• SOLVE The surface temperature of the filament of an incandescent lamp is 1300°C. If the electric power input is doubled, what will the temperature become? *Hint: Show that you can neglect the temperature of the surroundings.*

**50** •• Liquid helium is stored at its boiling point (4.2 K) in a spherical can that is separated by a vacuum space from a surrounding shield that is maintained at the temperature of liquid nitrogen (77 K). If the can is 30 cm in diameter and is blackened on the outside so that it acts as a blackbody radiator, how much helium boils away per hour?

## General Problems

**51** • SSM SOLVE A steel tape is placed around the earth at the equator when the temperature is 0°C. What will the clearance between the tape and the ground (assumed to be uniform) be if the temperature of the tape rises to 30°C? Neglect the expansion of the earth.

**52** •• Show that change in the density of an isotropic material due to an increase in temperature $\Delta T$ is given by $\Delta \rho = -\beta \rho \Delta T$.

**53** •• SOLVE The solar constant is the power received from the sun per unit area perpendicular to the sun's rays at the mean distance of the earth from the sun. Its value at the upper atmosphere of the earth is about 1.35 kW/m². Calculate the effective temperature of the sun if it radiates like a blackbody. (The radius of the sun is $6.96 \times 10^8$ m.)

**54** •• SOLVE✓ To determine the R value of insulating material that comes in sheets of $\frac{1}{2}$-in. thickness, you construct a cubical box of 12 in. per side and place a thermometer and a 100-W heater inside the box. After thermal equilibrium has been attained, the temperature inside the box is 90°C when the external temperature is 20°C. Determine the R value of this material.

**55** •• A 2-cm-thick copper sheet is pressed against a sheet of aluminum. What should be the thickness of the aluminum sheet so that the temperature of the copper–aluminum interface is $(T_1 + T_2)/2$, where $T_1$ and $T_2$ are the temperatures at the copper–air and aluminum–air interfaces?

**56** •• At a temperature of 20°C, a steel bar of radius 2.2 cm and length 60 cm is jammed horizontally perpendicular between two vertical concrete walls. With a blowtorch, the temperature of the bar is raised to 60°C. Find the force exerted by the bar on each wall.

**57** •• (a) From the definition of $\beta$, the coefficient of volume expansion (at constant pressure), show that $\beta = 1/T$ for an ideal gas. (b) The experimentally determined value of $\beta$ for $N_2$ gas at 0°C is 0.003673 K⁻¹. Compare this value with the theoretical value $\beta = 1/T$, assuming that $N_2$ is an ideal gas.

**58** •• One way to construct a device with two points whose separation remains the same in spite of temperature changes is to bolt together one end of two rods, both of which have different coefficients of linear expansion as in the arrangement shown in Figure 20-16. (a) Show that the distance L will not change with temperature if the lengths $L_A$ and $L_B$ are chosen such that $L_A/L_B = \alpha_B/\alpha_A$. (b) If material B is steel, material A is brass, and $L_A = 250$ cm at 0°C, what is the value of L?

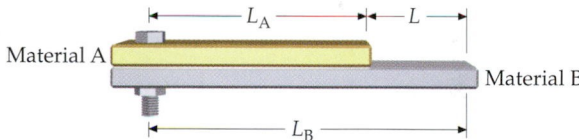

**FIGURE 20-16** Problem 58

**59** •• On the average, the temperature of the earth's crust increases 1.0 C° for every 30 m of depth. The average thermal conductivity of the earth's crust is 0.74 J/m·s·K. What is the heat loss of the earth per second due to conduction from the core? How does this heat loss compare with the average power received from the sun? (The solar constant is about 1.35 kW/m².)

**60** •• SOLVE A copper-bottomed saucepan containing 0.8 L of boiling water boils dry in 10 min. Assuming that all the heat flows through the flat copper bottom, which has a diameter of 15 cm and a thickness of 3.0 mm, calculate the temperature of the outside of the copper bottom while some water is still in the pan.

**61** •• SSM A hot-water tank of cylindrical shape has an inside diameter of 0.55 m and inside height of 1.2 m. The tank is enclosed with a 5-cm-thick insulating layer of glass wool whose thermal conductivity is 0.035 W/m·K. The metallic interior and exterior walls of the container have thermal conductivities that are much greater than that of the glass wool. How much power must be supplied to this tank in order to maintain the water temperature at 75°C when the external temperature is 1°C?

**62** ••• The diameter of a rod is given by $d = d_0(1 + ax)$, where $a$ is a constant and $x$ is the distance from one end. If the thermal conductivity of the material is $k$ what is the thermal resistance of the rod if its length is $L$?

**63** ••• A solid disk of radius $r$ and mass $m$ is spinning in a frictionless environment with angular velocity $\omega_1$ at temperature $T_1$. The temperature of the disk is then changed to $T_2$. Express the angular velocity $\omega_2$, rotational kinetic energy $E_2$, and angular momentum $L_2$ in terms of their values at the temperature $T_1$ and the linear expansion coefficient $\alpha$ of the disk.

**64** ••• Write a spreadsheet program to graph the temperature of the earth as a function of emissivity, using the results of Problem 18. How much does the emissivity have to change in order for the average temperature to rise by 1 K? This can be thought of as a model for the effect of increasing concentrations of greenhouse gases like methane and $CO_2$ in the earth's atmosphere.

**65** ••• A small pond has a layer of ice 1 cm thick floating on its surface. (a) If the air temperature is $-10°C$, find the rate in centimeters per hour at which ice is added to the bottom of the layer. The density of ice is $0.917 \text{ g/cm}^3$. (b) How long does it take for a 20-cm layer to be built up?

**66** ••• **SSM** A 200-g copper container holding 0.7 L of water is thermally isolated from its surroundings—except for a 10-cm-long copper rod of cross-sectional area 1.5 cm$^2$ connecting it to a second copper container filled with an ice and water mixture so its temperature remains at $0°C$. The initial temperature of the first container is $T_0 = 60°C$. (Assume the heat capacity of the rod to be negligible.) (a) Show that

the temperature $T$ of the first container changes over time $t$ according to

$$T = T_0 e^{-t/RC}$$

where $T$ is in degrees Celsius, $R$ is the thermal resistance of the rod, and $C$ is the total heat capacity of the container plus the water. (Neglect the heat capacity of the rod.) (b) Evaluate $R$, $C$, and the "time constant" $RC$. (c) Show that the total amount of heat $Q$ conducted in time $t$ is

$$Q = CT_0(1 - e^{-t/RC})$$

(d) Using a spreadsheet program, graph both $T(t)$ and $Q(t)$; from the graph, find the time it takes for the temperature of the first container to be reduced to $30°C$.

**67** ••• A blackened copper cube that is 1 cm along an edge is heated to a temperature of $300°C$, and then is placed in a vacuum chamber whose walls are at a temperature of $0°C$. In the vacuum chamber, the cube cools radiatively. (a) Show that the (absolute) temperature $T$ of the cube follows the differential equation:

$$\frac{dT}{dt} = -\frac{e\sigma A}{C}(T^4 - T_0^4)$$

where $C$ is the heat capacity of the cube, $A$ is its surface area, $e$ the emissivity, and $T_0$ the temperature of the vacuum chamber. (b) Using Euler's method, numerically solve the differential equation to find $T(t)$, and graph it. Assume $e = 1$. (The Euler method is discussed in Section 4 of Chapter 5.) How long does it take the cube to cool to a temperature of $15°C$?

# The Electric Field I: Discrete Charge Distributions

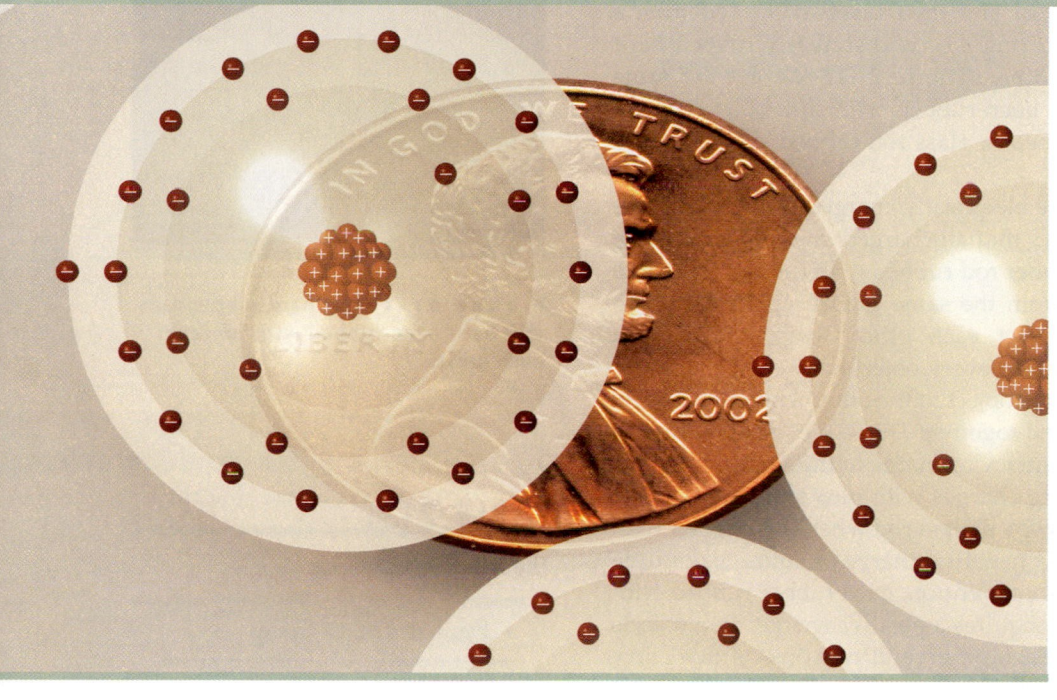

COPPER IS A CONDUCTOR, A MATERIAL WITH SPECIFIC PROPERTIES WE FIND USEFUL BECAUSE THESE PROPERTIES MAKE IT POSSIBLE TO TRANSPORT ELECTRICITY. THE ELECTRICITY WE HARNESS TO POWER MACHINES IS ALSO RESPONSIBLE FOR THE COPPER ATOM ITSELF: ATOMS ARE HELD TOGETHER BY ELECTRICAL FORCES.

**?** **What is the total charge of all the electrons in a penny?** **(See Example 21-1.)**

21-1    Electric Charge

21-2    Conductors and Insulators

21-3    Coulomb's Law

21-4    The Electric Field

21-5    Electric Field Lines

21-6    Motion of Point Charges in Electric Fields

21-7    Electric Dipoles in Electric Fields

**W**hile just a century ago we had nothing more than a few electric lights, we are now extremely dependent on electricity in our daily lives. Yet, although the use of electricity has only recently become widespread, the study of electricity has a history reaching long before the first electric lamp glowed. Observations of electrical attraction can be traced back to the ancient Greeks, who noticed that after amber was rubbed, it attracted small objects such as straw or feathers. Indeed, the word *electric* comes from the Greek word for amber, *elektron*.

➤ In this chapter, we begin our study of electricity with *electrostatics*, the study of electrical charges at rest. After introducing the concept of electric charge, we briefly look at conductors and insulators and how conductors can be given a net charge. We then study Coulomb's law, which describes the force exerted by one electric charge on another. Next, we introduce the electric field and show how it can be visualized by electric field lines that indicate the magnitude and direction of the field, just as we visualized the velocity field of a flowing fluid using streamlines (Chapter 13). Finally, we discuss the behavior of point charges and electric dipoles in electric fields.

# 21-1 Electric Charge

Suppose we rub a hard rubber rod with fur and then suspend the rod from a string so that it is free to rotate. Now we bring a second similarly rubbed hard rubber rod near it. The rods repel each other (Figure 21-1). We get the same results if we use two glass rods that have been rubbed with silk. But, when we place a hard rubber rod rubbed with fur near a glass rod rubbed with silk they attract each other.

Rubbing a rod causes the rod to become electrically charged. If we repeat the experiment with various materials, we find that all charged objects fall into one of just two groups—those like the hard rubber rod rubbed with fur and those like the glass rod rubbed with silk. Objects from the same group repel each other, while objects from different groups attract each other. Benjamin Franklin explained this by proposing a model in which every object has a *normal* amount of electricity that can be transferred from one object to the other when two objects are in close contact, as when they are rubbed together. This leaves one object with an excess charge and the other with a deficiency of charge in the same amount as the excess. Franklin described the resulting charges with plus and minus signs, choosing positive to be the charge acquired by a glass rod when it is rubbed with a piece of silk. The piece of silk acquires a negative charge of equal magnitude during the procedure. Based on Franklin's convention, hard rubber rubbed with fur acquires a negative charge and the fur acquires a positive charge. Two objects that carry the same type of charge repel each other, and two objects that carry opposite charges attract each other (Figure 21-2).

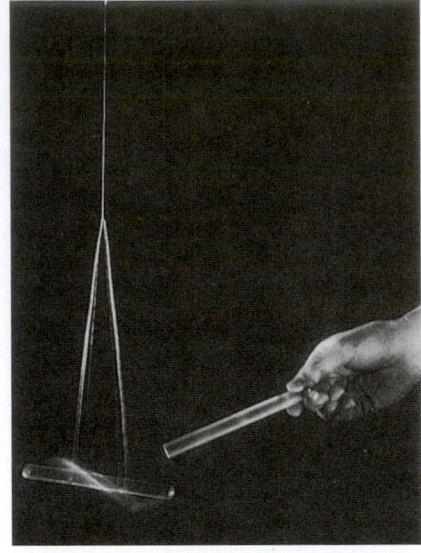

**FIGURE 21-1** Two hard rubber rods that have been rubbed with fur repel each other.

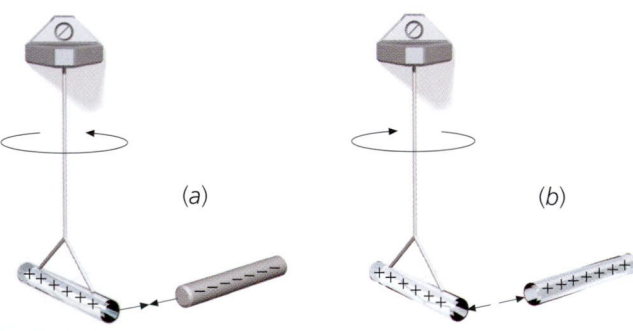

(a)     (b)

**FIGURE 21-2**
(*a*) Objects carrying charges of opposite sign attract each other. (*b*) Objects carrying charges of the same sign repel each other.

Today, we know that when glass is rubbed with silk, electrons are transferred from the glass to the silk. Because the silk is negatively charged (according to Franklin's convention, which we still use) electrons are said to carry a negative charge. Table 21-1 is a short version of the **triboelectric series.** (In Greek *tribos* means "a rubbing.") The further down the series a material is, the greater its affinity for electrons. If two of the materials are brought in contact, electrons are transferred from the material higher in the table to the one further down the table. For example, if Teflon is rubbed with nylon, electrons are transferred from the nylon to the Teflon.

## Charge Quantization

Matter consists of atoms that are electrically neutral. Each atom has a tiny but massive nucleus that contains protons and neutrons. Protons are positively charged, whereas neutrons are uncharged. The number of protons in the nucleus

## TABLE 21-1

**The Triboelectric Series**

| + Positive End of Series |
| --- |
| Asbestos |
| Glass |
| Nylon |
| Wool |
| Lead |
| Silk |
| Aluminum |
| Paper |
| Cotton |
| Steel |
| Hard rubber |
| Nickel and copper |
| Brass and silver |
| Synthetic rubber |
| Orlon |
| Saran |
| Polyethylene |
| Teflon |
| Silicone rubber |
| − Negative End of Series |

is the atomic number $Z$ of the element. Surrounding the nucleus is an equal number of negatively charged electrons, leaving the atom with zero net charge. The electron is about 2000 times less massive than the proton, yet the charges of these two particles are exactly equal in magnitude. The charge of the proton is $e$ and that of the electron is $-e$, where $e$ is called the **fundamental unit of charge.** The charge of an electron or proton is an intrinsic property of the particle, just as mass and spin are intrinsic properties of these particles.

All observable charges occur in integral amounts of the fundamental unit of charge $e$; that is, *charge is quantized*. Any charge $Q$ occurring in nature can be written $Q = \pm Ne$, where $N$ is an integer.[†] For ordinary objects, however, $N$ is usually very large and charge appears to be continuous, just as air appears to be continuous even though air consists of many discrete molecules. To give an everyday example of $N$, charging a plastic rod by rubbing it with a piece of fur typically transfers $10^{10}$ or more electrons to the rod.

## Charge Conservation

When objects are rubbed together, one object is left with an excess number of electrons and is therefore negatively charged; the other object is left lacking electrons and is therefore positively charged. The net charge of the two objects remains constant; that is, *charge is conserved*. The **law of conservation of charge** is a fundamental law of nature. In certain interactions among elementary particles, charged particles such as electrons are created or annihilated. However, in these processes, equal amounts of positive and negative charge are produced or destroyed, so the net charge of the universe is unchanged.

The SI unit of charge is the coulomb, which is defined in terms of the unit of electric current, the ampere (A).[‡] The **coulomb** (C) is the amount of charge flowing through a wire in one second when the current in the wire is one ampere. The fundamental unit of electric charge $e$ is related to the coulomb by

$$e = 1.602177 \times 10^{-19}\,\text{C} \approx 1.60 \times 10^{-19}\,\text{C} \qquad 21\text{-}1$$

FUNDAMENTAL UNIT OF CHARGE

**EXERCISE** A charge of magnitude 50 nC (1 nC = $10^{-9}$ C) can be produced in the laboratory by simply rubbing two objects together. How many electrons must be transferred to produce this charge?
(*Answer* $N = Q/e = (50 \times 10^{-9}\,\text{C})/(1.6 \times 10^{-19}\,\text{C}) = 3.12 \times 10^{11}$. Charge quantization cannot be detected in a charge of this size; even adding or subtracting a million electrons produces a negligibly small effect.)

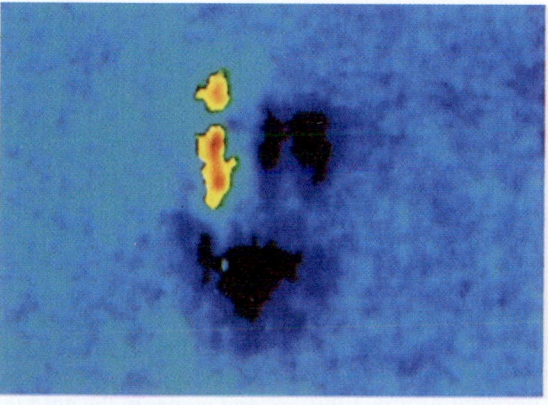

Charging by contact. A piece of plastic about 0.02 mm wide was charged by contact with a piece of nickel. Although the plastic carries a net positive charge, regions of negative charge (dark) as well as regions of positive charge (yellow) are indicated. The photograph was taken by sweeping a charged needle of width $10^{-7}$ m over the sample and recording the electrostatic force on the needle.

---

[†] In the standard model of elementary particles, protons, neutrons, and some other elementary particles are made up of more fundamental particles called quarks that carry charges of $\pm\frac{1}{3}e$ or $\pm\frac{2}{3}e$. Only combinations that result in a net charge of $\pm Ne$ or 0 are known.
[‡] The ampere (A) is the unit of current used in everyday electrical work.

A copper penny[†] ($Z = 29$) has a mass of 3 grams. What is the total charge of all the electrons in the penny?

**PICTURE THE PROBLEM**  The electrons have a total charge given by the number of electrons in the penny, $N_e$, times the charge of an electron, $-e$. The number of electrons is 29 (the atomic number of copper) times the number of copper atoms $N$. To find $N$, we use the fact that one mole of any substance has Avogadro's number ($N_A = 6.02 \times 10^{23}$) of molecules, and the number of grams in a mole is the molecular mass $M$, which is 63.5 g/mol for copper. Since each molecule of copper is just one copper atom, we find the number of atoms per gram by dividing $N_A$ (atoms/mole) by $M$ (grams/mole).

1. The total charge is the number of electrons times the electronic charge:

$$Q = N_e(-e)$$

2. The number of electrons is $Z$ times the number of copper atoms $N_a$:

$$N_e = ZN_a$$

3. Compute the number of copper atoms in 3 g of copper:

$$N_a = (3\text{ g})\frac{6.02 \times 10^{23}\text{ atoms/mol}}{63.5\text{ g/mol}} = 2.84 \times 10^{22}\text{ atoms}$$

4. Compute the number of electrons $N_e$:

$$N_e = ZN_a = (29\text{ electrons/atom})(2.84 \times 10^{22}\text{ atoms})$$
$$= 8.24 \times 10^{23}\text{ electrons}$$

5. Use this value of $N_e$ to find the total charge:

$$Q = N_e(-e)$$
$$= (8.24 \times 10^{23}\text{ electrons})(-1.6 \times 10^{-19}\text{ C/electron})$$
$$= \boxed{-1.32 \times 10^5\text{ C}}$$

**EXERCISE**  If one million electrons were given to each man, woman, and child in the United States (about 285 million people), what percentage of the number of electrons in a penny would this represent? (*Answer*   About $35 \times 10^{-9}$ percent)

# 21-2  Conductors and Insulators

In many materials, such as copper and other metals, some of the electrons are free to move about the entire material. Such materials are called **conductors**. In other materials, such as wood or glass, all the electrons are bound to nearby atoms and none can move freely. These materials are called **insulators**.

In a single atom of copper, 29 electrons are bound to the nucleus by the electrostatic attraction between the negatively charged electrons and the positively charged nucleus. The outer electrons are more weakly bound than the inner electrons because of their greater distance from the nucleus and because of the repulsive force exerted by the inner electrons. When a large number of copper atoms are combined in a piece of metallic copper, the binding of the electrons of each individual atom is reduced by interactions with neighboring atoms. One or more of the outer electrons in each atom is no longer bound

**FIGURE 21-3**  An electroscope. Two gold leaves are attached to a conducting post that has a conducting ball on top. The leaves are otherwise insulated from the container. When uncharged, the leaves hang together vertically. When the ball is touched by a negatively charged plastic rod, some of the negative charge from the rod is transferred to the ball and moves to the gold leaves, which then spread apart because of electrical repulsion between their negative charges. Touching the ball with a positively charged glass rod also causes the leaves to spread apart. In this case, the positively charged glass rod attracts electrons from the metal ball, leaving a net positive charge on the leaves.

---

† The penny was composed of 100 percent copper from 1793 to 1837. In 1982, the composition changed from 95 percent copper and 5 percent zinc to 2.5 percent copper and 97.5 percent zinc.

as a gas molecule is free to move about in a box. The number of free electrons depends on the particular metal, but it is typically about one per atom. An atom with an electron removed or added, resulting in a net charge on the atom, is called an **ion**. In metallic copper, the copper ions are arranged in a regular array called a *lattice*. A conductor is electrically neutral if for each lattice ion carrying a positive charge $+e$ there is a free electron carrying a negative charge $-e$. The net charge of the conductor can be changed by adding or removing electrons. A conductor with a negative net charge has an excess of free electrons, while a conductor with a positive net charge has a deficit of free electrons.

## Charging by Induction

The conservation of charge is illustrated by a simple method of charging a conductor called **charging by induction,** as shown in Figure 21-4. Two uncharged metal spheres are in contact. When a charged rod is brought near one of the spheres, free electrons flow from one sphere to the other, toward a positively charged rod or away from a negatively charged rod. The positively charged rod in Figure 21-4a attracts the negatively charged electrons, and the sphere nearest the rod acquires electrons from the sphere farther away. This leaves the near sphere with a net negative charge and the far sphere with an equal net positive charge. A conductor that has *separated* equal and opposite charges is said to be **polarized**. If the spheres are separated before the rod is removed, they will be left with equal amounts of opposite charges (Figure 21-4b). A similar result would be obtained with a negatively charged rod, which would drive electrons from the near sphere to the far sphere.

**EXERCISE** Two identical conducting spheres, one with an initial charge $+Q$, the other initially uncharged, are brought into contact. (a) What is the new charge on each sphere? (b) While the spheres are in contact, a negatively charged rod is moved close to one sphere, causing it to have a charge of $+2Q$. What is the charge on the other sphere? (*Answer* (a) $+\frac{1}{2}Q$. Since the spheres are identical, they must share the total charge equally. (b) $-Q$, which is necessary to satisfy the conservation of charge)

**EXERCISE** Two identical spheres are charged by induction and then separated; sphere 1 has charge $+Q$ and sphere 2 has charge $-Q$. A third identical sphere is initially uncharged. If sphere 3 is touched to sphere 1 and separated, then touched to sphere 2 and separated, what is the final charge on each of the three spheres? (*Answer* $Q_1 = +Q/2$, $Q_2 = -Q/4$, $Q_3 = -Q/4$)

For many purposes, the earth itself can be considered to be an infinitely large conductor with an abundant supply of free charge. If a conductor is connected to the earth, it is said to be **grounded** (indicated schematically in Figure 21-5b by a connecting wire ending in parallel horizontal lines). Figure 21-5 demonstrates

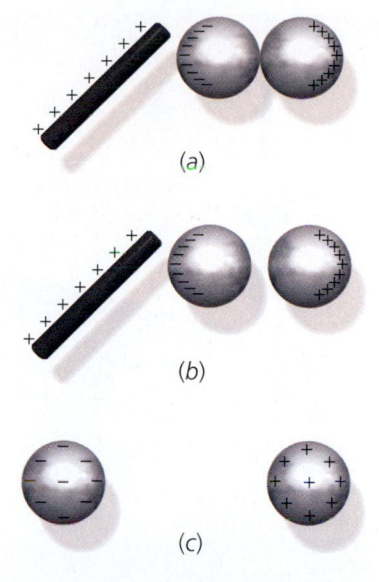

**FIGURE 21-4** Charging by induction. (*a*) Conductors in contact become oppositely charged when a charged rod attracts electrons to the left sphere. (*b*) If the spheres are separated before the rod is removed, they will retain their equal and opposite charges. (*c*) When the rod is removed and the spheres are far apart, the distribution of charge on each sphere approaches uniformity.

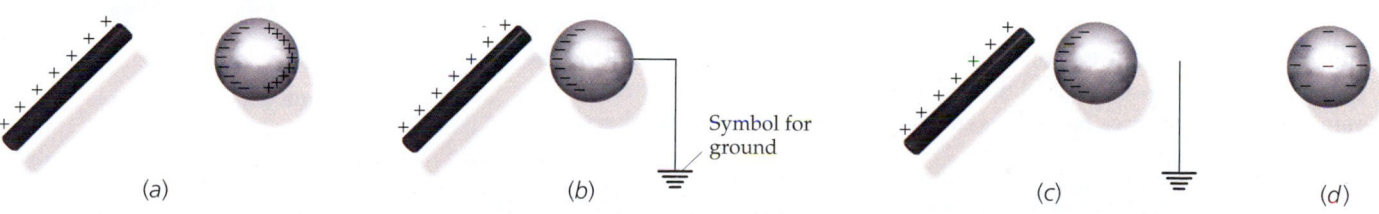

(a)          (b)          Symbol for ground          (c)          (d)

**FIGURE 21-5** Induction via grounding. (*a*) The free charge on the single conducting sphere is polarized by the positively charged rod, which attracts negative charges on the sphere. (*b*) When the conductor is grounded by connecting it with a wire to a very large conductor, such as the earth, electrons from the ground neutralize the positive charge on the far face. The conductor is then negatively charged. (*c*) The negative charge remains if the connection to the ground is broken before the rod is removed. (*d*) After the rod is removed, the sphere has a uniform negative charge.

The lightning rod on this building is grounded so that it can conduct electrons from the ground to the positively charged clouds, thus neutralizing them.

These fashionable ladies are wearing hats with metal chains that drag along the ground, which were supposed to protect them from lightning.

how we can induce a charge in a single conductor by transferring charge from the earth through the ground wire and then breaking the connection to the ground.

# 21-3 Coulomb's Law

Charles Coulomb (1736–1806) studied the force exerted by one charge on another using a torsion balance of his own invention.[†] In Coulomb's experiment, the charged spheres were much smaller than the distance between them so that the charges could be treated as point charges. Coulomb used the method of charging by induction to produce equally charged spheres and to vary the amount of charge on the spheres. For example, beginning with charge $q_0$ on each sphere, he could reduce the charge to $\frac{1}{2}q_0$ by temporarily grounding one sphere to discharge it and then placing the two spheres in contact. The results of the experiments of Coulomb and others are summarized in **Coulomb's law:**

The force exerted by one point charge on another acts along the line between the charges. It varies inversely as the square of the distance separating the charges and is proportional to the product of the charges. The force is repulsive if the charges have the same sign and attractive if the charges have opposite signs.

COULOMB'S LAW

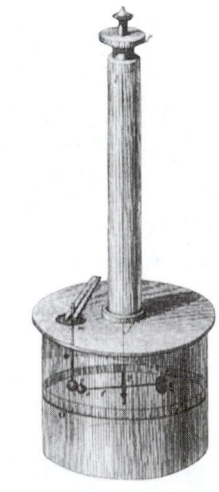

Coulomb's torsion balance.

---

[†] Coulomb's experimental apparatus was essentially the same as that described for the Cavendish experiment in Chapter 11, with the masses replaced by small charged spheres. For the magnitudes of charges easily transferred by rubbing, the gravitational attraction of the spheres is completely negligible compared with their electric attraction or repulsion.

The *magnitude* of the electric force exerted by a charge $q_1$ on another charge $q_2$ a distance $r$ away is thus given by

$$F = \frac{k|q_1 q_2|}{r^2}$$    21-2

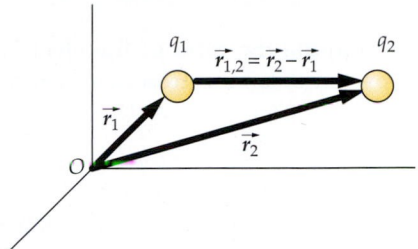

where $k$ is an experimentally determined constant called the **Coulomb constant,** which has the value

$$k = 8.99 \times 10^9 \, \text{N·m}^2/\text{C}^2$$    21-3

**FIGURE 21-6** Charge $q_1$ at position $\vec{r}_1$ and charge $q_2$ at $\vec{r}_2$ relative to the origin $O$. The force exerted by $q_1$ on $q_2$ is in the direction of the vector $\vec{r}_{1,2} = \vec{r}_2 - \vec{r}_1$ if both charges have the same sign, and in the opposite direction if they have opposite signs.

If $q_1$ is at position $\vec{r}_1$ and $q_2$ is at $\vec{r}_2$ (Figure 21-6), the force $\vec{F}_{1,2}$ exerted by $q_1$ on $q_2$ is

$$\vec{F}_{1,2} = \frac{kq_1 q_2}{r_{1,2}^2} \hat{r}_{1,2}$$    21-4

COULOMB'S LAW FOR THE FORCE EXERTED BY $q_1$ ON $q_2$

where $\vec{r}_{1,2} = \vec{r}_2 - \vec{r}_1$ is the vector pointing from $q_1$ to $q_2$, and $\hat{r}_{1,2} = \vec{r}_{1,2}/r_{1,2}$ is a unit vector pointing from $q_1$ to $q_2$.

By Newton's third law, the force $\vec{F}_{2,1}$ exerted by $q_2$ on $q_1$ is the negative of $\vec{F}_{1,2}$. Note the similarity between Coulomb's law and Newton's law of gravity. (See Equation 11-3.) Both are inverse-square laws. But the gravitational force between two particles is proportional to the masses of the particles and is always attractive, whereas the electric force is proportional to the charges of the particles and is repulsive if the charges have the same sign and attractive if they have opposite signs.

---

*ELECTRIC FORCE IN HYDROGEN*    **EXAMPLE 21-2**

In a hydrogen atom, the electron is separated from the proton by an average distance of about $5.3 \times 10^{-11}$ m. Calculate the magnitude of the electrostatic force of attraction exerted by the proton on the electron.

**PICTURE THE PROBLEM** Substitute the given values into Coulomb's law:

$$F = \frac{k|q_1 q_2|}{r^2} = \frac{ke^2}{r^2} = \frac{(8.99 \times 10^9 \, \text{N·m}^2)(1.6 \times 10^{-19} \, \text{C})^2}{(5.3 \times 10^{-11} \, \text{m})^2}$$

$$= \boxed{8.19 \times 10^{-8} \, \text{N}}$$

**REMARKS** Compared with macroscopic interactions, this is a very small force. However, since the mass of the electron is only about $10^{-30}$ kg, this force produces an enormous acceleration of $F/m = 8 \times 10^{22}$ m/s².

**EXERCISE** Two point charges of 0.05 $\mu$C each are separated by 10 cm. Find the magnitude of the force exerted by one point charge on the other. (*Answer* $2.25 \times 10^{-3}$ N)

Since the electrical force and the gravitational force between any two particles both vary inversely with the square of the separation between the particles, the ratio of these forces is independent of separation. We can therefore compare the relative strengths of the electrical and gravitational forces for elementary particles such as the electron and proton.

---

*RATIO OF ELECTRIC AND GRAVITATIONAL FORCES*      **EXAMPLE 21-3**

Compute the ratio of the electric force to the gravitational force exerted by a proton on an electron in a hydrogen atom.

**PICTURE THE PROBLEM** We use Coulomb's law with $q_1 = e$ and $q_2 = -e$ to find the electric force, and Newton's law of gravity with the mass of the proton, $m_p = 1.67 \times 10^{-27}$ kg, and the mass of the electron, $m_e = 9.11 \times 10^{-31}$ kg.

1. Express the magnitudes of the electric force $F_e$ and the gravitational force $F_g$ in terms of the charges, masses, separation distance $r$, and electrical and gravitational constants:

$$F_e = \frac{ke^2}{r^2}; F_g = \frac{Gm_p m_e}{r^2}$$

2. Take the ratio. Note that the separation distance $r$ cancels:

$$\frac{F_e}{F_g} = \frac{ke^2}{Gm_p m_e}$$

3. Substitute numerical values:

$$\frac{F_e}{F_g} = \frac{(8.99 \times 10^9 \text{ N·m}^2/\text{C}^2)(1.6 \times 10^{-19} \text{ C})^2}{(6.67 \times 10^{-11} \text{ N·m}^2/\text{kg}^2)(1.67 \times 10^{-27} \text{ kg})(9.11 \times 10^{-31} \text{ kg})}$$

$$= \boxed{2.27 \times 10^{39}}$$

**REMARKS** This result shows why the effects of gravity are not considered when discussing atomic or molecular interactions.

Although the gravitational force is incredibly weak compared with the electric force and plays essentially no role at the atomic level, it is the dominant force between large objects such as planets and stars. Because large objects contain almost equal numbers of positive and negative charges, the attractive and repulsive electrical forces cancel. The net force between astronomical objects is therefore essentially the force of gravitational attraction alone.

## Force Exerted by a System of Charges

In a system of charges, each charge exerts a force given by Equation 21-4 on every other charge. The net force on any charge is the vector sum of the individual forces exerted on that charge by all the other charges in the system. This follows from the principle of superposition of forces.

---

*NET FORCE*      **EXAMPLE 21-4**    **Try It Yourself**

Three point charges lie on the $x$ axis; $q_1$ is at the origin, $q_2$ is at $x = 2$ m, and $q_0$ is at position $x$ ($x > 2$m).

(a) Find the net force on $q_0$ due to $q_1$ and $q_2$ if $q_1 = +25$ nC, $q_2 = -10$ nC and $x = 3.5$ m (Figure 21-7).

(b) Find an expression for the net force on $q_0$ due to $q_1$ and $q_2$ throughout the region 2 m $< x < \infty$ (Figure 21-8).

**PICTURE THE PROBLEM** The net force on $q_0$ is the vector sum of the force $\vec{F}_{1,0}$ exerted by $q_1$, and the force $\vec{F}_{2,0}$ exerted by $q_2$. The individual forces are found using Coulomb's law. Note that $\hat{r}_{1,0} = \hat{r}_{2,0} = \hat{i}$ because both $\vec{r}_{1,0}$ and $\vec{r}_{2,0}$ are in the positive $x$ direction.

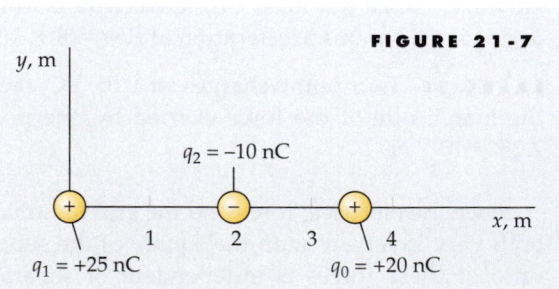

**FIGURE 21-7**

**Cover the column to the right and try these on your own before looking at the answers.**

**Steps**                                                    **Answers**

(a) 1. Draw a sketch of the system of charges.
       Label the distances $r_{1,0}$ and $r_{2,0}$.

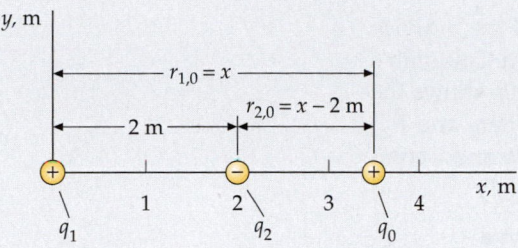

**FIGURE 21-8**

2. Find the force $\vec{F}_{1,0}$ due to $q_1$.                $\vec{F}_{1,0} = (0.367 \ \mu N)\hat{i}$

3. Find the force $\vec{F}_{2,0}$ due to $q_2$.                $\vec{F}_{2,0} = (-0.799 \ \mu N)\hat{i}$

4. Combine your results to obtain the net force.               $\vec{F}_{net} = \vec{F}_{1,0} + \vec{F}_{2,0} = \boxed{-(0.432 \ \mu N\hat{i})}$

(b) 1. Find an expression for the force due to $q_1$.          $\vec{F}_{1,0} = \dfrac{kq_1 q_0}{x^2}\hat{i}$

2. Find an expression for the force due to $q_2$.              $\vec{F}_{1,0} = \dfrac{kq_2 q_0}{(x - 2\,m)^2}\hat{i}$

3. Combine your results to obtain an expression for the        $\vec{F}_{net} = \vec{F}_{1,0} + \vec{F}_{2,0} = \left(\dfrac{kq_1 q_0}{x^2} + \dfrac{kq_2 q_0}{(x - 2\,m)^2}\right)\hat{i}$
   net force.

**REMARKS**  Figure 21-9 shows the $x$ component of the force $F_x$ on $q_0$ as a function of the position $x$ of $q_0$ throughout the region $2 \ m < x < \infty$. Near $q_2$ the force due to $q_2$ dominates, and because opposite charges attract the force on $q_2$ is in the negative $x$ direction. For $x \gg 2m$ the force is in the positive $x$ direction. This is because for large $x$ the distance between $q_1$ and $q_2$ is negligible so the force due to the two charges is almost the same as that for a single charge of $+15$ nC.

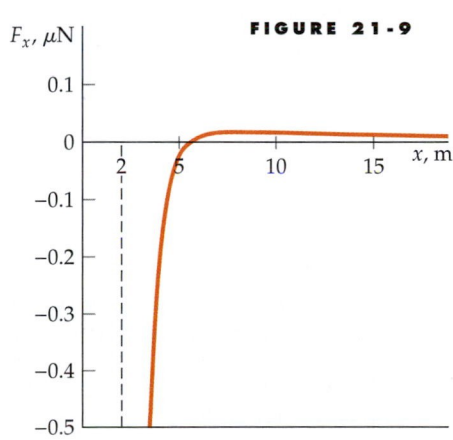

**FIGURE 21-9**

**EXERCISE**  If $q_0$ is at $x = 1$ m, find (a) $\hat{r}_{1,0}$, (b) $\hat{r}_{2,0}$, and (c) the net force acting on $q_0$. (Answer  (a) $\hat{i}$, (b) $-\hat{i}$, (c) $(6.29 \ \mu N)\hat{i}$)

If a system of charges is to remain stationary, then there must be other forces acting on the charges so that the net force from all sources acting on each charge is zero. In the preceding example, and those that follow throughout the book, we assume that there are such forces so that all the charges remain stationary.

NET FORCE IN TWO DIMENSIONS

**EXAMPLE 21-5**

Charge $q_1 = +25$ nC is at the origin, charge $q_2 = -15$ nC is on the $x$ axis at $x = 2$ m, and charge $q_0 = +20$ nC is at the point $x = 2$ m, $y = 2$ m as shown in Figure 21-10. Find the magnitude and direction of the resultant force $\Sigma\vec{F}$ on $q_0$.

**PICTURE THE PROBLEM** The resultant force is the vector sum of the individual forces exerted by each charge on $q_0$. We compute each force from Coulomb's law and write it in terms of its rectangular components. Figure 21-10a shows the resultant force on charge $q_0$ as the vector sum of the forces $\vec{F}_{1,0}$ due to $q_1$ and $\vec{F}_{2,0}$ due to $q_2$. Figure 21-10b shows the net force in Figure 21-10a and its $x$ and $y$ components.

**FIGURE 21-10**

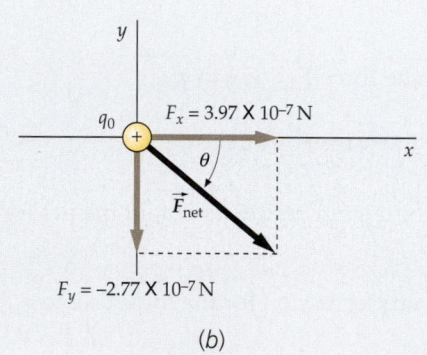

(a)

(b)

1. Draw the coordinate axes showing the positions of the three charges. Show the resultant force on charge $q_0$ as the vector sum of the forces $\vec{F}_{1,0}$ due to $q_1$ and $\vec{F}_{2,0}$ due to $q_2$.

2. The resultant force $\Sigma\vec{F}$ on $q_0$ is the sum of the individual forces:

$$\Sigma\vec{F} = \vec{F}_{1,0} + \vec{F}_{2,0}$$

$$\Sigma F_x = F_{1,0x} + F_{2,0x}$$

$$\Sigma F_y = F_{1,0y} + F_{2,0y}$$

3. The force $\vec{F}_{1,0}$ is directed along the line from $q_1$ to $q_0$. Use $r_{1,0} = 2\sqrt{2}$ for the distance between $q_1$ and $q_0$ to calculate its magnitude:

$$F_{1,0} = \frac{k|q_1 q_0|}{r_{1,0}^2}$$

$$= \frac{(8.99 \times 10^9 \text{ N·m}^2/\text{C}^2)(25 \times 10^{-9} \text{ C})(20 \times 10^{-9} \text{ C})}{(2\sqrt{2} \text{ m})^2}$$

$$= 5.62 \times 10^{-7} \text{ N}$$

4. Since $\vec{F}_{1,0}$ makes an angle of 45° with the $x$ and $y$ axes, its $x$ and $y$ components are equal to each other:

$$F_{1,0x} = F_{1,0y} = F_{1,0}\cos 45° = \frac{5.62 \times 10^{-7} \text{ N}}{\sqrt{2}} = 3.97 \times 10^{-7} \text{ N}$$

5. The force $\vec{F}_{2,0}$ exerted by $q_2$ on $q_0$ is attractive and in the negative $y$ direction as shown in Figure 21-10a:

$$\vec{F}_{2,0} = \frac{kq_2 q_0}{r_{2,0}^2}\hat{r}_{2,0}$$

$$= \frac{(8.99 \times 10^9 \text{ N·m}^2/\text{C}^2)(-15 \times 10^{-9} \text{ C})(20 \times 10^{-9} \text{ C})}{(2 \text{ m})^2}\hat{j}$$

$$= (-6.74 \times 10^{-7} \text{ N})\hat{j}$$

6. Calculate the components of the resultant force:

$$\Sigma F_x = F_{1,0x} + F_{2,0x} = (3.97 \times 10^{-7} \text{ N}) + 0 = 3.97 \times 10^{-7} \text{ N}$$

$$\Sigma F_y = F_{1,0y} + F_{2,0y} = (3.97 \times 10^{-7} \text{ N}) + (-6.74 \times 10^{-7} \text{ N})$$

$$= -2.77 \times 10^{-7} \text{ N}$$

7. Draw the resultant force along with its two components:

8. The magnitude of the resultant force is found from its components:

$$F = \sqrt{F_x^2 + F_y^2} = \sqrt{(3.97 \times 10^{-7}\,\text{N})^2 + (-2.77 \times 10^{-7}\,\text{N})^2}$$

$$= \boxed{4.84 \times 10^{-7}\,\text{N}}$$

9. The resultant force points to the right and downward as shown in Figure 21-10b, making an angle $\theta$ with the $x$ axis given by:

$$\tan\theta = \frac{F_y}{F_x} = \frac{-2.77}{3.97} = -0.698$$

$$\theta = \boxed{-34.9°}$$

**EXERCISE** Express $\hat{r}_{1,0}$ in Example 21-5 in terms of $\hat{i}$ and $\hat{j}$. [*Answer* $\hat{r}_{1,0} = (\hat{i} + \hat{j})/\sqrt{2}$]

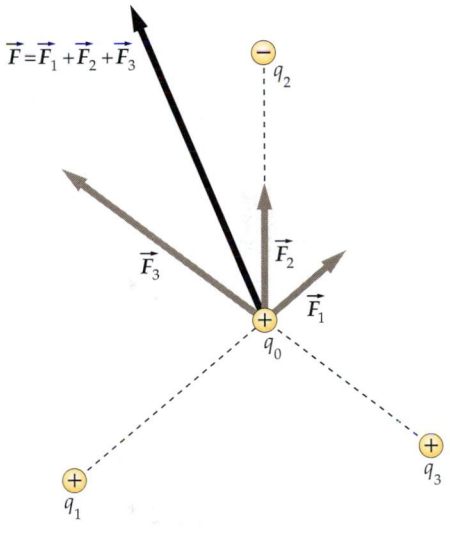

**FIGURE 21-11** A small test charge $q_0$ in the vicinity of a system of charges $q_1$, $q_2$, $q_3$, ... experiences a force $\vec{F}$ that is proportional to $q_0$. The ratio $\vec{F}/q_0$ is the electric field at that point.

## 21-4 The Electric Field

The electric force exerted by one charge on another is an example of an action-at-a-distance force, similar to the gravitational force exerted by one mass on another. The idea of action at a distance presents a difficult conceptual problem. What is the mechanism by which one particle can exert a force on another across the empty space between the particles? Suppose that a charged particle at some point is suddenly moved. Does the force exerted on the second particle some distance $r$ away change instantaneously? To avoid the problem of action at a distance, the concept of the **electric field** is introduced. One charge produces an electric field $\vec{E}$ everywhere in space, and this field exerts the force on the second charge. Thus, it is the *field* $\vec{E}$ at the position of the second charge that exerts the force on it, not the first charge itself which is some distance away. Changes in the field propagate through space at the speed of light, $c$. Thus, if a charge is suddenly moved, the force it exerts on a second charge a distance $r$ away does not change until a time $r/c$ later.

Figure 21-11 shows a set of point charges, $q_1$, $q_2$, and $q_3$, arbitrarily arranged in space. These charges produce an electric field $\vec{E}$ everywhere in space. If we place a small positive **test charge** $q_0$ at some point near the three charges, there will be a force exerted on $q_0$ due to the other charges.[†] The net force on $q_0$ is the vector sum of the individual forces exerted on $q_0$ by each of the other charges in the system. Because each of these forces is proportional to $q_0$, the net force will be proportional to $q_0$. The electric field $\vec{E}$ at a point is this force divided by $q_0$:[‡]

$$\vec{E} = \frac{\vec{F}}{q_0} \qquad (q_0 \text{ small}) \qquad\qquad 21\text{-}5$$

DEFINITION—ELECTRIC FIELD

The SI unit of the electric field is the newton per coulomb (N/C). Table 21-2 lists the magnitudes of some of the electric fields found in nature.

## TABLE 21-2

**Some Electric Fields in Nature**

| | $E$, N/C |
|---|---|
| In household wires | $10^{-2}$ |
| In radio waves | $10^{-1}$ |
| In the atmosphere | $10^2$ |
| In sunlight | $10^3$ |
| Under a thundercloud | $10^4$ |
| In a lightning bolt | $10^4$ |
| In an X-ray tube | $10^6$ |
| At the electron in a hydrogen atom | $6 \times 10^{11}$ |
| At the surface of a uranium nucleus | $2 \times 10^{21}$ |

---

† The presence of the charge $q_0$ will generally change the original distribution of the other charges, particularly if the charges are on conductors. However, we may choose $q_0$ to be small enough so that its effect on the original charge distribution is negligible.

‡ This definition is similar to that for the gravitational field of the earth, which was defined in Section 4-3 as the force per unit mass exerted by the earth on an object.

The electric field describes the condition in space set up by the system of point charges. By moving a test charge $q_0$ from point to point, we can find $\vec{E}$ at all points in space (except at any point occupied by a charge $q$). The electric field $\vec{E}$ is thus a vector function of position. The force exerted on a test charge $q_0$ at any point is related to the electric field at that point by

$$\vec{F} = q_0 \vec{E}$$

21-6

**EXERCISE** When a 5-nC test charge is placed at a certain point, it experiences a force of $2 \times 10^{-4}$ N in the direction of increasing $x$. What is the electric field $\vec{E}$ at that point? [*Answer* $\vec{E} = \vec{F}/q_0 = (4 \times 10^4 \text{ N/C})\hat{i}$]

**EXERCISE** What is the force on an electron placed at a point where the electric field is $\vec{E} = (4 \times 10^4 \text{ N/C})\hat{i}$? [*Answer* $(-6.4 \times 10^{-15} \text{ N})\hat{i}$]

The electric field due to a single point charge can be calculated from Coulomb's law. Consider a small, positive test charge $q_0$ at some point $P$ a distance $r_{i,P}$ away from a charge $q_i$. The force on it is

$$\vec{F}_{i,0} = \frac{kq_iq_0}{r_{i,P}^2}\hat{r}_{i,P}$$

The electric field at point $P$ due to charge $q_i$ (Figure 21-12) is thus

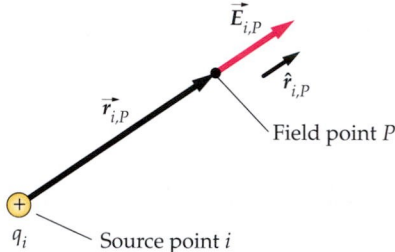

**FIGURE 21-12** The electric field $\vec{E}$ at a field point $P$ due to charge $q_i$ at a source point $i$.

$$\vec{E}_{i,P} = \frac{kq_i}{r_{i,P}^2}\hat{r}_{i,P}$$

21-7

COULOMB'S LAW FOR $\vec{E}$ DUE TO A POINT CHARGE

where $\hat{r}_{i,P}$ is the unit vector pointing from the **source point** $i$ to the **field point** $P$. The net electric field due to a distribution of point charges is found by summing the fields due to each charge separately:

$$\vec{E}_P = \sum_i \vec{E}_{i,P} = \sum_i \frac{kq_i}{r_{i,P}^2}\hat{r}_{i,P}$$

21-8

ELECTRIC FIELD $\vec{E}$ DUE TO A SYSTEM OF POINT CHARGES

---

*ELECTRIC FIELD ON A LINE THROUGH TWO POSITIVE CHARGES* **EXAMPLE 21-6**

A positive charge $q_1 = +8$ nC is at the origin, and a second positive charge $q_2 = +12$ nC is on the $x$ axis at $a = 4$ m (Figure 21-13). Find the net electric field (*a*) at point $P_1$ on the $x$ axis at $x = 7$ m, and (*b*) at point $P_2$ on the $x$ axis at $x = 3$ m.

**FIGURE 21-13**

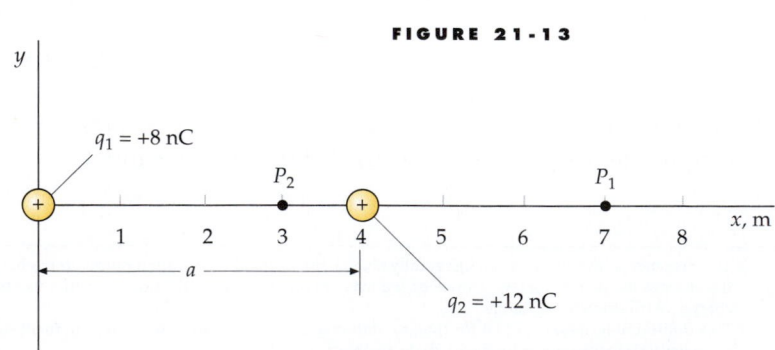

**PICTURE THE PROBLEM** Because point $P_1$ is to the right of both charges, each charge produces a field to the right at that point. At point $P_2$, which is between the charges, the 5-nC charge gives a field to the right and the 12-nC charge gives a field to the left. We calculate each field using

$$\vec{E} = \sum_i \frac{kq_i}{r_{i,P}^2}\hat{r}_{i,P}$$

At point $P_1$, both unit vectors point along the $x$ axis in the positive direction, so $\hat{r}_{1,P_1} = \hat{r}_{2,P_1} = \hat{i}$. At point $P_2$, $\hat{r}_{1,P_2} = \hat{i}$, but the unit vector from the 12-nC charge points along the negative $x$ direction, so $\hat{r}_{2,P_2} = -\hat{i}$.

1. Calculate $\vec{E}$ at point $P_1$, using $r_{1,P_1} = x = 7$ m and $r_{2,P_1} = (x - a) = 7$ m $- 4$ m $= 3$ m:

$$\vec{E} = \frac{kq_1}{r_{1,P_1}^2}\hat{r}_{1,P_1} + \frac{kq_2}{r_{2,P_1}^2}\hat{r}_{2,P_1} = \frac{kq_1}{x^2}\hat{i} + \frac{kq_2}{(x-a)^2}\hat{i}$$

$$= \frac{(8.99 \times 10^9 \text{ N·m}^2/\text{C}^2)(8 \times 10^{-9}\text{C})}{(7 \text{ m})^2}\hat{i}$$

$$+ \frac{(8.99 \times 10^9 \text{ N·m}^2/\text{C}^2)(12 \times 10^{-9}\text{C})}{(3 \text{ m})^2}\hat{i}$$

$$= (1.47 \text{ N/C})\hat{i} + (12.0 \text{ N/C})\hat{i} = \boxed{(13.5 \text{ N/C})\hat{i}}$$

2. Calculate $\vec{E}$ at point $P_2$, where $r_{1,P_2} = x = 3$ m and $r_{2,P_2} = a - x = 4$ m $- 3$ m $= 1$ m:

$$\vec{E} = \frac{kq_1}{r_{1,P_2}^2}\hat{r}_{1,P_2} + \frac{kq_2}{r_{2,P_2}^2}\hat{r}_{2,P_2} = \frac{kq_1}{x^2}\hat{i} + \frac{kq_2}{(a-x)^2}(-\hat{i})$$

$$= \frac{(8.99 \times 10^9 \text{ N·m}^2/\text{C}^2)(8 \times 10^{-9}\text{C})}{(3 \text{ m})^2}\hat{i}$$

$$+ \frac{(8.99 \times 10^9 \text{ N·m}^2/\text{C}^2)(12 \times 10^{-9}\text{C})}{(1 \text{ m})^2}(-\hat{i})$$

$$= (7.99 \text{ N/C})\hat{i} - (108 \text{ N/C})\hat{i} = \boxed{(-100 \text{ N/C})\hat{i}}$$

**REMARKS** The electric field at point $P_2$ is in the negative $x$ direction because the field due to the +12-nC charge, which is 1 m away, is larger than that due to the +8-nC charge, which is 3 m away. The electric field at source points close to the +8-nC charge is dominated by the field due to the +8-nC charge. There is one point between the charges where the net electric field is zero. At this point, a test charge would experience no net force. A sketch of $E_x$ versus $x$ for this system is shown in Figure 21-14.

**EXERCISE** Find the point on the $x$ axis where the electric field is zero. (*Answer* $x = 1.80$ m)

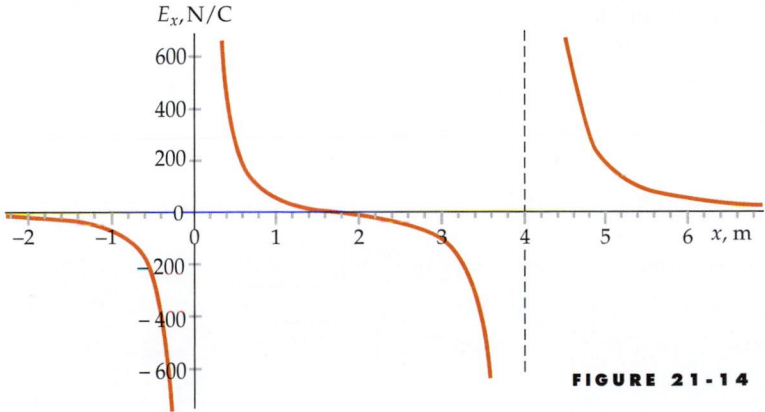

**FIGURE 21-14**

**EXAMPLE 21-7** Try It Yourself

Find the electric field on the $y$ axis at $y = 3$ m for the charges in Example 21-6.

**PICTURE THE PROBLEM** On the $y$ axis, the electric field $\vec{E}_1$ due to charge $q_1$ is directed along the $y$ axis, and the field $\vec{E}_2$ due to charge $q_2$ makes an angle $\theta$ with the $y$ axis (Figure 21-15a). To find the resultant field, we first find the $x$ and $y$ components of these fields, as shown in Figure 21-15b.

**FIGURE 21-15**

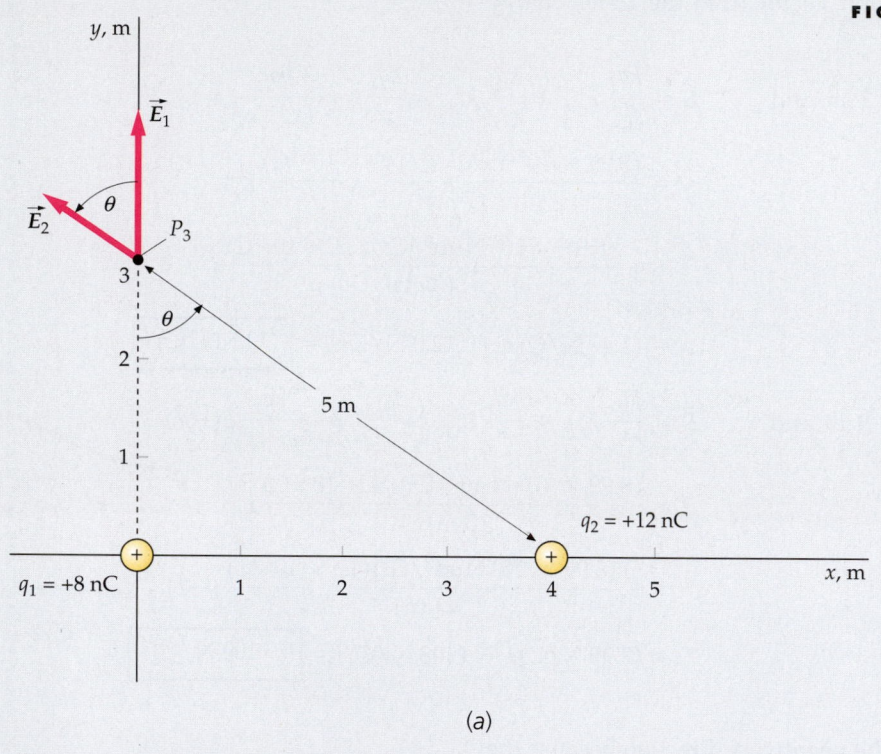

(a)

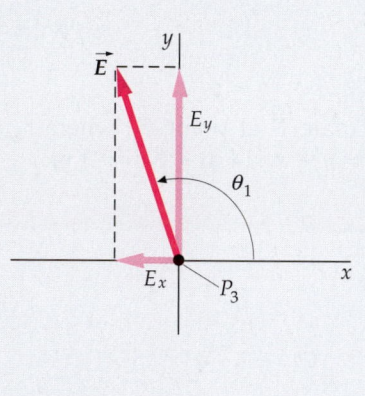

(b)

**Cover the column to the right and try these on your own before looking at the answers.**

| Steps | Answers |
|---|---|
| 1. Calculate the magnitude of the field $\vec{E}_1$ due to $q_1$. Find the $x$ and $y$ components of $\vec{E}_1$. | $E_1 = kq_1/y^2 = 7.99$ N/C $E_{1x} = 0, E_{1y} = 7.99$ N/C |
| 2. Calculate the magnitude of the field $\vec{E}_2$ due to $q_2$. | $E_2 = 4.32$ N/C |
| 3. Write the $x$ and $y$ components of $\vec{E}_2$ in terms of the angle $\theta$. | $E_x = -E_2 \sin \theta; E_y = E_2 \cos \theta$ |
| 4. Compute $\sin \theta$ and $\cos \theta$. | $\sin \theta = 0.8; \cos \theta = 0.6$ |
| 5. Calculate $E_{2x}$ and $E_{2y}$. | $E_{2x} = -3.46$ N/C; $E_{2y} = 2.59$ N/C |
| 6. Find the $x$ and $y$ components of the resultant field $\vec{E}$. | $E_x = -3.46$ N/C; $E_y = 10.6$ N/C |
| 7. Calculate the magnitude of $\vec{E}$ from its components. | $E = \sqrt{E_x^2 + E_y^2} = \boxed{11.2 \text{ N/C}}$ |
| 8. Find the angle $\theta_1$ made by $\vec{E}$ with the $x$ axis. | $\theta_1 = \tan^{-1}\left(\dfrac{E_y}{E_x}\right) = \boxed{108°}$ |

ELECTRIC FIELD DUE TO TWO EQUAL AND OPPOSITE CHARGES    **EXAMPLE   21-8**

A charge $+q$ is at $x = a$ and a second charge $-q$ is at $x = -a$ (Figure 21-16).
(a) Find the electric field on the $x$ axis at an arbitrary point $x > a$. (b) Find the
limiting form of the electric field for $x \gg a$.

**PICTURE THE PROBLEM** We calculate the electric field using

$$\vec{E} = \sum_i \frac{kq_i}{r_{i,P}^2}\hat{r}_{i,P}$$

(Equation 21-8). For $x > a$, the unit vector for each charge is $\hat{i}$.
The distances are $x - a$ to the plus charge and $x - (-a) = x + a$
to the minus charge.

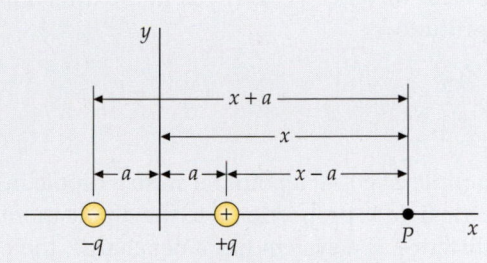

**FIGURE 21-16**

(a) 1. Draw the charge configuration on a coordinate axis and
label the distances from each charge to the field point:

2. Calculate $\vec{E}$ due to the two charges for $x > a$: (*Note:* The
equation on the right holds only for $x > a$. For
$x < a$, the signs of the two terms are reversed. For
$-a < x < a$, both terms have negative signs.)

$$\vec{E} = \frac{kq}{(x-a)^2}\hat{i} + \frac{k(-q)}{(x+a)^2}\hat{i}$$

$$= kq\left[\frac{1}{(x-a)^2} - \frac{1}{(x+a)^2}\right]\hat{i}$$

3. Put the terms in square brackets under a common
denominator and simplify:

$$\vec{E} = kq\left[\frac{(x+a)^2 - (x-a)^2}{(x+a)^2(x-a)^2}\right]\hat{i} = \boxed{kq\frac{4ax}{(x^2-a^2)^2}\hat{i}}$$

(b) In the limit $x \gg a$, we can neglect $a^2$ compared with $x^2$ in
the denominator:

$$\vec{E} = kq\frac{4ax}{(x^2-a^2)^2}\hat{i} \approx kq\frac{4ax}{x^4}\hat{i} = \boxed{\frac{4kqa}{x^3}\hat{i}}$$

**REMARKS** Figure 21-17 shows $E_x$ versus $x$ for all $x$, for $q = 1$ nC
and $a = 1$ m. Far from the charges, the field is given by

$$\vec{E} = \frac{4kqa}{|x|^3}\hat{i}$$

Between the charges, the contribution from each
charge is in the negative direction. An expression
that holds for all $x$ is

$$\vec{E} = \frac{kq}{(x-a)^2}\left[\frac{(x-a)\hat{i}}{|x-a|}\right] + \frac{k(-q)}{(x+a)^2}\left[\frac{(x+a)\hat{i}}{|x+a|}\right]$$

Note that the unit vectors (quantities in square
brackets in this expression) point in the proper direction for all $x$.

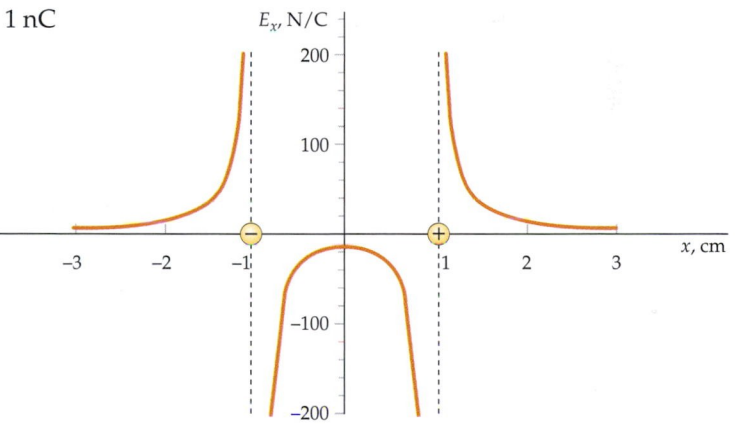

**FIGURE 21-17** A plot of $E_x$ versus $x$ on the
$x$ axis for the charge distribution in Example 21-8.

## Electric Dipoles

A system of two equal and opposite charges $q$ separated by a small distance
$L$ is called an **electric dipole**. Its strength and orientation are described by
the **electric dipole moment** $\vec{p}$, which is a vector that points from the negative charge to the positive charge and has the magnitude $qL$ (Figure 21-18).

$$\boxed{\vec{p} = q\vec{L}}$$

21-9

DEFINITION—ELECTRIC DIPOLE MOMENT

where $\vec{L}$ is the vector from the negative charge to the positive charge.

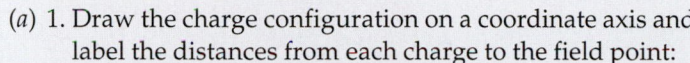

**FIGURE 21-18** An electric dipole consists of
a pair of equal and opposite charges. The dipole
moment is $\vec{p} = q\vec{L}$, where $q$ is the magnitude of
one of the charges and $\vec{L}$ is the relative position
vector from the negative to the positive charge.

For the system of charges in Figure 21-16, $\vec{L} = 2a\hat{i}$ and the electric dipole moment is

$$\vec{p} = 2aq\hat{i}$$

In terms of the dipole moment, the electric field on the axis of the dipole at a point a great distance $|x|$ away is in the direction of the dipole moment and has the magnitude

$$E = \frac{2kp}{|x|^3} \qquad\qquad \text{21-10}$$

(See Example 21-8). At a point far from a dipole in any direction, the magnitude of the electric field is proportional to the dipole moment and decreases with the cube of the distance. If a system has a net charge, the electric field decreases as $1/r^2$ at large distances. In a system with zero net charge, the electric field falls off more rapidly with distance. In the case of an electric dipole, the field falls off as $1/r^3$.

# 21-5 Electric Field Lines

We can picture the electric field by drawing lines to indicate its direction. At any given point, the field vector $\vec{E}$ is tangent to the lines. Electric field lines are also called **lines of force** because they show the direction of the force exerted on a positive test charge. At any point near a positive point charge, the electric field $\vec{E}$ points radially away from the charge. Consequently, the electric field lines near a positive charge also point away from the charge. Similarly, near a negative point charge the electric field lines point toward the negative charge.

Figure 21-19 shows the electric field lines of a single positive point charge. The spacing of the lines is related to the strength of the electric field. As we move away from the charge, the field becomes weaker and the lines become farther apart. Consider a spherical surface of radius $r$ with its center at the charge. Its area is $4\pi r^2$. Thus, as $r$ increases, the density of the field lines (the number of lines per unit area) decreases as $1/r^2$, the same rate of decrease as $E$. So, if we adopt the convention of drawing a fixed number of lines from a point charge, the number being proportional to the charge $q$, and if we draw the lines symmetrically about the point charge, the field strength is indicated by the density of the lines. The more closely spaced the lines, the stronger the electric field.

Figure 21-20 shows the electric field lines for two equal positive point charges $q$ separated by a small distance. Near each charge, the field is approximately due to that charge alone because the other charge is far away. Consequently, the field lines near either charge are radial and equally spaced. Because the charges are

(a)

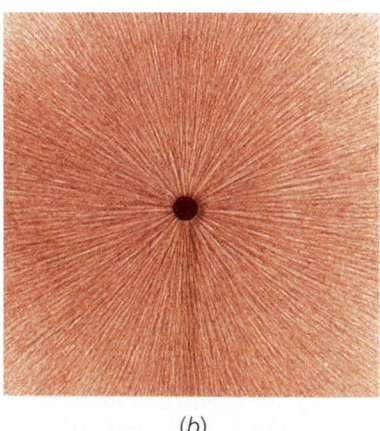

(b)

**FIGURE 21-19** (a) Electric field lines of a single positive point charge. If the charge were negative, the arrows would be reversed. (b) The same electric field lines shown by bits of thread suspended in oil. The electric field of the charged object in the center induces opposite charges on the ends of each bit of thread, causing the threads to align themselves parallel to the field.

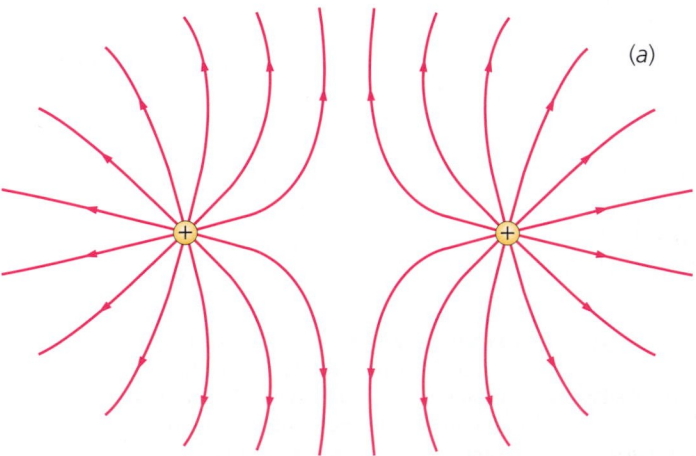

(a)

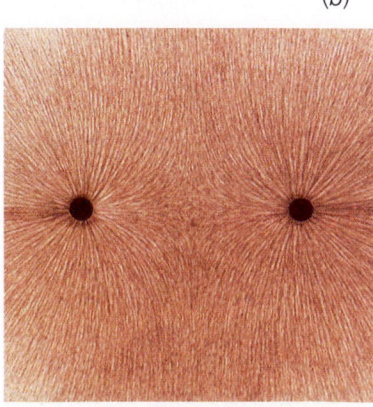

(b)

**FIGURE 21-20** (a) Electric field lines due to two positive point charges. The arrows would be reversed if both charges were negative. (b) The same electric field lines shown by bits of thread in oil.

equal, we draw an equal number of lines originating from each charge. At very large distances, the details of the charge configuration are not important and the system looks like a point charge of magnitude $2q$. (For example, if the two charges were 1 mm apart and we were looking at them from a point 100 km away, they would look like a single charge.) So at a large distance from the charges, the field is approximately the same as that due to a point charge $2q$ and the lines are approximately equally spaced. Looking at Figure 21-20, we see that the density of field lines in the region between the two charges is small compared to the density of lines in the region just to the left and just to the right of the charges. This indicates that the magnitude of the electric field is weaker in the region between the charges than it is in the region just to the right or left of the charges, where the lines are more closely spaced. This information can also be obtained by direct calculation of the field at points in these regions.

We can apply this reasoning to draw the electric field lines for any system of point charges. Very near each charge, the field lines are equally spaced and leave or enter the charge radially, depending on the sign of the charge. Very far from all the charges, the detailed structure of the system is not important so the field lines are just like those of a single point charge carrying the net charge of the system. The rules for drawing electric field lines can be summarized as follows:

1. Electric field lines begin on positive charges (or at infinity) and end on negative charges (or at infinity).

2. The lines are drawn uniformly spaced entering or leaving an isolated point charge.

3. The number of lines leaving a positive charge or entering a negative charge is proportional to the magnitude of the charge.

4. The density of the lines (the number of lines per unit area perpendicular to the lines) at any point is proportional to the magnitude of the field at that point.

5. At large distances from a system of charges with a net charge, the field lines are equally spaced and radial, as if they came from a single point charge equal to the net charge of the system.

6. Field lines do not cross. (If two field lines crossed, that would indicate two directions for $\vec{E}$ at the point of intersection.)

RULES FOR DRAWING ELECTRIC FIELD LINES

Figure 21-21 shows the electric field lines due to an electric dipole. Very near the positive charge, the lines are directed radially outward. Very near the negative charge, the lines are directed radially inward. Because the charges have equal magnitudes, the number of lines that begin at the positive charge equals the number that end at the negative charge. In this case, the field is strong in the region between the charges, as indicated by the high density of field lines in this region in the field.

Figure 21-22$a$ shows the electric field lines for a negative charge $-q$ at a small distance from a positive charge $+2q$. Twice as many lines leave the positive charge as enter the negative charge. Thus, half the lines beginning on the positive charge $+2q$ enter the negative charge $-q$; the rest leave the system. Very far from the charges (Figure 21-22$b$), the lines leaving the system are approximately symmetrically spaced and point radially outward, just as they would for a single positive charge $+q$.

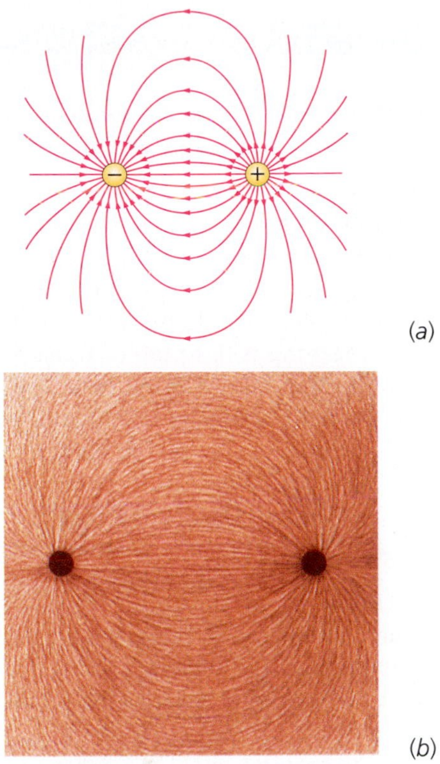

**FIGURE 21-21** (*a*) Electric field lines for an electric dipole. (*b*) The same field lines shown by bits of thread in oil.

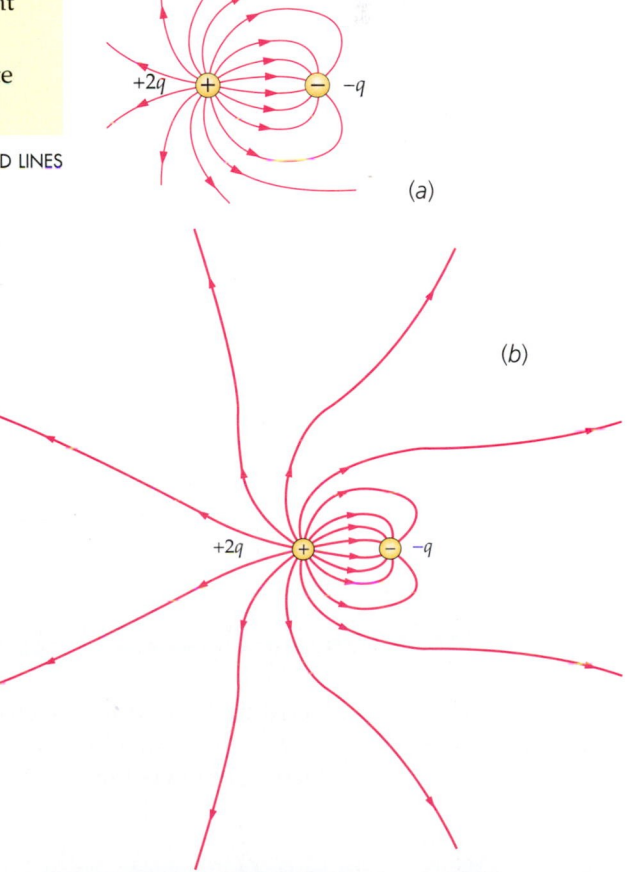

(a)

(b)

**FIGURE 21-22** (*a*) Electric field lines for a point charge $+2q$ and a second point charge $-q$. (*b*) At great distances from the charges, the field lines approach those for a single point charge $+q$ located at the center of charge.

**EXAMPLE 21-9**

The electric field lines for two conducting spheres are shown in Figure 21-23. What is the relative sign and magnitude of the charges on the two spheres?

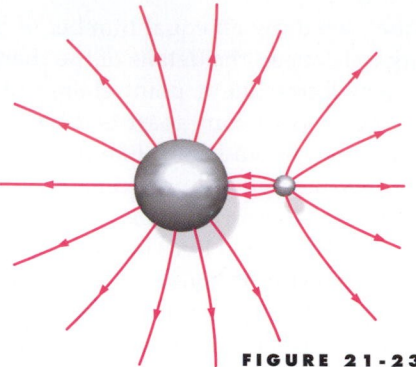

**FIGURE 21-23**

**PICTURE THE PROBLEM** The charge on a sphere is positive if more lines leave than enter and negative if more enter than leave. The ratio of the magnitudes of the charges equals the ratio of the net number of lines entering or leaving.

Since 11 electric field lines leave the large sphere on the left and 3 enter, the net number leaving is 8, so the charge on the large sphere is positive. For the small sphere on the right, 8 lines leave and none enter, so its charge is also positive. Since the net number of lines leaving each sphere is 8, the spheres carry equal positive charges. The charge on the small sphere creates an intense field at the nearby surface of the large sphere that causes a local accumulation of negative charge on the large sphere—indicated by the three entering field lines. Most of the large sphere's surface has positive charge, however, so its total charge is positive.

The convention relating the electric field strength to the electric field lines works because the electric field varies inversely as the square of the distance from a point charge. Because the gravitational field of a point mass also varies inversely as the square of the distance, field-line drawings are also useful for picturing the gravitational field. Near a point mass, the gravitational field lines converge on the mass just as electric field lines converge on a negative charge. However, unlike electric field lines near a positive charge, there are no points in space from which gravitational field lines diverge. That's because the gravitational force is always attractive, never repulsive.

# 21-6 Motion of Point Charges in Electric Fields

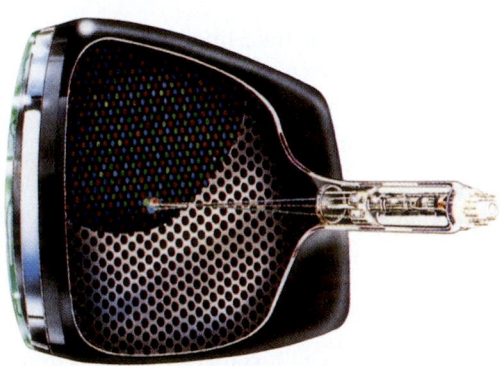

When a particle with a charge $q$ is placed in an electric field $\vec{E}$, it experiences a force $q\vec{E}$. If the electric force is the only significant force acting on the particle, the particle has acceleration

$$\vec{a} = \frac{\Sigma \vec{F}}{m} = \frac{q}{m}\vec{E}$$

where $m$ is the mass of the particle. (If the particle is an electron, its speed in an electric field is often a significant fraction of the speed of light. In such cases, Newton's laws of motion must be modified by Einstein's special theory of relativity.) If the electric field is known, the charge-to-mass ratio of the particle can be determined from the measured acceleration. J. J. Thomson used the deflection of electrons in a uniform electric field in 1897 to demonstrate the existence of electrons and to measure their charge-to-mass ratio. Familiar examples of devices that rely on the motion of electrons in electric fields are oscilloscopes, computer monitors, and television picture tubes.

Schematic drawing of a cathode-ray tube used for color television. The beams of electrons from the electron gun on the right activate phosphors on the screen at the left, giving rise to bright spots whose colors depend on the relative intensity of each beam. Electric fields between deflection plates in the gun (or magnetic fields from coils surrounding the gun) deflect the beams. The beams sweep across the screen in a horizontal line, are deflected downward, then sweep across again. The entire screen is covered in this way 30 times per second.

**FIGURE 21-24**

**EXAMPLE 21-10**

An electron is projected into a uniform electric field $\vec{E} = (1000 \text{ N/C})\hat{i}$ with an initial velocity $\vec{v}_0 = (2 \times 10^6 \text{ m/s})\hat{i}$ in the direction of the field (Figure 21-24). How far does the electron travel before it is brought momentarily to rest?

**PICTURE THE PROBLEM** Since the charge of the electron is negative, the force $\vec{F} = -e\vec{E}$ acting on the electron is in the direction opposite that of the field. Since $\vec{E}$ is constant, the force is constant and we can use constant acceleration formulas from Chapter 2. We choose the field to be in the positive $x$ direction.

1. The displacement $\Delta x$ is related to the initial and final velocities:

$$v_x^2 = v_{0x}^2 + 2a_x\,\Delta x$$

2. The acceleration is obtained from Newton's second law:

$$a_x = \frac{F_x}{m} = \frac{-eE}{m}$$

3. When $v_x = 0$, the displacement is:

$$\Delta x = \frac{v_x^2 - v_{0x}^2}{2a_x} = \frac{0 - v_{0x}^2}{2(-eE/m)} = \frac{mv_0^2}{2eE}$$

$$= \frac{(9.11 \times 10^{-31}\,\text{kg})(2 \times 10^6\,\text{m/s})^2}{2(1.6 \times 10^{-19}\,\text{C})(1000\,\text{N/C})}$$

$$= 1.14 \times 10^{-2}\,\text{m} = \boxed{1.14\,\text{cm}}$$

---

*ELECTRON MOVING PERPENDICULAR TO A UNIFORM ELECTRIC FIELD*    **EXAMPLE 21-11**

An electron enters a uniform electric field $\vec{E} = (-2000\,\text{N/C})\hat{j}$ with an initial velocity $\vec{v}_0 = (10^6\,\text{m/s})\hat{i}$ perpendicular to the field (Figure 21-25). (a) Compare the gravitational force acting on the electron to the electric force acting on it. (b) By how much has the electron been deflected after it has traveled 1 cm in the $x$ direction?

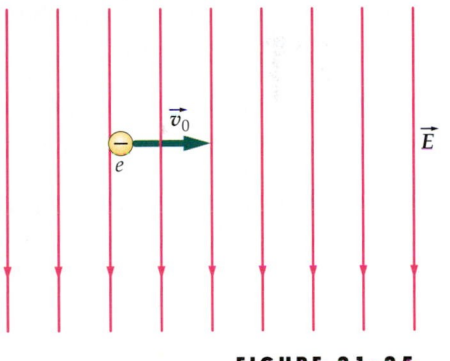

**FIGURE 21-25**

**PICTURE THE PROBLEM** (a) Calculate the ratio of the electric force $qE = -eE$ to the gravitational force $mg$. (b) Since $mg$ is negligible, the force on the electron is $-eE$ vertically upward. The electron thus moves with constant horizontal velocity $v_x$ and is deflected upward by an amount $y = \frac{1}{2}at^2$, where $t$ is the time to travel 1 cm in the $x$ direction.

(a) Calculate the ratio of the magnitude of the electric force, $F_e$, to the magnitude of the gravitational force, $F_g$:

$$\frac{F_e}{F_g} = \frac{eE}{mg} = \frac{(1.6 \times 10^{-19}\,\text{C})(2000\,\text{N/C})}{(9.11 \times 10^{-31}\,\text{kg})(9.81\,\text{N/kg})} = \boxed{3.6 \times 10^{13}}$$

(b) 1. Express the vertical deflection in terms of the acceleration $a$ and time $t$:

$$y = \frac{1}{2}a_y t^2$$

2. Express the time required for the electron to travel a horizontal distance $x$ with constant horizontal velocity $v_0$:

$$t = \frac{x}{v_0}$$

3. Use this result for $t$ and $eE/m$ for $a_y$ to calculate $y$:

$$y = \frac{1}{2}\frac{eE}{m}\left(\frac{x}{v_0}\right)^2$$

$$= \frac{1}{2}\frac{(1.6 \times 10^{-19}\,\text{C})(2000\,\text{N/C})}{9.11 \times 10^{-31}\,\text{kg}}\left(\frac{0.01\,\text{m}}{10^6\,\text{m/s}}\right)^2$$

$$= \boxed{1.76\,\text{cm}}$$

**REMARKS** (a) As is usually the case, the electric force is huge compared with the gravitational force. Thus, it is not necessary to consider gravity when designing a cathode-ray tube, for example, or when calculating the deflection in the problem above. In fact, a television picture tube works equally well upside down and right side up, as if gravity were not even present. (b) The path of an electron moving in a uniform electric field is a parabola, the same as the path of a neutron moving in a uniform gravitational field.

You've just finished printing out a long essay for your English professor, and you get to wondering about how the ink-jet printer knows where to place the ink. You search the Internet and find a picture (Figure 21-26) that shows that the ink drops are given a charge and pass between a pair of oppositely charged metal plates that provide a uniform electric field in the region between the plates. Since you've been studying the electric field in physics class, you wonder if you can determine how large a field is used in this type of printer. You do a bit more searching and find that the 40-$\mu$m-diameter ink drops have an initial velocity of 40 m/s, and that a drop with a 2-nC charge is deflected upward a distance of 3 mm as the drop transits the 1-cm-long region between the plates. Find the magnitude of the electric field. (Neglect any effects of gravity on the motion of the drops.)

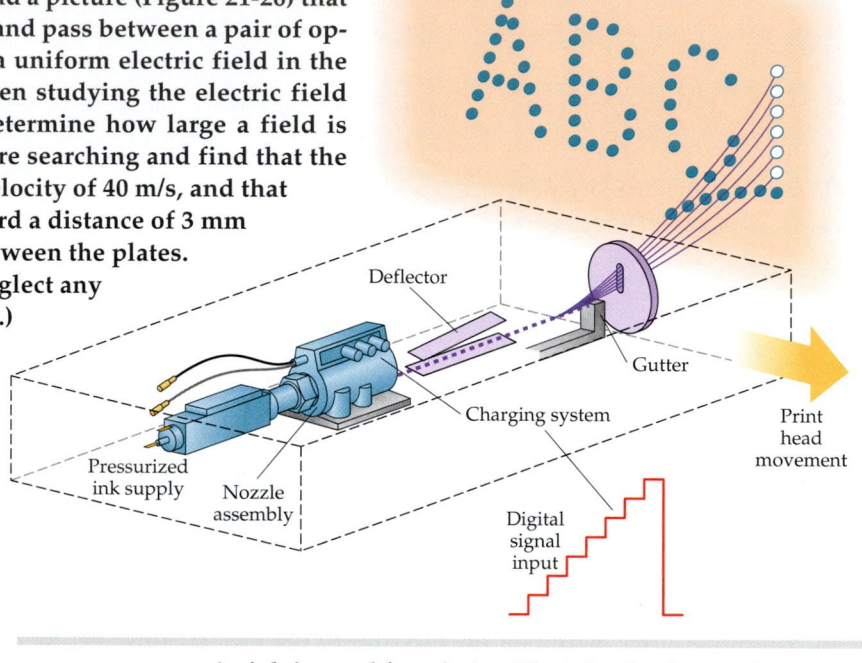

**PICTURE THE PROBLEM** The electric field $\vec{E}$ exerts a constant electric force $\vec{F}$ on the drop as it passes between the two plates, where $\vec{F} = q\vec{E}$. We are looking for $E$. We can get the force $\vec{F}$ by determining the mass and accelertion $\vec{F} = m\vec{a}$. The acceleration can be found from kinematics and mass can be found using the radius and assuming that the density $\rho$ of ink is 1000 kg/m$^3$ (the same as the density of water).

**FIGURE 21-26** An ink-jet used for printing. The ink exits the nozzle in discrete droplets. Any droplet destined to form a dot on the image is given a charge. The deflector consists of a pair of oppositely charged plates. The greater the charge a drop receives, the higher the drop is deflected as it passes between the deflector plates. Drops that do not receive a charge are not deflected upward. These drops end up in the gutter, and the ink is returned to the ink reservoir.

1. The electric field equals the force to charge ratio:

$$E = \frac{F}{q}$$

2. The force, which is in the $+y$ direction (upward), equals the mass times the acceleration:

$$F = ma$$

3. The vertical displacement is obtained using a constant-acceleration kinematic formula with $v_{0y} = 0$:

$$\Delta y = v_{0y}t + \tfrac{1}{2}at^2$$
$$= 0 + \tfrac{1}{2}at^2$$

4. The time is how long it takes for the drop to travel the $\Delta x = 1$ cm at $v_0 = 40$ m/s:

$$\Delta x = v_{0x}t = v_0 t, \text{ so } t = \Delta x/v_0$$

5. Solving for $a$ gives:

$$a = \frac{2\Delta y}{t^2} = \frac{2\Delta y}{(\Delta x/v_0)^2} = \frac{2v_0^2\Delta y}{(\Delta x)^2}$$

6. The mass equals the density times the volume:

$$m = \rho V = \rho \tfrac{4}{3}\pi r^3$$

7. Solve for $E$:

$$E = \frac{F}{q} = \frac{ma}{q} = \frac{\rho\tfrac{4}{3}\pi r^3}{q}\frac{2v_0^2\Delta y}{(\Delta x)^2}$$

$$= \frac{8\pi}{3}\frac{\rho r^3 v_0^2 \Delta y}{q(\Delta x)^2}$$

$$= \frac{8\pi}{3}\frac{(1000 \text{ kg/m}^3)(20 \times 10^{-6} \text{ m})^3(40 \text{ m/s})^2(3 \times 10^{-3} \text{ m})}{(2 \times 10^{-9} \text{ C})(0.01 \text{ m})^2}$$

$$= \boxed{1610 \text{ N/C}}$$

**REMARKS** The ink jet in this example is called a multiple-deflection continuous ink jet. It is used in some industrial printers. The ink-jet printers sold for use with home computers do not use charged droplets deflected by an electric field.

## 21-7 Electric Dipoles in Electric Fields

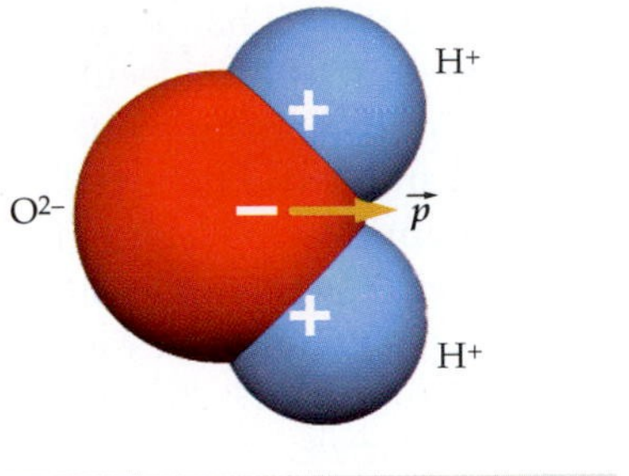

In Example 21-6 we found the electric field produced by a dipole, a system of two equal and opposite point charges that are close together. Here we consider the behavior of an electric dipole in an external electric field. Some molecules have permanent electric dipole moments due to a nonuniform distribution of charge within the molecule. Such molecules are called **polar molecules.** An example is HCl, which is essentially a positive hydrogen ion of charge $+e$ combined with a negative chlorine ion of charge $-e$. The center of charge of the positive ion does not coincide with the center of charge for the negative ion, so the molecule has a permanent dipole moment. Another example is water (Figure 21-27).

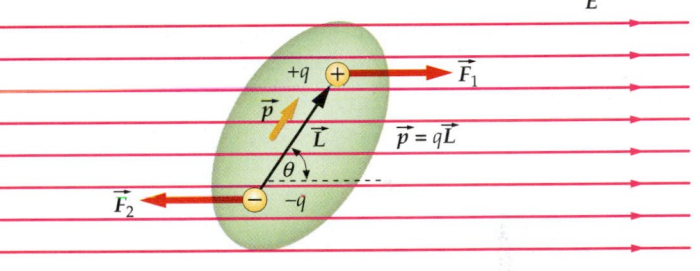

**FIGURE 21-27** An $H_2O$ molecule has a permanent electric dipole moment that points in the direction from the center of negative charge to the center of positive charge.

A uniform external electric field exerts no net force on a dipole, but it does exert a torque that tends to rotate the dipole into the direction of the field. We see in Figure 21-28 that the torque calculated about the position of either charge has the magnitude $F_1 L \sin \theta = qEL \sin \theta = pE \sin \theta$.[†] The direction of the torque is into the paper such that it rotates the dipole moment $\vec{p}$ into the direction of $\vec{E}$. The torque can be conveniently written as the cross product of the dipole moment $\vec{p}$ and the electric field $\vec{E}$.

$$\vec{\tau} = \vec{p} \times \vec{E} \qquad\qquad 21\text{-}11$$

When the dipole rotates through $d\theta$, the electric field does work:

$$dW = -\tau d\theta = -pE \sin \theta \, d\theta$$

(The minus sign arises because the torque opposes any increase in $\theta$.) Setting the negative of this work equal to the change in potential energy, we have

$$dU = -dW = +pE \sin \theta \, d\theta$$

Integrating, we obtain

$$U = -pE \cos \theta + U_0$$

If we choose the potential energy $U_0$ to be zero when $\theta = 90°$, then the potential energy of the dipole is

$$U = -pE \cos \theta = -\vec{p} \cdot \vec{E} \qquad\qquad 21\text{-}12$$

POTENTIAL ENERGY OF A DIPOLE IN AN ELECTRIC FIELD

**FIGURE 21-28** A dipole in a uniform electric field experiences equal and opposite forces that tend to rotate the dipole so that its dipole moment is aligned with the electric field.

Microwave ovens take advantage of the electric dipole moment of water molecules to cook food. Like all electromagnetic waves, microwaves have oscillating electric fields that exert torques on electric dipoles, torques that cause the water molecules to rotate with significant rotational kinetic energy. In this manner, energy is transferred from the microwave radiation to the water molecules throughout the food at a high rate, accounting for the rapid cooking times that make microwave ovens so convenient.

---

† The torque produced by two equal and opposite forces (an arrangement called a couple) is the same about any point in space.

**Nonpolar molecules** have no permanent electric dipole movement. However, all neutral molecules contain equal amounts of positive and negative charge. In the presence of an external electric field $\vec{E}$, the charges become separated in space. The positive charges are pushed in the direction of $\vec{E}$ and the negative charges are pushed in the opposite direction. The molecule thus acquires an induced dipole moment parallel to the external electric field and is said to be **polarized**.

In a nonuniform electric field, an electric dipole experiences a net force because the electric field has different magnitudes at the positive and negative poles. Figure 21-29 shows how a positive point charge polarizes a nonpolar molecule and then attracts it. A familiar example is the attraction that holds an electrostatically charged balloon against a wall. The nonuniform field produced by the charge on the balloon polarizes molecules in the wall and attracts them. An equal and opposite force is exerted by the wall molecules on the balloon.

The diameter of an atom or molecule is of the order of $10^{-10}$ m = 0.1 nm. A convenient unit for the electric dipole moment of atoms and molecules is the fundamental electronic charge $e$ times the distance 1 nm. For example, the dipole moment of $H_2O$ in these units has a magnitude of about $0.04e \cdot$nm.

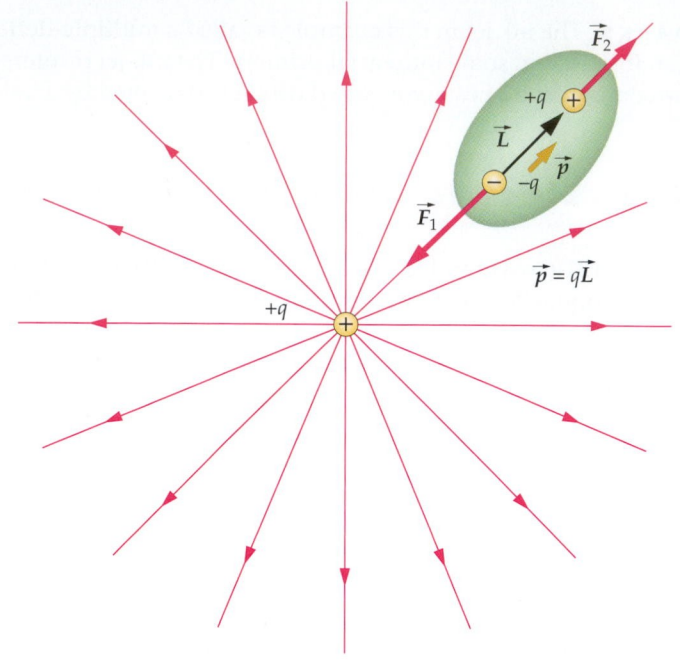

**FIGURE 21-29** A nonpolar molecule in the nonuniform electric field of a positive point charge. The induced electric dipole moment $\vec{p}$ is parallel to the field of the point charge. Because the point charge is closer to the center of negative charge than to the center of positive charge, there is a net force of attraction between the dipole and the point charge. If the point charge were negative, the induced dipole moment would be reversed, and the molecule would again be attracted to the point charge.

---

*TORQUE AND POTENTIAL ENERGY*                **EXAMPLE 21-13**

**A dipole with a moment of magnitude 0.02 $e \cdot$nm makes an angle of 20° with a uniform electric field of magnitude $3 \times 10^3$ N/C (Figure 21-30). Find (a) the magnitude of the torque on the dipole, and (b) the potential energy of the system.**

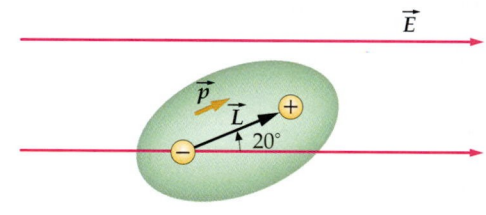

**FIGURE 21-30**

**PICTURE THE PROBLEM** The torque is found from $\vec{\tau} = \vec{p} \times \vec{E}$ and the potential energy is found from $U = -\vec{p} \cdot \vec{E}$.

1. Calculate the magnitude of the torque:

$$\tau = |\vec{p} \times \vec{E}| = pE\sin\theta = (0.02\ e \cdot \text{nm})(3 \times 10^3\ \text{N/C})(\sin 20°)$$

$$= (0.02)(1.6 \times 10^{-19}\ \text{C})(10^{-9}\ \text{m})(3 \times 10^3\ \text{N/C})(\sin 20°)$$

$$= \boxed{3.28 \times 10^{-27}\ \text{N} \cdot \text{m}}$$

2. Calculate the potential energy:

$$U = -\vec{p} \cdot \vec{E} = -pE\cos\theta$$

$$= -(0.02)(1.6 \times 10^{-19}\ \text{C})(10^{-9}\ \text{m})(3 \times 10^3\ \text{N/C})\cos 20°$$

$$= \boxed{-9.02 \times 10^{-27}\ \text{J}}$$

# SUMMARY

1. Quantization and conservation are fundamental properties of electric charge.
2. Coulomb's law is the fundamental law of interaction between charges at rest.
3. The electric field describes the condition in space set up by a charge distribution.

| Topic | Relevant Equations and Remarks | |
|---|---|---|
| **1. Electric Charge** | There are two kinds of electric charge, positive and negative. | |
| Quantization | Electric charge is quantized—it always occurs in integral multiples of the fundamental unit of charge $e$. The charge of the electron is $-e$ and that of the proton is $+e$. | |
| Magnitude | $e = 1.60 \times 10^{-19}$ C | 21-1 |
| Conservation | Charge is conserved. It is neither created nor destroyed in any process, but is merely transferred. | |
| **2. Conductors and Insulators** | In conductors, about one electron per atom is free to move about the entire material. In insulators, all the electrons are bound to nearby atoms. | |
| Ground | A very large conductor that can supply an unlimited amount of charge (such as the earth) is called a ground. | |
| **3. Charging by Induction** | A conductor can be charged by holding a charge near the conductor to attract or repel the free electrons and then grounding the conductor to drain off the faraway charges. | |
| **4. Coulomb's Law** | The force exerted by a charge $q_1$ on $q_2$ is given by | |
| | $$\vec{F}_{1,2} = \frac{kq_1 q_2}{r_{1,2}^2} \hat{r}_{1,2}$$ | 21-2 |
| | where $\hat{r}_{1,2}$ is a unit vector that points from $q_1$ to $q_2$. | |
| Coulomb constant | $k = 8.99 \times 10^9 \ \text{N}\cdot\text{m}^2/\text{C}^2$ | 21-3 |
| **5. Electric Field** | The electric field due to a system of charges at a point is defined as the net force exerted by those charges on a very small positive test charge $q_0$ divided by $q_0$: | |
| | $$\vec{E} = \frac{\vec{F}}{q_0}$$ | 21-5 |
| Due to a point charge | $$\vec{E}_{i,P} = \frac{kq_i}{r_{i,P}^2} \hat{r}_{i,P}$$ | 21-7 |
| Due to a system of point charges | The electric field due to several charges is the vector sum of the fields due to the individual charges: | |
| | $$\vec{E}_{i,P} = \sum_i \vec{E}_i = \sum_i \frac{kq_i}{r_{i,P}^2} \hat{r}_{i,P}$$ | 21-8 |
| **6. Electric Field Lines** | The electric field can be represented by electric field lines that originate on positive charges and end on negative charges. The strength of the electric field is indicated by the density of the electric field lines. | |

**7. Electric Dipole**

An electric dipole is a system of two equal but opposite charges separated by a small distance.

| | | |
|---|---|---|
| Dipole moment | $\vec{p} = q\vec{L}$ <br><br> where $\vec{L}$ points from the negative charge to the positive charge. | 21-9 |
| Field due to dipole | The electric field far from a dipole is proportional to the dipole moment and decreases with the cube of the distance. | |
| Torque on a dipole | In a uniform electric field, the net force on a dipole is zero, but there is a torque that tends to align the dipole in the direction of the field. <br><br> $\vec{\tau} = \vec{p} \times \vec{E}$ | 21-11 |
| Potential energy of a dipole | $U = -\vec{p} \cdot \vec{E}$ | 21-12 |

**8. Polar and Nonpolar Molecules**

Polar molecules, such as $H_2O$, have permanent dipole moments because their centers of positive and negative charge do not coincide. They behave like simple dipoles in an electric field. Nonpolar molecules do not have permanent dipole moments, but they acquire induced dipole moments in the presence of an electric field.

# PROBLEMS

- • Single-concept, single-step, relatively easy
- •• Intermediate-level, may require synthesis of concepts
- ••• Challenging
- **SSM** Solution is in the *Student Solutions Manual*
- **iSOLVE** Problems available on iSOLVE online homework service
- **iSOLVE✓** These "Checkpoint" online homework service problems ask students additional questions about their confidence level, and how they arrived at their answer.

In a few problems, you are given more data than you actually need; in a few other problems, you are required to supply data from your general knowledge, outside sources, or informed estimates.

## Conceptual Problems

**1** •• **SSM** Discuss the similarities and differences in the properties of electric charge and gravitational mass.

**2** • Can insulators be charged by induction?

**3** •• A metal rectangle $B$ is connected to ground through a switch $S$ that is initially closed (Figure 21-31). While the charge $+Q$ is near $B$, switch $S$ is opened. The charge $+Q$ is then removed. Afterward, what is the charge state of the metal rectangle $B$? (*a*) It is positively charged. (*b*) It is uncharged. (*c*) It is negatively charged. (*d*) It may be any of the above depending on the charge on $B$ before the charge $+Q$ was placed nearby.

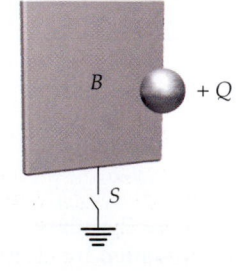

**FIGURE 21-31**
**Problem 3**

**4** •• Explain, giving each step, how a positively charged insulating rod can be used to give a metal sphere (*a*) a negative charge, and (*b*) a positive charge. (*c*) Can the same rod be used to simultaneously give one sphere a positive charge and another sphere a negative charge without the rod having to be recharged?

**5** •• **SSM** Two uncharged conducting spheres with their conducting surfaces in contact are supported on a large wooden table by insulated stands. A positively charged rod is brought up close to the surface of one of the spheres on the side opposite its point of contact with the other sphere. (*a*) Describe the induced charges on the two conducting spheres, and sketch the charge distributions on them. (*b*) The two spheres are separated far apart and the charged rod is removed. Sketch the charge distributions on the separated spheres.

**6** • Three charges, $+q$, $+Q$, and $-Q$, are placed at the corners of an equilateral triangle as shown in Figure 21-32. The net force on charge $+q$ due to the other two charges is (*a*) vertically up. (*b*) vertically down. (*c*) zero. (*d*) horizontal to the left. (*e*) horizontal to the right.

$+q$

$+Q$  $-Q$

**FIGURE 21-32** Problem 6

**7** • [SSM] A positive charge that is free to move but is at rest in an electric field $\vec{E}$ will

(*a*) accelerate in the direction perpendicular to $\vec{E}$.
(*b*) remain at rest.
(*c*) accelerate in the direction opposite to $\vec{E}$.
(*d*) accelerate in the same direction as $\vec{E}$.
(*e*) do none of the above.

**8** • [SSM] If four charges are placed at the corners of a square as shown in Figure 21-33, the field $\vec{E}$ is zero at

(*a*) all points along the sides of the square midway between two charges.
(*b*) the midpoint of the square.
(*c*) midway between the top two charges and midway between the bottom two charges.
(*d*) none of the above.

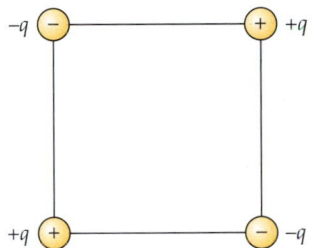

**FIGURE 21-33**
**Problem 8**

**9** •• At a particular point in space, a charge $Q$ experiences no net force. It follows that

(*a*) there are no charges nearby.
(*b*) if charges are nearby, they have the opposite sign of $Q$.
(*c*) if charges are nearby, the total positive charge must equal the total negative charge.
(*d*) none of the above need be true.

**10** • Two charges $+4q$ and $-3q$ are separated by a small distance. Draw the electric field lines for this system.

**11** • [SSM] Two charges $+q$ and $-3q$ are separated by a small distance. Draw the electric field lines for this system.

**12** • [SSM] Three equal positive point charges are situated at the corners of an equilateral triangle. Sketch the electric field lines in the plane of the triangle.

**13** • Which of the following statements are true?

(*a*) A positive charge experiences an attractive electrostatic force toward a nearby neutral conductor.
(*b*) A positive charge experiences no electrostatic force near a neutral conductor.
(*c*) A positive charge experiences a repulsive force, away from a nearby conductor.
(*d*) Whatever the force on a positive charge near a neutral conductor, the force on a negative charge is then oppositely directed.
(*e*) None of the above is correct.

**14** • [SSM] The electric field lines around an electrical dipole are best represented by which, if any, of the diagrams in Figure 21-34?

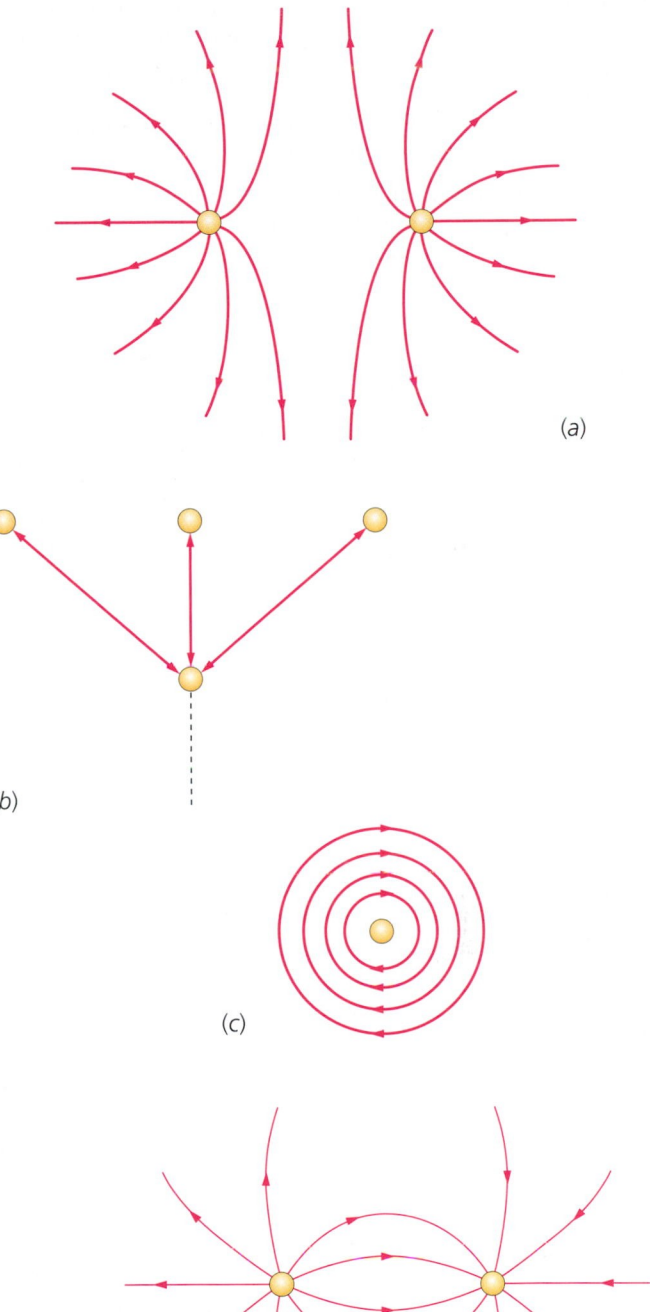

**FIGURE 21-34** Problem 14

**15** •• [SSM] A molecule with electric dipole moment $\vec{p}$ is oriented so that $\vec{p}$ makes an angle $\theta$ with a uniform electric field $\vec{E}$. The dipole is free to move in response to the force from the field. Describe the motion of the dipole. Suppose the electric field is nonuniform and is larger in the $x$ direction. How will the motion be changed?

**16** •• True or false:

(a) The electric field of a point charge always points away from the charge.

(b) All macroscopic charges $Q$ can be written as $Q = \pm Ne$, where $N$ is an integer and $e$ is the charge of the electron.

(c) Electric field lines never diverge from a point in space.

(d) Electric field lines never cross at a point in space.

(e) All molecules have electric dipole moments in the presence of an external electric field.

**17** •• Two metal balls have charges $+q$ and $-q$. How will the force on one of them change if (a) the balls are placed in water, the distance between them being unchanged, and (b) a third uncharged metal ball is placed between the first two? Explain.

**18** •• **SSM** A metal ball is positively charged. Is it possible for it to attract another positively charged ball? Explain.

**19** •• **SSM** A simple demonstration of electrostatic attraction can be done simply by tying a small ball of tinfoil on a hanging string, and bringing a charged wand near it. Initially, the ball will be attracted to the wand, but once they touch, the ball will be repelled violently from it. Explain this behavior.

### Estimation and Approximation

**20** •• Two small spheres are connected to opposite ends of a steel cable of length 1 m and cross-sectional area 1.5 cm². A positive charge $Q$ is placed on each sphere. Estimate the largest possible value $Q$ can have before the cable breaks, given that the tensile strength of steel is $5.2 \times 10^8$ N/m².

**21** •• The net charge on any object is the result of the surplus or deficit of only an extremely small fraction of the electrons in the object. In fact, a charge imbalance greater than this would result in the destruction of the object. (a) Estimate the force acting on a 0.5 cm × 0.5 cm × 4 cm rod of copper if the electrons in the copper outnumbered the protons by 0.0001%. Assume that half of the excess electrons migrate to opposite ends of the rod of the copper. (b) Calculate the largest possible imbalance, given that copper has a tensile strength of $2.3 \times 10^8$ N/m².

**22** ••• Electrical discharge (sparks) in air occur when free ions in the air are accelerated to a high enough velocity by an electric field to ionize other gas molecules on impact. (a) Assuming that the ion moves, on average, 1 mean free path through the gas before hitting a molecule, and that it needs to acquire an energy of approximately 1 eV to ionize it, estimate the field strength required for electrical breakdown in air at a pressure and temperature of $1 \times 10^5$ N/m² and 300 K. Assume that the cross-sectional area of a nitrogen molecule is about 0.1 nm². (b) How should the breakdown potential depend on temperature (all other things being equal)? On pressure?

**23** •• **SSM** A popular classroom demonstration consists of rubbing a "magic wand" made of plastic with fur to charge it, and then placing it near an empty soda can on its side (Figure 21-35.) The can will roll toward the wand, as it acquires a charge on the side nearest the wand by induction. Typically, if the wand is held about 10 cm away from the can, the can will have an initial acceleration of about 1 m/s². If the mass of the can is 0.018 kg, estimate the charge on the rod.

**FIGURE 21-35**
Problem 23

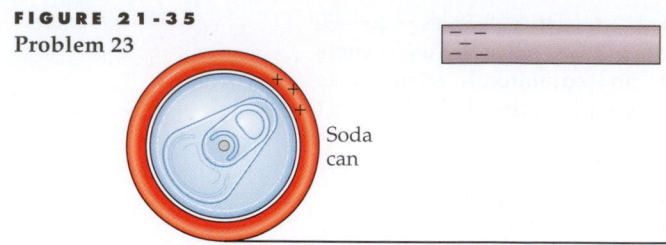

Soda can

**24** •• Estimate the force required to bind the He nucleus together, given that the extent of the nucleus is about $10^{-15}$ m and contains 2 protons.

### Electric Charge

**25** • **SOLVE** A plastic rod is rubbed against a wool shirt, thereby acquiring a charge of $-0.8$ $\mu$C. How many electrons are transferred from the wool shirt to the plastic rod?

**26** • A charge equal to the charge of Avogadro's number of protons ($N_A = 6.02 \times 10^{23}$) is called a *faraday*. Calculate the number of coulombs in a faraday.

**27** • **SSM** **SOLVE** How many coulombs of positive charge are there in 1 kg of carbon? Twelve grams of carbon contain Avogadro's number of atoms, with each atom having six protons and six electrons.

### Coulomb's Law

**28** • **SOLVE** A charge $q_1 = 4.0$ $\mu$C is at the origin, and a charge $q_2 = 6.0$ $\mu$C is on the $x$ axis at $x = 3.0$ m. (a) Find the force on charge $q_2$. (b) Find the force on $q_1$. (c) How would your answers for Parts (a) and (b) differ if $q_2$ were $-6.0$ $\mu$C?

**29** • **SOLVE**✓ Three point charges are on the $x$ axis: $q_1 = -6.0$ $\mu$C is at $x = -3.0$ m, $q_2 = 4.0$ $\mu$C is at the origin, and $q_3 = -6.0$ $\mu$C is at $x = 3.0$ m. Find the force on $q_1$.

**30** •• Three charges, each of magnitude 3 nC, are at separate corners of a square of edge length 5 cm. The two charges at opposite corners are positive, and the other charge is negative. Find the force exerted by these charges on a fourth charge $q = +3$ nC at the remaining corner.

**31** •• **SOLVE** A charge of 5 $\mu$C is on the $y$ axis at $y = 3$ cm, and a second charge of $-5$ $\mu$C is on the $y$ axis at $y = -3$ cm. Find the force on a charge of 2 $\mu$C on the $x$ axis at $x = 8$ cm.

**32** •• **SSM** A point charge of $-2.5$ $\mu$C is located at the origin. A second point charge of 6 $\mu$C is at $x = 1$ m, $y = 0.5$ m. Find the $x$ and $y$ coordinates of the position at which an electron would be in equilibrium.

**33** •• **SSM** A charge of $-1.0$ $\mu$C is located at the origin; a second charge of 2.0 $\mu$C is located at $x = 0$, $y = 0.1$ m; and a third charge of 4.0 $\mu$C is located at $x = 0.2$ m, $y = 0$. Find the forces that act on each of the three charges.

**34** •• A charge of 5.0 $\mu$C is located at $x = 0$, $y = 0$ and a charge $Q_2$ is located at $x = 4.0$ cm, $y = 0$. The force on a 2-$\mu$C charge at $x = 8.0$ cm, $y = 0$ is 19.7 N, pointing in the negative $x$ direction. When this 2-$\mu$C charge is positioned at $x = 17.75$ cm, $y = 0$, the force on it is zero. Determine the charge $Q_2$.

**35** •• Five equal charges $Q$ are equally spaced on a semicircle of radius $R$ as shown in Figure 21-36. Find the force on a charge $q$ located at the center of the semicircle.

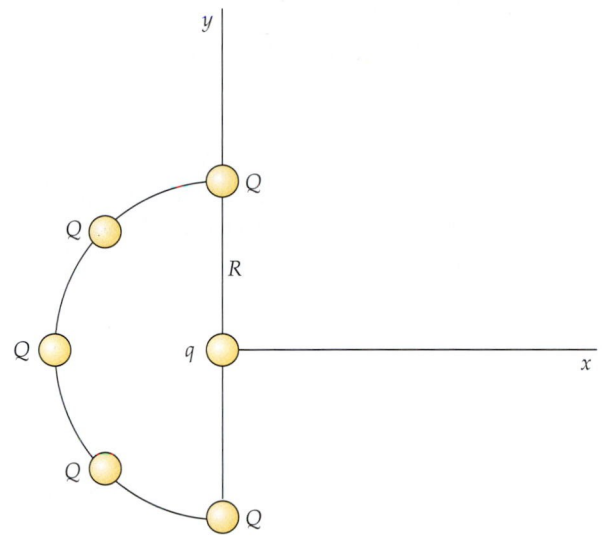

**FIGURE 21-36** Problem 35

**36** ••• The configuration of the $NH_3$ molecule is approximately that of a regular tetrahedron, with three $H^+$ ions forming the base and an $N^{3-}$ ion at the apex of the tetrahedron. The length of each side is $1.64 \times 10^{-10}$ m. Calculate the force that acts on each ion.

### The Electric Field

**37** • SSM ISOLVE A charge of 4.0 $\mu$C is at the origin. What is the magnitude and direction of the electric field on the $x$ axis at (a) $x = 6$ m, and (b) $x = -10$ m? (c) Sketch the function $E_x$ versus $x$ for both positive and negative values of $x$. (Remember that $E_x$ is negative when $E$ points in the negative $x$ direction.)

**38** • SSM ISOLVE✓ Two charges, each +4 $\mu$C, are on the $x$ axis, one at the origin and the other at $x = 8$ m. Find the electric field on the $x$ axis at (a) $x = -2$ m, (b) $x = 2$ m, (c) $x = 6$ m, and (d) $x = 10$ m. (e) At what point on the $x$ axis is the electric field zero? (f) Sketch $E_x$ versus $x$.

**39** • When a test charge $q_0 = 2$ nC is placed at the origin, it experiences a force of $8.0 \times 10^{-4}$ N in the positive $y$ direction. (a) What is the electric field at the origin? (b) What would be the force on a charge of $-4$ nC placed at the origin? (c) If this force is due to a charge on the $y$ axis at $y = 3$ cm, what is the value of that charge?

**40** • ISOLVE✓ The electric field near the surface of the earth points downward and has a magnitude of 150 N/C. (a) Compare the upward electric force on an electron with the downward gravitational force. (b) What charge should be placed on a penny of mass 3 g so that the electric force balances the weight of the penny near the earth's surface?

**41** •• ISOLVE✓ Two equal positive charges of magnitude $q_1 = q_2 = 6.0$ nC are on the $y$ axis at $y_1 = +3$ cm and $y_2 = -3$ cm. (a) What is the magnitude and direction of the electric field on the $x$ axis at $x = 4$ cm? (b) What is the force exerted on a third charge $q_0 = 2$ nC when it is placed on the $x$ axis at $x = 4$ cm?

**42** •• SSM ISOLVE✓ A point charge of +5.0 $\mu$C is located at $x = -3.0$ cm, and a second point charge of $-8.0$ $\mu$C is located at $x = +4.0$ cm. Where should a third charge of +6.0 $\mu$C be placed so that the electric field at $x = 0$ is zero?

**43** •• A point charge of $-5$ $\mu$C is located at $x = 4$ m, $y = -2$ m. A second point charge of 12 $\mu$C is located at $x = 1$ m, $y = 2$ m. (a) Find the magnitude and direction of the electric field at $x = -1$ m, $y = 0$. (b) Calculate the magnitude and direction of the force on an electron at $x = -1$ m, $y = 0$.

**44** •• Two equal positive charges $q$ are on the $y$ axis, one at $y = +a$ and the other at $y = -a$. (a) Show that the electric field on the $x$ axis is along the $x$ axis with $E_x = 2kqx(x^2 + a^2)^{-3/2}$. (b) Show that near the origin, when $x$ is much smaller than $a$, $E_x$ is approximately $2kqx/a^3$. (c) Show that for values of $x$ much larger than $a$, $E_x$ is approximately $2kq/x^2$. Explain why you would expect this result even before calculating it.

**45** •• SSM A 5-$\mu$C point charge is located at $x = 1$ m, $y = 3$ m; and a $-4$-$\mu$C point charge is located at $x = 2$ m, $y = -2$ m. (a) Find the magnitude and direction of the electric field at $x = -3$ m, $y = 1$ m. (b) Find the magnitude and direction of the force on a proton at $x = -3$ m, $y = 1$ m.

**46** •• (a) Show that the electric field for the charge distribution in Problem 44 has its greatest magnitude at the points $x = a/\sqrt{2}$ and $x = -a/\sqrt{2}$ by computing $dE_x/dx$ and setting the derivative equal to zero. (b) Sketch the function $E_x$ versus $x$ using your results for Part (a) of this problem and Parts (b) and (c) of Problem 44.

**47** ••• For the charge distribution in Problem 44, the electric field at the origin is zero. A test charge $q_0$ placed at the origin will therefore be in equilibrium. (a) Discuss the stability of the equilibrium for a positive test charge by considering small displacements from equilibrium along the $x$ axis and small displacements along the $y$ axis. (b) Repeat Part (a) for a negative test charge. (c) Find the magnitude and sign of a charge $q_0$ that when placed at the origin results in a net force of zero on each of the three charges. (d) What will happen if any of the charges is displaced slightly from equilibrium?

**48** ••• SSM Two positive point charges $+q$ are on the $y$ axis at $y = +a$ and $y = -a$ as in Problem 44. A bead of mass $m$ carrying a negative charge $-q$ slides without friction along a thread that runs along the $x$ axis. (a) Show that for small displacements of $x \ll a$, the bead experiences a restoring force that is proportional to $x$ and therefore undergoes simple harmonic motion. (b) Find the period of the motion.

### Motion of Point Charges in Electric Fields

**49** • ISOLVE The acceleration of a particle in an electric field depends on the ratio of the charge to the mass of the particle. (a) Compute $e/m$ for an electron. (b) What is the magnitude and direction of the acceleration of an electron in a uniform electric field with a magnitude of 100 N/C? (c) When the speed of an electron approaches the speed of light $c$, relativistic mechanics must be used to calculate its motion, but at speeds significantly less than $c$, Newtonian mechanics applies. Using Newtonian mechanics, compute the time it takes for an electron placed at rest in an electric field with a magnitude of 100 N/C to reach a speed of $0.01c$. (d) How far does the electron travel in that time?

**50** • SSM SOLVE (a) Compute $e/m$ for a proton, and find its acceleration in a uniform electric field with a magnitude of 100 N/C. (b) Find the time it takes for a proton initially at rest in such a field to reach a speed of $0.01c$ (where $c$ is the speed of light).

**51** • SOLVE✔ An electron has an initial velocity of $2 \times 10^6$ m/s in the $x$ direction. It enters a uniform electric field $\vec{E} = (400 \text{ N/C})\hat{j}$; which is in the $y$ direction. (a) Find the acceleration of the electron. (b) How long does it take for the electron to travel 10 cm in the $x$ direction in the field? (c) By how much, and in what direction, is the electron deflected after traveling 10 cm in the $x$ direction in the field?

**52** •• SOLVE✔ An electron, starting from rest, is accelerated by a uniform electric field of $8 \times 10^4$ N/C that extends over a distance of 5.0 cm. Find the speed of the electron after it leaves the region of uniform electric field.

**53** •• A 2-g object, located in a region of uniform electric field $\vec{E} = (300 \text{ N/C})\hat{i}$, carries a charge $Q$. The object, released from rest at $x = 0$, has a kinetic energy of 0.12 J at $x = 0.50$ m. Determine the charge $Q$.

**54** •• SSM SOLVE A particle leaves the origin with a speed of $3 \times 10^6$ m/s at 35° to the $x$ axis. It moves in a constant electric field $\vec{E} = E_y\hat{j}$. Find $E_y$ such that the particle will cross the $x$ axis at $x = 1.5$ cm if the particle is (a) an electron, and (b) a proton.

**55** •• An electron starts at the position shown in Figure 21-37 with an initial speed $v_0 = 5 \times 10^6$ m/s at 45° to the $x$ axis. The electric field is in the positive $y$ direction and has a magnitude of $3.5 \times 10^3$ N/C. On which plate and at what location will the electron strike?

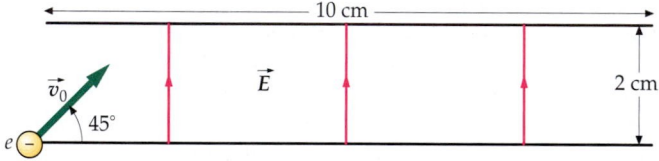

**FIGURE 21-37** Problem 55

**56** •• An electron with kinetic energy of $2 \times 10^{-16}$ J is moving to the right along the axis of a cathode-ray tube as shown in Figure 21-38. There is an electric field $\vec{E} = (2 \times 10^4 \text{ N/C})\hat{j}$ in the region between the deflection plates. Everywhere else, $\vec{E} = 0$. (a) How far is the electron from the axis of the tube when it reaches the end of the plates? (b) At what angle is the electron moving with respect to the axis? (c) At what distance from the axis will the electron strike the fluorescent screen?

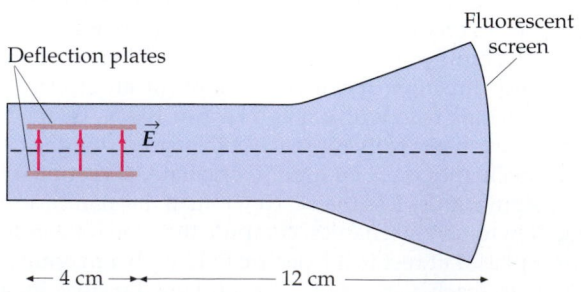

**FIGURE 21-38** Problem 56

**57** • SOLVE Two point charges, $q_1 = 2.0$ pC and $q_2 = -2.0$ pC, are separated by 4 μm. (a) What is the dipole moment of this pair of charges? (b) Sketch the pair, and show the direction of the dipole moment.

**58** • SSM SOLVE A dipole of moment 0.5 e·nm is placed in a uniform electric field with a magnitude of $4.0 \times 10^4$ N/C. What is the magnitude of the torque on the dipole when (a) the dipole is parallel to the electric field, (b) the dipole is perpendicular to the electric field, and (c) the dipole makes an angle of 30° with the electric field? (d) Find the potential energy of the dipole in the electric field for each case.

**59** •• SSM For a dipole oriented along the $x$ axis, the electric field falls off as $1/x^3$ in the $x$ direction and $1/y^3$ in the $y$ direction. Use dimensional analysis to prove that, in any direction, the field far from the dipole falls off as $1/r^3$.

**60** •• A water molecule has its oxygen atom at the origin, one hydrogen nucleus at $x = 0.077$ nm, $y = 0.058$ nm and the other hydrogen nucleus at $x = -0.077$ nm, $y = 0.058$ nm. If the hydrogen electrons are transferred completely to the oxygen atom so that it has a charge of $-2e$, what is the dipole moment of the water molecule? (Note that this characterization of the chemical bonds of water as totally ionic is simply an approximation that overestimates the dipole moment of a water molecule.)

**61** •• An electric dipole consists of two charges $+q$ and $-q$ separated by a very small distance $2a$. Its center is on the $x$ axis at $x = x_1$, and it points along the $x$ axis in the positive $x$ direction. The dipole is in a nonuniform electric field, which is also in the $x$ direction, given by $\vec{E} = Cx\hat{i}$, where $C$ is a constant. (a) Find the force on the positive charge and that on the negative charge, and show that the net force on the dipole is $Cp\hat{i}$. (b) Show that, in general, if a dipole of moment $\vec{p}$ lies along the $x$ axis in an electric field in the $x$ direction, the net force on the dipole is given approximately by $(dE_x/dx)p\hat{i}$.

**62** ••• A positive point charge $+Q$ is at the origin, and a dipole of moment $\vec{p}$ is a distance $r$ away ($r \gg L$) and in the radial direction as shown in Figure 21-29. (a) Show that the force exerted on the dipole by the point charge is attractive and has a magnitude $\approx 2kQp/r^3$ (see Problem 61). (b) Now assume that the dipole is centered at the origin and that a point charge $Q$ is a distance $r$ away along the line of the dipole. Using Newton's third law and your result for part (a), show that at the location of the positive point charge the electric field $\vec{E}$ due to the dipole is toward the dipole and has a magnitude of $\approx 2kp/r^3$.

## General Problems

**63** • SSM (a) What mass would a proton have if its gravitational attraction to another proton exactly balanced out the electrostatic repulsion between them? (b) What is the true ratio of these two forces?

**64** •• Point charges of $-5.0$ μC, $+3.0$ μC, and $+5.0$ μC are located along the $x$ axis at $x = -1.0$ cm, $x = 0$, and $x = +1.0$ cm, respectively. Calculate the electric field at $x = 3.0$ cm and at $x = 15.0$ cm. Is there some point on the $x$ axis where the magnitude of the electric field is zero? Locate that point.

**65** •• For the charge distribution of Problem 64, find the electric field at $x = 15.0$ cm as the vector sum of the electric field due to a dipole formed by the two 5.0-$\mu$C charges and a point charge of 3.0 $\mu$C, both located at the origin. Compare your result with the result obtained in Problem 64, and explain any difference between these two.

**66** •• **SSM** **iSOLVE** In copper, about one electron per atom is free to move about. A copper penny has a mass of 3 g. (a) What percentage of the free charge would have to be removed to give the penny a charge of 15 $\mu$C? (b) What would be the force of repulsion between two pennies carrying this charge if they were 25 cm apart? Assume that the pennies are point charges.

**67** •• Two charges $q_1$ and $q_2$ have a total charge of 6 $\mu$C. When they are separated by 3 m, the force exerted by one charge on the other has a magnitude of 8 mN. Find $q_1$ and $q_2$ if (a) both are positive so that they repel each other, and (b) one is positive and the other is negative so that they attract each other.

**68** •• Three charges, $+q$, $+2q$, and $+4q$, are connected by strings as shown in Figure 21-39. Find the tensions $T_1$ and $T_2$.

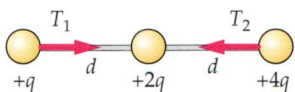

**FIGURE 21-39** Problem 68

**69** •• **SSM** A positive charge $Q$ is to be divided into two positive charges $q_1$ and $q_2$. Show that, for a given separation $D$, the force exerted by one charge on the other is greatest if $q_1 = q_2 = \frac{1}{2}Q$.

**70** •• **SSM** A charge $Q$ is located at $x = 0$, and a charge $4Q$ is at $x = 12.0$ cm. The force on a charge of $-2$ $\mu$C is zero if that charge is placed at $x = 4.0$ cm, and is 126.4 N in the positive $x$ direction if placed at $x = 8.0$ cm. Determine the charge $Q$.

**71** •• Two small spheres (point charges) separated by 0.60 m carry a total charge of 200 $\mu$C. (a) If the two spheres repel each other with a force of 80 N, what are the charges on each of the two spheres? (b) If the two spheres attract each other with a force of 80 N, what are the charges on the two spheres?

**72** •• **iSOLVE** ✓ A ball of known charge $q$ and unknown mass $m$, initially at rest, falls freely from a height $h$ in a uniform electric field $\vec{E}$ that is directed vertically downward. The ball hits the ground at a speed $v = 2\sqrt{gh}$. Find $m$ in terms of $E$, $q$, and $g$.

**73** •• **SSM** A rigid stick one meter long is pivoted about its center (Figure 21-40). A charge $q_1 = 5 \times 10^{-7}$ C is placed on one end of the rod, and an equal but opposite charge $q_2$ is placed a distance $d = 10$ cm directly below it. (a) What is the net force between the two charges? (b) What is the torque (measured from the center of the rod) due to that force? (c) To counterbalance the attraction between the two charges, we hang a block 25 cm from the pivot on the *opposite* side of the balance point. What value should we choose for the mass $m$ of the block? (See Figure 21-40.) (d) We now move the block and hang it a distance of 25 cm from the balance point on the *same* side of the balance as the charge. Keeping $q_1$ the same, and $d$ the same, what value should we choose for $q_2$ to keep this apparatus in balance?

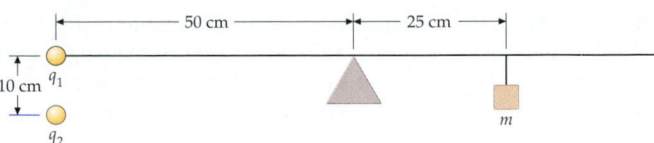

**FIGURE 21-40** Problem 73

**74** •• Charges of 3.0 $\mu$C are located at $x = 0$, $y = 2.0$ m, and at $x = 0$, $y = -2.0$ m. Charges $Q$ are located at $x = 4.0$ m, $y = 2.0$ m, and at $x = 4.0$ m, $y = -2.0$ m (Figure 21-41). The electric field at $x = 0$, $y = 0$ is $(4.0 \times 10^3 \text{ N/C})\hat{i}$. Determine $Q$.

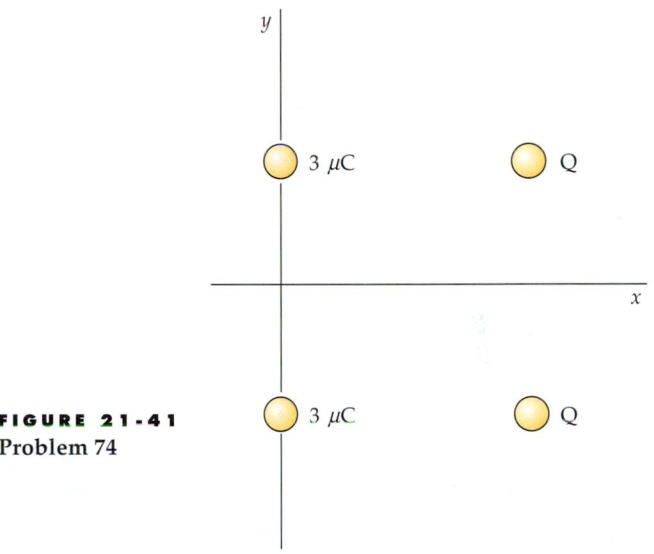

**FIGURE 21-41** Problem 74

**75** •• Two identical small spherical conductors (point charges), separated by 0.60 m, carry a total charge of 200 $\mu$C. They repel one another with a force of 120 N. (a) Find the charge on each sphere. (b) The two spheres are placed in electrical contact and then separated so that each carries 100 $\mu$C. Determine the force exerted by one sphere on the other when they are 0.60 m apart.

**76** •• Repeat Problem 75 if the two spheres initially attract one another with a force of 120 N.

**77** •• A charge of $-3.0$ $\mu$C is located at the origin; a charge of 4.0 $\mu$C is located at $x = 0.2$ m, $y = 0$; a third charge $Q$ is located at $x = 0.32$ m, $y = 0$. The force on the 4.0-$\mu$C charge is 240 N, directed in the positive $x$ direction. (a) Determine the charge $Q$. (b) With this configuration of three charges, where, along the $x$ direction, is the electric field zero?

**78** •• **SSM** Two small spheres of mass $m$ are suspended from a common point by threads of length $L$. When each sphere carries a charge $q$, each thread makes an angle $\theta$ with the vertical as shown in Figure 21-42. (a) Show that the charge $q$ is given by

$$q = 2L \sin \theta \sqrt{\frac{mg \tan \theta}{k}}$$

where $k$ is the Coulomb constant. (b) Find $q$ if $m = 10$ g, $L = 50$ cm, and $\theta = 10°$.

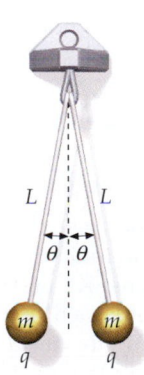

**FIGURE 21-42** Problem 78

**79** •• **SOLVE** (a) Suppose that in Problem 78 $L =$ 1.5 m, $m = 0.01$ kg, and $q = 0.75$ $\mu$C. What is the angle that each string makes with the vertical? (b) Find the angle that each string makes with the vertical if one mass carries a charge of 0.50 $\mu$C, the other a charge of 1.0 $\mu$C.

**80** •• Four charges of equal magnitude are arranged at the corners of a square of side $L$ as shown in Figure 21-43. (a) Find the magnitude and direction of the force exerted on the charge in the lower left corner by the other charges. (b) Show that the electric field at the midpoint of one of the sides of the square is directed along that side toward the negative charge and has a magnitude $E$ given by

$$E = k\frac{8q}{L^2}\left(1 - \frac{\sqrt{5}}{25}\right)$$

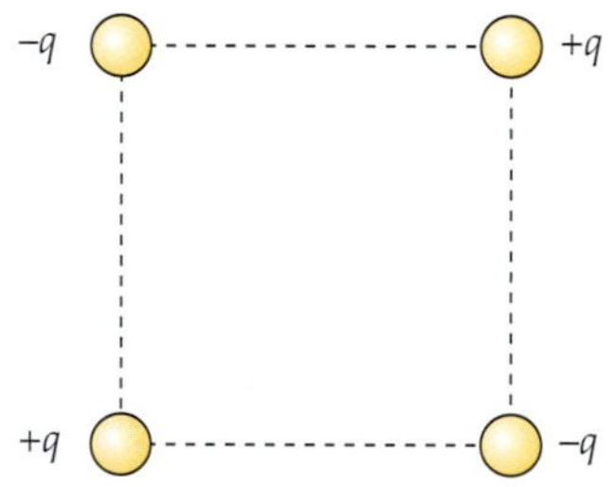

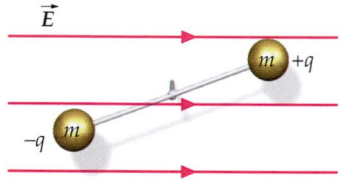

**FIGURE 21-43**
**Problem 80**

**81** •• Figure 21-44 shows a dumbbell consisting of two identical masses $m$ attached to the ends of a thin (massless) rod of length $a$ that is pivoted at its center. The masses carry charges of $+q$ and $-q$, and the system is located in a uniform electric field $\vec{E}$. Show that for small values of the angle $\theta$ between the direction of the dipole and the electric field, the system displays simple harmonic motion, and obtain an expression for the period of that motion.

**FIGURE 21-44** **Problems 81 and 82**

**82** •• For the dumbbell in Figure 21-44, let $m = 0.02$ kg, $a = 0.3$ m, and $\vec{E} = (600 \text{ N/C})\hat{i}$. Initially the dumbbell is at rest and makes an angle of 60° with the $x$ axis. The dumbbell is then released, and when it is momentarily aligned with the electric field, its kinetic energy is $5 \times 10^{-3}$ J. Determine the magnitude of $q$.

**83** •• **SSM** An electron (charge $-e$, mass $m$) and a positron (charge $+e$, mass $m$) revolve around their common center of mass under the influence of their attractive coulomb force. Find the speed of each particle $v$ in terms of $e$, $m$, $k$, and their separation $r$.

**84** •• The equilibrium separation between the nuclei of the ionic molecule KBr is 0.282 nm. The masses of the two ions, K$^+$ and Br$^-$, are very nearly the same, $1.4 \times 10^{-25}$ kg and each of the two ions carries a charge of magnitude $e$. Use the result of Problem 81 to determine the frequency of oscillation of a KBr molecule in a uniform electric field of 1000 N/C.

**85** ••• A small (point) mass $m$, which carries a charge $q$, is constrained to move vertically inside a narrow, frictionless cylinder (Figure 21-45). At the bottom of the cylinder is a point mass of charge $Q$ having the same sign as $q$. (a) Show that the mass $m$ will be in equilibrium at a height $y_0 = (kqQ/mg)^{1/2}$. (b) Show that if the mass $m$ is displaced by a small amount from its equilibrium position and released, it will exhibit simple harmonic motion with angular frequency $\omega = (2g/y_0)^{1/2}$.

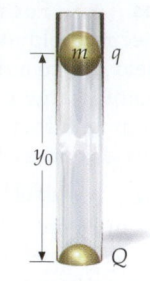

**FIGURE 21-45**
**Problem 85**

**86** ••• A small bead of mass $m$ and carrying a negative charge $-q$ is constrained to move along a thin, frictionless rod (Figure 21-46). A distance $L$ from this rod is a positive charge $Q$. Show that if the bead is displaced a distance $x$, where $x \ll L$, and released, it will exhibit simple harmonic motion. Obtain an expression for the period of this motion in terms of the parameters $L$, $Q$, $q$, and $m$.

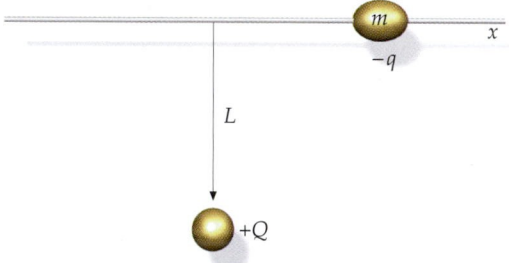

**FIGURE 21-46** **Problem 86**

**87** ••• Repeat Problem 79 with the system located in a uniform electric field of $1.0 \times 10^5$ N/C that points vertically downward.

**88** ••• Suppose that the two spheres of mass in Problem 78 are not equal. One mass is 0.01 kg, the other is 0.02 kg. The charges on the two masses are 2.0 $\mu$C and 1.0 $\mu$C, respectively. Determine the angle that each of the strings supporting the masses makes with the vertical.

**89** ••• **SOLVE** A simple pendulum of length $L = 1.0$ m and mass $M = 5.0 \times 10^{-3}$ kg is placed in a uniform, vertically directed electric field $\vec{E}$. The bob carries a charge of $-8.0$ $\mu$C. The period of the pendulum is 1.2 s. What is the magnitude and direction of $\vec{E}$?

**90** ••• **SSM** Two neutral polar molecules attract each other. Suppose that each molecule has a dipole moment $\vec{p}$, and that these dipoles are aligned along the $x$ axis and separated by a distance $d$. Derive an expression for the force of attraction in terms of $p$ and $d$.

**91** ••• Two equal positive charges $Q$ are on the $x$ axis at $x = \frac{1}{2}L$ and $x = -\frac{1}{2}L$. (a) Obtain an expression for the electric field as a function of $y$ on the $y$ axis. (b) A ring of mass $m$, which carries a charge $q$, moves on a thin, frictionless rod along the $y$ axis. Find the force that acts on the charge $q$ as a function of $y$; determine the sign of $q$ such that this force always points toward $y = 0$. (c) Show that for small values of $y$ the ring exhibits simple harmonic motion. (d) If $Q = 5$ $\mu$C, $|q| = 2$ $\mu$C, $L = 24$ cm, and $m = 0.03$ kg, what is the frequency of the oscillation for small amplitudes?

**92** ••• In the Millikan experiment used to determine the charge on the electron, a charged polystyrene microsphere is released in still air in a known vertical electric field. The charged microsphere will accelerate in the direction of the net force until it reaches terminal speed. The charge on the microsphere is determined by measuring the terminal speed. In one such experiment, the bead has radius $r = 5.5 \times 10^7$ m, and the field has a magnitude $E = 6 \times 10^4$ N/C. The magnitude of the drag force on the sphere is $F_D = 6\pi\eta rv$, where $v$ is the speed of the sphere and $\eta$ is the viscosity of air ($\eta = 1.8 \times 10^{-5}$ N·s/m²). The polystyrene has density $1.05 \times 10^3$ kg/m³. (a) If the electric field is pointing down so that the polystyrene microsphere rises with a terminal speed $v = 1.16 \times 10^{-4}$ m/s, what is the charge on the sphere? (b) How many excess electrons are on the sphere? (c) If the direction of the electric field is reversed but its magnitude remains the same, what is the terminal speed?

**93** ••• SSM In Problem 92, there was a description of the Millikan experiment used to determine the charge on the electron. In the experiment, a switchable power supply is used so that the electrical field can point both up and down, but with the same magnitude, so that one can measure the terminal speed of the microsphere as it is pushed up (against the force of gravity) and down. Let $v_u$ represent the terminal speed when the particle is moving up, and $v_d$ the terminal speed when moving down. (a) If we let $v = v_u + v_d$, show that $v = \dfrac{qE}{3\pi\eta r}$, where $q$ is the microsphere's net charge. What advantage does measuring both $v_u$ and $v_d$ give over measuring only one? (b) Because charge is quantized, $v$ can only change by steps of magnitude $\Delta v$. Using the data from Problem 92, calculate $\Delta v$.

# CHAPTER

# 22

# The Electric Field II: Continuous Charge Distributions

BY DESCRIBING CHARGE IN TERMS OF CONTINUOUS CHARGE DENSITY, IT BECOMES POSSIBLE TO CALCULATE THE CHARGE ON THE SURFACE OF OBJECTS AS LARGE AS CELESTIAL BODIES.

**?** **How would you calculate the charge on the surface of the Earth? (See Example 22-10.)**

22-1   Calculating $\vec{E}$ From Coulomb's Law

22-2   Gauss's Law

22-3   Calculating $\vec{E}$ From Gauss's Law

22-4   Discontinuity of $E_n$

22-5   Charge and Field at Conductor Surfaces

*22-6   Derivation of Gauss's Law From Coulomb's Law

O n a microscopic scale, electric charge is quantized. However, there are often situations in which many charges are so close together that they can be thought of as continuously distributed. The use of a continuous charge density to describe a large number of discrete charges is similar to the use of a continuous mass density to describe air, which actually consists of a large number of discrete molecules. In both cases, it is usually easy to find a volume element $\Delta V$ that is large enough to contain a multitude of individual charges or molecules and yet is small enough that replacing $\Delta V$ with a differential $dV$ and using calculus introduces negligible error.

We describe the charge per unit volume by the **volume charge density** $\rho$:

$$\rho = \frac{\Delta Q}{\Delta V} \qquad\qquad 22\text{-}1$$

Often charge is distributed in a very thin layer on the surface of an object. We define the **surface charge density** $\sigma$ as the charge per unit area:

$$\sigma = \frac{\Delta Q}{\Delta A} \qquad\qquad 22\text{-}2$$

Similarly, we sometimes encounter charge distributed along a line in space. We define the **linear charge density** $\lambda$ as the charge per unit length:

$$\lambda = \frac{\Delta Q}{\Delta L} \qquad\qquad 22\text{-}3$$

➤ In this chapter, we show how Coulomb's law is used to calculate the electric field produced by various types of continuous charge distributions. We then introduce Gauss's law, which relates the electric field on a closed surface to the net charge within the surface, and we use this relation to calculate the electric field for symmetric charge distributions.

# 22-1    Calculating $\vec{E}$ From Coulomb's Law

Figure 22-1 shows an element of charge $dq = \rho\, dV$ that is small enough to be considered a point charge. Coulomb's law gives the electric field $d\vec{E}$ at a field point $P$ due to this element of charge as:

$$d\vec{E} = \frac{k\, dq}{r^2}\, \hat{r}$$

where $\hat{r}$ is a unit vector that points from the source point to the field point $P$. The total field at $P$ is found by integrating this expression over the entire charge distribution. That is,

$$\vec{E} = \int_V \frac{k\, dq}{r^2}\, \hat{r} \qquad\qquad 22\text{-}4$$

ELECTRIC FIELD DUE TO A CONTINUOUS CHARGE DISTRIBUTION

where $dq = \rho\, dV$. If the charge is distributed on a surface or line, we use $dq = \sigma\, dA$ or $dq = \lambda\, dL$ and integrate over the surface or line.

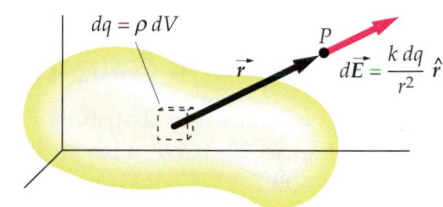

**FIGURE 22-1** An element of charge $dq$ produces a field $d\vec{E} = (k\, dq/r^2)\,\hat{r}$ at point $P$. The field at $P$ is found by integrating over the entire charge distribution.

## $\vec{E}$ on the Axis of a Finite Line Charge

A charge $Q$ is uniformly distributed along the $x$ axis from $x = -\frac{1}{2}L$ to $x = +\frac{1}{2}L$, as shown in Figure 22-2. The linear charge density for this charge is $\lambda = Q/L$. We wish to find the electric field produced by this line charge at some field point $P$ on the $x$ axis at $x = x_P$, where $x_P > \frac{1}{2}L$. In the figure, we have chosen the element of charge $dq$ to be the charge on a small element of length $dx$ at position $x$. Point $P$ is a distance $r = x_P - x$ from $dq$. Coulomb's law gives the electric field at $P$ due to the charge $dq$ on this length $dx$. It is directed along the $x$ axis and is given by

$$dE_x\hat{i} = \frac{k\, dq}{(x_P - x)^2}\hat{i} = \frac{k\lambda\, dx}{(x_P - x)^2}\hat{i}$$

We find the total field $\vec{E}$ by integrating over the entire line charge in the direction of increasing $x$ (from $x_1 = -\frac{1}{2}L$ to $x_2 = +\frac{1}{2}L$):

$$E_x = k\lambda \int_{-L/2}^{+L/2} \frac{dx}{(x_P - x)^2} = -k\lambda \int_{x_P + (L/2)}^{x_P - (L/2)} \frac{du}{u^2}$$

where $u = x_P - x$ (so $du = -dx$). Note that if $x = -\frac{1}{2}L$, $u = x_P + \frac{1}{2}L$, and if $x = +\frac{1}{2}L$, $u = x_P - \frac{1}{2}L$. Evaluating the integral gives

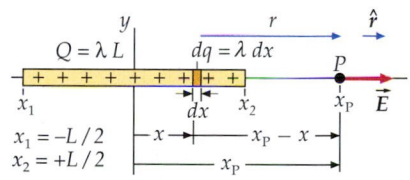

**FIGURE 22-2** Geometry for the calculation of the electric field on the axis of a uniform line charge of length $L$, charge $Q$, and linear charge density $\lambda = Q/L$. An element $dq = \lambda\, dx$ is treated as a point charge.

$$E_x = +k\lambda \frac{1}{u}\Big|_{x_P+(L/2)}^{x_P-(L/2)} = k\lambda \left\{\frac{1}{x_P - \frac{1}{2}L} - \frac{1}{x_P + \frac{1}{2}L}\right\} = \frac{k\lambda L}{x_P^2 - (\frac{1}{2}L)^2}$$

Substituting $Q$ for $\lambda L$, we obtain

$$E_x = \frac{kQ}{x_P^2 - (\frac{1}{2}L)^2}, \qquad x_P > \frac{1}{2}L \qquad\qquad 22\text{-}5$$

We can see that if $x_P$ is much larger than $L$, the electric field at $x_P$ is approximately $kQ/x_P^2$. That is, if we are sufficiently far away from the line charge, it approaches that of a point charge $Q$ at the origin.

**EXERCISE** The validity of Equation 22-5 is established for the region $x_P > \frac{1}{2}L$. Is it also valid in the region $-\frac{1}{2}L \leq x_P \leq \frac{1}{2}L$? Explain. (*Answer* No. Symmetry dictates that $E_x$ is zero at $x_P = 0$. However, Equation 22-5 gives a negative value for $E_x$ at $x_P = 0$. These contradictory results cannot both be valid.)

## $\vec{E}$ off the Axis of a Finite Line Charge

A charge $Q$ is uniformly distributed on a straight-line segment of length $L$, as shown in Figure 22-3. We wish to find the electric field at an arbitrarily positioned field point $P$. To calculate the electric field at $P$ we first choose coordinate axes. We choose the $x$ axis through the line charge and the $y$ axis through point $P$ as shown. The ends of the charged line segment are labeled $x_1$ and $x_2$. A typical charge element $dq = \lambda\,dx$ that produces a field $d\vec{E}$ is shown in the figure. The field at $P$ has both an $x$ and a $y$ component. Only the $y$ component is computed here. (The $x$ component is to be computed in Problem 22-27.)

The magnitude of the field produced by an element of charge $dq = \lambda\,dx$ is

$$|d\vec{E}| = \frac{k\,dq}{r^2} = \frac{k\lambda\,dx}{r^2}$$

and the $y$ component is

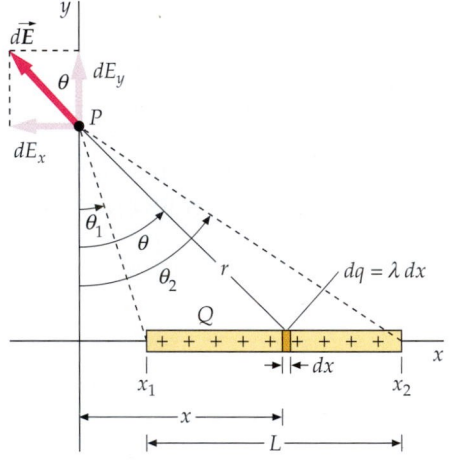

**FIGURE 22-3** Geometry for the calculation of the electric field at field point $P$ due to a uniform finite line charge.

$$dE_y = |d\vec{E}|\cos\theta = \frac{k\lambda\,dx}{r^2}\frac{y}{r} = \frac{k\lambda y\,dx}{r^3} \qquad\qquad 22\text{-}6$$

where $\cos\theta = y/r$ and $r = \sqrt{x^2 + y^2}$. The total $y$ component $E_y$ is computed by integrating from $x = x_1$ to $x = x_2$.

$$E_y = \int_{x=x_1}^{x=x_2} dE_y = k\lambda y \int_{x_1}^{x_2} \frac{dx}{r^3} \qquad\qquad 22\text{-}7$$

In calculating this integral $y$ remains fixed. One way to execute this calculation is to use trigonometric substitution. From the figure we can see that $x = y\tan\theta$, so $dx = y\sec^2\theta\,d\theta$.[†] We also can see that $y = r\cos\theta$, so $1/r = \cos\theta/y$. Substituting these into Equation 22-7 gives

$$E_y = k\lambda y\frac{1}{y^2}\int_{\theta_1}^{\theta_2}\cos\theta\,d\theta = \frac{k\lambda}{y}(\sin\theta_2 - \sin\theta_1) = \frac{kQ}{Ly}(\sin\theta_2 - \sin\theta_1) \qquad 22\text{-}8a$$

$E_Y$ DUE TO A UNIFORMLY CHARGED LINE SEGMENT

**EXERCISE** Show that for the line charge shown in Figure 22-3 $dE_x = -k\lambda x\,dx/r^3$.

---

[†] We have used the relation $d(\tan\theta)/d\theta = \sec^2\theta$.

The $x$ component for the finite line charge shown in Figure 22-3 (and computed in Problem 22-27) is

$$E_x = \frac{k\lambda}{y}(\cos\theta_2 - \cos\theta_1) \qquad\qquad 22\text{-}8b$$

$E_x$ DUE TO A UNIFORMLY CHARGED LINE SEGMENT

### $\vec{E}$ Due to an Infinite Line Charge

A line charge may be considered infinite if for any field point of interest $P$ (see Figure 22-3), $x_1 \rightarrow -\infty$ and $x_2 \rightarrow +\infty$. We compute $E_x$ and $E_y$ for an infinite line charge using Equations 22-8a and b in the limit that $\theta_1 \rightarrow -\pi/2$ and $\theta_2 \rightarrow \pi/2$. (From Figure 22-3 we can see that this is the same as the limit that $x_1 \rightarrow -\infty$ and $x_2 \rightarrow +\infty$.) Substituting $\theta_1 = -\pi/2$ and $\theta_2 = \pi/2$ into Equations 22-8a and b gives $E_x = 0$ and $E_y = \dfrac{2k\lambda}{y}$, where $y$ is the perpendicular distance from the line charge to the field point. Thus,

$$E_R = 2k\frac{\lambda}{R} \qquad\qquad 22\text{-}9$$

$\vec{E}$ AT A DISTANCE $R$ FROM AN INFINITE LINE CHARGE

where $R$ is the perpendicular distance from the line charge to the field point.

**EXERCISE** Show that Equation 22-9 has the correct units for the electric field.

Electric field lines near a long wire. The electric field near a high-voltage power line can be large enough to ionize air, making the air a conductor. The glow resulting from the recombination of free electrons with the ions is called corona discharge.

---

**FIGURE 22-4**

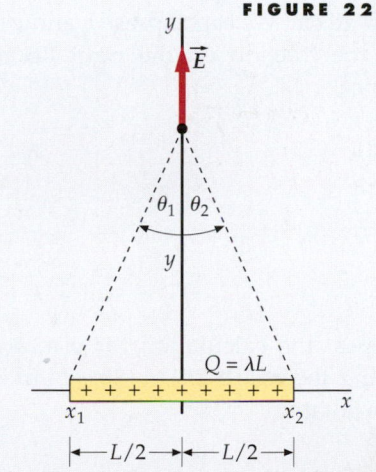

---

| *ELECTRIC FIELD ON THE AXIS OF A FINITE LINE CHARGE* | **E X A M P L E    2 2 - 1** |
| --- | --- |

**Using Equations 22-8a and b, obtain an expression for the electric field on the perpendicular bisector of a uniformly charged line segment with linear charge density $\lambda$ and length $L$.**

**PICTURE THE PROBLEM** Sketch the line charge on the $x$ axis with the $y$ axis as its perpendicular bisector. According to Figure 22-4 this means choosing $x_1 = -\frac{1}{2}L$ and $x_2 = \frac{1}{2}L$ so $\theta_1 = -\theta_2$. Then use Equations 22-8a and 22-8b to find the electric field.

1. Sketch the charge configuration with the line charge on the $x$ axis with the $y$ axis as its perpendicular bisector. Show the field point on the positive $y$ axis a distance $y$ from the origin:

2. Use Equation 22-8a to find an expression for $E_y$. Simplify using $\theta_2 = -\theta_1 = \theta$:

$$E_y = \frac{k\lambda}{y}(\sin\theta_2 - \sin\theta_1) = \frac{k\lambda}{y}[\sin\theta - \sin(-\theta)]$$

$$= \frac{2k\lambda}{y}\sin\theta$$

3. Express $\sin\theta$ in terms of $y$ and $L$ and substitute into the step 2 result:

$$\sin\theta = \frac{\frac{1}{2}L}{\sqrt{(\frac{1}{2}L)^2 + y^2}}$$

so

$$E_y = \frac{2k\lambda}{y}\frac{\frac{1}{2}L}{\sqrt{(\frac{1}{2}L)^2 + y^2}}$$

4. Use Equation 22-8b to determine $E_x$:

$$E_x = \frac{k\lambda}{y}(\cos\theta_2 - \cos\theta_1) = \frac{k\lambda}{y}[\cos\theta - \cos(-\theta)]$$

$$= \frac{k\lambda}{y}(\cos\theta - \cos\theta) = 0$$

5. Express the vector $\vec{E}$:

$$\vec{E} = E_x\hat{i} + E_y\hat{j} = \boxed{\frac{2k\lambda}{y}\frac{\frac{1}{2}L}{\sqrt{(\frac{1}{2}L)^2 + y^2}}\hat{j}}$$

*ELECTRIC FIELD NEAR AND FAR FROM A FINITE LINE CHARGE*    **EXAMPLE 22-2**

A line charge of linear density $\lambda = 4.5$ nC/m lies on the $x$ axis and extends from $x = -5$ cm to $x = 5$ cm. Using the expression for $E_y$ obtained in Example 22-1, calculate the electric field on the $y$ axis at (a) $y = 1$ cm, (b) $y = 4$ cm, and (c) $y = 40$ cm. (d) Estimate the electric field on the $y$ axis at $y = 1$ cm, assuming the line charge to be infinite. (e) Find the total charge and estimate the field at $y = 40$ cm, assuming the line charge to be a point charge.

**PICTURE THE PROBLEM** Use the result of Example 22-1 to obtain the electric field on the $y$ axis. In the expression for $\sin\theta_0$, we can express $L$ and $y$ in centimeters because the units cancel. (d) To find the field very near the line charge, we use $E_y = 2k\lambda/y$. (e) To find the field very far from the charge, we use $E_y = kQ/y^2$ with $Q = \lambda L$.

1. Calculate $E_y$ at $y = 1$ cm for $\lambda = 4.5$ nC/m and $L = 10$ cm. We can express $L$ and $y$ in centimeters in the fraction on the right because the units cancel:

$$E_y = \frac{2k\lambda}{y}\frac{\frac{1}{2}L}{\sqrt{(\frac{1}{2}L)^2 + y^2}}$$

$$= \frac{2(8.99 \times 10^9\ \text{N·m}^2/\text{C}^2)(4.5 \times 10^{-9}\ \text{C/m})}{0.01\ \text{m}}\frac{5\ \text{cm}}{\sqrt{(5\ \text{cm})^2 + (1\ \text{cm})^2}}$$

$$= \frac{80.9\ \text{N·m/C}}{0.01\ \text{m}}\frac{5\ \text{cm}}{\sqrt{(5\ \text{cm})^2 + (1\ \text{cm})^2}} = 7.93 \times 10^3\ \text{N/C}$$

$$= \boxed{7.93\ \text{kN/C}}$$

2. Repeat the calculation for $y = 4$ cm $= 0.04$ m using the result $2k\lambda = 80.9$ N·m/C to simplify the notation:

$$E_y = \frac{80.9\ \text{N·m/C}}{0.04\ \text{m}}\frac{5\ \text{cm}}{\sqrt{(5\ \text{cm})^2 + (4\ \text{cm})^2}} = 1.58 \times 10^3\ \text{N/C}$$

$$= \boxed{1.58\ \text{kN/C}}$$

3. Repeat the calculation for $y = 40$ cm:

$$E_y = \frac{80.9\ \text{N·m/C}}{0.40\ \text{m}}\frac{5\ \text{cm}}{\sqrt{(5\ \text{cm})^2 + (40\ \text{cm})^2}} = \boxed{25.1\ \text{N/C}}$$

4. Calculate the field at $y = 1$ cm $= 0.01$ m due to an infinite line charge:

$$E_y \approx \frac{2k\lambda}{y} = \frac{80.9\ \text{N·m/C}}{0.01\ \text{m}} = \boxed{8.09\ \text{kN/m}}$$

5. Calculate the total charge $\lambda L$ for $L = 0.1$ m and use it to find the field of a point charge at $y = 0.4$ m:

$$Q = \lambda L = (4.5\ \text{nC/m})(0.1\ \text{m}) = 0.45\ \text{nC}$$

$$E_y \approx \frac{k\lambda L}{y^2} = \frac{kQ}{y^2} = \frac{(8.99 \times 10^9\ \text{N·m}^2/\text{C}^2)(0.45 \times 10^{-9}\ \text{C})}{(0.40\ \text{m})^2}$$

$$= \boxed{25.3\ \text{N/C}}$$

**REMARKS** At 1 cm from the 10-cm-long line charge, the estimated value of 8.09 kN/C obtained by assuming an infinite line charge differs from the exact value of 7.93 calculated in (a) by about 2 percent. At 40 cm from the line charge, the approximate value of 25.3 N/C obtained by assuming the line charge to be a point charge differs from the exact value of 25.1 N/C obtained in (c) by about 1 percent. Figure 22-5 shows the exact result for this line segment of length 10 cm and charge density 4.5 nC/m, and for the limiting cases of an infinite line charge of the same charge density, and a point charge $Q = \lambda L$.

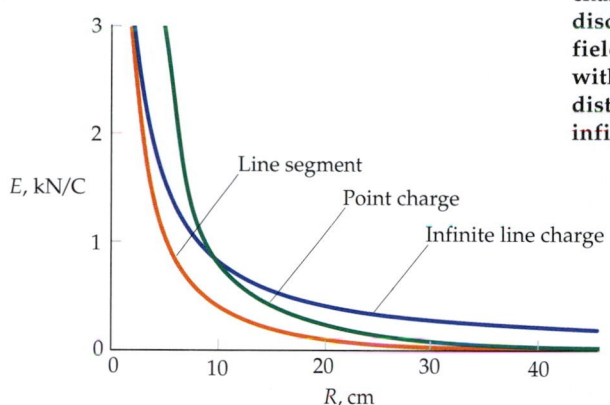

**FIGURE 22-5** The magnitude of the electric field is plotted versus distance for the 10-cm-long line charge, the point charge, and the infinite line charge discussed in Example 22-2. Note that the field of the finite line segment converges with the field of the point charge at large distances, and with the field of the infinite line charge at small distances.

---

*FIELD DUE TO A LINE CHARGE AND A POINT CHARGE*    **E X A M P L E    2 2 - 3**    **Try It Yourself**

An infinitely long line charge of linear charge density $\lambda = 0.6\ \mu\text{C/m}$ lies along the $z$ axis, and a point charge $q = 8\ \mu\text{C}$ lies on the $y$ axis at $y = 3$ m. Find the electric field at the point $P$ on the $x$ axis at $x = 4$ m.

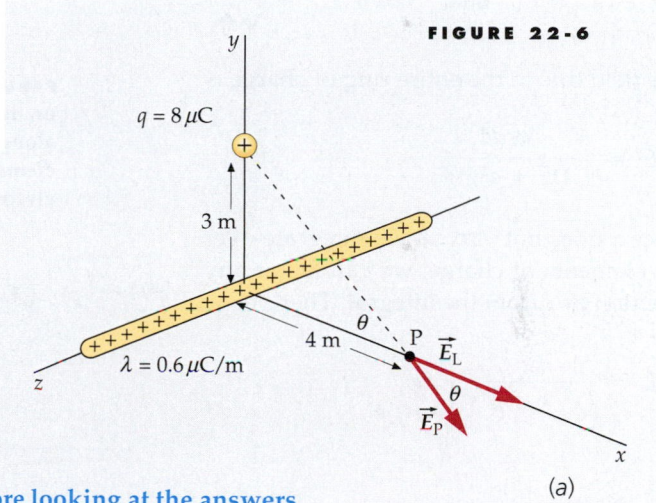

**FIGURE 22-6**

**PICTURE THE PROBLEM** The electric field for this system is the superposition of the fields due to the infinite line charge and the point charge. The field of the line charge, $\vec{E}_L$, points radially away from the $z$ axis (Figure 22-6). Thus, at point $P$ on the $x$ axis, $\vec{E}_L$ is in the positive $x$ direction. The point charge produces a field $\vec{E}_P$ along the line connecting $q$ and the point $P$. The distance from $q$ to $P$ is

$r = \sqrt{(3\ \text{m})^2 + (4\ \text{m})^2} = 5$ m.

(a)

**Cover the column to the right and try these on your own before looking at the answers.**

| Steps | Answers |
|---|---|
| 1. Calculate the field $\vec{E}_L$ at point $P$ due to the infinite line charge. | $\vec{E}_L = 2.70\ \text{kN/C}\,\hat{\imath}$ |
| 2. Find the field $\vec{E}_P$ at point $P$ due to the point charge. Express $\vec{E}_P$ in terms of the unit vector $\hat{r}$ that points from $q$ toward $P$. | $\vec{E}_P = 2.88\ \text{kN/C}\,\hat{r}$ |
| 3. Find the $x$ and $y$ components of $\vec{E}_P$. | $E_{Px} = E_P\,(0.8) = 2.30\ \text{kN/C}$ <br> $E_{Py} = E_P\,(-0.6) = -1.73\ \text{kN/C}$ |
| 4. Find the $x$ and $y$ components of the total field at point $P$. | $E_x = \boxed{5.00\ \text{kN/C}}$ , $E_y = \boxed{-1.73\ \text{kN/C}}$ |
| 5. Use your result in step 4 to calculate the magnitude of the total field. | $E = \sqrt{E_x^2 + E_y^2} = \boxed{5.29\ \text{kN/C}}$ |
| 6. Use your results in step 4 to find the angle $\phi$ between the field and the direction of increasing $x$. | $\phi = \tan^{-1}\dfrac{E_y}{E_x} = \boxed{-19.1°}$ |

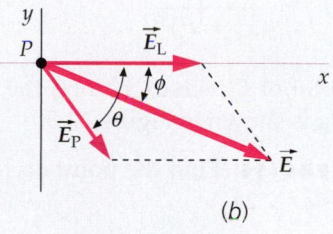

(b)

## $\vec{E}$ on the Axis of a Ring Charge

Figure 22-7a shows a uniform ring charge of radius $a$ and total charge $Q$. The field $d\vec{E}$ at point $P$ on the axis due to the charge element $dq$ is shown in the figure. This field has a component $dE_x$ directed along the axis of the ring and a component $dE_\perp$ directed perpendicular to the axis. The perpendicular components cancel in pairs, as can be seen in Figure 22-7b. From the symmetry of the charge distribution, we can see that the net field due to the entire ring must lie along the axis of the ring; that is, the perpendicular components sum to zero.

The axial component of the field due to the charge element shown is

$$dE_x = \frac{k\,dq}{r^2}\cos\theta = \frac{k\,dq}{r^2}\frac{x}{r} = \frac{k\,dq\,x}{(x^2+a^2)^{3/2}}$$

where

$$r^2 = x^2 + a^2 \quad \text{and} \quad \cos\theta = \frac{x}{r} = \frac{x}{\sqrt{x^2+a^2}}$$

The field due to the entire ring of charge is

$$E_x = \int \frac{kx\,dq}{(x^2+a^2)^{3/2}}$$

Since $x$ does not vary as we integrate over the elements of charge, we can factor any function of $x$ from the integral. Then

$$E_x = \frac{kx}{(x^2+a^2)^{3/2}} \int dq$$

or

$$E_x = \frac{kQx}{(x^2+a^2)^{3/2}} \qquad \text{22-10}$$

A plot of $E_x$ versus $x$ along the axis of the ring is shown in Figure 22-8.

**EXERCISE** Find the point on the axis of the ring where $E_x$ is maximum. (*Answer* $x = a/\sqrt{2}$)

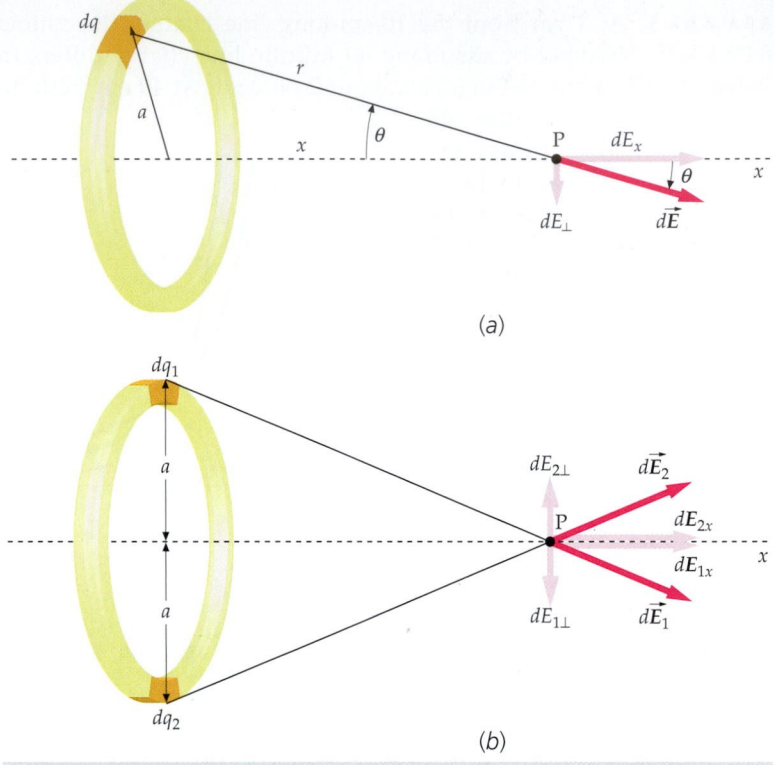

**FIGURE 22-7** (a) A ring charge of radius $a$. The electric field at point $P$ on the $x$ axis due to the charge element $dq$ shown has one component along the $x$ axis and one perpendicular to the $x$ axis. (b) For any charge element $dq_1$ there is an equal charge element $dq_2$ opposite it, and the electric-field components perpendicular to the $x$ axis sum to zero.

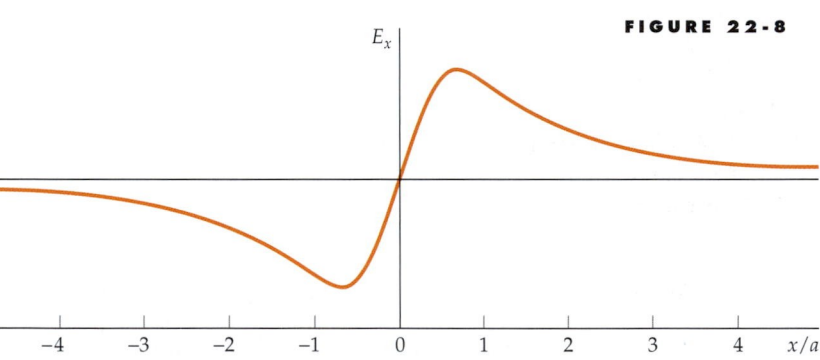

**FIGURE 22-8**

## $\vec{E}$ on the Axis of a Uniformly Charged Disk

Figure 22-9 shows a uniformly charged disk of radius $R$ and total charge $Q$. We can calculate the field on the axis of the disk by treating the disk as a set of concentric ring charges. Let the axis of the disk be the $x$ axis. $\vec{E}$ due to the charge on each ring is along the $x$ axis. A ring of radius $a$ and width $da$ is shown in the figure. The area of this ring is $dA = 2\pi a\,da$, and its charge is $dq = \sigma\,dA = 2\pi\sigma a\,da$, where $\sigma = Q/\pi R^2$ is the surface charge density (the charge per unit area). The field produced by this ring is given by Equation 22-10 if we replace $Q$ with $dq = 2\pi\sigma a\,da$.

$$dE_x = \frac{kx2\pi\sigma a\,da}{(x^2+a^2)^{3/2}}$$

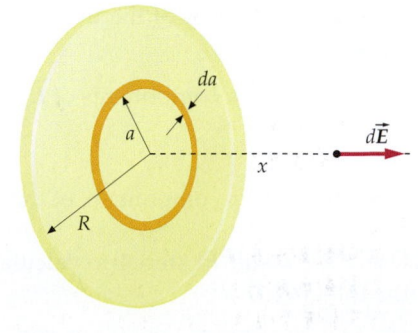

**FIGURE 22-9** A uniform disk of charge can be treated as a set of ring charges, each of radius $a$.

The total field is found by integrating from $a = 0$ to $a = R$:

$$E_x = \int_0^R \frac{kx2\pi\sigma a\,da}{(x^2 + a^2)^{3/2}} = kx\pi\sigma \int_0^R (x^2 + a^2)^{-3/2}2a\,da = kx\pi\sigma \int_{x^2+0^2}^{x^2+R^2} u^{-3/2}du$$

where $u = x^2 + a^2$, so $du = 2a\,da$. The integration thus gives

$$E_x = kx\pi\sigma \frac{u^{-1/2}}{-1/2}\bigg|_{x^2}^{x^2+R^2} = -2kx\pi\sigma\left(\frac{1}{\sqrt{x^2 + R^2}} - \frac{1}{\sqrt{x^2}}\right)$$

This can be expressed

$$E_x = 2\pi k\sigma\left(1 - \frac{1}{\sqrt{1 + \dfrac{R^2}{x^2}}}\right), \qquad x > 0 \qquad\qquad 22\text{-}11$$

$\vec{E}$ ON THE AXIS OF A DISK CHARGE

**EXERCISE** Find an expression for $E_x$ on the negative $x$ axis. (*Answer* $E_x = -2\pi k\sigma\left(1 - \dfrac{1}{\sqrt{1 + \dfrac{R^2}{x^2}}}\right)$ for $x < 0$)

For $x \gg R$ (on the positive $x$ axis far from the disk) we expect it to look like a point charge. If we merely replace $R^2/x^2$ with 0 for $x \gg R$, we get $E_x \to 0$. Although this is correct, it does not tell us anything about how $E_x$ depends on $x$ for large $x$. We can find this dependence by using the binomial expansion, $(1 + \epsilon)^n \approx 1 + n\epsilon$, for $|\epsilon| \ll 1$. Using this approximation on the second term in Equation 22-11, we obtain

$$\frac{1}{\left(1 + \dfrac{R^2}{x^2}\right)^{1/2}} = \left(1 + \frac{R^2}{x^2}\right)^{-1/2} \approx 1 - \frac{R^2}{2x^2}$$

Substituting this into Equation 22-11 we obtain

$$E_x \approx 2\pi k\sigma\left(1 - 1 + \frac{R^2}{2x^2}\right) = \frac{2k\pi R^2\sigma}{2x^2} = \frac{kQ}{x^2}, \qquad x \gg R \qquad\qquad 22\text{-}12$$

where $Q = \sigma\pi R^2$ is the total charge on the disk. For large $x$, the electric field of the charged disk approaches that of a point charge $Q$ at the origin.

## $\vec{E}$ Due to an Infinite Plane of Charge

The field of an infinite plane of charge can be obtained from Equation 22-11 by letting the ratio $R/x$ go to infinity. Then

$$E_x = 2\pi k\sigma, \qquad x > 0 \qquad\qquad 22\text{-}13a$$

$\vec{E}$ NEAR AN INFINITE PLANE OF CHARGE

Thus, the field due to an infinite-plane charge distribution is uniform; that is, the field does not depend on $x$. On the other side of the infinite plane, for negative values of $x$, the field points in the negative $x$ direction, so

$$E_x = -2\pi k\sigma, \qquad x < 0 \qquad\qquad 22\text{-}13b$$

As we move along the $x$ axis, the electric field jumps from $-2\pi k\sigma\,\hat{i}$ to $+2\pi k\sigma\,\hat{i}$ when we pass through an infinite plane of charge (Figure 22-10). There is thus a discontinuity in $E_x$ in the amount $4\pi k\sigma$.

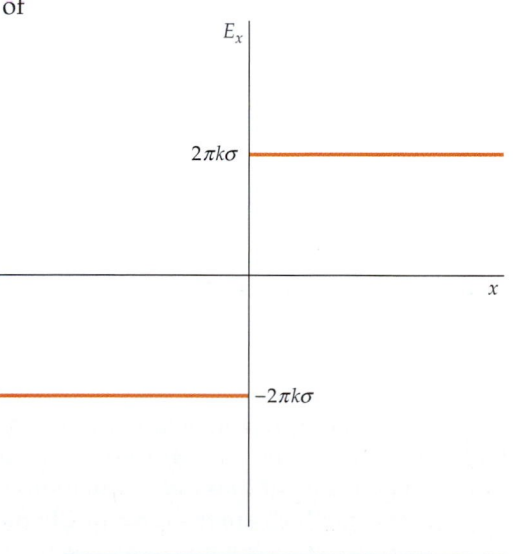

**FIGURE 22-10** Graph showing the discontinuity of $\vec{E}$ at a plane charge.

**E X A M P L E   2 2 - 4**

A disk of radius 5 cm carries a uniform surface charge density of 4 $\mu$C/m². Using appropriate approximations, find the electric field on the axis of the disk at distances of (a) 0.01 cm, (b) 0.03 cm, and (c) 6 m. (d) Compare the results for (a), (b), and (c) with the exact values arrived at by using Equation 22-11.

**PICTURE THE PROBLEM** For the comparisons in Part (d), we will carry out all calculations to five-figure accuracy. For (a) and (b), the field point is very near the disk compared with its radius, so we can approximate the disk as an infinite plane. For (c), the field point is sufficiently far from the disk ($x/R$ = 120) that we can approximate the disk as a point charge. (d) To compare, we find the percentage difference between the approximate values and the exact values.

(a) The electric field near the disk is approximately that due to an infinite plane charge:

$$E_x \approx 2\pi k\sigma$$

$$= 2\pi(8.98755 \times 10^9\ \text{N·m}^2/\text{C}^2)(4 \times 10^{-6}\ \text{C/m}^2)$$

$$= \boxed{225.88\ \text{kN/C}}$$

(b) Since 0.03 cm is still very near the disk, the disk still looks like an infinite plane charge:

$$E_x \approx 2\pi k\sigma = \boxed{225.88\ \text{kN/C}}$$

(c) Far from the disk, the field is approximately that due to a point charge:

$$E_x \approx \frac{kQ}{x^2} = \frac{k\sigma\pi R^2}{x^2} = 2\pi k\sigma \frac{R^2}{2x^2}$$

$$= (225.88\ \text{kN/C})\frac{(0.05\ \text{m})^2}{2(6\ \text{m})^2} = \boxed{7.8431\ \text{N/C}}$$

(d) Using the exact expression (Equation 22-11) for $E_x$, we calculate the exact values at the specified points:

$$E_x(\text{exact}) = 2\pi k\sigma\left(1 - \frac{1}{\sqrt{1 + \dfrac{R^2}{x^2}}}\right)$$

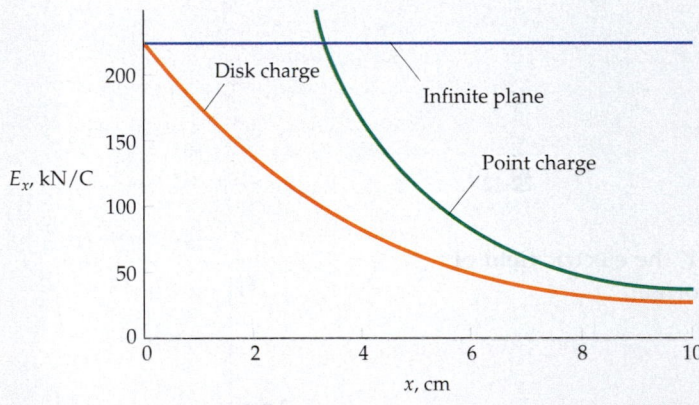

| $x$ (cm) | $E_x$ (exact) (N/C) | $E_x$ (approx) (N/C) | % diff |
|---|---|---|---|
| 0.01 | 225,430 | 225,880 | 0.2 |
| 0.03 | 224,530 | 225,880 | 0.6 |
| 600 | 7.8427 | 7.8431 | 0.005 |

**FIGURE 22-11** Note that the field of the disk charge converges with the field of the point charge at large distances, and equals the field of the infinite plane charge in the limit that $x$ approaches zero.

**REMARKS** Figure 22-11 shows $E_x$ versus $x$ for the disk charge in this example, for an infinite plane with the same charge density, and for a point charge.

## 22-2 Gauss's Law

In Chapter 21, the electric field is described visually via electric field lines. Here that description is put in rigorous mathematical language called Gauss's law. Gauss's law is one of Maxwell's equations—the fundamental equations of electromagnetism, which are the topic of Chapter 31. For static charges, Gauss's law and Coulomb's law are equivalent. Electric fields arising from some symmetrical charge distributions, such as a spherical shell of charge or an infinite line of charge, can be easily calculated using Gauss's law. In this section, we give an argument for the validity of Gauss's law based on the properties of electric field lines. A rigorous derivation of Gauss's law is presented in Section 22-6.

A closed surface is one that divides the universe into two distinct regions, the region inside the surface and the region outside the surface. Figure 22-12 shows a closed surface of arbitrary shape enclosing a dipole. The number of electric field lines beginning on the positive charge and penetrating the surface from the inside depends on where the surface is drawn, but any line penetrating the surface from the inside also penetrates it from the outside. To count the net number of lines out of any closed surface, count any line that penetrates from the inside as +1, and any penetration from the outside as −1. Thus, for the surface shown (Figure 22-12), the net number of lines out of the surface is zero. For surfaces enclosing other types of

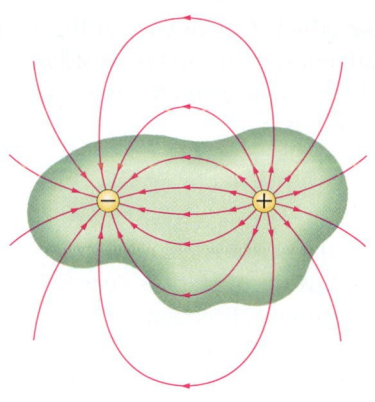

**FIGURE 22-12** A surface of arbitrary shape enclosing an electric dipole. As long as the surface encloses both charges, the number of lines penetrating the surface from the inside is exactly equal to the number of lines penetrating the surface from the outside no matter where the surface is drawn.

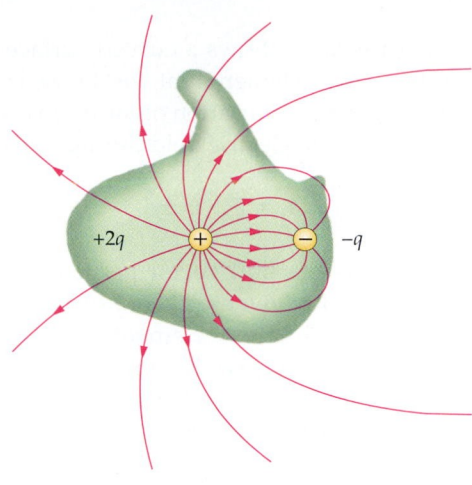

**FIGURE 22-13** A surface of arbitrary shape enclosing the charges +2q and −q. Either the field lines that end on −q do not pass through the surface or they penetrate it from the inside the same number of times as from the outside. The net number that exit, the same as that for a single charge of +q, is equal to the net charge enclosed by the surface.

charge distributions, such as that shown in Figure 22-13, *the net number of lines out of any surface enclosing the charges is proportional to the net charge enclosed by the surface.* This rule is a qualitative statement of Gauss's law.

## Electric Flux

The mathematical quantity that corresponds to the number of field lines penetrating a surface is called the **electric flux** $\phi$. For a surface perpendicular to $\vec{E}$ (Figure 22-14), the electric flux is the product of the magnitude of the field E and the area A:

$$\phi = EA$$

The units of flux are $N \cdot m^2/C$. Because E is proportional to the number of field lines per unit area, the flux is proportional to the number of field lines penetrating the surface.

In Figure 22-15, the surface of area $A_2$ is not perpendicular to the electric field $\vec{E}$. However, the number of lines that penetrate the surface of area $A_2$ is the same as the number that penetrate the surface of area $A_1$, which is perpendicular to $\vec{E}$. These areas are related by

$$A_2 \cos \theta = A_1 \qquad \text{22-14}$$

where $\theta$ is the angle between $\vec{E}$ and the unit vector $\hat{n}$ that is normal to the surface $A_2$, as shown in the figure. The electric flux through a surface is defined to be

$$\phi = \vec{E} \cdot \hat{n}A = EA \cos \theta = E_n A \qquad \text{22-15}$$

where $E_n = \vec{E} \cdot \hat{n}$ is the component of $\vec{E}$ normal (perpendicular) to the surface.

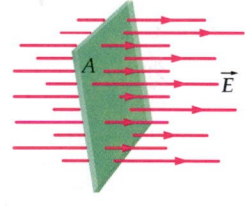

**FIGURE 22-14** Electric field lines of a uniform field penetrating a surface of area A that is oriented perpendicular to the field. The product EA is the electric flux through the surface.

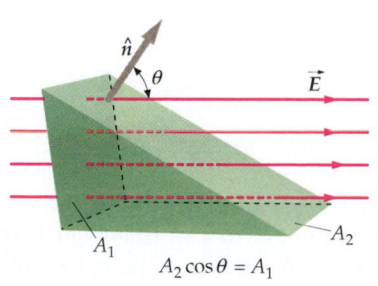

$A_2 \cos \theta = A_1$

**FIGURE 22-15** Electric field lines of a uniform electric field that is perpendicular to the surface of area $A_1$ but makes an angle $\theta$ with the unit vector $\hat{n}$ that is normal to the surface of area $A_2$. Where $\vec{E}$ is not perpendicular to the surface, the flux is $E_n A$, where $E_n = E \cos \theta$ is the component of $\vec{E}$ that is perpendicular to the surface. The flux through the surface of area $A_2$ is the same as that through the surface of area $A_1$.

Figure 22-16 shows a curved surface over which $\vec{E}$ may vary. If the area $\Delta A_i$ of the surface element that we choose is small enough, it can be considered to be a plane, and the variation of the electric field across the element can be neglected. The flux of the electric field through this element is

$$\Delta \phi_i = E_{ni} \Delta A_i = \vec{E}_i \cdot \hat{n}_i \Delta A_i$$

where $\hat{n}_i$ is the unit vector perpendicular to the surface element and $\vec{E}_i$ is the electric field anywhere on the surface element. If the surface is curved, the unit vectors for different elements will have different directions. The total flux through the surface is the sum of $\Delta\phi_i$ over all the elements making up the surface. In the limit, as the number of elements approaches infinity and the area of each element approaches zero, this sum becomes an integral. The general definition of electric flux is thus:

$$\phi = \lim_{\Delta A_i \to 0} \sum_i \vec{E}_i \cdot \hat{n}_i \Delta A_i = \int_S \vec{E} \cdot \hat{n}\, dA \qquad 22\text{-}16$$

DEFINITION—ELECTRIC FLUX

where the $S$ stands for the surface we are integrating over.

On a *closed* surface we are interested in the electric flux out of the surface, so we choose the unit vector $\hat{n}$ to be outward at each point. The integral over a closed surface is indicated by the symbol $\oint$. The total or net flux out of a closed surface is therefore written

$$\phi_{net} = \oint_S \vec{E} \cdot \hat{n}\, dA = \oint_S E_n\, dA \qquad 22\text{-}17$$

The net flux $\phi_{net}$ through the closed surface is positive or negative, depending on whether $\vec{E}$ is predominantly outward or inward at the surface. At points on the surface where $\vec{E}$ is inward, $E_n$ is negative.

## Quantitative Statement of Gauss's Law

Figure 22-17 shows a spherical surface of radius $R$ with a point charge $Q$ at its center. The electric field everywhere on this surface is normal to the surface and has the magnitude

$$E_n = \frac{kQ}{R^2}$$

The net flux of $\vec{E}$ out of this spherical surface is

$$\phi_{net} = \oint_S E_n\, dA = E_n \oint_S dA$$

where we have taken $E_n$ out of the integral because it is constant everywhere on the surface. The integral of $dA$ over the surface is just the total area of the surface, which for a sphere of radius $R$ is $4\pi R^2$. Using this and substituting $kQ/R^2$ for $E_n$, we obtain

$$\phi_{net} = \frac{kQ}{R^2} 4\pi R^2 = 4\pi k Q \qquad 22\text{-}18$$

Thus, the net flux out of a spherical surface with a point charge at its center is independent of the radius $R$ of the sphere and is equal to $4\pi k$ times $Q$ (the point charge). This is consistent with our previous observation that the net number of

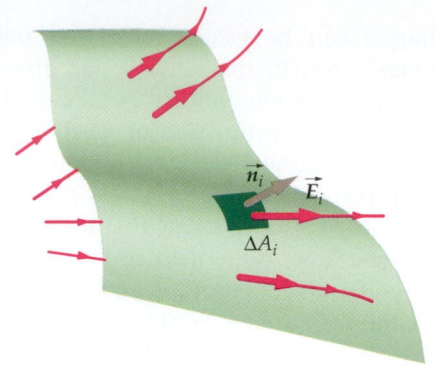

**FIGURE 22-16** If $E_n$ varies from place to place on a surface, either because $E$ varies or because the angle between $\vec{E}$ and $\hat{n}$ varies, the area of the surface is divided into small elements of area $\Delta A_i$. The flux through the surface is computed by summing $\vec{E}_i \cdot \hat{n}_i \Delta A_i$ over all the area elements.

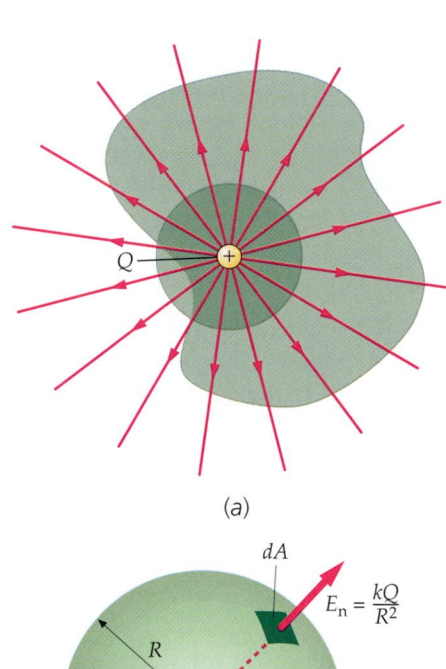

(a)

(b)

**FIGURE 22-17** A spherical surface enclosing a point charge $Q$. (a) The net number of electric field lines out of this surface and the net number out of any surface that also encloses $Q$ is the same. (b) The net flux is easily calculated for a spherical surface. It equals $E_n$ times the surface area, or $E_n 4\pi R^2$.

lines going out of a closed surface is proportional to the net charge inside the surface. *This number of lines is the same for all closed surfaces surrounding the charge, independent of the shape of the surface.* Thus, the net flux out of *any surface* surrounding a point charge $Q$ equals $4\pi k Q$.

We can extend this result to systems containing multiple charges. In Figure 22-18, the surface encloses two point charges, $q_1$ and $q_2$, and there is a third point charge $q_3$ outside the surface. Since the electric field at any point on the surface is the vector sum of the electric fields produced by each of the three charges, the net flux $\phi_{net} = \oint_s \vec{E} \cdot \hat{n} \, dA$ out of the surface is just the sum of the fluxes due to the individual charges. The flux due to charge $q_3$, which is outside the surface, is zero because every field line from $q_3$ that enters the surface at one point leaves the surface at some other point. The flux out of the surface due to charge $q_1$ is $4\pi k q_1$ and that due to charge $q_2$ is $4\pi k q_2$. The net flux out of the surface therefore equals $4\pi k(q_1 + q_2)$, which may be positive, negative, or zero depending on the signs and magnitudes of $q_1$ and $q_2$.

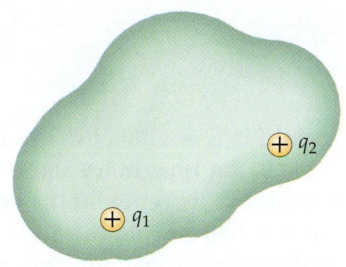

**FIGURE 22-18** A surface enclosing point charges $q_1$ and $q_2$, but not $q_3$. The net flux out of this surface is $4\pi k(q_1 + q_2)$.

> The net outward flux through any closed surface equals $4\pi k$ times the net charge inside the surface:
>
> $$\phi_{net} = \int_S E_n \, dA = 4\pi k Q_{inside} \qquad 22\text{-}19$$

GAUSS'S LAW

This is **Gauss's law.** Its validity depends on the fact that the electric field due to a single point charge varies inversely with the square of the distance from the charge. It was this property of the electric field that made it possible to draw a fixed number of electric field lines from a charge and have the density of lines be proportional to the field strength.

It is customary to write the Coulomb constant $k$ in terms of another constant $\epsilon_0$, which is called the **permittivity of free space:**

$$k = \frac{1}{4\pi \epsilon_0} \qquad 22\text{-}20$$

Using this notation, Coulomb's law for $\vec{E}$ is written

$$\vec{E} = \frac{1}{4\pi \epsilon_0} \frac{q}{r^2} \hat{r} \qquad 22\text{-}21$$

and Gauss's law is written

$$\phi_{net} = \oint_S E_n \, dA = \frac{Q_{inside}}{\epsilon_0} \qquad 22\text{-}22$$

The value of $\epsilon_0$ in SI units is

$$\epsilon_0 = \frac{1}{4\pi k} = \frac{1}{4\pi(8.99 \times 10^9 \, \text{N·m}^2/\text{C}^2)} = 8.85 \times 10^{-12} \, \text{C}^2/\text{N·m}^2 \qquad 22\text{-}23$$

Gauss's law is valid for all surfaces and all charge distributions. For charge distributions that have high degrees of symmetry, it can be used to calculate the electric field, as we illustrate in the next section. For static charge distributions, Gauss's law and Coulomb's law are equivalent. However, Gauss's law is more general in that it is always valid and Coulomb's law is valid only for static charge distributions.

**E X A M P L E   2 2 - 5**

An electric field is $\vec{E} = (200 \text{ N/C})\hat{i}$ in the region $x > 0$ and $\vec{E} = (-200 \text{ N/C})\hat{i}$ in the region $x < 0$. An imaginary soup-can shaped surface of length 20 cm and radius $R = 5$ cm has its center at the origin and its axis along the $x$ axis, so that one end is at $x = +10$ cm and the other is at $x = -10$ cm (Figure 22-19). (*a*) What is the net outward flux through the entire closed surface? (*b*) What is the net charge inside the closed surface?

**PICTURE THE PROBLEM** The closed surface described, which is piecewise continuous, consists of three pieces—two flat ends and a curved side. Separately calculate the flux of $\vec{E}$ out of each piece of the surface. To calculate the flux out of a piece draw the outward normal $\hat{n}$ at a randomly chosen point on the piece and draw the vector $\vec{E}$ at the same point. If $E_n = \vec{E} \cdot \hat{n}$ is the same everywhere on the piece, then the outward flux through it is $\phi = \vec{E} \cdot \hat{n}A$ (Equation 22-15). The net outward flux through the entire closed surface is obtained by summing the fluxes through the individual pieces. The net outward flux is related to the charge inside by Gauss's law (Equation 22-19).

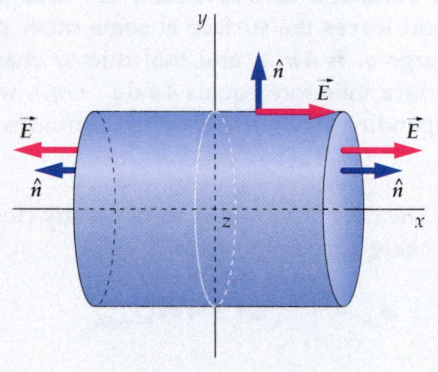

**FIGURE 22-19**

(*a*) 1. Sketch the soup-can shaped surface. On each piece of the surface draw the outward normal $\hat{n}$ and the vector $\vec{E}$:

2. Calculate the outward flux through the right circular flat surface where $\hat{n} = \hat{i}$:

$$\phi_{\text{right}} = \vec{E}_{\text{right}} \cdot \hat{n}_{\text{right}} A = \vec{E}_{\text{right}} \cdot \hat{i} \pi R^2$$
$$= (200 \text{ N/C})\hat{i} \cdot \hat{i}(\pi)(0.05 \text{ m})^2$$
$$= 1.57 \text{ N·m}^2/\text{C}$$

3. Calculate the outward flux through the left circular surface where $\hat{n} = -\hat{i}$:

$$\phi_{\text{left}} = \vec{E}_{\text{left}} \cdot \hat{n}_{\text{left}} A = \vec{E}_{\text{left}} \cdot (-\hat{i}) \pi R^2$$
$$= (-200 \text{ N/C})\hat{i} \cdot (-\hat{i})(\pi)(0.05 \text{ m})^2$$
$$= 1.57 \text{ N·m}^2/\text{C}$$

4. Calculate the outward flux through the curved surface where $\vec{E}$ is perpendicular to $\hat{n}$:

$$\phi_{\text{curved}} = \vec{E}_{\text{curved}} \cdot \hat{n}_{\text{curved}} A = 0$$

5. The net outward flux is the sum through all the individual surfaces:

$$\phi_{\text{net}} = \phi_{\text{right}} + \phi_{\text{left}} + \phi_{\text{curved}}$$
$$= 1.57 \text{ N·m}^2/\text{C} + 1.57 \text{ N·m}^2/\text{C} + 0$$
$$= \boxed{3.14 \text{ N·m}^2/\text{C}}$$

(*b*) Gauss's law relates the charge inside to the net flux:

$$Q_{\text{inside}} = \epsilon_0 \phi_{\text{net}}$$
$$= (8.85 \times 10^{-12} \text{ C}^2/\text{N·m}^2)(3.14 \times \text{N·m}^2/\text{C})$$
$$= \boxed{2.78 \times 10^{-11} \text{ C} = 27.8 \text{ pC}}$$

**REMARKS** The flux does not depend on the length of the can. This means the charge inside resides entirely on the *yz* plane.

# 22-3   Calculating $\vec{E}$ From Gauss's Law

Given a highly symmetrical charge distribution, the electric field can often be calculated more easily using Gauss's law than it can be using Coulomb's law. We first find an imaginary closed surface, called a **Gaussian surface** (the soup can in Example 22-5). Optimally, this surface is chosen so that on each of its pieces $\vec{E}$ is either zero, perpendicular to $\hat{n}$, or parallel to $\hat{n}$ with $E_n$ constant. Then the flux through each piece equals $E_n A$ and Gauss's law is used to relate the field to the charges inside the closed surface.

## Plane Symmetry

A charge distribution has **plane symmetry** if the views of it from all points on an infinite plain surface are the same. Figure 22-20 shows an infinite plane of charge of uniform surface charge density $\sigma$. By symmetry, $\vec{E}$ must be perpendicular to the plane and can depend only on the distance from it. Also, $\vec{E}$ must have the same magnitude but the opposite direction at points the same distance from the charged plane on either side of the plane. For our Gaussian surface, we choose a soup-can shaped cylinder as shown, with the charged plane bisecting the cylinder. On each piece of the cylinder is drawn both $\hat{n}$ and $\vec{E}$. Since $\vec{E} \cdot \hat{n}$ is zero everywhere on the curved piece of the Gaussian surface, there is no flux through it. The flux through each flat piece of the surface is $E_n A$, where $A$ is the area of each flat piece. Thus, the total outward flux through the closed surface is $2E_n A$. The net charge inside the surface is $\sigma A$. Gauss's law then gives

$$Q_{inside} = \epsilon_0 \, \phi_{net}$$

$$\sigma A = \epsilon_0 \, 2E_n A$$

(Can you see why $Q_{inside} = \sigma A$?) Solving for $E_n$ gives

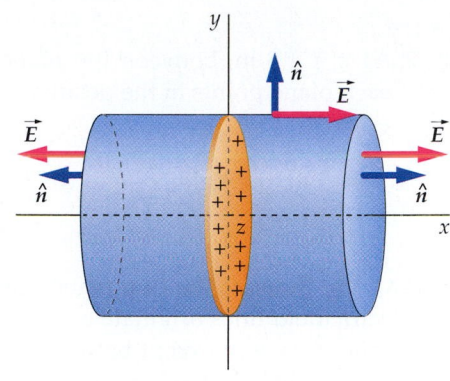

**FIGURE 22-20** Gaussian surface for the calculation of $\vec{E}$ due to an infinite plane of charge. (Only the part of the plane that is inside the Gaussian surface is shown.) On the flat faces of this soup can, $\vec{E}$ is perpendicular to the surface and constant in magnitude. On the curved surface $\vec{E}$ is parallel with the surface.

$$E_n = \frac{\sigma}{2\epsilon_0} = 2\pi k\sigma \qquad\qquad 22\text{-}24$$

$\vec{E}$ FOR AN INFINITE PLANE OF CHARGE

$E_n$ is positive if $\sigma$ is positive, and $E_n$ is negative if $\sigma$ is negative. This means if $\sigma$ is positive $\vec{E}$ is directed away from the charged plane, and if $\sigma$ is negative $\vec{E}$ points toward it. This is the same result that we obtained, with much more difficulty, using Coulomb's law (Equations 22-13a and b). Note that the field is discontinuous at the charged plane. If the charged plane is the $yz$ plane, the field is $\vec{E} = \sigma/(2\epsilon_0)\hat{i}$ in the region $x > 0$ and $\vec{E} = -\sigma/(2\epsilon_0)\hat{i}$ in the region $x < 0$. Thus, the field is discontinuous by $\Delta\vec{E} = \sigma/(2\epsilon_0)\hat{i} - [-\sigma/(2\epsilon_0)\hat{i}] = (\sigma/\epsilon_0)\hat{i}$.

---

*ELECTRIC FIELD DUE TO TWO INFINITE PLANES*   **EXAMPLE 22-6**

In Figure 22-21, an infinite plane of surface charge density $\sigma = +4.5$ nC/m$^2$ lies in the $x = 0$ plane, and a second infinite plane of surface charge density $\sigma = -4.5$ nC/m$^2$ lies in a plane parallel to the $x = 0$ plane at $x = 2$ m. Find the electric field at (a) $x = 1.8$ m and (b) $x = 5$ m.

**FIGURE 22-21**

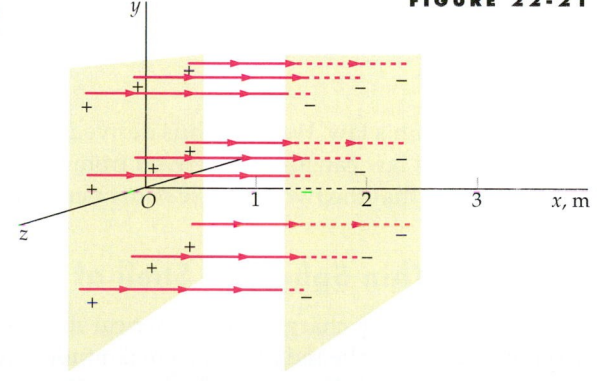

**PICTURE THE PROBLEM** Each plane produces a uniform electric field of magnitude $E = \sigma/(2\epsilon_0)$. We use superposition to find the resultant field. Between the planes the fields add, producing a net field of magnitude $\sigma/\epsilon_0$ in the positive $x$ direction. For $x > 2$ m and for $x < 0$, the fields point in opposite directions and cancel.

(a) 1. Calculate the magnitude of the field $E$ produced by each plane:

$$E = \frac{\sigma}{2\,\epsilon_0} = \frac{4.5 \times 10^{-9}\,\text{C/m}^2}{2(8.85 \times 10^{-12}\,\text{C}^2/\text{N·m}^2)}$$

$$= 254\,\text{N/C}$$

2. At $x = 1.8$ m, between the planes, the field due to each plane points in the positive $x$ direction:

$$E_{x,net} = E_1 + E_2 = 254 \text{ N/C} + 254 \text{ N/C}$$

$$= \boxed{508 \text{ N/C}}$$

(b) At $x = 5$ m, the fields due to the two planes are oppositely directed:

$$E_{x,net} = E_1 - E_2 = \boxed{0}$$

**REMARKS** Because the two planes carry equal and opposite charge densities, the electric field lines originate on the positive plane and terminate on the negative plane. $\vec{E}$ is zero except between the planes. Note that $E_{x,net} = 508$ N/C not just at $x = 1.8$ m but at any point in the region between the charged planes.

## Spherical Symmetry

Assume a charge distribution is concentric within a spherical surface. The charge distribution has **spherical symmetry** if the views of it from all points on the spherical surface are the same. To calculate the electric field due to spherically symmetric charge distributions, we use a spherical surface for our Gaussian surface. We illustrate this by first finding the electric field at a distance $r$ from a point charge $q$. We choose a spherical surface of radius $r$, centered at the point charge, for our Gaussian surface. By symmetry, $\vec{E}$ must be directed either radially outward or radially inward. It follows that the component of $\vec{E}$ normal to the surface equals the radial component of $E$ at each point on the surface. That is, $E_n = \vec{E} \cdot \hat{n} = E_r$, where $\hat{n}$ is the outward normal, has the same value everywhere on the spherical surface. Also, the magnitude of $\vec{E}$ can depend on the distance from the charge but not on the direction from the charge. The net flux through the spherical surface of radius $r$ is thus

$$\phi_{net} = \oint_S \vec{E} \cdot \hat{n} \, dA = \oint_S E_r \, dA = E_r \oint_S dA = E_r 4\pi r^2$$

where $\oint_S dA = 4\pi r^2$ the total area of the spherical surface. Since the total charge inside the surface is just the point charge $q$, Gauss's law gives

$$E_r 4\pi r^2 = \frac{q}{\epsilon_0}$$

Solving for $E_r$ gives

$$E_r = \frac{1}{4\pi\epsilon_0} \frac{q}{r^2}$$

which is Coulomb's law. We have thus derived Coulomb's law from Gauss's law. Because Gauss's law can also be derived from Coulomb's law (see Section 22-6), we have shown that the two laws are equivalent for static charges.

## $\vec{E}$ Due to a Thin Spherical Shell of Charge

Consider a uniformly charged thin spherical shell of radius $R$ and total charge $Q$. By symmetry, $\vec{E}$ must be radial, and its magnitude can depend only on the distance $r$ from the center of the sphere. In Figure 22-22, we have chosen a spherical Gaussian surface of radius $r > R$. Since $\vec{E}$ is normal to this surface, and has the same magnitude everywhere on the surface, the flux through the surface is

$$\phi_{net} = \oint_S E_r \, dA = E_r \oint_S dA = E_r 4\pi r^2$$

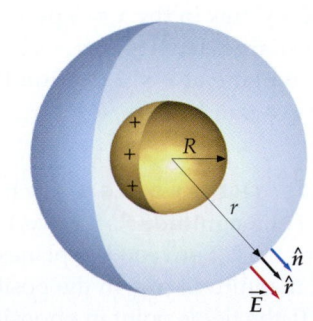

**FIGURE 22-22** Spherical Gaussian surface of radius $r > R$ for the calculation of the electric field outside a uniformly charged thin spherical shell of radius $R$.

Since the total charge inside the Gaussian surface is the total charge on the shell $Q$, Gauss's law gives

$$E_r 4\pi r^2 = \frac{Q}{\epsilon_0}$$

or

$$E_r = \frac{1}{4\pi\epsilon_0}\frac{Q}{r^2}, \qquad r > R \qquad\qquad\qquad 22\text{-}25a$$

$\vec{E}$ OUTSIDE A SPHERICAL SHELL OF CHARGE

Thus, the electric field outside a uniformly charged spherical shell is the same as if all the charge were at the center of the shell.

If we choose a spherical Gaussian surface inside the shell, where $r < R$, the net flux is again $E_r 4\pi r^2$, but the total charge inside the surface is zero. Therefore, for $r < R$, Gauss's law gives

$$\phi_{\text{net}} = E_r 4\pi r^2 = 0$$

so

$$E_r = 0, \qquad r < R \qquad\qquad\qquad 22\text{-}25b$$

$\vec{E}$ INSIDE A SPHERICAL SHELL OF CHARGE

These results can also be obtained by direct integration of Coulomb's law, but that calculation is much more difficult.

Figure 22-23 shows $E_r$ versus $r$ for a spherical-shell charge distribution. Again, note that the electric field is discontinuous at $r = R$, where the surface charge density is $\sigma = Q/4\pi R^2$. Just outside the shell at $r \approx R$, the electric field is $E_r = Q/4\pi\epsilon_0 R^2 = \sigma/\epsilon_0$, since $\sigma = Q/4\pi R^2$. Because the field just inside the shell is zero, the electric field is discontinuous by the amount $\sigma/\epsilon_0$ as we pass through the shell.

(a)

(b)

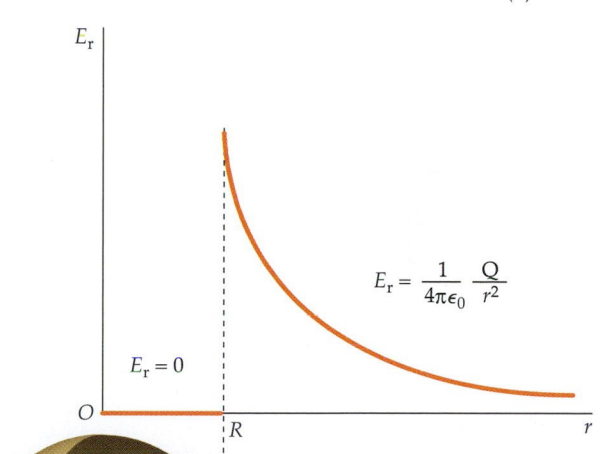

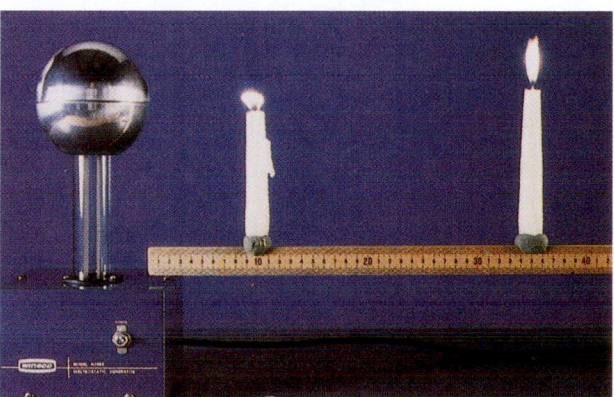

**FIGURE 22-23** (a) A plot of $E_r$ versus $r$ for a spherical-shell charge distribution. The electric field is discontinuous at $r = R$, where there is a surface charge of density $\sigma$. (b) The decrease in $E_r$ over distance due to a charged spherical shell is evident by the effect of the field on the flames of these two candles. The spherical shell at the left (part of a Van de Graaff generator, a device that is discussed in Chapter 24) carries a large negative charge that attracts the positive ions in the nearby candle flame. The flame at right, which is much farther away, is not noticeably affected.

**EXAMPLE 22-7**

A spherical shell of radius $R = 3$ m has its center at the origin and carries a surface charge density of $\sigma = 3$ nC/m². A point charge $q = 250$ nC is on the $y$ axis at $y = 2$ m. Find the electric field on the $x$ axis at (a) $x = 2$ m and (b) $x = 4$ m.

**PICTURE THE PROBLEM** We find the field due to the point charge and that due to the spherical shell and sum the field vectors. For (a), the field point is inside the shell, so the field is due only to the point charge (Figure 22-24a). For (b), the field point is outside the shell, so the shell can be considered as a point charge at the origin. We then find the field due to two point charges (Figure 22-24b).

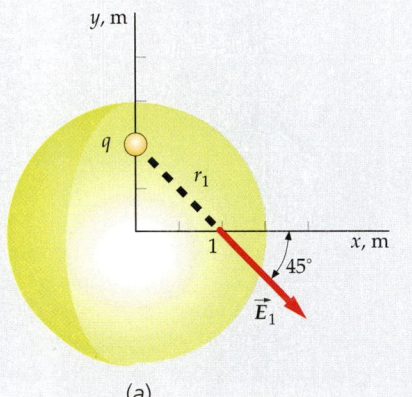

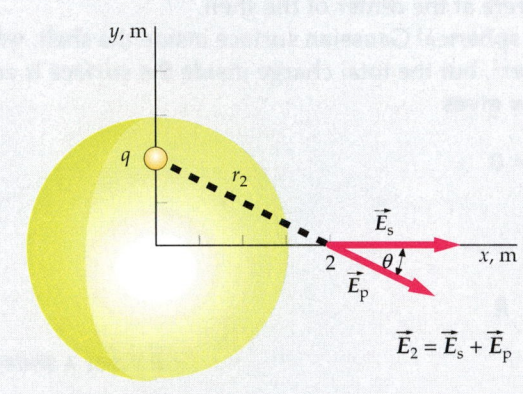

**FIGURE 22-24**

(a)  (b)

(a) 1. Inside the shell, $\vec{E}_1$ is due only to the point charge:

$$\vec{E}_1 = \frac{kq}{r_1^2}\hat{r}_1$$

2. Calculate the square of the distance $r_1$:

$$r_1^2 = (2\text{ m})^2 + (2\text{ m})^2 = 8\text{ m}^2$$

3. Use $r_1^2$ to calculate the magnitude of the field:

$$E_1 = \frac{kq}{r_1^2} = \frac{(8.99 \times 10^9\text{ N·m}^2/\text{C}^2)(250 \times 10^{-9}\text{ C})}{8\text{ m}^2}$$

$$= 281\text{ N/C}$$

4. From Figure 22-24a, we can see that the field makes an angle of 45° with the $x$ axis:

$$\theta_1 = 45°$$

5. Express $\vec{E}_1$ in terms of its components:

$$\vec{E}_1 = E_{1x}\hat{i} + E_{1y}\hat{j} = E_1\cos 45°\hat{i} - E_1\sin 45°\hat{j}$$

$$= (281\text{ N/C})\cos 45°\hat{i} - (281\text{ N/C})\sin 45°\hat{j}$$

$$= \boxed{199\,(\hat{i} - \hat{j})\text{ N/C}}$$

(b) 1. Outside of its perimeter, the shell can be treated as a point charge at the origin, and the field due to the shell $\vec{E}_s$ is therefore along the $x$ axis:

$$\vec{E}_s = \frac{kQ}{x_2^2}\hat{i}$$

2. Calculate the total charge $Q$ on the shell:

$$Q = \sigma 4\pi R^2 = (3\text{ nC/m}^2)4\pi(3\text{ m})^2 = 339\text{ nC}$$

3. Use $Q$ to calculate the field due to the shell:

$$E_s = \frac{kQ}{x_2^2} = \frac{(8.99 \times 10^9\text{ N·m}^2/\text{C}^2)(339 \times 10^{-9}\text{ C})}{(4\text{ m})^2}$$

$$= 190\text{ N/C}$$

4. The field due to the point charge is:

$$\vec{E}_p = \frac{kq}{r_2^2}\hat{r}_2$$

5. Calculate the square of the distance from the point charge $q$ on the $y$ axis to the field point at $x = 4$ m:

$$r_2^2 = (2\ \text{m})^2 + (4\ \text{m})^2 = 20\ \text{m}^2$$

6. Calculate the magnitude of the field due to the point charge:

$$E_p = \frac{kq}{r_2^2} = \frac{(8.99 \times 10^9\ \text{N·m}^2/\text{C}^2)(250 \times 10^{-9}\ \text{C})}{20\ \text{m}^2}$$

$$= 112\ \text{N/C}$$

7. This field makes an angle $\theta$ with the $x$ axis, where:

$$\tan \theta = \frac{2\ \text{m}}{4\ \text{m}} = \frac{1}{2} \Rightarrow \theta = \tan^{-1}\frac{1}{2} = 26.6°$$

8. The $x$ and $y$ components of the net electric field are thus:

$$E_x = E_{px} + E_{sx} = E_p \cos \theta + E_s$$

$$= (112\ \text{N/C}) \cos 26.6° + 190\ \text{N/C} = 290\ \text{N/C}$$

$$E_y = E_{py} + E_{sy} = -E_p \sin \theta + 0$$

$$= -(112\ \text{N/C}) \sin 26.6° = -50.0\ \text{N/C}$$

$$\boxed{\vec{E} = (290\hat{i} - 50.0\hat{j})\ \text{N/C}}$$

**REMARKS** Giving the $x$, $y$, and $z$ components of a vector completely specifies the vector. In these cases, the $z$ component is zero.

## $\vec{E}$ Due to a Uniformly Charged Sphere

| *ELECTRIC FIELD DUE TO A CHARGED SOLID SPHERE* | **EXAMPLE 22-8** |
| --- | --- |

Find the electric field (*a*) outside and (*b*) inside a uniformly charged solid sphere of radius $R$ carrying a total charge $Q$ that is uniformly distributed throughout the volume of the sphere with charge density $\rho = Q/V$, where $V = \frac{4}{3}\pi R^3$ is the volume of the sphere.

**FIGURE 22-25**

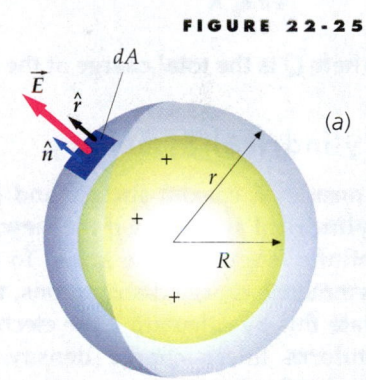

(*a*)

**PICTURE THE PROBLEM** By symmetry, the electric field must be radial. (*a*) To find $E_r$ outside the charged sphere, we choose a spherical Gaussian surface of radius $r > R$ (Figure 22-25*a*). (*b*) To find $E_r$ inside the charge we choose a spherical Gaussian surface of radius $r > R$ (Figure 22-25*b*). On each of these surfaces, $E_r$ is constant. Gauss's law then relates $E_r$ to the total charge inside the Gaussian surface.

(*a*) 1. (Outside) Draw a charged sphere of radius $R$ and draw a spherical Gaussian surface with radius $r > R$:

2. Relate the flux through the Gaussian surface to the electric field $E_r$ on it. At every point on this surface $\hat{n} = \hat{r}$ and $E_r$ has the same value:

$$\phi_{\text{net}} = \vec{E} \cdot \hat{n}A = \vec{E} \cdot \hat{r}A = E_r 4\pi r^2$$

3. Apply Gauss's law to relate the field to the total charge inside the surface, which is $Q$:

$$E_r 4\pi r^2 = \frac{Q_{\text{inside}}}{\epsilon_0} = \frac{Q}{\epsilon_0}$$

(*b*)

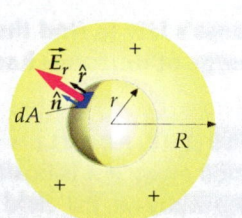

4. Solve for $E_r$:

$$\boxed{E_r = \frac{1}{4\pi\epsilon_0}\frac{Q}{r^2}, \quad r > R}$$

(*b*) 1. (Inside) Again draw the charged sphere of radius $R$. This time draw a spherical Gaussian surface with radius $r < R$:

2. Relate the flux through the Gaussian surface to the electric field $E_r$ on it. At every point on this surface $\hat{n} = \hat{r}$ and $E_r$ has the same value:

$$\phi_{\text{net}} = \vec{E} \cdot \hat{n}A = \vec{E} \cdot \hat{r}A = E_r 4\pi r^2$$

3. Apply Gauss's law to relate the field to the total charge inside the surface $Q_{inside}$:

$$E_r 4\pi r^2 = \frac{Q_{inside}}{\epsilon_0}$$

4. The total charge inside the surface is $\rho V'$, where $\rho = Q/V$, $V = \frac{4}{3}\pi R^3$ and $V' = \frac{4}{3}\pi r^3$. $V$ is the volume of the solid sphere and $V'$ is the volume inside the Gaussian surface:

$$Q_{inside} = \rho V' = \left(\frac{Q}{V}\right)V' = \left(\frac{Q}{\frac{4}{3}\pi R^3}\right)\left(\frac{4}{3}\pi r^3\right) = Q\frac{r^3}{R^3}$$

5. Substitute this value for $Q_{inside}$ and solve for $E_r$:

$$E_r 4\pi r^2 = \frac{Q_{inside}}{\epsilon_0} = \frac{1}{\epsilon_0}Q\frac{r^3}{R^3}$$

$$\boxed{E_r = \frac{1}{4\pi\,\epsilon_0}\frac{Q}{R^3}r, \quad r \leq R}$$

**REMARKS** Figure 22-26 shows $E_r$ versus $r$ for the charge distribution in this example. Inside a sphere of charge, $E_r$ increases with $r$. Note that $E_r$ is continuous at $r = R$. A uniformly charged sphere is sometimes used as a model to describe the electric field of an atomic nucleus.

We see from Example 22-8 that the electric field a distance $r$ from the center of a uniformly charged sphere of radius $R$ is given by

$$E_r = \frac{1}{4\pi\,\epsilon_0}\frac{Q}{r^2}, \qquad r \geq R \qquad 22\text{-}26a$$

$$E_r = \frac{1}{4\pi\,\epsilon_0}\frac{Q}{R^3}r, \qquad r \leq R \qquad 22\text{-}26b$$

where $Q$ is the total charge of the sphere.

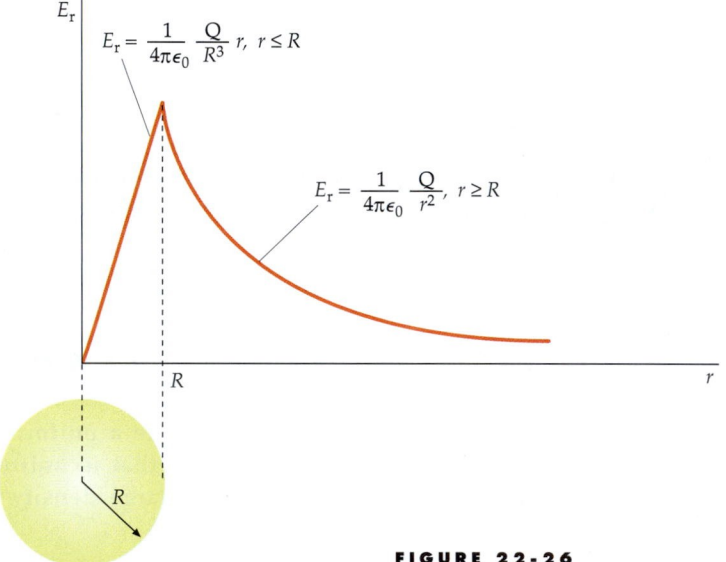

**FIGURE 22-26**

## Cylindrical Symmetry

Consider a coaxial surface and charge distribution. A charge distribution has **cylindrical symmetry** if the views of it from all points on a cylindrical surface of infinite length are the same. To calculate the electric field due to cylindrically symmetric charge distributions, we use a cylindrical Gaussian surface. We illustrate this by calculating the electric field due to an infinitely long line charge of uniform linear charge density, a problem we have already solved using Coulomb's law.

---

*ELECTRIC FIELD DUE TO INFINITE LINE CHARGE*  **EXAMPLE 22-9**

**Use Gauss's law to find the electric field everywhere due to an infinitely long line charge of uniform charge density $\lambda$.**

**PICTURE THE PROBLEM** Because of the symmetry, we know the electric field is directed away if $\lambda$ is positive (directly toward it if $\lambda$ is negative), and we know the magnitude of the field depends only on the radial distance from the line charge. We therefore choose a soup-can shaped Gaussian surface coaxial with the line. This surface consists of three pieces, the two flat ends and the curved side. We calculate the outward flux of $\vec{E}$ through each piece and, using Gauss's law, relate the net outward flux to the charge density $\lambda$.

1. Sketch the wire and a coaxial soup-can shaped Gaussian surface (Figure 22-27) with length $L$ and radius $R$. The closed surface consists of three pieces, the two flat ends and the curved side. At a randomly chosen point on each piece, draw the vectors $\vec{E}$ and $\hat{n}$. Because of the symmetry, we know that the direction of $\vec{E}$ is directly away from the line charge if $\lambda$ is positive (directly toward it if $\lambda$ is negative), and we know that the magnitude of $E$ depends only on the radial distance from the line charge.

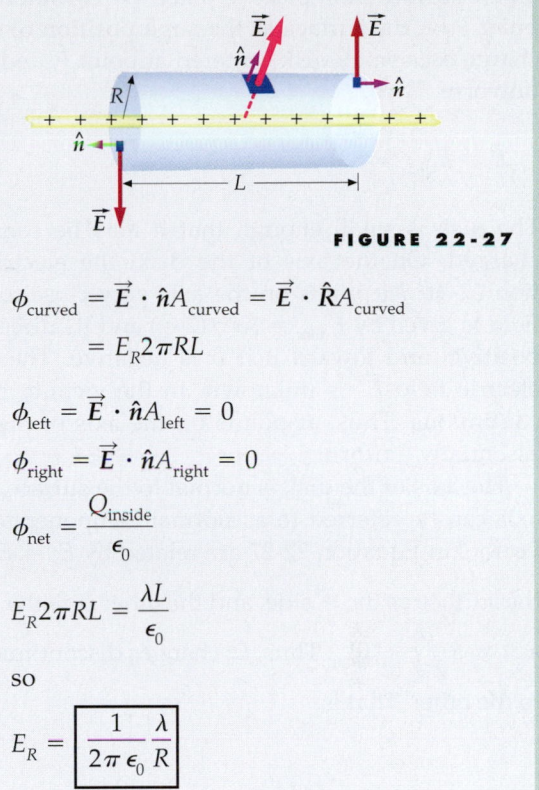

**FIGURE 22-27**

2. Calculate the outward flux through the curved piece of the Gaussian surface. At each point on the curved piece $\hat{R} = \hat{n}$, where $\hat{R}$ is the unit vector in the radial direction.

$$\phi_{\text{curved}} = \vec{E} \cdot \hat{n} A_{\text{curved}} = \vec{E} \cdot \hat{R} A_{\text{curved}}$$
$$= E_R 2\pi RL$$

3. Calculate the outward flux through each of the flat ends of the Gaussian surface. On these pieces the direction of $\hat{n}$ is parallel with the line charge (and thus perpendicular to $\vec{E}$):

$$\phi_{\text{left}} = \vec{E} \cdot \hat{n} A_{\text{left}} = 0$$
$$\phi_{\text{right}} = \vec{E} \cdot \hat{n} A_{\text{right}} = 0$$

4. Apply Gauss's law to relate the field to the total charge inside the surface $Q_{\text{inside}}$. The net flux out of the Gaussian surface is the sum of the fluxes out of the three pieces of the surface, and $Q_{\text{inside}}$ is the charge on a length $L$ of the line charge:

$$\phi_{\text{net}} = \frac{Q_{\text{inside}}}{\epsilon_0}$$

$$E_R 2\pi RL = \frac{\lambda L}{\epsilon_0}$$

so

$$\boxed{E_R = \frac{1}{2\pi\epsilon_0}\frac{\lambda}{R}}$$

■ **REMARKS** Since $1/(2\pi\epsilon_0) = 2k$, the field is $2k\lambda/R$, the same as Equation 22-9.

It is important to realize that although Gauss's law holds for any surface surrounding any charge distribution, it is very useful for calculating the electric fields of charge distributions that are highly symmetric. It is also useful doing calculations involving conductors in electrostatic equilibrium, as we shall see in Section 22.5. In the calculation of Example 22-9, we needed to assume that the field point was very far from the ends of the line charge so that $E_n$ would be constant everywhere on the cylindrical Gaussian surface. (This is equivalent to assuming that, at the distance $R$ from the line, the line charge appears to be infinitely long.) If we are near the end of a finite line charge, we cannot assume that $\vec{E}$ is perpendicular to the curved surface of the soup can, or that $E_n$ is constant everywhere on it, so we cannot use Gauss's law to calculate the electric field.

**FIGURE 22-28** (a) A surface carrying surface-charge. (b) The electric field $\vec{E}_{\text{disk}}$ due to the charge on a circular disk, plus the electric field $\vec{E}'$ due to all other charges. The right side of the disk is the + side, the left side the − side.

## 22-4 Discontinuity of $E_n$

We have seen that the electric field for an infinite plane of charge and a thin spherical shell of charge is discontinuous by the amount $\sigma/\epsilon_0$ on either side of a surface carrying charge density $\sigma$. We now show that this is a general result for the component of the electric field that is perpendicular to a surface carrying a charge density of $\sigma$.

Figure 22-28 shows an arbitrary surface carrying a surface charge density $\sigma$. The surface is arbitrary in that it is arbitrarily curved, although it does not have any sharp folds, and $\sigma$ may vary continuously

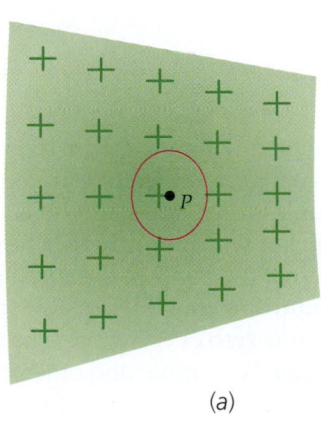

(a)

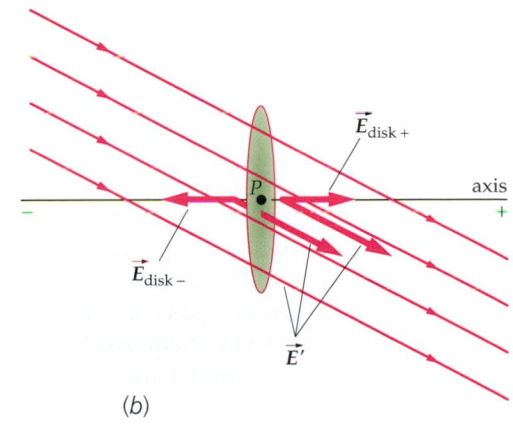

(b)

on the surface from place to place. We consider electric field $\vec{E}$ in the vicinity of a point $P$ on the surface as the superposition of electric field $\vec{E}_{disk}$, due just to the charge on a small disk centered at point $P$, and $\vec{E}$ due to all other charges in the universe. Thus,

$$\vec{E} = \vec{E}_{disk} + \vec{E}'$$ 22-27

The disk is small enough that it may be considered both flat and uniformly charged. On the axis of the disk, the electric field $\vec{E}_{disk}$ is given by Equation 22-11. At points on the axis very close to the disk, the magnitude of this field is given by $E_{disk} = |\sigma|/(2\,\epsilon_0)$ and its direction is away from the disk if $\sigma$ is positive, and toward it if $\sigma$ is negative. The magnitude and direction of the electric field $\vec{E}'$ is unknown. In the vicinity of point $P$, however, this field is continuous. Thus, at points on the axis of the disk and very close to it, $\vec{E}'$ is essentially uniform.

The axis of the disk is normal to the surface, so vector components along this axis can be referred to as normal components. The normal components of the vectors in Equation 22-27 are related by $E_n = E_{disk\,n} + E'_n$. If we refer one side of the surface as the $+$ side, and the other side the $-$ side, then $E_{n+} = \dfrac{\sigma}{2\,\epsilon_0} + E'_{n+}$ and $E_{n-} = -\dfrac{\sigma}{2\,\epsilon_0} + E'_{n+}$. Thus, $E_n$ changes discontinuously from one side of the surface to the other. That is:

$$\Delta E_n = E_{n+} - E_{n-} = \frac{\sigma}{2\,\epsilon_0} - \left(-\frac{\sigma}{2\,\epsilon_0}\right) = \frac{\sigma}{\epsilon_0}$$ 22-28

DISCONTINUITY OF $E_n$ AT A SURFACE CHARGE

where we have made use of the fact that near the disk $E'_{n+} = E'_{n-}$ (since $\vec{E}'$ is continuous and uniform).

Note that the discontinuity of $E_n$ occurs at a finite disk of charge, an infinite plane of charge (refer to Figure 22-10), and a thin spherical shell of charge (see Figure 22-23). However, it does not occur at the perimeter of a solid sphere of charge (see Figure 22-26). The electric field is discontinuous at any location with an infinite volume-charge density. These include locations with a finite point charge, locations with a finite line-charge density, and locations with a finite surface-charge density. At all locations with a finite surface-charge density, the normal component of the electric field is discontinuous—in accord with Equation 22-28.

## 22-5 Charge and Field at Conductor Surfaces

A conductor contains an enormous amount of mobile charge that can move freely within the conductor. If there is an electric field within a conductor, there will be a net force on this charge causing a momentary electric current (electric currents are discussed in Chapter 25). However, unless there is a source of energy to maintain this current, the free charge in a conductor will merely redistribute itself to create an electric field that cancels the external field within the conductor. The conductor is then said to be in **electrostatic equilibrium.** Thus, in electrostatic equilibrium, the electric field inside a conductor is zero everywhere. The time taken to reach equilibrium depends on the conductor. For copper and other metal conductors, the time is so small that in most cases electrostatic equilibrium is reached in a few nanoseconds.

We can use Gauss's law to show that any net electric charge on a conductor resides on the surface of the conductor. Consider a Gaussian surface completely inside the material of a conductor in electrostatic equilibrium (Figure 22-29). The size and shape of the Gaussian surface doesn't matter, as long as the entire surface is within the material of the conductor. The electric field is zero everywhere on the Gaussian surface because the surface is completely within the conductor where the field is everywhere zero. The net flux of the electric field through the surface must therefore be zero, and, by Gauss's law, the net charge inside the surface must be zero. Thus, there can be no net charge inside any surface lying completely within the material of the conductor. If a conductor carries a net charge, it must reside on the conductor's surface. At the surface of a conductor in electrostatic equilibrium, $\vec{E}$ must be perpendicular to the surface. We conclude this by reasoning that if the electric field had a tangential component at the surface, the free charge would be accelerated tangential to the surface until electrostatic equilibrium was reestablished.

Since $E_n$ is discontinuous at any charged surface by the amount $\sigma/\epsilon_0$, and since $\vec{E}$ is zero inside the material of a conductor, the field just outside the surface of a conductor is given by

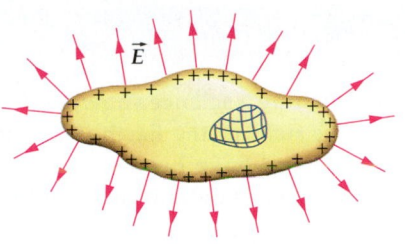

**FIGURE 22-29** A Gaussian surface completely within the material of a conductor. Since the electric field is zero inside a conductor in electrostatic equilibrium, the net flux through this surface must also be zero. Therefore, the net charge density $\rho$ within the material of a conductor must be zero.

$$E_n = \frac{\sigma}{\epsilon_0} \qquad\qquad 22\text{-}29$$

$E_n$ JUST OUTSIDE THE SURFACE OF A CONDUCTOR

This result is exactly twice the field produced by a uniform disk of charge. We can understand this result from Figure 22-30. The charge on the conductor consists of two parts: (1) the charge near point $P$ and (2) all the rest of the charge. The charge near point $P$ looks like a small, uniformly charged circular disk centered at $P$ that produces a field near $P$ of magnitude $\sigma/(2\,\epsilon_0)$ just inside and just outside the conductor. The rest of the charges in the universe must produce a field of magnitude $\sigma/(2\,\epsilon_0)$ that exactly cancels the field inside the conductor. This field due to the rest of the charge adds to the field due to the small charged disk just outside the conductor to give a total field of $\sigma/\epsilon_0$.

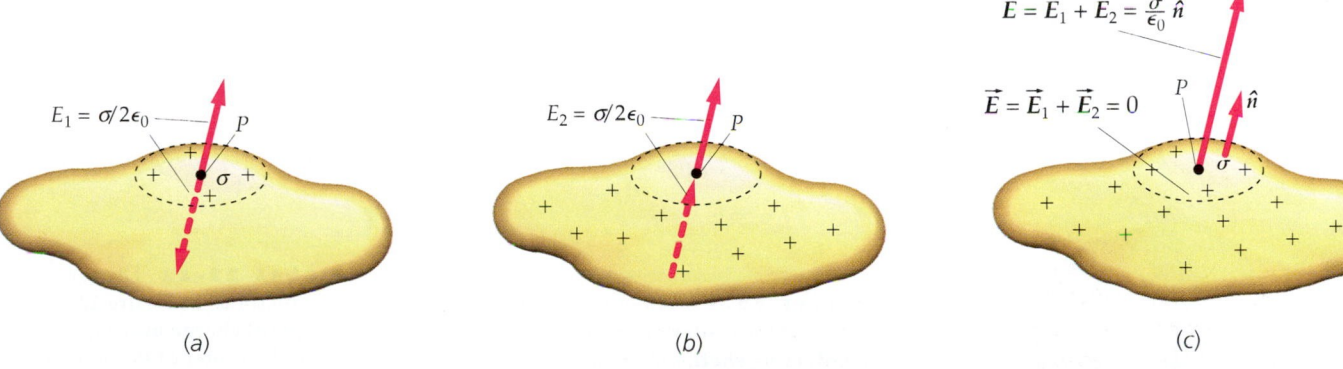

(a)  (b)  (c)

**FIGURE 22-30** An arbitrarily shaped conductor carrying a charge on its surface. (a) The charge in the vicinity of point $P$ near the surface looks like a small uniformly charged circular disk centered at $P$, giving an electric field of magnitude $\sigma/(2\epsilon_0)$ pointing away from the surface both inside and outside the surface. Inside the conductor, this field points away from point $P$ in the opposite direction. (b) Since the net field inside the conductor is zero, the rest of the charges in the universe must produce a field of magnitude $\sigma/(2\epsilon_0)$ in the outward direction. The field due to this charge is the same just inside the surface as it is just outside the surface. (c) Inside the surface, the fields shown in (a) and (b) cancel, but outside at point $P$ they add to give $E_n = \sigma/\epsilon_0$.

*THE CHARGE OF THE EARTH*                          **E X A M P L E     2 2 - 1 0**

While watching a science show on the atmosphere, you find out that on average the electric field of the Earth is about 100 N/C directed vertically downwards. Given that you have been studying electric fields in your physics class, you wonder if you can determine what the total charge on the Earth's surface is.

**PICTURE THE PROBLEM** The earth is a conductor, so any charge it carries resides on the surface of the earth. The surface charge density $\sigma$ is related to the normal component of the electric field $E_n$ by Equation 22-29. The total charge $Q$ equals the charge density $\sigma$ times the surface area $A$.

1. The surface charge density $\sigma$ is related to the normal component of the electric field $E_n$ by Equation 22-29:

$$E_n = \frac{\sigma}{\epsilon_0}$$

2. On the surface of the earth $\hat{n}$ is upward and $\vec{E}$ is downward, so $E_n$ is negative:

$$E_n = \vec{E} \cdot \hat{n} = E \times 1 \times \cos 180° = -E = -100 \text{ n/C}$$

3. The charge $Q$ is the charge per unit area. Combine this with the step 1 and 2 results to obtain an expression for $Q$:

$$Q = \sigma A = \epsilon_0 E_n A = -\epsilon_0 EA$$

4. The surface area of a sphere of radius $r$ is given by $A = 4\pi r^2$.

$$Q = -\epsilon_0 EA = -\epsilon_0 E 4\pi R_E^2 = -4\pi \epsilon_0 E R_E^2$$

5. The radius of the earth is $6.38 \times 10^6$ m:

$$Q = -4\pi \epsilon_0 E R_E^2$$
$$= -4\pi (8.85 \times 10^{-12} \text{ C}^2/\text{N·m}^2)(100 \text{ N/C})(6.38 \times 10^6 \text{ m})^2$$
$$= \boxed{-4.53 \times 10^5 \text{ C}}$$

Figure 22-31 shows a positive point charge $q$ at the center of a spherical cavity inside a spherical conductor. Since the net charge must be zero within any Gaussian surface drawn within the conductor, there must be a negative charge $-q$ induced in the inside surface. In Figure 22-32, the point charge has been moved so that it is no longer at the center of the cavity. The field lines in the cavity are altered, and the surface charge density of the induced negative charge on the inner surface is no longer uniform. However, the positive surface charge density on the outside surface is not disturbed—it is still uniform—because it is electrically shielded from the cavity by the conducting material.

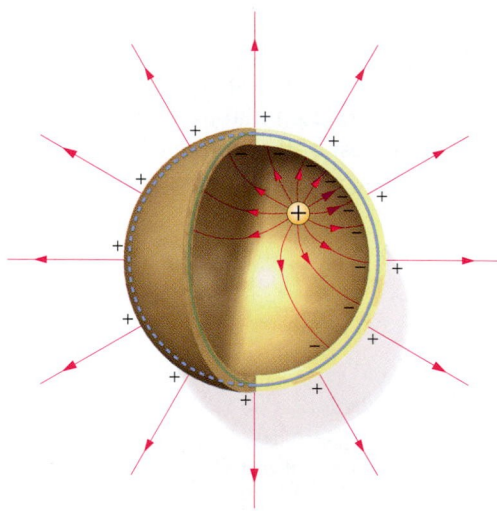

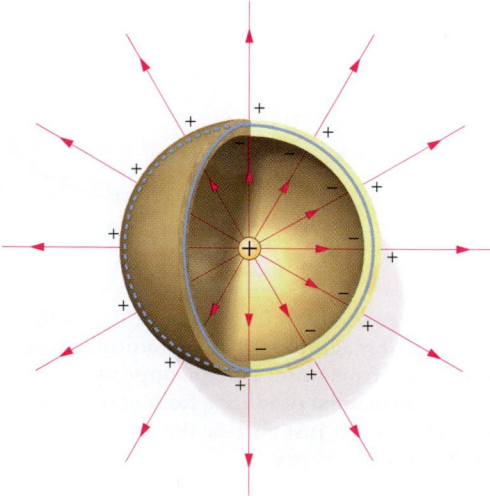

**FIGURE 22-31** A point charge $q$ in the cavity at the center of a thick spherical conducting shell. Since the net charge within the Gaussian surface (indicated in blue) must be zero, a surface charge $-q$ is induced on the inner surface of the shell, and since the conductor is neutral, an equal but opposite charge $+q$ is induced on the outer surface. Electric field lines begin on the point charge and end on the inner surface. Field lines begin again on the outer surface.

**FIGURE 22-32** The same conductor as in Figure 22-31 with the point charge moved away from the center of the sphere. The charge on the outer surface and the electric field lines outside the sphere are not affected.

An infinite, nonconducting, uniformly charged plane is located in the $x = -a$ plane, and a second such plane is located in the $x = +a$ plane (Figure 22-33a). The plane at $x = -a$ carries a positive charge density whereas the plane at $x = +a$ carries a negative charge density of the same magnitude. The electric field due to the charges on both planes is $\vec{E}_{\text{applied}} = (450 \text{ kN/C})\hat{i}$ in the region between them. A thin, uncharged 2-m diameter conducting disk is placed in the $x = 0$ plane and centered at the origin (Figure 22-33b). (a) Find the charge density on each face of the disk. Also, find the electric field just outside the disk at each face. (Assume that any charge on either face is uniformly distributed.) (b) A net charge of 96 $\mu$C is placed on the disk. Find the new charge density on each face and the electric field just outside each face but far from the edges of the sheet.

Electric field lines for an oppositely charged cylinder and plate, shown by bits of fine thread suspended in oil. Note that the field lines are perpendicular to the conductors and that there are no lines inside the cylinder.

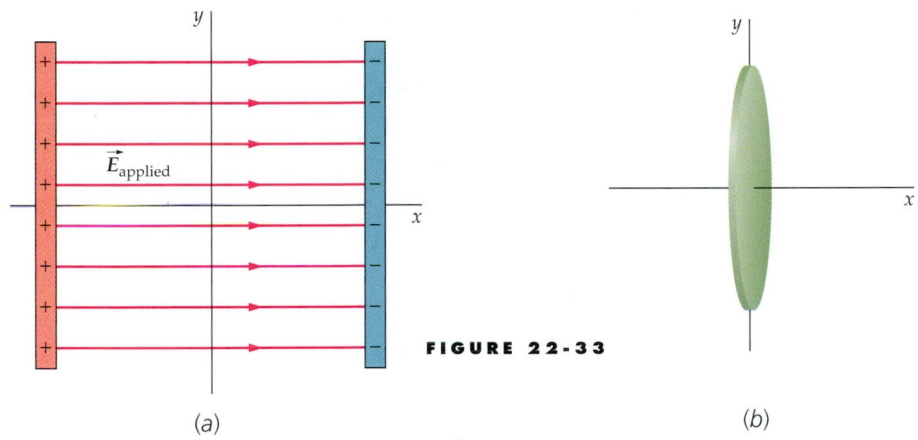

**FIGURE 22-33**

(a)                                    (b)

**PICTURE THE PROBLEM** (a) We find the charge density by using the fact that the total charge on the disk is zero and that there is no electric field inside the conducting material of the disk. The surface charges on the disk must produce an electric field inside it that exactly cancels $\vec{E}_{\text{applied}}$. (b) The additional charge of 96 $\mu$C must be distributed so that the electric field inside the conducting disk remains zero.

(a) 1. Let $\sigma_R$ and $\sigma_L$ be the charge densities on the right and left faces on the conducting sheet, respectively. Since the disk is uncharged, these densities must add to zero.

$$\sigma_R + \sigma_L = 0$$

so

$$\sigma_L = -\sigma_R$$

2. Inside the conducting sheet the electric field due to the charges on its surface must cancel $\vec{E}_{\text{applied}}$. Let $\vec{E}_R$ and $\vec{E}_L$ be the electric field due to the charges on the right and left faces, respectively.

$$\vec{E}_R + \vec{E}_L + \vec{E}_{\text{applied}} = 0$$

3. Using Equations 22-13a and b we can express the electric field due to the charge on each surface of the disk by the corresponding surface charge density. The field due to a disk of surface charge $\sigma$ next to the disk is given by $[\sigma/(2\,\epsilon_0)]\,\hat{u}$, where $\hat{u}$ is a unit vector directed away from the surface charge.

$$\vec{E}_R + \vec{E}_L + \vec{E}_{\text{applied}} = 0$$

$$\frac{\sigma_R}{2\,\epsilon_0}(-\hat{i}) + \frac{\sigma_L}{2\,\epsilon_0}\hat{i} + \vec{E}_{\text{applied}} = 0$$

4. Substituting $-\sigma_R$ for $\sigma_L$ and solving for the surface charge densities gives:

$$\frac{\sigma_R}{2\,\epsilon_0}(-\hat{i}) + \frac{-\sigma_R}{2\,\epsilon_0}\hat{i} + \vec{E}_{\text{applied}} = 0$$

$$-\frac{\sigma_R}{\epsilon_0}\hat{i} + \vec{E}_{\text{applied}} = 0$$

$$\sigma_R \hat{i} = \epsilon_0 \vec{E}_{\text{applied}}$$

$$= (8.85 \times 10^{-12}\,\text{C}^2/\text{N·m}^2)(450\,\text{kN/C})\hat{i}$$

$$\sigma_R = 3.98 \times 10^{-6}\,\text{C/m}^2 = \boxed{3.98\,\mu\text{C/m}^2}$$

$$\sigma_L = -\sigma_R = \boxed{-3.98\,\mu\text{C/m}^2}$$

5. Use Equation 22-29 ($E_n = \sigma/\epsilon_0$) to relate the electric field just outside a conductor to the surface charge density on it. Just outside the right side of the disk $\hat{n} = \hat{i}$, and just outside the left side $\hat{n} = -\hat{i}$:

$$E_{Rn} = \frac{\sigma_R}{\epsilon_0} = \frac{3.98\,\mu\text{C/m}^2}{8.85 \times 10^{-12}\,\text{C}^2/\text{N·m}^2}$$

$$= 450\,\text{kN/C}$$

$$\vec{E}_R = E_{Rn}\hat{n} = E_{Rn}\hat{i} = \boxed{450\,\text{kN/C}\,\hat{i}}$$

$$E_{Ln} = \frac{\sigma_L}{\epsilon_0} = \frac{-3.98\,\mu\text{C/m}^2}{8.85 \times 10^{-12}\,\text{C}^2/\text{N·m}^2}$$

$$\vec{E}_L = E_{Ln}\hat{n} = E_{Ln}(-\hat{i}) = \boxed{450\,\text{kN/C}\,\hat{i}}$$

(b) 1. The sum of the charges on the two faces of the disk must equal the net charge on the disk.

$$Q_R + Q_L = Q_{\text{net}}$$

$$\sigma_R A + \sigma_L A = Q_{\text{net}}$$

or

$$\sigma_L = \frac{Q_{\text{net}}}{A} - \sigma_R$$

2. Substitute for $\sigma_L$ in the Part (a), step 2 result and solve for the surface charge densities:

$$\frac{\sigma_R}{2\,\epsilon_0}(-\hat{i}) + \frac{(Q_{\text{net}}/A) - \sigma_R}{2\,\epsilon_0}\hat{i} + \vec{E}_{\text{applied}} = 0$$

$$\frac{(Q_{\text{net}}/A) - 2\sigma_R}{2\,\epsilon_0}\hat{i} + \vec{E}_{\text{applied}} = 0$$

$$\sigma_R \hat{i} = \epsilon_0 \vec{E}_{\text{applied}} + \frac{Q_{\text{net}}}{2A}\hat{i} = \epsilon_0(450\,\text{kN/C})\hat{i} + \frac{Q_{\text{net}}}{2A}\hat{i}$$

$$\sigma_R = (8.85 \times 10^{-12}\,\text{C}^2/\text{N·m}^2)(450\,\text{kN/C}) + \frac{Q_{\text{net}}}{2A}$$

$$= 3.98\,\mu\text{C/m}^2 + \frac{96\,\mu\text{C}}{2\pi(1\,\text{m})^2} = \boxed{19.3\,\mu\text{C/m}^2}$$

$$\sigma_L = \frac{Q_{\text{net}}}{A} - \sigma_R = \frac{Q_{\text{net}}}{A} - \left(\epsilon_0(450\,\text{kN/C}) + \frac{Q_{\text{net}}}{2A}\right)$$

$$= -\epsilon_0(450\,\text{kN/C}) + \frac{Q_{\text{net}}}{2A}$$

$$= -3.98\,\mu\text{C/m}^2 + \frac{96\,\mu\text{C}}{2\pi(1\,\text{m})^2} = \boxed{11.3\,\mu\text{C/m}^2}$$

3. Using Equation 22-29 ($E_n = \epsilon_0\sigma$), relate the electric field just outside a conductor to the surface charge density on it.

$$E_{Rn} = \frac{\sigma_R}{\epsilon_0} = \frac{19.3\,\mu\text{C/m}^2}{8.85 \times 10^{-12}\,\text{C}^2/\text{N·m}^2}$$

$$= 2.17 \times 10^6\,\text{N/C}$$

$$\vec{E}_R = E_{Rn}\hat{n} = E_{Rn}\hat{i} = \boxed{+2.17\,\text{MN/C}\,\hat{i}}$$

$$E_{Ln} = \frac{\sigma_L}{\epsilon_0} = \frac{11.3\,\mu\text{C/m}^2}{8.85 \times 10^{-12}\,\text{C}^2/\text{N·m}^2}$$

$$\vec{E}_L = E_{Ln}\hat{n} = E_{Ln}(-\hat{i}) = \boxed{-1.28\,\text{MN/C}\,\hat{i}}$$

**REMARKS**  The charge added to the disk was distributed equally, half on one side and half on the other. The electric field inside the disk due to this added charge is exactly zero. On each side of a real charged conducting thin disk the magnitude of the charge density is greatest near the edge of the disk.

**EXERCISE**  The electric field just outside the surface of a certain conductor points away from the conductor and has a magnitude of 2000 N/C. What is the surface charge density on the surface of the conductor? (*Answer*   17.7 nC/m²)

# *22-6  Derivation of Gauss's Law From Coulomb's Law

Gauss's law can be derived mathematically using the concept of the **solid angle.** Consider an area element $\Delta A$ on a spherical surface. The solid angle $\Delta\Omega$ subtended by $\Delta A$ at the center of the sphere is defined to be

$$\Delta\Omega = \frac{\Delta A}{r^2}$$

where $r$ is the radius of the sphere. Since $\Delta A$ and $r^2$ both have dimensions of length squared, the solid angle is dimensionless. The SI unit of the solid angle is the **steradian** (sr). Since the total area of a sphere is $4\pi r^2$, the total solid angle subtended by a sphere is

$$\frac{4\pi r^2}{r^2} = 4\pi \text{ steradians}$$

There is a close analogy between the solid angle and the ordinary plane angle $\Delta\theta$, which is defined to be the ratio of an element of arc length of a circle $\Delta s$ to the radius of the circle:

$$\Delta\theta = \frac{\Delta s}{r} \text{ radians}$$

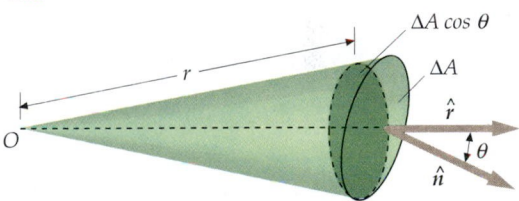

**FIGURE 22-34**  An area element $\Delta A$ whose normal is not parallel to the radial line from $O$ to the center of the element. The solid angle subtended by this element at $O$ is defined to be $(\Delta A \cos\theta)/r^2$.

The total plane angle subtended by a circle is $2\pi$ radians.

In Figure 22-34, the area element $\Delta A$ is not perpendicular to the radial lines from point $O$. The unit vector $\hat{n}$ normal to the area element makes an angle $\theta$ with the radial unit vector $\hat{r}$. In this case, the solid angle subtended by $\Delta A$ at point $O$ is

$$\Delta\Omega = \frac{\Delta A\,\hat{n}\cdot\hat{r}}{r^2} = \frac{\Delta A \cos\theta}{r^2} \qquad\qquad 22\text{-}30$$

Figure 22-35 shows a point charge $q$ surrounded by a surface $S$ of arbitrary shape. To calculate the flux of $\vec{E}$ through this surface, we want to find $\vec{E}\cdot\hat{n}\Delta A$ for each element of area on the surface and sum over the entire surface. The electric field at the area element shown is given by

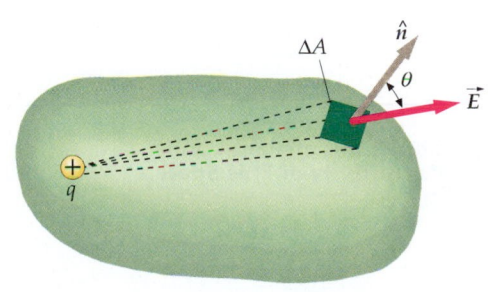

**FIGURE 22-35**  A point charge enclosed by an arbitrary surface $S$. The flux through an area element $\Delta A$ is proportional to the solid angle subtended by the area element at the charge. The net flux through the surface, found by summing over all the area elements, is proportional to the total solid angle $4\pi$ at the charge, which is independent of the shape of the surface.

$$\vec{E} = \frac{kq}{r^2} \hat{r}$$

so the flux through the element is

$$\Delta\phi = \vec{E} \cdot \hat{n} \, \Delta A = \frac{kq}{r^2} \hat{r} \cdot \hat{n} \, \Delta A = kq \, \Delta\Omega$$

The solid angle $\Delta\Omega$ is the same as that subtended by the corresponding area element of a spherical surface of any radius. The sum of the fluxes through the entire surface is $kq$ times the total solid angle subtended by the closed surface, which is $4\pi$ steradians:

$$\phi_{net} = \oint_S \vec{E} \cdot \hat{n} \, dA = kq \oint d\Omega = kq4\pi = 4\pi kq = \frac{q}{\epsilon_0} \qquad \text{22-31}$$

which is Gauss's law.

# SUMMARY

1. Gauss's law is a fundamental law of physics that is equivalent to Coulomb's law for static charges.
2. For highly symmetric charge distributions, Gauss's law can be used to calculate the electric field.

| Topic | Relevant Equations and Remarks | |
|---|---|---|
| 1. Electric Field for a Continuous Charge Distribution | $\vec{E} = \int_V \frac{k \, dq}{r^2} \hat{r} = \frac{1}{4\pi\epsilon_0} \int_V \frac{dq}{r^2} \hat{r}$ (Coulomb's law) | 22-4 |
| | where $dq = \rho \, dV$ for a charge distributed throughout a volume, $dq = \sigma \, dA$ for a charge distributed on a surface, and $dq = \lambda \, dL$ for a charge distributed along a line. | |
| 2. Electric Flux | $\phi = \lim\limits_{\Delta A_i \to 0} \sum_i \vec{E}_i \cdot \hat{n}_i \, \Delta A_i = \int_S \vec{E} \cdot \hat{n} dA$ | 22-16 |
| 3. Gauss's Law | $\phi_{net} = \int_S E_n \, dA = 4\pi k \, Q_{inside} = \frac{Q_{inside}}{\epsilon_0}$ | 22-19 |
| | The net outward flux through a closed surface equals $4\pi k$ times the net charge within the surface. | |
| 4. Coulomb Constant $k$ and Permittivity of Free Space $\epsilon_0$ | $k = \frac{1}{4\pi\epsilon_0} = 8.99 \times 10^9 \text{ N·m}^2/\text{C}^2$ | |
| | $\epsilon_0 = \frac{1}{4\pi k} = 8.85 \times 10^{-12} \text{ C}^2/\text{N·m}^2$ | 22-23 |

| 5. Coulomb's Law and Gauss's Law | $\vec{E} = \dfrac{1}{4\pi\epsilon_0}\dfrac{q}{r^2}\hat{r}$ | 22-21 |
|---|---|---|
| | $\phi_{net} = \oint_S E_n\, dA = \dfrac{Q_{inside}}{\epsilon_0}$ | 22-22 |

---

**6. Discontinuity of $E_n$**

At a surface carrying a surface charge density $\sigma$, the component of the electric field perpendicular to the surface is discontinuous by $\sigma/\epsilon_0$.

$$E_{n+} - E_{n-} = \frac{\sigma}{\epsilon_0}$$  22-28

---

**7. Charge on a Conductor**

In electrostatic equilibrium, the net electric charge on a conductor resides on the surface of the conductor.

---

**8. $\vec{E}$ Just Outside a Conductor**

The resultant electric field just outside the surface of a conductor is perpendicular to the surface and has the magnitude $\sigma/\epsilon_0$, where $\sigma$ is the local surface charge density at that point on the conductor:

$$E_n = \frac{\sigma}{\epsilon_0}$$  22-29

The force per unit area exerted on the charge on the surface of a conductor by all the other charges is called the electrostatic stress.

---

**9. Electric Fields for Various Uniform Charge Distributions**

| Of a line charge | $E_y = \dfrac{k\lambda}{y}(\sin\theta_2 - \sin\theta_1); E_x = \dfrac{k\lambda}{y}(\cos\theta_2 - \cos\theta_1)$ | 22-8 |
|---|---|---|
| Of a line charge of infinite length | $E_R = 2k\dfrac{\lambda}{R} = \dfrac{1}{2\pi\epsilon_0}\dfrac{\lambda}{R}$ | 22-9 |
| On the axis of a charged ring | $E_x = \dfrac{kQx}{(x^2 + a^2)^{3/2}}$ | 22-10 |
| On the axis of a charged disk | $E_x = \dfrac{\sigma}{2\epsilon_0}\left(1 - \dfrac{1}{\sqrt{1 + \dfrac{R^2}{x^2}}}\right),\quad x > 0$ | 22-11 |
| Of a charged plane | $E_x = \dfrac{\sigma}{2\epsilon_0},\quad x > 0$ | 22-24 |
| Of a charged spherical shell | $E_r = \dfrac{1}{4\pi\epsilon_0}\dfrac{Q}{r^2},\quad r > R$ | 22-25a |
| | $E_r = 0,\quad r < R$ | 22-25b |
| Of a charged solid sphere | $E_r = \dfrac{1}{4\pi\epsilon_0}\dfrac{Q}{r^2},\quad r \geq R$ | 22-26a |
| | $E_r = \dfrac{1}{4\pi\epsilon_0}\dfrac{Q}{R^3}r,\quad r \leq R$ | 22-26b |

## Conceptual Problems

**1** •• SSM True or false:

(a) Gauss's law holds only for symmetric charge distributions.

(b) The result that $E = 0$ inside a conductor can be derived from Gauss's law.

**2** •• What information, in addition to the total charge inside a surface, is needed to use Gauss's law to find the electric field?

**3** ••• Is the electric field $E$ in Gauss's law only that part of the electric field due to the charge inside a surface, or is it the total electric field due to all charges both inside and outside the surface?

**4** •• Explain why the electric field increases with $r$ rather than decreasing as $1/r^2$ as one moves out from the center inside a spherical charge distribution of constant volume charge density.

**5** • SSM True or false:

(a) If there is no charge in a region of space, the electric field on a surface surrounding the region must be zero everywhere.

(b) The electric field inside a uniformly charged spherical shell is zero.

(c) In electrostatic equilibrium, the electric field inside a conductor is zero.

(d) If the net charge on a conductor is zero, the charge density must be zero at every point on the surface of the conductor.

**6** • If the electric field $E$ is zero everywhere on a closed surface, is the net flux through the surface necessarily zero? What, then, is the net charge inside the surface?

**7** • A point charge $-Q$ is at the center of a spherical conducting shell of inner radius $R_1$ and outer radius $R_2$, as shown in Figure 22-36. The charge on the inner surface of the shell is (a) $+Q$. (b) zero. (c) $-Q$. (d) dependent on the total charge carried by the shell.

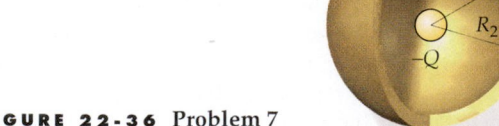

**FIGURE 22-36** Problem 7

**8** • For the configuration of Figure 22-36, the charge on the outer surface of the shell is (a) $+Q$. (b) zero. (c) $-Q$. (d) dependent on the total charge carried by the shell.

**9** •• SSM Suppose that the total charge on the conducting shell of Figure 22-36 is zero. It follows that the electric field for $r < R_1$ and $r > R_2$ points

(a) away from the center of the shell in both regions.

(b) toward the center of the shell in both regions.

(c) toward the center of the shell for $r < R_1$ and is zero for $r > R_2$.

(d) away from the center of the shell for $r < R_1$ and is zero for $r > R_2$.

**10** •• SSM If the conducting shell in Figure 22-36 is grounded, which of the following statements is then correct?

(a) The charge on the inner surface of the shell is $+Q$ and that on the outer surface is $-Q$.

(b) The charge on the inner surface of the shell is $+Q$ and that on the outer surface is zero.

(c) The charge on both surfaces of the shell is $+Q$.

(d) The charge on both surfaces of the shell is zero.

**11** •• For the configuration described in Problem 10, in which the conducting shell is grounded, the electric field for $r < R_1$ and $r > R_2$ points

(a) away from the center of the shell in both regions.

(b) toward the center of the shell in both regions.

(c) toward the center of the shell for $r < R_1$ and is zero for $r > R_2$.

(d) toward the center of the shell for $r < R_1$ and is zero for $r > R_1$.

**12** •• If the net flux through a closed surface is zero, does it follow that the electric field $E$ is zero everywhere on the surface? Does it follow that the net charge inside the surface is zero?

**13** •• True or false: The electric field is discontinuous at all points at which the charge density is discontinuous.

## Estimation and Approximation

**14** •• SSM Given that the maximum field sustainable in air without electrical discharge is approximately $3 \times 10^6$ N/C, estimate the total charge of a thundercloud. Make any assumptions that seem reasonable.

**15** •• If you rub a rubber balloon against dry hair, the resulting static charge will be enough to make the hair stand on end. Estimate the surface charge density on the balloon and its electric field.

**16** • A disk of radius 2.5 cm carries a uniform surface charge density of 3.6 $\mu C/m^2$. Using reasonable approximations, find the electric field on the axis at distances of (a) 0.01 cm, (b) 0.04 cm, (c) 5 m, and (d) 5 cm.

## Calculating $\vec{E}$ From Coulomb's Law

**17** • [SSM] [ISOLVE✓] A uniform line charge of linear charge density $\lambda = 3.5$ nC/m extends from $x = 0$ to $x = 5$ m. (a) What is the total charge? Find the electric field on the $x$ axis at (b) $x = 6$ m, (c) $x = 9$ m, and (d) $x = 250$ m. (e) Find the field at $x = 250$ m, using the approximation that the charge is a point charge at the origin, and compare your result with that for the exact calculation in Part (d).

**18** • Two infinite vertical planes of charge are parallel to each other and are separated by a distance $d = 4$ m. Find the electric field to the left of the planes, to the right of the planes, and between the planes (a) when each plane has a uniform surface charge density $\sigma = +3$ $\mu C/m^2$ and (b) when the left plane has a uniform surface charge density $\sigma = +3$ $\mu C/m^2$ and that of the right plane is $\sigma = -3$ $\mu C/m^2$. Draw the electric field lines for each case.

**19** • [ISOLVE✓] A 2.75-$\mu C$ charge is uniformly distributed on a ring of radius 8.5 cm. Find the electric field on the axis at (a) 1.2 cm, (b) 3.6 cm, and (c) 4.0 m from the center of the ring. (d) Find the field at 4.0 m using the approximation that the ring is a point charge at the origin, and compare your results with that for Part (c).

**20** • For the disk charge of Problem 16, calculate exactly the electric field on the axis at distances of (a) 0.04 cm and (b) 5 m, and compare your results with those for Parts (b) and (c) of Problem 16.

**21** • A uniform line charge extends from $x = -2.5$ cm to $x = +2.5$ cm and has a linear charge density of $\lambda = 6.0$ nC/m. (a) Find the total charge. Find the electric field on the $y$ axis at (b) $y = 4$ cm, (c) $y = 12$ cm, and (d) $y = 4.5$ m. (e) Find the field at $y = 4.5$ m, assuming the charge to be a point charge, and compare your result with that for Part (d).

**22** • [ISOLVE✓] A disk of radius $a$ lies in the $yz$ plane with its axis along the $x$ axis and carries a uniform surface charge density $\sigma$. Find the value of $x$ for which $E_x = \frac{1}{2}\sigma/2\epsilon_0$.

**23** • A ring of radius $a$ with its center at the origin and its axis along the $x$ axis carries a total charge $Q$. Find $E_x$ at (a) $x = 0.2a$, (b) $x = 0.5a$, (c) $x = 0.7a$, (d) $x = a$, and (e) $x = 2a$. (f) Use your results to plot $E_x$ versus $x$ for both positive and negative values of $x$.

**24** • Repeat Problem 23 for a disk of uniform surface charge density $\sigma$.

**25** •• [SSM] (a) Using a ==spreadsheet== program or graphing calculator, make a graph of the electric field on the axis of a disk of radius r = 30 cm carrying a surface charge density $\sigma = 0.5$ nC/m$^2$. (b) Compare the field to the approximation $E = 2\pi k\sigma$. At what distance does the approximation differ from the exact solution by 10 percent?

**26** •• Show that $E_x$ on the axis of a ring charge of radius $a$ has its maximum and minimum values at $x = +a/\sqrt{2}$ and $x = -a/\sqrt{2}$. Sketch $E_x$ versus $x$ for both positive and negative values of $x$.

**27** •• A line charge of uniform linear charge density $\lambda$ lies along the $x$ axis from $x = x_1$ to $x = x_2$ where $x_1 < x_2$. Show the $x$ component of the electric field at a point on the $y$ axis is given by

$$E_x = \frac{k\lambda}{y}(\cos\theta_2 - \cos\theta_1)$$

where $\theta_1 = \tan^{-1}(x_1/y)$ and $\theta_2 = \tan^{-1}(x_2/y)$.

**28** •• A ring of radius $R$ has a charge distribution on it that goes as $\lambda(\theta) = \lambda_0 \sin\theta$, as shown in the figure below. (a) In what direction does the field at the center of the ring point? (b) What is the magnitude of the field at the center of the ring?

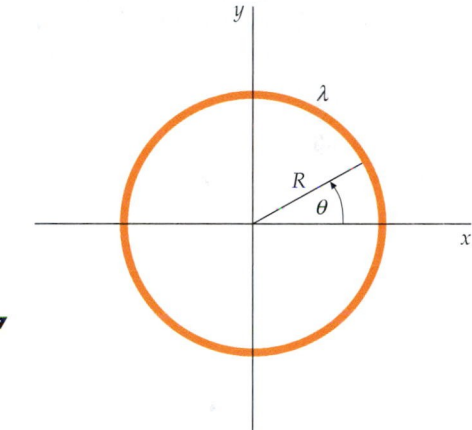

**FIGURE 22-37**
Problem 28

**29** •• A finite line charge of uniform linear charge density $\lambda$ lies on the $x$ axis from $x = 0$ to $x = a$. Show that the $y$ component of the electric field at a point on the $y$ axis is given by

$$E_y = \frac{k\lambda}{y}\frac{a}{\sqrt{y^2 + a^2}}$$

**30** ••• [SSM] A hemispherical thin shell of radius $R$ carries a uniform surface charge $\sigma$. Find the electric field at the center of the hemispherical shell ($r = 0$).

## Gauss's Law

**31** • Consider a uniform electric field $\vec{E} = 2$ kN/C$\hat{i}$. (a) What is the flux of this field through a square of side 10 cm in a plane parallel to the $yz$ plane? (b) What is the flux through the same square if the normal to its plane makes a 30° angle with the $x$ axis?

**32** • [SSM] A single point charge $q = +2$ $\mu C$ is at the origin. A spherical surface of radius 3.0 m has its center on the $x$ axis at $x = 5$ m. (a) Sketch electric field lines for the point charge. Do any lines enter the spherical surface? (b) What is the net number of lines that cross the spherical surface, counting those that enter as negative? (c) What is the net flux of the electric field due to the point charge through the spherical surface?

**33** • An electric field is $\vec{E} = 300$ N/C$\hat{i}$ for $x > 0$ and $\vec{E} = -300$ N/C$\hat{i}$ for $x < 0$. A cylinder of length 20 cm and radius 4 cm has its center at the origin and its axis along the $x$ axis such that one end is at $x = +10$ cm and the other is at $x = -10$ cm. (*a*) What is the flux through each end? (*b*) What is the flux through the curved surface of the cylinder? (*c*) What is the net outward flux through the entire cylindrical surface? (*d*) What is the net charge inside the cylinder?

**34** • Careful measurement of the electric field at the surface of a black box indicates that the net outward flux through the surface of the box is 6.0 kN·m²/C. (*a*) What is the net charge inside the box? (*b*) If the net outward flux through the surface of the box were zero, could you conclude that there were no charges inside the box? Why or why not?

**35** • A point charge $q = +2$ μC is at the center of a sphere of radius 0.5 m. (*a*) Find the surface area of the sphere. (*b*) Find the magnitude of the electric field at points on the surface of the sphere. (*c*) What is the flux of the electric field due to the point charge through the surface of the sphere? (*d*) Would your answer to Part (*c*) change if the point charge were moved so that it was inside the sphere but not at its center? (*e*) What is the net flux through a cube of side 1 m that encloses the sphere?

**36** • **SSM** Since Newton's law of gravity and Coulomb's law have the same inverse-square dependence on distance, an expression analogous in form to Gauss's law can be found for gravity. The gravitational field $\vec{g}$ is the force per unit mass on a test mass $m_0$. Then, for a point mass $m$ at the origin, the gravitational field $g$ at some position $r$ is

$$\vec{g} = -\frac{Gm}{r^2}\hat{r}$$

Compute the flux of the gravitational field through a spherical surface of radius $R$ centered at the origin, and show that the gravitational analog of Gauss's law is $\phi_{net} = -4\pi Gm_{inside}$.

**37** •• **ISOLVE** A charge of 2 μC is 20 cm above the center of a square of side length 40 cm. Find the flux through the square. (*Hint: Don't integrate.*)

**38** •• **ISOLVE**✓ In a particular region of the earth's atmosphere, the electric field above the earth's surface has been measured to be 150 N/C downward at an altitude of 250 m and 170 N/C downward at an altitude of 400 m. Calculate the volume charge density of the atmosphere assuming it to be uniform between 250 and 400 m. (You may neglect the curvature of the earth. Why?)

## Spherical Symmetry

**39** • A spherical shell of radius $R_1$ carries a total charge $q_1$ that is uniformly distributed on its surface. A second, larger spherical shell of radius $R_2$ that is concentric with the first carries a charge $q_2$ that is uniformly distributed on its surface. (*a*) Use Gauss's law to find the electric field in the regions $r < R_1$, $R_1 < r < R_2$, and $r > R_2$. (*b*) What should the ratio of the charges $q_1/q_2$ and their relative signs be for the electric field to be zero for $r > R_2$? (*c*) Sketch the electric field lines for the situation in Part (*b*) when $q_1$ is positive.

**40** • **ISOLVE**✓ A spherical shell of radius 6 cm carries a uniform surface charge density $\sigma = 9$ nC/m². (*a*) What is the total charge on the shell? Find the electric field at (*b*) $r = 2$ cm, (*c*) $r = 5.9$ cm, (*d*) $r = 6.1$ cm, and (*e*) $r = 10$ cm.

**41** •• A sphere of radius 6 cm carries a uniform volume charge density $\rho = 450$ nC/m³. (*a*) What is the total charge of the sphere? Find the electric field at (*b*) $r = 2$ cm, (*c*) $r = 5.9$ cm, (*d*) $r = 6.1$ cm, and (*e*) $r = 10$ cm. Compare your answers with Problem 40.

**42** •• **SSM** Consider two concentric conducting spheres (Figure 22-38). The outer sphere is hollow and initially has a charge $-7Q$ deposited on it. The inner sphere is solid and has a charge $+2Q$ on it. (*a*) How is the charge distributed on the outer sphere? That is, how much charge is on the outer surface and how much charge is on the inner surface? (*b*) Suppose a wire is connected between the inner and outer spheres. After electrostatic equilibrium is established, how much total charge is on the outside sphere? How much charge is on the outer surface of the outside sphere, and how much charge is on the inner surface? Does the electric field at the surface of the inside sphere change when the wire is connected? If so, how? (*c*) Suppose we return to the original conditions in Part (*a*), with $+2Q$ on the inner sphere and $-7Q$ on the outer. We now connect the outer sphere to ground with a wire and then disconnect it. How much total charge will be on the outer sphere? How much charge will be on the inner surface of the outer sphere and how much will be on the outer surface?

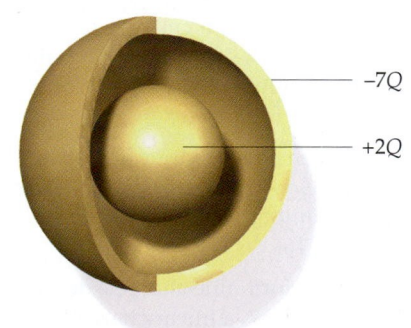

**FIGURE 22-38**
**Problem 42**

$-7Q$

$+2Q$

**43** •• **ISOLVE**✓ A nonconducting sphere of radius $R = 0.1$ m carries a uniform volume charge of charge density $\rho = 2.0$ nC/m³. The magnitude of the electric field at $r = 2R$ is 1883 N/C. Find the magnitude of the electric field at $r = 0.5R$.

**44** •• A nonconducting sphere of radius $R$ carries a volume charge density that is proportional to the distance from the center: $\rho = Ar$ for $r \le R$, where $A$ is a constant; $\rho = 0$ for $r > R$. (*a*) Find the total charge on the sphere by summing the charges on shells of thickness $dr$ and volume $4\pi r^2 \, dr$. (*b*) Find the electric field $E_r$ both inside and outside the charge distribution, and sketch $E_r$ versus $r$.

**45** •• Repeat Problem 44 for a sphere with volume charge density $\rho = B/r$ for $r < R$; $\rho = 0$ for $r > R$.

**46** •• **SSM** Repeat Problem 44 for a sphere with volume charge density $\rho = C/r^2$ for $r < R$; $\rho = 0$ for $r > R$.

**47** ••• A thick, nonconducting spherical shell of inner radius $a$ and outer radius $b$ has a uniform volume charge density $\rho$. Find (*a*) the total charge and (*b*) the electric field everywhere.

## Cylindrical Symmetry

**48** •• Show that the electric field due to an infinitely long, uniformly charged cylindrical shell of radius $R$ carrying a surface charge density $\sigma$ is given by

$$E_r = 0, \qquad r < R$$

$$E_r = \frac{\sigma R}{\epsilon_0 r} = \frac{\lambda}{2\pi\epsilon_0 r} \qquad r > R$$

where $\lambda = 2\pi R\sigma$ is the charge per unit length on the shell.

**49** •• **ISOLVE** A cylindrical shell of length 200 m and radius 6 cm carries a uniform surface charge density of $\sigma = 9 \text{ nC/m}^2$. (a) What is the total charge on the shell? Find the electric field at (b) $r = 2$ cm, (c) $r = 5.9$ cm, (d) $r = 6.1$ cm, and (e) $r = 10$ cm. (Use the results of Problem 48.)

**50** •• An infinitely long nonconducting cylinder of radius $R$ carries a uniform volume charge density of $\rho(r) = \rho_0$. Show that the electric field is given by

$$E_r = \frac{\rho R^2}{2\epsilon_0 r} = \frac{1}{2\pi\epsilon_0} \frac{\lambda}{r} \qquad r > R$$

$$E_r = \frac{\rho}{2\epsilon_0} r = \frac{\lambda}{2\pi\epsilon_0 R^2} r \qquad r < R$$

where $\lambda = \rho\pi R^2$ is the charge per unit length.

**51** •• A cylinder of length 200 m and radius 6 cm carries a uniform volume charge density of $\rho = 300 \text{ nC/m}^3$. (a) What is the total charge of the cylinder? Use the formulas given in Problem 50 to calculate the electric field at a point equidistant from the ends at (b) $r = 2$ cm, (c) $r = 5.9$ cm, (d) $r = 6.1$ cm, and (e) $r = 10$ cm. Compare your results with those in Problem 49.

**52** •• **SSM** Consider two infinitely long, concentric cylindrical shells. The inner shell has a radius $R_1$ and carries a uniform surface charge density of $\sigma_1$, and the outer shell has a radius $R_2$ and carries a uniform surface charge density of $\sigma_2$. (a) Use Gauss's law to find the electric field in the regions $r < R_1$, $R_1 < r < R_2$, and $r > R_2$. (b) What is the ratio of the surface charge densities $\sigma_2/\sigma_1$ and their relative signs if the electric field is zero at $r > R_2$? What would the electric field between the shells be in this case? (c) Sketch the electric field lines for the situation in Part (b) if $\sigma_1$ is positive.

**53** •• **ISOLVE** Figure 22-39 shows a portion of an infinitely long, concentric cable in cross section. The inner conductor carries a charge of 6 nC/m; the outer conductor is uncharged. (a) Find the electric field for all values of $r$, where $r$ is the distance from the axis of the cylindrical system. (b) What are the surface charge densities on the inside and the outside surfaces of the outer conductor?

**54** •• An infinitely long nonconducting cylinder of radius $R$ and carrying a nonuniform volume charge density of $\rho(r) = ar$. (a) Show that the charge per unit length of the cylinder is $\lambda = 2\pi a R^3/3$. (b) Find the expressions for the electric field due to this charged cylinder. You should find one expression for the electric field in the region $r < R$ and a second expression for the field in the region $r > R$, as in Problem 50.

**55** •• Repeat Problem 54 for a nonuniform volume charge density of $\rho = br^2$. In Part (a) show $\lambda = \pi b R^4/2$ (instead of the expression given for $\lambda$ in Problem 54).

**56** ••• An infinitely long, thick, nonconducting cylindrical shell of inner radius $a$ and outer radius $b$ has a uniform volume charge density $\rho$. Find the electric field everywhere.

**57** ••• Suppose that the inner cylinder of Figure 22-39 is made of nonconducting material and carries a volume charge distribution given by $\rho(r) = C/r$, where $C = 200 \text{ nC/m}^2$. The outer cylinder is metallic. (a) Find the charge per meter carried by the inner cylinder. (b) Calculate the electric field for all values of $r$.

## Charge and Field at Conductor Surfaces

**58** • **SSM** **ISOLVE** ✓ A penny is in an external electric field of magnitude 1.6 kN/C directed perpendicular to its faces. (a) Find the charge density on each face of the penny, assuming the faces are planes. (b) If the radius of the penny is 1 cm, find the total charge on one face.

**59** • **ISOLVE** ✓ An uncharged metal slab has square faces with 12-cm sides. It is placed in an external electric field that is perpendicular to its faces. The total charge induced on one of the faces is 1.2 nC. What is the magnitude of the electric field?

**60** • **ISOLVE** A charge of 6 nC is placed uniformly on a square sheet of nonconducting material of side 20 cm in the $yz$ plane. (a) What is the surface charge density $\sigma$? (b) What is the magnitude of the electric field just to the right and just to the left of the sheet? (c) The same charge is placed on a square conducting slab of side 20 cm and thickness 1 mm. What is the surface charge density $\sigma$? (Assume that the charge distributes itself uniformly on the large square surfaces.) (d) What is the magnitude of the electric field just to the right and just to the left of each face of the slab?

**61** • A spherical conducting shell with zero net charge has an inner radius $a$ and an outer radius $b$. A point charge $q$ is placed at the center of the shell. (a) Use Gauss's law and the properties of conductors in equilibrium to find the electric field in the regions $r < a$, $a < r < b$, and $b < r$. (b) Draw the electric field lines for this situation. (c) Find the charge density on the inner surface ($r = a$) and on the outer surface ($r = b$) of the shell.

**62** •• **ISOLVE** The electric field just above the surface of the earth has been measured to be 150 N/C downward. What total charge on the earth is implied by this measurement?

**63** •• **SSM** A positive point charge of magnitude 2.5 $\mu$C is at the center of an uncharged spherical conducting shell of inner radius 60 cm and outer radius 90 cm. (a) Find the charge densities on the inner and outer surfaces of the shell and the total charge on each surface. (b) Find the electric field everywhere. (c) Repeat Part (a) and Part (b) with a net charge of $+3.5 \mu$C placed on the shell.

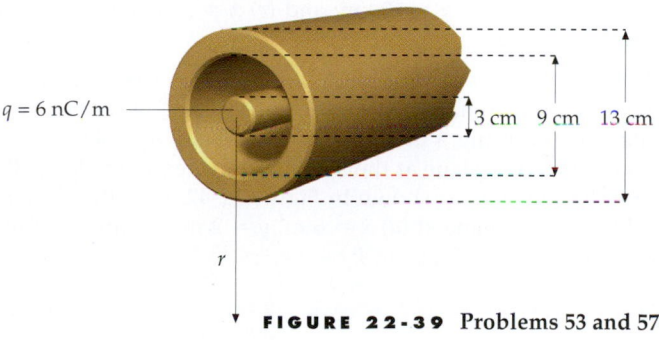

$q = 6 \text{ nC/m}$

3 cm  9 cm  13 cm

$r$

**FIGURE 22-39** Problems 53 and 57

**64 ••** 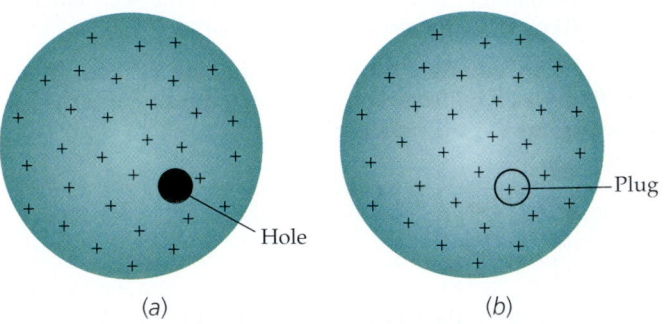✓ If the magnitude of an electric field in air is as great as $3 \times 10^6$ N/C, the air becomes ionized and begins to conduct electricity. This phenomenon is called dielectric breakdown. A charge of 18 $\mu$C is to be placed on a conducting sphere. What is the minimum radius of a sphere that can hold this charge without breakdown?

**65 ••** A square conducting slab with 5-m sides carries a net charge of 80 $\mu$C. (a) Find the charge density on each face of the slab and the electric field just outside one face of the slab. (b) The slab is placed to the right of an infinite charged nonconducting plane with charge density 2.0 $\mu$C/m² so that the faces of the slab are parallel to the plane. Find the electric field on each side of the slab far from its edges and the charge density on each face.

## General Problems

**66 ••** Consider the three concentric metal spheres shown in Figure 22-40. Sphere one is solid, with radius $R_1$. Sphere two is hollow, with inner radius $R_2$ and outer radius $R_3$. Sphere three is hollow, with inner radius $R_4$ and outer radius $R_5$. Initially, all three spheres have zero excess charge. Then a negative charge $-Q_0$ is placed on sphere one and a positive charge $+Q_0$ is placed on sphere three. (a) After the charges have reached equilibrium, will the electric field in the space between spheres one and two point *toward* the center, *away* from the center, or neither? (b) How much charge will be on the inner surface of sphere two? Give the correct sign. (c) How much charge will be on the outer surface of sphere two? (d) How much charge will be on the inner surface of sphere three? (e) How much charge will be on the outer surface of sphere three? (f) Plot E versus r.

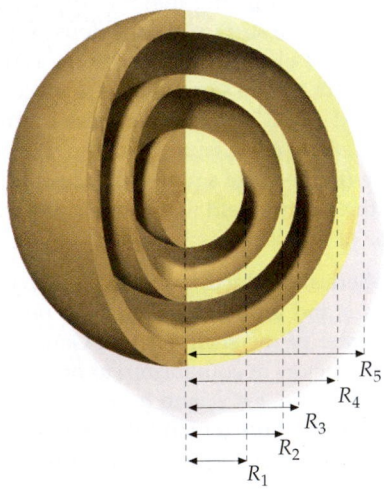

**FIGURE 22-40** Problem 66

**67 ••** ISOLVE A nonuniform surface charge lies in the yz plane. At the origin, the surface charge density is $\sigma = 3.10$ $\mu$C/m². Other charged objects are present as well. Just to the right of the origin, the x component of the electric field is $E_x = 4.65 \times 10^5$ N/C. What is $E_x$ just to the left of the origin?

**68 ••** An infinite line charge of uniform linear charge density $\lambda = -1.5$ $\mu$C/m lies parallel to the y axis at $x = -2$ m. A point charge of 1.3 $\mu$C is located at $x = 1$ m, $y = 2$ m. Find the electric field at $x = 2$ m, $y = 1.5$ m.

**69 ••** SSM A thin nonconducting uniformly charged spherical shell of radius $r$ (Figure 22-41a) has a total charge of $Q$. A small circular plug is removed from the surface. (a) What is the magnitude and direction of the electric field at the center of the hole? (b) The plug is put back in the hole (Figure 22-41b). Using the result of part $a$, calculate the force acting on the plug. (c) From this, calculate the "electrostatic pressure" (force/unit area) tending to expand the sphere.

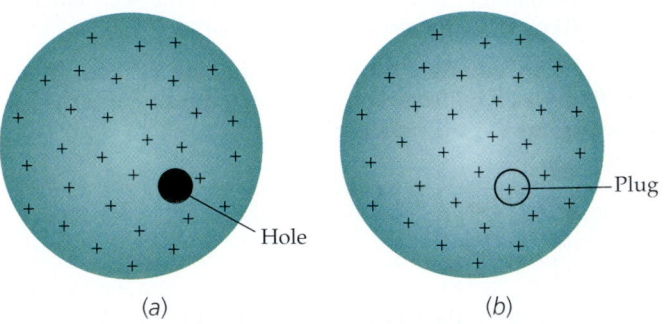

(a)                                           (b)

**FIGURE 22-41** Problem 69

**70 ••** A soap bubble of radius $R_1 = 10$ cm has a charge of 3 nC uniformly spread over it. Because of electrostatic repulsion, the soap bubble expands until it bursts at a radius $R_2 = 20$ cm. From the results of Problem 69, calculate the work done by the electrostatic force in expanding the soap bubble.

**71 ••** If the soap bubble of Problem 70 collapses into a spherical water droplet, estimate the electric field at its surface.

**72 ••** Two infinite planes of charge lie parallel to each other and to the yz plane. One is at $x = -2$ m and has a surface charge density of $\sigma = -3.5$ $\mu$C/m². The other is at $x = 2$ m and has a surface charge density of $\sigma = 6.0$ $\mu$C/m². Find the electric field for (a) $x < -2$ m, (b) $-2$ m $< x < 2$ m, and (c) $x > 2$ m.

**73 ••** SSM An infinitely long cylindrical shell is coaxial with the y axis and has a radius of 15 cm. It carries a uniform surface charge density $\sigma = 6$ $\mu$C/m². A spherical shell of radius 25 cm is centered on the x axis at $x = 50$ cm and carries a uniform surface charge density $\sigma = -12$ $\mu$C/m². Calculate the magnitude and direction of the electric field at (a) the origin; (b) $x = 20$ cm, $y = 10$ cm; and (c) $x = 50$ cm, $y = 20$ cm. (See Problem 48.)

**74 ••** ISOLVE An infinite plane in the xz plane carries a uniform surface charge density $\sigma_1 = 65$ nC/m². A second infinite plane carrying a uniform charge density $\sigma_2 = 45$ nC/m² intersects the xz plane at the z axis and makes an angle of 30° with the xz plane, as shown in Figure 22-42. Find the electric field in the xy plane at (a) $x = 6$ m, $y = 2$ m and (b) $x = 6$ m, $y = 5$ m.

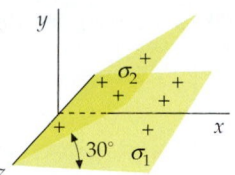

**FIGURE 22-42** Problem 74

**75** •• A quantum-mechanical treatment of the hydrogen atom shows that the electron in the atom can be treated as a smeared-out distribution of charge, which has the form: $\rho(r) = \rho_0 e^{-2r/a}$, where $r$ is the distance from the nucleus, and $a$ is the Bohr radius ($a = 0.0529$ nm). (a) Calculate $\rho_0$, from the fact that the atom is uncharged. (b) Calculate the electric field at any distance $r$ from the nucleus. Treat the proton as a point charge.

**76** •• **SSM** Using the results of Problem 75, if we placed a proton above the nucleus of a hydrogen atom, at what distance $r$ would the electric force on the proton balance the gravitational force $mg$ acting on it? From this result, explain why even though the electrostatic force is enormously stronger than the gravitational force, it is the gravitational force we notice more.

**77** •• A ring of radius $R$ carries a uniform, positive, linear charge density $\lambda$. Figure 22-43 shows a point $P$ in the plane of the ring but not at the center. Consider the two elements of the ring of lengths $s_1$ and $s_2$ shown in the figure at distances $r_1$ and $r_2$, respectively, from point $P$. (a) What is the ratio of the charges of these elements? Which produces the greater field at point $P$? (b) What is the direction of the field at point $P$ due to each element? What is the direction of the total electric field at point $P$? (c) Suppose that the electric field due to a point charge varied as $1/r$ rather than $1/r^2$. What would the electric field be at point $P$ due to the elements shown? (d) How would your answers to Parts (a), (b), and (c) differ if point $P$ were inside a spherical shell of uniform charge and the elements were of areas $s_1$ and $s_2$?

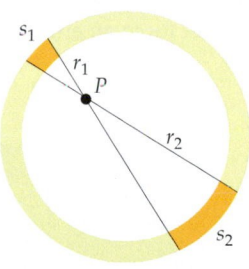

**FIGURE 22-43** Problem 77

**78** •• A uniformly charged ring of radius $R$ that lies in a horizontal plane carries a charge $Q$. A particle of mass $m$ carries a charge $q$, whose sign is opposite that of $Q$, is on the axis of the ring. (a) What is the minimum value of $|q|/m$ such that the particle will be in equilibrium under the action of gravity and the electrostatic force? (b) If $|q|/m$ is twice that calculated in Part (a), where will the particle be when it is in equilibrium?

**79** •• A long, thin, nonconducting plastic rod is bent into a loop with radius $R$. Between the ends of the rod, a small gap of length $l$ ($l \ll R$) remains. A charge $Q$ is equally distributed on the rod. (a) Indicate the direction of the electric field at the center of the loop. (b) Find the magnitude of the electric field at the center of the loop.

**80** •• A nonconducting sphere 1.2 m in diameter with its center on the $x$ axis at $x = 4$ m carries a uniform volume charge of density $\rho = 5\ \mu C/m^3$. Surrounding the sphere is a spherical shell with a diameter of 2.4 m and a uniform surface charge density $\sigma = -1.5\ \mu C/m^2$. Calculate the magnitude and direction of the electric field at (a) $x = 4.5$ m, $y = 0$; (b) $x = 4.0$ m, $y = 1.1$ m; and (c) $x = 2.0$ m, $y = 3.0$ m.

**81** •• An infinite plane of charge with surface charge density $\sigma_1 = 3\ \mu C/m^2$ is parallel to the $xz$ plane at $y = -0.6$ m. A second infinite plane of charge with surface charge density $\sigma_2 = -2\ \mu C/m^2$ is parallel to the $yz$ plane at $x = 1$ m. A sphere of radius 1 m with its center in the $xy$ plane at the intersection of the two charged planes ($x = 1$ m, $y = -0.6$ m) has a surface charge density $\sigma_3 = -3\ \mu C/m^2$. Find the magnitude and direction of the electric field on the $x$ axis at (a) $x = 0.4$ m and (b) $x = 2.5$ m.

**82** •• An infinite plane lies parallel to the $yz$ plane at $x = 2$ m and carries a uniform surface charge density $\sigma = 2\ \mu C/m^2$. An infinite line charge of uniform linear charge density $\lambda = 4\ \mu C/m$ passes through the origin at an angle of $45°$ with the $x$ axis in the $xy$ plane. A sphere of volume charge density $\rho = -6\ \mu C/m^3$ and radius 0.8 m is centered on the $x$ axis at $x = 1$ m. Calculate the magnitude and direction of the electric field in the $xy$ plane at $x = 1.5$ m, $y = 0.5$ m.

**83** •• **iSOLVE✓** An infinite line charge $\lambda$ is located along the $z$ axis. A particle of mass $m$ that carries a charge $q$ whose sign is opposite to that of $\lambda$ is in a circular orbit in the $xy$ plane about the line charge. Obtain an expression for the period of the orbit in terms of $m$, $q$, $R$, and $\lambda$, where $R$ is the radius of the orbit.

**84** •• **SSM** A ring of radius $R$ that lies in the $yz$ plane carries a positive charge $Q$ uniformly distributed over its length. A particle of mass $m$ that carries a negative charge of magnitude $q$ is at the center of the ring. (a) Show that if $x \ll R$, the electric field along the axis of the ring is proportional to $x$. (b) Find the force on the particle of mass $m$ as a function of $x$. (c) Show that if $m$ is given a small displacement in the $x$ direction, it will perform simple harmonic motion. Calculate the period of that motion.

**85** •• **iSOLVE** When the charges $Q$ and $q$ of Problem 84 are 5 $\mu C$ and $-5\ \mu C$, respectively, and the radius of the ring is 8.0 cm, the mass $m$ oscillates about its equilibrium position with an angular frequency of 21 rad/s. Find the angular frequency of oscillation of the mass if the radius of the ring is doubled to 16 cm and all other parameters remain unchanged.

**86** •• **iSOLVE** Given the initial conditions of Problem 85, find the angular frequency of oscillation of the mass if the radius of the ring is doubled to 16 cm while keeping the linear charge density on the ring constant.

**87** •• A uniformly charged nonconducting sphere of radius $a$ with center at the origin has volume charge density $\rho$. (a) Show that at a point within the sphere a distance $r$ from the center $\vec{E} = \dfrac{\rho}{3\,\epsilon_0} r\hat{r}$. (b) Material is removed from the sphere leaving a spherical cavity of radius $b = a/2$ with its center at $x = b$ on the $x$ axis (Figure 22-44). Calculate the electric field at points 1 and 2 shown in Figure 22-44. (*Hint: Replace the sphere-with-cavity with two uniform spheres of equal positive and negative charge densities.*)

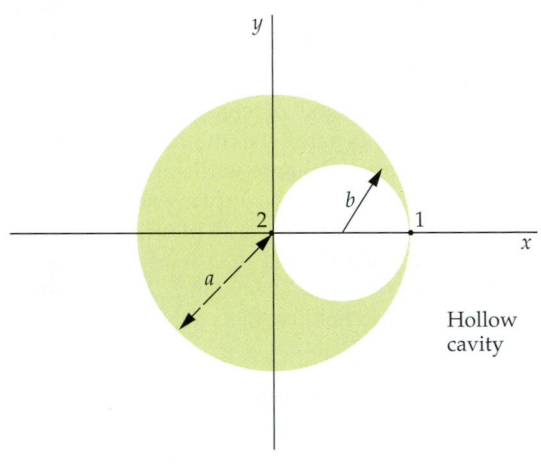

**FIGURE 22-44** Problem 87

**88** ••• Show that the electric field throughout the cavity of Problem 87 is uniform and is given by

$$\vec{E} = \frac{\rho}{3\,\epsilon_0}b\,\hat{i}$$

**89** •• Repeat Problem 87 assuming that the cavity is filled with a uniformly charged material wth a total charge of $Q$.

**90** •• A nonconducting cylinder of radius 1.2 m and length 2.0 m carries a charge of 50 $\mu$C uniformly distributed throughout the cylinder. Find the electric field *on the cylinder axis* at a distance of (a) 0.5 m, (b) 2.0 m, and (c) 20 m from the center of the cylinder.

**91** •• **ISOLVE** A uniform line charge of density $\lambda$ lies on the $x$ axis between $x = 0$ and $x = L$. Its total charge is $Q = 8$ nC. The electric field at $x = 2L$ is 600 N/C$\hat{i}$. Find the electric field at $x = 3L$.

**92** ••• A *small* gaussian surface in the shape of a cube with faces parallel to the $xy$, $xz$, and $yz$ planes (Figure 22-45) is in a region in which the electric field remains parallel with the $x$ axis. Using the Taylor series (and neglecting terms higher than first order), show that the net flux of the electric field out of the gaussian surface is given by

$$\phi_{\text{net}} = \frac{\partial E_x}{\partial x}\,\Delta V$$

where $\Delta V$ is the volume enclosed by the gaussian surface.

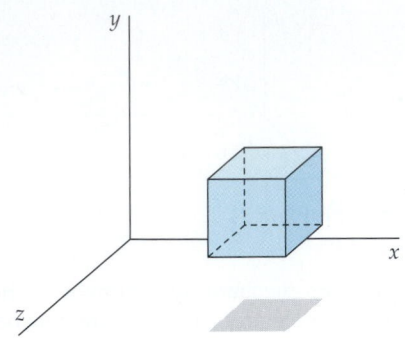

**FIGURE 22-45** Problem 92

*Remark:* The corresponding result for situations for which the direction of the electric field is not restricted to one dimension is

$$\phi_{\text{net}} = \left(\frac{\partial E_x}{\partial x} + \frac{\partial E_y}{\partial y} + \frac{\partial E_z}{\partial z}\right)\Delta V$$

where the combination of derivatives in the parentheses is commonly written $\vec{\nabla} \cdot \vec{E}$ and is called the *divergence* of $\vec{E}$.

**93** •• Using Gauss's law and the results of Problem 92 show that

$$\vec{\nabla} \cdot \vec{E} = \frac{\rho}{\epsilon_0}$$

where $\rho$ is the volume charge density. (This equation is known as the point form of Gauss's law.)

**94** ••• **SSM** A dipole $\vec{p}$ is located at a distance $r$ from an infinitely long line charge with a uniform linear charge density $\lambda$. Assume that the dipole is aligned with the field due to the line charge. Determine the force that acts on the dipole.

**95** •• Consider a simple but surprisingly accurate model for the Hydrogen molecule: two positive point charges, each with charge $+e$, are placed inside a sphere of radius $R$, which has uniform charge density $-2e$. The two point charges are placed symmetrically (Figure 22-46). Find the distance from the center, $a$, where the net force on either charge is 0.

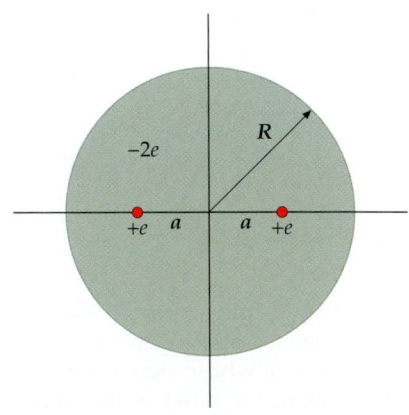

**FIGURE 22-46**
**Problem 95**

# Electric Potential

THIS GIRL HAS BEEN RAISED TO A HIGH POTENTIAL THROUGH CONTACT WITH THE DOME OF A VAN DE GRAAFF GENERATOR. SHE IS STANDING ON A PLATFORM THAT ELECTRICALLY INSULATES HER FROM THE FLOOR, SO SHE ACCUMULATES CHARGE FROM THE VAN DE GRAAFF. HER HAIR STANDS UP BECAUSE THE CHARGES ON HER HEAD AND THE CHARGES ON HER HAIR STRANDS HAVE THE SAME SIGN, AND LIKE CHARGES REPEL EACH OTHER.

 **Did you know that the maximum potential that the dome of a Van de Graaff generator can be raised to is determined by the radius of the dome? For a discussion of this, see Example 23-14.**

23-1 Potential Difference

23-2 Potential Due to a System of Point Charges

23-3 Computing the Electric Field From the Potential

23-4 Calculations of $V$ for Continuous Charge Distributions

23-5 Equipotential Surfaces

T he electric force between two charges is directed along the line joining the charges and varies inversely with the square of their separation, the same dependence as the gravitational force between two masses. Like the gravitational force, the electric force is conservative, so there is a potential energy function $U$ associated with it. If we place a test charge $q_0$ in an electric field, its potential energy is proportional to $q_0$. The potential energy per unit charge is a function of the position in space of the charge and is called the electric potential. As it is a scalar field, it is easier to manipulate than the electric field in many circumstances. ➤ **In this chapter, we will establish the relationship between the electric field and electric potential and calculate the electric potential of various continuous charge distributions. Then, we can use the electric potential to determine the electric field of these regions.**

## 23-1 Potential Difference

In general, when the point of application of a conservative force $\vec{F}$ undergoes a displacement $d\vec{\ell}$, the change in the potential energy function $dU$ is given by

$$dU = -\vec{F} \cdot d\vec{\ell}$$

The force exerted by an electric field $\vec{E}$ on a point charge $q_0$ is

$$\vec{F} = q_0 \vec{E}$$

Thus, when a charge undergoes a displacement $d\vec{\ell}$, the change in the electrostatic potential energy is

$$dU = -q_0 \vec{E} \cdot d\vec{\ell} \qquad 23\text{-}1$$

The potential energy change is proportional to the charge $q_0$. The potential energy change *per unit charge* is called the **potential difference** $dV$:

$$dV = \frac{dU}{q_0} = -\vec{E} \cdot d\vec{\ell} \qquad 23\text{-}2a$$

DEFINITION—POTENTIAL DIFFERENCE

For a finite displacement from point $a$ to point $b$, the change in potential is

$$\Delta V = V_b - V_a = \frac{\Delta U}{q_0} = -\int_a^b \vec{E} \cdot d\vec{\ell} \qquad 23\text{-}2b$$

DEFINITION—FINITE POTENTIAL DIFFERENCE

The potential difference $V_b - V_a$ is the negative of the work per unit charge done by the electric field on a small positive test charge when the test charge moves from point $a$ to point $b$. During this calculation, the positions of any and all other charges remain fixed.

The function $V$ is called the **electric potential,** often it is shortened to the **potential.** Like the electric field, the potential $V$ is a function of position. Unlike the electric field, $V$ is a scalar function, whereas $\vec{E}$ is a vector function. As with potential energy $U$, only *differences* in the potential $V$ are important. We are free to choose the potential to be zero at any convenient point, just as we are when dealing with potential energy. For convenience, the electric potential and the potential energy of a test charge are chosen to be zero at the same point. Under these conditions they are related by

$$U = q_0 V \qquad 23\text{-}3$$

RELATION BETWEEN POTENTIAL ENERGY AND POTENTIAL

## Continuity of V

In Chapter 22, we saw that the electric field is discontinuous by $\sigma/\epsilon_0$ at points where there is a surface charge density $\sigma$. The potential function, on the other hand, is continuous everywhere, except at points where the electric field is infinite (points where there is a point charge or a line charge). We can see this from its definition. Consider a region occupied by an electric field $\vec{E}$. The difference in potential between two nearby points separated by displacement $d\vec{\ell}$ is related to the electric field by $dV = -\vec{E} \cdot d\vec{\ell}$ (Equation 23-2a). The dot product can be expressed $\vec{E} \cdot d\vec{\ell} = E_{\parallel} d\ell$, where $E_{\parallel}$ is the component of $\vec{E}$ in the direction of $d\vec{\ell}$ and $d\ell$ is the magnitude of $d\vec{\ell}$. Substituting into Equation 23-2a gives $dV = -E_{\parallel} d\ell$. If $\vec{E}$ is finite at each of the two points and along the line segment of infinitesimal length $d\ell$ joining them, then $dV$ is infinitesimal. Thus, the potential function $V$ is continuous at any point not occupied by a point charge or a line charge.

## Units

Since electric potential is the potential energy per unit charge, the SI unit for potential and potential difference is the joule per coulomb, called the **volt** (V):

$$1\,V = 1\,J/C \qquad\qquad 23\text{-}4$$

The potential difference between two points (measured in volts) is sometimes called the **voltage.** In a 12-V car battery, the positive terminal has a potential 12 V higher than the negative terminal. If we attach an external circuit to the battery and one coulomb of charge is transferred from the positive terminal through the circuit to the negative terminal, the potential energy of the charge decreases by $Q\,\Delta V = (1\,C)(12\,V) = 12\,J$.

We can see from Equation 23-2 that the dimensions of potential are also those of electric field times distance. Thus, the unit of the electric field is equal to one volt per meter:

$$1\,N/C = 1\,V/m \qquad\qquad 23\text{-}5$$

so we may think of the electric field strength as either a force per unit charge or as a rate of change of $V$ with respect to distance. In atomic and nuclear physics, we often have elementary particles with charges of magnitude $e$, such as electrons and protons, moving through potential differences of several to thousands or even millions of volts. Since energy has dimensions of electric charge times electric potential, a unit of energy is the product of the fundamental charge unit $e$ times a volt. This unit is called an **electron volt** (eV). Energies in atomic and molecular physics are typically a few eV, making the electron volt a convenient-sized unit for atomic and molecular processes. The conversion between electron volts and joules is obtained by expressing the electronic charge in coulombs:

$$1\,eV = 1.60 \times 10^{-19}\,C \cdot V = 1.60 \times 10^{-19}\,J \qquad\qquad 23\text{-}6$$

THE ELECTRON VOLT

For example, an electron moving from the negative terminal to the positive terminal of a 12-V car battery loses 12 eV of potential energy.

## Potential and Electric Fields

If we place a positive test charge $q_0$ in an electric field $\vec{E}$ and release it, it accelerates in the direction of $\vec{E}$. As the kinetic energy of the charge increases, its potential energy decreases. The charge therefore accelerates toward a region of lower potential energy, just as a mass accelerates toward a region of lower gravitational potential energy (Figure 23-1). Thus, as illustrated in Figure 23-2,

> The electric field points in the direction in which the potential decreases most rapidly.

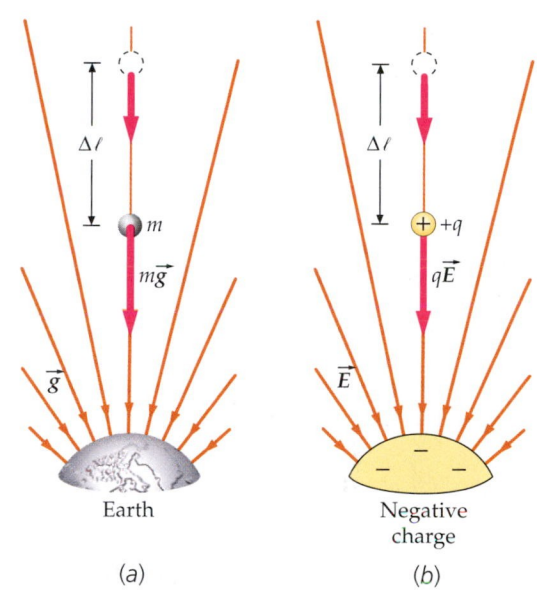

**FIGURE 23-1** (*a*) The work done by the gravitational field $\vec{g}$ on a mass $m$ is equal to the decrease in the gravitational potential energy. (*b*) The work done by the electric field $\vec{E}$ on a charge $q$ is equal to the decrease in the electric potential energy.

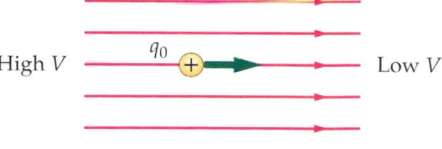

**FIGURE 23-2** The electric field points in the direction in which the potential decreases most rapidly. If a positive test charge $q_0$ is in an electric field, it accelerates in the direction of the field. If it is released from rest, its kinetic energy increases and its potential energy decreases.

FIND V FOR CONSTANT $\vec{E}$

## EXAMPLE 23-1

An electric field points in the positive $x$ direction and has a constant magnitude of $E = 10$ N/C $= 10$ V/m. Find the potential as a function of $x$, assuming that $V = 0$ at $x = 0$.

**PICTURE THE PROBLEM**

1. By definition, the change in potential $dV$ is related to the displacement $d\vec{\ell}$ and the electric field $\vec{E}$:

$$dV = -\vec{E} \cdot d\vec{\ell} = -E\hat{i} \cdot (dx\,\hat{i} + dy\,\hat{j} + dz\,\hat{k}) = -E\,dx$$

2. Integrate $dV$:

$$V(x) = \int dV = \int -E\,dx = -Ex + C$$

3. The constant of integration C is found by setting $V = 0$ at $x = 0$:

$$V(0) = C \quad \Rightarrow \quad 0 = C$$

4. The potential is then:

$$V(x) = -Ex = \boxed{-(10\text{ V/m})x}$$

**REMARKS** The potential is zero at $x = 0$ and decreases by 10 V for every 1-m increase in $x$.

**EXERCISE** Repeat this example for the electric field $\vec{E} = (10\text{ V/m}^2)x\hat{i}$ [*Answer* $V(x) = -(5\text{ V/m}^2)x^2$]

## 23-2 Potential Due to a System of Point Charges

The electric potential at a distance $r$ from a point charge $q$ at the origin can be calculated from the electric field:

$$\vec{E} = \frac{kq}{r^2}\hat{r}$$

For an infinitesimal displacement $d\vec{\ell}$ where we have replaced $r_P$ (the distance to the field point) with $r$ (Figure 23-3), the change in potential is

$$dV = -\vec{E} \cdot d\vec{\ell} = -\frac{kq}{r^2}\hat{r} \cdot d\vec{\ell} = -\frac{kq}{r^2}dr$$

Integrating along a path from an arbitrary reference point to an arbitrary field point gives

$$\int_{r_{\text{ref}}}^{P} dV = -\int_{r_{\text{ref}}}^{P} \vec{E} \cdot d\vec{\ell} = -kq\int_{r_{\text{ref}}}^{r_P} r^{-2}\,dr = -kq\frac{r^{-1}}{-1}\Big|_{r_{\text{ref}}}^{r_P} = \frac{kq}{r_P} - \frac{kq}{r_{\text{ref}}}$$

or

$$V = \frac{kq}{r} - \frac{kq}{r_{\text{ref}}} \qquad\qquad 23\text{-}7$$

POTENTIAL DUE TO A POINT CHARGE

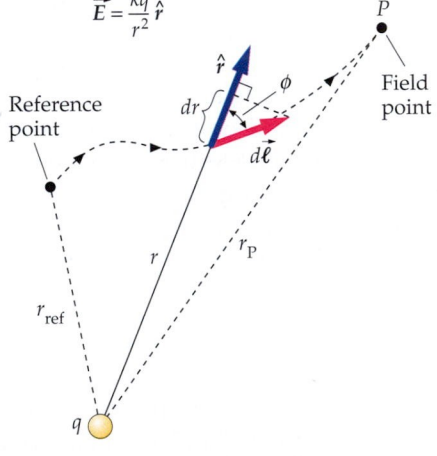

**FIGURE 23-3** The change in $r$ is $dr$. It is the component of $d\vec{\ell}$ in the direction of $\hat{r}$. It can be seen from the figure that $|d\vec{\ell}|\cos\phi = dr$. Since $\hat{r} \cdot d\vec{\ell} = |d\vec{\ell}|\cos\phi$, it follows that $dr = \hat{r} \cdot d\vec{\ell}$.

where we have replaced $r_P$ (the distance to the field point) with $r$. We are free to choose the reference point, so we choose it to give the potential the simplest algebraic form. Choosing the reference point infinitely far from the point charge ($r_{\text{ref}} = \infty$) accomplishes this. Thus,

$$V = \frac{kq}{r} \qquad \qquad 23\text{-}8$$

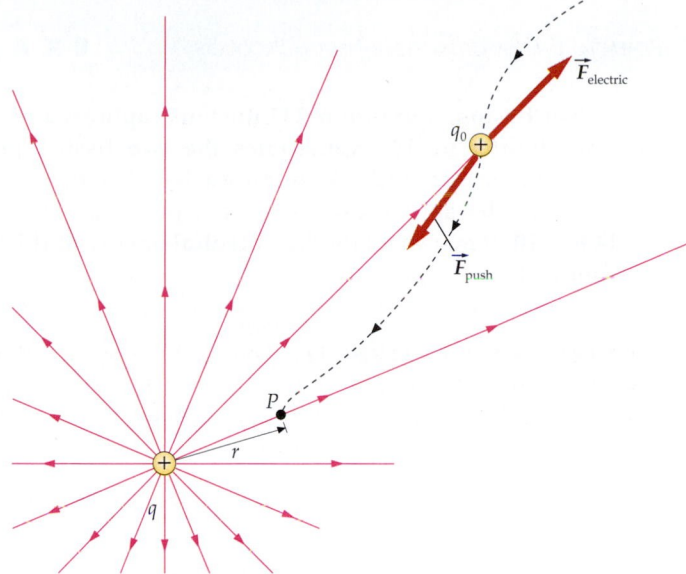

COULOMB POTENTIAL

The potential given by Equation 23-8 is called the **Coulomb potential.** It is positive or negative depending on whether $q$ is positive or negative.

The potential energy $U$ of a test charge $q_0$ placed a distance $r$ from the point charge $q$ is

$$U = q_0 V = \frac{kq_0 q}{r} \qquad \qquad 23\text{-}9$$

ELECTROSTATIC POTENTIAL ENERGY OF A TWO-CHARGE SYSTEM

This is the electric potential energy of the two-charge system relative to $U = 0$ at infinite separation. If we release a test charge $q_0$ from rest at a distance $r$ from $q$ (and hold $q$ fixed), the test charge will be accelerated away from $q$ (assuming that $q$ has the same sign as $q_0$). At a very great distance from $q$, its potential energy will be zero so its kinetic energy will be $kq_0 q/r$. Alternatively, if we move a test charge $q_0$ initially at rest at infinity to rest at a point a distance $r$ from $q$, the work we must do is $kq_0 q/r$ (Figure 23-4). The work per unit charge is $kq/r$, the electric potential at point $P$ relative to zero potential at infinity.

Choosing the electrostatic potential energy of two charges to be zero at an infinite separation is analogous to the choice we made in Chapter 11 when we chose the gravitational potential energy of two point masses to be zero at an infinite separation. If two charges (or two masses) are at infinite separation, we think of them as not interacting. It has a certain appeal that the potential energy is zero if the particles are not interacting.

**FIGURE 23-4** The work required to bring a test charge $q_0$ from infinity to a point $P$ is $kq_0 q/r$, where $r$ is the distance from $P$ to a charge $q$. The work per unit charge is $kq/r$, the electric potential at point $P$ relative to zero potential at infinity. If the test charge is released from point $P$, the electric field does work $kq_0 q/r$ on the charge as the charge moves out to infinity.

---

POTENTIAL ENERGY OF A HYDROGEN ATOM          **E X A M P L E    2 3 - 2**

(*a*) **What is the electric potential at a distance $r = 0.529 \times 10^{-10}$ m from a proton? (This is the average distance between the proton and electron in a hydrogen atom.)** (*b*) **What is the electric potential energy of the electron and the proton at this separation?**

**PICTURE THE PROBLEM**

(*a*) Use $V = kq/r$ to calculate the potential $V$ due to the proton:

$$V = \frac{kq}{r} = \frac{ke}{r} = \frac{(8.99 \times 10^9 \text{ N·m}^2/\text{C}^2)(1.6 \times 10^{-19} \text{ C})}{0.529 \times 10^{-10} \text{ m}}$$

$$= 27.2 \text{ N·m/C} = \boxed{27.2 \text{ V}}$$

(*b*) Use $U = q_0 V$, with $q_0 = -e$ to calculate the potential energy:

$$U = q_0 V = (-e)(27.2 \text{ V}) = \boxed{-27.2 \text{ eV}}$$

**REMARKS** If the electron were at rest at this distance from the proton, it would take a minimum of 27.2 eV to remove it from the atom. However, the electron has kinetic energy equal to 13.6 eV, so its total energy in the atom is 13.6 eV − 27.2 eV = −13.6 eV. The minimum energy needed to remove the electron from the atom is thus 13.6 eV. This energy is called the ionization energy.

**EXERCISE** What is the potential energy of the electron and proton in Example 23-2 in SI units? (*Answer* −4.35 × 10$^{-18}$ J)

**EXAMPLE 23-3**

In nuclear fission, a uranium-235 nucleus captures a neutron and splits apart into two lighter nuclei. Sometimes the two fission products are a barium nucleus (charge $56e$) and a krypton nucleus (charge $36e$). Assume that immediately after the split these nuclei are positive point charges separated by $r = 14.6 \times 10^{-15}$ m. Calculate the potential energy of this two-charge system in electron volts.

**PICTURE THE PROBLEM** The potential energy for two point charges separated by a distance $r$ is $U = kq_1q_2/r$. To find this energy in electron volts, we calculate the potential due to one of the charges $kq_1/r$ in volts and multiply by the other charge.

1. Equation 23-9 gives the potential energy of the two charges:

$$U = \frac{kq_1q_2}{r} = \frac{k(56e)(36e)}{r}$$

2. Factor out $e$ and substitute the given values:

$$U = \frac{k(56e)(36e)}{r} = e\frac{56 \cdot 36ke}{r}$$

$$= e\frac{56 \cdot 36 \cdot (8.99 \times 10^9 \text{ N·m}^2/\text{C}^2)(1.6 \times 10^{-19} \text{ C})}{14.6 \times 10^{-15} \text{ m}}$$

$$= e(1.99 \times 10^8 \text{ V}) = \boxed{199 \text{ MeV}}$$

**REMARKS** The separation distance $r$ was chosen to be the sum of the radii of the two nuclei. After the fission, the two nuclei repel because of their electrostatic repulsion. Their potential energy of 199 MeV is converted into kinetic energy and, upon colliding with surrounding atoms, thermal energy. Two or three neutrons are also released in the fission process. In a chain reaction, one or more of these neutrons produces a fission of another uranium nucleus. The average energy given off in chain reactions of this type is about 200 MeV per nucleus, as calculated in this example.

The potential at some point due to several point charges is the sum of the potentials due to each charge separately. (This follows from the superposition principle for the electric field.) The potential due to a system of point charges $q_i$ is thus given by

$$V = \sum_i \frac{kq_i}{r_i} \qquad \qquad 23\text{-}10$$

*POTENTIAL DUE TO A SYSTEM OF POINT CHARGES*

where the sum is over all the charges, and $r_i$ is the distance from the $i$th charge to the field point at which the potential is to be found.

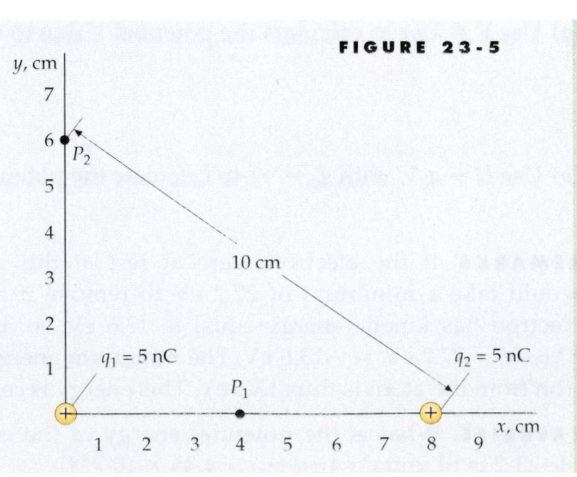

**FIGURE 23-5**

**EXAMPLE 23-4**

Two $+5$ nC point charges are on the $x$-axis, one at the origin and the other at $x = 8$ cm. Find the potential at ($a$) point $P_1$ on the $x$ axis at $x = 4$ cm and ($b$) point $P_2$ on the $y$-axis at $y = 6$ cm.

**PICTURE THE PROBLEM** The two positive point charges on the $x$-axis are shown in Figure 23-5, and the potential is to be found at points $P_1$ and $P_2$.

(a) 1. Use Equation 23-10 to write $V$ as a function of the distances $r_1$ and $r_2$ to the charges:

$$V = \sum_i \frac{kq_i}{r_i} = \frac{kq_1}{r_1} + \frac{kq_2}{r_2}$$

2. Point $P_1$ is 4 cm from each charge, and the charges are equal:

$$r_1 = r_2 = r = 0.04 \text{ m}$$
$$q_1 = q_2 = q = 5 \times 10^{-9} \text{ C}$$

3. Use these to find the potential at point $P_1$:

$$V = \frac{kq_1}{r_1} + \frac{kq_2}{r_2} = \frac{2kq}{r}$$

$$= \frac{2 \times (8.99 \times 10^9 \text{ N·m}^2/\text{C}^2)(5 \times 10^{-9} \text{ C})}{0.04 \text{ m}}$$

$$= 2250 \text{ V} = \boxed{2.25 \text{ kV}}$$

(b) Point $P_2$ is 6 cm from one charge and 10 cm from the other. Use these to find the potential at point $P_2$:

$$V = \frac{(8.99 \times 10^9 \text{ N·m}^2/\text{C}^2)(5 \times 10^{-9} \text{ C})}{0.06 \text{ m}}$$

$$+ \frac{(8.99 \times 10^9 \text{ N·m}^2/\text{C}^2)(5 \times 10^{-9} \text{ C})}{0.10 \text{ m}}$$

$$= 749 \text{ V} + 450 \text{ V} \approx \boxed{1.20 \text{ kV}}$$

**REMARKS** Note that in Part (a), the electric field is zero at the point midway between the charges but the potential is not. It takes work to bring a test charge to this point from a long distance away, because the electric field is zero only at the final position.

**FIGURE 23-6**

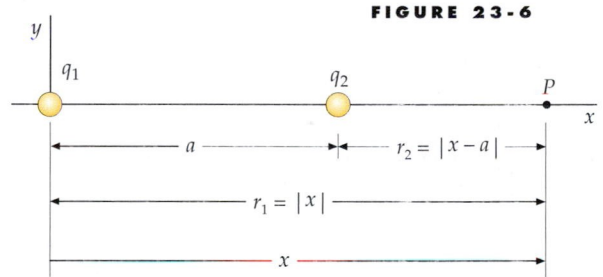

*POTENTIAL THROUGHOUT THE X-AXIS*    **EXAMPLE  23-5**

In Figure 23-6, a point charge $q_1$ is at the origin, and a second point charge $q_2$ is on the x-axis at $x = a$. **Find the potential everywhere on the x-axis.**

**PICTURE THE PROBLEM** The total potential is the sum of the potential due to each charge separately. The distance $r_1$ from $q_1$ to an arbitrary field point $P$ is $r_1 = |x|$, and the distance $r_2$ from $q_2$ to $P$ is $r_2 = |x - a|$.

Write the potential as a function of the distances to the two charges:

$$V = \frac{kq_1}{r_1} + \frac{kq_2}{r_2} = \boxed{\frac{kq_1}{|x|} + \frac{kq_2}{|x - a|}}$$

$$x \neq 0, \quad x \neq a$$

**REMARKS** Figure 23-7 shows $V$ versus $x$ for $q_1 = q_2 > 0$. The potential becomes infinite at each charge.

**FIGURE 23-7**

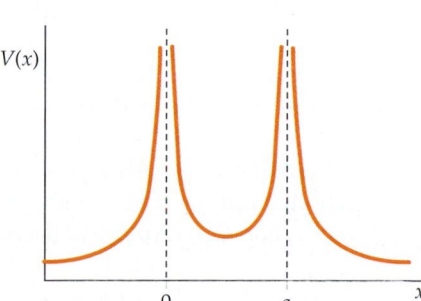

**E X A M P L E   2 3 - 6**

An electric dipole consists of a positive charge $+q$ on the x-axis at $x = +a$ and a negative charge $-q$ on the x-axis at $x = -a$, as shown in Figure 23-8. Find the potential on the x-axis for $x \gg a$ in terms of the dipole moment $p = 2qa$.

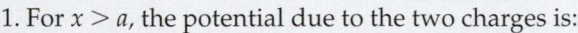
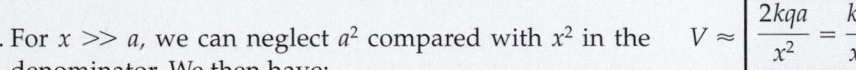

**FIGURE 23-8**

**PICTURE THE PROBLEM** The potential is the sum of the potentials for each charge. For $x > a$, the distance from the field point $P$ to the positive charge is $x - a$ and the distance to the negative charge is $x + a$.

1. For $x > a$, the potential due to the two charges is:

$$V = \frac{kq}{x - a} + \frac{k(-q)}{x + a} = \frac{2kqa}{x^2 - a^2}$$

2. For $x \gg a$, we can neglect $a^2$ compared with $x^2$ in the denominator. We then have:

$$V \approx \boxed{\frac{2kqa}{x^2} = \frac{kp}{x^2}, \quad x \gg a}$$  23-11

**REMARKS** Far from the dipole, the potential decreases as $1/r^2$ (compared to the potential of a point charge, which decreases as $1/r$).

## 23-3 Computing the Electric Field From the Potential

If we know the potential, we can use the potential to calculate the electric field. Consider a small displacement $d\vec{\ell}$ in an arbitrary electric field $\vec{E}$. The change in potential is

$$dV = -\vec{E} \cdot d\vec{\ell} = -E \cos\theta \, d\ell = -E_t \, d\ell$$  23-12

where $E_t = E \cos\theta$ is the component of $\vec{E}$ in the direction of $d\vec{\ell}$. Then

$$E_t = -\frac{dV}{d\ell}$$  23-13

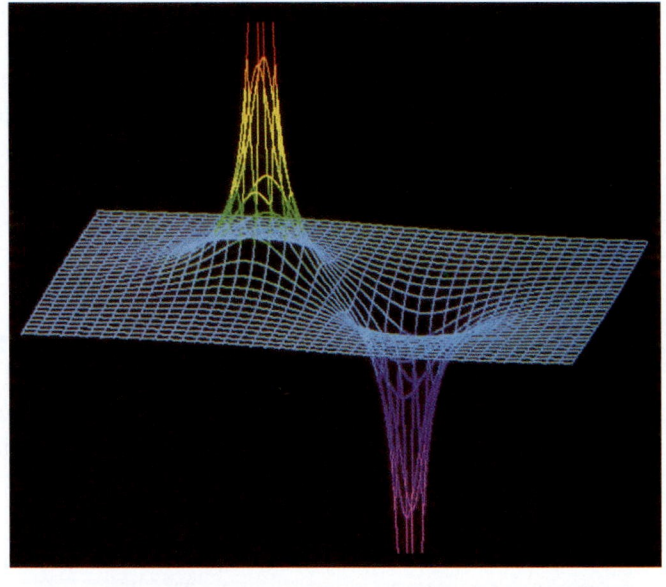

The elecrostatic potential in the plane of an electric dipole. The potential due to each charge is proportional to the charge and inversely proportional to the distance from the charge.

If the displacement $d\vec{\ell}$ is perpendicular to the electric field, then $dV = 0$ (the potential does not change). For a given $d\ell$, the maximum increase in $V$ occurs when the displacement $d\vec{\ell}$ is in the same direction as $-\vec{E}$. A vector that points in the direction of the greatest change in a scalar function and that has a magnitude equal to the derivative of that function with respect to the distance in that direction is called the **gradient** of the function. Thus, the electric field $\vec{E}$ is the negative gradient of the potential $V$. The electric field lines point in the direction of the greatest rate of decrease with respect to distance in the potential function.

If the potential $V$ depends only on $x$, there will be no change in $V$ for displacements in the $y$ or $z$ direction, it follows that $E_y$ and $E_z$ equal zero. For a displacement in the $x$ direction, $d\vec{\ell} = dx\hat{i}$, and Equation 23-12 becomes

$$dV(x) = -\vec{E} \cdot d\vec{\ell} = -\vec{E} \cdot dx\hat{i} = -\vec{E} \cdot \hat{i} dx = -E_x dx$$

Then

$$E_x = -\frac{dV(x)}{dx}$$  23-14

Similarly, for a spherically symmetric charge distribution, the potential can be a function only of the radial distance $r$. Displacements perpendicular to the radial direction give no change in $V(r)$, so the electric field must be radial. A displacement in the radial direction is written $d\vec{\ell} = dr\hat{r}$. Equation 23-12 is then

$$dV(r) = -\vec{E} \cdot d\vec{\ell} = -\vec{E} \cdot dr\hat{r} = -E_r dr$$

and

$$E_r = -\frac{dV(r)}{dr}$$  23-15

If we know either the potential or the electric field over some region of space, we can use one to calculate the other. The potential is often easier to calculate because it is a scalar function, whereas the electric field is a vector function. Note that we cannot calculate $\vec{E}$ if we know the potential $V$ at just a single point— we must know $V$ over a region of space to compute the derivative necessary to obtain $\vec{E}$.

---

**$\vec{E}$ FOR A POTENTIAL THAT VARIES WITH X**                    **E X A M P L E    2 3 - 7**

Find the electric field for the electric potential function $V$ given by $V = 100\ V - (25\ V/m)x$.

**PICTURE THE PROBLEM**  This potential function depends only on $x$. The electric field is found from Equation 23-14:

$$\vec{E} = -\frac{dV}{dx}\hat{i} = \boxed{+(25\ V/m)\hat{i}}$$

**REMARKS**  This electric field is uniform and in the $x$ direction. Note that the constant 100 V in the expression for $V(x)$ has no effect on the electric field. The electric field does not depend on the choice of zero for the potential function.

**EXERCISE**  (*a*) At what point does $V$ equal zero in this example? (*b*) Write the potential function corresponding to the same electric field with $V = 0$ at $x = 0$. [*Answer*   (*a*) $x = 4$ m, (*b*) $V = -(25\ V/m)x$]

## *General Relation Between $\vec{E}$ and $V$

In vector notation, the gradient of $V$ is written as either $\vec{grad}\ V$ or $\vec{\nabla}V$. Then

$$\vec{E} = -\vec{\nabla}V$$  23-16

In general, the potential function can depend on $x$, $y$, and $z$. The rectangular components of the electric field are related to the partial derivatives of the potential with respect to $x$, $y$, or $z$, while the other variables are held constant. For example, the $x$ component of the electric field is given by

$$E_x = -\frac{\partial V}{\partial x}$$  23-17a

Similarly, the $y$ and $z$ components of the electric field are related to the potential by

$$E_y = -\frac{\partial V}{\partial y}$$  23-17b

and

$$E_z = -\frac{\partial V}{\partial z}$$  23-17c

Thus, Equation 23-16 in rectangular coordinates is

$$\vec{E} = -\vec{\nabla}V = -\left(\frac{\partial V}{\partial x}\hat{i} + \frac{\partial V}{\partial y}\hat{j} + \frac{\partial V}{\partial z}\hat{k}\right)$$  23-18

# 23-4 Calculations of V for Continuous Charge Distributions

The potential due to a continuous distribution of charge can be calculated by choosing an element of charge $dq$, which we treat as a point charge, and, invoking superposition, changing the sum in Equation 23-10 to an integral:

$$V = \int \frac{k\,dq}{r}$$  23-19

POTENTIAL DUE TO A CONTINUOUS CHARGE DISTRIBUTION

This equation assumes that $V = 0$ at an infinite distance from the charges, so we cannot use it when there is charge at infinity, as is the case for artificial charge distributions like an infinite line charge or an infinite plane charge.

## V on the Axis of a Charged Ring

Figure 23-9 shows a uniformly charged ring of radius $a$ and charge $Q$. The distance from an element of charge $dq$ to the field point $P$ on the axis of the ring is $r = \sqrt{x^2 + a^2}$. Since this distance is the same for all elements of charge on the ring, we can remove this term from the integral in Equation 23-19. The potential at point $P$ due to the ring is thus

$$V = \int_0^Q \frac{k\,dq}{r} = \frac{k}{r}\int_0^Q dq = \frac{kQ}{r}$$

or

$$V = \frac{kQ}{\sqrt{x^2 + a^2}} = \frac{kQ}{|x|}\frac{1}{\sqrt{1 + (a^2/x^2)}}$$  23-20

POTENTIAL ON THE AXIS OF A CHARGED RING

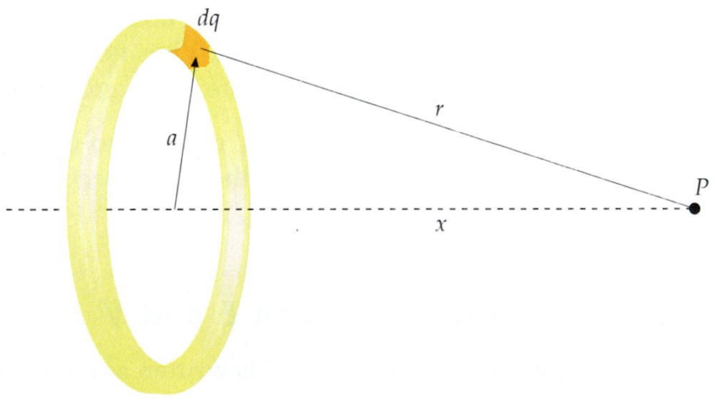

**FIGURE 23-9** Geometry for the calculation of the electric potential at a point on the axis of a charged ring of radius $a$.

Note that when $|x|$ is much greater than $a$, the potential approaches $kQ/|x|$, the same as for a point charge $Q$ at the origin.

*A RING AND A PARTICLE*                    **EXAMPLE 23-8** **Try It Yourself**

A ring of radius 4 cm is in the $yz$ plane with its center at the origin. The ring carries a uniform charge of 8 nC. A small particle of mass $m = 6$ mg $= 6 \times 10^{-6}$ kg
and charge $q_0 = 5$ nC is placed at $x = 3$ cm and released. Find the speed of the
particle when it is a great distance from the ring. Assume gravitational effects
are negligible.

**PICTURE THE PROBLEM** The particle is repelled by the ring. As the particle
moves along the $x$-axis, its potential energy decreases and its kinetic energy
increases. Use conservation of mechanical energy to find the kinetic energy of
the particle when it is far from the ring. The final speed is found from the final
kinetic energy.

**Cover the column to the right and try these on your own before looking at the answers.**

| Steps | Answers |
|---|---|
| 1. Write down the relation between the kinetic energy and the speed. | $K = \frac{1}{2}mv^2$ |
| 2. Use $U = q_0 V$, with $V$ given by Equation 23-20, to obtain an expression for the potential energy of the point charge $q_0$ as a function of its distance $x$ from the center of the ring. | $U = q_0 V = \dfrac{kq_0 Q}{\sqrt{x^2 + a^2}}$ |
| 3. Use conservation of mechanical energy to relate the speed to the position $x$ and solve for the speed when $x$ approaches infinity. | $U_f + K_f = U_i + K_i$ |

$$\frac{kq_0 Q}{\sqrt{x_f^2 + a^2}} + \frac{1}{2}mv_f^2 = \frac{kq_0 Q}{\sqrt{x_i^2 + a^2}} + \frac{1}{2}mv_i^2$$

so

$$v_f = \boxed{1.55 \text{ m/s}}$$

**EXERCISE** What is the potential energy of the particle when it is at
$x = 9$ cm? (*Answer* $3.65 \times 10^{-6}$ J)

## V on the Axis of a Uniformly Charged Disk

We can use our result for the potential on the axis of a ring charge to calculate the
potential on the axis of a uniformly charged disk.

*FIND V FOR A CHARGED DISK*                    **EXAMPLE 23-9**

Find the potential on the axis of a disk of radius $R$ that carries a
total charge $Q$ distributed uniformly on its surface.

**PICTURE THE PROBLEM** We take the axis of the disk to be the
$x$-axis, and we treat the disk as a set of ring charges. The ring of
radius $a$ and thickness $da$ in Figure 23-10 has an area of $2\pi a\, da$, and
its charge is $dq = \sigma\, dA = \sigma 2\pi a\, da$ where $\sigma = Q/(\pi R^2)$, the surface
charge density. The potential due to the charge on this ring at
point $P$ is given by Equation 23-20. We then integrate from $a = 0$ to
$a = R$ to find the total potential due to the charge on the disk.

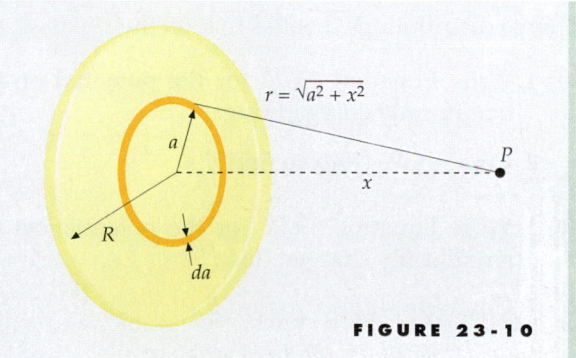

**FIGURE 23-10**

1. Write the potential $dV$ at point $P$ due to the charged ring of radius $a$:

$$dV = \frac{k\,dq}{(x^2 + a^2)^{1/2}} = \frac{k\sigma\,2\pi a\,da}{(x^2 + a^2)^{1/2}}$$

2. Integrate from $a = 0$ to $a = R$:

$$V = \int_0^R \frac{k\sigma\,2\pi a\,da}{(x^2 + a^2)^{1/2}} = k\sigma\pi\int_0^R (x^2 + a^2)^{-1/2} 2a\,da$$

3. The integral is of the form $\int u^n\,du$, with $u = x^2 + a^2$, $du = 2x\,dx$, and $n = -\frac{1}{2}$. When $a = 0$, $u = x^2$ and when $a = R$, $u = x^2 + R^2$:

$$V = k\sigma\pi\int_{x^2+0^2}^{x^2+R^2} u^{-1/2}\,du = k\sigma\pi\,\frac{u^{1/2}}{\frac{1}{2}}\Big|_{x^2}^{x^2+R^2} = 2k\sigma\pi\left(\sqrt{x^2 + R^2} - \sqrt{x^2}\right)$$

4. Rearranging this result to find $V$ gives:

$$\boxed{V = 2\pi k\sigma |x|\left(\sqrt{1 + \frac{R^2}{x^2}} - 1\right)}$$

---

**❶PLAUSIBILITY CHECK** For $|x| \gg R$, the potential function $V$ should approach that of a point charge $Q$ at the origin. We expect that for large $|x|$, $V \approx kQ/|x|$. To approximate our result for $|x| \gg R$, we use the binomial expansion:

$$\left(1 + \frac{R^2}{x^2}\right)^{1/2} \approx 1 + \frac{1}{2}\frac{R^2}{x^2} + \cdots$$

Then

$$V \approx 2\pi k\sigma |x|\left[\left(1 + \frac{1}{2}\frac{R^2}{x^2} + \cdots\right) - 1\right] = \frac{k(\sigma\pi R^2)}{|x|} = \frac{kQ}{|x|}$$

From Example 23-9, we see that the potential on the axis of a uniformly charged disk is

$$V = 2\pi k\sigma |x|\left(\sqrt{1 + \frac{R^2}{x^2}} - 1\right) \qquad\qquad 23\text{-}21$$

POTENTIAL ON THE AXIS OF A UNIFORMLY CHARGED DISK

---

*FIND $\vec{E}$ GIVEN V*  **E X A M P L E   2 3 - 1 0**

**Calculate the electric field on the axis of (a) a uniformly charged ring and (b) a uniformly charged disk using the potential functions previously given for these charge distributions.**

**PICTURE THE PROBLEM** Using $E_x = -dV/dx$, we can evaluate $E_x$ by direct differentiation. We cannot evaluate either $E_y$ or $E_z$ by direct differentiation because we do not know how $V$ varies in those directions. However, the symmetry of the charge distributions dictates that on the x-axis, $E_y = E_z = 0$.

(a) 1. Write Equation 23-20 for the potential on the axis of a uniformly charged ring:

$$V = \frac{kQ}{\sqrt{x^2 + a^2}} = kQ(x^2 + a^2)^{-1/2}$$

2. Compute $-dV/dx$ to find $E_x$:

$$E_x = -\frac{dV}{dx} = +\frac{1}{2}kQ(x^2 + a^2)^{-3/2}(2x) = \boxed{\frac{kQx}{(x^2 + a^2)^{3/2}}}$$

(b) 1. Write Equation 23-21 for the potential on the axis of a uniformly charged disk:

$$V = 2\pi k\sigma\left[(x^2 + a^2)^{1/2} - |x|\right]$$

2. Compute $-dV/dx$ to find $E_x$:

$$E_x = -\frac{dV}{dx} = -2\pi k\sigma\left[\frac{1}{2}(x^2 + a^2)^{-1/2}\,2x - \frac{d|x|}{dx}\right]$$

3. Evaluate $d|x|/dx$. It is the slope of a graph of $|x|$ versus $x$ (see Figure 23-11):

$$\frac{d|x|}{dx} = +1, \quad x > 0; \frac{d|x|}{dx} = -1, \quad x < 0$$

4. Substituting for $d|x|/dx$ in the Part (b), step 2 result gives:

$$E_x = -2\pi k\sigma\left(\frac{x}{\sqrt{x^2 + a^2}} - 1\right), \quad x > 0$$

and

$$E_x = -2\pi k\sigma\left(\frac{x}{\sqrt{x^2 + a^2}} + 1\right), \quad x < 0$$

5. A little rearranging puts these expressions in a form that better reveals that $E_x$ is an odd function (Figure 23-12). [A function $f$ is odd if $f(-x) = f(x)$ for all values of $x$]:

$$\boxed{E_x = +2\pi k\sigma\left(1 - \frac{1}{\sqrt{1 + (a^2/x^2)}}\right), \quad x > 0}$$

and

$$\boxed{E_x = -2\pi k\sigma\left(1 - \frac{1}{\sqrt{1 + (a^2/x^2)}}\right), \quad x < 0}$$

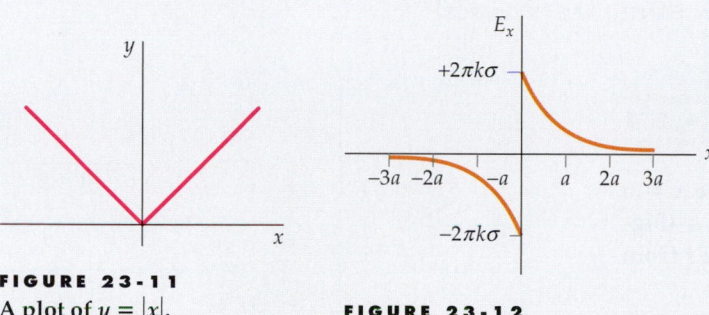

**FIGURE 23-11**
A plot of $y = |x|$.

**FIGURE 23-12**

**REMARKS** The results for Parts (a) and (b) are the same as Equations 22-10 and 22-11, which were calculated directly from Coulomb's law.

## V Due to an Infinite Plane of Charge

If we let $R$ become very large, our disk approaches an infinite plane. As $R$ approaches infinity, the potential function (Equation 23-21) approaches infinity. However, we obtained Equation 23-21 from Equation 23-19, which assumes that $V = 0$ at infinity, so Equation 23-21 cannot be used. For infinite charge distributions, we must choose $V = 0$ at some finite point rather than at infinity. For such cases, we first find the electric field $\vec{E}$ (by direct integration or from Gauss's law) and then calculate the potential function $V$ from its definition $dV = -\vec{E} \cdot d\vec{\ell}$. For an infinite plane of uniform charge of density $\sigma$ in the $yz$ plane, the electric field for positive $x$ is given by

$$\vec{E} = \frac{\sigma}{2\,\epsilon_0}\,\hat{i} = 2\pi k\sigma\hat{i}$$

The potential is then

$$dV = -\vec{E} \cdot d\vec{\ell} = -(2\pi k\sigma\hat{i}) \cdot (dx\hat{i} + dy\hat{j} + dz\hat{k}) = -2\pi k\sigma\,dx$$

where we have used $d\vec{\ell} = dx\hat{i} + dy\hat{j} + dz\hat{k}$. Integrating, we obtain

$$V = -2\pi k\sigma x + V_0$$

where the arbitrary integration constant $V_0$ is the potential at $x = 0$. Note that the potential decreases with distance from the plane and approaches $-\infty$ as $x$ approaches $+\infty$. Therefore, at $x = +\infty$ the potential equals negative infinity.

For negative $x$, the electric field is

$$\vec{E} = -2\pi k\sigma\hat{i}$$

so

$$dV = -\vec{E} \cdot d\vec{\ell} = +2\pi k\sigma \, dx$$

and the potential is

$$V = V_0 + 2\pi k\sigma x$$

Since $x$ is negative, the potential again decreases with distance from the plane and approaches $-\infty$ as $x$ approaches $-\infty$. For either positive or negative $x$, the potential can be written

$$V = V_0 - 2\pi k\sigma |x| \qquad 23\text{-}22$$

POTENTIAL NEAR AN INFINITE PLANE OF CHARGE

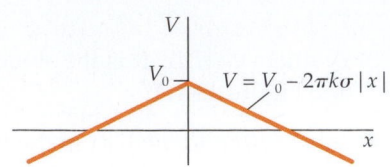

**FIGURE 23-13** Plot of $V$ versus $x$ for an infinite plane of charge in the $yz$ plane. Note that the potential is continuous at $x = 0$ even though $E_x = dV/dx$ is not.

---

*A PLANE AND A POINT CHARGE*  **EXAMPLE 23-11**

**An infinite plane of uniform charge density $\sigma$ is in the $x = 0$ plane, and a point charge $q$ is on the $x$ axis at $x = a$ (Figure 23-14). Find the potential at some point $P$ a distance $r$ from the point charge.**

**PICTURE THE PROBLEM** We can use the principle of superposition. The total potential $V$ is the sum of the individual potentials due to the plane and the point charge. We must add an arbitrary constant in our expression for $V$, which is determined by our choice of the reference point, where $V = 0$. We are free to choose the reference point, except at $x = \pm\infty$ or at $x = a$ on the $x$-axis. For this solution, we choose $V = 0$ at the origin.

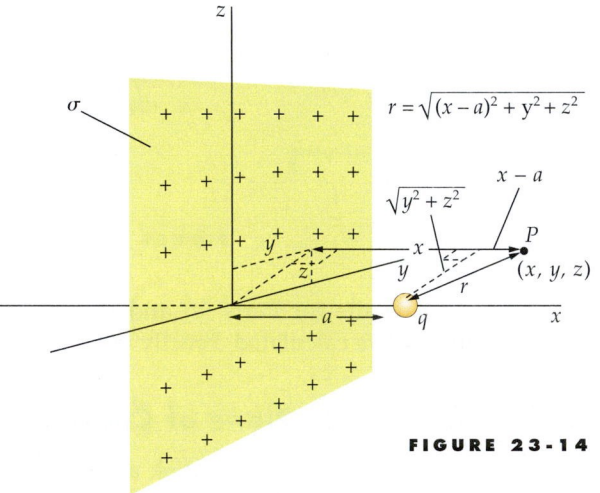

**FIGURE 23-14**

1. The potential due to the charged plane is given by Equation 23-22:

$$V_{\text{plane}} = -2\pi k\sigma |x|$$

2. Equation 23-7 gives the potential due to a point charge. The distance $r$ from the point charge to the field point equals $\sqrt{(x-a)^2 + y^2 + z^2}$:

$$V_{\text{point}} = \frac{kq}{r} = \frac{kq}{\sqrt{(x-a)^2 + y^2 + z^2}}$$

3. Sum the above results to find the total potential $V$. A constant is added to the sum so we can set the potential at the reference point to zero:

$$V = V_{\text{plane}} + V_{\text{point}}$$

$$= -2\pi k\sigma |x| + \frac{kq}{\sqrt{(x-a)^2 + y^2 + z^2}} + C$$

4. We choose to let $V = 0$ at the origin. To do that, set $V = 0$ at $x = y = z = 0$ and solve for the constant $C$:

$$0 = 0 + \frac{kq}{a} + C, \quad \text{so } C = -\frac{kq}{a}$$

5. Substitute $-kq/a$ for $C$ in the step 3 result:

$$V = -2\pi k\sigma |x| + \frac{kq}{\sqrt{(x-a)^2 + y^2 + z^2}} - \frac{kq}{a}$$

$$\boxed{\, = -2\pi k\sigma |x| + \frac{kq}{r} - \frac{kq}{a} \,}$$

**REMARKS** The answer is not unique. We could have specified the potential at any point other than at $x = a$ or at $x = \pm\infty$.

## V Inside and Outside a Spherical Shell of Charge

We find the potential due to a thin spherical shell of radius $R$ with charge $Q$ uniformly distributed on its surface next. We are interested in the potential at all points inside, outside, and on the shell. Unlike the infinite plane of charge, this charge distribution is confined to a finite region of space, so, in principle, we could calculate the potential by direct integration of Equation 23-19. However, there is a simpler way. Since the electric field for this charge distribution is easily obtained from Gauss's law, we will calculate the potential from the known electric field using $dV = -\vec{E} \cdot d\vec{\ell}$.

Outside the spherical shell, the electric field is radial and is the same as if all the charge $Q$ were a point charge at the origin:

$$\vec{E} = \frac{kQ}{r^2}\hat{r}$$

The change in the potential for some displacement $d\vec{r}$ outside the shell is then

$$dV = -\vec{E} \cdot d\vec{\ell} = -\frac{kQ}{r^2}\hat{r} \cdot d\vec{\ell}$$

The product $\hat{r} \cdot d\vec{\ell}$ is $dr$ (the component of $d\vec{\ell}$ in the direction of $\hat{r}$). Integrating along a path from the reference point at infinity, we obtain

$$V_P = -\int_\infty^{r_P} \vec{E} \cdot d\vec{\ell} = -\int_\infty^{r_P} \frac{kQ}{r^2}\,dr = -kQ\int_\infty^{r_P} r^{-2}\,dr = \frac{kQ}{r_P}$$

where $P$ is an arbitrary field point in the region $r \geq R$, and $r_P$ is the distance from the center of the shell to the field point $P$. The potential is chosen to be zero at infinity. Since $P$ is arbitrary, we let $r_P = r$ to obtain

$$V = \frac{kQ}{r}, \qquad r \geq R$$

Inside the spherical shell, the electric field is zero everywhere. Again integrating from the reference point at infinity, we obtain

$$V_P = -\int_\infty^{r_P} \vec{E} \cdot d\vec{r} = -\int_\infty^{R} \frac{kQ}{r^2}\,dr - \int_R^{r_P} (0)\,dr = \frac{kQ}{R}$$

where $P$ is an arbitrary field point in the region $r < R$, and $r_P$ is the distance from the center of the shell to the field point $P$. The potential at an arbitrary point inside the shell is $kQ/R$, where $R$ is the radius of the shell. Inside the shell $V$ is the same everywhere. It is the work per unit charge to bring a test charge from infinity to the shell. No additional work is required to bring it from the shell to any point inside the shell. Thus,

$$V = \frac{kQ}{r}, \qquad r \geq R$$

$$V = \frac{kQ}{R}, \qquad r \leq R \qquad\qquad \text{23-23}$$

POTENTIAL DUE TO A SPHERICAL SHELL

This potential function is plotted in Figure 23-15.

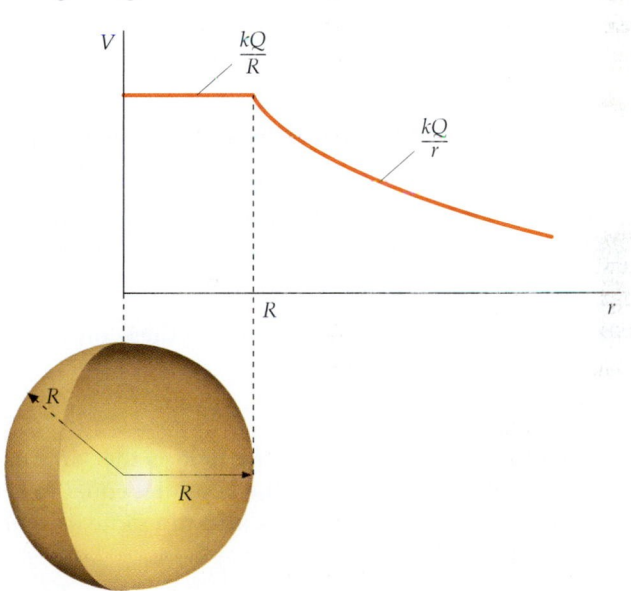

**FIGURE 23-15** Electric potential of a uniformly charged spherical shell of radius $R$ as a function of the distance $r$ from the center of the shell. Inside the shell, the potential has the constant value $kQ/R$. Outside the shell, the potential is the same as that due to a point charge at the center of the sphere.

❶    A common mistake is to think that the potential must be zero inside a spherical shell because the electric field is zero throughout that region. But a region of zero electric field merely implies that the potential is uniform. Consider a spherical shell with a small hole so that we can move a test charge in and out of the shell. If we move the test charge from an infinite distance to the shell, the work per charge we must do is $kQ/R$. Inside the shell there is no electric field, so it takes no work to move the test charge around inside the shell. The total amount of work per unit charge it takes to bring the test charge from infinity to any point inside the shell is just the work per charge it takes to bring the test charge up to the shell radius $R$, which is $kQ/R$. The potential is therefore $kQ/R$ everywhere inside the shell.

**EXERCISE** What is the potential of a spherical shell of radius 10 cm carrying a charge of 6 $\mu$C? (*Answer*   $5.39 \times 10^5$ V $= 539$ kV)

---

*FIND V FOR A UNIFORMLY CHARGED SPHERE*          **EXAMPLE  23-12**   **Try It Yourself**

In one model, a proton is considered to be a spherical ball of charge of uniform volume charge density with radius $R$ and total charge $Q$. The electric field inside the sphere is given by Equation 22-26b,

$$E_r = k\frac{Q}{R^3}r$$

Find the potential $V$ both inside and outside the sphere.

**PICTURE THE PROBLEM** Outside the sphere, the charge looks like a point charge, so the potential is given by $V = kQ/r$. Inside the sphere, $V$ can be found by integrating $dV = -\vec{E} \cdot d\vec{\ell}$.

**Cover the column to the right and try these on your own before looking at the answers.**

**Steps**

1. Outside the sphere, the electric field is the same as that of a point charge. If we set the potential equal to zero at infinity, the potential there is also the same as that of a point charge.

2. For $r \leq R$, find $dV$ from $dV = -\vec{E} \cdot d\vec{\ell}$.

3. Find the definite integral using your expression in step 2. Find the change in potential from infinity to an arbitrary field point $P$ in the region $r_P < R$, where $r_P$ is the distance of point $P$ from the center of the sphere:

4. Since the field point position is arbitrary, express the result in terms of $r = r_P$:

**Answers**

$$V(r) = \boxed{\frac{kQ}{r}}, \quad r \geq R$$

$$dV = -\vec{E} \cdot d\vec{\ell} = -E_r\,dr = -\frac{kQ}{R^3}r\,dr, \quad r \leq R$$

$$V_P = -\int_\infty^{r_P} E_r\,dr = -\int_\infty^R \frac{kQ}{r^2}\,dr - \int_R^{r_P} \frac{kQ}{R^3}r\,dr$$

$$= \frac{kQ}{R} - \frac{kQ}{2R^3}(r_P^2 - R^2) = \frac{kQ}{2R}\left(3 - \frac{r_P^2}{R^2}\right), \quad r \leq R$$

$$V(r) = \boxed{\frac{kQ}{2R}\left(3 - \frac{r^2}{R^2}\right), \quad r \leq R}$$

❶ **PLAUSIBILITY CHECK** Substituting $r = R$ in the step 4 result gives $V(R) = kQ/R$ as required. At $r = 0$, $V(0) = 3kQ/2R = 1.5\,kQ/R$, which is greater than $V(R)$, as it should be, because the electric field is in the positive radial direction for $r < R$, so positive work must be done to move a positive test charge against the field from $r = R$ to $r = 0$.

**REMARKS**  Figure 23-16 shows $V(r)$ as a function of $r$. Note that both $V(r)$ and $E_r = -dV/dr$ are continuous everywhere.

**EXERCISE**  What is $V(r)$, if we choose $V(R) = 0$? [Answer   $V(r) = kQ/r - kQ/R$ for $r \geq R$; $V(r) = \frac{1}{2}(kQ/R)(1 - r^2/R^2)$ for $r \leq R$]

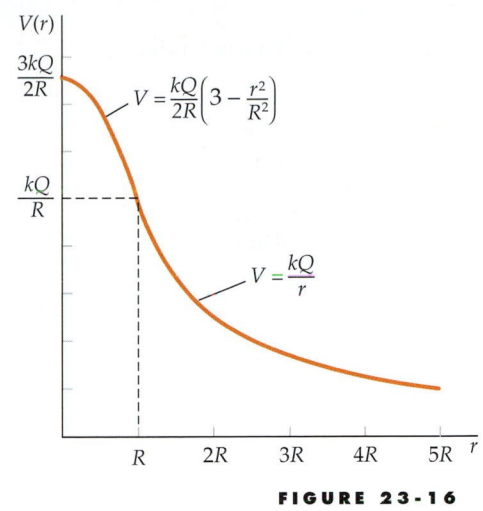

**FIGURE 23-16**

## V Due to an Infinite Line Charge

We will now calculate the potential due to a uniform infinite line charge. Let the charge per unit length be $\lambda$. Like the infinite plane of charge, this charge distribution is not confined to a finite region of space, so, in principle, we cannot calculate the potential by direct integration of $dV = kdq/r$ (Equation 23-19). Instead, we will find the potential by integrating the electric field directly. First, we must obtain the electric field of a uniformly charged infinite line. The field, a cylindrically symmetric charge distribution like this one, can be obtained using Gauss's law ($\phi_{net} = 4\pi kQ_{inside}$). The outward flux through a coaxial soup-can-shaped Gaussian surface of radius $R$ and length $L$ is $E_R (2\pi RL)$, and the charge inside is $\lambda L$. Substituting these expressions into the Gauss's-law equation and solving for $E_R$ gives $E_R = 2k\lambda/R$. The change in potential for a displacement $d\vec{\ell}$ is given by

$$dV = -\vec{E} \cdot d\vec{\ell} = -E_R \hat{R} \cdot d\vec{\ell}$$

where $\hat{R}$ is the radial direction. The product $\hat{R} \cdot d\vec{\ell}$ is $dR$ (the component of $d\vec{\ell}$ in the direction of $\hat{R}$), so $dV = -E_R dR$. Integrating from an arbitrary reference point to an arbitrary field point $P$ (Figure 23-17) gives

$$V_P - V_{ref} = -\int_{R_{ref}}^{R_P} E_R dR = -2k\lambda \int_{R_{ref}}^{R_P} \frac{dR}{R} = -2k\lambda \ln \frac{R_P}{R_{ref}}$$

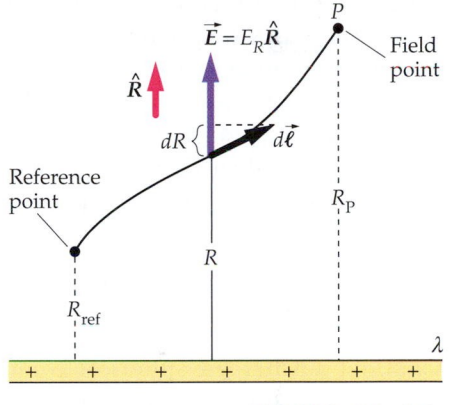

**FIGURE 23-17**

where $R_P$ and $R_{ref}$ are the radial distances of the field point $P$ and the reference point from the line charge, respectively. For convenience, we choose the potential to equal zero at the reference point ($V_{ref} = 0$). We cannot choose $R_{ref}$ to be zero because $\ln(0) = -\infty$, and we cannot choose $R_{ref}$ to be infinity because $\ln(\infty) = +\infty$. However, any other choice in the interval $0 < R_{ref} < \infty$ is acceptable, and the potential function is given by

$$V = 2k\lambda \ln \frac{R_{ref}}{R} \qquad\qquad 23\text{-}24$$

POTENTIAL DUE TO A LINE CHARGE

We do not encounter infinite planes or lines of charge, but these distributions make excellent models for some real situations. For example, the potential near a 500-m-long, nearly straight, high-voltage transmission power line.

## 23-5  Equipotential Surfaces

Since there is no electric field inside the material of a conductor that is in static equilibrium, the change in potential as we move about the region occupied by the conducting material is zero. Thus, the electric potential is the same throughout the material of the conductor; that is, the conductor is a three-dimensional **equipotential region** and its surface is an **equipotential surface**. Because the potential $V$ has the same value everywhere on an equipotential surface, the change in $V$ is zero. If a test charge on the surface is given a small displacement $d\vec{\ell}$ parallel to the surface, $dV = -\vec{E} \cdot d\vec{\ell} = 0$. Since $\vec{E} \cdot d\vec{\ell}$ is zero for *any* $d\vec{\ell}$ parallel to the surface, $\vec{E}$ must be perpendicular to any and every $d\vec{\ell}$ parallel to

**FIGURE 23-18** Equipotential surfaces and electric field lines outside a uniformly charged spherical conductor. The equipotential surfaces are spherical and the field lines are radial and perpendicular to the equipotential surfaces.

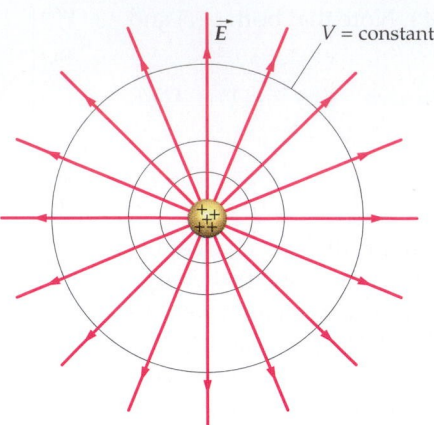

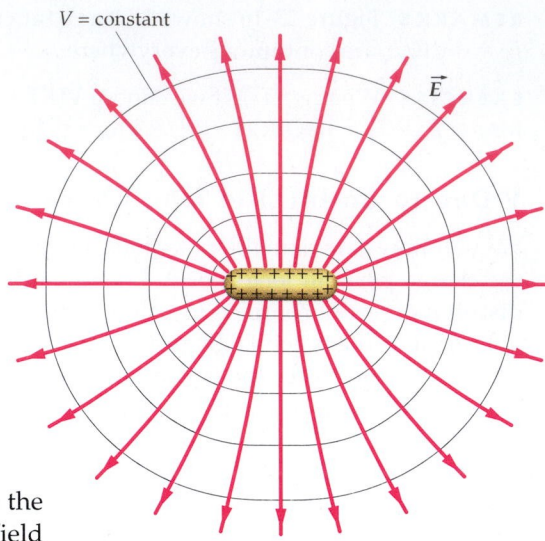

the surface. The only way $\vec{E}$ can be perpendicular to every $d\vec{\ell}$ parallel to the surface, however, is for $\vec{E}$ to be normal to the surface. Therefore, any electric field lines beginning or terminating on the equipotential surface must be normal to it. Figures 23-18 and 23-19 show equipotential surfaces near a spherical conductor and a nonspherical conductor. Note that anywhere a field line meets or penetrates an equipotential surface, shown in blue, the field line is normal to the equipotential surface. If we go from one equipotential surface to a neighboring equipotential surface by undergoing a displacement $d\vec{\ell}$ along a field line in the direction of the field, the potential changes by $dV = -\vec{E} \cdot d\vec{\ell} = -E d\ell$. It follows that equipotential surfaces that have a fixed potential difference between them are more closely spaced where the electric field strength $E$ is greater.

**FIGURE 23-19** Equipotential surfaces and electric field lines outside a nonspherical conductor. Electric field lines always intersect equipotential surfaces at right angles.

*A HOLLOW SPHERICAL SHELL*　　　　　　**EXAMPLE 23-13**

A hollow, uncharged spherical conducting shell has an inner radius $a$ and an outer radius $b$. A positive point charge $+q$ is in the cavity, at the center of the sphere. (*a*) Find the charge on each surface of the conductor. (*b*) Find the potential $V(r)$ everywhere, assuming that $V = 0$ at $r = \infty$.

**PICTURE THE PROBLEM** (*a*) The charge distribution has spherical symmetry, so applying Gauss's law should be a good method for finding the charges on the inner and outer surface of the shell. (*b*) Sum the individual potentials for the individual charges to obtain the resultant potential. The potential for a point charge and for a uniform thin spherical shell of charge have already been established (Equations 23-8 and 23-23).

(*a*) 1. The charge inside a closed surface is proportional to the outward flux of $\vec{E}$ through the surface:

$$\phi_{\text{net}} = 4\pi k Q_{\text{inside}}, \text{ where } \phi_{\text{net}} = \oint_S E_n \, dA$$

2. Sketch the point charge and the spherical shell. On a conductor, charge resides only on its surface. Label the charge on each surface of the shell. Include a Gaussian surface completely inside the material of the conductor:

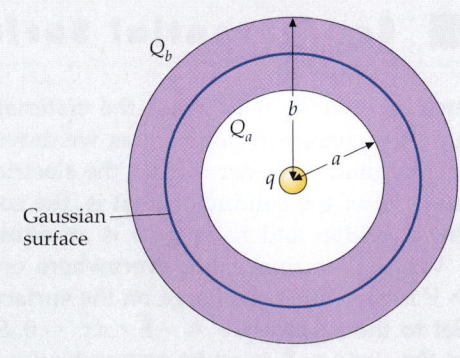

**FIGURE 23-20**

3. Apply Gauss's law (the step 1 result) to the Gaussian surface and solve for the charge on the inner surface of the shell:

$$E_n = 0 \quad \Rightarrow \quad Q_{inside} = q + Q_a = 0$$

so

$$\boxed{Q_a = -q}$$

4. The shell is neutral, so solve for the charge on its outer surface:

$$Q_a + Q_b = 0$$

so

$$\boxed{Q_b = +q}$$

(b) 1. The potential is the sum of the potentials due to the individual charges:

$$V = V_q + V_{Q_a} + V_{Q_b}$$

2. Add the potentials in the region outside the shell. The potential for a thin charged spherical shell is given in Equation 23-23:

$$V = \frac{kq}{r} - \frac{kq}{r} + \frac{kq}{r} = \boxed{\frac{kq}{r}, \quad r \geq b}$$

3. Add the potentials in the region inside the material of the conducting shell:

$$V = \frac{kq}{r} - \frac{kq}{r} + \frac{kq}{b} = \boxed{\frac{kq}{b}, \quad a \leq r \leq b}$$

4. Add the potentials in the region between the point charge and the shell:

$$V = \boxed{\frac{kq}{r} + \frac{kq}{b} - \frac{kq}{a}, \quad 0 < r \leq a}$$

**REMARKS** Each of the individual potential functions has its zero-potential reference point at $r = \infty$. Thus, the sum of these functions also has its zero-potential reference point at $r = \infty$. The potential arrived at in the example can be obtained by directly evaluating $-\int_{\infty}^{P} \vec{E} \cdot d\vec{\ell} = -\int_{\infty}^{r_p} E_r \, dr$. Yet a third way to obtain the potential is by evaluating the indefinite integral $-\int E_r \, dr$ in each region to find the integration constants by matching the potential functions at the boundaries. Matching the potential functions at the boundaries is valid because the potential must be continuous.

Figure 23-21 shows the electric potential as a function of the distance from the center of the cavity. Inside the conducting material, where $a \leq r \leq b$, the potential has the constant value $kq/b$. Outside the shell, the potential is the same as that of a point charge $q$ at the center of the shell. Note that $V(r)$ is continuous everywhere. The electric field is discontinuous at the conductor surfaces, as reflected in the discontinuous slope of $V(r)$ at $r = a$ and $r = b$.

In general, two conductors that are separated in space will not be at the same potential. The potential difference between such conductors depends on their geometrical shapes, their separation, and the net charge on each. When two conductors are brought into contact, the charge on the conductors distributes itself so that electrostatic equilibrium is established, and the electric field is zero inside both conductors. While in contact, the two conductors may be considered to be a single conductor with a single equipotential surface. If we put a spherical charged conductor in contact with a second spherical conductor that is uncharged, charge will flow between them until both conductors are at the same potential. If the spherical conductors are identical, they share the original charge equally. If the identical spherical conductors are now separated, each carries half the original charge.

**FIGURE 23-21**

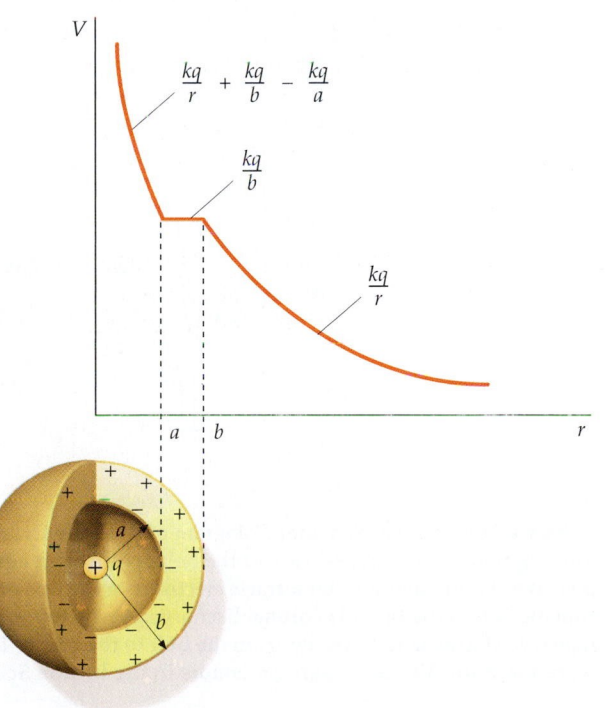

## The Van de Graaff Generator

In Figure 23-22, a small conductor carrying a positive charge $q$ is inside the cavity of a larger conductor. In equilibrium, the electric field is zero inside the conducting material of both conductors. The electric field lines that begin on the positive charge $q$ must terminate on the inner surface of the large conductor. This must occur no matter what the charge may be on the outside surface of the large conductor. Regardless of the charge on the large conductor, the small conductor in the cavity is at a greater potential because the electric field lines go from this conductor to the larger conductor. If the conductors are now connected, say, with a fine conducting wire, *all* the charge originally on the smaller conductor will flow to the larger conductor. When the connection is broken, there is no charge on the small conductor in the cavity, and there are no field lines between the conductors. The positive charge transferred from the smaller conductor resides completely on the outside surface of the larger conductor. If we put more positive charge on the small conductor in the cavity and again connect the conductors with a fine wire, all of the charge on the inner conductor will again flow to the outer conductor. The procedure can be repeated indefinitely. This method is used to produce large potentials in a device called the Van de Graaff generator, in which the charge is brought to the inner surface of a larger spherical conductor by a continuous charged belt (Figure 23-23). Work must be done by the motor driving the belt to bring the charge from the bottom to the top of the belt, where

**FIGURE 23-22** Small conductor carrying a positive charge inside a larger hollow conductor.

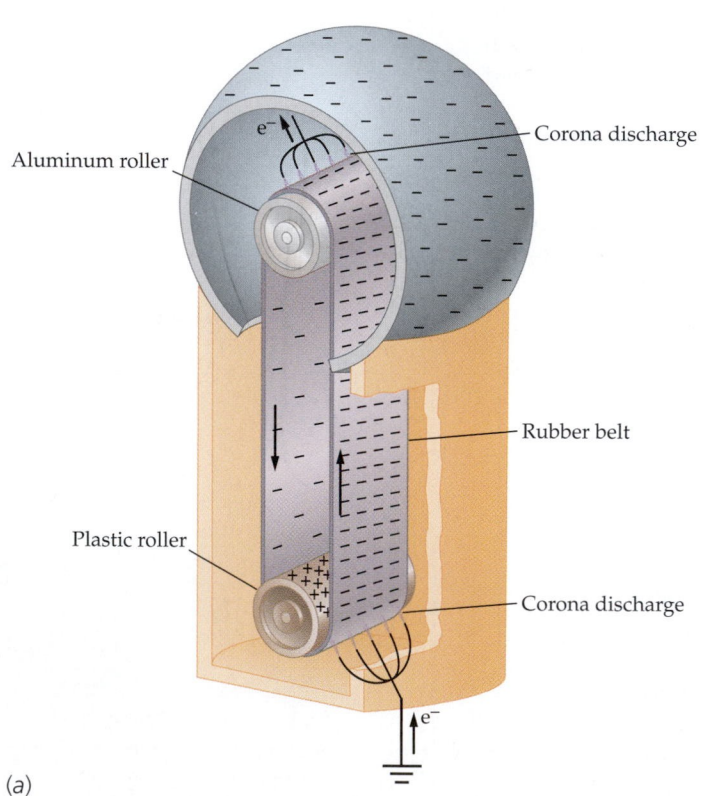

(a)

(b)

**FIGURE 23-23** (*a*) Schematic diagram of a Van de Graaff generator. The lower roller becomes positively charged due to contact with the moving belt. (The inner surface of the belt acquires an equal amount of negative charge that is distributed over a larger area.) The dense positive charge on the roller attracts electrons to the tips of the lower comb where dielectric breakdown takes place and negative charge is transported to the belt via corona discharge. At the top roller the negatively charged belt repels electrons from the tips of the comb and negative charge is transferred from the belt to the comb. The charge is then transferred to the outer surface of the dome. (*b*) These large demonstration Van de Graaff generators in the Boston Science Museum are discharging to the grounded wire cage housing the operator.

the potential is very high. One can often hear the motor speed decrease as the sphere charges. The greater the charge on the outer conductor, the greater its potential, and the greater the electric field just outside its outer surface. A Van de Graaff accelerator is a device that uses the intense electric field produced by a Van de Graaff generator to accelerate ions (charged atomic particles), such as protons.

## Dielectric Breakdown

Many nonconducting materials become ionized in very high electric fields and become conductors. This phenomenon, called **dielectric breakdown,** occurs in air at an electric field strength of $E_{max} \approx 3 \times 10^6$ V/m = 3 MN/C. In air, some of the existing ions are accelerated to greater kinetic energies before they collide with neighboring molecules. Dielectric breakdown occurs when these ions are accelerated to kinetic energies sufficient to result in a growth in ion concentration due to the collisions with neighboring molecules. The maximum potential that can be obtained in a Van de Graaff generator is limited by the dielecric break-down of the air. In a vacuum, Van de Graaff generators can achieve much higher potentials. The magnitude of the electric field for which dielectric breakdown occurs in a material is called the **dielectric strength** of that material. The dielec-tric strength of air is about 3 MV/m. The discharge through the conducting air resulting from dielectric breakdown is called **arc discharge.** The electric shock you receive when you touch a metal doorknob after walking across a rug on a dry day is a familiar example of arc discharge. These breakdowns occur more often on dry days because moist air can conduct the charge away before the breakdown condition is reached. Lightning is an example of arc discharge on a large scale.

---

*DIELECTRIC BREAKDOWN FOR A CHARGED SPHERE*     **EXAMPLE 23-14**

A spherical conductor has a radius of 30 cm ($\approx$1 ft). (*a*) **What is the maximum charge that can be placed on the sphere before dielectric breakdown of the surrounding air occurs?** (*b*) **What is the maximum potential of the sphere?**

**PICTURE THE PROBLEM** (*a*) We find the maximum charge by relating the charge to the electric field and setting the field equal to the dielectric strength of air, $E_{max}$. (*b*) The maximum potential is then found from the maximum charge calculated in Part (*a*).

(*a*) 1. The surface charge density on the conductor $\sigma$ is related to the electric field just outside the conductor:

$$E = \frac{\sigma}{\epsilon_0} = 4\pi k\sigma$$

2. Set this field equal to $E_{max}$:

$$E_{max} = 4\pi k\sigma_{max}$$

3. The maximum charge $Q_{max}$ is found from $\sigma_{max}$:

$$\sigma_{max} = \frac{Q_{max}}{4\pi R^2}$$

4. Solving for $Q_{max}$ gives:

$$Q_{max} = 4\pi R^2\sigma_{max} = 4\pi R^2 \frac{E_{max}}{4\pi k} = \frac{R^2 E_{max}}{k}$$

$$= \frac{(0.3\text{ m})^2(3\times10^6\text{ N/C})}{(8.99\times10^9\text{ N·m}^2/\text{C}^2)} = \boxed{3.00\times10^{-5}\text{ C}}$$

(*b*) Use the expression for the maximum charge to calculate the maximum potential of the sphere:

$$V_{max} = \frac{kQ_{max}}{R} = \frac{k}{R}\left(\frac{R^2 E_{max}}{k}\right) = RE_{max}$$

$$= (0.3\text{ m})(3\times10^6\text{ N/C}) = \boxed{9.00\times10^5\text{ V}}$$

**EXAMPLE 23-15**

Two charged spherical conductors of radius $R_1 = 6$ cm and $R_2 = 2$ cm (Figure 23-24) are separated by a distance much greater than 6 cm and are connected by a long, thin conducting wire. A total charge $Q = +80$ nC is placed on one of the spheres. (*a*) What is the charge on each sphere? (*b*) What is the electric field near the surface of each sphere? (*c*) What is the electric potential of each sphere? (Assume that the charge on the connecting wire is negligible.)

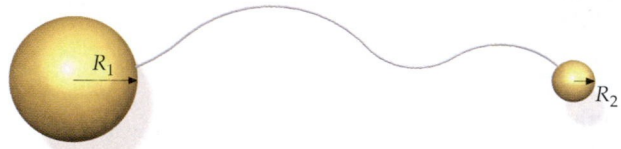

**FIGURE 23-24**

**PICTURE THE PROBLEM** The total charge will be distributed with $Q_1$ on sphere 1 and $Q_2$ on sphere 2 so that the spheres will be at the same potential. We can use $V = kQ/R$ for the potential of each sphere because they are far apart.

(*a*) 1. Conservation of charge gives us one relation between the charges $Q_1$ and $Q_2$:

$$Q_1 + Q_2 = Q$$

2. Equating the potential of the spheres gives us a second relation for the charges $Q_1$ and $Q_2$:

$$\frac{kQ_1}{R_1} = \frac{kQ_2}{R_2} \Rightarrow Q_2 = \frac{R_2}{R_1}Q_1$$

3. Combine the results from step 1 and step 2 and solve for $Q_1$ and $Q_2$:

$$Q_1 + \frac{R_2}{R_1}Q_1 = Q, \quad \text{so}$$

$$Q_1 = \frac{R_1}{R_1 + R_2}Q = \frac{6 \text{ cm}}{8 \text{ cm}}(80 \text{ nC}) = \boxed{60 \text{ nC}}$$

$$Q_2 = Q - Q_1 = \boxed{20 \text{ nC}}$$

(*b*) Use these results to calculate the electric fields at the surface of the spheres:

$$E_1 = \frac{kQ_1}{R_1^2} = \frac{(8.99 \times 10^9 \text{ N·m}^2/\text{C}^2)(60 \times 10^{-9} \text{ C})}{(0.06 \text{ m})^2}$$

$$= \boxed{150 \text{ kN/C}}$$

$$E_2 = \frac{kQ_2}{R_2^2} = \frac{(8.99 \times 10^9 \text{ N·m}^2/\text{C}^2)(20 \times 10^{-9} \text{ C})}{(0.02 \text{ m})^2}$$

$$= \boxed{450 \text{ kN/C}}$$

(*c*) Calculate the common potential from $kQ/R$ for either sphere:

$$V_1 = \frac{kQ_1}{R_1} = \frac{(8.99 \times 10^9 \text{ N·m}^2/\text{C}^2)(60 \times 10^{-9} \text{ C})}{0.06 \text{ m}}$$

$$= \boxed{8.99 \text{ kV}}$$

**PLAUSIBILITY CHECK** If we use sphere 2 to calculate $V$, we obtain $V_2 = kQ_2/R_2 = (8.99 \times 10^9 \text{ N·m}^2/\text{C}^2)(20 \times 10^{-9} \text{ C})/0.02 \text{ m} = 8.99 \times 10^3 \text{ V}$. An additional check is available, since the electric field at the surface of each sphere is proportional to its charge density. The radius of sphere 1 is three times that of sphere 2, so its surface area is nine times that of sphere 2. And since it carries three times the charge, its charge density is $\frac{1}{3}$ that of sphere 2. Therefore, the field of sphere 1 should be $\frac{1}{3}$ that of sphere 2, which is what we found.

When a charge is placed on a conductor of nonspherical shape, like that in Figure 23-25a, the surface of the conductor will be an equipotential surface, but the surface charge density and the electric field just outside the conductor will vary from point to point. Near a point where the radius of curvature is small, such as point A in the figure, the surface charge density and electric field will be large, whereas near a point where the radius of curvature is large, such as point B in the figure, the field and surface charge density will be small. We can understand this qualitatively by considering the ends of the conductor to be spheres of different radii. Let $\sigma$ be the surface charge density.

The potential of a sphere of radius R is

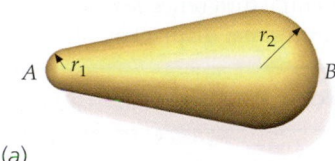

(a)

$$V = \frac{kQ}{R} = \frac{1}{4\pi\epsilon_0}\frac{Q}{R}$$  23-25

Since the area of a sphere is $4\pi R^2$, the charge on a sphere is related to the charge density by $Q = 4\pi R^2\sigma$. Substituting this expression for Q into Equation 23-25 we have

$$V = \frac{1}{4\pi\epsilon_0}\frac{4\pi R^2\sigma}{R} = \frac{R\sigma}{\epsilon_0}$$

Solving for $\sigma$, we obtain

$$\sigma = \frac{\epsilon_0 V}{R}$$  23-26

Since both *spheres* are at the same potential, the sphere with the smaller radius must have the greater surface charge density. And since $E = \sigma/\epsilon_0$ near the surface of a conductor, the electric field is greatest at points on the conductor where the radius of curvature is least.

For an arbitrarily shaped conductor, the potential at which dielectric breakdown occurs depends on the smallest radius of curvature of any part of the conductor. If the conductor has sharp points of very small radius of curvature, dielectric breakdown will occur at relatively low potentials. In the Van de Graaff generator (see Figure 23-23a), the charge is transferred onto the belt by sharp-edged conductors near the bottom of the belt. The charge is removed from the belt by sharp-edged conductors near the top of the belt. Lightning rods at the top of a tall building draw the charge off a nearby cloud before the potential of the cloud can build up to a destructively large value.

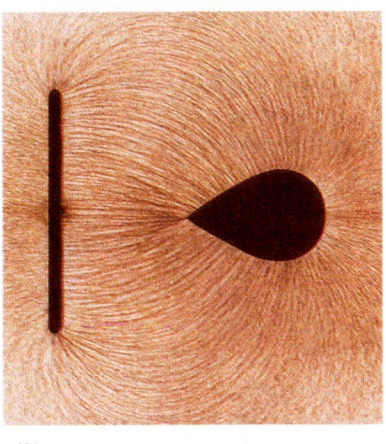

(b)

**FIGURE 23-25** (a) A nonspherical conductor. If a charge is placed on such a conductor, it will produce an electric field that is stronger near point A, where the radius of curvature is small, than near point B, where the radius of curvature is large. (b) Electric field lines near a nonspherical conductor and plate carrying equal and opposite charges. The lines are shown by small bits of thread suspended in oil. Note that the electric field is strongest near points of small radius of curvature, such as at the ends of the plate and at the pointed left side of the conductor. The equipotential surfaces are more closely spaced where the field strength is greater.

# SUMMARY

1. Electric potential, which is defined as the electrostatic potential energy per unit charge, is an important derived physical concept that is related to the electric field.

2. Because potential is a scalar quantity, it is often easier to calculate than the vector electric field. Once V is known, $\vec{E}$ can be calculated from V.

| Topic | Relevant Equations and Remarks |
|---|---|
| **1. Potential Difference** | The potential difference $V_b - V_a$ is defined as the negative of the work per unit charge done by the electric field when a test charge moves from point a to point b: $$\Delta V = V_b - V_a = \frac{\Delta U}{q_0} = -\int_a^b \vec{E} \cdot d\vec{\ell} \qquad 23\text{-}2b$$ |

| Potential difference for infinitesimal displacements | $dV = -\vec{E} \cdot d\vec{\ell}$ | **23-2a** |

**2. Electric Potential**

| Potential due to a point charge | $V = \dfrac{kq}{r} - \dfrac{kq}{r_{\text{ref}}}, \quad (V = 0 \text{ at } r = r_{\text{ref}})$ | **23-7** |

| Coulomb potential | $V = \dfrac{kq}{r}, \quad (V = 0 \text{ at } r = \infty)$ | **23-8** |

| Potential due to a system of point charges | $V = \sum_i \dfrac{kq_i}{r_i}, \quad (V = 0 \text{ at } r_i = \infty, i = 1, 2, \dots)$ | **23-10** |

| Potential due to a continuous charge distribution | $V = \displaystyle\int \dfrac{k\,dq}{r}, \quad (V = 0 \text{ at } r = \infty)$ | **23-19** |

This expression can be used only if the charge distribution is contained in a finite volume so that the potential can be chosen to be zero at infinity.

| Potential and electric field lines | Electric field lines point in the direction of decreasing electric potential. |

| Continuity of electric potential | The potential function $V$ is continuous everywhere in space. |

**3. Computing the Electric Field From the Potential**

The electric field points in the direction of the greatest decrease in the potential.

$$E_t = \frac{dV}{d\ell}$$

**23-13**

| Gradient | A vector that points in the direction of the greatest rate of change in a scalar function and that has a magnitude equal to the derivative of that function, with respect to the distance in that direction, is called the gradient of the function. $\vec{E}$ is the negative gradient of $V$. |

| Potential a function of $x$ alone | $E_x = -\dfrac{dV(x)}{dx}$ | **23-14** |

| Potential a function of $r$ alone | $E_r = -\dfrac{dV(r)}{dr}$ | **23-15** |

**4. *General Relation Between $\vec{E}$ and $V$**

$$\vec{E} = -\vec{\nabla}V = -\left(\frac{\partial V}{\partial x}\hat{i} + \frac{\partial V}{\partial y}\hat{j} + \frac{\partial V}{\partial z}\hat{k}\right) \quad \text{or} \quad V_b - V_a = -\int_a^b \vec{E} \cdot d\vec{\ell}$$

**23-18**

**5. Units**

| $V$ and $\Delta V$ | The SI unit of potential and potential difference is the volt (V): |
| | $1\,\text{V} = 1\,\text{J/C}$ | **23-4** |

| Electric field | $1\,\text{N/C} = 1\,\text{V/m}$ | **23-5** |

| Electron volt | The electron volt (eV) is the change in potential energy of a particle of charge $e$ as it moves from $a$ to $b$, where $\Delta V = V_b - V_a = 1$ volt: |
| | $1\,\text{eV} = 1.60 \times 10^{-19}\,\text{C} \cdot \text{V} = 1.60 \times 10^{-19}\,\text{J}$ | **23-6** |

**6. Potential Energy of Two Point Charges**

$$U = q_0 V = \frac{kq_0 q}{r}, \quad (U = 0 \text{ at } r = \infty)$$

**23-9**

## 7. Potential Functions

| | | | |
|---|---|---|---|
| On the axis of a uniformly charged ring | $V = \dfrac{kQ}{\sqrt{x^2 + a^2}}$, | $(V = 0 \text{ at } |x| = \infty)$ | 23-20 |
| On the axis of a uniformly charged disk | $V = 2\pi k\sigma|x|\left(\sqrt{1 + \dfrac{R^2}{x^2}} - 1\right)$, | $(V = 0 \text{ at } |x| = \infty)$ | 23-21 |
| Near an infinite plane of charge | $V = V_0 - 2\pi k\sigma|x|$, | $(V = V_0 \text{ at } x = 0)$ | 23-22 |
| For a spherical shell of charge | $V = \dfrac{kQ}{r}$, $r \geq R$ | $(V = 0 \text{ at } r = \infty)$ | |
| | $V = \dfrac{kQ}{R}$, $r \leq R$ | $(V = 0 \text{ at } r = \infty)$ | 23-23 |
| For an infinite line charge | $V = 2k\lambda \ln \dfrac{R_{\text{ref}}}{R}$, | $V = 0 \text{ at } r = R_{\text{ref}}$ | 23-24 |

## 8. Charge on a Nonspherical Conductor

On a conductor of arbitrary shape, the surface charge density $\sigma$ is greatest at points where the radius of curvature is smallest.

## 9. Dielectric Breakdown

The amount of charge that can be placed on a conductor is limited by the fact that molecules of the surrounding medium undergo dielectric breakdown at very high electric fields, causing the medium to become a conductor.

**Dielectric strength**

The dielectric strength is the magnitude of the electric field at which dielectric breakdown occurs. The dielectric strength of air is

$$E_{\text{max}} \approx 3 \times 10^6 \text{ V/m} = 3 \text{ MV/m}$$

# PROBLEMS

- Single-concept, single-step, relatively easy
- •• Intermediate-level, may require synthesis of concepts
- ••• Challenging, for advanced students
- **SSM** Solution is in the *Student Solutions Manual*
- **iSOLVE** Problems available on iSOLVE online homework service
- **iSOLVE✓** These "Checkpoint" online homework service problems ask students additional questions about their confidence level, and how they arrived at their answer.

In a few problems, you are given more data than you actually need; in a few other problems, you are required to supply data from your general knowledge, outside sources, or informed estimates.

## Conceptual Problems

1 • **SSM** A positive charge is released from rest in an electric field. Will it move toward a region of greater or smaller electric potential?

2 •• A lithium nucleus and an $\alpha$ particle are at rest. The lithium nucleus has a charge of $+3e$ and a mass of $7\,u$; the $\alpha$ particle has a charge of $+2e$ and a mass of $4\,u$. Which of the following methods would accelerate them both to the same kinetic energy? (a) Accelerate them through the same electrical potential difference. (b) Accelerate the $\alpha$ particle through potential $V_1$ and the lithium nucleus through $(2/3)V_1$. (c) Accelerate the $\alpha$ particle through potential $V_1$ and the lithium nucleus through $(7/4)V_1$. (d) Accelerate the $\alpha$ particle through potential $V_1$ and the lithium nucleus through $(2 \times 7)/(3 \times 4)V$. (e) None of the answers are correct.

3 • If the electric potential is constant throughout a region of space, what can you say about the electric field in that region?

**4** • If $E$ is known at just one point, can $V$ be found at that point?

**5** • In what direction can you move relative to an electric field so that the electric potential does not change?

**6** •• In the calculation of $V$ at a point $x$ on the axis of a ring of charge, does it matter whether the charge $Q$ is uniformly distributed around the ring? Would either $V$ or $E_x$ be different if it were not?

**7** •• SSM Figure 23-26 shows a metal sphere carrying a charge $-Q$ and a point charge $+Q$. Sketch the electric field lines and equipotential surfaces in the vicinity of this charge system.

$-Q$

$+Q$
$\oplus$

**FIGURE 23-26** Problems 7 and 8

**8** •• Repeat Problem 7 with the charge on the metal sphere changed to $+Q$.

**9** •• Sketch the electric field lines and the equipotential surfaces, both near and far, from the conductor shown in Figure 23-25a, assuming that the conductor carries some charge $Q$.

**10** •• Two equal positive charges are separated by a small distance. Sketch the electric field lines and the equipotential surfaces for this system.

**11** • SSM Two equal positive point charges $+Q$ are on the $x$-axis. One is at $x = -a$ and the other is at $x = +a$. At the origin, (a) $E = 0$ and $V = 0$, (b) $E = 0$ and $V = 2kQ/a$, (c) $\vec{E} = (2kQ^2/a^2)\hat{i}$ and $V = 0$, (d) $\vec{E} = (2kQ^2/a^2)\hat{i}$ and $V = 2kQ/a$, or (e) none of the answers are correct.

**12** • The electrostatic potential is measured to be $V(x, y, z) = 4|x| + V_0$, where $V_0$ is a constant. The charge distribution responsible for this potential is (a) a uniformly charged thread in the $xy$ plane, (b) a point charge at the origin, (c) a uniformly charged sheet in the $yz$ plane, or (d) a uniformly charged sphere of radius $1/\pi$ at the origin.

**13** • Two point charges of equal magnitude but opposite sign are on the $x$ axis; $+Q$ is at $x = -a$ and $-Q$ is at $x = +a$. At the origin, (a) $E = 0$ and $V = 0$, (b) $E = 0$ and $V = 2kQ/a$, (c) $\vec{E} = (2kQ^2/a^2)\hat{i}$ and $V = 0$, (d) $\vec{E} = (2kQ^2/a^2)\hat{i}$ and $V = 2kQ/a$, or (e) none of the answers are correct.

**14** •• True or false:

(a) If the electric field is zero in some region of space, the electric potential must also be zero in that region.

(b) If the electric potential is zero in some region of space, the electric field must also be zero in that region.

(c) If the electric potential is zero at a point, the electric field must also be zero at that point.

(d) Electric field lines always point toward regions of lower potential.

(e) The value of the electric potential can be chosen to be zero at any convenient point.

(f) In electrostatics, the surface of a conductor is an equipotential surface.

(g) Dielectric breakdown occurs in air when the potential is $3 \times 10^6$ V.

**15** •• (a) $V$ is constant on a conductor surface. Does this mean that $\sigma$ is constant? (b) If $E$ is constant on a conductor surface, does this mean that $\sigma$ is constant? Does it mean that $V$ is constant?

**16** • SSM Two charged metal spheres are connected by a wire, and sphere $A$ is larger than sphere $B$ (Figure 23-27). The magnitude of the electric potential of sphere $A$ is (a) greater than that at the surface of sphere $B$; (b) less than that at the surface of sphere $B$; (c) the same as that at the surface of sphere $B$; (d) greater than or less than that at the surface of sphere $B$, depending on the radii of the spheres; or (e) greater than or less than that at the surface of sphere $B$, depending on the charge on the spheres.

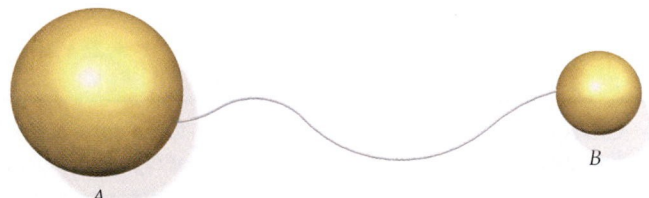

$B$

$A$

**FIGURE 23-27** Problem 16

## Estimation and Approximation

**17** • Estimate the potential difference between a thundercloud and the earth, given that the electrical breakdown of air occurs at fields of roughly $3 \times 10^6$ V/m.

**18** • SSM Estimate the potential difference across the spark gap in a typical automobile spark plug. Because of the high compression of the gas in the piston, the electric field at which the gas sparks is roughly $2 \times 10^7$ V/m.

**19** •• A proton can be thought of as having a "radius" of approximately $10^{-15}$ m. Two protons have a head-on collision with equal and opposite momenta. (a) Estimate the minimum kinetic energy (in MeV) required by each proton to allow the protons to overcome electrostatic repulsion and collide. Do this estimate without using relativity. (b) The rest energy of the proton is 938 MeV. If your value for the kinetic energy is much less than this, then a nonrelativistic calculation was justified. What fraction of the rest energy of the proton is the kinetic energy you calculated in Part (a)?

**20** •• SOLVE When you touch a friend after walking across a rug on a dry day, you typically draw a spark of about 2 mm. Estimate the potential difference between you and your friend before the spark.

## Potential Difference

**21** • SOLVE A uniform electric field of 2 kN/C is in the $x$ direction. A positive point charge $Q = 3$ $\mu$C is released from rest at the origin. (a) What is the potential difference $V(4 \text{ m}) - V(0)$? (b) What is the change in the potential energy of the charge from $x = 0$ to $x = 4$ m? (c) What is the kinetic energy of the charge when it is at $x = 4$ m? (d) Find the potential $V(x)$ if $V(x)$ is chosen to be zero at $x = 0$, (e) 4 kV at $x = 0$, and (f) zero at $x = 1$ m.

**22** • Two large parallel conducting plates separated by 10 cm carry equal and opposite surface charge densities so that the electric field between them is uniform. The difference in potential between the plates is 500 V. An electron is released from rest at the negative plate. (a) What is the magnitude of the electric field between the plates? Is the positive or negative plate at the higher potential? (b) Find the work done by the electric field on the electron as the electron moves from the negative plate to the positive plate. Express your answer in both electron volts and joules. (c) What is the change in potential energy of the electron when it moves from the negative plate to the positive plate? What is its kinetic energy when it reaches the positive plate?

**23** • A positive charge of magnitude 2 $\mu$C is at the origin. (a) What is the electric potential $V$ at a point 4 m from the origin relative to $V = 0$ at infinity? (b) How much work must be done by an outside agent to bring a 3-$\mu$C charge from infinity to $r = 4$ m, assuming that the 2-$\mu$C charge is held fixed at the origin? (c) How much work must be done by an outside agent to bring the 2-$\mu$C charge from infinity to the origin if the 3-$\mu$C charge is first placed at $r = 4$ m and is then held fixed?

**24** •• **iSOLVE** The distance between the $K^+$ and $Cl^-$ ions in KCl is $2.80 \times 10^{-10}$ m. Calculate the energy required to separate the two ions to an infinite distance apart, assuming them to be point charges initially at rest. Express your answer in eV.

**25** •• **iSOLVE** Protons from a Van de Graaff accelerator are released from rest at a potential of 5 MV and travel through a vacuum to a region at zero potential. (a) Find the final speed of the 5-MeV protons. (b) Find the accelerating electric field if the same potential change occurred *uniformly* over a distance of 2.0 m.

**26** •• **SSM** **iSOLVE** An electron gun fires electrons at the screen of a television tube. The electrons start from rest and are accelerated through a potential difference of 30,000 V. What is the energy of the electrons when they hit the screen (a) in electron volts and (b) in joules? (c) What is the speed of impact of electrons with the screen of the picture tube?

**27** •• (a) Derive an expression for the distance of closest approach of an $\alpha$ particle with kinetic energy $E$ to a massive nucleus of charge $Ze$. Assume that the nucleus is fixed in space. (b) Find the distance of closest approach of a 5.0- and a 9.0-MeV $\alpha$ particle to a gold nucleus; the charge of the gold nucleus is $79e$. (Neglect the recoil of the gold nucleus.)

### Potential Due to a System of Point Charges

**28** • Four 2-$\mu$C point charges are at the corners of a square of side 4 m. Find the potential at the center of the square (relative to zero potential at infinity) if (a) all the charges are positive, (b) three of the charges are positive and one is negative, and (c) two are positive and two are negative.

**29** • **iSOLVE** Three point charges are on the x-axis: $q_1$ is at the origin, $q_2$ is at $x = 3$ m, and $q_3$ is at $x = 6$ m. Find the potential at the point $x = 0$, $y = 3$ m if (a) $q_1 = q_2 = q_3 = 2\ \mu C$, (b) $q_1 = q_2 = 2\ \mu C$ and $q_3 = -2\ \mu C$, and (c) $q_1 = q_3 = 2\ \mu C$ and $q_2 = -2\ \mu C$.

**30** • Points $a$, $b$, and $c$ are at the corners of an equilateral triangle of side 3 m. Equal positive charges of 2 $\mu$C are at $a$ and $b$. (a) What is the potential at point $c$? (b) How much work is required to bring a positive charge of 5 $\mu$C from infinity to point $c$ if the other charges are held fixed? (c) Answer Parts (a) and (b) if the charge at $b$ is replaced by a charge of $-2\ \mu C$.

**31** • **iSOLVE** A sphere with radius 60 cm has its center at the origin. Equal charges of 3 $\mu$C are placed at 60° intervals along the equator of the sphere. (a) What is the electric potential at the origin? (b) What is the electric potential at the north pole?

**32** • **SSM** Two point charges $q$ and $q'$ are separated by a distance $a$. At a point $a/3$ from $q$ and along the line joining the two charges the potential is zero. Find the ratio $q/q'$.

**33** •• Two positive charges $+q$ are on the x-axis at $x = +a$ and $x = -a$. (a) Find the potential $V(x)$ as a function of $x$ for points on the x-axis. (b) Sketch $V(x)$ versus $x$. (c) What is the significance of the minimum on your curve?

**34** •• **SSM** A point charge of $+3e$ is at the origin and a second point charge of $-2e$ is on the x-axis at $x = a$. (a) Sketch the potential function $V(x)$ versus $x$ for all $x$. (b) At what point or points is $V(x)$ zero? (c) How much work is needed to bring a third charge $+e$ to the point $x = \frac{1}{2}a$ on the x-axis?

### Computing the Electric Field From the Potential

**35** • **iSOLVE** A uniform electric field is in the negative x direction. Points $a$ and $b$ are on the x-axis, $a$ at $x = 2$ m and $b$ at $x = 6$ m. (a) Is the potential difference $V_b - V_a$ positive or negative? (b) If the magnitude of $V_b - V_a$ is $10^5$ V, what is the magnitude $E$ of the electric field?

**36** • **SSM** The potential due to a particular charge distribution is measured at several points along the x-axis, as shown in Figure 23-28. For what value(s) in the range $0 < x < 10$ m is $E_x = 0$?

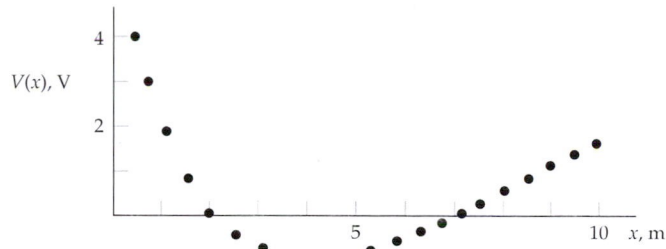

**FIGURE 23-28** Problem 36

**37** • A point charge $q = 3.00\ \mu C$ is at the origin. (a) Find the potential $V$ on the x-axis at $x = 3.00$ m and at $x = 3.01$ m. (b) Does the potential increase or decrease as $x$ increases? Compute $-\Delta V/\Delta x$, where $\Delta V$ is the change in potential from $x = 3.00$ m to $x = 3.01$ m and $x = 0.01$ m. (c) Find the electric field at $x = 3.00$ m, and compare its magnitude with $-\Delta V/\Delta x$ found in Part (b). (d) Find the potential (to three significant figures) at the point $x = 3.00$ m, $y = 0.01$ m, and compare your result with the potential on the x-axis at $x = 3.00$ m. Discuss the significance of this result.

**38** • A charge of $+3.00 \ \mu C$ is at the origin, and a charge of $-3.00 \ \mu C$ is on the $x$-axis at $x = 6.00$ m. (a) Find the potential on the $x$-axis at $x = 3.00$ m. (b) Find the electric field on the $x$-axis at $x = 3.00$ m. (c) Find the potential on the $x$-axis at $x = 3.01$ m, and compute $-\Delta V/\Delta x$, where $\Delta V$ is the change in potential from $x = 3.00$ m to $x = 3.01$ m and $x = 0.01$ m. Compare your result with your answer to Part (b).

**39** • A uniform electric field is in the positive $y$ direction. Points $a$ and $b$ are on the $y$-axis, $a$ at $y = 2$ m and $b$ at $x = 6$ m. (a) Is the potential difference $V_b - V_a$ positive or negative? (b) If the magnitude of $V_b - V_a$ is $2 \times 10^4$ V, what is the magnitude $E$ of the electric field?

**40** • In the following, $V$ is in volts and $x$ is in meters. Find $E_x$ when (a) $V(x) = 2000 + 3000x$, (b) $V(x) = 4000 + 3000x$, (c) $V(x) = 2000 - 3000x$, and (d) $V(x) = -2000$, independent of $x$.

**41** •• A charge $q$ is at $x = 0$ and a charge $-3q$ is at $x = 1$ m. (a) Find $V(x)$ for a general point on the $x$-axis. (b) Find the points on the $x$-axis where the potential is zero. (c) What is the electric field at these points? (d) Sketch $V(x)$ versus $x$.

**42** •• [SSM] [iSOLVE] An electric field is given by $E_x = 2.0x^3$ kN/C. Find the potential difference between the points on the $x$-axis at $x = 1$ m and $x = 2$ m.

**43** •• Three equal charges lie in the $xy$ plane. Two are on the $y$-axis at $y = -a$ and $y = +a$, and the third is on the $x$-axis at $x = a$. (a) What is the potential $V(x)$ due to these charges at a point on the $x$-axis? (b) Find $E_x$ along the $x$-axis from the potential function $V(x)$. Evaluate your answers to Parts (a) and (b) at the origin and at $x = \infty$ to see if they yield expected results.

## Calculations of V for Continuous Charge Distributions

**44** • A charge of $q = +10^{-8}$ C is uniformly distributed on a spherical shell of radius 12 cm. (a) What is the magnitude of the electric field just outside and just inside the shell? (b) What is the magnitude of the electric potential just outside and just inside the shell? (c) What is the electric potential at the center of the shell? What is the electric field at that point?

**45** • An infinite line charge of linear charge density $\lambda = 1.5 \ \mu C/m$ lies on the $z$-axis. Find the potential at distances from the line charge of (a) 2.0 m, (b) 4.0 m, and (c) 12 m, assuming that $V = 0$ at 2.5 m.

**46** •• Derive Equation 23-21 by integrating the electric field $E_x$ along the axis of the disk. (See Equation 22-11.)

**47** •• [SSM] A rod of length $L$ carries a charge $Q$ uniformly distributed along its length. The rod lies along the $y$-axis with its center at the origin. (a) Find the potential as a function of position along the $x$-axis. (b) Show that the result obtained in Part (a) reduces to $V = kQ/x$ for $x \gg L$.

**48** •• A disk of radius $R$ carries a surface charge distribution of $\sigma = \sigma_0 R/r$. (a) Find the total charge on the disk. (b) Find the potential on the axis of the disk a distance $x$ from its center.

**49** •• Repeat Problem 48 if the surface charge density is $\sigma = \sigma_0 r^2/R^2$.

**50** •• A rod of length $L$ carries a charge $Q$ uniformly distributed along its length. The rod lies along the $y$-axis with one end at the origin. Find the potential as a function of position along the $x$-axis.

**51** •• [SSM] A disk of radius $R$ carries a charge density $+\sigma_0$ for $r < a$ and an equal but opposite charge density $-\sigma_0$ for $a < r < R$. The total charge carried by the disk is zero. (a) Find the potential a distance $x$ along the axis of the disk. (b) Obtain an approximate expression for $V(x)$ when $x \gg R$.

**52** •• Use the result obtained in Problem 51a to calculate the electric field along the axis of the disk. Then calculate the electric field by direct integration using Coulomb's law.

**53** •• A rod of length $L$ has a charge $Q$ uniformly distributed along its length. The rod lies along the $x$-axis with its center at the origin. (a) What is the electric potential as a function of position along the $x$-axis for $x > L/2$? (b) Show that for $x \gg L/2$, your result reduces to that due to a point charge $Q$.

**54** •• A conducting spherical shell of inner radius $b$ and outer radius $c$ is concentric with a small metal sphere of radius $a < b$. The metal sphere has a positive charge $Q$. The total charge on the conducting spherical shell is $-Q$. (a) What is the potential of the spherical shell? (b) What is the potential of the metal sphere?

**55** •• Two very long, coaxial cylindrical shell conductors carry equal and opposite charges. The inner shell has radius $a$ and charge $+q$; the other shell has radius $b$ and charge $-q$. The length of each cylindrical shell is $L$. Find the potential difference between the shells.

**56** •• [iSOLVE✓] A uniformly charged sphere has a potential on its surface of 450 V. At a radial distance of 20 cm from this surface, the potential is 150 V. What is the radius of the sphere, and what is the charge of the sphere?

**57** •• Consider two infinite parallel planes of charge, one in the $yz$ plane and the other at distance $x = a$. (a) Find the potential everywhere in space when $V = 0$ at $x = 0$ if the planes carry equal positive charge densities $+\sigma$. (b) Repeat the problem with charge densities equal and opposite, and the charge in the $yz$ plane positive.

**58** •• [SSM] [iSOLVE] Show that for $x \gg R$ the potential on the axis of a disk charge approaches $kQ/x$, where $Q = \sigma \pi R^2$ is the total charge on the disk. [Hint: Write $(x^2 + R^2)^{1/2} = x(1 + R^2/x^2)^{1/2}$ and use the binomial expression.]

**59** •• In Example 23-12, you derived the expression

$$V(r) = \frac{kQ}{2R}\left(3 - \frac{r^2}{R^2}\right)$$

for the potential inside a solid sphere of constant charge density by first finding the electric field. In this problem you derive the same expression by direct integration. Consider a sphere of radius $R$ containing a charge $Q$ uniformly distributed. You wish to find $V$ at some point $r < R$. (a) Find the charge $q'$ inside a sphere of radius $r$ and the potential $V_1$ at $r$ due to this part of the charge. (b) Find the potential $dV_2$ at $r$ due to the charge in a shell of radius $r'$ and thickness $dr'$ at $r' > r$. (c) Integrate your expression in Part (b) from $r' = r$ to $r' = R$ to find $V_2$. (d) Find the total potential $V$ at $r$ from $V = V_1 + V_2$.

**60** • **SOLVE** ✓ An infinite plane of charge has a surface charge density 3.5 $\mu C/m^2$. How far apart are the equipotential surfaces whose potentials differ by 100 V?

**61** • A point charge $q = +\frac{1}{9} \times 10^{-8}$ C is at the origin. Taking the potential to be zero at $r = \infty$, locate the equipotential surfaces at 20-V intervals from 20 to 100 V, and sketch them to scale. Are these surfaces equally spaced?

**62** • **SOLVE** ✓ (a) Find the maximum net charge that can be placed on a spherical conductor of radius 16 cm before dielectric breakdown of the air occurs. (b) What is the potential of the sphere when it carries this maximum charge?

**63** • **SSM** **SOLVE** Find the greatest surface charge density $\sigma_{max}$ that can exist on a conductor before dielectric breakdown of the air occurs.

**64** •• Charge is placed on two conducting spheres that are very far apart and connected by a long thin wire. The radius of the smaller sphere is 5 cm and that of the larger sphere is 12 cm. The electric field at the surface of the larger sphere is 200 kV/m. Find the surface charge density on each sphere.

**65** •• Two concentric spherical shell conductors carry equal and opposite charges. The inner shell has radius $a$ and charge $+q$; the outer shell has radius $b$ and charge $-q$. Find the potential difference between the shells, $V_a - V_b$.

**66** ••• **SSM** Calculate the potential relative to infinity at the point a distance $R/2$ from the center of a uniformly charged thin spherical shell of radius $R$ and charge $Q$.

**67** •• Two identical uncharged metal spheres connected by a wire are placed close by two similar conducting spheres with equal and opposite charges, as shown in Figure 23-29. (a) Sketch the electric field lines between spheres 1 and 3 and between spheres 2 and 4. (b) What can be said about the potentials $V_1$, $V_2$, $V_3$, and $V_4$ of the spheres? (c) If spheres 3 and 4 are also connected by a wire, show that the final charge on each must be zero.

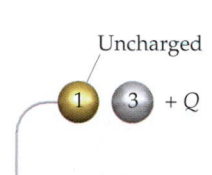

**FIGURE 23-29** Problem 67

## General Problems

**68** • An electric dipole has a positive charge of $4.8 \times 10^{-19}$ C separated from a negative charge of the same magnitude by $6.4 \times 10^{-10}$ m. What is the electric potential at a point $9.2 \times 10^{-10}$ m from each of the two charges? (a) 9.4 V. (b) Zero. (c) 4.2 V. (d) $5.1 \times 10^9$ V. (e) 1.7 V.

**69** • Two positive charges $+q$ are on the y-axis at $y = +a$ and $y = -a$. (a) Find the potential $V$ for any point on the x-axis. (b) Use your result in Part (a) to find the electric field at any point on the x-axis.

**70** • **SOLVE** ✓ If a conducting sphere is to be charged to a potential of 10,000 V, what is the smallest possible radius of the sphere so that the electric field will not exceed the dielectric strength of air?

**71** •• **SSM** Two infinitely long parallel wires carry a uniform charge per unit length $\lambda$ and $-\lambda$ respectively. The wires are in the $xz$ plane, parallel with the $z$ axis. The positively charged wire intersects the $x$ axis at $x = -a$, and the negatively charged wire intersects the $x$ axis at $x = +a$. (a) Choose the origin as the reference point where the potential is zero, and express the potential at an arbitrary point $(x, y)$ in the $xy$ plane in terms of $x$, $y$, $\lambda$, and $a$. Use this expression to solve for the potential everywhere on the $y$ axis. (b) Use a spreadsheet program to plot the equipotential curve in the $xy$ plane that passes through the point $x = \frac{1}{4}a$, $y = 0$. Use $a = 5$ cm and $\lambda = 5$ nC/m.

**72** •• The equipotential curve graphed in Problem 71 looks like a circle. (a) Show explicitly that it is a circle. (b) The equipotential circle in the $xy$ plane is the intersection of a three-dimensional equipotential surface and the $xy$ plane. Describe the three-dimensional surface in a sentence or two.

**73** •• The hydrogen atom can be modeled as a positive point charge of magnitude $+e$ (the proton) surrounded by a charge density (the electron) which has the formula $\rho = \rho_0 e^{-2r/a}$ (from quantum mechanics), where $a = 0.523$ nm. (a) Calculate the value of $\rho_0$ needed for charge neutrality. (b) Calculate the electrostatic potential (relative to infinity) at any distance $r$ from the proton.

**74** • **SOLVE** An isolated aluminum sphere of radius 5.0 cm is at a potential of 400 V. How many electrons have been removed from the sphere to raise it to this potential?

**75** • **SOLVE** A point charge $Q$ resides at the origin. A particle of mass $m = 0.002$ kg carries a charge of 4.0 $\mu C$. The particle is released from rest at $x = 1.5$ m. Its kinetic energy as it passes $x = 1.0$ m is 0.24 J. Find the charge $Q$.

**76** •• **SSM** **SOLVE** A Van de Graaff generator has a potential difference of 1.25 MV between the belt and the outer shell. Charge is supplied at the rate of 200 $\mu C/s$. What minimum power is needed to drive the moving belt?

**77** •• A positive point charge $+Q$ is located at $x = -a$. (a) How much work is required to bring a second equal positive point charge $+Q$ from infinity to $x = +a$? (b) With the two equal positive point charges at $x = -a$ and $x = +a$, how much work is required to bring a third charge $-Q$ from infinity to the origin? (c) How much work is required to move the charge $-Q$ from the origin to the point $x = 2a$ along the semicircular path shown (Figure 23-30)?

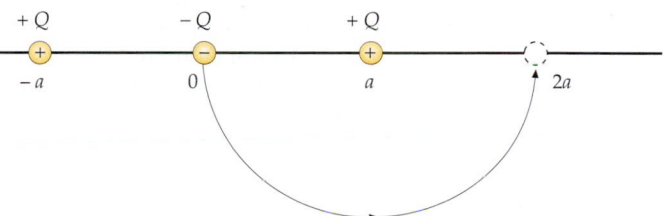

**FIGURE 23-30** Problem 77

**78** •• A charge of 2 nC is uniformly distributed around a ring of radius 10 cm that has its center at the origin and its axis along the x-axis. A point charge of 1 nC is located at $x = 50$ cm. Find the work required to move the point charge to the origin. Give your answer in both joules and electron volts.

**79** •• ✓ The centers of two metal spheres of radius 10 cm are 50 cm apart on the $x$-axis. The spheres are initially neutral, but a charge $Q$ is transferred from one sphere to the other, creating a potential difference between the spheres of 100 V. A proton is released from rest at the surface of the positively charged sphere and travels to the negatively charged sphere. At what speed does it strike the negatively charged sphere?

**80** •• (a) Using a spreadsheet program, graph $V(x)$ versus $x$ for the uniformly charged ring in the $yz$ plane given by Equation 23-20. (b) At what point is $V(x)$ a maximum? (c) What is $E_x$ at this point?

**81** •• A spherical conductor of radius $R_1$ is charged to 20 kV. When it is connected by a long fine wire to a second conducting sphere far away, its potential drops to 12 kV. What is the radius of the second sphere?

**82** •• SSM ✓ A metal sphere centered at the origin carries a surface charge of charge density $\sigma = 24.6 \text{ nC/m}^2$. At $r = 2.0$ m, the potential is 500 V and the magnitude of the electric field is 250 V/m. Determine the radius of the metal sphere.

**83** •• Along the axis of a uniformly charged disk, at a point 0.6 m from the center of the disk, the potential is 80 V and the magnitude of the electric field is 80 V/m; at a distance of 1.5 m, the potential is 40 V and the magnitude of the electric field is 23.5 V/m. Find the total charge residing on the disk.

**84** •• A radioactive $^{210}$Po nucleus emits an $\alpha$ particle of charge $+2e$ and energy 5.30 MeV. Assume that just after the $\alpha$ particle is formed and escapes from the nucleus, it is a distance $R$ from the center of the daughter nucleus $^{206}$Pb, which has a charge $+82e$. Calculate $R$ by setting the electrostatic potential energy of the two particles at this separation equal to 5.30 MeV. (Neglect the size of the $\alpha$ particle.)

**85** •• Two large, parallel, nonconducting planes carry equal and opposite charge densities of magnitude $\sigma$. The planes have area $A$ and are separated by a distance $d$. (a) Find the potential difference between the planes. (b) A conducting slab having thickness $a$ and area $A$, the same area as the planes, is inserted between the original two planes. The slab carries no net charge. Find the potential difference between the original two planes and sketch the electric field lines in the region between the original two planes.

**86** ••• A point charge $q_1$ is at the origin and a second point charge $q_2$ is on the $x$-axis at $x = a$, as in Example 23-5. (a) Calculate the electric field everywhere on the $x$-axis from the potential function given in that example. (b) Find the potential at a general point on the $y$-axis. (c) Use your result from Part (b) to calculate the $y$ component of the electric field on the $y$-axis. Compare your result with that obtained directly from Coulomb's law.

**87** ••• SSM A point charge $q$ is a distance $d$ from a grounded conducting plane of infinite extent (Figure 23-31a). For this configuration the potential $V$ is zero, both at all points infinitely far from the particle in all directions, and at all points on the conducting plane. Consider a set of coordinate axes with the particle located on the $x$ axis at $x = d$. A second configuraton (Figure 23-31b) has the conducting plane replaced by a particle of charge $-q$ located on the $x$ axis at $x = -d$. (a) Show

that for the second configuration the potential function is zero at all points infinitely far from the particle in all directions, and at all points on the $yz$ plane—just as was the case for the first configuration. (b) A theorem, called the uniqueness theorem, shows that throughout the half-space $x > 0$ the potential function $V$—and thus the electric field $\vec{E}$—for the two configurations are identical. Using this result, obtain the electric field $\vec{E}$ at every point in the $yz$ plane in the second configuration. (The uniqueness theorem tells us that in the first configuration the electric field at each point in the $yz$ plane is the same as it is in the second configuration.) Use this result to find the surface charge density $\sigma$ at each point in the conducting plane (in the first configuration).

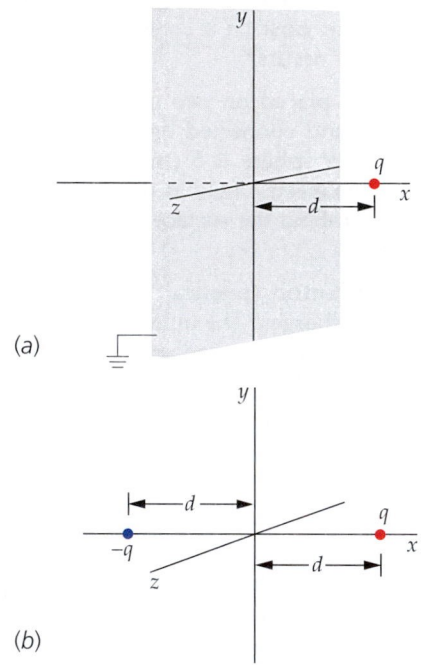

(a)

(b)

**FIGURE 23-31** Problem 87

**88** ••• A particle of mass $m$ carrying a positive charge $q$ is constrained to move along the $x$-axis. At $x = -L$ and $x = L$ are two ring charges of radius $L$ (Figure 23-32). Each ring is centered on the $x$-axis and lies in a plane perpendicular to it. Each carries a positive charge $Q$. (a) Obtain an expression for the potential due to the ring charges as a function of $x$. (b) Show that $V(x)$ is a minimum at $x = 0$. (c) Show that for $x << L$, the potential is of the form $V(x) = V(0) + \alpha x^2$. (d) Derive an expression for the angular frequency of oscillation of the mass $m$ if it is displaced slightly from the origin and released.

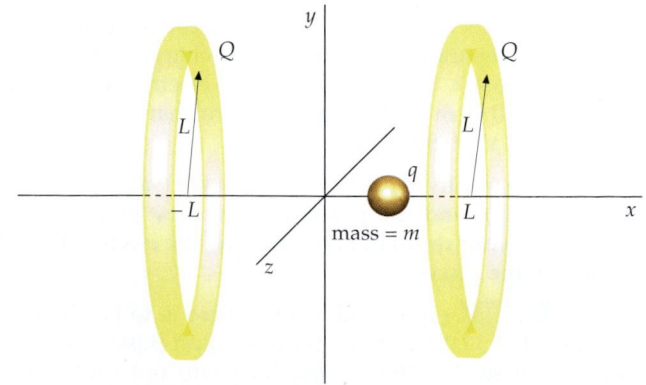

**FIGURE 23-32** Problem 88

**89** ••• Three concentric conducting spherical shells have radii $a$, $b$, and $c$ so that $a < b < c$. Initially, the inner shell is uncharged, the middle shell has a positive charge $Q$, and the outer shell has a negative charge $-Q$. (a) Find the electric potential of the three shells. (b) If the inner and outer shells are now connected by a wire that is insulated as it passes through the middle shell, what is the electric potential of each of the three shells, and what is the final charge on each shell?

**90** ••• **SSM** Consider two concentric spherical metal shells of radii $a$ and $b$, where $b > a$. The outer shell has a charge $Q$, but the inner shell is grounded. This means that the inner shell is at zero potential and that electric field lines leave the outer shell and go to infinity, but other electric field lines leave the outer shell and end on the inner shell. Find the charge on the inner shell.

**91** ••• Show that the total work needed to assemble a uniformly charged sphere with charge $Q$ and radius $R$ is given by $W = U = \dfrac{3}{5}\dfrac{Q^2}{4\pi\epsilon_0 R}$, where $U$ is the electrostatic potential energy of the sphere. *Hint: Let $\rho$ be the charge density of the sphere with charge $Q$ and radius $R$. Calculate the work $dW$ to bring in charge $dq$ from infinity to the surface of a uniformly charged sphere of radius $r$ ($r < R$) and charge density $\rho$. (No additional work is required to smear $dq$ throughout a spherical shell of radius $r$, thickness $dr$, and charge density $\rho$.)*

**92** •• Use the result of Problem 91 to calculate the *classical electron radius*, the radius of a uniform sphere of charge $-e$ that has electrostatic potential energy equal to the rest energy of the electron ($5.11 \times 10^5$ eV). Comment on the shortcomings of this model for the electron.

**93** •• (a) Consider a uniformly charged sphere of radius $R$ and total charge $Q$ which is composed of an incompressible fluid, such as water. If the sphere fissions (splits) into two halves of equal volume and equal charge, and if these halves stabilize into spheres, what is the radius $R'$ of each? (b) Using the expression for potential energy shown in Problem 91, calculate the change in the total electrostatic potential energy of the charged fluid. Assume that the spheres are separated by a large distance.

**94** ••• **SSM** Problem 93 can be modified to be used as a very simple model for nuclear fission. When a $^{235}$U nucleus absorbs a neutron, it can fission into the fragments $^{140}$Xe and $^{94}$Sr, plus 2 neutrons ejected. The $^{235}$U has 92 protons, while $^{140}$Xe has 54 and $^{94}$Sr has 38. Estimate the energy liberated by this fission process (in MeV), assuming that the mass density of the nucleus is constant and has a value $\rho \sim 4 \times 10^{17}$ kg/m$^3$.

**95** ••• (a) Consider an imaginary spherical surface and a point charge $q$ that is located outside the surface. Show by direct integration that the potential at the center of the spherical surface due to the presence of point charge is the average of the potential over the surface of the sphere. (b) Argue from the superposition principle that this result must hold for any spherical surface and any configuration of charges outside the surface.

# Electrostatic Energy and Capacitance

THE ENERGY FOR THE ELECTRONIC FLASH OF THE CAMERA WAS STORED IN A CAPACITOR IN THE FLASH UNIT.

**?** **How is energy stored in a capacitor? (See Section 24-3.)**

24-1    Electrostatic Potential Energy

24-2    Capacitance

24-3    The Storage of Electrical Energy

24-4    Capacitors, Batteries, and Circuits

24-5    Dielectrics

24-6    Molecular View of a Dielectric

**W**hen we bring a point charge $q$ from far away to a region where other charges are present, we must do work $qV$, where $V$ is the potential at the final position due to the other charges in the vicinity. The work done is stored as electrostatic potential energy. The electrostatic potential energy of a system of charges is the total work needed to assemble the system.

When positive charge is placed on an isolated conductor, the potential of the conductor increases. The ratio of the charge to the potential is called the **capacitance** of the conductor. A useful device for storing charge and energy is the capacitor, which consists of two conductors, closely spaced but insulated from each other. When attached to a source of potential difference, such as a battery, the conductors acquire equal and opposite charges. The ratio of the magnitude of the charge on either conductor to the potential difference between the conductors is the capacitance of the capacitor. Capacitors have many uses. The flash attachment for your camera uses a capacitor to store the energy needed to provide the sudden flash of light. Capacitors are also used in the tuning circuits of devices such as radios, televisions, and cellular phones, allowing them to operate at specific frequencies. ➤ **Circuits containing batteries and capacitors are presented in this chapter. In the next few chapters, these techniques and concepts will be further developed in circuits containing resistors, inductors, and other devices.**

The first capacitor was the Leyden jar, a glass container lined inside and out with gold foil. It was invented at the University of Leyden in the Netherlands by eighteenth-century experimenters who, while studying the effects of electric charges on people and animals, got the idea of trying to store a large amount of charge in a bottle of water. An experimenter held up a jar of water in one hand while charge was conducted to the water by a chain from a static electric generator. When the experimenter reached over to lift the chain out of the water with his other hand, he was knocked unconscious. Benjamin Franklin realized that the device for storing charge did not have to be jar shaped and used foil-covered window glass, called Franklin panes. With several of these connected in parallel, Franklin stored a large charge and attempted to kill a turkey with it. Instead, he knocked himself out. Franklin later wrote, "I tried to kill a turkey but nearly succeeded in killing a goose."

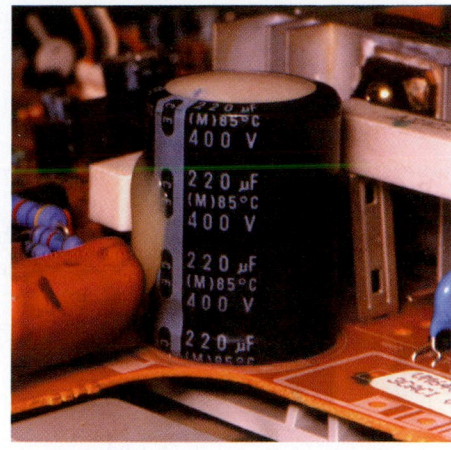

Capacitors are used in large numbers in common electronic devices such as television sets. Some capacitors are used to store energy, but the majority of them are used to filter unwanted electrical frequencies.

## 24-1 Electrostatic Potential Energy

If we have a point charge $q_1$ at point 1, the potential $V_2$ at point 2 a distance $r_{1,2}$ away is given by

$$V_2 = \frac{kq_1}{r_{1,2}}$$

To bring a second point charge $q_2$ in from rest at infinity to rest at point 2 requires that we do work:

$$W_2 = q_2 V_2 = \frac{kq_2 q_1}{r_{1,2}}$$

The potential at point 3, a distance $r_{1,3}$ from $q_1$ and a distance $r_{2,3}$ from $q_2$, is given by

$$V_3 = \frac{kq_1}{r_{1,3}} + \frac{kq_2}{r_{2,3}}$$

To bring in an additional point charge $q_3$ from rest at infinity to rest at point 3 requires that we must do additional work:

$$W_3 = q_3 V_3 = \frac{kq_3 q_1}{r_{1,3}} + \frac{kq_3 q_2}{r_{2,3}}$$

The total work required to assemble the three charges is the **electrostatic potential energy** $U$ of the system of three point charges:

$$U = \frac{kq_2 q_1}{r_{1,2}} + \frac{kq_3 q_1}{r_{1,3}} + \frac{kq_3 q_2}{r_{2,3}} \qquad \text{24-1}$$

This quantity of work is independent of the order in which the charges are brought to their final positions. In general,

> The electrostatic potential energy of a system of point charges is the work needed to bring the charges from an infinite separation to their final positions.

ELECTROSTATIC POTENTIAL ENERGY OF A SYSTEM

The first two terms on the right-hand side of Equation 24-1 can be written

$$\frac{kq_2q_1}{r_{1,2}} + \frac{kq_3q_1}{r_{1,3}} = q_1\left(\frac{kq_2}{r_{1,2}} + \frac{kq_3}{r_{1,3}}\right) = q_1V_1$$

where $V_1$ is the potential at the location of $q_1$ due to charges $q_2$ and $q_3$. Similarly, the second and third terms represent the charge $q_3$ times the potential due to charges $q_1$ and $q_2$, and the first and third terms equal the charge $q_2$ times the potential due to charges $q_1$ and $q_2$. We can thus rewrite Equation 24-1 as

$$U = \frac{kq_2q_1}{r_{1,2}} + \frac{kq_3q_1}{r_{1,3}} + \frac{kq_3q_2}{r_{2,3}} = \frac{1}{2}(U + U)$$

$$= \frac{1}{2}\left(\frac{kq_2q_1}{r_{1,2}} + \frac{kq_3q_1}{r_{1,3}} + \frac{kq_3q_2}{r_{2,3}} + \frac{kq_2q_1}{r_{1,2}} + \frac{kq_3q_1}{r_{1,3}} + \frac{kq_3q_2}{r_{2,3}}\right)$$

$$= \frac{1}{2}\left[q_1\left(\frac{kq_2}{r_{1,2}} + \frac{kq_3}{r_{1,3}}\right) + q_2\left(\frac{kq_3}{r_{2,3}} + \frac{kq_1}{r_{1,2}}\right) + q_3\left(\frac{kq_1}{r_{1,3}} + \frac{kq_2}{r_{2,3}}\right)\right]$$

The electrostatic potential energy $U$ of a system of $n$ point charges is thus

$$U = \frac{1}{2}\sum_{i=1}^{n} q_iV_i \qquad\qquad 24\text{-}2$$

ELECTROSTATIC POTENTIAL ENERGY OF A SYSTEM OF POINT CHARGES

where $V_i$ is the potential at the location of the $i$th charge due to all of the other charges.

Equation 24-2 also describes the electrostatic potential energy of a continuous charge distribution. Consider a spherical conductor of radius $R$. When the sphere carries a charge $q$, its potential relative to $V = 0$ at infinity is

$$V = \frac{kq}{R}$$

The work we must do to bring an additional amount of charge $dq$ from infinity to the conductor is $V\,dq$. This work equals the increase in the potential energy of the conductor:

$$dU = V\,dq = \frac{kq}{R}\,dq$$

The total potential energy $U$ is the integral of $dU$ as $q$ increases from zero to its final value $Q$. Integrating, we obtain

$$U = \frac{k}{R}\int_0^Q q\,dq = \frac{kQ^2}{2R} = \frac{1}{2}QV \qquad\qquad 24\text{-}3$$

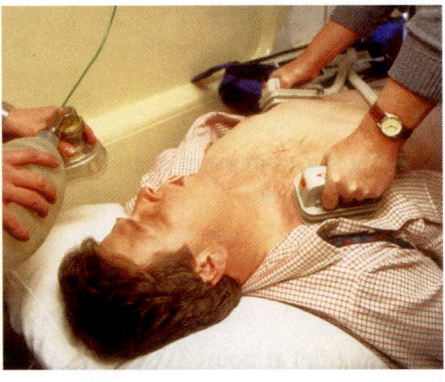

In about two thirds of the people that go into cardiac arrest the heart goes into a state called ventricular fribrillation. In this state the heart quivers, spasms chaotically, and does not pump. To defribrillate the heart a significant current is passed through it, which causes it to stop. Then the pacemaker cells in the heart can again establish a regular heartbeat. An external defibrillator applies a large voltage across the chest.

where $V = kQ/R$ is the potential on the surface of the fully charged sphere. We can interpret Equation 24-3 as $U = Q \times \frac{1}{2}V$ where $\frac{1}{2}V$ is the average potential of the sphere during the charging process. During the charging process, bringing the first element of charge in from infinity to the uncharged sphere requires no work because the sphere is initially uncharged. Therefore, the charge being brought in is not repelled by the charge on the sphere. As the charge on the sphere accumulates, bringing in additional elements of charge to the sphere requires more and more work; when the sphere is almost fully charged, bringing the last element of charge in against the repulsive force of the charge on the sphere requires the most work. The average potential of the sphere during the charging process is one-half its final potential $V$, so the total work required to bring in the entire charge equals $\frac{1}{2}QV$. Although we derived Equation 24-3 for a spherical conductor, it holds for any conductor. The potential of any conductor is proportional to its charge $q$, so we can write $V = \alpha q$, where $\alpha$ is a constant. The work needed to bring an additional charge $dq$ from infinity to the conductor is $V\,dq = \alpha q\,dq$, and the total work needed to put a charge $Q$ on the conductor is $\frac{1}{2}\alpha Q^2 = \frac{1}{2}QV$. If we have a set of $n$ conductors with the $i$th conductor at potential $V_i$ and carrying a charge $Q_i$, the electrostatic potential energy is

$$U = \frac{1}{2}\sum_{i=1}^{n} Q_i V_i \qquad\qquad 24\text{-}4$$

ELECTROSTATIC POTENTIAL ENERGY OF A SYSTEM OF CONDUCTORS

---

*WORK REQUIRED TO MOVE POINT CHARGES*    **EXAMPLE 24-1**

**Points $A$, $B$, $C$, and $D$ are at the corners of a square of side $a$, as shown in Figure 24-1. Four identical positive point charges, each with charge $q$, are initially at rest at infinite separation. ($a$) Calculate the total work required to place the point charges at each corner of the square by separately calculating the work required to move each charge to its final position. ($b$) Show that Equation 24-2 gives the total work.**

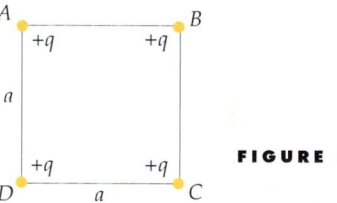

**FIGURE 24-1**

**PICTURE THE PROBLEM**   No work is needed to place the first charge at point $A$ because the potential is zero when the other three charges are at infinity. As each additional charge is brought into place, work must be done because of the presence of the previous charges.

($a$) 1. Place the first charge at point $A$. To accomplish this step, the work $W_A$ that is needed is zero:

$$W_A = 0$$

2. Bring the second charge to point $B$. The work required is $W_B = qV_A$, where $V_A$ is the potential at point $B$ due to the first charge at point $A$ a distance $a$ away:

$$W_B = qV_A = q\left(\frac{kq}{a}\right) = \frac{kq^2}{a}$$

3. $W_C = qV_C$, where $V_C$ is the potential at point $C$ due to $q$ at point $A$ a distance $\sqrt{2}a$ away and $q$ at point $B$ a distance $a$ away:

$$W_C = qV_C = q\left(\frac{kq}{a} + \frac{kq}{\sqrt{2}a}\right) = \left(1 + \frac{1}{\sqrt{2}}\right)\frac{kq^2}{a}$$

4. Similar considerations give $W_D$, the work needed to bring the fourth charge to point $D$:

$$W_D = qV_D = q\left(\frac{kq}{a} + \frac{kq}{\sqrt{2}a} + \frac{kq}{a}\right)$$

$$= \left(2 + \frac{1}{\sqrt{2}}\right)\frac{kq^2}{a}$$

5. Summing the individual contributions gives the total work required to assemble the four charges:

$$W_{\text{total}} = W_A + W_B + W_C + W_D = \boxed{\left(4 + \sqrt{2}\right)\frac{kq^2}{a}}$$

($b$) 1. Calculate $W_{\text{total}}$ from Equation 24-2. Use $V_D$ from Part ($a$), step 4 for the potential at the location of each charge. There are four identical terms, one from each charge:

$$W_{\text{total}} = U = \frac{1}{2}\sum_{i=1}^{4} q_i V_i$$

2. The potential at the location of each charge is $V_D$ from step 4. Substitute $V_D$ for $V_i$ and solve for $W_{\text{total}}$:

$$W_{\text{total}} = \frac{1}{2}\sum_{i=1}^{4}\left[q_i\left(2 + \frac{1}{\sqrt{2}}\right)\frac{kq}{a}\right] = \frac{1}{2}\left(2 + \frac{1}{\sqrt{2}}\right)\frac{kq}{a}\sum_{i=1}^{4} q_i$$

$$= \frac{1}{2}\left(2 + \frac{1}{\sqrt{2}}\right)\frac{kq}{a}4q = \boxed{\left(4 + \sqrt{2}\right)\frac{kq^2}{a}}$$

**REMARKS** $W_{total}$ equals the total electrostatic energy of the charge distribution.

**EXERCISE** (a) How much additional work is required to bring a fifth positive charge $q$ from infinity to the center of the square? (b) What is the total work required to assemble the five-charge system? [*Answer* (a) $4\sqrt{2}\, kq^2/a$, (b) $(4 + 5\sqrt{2})\, kq^2/a$]

## **24-2 Capacitance**

The potential $V$ due to the charge $Q$ on a single isolated conductor is proportional to $Q$ and depends on the size and shape of the conductor. Typically, the larger the surface area of a conductor the more charge it can carry for a given potential. For example, the potential of a spherical conductor of radius $R$ carrying a charge $Q$ is

$$V = \frac{kQ}{R}$$

The ratio of charge $Q$ to the potential $V$ of an isolated conductor is called its capacitance $C$:

$$C = \frac{Q}{V} \qquad \text{24-5}$$

DEFINITION—CAPACITANCE

Capacitance is a measure of the capacity to store charge for a given potential difference. Since the potential is proportional to the charge, this ratio does not depend on either $Q$ or $V$, but only on the size and shape of the conductor. The self-capacitance of a spherical conductor is

$$C = \frac{Q}{V} = \frac{Q}{kQ/R} = \frac{R}{k} = 4\pi \epsilon_0 R \qquad \text{24-6}$$

The SI unit of capacitance is the coulomb per volt, which is called a **farad** (F) after the great English experimentalist Michael Faraday:

$$1\,\text{F} = 1\,\text{C/V} \qquad \text{24-7}$$

The farad is a rather large unit, so submultiples such as the microfarad (1 $\mu$F = $10^{-6}$ F) or the picofarad (1 pF = $10^{-12}$ F) are often used. Since capacitance is in farads and $R$ is in meters, we can see from Equation 24-6 that the SI unit for the permittivity of free space, $\epsilon_0$, can also be written as a farad per meter:

$$\epsilon_0 = 8.85 \times 10^{-12}\,\text{F/m} = 8.85\,\text{pF/m} \qquad \text{24-8}$$

**EXERCISE** Find the radius of a spherical conductor that has a capacitance of 1 F. (*Answer* $8.99 \times 10^9$ m, which is about 1400 times the radius of the earth)

We see from the above exercise that the farad is indeed a very large unit.

**EXERCISE** A sphere of capacitance $C_1$ carries a charge of 20 $\mu$C. If the charge is increased to 60 $\mu$C, what is the new capacitance $C_2$? (*Answer* $C_2 = C_1$. The capacitance does not depend on the charge. If the charge is tripled, the potential of the sphere will be tripled and the ratio $Q/V$, which depends only on the radius of the sphere, remains unchanged.)

## Capacitors

A device consisting of two conductors carrying equal but opposite charges is called a **capacitor.** A capacitor is usually charged by transferring a charge $Q$ from one conductor to the other conductor, which leaves one of the conductors with a charge $+Q$ and the other conductor with a charge $-Q$. The capacitance of the device is defined to be $Q/V$, where $Q$ is the magnitude of the charge on either conductor and $V$ is the magnitude of the potential difference between the conductors.[†] To calculate the capacitance, we place equal and opposite charges on the conductors and then find the potential difference $V$ by first finding the electric field $\vec{E}$ due to the charges.

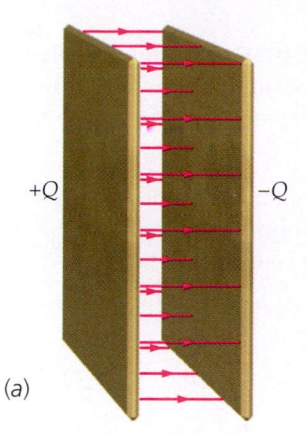

(a)

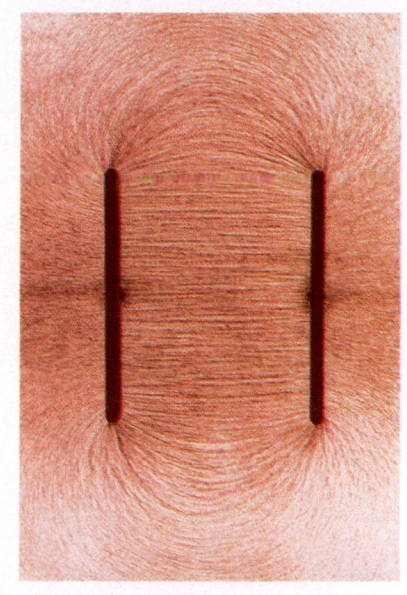

(b)

**FIGURE 24-2** (*a*) Electric field lines between the plates of a parallel-plate capacitor. The lines are equally spaced between the plates, indicating that the field is uniform. (*b*) Electric field lines in a parallel-plate capacitor shown by small bits of thread suspended in oil.

## Parallel-Plate Capacitors

A common capacitor is the **parallel-plate capacitor,** which utilizes two parallel conducting plates. In practice, the plates are often thin metallic foils that are separated and insulated from one another by a thin plastic film. This "sandwich" is then rolled up, which allows for a large surface area in a relatively small space. Let $A$ be the area of the surface (the area of the side of each plate that faces the other plate), and let $d$ be the separation distance, which is small compared to the length and width of the plates. We place a charge $+Q$ on one plate and $-Q$ on the other plate. These charges attract each other and become uniformly distributed on the inside surfaces of the plates. Since the plates are close together, the electric field between them is approximately the same as the field between two infinite planes of equal and opposite charge. Each plate contributes a uniform field of magnitude $E = \sigma/(2\epsilon_0)$; Equation 22-24 giving a total field strength $E = \sigma/\epsilon_0$, where $\sigma = Q/A$ is the magnitude of the charge per unit area on either plate. Since $\vec{E}$ is uniform between the plates (Figure 24-2), the potential difference between the plates equals the field strength $E$ times the plate separation $d$:

$$V = Ed = \frac{\sigma}{\epsilon_0} d = \frac{Qd}{\epsilon_0 A} \qquad \text{24-9}$$

The capacitance of the parallel-plate capacitor is thus

$$C = \frac{Q}{V} = \frac{\epsilon_0 A}{d} \qquad \text{24-10}$$

CAPACITANCE OF A PARALLEL-PLATE CAPACITOR

Note that because $V$ is proportional to $Q$, the capacitance does not depend on either $Q$ or $V$. For a parallel-plate capacitor, the capacitance is proportional to the area of the plates and is inversely proportional to the gap width (separation distance). In general, capacitance depends on the size, shape, and geometrical arrangement of the conductors and capacitance also depends on the properties of the insulating medium between the conductors.

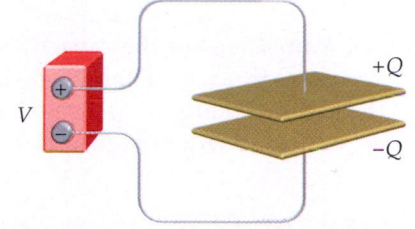

**FIGURE 24-3** When the conductors of an uncharged capacitor are connected to the terminals of a battery, the battery "pumps" charge from one conductor to the other until the potential difference between the conductors equals that between the battery terminals.[‡] The amount of charge transferred through the battery is $Q = CV$.

---

[†] When we speak of the charge on a capacitor, we mean the magnitude of the charge on either conductor. The use of $V$ rather than $\Delta V$ for the magnitude of the potential difference between the plates is standard and simplifies many of the equations relating to capacitance.

[‡] We will discuss batteries more fully in Chapter 25. Here, all we need to know is that a battery is a device that stores energy, supplies electrical energy, and maintains a constant potential difference $V$ between its terminals.

**EXAMPLE 24-2**

A parallel-plate capacitor has square plates of edge length 10 cm separated by 1 mm. (*a*) Calculate the capacitance of this device. (*b*) If this capacitor is charged to 12 V, how much charge is transferred from one plate to another?

**PICTURE THE PROBLEM** The capacitance $C$ is determined by the area and the separation of the plates. Once $C$ is found, the charge for a given voltage $V$ is found from the definition of capacitance $C = Q/V$.

1. We find the capacitance using Equation 24-10:

$$C = \frac{\epsilon_0 A}{d} = \frac{(8.85 \text{ pF/m})(0.1 \text{ m})^2}{0.001 \text{ m}} = \boxed{88.5 \text{ pF}}$$

2. The charge transferred is found from the definition of capacitance:

$$Q = CV = (88.5 \text{ pF})(12 \text{ V}) = 1.06 \times 10^{-9} \text{ C}$$

$$= \boxed{1.06 \text{ nC}}$$

**REMARKS** $Q$ is the magnitude of the charge on each plate of the capacitor. In this case, $Q$ corresponds to roughly $6.6 \times 10^9$ electrons.

**EXERCISE** How large would the plate area have to be for the capacitance to be 1 F? (*Answer* $A = 1.13 \times 10^8 \text{ m}^2$, which corresponds to a square 10.6 km on a side)

## Cylindrical Capacitors

A cylindrical capacitor consists of a small conducting cylinder or wire of radius $R_1$ and a larger, concentric cylindrical conducting shell of radius $R_2$. A coaxial cable, such as that used for cable television, can be thought of as a cylindrical capacitor. The capacitance per unit length of a coaxial cable is important in determining the transmission characteristics of the cable.

**FIGURE 24-4**

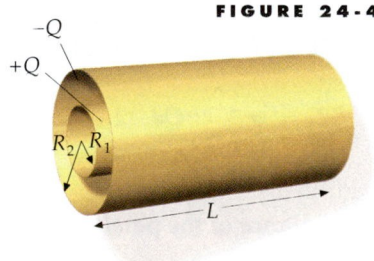

**EXAMPLE 24-3**

Find an expression for the capacitance of a cylindrical capacitor consisting of two conductors, each of length $L$. One conductor is a cylinder of radius $R_1$ and the second conductor is a coaxial cylindrical shell of inner radius $R_2$, with $R_1 < R_2 \ll L$ as shown in Figure 24-4.

**PICTURE THE PROBLEM** We place charge $+Q$ on the inner conductor and charge $-Q$ on the outer conductor and calculate the potential difference $V = V_b - V_a$ from the electric field between the conductors, which is found from Gauss's law. Since the electric field is not uniform (it depends on $R$) we must integrate to find the potential difference.

1. The capacitance is defined as the ratio $Q/V$:           $C = Q/V$

2. $V$ is related to the electric field:           $dV = -\vec{E} \cdot d\vec{\ell}$

3. To find $E_R$ we choose a soup-can shaped Gaussian surface of radius $R$ and length $\ell$, where ($R_1 < R < R_2$) and $\ell \ll L$. The Gaussian surface is located far from the ends of the cylindrical shells (Figure 24-5):

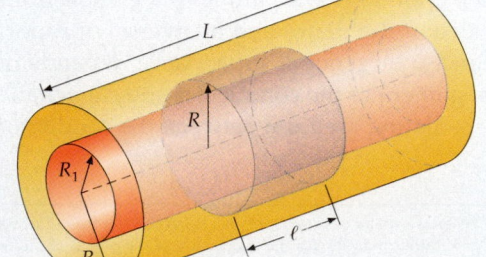

**FIGURE 24-5**

4. Far from the ends of the shells $\vec{E}$ is radial, so there is no flux of $\vec{E}$ through the flat ends of the can. The area of the curved part of the can is $2\pi R\ell$, so Gauss's law gives:

$$\phi_{net} = \oint_S E_n \, dA = \frac{1}{\epsilon_0} Q_{inside}$$

$$= E_R 2\pi R\ell = \frac{1}{\epsilon_0} Q_{inside}$$

5. Assuming the charge per unit length on the inner shell is uniformly distributed, find $Q_{inside}$:

$$Q_{inside} = \frac{\ell}{L} Q$$

6. Substitute for $Q_{inside}$ and solve for $E_R$:

$$E_R 2\pi R\ell = \frac{1}{\epsilon_0} \frac{\ell}{L} Q$$

so

$$E_R = \frac{Q}{2\pi L \epsilon_0 R}$$

7. Integrate to find $V = |V_{R_2} - V_{R_1}|$:

$$V_{R_2} - V_{R_1} = \int_{V_{R_1}}^{V_{R_2}} dV = -\int_{R_1}^{R_2} E_R \, dR$$

$$= -\frac{Q}{2\pi L \epsilon_0} \int_{R_1}^{R_2} \frac{dR}{R} = -\frac{Q}{2\pi L \epsilon_0} \ln \frac{R_2}{R_1}$$

so

$$V = |V_{R_2} - V_{R_1}| = \frac{Q}{2\pi L \epsilon_0} \ln \frac{R_2}{R_1}$$

8. Substitute this result to find $C$:

$$C = \frac{Q}{V} = \boxed{\frac{2\pi \epsilon_0 L}{\ln(R_2/R_1)}}$$

**REMARKS** The capacitance of a cylindrical capacitor is proportional to the length of the conductors.

**EXERCISE** How is the capacitance affected if the potential across a cylindrical capacitor is increased from 20 V to 80 V? (*Answer* The capacitance of any capacitor does not depend on the potential. To increase $V$, you must increase the charge $Q$. The ratio $Q/V$ depends only on the geometry of the capacitor and the nature of the insulators.)

From Example 24-3 we see that the capacitance of a cylindrical capacitor is given by

$$C = \frac{2\pi \epsilon_0 L}{\ln(R_2/R_1)} \qquad\qquad 24\text{-}11$$

CAPACITANCE OF A CYLINDRICAL CAPACITOR

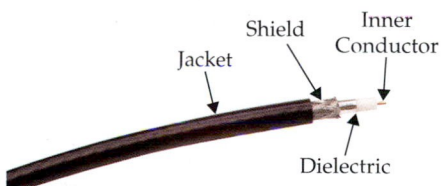

A coaxial cable is a long cylindrical capacitor with a solid wire for the inner conductor and a braided-wire shield for the outer conductor. The outer rubber coating has been peeled back from the cable to show the conductors and the white plastic insulator that separates the conductors.

Cutaway of a 200-$\mu$F capacitor used in an electronic strobe light.

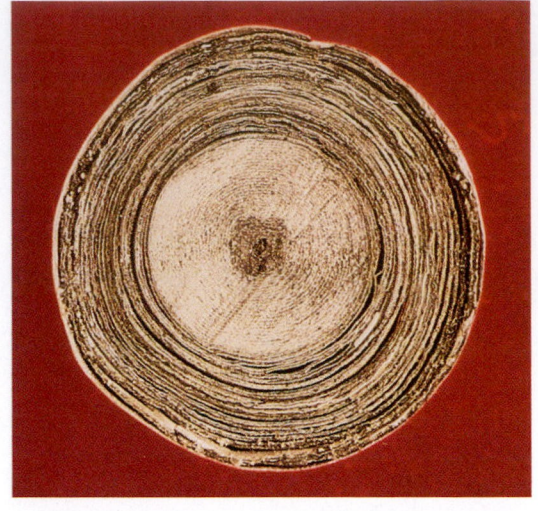

Cross section of a foil-wound capacitor.

A variable air-gap capacitor like those that were used in the tuning circuits of old radios. The semicircular plates rotate through the fixed plates, which changes the amount of surface area between the plates, and hence the capacitance.

Ceramic capacitors for use in electronic circuits.

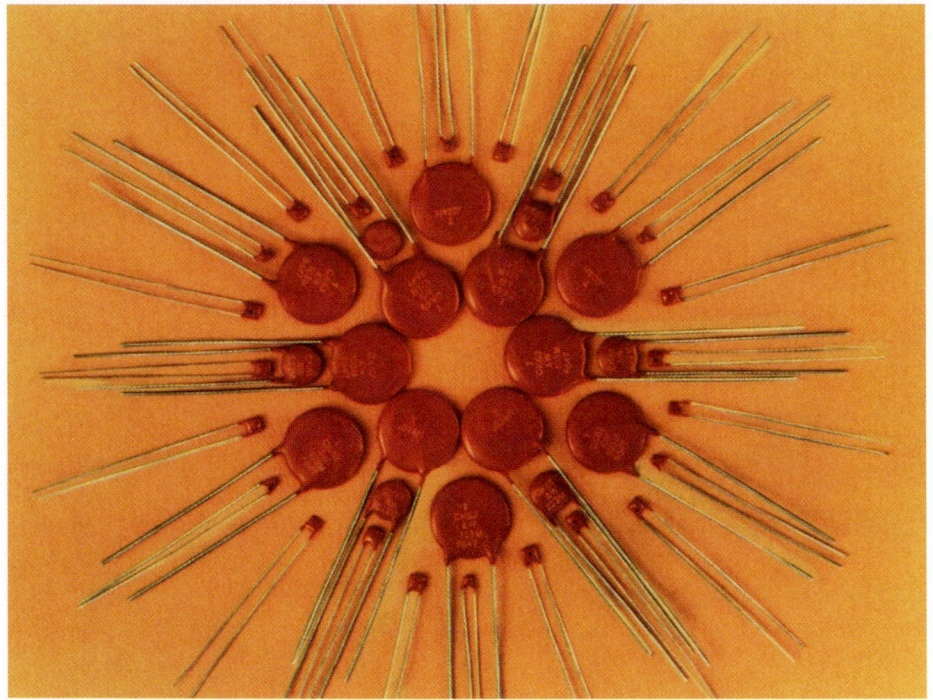

# 24-3 The Storage of Electrical Energy

When a capacitor is being charged, typically electrons are transferred from the positively charged conductor to the negatively charged conductor. This leaves the positive conductor with an electron deficit and the negative conductor with an electron surplus. Alternatively, transferring positive charges from the negative to the positive conductor can also charge capacitors. Either way, work must be done to charge a capacitor, and at least some of this work is stored as electrostatic potential energy.

Let $q$ be the positive charge that has been transferred at some time during the charging process. The potential difference is then $V = q/C$. If a small amount of additional positive charge $dq$ is now transferred from the negative conductor to the positive conductor through a potential increase of $V$ (Figure 24-6), the potential energy of the charge, and thus the capacitor, is increased by

$$dU = V\,dq = \frac{q}{C}\,dq$$

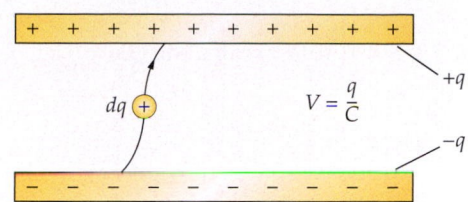

The total increase in potential energy $U$ is the integral of $dU$ as $q$ increases from zero to its final value $Q$ (Figure 24-7):

$$U = \int dU = \int_0^Q \frac{q}{C}\,dq = \frac{1}{2}\frac{Q^2}{C}$$

This potential energy is the energy stored in the capacitor. Using $C = Q/V$, we can express this energy in a variety of ways:

$$U = \frac{1}{2}\frac{Q^2}{C} = \frac{1}{2}QV = \frac{1}{2}CV^2 \qquad\qquad 24\text{-}12$$

ENERGY STORED IN A CAPACITOR

**FIGURE 24-6** When a small amount of positive charge $dq$ is moved from the negative conductor to the positive conductor, its potential energy is increased by $dU = V\,dq$, where $V$ is the potential difference between the conductors.

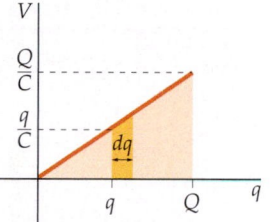

**EXERCISE** A 15-$\mu$F capacitor is charged to 60 V. How much energy is stored in the capacitor? (*Answer* 0.027 J)

**EXERCISE** Obtain the expression for the electrostatic energy stored in a capacitor (Equation 24-12) from Equation 24-4, using $Q_1 = +Q$, $Q_2 = -Q$, $n = 2$, and $V = V_1 - V_2$.

Suppose we charge a capacitor by connecting it to a battery. The potential difference $V$ when the capacitor is fully charged with charge $Q$ is just the potential difference between the terminals of the battery before they were connected to the capacitor. The total work done *by the battery* in charging the capacitor is $QV$, which is twice the energy stored in the capacitor. The additional work done by the battery is either dissipated as thermal energy in the battery and in the connecting wires[†] or radiated as electromagnetic energy via an electromagnetic wave.[‡]

**FIGURE 24-7** The work needed to charge a capacitor is the integral of $V\,dq$ from the original charge of $q = 0$ to the final charge of $q = Q$. This work is the triangular area under the curve $\frac{1}{2}(Q/C)Q$.

---

[†] We will show in Section 25-6 that if the capacitor is connected to a battery by wires of some resistance R, half the energy supplied by the battery in charging the capacitor is dissipated as thermal energy in the wires.
[‡] We will show in Section 30-3 that under certain circumstances the circuit will act as a broadcast antenna and a significant portion of the work will be broadcast as electromagnetic radiation.

---

*CHARGING A PARALLEL-PLATE CAPACITOR WITH A BATTERY* **EXAMPLE 24-4**

A parallel-plate capacitor with square plates 14 cm on a side and separated by 2.0 mm is connected to a battery and charged to 12 V. (*a*) What is the charge on the capacitor? (*b*) How much energy is stored in the capacitor? (*c*) The battery is then disconnected from the capacitor and the plate separation is then increased to 3.5 mm. By how much is the energy increased when the plate separation is changed?

**PICTURE THE PROBLEM** (*a*) The charge on the capacitor can be calculated from the capacitance and then used to calculate the energy stored in Part (*b*). (*c*) Since the capacitor is no longer connected to the battery, the charge remains constant as the plates are separated. The energy increase is found by using the charge and new potential to calculate the new energy, from which we subtract the original energy.

(*a*) 1. The charge $Q$ on the capacitor equals the product of $C_0$ and $V_0$, where $C_0$ is the capacitance and $V_0 = 12$ V is the battery voltage:

$$Q = C_0 V_0$$

2. Calculate the capacitance of the parallel-plate capacitor:

$$C_0 = \frac{\epsilon_0 A}{d_0}$$

3. Substitute to calculate $Q$:

$$Q = C_0 V_0 = \frac{\epsilon_0 A}{d_0} V_0$$

$$= \frac{(8.85 \text{ pF/m})(0.14 \text{ m})^2}{0.002 \text{ m}} (12 \text{ V}) = \boxed{1.04 \text{ nC}}$$

(b) Calculate the energy stored:

$$U_0 = \tfrac{1}{2} Q V_0 = \tfrac{1}{2}(1.04 \text{ nC})(12 \text{ V}) = \boxed{6.24 \text{ nJ}}$$

(c) 1. The battery is disconnected. The potential difference $V$ between the plates is the field strength $E$ times the separation distance $d$:

$$V = Ed$$

2. At the surface of a conductor, $E$ is proportional to the surface charge density $\sigma = Q/A$. Since $Q$ is constant, so is $\sigma$, and thus $E$:

$$E = \frac{\sigma}{\epsilon_0} = \frac{Q}{A \epsilon_0}$$

3. Combining the last two steps reveals that $V$ is proportional to $d$:

$$V = Ed = \frac{Q}{A \epsilon_0} d$$

so

$$\frac{V}{d} = \frac{V_0}{d_0}, \quad \text{or} \quad \left( V = \frac{d}{d_0} V_0 \right)$$

4. Calculate $U$ and $\Delta U$, obtaining $U_0$ from Part (b):

$$U = \frac{1}{2} QV = \frac{1}{2} Q \frac{d}{d_0} V_0 = \frac{d}{d_0} \frac{1}{2} QV_0 = \frac{d}{d_0} U_0$$

so

$$\Delta U = U - U_0 = \frac{d}{d_0} U_0 - U_0 = \left( \frac{d}{d_0} - 1 \right) U_0$$

$$= \left( \frac{3.5 \text{ mm}}{2.0 \text{ mm}} - 1 \right)(6.24 \text{ nJ}) = \boxed{4.68 \text{ nJ}}$$

**REMARKS** The additional energy calculated in Part (c) comes from work done by the agent responsible for increasing the separation between the plates, which attract each other. An application of the dependence of capacitance on separation distance is shown in Figure 24-8.

**EXERCISE** Find the final voltage $V$ between the capacitor plates. (*Answer* 21.0 V)

**EXERCISE** (a) Find the initial capacitance $C_0$ in this example when separation of the plates is 2.0 mm. (b) Find the final capacitance $C$ when separation of the plates is 3.5 mm. (*Answer* (a) $C_0 = 86.7$ pF (b) $C = 49.6$ pF)

It is instructive to work Part (c) of Example 24-4 in another way. The oppositely charged plates of a capacitor exert attractive forces on one another. Work must be done against these forces to increase the plate separation. Assume that the lower plate is held fixed and the upper plate is moved. The force on the upper plate is the charge $Q$ on the plate times the electric field $\vec{E}'$ *due to the charge* $-Q$ *on the lower plate*. This field is half the total field $\vec{E}$ between the plates (because the charge on the upper plate and the charge on the lower plate contribute equally to the field). When the potential difference is 12 V and the separation is 2 mm, the total field strength between the plates is

$$E = \frac{V}{d} = \frac{12 \text{ V}}{2 \text{ mm}} = 6 \text{ V/mm} = 6 \text{ kV/m}$$

The magnitude of the force exerted on the upper plate by the bottom plate is thus

$$F = QE' = Q(\tfrac{1}{2}E) = (1.04 \text{ nC})(3 \text{ kV/m}) = 3.12 \ \mu\text{N}$$

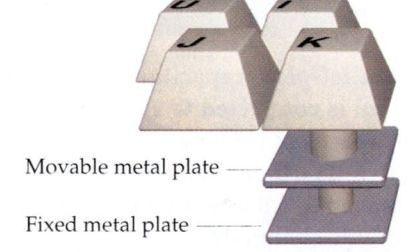

Movable metal plate

Fixed metal plate

**FIGURE 24-8** Capacitance switching in computer keyboards. A metal plate attached to each key acts as the top plate of a capacitor. Depressing the key decreases the separation between the top and bottom plates and increases the capacitance, which triggers the electronic circuitry of the computer to acknowledge the keystroke.

The work that must be done to move the upper plate a distance of $\Delta d = 1.5$ mm is then

$$W = F \, \Delta d = (3.12 \, \mu N)(1.5 \text{ mm}) = 4.68 \text{ nJ}$$

This is the same number of joules calculated in Part (c) of Example 24-4. This work equals the increase in the energy stored.

## Electrostatic Field Energy

In the process of charging a capacitor, an electric field is produced between the plates. The work required to charge the capacitor can be thought of as the work required to create the electric field. That is, we can think of the energy stored in a capacitor as energy stored in the electric field, called **electrostatic field energy.**

Consider a parallel-plate capacitor. We can relate the energy stored in the capacitor to the electric field strength $E$ between the plates. The potential difference between the plates is related to the electric field by $V = Ed$, where $d$ is the plate separation distance. The capacitance is given by $C = \epsilon_0 A/d$ (Equation 24-10). The energy stored is

$$U = \frac{1}{2}CV^2 = \frac{1}{2}\left(\frac{\epsilon_0 A}{d}\right)(Ed)^2 = \frac{1}{2}\epsilon_0 E^2(Ad)$$

The quantity $Ad$ is the volume of the space between the plates of the capacitor containing the electric field. The energy-per-unit volume is called the **energy density** $u_e$. The energy density in an electric field strength $E$ is thus

$$u_e = \frac{energy}{volume} = \frac{1}{2}\epsilon_0 E^2 \qquad\qquad\qquad 24\text{-}13$$

ENERGY DENSITY OF AN ELECTROSTATIC FIELD

Thus, the energy per unit volume of the electrostatic field is proportional to the square of the electric field strength. *Although we obtained Equation 24-13 by considering the electric field between the plates of a parallel-plate capacitor, the result applies to any electric field.* Whenever there is an electric field in space, the electrostatic energy per unit volume is given by Equation 24-13.

**EXERCISE** (a) Calculate the energy density $u_e$ for Example 24-4 when the plate separation is 2.0 mm. (b) Show that the increase in energy in Example 24-4 is equal to $u_e$ times the increase in volume ($\Delta$ vol) between the plates. (*Answer* (a) $u_e = \frac{1}{2}\epsilon_0 E^2 = 159.3 \, \mu J/m^3$, (b) $\Delta vol = A \, \Delta d = 2.94 \times 10^{-5} \text{ m}^3$, $u_e \, \Delta vol = 4.68$ nJ, in agreement with Example 24-4)

We can illustrate the generality of Equation 24-13 by calculating the electrostatic field energy of a spherical conductor of radius $R$ that carries a charge $Q$. The electrostatic potential energy in terms of the charge $Q$ and potential $V$ is given by Equation 24-12:

$$U = \frac{kQ^2}{2R} = \frac{1}{2}QV \qquad\qquad\qquad 24\text{-}14$$

We now obtain the same result by considering the energy density of an electric field given by Equation 24-13. When the conductor carries a charge $Q$, the electric field is radial and is given by

$$E_r = 0, \quad r < R \text{ (inside the conductor)}$$

$$E_r = \frac{kQ}{r^2}, \quad r > R \text{ (outside the conductor)}$$

Since the electric field is spherically symmetric, we choose a spherical shell for our volume element. If the radius of the shell is $r$ and its thickness is $dr$, the volume is $d\mathcal{V} = 4\pi r^2\,dr$ (Figure 24-9). The energy $dU$ in this volume element is

$$dU = u_e\,d\mathcal{V} = \frac{1}{2}(\epsilon_0 E^2)4\pi r^2 dr$$

$$= \frac{1}{2}\epsilon_0\left(\frac{kQ}{r^2}\right)^2(4\pi r^2\,dr) = \frac{1}{2}(4\pi\epsilon_0 k^2)Q^2\frac{dr}{r^2} = \frac{1}{2}kQ^2 r^{-2}\,dr$$

where we have used $4\pi\epsilon_0 = 1/k$. Since the electric field is zero for $r < R$, we obtain the total energy in the electric field by integrating from $r = R$ to $r = \infty$:

$$U = \int u_e\,d\mathcal{V} = \frac{1}{2}kQ^2\int_R^\infty r^{-2}dr = \frac{1}{2}k\frac{Q^2}{R} = \frac{1}{2}Q\left(\frac{kQ}{R}\right) = \frac{1}{2}QV \qquad \text{24-15}$$

which is the same as Equation 24-12.

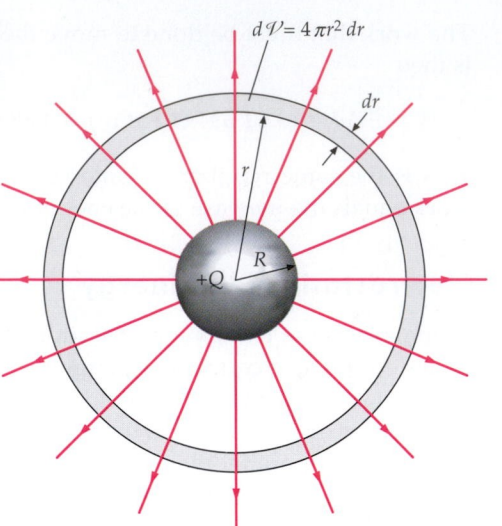

**FIGURE 24-9** Geometry for the calculation of the electrostatic energy of a spherical conductor carrying a charge $Q$. The volume of the space between $r$ and $r + dr$ is $d\mathcal{V} = 4\pi r^2\,dr$. The electrostatic field energy in this volume element is $u_e d\mathcal{V}$, where $u_e = \frac{1}{2}\epsilon_0 E^2$ is the energy density.

# 24-4 Capacitors, Batteries, and Circuits

Next, we examine what happens when an initially uncharged capacitor is connected to the terminals of a battery. The potential difference between the two terminals of a battery is called its **terminal voltage.** Typically, one terminal of a battery is positively charged and the other terminal is negatively charged; this charge separation is maintained by chemical action within the battery. Within the battery, there is an electric field directed away from the positive terminal toward the negative terminal.[†] When a plate of an uncharged capacitor is connected to the negative terminal of the battery, the negative charge on that terminal is shared with the plate. This gives the plate a small negative charge and momentarily reduces the amount of negative charge on that battery terminal. If the second capacitor plate is then connected to the positive battery terminal, the charge on the positive battery terminal is then shared with it—momentarily reducing the positive charge on that battery terminal. These charge reductions on the battery terminals result in a decrease in the terminal voltage of the battery. This decrease in terminal voltage triggers the chemical activity within the battery that transfers charge from one terminal to the other terminal in an effort to maintain the terminal voltage at its initial level, which is called the **open-circuit terminal voltage.** This chemical action ceases when the battery has transferred sufficient charge from one capacitor plate to the other capacitor plate to raise the potential difference between the plates to the open-circuit terminal voltage of the battery.

It is useful to think of a battery as a charge pump. When we connect the plates of an uncharged capacitor to the terminals of a battery, the terminal voltage drops causing the battery to pump charge from one plate to the other plate until the open circuit terminal voltage is again reached.

In electric circuit diagrams, the symbol representing a battery is ⊣⊢, where the longer, thinner vertical line represents the positive terminal and the shorter, thicker vertical line represents the negative terminal. The symbol representing a capacitor is ⊣⊢.

**EXERCISE** A 6-$\mu$F capacitor, initially uncharged, is connected to the terminals of a 9-V battery. What total amount of charge flows through the battery? (*Answer* 54 $\mu$C)

---

[†] This electric field from the positive to the negative terminal exists outside the battery also.

## Combinations of Capacitors

CAPACITORS CONNECTED IN PARALLEL                    **EXAMPLE  24 - 5**

**FIGURE 24-10**

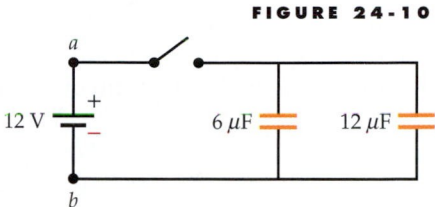

A circuit consists of a 6-$\mu$F capacitor, a 12-$\mu$F capacitor, a 12-V battery, and a switch, connected as shown in Figure 24-10. Initially, the switch is open and the capacitors are uncharged. The switch is then closed and the capacitors charge. When the capacitors are fully charged and open-circuit terminal voltage is restored (*a*) what is the potential of each conductor in the circuit? (Choose the zero-potential reference point on the negative battery terminal.) (*b*) What is the charge on each capacitor plate? (*c*) What total charge passed through the battery?

**PICTURE THE PROBLEM** The potential is the same throughout a conductor in electrostatic equilibrium. After the charges stop moving, all of the conductors connected by a conducting wire are at the same potential. The charge on a capacitor (step 2 and step 3) is related to the potential difference across the capacitor by $Q = CV$. The charges on the plates of a single capacitor are equal but opposite.

**FIGURE 24-11**

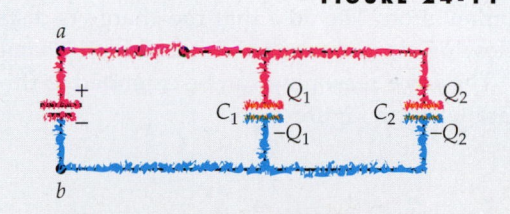

(*a*) Use a red marker to color the positive ($+$) battery terminal and all the conductors connected to it (Figure 24-11), and use a blue marker to color the negative ($-$) battery terminal and all the conductors connected to it:

All points colored red are at potential $\boxed{V_a = 12 \text{ V}}$

All points colored blue are at potential $\boxed{V_b = 0}$

(*b*) Use $Q = CV$ to find the magnitude of the charge on the plates. The capacitor plate at the higher potential carries a positive charge:

$Q_1 = C_1 V = (6 \ \mu\text{F})(12 \text{ V}) = \boxed{72 \ \mu\text{C}}$

$Q_2 = C_2 V = (12 \ \mu\text{F})(12 \text{ V}) = \boxed{144 \ \mu\text{C}}$

(*c*) The plates become charged because the battery acts as a charge pump:

$Q = Q_1 + Q_2 = \boxed{216 \ \mu\text{C}}$

**REMARKS** The equivalent capacitance of the two-capacitor combination is $Q/V$, where $Q$ is the charge passing through the battery and $V$ is the open-circuit terminal voltage of the battery. For this example $C_{\text{eq}} = (216 \ \mu\text{C})/(12 \text{ V}) = 18 \ \mu\text{F}$.

When two capacitors are connected, as shown in Figure 24-12, so that the upper plates of the two capacitors are connected by a conducting wire and are therefore at a common potential, and the lower plates are also connected together and are at a common potential, just like the capacitors in Example 24-5, the capacitors are said to be connected in **parallel.** Devices connected in parallel share a common potential difference across each device *due solely to the way they are connected.*

In Figure 24-12, assume that points $a$ and $b$ are connected to a battery or some other device that provides a potential difference $V = V_a - V_b$ between the plates of each capacitor. If the capacitances are $C_1$ and $C_2$, the charges $Q_1$ and $Q_2$ stored on the plates are given by

$Q_1 = C_1 V$

and

$Q_2 = C_2 V$

The total charge stored is

$Q = Q_1 + Q_2 = C_1 V + C_2 V = (C_1 + C_2)V$

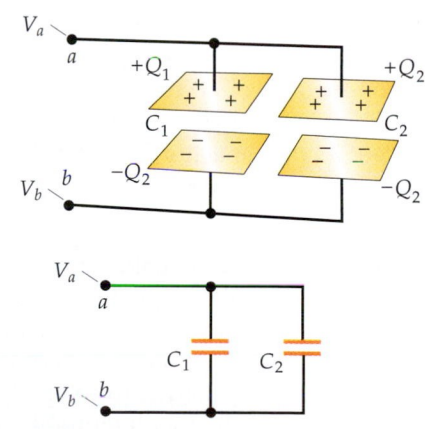

**FIGURE 24-12** Two capacitors in parallel. The upper plates are connected together and are therefore at a common potential $V_a$; the lower plates are similarly connected together and therefore at a common potential $V_b$.

A combination of capacitors in a circuit can sometimes be substituted with a single capacitor that is operationally equivalent to the combination. The substitute capacitor is said to have an **equivalent capacitance.** That is, if a combination of initially uncharged capacitors is connected to a battery, the charge $Q$ that flows through the battery as the capacitor combination becomes charged is the same as the charge that flows through the same battery if connected to a single uncharged capacitor of equivalent capacitance. Therefore, the equivalent capacitance of two capacitors in parallel is the ratio of the charge $Q_1 + Q_2$ to the potential difference:

$$C_{eq} = \frac{Q}{V} = \frac{Q_1 + Q_2}{V} = \frac{Q_1}{V} + \frac{Q_2}{V} = C_1 + C_2 \qquad \text{24-16}$$

Thus, for two capacitors in parallel, $C_{eq}$ is the sum of the individual capacitances. When we add a second capacitor in parallel, we increase the capacitance of the combination. The area that the charge is distributed on is effectively increased, allowing more charge to be stored for the same potential difference.

The same reasoning can be extended to three or more capacitors connected in parallel, as in Figure 24-13:

$$C_{eq} = C_1 + C_2 + C_3 + \dots \qquad \text{24-17}$$

EQUIVALENT CAPACITANCE FOR CAPACITORS IN PARALLEL

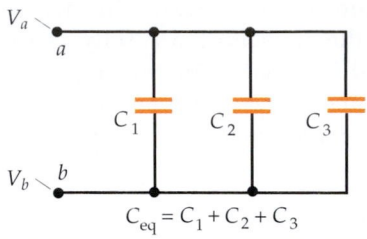

**FIGURE 24-13** Three capacitors in parallel. The effect of adding a parallel capacitor to a circuit is an increase in the equivalent capacitance.

---

*CAPACITORS CONNECTED IN SERIES*   **EXAMPLE 24-6**

A circuit consists of a 6-$\mu$F capacitor, a 12 $\mu$-F capacitor, a 12-V battery, and a switch, connected as shown in Figure 24-14. Initially, the switch is open and the capacitors are uncharged. The switch is then closed and the capacitors charge. When the capacitors are fully charged and open-circuit terminal voltage is restored, (*a*) what is the potential of each conductor in the circuit? (Choose the zero-potential reference point on the negative battery terminal.) If the potential of a conductor is not known, represent its potential symbolically. (*b*) What is the charge on each capacitor plate? (*c*) What total charge passed through the battery?

**FIGURE 24-14**

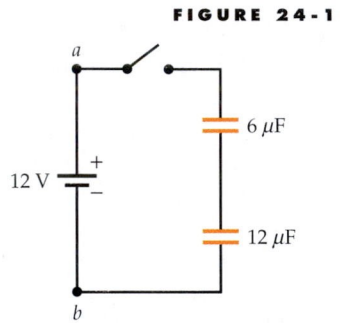

**PICTURE THE PROBLEM** (*a*) The potential is the same throughout a conductor in electrostatic equilibrium. After the charges stop moving, all of the conductors connected by a conducting wire are at the same potential. The charge on a capacitor, Parts (*b*) and (*c*), is related to the potential difference across the capacitor by $Q = CV$. Charge does not travel from one plate of a capacitor to the other.

**FIGURE 24-15**

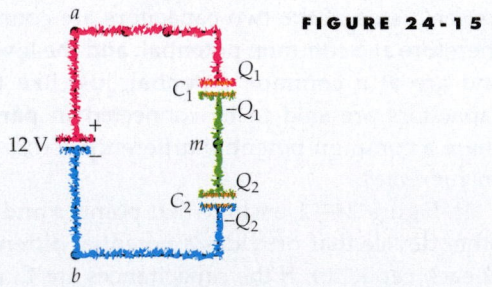

(*a*) Use a red marker to color the positive (+) battery terminal and all conductors connected to it, use a blue marker to color the negative (−) battery terminal and all the conductors connected to it, and use a green marker to color all other mutually connected conductors (Figure 24-15):

All points colored red are at potential $\boxed{V_a = 12 \text{ V}}$

All points colored blue are at potential $\boxed{V_b = 0}$

All points colored green are at the yet unknown potential $\boxed{V_m}$

(*b*) 1. Express the potential difference across each capacitor in terms of the Part (*a*) results:

$V_1 = V_a - V_m$

and

$V_2 = V_m - V_b$

2. Use $Q = CV$ to relate the charge on each capacitor to the potential difference:

$$Q_1 = C_1 V_1 = C_1(V_a - V_m)$$

and

$$Q_2 = C_2 V_2 = C_2(V_m - V_b)$$

3. Eliminating $V_m$ gives:

$$\left. \begin{array}{l} V_a - V_m = \dfrac{Q_1}{C_1} \\[2mm] V_m - V_b = \dfrac{Q_2}{C_2} \end{array} \right\} \Rightarrow V_a - V_b = \dfrac{Q_1}{C_1} + \dfrac{Q_2}{C_2}$$

4. During charging, there is no charge transferred either to or from the green region in Figure 24-15, so its net charge remains zero:

$$(-Q_1) + Q_2 = 0$$

so

$$Q_1 = Q_2$$

5. Let $Q = Q_1 = Q_2$. Substitute $Q$ for $Q_1$ and $Q_2$ and solve for $Q$:

$$V_a - V_b = \frac{Q}{C_1} + \frac{Q}{C_2}$$

so

$$Q = \frac{V_a - V_b}{\dfrac{1}{C_1} + \dfrac{1}{C_2}} = \frac{12\ \text{V} - 0}{\dfrac{1}{6\ \mu\text{F}} + \dfrac{1}{12\ \mu\text{F}}} = 48\ \mu\text{C}$$

$$Q_1 = Q_2 = \boxed{48\ \mu\text{C}}$$

(c) All the charge passing through the battery ends up on the upper plate of $C_1$:

$$Q_1 = Q = \boxed{48\ \mu\text{C}}$$

**REMARKS** The equivalent capacitance of the two-capacitor combination is $Q/V$, where $Q$ is the charge passing through the battery and $V$ is the open-circuit terminal voltage of the battery. For this example $C_{eq} = (48\ \mu\text{C})/(12\ \text{V}) = 4\ \mu\text{F}$.

**EXERCISE** Find the potential $V_m$ on the conductors colored green in Figure 24-15. (*Answer* 4.0 V)

In Figure 24-16, two capacitors are connected so that the potential difference across the pair is the sum of the potential differences across the individual capacitors, just like those in Example 24-6. Devices connected in this manner are connected in **series**.

Capacitors $C_1$ and $C_2$ in Figure 24-16 are connected in series and initially they are without charge. If points $a$ and $b$ are then connected to the terminals of a battery, electrons will be pumped from the upper plate of $C_1$ to the lower plate of $C_2$. This leaves the upper plate of $C_1$ with a charge $+Q$ and the lower plate of $C_2$ with a charge $-Q$. When a charge $+Q$ appears on the upper plate of $C_1$, the electric field produced by that charge induces an equal negative charge, $-Q$, on the lower plate of $C_1$. This charge comes from electrons drawn from the upper plate of $C_2$. Thus, there will be an equal charge $+Q$ on the upper plate of the second capacitor and a corresponding charge $-Q$ on its lower plate. The potential difference across the first capacitor is

$$V_1 = \frac{Q}{C_1}$$

Similarly, the potential difference across the second capacitor is

$$V_2 = \frac{Q}{C_2}$$

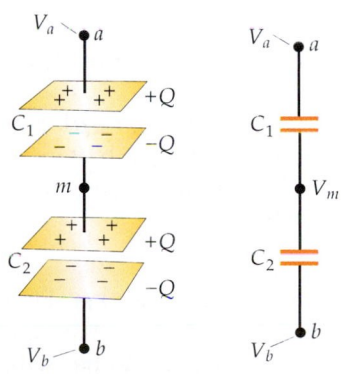

**FIGURE 24-16** The total charge on the two interconnected capacitor plates equals zero. The potential difference across the pair equals the sum of the potential differences across the individual capacitors. The two capacitors are connected in series.

The potential difference across the two capacitors in series is the sum of these potential differences:

$$V = V_a - V_b = V_1 + V_2 = \frac{Q}{C_1} + \frac{Q}{C_2} = Q\left(\frac{1}{C_1} + \frac{1}{C_2}\right) \qquad 24\text{-}18$$

The equivalent capacitance of the two capacitors in series is defined as

$$C_{eq} = \frac{Q}{V} \qquad 24\text{-}19$$

Substituting $Q/C_{eq}$ for $V$ in Equation 24-18, and then dividing both sides by $Q$, gives

$$\frac{1}{C_{eq}} = \frac{1}{C_1} + \frac{1}{C_2} \qquad 24\text{-}20$$

Note that in the preceding exercise, the equivalent capacitance of the two capacitors in series is less than the capacitance of either capacitor. Adding a capacitor in series increases $1/C_{eq}$, which means the equivalent capacitance $C_{eq}$ decreases. When we add a second capacitor in series, we decrease the capacitance of the combination. The plate separation is essentially increased, requiring a greater potential difference to store the same charge.

Equation 24-20 can be generalized to three or more capacitors connected in series:

$$\frac{1}{C_{eq}} = \frac{1}{C_1} + \frac{1}{C_2} + \frac{1}{C_3} + \ldots \qquad 24\text{-}21$$

EQUIVALENT CAPACITANCE FOR EQUALLY CHARGED CAPACITORS IN SERIES

A capacitor bank for storing energy to be used by the pulsed Nova laser at Lawrence Livermore Laboratories. The laser is used in fusion studies.

❗ **This formula is valid only if the capacitors are in series *and* the total charge on each pair of capacitor plates connected by a wire is zero.**

**EXERCISE** Two capacitors have capacitances of 20 $\mu$F and 30 $\mu$F. Find the equivalent capacitance if the capacitors are connected (*a*) in parallel and (*b*) in series. (*Answer*   (*a*) 50 $\mu$F, (*b*) 12 $\mu$F)

---

*USING THE EQUIVALENCE FORMULA*      **EXAMPLE 24-7**

A 6-$\mu$F capacitor and a 12-$\mu$F capacitor, each initially uncharged, are connected in series across a 12-V battery. Using the equivalence formula for capacitors in series, find the charge on each capacitor and the potential difference across each.

**PICTURE THE PROBLEM** Figure 24-17*a* shows the circuit in this example and Figure 24-17*b* shows an equivalent capacitor that carries the same charge $Q = C_{eq}V$. After finding the charge, we can find the potential drop across each capacitor.

1. The charge on each capacitor equals the charge on the equivalent capacitor:  $\qquad Q = C_{eq}V$

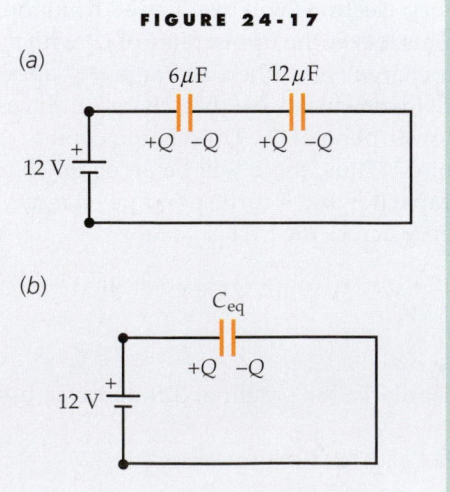

**FIGURE 24-17**

2. The equivalent capacitance of the series combination is found from:

$$\frac{1}{C_{eq}} = \frac{1}{C_1} + \frac{1}{C_2} = \frac{1}{6\ \mu F} + \frac{1}{12\ \mu F} = \frac{3}{12\ \mu F}$$

$$C_{eq} = 4\ \mu F$$

3. Use this value to find the charge $Q$. This is the charge that went through the battery. It is the charge on each capacitor:

$$Q = C_{eq}V = (4\ \mu F)(12\ V) = \boxed{48\ \mu C}$$

4. Use the result for $Q$ to find the potential across the 6-$\mu$F capacitor:

$$V_1 = \frac{Q}{C_1} = \frac{48\ \mu C}{6\ \mu F} = \boxed{8\ V}$$

5. Again, use the result for $Q$ to find the potential across the 12-$\mu$F capacitor:

$$V_2 = \frac{Q}{C_2} = \frac{48\ \mu C}{12\ \mu F} = \boxed{4\ V}$$

**⊘ PLAUSIBILITY  CHECK**  The sum of these potential differences is 12 V, as required.

**■ REMARKS**  The results are the same as those obtained in Example 24-6.

---

*CAPACITORS IN SERIES REARRANGED IN PARALLEL*    **EXAMPLE  24-8**    **Try It Yourself**

The two capacitors in Example 24-7 are removed from the battery and carefully disconnected from each other so that the charge on the plates is not disturbed (Figure 24-18a). They are then reconnected in a circuit containing open switches, positive plate to positive plate and negative plate to negative plate (Figure 24-18b). Find the potential difference across the capacitors and the charge on each capacitor after the switches are closed and the charges have stopped flowing.

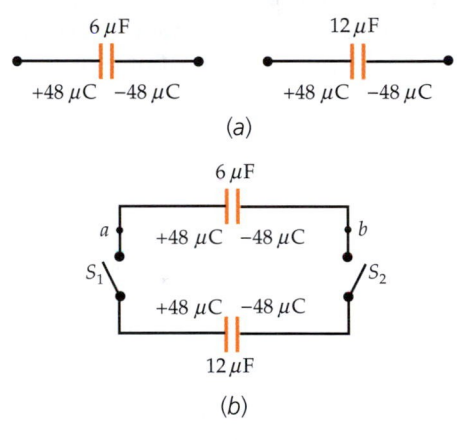

(a)

(b)

**FIGURE 24-18**

**PICTURE  THE  PROBLEM**  Just after the two capacitors are disconnected from the battery, they carry equal charges of 48 $\mu$C. After switches $S_1$ and $S_2$ in the new circuit are closed, the capacitors are in parallel between points $a$ and $b$. The potential across each of them is the same, and the equivalent capacitance of the system is $C_{eq} = C_1 + C_2$. The two positive plates form a single conductor with charge $Q = 48\ \mu C$, and the negative plates form a conductor with charge $-Q = -48\ \mu C$. Therefore, the potential difference is $V = Q/C_{eq}$, and the charges on the two capacitors are $Q_1 = C_1V$ and $Q_2 = C_2V$.

**Cover the column to the right and try these on your own before looking at the answers.**

| Steps | Answers |
|---|---|
| 1. The wiring is such that after the switches are closed the potential difference is the same across each capacitor. | $V = V_1 = V_2$ |
| 2. For each capacitor $V = Q/C$. Substitute this into the step 1 result. Let $C_1$ be the 6-$\mu$F capacitor. | $\dfrac{Q_1}{C_1} = \dfrac{Q_2}{C_2}$ |
| 3. The sum of the charges on the two capacitor plates on the left remains 96 $\mu$C. | $Q_1 + Q_2 = 96\ \mu C$ |
| 4. Solve for the charge on each capacitor. | $Q_1 = \boxed{32\ \mu C}$,   $Q_2 = \boxed{64\ \mu C}$ |
| 5. Calculate the potential difference. | $V = \dfrac{Q_1}{C_1} = \boxed{5.33\ V}$ |

**⊘ PLAUSIBILITY  CHECK**  Note that $Q = Q_1 + Q_2 = 96\ \mu C$, and that $Q_2/C_2 = 5.33\ V$ as required.

**REMARKS** After the switches are closed, the two capacitors are connected in parallel with the potential difference between point $a$ and point $b$ being the potential difference across the pair. Thus, $C_{eq} = C_1 + C_2 = 18 \ \mu F$, $Q = Q_1 + Q_2 = 96 \ \mu C$, and $V = Q/C_{eq} = 5.33 \ V$.

**EXERCISE** Find the energy stored in the capacitors before and after they are connected. [*Answer* $U_i = q^2/(2C_1) + q^2/(2C_2)$, where $q = 48 \ \mu C$. Thus, $U_i = 288 \ \mu J$. $U_f = Q_1^2/(2C_1) + Q_2^2/(2C_2) = 256 \ \mu J$. Note that 32 $\mu J$ is *lost* to thermal energy in the wires or radiated away.]

**FIGURE 24-19**

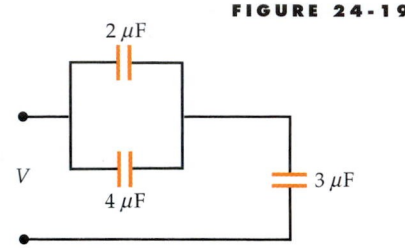

**CAPACITORS IN SERIES AND IN PARALLEL**     **EXAMPLE 24-9**

(*a*) Find the equivalent capacitance of the network of three capacitors in Figure 24-19. (*b*) The capacitors are initially uncharged. Find the charge on each capacitor and the voltage drop across it after the capacitor combination is connected to a 6-V battery.

**PICTURE THE PROBLEM** (*a*) The 2-$\mu F$ capacitor and the 4-$\mu F$ capacitor are connected in parallel, and the parallel combination is connected in series with the 3-$\mu F$ capacitor. We first find the equivalent capacitance of the 2-$\mu F$ capacitor and the 4-$\mu F$ capacitor (Figure 24-20*a*), then combine this equivalent capacitance with the 3-$\mu F$ capacitor to reach a final equivalent capacitance (Figure 24-20*b*). (*b*) The charge on the 3-$\mu F$ capacitor is the charge passing through the battery $Q = C_{eq} V$ as shown in Figure 24-20*a*.

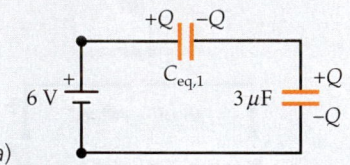 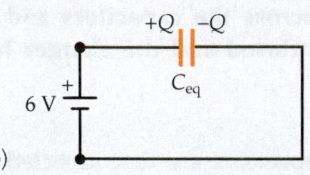

**FIGURE 24-20**

(*a*)    (*b*)

(*a*) 1. The equivalent capacitance of the two capacitors in parallel is the sum of the capacitances:

$$C_{eq,1} = C_1 + C_2 = 2 \ \mu F + 4 \ \mu F = 6 \ \mu F$$

2. Find the equivalent capacitance of a 6-$\mu F$ capacitor in series with a 3-$\mu F$ capacitor:

$$\frac{1}{C_{eq}} = \frac{1}{C_{eq,1}} + \frac{1}{C_3} = \frac{1}{6 \ \mu F} + \frac{1}{3 \ \mu F} = \frac{1}{2 \ \mu F}$$

$$C_{eq} = \boxed{2 \ \mu F}$$

(*b*) 1. Calculate the charge $Q$ delivered by the battery. This is also the charge on the 3-$\mu F$ capacitor:

$$Q = C_{eq} V = (2 \ \mu F)(6 \ V) = 12 \ \mu C$$

2. The potential drop across the 3-$\mu F$ capacitor is $Q/C_3$:

$$V_3 = \frac{Q_3}{C_3} = \frac{Q}{C_3} = \frac{12 \ \mu C}{3 \ \mu F} = \boxed{4 \ V}$$

3. The potential drop across the parallel combination $V_{2,4}$ is $Q/C_{eq,1}$:

$$V_{2,4} = \frac{Q}{C_{eq,1}} = \frac{12 \ \mu C}{6 \ \mu F} = \boxed{2 \ V}$$

4. The charge on each of the parallel capacitors is found from $Q_i = C_i V_{2,4}$, where $V_{2,4} = 2 \ V$:

$$Q_2 = C_2 V_{2,4} = (2 \ \mu F)(2 \ V) = \boxed{4 \ \mu C}$$

$$Q_4 = C_4 V_{2,4} = (4 \ \mu F)(2 \ V) = \boxed{8 \ \mu C}$$

**PLAUSIBILITY CHECK** The voltage drop across the parallel combination (2 V) plus that across the 3-$\mu F$ capacitor (4 V) equals the voltage of the battery. Also, the sum of the charges on the parallel capacitors (4 $\mu C$ + 8 $\mu C$) equals the total charge (12 $\mu C$) on the 3-$\mu F$ capacitor.

**EXERCISE** Find the energy stored in each capacitor. (*Answer* $U_2 = 4 \ \mu J$, $U_3 = 24 \ \mu J$, $U_4 = 8 \ \mu J$. Note that $U_2 + U_3 + U_4 = 36 \ \mu J = \frac{1}{2} QV = \frac{1}{2} Q^2/C_{eq} = \frac{1}{2} C_{eq} V^2$.)

## 24-5 Dielectrics

A nonconducting material (e.g., air, glass, paper, or wood) is called a **dielectric**. When the space between the two conductors of a capacitor is occupied by a dielectric, the capacitance is increased by a factor $\kappa$ that is characteristic of the dielectric, a fact discovered experimentally by Michael Faraday. The reason for this increase is that the electric field between the plates of a capacitor is weakened by the dielectric. Thus, for a given charge on the plates, the potential difference is reduced and the capacitance ($Q/V$) is increased.

Consider an isolated charged capacitor without a dielectric between its plates. A dielectric slab is then inserted between the plates, completely filling the space between the plates. If the electric field is $E_0$ before the dielectric slab is inserted, after the dielectric slab is inserted between the plates the field is

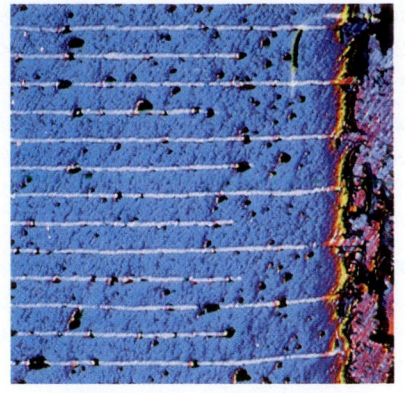

A cut section of a multilayer capacitor with a ceramic dielectric. The white lines are the edges of the conducting plates.

$$E = \frac{E_0}{\kappa} \qquad\qquad 24\text{-}22$$

ELECTRIC FIELD INSIDE A DIELECTRIC

where $\kappa$ (kappa) is called the **dielectric constant**. For a parallel-plate capacitor of separation $d$, the potential difference $V$ between the plates is

$$V = Ed = \frac{E_0 d}{\kappa} = \frac{V_0}{\kappa}$$

where $V$ is the potential difference with the dielectric and $V_0 = E_0 d$ is the original potential difference without the dielectric. The new capacitance is

$$C = \frac{Q}{V} = \frac{Q}{V_0/\kappa} = \kappa \frac{Q}{V_0}$$

or

$$C = \kappa C_0 \qquad\qquad 24\text{-}23$$

EFFECT OF A DIELECTRIC ON CAPACITANCE

where $C_0 = Q/V_0$ is the capacitance without the dielectric. The capacitance of a parallel-plate capacitor filled with a dielectric of constant $\kappa$ is thus

$$C = \frac{\kappa \epsilon_0 A}{d} = \frac{\epsilon A}{d} \qquad\qquad 24\text{-}24$$

where

$$\epsilon = \kappa \epsilon_0 \qquad\qquad 24\text{-}25$$

is called the **permittivity** of the dielectric.

In the preceding discussion, the capacitor was isolated so we assumed that the charge on its plates did not change as the dielectric was inserted. This is the case if the capacitor is charged and then removed from the charging source (the battery) before the insertion of the dielectric. If the dielectric is inserted while the battery remains connected, the battery pumps additional charge to maintain the original potential difference. The total charge on the plates is then $Q = \kappa Q_0$. In either case, the capacitance ($Q/V$) is increased by the factor $\kappa$.

**EXERCISE** The 88.5-pF capacitor of Example 24-2 is filled with a dielectric of constant $\kappa = 2$. (*a*) Find the new capacitance. (*b*) Find the charge on the capacitor with the dielectric in place if the capacitor is attached to a 12-V battery. (*Answer* (*a*) 177 pF, (*b*) 2.12 nC)

**EXERCISE** The capacitor in the previous exercise is charged to 12 V without the dielectric and is then disconnected from the battery. The dielectric of constant $\kappa = 2$ is then inserted. Find the new values for (*a*) the charge $Q$, (*b*) the voltage $V$, and (*c*) the capacitance $C$. (*Answer* (*a*) $Q = 1.06$ nC, which is unchanged; (*b*) $V = 6$ V; (*c*) $C = 177$ pF)

Dielectrics not only increase the capacitance of a capacitor, they also provide a means for keeping parallel conducting plates apart and they raise the potential difference at which dielectric breakdown occurs.[†] Consider a parallel-plate capacitor made from two sheets of metal foil that are separated by a thin plastic sheet. The plastic sheet allows the metal sheets to be very close together without actually being in electrical contact, and because the dielectric strength of plastic is greater than that of air, a greater potential difference can be attained before dielectric breakdown occurs. Table 24-1 lists the dielectric constants and dielectric strengths of some dielectrics. Note that for air $\kappa \approx 1$; so, for most situations we do not need to distinguish between air and a vacuum.

## TABLE 24-1

**Dielectric Constants and Dielectric Strengths of Various Materials**

| Material | Dielectric Constant $\kappa$ | Dielectric Strength, kV/mm |
|---|---|---|
| Air | 1.00059 | 3 |
| Bakelite | 4.9 | 24 |
| Glass (Pyrex) | 5.6 | 14 |
| Mica | 5.4 | 10–100 |
| Neoprene | 6.9 | 12 |
| Paper | 3.7 | 16 |
| Paraffin | 2.1–2.5 | 10 |
| Plexiglas | 3.4 | 40 |
| Polystyrene | 2.55 | 24 |
| Porcelain | 7 | 5.7 |
| Transformer oil | 2.24 | 12 |

*USING A DIELECTRIC IN A PARALLEL-PLATE CAPACITOR* **EXAMPLE 24-10**

A parallel-plate capacitor has square plates of edge length 10 cm and a separation of $d = 4$ mm. A dielectric slab of constant $\kappa = 2$ has dimensions 10 cm × 10 cm × 4 mm. (*a*) What is the capacitance without the dielectric? (*b*) What is the capacitance if the dielectric slab fills the space between the plates? (*c*) What is the capacitance if a dielectric slab with dimensions 10 cm × 10 cm × 3 mm is inserted into the 4-mm gap?

[†] Recall from Chapter 23 that for electric fields greater than about $3 \times 10^6$ V/m, air breaks down; that is, it becomes ionized and begins to conduct.

**PICTURE THE PROBLEM** The capacitance without the dielectric, $C_0$, is found from the area and spacing of the plates (Figure 24-21a). When the capacitor is filled with a dielectric $\kappa$, (Figure 24-21b), the capacitance is $C = \kappa C_0$ (Equation 24-23). If the dielectric only partially fills the capacitor (Figure 24-21c), we calculate the potential difference $V$ for a given charge $Q$, then apply the definition of capacitance, $C = Q/V$.

**FIGURE 24-21**

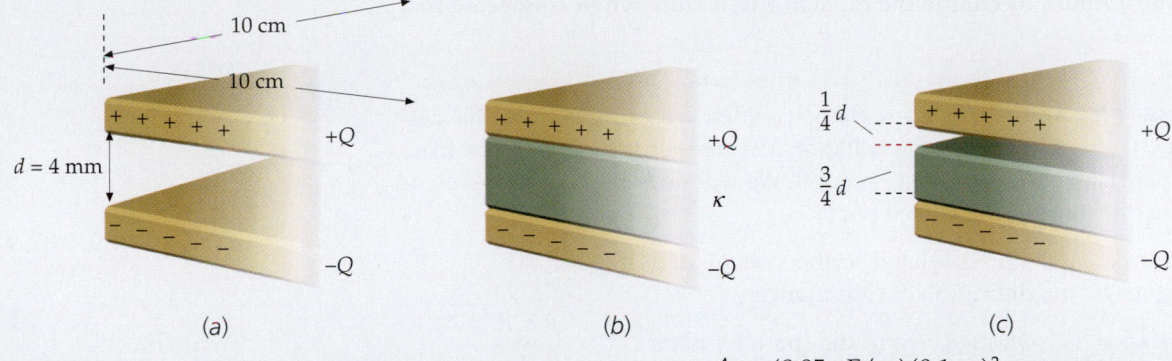

(a)            (b)            (c)

(a) If there is no dielectric, the capacitance $C_0$ is given by Equation 24-10:

$$C_0 = \frac{\epsilon_0 A}{d} = \frac{(8.85 \text{ pF/m})(0.1 \text{ m})^2}{0.004 \text{ m}} = \boxed{22.1 \text{ pF}}$$

(b) When the capacitor is filled with a dielectric $\kappa$, its capacitance $C$ is increased by the factor $\kappa$:

$$C = \kappa C_0 = (2)(22.1 \text{ pF}) = \boxed{44.2 \text{ pF}}$$

(c) 1. The new capacitance is related to the original charge $Q$ and the new potential difference $V$:

$$C = \frac{Q}{V}$$

2. The potential difference $V$ between the plates is the sum of the potential difference for the empty gap plus the potential difference for the dielectric slab:

$$V = V_{\text{gap}} + V_{\text{slab}} = E_{\text{gap}}(\tfrac{1}{4}d) + E_{\text{slab}}(\tfrac{3}{4}d)$$

3. The field in the gap just outside the conductor is the original field $E_0$:

$$E_{\text{gap}} = E_0 = \frac{Q}{\epsilon_0 A}$$

4. The field in the dielectric slab is reduced by the factor $\kappa$:

$$E_{\text{slab}} = \frac{E_0}{\kappa}$$

5. Combining the previous two results yields $V$ in terms of $\kappa$. Note that the original potential difference is $V_0 = E_0 d$:

$$V = E_0\left(\frac{1}{4}d\right) + \frac{E_0}{\kappa}\left(\frac{3}{4}d\right) = E_0 d\left(\frac{1}{4} + \frac{3}{4\kappa}\right)$$

$$= V_0\left(\frac{\kappa + 3}{4\kappa}\right)$$

6. Using $C = Q/V$, we find the new capacitance in terms of the original capacitance, $C_0 = Q/V_0$:

$$C = \frac{Q}{V} = \frac{Q}{V_0\dfrac{\kappa + 3}{4\kappa}} = \frac{Q}{V_0}\left(\frac{4\kappa}{\kappa + 3}\right) = C_0\left(\frac{4\kappa}{\kappa + 3}\right)$$

$$= (22.1 \text{ pF})\left(\frac{8}{5}\right) = \boxed{35.4 \text{ pF}}$$

**PLAUSIBILITY CHECK** The absence of a dielectric corresponds to $\kappa = 1$. In this case, our result for the final step in Part (c) would reduce to $C = C_0$ as expected. Suppose that the dielectric slab were a conducting slab. In a conductor, $E = 0$; so, according to Equation 24-22, $\kappa$ for a conductor would equal infinity. As $\kappa$ approaches infinity, the quantity $4\kappa/(\kappa + 3)$ approaches 4, so the result for the final step in Part (c) approaches $4C_0$. A conducting slab simply extends the capacitor plate, hence the plate separation with the conducting dielectric in place would be $\tfrac{1}{4}d$. This means that $C$ should be $4C_0$, as it is for very large $\kappa$.

**REMARKS** Note that the results of this example are independent of the vertical position of the dielectric (or conducting) slab in the space between the plates.

A HOMEMADE CAPACITOR               **EXAMPLE  24-11**   **Put It in Context**

When studying capacitors in physics class, your professor claims that you could build a parallel-plate capacitor from waxed paper and aluminum foil. You decide to try it, and build one about the size of a piece of notebook paper. Before testing its charge-storing power on your gullible roommate, you decide to calculate the amount of charge the capacitor will store when connected to a 9-V battery.

**PICTURE THE PROBLEM** We want charge, which we can get from the definition $C = Q/V$ if we know the capacitance. We can get the capacitance from the parallel-plate capacitor formula $C = \epsilon_0 A/d$. We will need to either measure or to estimate the thickness of the waxed paper.

1. The charge on a capacitor is related to the voltage and the capacitance by the definition of capacitance:

$$Q = CV$$

2. The capacitance is obtained from the parallel-plate capacitance formula:

$$C = \frac{\kappa \epsilon_0 A}{d}$$

3. Substituting for $C$ and solving for $Q$ gives:

$$Q = CV = \frac{\kappa \epsilon_0 VA}{d}$$

4. A sheet of notebook paper is approximately 8.5-by-11 in.:

$$A = 8.5 \text{ in.} \times 11 \text{ in.} = 93.5 \text{ in.}^2 = 0.0603 \text{ m}^2$$

5. We assume a sheet of wax paper is the same thickness as a sheet of the paper your physics textbook is made of. Measure the thickness of 300 sheets of paper in your book (from page 1 through page 600):

300 sheets of paper are 2.0 cm (0.020 m) thick. So, the thickness of a single sheet of paper is

$$0.020 \text{ m}/300 = 66.7 \ \mu m$$

6. Using the step 3 result, solve for the charge. Assume the dielectric constant of wax paper is 2.3 (the same as that of paraffin):

$$Q = \frac{\kappa \epsilon_0 AV}{d} = \frac{2.3(88.6 \text{ pF/m})(0.0603 \text{ m}^2)(9 \text{ V})}{66.7 \times 10^{-6} \text{ m}}$$

$$= 1.66 \times 10^6 \text{ pC} = \boxed{1.66 \ \mu C}$$

## Energy Stored in the Presence of a Dielectric

The energy stored in a parallel-plate capacitor with dielectric is

$$U = \tfrac{1}{2}QV = \tfrac{1}{2}CV^2$$

We can express the capacitance $C$ in terms of the area and separation of the plates, and the voltage difference $V$ in terms of the electric field and plate separation, to obtain

$$U = \frac{1}{2}CV^2 = \frac{1}{2}\left(\frac{\epsilon A}{d}\right)(Ed)^2 = \frac{1}{2}\epsilon E^2 \,(Ad)$$

The quantity $Ad$ is the volume between the plates containing the electric field. The energy per unit volume is thus

$$u_e = \tfrac{1}{2}\epsilon E^2 = \tfrac{1}{2}\kappa \epsilon_0 E^2 \qquad\qquad 24\text{-}26$$

Part of this energy is the energy associated with the electric field (Equation 24-13) and the rest is the energy associated with the polarization of the dielectric (discussed in Section 24-6).

**EXAMPLE 24-12**

Two parallel-plate capacitors, each having a capacitance of $C_1 = C_2 = 2\ \mu F$, are connected in parallel across a 12-V battery. (*a*) Find the charge on each capacitor. (*b*) Find the total energy stored in the capacitors.

The parallel combination is then disconnected from the battery and a dielectric slab of constant $\kappa = 2.5$ is inserted between the plates of the capacitor $C_2$, completely filling the gap. After the dielectric is inserted, find (*c*) the potential difference across each capacitor, (*d*) the charge on each capacitor, and (*e*) the total energy stored in the capacitors.

**PICTURE THE PROBLEM** (*a*) The charge $Q$ and (*b*) total energy $U$ can be found for each capacitor from its capacitance $C$ and voltage $V$. (*c*) After the capacitors are removed from the battery, the total charge on the pair remains the same. When the dielectric is inserted into one of the capacitors, its capacitance $C_2$ changes. The potential across the parallel combination can be found from the total charge and the equivalent capacitance.

(*a*) The charge on each capacitor is found from its capacitance $C$ and voltage $V$:

$$Q = CV = (2\ \mu F)(12\ V) = \boxed{24\ \mu C}$$

(*b*) 1. The energy stored in each capacitor is found from its charge $Q$ and its voltage $V$:

$$U = \tfrac{1}{2}QV = \tfrac{1}{2}(24\ \mu C)(12\ V) = 144\ \mu J$$

2. The total energy is twice that stored in each capacitor:

$$U_{total} = 2U = \boxed{288\ \mu J}$$

(*c*) 1. The potential across the parallel combination is related to the total charge $Q_{total}$ and the equivalent capacitance $C_{eq}$:

$$V = \frac{Q_{total}}{C_{eq}}$$

2. The capacitance $C_2$ of the capacitor with the dielectric is increased by the factor $\kappa$. The equivalent capacitance is the sum of the capacitances:

$$C_{eq} = C_1 + C_2 = C_1 + \kappa C_2 = (2\ \mu F) + (2.5)(2\ \mu F)$$
$$= 2\ \mu F + 5\ \mu F = 7\ \mu F$$

3. The total charge remains 48 $\mu C$. Substitute for $Q_{total}$ and $C_{eq}$ to calculate $V$:

$$V = \frac{Q_{total}}{C_{eq}} = \frac{48\ \mu C}{7\ \mu F} = \boxed{6.86\ V}$$

(*d*) The charge on each capacitor is again derived from its capacitance and the voltage $V$:

$$Q_1 = C_1 V = (2\ \mu F)(6.86\ V) = \boxed{13.7\ \mu C}$$

$$Q_2 = C_2 V = (5\ \mu F)(6.86\ V) = \boxed{34.3\ \mu C}$$

(*e*) The energy stored in each capacitor is found from its new charge and new voltage:

$$U = U_1 + U_2 = \tfrac{1}{2}Q_1 V + \tfrac{1}{2}Q_2 V = \tfrac{1}{2}(Q_1 + Q_2)V$$
$$= \tfrac{1}{2}(13.7\ \mu C + 34.3\ \mu C)(6.86\ V) = \boxed{165\ \mu J}$$

**PLAUSIBILITY CHECK** When the dielectric is inserted into one of the capacitors, the field is weakened and the potential difference is lowered. Since the two capacitors are connected in parallel, charge must flow from the other capacitor so that the potential difference is the same across both capacitors. Note that the capacitor with the dielectric has the greater charge, and that when the charges calculated for each capacitor in Part (*d*) are added, $Q_1 + Q_2 = 13.7\ \mu C + 34.3\ \mu C = 48\ \mu C$, the result is the same as the original sum.

**REMARKS** The total energy of 165 $\mu J$ is less than the original energy of 288 $\mu J$. When the dielectric is inserted, it is pulled in and work is done on whatever was holding it. To remove the dielectric, work $W = 288\ \mu J - 165\ \mu J = 123\ \mu J$ must be done, and this work is stored as electrostatic potential energy.

Find (a) the charge on each capacitor and (b) the total energy stored in the capacitors of Example 24-12, if the dielectric is inserted into one of the capacitors while the battery is still connected.

**PICTURE THE PROBLEM** Since the battery is still connected, the potential difference across the capacitors remains 12 V. This condition determines the charge and energy stored in each capacitor. Let subscript 1 refer to the capacitor without the dielectric and subscript 2 refer to the capacitor with the dielectric.

**Cover the column to the right and try these on your own before looking at the answers.**

Steps | Answers
--- | ---

(a) Calculate the charge on each capacitor from $Q = CV$ using the result that $C_1 = 2\ \mu F$ and $C_2 = 5\ \mu F$ as found in Example 24-12.

$Q_1 = C_1 V = \boxed{24\ \mu C}$

$Q_2 = C_2 V = \boxed{60\ \mu C}$

(b) 1. Calculate the energy stored in each capacitor from $U = \frac{1}{2} CV^2$. Check your results by using $U = \frac{1}{2} QV$.

$U_1 = 144\ \mu J, U_2 = 360\ \mu J$

2. Add your results for $U_1$ and $U_2$ to obtain the final energy.

$U_{total} = \boxed{504\ \mu J}$

**REMARKS** Note that $Q_2$ is two and a half times its value before the dielectric was inserted (since $\kappa = 2.5$). The battery supplies this additional charge in order to maintain a fixed potential difference. Because of the work done by the battery to supply this charge, the total energy of the system is higher with the dielectric in place (504 $\mu J$) than without the dielectric (288 $\mu J$).

# 24-6 Molecular View of a Dielectric

A dielectric weakens the electric field between the plates of a capacitor, because the molecules in the dielectric produce an electric field within the dielectric in a direction opposite to the field produced by the charges on the plates. The electric field produced by the dielectric is due to the electric dipole moments of the molecules of the dielectric.

Although atoms and molecules are electrically neutral, they are affected by electric fields because they contain positive and negative charges that can respond to external fields. We can think of an atom as a very small, positively charged nucleus surrounded by a negatively charged electron cloud. In some atoms and molecules, the electron cloud is spherically symmetric, so its "center of negative charge" is at the center of the atom or molecule, coinciding with the center of positive charge. An atom or molecule like this has zero dipole moment and is said to be nonpolar. But in the presence of an external electric field, the positive and negative charges experience forces in opposite directions, so the positive and negative charges then separate until the attractive force they exert on each other balances the forces due to the external electric field (Figure 24-22). The molecule is then said to be polarized and it behaves like an electric dipole.

In some molecules (e.g., HCl and $H_2O$), the centers of positive and negative charge do not coincide, even in the absence of an external electric field. As we noted in Chapter 21, these polar molecules have a permanent electric dipole moment.

When a dielectric is placed in the field of a capacitor, its molecules are polarized in such a way that there is a net dipole moment parallel to the field. If the

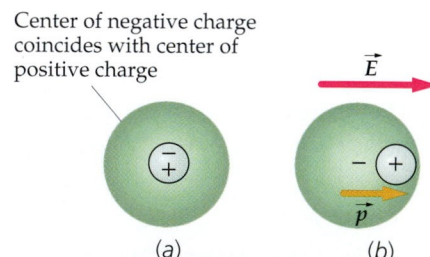

(a)    (b)

**FIGURE 24-22** Schematic diagrams of the charge distributions of an atom or nonpolar molecule. (a) In the absence of an external electric field, the center of positive charge coincides with the center of negative charge. (b) In the presence of an external electric field, the centers of positive and negative charge are displaced, producing an induced dipole moment in the direction of the external field.

molecules are polar their dipole moments, originally oriented at random, tend to become aligned due to the torque exerted by the field.[†] If the molecules are nonpolar, the field induces dipole moments that are parallel to the field. In either case, the molecules in the dielectric are polarized in the direction of the external field (Figure 24-23).

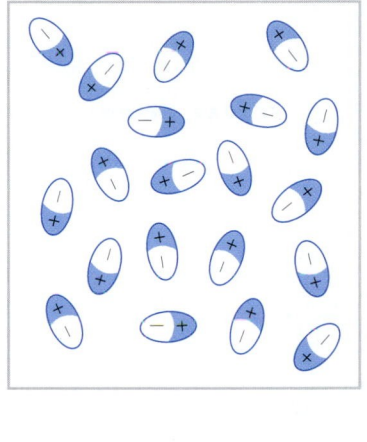

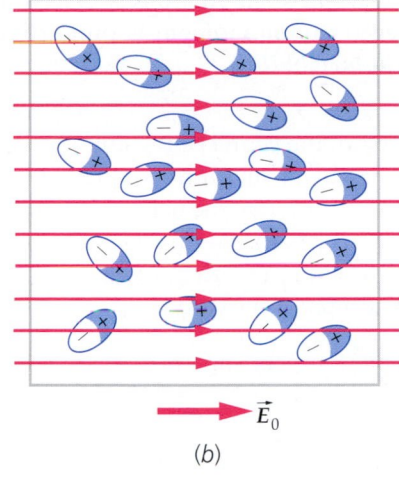

(a)                                         (b)

**FIGURE 24-23** (*a*) The randomly oriented electric dipoles of a polar dielectric in the absence of an external electric field. (*b*) In the presence of an external electric field, the dipoles are partially aligned parallel to the field.

The net effect of the polarization of a homogeneous dielectric in a parallel-plate capacitor is the creation of a surface charge on the dielectric faces near the plates, as shown in Figure 24-24. The surface charge on the dielectric is called a **bound charge,** because the surface charge is bound to the molecules of the dielectric and cannot move about like the free charge on the conducting capacitor plates. This bound charge produces an electric field opposite in direction to the electric field produced by the free charge on the conductors. Thus, the net electric field between the plates is reduced, as illustrated in Figure 24-25.

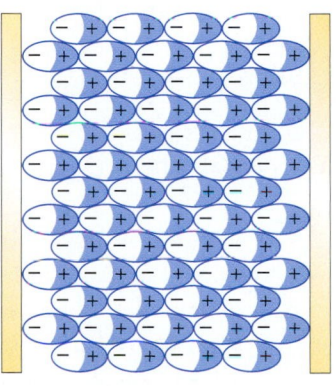

**FIGURE 24-24** When a dielectric is placed between the plates of a capacitor, the electric field of the capacitor polarizes the molecules of the dielectric. The result is a bound charge on the surface of the dielectric that produces its own electric field; this field opposes the external field. The field of the bound surface charges thus weakens the electric field within the dielectric.

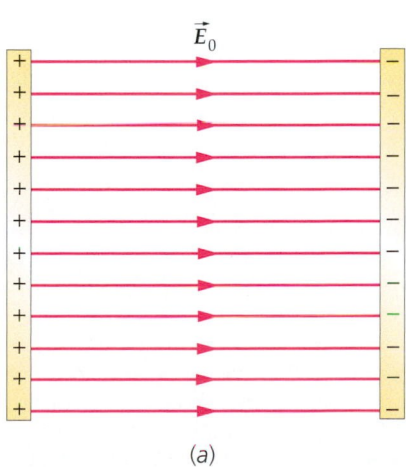

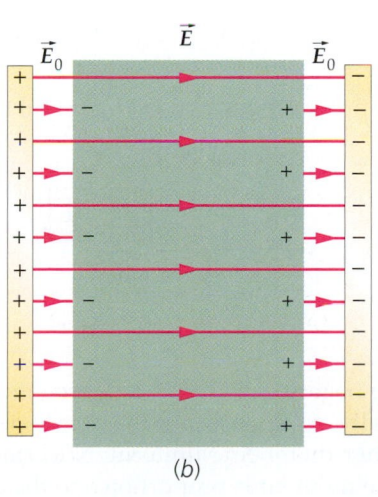

(a)                                         (b)

**FIGURE 24-25** The electric field between the plates of a capacitor (*a*) with no dielectric and (*b*) with a dielectric. The surface charge on the dielectric weakens the original field between the plates.

---

[†] The degree of alignment depends on the external field and on the temperature. It is approximately proportional to $pE/kT$, where $pE$ is the maximum energy of a dipole in a field $E$, and $kT$ is the characteristic thermal energy.

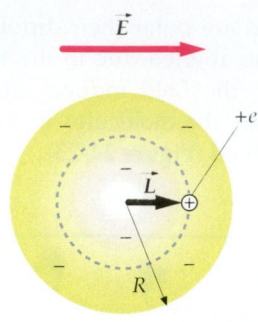

**FIGURE 24-26**

A hydrogen atom consists of a proton nucleus of charge $+e$ and an electron of charge $-e$. The charge distribution of the atom is spherically symmetric, so the atom is nonpolar. Consider a model in which the hydrogen atom consists of a positive point charge $+e$ at the center of a uniformly charged spherical cloud of radius $R$ and total charge $-e$. Show that when such an atom is placed in a uniform external electric field $\vec{E}$, the induced dipole moment is proportional to $\vec{E}$; that is, $\vec{p} = \alpha\vec{E}$, where $\alpha$ is called the *polarizability*.

**PICTURE THE PROBLEM** In the external field, the center of the uniform negative cloud is displaced from the positive charge by an amount $L$ so that the force exerted by the field $e\vec{E}$ is balanced by the force exerted by the negative cloud $e\vec{E}'$, where $\vec{E}'$ is the field due to the cloud (Figure 24-26). We use Gauss's law to find $E'$, and then we calculate the induced dipole moment $\vec{p} = e\vec{L}$, where $\vec{L}$ is the position of the positive charge relative to the center of the cloud.

1. Write the magnitude of the induced dipole moment in terms of $e$ and $L$:

$$p = eL$$

2. We can find $L$ by calculating the field $E'_n$ due to the negatively charged cloud at a distance $L$ from the center. We use Gauss's law to compute $E'_n$. Choose a spherical Gaussian surface of radius $L$ concentric with the cloud. Then $E'_n$ is constant on this surface:

$$\phi_{net} = \oint E_n \, dA = \frac{Q_{inside}}{\epsilon_0}$$

$$E'_n (4\pi L^2) = \frac{Q_{inside}}{\epsilon_0}$$

$$E'_n = \frac{Q_{inside}}{4\pi \epsilon_0 L^2}$$

3. The charge inside the sphere of radius $L$ equals the charge density times the volume:

$$Q_{inside} = \rho \frac{4}{3}\pi L^3 = \frac{-e}{\frac{4}{3}\pi R^3}\frac{4}{3}\pi L^3 = -e\frac{L^3}{R^3}$$

4. Substitute this value of $Q_{inside}$ to calculate $E'_n$:

$$E'_n = \frac{Q_{inside}}{4\pi \epsilon_0 L^2} = \frac{-eL^3/R^3}{4\pi \epsilon_0 L^2} = -\frac{e}{4\pi \epsilon_0 R^3}L$$

5. Solve for $L$:

$$L = -\frac{4\pi \epsilon_0 R^3}{e}E'_n$$

6. $E'_n$ is negative because it points inward on the Gaussian surface. At the positive charge, $E'_n$ points to the left, so $E'_n = -E$:

$$E'_n = -E$$

so

$$L = \frac{4\pi \epsilon_0 R^3}{e}E$$

7. Substitute these results for $L$ and $E'_n$ to express $p$ in terms of the external field $E$:

$$p = eL = 4\pi \epsilon_0 R^3 E = \alpha E$$

so

$$\boxed{\vec{p} = \alpha\vec{L}}$$

where

$$\alpha = 4\pi \epsilon_0 R^3$$

**REMARKS** The charge distribution of the negative charge in a hydrogen atom, obtained from quantum theory, is spherically symmetric, but the charge density decreases exponentially with distance rather than being uniform. Nevertheless, the above calculation shows that the dipole moment is proportional to the external field $p = \alpha E$, and the polarizability $\alpha$ is of the order of $4\pi \epsilon_0 R^3$ where $R$ is the radius of the atom or molecule. The dielectric constant $\kappa$ can be related to the polarizability and to the number of molecules per unit volume.

## Magnitude of the Bound Charge

The bound charge density $\sigma_b$ on the surfaces of the dielectric is related to the dielectric constant $\kappa$ and to the free charge density $\sigma_f$ on the plates. Consider a dielectric slab between the plates of a parallel-plate capacitor, as shown in Figure 24-27. If the dielectric is a very thin slab between plates that are close together, the electric field inside the dielectric slab due to the bound charge densities, $+\sigma_b$ on the right and $-\sigma_b$ on the left, is just the field due to two infinite-plane charge densities. Thus, the field $E_b$ has the magnitude

$$E_b = \frac{\sigma_b}{\epsilon_0}$$

This field is directed to the left and subtracts from the electric field $E_0$ due to the free charge density on the capacitor plates, which has the magnitude

$$E_0 = \frac{\sigma_f}{\epsilon_0}$$

The magnitude of the net field $E = E_0/\kappa$ is the difference between these magnitudes:

$$E = E_0 - E_b = \frac{E_0}{\kappa}$$

or

$$E_b = \left(1 - \frac{1}{\kappa}\right)E_0$$

Writing $\sigma_b/\epsilon_0$ for $E_b$ and $\sigma_f/\epsilon_0$ for $E_0$, we obtain

$$\sigma_b = \left(1 - \frac{1}{\kappa}\right)\sigma_f \qquad\qquad 24\text{-}27$$

The bound charge density $\sigma_b$ is always less than the free charge density $\sigma_f$ on the capacitor plates, and it is zero if $\kappa = 1$, which is the case when there is no dielectric. For a conducting slab, $\kappa = \infty$ and $\sigma_b = \sigma_f$.

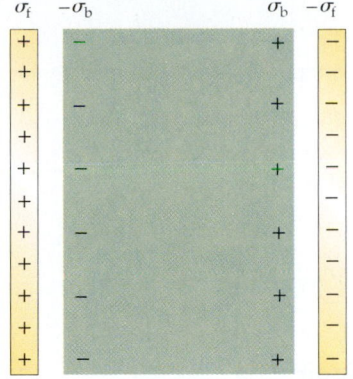

**FIGURE 24-27** A parallel-plate capacitor with a dielectric slab between the plates. If the plates are closely spaced, each of the surface charges can be considered as an infinite plane charge. The electric field due to the free charge on the plates is directed to the right and has a magnitude $E_0 = \sigma_f/\epsilon_0$. That due to the bound charge is directed to the left and has a magnitude $E_b = \sigma_b/\epsilon_0$.

## *The Piezoelectric Effect

In certain crystals that contain polar molecules (e.g., quartz, tourmaline, and topaz), a mechanical stress applied to the crystal produces polarization of the molecules. This is known as the **piezoelectric effect.** The polarization of the stressed crystal causes a potential difference across the crystal, which can be used to produce an electric current. Piezoelectric crystals are used in transducers (e.g., microphones, phonograph pickups, and vibration-sensing devices) to convert mechanical strain into electrical signals. The converse piezoelectric effect, in which a voltage applied to such a crystal induces mechanical strain (deformation), is used in headphones and many other devices.

Because the natural frequency of vibration of quartz is in the range of radio frequencies, and because its resonance curve is very sharp,[†] quartz is used extensively to stabilize radio-frequency oscillators and to make accurate clocks.

---

† Resonance in AC circuits, which will be discussed in Chapter 29, is analogous to mechanical resonance, which was discussed in Chapter 14.

1. Capacitance is an important defined quantity that relates charge to potential difference.
2. Devices connected in *parallel* share a common potential difference across each device *due solely to the way they are connected*.

| Topic | Relevant Equations and Remarks |
|---|---|
| **1. Electrostatic Potential Energy** | The electrostatic potential energy of a system of point charges is the work needed to bring the charges from an infinite separation to their final positions. |
| Of point charges | $U = \dfrac{1}{2} \sum\limits_{i=1}^{n} q_i V_i$     24-2 |
| Of a conductor with charge $Q$ at potential $V$ | $U = \frac{1}{2} QV$     24-3 |
| Of a system of conductors | $U = \dfrac{1}{2} \sum\limits_{i=1}^{n} Q_i V_i$     24-4 |
| Energy stored in a capacitor | $U = \dfrac{1}{2} \dfrac{Q^2}{C} = \dfrac{1}{2} QV = \dfrac{1}{2} CV^2$     24-12 |
| Energy density of an electric field | $u_e = \frac{1}{2} \epsilon_0 E^2$     24-13 |
| **2. Capacitor** | A capacitor is a device for storing charge and energy. It consists of two conductors insulated from each other that carry equal and opposite charges. |
| **3. Capacitance** | Definition of capacitance. $C = \dfrac{Q}{V}$     24-5 |
| Isolated conductor | $Q$ is the conductor's total charge, $V$ is the conductor's potential relative to infinity. |
| Capacitor | $Q$ is the magnitude of the charge on either conductor, $V$ is the magnitude of the potential difference between the conductors. |
| Of an isolated spherical conductor | $C = 4\pi \epsilon_0 R$     24-6 |
| Of a parallel-plate capacitor | $C = \dfrac{\epsilon_0 A}{d}$     24-10 |
| Of a cylindrical capacitor | $C = \dfrac{2\pi \epsilon_0 L}{\ln(R_2/R_1)}$     24-11 |
| **4. Equivalent Capacitance** | |
| Parallel capacitors | When devices are connected in parallel, the voltage drop is the same across each. $C_{eq} = C_1 + C_2 + C_3 + \dots$     24-17 |

| Series capacitors | When capacitors are in series, the voltage drops add. If the net charge on each connected pair of plates is zero, then: |
|---|---|

$$\frac{1}{C_{eq}} = \frac{1}{C_1} + \frac{1}{C_2} + \frac{1}{C_3} + \dots$$

24-21

## 5. Dielectrics

| Macroscopic behavior | A nonconducting material is called a dielectric. When a dielectric is inserted between the plates of a capacitor, the electric field within the dielectric is weakened and the capacitance is thereby increased by the factor $\kappa$, which is the dielectric constant. |
|---|---|
| Microscopic view | The field in the dielectric of a capacitor is weakened because the dipole moments of the molecules (either preexisting or induced) tend to align with the field and thereby produce an electric field inside the dielectric that opposes the applied field. The aligned dipole moment of the dielectric is proportional to the applied field. |
| Electric field inside | $E = \dfrac{E_0}{\kappa}$ |
| Effect on capacitance | $C = \kappa C_0$ |
| Permittivity $\epsilon$ | $\epsilon = \kappa \epsilon_0$ |
| Uses of a dielectric | 1. Increases capacitance<br>2. Increases dielectric strength<br>3. Physically separates conductors |

24-22

24-23

24-25

| *6. Piezoelectric Effect | In certain crystals containing polar molecules, a mechanical stress polarizes the molecules, which induces a voltage across the crystal. Conversely, an applied voltage induces mechanical strain (deformation) in the crystal. |
|---|---|

# PROBLEMS

• Single-concept, single-step, relatively easy

•• Intermediate-level, may require synthesis of concepts

••• Challenging, for advanced students

SSM Solution is in the *Student Solutions Manual*

 Problems available on iSOLVE online homework service

✓ These "Checkpoint" online homework service problems ask students additional questions about their confidence level, and how they arrived at their answer.

In a few problems, you are given more data than you actually need; in a few other problems, you are required to supply data from your general knowledge, outside sources, or informed estimates.

## Conceptual Problems

**1** • SSM If the voltage across a parallel-plate capacitor is doubled, its capacitance (*a*) doubles. (*b*) drops by half. (*c*) remains the same.

**2** • If the charge on an isolated spherical conductor is doubled, its capacitance (*a*) doubles. (*b*) drops by half. (*c*) remains the same.

**3** • True or false: The electrostatic energy per unit volume at some point is proportional to the square of the electric field at that point.

**4** • If the potential difference of a parallel-plate capacitor is doubled by changing the plate separation without changing the charge, by what factor does its stored electric energy change?

**5** •• [SSM] A parallel-plate air capacitor is connected to a constant-voltage battery. If the separation between the capacitor plates is doubled while the capacitor remains connected to the battery, the energy stored in the capacitor (a) quadruples. (b) doubles. (c) remains unchanged. (d) drops to half its previous value. (e) drops to one-fourth its previous value.

**6** •• If the capacitor of Problem 5 is disconnected from the battery before the separation between the plates is doubled, the energy stored in the capacitor upon separation of the plates (a) quadruples. (b) doubles. (c) remains unchanged. (d) drops to half its previous value. (e) drops to one-fourth its previous value.

**7** • True or false:

(a) The equivalent capacitance of two capacitors in parallel equals the sum of the individual capacitances.

(b) The equivalent capacitance of two capacitors in series is less than the capacitance of either capacitor alone.

**8** •• Two initially uncharged capacitors of capacitance $C_0$ and $2C_0$, respectively, are connected in series across a battery. Which of the following is true?

(a) The capacitor $2C_0$ carries twice the charge of the other capacitor.

(b) The voltage across each capacitor is the same.

(c) The energy stored by each capacitor is the same.

(d) None of the above statements is correct.

**9** • True or false: A dielectric inserted into a capacitor increases the capacitance.

**10** •• [SSM] Two capacitors half-filled with a dielectric are shown in Figure 24-28. The area and separation of each capacitor is the same. Which has the higher capacitance, that shown in Figure (a) or in Figure (b)?

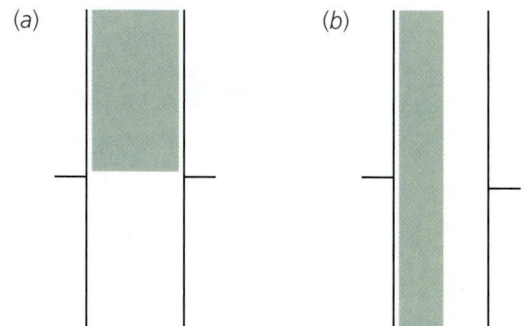

**FIGURE 24-28** Problem 10

**11** • True or false:

(a) The capacitance of a capacitor is defined as the total amount of charge the capacitor can hold.

(b) The capacitance of a parallel-plate capacitor depends on the voltage difference between the plates.

(c) The capacitance of a parallel-plate capacitor is proportional to the charge on its plates.

**12** •• Two identical capacitors are connected in series to a 100-V battery. When only one capacitor is connected to this battery, the energy stored is $U_0$. What is the total energy stored in the two capacitors when the series combination is connected to the battery? (a) $4U_0$. (b) $2U_0$. (c) $U_0$. (d) $U_0/2$. (e) $U_0/4$.

## Estimation and Approximation

**13** •• Disconnect the coaxial cable from a television or other device and measure (estimate) the diameter of the center conductor and the braided conductor, shown in the photo on page 755. Assume a plausible value (see Table 24-1) for the dielectric constant of the material separating the two conductors and estimate the capacitance per unit length of the cable.

**14** •• [SSM] To create the high-energy densities needed to operate a pulsed nitrogen laser, the discharge from a high-capacitance capacitor is used. Typically, the energy requirement per pulse (i.e., per discharge) is 100 J. Estimate the capacitance required if the discharge is applied through a spark gap of 1 cm width. Assume that the dielectric breakdown of nitrogen occurs at $E \approx 3 \times 10^6$ V/m.

**15** •• Measurements reveal that the earth's electric field extends upward for 1000 m and has an average magnitude of 200 V/m. Estimate the electrical energy stored in the atmosphere. (*Hint: You may treat the atmosphere as a flat slab with an area equal to the surface area of the earth. Why?*)

**16** •• Estimate the capacitance of a typical hot-air balloon.

## Electrostatic Potential Energy

**17** • Three point charges are on the x axis: $q_1$ at the origin, $q_2$ at $x = 3$ m, and $q_3$ at $x = 6$ m. Find the electrostatic potential energy for (a) $q_1 = q_2 = q_3 = 2 \mu C$; (b) $q_1 = q_2 = 2 \mu C$, and $q_3 = -2 \mu C$; and (c) $q_1 = q_3 = 2 \mu C$, and $q_2 = -2 \mu C$.

**18** • Point charges $q_1$, $q_2$, and $q_3$ are at the corners of an equilateral triangle of side 2.5 m. Find the electrostatic potential energy of this charge distribution if (a) $q_1 = q_2 = q_3 = 4.2 \mu C$; (b) $q_1 = q_2 = 4.2 \mu C$, and $q_3 = -4.2 \mu C$; and (c) $q_1 = q_2 = -4.2 \mu C$, and $q_3 = +4.2 \mu C$.

**19** • [SSM] [SOLVE] What is the electrostatic potential energy of an isolated spherical conductor of 10 cm radius that is charged to 2 kV?

**20** •• [SOLVE] Four point charges of magnitude 2 $\mu C$ are at the corners of a square of side 4 m. Find the electrostatic potential energy if (a) all of the charges are negative, (b) three of the charges are positive and one of the charges is negative, and (c) two of the charges are positive and two of the charges are negative.

**21** •• [SOLVE] Four charges are at the corners of a square centered at the origin as follows: q at $(-a, +a)$; 2q at $(+a, +a)$; $-3q$ at $(+a, -a)$; and 6q at $(-a, -a)$. A fifth charge $+q$ is placed at the origin and released from rest. Find its speed when it is a great distance from the origin.

## Capacitance

**22** • [SSM] An isolated spherical conductor of 10 cm radius is charged to 2 kV. (a) How much charge is on the conductor? (b) What is the capacitance of the sphere? (c) How does the capacitance change if the sphere is charged to 6 kV?

**23** • **i SOLVE** A capacitor has a charge of 30 $\mu$C. The potential difference between the conductors is 400 V. What is the capacitance?

**24** •• Two isolated conducting spheres of equal radius have charges $+Q$ and $-Q$, respectively. If they are separated by a large distance compared to their radius, what is the capacitance of this unusual capacitor?

## The Storage of Electrical Energy

**25** • **i SOLVE** (a) A 3-$\mu$F capacitor is charged to 100 V. How much energy is stored in the capacitor? (b) How much additional energy is required to charge the capacitor from 100 V to 200 V?

**26** • **i SOLVE** A 10-$\mu$F capacitor is charged to $Q = 4$ $\mu$C. (a) How much energy is stored in the capacitor? (b) If half the charge is removed from the capacitor, how much energy remains?

**27** • **i SOLVE** (a) Find the energy stored in a 20-pF capacitor when it is charged to 5 $\mu$C. (b) How much additional energy is required to increase the charge from 5 $\mu$C to 10 $\mu$C?

**28** • **SSM** Find the energy per unit volume in an electric field that is equal to 3 MV/m, which is the dielectric strength of air.

**29** • A parallel-plate capacitor with a plate area of 2 m² and a separation of 1.0 mm is charged to 100 V. (a) What is the electric field between the plates? (b) What is the energy per unit volume in the space between the plates? (c) Find the total energy by multiplying your answer from Part (b) by the total volume between the plates. (d) Find the capacitance C. (e) Calculate the total energy from $U = \frac{1}{2} CV^2$, and compare your answer with your result from Part (c).

**30** •• **i SOLVE** Two concentric metal spheres have radii $r_1 = 10$ cm and $r_2 = 10.5$ cm, respectively. The inner sphere has a charge $Q = 5$ nC spread uniformly on its surface, and the outer sphere has charge $-Q$ on its surface. (a) Calculate the total energy stored in the electric field inside the spheres. *Hint: You can treat the spheres essentially as parallel flat slabs separated by 0.5 cm—why?* (b) Find the capacitance of this two-sphere system and show that the total energy stored in the field is equal to $\frac{1}{2}Q^2/C$.

**31** •• **SSM** **i SOLVE** ✓ A parallel-plate capacitor with plates of area 500 cm² is charged to a potential difference V and is then disconnected from the voltage source. When the plates are moved 0.4 cm farther apart, the voltage between the plates increases by 100 V. (a) What is the charge Q on the positive plate of the capacitor? (b) How much does the energy stored in the capacitor increase due to the movement of the plates?

**32** ••• A ball of charge of radius R has a uniform charge density $\rho$ and a total charge $Q = \frac{4}{3} \pi R^3 \rho$. (a) Find the electrostatic energy density at a distance $r$ from the center of the ball for $r < R$ and for $r > R$. (b) Find the energy in a spherical shell of volume $4\pi r^2 \, dr$ for $r < R$ and for $r > R$. (c) Compute the total electrostatic energy by integrating your expressions from Part (b), and show that your result can be written $U = kQ^2/R$. Explain why this result is greater than that for a spherical conductor of radius R carrying a total charge Q.

## Combinations of Capacitors

**33** • (a) How many 1-$\mu$F capacitors connected in parallel would it take to store a total charge of 1 mC with a potential difference of 10 V across each capacitor? (b) What would be the potential difference across the combination? (c) If the number of 1-$\mu$F capacitors found in Part (a) is connected in series and the potential difference across each is 10 V, find the charge on each capacitor and the potential difference across the combination.

**34** • **i SOLVE** ✓ A 3-$\mu$F capacitor and a 6-$\mu$F capacitor are connected in series, and the combination is connected in parallel with an 8-$\mu$F capacitor. What is the equivalent capacitance of this combination?

**35** • **SSM** Three capacitors are connected in a triangle as shown in Figure 24-29. Find the equivalent capacitance between points $a$ and $c$.

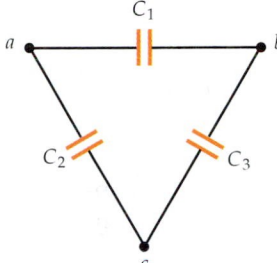

**FIGURE 24-29** Problem 35

**36** • A 10-$\mu$F capacitor and a 20-$\mu$F capacitor are connected in parallel across a 6-V battery. (a) What is the equivalent capacitance of this combination? (b) What is the potential difference across each capacitor? (c) Find the charge on each capacitor.

**37** •• A 10-$\mu$F capacitor is connected in series with a 20-$\mu$F capacitor across a 6-V battery. (a) Find the charge on each capacitor. (b) Find the potential difference across each capacitor.

**38** •• **SSM** Three identical capacitors are connected so that their maximum equivalent capacitance is 15 $\mu$F. (a) Describe how the capacitors are combined. (b) There are three other ways to combine all three capacitors in a circuit. What are the equivalent capacitances for each arrangement?

**39** •• For the circuit shown in Figure 24-30, find (a) the total equivalent capacitance between the terminals, (b) the charge stored on each capacitor, and (c) the total stored energy.

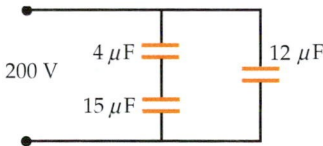

**FIGURE 24-30** Problem 39

**40** •• (a) Show that the equivalent capacitance of two capacitors in series can be written

$$C_{eq} = \frac{C_1 C_2}{C_1 + C_2}$$

(b) Use this expression to show that $C_{eq} < C_1$ and $C_{eq} < C_2$.
(c) Show that the correct expression for the equivalent capacitance of three capacitors in series is

$$C_{eq} = \frac{C_1 C_2 C_3}{C_1 C_2 + C_2 C_3 + C_1 C_3}$$

**41** •• For the circuit shown in Figure 24-31, find (*a*) the total equivalent capacitance between the terminals, (*b*) the charge stored on each capacitor, and (*c*) the total stored energy.

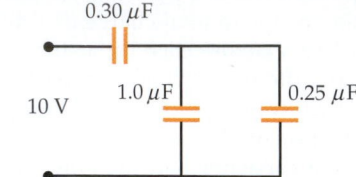

**FIGURE 24-31**
**Problem 41**

**42** •• Five identical capacitors of capacitance $C_0$ are connected in a bridge network, as shown in Figure 24-32. (*a*) What is the equivalent capacitance between points *a* and *b*? (*b*) Find the equivalent capacitance between points *a* and *b* if the capacitor at the center is replaced by a capacitor with a capacitance of $10 C_0$.

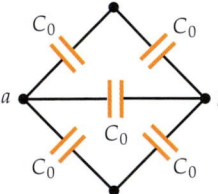

**FIGURE 24-32** **Problem 42**

**43** •• Design a network of capacitors that has a capacitance of 2 $\mu$F and breakdown voltage of 400 V, using only 2-$\mu$F capacitors that have individual breakdown voltages of 100 V.

**44** •• **SSM** Find all the different possible equivalent capacitances that can be obtained using a 1-$\mu$F, a 2-$\mu$F, and a 4-$\mu$F capacitor in any combination that includes all three, or any two, of the capacitors.

**45** ••• (*a*) What is the capacitance of the infinite ladder of capacitors shown in Figure 24-33*a*? (*b*) If we were to replace the ladder with a single capacitor (as shown in Figure 24-33*b*), what capacitance *C* would we need so that the combination had the same capacitance as the infinite ladder?

(*a*)

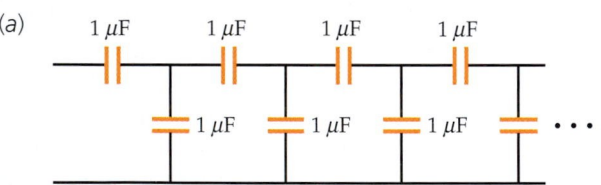

(*b*)

**FIGURE 24-33** **Problem 45**

## Parallel-Plate Capacitors

**46** • **iSOLVE✓** A parallel-plate capacitor has a capacitance of 2 $\mu$F and a plate separation of 1.6 mm. (*a*) What is the maximum potential difference between the plates, so that dielectric breakdown of the air between the plates does not occur? (Use $E_{max} = 3$ MV/m.) (*b*) How much charge is stored at this maximum potential difference?

**47** • **iSOLVE✓** An electric field of $2 \times 10^4$ V/m exists between the plates of a circular parallel-plate capacitor that has a plate separation of 2 mm. (*a*) What is the voltage across the capacitor? (*b*) What plate radius is required if the stored charge is 10 $\mu$C?

**48** •• A parallel-plate, air-gap capacitor has a capacitance of 0.14 $\mu$F. The plates are 0.5 mm apart. (*a*) What is the area of each plate? (*b*) What is the potential difference if the capacitor is charged to 3.2 $\mu$C? (*c*) What is the stored energy? (*d*) How much charge can the capacitor carry before dielectric breakdown of the air between the plates occurs?

**49** •• **SSM** **iSOLVE✓** Design a 0.1-$\mu$F parallel-plate capacitor with air between the plates that can be charged to a maximum potential difference of 1000 V. (*a*) What is the minimum possible separation between the plates? (*b*) What minimum area must the plates of the capacitor have?

## Cylindrical Capacitors

**50** • A Geiger tube consists of a wire of radius $R = 0.2$ mm, length $L = 12$ cm, and a coaxial cylindrical shell conductor of the same length $L = 12$ cm with a radius of 1.5 cm. (*a*) Find the capacitance, assuming that the gas in the tube has a dielectric constant of $\kappa = 1$. (*b*) Find the charge per unit length on the wire, when the potential difference between the wire and shell is 1.2 kV.

**51** •• A cylindrical capacitor consists of a long wire of radius $R_1$ and length $L$ with a charge $+Q$ and a concentric outer cylindrical shell of radius $R_2$, length $L$, and charge $-Q$. (*a*) Find the electric field and energy density at any point in space. (*b*) How much energy resides in a cylindrical shell between the conductors of radius $R$, thickness $dr$, and volume $2\pi rL\, dr$? (*c*) Integrate your expression from Part (*b*) to find the total energy stored in the capacitor and compare your result with that obtained, using $U = \frac{1}{2}CV^2$.

**52** ••• Three concentric, thin conducting cylindrical shells have radii of 0.2, 0.5, and 0.8 cm. The space between the shells is filled with air. The innermost and outermost cylinders are connected at one end by a conducting wire. Find the capacitance per unit length of this system.

**53** •• **SSM** A goniometer is a precise instrument for measuring angles. A capacitive goniometer is shown in Figure 24-34*a*. Each plate of the variable capacitor (Figure 24-34*b*) consists of a flat metal semicircle with inner radius $R_1$ and the outer radius $R_2$. The plates share a common rotation axis, and the width of the air gap separating the plates is *d*. Calculate the capacitance as a function of the angle $\theta$ and the parameters given.

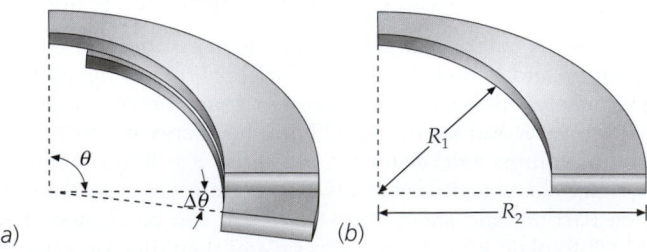

(*a*)　(*b*)

**FIGURE 24-34** **Problem 53**

**54** •• A capacitive pressure gauge is shown in Figure 24-35. Two plates of area $A$ are separated by a material with dielectric constant $\kappa$, thickness $d$, and Young's modulus $Y$. If a pressure increase $\Delta P$ is applied to the plates, what is the change in their capacitance?

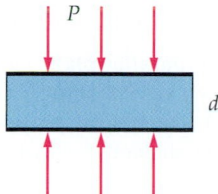

**FIGURE 24-35** Problem 54

## Spherical Capacitors

**55** •• SSM A spherical capacitor consists of two thin, concentric spherical shells of radii $R_1$ and $R_2$. (a) Show that the capacitance is given by $C = 4\pi\epsilon_0 R_1 R_2/(R_2 - R_1)$. (b) Show that when the radii of the shells are nearly equal, the capacitance is given approximately by the expression for the capacitance of a parallel-plate capacitor, $C = \epsilon_0 A/d$, where $A$ is the area of the sphere and $d = R_2 - R_1$.

**56** •• A spherical capacitor has an inner sphere of radius $R_1$ with a charge of $+Q$ and an outer concentric spherical thin shell of radius $R_2$ with a charge of $-Q$. (a) Find the electric field and the energy density at any point in space. (b) Calculate the energy in the electrostatic field in a spherical shell of radius $r$, thickness $dr$, and volume $4\pi r^2\, dr$ between the conductors? (c) Integrate your expression from Part (b) to find the total energy stored in the capacitor, and compare your result with that obtained using $U = \frac{1}{2}QV$.

**57** ••• A spherical shell of radius $R$ carries a charge $Q$ distributed uniformly over its surface. Find the distance from the center of the sphere such that half the total electrostatic field energy of the system is within that distance.

## Disconnected and Reconnected Capacitors

**58** •• ISOLVE✓ A 2-$\mu$F capacitor is charged to a potential difference of 12 V. The wires connecting the capacitor to the battery are then disconnected from the battery and connected across a second, initially uncharged, capacitor. The potential difference across the 2-$\mu$F capacitor then drops to 4 V. What is the capacitance of the second capacitor?

**59** •• ISOLVE A 100-pF capacitor and a 400-pF capacitor are both charged to 2 kV. They are then disconnected from the voltage source and are connected together, positive plate to positive plate and negative plate to negative plate. (a) Find the resulting potential difference across each capacitor. (b) Find the energy lost when the connections are made.

**60** •• SSM ISOLVE Two capacitors, $C_1 = 4\ \mu$F and $C_2 = 12\ \mu$F, are connected in series across a 12-V battery. They are carefully disconnected so that they are not discharged and they are then reconnected to each other, with positive plate to positive plate and negative plate to negative plate. (a) Find the potential difference across each capacitor after they are connected. (b) Find the initial energy stored and the final energy stored in the capacitors.

**61** •• A 1.2-$\mu$F capacitor is charged to 30 V. After charging, the capacitor is disconnected from the voltage source and is connected to another uncharged capacitor. The final voltage is 10 V. (a) What is the capacitance of the first capacitor? (b) How much energy was lost when the connection was made?

**62** •• Rework Problem 59, imagining that the capacitors are connected positive plate to negative plate, after they have been charged to 2 kV.

**63** •• Rework Problem 60, imagining that the two capacitors are first connected in parallel across the 12-V battery and are then connected, with the positive plate of each capacitor connected to the negative plate of the other.

**64** •• SSM ISOLVE A 20-pF capacitor is charged to 3 kV and then removed from the battery and connected to an uncharged 50-pF capacitor. (a) What is the new charge on each capacitor? (b) Find the initial energy stored in the 20-pF capacitor, and find the final energy stored in the two capacitors. Is electrostatic potential energy gained or lost when the two capacitors are connected?

**65** ••• A parallel combination of three capacitors, $C_1 = 2\ \mu$F, $C_2 = 4\ \mu$F, and $C_3 = 6\ \mu$F is charged with a 200-V source. The capacitors are then disconnected from the voltage source and from each other and are reconnected positive plates to negative plates, as shown in Figure 24-36. (a) What is the voltage across each capacitor with switches $S_1$ and $S_2$ closed but switch $S_3$ open? (b) After switch $S_3$ is closed, what is the final charge on each capacitor? (c) Give the voltage across each capacitor after switch $S_3$ is closed.

**FIGURE 24-36**
Problem 65

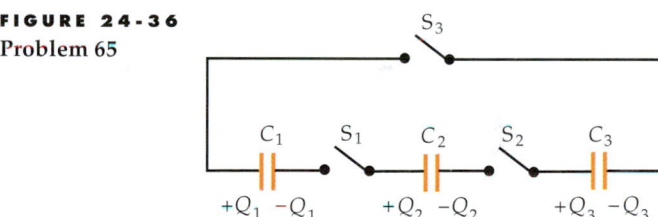

**66** •• SSM A capacitor of capacitance $C$ has a charge $Q$. A student connects one terminal of the capacitor to a terminal of an identical uncharged capacitor. When the remaining two terminals are connected, charge flows until electrostatic equilibrium is reestablished and both capacitors have charge $Q/2$ on them. Compare the total energy initially stored in the one capacitor to the total energy stored in the two after the second electrostatic equilibrium. Where did the missing energy go? This energy was dissipated in the connecting wires via Joule heating, which is discussed in Chapter 25.

## Dielectrics

**67** • ISOLVE✓ A parallel-plate capacitor is made by placing polyethylene ($\kappa = 2.3$) between two sheets of aluminum foil. The area of each sheet of aluminum foil is 400 cm$^2$, and the thickness of the polyethylene is 0.3 mm. Find the capacitance.

**68** •• Suppose the Geiger tube of Problem 50 is filled with a gas of dielectric constant $\kappa = 1.8$ and breakdown field of $2 \times 10^6$ V/m. (a) What is the maximum potential difference that can be maintained between the wire and shell? (b) What is the charge per unit length on the wire?

**69** •• Repeat Problem 56, with the space between the two spherical shells filled with a dielectric of dielectric constant $\kappa$.

**70** •• [SOLVE] ✓ A certain dielectric, with a dielectric constant $\kappa = 24$, can withstand an electric field of $4 \times 10^7$ V/m. Suppose we want to use this dielectric to construct a 0.1-$\mu$F capacitor that can withstand a potential difference of 2000 V. (a) What is the minimum plate separation? (b) What must the area of the plates be?

**71** •• A parallel-plate capacitor has plates separated by a distance $d$. The space between the plates is filled with two dielectrics, one of thickness $\frac{1}{4}d$ and dielectric constant $\kappa_1$, and the other with thickness $\frac{3}{4}d$ and dielectric constant $\kappa_2$. Find the capacitance of this capacitor in terms of $C_0$, the capacitance with no dielectrics.

**72** •• [SSM] Two capacitors, each consisting of two conducting plates of surface area $A$, with an air gap of width $d$. They are connected in parallel, as shown in Figure 24-37, and each has a charge $Q$. A slab of width $d$ and area $A$ with dielectric constant $\kappa$ is inserted between the plates of one of the capacitors. Calculate the new charge $Q'$ on that capacitor.

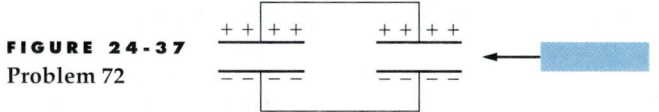

**FIGURE 24-37**
**Problem 72**

**73** •• A parallel-plate capacitor with no dielectric has a capacitance $C_0$. If the separation distance between the plates is $d$, and a slab with dielectric constant $\kappa$ and thickness $t < d$ is placed in the capacitor, find the new capacitance.

**74** •• The membrane of the axon of a nerve cell is a thin cylindrical shell of radius $R = 10^{-5}$ m, length $L = 0.1$ m, and thickness $d = 10^{-8}$ m. The membrane has a positive charge on one side and a negative charge on the other, and the membrane acts as a parallel-plate capacitor of area $A = 2\pi rL$ and separation $d$. The membrane's dielectric constant is approximately $\kappa = 3$. (a) Find the capacitance of the membrane. If the potential difference across the membrane is 70 mV, find (b) the charge on each side of the membrane, and (c) the electric field through the membrane.

**75** •• [SSM] What is the dielectric constant of a dielectric on which the induced bound charge density is (a) 80 percent of the free-charge density on the plates of a capacitor filled by the dielectric, (b) 20 percent of the free charge density, and (c) 98 percent of the free charge density?

**76** •• Two parallel plates have charges $Q$ and $-Q$. When the space between the plates is devoid of matter, the electric field is $2.5 \times 10^5$ V/m. When the space is filled with a certain dielectric, the field is reduced to $1.2 \times 10^5$ V/m. (a) What is the dielectric constant of the dielectric? (b) If $Q = 10$ nC, what is the area of the plates? (c) What is the total induced charge on either face of the dielectric?

**77** •• [SSM] Find the capacitance of the parallel-plate capacitor shown in Figure 24-38.

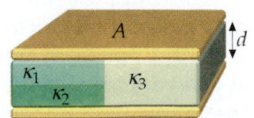

**FIGURE 24-38** Problem 77

**78** •• A parallel-plate capacitor has plates of area 600 cm² and a separation of 4 mm. The capacitor is charged to 100 V and is then disconnected from the battery. (a) Find the electric field $E_0$ and the electrostatic energy $U$. A dielectric of constant $\kappa = 4$ is then inserted, completely filling the space between the plates. Find (b) the new electric field $E$, (c) the potential difference $V$, and (d) the new electrostatic energy.

**79** ••• A parallel-plate capacitor is constructed using a dielectric whose constant varies with position. The plates have area $A$. The bottom plate is at $y = 0$ and the top plate is at $y = y_0$. The dielectric constant is given as a function of $y$ according to $\kappa = 1 + (3/y_0)y$. (a) What is the capacitance? (b) Find $\sigma_b/\sigma_f$ on the surfaces of the dielectric. (c) Use Gauss's law to find the induced volume charge density $\rho(y)$ within this dielectric. (d) Integrate the expression for the volume charge density found in Part (c) over the dielectric, and show that the total induced bound charge, including that on the surfaces, is zero.

## General Problems

**80** •• You are given 4 identical capacitors and a 100-V battery. When only one capacitor is connected to this battery the energy stored is $U_0$. Can you find a combination of the four capacitors so that the total energy stored in all four capacitors is $U_0$?

**81** • [SSM] Three capacitors have capacitances of 2 $\mu$F, 4 $\mu$F, and 8 $\mu$F. Find the equivalent capacitance if (a) the capacitors are connected in parallel and (b) if the capacitors are connected in series.

**82** • A 1-$\mu$F capacitor is connected in parallel with a 2-$\mu$F capacitor, and the combination is connected in series with a 6-$\mu$F capacitor. What is the equivalent capacitance of this combination?

**83** • The voltage across a parallel-plate capacitor with plate separation 0.5 mm is 1200 V. The capacitor is disconnected from the voltage source and the separation between the plates is increased until the energy stored in the capacitor has been doubled. Determine the final separation between the plates.

**84** •• [SOLVE] ✓ Determine the capacitance of each of the networks shown in Figure 24-39.

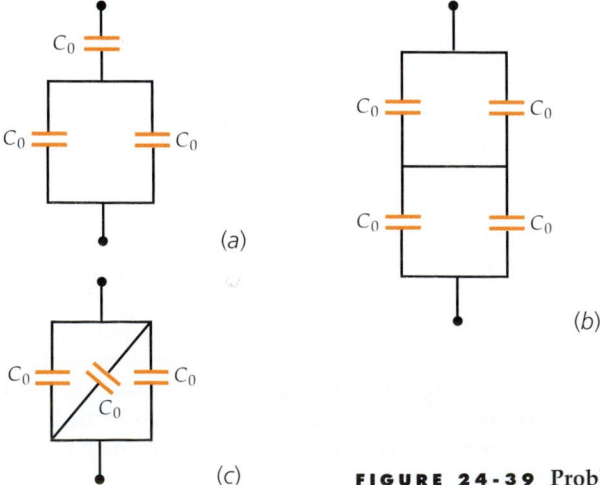

**FIGURE 24-39** Problem 84

**85** •• SSM Figure 24-40 shows four capacitors connected in the arrangement known as a capacitance bridge. The capacitors are initially uncharged. What must the relation between the four capacitances be so that the potential between points $c$ and $d$ is zero when a voltage V is applied between points $a$ and $b$?

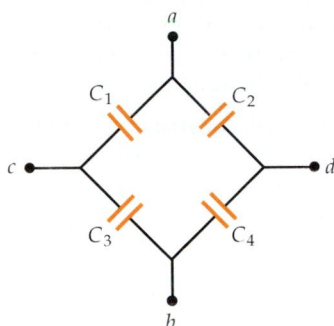

**FIGURE 24-40** Problem 85

**86** •• Two conducting spheres of radius $R$ are separated by a large distance, compared to their size. One sphere initially has a charge $Q$, and the other sphere is uncharged. A thin wire is then connected between the spheres. What fraction of the initial energy is dissipated?

**87** •• A parallel-plate capacitor of area $A$ and separation distance $d$ is charged to a potential difference $V$ and then disconnected from the charging source. The plates are then pulled apart until the separation is $2d$. find expressions in terms of $A$, $d$, and $V$ for (a) the new capacitance, (b) the new potential difference, and (c) the new stored energy. (d) How much work was required to change the plate separation from $d$ to $2d$?

**88** •• A parallel-plate capacitor has capacitance $C_0$ with no dielectric. The capacitor is then filled with dielectric of constant $\kappa$. When a second capacitor of capacitance $C'$ is connected in series with the first capacitor, the capacitance of the series combination is $C_0$. Find $C'$.

**89** •• A Leyden jar, the earliest type of capacitor, is a glass jar coated inside and out with metal foil. Suppose that a Leyden jar is a cylinder 40-cm high, with 2.0-mm-thick walls, and an inner diameter of 8 cm. Ignore any field fringing. (a) Find the capacitance of this Leyden jar, if the dielectric constant $\kappa$ of the glass is 5. (b) If the dielectric strength of the glass is 15 MV/m, what maximum charge can the Leyden jar carry without undergoing dielectric breakdown? (*Hint: Treat the device as a parallel-plate capacitor.*)

**90** •• SSM iSOLVE✓ A parallel-plate capacitor is constructed from a layer of silicon dioxide of thickness $5 \times 10^{-6}$ m between two conducting films. The dielectric constant of silicon dioxide is 3.8 and its dielectric strength is $8 \times 10^6$ V/m. (a) What voltage can be applied across this capacitor without dielectric breakdown? (b) What should the surface area of the layer of silicon dioxide be for a 10-pF capacitor? (c) Estimate the number of these capacitors that can fit into a square 1 cm by 1 cm.

**91** •• A parallel combination of two identical 2-$\mu$F parallel-plate capacitors is connected to a 100-V battery. The battery is then removed and the separation between the plates of one of the capacitors is doubled. Find the charge on each of the capacitors.

**92** •• A parallel-plate capacitor has a capacitance $C_0$ and a plate separation $d$. Two dielectric slabs of constants $\kappa_1$ and $\kappa_2$, each of thickness $\frac{1}{2}d$ and having the same area as the plates, are inserted between the plates as shown in Figure 24-41. When the charge on the plates is $Q$, find (a) the electric field in each dielectric, and (b) the potential difference between the plates. (c) Show that the new capacitance is given by $C = 2\kappa_1\kappa_2/(\kappa_1 + \kappa_2)C_0$. (d) Show that this system can be considered to be a series combination of two capacitors of thickness $\frac{1}{2}d$ filled with dielectrics of constant $\kappa_1$ and $\kappa_2$.

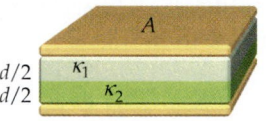

**FIGURE 24-41** Problem 92

**93** •• A parallel-plate capacitor has a plate area $A$ and a separation distance $d$. A metal slab of thickness $t$ and area $A$ is inserted between the plates. (a) Show that the capacitance is given by $C = \epsilon_0 A/(d - t)$, regardless of where the metal slab is placed. (b) Show that this arrangement can be considered to be a capacitor of separation $a$ in series with one of separation $b$, where $a + b + t = d$.

**94** •• SSM A parallel-plate capacitor is filled with two dielectrics of equal size, as shown in Figure 24-42. (a) Show that this system can be considered to be two capacitors of area $\frac{1}{2}A$ connected in parallel. (b) Show that the capacitance is increased by the factor $(\kappa_1 + \kappa_2)/2$.

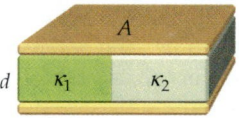

**FIGURE 24-42** Problem 94

**95** •• A parallel-plate capacitor of plate area $A$ and separation $x$ is given a charge $Q$ and is then removed from the charging source. (a) Find the stored electrostatic energy as a function of $x$. (b) Find the increase in energy $dU$ due to an increase in plate separation $dx$ from $dU = (dU/dx)\,dx$. (c) If $F$ is the force exerted by one plate on the other, the work needed to move one plate a distance $dx$ is $F\,dx = dU$. Show that $F = Q^2/2\epsilon_0 A$. (d) Show that the force in Part (c) equals $\frac{1}{2}EQ$, where $Q$ is the charge on one plate and $E$ is the electric field between the plates. Discuss the reason for the factor $\frac{1}{2}$ in this result.

**96** •• A rectangular parallel-plate capacitor of length $a$ and width $b$ has a dielectric of width $b$ partially inserted a distance $x$ between the plates, as shown in Figure 24-43. (a) Find the capacitance as a function of $x$. Neglect edge effects. (b) Show that your answer gives the expected results for $x = 0$ and $x = a$.

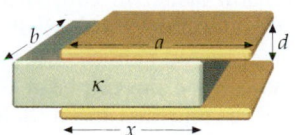

**FIGURE 24-43** Problem 96

**97** ••• **SSM** An electrically isolated capacitor with charge $Q$ is partly filled with a dielectric substance as shown in Figure 24-43. The capacitor consists of two rectangular plates of edge lengths $a$ and $b$ separated by distance $d$. The distance which the dielectric is inserted is $x$. (*a*) What is the energy stored in the capacitor? (*Hint: the capacitor can be thought of as two capacitors connected in parallel.*) (*b*) Because the energy of the capacitor decreases as $x$ increases, the electric field must be doing positive work on the dielectric, meaning that there must be an electric force pulling it in. Calculate the force by examining how the stored energy varies with $x$. (*c*) Express the force in terms of the capacitance and voltage. (*d*) Where does this force originate from?

**98** •• Two identical, 4-$\mu$F parallel-plate capacitors are connected in series across a 24-V battery. (*a*) What is the charge on each capacitor? (*b*) What is the total stored energy of the capacitors? A dielectric that has a dielectric constant of 4.2 is inserted between the plates of one of the capacitors, while the battery is still connected. (*c*) After the dielectric is inserted, what is the charge on each capacitor? (*d*) What is the potential difference across each capacitor? (*e*) What is the total stored energy of the capacitors?

**99** •• A parallel-plate capacitor has a plate area $A$ of 1 m$^2$ and a plate separation distance $d$ of 0.5 cm. Completely filling the space between the conducting plates is a glass plate that has a dielectric constant of $\kappa = 5$. The capacitor is charged to a potential difference of 12 V and the capacitor is then removed from its charging source. How much work is required to pull the glass plate out of the capacitor?

**100** •• A capacitor carries a charge of 15 $\mu$C, when the potential between its plates is $V$. When the charge on the capacitor is increased to 18 $\mu$C, the potential between the plates increases by 6 V. Find the capacitance of the capacitor and the initial and final voltages.

**101** •• A capacitance balance is shown in Figure 24-44. On one side of the balance, a weight is attached, while on the other side is a capacitor whose two plates are separated by a gap of variable width. When the capacitor is charged to a voltage $V$, the attractive force between the plates balances the weight of the hanging mass. (*a*) Is the balance stable? That is, if we balance it out, and then move the plates a little closer together, will they snap shut or move back to the equilibrium point? (*b*) Calculate the voltage required to balance a mass $M$, assuming the plates are separated by distance $d$ and have area $A$. The force between the plates is given by the derivative of the stored energy with respect to the plate separation. Why is this?

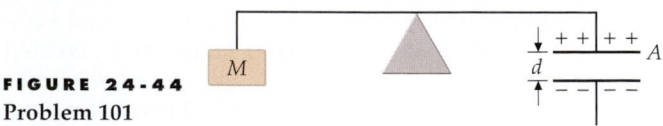

**FIGURE 24-44**
Problem 101

**102** ••• **SSM** You are asked to construct a parallel-plate, air-gap capacitor that will store 100 kJ of energy. (*a*) What minimum volume is required between the plates of the capacitor? (*b*) Suppose you have developed a dielectric that can withstand 3 × 10$^8$ V/m and that has a dielectric constant of $\kappa = 5$. What volume of this dielectric, between the plates of the capacitor, is required for it to be able to store 100 kJ of energy?

**103** ••• Consider two parallel-plate capacitors, $C_1$ and $C_2$, that are connected in parallel. The capacitors are identical except that $C_2$ has a dielectric inserted between its plates. A voltage source of 200 V is connected across the capacitors to charge them and, the voltage source is then disconnected. (*a*) What is the charge on each capacitor? (*b*) What is the total stored energy of the capacitors? (*c*) The dielectric is removed from $C_2$. What is the final stored energy of the capacitors? (*d*) What is the final voltage across the two capacitors?

**104** ••• A capacitor is constructed of two concentric cylinders of radii $a$ and $b$ ($b > a$), which has a length $L \gg b$. A charge of $+Q$ is on the inner cylinder, and a charge of $-Q$ is on the outer cylinder. The region between the two cylinders is filled with a dielectric that has a dielectric constant $\kappa$. (*a*) Find the potential difference between the cylinders. (*b*) Find the density of the free charge $\sigma_f$ on the inner cylinder and the outer cylinder. (*c*) Find the bound charge density $\sigma_b$ on the inner cylindrical surface of the dielectric and on the outer cylindrical surface of the dielectric. (*d*) Find the total stored electrostatic energy. (*e*) If the dielectric will move without friction, how much mechanical work is required to remove the dielectric cylindrical shell?

**105** ••• Two parallel-plate capacitors have the same separation distance and plate area. The capacitance of each is initially 10 $\mu$F. When a dielectric is inserted, so that it completely fills the space between the plates of one of the capacitors, the capacitance of that capacitor increases to 35 $\mu$F. The 35-$\mu$F and 10-$\mu$F capacitors are connected in parallel and are charged to a potential difference of 100 V. The voltage source is then disconnected. (*a*) What is the stored energy of this system? (*b*) What are the charges on the two capacitors? (*c*) The dielectric is removed from the capacitor. What are the new charges on the plates of the capacitors? (*d*) What is the final stored energy of the system?

**106** ••• **SSM** The two capacitors shown in Figure 24-45 have capacitances $C_1 = 0.4$ $\mu$F and $C_2 = 1.2$ $\mu$F. The voltages across the two capacitors are $V_1$ and $V_2$, respectively, and the total stored energy in the two capacitors is 1.14 mJ. If terminals $b$ and $c$ are connected together, the voltage is $V_a - V_d = 80$ V; if terminal $a$ is connected to terminal $b$, and terminal $c$ is connected to terminal $d$, the voltage $V_a - V_d = 20$ V. Find the initial voltages $V_1$ and $V_2$.

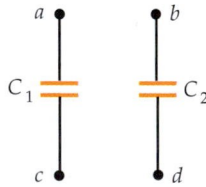

**FIGURE 24-45** Problem 106

**107** ••• Before Switch S is closed, as shown in Figure 24-46, the voltage across the terminals of the switch is 120 V and the voltage across the 0.2 $\mu$F capacitor is 40 V. The total energy stored in the two capacitors is 1440 $\mu$J. After closing the switch, the voltage across each capacitor is 80 V, and the energy stored by the two capacitors has dropped to 960 $\mu$J. Determine the capacitance of $C_2$ and the charge on that capacitor before the switch was closed.

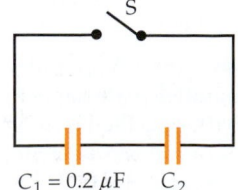

**FIGURE 24-46** Problem 107

**108 •••** A parallel-plate capacitor of area $A$ and separation distance $d$ is charged to a potential difference $V$ and is then removed from the charging source. A dielectric slab of constant $\kappa = 2$, thickness $d$, and area $\frac{1}{2}A$ is inserted, as shown in Figure 24-47. Let $\sigma_1$ be the free charge density at the conductor–dielectric surface, and let $\sigma_2$ be the free charge density at the conductor–air surface. (a) Why must the electric field have the same value inside the dielectric as in the free space between the plates? (b) Show that $\sigma_1 = 2\,\sigma_2$. (c) Show that the new capacitance is $3\epsilon_0 A/2d$, and that the new potential difference is $\frac{2}{3}V$.

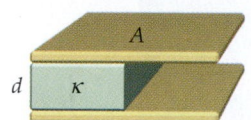

**FIGURE 24-47** Problem 108

**109 •••** Two identical, 10-$\mu$F parallel-plate capacitors are given equal charges of 100 $\mu$C each and are then removed from the charging source. The charged capacitors are connected by a wire between their positive plates and by another wire between their negative plates. (a) What is the stored energy of the system? A dielectric that has a dielectric constant of $\kappa = 3.2$ is inserted between the plates of one of the capacitors, so that it completely fills the region between the plates. (b) What is the final charge on each capacitor? (c) What is the final stored energy of the system?

**110 •••** **SSM** A capacitor has rectangular plates of length $a$ and width $b$. The top plate is inclined at a small angle, as shown in Figure 24-48. The plate separation varies from $d = y_0$ at the left to $d = 2y_0$ at the right, where $y_0$ is much less than $a$ or $b$. Calculate the capacitance using strips of width $dx$ and length $b$ to approximate differential capacitors of area $b\, dx$ and separation $d = y_0 + (y_0/a)x$ that are connected in parallel.

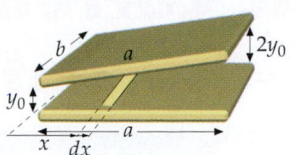

**FIGURE 24-48** Problem 110

**111 •••** Not all dielectrics that separate the plates of a capacitor are rigid. For example, the membrane of a nerve axon is a bilipid layer that has a finite compressibility. Consider a parallel-plate capacitor whose plate separation is maintained by a dielectric of dielectric constant $\kappa = 3.0$ and thickness $d = 0.2$ mm, when the potential across the capacitor is zero. The dielectric, which has a dielectric strength of 40 kV/mm, is highly compressible, with a Young's modulus[†] for compressive stress of $5 \times 10^6$ N/m$^2$. The capacitance of the capacitor in the limit $V \to 0$ is $C_0$. (a) Derive an expression for the capacitance, as a function of voltage across the capacitor. (b) What is the maximum voltage that can be applied to the capacitor? (Assume that $\kappa$ does not change under compression.) (c) What fraction of the total energy of the capacitor is electrostatic field energy and what fraction is mechanical stress energy stored in the compressed dielectric when the voltage across the capacitor is just below the breakdown voltage?

**112 •••** A conducting sphere of radius $R_1$ is given a free charge $Q$. The sphere is surrounded by an uncharged, concentric spherical dielectric shell that has an inner radius $R_1$, an outer radius $R_2$, and a dielectric constant $\kappa$. The system is far removed from other objects. (a) Find the electric field everywhere in space. (b) What is the potential of the conducting sphere relative to $V = 0$ at infinity? (c) Find the total electrostatic potential energy of the system.

---

† Young's modulus is discussed in Section 12-8.

# Electric Current and Direct-Current Circuits

UNDERSTANDING DIRECT CURRENT CIRCUITS CAN HELP YOU PERFORM POTENTIALLY DANGEROUS TASKS LIKE JUMP-STARTING A VEHICLE.

**?** **When jump-starting your car, which battery terminals should be connected? (See Example 25-15.)**

25-1   Current and the Motion of Charges

25-2   Resistance and Ohm's Law

25-3   Energy in Electric Circuits

25-4   Combinations of Resistors

25-5   Kirchhoff's Rules

25-6   *RC* Circuits

When we turn on a light, we connect the wire filament in the lightbulb across a potential difference that causes electric charge to flow through the wire, which is similar to the way a pressure difference in a garden hose causes water to flow through the hose. The flow of electric charge constitutes an electric current. Usually we think of currents as being in conducting wires, but the electron beam in a video monitor and a beam of charged ions from a particle accelerator also constitute electric currents.

➤ In Chapter 25, we will look at direct current (dc) circuits, which are circuits where the direction of the current in a circuit element does not vary. Direct currents can be produced by batteries connected to resistors and capacitors. In Chapter 29, we discuss alternating current (ac) circuits, in which the direction of the current alternates.

When a switch is thrown to turn on a circuit, a very small amount of charge accumulates along the surfaces of the wires and other conducting elements of the circuit and these charges produce an electric field that, within the material of the conductors, drives the motion of mobile charges throughout the conducting materials in the circuit. Many complicated changes take place as the current builds up and small charges accumulate at various points in the circuit, but an equilibrium or steady state is quickly established. The time for steady state to be established depends on the size and the conductivity of the elements in the circuit, but the time is practically instantaneous as far as our perceptions are concerned. In steady state, charge no longer continues to accumulate at points along the circuit and the current is steady. (For circuits containing capacitors and resistors, the current may increase or decrease slowly, but appreciable changes occur only over a period of time that is much longer than the time needed to reach the steady state.)

# 25-1 Current and the Motion of Charges

Electric **current** is defined as the rate of flow of electric charge through a cross-sectional area. Figure 25-1 shows a segment of a current-carrying wire in which charge carriers are moving. If $\Delta Q$ is the charge that flows through the cross-sectional area $A$ in time $\Delta t$, the current $I$ is

$$I = \frac{\Delta Q}{\Delta t} \qquad\qquad 25\text{-}1$$

DEFINITION—ELECTRIC CURRENT

The SI unit of current is the **ampere** (A)[†]:

$$1\,\text{A} = 1\,\text{C/s} \qquad\qquad 25\text{-}2$$

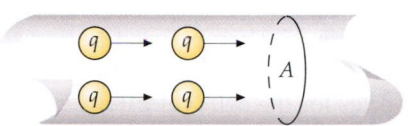

**FIGURE 25-1** A segment of a current-carrying wire. If $\Delta Q$ is the amount of charge that flows through the cross-sectional area $A$ in time $\Delta t$, the current through $A$ is $I = \Delta Q/\Delta t$.

By convention, the direction of current is considered to be the direction of flow of positive charge. This convention was established before it was known that free electrons are the particles that actually travel in current-carrying metal wires. Thus, electrons move in the direction *opposite* to the direction of the conventional current. (In an accelerator that produces a proton beam, both the direction of the current and the direction of motion of the positively charged protons are the same.)

In a conducting metal wire, the motion of negatively charged free electrons is quite complex. When there is no electric field in the wire, the free electrons move in random directions with relatively large speeds of the order of $10^6$ m/s.[‡] In addition, the electrons collide repeatedly with the lattice ions in the wire. Since the velocity vectors of the electrons are randomly oriented, the *average* velocity is zero. When an electric field is applied, the field exerts a force $-e\vec{E}$ on each free electron, giving it a change in velocity in the direction opposite the field. However, any additional kinetic energy acquired is quickly dissipated by collisions with the lattice ions in the wire. During the time between collisions with the lattice ions, the free electrons, on average, acquire an additional velocity in the direction opposite to the field. The net result of this repeated acceleration and dissipation of energy is that the electrons drift along the wire with a small average velocity, directed opposite to the electric-field direction, called their **drift velocity**. The **drift speed** is the magnitude of the drift velocity.

The motion of the free electrons in a metal is similar to the motion of the molecules of a gas, such as air. In still air, the gas molecules move with large instantaneous velocities (due to their thermal energy), but their average velocity is zero. When there is a breeze, the air molecules have a small average velocity or drift velocity in the direction of the breeze superimposed on their much larger instantaneous velocities. Similarly, when there is no applied electric field, the *electron gas* in a metal has a zero average velocity, but when there is an applied electric field, the electron gas acquires a small drift velocity.

---

[†] The ampere is operationally defined (see Chapter 26) in terms of the magnetic force that current-carrying wires exert on one another. The coulomb is then defined as the ampere·second.

[‡] The average energy of the free electrons in a metal is quite large, even at very low temperatures. These electrons do not have the classical Maxwell–Boltzmann energy distribution and do not obey the classical equipartition theorem. We discuss the energy distribution of these electrons and calculate their average speed in Chapter 38.

Let $n$ be the number of free charge-carrying particles per unit volume in a conducting wire of cross-sectional area $A$. We call $n$ the **number density** of charge carriers. Assume that each particle carries a charge $q$ and moves with a drift velocity $v_d$. In a time $\Delta t$, all the particles in the volume $Av_d \Delta t$, shown in Figure 25-2 as a shaded region, pass through the area element. The number of particles in this volume is $nAv_d \Delta t$, and the total free charge is

$$\Delta Q = qnAv_d \Delta t$$

The current is thus

$$I = \frac{\Delta Q}{\Delta t} = qnAv_d \qquad \text{25-3}$$

RELATION BETWEEN CURRENT AND DRIFT VELOCITY

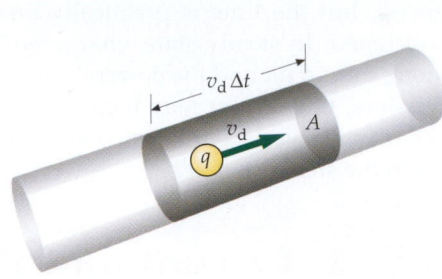

**FIGURE 25-2** In time $\Delta t$, all the free charges in the shaded volume pass through $A$. If there are $n$ charge carriers per unit volume, each with charge $q$, the total free charge in this volume is $\Delta Q = qnAv_d \Delta t$, where $v_d$ is the drift velocity of the charge carriers.

Equation 25-3 can be used to find the current due to the flow of any species of charged particle, simply by substituting the average velocity of the particle species for the drift velocity $v_d$.

The number density of charge carriers in a conductor can be measured by the Hall effect, which is discussed in Chapter 26. The result is that, in most metals, there is about one free electron per atom.

---

*FINDING THE DRIFT SPEED*                 **EXAMPLE 25-1**

**A typical wire for laboratory experiments is made of copper and has a radius 0.815 mm. Calculate the drift speed of electrons in such a wire carrying a current of 1 A, assuming one free electron per atom.**

**PICTURE THE PROBLEM** Equation 25-3 relates the drift speed to the number density of charge carriers, which equals the number density of copper atoms $n_a$. We can find $n_a$ from the mass density of copper, its molecular mass, and Avogadro's number.

1. The drift velocity is related to the current and number density of charge carriers:

$$I = nqv_d A$$

2. If there is one free electron per atom, the number density of free electrons equals the number density of atoms $n_a$:

$$n = n_a$$

3. The number density of atoms $n_a$ is related to the mass density $\rho_m$, Avogadro's number $N_A$, and the molar mass $M$. For copper, $\rho = 8.93 \text{ g/cm}^3$ and $M = 63.5 \text{ g/mol}$:

$$n_a = \frac{\rho_m N_A}{M}$$

$$= \frac{(8.93 \text{ g/cm}^3)(6.02 \times 10^{23} \text{ atoms/mol})}{63.5 \text{ g/mol}}$$

$$= 8.47 \times 10^{22} \text{ atoms/cm}^3 = 84.7 \text{ atoms/nm}^3$$

$$= 8.47 \times 10^{28} \text{ atoms/m}^3$$

4. The magnitude of the charge is $e$, and the area is related to the radius $r$ of the wire:

$$q = e$$
$$A = \pi r^2$$

5. Substituting numerical values yields $v_d$:

$$v_d = \frac{I}{nqA} = \frac{I}{n_a e \pi r^2}$$

$$= \frac{1 \text{ C/s}}{(8.47 \times 10^{28} \text{ m}^{-3})(1.6 \times 10^{-19} \text{ C})\pi(8.15 \times 10^{-4} \text{ m})^2}$$

$$= 3.54 \times 10^{-5} \text{ m/s} = \boxed{3.54 \times 10^{-2} \text{ mm/s}}$$

**REMARKS** Typical drift speeds are of the order of a few hundredths of a millimeter per second, quite small by macroscopic standards.

**EXERCISE** How long would it take for an electron to drift from your car battery to the starter motor, a distance of about 1 m, if its drift speed is $3.5 \times 10^{-5}$ m/s? (*Answer* 7.9 h)

If electrons drift down a wire at such low speeds, why does an electric light come on instantly when the switch is thrown? A comparison with water in a hose may prove useful. If you attach an empty 100-ft-long hose to a water faucet and turn on the water, it typically takes several seconds for the water to travel the length of the hose to the nozzle. However, if the hose is already full of water when the faucet is opened, the water emerges from the nozzle almost instantaneously. Because of the water pressure at the faucet, the segment of water near the faucet pushes on the water immediately next to it, which pushes on the next segment of water, and so on, until the last segment of water is pushed out the nozzle. This pressure wave moves down the hose at the speed of sound in water, and the water quickly reaches a steady flow rate.

Unlike a water hose, a metal wire is never empty. That is, there are always a very large number of conduction electrons throughout the metal wire. Thus, charge starts moving along the entire length of the wire (including the wire inside the lightbulb) almost immediately after the light switch is thrown. The transport of a significant amount of charge in a wire is accomplished not by a few charges moving rapidly down the wire, but by a very large number of charges slowly drifting down the wire. Surface charges on the wires produce an electric field, and this electric field drives the conduction electrons through the wire.

---

*FINDING THE NUMBER DENSITY*                    **EXAMPLE 25-2**

In a certain particle accelerator, a current of 0.5 mA is carried by a 5-MeV proton beam that has a radius of 1.5 mm. (*a*) Find the number density of protons in the beam. (*b*) If the beam hits a target, how many protons hit the target in 1 s?

**PICTURE THE PROBLEM** To find the number density, we use the relation $I = qnAv$ (Equation 25-3), where $v$ is the drift speed of the charge carriers. (The drift speed is the magnitude of the average velocity.) We can find $v$ from the energy. The amount of charge $Q$ that hits the target in time $\Delta t$ is $I\Delta t$, and the number $N$ of protons that hits the target is $Q$ divided by the charge per proton.

(*a*) 1. The number density is related to the current, the charge, the cross-sectional area, and the speed:

$$I = qnAv$$

2. We find the speed of the protons from their kinetic energy:

$$K = \tfrac{1}{2}mv^2 = 5 \text{ MeV}$$

3. Use $m = 1.67 \times 10^{-27}$ kg for the mass of a proton, and solve for the speed:

$$v = \sqrt{\frac{2K}{m}} = \sqrt{\frac{(2)(5 \times 10^6 \text{ eV})}{1.6 \times 10^{-27} \text{ kg}}} \times \frac{1.6 \times 10^{-19} \text{ J}}{1 \text{ eV}}$$

$$= \boxed{3.10 \times 10^7 \text{ m/s}}$$

4. Substitute to calculate $n$:

$$n = \frac{I}{qAv}$$

$$= \frac{0.5 \times 10^{-3} \text{ A}}{(1.6 \times 10^{-19} \text{ C/proton})\,\pi(1.5 \times 10^{-3} \text{ m})^2(3.10 \times 10^7 \text{ m/s})}$$

$$= \boxed{1.43 \times 10^{13} \text{ protons/m}^3}$$

(b) 1. The number of protons $N$ that hit the target in 1 s is related to the total charge $\Delta Q$ that hits in 1 s and the proton charge $q$:

$$\Delta Q = Nq$$

2. The charge $\Delta Q$ that strikes the target in time $\Delta t$ is the current times the time:

$$\Delta Q = I \Delta t$$

3. The number of protons is then:

$$N = \frac{\Delta Q}{q} = \frac{I \Delta t}{q} = \frac{(0.5 \times 10^{-3}\,\text{A})(1\,\text{s})}{1.6 \times 10^{-19}\,\text{C/proton}}$$

$$= \boxed{3.13 \times 10^{15}\ \text{protons}}$$

**✓ PLAUSIBILITY CHECK** The number of protons $N$ hitting the target in time $\Delta t$ is the number in the volume $Av\,\Delta t$. Then $N = nAv\,\Delta t$. Substituting $n = I/(qAv)$ then gives $N = nAv\,\Delta t = [I/(qAv)](Av)\,\Delta t = I\,\Delta t/q = \Delta Q/q$, which is what we used in Part (b).

**REMARKS** We were able to use the classical expression for kinetic energy in step 2 without taking relativity into consideration, because the proton kinetic energy of 5 MeV is much less than the proton rest energy (about 931 MeV). The speed found, $3.1 \times 10^7$ m/s, is about one-tenth the speed of light.

**EXERCISE** Using the number density found in Part (a), how many protons are there in a volume of 1 mm$^3$ of the space containing the beam? (*Answer* 14,300)

# 25-2 Resistance and Ohm's Law

Current in a conductor is driven by an electric field $\vec{E}$ inside the conductor that exerts a force $q\vec{E}$ on the free charges. (In electrostatic equilibrium, the electric field must be zero inside a conductor, but when a conductor carries a current, it is no longer in electrostatic equilibrium and the free charge drifts down the conductor, driven by the electric field.) Since the direction of the force on a positive charge is also the direction of the electric field, $\vec{E}$ is in the direction of the current.

Figure 25-3 shows a wire segment of length $\Delta L$ and cross-sectional area $A$ carrying a current $I$. Since the electric field points in the direction of decreasing potential, the potential at point $a$ is greater than the potential at point $b$. If we think of the current as the flow of positive charge, these positive charges move in the direction of decreasing potential. Assuming the electric field $\vec{E}$ to be uniform throughout the segment, the **potential drop** $V$ between points $a$ and $b$ is

$$V = V_a - V_b = E\,\Delta L \qquad\qquad 25\text{-}4$$

The ratio of the potential drop to the current is called the **resistance** of the segment.

$$R = \frac{V}{I} \qquad\qquad 25\text{-}5$$

DEFINITION—RESISTANCE

The SI unit of resistance, the volt per ampere, is called an **ohm** ($\Omega$):

$$1\,\Omega = 1\,\text{V/A} \qquad\qquad 25\text{-}6$$

For many materials, the resistance does not depend on the potential drop or the current. Such materials, which include most metals, are called **ohmic materials.**

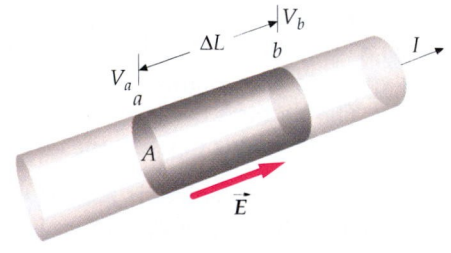

**FIGURE 25-3** A segment of wire carrying a current $I$. The potential drop is related to the electric field by $V_a - V_b = E\,\Delta L$.

For ohmic materials, the potential drop across a segment is proportional to the current:

$$V = IR, \quad R \text{ constant} \qquad 25\text{-}7$$

OHM'S LAW

For **nonohmic materials,** the resistance depends on the current $I$, so $V$ is not proportional to $I$. Figure 25-4 shows the potential difference $V$ versus the current $I$ for ohmic and nonohmic materials. For ohmic materials (Figure 25-4$a$), the relation is linear, but for nonohmic materials (Figure 25-4$b$), the relation is not linear. Ohm's law is not a fundamental law of nature, like Newton's laws or the laws of thermodynamics, but rather is an empirical description of a property shared by many materials.

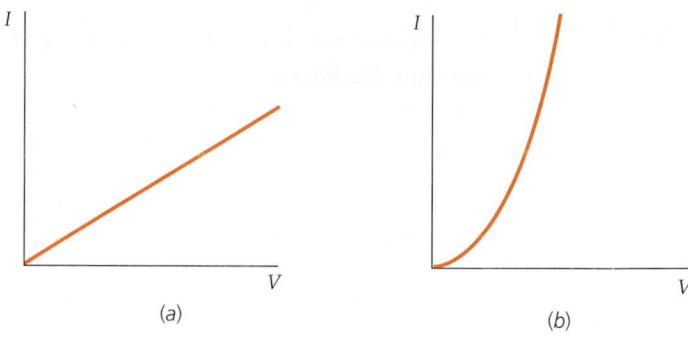

**FIGURE 25-4** Plots of $V$ versus $I$ for ($a$) ohmic and ($b$) nonohmic materials. The resistance $R = V/I$ is independent of $I$ for ohmic materials, as indicated by the constant slope of the line in Figure 25-4$a$.

**EXERCISE**   A wire of resistance 3 Ω carries a current of 1.5 A. What is the potential drop across the wire? (*Answer*   4.5 V)

The resistance of a conducting wire is found to be proportional to the length of the wire and inversely proportional to its cross-sectional area:

$$R = \rho \frac{L}{A} \qquad 25\text{-}8$$

where the proportionality constant $\rho$ is called the **resistivity** of the conducting material.[†] The unit of resistivity is the ohm-meter (Ω·m). Note that Equation 25-7 and Equation 25-8 for electrical conduction and electrical resistance are of the same form as Equation 20-9 ($\Delta T = IR$) and Equation 20-10 [$R = \Delta x/(kA)$] for thermal conduction and thermal resistance. For the electrical equations, the potential difference $V$ replaces the temperature difference $\Delta T$ and $1/\rho$ replaces the thermal conductivity $k$. (In fact, $1/\rho$ is called the electrical conductivity.[‡]) Ohm was led to his law by the similarity between the conduction of electricity and the conduction of heat.

---

*THE LENGTH OF A 2-Ω RESISTOR*                    **EXAMPLE   25-3**

A Nichrome wire ($\rho = 10^{-6}$ Ω·m) has a radius of 0.65 mm. What length of wire is needed to obtain a resistance of 2.0 Ω?

Solve $R = \rho L/A$
(Equation 25-8) for $L$:
$$L = \frac{RA}{\rho} = \frac{(2\ \Omega)\pi(6.5 \times 10^{-4}\ \text{m})^2}{10^{-6}\ \Omega \cdot \text{m}} = \boxed{2.65\ \text{m}}$$

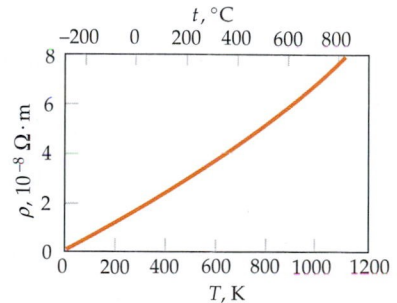

The resistivity of any given metal depends on the temperature. Figure 25-5 shows the temperature dependence of the resistivity of copper. This graph is nearly a straight line, which means that the resistivity varies nearly linearly with temperature.[§] In tables, the resistivity is usually given in terms of its value at 20°C, $\rho_{20}$, along with the **temperature coefficient of resistivity,** $\alpha$, which is defined by

$$\alpha = \frac{(\rho - \rho_{20})/\rho_{20}}{t_{\text{C}} - 20°\text{C}} \qquad 25\text{-}9$$

**FIGURE 25-5** Plot of resistivity $\rho$ versus temperature for copper. Since the Celsius and absolute temperatures differ only in the choice of zero, the resistivity has the same slope whether it is plotted against $t$ or $T$.

---

[†] The symbol $\rho$ used here for the resistivity was used in previous chapters for volume charge density. Care must be taken to distinguish what quantity $\rho$ refers to. Usually this will be clear from the context.
[‡] The unit of conductivity is the siemens (S), 1 siemens = 1 Ω$^{-1}$·m$^{-1}$.
[§] There is a breakdown in this linearity for all metals at very low temperatures that is not shown in Figure 25-5.

## TABLE 25-1

**Resistivities and Temperature Coefficients**

| Material | Resistivity $\rho$ at 20°C, $\Omega \cdot$m | Temperature Coefficient $\alpha$ at 20°C, $K^{-1}$ |
|---|---|---|
| Silver | $1.6 \times 10^{-8}$ | $3.8 \times 10^{-3}$ |
| Copper | $1.7 \times 10^{-8}$ | $3.9 \times 10^{-3}$ |
| Aluminum | $2.8 \times 10^{-8}$ | $3.9 \times 10^{-3}$ |
| Tungsten | $5.5 \times 10^{-8}$ | $4.5 \times 10^{-3}$ |
| Iron | $10 \times 10^{-8}$ | $5.0 \times 10^{-3}$ |
| Lead | $22 \times 10^{-8}$ | $4.3 \times 10^{-3}$ |
| Mercury | $96 \times 10^{-8}$ | $0.9 \times 10^{-3}$ |
| Nichrome | $100 \times 10^{-8}$ | $0.4 \times 10^{-3}$ |
| Carbon | $3500 \times 10^{-8}$ | $-0.5 \times 10^{-3}$ |
| Germanium | $0.45$ | $-4.8 \times 10^{-2}$ |
| Silicon | $640$ | $-7.5 \times 10^{-2}$ |
| Wood | $10^8 - 10^{14}$ | |
| Glass | $10^{10} - 10^{14}$ | |
| Hard rubber | $10^{13} - 10^{16}$ | |
| Amber | $5 \times 10^{14}$ | |
| Sulfur | $1 \times 10^{15}$ | |

## TABLE 25-2

**Wire Diameters and Cross-Sectional Areas for Commonly Used Copper Wires**

| Gauge Number | Diameter at 20°C, mm | Area, mm$^2$ |
|---|---|---|
| 4 | 5.189 | 21.15 |
| 6 | 4.115 | 13.30 |
| 8 | 3.264 | 8.366 |
| 10 | 2.588 | 5.261 |
| 12 | 2.053 | 3.309 |
| 14 | 1.628 | 2.081 |
| 16 | 1.291 | 1.309 |
| 18 | 1.024 | 0.8235 |
| 20 | 0.8118 | 0.5176 |
| 22 | 0.6438 | 0.3255 |

Table 25-1 gives the resistivity $\rho$ at 20°C and the temperature coefficient $\alpha$ at 20°C for various materials. Note the tremendous range of values for $\rho$.

Electrical wires are manufactured in standard sizes. The diameter of the circular cross section is indicated by a *gauge number,* with higher numbers corresponding to smaller diameters, as can be seen from Table 25-2.

*RESISTANCE PER UNIT LENGTH*                    **EXAMPLE 25-4**

**Calculate the resistance per unit length of a 14-gauge copper wire.**

1. From Equation 25-8, the resistance per unit length equals the resistivity per unit area:

$$R = \rho \frac{L}{A}$$

so

$$\frac{R}{L} = \frac{\rho}{A}$$

2. Find the resistivity of copper from Table 25-1 and the area from Table 25-2:

$$\rho = 1.7 \times 10^{-8} \ \Omega \cdot m$$

$$A = 2.08 \ \text{mm}^2$$

3. Use these values to find $R/L$:

$$\frac{R}{L} = \frac{\rho}{A} = \frac{1.7 \times 10^{-8} \ \Omega \cdot m}{2.08 \times 10^{-6} \ m^2} = \boxed{8.17 \times 10^{-3} \ \Omega/m}$$

**REMARKS** 14-gauge copper wire is commonly used for household lighting circuits. The resistance of a 100-W, 120-V lightbulb filament is 144 $\Omega$ and the resistance of a 100 m of the wire is 0.817 $\Omega$, so the resistance of the wire is negligible compared to the resistance of the lightbulb filament.

Carbon, which has a relatively high resistivity, is used in resistors found in electronic equipment. Resistors are often marked with colored stripes that indicate their resistance value. The code for interpreting these colors is given in Table 25-3.

## TABLE 25-3

**The Color Code for Resistors and Other Devices**

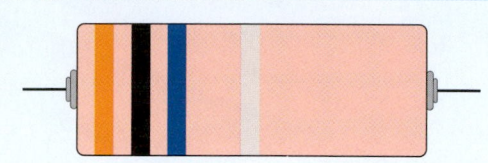

| Colors | Numeral | | Tolerance | |
|--------|---------|--------|-----------|------|
| Black  | = 0 | Brown | = | 1 % |
| Brown  | = 1 | Red   | = | 2 % |
| Red    | = 2 | Gold  | = | 5 % |
| Orange | = 3 | Silver | = | 10 % |
| Yellow | = 4 | None  | = | 20 % |
| Green  | = 5 |       |   |     |
| Blue   | = 6 |       |   |     |
| Violet | = 7 |       |   |     |
| Gray   | = 8 |       |   |     |
| White  | = 9 |       |   |     |

The color bands are read starting with the band closest to the end of the resistor. The first two bands represent an integer between 1 and 99. The third band represents the number of zeros that follow. For the resistor shown, the colors of the first three bands are, respectively, orange, black, and blue. Thus, the number is 30,000,000 and the resistance is 30 MΩ. The fourth band is the tolerance band. If the fourth band is silver, as shown here, the tolerance is 10 percent. Ten percent of 30 is 3, so the resistance is $(30 \pm 3)$ MΩ.

Color-coded carbon resistors on a circuit board.

---

*THE ELECTRIC FIELD THAT DRIVES THE CURRENT*    **E X A M P L E    2 5 - 5**

**Find the electric field strength $E$ in the 14-gauge copper wire of Example 25-4 when the wire is carrying a current of 1.3 A.**

**PICTURE THE PROBLEM** We find the electric field strength as the potential drop for a given length of wire, $E = V/L$. The potential drop is found using Ohm's law, $V = IR$, and the resistance per length is given in Example 25-4.

1. The electric field strength equals the potential drop per unit length:

$$E = \frac{V}{L}$$

2. Write Ohm's law for the potential drop:

$$V = IR$$

3. Substitute this expression into the equation for $E$:

$$E = \frac{V}{L} = \frac{IR}{L} = I\frac{R}{L}$$

4. Substitute the value of $R/L$ found in Example 25-4 to calculate $E$:

$$E = I\frac{R}{L} = (1.3 \text{ A})(8.17 \times 10^{-3} \, \Omega/\text{m}) = \boxed{1.06 \times 10^{-2} \text{ V/m}}$$

**REMARKS** Since $R/L = \rho/A$, $E = I\rho/A$, which is the same throughout the length of the wire. Thus, $E$ is uniform throughout the length of the wire.

## **25-3** Energy in Electric Circuits

When there is an electric field in a conductor, the *electron gas* gains kinetic energy due to the work done on the free electrons by the field. However, steady state is soon achieved as the kinetic energy gain is continuously dissipated into the thermal energy of the conductor by collisions between the electrons and the lattice ions of the conductor. This mechanism for increasing the thermal energy of a conductor is called **Joule heating.**

Consider the segment of wire of length $L$ and cross-sectional area $A$ shown in Figure 25-6a. The wire is carrying a steady current to the right. Consider the free charge $Q$ initially in the segment. During time $\Delta t$, this free charge undergoes a small displacement to the right (Figure 25-6b). This displacement is equivalent to an amount of charge $\Delta Q$ (Figure 25-6c) being moved from its left end, where it had potential energy $\Delta Q\,V_a$, to its right end, where it has potential energy $\Delta Q\,V_b$. The net change in the potential energy of $Q$ is thus

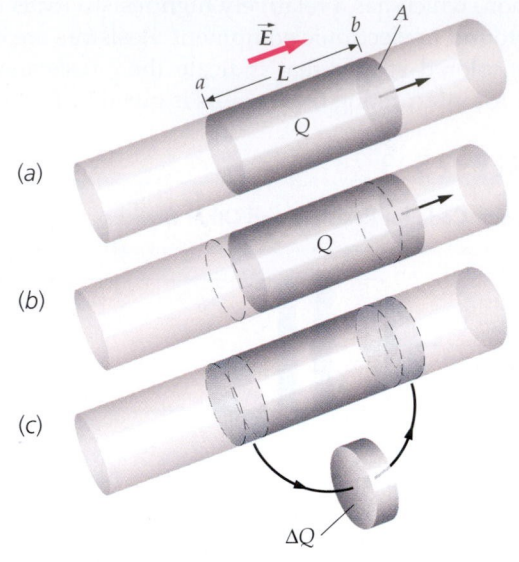

**FIGURE 25-6** During a time $\Delta t$, an amount of charge $\Delta Q$ passes point $a$, where the potential is $V_a$. During the same time interval, an equal amount of charge leaves the segment, passing point $b$, where the potential is $V_b$. The net effect during time $\Delta t$ is: the charge $Q$ that was initially in the segment both loses an amount of potential energy equal to $\Delta Q\,V_a$, and gains an amount equal to $\Delta Q\,V_b$. This amounts to a net decrease in potential energy since $V_a > V_b$.

$$\Delta U = \Delta Q(V_b - V_a)$$

since $V_a > V_b$, this represents a net loss in the potential energy of $Q$. The potential energy lost is then

$$-\Delta U = \Delta Q\,V$$

where $V = V_a - V_b$ is the potential drop across the segment. The rate of potential energy loss is

$$-\frac{\Delta U}{\Delta t} = \frac{\Delta Q}{\Delta t}V = IV$$

where $I = \Delta Q/\Delta t$ is the current. The potential energy loss per unit time is the power $P$ dissipated in the conducting segment:

$$P = IV \qquad\qquad\qquad 25\text{-}10$$

POTENTIAL ENERGY LOSS PER UNIT TIME

If $V$ is in volts and $I$ is in amperes, the power is in watts. The power loss is the product of the decrease in potential energy per unit charge, $V$, and the charge flowing per unit time, $I$. Equation 25-10 applies to any device in a circuit. The rate at which potential energy is delivered to the device is the product of the potential drop across the device and the current through the device. In a conductor, the potential energy is dissipated as thermal energy in the conductor. Using $V = IR$, or $I = V/R$, we can write Equation 25-10 in other useful forms

$$P = IV = I^2 R = \frac{V^2}{R} \qquad\qquad 25\text{-}11$$

POWER DISSIPATED IN A RESISTOR

| | |
|---|---|
| *POWER DISSIPATED IN A RESISTOR* | **E X A M P L E   2 5 - 6** |

**A 12-$\Omega$ resistor carries a current of 3 A. Find the power dissipated in this resistor.**

**PICTURE THE PROBLEM** Since we are given the current and the resistance, but not the potential drop, $P = I^2R$ is the most convenient equation to use. Alternatively, we could find the potential drop from $V = IR$, then use $P = IV$.

Compute $I^2R$:
$$P = I^2R = (3\text{ A})^2(12\ \Omega) = \boxed{108\text{ W}}$$

**PLAUSIBILITY CHECK** The potential drop across the resistor is $V = IR = (3\text{ A})(12\ \Omega) = 36\text{ V}$. We can use this to find the power from $P = IV = (3\text{ A})(36\text{ V}) = 108\text{ W}$.

**EXERCISE** A wire of resistance 5 $\Omega$ carries a current of 3 A for 6 s. (*a*) How much power is put into the wire? (*b*) How much thermal energy is produced? (*Answer* (*a*) 45 W, (*b*) 270 J)

## EMF and Batteries

To maintain a steady current in a conductor, we need a constant supply of electrical energy. A device that supplies electrical energy to a circuit is called a **source of emf.** (The letters *emf* stand for *electromotive force*, a term that is now rarely used. The term is something of a misnomer because it is definitely not a force.) Examples of emf sources are a battery, which converts chemical energy into electrical energy, and a generator, which converts mechanical energy into electrical energy. A source of emf does work on the charge passing through it, raising the

The electric ray has two large electric organs on each side of its head, where current passes from the lower to the upper surface of the body. These organs are composed of columns, with each column consisting of one hundred forty to half a million gelatinous plates. In saltwater fish, these batteries are connected in parallel, whereas in freshwater fish the batteries are connected in series, transmitting discharges of higher voltage. Fresh water has a higher resistivity than salt water, so to be effective a higher voltage is required. It is with such a battery that an average electric ray can electrocute a fish, delivering 50 A at 50 V.

potential energy of the charge. The work per unit charge is called the **emf** $\mathcal{E}$ of the source. The unit of emf is the volt, the same as the unit of potential difference. An **ideal battery** is a source of emf that maintains a constant potential difference between its two terminals, independent of the current through the battery. The potential difference between the terminals of an ideal battery is equal in magnitude to the emf of the battery.

Figure 25-7 shows a simple circuit consisting of a resistance $R$ connected to an ideal battery. The resistance is indicated by the symbol -$\wedge\wedge\wedge$-. The straight lines indicate connecting wires of negligible resistance. The source of emf ideally maintains a constant potential difference equal to $\mathcal{E}$ between points $a$ and $b$, with point $a$ being at the higher potential. There is negligible potential difference between points $a$ and $c$ and between points $d$ and $b$, because the connecting wire is assumed to have negligible resistance. The potential drop from points $c$ to $d$ is therefore equal in magnitude to the emf $\mathcal{E}$, and the current through the resistor is given by $I = \mathcal{E}/R$. The direction of the current in this circuit is clockwise, as shown in the figure.

Note that *inside* the source of emf, the charge flows from a region of low potential to a region of high potential, so it gains potential energy.[†] When charge $\Delta Q$ flows through the source of emf $\mathcal{E}$, its potential energy is increased by the amount $\Delta Q\,\mathcal{E}$. The charge then flows through the resistor, where this potential energy is

**FIGURE 25-7** A simple circuit consisting of an ideal battery of emf $\mathcal{E}$, a resistance $R$, and connecting wires that are assumed to be of negligible resistance.

---

† When a battery is being charged by a generator or by another battery, the charge flows from a high-potential to a low-potential region within the battery being charged, thus losing electrostatic potential energy. The energy lost is converted to chemical energy and stored in the battery being charged.

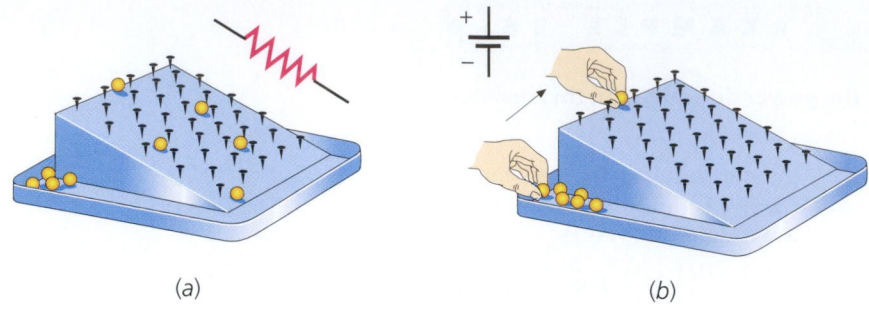

(a)                                 (b)

dissipated as thermal energy. The rate at which energy is supplied by the source of emf is the power output:

$$P = \frac{\Delta Q \mathcal{E}}{\Delta t} = I\mathcal{E} \qquad\qquad 25\text{-}12$$

POWER SUPPLIED BY AN EMF SOURCE

In the simple circuit of Figure 25-7, the power output by the source of emf equals that dissipated in the resistor.

A source of emf can be thought of as a charge pump that pumps the charge from a region of low potential energy to a region of higher potential energy. Figure 25-8 shows a mechanical analog of the simple electric circuit just discussed.

In a **real battery,** the potential difference across the battery terminals, called the **terminal voltage,** is not simply equal to the emf of the battery. Consider the circuit consisting of a real battery and a resistor in Figure 25-9. If the current is varied by varying the resistance $R$ and the terminal voltage is measured, the terminal voltage is found to decrease slightly as the current increases (Figure 25-10), just as if there were a small resistance within the battery.

Thus, we can consider a real battery to consist of an ideal battery of emf $\mathcal{E}$ plus a small resistance $r$, called the **internal resistance** of the battery.

The circuit diagram for a real battery and resistor is shown in Figure 25-11. If the current in the circuit is $I$, the potential at point $a$ is related to the potential at point $b$ by

$$V_a = V_b + \mathcal{E} - Ir$$

The terminal voltage is thus

$$V_a - V_b = \mathcal{E} - Ir \qquad\qquad 25\text{-}13$$

The terminal voltage of the battery decreases linearly with current, as we saw in Figure 25-10. The potential drop across the resistor $R$ is $IR$ and is equal to the terminal voltage:

$$IR = V_a - V_b = \mathcal{E} - Ir$$

Solving for the current $I$, we obtain

$$I = \frac{\mathcal{E}}{R + r} \qquad\qquad 25\text{-}14$$

**FIGURE 25-8** A mechanical analog of a simple circuit consisting of a resistance and source of emf. (*a*) The marbles start at some height *h* above the bottom and are accelerated between collisions with the nails by the gravitational field. The nails are analogous to the lattice ions in the resistor. During the collisions, the marbles transfer the kinetic energy they obtained between collisions to the nails. Because of the many collisions, the marbles have only a small, approximately constant, drift velocity toward the bottom. (*b*) When the marbles reach the bottom, a child picks them up, lifts them to their original height *h*, and starts them again. The child, who does work *mgh* on each marble, is analogous to the source of emf. The energy source in this case is the internal chemical energy of the child.

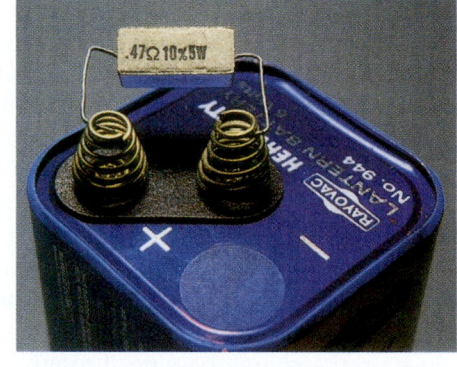

**FIGURE 25-9** A simple circuit consisting of a real battery, a resistor, and connecting wires.

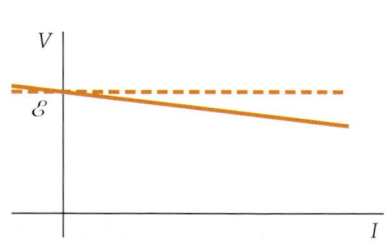

**FIGURE 25-10** Terminal voltage *V* versus *I* for a real battery. The dashed line shows the terminal voltage of an ideal battery, which has the same magnitude as $\mathcal{E}$.

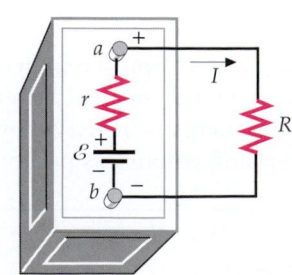

**FIGURE 25-11** Circuit diagram for the circuit shown in Figure 25-9. A real battery can be represented by an ideal battery of emf $\mathcal{E}$ and a small resistance *r*.

If a battery is connected as shown in Figure 25-11, the terminal voltage given by Equation 25-13 is less than the emf of the battery because of the potential drop across the internal resistance of the battery. Real batteries, such as a good car battery, usually have an internal resistance of the order of a few hundredths of an ohm, so the terminal voltage is nearly equal to the emf unless the current is very large. One sign of a bad battery is an unusually high internal resistance. If you suspect that your car battery is bad, checking the terminal voltage with a voltmeter, which draws very little current, is not always sufficient. You need to check the terminal voltage while current is being drawn from the battery, such as while you are trying to start your car. Then the terminal voltage may drop considerably, indicating a high internal resistance and a bad battery.

Batteries are often rated in ampere-hours (A·h), which is the total charge that batteries can deliver:

$$1 \text{ A·h} = (1 \text{ C/s}) (3600 \text{ s}) = 3600 \text{ C}$$

The total energy stored in the battery is the product of the emf and the total charge it can deliver:

$$W = Q\mathcal{E} \qquad\qquad 25\text{-}15$$

---

*TERMINAL VOLTAGE, POWER, AND STORED ENERGY*      **EXAMPLE 25-7**

**An 11-$\Omega$ resistor is connected across a battery of emf 6 V and internal resistance 1 $\Omega$. Find (a) the current, (b) the terminal voltage of the battery, (c) the power delivered by the emf source, (d) the power delivered to the external resistor, and (e) the power dissipated by the battery's internal resistance. (f) If the battery is rated at 150 A·h, how much energy does the battery store?**

**PICTURE THE PROBLEM** The circuit diagram is the same as the circuit diagram shown in Figure 25-11. We find the current from Equation 25-14 and then use it to find the terminal voltage and power delivered to the resistors.

1. Equation 25-14 gives the current:
$$I = \frac{\mathcal{E}}{R + r} = \frac{6 \text{ V}}{11 \ \Omega + 1 \ \Omega} = \boxed{0.5 \text{ A}}$$

2. Use the current to calculate the terminal voltage of the battery:
$$V_a - V_b = \mathcal{E} - Ir = 6 \text{ V} - (0.5 \text{ A})(1 \ \Omega) = \boxed{5.5 \text{ V}}$$

3. The power delivered by the source of emf equals $\mathcal{E}I$:
$$P = \mathcal{E}I = (6 \text{ V})(0.5 \text{ A}) = \boxed{3 \text{ W}}$$

4. The power delivered to and dissipated by the external resistance equals $I^2R$:
$$I^2R = (0.5 \text{ A})^2(11 \ \Omega) = \boxed{2.75 \text{ W}}$$

5. The power dissipated in the internal resistance is $I^2r$.
$$I^2r = (0.5 \text{ A})^2(1 \ \Omega) = \boxed{0.25 \text{ W}}$$

6. The total energy stored is the emf times the total charge it can deliver:
$$W = Q\mathcal{E} = 150 \text{ A·h} \times \frac{3600 \text{ C}}{1 \text{ A·h}} \times 6 \text{ V} = \boxed{3.24 \text{ MJ}}$$

**REMARKS** The value of the internal resistance is exaggerated in this example to simplify calculations. In other examples, we may simply ignore the internal resistance. Of the 3 W of power delivered by the battery, 2.75 W is dissipated in the resistor and 0.25 W is dissipated in the internal resistance of the battery.

**EXAMPLE 25-8** Try It Yourself

For a battery of given emf $\mathcal{E}$ and internal resistance $r$, what value of external resistance $R$ should be placed across the terminals to obtain the maximum power delivered to the resistor?

**PICTURE THE PROBLEM** The circuit diagram is the same as the circuit diagram shown in Figure 25-11. The power input to $R$ is $I^2R$, where $I = \mathcal{E}/(R + r)$. To find the maximum power, we compute $dP/dR$ and set it equal to zero.

**Cover the column to the right and try these on your own before looking at the answers.**

| Steps | Answers |
|---|---|
| 1. Use Equation 25-14 to eliminate $I$ from $P = I^2R$ so that $P$ is written as a function of $R$ and the constants $\mathcal{E}$ and $r$ only. | $P = \dfrac{\mathcal{E}^2 R}{(R + r)^2} = \mathcal{E}^2 R(R + r)^{-2}$ |
| 2. Calculate the derivative $dP/dR$. Use the product rule. | $\dfrac{dP}{dR} = \mathcal{E}^2(R + r)^{-2} - 2\mathcal{E}^2 R(R + r)^{-3}$ |
| 3. Set $dP/dR = 0$ and solve for $R$ in terms of $r$. | $R = r$ |

**REMARKS** The maximum value of $P$ occurs when $R = r$, that is, when the load resistance equals the internal resistance. A similar result holds for alternating current circuits. Choosing $R = r$ to maximize the power delivered to the load is known as *impedance matching*. A graph of $P$ versus $R$ is shown in Figure 25-12.

# 25-4 Combinations of Resistors

The analysis of a circuit can often be simplified by replacing two or more resistors with a single equivalent resistor that carries the same current with the same potential drop as the original resistors. The replacement of a set of resistors by an equivalent resistor is similar to the replacement of a set of capacitors by an equivalent capacitor, discussed in Chapter 24.

**FIGURE 25-12** The power delivered to the external resistor is maximum if $R = r$.

## Resistors in Series

When two or more resistors are connected like $R_1$ and $R_2$ in Figure 25-13 so that they carry the same current $I$, the resistors are said to be connected in series. The potential drop across $R_1$ is $IR_1$ and the potential drop across $R_2$ is $IR_2$. The potential drop across the two resistors is the sum of the potential drops across the individual resistors:

$$V = IR_1 + IR_2 = I(R_1 + R_2) \qquad 25\text{-}16$$

The single equivalent resistance $R_{eq}$ that gives the same total potential drop $V$ when carrying the same current $I$ is found by setting $V$ equal to $IR_{eq}$ (Figure 25-13b). Then $R_{eq}$ is given by

$$R_{eq} = R_1 + R_2$$

**FIGURE 25-13** (*a*) Two resistors in series carry the same current. (*b*) The resistors in Figure 25-13*a* can be replaced by a single equivalent resistance $R_{eq} = R_1 + R_2$ that gives the same total potential drop when carrying the same current as in Figure 25-13*a*.

(a)

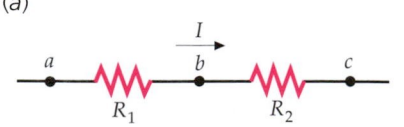

(b)

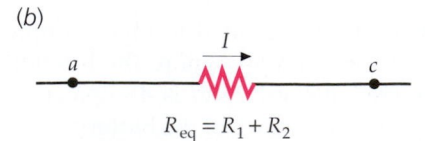

When there are more than two resistors in series, the equivalent resistance is

$$R_{eq} = R_1 + R_2 + R_3 + \ldots \qquad 25\text{-}17$$

EQUIVALENT RESISTANCE FOR RESISTORS IN SERIES

### Resistors in Parallel

Two resistors that are connected, as in Figure 25-14a, so that they have the same potential difference across them, are in parallel. Note that the resistors are connected at both ends by wires. Let $I$ be the current leading to point $a$. At point $a$ the circuit branches out into two branches, and the current $I$ divides into two parts, with current $I_1$ in the upper branch containing resistor $R_1$, and with current $I_2$ in the lower branch containing $R_2$. The two **branch currents** sum to the current in the wire leading into point $a$:

$$I = I_1 + I_2 \qquad 25\text{-}18$$

At point $b$ the branch currents recombine so the current in the wire following point $b$ is also equal to $I = I_1 + I_2$. The potential drop across either resistor, $V = V_a - V_b$, is related to the currents by

$$V = I_1 R_1 = I_2 R_2 \qquad 25\text{-}19$$

The equivalent resistance for parallel resistors is the resistance $R_{eq}$ for which the same total current $I$ requires the same potential drop $V$ (Figure 25-14b):

$$R_{eq} = \frac{V}{I}$$

Solving this equation for $I$ and using $I = I_1 + I_2$, we have

$$I = \frac{V}{R_{eq}} = I_1 + I_2 = \frac{V}{R_1} + \frac{V}{R_2} = V\left(\frac{1}{R_1} + \frac{1}{R_2}\right) \qquad 25\text{-}20$$

where we have used Equation 25-19 for $I_1$ and $I_2$. The equivalent resistance for two resistors in parallel is therefore given by

$$\frac{1}{R_{eq}} = \frac{1}{R_1} + \frac{1}{R_2}$$

This result can be generalized for combinations, such as that in Figure 25-15, in which three or more resistors are connected in parallel:

$$\frac{1}{R_{eq}} = \frac{1}{R_1} + \frac{1}{R_2} + \frac{1}{R_3} + \ldots \qquad 25\text{-}21$$

EQUIVALENT RESISTANCE FOR RESISTORS IN PARALLEL

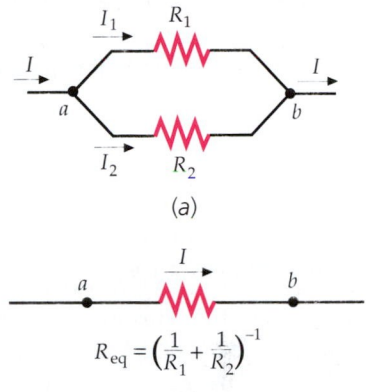

**FIGURE 25-14** (a) Two resistors are in parallel when they are connected together at both ends so that the potential drop is the same across each. (b) The two resistors in Figure 25-14a can be replaced by an equivalent resistance $R_{eq}$ that is related to $R_1$ and $R_2$ by $1/R_{eq} = 1/R_1 + 1/R_2$.

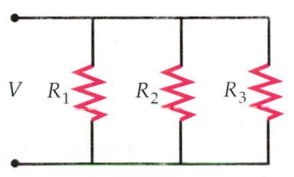

**FIGURE 25-15** Three resistors in parallel.

**EXERCISE** A 2-$\Omega$ resistor and a 4-$\Omega$ resistor are connected (a) in series and (b) in parallel. Find the equivalent resistances for both cases. (*Answer* (a) 6 $\Omega$, (b) 1.33 $\Omega$)

*RESISTORS IN PARALLEL*　　　　　**EXAMPLE 25-9**

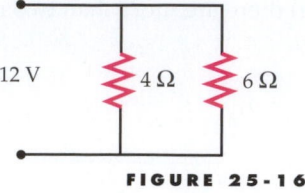

**FIGURE 25-16**

A battery applies a potential difference of 12 V across the parallel combination of 4-$\Omega$ and 6-$\Omega$ resistors shown in Figure 25-16. Find (*a*) the equivalent resistance, (*b*) the total current, (*c*) the current through each resistor, (*d*) the power dissipated in each resistor, and (*e*) the power delivered by the battery.

**PICTURE THE PROBLEM** Choose symbols and directions for the currents in Figure 25-17.

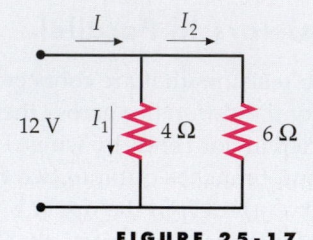

**FIGURE 25-17**

(*a*) Calculate the equivalent resistance:

$$\frac{1}{R_{eq}} = \frac{1}{4\ \Omega} + \frac{1}{6\ \Omega} = \frac{3}{12\ \Omega} + \frac{2}{12\ \Omega} = \frac{5}{12\ \Omega}$$

$$R_{eq} = \frac{12\ \Omega}{5} = \boxed{2.4\ \Omega}$$

(*b*) The total current is the potential drop divided by the equivalent resistance:

$$I = \frac{V}{R_{eq}} = \frac{12\ V}{2.4\ \Omega} = \boxed{5\ A}$$

(*c*) We obtain the current through each resistor using Equation 25-19 and the fact that the potential drop is 12 V across the parallel combination:

$$V = IR$$

$$I_1 = \frac{12\ V}{4\ \Omega} = \boxed{3\ A}$$

$$I_2 = \frac{12\ V}{6\ \Omega} = \boxed{2\ A}$$

(*d*) Use these currents to find the power dissipated in each resistor:

$$P_1 = I_1^2 R = (3\ A)^2 (4\ \Omega) = \boxed{36\ W}$$

$$P_2 = I_2^2 R = (2\ A)^2 (6\ \Omega) = \boxed{24\ W}$$

(*e*) Use $P = VI$ to find the power delivered by the battery:

$$P = VI = (12\ V)(5\ A) = \boxed{60\ W}$$

**PLAUSIBILITY CHECK** The power delivered by the battery equals the power dissipated in the two resistors $P = 60\ W = 36\ W + 24\ W$. In part (*d*), we could have calculated the power dissipated in each resistor from $P_4 = VI_4 = (12\ V)(3\ A) = 36\ W$ and $P_6 = VI_6 = (12\ V)(2\ A) = 24\ W$.

*RESISTORS IN SERIES*　　　　**EXAMPLE 25-10**　**Try It Yourself**

A 4-$\Omega$ resistor and a 6-$\Omega$ resistor are connected in series to a battery of emf 12 V with negligible internal resistance. Find (*a*) the equivalent resistance of the two resistors, (*b*) the current in the circuit, (*c*) the potential drop across each resistor, (*d*) the power dissipated in each resistor, and (*e*) the total power dissipated.

**Cover the column to the right and try these on your own before looking at the answers.**

| Steps | Answers |
|---|---|
| (*a*) 1. Draw a circuit diagram (Figure 25-18). | |
| 2. Calculate $R_{eq}$ for the two series resistors. | $R_{eq} = \boxed{10\ \Omega}$ |
| (*b*) Use $V = IR_{eq}$ to find the current through the battery. | $I = \boxed{1.2\ A}$ |

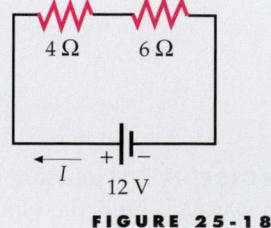

**FIGURE 25-18**

(c) Use Ohm's law to find the potential drop across each resistor.    $V_4 =$ 4.8 V ,    $V_6 =$ 7.2 V

(d) Find the power dissipated in each resistor using $P = I^2R$. Check your result using $P = IV$ for each resistor.    $P_4 =$ 5.76 W ,    $P_6 =$ 8.64 W

(e) Add your results from Part (d) to find the total power. Check your result, using $P = IV$ and $P = I^2R_{eq}$.    $P =$ 14.4 W

**REMARKS**  Note that much less power is dissipated in the series circuit than in the corresponding parallel circuit of Example 25-9.

Note from Example 25-9 that the equivalent resistance of two parallel resistances is less than the resistance of either resistor alone. This is a general result. Suppose we have a single resistor $R_1$ carrying current $I_1$ with potential drop $V = I_1R_1$. If we add a second resistor in parallel, it will carry some additional current $I_2$ without affecting $I_1$. The equivalent resistance is $V/(I_1 + I_2)$, which is less than $R_1 = V/I_1$. Note also from Example 25-9 that the ratio of the currents in the two parallel resistors equals the inverse ratio of the resistances. This general result follows from Equation 25-19:

$$I_1R_1 = I_2R_2$$

$$\frac{I_1}{I_2} = \frac{R_2}{R_1} \text{ (parallel resistors)}\qquad\qquad 25\text{-}22$$

---

*SERIES AND PARALLEL COMBINATIONS*          **EXAMPLE  25-11**    **Try It Yourself**

Consider the circuit in Figure 25-19. When the switch $S_1$ is open and switch $S_2$ is closed, find (a) the equivalent resistance of the circuit, (b) the total current in the source of emf, (c) the potential drop across each resistor, and (d) the current carried by each resistor. (e) If switch $S_1$ is now closed, find the current in the 2-$\Omega$ resistor. (f) If switch $S_2$ is now opened (while switch $S_1$ remains closed), find the potential drops across the 6-$\Omega$ resistor and across switch $S_2$.

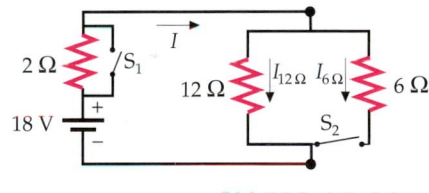

**FIGURE  25-19**

**PICTURE THE PROBLEM**  (a) To find the equivalent resistance of the circuit, first replace the two parallel resistors by their equivalent resistance. Ohm's law can then be used to find the current and potential drops. For Part (b) and Part (c), use Ohm's law.

**Cover the column to the right and try these on your own before looking at the answers.**

| **Steps** | **Answers** |
|---|---|
| (a) 1. Find the equivalent resistance of the 6- and 12-$\Omega$ parallel combination. | $R_{eq} = 4\ \Omega$ |
| 2. Combine your result in step 1 with the 2-$\Omega$ resistor in series to find the total equivalent resistance of the circuit. | $R'_{eq} =$ 6 $\Omega$ |
| (b) Find the total current using Ohm's law. This is the current in both the battery and in the 2-$\Omega$ resistor. | $I =$ 3 A |
| (c) 1. Find the potential drop across the 2-$\Omega$ resistor from $V_2 = IR$. | $V_{2\Omega} =$ 6 V |
| 2. Find the potential drop across each resistor in the parallel combination using $V_p = IR_{eq}$. | $V_{6\Omega} = V_{12\Omega} =$ 12 V |

(d) Find the current in the 6-$\Omega$ and 12-$\Omega$ resistors from $I = V_p/R$.

$$I_{6\Omega} = \boxed{2\text{ A}}, \quad I_{12\Omega} = \boxed{1\text{ A}}$$

(e) With $S_1$ closed the potential drop across the 2-$\Omega$ resistor is zero. Using Ohm's law, calculate the current through the 2-$\Omega$ resistor.

$$I_{2\Omega} = \boxed{0}$$

(f) With $S_2$ open, the current through the 6-$\Omega$ resistor is zero. Using Ohm's law, calculate the potential drop across the 6-$\Omega$ resistor. The potential drop across the 6-$\Omega$ resistor plus the potential drop across switch $S_2$ equals the potential drop across the 12-$\Omega$ resistor.

$$V_{6\Omega} = \boxed{0}, \quad V_{S_2} = \boxed{18\text{ V}}$$

**PLAUSIBILITY CHECK** The current in the 6-$\Omega$ resistor is twice that in the 12-$\Omega$ resistor, as we should expect. Also, these two currents sum to give $I$, the total current in the circuit, as they must. Finally, note that the potential drops across the 2-$\Omega$ resistor and the parallel combination sum to the emf of the battery; $V_2 + V_p = 6\text{ V} + 12\text{ V} = 18\text{ V}$.

**EXERCISE** Repeat Part (a) through Part (d) of this example with the 6-$\Omega$ resistor replaced by a wire of negligible resistance. (*Answer* (a) $R'_{eq} = 2\ \Omega$; (b) $I = 9$ A; (c) $V_2 = 18$ V, $V_0 = 0$, $V_{12} = 0$; (d) $I_2 = 9$ A, $I_0 = 9$ A, $I_{12} = 0$)

---

Find the equivalent resistance of the combination of resistors shown in Figure 25-20.

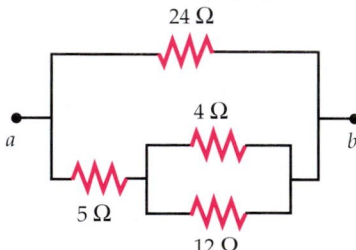

**FIGURE 25-20**

**PICTURE THE PROBLEM** You can analyze this complicated combination step by step. First, find the equivalent resistance $R_{eq}$ of the 4-$\Omega$ and 12-$\Omega$ parallel combination; next, find the equivalent resistance $R'_{eq}$ of the series combination of the 5-$\Omega$ resistor and $R_{eq}$; and finally, find the equivalent resistance $R''_{eq}$ of the parallel combination of the 24-$\Omega$ resistor and $R'_{eq}$.

**Cover the column to the right and try these on your own before looking at the answers.**

**Steps**

1. Find the equivalent resistance $R_{eq}$ of the 4-$\Omega$ and 12-$\Omega$ resistors in parallel.

2. Find the equivalent resistance $R'_{eq}$ of $R_{eq}$ in series with the 5-$\Omega$ resistor.

3. Find the equivalent resistance of $R'_{eq}$ in parallel with the 24-$\Omega$ resistor.

**Answers**

$R_{eq} = 3\ \Omega$

$R'_{eq} = 8\ \Omega$

$R''_{eq} = \boxed{6\ \Omega}$

**BLOWING THE FUSE**          **E X A M P L E   2 5 - 1 3**   **Put It in Context**

You are making a snack for some friends to help you get ready for a full night of studying. You decide that coffee, toast, and popcorn would be a good start. You start the toaster and get some popcorn going in the microwave. Since your apartment is in an older building, you know you have problems with the fuse blowing when you turn too many things on. Should you start the coffeemaker? You look on the appliances and find that the toaster has a rating of 900 W, the microwave is rated at 1200 W, and the coffeemaker is rated at 600 W. Past experience with replacing fuses has shown that your house has 20-A fuses.

**PICTURE THE PROBLEM** We can assume that household circuits are wired in parallel, since plugging in one device usually does not affect others that are in the circuit. Household voltage in the United States is 120 V. (We can neglect the fact that it is not dc.) If we can determine the current through each device, we can add up the total current in the circuit and see how it compares to the fuse current.

1. The power delivered to a device is the current times the potential drop. That is, $P = IV$. Solve for the current for each device:

$$I_{toaster} = \frac{P_{toaster}}{V} = \frac{900 \text{ W}}{120 \text{ V}} = 7.5 \text{ A}$$

$$I_{m\text{-}wave} = \frac{P_{m\text{-}wave}}{V} = \frac{1200 \text{ W}}{120 \text{ V}} = 10 \text{ A}$$

$$I_{c\text{-}maker} = \frac{P_{c\text{-}maker}}{V} = \frac{600 \text{ W}}{120 \text{ V}} = 5 \text{ A}$$

2. The current through the fuse is the sum of these currents: $I_{fuse} = 22.5 \text{ A}$

3. A current this large is above the 20-A rating of the fuse:

> Your guests will have to wait on the coffee.

**REMARKS** We have assumed that the apartment has only one circuit, and thus only one fuse. Typically, there are several circuits, each fused separately. The coffeemaker can be plugged into an outlet that is on a different circuit than the outlet for the toaster and microwave without a fuse blowing.

# 25-5 Kirchhoff's Rules

There are many simple circuits, such as the simple circuit shown in Figure 25-21, that cannot be analyzed by merely replacing combinations of resistors by an equivalent resistance. The two resistors $R_1$ and $R_2$ in this circuit look as if they might be in parallel, but they are not. The potential drop is not the same across both resistors because of the presence of the emf source $\mathcal{E}_2$ in series with $R_2$. Nor are $R_1$ and $R_2$ in series, because they do not carry the same current.

Two rules, called **Kirchhoff's rules**, apply to this circuit and to any other circuit:

1. When any closed-circuit loop is traversed, the algebraic sum of the changes in potential must equal zero.

2. At any junction (branch point) in a circuit where the current can divide, the sum of the currents into the junction must equal the sum of the currents out of the junction.

KIRCHHOFF'S RULES

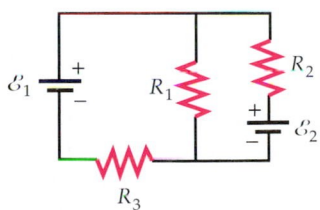

**FIGURE 25-21** An example of a simple circuit that cannot be analyzed by replacing combinations of resistors in series or parallel with their equivalent resistances. The potential drops across $R_1$ and $R_2$ are not equal because of the emf source $\mathcal{E}_2$, so these resistors are not in parallel. (Parallel resistors would be connected together at both ends.) The resistors do not carry the same current, so they are not in series.

Kirchhoff's first rule, called the **loop rule,** follows directly from the presence of a conservative field $\vec{E}$.[†] To say $\vec{E}$ is conservative means that

$$\oint_C \vec{E} \cdot d\vec{r} = 0 \qquad\qquad \text{25-23}$$

where the integral is taken around any closed curve $C$. Changes in potential $\Delta V$ and $\vec{E}$ are related by $\Delta V = V_b - V_a = -\int_a^b \vec{E} \cdot d\vec{r}$. Thus, Equation 25-23 implies that the sum of the changes in potential (the sum of the $\Delta V$s) around any closed path equals zero.

Kirchhoff's second rule, called the **junction rule,** follows from the conservation of charge. Figure 25-22 shows the junction of three wires carrying currents $I_1$, $I_2$, and $I_3$. Since charge does not originate or accumulate at this point, the conservation of charge implies the junction rule, which for this case gives

$$I_1 = I_2 + I_3 \qquad\qquad \text{25-24}$$

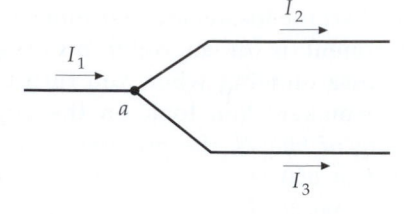

**FIGURE 25-22** Illustration of Kirchhoff's junction rule. The current $I_1$ into point $a$ equals the sum $I_2 + I_3$ of the currents out of point $a$.

## Single-Loop Circuits

As an example of using Kirchhoff's loop rule, consider the circuit shown in Figure 25-23, which contains two batteries with internal resistances $r_1$ and $r_2$ and three external resistors. We wish to find the current in terms of the emfs and resistances.

We choose clockwise as positive, as indicated in Figure 25-23. We then apply Kirchhoff's loop rule as we traverse the circuit in the positive direction, beginning at point $a$. Note that we encounter a potential drop as we traverse the source of emf between points $c$ and $d$ and we encounter a potential increase as we traverse the source of emf between $e$ and $a$. Assuming that $I$ is positive, we encounter a potential drop as we traverse each resistor. Beginning at point $a$, we obtain from Kirchhoff's loop rule

$$-IR_1 - IR_2 - \mathcal{E}_2 - Ir_2 - IR_3 + \mathcal{E}_1 - Ir_1 = 0$$

Solving for the current $I$, we obtain

$$I = \frac{\mathcal{E}_1 - \mathcal{E}_2}{R_1 + R_2 + R_3 + r_1 + r_2} \qquad\qquad \text{25-25}$$

If $\mathcal{E}_2$ is greater than $\mathcal{E}_1$, we get a negative value for the current $I$, indicating that the current is in the negative direction (counterclockwise).

For this example, suppose that $\mathcal{E}_1$ is the greater emf. In battery 2, the charge flows from high potential to low potential. Therefore, a charge $\Delta Q$ moving through battery 2 from point $c$ to point $d$ loses potential energy $\Delta Q \, \mathcal{E}_2$ (plus any energy dissipated within the battery via Joule heating). If battery 2 is a rechargeable battery, much of this potential energy is stored in the battery as chemical energy, which means that battery 2 is *charging*.

The analysis of a circuit is usually simplified if we choose one point to be at zero potential and then find the potentials of the other points relative to it. Since only potential differences are important, any point in a circuit can be chosen to have zero potential. In the following example, we choose point $e$ in the figure to be at zero potential. This is indicated by the ground symbol $\perp$ at point $e$.[‡]

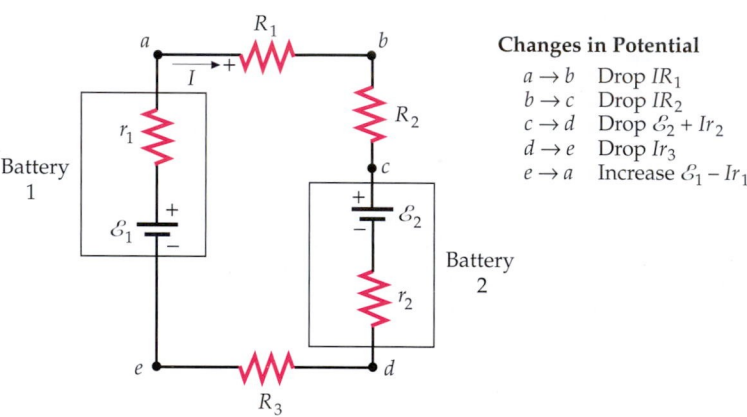

**Changes in Potential**

| | |
|---|---|
| $a \rightarrow b$ | Drop $IR_1$ |
| $b \rightarrow c$ | Drop $IR_2$ |
| $c \rightarrow d$ | Drop $\mathcal{E}_2 + Ir_2$ |
| $d \rightarrow e$ | Drop $Ir_3$ |
| $e \rightarrow a$ | Increase $\mathcal{E}_1 - Ir_1$ |

**FIGURE 25-23** Circuit containing two batteries and three external resistors.

---

[†] There is also a nonconservative electric field that is discussed in Chapter 28. The resultant electric field is the superposition of the conservative electric field and the nonconservative electric field.

[‡] As we saw in Section 21-2, the earth can be considered to be a very large conductor with a nearly unlimited supply of charge, which means that the potential of the earth remains essentially constant. In practice, electrical circuits are often grounded by connecting one point to the earth. The outside metal case of a washing machine, for example, is usually grounded by connecting it by a wire to a water pipe that is in contact with the earth. Since everything so grounded is at the same potential, it is convenient to designate this potential as zero.

**E X A M P L E   2 5 - 1 4**

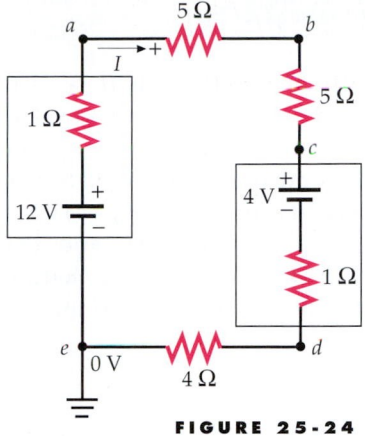

**FIGURE 25-24**

Suppose the elements in the circuit in Figure 25-23 have the values $\mathcal{E}_1 = 12$ V, $\mathcal{E}_2 = 4$ V, $r_1 = r_2 = 1\ \Omega$, $R_1 = R_2 = 5\ \Omega$, and $R_3 = 4\ \Omega$, as shown in Figure 25-24. (*a*) Find the potentials at points *a* through *e* in the figure, assuming that the potential at point *e* is zero. (*b*) Find the power input and output in the circuit.

**PICTURE THE PROBLEM** To find the potential differences, we first need to find the current *I* in the circuit. The potential drop across each resistor is then *IR*. To discuss the energy balance, we calculate the power into or out of each element using Equations 25-11 and 25-12.

(*a*) 1. The current *I* in the circuit is found using Equation 25-25:

$$I = \frac{12\ \text{V} - 4\ \text{V}}{5\ \Omega + 5\ \Omega + 4\ \Omega + 1\ \Omega + 1\ \Omega} = \frac{8\ \text{V}}{16\ \Omega} = 0.5\ \text{A}$$

2. We now find the potential at each labeled point in the circuit:

$$V_a = V_e + \mathcal{E}_1 - Ir_1 = 0 + 12\ \text{V} - (0.5\ \text{A})(1\ \Omega) = \boxed{11.5\ \text{V}}$$

$$V_b = V_a - IR_1 = 11.5\ \text{V} - (0.5\ \text{A})(5\ \Omega) = \boxed{9\ \text{V}}$$

$$V_c = V_b - IR_2 = 9\ \text{V} - (0.5\ \text{A})(5\ \Omega) = \boxed{6.5\ \text{V}}$$

$$V_d = V_c - \mathcal{E}_2 - Ir_2 = 6.5\ \text{V} - 4\ \text{V} - (0.5\ \text{A})(1\ \Omega) = \boxed{2.0\ \text{V}}$$

$$V_e = V_d - IR_3 = 2.0\ \text{V} - (0.5\ \text{A})(4\ \Omega) = \boxed{0}$$

(*b*) 1. First, calculate the power supplied by the emf source $\mathcal{E}_1$:

$$P_{\mathcal{E}_1} = \mathcal{E}_1 I = (12\ \text{V})(0.5\ \text{A}) = \boxed{6\ \text{W}}$$

2. Part of this power is dissipated in the resistors, both internal and external:

$$P_R = I^2R_1 + I^2R_2 + I^2R_3 + I^2r_1 + I^2r_2$$

$$= (0.5\ \text{A})^2(5\ \Omega + 5\ \Omega + 4\ \Omega + 1\ \Omega + 1\ \Omega) = 4.0\ \text{W}$$

3. The remaining 2 W of power goes into charging battery 2:

$$P_{\mathcal{E}_2} = \mathcal{E}_2 I = (4\ \text{V})(0.5\ \text{A}) = 2\ \text{W}$$

4. The rate at which potential energy being taken out of the circuit is:

$$P = P_R + P_{\mathcal{E}_1} = \boxed{6\ \text{W}}$$

Note that the terminal voltage of the battery that is being charged in Example 25-14 is $V_c - V_d = 4.5$ V, which is greater than the emf of the battery. If the same 4-V battery were to deliver 0.5 A to an external circuit, its terminal voltage would be 3.5 V (again assuming that its internal resistance is 1 $\Omega$). If the internal resistance is very small, the terminal voltage of a battery is nearly equal to its emf, whether the battery is delivering energy to an external circuit or is being charged. Some real batteries, such as those used in automobiles, are nearly reversible and can easily be recharged. Other types of batteries are not reversible. If you attempt to recharge one of these by driving current from its positive to its negative terminal, most, if not all, of the energy will be dissipated into thermal energy rather than being transformed into the chemical energy of the battery.

**E X A M P L E   2 5 - 1 5**

A fully charged[†] car battery is to be connected by jumper cables to a discharged car battery in order to charge it. (*a*) To which terminal of the discharged battery should the positive terminal of the charged battery be connected? (*b*) Assume that the charged battery has an emf of $\mathcal{E}_1 = 12$ V and the discharged battery has an emf of $\mathcal{E}_2 = 11$ V, that the internal resistances of the batteries are $r_1 = r_2 = 0.02$ Ω, and that the resistance of the jumper cables is $R = 0.01$ Ω. What will the charging current be? (*c*) What will the current be if the batteries are connected incorrectly?

**PICTURE THE PROBLEM**

**FIGURE 25-25**

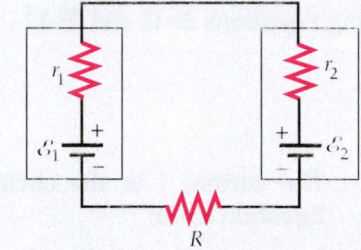

1. To charge the discharged battery, we connect the terminals positive to positive and negative to negative, to drive current through the discharged battery from the positive terminal to the negative terminal (Figure 25-25):

2. Use Kirchhoff's loop rule to find the charging current:

$$\mathcal{E}_1 - Ir_1 - Ir_2 - \mathcal{E}_2 - IR = 0$$

so

$$I = \frac{\mathcal{E}_1 - \mathcal{E}_2}{R + r_1 + r_2} = \frac{12\ V - 11\ V}{0.05\ \Omega} = \boxed{20\ A}$$

3. When the batteries are connected incorrectly, positive terminals to negative terminals, the emfs add:

$$\mathcal{E}_1 - Ir_1 + \mathcal{E}_2 - Ir_2 - IR = 0$$

so

$$I = \frac{\mathcal{E}_1 + \mathcal{E}_2}{R + r_1 + r_2} = \frac{12\ V + 11\ V}{0.05\ \Omega} = \boxed{460\ A}$$

**REMARKS** If the batteries are connected incorrectly, as shown in Figure 25-26, the total resistance of the circuit is of the order of hundredths of an ohm, the current is very large, and the batteries could explode in a shower of boiling battery acid.

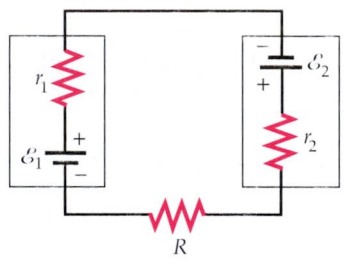

**FIGURE 25-26** Two batteries connected incorrectly—dangerous!

## Multiloop Circuits

In multiloop circuits, often the directions of the currents in the different branches of the circuit are unknown. Fortunately, Kirchhoff's rules do not require that we know the directions of the current initially. In fact, these rules allow us to solve for the directions of the currents. To accomplish this, for each branch we arbitrarily assign a positive direction along the branch, and we indicate this assignment by placing a corresponding arrow on the circuit diagram (Figure 25-27). If the actual current in the branch is in the positive direction, when we solve for it we will get a positive value, and if the actual current is opposite to the positive direction, when we solve for it we will get a negative value. The current through a resistor always goes from high potential to low potential. Therefore, any time we traverse a resistor in the direction of the current, the change in potential is negative, and vice versa. Here is the rule:

> For each branch of a circuit, we draw an arrow to indicate the positive direction for that branch. Then, if we traverse a resistor in the direction of the arrow, the change in potential $\Delta V$ is equal to $-IR$ (and if we traverse a resistor in the opposite direction, $\Delta V$ is equal to $+IR$).
>
> SIGN RULE FOR THE CHANGE IN POTENTIAL ACROSS A RESISTOR

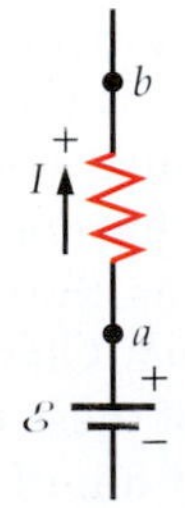

**FIGURE 25-27** It is not known whether or not the current $I$ has a positive or a negative value. Whether it is positive or negative, $V_b - V_a = -IR$. If the current is upward, then $I$ is positive and $-IR$ is negative. However, if the current is downward, then $I$ is negative and $-IR$ is positive.

---

† Batteries do not store charge. A *fully charged* battery is one with a maximum amount of stored chemical energy.

If we traverse a resistor in the positive direction, and if $I$ is positive, then $-IR$ is negative. This is as expected, since the current is always in the direction of decreasing potential. If we traverse a resistor in the positive direction, and if $I$ is negative, then $-IR$ is positive. Similarly, if we traverse a resistor in the negative direction, and if $I$ is positive, then $+IR$ is positive, and if we traverse a resistor in the negative direction and if $I$ is negative, then $+IR$ is negative.

To analyze circuits containing more than one loop, we need to use both of Kirchhoff's rules, with Kirchhoff's junction rule applied to points where the current splits into two or more parts.

**FIGURE 25-28**

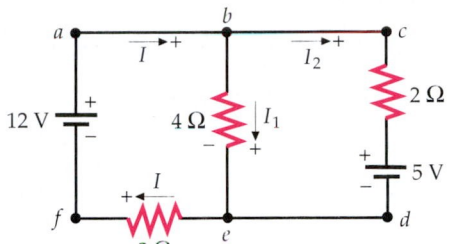

*APPLYING KIRCHHOFF'S RULES*    **E X A M P L E    2 5 - 1 6**

(*a*) **Find the current in each branch of the circuit shown in Figure 25-28.**
(*b*) **Find the energy dissipated in the 4-$\Omega$ resistor in 3 s.**

**PICTURE THE PROBLEM** There are three branch currents, $I$, $I_1$, and $I_2$, to be determined, so we need three relations. One relation comes from applying the junction rule to point *b*. (We can also apply the junction rule to point *e*, the only other junction in the circuit, but it gives exactly the same information.) The other two relations are obtained by applying the loop rule. There are three loops in the circuit: the two interior loops, *abefa* and *bcdeb*, and the exterior loop, *abcdefa*. We can use any two of these loops—the third will give redundant information. There is a direction arrow on each branch in Figure 25-28. Each direction arrow indicates the positive direction for that branch. If our analysis results in a negative value for a branch current, then that current is in the direction opposite to the direction arrow for that branch.

(*a*) 1. Apply the junction rule to point *b*:

$I = I_1 + I_2$

2. Apply the loop rule to the outer loop, *abcdefa*:

$12\text{ V} - (2\ \Omega)I_2 - 5\text{ V} - (3\ \Omega)(I_1 + I_2) = 0$

3. Divide the above equation by 1 $\Omega$, recalling that $(1\text{ V})/(1\ \Omega) = 1$ A, then simplify:

$7\text{ A} - 3I_1 - 5I_2 = 0$

4. For the third condition, apply the loop rule to the loop on the right, *bcdeb*:

$-(2\ \Omega)I_2 - 5\text{ V} + (4\ \Omega)I_1 = 0$

$-5\text{ A} + 4I_1 - 2I_2 = 0$

5. The results for steps 3 and 4 can be combined to solve for $I_1$ and $I_2$. To do so, first multiply the result for step 3 by 2, and then multiply the result for step 4 by $-5$:

$14\text{ A} - 6I_1 - 10I_2 = 0$

$25\text{ A} - 20I_1 + 10I_2 = 0$

6. Add the equations in step 5 to eliminate $I_2$, then solve for $I_1$:

$39\text{ A} - 26I_1 = 0$

$I_1 = \dfrac{39\text{ A}}{26} = \boxed{1.5\text{ A}}$

7. Substitute $I_1$ in the results for step 3 or 4 to solve for $I_2$:

$7\text{ A} - 3(1.5\text{ A}) - 5I_2 = 0$

$I_2 = \dfrac{2.5\text{ A}}{5} = \boxed{0.5\text{ A}}$

8. Finally, $I_1$ and $I_2$ determine $I$ using the equation in step 1:

$I = I_1 + I_2 = 1.5\text{ A} + 0.5\text{ A} = \boxed{2.0\text{ A}}$

(*b*) 1. The power dissipated in the 4-$\Omega$ resistor is found using $P = I_1^2 R$:

$P = I_1^2 R = (1.5\text{ A})^2(4\ \Omega) = 9\text{ W}$

2. The total energy dissipated in a time $\Delta t$ is $W = P\Delta t$. In this case, $t = 3$ s:

$W = P\Delta t = (9\text{ W})(3\text{ s}) = \boxed{27\text{ J}}$

**⊘ PLAUSIBILITY CHECK** In Figure 25-29, we have chosen the potential to be zero at point $f$, and we have labeled the currents and the potentials at the other points. Note that $V_b - V_e = 6$ V and $V_e - V_f = 6$ V.

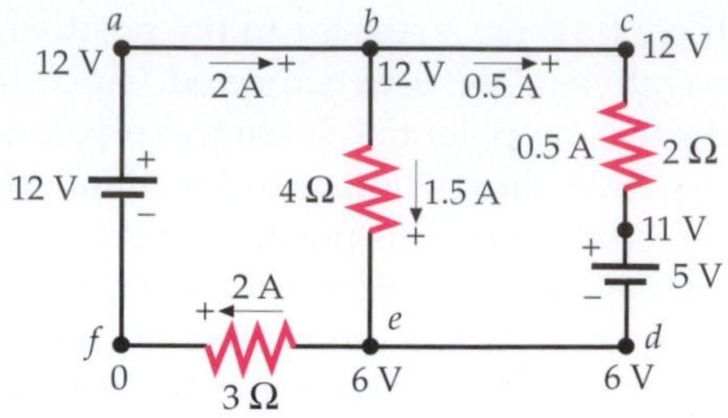

**REMARKS** Applying the loop rule to the loop on the left, *abefa*, gives $12\text{ V} - (4\ \Omega)\ I_1 - (3\ \Omega)(I_1 + I_2) = 0$, or $12\text{ A} - 7I_1 - 3I_2 = 0$. Note that this is just the result for step 3 minus the result for step 4 and hence contains no new information, as expected.

**EXERCISE** Find $I_1$ for the case in which the 3-$\Omega$ resistor approaches (*a*) zero resistance and (*b*) infinite resistance. [*Answer* (*a*) The potential drop across the 4-$\Omega$ resistor is 12 V; thus, $I_1 = 3$ A. (*b*) In this case, the loop on the left is an open circuit, so $I = 0$ and $I_2 = -I_1$. Thus, $I_1 = (5\text{ V})/(2\ \Omega + 4\ \Omega) = 0.833$ A.]

Example 25-16 illustrates the general methods for the analysis of multiloop circuits:

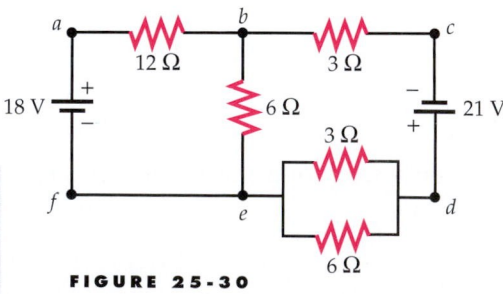

**FIGURE 25-29**

1. Draw a sketch of the circuit.

2. Replace any series or parallel resistor combinations or capacitor combinations by their equivalent values.

3. Choose the positive direction for each branch of the circuit and indicate the positive direction with a direction arrow. Label the current in each branch. Add plus and minus signs to indicate the high-potential terminal and low-potential terminal of each source of emf.

4. Apply the junction rule to all but one of the branch points (junctions).

5. Apply the loop rule to each loop until you obtain as many independent equations as there are unknowns. When traversing a resistor in the positive direction, the change in potential equals $-IR$. When traversing a battery from the negative terminal to the positive terminal, the change in potential equals $\mathcal{E} - IR$.

6. Solve the equations to obtain the desired values.

7. Check your results by assigning a potential of zero to one point in the circuit and use the values of the currents found to determine the potentials at other points in the circuit.

GENERAL METHOD FOR ANALYZING MULTILOOP CIRCUITS

---

*A THREE-BRANCH CIRCUIT*                **EXAMPLE   2 5 - 1 7   Try It Yourself**

(*a*) **Find the current in each part of the circuit shown in Figure 25-30. Draw the circuit diagram with the correct magnitudes and directions for the current in each part. (*b*) Assign $V = 0$ to point $c$ and then label the potential at each other point $a$ through $f$.**

**PICTURE THE PROBLEM** First, replace the two parallel resistors by an equivalent resistance. Let $I$ be the current through the 18-V battery, and let $I_1$ be the current from point $b$ to point $e$. The currents can then be found by applying the junction rule at branch points $b$ and $c$ and by applying the loop rule to each loop.

**FIGURE 25-30**

**Cover the column to the right and try these on your own before looking at the answers.**

| Steps | Answers |
|---|---|
| (*a*)  1. Find the equivalent resistance of the 3-$\Omega$ and 6-$\Omega$ parallel resistors. | $R_{eq} = 2\ \Omega$ |

**FIGURE 25-31**

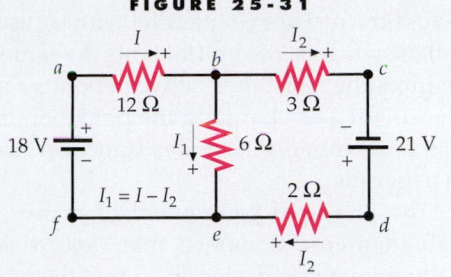

2. Apply the junction rule at points *b* and *e* and redraw the circuit diagram with the positive branch directions indicated (Figure 25-31).

$I = I_1 + I_2$   or   $(I_1 = I - I_2)$

3. Apply Kirchhoff's loop rule to loop *abefa* to obtain an equation involving *I* and $I_2$.

$18\,\text{V} - (12\,\Omega)I - (6\,\Omega)(I - I_2) = 0$

4. Simplify your equation from step 3.

$3\,\text{A} - 3I + I_2 = 0$

5. Apply Kirchhoff's loop rule to loop *bcdeb* to obtain an equation involving *I* and $I_2$.

$-(3\,\Omega)I_2 + 21\,\text{V} - (2\,\Omega)I_2 + (6\,\Omega)(I - I_2) = 0$

6. Simplify your equation in step 5.

$21\,\text{A} + 6I - 11I_2 = 0$

7. Solve your simultaneous equations from step 4 and step 6 for *I* and $I_2$. One way to do this is to multiply the equation in step 4 by 11 and then add the equations to eliminate $I_2$.

$I = \boxed{2\,\text{A}}, \quad I_2 = \boxed{3\,\text{A}}$

8. Find the current through the 6-$\Omega$ resistor.

$I_1 = I - I_2 = \boxed{-1\,\text{A}}$

9. Use $V = I_2 R_{eq}$ to find the potential drop across the parallel 3-$\Omega$ and 6-$\Omega$ resistors.

$V = 6\,\text{V}$

10. Use the result of step 9 to find the current in each of the parallel resistors.

$I_{3\Omega} = \boxed{2\,\text{A}}, \quad I_{6\Omega} = \boxed{1\,\text{A}}$

(*b*) Redraw Figure 25-31 showing the current through each part of the circuit (Figure 25-32). Begin with $V = 0$ at point *c* and calculate the potential at points *d, e, f, a*, and *b*.

$V_d = V_c + 21\,\text{V} = 0 + 21\,\text{V} = \boxed{21\,\text{V}}$

$V_e = V_d - (3\,\text{A})(2\,\Omega) = 21\,\text{V} - 6\,\text{V} = \boxed{15\,\text{V}}$

$V_f = V_e = \boxed{15\,\text{V}}$

$V_a = V_f + 18\,\text{V} = 15\,\text{V} + 18\,\text{V} = \boxed{33\,\text{V}}$

$V_b = V_a - (2\,\text{A})(12\,\Omega) = 33\,\text{V} - 24\,\text{V} = \boxed{9\,\text{V}}$

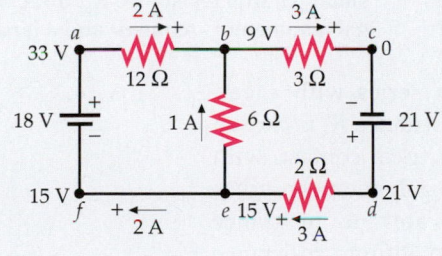

**FIGURE 25-32**

**PLAUSIBILITY CHECK** From point *b* to point *c* the potential drops by $(3\,\text{A})(3\,\Omega) = 9\,\text{V}$, which gives $V_c = 0$, as assumed. From point *e* to point *b* the potential drops by $(1\,\text{A})(6\,\Omega) = 6\,\text{V}$, so $V_b = V_e - 6\,\text{V} = 15\,\text{V} - 6\,\text{V} = 9\,\text{V}$.

## Ammeters, Voltmeters, and Ohmmeters

The devices that measure current, potential difference, and resistance are called **ammeters, voltmeters,** and **ohmmeters,** respectively. Often, all three of these meters are included in a single *multimeter* that can be switched from one use to another. You might use a voltmeter to measure the terminal voltage of your car battery and an ohmmeter to measure the resistance of some electrical device at home (e.g., a toaster or lightbulb) when you suspect a short circuit or a broken wire.

To measure the current through a resistor in a simple circuit, we place an ammeter in series with the resistor, as shown in Figure 25-33, so that the ammeter and the resistor carry the same current. Since the ammeter has a very low (but finite) resistance, the current in the circuit decreases very slightly when the ammeter is inserted. Ideally, the ammeter should have a negligibly small resistance so that the current to be measured is only negligibly affected.

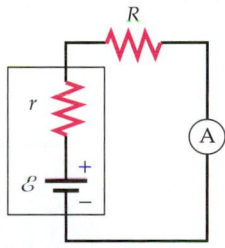

**FIGURE 25-33** To measure the current in a resistor *R*, an ammeter A (circled) is placed in series with the resistor so that it carries the same current as the resistor.

The potential difference across a resistor is measured by placing a voltmeter across the resistor (in parallel with it), as shown in Figure 25-34, so that the potential drop across the voltmeter is the same as that across the resistor. The voltmeter reduces the resistance between points $a$ and $b$, thus increasing the total current in the circuit and changing the potential drop across the resistor. A good voltmeter has an extremely large resistance so that its effect on the current in the circuit is negligible.

The principal component of many common ammeters and voltmeters is a **galvanometer,** a device that detects small currents passing through it. The galvanometer is designed so that the scale reading is proportional to the current passing through. A typical galvanometer used in many student laboratories consists of a coil of wire in the magnetic field of a permanent magnet. When the coil carries a current, the magnetic field exerts a torque on the coil, which causes the coil to rotate. A pointer attached to the coil indicates the reading on a scale. The coil itself contributes a small amount of resistance when the galvanometer is placed within a circuit.

To construct an ammeter from a galvanometer, we place a small resistor called a **shunt resistor** in *parallel* with the galvanometer. The shunt resistance is usually much smaller than the resistance of the galvanometer so that most of the current is carried by the shunt resistor. The equivalent resistance of the ammeter is then approximately equal to the shunt resistance, which is much smaller than the internal resistance of the galvanometer alone. To construct a voltmeter, we place a resistor with a large resistance in *series* with the galvanometer so that the equivalent resistance of the voltmeter is much larger than that of the galvanometer alone. Figure 25-35 illustrates the construction of an ammeter and voltmeter from a galvanometer. The resistance of the galvanometer $R_g$ is shown separately in these schematic drawings, but it is actually part of the galvanometer.

A simple ohmmeter consists of a battery connected in series with a galvanometer and a resistor, as shown in Figure 25-36a. The resistance $R_s$ is chosen so that when the terminals $a$ and $b$ are shorted (put in electrical contact, with negligible resistance between them), the current through the galvanometer gives a full-scale deflection. Thus, a full-scale deflection indicates no resistance between terminals $a$ and $b$. A zero deflection indicates an infinite resistance between the terminals. When the terminals are connected across an unknown resistance $R$, the current through the galvanometer depends on $R$, so the scale can be calibrated to give a direct reading of $R$, as shown in Figure 25-36b. Because an ohmmeter sends a current through the resistance to be measured, some caution must be exercised when using this instrument. For example, you would not want to try to measure the resistance of a sensitive galvanometer with an ohmmeter, because the current provided by the battery in the ohmmeter would probably damage the galvanometer.

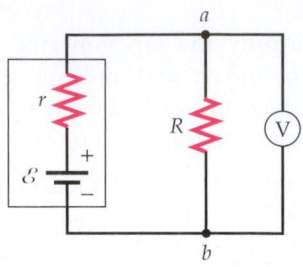

**FIGURE 25-34** To measure the potential drop across a resistor, a voltmeter V (circled) is placed in parallel with the resistor so that the potential drops across the voltmeter and the resistor are the same.

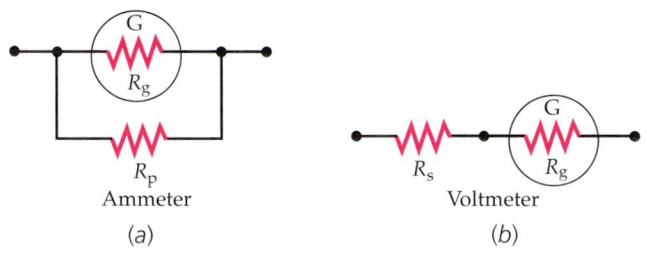

Ammeter
(a)

Voltmeter
(b)

**FIGURE 25-35** (*a*) An ammeter consists of a galvanometer G (circled) whose resistance is $R_g$ and a small parallel resistance $R_p$. (*b*) A voltmeter consists of a galvanometer G (circled) and a large series resistance $R_s$.

(a)

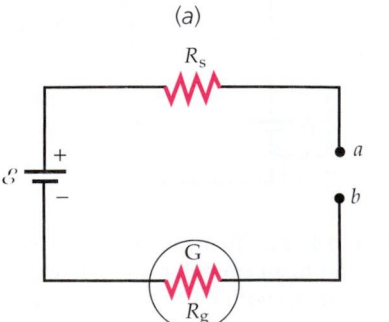

(b)

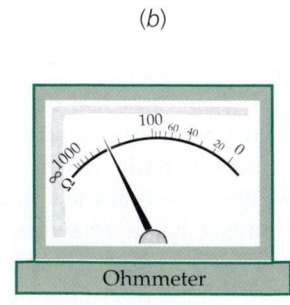

Ohmmeter

**FIGURE 25-36** (*a*) An ohmmeter consists of a battery connected in series with a galvanometer and a resistor $R_s$, which is chosen so that the galvanometer gives full-scale deflection when points $a$ and $b$ are shorted. (*b*) When a resistor $R$ is placed across $a$ and $b$, the galvanometer needle deflects by an amount that depends on the value of $R$. The galvanometer scale is calibrated to give a readout in ohms.

# 25-6 *RC Circuits*

A circuit containing a resistor and a capacitor is called an **RC circuit.** The current in an *RC* circuit flows in a single direction, as in all dc circuits, but the magnitude of the current varies with time. A practical example of an *RC* circuit is the circuit in the flash attachment of a camera. Before a flash photograph is taken, a battery in the flash attachment charges the capacitor through a resistor. When the charge is accomplished, the flash is ready. When the picture is taken, the capacitor discharges through the flashbulb. The battery then recharges the capacitor, and a short time later the flash is ready for another picture. Using Kirchhoff's rules, we can obtain equations for the charge *Q* and the current *I* as functions of time for both the charging and discharging of a capacitor through a resistor.

## Discharging a Capacitor

Figure 25-37 shows a capacitor with initial charges of $+Q_0$ on the upper plate and $-Q_0$ on the lower plate. The capacitor is connected to a resistor *R* and a switch *S*, which is initially open. The potential difference across the capacitor is initially $V_0 = Q_0/C$, where *C* is the capacitance.

We close the switch at time $t = 0$. Since there is now a potential difference across the resistor, there must be a current in it. The initial current is

$$I_0 = \frac{V_0}{R} = \frac{Q_0}{RC} \qquad\qquad 25\text{-}26$$

The current is due to the flow of charge from the positive plate of the capacitor to the negative plate through the resistor. After a time, the charge on the capacitor is reduced. If we choose the positive direction to be clockwise, then the current equals the rate of decrease of that charge. If *Q* is the charge on the upper plate of the capacitor at time *t*, the current at that time is

$$I = -\frac{dQ}{dt} \qquad\qquad 25\text{-}27$$

(The minus sign is needed because while *Q* decreases, $dQ/dt$ is negative.)[†] Traversing the circuit in the clockwise direction, we encounter a potential drop $IR$ across the resistor and a potential increase $Q/C$ across the capacitor. Thus, Kirchhoff's loop rule gives

$$\frac{Q}{C} - IR = 0 \qquad\qquad 25\text{-}28$$

where *Q* and *I*, both functions of time, are related by Equation 25-27. Substituting $-dQ/dt$ for *I* in Equation 25-28, we have

$$\frac{Q}{C} + R\frac{dQ}{dt} = 0$$

or

$$\frac{dQ}{dt} = -\frac{1}{RC}Q \qquad\qquad 25\text{-}29$$

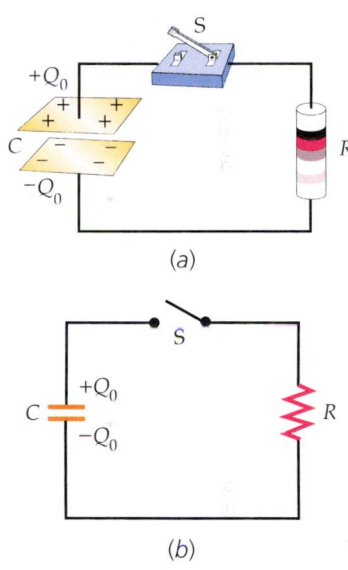

(a)

(b)

**FIGURE 25-37** (*a*) A parallel-plate capacitor in series with a switch *S* and a resistor *R*. (*b*) A circuit diagram for Figure 25-37*a*.

---

[†] If the positive direction were chosen to be counterclockwise, then the sign in Equation 25-27 would be a positive sign.

To solve this equation, we first separate the variables $Q$ and $t$ by multiplying both sides by $dt/Q$, and then integrate. Multiplying both sides by $dt/Q$, we obtain

$$\frac{dQ}{Q} = -\frac{1}{RC}\,dt \qquad\qquad 25\text{-}30$$

The variables $Q$ and $t$ are now in separate terms. Integrating from $Q_0$ at $t = 0$ to $Q'$ at time $t'$ gives

$$\int_{Q_0}^{Q'} \frac{dQ}{Q} = -\frac{1}{RC}\int_0^{t'} dt$$

so

$$\ln\frac{Q'}{Q_0} = -\frac{t'}{RC}$$

Since $t'$ is arbitrary, we can replace $t'$ with $t$, and then $Q' = Q(t)$. Solving for $Q(t)$ gives

$$Q(t) = Q_0 e^{-t/(RC)} = Q_0 e^{-t/\tau} \qquad\qquad 25\text{-}31$$

where $\tau$, called the **time constant,** is the time it takes for the charge to decrease by a factor of $e^{-1}$:

$$\tau = RC \qquad\qquad 25\text{-}32$$

DEFINITION—TIME CONSTANT

Figure 25-38 shows the charge on the capacitor in the circuit of Figure 25-37 as a function of time. After a time $t = \tau$, the charge is $Q = e^{-1}Q_0 = 0.37\,Q_0$, after a time $t = 2\tau$, the charge is $Q = e^{-2}Q_0 = 0.135Q_0$, and so forth. After a time equal to several time constants, the charge $Q$ is negligible. This type of decrease, which is called an **exponential decrease,** is very common in nature. It occurs whenever the rate at which a quantity decreases is proportional to the quantity itself.[†]

The decrease in the charge on a capacitor can be likened to the decrease in the amount of water in a bucket with vertical sides that has a small hole in the bottom. The rate at which the water flows out of the bucket is proportional to the pressure of the water, which is in turn proportional to the amount of water still in the bucket.

The current is obtained by differentiating Equation 25-31

$$I = -\frac{dQ}{dt} = \frac{Q_0}{RC}\,e^{-t/(RC)}$$

Substituting, using Equation 25-26, we obtain

$$I = I_0 e^{-t/\tau} \qquad\qquad 25\text{-}33$$

where $I_0 = V_0/R = Q_0/(RC)$ is the initial current. The current as a function of time is shown in Figure 25-39. As with the charge, the current decreases exponentially with time constant $\tau = RC$.

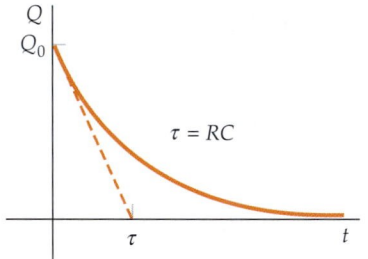

**FIGURE 25-38** Plot of the charge on the capacitor versus time for the circuit shown in Figure 25-37 when the switch is closed at time $t = 0$. The time constant $\tau = RC$ is the time it takes for the charge to decrease by a factor of $e^{-1}$. (The time constant is also the time it would take the capacitor to discharge fully if its discharge rate remains constant, as indicated by the dashed line.)

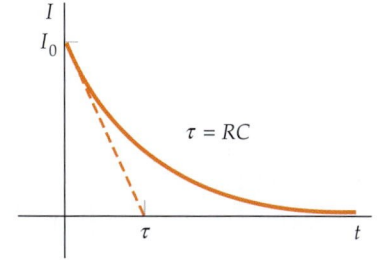

**FIGURE 25-39** Plot of the current versus time for the circuit in Figure 25-37. The curve has the same shape as that in Figure 25-38. If the rate of decrease of the current remains constant, the current would reach zero after one time constant, as indicated by the dashed line.

---

† We encountered exponential decreases in Chapter 14 when we studied the damped oscillator.

---

*DISCHARGING A CAPACITOR*  **EXAMPLE 25-18**

A 4-$\mu$F capacitor is charged to 24 V and then connected across a 200-$\Omega$ resistor. Find (*a*) the initial charge on the capacitor, (*b*) the initial current through the 200-$\Omega$ resistor, (*c*) the time constant, and (*d*) the charge on the capacitor after 4 ms.

**PICTURE THE PROBLEM** The circuit diagram is the same as the circuit diagram shown in Figure 25-37.

(*a*) The initial charge is related to the capacitance and voltage: $\qquad Q_0 = CV_0 = (4\ \mu\text{F})(24\ \text{V}) = \boxed{96\ \mu\text{C}}$

(*b*) The initial current is the initial voltage divided by the resistance: $\qquad I_0 = \dfrac{V_0}{R} = \dfrac{24\ \text{V}}{200\ \Omega} = \boxed{0.12\ \text{A}}$

(*c*) The time constant is $RC$: $\qquad \tau = RC = (200\ \Omega)(4\ \mu\text{F}) = 800\ \mu\text{s} = \boxed{0.8\ \text{ms}}$

(*d*) Substitute $t = 4$ ms into Equation 25-31 to find the charge on the capacitor at that time: $\qquad Q = Q_0 e^{-t/\tau} = (96\ \mu\text{C})e^{-(4\ \text{ms})/(0.8\ \text{ms})}$

$$= (96\ \mu\text{C})e^{-5} = \boxed{0.647\ \mu\text{C}}$$

**REMARKS** After five time constants, the $Q$ is less than 1 percent of its initial value.

**EXERCISE** Find the current through the 200-$\Omega$ resistor at $t = 4$ ms. (*Answer* 0.809 mA)

## Charging a Capacitor

Figure 25-40*a* shows a circuit for charging a capacitor. The capacitor is initially uncharged. The switch S, originally open, is closed at time $t = 0$. Charge immediately begins to flow through the battery (Figure 25-40*b*). If the charge on the rightmost plate of the capacitor at time $t$ is $Q$, the current in the circuit is $I$, and clockwise is positive, then Kirchhoff's loop rule gives

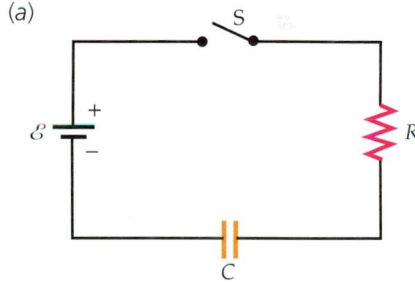

(*a*)

$$\mathcal{E} - IR - \frac{Q}{C} = 0 \qquad\qquad \text{25-34}$$

By inspecting this equation we can see that at time $t = 0$, the charge on the capacitor is zero and the current is $I_0 = \mathcal{E}/R$. The charge then increases and the current decreases. The charge reaches a maximum value of $Q_f = C\mathcal{E}$ when the current $I$ equals zero, as can also be seen from Equation 25-34.

In this circuit, we have chosen the positive direction so if $I$ is positive $Q$ is increasing. Thus,

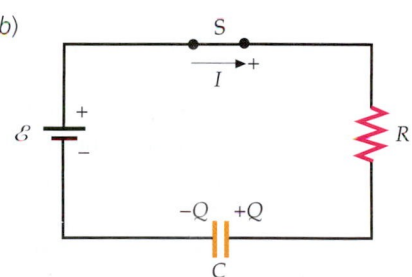

(*b*)

$$I = +\frac{dQ}{dt}$$

Substituting $dQ/dt$ for $I$ in Equation 25-34 gives

$$\mathcal{E} - R\frac{dQ}{dt} - \frac{Q}{C} = 0 \qquad\qquad \text{25-35}$$

**FIGURE 25-40** (*a*) A circuit for charging a capacitor to a potential difference $\mathcal{E}$. (*b*) After the switch is closed, there is current through and a potential drop across the resistor and a charge on and a potential drop across the capacitor.

Equation 25-35 can be solved in the same way as Equation 25-29. The details are left as a problem (see Problem 119). The result is

$$Q = C\mathcal{E}(1 - e^{-t/(RC)}) = Q_f(1 - e^{-t/\tau}) \qquad\qquad \text{25-36}$$

where $Q_f = C\mathcal{E}$ is the final charge. The current is obtained from $I = dQ/dt$:

$$I = \frac{dQ}{dt} = C\mathcal{E}\left(-\frac{-1}{RC}e^{-t/(RC)}\right) = \frac{\mathcal{E}}{R}e^{-t/(RC)}$$

or

$$I = \frac{\mathcal{E}}{R}e^{-t/(RC)} = I_0 e^{-t/\tau} \qquad\qquad 25\text{-}37$$

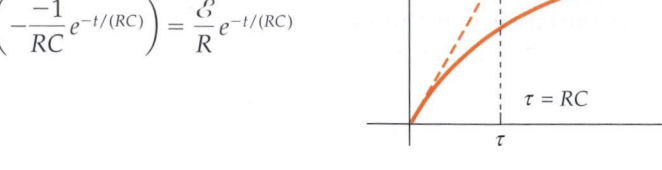

**FIGURE 25-41** Plot of the charge on the capacitor versus time for the charging circuit of Figure 25-40 after the switch is closed (at $t = 0$). After a time $t = \tau = RC$, the charge on the capacitor is 0.63 $C\mathcal{E}$, where $C\mathcal{E}$ is its final charge. If the charging rate were constant, the capacitor would be fully charged after a time $t = \tau$.

where the initial current in this case is $I_0 = \mathcal{E}/R$.

Figure 25-41 and Figure 25-42 show the charge and the current as functions of time.

**EXERCISE** Show that Equation 25-36 does indeed satisfy Equation 25-35 by substituting $Q(t)$ and $dQ/dt$ into Equation 25-35.

**EXERCISE** What fraction of the maximum charge is on the charging capacitor after a time $t = 2\tau$? (*Answer*  0.86)

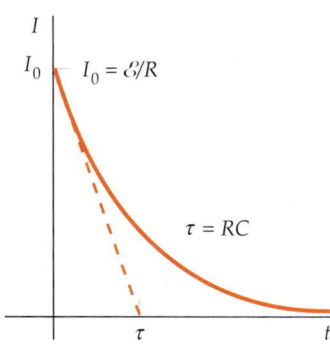

**FIGURE 25-42** Plot of the current versus time for the charging circuit of Figure 25-40. The current is initially $\mathcal{E}/R$, and the current decreases exponentially with time.

---

*CHARGING A CAPACITOR*      **EXAMPLE 25-19**    **Try It Yourself**

A 6-V battery of negligible internal resistance is used to charge a 2-$\mu$F capacitor through a 100-$\Omega$ resistor. Find (*a*) the initial current, (*b*) the final charge on the capacitor, (*c*) the time required for the charge to reach 90 percent of its final value, and (*d*) the charge when the current is half its initial value.

**PICTURE THE PROBLEM**

**Cover the column to the right and try these on your own before looking at the answers.**

| Steps | Answers |
|---|---|
| (*a*) Find the initial current from $I_0 = \mathcal{E}/R$. | $I_0 = \boxed{0.06\ \text{A}}$ |
| (*b*) Find the final charge from $Q = C\mathcal{E}$. | $Q_f = \boxed{12\ \mu\text{C}}$ |
| (*c*) Set $Q = 0.9\ Q_f$ in Equation 25-36 and solve for $t$. (First solve for $e^{t/\tau}$, then take the natural log of both sides, then solve for $t$.) | $t = 2.3\ \tau = \boxed{460\ \mu\text{s}}$ |
| (*d*) 1. Apply Kirchhoff's loop rule to the circuit using Figure 25-40*b*. | $\mathcal{E} - IR - \dfrac{Q}{C} = 0$ |
| 2. Set $I = I_0/2$ and solve for $Q$. | $Q = \dfrac{Q_f}{2} = \boxed{6\ \mu\text{C}}$ |

**REMARKS** The answer to Part (*d*) can be obtained by first solving for $t$ using Equation 25-37, then substituting that time into Equation 25-36 and solving for $Q$. However, using the loop rule is certainly the more direct approach.

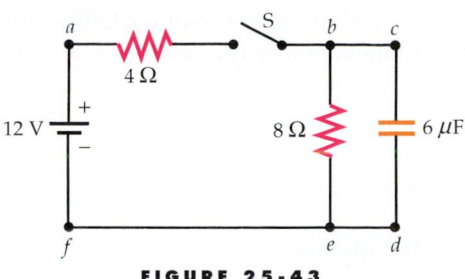

**FIGURE 25-43**

The 6-$\mu$F capacitor in the circuit shown in Figure 25-43 is initially uncharged. Find the current through the 4-$\Omega$ resistor and the current through the 8-$\Omega$ resistor (a) immediately after the switch is closed, and (b) a long time after the switch is closed. (c) Find the charge on the capacitor a long time after the switch is closed.

**PICTURE THE PROBLEM** Since the capacitor is initially uncharged, and since the 4-$\Omega$ resistor limits the current through the battery, the initial potential difference across the capacitor is zero. The capacitor and the 8-$\Omega$ resistor are connected in parallel, and the difference in potential across each is the same. Thus, the initial potential difference across the 8-$\Omega$ resistor is also zero.

(a) Apply the loop rule to the outer loop and solve for the current through the 4-$\Omega$ resistor. The potential difference across the 8-$\Omega$ resistor and the capacitor are equal. Set the initial charge on the capacitor equal to zero and solve for the current through the 8-$\Omega$ resistor:

$$12 \text{ V} - (4 \text{ }\Omega)I_{4\Omega,0} + 0 = 0, I_{4\Omega,0} = \boxed{3 \text{ A}}$$

$$I_{8\Omega,0}(8 \text{ }\Omega) = \frac{Q_0}{C}, \quad I_{8\Omega,0} = \boxed{0}$$

(b) After a long time, the capacitor is fully charged (no more charge flows onto its the plates) and the current through both resistors is the same. Apply the loop rule to the left loop and solve for the current:

$$12 \text{ V} - (4 \text{ }\Omega)I_f - (8 \text{ }\Omega)I_f = 0$$

$$I_f = \boxed{1 \text{ A}}$$

(c) The potential difference across the 8-$\Omega$ resistor and the capacitor are equal. Use this to solve for $Q_f$:

$$I_f(8 \text{ }\Omega) = \frac{Q_f}{C}$$

$$Q_f = (1 \text{ A})(8 \text{ }\Omega)(6 \text{ }\mu\text{F}) = \boxed{48 \text{ }\mu\text{C}}$$

**REMARKS** The analysis of this circuit at the extreme times when the capacitor is either uncharged or fully charged is simple. When the capacitor is uncharged, it acts like a short circuit between points $c$ and $d$; that is, the circuit is the same as the one shown in Figure 25-44$a$, where we have replaced the capacitor by a wire of zero resistance. When the capacitor is fully charged, it acts like an open circuit, as shown in Figure 25-44$b$.

## Energy Conservation in Charging a Capacitor

During the charging process, a total charge $Q_f = \mathscr{E}C$ flows through the battery. The battery therefore does work

$$W = Q_f\mathscr{E} = C\mathscr{E}^2$$

Half of this work is accounted for by the energy stored in the capacitor (see Equation 24-12):

$$U = \tfrac{1}{2}Q_f\mathscr{E}$$

(a)

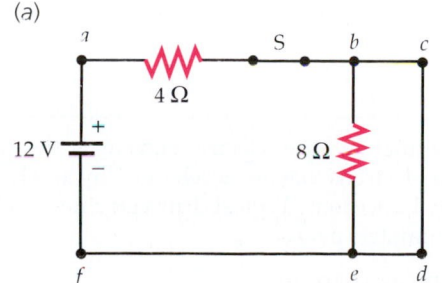

(b)

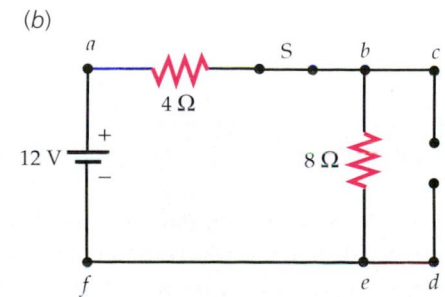

**FIGURE 25-44**

We now show that the other half of work done by the battery is dissipated as thermal energy by the resistance of the circuit. The rate at which energy is dissipated by the resistance $R$ is

$$\frac{dW_R}{dt} = I^2 R$$

Using Equation 25-37 for the current, we have

$$\frac{dW_R}{dt} = \left(\frac{\mathscr{E}}{R} e^{-t/(RC)}\right)^2 R = \frac{\mathscr{E}^2}{R} e^{-2t/(RC)}$$

We find the total energy dissipated by integrating from $t = 0$ to $t = \infty$:

$$W_R = \int_0^\infty \frac{\mathscr{E}^2}{R} e^{-2t/(RC)} \, dt = \frac{\mathscr{E}^2}{R} \int_0^\infty e^{-at} \, dt$$

where $a = 2/RC$. Thus,

$$W_R = \frac{\mathscr{E}^2}{R} \frac{e^{-at}}{-a} \bigg|_0^\infty = -\frac{\mathscr{E}^2}{Ra}(0 - 1) = \frac{\mathscr{E}^2}{R}\frac{1}{a} = \frac{\mathscr{E}^2}{R}\frac{RC}{2}$$

The total amount of Joule heating is thus

$$W_R = \frac{1}{2}\mathscr{E}^2 C = \frac{1}{2}Q_f \mathscr{E}$$

where $Q_f = \mathscr{E}C$. This result is independent of the resistance $R$. Thus, when a capacitor is charged through a resistor by a constant source of emf, half the energy provided by the source of emf is stored in the capacitor and half goes into thermal energy. This thermal energy includes the energy that goes into the internal resistance of the source of emf.

# SUMMARY

1. Ohm's law is an empirical law that holds only for certain materials.
2. Current, resistance, and emf are important *defined* quantities.
3. Kirchhoff's rules follow from the conservation of charge and the conservative nature of the electric field.

| Topic | Relevant Equations and Remarks |
|---|---|
| **1. Electric Current** | Electric current is the rate of flow of electric charge through a cross-sectional area. |
| | $$I = \frac{\Delta Q}{\Delta t} \qquad \text{25-1}$$ |
| Drift velocity | In a conducting wire, electric current is the result of the slow drift of negatively charged electrons that are accelerated by an electric field in the wire and then collide with the lattice ions. Typical drift velocities of electrons in wires are of the order of a few millimeters per second. |
| | $$I = qnAv_d \qquad \text{25-3}$$ |

## 2. Resistance

| | | |
|---|---|---|
| Definition of resistance | $R = \dfrac{V}{I}$ | 25-5 |
| Resistivity, $\rho$ | $R = \rho \dfrac{L}{A}$ | 25-8 |
| Temperature coefficient of resistivity, $\alpha$ | $\alpha = \dfrac{(\rho - \rho_{20})/\rho_{20}}{t_C - 20°C}$ | 25-9 |

## 3. Ohm's Law

For ohmic materials, the resistance does not depend on the current or the potential drop:

$$V = IR, \quad R \text{ constant} \qquad \text{25-7}$$

## 4. Power

| | | |
|---|---|---|
| Supplied to a device or segment | $P = IV$ | 25-10 |
| Dissipated in a resistor | $P = IV = I^2 R = \dfrac{V^2}{R}$ | 25-11 |

## 5. EMF

| | |
|---|---|
| Source of emf | A device that supplies electrical energy to a circuit. |

| | | |
|---|---|---|
| Power supplied by an emf source | $P = \mathcal{E}I$ | 25-12 |

## 6. Battery

| | |
|---|---|
| Ideal | An ideal battery is a source of emf that maintains a constant potential difference between its two terminals, independent of the current through the battery. |
| Real | A real battery can be considered as an ideal battery in series with a small resistance called its internal resistance. |

| | | |
|---|---|---|
| Terminal voltage | $V_a - V_b = \mathcal{E} - Ir$ | 25-13 |
| Total energy stored | $W = Q\mathcal{E}$ | 25-15 |

## 7. Equivalent Resistance

| | | |
|---|---|---|
| Resistors in series | $R_{eq} = R_1 + R_2 + R_3 + \ldots$ | 25-17 |
| Resistors in parallel | $\dfrac{1}{R_{eq}} = \dfrac{1}{R_1} + \dfrac{1}{R_2} + \dfrac{1}{R_3} + \ldots$ | 25-21 |

## 8. Kirchhoff's Rules

1. When any closed-circuit loop is traversed, the algebraic sum of the changes in potential must equal zero.
2. At any junction (branch point) in a circuit where the current can divide, the sum of the currents into the junction must equal the sum of the currents out of the junction.

## 9. Measuring Devices

| | |
|---|---|
| Ammeter | An ammeter is a very low resistance device that is placed in series with a circuit element to measure the current in the element. |

| Voltmeter | A voltmeter is a very high resistance device that is placed in parallel with a circuit element to measure the potential drop across the element. |
|---|---|
| Ohmmeter | An ohmmeter is a device containing a battery connected in series with a galvanometer and a resistor that is used to measure the resistance of a circuit element placed across its terminals. |

**10. Discharging a Capacitor**

| Charge on the capacitor | $Q(t) = Q_0 e^{-t/(RC)} = Q_0 e^{-t/\tau}$ | 25-31 |
|---|---|---|
| Current in the circuit | $I = -\dfrac{dQ}{dt} = \dfrac{V_0}{R} e^{-t/(RC)} = I_0 e^{-t/\tau}$ | 25-33 |
| Time constant | $\tau = RC$ | 25-32 |

**11. Charging a Capacitor**

| Charge on the capacitor | $Q = C\mathcal{E}(1 - e^{-t/(RC)}) = Q_f(1 - e^{-t/\tau})$ | 25-36 |
|---|---|---|
| Current in the circuit | $I = +\dfrac{dQ}{dt} = \dfrac{\mathcal{E}}{R} e^{-t/(RC)} = I_0 e^{-t/\tau}$ | 25-37 |

# PROBLEMS

- • Single-concept, single-step, relatively easy
- •• Intermediate-level, may require synthesis of concepts
- ••• Challenging
- **SSM** Solution is in the *Student Solutions Manual*
- **iSOLVE** Problems available on iSOLVE online homework service
- **iSOLVE✔** These "Checkpoint" online homework service problems ask students additional questions about their confidence level, and how they arrived at their answer.

In a few problems, you are given more data than you actually need; in a few other problems, you are required to supply data from your general knowledge, outside sources, or informed estimates.

## Conceptual Problems

**1** • **SSM** In our study of electrostatics, we concluded that there is no electric field within a conductor in electrostatic equilibrium. How is it that we can now discuss electric fields inside a conductor?

**2** • Figure 25-8 illustrates a mechanical analog of a simple electric circuit. Devise another mechanical analog in which the current is represented by a flow of water instead of marbles.

**3** • **iSOLVE✔** Two wires of the same material with the same length have different diameters. Wire A has twice the diameter of wire B. If the resistance of wire B is $R$, then what is the resistance of wire A? (a) $R$ (b) $2R$ (c) $R/2$ (d) $4R$ (e) $R/4$

**4** •• Discuss the difference between an emf and a potential difference.

**5** •• **SSM** A metal bar is to be used as a resistor. Its dimensions are 2 by 4 by 10 units. To get the smallest resistance from this bar, one should attach leads to the opposite sides that have the dimensions of

(a) 2 by 4 units.
(b) 2 by 10 units.
(c) 4 by 10 units.
(d) All connections will give the same resistance.
(e) None of the above is correct.

**6** •• Two cylindrical copper wires have the same mass. Wire A is twice as long as wire B. Their resistances are related by (a) $R_A = 8R_B$. (b) $R_A = 4R_B$. (c) $R_A = 2R_B$. (d) $R_A = R_B$.

**7** • A resistor carries a current $I$. The power dissipated in the resistor is $P$. What is the power dissipated if the same resistor carries current $3I$? (Assume no change in resistance.)
(a) $P$ (b) $3P$ (c) $P/3$ (d) $9P$ (e) $P/9$

**8** • The power dissipated in a resistor is $P$ when the potential drop across it is $V$. If the voltage drop is increased to 2 V (with no change in resistance), what is the power dissipated? (a) $P$ (b) $2P$ (c) $4P$ (d) $P/2$ (e) $P/4$

**9** • A heater consists of a variable resistance connected across a constant voltage supply. To increase the heat output, should you decrease the resistance or increase the resistance?

**10** • **SSM** Two resistors with resistances $R_1$ and $R_2$ are connected in parallel. If $R_1 \gg R_2$, the equivalent resistance of the combination is approximately (a) $R_1$. (b) $R_2$. (c) 0. (d) infinity.

**11** • Answer Problem 10 with resistors $R_1$ and $R_2$ connected in series.

**12** • Two resistors are connected in parallel across a potential difference. The resistance of resistor A is twice that of resistor B. If the current carried by resistor A is $I$, then what is the current carried by resistor B? (a) $I$ (b) $2I$ (c) $I/2$ (d) $4I$ (e) $I/4$

**13** • **SSM** Two resistors are connected in series across a potential difference. Resistor A has twice the resistance of resistor B. If the current carried by resistor A is $I$, then what is the current carried by resistor B? (a) $I$ (b) $2I$ (c) $I/2$ (d) $4I$ (e) $I/4$

**14** •• When two identical resistors are connected in series across the terminals of a battery, the power delivered by the battery is 20 W. If these resistors are connected in parallel across the terminals of the same battery, what is the power delivered by the battery? (a) 5 W (b) 10 W (c) 20 W (d) 40 W (e) 80 W

**15** • Kirchhoff's loop rule follows from (a) conservation of charge. (b) conservation of energy. (c) Newton's laws. (d) Coulomb's law. (e) quantization of charge.

**16** • An ideal voltmeter should have _____ internal resistance.

(a) infinite
(b) zero

**17** • **SSM** An ideal ammeter should have _____ internal resistance.

(a) infinite
(b) zero

**18** • An ideal voltage source should have _____ internal resistance.

(a) infinite
(b) zero

**19** • The capacitor $C$ in Figure 25-45 is initially uncharged. Just after the switch S is closed, (a) the voltage across $C$ equals $\mathcal{E}$. (b) the voltage across $R$ equals $\mathcal{E}$. (c) the current in the circuit is zero. (d) both (a) and (c) are correct.

**20** •• During the time it takes to fully charge the capacitor of Figure 25-45, (a) the energy supplied by the battery is $\frac{1}{2}C\mathcal{E}^2$. (b) the energy dissipated in the resistor is $\frac{1}{2}C\mathcal{E}^2$. (c) energy in the resistor is dissipated at a constant rate. (d) the total charge flowing through the resistor is $\frac{1}{2}C\mathcal{E}$.

**21** •• **SSM** A battery is connected to a series combination of a switch, a resistor, and an initially uncharged capacitor. The switch is closed at $t = 0$. Which of the following statements is true?

(a) As the charge on the capacitor increases, the current increases.
(b) As the charge on the capacitor increases, the voltage drop across the resistor increases.
(c) As the charge on the capacitor increases, the current remains constant.
(d) As the charge on the capacitor increases, the voltage drop across the capacitor decreases.
(e) As the charge on the capacitor increases, the voltage drop across the resistor decreases.

**22** •• A capacitor is discharging through a resistor. If it takes a time $T$ for the charge on a capacitor to drop to half its initial value, how long does it take for the energy to drop to half its initial value?

**23** • Which will produce more thermal energy when connected across an ideal battery, a small resistance or a large resistance?

**24** • **SSM** All voltage sources have some internal resistance, usually on the order of 100 Ω or less. From this fact, explain the following statement that appears in some electronics textbooks: "A voltage source likes to see a high resistance."

**25** • Do Kirchhoff's rules apply to circuits containing capacitors?

**26** •• In Figure 25-46, all three resistors are identical. The power dissipated is (a) the same in $R_1$ as in the parallel combination of $R_2$ and $R_3$. (b) the same in $R_1$ and $R_2$. (c) greatest in $R_1$. (d) smallest in $R_1$.

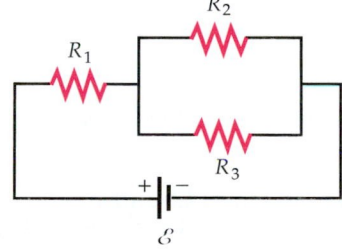

**FIGURE 25-46** Problem 26

### Estimation and Approximation

**27** •• A 16-gauge copper wire insulated with rubber can safely carry a maximum current of 6 A. (a) How great a potential difference can be applied across 40 m of this wire? (b) Find the electric field in the wire when it carries a current of 6 A. (c) Find the power dissipated in the wire when it carries a current of 6 A.

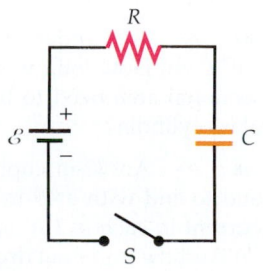

**FIGURE 25-45** Problems 19 and 20

**28** •• An automobile jumper cable 3 m long is constructed of multiple strands of copper wire that has an equivalent cross-sectional area of 10 mm². (*a*) What is the resistance of the jumper cable? (*b*) When the cable is used to start a car, it carries a current of 90 A. What is the potential drop that occurs across the jumper cable? (*c*) How much power is dissipated in the jumper cable?

**29** •• A coil of Nichrome wire is to be used as the heating element in a water boiler that is required to generate 8 g of steam per second. The wire has a diameter of 1.80 mm and is connected to a 120-V power supply. Find the length of wire required.

**30** •• SSM Compact fluorescent lightbulbs cost $6 each and have an expected lifetime of 8000 h. These bulbs consume 20 W of power, but produce the illumination equivalent to 75-W incandescent bulbs. Incandescent bulbs cost approximately $1.50 each and have an expected lifetime of 1200 h. If the average household has, on the average, six 75-W incandescent lightbulbs on constantly, and if energy costs 11.5 cents per kilowatt-hour, how much money would a consumer save each year by installing the energy-efficient fluorescent lightbulbs?

**31** •• The wires in a house must be large enough in diameter so that they do not get hot enough to start a fire. Suppose a certain wire is to carry a current of 20 A, and it is determined that the joule heating of the wire should not exceed 2 W/m. What diameter must a copper wire have to be safe for this current?

**32** •• SSM A laser diode used in making a laser pointer is a highly nonlinear circuit element. For a voltage drop across it less than approximately 2.3 V, it behaves as if it has effectively infinite internal resistance, but for voltages across it higher than this it has a very low internal resistance—effectively zero. (*a*) A laser pointer is made by putting two 1.55 V watch batteries in series across the laser diode. If the batteries each have an internal resistance between 100 Ω and 150 Ω, estimate the current in the laser diode. (*b*) About half of the power delivered to the laser diode goes into radiant energy. Using this fact, estimate the power of the laser diode, and compare this to typical quoted values of about 3 mW. (*c*) If the batteries each have a capacity of 20-mA hours (i.e., they can deliver a constant current of 20 mA for approximately one hour before discharging), estimate how long one can continuously operate the laser pointer before replacing the batteries.

## Current and the Motion of Charges

**33** • ISOLVE A 10-gauge copper wire carries a current of 20 A. Assuming one free electron per copper atom, calculate the drift velocity of the electrons.

**34** • ISOLVE In a fluorescent tube of diameter 3 cm, $2.0 \times 10^{18}$ electrons and $0.5 \times 10^{18}$ positive ions (with a charge of $+e$) flow through a cross-sectional area each second. What is the current in the tube?

**35** • In a certain electron beam, there are $5.0 \times 10^6$ electrons per cubic centimeter. Suppose the kinetic energy of each electron is 10 keV and the beam is cylindrical with a diameter of 1 mm. (*a*) What is the velocity of an electron in the beam? (*b*) Find the beam current.

**36** •• A ring of radius $a$ with a linear charge density $\lambda$ rotates about its axis with angular velocity $\omega$. Find an expression for the current.

**37** •• SSM ISOLVE A 10-gauge copper wire and a 14-gauge copper wire are welded together end to end. The wires carry a current of 15 A. If there is one free electron per copper atom in each wire, find the drift velocity of the electrons in each wire.

**38** •• In a certain particle accelerator, a proton beam with a diameter of 2 mm constitutes a current of 1 mA. The kinetic energy of each proton is 20 MeV. The beam strikes a metal target and is absorbed by it. (*a*) What is the number $n$ of protons per unit volume in the beam? (*b*) How many protons strike the target in 1 minute? (*c*) If the target is initially uncharged, express the charge of the target as a function of time.

**39** •• SSM In a proton supercollider, the protons in a 5-mA beam move with nearly the speed of light. (*a*) How many protons are there per meter of the beam? (*b*) If the cross-sectional area of the beam is $10^{-6}$ m², what is the number density of protons?

## Resistance and Ohm's Law

**40** • A 10-m-long wire of resistance 0.2 Ω carries a current of 5 A. (*a*) What is the potential difference across the wire? (*b*) What is the magnitude of the electric field in the wire?

**41** • ISOLVE A potential difference of 100 V produces a current of 3 A in a certain resistor. (*a*) What is the resistance of the resistor? (*b*) What is the current when the potential difference is 25 V?

**42** • ISOLVE A block of carbon is 3.0 cm long and has a square cross-sectional area with sides of 0.5 cm. A potential difference of 8.4 V is maintained across its length. (*a*) What is the resistance of the block? (*b*) What is the current in this resistor?

**43** • ISOLVE A carbon rod with a radius of 0.1 mm is used to make a resistor. The resistivity of this material is $3.5 \times 10^{-5}$ Ω·m. What length of the carbon rod will make a 10-Ω resistor?

**44** • SSM The third (current-carrying) rail of a subway track is made of steel and has a cross-sectional area of about 55 cm². The resistivity of steel is $10^{-7}$ Ω·m. What is the resistance of 10 km of this track?

**45** • ISOLVE What is the potential difference across one wire of a 30-m extension cord made of 16-gauge copper wire carrying a current of 5 A?

**46** • ISOLVE How long is a 14-gauge copper wire that has a resistance of 2 Ω?

**47** •• A cylinder of glass 1 cm long has a resistivity of $10^{12}$ Ω·m. How long would a copper wire of the same cross-sectional area need to be to have the same resistance as the glass cylinder?

**48** •• An 80-m copper wire 1 mm in diameter is joined end to end with a 49-m iron wire of the same diameter. The current in each is 2 A. (*a*) Find the electric field in each wire. (*b*) Find the potential drop across each wire.

**49** •• [SSM] A copper wire and an iron wire with the same length and diameter carry the same current $I$. (a) Find the ratio of the potential drops across these wires. (b) In which wire is the electric field greater?

**50** •• A rubber tube 1 m long with an inside diameter of 4 mm is filled with a salt solution that has a resistivity of $10^{-3}$ $\Omega \cdot$m. Metal plugs form electrodes at the ends of the tube. (a) What is the resistance of the filled tube? (b) What is the resistance of the filled tube if it is uniformly stretched to a length of 2 m?

**51** •• A wire of length 1 m has a resistance of 0.3 $\Omega$. It is uniformly stretched to a length of 2 m. What is its new resistance?

**52** •• Currents up to 30 A can be carried by 10-gauge copper wire. (a) What is the resistance of 100 m of 10-gauge copper wire? (b) What is the electric field in the wire when the current is 30 A? (c) How long does it take for an electron to travel 100 m in the wire when the current is 30 A?

**53** •• A cube of copper has sides of 2 cm. If it is drawn out to form a 14-gauge wire, what will its resistance be?

**54** •• [SSM] A diode is a circuit element with a very nonlinear $IV$ curve. In a diode, $I = I_0(e^{V/(25\ \text{mV})} - 1)$, where $I_0 \sim 2 \times 10^{-9}$ A. Using a spreadsheet program, make a graph of $I$ versus $V$ for a typical diode, for both forward biasing ($V > 0$) and back-biasing ($V < 0$). Show that a plot $\ln(I)$ versus $V$ for forward biasing (using $V > 0.3$ V), is nearly a straight line. What is the slope of the line?

**55** •• (a) From the results of Problem 54, show that a diode effectively behaves like a resistor with infinite resistance if the voltage $V$ applied across the diode is less than approximately 0.6 V, and behaves like a resistor with zero resistance if $V > 0.6$ V. (b) Estimate the current flowing through the forward biased diode in the circuit shown in Figure 25-47.

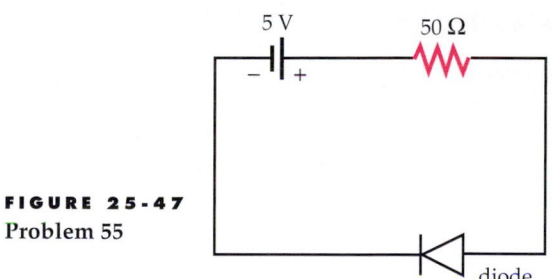

**FIGURE 25-47**
**Problem 55**

**56** ••• Find the resistance between the ends of the half ring shown in Figure 25-48. The resistivity of the material of the ring is $\rho$.

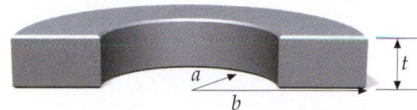

**FIGURE 25-48** Problem 56

**57** ••• The radius of a wire of length $L$ increases linearly along its length according to $r = a + [(b - a)/L]x$, where $x$ is the distance from the small end of radius $a$. What is the resistance of this wire in terms of its resistivity $\rho$, length $L$, radius $a$, and radius $b$?

**58** ••• [SSM] The space between two concentric spherical-shell conductors is filled with a material that has a resistivity of $10^9$ $\Omega \cdot$m. If the inner shell has a radius of 1.5 cm and the outer shell has a radius of 5 cm, what is the resistance between the conductors? (Hint: Find the resistance of a spherical-shell element of the material of area $4\pi r^2$ and length $dr$, and integrate to find the total resistance of the set of shells in series.)

**59** ••• The space between two metallic coaxial cylinders of length $L$ and radii $a$ and $b$ is completely filled with a material having a resistivity $\rho$. (a) What is the resistance between the two cylinders? (See the hint in Problem 58.) (b) Find the current between the two cylinders if $\rho = 30$ $\Omega \cdot$m, $a = 1.5$ cm, $b = 2.5$ cm, $L = 50$ cm, and a potential difference of 10 V is maintained between the two cylinders.

## Temperature Dependence of Resistance

**60** • [SSM] [ISOLVE ✓] A tungsten rod is 50 cm long and has a square cross-sectional area with sides of 1 mm. (a) What is its resistance at 20°C? (b) What is its resistance at 40°C?

**61** • At what temperature will the resistance of a copper wire be 10 percent greater than it is at 20°C?

**62** •• A toaster with a Nichrome heating element has a resistance of 80 $\Omega$ at 20°C and an initial current of 1.5 A. When the heating element reaches its final temperature, the current is 1.3 A. What is the final temperature of the heating element?

**63** •• An electric space heater has a Nichrome heating element with a resistance of 8 $\Omega$ at 20°C. When 120 V are applied, the electric current heats the Nichrome wire to 1000°C. (a) What is the initial current drawn by the cold heating element? (b) What is the resistance of the heating element at 1000°C? (c) What is the operating wattage of this heater?

**64** •• A 10-$\Omega$ Nichrome resistor is wired into an electronic circuit using copper leads (wires) of diameter 0.6 mm, with a total length of 50 cm. (a) What additional resistance is due to the copper leads? (b) What percentage error in the total added resistance is produced by neglecting the resistance of the copper leads? (c) What change in temperature would produce a change in resistance of the Nichrome wire equal to the resistance of the copper leads?

**65** ••• [SSM] A wire of cross-sectional area $A$, length $L_1$, resistivity $\rho_1$, and temperature coefficient $\alpha_1$ is connected end to end to a second wire of the same cross-sectional area, length $L_2$, resistivity $\rho_2$, and temperature coefficient $\alpha_2$, so that the wires carry the same current. (a) Show that if $\rho_1 L_1 \alpha_1 + \rho_2 L_2 \alpha_2 = 0$, the total resistance $R$ is independent of temperature for small temperature changes. (b) If one wire is made of carbon and the other wire is made of copper, find the ratio of their lengths for which $R$ is approximately independent of temperature.

**66** ••• The resistivity of tungsten increases approximately linearly from 56 n$\Omega \cdot$m at 293 K and 1.1 $\mu\Omega \cdot$m at 3500 K. Estimate (a) the resistance and (b) the diameter of a tungsten filament used in a 40-W bulb, assuming that the filament temperature is about 2500 K and that a 100-V dc supply is used to power the lightbulb. Assume that the length of the filament is constant and equal to 0.5 cm.

**67** ••• A small light bulb used in an electronics class has a carbon filament in the form of a cylinder with a length of 3 cm and a diameter of $d = 40\mu$m. At temperatures between 500K and 700K, the resistivity of the carbon used in making small light bulb filaments is about $3 \times 10^{-5}$ $\Omega \cdot$m. (a) Assuming that the bulb is a perfect blackbody radiator, calculate the temperature of the filament when a voltage $V = 5$ V is placed across it. (b) One problem with carbon filament bulbs, unlike tungsten filament bulbs, is that the resistivity of carbon decreases with increasing temperature. Explain why this is a problem.

## Energy in Electric Circuits

**68** • [SSM] Find the power dissipated in a resistor connected across a constant potential difference of 120 V if its resistance is (a) 5 $\Omega$ and (b) 10 $\Omega$.

**69** • [SOLVE] A 10,000-$\Omega$ carbon resistor used in electronic circuits is rated at 0.25 W. (a) What maximum current can this resistor carry? (b) What maximum voltage can be placed across this resistor?

**70** • A 1-kW heater is designed to operate at 240 V. (a) What is the heater's resistance and what current does the heater draw? (b) What is the power dissipated in this resistor if it operates at 120 V? Assume that its resistance is constant.

**71** • A battery has an emf of 12 V. How much work does it do in 5 s if it delivers a current of 3 A?

**72** • [SOLVE] A battery with 12-V emf has a terminal voltage of 11.4 V when it delivers a current of 20 A to the starter of a car. What is the internal resistance $r$ of the battery?

**73** • [SSM] (a) How much power is delivered by the emf of the battery in Problem 72 when it delivers a current of 20 A? (b) How much of this power is delivered to the starter? (c) By how much does the chemical energy of the battery decrease when it delivers a current of 20 A to the starter for 3 min? (d) How much heat is developed in the battery when it delivers a current of 20 A for 3 min?

**74** • A battery with an emf of 6 V and an internal resistance of 0.3 $\Omega$ is connected to a variable resistance $R$. Find the current and power delivered by the battery when $R$ is (a) 0, (b) 5 $\Omega$, (c) 10 $\Omega$, and (d) infinite.

**75** •• A 12-V automobile battery with negligible internal resistance can deliver a total charge of 160 A·h. (a) What is the total stored energy in the battery? (b) How long could this battery provide 150 W to a pair of headlights?

**76** •• A space heater in an old home draws a 12.5-A current. A pair of 12-gauge copper wires carries the current from the fuse box to the wall outlet, a distance of 30 m. The voltage at the fuse box is exactly 120 V. (a) What is the voltage delivered to the space heater? (b) If the fuse will blow at a current of 20 A, how many 60-W bulbs can be supplied by this line when the space heater is on? (Assume that the wires from the wall to the space heater and to the light fixtures have negligible resistance.)

**77** •• [SSM] [SOLVE] A lightweight electric car is powered by ten 12-V batteries. At a speed of 80 km/h, the average frictional force is 1200 N. (a) What must be the power of the electric motor if the car is to travel at a speed of 80 km/h? (b) If each battery can deliver a total charge of 160 A·h before recharging, what is the total charge in coulombs that can be delivered by the ten batteries before charging? (c) What is the total electrical energy delivered by the ten batteries before recharging? (d) How far can the car travel at 80 km/h before the batteries must be recharged? (e) What is the cost per kilometer if the cost of recharging the batteries is 9 cents per kilowatt-hour?

**78** ••• [SOLVE]✓ A 100-W heater is designed to operate with an applied voltage of 120 V. (a) What is the heater's resistance, and what current does the heater draw? (b) Show that if the potential difference $V$ across the heater changes by a small amount $\Delta V$, the power $P$ changes by a small amount $\Delta P$, where $\Delta P/P \approx 2\Delta V/V$. (Hint: Approximate the changes with differentials and assume the resistance is constant.) (c) Find the approximate power dissipated in the heater, if the potential difference is decreased to 115 V.

## Combinations of Resistors

**79** • [SSM] (a) Find the equivalent resistance between point $a$ and point $b$ in Figure 25-49. (b) If the potential drop between point $a$ and point $b$ is 12 V, find the current in each resistor.

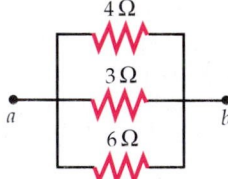

**FIGURE 25-49** Problem 79

**80** • Repeat Problem 79 for the resistor network shown in Figure 25-50.

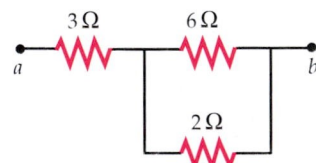

**FIGURE 25-50** Problem 80

**81** • (a) Show that the equivalent resistance between point $a$ and point $b$ in Figure 25-51 is $R$. (b) What would be the effect of adding a resistance $R$ between point $c$ and point $d$?

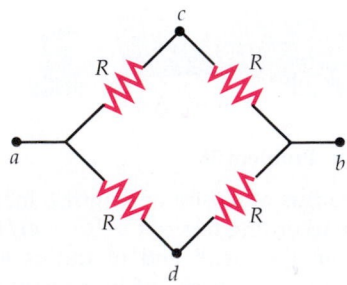

**FIGURE 25-51** Problem 81

**82** •• The battery in Figure 25-52 has negligible internal resistance. Find (*a*) the current in each resistor and (*b*) the power delivered by the battery.

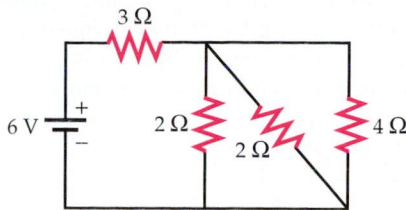

**FIGURE 25-52 Problem 82**

**83** •• **SSM** A 5-V power supply has an internal resistance of 50 Ω. What is the smallest resistor that we can put in series with the power supply so that the voltage drop across the resistor is larger than 4.5 V?

**84** •• A battery has an emf $\mathcal{E}$ and an internal resistance *r*. When a 5-Ω resistor is connected across the terminals, the current is 0.5 A. When this resistor is replaced by an 11-Ω resistor, the current is 0.25 A. Find (*a*) the emf $\mathcal{E}$ and (*b*) the internal resistance *r*.

**85** •• Consider the equivalent resistance of two resistors $R_1$ and $R_2$ connected in parallel as a function $R_1$ and *x*, where *x* is the ratio $R_2/R_1$. (*a*) Show that $R_{eq} = R_1 x/(1 + x)$. (*b*) Sketch a plot of $R_{eq}/R_1$ as a function of *x*.

**86** •• An ideal current source supplies a constant current regardless of the *load* that it is attached to. An almost-ideal current source can be made by putting a large resistor in series with an ideal voltage source. (*a*) What resistance is needed to turn an ideal 5-V voltage source into an almost-ideal 10-mA current source? (*b*) If we wish the current to drop by less than 10 percent when we load this current source, what is the largest resistance we can place in series with this current source?

**87** •• Repeat Problem 79 for the resistor network shown in Figure 25-53.

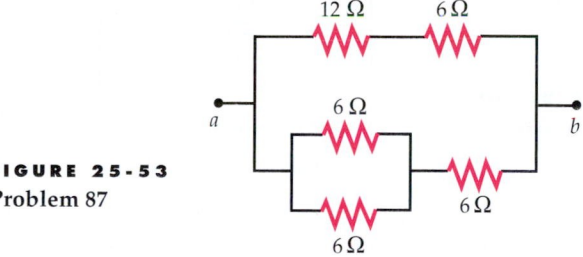

**FIGURE 25-53
Problem 87**

**88** •• Repeat Problem 79 for the resistor network shown in Figure 25-54.

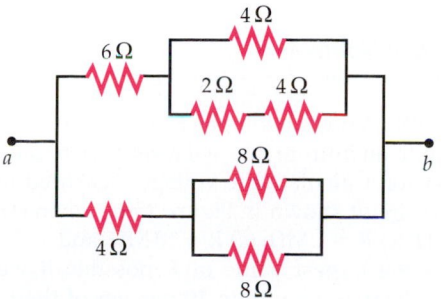

**FIGURE 25-54 Problem 88**

**89** •• **SSM** A length of wire has a resistance of 120 Ω. The wire is cut into *N* identical pieces that are then connected in parallel. The resistance of the parallel arrangement is 1.875 Ω. Find *N*.

**90** •• A parallel combination of an 8-Ω resistor and an unknown resistor *R* is connected in series with a 16-Ω resistor and a battery. This circuit is then disassembled and the three resistors are then connected in series with each other and the same battery. In both arrangements, the current through the 8-Ω resistor is the same. What is the unknown resistance *R*?

**91** •• **iSOLVE** For the resistance network shown in Figure 25-55, find (*a*) $R_3$, so that $R_{ab} = R_1$, (*b*) $R_2$, so that $R_{ab} = R_3$; and (*c*) $R_1$, so that $R_{ab} = R_1$.

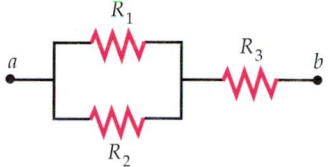

**FIGURE 25-55 Problems 91 and 92**

**92** •• Check your results for Problem 91 using (*a*) $R_1 = 4\,\Omega$, $R_2 = 6\,\Omega$; (*b*) $R_1 = 4\,\Omega$, $R_3 = 3\,\Omega$; and (*c*) $R_2 = 6\,\Omega$, $R_3 = 3\,\Omega$.

## Kirchhoff's Rules

**93** • **SSM** In Figure 25-56, the emf is 6 V and $R = 0.5\,\Omega$. The rate of joule heating in *R* is 8 W. (*a*) What is the current in the circuit? (*b*) What is the potential difference across *R*? (*c*) What is *r*?

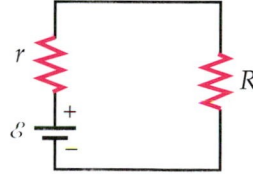

**FIGURE 25-56 Problem 93**

**94** • For the circuit in Figure 25-57, find (*a*) the current, (*b*) the power delivered or absorbed by each source of emf, and (*c*) the rate of joule heating in each resistor. (Assume that the batteries have negligible internal resistance.)

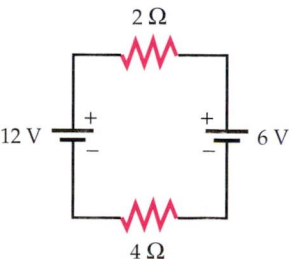

**FIGURE 25-57 Problem 94**

**95** •• A sick car battery with an emf of 11.4 V and an internal resistance of 0.01 Ω is connected to a load of 2 Ω. To help the ailing battery, a second battery with an emf of 12.6 V and an internal resistance of 0.01 Ω is connected by jumper cables to the terminals of the first battery. (*a*) Draw a diagram of this circuit. (*b*) Find the current in each part of the circuit. (*c*) Find the power delivered by the second battery and discuss where this power goes, assuming that the emfs and internal resistances of both batteries remain constant.

**96** •• In the circuit in Figure 25-58, the reading of the ammeter is the same with both switches open and both switches closed. Find the resistance R.

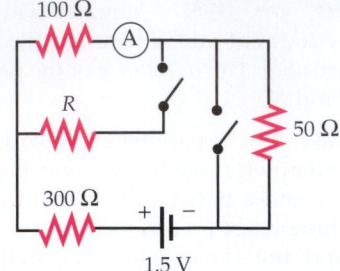

**FIGURE 25-58** Problem 96

**97** •• SSM In the circuit shown in Figure 25-59, the batteries have negligible internal resistance. Find (*a*) the current in each resistor, (*b*) the potential difference between point *a* and point *b*, and (*c*) the power supplied by each battery.

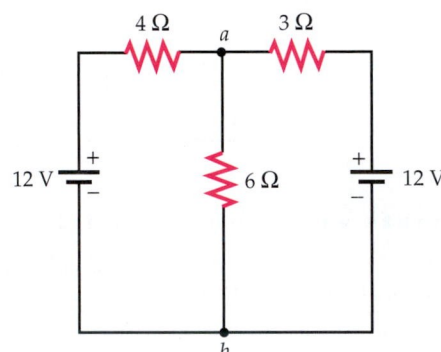

**FIGURE 25-59**
Problem 97

**98** •• Repeat Problem 97 for the circuit in Figure 25-60.

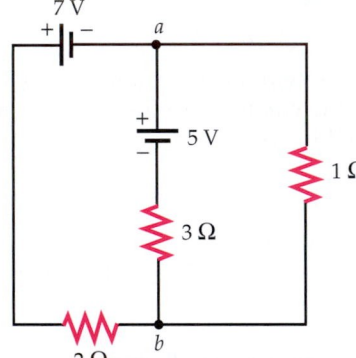

**FIGURE 25-60**
Problem 98

**99** •• Two identical batteries, each with an emf $\mathcal{E}$ and an internal resistance $r$, can be connected across a resistance $R$ either in series or in parallel. Is the power supplied to $R$ greater when $R < r$ or when $R > r$?

**100** •• SSM The circuit fragment shown in Figure 25-61 is called a *voltage divider*. (*a*) If $R_{\text{load}}$ is not attached, show that $V_{\text{out}} = VR_2/(R_1 + R_2)$. (*b*) If $R_1 = R_2 = 10 \text{ k}\Omega$, what is the smallest value of $R_{\text{load}}$ that can be used so that $V_{\text{out}}$ drops by less than 10 percent from its unloaded value? ($V_{\text{out}}$ is measured with respect to ground.)

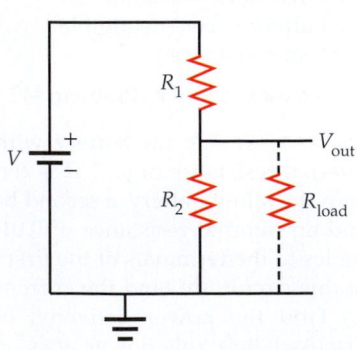

**FIGURE 25-61** Problem 100

**101** •• Thevenin's theorem states that the voltage divider circuit of Problem 100 can be replaced by a constant voltage source with voltage $V'$ in series with a Thevenin resistance $R'$ in series with the load resistor $R_{\text{load}}$. $V'$ and $R'$ depend only on $V$, $R_1$ and $R_2$. In this arrangement, the voltage drop across $R_{\text{load}}$ will be the same as if the load resistor were placed in parallel with $R_2$ in the voltage divider from Problem 100.

(*a*) Show that $R' = \dfrac{R_1 R_2}{R_1 + R_2}$.

(*b*) Show that $V' = V\dfrac{R_2}{R_1 + R_2}$.

**102** •• For the circuit shown in Figure 25-62, find (*a*) the current in each resistor, (*b*) the power supplied by each source of emf, and (*c*) the power dissipated in each resistor.

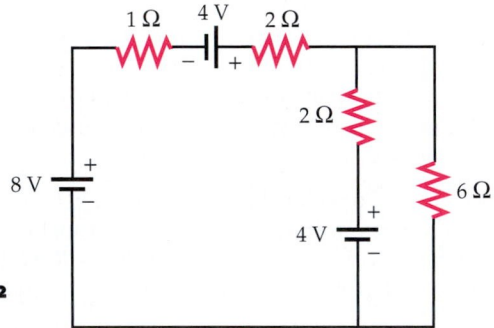

**FIGURE 25-62**
Problem 102

**103** •• For the circuit shown in Figure 25-63, find the potential difference between point *a* and point *b*.

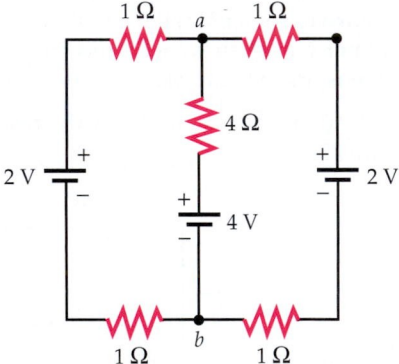

**FIGURE 25-63**
Problem 103

**104** •• You have two batteries, one with $\mathcal{E} = 9$ V and $r = 0.8 \ \Omega$ and the other battery with $\mathcal{E} = 3$ V and $r = 0.4 \ \Omega$. (*a*) Show how you would connect the batteries to give the largest current through a resistor $R$. Find the current for (*b*) $R = 0.2 \ \Omega$, (*c*) $R = 0.6 \ \Omega$, (*d*) $R = 1.0 \ \Omega$, and (*e*) $R = 1.5 \ \Omega$.

## Ammeters and Voltmeters

**105** •• SSM A digital voltmeter can be modeled as an ideal voltmeter with an infinite internal resistance in parallel with a 10 M·$\Omega$ resistor. Calculate the voltage measured by the voltmeter in the circuit shown in Figure 25-64 when (*a*) $R = 1 \text{ k}\Omega$, (*b*) $R = 10 \text{ k}\Omega$, (*c*) $R = 1 \text{ M}\Omega$, (*d*) $R = 10 \text{ M}\Omega$, and (*e*) $R = 100 \text{ M}\Omega$. (*f*) What is the largest value of $R$ possible if we wish the measured voltage to be within 10 percent of the *true* voltage (i.e., the voltage drop without the voltmeter in place)?

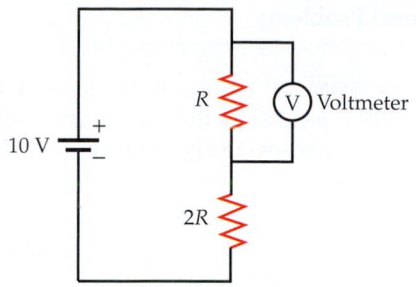

**FIGURE 25-64**
Problem 105

**106** •• You are given a galvanometer meter movement that will deflect full scale if a current of 50 μA runs through the galvanometer. At this current, there is a voltage drop of 0.25 V across the meter. What is the meter's internal resistance?

**107** •• We wish to change the meter in Problem 106 into an ammeter that can measure currents up to 100 mA. Show that this can be done by placing a resistor in parallel with the meter, and find the value of its resistance.

**108** •• (a) If the ammeter from Problem 107 is used to measure the current through a 100-Ω resistor that is hooked up to a 10-V power supply, what current will the meter read? (The question is not as simple as it sounds.) (b) What if the ammeter is used to measure the current flowing through a 10-Ω resistor that is hooked up to a 1-V power supply?

**109** •• [SSM] Show that the meter movement in Problem 106 can be converted into a voltmeter by placing a large resistance in series with the meter movement, and find the resistance needed for a full-scale deflection when 10 V are placed across it.

**110** •• If the voltmeter described in Problem 109 is used to measure the voltage drop across $R_1$ in the circuit shown in Figure 25-65, what voltage will the voltmeter read?

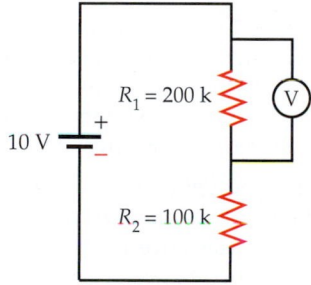

**FIGURE 25-65** Problem 110

## RC Circuits

**111** • [SOLVE] A 6-μF capacitor is charged to 100 V and is then connected across a 500-Ω resistor. (a) What is the initial charge on the capacitor? (b) What is the initial current just after the capacitor is connected to the resistor? (c) What is the time constant of this circuit? (d) How much charge is on the capacitor after 6 ms?

**112** • (a) Find the initial energy stored in the capacitor of Problem 111. (b) Show that the energy stored in the capacitor is given by $U = U_0 e^{-2t/\tau}$, where $U_0$ is the initial energy and $\tau = RC$ is the time constant. (c) Sketch a plot of the energy $U$ in the capacitor versus time $t$.

**113** •• [SSM] In the circuit previously shown in Figure 25-40, emf $\mathcal{E} = 50$ V and $C = 2.0$ μF; the capacitor is initially uncharged. At 4 s after switch S is closed, the voltage drop across the resistor is 20 V. Find the resistance of the resistor.

**114** •• [SSM] [SOLVE✓] A 0.12-μF capacitor is given a charge $Q_0$. After 4 s, the capacitor's charge is $\frac{1}{2}Q_0$. What is the effective resistance across this capacitor?

**115** •• [SOLVE✓] A 1.6-μF capacitor, initially uncharged, is connected in series with a 10-kΩ resistor and a 5-V battery of negligible internal resistance. (a) What is the charge on the capacitor after a very long time? (b) How long does it take the capacitor to reach 99 percent of its final charge?

**116** •• Consider the circuit shown in Figure 25-66. From your knowledge of how capacitors behave in circuits, find (a) the initial current through the battery just after the switch is closed, (b) the steady-state current through the battery when the switch has been closed for a long time, and (c) the maximum voltage across the capacitor.

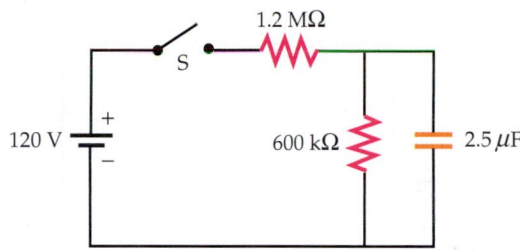

**FIGURE 25-66** Problem 116

**117** •• A 2-MΩ resistor is connected in series with a 1.5-μF capacitor and a 6.0-V battery of negligible internal resistance. The capacitor is initially uncharged. After a time $t = \tau = RC$, find (a) the charge on the capacitor, (b) the rate at which the charge is increasing, (c) the current, (d) the power supplied by the battery, (e) the power dissipated in the resistor, and (f) the rate at which the energy stored in the capacitor is increasing.

**118** •• In the steady state, the charge on the 5-μF capacitor in the circuit shown in Figure 25-67 is 1000 μC. (a) Find the battery current. (b) Find the resistances $R_1$, $R_2$, and $R_3$.

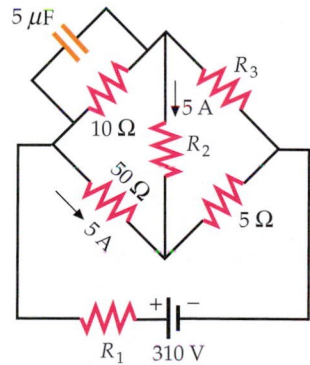

**FIGURE 25-67** Problem 118

**119** •• Show that Equation 25-35 can be written

$$\frac{dQ}{\mathcal{E}C - Q} = \frac{dt}{RC}$$

Integrate this equation to derive the solution given by Equation 25-36.

**120 •••** [SSM] A photojournalist's flash unit uses a 9-V battery pack to charge a 0.15-$\mu$F capacitor, which is then discharged through the flash lamp of 10.5-$\Omega$ resistance when a switch is closed. The minimum voltage necessary for the flash discharge is 7 V. The capacitor is charged through an 18-k$\Omega$ resistor. (a) How much time is required to charge the capacitor to the required 7 V? (b) How much energy is released when the lamp flashes? (c) How much energy is supplied by the battery during the charging cycle and what fraction of that energy is dissipated in the resistor?

**121 •••** For the circuit shown in Figure 25-68, (a) what is the initial battery current immediately after switch S is closed? (b) What is the battery current a long time after switch S is closed? (c) What is the current in the 600-$\Omega$ resistor as a function of time?

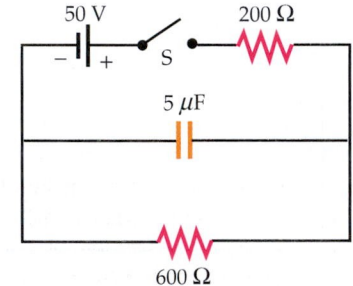

**FIGURE 25-68**
**Problem 121**

**122 •••** For the circuit shown in Figure 25-69, (a) what is the initial battery current immediately after switch S is closed? (b) What is the battery current a long time after switch S is closed? (c) If the switch has been closed for a long time and is then opened, find the current through the 600-k$\Omega$ resistor as a function of time.

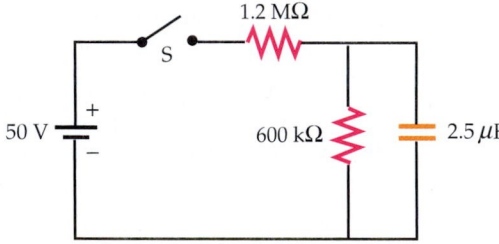

**FIGURE 25-69** **Problem 122**

**123 •••** In the circuit shown in Figure 25-70, the capacitor has a capacitance of 2.5 $\mu$F and the resistor has a resistance of 0.5 M$\Omega$. Before the switch is closed, the potential drop across the capacitor is 12 V, as shown. Switch S is closed at $t = 0$. (a) What is the current in R immediately after switch S is closed? (b) At what time $t$ is the voltage across the capacitor 24 V?

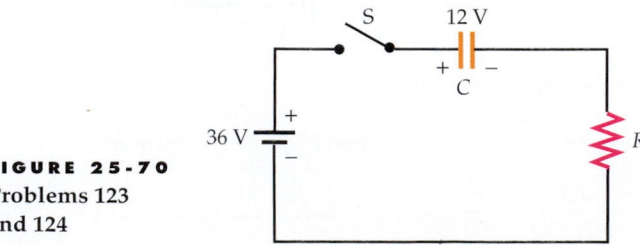

**FIGURE 25-70**
**Problems 123**
**and 124**

**124 •••** Repeat Problem 123 if the capacitor is connected with reversed polarity.

## General Problems

**125 ••** [SSM] In Figure 25-71, $R_1 = 4\ \Omega$, $R_2 = 6\ \Omega$, and $R_3 = 12\ \Omega$. If we denote the currents through these resistors by $I_1$, $I_2$, and $I_3$, respectively, then (a) $I_1 > I_2 > I_3$. (b) $I_2 = I_3$. (c) $I_3 > I_2$. (d) none of the above is correct.

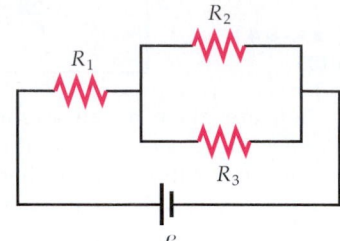

**FIGURE 25-71**
**Problems 125 and 127**

**126 ••** A 25-W lightbulb is connected in series with a 100-W lightbulb and a voltage V is placed across the combination. Which lightbulb is brighter? Explain.

**127 •** If the battery emf in Figure 25-71 is 24 V and $R_1 = 4\ \Omega$, $R_2 = 6\ \Omega$, and $R_3 = 12\ \Omega$, then (a) $I_2 = 4$ A. (b) $I_2 = 2$ A. (c) $I_2 = 1$ A. (d) none of the above is correct.

**128 •** A 10-$\Omega$ resistor is rated as being capable of dissipating 5 W of power. (a) What maximum current can this resistor tolerate? (b) What voltage across this resistor will produce the maximum current?

**129 •** A 12-V car battery has an internal resistance of 0.4 $\Omega$. (a) What is the current if the battery is shorted momentarily? (b) What is the terminal voltage when the battery delivers a current of 20 A to start the car?

**130 ••** The current drawn from a battery is 1.80 A when a 7-$\Omega$ resistor is connected across the battery terminals. If a second 12-$\Omega$ resistor is connected in parallel with the 7-$\Omega$ resistor, the battery delivers a current of 2.20 A. What are the emf and internal resistance of the battery?

**131 ••** [SSM] A closed box has two metal terminals $a$ and $b$. The inside of the box contains an unknown emf $\mathcal{E}$ in series with a resistance R. When a potential difference of 21 V is maintained between terminal $a$ and terminal $b$, there is a current of 1 A between the terminals $a$ and $b$. If this potential difference is reversed, a current of 2 A in the reverse direction is observed. Find $\mathcal{E}$ and R.

**132 ••** The capacitors in the circuit shown in Figure 25-72 are initially uncharged. (a) What is the initial value of the battery current when switch S is closed? (b) What is the battery current after a long time? (c) What are the final charges on the capacitors?

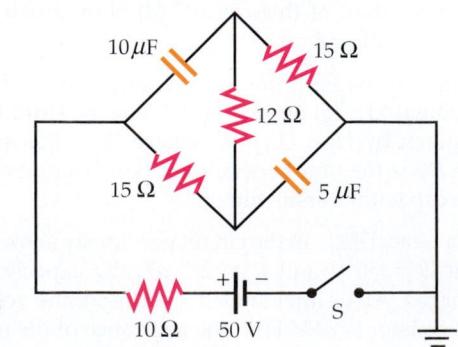

**FIGURE 25-72**
**Probles 132**

**133** •• **SSM** **iSOLVE✓** The circuit shown in Figure 25-73 is a slide-type *Wheatstone bridge*. This bridge is used to determine an unknown resistance $R_x$, in terms of the known resistances $R_1$, $R_2$, and $R_0$. The resistances $R_1$ and $R_2$ comprise a wire 1 m long. Point $a$ is a sliding contact that is moved along the wire to vary these resistances. Resistance $R_1$ is proportional to the distance from the left end of the wire (labeled 0 cm) to point $a$, and $R_2$ is proportional to the distance from point $a$ to the right end of the wire (labeled 100 cm). The sum of $R_1$ and $R_2$ remains constant. When points $a$ and $b$ are at the same potential, there is no current in the galvanometer and the bridge is said to be balanced. (Because the galvanometer is used to detect the absence of a current, it is called a *null detector*.) If the fixed resistance $R_0 = 200\ \Omega$, find the unknown resistance $R_x$ if ($a$) the bridge balances at the 18-cm mark, ($b$) the bridge balances at the 60-cm mark, and ($c$) the bridge balances at the 95-cm mark.

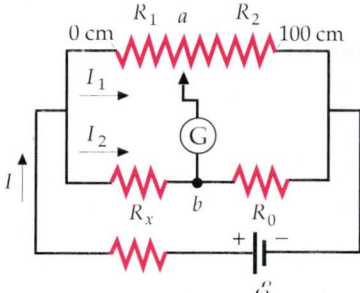

**FIGURE 25-73**
**Problems 133 and 134**

**134** •• For the Wheatstone bridge presented in Problem 133, the bridge balances at the 98-cm mark when $R_0 = 200\ \Omega$. ($a$) What is the unknown resistance? ($b$) What effect would an error of 2 mm in the location of the balance point have on the measured value of the unknown resistance? ($c$) How should $R_0$ be changed so that the balance point for this unknown resistor will be nearer the 50-cm mark?

**135** •• A cyclotron produces a 3.50-$\mu$A proton beam of 60-MeV energy. The protons impinge and come to rest inside a 50-g copper target within the vacuum chamber. ($a$) Determine the number of protons that strike the target per second. ($b$) Find the energy deposited in the target per second. ($c$) How much time elapses before the target temperature rises 300°C? (Neglect cooling by radiation.)

**136** •• The belt of a Van de Graaff generator carries a surface charge density of 5 mC/m². The belt is 0.5 m wide and moves at 20 m/s. ($a$) What current does it carry? ($b$) If this charge is raised to a potential of 100 kV, what is the minimum power of the motor needed to drive the belt?

**137** •• Conventional large electromagnets use water cooling to prevent excessive heating of the magnet coils. A large laboratory electromagnet draws 100 A when a voltage of 240 V is applied to the terminals of the energizing coils. To cool the coils, water at an initial temperature of 15°C is circulated through the coils. How many liters per second must pass through the coils if their temperature should not exceed 50°C?

**138** •• A parallel-plate capacitor is made from plates of area $A$ separated by a distance $d$, and are filled with a dielectric with dielectric constant $\kappa$ and resistivity $\rho$, show that the product of the resistance $R$ of this dielectric, with the capacitance of the capacitor, is $RC = \epsilon_0 \rho \kappa$.

**139** •• Show that the result of Problem 138 is true for a cylindrical capacitor or resistor. Should it be true for a capacitor or resistor of any shape?

**140** •• **SSM** ($a$) Show that a leaky capacitor (one for which the resistance of the dielectric is finite) can be modeled as a capacitor with infinite resistance in parallel with a resistor. ($b$) Show that the time constant for discharging this capacitor is $\tau = \epsilon_0 \rho \kappa$. ($c$) Mica has a dielectric constant $\kappa = 5$ and a resistivity $\rho = 9 \times 10^{13}\ \Omega\cdot$m. Calculate the time it takes for the charge of a mica-filled capacitor to decrease to 10 percent of its initial value.

**141** ••• Figure 25-74 shows the basis of the sweep circuit used in an oscilloscope. Switch S is an electronic switch that closes whenever the potential across the terminals switches reaches a value $V_c$; switch S opens when the potential has dropped to 0.2 V. The emf $\mathcal{E}$, which is much greater than $V_c$, charges the capacitor C through a resistor $R_1$. The resistor $R_2$ represents the small but finite resistance of the electronic switch. In a typical circuit, $\mathcal{E} = 800$ V, $V_c = 4.2$ V, $R_2 = 0.001\ \Omega$, $R_1 = 0.5$ M$\Omega$, and $C = 0.02\ \mu$F. ($a$) What is the time constant for charging of the capacitor C? ($b$) Show that in the time required to bring the potential across switch S to the critical potential $V_c = 4.2$ V, the voltage across the capacitor increases almost linearly with time. (*Hint:* Use the expansion of the exponential for small values of exponent.) ($c$) What should the value of $R_1$ be so that C charges from 0.2 V to 4.2 V in 0.1 s? ($d$) How much time elapses during the discharge of C through switch S? ($e$) At what rate is energy dissipated in the resistor $R_1$ and in the switch resistance?

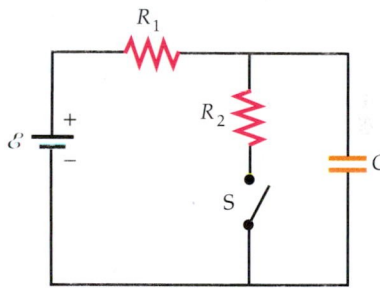

**FIGURE 25-74** Problem 141

**142** ••• In the circuit shown in Figure 25-75, $R_1 = 2$ M$\Omega$, $R_2 = 5$ M$\Omega$, and $C = 1\ \mu$F. At $t = 0$, switch S is closed, and at $t = 2.0$ s switch S is opened. ($a$) Sketch the voltage across C and the current through $R_2$ between $t = 0$ and $t = 10$ s. ($b$) Find the voltage across the capacitor at $t = 2$ s and at $t = 8$ s.

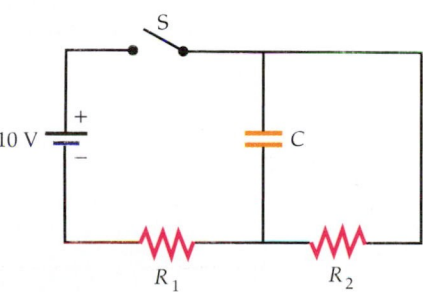

**FIGURE 25-75** Problem 142

**143 •••** Two batteries with emfs $\mathcal{E}_1$ and $\mathcal{E}_2$ and internal resistances $r_1$ and $r_2$ are connected in parallel. Prove that if a resistor is connected in parallel with this combination the optimal load resistance (the resistance at which maximum power is delivered) is $R = r_1 r_2/(r_1 + r_2)$.

**144 •••** **SSM** Capacitors $C_1$ and $C_2$ are connected in parallel by a resistor and two switches, as shown in Figure 25-76. Capacitor $C_1$ is initially charged to a voltage $V_0$, and capacitor $C_2$ is uncharged. The switches $S_1$ and $S_2$ are then closed. (*a*) What are the final charges on $C_1$ and $C_2$? (*b*) Compare the initial and final stored energies of the system. (*c*) What caused the decrease in the capacitor-stored energy?

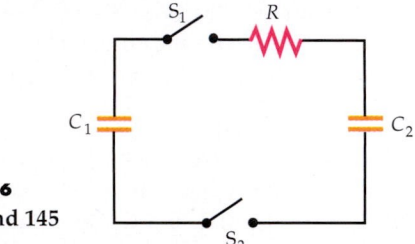

**FIGURE 25-76**
**Problems 144 and 145**

**145 •••** (*a*) In Problem 144, find the current through $R$ after the switches $S_1$ and $S_2$ are closed as a function of time. (*b*) Find the energy dissipated in the resistor as a function of time. (*c*) Find the total energy dissipated in the resistor and compare it with the loss of stored energy found in Part (*b*) of Problem 144.

**146 •••** In the circuit shown in Figure 25-77, the capacitors are initially uncharged. Switch $S_2$ is closed and then switch $S_1$ is closed. (*a*) What is the battery current immediately after $S_1$ is closed? (*b*) What is the battery current a long time after both switches are closed? (*c*) What is the final voltage across $C_1$? (*d*) What is the final voltage across $C_2$? (*e*) Switch $S_2$ is opened again after a long time. Give the current in the 150-$\Omega$ resistor as a function of time.

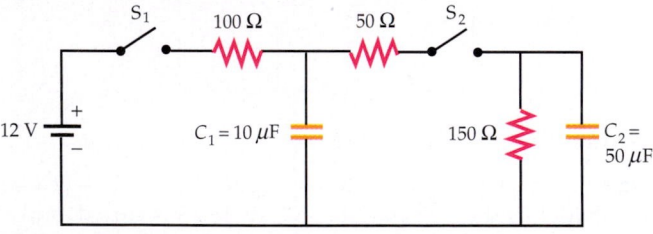

**FIGURE 25-77** Problem 146

**147 •••** **SSM** The differential resistance[†] of a nonohmic circuit element is defined as $R_d = dV/dI$, where $V$ is the voltage across the element, and $I$ is the current through the element. Show that for $V > 0.6$ V, the differential resistance of a diode (Problem 54) is approximately $R_d = (25\ \text{mV})/I$, and for $V < 0$, $R_d$ increases exponentially with $|V|$. Use this result to justify the approximation given in Problem 55.

**148 ••** A graph of the voltage as a function of current for an Esaki diode is shown in Figure 25-78. Make a graph of the differential resistance of the diode as a function of voltage. (See Problem 147 for a definition of differential resistance.) At what value of the voltage does the differential resistance become negative?

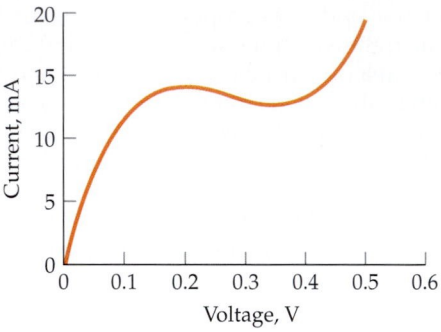

**FIGURE 25-78** Problem 148

**149 •••** A linear accelerator produces a pulsed beam of electrons. The current is 1.6 A for the 0.1-$\mu$s duration of each pulse. (*a*) How many electrons are in each pulse? (*b*) What is the average current of the beam if there are 1000 pulses per second? (*c*) If each electron acquires an energy of 400 MeV, what is the average power output of the accelerator? (*d*) What is the peak power output? (*e*) What fraction of the time is the accelerator actually accelerating electrons? (This is called the *duty factor* of the accelerator.)

**150 •••** Calculate the equivalent resistance between points $a$ and $b$ for the infinite ladder of resistors shown in Figure 25-79.

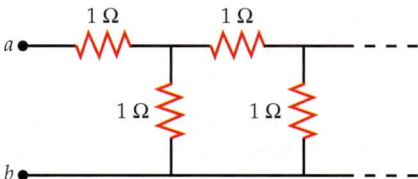

**FIGURE 25-79** Problem 150

**151 •••** **SSM** Calculate the equivalent resistance between points $a$ and $b$ for the infinite ladder of resistors shown in Figure 25-80, where $R_1$ and $R_2$ can take any value.

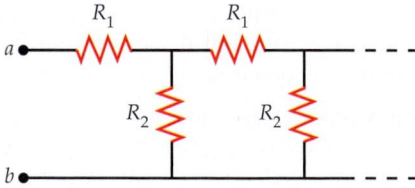

**FIGURE 25-80** Problem 151

---

[†] Differential resistance ($dV/dI$) is also called dynamic resistance or dynamic impedance.

# The Magnetic Field

THE AURORA BOREALIS APPEARS WHEN "SOLAR WIND," CHARGED PARTICLES PRODUCED BY NUCLEAR FUSION REACTIONS IN THE SUN, BECOME TRAPPED IN THE EARTH'S MAGNETIC FIELD.

**?** **How does the earth's magnetic field act on subatomic particles? (See Example 26-1.)**

26-1    The Force Exerted by a Magnetic Field

26-2    Motion of a Point Charge in a Magnetic Field

26-3    Torques on Current Loops and Magnets

26-4    The Hall Effect

**M**ore than 2000 years ago, the Greeks were aware that a certain type of stone (now called magnetite) attracts pieces of iron, and there are written references to the use of magnets for navigation dating from the twelfth century.

In 1269, Pierre de Maricourt discovered that a needle laid at various positions on a spherical natural magnet orients itself along lines that pass through points at opposite ends of the sphere. He called these points the poles of the magnet. Subsequently, many experimenters noted that every magnet of any shape has two poles, designated the north and south poles, where the force exerted by the magnet is strongest. It was also noted that the like poles of two magnets repel each other and the unlike poles of two magnets attract each other.

In 1600, William Gilbert discovered that the earth is a natural magnet with magnetic poles near the north and south geographic poles. Since the north pole of a compass needle points toward the south pole of a given magnet, what we call the north pole of the earth is actually a south magnetic pole, as illustrated in Figure 26-1.

Although electric charges and magnetic poles are similar in many respects, there is an important difference: Magnetic poles always occur in pairs. When a magnet is broken in half, equal and opposite poles appear at either side of the break point. The result is two magnets, each with a north and south pole. There has long been speculation about the existence of an isolated magnetic pole, and in recent years considerable experimental effort has been made to find such an object. Thus far, there is no conclusive evidence that an isolated magnetic pole exists.

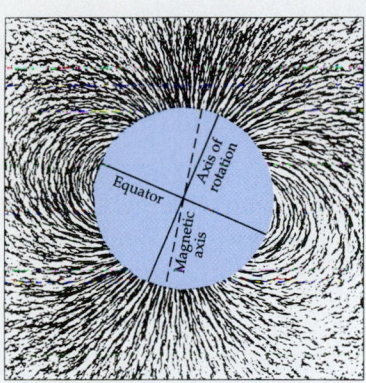

**FIGURE 26-1** Magnetic field lines of the earth depicted by iron filings around a uniformly magnetized sphere. The field lines exit from the north magnetic pole, which is near the south geographic pole, and enter the south magnetic pole, which is near the north geographic pole.

➤ In this chapter, we consider the effects of a given magnetic field on moving charges and on wires carrying currents. The sources of magnetic fields are discussed in the next chapter.

# 26-1 The Force Exerted by a Magnetic Field

The existence of a magnetic field $\vec{B}$ at some point in space can be demonstrated with a compass needle. If there is a magnetic field, the needle will align itself in the direction of the field.

Experimentally it is observed that, when a charge $q$ has velocity $\vec{v}$ in a magnetic field, there is a force on the magnetic field that is proportional to $q$ and to $v$, and to the sine of the angle between the directions of $\vec{v}$ and $\vec{B}$. Surprisingly, the force is perpendicular to both the velocity and the field. These experimental results can be summarized as follows: When a charge $q$ moves with velocity $\vec{v}$ in a magnetic field $\vec{B}$, the magnetic force $\vec{F}$ on the charge is

$$\vec{F} = q\vec{v} \times \vec{B}$$
26-1

MAGNETIC FORCE ON A MOVING CHARGE

Since $\vec{F}$ is perpendicular to both $\vec{v}$ and $\vec{B}$, $\vec{F}$ is perpendicular to the plane defined by these two vectors. The direction of $\vec{v} \times \vec{B}$ is given by the right-hand rule as $\vec{v}$ is rotated into $\vec{B}$, as illustrated in Figure 26-2. If $q$ is positive, then $\vec{F}$ is in the same direction as $\vec{v} \times \vec{B}$.

Examples of the direction of the forces exerted on moving charges when the magnetic field vector $\vec{B}$ is in the vertical direction are shown in Figure 26-3. Note that the direction of any particular magnetic field $\vec{B}$ can be found experimentally by measuring $\vec{F}$ and $\vec{v}$ for several velocities in different directions and then applying Equation 26-1.

Equation 26-1 defines the **magnetic field** $\vec{B}$ in terms of the force exerted on a moving charge. The SI unit of magnetic field is the **tesla** (T). A charge of one coulomb moving with a velocity of one meter per second perpendicular to a magnetic field of one tesla experiences a force of one newton:

$$1\,\text{T} = 1\frac{\text{N}}{\text{C}\cdot\text{m/s}} = 1\,\text{N}/(\text{A}\cdot\text{m})$$
26-2

This unit is rather large. The magnetic field of the earth has a magnitude somewhat less than $10^{-4}$ T on the earth's surface. The magnetic fields near powerful

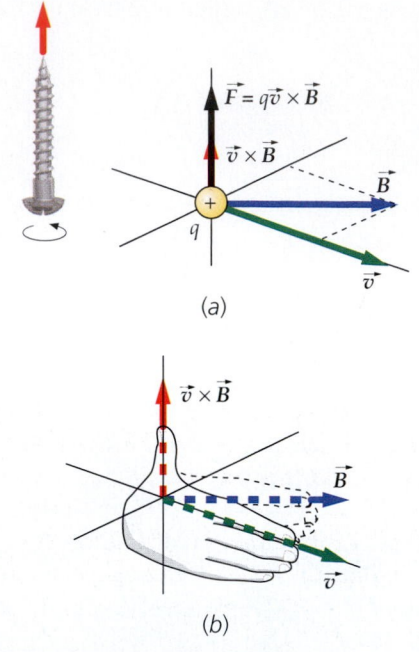

**FIGURE 26-2** Right-hand rule for determining the direction of a force exerted on a charge moving in a magnetic field. If $q$ is positive, then $\vec{F}$ is in the same direction as $\vec{v} \times \vec{B}$. (*a*) The cross product $\vec{v} \times \vec{B}$ is perpendicular to both $\vec{v}$ and $\vec{B}$ and is in the direction of the advance of a right-hand-threaded screw if turned in the same direction as to rotate $\vec{v}$ into $\vec{B}$. (*b*) If the fingers of the right hand are in the direction of $\vec{v}$ so that they can be curled toward $\vec{B}$, the thumb points in the direction of $\vec{v} \times \vec{B}$.

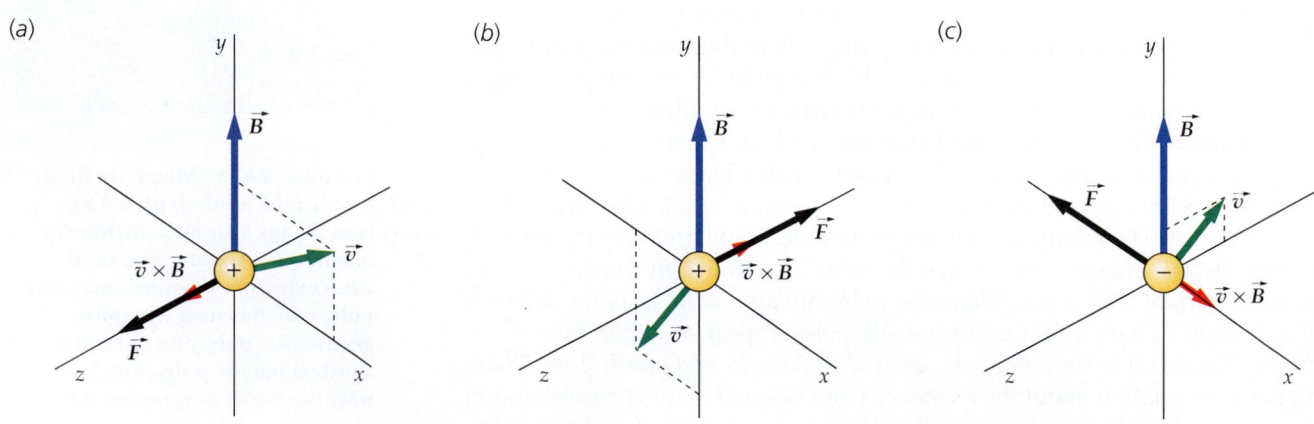

**FIGURE 26-3** Direction of the magnetic force on a charged particle moving with velocity $\vec{v}$ in a magnetic field $\vec{B}$.

permanent magnets are about 0.1 T to 0.5 T, and powerful laboratory and industrial electromagnets produce fields of 1 T to 2 T. Fields greater than 10 T are difficult to produce because the resulting magnetic forces will either tear the magnets apart or crush the magnets. A commonly used unit, derived from the cgs system, is the **gauss** (G), which is related to the tesla as follows

$$1\,G = 10^{-4}\,T \qquad\qquad 26\text{-}3$$

DEFINITION—GAUSS

Since magnetic fields are often given in gauss, which is not an SI unit, remember to convert from gauss to teslas when making calculations.

**FIGURE 26-4**

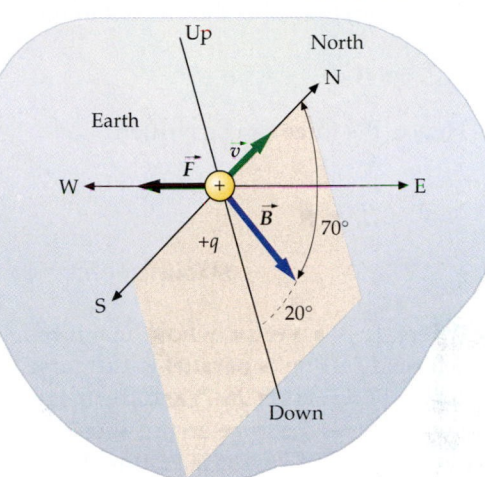

*FORCE ON A PROTON GOING NORTH* **EXAMPLE 26-1**

The magnetic field of the earth is measured at a point on the surface to have a magnitude of 0.6 G and is directed downward and, in the northern hemisphere, northward, making an angle of about 70° with the horizontal, as shown in Figure 26-4. (The earth's magnetic field varies from place to place. These data are approximately correct for the central United States.) A proton ($q = +e$) is moving horizontally in the northward direction with speed $v = 10$ Mm/s $= 10^7$ m/s. Calculate the magnetic force on the proton (*a*) using $F = qvB \sin \theta$ and (*b*) by expressing $\vec{v}$ and $\vec{B}$ in terms of the unit vectors $\hat{i}, \hat{j}, \hat{k}$, and computing $\vec{F} = q\vec{v} \times \vec{B}$.

**PICTURE THE PROBLEM** Let the $x$ and $y$ directions be east and north, respectively, and let the $z$ direction be upward (Figure 26-5). The velocity vector is then in the $y$ direction.

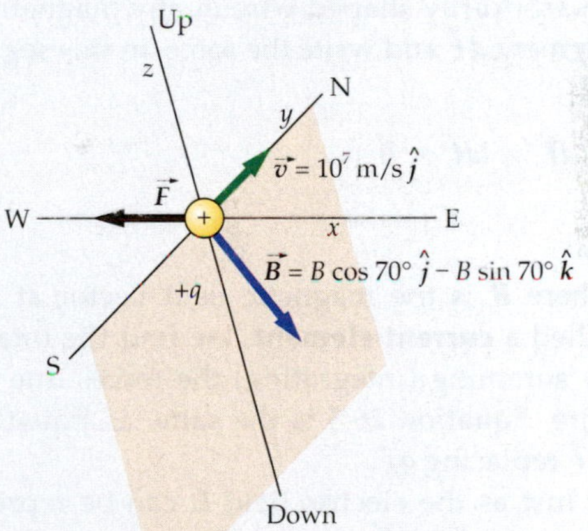

(*a*) Calculate $F = qvB \sin \theta$ using $\theta = 70°$. From Figure 26-4 we see that the direction of the force is westward:

$$F = qvB \sin 70°$$

$$= (1.6 \times 10^{-19}\,\text{C})(10^7\,\text{m/s})(0.6 \times 10^{-4}\,\text{T})(0.94)$$

$$= \boxed{9.02 \times 10^{-17}\,\text{N}}$$

(*b*) 1. The magnetic force is the vector product of $q\vec{v}$ and $\vec{B}$:  $\vec{F} = q\vec{v} \times \vec{B}$

2. Express $\vec{v}$ and $\vec{B}$ in terms of their components:  $\vec{v} = v_y \hat{j}$

$$\vec{B} = B_y \hat{j} + B_z \hat{k}$$

3. Write $\vec{F} = q\vec{v} \times \vec{B}$ in terms of these components:  $\vec{F} = q\vec{v} \times \vec{B} = q(v_y \hat{j}) \times (B_y \hat{j} + B_z \hat{k})$

$$= qv_y B_y (\hat{j} \times \hat{j}) + qv_y B_z (\hat{j} \times \hat{k}) = 0 + qv_y B_z \hat{i}$$

4. Evaluate $\vec{F}$:  $\vec{F} = qv\,(-B \sin \theta)\hat{i}$

$$= -(1.6 \times 10^{-19}\,\text{C})(10^7\,\text{m/s})(0.6 \times 10^{-4}\,\text{T})\sin 70°\,\hat{i}$$

$$= \boxed{-9.02 \times 10^{-17}\,\text{N}\hat{i}}$$

**FIGURE 26-5**

**REMARKS** Note that the direction of $\hat{i}$ is eastward, so the force is directed westward as shown in Figure 26-5.

**EXERCISE** Find the force on a proton moving with velocity $\vec{v} = 4 \times 10^6$ m/s$\hat{i}$ in a magnetic field $\vec{B} = 2.0$ T$\hat{k}$. (*Answer* $-1.28 \times 10^{-12}$ N$\hat{j}$)

When a wire carries a current in a magnetic field, there is a force on the wire that is equal to the sum of the magnetic forces on the charged particles whose motion produces the current. Figure 26-6 shows a short segment of wire of cross-sectional area $A$ and length $L$ carrying a current $I$. If the wire is in a magnetic field $\vec{B}$, the magnetic force on each charge is $q\vec{v}_d \times \vec{B}$, where $\vec{v}_d$ is the drift velocity of the charge carriers (the drift velocity is the same as the average velocity). The number of charges in the wire segment is the number $n$ per unit volume times the volume $AL$. Thus, the total force on the wire segment is

$$\vec{F} = (q\vec{v}_d \times \vec{B})nAL$$

From Equation 25-3, the current in the wire is

$$I = nqv_d A$$

Hence, the force can be written

$$\vec{F} = I\vec{L} \times \vec{B} \qquad 26\text{-}4$$

MAGNETIC FORCE ON A SEGMENT OF CURRENT-CARRYING WIRE

where $\vec{L}$ is a vector whose magnitude is the length of the wire and whose direction is parallel to the current. For the current in the positive $x$ direction (Figure 26-7) and the magnetic field vector at the segment in the $xy$ plane, the force on the wire is directed along the $z$ axis.

In Equation 26-4 it is assumed that the wire segment is straight and that the magnetic field does not vary over its length. The equation can be generalized for an arbitrarily shaped wire in any magnetic field. If we choose a very small wire segment $d\vec{\ell}$ and write the force on this segment as $d\vec{F}$, we have

$$d\vec{F} = I d\vec{\ell} \times \vec{B} \qquad 26\text{-}5$$

MAGNETIC FORCE ON A CURRENT ELEMENT

where $\vec{B}$ is the magnetic field vector at the segment. The quantity $I d\vec{\ell}$ is called a **current element**. We find the total force on a current-carrying wire by summing (integrating) the forces due to all the current elements in the wire. Equation 26-5 is the same as Equation 26-1 with the current element $I d\vec{\ell}$ replacing $q\vec{v}$.

Just as the electric field $\vec{E}$ can be represented by electric field lines, the magnetic field $\vec{B}$ can be represented by **magnetic field lines**. In both cases, the direction of the field is indicated by the direction of the field lines and the magnitude of the field is indicated by their density. There are, however, two important differences between electric field lines and magnetic field lines:

1. Electric field lines are in the direction of the electric force on a positive charge, but the magnetic field lines are perpendicular to the magnetic force on a moving charge.

2. Electric field lines begin on positive charges and end on negative charges; magnetic field lines neither begin nor end.

Figure 26-8 shows the magnetic field lines both inside and outside a bar magnet.

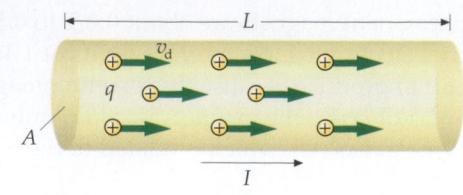

**FIGURE 26-6** Wire segment of length $L$ carrying current $I$. If the wire is in a magnetic field, there will be a force on each charge carrier resulting in a force on the wire.

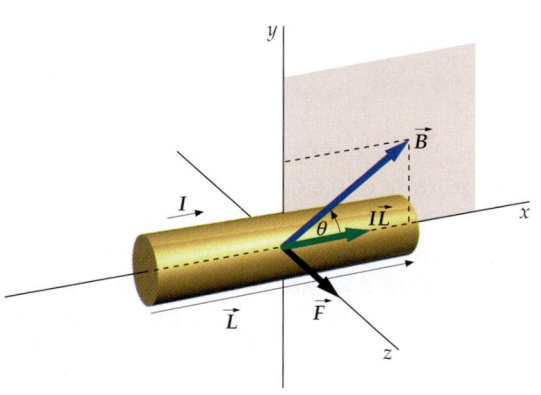

**FIGURE 26-7** Magnetic force on a current-carrying segment of wire in a magnetic field. The current is in the $x$ direction, and the magnetic field is in the $xy$ plane and makes an angle $\theta$ with the $+x$ direction. The force $\vec{F}$ is in the $+z$ direction, perpendicular to both $\vec{B}$ and $\vec{L}$, and has magnitude $ILB \sin \theta$.

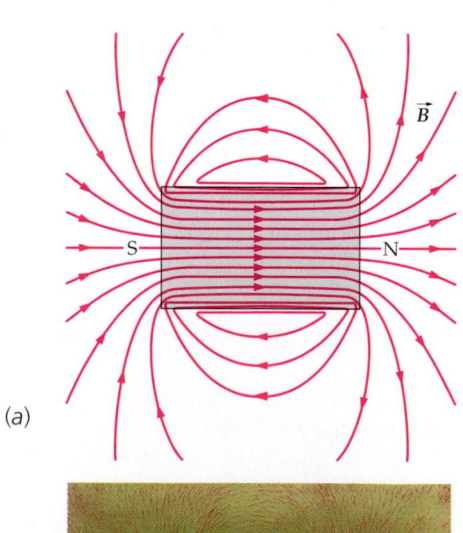

(a)

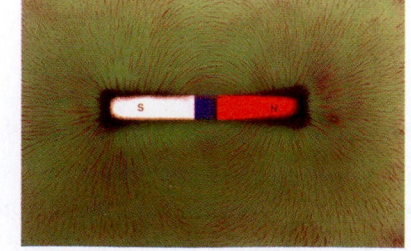

(b)

**FIGURE 26-8** (a) Magnetic field lines inside and outside a bar magnet. The lines emerge from the north pole and enter the south pole, but they have no beginning or end. Instead, they form closed loops. (b) Magnetic field lines outside a bar magnet as indicated by iron filings.

**EXAMPLE  26-2**

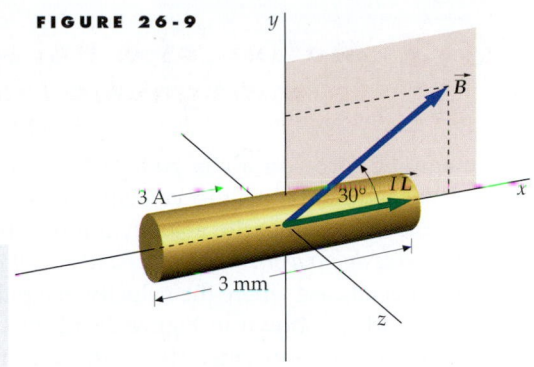

**FIGURE 26-9**

A wire segment 3 mm long carries a current of 3 A in the $+x$ direction. It lies in a magnetic field of magnitude 0.02 T that is in the $xy$ plane and makes an angle of 30° with the $+x$ direction, as shown in Figure 26-9. What is the magnetic force exerted on the wire segment?

**PICTURE THE PROBLEM**  The magnetic force is in the direction of $\vec{L} \times \vec{B}$, which we see from Figure 26-9 is in the positive $z$ direction.

The magnetic force is given by Equation 26-4:

$$\vec{F} = I\vec{L} \times \vec{B} = ILB \sin 30° \, \hat{k}$$

$$= (3 \text{ A})(0.003 \text{ m})(0.02 \text{ T})(\sin 30°)\hat{k}$$

$$= \boxed{9 \times 10^{-5} \text{ N}\hat{k}}$$

*FORCE ON A BENT WIRE*    **EXAMPLE  26-3**

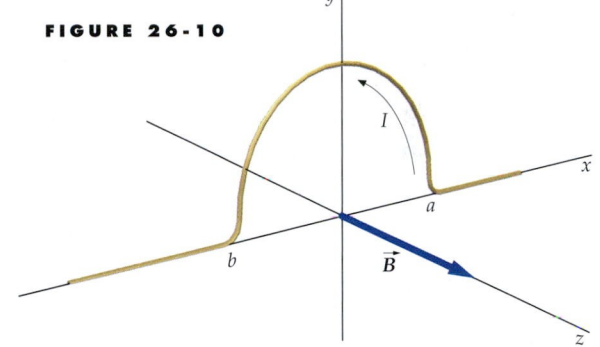

**FIGURE 26-10**

A wire bent into a semicircular loop of radius $R$ lies in the $xy$ plane. It carries a current $I$ from point $a$ to point $b$, as shown in Figure 26-10. There is a uniform magnetic field $\vec{B} = B\hat{k}$ perpendicular to the plane of the loop. Find the force acting on the semicircular loop part of the wire.

**PICTURE THE PROBLEM**  The force $d\vec{F}$ exerted on a segment of the semicircular wire lies in the $xy$ plane, as shown in Figure 26-11. We find the total force by expressing the $x$ and $y$ components of $d\vec{F}$ in terms of $\theta$ and integrating them separately from $\theta = 0$ to $\theta = \pi$.

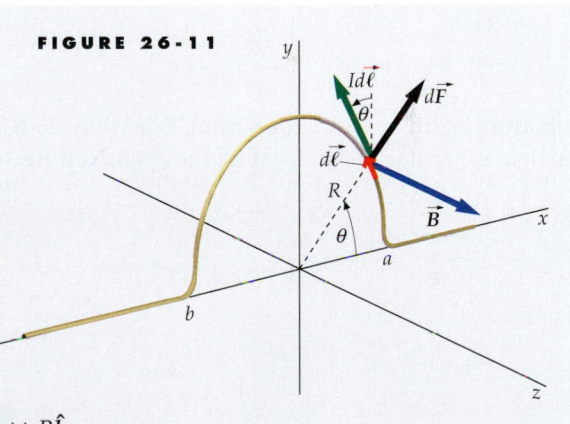

**FIGURE 26-11**

1. Write the force $d\vec{F}$ on a current element $d\vec{\ell}$.
$$d\vec{F} = Id\vec{\ell} \times \vec{B}$$

2. Express $d\vec{\ell}$ in terms of the unit vectors $\hat{i}$ and $\hat{j}$:
$$d\vec{\ell} = -d\ell \sin \theta \hat{i} + d\ell \cos \theta \hat{j}$$

3. Compute $Id\vec{\ell}$ using $d\ell = Rd\theta$ and $\vec{B} = B\hat{k}$:
$$d\vec{F} = Id\vec{\ell} \times \vec{B}$$
$$= I(-R \sin \theta d\theta \hat{i} + R \cos \theta d\theta \hat{j}) \times B\hat{k}$$
$$= IRB \sin \theta d\theta \hat{j} + IRB \cos \theta d\theta \hat{i}$$

4. Integrate each component of $d\vec{F}$ from $\theta = 0$ to $\theta = \pi$:
$$\vec{F} = \int d\vec{F} = IRB\hat{i} \int_0^\pi \cos \theta d\theta + IRB\hat{j} \int_0^\pi \sin \theta d\theta$$
$$= IRB\hat{i}\,(0) + IRB\,\hat{j}\,(2) = \boxed{2IRB\hat{j}}$$

**PLAUSIBILITY CHECK**  The result that the $x$ component of $\vec{F}$ is zero can be seen from symmetry. For the right half of the loop, $d\vec{F}$ points to the right; for the left half of the loop, $d\vec{F}$ points to the left.

**REMARKS**  The net force on the semicircular wire is the same as if the semicircle were replaced by a straight-line segment of length $2R$ connecting points $a$ and $b$. (This is a general result, as shown in Problem 30.)

# 26-2 Motion of a Point Charge in a Magnetic Field

The magnetic force on a charged particle moving through a magnetic field is always perpendicular to the velocity of the particle. The magnetic force thus changes the direction of the velocity but not the velocity's magnitude. Therefore, *magnetic fields do no work on particles and do not change their kinetic energy.*

In the special case where the velocity of a particle is perpendicular to a uniform magnetic field, as shown in Figure 26-12, the particle moves in a circular orbit. The magnetic force provides the centripetal force necessary for the centripetal acceleration $v^2/r$ in circular motion. We can use Newton's second law to relate the radius of the circle to the magnetic field and the speed of the particle. If the velocity is $\vec{v}$, the magnitude of the net force is $qvB$, since $\vec{v}$ and $\vec{B}$ are perpendicular. Newton's second law gives

$$F = ma$$

$$qvB = m\frac{v^2}{r}$$

or

$$r = \frac{mv}{qB} \qquad\qquad 26\text{-}6$$

The period of the circular motion is the time it takes the particle to travel once around the circumference of the circle. The period is related to the speed by

$$T = \frac{2\pi r}{v}$$

Substituting in $r = mv/(qB)$ from Equation 26-6, we obtain the period of the particle's circular motion, called the **cyclotron period:**

$$T = \frac{2\pi(mv/qB)}{v} = \frac{2\pi m}{qB} \qquad\qquad 26\text{-}7$$

CYCLOTRON PERIOD

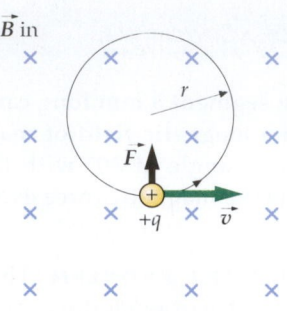

**FIGURE 26-12** Charged particle moving in a plane perpendicular to a uniform magnetic field. The magnetic field is into the page as indicated by the crosses. (Each cross represents the tail feathers of an arrow. A field out of the plane of the page would be indicated by dots, each dot representing the point of an arrow.) The magnetic force is perpendicular to the velocity of the particle, causing it to move in a circular orbit.

(*a*) Circular path of electrons moving in the magnetic field produced by two large coils. The electrons ionize the gas in the tube, causing it to give off a bluish glow that indicates the path of the beam. (*b*) False-color photograph showing tracks of a 1.6-MeV proton (red) and a 7-MeV $\alpha$ particle (yellow) in a cloud chamber. The radius of curvature is proportional to the momentum and inversely proportional to the charge of the particle. For these energies, the momentum of the $\alpha$ particle, which has twice the charge of the proton, is about four times that of the proton and so its radius of curvature is greater.

(b)

(a)

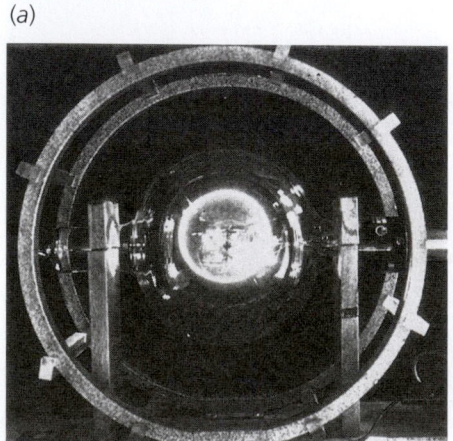

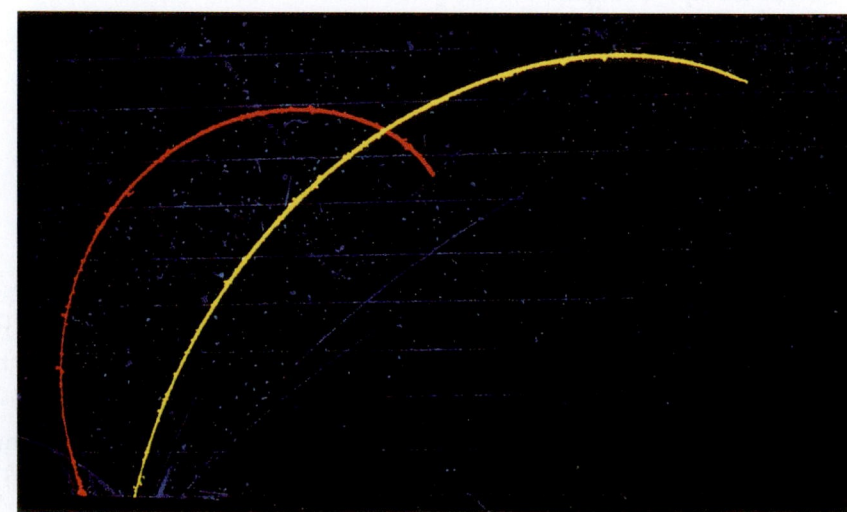

The frequency of the circular motion, called the **cyclotron frequency,** is the reciprocal of the period:

$$f = \frac{1}{T} = \frac{qB}{2\pi m}, \quad \text{so} \quad \omega = 2\pi f = \frac{q}{m}B \qquad \qquad 26\text{-}8$$

CYCLOTRON FREQUENCY

Note that the period and the frequency given by Equations 26-7 and 26-8 depend on the charge-to-mass ratio $q/m$, but the period and the frequency are independent of the velocity $v$ or the radius $r$. Two important applications of the circular motion of charged particles in a uniform magnetic field, the mass spectrometer and the cyclotron, are discussed later in this section.

---

CYCLOTRON PERIOD                                 **E X A M P L E    2 6 - 4**

A proton of mass $m = 1.67 \times 10^{-27}$ kg and charge $q = e = 1.6 \times 10^{-19}$ C moves in a circle of radius $r = 21$ cm perpendicular to a magnetic field $B = 4000$ G. Find (a) the period of the motion and (b) the speed of the proton.

1. Calculate the period $T$ from Equation 26-7 with $B = 4000$ G $= 0.4$ T:

$$T = \frac{2\pi m}{qB} = \frac{2\pi(1.67 \times 10^{-27} \text{ kg})}{(1.6 \times 10^{-19} \text{ C})(0.4 \text{ T})}$$

$$= \boxed{1.64 \times 10^{-7} \text{ s} = 164 \text{ ns}}$$

2. Calculate the speed $v$ from Equation 26-6:

$$v = \frac{rqB}{m} = \frac{(0.21 \text{ m})(1.6 \times 10^{-19} \text{ C})(0.4 \text{ T})}{1.67 \times 10^{-27} \text{ kg}}$$

$$= \boxed{8.05 \times 10^6 \text{ m/s} = 8.05 \text{ m/}\mu\text{s}}$$

**REMARKS** The radius of the circular motion is proportional to the speed, but the period is independent of both the speed and radius.

**PLAUSIBILITY CHECK** Note that the product of the speed $v$ and the period $T$ equals the circumference of the circle $2\pi r$ as expected: $vT = (8.05 \times 10^6 \text{ m/s})(1.64 \times 10^{-7} \text{ s}) = 1.32$ m; $2\pi r = 2\pi(0.21 \text{ m}) = 1.32$ m.

Suppose that a charged particle enters a uniform magnetic field with a velocity that is not perpendicular to $\vec{B}$. There is no force component, and thus no acceleration component, parallel to $\vec{B}$, so the component of the velocity parallel to $\vec{B}$ remains constant. The magnetic force on the particle is perpendicular to $\vec{B}$, so the change in motion of the particle due to this force is the same as that just discussed. The path of the particle is thus a helix, as shown in Figure 26-13.

(a)

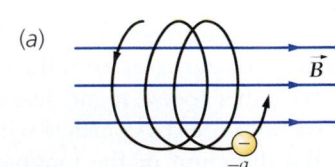

(b)

**FIGURE 26-13** (a) When a particle has a velocity component parallel to a magnetic field as well as a velocity component perpendicular to the magnetic field the particle moves in a helical path around the field lines. (b) Cloud-chamber photograph of the helical path of an electron moving in a magnetic field. The path of the electron is made visible by the condensation of water droplets in the cloud chamber.

The motion of charged particles in nonuniform magnetic fields can be quite complex. Figure 26-14 shows a **magnetic bottle,** an interesting magnetic field configuration in which the field is weak at the center and strong at both ends. A detailed analysis of the motion of a charged particle in such a field shows that the particle spirals around the field lines and becomes trapped, oscillating back and forth between points $P_1$ and $P_2$ in the figure. Such magnetic field configurations are used to confine dense beams of charged particles, called *plasmas,* in nuclear fusion research. A similar phenomenon is the oscillation of ions back and forth between the earth's magnetic poles in the Van Allen belts (Figure 26-15).

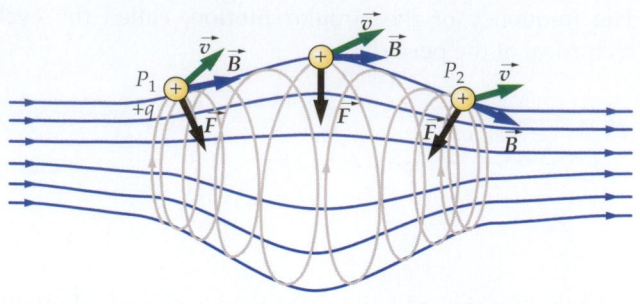

**FIGURE 26-14** Magnetic bottle. When a charged particle moves in such a field, which is strong at both ends and weak in the middle, the particle becomes trapped and moves back and forth, spiraling around the field lines.

## *The Velocity Selector

The magnetic force on a charged particle moving in a uniform magnetic field can be balanced by an electric force if the magnitudes and directions of the magnetic field and the electric field are properly chosen. Since the electric force is in the direction of the electric field (for positive particles) and the magnetic force is perpendicular to the magnetic field, the electric and magnetic fields in the region through which the particle is moving must be perpendicular to each other if the forces are to balance. Such a region is said to have **crossed fields.**

Figure 26-16 shows a region of space between the plates of a capacitor where there is an electric field and a perpendicular magnetic field (produced by a magnet with poles above and below the paper). Consider a particle of charge $q$ entering this space from the left. The net force on the particle is

$$\vec{F} = q\vec{E} + q\vec{v} \times \vec{B}$$

If $q$ is positive, the electric force of magnitude $qE$ is down and the magnetic force of magnitude $qvB$ is up. If the charge is negative, each of these forces is reversed. The two forces balance if $qE = qvB$ or

$$v = \frac{E}{B} \qquad\qquad 26\text{-}9$$

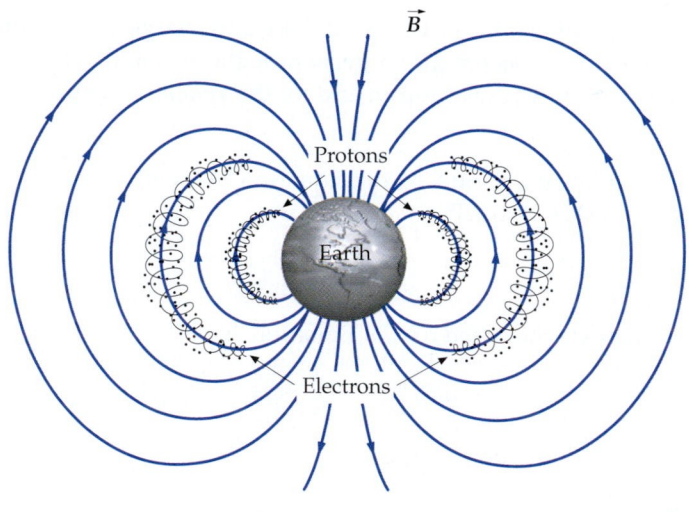

**FIGURE 26-15** Van Allen belts. Protons (inner belts) and electrons (outer belts) are trapped in the earth's magnetic field and spiral around the field lines between the north and south poles.

For given magnitudes of the electric and magnetic fields, the forces balance only for particles with the speed given by Equation 26-9. Any particle with this speed, regardless of its mass or charge, will traverse the space undeflected. A particle with a greater speed will be deflected toward the direction of the magnetic force, and a particle with less speed will be deflected in the direction of the electric force. This arrangement of fields is often used as a **velocity selector,** which is a device that allows only particles with speed, given by Equation 26-9, to pass.

**EXERCISE** A proton is moving in the $x$ direction in a region of crossed fields where $\vec{E} = 2 \times 10^5$ N/C $\hat{k}$ and $\vec{B} = -3000$ G $\hat{j}$. (a) What is the speed of the proton if it is not deflected? (b) If the proton moves with twice this speed, in which direction will it be deflected? (*Answer* (a) 667 km/s (b) in the negative z direction)

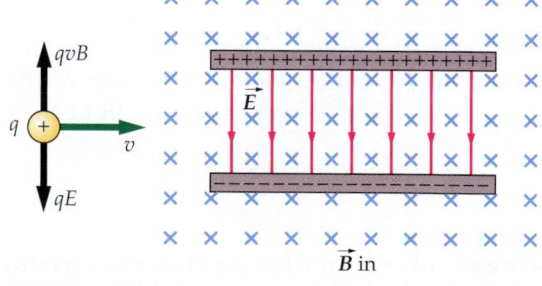

**FIGURE 26-16** Crossed electric and magnetic fields. When a positive particle moves to the right, the particle experiences a downward electric force and an upward magnetic force. These forces balance if the speed of the particle is related to the field strengths by $vB = E$.

## *Thomson's Measurement of $q/m$ for Electrons

An example of the use of crossed electric and magnetic fields is the famous experiment performed by J. J. Thomson in 1897 where he showed that the rays of a cathode-ray tube can be deflected by electric and magnetic fields, indicating that they must consist of charged particles. By measuring the deflections of these particles Thomson showed that all the particles have the same charge-to-mass ratio $q/m$. He also showed that particles with this charge-to-mass ratio can be obtained using any material for a source, which means that these particles, now called electrons, are a fundamental constituent of all matter.

Figure 26-17 shows a schematic diagram of the cathode-ray tube Thomson used. Electrons are emitted from the cathode C, which is at a negative potential relative to the slits A and B. An electric field in the direction from A to C accelerates the electrons, and the electrons pass through slits A and B into a field-free region. The electrons then enter the electric field between the capacitor plates D and F that is perpendicular to the velocity of the electrons. This field accelerates the electrons vertically for the short time that they are between the plates. The electrons are deflected and strike the phosphorescent screen S at the far right side of the tube at some deflection $\Delta y$ from the point at which they strike when there is no field between the plates. The screen glows where the electrons strike the screen, indicating the location of the beam. The initial speed of the electrons $v_0$ is determined by introducing a magnetic field $\vec{B}$ between the plates in a direction that is perpendicular to both the electric field and the initial velocity of the electrons. The magnitude of $\vec{B}$ is adjusted until the beam is not deflected. The speed is then found from Equation 26-9.

With the magnetic field turned off, the beam is deflected by an amount $\Delta y$, which consists of two parts: the deflection $\Delta y_1$, which occurs while the electrons are between the plates, and the deflection $\Delta y_2$, which occurs after the electrons leave the region between the plates (Figure 26-18).

Let $x_1$ be the horizontal distance across the deflection plates D and F. If the electron is moving horizontally with speed $v_0$ when it enters region between the plates, the time spent between the plates is $t_1 = x_1/v_0$, and the vertical velocity when it leaves the plates is

$$v_y = a_y t_1 = \frac{qE_y}{m} t_1 = \frac{qE_y}{m} \frac{x_1}{v_0}$$

where $E_y$ is the upward component of the electric field between the plates. The deflection in this region is

$$\Delta y_1 = \frac{1}{2} a_y t_1^2 = \frac{1}{2} \frac{qE_y}{m} \left( \frac{x_1}{v_0} \right)^2$$

The electron then travels an additional horizontal distance $x_2$ in the field-free region from the deflection plates to the screen. Since the velocity of the electron is constant in this region, the time to reach the screen is $t_2 = x_2/v_0$, and the additional vertical deflection is

$$\Delta y_2 = v_y t_2 = \frac{qE_y}{m} \frac{x_1}{v_0} \frac{x_2}{v_0}$$

The total deflection at the screen is therefore

$$\Delta y = \Delta y_1 + \Delta y_2 = \frac{1}{2} \frac{qE_y}{mv_0^2} x_1^2 + \frac{qE_y}{mv_0^2} x_1 x_2 \qquad \text{26-10}$$

The measured deflection $\Delta y$ can be used to determine the charge-to-mass ratio, $q/m$, from Equation 26-10.

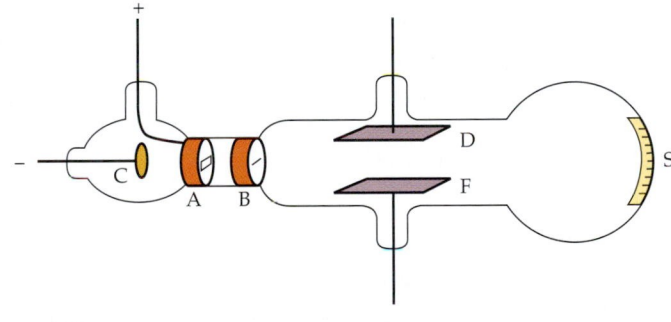

**FIGURE 26-17** Thomson's tube for measuring $q/m$ for the particles of cathode rays (electrons). Electrons from the cathode C pass through the slits at A and B and strike a phosphorescent screen S. The beam can be deflected by an electric field between plates D and F or by a magnetic field (not shown).

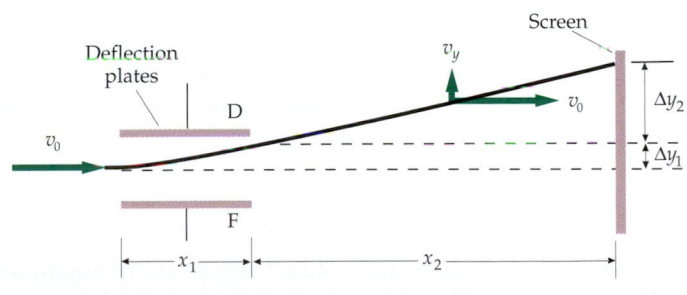

**FIGURE 26-18** The total deflection of the beam in the J. J. Thomson experiments consists of the deflection $\Delta y_1$ while the electrons are between the plates plus the deflection $\Delta y_2$ that occurs in the field-free region between the plates and the screen.

**EXAMPLE 26-5**

Electrons pass undeflected through the plates of Thomson's apparatus when the electric field is 3000 V/m and there is a crossed magnetic field of 1.40 G. If the plates are 4-cm long and the ends of the plates are 30 cm from the screen, find the deflection on the screen when the magnetic field is turned off.

**PICTURE THE PROBLEM** The mass and charge of the electron are known: $m = 9.11 \times 10^{-31}$ kg and $q = -e = -1.6 \times 10^{-19}$ C. The speed of the electron can be found from the ratio of the magnetic and electric fields.

1. The total deflection of the electron is given by Equation 26-10:
$$\Delta y = \Delta y_1 + \Delta y_2 = \frac{1}{2}\frac{qE_y}{mv_0^2}x_1^2 + \frac{qE_y}{mv_0^2}x_1x_2$$

2. The speed $v_0$ equals $E/B$:
$$v_0 = \frac{E}{B} = \frac{3000 \text{ V/m}}{1.40 \times 10^{-4} \text{ T}} = 2.14 \times 10^7 \text{ m/s}$$

3. Substitute this value for $v_0$, the given value of $E$, and the known values for $m$ and $q$ to find $\Delta y$:
$$\Delta y_1 = \frac{1}{2}\frac{(-1.6 \times 10^{-19}\text{ C})(-3000 \text{ V/m})}{(9.11 \times 10^{-31}\text{ kg})(2.14 \times 10^7\text{ m/s})^2}(0.04 \text{ m})^2$$
$$= 9.20 \times 10^{-4} \text{ m}$$

$$\Delta y_2 = \frac{(-1.6 \times 10^{-19}\text{ C})(-3000 \text{ V/m})}{(9.11 \times 10^{-31}\text{ kg})(2.14 \times 10^7\text{ m/s})^2}(0.04 \text{ m})(0.30 \text{ m})$$
$$= 1.38 \times 10^{-2} \text{ m}$$

$$\Delta y = \Delta y_1 + \Delta y_2$$
$$= 9.20 \times 10^{-4}\text{ m} + 1.38 \times 10^{-2}\text{ m}$$
$$= 0.92 \text{ mm} + 13.8 \text{ mm} = \boxed{14.7 \text{ mm}}$$

## *The Mass Spectrometer

The **mass spectrometer,** first designed by Francis William Aston in 1919, was developed as a means of measuring the masses of isotopes. Such measurements are important in determining both the presence of isotopes and their abundance in nature. For example, natural magnesium has been found to consist of 78.7 percent $^{24}$Mg, 10.1 percent $^{25}$Mg, and 11.2 percent $^{26}$Mg. These isotopes have masses in the approximate ratio 24:25:26.

Figure 26-19 shows a simple schematic drawing of a mass spectrometer. Positive ions are formed by bombarding neutral atoms with X rays or a beam of electrons. (Electrons are knocked out of the atoms by the X rays or bombarding electrons.) These ions are accelerated by an electric field and enter a uniform magnetic field. If the positive ions start from rest and move through a potential difference $\Delta V$, the ions kinetic energy when they enter the magnetic field equals their loss in potential energy, $q|\Delta V|$:

$$\tfrac{1}{2}mv^2 = q|\Delta V| \qquad\qquad 26\text{-}11$$

The ions move in a semicircle of radius $r$ given by Equation 26-6, $r = mv/qB$, and strike a photographic plate at point $P_2$, a distance $2r$ from the point $P_1$ where the ions entered the magnetic field.

The speed $v$ can be eliminated from Equations 26-6 and 26-11 to find $m/q$ in terms of the known quantities $\Delta V$, $B$, and $r$. We first solve Equation 26-6 for $v$ and square each term, which gives

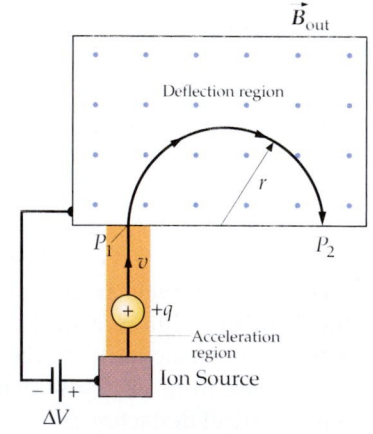

**FIGURE 26-19** Schematic drawing of a mass spectrometer. Positive ions from an ion source are accelerated through a potential difference $\Delta V$ and enter a uniform magnetic field. The magnetic field is out of the plane of the page as indicated by the dots. The ions are bent into a circular arc and emerge at $P_2$. The radius of the circle varies with the mass of the ion.

$$v^2 = \frac{r^2 q^2 B^2}{m^2}$$

Substituting this expression for $v^2$ into Equation 26-11, we obtain

$$\frac{1}{2} m \left( \frac{r^2 q^2 B^2}{m^2} \right) = q |\Delta V|$$

Simplifying this equation and solving for $m/q$, we obtain

$$\frac{m}{q} = \frac{B^2 r^2}{2 |\Delta V|} \qquad\qquad\qquad 26\text{-}12$$

In Aston's original mass spectrometer, mass differences could be measured to a precision of about 1 part in 10,000. The precision has been improved by introducing a velocity selector between the ion source and the magnet, which increases the degree of accuracy with which the velocities of the incoming ions can be determined.

---

*SEPARATING ISOTOPES OF NICKEL*                    **E X A M P L E   2 6 - 6**

A $^{58}$Ni ion of charge $+e$ and mass $9.62 \times 10^{-26}$ kg is accelerated through a potential drop of 3 kV and deflected in a magnetic field of 0.12 T. (*a*) Find the radius of curvature of the orbit of the ion. (*b*) Find the difference in the radii of curvature of $^{58}$Ni ions and $^{60}$Ni ions. (Assume that the mass ratio is 58:60.)

**PICTURE THE PROBLEM**   The radius of curvature $r$ can be found using Equation 26-12. Using the mass dependence of $r$, we can find the radius for $^{60}$Ni ions from the radius for $^{58}$Ni ions and then take the difference.

(*a*) Solve Equation 26-12 for $r$:

$$r = \sqrt{\frac{2m |\Delta V|}{qB^2}} = \left[ \frac{2(9.62 \times 10^{-26}\ \text{kg})(3000\ \text{V})}{(1.6 \times 10^{-19}\ \text{C})(0.12\ \text{T})^2} \right]^{1/2}$$

$$= \boxed{0.501\ \text{m}}$$

(*b*) 1. Let $r_1$ and $r_2$ be the radius of the orbit of the $^{58}$Ni ion and the $^{60}$Ni ion, respectively. Use the result in Part (*a*) to find the ratio of $r_2$ to $r_1$:

$$\frac{r_2}{r_1} = \sqrt{\frac{m_2}{m_1}} = \sqrt{\frac{60}{58}} = 1.017$$

2. Use the result of the previous step to calculate $r_2$ for $^{60}$Ni:

$$r_2 = 1.017\, r_1 = (1.017)(0.501\ \text{m}) = 0.510\ \text{m}$$

3. The difference in orbital radii is $r_2 - r_1$:

$$r_2 - r_1 = 0.510\ \text{m} - 0.501\ \text{m} = \boxed{9\ \text{mm}}$$

---

## The Cyclotron

The cyclotron was invented by E. O. Lawrence and M. S. Livingston in 1934 to accelerate particles, such as protons or deuterons, to high kinetic energies.[†] The high-energy particles are used to bombard atomic nuclei, causing nuclear reactions that are then studied to obtain information about the nucleus. High-energy protons and deuterons are also used to produce radioactive materials and for medical purposes.

---

† A deuteron is the nucleus of heavy hydrogen, $^2$H, which consists of a proton and neutron tightly bound together.

Figure 26-20 is a schematic drawing of a cyclotron. The particles move in two semicircular metal containers called *dees*, after their shape. The dees are housed in a vacuum chamber that is in a uniform magnetic field provided by an electromagnet. The region in which the particles move must be evacuated so that the particles will not be scattered in collisions with air molecules and lose energy. A potential difference $\Delta V$, which alternates in time with a period $T$, is maintained between the dees. The period is chosen to be the cyclotron period $T = 2\pi m/(qB)$ (Equation 26-7). The potential difference creates an electric field across the gap between the dees. At the same time, there is no electric field within each dee because the metal dees act as shields.

Positively charged particles are initially injected into $\text{dee}_1$ with a small velocity from an ion source S near the center of the dees. They move in a semicircle in $\text{dee}_1$ and arrive at the gap between $\text{dee}_1$ and $\text{dee}_2$ after a time $\frac{1}{2}T$. The potential is adjusted so that $\text{dee}_1$ is at a higher potential than $\text{dee}_2$ when the particles arrive at the gap between them. Each particle is therefore accelerated across the gap by the electric field and gains kinetic energy equal to $q\,\Delta V$.

Because the particle now has more kinetic energy, the particle moves in a semicircle of larger radius in $\text{dee}_2$. It arrives at the gap again after a time $\frac{1}{2}T$, because the period is independent of the particle's speed. By this time, the potential between the dees has been reversed so that $\text{dee}_2$ is now at the higher potential. Once more the particle is accelerated across the gap and gains additional kinetic energy equal to $q\,\Delta V$. Each time the particle arrives at the gap, it is accelerated and gains kinetic energy equal to $q\,\Delta V$. Thus, the particle moves in larger and larger semicircular orbits until it eventually leaves the magnetic field. In the typical cyclotron, each particle may make 50 to 100 revolutions and exit with energies of up to several hundred mega-electron volts.

The kinetic energy of a particle leaving a cyclotron can be calculated by setting $r$ in Equation 26-6 equal to the maximum radius of the dees and solving the equation for $v$:

$$r = \frac{mv}{qB}, \quad v = \frac{qBr}{m}$$

Then

$$K = \frac{1}{2}mv^2 = \frac{1}{2}\left(\frac{q^2B^2}{m}\right)r^2 \qquad\qquad 26\text{-}13$$

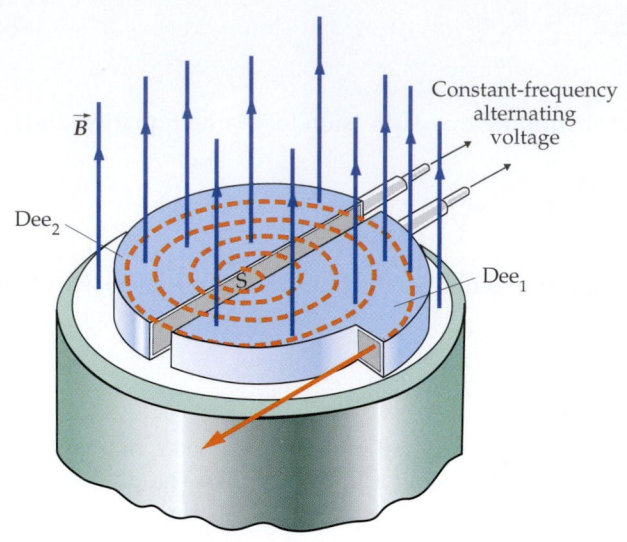

**FIGURE 26-20** Schematic drawing of a cyclotron. The upper-pole face of the magnet has been omitted. Charged particles, such as protons, are accelerated from a source at the center by the potential difference across the gap between the dees. When the charged particles arrive at the gap again the potential difference has changed sign so they are again accelerated across the gap and move in a larger circle. The potential difference across the gap alternates with the cyclotron frequency of the particle, which is independent of the radius of the circle.

---

*ENERGY OF ACCELERATED PROTON* **EXAMPLE 26-7**

**A cyclotron for accelerating protons has a magnetic field of 1.5 T and a maximum radius of 0.5 m. (*a*) What is the cyclotron frequency? (*b*) What is the kinetic energy of the protons when they emerge?**

(*a*) The cyclotron frequency is given by Equation 26-8:

$$f = \frac{qB}{2\pi m} = \frac{(1.6 \times 10^{-19}\,\text{C})(1.5\,\text{T})}{2\pi(1.67 \times 10^{-27}\,\text{kg})} = 2.29 \times 10^7\,\text{Hz}$$

$$= \boxed{22.9\,\text{MHz}}$$

(b) 1. The kinetic energy of the emerging protons is given by Equation 26-13:

$$K = \frac{1}{2}\left[\frac{(1.6 \times 10^{-19}\,\text{C})^2(1.5\,\text{T})^2}{1.67 \times 10^{-27}\,\text{kg}}\right](0.5\,\text{m})^2$$

$$= 4.31 \times 10^{-12}\,\text{J}$$

2. The energies of protons and other elementary particles are usually expressed in electron volts. Use $1\,\text{eV} = 1.6 \times 10^{-19}\,\text{J}$ to convert to eV:

$$K = 4.31 \times 10^{-12}\,\text{J} \times \frac{1\,\text{eV}}{1.6 \times 10^{-19}\,\text{J}} = \boxed{26.9\,\text{MeV}}$$

# 26-3 Torques on Current Loops and Magnets

A current-carrying loop experiences no net force in a uniform magnetic field, but it does experience a torque that tends to twist the current-carrying loop. The orientation of the loop can be described conveniently by a unit vector $\hat{n}$ that is perpendicular to the plane of the loop, as illustrated in Figure 26-21. If the fingers of the right hand curl around the loop in the direction of the current, the thumb points in the direction of $\hat{n}$.

Figure 26-22 shows the forces exerted by a uniform magnetic field on a rectangular loop whose normal unit vector $\hat{n}$ makes an angle $\theta$ with the magnetic field $\vec{B}$. The net force on the loop is zero. The forces $\vec{F}_1$ and $\vec{F}_2$ have the magnitude

$$F_1 = F_2 = IaB$$

These forces form a couple, so the torque is the same about any point. Point $P$ in Figure 26-22 is a convenient point about which to compute the torque. The magnitude of the torque is

$$\tau = F_2 b \sin\theta = IaBb \sin\theta = IAB \sin\theta$$

where $A = ab$ is the area of the loop. For a loop with $N$ turns, the torque has the magnitude

$$\tau = NIAB \sin\theta$$

This torque tends to twist the loop so that $\hat{n}$ is in the same direction as $\vec{B}$ (i.e., so that its plane is perpendicular to $\vec{B}$).

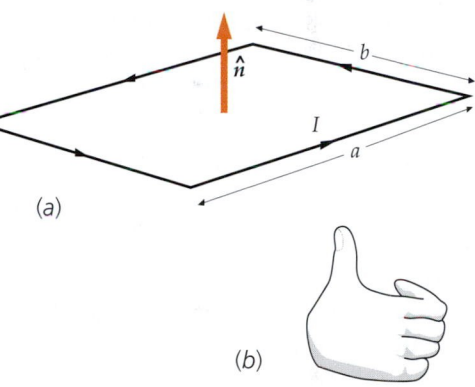

(a)

(b)

**FIGURE 26-21** (*a*) The orientation of a current loop is described by the unit vector $\hat{n}$ perpendicular to the plane of the loop. (*b*) Right-hand rule for determining the direction of $\hat{n}$. If the fingers of the right hand curl around the loop in the direction of the current, the thumb points in the direction of $\hat{n}$.

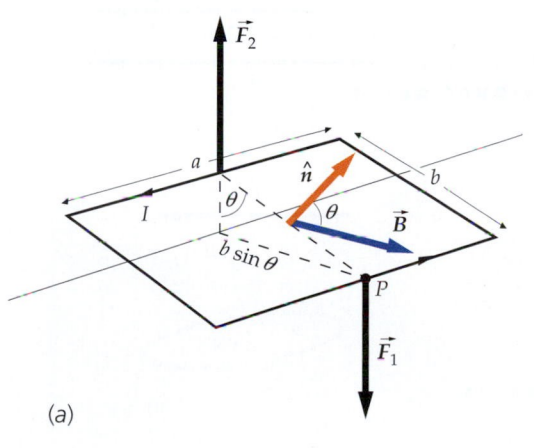

(a)

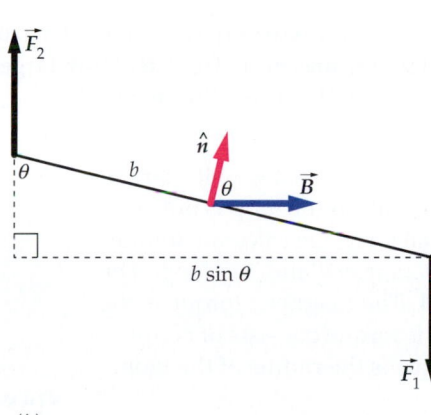

(b)

**FIGURE 26-22** (*a*) Rectangular current loop whose unit normal $\hat{n}$ makes an angle $\theta$ with a uniform magnetic field $\vec{B}$. (*b*) An edge-on view of the current loop. The torque on the loop has magnitude $IAB \sin\theta$ and is in the direction such that $\hat{n}$ tends to rotate into $\vec{B}$.

The torque can be written conveniently in terms of the **magnetic dipole moment** $\vec{\mu}$ (also referred to simply as the **magnetic moment**) of the current loop, which is defined as

$$\vec{\mu} = NIA\hat{n} \qquad\qquad 26\text{-}14$$

MAGNETIC DIPOLE MOMENT OF A CURRENT LOOP

The SI unit of magnetic moment is the ampere-meter$^2$ (A·m$^2$). In terms of the magnetic dipole moment, the torque on the current loop is given by

$$\vec{\tau} = \vec{\mu} \times \vec{B} \qquad\qquad 26\text{-}15$$

TORQUE ON A CURRENT LOOP

Equation 26-15, which we have derived for a rectangular loop, holds in general for a loop of any shape that is in a single plane. The torque on any loop is the cross product of the magnetic moment $\vec{\mu}$ of the loop and the magnetic field $\vec{B}$, where the magnetic moment is defined as a vector that is perpendicular to the plane of the loop (Figure 26-23), has magnitude equal to $NIA$, and has the same direction as $\hat{n}$. Comparing Equation 26-15 with Equation 21-11 ($\vec{\tau} = \vec{p} \times \vec{E}$) for the torque on an electric dipole, we see that the expression for the torque on a current loop in a magnetic field has the same form as that for the torque on an electric dipole in an electric field.

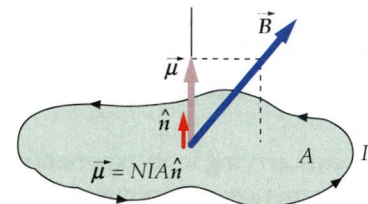

**FIGURE 26-23** A flat current loop of arbitrary shape is described by its magnetic moment $\vec{\mu} = NIA\hat{n}$. In a magnetic field $\vec{B}$, the loop experiences a torque $\vec{\mu} \times \vec{B}$.

---

TORQUE ON A CURRENT LOOP | **EXAMPLE 26-8**

**A circular loop of radius 2 cm has 10 turns of wire and carries a current of 3 A. The axis of the loop makes an angle of 30° with a magnetic field of 8000 G. Find the magnitude of the torque on the loop.**

The magnitude of the torque is given by Equation 26-15:

$$\tau = |\vec{\mu} \times \vec{B}| = \mu B \sin\theta = NIAB \sin\theta$$
$$= (10)(3\text{ A})\pi(0.02\text{ m})^2(0.8\text{ T}) \sin 30°$$
$$= \boxed{1.51 \times 10^{-2}\text{ N·m}^{-2}}$$

---

TILTING A LOOP | **EXAMPLE 26-9** **Try It Yourself**

**A circular wire loop of radius $R$, mass $m$, and current $I$ lies on a horizontal surface (Figure 26-24). There is a horizontal magnetic field $\vec{B}$. How large can the current $I$ be before one edge of the loop will lift off the surface?**

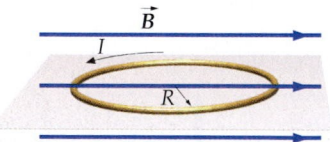

**FIGURE 26-24**

**PICTURE THE PROBLEM** The loop (Figure 26-25) will start to rotate when the magnitude of the net torque on the loop is not zero. To eliminate the torque due to the normal force, we calculate torques about the point of contact between the surface and the loop. The magnetic torque is given by $\vec{\tau} = \vec{\mu} \times \vec{B}$. The magnetic torque is the same about any point since the magnetic torque consists of couples. The lever arm for the gravitational torque is the radius of the loop.

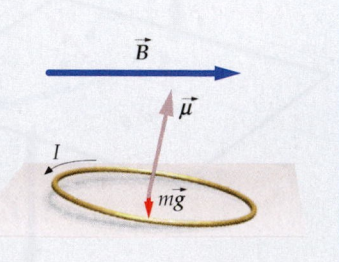

**FIGURE 26-25**

| Steps | Answers |
|---|---|
| 1. Find the magnitude of the magnetic torque acting on the loop. | $\tau_m = \mu B = I\pi R^2 B$ |
| 2. Find the magnitude of the gravitational torque exerted on the loop. | $\tau_g = mgR$ |
| 3. Equate the magnitudes of the torques and solve for the current $I$. | $I = \boxed{\dfrac{mg}{\pi RB}}$ |

■ **REMARKS** The torque vectors are equal and opposite.

## Potential Energy of a Magnetic Dipole in a Magnetic Field

When a torque is exerted through an angle, work is done. When a dipole is rotated through an angle $d\theta$, the work done is

$$dW = -\tau d\theta = -\mu B \sin\theta d\theta$$

The minus sign arises because the torque tends to decrease $\theta$. Setting this work equal to the decrease in potential energy, we have

$$dU = -dW = +\mu B \sin\theta d\theta$$

Integrating, we obtain

$$U = -\mu B \cos\theta + U_0$$

We choose the potential energy to be zero when $\theta = 90°$. Then $U_0 = 0$ and the potential energy of the dipole is

$$U = -\mu B \cos\theta = -\vec{\mu} \cdot \vec{B} \qquad\qquad 26\text{-}16$$

POTENTIAL ENERGY OF A MAGNETIC DIPOLE

Equation 26-16 gives the potential energy of a magnetic dipole at an angle $\theta$ to the direction of a magnetic field.

---

*TORQUE ON A COIL*        **E X A M P L E    2 6 - 1 0**

A square 12-turn coil with edge-length 40 cm carries a current of 3 A. It lies in the $xy$ plane, as shown in a uniform magnetic field $\vec{B} = 0.3$ T $\hat{i} + 0.4$ T $\hat{k}$. Find (*a*) the magnetic moment of the coil and (*b*) the torque exerted on the coil. (*c*) Find the potential energy of the coil.

**PICTURE THE PROBLEM** From Figure 26-26, we see that the magnetic moment of the loop is in the positive $z$ direction.

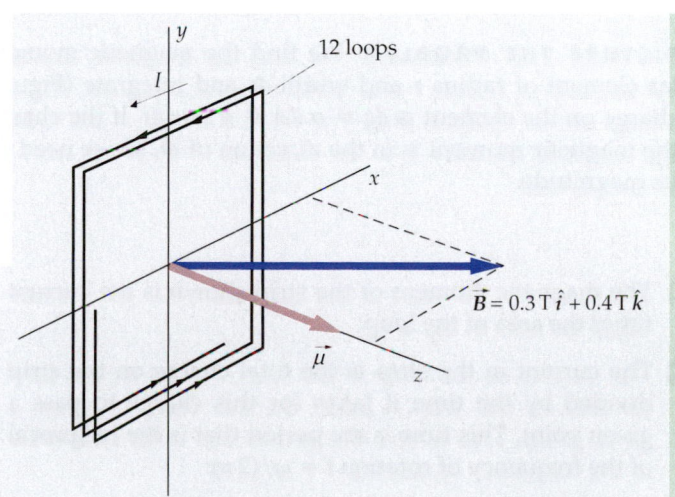

**FIGURE 26-26**

(a) Calculate the magnetic moment of the loop:

$$\vec{\mu} = NIA\,\hat{k} = (12)(3\text{ A})(0.40\text{ m})^2\,\hat{k}$$

$$= \boxed{5.76\text{ A}\cdot\text{m}^2\,\hat{k}}$$

(b) The torque on the current loop is given by Equation 26-15:

$$\vec{\tau} = \vec{\mu} \times \vec{B}$$

$$= (5.76\text{ A}\cdot\text{m}^2\,\hat{k}) \times (0.3\text{ T}\,\hat{i} + 0.4\text{ T}\,\hat{k})$$

$$= \boxed{1.73\text{ N}\cdot\text{m}\,\hat{j}}$$

(c) The potential energy is the negative dot product of $\vec{\mu}$ and $\vec{B}$:

$$U = -\vec{\mu} \cdot \vec{B}$$

$$= -(5.76\text{ A}\cdot\text{m}^2\,\hat{k}) \cdot (0.3\text{ T}\,\hat{i} + 0.4\text{ T}\,\hat{k})$$

$$= \boxed{-2.30\text{ J}}$$

**REMARKS** We have used $\hat{k} \times \hat{k} = 0$ and $\hat{k} \times \hat{i} = \hat{j}$, $\hat{k} \cdot \hat{i} = 0$ and $\hat{k} \cdot \hat{k} = 1$. The torque is in the $y$ direction.

**EXERCISE** Calculate $U$ if $\vec{B}$ and the magnetic moment $\vec{\mu}$ are in the same direction. (*Answer* $U = -\mu B = -(5.76\text{ A}\cdot\text{m}^2)(0.5\text{ T}) = -2.88$ J. Note that this potential energy is lower than that found in the example. The potential energy is lowest when $\vec{\mu}$ and $\vec{B}$ are in the same direction.)

When a small permanent magnet, such as a compass needle, is placed in a magnetic field $\vec{B}$, the field exerts a torque on the magnet that tends to rotate the magnet so that it lines up with the field. This effect also occurs with previously unmagnetized iron filings, which become magnetized in the presence of a $\vec{B}$ field. The bar magnet is characterized by a magnetic moment $\vec{\mu}$, a vector that points in the same direction as an arrow drawn from the south pole to the north pole. A small bar magnet thus behaves like a current loop. This is not a coincidence. The origin of the magnetic moment of a bar magnet is, in fact, microscopic current loops that result from the motion of electrons in the atoms of the magnet.

---

$\vec{\mu}$ OF A ROTATING DISK          **EXAMPLE 26-11**

**FIGURE 26-27**

**A thin nonconducting disk of mass $m$ and radius $R$ has a uniform surface charge per unit area $\sigma$ and rotates with angular velocity $\vec{\omega}$ about its axis. Find the magnetic moment of the rotating disk.**

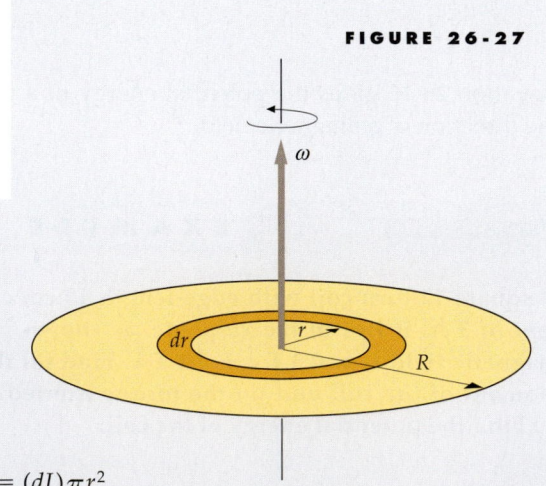

**PICTURE THE PROBLEM** We find the magnetic moment of a circular element of radius $r$ and width $dr$ and integrate (Figure 26-27). The charge on the element is $dq = \sigma\,dA = \sigma 2\pi r\,dr$. If the charge is positive, the magnetic moment is in the direction of $\vec{\omega}$, so we need only calculate its magnitude.

1. The magnetic moment of the strip shown is the current times the area of the loop:

$$d\mu = (dI)A = (dI)\pi r^2$$

2. The current in the strip is the total charge on the strip divided by the time it takes for this charge to pass a given point. This time is the period that is the reciprocal of the frequency of rotation $f = \omega/(2\pi)$:

$$dI = \frac{dq}{T} = (dq)f = (\sigma\,dA)\frac{\omega}{2\pi}$$

$$= (\sigma 2\pi r\,dr)\frac{\omega}{2\pi} = \sigma\omega r\,dr$$

3. Substitute to obtain the magnetic moment of the strip $d\mu$ in terms of $r$ and $dr$:

$$d\mu = (dI)\pi r^2 = (\sigma\omega r\, dr)\pi r^2 = \pi\sigma\omega r^3\, dr$$

4. Integrate from $r = 0$ to $r = R$:

$$\mu = \int d\mu = \int_0^R \pi\sigma\omega r^3\, dr = \frac{1}{4}\pi\sigma\omega R^4$$

5. Use the fact that $\vec{\mu}$ is parallel to $\vec{\omega}$ if $\sigma$ is positive to write the magnetic moment as a vector:

$$\boxed{\vec{\mu} = \tfrac{1}{4}\,\pi\sigma R^4\,\vec{\omega}}$$

**REMARKS** In terms of the total charge $Q = \sigma\pi R^2$, the magnetic moment is $\vec{\mu} = \frac{1}{4}QR^2\vec{\omega}$. The angular momentum of the disk is $\vec{L} = (\frac{1}{2}mR^2)\vec{\omega}$, so the magnetic moment can be written $\vec{\mu} = [Q/(2m)]\vec{L}$, which is a more general result. (See Problem 63.)

# 26-4 The Hall Effect

As we have seen, charges moving in a magnetic field experience a force perpendicular to their motion. When these charges are traveling in a conducting wire, they will be pushed to one side of the wire. This results in a separation of charge in the wire called the **Hall effect.** This phenomenon allows us to determine the sign of the charge on the charge carriers and the number of charge carriers per unit volume $n$ in a conductor. The Hall effect also provides a convenient method for measuring magnetic fields.

Figure 26-28 shows two conducting strips; each conducting strip carries a current $I$ to the right because the left sides of the strips are connected to the positive terminal of a battery and the right sides are connected to the negative terminal. The strips are in a magnetic field that is directed into the paper. Let us assume for the moment that the current in the strip consists of positively charged particles moving to the right, as shown in Figure 26-28a. The magnetic force on these particles is $q\vec{v}_d \times \vec{B}$ (where $\vec{v}_d$ is the drift velocity of the charge carriers). This force is directed upward. The positive particles therefore move up to the top of the strip, leaving the bottom of the strip with an excess negative charge. This separation of charge produces an electric field in the strip that opposes the magnetic force on the charge carriers. When the electric forces and magnetic forces balance, the charge carriers no longer move upward. Since the electric field points in the direction of decreasing potential, the upper part of the strip is at a higher potential than is the lower part of the strip. This potential difference can be measured using a sensitive voltmeter. On the other hand, if the current consists of moving negatively charged particles, as shown in Figure 26-28b, the charge carriers in the strip must move to the left (since the current is still to the right). The magnetic force $q\vec{v}_d \times \vec{B}$ is again up, because the signs of both $q$ and $\vec{v}_d$ have been reversed. Again the carriers are forced to the upper part of the strip, but the upper part of the strip now carries a negative charge (because the charge carriers are negative) and the lower part of the strip now carries a positive charge.

A measurement of the sign of the potential difference between the upper and lower parts of the strip tells us the sign of the charge carriers. In semiconductors, the charge carriers may be negative electrons or positive holes. A measurement of the sign of the potential difference tells us which are dominant for a particular semiconductor. For a normal metallic conductor, we find that the upper part of the strip in Figure 26-28b is at a lower potential than is the lower part of the strip—which means that the upper part must carry a negative charge. Thus, Figure 26-28b is the correct illustration of the current in a normal metallic conductor. It was this type of experiment that led to the discovery that the charge carriers in metallic conductors are negative.

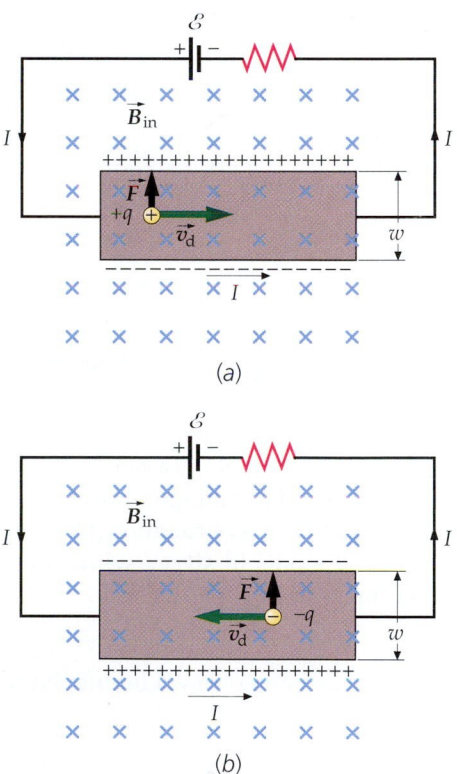

**FIGURE 26-28** The Hall effect. The magnetic field is directed into the plane of the page as indicated by the crosses. The magnetic force on a charged particle is upward for a current to the right whether the current is due to (a) positive particles moving to the right or (b) negative particles moving to the left.

The potential difference between the top of the strip and the bottom of the strip is called the **Hall voltage.** We can calculate the magnitude of the Hall voltage in terms of the drift velocity. The magnitude of the magnetic force on the charge carriers in the strip is $qv_dB$. This magnetic force is balanced by the electrostatic force of magnitude $qE_H$, where $E_H$ is the electric field due to the charge separation. Thus, we have $E_H = v_dB$. If the width of the strip is $w$, the potential difference is $E_Hw$. The Hall voltage is therefore

$$V_H = E_Hw = v_dBw \qquad\qquad 26\text{-}17$$

**EXERCISE** A conducting strip of width $w = 2.0$ cm is placed in a magnetic field of 0.8 T. The Hall voltage is measured to be 0.64 $\mu$V. Calculate the drift velocity of the electrons. (*Answer* $4.0 \times 10^{-5}$ m/s)

Since the drift velocity for ordinary currents is very small, we can see from Equation 26-17 that the Hall voltage is very small for ordinary-sized strips and magnetic fields. From measurements of the Hall voltage for a strip of a given size, we can determine the number of charge carriers per unit volume in the strip. The current is given by Equation 25-3:

$$I = nqv_dA$$

where $A$ is the cross-sectional area of the strip. For a strip of width $w$ and thickness $t$, the cross-sectional area is $A = wt$. Since the charge carriers are electrons, the quantity $q$ is the charge on one electron $e$. The number density of charge carriers $n$ is thus given by

$$n = \frac{I}{Aqv_d} = \frac{I}{wtev_d} \qquad\qquad 26\text{-}18$$

Substituting $V_H/B$ for $v_dw$ (Equation 26-17), we have

$$n = \frac{IB}{teV_H} \qquad\qquad 26\text{-}19$$

---

*CHARGE CARRIER NUMBER DENSITY IN SILVER* **EXAMPLE 26-12**

A silver slab of thickness 1 mm and width 1.5 cm carries a current of 2.5 A in a region in which there is a magnetic field of magnitude 1.25 T perpendicular to the slab. The Hall voltage is measured to be 0.334 $\mu$V. (*a*) Calculate the number density of the charge carriers. (*b*) Compare your answer in step 1 to the number density of atoms in silver, which has a mass density of $\rho = 10.5$ g/cm$^3$ and a molar mass of $M = 107.9$ g/mol.

1. Substitute numerical values into Equation 26-19 to find $n$:

$$n = \frac{IB}{teV_H} = \frac{(2.5\text{ A})(1.25\text{ T})}{(0.001\text{ m})(1.6 \times 10^{-19}\text{ C})(3.34 \times 10^{-7}\text{ V})}$$

$$= \boxed{5.85 \times 10^{28}\text{ electrons/m}^3}$$

2. The number of atoms per unit volume is $\rho N_A/M$:

$$n_a = \rho\frac{N_A}{M} = (10.5\text{ g/cm}^3)\frac{6.02 \times 10^{23}\text{ atoms/mol}}{107.9\text{ g/mol}}$$

$$= \boxed{5.86 \times 10^{22}\text{ atoms/cm}^3 = 5.86 \times 10^{28}\text{ atoms/m}^3}$$

**REMARKS** These results indicate that the number of charge carriers in silver is very nearly one per atom.

The Hall voltage provides a convenient method for measuring magnetic fields. If we rearrange Equation 26-19, we can write for the Hall voltage

$$V_H = \frac{I}{nte}B \qquad\qquad 26\text{-}20$$

A given strip can be calibrated by measuring the Hall voltage for a given current in a known magnetic field. The strip can then be used to measure an unknown magnetic field $B$ by measuring the Hall voltage for a given current.

## *The Quantum Hall Effects

According to Equation 26-20, the Hall voltage should increase linearly with magnetic field $B$ for a given current in a given slab. In 1980, while studying the Hall effect in semiconductors at very low temperatures and very large magnetic fields, the German physicist Klaus von Klitzing discovered that a plot of $V_H$ versus $B$ resulted in a series of plateaus, as shown in Figure 26-29, rather than a straight line. That is, the Hall voltage is quantized. For the discovery of the integer quantum Hall effect, von Klitzing won the Nobel Prize in physics in 1985.

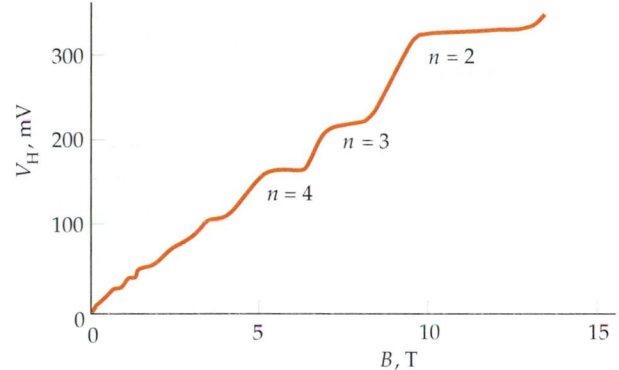

**FIGURE 26-29** A plot of the Hall voltage versus applied magnetic field shows plateaus, indicating that the Hall voltage is quantized. These data were taken at a temperature of 1.39 K with the current $I$ held fixed at 25.52 $\mu$A.

In the theory of the integer quantum Hall effect, the Hall resistance, defined as $R_H = V_H/I$, can take on only the values

$$R_H = \frac{V_H}{I} = \frac{R_K}{n}, \quad n = 1, 2, 3, \dots \qquad\qquad 26\text{-}21$$

where $n$ is an integer, and $R_K$, called the **von Klitzing constant,** is related to the fundamental electronic charge $e$ and Planck's constant $h$ by

$$R_K = \frac{h}{e^2} \qquad\qquad 26\text{-}22$$

Because the von Klitzing constant can be measured to an accuracy of a few parts per billion, the quantum Hall effect is now used to define a standard of resistance. As of January 1990, the **ohm** is defined in terms of the conventional value[†] of the von Klitzing constant $R_{K-90}$, which has the value

$$R_{K-90} = 25{,}812.807 \ \Omega \ (\text{exact}) \qquad\qquad 26\text{-}23$$

In 1982 it was observed that under certain special conditions the Hall resistance is given by Equation 26-22, but with the integer $n$ replaced by a series of rational fractions. This is called the fractional quantum Hall effect. For the discovery and explanation of the fractional quantum Hall effect, American professors Laughlin, Stormer, and Tsui won the Nobel Prize in physics in 1998.

---

[†] The value of $R_{K-90}$ differs only slightly from that of $R_K$. The currently used value of the von Klitzing constant is $R_K = (25\,812.807\,572 \pm 0.000\,095)\ \Omega$

1. The magnetic field describes the condition in space in which moving charges experience a force perpendicular to their velocity.

2. The magnetic force is part of the electromagnetic force, one of the four fundamental forces of nature.

3. The magnitude and direction of a magnetic field $\vec{B}$ are defined by the force $\vec{F} = q\vec{v} \times \vec{B}$ exerted on moving charges.

| Topic | Relevant Equations and Remarks | |
|---|---|---|
| **1. Magnetic Force** | | |
| On a moving charge | $\vec{F} = q\vec{v} \times \vec{B}$ | 26-1 |
| On a current element | $d\vec{F} = I\,d\vec{\ell} \times \vec{B}$ | 26-5 |
| Unit of the magnetic field | The SI unit of magnetic fields is the tesla (T). A commonly used unit is the gauss (G), which is related to the tesla by | |
| | $1\,\text{G} = 10^{-4}\,\text{T}$ | 26-3 |
| **2. Motion of Point Charges** | A particle of mass $m$ and charge $q$ moving with speed $v$ in a plane perpendicular to a uniform magnetic field moves in a circular orbit. The period and frequency of this circular motion are independent of the radius of the orbit and of the speed of the particle. | |
| Newton's second law | $qvB = m\dfrac{v^2}{r}$ | 26-6 |
| Cyclotron period | $T = \dfrac{2\pi m}{qB}$ | 26-7 |
| Cyclotron frequency | $f = \dfrac{1}{T} = \dfrac{qB}{2\pi m}$ | 26-8 |
| *Velocity selector | A velocity selector consists of crossed electric and magnetic fields so that the electric and magnetic forces balance for a particle moving with speed $v$. | |
| | $E = vB$ | 26-9 |
| *Thomson's measurement of $q/m$ | The deflection of a charged particle in an electric field depends on the speed of the particle and is proportional to the charge-to-mass ratio $q/m$ of the particle. J. J. Thomson used crossed electric and magnetic fields to measure the speed of cathode rays and then measured $q/m$ for these particles by deflecting them in an electric field. He showed that all cathode rays consist of particles that all have the same charge-to-mass ratio. These particles are now called electrons. | |
| *Mass spectrometer | The mass-to-charge ratio of an ion of known speed can be determined by measuring the radius of the circular path taken by the ion in a known magnetic field. | |
| **3. Current Loops** | | |
| Magnetic dipole moment | $\vec{\mu} = NIA\hat{n}$ | 26-14 |
| Torque | $\vec{\tau} = \vec{\mu} \times \vec{B}$ | 26-15 |

| Potential energy of a magnetic dipole | $U = -\vec{\mu} \cdot \vec{B}$ | 26-16 |

| Net force | The net force on a current loop in a *uniform* magnetic field is zero. | |

## 4. The Hall Effect

When a conducting strip carrying a current is placed in a magnetic field, the magnetic force on the charge carriers causes a separation of charge called the Hall effect. This results in a voltage $V_H$, called the Hall voltage. The sign of the charge carriers can be determined from a measurement of the sign of the Hall voltage, and the number of carriers per unit volume can be determined from the magnitude of $V_H$.

| Hall voltage | $V_H = E_H w = v_d B w = \dfrac{I}{nte}B$ | 26-17, 26-20 |

| *Quantum Hall effects | Measurements at very low temperatures in very large magnetic fields indicate that the Hall resistance $R_H = V_H/I$ is quantized and can take on only the values given by | |

$$R_H = \frac{V_H}{I} = \frac{R_K}{n}, \quad n = 1, 2, 3, \ldots \qquad \text{26-21}$$

| *Conventional von Klitzing constant (definition of ohm) | $R_{K-90} = 25{,}812.807 \ \Omega$ (exact) | 26-23 |

# PROBLEMS

- • Single-concept, single-step, relatively easy
- •• Intermediate-level, may require synthesis of concepts
- ••• Challenging
- SSM Solution is in the *Student Solutions Manual*
- iSOLVE Problems available on iSOLVE online homework service
- iSOLVE✓ These "Checkpoint" online homework service problems ask students additional questions about their confidence level, and how they arrived at their answer.

In a few problems, you are given more data than you actually need; in a few other problems, you are required to supply data from your general knowledge, outside sources, or informed estimates.

## Conceptual Problems

**1** • SSM When a cathode-ray tube is placed horizontally in a magnetic field that is directed vertically upward, the electrons emitted from the cathode follow one of the dashed paths to the face of the tube in Figure 26-30. The correct path is (a) 1. (b) 2. (c) 3. (d) 4. (e) 5.

FIGURE 26-30
Problem 1

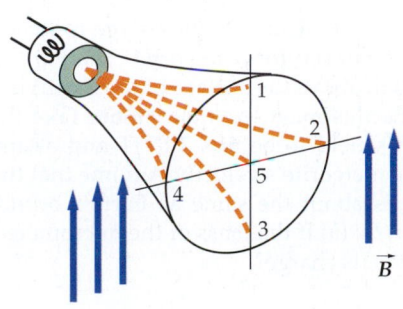

**2** • Why not define $\vec{B}$ to be in the direction of $\vec{F}$, as we do for $\vec{E}$?

**3** • True or false: The magnetic force does not accelerate a charged particle because the magnetic force is perpendicular to the velocity of the particle.

**4** • A beam of positively charged particles passes undeflected from left to right through a velocity selector in which the electric field is up. The beam is then reversed so that it travels from right to left. Will the beam now be deflected in the velocity selector? If so, in which direction?

**5** • SSM A *flicker bulb* is a lightbulb with a long, thin filament. When it is plugged in and a magnet is brought near the lightbulb, the filament is seen to oscillate rapidly back and forth. Why does the filament oscillate, and what is the frequency of oscillation?

**6** • What orientation of a current loop relative to the direction of the magnetic field gives maximum torque?

**7** • True or false:

(a) The magnetic force on a moving charged particle is always perpendicular to the velocity of the particle.

(b) The torque on a magnet by a magnetic field tends to align the magnet's magnetic moment in the direction of the magnetic field.

(c) A current loop in a uniform magnetic field responds to the field in the same manner as a small permanent magnet.

(d) The period of a particle moving in a circle in a magnetic field is proportional to the radius of the circle.

(e) The drift velocity of electrons in a wire can be determined from the Hall effect.

**8** • [SSM] The north-seeking pole of a compass needle located on the magnetic equator is the end of the needle that points toward the north, and the direction of any magnetic field $\vec{B}$ is specified as the direction that the north-seeking pole of a compass needle points when the needle is aligned in the field. Suppose that the direction of the magnetic field $\vec{B}$ were instead specified as the direction of a south-seeking pole of a compass needle aligned in the field. Would the right-hand rule shown in Figure 26-2 then give the direction of the magnetic force on the moving positive charge, or would a left-hand rule be required? Explain.

**9** • If the magnetic field is directed toward the north and a positively charged particle is moving toward the east, what is the direction of the magnetic force on the particle?

**10** • A positively charged particle is moving northward in a magnetic field. The magnetic force on the particle is toward the northeast. What is the direction of the magnetic field? (a) Up (b) West (c) South (d) Down (e) The force cannot be directed toward the northeast.

**11** • A $^7$Li nucleus with a charge of $+3e$ and a mass of 7 u (1 u $= 1.66 \times 10^{-27}$ kg) and a proton with charge $+e$ and mass 1 u are both moving in a plane perpendicular to a magnetic field $\vec{B}$. The magnitude of the momenta of the two particles are equal. The ratio of the radius of curvature of the path of the proton, $R_p$, to that of the $^7$Li nucleus, $R_{Li}$, is (a) $R_p/R_{Li} = 3$. (b) $R_p/R_{Li} = 1/3$. (c) $R_p/R_{Li} = 1/7$. (d) $R_p/R_{Li} = 3/7$. (e) none of these.

**12** • [SSM] An electron moving with speed $v$ to the right enters a region of uniform magnetic field directed out of the paper. When the electron enters this region, it will be (a) deflected out of the plane of the paper. (b) deflected into the plane of the paper. (c) deflected upward. (d) deflected downward. (e) undeviated in its motion.

**13** ••• The theory of relativity tells us that none of the laws of physics can depend on the absolute velocity of an object, which is in fact impossible to define. Instead, the behavior of physical objects can only depend on the relative velocity between the objects. The development of new physical insights can come from this idea. For example, in Figure 26-31, a magnet moving at high speed flies by an electron that is at rest relative to a physicist observing it in a laboratory. Explain why you are sure that a force must be acting on it. What direction will the force point when the north pole of the magnet passes directly underneath the electron? Explain.

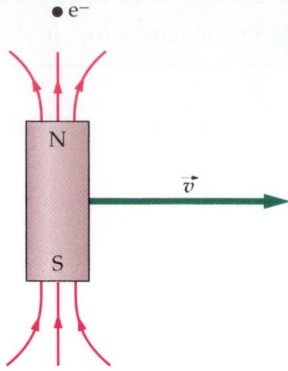

FIGURE 26-31 Problem 13

**14** • How are magnetic field lines similar to electric field lines? How are they different?

**15** • If a current $I$ in a given wire and a magnetic field $\vec{B}$ are known, the force $\vec{F}$ on the current is uniquely determined. Show that knowing $\vec{F}$ and $I$ does not provide complete knowledge of $\vec{B}$.

## Estimation and Approximation

**16** •• [SSM] CRT's used in monitors and televisions commonly use magnetic deflection to steer the electron beams. A schematic diagram is shown in Figure 26-32. The electron beam is accelerated through a potential difference and the electron beam is then accelerated through a magnetic field that deflects the electron beam, as shown in the figure. Given the following parameters, estimate the magnitude of the magnetic field needed for maximum deflection: accelerating voltage, $V = 15$ kV; distance over which electron is in magnetic field, $d = 5$ cm; length, $L = 50$ cm; diagonal of CRT, $r = 19$ in.

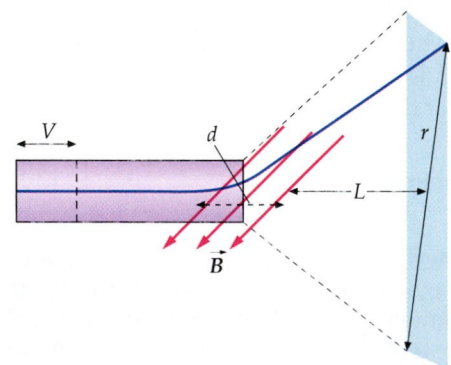

FIGURE 26-32 Problem 16

**17** •• (a) Estimate the charge-to-mass ratio of a micrometeorite needed for it to "orbit" the Earth in a low-earth orbit (400 km above the surface of the Earth) under the influence of the Earth's magnetic field alone. Take the magnitude of the Earth's field to be $5 \times 10^{-5}$ T and assume it perpendicular to the meteorite's velocity. Assume that the speed of the meteorite is about the same as Earth's orbital speed of roughly 30 km/s. (b) If the mass of the micrometeorite is $3 \times 10^{-10}$ kg, what is its charge?

## The Force Exerted by a Magnetic Field

**18** • Find the magnetic force on a proton moving with velocity 4.46 Mm/s in the positive $x$ direction in a magnetic field of 1.75 T in the positive $z$ direction.

**19** • **ISOLVE** ✔ A charge $q = -3.64$ nC moves with a velocity of $2.75 \times 10^6$ m/s $\hat{\imath}$. Find the force on the charge if the magnetic field is (a) $\vec{B} = 0.38$ T $\hat{\jmath}$, (b) $\vec{B} = 0.75$ T $\hat{\imath} + 0.75$ T $\hat{\jmath}$, (c) $\vec{B} = 0.65$ T $\hat{\imath}$, and (d) $\vec{B} = 0.75$ T $\hat{\imath} + 0.75$ T $\hat{k}$.

**20** • **ISOLVE** A uniform magnetic field of magnitude 1.48 T is in the positive $z$ direction. Find the force exerted by the field on a proton if the proton's velocity is (a) $\vec{v} = 2.7$ Mm/s $\hat{\imath}$, (b) $\vec{v} = 3.7$ Mm/s $\hat{\jmath}$, (c) $\vec{v} = 6.8$ Mm/s $\hat{k}$, and (d) $\vec{v} = 4.0$ Mm/s $\hat{\imath} + 3.0$ Mm/s $\hat{\jmath}$.

**21** • **ISOLVE** A straight wire segment 2 m long makes an angle of 30° with a uniform magnetic field of 0.37 T. Find the magnitude of the force on the wire if it carries a current of 2.6 A.

**22** • **SSM** A straight wire segment $I\vec{L} = (2.7 \text{ A})(3 \text{ cm } \hat{\imath} + 4 \text{ cm } \hat{\jmath})$ is in a uniform magnetic field $\vec{B} = 1.3$ T $\hat{\imath}$. Find the force on the wire.

**23** • **ISOLVE** ✔ What is the force (magnitude and direction) on an electron with velocity $\vec{v} = (2\hat{\imath} - 3\hat{\jmath} \times 10^6)$ m/s in a magnetic field $\vec{B} = (0.8 \hat{\imath} + 0.6 \hat{\jmath} - 0.4 \hat{k})$T?

**24** •• The wire segment shown in Figure 26-33 carries a current of 1.8 A from $a$ to $b$. There is a magnetic field $\vec{B} = 1.2$ T $\hat{k}$. Find the total force on the wire and show that the total force is the same as if the wire were a straight segment from $a$ to $b$.

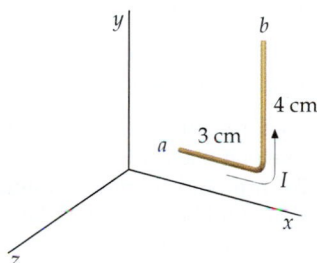

**FIGURE 26-33**
**Problem 24**

**25** •• **ISOLVE** A straight, stiff, horizontal wire of length 25 cm and mass 50 g is connected to a source of emf by light, flexible leads. A magnetic field of 1.33 T is horizontal and perpendicular to the wire. Find the current necessary to float the wire; that is, find the current so the magnetic force balances the weight of the wire.

**26** •• **SSM** A simple gaussmeter for measuring horizontal magnetic fields consists of a stiff 50-cm wire that hangs vertically from a conducting pivot so that its free end makes contact with a pool of mercury in a dish below. The mercury provides an electrical contact without constraining the movement of the wire. The wire has a mass of 5 g and conducts a current downward. (a) What is the equilibrium angular displacement of the wire from vertical if the horizontal magnetic field is 0.04 T and the current is 0.20 A? (b) If the current is 20 A and a displacement from vertical of 0.5 mm can be detected for the free end, what is the horizontal magnetic field sensitivity of this gaussmeter?

**27** •• A current-carrying wire is bent into a semicircular loop of radius $R$ that lies in the $xy$ plane. There is a uniform magnetic field $\vec{B} = B\hat{k}$ perpendicular to the plane of the loop (Figure 26-34). Verify that the force acting on the loop is 0.

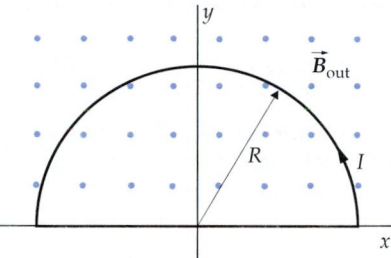

**FIGURE 26-34** Problem 27

**28** •• A 10-cm length of wire carries a current of 4.0 A in the positive $z$ direction. The force on this wire due to a uniform magnetic field $\vec{B}$ is $\vec{F} = (-0.2 \hat{\imath} + 0.2 \hat{\jmath})$N. If this wire is rotated so that the current flows in the positive $x$ direction, the force on the wire is $\vec{F} = 0.2 \hat{k}$N. Find the magnetic field $\vec{B}$.

**29** •• **ISOLVE** A 10-cm length of wire carries a current of 2.0 A in the positive $x$ direction. The force on this wire due to the presence of a magnetic field $\vec{B}$ is $\vec{F} = (3.0 \hat{\jmath} + 2.0 \hat{k})$N. If this wire is now rotated so that the current flows in the positive $y$ direction, the force on the wire is $\vec{F} = (-3.0 \hat{\imath} - 2.0 \hat{k})$N. Determine the magnetic field $\vec{B}$.

**30** ••• A wire bent in some arbitrary shape carries a current $I$ in a uniform magnetic field $\vec{B}$. Show explicitly that the total force on the part of the wire from some point $a$ to some point $b$ is $\vec{F} = I\vec{L} \times \vec{B}$, where $\vec{L}$ is the vector from point $a$ to point $b$.

## Motion of a Point Charge in a Magnetic Field

**31** • **SSM** A proton moves in a circular orbit of radius 65 cm perpendicular to a uniform magnetic field of magnitude 0.75 T. (a) What is the period for this motion? (b) Find the speed of the proton. (c) Find the kinetic energy of the proton.

**32** • An electron of kinetic energy 45 keV moves in a circular orbit perpendicular to a magnetic field of 0.325 T. (a) Find the radius of the orbit. (b) Find the frequency and period of the motion.

**33** • **ISOLVE** ✔ An electron from the sun with a speed of $1 \times 10^7$ m/s enters the earth's magnetic field high above the equator where the magnetic field is $4 \times 10^{-7}$ T. The electron moves nearly in a circle, except for a small drift along the direction of the earth's magnetic field that will take the electron toward the north pole. (a) What is the radius of the circular motion? (b) What is the radius of the circular motion near the north pole where the magnetic field is $2 \times 10^{-5}$ T?

**34** •• Protons and deuterons (each with charge $+e$) and alpha particles (with charge $+2e$) of the same kinetic energy enter a uniform magnetic field $\vec{B}$ that is perpendicular to their velocities. Let $R_p$, $R_d$, and $R_\alpha$ be the radii of their circular orbits. Find the ratios $R_d/R_p$ and $R_\alpha/R_p$. Assume that $m_\alpha = 2m_d = 4m_p$.

**35** •• A proton and an alpha particle move in a uniform magnetic field in circles of the same radii. Compare (a) their velocities, (b) their kinetic energies, and (c) their angular momenta. (See Problem 34.)

**36** •• A particle of charge $q$ and mass $m$ has momentum $p = mv$ and kinetic energy $K = p^2/2m$. If the particle moves in a circular orbit of radius $R$ perpendicular to a uniform magnetic field $\vec{B}$, show that (a) $p = BqR$ and (b) $K = \frac{1}{2}B^2q^2R^2/m$.

**37** •• [SSM] A beam of particles with velocity $\vec{v}$ enters a region of uniform magnetic field $\vec{B}$ that makes a small angle $\theta$ with $\vec{v}$. Show that after a particle moves a distance $2\pi(m/qB)v\cos\theta$, measured along the direction of $\vec{B}$, the velocity of the particle is in the same direction as it was when the particle entered the field.

**38** •• [SOLVE] A proton with speed $v = 10^7$ m/s enters a region of uniform magnetic field $B = 0.8$ T, which is into the page, as shown in Figure 26-35. The angle $\theta = 60°$. Find the angle $\phi$ and the distance $d$.

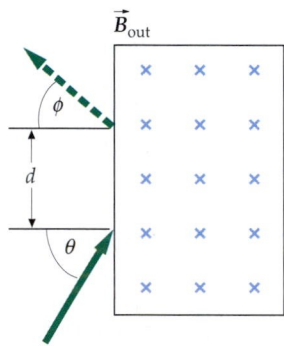

$\vec{B}_{\text{out}}$

**FIGURE 26-35** Problems 38 and 39

**39** •• Suppose that in Figure 26-35 $B = 0.6$ T, the distance $d = 0.4$ m, and $\theta = 24°$. Find the speed $v$ and the angle $\phi$ if the particles are (a) protons and (b) deuterons.

**40** •• The galactic magnetic field in some region of interstellar space has a magnitude of $10^{-9}$ T. A particle of interstellar dust has a mass of 10 $\mu$g and a total charge of 0.3 nC. How many years does it take to complete a circular orbit in the magnetic field?

## The Velocity Selector

**41** • [SSM] [SOLVE] A velocity selector has a magnetic field of magnitude 0.28 T perpendicular to an electric field of magnitude 0.46 MV/m. (a) What must the speed of a particle be for the particle to pass through undeflected? What energy must (b) protons and (c) electrons have to pass through undeflected?

**42** • A beam of protons moves along the $x$ axis in the positive $x$ direction with a speed of 12.4 km/s through a region of crossed fields balanced for zero deflection. (a) If there is a magnetic field of magnitude 0.85 T in the positive $y$ direction, find the magnitude and direction of the electric field. (b) Would electrons of the same velocity be deflected by these fields? If so, in what direction?

## Thomson's Measurement of q/m for Electrons and the Mass Spectrometer

**43** •• [SSM] [SOLVE ✔] The plates of a Thomson $q/m$ apparatus are 6.0 cm long and are separated by 1.2 cm. The end of the plates is 30.0 cm from the tube screen. The kinetic energy of the electrons is 2.8 keV. (a) If a potential of 25 V is applied across the deflection plates, by how much will the beam deflect? (b) Find the magnitude of the crossed magnetic field that will allow the beam to pass between the plates undeflected.

**44** •• Chlorine has two stable isotopes, $^{35}$Cl and $^{37}$Cl, whose natural abundances are about 76 percent and 24 percent, respectively. Singly ionized chlorine gas is to be separated into its isotopic components using a mass spectrometer. The magnetic field in the spectrometer is 1.2 T. What is the minimum value of the potential through which these ions must be accelerated so that the separation between them is 1.4 cm?

**45** •• [SOLVE] A singly ionized $^{24}$Mg ion (mass $3.983 \times 10^{-26}$ kg) is accelerated through a 2.5-kV potential difference and deflected in a magnetic field of 557 G in a mass spectrometer. (a) Find the radius of curvature of the orbit for the ion. (b) What is the difference in radius for $^{26}$Mg ions and for $^{24}$Mg ions? (Assume that their mass ratio is 26:24.)

**46** •• [SSM] A beam of $^6$Li and $^7$Li ions passes through a velocity selector and enters a magnetic spectrometer. If the diameter of the orbit of the $^6$Li ions is 15 cm, what is the diameter of the orbit for $^7$Li ions?

## The Cyclotron

**47** •• In Example 26-6, determine the time required for a $^{58}$Ni ion and a $^{60}$Ni ion to complete the semicircular path.

**48** •• Before entering a mass spectrometer, ions pass through a velocity selector consisting of parallel plates separated by 2.0 mm and having a potential difference of 160 V. The magnetic field between the plates is 0.42 T. The magnetic field in the mass spectrometer is 1.2 T. Find (a) the speed of the ions entering the mass spectrometer and (b) the difference in the diameters of the orbits of singly ionized $^{238}$U and $^{235}$U. (The mass of a $^{235}$U ion is $3.903 \times 10^{-25}$ kg.)

**49** •• [SSM] [SOLVE] A cyclotron for accelerating protons has a magnetic field of 1.4 T and a radius of 0.7 m. (a) What is the cyclotron frequency? (b) Find the maximum energy of the protons when they emerge. (c) How will your answers change if deuterons, which have the same charge but twice the mass, are used instead of protons?

**50** •• A certain cyclotron with a magnetic field of 1.8 T is designed to accelerate protons to 25 MeV. (a) What is the cyclotron frequency? (b) What must the minimum radius of the magnet be to achieve a 25-MeV emergence energy? (c) If the alternating potential applied to the dees has a maximum value of 50 kV, how many revolutions must the protons make before emerging with an energy of 25 MeV?

**51** •• [SOLVE] Show that for a certain cyclotron the cyclotron frequencies of deuterons and alpha particles are the same and are half that of a proton in the same magnetic field. (See Problem 34.)

**52** •• Show that the radius of the orbit of a charged particle in a cyclotron is proportional to the square root of the number of orbits completed.

## Torques on Current Loops and Magnets

**53** • A small circular coil of 20 turns of wire lies in a uniform magnetic field of 0.5 T, so that the normal to the plane of the coil makes an angle of 60° with the direction of $\vec{B}$. The radius of the coil is 4 cm, and it carries a current of 3 A. (a) What is the magnitude of the magnetic moment of the coil? (b) What is the magnitude of the torque exerted on the coil?

**54** • **iSOLVE** ✓ What is the maximum torque on a 400-turn circular coil of radius 0.75 cm that carries a current of 1.6 mA and resides in a uniform magnetic field of 0.25 T?

**55** • **SSM** **iSOLVE** ✓ A current-carrying wire is bent into the shape of a square of edge-length $L = 6$ cm and is placed in the $xy$ plane. It carries a current $I = 2.5$ A. What is the magnitude of the torque on the wire if there is a uniform magnetic field of 0.3 T (a) in the $z$ direction and (b) in the $x$ direction?

**56** • Repeat Problem 55 if the wire is bent into an equilateral triangle of edge-length 8 cm.

**57** •• A rigid, circular loop of radius $R$ and mass $m$ carries a current $I$ and lies in the $xy$ plane on a rough, flat table. There is a horizontal magnetic field of magnitude $B$. What is the minimum value of $B$ so that one edge of the loop will lift off the table?

**58** •• **iSOLVE** ✓ A rectangular, 50-turn coil has sides 6-cm long and 8-cm long and carries a current $I$ of 1.75 A. It is oriented and pivoted about the $z$ axis, as shown in Figure 26-36. (a) If the wire in the $xy$ plane makes an angle $\theta = 37°$ with the $y$ axis as shown, what angle does the unit normal $\hat{n}$ make with the $x$ axis? (b) Write an expression for $\hat{n}$ in terms of the unit vectors $\hat{i}$ and $\hat{j}$. (c) What is the magnetic moment of the coil? (d) Find the torque on the coil when there is a uniform magnetic field $\vec{B} = 1.5$ T $\hat{j}$. (e) Find the potential energy of the coil in this field.

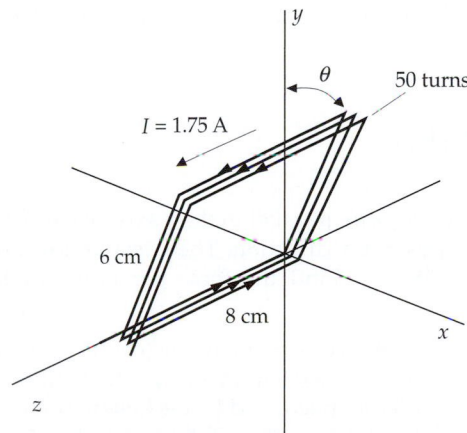

**FIGURE 26-36** Problems 58 and 59

**59** •• The coil in Problem 58 is pivoted about the $z$ axis and held at various positions in a uniform magnetic field $\vec{B} = 2.0$ T $\hat{j}$. Sketch the position of the coil and find the torque exerted when the unit normal is (a) $\hat{n} = \hat{i}$, (b) $\hat{n} = \hat{j}$, (c) $\hat{n} = -\hat{j}$, and (d) $\hat{n} = (\hat{i} + \hat{j})/\sqrt{2}$.

## Magnetic Moments

**60** •• **SSM** A small magnet of length 6.8 cm is placed at an angle of 60° to the direction of a uniform magnetic field of magnitude 0.04 T. The observed torque has a magnitude of 0.10 N·m. Find the magnetic moment of the magnet.

**61** •• **iSOLVE** ✓ A wire loop consists of two semicircles connected by straight segments (Figure 26-37). The inner and outer radii are 0.3 m and 0.5 m, respectively. A current $I$ of 1.5 A flows in this loop with the current in the outer semicircle in the clockwise direction. What is the magnetic moment of this current loop?

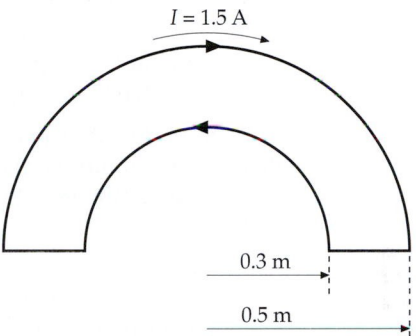

**FIGURE 26-37** Problem 61

**62** •• A wire of length $L$ is wound into a circular coil of $N$ loops. Show that when this coil carries a current $I$, its magnetic moment has the magnitude $IL^2/4\pi N$.

**63** •• A particle of charge $q$ and mass $m$ moves in a circle of radius $R$ and with angular velocity $\omega$. (a) Show that the average current is $I = q\omega/(2\pi)$ and that the magnetic moment has the magnitude $\mu = \frac{1}{2}q\omega r^2$. (b) Show that the angular momentum of this particle has the magnitude $L = mr^2\omega$ and that the magnetic moment and angular momentum vectors are related by $\vec{\mu} = (\frac{1}{2}q/m)\vec{L}$.

**64** ••• **SSM** A hollow cylinder has length $L$ and inner and outer radii $R_i$ and $R_o$, respectively (Figure 26-38). The cylinder carries a uniform charge density $\rho$. Derive an expression for the magnetic moment as a function of $\omega$, the angular velocity of rotation of the cylinder about its axis.

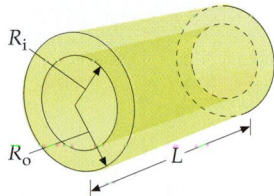

**FIGURE 26-38** Problem 64

**65** ••• A nonconducting rod of mass $m$ and length $L$ has a uniform charge per unit length $\lambda$ and rotates with angular velocity $\omega$ about an axis through one end and perpendicular to the rod. (a) Consider a small segment of the rod of length $dx$ and charge $dq = \lambda\,dx$ at a distance $x$ from the pivot (Figure 26-39). Show that the magnetic moment of this segment is $\frac{1}{2}\lambda\omega x^2 dx$. (b) Integrate your result to show that the total magnetic moment of the rod is $\mu = \frac{1}{6}\lambda\omega L^3$. (c) Show that the magnetic moment $\vec{\mu}$ and angular momentum $\vec{L}$ are related by $\vec{\mu} = (\frac{1}{2}Q/m)\vec{L}$, where $Q$ is the total charge on the rod.

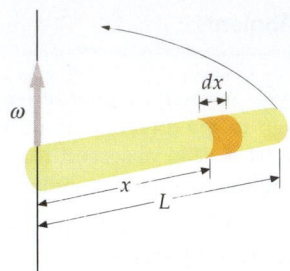

**FIGURE 26-39**
Problem 65

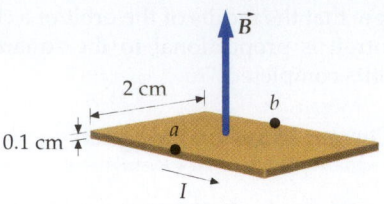

**FIGURE 26-41** Problems 70 and 71

**66** ••• A nonuniform, nonconducting disk of mass $m$, radius $R$, and total charge $Q$ has a surface charge density $\sigma = \sigma_0 r/R$ and a mass per unit area $\sigma_m = (m/Q)\sigma$. The disk rotates with angular velocity $\omega$ about its axis. (a) Show that the magnetic moment of the disk has a magnitude $\mu = \frac{1}{5}\pi\omega\sigma_0 R^4 = \frac{3}{10}Q\omega R^2$. (b) Show that the magnetic moment $\vec{\mu}$ and angular momentum $\vec{L}$ are related by $\vec{\mu} = (\frac{1}{2}Q/m)\vec{L}$.

**67** ••• A spherical shell of radius $R$ carries a surface charge density $\sigma$. The sphere rotates about its diameter with angular velocity $\omega$. Find the magnetic moment of the rotating sphere.

**68** ••• A solid sphere of radius $R$ carries a uniform volume charge density $\rho$. The sphere rotates about its diameter with angular velocity $\omega$. Find the magnetic moment of this rotating sphere.

**69** ••• **SSM** A uniform disk of mass $m$, radius $R$, and surface charge $\sigma$ rotates about its center with angular velocity $\omega$ in Figure 26-40. A uniform magnetic field of magnitude $\vec{B}$ threads the disk, making an angle $\theta$ with respect to the rotation axis of the disk. Calculate (a) the net torque acting on the disk and (b) the precession frequency of the disk in the magnetic field. (See pp. 316–317 for a discussion of precession.)

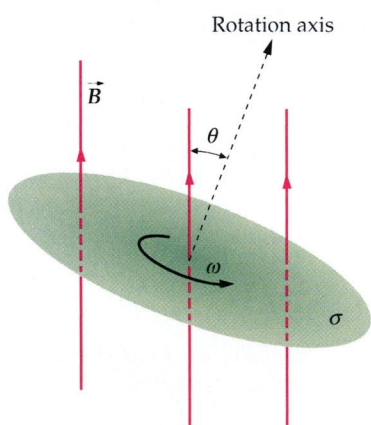

**FIGURE 26-40** Problem 69

## The Hall Effect

**70** • A metal strip 2-cm wide and 0.1-cm thick carries a current of 20 A in a uniform magnetic field of 2 T, as shown in Figure 26-41. The Hall voltage is measured to be 4.27 $\mu$V. (a) Calculate the drift velocity of the electrons in the strip. (b) Find the number density of the charge carriers in the strip. (c) Is point $a$ or point $b$ at the higher potential?

**71** •• **ISOLVE** ✓ The number density of free electrons in copper is $8.47 \times 10^{22}$ electrons per cubic centimeter. If the metal strip in Figure 26-41 is copper and the current is 10 A, find (a) the drift velocity $v_d$ and (b) the Hall voltage. (Assume that the magnetic field is 2.0 T.)

**72** •• **SSM** **ISOLVE** A copper strip ($n = 8.47 \times 10^{22}$ electrons per cubic centimeter) 2-cm wide and 0.1-cm thick is used to measure the magnitudes of unknown magnetic fields that are perpendicular to the strip. Find the magnitude of $B$ when $I = 20$ A and the Hall voltage is (a) 2.00 $\mu$V, (b) 5.25 $\mu$V, and (c) 8.00 $\mu$V.

**73** •• **ISOLVE** Because blood contains charged ions, moving blood develops a Hall voltage across the diameter of an artery. A large artery with a diameter of 0.85 cm has a flow speed of 0.6 m/s. If a section of this artery is in a magnetic field of 0.2 T, what is the maximum possible potential difference across the diameter of the artery?

**74** •• **ISOLVE** The Hall coefficient $R$ is defined as $R = E_y/(J_x B_z)$, where $J_x$ is the current per unit area in the $x$ direction in the slab, $B_z$ is the magnetic field in the $z$ direction, and $E_y$ is the resulting Hall field in the $y$ direction. Show that the Hall coefficient is $1/(nq)$, where $q$ is the charge of the charge carriers, $-1.6 \times 10^{-19}$ C if they are electrons. (The Hall coefficients of monovalent metals, such as copper, silver, and sodium are therefore negative.)

**75** •• **SSM** Aluminum has a density of $2.7 \times 10^3$ kg/m$^3$ and a molar mass of 27 g/mol. The Hall coefficient of aluminum is $R = -0.3 \times 10^{-10}$ m$^3$/C. (See Problem 74 for the definition of $R$.) Find the number of conduction electrons per aluminum atom.

## General Problems

**76** • A long wire parallel to the $x$ axis carries a current of 6.5 A in the positive $x$ direction. There is a uniform magnetic field $\vec{B} = 1.35$ T $\hat{j}$. Find the force per unit length on the wire.

**77** • **ISOLVE** ✓ An alpha particle (charge $+2e$) travels in a circular path of radius 0.5 m in a magnetic field of 1 T. Find (a) the period, (b) the speed, and (c) the kinetic energy (in electron volts) of the alpha particle. Take $m = 6.65 \times 10^{-27}$ kg for the mass of the alpha particle.

**78** •• The pole strength $q_m$ of a bar magnet is defined by $q_m = |\vec{\mu}|/L$, where $L$ is the length of the magnet. Show that the torque exerted on a bar magnet in a uniform magnetic field $\vec{B}$ is the same as if a force $+q_m\vec{B}$ is exerted on the north pole and a force $-q_m\vec{B}$ is exerted on the south pole.

**79** •• [SSM] A particle of mass $m$ and charge $q$ enters a region where there is a uniform magnetic field $\vec{B}$ along the $x$ axis. The initial velocity of the particle is $\vec{v} = v_{0x}\hat{i} + v_{0y}\hat{j}$, so the particle moves in a helix. (a) Show that the radius of the helix is $r = mv_{0y}/qB$. (b) Show that the particle takes a time $t = 2\pi m/qB$ to make one orbit around the helix.

**80** •• [SSM] [iSOLVE] A metal crossbar of mass $m$ rides on a pair of long, horizontal conducting rails separated by a distance $L$ and connected to a device that supplies constant current $I$ to the circuit, as shown in Figure 26-42. A uniform magnetic field $\vec{B}$ is established, as shown. (a) If there is no friction and the bar starts from rest at $t = 0$, show that at time $t$ the bar has velocity $v = (BIL/m)t$. (b) In which direction will the bar move? (c) If the coefficient of static friction is $\mu_S$, find the minimum field $B$ necessary to start the bar moving.

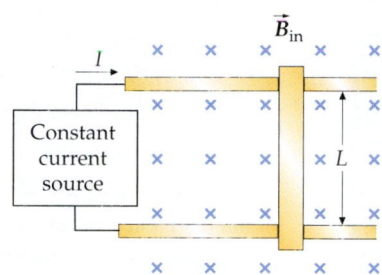

**FIGURE 26-42** Problems 80 and 81

**81** •• Assume that the rails in Figure 26-42 are frictionless but tilted upward so that they make an angle $\theta$ with the horizontal. (a) What vertical magnetic field $\vec{B}$ is needed to keep the bar from sliding down the rails? (b) What is the acceleration of the bar if $B$ has twice the value found in Part (a)?

**82** •• A long, narrow bar magnet that has magnetic moment $\vec{\mu}$ parallel to its long axis is suspended at its center as a frictionless compass needle. When placed in a horizontal magnetic field $\vec{B}$, the needle lines up with the field. If it is displaced by a small angle $\theta$, show that the needle will oscillate about its equilibrium position with frequency $f = \frac{1}{2\pi}\sqrt{\mu B/I}$, where $I$ is the moment of inertia about the point of suspension.

**83** •• [iSOLVE] A conducting wire is parallel to the $y$ axis. It moves in the positive $x$ direction with a speed of 20 m/s in a magnetic field $\vec{B} = 0.5$ T $\hat{k}$. (a) What are the magnitude and direction of the magnetic force on an electron in the conductor? (b) Because of this magnetic force, electrons move to one end of the wire leaving the other end positively charged, until the electric field due to this charge separation exerts a force on the electrons that balances the magnetic force. Find the magnitude and direction of this electric field in the steady state. (c) Suppose the moving wire is 2-m long. What is the potential difference between its two ends due to this electric field?

**84** ••• The rectangular frame shown in Figure 26-43 is free to rotate about the axis A–A on the horizontal shaft. The frame is 10-cm long and 6-cm wide, and the rods that make up the frame have a mass per unit length of 20 g/cm. A uniform magnetic field $B = 0.2$ T is directed, as shown. A current may be sent around the frame by means of the wires attached at the top. (a) If no current passes through the frame, what is the period of this physical pendulum for small oscillations? (b) If a current of 8 A passes through the frame in the direction indicated by the arrow, what is then the period of this physical

pendulum? (c) Suppose the direction of the current is opposite to the direction shown. The frame is displaced from the vertical by some angle $\theta$. What must be the magnitude of the current so that this frame will be in equilibrium?

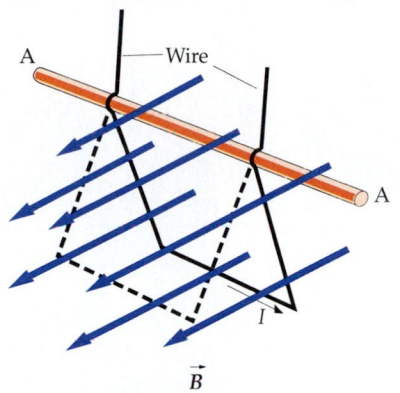

**FIGURE 26-43** Problem 84

**85** ••• [SSM] A stiff, straight horizontal wire of length 25 cm and mass 20 g is supported by electrical contacts at its ends, but is otherwise free to move vertically upward. The wire is in a uniform, horizontal magnetic field of magnitude 0.4 T perpendicular to the wire. A switch connecting the wire to a battery is closed and the wire flies upward, rising to a maximum height $h$. The battery delivers a total charge of 2 C during the short time it makes contact with the wire. Find the height $h$.

**86** ••• A circular loop of wire with mass $m$ carries a current $I$ in a uniform magnetic field. It is initially in equilibrium with its magnetic moment vector aligned with the magnetic field. The loop is given a small twist about a diameter and then released. What is the period of the motion? (Assume that the only torque exerted on the loop is due to the magnetic field.)

**87** ••• A small bar magnet has a magnetic moment $\vec{\mu}$ that makes an angle $\theta$ with the $x$ axis and lies in a nonuniform magnetic field given by $\vec{B} = B_x(x)\hat{i} + B_y(y)\hat{j}$. Use $F_x = -dU/dx$ and $F_y = -dU/dy$ to show that there is a net force on the magnet that is given by

$$\vec{F} = \mu_x \frac{\partial B_x}{\partial x}\hat{i} + \mu_y \frac{\partial B_y}{\partial y}\hat{j}$$

**88** ••• [SSM] The special theory of relativity tells us that a particle's mass depends on its speed through the formula:

$$m(v) = \frac{m_0}{\sqrt{1 - \dfrac{v^2}{c^2}}} = \gamma(v)m_0$$

where $m_0$ is the particle's rest mass and $\gamma(v) = 1/\sqrt{1 - (v^2/c^2)}$ (a) *Taking into account the special theory of relativity,* what is the radius and period of a particle's orbit if it has speed $v$ and is moving in a magnetic field with magnitude $B$ that is perpendicular to the direction of the velocity? Assume the force on the particle is given by $\vec{F} = q(\vec{v} \times \vec{B})$. The particle has rest mass $m_0$ and charge $q$. (b) Using a spreadsheet program, make graphs of the radius and period of the orbit of an electron in a 10-T magnetic field versus $\gamma(v)$ for speeds between $v = 0.1c$ and $v = 0.999c$. Use a logarithmic scale to display $\gamma(v)$.

# Sources of the Magnetic Field

THESE COILS AT THE KETTERING MAGNETICS LABORATORY AT OAKLAND UNIVERSITY ARE CALLED HELMHOLTZ COILS. THEY ARE USED TO CANCEL THE EARTH'S MAGNETIC FIELD AND TO PROVIDE A UNIFORM MAGNETIC FIELD IN A SMALL REGION OF SPACE FOR STUDYING THE MAGNETIC PROPERTIES OF MATTER.

**?** Have you any idea what the magnetic field of a current-carrying coil looks like? There are illustrations of the magnetic field of a coil in Section 27-2.

27-1    The Magnetic Field of Moving Point Charges

27-2    The Magnetic Field of Currents: The Biot–Savart Law

27-3    Gauss's Law for Magnetism

27-4    Ampère's Law

27-5    Magnetism in Matter

The earliest known sources of magnetism were permanent magnets. One month after Oersted announced his discovery that a compass needle is deflected by an electric current, Jean-Baptiste Biot and Félix Savart announced the results of their measurements of the torque on a magnet near a long, current-carrying wire and they analyzed these results in terms of the magnetic field produced by each element of the current. André-Marie Ampère extended these experiments and showed that current elements also experience a force in the presence of a magnetic field and that two currents exert forces on each other.

➤ In this chapter, we begin by considering the magnetic field produced by a single moving charge and by the moving charges in a current element. We then calculate the magnetic fields produced by some common current configurations, such as a straight wire segment; a long, straight wire; a current loop; and a solenoid. Next we discuss Ampère's law, which relates the line integral of the magnetic field around a closed loop to the total current that passes through the loop. Finally, we consider the magnetic properties of matter.

# 27-1 The Magnetic Field of Moving Point Charges

When a point charge $q$ moves with velocity $\vec{v}$, the moving point charge produces a magnetic field $\vec{B}$ in space, given by[†]

$$\vec{B} = \frac{\mu_0}{4\pi} \frac{q\vec{v} \times \hat{r}}{r^2} \qquad \qquad 27\text{-}1$$

MAGNETIC FIELD OF A MOVING POINT CHARGE

where $\hat{r}$ is a unit vector (see Figure 27-1) that points to the field point $P$ from the charge $q$ moving with velocity $\vec{v}$, and $\mu_0$ is a constant of proportionality called the **permeability of free space**,[‡] which has the exact value

$$\mu_0 = 4\pi \times 10^{-7}\,\text{T} \cdot \text{m/A} = 4\pi \times 10^{-7}\,\text{N/A}^2 \qquad \qquad 27\text{-}2$$

The units of $\mu_0$ are such that $B$ is in teslas when $q$ is in coulombs, $v$ is in meters per second, and $r$ is in meters. The unit $\text{N/A}^2$ comes from the fact that $1\,\text{T} = 1\,\text{N/(A·m)}$. The constant $1/(4\pi)$ is arbitrarily included in Equation 27-1 so that the factor $4\pi$ will not appear in Ampère's law (Equation 27-15), which we will study in Section 27-4.

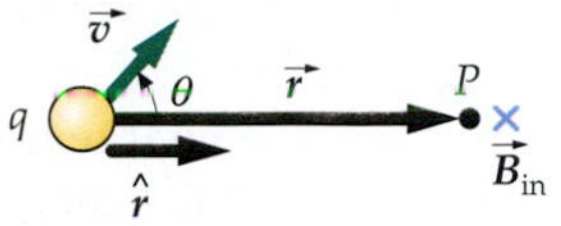

**FIGURE 27-1** A positive point charge $q$ moving with velocity $\vec{v}$ produces a magnetic field $\vec{B}$ at a field point $P$ that is in the direction $\vec{v} \times \hat{r}$, where $\hat{r}$ is the unit vector pointing from the charge to the field point. The field varies inversely as the square of the distance from the charge to the field point and is proportional to the sine of the angle between $\vec{v}$ and $\hat{r}$. (The blue × at the field point indicates that the direction of the field is into the page.)

---

MAGNETIC FIELD OF A MOVING POINT CHARGE

### EXAMPLE 27-1

A point particle with charge $q = 4.5$ nC is moving with velocity $\vec{v} = 3 \times 10^3$ m/s$\hat{i}$ parallel to the $x$ axis along the line $y = 3$ m. Find the magnetic field at the origin produced by this charge when the charge is at the point $x = -4$ m, $y = 3$ m, as shown in Figure 27-2.

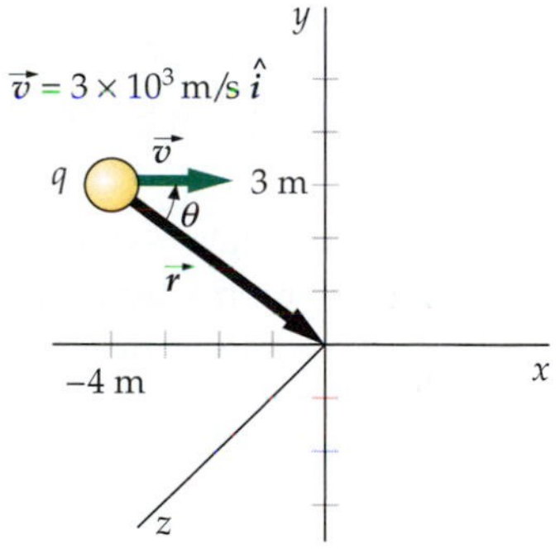

**FIGURE 27-2**

1. The magnetic field is given by Equation 27-1:

$$\vec{B} = \frac{\mu_0}{4\pi} \frac{q\vec{v} \times \hat{r}}{r^2}, \quad \text{with } \vec{v} = v\hat{i}$$

2. Find $\vec{r}$ and $r$ from Figure 27-2 and write $\hat{r}$ in terms of $\hat{i}$ and $\hat{j}$:

$$\vec{r} = 4\,\text{m}\hat{i} - 3\,\text{m}\hat{j}$$

$$r = \sqrt{4^2 + 3^2}\,\text{m} = 5\,\text{m}$$

$$\hat{r} = \frac{\vec{r}}{r} = \frac{4\,\text{m}\hat{i} - 3\,\text{m}\hat{j}}{5\,\text{m}} = 0.8\,\hat{i} - 0.6\,\hat{j}$$

3. Substitute the above results in Equation 27-1 to obtain:

$$\vec{B} = \frac{\mu_0}{4\pi} \frac{q\vec{v} \times \hat{r}}{r^2} = \frac{\mu_0}{4\pi} \frac{q(v\hat{i}) \times (0.8\,\hat{i} - 0.6\,\hat{j})}{r^2} = \frac{\mu_0}{4\pi} \frac{q(-0.6\,v\hat{k})}{r^2}$$

$$= -(10^{-7}\,\text{T·m/A}) \frac{(4.5 \times 10^{-9}\,\text{C})(0.6)(3 \times 10^3\,\text{m/s})}{(5\,\text{m})^2}\,\hat{k}$$

$$= \boxed{-3.24 \times 10^{-14}\,\text{T}\hat{k}}$$

---

[†] This expression is used for speeds much less than the speed of light.
[‡] Some care must be taken not to confuse the constant $\mu_0$ with the magnitude of the magnetic moment vector $\vec{\mu}$.

**REMARKS** It is also possible to obtain $\vec{B}$ without finding an explicit expression for the unit vector $\hat{r}$. From Figure 27-2 we note that $\vec{v} \times \hat{r}$ is in the negative $z$ direction. In addition, the magnitude of $\vec{v} \times \hat{r}$ is $v \sin \theta$, where $\sin \theta = (3 \text{ m})/(5 \text{ m}) = 0.6$. Combining these results, we have $\vec{v} \times \hat{r} = v \sin \theta(-\hat{k}) = -v(0.6)\hat{k}$, in agreement with our result in line 1 of step 3. Finally, this example shows that the magnetic field due to a moving charge is quite small. For comparison, the earth's magnetic field near its surface has a magnitude of about $10^{-4}$ T.

**EXERCISE** At the same instant, find the magnetic field on the $y$ axis both at $y = 3$ m and at $y = 6$ m. (*Answer* $\vec{B} = 0, \vec{B} = 3.24 \times 10^{-14}$ T$\hat{k}$)

(a)

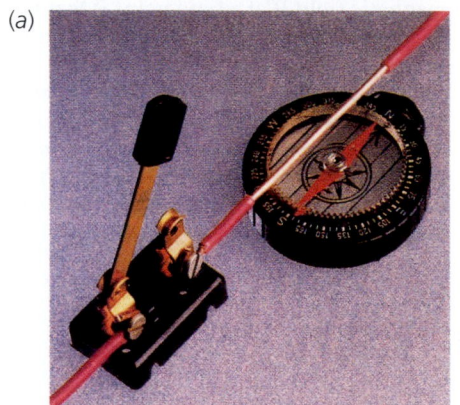

(b)

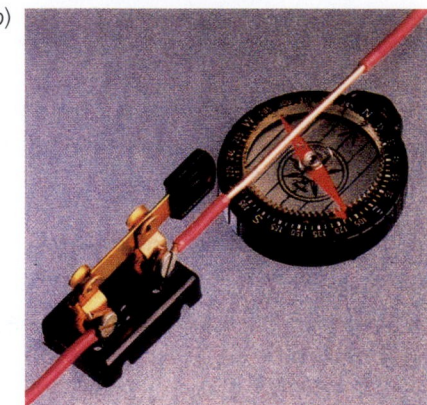

Oersted's experiment. (*a*) With no current in the wire, the compass needle points north. (*b*) When the wire carries a current, the compass needle is deflected in the direction of the resultant magnetic field. The current in the wire is directed upward, from left to right. The insulation has been stripped from the wire to improve the contrast of the photograph.

# 27-2 The Magnetic Field of Currents: The Biot–Savart Law

In the previous chapter we extended our discussion of forces on point charges to forces on current elements by replacing $q\vec{v}$ with the current element $I \, d\vec{\ell}$. We do the same for the magnetic field produced by a current element. The magnetic field $d\vec{B}$ produced by a current element $I \, d\vec{\ell}$ is given by Equation 27-1, with $q\vec{v}$ replaced by $I \, d\vec{\ell}$:

$$d\vec{B} = \frac{\mu_0}{4\pi} \frac{I \, d\vec{\ell} \times \hat{r}}{r^2} \qquad \qquad 27\text{-}3$$

BIOT–SAVART LAW

Equation 27-3, known as the **Biot–Savart law,** was also deduced by Ampère. The Biot–Savart law and Equation 27-1 are analogous to Coulomb's law for the electric field of a point charge. The source of the magnetic field is a moving charge $q\vec{v}$ or a current element $I \, d\vec{\ell}$, just as the charge $q$ is the source of the electrostatic field. The magnetic field decreases with the square of the distance from the moving charge or current element, just as the electric field decreases with the square of the distance from a point charge. However, the directional aspects of the electric and magnetic fields are quite different. Whereas the electric field points in the radial direction $\hat{r}$ from the point charge to the field point (for a positive charge), the magnetic field is perpendicular to both $\hat{r}$ and to $\vec{v}$, in the case of a point charge, or to $d\vec{\ell}$ in the case of a current element. At a point along the line of a current element, such as point $P_2$ in Figure 27-3, the magnetic field due to that element is zero. (Equation 27-3 gives $d\vec{B} = 0$ if $d\vec{\ell}$ and $\hat{r}$ are either parallel or antiparallel.)

The magnetic field due to the total current in a circuit can be calculated by using the Biot–Savart law to find the field due to each current element, and then summing (integrating) over all the current elements in the circuit. This calculation is difficult for all but the simplest circuit geometries.

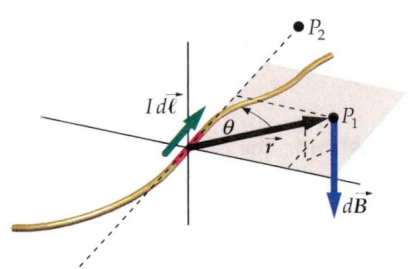

**FIGURE 27-3** The current element $I \, d\vec{\ell}$ produces a magnetic field at point $P_1$ that is perpendicular to both $d\vec{\ell}$ and $\hat{r}$. The current element produces no magnetic field at point $P_2$, which is along the line of $d\vec{\ell}$.

## $\vec{B}$ Due to a Current Loop

Figure 27-4 shows a current element $I\,d\vec{\ell}$ of a current loop of radius $R$ and the unit vector $\hat{r}$ that is directed from the element to the center of the loop. The magnetic field at the center of the loop due to this element is directed along the axis of the loop, and its magnitude is given by

$$dB = \frac{\mu_0}{4\pi} \frac{I\,d\ell\,\sin\theta}{R^2}$$

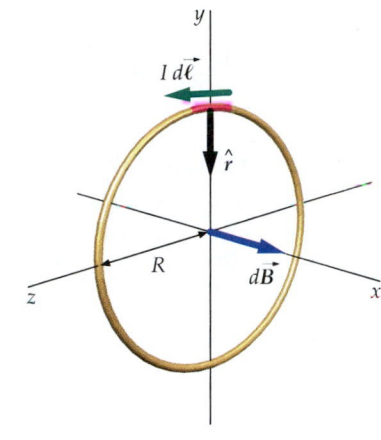

where $\theta$ is the angle between $d\vec{\ell}$ and $\hat{r}$, which is 90° for each current element, so $\sin\theta = 1$. The magnetic field due to the entire current is found by integrating over all the current elements in the loop. Since $R$ is the same for all elements, we obtain

$$B = \int dB = \frac{\mu_0}{4\pi}\frac{I}{R^2}\oint d\ell$$

The integral of $d\ell$ around the complete loop gives the total length $2\pi R$, the circumference of the loop. The magnetic field due to the entire loop is thus

**FIGURE 27-4** Current element for calculating the magnetic field at the center of a circular current loop. Each element produces a magnetic field that is directed along the axis of the loop.

$$B = \frac{\mu_0}{4\pi}\frac{I}{R^2}2\pi R = \frac{\mu_0 I}{2R} \qquad\qquad 27\text{-}4$$

$B$ AT THE CENTER OF A CURRENT LOOP

**EXERCISE** Find the current in a circular loop of radius 8 cm that will give a magnetic field of 2 G at the center of the loop. (*Answer*  25.5 A)

Figure 27-5 shows the geometry for calculating the magnetic field at a point on the axis of a circular current loop a distance $x$ from the circular loop's center. We first consider the current element at the top of the loop. Here, as everywhere on the loop, $I\,d\vec{\ell}$ is tangent to the loop and perpendicular to the vector $\vec{r}$ from the current element to the field point $P$. The magnetic field $d\vec{B}$ due to this element is in the direction shown in the figure, perpendicular to $\hat{r}$ and also perpendicular to $I\,d\vec{\ell}$. The magnitude of $d\vec{B}$ is

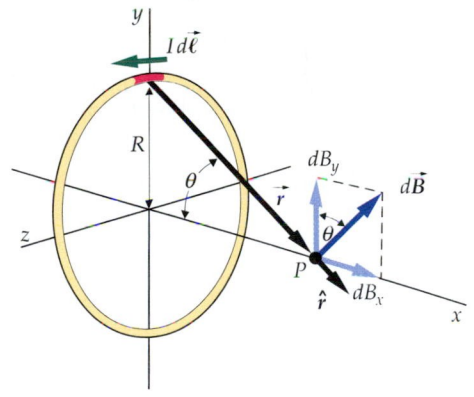

$$|d\vec{B}| = \frac{\mu_0}{4\pi}\frac{I|d\vec{\ell}\times\hat{r}|}{r^2} = \frac{\mu_0}{4\pi}\frac{I\,d\ell}{(x^2+R^2)}$$

where we have used the facts that $r^2 = x^2 + R^2$ and that $d\vec{\ell}$ and $\hat{r}$ are perpendicular, so $|d\vec{\ell}\times\hat{r}| = d\ell$.

When we sum around all the current elements in the loop, the components of $d\vec{B}$ perpendicular to the axis of the loop, such as $dB_y$ in Figure 27-5, sum to zero, which leave only the components $dB_x$ that are parallel to the axis. We thus compute only the $x$ component of the field. From Figure 27-5, we have

**FIGURE 27-5** Geometry for calculating the magnetic field at a point on the axis of a circular current loop.

$$dB_x = dB\sin\theta = \left(\frac{\mu_0}{4\pi}\frac{I\,d\ell}{(x^2+R^2)}\right)\left(\frac{R}{\sqrt{x^2+R^2}}\right) = \frac{\mu_0}{4\pi}\frac{IR\,d\ell}{(x^2+R^2)^{3/2}}$$

To find the field due to the entire loop of current, we integrate $dB_x$ around the loop:

$$B_x = \oint dB_x = \oint \frac{\mu_0}{4\pi}\frac{IR}{(x^2+R^2)^{3/2}}\,d\ell$$

Since neither $x$ nor $R$ varies as we sum over the elements in the loop, we can remove these quantities from the integral. Then,

$$B_x = \frac{\mu_0}{4\pi} \frac{IR}{(x^2 + R^2)^{3/2}} \oint d\ell$$

The integral of $d\ell$ around the loop gives $2\pi R$. Thus,

$$B_x = \frac{\mu_0}{4\pi} \frac{IR}{(x^2 + R^2)^{3/2}} 2\pi R = \frac{\mu_0}{4\pi} \frac{2\pi R^2 I}{(x^2 + R^2)^{3/2}} \qquad \text{27-5}$$

$B$ ON THE AXIS OF A CURRENT LOOP

**EXERCISE** Show that Equation 27-5 reduces to $B_x = \mu_0 I/2R$ (Equation 27-4) at the center of the loop.

At great distances from the loop, $|x|$ is much greater than $R$, so $(x^2 + R^2)^{3/2} \approx (x^2)^{3/2} = |x|^3$. Then,

$$B_x = \frac{\mu_0}{4\pi} \frac{2I\pi R^2}{|x|^3}$$

or

$$B_x = \frac{\mu_0}{4\pi} \frac{2\mu}{|x|^3} \qquad \text{27-6}$$

Magnetic-dipole field on the axis of the dipole

where $\mu = I\pi R^2$ is the magnitude of the magnetic moment of the loop. Note the similarity of this expression and the electric field on the axis of an electric dipole of moment $\rho$ (Equation 21-10):

$$E_x = \frac{1}{4\pi\epsilon_0} \frac{2\rho}{|x|^3}$$

Although it has not been demonstrated, our result that a current loop produces a magnetic dipole field far away holds in general for any point whether it is on the axis of the loop or off of the axis of the loop. Thus, a current loop behaves as a magnetic dipole because it experiences a torque $\vec{\mu} \times \vec{B}$ when placed in an external magnetic field (as was shown in Chapter 26) and it also produces a magnetic dipole field at a great distance from the current loop. Figure 27-6 shows the magnetic field lines for a current loop.

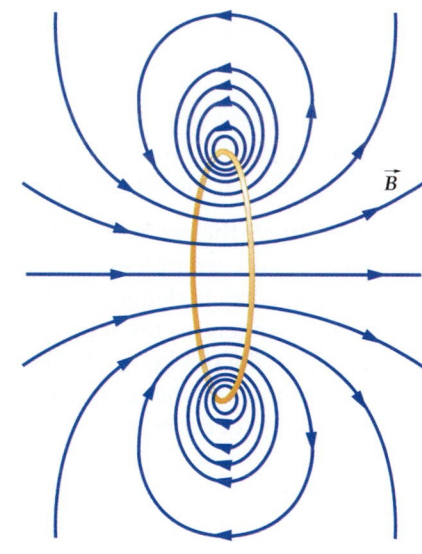

(a)

(b)

**FIGURE 27-6** (a) The magnetic field lines of a circular current loop. (b) The magnetic field lines of a circular current loop indicated by iron filings.

A circular coil of radius 5.0 cm has 12 turns and lies in the $x = 0$ plane and is centered at the origin. It carries a current of 4 A so that the direction of the magnetic moment of the coil is along the $x$ axis. Using Equation 27-5, find the magnetic field on the $x$ axis at (*a*) $x = 0$, (*b*) $x = 15$ cm, and (*c*) $x = 3$ m. (*d*) Using Equation 27-6, find the magnetic field on the $x$ axis at $x = 3$ m.

**PICTURE THE PROBLEM** The magnetic field due to a loop with $N$ turns is $N$ times that due to a single turn. (*a*) At $x = 0$ (center of the loops) $B = \mu_0 N/(2R)$ (from Equation 27-4). Equation 27-5 gives the magnetic field on axis due to the current in a single turn. Far from the loop, as in Part (*c*), the field can be found using Equation 27-6. In this case, since we have $N$ loops, the magnetic moment is $\mu = NI\pi R^2$.

(*a*) $B_x$ at the center is $N$ times that given by Equation 27-4 for a single loop:

$$B_x = \frac{\mu_0 NI}{2R}$$

$$= (4\pi \times 10^{-7}\ \text{T·m/A})\,\frac{(12)(4\ \text{A})}{2(0.05\ \text{m})} = \boxed{6.03 \times 10^{-4}\ \text{T}}$$

(*b*) $B_x$ on the axis is $N$ times that given by Equation 27-5:

$$B_x = \frac{\mu_0}{4\pi}\,\frac{2\pi R^2 NI}{(x^2 + R^2)^{3/2}}$$

$$= (10^{-7}\ \text{T·m/A})\,\frac{2\pi(0.05\ \text{m})^2(12)(4\ \text{A})}{[(0.15\ \text{m})^2 + (0.05\ \text{m})^2]^{3/2}}$$

$$= \boxed{1.91 \times 10^{-5}\ \text{T}}$$

(*c*) Use Equation 27-5 again:

$$B_x = \frac{\mu_0}{4\pi}\,\frac{2\pi R^2 NI}{(x^2 + R^2)^{3/2}}$$

$$= (10^{-7}\ \text{T·m/A})\,\frac{2\pi(0.05\ \text{m})^2(12)(4\ \text{A})}{[(3\ \text{m})^2 + (0.05\ \text{m})^2]^{3/2}}$$

$$= \boxed{2.791 \times 10^{-9}\ \text{T}}$$

(*d*) 1. Since 3 m is much greater than the radius $R = 0.05$ m, we can use Equation 27-6 for the magnetic field far from the loop:

$$B_x = \frac{\mu_0}{4\pi}\,\frac{2\mu}{|x|^3}$$

2. The magnitude of the magnetic moment of the loop is N/A:

$$\mu = NI\pi R^2 = (12)(4\ \text{A})\pi(0.05\ \text{m})^2 = 0.377\ \text{A·m}^2$$

3. Substitute $\mu$ and $x = 3$ m into $B_x$ in step 1:

$$B_x = \frac{\mu_0}{4\pi}\,\frac{2\mu}{|x|^3} = (10^{-7}\ \text{T·m/A})\,\frac{2(0.377\ \text{A·m}^2)}{(3\ \text{m})^3}$$

$$= \boxed{2.793 \times 10^{-9}\ \text{T}}$$

**REMARKS** In Part (*d*) $x = 60R$, so we were able to use an approximation that is valid for $x \gg R$. The result differs from the exact value, calculated in Part (*c*), by less than one tenth of one percent.

CIRCULATING THE AMOUNT OF MOBILE CHARGE

**EXAMPLE 27-3**

In the coil described in Example 27-2 the current is 4 A. Assuming the drift speed is $1.4 \times 10^{-4}$ m/s, find the number of coulombs of mobile charge in the wire. (The drift speed for a wire carrying a current of 1 A was found to be $3.4 \times 10^{-5}$ m/s in Example 25-1.)

**PICTURE THE PROBLEM** The amount of moving charge $Q$ in the wire is the product of the rate at which charge enters one end of the wire and the time it takes the charge to travel the length of the wire. The rate at which charge enters one end of the wire is the current $I$, and the time for the charge to travel the length $L$ of the wire is $L/v_d$, which is the drift speed.

1. The amount of moving charge is the product of the current and the time for a charge carrier to travel the length of the wire:

$$Q = I\,\Delta t$$

2. The drift speed is the length of the wire divided by the time:

$$v_d = \frac{L}{\Delta t}$$

3. The length $L$ is the number of turns times the length per turn. Also, we solve the step 2 result for the time:

$$L = N2\pi R = (12)2\pi(.05\text{ m}) = 3.77\text{ m}$$

and

$$\Delta t = \frac{L}{v_d} = \frac{3.77\text{ m}}{1.4 \times 10^{-4}\text{ m/s}} = 2.69 \times 10^4\text{ s}$$

4. Solve the step 1 result for the amount of moving charge in the wire:

$$Q = I\,\Delta t = (4\text{ A})(2.69 \times 10^4\text{ s})$$

$$= \boxed{1.08 \times 10^5\text{ C}}$$

**REMARKS** The current consists of more than $10^5$ C of moving charges. This is an enormous amount of charge, in comparison to the amount of charge stored in an ordinary capacitor.

---

TORQUE ON A BAR MAGNET

**EXAMPLE 27-4** **Try It Yourself**

A small bar magnet of magnetic moment $\mu = 0.03$ A·m² is placed at the center of the coil of Example 27-2 so that its magnetic moment vector lies in the $xy$ plane and makes an angle of 30° with the $x$ axis. Neglecting any variation in $\vec{B}$ over the region of the magnet, find the torque on the magnet.

**PICTURE THE PROBLEM** The torque on a magnetic moment is given by $\vec{\tau} = \vec{\mu} \times \vec{B}$. Since $\vec{B}$ is in the positive $x$ direction, you can see from Figure 27-7 that $\vec{\mu} \times \vec{B}$ is in the negative $z$ direction.

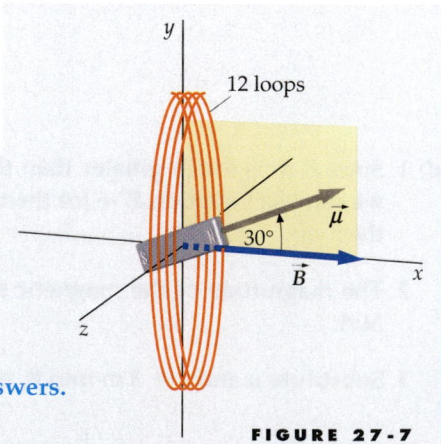

**FIGURE 27-7**

**Cover the column to the right and try these on your own before looking at the answers.**

| Steps | Answers |
|---|---|
| 1. Compute the magnitude of the torque from $\vec{\tau} = \vec{\mu} \times \vec{B}$. | $\tau = 9.04 \times 10^{-6}$ N·m |
| 2. Indicate the direction with a unit vector. | $\vec{\tau} = \boxed{-(9.04 \times 10^{-6}\text{ N·m})\hat{k}}$ |

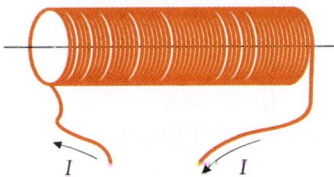

**FIGURE 27-8** A tightly wound solenoid can be considered as a set of circular current loops placed side by side that carry the same current. The solenoid produces a uniform magnetic field inside the loops.

## $\vec{B}$ Due to a Current in a Solenoid

A **solenoid** is a wire tightly wound into a helix of closely spaced turns, as illustrated in Figure 27-8. A solenoid is used to produce a strong, uniform magnetic field in the region surrounded by its loops. The solenoid's role in magnetism is analogous to that of the parallel-plate capacitor, which produces a strong, uniform electric field between its plates. The magnetic field of a solenoid is essentially that of a set of $N$ identical current loops placed side by side. Figure 27-9 shows the magnetic field lines for two such loops.

Figure 27-10 shows the magnetic field lines for a long, tightly wound solenoid. Inside the solenoid, the field lines are approximately parallel to the axis and are closely and uniformly spaced, indicating a strong, uniform magnetic field. Outside the solenoid, the lines are much less dense. The field lines diverge from one end and converge at the other end. Comparing this figure with Figure 27-8, we see that the field lines of a solenoid, both inside and outside the solenoid, are identical to those of a bar magnet of the same shape as the solenoid.

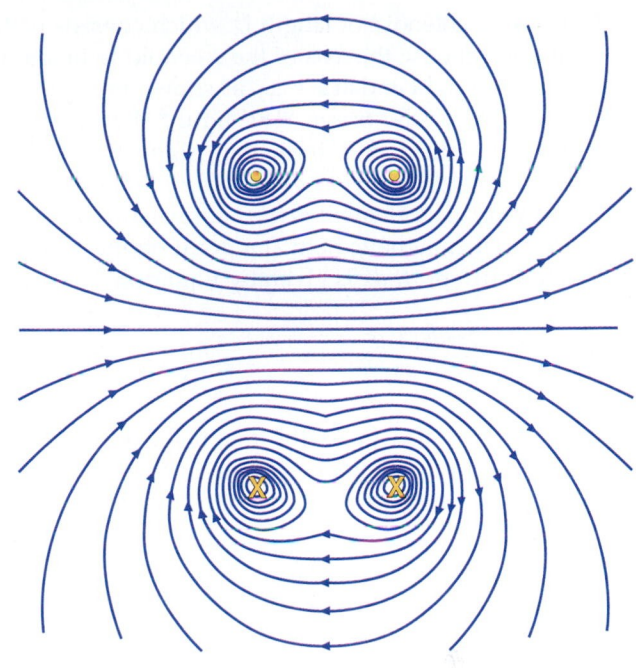

**FIGURE 27-9** Magnetic field lines due to two coaxial loops carrying the same current. The points where the loops intersect the plane of the page are each marked by an × where the current enters and by a dot where the current emerges. In the region between the loops near the axis the magnetic fields of the individual loops superpose, so the resultant field is strong and surprisingly uniform. In the regions away from the loops, the resultant field is relatively weak.

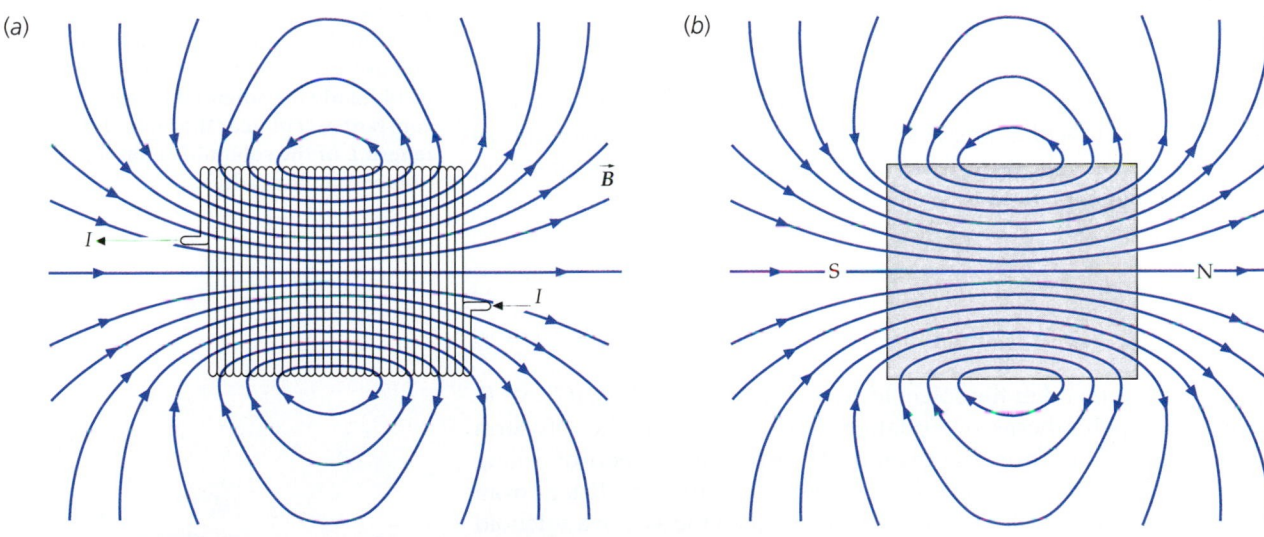

(a)

(b)

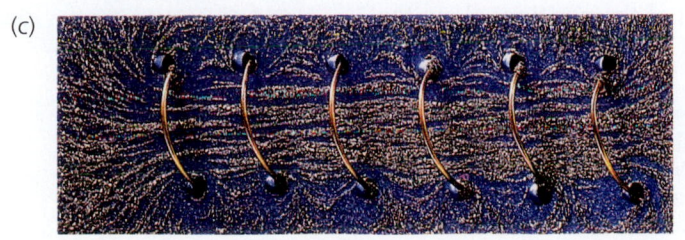

(c)

**FIGURE 27-10** (a) Magnetic field lines of a solenoid. The lines are identical to those of a bar magnet of the same shape, as in Figure 27-10 (b). (c) Magnetic field lines of a solenoid shown by iron filings.

Consider a solenoid of length $L$, which consists of $N$ turns of wire carrying a current $I$. We choose the axis of the solenoid to be the $x$ axis, with the left end at $x = x_1$ and the right end at $x = x_2$, as shown in Figure 27-11. We will calculate the magnetic field at the origin. The figure shows an element of the solenoid of length $dx$ at a distance $x$ from the origin. If $n = N/L$ is the number of turns per unit length, there are $n\,dx$ turns of wire in this element, with each turn carrying a current $I$. The element is thus equivalent to a single loop carrying a current $di = nI\,dx$. The magnetic field at a point on the $x$ axis due to a loop at the origin carrying a current $nI\,dx$ is given by Equation 27-5 with $I$ replaced by $di = nI\,dx$:

$$dB_x = \frac{\mu_0}{4\pi} \frac{2\pi R^2 nI\,dx}{(x^2 + R^2)^{3/2}}$$

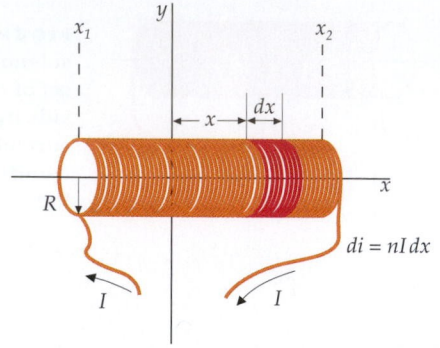

**FIGURE 27-11** Geometry for calculating the magnetic field inside a solenoid on its axis. The number of turns in the element $dx$ is $n\,dx$, where $n = N/L$ is the number of turns per unit length. The element $dx$ is treated as a current loop carrying a current $di = nI\,dx$.

This expression also gives the magnetic field at the origin due to a current loop at $x$. We find the magnetic field at the origin due to the entire solenoid by integrating this expression from $x = x_1$ to $x = x_2$:

$$B_x = \frac{\mu_0}{4\pi} 2\pi R^2 nI \int_{x_1}^{x_2} \frac{dx}{(x^2 + R^2)^{3/2}} \qquad 27\text{-}7$$

The integral in Equation 27-7 can be evaluated using trigonometric substitution with $x = R \tan \theta$. Also, the integral can be looked up in standard tables of integrals. The integral's value is

$$\int_{x_1}^{x_2} \frac{dx}{(x^2 + R^2)^{3/2}} = \frac{x}{R^2 \sqrt{x^2 + R^2}}\bigg|_{x_1}^{x_2} = \frac{1}{R^2}\left(\frac{x_2}{\sqrt{x_2^2 + R^2}} - \frac{x_1}{\sqrt{x_1^2 + R^2}}\right)$$

Substituting this into Equation 27-7, we obtain

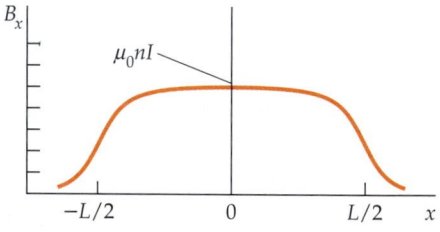

$$B_x = \frac{1}{2}\mu_0 nI\left(\frac{x_2}{\sqrt{x_2^2 + R^2}} - \frac{x_1}{\sqrt{x_1^2 + R^2}}\right) \qquad 27\text{-}8$$

$B_x$ ON THE AXIS OF A SOLENOID AT $X = 0$

**FIGURE 27-12** Graph of the magnetic field on the axis inside a solenoid versus the position $x$ on the axis. The field inside the solenoid is nearly constant except near the ends. The length $L$ of the solenoid is ten times longer than the radius.

A solenoid is called a long solenoid if its length $L$ is much greater than its radius $R$. Inside and far from the ends of a long solenoid, the left term in the parentheses tends toward $+1$ and the right term tends toward $-1$. In the region satisfying these conditions, the magnetic field is

$$B_x = \mu_0 nI \qquad 27\text{-}9$$

$B_x$ INSIDE A LONG SOLENOID

If the origin is at the left end of the solenoid, $x_1 = 0$ and $x_2 = L$. Then, if $L \gg R$, the right term in the parentheses of Equation 27-8 is zero and the left term approaches 1, so $B \approx \frac{1}{2}\mu_0 nI$. Thus, the magnitude of $\vec{B}$ at either end of a long solenoid is half the magnitude at points within the solenoid that are distant from either end. Figure 27-12 gives a plot of the magnetic field on the axis of a solenoid versus position $x$ on the axis (with the origin at the center of the solenoid). The approximation that the field is uniform (independent of the position) along the axis is good, except for very near the ends.

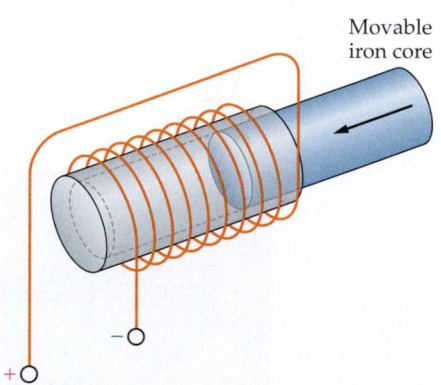

Movable iron core

**FIGURE 27-13** An automotive starter solenoid. When the solenoid is energized, its magnetic field pulls in the iron core. This engages gears that connect the starter motor to the flywheel of the engine. Once the current to the solenoid is interrupted, a spring disengages the gears and pushes the iron core to the right.

**$\vec{B}$ AT CENTER OF A SOLENOID**    **EXAMPLE 27-5**

**Find the magnetic field at the center of a solenoid of length 20 cm, radius 1.4 cm, and 600 turns that carries a current of 4 A.**

**PICTURE THE PROBLEM**

1. We will calculate the field exactly, using Equation 27-8:

$$B_x = \frac{1}{2}\mu_0 nI \left( \frac{x_2}{\sqrt{x_2^2 + R^2}} - \frac{x_1}{\sqrt{x_1^2 + R^2}} \right)$$

2. For a point at the center of the solenoid, $x_1 = -10$ cm and $x_2 = +10$ cm. Thus, the terms in the parentheses in Equation 27-8 have values of:

$$\frac{x_2}{\sqrt{x_2^2 + R^2}} = \frac{10\text{ cm}}{\sqrt{(10\text{ cm})^2 + (1.4\text{ cm})^2}} = 0.990$$

$$\frac{x_1}{\sqrt{x_1^2 + R^2}} = \frac{-10\text{ cm}}{\sqrt{(-10\text{ cm})^2 + (1.4\text{ cm})^2}} = -0.990$$

3. Substitute these results into $B_x$ in step 1:

$$B_x = \frac{1}{2}(4\pi \times 10^{-7}\text{ T·m/A})[(600\text{ turns})/(0.2\text{ m})](4\text{ A})(0.990 + 0.990)$$

$$= \boxed{1.50 \times 10^{-2}\text{ T}}$$

**REMARKS** Note that the approximation obtained using Equation 27-9 amounts to replacing 0.99 by 1.00, which differs by only one percent. Note also that the magnitude of the magnetic field inside this solenoid is fairly large—about 250 times the magnetic field of the earth.

**EXERCISE** Calculate $B_x$ using the long-solenoid approximation. (*Answer*  $1.51 \times 10^{-2}$ T)

### $\vec{B}$ Due to a Current in a Straight Wire

Figure 27-14 shows the geometry for calculating the magnetic field $\vec{B}$ at a point $P$ due to the current in the straight wire segment shown. We choose $R$ to be the perpendicular distance from the wire to point $P$, and we choose the $x$ axis to be along the wire with $x = 0$ at the projection of $P$ onto the $x$ axis.

A typical current element $I\,d\vec{\ell}$ at a distance $x$ from the origin is shown. The vector $\vec{r}$ points from the element to the field point $P$. The direction of the magnetic field at $P$ due to this element is the direction of $I\,d\vec{\ell} \times \hat{r}$, which is out of the paper. Note that the magnetic fields due to all the current elements of the wire are in this same direction. Thus, we need to compute only the magnitude of the field. The field due to the current element shown has the magnitude (Equation 27-3)

$$dB = \frac{\mu_0}{4\pi}\frac{I\,dx}{r^2}\sin\phi$$

It is more convenient to write this in terms of $\theta$ rather than $\phi$:

$$dB = \frac{\mu_0}{4\pi}\frac{I\,dx}{r^2}\cos\theta \qquad\qquad 27\text{-}10$$

**FIGURE 27-14** Geometry for calculating the magnetic field at point $P$ due to a straight current segment. Each element of the segment contributes to the total magnetic field at point $P$, which is directed out of the paper. The result is expressed in terms of the angles $\theta_1$ and $\theta_2$.

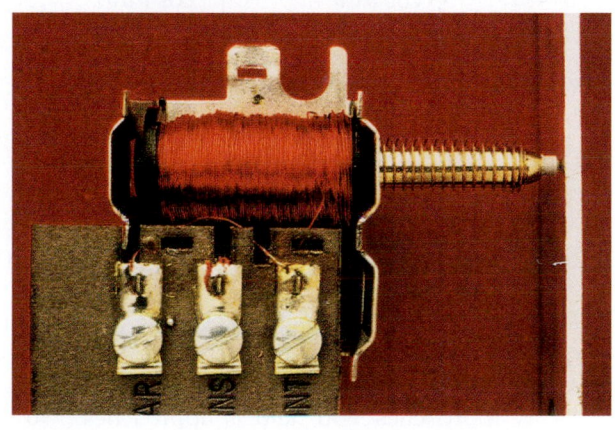

A cross section of a doorbell. When the solenoid is energized, its magnetic field pulls on the plunger, causing it to strike the bell (not shown). The spring returns the plunger to its normal position.

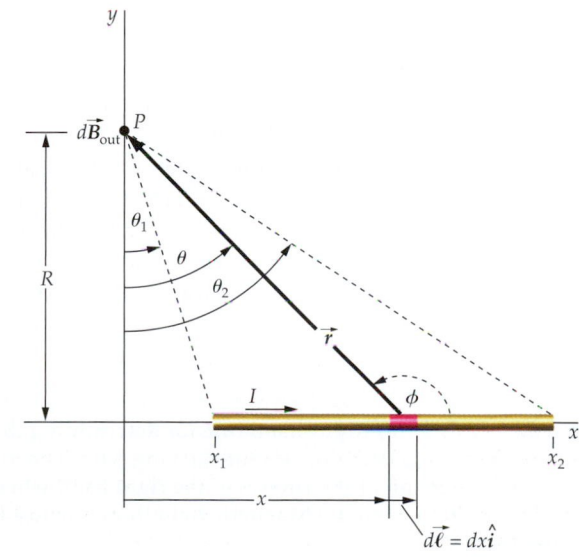

To sum over all the current elements, we need to relate the variables $\theta$, $r$, and $x$. It turns out to be easiest to express $x$ and $r$ in terms of $\theta$. We have

$$x = R \tan \theta$$

Then, taking the differential of each side with $R$ as a constant gives

$$dx = R \sec^2 \theta \, d\theta = R\frac{r^2}{R^2} \, d\theta = \frac{r^2}{R} \, d\theta$$

where we have used $\sec \theta = r/R$. Substituting this expression for $dx$ into Equation 27-10, we obtain

$$dB = \frac{\mu_0}{4\pi} \frac{I}{r^2} \frac{r^2 \, d\theta}{R} \cos \theta = \frac{\mu_0}{4\pi} \frac{I}{R} \cos \theta \, d\theta$$

We sum over these elements by integrating from $\theta = \theta_1$ to $\theta = \theta_2$, where $\theta_1$ and $\theta_2$ are shown in Figure 27-14. This gives

$$B = \int_{\theta_1}^{\theta_2} \frac{\mu_0}{4\pi} \frac{I}{R} \cos \theta \, d\theta = \frac{\mu_0}{4\pi} \frac{I}{R} \int_{\theta_1}^{\theta_2} \cos \theta \, d\theta$$

Evaluating the integral, we obtain

$$B = \frac{\mu_0}{4\pi} \frac{I}{R} (\sin \theta_2 - \sin \theta_1) \qquad\qquad 27\text{-}11$$

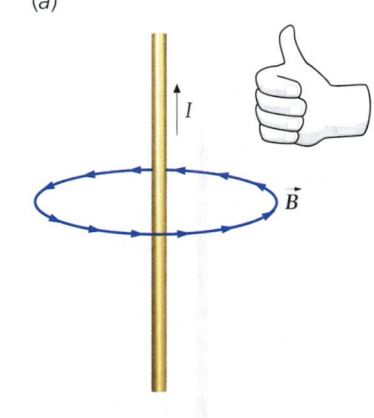

(a)

**B DUE TO A STRAIGHT WIRE SEGMENT**

This result gives the magnetic field due to any wire segment in terms of the perpendicular distance $R$ and $\theta_1$ and $\theta_2$ are the angles subtended at the field point by the ends of the wire. If the length of the wire approaches infinity in both directions, $\theta_2$ approaches $+90°$ and $\theta_1$ approaches $-90°$. The result for such a very long wire is obtained from Equation 27-11, by setting $\theta_1 = -90°$ and $\theta_2 = +90°$:

$$B = \frac{\mu_0}{4\pi} \frac{2I}{R} \qquad\qquad 27\text{-}12$$

**B DUE TO AN INFINITELY LONG, STRAIGHT WIRE**

At any point in space, the magnetic field lines of a long, straight, current-carrying wire are tangent to a circle of radius $R$ about the wire, where $R$ is the perpendicular distance from the wire to the field point. The direction of $\vec{B}$ can be determined by applying the right-hand rule, as shown in Figure 27-15a. The magnetic field lines thus encircle the wire, as shown in Figure 27-15b.

The result expressed by Equation 27-12 was found experimentally by Biot and Savart in 1820. From their analysis, Biot and Savart were able to discover the expression given in Equation 27-3 for the magnetic field due to a current element.

(b)

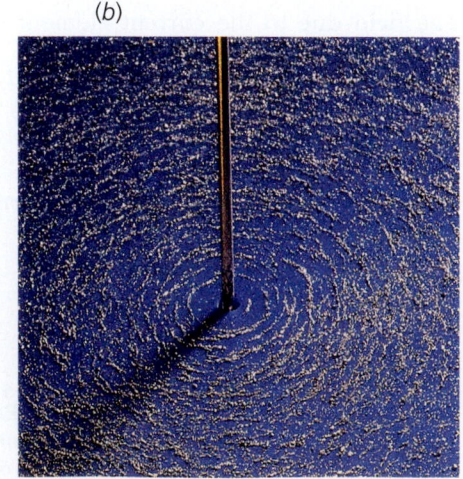

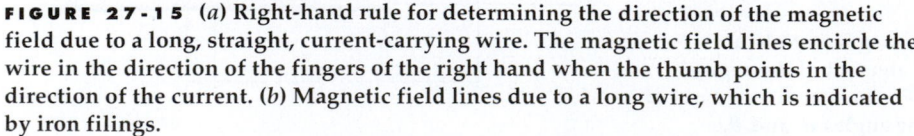

**FIGURE 27-15** (a) Right-hand rule for determining the direction of the magnetic field due to a long, straight, current-carrying wire. The magnetic field lines encircle the wire in the direction of the fingers of the right hand when the thumb points in the direction of the current. (b) Magnetic field lines due to a long wire, which is indicated by iron filings.

$\vec{B}$ AT CENTER OF SQUARE CURRENT LOOP          **E X A M P L E   2 7 - 6**

Find the magnetic field at the center of a square current loop of edge length $L = 50$ cm, which carries a current of 1.5 A.

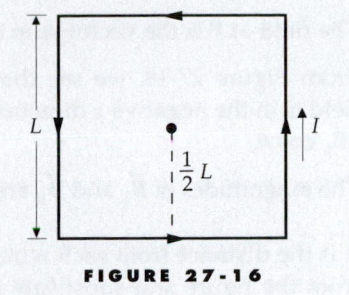

**FIGURE 27-16**

**PICTURE THE PROBLEM**  The magnetic field at the center of the loop is the sum of the contributions from each of the four sides of the loop. From Figure 27-16, we can see that each side of the loop produces a field of equal magnitude pointing out of the page. Thus, we use Equation 27-11 for a given side, then multiply by 4 for the total field.

1. The total field is 4 times the field $B_s$ due to a side:

$$B = 4B_s$$

2. Calculate the magnetic field $B_s$ due to a given side of the loop. Note from the figure that $R = \frac{1}{2}L$ and $\theta_1 = -45°$ and $\theta_2 = +45°$:

$$B_s = \frac{\mu_0}{4\pi}\frac{I}{R}(\sin\theta_2 - \sin\theta_1) = \frac{\mu_0}{4\pi}\frac{I}{\frac{1}{2}L}\left[\sin(+45°) - \sin(-45°)\right]$$

$$= (10^{-7}\,\text{T·m/A})\frac{1.5\,\text{A}}{0.25\,\text{m}}2\sin 45° = 8.49 \times 10^{-7}\,\text{T}$$

3. Multiply this value by 4 to find the total field:

$$B = 4B_s = 4(8.49 \times 10^{-7}\,\text{T}) = \boxed{3.39 \times 19^{-6}\,\text{T}}$$

**EXERCISE**  Compare the magnetic field at the center of a circular current loop of radius R with the magnetic field at the center of a square current loop of side $L = 2R$ carrying the same current. Which is larger? (*Answer*   B at the center is larger for the circle, by about 10 percent)

**EXERCISE**  Find the distance from a long, straight wire carrying a current of 12 A, where the magnetic field due to the current in the wire is equal in magnitude to 0.6 G (the magnitude of the earth's magnetic field). (*Answer*   R = 4.00 cm)

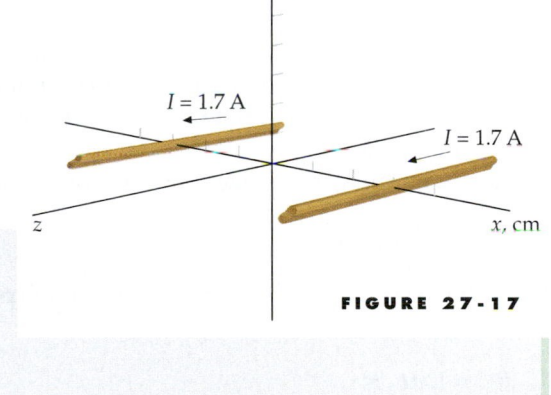

A current gun used to measure electric current. The jaws of the current gun clamp around a current-carrying wire without touching the wire. The magnetic field produced by the wire is measured with a Hall-effect device mounted in the current gun. The Hall-effect device puts out a voltage proportional to the magnetic field, which in turn is proportional to the current in the wire.

$\vec{B}$ DUE TO TWO PARALLEL WIRES          **E X A M P L E   2 7 - 7**

A long, straight wire carrying a current of 1.7 A in the positive z direction lies along the line $x = -3$ cm, $y = 0$. A second such wire carrying a current of 1.7 A in the positive z direction lies along the line $x = +3$ cm, $y = 0$, as shown in Figure 27-17. Find the magnetic field at a point $P$ on the y axis at $y = 6$ cm.

**FIGURE 27-17**

**PICTURE THE PROBLEM**  The magnetic field at point $P$ is the vector sum of the field $\vec{B}_L$ due to the wire on the left in Figure 27-18, and the field $\vec{B}_R$ due to the wire on the right. Since each wire carries the same current, and each wire is the same distance from point $P$, the magnitudes $B_L$ and $B_R$ are equal. $\vec{B}_L$ is perpendicular to the radius from the left wire to point $P$, and $\vec{B}_R$ is perpendicular to the radius from the right wire to the point $P$.

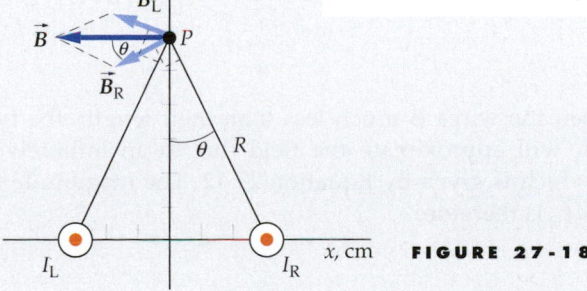

**FIGURE 27-18**

1. The field at $P$ is the vector sum of the fields $\vec{B}_L$ and $\vec{B}_R$:

$$\vec{B} = \vec{B}_L + \vec{B}_R$$

2. From Figure 27-18, we see that the resultant magnetic field is in the negative $x$ direction and has the magnitude $2B_L \cos \theta$.

$$\vec{B} = -2B_L \cos \theta \, \hat{i}$$

3. The magnitudes of $\vec{B}_L$ and $\vec{B}_R$ are given by Equation 27-12:

$$B_L = B_R = \frac{\mu_0}{4\pi} \frac{2I}{R}$$

4. $R$ is the distance from each wire to the point $P$. We find $R$ from the figure and substitute $R$ into the expression for $B_L$ and $B_R$:

$$R = \sqrt{(3 \text{ cm})^2 + (6 \text{ cm})^2} = 6.71 \text{ cm}$$

so

$$B_L = B_R = (10^{-7} \text{ T·m/A}) \frac{2(1.7 \text{ A})}{0.0671 \text{ m}} = 5.07 \times 10^{-6} \text{ T}$$

5. We obtain $\cos \theta$ from the figure:

$$\cos \theta = \frac{6 \text{ cm}}{R} = \frac{6 \text{ cm}}{6.71 \text{ cm}} = 0.894$$

6. Substitute the values of $\cos \theta$ and $B_L$ into the equation in step 2 for $\vec{B}$:

$$\vec{B} = -2(5.07 \times 10^{-6} \text{ T})(0.894)\hat{i} = \boxed{-9.07 \times 10^{-6} \text{ T } \hat{i}}$$

**EXERCISE** Find $\vec{B}$ at the origin. (*Answer* 0)

**EXERCISE** Find $\vec{B}$ at the origin assuming that $I_R$ goes into the page. (*Answer* $\vec{B} = 2.27 \times 10^{-5} \text{ T } \hat{j}$)

## Magnetic Force Between Parallel Wires

We can use Equation 27-12 for the magnetic field due to a long, straight, current-carrying wire and $d\vec{F} = I \, d\vec{\ell} \times \vec{B}$ (Equation 26-5) for the force exerted by a magnetic field on a segment of a current-carrying wire to find the force exerted by one long straight current on another. Figure 27-19 shows two long parallel wires carrying currents in the same direction. We consider the force on a segment $d\vec{\ell}_2$ carrying current $I_2$, as shown. The magnetic field $\vec{B}_1$ at this segment due to current $I_1$ is perpendicular to the segment $I_2 \, d\vec{\ell}_2$, as shown. This is true for all current elements along the wire. The magnetic force $d\vec{F}_2$ on current segment $I_2 \, d\vec{\ell}_2$ is directed toward current $I_1$, since $d\vec{F}_2 = I_2 \, d\vec{\ell}_2 \times \vec{B}_2$. Similarly, a current segment $I_1 \, d\vec{\ell}_1$ will experience a magnetic force directed toward current $I_2$ due to a magnetic field arising from current $I_2$. Thus, two parallel currents attract each other. If one of the currents is reversed the force will be reversed, so two antiparallel currents will repel each other. The attraction or repulsion of parallel or antiparallel currents was discovered experimentally by Ampère one week after he heard of Oersted's discovery of the effect of a current on a compass needle.

The magnitude of the magnetic force on the segment $I_2 \, d\vec{\ell}_2$ is

$$dF_2 = |I_2 \, d\vec{\ell}_2 \times \vec{B}_1|$$

Since the magnetic field at segment $I_2 \, d\vec{\ell}_2$ is perpendicular to the current segment, we have

$$dF_2 = I_2 \, d\ell_2 \, B_1$$

If the distance $R$ between the wires is much less than their length, the field at $I_2 \, d\vec{\ell}_2$ due to current $I_1$ will approximate the field due to an infinitely long, current-carrying wire, which is given by Equation 27-12. The magnitude of the force on the segment $I_2 \, d\vec{\ell}_2$ is therefore

$$dF_2 = I_2 \, d\ell_2 \frac{\mu_0 I_1}{2\pi R}$$

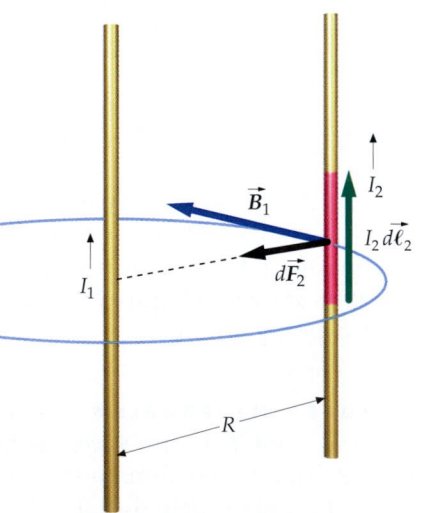

**FIGURE 27-19** Two long straight wires carrying parallel currents. The magnetic field $\vec{B}_1$ due to current $I_1$ is perpendicular to current $I_2$. The force on current $I_2$ is toward current $I_1$. There is an equal and opposite force exerted by current $I_2$ on $I_1$. The current-carrying wires thus attract each other.

The force per unit length is

$$\frac{dF_2}{d\ell_2} = I_2 \frac{\mu_0 I_1}{2\pi R} = 2\frac{\mu_0}{4\pi}\frac{I_1 I_2}{R}$$    27-13

In Chapter 21, the coulomb was defined in terms of the ampere, but the definition of the ampere was deferred. The ampere is defined as follows:

> The ampere is the constant electric current that, when maintained in two straight parallel conductors of infinite length and of negligible circular cross sections placed one meter apart in a vacuum, would produce a force between the conductors equal to $2 \times 10^{-7}$ newtons per meter of length.

DEFINITION—AMPERE

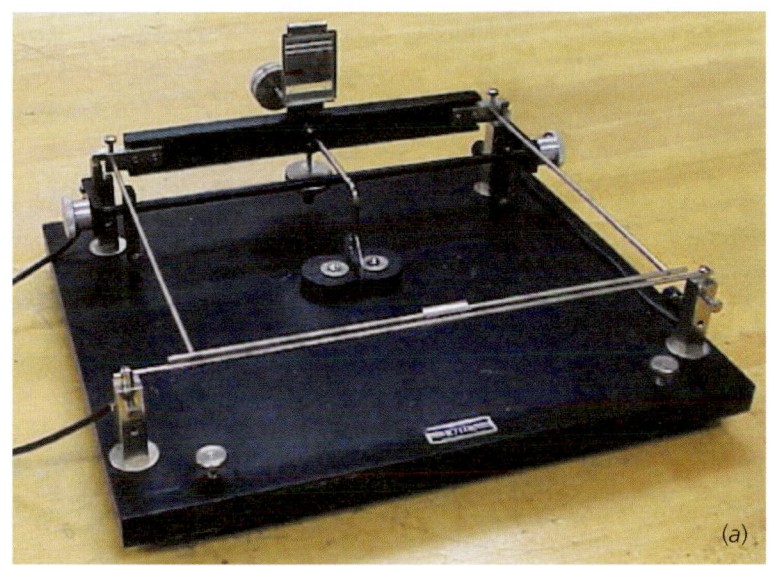

(a)

This definition of the ampere makes the permeability of free space $\mu_0$ equal to exactly $4\pi \times 10^{-7}$ N/A$^2$. It also allows the unit of current (and therefore the unit of electric charge) to be determined by a mechanical measurement. In practice, currents much closer together than 1 m are used so that the force can be measured accurately with long but finite wires.

Figure 27-20 shows a **current balance,** which is a device that can be used to calibrate an ammeter from the definition of the ampere. The upper conductor, directly above the lower conductor, is free to rotate about knife-edge contacts and is balanced so that the wires (or conducting rods) are a small distance apart. The conductors are connected in series to carry the same current but in opposite directions so that the currents will repel each other. Weights are placed on the upper conductor until it balances again at the original separation. The force of repulsion is thus determined by measuring the total weight required to balance the upper conductor.

**FIGURE 27-20** (a) A picture of a current balance used in a general physics lab. (b) A schematic diagram of a current balance. The two parallel rods in front carry equal but oppositely directed currents and therefore repel each other. The force of repulsion is balanced by weights placed on the upper rod, which is part of a rectangle that is balanced on knife edges at the back. The mirror on top is used to reflect a beam of laser light to accurately determine the position of the upper rod.

Mirror

Knife–edge contacts

Beam deflected upward

Laser beam

(b)

BALANCING THE MAGNETIC FORCE **EXAMPLE 27-8** Try It Yourself

Two straight rods 50-cm long with axes 1.5-mm apart in a current balance carry currents of 15 A each in opposite directions. What mass must be placed on the upper rod to balance the magnetic force of repulsion?

**PICTURE THE PROBLEM** Equation 27-13 gives the magnitude of the magnetic force per unit length exerted by the lower rod on the upper rod. Find this force for a rod of length $L$ and set it equal to the weight $mg$.

**Cover the column to the right and try these on your own before looking at the answers.**

| Steps | Answers |
|---|---|
| 1. Set the weight $mg$ equal to the magnetic force of repulsion of the rods. | $mg = 2\dfrac{\mu_0}{4\pi}\dfrac{I_1 I_2}{R}L$ |
| 2. Solve for the mass $m$. | $m = 1.53 \times 10^{-3}\,\text{kg} = \boxed{1.53\,\text{g}}$ |

**REMARKS** Since only 1.53 g are required to balance the system, we see that the magnetic force between two straight current-carrying wires is relatively small, even for currents as large as 15 A separated by only 1.5 mm.

# 27-3 Gauss's Law for Magnetism

The magnetic field lines shown in Figure 27-6, Figure 27-9, and Figure 27-10 differ from electric field lines because the lines of $\vec{B}$ form closed curves, whereas lines of $\vec{E}$ begin and end on electric charges. The magnetic equivalent of an electric charge is a magnetic pole, such as appears to be at the ends of a bar magnet. Magnetic field lines appear to diverge from the north-pole end of a bar magnet (Figure 27-10$b$) and appear to converge on the south-pole end. However, inside the magnet the magnetic field lines neither diverge from a point near the north-pole end, nor do they converge on a point near the south-pole end. Instead, the magnetic field lines pass through the bar magnet from the south-pole end to the north-pole end, as shown in Figure 27-10$b$. If a Gaussian surface encloses one end of a bar magnet, the number of magnetic field lines that leave through the surface is exactly equal to the number of magnetic field lines that enter through the surface. That is, the net flux $\phi_{m,\,net}$ of the magnetic field through any closed surface $S$ is always zero.[†]

$$\phi_{m,\,net} = \oint_S B_n\,dA = 0 \qquad\qquad 27\text{-}14$$

GAUSS'S LAW FOR MAGNETISM

where $B_n$ is the component of $\vec{B}$ normal to surface $S$ at area element $dA$. The definition of the magnetic flux $\phi_m$ is exactly analogous to the electric flux, with $\vec{B}$ replacing $\vec{E}$. This result is called Gauss's law for magnetism. It is the mathematical statement that there exist no points in space from which magnetic field lines diverge, or to which magnetic field lines converge. That is, isolated magnetic poles do not exist.[‡] The fundamental unit of magnetism is the magnetic

---

[†] Recall that the net flux of the electric field is a measure of the net number of field lines that leave a closed surface and is equal to $Q_{inside}/\epsilon_0$.

[‡] The existence of magnetic monopoles is a subject of great debate, and the search for magnetic monopoles remains active. To date, however, none have been discovered.

dipole. Figure 27-21 compares the field lines of $\vec{B}$ for a magnetic dipole with the field lines of $\vec{E}$ for an electric dipole. Note that far from the dipoles the field lines are identical. But inside the dipole, the field lines of $\vec{E}$ are directed opposite to the field lines of $\vec{B}$. The field lines of $\vec{E}$ diverge from the positive charge and converge to the negative charge, whereas the field lines of $\vec{B}$ are continuous loops.

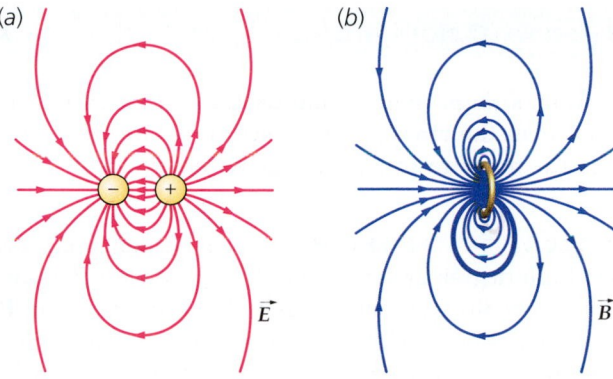

# 27-4 Ampère's Law

In Chapter 22, we found that for highly symmetric charge distributions we could calculate the electric field more easily using Gauss's law than Coulomb's law. A similar situation exists in magnetism. Ampère's law relates the tangential component $B_t$ of the magnetic field summed (integrated) around a closed curve $C$ to the current $I_C$ that passes through any surface bounded by $C$. It can be used to obtain an expression for the magnetic field in situations that have a high degree of symmetry. In mathematical form, **Ampère's law** is

$$\oint_C B_t \, d\ell = \oint_C \vec{B} \cdot d\vec{\ell} = \mu_0 I_C, \quad C \text{ is any closed curve} \qquad 27\text{-}15$$

AMPÈRE'S LAW

where $I_C$ is the net current that penetrates any surface $S$ bounded by the curve $C$. The positive tangential direction for the path integral is related to the choice for the positive direction for the current $I_C$ through $S$ by the right-hand rule shown in Figure 27-22. Ampère's law holds for any curve $C$, as long as the currents are steady and continuous. This means the current does not change in time and that charge is not accumulating anywhere. Ampère's law is useful in calculating the magnetic field $\vec{B}$ in situations that have a high degree of symmetry so that the line integral $\oint_C \vec{B} \cdot d\vec{\ell}$ can be written as $B \oint_C d\ell$ (the product of $B$ and some distance). The integral $\oint_C \vec{B} \cdot d\vec{\ell}$ is called a **circulation integral**. More specifically, $\oint_C \vec{B} \cdot d\vec{\ell}$ is called the circulation of $\vec{B}$ around curve $C$. Ampère's law and Gauss's law are both of considerable theoretical importance, and both laws hold whether there is symmetry or there is no symmetry. If there is no symmetry, neither law is very useful in calculating electric or magnetic fields.

The simplest application of Ampère's law is to find the magnetic field of an infinitely long, straight, current-carrying wire. Figure 27-23 shows a circular curve around a long wire with its center at the wire. We know the direction of the magnetic field due to each current element is tangent to this circle from the Biot–Savart law. Assuming that the magnetic field is tangent to this circle, that the magnetic field is in the same direction as $d\vec{\ell}$, and that the magnetic field has the same magnitude $B$ at any point on the circle, Ampère's law ($\oint_C B_t \, d\vec{\ell} = \mu_0 I_C$) then gives

$$B \oint_C d\ell = \mu_0 I_C$$

where $B = B_t$. We can factor $B$ out of the integral because $B$ has the same value everywhere on the circle. The integral of $d\ell$ around the circle equals $2\pi R$ (the circumference of the circle). The current $I_C$ is the current $I$ in the wire. We thus obtain $B 2\pi R = \mu_0 I$

$$B = \frac{\mu_0 I}{2\pi R}$$

which is Equation 27-12.

**FIGURE 27-21** (*a*) Electric field lines of an electric dipole. (*b*) Magnetic field lines of a magnetic dipole. Far from the dipoles, the field lines are identical. In the region between the charges in Figure 27-21(*a*), the electric field lines are opposite the direction of the dipole moment, whereas inside the loop in Figure 27-21(*b*), the magnetic field lines are parallel to the direction of the dipole moment.

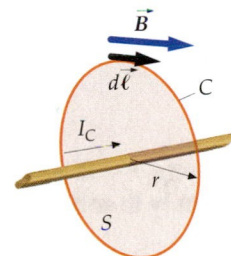

**FIGURE 27-22** The positive direction for the path integral for Ampère's law is related to the positive direction for the current passing through the surface by a right-hand rule.

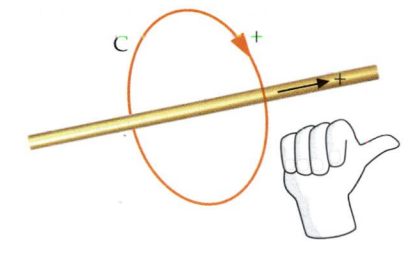

**FIGURE 27-23** Geometry for calculating the magnetic field of a long, straight, current-carrying wire using Ampère's law. On a circle around the wire, the magnetic field is constant and tangent to the circle.

**B** *INSIDE AND OUTSIDE A WIRE*

### EXAMPLE 27-9

A long, straight wire of radius $R$ carries a current $I$ that is uniformly distributed over the circular cross section of the wire. Find the magnetic field both outside the wire and inside the wire.

**PICTURE THE PROBLEM** We can use Ampère's law to calculate $\vec{B}$ because of the high degree of symmetry. At a distance $r$ (Figure 27-24), we know that $\vec{B}$ is tangent to the circle of radius $r$ about the wire and $\vec{B}$ is constant in magnitude everywhere on the circle. The current through the surface $S$ bounded by $C$ depends on whether $r$ is less than or greater than the radius of the wire $R$.

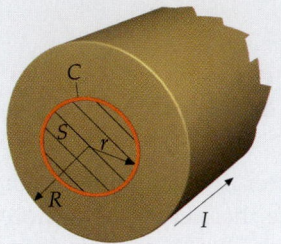

**FIGURE 27-24**

1. Ampère's law is used to relate the circulation of $\vec{B}$ around curve $C$ to the current passing through the surface $S$ bounded by $C$:

$$\oint_C \vec{B} \cdot d\vec{\ell} = \mu_0 I_C$$

2. Evaluate the circulation of $\vec{B}$ around a circle of radius $r$ that is coaxial with the wire:

$$\oint_C \vec{B} \cdot d\vec{\ell} = B \oint_C d\ell = B2\pi r$$

3. Substitute into Ampère's law and solve for $B$:

$$B2\pi r = \mu_0 I_C$$

so

$$B = \frac{\mu_0 I_C}{2\pi r}$$

4. Outside the wire, $r > R$, and the total current passes through the surface bounded by $C$:

$$I_C = I$$

$$\boxed{B = \frac{\mu_0 I}{2\pi r} \quad (r \geq R)}$$

5. Inside the wire, $r < R$. Assume that the current is distributed uniformly to solve for $I_C$. Solve for $B$:

$$\frac{I_C}{\pi r^2} = \frac{I}{\pi R^2}$$

or

$$\left( I_C = \frac{r^2}{R^2} I \right)$$

so

$$B = \frac{\mu_0}{2\pi} \frac{I_C}{r} = \frac{\mu_0}{2\pi} \frac{(r^2/R^2)I}{r} = \boxed{\frac{\mu_0}{2\pi} \frac{I}{R^2} r \quad r \leq R}$$

**REMARKS** Inside the wire, the field increases with distance from the center of the wire. Figure 27-25 shows the graph of $B$ versus $r$ for this example.

We see from Example 27-9 that the magnetic field due to a current uniformly distributed over a wire of radius $R$ is given by

$$B = \frac{\mu_0 I}{2\pi R^2} r \quad (r \leq R)$$

$$B = \frac{\mu_0 I}{2\pi r} \quad (r \geq R)$$ 27-16

For the next application of Ampère's law, we calculate the magnetic field of a tightly wound **toroid**, which consists of loops of wire wound around a doughnut-

**FIGURE 27-25**

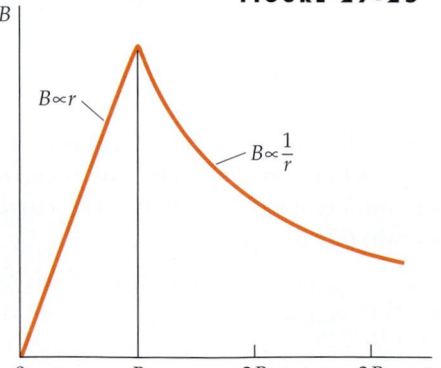

shaped form, as shown in Figure 27-26. There are $N$ turns of wire, each carrying a current $I$. To calculate $B$, we evaluate the line integral $\oint_C \vec{B} \cdot d\vec{\ell}$ around a circle of radius $r$ centered in the middle of the toroid. By symmetry, $\vec{B}$ is tangent to this circle and constant in magnitude at every point on the circle. Then,

$$\oint_C \vec{B} \cdot d\vec{\ell} = B2\pi r = \mu_0 I_C$$

Let $a$ and $b$ be the inner and outer radii of the toroid, respectively. The total current through the surface $S$ bounded by a circle of radius $r$ for $a < r < b$ is $NI$. Ampère's law then gives

$$\oint_C \vec{B} \cdot d\vec{\ell} = \mu_0 I_C, \quad \text{or} \quad (B2\pi r = \mu_0 NI)$$

or

$$B = \frac{\mu_0 NI}{2\pi r}, \quad a < r < b \qquad \text{27-17}$$

B INSIDE A TIGHTLY WOUND TOROID

If $r$ is less than $a$, there is no current through the surface $S$. If $r$ is greater than $b$, the total current through $S$ is zero because for each turn of the wire the current penetrates the surface twice (Figure 27-27), once going into the page and once coming out of the page. Thus, the magnetic field is zero for both $r < a$ and $r > b$:

$$B = 0, \quad r < a \quad \text{or} \quad r > b$$

The magnetic field intensity inside the toroid is not uniform but decreases with increasing $r$. However, if the radius of the loops of the coil, $\frac{1}{2}(b - a)$ is much less than the radius $\frac{1}{2}(b + a)$ of the center of the loops, the variation in $r$ from $r = a$ to $r = b$ is small and $B$ is approximately uniform, as it is in a solenoid.

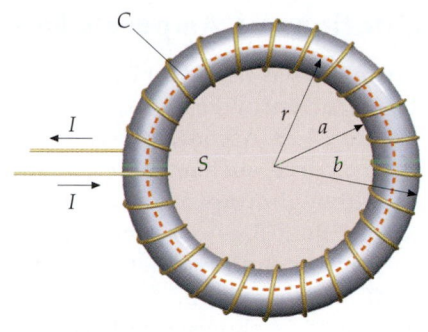

**FIGURE 27-26** A toroid consists of loops of wire wound around a doughnut-shaped form. The magnetic field at any distance $r$ can be found by applying Ampère's law to the circle of radius $r$. The surface $S$ is bounded by curve $C$. The wire penetrates $S$ once for each turn.

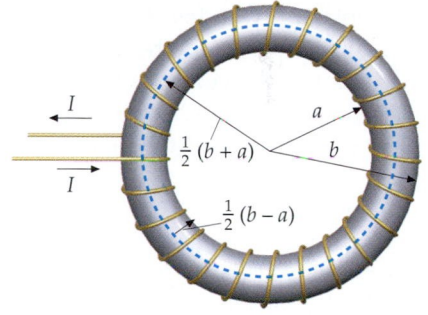

**FIGURE 27-27** The toroid has mean radius $r = \frac{1}{2}(b + a)$, where $a$ and $b$ are the inner and outer radii of the toroid. Each turn of the wire is a circle of radius $\frac{1}{2}(b - a)$.

(a)

(b)

(*a*) The Tokamak fusion-test reactor is a large toroid that produces a magnetic field for confining charged particles. Coils containing over 10 km of water-cooled copper wire carry a pulsed current, which has a peak value of 73,000 A and produces a magnetic field of 5.2 T for about 3 s. (*b*) Inspection of the assembly of the Tokamak reactor from inside the toroid.

## Limitations of Ampère's Law

Ampère's law is useful for calculating the magnetic field only when there is both a steady current and a high degree of symmetry. Consider the current loop shown in Figure 27-28. According to Ampère's law, the line integral $\oint_C \vec{B} \cdot d\vec{\ell} = \oint_C B_t \, d\ell$ around a curve, such as curve $C$ in the figure, equals $\mu_0$ times the current $I$ in the loop. Although Ampère's law is valid for this curve, the tangential component of magnetic field $B_t$ is not constant along any curve encircling the current. Thus, there is not enough symmetry in this situation to allow us to evaluate the integral $\oint_C B_t \, d\ell$ and solve for $B_t$.

Figure 27-29 shows a finite current segment of length $\ell$. We wish to find the magnetic field at point $P$, which is equidistant from the ends of the segment and at a distance $r$ from the center of the segment. A direct application of Ampère's law gives

$$B = \frac{\mu_0}{2\pi} \frac{I}{r}$$

This result is the same as for an infinitely long wire, since the same symmetry arguments apply. It does not agree with the result obtained from the Biot–Savart law, which depends on the length of the current segment and which agrees with experiment. If the current segment is just one part of a continuous circuit carrying a current, as shown in Figure 27-30, Ampère's law for curve $C$ is valid, but it cannot be used to find the magnetic field at point $P$ because there is insufficient symmetry.

In Figure 27-31, the current in the segment arises from a small spherical conductor with initial charge $+Q$ at the left of the segment and another small spherical conductor at the right with charge $-Q$. When they are connected, a current $I = -dQ/dt$ exists in the segment for a short time, until the spheres are uncharged. For this case, we *do* have the symmetry needed to assume that $\vec{B}$ is tangential to the curve and $\vec{B}$ is constant in magnitude along the curve. For a situation like this, in which the current is discontinuous in space, Ampère's law is not valid. In Chapter 30, we will see how Maxwell was able to modify Ampère's law so that it holds for all currents. When Maxwell's generalized form of Ampère's law is used to calculate the magnetic field for a current segment, such as the current segment shown in Figure 27-31, the result agrees with the result found from the Biot–Savart law.

## 27-5 Magnetism in Matter

Atoms have magnetic dipole moments due to the motion of their electrons and due to the intrinsic magnetic dipole moment associated with the spin of the electrons. Unlike the situation with electric dipoles, the alignment of magnetic dipoles parallel to an external magnetic field tends to *increase* the field. We can see this difference by comparing the electric field lines of an electric dipole with the magnetic field lines of a magnetic dipole, such as a small current loop, as was shown in Figure 27-21. Far from the dipoles, the field lines are identical. However, between the charges of the electric dipole, the electric field lines are opposite the direction of the dipole moment, whereas inside the current loop, the magnetic field lines are parallel to the magnetic dipole moment. Thus, inside a magnetically polarized material, the magnetic dipoles create a magnetic field that is parallel to the magnetic dipole moment vectors.

Materials fall into three categories—**paramagnetic, diamagnetic,** and **ferromagnetic**—according to the behavior of their magnetic moments in an external magnetic field. Paramagnetism arises from the partial alignment of the electron spins (in metals) or from the atomic or molecular magnetic moments by an

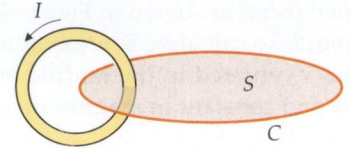

**FIGURE 27-28** Ampère's law holds for the curve $C$ encircling the current in the circular loop, but it is not useful for finding $B_t$, because $B_t$ cannot be factored out of the circulation integral.

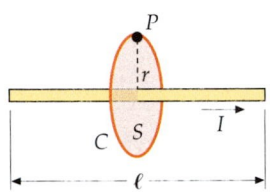

**FIGURE 27-29** The application of Ampère's law to find the magnetic field on the bisector of a finite current segment gives an incorrect result.

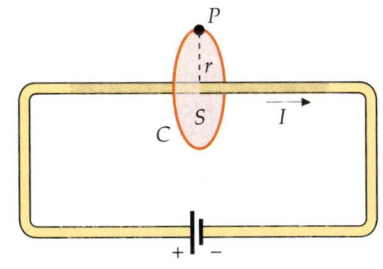

**FIGURE 27-30** If the current segment in Figure 27-28 is part of a complete circuit, Ampère's law for the curve $C$ is valid, but there is not enough symmetry to use Ampère's law to find the magnetic field at point $P$.

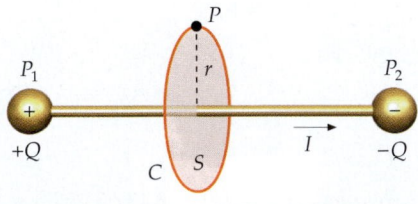

**FIGURE 27-31** If the current segment in Figure 27-29 is due to a momentary flow of charge from a small conductor on the left to a small conductor on the right, there is enough symmetry to use Ampère's law to compute the magnetic field at $P$, but Ampère's law is not valid because the current is not continuous in space.

applied magnetic field in the direction of the field. In paramagnetic materials, the magnetic dipoles do not interact strongly with each other and are normally randomly oriented. In the presence of an applied magnetic field, the dipoles are partially aligned in the direction of the field, thereby increasing the field. However, in external magnetic fields of ordinary strength at ordinary temperatures, only a very small fraction of the molecules are aligned because thermal motion tends to randomize their orientation. The increase in the total magnetic field is therefore very small. Ferromagnetism is much more complicated. Because of a strong interaction between neighboring magnetic dipoles, a high degree of alignment occurs even in weak external magnetic fields, which causes a very large increase in the total field. Even when there is no external magnetic field, a ferromagnetic material may have its magnetic dipoles aligned, as in permanent magnets. Diamagnetism arises from the orbital magnetic dipole moments induced by an applied magnetic field. These magnetic moments are opposite the direction of the applied magnetic field, thereby decreasing the field. This effect actually occurs in all materials; however, because the induced magnetic moments are very small compared to the permanent magnetic moments, diamagnetism is often masked by paramagnetic or ferromagnetic effects. Diamagnetism is thus observed only in materials whose molecules have no permanent magnetic moments.

## Magnetization and Magnetic Susceptibility

When some material is placed in a strong magnetic field, such as that of a solenoid, the magnetic field of the solenoid tends to align the magnetic dipole moments (either permanent or induced) inside the material and the material is said to be magnetized. We describe a magnetized material by its **magnetization $\vec{M}$,** which is defined as the net magnetic dipole moment per unit volume of the material:

$$\vec{M} = \frac{d\vec{\mu}}{dV} \qquad \text{27-18}$$

Long before we had any understanding of atomic or molecular structure, Ampère proposed a model of magnetism in which the magnetization of materials is due to microscopic current loops inside the magnetized material. We now know that these current loops are a classical model for the orbital motion and spin of the electrons in atoms. Consider a cylinder of magnetized material. Figure 27-32 shows atomic current loops in the cylinder aligned with their magnetic moments along the axis of the cylinder. Because of cancellation of neighboring current loops, the net current at any point inside the material is zero, leaving a net current on the surface of the material (Figure 27-33). This surface current, called an **amperian current,** is similar to the real current in the windings of the solenoid.

Figure 27-34 shows a small disk of cross-sectional area $A$, length $d\ell$, and volume $dV = A\,d\ell$. Let $di$ be the amperian current on the surface of the disk. The magnitude of the magnetic dipole moment of the disk is the same as that of a current loop of area $A$ carrying a current $di$:

$$d\mu = A\,di$$

The magnitude of the magnetization of the disk is the magnetic moment per unit volume:

$$M = \frac{d\mu}{dV} = \frac{A\,di}{A\,d\ell} = \frac{di}{d\ell} \qquad \text{27-19}$$

Thus, the magnitude of the magnetization vector is the amperian current per unit length along the surface of the magnetized material. We see from this result that the units of $M$ are amperes per meter.

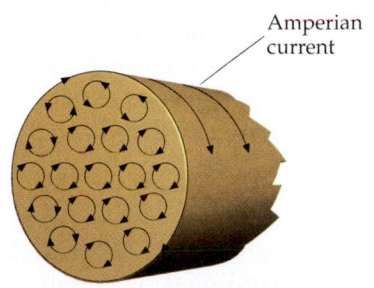

**FIGURE 27-32** A model of atomic current loops in which all the atomic dipoles are parallel to the axis of the cylinder. The net current at any point inside the material is zero due to cancellation of neighboring atoms. The result is a surface current similar to that of a solenoid.

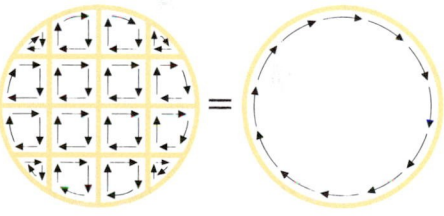

**FIGURE 27-33** The currents in the adjacent current loops in the interior of a uniformly magnetized material cancel, leaving only a surface current. Cancellation occurs at every interior point independent of the shape of the loops.

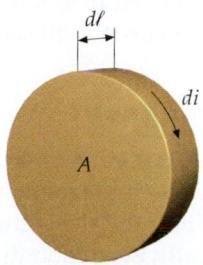

**FIGURE 27-34** A disk element for relating the magnetization $M$ to the surface current per unit length.

Consider a cylinder that has a uniform magnetization $\vec{M}$ parallel to its axis. The effect of the magnetization is the same as if the cylinder carried a surface current per unit length of magnitude $M$. This current is similar to the current carried by a tightly wound solenoid. For a solenoid, the current per unit length is $nI$, where $n$ is the number of turns per unit length and $I$ is the current in each turn. The magnitude of the magnetic field $B_m$ inside the cylinder and far from its ends is thus given by Equation 27-9 for a solenoid with $nI$ replaced by $M$:

$$B_m = \mu_0 M \qquad\qquad 27\text{-}20$$

Suppose we place a cylinder of magnetic material inside a long solenoid with $n$ turns per unit length that carries a current $I$. The applied field of the solenoid $\vec{B}_{app}$ ($B_{app} = \mu_0 nI$) magnetizes the material so that it has a magnetization $\vec{M}$. The resultant magnetic field at a point inside the solenoid and far from its ends due to the current in the solenoid plus the magnetized material is

$$\vec{B} = \vec{B}_{app} + \mu_0 \vec{M} \qquad\qquad 27\text{-}21$$

For paramagnetic and ferromagnetic materials, $\vec{M}$ is in the same direction as $\vec{B}_{app}$; for diamagnetic materials, $\vec{M}$ is opposite to $\vec{B}_{app}$. For paramagnetic and diamagnetic materials, the magnetization is found to be proportional to the applied magnetic field that produces the alignment of the magnetic dipoles in the material. We can thus write

$$\vec{M} = \chi_m \frac{\vec{B}_{app}}{\mu_0} \qquad\qquad 27\text{-}22$$

where $\chi_m$ is a dimensionless number called the **magnetic susceptibility**. Equation 27-21 is then

$$\vec{B} = \vec{B}_{app} + \mu_0 \vec{M} = \vec{B}_{app}(1 + \chi_m) = K_m \vec{B}_{app} \qquad\qquad 27\text{-}23$$

where

$$K_m = 1 + \chi_m \qquad\qquad 27\text{-}24$$

is called the **relative permeability** of the material. For paramagnetic materials, $\chi_m$ is a small positive number that depends on temperature. For diamagnetic materials (other than superconductors), it is a small negative constant independent of temperature. Table 27-1 lists the magnetic susceptibility of various paramagnetic and diamagnetic materials. We see that the magnetic susceptibility for the solids listed is of the order of $10^{-5}$, and $K_m \approx 1$.

The magnetization of ferromagnetic materials, which we discuss shortly, is much more complicated. The relative permeability $K_m$ defined as the ratio $B/B_{app}$ is not constant and has maximum values ranging from 5000 to 100,000. In the case of permanent magnets, $K_m$ is not even defined since such materials exhibit magnetization even in the absence of an applied field.

## Atomic Magnetic Moments

The magnetization of a paramagnetic or ferromagnetic material can be related to the permanent magnetic moments of the individual atoms or electrons of the material. The orbital magnetic moment of an atomic electron can be derived semiclassically, even though it is quantum mechanical in origin. Consider a particle of mass $m$ and charge $q$ moving with speed $v$ in a circle of radius $r$, as shown in Figure 27-35. The magnitude of the angular momentum of the particle is

$$L = mvr \qquad\qquad 27\text{-}25$$

### TABLE 27-1

**Magnetic Susceptibility of Various Materials at 20°C**

| Material | $\chi_m$ |
| --- | --- |
| Aluminum | $2.3 \times 10^{-5}$ |
| Bismuth | $-1.66 \times 10^{-5}$ |
| Copper | $-0.98 \times 10^{-5}$ |
| Diamond | $-2.2 \times 10^{-5}$ |
| Gold | $-3.6 \times 10^{-5}$ |
| Magnesium | $1.2 \times 10^{-5}$ |
| Mercury | $-3.2 \times 10^{-5}$ |
| Silver | $-2.6 \times 10^{-5}$ |
| Sodium | $-0.24 \times 10^{-5}$ |
| Titanium | $7.06 \times 10^{-5}$ |
| Tungsten | $6.8 \times 10^{-5}$ |
| Hydrogen (1 atm) | $-9.9 \times 10^{-9}$ |
| Carbon dioxide (1 atm) | $-2.3 \times 10^{-9}$ |
| Nitrogen (1 atm) | $-5.0 \times 10^{-9}$ |
| Oxygen (1 atm) | $2090 \times 10^{-9}$ |

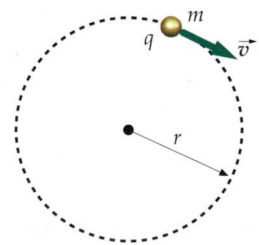

**FIGURE 27-35** A particle of charge $q$ and mass $m$ moving with speed $v$ in a circle of radius $r$. The angular momentum is into the paper and has a magnitude $mvr$ and the magnetic moment is into the paper (if $q$ is positive) and has a magnitude $\frac{1}{2} qvr$.

The magnitude of the magnetic moment is the product of the current and the area of the circle:

$$\mu = IA = I\pi r^2$$

If $T$ is the time for the charge to complete one revolution, the current (charge passing a point per unit time) is $q/T$. Since the period $T$ is the distance $2\pi r$ divided by the velocity $v$, the current is

$$I = \frac{q}{T} = \frac{qv}{2\pi r}$$

The magnetic moment is then

$$\mu = IA = \frac{qv}{2\pi r}\pi r^2 = \frac{1}{2}qvr \qquad\qquad 27\text{-}26$$

Using $vr = L/m$ from Equation 27-25, we have for the magnetic moment

$$\mu = \frac{q}{2m}L$$

If the charge $q$ is positive, the angular momentum and magnetic moment are in the same direction. We can therefore write

$$\vec{\mu} = \frac{q}{2m}\vec{L} \qquad\qquad 27\text{-}27$$

CLASSICAL RELATION BETWEEN MAGNETIC MOMENT AND ANGULAR MOMENTUM

Equation 27-27 is the general classical relation between magnetic moment and angular momentum. It also holds in the quantum theory of the atom for orbital angular momentum, but the equation does not hold for the intrinsic spin angular momentum of the electron. For electron spin, the magnetic moment is twice that predicted by this equation.[†] The extra factor of 2 is a result from quantum theory that has no analog in classical mechanics.

Since angular momentum is quantized, the magnetic moment of an atom is also quantized. The quantum of angular momentum is $\hbar = h/(2\pi)$, where $h$ is Planck's constant, so we express the magnetic moment in terms of $\vec{L}/\hbar$

$$\vec{\mu} = \frac{q\hbar}{2m}\frac{\vec{L}}{\hbar}$$

For an electron, $m = m_e$ and $q = -e$, so the magnetic moment of the electron due to its orbital motion is

$$\vec{\mu}_\ell = -\frac{e\hbar}{2m_e}\frac{\vec{L}}{\hbar} = -\mu_B\frac{\vec{L}}{\hbar} \qquad\qquad 27\text{-}28$$

MAGNETIC MOMENT DUE TO THE ORBITAL MOTION OF AN ELECTRON

---

† This result, and the phenomenon of electron spin itself, was predicted in 1927 by Paul Dirac, who combined special relativity and quantum mechanics into a relativistic wave equation called the Dirac equation. Precise measurements indicate that the magnetic moment of the electron due to its spin is 2.00232 times that predicted by Equation 27-27. The fact that the intrinsic magnetic moment of the electron is approximately twice what we would expect makes it clear that the simple model of the electron as a spinning ball is not to be taken literally.

where

$$\mu_B = \frac{e\hbar}{2m_e} = 9.27 \times 10^{-24} \text{ A} \cdot \text{m}^2 = 9.27 \times 10^{-24} \text{ J/T}$$

$$= 5.79 \times 10^{-5} \text{ eV/T} \qquad\qquad 27\text{-}29$$

BOHR MAGNETON

is the quantum unit of magnetic moment called a **Bohr magneton.** The magnetic moment of an electron due to its intrinsic spin angular momentum $\vec{S}$ is

$$\vec{\mu}_s = -2 \times \frac{e\hbar}{2m_e} \frac{\vec{S}}{\hbar} = -2\mu_B \frac{\vec{S}}{\hbar} \qquad\qquad 27\text{-}30$$

MAGNETIC MOMENT DUE TO ELECTRON SPIN

Although the calculation of the magnetic moment of any atom is a complicated problem in quantum theory, the result for all electrons, according to both theory and experiment, is that the magnetic moment is of the order of a few Bohr magnetons. For atoms with zero net angular momentum, the net magnetic moment is zero. (The shell structure of atoms is discussed in Chapter 36.)

If all the atoms or molecules in some material have their magnetic moments aligned, the magnetic moment per unit volume of the material is the product of the number of molecules per unit volume $n$ and the magnetic moment $\mu$ of each molecule. For this extreme case, the **saturation magnetization** $M_s$ is

$$M_s = n\mu \qquad\qquad 27\text{-}31$$

The number of molecules per unit volume can be found from the molecular mass $M$, the density $\rho$ of the material, and Avogadro's number $N_A$:

$$n = \frac{N_A \text{ (atoms/mol)}}{M \text{ (kg/mol)}} \rho(\text{kg/m}^3) \qquad\qquad 27\text{-}32$$

*SATURIZATION MAGNETIZATION FOR IRON*                **EXAMPLE 27-10**

**Find the saturation magnetization and the magnetic field it produces for iron, assuming that each iron atom has a magnetic moment of 1 Bohr magneton.**

**PICTURE THE PROBLEM** We find the number of molecules per unit volume from the density of iron, $\rho = 7.9 \times 10^3$ kg/m³, and its molecular mass $M = 55.8 \times 10^{-3}$ kg/mol.

1. The saturation magnetization is the product of the number of molecules per unit volume and the magnetic moment of each molecule:

$$M_s = n\mu$$

2. Calculate the number of molecules per unit volume from Avogadro's number, the molecular mass, and the density:

$$n = \frac{N_A}{M}\rho = \frac{6.02 \times 10^{23} \text{ atoms/mol}}{55.8 \times 10^{-3} \text{ kg/mol}} (7.9 \times 10^3 \text{ kg/m}^3)$$

$$= 8.52 \times 10^{28} \text{ atoms/m}^3$$

3. Substitute this result and $\mu = 1$ Bohr magneton to calculate the saturation magnetization:

$$M_s = n\mu$$

$$= (8.52 \times 10^{28} \text{ atoms/m}^3)(9.27 \times 10^{-24} \text{ A·m}^2)$$

$$= \boxed{7.90 \times 10^5 \text{ A/m}}$$

4. The magnetic field on the axis inside a long iron cylinder resulting from this maximum magnetization is given by $B = \mu_0 M_s$:

$$B = \mu_0 M_s$$

$$= (4\pi \times 10^{-7} \text{ T·A})(7.90 \times 10^5 \text{ A/m})$$

$$= \boxed{0.993 \text{ T} \approx 1 \text{ T}}$$

**REMARKS** The measured saturation magnetic field of annealed iron is about 2.16 T, indicating that the magnetic moment of an iron atom is slightly greater than 2 Bohr magnetons. This magnetic moment is due mainly to the spins of two unpaired electrons in the iron atom.

## *Paramagnetism

Paramagnetism occurs in materials whose atoms have permanent magnetic moments that interact with each other only very weakly, resulting in a very small, positive magnetic susceptibility $\chi_m$. When there is no external magnetic field, these magnetic moments are randomly oriented. In the presence of an external magnetic field, the magnetic moments tend to line up parallel to the field, but this is counteracted by the tendency for the magnetic moments to be randomly oriented due to thermal motion. The degree to which the moments line up with the field depends on the strength of the field and on the temperature. This degree of alignment usually is small because the energy of a magnetic moment in an external magnetic field is typically much smaller than the thermal energy of an atom of the material, which is of the order of $kT$, where $k$ is Boltzmann's constant and $T$ is the absolute temperature.

The potential energy of a magnetic dipole of moment $\vec{\mu}$ in an external magnetic field $\vec{B}$ is given by Equation 26-16:

$$U = -\mu B \cos\theta = -\vec{\mu} \cdot \vec{B}$$

The potential energy when the moment is parallel with the field ($\theta = 0$) is thus lower than when the moment is antiparallel ($\theta = 180°$) by the amount $2\mu B$. For a typical atomic magnetic moment of 1 Bohr magneton and a typical strong magnetic field of 1 T, the difference in potential energy is

$$\Delta U = 2\mu_B B = 2(5.79 \times 10^{-5} \text{ eV/T})(1 \text{ T}) = 1.16 \times 10^{-4} \text{ eV}$$

At a normal temperature of $T = 300$ K, the typical thermal energy $kT$ is

$$kT = (8.62 \times 10^{-5} \text{ eV/K})(300 \text{ K}) = 2.59 \times 10^{-2} \text{ eV}$$

which is more than 200 times greater than $2\mu_B B$. Thus, even in a very strong magnetic field of 1 T, most of the magnetic moments will be randomly oriented because of thermal motions (unless the temperature is very low).

Figure 27-36 shows a plot of the magnetization $M$ versus an applied external magnetic field $B_{app}$ at a given temperature. In very strong fields, nearly all the magnetic moments are aligned with the field and $M \approx M_s$. (For magnetic fields attainable in the laboratory, this can occur only for very low temperatures.) When $B_{app} = 0$, $M = 0$, indicating that the orientation of the moments is completely random. In weak fields, the magnetization is approximately proportional to the

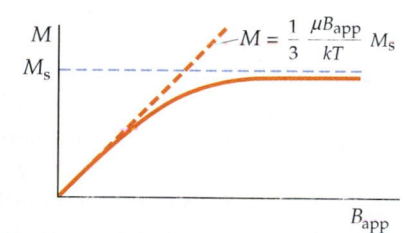

**FIGURE 27-36** Plot of magnetization $M$ versus an applied magnetic field $B_{app}$. In very strong fields, the magnetization approaches the saturation value $M_s$. This can be achieved only at very low temperatures. In weak fields, the magnetization is approximately proportional to $B_{app}$, a result known as Curie's law.

applied field, as indicated by the orange dashed line in the figure. In this region, the magnetization is given by

$$M = \frac{1}{3} \frac{\mu B_{app}}{kT} M_s$$   27-33

CURIE'S LAW

Note that $\mu B_{app}/(kT)$ is the ratio of the maximum energy of a dipole in the magnetic field to the characteristic thermal energy. The result that the magnetization varies inversely with the absolute temperature was discovered experimentally by Pierre Curie and is known as **Curie's law.**

Liquid oxygen, which is paramagnetic, is attracted by the magnetic field of a permanent magnet. A net force is exerted on the magnetic dipoles because the magnetic field is not uniform.

APPLYING CURIE'S LAW          **E X A M P L E  2 7 - 1 1**

If $\mu = \mu_B$, at what temperature will the magnetization be 1 percent of the saturation magnetization in an applied magnetic field of 1 T?

**PICTURE THE PROBLEM**

1. Curie's law relates $M$, $T$, $M_s$, and $B_{app}$:

$$M = \frac{1}{3} \frac{\mu B_{app}}{kT} M_s$$

2. Solve for $T$ using $\mu = \mu_B$ and $M/M_s = 0.01$:

$$T = \frac{\mu_B B_{app}}{3k} \frac{M_s}{M} = \frac{(5.79 \times 10^{-5} \text{ eV/T})(1 \text{ T})}{3(8.62 \times 10^{-5} \text{ eV/K})} 100$$

$$= \boxed{22.4 \text{ K}}$$

**REMARKS** From this example, we see that even in a strong applied magnetic field of 1 T, the magnetization is less than 1 percent of saturation at temperatures above 22.4 K.

**EXERCISE** If $\mu = \mu_B$, what fraction of the saturation magnetization is $M$ at 300 K for an external magnetic field of 1.5 T? (*Answer* $M/M_s = 1.12 \times 10^{-3}$)

## *Ferromagnetism

Ferromagnetism occurs in pure iron, cobalt, and nickel as well as in alloys of these metals with each other. It also occurs in gadolinium, dysprosium, and a few compounds. Ferromagnetism arises from a strong interaction between the electrons in a partially full band in a metal or between the localized electrons that form magnetic moments on neighboring atoms or molecules. This interaction, called the **exchange interaction,** lowers the energy of a pair of electrons with parallel spins.

Ferromagnetic materials have very large positive values of magnetic susceptibility $\chi_m$ (as measured under conditions described, which follow). In these substances, a small external magnetic field can produce a very large degree of alignment of the atomic magnetic dipole moments. In some cases, the alignment can persist even when the external magnetizing field is removed. This alignment persists because the magnetic dipole moments exert strong forces on their neighbors so that over a small region of space the moments are aligned with each other even when there is no external field. The region of space over which the magnetic dipole moments are aligned is called a **magnetic domain.** The size of a domain is usually microscopic. Within the domain, all the permanent atomic magnetic moments are aligned, but the direction of alignment varies from domain to domain so that the net magnetic moment of a macroscopic piece of ferromagnetic

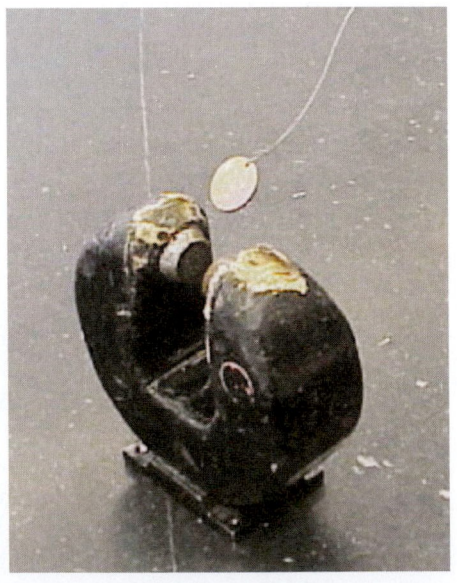

A Canadian quarter that is attracted by a magnet. Canadian coins often contain significant amounts of nickel, which is ferromagnetic.

(a)

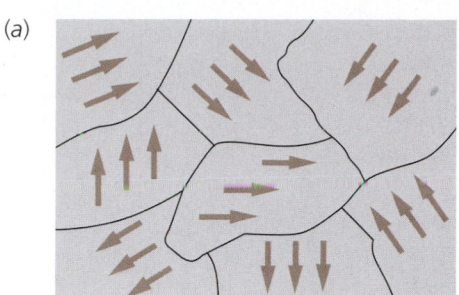

(b)

**FIGURE 27-37** (a) Schematic illustration of ferromagnetic domains. Within a domain, the magnetic dipoles are aligned, but the direction of alignment varies from domain to domain so that the net magnetic moment is zero. A small external magnetic field may cause the enlargement of those domains that are aligned parallel to the field, or it may cause the alignment within a domain to rotate. In either case, the result is a net magnetic moment parallel to the field. (b) Magnetic domains on the surface of an FE−3 percent Si crystal observed using a scanning electron microscope with polarization analysis. The four colors indicate four possible domain orientations.

material is zero in the normal state. Figure 27-37 illustrates this situation. The dipole forces that produce this alignment are predicted by quantum theory but cannot be explained with classical physics. At temperatures above a critical temperature, called the **Curie temperature,** thermal agitation is great enough to break up this alignment and ferromagnetic materials become paramagnetic.

When an external magnetic field is applied, the boundaries of the domains may shift or the direction of alignment within a domain may change so that there is a net macroscopic magnetic moment in the direction of the applied field. Since the degree of alignment is large for even a small external field, the magnetic field produced in the material by the dipoles is often much greater than the external field.

Let us consider what happens when we magnetize a long iron rod by placing it inside a solenoid and gradually increase the current in the solenoid windings. We assume that the rod and the solenoid are long enough to permit us to neglect end effects. Since the induced magnetic moments are in the same direction as the applied field, $\vec{B}_{app}$ and $\vec{M}$ are in the same direction. Then,

$$B = B_{app} + \mu_0 M = \mu_0 n I + \mu_0 M \qquad \text{27-34}$$

In ferromagnetic materials, the magnetic field $\mu_0 M$ due to the magnetic moments is often greater than the magnetizing field $B_{app}$ by a factor of several thousand.

**A chunk of magnetite (lodestone) attracts the needle of a compass.**

(a)

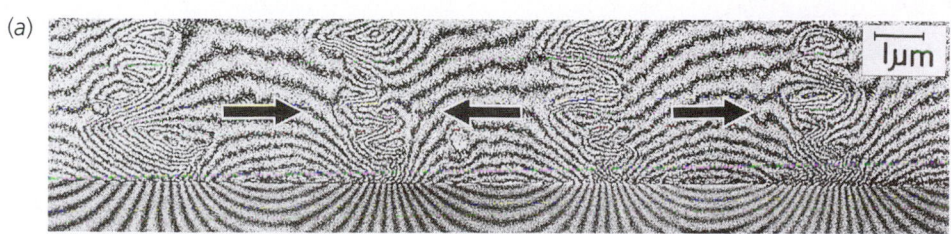

1μm

(b)

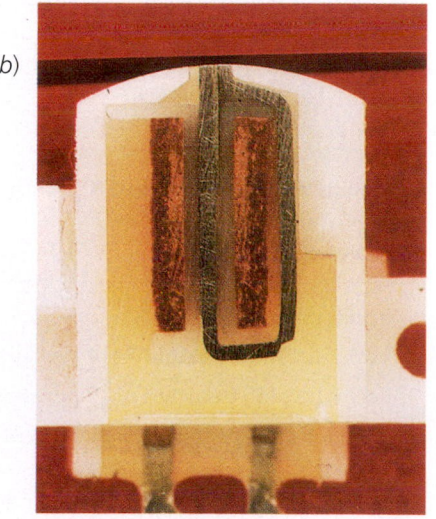

(a) Magnetic field lines on a cobalt magnetic recording tape. The solid arrows indicate the encoded magnetic bits. (b) Cross section of a magnetic tape recording head. Current from an audio amplifier is sent to wires around a magnetic core in the recording head where it produces a magnetic field. When the tape passes over a gap in the core of the recording head, the fringing magnetic field encodes information on the tape.

Figure 27-38 shows a plot of $B$ versus the magnetizing field $B_{app}$. As the current is gradually increased from zero, $B$ increases from zero along the part of the curve from the origin $O$ to point $P_1$. The flattening of this curve near point $P_1$ indicates that the magnetization $M$ is approaching its saturation value $M_s$, at which all the atomic magnetic moments are aligned. Above saturation, $B$ increases only because the magnetizing field $B_{app} = \mu_0 nI$ increases. When $B_{app}$ is gradually decreased from point $P_1$, there is not a corresponding decrease in the magnetization. The shift of the domains in a ferromagnetic material is not completely reversible, and some magnetization remains even when $B_{app}$ is reduced to zero, as indicated in the figure. This effect is called **hysteresis,** from the Greek word *hysteros* meaning later or behind, and the curve in Figure 27-38 is called a **hysteresis curve.** The value of the magnetic field at point $P_4$ when $B_{app}$ is zero is called the **remnant field** $B_{rem}$. At this point, the iron rod is a permanent magnet. If the current in the solenoid is now reversed so that $B_{app}$ is in the opposite direction, the magnetic field $B$ is gradually brought to zero at point $c$. The remaining part of the hysteresis curve is obtained by further increasing the current in the opposite direction until point $P_2$ is reached, which corresponds to saturation in the opposite direction, and then decreasing the current to zero at point $P_3$ and increasing it again in its original direction.

Since the magnetization $M$ depends on the previous history of the material, and since it can have a large value even when the applied field is zero, it is not simply related to the applied field $B_{app}$. However, if we confined ourselves to that part of the magnetization curve from the origin to point $P_1$ in Figure 27-38, $\vec{B}_{app}$ and $\vec{M}$ are parallel and $M$ is zero when $B_{app}$ is zero. We can then define the magnetic susceptibility as in Equation 27-22,

$$M = \chi_m \frac{B_{app}}{\mu_0}$$

and

$$B = B_{app} + \mu_0 M = B_{app}(1 + \chi_m) = K_m \mu_0 nI = \mu nI \qquad 27\text{-}35$$

where

$$\mu = (1 + \chi_m)\mu_0 = K_m \mu_0 \qquad 27\text{-}36$$

is called the **permeability** of the material. (For paramagnetic and diamagnetic materials, $\chi_m$ is much less than 1 so the permeability $\mu$ and the permeability of free space $\mu_0$ are very nearly equal.)

Since $B$ does not vary linearly with $B_{app}$, as can be seen from Figure 27-38, the relative permeability is not constant. The maximum value of $K_m$ occurs at a magnetization that is considerably less than the saturation magnetization. Table 27-2 lists the saturation magnetic field $\mu_0 M_s$ and the maximum values of $K_m$ for some ferromagnetic materials. Note that the maximum values of $K_m$ are much greater than 1.

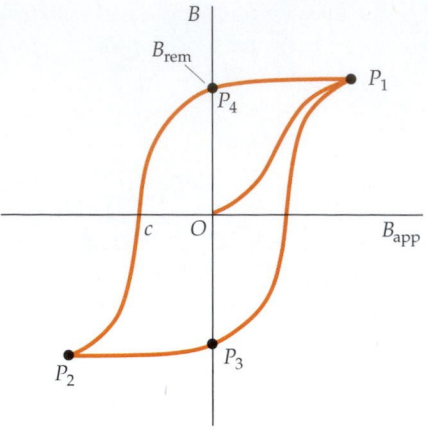

**FIGURE 27-38** Plot of $B$ versus the applied magnetizing field $B_{app}$. The outer curve is called a hysteresis curve. The field $B_{rem}$ is called the remnant field. It remains when the applied field returns to zero.

## TABLE 27-2

**Maximum Values of $\mu_0 M$ and $K_m$ for Some Ferromagnetic Materials**

| Material | $\mu_0 M_s$, T | $K_m$ |
|---|---|---|
| Iron (annealed) | 2.16 | 5,500 |
| Iron-silicon (96 percent Fe, 4 percent Si) | 1.95 | 7,000 |
| Permalloy (55 percent Fe, 45 percent Ni) | 1.60 | 25,000 |
| Mu-metal (77 percent Ni, 16 percent Fe, 5 percent Cu, 2 percent Cr) | 0.65 | 100,000 |

The area enclosed by the hysteresis curve is proportional to the energy dissipated as heat in the irreversible process of magnetizing and demagnetizing. If the hysteresis effect is small, so that the area inside the curve is small, indicating a small energy loss, the material is called **magnetically soft.** Soft iron is an example. The hysteresis curve for a magnetically soft material is shown in Figure 27-39. Here the remnant field $B_{rem}$ is nearly zero, and the energy loss per cycle is small. Magnetically soft materials are used for transformer cores to allow the magnetic field $B$ to change without incurring large energy losses as the field alternates. On the other hand, a large remnant field is desirable in a permanent magnet. **Magnetically hard** materials, such as carbon steel and the alloy Alnico 5, are used for permanent magnets.

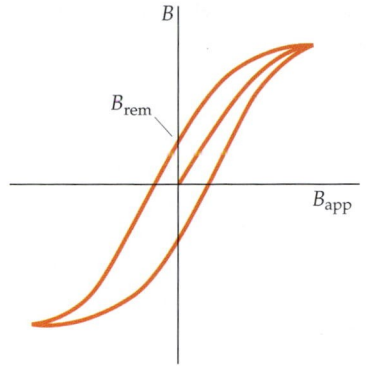

**FIGURE 27-39** Hysteresis curve for a magnetically soft material. The remnant field is very small compared with the remnant field for a magnetically hard material such as that shown in Figure 27-38.

(a) An extremely high-capacity, hard-disk drive for magnetic storage of information, capable of storing over 250 gigabytes of information. (b) A magnetic test pattern on a hard disk, magnified 2400 times. The light and dark regions correspond to oppositely directed magnetic fields. The smooth region just outside the pattern is a region of the disk that has been erased just prior to writing.

(a)

(b)

SOLENOID WITH IRON CORE        **EXAMPLE 27-12**

A long solenoid with 12 turns per centimeter has a core of annealed iron. When the current is 0.50 A, the magnetic field inside the iron core is 1.36 T. Find (a) the applied field $B_{app}$, (b) the relative permeability $K_m$, and (c) the magnetization $M$.

**PICTURE THE PROBLEM** The applied field is just that of a long solenoid given by $B_{app} = \mu_0 nI$. Since the total magnetic field is given, we can find the relative permeability from its definition ($K_m = B/B_{app}$) and we can find $M$ from $B = B_{app} + \mu_0 M$.

1. The applied field is given by Equation 27-9:

$$B_{app} = \mu_0 nI$$

$$= (4\pi \times 10^{-7}\ \text{T·m/A})(1200\ \text{turns/m})(0.5\ \text{A})$$

$$= 7.54 \times 10^{-4}\ \text{T}$$

2. The relative permeability is the ratio of $B$ to $B_{app}$:

$$K_m = \frac{B}{B_{app}} = \frac{1.36\ \text{T}}{7.54 \times 10^{-4}\ \text{T}} = 1.80 \times 10^3$$

3. The magnetization $M$ is found from Equation 27-34:

$$\mu_0 M = B - B_{app}$$

$$= 1.36\ \text{T} - 7.54 \times 10^{-4}\ \text{T} \approx B = 1.36\ \text{T}$$

$$M = \frac{B}{\mu_0} = \frac{1.36\ \text{T}}{4\pi \times 10^{-7}\ \text{T·m/A}} = \boxed{1.08 \times 10^6\ \text{A/m}}$$

**REMARKS** The applied magnetic field of $7.54 \times 10^{-4}$ T is a negligible fraction of the total field of 1.36 T. Note that the value for $K_m$ of 1800 is considerably smaller than the maximum value of 5500 in Table 27-2. Note also that the susceptibility $\chi_m = K_m - 1 \approx K_m$ to the three-place accuracy with which we calculated $K_m$.

## *Diamagnetism

Diamagnetic materials are those materials that have very small negative values of magnetic susceptibility $\chi_m$. Diamagnetism was discovered by Michael Faraday in 1845 when Faraday found that a piece of bismuth is repelled by either pole of a magnet, indicating that the external field of the magnet induces a magnetic moment in bismuth in the direction opposite the field.

We can understand this effect qualitatively from Figure 27-40, which shows two positive charges moving in circular orbits with the same speed but in opposite directions. Their magnetic moments are in opposite directions and therefore cancel.[†] In the presence of an external magnetic field $\vec{B}$ directed into the paper, the charges experience an extra force $q\vec{v} \times \vec{B}$, which is along the radial direction. For the charge on the left, this extra force is inward, increasing the centripetal force. If the charge is to remain in the same circular orbit, it must speed up so that $mv^2/r$ equals the total centripetal force.[‡] Its magnetic moment, which is outward, is thus increased. For the charge on the right, the additional force is outward, so the particle must slow down to maintain its circular orbit. Its magnetic moment, which is

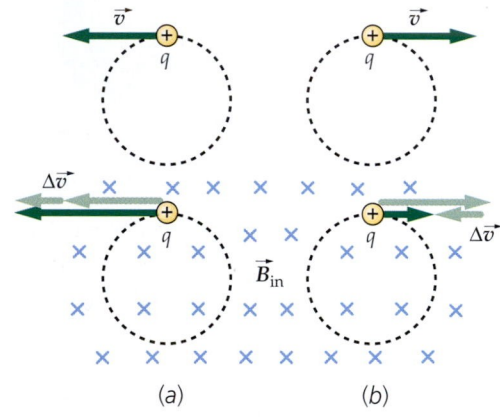

**FIGURE 27-40** (a) A positive charge moving counterclockwise in a circle has its magnetic moment directed out of the page. When an external, magnetic field directed into the page is turned on, the magnetic force increases the centripetal force so the speed of the particle must increase. The change in the magnetic moment is out of the page. (b) A positive charge moving clockwise in a circle has its magnetic moment directed into the page. When an external, magnetic field directed into the page is turned on, the magnetic force decreases the centripetal force so the speed of the particle must decrease. As in (a), the change in the magnetic moment is directed out of the page.

---

[†] It is simpler to consider positive charges even though it is the negatively charged electrons that provide the magnetic moments in matter.

[‡] The electron speeds up because of an electric field induced by the changing magnetic field, an effect called induction, which we discuss in Chapter 28.

inward, is decreased. In each case, the *change* in the magnetic moment of the charges is in the direction out of the page, opposite that of the external applied field. Since the permanent magnetic moments of the two charges are equal and oppositely directed they add to zero, leaving only the induced magnetic moments which are both opposite the direction of the applied magnetic field.

A material will be diamagnetic if its atoms have zero net angular momentum and therefore no permanent magnetic moment. (The net angular momentum of an atom depends on the electronic structure of the atom, which is a subject that we will study in Chapter 35.) The induced magnetic moments that cause diamagnetism have magnitudes of the order of $10^{-5}$ Bohr magnetons. Since this is much smaller than the permanent magnetic moments of the atoms of paramagnetic or ferromagnetic materials, the diamagnetic effect in these atoms is masked by the alignment of their permanent magnetic moments. However, since this alignment decreases with temperature, all materials are theoretically diamagnetic at sufficiently high temperatures.

When a superconductor is placed in an external magnetic field, electric currents are induced on the superconductor's surface so that the net magnetic field in the superconductor is zero. Consider a superconducting rod inside a solenoid of $n$ turns per unit length. When the solenoid is connected to a source of emf so that it carries a current $I$, the magnetic field due to the solenoid is $\mu_0 nI$. A surface current of $-nI$ per unit length is induced on the superconducting rod that cancels out the field due to the solenoid so that the net field inside the superconductor is zero. From Equation 27-23,

$$\vec{B} = \vec{B}_{app}(1 + \chi_m) = 0$$

so

$$\chi_m = -1$$

A superconductor is thus a perfect diamagnet with a magnetic susceptibility of $-1$.

A superconductor is a perfect diamagnet. Here, the superconducting pendulum bob is repelled by the permanent magnet.

## SUMMARY

1. Magnetic fields arise from moving charges, and therefore from currents.
2. The Biot–Savart law describes the magnetic field produced by a current element.
3. Ampère's law relates the line integral of the magnetic field along some closed curve to the current that passes through any surface bounded by the curve.
4. The magnetization vector $\vec{M}$ describes the magnetic moment per unit volume of matter.
5. The classical relation $\vec{\mu} = [q/(2m)]\,\vec{L}$ is derived from the definitions of angular momentum and magnetic moment.
6. The Bohr magneton is a convenient unit for atomic and nuclear magnetic moments.

| Topic | Relevant Equations and Remarks | |
|---|---|---|

### 1. Magnetic Field $\vec{B}$

| | | |
|---|---|---|
| Due to a moving point charge | $$\vec{B} = \frac{\mu_0}{4\pi}\frac{q\vec{v}\times\hat{r}}{r^2}$$ | 27-1 |
| | where $\hat{r}$ is a unit vector that points to the field point $P$ from the charge $q$ moving with velocity $\vec{v}$, and $\mu_0$ is a constant of proportionality called the permeability of free space: | |
| | $$\mu_0 = 4\pi\times 10^{-7}\ \text{T}\cdot\text{m/A} = 4\pi\times 10^{-7}\ \text{N/A}^2$$ | 27-2 |
| Due to a current element (Biot–Savart law) | $$d\vec{B} = \frac{\mu_0}{4\pi}\frac{I\,d\vec{\ell}\times\hat{r}}{r^2}$$ | 27-3 |
| On the axis of a current loop | $$B_x = \frac{\mu_0}{4\pi}\frac{2\pi R^2 I}{(x^2 + R^2)^{3/2}}$$ | 27-5 |
| On the axis of a current loop, at great distances from the loop | $$B_x = \frac{\mu_0}{4\pi}\frac{2\mu}{|x|^3}$$ <br> where $\mu$ is the magnitude of the magnetic moment of the loop. | 27-6 |
| Inside a long solenoid, far from the ends | $B_x = \mu_0 n I$ <br> where $n$ is the number of turns per unit length. | 27-9 |
| Due to a straight wire segment | $$B = \frac{\mu_0}{4\pi}\frac{I}{R}(\sin\theta_2 - \sin\theta_1)$$ <br> where $R$ is the perpendicular distance to the wire and $\theta_1$ and $\theta_2$ are the angles subtended at the field point by the ends of the wire. | 27-11 |
| Due to a long, straight wire | $$B = \frac{\mu_0}{4\pi}\frac{2I}{R}$$ <br> The direction of $\vec{B}$ is such that the magnetic field lines of $\vec{B}$ encircle the wire in the direction of the fingers of the right hand if the thumb points in the direction of the current. | 27-12 |
| Inside a tightly wound toroid | $$B = \frac{\mu_0}{2\pi}\frac{NI}{r},\quad a < r < b$$ | 27-17 |

### 2. Magnetic Field Lines

The magnetic field is indicated by lines parallel to $\vec{B}$ at any point whose density is proportional to the magnitude of $\vec{B}$. Magnetic lines do not begin or end at any point in space. Instead, they form continuous loops.

| | | |
|---|---|---|
| **3. Gauss's Law for Magnetism** | $\phi_{m, \text{net}} = \oint_S B_n \, dA = 0$ | 27-14 |

| | |
|---|---|
| **4. Magnetic Poles** | Magnetic poles always occur in pairs. Isolated magnetic poles have not been found. |

**5. Ampère's Law**

$$\oint_C \vec{B} \cdot d\vec{\ell} = \mu_0 I_C$$

where $C$ is any closed curve. **27-15**

| | |
|---|---|
| Validity of Ampère's law | Ampère's law is valid only if the currents are steady and continuous. It can be used to derive expressions for the magnetic field for situations with a high degree of symmetry, such as a long, straight, current-carrying wire or a long, tightly wound solenoid. |

| | |
|---|---|
| **6. Magnetism in Matter** | Matter can be classified as paramagnetic, ferromagnetic, or diamagnetic. |

| | |
|---|---|
| Magnetization | A magnetized material is described by its magnetization vector $\vec{M}$, which is defined as the net magnetic dipole moment per unit volume of the material: |

$$\vec{M} = \frac{d\vec{\mu}}{dV}$$

**27-18**

The magnetic field due to a uniformly magnetized cylinder is the same as if the cylinder carried a current per unit length of magnitude $M$ on its surface. This current, which is due to the intrinsic motion of the atomic charges in the cylinder, is called an amperian current.

**7. $\vec{B}$ in Magnetic Materials** $\qquad \vec{B} = \vec{B}_{\text{app}} + \mu_0 \vec{M}$ **27-21**

Magnetic susceptibility $\chi_m$ $\qquad \vec{M} = \chi_m \dfrac{\vec{B}_{\text{app}}}{\mu_0}$ **27-22**

For paramagnetic materials, $\chi_m$ is a small positive number that depends on temperature. For diamagnetic materials (other than superconductors), it is a small negative constant independent of temperature. For superconductors, $\chi_m = -1$. For ferromagnetic materials, the magnetization depends not only on the magnetizing current but also on the past history of the material.

Relative permeability $\qquad \vec{B} = K_m \vec{B}_{\text{app}}$ **27-23**

where

$$K_m = 1 + \chi_m$$

**27-24**

**8. Atomic Magnetic Moments** $\qquad \vec{\mu} = \dfrac{q}{2m} \vec{L}$ **27-27**

where $\vec{L}$ is the orbital angular momentum of the particle.

| | | |
|---|---|---|
| Due to the orbital motion of an electron | $\vec{\mu}_\ell = -\dfrac{e\hbar}{2m_e} \dfrac{\vec{L}}{\hbar} = -\mu_B \dfrac{\vec{L}}{\hbar}$ | 27-28 |

| | | |
|---|---|---|
| Due to electron spin | $\vec{\mu}_s = -2 \times \dfrac{e\hbar}{2m_e} \dfrac{\vec{S}}{\hbar} = -2\mu_B \dfrac{\vec{S}}{\hbar}$ | 27-30 |

Bohr magneton $\qquad \mu_B = \dfrac{e\hbar}{2m_e} = 9.27 \times 10^{-24} \text{ A·m}^2$

$$= 9.27 \times 10^{-24} \text{ J/T} = 5.79 \times 10^{-5} \text{ eV/T}$$

**27-29**

where

$$\hbar = \frac{h}{2\pi} = 1.05 \times 10^{-34} \text{ J·s}$$

and $h = 6.626 \times 10^{-34}$ J·s is Planck's constant.

| | |
|---|---|
| **\*9. Paramagnetism** | Paramagnetic materials have permanent atomic magnetic moments that have random directions in the absence of an applied magnetic field. In an applied field these dipoles are aligned with the field to some degree, producing a small contribution to the total field that adds to the applied field. The degree of alignment is small except in very strong fields and at very low temperatures. At ordinary temperatures, thermal motion tends to maintain the random directions of the magnetic moments. |
| Curie's law | In weak fields, the magnetization is approximately proportional to the applied field and inversely proportional to the absolute temperature. |

$$M = \frac{1}{3}\frac{\mu B_{\text{app}}}{kT}M_s \qquad\qquad 27\text{-}33$$

| | |
|---|---|
| **\*10. Ferromagnetism** | Ferromagnetic materials have small regions of space called magnetic domains in which all the permanent atomic magnetic moments are aligned. When the material is unmagnetized, the direction of alignment in one domain is independent of that in another domain so that no net magnetic field is produced. When the material is magnetized, the domains of a ferromagnetic material are aligned, producing a very strong contribution to the magnetic field. This alignment can persist even when the external field is removed, thus leading to permanent magnetism. |
| **\*11. Diamagnetism** | Diamagnetic materials are those materials in which the magnetic moments of all electrons in each atom cancel, leaving each atom with a zero magnetic moment in the absence of an external field. In an external field, a very small magnetic moment is induced that tends to weaken the field. This effect is independent of temperature. Superconductors are diamagnetic with a magnetic susceptibility equal to $-1$. |

# PROBLEMS

- Single-concept, single-step, relatively easy
- •• Intermediate-level, may require synthesis of concepts
- ••• Challenging
- **SSM** Solution is in the *Student Solutions Manual*
- **iSOLVE** Problems available on iSOLVE online homework service
- **iSOLVE ✓** These "Checkpoint" online homework service problems ask students additional questions about their confidence level, and how they arrived at their answer.

In a few problems, you are given more data than you actually need; in a few other problems, you are required to supply data from your general knowledge, outside sources, or informed estimates.

## Conceptual Problems

**1** • **SSM** Compare the directions of the electric force and the magnetic force between two positive charges, which move along parallel paths (*a*) in the same direction and (*b*) in opposite directions.

**2** • Is $\vec{B}$ uniform everywhere within a current loop? Explain.

**3** • Sketch the field lines for the electric dipole and the magnetic dipole shown in Figure 27-41. How do they differ in appearance close to the center of each dipole?

 +q

 −q

I

Electric dipole      Magnetic dipole

**FIGURE 27-41** Problem 3

**4** • Two wires lie in the plane of the paper and carry equal currents in opposite directions, as shown in Figure 27-42. At a point midway between the wires, the magnetic field is (a) zero. (b) into the page. (c) out of the page. (d) toward the top or bottom of the page. (e) toward one of the two wires.

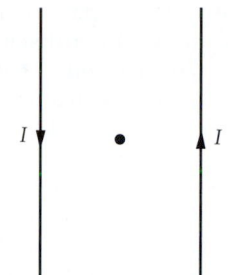

**FIGURE 27-42** Problem 4

**5** • Two parallel wires carry currents $I_1$ and $I_2 = 2I_1$ in the same direction. The forces $F_1$ and $F_2$ on the wires are related by (a) $F_1 = F_2$. (b) $F_1 = 2F_2$. (c) $2F_1 = F_2$. (d) $F_1 = 4F_2$. (e) $4F_1 = F_2$.

**6** • [SSM] A wire carries an electrical current straight up. What is the direction of the magnetic field due to the wire a distance of 2 m north of the wire? (a) North (b) East (c) West (d) South (e) Upward

**7** • Two current-carrying wires are perpendicular to each other. The current in one wire flows vertically upward and the current in the other wire flows horizontally toward the east. The horizontal wire is 1 m south of the vertical wire. What is the direction of the net magnetic force on the horizontal wire? (a) North (b) East (c) West (d) South (e) There is no net magnetic force on the horizontal wire.

**8** • Make a field-line sketch of the magnetic field due to the currents in the pair of coaxial coils (Figure 27-43). The currents in the coils have the same magnitude and are in the same direction in each coil.

**FIGURE 27-43** Problems 8 and 9

**9** • [SSM] Make a field-line sketch of the magnetic field due to the currents in the pair of coaxial coils (Figure 27-43). The currents in the coils have the same magnitude but are opposite in direction in each coil.

**10** • Ampère's law is valid (a) when there is a high degree of symmetry. (b) when there is no symmetry. (c) when the current is constant. (d) when the magnetic field is constant. (e) in all of these situations if the current is continuous.

**11** • True or false:

(a) Diamagnetism is the result of induced magnetic dipole moments.

(b) Paramagnetism is the result of the partial alignment of permanent magnetic dipole moments.

**12** • [SSM] If the magnetic susceptibility is positive, (a) paramagnetic effects or ferromagnetic effects must be greater than diamagnetic effects. (b) diamagnetic effects must be greater than paramagnetic effects. (c) diamagnetic effects must be greater than ferromagnetic effects. (d) ferromagnetic effects must be greater than paramagnetic effects. (e) paramagnetic effects must be greater than ferromagnetic effects.

**13** • True or false:

(a) The magnetic field due to a current element is parallel to the current element.

(b) The magnetic field due to a current element varies inversely with the square of the distance from the element.

(c) The magnetic field due to a long wire varies inversely with the square of the distance from the wire.

(d) Ampère's law is valid only if there is a high degree of symmetry.

(e) Ampère's law is valid only for continuous currents.

**14** • Can a particle have angular momentum and not have a magnetic moment?

**15** • Can a particle have a magnetic moment and not have angular momentum?

**16** • A circular loop of wire carries a current $I$. Is there angular momentum associated with the magnetic moment of the loop? If so, why is it not noticed?

**17** • A hollow tube carries a current. Inside the tube, $\vec{B} = 0$. Why is this the case, because $\vec{B}$ is strong inside a solenoid?

**18** • [SSM] When a current is passed through the wire in Figure 27-44, will the wire tend to bunch up or form a circle?

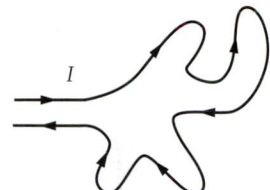

**FIGURE 27-44** Problem 18

**19** • Which of the four gases listed in Table 27-1 are diamagnetic and which of the four gases are paramagnetic?

## Estimation and Approximation

**20** •• The magnetic moment of the earth is about $9 \times 10^{22}$ A·m². (a) If the magnetization of the earth's core were $1.5 \times 10^9$ A/m, what is the core volume? (b) What is the radius of such a core if it were spherical and centered with the earth?

**21** •• [SSM] Estimate the transient magnetic field 100 m away from a lightning bolt if a charge of about 30 C is transferred from cloud to ground and the average speed of the charges is $10^6$ m/s.

**22** •• [SSM] The rotating disk of Problem 125 (page 896) can be used as a model for the magnetic field due to a sunspot. If the sunspot radius is approximately $10^7$ m rotating at an angular velocity of about $10^{-2}$ rad/s, calculate the total charge $Q$ on the sunspot needed to create a magnetic field of order 0.1 T at the center of the sunspot. What is the electrical field magnitude just above the center of the sunspot due to this charge?

## The Magnetic Field of Moving Point Charges

**23** • At time $t = 0$, a particle with charge $q = 12$ $\mu$C is located at $x = 0$, $y = 2$ m; the particle's velocity at that time is $\vec{v} = 30$ m/s$\hat{i}$. Find the magnetic field at (a) the origin; (b) $x = 0$, $y = 1$ m; (c) $x = 0$, $y = 3$ m; and (d) $x = 0$, $y = 4$ m.

**24** • For the particle in Problem 23, find the magnetic field at (a) $x = 1$ m, $y = 3$ m; (b) $x = 2$ m, $y = 2$ m; and (c) $x = 2$ m, $y = 3$ m.

**25** • A proton (charge $+e$) traveling with a velocity of $\vec{v} = 1 \times 10^4$ m/s$\hat{i} + 2 \times 10^4$ m/s$\hat{j}$ is located at $x = 3$ m, $y = 4$ m at some time $t$. Find the magnetic field at (a) $x = 2$ m, $y = 2$ m; (b) $x = 6$ m, $y = 4$ m; and (c) $x = 3$ m, $y = 6$ m.

**26** • **iSOLVE** An electron orbits a proton at a radius of $5.29 \times 10^{-11}$ m. What is the magnetic field at the proton due to the orbital motion of the electron?

**27** •• **SSM** Two equal charges $q$ located at $(0, 0, 0)$ and at $(0, b, 0)$ at time zero are moving with speed $v$ in the positive $x$ direction ($v \ll c$). Find the ratio of the magnitudes of the magnetic force and electrostatic force on each charge.

## The Magnetic Field of Currents: The Biot–Savart Law

**28** • A small current element $I\, d\vec{\ell}$ with $d\vec{\ell} = 2$ mm$\hat{k}$ and $I = 2$ A, is centered at the origin. Find the magnetic field $d\vec{B}$ at the following points: (a) on the $x$ axis at $x = 3$ m, (b) on the $x$ axis at $x = -6$ m, (c) on the $z$ axis at $z = 3$ m, and (d) on the $y$ axis at $y = 3$ m.

**29** • For the current element in Problem 28, find the magnitude and direction of $d\vec{B}$ at $x = 0, y = 3$ m, $z = 4$ m.

**30** • **SSM** For the current element in Problem 28, find the magnitude of $d\vec{B}$ and indicate its direction on a diagram at (a) $x = 2$ m, $y = 4$ m, $z = 0$ and (b) $x = 2$ m, $y = 0$, $z = 4$ m.

## $\vec{B}$ Due to a Current Loop

**31** • A single loop of wire with radius 3 cm carries a current of 2.6 A. What is the magnitude of $B$ on the axis of the loop at (a) the center of the loop, (b) 1 cm from the center, (c) 2 cm from the center, and (d) 35 cm from the center?

**32** • **SSM** **iSOLVE** A single-turn circular loop of radius 10.0 cm is to produce a field at its center that will just cancel the earth's magnetic field at the equator, which is 0.7 G directed north. Find the current in the loop and make a sketch that shows the orientation of the loop and the current.

**33** •• For the loop of wire in Problem 32, at what point along the axis of the loop is the magnetic field (a) 10 percent of the field at the center, (b) 1 percent of the field at the center, and (c) 0.1 percent of the field at the center?

**34** •• **iSOLVE** A single-turn circular loop of radius 8.5 cm is to produce a field at its center that will just cancel the earth's field of magnitude 0.7 G directed at 70° below the horizontal north direction. Find the current in the loop and make a sketch that shows the orientation of the loop and the current in the loop.

**35** •• A circular current loop of radius $R$ carrying a current $I = 10$ A is centered at the origin with its axis along the $x$ axis. Its current is such that it produces a magnetic field in the positive $x$ direction. (a) Using a spreadsheet program or graphing calculator, construct a graph of $B_x$ versus $x/R$ for points on the $x/R$ axis $-5 < x/R < +5$. Compare this graph with that for $E_x$ due to a charged ring of the same size. (b) A second, identical current loop, carrying an equal current in the same sense, is in a plane parallel to the $yz$ plane with its center at $x = R$. Make separate graphs of $B_x$ on the $x$ axis due to each loop and also graph the resultant field due to the two loops. Show from your sketch that $dB_x/dx$ is zero midway between the two loops.

**36** •• A pair of identical coils, each of radius $r$, are separated by a distance $r$. Called *Helmholtz coils*, the coils are coaxial and carry equal currents such that their axial fields add. A feature of Helmholtz coils is that the resultant magnetic field in the region between the coils is very uniform. Let $r = 30$ cm, $I = 15$ A and $N = 250$ turns for each coil. Using a spreadsheet program calculate and graph the magnetic field as a function of $x$, the distance from the center of the coils along the common axis, for $-r < x < r$. Over what range of $x$ does the field vary by less than 20%?

**37** ••• Two Helmholtz coils with radii $R$ have their axes along the $x$ axis (see Problem 36). One coil is in the $yz$ plane and the second coil is in a parallel plane at $x = R$. Show that at the midpoint of the coils $dB_x/dx = 0$, $d^2B_x/dx^2 = 0$, and $d^3B_x/dx^3 = 0$. (*Note:* This shows that the magnetic field at points near the midpoint is approximately equal to that at the midpoint.)

**38** ••• **SSM** *Anti-Helmholtz* coils are used in many physics applications, such as laser cooling and trapping, where a spatially inhomogeneous field with a uniform gradient is desired. These coils have the same construction as a Helmholtz coil, except that the currents flow in opposite directions, so that the axial fields subtract, and the coil separation is $r\sqrt{3}$ rather than $r$. Graph the magnetic field as a function of $x$, the axial distance from the center of the coils, for an anti-Helmholtz coil using the same parameters as in Problem 36.

**39** •• Two concentric coplanar conducting circular loops have radii $r_1 = 10$ cm and $r_2 > r_1$ are in a horizontal plane. A current $I = 1$ A flows in each coil, but in opposite directions, with the current in the inner coil being counterclockwise as viewed from above. Using a spreadsheet program, calculate and graph the magnetic field as a function of the height $x$ above the center of the coils for $r_2 = $ (a) 10.1 cm, (b) 11 cm, (c) 15 cm and (d) 20 cm.

**40** ••• Two concentric circular loops of wire in the same plane have radii $r_1 = 10$ cm and $r_2 > r_1$. A current $I = 1$ A flows in each loop but in the opposite direction. Using a spreadsheet program, calculate and graph the magnetic field component $B_x$ on the axis of the loops as a function of the distance $x$ from the center of the coils. Construct a separate curve for $r_2 = $ (a) 10.1 cm, (b) 11 cm, (c) 15 cm, and (d) 20 cm.

**41** ••• For the coils considered in Problem 40, show that if $r_2 = r_1 + \Delta r$, where $\Delta r \ll r_1$, then

$$B(x) \approx \left(\frac{\mu_0 I \Delta r}{2}\right)\left(\frac{2rx^2 - r^3}{(x^2 + r_1^2)^{5/2}}\right).$$

## Straight-Line Current Segments

**42** •• For the coils considered in Problem 41, show that when $x \gg r_1$, then

$$B(x) \approx -\left(\frac{\mu_0 I \Delta r}{2}\right)\left(\frac{2r_1}{x^3}\right).$$

Compare this to the results of Problem 39 (a).

Problems 43 to 48 refer to Figure 27-45, which shows two long straight wires in the xy plane and parallel to the x axis. One wire is at y = −6 cm and the other wire is at y = +6 cm. The current in each wire is 20 A.

**FIGURE 27-45**
**Problems 43–48**

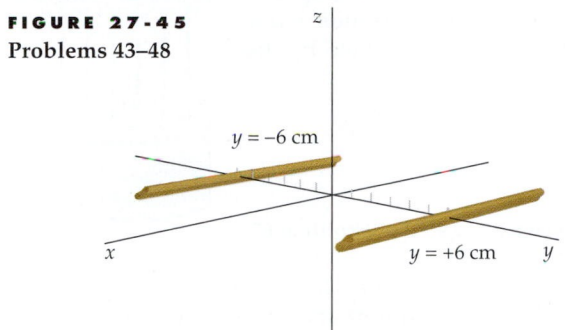

**43** • **SSM** If the currents in Figure 27-45 are in the negative x direction, find $\vec{B}$ at the points on the y axis at (a) y = −3 cm, (b) y = 0, (c) y = +3 cm, and (d) y = +9 cm.

**44** •• Using a spreadsheet program or graphing calculator, graph $B_z$ versus y for points on the y axis when both currents are in the negative x direction.

**45** • Find $\vec{B}$ at points on the y axis, as in Problem 43, when the current in the wire at y = −6 cm is in the negative x direction and the current in the wire at y = +6 cm is in the positive x direction.

**46** •• Using a spreadsheet program or graphing calculator, graph $B_z$ versus y for points on the y axis when the directions of the currents are opposite to those in Problem 45.

**47** • Find $\vec{B}$ on the z axis at z = +8 cm if (a) the currents are parallel, as in Problem 43 and (b) the currents are antiparallel, as in Problem 45.

**48** • **ISOLVE** Find the magnitude of the force per unit length exerted by one wire on the other.

**49** • **ISOLVE** Two long, straight parallel wires 8.6 cm apart carry currents of equal magnitude I. The parallel wires repel each other with a force per unit length of 3.6 nN/m. (a) Are the currents parallel or antiparallel? (b) Find I.

**50** •• **ISOLVE** The current in the wire shown in Figure 27-46 is 8 A. Find B at point P due to each wire segment and sum to find the resultant B.

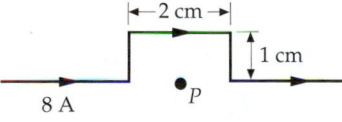

**FIGURE 27-46** Problem 50

**51** •• **ISOLVE** A wire of length 16 cm is suspended by flexible leads above a long straight wire. Equal but opposite currents are established in the wires so that the 16-cm wire floats 1.5 mm above the long wire with no tension in its suspension leads. If the mass of the 16-cm wire is 14 g, what is the current?

**52** •• **SSM** Three long, parallel straight wires pass through the corners of an equilateral triangle of sides 10 cm, as shown in Figure 27-47, where a dot means that the current is out of the paper and a cross means that the current is into the paper. If each current is 15 A, find

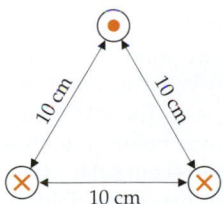

**FIGURE 27-47** Problems 52 and 53

(a) the force per unit length on the upper wire and (b) the magnetic field B at the upper wire due to the two lower wires.

**53** •• Rework Problem 52, with the current in the lower right corner of Figure 27-47 reversed.

**54** •• An infinitely long insulated wire lies along the x axis and carries current I in the positive x direction. A second infinitely long insulated wire lies along the y axis and carries current I in the positive y direction. Where in the xy plane is the resultant magnetic field zero?

**55** •• An infinitely long wire lies along the z axis and carries a current of 20 A in the positive z direction. A second infinitely long wire is parallel to the z axis at x = 10 cm. (a) Find the current in the second wire if the magnetic field at x = 2 cm is zero. (b) What is the magnetic field at x = 5 cm?

**56** •• Three very long parallel wires are at the corners of a square, as shown in Figure 27-48. The wires each carry a current of magnitude I. Find the magnetic field B at the unoccupied corner of the square when (a) all the currents are into the paper, (b) $I_1$ and $I_3$ are into the paper and $I_2$ is out, and (c) $I_1$ and $I_2$ are into the paper and $I_3$ is out.

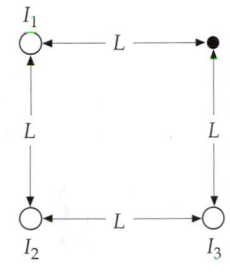

**FIGURE 27-48** Problem 56

**57** •• **SSM** Four long, straight parallel wires each carry current I. In a plane perpendicular to the wires, the wires are at the corners of a square of side a. Find the force per unit length on one of the wires if (a) all the currents are in the same direction and (b) the currents in the wires at adjacent corners are oppositely directed.

**58** •• An infinitely long nonconducting cylinder of radius R lies along the z axis. Five long conducting wires are parallel to the cylinder and are spaced equally on the upper half of the cylinder's surface. Each wire carries a current I in the positive z direction. Find the magnetic field on the z axis.

### $\vec{B}$ Due to a Current in a Solenoid

**59** • A solenoid with length 30 cm, radius 1.2 cm, and 300 turns carries a current of 2.6 A. Find B on the axis of the solenoid (a) at the center, (b) inside the solenoid at a point 10 cm from one end, and (c) at one end.

**60** • **SSM** **ISOLVE** A solenoid 2.7-m long has a radius of 0.85 cm and 600 turns. It carries a current I of 2.5 A. What is the approximate magnetic field B on the axis of the solenoid?

**61** ••• A solenoid has n turns per unit length and radius R and carries a current I. Its axis is along the x axis with one end at $x = -\frac{1}{2}\ell$ and the other end at $x = +\frac{1}{2}\ell$, where $\ell$ is the total length of the solenoid. Show that the magnetic field B at a point on the axis outside the solenoid is given by

$$B = \tfrac{1}{2}\mu_0 nI(\cos\theta_1 - \cos\theta_2)$$

where $\cos\theta_1 = \dfrac{x + \frac{1}{2}\ell}{[R^2 + (x + \frac{1}{2}\ell)^2]^{1/2}}$

and $\cos\theta_2 = \dfrac{x - \frac{1}{2}\ell}{[R^2 + (x - \frac{1}{2}\ell)^2]^{1/2}}$

**62 •••** In Problem 61, a formula for the magnetic field along the axis of a solenoid is given. For $x \gg \ell$ and $\ell > R$, the angles $\theta_1$ and $\theta_2$ are very small, so the small-angle approximation $\cos \theta \approx 1 - \theta^2/2$ is valid. (a) Draw a diagram and show that

$$\theta_1 \approx \frac{R}{x + \frac{1}{2}\ell}$$

and

$$\theta_2 \approx \frac{R}{x - \frac{1}{2}\ell}$$

(b) Show that the magnetic field at a point far from either end of the solenoid can be written

$$B = \frac{\mu_0}{4\pi}\left(\frac{q_m}{r_1^2} - \frac{q_m}{r_2^2}\right)$$

where $r_1 = x - \frac{1}{2}\ell$ is the distance to the near end of the solenoid, $r_1 = x + \frac{1}{2}\ell$ is the distance to the far end, and $q_m = nI\pi R^2 = \mu/\ell$, where $\mu = NI\pi R^2$ is the magnetic moment of the solenoid.

## Ampère's Law

**63 •** [SSM] [SOLVE] A long, straight, thin-walled cylindrical shell of radius $R$ carries a current $I$. Find $B$ inside the cylinder and outside the cylinder.

**64 •** In Figure 27-49, one current is 8 A into the paper, the other current is 8 A out of the paper, and each curve is a circular path. (a) Find $\oint_C \vec{B} \cdot d\vec{\ell}$ for each path indicated, where each integral is taken with $d\vec{\ell}$ counterclockwise. (b) Which path, if any, can be used to find $B$ at some point due to these currents?

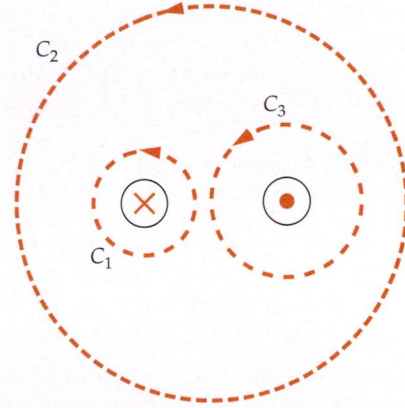

**FIGURE 27-49**
**Problem 64**

**65 •** A very long coaxial cable consists of an inner wire and a concentric outer cylindrical conducting shell of radius $R$. At one end, the wire is connected to the shell. At the other end, the wire and shell are connected to opposite terminals of a battery, so there is a current down the wire and back up the shell. Assume that the cable is straight. Find $B$ (a) at points between the wire and the shell far from the ends and (b) outside the cable.

**66 ••** [SOLVE] A wire of radius 0.5 cm carries a current of 100 A that is uniformly distributed over its cross-sectional area. Find $B$ (a) 0.1 cm from the center of the wire, (b) at the surface of the wire, and (c) at a point outside the wire 0.2 cm from the surface of the wire. (d) Sketch a graph of $B$ versus the distance from the center of the wire.

**67 ••** [SSM] Show that a uniform magnetic field with no fringing field, such as that shown in Figure 27-50, is impossible because it violates Ampère's law. Do this by applying Ampère's law to the rectangular curve shown by the dashed lines.

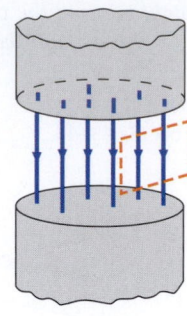

**FIGURE 27-50** Problem 67

**68 ••** A coaxial cable consists of a solid inner cylindrical conductor of radius 1.00 mm and an outer cylindrical shell conductor of inner radius 2.00 mm with an outer radius of 3.00 mm. There is a current of 15 A down the inner wire and an equal return current in the outer wire. The currents are uniform over the cross section of each conductor. Using a spreadsheet program or graphing calculator, graph the magnitude of the magnetic field $B$ as a function of the distance $r$ from the cable axis for 0 mm $< r <$ 3.00 mm. What is the field outside the wire?

**69 ••** An infinitely long, thick cylindrical shell of inner radius $a$ and outer radius $b$ carries a current $I$ uniformly distributed across a cross section of the shell. Find the magnetic field for (a) $r < a$, (b) $a < r < b$, and (c) $r > b$.

**70 ••** Figure 27-51 shows a solenoid carrying a current $I$ with $n$ turns per unit length. Apply Ampère's law to the rectangular curve shown in the figure to derive an expression for $B$, assuming that $B$ is uniform inside the solenoid and that $B$ is zero outside the solenoid.

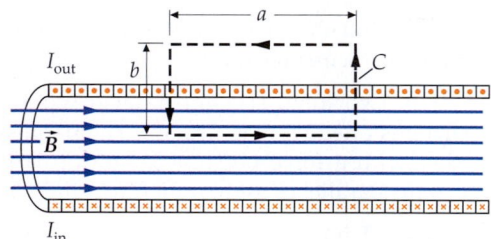

**FIGURE 27-51** Problem 70

**71 ••** [SOLVE] A tightly wound toroid of inner radius 1 cm and outer radius 2 cm has 1000 turns of wire and carries a current of 1.5 A. (a) What is the magnetic field at a distance of 1.1 cm from the center? (b) What is the magnetic field at a distance of 1.5 cm from the center?

**72 ••** [SSM] The $xz$ plane contains an infinite sheet of current in the positive $z$ direction. The current per unit length (along the $x$ direction) is $\lambda$. Figure 27-52$a$ shows a point $P$ above the sheet ($y > 0$) and two portions of the current sheet labeled $I_1$ and $I_2$. (a) What is the direction of the magnetic field $\vec{B}$ at point $P$ due to the two portions of the current shown? (b) What is the direction of the magnetic field $\vec{B}$ at point $P$ due to the entire sheet? (c) What is the direction of $\vec{B}$ at a point below the sheet ($y < 0$)? (d) Apply Ampère's law to the rectangular curve shown in Figure 27-52$b$ to show that the magnetic field at any point above the sheet is given by $\vec{B} = -\frac{1}{2}\mu_0\lambda\hat{i}$

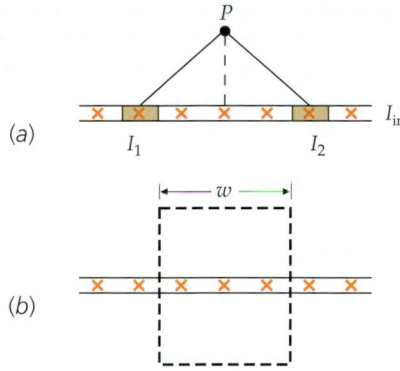

**FIGURE 27-52** Problem 72

## Magnetization and Magnetic Susceptibility

**73** • A tightly wound solenoid 20-cm long has 400 turns and carries a current of 4 A so that its axial field is in the $z$ direction. Neglecting end effects, find $B$ and $B_{app}$ at the center when ($a$) there is no core in the solenoid and ($b$) there is an iron core with a magnetization $M = 1.2 \times 10^6$ A/m.

**74** • If the solenoid of Problem 73 has an aluminum core, find $B_{app}$, $M$, and $B$ at the center, neglecting end effects.

**75** • Repeat Problem 74 for a tungsten core.

**76** • A long solenoid is wound around a tungsten core and carries a current. ($a$) If the core is removed while the current is held constant, does the magnetic field inside the solenoid decrease or increase? ($b$) By what percentage does the magnetic field inside the solenoid decrease or increase?

**77** • **SOLVE** When a sample of liquid is inserted into a solenoid carrying a constant current, the magnetic field inside the solenoid decreases by 0.004 percent. What is the magnetic susceptibility of the liquid?

**78** • **SOLVE** A long solenoid carrying a current of 10 A has 50 turns/cm. What is the magnetic field in the interior of the solenoid when the interior is ($a$) a vacuum, ($b$) filled with aluminum, and ($c$) filled with silver?

**79** •• **SSM** A cylinder of magnetic material is placed in a long solenoid of $n$ turns per unit length and current $I$. The values for magnetic field $B$ within the material versus $nI$ is given below. Use these values to plot $B$ versus $B_{app}$ and $K_m$ versus $nI$.

| $nI$, A/m | 0 | 50 | 100 | 150 | 200 | 500 | 1000 | 10,000 |
|---|---|---|---|---|---|---|---|---|
| $B$, T | 0 | 0.04 | 0.67 | 1.00 | 1.2 | 1.4 | 1.6 | 1.7 |

**80** •• A small magnetic sample is in the form of a disk that has a radius of 1.4 cm, a thickness of 0.3 cm, and a uniform magnetization along its axis throughout its volume. The magnetic moment of the sample is $1.5 \times 10^{-2}$ A·m². ($a$) What is the magnetization $\vec{M}$ of the sample? ($b$) If this magnetization is due to the alignment of $N$ electrons, each with a magnetic moment of 1 $\mu_B$, what is $N$? ($c$) If the magnetization is along the axis of the disk, what is the magnitude of the amperian surface current?

**81** •• A cylindrical shell in the shape of a flat washer has inner radius $r$, outer radius $R$, and thickness (length) $t$, where $t \ll R$. The material of the shell has a uniform magnetization of magnitude $M$ parallel to its axis. Show that the magnetic field due to the cylinder can be modeled using the concentric conducting loops model of Problem 39. What is the amperian current $I$ which we must use to model this field?

## Atomic Magnetic Moments

**82** •• **SSM** **SOLVE** Nickel has a density of 8.7 g/cm³ and a molecular mass of 58.7 g/mol. Nickel's saturation magnetization is given by $\mu_0 M_s = 0.61$ T. Calculate the magnetic moment of a nickel atom in Bohr magnetons.

**83** •• **SOLVE** Repeat Problem 82 for cobalt, which has a density of 8.9 g/cm³, a molecular mass of 58.9 g/mol, and a saturation magnetization given by $\mu_0 M_s = 1.79$ T.

## *Paramagnetism

**84** • Show that Curie's law predicts that the magnetic susceptibility of a paramagnetic substance is $\chi_m = \mu_0 M_s / 3kT$.

**85** •• In a simple model of paramagnetism, we can consider that some fraction $f$ of the molecules have their magnetic moments aligned with the external magnetic field and that the rest of the molecules are randomly oriented and therefore do not contribute to the magnetic field. ($a$) Use this model and Curie's law to show that at temperature $T$ and external magnetic field $B$ the fraction of aligned molecules is $f = \mu B / 3kT$. ($b$) Calculate this fraction for $T = 300$ K, $B = 1$ T, assuming $\mu$ to be 1 Bohr magneton.

**86** •• **SSM** Assume that the magnetic moment of an aluminum atom is 1 Bohr magneton. The density of aluminum is 2.7 g/cm³, and its molecular mass is 27 g/mol. ($a$) Calculate $M_s$ and $\mu_0 M_s$ for aluminum. ($b$) Use the results of Problem 84 to calculate $\chi_m$ at $T = 300$ K. ($c$) Explain why the result for Part ($b$) is larger than the value listed in Table 27-1.

**87** •• A toroid with $N$ turns carrying a current $I$ has a mean radius $R$ and a cross-sectional radius $r$, where $r \ll R$ (Figure 27-53). When the toroid is filled with material, it is called a *Rowland ring*. Find $B_{app}$ and $B$ in such a ring, assuming a magnetization $\vec{M}$ everywhere parallel to $\vec{B}_{app}$.

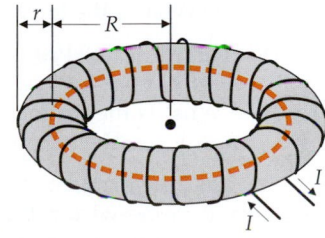

**FIGURE 27-53** Problem 87

**88** •• A toroid is filled with liquid oxygen that has a susceptibility of $4 \times 10^{-3}$. The toroid has 2000 turns and carries a current of 15 A. Its mean radius is 20 cm, and the radius of its cross section is 0.8 cm. ($a$) What is the magnetization $M$? ($b$) What is the magnetic field $B$? ($c$) What is the percentage increase in $B$ produced by the liquid oxygen?

**89 ••** A toroid has an average radius of 14 cm and a cross-sectional area of 3 cm². It is wound with fine wire, 60 turns/cm measured along its mean circumference, and the wire carries a current of 4 A. The core is filled with a paramagnetic material of magnetic susceptibility $2.9 \times 10^{-4}$. (*a*) What is the magnitude of the magnetic field within the substance? (*b*) What is the magnitude of the magnetization? (*c*) What would the magnitude of the magnetic field be if there were no paramagnetic core present?

## *Ferromagnetism

**90 •** **SSM** For annealed iron, the relative permeability $K_m$ has its maximum value of approximately 5500 at $B_{app} = 1.57 \times 10^{-4}$ T. Find $M$ and $B$ when $K_m$ is maximum.

**91 ••** The saturation magnetization for annealed iron occurs when $B_{app} = 0.201$ T. Find the permeability $\mu$ and the relative permeability $K_m$ of annealed iron at saturation. (See Table 27-2.)

**92 ••** **iSOLVE** The coercive force is defined to be the applied magnetic field needed to bring $B$ back to zero along the hysteresis curve (which is point $c$ in Figure 27-38). For a certain permanent bar magnet, the coercive force $B_{app} = 5.53 \times 10^{-2}$ T. The bar magnet is to be demagnetized by placing it inside a 15-cm-long solenoid with 600 turns. What minimum current is needed in the solenoid to demagnetize the magnet?

**93 ••** A long solenoid with 50 turns/cm carries a current of 2 A. The solenoid is filled with iron and $B$ is measured to be 1.72 T. (*a*) Neglecting end effects, what is $B_{app}$? (*b*) What is $M$? (*c*) What is the relative permeability $K_m$?

**94 ••** When the current in Problem 93 is 0.2 A, the magnetic field is measured to be 1.58 T. (*a*) Neglecting end effects, what is $B_{app}$? (*b*) What is $M$? (*c*) What is the relative permeability $K_m$?

**95 ••** A long, iron-core solenoid with 2000 turns/m carries a current of 20 mA. At this current, the relative permeability of the iron core is 1200. (*a*) What is the magnetic field within the solenoid? (*b*) With the iron core removed, what current will produce the same field within the solenoid?

**96 ••** **SSM** Two long straight wires 4-cm apart are embedded in a uniform insulator that has a relative permeability of $K_m = 120$. The wires carry 40 A in opposite directions. (*a*) What is the magnetic field at the midpoint of the plane of the wires? (*b*) What is the force per unit length on the wires?

**97 ••** The toroid of Problem 88 has its core filled with iron. When the current is 10 A, the magnetic field in the toroid is 1.8 T. (*a*) What is the magnetization $M$? (*b*) Find the values for $K_m$, $\mu$, and $\chi_m$ for the iron sample.

**98 ••** Find the magnetic field in the toroid of Problem 89 if the current in the wire is 0.2 A and soft iron, which has a relative permeability of 500, is substituted for the paramagnetic core?

**99 ••** A long straight wire with a radius of 1.0 mm is coated with an insulating ferromagnetic material that has a thickness of 3.0 mm and a relative magnetic permeability of $K_m = 400$. The coated wire is in air and the wire itself is nonmagnetic. The wire carries a current of 40 A. (*a*) Find the

magnetic field inside the wire as a function of radius $R$. (*b*) Find the magnetic field inside the ferromagnetic material as a function of radius $R$. (*c*) Find the magnetic field outside the ferromagnetic material as a function of $R$. (*d*) What must the magnitudes and directions of the amperian currents be on the surfaces of the ferromagnetic material to account for the magnetic fields observed?

## General Problems

**100 •** Find the magnetic field at point $P$ in Figure 27-54.

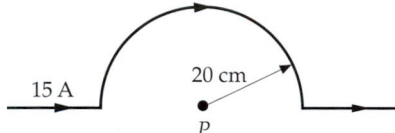

**FIGURE 27-54** Problem 100

**101 •** **SSM** In Figure 27-55, find the magnetic field at point $P$, which is at the common center of the two semicircular arcs.

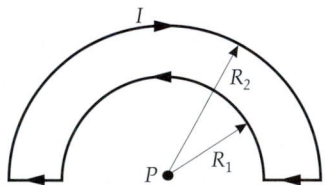

**FIGURE 27-55** Problem 101

**102 ••** A wire of length $\ell$ is wound into a circular coil of $N$ loops and carries a current $I$. Show that the magnetic field at the center of the coil is given by $B = \mu_0 \pi N^2 I / \ell$.

**103 ••** A very long wire carrying a current $I$ is bent into the shape shown in Figure 27-56. Find the magnetic field at point $P$.

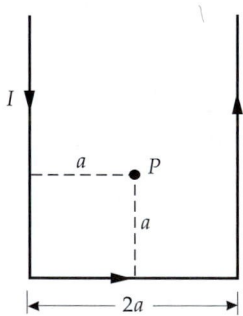

**FIGURE 27-56** Problem 103

**104 ••** **SSM** A power cable carrying 50 A is 2 m below the earth's surface, but the cable's direction and precise position are unknown. Show how you could locate the cable using a compass. Assume that you are at the equator, where the earth's magnetic field is 0.7 G north.

**105 ••** A long straight wire carries a current of 20 A, as shown in Figure 27-57. A rectangular coil with two sides parallel to the straight wire has sides 5 cm and 10 cm with the near side a distance 2 cm from the wire. The coil carries a current of 5 A. (*a*) Find the force on each segment of the rectangular coil due to the current in the long straight wire.

(b) What is the net force on the coil?

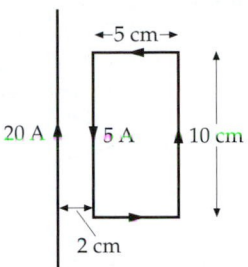

FIGURE 27-57 Problem 105

**106** •• SOLVE The closed loop shown in Figure 27-58 carries a current of 8 A in the counterclockwise direction. The radius of the outer arc is 60 cm, that of the inner arc is 40 cm. Find the magnetic field at point P.

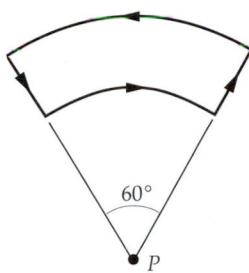

FIGURE 27-58 Problem 106

**107** •• SOLVE A closed circuit consists of two semicircles of radii 40 cm and 20 cm that are connected by straight segments, as shown in Figure 27-59. A current of 3 A flows around this circuit in the clockwise direction. Find the magnetic field at point P.

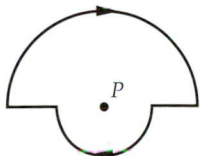

FIGURE 27-59 Problem 107

**108** •• SSM SOLVE A very long straight wire carries a current of 20 A. An electron 1 cm from the center of the wire is moving with a speed of $5.0 \times 10^6$ m/s. Find the force on the electron when it moves (a) directly away from the wire, (b) parallel to the wire in the direction of the current, and (c) perpendicular to the wire and tangent to a circle around the wire.

**109** •• A current $I$ of 5 A is uniformly distributed over the cross section of a long straight wire of radius $r_0 = 2.55$ mm. Using a spreadsheet program, graph the magnitude of the magnetic field as a function of $r$ and the distance from the center of the wire for $0 \le r \le 10r_0$.

**110** •• A large, 50-turn circular coil of radius 10 cm carries a current of 4 A. At the center of the large coil is a small 20-turn coil of radius 0.5 cm carrying a current of 1 A. The planes of the two coils are perpendicular. Find the torque exerted by the large coil on the small coil. (Neglect any variation in $B$ due to the large coil over the region occupied by the small coil.)

**111** •• SSM Figure 27-60 shows a bar magnet suspended by a thin wire that provides a restoring torque $-\kappa\theta$. The magnet is 16-cm long, has a mass of 0.8 kg, a dipole moment of $\mu = 0.12$ A·m², and it is located in a region where a uniform magnetic field B can be established. When the external magnetic field is 0.2 T and the magnet is given a small angular displacement $\Delta\theta$, the bar magnet oscillates about its equilibrium position with a period of 0.500 s. Determine the constant $\kappa$ and the period of this torsional pendulum when $B = 0$.

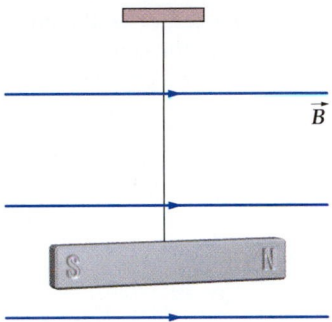

FIGURE 27-60 Problem 111

**112** •• A long, narrow bar magnet that has magnetic moment $\mu$ parallel to its long axis is suspended at its center as a frictionless compass needle. When placed in a magnetic field $\vec{B}$, the needle lines up with the field. If it is displaced by a small angle $\theta$, show that the needle will oscillate about its equilibrium position with frequency $f = (\frac{1}{2})\pi\sqrt{\mu B/I}$, where $I$ is the moment of inertia about the point of suspension.

**113** •• A small bar magnet of mass 0.1 kg, length 1 cm, and magnetic moment $\mu = 0.04$ A·m² is located at the center of a 100-turn loop of 0.2 m diameter. The loop carries a current of 5.0 A. At equilibrium, the bar magnet is aligned with the field due to the current loop. The bar magnet is given a displacement along the axis of the loop and released. Show that if the displacement is small, the bar magnet executes simple harmonic motion, and find the period of this motion.

**114** •• Suppose the needle in Problem 112 is a uniformly magnetized iron rod that is 8 cm long and has a cross-sectional area of 3 mm². Assume that the magnetic dipole moment for each iron atom is 2.2 $\mu_B$ and that all the iron atoms have their dipole moments aligned. Calculate the frequency of small oscillations about the equilibrium position when the magnetic field is 0.5 G.

**115** •• The needle of a magnetic compass has a length of 3 cm, a radius of 0.85 mm, and a density of $7.96 \times 10^3$ kg/m³. The needle is free to rotate in a horizontal plane, where the horizontal component of the earth's magnetic field is 0.6 G. When disturbed slightly, the compass executes simple harmonic motion about its midpoint with a frequency of 1.4 Hz. (a) What is the magnetic dipole moment of the needle? (b) What is the magnetization $M$? (c) What is the amperian current on the surface of the needle? (See Problem 112.)

**116** •• SSM An iron bar of length 1.4 m has a diameter of 2 cm and a uniform magnetization of $1.72 \times 10^6$ A/m directed along the bar's length. The bar is stationary in space and is suddenly demagnetized so that its magnetization disappears. What is the rotational angular velocity of the bar if its angular momentum is conserved? (Assume that Equation 27-27 holds where $m$ is the mass of an electron and $q = -e$.)

**117** •• The magnetic dipole moment of an iron atom is $2.219\ \mu_B$. (a) If all the atoms in an iron bar of length 20 cm and cross-sectional area 2 cm² have their dipole moments aligned, what is the dipole moment of the bar? (b) What torque must be supplied to hold the iron bar perpendicular to a magnetic field of 0.25 T?

**118** •• SSM A relatively inexpensive ammeter, called a *tangent galvanometer*, can be made using the earth's field. A plane circular coil of N turns and radius R is oriented so the field $B_c$ it produces in the center of the coil is either east or west. A compass is placed at the center of the coil. When there is no current in the coil, the compass needle points north. When there is a current I, the compass needle points in the direction of the resultant magnetic field $\vec{B}$ at an angle $\theta$ to the north. Show that the current I is related to $\theta$ and to the horizontal component of the earth's field $B_e$ by

$$I = \frac{2RB_e}{\mu_0 N}\tan\theta$$

**119** •• SOLVE An infinitely long straight wire is bent, as shown in Figure 27-61. The circular portion has a radius of 10 cm with its center a distance r from the straight part. Find r so that the magnetic field at the center of the circular portion is zero.

**FIGURE 27-61**
**Problem 119**

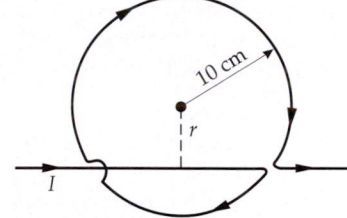

**120** •• (a) Find the magnetic field at point P for the wire carrying current I, as shown in Figure 27-62. (b) Use your result from Part (a) to find the field at the center of a polygon of N sides. Show that when N is very large, your result approaches that for the magnetic field at the center of a circle.

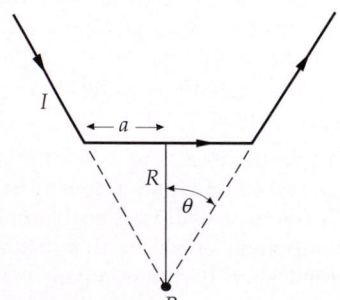

**FIGURE 27-62**
**Problem 120**

**121** •• The current in a long cylindrical conductor of radius R = 10 cm varies with distance from the axis of the cylinder according to the relation $I(r) = (50\ \text{A/m})r$. Find the magnetic field at (a) r = 5 cm, (b) r = 10 cm, and (c) r = 20 cm.

**122** •• Figure 27-63 shows a square loop, 20 cm per side, in the xy plane with its center at the origin. The loop carries a current of 5 A. Above it at y = 0, z = 10 cm is an infinitely long wire parallel to the x axis carrying a current of 10 A. (a) Find the torque on the loop. (b) Find the net force on the loop.

**FIGURE 27-63**
**Problem 122**

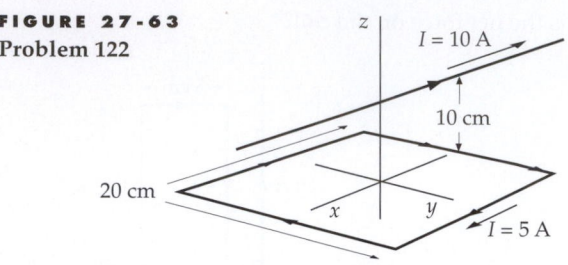

**123** •• A current balance is constructed in the following way: A 10-cm-long section of wire is placed on top of the pan of an electronic balance used in a chemistry lab. Leads are clipped to it running into a power supply and through the supply to another segment of wire that is suspended directly above it, parallel with it. (See figure below.) The distance between the two wires is L = 2.0 cm. The power supply provides a current I running through the wires. When the power supply is switched on, the reading on the balance increases by 5.0 mg (1 mg = $10^{-6}$ kg). What is the current running through the wire?

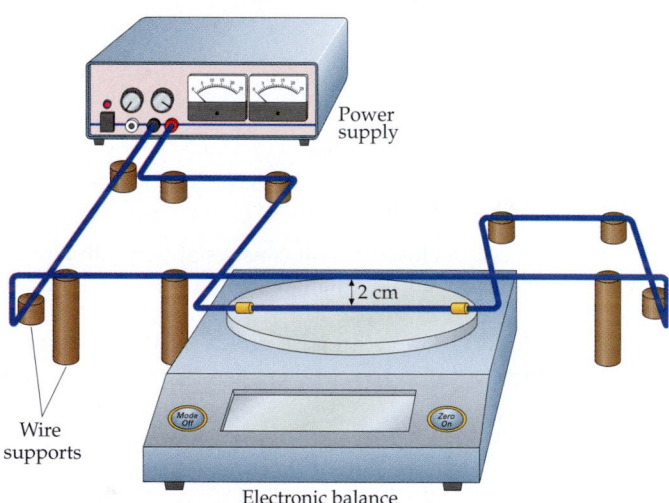

**FIGURE 27-64** **Problem 123**

**124** •• Consider the current balance of Problem 123. If the sensitivity of the balance is 0.1 mg, what is the minimum current detectable using this current balance? Discuss any advantages or disadvantages of this type of current balance versus the "standard" current balance discussed in the chapter.

**125** ••• SSM A disk of radius R carries a fixed charge density $\sigma$ and rotates with angular velocity $\omega$. (a) Consider a circular strip of radius r and width dr with charge dq. Show that the current produced by this strip $dI = (\omega/2\pi)\ dq = \omega\sigma r\ dr$. (b) Use your result from Part (a) to show that the magnetic field at the center of the disk is $B = \frac{1}{2}\mu_0\sigma\omega R$. (c) Use your result from Part (a) to find the magnetic field at a point on the axis of the disk a distance x from the center.

**126** ••• A square loop of side $\ell$ lies in the yz plane with its center at the origin. It carries a current I. Find the magnetic field B at any point on the x axis and show from your expression that for x much larger than $\ell$,

$$B \approx \frac{\mu_0\ 2\mu}{4\pi x^3}$$

where $\mu = I\ell^2$ is the magnetic moment of the loop.

# Magnetic Induction

DEMONSTRATION OF INDUCED EMF. WHEN THE MAGNET IS MOVING TOWARD OR AWAY FROM THE COIL, AN EMF IS INDUCED IN THE COIL, AS SHOWN BY THE GALVANOMETER'S DEFLECTION. NO DEFLECTION IS OBSERVED WHEN THE MAGNET IS STATIONARY.

**?** **What causes the current when the magnet moves? This is discussed in Section 28-2.**

28-1   Magnetic Flux

28-2   Induced EMF and Faraday's Law

28-3   Lenz's Law

28-4   Motional EMF

28-5   Eddy Currents

28-6   Inductance

28-7   Magnetic Energy

*28-8   *RL* Circuits

*28-9   Magnetic Properties of Superconductors

In the early 1830s, Michael Faraday in England and Joseph Henry in America independently discovered that in a *changing* magnetic field a changing magnetic flux through a surface bounded by a closed stationary loop of wire induces a current in the wire. The emfs and currents caused by such changing magnetic fluxes are called **induced emfs** and **induced currents.** The process itself is referred to as **induction.** Faraday and Henry also discovered that in a *static* magnetic field a changing magnetic flux through a surface bounded by a moving loop of wire induces an emf in the wire. An emf caused by the motion of a conductor in a region with a magnetic field is called a **motional emf.**

When you pull the plug of an electric cord from its socket, you sometimes observe a small spark. Before the cord is disconnected, the cord carries a current that produces a magnetic field encircling the current. When the cord is disconnected, the current abruptly ceases and the magnetic field around the cord collapses. This changing magnetic field induces an emf that tends to maintain the original current, resulting in a spark at the points of the disconnect. Once the magnetic field collapses to zero it is no longer changing, and the induced emf is zero.

Changing magnetic fields can result from changing currents or from moving magnets. The chapter-opening photo illustrates a simple classroom demonstration of emf induced by a changing magnetic field. The ends of a coil are attached to a galvanometer and a strong magnet is moved toward or away from the coil. The momentary deflection shown by the galvanometer *during* the motion indicates that there is an induced electric current in the coil–galvanometer circuit. A current is also induced if the coil is moved toward a stationary magnet, away from a stationary magnet, or if the coil is rotated in a region with a static magnetic field. A coil rotating in a static magnetic field is the basic element of a generator, which converts mechanical energy into electrical energy.

➤ **This chapter will explore the various methods of magnetic induction, all of which can be summarized by a single relation known as Faraday's law. Faraday's law relates the induced emf in a circuit to the rate of change in magnetic flux through the circuit. (The *magnetic flux through the circuit* refers to the flux of the magnetic field through any surface bounded by the circuit.)**

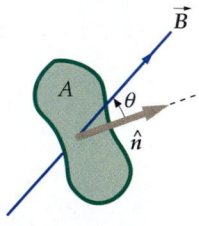

**FIGURE 28-1** When $\vec{B}$ makes an angle $\theta$ with the normal to the area of a loop, the flux through the loop is $\vec{B} \cdot \hat{n}A = BA \cos \theta$.

# 28-1 Magnetic Flux

The flux of any vector field through a surface is calculated in the same way as the flux of an electric field through a surface (Section 22-2). Let $dA$ be an element of area on the surface $S$, and let $\hat{n}$ be a unit normal, a unit vector normal to the area element (Figure 28-1). There are two directions normal to any area element, and which of the two directions is selected for the direction of $\hat{n}$ is a matter of choice. However, the sign of the flux does depend on this choice. The magnetic flux $\phi_m$ through $S$ is

$$\phi_m = \int_S \vec{B} \cdot \hat{n} \, dA = \int_S B_n \, dA \qquad 28\text{-}1$$

MAGNETIC FLUX

The unit of magnetic flux is that of magnetic field intensity times area, teslameter squared, which is called a **weber** (Wb):

$$1 \text{ Wb} = 1 \text{ T·m}^2 \qquad 28\text{-}2$$

Since $B$ is proportional to the number of field lines per unit area, the magnetic flux is proportional to the number of lines through an element of area.

**EXERCISE** Show that a weber per second is a volt.

If the surface is flat with area $A$, and if $\vec{B}$ is uniform (has the same magnitude and direction) over the surface, the magnetic flux through the surface is

$$\phi_m = \vec{B} \cdot \hat{n}A = BA \cos \theta = B_n A$$

where $\theta$ is the angle between the direction of $\vec{B}$ and the positive normal direction. We are often interested in the flux through a surface bounded by a coil that contains several turns of wire. If the coil contains $N$ turns, the flux through the surface is $N$ times the flux through each turn (Figure 28-2). That is,

$$\phi_m = NBA \cos \theta \qquad 28\text{-}3$$

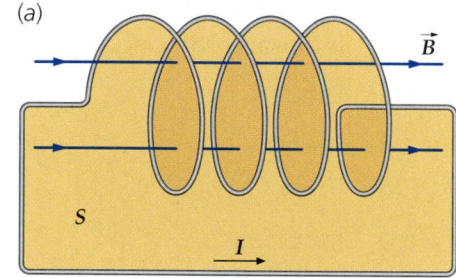

(a)

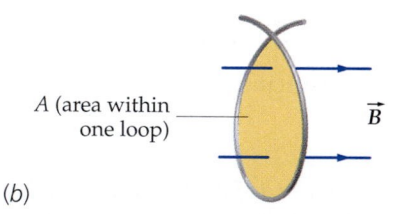

(b)

**FIGURE 28-2** (a) The flux through the surface $S$ bounded by a coil with $N$ turns is proportional to the number of field lines penetrating the surface. The coil shown has 4 turns. For the two field lines shown, each line penetrates the surface four times, once for each turn, so the flux through $S$ is four times greater than the flux through the surface "bounded" by a single turn of the coil. The coil shown is not tightly wound so the surface $S$ can better be observed. (b) The area $A$ of the flat surface that is (almost) bounded by a single turn.

where $A$ is the area of the flat surface bounded by a single turn. (*Note:* Only a closed curve can actually bound a surface.) A single turn of a multiturn coil is not closed, so a single turn can not actually bound a surface. However, if a coil is tightly wound a single turn is almost closed, and $A$ is the area of the flat surface that it (almost) bounds.

---

*FLUX THROUGH A SOLENOID*                                    **E X A M P L E    2 8 - 1**

**Find the magnetic flux through a solenoid that is 40-cm long, has a radius of 2.5 cm, has 600 turns, and carries a current of 7.5 A.**

**PICTURE THE PROBLEM** The magnetic field $\vec{B}$ inside the solenoid is uniform and parallel with the axis of the solenoid. It is therefore perpendicular to the plane of each coil. Therefore, we need to find $B$ inside the solenoid and then multiply $B$ by $NA$.

1. The magnetic flux is the product of the number of turns, the magnetic field strength, and the area bounded by one turn:

$$\phi_m = NBA$$

2. The magnetic field inside the solenoid is given by $B = \mu_0 nI$, where $n = N/\ell$ is the number of turns per unit length:

$$\phi_m = N\mu_0 nIA = N\mu_0 \frac{N}{\ell}IA = \frac{\mu_0 N^2 IA}{\ell}$$

3. Express the area $A$ in terms of its radius:

$$A = \pi r^2$$

4. Substitute the given values to calculate the flux:

$$\phi_m = \frac{\mu_0 N^2 I \pi r^2}{\ell}$$

$$= \frac{(4\pi \times 10^{-7}\ \text{T·m/A})(600\ \text{turns})^2 (7.5\ \text{A})\pi(0.025\ \text{m})^2}{0.40\ \text{m}}$$

$$= \boxed{1.66 \times 10^{-2}\ \text{Wb}}$$

**REMARKS** Note that since $\phi_m = NBA$ and $B$ is proportional to the number of turns $N$, $\phi_m$ is proportional to $N^2$.

---

# 28-2 Induced EMF and Faraday's Law

Experiments by Faraday, Henry, and others showed that if the magnetic flux through a surface bounded by a circuit is changed by any means, an emf equal in magnitude to the rate of change of the flux is induced in the circuit. We usually detect the emf by observing a current in the circuit, but the emf is present even if the circuit is nonexistent or incomplete (not closed) and there is no current. Previously we considered emfs that were localized in a specific part of the circuit, such as between the terminals of the battery. However, induced emfs can be distributed throughout the circuit.

The magnetic flux through a surface bounded by a circuit can be changed in several ways. The current producing the magnetic field may be increased or decreased, permanent magnets may be moved toward the surface or away from the surface, the circuit itself may be rotated in a region with a static magnetic field or translated in a region with a nonuniform static magnetic field $\vec{B}$, the orientation of the circuit may be changed, or the area of the surface in a region with a uniform static magnetic field may be increased or decreased. In every case, an emf $\mathcal{E}$ is induced in the circuit that is equal in magnitude to the rate of change of the magnetic flux through (a surface bounded by) the circuit. That is

$$\mathcal{E} = -\frac{d\phi_m}{dt} \qquad\qquad 28\text{-}4$$

FARADAY'S LAW

This result is known as **Faraday's law.** The negative sign in Faraday's law has to do with the direction of the induced emf, which is addressed shortly.

Figure 28-3 shows a single stationary loop of wire in a magnetic field. The flux through the loop is changing because the magnetic field strength is increasing, so an emf is induced in the loop. Since emf is the work done per unit charge, we know there must be forces exerted on the mobile charges doing work on them. Magnetic forces can do no work, therefore, we cannot attribute the emf to the work done by magnetic forces. It is electric forces associated with a nonconservative electric field $\vec{E}_{nc}$ doing the work on the mobile charges. The line integral of this electric field around a complete circuit equals the work done per unit charge, which is the induced emf in the circuit.

The electric fields that we studied in earlier chapters resulted from static electric charges. Such electric fields are conservative, meaning that their circulation about any curve $C$ is zero. (The circulation of a vector field $\vec{A}$ about a closed curve $C$ is defined as $\oint_C \vec{A} \cdot d\vec{\ell}$.) However, the electric field associated with a changing magnetic field is nonconservative. Its circulation about $C$ is an induced emf, equal to the negative of the rate of change of the magnetic flux through any surface $S$ bounded by $C$:

$$\mathcal{E} = \oint_C \vec{E}_{nc} \cdot d\vec{\ell} = -\frac{d}{dt}\int_S \vec{B} \cdot \hat{n}\, dA = -\frac{d\phi_m}{dt} \qquad\qquad 28\text{-}5$$

INDUCED EMF IN A STATIONARY CIRCUIT IN A CHANGING MAGNETIC FIELD

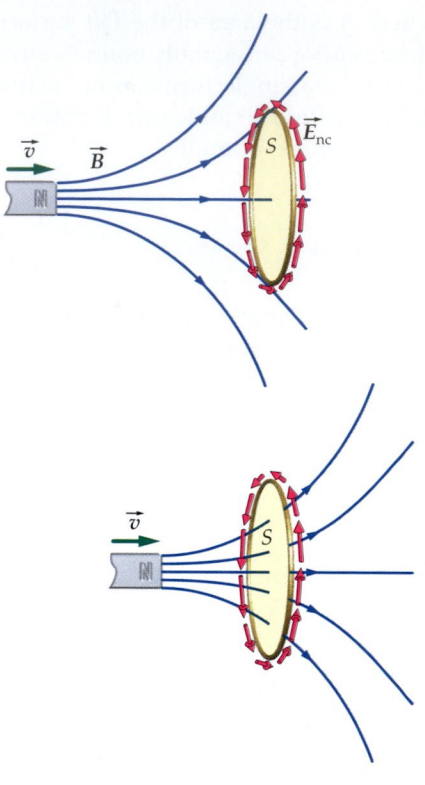

**FIGURE 28-3** If the magnetic flux through the stationary wire loop is changing, an emf is induced in the loop. The emf is distributed throughout the loop, which is due to a nonconservative electric field $\vec{E}_{nc}$ tangent to the wire.

---

*INDUCED EMF IN A CIRCULAR COIL I*  **EXAMPLE 28-2**

A uniform magnetic field makes an angle of 30° with the axis of a circular coil of 300 turns and a radius of 4 cm. The magnitude of the magnetic field increases at a rate of 85 T/s while its direction remains fixed. Find the magnitude of the induced emf in the coil.

**PICTURE THE PROBLEM** The induced emf equals $N$ times the rate of change of the flux through a single turn. Since $\vec{B}$ is uniform, the flux through each turn is simply $\phi_m = BA \cos\theta$, where $A = \pi r^2$ is the area of the coil.

1. The magnitude of the induced emf is given by Faraday's law:

$$\mathcal{E} = -\frac{d\phi_m}{dt}$$

2. For a uniform field, the flux is:

$$\phi_m = N\vec{B} \cdot \hat{n}A = NBA \cos\theta$$

3. Substitute this expression for $\phi_m$ and calculate $\mathcal{E}$:

$$\mathcal{E} = -\frac{d\phi_m}{dt} = -\frac{d}{dt}(NBA \cos\theta) = -N\pi r^2 \cos\theta\frac{dB}{dt}$$

$$= -(300)\pi(0.04\text{ m})^2 \cos 30°(85\text{ T/s}) = -111\text{ V}$$

$$|\mathcal{E}| = \boxed{111\text{ V}}$$

**EXERCISE** If the resistance of the coil is 200 Ω, what is the induced current? (*Answer* 0.555 A)

*INDUCED EMF IN A CIRCULAR COIL II*          **E X A M P L E   2 8 - 3    Try It Yourself**

An 80-turn coil of radius 5 cm and resistance of 30 $\Omega$ sits in a region with a uniform magnetic field normal to the plane of the coil. At what rate must the magnitude of the magnetic field change to produce a current of 4 A in the coil?

**PICTURE THE PROBLEM**   The rate of change of the magnetic field is related to the rate of change of the flux, which is related to the induced emf by Faraday's law. The emf in the coil equals *IR*.

**Cover the column to the right and try these on your own before looking at the answers.**

| Steps | Answers |
|---|---|
| 1. Write the magnetic flux in terms of *B*, *N*, and the radius *r*, and solve for *B*. | $\phi_m = NBA = NB\pi r^2$ $$B = \frac{\phi_m}{N\pi r^2}$$ |
| 2. Take the time derivative of *B*. | $$\frac{dB}{dt} = \frac{1}{N\pi r^2}\frac{d\phi_m}{dt}$$ |
| 3. Use Faraday's law to relate the rate of change of the flux to the emf. | $$\mathcal{E} = -\frac{d\phi_m}{dt}$$ |
| 4. Calculate the magnitude of the emf in the coil from the current and resistance of the coil. | $|\mathcal{E}| = IR = 120\text{ V}$ |
| 5. Substitute numerical values of *E*, *N*, and *r* to calculate *dB/dt*. | $$\left|\frac{dB}{dt}\right| = \frac{1}{N\pi r^2}|\mathcal{E}| = \boxed{191\text{ T/s}}$$ |

A sign convention allows us to use Equation 28-5 to find the direction of both the induced electric field and the induced emf. According to this convention, the positive tangential direction along the integration path *C* is related to the direction of the unit normal $\hat{n}$ on the surface *S* bounded by *C* by a right-hand rule (Figure 28-4). By placing your right thumb in the direction of $\hat{n}$, the fingers of your hand curl in the positive tangential direction on *C*. If $d\phi_m/dt$ is positive, then in accord with Faraday's law (Equation 28-5), both $\vec{E}_{nc}$ and $\mathcal{E}$ are in the negative tangential direction. (The direction of both $\vec{E}_{nc}$ and $\mathcal{E}$ can be determined via Lenz's law, which is discussed in Section 28-3.)

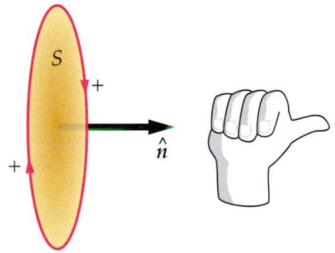

**FIGURE 28-4**  By placing your right thumb in the direction of $\hat{n}$ on the surface *S*, the fingers of your hand curl in the positive tangential direction on *C*.

*INDUCED NONCONSERVATIVE ELECTRIC FIELD*          **E X A M P L E   2 8 - 4**

A magnetic field $\vec{B}$ is perpendicular to the plane of the page. $\vec{B}$ is uniform throughout a circular region of radius *R*, as shown in Figure 28-5. Outside this region, *B* equals zero. The direction of $\vec{B}$ remains fixed and rate of change of *B* is *dB/dt*. What are the magnitude and direction of the induced electric field in the plane of the page (*a*) a distance *r < R* from the center of the circular region and (*b*) a distance *r > R* from the center, where *B* = 0.

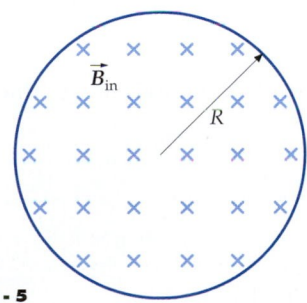

**FIGURE 28-5**

**PICTURE THE PROBLEM** The magnetic field $\vec{B}$ is into the page and uniform over a circular region of radius $R$, as shown in Figure 28-6. As $B$ increases or decreases, the magnetic flux through a surface bounded by closed curve $C$ also changes, and an emf $\mathcal{E} = \oint_C \vec{E} \cdot d\vec{\ell}$ is induced around $C$. The induced electric field is found by applying $\oint_C \vec{E} \cdot d\vec{\ell} = -d\phi_m/dt$ (Equation 28-5). To take advantage of the system's symmetry, we choose $C$ to be a circular curve of radius $r$ and then evaluate the line integral. By symmetry, $\vec{E}$ is tangent to circle $C$ and has the same magnitude at any point on the circle. We will assign into the page as the direction of $\hat{n}$. The sign convention then tells us that the positive tangential direction is clockwise. We then calculate the magnetic flux $\phi_m$, take its time derivative, and solve for $E_t$.

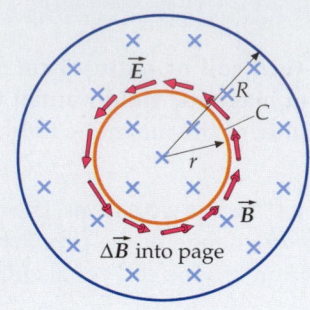

**FIGURE 28-6**

(a) 1. The $\vec{E}$ and $\vec{B}$ fields are related by Equation 28-5:

$$\oint_C \vec{E} \cdot d\vec{\ell} = -\frac{d\phi_m}{dt}$$

where

$$\phi_m = \int_S \vec{B} \cdot \hat{n}\, dA$$

2. $E_t$ (the tangential component of $\vec{E}$) is found from the line integral for a circle of radius $r < R$. $\vec{E}$ is tangent to the circle and has a constant magnitude:

$$\oint_C \vec{E} \cdot d\vec{\ell} = \oint_C E_t\, d\ell = E_t \oint_C d\ell = E_t\, 2\pi r$$

3. For $r < R$, $\vec{B}$ is uniform on the flat surface $S$ bounded by the circle $C$. We choose into the page as the direction of $\hat{n}$. Because $\vec{B}$ is also into the page, the flux through $S$ is simply $BA$:

$$\phi_m = \int_S \vec{B} \cdot \hat{n}\, dA = \int_S B_n\, dA = B_n \int_S dA$$

$$= BA = B\pi r^2$$

4. Calculate the time derivative of $\phi_m$:

$$\frac{d\phi_m}{dt} = \frac{d}{dt}(B\pi r^2) = \frac{dB}{dt}\pi r^2$$

5. Substitute the step 2 and step 4 results into the step 1 result and solve for $E_t$. The positive tangential direction is clockwise.

$$E_t\, 2\pi r = -\frac{dB}{dt}\pi r^2$$

so

$$\boxed{E_t = -\frac{r}{2}\frac{dB}{dt},\quad r < R}$$

6. For the choice for the direction of $\hat{n}$ in step 3, the positive tangential direction is clockwise:

$E_t$ is negative, so $\vec{E}$ is $\boxed{\text{counterclockwise}}$.

(b) 1. For a circle of radius $r > R$ (the region where the magnetic field is zero), the line integral is the same as before:

$$\oint_C \vec{E} \cdot d\vec{\ell} = E_t\, 2\pi r$$

2. Since $B = 0$ for $r > R$, the magnetic flux through $S$ is $B\pi R^2$:

$$\phi_m = B\pi R^2$$

3. Apply Faraday's law to find $E_t$:

$$E_t\, 2\pi r = -\frac{dB}{dt}\pi R^2$$

$$E_t = -\frac{R^2}{2r}\frac{dB}{dt},\quad r > R$$

$E_t$ is negative, so $\vec{E}$ is $\boxed{\text{counterclockwise}}$.

**REMARKS** The positive tangential direction is clockwise. When $d\phi_m/dt$ is positive, $E_t$ is negative and the electric field direction is counterclockwise, as shown in Figure 28-7. Note that the electric field in this example is produced by a changing magnetic field rather than by electric charges. Note also that $\vec{E}$, and thus the emf, exists along any closed curve bounding the area through which the magnetic flux is changing, whether there is a wire or circuit along the curve or there is not.

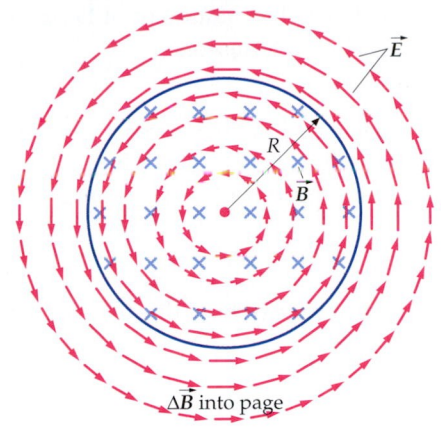

**FIGURE 28-7** The magnetic field is into the page and increasing in magnitude. The induced electric field is counterclockwise.

## 28-3 Lenz's Law

The negative sign in Faraday's law has to do with the direction of the induced emf. This can be obtained by the sign convention described in the previous section, or by a general physical principle known as **Lenz's law**:

> The induced emf is in such a direction as to oppose, or tend to oppose, the change that produces it.

LENZ'S LAW

Note that Lenz's law does not specify just what kind of change causes the induced emf and current. The statement of Lenz's law is purposely left vague to cover a variety of conditions, which we will now illustrate.

Figure 28-8 shows a bar magnet moving toward a loop that has a resistance $R$. It is the motion of the bar magnetic to the right that induces an emf and current in the loop. Lenz's law tells us that this induced emf and current must be in a direction to oppose the motion of the bar magnet. That is, the current induced in the loop produces a magnetic field of its own, and this magnetic field must exert a force to the left on the approaching bar magnet. Figure 28-9 shows the induced magnetic moment of the current loop when the magnet is moving toward it. The loop acts like a small magnet with its north pole to the left and its south pole to the right. Since like poles repel, the induced magnetic moment of the loop repels the bar magnet; that is, it opposes its motion toward the loop. This means the direction of the induced current in the loop must be as shown in Figure 28-9.

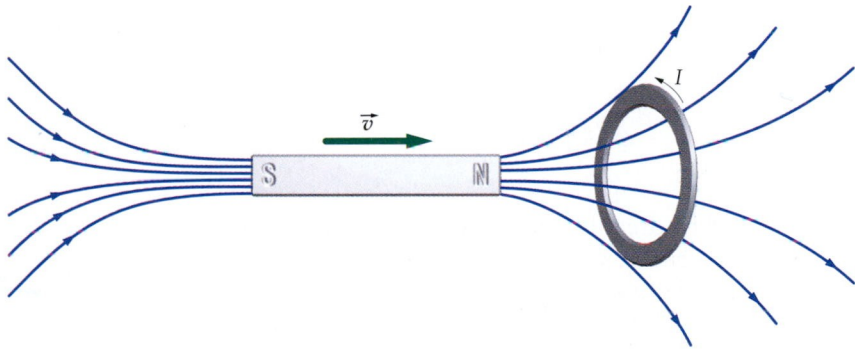

**FIGURE 28-8** When the bar magnet is moving to the right, toward the loop, the emf induced in the loop produces an induced current in the direction shown. The magnetic field due to this induced current in the loop produces a magnetic field that exerts a force on the bar magnet opposing its motion to the right.

Suppose the induced current in the loop shown in Figure 28-9 was opposite to the direction shown. Then there would be a magnetic force on the approaching bar magnet to the right, causing it to gain speed. This gain in speed would cause an increase in the induced current, which in turn would cause the force on the bar magnet to increase, and so forth. This is too good to be true. Any time we nudge a bar magnetic toward a conducting loop it would move toward the loop with ever increasing speed and with no significant effort on our part. Were this to occur, it would be a violation of energy conservation. However, the reality is that energy is conserved, and the statement called Lenz's law is consistent with this reality.

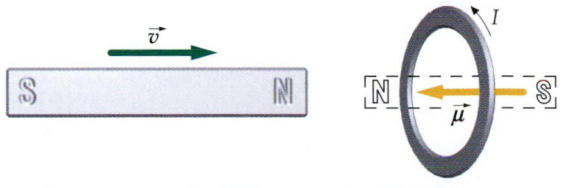

**FIGURE 28-9** The magnetic moment of the loop $\vec{\mu}$ (shown in outline as if it were a bar magnet) due to the induced current is such as to oppose the motion of the bar magnet. The bar magnet is moving toward the loop, so the induced magnetic moment repels the bar magnet.

An alternative statement of Lenz's law in terms of magnetic flux is frequently of use. This statement is:

> When a magnetic flux through a surface changes, the magnetic field due to any induced current produces a flux of its own—through the same surface and in opposition to the change.

ALTERNATIVE STATEMENT OF LENZ'S LAW

For an example of how this alternative statement is applied, see Example 28-5.

---

LENZ'S LAW AND INDUCED CURRENT          **E X A M P L E    2 8 - 5**

**Using the alternative statement of Lenz's law, find the direction of the induced current in the loop shown in Figure 28-8.**

**PICTURE THE PROBLEM** Use the alternative statement of Lenz's law to determine the direction of the magnetic field due to the current induced in the loop. Then use a right-hand rule to determine the direction of the induced current.

1. Draw a sketch of the loop bounding the flat surface $S$ (Figure 28-10). On surface $S$ draw the vector $\Delta \vec{B}_1$, which is the change in the magnetic field $\vec{B}_1$ of the approaching bar magnet:

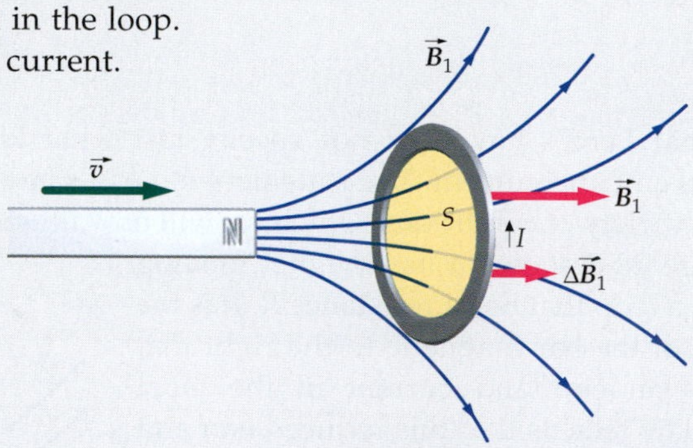

FIGURE 28-10

2. On the sketch draw the vector $\vec{B}_2$, which is the magnetic field of the current induced in the loop (Figure 28-11). Use the alternative statement of Lenz's law to determine the direction of $\vec{B}_2$:

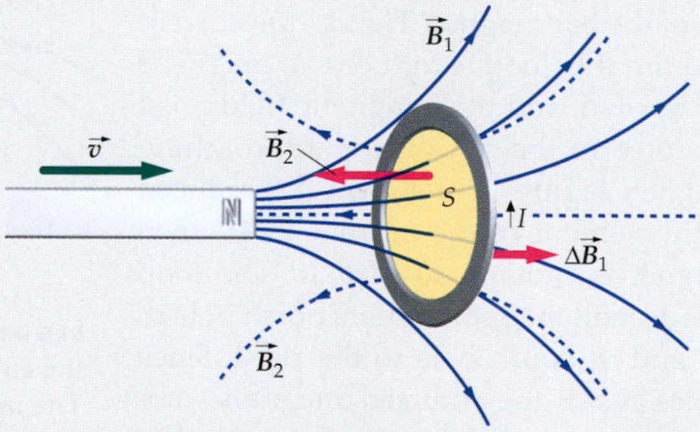

FIGURE 28-11

3. Using the right-hand rule and the direction of $\vec{B}_2$, determine the direction of the current induced in the loop (Figure 28-12):

FIGURE 28-12

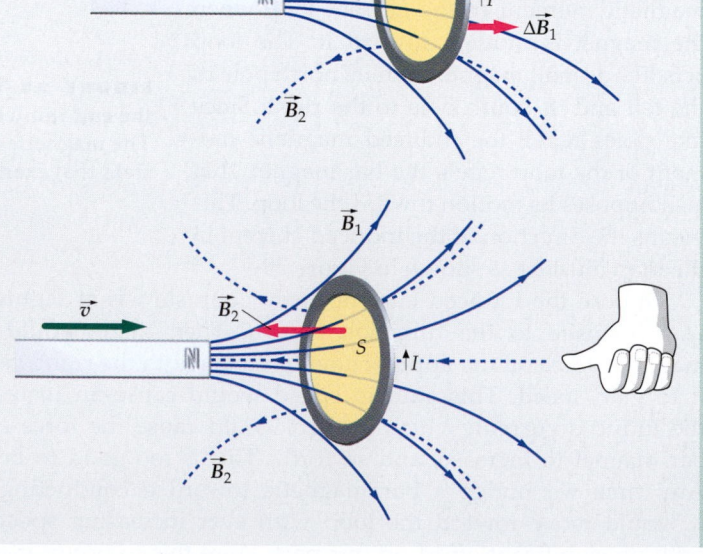

**EXERCISE** Using the alternative statement of Lenz's law, find the direction of the induced current in the loop shown in Figure 28-8 if the magnet is moving to the left (away from the loop). (*Answer* Opposite to the direction shown in Figure 28-12)

In Figure 28-13, the bar magnet is at rest and the loop is moving away from the magnet. The induced current and magnetic moment are shown in the figure. In this case, the bar magnet attracts the loop, thus opposing the motion of the loop as required by Lenz's law.

In Figure 28-14, when the current in circuit 1 is changing, there is a changing flux through circuit 2. Suppose that the switch S in circuit 1 is initially open so that there is no current in the circuit (Figure 28-14a). When we close the switch (Figure 28-14b), the current in circuit 1 does not reach its steady value $\mathcal{E}_1/R_1$ instantaneously but takes some time to change from zero to this value. During the time the current is increasing, the flux through circuit 2 is changing and a current is induced in circuit 2 in the direction shown. When the current in circuit 1 reaches its steady value, the flux through circuit 2 is no longer changing, so there is no longer an induced current in circuit 2. An induced current in circuit 2 in the opposite direction appears momentarily when the switch in circuit 1 is opened (Figure 28-14c) and the current in circuit 1 is decreasing to zero. It is important to understand that there is an induced emf *only while the flux is changing*. The emf does not depend on the magnitude of the flux itself, but only on its rate of change. If there is a large steady flux through a circuit, there is no induced emf.

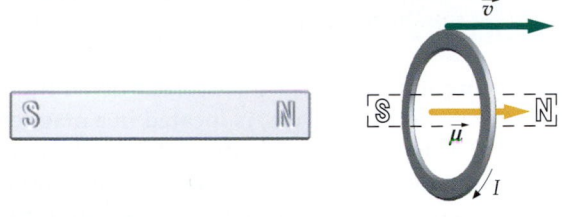

**FIGURE 28-13** When the loop is moving away from the stationary bar magnet, the bar magnet attracts the magnetic moment of the loop, again opposing the relative motion.

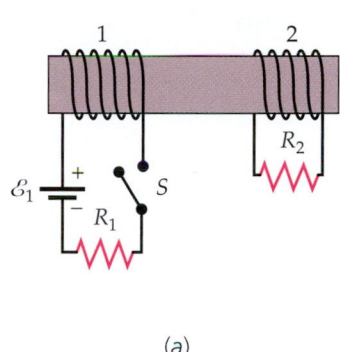

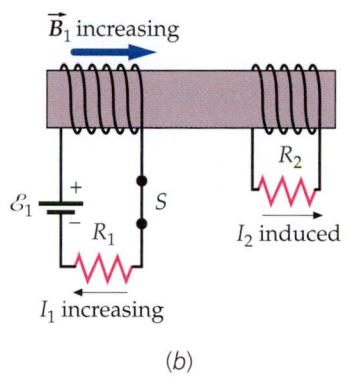

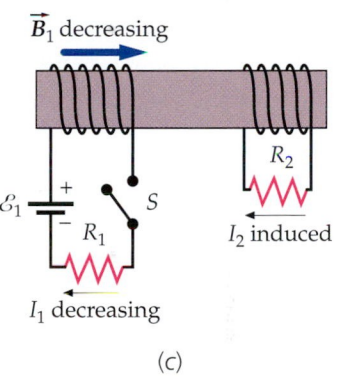

(a)    (b)    (c)

For our next example, we consider the single isolated circuit shown in Figure 28-15. If there is a current in the circuit, there is a magnetic flux through the coil due to its own current. If the current is changing, the flux in the coil is changing and there is an induced emf in the circuit while the flux is changing. This *self-induced emf* opposes the change in the current. It is therefore called a **back emf**. Because of this self-induced emf, the current in a circuit cannot jump instantaneously from zero to some finite value or from some finite value to zero. Henry first noticed this effect when he was experimenting with a circuit consisting of many turns of a wire like that in Figure 28-15. This arrangement gives a large flux through the circuit for even a small current. Joseph Henry noticed a spark across the switch when he tried to break the circuit. Such a spark is due to the large induced emf that occurs when the current varies rapidly, as during the opening of the switch. In this case, the induced emf is directed so as to maintain the original current. The large induced emf produces a large potential difference across the switch as it is opened. The electric field between the contacts of the switch is large enough to produce dielectric breakdown in the surrounding air. When dielectric breakdown occurs, the air conducts electric current in the form of a spark.

**FIGURE 28-14** (a) Two adjacent circuits. (b) Just after the switch is closed, $I_1$ is increasing in the direction shown. The changing flux through circuit 2 induces the current $I_2$. The flux through circuit 2 due to $I_2$ opposes the change in flux due to $I_1$. (c) As the switch is opened, $I_1$ decreases and the flux through circuit 2 changes. The induced current $I_2$ then tends to maintain the flux through circuit 2.

**FIGURE 28-15** The coil with many turns of wire gives a large flux for a given current in the circuit. Thus, when the current changes, there is a large emf induced in the coil opposing the change.

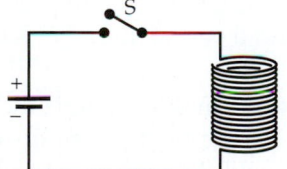

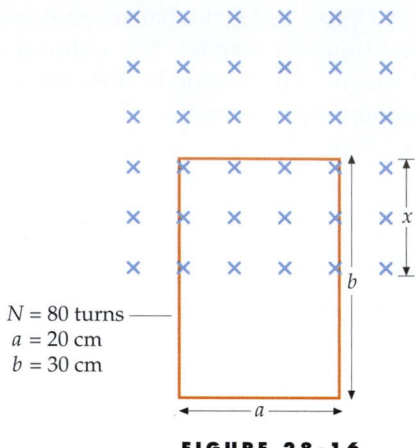

**FIGURE 28-16**

**EXAMPLE 28-6**

A rectangular coil of $N$ turns, each of width $a$ and length $b$; where $N = 80$, $a = 20$ cm, and $b = 30$ cm; is located in a magnetic field of magnitude $B = 0.8$ T directed into the page (Figure 28-16), with only half of the coil in the region of the magnetic field. The resistance $R$ of the coil is 30 $\Omega$. Find the magnitude and direction of the induced current if the coil is moved with a speed of 2 m/s (a) to the right, (b) up, and (c) down.

**PICTURE THE PROBLEM** The induced current equals the induced emf divided by the resistance. We can calculate the emf induced in the circuit as the coil moves by calculating the rate of change of the flux through the coil. The flux is proportional to the distance $x$. The direction of the induced current is found from Lenz's law.

(a) 1. The induced current equals the emf divided by the resistance:

$$I = \frac{\mathcal{E}}{R}$$

2. The induced emf and the magnetic flux are related by Faraday's law:

$$\mathcal{E} = -\frac{d\phi_m}{dt}$$

3. The flux through the coil is $N$ times the flux through each turn of the coil. We choose into the page as the direction of $\hat{n}$. The flux through the surface $S$ bounded by a single turn is $Bax$:

$$\phi_m = N\vec{B} \cdot \hat{n}A = NBax$$

4. When the coil is moving to the right (or to the left), the flux does not change (until the coil leaves the region of magnetic field). The current is therefore zero:

$$\mathcal{E} = -\frac{d\phi_m}{dt} = 0$$

so

$$I = \boxed{0}$$

(b) 1. Compute the rate of change of the flux when the coil is moving up. In this case $x$ is increasing, so $dx/dt$ is positive:

$$\frac{d\phi_m}{dt} = \frac{d}{dt}(NBax) = NBa\frac{dx}{dt}$$

2. Calculate the magnitude of the current:

$$I = \frac{\mathcal{E}}{R} = \frac{NBa(dx/dt)}{R}$$

$$= \frac{(80)(0.8\text{ T})(0.20\text{ m})(2\text{ m/s})}{30\ \Omega} = 0.853\text{ A}$$

3. As the coil moves upward, the flux of $\vec{B}$ through $S$ is increasing. The induced current must produce a magnetic field whose flux through $S$ decreases as $x$ increases. That would be a magnetic field whose dot product with $\hat{n}$ is negative. Such a magnetic field is directed out of the page on $S$. To produce a magnetic field in this direction the induced current must be counterclockwise:

$$\boxed{I = 0.853\text{ A, counterclockwise}}$$

(c) As the coil moves downward, the flux of $\vec{B}$ through $S$ is decreasing. The induced current must produce a magnetic field whose flux through $S$ increases as $x$ decreases. That would be a magnetic field whose dot product with $\hat{n}$ is positive. Such a magnetic field is directed into the page on $S$. To produce a magnetic field in this direction the induced current must be clockwise:

$$\boxed{I = 0.853\text{ A, clockwise}}$$

**REMARKS** In this example the magnetic field is static, so there is no nonconservative electric field. Thus, the emf is not the work done by a nonconservative electric field. This issue is examined in the next section.

# 28-4 Motional EMF

The emf induced in a conductor moving through a magnetic field is called **motional emf.** More generally,

> Motional emf is any emf induced by the motion of a conductor in a magnetic field.

DEFINITION—MOTIONAL EMF

---

*TOTAL CHARGE THROUGH A FLIPPED COIL*                    **EXAMPLE 28-7**

A small coil of $N$ turns has its plane perpendicular to a uniform static magnetic field $\vec{B}$, as shown in Figure 28-17. The coil is connected to a current integrator (C.I.), which is a device used to measure the total charge passing through the coil. Find the charge passing through the coil if the coil is rotated through 180° about the axis shown.

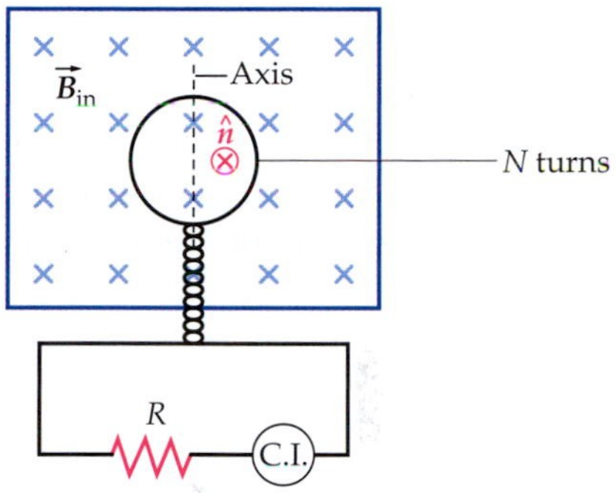

**FIGURE 28-17**

**PICTURE THE PROBLEM** When the coil in Figure 28-17 is rotated, the magnetic flux through the coil changes, causing an induced emf $\mathcal{E}$. The emf in turn causes a current $I = \mathcal{E}/R$, where $R$ is the total resistance of the circuit. Since $I = dq/dt$, we can find the charge $Q$ passing through the integrator by integrating $I$; that is, $Q = \int dq = \int I\,dt$.

1. The increment of charge $dq$ equals the current $I$ times the increment of time $dt$:

$$dq = I\,dt$$

2. The emf $\mathcal{E}$ is related to $I$ by Ohm's law:

$$\mathcal{E} = RI$$

so

$$\mathcal{E}\,dt = RI\,dt$$

3. The emf is related to the flux $\phi_m$ by Faraday's law:

$$\mathcal{E} = -\frac{d\phi_m}{dt}$$

or

$$\mathcal{E}\,dt = -d\phi_m$$

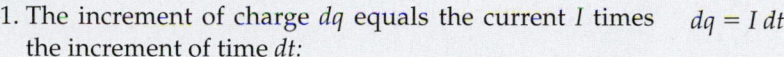

Before rotation          After rotation

**FIGURE 28-18**

4. Substitute $-d\phi_m$ for $\mathcal{E}\,dt$ and $dq$ for $I\,dt$ in the step 2 result and solve for $dq$:

$$-d\phi_m = R\,dq$$

so

$$dq = -\frac{1}{R}\,d\phi_m$$

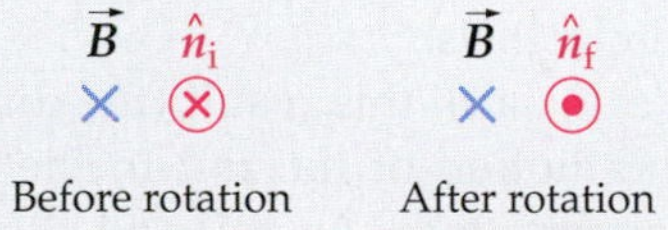

5. Integrate to find the total charge $Q$:

$$Q = \int_0^Q dq = -\frac{1}{R}\int_{\phi_{m,i}}^{\phi_{m,f}} d\phi_m = -\frac{1}{R}(\phi_{m,f} - \phi_{m,i}) = -\frac{\Delta\phi_m}{R}$$

6. The flux through the coil is $\phi_m = N\vec{B} \cdot \hat{n}A$, where $\hat{n}$ is the normal to the flat surface bounded by the coil (Figure 28-18). Initially, the normal is directed into the page. When the coil rotates, so does the surface and its normal. Find the change in $\phi_m$ when the coil rotates 180°:

$$\Delta\phi_m = \phi_{m,f} - \phi_{m,i} = N\vec{B} \cdot \hat{n}_f A - N\vec{B} \cdot \hat{n}_i A$$
$$= NA(\vec{B} \cdot \hat{n}_f - \vec{B} \cdot \hat{n}_i) = NA[(-B) - (+B)] = -2NBA$$

7. Combining the previous two results yields $Q$:

$$\boxed{Q = \frac{2NBA}{R}}$$

**REMARKS** Note that the charge $Q$ does not depend on whether or not the coil is rotated slowly or quickly—all that matters is the change in magnetic flux through the coil. A coil used in this way is called a *flip coil*. It is used to measure magnetic fields. For example, if the current integrator (C.I.) measures a total charge $Q$ passing through the coil when it is flipped, the magnetic field strength can be found from $B = RQ/(2NA)$.

**EXERCISE** A flip coil of 40 turns has a radius of 3 cm, a resistance of 16 $\Omega$, and the plane of the coil is initially perpendicular to a static, uniform 0.50-T magnetic field. If the coil is flipped through 90°, how much charge passes through the coil? (*Answer* 3.53 mC)

Figure 28-19 shows a thin conducting rod sliding to the right along conducting rails that are connected by a resistor. A uniform magnetic field $\vec{B}$ is directed into the page.

Consider the magnetic flux through the flat surface $S$ bounded by the circuit. Let the normal $\hat{n}$ to the surface be into the page. As the rod moves to the right the surface $S$ increases, as does the magnetic flux through the surface $S$. Thus, an emf is induced in the circuit. Let $\ell$ be the separation of the rails and $x$ be the distance from the left end of the rails to the rod. The area of surface $S$ is then $\ell x$, and the magnetic flux through $S$ is

$$\phi_m = \vec{B} \cdot \hat{n} A = B_n A = B\ell x$$

When $x$ increases by $dx$, the area of surface $S$ increases by $dA = \ell\,dx$ and the flux $\phi_m$ increases by $d\phi_m = B\ell\,dx$. The rate of change of the flux is

$$\frac{d\phi_m}{dt} = B\ell\frac{dx}{dt} = B\ell v$$

where $v = dx/dt$ is the speed of the rod. The emf induced in this circuit is therefore

$$\mathcal{E} = -\frac{d\phi_m}{dt} = -B\ell v$$

where the negative sign tells us that the emf is in the negative tangential direction. Put your right thumb in the direction of $\hat{n}$ (into the page) and your fingers will curl in the positive tangential direction (clockwise). Thus, the induced emf is counterclockwise.

We can check this result (the direction of the induced emf) using Lenz's law. It is the motion of the rod to the right that produces the induced current, so the magnetic force on this rod due to the induced current must be to the left. The magnetic force on a current-carrying conductor is given by $I\vec{L} \times \vec{B}$ (Equation 26-4), where $\vec{L}$ is in the direction of the current. If $\vec{L}$ is upward the force is to the left, which affirms our previous result (that the induced emf is counterclockwise). If the rod is given some initial velocity $\vec{v}$ to the right and is then released, the force due to the induced current slows the rod until it stops. To maintain the motion of the rod, an external force pushing the rod to the right must be maintained.

A second check on the direction of the induced emf and current is implemented by considering the direction of the magnetic force on the charge carriers moving to the right with the rod. The charge carriers move rightward with the same velocity $\vec{v}$ as the rod, so the charge carriers experience a magnetic force $\vec{F} = q\vec{v} \times \vec{B}$. If $q$ is positive this force is upward, which means the induced emf is counterclockwise.

Emf is the work per unit charge on the charge carriers, but what is the force that is doing this work in the circuit shown in Figure 28-19? It turns out this work is

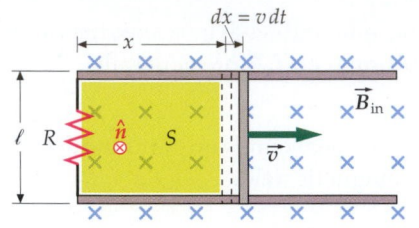

**FIGURE 28-19** A conducting rod sliding on conducting rails in a magnetic field. As the rod moves to the right, the area of the surface $S$ increases, so the magnetic flux through $S$ into the paper increases. An emf of magnitude $B\ell v$ is induced in the circuit, inducing a counterclockwise current that produces flux through the surface $S$ directed out of the paper opposing the change in flux due to the motion of the rod.

done by the superposition of a magnetic force and an electric force (Figure 28-20). To see how this comes about, consider that the current in the rod is upward, so the drift velocity $\vec{v}_d$ of the assumed positive charge carriers is upward. Thus, a magnetic force ($\vec{F}_L = q\vec{v}_d \times \vec{B}$) toward the left acts on the charge carriers and, as a result, the rod becomes polarized—its left side positively charged and its right side negatively charged. These surface charges produce an electric field $\vec{E}_\perp$ inside the rod toward the right, and this field exerts an electric force ($\vec{F}_R = q\vec{E}_\perp$) toward the right on the charge carriers. The sum $\vec{F}_L + q\vec{E}_\perp = 0$, since the net horizontal force on the charge carriers is zero. In addition, an upward magnetic force $\vec{F}_U = q\vec{v} \times \vec{B}$, where $\vec{v}$ is the velocity of both the charge carriers and the rod to the right. The total work done by all three of these forces on a charge carrier traversing the rod is just the work done by $\vec{F}_U$, and this work is $F_U\ell = qvB\ell$. Thus, the work per unit charge is $vB\ell$, which is obtained by dividing the total work by the charge $q$. The magnitude of the emf equals this work.

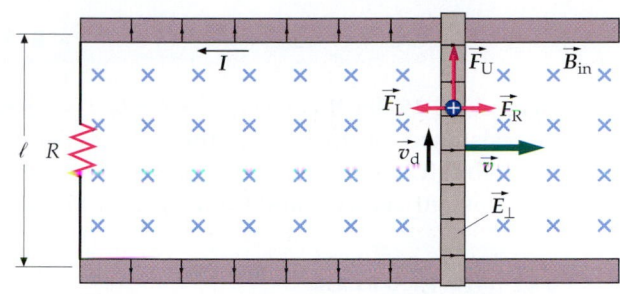

**FIGURE 28-20** As a positive charge carrier moves along the moving rod, electric forces and magnetic forces act on the charge carrier. The net electromagnetic force on the charge carrier is directed upward, in the direction of the drift velocity. The work per unit charge done by this force on the charge carrier as it transverses the rod is the motional emf.

$$\mathcal{E} = vB\ell \qquad\qquad 28\text{-}6$$

MAGNITUDE OF EMF FOR A ROD MOVING PERPENDICULAR TO BOTH THE ROD AND $\vec{B}$

The magnitude of the emf is the total work per unit charge done by all three forces $\vec{F}_L$, $\vec{F}_R$, and $\vec{F}_U$. Taken together, $\vec{F}_L$ and $\vec{F}_U$ constitute the total magnetic force. The total magnetic force, however, is perpendicular to the velocity of the charge carriers and thus does no work. Therefore, the total work done by all three forces is done solely by the electric force $\vec{F}_R$.

Figure 28-21 shows a positive charge carrier in a conducting rod that is moving at constant speed through a uniform magnetic field directed into the paper. Because the charge carrier is moving horizontally with the rod, there is an upward magnetic force on the charge carrier of magnitude $qvB$. Responding to this force, the charge carriers in the rod move upward, producing a net positive charge at the top of the rod and leaving a net negative charge at the bottom of the rod. The charge carriers continue to move upward until the electric field $\vec{E}_\parallel$ produced by the separated charges exerts a downward force of magnitude $qE_\parallel$ on the separated charges, which balances the upward magnetic force $qvB$. In equilibrium, the magnitude of this electric field in the rod is

$$E_\parallel = vB$$

The direction of this electric field is parallel to the rod, directed downward. The associated potential difference across the length $\ell$ of the rod is

$$\Delta V = E_\parallel \ell = vB\ell$$

with the potential being higher at the top. That is, when there is no current through the rod, the potential difference across the rod equals $vB\ell$ (the motional emf). When there is a current $I$ through the rod, the potential difference is

$$\Delta V = vB\ell - Ir \qquad\qquad 28\text{-}7$$

POTENTIAL DIFFERENCE ACROSS A MOVING ROD

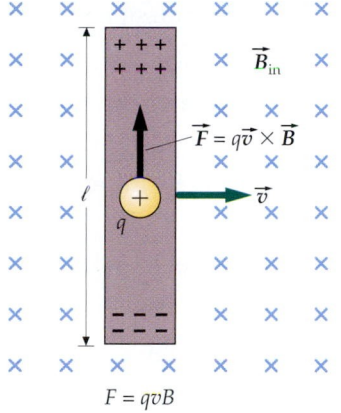

$$F = qvB$$

**FIGURE 28-21** A positive charge carrier in a conducting rod that is moving through a magnetic field experiences a magnetic force that has an upward component. Some of these charge carriers move to the top of the rod, leaving the bottom of the rod negative. The charge separation produces a downward electric field of magnitude $E_\parallel = vB$ in the rod. Thus, the potential at the top of the rod is greater than the potential at the bottom of the rod by $E_\parallel \ell = vB\ell$.

where $r$ is the resistance of the rod.

**EXERCISE** A rod 40-cm long moves at 12 m/s in a plane perpendicular to a magnetic field of 0.30 T. The rod's velocity is perpendicular to its length. Find the emf induced in the rod. (*Answer* 1.44 V)

---

*A U-Shaped Conductor and a Sliding Rod* **EXAMPLE 28-8 Try It Yourself**

Using Figure 28-19, let $B = 0.6$ T, $v = 8$ m/s, $\ell = 15$ cm, and $R = 25$ $\Omega$; assume that the resistances of the rod and the rails are negligible. Find (*a*) the induced emf in the circuit, (*b*) the current in the circuit, (*c*) the force needed to move the rod with constant velocity, and (*d*) the power dissipated in the resistor.

**PICTURE THE PROBLEM**

Cover the column to the right and try these on your own before looking at the answers.

| Steps | Answers |
|---|---|
| 1. Calculate the induced emf from Equation 28-6. | $\mathcal{E} = Bv\ell = \boxed{0.720 \text{ V}}$ |
| 2. Find the current from Ohm's law. | $I = \dfrac{\mathcal{E}}{R} = \boxed{28.8 \text{ mA}}$ |
| 3. The force needed to move the rod with constant velocity is equal and opposite to the force exerted by the magnetic field on the rod, which has the magnitude $I\ell B$ (Equation 26-4). Calculate the magnitude of this force. | $F = IB\ell = \boxed{2.59 \text{ mN}}$ |
| 4. Find the power dissipated in the resistor. | $P = I^2 R = \boxed{20.7 \text{ mW}}$ |

**PLAUSIBILITY CHECK** Using $P = Fv$, we confirm that the power is 20.7 mW.

**REMARKS** The potential at the top of the rod is greater than the potential at the bottom of the rod by the emf.

---

*Magnetic Drag* **EXAMPLE 28-9**

A rod of mass $m$ slides on frictionless conducting rails in a region of static uniform magnetic field $\vec{B}$ directed into the page (Figure 28-22). An external agent is pushing the rod, maintaining its motion to the right at constant speed $v_0$. At time $t = 0$, the agent abruptly stops pushing and the rod continues forward, being slowed by the magnetic force. Find the speed $v$ of the rod as a function of time.

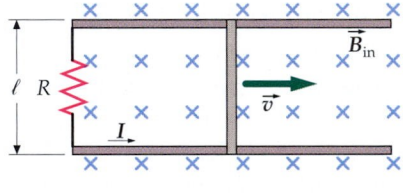

**FIGURE 28-22**

**PICTURE THE PROBLEM** The speed of the rod changes because a magnetic force acts on the induced current. The motion of the rod through a magnetic field induces an emf $\mathcal{E} = B\ell v$ and, therefore, a current in the rod, $I = \mathcal{E}/R$. This causes a magnetic force to act on the rod, $F = IB\ell$. With the force known, we apply Newton's second law to find the speed as a function of time. Take the positive $x$ direction as being to the right.

| | |
|---|---|
| 1. Apply Newton's second law to the rod: | $F_x = ma_x = m\dfrac{dv}{dt}$ |
| 2. The force exerted on the rod is the magnetic force, which is proportional to the current and in the negative $x$ direction, as shown in Figure 28-22: | $F_x = -IB\ell$ |
| 3. The current equals the motional emf divided by the resistance of the rod: | $I = \dfrac{\mathcal{E}}{R} = \dfrac{B\ell v}{R}$ |
| 4. Combining these results, we find the magnitude of the magnetic force exerted on the rod: | $F_x = -IB\ell = -\dfrac{B\ell v}{R}B\ell = -\dfrac{B^2\ell^2 v}{R}$ |
| 5. Newton's second law then gives: | $-\dfrac{B^2\ell^2 v}{R} = m\dfrac{dv}{dt}$ |

6. Separate the variables, then integrate the velocity from $v_0$ to $v_f$ and integrate the time from 0 to $t_f$:

$$\frac{dv}{v} = -\frac{B^2\ell^2}{mR}\,dt$$

$$\int_{v_0}^{v_f}\frac{dv}{v} = -\frac{B^2\ell^2}{mR}\int_0^{t_f}dt$$

$$\ln\frac{v_f}{v_0} = -\frac{B^2\ell^2}{mR}\,t_f$$

7. Let $v = v_f$ and $t = t_f$, then solve for $v$:

$$\boxed{v = v_0 e^{-t/\tau}, \text{ where } \tau = \frac{mR}{B^2\ell^2}}$$

**REMARKS** If the force were constant, the rod's speed would decrease linearly with time. However, because the force is proportional to the rod's speed, as found in step 4, the force is large initially but the force decreases as the speed decreases. In principle, the rod never stops moving. Even so, the rod travels only a finite distance. (See Problem 37.)

The general equation for motional emf is

$$\boxed{\mathcal{E} = \oint_C (\vec{v} \times \vec{B}) \cdot d\vec{\ell} = -\frac{d\phi_m}{dt}}$$
28-8

GENERAL EQUATION FOR MOTIONAL EMF

where $\vec{v}$ is the velocity of the wire at the element $d\vec{\ell}$. The integral is taken at an instant in time.

**FIGURE 28-23** The positive $x$, $y$, and $z$ directions are to the right, into the page, and up the page respectively. The rod moves to the right with the velocity $\vec{v}_r$, and there is a uniform static magnetic field directed into the page.

---

*VERIFYING* $\mathcal{E} = vB\ell$ **EXAMPLE 28-10**

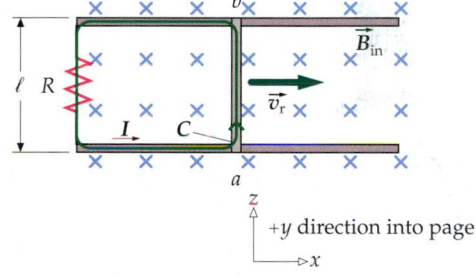

+y direction into page

Integrate $\mathcal{E} = \oint_C(\vec{v} \times \vec{B})\cdot d\vec{\ell}$ to show that the emf in the circuit in Figure 28-23 is given by Equation 28-6.

**PICTURE THE PROBLEM** The circuit $C$ can be divided into two parts: part $C_1$, which is moving, and part $C_2$, which is stationary.

1. Divide the circuit into two parts, $C_1$ and $C_2$. On $C_1$, $\vec{v} = \vec{v}_r$ and on $C_2$, $\vec{v} = 0$:

$$\oint_C (\vec{v} \times \vec{B}) \cdot d\vec{\ell} = \int_{a\,C_1}^b (\vec{v} \times \vec{B}) \cdot d\vec{\ell} + \int_{b\,C_2}^a (\vec{v} \times \vec{B}) \cdot d\vec{\ell}$$

$$= \int_{a\,C_1}^b (\vec{v}_r \times \vec{B}) \cdot d\vec{\ell} + 0$$

2. Evaluate $(\vec{v}_r \times \vec{B})\cdot d\vec{\ell}$ on $C_1$:

$$\vec{v}_r \times \vec{B} = v_r\hat{i} \times B\hat{j} = v_r B\,\hat{k}$$

and

$$d\vec{\ell} = d\ell\,\hat{k}$$

so

$$(\vec{v}_r \times \vec{B}) \cdot d\vec{\ell} = v_r B\hat{k} \cdot d\ell\,\hat{k} = v_r B\,d\ell$$

3. Evaluate the integral and find the emf:

$$\mathcal{E} = \int_{a\,C_1}^b (\vec{v}_r \times \vec{B}) \cdot d\vec{\ell} = \int_{a\,C_1}^b v_r B\,d\ell = v_r B \int_{a\,C_1}^b d\ell$$

$$= \boxed{v_r B\ell}$$

## 28-5 Eddy Currents

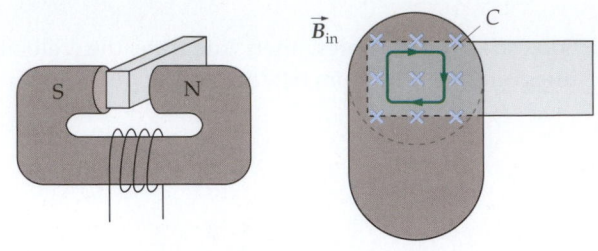

In the examples we have discussed, currents were induced in thin wires or rods. Often a changing flux sets up circulating currents, which are called *eddy currents,* in a piece of bulk metal like the core of a transformer. The heat produced by such current constitutes a power loss in the transformer. Consider a conducting slab between the pole faces of an electromagnet (Figure 28-24). If the magnetic field $\vec{B}$ between the pole faces is changing with time (as it will if the current in the magnet windings is alternating current), the flux through any closed loop in the slab, such as through the curve C indicated in the figure, will be changing. Since path C is in a conductor, there will be an induced emf around C.

The existence of eddy currents can be demonstrated by pulling a copper or aluminum sheet between the poles of a strong permanent magnet (Figure 28-25). Part of the area enclosed by curve C in this figure is in the magnetic field, and part of the area enclosed by curve C is outside the magnetic field. As the sheet is pulled to the right, the flux through this curve decreases (assuming that into the paper is the positive normal direction). A clockwise emf is induced around this curve. This emf drives a current that is directed upward in the region between the pole faces, and the magnetic field exerts a force on this current to the left opposing motion of the sheet. You can feel this drag force on the sheet if you pull a conducting sheet rapidly through a strong magnetic field.

Eddy currents are usually unwanted because power is lost due to joule heating by the current, and this dissipated energy must be transferred to the environment. The power loss can be reduced by increasing the resistance of the possible paths for the eddy currents, as shown in Figure 28-26a. Here the conducting slab is laminated; that is, the conducting slab is made up of small strips glued together. Because insulating glue separates the strips, the eddy currents are essentially confined to the strips. The large eddy-current loops are broken up, and the power loss is greatly reduced. Similarly, if there are cuts in the sheet, as shown in Figure 28-26b, the eddy currents are lessened and the magnetic force is greatly reduced.

Eddy currents are not always undesirable. For example, eddy currents are often used to damp unwanted oscillations. With no damping present, sensitive mechanical balance scales that are used to weigh small masses might oscillate back and forth around their equilibrium reading many times. Such scales are usually designed so that a small sheet of aluminum (or some other metal) moves between the poles of a magnet as the scales oscillate. The resulting eddy currents dampen the oscillations so that equilibrium is quickly reached. Eddy currents also play a role in the magnetic braking systems of some rapid transit cars. A large electromagnet is positioned in the vehicle over the rails. If the magnet is energized by a current in its windings, eddy currents are induced in the rails by the motion of the magnet and the magnetic forces provide a drag force on the magnet that slows the car.

**FIGURE 28-24** Eddy currents. When the magnetic field through a metal slab is changing, and emf is induced in any closed loop in the metal, such as loop C. The induced emfs drive currents, which are called eddy currents.

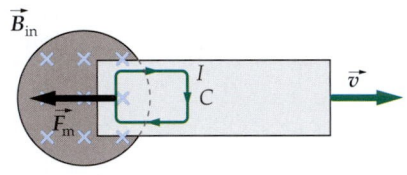

**FIGURE 28-25** Demonstration of eddy currents. When the metal sheet is pulled to the right, there is a magnetic force to the left on the induced current opposing the motion.

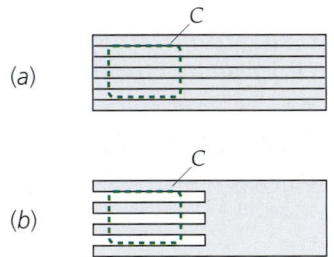

(a)

(b)

**FIGURE 28-26** Disrupting the conduction paths in the metal slab can reduce the eddy current. (*a*) If the slab is constructed from strips of metal glued together, the insulating glue between the slabs increases the resistance of the closed loop C. (*b*) Slots cut into the metal slab also reduce the eddy current.

## 28-6 Inductance

### Self-Inductance

The magnetic flux through a circuit is related to the current in that circuit and the currents in other nearby circuits.[†] Consider a coil carrying a current $I$. The current in the coil produces a magnetic field $\vec{B}$ that varies from point to point, but the value of $\vec{B}$ at each point is proportional to $I$. The magnetic flux through the coil is therefore also proportional to $I$:

---

[†] We are assuming that there are no permanent magnets around.

$$\phi_{\mathrm{m}} = LI \qquad\qquad\qquad 28\text{-}9$$

DEFINITION—SELF-INDUCTANCE

where $L$, the proportionality constant, is called the self-inductance of the coil. The self-inductance depends on the geometric shape of the coil. The SI unit of inductance is the **henry** (H). From Equation 28-9, we can see that the unit of inductance equals the unit of flux divided by the unit of current:

$$1\,\mathrm{H} = 1\,\frac{\mathrm{Wb}}{\mathrm{A}} = 1\,\frac{\mathrm{T \cdot m^2}}{\mathrm{A}}$$

In principle, the self-inductance of any coil or circuit can be calculated by assuming a current $I$, calculating $\vec{B}$ at every point on a surface bounded by the coil, calculating the flux $\phi_{\mathrm{m}}$, and using $L = \phi_{\mathrm{m}}/I$. In actual practice, the calculation is often very challenging. However, the self-inductance of a long, tightly wound solenoid can be calculated directly. The magnetic flux through a solenoid of length $\ell$ and $N$ turns carrying a current $I$ was calculated in Example 28-1:

$$\phi_{\mathrm{m}} = \frac{\mu_0 N^2 I A}{\ell} = \mu_0 n^2 I A \ell \qquad\qquad 28\text{-}10$$

where $n = N/\ell$ is the number of turns per unit length. As expected, the flux is proportional to the current $I$. The proportionality constant is the self-inductance:

$$L = \frac{\phi_{\mathrm{m}}}{I} = \mu_0 n^2 A \ell \qquad\qquad 28\text{-}11$$

SELF-INDUCTANCE OF A SOLENOID

The self-inductance of a solenoid is proportional to the square of the number of turns per unit length $n$ and to the volume $A\ell$. Thus, like capacitance, self-inductance depends only on geometric factors.[†] From the dimensions of Equation 28-11, we can see that $\mu_0$ can be expressed in henrys per meter:

$$\mu_0 = 4\pi \times 10^{-7}\,\mathrm{H/m}$$

[†] If the inductor has an iron core, the self-inductance also depends on properties of the core.

---

*SELF-INDUCTANCE OF A SOLENOID*                    **EXAMPLE 28-11**

**Find the self-inductance of a solenoid of length 10 cm, area 5 cm², and 100 turns.**

**PICTURE THE PROBLEM** We can calculate the self-inductance in henrys from Equation 28-11.

1. $L$ is given by Equation 28-11:

$$L = \mu_0 n^2 A \ell$$

2. Convert the given quantities to SI units:

$$\ell = 10\,\mathrm{cm} = 0.1\,\mathrm{m}$$

$$A = 5\,\mathrm{cm^2} = 5 \times 10^{-4}\,\mathrm{m^2}$$

$$n = N/\ell = (100\ \text{turns})/(0.1\,\mathrm{m}) = 1000\ \text{turns/m}$$

$$\mu_0 = 4\pi \times 10^{-7}\,\mathrm{H/m}$$

3. Substitute the given quantities:

$$L = \mu_0 n^2 A \ell$$

$$= (4\pi \times 10^{-7}\,\mathrm{H/m})(10^3\ \text{turns/m})^2(5 \times 10^{-4}\,\mathrm{m^2})(0.1\,\mathrm{m})$$

$$= \boxed{6.28 \times 10^{-5}\,\mathrm{H}}$$

When the current in a circuit is changing, the magnetic flux due to the current is also changing, so an emf is induced in the circuit. Because the self-inductance of a circuit is constant, the change in flux is related to the change in current by

$$\frac{d\phi_m}{dt} = \frac{d(LI)}{dt} = L\frac{dI}{dt}$$

According to Faraday's law, we have

$$\mathcal{E} = -\frac{d\phi_m}{dt} = -L\frac{dI}{dt} \qquad\qquad 28\text{-}12$$

Thus, the self-induced emf is proportional to the rate of change of the current. A coil or solenoid with many turns has a large self-inductance and is called an **inductor**. In circuits, it is denoted by the symbol ⌒⌒⌒. Typically, we can neglect the self-inductance of the rest of the circuit compared with that of an inductor. The potential difference across an inductor is given by

$$\Delta V = \mathcal{E} - Ir = -L\frac{dI}{dt} - Ir \qquad\qquad 28\text{-}13$$

POTENTIAL DIFFERENCE ACROSS AN INDUCTOR

where $r$ is the internal resistance of the inductor.[†] For an ideal inductor, $r = 0$.

**EXERCISE** At what rate must the current in the solenoid of Example 28-11 change to induce a back emf of 20 V? (*Answer*   $3.18 \times 10^5$ A/s)

## Mutual Inductance

When two or more circuits are close to each other, as in Figure 28-27, the magnetic flux through one circuit depends not only on the current in that circuit but also on the current in the nearby circuits. Let $I_1$ be the current in circuit 1, on the left in Figure 28-27, and let $I_2$ be the current in circuit 2, on the right in Figure 28-27. The magnetic field $\vec{B}$ at surface $S_2$ is the superposition of $\vec{B}_1$ due to $I_1$, and $\vec{B}_2$ due to $I_2$, where $\vec{B}_1$ is proportional to $I_1$ (and $\vec{B}_2$ is proportional to $I_2$). We can therefore write the flux of $\vec{B}_1$ through circuit 2, $\phi_{m2,1}$ as:

$$\phi_{m2,1} = M_{2,1}I_1 \qquad\qquad 28\text{-}14a$$

DEFINITION—MUTUAL INDUCTANCE

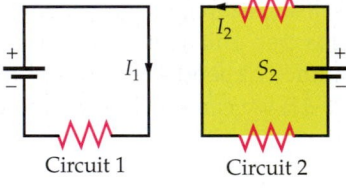

**FIGURE 28-27** Two adjacent circuits. The magnetic field on $S_2$ is partly due to current $I_1$ and partly due to current $I_2$. The flux through the magnetic field is the sum of two terms, one proportional to $I_1$ and the other to $I_2$.

where $M_{2,1}$ is called the **mutual inductance** of the two circuits. The mutual inductance depends on the geometrical arrangement of the two circuits. For instance, if the circuits are far apart, the flux of $\vec{B}_1$ through circuit 2 will be small and the mutual inductance will be small. (The net flux $\phi_{m2}$ of $\vec{B} = \vec{B}_1 + \vec{B}_2$ through circuit 2 is given by $\phi_{m2} = \phi_{m2,2} + \phi_{m2,1}$.) An equation similar to Equation 28-14a can be written for the flux of $\vec{B}_2$ through circuit 1:

$$\phi_{m1,2} = M_{1,2}I_2 \qquad\qquad 28\text{-}14b$$

We can calculate the mutual inductance for two tightly wound concentric solenoids like the solenoids shown in Figure 28-28. Let $\ell$ be the length of both solenoids, and let the inner solenoid have $N_1$ turns and radius $r_1$ and the outer

---

† If the inductor has an iron core, the internal resistance includes properties of the core.

solenoid have $N_2$ turns and radius $r_2$. We will first calculate the mutual inductance $M_{2,1}$ by assuming that the inner solenoid carries a current $I_1$ and finding the magnetic flux $\phi_{m2}$ due to this current through the outer solenoid.

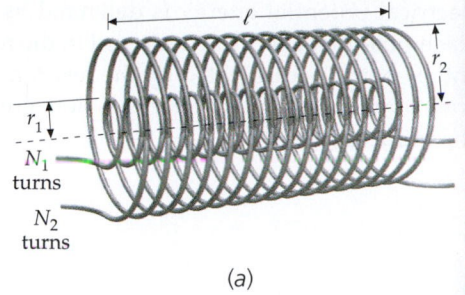

The magnetic field $\vec{B}_1$ due to the current in the inner solenoid is constant in the space within the inner solenoid and has magnitude

$$B_1 = \mu_0(N_1/\ell)I_1 = \mu_0 n_1 I_1, \quad r < r_1 \qquad \text{28-15}$$

and outside the inner solenoid this magnetic field $B_1$ is negligible. The flux of $\vec{B}_1$ through the outer solenoid is therefore

$$\phi_{m2} = N_2 B_1 (\pi r_1^2) = n_2 \ell B_1 (\pi r_1^2) = \mu_0 n_2 n_1 \ell (\pi r_1^2) I_1$$

Note that the area used to compute the flux through the outer solenoid is not the area of that solenoid, $\pi r_2^2$, but rather is the area of the inner solenoid, $\pi r_1^2$, because the magnetic field due to the inner solenoid is zero outside the inner solenoid. The mutual inductance $M_{1,2}$ is thus

$$M_{2,1} = \frac{\phi_{m2,1}}{I_1} = \mu_0 n_2 n_1 \ell \pi r_1^2 \qquad \text{28-16}$$

**EXERCISE**  Calculate the mutual inductance $M_{1,2}$ of the concentric solenoids of Figure 28-28 by finding the flux through the inner solenoid due to a current $I_2$ in the outer solenoid. (*Answer*  $M_{1,2} = M_{2,1} = \mu_0 n_2 n_1 \ell \pi r_1^2$)

Note from the exercise that $M_{1,2} = M_{2,1}$. It can be shown that this is a general result. We will therefore drop the subscripts for mutual inductance and simply write $M$.

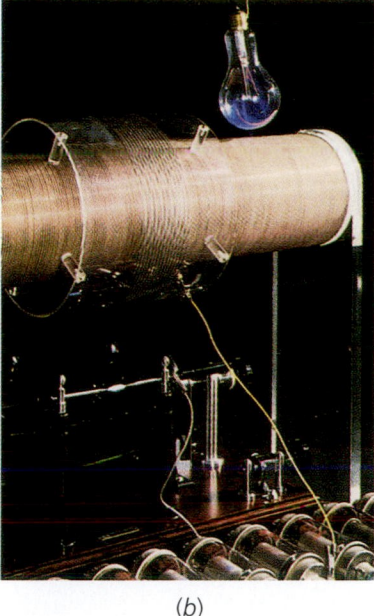

(b)

**FIGURE 28-28**  (a) A long narrow solenoid inside a second solenoid of the same length. A current in either solenoid produces magnetic flux in the other. (b) A tesla coil illustrating the geometry of the wires in Figure 28-28a. Such a device functions as a transformer.[†] Here, low-voltage alternating current in the outer winding is transformed into a higher-voltage alternating current in the inner winding. The emf induced in the inner coil by the field of the charging current in the outer coil is high enough to light the bulb above the coils.

## 28-7  Magnetic Energy

An inductor stores magnetic energy, just as a capacitor stores electrical energy. Consider the circuit shown in Figure 28-29, which consists of an inductance $L$ and a resistance $R$ in series with a battery of emf $\mathcal{E}_0$ and a switch S. We assume that $R$ and $L$ are the resistance and inductance of the entire circuit. The switch is initially open, so there is no current in the circuit. A short time after the switch is closed there is a current $I$ in the circuit, a potential difference $-IR$ across the resistor, and a potential difference $-L\,dI/dt$ across the inductor. (For an inductor with negligible resistance, the difference in potential across the inductor equals the back emf, which was given in Equation 28-12.) Applying Kirchhoff's loop rule to this circuit gives

$$\mathcal{E}_0 - IR - L\frac{dI}{dt} = 0 \qquad \text{28-17}$$

If we multiply each term by the current $I$ and rearrange, we obtain

$$\mathcal{E}_0 I = I^2 R + LI\frac{dI}{dt} \qquad \text{28-18}$$

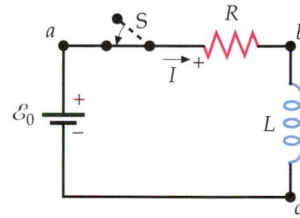

**FIGURE 28-29**  Just after the switch S is closed in this circuit, the current begins to increase and a back emf of magnitude $L\,dI/dt$ is induced in the inductor. The potential drop across the resistor $IR$ plus the potential drop across the inductor $L\,dI/dt$ equals the emf of the battery.

---

† The transformer is discussed in Chapter 29.

The term $\mathcal{E}_0 I$ is the rate at which electrical potential energy is delivered by the battery. The term $I^2R$ is the rate at which potential energy is delivered to the resistor. (It is also the rate at which potential energy is dissipated by the resistance in the circuit.) The term $LI\, dI/dt$ is the rate at which potential energy is delivered to the inductor. If $U_m$ is the energy in the inductor, then

$$\frac{dU_m}{dt} = LI\frac{dI}{dt}$$

which implies

$$dU_m = LI\, dI$$

Integrating this equation from time $t = 0$, when the current is zero, to $t = \infty$, when the current has reached its final value $I_f$, we obtain

$$U_m = \int dU_m = \int_0^{I_f} LI\, dI = \frac{1}{2}LI_f^2$$

The energy stored in an inductor carrying a current $I$ is thus given by

$$U_m = \frac{1}{2}LI^2 \qquad\qquad 28\text{-}19$$

ENERGY STORED IN AN INDUCTOR

When a current is produced in an inductor, a magnetic field is created in the space within the inductor coil. We can think of the energy stored in an inductor as energy stored in the magnetic field. For the special case of a long solenoid, the magnetic field is related to the current $I$ and the number of turns per unit length $n$ by

$$B = \mu_0 nI$$

and the self-inductance is given by Equation 28-11:

$$L = \mu_0 n^2 A\ell$$

where $A$ is the cross-sectional area and $\ell$ is the length. Substituting $B/(\mu_0 n)$ for $I$ and $\mu_0 n^2 A\ell$ for $L$ in Equation 28-19, we obtain

$$U_m = \frac{1}{2}LI^2 = \frac{1}{2}\mu_0 n^2 A\ell\left(\frac{B}{\mu_0 n}\right)^2 = \frac{B^2}{2\mu_0}A\ell$$

The quantity $A\ell$ is the volume of the space within the solenoid containing the magnetic field. The energy per unit volume is the **magnetic energy density** $u_m$:

$$u_m = \frac{B^2}{2\mu_0} \qquad\qquad 28\text{-}20$$

MAGNETIC ENERGY DENSITY

Although we derived this by considering the special case of the magnetic field in a long solenoid, it is a general result. Whenever there is a magnetic field in space, the magnetic energy per unit volume is given by Equation 28-20. Note the similarity to the energy density in an electric field (Equation 24-13):

$$u_e = \frac{1}{2}\epsilon_0 E^2$$

_Electromagnetic Energy Density_                    **E X A M P L E    2 8 - 1 2**

A certain region of space contains a uniform magnetic field of 0.020 T and a uniform electric field of $2.5 \times 10^6$ N/C. Find (a) the total electromagnetic energy density and (b) the energy in a cubical box of edge length $\ell = 12$ cm.

**PICTURE THE PROBLEM** The total energy density $u$ is the sum of the electrical and magnetic energy densities, $u = u_e + u_m$. The energy in a volume $\mathcal{V}$ is given by $U = u\mathcal{V}$.

(a) 1. Calculate the electrical energy density:

$$u_e = \frac{1}{2}\,\epsilon_0\,E^2$$

$$= \frac{1}{2}\,(8.85 \times 10^{-12}\,\text{C}^2/\text{N·m}^2)(2.5 \times 10^6\,\text{N/C})^2$$

$$= 27.7\,\text{J/m}^3$$

2. Calculate the magnetic energy density:

$$u_m = \frac{B^2}{2\mu_0} = \frac{(0.02\,\text{T})^2}{2(4\pi \times 10^{-7}\,\text{N/A}^2)} = 159\,\text{J/m}^3$$

3. The total energy density is the sum of the above two contributions:

$$u = u_e + u_m = 27.7\,\text{J/m}^3 + 159\,\text{J/m}^3 = \boxed{187\,\text{J/m}^3}$$

(b) The total energy in the box is $U = u\mathcal{V}$, where $\mathcal{V} = \ell^3$ is the volume of the box:

$$U = u\mathcal{V} = u\ell^3 = (187\,\text{J/m}^3)(0.12\,\text{m})^3 = \boxed{0.323\,\text{J}}$$

## *28-8  RL Circuits

A circuit containing a resistor and an inductor, such as that shown in Figure 28-29, is called an **RL circuit.** Because all circuits have resistance and self-inductance at room temperature, the analysis of an RL circuit can be applied to some extent to all circuits.[†]

For the circuit shown in Figure 28-29, application of Kirchhoff's loop rule (Equation 28-17) gave us

$$\mathcal{E}_0 - IR - L\frac{dI}{dt} = 0$$

Let us look at some general features of the current before we solve this equation. Just after we close the switch in the circuit the current is still zero, so $IR$ is zero, and $L\,dI/dt$ equals the emf of the battery, $\mathcal{E}_0$. Setting $I = 0$ in Equation 28-17, we get

$$\left.\frac{dI}{dt}\right|_{I=0} = \frac{\mathcal{E}_0}{L} \qquad\qquad 28\text{-}21$$

As the current increases $IR$ increases, and $dI/dt$ decreases. Note that the current cannot abruptly jump from zero to some finite value as it would if there were no inductance. When the inductance $L$ is not negligible $dI/dt$ is finite, and therefore the current must be continuous in time. After a short time, the current has reached a positive value $I$, and the rate of change of the current is

$$\frac{dI}{dt} = \frac{\mathcal{E}_0}{L} - \frac{IR}{L}$$

_____

[†] All circuits also have some capacitance between parts of the circuits at different potentials. We will consider the effects of capacitance in Chapter 29 when we study ac circuits. Here we will neglect capacitance to simplify the analysis and to focus on the effects of inductance.

At this time the current is still increasing, but its rate of increase is less than at $t = 0$. The final value of the current can be obtained by setting $dI/dt$ equal to zero:

$$I_f = \frac{\mathcal{E}_0}{R} \qquad\qquad 28\text{-}22$$

Figure 28-30 shows the current in this circuit as a function of time. This figure is the same as that for the charge on a capacitor as a function of time when the capacitor is charged in an $RC$ circuit (Figure 25-41).

Equation 28-17 is of the same form as Equation 25-36 for the charging of a capacitor and can be solved in the same way—by separating variables and integrating. The result is

$$I = \frac{\mathcal{E}_0}{R}(1 - e^{-(R/L)t}) = I_f(1 - e^{-t/\tau}) \qquad\qquad 28\text{-}23$$

where $I_f = \mathcal{E}_0/R$ is the current as $t \to \infty$, and

$$\tau = \frac{L}{R} \qquad\qquad 28\text{-}24$$

is the **time constant** of the circuit. The larger the self-inductance $L$ or the smaller the resistance $R$, the longer it takes for the current to reach any specified fraction of its final current $I_f$.

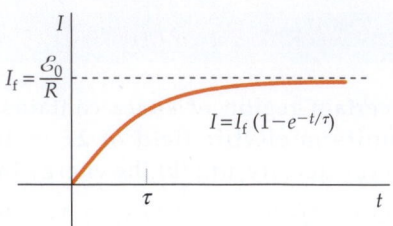

**FIGURE 28-30** Current versus time in an $RL$ circuit. At a time $t = \tau = L/R$, the current is at 63 percent of its maximum value $\mathcal{E}_0/R$.

---

*ENERGIZING A COIL*                    **EXAMPLE 28-13**

**A coil of self-inductance 5 mH and a resistance of 15 $\Omega$ is placed across the terminals of a 12-V battery of negligible internal resistance. (*a*) What is the final current? (*b*) What is the time constant? (*c*) How many time constants does it take for the current to reach 99 percent of its final value?**

**PICTURE THE PROBLEM** The final current is the current when $dI/dt = 0$, as given in Equation 28-22. The current as a function of time is given by Equation 28-23, $I = I_f(1 - e^{-t/\tau})$, where $\tau = L/R$.

1. Use Equation 28-22 to find the final current, $I_f$:

$$I_f = \frac{\mathcal{E}_0}{R} = \frac{12\ \text{V}}{15\ \Omega} = \boxed{0.800\ \text{A}}$$

2. Calculate the time constant $\tau$.

$$\tau = \frac{L}{R} = \frac{5 \times 10^{-3}\ \text{H}}{15\ \Omega} = \boxed{333\ \mu\text{s}}$$

3. Use Equation 28-23 and calculate the time $t$ for $I = 0.99 I_f$:

$$I = I_f(1 - e^{-t/\tau})$$

so

$$e^{-t/\tau} = \left(1 - \frac{I}{I_f}\right)$$

and

$$-\frac{t}{\tau} = \ln\left(1 - \frac{I}{I_f}\right)$$

Thus,

$$t = -\tau \ln\left(1 - \frac{I}{I_f}\right) = -\tau \ln(1 - 0.99)$$

$$= -\tau \ln(0.01) = +\tau \ln 100 = \boxed{4.61\ \tau}$$

**REMARKS** In five time constants the current is within one percent of its final value.

**EXERCISE** How much energy is stored in this inductor when the final current has been attained? (*Answer* $U_\text{m} = \frac{1}{2}LI_\text{f}^2 = 1.6 \times 10^{-3}\,\text{J}$)

In Figure 28-31, the circuit has a make-before-break switch (shown in Figure 28-32) that allows us to remove the battery from the circuit without interrupting the current through the inductor. The resistor $R_1$ protects the battery so that the battery is not shorted when the switch is thrown. If the switch pole is in position $e$, the battery, the inductor, and the two resistors are connected in series and the current builds up in the circuit as just discussed, except that the total resistance is now $R_1 + R$ and the final current is $\mathscr{E}_0/(R + R_1)$. Suppose that the pole has been in position $e$ for a long time, so that the current remains at its final value, which we will call $I_0$. At time $t = 0$ we rapidly move the pole to position $f$ (to remove the battery from consideration completely). We now have a circuit (loop $abcda$) with just a resistor and an inductor carrying an initial current $I_0$. Applying Kirchoff's loop rule to this circuit gives

$$-IR - L\frac{dI}{dt} = 0$$

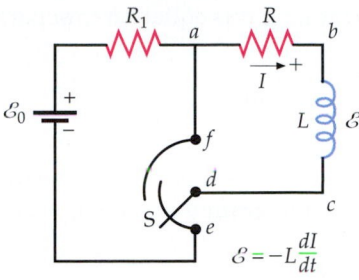

**FIGURE 28-31** An *RL* circuit with a make-before-break switch so that the battery can be removed from the circuit without interrupting the current through the inductor. The current in the inductor reaches its maximum value with the switch pole in position $e$. The pole is then rapidly moved to position $f$.

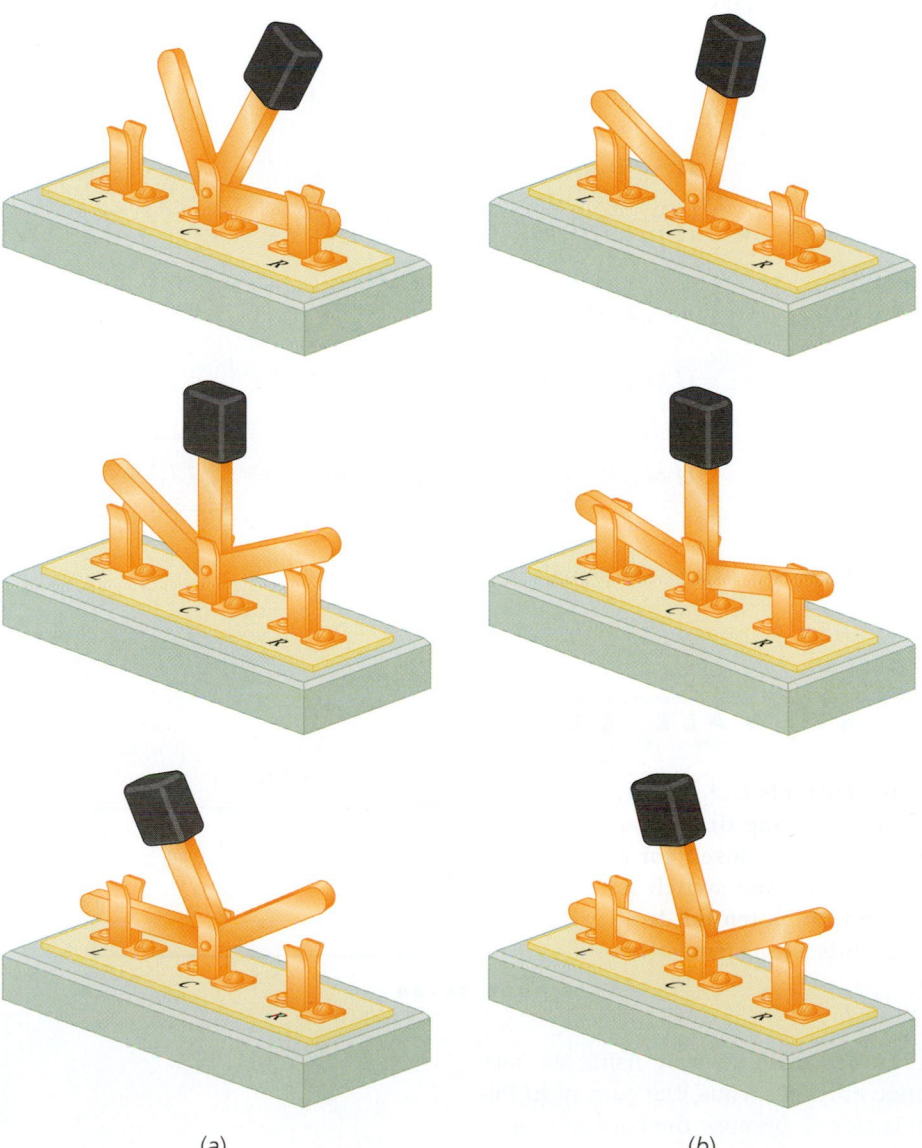

(a)                      (b)

**FIGURE 28-32** (*a*) The standard single-pole, double-throw switch is a break-before-make switch. That is, it breaks the first contact before making the second contact. (*b*) In a make-before-break, single-pole, double-throw switch the throw makes the second contact before breaking the first contact. With the throw in the middle position, the throw is in electrical contact with contact *L* and contact *R*.

Rearranging this equation to separate the variables $I$ and $t$ gives

$$\frac{dI}{I} = -\frac{R}{L} dt \qquad\qquad 28\text{-}25$$

Equation 28-25 is of the same form as Equation 25-31 for the discharge of a capacitor. Integrating and then solving for $I$ gives

$$I = I_0 e^{-t/\tau} \qquad\qquad 28\text{-}26$$

where $\tau = L/R$ is the time constant. Figure 28-33 shows the current as a function of time.

**EXERCISE** What is the time constant of a circuit of resistance 85 $\Omega$ and inductance 6 mH? (*Answer* 70.6 $\mu$s)

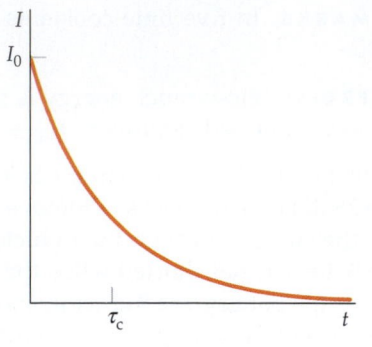

**FIGURE 28-33** Current versus time for the circuit in Figure 28-31. The current decreases exponentially with time.

---

*ENERGY DISSIPATED*  **EXAMPLE 28-14**

**Find the total energy dissipated in the resistor $R$, as shown in Figure 28-31, when the current in the inductor decreases from its initial value of $I_0$ to 0.**

**PICTURE THE PROBLEM** The rate of energy dissipation $I^2R$ varies with time so to calculate the total energy dissipated requires that we integrate.

1. The rate of heat production is $I^2R$: $\qquad\qquad P = I^2R$

2. The total energy $U$ dissipated in the resistor is the integral of $P\,dt$ from $t = 0$ to $t = \infty$: $\qquad U = \int_0^\infty I^2R\,dt$

3. The current $I$ is given by Equation 28-26: $\qquad I = I_0 e^{-(R/L)t}$

4. Substitute this current into the integral: $\qquad U = \int_0^\infty I^2R\,dt = \int_0^\infty I_0^2 e^{-2(R/L)t}R\,dt = I_0^2 R \int_0^\infty e^{-2(R/L)t}\,dt$

5. The integration can be done by substituting $x = 2Rt/L$: $\qquad U = I_0^2 R \frac{e^{-2(R/L)t}}{-2(R/L)}\Bigg|_0^\infty = I_0^2 R \frac{-L}{2R}(0 - 1) = \boxed{\frac{1}{2}LI_0^2}$

**PLAUSIBILITY CHECK** The total amount of energy dissipated equals the energy $\frac{1}{2}LI_0^2$ originally stored in the inductor.

---

*INITIAL CURRENTS AND FINAL CURRENTS*  **EXAMPLE 28-15**

**For the circuit shown in Figure 28-34, find the currents $I_1$, $I_2$, and $I_3$ (a) immediately after switch S is closed and (b) a long time after switch S has been closed. After the switch has been closed for a long time the switch is opened. Immediately after the switch is opened (c) find the three currents and (d) find the potential drop across the 20-$\Omega$ resistor. (e) Find all three currents a long time after switch S was opened.**

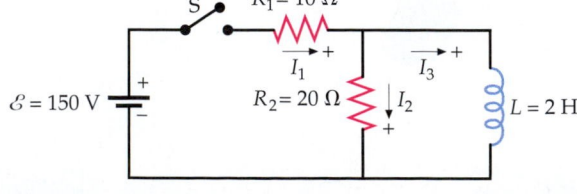

**FIGURE 28-34**

**PICTURE THE PROBLEM** (a) We simplify our calculations by using the fact that the current in an inductor cannot change abruptly. Thus, the current in the inductor must be zero just after the switch is closed, because the current is zero before the switch is closed. (*b*) When the current reaches its final value $dI/dt$ equals zero, so there is no potential drop across the inductor. The inductor thus acts like a short circuit; that is, the inductor acts like a wire with zero resistance. (*c*) Immediately after the switch is opened, the current in the inductor is the same as it was before. (*d*) A long time after the switch is opened, all the currents must be zero.

(*a*) 1. The switch is just opened. The current through the inductor is zero, just as it was before the switch was closed. Apply the junction rule to relate $I_1$ and $I_2$:

$$I_3 = \boxed{0}$$

$$I_1 = I_2 + I_3$$

so

$$I_1 = I_2$$

2. The current in the left loop is obtained by applying the loop rule to the loop on the left:

$$\mathcal{E} - I_1 R_1 - I_2 R_2 = 0$$

so

$$I_1 = \frac{\mathcal{E}}{R_1 + R_2} = \frac{150\ \text{V}}{10\ \Omega + 20\ \Omega} = \boxed{5\ \text{A}} = I_2$$

(*b*) 1. After a long time, the currents are steady and the inductor acts like a short circuit, so the potential drop across $R_2$ is zero. Apply the loop rule to the right loop and solve for $I_2$:

$$-L\frac{dI_3}{dt} + I_2 R_2 = 0$$

$$0 + I_2 R_2 = 0 \quad \Rightarrow \quad I_2 = \boxed{0}$$

2. Apply the loop rule to the left loop and solve for $I_1$:

$$\mathcal{E} - I_1 R_1 - I_2 R_2 = 0$$

$$\mathcal{E} - I_1 R_1 - 0 = 0$$

so

$$I_1 = \frac{\mathcal{E}}{R_1} = \frac{150\ \text{V}}{10\ \Omega} = \boxed{15\ \text{A}}$$

3. Apply the junction rule and solve for $I_3$:

$$I_1 = I_2 + I_3$$

$$15\ \text{A} = 0 + I_3$$

so

$$I_3 = \boxed{15\ \text{A}}$$

(*c*) When the switch is reopened, $I_1$ *instantly* becomes zero. The current $I_3$ in the inductor changes continuously, so at that instant $I_3 = 15$ A. Apply the junction rule and solve for $I_2$:

$$I_3 = \boxed{15\ \text{A}}$$

$$I_1 = I_2 + I_3$$

so

$$I_2 = I_1 - I_3 = 0 - 15\ \text{A} = \boxed{-15\ \text{A}}$$

(*d*) Apply Ohm's law to find the potential drop across $R_2$:

$$V = I_2 R_2 = (15\ \text{A})(20\ \Omega) = \boxed{300\ \text{V}}$$

(*e*) A long time after the switch is opened, all the currents must equal zero.

$$I_1 = I_2 = I_3 = \boxed{0}$$

**REMARKS** Were you surprised to find the potential drop across $R_2$ in Part (*d*) to be larger than the emf of the battery? This potential drop is equal to the emf of the inductor.

**EXERCISE** Suppose $R_2 = 200\ \Omega$ and the switch has been closed for a long time. What is the potential drop across it immediately after the switch is then opened? (*Answer* 3000 V)

# *28-9 Magnetic Properties of Superconductors

Superconductors have resistivities of zero below a critical temperature $T_c$, which varies from material to material. In the presence of a magnetic field $\vec{B}$, the critical temperature is lower than the critical temperature is when there is no field. As the magnetic field increases, the critical temperature decreases. If the magnetic field magnitude is greater than some critical field $B_c$, superconductivity does not exist at any temperature.

## *Meissner Effect

As a superconductor is cooled below the critical temperature in an applied magnetic field, the magnetic field inside the superconductor becomes zero (Figure 28-35). This effect was discovered by Walter Meissner and Robert Ochsenfeld in 1933 and is now known as the **Meissner effect**. The magnetic field becomes zero because superconducting currents induced on the surface of the superconductor produce a second magnetic field that cancels out the applied one. The magnetic levitation (see the following photo) results from the repulsion between the permanent magnet producing the applied field and the magnetic field produced by the currents induced in the superconductor. Only certain superconductors, called **type I superconductors**, exhibit the complete Meissner effect. Figure 28-36a shows a plot of the magnetization $M$ times $\mu_0$ versus the applied magnetic field $B_{app}$ for a type I superconductor. For a magnetic field less than the critical field $B_c$, the magnetic field $\mu_0 M$ induced in the superconductor is equal and opposite to the applied magnetic field. The values of $B_c$ for type I superconductors are always too small for such materials to be useful in the coils of a superconducting magnet.

Other materials, known as **type II superconductors**, have a magnetization curve similar to that in Figure 28-36b. Such materials are usually alloys or metals that have large resistivities in the normal state. Type II superconductors exhibit the electrical properties of superconductors except for the Meissner effect up to the critical field $B_{c2}$, which may be several hundred times the typical values of critical fields for type I superconductors. For example, the alloy $Nb_3Ge$ has a critical field $B_{c2} = 34$ T. Such materials can be used for high-field superconducting magnets. Below the critical field $B_{c1}$, the behavior of a type II superconductor is the same as that of a type I superconductor. In the region between fields $B_{c1}$ and $B_{c2}$, the superconductor is said to be in a vortex state.

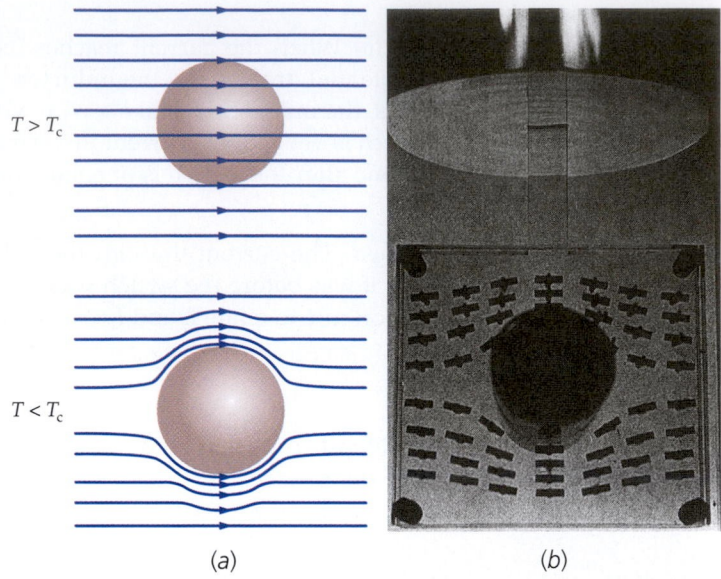

(a)      (b)

**FIGURE 28-35** (a) The Meissner effect in a superconducting solid sphere cooled in a constant applied magnetic field. As the temperature drops below the critical temperature $T_c$, the magnetic field inside the sphere becomes zero. (b) Demonstration of the Meissner effect. A superconducting tin cylinder is situated with its axis perpendicular to a horizontal magnetic field. The directions of the field lines are indicated by weakly magnetized compass needles mounted in a Lucite sandwich so that they are free to turn.

The cube is a superconductor. The magnetic levitation results from the repulsion between the permanent magnet producing the applied field and the magnetic field produced by the currents induced in the superconductor.

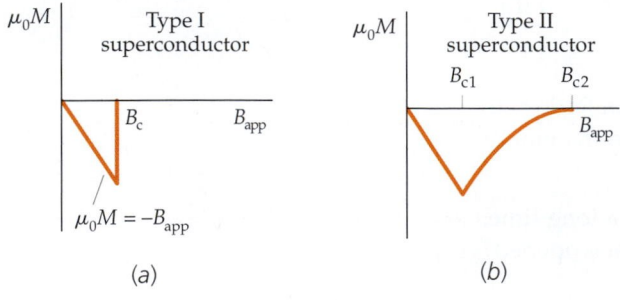

(a)          (b)

**FIGURE 28-36** Plots of $\mu_0$ times the magnetization $M$ versus applied magnetic field for type I and type II superconductors. (a) In a type I superconductor, the resultant magnetic field is zero below a critical applied field $B_c$ because the field due to induced currents on the surface of the superconductor exactly cancels the applied field. Above the critical field, the material is a normal conductor and the magnetization is too small to be seen on this scale. (b) In a type II superconductor, the magnetic field starts to penetrate the superconductor at a field $B_{c1}$, but the material remains superconducting up to a field $B_{c2}$, after which the material becomes a normal conductor.

## *Flux Quantization

Consider a superconducting ring of area $A$ carrying a current. There can be a magnetic flux $\phi_m = B_n A$ through the flat surface $S$ bounded by the ring due to the current in the ring and due also perhaps to other currents external to the ring. According to Equation 28-5, if the flux through $S$ changes, an electric field will be induced in the ring whose circulation is proportional to the rate of change of the flux. But there can be no electric field in a superconducting ring because it has no resistance, so a finite electric field would drive an infinite current. The flux through the ring is thus frozen and cannot change.

Another effect, which results from the quantum-mechanical treatment of superconductivity, is that the total flux through surface $S$ is quantized and is given by

$$\phi_m = n\frac{h}{2e}, \quad n = 1, 2, 3, \ldots \qquad 28\text{-}27$$

The smallest unit of flux, called a **fluxon**, is

$$\phi_0 = \frac{h}{2e} = 2.0678 \times 10^{-15} \text{ T·m}^2 \qquad 28\text{-}28$$

# SUMMARY

1. Faraday's law and Lenz's law are fundamental laws of physics.
2. Self-inductance is a property of a circuit element that relates the flux through the element to the current.

| Topic | Relevant Equations and Remarks | |
|---|---|---|
| **1. Magnetic Flux $\phi_m$** | | |
| General definition | $\phi_m = \displaystyle\int_S \vec{B} \cdot \hat{n}\, dA$ | 28-1 |
| Uniform field, flat surface bounded by coil of $N$ turns | $\phi_m = NBA \cos\theta$ <br> where $A$ is the area of the flat surface bounded by a single turn. | 28-3 |
| Units | $1 \text{ Wb} = 1 \text{ T·m}^2$ | 28-2 |
| Due to current in a circuit | $\phi_m = LI$ | 28-9 |
| Due to current in two circuits | $\phi_{m1} = L_1 I_1 + M I_2$ <br> $\phi_{m2} = L_2 I_2 + M I_1$ | 28-14 |
| *Quantization | $\phi_m = n\dfrac{h}{2e}, \quad n = 1, 2, 3, \cdots$ | 28-27 |
| *Fluxon | $\phi_0 = \dfrac{h}{2e} = 2.0678 \times 10^{-15} \text{ T·m}^2$ | 28-28 |

## 2. EMF

| | | |
|---|---|---|
| Faraday's law (includes both induction and motional emf) | $\mathcal{E} = -\dfrac{d\phi_m}{dt}$ | 28-4 |
| Induction (time varying magnetic field, $C$ stationary) | $\mathcal{E} = \oint_C \vec{E} \cdot d\vec{\ell}$ | 28-5 |
| Motional (static magnetic field, $C$ not stationary) | $\mathcal{E} = \oint_C (\vec{v} \times \vec{B}) \cdot d\vec{\ell}$ | 28-8 |

where $\vec{v}$ is the velocity of the conducting path.

| | | |
|---|---|---|
| Rod moving perpendicular to both itself and $\vec{B}$ | $\mathcal{E} = vB\ell$ | 28-6 |
| Self-induced (back emf) | $\mathcal{E} = -L\dfrac{dI}{dt}$ | 28-12 |

## 3. Faraday's Law

$$\mathcal{E} = -\frac{d\phi_m}{dt} \qquad \text{28-4}$$

## 4. Lenz's Law

The induced emf and induced current are in such a direction as to oppose, or tend to oppose, the change that produces them.

**Alternative statement** — When a magnetic flux through a surface changes, the magnetic field due to any induced current produces a flux of its own—through the same surface and in opposition to the change.

## 5. Inductance

| | | |
|---|---|---|
| Self-inductance | $L = \dfrac{\phi_m}{I}$ | 28-9 |
| Self-inductance of a solenoid | $L = \mu_0 n^2 A\ell$ | 28-11 |
| Mutual inductance | $M = \dfrac{\phi_{m2,1}}{I_1} = \dfrac{\phi_{m1,2}}{I_2}$ | 28-16 |
| Units | $1\,\text{H} = 1\dfrac{\text{Wb}}{\text{A}} = 1\dfrac{\text{T·m}^2}{\text{A}}$ $\mu_0 = 4\pi \times 10^{-7}\,\text{H/m}$ | |

## 6. Magnetic Energy

| | | |
|---|---|---|
| Energy stored in an inductor | $U_m = \dfrac{1}{2}LI^2$ | 28-19 |
| Energy density in a magnetic field | $u_m = \dfrac{B^2}{2\mu_0} = \dfrac{1}{2}\mu_0^{-1}B^2$ | 28-20 |

## *7. *RL* Circuits

| | | |
|---|---|---|
| Potential difference across an inductor | $\Delta V = \mathcal{E} - Ir = -L\dfrac{dI}{dt} - Ir$ | 28-13 |

where $r$ is the internal resistance of the inductor. For an ideal inductor $r = 0$.

| Energizing an inductor with a battery | In a circuit consisting of a resistance $R$, an inductance $L$, and a battery of emf $\mathcal{E}_0$ in series, the current does not reach its maximum value $I_f$ instantaneously, but rather takes some time to build up. If the current is initially zero, its value at some later time $t$ is given by | |
|---|---|---|
| | $$I = \frac{\mathcal{E}_0}{R}(1 - e^{-t/\tau}) = I_f(1 - e^{-t/\tau})$$ | 28-23 |
| Time constant $\tau$ | $$\tau = \frac{L}{R}$$ | 28-24 |
| De-energizing an inductor through a resistor | In a circuit consisting of a resistance $R$ and an inductance $L$, the current does not drop to zero instantaneously, but rather takes some time to decrease. If the current is initially $I_0$, its value at some later time $t$ is given by | |
| | $$I = I_0 e^{-t/\tau}$$ | 28-26 |

# PROBLEMS

- • Single-concept, single-step, relatively easy
- •• Intermediate-level, may require synthesis of concepts
- ••• Challenging
- SSM Solution is in the *Student Solutions Manual*
- ISOLVE Problems available on iSOLVE online homework service
- ISOLVE✓ These "Checkpoint" online homework service problems ask students additional questions about their confidence level, and how they arrived at their answer.

In a few problems, you are given more data than you actually need; in a few other problems, you are required to supply data from your general knowledge, outside sources, or informed estimates.

## Conceptual Problems

1 • SSM ISOLVE A conducting loop lies in the plane of this page and carries a clockwise induced current. Which of the following statements could be true? (a) A constant magnetic field is directed into the page. (b) A constant magnetic field is directed out of the page. (c) An increasing magnetic field is directed into the page. (d) A decreasing magnetic field is directed into the page. (e) A decreasing magnetic field is directed out of the page.

2 • Give the direction of the induced current in the circuit, shown on the right in Figure 28-37, when the resistance in the circuit on the left is suddenly (a) increased and (b) decreased.

3 •• The two circular loops in Figure 28-38 have their planes parallel to each other. As viewed from the left, there is a counterclockwise current in loop A. Give the direction of the current in loop B and state whether the loops attract or repel each other if the current in loop A is (a) increasing and (b) decreasing.

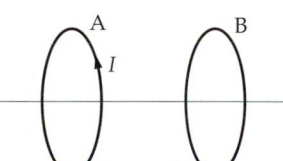

**FIGURE 28-38** Problem 3

4 •• A bar magnet moves with constant velocity along the axis of a loop, as shown in Figure 28-39. (a) Make a qualitative graph of the flux $\phi_m$ through the loop as a function of time. Indicate the time $t_1$ when the magnet is halfway through the loop. (b) Sketch a graph of the current $I$ in the loop versus time, choosing $I$ to be positive when it is clockwise as viewed from the left.

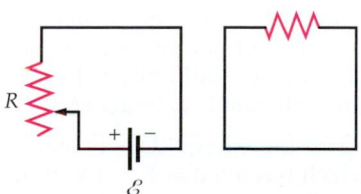

**FIGURE 28-37** Problem 2

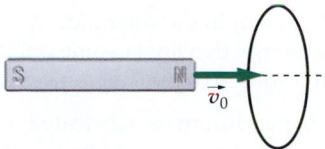

**FIGURE 28-39** Problem 4

**5** •• A bar magnet is mounted on the end of a coiled spring in such a way that it moves with simple harmonic motion along the axis of a loop, as shown in Figure 28-40. (a) Make a qualitative graph of the flux $\phi_m$ through the loop as a function of time. Indicate the time $t_1$ when the magnet is halfway through the loop. (b) Sketch the current $I$ in the loop versus time, choosing $I$ to be positive when it is clockwise as viewed from above.

**FIGURE 28-40** Problem 5

**6** • [SSM] If the current through an inductor were doubled, the energy stored in the inductor would be (a) the same. (b) doubled. (c) quadrupled. (d) halved. (e) quartered.

**7** • Inductors in circuits that are switched on and off are often protected by being placed in parallel with a diode, as shown in Figure 28-41. (A diode is a one-way valve for current; current can flow in the direction of the arrow, but not opposite to the direction of the arrow.) Why is such protection needed? Explain how the voltage across the inductor changes with no diode protection if a switch is opened suddenly while current is flowing through the inductor.

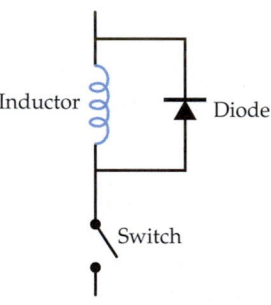

**FIGURE 28-41** Problem 7

**8** • Two inductors are made from identical lengths of wire wrapped around identical circular cores of the same radius. However, one inductor has three times the number of coils per unit length as the other. Which coil has the higher self-inductance? What is the ratio of the self-inductance of the two coils?

**9** • True or false:

(a) The induced emf in a circuit is proportional to the magnetic flux through the circuit.
(b) There can be an induced emf at an instant when the flux through the circuit is zero.
(c) Lenz's law is related to the conservation of energy.
(d) The inductance of a solenoid is proportional to the rate of change of the current in the solenoid.
(e) The magnetic energy density at some point in space is proportional to the square of the magnetic field at that point.

**10** • [SSM] A pendulum is fabricated from a thin, flat piece of aluminum. At the bottom of its arc, it passes between the poles of a strong permanent magnet. In Figure 28-42a, the

metal sheet is continuous, whereas in Figure 28-42b, there are slots in it. The pendulum with slots swings back and forth many times, but the pendulum without slots comes to a stop in no more than one complete oscillation. Explain why.

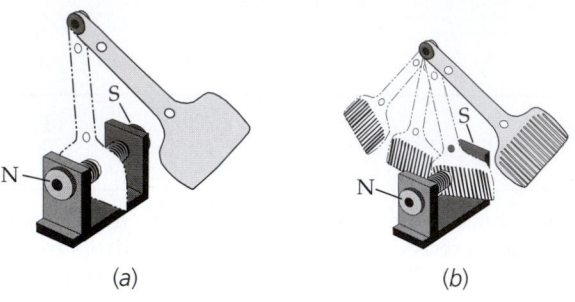

(a)                                    (b)

**FIGURE 28-42** Problem 10

**11** • A bar magnet is dropped inside a long vertical tube. If the tube is made of metal, the magnet quickly approaches a terminal speed, but if the tube is made of cardboard, the magnet does not. Explain.

**12** •• An experimental setup for a jumping ring demonstration is shown in Figure 28-43. A metal ring is placed on top of a large coil, with an iron rod threading the center of the ring and the coil. When a current is suddenly started in the coil, the ring jumps several feet into the air. Explain how the demonstration works. Will the demonstration work if a slot is cut into the ring?

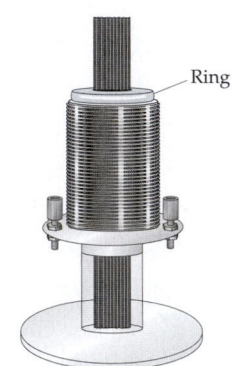

**FIGURE 28-43** Problem 12

## Estimation and Approximation

**13** •• [SSM] A physics teacher attempts the following emf demonstration. She has two of her students hold a long wire connected to a voltmeter. The wire is held slack, so that there is a large arc in it. When she says "start," the students begin rotating the wire in a large vertical arc, as if they were playing jump rope. The students stand 3.0 m apart, and the sag in the wire is about 1.5 m. (You may idealize the shape of the wire as a perfect semicircular arc of diameter $d = 1.5$ m.) The induced emf from the jump rope is then measured on the voltmeter. (a) Estimate a reasonable value for the maximum angular velocity that the students can rotate the wire. (b) From this, estimate the maximum emf induced in the wire. The magnitude of the Earth's magnetic field is approximately 0.7 G. (c) Can the students rotate the jump rope fast enough to generate an emf of 1 V? (d) Suggest modifications to the demonstration that would allow higher emfs to be generated.

**14** • Compare the energy density stored in the earth's electric field, which has a value $E \sim 100$ V/m at the surface of the earth, to that of the earth's magnetic field, where $B \sim 5 \times 10^{-5}$ T.

**15** •• A lightning strike transfers roughly 30 C of charge from the sky to the ground in approximately 1 $\mu s$. Estimate the maximum emf induced by the lightning strike in an antenna consisting of a single loop of wire with cross-sectional area 0.1 m² a distance 300 m away from the lightning strike.

## Magnetic Flux

**16** • A uniform magnetic field of magnitude 2000 G is parallel to the $x$ axis. A square coil of side 5 cm has a single turn and makes an angle $\theta$ with the $z$ axis, as shown in Figure 28-44. Find the magnetic flux through the coil when (a) $\theta = 0°$, (b) $\theta = 30°$, (c) $\theta = 60°$, and (d) $\theta = 90°$.

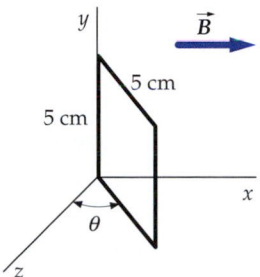

**FIGURE 28-44** Problem 16

**17** • **SSM** A circular coil has 25 turns and a radius of 5 cm. It is at the equator, where the earth's magnetic field is 0.7 G north. Find the magnetic flux through the coil when its plane is (a) horizontal, (b) vertical with its axis pointing north, (c) vertical with its axis pointing east, and (d) vertical with its axis making an angle of 30° with north.

**18** • **ISOLVE** A magnetic field of 1.2 T is perpendicular to a square coil of 14 turns. The length of each side of the coil is 5 cm. (a) Find the magnetic flux through the coil. (b) Find the magnetic flux through the coil if the magnetic field makes an angle of 60° with the normal to the plane of the coil.

**19** • A uniform magnetic field $\vec{B}$ is perpendicular to the base of a hemisphere of radius $R$. Calculate the magnetic flux through the spherical surface of the hemisphere.

**20** •• **ISOLVE** Find the magnetic flux through a 400-turn solenoid of length 25 cm and radius 1 cm that carries a current of 3 A.

**21** •• **ISOLVE** Rework Problem 20 for an 800-turn solenoid of length 30 cm and radius 2 cm that carries a current of 2 A.

**22** •• A circular coil of 15 turns of radius 4 cm is in a uniform magnetic field of 4000 G in the positive $x$ direction. Find the flux through the coil when the unit vector perpendicular to the plane of the coil is (a) $\hat{n} = \hat{i}$, (b) $\hat{n} = \hat{j}$, (c) $\hat{n} = (\hat{i} + \hat{j})/\sqrt{2}$, (d) $\hat{n} = \hat{k}$, and (e) $\hat{n} = 0.6\hat{i} + 0.8\hat{j}$.

**23** •• A solenoid has $n$ turns per unit length, radius $R_1$, and carries a current $I$. (a) A large circular loop of radius $R_2 > R_1$ and $N$ turns encircles the solenoid at a point far away from the ends of the solenoid. Find the magnetic flux through the loop. (b) A small circular loop of $N$ turns and radius $R_3 < R_1$ is completely inside the solenoid, far from its ends, with its axis parallel to that of the solenoid. Find the magnetic flux through this small loop.

**24** •• **SSM** A long straight wire carries a current $I$. A rectangular loop with two sides parallel to the straight wire has sides $a$ and $b$, with its near side a distance $d$ from the straight wire, as shown in Figure 28-45. (a) Compute the magnetic flux through the rectangular loop. (Hint: Calculate the flux through a strip of area $dA = b\,dx$ and integrate from $x = d$ to $x = d + a$.) (b) Evaluate your answer for $a = 5$ cm, $b = 10$ cm, $d = 2$ cm, and $I = 20$ A.

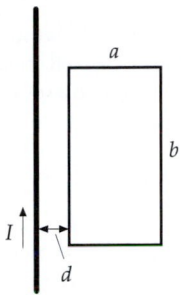

**FIGURE 28-45** Problem 24

**25** ••• **ISOLVE** A long cylindrical conductor of radius $R$ carries a current $I$ that is uniformly distributed over its cross-sectional area. Find the magnetic flux per unit length through the area indicated in Figure 28-46.

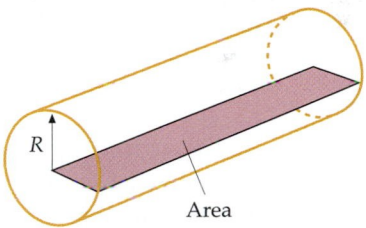

**FIGURE 28-46** Problem 25

**26** ••• A rectangular coil in the plane of the page has dimensions $a$ and $b$. A long wire that carries a current $I$ is placed directly above the coil (Figure 28-47). (a) Obtain an expression for the magnetic flux through the coil as a function of $x$ for $0 \le x \le 2b$. (b) For what value of $x$ is the flux through the coil a maximum? For what value of $x$ is the flux a minimum?

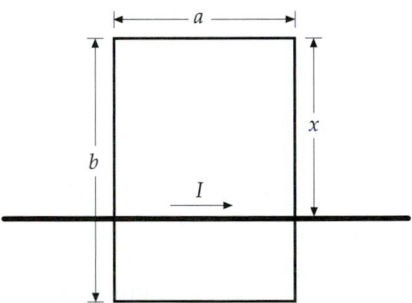

**FIGURE 28-47** Problem 26

## Induced EMF and Faraday's Law

**27** • **SSM** A uniform magnetic field $\vec{B}$ is established perpendicular to the plane of a loop of radius 5 cm, resistance 0.4 Ω, and negligible self-inductance. The magnitude of $\vec{B}$ is increasing at a rate of 40 mT/s. Find (a) the induced emf $\mathcal{E}$ in the loop, (b) the induced current in the loop, and (c) the rate of joule heating in the loop.

**28** • The flux through a loop is given by $\phi_m = (t^2 - 4t) \times 10^{-1}$ Wb, where $t$ is in seconds. (a) Find the induced emf $\mathcal{E}$ as a function of time. (b) Find both $\phi_m$ and $\mathcal{E}$ at $t = 0$, $t = 2$ s, $t = 4$ s, and $t = 6$ s.

**29** • (a) For the flux given in Problem 28, sketch graphs of $\phi_m$ and $\mathcal{E}$ versus $t$. (b) At what time is the flux minimum? What is the emf at this time? (c) At what times is the flux zero? What is the emf at these times?

**30** • A solenoid of length 25 cm and radius 0.8 cm with 400 turns is in an external magnetic field of 600 G that makes an angle of 50° with the axis of the solenoid. (a) Find the magnetic flux through the solenoid. (b) Find the magnitude of the emf induced in the solenoid if the external magnetic field is reduced to zero in 1.4 s.

**31** •• SSM ISOLVE✓ A 100-turn circular coil has a diameter of 2 cm and resistance of 50 Ω. The plane of the coil is perpendicular to a uniform magnetic field of magnitude 1 T. The direction of the field is suddenly reversed. (a) Find the total charge that passes through the coil. If the reversal takes 0.1 s, find (b) the average current in the coil and (c) the average emf in the coil.

**32** •• At the equator, a 1000-turn coil with a cross-sectional area of 300 cm² and a resistance of 15 Ω is aligned with its plane perpendicular to the earth's magnetic field of 0.7 G. If the coil is flipped over, how much charge flows through the coil?

**33** •• ISOLVE✓ A circular coil of 300 turns and radius 5 cm is connected to a current integrator. The total resistance of the circuit is 20 Ω. The plane of the coil is originally aligned perpendicular to the earth's magnetic field at some point. When the coil is rotated through 90°, the charge that passes through the current integrator is measured to be 9.4 μC. Calculate the magnitude of the earth's magnetic field at that point.

**34** •• The wire in Problem 26 is placed at $x = b/4$. (a) Obtain an expression for the emf induced in the coil if the current varies with time according to $I = 2t$. (b) If $a = 1.5$ m and $b = 2.5$ m, what should be the resistance of the coil so that the induced current is 0.1 A? What is the direction of this current?

**35** •• Repeat Problem 34 if the wire is placed at $x = b/3$.

## Motional EMF

**36** • SSM ISOLVE A rod 30 cm long moves at 8 m/s in a plane perpendicular to a magnetic field of 500 G. The velocity of the rod is perpendicular to its length. Find (a) the magnetic force on an electron in the rod, (b) the electrostatic field $\vec{E}$ in the rod, and (c) the potential difference $V$ between the ends of the rod.

**37** • ISOLVE Find the speed of the rod in Problem 36 if the potential difference between the ends is 6 V.

**38** • In Figure 28-22, let $B$ be 0.8 T, $v = 10$ m/s, $\ell = 20$ cm, and $R = 2$ Ω. Find (a) the induced emf in the circuit, (b) the current in the circuit, and (c) the force needed to move the rod with constant velocity assuming negligible friction. Find (d) the power input by the force found in Part (c), and (e) the rate of joule heat production $I^2R$.

**FIGURE 28-22**
Problem 38

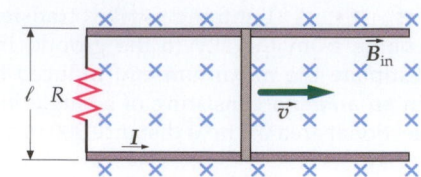

**39** •• A 10-cm by 5-cm rectangular loop with resistance 2.5 Ω is pulled through a region of uniform magnetic field $B = 1.7$ T (Figure 28-48) with constant speed $v = 2.4$ cm/s. The front of the loop enters the region of the magnetic field at time $t = 0$. (a) Find and graph the flux through the loop as a function of time. (b) Find and graph the induced emf and the current in the loop as functions of time. Neglect any self-inductance of the loop and extend your graphs from $t = 0$ to $t = 16$ s.

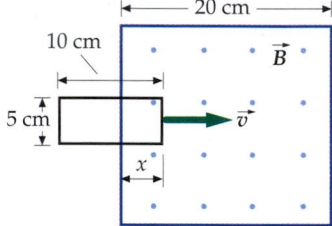

**FIGURE 28-48** Problem 39

**40** •• A uniform magnetic field of magnitude 1.2 T is in the z direction. A conducting rod of length 15 cm lies parallel to the y axis and oscillates in the x direction with displacement given by $x = (2 \text{ cm}) \cos 120 \pi t$. What is the emf induced in the rod?

**41** •• ISOLVE✓ In Figure 28-49, the rod has a resistance $R$ and the rails are horizontal and have negligible resistance. A battery of emf $\mathcal{E}$ and negligible internal resistance is connected between points $a$ and $b$ so that the current in the rod is downward. The rod is placed at rest at $t = 0$. (a) Find the force on the rod as a function of the speed $v$ and write Newton's second law for the rod when it has speed $v$. (b) Show that the rod moves at a terminal speed and find an expression for it. (c) What is the current when the rod will approach its terminal speed?

**FIGURE 28-49**
Problems 41 and 44

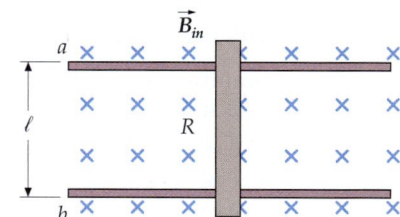

**42** •• SSM In Example 28-9, find the total energy dissipated in the resistance and show that it is equal to $mv_0^2$.

**43** •• Find the total distance traveled by the rod in Example 28-9.

**44** •• In Figure 28-49, the rod has a resistance $R$ and the rails have negligible resistance. A capacitor with charge $Q_0$ and capacitance $C$ is connected between points $a$ and $b$ so that the current in the rod is downward. The rod is placed at rest at $t = 0$. (a) Write the equation of motion for the rod on the rails. (b) Show that the terminal speed of the rod down the rails is related to the final charge on the capacitor.

**45** •• SSM In Figure 28-50, a conducting rod of mass $m$ and negligible resistance is free to slide without friction along two parallel rails of negligible resistance separated by a distance $\ell$ and connected by a resistance $R$. The rails are attached to a long inclined plane that makes an angle $\theta$ with the horizontal. There is a magnetic field $B$ directed upward. (a) Show that there is a retarding force directed up the incline given by $F = (B^2\ell^2v \cos^2 \theta)/R$. (b) Show that the terminal speed of the rod is $v_t = (mgR \sin \theta)/(B^2\ell^2 \cos^2 \theta)$.

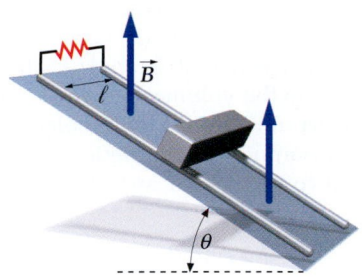

**FIGURE 28-50** Problems 45 and 49

**46** •• A square loop of a conducting wire (area $A$) is pulled out of a region of constant, very high magnetic field $B$ that is directed perpendicular to the plane of the wire. Half of the wire is in the field and half of the wire is out of the field when the wire is pulled out. A constant force $F$ is exerted on the wire to pull the wire out. The wire is pulled out in time $t$. All else being equal, if the force were doubled, approximately how long would it take to pull the wire out? (a) $t$ (b) $t/\sqrt{2}$ (c) $t/2$ (d) $t/4$

**47** •• If instead of doubling the force the resistance of the wire in Problem 46 were halved (all else being equal), what would the new time be? (a) $t$ (b) $2t$ (c) $t/2$ (d) $t\sqrt{2}$

**48** •• A wire lies along the $z$ axis and carries current $I = 20$ A in the positive $z$ direction. A small conducting sphere of radius $R = 2$ cm is initially at rest on the $y$ axis at a distance $h = 45$ m above the wire. The sphere is dropped at time $t = 0$. (a) What is the electric field at the center of the sphere at $t = 3$ s? Assume that the only magnetic field is the magnetic field produced by the wire. (b) What is the voltage across the sphere at $t = 3$ s?

**49** •• **iSOLVE** In Figure 28-50, let $\theta = 30°$; $m = 0.4$ kg, $\ell = 15$ m, and $R = 2$ Ω. The rod starts from rest at the top of the inclined plane at $t = 0$. The rails have negligible resistance. There is a constant, vertically directed magnetic field of magnitude $B = 1.2$ T. (a) Find the emf induced in the rod as a function of its velocity down the rails. (b) Write Newton's law of motion for the rod; show that the rod will approach a terminal speed and determine its value.

**50** ••• A solid conducting cylinder of radius 0.1 m and mass 4 kg rests on horizontal conducting rails (Figure 28-51). The rails, separated by a distance $a = 0.4$ m, have a rough surface, so the cylinder rolls rather than slides. A 12-V battery is connected to the rails as shown. The only significant resistance in the circuit is the contact resistance of 6 Ω between the cylinder and rails. The system is in a uniform vertical magnetic field. The cylinder is initially at rest next to the battery. (a) What must be the magnitude and the direction of $\vec{B}$ so that the cylinder has an initial acceleration of 0.1 m/s² to the right? (b) Find the force on the cylinder as a function of its speed $v$. (c) Find the terminal velocity of the cylinder. (d) What is the kinetic energy of the cylinder when it has reached its terminal velocity? (Neglect the magnetic field due to the current in

the battery–rails–cylinder loop, and assume that the current density in the cylinder is uniform.)

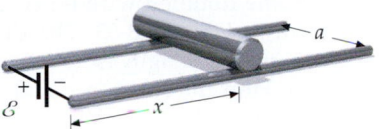

**FIGURE 28-51** Problem 50

**51** ••• **SSM** The loop in Problem 24 moves away from the wire with a constant speed $v$. At time $t = 0$, the left side of the loop is a distance $d$ from the long straight wire. (a) Compute the emf in the loop by computing the motional emf in each segment of the loop that is parallel to the long wire. Explain why you can neglect the emf in the segments that are perpendicular to the wire. (b) Compute the emf in the loop by first computing the flux through the loop as a function of time and then using $\mathcal{E} = -d\phi_m/dt$. Compare your answer with that obtained in Part (a).

**52** ••• A conducting rod of length $\ell$ rotates at constant angular velocity about one end, in a plane perpendicular to a uniform magnetic field $B$ (Figure 28-52). (a) Show that the magnetic force on a body whose charge is $q$ at a distance $r$ from the pivot is $Bqr\omega$. (b) Show that the potential difference between the ends of the rod is $V = \frac{1}{2}B\omega\ell^2$. (c) Draw any radial line in the plane from which to measure $\theta = \omega t$. Show that the area of the pie-shaped region between the reference line and the rod is $A = \frac{1}{2}\ell^2 \theta$. Compute the flux through this area, and show that $\mathcal{E} = \frac{1}{2}B\omega\ell^2$ follows when Faraday's law is applied to this area.

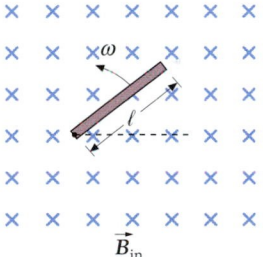

**FIGURE 28-52** Problem 52

## Inductance

**53** • A coil with a self-inductance of 8 H carries a current of 3 A that is changing at a rate of 200 A/s. Find (a) the magnetic flux through the coil and (b) the induced emf in the coil.

**54** • **SSM** A coil with self-inductance $L$ carries a current $I$, given by $I = I_0 \sin 2\pi ft$. Find and graph the flux $\phi_m$ and the self-induced emf as functions of time.

**55** •• A solenoid has a length of 25 cm, a radius of 1 cm, 400 turns, and carries a 3-A current. Find (a) $B$ on the axis at the center of the solenoid; (b) the flux through the solenoid, assuming $B$ to be uniform; (c) the self-inductance of the solenoid; and (d) the induced emf in the solenoid when the current changes at 150 A/s.

**56** •• **iSOLVE** Two solenoids of radii 2 cm and 5 cm are coaxial. They are each 25 cm long and have 300 turns and 1000 turns, respectively. Find their mutual inductance.

**57** •• SSM A long insulated wire with a resistance of 18 Ω/m is to be used to construct a resistor. First, the wire is bent in half and then the doubled wire is wound in a cylindrical form, as shown in Figure 28-53. The diameter of the cylindrical form is 2 cm, its length is 25 cm, and the total length of wire is 9 m. Find the resistance and inductance of this wire-wound resistor.

**FIGURE 28-53** Problem 57

**58** ••• ISOLVE✓ In Figure 28-54, circuit 2 has a total resistance of 300 Ω. A total charge of $2 \times 10^{-4}$ C flows through the galvanometer in circuit 2 when switch S in circuit 1 is closed. After a long time, the current in circuit 1 is 5 A. What is the mutual inductance between the two coils?

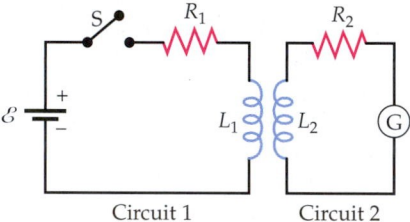

**FIGURE 28-54** Problem 58

**59** ••• Show that the inductance of a toroid of rectangular cross section, as shown in Figure 28-55, is given by

$$L = \frac{\mu_0 N^2 H \ln(b/a)}{2\pi}$$

where $N$ is the total number of turns, $a$ is the inside radius, $b$ is the outside radius, and $H$ is the height of the toroid.

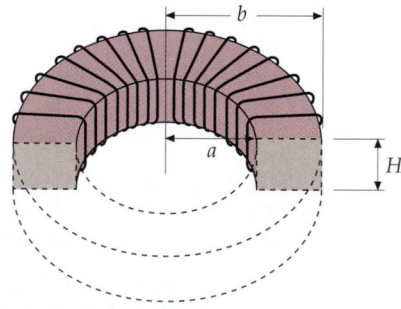

**FIGURE 28-55** Problem 59

## Magnetic Energy

**60** • A coil with a self-inductance of 2 H and a resistance of 12 Ω is connected across a 24-V battery of negligible internal resistance. (a) What is the final current? (b) How much energy is stored in the inductor when the final current is attained?

**61** •• SSM In a plane electromagnetic wave, such as a light wave, the magnitudes of the electric fields and magnetic fields are related by $E = cB$, where $c = 1/\sqrt{\epsilon_0\mu_0}$ is the speed of light. Show that in this case the electric energy and the magnetic energy densities are equal.

**62** •• A solenoid of 2000 turns, area 4 cm², and length 30 cm carries a current of 4 A. (a) Calculate the magnetic energy stored in the solenoid from $\frac{1}{2}LI^2$. (b) Divide your answer in Part (a) by the volume of the solenoid to find the magnetic energy per unit volume in the solenoid. (c) Find $B$ in the solenoid. (d) Compute the magnetic energy density from $u_m = B^2/2\mu_0$, and compare your answer with your result for Part (b).

**63** •• ISOLVE A long cylindrical wire of radius $a = 2$ cm carries a current $I = 80$ A uniformly distributed over its cross-sectional area. Find the magnetic energy per unit length within the wire.

**64** •• SSM You are given a length $d$ of wire that has radius $a$ and are told to wind it into an inductor in the shape of a cylinder with a circular cross section of radius $r$. The windings are to be as close together as possible without overlapping. Show that the self-inductance of this inductor is

$$L = \mu_0\left(\frac{rd}{4a}\right)$$

**65** • Using the result of Problem 64, calculate the self-inductance of an inductor wound from 10 cm of wire with a diameter of 1 mm into a coil with radius $R = 0.25$ cm.

**66** •• A toroid of mean radius 25 cm and circular cross section of radius 2 cm is wound with a superconducting wire of length 1000 m that carries a current of 400 A. (a) What is the number of turns on the coil? (b) What is the magnetic field at the mean radius? (c) Assuming that $B$ is constant over the area of the coil, calculate the magnetic energy density and the total energy stored in the toroid.

## *RL Circuits

**67** • A coil of resistance 8 Ω and self-inductance 4 H is suddenly connected across a constant potential difference of 100 V. Let $t = 0$ be the time of connection, at which the current is zero. Find the current $I$ and its rate of change $dI/dt$ at times (a) $t = 0$, (b) $t = 0.1$ s, (c) $t = 0.5$ s, and (d) $t = 1.0$ s.

**68** • The current in a coil with a self-inductance of 1 mH is 2 A at $t = 0$, when the coil is shorted through a resistor. The total resistance of the coil plus the resistor is 10 Ω. Find the current after (a) 0.5 ms and (b) 10 ms.

**69** •• SSM In the circuit shown Figure 28-29, let $\mathcal{E}_0 = 12$ V, $R = 3$ Ω, and $L = 0.6$ H. The switch is closed at time $t = 0$. At time $t = 0.5$ s, find (a) the rate at which the battery supplies power, (b) the rate of joule heating, and (c) the rate at which energy is being stored in the inductor.

**70** •• Rework Problem 69 for the times $t = 1$ s and $t = 100$ s.

**71** •• The current in an $RL$ circuit is zero at time $t = 0$ and increases to half its final value in 4 s. (a) What is the time constant of this circuit? (b) If the total resistance is 5 Ω, what is the self-inductance?

**72** •• How many time constants must elapse before the current in an *RL* circuit that is initially zero reaches (*a*) 90 percent, (*b*) 99 percent, and (*c*) 99.9 percent of its final value?

**73** •• A coil with inductance 4 mH and resistance 150 $\Omega$ is connected across a battery of emf 12 V and negligible internal resistance. (*a*) What is the initial rate of increase of the current? (*b*) What is the rate of increase when the current is half its final value? (*c*) What is the final current? (*d*) How long does it take for the current to reach 99 percent of its final value?

**74** •• **ISOLVE** A large electromagnet has an inductance of 50 H and a resistance of 8 $\Omega$. It is connected to a dc power source of 250 V. Find the time for the current to reach (*a*) 10 A and (*b*) 30 A.

**75** ••• **SSM** Given the circuit shown in Figure 28-56, assume that the switch S has been closed for a long time so that steady currents exist in the inductor, and that the inductor *L* has negligible resistance. (*a*) Find the battery current, the current in the 100 $\Omega$ resistor, and the current through the inductor. (*b*) Find the initial voltage across the inductor when the switch S is opened. (*c*) Using a spreadsheet program, make graphs of the current and voltage across the inductor as a function of time.

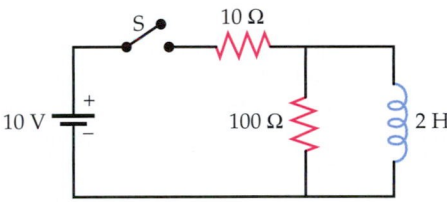

**FIGURE 28-56** Problem 75

**76** •• Compute the initial slope *dI/dt* at *t* = 0 from Equation 28-26, and show that if the current decreased steadily at this rate the current would be zero after one time constant.

**77** •• An inductance *L* and resistance *R* are connected in series with a battery, as shown in Figure 28-31. A long time after switch S$_1$ is closed, the current is 2.5 A. When the battery is switched out of the circuit by opening switch S$_1$ and closing S$_2$, the current drops to 1.5 A in 45 ms. (*a*) What is the time constant for this circuit? (*b*) If *R* = 0.4 $\Omega$, what is *L*?

**78** • **ISOLVE** When the current in a certain coil is 5 A and the current is increasing at the rate of 10 A/s, the potential difference across the coil is 140 V. When the current is 5 A and the current is decreasing at the rate of 10 A/s, the potential difference is 60 V. Find the resistance and self-inductance of the coil.

**79** •• For the circuit shown in Figure 28-57, (*a*) find the rate of change of the current in each inductor and in the resistor just after the switch is closed. (*b*) What is the final current? (Use the result from Problem 88.)

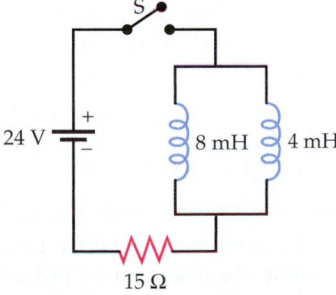

**FIGURE 28-57** Problem 79

**80** •• **SSM** For the circuit of Example 28-11, find the time at which the power dissipation in the resistor equals the rate at which magnetic energy is stored in the inductor.

**81** ••• In the circuit shown in Figure 28-29, let $\mathcal{E}_0$ = 12 V, *R* = 3 $\Omega$, and *L* = 0.6 H. The switch is closed at time *t* = 0. From time *t* = 0 to *t* = $\tau$, find (*a*) the total energy that has been supplied by the battery, (*b*) the total energy that has been dissipated in the resistor, and (*c*) the energy that has been stored in the inductor. (*Hint:* Find the rates as functions of time and integrate from *t* = 0 to *t* = $\tau$ = *L/R*.)

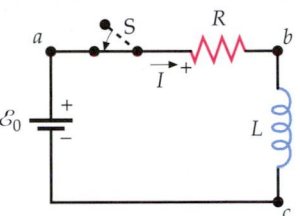

**FIGURE 28-29** Problem 81

## General Problems

**82** • A circular coil of radius 3 cm has 6 turns. A magnetic field *B* = 5000 G is perpendicular to the coil. (*a*) Find the magnetic flux through the coil. (*b*) Find the magnetic flux through the coil if the coil makes an angle of 20° with the magnetic field.

**83** • The magnetic field in Problem 82 is steadily reduced to zero in 1.2 s. Find the emf induced in the coil when (*a*) the magnetic field is perpendicular to the coil and (*b*) the magnetic field makes an angle of 20° with the normal to the coil.

**84** • A 100-turn coil has a radius of 4 cm and a resistance of 25 $\Omega$. At what rate must a perpendicular magnetic field change to produce a current of 4 A in the coil?

**85** •• **SSM** **ISOLVE** ✔ Figure 28-58 shows an ac generator. The generator consists of a rectangular loop of dimensions *a* and *b* with *N* turns connected to slip rings. The loop rotates with an angular velocity $\omega$ in a uniform magnetic field $\vec{B}$. (*a*) Show that the potential difference between the two slip rings is $\mathcal{E} = NBab\omega \sin \omega t$. (*b*) If *a* = 1 cm, *b* = 2 cm, *N* = 1000, and *B* = 2 T, at what angular frequency $\omega$ must the coil rotate to generate an emf whose maximum value is 110 V?

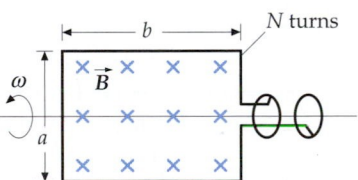

**FIGURE 28-58** Problems 85 and 86

**86** •• Prior to 1960, magnetic field strength was measured by means of a rotating coil gaussmeter. This device used a small loop of many turns rotating on an axis perpendicular to the magnetic field at fairly high speed, which was connected to an ac voltmeter by means of slip rings, like those shown in Figure 28-58. The sensing coil for a rotating coil gaussmeter has 400 turns and an area of 1.4 cm². The coil rotates at 180 rpm. If the magnetic field strength is 0.45 T, find the maximum induced emf in the coil and the orientation of the coil relative to the field for which this maximum induced emf occurs.

**87** •• **iSOLVE**✓ Show that the effective inductance for two inductors $L_1$ and $L_2$ connected in series, so that none of the flux from either passes through the other, is given by $L_{eff} = L_1 + L_2$.

**88** •• **SSM** Show that the effective inductance for two inductors $L_1$ and $L_2$ connected in parallel, so that none of the flux from either passes through the other, is given by

$$\frac{1}{L_{eff}} = \frac{1}{L_1} + \frac{1}{L_2}$$

**89** •• **SSM** Figure 28-59(a) shows an experiment designed to measure the acceleration of gravity. A large plastic tube is encircled by a wire, which is arranged in single loops separated by a distance of 10 cm. A strong magnet is dropped through the top of the loop. As the magnet falls through each loop the voltage rises and then the voltage rapidly falls through 0 to a large negative value as the magnet passes through the loop and then returns to 0. The shape of the voltage signal is shown in Figure 28-59(b). (a) Explain how this experiment works. (b) Explain why the tube cannot be made of a conductive material. (c) Qualitatively explain the shape of the voltage signal in Figure 28-59(b). (d) The times at which the voltage crosses 0 as the magnet falls through each loop in succession are given in the table in the next column. Use these data to calculate a value for $g$.

| Loop Number | Zero Crossing Time(s) |
|---|---|
| 1 | 0.011189 |
| 2 | 0.063133 |
| 3 | 0.10874 |
| 4 | 0.14703 |
| 5 | 0.18052 |
| 6 | 0.21025 |
| 7 | 0.23851 |
| 8 | 0.26363 |
| 9 | 0.28853 |
| 10 | 0.31144 |
| 11 | 0.33494 |
| 12 | 0.35476 |
| 13 | 0.37592 |
| 14 | 0.39107 |

**90** •• The rectangular coil shown in Figure 28-60 has 80 turns, is 25 cm wide, is 30 cm long, and is located in a magnetic field $B = 1.4$ T directed out of the page, as shown, with only half of the coil in the region of the magnetic field. The resistance of the coil is 24 $\Omega$. Find the magnitude and the direction of the induced current if the coil is moved with a speed of 2 m/s (a) to the right, (b) up, (c) to the left, and (d) down.

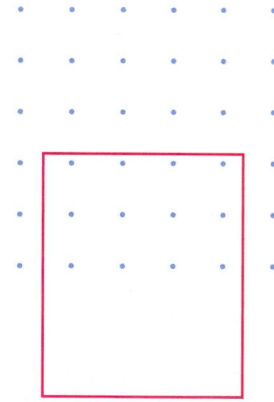

**FIGURE 28-60** Problem 90

**91** •• **SSM** Suppose the coil of Problem 90 is rotated about its vertical centerline at constant angular velocity of 2 rad/s. Find the induced current as a function of time.

**92** •• Show that if the flux through each turn of an $N$-turn coil of resistance $R$ changes from $\phi_{m1}$ to $\phi_{m2}$, the total charge passing through the coil is given by $Q = N(\phi_{m1} - \phi_{m2})/R$.

**93** •• A long solenoid has $n$ turns per unit length and carries a current given by $I = I_0 \sin \omega t$. The solenoid has a circular cross section of radius $R$. Find the induced electric field at a radius $r$ from the axis of the solenoid for (a) $r < R$ and (b) $r > R$.

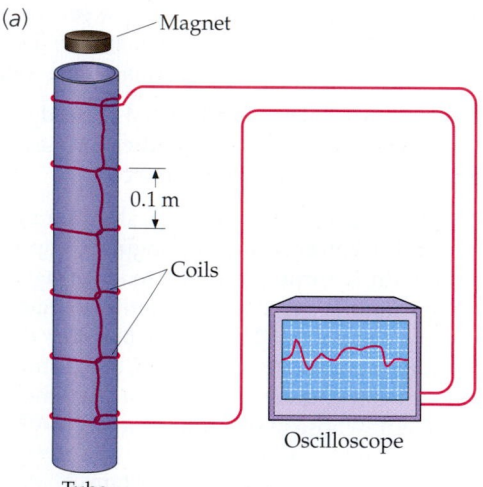

(a)

Magnet

0.1 m

Coils

Oscilloscope

Tube

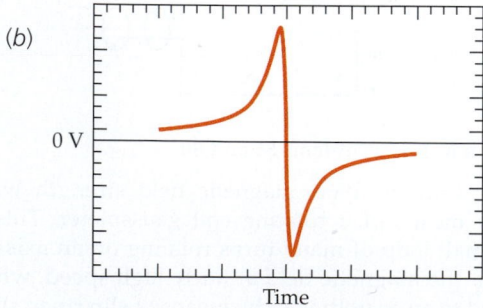

(b)

0 V

Time

**FIGURE 28-59** Problem 89

**94 •••** A coaxial cable consists of two very thin-walled conducting cylinders of radii $r_1$ and $r_2$ (Figure 28-61). Current $I$ goes in one direction down the inner cylinder and in the opposite direction in the outer cylinder. (a) Use Ampère's law to find $B$. Show that $B = 0$, except in the region between the conductors. (b) Show that the magnetic energy density in the region between the cylinders is

$$u_m = \frac{\mu_0 I^2}{8\pi^2 r^2}$$

(c) Find the magnetic energy in a cylindrical shell volume element of length $\ell$ and volume $dV = \ell 2\pi r\, dr$, and integrate your result to show that the total magnetic energy in the volume of length $\ell$ is

$$U_m = \frac{\mu_0}{4\pi} I^2 \ell \ln \frac{r_2}{r_1}$$

(d) Use the result in Part (c) and $U_m = \frac{1}{2}LI^2$ to show that the self-inductance per unit length is

$$\frac{L}{\ell} = \frac{\mu_0}{2\pi} \ln \frac{r_2}{r_1}$$

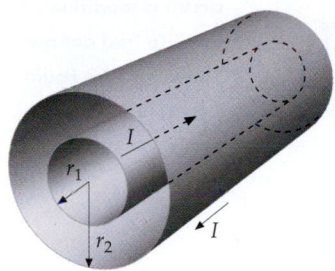

FIGURE 28-61 Problems 94 and 95

**95 •••** Using Figure 28-61, compute the flux through a rectangular area of sides $\ell$ and $r_2 - r_1$ between the conductors. Show that the self-inductance per unit length can be found from $\phi_m = LI$ (see Part (d) of Problem 94).

**96 •••** SSM Figure 28-62 shows a rectangular loop of wire, 0.30 m wide and 1.50 m long, in the vertical plane and perpendicular to a uniform magnetic field $B = 0.40$ T, directed inward as shown. The portion of the loop not in the magnetic field is 0.10 m long. The resistance of the loop is 0.20 $\Omega$ and its mass is 0.05 kg. The loop is released from rest at $t = 0$. (a) What is the magnitude and direction of the induced current when the loop has a downward velocity $v$? (b) What is the force that acts on the loop as a result of this current? (c) What is the net force acting on the loop? (d) Write the equation of motion of the loop. (e) Obtain an expression for the velocity of the loop as a function of time. (f) Integrate the

expression obtained in Part (e) to find the displacement $y$ as a function of time. (g) Using a spreadsheet program, make a graph of the position $y$ of the loop as a function of time for values of $y$ between 0 m and 1.4 m (i.e., when the loop leaves the magnetic field). At what time $t$ does $y = 1.4$ m? Compare this to the time it would have taken if $B = 0$.

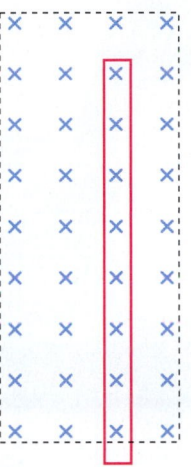

FIGURE 28-62 Problems 96 and 97

**97 •••** The loop of Problem 96 is attached to a plastic spring of spring constant $\kappa$ (see Figure 28-62). (a) When $B = 0$, the period of small-amplitude vertical oscillations of the mass–spring system is 0.8 s. Find the spring constant $\kappa$. (b) When $B \neq 0$, a current is induced in the loop as a result of its up and down motion. Obtain an expression for the induced current as a function of time when $B = 0.40$ T, and the displacement of the center of the loop is $y = 0.05$ m downward. (c) Show that the force on the loop is of the form $-\beta v$, where $v = dy/dt$, and find an expression for $\beta$ in terms of $B$, $w$, and $R$, where $w$ is the width of the wire loop and $R$ is its resistance. (d) Using a spreadsheet program, make graphs of the position $y$ and the velocity $v$ of the center of the loop as a function of time, use the parameters given.

**98 •••** A coil of $N$ turns and area $A$ hangs from a wire that provides a linear restoring torque with torsion constant $\kappa$. The two ends of the coil are connected to each other, the coil has resistance $R$, and the moment of inertia of the coils is $I$. The plane of the coil is vertical, and parallel to a uniform horizontal magnetic field $B$ when the wire is not twisted (i.e., $\theta = 0$). The coil is twisted and released from a small angle $\theta = \theta_0$. Show that the orientation of the coil will undergo damped harmonic oscillation according to $\theta(t) = \theta_0 e^{-\beta t} \cos \omega t$, where

$$\omega = \sqrt{\kappa/I} \quad \text{and} \quad \beta = \frac{N^2 B^2 A^2}{RI}.$$

# Alternating Current Circuits

THIS HIP-LOOKING LISTENER DIALS IN HER FAVORITE RADIO STATION. THIS CHANGES THE RESONANT FREQUENCY OF AN OSCILLATING ELECTRIC CIRCUIT WITHIN THE TUNER, SO ONLY THE STATION SHE SELECTS IS AMPLIFIED.

**?** **What component of the circuit is modified as she turns the dial? To find out more about the workings of a radio turner, see Example 29-9.**

29-1    Alternating Current Generators

29-2    Alternating Current in a Resistor

29-3    Alternating Current Circuits

*29-4    Phasors

*29-5    *LC* and *RLC* Circuits Without a Generator

*29-6    Driven *RLC* Circuits

*29-7    The Transformer

**M**ore than 99 percent of the electrical energy used today is produced by electrical generators in the form of alternating current, which has a great advantage over direct current, because electrical energy can be transported over long distances at very high voltages and low currents to reduce energy losses due to Joule heating. Electrical energy can then be transformed, with almost no energy loss, to lower and safer voltages and correspondingly higher currents for everyday use. The transformer that accomplishes these changes in potential difference and current works on the basis of magnetic induction. In North America, power is delivered by a sinusoidal current of frequency 60 Hz. Devices such as radios, television sets, and microwave ovens detect or generate alternating currents of much higher frequencies.

Alternating current is produced by motional emf or magnetic induction in an ac generator, which is designed to provide a sinusoidal emf.

➤ **In this chapter, we will see that when the generator output is sinusoidal, the current in an inductor, a capacitor, or a resistor is also sinusoidal, although it is generally not in phase with the generator's emf. When the emf and current are both sinusoidal, their maximum values are related. The study of sinusoidal** currents is particularly important because even currents that are not sinusoidal can be analyzed in terms of sinusoidal components using Fourier analysis.

# 29-1 Alternating Current Generators

Figure 29-1 shows a simple **ac generator** that consists of a coil of area $A$ and $N$ turns rotating in a uniform magnetic field. The ends of the coil are connected to rings, called slip rings, that rotate with the coil. They make electrical contact through stationary conducting brushes that are in contact with the rings.

When the normal to the plane of the coil makes an angle $\theta$ with a uniform magnetic field $\vec{B}$, as shown in the figure, the magnetic flux through the coil is

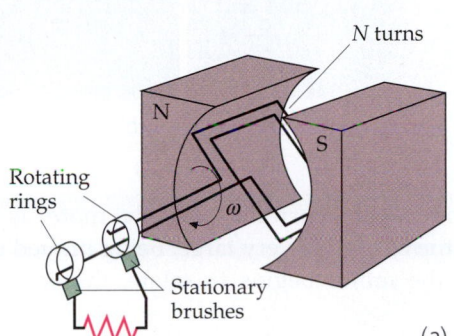
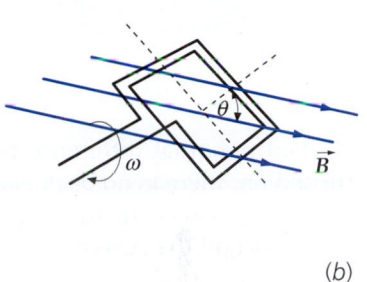

$$\phi_\mathrm{m} = NBA \cos \theta \qquad \text{29-1}$$

where $A$ is the area of the flat surface bounded by a single turn of the coil and $N$ is the number of turns. When the coil is mechanically rotated, the flux through the coil will change, and an emf will be induced. If $\omega$ is the angular velocity of rotation and the initial angle is $\delta$, the angle at some later time $t$ is given by

$$\theta = \omega t + \delta$$

Then

$$\phi_\mathrm{m} = NBA \cos(\omega t + \delta) = NBA \cos(2\pi f t + \delta)$$

The emf in the coil will then be

$$\mathcal{E} = -\frac{d\phi_\mathrm{m}}{dt} = -NBA\frac{d}{dt}\cos(\omega t + \delta) = +NBA\omega \sin(\omega t + \delta) \qquad \text{29-2}$$

where $NBA\omega$ is the peak (maximum) emf. Thus,

$$\mathcal{E} = \mathcal{E}_\mathrm{peak} \sin(\omega t + \delta) \qquad \text{29-3}$$

where the emf amplitude is given by

$$\mathcal{E}_\mathrm{peak} = NBA\omega \qquad \text{29-4}$$

We can thus produce a sinusoidal emf in a coil by rotating the coil with constant angular velocity in a magnetic field. Although practical generators are considerably more complicated, they produce a sinusoidal emf either via induction or via motional emf. In circuit diagrams, an ac generator is represented by the symbol ⊝.

The same coil in a static magnetic field that can be used to generate an alternating emf can also be used as an **ac motor**. Instead of mechanically rotating the coil to generate an emf, we apply an ac potential difference generated by another ac generator to the coil. This produces an ac current in the coil, and the magnetic field exerts forces on the wires producing a torque that rotates the coil. As the coil rotates in the magnetic field, a back emf is generated that tends to counter the

**FIGURE 29-1** (*a*) An ac generator. A coil rotating with constant angular frequency $\omega$ in a static magnetic field $\vec{B}$ generates a sinusoidal emf. Energy from a waterfall or a steam turbine is used to rotate the coil to produce electrical energy. The emf is supplied to an external circuit by the brushes that are in contact with the rings. (*b*) At this instant, the normal to the plane of the coil makes an angle $\theta$ with the magnetic field, and the flux through the flat surface bounded by the coil is $BA \cos \theta$.

(a)

(b)

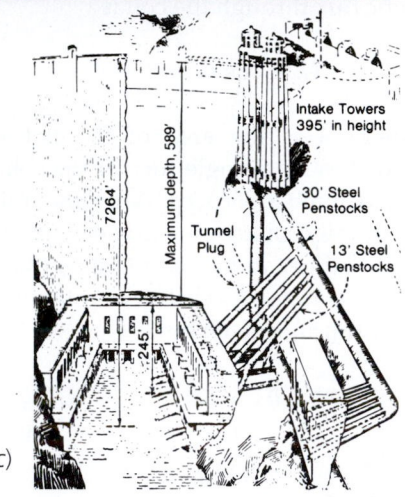

(c)

applied potential difference that produces the current. When the motor is first turned on, there is no back emf and the current is very large, being limited only by the resistance in the circuit. As the motor begins to rotate, the back emf increases and the current decreases.

**EXERCISE** A 250-turn coil has an area per turn of 3 cm². If it rotates in a magnetic field of 0.4 T at 60 Hz, what is $\mathcal{E}_{peak}$? (*Answer* $\mathcal{E}_{peak} = 11.3$ V)

# 29-2 Alternating Current in a Resistor

Figure 29-2 shows a simple ac circuit that consists of an ideal generator and a resistor. (A generator is ideal if its internal resistance, self-inductance, and capacitance are negligible.) The voltage drop across the resistor $V_R$ is equal to the emf $\mathcal{E}$ of the generator. If the generator produces an emf given by Equation 29-3, we have

$$V_R = \mathcal{E} = \mathcal{E}_{peak} \sin(\omega t + \delta) = V_{R,peak} \sin(\omega t + \delta)$$

where $V_{R,peak} = \mathcal{E}_{peak}$. In this equation, the phase constant $\delta$ is arbitrary. It is convenient to choose $\delta = \pi/2$ so that

$$V_R = V_{R,peak} \sin\left(\omega t + \frac{\pi}{2}\right) = V_{R,peak} \cos \omega t$$

Applying Ohm's law, we have

$$V_R = IR \qquad\qquad 29\text{-}5$$

Thus,

$$V_{R,peak} \cos \omega t = IR \qquad\qquad 29\text{-}6$$

so the current in the resistor is

$$I = \frac{V_{R,peak}}{R} \cos \omega t = I_{peak} \cos \omega t \qquad\qquad 29\text{-}7$$

(a) The mechanical energy of falling water drives turbines (b) for the generation of electricity. (c) Schematic drawing of the Hoover Dam showing the intake towers and pipes (penstocks) that carry the water to the generators below.

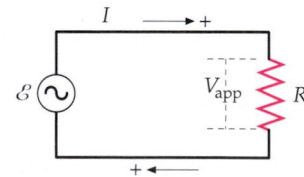

**FIGURE 29-2** An ac generator in series with a resistor $R$.

where

$$I_{peak} = \frac{V_{R,peak}}{R} \qquad \text{29-8}$$

Note that the current through the resistor is in phase with the potential drop across the resistor, as shown in Figure 29-3.

The power dissipated in the resistor varies with time. Its instantaneous value is

$$P = I^2 R = (I_{peak} \cos \omega t)^2 R = I_{peak}^2 R \cos^2 \omega t \qquad \text{29-9}$$

Figure 29-4 shows the power as a function of time. The power varies from zero to its peak value $I_{peak}^2 R$, as shown. We are usually interested in the average power over one or more complete cycles:

$$P_{av} = (I^2 R)_{av} = I_{peak}^2 R (\cos^2 \omega t)_{av}$$

The average value of $\cos^2 \omega t$ over one or more periods is $\frac{1}{2}$. This can be seen from the identity $\cos^2 \omega t + \sin^2 \omega t = 1$. A plot of $\sin^2 \omega t$ looks the same as a plot of $\cos^2 \omega t$ except that the plot is shifted by 90°. Both have the same average value over one or more periods, and since their sum is 1, the average value of each must be $\frac{1}{2}$. The average power dissipated in the resistor is thus

$$P_{av} = (I^2 R)_{av} = \frac{1}{2} I_{peak}^2 R \qquad \text{29-10}$$

### Root-Mean-Square Values

Most ac ammeters and voltmeters are designed to measure the **root-mean-square (rms) values** of current and potential difference rather than the peak values. The rms value of a current $I_{rms}$ is defined by

$$I_{rms} = \sqrt{(I^2)_{av}} \qquad \text{29-11}$$

DEFINITION—RMS CURRENT

For a sinusoidal current, the average value of $I^2$ is

$$(I^2)_{av} = [(I_{peak} \cos \omega t)^2]_{av} = \frac{1}{2} I_{peak}^2$$

Substituting $\frac{1}{2} I_{peak}^2$ for $(I^2)_{av}$ in Equation 29-11, we obtain

$$I_{rms} = \frac{1}{\sqrt{2}} I_{peak} \approx 0.707 I_{peak} \qquad \text{29-12}$$

RMS VALUE RELATED TO PEAK VALUE

The rms value of any quantity that varies sinusoidally equals the peak value of that quantity divided by $\sqrt{2}$.

Substituting $I_{rms}^2$ for $\frac{1}{2} I_{peak}^2$ in Equation 29-10, we obtain for the average power dissipated in the resistor

$$P_{av} = I_{rms}^2 R \qquad \text{29-13}$$

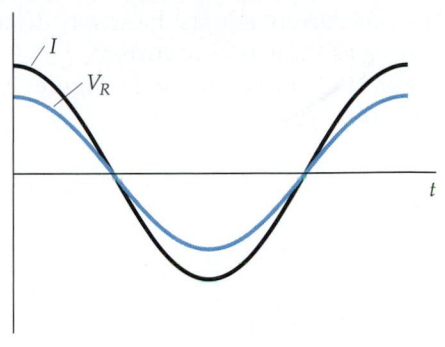

**FIGURE 29-3** The voltage drop across a resistor is in phase with the current.

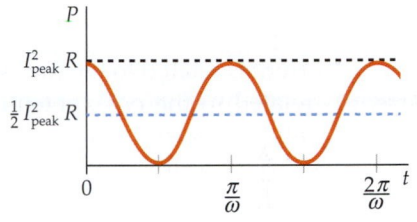

**FIGURE 29-4** Plot of the power dissipated in the resistor shown in Figure 29-2 versus time. The power varies from zero to a peak value $I_{peak}^2 R$. The average power is half the peak power.

The rms current equals the steady dc current that would produce the same Joule heating as the actual ac current.

For the simple circuit in Figure 29-2, the average power delivered by the generator is:

$$P_{av} = (\mathcal{E}I)_{av} = [(\mathcal{E}_{peak} \cos \omega t)(I_{peak} \cos \omega t)]_{av} = \mathcal{E}_{peak} I_{peak} (\cos^2 \omega t)_{av}$$

or

$$P_{av} = \tfrac{1}{2}\mathcal{E}_{peak} I_{peak}$$

Using $I_{rms} = I_{peak}/\sqrt{2}$ and $\mathcal{E}_{rms} = \mathcal{E}_{peak}/\sqrt{2}$, this can be written

$$P_{av} = \mathcal{E}_{rms} I_{rms} \qquad\qquad 29\text{-}14$$

AVERAGE POWER DELIVERED BY A GENERATOR

The rms current is related to the rms potential drop in the same way that the peak current is related to the peak potential drop. We can see this by dividing each side of Equation 29-8 by $\sqrt{2}$ and using $I_{rms} = I_{peak}/\sqrt{2}$ and $V_{R,rms} = V_{R,peak}/\sqrt{2}$.

$$I_{rms} = \frac{V_{R,rms}}{R} \qquad\qquad 29\text{-}15$$

Equations 29-13, 29-14, and 29-15 are of the same form as the corresponding equations for direct-current circuits with $I$ replaced by $I_{rms}$ and $V_R$ replaced by $V_{R,rms}$. We can therefore calculate the power input and the heat generated using the same equations that we used for direct current, if we use rms values for the current and potential drop.

**EXERCISE** The sinusoidal potential drop across a 12-$\Omega$ resistor has a peak value of 48 V. Find (a) the rms current, (b) the average power, and (c) the maximum power. (*Answer* (a) 2.83 A, (b) 96 W, (c) 192 W)

The ac power supplied to domestic wall outlets and light fixtures in the United States has an rms potential difference of 120 V at a frequency of 60 Hz. This potential difference is maintained, independent of the current. If you plug a 1600-W space heater into a wall outlet it will draw a current of

$$I_{rms} = \frac{P_{av}}{V_{rms}} = \frac{1600 \text{ W}}{120 \text{ V}} = 13.3 \text{ A}$$

All appliances plugged into the outlets of a single 120-V circuit are connected in parallel. If you plug a 500-W toaster into another outlet of the same circuit, it will draw a current of 500 W/120 V = 4.17 A, and the total current through the parallel combination will be 17.5 A. Typical household wall outlets are rated at 15 A and are part of a circuit using wires rated at either 15 A or 20 A, with each circuit having several outlets. The wire in each circuit is rated at 15 A or 20 A, correspondingly. A total current greater than the rated current for the wiring is likely to overheat the wiring and is a fire hazard. Each circuit is therefore equipped with a circuit breaker (or a fuse in older houses) that trips (or blows) when the total current exceeds the 15-A or 20-A rating.

High-power domestic appliances, such as electric clothes dryers, kitchen ranges, and hot water heaters, typically require power delivered at 240-V rms. For a given power requirement, only half as much current is required at 240 V as at 120 V, but 240 V is more likely to deliver a fatal shock or to start a fire than 120 V.

*SAWTOOTH WAVEFORM*    **EXAMPLE   29-1**

Find (*a*) the average current and (*b*) the rms current for the sawtooth waveform shown in Figure 29-5. In the region $0 < t < T$, the current is given by $I = (I_0/T)t$.

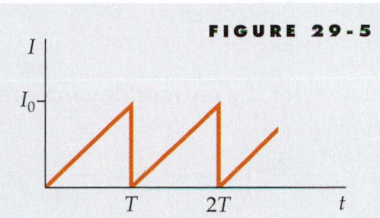

**FIGURE 29-5**

**PICTURE THE PROBLEM**   The average of any quantity over a time interval $T$ is the integral of the quantity over the interval divided by $T$. We use this to find both the average current, $I_{av}$, and the average of the current squared, $(I^2)_{av}$.

(*a*) Calculate $I_{av}$ by integrating $I$ from $t = 0$ to $t = T$ and dividing by $T$:

$$I_{av} = \frac{1}{T}\int_0^T I\,dt = \frac{1}{T}\int_0^T \frac{I_0}{T}t\,dt = \frac{I_0}{T^2}\frac{T^2}{2} = \frac{1}{2}I_0$$

(*b*) 1. Find $(I^2)_{av}$ by integrating $I^2$:

$$(I^2)_{av} = \frac{1}{T}\int_0^T I^2\,dt = \frac{1}{T}\left(\frac{I_0}{T}\right)^2\int_0^T t^2\,dt = \frac{I_0^2}{T^3}\frac{T^3}{3} = \frac{1}{3}I_0^2$$

2. The rms current is the square root of $(I^2)_{av}$:

$$I_{rms} = \sqrt{(I^2)_{av}} = \boxed{\frac{I_0}{\sqrt{3}}}$$

# 29-3  Alternating Current Circuits

Alternating current behaves differently than direct current in inductors and capacitors. When a capacitor becomes fully charged in a dc circuit, the capacitor blocks the current; that is, the capacitor acts like an open circuit. However, if the current alternates, charge continually flows onto the plates or off the plates of the capacitor. We will see that at high frequencies, a capacitor hardly impedes the current at all. That is, the capacitor acts like a short circuit. Conversely, an induction coil usually has a low resistance and is essentially a short circuit for direct current; however, when the current is changing, a back emf is generated in an inductor that is proportional to $dI/dt$. At high frequencies, the back emf is large and the inductor acts like an open circuit.

## Inductors in Alternating Current Circuits

Figure 29-6 shows an inductor coil in series with an ac generator. When the current changes in the inductor, a back emf of magnitude $L\,dI/dt$ is generated due to the changing flux. Usually this back emf is much greater than the $IR$ drop due to the resistance of the coil, so we normally neglect the resistance of the coil. The potential drop across the inductor $V_L$ is then given by

$$V_L = L\frac{dI}{dt} \tag{29-16}$$

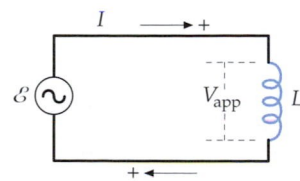

**FIGURE 29-6**  An ac generator in series with an inductor $L$. The arrow indicates the positive direction along the wire. Note that for a positive value of $dI/dt$, the voltage drop $V_L$ across the inductor is positive. That is, if you traverse the inductor in the direction of the direction arrow you go in the direction of decreasing potential.

In this circuit, the potential drop $V_L$ across the inductor equals the emf $\mathcal{E}$ of the generator. That is,

$$V_L = \mathcal{E} = \mathcal{E}_{max}\cos\omega t = V_{L,peak}\cos\omega t$$

where $V_{L,peak} = \mathcal{E}_{peak}$. Substituting for $V_L$ in Equation 29-16 gives

$$V_{L,peak}\cos\omega t = L\frac{dI}{dt} \tag{29-17}$$

Rearranging, we obtain

$$dI = \frac{V_{L,\text{peak}}}{L} \cos \omega t \, dt \qquad\qquad 29\text{-}18$$

We solve for the current $I$ by integrating both sides of the equation:

$$I = \frac{V_{L,\text{peak}}}{L} \int \cos \omega t \, dt = \frac{V_{L,\text{peak}}}{\omega L} \sin \omega t + C \qquad\qquad 29\text{-}19$$

where the constant of integration $C$ is the dc component of the current. Setting the dc component of the current to be zero, we have

$$I = \frac{V_{L,\text{peak}}}{\omega L} \sin \omega t = I_{\text{peak}} \sin \omega t \qquad\qquad 29\text{-}20$$

where

$$I_{\text{peak}} = \frac{V_{L,\text{peak}}}{\omega L} \qquad\qquad 29\text{-}21$$

The potential drop $V_L = V_{L,\text{peak}} \cos \omega t$ across the inductor is 90° out of phase with the current $I = I_{\text{peak}} \sin \omega t$. From Figure 29-7, which shows $I$ and $V_L$ as functions of time, we can see that the peak value of the potential drop occurs 90° or one-fourth period prior to the corresponding peak value of the current. The potential drop across an inductor is said to *lead the current by 90°*. We can understand this physically. When $I$ is zero but increasing, $dI/dt$ is maximum, so the back emf induced in the inductor is at its maximum. One-quarter cycle later, $I$ is maximum. At this time, $dI/dt$ is zero, so $V_L$ is zero. Using the trigonometric identity $\sin \theta = \cos\left(\theta - \dfrac{\pi}{2}\right)$, where $\theta = \omega t$, Equation 29-20 for the current can be written

$$I = I_{\text{peak}} \cos\left(\omega t - \frac{\pi}{2}\right) \qquad\qquad 29\text{-}22$$

The relation between the peak current and the peak potential drop (or between the rms current and rms potential drop) for an inductor can be written in a form similar to Equation 29-15 for a resistor. From Equation 29-21, we have

$$I_{\text{peak}} = \frac{V_{L,\text{peak}}}{\omega L} = \frac{V_{L,\text{peak}}}{X_L} \qquad\qquad 29\text{-}23$$

where

$$X_L = \omega L \qquad\qquad 29\text{-}24$$

DEFINITION—INDUCTIVE REACTANCE

is called the **inductive reactance**. Since $I_{\text{rms}} = I_{\text{peak}}/\sqrt{2}$ and $V_{L,\text{rms}} = V_{L,\text{peak}}/\sqrt{2}$ the rms current is given by

$$I_{\text{rms}} = \frac{V_{L,\text{rms}}}{X_L} \qquad\qquad 29\text{-}25$$

Like resistance, inductive reactance has units of ohms. As we can see from Equation 29-25, the larger the reactance for a given potential drop, the smaller

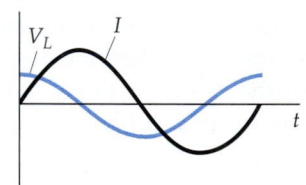

**FIGURE 29-7** Current and potential drop across the inductor shown in Figure 29-6 as functions of time. The maximum potential drop occurs one-fourth period before the maximum current. Thus, the potential drop is said to lead the current by one-fourth period or 90°.

the peak current. Unlike resistance, the inductive reactance depends on the frequency of the current—the greater the frequency, the greater the reactance.

The *instantaneous* power delivered to the inductor from the generator is

$$P = V_L I = (V_{L,\text{peak}} \cos \omega t)(I_{\text{peak}} \sin \omega t) = V_{L,\text{peak}} I_{\text{peak}} \cos \omega t \sin \omega t$$

The *average* power delivered to the inductor is zero. We can see this by using the trigonometric identity

$$2 \cos \omega t \sin \omega t = \sin 2\omega t$$

The value of $\sin 2\omega t$ oscillates twice during each cycle and is negative as often as it is positive. Thus, on the average, no energy is dissipated in an inductor. (This is the case only if the resistance of the inductor is negligible.)

---

*INDUCTIVE REACTANCE*  **EXAMPLE 29-2**

**The potential drop across a 40-mH inductor is sinusoidal with a peak potential drop of 120 V. Find the inductive reactance and the peak current when the frequency is (a) 60 Hz and (b) 2000 Hz.**

**PICTURE THE PROBLEM** We calculate the inductive reactance at each frequency and use Equation 29-23 to find the peak current.

(a) 1. The peak current equals the peak potential drop divided by the inductive reactance. The peak potential drop equals the emf:

$$I_{\text{peak}} = \frac{V_{L,\text{peak}}}{X_L}$$

2. Compute the inductive reactance at 60 Hz:

$$X_{L1} = \omega_1 L = 2\pi f_1 L = (2\pi)(60 \text{ Hz})(40 \times 10^{-3} \text{ H})$$

$$= \boxed{15.1 \ \Omega}$$

3. Use this value of $X_L$ to compute the peak current at 60 Hz:

$$I_{1,\text{peak}} = \frac{120 \text{ V}}{15.1 \ \Omega} = \boxed{7.95 \text{ A}}$$

(b) 1. Compute the inductive reactance at 2000 Hz:

$$X_{L2} = \omega_2 L = 2\pi f_2 L$$

$$= (2\pi)(2000 \text{ Hz})(40 \times 10^{-3} \text{ H}) = \boxed{503 \ \Omega}$$

2. Use this value of $X_L$ to compute the peak current at 2000 Hz:

$$I_{2,\text{peak}} = \frac{120 \text{ V}}{503 \ \Omega} = \boxed{0.239 \text{ A}}$$

---

## Capacitors in Alternating Current Circuits

When a capacitor is connected across the terminals of an ac generator (Figure 29-8), the voltage drop across the capacitor is

$$V_C = \frac{Q}{C} \qquad\qquad 29\text{-}26$$

where $Q$ is the charge on the upper plate of the capacitor.

In this circuit, the potential drop $V_C$ across the capacitor equals the emf $\mathcal{E}$ of the generator. That is,

$$V_C = \mathcal{E} = \mathcal{E}_{\text{peak}} \cos \omega t = V_{C,\text{peak}} \cos \omega t$$

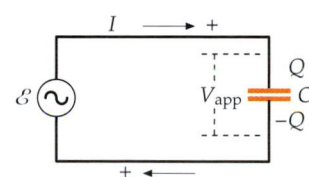

**FIGURE 29-8** An ac generator in series with a capacitor $C$. The positive direction along the circuit is such that when the current is positive the charge $Q$ on the upper capacitor plate is increasing, so the current is related to the charge by $I = +dQ/dt$.

where $V_{C,peak} = \mathcal{E}_{peak}$. Substituting for $V_C$ in Equation 29-26 and solving for $Q$ gives

$$Q = V_C C = V_{C,peak} C \cos \omega t$$

The current is

$$I = \frac{dQ}{dt} = -\omega V_{C,peak} C \sin \omega t = -I_{peak} \sin \omega t$$

where

$$I_{peak} = \omega V_{C,peak} C \qquad\qquad 29\text{-}27$$

Using the trigonometric identity $\sin \theta = -\cos\left(\theta + \frac{\pi}{2}\right)$, where $\theta = \omega t$, we obtain

$$I = -\omega C V_{C,peak} \sin \omega t = I_{peak} \cos\left(\omega t + \frac{\pi}{2}\right) \qquad\qquad 29\text{-}28$$

As with the inductor, the voltage drop $V_C = V_{C,peak} \cos \omega t$ across the capacitor is out of phase with the current

$$I = I_{peak} \cos\left(\omega t + \frac{\pi}{2}\right)$$

in the circuit. From Figure 29-9, we see that the maximum value of the potential drop occurs 90° or one-fourth period *after* the maximum value of the current. Thus, *the potential drop across a capacitor lags the current by 90°.* Again, we can understand this physically. The charge $Q$ is proportional to the potential drop $V_C$. The maximum value of $dQ/dt = I$ occurs when the charge $Q$, and therefore when $V_C$, is zero. As the charge on the capacitor plate increases the current decreases until, one-fourth period later, the charge $Q$, and therefore $V_C$, is a maximum and the current is zero. The current then becomes negative as the charge $Q$ decreases.

Again, we can relate the current to the potential drop in a form similar to Equation 29-8 for a resistor. From Equation 29-27, we have

$$I_{peak} = \omega C V_{C,peak} = \frac{V_{C,peak}}{1/(\omega C)} = \frac{V_{C,peak}}{X_C}$$

and, similarly,

$$I_{rms} = \frac{V_{C,rms}}{X_C} \qquad\qquad 29\text{-}29$$

where

$$X_C = \frac{1}{\omega C} \qquad\qquad 29\text{-}30$$

DEFINITION—CAPACITIVE REACTANCE

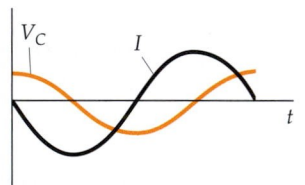

**FIGURE 29-9** Current and potential drop across the capacitor shown in Figure 29-8 versus time. The maximum potential drop occurs one-fourth period after the maximum current. Thus, the potential drop is said to lag the current by 90°.

is called the **capacitive reactance** of the circuit. Like resistance and inductive reactance, capacitive reactance has units of ohms and, like inductive reactance, capacitive reactance depends on the frequency of the current. In this case, the greater the frequency, the smaller the reactance. The average power delivered to a capacitor from an ac generator is zero, as it is for an inductor. This occurs because the potential drop is proportional to $\cos \omega t$ and the current is proportional to $\sin \omega t$ and $(\cos \omega t \sin \omega t)_{av} = 0$. Thus, like inductors with no resistance, capacitors dissipate no energy.

Since charge cannot pass across the space between the plates of a capacitor, it may seem strange that there is a continuing alternating current in the circuit shown in Figure 29-8. Suppose we choose the time to be zero at the instant that the voltage drop $V_C$ across the capacitor is both zero and increasing. (At this same instant, the charge $Q$ on the upper plate of the capacitor is also both zero and increasing.) As $V_C$ then increases, positive charge flows off the lower plate and onto the upper plate, and $Q$ reaches its maximum value $Q_{peak}$ a quarter period later. After $Q$ reaches its maximum value $Q$ continues to change, reaching zero at the half-period point, $-Q_{peak}$ at the three-quarter-period point, and zero (again) at the completion of the cycle at the full-period point. The charge $Q_{peak}$ flows through the generator each quarter period. If we double the frequency, we halve the period. Thus, if we double the frequency we halve the time for the charge $Q_{peak}$ to flow through the generator, so we have doubled the current amplitude $I_{peak}$. Hence, the greater the frequency, the less the capacitor impedes the flow of charge.

---

| *CAPACITIVE REACTANCE* | **EXAMPLE 29-3** |
|---|---|

**A 20-$\mu$F capacitor is placed across an ac generator that applies a potential drop with an amplitude (peak value) of 100 V. Find the capacitive reactance and the current amplitude when the frequency is 60 Hz and when the frequency is 6000 Hz.**

**PICTURE THE PROBLEM** The capacitive reactance is $X_C = 1/(\omega C)$ and the peak current is $I_{peak} = V_{C,peak}/X_C$.

1. Calculate the capacitive reactance at 60 Hz and at 6000 Hz:

$$X_{C1} = \frac{1}{\omega_1 C} = \frac{1}{2\pi f_1 C} = \frac{1}{2\pi (60\ \text{Hz})(20 \times 10^{-6}\ \text{F})} = \boxed{133\ \Omega}$$

$$X_{C2} = \frac{1}{\omega_2 C} = \frac{1}{2\pi f_2 C} = \frac{1}{2\pi (6000\ \text{Hz})(20 \times 10^{-6}\ \text{F})} = \boxed{1.33\ \Omega}$$

2. Use these values of $X_C$ to find the peak currents:

$$I_{1,peak} = \frac{V_{C,peak}}{X_{C1}} = \frac{100\ \text{V}}{133\ \Omega} = \boxed{0.752\ \text{A}}$$

$$I_{2,peak} = \frac{V_{C,peak}}{X_{C2}} = \frac{100\ \text{V}}{1.33\ \Omega} = \boxed{75.2\ \text{A}}$$

**REMARKS** Note that the capacitive reactance is inversely proportional to the frequency, so increasing the frequency by two orders of magnitude decreases the reactance by two orders of magnitude. The current is directly proportional to the frequency, as expected.

---

## *29-4 Phasors

Until this point, the circuits considered contained an ideal ac generator and only a single passive element (i.e., resistor, inductor, or capacitor). In these circuits, the potential drop across the passive element equaled the emf of the generator. In circuits that contain an ideal ac generator and two or more additional elements connected in series, the sum of the potential drops across the elements is equal to the generator emf; which is the same as with dc circuits. However, in ac circuits these potential drops typically are not in phase, so the sum of their rms values does not equal the rms value of the generator emf.

Two-dimensional vectors, which are called phasors, can represent the phase relations between the current and the potential drops across resistors, capacitors, or inductors. In Figure 29-10, the potential drop across a resistor $V_R$ is represented by a vector $\vec{V}_R$ that has magnitude $I_{peak}R$ and makes an angle $\theta$ with the $x$ axis. This potential drop is in phase with the current. In general, the current in a steady-state ac circuit varies with time, as

$$I = I_{peak} \cos \theta = I_{peak} \cos (\omega t - \delta) \qquad \text{29-31}$$

where $\omega$ is the angular frequency and $\delta$ is some phase constant. The potential drop across a resistor is then given by

$$V_R = IR = I_{peak}R \cos (\omega t - \delta) \qquad \text{29-32}$$

The potential drop across a resistor is thus equal to the $x$ component of the phasor vector $\vec{V}_R$, which rotates counterclockwise with an angular frequency $\omega$. The current $I$ may be written as the $x$ component of a phasor $\vec{I}$ having the same direction as $\vec{V}_R$.

When several components are connected together in a series combination, their potential drops add. When several components are connected in parallel, their currents add. Unfortunately, adding sines or cosines of different amplitudes and phases algebraically is awkward. It is much easier to do this by vector addition.[†]

Let us look at how phasors are used. Any ac current or potential drop is written in the form $A \cos(\omega t - \delta)$, which in turn is treated as $A_x$, the $x$ component of a phasor that makes an angle $(\omega t - \delta)$ with the positive $x$ direction. Instead of adding two potential drops or currents algebraically, as $A \cos(\omega t - \delta_1) + B \cos(\omega t - \delta_2)$, we represent these quantities as phasors $\vec{A}$ and $\vec{B}$ and find the phasor sum $\vec{C} = \vec{A} + \vec{B}$ geometrically. The resultant potential drop or current is then the $x$ component of the resultant phasor, $C_x = A_x + B_x$. The geometric representation conveniently shows the relative amplitudes and phases of the phasors.

Consider a circuit that contains an inductor $L$, a capacitor $C$, and a resistor $R$ connected in series. They all carry the same current, which is represented as the $x$ component of the current phasor $\vec{I}$. The potential drop across the inductor $V_L$ is represented by a phasor $\vec{V}_L$ that has magnitude $I_{peak}X_L$ and leads the current phasor $\vec{I}$ by 90°. Similarly, the potential drop across the capacitor $V_C$ is represented by a phasor $\vec{V}_C$ that has magnitude $I_{peak}X_C$ and lags the current by 90°. Figure 29-11 shows the phasors $\vec{V}_R$, $\vec{V}_L$, and $\vec{V}_C$. As time passes, the three phasors rotate counterclockwise with an angular frequency $\omega$, so the relative positions of the vectors do not change. At any time, the instantaneous value of the potential drop across any of these elements equals the $x$ component of the corresponding phasor.

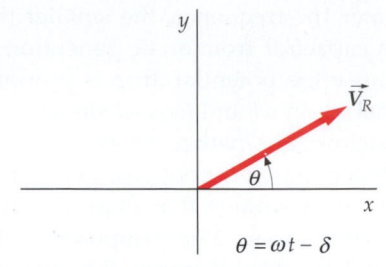

**FIGURE 29-10** The potential drop across a resistor can be represented by a vector $\vec{V}_R$, which is called a phasor, that has magnitude $I_{peak}R$ and makes an angle $\theta = \omega t - \delta$ with the $x$ axis. The phasor rotates with an angular frequency $\omega$. The potential drop $V_R = IR$ is the $x$ component of $\vec{V}_R$.

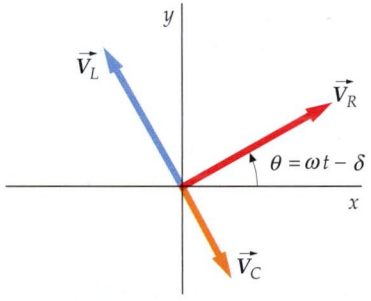

**FIGURE 29-11** Phasor representations of the potential drops $V_R$, $V_L$, and $V_C$. Each vector rotates in the counterclockwise direction with an angular frequency $\omega$. At any instant, the potential drop across an element equals the $x$ component of the corresponding phasor, and the potential drop across the *RLC*-series combination, which equals the sum of the potential drops, equals the $x$ component of the vector sum $\vec{V}_R + \vec{V}_L + \vec{V}_C$.

## *29-5 *LC* and *RLC* Circuits Without a Generator

Figure 29-12 shows a simple circuit with inductance and capacitance but with no resistance. Such a circuit is called an *LC* circuit. We assume that the upper capacitor plate carries an initial positive charge $Q_0$ and that the switch is initially open.

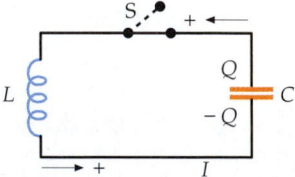

**FIGURE 29-12** An *LC* circuit. When the switch is closed, the initially charged capacitor discharges through the inductor, producing a back emf.

---

[†] It is also easier to do using complex numbers.

After the switch is closed at $t = 0$, the charge begins to flow through the inductor. Let $Q$ be the charge on the upper plate of the capacitor and let the positive direction around the circuit be counterclockwise, as shown. Then,

$$I = -\frac{dQ}{dt}$$

Applying Kirchhoff's loop rule to the circuit, we have

$$-L\frac{dI}{dt} + \frac{Q}{C} = 0 \qquad\qquad 29\text{-}33$$

Substituting $-dQ/dt$ for $I$ and multiplying both sides by $-1$ gives

$$L\frac{d^2Q}{dt^2} + \frac{Q}{C} = 0 \qquad\qquad 29\text{-}34$$

This equation is of the same form as Equation 14-2 for the acceleration of a mass on a spring:

$$m\frac{d^2x}{dt^2} + kx = 0$$

The behavior of an *LC* circuit is thus analogous to that of a mass on a spring, with $L$ analogous to the mass $m$, $Q$ analogous to the position $x$, and $1/C$ analogous to the spring constant $k$. Also, the current $I$ is analogous to the velocity $v$, since $v = dx/dt$ and $I = dQ/dt$. In mechanics, the mass of an object describes the inertia of the object. The greater the mass, the more difficult it is to change the velocity of the object. Similarly, the inductance $L$ can be thought of as the inertia of an ac circuit. The greater the inductance, the more opposition there is to changes in the current $I$.

If we divide each term in Equation 29-34 by $L$ and rearrange, we obtain

$$\frac{d^2Q}{dt^2} = -\frac{1}{LC}Q \qquad\qquad 29\text{-}35$$

which is analogous to

$$\frac{d^2x}{dt^2} = -\frac{k}{m}x \qquad\qquad 29\text{-}36$$

In Chapter 14, we found that we could write the solution of Equation 29-36 for simple harmonic motion in the form

$$x = A\cos(\omega t - \delta)$$

where $\omega = \sqrt{k/m}$ is the angular frequency, $A$ is the displacement amplitude, and $\delta$ is the phase constant, which depends on the initial conditions. The solution to Equation 29-35 is thus

$$Q = A\cos(\omega t - \delta)$$

with

$$\omega = \frac{1}{\sqrt{LC}} \qquad\qquad 29\text{-}37$$

The current $I$ is found by differentiating:

$$I = \frac{dQ}{dt} = -\omega A \sin(\omega t - \delta)$$

If we choose our initial conditions to be $Q = Q_{peak}$ and $I = 0$ at $t = 0$, the phase constant $\delta$ is zero and $A = Q_{peak}$. Our solutions are then

$$Q = Q_{peak} \cos \omega t \qquad\qquad 29\text{-}38$$

and

$$I = -\omega Q_{peak} \sin \omega t = -I_{peak} \sin \omega t \qquad\qquad 29\text{-}39$$

where $I_{peak} = \omega Q_{peak}$.

Figure 29-13 shows graphs of $Q$ and $I$ versus time. The charge oscillates between the values $+Q_{peak}$ and $-Q_{peak}$ with angular frequency $\omega = 1/\sqrt{(LC)}$. The current oscillates between $+\omega Q_{peak}$ and $-\omega Q_{peak}$ with the same frequency. Also, the current leads the charge by 90° (see Problem 29-37). The current is maximum when the charge is zero and the current is zero when the charge is maximum.

In our study of the oscillations of a mass on a spring, we found that the total energy is constant, and that the total energy oscillates between potential energy and kinetic energy. We also have two kinds of energy in the $LC$ circuit, electric energy and magnetic energy. The electric energy stored in the capacitor is

$$U_e = \frac{1}{2}QV_C = \frac{1}{2}\frac{Q^2}{C}$$

Substituting $Q_{peak} \cos \omega t$ for $Q$, we have for the electric energy

$$U_e = \frac{1}{2}\frac{Q_{peak}^2}{C} \cos^2 \omega t \qquad\qquad 29\text{-}40$$

The electric energy oscillates between its maximum value $Q_0^2/(2C)$ and zero at an angular frequency of $2\omega$ (see Problem 29-37). The magnetic energy stored in the inductor is

$$U_m = \frac{1}{2}LI^2 \qquad\qquad 29\text{-}41$$

Substituting $I = -\omega Q_{peak} \sin \omega t$ (Equation 29-39), we get

$$U_m = \frac{1}{2}L\omega^2 Q_{peak}^2 \sin^2 \omega t = \frac{1}{2}\frac{Q_{peak}^2}{C} \sin^2 \omega t \qquad\qquad 29\text{-}42$$

where we have used $\omega^2 = 1/LC$. The magnetic energy also oscillates between its maximum value of $Q_{peak}^2/2C$ and zero at an angular frequency of $2\omega$. The sum of the electrostatic energy and the magnetic energy is the total energy, which is constant in time:

$$U_{total} = U_e + U_m = \frac{1}{2}\frac{Q_{peak}^2}{C} \cos^2 \omega t + \frac{1}{2}\frac{Q_{peak}^2}{C} \sin^2 \omega t = \frac{1}{2}\frac{Q_{peak}^2}{C}$$

This sum equals the energy initially stored on the capacitor.

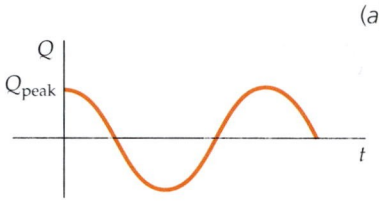

(a)

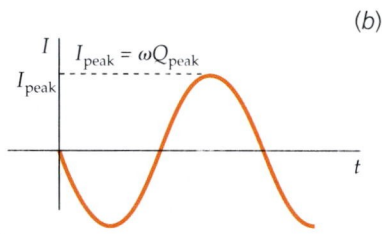

(b)

**FIGURE 29-13** Graphs of (a) $Q$ versus $t$ and (b) $I$ versus $t$ for the $LC$ circuit shown in Figure 29-12.

LC OSCILLATOR **EXAMPLE 29-4**

A 2-$\mu$F capacitor is charged to 20 V and the capacitor is then connected across a 6-$\mu$H inductor. (*a*) What is the frequency of oscillation? (*b*) What is the peak value of the current?

**PICTURE THE PROBLEM** In (*b*), the current is maximum when $dQ/dt$ is maximum, so the current amplitude is $\omega Q_{peak}$. $Q = Q_{peak}$ when $V = V_{peak}$, where $V$ is the voltage across the capacitor.

(*a*) The frequency of oscillation depends only on the values of the capacitance and the inductance:

$$f = \frac{\omega}{2\pi} = \frac{1}{2\pi\sqrt{LC}} = \frac{1}{2\pi\sqrt{(6 \times 10^{-6}\,\text{H})(2 \times 10^{-6}\,\text{F})}}$$

$$= \boxed{4.59 \times 10^4\,\text{Hz}}$$

(*b*) 1. The peak value of the current is related to the peak value of the charge:

$$I_{peak} = \omega Q_{peak} = \frac{Q_{peak}}{\sqrt{LC}}$$

2. The peak charge on the capacitor is related to the peak potential drop across the capacitor:

$$Q_{peak} = CV_{peak}$$

3. Substitute $CV_{peak}$ for $Q_{peak}$ and calculate $I_{peak}$:

$$I_{peak} = \frac{CV_{peak}}{\sqrt{LC}} = \frac{(2\,\mu\text{F})(20\,\text{V})}{\sqrt{(6\,\mu\text{H})(2\,\mu\text{F})}} = \boxed{11.5\,\text{A}}$$

**EXERCISE** A 5-$\mu$F capacitor is charged and is then discharged through an inductor. What should the value of the inductance be so that the current oscillates with frequency 8 kHz? (*Answer*   79.2 $\mu$H)

If we include a resistor in series with the capacitor and the inductor, as in Figure 29-14, we have an **RLC circuit.** Kirchhoff's loop rule gives

$$L\frac{dI}{dt} + IR + \frac{Q}{C} = 0 \qquad\qquad 29\text{-}43a$$

or

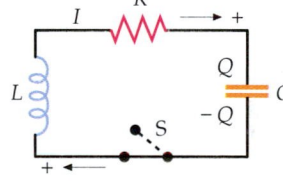

**FIGURE 29-14** An *RLC* circuit.

$$L\frac{d^2Q}{dt^2} + R\frac{dQ}{dt} + \frac{1}{C}Q = 0 \qquad\qquad 29\text{-}43b$$

where we have used $I = dQ/dt$ as before. Equations 29-43*a* and 29-43*b* are analogous to the equation for a damped harmonic oscillator (see Equation 14-35):

$$m\frac{d^2x}{dt^2} + b\frac{dx}{dt} + kx = 0$$

The first term, $L\,dI/dt = L\,d^2Q/dt^2$, is analogous to the mass times the acceleration, $m\,dv/dt = m\,d^2x/dt^2$; the second term, $IR = R\,dQ/dt$, is analogous to the damping term, $bv = b\,dx/dt$; and the third term, $Q/C$, is analogous to the restoring force $kx$. In the oscillation of a mass on a spring, the damping constant $b$ leads to a dissipation of mechanical energy. In an *RLC* circuit, the resistance $R$ is analogous to the damping constant $b$ and leads to a dissipation of electrical energy.

If the resistance is small, the charge and the current oscillate with (angular) frequency[†] that is very nearly equal to $\omega_0 = 1/\sqrt{LC}$, which is called the

[†] As in Chapter 14 when we discussed mechanical oscillations, we usually omit the word *angular* when the omission will not cause confusion.

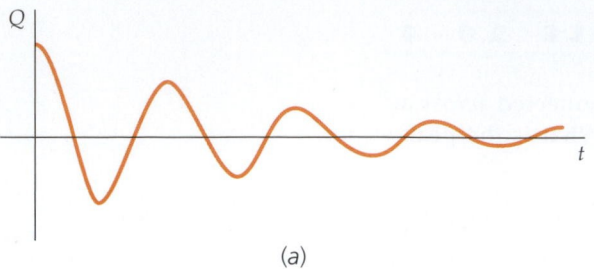

(a)

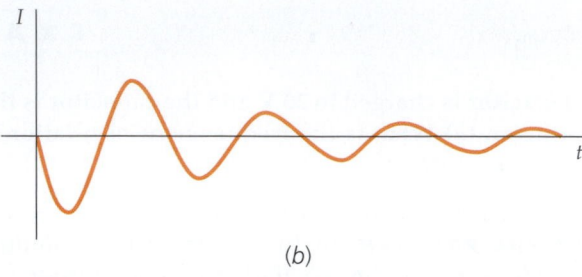

(b)

**natural frequency** of the circuit, but the oscillations are damped. We can understand this qualitatively from energy considerations. If we multiply each term in Equation 29-43$a$ by the current $I$, we obtain

$$LI\frac{dI}{dt} + I^2R + I\frac{Q}{C} = 0 \qquad 29\text{-}44$$

The magnetic energy in the inductor is given by $\frac{1}{2}LI^2$ (see Equation 28-20). Note that

$$\frac{d(\frac{1}{2}LI^2)}{dt} = LI\frac{dI}{dt}$$

where $LI\,dI/dt$ is the first term in Equation 29-44. If $LI\,dI/dt$ is positive, it equals the rate at which electrical potential energy is transformed into magnetic energy. If $LI\,dI/dt$ is negative, it equals the rate at which magnetic energy is transformed back into electrical potential energy. Note that $LI\,dI/dt$ is positive or negative depending on whether $I$ and $dI/dt$ have the same sign or different signs. The second term in Equation 29-44 is $I^2R$, the rate at which electrical potential energy is dissipated in the resistor. $I^2R$ is never negative. Note that

$$\frac{d(\frac{1}{2}Q^2/C)}{dt} = \frac{Q}{C}\frac{dQ}{dt} = I\frac{Q}{C}$$

where $IQ/C$ is the third term in Equation 29-44. This is the rate of change of the electric potential energy of the capacitor, which may be positive or negative. The sum of the electric and magnetic energies is not constant for this circuit because energy is continually dissipated in the resistor. Figure 29-15 shows graphs of $Q$ versus $t$ and $I$ versus $t$ for a small resistance $R$ in an $RLC$ circuit. If we increase $R$, the oscillations become more heavily damped until a critical value of $R$ is reached for which there is not even one oscillation. Figure 29-16 shows a graph of $Q$ versus $t$ in an $RLC$ circuit when the value of $R$ is greater than the critical damping value.

**FIGURE 29-15** Graphs of (a) $Q$ versus $t$ and (b) $I$ versus $t$ for the $RLC$ circuit shown in Figure 29-14 when the value of $R$ is small enough so that the oscillations are underdamped.

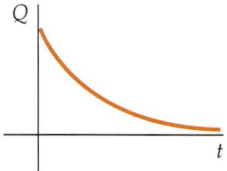

**FIGURE 29-16** A graph of $Q$ versus $t$ for the $RLC$ circuit shown in Figure 29-14 when the value of $R$ is so large that the oscillations are overdamped.

## *29-6 Driven *RLC* Circuits

### Series *RLC* Circuit

Figure 29-17 shows a series $RLC$ circuit being sinusoidally driven by an ac generator. If the potential drop applied by the generator to the series $RLC$ combination is $V_{app} = V_{app,peak} \cos \omega t$, applying Kirchhoff's loop rule gives

$$V_{app,peak} \cos \omega t - L\frac{dI}{dt} - IR - \frac{Q}{C} = 0$$

**FIGURE 29-17** A series $RLC$ circuit with an ac generator.

Using $I = dQ/dt$ and rearranging, we obtain

$$L\frac{d^2Q}{dt^2} + R\frac{dQ}{dt} + \frac{1}{C}Q = V_{app,peak}\cos\omega t \qquad\qquad 29\text{-}45$$

This equation is analogous to Equation 14-51 for the forced oscillation of a mass on a spring:

$$m\frac{d^2x}{dt^2} + b\frac{dx}{dt} + m\omega_0^2 x = F_0\cos\omega t$$

(In Equation 14-51, the force constant $k$ was written in terms of the mass $m$ and the natural angular frequency $\omega_0$ using $k = m\omega_0^2$. The capacitance in Equation 29-45 could be similarly written in terms of $L$ and the natural angular frequency using $1/C = L\omega_0^2$.)

We will discuss the solution of Equation 29-45 qualitatively as we did with Equation 14-51 for the forced oscillator. The current in the circuit consists of a transient current that depends on the initial conditions (e.g., the initial phase of the generator and the initial charge on the capacitor) and a steady-state current that does not depend on the initial conditions. We will ignore the transient current, which decreases exponentially with time and is eventually negligible, and concentrate on the steady-state current. The steady-state current obtained by solving Equation 29-45 is

$$I = I_{peak}\cos(\omega t - \delta) \qquad\qquad 29\text{-}46$$

where the phase angle $\delta$ is given by

$$\tan\delta = \frac{X_L - X_C}{R} \qquad\qquad 29\text{-}47$$

PHASE CONSTANT FOR A SERIES *RLC* CIRCUIT

The peak current is

$$I_{peak} = \frac{V_{app,peak}}{\sqrt{R^2 + (X_L - X_C)^2}} = \frac{V_{app,peak}}{Z} \qquad\qquad 29\text{-}48$$

PEAK CURRENT IN A SERIES *RLC* CIRCUIT

where

$$Z = \sqrt{R^2 + (X_L - X_C)^2} \qquad\qquad 29\text{-}49$$

IMPEDANCE OF A SERIES *RLC* CIRCUIT

The quantity $X_L - X_C$ is called the **total reactance,** and $Z$ is called the **impedance.** Combining these results, we have

$$I = \frac{V_{app,peak}}{Z}\cos(\omega t - \delta) \qquad\qquad 29\text{-}50$$

Equation 29-50 can also be obtained from a simple diagram using the phasor representations. Figure 29-18 shows the phasors representing the potential drops across the resistance, the inductance, and the capacitance. The $x$ component of each of these vectors equals the instantaneous potential drop across the corresponding element. Since the sum of the $x$ components equals the $x$ component of the sum, the sum of the $x$ components equals the sum of the potential drops across these elements, which by Kirchhoff's loop rule equals the instantaneous applied potential drop.

If we represent the potential drop applied across the series combination $V_{app} = V_{app,peak} \cos \omega t$ as a phasor $\vec{V}_{app}$ that has the magnitude $V_{app,peak}$, we have

$$\vec{V}_{app} = \vec{V}_R + \vec{V}_L + \vec{V}_C \qquad\qquad 29\text{-}51$$

In terms of the magnitudes,

$$V_{app,peak} = |\vec{V}_R + \vec{V}_L + \vec{V}_C| = \sqrt{V_{R,peak}^2 + (V_{L,peak} - V_{C,peak})^2}$$

But $V_R = I_{peak}R$, $V_L = I_{peak}X_L$, and $V_C = I_{peak}X_C$. Thus,

$$V_{app,peak} = I_{peak}\sqrt{R^2 + (X_L - X_C)^2} = I_{peak}Z$$

The phasor $\vec{V}_{app}$ makes an angle $\delta$ with $\vec{V}_R$, as shown in Figure 29-18. From the figure, we can see that

$$\tan \delta = \frac{|\vec{V}_L + \vec{V}_C|}{|\vec{V}_R|} = \frac{I_{peak}X_L - I_{peak}X_C}{I_{peak}R} = \frac{X_L - X_C}{R}$$

in agreement with Equation 29-47. Since $\vec{V}_{app}$ makes an angle $\omega t$ with the $x$ axis, $\vec{V}_R$ makes an angle $\omega t - \delta$ with the $x$ axis. This applied potential drop is in phase with the current, which is therefore given by

$$I = I_{peak}\cos(\omega t - \delta) = \frac{V_{app,peak}}{Z}\cos(\omega t - \delta)$$

This is Equation 29-50. The relation between the impedance $Z$, the resistance $R$, and the total reactance $X_L - X_C$ is best remembered by using the right triangle shown in Figure 29-19.

## Resonance

When $X_L$ and $X_C$ are equal, the total reactance is zero, and the impedance $Z$ has its smallest value $R$. Then $I_{peak}$ has its greatest value and the phase angle $\delta$ is zero, which means that the current is in phase with the applied potential drop. Let $\omega_{res}$ be the value of $\omega$ for which $X_L$ and $X_C$ are equal. It is obtained from

$$X_L = X_C$$

$$\omega_{res}L = \frac{1}{\omega_{res}C}$$

or

$$\omega_{res} = \frac{1}{\sqrt{LC}}$$

which equals the natural frequency $\omega_0$. When the frequency of the applied potential drop equals the natural frequency $\omega_0$, the impedance is smallest, $I_{peak}$ is

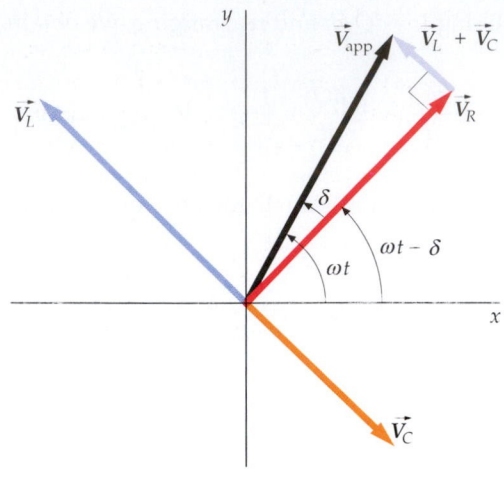

**FIGURE 29-18** Phase relations among potential drops in a series $RLC$ circuit. The potential drop across the resistor is in phase with the current. The potential drop across the inductor $V_L$ leads the current by 90°. The potential drop across the capacitor lags the current by 90°. The sum of the vectors representing these potential drops gives a vector at an angle $\delta$ with the current representing the applied emf. For the case shown here, $V_L$ is greater than $V_C$, and the current lags the applied potential drop by $\delta$.

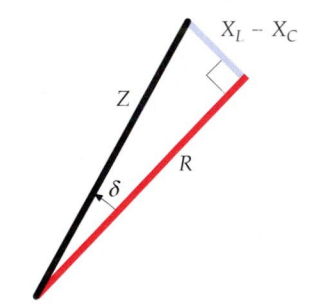

**FIGURE 29-19** A right triangle relating capacitive and inductive reactance, resistance, impedance, and the phase angle in an $RLC$ circuit.

greatest, and the circuit is said to be at **resonance.** The natural frequency $\omega_0$ is therefore also called the **resonance frequency.** This resonance condition in a driven RLC circuit is similar to that in a driven simple harmonic oscillator.

Since neither an inductor nor a capacitor dissipates energy, the average power delivered to a series RLC circuit is the average power supplied to the resistor. The instantaneous power supplied to the resistor is

$$P = I^2R = [I_{\text{peak}} \cos(\omega t - \delta)]^2 R$$

Averaging over one or more cycles and using $(\cos^2 \theta)_{\text{av}} = \frac{1}{2}$, we obtain for the average power

$$P_{\text{av}} = \frac{1}{2}I^2_{\text{peak}}R = I^2_{\text{rms}}R \qquad\qquad 29\text{-}52$$

Using $R/Z = \cos \delta$ from Figure 29-19 and $I_{\text{peak}} = V_{\text{app,peak}}/Z$, this can be written

$$P_{\text{av}} = \frac{1}{2}V_{\text{app,peak}}I_{\text{peak}} \cos \delta = V_{\text{app,rms}}I_{\text{rms}} \cos \delta \qquad\qquad 29\text{-}53$$

The quantity $\cos \delta$ is called the **power factor** of the RLC circuit. At resonance, $\delta$ is zero, and the power factor is 1.

The power can also be expressed as a function of the angular frequency $\omega$. Using $I_{\text{rms}} = V_{\text{app,rms}}/Z$ Equation 29-52 becomes

$$P_{\text{av}} = I^2_{\text{rms}}R = V^2_{\text{app,rms}}\frac{R}{Z^2}$$

From the definition of impedance Z, we have

$$Z^2 = (X_L - X_C)^2 + R^2 = \left(\omega L - \frac{1}{\omega C}\right)^2 + R^2$$

$$= \frac{L^2}{\omega^2}\left(\omega^2 - \frac{1}{LC}\right)^2 + R^2$$

$$= \frac{L^2}{\omega^2}(\omega^2 - \omega_0^2)^2 + R^2$$

where we have used $\omega_0 = 1/\sqrt{LC}$. Using this expression for $Z^2$, we obtain the average power as a function of $\omega$:

$$P_{\text{av}} = \frac{V^2_{\text{app,rms}}R\omega^2}{L^2(\omega^2 - \omega_0^2)^2 + \omega^2R^2} \qquad\qquad 29\text{-}54$$

Figure 29-20 shows the average power supplied by the generator to the series combination as a function of generator frequency for two different values of the resistance R. These curves, called **resonance curves,** are the same as the power-versus-frequency curves for a driven damped oscillator (see Section 14-5). The average power is greatest when the generator frequency equals the resonance frequency. When the resistance is small, the resonance curve is narrow; when the resistance is large, the resonance curve is broad. A resonance curve can be characterized by the **resonance width** $\Delta\omega$. As shown in Figure 29-20, the resonance width is the frequency difference between the two points on the curve where the power is half its maximum value. When the width is small compared with the resonance frequency, the resonance is sharp; that is, the resonance curve is narrow.

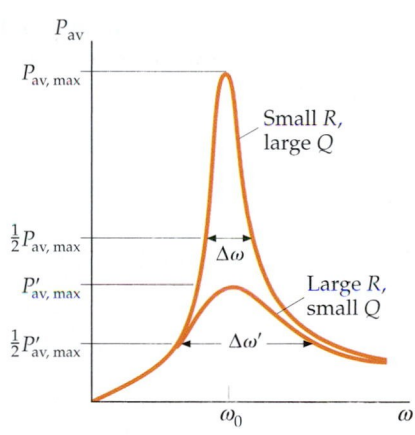

**FIGURE 29-20** Plot of average power versus frequency for a series RLC circuit. The power is maximum when the frequency of the generator $\omega$ equals the natural frequency of the circuit $\omega_0 = 1/\sqrt{LC}$. If the resistance is small, the Q factor is large and the resonance is sharp. The resonance width $\Delta\omega$ of the curves is measured between points where the power is half its maximum value.

In Chapter 14, the $Q$ factor for a mechanical oscillator is defined as $Q = \omega_0 m/b$ where $m$ is the mass and $b$ is the damping constant. We then saw that for an underdamped oscillator $Q = 2\pi E/|\Delta E|$, where $E$ is the total energy of the system at the beginning of a cycle and $\Delta E$ is the energy dissipated during the cycle. The **Q factor** for an $RLC$ circuit can be defined in a similar way. Since $L$ is analogous to the mass $m$ and $R$ is analogous to the damping constant $b$, the $Q$ factor for an $RLC$ circuit is given by

$$Q = 2\pi \frac{E}{|\Delta E|} = \frac{\omega_0 L}{R} \qquad\qquad 29\text{-}55$$

When the resonance curve is reasonably narrow (that is, when $Q$ is greater than about 2 or 3), the $Q$ factor can be approximated by

$$Q = \frac{\omega_0}{\Delta\omega} = \frac{f_0}{\Delta f} \qquad\qquad 29\text{-}56$$

$Q$ FACTOR FOR AN $RLC$ CIRCUIT

Resonance circuits are used in radio receivers, where the resonance frequency of the circuit is varied either by varying the capacitance or the inductance. Resonance occurs when the natural frequency of the circuit equals one of the frequencies of the radio waves picked up at the antenna. At resonance, there is a relatively large current in the antenna circuit. If the $Q$ factor of the circuit is sufficiently high, currents due to other station frequencies off resonance will be negligible compared with those currents due to the station frequency to which the circuit is tuned.

---

*Driven Series RLC Circuit*                    **EXAMPLE 29-5**

**A series $RLC$ combination with $L = 2$ H, $C = 2$ $\mu$F, and $R = 20$ $\Omega$ is driven by an ideal generator with a peak emf of 100 V and a frequency that can be varied. Find (a) the resonance frequency $f_0$, (b) the $Q$ value, (c) the width of the resonance $\Delta f$, and (d) the current amplitude at resonance.**

**PICTURE THE PROBLEM** The resonance frequency is found from $\omega_0 = 1/\sqrt{LC}$ and the $Q$ value is found from $Q = \omega_0 L/R$.

1. The resonance frequency is $f_0 = \omega_0/2\pi$:

$$f_0 = \frac{\omega_0}{2\pi} = \frac{1}{2\pi\sqrt{LC}}$$

$$= \frac{1}{2\pi\sqrt{(2\text{ H})(2\times 10^{-6}\text{ F})}} = \boxed{79.6\text{ Hz}}$$

2. Use this result to calculate $Q$:

$$Q = \frac{\omega_0 L}{R} = \frac{2\pi(79.6\text{ Hz})(2\text{ H})}{20\ \Omega} = \boxed{50}$$

3. Use the value of $Q$ to find the width of the resonance $\Delta f$:

$$\Delta f = \frac{f_0}{Q} = \frac{79.6\text{ Hz}}{50} = \boxed{1.59\text{ Hz}}$$

4. At resonance, the impedance is $R$ and $I_{\text{peak}}$ is $V_{\text{app,peak}}/R$:

$$I_{\text{max}} = \frac{V_{\text{app,peak}}}{R} = \frac{\mathcal{E}_{\text{peak}}}{R} = \frac{100\text{ V}}{20\ \Omega} = \boxed{5\text{ A}}$$

**REMARKS** The width of 1.59 Hz is less than 2 percent of the resonance frequency of 79.6 Hz, so the resonance peak is quite sharp.

*DRIVEN SERIES RLC CIRCUIT CURRENT, PHASE, AND POWER* **EXAMPLE 29-6 Try It Yourself**

If the generator in Example 29-5 has a frequency of 60 Hz, find (*a*) the current amplitude, (*b*) the phase constant $\delta$, (*c*) the power factor, and (*d*) the average power delivered.

**PICTURE THE PROBLEM** The current amplitude is the amplitude of the applied potential drop divided by the total impedance of the series combination. The phase angle $\delta$ is found from $\tan \delta = (X_L - X_C)/R$. You can use either Equation 29-52 or Equation 29-53 to find the average power delivered.

**Cover the column to the right and try these on your own before looking at the answers.**

| Steps | Answers |
|---|---|
| (*a*) 1. Write the peak current in terms of $V_{app,peak}$ and the impedance. | $I_{peak} = \dfrac{V_{app,peak}}{Z} = \dfrac{\mathcal{E}_{peak}}{Z}$ |
| 2. Calculate the capacitive and inductive reactances and the total reactance. | $X_C = 1326\ \Omega,\ X_L = 754\ \Omega$ <br> so <br> $X_L - X_C = -572\ \Omega$ |
| 3. Calculate the total impedance Z. | $Z = 573\ \Omega$ |
| 4. Use the results of steps 2 and 3 to calculate $I_{peak}$. | $I_{peak} = \boxed{0.175\ A}$ |
| (*b*) Use the results of Part (*a*) steps 2 and 3 to calculate $\delta$. | $\delta = \tan^{-1}\dfrac{X_L - X_C}{R} = \boxed{-88.0°}$ |
| (*c*) Use your value of $\delta$ to compute the power factor. | $\cos \delta = 0.0349$ |
| (*d*) Calculate the average power delivered from Equation 29-52. | $P_{av} = \frac{1}{2} I_{peak}^2 R = \boxed{0.305\ W}$ |

**PLAUSIBILITY CHECK** To check our result for the average power using the power factor found in Part (*c*), we have $P_{av} = \frac{1}{2} V_{app,peak} I_{peak} \cos \delta = \frac{1}{2} \mathcal{E}_{peak} I_{peak} \cos \delta = 0.305$ W. This is in agreement with our result for Part (*d*).

**REMARKS** The frequency of 60 Hz is well below the resonance frequency of 79.6 Hz. (Recall that the width as calculated in Example 29-5 is only 1.59 Hz.) As a result, the total reactance is much greater in magnitude than the resistance. This is always the case far from resonance. Similarly, an $I_{peak}$ of 0.175 A is much less than an $I_{peak}$ at resonance, which was found to be 5 A. Finally, we see from Figure 29-18 that a negative phase angle $\delta$ means that the current leads the applied potential drop.

*DRIVEN SERIES RLC CIRCUIT AT RESONANCE* **EXAMPLE 29-7 Try It Yourself**

Find the peak potential drop across the resistor, the inductor, and the capacitor at resonance for the circuit in Example 29-5.

**PICTURE THE PROBLEM** The peak potential drop across the resistor is $I_{peak}$ times R. Similarly, the peak potential drop across the inductor or capacitor is $I_{peak}$ times the appropriate reactance. We found that at resonance $I_{peak} = 5$ A and $f_0 = 79.6$ Hz in Example 29-5.

**Cover the column to the right and try these on your own before looking at the answers.**

| Steps | Answers |
|---|---|
| 1. Calculate $V_{R,\text{peak}} = I_{\text{peak}}R$. | $V_{R,\text{peak}} = I_{\text{peak}}R = \boxed{100 \text{ V}}$ |
| 2. Express $V_{L,\text{peak}}$ in terms of $I_{\text{peak}}$ and $X_L$. | $V_{L,\text{peak}} = I_{\text{peak}}X_L = I_{\text{peak}}\omega_0 L = \boxed{5000 \text{ V}}$ |
| 3. Express $V_{C,\text{peak}}$ in terms of $I_{\text{peak}}$ and $X_C$. | $V_{C,\text{peak}} = I_{\text{peak}}X_C = \dfrac{I_{\text{peak}}}{\omega_0 C} = \boxed{5000 \text{ V}}$ |

**REMARKS** The inductive and capacitive reactances are equal, as we would expect, since we found the resonance frequency by setting them equal. The phasor diagram for the potential drops across the resistor, capacitor, and inductor is shown in Figure 29-21. The peak potential drop across the resistor is a relatively safe 100 V, equal to the peak emf of the generator. However, the peak potential drops across the inductor and the capacitor are a dangerously high 5000 V. These potential drops are 180° out of phase. At resonance, the potential drop across the inductor at any instant is the negative of that across the capacitor, so they always sum to zero, leaving the potential drop across the resistor equal to the emf in the circuit.

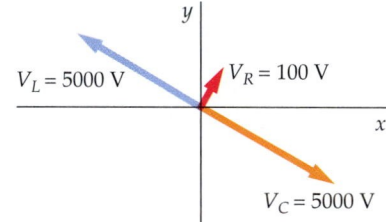

**FIGURE 29-21**

$V_L = 5000$ V  $V_R = 100$ V  $V_C = 5000$ V

---

RC Low-Pass Filter   **EXAMPLE 29-8**

**A resistor $R$ and capacitor $C$ are in series with a generator, as shown in Figure 29-22. The generator applies a potential drop across the $RC$ combination given by $V_{\text{app}} = \sqrt{2}V_{\text{app,rms}} \cos \omega t$. Find the rms potential drop across the capacitor $V_{\text{out,rms}}$ as a function of frequency $\omega$.**

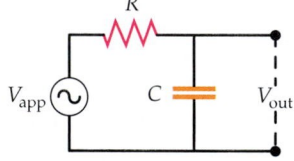

**FIGURE 29-22** The peak output voltage decreases as frequency increases.

**PICTURE THE PROBLEM** The rms potential drop across the capacitor is the product of the rms current and the capacitive reactance. The rms current is found from the potential drop applied by the generator and the impedance of the series $RC$ combination.

1. The potential drop across the capacitor is $I_{\text{rms}}$ times $X_C$:

$$V_{\text{out,rms}} = I_{\text{rms}}X_C$$

2. The rms current depends on the applied rms potential drop and the impedance:

$$I_{\text{rms}} = \frac{V_{\text{app,rms}}}{Z}$$

3. In this circuit, only $R$ and $X_C$ contribute to the total impedance:

$$Z = \sqrt{R^2 + X_C^2}$$

4. Substitute these values and $X_C = 1/(\omega C)$ to find the output rms potential drop:

$$V_{\text{out,rms}} = I_{\text{rms}}X_C = \frac{V_{\text{app,rms}}}{Z}X_C = \frac{V_{\text{app,rms}}X_C}{\sqrt{R^2 + X_C^2}}$$

$$= \frac{V_{\text{app,rms}}\left(\dfrac{1}{\omega C}\right)}{\sqrt{R^2 + \left(\dfrac{1}{\omega C}\right)^2}} = \boxed{\frac{V_{\text{app,rms}}}{\sqrt{1 + \omega^2(RC)^2}}}$$

**REMARKS** This circuit is called an *RC low-pass filter,* since it transmits low frequencies with greater amplitude than high frequencies. In fact, the output potential drop equals the potential drop applied by the generator in the limit that $\omega \to 0$, but approaches zero for $\omega \to \infty$, as shown in the graph of the ratio of output potential drop to applied potential drop in Figure 29-23.

**EXERCISE** Find the output potential drop for this circuit if the capacitor is replaced by an inductor $L$. (*Answer* $V_{\text{out,rms}} = V_{\text{in,rms}} / \sqrt{1 + (R/L)^2 / \omega^2}$. This circuit is a *high-pass filter.*)

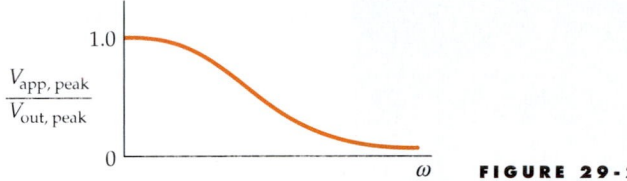

FIGURE 29-23

---

AN FM TUNER                    **EXAMPLE 29-9**    **Put It in Context**

You have been tinkering with building a radio tuner using your new knowledge of physics. You know that the FM dial gives its frequencies in megahertz, and you would like to determine what percentage of change in an inductor would allow you to tune for the whole FM range. You decide to start at midrange and determine a percent increase and decrease needed for inductance. A variable inductor is usually an iron-core solenoid, and the inductance is increased by further inserting the core. The FM dial goes from 88 MHz to 108 MHz.

**PICTURE THE PROBLEM** We can relate inductance to the resonant frequency with $\omega = 2\pi f$ and $\omega = 1/\sqrt{LC}$. Then, if we find the percent change in frequency, we can determine the percent change in inductance. The capacitance $C$ does not vary.

1. The resonant angular frequency $\omega$ is related to the inductance $L$:

$$\omega = 1/\sqrt{LC}$$

and

$$\omega = 2\pi f$$

so

$$f = \frac{1}{2\pi \sqrt{LC}}$$

2. $L$ is inversely proportional to $f^2$:

$$L = af^{-2}$$

where

$$a = (4\pi^2 C)^{-1}$$

3. Express the fractional change in $L$ in terms of the frequencies: When $L$ is maximum, $f$ is minimum and vice versa. The middle frequency $f_{\text{mid}}$ is halfway between the maximum and minimum frequency, and $L_{\text{mid}}$ is the inductance when $f = f_{\text{mid}}$:

$$\frac{\Delta L}{L} = \frac{L_{\text{max}} - L_{\text{min}}}{L_{\text{mid}}} = \frac{af_{\text{max}}^{-2} - af_{\text{min}}^{-2}}{af_{\text{mid}}^{-2}}$$

$$= f_{\text{mid}}^2 \left( \frac{1}{f_{\text{max}}^2} - \frac{1}{f_{\text{min}}^2} \right) = 98^2 \left( \frac{1}{108^2} - \frac{1}{88^2} \right)$$

$$= -0.417$$

4. The negative sign is not relevant, except as an indication that when the inductance increases the resonant frequency decreases. Express the step 3 result as a percentage:

The inductance varies by about $\boxed{42 \text{ percent}}$

A shipboard radio, circa 1920. Exposed at the operator's left are the inductance coils and capacitor plates of the tuning circuit.

## Parallel *RLC* Circuit

Figure 29-24 shows a resistor $R$, a capacitor $C$, and an inductor $L$ connected in parallel across an ac generator. The total current $I$ from the generator divides into three currents. The current $I_R$ in the resistor, the current $I_C$ in the capacitor, and the current $I_L$ in the inductor. The instantaneous potential drop $V_{app}$ is the same across each element. The current in the resistor is in phase with the potential drop and the phasor $\vec{I}_R$ has magnitude $V_{peak}/R$. Since the potential drop across an inductor *leads* the current in the inductor by 90°, $I_L$ *lags* the potential drop by 90°, and the phasor $\vec{I}_L$ has magnitude $V_{peak}/X_L$. Similarly, the $I_C$ leads the potential drop by 90° and the phasor $\vec{I}_C$ has magnitude $V_{peak}/X_C$. These currents are represented by phasors in Figure 29-25. The total current $I$ is the $x$ component of the vector sum of the individual currents as shown in the figure. The magnitude of the total current is

$$I = \sqrt{I_R^2 + (I_L - I_C)^2} = \sqrt{\left(\frac{V_{peak}}{R}\right)^2 + \left(\frac{V_{peak}}{X_L} - \frac{V_{peak}}{X_C}\right)^2} = \frac{V_{peak}}{Z} \qquad 29\text{-}57$$

where the total impedance $Z$ is related to the resistance and the capacitive and inductive reactances by

$$\frac{1}{Z} = \sqrt{\left(\frac{1}{R}\right)^2 + \left(\frac{1}{X_L} - \frac{1}{X_C}\right)^2} \qquad 29\text{-}58$$

At resonance, the currents in the inductor and capacitor are 180° out of phase, so the total current is a minimum and is just the current in the resistor. We see from Equation 29-57 that this occurs if $Z$ is maximum, so $1/Z$ is minimum. Then, we see from Equation 29-58 that if $X_L = X_C$, $1/Z$ has its minimum value $1/R$. Equating $X_L$ with $X_C$ and solving for $\omega$ obtains the resonant frequency, which equals the natural frequency $\omega_0 = 1/\sqrt{LC}$.

## *29-7 The Transformer

A transformer is a device used to raise or lower the voltage in a circuit without an appreciable loss of power. Figure 29-26 shows a simple transformer consisting of two wire coils around a common iron core. The coil carrying the input power is

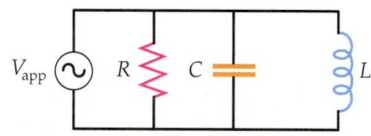

**FIGURE 29-24** A parallel *RLC* circuit.

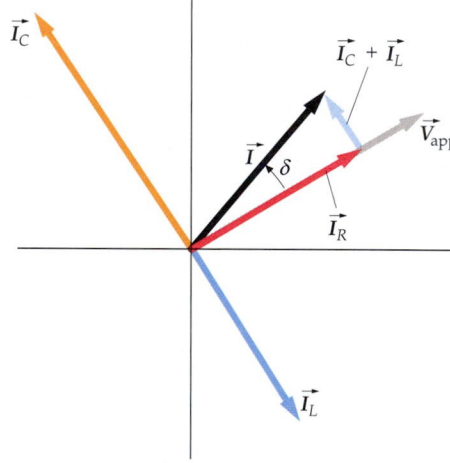

**FIGURE 29-25** A phasor diagram for the currents in the parallel *RLC* circuit shown in Figure 29-24. The potential drop is the same across each element. The current in the resistor is in phase with the potential drop. The current in the capacitor leads the potential drop by 90° and the current in the inductor lags the potential drop by 90°. The phase difference $\delta$ between the total current and the potential drop depends on the relative magnitudes of the currents, which depend on the values of the resistance and of the capacitive and inductive reactances.

called the primary, and the other coil is called the secondary. Either coil of a transformer can be used for the primary or secondary. The transformer operates on the principle that an alternating current in one circuit induces an alternating emf in a nearby circuit due to the mutual inductance of the two circuits. The iron core increases the magnetic field for a given current and guides it so that nearly all the magnetic flux through one coil goes through the other coil. If no power were lost, the product of the potential difference across and the current in the secondary windings would equal the product of the potential drop across and the current in the primary windings. Thus, if the potential difference across the secondary coil is higher than the potential drop across the primary circuit, the current in the secondary coil is lower than the current in the primary coil, and vice versa. Power losses arise because of the Joule heating in the small resistances in both coils, or in current loops within the core,[†] and from hysteresis in the iron cores. We will neglect these losses and consider an ideal transformer of 100 percent efficiency, for which all of the power supplied to the primary coil appears in the secondary coil. Actual transformers are often 90 percent to 95 percent efficient.

Consider a transformer with a potential drop $V_1$ across the primary coil of $N_1$ turns; the secondary coil of $N_2$ turns is an open circuit. Because of the iron core, there is a large flux through each coil even when the magnetizing current $I_m$ in the primary circuit is very small. We can ignore the resistances of the coils, which are negligible in comparison with their inductive reactances. The primary circuit is then a simple circuit consisting of an ac generator and a pure inductance, like that discussed in Section 29-3. The current magnetizing in the primary coil and the voltage drop across the primary coil are out of phase by 90°, and the average power dissipated in the primary coil is zero. If $\phi_{turn}$ is the magnetic flux through a single turn of the primary coil, the potential drop across the primary coil is equal to the back emf, so

$$V_1 = N_1 \frac{d\phi_{turn}}{dt} \qquad \text{29-59}$$

If there is no flux leakage out of the iron core, the flux through each turn is the same for both coils. Thus, the total flux through the secondary coil is $N_2 \phi_{turn}$, and the potential difference across the secondary coil is

$$V_2 = N_2 \frac{d\phi_{turn}}{dt} \qquad \text{29-60}$$

Comparing Equations 29-59 and 29-60, we can see that

$$V_2 = \frac{N_2}{N_1} V_1 \qquad \text{29-61}$$

If $N_2$ is greater than $N_1$, the potential difference across the secondary coil is greater than the potential drop across the primary coil, and the transformer is called a step-up transformer. If $N_2$ is less than $N_1$, the potential difference across the secondary coil is less than the potential drop across the primary coil, and the transformer is called a step-down transformer.

When we put a resistance $R$, called a load resistance, across the secondary coil, there will then be a current $I_2$ in the secondary circuit that is in phase with the potential drop $V_2$ across the resistance. This current sets up an additional

---

[†] The induced currents, called eddy currents, can be greatly reduced by using a core of laminated metal to break up current paths.

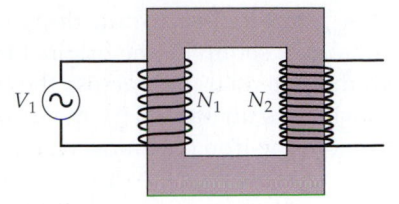

**FIGURE 29-26** A transformer with $N_1$ turns in the primary and $N_2$ turns in the secondary.

(a)

(b)

(a) A power box with a transformer for stepping down voltage for distribution to homes. (b) A suburban power substation where transformers step down voltage from high-voltage transmission lines.

flux $\phi'_{turn}$ through each turn that is proportional to $N_2 I_2$. This flux opposes the original flux set up by the original magnetizing current $I_m$ in the primary. However, the potential drop across the primary coil is determined by the generator emf, which is unaffected by the secondary circuit. According to Equation 29-60, the flux in the iron core must change at the original rate; that is, the total flux in the iron core must be the same as when there is no load across the secondary. The primary coil thus draws an additional current $I_1$ to maintain the original flux $\phi_{turn}$. The flux through each turn produced by this additional current is proportional to $N_1 I_1$. Since this flux equals $-\phi'_{turn}$, the additional current $I_1$ in the primary is related to the current $I_2$ in the secondary by

$$N_1 I_1 = -N_2 I_2 \qquad\qquad 29\text{-}62$$

These currents are 180° out of phase and produce counteracting fluxes. Since $I_2$ is in phase with $V_2$, the additional current $I_1$ is in phase with the potential drop across the primary circuit. The power input from the generator is $V_{1,rms} I_{1,rms}$, and the power output is $V_{2,rms} I_{2,rms}$. (The magnetizing current does not contribute to the power input because it is 90° out of phase with the generator voltage.) If there are no losses,

$$V_{1,rms} I_{1,rms} = V_{2,rms} I_{2,rms} \qquad\qquad 29\text{-}63$$

In most cases, the additional current in the primary $I_1$ is much greater than the original magnetizing current $I_m$ that is drawn from the generator when there is no load. This can be demonstrated by putting a lightbulb in series with the primary coil. The lightbulb is much brighter when there is a load across the secondary circuit than when the secondary circuit is open. If $I_m$ can be neglected, Equation 29-63 relates the total currents in the primary and secondary circuits.

---

*DOORBELL TRANSFORMER*       **EXAMPLE 29-10**

**A doorbell requires 0.4 A at 6 V. It is connected to a transformer whose primary, containing 2000 turns, is connected to a 120-V ac line. (*a*) How many turns should there be in the secondary? (*b*) What is the current in the primary?**

**PICTURE THE PROBLEM** We can find the number of turns from the turns ratio, which equals the voltage ratio. The primary current can be found by equating the power out to the power in.

1. The turns ratio can be obtained from Equation 29-61. Solve for the number of turns in the secondary, $N_2$:

$$\frac{N_2}{N_1} = \frac{V_2}{V_1}$$

so

$$N_2 = \frac{V_{2,rms}}{V_{1,rms}} N_1 = \frac{6\ V}{120\ V}\,2000\ \text{turns} = \boxed{100\ \text{turns}}$$

2. Since we are assuming 100 percent efficiency in power transmission, the input and output currents are related by Equation 29-62. Solve for the current in the primary, $I_1$:

$$V_2 I_2 = V_1 I_1$$

so

$$I_1 = \frac{V_2}{V_1} I_2 = \frac{6\ V}{120\ V}\,(0.4\ A) = \boxed{0.02\ A}$$

---

An important use of transformers is in the transport of electrical power. To minimize the $I^2 R$ heat loss (Joule heating) in transmission lines, it is economical to use a high voltage and a low current. On the other hand, safety and other considerations require that power be delivered to consumers at lower voltages and therefore with higher currents. Suppose, for example, that each person in a city with a population of 50,000 uses 1.2 kW of electric power. (The per capita consumption of power in the United States is actually somewhat higher than this.) At 120 V, the current required for each person would be

$$I = \frac{1200 \text{ W}}{120 \text{ V}} = 10 \text{ A}$$

The total current for 50,000 people would then be 500,000 A. The transport of such a current from a power-plant generator to a city many kilometers away would require conductors of enormous thickness, and the $I^2R$ power loss would be substantial. Rather than transmit the power at 120 V, step-up transformers are used at the power plant to step up the voltage to some very large value, such as 600,000 V. For this voltage, the current needed is only

$$I = \frac{120 \text{ V}}{600,000 \text{ V}} (500,000 \text{ A}) = 100 \text{ A}$$

To reduce the voltage to a safer level for transport within a city, power substations are located just outside the city to step down the voltage to a safer value, such as 10,000 V. Transformers in boxes attached to the power poles outside each house again step down the voltage to 120 V (or 240 V) for distribution to the house. Because of the ease of stepping the voltage up or down with transformers, alternating current rather than direct current is in common use.

---

*TRANSMISSION LOSSES*                          **EXAMPLE  29-11**

**A transmission line has a resistance of 0.02 $\Omega$/km. Calculate the $I^2R$ power loss if 200 kW of power is transmitted from a power generator to a city 10 km away at (*a*) 240 V and (*b*) 4.4 kV.**

**PICTURE THE PROBLEM** First, note that the total resistance of 10 km of wire is $R = (0.02 \ \Omega/\text{km})(10 \text{ km}) = 0.2 \ \Omega$. In each case, begin by finding the current needed to transmit 200 kW using $P = IV$, then find the power loss using $I^2R$.

(*a*) 1. Find the current needed to transmit 200 kW of power at 240 V:

$$I = \frac{P}{V} = \frac{200 \text{ kW}}{240 \text{ V}} = 833 \text{ A}$$

2. Calculate the power loss:

$$I^2R = (833 \text{ A})^2(0.2 \ \Omega) = \boxed{139,000 \text{ W}}$$

(*b*) 1. Now, find the current needed to transmit 200 kW of power at 4.4 kV:

$$I = \frac{P}{V} = \frac{200 \text{ kW}}{4.4 \text{ kV}} = 45.5 \text{ A}$$

2. Calculate the power loss:

$$I^2R = (45.5 \text{ A})^2(0.2 \ \Omega) = \boxed{414 \text{ W}}$$

**REMARKS** Note that with a transmission voltage of 240 V almost 70 percent of the power is wasted through heat loss, and there is an $IR$ (voltage) drop across the transmission line of 167 V, so the power is delivered at only 73 V. However, with transmission at 4.4 kV only about 0.2 percent of the power is lost in transmission, and there is an $IR$ drop across the transmission line of only 9 V, so the power is delivered with only a 0.2 percent voltage drop. This illustrates the advantages of high-voltage power transmission.

# SUMMARY

1. Reactance is a frequency-dependent property of capacitors and inductors that is analogous to the resistance of a resistor.

2. Impedance is a frequency-dependent property of an ac circuit or circuit loop that is analogous to the resistance in a dc circuit.

3. Phasors are two-dimensional vectors that allow us to picture the phase relations in a circuit.

4. Resonance occurs when the frequency of the generator equals the natural frequency of the oscillating circuit.

| Topic | Relevant Equations and Remarks | |
|---|---|---|
| **1. Alternating Current Generators** | An ac generator is a device that transforms mechanical energy into electrical energy. This transformation can be accomplished by using the mechanical energy to either rotate a conducting coil in a magnetic field or rotating a magnet in a conducting coil. | |
| EMF generated | $\mathcal{E} = \mathcal{E}_{peak} \sin(\omega t + \delta) = NBA\omega \sin(\omega t + \delta)$ | **29-3, 29-4** |
| **2. Current** | | |
| RMS current | $I_{rms} = \sqrt{(I^2)_{av}}$ | **29-11** |
| RMS current and peak current | $I_{rms} = \dfrac{1}{\sqrt{2}} I_{peak}$ | **29-12** |
| For a resistor | $I_{rms} = \dfrac{V_{R,rms}}{R}$ <br><br> potential drop and current in phase | **29-15** |
| For an inductor | $I_{rms} = \dfrac{V_{L,rms}}{\omega L} = \dfrac{V_{L,rms}}{X_L}$ <br><br> potential drop leads current by 90° | **29-25** |
| For a capacitor | $I_{rms} = \dfrac{V_{C,rms}}{1/\omega L} = \dfrac{V_{C,rms}}{X_C}$ <br><br> potential drop lags current by 90° | **29-29** |
| **3. Reactance** | | |
| Inductive reactance | $X_L = \omega L$ | **29-24** |
| Capacitive reactance | $X_C = \dfrac{1}{\omega C}$ | **29-30** |
| **4. Average Power Dissipation** | | |
| By a resistor | $P_{av} = V_{R,rms}I_{rms} = I_{rms}^2 R$ | **29-13, 29-15** |
| By an inductor or by a capacitor | $P_{av} = 0$ | |

5. **\*Phasors**

Phasors are two-dimensional vectors that represent the current $\vec{I}$, the potential drop across a resistor $\vec{V}_R$, the potential drop across a capacitor $\vec{V}_C$, and the potential drop across an inductor $\vec{V}_L$ in an ac circuit. These phasors rotate in the counterclockwise direction with an angular velocity that is equal to the angular frequency $\omega$ of the current. $\vec{V}_R$ is in phase with the current, $\vec{V}_L$ leads the current by 90°, and $\vec{V}_C$ lags the current by 90°. The $x$ component of each phasor equals the magnitude of the current or the corresponding potential drop at any instant.

6. **\*LC and RLC Series Circuits**

If a capacitor is discharged through an inductor, the charge and the voltage on the capacitor oscillate with angular frequency

$$\omega = \frac{1}{\sqrt{LC}} \qquad \text{29-37}$$

The current in the inductor oscillates with the same frequency, but it is out of phase with the charge by 90°. The energy oscillates between electric energy in the capacitor and magnetic energy in the inductor. If the circuit also has resistance, the oscillations are damped because energy is dissipated in the resistor.

7. **Series RLC Circuit Driven by an Applied Potential Drop of Frequency $\omega$**

| | | |
|---|---|---|
| Applied potential drop | $V_{app} = V_{app,peak} \cos \omega t$ | |
| Current | $I = \dfrac{V_{app,peak}}{Z} \cos(\omega t - \delta)$ | 29-50 |
| Impedance Z | $Z = \sqrt{R^2 + (X_L - X_C)^2}$ | 29-49 |
| Phase angle $\delta$ | $\tan \delta = \dfrac{X_L - X_C}{R}$ | 29-47 |
| Average power | $P_{av} = I_{rms}^2 R = V_{app,rms} I_{rms} \cos \delta = \dfrac{V_{app,rms}^2 R \omega^2}{L^2(\omega^2 - \omega_0^2)^2 + \omega^2 R^2}$ | 29-52, 29-53, 29-54 |

Power factor

The quantity $\cos \delta$ in Equation 29-53 is called the power factor of the RLC circuit. At resonance, $\delta$ is zero, the power factor is 1, and

$$P_{av} = V_{app,rms} I_{rms}$$

Resonance

When the rms current is maximum, the circuit is said to be at resonance. The conditions for resonance are

$$X_L = X_C, \quad \text{so} \quad Z = \sqrt{R^2 + (X_L - X_C)^2} = R$$

$$\omega = \omega_0 = \frac{1}{\sqrt{LC}} \quad \text{and} \quad \delta = 0$$

8. **Q Factor**

The sharpness of the resonance curve is described by the Q factor

$$Q = \frac{\omega_0 L}{R} \qquad \text{29-55}$$

When the resonance curve is reasonably narrow, the Q factor can be approximated by

$$Q = \frac{\omega_0}{\Delta \omega} = \frac{f_0}{\Delta f} \qquad \text{29-56}$$

**9. Transformers**

A transformer is a device used to raise or lower the voltage in a circuit without an appreciable loss in power. For a transformer with $N_1$ turns in the primary and $N_2$ turns in the secondary, the potential difference across the secondary coil is related to the potential drop across the primary coil by

$$V_2 = \frac{N_2}{N_1} V_1 \qquad\qquad \textbf{29-61}$$

If there are no power losses,

$$V_{1,\mathrm{rms}} I_{1,\mathrm{rms}} = V_{2,\mathrm{rms}} I_{2,\mathrm{rms}} \qquad\qquad \textbf{29-63}$$

# PROBLEMS

- Single-concept, single-step, relatively easy
- •• Intermediate-level, may require synthesis of concepts
- ••• Challenging
- **SSM** Solution is in the *Student Solutions Manual*
-  Problems available on iSOLVE online homework service
- ✓ These "Checkpoint" online homework service problems ask students additional questions about their confidence level, and how they arrived at their answer.

In a few problems, you are given more data than you actually need; in a few other problems, you are required to supply data from your general knowledge, outside sources, or informed estimates.

## Conceptual Problems

**1** • **SSM** As the frequency in the simple ac circuit in Figure 29-27 increases, the rms current through the resistor (*a*) increases. (*b*) does not change. (*c*) may increase or decrease depending on the magnitude of the original frequency. (*d*) may increase or decrease depending on the magnitude of the resistance. (*e*) decreases.

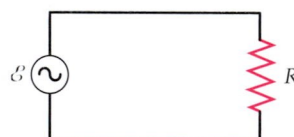

**FIGURE 29-27** Problem 1

**2** • If the rms voltage in an ac circuit is doubled, the peak voltage is (*a*) increased by a factor of 2. (*b*) decreased by a factor of 2. (*c*) increased by a factor of $\sqrt{2}$. (*d*) decreased by a factor of $\sqrt{2}$. (*e*) not changed.

**3** • If the frequency in the circuit shown in Figure 29-28 is doubled, the inductance of the inductor will (*a*) increase by a factor of 2. (*b*) not change. (*c*) decrease by a factor of 2. (*d*) increase by a factor of 4. (*e*) decrease by a factor of 4.

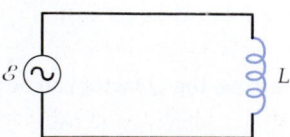

**FIGURE 29-28** Problems 3 and 4

**4** • If the frequency in the circuit shown in Figure 29-28 is doubled, the inductive reactance of the inductor will (*a*) increase by a factor of 2. (*b*) not change. (*c*) decrease by a factor of 2. (*d*) increase by a factor of 4. (*e*) decrease by a factor of 4.

**5** • **SSM** If the frequency in the circuit in Figure 29-29 is doubled, the capacitive reactance of the circuit will (*a*) increase by a factor of 2. (*b*) not change. (*c*) decrease by a factor of 2. (*d*) increase by a factor of 4. (*e*) decrease by a factor of 4.

**FIGURE 29-29** Problem 5

**6** • In a circuit consisting of a generator and an inductor, are there any times when the inductor absorbs power from the generator? Are there any times when the inductor supplies power to the generator?

**7** • In a circuit consisting of a generator and a capacitor, are there any times when the capacitor absorbs power from the generator? Are there any times when the capacitor supplies power to the generator?

**8** • The SI units of inductance times capacitance are (*a*) seconds squared. (*b*) hertz. (*c*) volts. (*d*) amperes. (*e*) ohms.

**9** •• **SSM** Making *LC* circuits with oscillation frequencies of thousands of hertz or more is easy, but making *LC* circuits that have small frequencies is difficult. Why?

**10** • True or false:

(a) An *RLC* circuit with a high *Q* factor has a narrow resonance curve.

(b) At resonance, the impedance of an *RLC* circuit equals the resistance *R*.

(c) At resonance, the current and generator voltage are in phase.

**11** • Does the power factor depend on the frequency?

**12** • **SSM** Are there any disadvantages to having a radio tuning circuit with an extremely large *Q* factor?

**13** • What is the power factor for a circuit that has inductance and capacitance but no resistance?

**14** • A transformer is used to change (a) capacitance, (b) frequency, (c) voltage, (d) power, (e) none of these.

**15** • True or false: If a transformer increases the current, it must decrease the voltage.

**16** •• An ideal transformer has $N_1$ turns on its primary and $N_2$ turns on its secondary. The power dissipated in a load resistance *R* connected across the secondary is $P_2$ when the primary voltage is $V_1$. The current in the primary windings is then (a) $P_2/V_1$. (b) $(N_1/N_2)(P_2/V_1)$. (c) $(N_2/N_1)(P_2/V_1)$. (d) $(N_2/N_1)^2(P_2/V_1)$.

**17** • True or false:

(a) Alternating current in a resistance dissipates no power because the current is negative as often as the current is positive.

(b) At very high frequencies, a capacitor acts like a short circuit.

## Estimation and Approximation

**18** •• **SSM** The impedances of motors, transformers, and electromagnets have inductive reactance. Suppose that the phase angle of the total impedance of a large industrial plant is 25° when the plant is under full operation and using 2.3 MW of power. The power is supplied to the plant from a substation 4.5 km from the plant; the 60 Hz rms line voltage at the plant is 40,000 V. The resistance of the transmission line from the substation to the plant is 5.2 Ω. The cost per kilowatt-hour is 0.07 dollars. The plant pays only for the actual energy used. (a) What are the resistance and inductive reactance of the plant's total load? (b) What is the current in the power lines and what must be the rms voltage at the substation to maintain the voltage at the plant at 40,000 V? (c) How much power is lost in transmission? (d) Suppose that the phase angle of the plant's impedance were reduced to 18° by adding a bank of capacitors in series with the load. How much money would be saved by the electric utility during one month of operation, assuming the plant operates at full capacity for 16 h each day? (e) What must be the capacitance of this bank of capacitors?

## Alternating Current Generators

**19** • A 200-turn coil has an area of 4 cm² and rotates in a magnetic field of 0.5 T. (a) What frequency will generate a maximum emf of 10 V? (b) If the coil rotates at 60 Hz, what is the maximum emf?

**20** • In what magnetic field must the coil of Problem 19 be rotating to generate a maximum emf of 10 V at 60 Hz?

**21** • **SSM** A 2-cm by 1.5-cm rectangular coil has 300 turns and rotates in a magnetic field of 4000 G. (a) What is the maximum emf generated when the coil rotates at 60 Hz? (b) What must its frequency be to generate a maximum emf of 110 V?

**22** • The coil of Problem 21 rotates at 60 Hz in a magnetic field *B*. What value of *B* will generate a maximum emf of 24 V?

## Alternating Current in a Resistor

**23** • **SSM** A 100-W lightbulb is plugged into a standard 120-V (rms) outlet. Find (a) $I_{rms}$, (b) $I_{max}$, and (c) the maximum power.

**24** • **SOLVE** A circuit breaker is rated for a current of 15 A rms at a voltage of 120 V rms. (a) What is the largest value of $I_{max}$ that the breaker can carry? (b) What average power can be supplied by this circuit?

## Alternating Current in Inductors and Capacitors

**25** • What is the reactance of a 1-mH inductor at (a) 60 Hz, (b) 600 Hz, and (c) 6 kHz?

**26** • **SOLVE** An inductor has a reactance of 100 Ω at 80 Hz. (a) What is its inductance? (b) What is its reactance at 160 Hz?

**27** • **SOLVE** At what frequency would the reactance of a 10-μF capacitor equal that of a 1-mH inductor?

**28** • What is the reactance of a 1-nF capacitor at (a) 60 Hz, (b) 6 kHz, and (c) 6 MHz?

**29** • **SSM** An emf of 10 V maximum and frequency 20 Hz is applied to a 20-μF capacitor. Find (a) $I_{max}$ and (b) $I_{rms}$.

**30** • **SOLVE** At what frequency is the reactance of a 10-μF capacitor (a) 1 Ω, (b) 100 Ω, and (c) 0.01 Ω?

**31** •• **SOLVE** Two ac voltage sources are connected in series with a resistor R = 25 Ω. One source is given by

$V_1 = (5\text{ V}) \cos(\omega t - \alpha),$

and the other source is

$V_2 = (5\text{ V}) \cos(\omega t + \alpha),$

with $\alpha = \pi/6$. (a) Find the current in R using a trigonometric identity for the sum of two cosines. (b) Use phasor diagrams to find the current in R. (c) Find the current in R if $\alpha = \pi/4$ and the amplitude of $V_2$ is increased from 5 V to 7 V.

## *LC and RLC Circuits Without a Generator

**32** • **SSM** Show from the definitions of the henry and the farad that $1/\sqrt{LC}$ has the unit s⁻¹.

**33** • (a) What is the period of oscillation of an *LC* circuit consisting of a 2-mH coil and a 20-μF capacitor? (b) What inductance is needed with an 80-μF capacitor to construct an *LC* circuit that oscillates with a frequency of 60 Hz?

**34** •• An $LC$ circuit has capacitance $C_1$ and inductance $L_1$. A second circuit has capacitance $C_2 = \frac{1}{2}C_1$ and $L_2 = 2L_1$, and a third circuit has capacitance $C_3 = 2C_1$ and $L_3 = \frac{1}{2}L_1$. (a) Show that each circuit oscillates with the same frequency. (b) In which circuit would the maximum current be greatest if the capacitor in each were charged to the same potential $V$?

**35** •• **ISOLVE** A 5-$\mu$F capacitor is charged to 30 V and is then connected across a 10-mH inductor. (a) How much energy is stored in the system? (b) What is the frequency of oscillation of the circuit? (c) What is the maximum current in the circuit?

**36** • **ISOLVE✓** A coil can be considered to be a resistance and an inductance in series. Assume that $R = 100\ \Omega$ and $L = 0.4$ H. The coil is connected across a 120-V rms, 60-Hz line. Find (a) the power factor, (b) the rms current, and (c) the average power supplied.

**37** •• **SSM** An inductor and a capacitor are connected, as shown in Figure 29-30. With the switch open, the left plate of the capacitor has charge $Q_0$. The switch is closed and the charge and current vary sinusoidally with time. (a) Plot both $Q$ versus $t$ and $I$ versus $t$ and explain how to interpret these two plots to illustrate that the current leads the charge by 90°. (b) Using a trig identity, show the expression for the current (Equation 29-38) leads the expression for the charge (Equation 29-39) by 90°. That is, show

$$I = -I_{peak} \sin \omega t = I_{peak} \cos\left(\omega t + \frac{\pi}{2}\right).$$

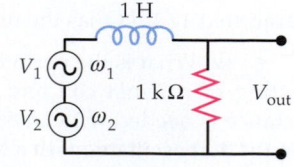

**FIGURE 29-30** Problem 37

### RL Circuits With a Generator

**38** •• **ISOLVE✓** A resistance $R$ and a 1.4-H inductance are in series across a 60-Hz ac voltage. The voltage across the resistor is 30 V and the voltage across the inductor is 40 V. (a) What is the resistance $R$? (b) What is the ac input voltage?

**39** •• **ISOLVE✓** A coil has a dc resistance of 80 $\Omega$ and an impedance of 200 $\Omega$ at a frequency of 1 kHz. Neglect the wiring capacitance of the coil at this frequency. What is the inductance of the coil?

**40** •• A single transmission line carries two voltage signals given by $V_1 = 10$ V cos $100t$ and $V_2 = 10$ V cos $10,000\,t$, where $t$ is in seconds. A series inductor of 1 H and a shunting resistor of 1 k$\Omega$ are inserted into the transmission line, as indicated in Figure 29-31. (a) What is the voltage signal observed at the output side of the transmission line? (b) What is the ratio of the low-frequency amplitude to the high-frequency amplitude?

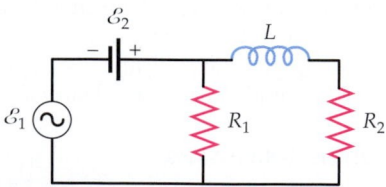

**FIGURE 29-31** Problem 40

**41** •• **ISOLVE** A coil with resistance and inductance is connected to a 120-V rms, 60-Hz line. The average power supplied to the coil is 60 W, and the rms current is 1.5 A. Find (a) the power factor, (b) the resistance of the coil, and (c) the inductance of the coil. (d) Does the current lag or lead the voltage? What is the phase angle $\delta$?

**42** •• **ISOLVE✓** A 36-mH inductor with a resistance of 40 $\Omega$ is connected to a source whose voltage is $\mathcal{E} = 345$ V cos $150\pi t$, where $t$ is in seconds. Determine the maximum current in the circuit, the maximum and rms voltages across the inductor, the average power dissipation, and the maximum and average energy stored in the magnetic field of the inductor.

**43** •• A coil of resistance $R$, inductance $L$, and negligible capacitance has a power factor of 0.866 at a frequency of 60 Hz. What is the power factor for a frequency of 240 Hz?

**44** •• **SSM** A resistor and an inductor are connected in parallel across an emf $\mathcal{E} = \mathcal{E}_{max}$ as shown in Figure 29-32. Show that (a) the current in the resistor is $I_R = \mathcal{E}_{max}/R \cos \omega t$, (b) the current in the inductor is $I_L = \mathcal{E}_{max}/X_L \cos(\omega t - 90°)$, and (c) $I = I_R + I_L = I_{max} \cos(\omega t - \delta)$, where $\tan \delta = R/X_L$ and $I_{max} = \mathcal{E}_{max}/Z$ with $Z^{-2} = R^{-2} + X_L^{-2}$.

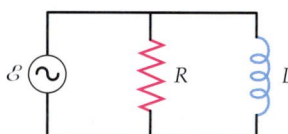

**FIGURE 29-32** Problem 44

**45** •• Figure 29-33 shows a load resistor $R_L = 20\ \Omega$ connected to a high-pass filter consisting of an inductor $L = 3.2$ mH and a resistor $R = 4\ \Omega$. The input voltage is $\mathcal{E} = 100$ V cos $2\pi ft$. Find the rms currents in $R$, $L$, and $R_L$ if (a) $f = 500$ Hz and (b) $f = 2000$ Hz. (c) What fraction of the total power delivered by the voltage source is dissipated in the load resistor if the frequency is 500 Hz and if the frequency is 2000 Hz?

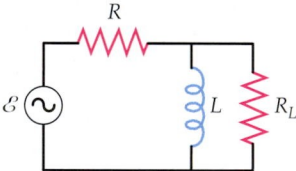

**FIGURE 29-33** Problem 45

**46** •• An ac source $\mathcal{E}_1 = 20$ V cos $2\pi ft$ in series with a battery whose emf is $\mathcal{E}_2 = 16$ V is connected to a circuit consisting of resistors $R_1 = 10\ \Omega$ and $R_2 = 8\ \Omega$ and an inductor $L = 6$ mH (Figure 29-34). Find the power dissipated in $R_1$ and $R_2$ if (a) $f = 100$ Hz, (b) $f = 200$ Hz, and (c) $f = 800$ Hz.

**FIGURE 29-34** Problem 46

**47** •• A 100-V rms voltage is applied to a series $RC$ circuit. The rms voltage across the capacitor is 80 V. What is the voltage across the resistor?

## Filters and Rectifiers

**48** •• **SSM** The circuit shown in Figure 29-35 is called an *RC* high-pass filter because it transmits signals with a high-input frequency with greater amplitude than low-frequency signals. If the input voltage is $V_{in} = V_{peak} \cos \omega t$, show that the output voltage is $V_{out} = V_H \cos(\omega t - \delta)$ where

$$V_H = \frac{V_{peak}}{\sqrt{1 + \left(\dfrac{1}{\omega RC}\right)^2}}$$

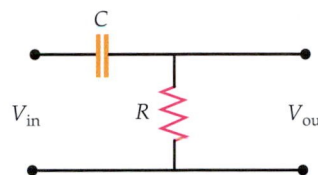

**FIGURE 29-35** Problem 48

**49** •• (*a*) Show that the phase constant $\delta$ in Problem 48 is given by

$$\tan \delta = -\left(\frac{1}{\omega RC}\right)$$

(*b*) What is the value of $\delta$ in the limit as $\omega \to 0$? (*c*) What is the value of $\delta$ in the limit $\omega \to \infty$?

**50** •• Assume that the resistor of Problem 48 has value $R = 20 \text{ k}\Omega$ and the capacitor has value $C = 15$ nF. (*a*) At what frequency $f$ is $V_{out} = V_{in}/\sqrt{2}$? (This is known as the 3 dB frequency, or $f_{3dB}$ for the circuit.) (*b*) Using a spreadsheet program, make a graph of $V_{out}$ versus $f$. Use a logarithmic scale for each variable. Make sure that the scale extends from at least $0.1 f_{3dB}$ to $10 f_{3dB}$ (*c*) Make a graph of $\delta$ versus $f$ and graph $f$ on a logarithmic scale. What value does $\delta$ have at $f = f_{3dB}$?

**51** ••• Show that if an arbitrary voltage signal is fed into the high-pass filter of Problem 48, in which the time variance of the signal is much slower than $1/(RC)$, the output of the circuit will be proportional to the time derivative of the input.

**52** •• We define the output from the high-pass filter from Problem 48 in the decibel scale as

$$\beta = 20 \log_{10} \frac{V_H}{V_{peak}}$$

Show that for $f \ll f_{3dB}$, where $f_{3dB}$ is defined in Problem 50, the output drops at a rate of 6 dB per octave. That is, every time the frequency is halved, the output drops by 6 dB.

**53** •• **SSM** Show that the average power dissipated in the resistor of the high-pass filter of Problem 48 is given by

$$P_{ave} = \frac{V_{peak}^2}{2R}\left(\frac{(\omega RC)^2}{1 + (\omega RC)^2}\right)$$

**54** •• One application of the high-pass filter of Problem 48 is that of a noise filter for electronic circuits (i.e., one that blocks out low-frequency noise). Using $R = 20 \text{ k}\Omega$, pick a value for $C$ for a high-pass filter that attenuates an input voltage signal at $f = 60$ Hz by a factor of 10.

**55** •• The circuit shown in Figure 29-36 is a low-pass filter. If the input voltage is

$$V_{in} = V_{peak} \cos \omega t \text{ show that the output voltage is}$$
$$V_{out} = V_L \cos(\omega t - \delta) \text{ where}$$
$$V_L = \frac{V_{peak}}{\sqrt{1 + (\omega RC)^2}}$$

Discuss the behavior of the output voltage in the limiting cases $\omega \to 0$ and $\omega \to \infty$.

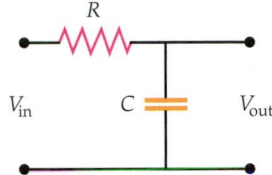

**FIGURE 29-36** Problem 55

**56** •• Show that $\delta$ for the low-pass filter of Problem 55 is given by the expression $\tan \delta = \omega RC$. Find the value of $\delta$ in the limit $\omega \to 0$ and $\omega \to \infty$.

**57** •• **SSM** Using a spreadsheet program, make a graph of $V_L$ versus $f = \omega/2\pi$ and $\delta$ versus $f$ for the low-pass filter of Problem 55. Use $R = 10 \text{ k}\Omega$ and $C = 5$ nF.

**58** ••• Show that if an arbitrary voltage signal is fed into the low-pass filter of Problem 55, in which the time variance of the signal is much faster than $1/(RC)$, the output of the circuit will be proportional to the integral of the input.

**59** ••• **SSM** Show the *trap* filter, shown in Figure 29-37, acts to reject signals at a frequency $\omega = 1/\sqrt{LC}$. How does the width of the frequency band rejected depend on the resistance $R$?

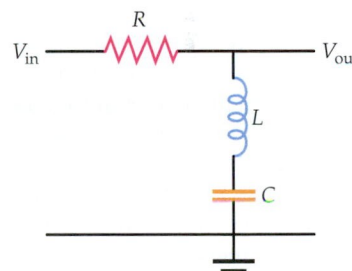

**FIGURE 29-37** Problem 59

**60** •• A half-wave rectifier for transforming an ac voltage into a dc voltage is shown in Figure 29-38. The diode in the figure can be thought of as a one-way valve for current, allowing current to pass in the forward (upward) direction when the voltage between points $A$ and $B$ is greater than $+0.6$ V. The resistance of the diode is effectively infinite when the voltage is less than $+0.6$ V. Using the same axes plot two cycles of both $V_{in}$ and $V_{out}$ versus $t$ when $V_{in} = V_{peak} \cos \omega t$.

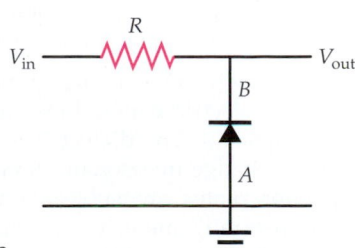

**FIGURE 29-38** Problem 60

**61** •• (a) The output of rectifier of Problem 60, Figure 29-39a, can be smoothed by putting its output through a low-pass filter. The resulting output is a dc voltage with a small amount of ripple on it, as shown in Figure 29-39b. If the input frequency $f = \omega/2\pi = 60$ Hz and the resistance is $R = 1$ k$\Omega$, find an approximate value for $C$, so that the output voltage varies by less than 50 percent of the mean value over one cycle.

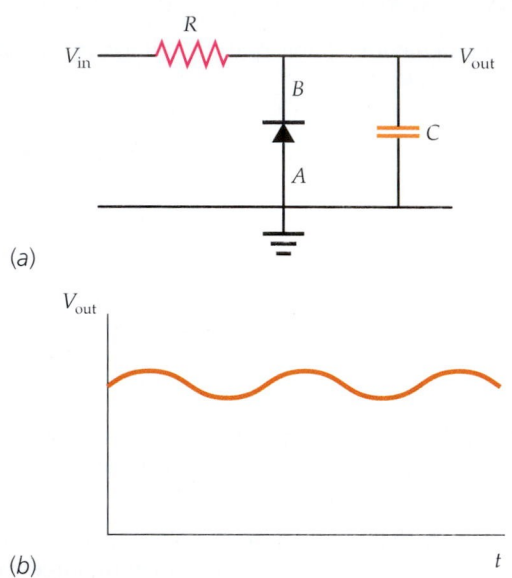

(a)

(b)                                            $t$

**FIGURE 29-39** Problem 61

## LC Circuits With a Generator

**62** •• The generator voltage in Figure 29-40 is given by $\mathcal{E} = (100 \text{ V}) \cos 2\pi ft$. (a) For each branch, what is the amplitude of the current and what is its phase relative to the applied voltage? (b) What is the angular frequency $\omega$ so that the current in the generator vanishes? (c) At this resonance, what is the current in the inductor? What is the current in the capacitor? (d) Draw a phasor diagram showing the general relationships between the applied voltage, the generator current, the capacitor current, and the inductor current for the case where the inductive reactance is larger than the capacitive reactance.

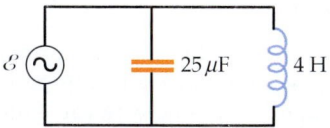

**FIGURE 29-40** Problem 62

**63** •• **iSOLVE** The charge on the capacitor of a series $LC$ circuit is given by $Q = (15 \ \mu\text{C}) \cos(1250t + \frac{\pi}{4})$, where $t$ is in seconds. (a) Find the current as a function of time. (b) Find $C$ if $L = 28$ mH. (c) Write expressions for the electrical energy $U_e$, the magnetic energy $U_m$, and the total energy $U$.

**64** ••• **SSM** One method for measuring the compressibility of a dielectric material uses an $LC$ circuit with a parallel-plate capacitor. The dielectric is inserted between the plates and the change in resonance frequency is determined as the capacitor plates are subjected to a compressive stress. In such an arrangement, the resonance frequency is 120 MHz when a dielectric of thickness 0.1 cm and dielectric constant

$\kappa = 6.8$ is placed between the capacitor plates. Under a compressive stress of 800 atm, the resonance frequency decreases to 116 MHz. Find Young's modulus of the dielectric material.

**65** ••• Figure 29-41 shows an inductance $L$ and a parallel plate capacitor of width $w = 20$ cm and thickness 0.2 cm. A dielectric with dielectric constant $\kappa = 4.8$ that can completely fill the space between the capacitor plates can be slid between the plates. The inductor has an inductance $L = 2$ mH. When half the dielectric is between the capacitor plates (i.e., when $x = \frac{1}{2} w$), the resonant frequency of this $LC$ combination is 90 MHz. (a) What is the capacitance of the capacitor without the dielectric? (b) Find the resonance frequency as a function of $x$.

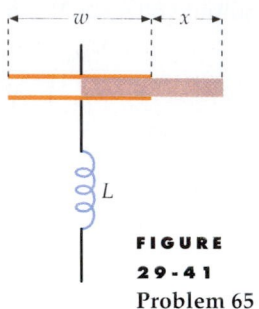

**FIGURE 29-41** Problem 65

## RLC Circuits With a Generator

**66** • A series $RLC$ circuit in a radio receiver is tuned by a variable capacitor, so that it can resonate at frequencies from 500 to 1600 kHz. If $L = 1 \ \mu$H, find the range of capacitances necessary to cover this range of frequencies.

**67** • (a) Find the power factor for the circuit in Example 29-5 when $\omega = 400$ rad/s. (b) At what angular frequency is the power factor 0.5?

**68** • **iSOLVE** An ac generator with a maximum emf of 20 V is connected in series with a 20-$\mu$F capacitor and an 80-$\Omega$ resistor. There is no inductance in the circuit. Find (a) the power factor, (b) the rms current, and (c) the average power if the angular frequency of the generator is 400 rad/s.

**69** •• **SSM** Show that the formula $P_{av} = R\mathcal{E}_{rms}^2/Z^2$ gives the correct result for a circuit containing only a generator and (a) a resistor, (b) a capacitor, and (c) an inductor.

**70** •• **iSOLVE** A series $RLC$ circuit with $L = 10$ mH, $C = 2 \ \mu$F, and $R = 5 \ \Omega$ is driven by a generator with a maximum emf of 100 V and a variable angular frequency $\omega$. Find (a) the resonant frequency $\omega_0$ and (b) $I_{rms}$ at resonance. When $\omega = 8000$ rad/s, find (c) $X_C$ and $X_L$, (d) $Z$ and $I_{rms}$, and (e) the phase angle $\delta$.

**71** •• For the circuit in problem 70, let the generator frequency be $f = \omega/2\pi = 1$ kHz. Find (a) the resonance frequency $f_0 = \omega_0/2\pi$, (b) $X_C$ and $X_L$, (c) the total impedance $Z$ and $I_{rms}$, and (d) the phase angle $\delta$.

**72** •• Find the power factor and the phase angle $\delta$ for the circuit in Problem 70 when the generator frequency is (a) 900 Hz, (b) 1.1 kHz, and (c) 1.3 kHz.

**73** •• Find (a) the $Q$ factor and (b) the resonance width for the circuit in Problem 70. (c) What is the power factor when $\omega = 8000$ rad/s?

**74** •• **SSM** **iSOLVE** ✓ FM radio stations have carrier frequencies that are separated by 0.20 MHz. When the radio is tuned to a station, such as 100.1 MHz, the resonance width of the receiver circuit should be much smaller than 0.2 MHz, so that adjacent stations are not received. If $f_0 = 100.1$ MHz and $\Delta f = 0.05$ MHz, what is the $Q$ factor for the circuit?

**75** •• **ISOLVE** A coil is connected to a 60-Hz, 100-V ac generator. At this frequency, the coil has an impedance of 10 Ω and a reactance of 8 Ω. (a) What is the current in the coil? (b) What is the phase angle between the current and the applied voltage? (c) What series capacitance is required so that the current and voltage are in phase? (d) What is the voltage measured across the capacitor?

**76** •• An 0.25-H inductor and a capacitor C are connected in series with a 60-Hz ac generator. An ac voltmeter is used to measure the rms voltages across the inductor and capacitor separately. The rms voltage across the capacitor is 75 V and that across the inductor is 50 V. (a) Find the capacitance C and the rms current in the circuit. (b) What would be the measured rms voltage across both the capacitor and inductor together?

**77** •• (a) Show that Equation 29-47 can be written as

$$\tan\delta = \frac{L(\omega^2 - \omega_0^2)}{\omega R}$$

Find δ approximately at (b) very low frequencies and (c) very high frequencies.

**78** •• (a) Show that in a series RC circuit with no inductance, the power factor is given by

$$\cos\delta = \frac{RC\omega}{\sqrt{1 + (RC\omega)^2}}$$

(b) Using a spreadsheet program, graph the power factor versus ω.

**79** •• **SSM** In the circuit shown in Figure 29-42, the ac generator produces an rms voltage of 115 V when operated at 60 Hz. What is the rms voltage across points (a) AB, (b) BC, (c) CD, (d) AC, and (e) BD?

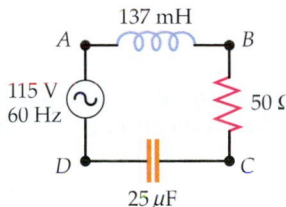

**FIGURE 29-42** Problem 79

**80** •• When an RLC series circuit is connected to a 120-V rms, 60-Hz line, the current is $I_{rms} = 11$ A and the current leads the voltage by 45°. (a) Find the power supplied to the circuit. (b) What is the resistance? (c) If the inductance L = 0.05 H, find the capacitance C. (d) What capacitance or inductance should you add to make the power factor 1?

**81** •• **ISOLVE** ✔ A series RLC circuit is driven at a frequency of 500 Hz. The phase angle between the applied voltage and current is determined from an oscilloscope measurement to be δ = 75°. If the total resistance is known to be 35 Ω and the inductance is 0.15 H, what is the capacitance of the circuit?

**82** •• A series RLC circuit with R = 400 Ω, L = 0.35 H, and C = 5 μF is driven by a generator of variable frequency f. (a) What is the resonance frequency $f_0$? Find f and $f/f_0$ when the phase angle δ is (b) 60° and (c) −60°.

**83** •• Sketch the impedance Z versus ω for (a) a series LR circuit, (b) a series RC circuit, and (c) a series RLC circuit.

**84** •• **SSM** Show that Equation 29-48 can be written as

$$I_{max} = \frac{\omega \mathscr{E}_{max}}{\sqrt{L^2(\omega^2 - \omega_0^2)^2 + \omega^2 R^2}}$$

**85** •• In a series RLC circuit, $X_C = 16$ Ω and $X_L = 4$ Ω at some frequency. The resonance frequency is $\omega_0 = 10^4$ rad/s. (a) Find L and C. If R = 5 Ω and $\mathscr{E}_{max} = 26$ V, find (b) the Q factor, and (c) the maximum current.

**86** •• In a series RLC circuit connected to an ac generator whose maximum emf is 200 V, the resistance is 60 Ω and the capacitance is 8 μF. The inductance can be varied from 8 mH to 40 mH, by the insertion of an iron core in the solenoid. The angular frequency of the generator is 2500 rad/s. If the capacitor voltage is not to exceed 150 V, find (a) the maximum current and (b) the range of inductance that is safe to use.

**87** •• A certain electrical device draws 10 A rms and has an average power of 720 W when connected to a 120-V rms, 60-Hz power line. (a) What is the impedance of the device? (b) What series combination of resistance and reactance is this device equivalent to? (c) If the current leads the emf, is the reactance inductive or capacitive?

**88** •• **SSM** A method for measuring inductance is to connect the inductor in series with a known capacitance, a known resistance, an ac ammeter, and a variable-frequency signal generator. The frequency of the signal generator is varied and the emf is kept constant until the current is maximum. (a) If C = 10 μF, $\mathscr{E}_{max} = 10$ V, R = 100 Ω, and I is maximum at ω = 5000 rad/s, what is L? (b) What is $I_{max}$?

**89** •• A resistor and a capacitor are connected in parallel across a sinusoidal emf $\mathscr{E} = \mathscr{E}_{max} \cos\omega t$, as shown in Figure 29-43. (a) Show that the current in the resistor is $I_R = (\mathscr{E}_{max}/R) \cos\omega t$. (b) Show that the current in the capacitor branch is $I_C = (\mathscr{E}_{max}/X_C) \cos(\omega t + 90°)$. (c) Show that the total current is given by $I = I_R + I_C = I_{max} \cos(\omega t + \delta)$, where $\tan\delta = R/X_C$ and $I_{max} = \mathscr{E}_{max}/Z$ with $Z^{-2} = R^{-2} + X_C^{-2}$.

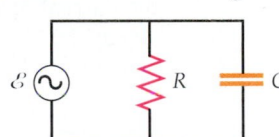

**FIGURE 29-43** Problem 89

**90** •• **SSM** In the circuit shown in Figure 29-44, R = 10 Ω, $R_L = 30$ Ω, L = 150 mH, and C = 8 μF; the frequency of the ac source is 10 Hz and its amplitude is 100 V. (a) Using phasor diagrams, determine the impedance of the circuit when switch S is closed. (b) Determine the impedance of the circuit when switch S is open. (c) What are the voltages across the load resistor $R_L$ when switch S is closed and when it is open? (d) Repeat Parts (a), (b), and (c) with the frequency of the source changed to 1000 Hz. (e) Which arrangement is a better low-pass filter, S open or S closed?

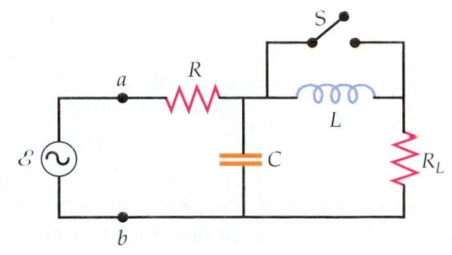

**FIGURE 29-44**
Problem 90

**91** •• In the circuit shown in Figure 29-45, $R_1 = 2\ \Omega$, $R_2 = 4\ \Omega$, $L = 12$ mH, $C = 30\ \mu$F, and $\mathcal{E} = (40\ \text{V})\cos \omega t$. (a) Find the resonance frequency. (b) At the resonance frequency, what are the rms currents in each resistor and the rms current supplied by the source emf?

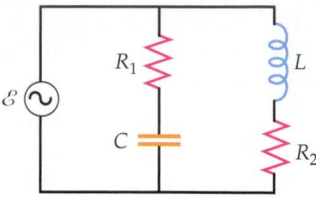

**FIGURE 29-45** Problems 91, 103, and 104

**92** •• For the circuit in Figure 29-24, derive an expression for the $Q$ of the circuit, assuming the resonance is sharp.

**93** •• **iSOLVE**✓ For the circuit in Figure 29-24, $L = 4$ mH. (a) What capacitance $C$ will result in a resonance frequency of 4 kHz? (b) When $C$ has the value found in Part (a), what should be the resistance $R$, so that the $Q$ of the circuit is 8?

**94** •• If the capacitance of $C$ in Problem 93 is reduced to half the value found in Problem 93, what then are the resonance frequency and the $Q$ of the circuit? What should be the resistance $R$ to give $Q = 8$?

**95** •• **iSOLVE** A series circuit consists of a 4.0-nF capacitor, a 36-mH inductor, and a 100-$\Omega$ resistor. The circuit is connected to a 20-V ac source whose frequency can be varied over a wide range. (a) Find the resonance frequency $f_0$ of the circuit. (b) At resonance, what is the rms current in the circuit and what are the rms voltages across the inductor and capacitor? (c) What is the rms current and what are the rms voltages across the inductor and capacitor at $f = f_0 + \frac{1}{2}\Delta f$, where $\Delta f$ is the width of the resonance?

**96** ••• In the parallel circuit shown in Figure 29-46, $V_{max} = 110$ V. (a) What is the impedance of each branch? (b) For each branch, what is the current amplitude and its phase relative to the applied voltage? (c) Give the current phasor diagram, and use it to find the total current and its phase relative to the applied voltage.

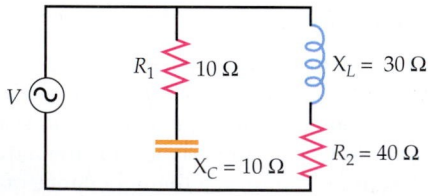

**FIGURE 29-46** Problem 96

**97** ••• **SSM** (a) Show that Equation 29-47 can be written as

$$\tan \delta = \frac{Q(\omega^2 - \omega_0^2)}{\omega \omega_0}$$

(b) Show that near resonance

$$\tan \delta \approx \frac{2Q(\omega - \omega)}{\omega}$$

(c) Sketch a plot of $\delta$ versus $x$, where $x = \omega/\omega_0$, for a circuit with high $Q$ and for one with low $Q$.

**98** ••• Show by direct substitution that the current given by Equation 29-46 with $\delta$ and $I_{max}$ given by Equations 29-47 and 29-48, respectively, satisfies Equation 29-45. (*Hint:* Use trigonometric identities for the sine and cosine of the sum of two angles, and write the equation in the form $A \sin \omega t + B \cos \omega t = 0$. Because this equation must hold for all times, $A = 0$ and $B = 0$.)

**99** ••• An ac generator is in series with a capacitor and an inductor in a circuit with negligible resistance. (a) Show that the charge on the capacitor obeys the equation

$$L\frac{d^2Q}{dt^2} + \frac{Q}{C} = \mathcal{E}_{max}\cos \omega t$$

(b) Show by direct substitution that this equation is satisfied by $Q = Q_{max}\cos \omega t$ if,

$$Q_{max} = -\frac{\mathcal{E}_{max}}{L(\omega^2 - \omega_0^2)}$$

(c) Show that the current can be written as $I = I_{max}\cos(\omega t - \delta)$, where

$$I_{max} = \frac{\omega \mathcal{E}_{max}}{L|\omega^2 - \omega_0^2|} = \frac{\mathcal{E}_{max}}{|X_L - X_C|}$$

and $\delta = -90°$ for $\omega < \omega_0$ and $\delta = 90°$ for $\omega > \omega_0$.

**100** ••• Figure 29-20 shows a plot of average power $P_{av}$ versus generator frequency $\omega$ for an $RLC$ circuit with a generator. The average power $P_{av}$ is given by Equation 29-54. The full width at half-maximum, $\Delta\omega$, is the width of the resonance curve between the two points, where $P_{av}$ is one-half its maximum value. Show that for a sharply peaked resonance, $\Delta\omega \approx R/L$ and, hence, that $Q \approx \omega_0/\Delta\omega$ in this case (Equation 29-56). (*Hint:* At resonance, the denominator of the expression on the right of Equation 29-54 is $\omega^2 R^2$. The half-power points will occur when the denominator is twice the value near resonance; that is, when $L^2(\omega^2 - \omega_0^2)^2 = \omega^2 R^2 \approx \omega_0^2 R^2$. Let $\omega_1$ and $\omega_2$ be the solutions of this equation. For a sharply peaked resonance, $\omega_1 \approx \omega_0$ and $\omega_2 \approx \omega_0$. Then, using the fact that $\omega + \omega_0 \approx 2\omega_0$, one finds that $\Delta\omega = \omega_2 - \omega_1 \approx R/L$.)

**101** • Show by direct substitution that

$$L\frac{d^2Q}{dt^2} + R\frac{dQ}{dt} + \frac{1}{C}Q = 0$$

(Equation 29-43b) is satisfied by

$$Q = Q_0 e^{-Rt/2L}\cos \omega' t$$

where

$$\omega' = \sqrt{(1/LC - (R/2L)^2}$$

and $Q_0$ is the charge on the capacitor at $t = 0$.

**102** ••• **SSM** One method for measuring the magnetic susceptibility of a sample uses an $LC$ circuit consisting of an air-core solenoid and a capacitor. The resonant frequency of the circuit without the sample is determined and then measured again with the sample inserted in the solenoid. Suppose the solenoid is 4 cm long, 0.3 cm in diameter, and has 400 turns of fine wire. Assume that the sample that is inserted in the solenoid is also 4 cm long and fills the air space. Neglect end effects. (In practice, a test sample of known susceptibility of the same shape as the unknown is used to calibrate the instrument.) (a) What is the inductance of the empty solenoid? (b) What should be the capacitance of the capacitor so that the resonance frequency of the circuit without a sample is 6.0000 MHz? (c) When a sample is inserted in the solenoid, the resonance frequency drops to 5.9989 MHz. Determine the sample's susceptibility.

**103 •••** (a) Find the angular frequency $\omega$ for the circuit in Problem 91 so that the magnitude of the reactance of the two parallel branches are equal. (b) At that frequency, what is the power dissipation in each of the two resistors?

**104 •••** (a) For the circuit of Problem 91, find the angular frequency $\omega$ for which the power dissipation in the two resistors is the same. (b) At that angular frequency, what is the reactance of each of the two parallel branches? (c) Draw a phasor diagram showing the current through each of the two parallel branches. (d) What is the impedance of the circuit?

### *The Transformer

**105 •** SSM An ac voltage of 24 V is required for a device whose impedance is 12 $\Omega$. (a) What should the turn ratio of a transformer be, so that the device can be operated from a 120-V line? (b) Suppose the transformer is accidentally connected reversed (i.e., with the secondary winding across the 120-V line and the 12-$\Omega$ load across the primary). How much current will then flow in the primary winding?

**106 •** A transformer has 400 turns in the primary and 8 turns in the secondary. (a) Is this a step-up or a step-down transformer? (b) If the primary is connected across 120 V rms, what is the open-circuit voltage across the secondary? (c) If the primary current is 0.1 A, what is the secondary current, assuming negligible magnetization current and no power loss?

**107 •** SOLVE The primary of a step-down transformer has 250 turns and is connected to a 120-V rms line. The secondary is to supply 20 A at 9 V. Find (a) the current in the primary and (b) the number of turns in the secondary, assuming 100 percent efficiency.

**108 •** A transformer has 500 turns in its primary, which is connected to 120 V rms. Its secondary coil is tapped at three places to give outputs of 2.5 V, 7.5 V, and 9 V. How many turns are needed for each part of the secondary coil?

**109 •** The distribution circuit of a residential power line is operated at 2000 V rms. This voltage must be reduced to 240 V rms for use within residences. If the secondary side of the transformer has 400 turns, how many turns are in the primary?

**110 ••** SSM An audio oscillator (ac source) with an internal resistance of 2000 $\Omega$ and an open-circuit rms output voltage of 12 V is to be used to drive a loudspeaker with a resistance of 8 $\Omega$. What should be the ratio of primary to secondary turns of a transformer, so that maximum power is transferred to the speaker? Suppose a second identical speaker is connected in parallel with the first speaker. How much power is then supplied to the two speakers combined?

**111 ••** One use of a transformer is for *impedance matching*. For example, the output impedance of a stereo amplifier is matched to the impedance of a speaker by a transformer. In Equation 29-63, the currents $I_1$ and $I_2$ can be related to the impedance $Z$ in the secondary because $I_2 = V_2/Z$. Using Equations 29-61 and 29-62, show that

$$I_1 = \mathcal{E}/[(N_1/N_2)^2 Z]$$

and, therefore, $Z_{\text{eff}} = (N_1/N_2)^2 Z$.

### General Problems

**112 •** A 5-kW electric clothes dryer runs on 240 V rms. Find (a) $I_{\text{rms}}$ and (b) $I_{\text{max}}$. (c) Find the same quantities for a dryer of the same power that operates at 120 V rms.

**113 •** Find the reactance of a 10.0-$\mu$F capacitor at (a) 60 Hz, (b) 6 kHz, and (c) 6 MHz.

**114 ••** SOLVE A resistance $R$ carries a current $I = 5$ A sin $120\pi t + 7$ A sin $240\pi t$. (a) What is the rms current? (b) If the resistance $R$ is 12 $\Omega$, what is the power dissipated in the resistor? (c) What is the rms voltage across the resistor?

**115 ••** SSM Figure 29-47 shows the voltage $V$ versus time $t$ for a *square-wave* voltage. If $V_0 = 12$ V, (a) what is the rms voltage of this waveform? (b) If this alternating waveform is rectified by eliminating the negative voltages, so that only the positive voltages remain, what now is the rms voltage of the rectified waveform?

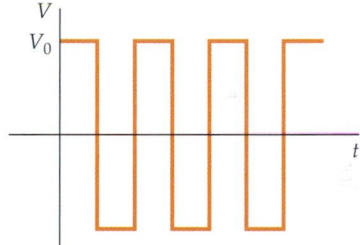

**FIGURE 29-47** Problem 115

**116 ••** SOLVE A pulsed current has a constant value of 15 A for the first 0.1 s of each second and is then 0 for the next 0.9 s of each second. (a) What is the rms value for this current waveform? (b) Each current pulse is generated by a voltage pulse of maximum value 100 V. What is the average power delivered by the pulse generator?

**117 ••** A circuit consists of two capacitors, a 24-V battery, and an ac voltage connected, as shown in Figure 29-48. The ac voltage is given by $\mathcal{E} = 20$ V cos $120\pi t$, where $t$ is in seconds. (a) Find the charge on each capacitor as a function of time. Assume transient effects have had sufficient time to decay. (b) What is the steady-state current? (c) What is the maximum energy stored in the capacitors? (d) What is the minimum energy stored in the capacitors?

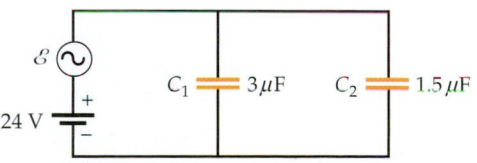

**FIGURE 29-48** Problem 117

**118** •• What are the average values and rms values of current for the two current waveforms shown in Figure 29-49?

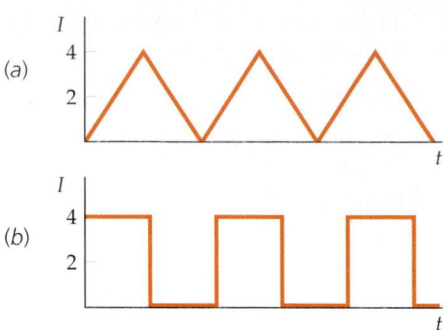

**FIGURE 29-49** Problem 118

**119** •• **iSOLVE** In the circuit shown in Figure 29-50, $\mathcal{E}_1 =$ (20 V) cos $2\pi ft$, $f = 180$ Hz; $\mathcal{E}_2 = 18$ V, and $R = 36$ Ω. Find the maximum, minimum, average, and rms values of the current through the resistor.

**FIGURE 29-50**
**Problems 119 through 121**

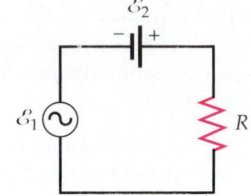

**120** •• **SSM** Repeat Problem 119 if the resistor $R$ is replaced by a 2-$\mu$F capacitor.

**121** •• Repeat Problem 119 if the resistor $R$ is replaced by a 12-mH inductor.

# Maxwell's Equations and Electromagnetic Waves

THE 70-M ANTENNA AT GOLDSTONE, CALIFORNIA. THE GOLDSTONE DEEP SPACE COMMUNICATIONS COMPLEX, LOCATED IN THE MOJAVE DESERT IN CALIFORNIA, IS ONE OF THREE COMPLEXES THAT COMPRISE NASA'S DEEP SPACE NETWORK. THIS NETWORK PROVIDES RADIO COMMUNICATIONS FOR ALL OF NASA'S INTERPLANETARY SPACECRAFT AND IS ALSO UTILIZED FOR RADIO ASTRONOMY AND RADAR OBSERVATIONS OF THE SOLAR SYSTEM AND THE UNIVERSE.

30-1    Maxwell's Displacement Current

30-2    Maxwell's Equations

30-3    Electromagnetic Waves

*30-4   The Wave Equation for Electromagnetic Waves

  **Did you ever wonder whether a radio antenna generates a wave equally in all directions? This topic is discussed in Section 30-3.**

**M**axwell's equations, first proposed by the great Scottish physicist James Clerk Maxwell, relate the electric and magnetic field vectors $\vec{E}$ and $\vec{B}$ and their sources, which are electric charges and currents. These equations summarize the experimental laws of electricity and magnetism—the laws of Coulomb, Gauss, Biot–Savart, Ampère, and Faraday. These experimental laws hold in general except for Ampère's law, which applies only to steady continuous currents.

➤ **In this chapter, we will see how Maxwell was able to generalize Ampère's law with the invention of the displacement current (Section 30-1). Maxwell was then able to show that the generalized laws of electricity and magnetism imply the existence of electromagnetic waves.**

Maxwell's equations play a role in classical electromagnetism analogous to that of Newton's laws in classical mechanics. In principle, all problems in classical electricity and magnetism can be solved using Maxwell's equations, just as all problems in classical mechanics can be solved using Newton's laws. Maxwell's equations are considerably more complicated than Newton's laws, however, and their application to most problems involves mathematics beyond the scope of this book. Nevertheless, Maxwell's equations are of great theoretical importance. For example, Maxwell showed that these equations can be combined to yield a wave equation for the electric and magnetic field vectors $\vec{E}$ and $\vec{B}$. Such **electromagnetic waves** are caused by accelerating charges, (e.g., the charges in an alternating current in an antenna). These electromagnetic waves were first produced in the laboratory by Heinrich Hertz in 1887. Maxwell showed that his equations predicted the speed of electromagnetic waves in free space to be

$$c = \frac{1}{\sqrt{\mu_0 \epsilon_0}}$$  30-1

THE SPEED OF ELECTROMAGNETIC WAVES

where $\epsilon_0$, the permittivity of free space, is the constant appearing in Coulomb's and Gauss's laws and $\mu_0$, the permeability of free space, is the constant appearing in the Biot–Savart law and Ampère's law. Maxwell noticed with great excitement the coincidence that the measure for the speed of light equaled $1/\sqrt{\mu_0 \epsilon_0}$, and Maxwell correctly surmised that light itself is an electromagnetic wave. Today, the value of $c$ is defined as $2.99792458 \times 10^8$ m/s, the value of $\mu_0$ is defined as $4\pi \times 10^7$ N/A$^2$, and the value of $\epsilon_0$ is defined by Equation 30-1.

# 30-1 Maxwell's Displacement Current

Ampère's law (Equation 27-15) relates the line integral of the magnetic field around some closed curve $C$ to the current that passes through any surface bounded by that curve:

$$\oint_C \vec{B} \cdot d\vec{\ell} = \mu_0 I_s, \text{ for any closed curve } C$$  30-2

Maxwell recognized a flaw in Ampère's law. Figure 30-1 shows two different surfaces, $S_1$ and $S_2$, bounded by the same curve $C$, which encircles a wire carrying current to a capacitor plate. The current through surface $S_1$ is $I$, but there is no current through surface $S_2$ because the charge stops on the capacitor plate. Thus, there is ambiguity in the phrase "the current through any surface bounded by the curve." Such a problem arises when the current is not continuous.

Maxwell showed that the law can be generalized to include all situations if the current $I$ in the equation is replaced by the sum of the current $I$ and another term $I_d$, called **Maxwell's displacement current,** defined as

$$I_d = \epsilon_0 \frac{d\phi_e}{dt}$$  30-3

DEFINITION—DISPLACEMENT CURRENT

where $\phi_e$ is the flux of the electric field through the same surface bounded by the curve $C$. The generalized form of Ampère's law is then

$$\oint_C \vec{B} \cdot d\vec{\ell} = \mu_0(I + I_d) = \mu_0 I + \mu_0 \epsilon_0 \frac{d\phi_e}{dt}$$  30-4

GENERALIZED FORM OF AMPÈRE'S LAW

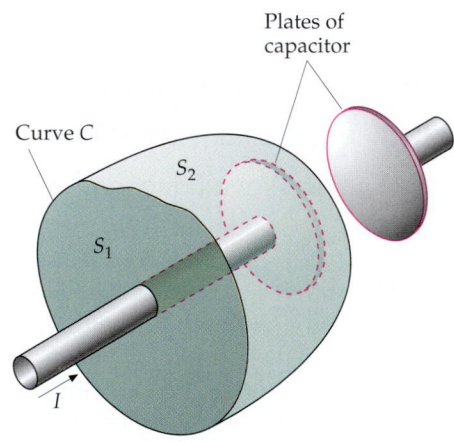

**FIGURE 30-1** Two surfaces $S_1$ and $S_2$ bounded by the same curve $C$. The current $I$ passes through surface $S_1$ but not through surface $S_2$. Ampère's law, which relates the line integral of the magnetic field around the curve $C$ to the total current passing through any surface bounded by $C$, is not valid when the current is not continuous, as when it stops at the capacitor plate here.

We can understand this generalization by considering Figure 30-1 again. Let us call the sum $I + I_d$ the generalized current. According to the argument just stated, the same generalized current must cross any surface bounded by the curve C. Thus, there can be no net generalized current into or out of the volume bounded by the two surfaces $S_1$ and $S_2$, which together form a closed surface. If there is a net current $I$ into the volume, there must be an equal net displacement current $I_d$ out of the volume. In the volume in the figure, there is a net current $I$ into the volume that increases the charge $Q_{inside}$ within the volume:

$$I = \frac{dQ_{inside}}{dt}$$

The flux of the electric field out of the volume is related to the charge by Gauss's law:

$$\phi_{e,\,net} = \oint_S E_n\, dA = \frac{1}{\epsilon_0} Q_{inside}$$

Solving for the charge gives

$$Q_{inside} = \epsilon_0\, \phi_{e,\,net}$$

and taking the derivative of each side gives

$$\frac{dQ_{inside}}{dt} = \epsilon_0 \frac{d\phi_{e,\,net}}{dt}$$

The rate of increase of the charge is thus proportional to the rate of increase of the net flux out of the volume:

$$\frac{dQ_{inside}}{dt} = \epsilon_0 \frac{d\phi_{e,\,net}}{dt} = I_d$$

Thus, the net conduction current into the volume equals the net displacement current out of the volume. The generalized current is thus continuous, and this is *always* the case.

It is interesting to compare Equation 30-4 to Equation 28-5:

$$\mathcal{E} = \oint_C \vec{E} \cdot d\vec{\ell} = -\frac{d\phi_m}{dt} = -\int_S \frac{\partial B_n}{\partial t}\, dA \qquad\qquad 30\text{-}5$$

which in this chapter will be referred to as Faraday's law. (Equation 30-5 is a restricted form of Faraday's law, a form that does not include motional emfs. Equation 30-5 does include emfs associated with a time varying magnetic field.) According to Faraday's law, a changing magnetic flux produces an electric field whose line integral around a closed curve is proportional to the rate of change of magnetic flux through any surface bounded by the curve. Maxwell's modification of Ampère's law shows that a changing electric flux produces a magnetic field whose line integral around a curve is proportional to the rate of change of the electric flux. We thus have the interesting reciprocal result that a changing magnetic field produces an electric field (Faraday's law) and a changing electric field produces a magnetic field (generalized form of Ampère's law). Note, there is no magnetic analog of a current $I$. This is because the magnetic monopole, the magnetic analog of an electric charge, does not exist.[†]

† The question of the existence of magnetic monopoles has theoretical importance. There have been numerous attempts to observe magnetic monopoles but to date no one has been successful at doing so.

**EXAMPLE 30-1**

A parallel-plate capacitor has closely spaced circular plates of radius $R$. Charge is flowing onto the positive plate and off the negative plate at the rate $I = dQ/dt = 2.5$ A. Compute the displacement current through surface $S$ passing between the plates (Figure 30-2) by directly computing the rate of change of the flux of $\vec{E}$ through surface $S$.

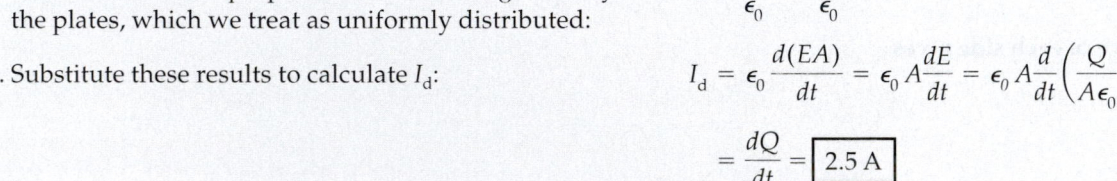

**PICTURE THE PROBLEM** The displacement current is $I_d = \epsilon_0 \, d\phi_e/dt$, where $\phi_e$ is the electric flux through the surface between the plates. Since the parallel plates are closely spaced, in the region between the plates the electric field is uniform and perpendicular to the plates. Outside the capacitor the electric field is negligible. Thus, the electric flux is simply $\phi_e = EA$, where $E$ is the electric field between the plates and $A$ is the plate area.

**FIGURE 30-2** The surface $S$ passes between the capacitor plates. The charge $Q$ is increasing at 2.5 C/s = 2.5 A. The distance between the plates is not drawn to scale. The plates are much closer together than the plates shown in the figure.

1. The displacement current is found by taking the time derivative of the electric flux:

$$I_d = \epsilon_0 \frac{d\phi_e}{dt}$$

2. The flux equals the electric field magnitude times the plate area:

$$\phi_e = EA$$

3. The electric field is proportional to the charge density on the plates, which we treat as uniformly distributed:

$$E = \frac{\sigma}{\epsilon_0} = \frac{Q/A}{\epsilon_0}$$

4. Substitute these results to calculate $I_d$:

$$I_d = \epsilon_0 \frac{d(EA)}{dt} = \epsilon_0 A \frac{dE}{dt} = \epsilon_0 A \frac{d}{dt}\left(\frac{Q}{A\epsilon_0}\right)$$

$$= \frac{dQ}{dt} = \boxed{2.5 \text{ A}}$$

**REMARKS** Note that the displacement current through the surface passing between the plates of the capacitor is equal to the current in the wires carrying charge to and from the capacitor.

**EXAMPLE 30-2**

The circular plates in Example 30-1 have a radius of $R = 3.0$ cm. Find the magnetic field strength $B$ at a point between the plates a distance $r = 2.0$ cm from the axis of the plates when the current into the positive plate is 2.5 A.

**PICTURE THE PROBLEM** We find $B$ from the generalized form of Ampère's law (Equation 30-4). We chose a circular path $C$ of radius $r = 2.0$ cm about the centerline joining the plates, as shown in Figure 30-3. We then calculate the displacement current through the surface $S$ bounded by $C$. By symmetry, $\vec{B}$ is tangent to $C$ and has the same magnitude everywhere on $C$.

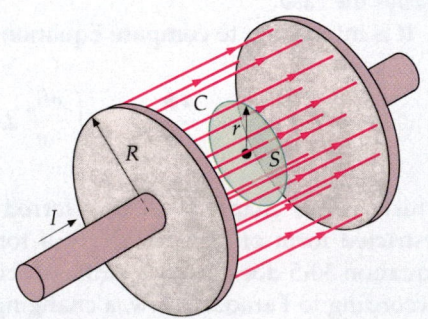

1. We find $B$ from the generalized form of Ampère's law:

$$\oint_C \vec{B} \cdot d\vec{\ell} = \mu_0(I + I_d)$$

where

$$I_d = \epsilon_0 \frac{d\phi_e}{dt}$$

**FIGURE 30-3** The space distance between the plates is not drawn to scale. The plates are much closer together than they appear.

2. The line integral is $B$ times the circumference of the circle:

$$\oint_C \vec{B} \cdot d\vec{\ell} = B(2\pi r)$$

3. Since there are no moving charges between the plates of the capacitor, $I = 0$. The generalized current through $S$ is just the displacement current:

$$\oint_C \vec{B} \cdot d\vec{\ell} = \mu_0 I + \mu_0 \epsilon_0 \frac{d\phi_e}{dt}$$

$$B(2\pi r) = 0 + \mu_0 \epsilon_0 \frac{d\phi_e}{dt}$$

4. The electric flux equals the product of the uniform field $E$ and the area of the flat surface $S$ bounded by the curve $C$:

$$\phi_e = \pi r^2 E = \pi r^2 \frac{\sigma}{\epsilon_0} = \pi r^2 \frac{Q}{\epsilon_0 \pi R^2}$$

$$= \frac{Q r^2}{\epsilon_0 R^2}$$

5. Substitute these results into step 3 and solve for $B$:

$$B(2\pi r) = \mu_0 \epsilon_0 \frac{d}{dt}\left(\frac{Q r^2}{\epsilon_0 R^2}\right) = \mu_0 \frac{r^2}{R^2} \frac{dQ}{dt}$$

$$B = \frac{\mu_0}{2\pi} \frac{r}{R^2} \frac{dQ}{dt} = \frac{\mu_0}{2\pi} \frac{r}{R^2} I$$

$$= (2 \times 10^{-7}\ \text{T·m/A}) \frac{0.02\ \text{m}}{(0.03\ \text{m})^2} (2.5\ \text{A})$$

$$= \boxed{1.11 \times 10^{-5}\ \text{T}}$$

## 30-2 Maxwell's Equations

Maxwell's equations are

$$\oint_S E_n\, dA = \frac{1}{\epsilon_0} Q_{\text{inside}}$$

30-6a

$$\oint_S B_n\, dA = 0$$

30-6b

$$\oint_C \vec{E} \cdot d\vec{\ell} = -\frac{d}{dt}\int_S B_n\, dA = -\int_S \frac{\partial B_n}{\partial t}\, dA$$

30-6c

$$\oint_C \vec{B} \cdot d\vec{\ell} = \mu_0(I + I_d)$$

$$= \mu_0 I + \mu_0 \epsilon_0 \frac{d}{dt}\int_S E_n\, dA = \mu_0 I + \mu_0 \epsilon_0 \int_S \frac{\partial E_n}{\partial t}\, dA$$

30-6d

MAXWELL'S EQUATIONS[†]

Equation 30-6a is Gauss's law; it states that the flux of the electric field through any closed surface equals $1/\epsilon_0$ times the net charge inside the surface. As discussed in Chapter 22, Gauss's law implies that the electric field due to a point charge varies inversely as the square of the distance from the charge. This law describes how electric field lines diverge from a positive charge and converge on a negative charge. Its experimental basis is Coulomb's law.

[†] In all four equations, the integration paths $C$ and the integration surfaces $S$ are at rest and the integrations take place at an instant in time.

Equation 30-6b, sometimes called Gauss's law for magnetism, states that the flux of the magnetic field vector $\vec{B}$ is zero through *any* closed surface. This equation describes the experimental observation that magnetic field lines do not diverge from any point in space or converge on any point; that is, it implies that isolated magnetic poles do not exist.

Equation 30-6c is Faraday's law; it states that the integral of the electric field around any closed curve C, which is the emf, equals the (negative) rate of change of the magnetic flux through any surface S bounded by the curve. (S is not a closed surface, so the magnetic flux through S is not necessarily zero.) Faraday's law describes how electric field lines encircle any area through which the magnetic flux is changing, and it relates the electric field vector $\vec{E}$ to the rate of change of the magnetic field vector $\vec{B}$.

Equation 30-6d, which is Ampère's law modified to include Maxwell's displacement current, states that the line integral of the magnetic field $\vec{B}$ around any closed curve C equals $\mu_0$ times the current through any surface S bounded by the curve plus $\mu_0 \epsilon_0$ times the rate of change of the electric flux through the same surface S. This law describes how the magnetic field lines encircle an area through which a current is passing or through which the electric flux is changing.

In Section 30-4, we show how wave equations for both the electric field $\vec{E}$ and the magnetic field $\vec{B}$ can be derived from Maxwell's equations.

# 30-3 Electromagnetic Waves

Figure 30-4 shows the electric and magnetic field vectors of an electromagnetic wave. The electric and magnetic fields are perpendicular to each other and perpendicular to the direction of propagation of the wave. Electromagnetic waves are thus transverse waves. The electric and magnetic fields are in phase and, at each point in space and at each instant in time, their magnitudes are related by

$$E = cB \qquad\qquad 30\text{-}7$$

where $c = 1/\sqrt{\mu_0 \epsilon_0}$ is the speed of the wave. The direction of propagation of an electromagnetic wave is the direction of the cross product $\vec{E} \times \vec{B}$.

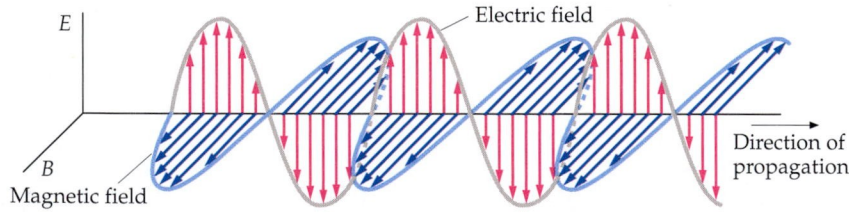

**FIGURE 30-4** The electric and magnetic field vectors in an electromagnetic wave. The fields are in phase, perpendicular to each other, and perpendicular to the direction of propagation of the wave.

## The Electromagnetic Spectrum

The various types of electromagnetic waves—light, radio waves, X rays, gamma rays, microwaves, and others—differ only in wavelength and frequency, which are related to the speed c in the usual way, $f\lambda = c$. Table 30-1 gives the **electromagnetic spectrum** and the names usually associated with the various frequency and wavelength ranges. These ranges are often not well defined and sometimes overlap. For example, electromagnetic waves with wavelengths of approximately 0.1 nm are usually called X rays, but if the electromagnetic waves originate from nuclear radioactivity, they are called gamma rays.

The human eye is sensitive to electromagnetic radiation with wavelengths from approximately 400 nm to 700 nm, which is the range called **visible light.** The shortest wavelengths in the visible spectrum correspond to violet light and

## TABLE 30-1

**The Electromagnetic Spectrum**

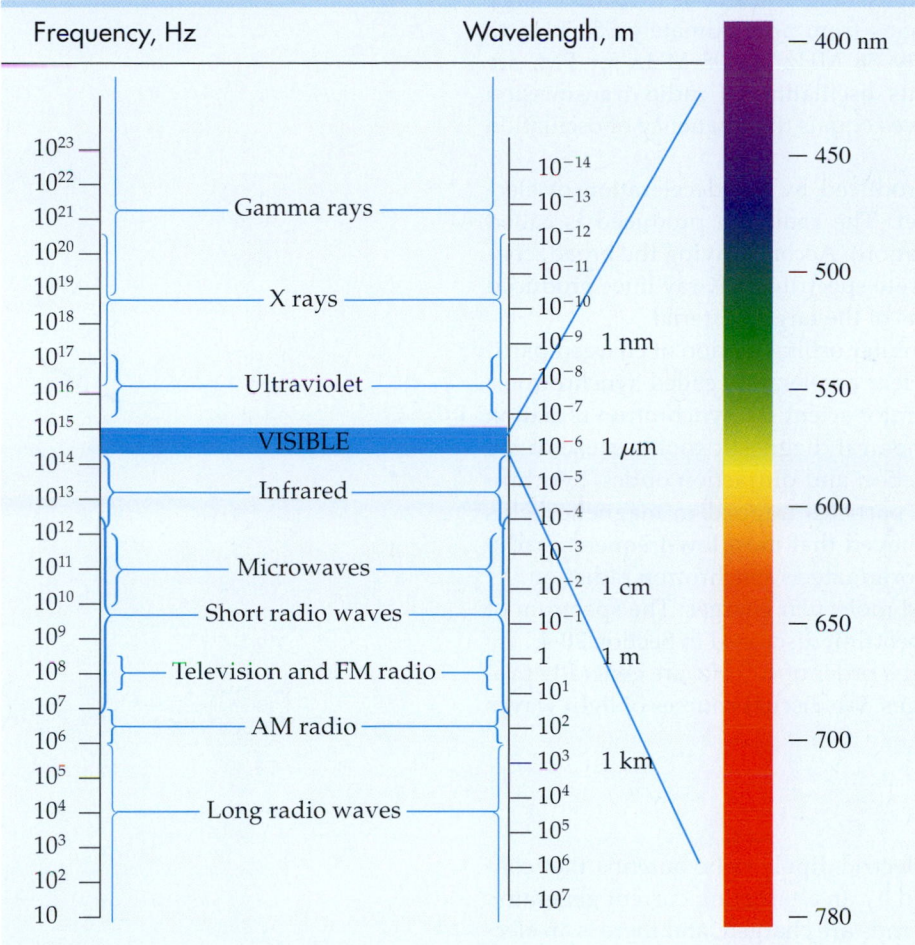

the longest wavelengths to red light, with all the colors of the rainbow falling between these extremes. Electromagnetic waves with wavelengths just beyond the visible spectrum on the short-wavelength side are called **ultraviolet rays,** and those with wavelengths just beyond the visible spectrum on the long-wavelength side are called **infrared waves.** Heat radiation given off by bodies at ordinary temperatures is in the infrared region of the electromagnetic spectrum. There are no limits on the wavelengths of electromagnetic radiation; that is, all wavelengths (or frequencies) are theoretically possible.

The differences in wavelengths of the various kinds of electromagnetic waves have important physical consequences. As we know, the behavior of waves depends strongly on the relative sizes of the wavelengths and the physical objects or apertures the waves encounter. Since the wavelengths of light are in the rather narrow range from approximately 400 nm to 700 nm, they are much smaller than most obstacles, so the ray approximation (introduced in Section 15-4) is often valid. The wavelength and frequency are also important in determining the kinds of interactions between electromagnetic waves and matter. X rays, for example, have very short wavelengths and high frequencies. They easily penetrate many materials that are opaque to lower-frequency light waves, which are absorbed by the materials. Microwaves have wavelengths of the order of a few centimeters and frequencies that are close to the natural resonance frequencies of water molecules in solids and liquids. Microwaves are therefore readily absorbed by the water molecules in foods, which is the mechanism by which food is heated in microwave ovens.

## Production of Electromagnetic Waves

Electromagnetic waves are produced when free electric charges accelerate or when electrons bound to atoms and molecules make transitions to lower energy states. Radio waves, which have frequencies from approximately 550 kHz to 1600 kHz for AM and from approximately 88 MHz to 108 MHz for FM, are produced by macroscopic electric currents oscillating in radio transmission antennas. The frequency of the emitted waves equals the frequency of oscillation of the charges.

A continuous spectrum of X rays is produced by the deceleration of electrons when they crash into a metal target. The radiation produced is called **bremsstrahlung** (German for braking radiation). Accompanying the broad, continuous bremsstrahlung spectrum is a discrete spectrum of X-ray lines produced by transitions of inner electrons in the atoms of the target material.

Synchrotron radiation arises from the circular orbital motion of charged particles (usually electrons or positrons) in nuclear accelerators called synchrotrons. Originally considered a nuisance by accelerator scientists, synchrotron radiation X rays are now produced and used as a medical diagnostic tool because of the ease of manipulating the beams with reflection and diffraction optics. Synchrotron radiation is also emitted by charged particles trapped in magnetic fields associated with stars and galaxies. It is believed that most low-frequency radio waves reaching the earth from outer space originate as synchrotron radiation.

Heat is radiated by the thermally excited molecular charges. The spectrum of heat radiation is the blackbody radiation spectrum discussed in Section 20-4.

Light waves, which have frequencies of the order of $10^{14}$ Hz, are generally produced by transitions of bound atomic charges. We discuss sources of light waves in Chapter 31.

## Electric Dipole Radiation

Figure 30-5 is a schematic drawing of an electric-dipole radio antenna that consists of two conducting rods along a line fed by an alternating current generator. At time $t = 0$ (Figure 30-5a), the ends of the rods are charged, and there is an electric field near the rod parallel to the rod. There is also a magnetic field, which is not shown, encircling the rods due to the current in the rods. The fluctuations in these fields move out away from the rods with the speed of light. After one-fourth period, at $t = T/4$ (Figure 30-5b), the rods are uncharged, and the electric field near the rod is zero. At $t = T/2$ (Figure 30-5c), the rods are again charged, but the charges are opposite those at $t = 0$. The electric and magnetic fields at a great distance from the antenna are quite different from the fields near the antenna. Far from the antenna, the electric and magnetic fields oscillate in phase with simple harmonic motion, perpendicular to each other and to the direction of propagation of the wave. Figure 30-6 shows the electric and magnetic fields far from an electric dipole antenna.

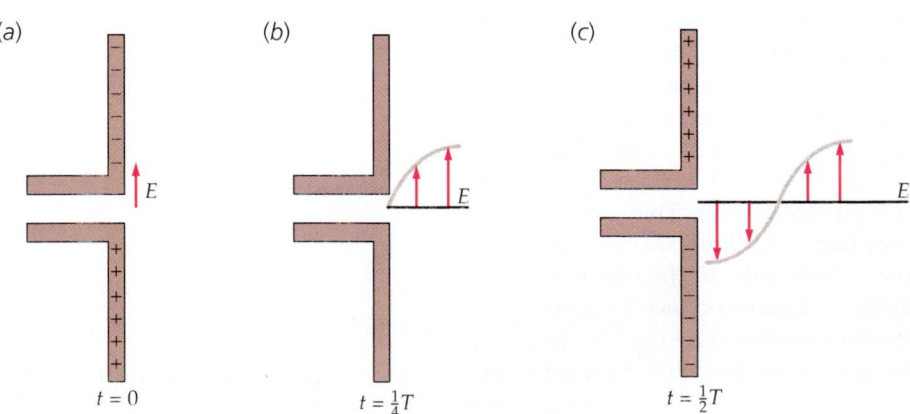

(a)  (b)  (c)

$t = 0$  $t = \frac{1}{4}T$  $t = \frac{1}{2}T$

**FIGURE 30-5** An electric dipole radio antenna for radiating electromagnetic waves. Alternating current is supplied to the antenna by a generator (not shown). The fluctuations in the electric field due to the fluctuations in the charges in the antenna propagates outward at the speed of light. There is also a fluctuating magnetic field (not shown) perpendicular to the paper due to the current in the antenna.

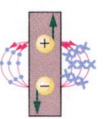 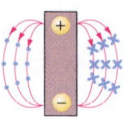 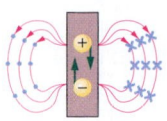 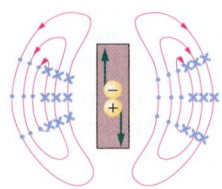

**FIGURE 30-6** Electric field lines (in red) and magnetic field lines (in blue) produced by an oscillating electric dipole. Each magnetic field line is a circle with the dipole along its axis. The cross product $\vec{E} \times \vec{B}$ is directed away from the dipole at all points.

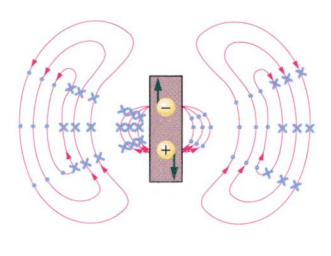 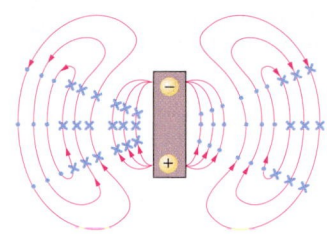

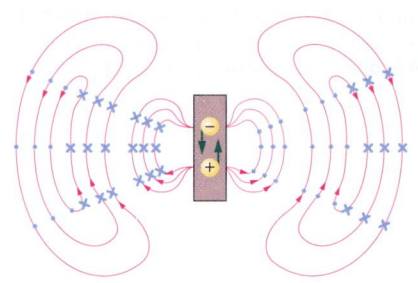 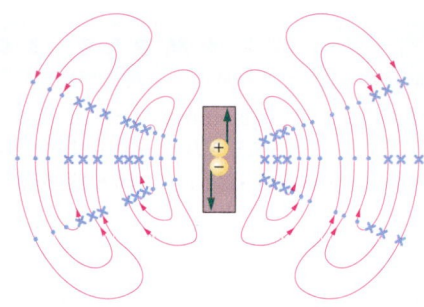

Electromagnetic waves of radio or television frequencies can be detected by an electric dipole antenna placed parallel to the electric field of the incoming wave, so that it induces an alternating current in the antenna (Figure 30-7). These electromagnetic waves can also be detected by a loop antenna placed perpendicular to the magnetic field, so that the changing magnetic flux through the loop induces a current in the loop (Figure 30-8). Electromagnetic waves of frequency in the visible light range are detected by the eye or by photographic film, both of which are mainly sensitive to the electric field.

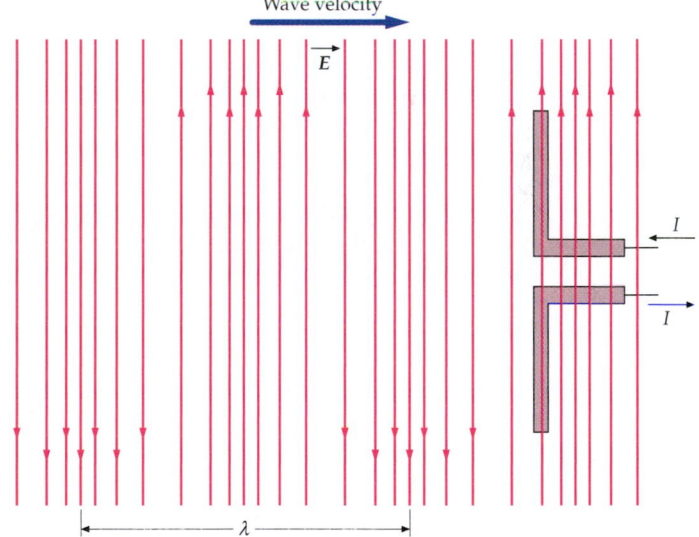

**FIGURE 30-7** An electric dipole antenna for detecting electromagnetic waves. The alternating electric field of the incoming wave produces an alternating current in the antenna.

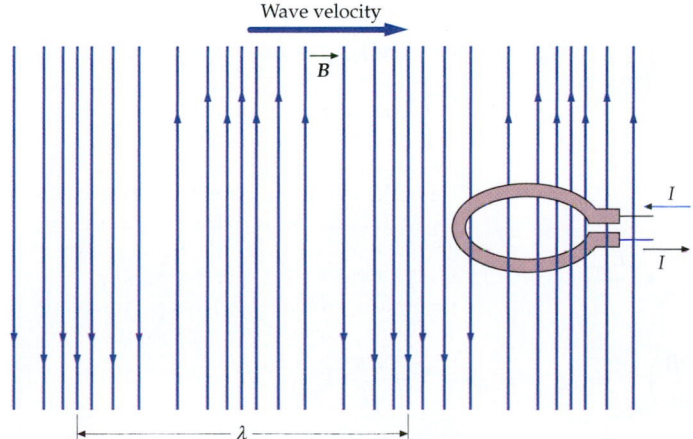

**FIGURE 30-8** Loop antenna for detecting electromagnetic radiation. The alternating magnetic flux through the loop due to the magnetic field of the radiation induces an alternating current in the loop.

The radiation from a dipole antenna, such as that shown in Figure 30-5, is called electric dipole radiation. Many electromagnetic waves exhibit the characteristics of electric dipole radiation. An important feature of this type of radiation is that the intensity of the electromagnetic waves radiated by a dipole antenna is zero along the axis of the antenna and maximum in the radial direction (away from the axis). If the dipole is in the $y$ direction with its center at the origin, as in Figure 30-9, the intensity is zero along the $y$ axis and maximum in the $xz$ plane. In the direction of a line making an angle $\theta$ with the $y$ axis, the intensity is proportional to $\sin^2 \theta$.

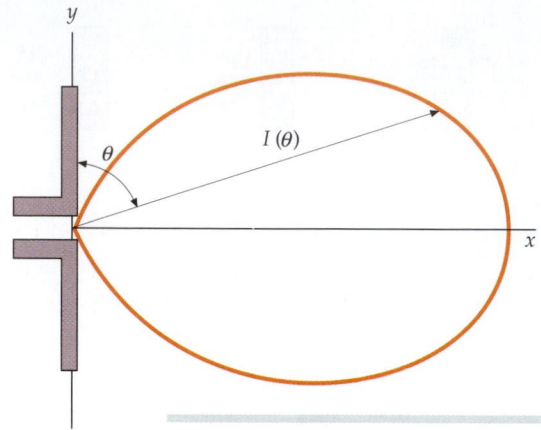

**FIGURE 30-9** Polar plot of the intensity of electromagnetic radiation from an electric dipole antenna versus angle. The intensity $I(\theta)$ is proportional to the length of the arrow. The intensity is maximum perpendicular to the antenna at $\theta = 90°$ and minimum along the antenna at $\theta = 0°$ or $\theta = 180°$.

*EMF INDUCED IN A LOOP ANTENNA*                    **EXAMPLE 30-3**

A loop antenna consisting of a single 10-cm radius loop of wire is used to detect electromagnetic waves for which $E_{rms} = 0.15$ V/m. Find the rms emf induced in the loop if the wave frequency is (a) 600 kHz and (b) 60 MHz.

**PICTURE THE PROBLEM** The induced emf in the wire is related to the rate of change of the magnetic flux through the loop by Faraday's law (Equation 30-5). Using Equation 30-7, we can obtain the rms value of the magnetic field from the given rms value of the electric field.

(a) 1. Faraday's law relates the magnitude of the emf to the rate of change of the magnetic flux through the flat stationary surface bounded by the loop:

$$|\mathcal{E}| = \frac{d\phi_m}{dt}$$

2. The wavelength of a 600 kHz wave traveling at speed $c$ is $\lambda = c/f = 500$ m. Over the flat surface bounded by the 10-cm radius loop, $\vec{B}$ is quite uniform.

$$\phi_m = BA = \pi r^2 B, \text{ so } |\mathcal{E}| = \frac{d\phi_m}{dt} = \pi r^2 \frac{\partial B}{\partial t}$$

and

$$\mathcal{E}_{rms} = \pi r^2 \left(\frac{\partial B}{\partial t}\right)_{rms}$$

3. Compute $dB_{rms}/dt$ from a sinusoidal $B$:

$$B = B_0 \sin(kx - \omega t)$$

$$\frac{\partial B}{\partial t} = -\omega B_0 \cos(kx - \omega t)$$

4. Calculate the rms value of $\partial B/\partial t$. The rms value of any sinusoidal function of time equals $1/\sqrt{2}$, and the peak value divided by $\sqrt{2}$ equals the rms value:

$$\left(\frac{\partial B}{\partial t}\right)_{rms} = \omega B_0 [-\cos(kx - \omega t)]_{rms} = \omega B_0/\sqrt{2} = \omega B_{rms}$$

5. Using Equation 30-7 ($E = cB$), relate the rms value of $\partial B/\partial t$ to $E_{rms}$:

$$E = cB$$

so

$$B_{rms} = \frac{E_{rms}}{c}$$

6. Substituting into the step 3 result gives:

$$\left(\frac{\partial B}{\partial t}\right)_{rms} = \omega B_{rms} = \omega \frac{E_{rms}}{c} = \frac{2\pi f}{c} E_{rms}$$

7. Substituting the step 6 result into the step 2 result, calculate $\mathscr{E}_{rms}$ at $f = 600$ kHz:

$$\mathscr{E}_{rms} = \pi r^2 \left(\frac{\partial B}{\partial t}\right)_{rms} = \pi r^2 \frac{2\pi f}{c} E_{rms}$$

$$= \pi (0.1 \text{ m})^2 \frac{2\pi (6 \times 10^5 \text{ Hz})}{3 \times 10^8 \text{ m/s}} (0.15 \text{ V/m})$$

$$= \boxed{5.92 \times 10^{-5} \text{ V} = 59.2 \ \mu\text{V}}$$

(b) The induced emf is proportional to the frequency (step 4), so at 60 MHz it will be 100 times greater than at 600 kHz:

$$\mathscr{E}_{rms} = (100)(5.92 \times 10^{-5} \text{ V}) = 0.00592 \text{ V}$$

$$= \boxed{5.92 \text{ mV}}$$

**REMARKS** For part (b) the frequency is 60 MHz, so $\lambda = c/f = 5$ m. $\vec{B}$ is not as uniform over the surface bounded by the 10-cm radius loop when $\lambda = 5$ m as it is when $\lambda = 500$ m, as in part (a). Thus, $\vec{B}$ on the surface when $\lambda = 5$ m is uniform enough that the part (b) result is sufficiently accurate for most purposes.

## Energy and Momentum in an Electromagnetic Wave

Like other waves, electromagnetic waves carry energy and momentum. The energy carried is described by the intensity, which is the average power per unit area incident on a surface perpendicular to the direction of propagation. The momentum per unit time per unit area carried by an electromagnetic wave is called the **radiation pressure.**

**Intensity**  Consider an electromagnetic wave traveling toward the right and a cylindrical region of length $L$ and cross-sectional area $A$ with its axis oriented from left to right. The average amount of electromagnetic energy $U_{av}$ within this region equals $u_{av}V$, where $u_{av}$ is the average energy density and $V = LA$ is the volume of the region. In the time it takes the electromagnetic wave to travel the distance $L$, all of this energy passes through the right end of the region. The time $\Delta t$ for the wave to travel the distance $L$ is $L/c$, so the power $P_{av}$ (the energy per unit time) passing out the right end of the region is

$$P_{av} = U_{av}/\Delta t = u_{av}LA/(L/c) = u_{av}Ac$$

and the intensity $I$ (the average power per unit area) is

$$I = P_{av}/A = u_{av}c$$

The total energy density in the wave $u$ is the sum of the electric and magnetic energy densities. The electric energy density $u_e$ (Equation 24-13) and magnetic energy density $u_m$ (Equation 28-20) are given by

$$u_e = \frac{1}{2}\epsilon_0 E^2 \quad \text{and} \quad u_m = \frac{B^2}{2\mu_0}$$

In an electromagnetic wave in free space, $E$ equals $cB$, so we can express the magnetic energy density in terms of the electric field:

$$u_m = \frac{B^2}{2\mu_0} = \frac{(E/c)^2}{2\mu_0} = \frac{E^2}{2\mu_0 c^2} = \frac{1}{2}\epsilon_0 E^2$$

where we have used $c^2 = 1/(\epsilon_0\mu_0)$. Thus, the electric and magnetic energy densities are equal. Using $E = cB$, we may express the total energy density in several useful ways:

$$u = u_e + u_m = \epsilon_0 E^2 = \frac{B^2}{\mu_0} = \frac{EB}{\mu_0 c} \qquad \text{30-8}$$

ENERGY DENSITY IN AN ELECTROMAGNETIC WAVE

To compute the average energy density, we replace the instantaneous fields $E$ and $B$ by their rms values $E_{rms} = E_0/\sqrt{2}$ and $B_{rms} = B_0/\sqrt{2}$, where $E_0$ and $B_0$ are the maximum values of the fields. The intensity is then

$$I = u_{av} c = \frac{E_{rms} B_{rms}}{\mu_0} = \frac{1}{2} \frac{E_0 B_0}{\mu_0} = |\vec{S}|_{av} \qquad \text{30-9}$$

INTENSITY OF AN ELECTROMAGNETIC WAVE

where the vector

$$\vec{S} = \frac{\vec{E} \times \vec{B}}{\mu_0} \qquad \text{30-10}$$

DEFINITION—POYNTING VECTOR

is called the **Poynting vector** after its discoverer, John Poynting. The average magnitude of $\vec{S}$ is the intensity of the wave, and the direction of $\vec{S}$ is the direction of propagation of the wave.

**Radiation Pressure**  We now show by a simple example that an electromagnetic wave carries momentum. Consider a wave moving along the $x$ axis that is incident on a stationary charge, as shown in Figure 30-10. For simplicity, we assume that $\vec{E}$ is in the $y$ direction and $\vec{B}$ is in the $z$ direction, and we neglect the time dependence of the fields. The particle experiences a force $q\vec{E}$ in the $y$ direction and is thus accelerated by the electric field. At any time $t$, the velocity in the $y$ direction is

$$v_y = at = \frac{qE}{m} t$$

After a short time $t_1$, the charge has acquired kinetic energy equal to

$$K = \frac{1}{2} m v_y^2 = \frac{1}{2} \frac{m q^2 E^2 t_1^2}{m^2} = \frac{1}{2} \frac{q^2 E^2}{m} t_1^2 \qquad \text{30-11}$$

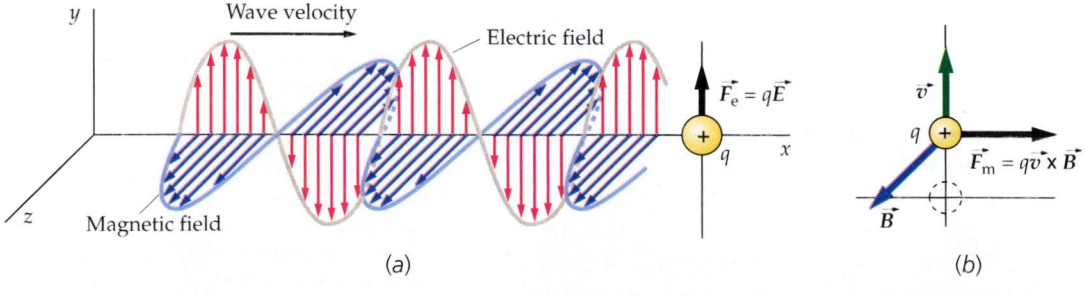

(a)

(b)

**FIGURE 30-10**  An electromagnetic wave incident on a point charge that is initially at rest on the $x$ axis. (a) The electric force $q\vec{E}$ accelerates the charge in the upward direction. (b) When the velocity $\vec{v}$ of the charge is upward, the magnetic force $q\vec{v} \times \vec{B}$ accelerates the charge in the direction of the wave.

When the charge is moving in the $y$ direction, it experiences a magnetic force

$$\vec{F}_m = q\vec{v} \times \vec{B} = qv_y\hat{j} \times B\hat{k} = qv_yB\hat{i} = \frac{q^2EB}{m}t\hat{i}$$

Note that this force is in the direction of propagation of the wave. Using $dp_x = F_x\,dt$, we find for the momentum $p_x$ transferred by the wave to the particle in time $t_1$:

$$p_x = \int_0^{t_1} F_x\,dt = \int_0^{t_1}\frac{q^2EB}{m}t\,dt = \frac{1}{2}\frac{q^2EB}{m}t_1^2$$

If we use $B = E/c$, this becomes

$$p_x = \frac{1}{c}\left(\frac{1}{2}\frac{q^2E^2}{m}t_1^2\right) \qquad\qquad 30\text{-}12$$

Comparing Equations 30-11 and 30-12, we see that the momentum acquired by the charge in the direction of the wave is $1/c$ times the energy. Although our simple calculation was not rigorous, the results are correct. The magnitude of the momentum carried by an electromagnetic wave is $1/c$ times the energy carried by the wave:

$$p = \frac{U}{c} \qquad\qquad 30\text{-}13$$

MOMENTUM AND ENERGY IN AN ELECTROMAGNETIC WAVE

Since the intensity is the energy per unit area per unit time, the intensity divided by $c$ is the momentum carried by the wave per unit area per unit time. The momentum carried per unit time is a force. The intensity divided by $c$ is thus a force per unit area, which is a pressure. This pressure is the radiation pressure $P_r$:

$$P_r = \frac{I}{c} \qquad\qquad 30\text{-}14$$

RADIATION PRESSURE AND INTENSITY

We can relate the radiation pressure to the electric or magnetic fields by using Equation 30-9 to relate $I$ to $E$ and $B$, and Equation 30-7 to eliminate either $E$ or $B$:

$$P_r = \frac{I}{c} = \frac{E_0B_0}{2\mu_0c} = \frac{E_{rms}B_{rms}}{\mu_0c} = \frac{E_0^2}{2\mu_0c^2} = \frac{B_0^2}{2\mu_0} \qquad\qquad 30\text{-}15$$

RADIATION PRESSURE IN TERMS OF $E$ AND $B$

Consider an electromagnetic wave incident normally on some surface. If the surface absorbs energy $U$ from the electromagnetic wave, it also absorbs momentum $p$ given by Equation 30-13, and the pressure exerted on the surface equals the radiation pressure. If the wave is reflected, the momentum transferred is $2p$ because the wave now carries momentum in the opposite direction. The pressure exerted on the surface by the wave is then twice that given by Equation 30-15.

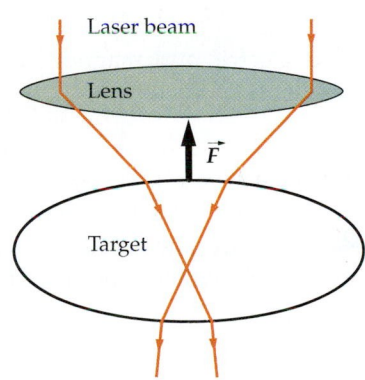

"Laser tweezers" make use of the momentum carried by electromagnetic waves to manipulate targets on a molecular scale. The two rays shown are refracted as they pass through a transparent target, such as a biological cell, or on an even smaller scale, as a tiny transparent bead attached to a large molecule within a cell. At each refraction, the rays are bent downward, which increases the downward component of momentum of the rays. The target thus exerts a downward force on the laser beams, and the laser beams exert an upward force on the target, which pulls the target toward the laser source. The force is typically of the order of piconewtons. Laser tweezers have been used to accomplish such astonishing feats as stretching out coiled DNA.

*RADIATION PRESSURE 3 M FROM A LIGHTBULB*　　　　**EXAMPLE 30-4**

**A lightbulb emits spherical electromagnetic waves uniformly in all directions. Find (a) the intensity, (b) the radiation pressure, and (c) the electric and magnetic field magnitudes at a distance of 3 m from the lightbulb, assuming that 50 W of electromagnetic radiation is emitted.**

**PICTURE THE PROBLEM** At a distance $r$ from the lightbulb, the energy is spread uniformly over an area $4\pi r^2$. The intensity is the power divided by the area. The radiation pressure can then be found from $P_r = I/c$.

(a) 1. Divide the power output by the area to find the intensity:

$$I = \frac{50 \text{ W}}{4\pi r^2}$$

2. Substitute $r = 3$ m:

$$I = \frac{50 \text{ W}}{4\pi(3 \text{ m})^2} = \boxed{0.442 \text{ W/m}^2}$$

(b) The radiation pressure is the intensity divided by the speed of light:

$$P_r = \frac{I}{c} = \frac{0.442 \text{ W/m}^2}{3 \times 10^8 \text{ m/s}} = \boxed{1.47 \times 10^{-9} \text{ Pa}}$$

(c) 1. $B_0$ is related to $P_r$ by Equation 30-15:

$$B_0 = \sqrt{2\mu_0 P_r}$$
$$= [2(4\pi \times 10^{-7} \text{ T·m/A})(1.47 \times 10^{-9} \text{ Pa})]^{1/2}$$
$$= 6.08 \times 10^{-8} \text{ T}$$

2. The maximum value of the electric field $E_0$ is $c$ times $B_0$:

$$E_0 = cB_0 = (3 \times 10^8 \text{ m/s})(6.08 \times 10^{-8} \text{ T})$$
$$= 18.2 \text{ V/m}$$

3. The electric and magnetic field magnitudes at that point are of the form:

$$\boxed{\begin{array}{l} E = E_0 \sin \omega t \quad \text{and} \quad B = B_0 \sin \omega t \\ \text{with } E_0 = 18.2 \text{ V/m} \quad \text{and} \quad B_0 = 6.08 \times 10^{-8} \text{ T} \end{array}}$$

 **REMARKS** Only about 2 percent of the power consumed by incandescent bulbs is transformed into visible light. Note that the radiation pressure calculated in Part (b) is very small compared with the atmospheric pressure, which is of the order of $10^5$ Pa.

---

*A LASER ROCKET*　　　　**EXAMPLE 30-5**

**You are stranded in space a distance of 20 m from your spaceship. You carry a 1-kW laser. If your total mass, including your space suit and laser, is 95 kg, how long will it take you to reach the spaceship if you point the laser directly away from it?**

**PICTURE THE PROBLEM** The laser emits light, which carries with it momentum. By momentum conservation, you are given an equal and opposite momentum toward the spaceship. The momentum carried by light is $p = U/c$, where $U$ is the energy of the light. If the power of the laser is $P = dU/dt$, then the rate of change of momentum produced by the laser is $dp/dt = (dU/dt)/c = P/c$. This is the force exerted on you, which is constant.

1. The time taken is related to the distance and the acceleration. We assume that you are initially at rest relative to the spaceship:

$$x = \frac{1}{2}at^2; \quad t = \sqrt{\frac{2x}{a}}$$

2. Your acceleration is the force divided by your mass, and the force is the power divided by $c$:

$$a = \frac{F}{m} = \frac{P/c}{m} = \frac{P}{mc}$$

3. Use this acceleration to calculate the time $t$:

$$t = \sqrt{\frac{2x}{a}} = \sqrt{\frac{2xmc}{P}}$$

$$= \sqrt{\frac{2(20 \text{ m})(95 \text{ kg})(3 \times 10^8 \text{ m/s})}{1000 \text{ W}}}$$

$$= 3.38 \times 10^4 \text{ s} = \boxed{9.38 \text{ h}}$$

**REMARKS** Note that the acceleration is extremely small—only about $10^{-9}$ g. Your speed when you reach the spaceship would be $v = at = 1.19$ mm/s, which is practically imperceptible.

**EXERCISE** How long would it take you to reach the spaceship if you took off one of your shoelaces and threw it as fast as you could in the direction opposite the ship? (To answer this, you must first estimate the mass of the shoelace and the maximum speed that you can throw the shoelace.) (*Answer* About 5 h for a 10-g shoelace thrown at 10 m/s)

## *30-4 The Wave Equation for Electromagnetic Waves

In Section 15-1, we saw that waves on a string obey a partial differential equation called the **wave equation:**

$$\frac{\partial^2 y(x, t)}{\partial x^2} = \frac{1}{v^2} \frac{\partial^2 y(x, t)}{\partial t^2} \qquad \text{30-16}$$

where $y(x, t)$ is the wave function, which for string waves is the displacement of the string. The velocity of the wave is given by $v = \sqrt{F/\mu}$, where $F$ is the tension and $\mu$ is the linear mass density. The general solution to this equation is

$$y(x, t) = f_1(x - vt) + f_2(x + vt)$$

The general solution functions can be expressed as a superposition of harmonic wave functions of the form

$$y(x, t) = y_0 \sin(kx - \omega t) \quad \text{and} \quad y(x, t) = y_0 \sin(kx + \omega t)$$

where $k = 2\pi/\lambda$ is the wave number and $\omega = 2\pi f$ is the angular frequency.

Maxwell's equations imply that $\vec{E}$ and $\vec{B}$ obey wave equations similar to Equation 30-16. We consider only free space, in which there are no charges or currents, and we assume that the electric and magnetic fields $\vec{E}$ and $\vec{B}$ are functions of time and one space coordinate only, which we will take to be the $x$ coordinate. Such a wave is called a **plane wave,** because $\vec{E}$ and $\vec{B}$ are uniform throughout any plane perpendicular to the $x$ axis. For a plane electromagnetic wave traveling parallel to the $x$ axis, the $x$ components of the fields are zero, so the vectors $\vec{E}$ and $\vec{B}$ are perpendicular to the $x$ axis and each obeys the wave equation:

$$\frac{\partial^2 \vec{E}}{\partial x^2} = \frac{1}{c^2} \frac{\partial^2 \vec{E}}{\partial t^2} \qquad \text{30-17a}$$

WAVE EQUATION FOR $\vec{E}$

$$\frac{\partial^2 \vec{B}}{\partial x^2} = \frac{1}{c^2}\frac{\partial^2 \vec{B}}{\partial t^2}$$

30-17b

WAVE EQUATION FOR $\vec{B}$

where $c = 1/\sqrt{\mu_0 \epsilon_0}$ is the speed of the waves. (*Note:* Dimensional reasoning helps in remembering these equations. For each equation, the numerators on both sides are the same and the denominators on both sides have the dimension of length squared.)

## *Derivation of the Wave Equation

We can relate the space derivative of one of the field vectors to the time derivative of the other field vector by applying Faraday's law (Equation 30-6c) and the modified version of Ampère's law (Equation 30-6d) to appropriately chosen curves in space. We first relate the space derivative of $E_y$ to the time derivative of $B_z$ by applying Equation 30-6c (Faraday's law) to the rectangular curve of sides $\Delta x$ and $\Delta y$ lying in the $xy$ plane (Figure 30-11). The circulation of $\vec{E}$ around C (the line integral of $\vec{E}$ around curve C) is

$$\oint_C \vec{E} \cdot d\vec{\ell} = E_y(x_2)\Delta y - E_y(x_1)\Delta y$$

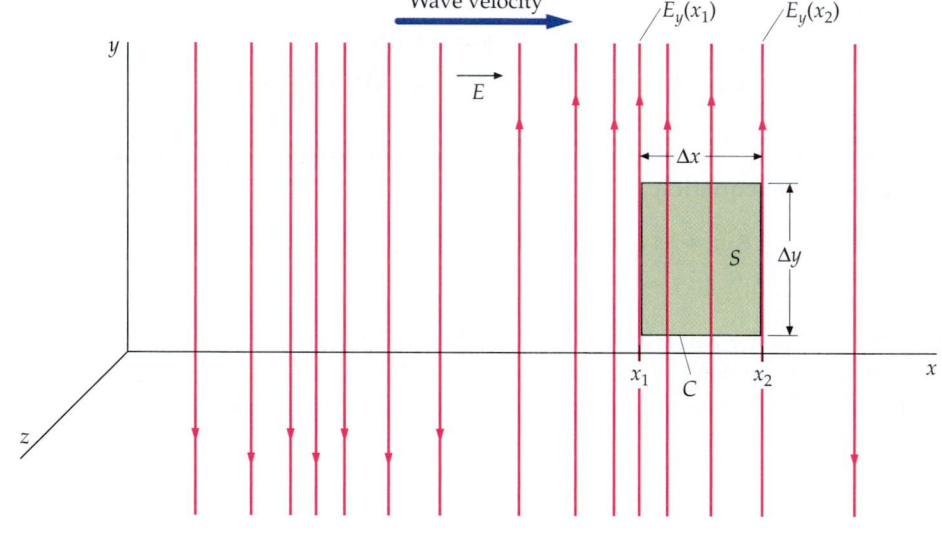

where $E_y(x_1)$ is the value of $E_y$ at the point $x_1$ and $E_y(x_2)$ is the value of $E_y$ at the point $x_2$. The contributions of the type $E_x\Delta x$ from the top and bottom of this curve are zero because $E_x = 0$. Since $\Delta x$ is very small (compared to the wavelength), we can approximate the difference in $E_y$ on the left and right sides of this curve (at $x_1$ and at $x_2$) by

$$E_y(x_2) - E_y(x_1) = \Delta E_y \approx \frac{\partial E_y}{\partial x}\Delta x$$

Then

$$\oint_C \vec{E} \cdot d\vec{\ell} \approx \frac{\partial E_y}{\partial x}\Delta x\,\Delta y$$

**FIGURE 30-11** A rectangular curve in the $xy$ plane for the derivation of Equation 30-18.

Faraday's law is

$$\oint_C \vec{E} \cdot d\vec{\ell} = -\int_S \frac{\partial B_n}{\partial t}\,dA$$

The flux of $\partial B_n/\partial t$ through the rectangular surface bounded by this curve is approximately

$$\int_S B_n\,dA \approx \frac{\partial B_z}{\partial t}\Delta x\,\Delta y$$

Faraday's law then gives

$$\frac{\partial E_y}{\partial x}\Delta x\,\Delta y = -\frac{\partial B_z}{\partial t}\Delta x\,\Delta y$$

or

$$\frac{\partial E_y}{\partial x} = -\frac{\partial B_z}{\partial t}$$

30-18

Equation 30-18 implies that if there is a component of the electric field $E_y$ that depends on $x$, there must be a component of the magnetic field $B_z$ that depends on time or, conversely, if there is a component of the magnetic field $B_z$ that depends on time, there must be a component of the electric field $E_y$ that depends on $x$. We can get a similar equation relating the space derivative of the magnetic field $B_z$ to the time derivative of the electric field $E_y$ by applying Ampère's law (Equation 30-6d) to the curve of sides $\Delta x$ and $\Delta z$ in the $xz$ plane shown in Figure 30-12.

For the case of no conduction currents ($I = 0$), Equation 30-6d is

$$\oint_C \vec{B} \cdot d\vec{\ell} = \mu_0 \epsilon_0 \int_S \frac{\partial E_n}{\partial t} \, dA$$

The details of this calculation are similar to those for Equation 30-18. The result is

$$\frac{\partial B_z}{\partial x} = -\mu_0 \epsilon_0 \frac{\partial E_y}{\partial t}$$

30-19

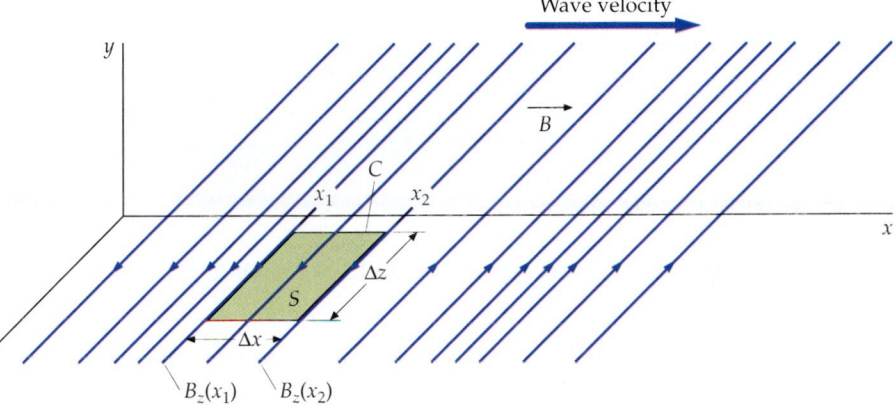

We can eliminate either $B_z$ or $E_y$ from Equations 30-18 and 30-19 by differentiating both sides of either equation with respect to either $x$ or $t$. If we differentiate both sides of Equation 30-18 with respect to $x$, we obtain

**FIGURE 30-12** A rectangular curve in the $xz$ plane for the derivation of Equation 30-19.

$$\frac{\partial}{\partial x}\left(\frac{\partial E_y}{\partial x}\right) = -\frac{\partial}{\partial x}\left(\frac{\partial B_z}{\partial t}\right)$$

Interchanging the order of the time and space derivatives on the term to the right of the equal sign gives

$$\frac{\partial^2 E_y}{\partial x^2} = -\frac{\partial}{\partial t}\left(\frac{\partial B_z}{\partial x}\right)$$

Using Equation 30-19, we substitute for $\partial B_z / \partial x$ to obtain

$$\frac{\partial^2 E_y}{\partial x^2} = -\frac{\partial}{\partial t}\left(-\mu_0 \epsilon_0 \frac{\partial E_y}{\partial t}\right)$$

which yields the wave equation

$$\frac{\partial^2 E_y}{\partial x^2} = \mu_0 \epsilon_0 \frac{\partial^2 E_y}{\partial t^2}$$

30-20

Comparing Equation 30-20 with Equation 30-16, we see that $E_y$ obeys a wave equation for waves with speed $c = 1/\sqrt{\mu_0 \epsilon_0}$, which is Equation 30-1.

If we had instead chosen to eliminate $E_y$ from Equations 30-18 and 30-19 (by differentiating Equation 30-18 with respect to $t$, for example), we would have obtained an equation identical to Equation 30-20 except with $B_z$ replacing $E_y$. We can thus see that both the electric field $E_y$ and the magnetic field $B_z$ obey a wave equation for waves traveling with the velocity $1/\sqrt{\mu_0 \epsilon_0}$, which is the velocity of light.

By following the same line of reasoning as used above, and applying Equation 30-6c (Faraday's law) to the curve in the $xz$ plane (Figure 30-12), we would obtain

$$\frac{\partial E_z}{\partial x} = \frac{\partial B_y}{\partial t} \qquad\qquad 30\text{-}21$$

Similarly, the application of Equation 30-6d to the curve in the $xy$ plane (Figure 30-11) gives

$$\frac{\partial B_y}{\partial x} = \mu_0 \epsilon_0 \frac{\partial E_z}{\partial t} \qquad\qquad 30\text{-}22$$

We can use these results to show that, for a wave propagating in the $x$ direction, the components $E_z$ and $B_y$ also obey the wave equation.

To show that the magnetic field $B_z$ is in phase with the electric field $E_y$, consider the harmonic wave function of the form

$$E_y = E_{y0} \sin(kx - \omega t) \qquad\qquad 30\text{-}23$$

If we substitute this solution into Equation 30-18, we have

$$\frac{\partial B_z}{\partial t} = -\frac{\partial E_y}{\partial x} = -k E_{y0} \cos(kx - \omega t)$$

To solve for $B_z$, we take the integral of $\partial B_z / \partial t$ with respect to time. Doing so yields

$$B_z = \int \frac{\partial B_z}{\partial t}\, dt = \frac{k}{\omega} E_{y0} \sin(kx - \omega t) + f(x) \qquad\qquad 30\text{-}24$$

where $f(x)$ is an arbitrary function of $x$.

**EXERCISE** Verify Equation 30-24 by taking $\partial B_z / \partial t$, where $B_z = (k/\omega) E_{y0} \sin(kx - \omega t) + f(x)$.

The result should be $-k E_{y0} \cos(kx - \omega t)$, which is the right hand side of the previous equation.

We next substitute the solution (Equation 30-23) into Equation 30-19 and obtain

$$\frac{\partial B_z}{\partial x} = -\mu_0 \epsilon_0 \frac{\partial E_y}{\partial t} = \omega \mu_0 \epsilon_0 E_{y0} \cos(kx - \omega t)$$

Solving for $B_z$ gives

$$B_z = \int \frac{\partial B_z}{\partial x}\, dx = \frac{\omega \mu_0 \epsilon_0}{k} E_{y0} \sin(kx - \omega t) + g(t) \qquad\qquad 30\text{-}25$$

where $g(t)$ is an arbitrary function of time. Equating the right sides of Equations 30-24 and 30-25 gives

$$\frac{k}{\omega} E_{y0} \sin(kx - \omega t) + f(x) = \frac{\omega \mu_0 \epsilon_0}{k} E_{y0} \sin(kx - \omega t) + g(t)$$

Substituting $c$ for $\omega/k$ and $1/c^2$ for $\mu_0 \epsilon_0$ gives

$$\frac{1}{c} E_{y0} \sin(kx - \omega t) + f(x) = \frac{1}{c} E_{y0} \sin(kx - \omega t) + g(t)$$

which implies $f(x) = g(t)$ for all values of $x$ and $t$. These remain equal only if $f(x) = g(t) = $ constant (independent of both $x$ and $t$). Thus, Equation 30-24 becomes

$$B_z = \frac{k}{\omega} E_{y0} \sin(kx - \omega t) + \text{constant} = B_{z0} \sin(kx - \omega t) \qquad \text{30-26}$$

where $B_{z0} = (k/\omega)E_{y0} = (1/c)E_{y0}$. The integration constant was dropped because it plays no part in the wave. It merely allows for the presence of a static uniform magnetic field. Since the electric and magnetic fields oscillate in phase with the same frequency, we have the general result that the magnitude of the electric field is $c$ times the magnitude of the magnetic field for an electromagnetic wave:

$$E = cB$$

which is Equation 30-7.

We see that Maxwell's equations imply wave equations 30-17$a$ and 30-17$b$ for the electric and magnetic fields; and that if $E_y$ varies harmonically, as in Equation 30-23, the magnetic field $B_z$ is in phase with $E_y$ and has an amplitude related to the amplitude of $E_y$ by $B_z = E_y/c$. The electric and magnetic fields are perpendicular to each other and to the direction of the wave propagation, as shown in Figure 30-4.

---

**$\vec{B}(x, t)$ FOR A LINEARLY POLARIZED PLANE WAVE**    **EXAMPLE 30-6**

The electric field of an electromagnetic wave is given by $\vec{E}(x, t) = E_0 \cos(kx - \omega t)\hat{k}$. (a) What is the direction of propagation of the wave? (b) What is the direction of the magnetic field in the $x = 0$ plane at time $t = 0$? (c) Find the magnetic field of the same wave. (d) Compute $\vec{E} \times \vec{B}$.

**PICTURE THE PROBLEM**  The argument of the cosine gives the direction of propagation. $\vec{B}$ is perpendicular to both $\vec{E}$ and to the direction of propagation. $\vec{B}$ and $\vec{E}$ are in phase.

(a) The argument of the cosine function $(kx - \omega t)$ tells us the direction of propagation:

> The direction of propagation is the direction of increasing $x$, which is the direction of $\hat{i}$.

(b) 1. $\vec{B}$ is in phase with $\vec{E}$ and is perpendicular to both $\vec{E}$ and the direction of propagation $\hat{k}$. (That is, $\vec{B}$ is perpendicular to both $\hat{i}$ and $\hat{k}$.) That means:

$$\vec{B}(x, t) = \pm B_0 \cos(kx - \omega t)\hat{j}$$

2. $\vec{E} \times \vec{B}$ is in the direction of propagation $\hat{i}$. Use the expressions for $\vec{E}$ and $\vec{B}$ and take the cross product:

$$\vec{E} \times \vec{B} = E_0 \cos(kx - \omega t)\hat{k} \times (\pm B_0 \cos(kx - \omega t)\hat{j})$$
$$= E_0(\pm B_0) \cos^2(kx - \omega t)(\hat{k} \times \hat{j})$$
$$= E_0(\pm B_0) \cos^2(kx - \omega t)(-\hat{i})$$

3. Choose the sign so that $\vec{E} \times \vec{B}$ is in the $\hat{i}$ direction:

$$\vec{E} \times \vec{B} = E_0(-B_0) \cos^2(kx - \omega t)(-\hat{i})$$

so

$$\vec{B}(x, t) = -B_0 \cos(kx - \omega t)\hat{j}$$

4. Evaluate $\vec{B}$ when both $x$ and $t$ equal zero.

$$\vec{B}(0, 0) = -B_0 \cos[k(0) - \omega(0)]\hat{j} = -B_0\hat{j}$$

$$\therefore \boxed{\vec{B}(0, 0) \text{ is in the negative } y \text{ direction.}}$$

(c) In an electromagnetic wave, $E_0 = cB_0$ and $\vec{B}$ and $\vec{E}$ are in phase. Thus:

$$\vec{B}(x, t) = \boxed{-B_0 \cos(kx - \omega t)\,\hat{j}}, \text{ where } B_0 = E_0/c$$

(d) Calculate $\vec{E} \times \vec{B}$. Let $\theta = kx - \omega t$ and do the calculation:

$$\vec{E} \times \vec{B} = (E_0 \cos\theta\,\hat{k}) \times (-B_0 \cos\theta\,\hat{j})$$

$$= -E_0 B_0 \cos^2\theta\,(\hat{k} \times \hat{j})$$

$$= \boxed{E_0 B_0 \cos^2\theta\,\hat{i}}, \text{ where } \theta = kx - \omega t$$

**REMARKS** The Part (d) result confirms the Part (a) result, because for an electromagnetic wave $\vec{E} \times \vec{B}$ is always in the direction of propagation.

---

■ $\vec{B}(x, t)$ FOR A CIRCULAR POLARIZED PLANE WAVE             **EXAMPLE 30-7**

The electric field of an electromagnetic wave is given by $\vec{E}(x, t) = E_0 \sin(kx - \omega t)\,\hat{j} + E_0 \cos(kx - \omega t)\,\hat{k}$. (a) Find the magnetic field of the same wave. (b) Compute $\vec{E} \cdot \vec{B}$ and $\vec{E} \times \vec{B}$.

**PICTURE THE PROBLEM** We can solve this using the principle of superposition. The given electric field is the superposition of two fields, the one given in Equation 30-23 and the one given in the problem statement of Example 30-6.

(a) 1. From the phase (the argument of the trig functions) we can see that the direction of propagation is the positive $x$ direction:

The phase is for a wave traveling in the positive $x$ direction.

2. The given electric field can be considered as the superposition of $\vec{E}_1 = E_0 \sin(kx - \omega t)\,\hat{j}$ and $\vec{E}_2 = E_0 \cos(kx - \omega t)\,\hat{k}$. Find the magnetic fields $\vec{B}_1$ and $\vec{B}_2$ associated with these electric fields, respectively. Use the procedure followed in Example 30-6:

For $\vec{E}_1 = E_0 \sin(kx - \omega t)\,\hat{j}$, $\vec{B}_1 = B_0 \sin(kx - \omega t)\,\hat{k}$

and

For $\vec{E}_2 = E_0 \cos(kx - \omega t)\,\hat{k}$, $\vec{B}_2 = -B_0 \cos(kx - \omega t)\,\hat{j}$

where

$$E_0 = cB_0$$

3. The superposition of magnetic fields gives the resultant magnetic fields:

$$\vec{B}(x, t) = \vec{B}_1 + \vec{B}_2$$

$$= B_0 \sin(kx - \omega t)\,\hat{k} - B_0 \cos(kx - \omega t)\,\hat{j}$$

where

$$B_0 = E_0/c$$

(b) 1. Let $\theta = kx - \omega t$ to simplify the notation and calculate $\vec{E} \cdot \vec{B}$:

$$\vec{E} \cdot \vec{B} = (E_0 \sin\theta\,\hat{j} + E_0 \cos\theta\,\hat{k}) \cdot (B_0 \sin\theta\,\hat{k} - B_0 \cos\theta\,\hat{j})$$

$$= E_0 B_0 \sin^2\theta\,\hat{j} \cdot \hat{k} - E_0 B_0 \sin\theta\cos\theta\,\hat{j} \cdot \hat{j}$$

$$+ E_0 B_0 \cos\theta\sin\theta\,\hat{k} \cdot \hat{k} - E_0 B_0 \cos^2\theta\,\hat{k} \cdot \hat{j}$$

$$= 0 - E_0 B_0 \sin\theta\cos\theta + E_0 B_0 \cos\theta\sin\theta - 0 = \boxed{0}$$

2. Calculate $\vec{E} \times \vec{B}$:

$$\vec{E} \times \vec{B} = (E_0 \sin\theta\,\hat{j} + E_0 \cos\theta\,\hat{k}) \times (-B_0 \cos\theta\,\hat{j} + B_0 \sin\theta\,\hat{k})$$

$$= -E_0 B_0 \sin\theta\cos\theta\,(\hat{j} \times \hat{j}) + E_0 B_0 \sin^2\theta\,(\hat{j} \times \hat{k})$$

$$- E_0 B_0 \cos^2\theta\,(\hat{k} \times \hat{j}) + E_0 B_0 \cos\theta\sin\theta\,(\hat{k} \times \hat{k})$$

$$= 0 + E_0 B_0 \sin^2\theta\,\hat{i} + E_0 B_0 \cos^2\theta\,\hat{i} + 0 = \boxed{E_0 B_0\,\hat{i}}$$

**REMARKS** We see that $\vec{E}$ and $\vec{B}$ are perpendicular to one another, and that $\vec{E} \times \vec{B}$ is in the direction of propagation of the wave. This type of electromagnetic wave is said to be *circularly polarized*. At a fixed value of $x$, both $\vec{E}$ and $\vec{B}$ rotate in a circle in a plane perpendicular to $\hat{i}$ with angular frequency $\omega$.

**EXERCISE** Calculate $\vec{E} \cdot \vec{E}$ and $\vec{B} \cdot \vec{B}$. [*Answer* $\vec{E} \cdot \vec{E} = E_y^2 + E_z^2 = E_0^2 \sin^2(kx - \omega t) + E_0^2 \cos^2(kx - \omega t) = E_0^2$ and $\vec{B} \cdot \vec{B} = B_y^2 + B_z^2 = B_0^2 \cos^2(kx - \omega t) + B_0^2 \sin^2(kx - \omega t) = B_0^2$]

■ **REMARKS** The fields $\vec{E}$ and $\vec{B}$ are constant in magnitude.

# SUMMARY

1. Maxwell's equations summarize the fundamental laws of physics that govern electricity and magnetism.

2. Electromagnetic waves include light, radio and television waves, X rays, gamma rays, microwaves, and others.

| Topic | Relevant Equations and Remarks | |
|---|---|---|
| **1. Maxwell's Displacement Current** | Ampère's law can be generalized to apply to currents that are not steady (and not continuous) if the current $I$ is replaced by $I + I_d$, where $I_d$ is Maxwell's displacement current: | |
| | $$I_d = \epsilon_0 \frac{d\phi_e}{dt}$$ | 30-3 |
| Generalized form of Ampère's law | $$\oint_C \vec{B} \cdot d\vec{\ell} = \mu_0(I + I_d) = \mu_0 I + \mu_0 \epsilon_0 \frac{d\phi_e}{dt}$$ | 30-4 |
| **2. Maxwell's Equations** | The laws of electricity and magnetism are summarized by Maxwell's equations. | |
| Gauss's law | $$\oint_S E_n \, dA = \frac{1}{\epsilon_0} Q_{inside}$$ | 30-6a |
| Gauss's law for magnetism (isolated magnetic poles do not exist) | $$\oint_S B_n \, dA = 0$$ | 30-6b |
| Faraday's law | $$\oint_C \vec{E} \cdot d\vec{\ell} = -\frac{d}{dt} \int_S B_n \, dA = -\int_S \frac{\partial B_n}{\partial t} \, dA$$ | 30-6c |
| Ampère's law modified | $$\oint_C \vec{B} \cdot d\vec{\ell} = \mu_0(I + I_d)$$ $$= \mu_0 I + \mu_0 \epsilon_0 \frac{d}{dt} \int_S E_n \, dA = \mu_0 I + \mu_0 \epsilon_0 \int_S \frac{\partial E_n}{\partial t} \, dA$$ | 30-6d |
| **3. Electromagnetic Waves** | In an electromagnetic wave, the electric and magnetic field vectors are perpendicular to each other and to the direction of propagation. Their magnitudes are related by | |
| | $$E = cB$$ | 30-7 |
| Wave speed | $$c = \frac{1}{\sqrt{\mu_0 \epsilon_0}} \approx 3 \times 10^8 \text{ m/s}$$ | 30-1 |

| | |
|---|---|
| Electromagnetic spectrum | The various types of electromagnetic waves—light, radio waves, X rays, gamma rays, microwaves, and others—differ only in wavelength and frequency. The human eye is sensitive to the range from about 400 nm to 700 nm. |
| Electric dipole radiation | Electromagnetic waves are produced when free electric charges accelerate. Oscillating charges in an electric dipole antenna radiate electromagnetic waves with an intensity that is greatest in directions perpendicular to the antenna. There is no radiated intensity along the axis of the antenna. Perpendicular to the antenna and far away from it, the electric field of the electromagnetic wave is parallel to the antenna. |

| | | |
|---|---|---|
| Energy density in an electromagnetic wave | $u = u_e + u_m = \epsilon_0 E^2 = \dfrac{B^2}{\mu_0} = \dfrac{EB}{\mu_0 c}$ | 30-8 |
| Intensity of an electromagnetic wave | $I = u_{av} c = \dfrac{E_{rms} B_{rms}}{\mu_0} = \dfrac{1}{2} \dfrac{E_0 B_0}{\mu_0} = \left| \vec{S} \right|_{av}$ | 30-9 |
| Poynting vector | $\vec{S} = \dfrac{\vec{E} \times \vec{B}}{\mu_0}$ | 30-10 |
| Momentum and energy in an electromagnetic wave | $p = \dfrac{U}{c}$ | 30-13 |
| Radiation pressure and intensity | $P_r = \dfrac{I}{c}$ | 30-14 |

**4. *The Wave Equation for Electromagnetic Waves**

Maxwell's equations imply that the electric and magnetic field vectors in free space obey a wave equation.

$$\frac{\partial^2 \vec{E}}{\partial x^2} = \frac{1}{c^2} \frac{\partial^2 \vec{E}}{\partial t^2} \qquad \text{30-17}a$$

$$\frac{\partial^2 \vec{B}}{\partial x^2} = \frac{1}{c^2} \frac{\partial^2 \vec{B}}{\partial t^2} \qquad \text{30-17}b$$

# PROBLEMS

- Single-concept, single-step, relatively easy
- •• Intermediate-level, may require synthesis of concepts
- ••• Challenging
- **SSM** Solution is in the *Student Solutions Manual*
- **ISOLVE** Problems available on iSOLVE online homework service
- **ISOLVE✓** These "Checkpoint" online homework service problems ask students additional questions about their confidence level, and how they arrived at their answer.

In a few problems, you are given more data than you actually need; in a few other problems, you are required to supply data from your general knowledge, outside sources, or informed estimates.

## Conceptual Problems

1 • **SSM** True or false:

(a) Maxwell's equations apply only to fields that are constant over time.

(b) The wave equation can be derived from Maxwell's equations.

(c) Electromagnetic waves are transverse waves.

(d) In an electromagnetic wave in free space, the electric and magnetic fields are in phase.

(e) In an electromagnetic wave in free space, the electric and magnetic field vectors $\vec{E}$ and $\vec{B}$ are equal in magnitude.

(f) In an electromagnetic wave in free space, the electric and magnetic energy densities are equal.

**2** •• Theorists have speculated about the possible existence of magnetic monopoles, and there have been several, as yet unsuccessful, experimental searches for such monopoles. Suppose magnetic monopoles were found and that the magnetic field at a distance $r$ from a monopole of strength $q_m$ is given by $B = (\mu_0/4\pi)q_m/r^2$. How would Maxwell's equations have to be modified to be consistent with such a discovery?

**3** • Which waves have greater frequencies, light waves or X rays?

**4** • SSM Are the frequencies of ultraviolet radiation greater or less than those of infrared radiation?

**5** • What kind of waves have wavelengths of the order of a few meters?

**6** • The detection of radio waves can be accomplished with either a dipole antenna or a loop antenna. The dipole antenna detects the (pick one) *electric/magnetic* field of the wave, and the loop antenna detects the *electric/magnetic* field of the wave.

**7** • A transmitter uses a loop antenna with the loop in the horizontal plane. What should be the orientation of a dipole antenna at the receiver for optimum signal reception?

**8** • SSM A helium-neon laser has a red beam. It is shone in turn on a red plastic filter (of the kind used for theater lighting) and a green plastic filter. (A red theater-lighting filter transmitts only red light.) On which filter will the laser exert a larger force?

## Estimation and Approximation

**9** •• Estimate the intensity and total power needed in a laser beam to lift a 15-$\mu$m diameter plastic bead against the force of gravity. Make any assumptions you think reasonable.

**10** ••• Some science fiction writers have used solar sails to propel interstellar spaceships. Imagine a giant sail erected on a spacecraft subjected to the solar radiation pressure. (a) Show that the spacecraft's acceleration is given by

$$a = \frac{P_S A}{4\pi r^2 cm}$$

where $P_S$ is the power output of the sun and is equal to $3.8 \times 10^{26}$ W, $A$ is the surface area of the sail, $m$ is the total mass of the spacecraft, $r$ is the distance from the sun, and $c$ is the speed of light. (b) Show that the velocity of the spacecraft at a distance $r$ from the sun is found from

$$v^2 = v_0^2 + \left(\frac{P_S A}{2\pi mc}\right)\left(\frac{1}{r_0} - \frac{1}{r}\right)$$

where $v_0$ is the initial velocity at $r_0$. (c) Compare the relative accelerations due to the radiation pressure and the gravitational force. Use reasonable values for $A$ and $m$. Will such a system work?

**11** •• The intensity of sunlight striking the earth's upper atmosphere (called the solar constant) is 1.37 kW/m$^2$. (a) Find

$E_{rms}$ and $B_{rms}$ due to the sun at the upper atmosphere of the earth. (b) Find the average power output of the sun. (c) Find the intensity and the radiation pressure at the surface of the sun.

**12** •• SSM Estimate the radiation pressure force exerted on the earth by the sun, and compare the radiation pressure force to the gravitational attraction of the sun. At the earth's orbit the intensity of sunlight is 1.37 kW/m$^2$

**13** •• SSM Repeat Problem 12 for the planet Mars. Which planet has the larger ratio of radiation pressure to gravitational attraction. Why?

**14** •• SSM In the new field of laser cooling and trapping, the forces associated with radiation pressure are used to slow down atoms from thermal speeds of hundreds of meters per second at room temperature to speeds of just a few meters per second or slower. An isolated atom will absorb radiation only at specific resonant frequencies. If the frequency of the laser-beam radiation is one of the resonant frequencies of the target atom, then the radiation is absorbed via a process called resonant absorption. The effective cross-sectional area of the atom for resonant absorption is approximately equal to $\lambda^2$, where $\lambda$ is the wavelength of the laser beam. (a) Estimate the acceleration of a rubidium atom (atomic mass 85 g/mol) in a laser beam whose wavelength is 780 nm and intensity is 10 W/m$^2$. (b) About how long would it take such a light beam to slow a rubidium atom in a gas at room temperature (300 K) down to near-zero velocity?

## Maxwell's Displacement Current

**15** • ISOLVE✓ A parallel-plate capacitor with no material between the plates has circular plates of radius 2.3 cm separated by 1.1 mm. Charge is flowing onto the upper plate and off the lower plate at a rate of 5 A. (a) Find the time rate of change of the electric field between the plates. (b) Compute the displacement current between the plates and show that the displacement current equals 5 A.

**16** • ISOLVE In a region of space, the electric field varies according to $E = (0.05 \text{ N/C}) \sin 2000t$, where $t$ is in seconds. Find the maximum displacement current through a 1-m$^2$ area perpendicular to $\vec{E}$.

**17** •• For Problem 15, show that at a distance $r$ from the axis of the plates the magnetic field between the plates is given by $B = (1.89 \times 10^{-3} \text{ T/m})r$, if $r$ is less than the radius of the plates.

**18** •• (a) Show that for a parallel-plate capacitor with no material between the plates the displacement current is given by $I_d = C \, dV/dt$, where $C$ is the capacitance and $V$ is the voltage across the capacitor. (b) A 5-nF parallel-plate capacitor is connected to an emf $\mathcal{E} = \mathcal{E}_0 \cos \omega t$, where $\mathcal{E}_0 = 3$ V and $\omega = 500\pi$. Find the displacement current between the plates as a function of time. Neglect any resistance in the circuit.

**19** •• SSM ISOLVE✓ Current of 10 A flows into a capacitor having plates with areas of 0.5 m$^2$. There is no material between the plates. (a) What is the displacement current between the plates? (b) What is $dE/dt$ between the plates for this current? (c) What is the line integral of $\vec{B} \cdot d\vec{\ell}$ around a circle of radius 10 cm that lies within the plates and parallel to the plates?

**20** •• A parallel-plate capacitor with circular plates is given a charge $Q_0$. Between the plates is a leaky dielectric having a dielectric constant of $\kappa$ and a resistivity $\rho$. (a) Find the conduction current between the plates as a function of time. (b) Find the displacement current between the plates as a function of time. What is the total (conduction plus displacement) current? (c) Find the magnetic field produced between the plates by the leakage discharge current as a function of time. (d) Find the magnetic field between the plates produced by the displacement current as a function of time. (e) What is the total magnetic field between the plates during discharge of the capacitor?

**21** •• The leaky capacitor of Problem 20 has plate separation $d$. It is being charged such that the voltage across the capacitor is given by $V(t) = (0.01 \text{ V/s})t$. (a) Find the conduction current as a function of time. (b) Find the displacement current. (c) Find the time for which the displacement current is equal to the conduction current.

**22** •• The space between the plates of a capacitor is filled with a material of resistivity $\rho = 10^4 \ \Omega \cdot m$ and dielectric constant $\kappa = 2.5$. The parallel plates are circular with a radius of 20 cm and are separated by 1 mm. The voltage across the plates is given by $V_0 \cos \omega t$, with $V_0 = 40$ V and $\omega = 120\pi$ rad/s. (a) What is the displacement current density? (b) What is the conduction current between the plates? (c) At what angular frequency is the total current 45° out of phase with the applied voltage?

**23** ••• [SSM] Show that the generalized form of Ampère's law (Equation 30-4) and the Biot–Savart law give the same result in a situation in which they both can be used. Figure 30-13 shows two charges $+Q$ and $-Q$ on the x axis at $x = -a$ and $x = +a$, with a current $I = -dQ/dt$ along the line between them. Point $P$ is on the y axis at $y = R$. (a) Use the Biot–Savart law to show that the magnitude of $B$ at point $P$ is

$$B = \frac{\mu_0 I a}{2\pi R} \frac{1}{\sqrt{R^2 + a^2}}$$

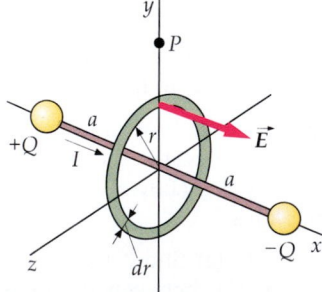

**FIGURE 30-13** Problem 23

(b) Consider a circular strip of radius $r$ and width $dr$ in the $yz$ plane with its center at the origin. Show that the flux of the electric field through this strip is

$$E_x \, dA = \frac{Q}{\epsilon_0} a(r^2 + a^2)^{-3/2} r \, dr$$

(c) Use your result from Part (b) to find the total flux $\phi_e$ through a circular area of radius $R$. Show that

$$\epsilon_0 \phi_e = Q\left(1 - \frac{a}{\sqrt{a^2 + R^2}}\right)$$

(d) Find the displacement current $I_d$, and show that

$$I + I_d = I\frac{a}{\sqrt{a^2 + R^2}}$$

(e) Finally, show that Equation 30-4 gives the same result for B as the result found in Part (a).

## Maxwell's Equations and the Electromagnetic Spectrum

**24** •• Show that the normal component of the magnetic field $\vec{B}$ is continuous across a surface, by applying Gauss's law for $\vec{B}$ ($\int B_n \, dA = 0$) to a pillbox Gaussian surface that has a face on each side of the surface.

**25** • [SSM] [ISOLVE]✓ Find the wavelength for (a) a typical AM radio wave with a frequency of 1000 kHz and (b) a typical FM radio wave with a frequency of 100 MHz.

**26** • [SSM] [ISOLVE] What is the frequency of a 3-cm microwave?

**27** • [ISOLVE] What is the frequency of an X ray with a wavelength of 0.1 nm?

## Electric Dipole Radiation

**28** •• [ISOLVE] The intensity of radiation from an electric dipole is proportional to $\sin^2 \theta / r^2$, where $\theta$ is the angle between the electric dipole moment and the position vector $\vec{r}$. A radiating electric dipole lies along the z axis (its dipole moment is in the z direction). Let $I_1$ be the intensity of the radiation at a distance $r = 10$ m and at angle $\theta = 90°$. Find the intensity (in terms of $I_1$) at (a) $r = 30$ m, $\theta = 90°$; (b) $r = 10$ m, $\theta = 45°$; and (c) $r = 20$ m, $\theta = 30°$.

**29** •• (a) For the situation described in Problem 28, at what angle is the intensity at $r = 5$ m equal to $I_1$? (b) At what distance is the intensity equal to $I_1$ at $\theta = 45°$?

**30** •• [ISOLVE] The transmitting antenna of a station is a dipole located atop a mountain 2000 m above sea level. The intensity of the signal on a nearby mountain 4 km distant and also 2000 m above sea level is $4 \times 10^{-12}$ W/m$^2$. What is the intensity of the signal at sea level and 1.5 km from the transmitter? (See Problem 28.)

**31** ••• A radio station that uses a vertical dipole antenna broadcasts at a frequency of 1.20 MHz with total power output of 500 kW. The radiation pattern is as shown in Figure 30-8 (i.e., the intensity of the signal varies as $\sin^2 \theta$, where $\theta$ is the angle between the direction of propagation and the vertical and is independent of azimuthal angle). Calculate the intensity of the signal at a horizontal distance of 120 km from the station. What is the intensity at that point as measured in photons per square centimeter per second?

**32** ••• [SSM] At a distance of 30 km from a radio station broadcasting at a frequency of 0.8 MHz, the intensity of the electromagnetic wave is $2 \times 10^{-13}$ W/m$^2$. The transmitting antenna is a vertical dipole. What is the total power radiated by the station?

**33** ••• [ISOLVE]✓ A small private plane approaching an airport is flying at an altitude of 2500 m above ground. The airport's flight control system transmits 100 W at 24 MHz, using a vertical dipole antenna. What is the intensity of the signal at the plane's receiving antenna when the plane's position on a map is 4 km from the airport?

## Energy and Momentum in an Electromagnetic Wave

**34** • ✓ISOLVE✓ An electromagnetic wave has an intensity of 100 W/m². Find (a) the radiation pressure $P_r$, (b) $E_{rms}$, and (c) $B_{rms}$.

**35** • ✓ISOLVE✓ The amplitude of an electromagnetic wave is $E_0 = 400$ V/m. Find (a) $E_{rms}$, (b) $B_{rms}$, (c) the intensity $I$, and (d) the radiation pressure $P_r$.

**36** • ISOLVE The rms value of the electric field in an electromagnetic wave is $E_{rms} = 400$ V/m. (a) Find $B_{rms}$, (b) the average energy density, and (c) the intensity.

**37** • Show that the units of $E = cB$ are consistent; that is, show that when $B$ is in teslas and $c$ is in meters per second, the units of $cB$ are volts per meter or newtons per coulomb.

**38** • SSM ISOLVE The rms value of the magnitude of the magnetic field in an electromagnetic wave is $B_{rms} = 0.245\ \mu T$. Find (a) $E_{rms}$, (b) the average energy density, and (c) the intensity.

**39** •• ISOLVE (a) An electromagnetic wave of intensity 200 W/m² is incident normally on a rectangular black card with sides of 20 cm and 30 cm that absorbs all the radiation. Find the force exerted on the card by the radiation. (b) Find the force exerted by the same wave if the card reflects all the radiation incident on it.

**40** •• Find the force exerted by the electromagnetic wave on the reflecting card in Part (b) of Problem 39 if the radiation is incident at an angle of 30° to the normal.

**41** •• SSM An AM radio station radiates an isotropic sinusoidal wave with an average power of 50 kW. What are the amplitudes of $E_{max}$ and $B_{max}$ at a distance of (a) 500 m, (b) 5 km, and (c) 50 km?

**42** •• ✓ISOLVE✓ A laser beam has a diameter of 1.0 mm and average power of 1.5 mW. Find (a) the intensity of the beam, (b) $E_{rms}$, (c) $B_{rms}$, and (d) the radiation pressure.

**43** •• SSM ISOLVE Instead of sending power by a 750-kV, 1000-A transmission line, one desires to beam this energy via an electromagnetic wave. The beam has a uniform intensity within a cross-sectional area of 50 m². What are the rms values of the electric and the magnetic fields?

**44** •• ✓ISOLVE✓ A laser pulse has an energy of 20 J and a beam radius of 2 mm. The pulse duration is 10 ns and the energy density is constant within the pulse. (a) What is the spatial length of the pulse? (b) What is the energy density within the pulse? (c) Find the electric and magnetic amplitudes of the laser pulse.

**45** •• SSM The electric field of an electromagnetic wave oscillates in the $y$ direction and the Poynting vector is given by

$$\vec{S}(x, t) = (100\ W/m^2)\cos^2[10x - (3 \times 10^9)t]\hat{i}$$

where $x$ is in meters and $t$ is in seconds. (a) What is the direction of propagation of the wave? (b) Find the wavelength and the frequency. (c) Find the electric and magnetic fields.

**46** •• A parallel-plate capacitor is being charged. The capacitor consists of two circular parallel plates of area A and separation $d$. (a) Show that the displacement current in the capacitor gap has the same value as the conduction current in the capacitor leads. (b) What is the direction of the Poynting vector in the region of space between the capacitor plates? (c) Calculate the Poynting vector S in this region and show that the flux of $S$ into this region is equal to the rate of change of the energy stored in the capacitor.

**47** •• ✓ISOLVE✓ A pulsed laser fires a 1000-MW pulse of 200-ns duration at a small object of mass 10 mg suspended by a fine fiber 4 cm long. If the radiation is completely absorbed without other effects, what is the maximum angle of deflection of this pendulum?

**48** •• The mirrors used in a particular type of laser are 99.99% reflecting. (a) If the laser has an average output power of 15 W, what is the average power of the radiation incident on one of the mirrors? (b) What is the force due to radiation pressure on one of the mirrors?

**49** •• A 10-cm by 15-cm card has a mass of 2 g and is perfectly reflecting. The card hangs in a vertical plane and is free to rotate about a horizontal axis through the top edge. The card is illuminated uniformly by an intense light that causes the card to make an angle of 1° with the vertical. Find the intensity of the light.

## *The Wave Equation for Electromagnetic Waves

**50** • Show by direct substitution that Equation 30-17a is satisfied by the wave function

$$E_y = E_0 \sin(kx - \omega t) = E_0 \sin k(x - ct)$$

where $c = \omega/k$.

**51** • Use the known values of $\mu_0$ and $\epsilon_0$ in SI units to compute $c = 1/\sqrt{\epsilon_0\mu_0}$, and show that it is approximately $3 \times 10^8$ m/s.

**52** ••• SSM (a) Using arguments similar to those given in the text, show that for a plane wave, in which E and B are independent of $y$ and $z$,

$$\frac{\partial E_z}{\partial x} = \frac{\partial B_y}{\partial t} \quad \text{and} \quad \frac{\partial B_y}{\partial x} = \mu_0\epsilon_0\frac{\partial E_z}{\partial t}$$

(b) Show that $E_z$ and $B_y$ also satisfy the wave equation.

**53** ••• Show that any function of the form $y(x, t) = f(x - vt)$ or $y(x, t) = g(x + vt)$ satisfies the wave Equation 30-16.

## General Problems

**54** • (a) Show that if $E$ is in volts per meter and $B$ is in teslas, the units of the Poynting vector $\vec{S} = (\vec{E} \times \vec{B})/\mu_0$ are watts per square meter. (b) Show that if the intensity $I$ is in watts per square meter, the units of radiation pressure $P_r = I/c$ are newtons per square meter.

**55** •• A loop antenna that may be rotated about a vertical axis is used to locate an unlicensed amateur radio transmitter. If the output of the receiver is proportional to the intensity of the received signal, how does the output of the receiver vary with the orientation of the loop antenna?

**56** •• An electromagnetic wave has a frequency of 100 MHz and is traveling in a vacuum. The magnetic field is given by $\vec{B}\,(z, t) = (10^{-8}\,\text{T})\cos(kz - \omega t)\hat{i}$. (a) Find the wavelength and the direction of propagation of this wave. (b) Find the electric vector $\vec{E}\,(z, t)$. (c) Give Poynting's vector, and find the intensity of this wave.

**57** •• SSM A circular loop of wire can be used to detect electromagnetic waves. Suppose a 100-MHz FM radio station radiates 50 kW uniformly in all directions. What is the maximum rms voltage induced in a loop of radius 30 cm at a distance of $10^5$ m from the station?

**58** •• SOLVE The electric field from a radio station some distance from the transmitter is given by $E = (10^{-4}\,\text{N/C})\cos 10^6 t$, where $t$ is in seconds. (a) What voltage is picked up on a 50-cm wire oriented along the electric field direction? (b) What voltage can be induced in a loop of radius 20 cm?

**59** •• A circular capacitor of radius $a$ has a thin wire of resistance $R$ connecting the centers of the two plates. A voltage $V_0 \sin \omega t$ is applied between the plates. (a) What is the current drawn by this capacitor? (b) What is the magnetic field as a function of radial distance $r$ from the centerline within the plates of this capacitor? (c) What is the phase angle between the current and the applied voltage?

**60** •• SOLVE A 20-kW beam of radiation is incident normally on a surface that reflects half of the radiation. What is the force on this surface?

**61** •• SSM The electric fields of two harmonic waves of angular frequency $\omega_1$ and $\omega_2$ are given by $\vec{E}_1 = E_{1,0}\cos(k_1 x - \omega_1 t)\hat{j}$ and by $\vec{E}_2 = E_{2,0}\cos(k_2 x - \omega_2 t + \delta)\hat{j}$. Find (a) the instantaneous Poynting vector for the resultant wave motion and (b) the time-average Poynting vector. If the direction of propagation of the second wave is reversed so $\vec{E}_2 = E_{2,0}\cos(k_2 x + \omega_2 t + \delta)\hat{j}$, find (c) the instantaneous Poynting vector for the resultant wave motion and (d) the time-average Poynting vector.

**62** •• SSM SOLVE At the surface of the earth, there is an approximate average solar flux of $0.75\,\text{kW/m}^2$. A family wishes to construct a solar energy conversion system to power their home. If the conversion system is 30 percent efficient and the family needs a maximum of 25 kW, what effective surface area is needed for perfectly absorbing collectors?

**63** •• SOLVE✓ Suppose one has an excellent radio capable of detecting a signal as weak as $10^{-14}\,\text{W/m}^2$. This radio has a 2000-turn coil antenna that has a radius of 1 cm wound on an iron core that increases the magnetic field by a factor of 200. The radio frequency is 140 KHz. (a) What is the amplitude of the magnetic field in this wave? (b) What is the emf induced in the antenna? (c) What would be the emf induced in a 2-m wire oriented in the direction of the electric field?

**64** •• Show that
$$\frac{\partial B_z}{\partial x} = -\mu_0 \epsilon_0 \frac{\partial E_y}{\partial t}$$

(Equation 30-19) follows from

$$\oint_C \vec{B} \cdot d\vec{\ell} = \mu_0 \epsilon_0 \int_S \frac{\partial E_n}{\partial t}\, dA$$

(Equation 30-6d with $I = 0$) by integrating along a suitable curve $C$ and over a suitable surface $S$ in a manner that parallels the derivation of Equation 30-18.

**65** ••• SSM A long cylindrical conductor of length $L$, radius $a$, and resistivity $\rho$ carries a steady current $I$ that is uniformly distributed over its cross-sectional area. (a) Use Ohm's law to relate the electric field $E$ in the conductor to $I$, $\rho$, and $a$. (b) Find the magnetic field $B$ just outside the conductor. (c) Use the results from Part (a) and Part (b) to compute the Poynting vector $\vec{S} = (\vec{E} \times \vec{B})/\mu_0$ at $r = a$ (the edge of the conductor). In what direction is $\vec{S}$? (d) Find the flux $\oint S_n\, dA$ through the surface of the conductor into the conductor, and show that the rate of energy flow into the conductor equals $I^2 R$, where $R$ is the resistance of the cylinder. (Here, $S_n$ is the *inward* component of $\vec{S}$ perpendicular to the surface of the conductor.)

**66** ••• A long solenoid of $n$ turns per unit length has a current that slowly increases with time. The solenoid has radius $R$, and the current in the windings has the form $I(t) = at$. (a) Find the induced electric field at a distance $r < R$ from the solenoid axis. (b) Find the magnitude and direction of the Poynting vector $\vec{S}$ at the cylindrical surface $r = R$ just inside the solenoid windings. (c) Calculate the flux $\oint S_n\, dA$ into the solenoid, and show that the flux equals the rate of increase of the magnetic energy inside the solenoid. (Here, $S_n$ is the *inward* component of $\vec{S}$ perpendicular to the surface of the solenoid.)

**67** ••• SSM SOLVE Small particles might be blown out of solar systems by the radiation pressure of sunlight. Assume that the particles are spherical with a radius $r$ and a density of $1\,\text{g/cm}^3$ and that they absorb all the radiation in a cross-sectional area of $\pi r^2$. The particles are a distance $R$ from the sun, which has a power output of $3.83 \times 10^{26}$ W. What is the radius $r$ for which the radiation force of repulsion just balances the gravitational force of attraction to the sun?

**68** ••• When an electromagnetic wave is reflected at normal incidence on a perfectly conducting surface, the electric field vector of the reflected wave at the reflecting surface is the negative of that of the incident wave. (a) Explain why this should be. (b) Show that the superposition of incident and reflected waves results in a standing wave. (c) What is the relationship between the magnetic field vector of the incident waves and reflected waves at the reflecting surface?

**69** ••• SSM An intense point source of light radiates 1 MW isotropically. The source is located 1 m above an infinite, perfectly reflecting plane. Determine the force that acts on the plane.

# Properties of Light

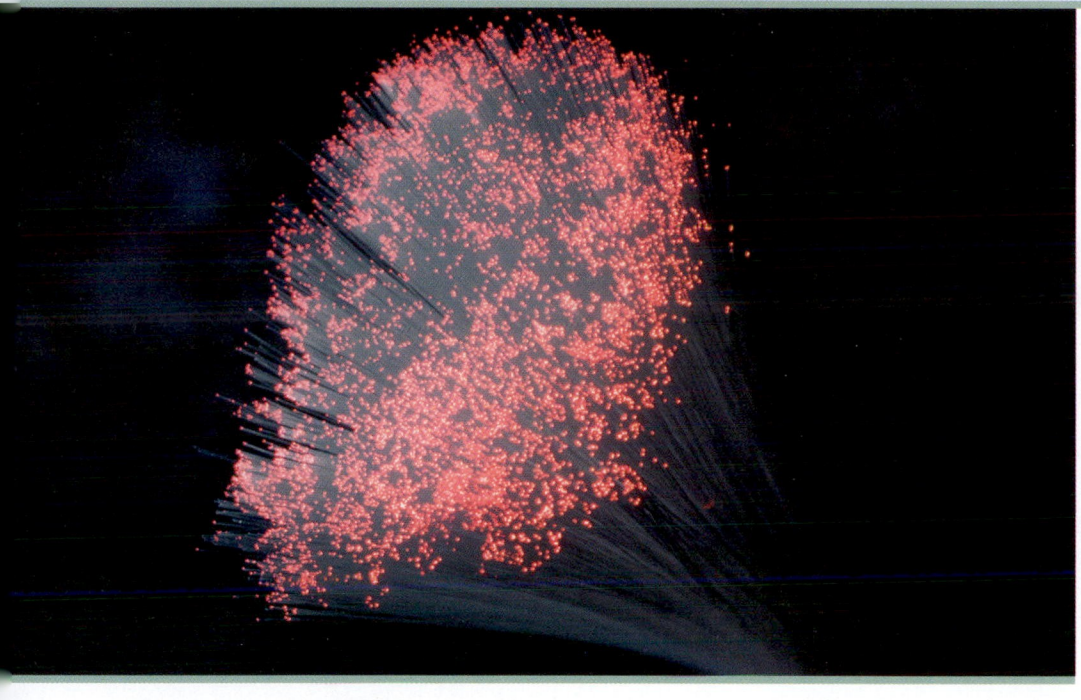

LIGHT IS TRANSMITTED BY TOTAL
INTERNAL REFLECTION THROUGH
TINY GLASS FIBERS.

 **How large must the angle
of incidence of the light on the
wall of the tube be so that no light
escapes? (See Example 31-5.)**

31-1   Wave–Particle Duality

31-2   Light Spectra

31-3   Sources of Light

31-4   The Speed of Light

31-5   The Propagation of Light

31-6   Reflection and Refraction

31-7   Polarization

31-8   Derivation of the Laws of Reflection and Refraction

The human eye is sensitive to electromagnetic radiation with wavelengths from approximately 400 nm to 700 nm. The shortest wavelengths in the visible spectrum correspond to violet light and the longest to red light. The perceived colors of light are the result of the physiological and psychological response of the eye–brain sensing system to the different frequencies of visible light. Although the correspondence between perceived color and frequency is quite good, there are many interesting deviations. For example, a mixture of red light and green light is perceived by the eye–brain sensing system as yellow even in the absence of light in the yellow region of the spectrum.

➤ In this chapter, we study how light is produced; how its speed is measured; and how light is scattered, reflected, refracted, and polarized.

# 31-1 Wave–Particle Duality

The wave nature of light was first demonstrated by Thomas Young, who observed the interference pattern of two coherent light sources produced by illuminating a pair of narrow, parallel slits with a single source. The wave theory of light culminated in 1860 with Maxwell's prediction of electromagnetic waves. The particle nature of light was first proposed by Albert Einstein in 1905 in his explanation of the photoelectric effect.[†] A particle of light called a **photon** has energy $E$ that is related to the frequency $f$ and wavelength $\lambda$ of the light wave by the Einstein equation

$$E = hf = \frac{hc}{\lambda} \qquad \text{31-1}$$

EINSTEIN'S EQUATION FOR PHOTON ENERGY

where $c$ is the speed of light and $h$ is Planck's constant:

$$h = 6.626 \times 10^{-34} \text{ J·s} = 4.136 \times 10^{-15} \text{ eV·s}$$

Since energies are often given in electron volts and wavelengths are given in nanometers, it is convenient to express the combination $hc$ in eV·nm. We have

$$hc = (4.136 \times 10^{-15} \text{ eV·s})(2.998 \times 10^8 \text{ m/s}) = 1.240 \times 10^{-6} \text{ eV·m}$$

or

$$hc = 1240 \text{ eV·nm} \qquad \text{31-2}$$

The propagation of light is governed by its wave properties, whereas the exchange of energy between light and matter is governed by its particle properties. This wave–particle duality is a general property of nature. For example, the propagation of electrons (and other so-called particles) is also governed by wave properties, whereas the exchange of energy between the electrons and other particles is governed by particle properties.

# 31-2 Light Spectra

Newton was the first to recognize that white light is a mixture of light of all colors of approximately equal intensity. He demonstrated this by letting sunlight fall on a glass prism and observing the spectrum of the refracted light (Figure 31-1). Because the angle of refraction produced by a glass prism depends slightly on wavelength, the refracted beam is spread out in space into its component colors or wavelengths, like a rainbow.

**FIGURE 31-1** Newton demonstrating the spectrum of sunlight with a glass prism.

---

[†] The photoelectric effect is discussed in Chapter 34.

Figure 31-2 shows a spectroscope, which is a device for analyzing the spectra of a light source. Light from the source passes through a narrow slit, traverses a lens to make the beam parallel, and falls on a glass prism. The refracted beam is viewed with a telescope, which is mounted on a rotating platform so that the angle of the refracted beam, which depends on the wavelength, can be measured. The spectrum of the light source can thus be analyzed in terms of its component wavelengths. The spectrum of sunlight contains a continuous range of wavelengths and is therefore called a **continuous spectrum.** The light emitted by the atoms in low-pressure gases, such as mercury atoms in a fluorescent lamp, contains only a discrete set of wavelengths. Each wavelength emitted by the source produces a separate image of the collimating slit in the spectroscope. Such a spectrum is called a **line spectrum.** The continuous visible spectrum and the line spectra from several elements are shown in the photograph.

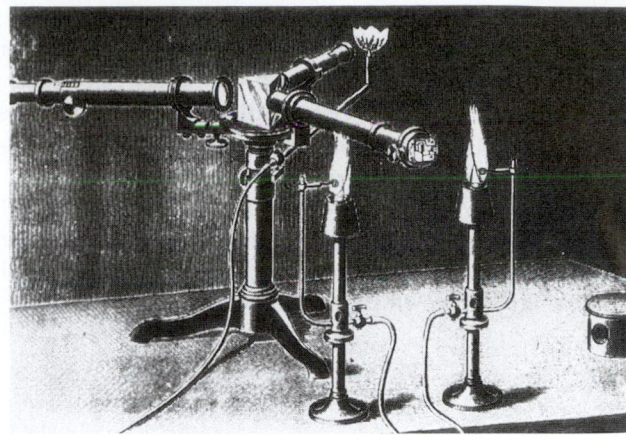

**FIGURE 31-2** A late nineteenth-century spectroscope belonging to Gustav Kirchhoff. Modern student spectroscopes usually share the same general design.

# 31-3 Sources of Light

## Line Spectra

The most common sources of visible light are transitions of the outer electrons in atoms. Normally an atom is in its ground state with its electrons at their lowest allowed energy levels consistent with the exclusion principle. (The exclusion principle, which was first enunciated by Wolfgang Pauli in 1925 to explain the electronic structure of atoms, states that no two electrons in an atom can be in the same quantum state.) The lowest energy electrons are closest to the nucleus and are tightly bound, forming a stable inner core. The one or two electrons in the highest energy states are much farther from the nucleus and are relatively easily excited to vacant higher energy states. These outer electrons are responsible for the energy changes in the atom that result in the emission or absorption of visible light.

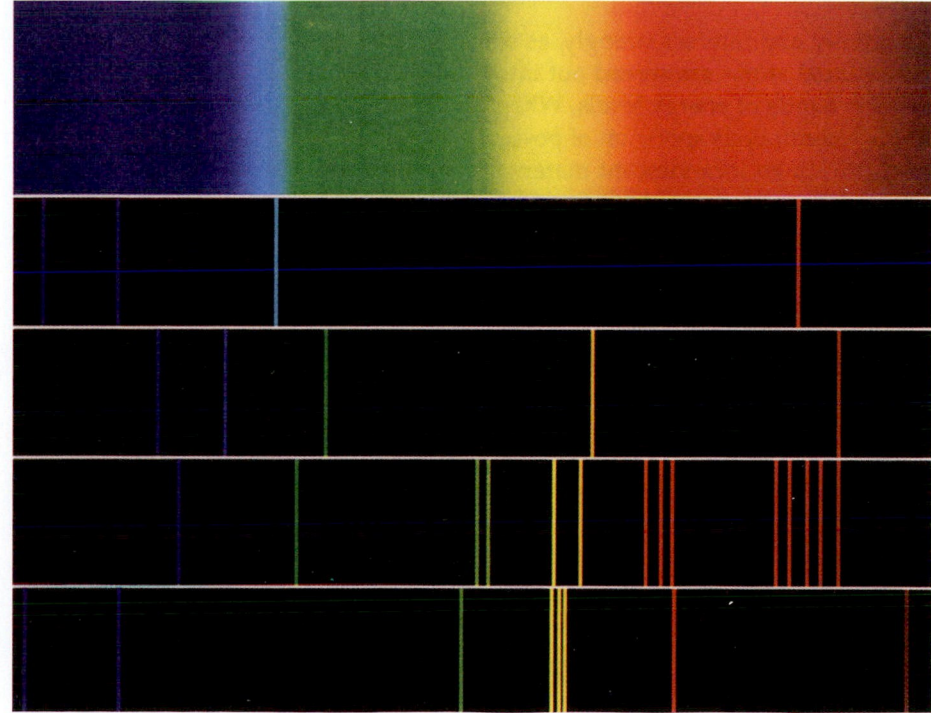

The continuous visible spectrum (top) and the line spectra of (from top to bottom) hydrogen, helium, barium, and mercury.

When an atom collides with another atom or with a free electron, or when the atom absorbs electromagnetic energy, the outer electrons can be excited to higher energy states. After a time of approximately 10 ns (1 ns = $10^{-9}$ s), these outer electrons spontaneously make transitions to lower energy states with the emission of a photon. This process, called **spontaneous emission,** is random; the photons emitted from two different atoms are not correlated. The emitted light is thus incoherent. By conservation of energy, the energy of an emitted photon is the energy difference $|\Delta E|$ between the initial state and the final state. The frequency of the light wave is related to the energy by the Einstein equation, $|\Delta E| = hf$. The wavelength of the emitted light is then

$$\lambda = \frac{c}{f} = \frac{hc}{hf} = \frac{hc}{|\Delta E|}$$
31-3

The photon energies corresponding to shortest wavelengths (400 nm) and longest (700 nm) wavelengths in the visible spectrum are

$$E_{400\,nm} = \frac{hc}{\lambda} = \frac{1240\ eV \cdot nm}{400\ nm} = 3.10\ eV$$
31-4a

and

$$E_{700\,nm} = \frac{hc}{\lambda} = \frac{1240\ eV \cdot nm}{700\ nm} = 1.77\ eV$$
31-4b

Because the energy levels in atoms form a discrete set, the emission spectrum of light from single atoms or from atoms in low-pressure gases consists of a set of sharp discrete lines that are characteristic of the element. These narrow lines are broadened somewhat by Doppler shifts, due to the motion of the atom relative to the observer and by collisions with other atoms; but, generally, if the gas density is low enough, the lines are narrow and well separated from one another. The study of the line spectra of hydrogen and other atoms led to the first understanding of the energy levels of atoms.

**Continuous Spectra** When atoms are close together and interact strongly, as in liquids and solids, the energy levels of the individual atoms are spread out into energy bands, resulting in essentially continuous bands of energy levels. When the bands overlap, as they often do, the result is a continuous spectrum of possible energies and a continuous emission spectrum. In an incandescent material such as a hot metal filament, electrons are randomly accelerated by frequent collisions, resulting in a broad spectrum of thermal radiation. The rate at which an object radiates thermal energy is proportional to the fourth power of its absolute temperature.[†] The radiation emitted by an object at temperatures below approximately 600°C is concentrated in the infrared and is not visible. As an object is heated, the energy radiated extends to shorter and shorter wavelengths. Between approximately 600°C and 700°C, enough of the radiated energy is in the visible spectrum for the object to glow a dull red. At higher and higher temperatures, the object becomes bright red and then white. For a given temperature, the wavelength $\lambda_{peak}$ at which the emitted power is a maximum varies inversely with the temperature, a result known as Wien's displacement law. The surface of the sun at $T = 6000$ K emits a continuous spectrum of approximately constant intensity over the visible range of wavelengths.

---

† This is known as the Stefan–Boltzmann law. This and other properties of thermal radiation, such as Wien's displacement law, are discussed more fully in Section 20-4.

# Absorption, Scattering, Spontaneous Emission, and Stimulated Emission

When radiation is emitted, an atom makes a transition from an excited state to a state of lower energy; when radiation is absorbed, an atom makes a transition from a lower state to a higher state. When atoms are irradiated with a continuous spectrum of radiation, the transmitted spectrum shows dark lines corresponding to the absorption of light at discrete wavelengths. The absorption spectra of atoms were the first line spectra observed. Since atoms and molecules at normal temperatures are in either their ground states or low-lying excited states, only transitions from a ground state (or a near ground state) to a more highly excited state are observed. Thus, absorption spectra usually have far fewer lines than emission spectra have.

Figure 31-3 illustrates several interesting phenomena that can occur when a photon is incident on an atom. In Figure 31-3a, the energy of the incoming photon is too small to excite the atom to an excited state, so the atom remains in its ground state and the photon is said to be scattered. Since the incoming and outgoing or scattered photons have the same energy, the scattering is said to be elastic. If the wavelength of the incident light is large compared with the size of the atom, the scattering can be described in terms of classical electromagnetic theory and is called **Rayleigh scattering** after Lord Rayleigh, who worked out the theory in 1871. The probability of Rayleigh scattering varies as $1/\lambda^4$. This means that blue light is scattered much more readily than red light, which accounts for the bluish color of the sky. The removal of blue light by Rayleigh scattering also accounts for some of the reddish color of the transmitted light seen in sunsets.

**Inelastic scattering,** also called **Raman scattering,** occurs when an incident photon with just the right amount of energy is absorbed and the molecule undergoes a transition to a more energetic state. Then the molecule emits a photon as it undergoes a transition to a less energetic state, whose energy differs from that of the initial state. If the energy of the scattered photon $hf'$ is less than that of the incident photon $hf$ (Figure 31-3b), it is called **Stokes Raman scattering.** If the energy of the scattered photon is greater than that of the incident photon (Figure 31-3c), it is called **anti-Stokes Raman scattering.**

In Figure 31-3d, the energy of the incident photon is just equal to the difference in energy between the initial state and a more energetic state. The atom absorbs the photon and makes a transition to the more excited state in a process called **resonance absorption.**

In Figure 31-3e, an atom in an excited state spontaneously undergoes a transition to a less energetic state, in a process called **spontaneous emission.** Often an atom in an excited state undergoes transitions to one or more intermediate states as it returns to the ground state. A common example occurs when an atom is excited by ultraviolet light and emits visible light as it returns via multiple transitions to its ground state. This process, often called **fluorescence,** occurs in a thin film lining the inside of the glass tubes of fluorescent light bulbs. Since the lifetime of a typical excited atomic energy state is of the order of 10 ns, this process appears to occur instantaneously. However, some excited states have much longer lifetimes—of the order of milliseconds or occasionally seconds or even minutes. Such a state is called a **metastable state. Phosphorescent materials** have very long-lived metastable states and emit light long after the original excitation.

Figure 31-3f illustrates the photoelectric effect, in which the absorption of the photon ionizes the atom by causing the emission of an electron. Figure 31-3g illustrates **stimulated emission.** This process occurs if the atom or molecule is initially in an excited state of energy $E_H$, and the energy of the incident photon is equal to $E_H - E_L$, where $E_L$ is the energy of a lower state. In this case, the oscillating electromagnetic field associated with the incident photon can stimulate the excited atom or molecule, which then emits a photon in the same direction as the incident photon and in phase with it. The photons from the stimulated atoms or molecules can stimulate the emission of additional photons propagating in the same direction

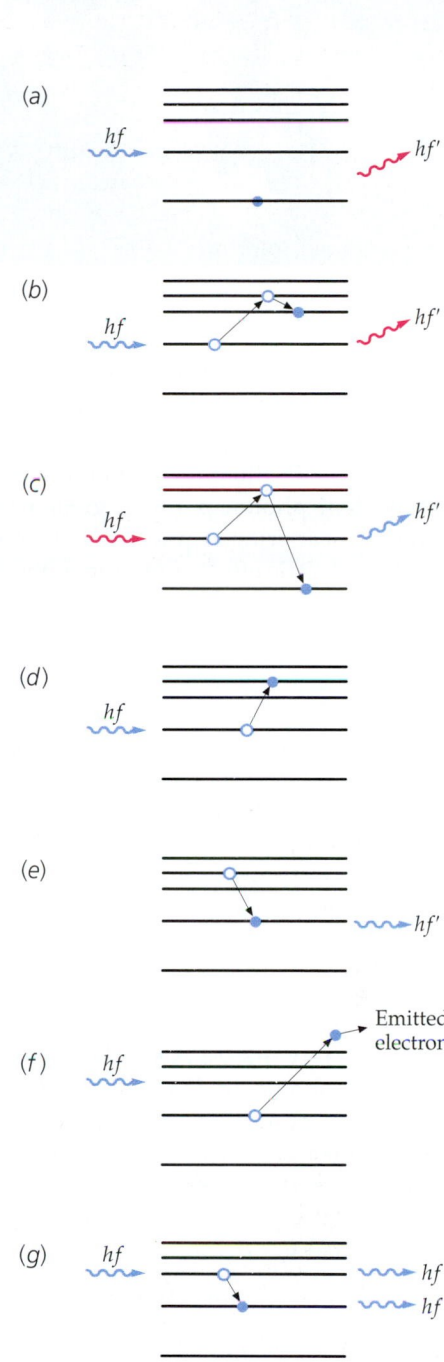

**FIGURE 31-3** Photon-atom and photon-molecule interactions. (*a*) Elastic scattering (*b*) Stokes Raman scattering (*c*) Anti-Stokes Raman scattering (*d*) Resonance absorption (*e*) Spontaneous emission (*f*) Photoelectric effect (*g*) Stimulated emission (*h*) Compton scattering.

(a)

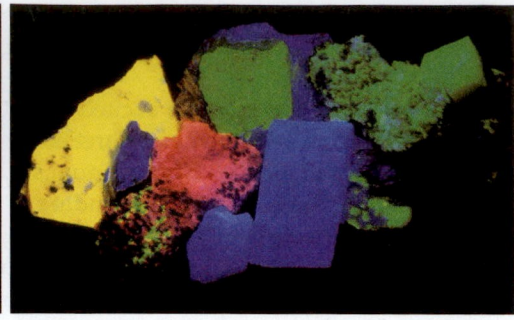

(b)

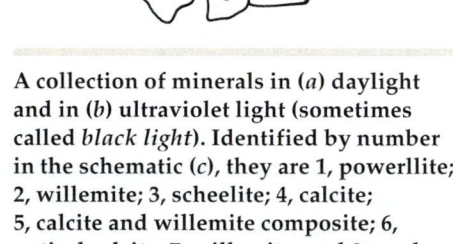

(c)

with the same phase. This process amplifies the initially emitted photon, yielding a beam of light originating from different atoms that is coherent. As a result, interference of the light from a large number of atoms can easily be observed.

Figure 31-3*h* illustrates **Compton scattering,** which occurs if the energy of the incident photon is much greater than the ionization energy. Note that in Compton scattering, a photon is absorbed and a photon is emitted, whereas in the photoelectric effect, a photon is absorbed with none emitted.

A collection of minerals in (*a*) daylight and in (*b*) ultraviolet light (sometimes called *black light*). Identified by number in the schematic (*c*), they are 1, powerllite; 2, willemite; 3, scheelite; 4, calcite; 5, calcite and willemite composite; 6, optical calcite; 7, willemite; and 8, opal. The change in color is due to the minerals fluorescing under the ultraviolet light. In optical calcite, both fluorescence and phosphorescence occur.

---

*RESONANT ABSORPTION AND EMISSION*　　　　**EXAMPLE 31-1**

The first excited state of potassium is $E_1 = 1.62$ eV above the ground state $E_0$, which we take to be zero. The second and third excited states of potassium have energy levels at $E_2 = 2.61$ eV and $E_3 = 3.07$ eV above the ground state. (*a*) What is the maximum wavelength of radiation that can be absorbed by potassium in its ground state? Calculate the wavelength of the emitted photon when the atom makes a transition from (*b*) the second excited state ($E_2$) to the ground state and from (*c*) the third excited state ($E_3$) to the second excited state ($E_2$).

**PICTURE THE PROBLEM** The ground state and the first three excited energy levels are shown in Figure 31-4. (*a*) Since the wavelength is related to the energy of a photon by $\lambda = hc/\Delta E$, longer wavelengths correspond to smaller energy differences. The smallest energy difference for a transition originating at the ground state is from the ground state to the first excited state. (*b*) The wavelengths of the photons given off when the atom de-excites are related to the energy differences by $\lambda = hc/|\Delta E|$.

$$\begin{aligned} &\underline{\phantom{xxxxxxxxxx}}\quad E_3 = 3.07 \text{ eV} \\ &\underline{\phantom{xxxxxxxxxx}}\quad E_2 = 2.61 \text{ eV} \\[1em] &\underline{\phantom{xxxxxxxxxx}}\quad E_1 = 1.62 \text{ eV} \\[2em] &\underline{\phantom{xxxxxxxxxx}}\quad E_0 = 0 \end{aligned}$$

**FIGURE 31-4**

(*a*) Calculate the wavelength of radiation absorbed in a transition from the ground state to the first excited state:

$$\lambda = \frac{hc}{\Delta E} = \frac{hc}{E_1 - E_0} = \frac{1240 \text{ eV·nm}}{1.62 \text{ eV} - 0} = \boxed{765 \text{ nm}}$$

(*b*) For the transition from $E_3$ to the ground state, the photon energy is $E_3 - E_0 = E_3$. Calculate the wavelength of radiation emitted in this transition:

$$\lambda = \frac{hc}{|\Delta E|} = \frac{hc}{E_3 - E_0} = \frac{1240 \text{ eV·nm}}{3.07 \text{ eV} - 0} = \boxed{404 \text{ nm}}$$

(*c*) For the transition from $E_3$ to $E_2$, the photon energy is $E_3 - E_2$. Calculate the wavelength of radiation emitted in this transition:

$$\lambda = \frac{hc}{|\Delta E|} = \frac{hc}{E_3 - E_2} = \frac{1240 \text{ eV·nm}}{3.07 \text{ eV} - 2.61 \text{ eV}}$$

$$= \boxed{2700 \text{ nm}}$$

**REMARKS** The wavelength of radiation emitted in the transition from $E_1$ to the ground state $E_0$ is 765 nm, the same as that for radiation absorbed in the transition from the ground state to $E_1$. This transition and the transmission from $E_3$ to the ground state both result in photons in the visible spectrum.

## Lasers

The *laser* (*l*ight *a*mplification by *s*timulated *e*mission of *r*adiation) is a device that produces a strong beam of coherent photons by stimulated emission. Consider a system consisting of atoms that have a ground state of energy $E_1$ and an excited metastable state of energy $E_2$. If these atoms are irradiated by photons of energy $E_2 - E_1$, those atoms in the ground state can absorb a photon and make the transition to state $E_2$, whereas those atoms already in the excited state may be stimulated to decay back to the ground state. The relative probabilities of absorption and stimulated emission, first worked out by Einstein, are equal. Ordinarily, nearly all the atoms of the system at normal temperature will initially be in the ground state, so absorption will be the main effect. To produce more stimulated-emission transitions than absorption transitions, we must arrange to have more atoms in the excited state than in the ground state. This condition, called population inversion, can be achieved by a method called optical pumping in which atoms are *pumped* up to levels of energy greater than $E_2$ by the absorption of an intense auxiliary radiation. The atoms then decay down to state $E_2$ either by spontaneous emission or by nonradiative transitions, such as those due to collisions.

Figure 31-5 shows a schematic diagram of the first laser, a ruby laser built by Theodore Maiman in 1960. The laser consists of a ruby rod a few centimeters long surrounded by a helical gaseous flashtube that emits a broad spectrum of light. The ends of the ruby rod are flat and perpendicular to the axis of the rod. Ruby is a transparent crystal of $Al_2O_3$ with a small amount (about 0.05 percent) of chromium. It appears red because the chromium ions ($Cr^{3+}$) have strong absorption bands in the blue and green regions of the visible spectrum, as shown in Figure 31-6. The energy levels of chromium—important for the operation of a ruby laser—are shown in Figure 31-7. When the flashtube is fired, there is an intense burst of light that lasts several milliseconds. Photon absorption excites many of the chromium ions to the bands of energy levels indicated by the shading in Figure 31-7. The excited chromium ions then rapidly drop down to a closely spaced pair of metastable states labeled $E_2$ in the figure. These metastable states are approximately 1.79 eV above the ground state. The expected lifetime for a chromium ion to remain in one of these metastable states is about 5 ms, after which the chromium ion spontaneously emits a photon and decays to the ground state. A millisecond is a long time for an atomic process. Consequently, if the flash is intense enough, the number of chromium ions populating the two metastable states will exceed the population of chromium ions in the ground state. It follows that during the time that the flashtube is firing, the populations of ions in the ground state and the metastable states are inverted. When the chromium ions in the state $E_2$ decay to the ground state by spontaneous emission, they emit photons of energy 1.79 eV and wavelength 694.3 nm. These photons have just the right energy to stimulate chromium ions in the metastable states to emit photons of the same energy (and wavelength) as they undergo the transition to the ground state. The photons also have just the right energy to stimulate chromium ions in the ground state to absorb a photon as they undergo the transition to one of the metastable states. These are competing processes, and the stimulated emission process dominates as long as the population of chromium ions in the metastable states exceeds the population in the ground state.

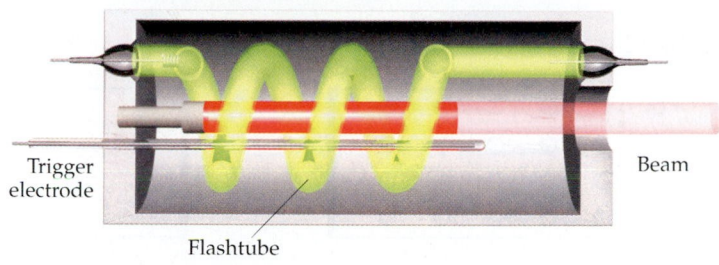

**FIGURE 31-5** Schematic diagram of the first ruby laser.

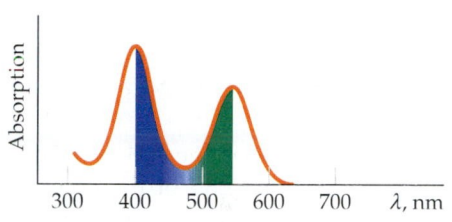

**FIGURE 31-6** Absorption versus wavelength for $Cr^{3+}$ in ruby. Ruby appears red because of the strong absorption of green and blue light by the chromium ions.

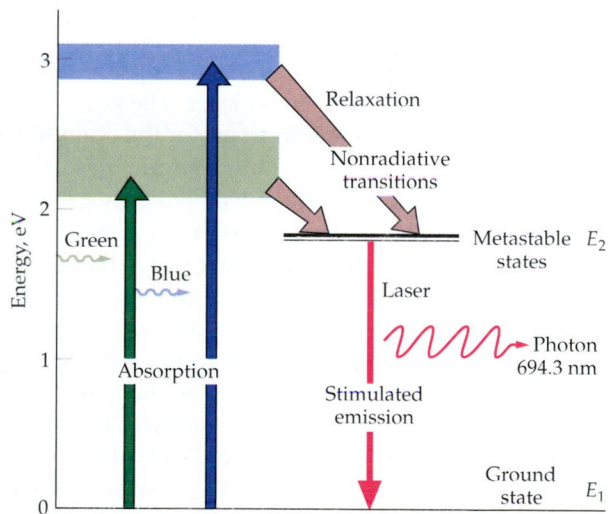

**FIGURE 31-7** Energy levels in a ruby laser. To make the population of the metastable states greater than that of the ground state, the ruby crystal is subjected to intense radiation that contains energy in the green and blue wavelengths. This excites atoms from the ground state to the bands of energy levels indicated by the shading, from which the atoms decay to the metastable states by nonradiative transitions. Then, by stimulated emission, the atoms undergo the transition from the metastable states to the ground state.

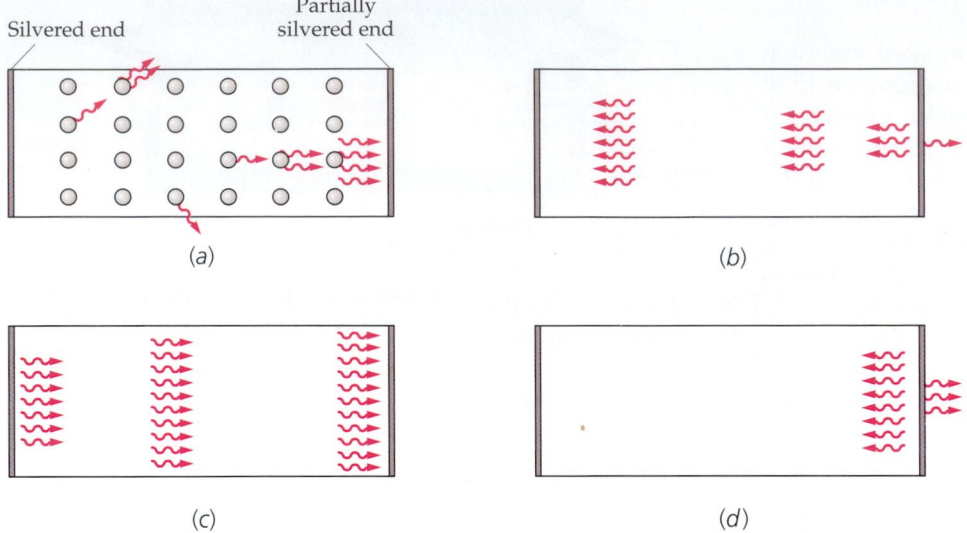

Silvered end / Partially silvered end

(a)   (b)   (c)   (d)

**FIGURE 31-8** Buildup of a photon beam in a laser. (*a*) When irradiated, some atoms spontaneously emit photons, some of which travel to the right and stimulate other atoms to emit photons parallel to the axis of the crystal. (*b*) Of the four photons that strike the right face, one is transmitted and three are reflected. As the reflected photons traverse the laser crystal, they stimulate other atoms to emit photons, and the beam builds up. By the time the beam reaches the right face again (*c*), it is comprised of many photons. (*d*) Some of these photons are transmitted, the rest of the photons are reflected.

In the ruby laser, one end of the crystal is fully silvered, so it is 100 percent reflecting; the other end of the crystal, called the output coupler, is partially silvered, leaving it about 85 percent reflecting. When photons traveling parallel to the axis of the crystal strike the silvered ends, all are reflected from the back face and 85 percent are reflected from the front face, with 15 percent of the photons escaping through the partially silvered front face. During each pass through the crystal, the photons stimulate more and more atoms so that an intense beam is emitted from the partially silvered end (Figure 31-8). Because the duration of each flash of the flashtube is between two and three seconds, the laser beam is produced in pulses lasting a few milliseconds. Modern ruby lasers generate intense light beams with energies ranging from 50 J to 100 J. The beam can have a diameter as small as 1 mm and an angular divergence as small as 0.25 milliradian to about 7 milliradians.

Population inversion is achieved somewhat differently in the continuous helium–neon laser. The energy levels of helium and neon that are important for the operation of the laser are shown in Figure 31-9. Helium has an excited energy state $E_{2,\text{He}}$ that is 20.61 eV above its ground state. Helium atoms are excited to state $E_{2,\text{He}}$ by an electric discharge. Neon has an excited state $E_{3,\text{Ne}}$ that is 20.66 eV above its ground state. This is just 0.05 eV above the first excited state of helium. The neon atoms are excited to state $E_{3,\text{Ne}}$ by collisions with excited helium atoms. The kinetic energy of the helium atoms provides the extra 0.05 eV of energy needed to excite the neon atoms. There is another excited state of neon $E_{2,\text{Ne}}$ that is 18.70 eV above its ground state and 1.96 eV below state $E_{3,\text{Ne}}$. Since state $E_{2,\text{Ne}}$ is normally unoccupied, population inversion between states $E_{3,\text{Ne}}$ and $E_{2,\text{Ne}}$ is obtained immediately. The stimulated emission that occurs between these states results in photons of energy 1.96 eV and wavelength 632.8 nm, which produces a bright red light. After stimulated emission, the atoms in state $E_{2,\text{Ne}}$ decay to the ground state by spontaneous emission.

Note that there are four energy levels involved in the helium–neon laser, whereas the ruby laser involved only three levels. In a three-level laser, population inversion is

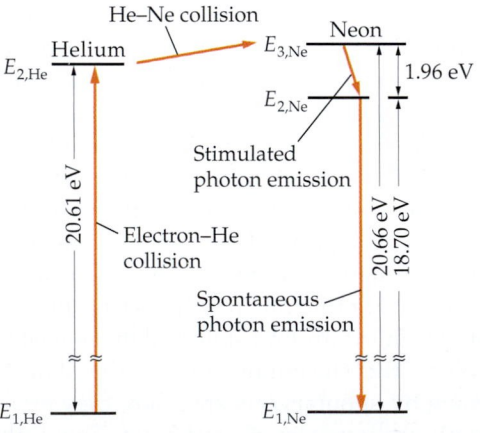

**FIGURE 31-9** Energy levels of helium and neon that are important for the helium–neon laser. The helium atoms are excited by electrical discharge to an energy state 20.61 eV above the ground state. They collide with neon atoms, exciting some neon atoms to an energy state 20.66 eV above the ground state. Population inversion is thus achieved between this level and one 1.96 eV below it. The spontaneous emission of photons of energy 1.96 eV stimulates other atoms in the upper state to emit photons of energy 1.96 eV.

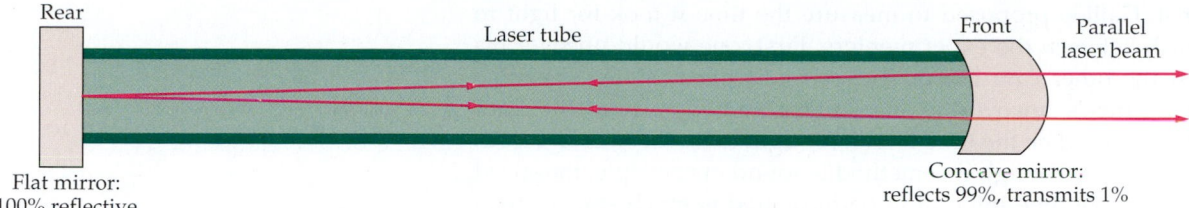

Rear

Laser tube

Front   Parallel
laser beam

Flat mirror:
100% reflective

Concave mirror:
reflects 99%, transmits 1%

**FIGURE 31-10** Schematic drawing of a helium–neon laser. The use of a concave mirror rather than a second plane mirror makes the alignment of the mirrors less critical than it is for the ruby laser. The concave mirror on the right also serves as a lens that focuses the emitted light into a parallel beam.

difficult to achieve because more than half the atoms in the ground state must be excited. In a four-level laser, population inversion is easily achieved because the state after stimulated emission is not the ground state but an excited state that is normally unpopulated.

Figure 31-10 shows a schematic diagram of a helium–neon laser commonly used for physics demonstrations. The helium–neon laser consists of a gas tube that contains 15 percent helium gas and 85 percent neon gas. A totally reflecting flat mirror is mounted at one end of the gas tube and a 99 percent reflecting concave mirror is placed at the other end of the gas tube. The concave mirror focuses parallel light at the flat mirror and also acts as a lens that transmits part of the light, so that the light emerges as a parallel beam.

A laser beam is coherent, very narrow, and intense. Its coherence makes the laser beam useful in the production of holograms, which we discuss in Chapter 33. The precise direction and small angular spread of the laser beam make it useful as a surgical tool for destroying cancer cells or reattaching a detached retina. Lasers are also used by surveyors for precise alignment over large distances. Distances can be accurately measured by reflecting a laser pulse from a mirror and measuring the time the pulse takes to travel to the mirror and back. The distance to the moon has been measured to within a few centimeters using a mirror placed on the moon for that purpose. Laser beams are also used in fusion research. An intense laser pulse is focused on tiny pellets of deuterium–tritium in a combustion chamber. The beam heats the pellets to temperatures of the order of $10^8$ K in a very short time, causing the deuterium and tritium to fuse and release energy.

Laser technology is advancing so quickly that it is possible to mention only a few of the recent developments. In addition to the ruby laser, there are many other solid-state lasers with output wavelengths that range from approximately 170 nm to 3900 nm. Lasers that generate more than 1 kW of continuous power have been constructed. Pulsed lasers can now deliver nanosecond pulses of power exceeding $10^{14}$ W. Various gas lasers can now produce beams of wavelengths that range from the far infrared to the ultraviolet. Semiconductor lasers (also known as diode lasers or junction lasers) have shrunk in just 10 years from the size of a pinhead to mere billionths of a meter. Liquid lasers that use chemical dyes can be tuned over a range of wavelengths (approximately 70 nm for continuous lasers and more than 170 nm for pulsed lasers). A relatively new laser, the free-electron laser, extracts light energy from a beam of free electrons moving through a spatially varying magnetic field. The free-electron laser has the potential for very high power and high efficiency and can be tuned over a large range of wavelengths. There appears to be no limit to the variety and uses of modern lasers.

## 31-4 The Speed of Light

Prior to the seventeenth century the speed of light was thought by many to be infinite, and an effort to measure the speed of light was made by Galileo. He and a partner stood on hilltops about three kilometers apart, each with a lantern and a shutter to cover it. Galileo proposed to measure the time it took for light to travel back and forth between the experimenters. First, one would uncover his lantern, and when the other saw the light, he would uncover his. The time between the first partner's uncovering his lantern and his seeing the light from the other lantern would be the time it took for light to travel back and forth between the experimenters. Though this method is sound in principle, the speed of light is so great that the time interval to be measured is much smaller than fluctuations in human response time, so Galileo was unable to obtain a value for the speed of light.

The first indication of the true magnitude of the speed of light came from astronomical observations of the period of Io, one of the moons of Jupiter. This period is determined by measuring the time between eclipses of Io behind Jupiter. The eclipse period is about 42.5 h, but measurements made when the earth is moving away from Jupiter along path $ABC$ in Figure 31-11 give a greater time for this period than do measurements made when the earth is moving toward Jupiter along path $CDA$ in the figure. Since these measurements differ from the average value by only about 15 s, the discrepancies were difficult to measure accurately. In 1675, the astronomer Ole Römer attributed these discrepancies to the fact that the speed of light is finite, and that during the 42.5 h between eclipses of Jupiter's moon, the distance between the earth and Jupiter changes, making the path for the light longer or shorter. Römer devised the following method for measuring the cumulative effect of these discrepancies. Jupiter is moving much more slowly than the earth, so we can neglect its motion. When the earth is at point $A$, nearest to Jupiter, the distance between the earth and Jupiter is changing negligibly. The period of Io's eclipse is measured, providing the time between the beginnings of successive eclipses. Based on this measurement, the number of occultations during 6 months is computed, and the time when an eclipse should begin a half-year later when the earth is at point $C$ is predicted. When the earth is actually at point $C$, the observed beginning of the eclipse is about 16.6 min later than predicted. This is the time it takes light to travel a distance equal to the diameter of the earth's orbit. This calculation neglects the distance traveled by Jupiter toward the earth. However, because the orbital speed of Jupiter is so much slower than that of the earth, the distance Jupiter moves toward (or away from) the earth during the 6 months is much less than the diameter of the earth's orbit.

**EXERCISE** Calculate (a) the distance traveled by the earth between successive eclipses of Io and (b) the speed of light, given that the time between successive eclipses is 15 s longer than average when the earth is moving directly away from Jupiter. (*Answer* (a) $4.59 \times 10^6$ km (b) $3.06 \times 10^8$ m/s)

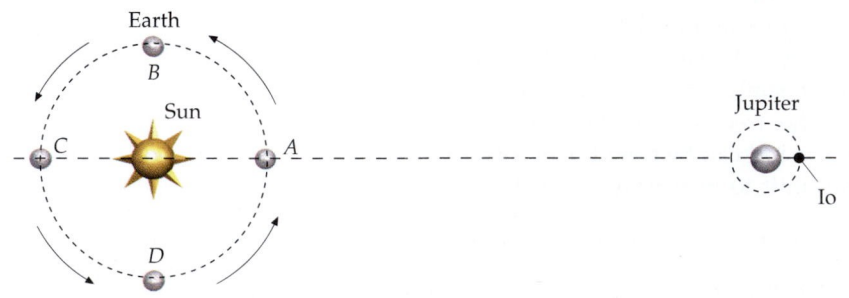

**FIGURE 31-11** Römer's method of measuring the speed of light. The time between eclipses of Jupiter's moon Io appears to be greater when the earth is moving along path $ABC$ than when the earth is moving along path $CDA$. The difference is due to the time it takes light to travel the distance traveled by the earth along the line of sight during one period of Io. (The distance traveled by Jupiter in one earth year is negligible.)

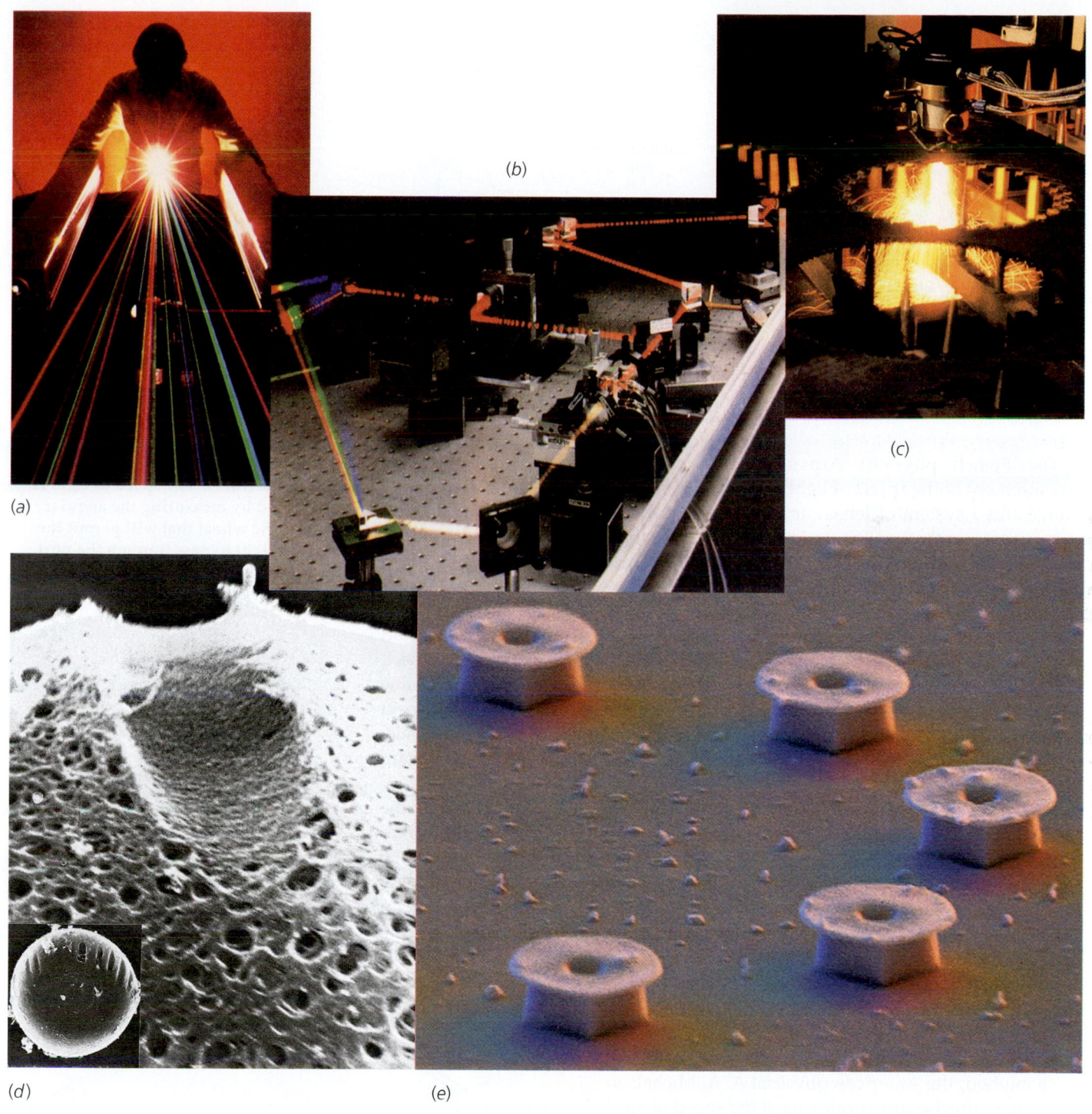

(*a*) Beams from a krypton laser and an argon laser, split into their component wavelengths. In these gas lasers, krypton and argon atoms have been stripped of multiple electrons, forming positive ions. The light-emitting energy transitions occur when excited electrons in the ions decay from one upper energy level to another. Here, several energy transitions are occurring at once, each corresponding to emitted light of a different wavelength. (*b*) A femtosecond pulsed laser. By a technique known as *modelocking*, different excited modes within a laser's cavity can be made to interfere with one another and create a series of ultrashort pulses, which are picoseconds long, that correspond to the time it takes light to bounce back and forth once within the cavity. Ultrashort pulses have been used as probes to study the behavior of molecules during chemical reactions. (*c*) A carbon dioxide laser takes just 2 minutes to cut out a steel saw blade. (*d*) A groove etched in the zona pellucida (protective outer covering) of a mouse egg by a *laser scissor* facilitates implantation. This technique has already been applied in human fertility therapies. Several effects contribute to the ability of the finely focused laser to cut on such a delicate scale—photon absorption may heat the target, break molecular bonds, or drive chemical reactions. (*e*) The so-called nanolasers shown are semiconductor disks mere microns in diameter and fractions of a micron in width. These tiny lasers work like their larger counterparts. Exploiting quantum effects that prevail on this microscopic scale, nanolasers promise great efficiency and they are being explored as ultrafast, low-energy switching devices.

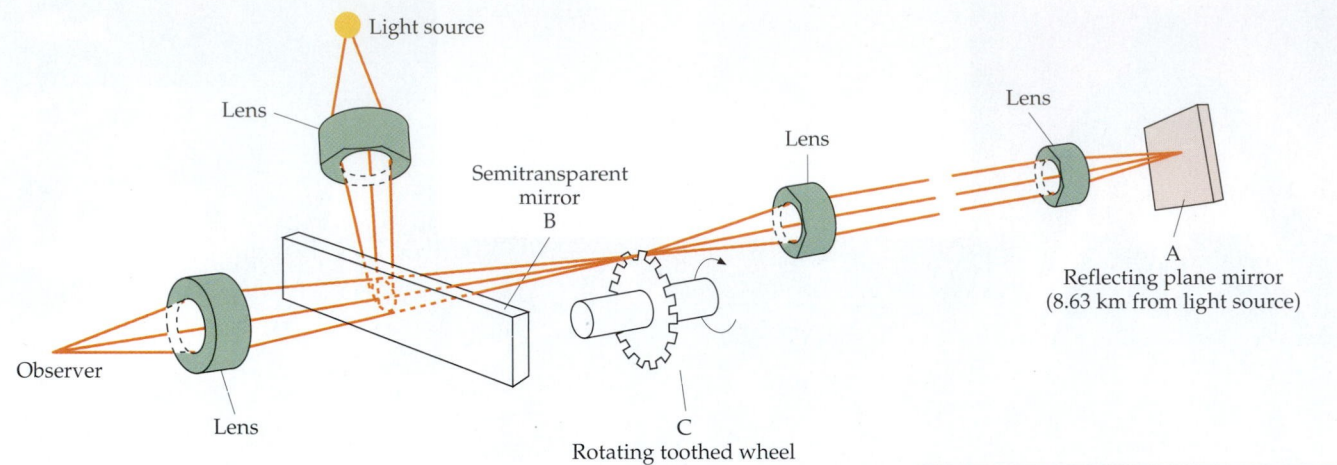

The French physicist Armand Fizeau made the first nonastronomical measurement of the speed of light in 1849. On a hill in Paris, Fizeau placed a light source and a system of lenses arranged so that the light reflected from a semitransparent mirror was focused on a gap in a toothed wheel, as shown in Figure 31-12. On a distant hill (about 8.63 km away) Fizeau placed a mirror to reflect the light back, to be viewed by an observer as shown. The toothed wheel was rotated, and the speed of rotation was varied. At low speeds of rotation, no light was visible because the light that passed through a gap in the rotating wheel and was reflected back by the mirror was obstructed by the next tooth of the wheel. The speed of rotation was then increased. The light suddenly became visible when the rotation speed was such that the reflected light passed through the next gap in the wheel. The time for the wheel to rotate through the angle between successive gaps equals the time for the light to make the round trip to the distant mirror and back.

Fizeau's method was improved upon by Jean Foucault, who replaced the toothed wheel with a rotating mirror, as shown in Figure 31-13. Light strikes the rotating mirror, the light is reflected to a distant fixed mirror, the light is then reflected back to the rotating mirror, and then to an observing telescope. During the time taken for the light to travel from the rotating mirror to the distant fixed mirror and back, the mirror rotates through a small angle. By measuring the angle $\theta$, the time for the light to travel to the distant mirror and back is determined. In approximately 1850, Foucault measured the speed of light in air and in water, and he showed that the speed of light is less in water than the speed of light in air. Using essentially the same method, the American physicist A. A. Michelson made more precise measurements of the speed of light in approximately 1880. A half-century later, Michelson made even more precise measurements of the speed of light, using an octagonal rotating mirror (Figure 31-14). In these measurements, the mirror rotates through one-eighth of a turn during the time it takes for the light to travel to the fixed mirror and back. The rotation rate is varied until another face of the mirror is in the right position for the reflected light to enter the telescope.

Another method of determining the speed of light involves the measurement of the electrical constants $\epsilon_0$ and $\mu_0$ to determine $c$ from $c = 1/\sqrt{\epsilon_0 \mu_0}$.

**FIGURE 31-12** Fizeau's method of measuring the speed of light. Light from the source is reflected by mirror B and is transmitted through a gap in the toothed wheel to mirror A. The speed of light is determined by measuring the angular speed of the wheel that will permit the reflected light to pass through the next gap in the toothed wheel so that an image of the source is observed.

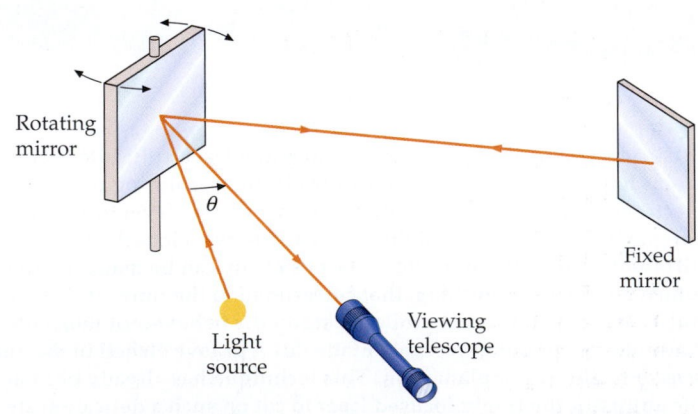

**FIGURE 31-13** Simplified drawing of Foucault's method of measuring the speed of light.

The various methods we have discussed for measuring the speed of light are all in general agreement. Today, the speed of light is defined to be exactly

$$c = 299{,}792{,}458 \text{ m/s} \qquad 31\text{-}5$$

DEFINITION—SPEED OF LIGHT

and the standard unit of length, the meter, is defined in terms of this speed and the standard unit of time. The meter is the distance light travels (in a vacuum) in $1/299{,}792{,}458$ s. The value $3 \times 10^8$ m/s for the speed of light is accurate enough for nearly all calculations. The speed of radio waves and all other electromagnetic waves (in a vacuum) is the same as the speed of light.

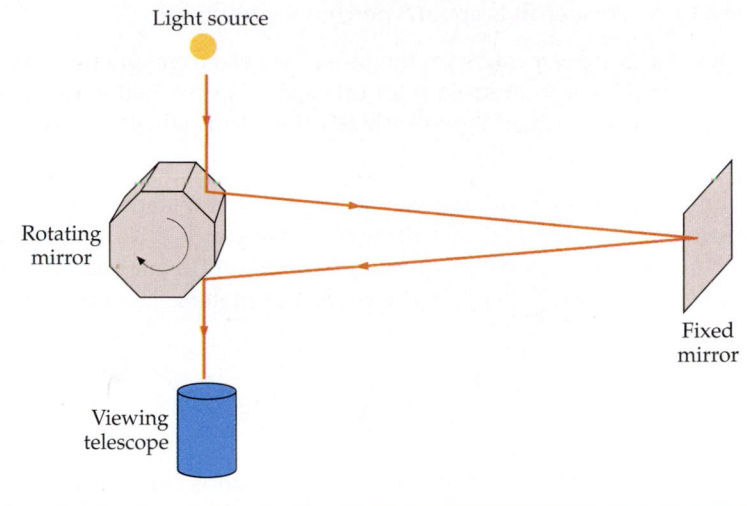

Light source

Rotating mirror

Viewing telescope

Fixed mirror

**FIGURE 31-14** Simplified drawing of Michelson's method of measuring the speed of light at Mt. Wilson in the late 1920s.

---

*THE SPEED OF LIGHT*                    **EXAMPLE 31-2**

**What is the speed of light in feet per nanosecond?**

**PICTURE THE PROBLEM** This is an exercise in unit conversions. There are $\sim 30$ cm $= 0.3$ m in 1 ft.

1. Convert m/s to ft/ns:

$$c = 3 \times 10^8 \text{ m/s} \times \left(\frac{1 \text{ ft}}{0.3 \text{ m}}\right) \times \left(\frac{1 \text{ s}}{10^9 \text{ ns}}\right) = \boxed{1 \text{ ft/ns}}$$

---

*FIZEAU'S DETERMINATION OF C*          **EXAMPLE 31-3**

**In Fizeau's experiment, his wheel had 720 teeth, and light was observed when the wheel rotated at 25.2 revolutions per second. If the distance from the wheel to the distant mirror was 8.63 km, what was Fizeau's value for the speed of light?**

**PICTURE THE PROBLEM** The time taken for the light to travel from the wheel to the mirror and back is the time for the wheel to rotate one $N$th of a revolution, where $N = 720$ is the total number of teeth.

1. The speed is the distance divided by the time. The distance from the wheel to the mirror is $L$:

$$c = \frac{2L}{\Delta t}$$

2. The angular displacement equals the angular speed times the time:

$$\Delta\theta = \omega\,\Delta t$$

3. Solve for the time:

$$\Delta t = \frac{\Delta\theta}{\omega}$$

4. Substitute for $\Delta t$ and solve for $c$:

$$c = \frac{2L\omega}{\Delta\theta} = \frac{2(8.63 \times 10^3 \text{ m})(25.2 \text{ rev/s})}{\dfrac{1}{720}\text{ rev}}$$

$$= \boxed{3.13 \times 10^8 \text{ m/s}}$$

**REMARKS** This result is about 5 percent too high.

**EXERCISE** Space travelers on the moon use electromagnetic waves to communicate with the space control center on earth. Use $c = 3.00 \times 10^8$ m/s to calculate the time delay for their signal to reach the earth, which is $3.84 \times 10^8$ m away. (*Answer*   1.28 s each way)

Large distances are often given in terms of the distance traveled by light in a given time. For example, the distance to the sun is 8.33 light-minutes, written 8.33 *c*-min. A light-year is the distance light travels in one year. We can easily find a conversion factor between light-years and meters. The number of seconds in one year is

$$1 \text{ y} = 1 \text{ y} \times \frac{365.24 \text{ d}}{1 \text{ y}} \times \frac{24 \text{ h}}{1 \text{ d}} \times \frac{3600 \text{ s}}{1 \text{ h}} = 3.156 \times 10^7 \text{ s}$$

(*Note:* There are approximately $\pi$ times $10^7$ seconds per year, which is how some individuals remember the approximate value of the conversion.) The number of meters in one light-year is thus

$$1c\text{-year} = (2.998 \times 10^8 \text{ m/s})(3.156 \times 10^7 \text{ s}) = 9.46 \times 10^{15} \text{ m} \qquad \text{31-6}$$

# 31-5  The Propagation of Light

The propagation of light is governed by the wave equation discussed in Chapter 30. But long before Maxwell's theory of electromagnetic waves, the propagation of light and other waves was described empirically by two interesting and very different principles attributed to the Dutch physicist Christian Huygens (1629–1695) and the French mathematician Pierre de Fermat (1601–1665).

## Huygens's Principle

Figure 31-15 shows a portion of a spherical wavefront emanating from a point source. The wavefront is the locus of points of constant phase. If the radius of the wavefront is *r* at time *t*, its radius at time $t + \Delta t$ is $r + c \Delta t$, where *c* is the speed of the wave. However, if a part of the wave is blocked by some obstacle or if the wave passes through a different medium, as in Figure 31-16, the determination of the new wavefront position at time $t + \Delta t$ is much more difficult. The propagation of any wavefront through space can be described using a geometric method discovered by Huygens in approximately 1678, which is now known as **Huygens's principle** or **Huygens's construction:**

> Each point on a primary wavefront serves as the source of spherical secondary wavelets that advance with a speed and frequency equal to those of the primary wave. The primary wavefront at some later time is the envelope of these wavelets.

HUYGENS'S PRINCIPLE

Figure 31-17 shows the application of Huygens's principle to the propagation of a plane wave and the propagation of a spherical wave. Of course, if each point on a wavefront were really a point source, there would be waves in the backward direction as well. Huygens ignored these back waves.

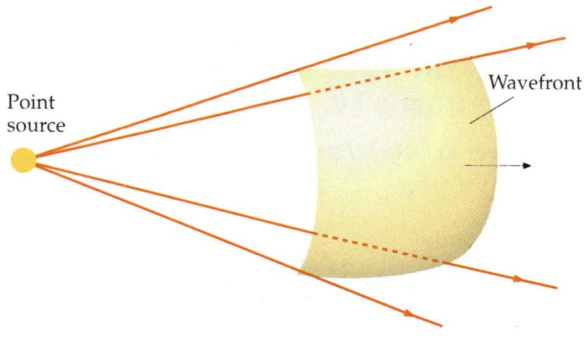

**FIGURE 31-15** Spherical wavefront from a point source.

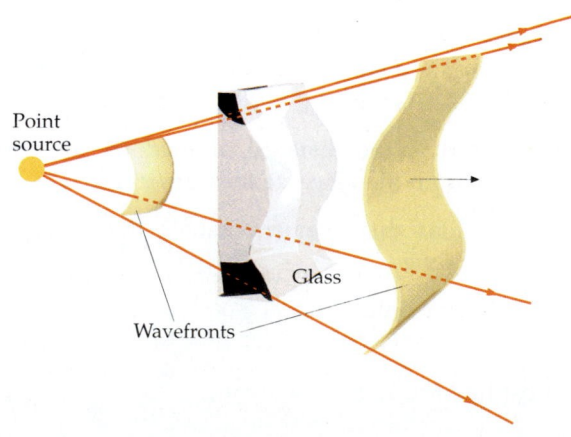

**FIGURE 31-16** Wavefront from a point source before and after passing through a piece of glass of varied thickness.

Huygens's principle was later modified by Augustin Fresnel, so that the new wavefront was calculated from the old wavefront by superposition of the wavelets considering their relative amplitudes and phases. Kirchhoff later showed that the Huygens–Fresnel principle was a consequence of the wave equation (Equation 30-17), thus putting it on a firm mathematical basis. Kirchhoff showed that the intensity of each wavelet depends on the angle and is zero at 180° (the backward direction).

We will use Huygens's principle to derive the laws of reflection and refraction in Section 31-8. In Chapter 33, we apply Huygens's principle with Fresnel's modification to calculate the diffraction pattern of a single slit. Because the wavelength of light is so small, we can often use the ray approximation to describe its propagation.

### Fermat's Principle

The propagation of light can also be described by Fermat's principle:

> The path taken by light traveling from one point to another is such that the time of travel is a minimum.[†]

FERMAT'S PRINCIPLE

In Section 31-8, we will use Fermat's principle to derive the laws of reflection and refraction.

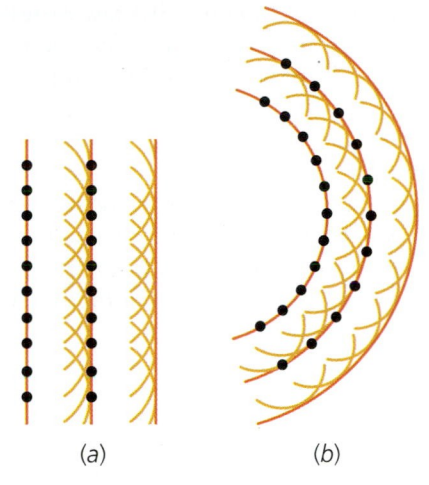

(a)                    (b)

**FIGURE 31-17** Huygens's construction for the propagation to the right of (a) a plane wave and (b) an outgoing spherical, or circular, wave.

## 31-6   Reflection and Refraction

The speed of light in a transparent medium such as air, water, or glass is less than the speed $c = 3 \times 10^8$ m/s in vacuum. A transparent medium is characterized by the **index of refraction,** $n$, which is defined as the ratio of the speed of light in a vacuum, $c$, to the speed in the medium, $v$:

$$n = \frac{c}{v} \qquad\qquad 31\text{-}7$$

DEFINITION—INDEX OF REFRACTION

For water, $n = 1.33$, whereas for glass $n$ ranges from approximately 1.50 to 1.66, depending on the type of glass. Diamond has a very high index of refraction— approximately 2.4. The index of refraction of air is approximately 1.0003, so for most purposes we can assume that the speed of light in air is the same as the speed of light in vacuum.

When a beam of light strikes a boundary surface separating two different media, such as an air–glass interface, part of the light energy is reflected and part of the light energy enters the second medium. If the incident light is not perpendicular to the surface, then the transmitted beam is not parallel to the incident beam. The change in direction of the transmitted ray is called **refraction.** Figure 31-18 shows a light ray striking a smooth air–glass interface. The angle $\theta_1$ between the incident ray and the normal (the line perpendicular to the surface) is called the **angle of incidence,** and the plane defined by these two lines is called the **plane of incidence.** The reflected ray lies in the plane of incidence and makes an angle $\theta_1'$ with the normal that is equal to the angle of incidence as shown in the figure:

$$\theta_1' = \theta_1 \qquad\qquad 31\text{-}8$$

LAW OF REFLECTION

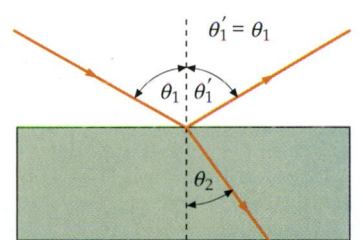

**FIGURE 31-18** The angle of reflection $\theta_1'$ equals the angle of incidence $\theta_1$. The angle of refraction $\theta_2$ is less than the angle of incidence if the light speed in the second medium is less than that in the incident medium.

---

[†] A more complete and general statement is that the time of travel is stationary with respect to variations in path; that is, if $t$ is expressed in terms of some parameter $x$, the path taken will be such that $dt/dx = 0$. The important characteristic of a stationary path is that the time taken along nearby paths will be approximately the same as that along the true path.

This result is known as the **law of reflection.** The law of reflection holds for any type of wave. Figure 31-19 illustrates the law of reflection for rays of light and for wavefronts of ultrasonic waves.

The ray that enters the glass in Figure 31-18 is called the refracted ray, and the angle $\theta_2$ is called the angle of refraction. When a wave crosses a boundary where the wave speed is reduced, as in the case of light entering glass from air, the angle of refraction is less than the angle of incidence $\theta_1$, as shown in Figure 31-18; that is, the refracted ray is bent toward the normal. If, on the other hand, the light beam originates in the glass and is refracted into the air, then the refracted ray is bent away from the normal.

The angle of refraction $\theta_2$ depends on the angle of incidence and on the relative speed of light waves in the two mediums. If $v_1$ is the wave speed in the incident medium and $v_2$ is the wave speed in the transmission medium, the angles of incidence and refraction are related by

$$\frac{1}{v_1} \sin \theta_1 = \frac{1}{v_2} \sin \theta_2$$ 31-9a

Equation 31-9a holds for the refraction of any kind of wave incident on a boundary separating two media.

In terms of the indexes of refraction of the two media $n_1$ and $n_2$, Equation 31-9a is

$$n_1 \sin \theta_1 = n_2 \sin \theta_2$$ 31-9b

SNELL'S LAW OF REFRACTION

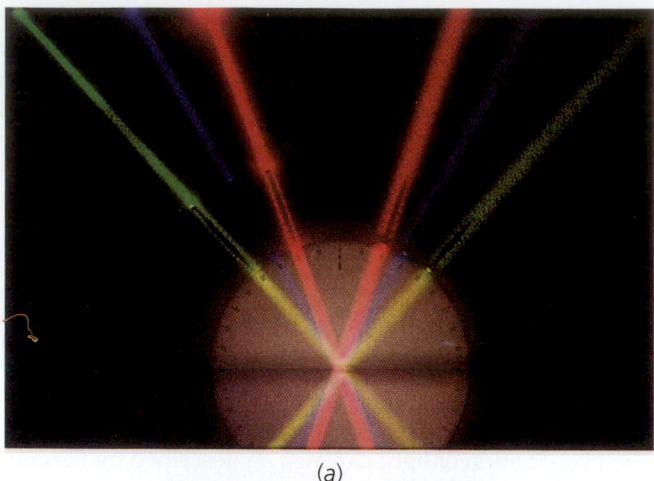
(a)

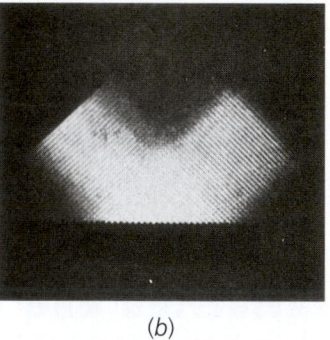
(b)

**FIGURE 31-19** (*a*) Light rays reflecting from an air–glass interface showing equal angles of incidence and reflection. (*b*) Ultrasonic plane waves in water reflecting from a steel plate.

This result was discovered experimentally in 1621 by the Dutch scientist Willebrord Snell and is known as **Snell's law** or the **law of refraction.** It was independently discovered a few years later by the French mathematician and philosopher René Descartes.

Reflection and refraction of a beam of light incident on a glass slab.

## Physical Mechanisms for Reflection and Refraction

The physical mechanism of the reflection and refraction of light can be understood in terms of the absorption and reradiation of the light by the atoms in the reflecting or refracting medium. When light traveling in air strikes a glass surface, the atoms in the glass absorb the light and reradiate it at the same frequency in all directions. The waves radiated backward by the glass atoms interfere constructively at an angle equal to the angle of incidence to produce the reflected wave.

The transmitted wave is the result of the interference of the incident wave and the wave produced by the absorption and reradiation of light energy by the atoms in the medium. For light entering glass from air, there is a phase lag between the reradiated wave and the incident wave. There is, therefore, also a phase lag between the resultant wave and the incident wave. This phase lag means that the position of a wave crest of the transmitted wave is retarded relative to the position of a wave crest of the incident wave in the medium. As a result, a transmitted wave crest does not travel as far in a given time as the original incident wave crest; that is, the velocity of the transmitted wave is less than that of the incident wave. The index of refraction is therefore greater than 1. The frequency of the light in the second medium is the same as the frequency of the incident light—the atoms absorb and reradiate the light at the same frequency—but the wave speed is different, so the wavelength of the transmitted light is different from that of the incident light. If $\lambda$ is the wavelength of light in a vacuum, then $\lambda f = c$, and if $\lambda'$ is the wavelength in a medium in which it has speed $v$, then $\lambda' f = v$. Combining these two relations gives $\lambda / \lambda' = c/v$, or

$$\lambda' = \frac{\lambda}{c/v} = \frac{\lambda}{n} \qquad\qquad 31\text{-}10$$

where $n = c/v$ is the index of refraction of the medium.

## Specular Reflection and Diffuse Reflection

Figure 31-20a shows a bundle of light rays from a point source $P$ that are reflected from a flat surface. After reflection, the rays diverge exactly as if they came from a point $P'$ behind the surface. (This point is called the *image point*. We will study the formation of images by reflecting and refracting surfaces in the next chapter.) When these rays enter the eye, they cannot be distinguished from rays actually diverging from a source at $P'$.

**FIGURE 31-20** (a) Specular reflection from a smooth surface. (b) Specular reflection of trees from water.

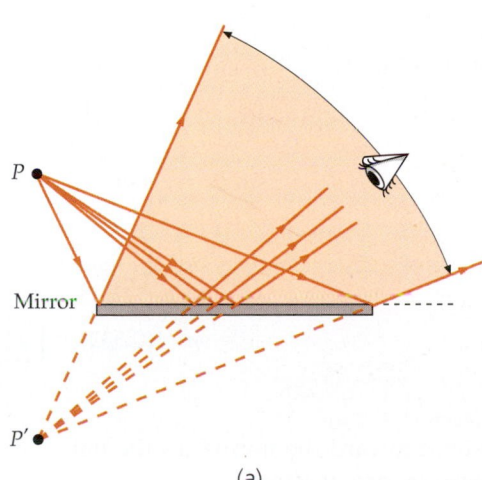

(a)

(b)

Reflection from a smooth surface is called **specular reflection.** It differs from **diffuse reflection,** which is illustrated in Figure 31-21. Here, because the surface is rough, the rays from a point reflect in random directions and do not diverge from any point, so there is no image. The reflection of light from the page of this book is diffuse reflection. The glass used in picture frames is sometimes ground slightly to give diffuse reflection and thereby cut down on glare from the light used to illuminate the picture. Diffuse reflection from the surface of the road allows you to see the road when you are driving at night because some of the light from your headlights reflects back toward you. In wet weather the reflection is mostly specular; therefore, little light is reflected back toward you, which makes the road difficult to see.

### Relative Intensity of Reflected and Transmitted Light

The fraction of light energy reflected at a boundary, such as an air–glass interface, depends in a complicated way on the angle of incidence, the orientation of the electric field vector associated with the wave, and the indexes of refraction of the two media. For the special case of normal incidence ($\theta_1 = \theta_1' = 0$), the reflected intensity can be shown to be

$$I = \left(\frac{n_1 - n_2}{n_1 + n_2}\right)^2 I_0 \qquad \text{31-11}$$

where $I_0$ is the incident intensity and $n_1$ and $n_2$ are the indexes of refraction of the two media. For a typical case of reflection from an air–glass interface for which $n_1 = 1$ and $n_2 = 1.5$, Equation 31-11 gives $I = I_0/25$. Only about 4 percent of the energy is reflected; the remainder of the energy is transmitted.

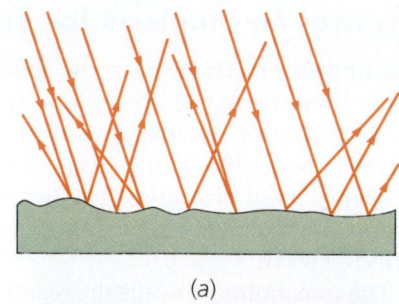

(a)

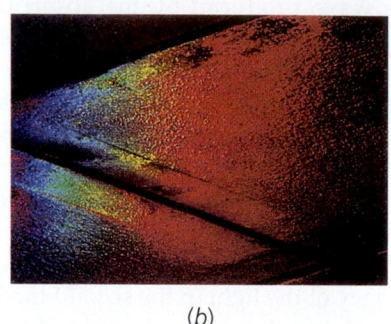

(b)

**FIGURE 31-21** (a) Diffuse reflection from a rough surface. (b) Diffuse reflection of colored lights from a sidewalk.

---

*REFRACTION FROM AIR TO WATER*      **EXAMPLE 31-4**

**Light traveling in air enters water with an angle of incidence of 45°. If the index of refraction of water is 1.33, what is the angle of refraction?**

**PICTURE THE PROBLEM** The angle of refraction is found using Snell's law. Let subscripts 1 and 2 refer to the air and water, respectively. Then $n_1 = 1$, $\theta_1 = 45°$, $n_2 = 1.33$, and $\theta_2$ is the angle of refraction (Figure 31-22).

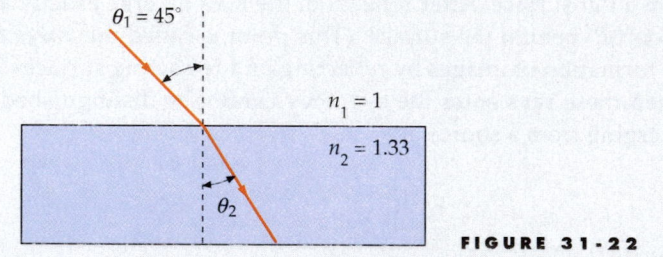

**FIGURE 31-22**

1. Use Snell's law to solve for $\sin\theta_2$, the sine of the angle of refraction:

$$n_1 \sin\theta_1 = n_2 \sin\theta_2$$

so

$$\sin\theta_2 = \frac{n_1}{n_2}\sin\theta_1$$

2. Find the angle whose sine is 0.532:

$$\theta_2 = \sin^{-1}\left(\frac{n_1}{n_2}\sin\theta_1\right) = \sin^{-1}\left(\frac{1.00}{1.33}\sin 45°\right)$$

$$= \sin^{-1}(0.532) = \boxed{32.1°}$$

**REMARKS** Note that the light is bent toward the normal as the light travels into the medium with the larger index of refraction.

(a)

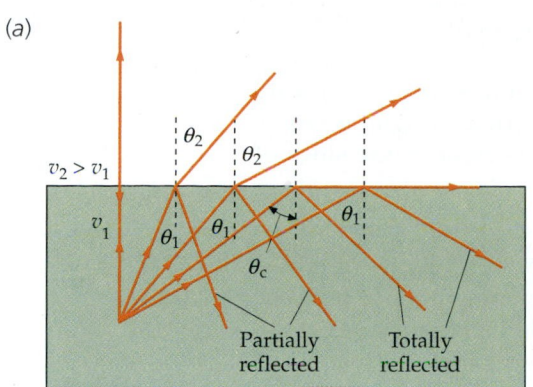

(b)

## Total Internal Reflection

Figure 31-23 shows a point source in glass with rays striking the glass–air interface at various angles. All the rays not perpendicular to the interface are bent away from the normal. As the angle of incidence is increased, the angle of refraction increases until a critical angle of incidence $\theta_c$ is reached for which the angle of refraction is 90°. For incident angles greater than this critical angle, there is no refracted ray. All the energy is reflected. This phenomenon is called **total internal reflection**. The critical angle can be found in terms of the indexes of refraction of the two media by solving Equation 31-9b ($n_1 \sin \theta_1 = n_2 \sin \theta_2$) for $\sin \theta_1$ and setting $\theta_2$ equal to 90°. That is,

$$\sin \theta_c = \frac{n_2}{n_1} \sin 90° = \frac{n_2}{n_1} \qquad 31\text{-}12$$

CRITICAL ANGLE FOR TOTAL INTERNAL REFLECTION

Note that total internal reflection occurs only when the incident light is in the medium with the higher index of refraction. Mathematically, if $n_2$ is greater than $n_1$, Snell's law of refraction cannot be satisfied because there is no real-valued angle whose sine is greater than 1.

**FIGURE 31-23** (*a*) Total internal reflection. As the angle of incidence is increased, the angle of refraction is increased until, at a critical angle of incidence $\theta_c$, the angle of refraction is 90°. For angles of incidence greater than the critical angle, there is no refracted ray. (*b*) A photograph of refraction and total internal reflection from a water–air interface.

---

*TOTAL INTERNAL REFLECTION*          **EXAMPLE 31-5**   **Try It Yourself**

A particular glass has an index of refraction of $n = 1.50$. What is the critical angle for total internal reflection for light leaving this glass and entering air, for which $n = 1.00$ (Figure 31-24)?

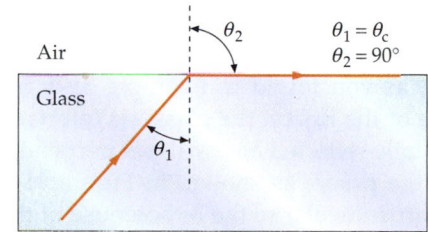

**FIGURE 31-24**

---

**Cover the column to the right and try these on your own before looking at the answers.**

**Steps**                                              **Answers**

1. Make a diagram showing the incident and refracted rays. For the critical angle the angle of refraction is 90°.

2. Apply the law of refraction Equation 31-9b. The critical angle is the angle of incidence.          $\theta_c = \boxed{41.8°}$

You are enjoying a nice break at the pool. While under the water, you look up and notice that you see objects above water level in a circle of light of radius approximately 2.0 m, and the rest of your vision is the color of the sides of the pool. How deep are you in the pool?

**PICTURE THE PROBLEM** We can determine the depth of the pool from the radius of the light and the angle at which the light is entering our eye from the edge of the circle. At the edge of the circle the light is entering the water at 90°, so the angle of refraction at the air–water surface is the critical angle for total internal refraction at the water–air surface. From Figure 31-25, we see that the depth $y$ is related to this angle and the radius of the circle $R$ by $\tan \theta_c = R/y$. The critical angle is found from Equation 31-12 with $n_2 = 1$ and $n_1 = 1.33$.

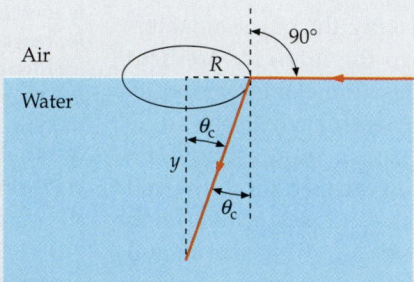

**FIGURE 31-25**

1. The depth $y$ is related to the radius of the circle $R$ and the critical angle $\theta_c$:

$$\tan \theta_c = R/y$$

2. Solve for the depth $y$:

$$y = \frac{R}{\tan \theta_c}$$

3. Find the critical angle for total internal refraction at a water–air surface:

$$\sin \theta_c = \frac{n_2}{n_1} = \frac{1}{1.33} = 0.752$$

$$\theta_c = 48.8°$$

4. Solve for the depth $y$:

$$y = \frac{R}{\tan \theta_c} = \frac{2.0 \text{ m}}{\tan 48.8°} = \boxed{1.75 \text{ m}}$$

Figure 31-26*a* shows light incident normally on one of the short sides of a 45–45–90° glass prism. If the index of refraction of the prism is 1.5, the critical angle for total internal reflection is 41.8°, as you found in Example 31-5. Since the angle of incidence of the ray on the glass–air interface is 45°, the light will be totally reflected and will exit perpendicular to the other face of the prism, as shown. In Figure 31-26*b*, the light is incident perpendicular to the hypotenuse of the prism and is totally reflected twice so that it emerges at 180° to its original direction. Prisms are used to change the directions of light rays. In binoculars, two prisms are used on each side. These prisms reflect the light, thus shortening the required length, and reinvert the image (first inverted by a lens).[†] Diamonds have a very high index of refraction ($n \approx 2.4$), so nearly all the light that enters a diamond is eventually reflected back out, giving the diamond its sparkle.

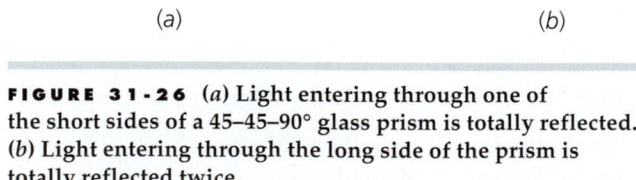

**FIGURE 31-26** (*a*) Light entering through one of the short sides of a 45–45–90° glass prism is totally reflected. (*b*) Light entering through the long side of the prism is totally reflected twice.

---

† The image produced by the objective lens of a telescope is discussed in Section 32-4.

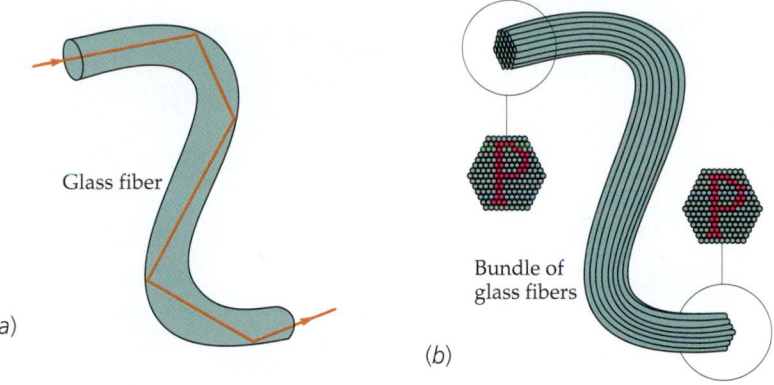

(a)

(b)

Bundle of glass fibers

(c)

### Fiber Optics

An interesting application of total internal reflection is the transmission of a beam of light down a long, narrow, transparent glass fiber (Figure 31-27a). If the beam begins approximately parallel to the axis of the fiber, it will strike the walls of the fiber at angles greater than the critical angle (if the bends in the fiber are not too sharp) and no light energy will be lost through the walls of the fiber. A bundle of such fibers can be used for imaging, as illustrated in Figure 31-27b. Fiber optics has many applications in medicine and in communications. In medicine, light is transmitted along tiny fibers to visually probe various internal organs without surgery. In communications, the rate at which information can be transmitted is related to the signal frequency. A transmission system using light of frequencies of the order of $10^{14}$ Hz can transmit information at a much greater rate than one using radio waves, which have frequencies of the order of $10^6$ Hz. In telecommunication systems, a single glass fiber the thickness of a human hair can transmit audio or video information equivalent to 32,000 voices speaking simultaneously.

### Mirages

When the index of refraction of a medium changes gradually, the refraction is continuous, leading to a gradual bending of the light. An interesting example of this is the formation of a mirage. On a hot and sunny day, the surface of exposed rocks, pavement, and sand often gets very hot. In this case there is often a layer of air near the ground that is warmer, and therefore less dense, than the air just above it. The speed

**FIGURE 31-27** (*a*) A light pipe. Light inside the pipe is always incident at an angle greater than the critical angle, so no light escapes the pipe by refraction. (*b*) Light from the object is transported by a bundle of glass fibers to form an image of the object at the other end of the pipe. (*c*) Light emerging from a bundle of glass fibers.

(a)

(b)

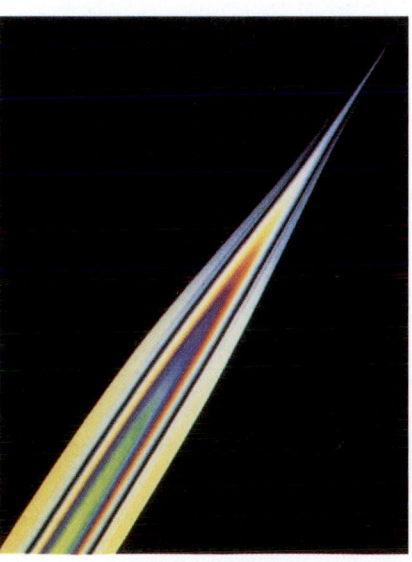

(*a*) In this demonstration at the Naval Research Laboratory, a combination of laser sources generates different colors that excite adjacent fiber sensor elements, leading to a separation of the information as indicated by the separation of the colors. (*b*) The tip of a light guide preform is softened by heat and drawn into a long, tiny fiber. The colors in the preform indicate a layered structure of differing compositions, which is retained in the fiber.

(a)

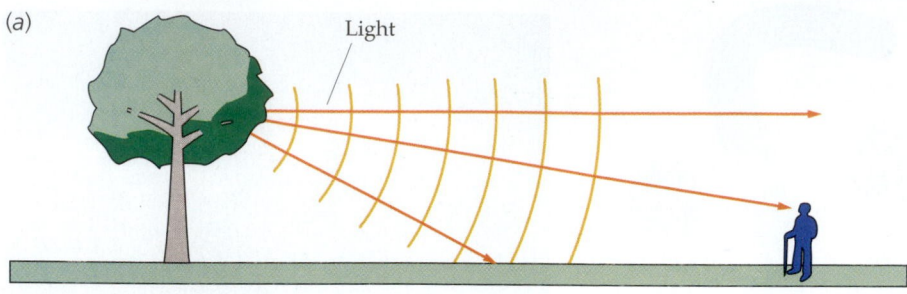

(b)

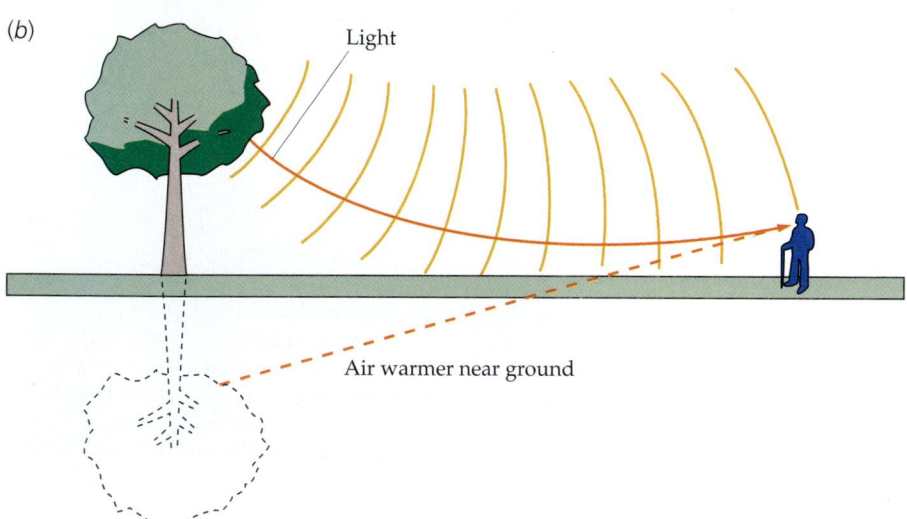

(c)

**FIGURE 31-28** A mirage. (*a*) When the air is at a uniform temperature, the wavefronts of the light from the tree are spherical. (*b*) When the air near the ground is warmer, the wavefronts are not spherical and the light from the tree is continuously refracted into a curved path. (*c*) Apparent reflections of motorcycles on a hot road.

of any light wave is slightly greater in this less dense layer, so a light beam passing from the cooler layer into the warmer layer is bent. Figure 31-28*a* shows the light from a tree when all the surrounding air is at the same temperature. The wavefronts are spherical, and the rays are straight lines. In Figure 31-28*b*, the air near the ground is warmer, resulting in the wavefronts traveling faster there. The portions of the wavefronts near the hot ground get ahead of the higher portions, creating a nonspherical wavefront and causing a curving of the rays. Thus, the two rays shown initially heading for the ground are bent upward. As a result, the viewer sees an image of the tree looking as if it were reflected off a water surface on the ground. When driving on a hot sunny day, you may have noticed apparent wet spots on the highway ahead that disappear as you approach them. These mirages are due to the refraction of light from the sky by a layer of air that has been heated due to its proximity to the hot pavement.

## Dispersion

The index of refraction of a material has a slight dependence on wavelength. For many materials, $n$ decreases slightly as the wavelength increases, as shown in Figure 31-29. The dependence of the index of refraction on wavelength (and therefore on frequency) is called **dispersion**. When a beam of white light is incident at some angle on the surface of a glass prism, the angle of refraction (which is measured relative to the normal) for the shorter wavelengths is slightly smaller than the angle of refraction for the longer wavelengths. The light of shorter wavelength (toward the violet end of the spectrum) is therefore bent more toward the normal than that of longer wavelength. The beam of white light is thus spread out or dispersed into its component colors or wavelengths (Figure 31-30).

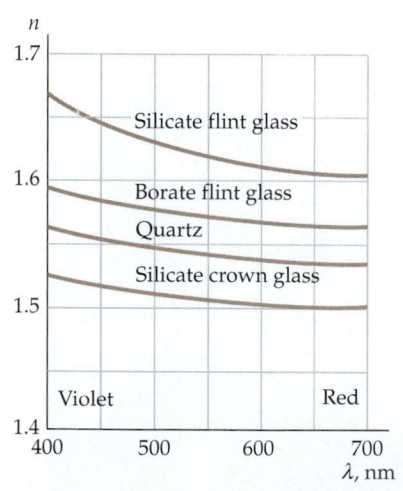

**FIGURE 31-29** Index of refraction versus wavelength for various materials.

**FIGURE 31-30** A beam of white light incident on a glass prism is dispersed into its component colors. The index of refraction decreases as the wavelength increases so that the longer wavelengths (red) are bent less than the shorter wavelengths (violet).

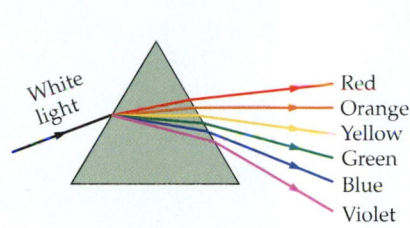

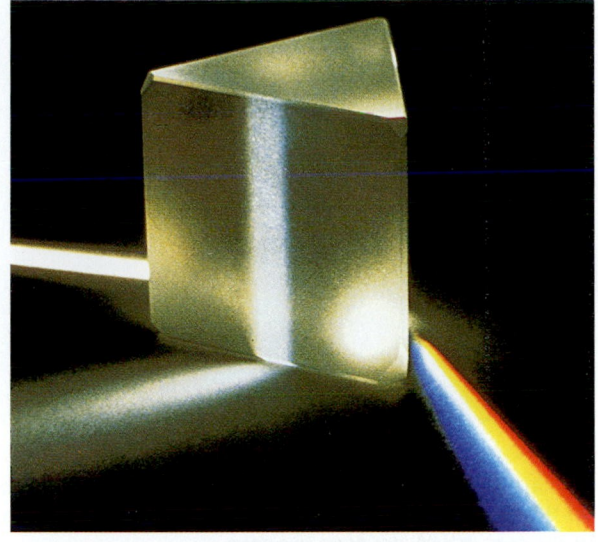

**Rainbows**   The rainbow is a familiar example of dispersion, in this case the dispersion of sunlight. Figure 31-31 is a diagram originally drawn by Descartes, showing parallel rays of light from the sun entering a spherical water drop. First, the rays are refracted as they enter the drop. The rays are then reflected from the water–air interface on the other side of the drop and finally are refracted again as they leave the drop.

From Figure 31-31, we can see that the angle made by the emerging rays and the diameter (along ray 1) reaches a maximum around ray 7 and then decreases. The concentration of rays emerging at approximately the maximum angle gives rise to the rainbow. By construction, using the law of refraction, Descartes showed that the maximum angle is about 42°. To observe a rainbow, we must therefore look at the water drops at an angle of 42° relative to the line back to the sun, as shown in Figure 31-32. The angular radius of the rainbow is therefore 42°.

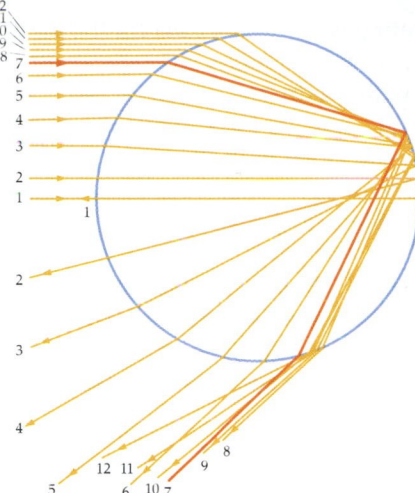

**FIGURE 31-31** Descartes's construction of parallel rays of light entering a spherical water drop. Ray 1 enters the drop along a diameter and is reflected back along its incident path. Ray 2 enters slightly above the diameter and emerges below the diameter at a small angle with the diameter. The rays entering farther and farther away from the diameter emerge at greater and greater angles up to ray 7, shown as the heavy line. Rays entering above ray 7 emerge at smaller and smaller angles with the diameter.

The separation of the colors in the rainbow results from the fact that the index of refraction of water depends slightly on the wavelength of light. The angular radius of the bow will therefore depend slightly on the wavelength of the light. The observed rainbow is made up of light rays from many different droplets of water (Figure 31-33). The color seen at a particular angular radius corresponds to the wavelength of light that allows the light to reach the eye from the droplets at that angular radius. Because $n_{water}$ is smaller for red light than for blue light, the red part of the rainbow is at a slightly greater angular radius than the blue part of the rainbow, so red is at the outer side of the rainbow.

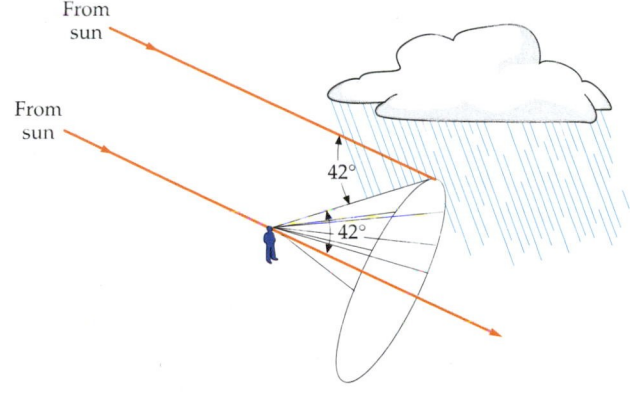

**FIGURE 31-32** A rainbow is viewed at an angle of 42° from the line to the sun, as predicted by Descartes's construction, as shown in Figure 31-31.

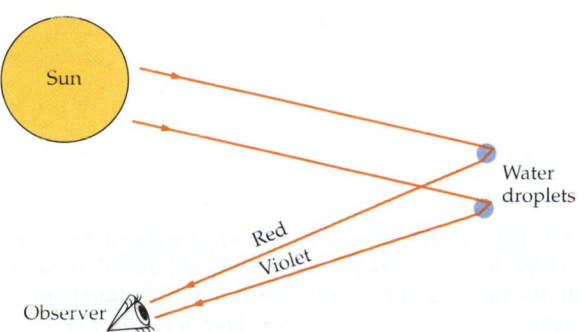

**FIGURE 31-33** The rainbow results from light rays from many different water droplets.

(a)

(b)

(a) This 22° halo around the sun results from refraction by hexagonal ice crystals that are randomly oriented in the upper atmosphere. (b) When the ice crystals are not randomly oriented but are falling with their flat bases horizontal, only parts of the halo on each side of the sun, called *sun dogs*, are seen.

When a light ray strikes a surface separating water and air, part of the light is reflected and part of the light is refracted. A secondary rainbow results from the light rays that are reflected twice within a droplet (Figure 31-34). The secondary bow has an angular radius of 51°, and its color sequence is the reverse of that of the primary bow; that is, the violet is on the outside in the secondary bow. Because of the small fraction of light reflected from a water–air interface, the secondary bow is considerably fainter than the primary bow.

**\*Calculating the Angular Radius of the Rainbow**   We can calculate the angular radius of the rainbow from the laws of reflection and refraction. Figure 31-35 shows a ray of light incident on a spherical water droplet at point $A$. The angle of refraction $\theta_2$ is related to the angle of incidence $\theta_1$ by Snell's law of refraction:

$$n_{air} \sin \theta_1 = n_{water} \sin \theta_2 \qquad 31\text{-}13$$

Point $P$ in Figure 31-35 is the intersection of the line of the incident ray and the line of the emerging ray. The angle $\phi_d$ is called the angle of deviation of the ray, and $\phi_d$ and $2\beta$ form a straight angle. Thus,

$$\phi_d + 2\beta = \pi \qquad 31\text{-}14$$

We wish to relate the angle of deviation $\phi_d$ to the angle of incidence $\theta_1$. From the triangle $AOB$, we have

$$2\theta_2 + \alpha = \pi \qquad 31\text{-}15$$

Similarly, from the triangle $AOP$, we have

$$\theta_1 + \beta + \alpha = \pi \qquad 31\text{-}16$$

Eliminating $\alpha$ from Equations 31-15 and 31-16 and solving for $\beta$ gives

$$\beta = \pi - \theta_1 - \alpha = \pi - \theta_1 - (\pi - 2\theta_2) = 2\theta_2 - \theta_1$$

Substituting this value for $\beta$ into Equation 31-14 gives the angle of deviation:

$$\phi_d = \pi - 2\beta = \pi - 4\theta_2 + 2\theta_1 \qquad 31\text{-}17$$

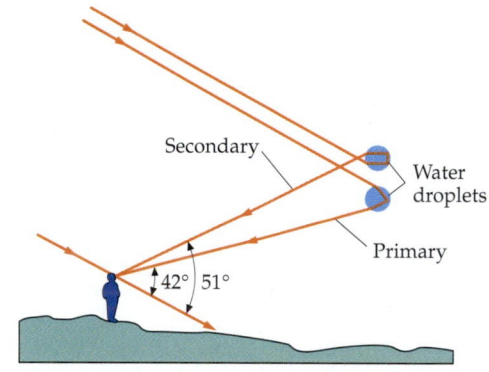

**FIGURE 31-34** The secondary rainbow results from light rays that are reflected twice within a water droplet.

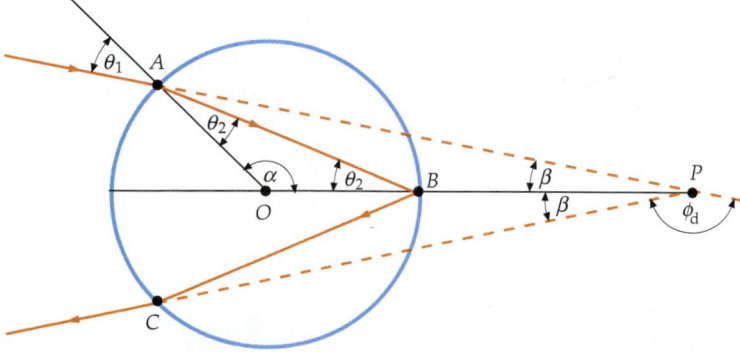

**FIGURE 31-35** Light ray incident on a spherical water drop. The refracted ray strikes the back of the water droplet at point $B$. It makes an angle $\theta_2$ with the radial line $OB$ and is reflected at an equal angle. The ray is refracted again at point $C$, where it leaves the droplet.

Equation 31-17 can be combined with Equation 31-13 to eliminate $\theta_2$ and give the angle of deviation $\phi_d$ in terms of the angle of incidence $\theta_1$:

$$\phi_d = \pi + 2\theta_1 - 4\sin^{-1}\left(\frac{n_{air}}{n_{water}}\sin\theta_1\right) \qquad 31\text{-}18$$

Figure 31-36 shows a plot of $\phi_d$ versus $\theta_1$. The angle of deviation $\phi_d$ has its minimum value when $\theta_1 \approx 60°$. At an angle of incidence of 60°, the angle of deviation is $\phi_{d,min} = 138°$. This angle is called the **angle of minimum deviation.** At incident angles that are slightly greater or slightly smaller than 60°, the angle of deviation is approximately the same. Therefore, the intensity of the light reflected by the water droplet will be a maximum at the angle of minimum deviation. We can see from Figure 31-35 that the maximum value of $\beta$ corresponds to the minimum value of $\phi_d$. Thus, angular radius of the intensity maximum, given by $2\beta_{max}$, is

$$2\beta_{max} = \pi - \phi_{d,min} = 180° - 138° = 42° \qquad 31\text{-}19$$

The index of refraction of water varies slightly with wavelength. Thus, for each wavelength (color), the intensity maxima occurs at an angular radius slightly different than that of neighboring wavelengths.

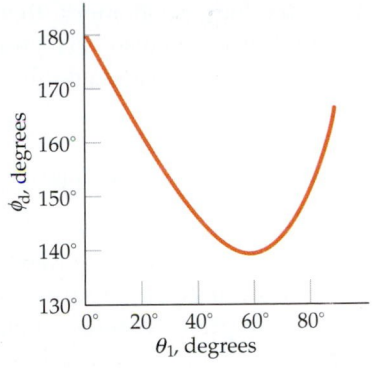

**FIGURE 31-36** Plot of the angle of deviation $\phi_d$ as a function of incident angle $\theta_1$. The angle of deviation has its minimum value of 138° when the angle of incidence is 60°. Since $d\phi_d/d\theta_1 = 0$ at minimum deviation, the deviation of rays with incident angles slightly less or slightly greater than 60° will be approximately the same.

# 31-7  Polarization

In a transverse mechanical wave, the vibration is perpendicular to the direction of propagation of the wave. If the vibration remains parallel to a plane, the wave is said to be **plane polarized** or **linearly polarized.** We can visualize polarization most easily by considering mechanical waves on a taut horizontal string. Let the x axis be along the string, let the z axis be vertical, and let the y axis be horizontal, perpendicular to both the x and z axes. If one end of the string is shaken up and down, the resulting waves on the string are linearly polarized with each element of the string vibrating up and down. Similarly, if one end is shaken back and forth along the y axis, the displacements of the string are linearly polarized with each element vibrating parallel with the y axis. If one end of the string is moved with constant speed in an ellipse in the x = 0 plane, the resulting wave is said to be **elliptically polarized.** In this case, each element of the string moves in an ellipse in a plane of constant x. Unpolarized waves can be produced by moving the end of the string in the x = 0 plane in a random way. Then the vibrations will have both y and z components that vary randomly. A linearly polarized electromagnetic wave is one in which the electric field remains parallel to a line. A wave produced by an electric dipole antenna is polarized with the electric field vector at any field point remaining parallel with the plane containing the field point and the antenna axis. Waves produced by numerous sources are usually unpolarized. A typical light source, for example, contains millions of atoms acting independently. The electric field for such a wave can be resolved into x and y components that vary randomly, because there is no correlation between the individual atoms producing the light.

The polarization of electromagnetic waves can be demonstrated with microwaves, which have wavelengths on the order of centimeters. In a typical microwave generator, polarized waves are radiated by an electric dipole antenna. In Figure 31-37, the electric dipole antenna is vertical, so the electric field vector $\vec{E}$ of the horizontally radiated waves is also vertical. An absorber can be made of a screen of parallel straight wires. When the wires are vertical, as in Figure 31-37a, the electric field parallel to the wires sets up currents in the wires and energy is absorbed. When the wires are horizontal and therefore perpendicular to $\vec{E}$, as in Figure 31-37b, no currents are set up and the waves are transmitted.

(a)

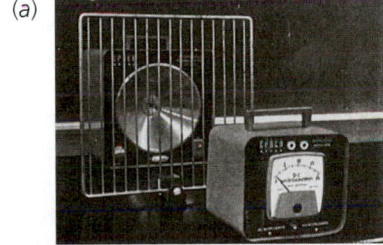

(b)

**FIGURE 31-37** Demonstration showing the polarization of microwaves. The electric field of the microwaves is vertical, parallel to the vertical dipole antenna. (a) When the metal wires of the absorber are vertical, electric currents are set up in the wires and energy is absorbed, as indicated by the low reading on the microwave detector. (b) When the wires are horizontal, no currents are set up and the microwaves are transmitted, as indicated by the high reading on the detector.

There are four phenomena that produce polarized electromagnetic waves from unpolarized waves: (1) absorption, (2) reflection, (3) scattering, and (4) birefringence (also called double refraction), each of which is examined in the upcoming sections.

## Polarization by Absorption

Several naturally occurring crystals, when cut into appropriate shapes, absorb and transmit light differently depending on the polarization of the light. These crystals can be used to produce linearly polarized light. In 1938, E. H. Land invented a simple commercial polarizing sheet called Polaroid. This material contains long-chain hydrocarbon molecules that are aligned when the sheet is stretched in one direction during the manufacturing process. These chains become conducting at optical frequencies when the sheet is dipped in a solution containing iodine. When light is incident with its electric field vector parallel to the chains, electric currents are set up along the chains, and the light energy is absorbed, just as the microwaves are absorbed by the wires in Figure 31-37. If the electric field is perpendicular to the chains, the light is transmitted. The direction perpendicular to the chains is called the **transmission axis.** We will make the simplifying assumption that all the light is transmitted when the electric field is parallel to the transmission axis and all the light is absorbed when it is perpendicular to the transmission axis. In reality, Polaroid absorbs some of the light, even when the electric field is parallel to the transmission axis.

Consider an unpolarized light beam incident on a polarizing sheet with its transmission axis along the $x$ direction, as shown in Figure 31-38. The beam is incident on a second polarizing sheet, the analyzer, whose transmission axis makes an angle $\theta$ with the $x$ axis. If $E$ is the electric field amplitude of the incident beam, the component parallel with the transmission axis is $E_\parallel = E \cos\theta$, and the component perpendicular to the transmission axis is $E_\perp = E \sin\theta$. The sheet absorbs $E_\perp$ and transmits $E_\parallel$, so the transmitted beam has an electric field amplitude of $E_\parallel = E \cos\theta$ and is linearly polarized in the direction of the transmission axis. Because the intensity of light is proportional to the square of the magnitude of the electric field amplitude, the intensity $I$ of light transmitted by the sheet is given by

$$I = I_0 \cos^2\theta \qquad \text{31-20}$$

LAW OF MALUS

where $I_0$ is the intensity of the incident beam. If we have an incident beam of unpolarized light of intensity $I_0$ incident on a polarizing sheet, the direction of the incident electric field varies from location to location on the sheet, and at each location it fluctuates in time. At each location the angle between the electric field and the transmission axis is, on average, 45°, so applying Equation 31-20 gives $I = I_0 \cos^2 45° = \frac{1}{2}I_0$, where $I$ is the intensity of the transmitted beam.

When two polarizing elements are placed in succession in a beam of unpolarized light, the first polarizing element is called the **polarizer** and the second polarizing element is called the **analyzer.** If the polarizer and the analyzer are crossed, that is, if their transmission axes are perpendicular to each other, no light gets through. Equation 31-20 is known as the **law of Malus** after its discoverer, E. L. Malus (1775–1812). It applies to any two polarizing elements whose transmission axes make an angle $\theta$ with each other.

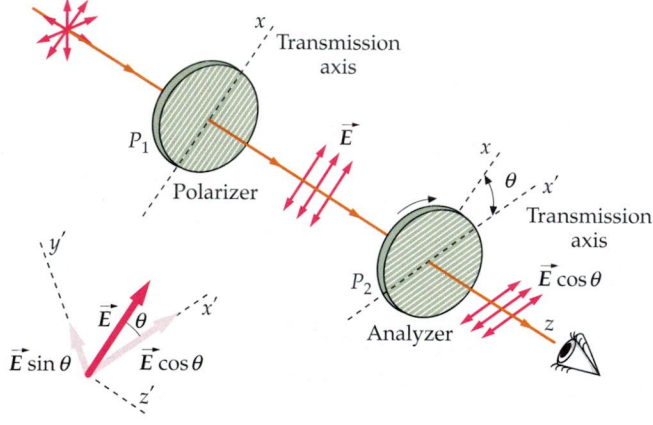

**FIGURE 31-38** A vertically polarized beam is incident on a polarizing sheet with its transmission axes making an angle $\theta$ with the vertical. Only the component $E \cos\theta$ is transmitted through the second sheet, and the transmitted beam is linearly polarized in the direction of the transmission axis. If the intensity between the sheets is $I_0$, the intensity transmitted by both sheets is $I_0 \cos^2\theta$.

(a)

(b)

(*a*) Cross polarizers block out all of the light. (*b*) In a liquid crystal display, the crystal is between crossed polarizers. Light incident on the crystal is transmitted because the crystal rotates the direction of polarization of the light 90°. The light is reflected back out through the crystal by a mirror behind the crystal, and a uniform background is seen. When a voltage is applied across a small segment of the crystal, the polarization is not rotated, so no light is transmitted and the segment appears black.

**FIGURE 31-39**

*INTENSITY TRANSMITTED*          **EXAMPLE 31-7**

Unpolarized light of intensity 3.0 W/m² is incident on two polarizing sheets whose transmission axes make an angle of 60° (Figure 31-39). What is the intensity of light transmitted by the second sheet?

**PICTURE THE PROBLEM**  The incident light is unpolarized, so the intensity transmitted by the first polarizing sheet is half the incident intensity. The second sheet further reduces the intensity by a factor of $\cos^2 \theta$, with $\theta = 60°$.

1. The intensity $I_1$ transmitted by the first sheet is half the intensity $I_0$ of unpolarized light incident on the first sheet: $\qquad I_1 = \frac{1}{2} I_0$

2. The intensity $I_2$ transmitted by the second sheet is related to the intensity $I_1$ of the light incident on the second sheet by Equation 31-20: $\qquad I_2 = I_1 \cos^2 \theta$

3. Combine these results and substitute the given data: $\qquad I_2 = \frac{1}{2} I_0 \cos^2 60° = \frac{1}{2} (3.0\ \text{W/m}^2)(0.500)^2$

$$= \boxed{0.375\ \text{W/m}^2}$$

**REMARKS**  Half the intensity passes through the first sheet no matter what the orientation of that sheet's transmission axis. Note that the second sheet rotates the plane of polarization by 60°.

## Polarization by Reflection

When unpolarized light is reflected from a plane surface boundary between two transparent media, such as air and glass or air and water, the reflected light is partially polarized. The degree of polarization depends on the angle of incidence and on the ratio of the wave speeds in the two media. For a certain angle of incidence called the polarizing angle $\theta_p$, the reflected light is completely polarized. At the polarizing angle, the reflected and refracted rays are perpendicular to each other. David Brewster (1781–1868), a Scottish scientist and an inventor of numerous instruments (including the kaleidoscope), discovered this experimentally in 1812. The polarizing angle is also referred to as Brewster's angle.

Figure 31-40 shows light incident at the polarizing angle $\theta_p$ for which the reflected light is completely polarized. The electric field of the incident light can be resolved into components parallel and perpendicular to the plane of incidence. The reflected light is linearly polarized with its electric field perpendicular to the plane of incidence. We can relate the polarizing angle to the indexes of refraction of the media using Snell's law (the law of refraction). If $n_1$ is the index of refraction of the first medium and $n_2$ is the index of refraction of the second medium, the law of refraction gives

$$n_1 \sin \theta_p = n_2 \sin \theta_2$$

where $\theta_2$ is the angle of refraction. From Figure 31-40, we can see that the sum of the angle of reflection and the angle of refraction is 90°. Since the angle of reflection equals the angle of incidence, we have

$$\theta_2 = 90° - \theta_p$$

Then

$$n_1 \sin \theta_p = n_2 \sin(90° - \theta_p) = n_2 \cos \theta_p$$

or

$$\tan \theta_p = \frac{n_2}{n_1} \qquad \text{31-21}$$

POLARIZING ANGLE

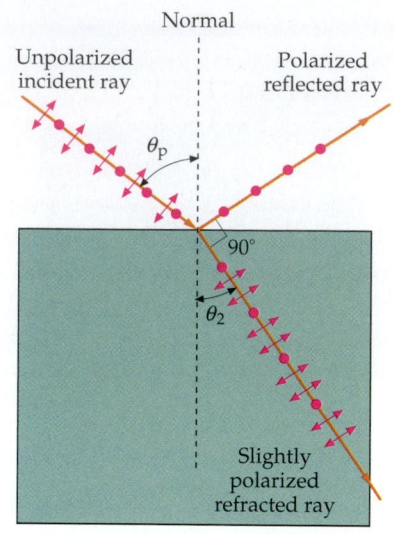

**FIGURE 31-40** Polarization by reflection. The incident wave is unpolarized and has components of the electric field parallel to the plane of incidence (arrows) and components perpendicular to this plane (dots). For incidence at the polarizing angle, the reflected wave is completely polarized, with its electric field perpendicular to the plane of incidence.

Although the reflected light is completely polarized for this angle of incidence, the transmitted light is only partially polarized because only a small fraction of the incident light is reflected. If the incident light itself is polarized with the electric field in the plane of incidence, there is no reflected light when the angle of incidence is $\theta_p$. We can understand this qualitatively from Figure 31-41. If we consider the molecules next to the surface of the second medium to be oscillating parallel to the electric field of the refracted ray, there can be no reflected ray because for an electric dipole antenna no energy is radiated along the line of oscillation. (Each of the oscillating molecules are a small electric dipole antenna.)

Because of the polarization of reflected light, sunglasses that contain a polarizing sheet can be very effective in cutting out glare. If light is reflected from a horizontal surface, such as a lake surface or snow on the ground, the electric field of the reflected light will be predominantly horizontal and the plane of incidence on the glasses will be predominantly vertical. Polarized sunglasses with a vertical transmission axis will then reduce glare by absorbing much of the reflected light. If you have polarized sunglasses, you can observe this effect by looking through the glasses at reflected light and then rotating the glasses 90°; much more of the light will be transmitted.

## Polarization by Scattering

The phenomenon of absorption and reradiation is called **scattering**. Scattering can be demonstrated by passing a light beam through a container of water to which a small amount of powdered milk has been added. The milk particles absorb light and reradiate it, making the light beam visible. Similarly, laser beams can be made visible by introducing chalk or smoke particles into the air to scatter the light. A familiar example of light scattering is that from air molecules, which tend to scatter short wavelengths more than long wavelengths, thereby giving the sky its blue color.

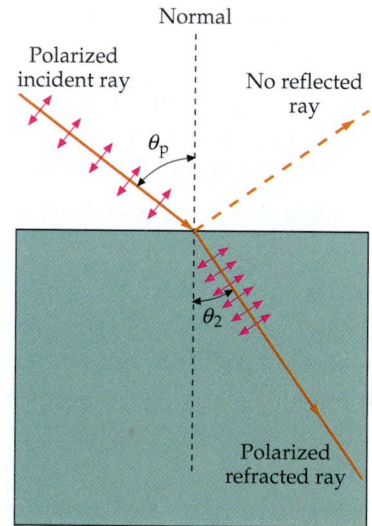

**FIGURE 31-41** Polarized light incident at the polarizing angle. When the incident light is polarized with $\vec{E}$ in the plane of incidence, there is no reflected ray.

We can understand polarization by scattering if we think of a scattering molecule as an electric dipole antenna that radiates waves with a maximum intensity in directions perpendicular to the antenna axis and zero intensity in the direction along the antenna axis. The electric field vector of the scattered light perpendicular to the direction of propagation is in the plane of the antenna axis and the field point. Figure 31-42 shows a beam of unpolarized light that initially travels along the z axis, striking a molecule at the origin. The electric field in the light beam has components in both the x and y directions perpendicular to the direction of motion of the light beam. These fields set up oscillations of the charges within the molecule in the z = 0 plane, and there is no oscillation along the z direction. These oscillations can be thought of as a superposition of an oscillation along the x axis and another along the y axis, with each of these oscillations producing dipole radiation. Thus, the oscillation along the x axis produces no radiation along the x axis, which means the light radiated along the x axis is produced only by the oscillation along the y axis. It follows that the light radiated along the x axis is polarized with its electric field parallel with the y axis. There is nothing special about the choice of axes for this discussion, so the result can be generalized. That is, the light scattered in a direction perpendicular to the incident light beam is polarized with its electric field perpendicular to both the incident beam and the direction of propagation of the scattered light. This can be seen easily by examining the scattered light with a piece of polarizing sheet.

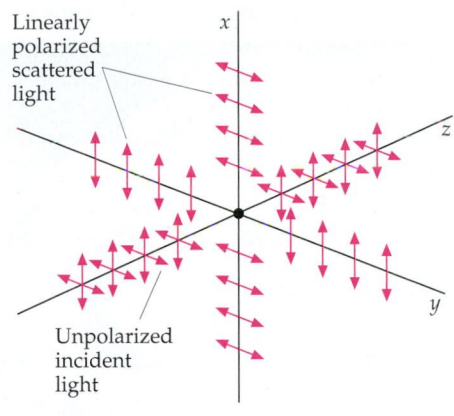

**FIGURE 31-42** Polarization by scattering. Unpolarized light propagating in the z direction is incident on a scattering center at the origin. The light scattered in the z = 0 plane along the x direction is polarized parallel with the y axis (and the light scattered in the y direction is polarized parallel with the x axis).

## Polarization by Birefringence

**Birefringence** is a complicated phenomenon that occurs in calcite and other noncubic crystals and in some stressed plastics, such as cellophane. Most materials are **isotropic,** that is, the speed of light passing through the material is the same in all directions. Because of their atomic structure, birefringent materials are **anisotropic.** The speed of light depends on the plane of polarization and on the direction of propagation of the light. When a light ray is incident on such materials, it may be separated into two rays called the *ordinary ray* and the *extraordinary ray.* These rays are polarized in mutually perpendicular directions, and they travel with different speeds. Depending on the relative orientation of the material and the incident light beam, the rays may also travel in different directions.

There is one particular direction in a birefringent material in which both rays propagate with the same speed. This direction is called the **optic axis** of the material. (The optic axis is actually a *direction* rather than a line in the material.) Nothing unusual happens when light travels in the direction of the optic axis. However, when light is incident at an angle to the optic axis, as shown in Figure 31-43, the rays travel in different directions and emerge separated in space. If the material is rotated, the extraordinary ray (the e ray in the figure) revolves in space around the ordinary ray (o ray).

If light is incident on a birefringent plate perpendicular to its crystal face and perpendicular to the optic axis, the two rays travel in the same direction but at different speeds. The number of wavelengths in the two rays in the plate is different because the wavelengths ($\lambda = v/f$) of the rays differ. The rays emerge with a phase difference that depends on the thickness of the plate and on the wavelength of the incident light. In a **quarter-wave plate,** the thickness is such that there is a 90° phase difference between the waves of a particular wavelength when they emerge. In a **half-wave plate,** the rays emerge with a phase difference of 180°.

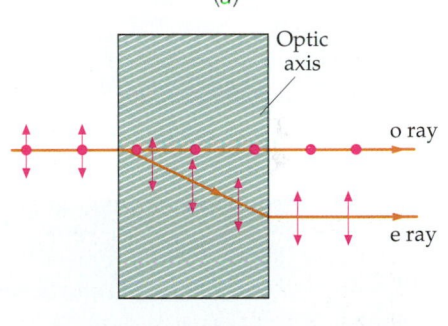

(a)

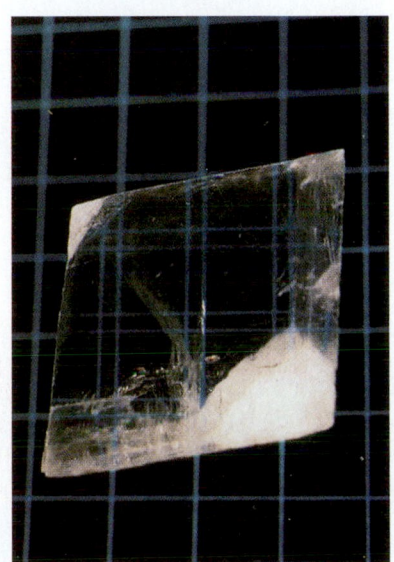

(b)

**FIGURE 31-43** (a) A narrow beam of light incident on a birefringent crystal such as calcite is split into two beams, called the ordinary ray (o ray) and the extraordinary ray (e ray), that have mutually perpendicular polarizations. If the crystal is rotated, the extraordinary ray rotates in space. (b) A double image of the cross hatching is produced by this birefringent crystal of calcium carbonate.

(a)

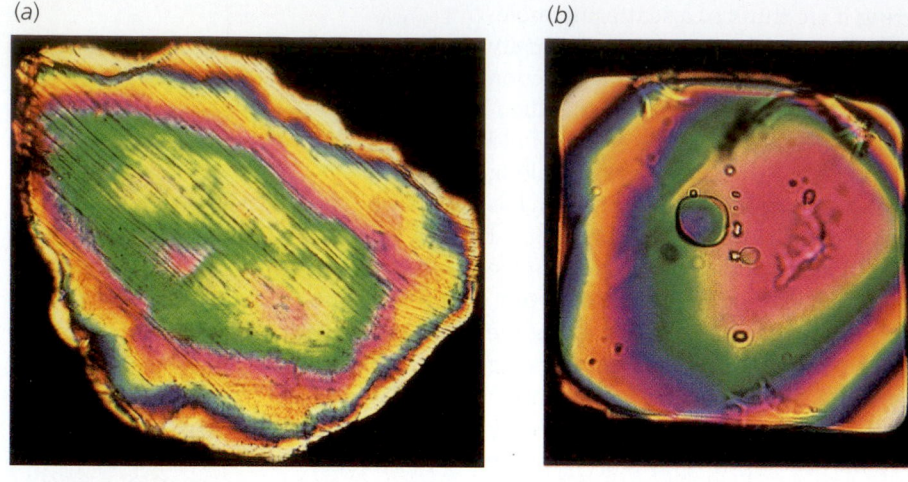

(b)

(c)

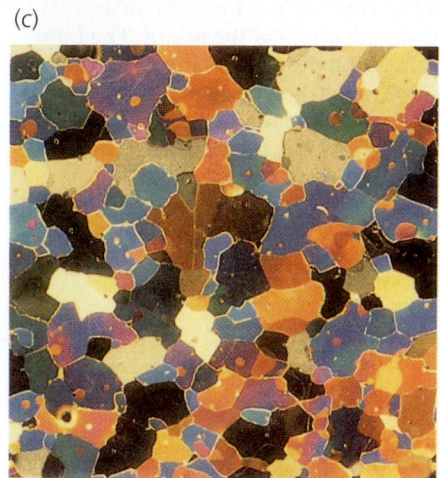

(d)

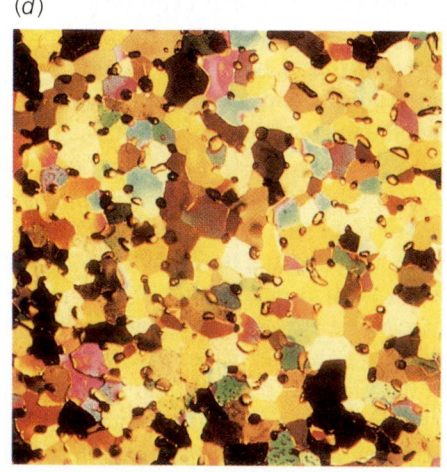

When the transmission axes of two polarizing sheets are perpendicular, the polarizers are said to be crossed and no light is transmitted. However, many materials are birefringent or become so under stress. Such materials rotate the direction of polarization of the light so that light of a particular wavelength is transmitted through both polarizers. When a birefringent material is viewed between crossed polarizers, information about its internal structure is revealed. (a) A shocked quartz grain from the site of a meteorite crater. The layered structure, evidenced by the parallel lines, arises from the shock of the impact of the meteor. (b) A grain of quartz typically found in silicic volcanic rocks. No shock lines are seen. (c) Thin sections of ice core from the antarctic ice sheet reveal bubbles of trapped $CO_2$, which appear amber-colored. This sample was taken from a depth of 194 m, corresponding to air trapped 1600 years ago, whereas the sample in (d) is from a depth of 56 m, corresponding to air trapped 450 years ago. Ice core measurements have replaced the less reliable technique of analyzing carbon in tree rings to compare current atmospheric $CO_2$ levels with those of the recent past. (e) Robert Mark of the Princeton School of Architecture examines the stress patterns in a plastic model of the nave structure of Chartres Cathedral.

(e)

Suppose that the incident light is linearly polarized so that the electric field vector is at 45° to the optic axis, as illustrated in Figure 31-44. The ordinary and extraordinary rays start out in phase and have equal amplitudes. With a quarter-wave plate, the waves emerge with a phase difference of 90°, so the resultant electric field has components $E_x = E_0 \sin \omega t$ and $E_y = E_0 \sin(\omega t + 90°) = E_0 \cos \omega t$. The electric field vector thus rotates in a circle and the wave is circularly polarized.

With a half-wave plate, the waves emerge with a phase difference of 180°, so the resultant electric field is linearly polarized with components $E_x = E_0 \sin \omega t$ and $E_y = E_0 \sin(\omega t + 180°) = -E_0 \sin \omega t$. The net effect is that the direction of polarization of the wave is rotated by 90° relative to that of the incident light, as shown in Figure 31-45.

Interesting and beautiful patterns can be observed by placing birefringent materials, such as cellophane or stressed plastic, between two polarizing sheets with their transmission axes perpendicular to each other. Ordinarily, no light is transmitted through crossed polarizing sheets. However, if we place a birefringent material between the crossed polarizing sheets, the material acts as a half-wave plate for light of a certain color depending on the material's thickness. The direction of polarization is rotated and some light gets through both sheets. Various glasses and plastics become birefringent when under stress. The stress patterns can be observed when the material is placed between crossed polarizing sheets.

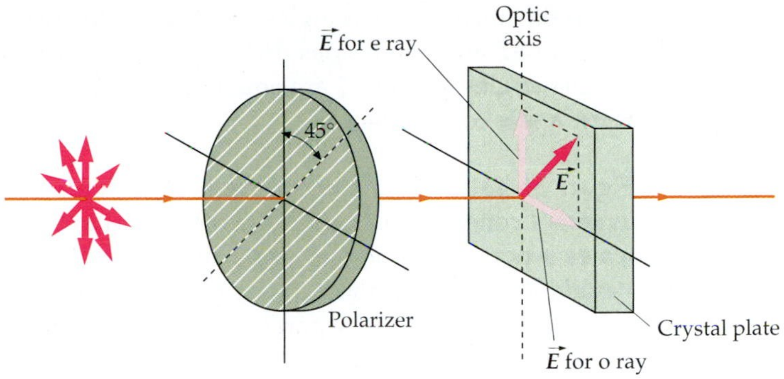

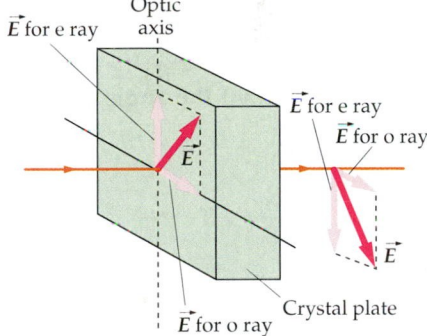

**FIGURE 31-44** Polarized light emerging from the polarizer is incident on a birefringent crystal so that the electric field vector makes a 45° angle with the optic axis, which is perpendicular to the light beam. The ordinary and extraordinary rays travel in the same direction but at different speeds. The polarization of the emerging light depends on the thickness of the crystal and the wavelength of the light.

# 31-8 Derivation of the Laws of Reflection and Refraction

The laws of reflection and refraction can be derived from either Huygens's principle or Fermat's principle.

## Huygens's Principle

**Reflection**   Figure 31-46 shows a plane wavefront $AA'$ striking a mirror at point $A$. As can be seen from the figure, the angle $\phi_1$ between the wavefront and the mirror is the same as the angle of incidence $\theta_1$, which is the angle between the normal to the mirror and the rays (which are perpendicular to the wavefronts). According to Huygens's principle, each point on a given wavefront can be considered a point source of secondary wavelets. The position of the wavefront after a time $t$ is found by constructing wavelets of radius $ct$ with their centers on the wavefront $AA'$. Wavelets that have not yet reached the mirror form the portion of the new wavefront $BB'$. Wavelets that have already reached the mirror are reflected and form the portion of the new wavefront $B''B$. By a similar construction, the wavefront $C''C$ is obtained from the Huygens's wavelets originating on the wavefront $B''B$. Figure 31-47 is an enlargement of a portion of Figure 31-46 showing $AP$, which is part of the initial position of the wavefront. During the time $t$, the wavelet from point $P$ reaches the mirror at point $B$, and the wavelet from point $A$ reaches point $B''$. The reflected wavefront $B''B$ makes an angle $\phi_1'$ with the mirror that is equal to the angle of reflection $\theta_1'$ between the reflected ray

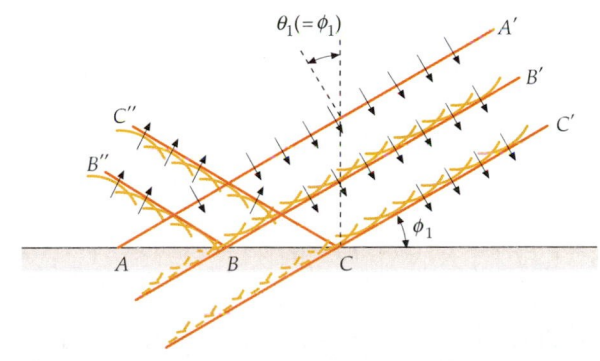

**FIGURE 31-45** If the birefringent crystal in Figure 31-44 is a half-wave plate, and if the electric field vector of the incident light makes an angle of 45° with the optic axis, then the direction of polarization of the emerging light is rotated by 90°.

**FIGURE 31-46** Plane wave reflected at a plane mirror. The angle $\theta_1$ between the incident ray and the normal to the mirror is the angle of incidence. It is equal to the angle $\phi_1$ between the incident wavefront and the mirror.

and the normal to the mirror. The triangles $AB''B$ and $APB$ are both right triangles with a common side $AB$ and equal sides $AB'' = BP = ct$. Hence, these triangles are congruent, and the angles $\phi_1$ and $\phi_1'$ are equal, implying that the angle of reflection $\theta_1'$ equals the angle of incidence $\theta_1$.

**Refraction** Figure 31-48 shows a plane wave incident on an air–glass interface. We apply Huygens's construction to find the wavefront in the transmitted wave. Line $AP$ indicates a portion of the wavefront in medium 1 that strikes the glass surface at an angle $\phi_1$. In time $t$, the wavelet from $P$ travels the distance $v_1t$ and reaches the point $B$ on the line $AB$ separating the two media, while the wavelet from point $A$ travels a shorter distance $v_2t$ into the second medium. The new wavefront $BB'$ is not parallel to the original wavefront $AP$ because the speeds $v_1$ and $v_2$ are different. From the triangle $APB$,

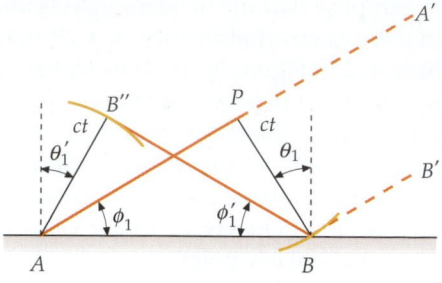

**FIGURE 31-47** Geometry of Huygens's construction for the calculation of the law of reflection. The wavefront $AP$ initially strikes the mirror at point $A$. After a time $t$, the Huygens wavelet from $P$ strikes the mirror at point $B$, and the Huygens wavelet from point $A$ reaches point $B$.

$$\sin \phi_1 = \frac{v_1t}{AB}$$

or

$$AB = \frac{v_1t}{\sin \phi_1} = \frac{v_1t}{\sin \theta_1}$$

using the fact that the angle $\phi_1$ equals the angle of incidence $\theta_1$. Similarly, from triangle $AB'B$,

$$\sin \phi_2 = \frac{v_2t}{AB}$$

or

$$AB = \frac{v_2t}{\sin \phi_2} = \frac{v_2t}{\sin \theta_2}$$

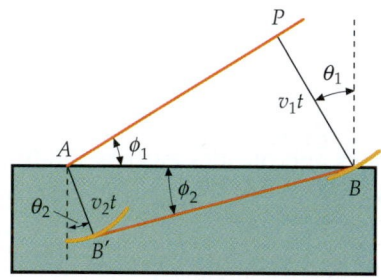

**FIGURE 31-48** Application of Huygens's principle to the refraction of plane waves at the surface separating a medium in which the wave speed is $v_1$ from a medium in which the wave speed $v_2$ is less than $v_1$. The angle of refraction in this case is less than the angle of incidence.

where $\theta_2 = \phi_2$ is the angle of refraction. Equating the two values for $AB$, we obtain

$$\frac{\sin \theta_1}{v_1} = \frac{\sin \theta_2}{v_2} \tag{31-22}$$

Substituting $v_1 = c/n_1$ and $v_2 = c/n_2$ in this equation and multiplying by $c$, we obtain $n_1 \sin \theta_1 = n_2 \sin \theta_2$, which is Snell's law.

## Fermat's Principle

**Reflection** Figure 31-49 shows two paths in which light leaves point $A$, strikes the plane surface, which we can consider to be a mirror, and travels to point $B$. The problem for the application of Fermat's principle to reflection can be stated as follows: At what point $P$ in the figure must the light strike the mirror so that it will travel from point $A$ to point $B$ in the least time? Since the light is traveling in the same medium for this problem, the time will be minimum when the distance is minimum. In Figure 31-49 the distance $APB$ is the same as the distance $A'PB$, where point $A'$ lies along the perpendicular from $A$ to the mirror and is equidistant behind the mirror. As we vary point $P$, the distance $A'PB$ is least when the points $A'$, $P$, and $B$ lie on a straight line. We can see from the figure that this occurs when the angle of incidence equals the angle of reflection.

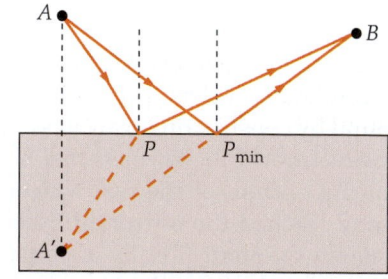

**FIGURE 31-49** Geometry for deriving the law of reflection from Fermat's principle. The time it takes for the light to travel from point $A$ to point $B$ is a minimum for light striking the surface at point $P$.

**Refraction** The derivation of Snell's law of refraction from Fermat's principle is slightly more complicated. Figure 31-50 shows the possible paths for light traveling from point $A$ in air to point $B$ in glass. Point $P_1$ is on the straight line between $A$ and $B$, but this path is not the one for the shortest travel time because light travels with a smaller speed in the glass. If we move slightly to the right of $P_1$, the total path length is greater, but the distance traveled in the slower medium is less than for the path through $P_1$. It is not apparent from the figure which path is the path of least time, but it is not surprising that a path slightly to the right of the straight-line path takes less time because the time gained by traveling a shorter distance in the glass more than compensates for the time lost traveling a longer distance in the air. As we move the point of intersection of the possible path to the right of point $P_1$, the total time of travel from point $A$ to point $B$ decreases until we reach a minimum at point $P_{min}$. Beyond this point, the time saved by traveling a shorter distance in the glass does not compensate for the greater time required for the greater distance traveled in the air.

Figure 31-51 shows the geometry for finding the path of least time. If $L_1$ is the distance traveled in medium 1 with index of refraction $n_1$, and $L_2$ is the distance traveled in medium 2 with index of refraction $n_2$, the time for light to travel the total path $AB$ is

$$t = \frac{L_1}{v_1} + \frac{L_2}{v_2} = \frac{L_1}{c/n_1} + \frac{L_2}{c/n_2} = \frac{n_1 L_1}{c} + \frac{n_2 L_2}{c} \qquad 31\text{-}23$$

We wish to find the point $P_{min}$ for which this time is a minimum. We do this by expressing the time in terms of a single parameter $x$, as shown in the figure, indicating the position of point $P_{min}$. In terms of the distance $x$,

$$L_1^2 = a^2 + x^2 \quad \text{and} \quad L_2^2 = b^2 + (d - x)^2 \qquad 31\text{-}24$$

Figure 31-52 shows the time $t$ as a function of $x$. At the value of $x$ for which the time is a minimum, the slope of the graph of $t$ versus $x$ is zero:

$$\frac{dt}{dx} = 0$$

Differentiating each term in Equation 31-23 with respect to $x$ and setting the result equal to zero, we obtain

$$\frac{dt}{dx} = \frac{1}{c}\left(n_1 \frac{dL_1}{dx} + n_2 \frac{dL_2}{dx}\right) = 0 \qquad 31\text{-}25$$

We can compute these derivatives from Equations 31-24. We have

$$2L_1 \frac{dL_1}{dx} = 2x \quad \text{or} \quad \frac{dL_1}{dx} = \frac{x}{L_1}$$

where $x/L_1$ is just $\sin \theta_1$ and $\theta_1$ is the angle of incidence. Thus,

$$\frac{dL_1}{dx} = \sin \theta_1 \qquad 31\text{-}26$$

Similarly,

$$2L_2 \frac{dL_2}{dx} = 2(d - x)(-1)$$

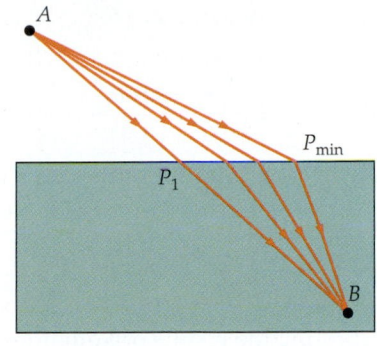
**FIGURE 31-50** Geometry for deriving Snell's law from Fermat's principle. The point $P_{min}$ is the point at which light must strike the glass in order that the travel time from point $A$ to point $B$ is a minimum.

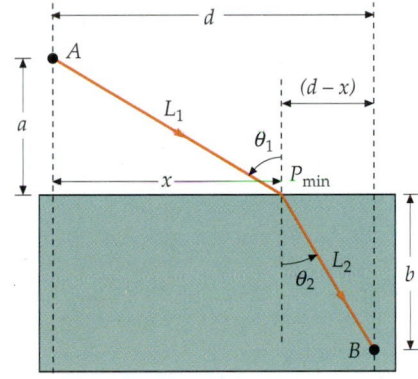
**FIGURE 31-51** Geometry for calculating the minimum time in the derivation of Snell's law from Fermat's principle.

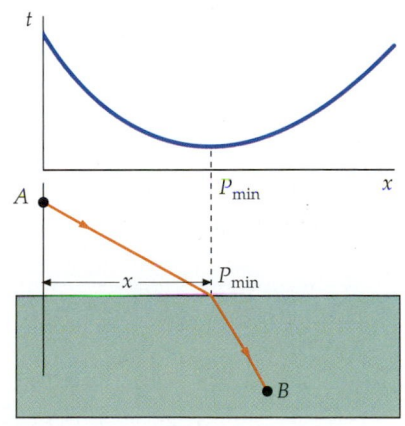
**FIGURE 31-52** Graph of the time it takes for light to travel from point $A$ to point $B$ versus $x$, measured along the refracting surface. The time is a minimum at the point at which the angles of incidence and refraction obey Snell's law.

or

$$\frac{dL_2}{dx} = -\frac{d-x}{L_2} = -\sin\theta_2 \qquad\qquad 31\text{-}27$$

where $\theta_2$ is the angle of refraction. From Equation 31-25,

$$n_1\frac{dL_1}{dx} + n_2\frac{dL_2}{dx} = 0 \qquad\qquad 31\text{-}28$$

Substituting the results of Equations 31-26 and 31-27 for $dL_1/dx$ and $dL_2/dx$ gives

$$n_1\sin\theta_1 + n_2(-\sin\theta_2) = 0$$

or

$$n_1\sin\theta_1 = n_2\sin\theta_2$$

which is Snell's law.

# SUMMARY

| Topic | Relevant Equations and Remarks |
|---|---|
| 1. **Visible Light** | The human eye is sensitive to electromagnetic radiation with wavelengths from approximately 400 nm (violet) to 700 nm (red). The photon energies range from approximately 1.8 eV to 3.1 eV. A uniform mixture of wavelengths, such as the wavelengths emitted by the sun, appears white to our eyes. |
| 2. **Wave–Particle Duality** | Light propagates like a wave, but interacts with matter like a particle. |
|   Photon energy | $E = hf = \dfrac{hc}{\lambda}$        31-1 |
|   Planck's constant | $h = 6.626\times10^{-34}\,\text{J·s} = 4.136\times10^{-15}\,\text{eV·s}$ |
|   $hc$ | $hc = 1240\,\text{eV·nm}$        31-2 |
| 3. **Emission of Light** | Light is emitted when an outer atomic electron makes a transition from an excited state to a state of lower energy. |
|   Line spectra | Atoms in dilute gases emit a discrete set of wavelengths called a line spectra. The photon energy $E = hf = hc/\lambda$ equals the difference in energy of the initial and final states of the atom. |
|   Continuous spectra | Atoms in high-density gases, liquids, or solids have continuous bands of energy levels, so they emit a continuous spectrum of light. Thermal radiation is visible if the temperature of the emitting object is above approximately 600°C. |
|   Spontaneous emission | An atom in an excited state will spontaneously make a transition to a lower state with the emission of a photon. This process is random, with a characteristic lifetime of about $10^{-8}$ s. The photons from two or more atoms are not correlated, so the light is incoherent. |

| | |
|---|---|
| Stimulated emission | Stimulated emission occurs if an atom is initially in an excited state and a photon of energy equal to the energy difference between that state and a lower state is incident on the atom. The oscillating electromagnetic field of the incident photon stimulates the excited atom to emit another photon in the same direction and in phase with the incident photon. The emitted light is coherent. |
| **4. Lasers** | A laser produces an intense, coherent, and narrow beam of photons as the result of stimulated emission. The operation of a laser depends on population inversion, in which there are more atoms in an excited state than in the ground state or a lower state. |
| **5. Speed of Light** | The SI unit of length, the meter, is defined so that the speed of light in vacuum is exactly |

$$c = 299{,}792{,}458 \text{ m/s} \qquad \text{31-5}$$

| | |
|---|---|
| $v$ in a transparent medium | $$v = \frac{c}{n} \qquad \text{31-7}$$ |

where $n$ is the index of refraction.

| | |
|---|---|
| **6. Huygens's Principle** | Each point on a primary wavefront serves as the source of spherical secondary wavelets that advance with a speed and frequency equal to that of the primary wave. The primary wavefront at some later time is the envelope of these wavelets. |
| **7. Reflection and Refraction** | When light is incident on a surface separating two media in which the speed of light differs, part of the light energy is transmitted and part of the light energy is reflected. |
| Law of reflection | The reflected ray lies in the plane of incidence and makes an angle $\theta_1'$ with the normal that is equal to the angle of incidence. |

$$\theta_1' = \theta_1 \qquad \text{31-8}$$

| | |
|---|---|
| Reflected intensity, normal incidence | $$I = \left(\frac{n_1 - n_2}{n_1 + n_2}\right)^2 I_0 \qquad \text{31-11}$$ |
| Index of refraction | $$n = \frac{c}{v} \qquad \text{31-7}$$ |
| Law of refraction (Snell's law) | $$n_1 \sin \theta_1 = n_2 \sin \theta_2 \qquad \text{31-9}b$$ |
| Total internal reflection | When light is traveling in a medium with an index of refraction $n_1$ and is incident on the boundary of a second medium with a lower index of refraction $n_2 < n_1$, the light is totally reflected if the angle of incidence is greater than the critical angle $\theta_c$ given by |
| Critical angle | $$\sin \theta_c = \frac{n_2}{n_1} \qquad \text{31-12}$$ |
| Dispersion | The speed of light in a medium, and therefore the index of refraction of that medium, depends on the wavelength of light. Because of dispersion, a beam of white light incident on a refracting prism is dispersed into its component colors. Similarly, the reflection and refraction of sunlight by raindrops produces a rainbow. |
| **8. Polarization** | Transverse waves can be polarized. The four phenomena that produce polarized electromagnetic waves from unpolarized waves are: (1) absorption, (2) scattering, (3) reflection, and (4) birefringence. |
| Malus's law | When two polarizers have their transmission axes at an angle $\theta$, the intensity transmitted by the second polarizer is reduced by the factor $\cos^2 \theta$. |

$$I = I_0 \cos^2 \theta \qquad \text{31-20}$$

# PROBLEMS

- Single-concept, single-step, relatively easy
- •• Intermediate-level, may require synthesis of concepts
- ••• Challenging
- **SSM** Solution is in the *Student Solutions Manual*
- **iSOLVE** Problems available on iSOLVE online homework service
- **iSOLVE✓** These "Checkpoint" online homework service problems ask students additional questions about their confidence level, and how they arrived at their answer.

In a few problems, you are given more data than you actually need; in a few other problems, you are required to supply data from your general knowledge, outside sources, or informed estimates.

## Conceptual Problems

**1** • Why is helium needed in a helium–neon laser? Why not just use neon?

**2** •• When a beam of visible white light passes through a gas of atomic hydrogen and is viewed with a spectroscope, dark lines are observed at the wavelengths of the emission series. The atoms that participate in the resonance absorption then emit this same wavelength light as they return to the ground state. Explain why the observed spectrum nevertheless exhibits pronounced dark lines.

**3** • How does a thin layer of water on the road affect the light you see reflected off the road from your own headlights? How does it affect the light you see reflected from the headlights of an oncoming car?

**4** • A ray of light passes from air into water, striking the surface of the water with an angle of incidence of 45°. Which of the following four quantities change as the light enters the water: (1) wavelength, (2) frequency, (3) speed of propagation, (4) direction of propagation? (*a*) 1 and 2 only; (*b*) 2, 3, and 4 only; (*c*) 1, 3, and 4 only; (*d*) 3 and 4 only; or (*e*) 1, 2, 3, and 4.

**5** •• **SSM** The density of the atmosphere decreases with height, as does the index of refraction. Explain how one can see the sun after it has set. Why does the setting sun appear flattened?

**6** • A physics student playing pocket billiards wants to strike her cue ball so that it hits a cushion and then hits the eight ball squarely. She chooses several points on the cushion and for each point measures the distance from it to the cue ball and to the eight ball. She aims at the point for which the sum of these distances is least. (*a*) Will her cue ball hit the eight ball? (*b*) How is her method related to Fermat's principle?

**7** • A swimmer at point *S* in Figure 31-53 develops a leg cramp while swimming near the shore of a calm lake and calls for help. A lifeguard at point *L* hears the call. The lifeguard can run 9 m/s and swim 3 m/s. She knows physics and chooses a path that will take the least time to reach the swimmer. Which of the paths shown in Figure 31-53 does the lifeguard take?

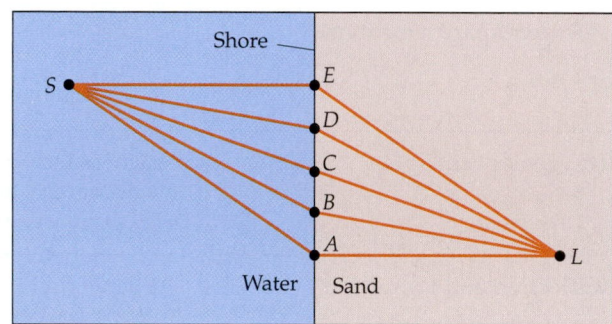

**FIGURE 31-53** Problem 7

**8** • Two polarizers have their transmission axes at an angle $\theta$. Unpolarized light of intensity $I$ is incident upon the first polarizer. What is the intensity of the light transmitted by the second polarizer? (*a*) $I \cos^2 \theta$, (*b*) $(I \cos^2 \theta)/2$, (*c*) $(I \cos^2 \theta)/4$, (*d*) $I \cos \theta$, (*e*) $(I \cos \theta)/4$, or (*f*) none of the answers are correct.

**9** • Which of the following is *not* a phenomenon whereby polarized light can be produced from unpolarized light? (*a*) absorption (*b*) reflection (*c*) birefringence (*d*) diffraction (*e*) scattering

**10** •• **SSM** We learned in Chapter 30, Section 30-3, that an oscillating electric dipole produces electromagnetic radiation (see Figure 30-8). Assuming that the light reflected off and refracted into the surface of a piece of transparent material is caused by such dipoles, show that the condition for Brewster's angle (Equation 31-21) is exactly the same as saying that the refracted ray is perpendicular to the axis of the radiating dipoles for light polarized in the plane of incidence.

**11** •• Draw a diagram to explain how Polaroid sunglasses reduce glare from sunlight reflected from a smooth horizontal surface, such as the surface found on a pool of water. Your diagram should clearly indicate the direction of polarization of the light as it propagates from the sun to the reflecting surface and then through the sunglasses into the eye.

**12** • **iSOLVE** True or false:

(*a*) Light and radio waves travel with the same speed through a vacuum.
(*b*) Most of the light incident normally on an air–glass interface is reflected.

(c) The angle of refraction of light is always less than the angle of incidence.

(d) The index of refraction of water is the same for all wavelengths in the visible spectrum.

(e) Longitudinal waves cannot be polarized.

**13** •• **[SOLVE]** ✔ Of the following statements about the speeds of the various colors of light in glass, which are true?

(a) All colors of light have the same speed in glass.

(b) Violet has the highest speed, red the lowest.

(c) Red has the highest speed, violet the lowest.

(d) Green has the highest speed, red and violet the lowest.

(e) Red and violet have the highest speed, green the lowest.

**14** •• **[SSM]** It is a common experience that on a calm, sunny day one can hear voices of persons in a boat over great distances. Explain this phenomenon, keeping in mind that sound is reflected from the surface of the water and that the temperature of the air just above the water's surface is usually less than that at a height of 10 m or 20 m above the water.

**15** • The human eye perceives color using a structure called a cone, located on the retina. The molecules of the cones come in three types that respond in a process similar to resonance absorption to red, green, and blue light, respectively. Use this fact to explain why the color of a blue object (450 nm in air) does not appear to change when immersed in clear colorless water, in spite of the fact that the wavelength of the light is shortened in accordance with Equation 31-10.

### Estimation and Approximation

**16** • Estimate the time required for light to make the round trip in Galileo's experiment to determine the speed of light.

**17** • Ole Römer's method for measuring the speed of light requires the precise prediction of the time of occurrence for the eclipse of Jupiter's moon Io. Assuming an eclipse took place on June 1 at midnight when the earth was in location A, as shown in Figure 31-11, predict the expected time of the eclipse one-quarter year later at location B, assuming (a) the speed of light is infinite and (b) the speed of light is the presently defined value of $2.998 \times 10^8$ m/s.

**18** •• If the angle of incidence is small enough, the approximation $\sin \theta \approx \theta$ may be used to simplify Snell's law. Calculate the angle of incidence that would make the error in calculating the angle of refraction using this small angle approximation no worse than 1 percent when compared to the exact formula. This approximation will be used in connection with image formation by spherical surfaces in Chapter 32.

### Sources of Light

**19** • **[SOLVE]** A pulse from a ruby laser has an average power of 10 MW and lasts 1.5 ns. (a) What is the total energy of the pulse? (b) How many photons are emitted in this pulse?

**20** • **[SOLVE]** A helium–neon laser emits light of wavelength 632.8 nm and has a power output of 4 mW. How many photons are emitted per second by this laser?

**21** • **[SOLVE]** ✔ The first excited state of an atom of a gas is 2.85 eV above the ground state. (a) What is the wavelength of radiation for resonance absorption? (b) If the gas is irradiated with monochromatic light of 320 nm wavelength, what is the wavelength of the Raman scattered light?

**22** •• A gas is irradiated with monochromatic ultraviolet light of 368 nm wavelength. Scattered light of the same wavelength and of 658 nm wavelength is observed. Assuming that the gas atoms were in their ground state prior to irradiation, find the energy difference between the ground state and the atomic state excited by the irradiation.

**23** •• Sodium has excited states 2.11 eV, 3.2 eV, and 4.35 eV above the ground state. (a) What is the maximum wavelength of radiation that will result in resonance fluorescence? What is the wavelength of the fluorescent radiation? (b) What wavelength will result in excitation of the state 4.35 eV above the ground state? If that state is excited, what are the possible wavelengths of resonance fluorescence that might be observed?

**24** •• **[SSM]** Singly ionized helium is a hydrogen-like atom with a nuclear charge of 2e. Its energy levels are given by $E_n = -4E_0/n^2$, where $E_0 = 13.6$ eV. If a beam of visible white light is sent through a gas of singly ionized helium, at what wavelengths will dark lines be found in the spectrum of the transmitted radiation?

### The Speed of Light

**25** • Mission Control sends a brief wake-up call to astronauts in a far away spaceship. Five seconds after the call is sent, Mission Control can hear the groans of the astronauts. How far away (at most) from the earth is the spaceship? (a) $7.5 \times 10^8$ m. (b) $15 \times 10^8$ m. (c) $30 \times 10^8$ m. (d) $45 \times 10^8$ m. (e) The spaceship is on the moon.

**26** • **[SOLVE]** The spiral galaxy in the Andromeda constellation is about $2 \times 10^{19}$ km away from us. How many light-years is this?

**27** • **[SOLVE]** On a spacecraft sent to Mars to take pictures, the camera is triggered by radio waves, which like all electromagnetic waves, travel with the speed of light. What is the time delay between sending the signal from the earth and receiving the signal on Mars? (Take the distance to Mars to be $9.7 \times 10^{10}$ m.)

**28** • The distance from a point on the surface of the earth to one on the surface of the moon is measured by aiming a laser light beam at a reflector on the surface of the moon and measuring the time required for the light to make a round trip. The uncertainty in the measured distance $\Delta x$ is related to the uncertainty in the time $\Delta t$ by $\Delta x = c \Delta t$. If the time intervals can be measured to $\pm 1$ ns, find the uncertainty of the distance in meters.

**29** •• **[SSM]** **[SOLVE]** In Galileo's attempt to determine the speed of light, he and his assistant were located on hilltops about 3 km apart. Galileo flashed a light and received a return flash from his assistant. (a) If his assistant had an instant reaction, what time difference would Galileo need to be able to measure for this method to be successful? (b) How does this time compare with human reaction time, which is about 0.2 s?

## Reflection and Refraction

**30** • **iSOLVE✓** Calculate the fraction of light energy reflected from an air–water interface at normal incidence.

**31** •• **SSM** A ray of light is incident on one of a pair of mirrors set at right angles to each other. The plane of incidence is perpendicular to both mirrors. Show that after reflecting off each mirror the ray will emerge in the opposite direction, regardless of the angle of incidence.

**32** •• (a) A beam of light in air is incident on an air–water interface. Using a spreadsheet or graphing program, plot the angle of refraction as a function of the angle of incidence from 0° to 90°. (b) Repeat Part (a), but for a beam of light initially in water, incident on a water–air interface. For Part (b), what is the meaning of your graph for angles of incidence that are greater than the critical angle?

**33** • **iSOLVE✓** Find the speed of light in water and in glass.

**34** • **iSOLVE** The index of refraction for silicate flint glass is 1.66 for light with a wavelength of 400 nm and 1.61 for light with a wavelength of 700 nm. Find the angles of refraction for light of these wavelengths that is incident on this glass at an angle of 45°.

**35** •• **iSOLVE✓** A slab of glass with an index of refraction of 1.5 is submerged in water with an index of refraction of 1.33. Light in the water is incident on the glass. Find the angle of refraction if the angle of incidence is (a) 60°, (b) 45°, and (c) 30°.

**36** •• Repeat Problem 35 for a beam of light initially in the glass that is incident on the glass–water interface at the same angles.

**37** •• **SSM** **iSOLVE✓** Light is incident normally on a slab of glass with an index of refraction $n = 1.5$. Reflection occurs at both surfaces of the slab. Approximately what percentage of the incident light energy is transmitted by the slab?

**38** •• This problem is a refraction analogy. A band is marching down a football field with a constant speed $v_1$. About midfield, the band comes to a section of muddy ground that has a sharp boundary making an angle of 30° with the 50-yd line, as shown in Figure 31-54. In the mud, the marchers move with speed $v_2 = v_1/2$. Diagram how each line of marchers is bent as it encounters the muddy section of the field so that the band is eventually marching in a different direction. Indicate the original direction by a ray, the final direction by a second ray, and find the angles between the rays and the line perpendicular to the boundary. Is their direction of motion bent toward the perpendicular to the boundary or away from it?

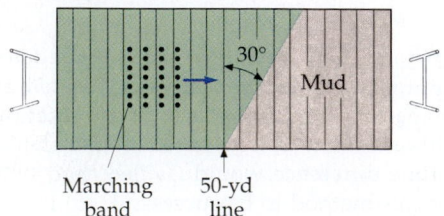

**FIGURE 31-54** Problem 38

**39** •• In Figure 31-55, light is initially in a medium (e.g., air) of index of refraction $n_1$. It is incident at angle $\theta_1$ on the surface of a liquid (e.g., water) of index of refraction $n_2$. The light passes through the layer of water and enters glass of index of refraction $n_3$. If $\theta_3$ is the angle of refraction in the glass, show that $n_1 \sin \theta_1 = n_3 \sin \theta_3$. That is, show that the second medium can be neglected when finding the angle of refraction in the third medium.

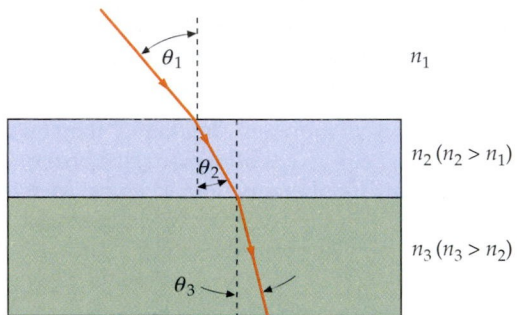

**FIGURE 31-55** Problem 39

**40** ••• **SSM** Figure 31-56 shows a beam of light incident on a glass plate of thickness $d$ and index of refraction $n$. (a) Find the angle of incidence so that the perpendicular separation between the ray reflected from the top surface and the ray reflected from the bottom surface and exiting the top surface is a maximum. (b) What is this angle of incidence if the index of refraction of the glass is 1.60? What is the separation of the two beams if the thickness of the glass plate is 4.0 cm?

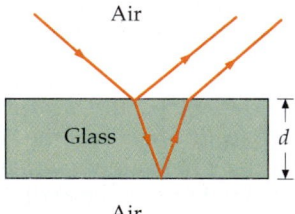

**FIGURE 31-56** Problem 40

## Total Internal Reflection

**41** • What is the critical angle for total internal reflection for light traveling initially in water that is incident on a water–air interface?

**42** •• **iSOLVE** A glass surface ($n = 1.50$) has a layer of water ($n = 1.33$) on it. Light in the glass is incident on the glass–water interface. Find the critical angle for total internal reflection.

**43** •• **iSOLVE✓** A point source of light is located 5 m below the surface of a large pool of water. Find the area of the largest circle on the pool's surface through which light coming directly from the source can emerge.

**44** •• Light is incident normally on the largest face of an isosceles-right-triangle prism. What is the speed of light in this prism if the prism is just barely able to produce total internal reflection?

**45** •• ISOLVE ✓ A point source of light is located at the bottom of a steel tank, and an opaque circular card of radius 6 cm is placed over it. A transparent fluid is gently added to the tank so that the card floats on the surface with its center directly above the light source. No light is seen by an observer above the surface until the fluid is 5 cm deep. What is the index of refraction of the fluid?

**46** •• SSM An optical fiber allows rays of light to propagate long distances through total internal reflection. As shown in Figure 31-57, the fiber consists of a core material with index of refraction $n_2$ and radius $b$, surrounded by a cladding material of index $n_3 < n_2$. The numerical aperture of the fiber is defined as $\sin\theta_1$, where $\theta_1$ is the angle of incidence of a ray of light impinging the end of the fiber that reflects off the core-cladding interface at the critical angle. Using the figure as a guide, show that the numerical aperture is given by

$$\sqrt{n_2^2 - n_3^2}$$

assuming the ray is incident from air. (*Hint: Use of the Pythagorean theorem may be required.*)

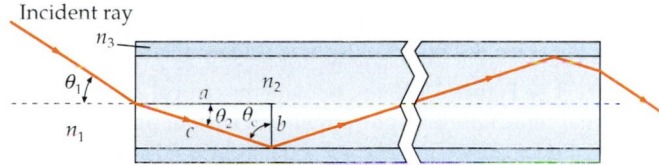

**FIGURE 31-57** Problems 46, 47, and 48

**47** • Find the maximum angle of incidence of a ray that would propagate through an optical fiber with a core index of refraction of 1.492, a core radius of 50 $\mu$m, and a cladding index of 1.489. See Problem 46 and Figure 31-57.

**48** •• Calculate the difference in time needed for two pulses of light to travel down 15 km of the fiber of Problem 47. Assume that one pulse enters the fiber at normal incidence, and the second pulse enters the fiber at the maximum angle of incidence calculated in Problem 47 (see Figure 31-57). In fiber optics, this effect is known as modal dispersion.

**49** ••• Investigate how a thin film of water on a glass surface affects the critical angle for total reflection. Take $n = 1.5$ for glass and $n = 1.33$ for water. (a) What is the critical angle for total internal reflection at the glass–water interface? (b) Is there any range of incident angles that are greater than $\theta_c$ for glass-to-air refraction and for which light rays will leave the glass and the water and pass into the air?

**50** ••• A laser beam is incident on a plate of glass of thickness 3 cm. The glass has an index of refraction of 1.5 and the angle of incidence is 40°. The top and bottom surfaces of the glass are parallel and both produce reflected beams of nearly the same intensity. What is the perpendicular distance $d$ between the two adjacent reflected beams?

## Dispersion

**51** •• SSM ISOLVE A beam of light strikes the plane surface of silicate flint glass at an angle of incidence of 45°. The index of refraction of the glass varies with wavelength, as shown in the graph in Figure 31-26. How much smaller is the

angle of refraction for violet light of wavelength 400 nm than that for red light of wavelength 700 nm?

**52** •• Different colors (frequencies) of light travel at different speeds (a phenomena referred to as dispersion). This can cause problems in fiber-optic communications systems where pulses of light must travel very long distances in glass. Assuming a fiber is made of silicate crown glass, calculate the difference in time needed for two short pulses of light to travel 15 km of fiber if the first pulse has a wavelength of 700 nm and the second pulse has a wavelength of 500 nm.

## Polarization

**53** • ISOLVE ✓ What is the polarizing angle for (a) water with $n = 1.33$ and (b) glass with $n = 1.5$?

**54** • Light known to be polarized in the horizontal direction is incident on a polarizing sheet. It is observed that only 15 percent of the intensity of the incident light is transmitted through the sheet. What angle does the transmission axis of the sheet make with the horizontal? (a) 8.6°, (b) 21°, (c) 23°, (d) 67°, or (e) 81°.

**55** • Two polarizing sheets have their transmission axes crossed so that no light gets through. A third sheet is inserted between the first two so that its transmission axis makes an angle $\theta$ with that of the first sheet. Unpolarized light of intensity $I_0$ is incident on the first sheet. Find the intensity of the light transmitted through all three sheets if (a) $\theta = 45°$ and (b) $\theta = 30°$.

**56** •• A horizontal 5 mW laser beam polarized in the vertical direction is incident on a pair of polarizers. The first is oriented so that its transmission axis is vertical and the second is oriented with its transmission axis at an angle of 27° with respect to the first. What is the power of the transmitted beam?

**57** •• The polarizing angle for a certain substance is 60°. (a) What is the angle of refraction of light incident at this angle? (b) What is the index of refraction of this substance?

**58** •• Two polarizing sheets have their transmission axes crossed and a third sheet is inserted so that its transmission axis makes an angle $\theta$ with that of the first sheet, as in Problem 55. Find the intensity of the transmitted light as a function of $\theta$. Show that the intensity transmitted through all three sheets is maximum when $\theta = 45°$.

**59** •• If the middle polarizing sheet in Problem 58 is rotating at an angular velocity $\omega$ about an axis parallel to the light beam, find the intensity transmitted through all three sheets as a function of time. Assume that $\theta = 0$ at time $t = 0$.

**60** •• SSM A stack of $N + 1$ ideal polarizing sheets is arranged with each sheet rotated by an angle of $\pi/(2N)$ rad with respect to the preceding sheet. A plane, linearly polarized light wave of intensity $I_0$ is incident normally on the stack. The incident light is polarized along the transmission axis of the first sheet and is therefore perpendicular to the transmission axis of the last sheet in the stack. (a) Show that the transmitted intensity through the stack is given by the expression

$$I_0 \cos^{2N}\left(\frac{\pi}{2N}\right)$$

(b) Using a spreadsheet or graphing program, plot the transmitted intensity as a function of $N$ for values of $N$ from 2 to 100. (c) What is the direction of polarization of the transmitted beam in each case?

**61** •• The device described in Problem 60 could serve as a polarization *rotator,* one that takes the linear plane of polarization from one direction to another. The efficiency of such a device is measured by taking the ratio of the output intensity at the desired polarization to the input intensity. The result of the previous problem suggests that the best way to do this would be to use a large number $N$. But in a real polarizer, a small amount of intensity is lost regardless of the input polarization. For each polarizer, assume the transmitted intensity is 98 percent of the amount predicted by the law of Malus and use a spreadsheet or graphing program to determine the optimum number of sheets you should use to rotate the polarization 90°.

**62** •• SSM Show that a linearly polarized wave can be thought of as a superposition of a right and a left circularly polarized wave.

**63** •• Suppose that the middle sheet in Problem 55 is replaced by two polarizing sheets. If the angles between the directions of polarization of adjacent sheets is 30°, what is the intensity of the transmitted light? How does this compare with the intensity obtained in Problem 55, Part (a)?

**64** •• SSM Show that the electric field of a circularly polarized wave propagating in the $x$ direction can be expressed by

$$\vec{E} = E_0 \sin(kx - \omega t)\hat{j} + E_0 \cos(kx - \omega t)\hat{k}$$

**65** •• A circularly polarized wave is said to be *right circularly polarized* if the electric and magnetic fields rotate clockwise when viewed along the direction of propagation and *left circularly polarized* if the fields rotate counterclockwise. What is the sense of the circular polarization for the wave described by the expression in Problem 64? What would be the corresponding expression for a circularly polarized wave of the opposite sense?

## General Problems

**66** • ISOLVE A beam of monochromatic red light with a wavelength of 700 nm in air travels in water. (a) What is the wavelength in water? (b) Does a swimmer underwater observe the same color or a different color for this light?

**67** •• The critical angle for total internal reflection for a substance is 45°. What is the polarizing angle for this substance?

**68** •• SSM Figure 31-58 shows two plane mirrors that make an angle $\theta$ with each other. Show that the angle between the incident and reflected rays is $2\theta$.

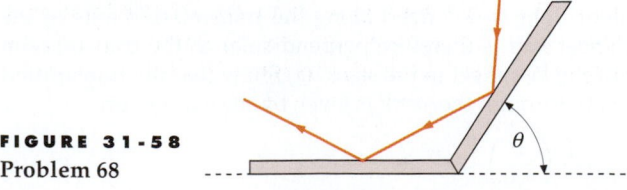

**FIGURE 31-58**
Problem 68

**69** •• A silver coin sits on the bottom of a swimming pool that is 4 m deep. A beam of light reflected from the coin emerges from the pool making an angle of 20° with respect to the water's surface and enters the eye of an observer. Draw a ray from the coin to the eye of the observer. Extend this ray, which goes from the water–air interface to the eye, straight back until it intersects with the vertical line drawn through the coin. What is the apparent depth of the swimming pool to this observer?

**70** •• Fishermen always insist on silence because noise on shore will scare fish away. Two gregarious fishermen purposely stand on shore at least 1.00 m back from the edge of a calm lake. The shore rises gradually, so their heads are never more than 2.00 m above the level of the surface of the water. Show that the sounds emanating from their mouths cannot be heard by any fish submerged in the lake. (Assume that sound does not travel through the ground and into the water.) *Hint:* The speed of sound in air is 330 m/s and the speed of sound in water is 1450 m/s.

**71** •• SSM ISOLVE✓ A swimmer at the bottom of a pool 3 m deep looks up and sees a circle of light. If the index of refraction of the water in the pool is 1.33, find the radius of the circle.

**72** •• Show that when a mirror is rotated through an angle $\theta$, the reflected beam of light is rotated through $2\theta$.

**73** •• Use Figure 31-25 to calculate the critical angles for total internal reflection for light initially in silicate flint glass that is incident on a glass–air interface if the light is (a) violet light of wavelength 400 nm and (b) red light of wavelength 700 nm.

**74** •• Show that for normally incident light, the intensity transmitted through a glass slab with an index of refraction of $n$ is approximately given by

$$I_T = I_0 \left[ \frac{4n}{(n + 1)^2} \right]^2$$

**75** •• A ray of light begins at the point $x = -2$ m, $y = 2$ m, strikes a mirror in the $xz$ plane at some point $x$, and reflects through the point $x = 2$ m, $y = 6$ m. (a) Find the value of $x$ that makes the total distance traveled by the ray a minimum. (b) What is the angle of incidence on the reflecting plane? What is the angle of reflection?

**76** •• SSM A Brewster window is used in lasers to preferentially transmit light of one polarization, as shown in Figure 31-59. Show that if $\theta_{P1}$ is the polarizing angle for the $n_1/n_2$ interface, then $\theta_{P2}$ is the polarizing angle for the $n_2/n_1$ interface.

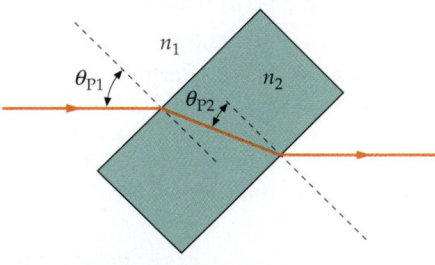

**FIGURE 31-59** Problem 76

**77** •• From the data provided in Figure 31-29, calculate the polarization angle for an air–glass interface, using light of wavelength 550 nm in each of the four types of glass shown in the figure.

**78** ••• Light passes symmetrically through a prism with an apex angle of $\alpha$, as shown in Figure 31-60. (a) Show that the angle of deviation $\delta$ is given by

$$\sin\frac{\alpha + \delta}{2} = n \sin\frac{\alpha}{2}$$

(b) If the refractive index for red light is 1.48 and the refractive index for violet light is 1.52, what is the angular separation of visible light for a prism with an apex angle of 60°?

**FIGURE 31-60**
Problems 78 and 89

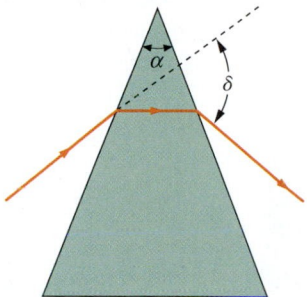

**79** •• **SSM** (a) For a light ray inside a transparent medium that has a planar interface with a vacuum, show that the polarizing angle and the critical angle for internal reflection satisfy $\tan\theta_p = \sin\theta_c$. (b) Which angle is larger?

**80** •• **SOLVE** Light is incident from air on a transparent substance at an angle of 58° with the normal. The reflected and refracted rays are observed to be mutually perpendicular. (a) What is the index of refraction of the transparent substance? (b) What is the critical angle for total internal reflection in this substance?

**81** •• **SOLVE** A light ray in dense flint glass with an index of refraction of 1.655 is incident on the glass surface. An unknown liquid condenses on the surface of the glass. Total internal reflection on the glass–liquid interface occurs for an angle of incidence on the glass–liquid interface of 53.7°. (a) What is the refractive index of the unknown liquid? (b) If the liquid is removed, what is the angle of incidence for total internal reflection? (c) For the angle of incidence found in Part (b), what is the angle of refraction of the ray into the liquid film? Does a ray emerge from the liquid film into the air above? Assume the glass and liquid have perfect planar surfaces.

**82** •• Given that the index of refraction for red light in water is 1.3318 and that the index of refraction for blue light in water is 1.3435, find the angular separation of these colors in the primary rainbow. (Use the equation given in Problem 86.)

**83** •• (a) Use the result from Problem 74 to find the ratio of the transmitted intensity to the incident intensity through $N$ parallel slabs of glass for light of normal incidence. (b) Find this ratio for three slabs of glass with $n = 1.5$. (c) How many slabs of glass with $n = 1.5$ will reduce the intensity to 10 percent of the incident intensity?

**84** •• Light is incident on a slab of transparent material at an angle $\theta_1$, as shown in Figure 31-61. The slab has a thickness $t$ and an index of refraction $n$. Show that

$$n = \frac{\sin\theta_1}{\sin[\arctan(d/t)]}$$

where $d$ is the distance shown in the figure and $\arctan(d/t)$ is the angle whose tangent is $d/t$.

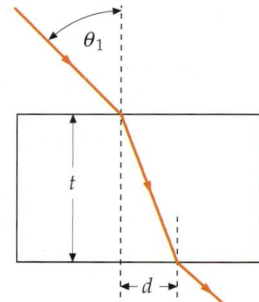

**FIGURE 31-61** Problem 84

**85** •• **SSM** Suppose rain falls vertically from a stationary cloud 10,000 m above a confused marathoner running in a circle with constant speed of 4 m/s. The rain has a terminal speed of 9 m/s. (a) What is the angle that the rain appears to make with the vertical to the marathoner? (b) What is the apparent motion of the cloud as observed by the marathoner? (c) A star on the axis of the earth's orbit appears to have a circular orbit of angular diameter of 41.2 seconds of arc. How is this angle related to the earth's speed in its orbit and the velocity of photons received from this distant star? (d) What is the speed of light as determined from the data in Part (c)?

**86** ••• Equation 31-18 gives the relation between the angle of deviation $\phi_d$ of a light ray incident on a spherical drop of water in terms of the incident angle $\theta_1$ and the index of refraction of water. (a) Assume that $n_{air} = 1$, and differentiate $\phi_d$ with respect to $\theta_1$. [Hint: If $y = \arcsin x$, then $dy/dx = (1 - x^2)^{-1/2}$.] (b) Set $d\phi_d/d\theta_1 = 0$ and show that the angle of incidence $\theta_{1\,m}$ for minimum deviation is given by

$$\cos\theta_{1m} = \sqrt{\frac{n^2 - 1}{3}}$$

and find $\theta_{1m}$ for water, where the index of refraction for water is 1.33.

**87** ••• **SOLVE** Show that a light ray incident on a rectangular glass slab of thickness $t$ at angle of incidence $\theta_1$ emerges parallel to the incident ray but displaced from it by an amount $s$ given by

$$s = \frac{t \sin(\theta_1 - \theta_2)}{\cos(\theta_2)}$$

where $\theta_2$ is the angle of refraction.

**88** •• Use the result of Problem 87 to find the lateral displacement of a laser beam incident at an angle of 30° on a 15 mm thick rectangular slab of glass with index of refraction 1.5.

**89** ••• Show that the angle of deviation $\delta$ is a minimum if the angle of incidence is such that the ray passes through the prism symmetrically, as shown in Figure 31-60.

# Optical Images

NOTE THAT THE PHOTOGRAPH SHOWS EVIDENCE THAT THE HANDS ON THE MIRROR IMAGE OF AN ORDINARY CLOCK ROTATE NOT CLOCKWISE, BUT COUNTERCLOCKWISE. IT IS ALSO TRUE THAT IF YOU COULD LOOK AT A CLOCK FACE FROM BEHIND THE CLOCK IT WOULD LOOK JUST LIKE THE MIRROR IMAGE OF THE CLOCK FACE.

**?** **When looked at from behind, do the hands of a clock rotate clockwise or counterclockwise? A number of observations concerning mirror images are discussed in Section 32-1.**

32-1  Mirrors

32-2  Lenses

*32-3  Aberrations

*32-4  Optical Instruments

**B**ecause the wavelength of light is very small compared with most obstacles and openings, diffraction—the bending of waves around corners—is often negligible, and the ray approximation, in which waves are considered to propagate in straight lines, accurately describes observations.
➤ **In this chapter, we apply the laws of reflection and refraction to the formation of images by mirrors and lenses.**

## 32-1 Mirrors

### Plane Mirrors

Figure 32-1 shows a bundle of light rays emanating from a point source $P$ and reflected from a plane mirror. After reflection, the rays diverge exactly as if they came from a point $P'$ behind the plane of the mirror. The point $P'$ is called the **image** of the **object** $P$. When these reflected rays enter the eye, they cannot be distinguished from rays diverging from a source at $P'$ with no mirror present. This image at $P'$ is called a **virtual image** because the light does not actually emanate from it. The plane of the mirror is the perpendicular bisector of the line from the object point $P$ to the image point $P'$ as shown. The image can be seen by an eye located anywhere in the shaded region indicated, in which a straight line from

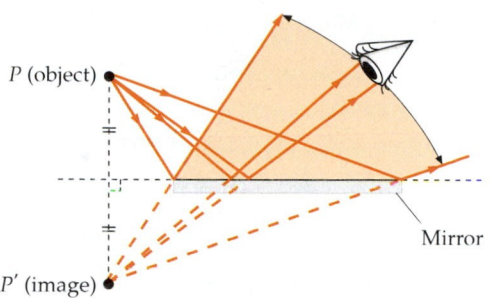

**FIGURE 32-1** Image formed by a plane mirror. The rays from point $P$ that strike the mirror and enter the eye appear to come from the image point $P'$ behind the mirror. The image can be seen by an eye located anywhere in the shaded region.

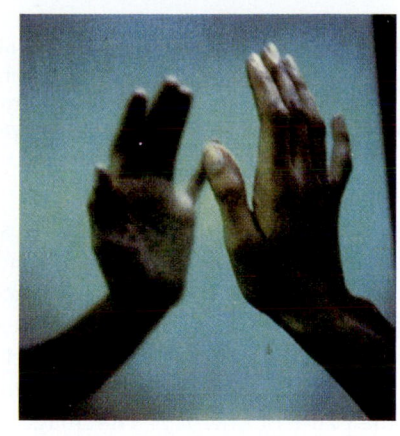

**FIGURE 32-2** The image of a right hand in a plane mirror is transformed to a left hand. This right-to-left reversal is a result of depth inversion.

the image to the eye passes through the mirror. The object need not be directly in front of the mirror. As long as the object is not behind the plane of the mirror, there is some location at which the eye can be located to view the image.

If you hold up your right hand and look in the mirror, the image you see is neither magnified nor reduced, but it looks like a left hand (Figure 32-2). This right-to-left reversal is a result of **depth inversion**—the hand is transformed from a right hand to a left hand because the front and the back of the hand are reversed by the mirror. Depth inversion is also illustrated in Figure 32-3. Figure 32-4 shows the image of a simple rectangular coordinate system. The mirror transforms a right-handed coordinate system, for which $\hat{i} \times \hat{j} = \hat{k}$, into a left-handed coordinate system, for which $\hat{i} \times \hat{j} = -\hat{k}$.

Figure 32-5 shows an arrow of height $y$ standing parallel to a plane mirror a distance $s$ from the mirror. We can locate the image of the arrowhead (and of any other point on the arrow) by drawing two rays. One ray, drawn perpendicular to the mirror, hits the mirror at point $A$ and is reflected back onto itself. The other ray, making an angle $\theta$ with the normal to the mirror, is reflected, making an equal angle $\theta$ with the $x$ axis. The extension of these two rays back behind the mirror locates the image of the arrowhead, as shown by the dashed lines in the figure. We can see from this figure that the image is the same distance behind the mirror as the object is in front of the mirror, and that the image is upright (points in the same direction as the object) and the image is the same size as the object.

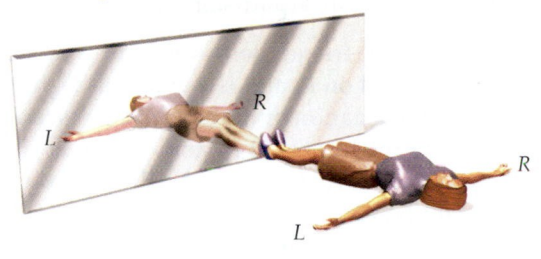

**FIGURE 32-3** A person lying down with her feet against the mirror. The image is depth inverted.

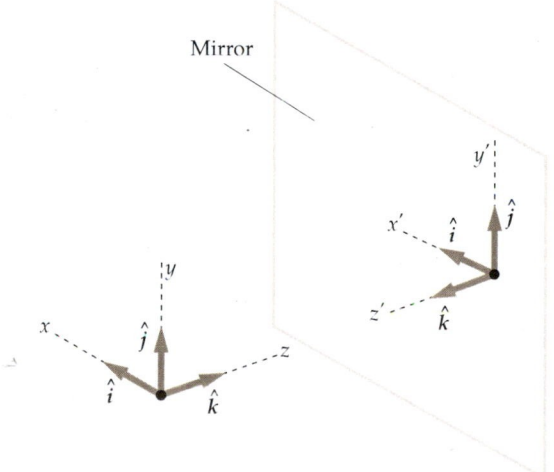

**FIGURE 32-4** Image of a rectangular coordinate system in a plane mirror. The arrow along the $z$ axis is reversed in the image. The image of the original right-handed coordinate system, for which $\hat{i} \times \hat{j} = \hat{k}$, is a left-handed coordinate system, for which $\hat{i} \times \hat{j} = -\hat{k}$.

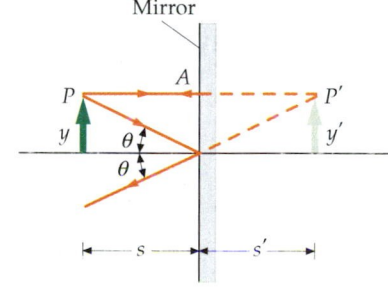

**FIGURE 32-5** Ray diagram for locating the image of an arrow in a plane mirror.

**FIGURE 32-6** Images formed by two plane mirrors. $P'_1$ is the image of the object $P$ in mirror 1, and $P'_2$ is the image of the object in mirror 2. Point $P''_{1,2}$ is the image of $P'_1$ in mirror 2, which is seen when light rays from the object reflect first from mirror 1 and then from mirror 2. The image $P'_2$ does not have an image in mirror 1 because it is behind that mirror.

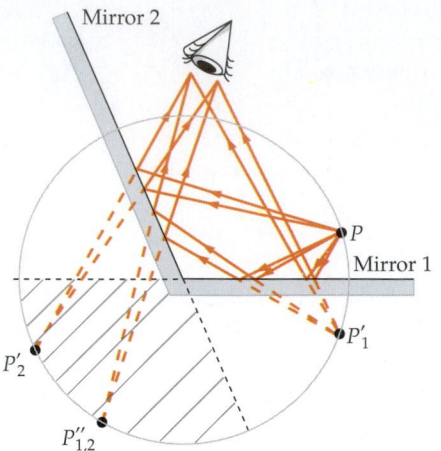

The formation of multiple images by two plane mirrors that make an angle with each other is illustrated in Figure 32-6. We frequently see this phenomenon in clothing stores that provide adjacent mirrors. The light from source point $P$ that is reflected from mirror 1 strikes mirror 2 just as if it came from the image point $P'_1$. The image $P'_1$ is the object for mirror 2. Its image is behind mirror 2 at point $P''_{1,2}$. This image will be formed whenever the image point $P'_1$ is in front of the plane of mirror 2. The image at point $P'_2$ is due to rays from $P$ that reflect directly from mirror 2. Since $P'_2$ is behind the plane of mirror 1, it cannot serve as an object point for a further image in mirror 1. The image at point $P'_2$ cannot serve as an object for mirror 1 because the geometry dictates that none of the rays from $P$ that reflect directly from mirror 2 can then strike mirror 1. An alternative way of stating this is that since $P'_2$ is behind the plane of mirror 1, the image at $P'_2$ cannot serve as an object for mirror one. The number of multiple images formed by two mirrors depends on the angle between the mirrors and the position of the object.

**EXERCISE** Show that a source point and all consequent image points formed by two plane mirrors are equidistant from the intersection of the two planes.

Suppose your friend Ben is standing at point $P$, is wearing a sweatshirt with BEN printed on it, and is waving his right hand. Also, suppose that you are standing at the location of the eye. You can see an image of Ben at all three image locations. For the images at $P'_1$ and $P'_2$, Ben is waving his left hand and the printing on his sweatshirt appears as **NƎƎ**. However, for the image at $P''_{1,2}$, Ben is waving his right hand and the printing appears as BEN. For the image at $P''_{1,2}$ depth inversion occurs twice, once for each reflection, so the result is as if no depth inversion occurs.

**EXERCISE** Which of the images of himself can Ben see? (*Answer* Ben can see only the image at $P'_1$.)

Figure 32-7 illustrates the fact that a horizontal ray reflected from two perpendicular vertical mirrors is reflected back along a parallel path no matter what angle the ray makes with the mirrors. If three mirrors are placed perpendicular to each other, like the sides of an inside corner of a box, any ray incident on any of the mirrors from any direction is reflected back on a path parallel to that of the incident

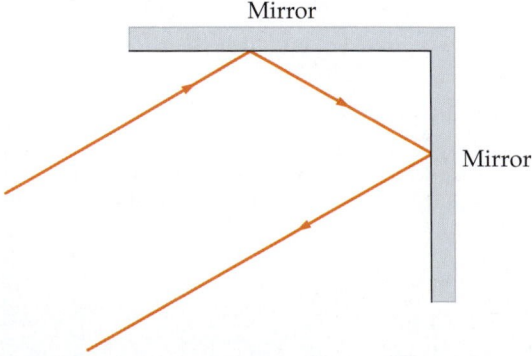

**FIGURE 32-7** A horizontal ray striking one of two perpendicular plane mirrors is reflected from the second mirror in the direction opposite the original direction for any angle of incidence.

ray. A set of three mirrors arranged in this manner is called a corner-cube reflector. An array of corner-cube reflectors was placed on the moon in the Sea of Tranquility by the Apollo 11 astronauts in 1969. A laser beam from the earth that is directed at the mirrors is reflected back to the same place on the earth. Such a beam has been used to measure the distance to the mirrors to within a few centimeters by measuring the time it takes for the light to reach the mirrors and return.

## Spherical Mirrors

Figure 32-8 shows a bundle of rays from a point source $P$ on the axis of a concave spherical mirror reflecting from the mirror and converging at point $P'$. (A concave mirror is shaped like a cave when you look into it.) The rays then diverge from this point, just as if there were an object at the point. This image is called a **real image,** because light actually does emanate from the image point. The image can be seen by an eye at the left of the image looking into the mirror. It could also be observed on a small viewing screen[†] or on a small piece of photographic film placed at the image point. A virtual image, such as that formed by a plane mirror as discussed in the previous section, cannot be observed on a screen at the image point because there is no light at the image point. Despite this distinction between real and virtual images, the eye makes no distinction between them. The light rays diverging from a real image and those appearing to diverge from a virtual image are the same to the eye.

From Figure 32-9, we can see that only rays that strike the spherical mirror at points near the axis $AV$ are reflected through the image point. Rays almost parallel with the axis and near to it are called **paraxial rays.** Rays that strike the mirror at points far from the axis upon reflection pass near the image point, but not through it. Such rays cause the image to appear blurred, an effect called **spherical aberration.** The image can be sharpened by blocking off all but the central part of the mirror, so that rays far from the axis do not strike it. The image is then sharper, but its brightness is reduced because less light is reflected to the image point.

We wish to obtain an equation relating the position of the image point to the position of the object point. To do this, we draw two rays (Figure 32-10a) from an arbitrarily positioned object point $P$. One ray passes through point $C$, the center of curvature of the mirror, and the other ray strikes point $A$, an arbitrarily positioned point on the mirror. The image point $P'$ is where these two rays intersect after reflecting off the mirror. Using the law of reflection, we obtain the location of $P'$. The ray passing through point $C$ strikes the mirror at normal incidence, so the ray reflects back upon itself. The ray striking the mirror at $A$ makes angle $\theta$ with the normal, so, as shown, the reflected ray also makes angle $\theta$ with the normal. (Any line normal to a spherical surface passes through the center of curvature.) The image distance $s'$ and object distance $s$ are measured from the plane tangent to the mirror at its vertex $V$. The angle $\beta$ is an exterior angle to the triangle $PAC$, therefore, $\beta = \alpha + \theta$. Similarly, from the triangle $PAP'$, $\gamma = \alpha + 2\theta$. Eliminating $\theta$ from these equations gives $2\beta = \alpha + \gamma$. By assuming all rays are paraxial, we can substitute using the small-angle approximations: $\alpha \approx \ell/s$, $\beta \approx \ell/r$, and $\gamma \approx \ell/s'$. Equation 32-1 follows directly:

$$\frac{1}{s} + \frac{1}{s'} = \frac{2}{r}$$

32-1

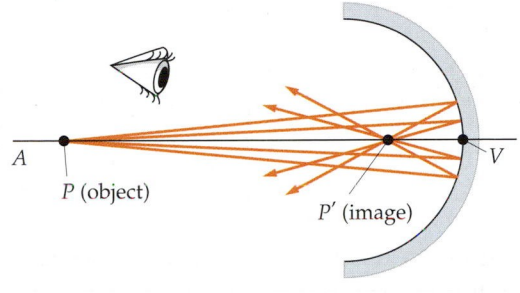

**FIGURE 32-8** Rays from a point object $P$ on the axis $AV$ of a concave spherical mirror form an image at $P'$. The image is sharp if the rays strike the mirror near the axis and if the rays are almost parallel with the axis.

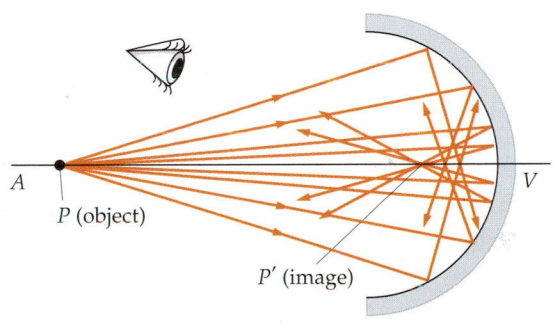

**FIGURE 32-9** Spherical aberration of a mirror. Nonparaxial rays that strike the mirror at points far from the axis $AV$ are not reflected through the image point $P'$ formed by the paraxial rays. The nonparaxial rays blur the image.

---

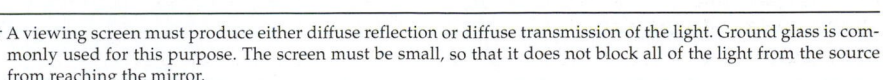

† A viewing screen must produce either diffuse reflection or diffuse transmission of the light. Ground glass is commonly used for this purpose. The screen must be small, so that it does not block all of the light from the source from reaching the mirror.

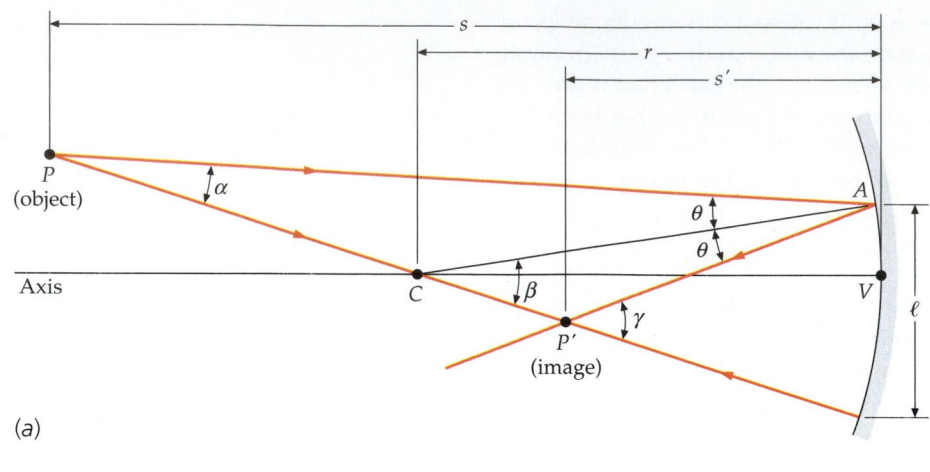

(a)

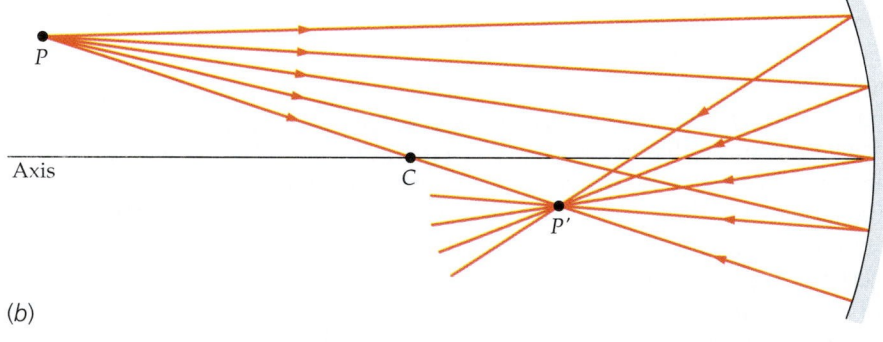

(b)

**FIGURE 32-10** (*a*) Geometry for calculating the image distance $s'$ from the object distance $s$ and the radius of curvature $r$. The angle $\beta$ is an exterior angle to the triangle $PAC$; therefore, $\beta = \alpha + \theta$. Similarly, from the triangle $PAP'$, $\gamma = \alpha + 2\theta$. Eliminating $\theta$ from these equations gives $2\beta = \alpha + \gamma$. Equation 32-1 follows directly, if we assume the following small-angle approximations: $\alpha \approx \ell/s$, $\beta \approx \ell/r$, and $\gamma \approx \ell/s'$. (*b*) All paraxial rays from object point $P$ pass through image point $P'$ after reflecting off the mirror.

This equation relates the object and image distances with the radius of curvature. The striking thing about this equation is that it contains absolutely nothing about the location of point $A$. Therefore, the equation is valid for *any* choice for the location of point $A$, as long as point $A$ is on the surface of the mirror and all rays are paraxial. That is, as shown in Figure 32-10*b*, *all* paraxial rays emanating from an object point will, upon reflection, pass through a *single* image point.

Equation 32-1 specifies the image position in terms of its distance from the mirror. We now specify the image position in terms of its distance from the axis. We first draw a single ray (Figure 32-11) that reflects off the mirror at its vertex. The two right triangles formed are similar. Corresponding sides of similar triangles are equal, so

$$\frac{y'}{y} = -\frac{s'}{s} \qquad\qquad 32\text{-}2$$

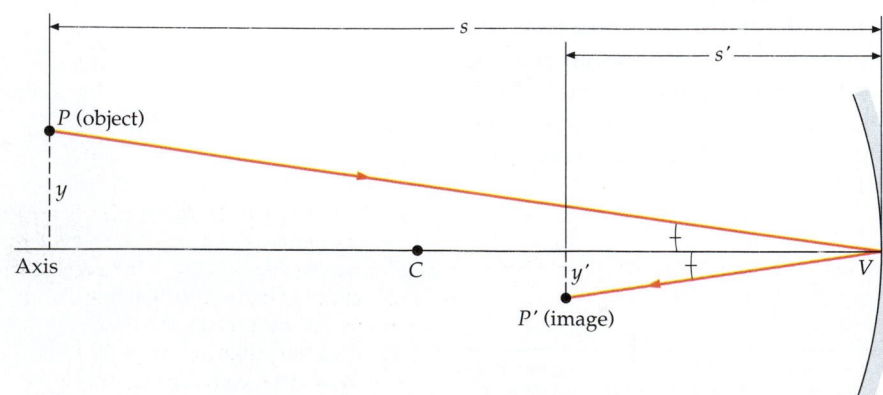

**FIGURE 32-11** Geometry for calculating the position $y'$ of the image point with respect to its distance from the axis.

The negative sign takes into account that $y'/y$ is negative as $P$ and $P'$ are on opposite sides of the axis. Thus, if $y$ is positive, $y'$ is negative and if $y$ is negative, $y'$ is positive.

**EXERCISE** For the image point and object point shown in Figure 32-11, show that

$$\frac{y'}{y} = -\frac{r/2}{s - (r/2)}$$

(*Hint:* Solve Equation 32-1 for $s'$ and substitute your result into Equation 32-2.)

When the object distance is large compared with the radius of curvature of the mirror, the term $1/s$ in Equation 32-1 is much smaller than $2/r$ and can be neglected. That is, as $s \to \infty$, $s' \to \frac{1}{2}r$, where $s'$ is the image distance. This distance is called the **focal length** $f$ of the mirror, and the plane on which parallel rays incident on the mirror are focused is called the **focal plane.** The intersection of the axis with the focal plane is called the **focal point** $F$, as illustrated in Figure 32-12a. (Again, only paraxial rays are focused at a single point.)

$$f = \tfrac{1}{2}r \qquad\qquad\qquad 32\text{-}3$$

FOCAL LENGTH FOR A MIRROR

**EXERCISE** Show that solving Equation 32-1 for $s'$ gives

$$s' = \frac{r}{2 - \dfrac{r}{s}}$$

Then show that as $s \to \infty$, $s' \to \frac{1}{2}r$.

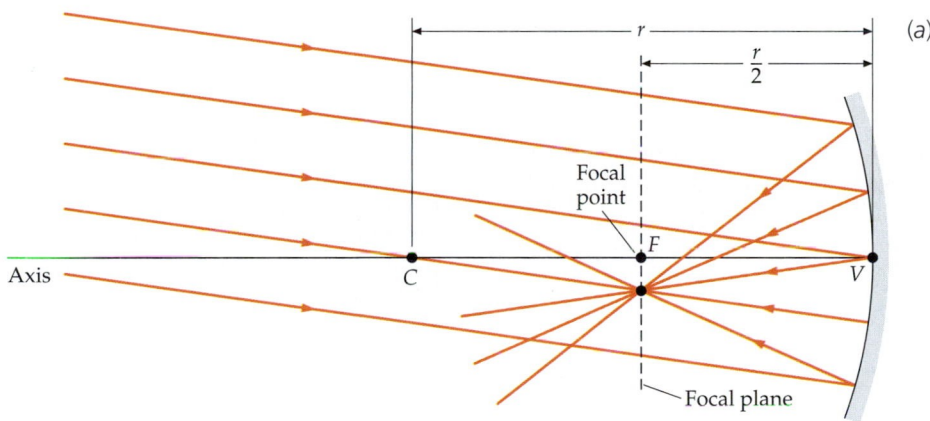

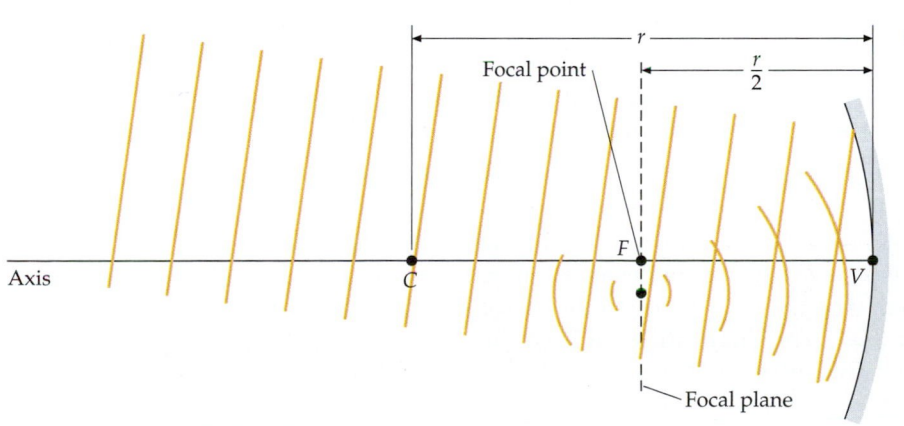

**FIGURE 32-12** (*a*) Parallel rays strike a concave mirror and are reflected to a point on the focal plane a distance $r/2$ to the left of the mirror. (*b*) The incoming wavefronts are plane waves; upon reflection, they become spherical wavefronts that converge to, and then diverge from, the focal point.

The focal length of a spherical mirror is half the radius of curvature. In terms of the focal length $f$, Equation 32-1 is

$$\frac{1}{s} + \frac{1}{s'} = \frac{1}{f} \qquad\qquad 32\text{-}4$$

MIRROR EQUATION

Equation 32-4 is called the **mirror equation**.

When an object point is very far from the mirror, the rays are parallel, and the wavefronts are approximately planes (Figure 32-12$b$). In Figure 32-12$b$, note that the last part of each wavefront to reflect off the concave mirror surface is the part just below the vertex $V$. This results in a spherical wavefront upon reflection. Figure 32-13 shows both the wavefronts and the rays for plane waves striking a convex mirror. In this case, the central part of the wavefront strikes the mirror first, and the reflected waves appear to come from the focal point behind the mirror.

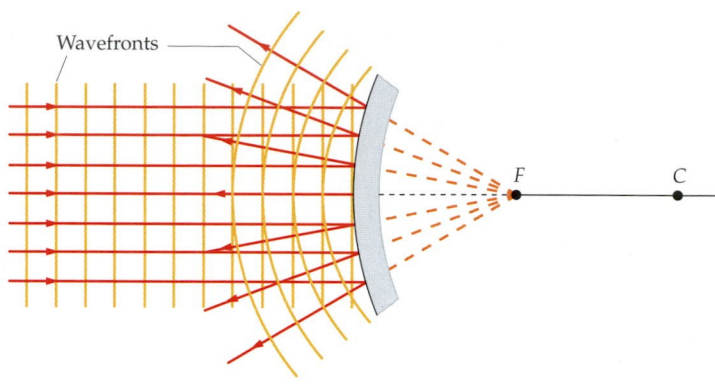

Figure 32-14 illustrates a property of waves called **reversibility.** If we reverse the direction of a reflected ray, the law of reflection assures us that the reflected ray will be along the original incoming ray, but in the opposite direction. (Reversibility holds also for refracted rays, which are discussed in later sections.) Thus, if we have a real image of an object formed by a reflecting (or refracting) surface, we can place an object at the image point and a new, real image will be formed at the position of the original object.

**FIGURE 32-13** Reflection of plane waves from a convex mirror. The outgoing wavefronts are spherical, as if emanating from the focal point $F$ behind the mirror. The rays are normal to the wavefronts and appear to diverge from $F$.

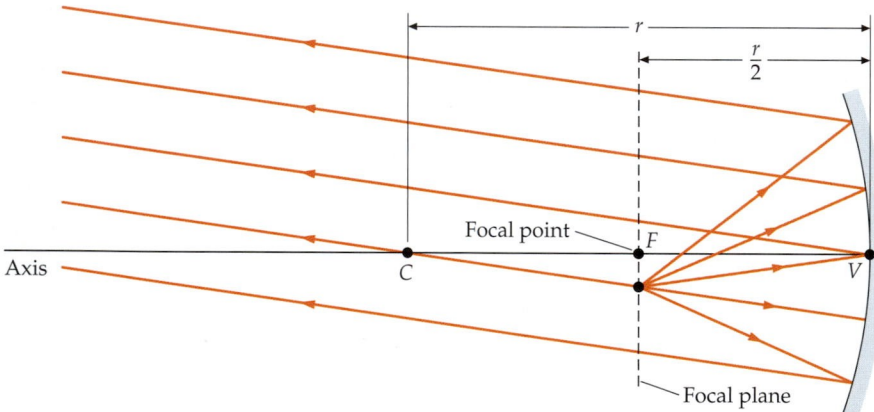

**FIGURE 32-14** Reversibility. Rays diverging from a point source on the focal plane of a concave mirror are reflected from the mirror as parallel rays. The rays follow the same paths as in Figure 32-12$a$ but in the reverse direction.

---

IMAGE IN A CONCAVE MIRROR          **EXAMPLE 32-1**

A point object is 12 cm from a concave mirror and 3 cm above the axis of the mirror. The radius of curvature of the mirror is 6 cm. Find (*a*) the focal length of the mirror and (*b*) the image distance. (*c*) Find the position of the image relative to the axis.

**PICTURE THE PROBLEM** The focal length of a spherical mirror is half the radius of curvature. Once the focal length is known, the image distance can be found using the mirror equation (Equation 32-4), and the distance of the image from the axis can be found using Equation 32-2. The image distance from the mirror is the distance from the plane tangent to the mirror at its vertex.

(*a*) The focal length is half the radius of curvature:

$$f = \tfrac{1}{2}r = \tfrac{1}{2}(6 \text{ cm}) = \boxed{3 \text{ cm}}$$

(*b*) 1. Use the mirror equation to find a relation for the image distance *s'*:

$$\frac{1}{s} + \frac{1}{s'} = \frac{1}{f}$$

or

$$\frac{1}{12 \text{ cm}} + \frac{1}{s'} = \frac{1}{3 \text{ cm}}$$

2. Solve for *s'*:

$$\frac{1}{s'} = \frac{4}{12 \text{ cm}} - \frac{1}{12 \text{ cm}} = \frac{3}{12 \text{ cm}}$$

$$s' = \boxed{4 \text{ cm}}$$

(*c*) 1. Use Equation 32-2 to find the distance *y'* of image from the axis:

$$\frac{y'}{y} = -\frac{s'}{s}$$

2. Solve for *y'*:

$$y' = -\frac{s'}{s}y = -\frac{4 \text{ cm}}{12 \text{ cm}}(3 \text{ cm}) = \boxed{-1 \text{ cm}}$$

**EXERCISE** A concave mirror has a focal length of 4 cm. (*a*) What is the mirror's radius of curvature? (*b*) Find the image distance for an object 2 cm from the mirror. (*Answer* (*a*) 8 cm (*b*) *s'* = −4 cm)

**REMARKS** A negative image distance means the image is on the opposite side of the mirror from the reflected light, as is the case with a plane mirror.

**EXERCISE** What is the radius of curvature of a plane mirror? (*Answer* Infinity)

## Ray Diagrams for Mirrors

A useful method to locate images is by geometric construction of a **ray diagram,** as illustrated in Figure 32-15, where the object is a human figure perpendicular to the axis a distance *s* from the mirror. By the judicious choice of rays from the head of the figure, we can quickly locate the image. There are three **principal rays** that are convenient to use:

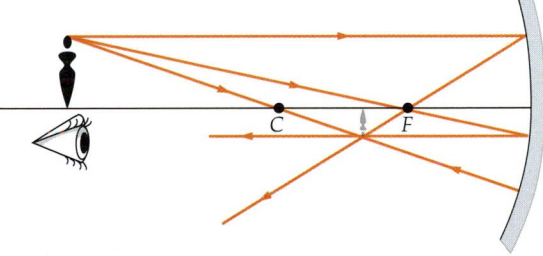

1. The **parallel ray,** drawn parallel to the axis. This ray is reflected through the focal point.

2. The **focal ray,** drawn through the focal point. This ray is reflected parallel to the axis.

3. The **radial ray,** drawn through the center of curvature. This ray strikes the mirror perpendicular to its surface and is thus reflected back on itself.

**FIGURE 32-15** Ray diagram for the location of the image by geometric construction.

PRINCIPAL RAYS FOR A MIRROR

These rays are shown in Figure 32-15. The intersection of any two paraxial rays locates the image point of the head. The three principal rays are easier to draw than any of the other rays. Typically, you draw two of the principal rays to locate the image, and then draw the third principal ray as a check to verify the result. Ray diagrams are best drawn with the mirror replaced by a straight line that extends as far as necessary to intercept the rays, as shown in Figure 32-16. Note that the image in this case is inverted and smaller than the object.

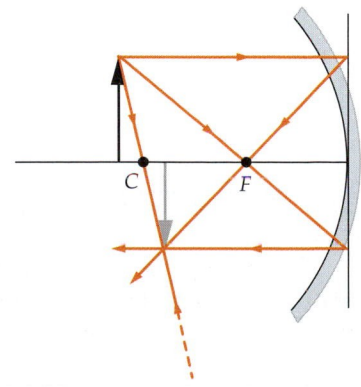

**FIGURE 32-16** Ray diagrams are easier to construct if the curved surface is replaced by a plane tangent to the surface at the vertex.

When the object is between the mirror and its focal point, the rays reflected from the mirror do not converge but appear to diverge from a point behind the mirror, as illustrated in Figure 32-17. In this case, the image is virtual and upright (*upright* meaning not inverted relative to the object). For an object between the mirror and the focal point, $s$ is less than $r/2$, so the image distance $s'$ calculated from Equation 32-1 turns out to be negative. We can apply Equations 32-1, 32-2, 32-3, and 32-4 to this case and to convex mirrors if we adopt a convenient sign convention. Whether the mirror is convex or concave, real images can be formed only in front of the mirror, that is, on the same side of the mirror as the reflected light (and the object). Virtual images are formed behind the mirror where there is no actual light from the object. Our sign convention is as follows:

1. $s$ is positive if the object is on the incident-light side of the mirror.

2. $s'$ is positive if the image is on the reflected-light side of the mirror.

3. $r$ (and $f$) is positive if the mirror is concave so the center of curvature is on the reflected-light side of the mirror.

SIGN CONVENTIONS FOR REFLECTION

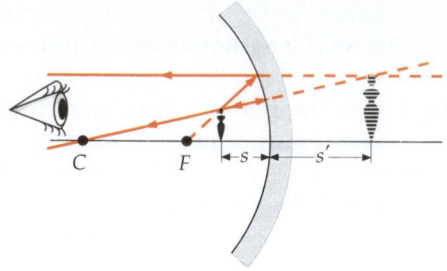

**FIGURE 32-17** A virtual image is formed by a concave mirror when the object is inside the focal point. Here the image is located by the radial ray, which is reflected back on itself, and the focal ray, which is reflected parallel to the axis. The two reflected rays appear to diverge from an image point behind the mirror. This image point is found by constructing extensions to the reflected rays.

The incident-light side and the reflected-light side are, of course, the same. The parameters $s$, $s'$, $r$, and $f$ are all positive if a real object[†] is in front of a concave mirror that forms a real image. A parameter is negative if it does not meet the stated condition for being positive.

The ratio of the image size to the object size is defined as the **lateral magnification** $m$ of the image. From Figure 32-18 and Equation 32-2, we see that the lateral magnification is

$$m = \frac{y'}{y} = -\frac{s'}{s} \tag{32-5}$$

LATERAL MAGNIFICATION

A negative magnification, which occurs when both $s$ and $s'$ are positive, indicates that the image is inverted.

For plane mirrors, the radius of curvature is infinite. The focal length given by Equation 32-3 is then also infinite. Equation 32-4 then gives $s' = -s$, indicating that the image is behind the mirror at a distance equal to the object distance. The magnification given by Equation 32-5 is then $+1$, indicating that the image is upright and the same size as the object.

Although the preceding equations, coupled with our sign conventions, are relatively easy to use, we often need to know only the approximate location and magnification of the image and whether it is real or virtual and upright or inverted. This knowledge is usually easiest to obtain by constructing a ray diagram. It is always a good idea to use both the graphical method and the algebraic method to locate an image, so that one method serves as a check on the results of the other.

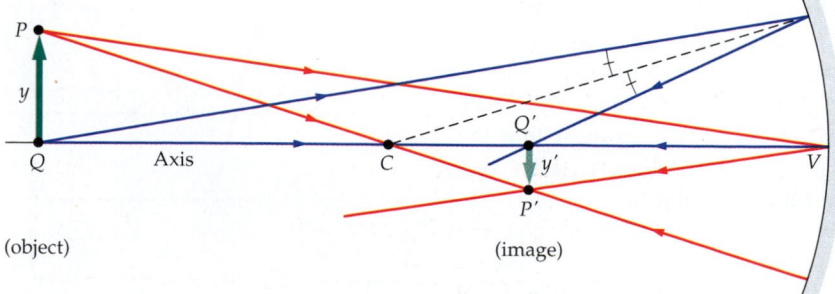

(object)          (image)

**FIGURE 32-18** Geometry for showing the lateral magnification. Rays from the top of the object at $P$, upon reflection, intersect at $P'$; and rays from the bottom of the object at $Q$ intersect at $Q'$, where points $P$ and $P'$ have vertical positions $y$ and $y'$, respectively. The lateral magnification $m$ is given by the ratio $y'/y$. In accord with Equation 32-2, $y'/y = -s'/s$. The minus sign results from the fact that $y'/y$ is negative when $s$ and $s'$ are both positive. A negative $m$ means the image is inverted.

---

† An object is real if it is on the same side of the mirror as the incident light.

**Convex Mirrors** Figure 32-19 shows a ray diagram for an object in front of a convex mirror. The central ray heading toward the center of curvature C is perpendicular to the mirror and is reflected back on itself. The parallel ray is reflected as if it came from the focal point F behind the mirror. The focal ray (not shown) would be drawn toward the focal point and would be reflected parallel to the axis. We can see from the figure that the image is behind the mirror and is therefore virtual. The image is also upright and smaller than the object.

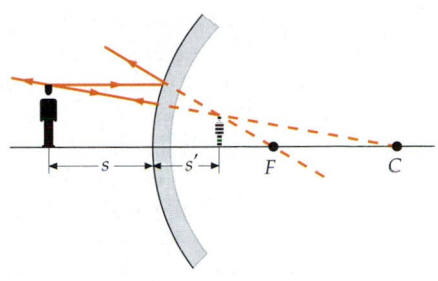

**FIGURE 32-19** Ray diagram for an object in front of a convex mirror.

| IMAGE IN A CONVEX MIRROR | **EXAMPLE 32-2** |
|---|---|

An object 2 cm high is 10 cm from a convex mirror with a radius of curvature of 10 cm. (*a*) Locate the image and (*b*) find the height of the image.

**PICTURE THE PROBLEM** The ray diagram for this problem is the same as shown in Figure 32-19. From this figure, we see that the image is upright, virtual, and smaller than the object. To find the exact location and height of the image, we use the mirror equation, with $s = 10$ cm and $r = -10$ cm.

(*a*) 1. The image distance $s'$ is related to the object distance $s$ and the focal length $f$ by the mirror equation:

$$\frac{1}{s} + \frac{1}{s'} = \frac{1}{f}$$

2. Calculate the focal length of the mirror:

$$f = \tfrac{1}{2}r = \tfrac{1}{2}(-10 \text{ cm}) = -5 \text{ cm}$$

3. Substitute $s = 10$ cm and $f = -5$ cm into the mirror equation to find the image distance:

$$\frac{1}{10 \text{ cm}} + \frac{1}{s'} = \frac{1}{-5 \text{ cm}}$$

4. Solve for $s'$:

$$s' = \boxed{-3.33 \text{ cm}}$$

(*b*) 1. The height of the image is $m$ times the height of the object:

$$y' = my$$

2. Calculate the magnification $m$:

$$m = -\frac{s'}{s} = -\frac{-3.33 \text{ cm}}{10 \text{ cm}} = +0.333$$

3. Use $m$ to find the height of the image:

$$y' = my = (0.333)(2 \text{ cm}) = \boxed{0.666 \text{ cm}}$$

**REMARKS** The image distance is negative, indicating a virtual image behind the mirror. The magnification is positive and less than one, indicating that the image is upright and smaller than the object.

**EXERCISE** Find the image distance and magnification for an object 5 cm away from the mirror in Example 32-2, and draw a ray diagram. (*Answer* $s' = -2.5$ cm, $m = +0.5$; the image is upright, virtual, and reduced in size. The ray diagram is shown in Figure 32-20.)

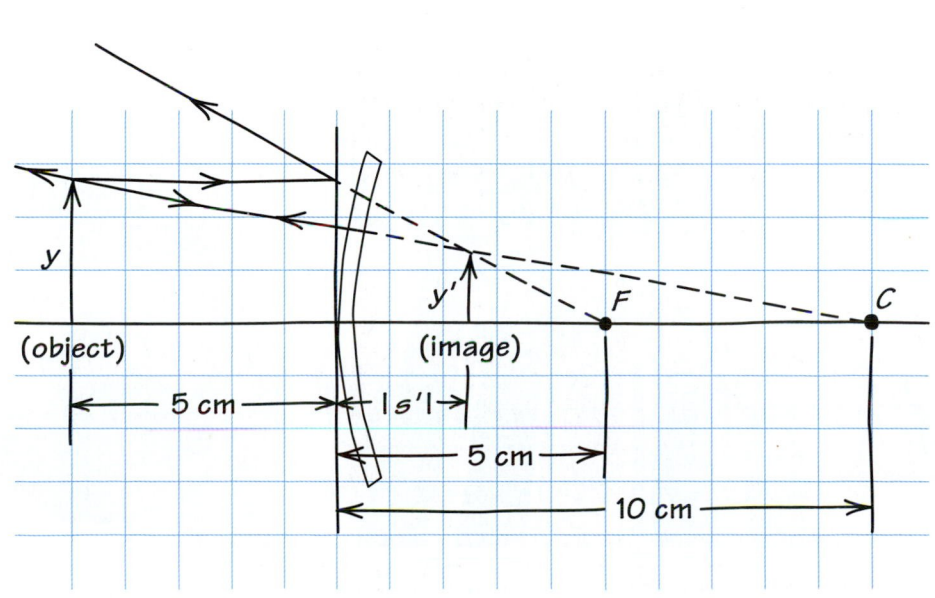

(object)  (image)

5 cm   |s'|

5 cm

10 cm

**FIGURE 32-20**

*DETERMINING THE RANGE*       **EXAMPLE 32-3**    **Put It in Context**

You have a part-time job at Pleasant Hills Golf Course. The fairway of the 16th hole is horizontal for the first 50 yd and then goes down a not-too-steep hill (Figure 32-21), so the people on the tee cannot see the party in front. To prevent people from driving off the tee into the party in front, a convex mirror is mounted on a pole, enabling golfers on the tee to see if the party in front is out of range. Your boss says that a range finder that works by triangulation could be placed facing the mirror, so the golfers could measure how the image of the party in front is behind the mirror. Then the golfers could be given a chart telling them how far the next party is from the tee. Your boss knows you are taking a physics course, so he asks you to calculate the distance of the image behind the mirror if the next party is 250 yd from the tee. The radius of curvature of the mirror has a magnitude of 20 yd.

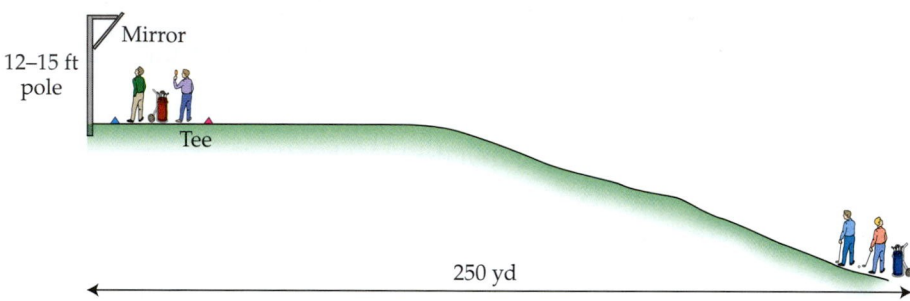

**FIGURE 32-21**

**PICTURE THE PROBLEM** The image distance is related to the object distance by the mirror formula, and the focal length of the lens is half the radius of curvature.

1. Use the mirror equation. For a convex mirror, the radius of curvature is negative:

$$\frac{1}{s} + \frac{1}{s'} = \frac{1}{f}$$

and

$$f = \frac{2}{r}$$

so

$$\frac{1}{250\ \text{yd}} + \frac{1}{s'} = \frac{2}{-20\ \text{yd}}$$

2. The image is 9.62 yd behind the mirror:

$$s' = \boxed{-9.62\ \text{yd}}$$

**EXERCISE** What is the distance to the party in front if the image is 9.75 yd behind the mirror? (*Answer* 390 yd)

(a)

(b)

(*a*) A convex mirror resting on paper with equally spaced parallel stripes. Note the large number of lines imaged in a small space and the reduction in size and distortion in shape of the image. (*b*) A convex mirror is used for security in a store.

# 32-2 Lenses

## Images Formed by Refraction

One end of a long transparent cylinder is machined and polished to form a convex spherical surface. Figure 32-22 illustrates the formation of an image by refraction at such a surface. Suppose the cylinder is submerged in a transparent liquid with index of refraction $n_1$, and suppose the cylinder is made of a plastic material with index of refraction $n_2$, where $n_2$ is greater than $n_1$. Again, only paraxial rays converge to one point. An equation relating the image distance to the object distance, the radius of curvature, and the indexes of refraction can be derived by applying the law of refraction (Snell's law) to these rays and using small-angle approximations. The geometry is shown in Figure 32-23. The result is

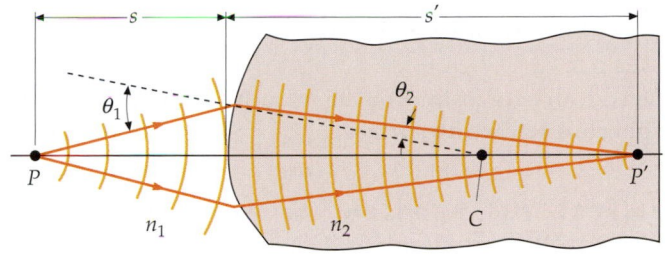

**FIGURE 32-22** Image formed by refraction at a spherical surface between two media where the waves move slower in the second medium.

$$\frac{n_1}{s} + \frac{n_2}{s'} = \frac{n_2 - n_1}{r} \qquad\qquad\qquad 32\text{-}6$$

REFRACTION AT A SINGLE SURFACE

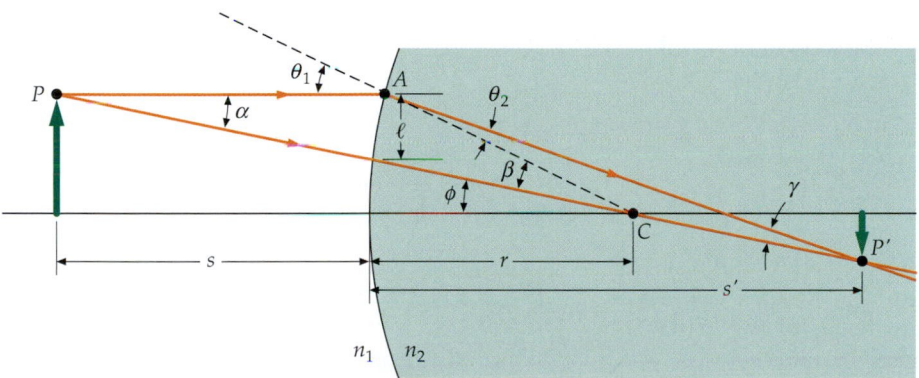

**FIGURE 32-23** Geometry for relating the image position to the object position for refraction at a single spherical surface. The angles $\theta_1$ and $\theta_2$ are related by Snell's law of refraction: $n_1 \sin \theta_1 = n_2 \sin \theta_2$. The small-angle approximation $\sin \theta = \theta$ gives $n_1 \theta_1 = n_2 \theta_2$. From triangle $ACP'$, we have $\beta = \theta_2 + \gamma = (n_1/n_2)\theta_1 + \gamma$. We can obtain another relation for $\theta_1$ from triangle $PAC$: $\theta_1 = \alpha + \beta$. Eliminate $\theta_1$ from these two equations: $n_1\alpha + n_1\beta + n_2\gamma = n_2\beta$. Simplify: $n_1\alpha + n_2\gamma = (n_2 - n_1)\beta$. Using the small-angle approximations $\alpha \approx \ell/s$, $\beta \approx \ell/r$, and $\gamma \approx \ell/s'$ gives Equation 32-6.

In refraction, real images are formed in back of the surface, which we will call the refracted-light side, whereas virtual images occur on the incident-light side, in front of the surface. The sign conventions we use for refraction are similar to those for reflection:

1. $s$ is positive for objects on the incident-light side of the surface.

2. $s'$ is positive for images on the refracted-light side of the surface.

3. $r$ is positive if the center of curvature is on the refracted-light side of the surface.

SIGN CONVENTIONS FOR REFRACTION[†]

---

[†] The sign convention of choice for advanced work on optical design is the Cartesian sign convention. It can readily be found on the Internet.

We see that parameters $s$, $s'$, and $r$ are all positive if a real object is in front of a convex refracting surface that forms a real image. A parameter is negative if it does not meet the stated condition for being positive.

---

*MAGNIFICATION BY A REFRACTING SURFACE*     **EXAMPLE 32-4**    **Try It Yourself**

**Derive an expression for the magnification $m = y'/y$ of an image formed by a spherical refracting surface.**

**PICTURE THE PROBLEM** The magnification is the ratio of $y'$ to $y$. Using Figure 32-18 and Figure 32-23 as guides, draw a ray diagram suitable for this derivation. These heights are related to the tangents of the angles $\theta_1$ and $\theta_2$, as shown in Figure 32-24. The angles are related by Snell's law. For paraxial rays, you can use the approximations $\tan \theta \approx \sin \theta \approx \theta$, and $\cos \theta \approx 1$.

**Cover the column to the right and try these on your own before looking at the answers.**

**Steps**

**Answers**

1. Using Figure 32-18 and Figure 32-23 as guides, draw a ray diagram suitable for this derivation. This drawing should include an object, a real image, a refracting surface, and an axis. Then draw an incident ray from the top of the object to the intersection of the axis with the refracting surface, and draw the refracted ray to the corresponding image point.

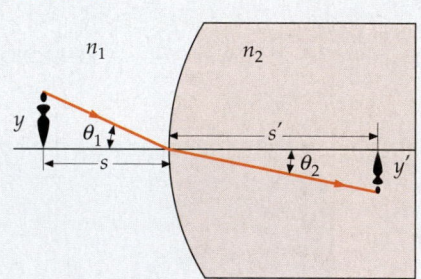

**FIGURE 32-24**

2. Write expressions for $\tan \theta_1$ and $\tan \theta_2$ in terms of the heights $y$ and $-y'$ and the object and image distances $s$ and $s'$. (Since $y'$ is negative, use $-y'$, so that $\tan \theta_2$ is positive.)

$$\tan \theta_1 = \frac{y}{s}; \tan \theta_2 = \frac{-y'}{s'}$$

3. Apply the small-angle approximation $\tan \theta \approx \theta$ to your expressions.

$$\theta_1 = \frac{y}{s}; \theta_2 = \frac{-y'}{s'}$$

4. Write Snell's law of refraction relating the angles $\theta_1$ and $\theta_2$ using the small-angle approximation $\sin \theta \approx \theta$.

$$n_1 \sin \theta_1 = n_2 \sin \theta_2$$
$$n_1 \theta_1 = n_2 \theta_2$$

5. Substitute the expressions for $\theta_1$ and $\theta_2$ found in step 3.

$$n_1 \left( \frac{y}{s} \right) = n_2 \left( \frac{-y'}{s'} \right)$$

6. Solve for the magnification $m = y'/y$.

$$m = \frac{y'}{y} = -\frac{n_1 s'}{n_2 s}$$

---

We see from Example 32-4 that the magnification due to refraction at a spherical surface is

$$m = \frac{y'}{y} = -\frac{n_1 s'}{n_2 s} \qquad\qquad 32\text{-}7$$

MAGNIFICATION FOR A REFRACTING BOUNDARY

IMAGE SEEN FROM A GOLDFISH BOWL                    EXAMPLE 32-5

Goldie the goldfish is in a 15-cm-radius spherical bowl of water with an index of refraction of 1.33. Fluffy the cat is sitting on the table with her nose 10 cm from the surface of the bowl (Figure 32-25). The light from Fluffy's nose is refracted by the air–water boundary to form an image. Find (a) the image distance and (b) the magnification of the image of Fluffy's nose. Neglect any effect of the bowl's thin glass wall.

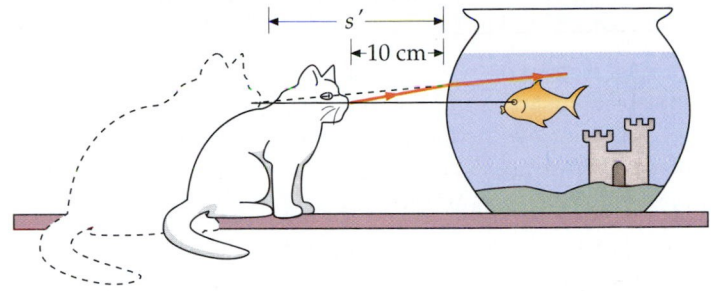

FIGURE 32-25

PICTURE THE PROBLEM We find the image distance $s'$ using Equation 32-6 and the magnification using Equation 32-7. Since we are interested in light that goes from Fluffy's nose to the bowl, it follows that the air–water boundary is convex, and that air is the incident-light side of boundary and water is the refracted-light side of boundary. With these identifications, we have $n_1 = 1$, $n_2 = 1.33$, $s = +10$ cm, and $r = +15$ cm.

(a) 1. The equation relating the object distance to the image distance is Equation 32-6:

$$\frac{n_1}{s} + \frac{n_2}{s'} = \frac{n_2 - n_1}{r}$$

2. Identify and assign signs to the parameters in the previous step:

$n_1 = 1$, $n_2 = 1.33$, $s = +10$ cm, and $r = +15$ cm

3. Substitute numerical values and solve for $s'$:

$$\frac{1}{10 \text{ cm}} + \frac{1.33}{s'} = \frac{1.33 - 1}{15 \text{ cm}}$$

so

$$s' = \boxed{-17.1 \text{ cm}}$$

(b) Substitute numerical values into Equation 32-7 to find the magnification $m$:

$$m = -\frac{n_1 s'}{n_2 s} = -\frac{(1)(-17.1 \text{ cm})}{(1.33)(10 \text{ cm})} = \boxed{1.29}$$

REMARKS Since $s'$ is negative, the image is virtual; that is, the image is on the opposite side of the refracting surface as the refracted light, as shown in Figure 32-25. The fish, Goldie, would see Fluffy to be slightly farther away ($|s'| > s$) than she actually is, and larger ($|m| > 1$) than she actually is. That $m$ is positive indicates the image is upright.

EXERCISE If Goldie is 7.5 cm from the side of the bowl nearest Fluffy, find (a) the location and (b) the magnification of Goldie's image, as seen by Fluffy. (Answer   $n_1 = 1.33$, $n_2 = 1$, $s = 7.5$ cm, $r = -15$ cm; thus, (a) $s' = -6.44$ cm and (b) $m = 1.14$. Fluffy sees Goldie to be slightly closer and slightly larger than she actually is.)

IMAGE SEEN FROM AN OVERHEAD BRANCH                EXAMPLE 32-6

During the summer months, Goldie the fish spends much of her time in a small pond in her owner's backyard. While enjoying a rest at the bottom of the 1-m deep pond, Goldie is being watched by Fluffy the cat, who is perched on a tree limb 3 m above the surface of the pond. How far below the surface is the image of the fish that Fluffy sees? (The index of refraction of water is 1.33.)

**PICTURE THE PROBLEM** The surface of the pond is a spherical refracting surface with an infinite radius of curvature. Thus, Equation 32-6 applies. Since the light reaching Fluffy originates in the water, use $n_1 = \frac{4}{3}$ and $n_2 = 1$.

1. Draw a picture of the situation. Label the object distance and the indexes of refraction of the media. Goldie is the object (Figure 32-26):

2. Using Equation 32-6, relate the image position $s'$ to the other relevant parameters:

$$\frac{n_1}{s} + \frac{n_2}{s'} = \frac{n_2 - n_1}{r}$$

3. The refracting surface is flat. Using $r = \infty$, solve for $s'$:

$$s' = -\frac{n_2}{n_1} s$$

4. Using the given values $n_1 = 1.33$, $n_2 = 1$, and $s = 1$ m, substitute to obtain $s'$:

$$s' = -\frac{1}{1.33}(1\text{ m}) = -0.75\text{ m}$$

That the image is negative means that the image is on the side of the surface opposite the refracted light. That is, it is 0.75 m below the surface.

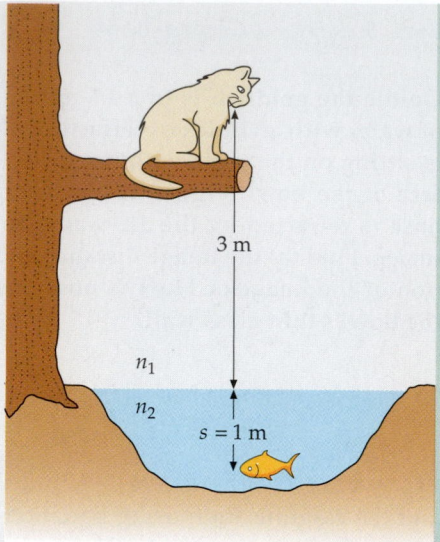

**FIGURE 32-26**

**REMARKS** (1) This image can be seen at the calculated position only when the object is viewed from directly overhead, or nearly so. From that observation point the rays are paraxial, a condition necessary for Equation 32-6 to be valid. If Fluffy is standing on the edge of the pond, the rays will not satisfy the paraxial approximation and Equation 32-6 will not correctly predict the location of the image. (2) The distance $(n_2/n_1)s$ is called the apparent depth of the submerged object. If $n_2 = 1$, the apparent depth equals $s/n_1$.

**EXERCISE** Draw a ray diagram for the image of Goldie, as described in Example 32-6. That is, draw several rays diverging from an object point $P$ on Goldie, and show that after refracting the rays appear to diverge from an image point $P'$ somewhat above the object point (Figure 32-27).

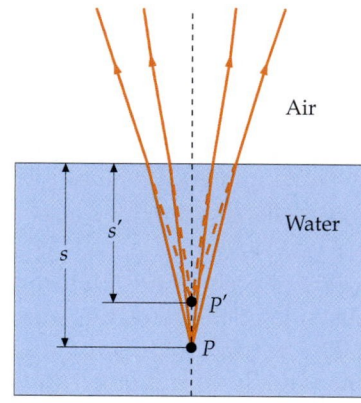

**FIGURE 32-27** Ray diagram for the image of an object in water as viewed from directly overhead. The depth of the image is less than the depth of the object.

## Thin Lenses

The most important application of Equation 32-6 for refraction at a single surface is finding the position of the image formed by a lens. This is done by considering the refraction at each surface of the lens separately to derive an equation relating the image distance to the object distance, the radius of curvature of each surface of the lens, and the index of refraction of the lens.

We will consider a thin lens of index of refraction $n$ with air on both sides. Let the radii of curvature of the surfaces of the lens be $r_1$ and $r_2$. If an object is at a distance $s$ from the first surface (and therefore from the lens), the distance $s_1'$ of the image due to refraction at the first surface can be found using Equation 32-6:

$$\frac{1}{s} + \frac{n}{s_1'} = \frac{n - 1}{r_1}$$

32-8

The light refracted at the first surface is again refracted at the second surface. Figure 32-28 shows the case when the image distance $s_1'$ for the first surface is negative, indicating a virtual image to the left of the surface. Rays in the glass refracted from the first surface diverge as if they came from the image point $P_1'$. The rays strike the second surface at the same angles as if there were an object at image point $P_1'$. The image for the first surface therefore becomes the object for the second surface. Since the lens is of negligible thickness, the object distance is equal in magnitude to $s_1'$. Object distances for objects on the incident-light side of a surface are positive, whereas image distances for images located there are negative. Thus, the object distance for the second surface is $s_2' = -s_1'$.[†] We now write Equation 32-6 for the second surface with $n_1 = n$, $n_2 = 1$, and $s = -s_1'$. The image distance for the second surface is the final image distance $s'$ for the lens:

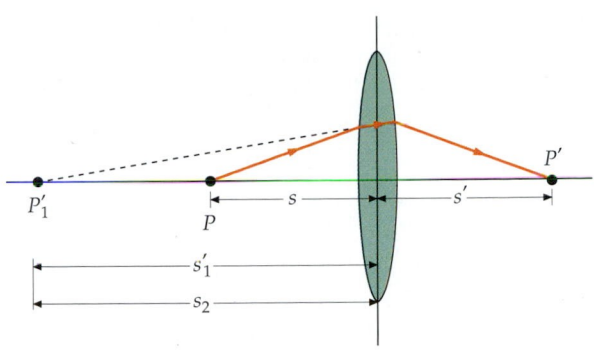

**FIGURE 32-28** Refraction occurs at both surfaces of a lens. Here, the refraction at the first surface leads to a virtual image at $P_1'$. The rays strike the second surface as if they came from $P_1'$. Image distances are negative when the image is on the incident-light side of the surface, whereas object distances are positive for objects located there. Thus, $s_2 = -s_1'$ is the object distance for the second surface of the lens.

$$\frac{n}{-s_1'} + \frac{1}{s'} = \frac{1-n}{r_2}$$

32-9

We can eliminate the image distance for the first surface $s_1'$ by adding Equations 32-8 and 32-9. We obtain

$$\frac{1}{s} + \frac{1}{s'} = (n-1)\left(\frac{1}{r_1} - \frac{1}{r_2}\right)$$

32-10

Equation 32-10 gives the image distance $s'$ in terms of the object distance $s$ and the properties of the thin lens—$r_1$, $r_2$, and $n$. As with mirrors, the focal length $f$ of a thin lens is defined as the image distance when the object distance is infinite. Setting $s$ equal to infinity and writing $f$ for the image distance $s'$, we obtain

$$\frac{1}{f} = (n-1)\left(\frac{1}{r_1} - \frac{1}{r_2}\right)$$

32-11

LENS-MAKER'S EQUATION

Equation 32-11 is called the **lens-maker's equation;** it gives the focal length of a thin lens in terms of the properties of the lens. Substituting $1/f$ for the right side of Equation 32-10, we obtain

$$\frac{1}{s} + \frac{1}{s'} = \frac{1}{f}$$

32-12

THIN-LENS EQUATION

This **thin-lens equation** is the same as the mirror equation (Equation 32-4). Recall, however, that the sign conventions for refraction are somewhat different from those for reflection. For lenses, the image distance $s'$ is positive when the image is on the refracted-light side of the lens, that is, when it is on the side opposite the incident-light side. The sign of the focal length (see Equation 32-11) is determined by the sign convention for a single refracting boundary. That is, $r$ is positive if the center of curvature is on the same side of the surface as the refracted light. For a lens like that shown in Figure 32-28, $r_1$ is positive and $r_2$ is negative, so $f$ is positive.

---

† If $s_1'$ were positive, the rays would be converging as they strike the second surface. The object for the second surface would then be a virtual object located to the right of the second surface. This object would be a virtual object. Again, $s_2 = -s_1'$.

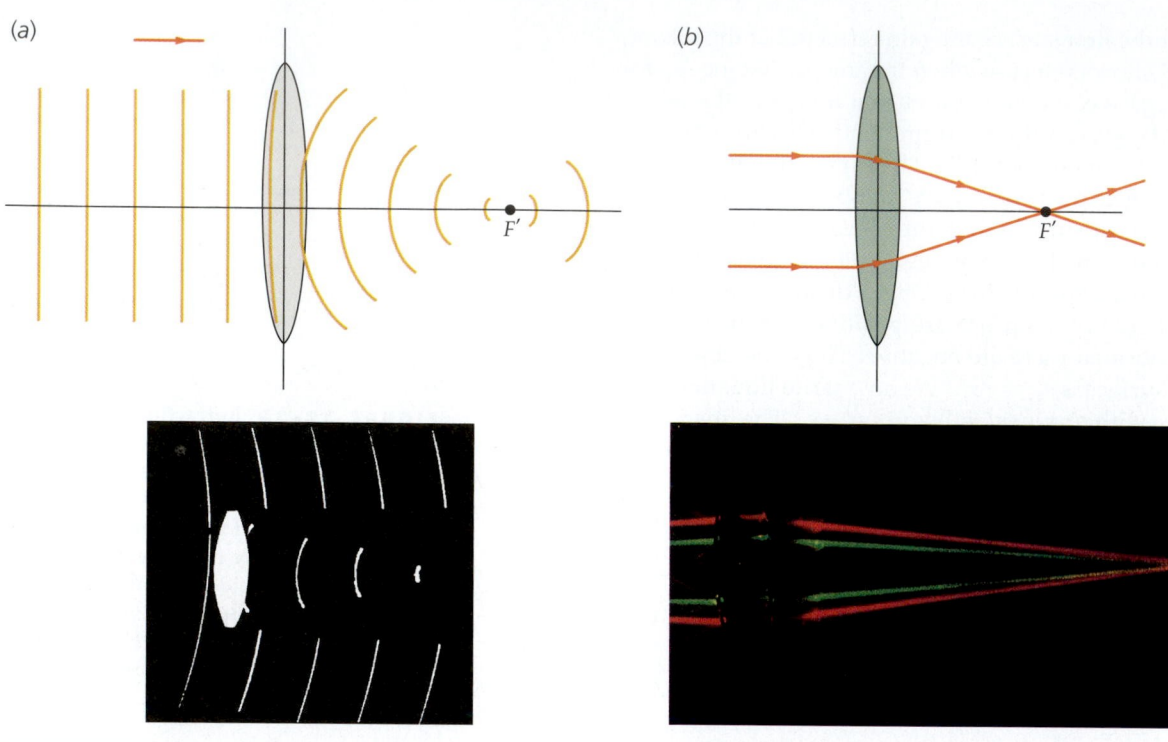

(a)

(b)

**FIGURE 32-29** (a) *Top:* Wavefronts for plane waves striking a converging lens. The central part of the wavefront is retarded more by the lens than the outer part, resulting in a spherical wave that converges at the focal point $F'$. *Bottom:* Wavefronts passing through a lens, shown by a photographic technique called *light-in-flight recording* that uses a pulsed laser to make a hologram of the wavefronts of light. (b) *Top:* Rays for plane waves striking a converging lens. The rays are bent at each surface and converge at the focal point. *Bottom:* A photograph of rays focused by a converging lens.

Figure 32-29a shows the wavefronts of plane waves incident on a double convex lens. The central part of the wavefront strikes the lens first. Since the wave speed in the lens is less than that in air (assuming $n > 1$), the central part of the wavefront lags behind the outer parts, resulting in a spherical wavefront that converges at the focal point $F'$. The rays for this situation are shown in Figure 32-29b. Such a lens is called a **converging lens.** Since its focal length as calculated from Equation 32-11 is positive, it is also called a **positive lens.** Any lens that is thicker in the middle than at the edges is a converging lens (providing that the index of refraction of the lens is greater than that of the surrounding medium). Figure 32-30 shows the wavefronts and rays for plane waves incident on a double concave lens. In this case, the outer part of the wavefronts lag behind the central parts, resulting in outgoing spherical waves that diverge from a focal point on the incident-light side of the lens. The focal length of this lens is negative. Any lens that is thinner in the middle than at the edges is a **diverging,** or **negative,** lens (providing that the index of refraction of the lens is greater than that of the surrounding medium).

**FIGURE 32-30** (a) Wavefronts for plane waves striking a diverging lens. Here, the outer part of the wavefront is retarded more than the central part, resulting in a spherical wave that diverges as it moves out, as if it came from the focal point $F'$ to the left of the lens. (b) Rays for plane waves striking the same diverging lens. The rays are bent outward and diverge, as if they came from the focal point $F'$. (c) A photograph of rays passing through a diverging lens.

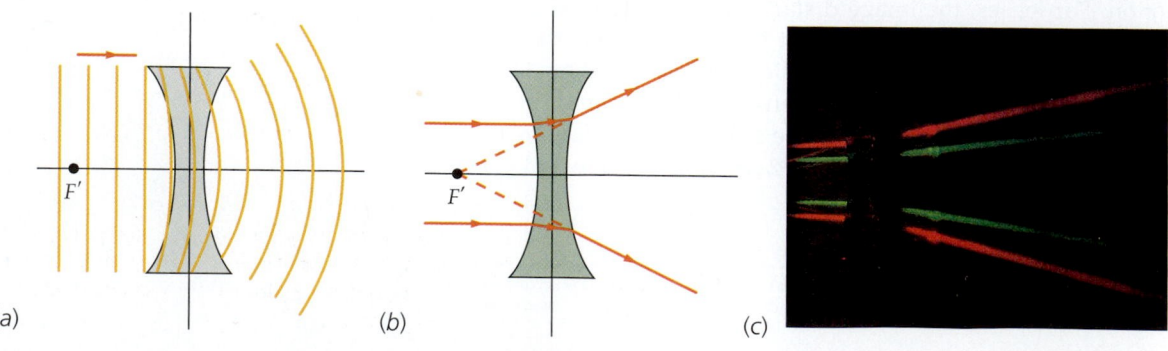

(a)

(b)

(c)

**EXAMPLE  3 2 - 7**

A double convex, thin glass lens with index of refraction $n = 1.5$ has radii of curvature of magnitude 10 cm and 15 cm, as shown in Figure 32-31. Find its focal length.

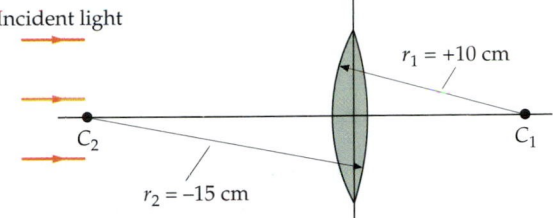

**FIGURE 32-31**

**PICTURE THE PROBLEM** We can find the focal length using the lens-maker's equation (Equation 32-11). Here, light is incident on the surface with the smaller radius of curvature. The center of curvature of this surface, $C_1$, is on the refracted-light side of the lens; thus, $r_1 = +10$ cm. For the second surface, the center of curvature, $C_2$, is on the incident-light side; therefore, $r_2 = -15$ cm.

Numerical substitution in Equation 32-11 yields the focal length $f$:

$$\frac{1}{f} = (n - 1)\left(\frac{1}{r_1} - \frac{1}{r_2}\right)$$

$$= (1.5 - 1)\left(\frac{1}{10 \text{ cm}} - \frac{1}{-15 \text{ cm}}\right) = 0.5\left(\frac{1}{6 \text{ cm}}\right)$$

$$f = \boxed{12 \text{ cm}}$$

**REMARKS** Note that both surfaces tend to converge the light rays; therefore, they both make a positive contribution to the focal length of the lens.

**EXERCISE** A double convex thin lens has an index of refraction $n = 1.6$ and radii of curvature of equal magnitude. If its focal length is 15 cm, what is the magnitude of the radius of curvature of each surface? (*Answer* 18 cm)

**EXERCISE** Show that if you reverse the direction of the incoming light for the lens shown in Example 32-7, so that it is incident on the surface with the greater radius of curvature, you get the same result for the focal length.

If parallel light strikes the lens of Example 32-7 from the left, it is focused at a point 12 cm to the right of the lens; whereas if parallel light strikes the lens from the right, it is focused at 12 cm to the left of the lens. Both of these points are focal points of the lens. Using the reversibility property of light rays, we can see that light diverging from a focal point and striking a lens will leave the lens as a parallel beam, as shown in Figure 32-32. In a particular lens problem in which the direction of the incident light is specified, the object point for which light emerges as a parallel beam is called the **first focal point** $F$, and the point at which parallel light is focused is called the **second focal point** $F'$. For a positive lens, the first focal point is on the incident-light side and the second focal point is on the refracted-light side. If parallel light is incident on the lens at a small angle with the axis, as in Figure 32-33, it is focused at a point in the **focal plane** a distance $f$ from the lens.

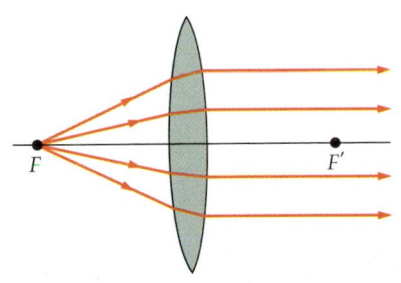

**FIGURE 32-32** Light rays diverging from the focal point of a positive lens emerge parallel to the axis.

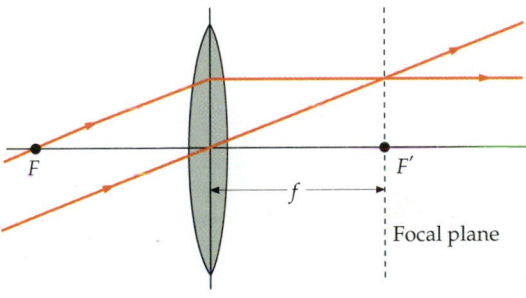

**FIGURE 32-33** Parallel rays incident on the lens at an angle to its axis are focused at a point in the focal plane of the lens.

The reciprocal of the focal length is called the **power of a lens.** When the focal length is expressed in meters, the power is given in reciprocal meters, called **diopters** (D):

$$P = \frac{1}{f} \qquad\qquad\qquad 32\text{-}13$$

The power of a lens measures its ability to focus parallel light at a short distance from the lens. The shorter the focal length, the greater the power. For example, a lens with a focal length of 25 cm = 0.25 m has a power of 4 D. A lens with a focal length of 10 cm = 0.10 m has a power of 10 D. Since the focal length of a diverging lens is negative, its power is negative.

---

*POWER OF A LENS*                         **EXAMPLE 32-8**

The lens shown in Figure 32-34 has an index of refraction of 1.5 and radii of curvature of magnitude 10 cm and 13 cm. Find (*a*) its focal length and (*b*) its power.

**PICTURE THE PROBLEM** For the orientation of the lens relative to the incident light shown in Figure 32-34, the radius of curvature of the first surface is $r_1 = +10$ cm and that of the second surface is $r_2 = +13$ cm.

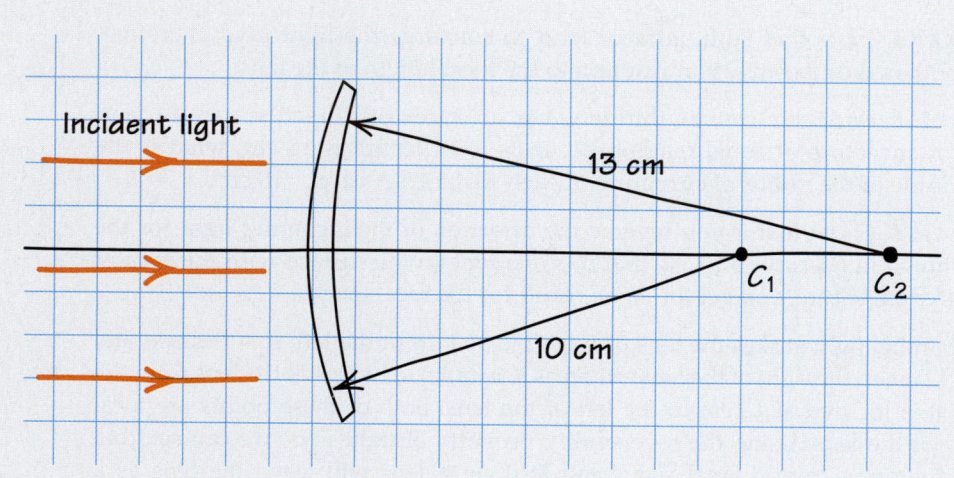

**FIGURE 32-34**

Incident light

13 cm

10 cm

$C_1$   $C_2$

1. Calculate $f$ from the lens-maker's equation using the given value of $n$ and the values of $r_1$ and $r_2$ for the orientation shown:

$$\frac{1}{f} = (n - 1)\left(\frac{1}{r_1} - \frac{1}{r_2}\right)$$

$$= (1.5 - 1)\left(\frac{1}{10\text{ cm}} - \frac{1}{13\text{ cm}}\right)$$

$$f = \boxed{86.7\text{ cm}}$$

2. The power is the reciprocal of the focal length expressed in meters:

$$P = \frac{1}{f} = \frac{1}{0.867\text{ m}} = \boxed{1.15\text{ D}}$$

**REMARKS** We obtain the same result no matter which surface the light strikes first.

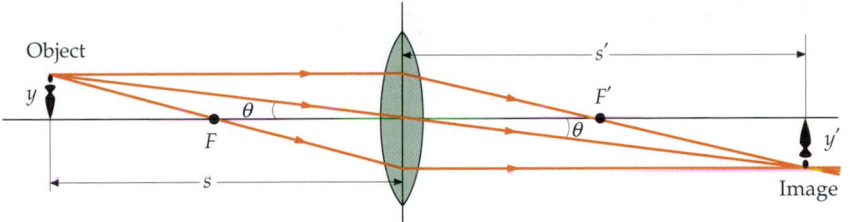

Object
$y$
$\theta$
$F$
$s$
$s'$
$F'$
$\theta$
$y'$
Image

**FIGURE 32-35** Ray diagram for a thin converging lens. We assume that all the bending of light takes place at the central plane. The ray through the center is undeflected because the lens surfaces there are parallel and close together.

In laboratory experiments involving lenses, it is usually much easier to measure the focal length than to calculate the focal length from the radii of curvature of the surfaces.

## Ray Diagrams for Lenses

As with images formed by mirrors, it is convenient to locate the images of lenses by graphical methods. Figure 32-35 illustrates the graphical method for a thin converging lens. We consider the rays to bend at the plane through the center of the lens. The three principal rays are as follows:

1. The **parallel ray,** drawn parallel to the axis. The emerging ray is directed toward (or away from) the second focal point of the lens.

2. The **central ray,** drawn through the center (the vertex) of the lens. This ray is undeflected. (The faces of the lens are parallel at this point, so the ray emerges in the same direction but displaced slightly. Since the lens is thin, the displacement is negligible.)

3. The **focal ray,** drawn through the first focal point.[†] This ray emerges parallel to the axis.

PRINCIPAL RAYS FOR A THIN LENS

The weight and bulk of a large-diameter lens can be reduced by constructing the lens from annular segments at different angles so that light from a point is refracted by the segments into a parallel beam. Such an arrangement is called a Fresnel lens. Several Fresnel lenses are used in this lighthouse to produce intense parallel beams of light from a source at the focal point of the lenses. The illuminated surface of an overhead projector is a Fresnel lens.

These three rays converge to the image point, as shown in Figure 32-35. In this case, the image is real and inverted. From the figure, we have $\tan \theta = y/s = -y'/s'$. The lateral magnification is then

$$m = \frac{y'}{y} = -\frac{s'}{s} \qquad \qquad 32\text{-}14$$

This expression is the same as the expression for mirrors. Again, a negative magnification indicates that the image is inverted. The ray diagram for a diverging lens is shown in Figure 32-36.

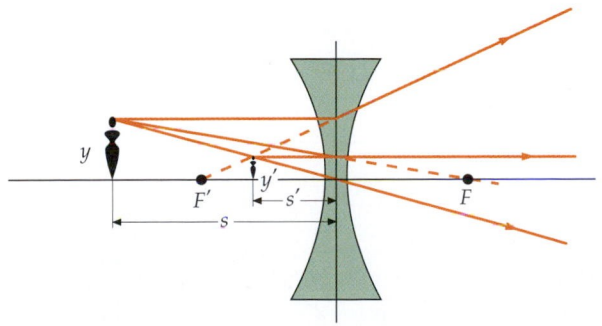

$y$
$F'$
$y'$
$s'$
$F$
$s$

**FIGURE 32-36** Ray diagram for a diverging lens. The parallel ray is bent away from the axis, as if it came from the second focal point $F'$. The ray toward the first focal point $F$ emerges parallel to the axis.

† The focal ray is drawn toward the first focal point for a diverging lens.

## EXAMPLE 32-9

An object 1.2 cm high is placed 4 cm from a double convex lens with a focal length of 12 cm. Locate the image both graphically and algebraically, state whether the image is real or virtual, and find its height. Place an eye on the figure positioned and oriented so as to view the image.

1. Draw the parallel ray. This ray leaves the object parallel to the axis, then is bent by the lens to pass through the second focal point, *F′* (Figure 32-37):

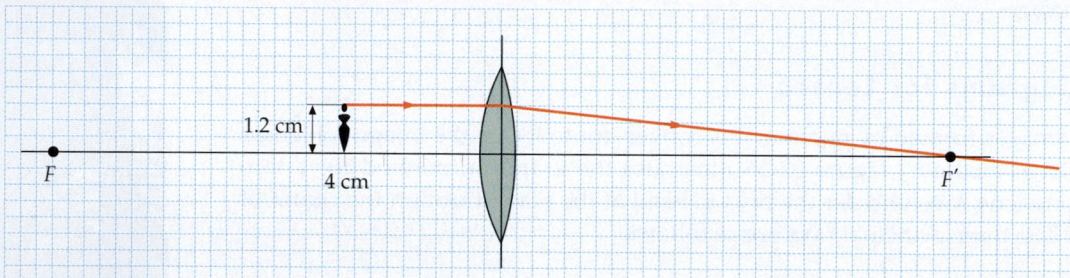

**FIGURE 32-37**

2. Draw the central ray, which passes undeflected through the center of the lens. Since the two rays are diverging on the refracted-light side, we extend them back to the incident-light side to find the image (Figure 32-38):

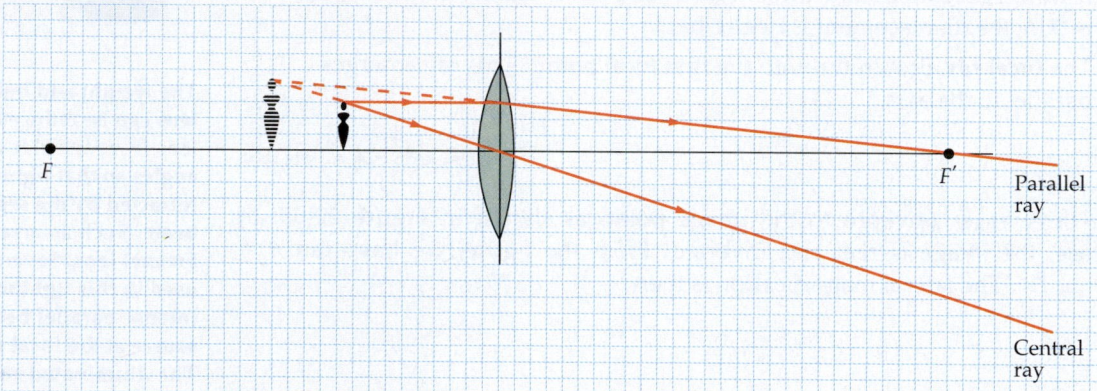

**FIGURE 32-38**

3. As a check, we also draw the focal ray. This ray leaves the object on a line passing through the first focal point, then emerges parallel to the axis. Note that the image is virtual, upright, and enlarged (Figure 32-39):

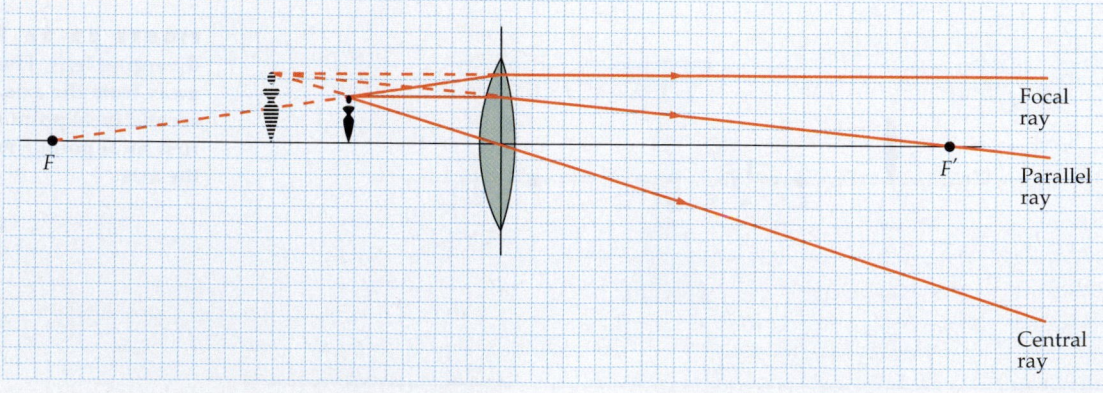

**FIGURE 32-39**

4. The eye must be positioned so the light from the image enters the eye.

5. We now verify the results of the ray diagram algebraically. First, find the image distance using Equation 32-12:

$$\frac{1}{4 \text{ cm}} + \frac{1}{s'} = \frac{1}{12 \text{ cm}}$$

$$\frac{1}{s'} = \frac{1}{12 \text{ cm}} - \frac{1}{4 \text{ cm}} = -\frac{1}{6 \text{ cm}}$$

$$s' = -6 \text{ cm}$$

6. The height of the image is found from the height of the object and the magnification:

$$h' = mh$$

7. The magnification $m$ is given by Equation 32-14:

$$m = -\frac{s'}{s} = -\frac{-6 \text{ cm}}{4 \text{ cm}} = \boxed{+1.5}$$

8. Using this result we find the height of the image, $h'$:

$$h' = mh = (1.5)(1.2 \text{ cm}) = \boxed{1.8 \text{ cm}}$$

**REMARKS** Note the agreement between the algebraic and ray diagram results. Algebraically, we find that the image is 6 cm from the lens on the incident-light side (since $s' < 0$); that is, the image is 2 cm to the left of the object. Since $m > 0$, it follows that the image is upright, and because $m > 1$, the image is enlarged. It is good practice to process lens problems both graphically and algebraically and to compare the results.

**EXERCISE** An object is placed 15 cm from a double convex lens of focal length 10 cm. Find the image distance and the magnification. Draw a ray diagram. Is the image real or virtual? Is the image upright or inverted? (*Answer* $s' = 30$ cm, $m = -2$; real, inverted)

**EXERCISE** Work the previous exercise for an object placed 5 cm from a lens with a focal length of 10 cm. (*Answer* $s' = -10$ cm, $m = 2$; virtual, upright)

## Combinations of Lenses

If we have two or more thin lenses, we can find the final image produced by the system by finding the image distance for the first lens and then using it, along with the distance between the lenses, to find the object distance for the second lens. That is, we consider each image, whether it is real or virtual—and whether it is actually formed or not—as the object for the next lens.

---

*IMAGE FORMED BY A SECOND LENS*                    **EXAMPLE 32-10**

A second lens of focal length +6 cm is placed 12 cm to the right of the lens in Example 32-9. Locate the final image.

**PICTURE THE PROBLEM** The principal rays used to locate the image of the first lens will not necessarily be principal rays for the second lens. In this example, however, we have chosen the position of the second lens (Figure 32-40a) so that the parallel ray for the first lens turns out to be the central ray for the second lens. Also, the focal ray for the first lens emerges parallel to the axis and is therefore the parallel ray for the second lens. If additional principal rays for the second lens are needed, we simply draw them from the image formed by the first lens. For example, in Figure 32-40*b* we added such a ray, drawn from the first image through the first focal point $F_2$ of the second lens.

**FIGURE 32-40**

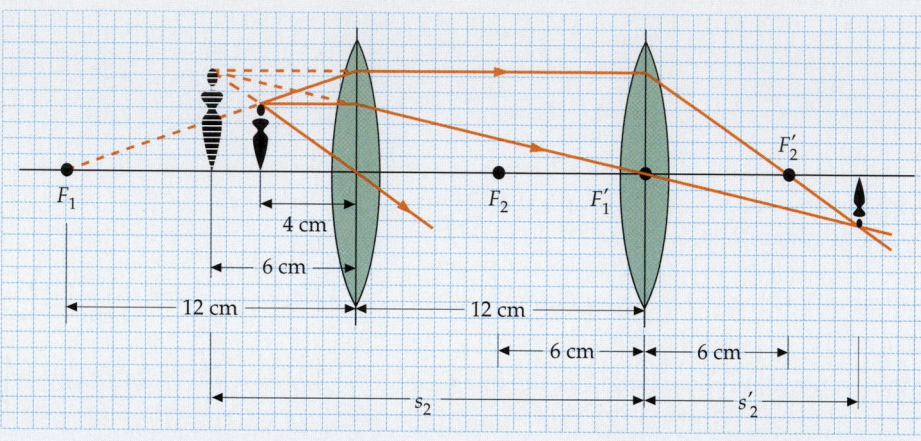

(a)

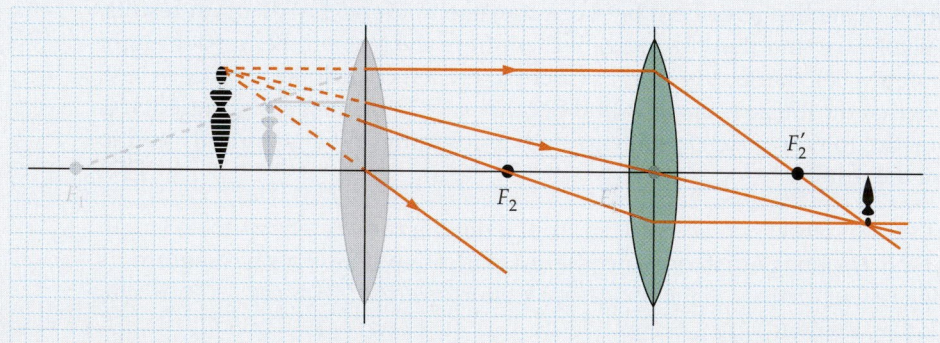

(b)

Algebraically we use $s_2 = 18$ cm, because the first image is 6 cm to the left of the first lens and therefore 18 cm to the left of the second lens.

Use $s_2 = 18$ cm and $f = 6$ cm to calculate $s_2'$:

$$\frac{1}{s_2} + \frac{1}{s_2'} = \frac{1}{f_2}$$

$$\frac{1}{18 \text{ cm}} + \frac{1}{s_2'} = \frac{1}{6 \text{ cm}}$$

$$s_2' = \boxed{9 \text{ cm}}$$

---

*A Combination of Two Lenses*  **EXAMPLE 32-11** **Try It Yourself**

**Two lenses, each of focal length 10 cm, are 15 cm apart. Find the final image of an object 15 cm from one of the lenses.**

**PICTURE THE PROBLEM** Use a ray diagram to find the location of the image formed by lens 1. When these rays strike lens 2 they are further refracted, leading to the final image. More accurate results are obtained algebraically using the thin-lens equation for both lens 1 and lens 2.

Cover the column to the right and try these on your own before looking at the answers.

**Steps**

**Answers**

1. Draw the (*a*) parallel, (*b*) central, and (*c*) focal rays for lens 1 (Figure 32-41). If lens 2 did not alter these rays, they would form an image at $I_1$.

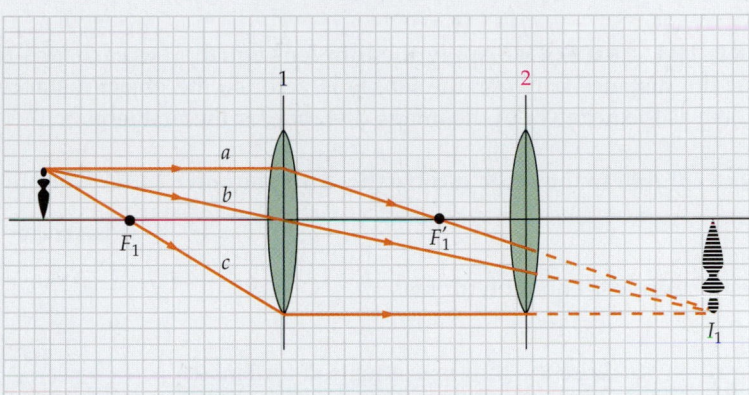

FIGURE 32-41

2. To locate the final image, add three principal rays (*d*, *e*, and *f*) for lens 2. The intersection of these rays gives the image location (Figure 32-42).

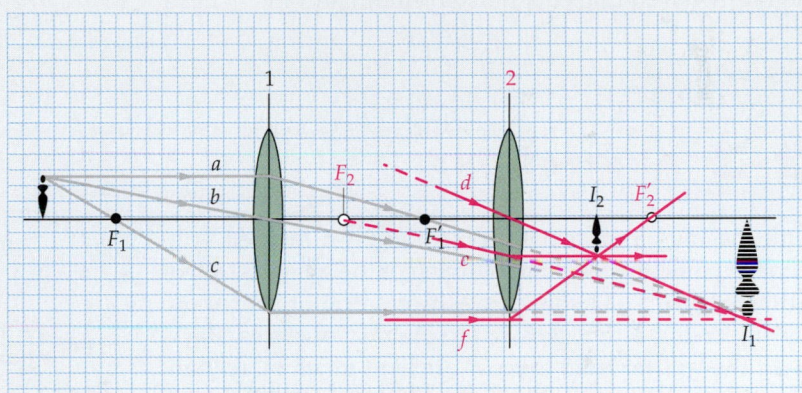

FIGURE 32-42

3. To proceed algebraically, use the thin-lens equation to find the image distance $s_1'$ produced by lens 1.

$s_1' = 30$ cm

4. For lens 2, the image, $I_1$ is 15 cm from the lens on the refracted-light side; hence, $s_2 = -15$ cm. Use this to find the final image distance $s_2'$.

$s_2' = \boxed{6 \text{ cm}}$

**REMARKS** From the ray diagram we see that the final image is real, inverted, and slightly reduced.

## Compound Lenses

When two thin lenses of focal lengths $f_1$ and $f_2$ are placed together, the effective focal length of the combination $f_{eff}$ is given by

$$\frac{1}{f_{eff}} = \frac{1}{f_1} + \frac{1}{f_2}$$

32-15

as is shown in the following Example 32-12. The power of two lenses in contact is given by

$$P_{eff} = P_1 + P_2$$

32-16

For two lenses very close together, derive the relation $\dfrac{1}{f_{eff}} = \dfrac{1}{f_1} + \dfrac{1}{f_2}$.

**PICTURE THE PROBLEM**  Apply the thin-lens equation to each lens using the fact that the distance between the lenses is zero, so the object distance for the second lens is the negative of the image distance for the first lens.

Cover the column to the right and try these on your own before looking at the answers.

| Steps | Answers |
|---|---|
| 1. Write the thin-lens equation for lens 1. | $\dfrac{1}{s} + \dfrac{1}{s_1'} = \dfrac{1}{f_1}$ |
| 2. Using $s_2 = -s_1'$, write the thin-lens equation for lens 2. | $\dfrac{1}{-s_1'} + \dfrac{1}{s'} = \dfrac{1}{f_2}$ |
| 3. Add your two resulting equations to eliminate $s_1'$. | $\boxed{\dfrac{1}{s} + \dfrac{1}{s'} = \dfrac{1}{f_1} + \dfrac{1}{f_2} = \dfrac{1}{f_{eff}}}$ |

# *32-3  Aberrations

When all the rays from a point object are not focused at a single image point, the resultant blurring of the image is called **aberration.** Figure 32-43 shows rays from a point source on the axis traversing a thin lens with spherical surfaces. Rays that strike the lens far from the axis are bent much more than are the rays near the axis, with the result that not all the rays are focused at a single point. Instead, the image appears as a circular disk. The **circle of least confusion** is at point $C$, where the diameter is minimum. This type of aberration in a lens is called **spherical aberration;** it is the same as the spherical aberration of mirrors discussed in Section 32-1. Similar but more complicated aberrations called *coma* (for the comet-shaped image) and *astigmatism* occur when objects are off axis. The aberration in the shape of the image of an extended object that occurs, because the magnification depends on the distance of the object point from the axis, is called **distortion.** We will not discuss these aberrations further, except to point out that they do not arise from any defect in the lens or mirror but instead result from the application of the laws of refraction and reflection to spherical surfaces. These aberrations are not evident in our simple equations, because we used small-angle approximations in the derivation of these equations.

Some aberrations can be eliminated or partially corrected by using nonspherical surfaces for mirrors or lenses, but nonspherical surfaces are usually much more difficult and costly to produce than spherical surfaces. One example of a nonspherical reflecting surface is the parabolic mirror illustrated in Figure 32-44. Rays that are parallel to the axis of a parabolic surface are reflected and focused at a common point, no matter how far the rays are from the axis. Parabolic reflecting surfaces are sometimes used in large astronomical telescopes, which need a large reflecting surface to gather as much light as possible to make the image as intense as possible (reflecting

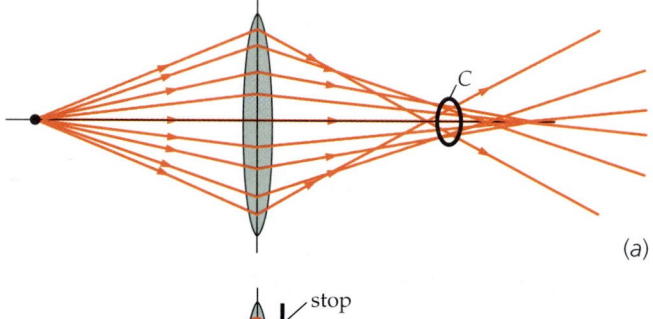

(a)

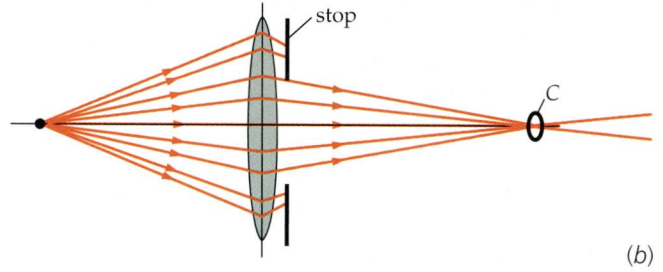

(b)

**FIGURE 32-43**  Spherical aberration in a lens. (*a*) Rays from a point object on the axis are not focused at a point. (*b*) Spherical aberration can be reduced by using a stop to block off the outer parts of the lens, but this also reduces the amount of light reaching the image.

telescopes are described in the upcoming optional Section 32-4). Satellite dishes use parabolic surfaces to focus microwaves from communications satellites. A parabolic surface can also be used in a searchlight to produce a parallel beam of light from a small source placed at the focal point of the surface.

An important aberration found with lenses but not found with mirrors is **chromatic aberration,** which is due to variations in the index of refraction with wavelength. From Equation 32-11, we can see that the focal length of a lens depends on its index of refraction and is therefore different for different wavelengths. Since $n$ is slightly greater for blue light than for red light, the focal length for blue light will be shorter than the focal length for red light. Because chromatic aberration does not occur for mirrors, many large telescopes use a large mirror instead of the large, light-gathering (objective) lens.

Chromatic aberration and other aberrations can be partially corrected by using combinations of lenses instead of a single lens. For example, a positive lens and a negative lens of greater focal length can be used together to produce a converging lens system that has much less chromatic aberration than a single lens of the same focal length. The lens of a good camera typically contains six elements to correct the various aberrations that are present.

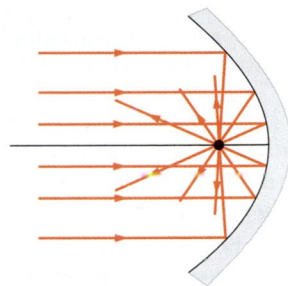

**FIGURE 32-44** A parabolic mirror focuses all rays parallel to the axis to a single point with no spherical aberration.

# *32-4 Optical Instruments

## *The Eye

The optical system of prime importance is the eye, which is shown in Figure 32-45. Light enters the eye through a variable aperture, the pupil. The light is focused by the cornea, with assistance from the lens, on the retina, which has a film of nerve fibers covering the back surface. The retina contains tiny sensing structures called *rods* and *cones,* which detect the light and transmit the information along the optic nerve to the brain. The shape of the crystalline lens can be altered slightly by the action of the ciliary muscle. When the eye is focused on an object far away, the muscle is relaxed and the cornea–lens system has its maximum focal length, about 2.5 cm, which is the distance from the cornea to the retina. When the object is brought closer to the eye, the ciliary muscle increases the curvature of the lens slightly, thereby decreasing its focal length, so that the image is again focused on the retina. This process is called *accommodation.* If the object is too close to the eye, the lens cannot focus the light on the retina and the image is blurred. The closest point for which the lens can focus the image on the retina is called the **near point.** The distance from the eye to the near point varies greatly from one person to another and changes with age. At 10 years, the near point may be as close as 7 cm, whereas at 60 years it may recede to 200 cm because of the loss of flexibility of the lens. The standard value taken for the near point is 25 cm.

If the eye underconverges, resulting in the images being focused behind the retina, the person is said to be farsighted. A farsighted person can see distant objects where little convergence is required, but has trouble seeing close objects. Farsightedness is corrected with a converging (positive) lens (Figure 32-46).

On the other hand, the eye of a nearsighted person overconverges and focuses light from distant objects in front of the retina. A nearsighted person can see nearby objects for which the widely diverging incident rays can be focused on

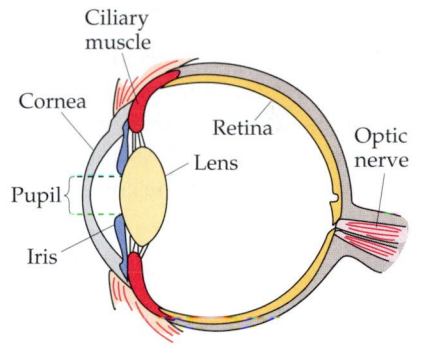

**FIGURE 32-45** The human eye. The amount of light entering the eye is controlled by the iris, which regulates the size of the pupil. The lens thickness is controlled by the ciliary muscle. The cornea and lens together focus the image on the retina, which contains approximately 125 million receptors, called rods and cones, and approximately 1 million optic-nerve fibers.

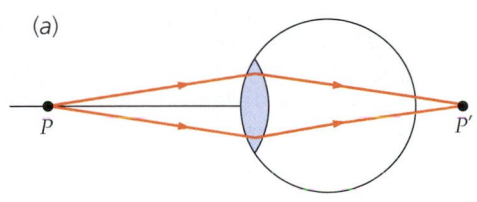

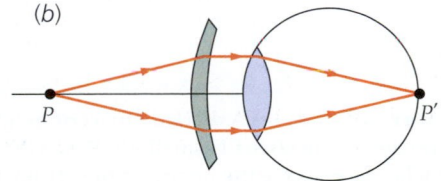

**FIGURE 32-46** (*a*) A farsighted eye focuses rays from a nearby object to a point behind the retina. (*b*) A converging lens corrects this defect by bringing the image onto the retina. These diagrams, and those following, are drawn as if all the focusing of the eye is done at the lens; whereas, in fact, the lens and cornea system act more like a spherical refracting surface than a thin lens.

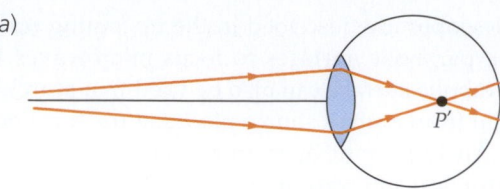

(a)

**FIGURE 32-47** (*a*) A nearsighted eye focuses rays from a distant object to a point in front of the retina. (*b*) A diverging lens corrects this defect.

the retina, but has trouble seeing distant objects clearly. Nearsightedness is corrected with a diverging (negative) lens (Figure 32-47).

Another common defect of vision is astigmatism, which is caused by the cornea being not quite spherical but having a different curvature in one plane than in another. This results in a blurring of the image of a point object into a short line. Astigmatism is corrected by glasses using lenses of cylindrical rather than spherical shape.

(b)

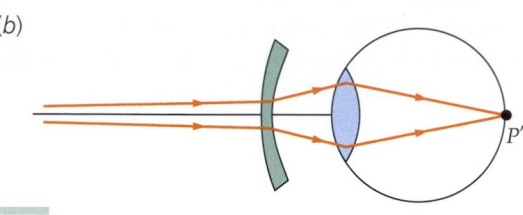

---

*FOCAL LENGTH OF THE CORNEA–LENS SYSTEM*      **EXAMPLE 32-13**

Both a thin lens and a spherical mirror have a focal length given by the formula $\frac{1}{s} + \frac{1}{s'} = \frac{1}{f}$, where $f$ is a constant. Using the same formula, we define the focal length of a spherical refracting surface. However, in this case, the focal length is not constant but depends upon $s$. By how much does the focal length of the cornea–lens system of the eye change if the object is moved from infinity to the near point at 25 cm? Assume that all the focusing is done at the cornea, and that the distance from the cornea to the retina is 2.5 cm.

**PICTURE THE PROBLEM** With the object at infinity, the focal length is 2.5 cm. We use the thin-lens equation to calculate the focal length when $s = 25$ cm and $s' = 2.5$ cm.

1. Use the thin-lens equation to calculate $f$:

$$\frac{1}{f} = \frac{1}{s} + \frac{1}{s'} = \frac{1}{25 \text{ cm}} + \frac{1}{2.5 \text{ cm}}$$

$$= \frac{1}{25 \text{ cm}} + \frac{10}{25 \text{ cm}} = \frac{11}{25 \text{ cm}}$$

so

$$f = 2.27 \text{ cm}$$

2. Subtract the original focal length of 2.5 cm to find the change:    $\Delta f = 2.27 \text{ cm} - 2.5 \text{ cm} = \boxed{-0.23 \text{ cm}}$

**REMARKS** In terms of the power of the cornea–lens system, when the focal length is 2.5 cm = 0.025 m for distant objects, the power is $P = 1/f = 40$ D. When the focal length is 2.27 cm, the power is 44 D.

**EXERCISE** Find the change in the focal length of the eye when an object originally at 4 m is brought to 40 cm from the eye. (Assume that the distance from the cornea to the retina is 2.5 cm.) (*Answer* −0.13 cm)

(a)

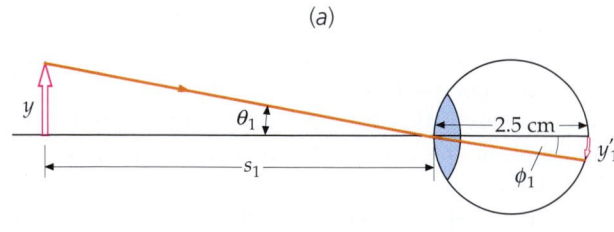

The apparent size of an object is determined by the actual size of the image on the retina. The larger the image on the retina, the greater the number of rods and cones activated. From Figure 32-48, we see that the size of the image on the retina is greater when the

(b)

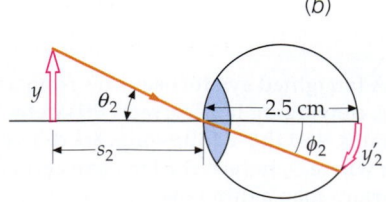

**FIGURE 32-48** (*a*) A distant object of height $y$ looks small because the image on the retina is small. (*b*) When the same object is closer, it looks larger because the image on the retina is larger.

object is close than it is when the object is far away. The apparent size of an object is thus greater when it is closer to the eye. The image size is proportional to the angle $\theta$ subtended by the object at the eye. For Figure 32-48,

$$\phi \approx \frac{y'}{2.5 \text{ cm}} \quad \text{and} \quad \theta \approx \frac{y}{s} \qquad \qquad 32\text{-}17$$

for small angles. Applying the law of refraction gives $n_{\text{Air}} \sin \theta = n \sin \phi$, where $n_{\text{Air}} = 1.00$ and $n$ is the refractive index inside the eye. For small angles this becomes

$$\theta \approx n\phi \qquad \qquad 32\text{-}18$$

Combining Equations 32-17 and 32-18 gives

$$\frac{y}{s} \approx n\frac{y'}{2.5 \text{ cm}}, \quad \text{or} \quad y' \approx \frac{2.5 \text{ cm}}{n}\frac{y}{s} \qquad \qquad 32\text{-}19$$

The size of the image on the retina is proportional to the size of the object and inversely proportional to its distance from the eye. Since the near point is the closest point to the eye for which a sharp image can be formed on the retina, the distance to the near point is called the *distance of most distinct vision.*

---

*READING GLASSES*                                   **E X A M P L E    3 2 - 1 4**

The near-point distance of a person's eye is 75 cm. With a reading glasses lens placed a negligible distance from the eye, the near-point distance of the lens–eye system is 25 cm. That is, if an object is placed 25 cm in front of the lens, then the lens forms a virtual image of the object a distance 75 cm in front of the lens. (*a*) What power is the reading glasses lens and (*b*) what is the lateral magnification of the image formed by the lens? (*c*) Which produces the bigger image on the retina, (1) the object at the near point of, and viewed by, the unaided eye or (2) the object at the near point of the lens–eye system and viewed through the lens that is immediately in front of the eye?

**PICTURE THE PROBLEM**  A near-point distance of the lens–eye system of 25 cm means the lens forms a virtual image 75 cm in front of the lens if an object is placed 25 cm in front of the lens. Figure 32-49*a* shows a diagram of an object 25 cm from a converging lens that produces a virtual, upright image at $s' = -75$ cm. Figure 32-49*b* shows the image on the retina formed by the focusing power of the eye.

**FIGURE 32-49**

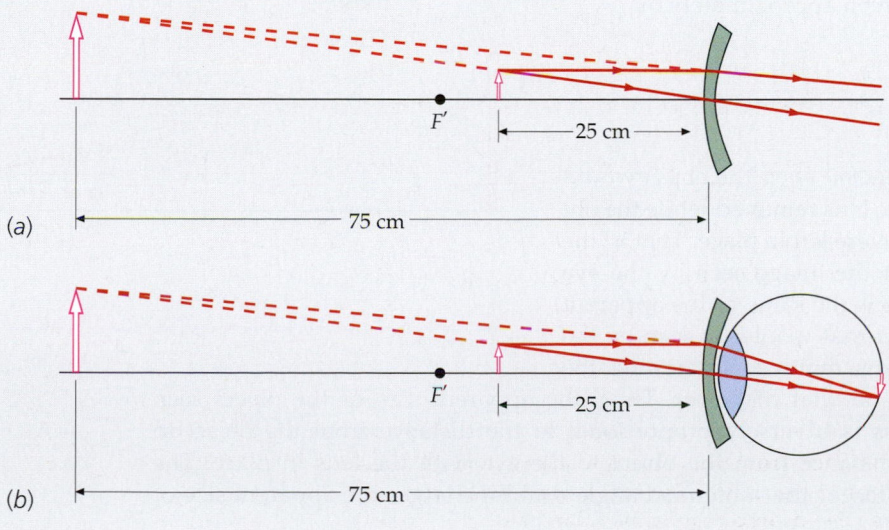

(a)

(b)

1. Use the thin-lens equation with $s = 25$ cm and $s' = -75$ cm to calculate the power, $1/f$:

$$\frac{1}{f} = \frac{1}{s} + \frac{1}{s'} = \frac{1}{25 \text{ cm}} + \frac{1}{-75 \text{ cm}}$$

$$= \frac{2}{75 \text{ cm}} = \frac{2}{0.75 \text{ m}} = \boxed{2.67 \text{ D}}$$

2. Using $m = -s'/s$, find $m$:

$$m = -\frac{s'}{s} = -\frac{-75 \text{ cm}}{25 \text{ cm}} = \boxed{3}$$

3. In both cases, the rays entering the eye appear to diverge from an image 75 cm in front of the eye. However, with the lens in place, the image there is larger by a factor of 3:

$\boxed{\text{Option 2}}$

**REMARKS** (1) If your near point is 75 cm, you are farsighted. To read a book you must hold it at least 75 cm from your eye to be able to focus on the print. The image of the print on your retina is then very small. The reading glasses lens produces an image also 75 cm from your eye, and this image is three times larger than the actual print. Thus, looking through the lens, the image of the print on the retina is larger by a factor of 3. (2) In this example, the distance from the lens to the eye was negligible. The results are slightly different if this distance is not negligible and is factored into the calculations.

**EXERCISE** Calculate the power of the eye for which the near point is 75 cm and the cornea–retina distance is 2.5 cm, and calculate the combined power of the lens and eye when they are in contact. Compare this with the power of a lens for which $s' = 2.5$ cm, when $s = 25$ cm. (*Answer* $P_{\text{eye}} = 41.33$ D; $P_c = 41.33$ D $+ 2.67$ D $= 44$ D; $P = 44$ D)

## *The Simple Magnifier

We saw in Example 32-14 that the *apparent* size of an object can be increased by using a converging lens placed next to the eye. A converging lens is called a **simple magnifier** if it is placed next to the eye and if the object is placed closer to the lens than its focal length, as was the case for the lens in Example 32-14. In that example, the lens formed a virtual image at the near point of the eye, the same location that the object must be placed for best viewing by the unaided eye. So, with the lens in place, the magnitude of the image distance $|s'|$ was greater than the object distance $s$, so the image seen by the eye is magnified by $m = |s'|/s$. If the actual height of the object was $y$, then the height $y'$ of the image formed by the lens would have been $my$. To the eye, this image subtended an angle $\theta$ (Figure 32-50) given approximately by

**FIGURE 32-50**

$$\theta = \frac{my}{|s'|} = m\frac{y}{|s'|} = \frac{|s'|}{s}\frac{y}{|s'|} = \frac{y}{s}$$

$$m = \frac{y'}{y} = \frac{|s'|}{s}$$

which is the *very same angle* the object would subtend were the lens removed while the object and the eye were left in place. That is, the apparent size of the image seen by the eye through the lens is the same as the apparent size of the object that would be seen by the eye were the lens removed (assuming the eye could focus at that distance). Thus, the apparent size of the object seen through the lens is inversely proportional to the distance from the object to the eye to the distance from the object to the eye with the lens in place. The smaller $s$ is, the larger the subtended angle $\theta$ and the larger the apparent size of the object.

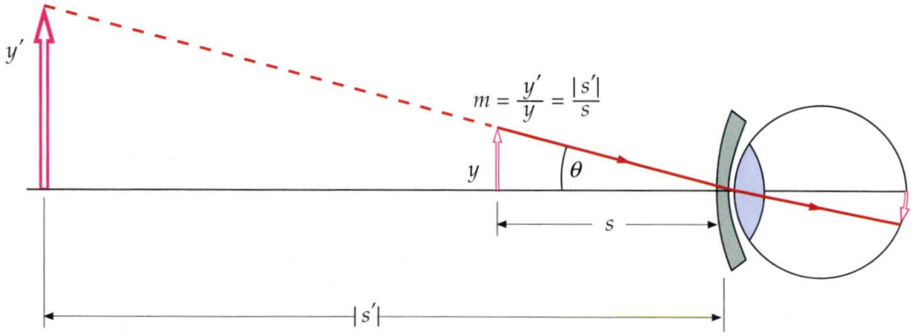

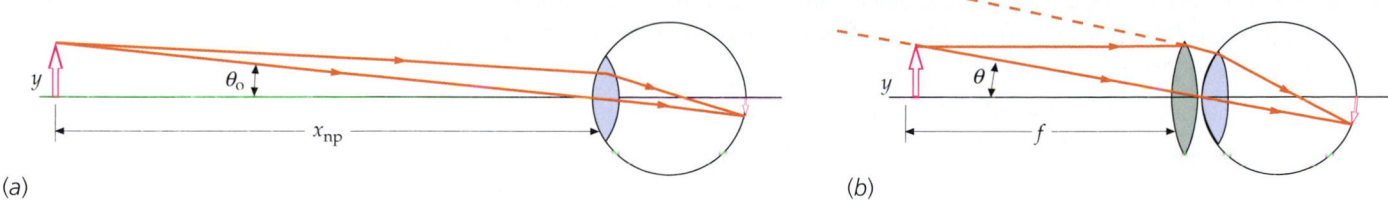

(a)                                                                (b)

**FIGURE 32-51** (*a*) An object at the near point subtends an angle $\theta_o$ at the naked eye. (*b*) When the object is at the focal point of the converging lens, the rays emerge from the lens parallel and enter the eye as if they came from an object a very large distance away. The image can thus be viewed at infinity by the relaxed eye. When *f* is less than the near-point distance, the converging lens allows the object to be brought closer to the eye. This increases the angle subtended by the object to $\theta$, thereby increasing the size of the image on the retina.

In Figure 32-51*a*, a small object of height $y$ is at the near point of the eye at a distance $x_{np}$. The angle subtended, $\theta_o$, is given approximately by

$$\theta_o = \frac{y}{x_{np}}$$

In Figure 32-51*b*, a converging lens of focal length $f$ that is smaller than $x_{np}$ is placed a negligible distance in front of the eye, and the object is placed in the focal plane of the lens. The rays emerge from the lens parallel, indicating that the image is located an infinite distance in front of the lens. The parallel rays are focused by the relaxed eye on the retina. The angle subtended by this image is equal to the angle subtended by the object (assuming that the lens is a negligible distance from the eye). The angle subtended by the object is approximately

$$\theta = \frac{y}{f}$$

The ratio $\theta/\theta_o$ is called the *angular magnification* or *magnifying power M* of the lens:

$$M = \frac{\theta}{\theta_o} = \frac{x_{np}}{f} \qquad\qquad 32\text{-}20$$

Simple magnifiers are used as eyepieces (called oculars) in microscopes and telescopes to view the image formed by another lens or lens system. To correct aberrations, combinations of lenses that result in a short positive focal length may be used in place of a single lens, but the principle of the simple magnifier is the same.

---

*ANGULAR MAGNIFICATION OF A SIMPLE MAGNIFIER*    **E X A M P L E   3 2 - 1 5**    **Try It Yourself**

**A person with a near point of 25 cm uses a 40-D lens as a simple magnifier. What angular magnification is obtained?**

**PICTURE THE PROBLEM** The angular magnification is found from the focal length $f$ (Equation 32-20), which is the reciprocal of the power.

**Cover the column to the right and try these on your own before looking at the answers.**

| Steps | Answers |
|---|---|
| 1. Calculate the focal length of the lens. | $f = 2.5$ cm |
| 2. Use your result from step 1 and incorporate the result into Equation 32-20 to calculate the angular magnification. | $M = \boxed{10}$ |

**REMARKS** Looking through the lens, the object appears 10 times larger because it can be placed at 2.5 cm rather than at 25 cm from the eye, thus increasing the size of the image on the retina tenfold.

**EXERCISE** What is the magnification in this example if the near point of the person is 30 cm rather than 25 cm? (*Answer* $M = 12$)

## *The Compound Microscope

The compound microscope (Figure 32-52) is used to look at very small objects at short distances. In its simplest form, it consists of two converging lenses. The lens nearest the object, called the **objective,** forms a real image of the object. This image is enlarged and inverted. The lens nearest the eye, called the **eyepiece** or **ocular,** is used as a simple magnifier to view the image formed by the objective.

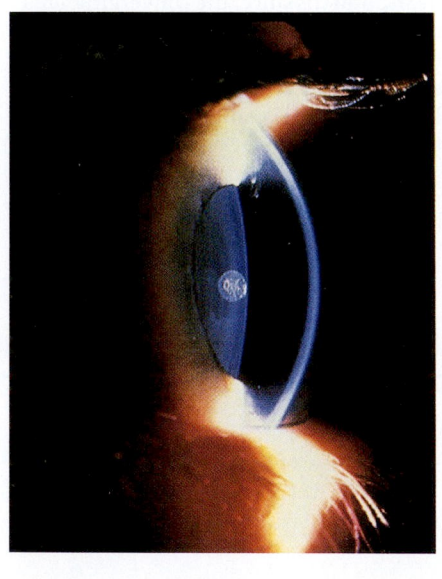

(a)

(b)

(*a*) The human eye in profile. (*b*) The lens of the eye is kept in place by the ciliary muscle (shown here in the upper left), which rings the lens. When the ciliary muscle contracts, the lens tends to bulge. The greater lens curvature enables the eye to focus on nearby objects. (*c*) Some of the 120 million rods and 7 million cones in the eye, magnified approximately 5000 times. The rods (the more slender of the two) are more sensitive in dim light, whereas the cones are more sensitive to color. The rods and cones form the bottom layer of the retina and are covered by nerve cells, blood vessels, and supporting cells. Most of the light entering the eye is reflected or absorbed before reaching the rods and cones. The light that does reach them triggers electrical impulses along nerve fibers that ultimately reach the brain. (*d*) A neural net used in the vision system of certain robots. Loosely modeled on the human eye, it contains 1920 sensors.

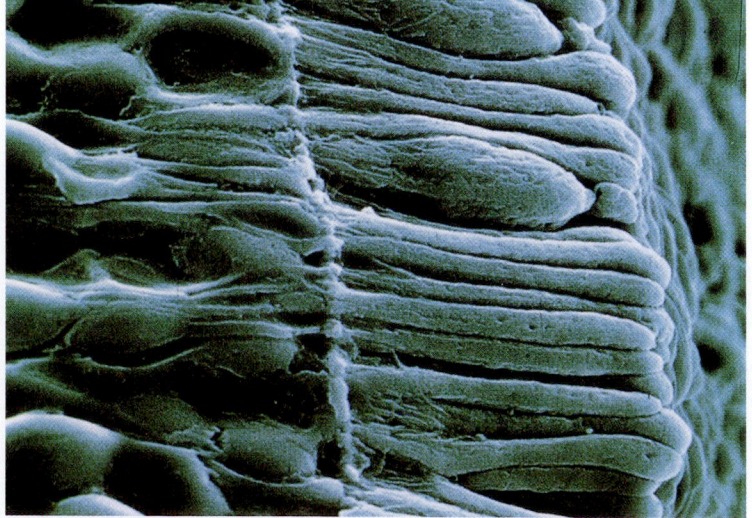

(c)

(d)

The eyepiece is placed so that the image formed by the objective falls at the first focal point of the eyepiece. The light from each point on the object thus emerges from the eyepiece as a parallel beam, as if it were coming from a point a great distance in front of the eye. (This is commonly called *viewing the image at infinity*.)

The distance between the second focal point of the objective and the first focal point of the eyepiece is called the **tube length** $L$. It is typically fixed at approximately 16 cm. The object is placed just outside the first focal point of the objective so that an enlarged image is formed at the first focal point of the eyepiece a distance $L + f_o$ from the objective, where $f_o$ is the focal length of the objective. From Figure 32-52, $\tan \beta = y/f_o = -y'/L$. The lateral magnification of the objective is therefore

$$m_o = \frac{y'}{y} = -\frac{L}{f_o} \qquad\qquad 32\text{-}21$$

The angular magnification of the eyepiece (from Equation 32-20) is

$$M_e = \frac{x_{np}}{f_e}$$

where $x_{np}$ is the near-point distance of the viewer, and $f_e$ is the focal length of the eyepiece. The magnifying power of the compound microscope is the product of the lateral magnification of the objective and the angular magnification of the eyepiece:

$$M = m_o M_e = -\frac{L}{f_o}\frac{x_{np}}{f_e} \qquad\qquad 32\text{-}22$$

MAGNIFYING POWER OF A MICROSCOPE

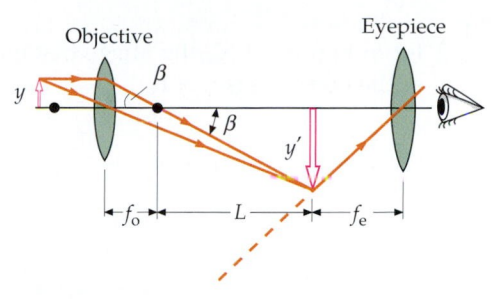

**FIGURE 32-52** Schematic diagram of a compound microscope consisting of two positive lenses, the objective of focal length $f_o$ and the ocular, or eyepiece, of focal length $f_e$. The real image of the object formed by the objective is viewed by the eyepiece, which acts as a simple magnifier. The final image is at infinity.

---

*THE COMPOUND MICROSCOPE*          **E X A M P L E    3 2 - 1 6**

**A microscope has an objective lens of focal length 1.2 cm and an eyepiece of focal length 2 cm. These lenses are separated by 20 cm. (a) Find the magnifying power if the near point of the viewer is 25 cm. (b) Where should the object be placed if the final image is to be viewed at infinity?**

(a) 1. The magnifying power is given by Equation 32-22:

$$M = -\frac{L}{f_o}\frac{x_{np}}{f_e}$$

2. The tube length $L$ is the distance between the lenses minus the focal distances:

$$L = 20 \text{ cm} - 2 \text{ cm} - 1.2 \text{ cm} = 16.8 \text{ cm}$$

3. Substitute this value for $L$ and the given values of $x_{np}, f_o,$ and $f_e$ to calculate $M$:

$$M = -\frac{L}{f_o}\frac{x_{np}}{f_e} = -\frac{16.8 \text{ cm}}{1.2 \text{ cm}}\frac{25 \text{ cm}}{2 \text{ cm}}$$

$$= \boxed{-175}$$

(b) 1. Calculate the object distance $s$ in terms of the image distance for the objective $s'$ and the focal length $f_o$:

$$\frac{1}{s} + \frac{1}{s'} = \frac{1}{f_o}$$

2. From Figure 32-52, the image distance for the image of the objective is $f_o + L$:

$$s' = f_o + L = 1.2 \text{ cm} + 16.8 \text{ cm}$$
$$= 18 \text{ cm}$$

3. Substitute to calculate $s$:

$$\frac{1}{s} + \frac{1}{18 \text{ cm}} = \frac{1}{1.2 \text{ cm}}$$

$$s = \boxed{1.29 \text{ cm}}$$

**REMARKS** The object should thus be placed at 1.29 cm from the objective or 0.09 cm outside its first focal point.

## *The Telescope

A telescope is used to view objects that are far away and are often large. The telescope works by creating a real image of the object that is much closer than the object. The astronomical telescope, illustrated schematically in Figure 32-53, consists of two positive lenses—an objective lens that forms a real, inverted image and an eyepiece that is used as a simple magnifier to view that image. Because the object is very far away, the image of the objective lies in the focal plane of the objective, and the image distance equals the focal length $f_o$. The image formed by the objective is much smaller than the object because the object distance is much larger than the focal length of the objective. For example, if we are looking at the moon, the image of the moon formed by the objective is much smaller than the moon itself. The purpose of the objective is not to magnify the object, but to produce an image that is close to us so it can be viewed by the eyepiece. The eyepiece is placed a distance $f_e$ from the image, where $f_e$ is the focal length of the eyepiece, so the final image can be viewed at infinity. Since this image is at the second focal plane of the objective and at the first focal plane of the eyepiece, the objective and eyepiece must be separated by the sum of the focal lengths of the objective and eyepiece, $f_o + f_e$.

The magnifying power of the telescope is the angular magnification $\theta_e / \theta_o$, where $\theta_e$ is the angle subtended by the final image as viewed through the eyepiece and $\theta_o$ is the angle subtended by the object when it is viewed directly by the unaided eye. The angle $\theta_o$ is the same as that subtended by the object at the objective shown in Figure 32-53. (The distance from a distant object, such as the moon, to the objective is essentially the same as the distance to the eye.) From this figure, we can see that

$$\tan \theta_o = \frac{y}{s} = -\frac{y'}{f_o} \approx \theta_o$$

where we have used the small-angle approximation $\tan \theta \approx \theta$. The angle $\theta_e$ in the figure is that subtended by the image at infinity formed by the eyepiece:

$$\tan \theta_e = \frac{y'}{f_e} \approx \theta_e$$

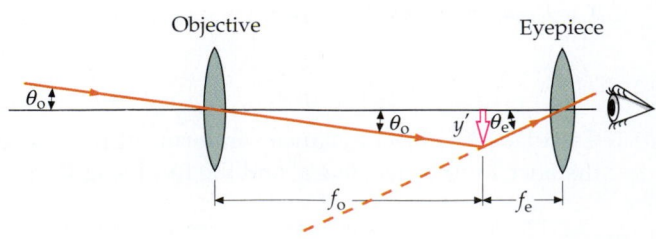

**FIGURE 32-53** Schematic diagram of an astronomical telescope. The objective lens forms a real, inverted image of a distant object near its second focal point, which coincides with the first focal point of the eyepiece. The eyepiece serves as a simple magnifier to view the image.

Since $y'$ is negative, $\theta_e$ is negative, indicating that the image is inverted. The magnifying power of the telescope is then

$$M = \frac{\theta_e}{\theta_o} = -\frac{f_o}{f_e} \qquad\qquad 32\text{-}23$$

MAGNIFYING POWER OF A TELESCOPE

From Equation 32-23, we can see that a large magnifying power is obtained with an objective of large focal length and an eyepiece of short focal length.

**EXERCISE** The world's largest refracting telescope is at the Yerkes Observatory of the University of Chicago at Williams Bay, Wisconsin. The telescope's objective has a diameter of 102 cm and a focal length of 19.5 m. The focal length of the eyepiece is 10 cm. What is its magnifying power? (*Answer* −195)

The main consideration with an astronomical telescope is not its magnifying power but its light-gathering power, which depends on the size of the objective. The larger the objective, the brighter the image. Very large lenses without aberrations are difficult to produce. In addition, there are mechanical problems in supporting very large, heavy lenses by their edges. A reflecting telescope (Figure 32-54 and Figure 32-55) uses a concave mirror instead of a lens for its objective. This offers several advantages. For one, a mirror does not produce chromatic aberration. In addition, mechanical support is much simpler, since the mirror weighs far less than a lens of equivalent optical quality, and the mirror can be supported over its entire back surface. In modern earth-based telescopes, the objective mirror consists of several dozen adaptive mirror segments that can be adjusted individually to correct for minute variations in gravitational stress when the telescope is tilted, and to compensate for thermal expansions and contractions and other changes caused by climatic conditions. In addition, they can adjust to nullify the distortions produced by atmospheric fluctuations.

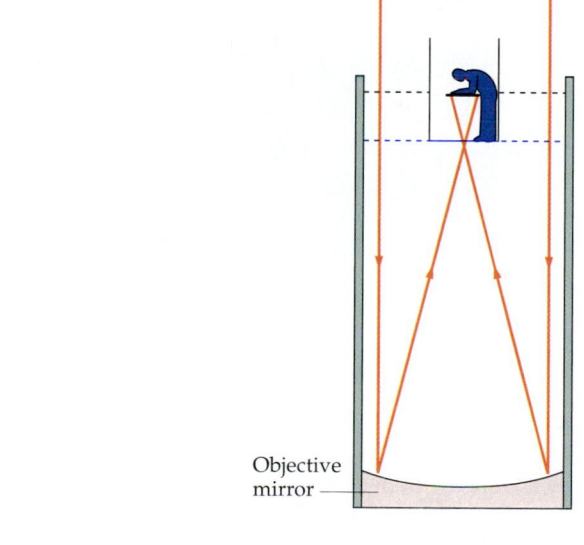

Objective mirror

**FIGURE 32-54** A reflecting telescope uses a concave mirror instead of a lens for its objective. Because the viewer compartment blocks off some of the incoming light, the arrangement shown here is used only in telescopes with very large objective mirrors.

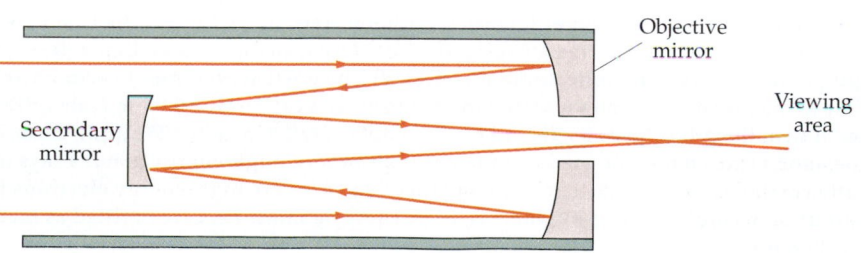

Objective mirror

Secondary mirror

Viewing area

**FIGURE 32-55** This reflecting telescope has a secondary mirror to redirect the light through a small hole in the objective mirror, thus providing more room for auxiliary instruments in the viewing area.

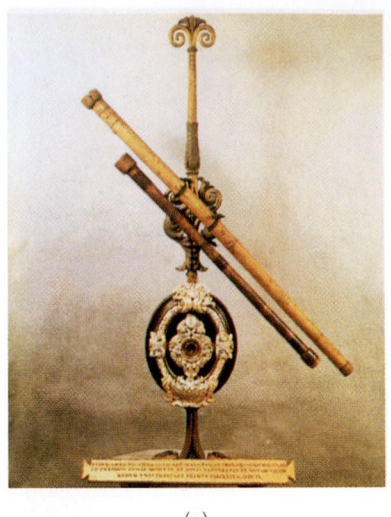

(a)

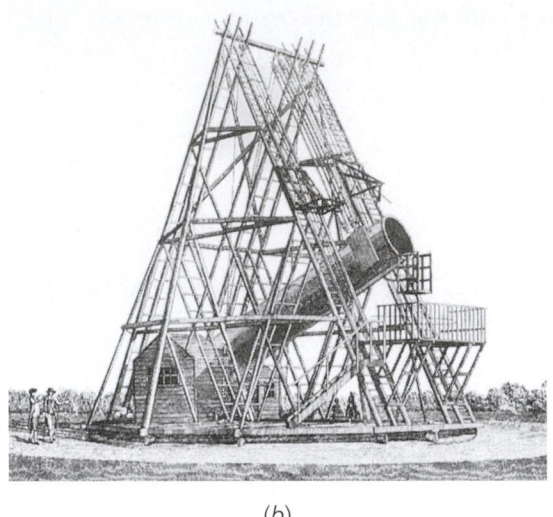

(b)

(c)

(d)

(e)

Astronomy at optical wavelengths began with Galileo approximately 400 years ago. In this century, astronomers began to explore the electromagnetic spectrum at other wavelengths; beginning with radio astronomy in the 1940s, satellite-based X-ray astronomy in the early 1960s, and more recently, ultraviolet, infrared, and gamma-ray astronomy. (*a*) Galileo's seventeenth-century telescope, with which he discovered mountains on the moon, sunspots, Saturn's rings, and the bands and moons of Jupiter. (*b*) An engraving of the reflector telescope built in the 1780s and used by the great astronomer Friedrich Wilhelm Herschel, who was the first to observe galaxies outside our own. (*c*) Because it is difficult to make large, flaw-free lenses, refractor telescopes like this 91.4-cm telescope at Lick Observatory have been superseded in light-gathering power by reflector telescopes. (*d*) The great astronomer Edwin Powell Hubble, who discovered the apparent expansion of the universe, is shown seated in the observer's cage of the 5.08-m Hale reflecting telescope, which is large enough for the observer to sit at the prime focus itself. (*e*) This 10-m optical reflector at the Whipple Observatory in southern Arizona is the largest instrument designed exclusively for use in gamma-ray astronomy. High-energy gamma rays of unknown origin strike the upper atmosphere and create cascades of particles. Among these particles are high-energy electrons that emit Cerenkov radiation observable from the ground. According to one hypothesis, high-energy gamma rays are emitted as matter is accelerated toward ultradense rotating stars called pulsars.

(a)

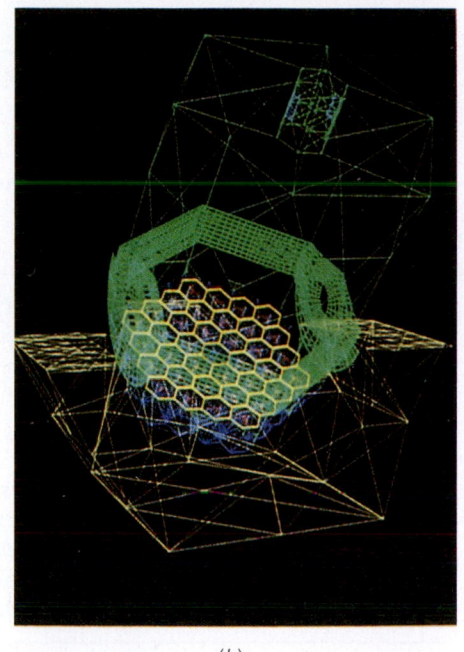

(b)

(c)

(a) The Keck Observatory, atop the inactive volcano of Mauna Kea, Hawaii, houses the world's largest optical telescope. The clear, dry air and lack of light pollution make the remote heights of Mauna Kea an ideal site for astronomical observations. (b) The Keck telescope is composed of 36 hexagonal mirror segments performing together as if they were a single mirror 10-m wide—roughly twice as large as the largest single-mirror telescope presently in operation. (c) Beneath each Keck mirror is a system of computer-controlled sensors and motor-driven actuators that can continuously vary the mirror's shape. These variations, which are sensitive to within 100 nm, enable the system to correct for variations in the alignments of the segments due to minute variations in gravitational stress when the telescope is tilted and to compensate for thermal expansions and contractions and fluctuations caused by gusts of wind on the mountaintop.

The Hubble Space Telescope is high above the atmospheric turbulence that limits the ability of ground-based telescopes to resolve images at optical wavelengths.

| Topic | Relevant Equations and Remarks |
|---|---|
| **1. Virtual and Real Images and Objects** | |
| Images | An image is real if actual light rays converge to each image point. This can occur in front of a mirror, or on the refracted-light side of a thin lens or refracting surface. An image is virtual if only extensions of the actual light rays converge to each image point. This can occur behind a mirror or on the incident-light side of a lens or refracting surface. |
| Virtual object | An object is real if actual light rays diverge from each object point. This can occur only on the incident-light side of a mirror, lens, or refracting surface. A real object is either an actual object or a real image. An object is virtual if only extensions of actual light rays diverge from each object point. This can occur only behind a mirror or on the refracted-light side of a lens or refracting surface. |

| **2. Spherical Mirrors** | | |
|---|---|---|
| Focal length | The focal length is the image distance when the object is at infinity, so the incident light is parallel to the axis: | |
| Mirror equation (for locating an image) | $$\frac{1}{s} + \frac{1}{s'} = \frac{1}{f}$$ where $$f = \frac{r}{2}$$ | 32-4  32-3 |
| Lateral magnification | $$m = \frac{y'}{y} = -\frac{s'}{s}$$ | 32-5 |
| Ray diagrams | Images can be located by a ray diagram using any two paraxial rays. The parallel, focal, and radial rays are the easiest to draw: 1. The parallel ray, drawn parallel to the axis, is reflected through the focal point. 2. The focal ray, drawn through the focal point, is reflected parallel to the axis. 3. The radial ray, drawn through the center of curvature, strikes the mirror perpendicular to its surface and is thus reflected back on itself. | |
| Sign conventions for reflection | 1. $s$ is positive if the object is on the incident-light side of the mirror. 2. $s'$ is positive if the image is on the reflected-light side of the mirror. 3. $r$ (and $f$) is positive if the mirror is concave so the center of curvature is on the reflected-light side of the mirror. | |

| **3. Images Formed by Refraction** | | |
|---|---|---|
| Refraction at a single surface | $$\frac{n_1}{s} + \frac{n_2}{s'} = \frac{n_2 - n_1}{r}$$ where $n_1$ is the index of refraction of the medium on the incident-light side of the surface. | 32-6 |
| Magnification | $$m = \frac{y'}{y} = -\frac{n_1 s'}{n_2 s}$$ | 32-7 |

| | |
|---|---|
| Sign conventions for refraction | 1. $s$ is positive for objects on the incident-light side of the surface. |
| | 2. $s'$ is positive for images on the refracted-light side of the surface. |
| | 3. $r$ is positive if the center of curvature is on the refracted-light side of the surface. |

**4. Thin Lenses**

| | | |
|---|---|---|
| Focal length (lens-maker's equation) | $$\frac{1}{f} = (n-1)\left(\frac{1}{r_1} - \frac{1}{r_2}\right)$$ | 32-11 |
| | A positive lens ($f > 0$) is a converging lens (like a double convex lens). A negative lens ($f < 0$) is a diverging lens (like a double concave lens). | |
| First and second focal points | Incident rays parallel to the axis emerge directed either toward or away from the *first focal point F'*. Incident rays directed either toward or away from the *second focal point F* emerge parallel with the axis. | |
| Power | $$P = \frac{1}{f}$$ | 32-13 |
| Thin-lens equation (for locating image) | $$\frac{1}{s} + \frac{1}{s'} = \frac{1}{f}$$ | 32-12 |
| Magnification | $$m = \frac{y'}{y} = -\frac{s'}{s}$$ | 32-14 |
| Ray diagrams | Images can be located by a ray diagram using any two paraxial rays. The parallel, central, and focal rays are the easiest to draw: | |
| | 1. The parallel ray, drawn parallel to the axis, emerges directed toward (or away from) the second focal point of the lens. | |
| | 2. The central ray, drawn through the center of the lens, is not deflected. | |
| | 3. The focal ray, drawn through (or toward) the first focal point, emerges parallel to the axis. | |
| Sign conventions for lenses | The sign conventions are the same as for refraction at a spherical surface. | |

**5. *Aberrations**

Blurring of the image of a single object point is called aberration. Spherical aberration results from the fact that a spherical surface focuses only paraxial rays (those that travel close to the axis) at a single point. Nonparaxial rays are focused at nearby points depending on the angle made with the axis. Spherical aberration can be reduced by blocking the rays farthest from the axis. This, of course, reduces the amount of light reaching the image.

Chromatic aberration, which occurs with lenses but not mirrors, results from the variation in the index of refraction with wavelength. Lens aberrations are most commonly reduced by using a series of lens elements.

**6. *The Eye**

The cornea–lens system of the eye focuses light on the retina, where it is sensed by the rods and cones that send information along the optic nerve to the brain. When the eye is relaxed, the focal length of the cornea–lens system is about 2.5 cm, which is the distance to the retina. When objects are brought near the eye, the lens changes shape to decrease the overall focal length so that the image remains focused on the retina. The closest distance for which the image can be focused on the retina is called the near point, typically about 25 cm. The apparent size of an object depends on the size of the image on the retina. The closer the object, the larger the image on the retina and therefore the larger the apparent size of the object.

**7. *The Simple Magnifier**

A simple magnifier consists of a lens with a positive focal length that is smaller than the near point.

| | | |
|---|---|---|
| Magnifying power (angular magnification) | $$M = \frac{\theta}{\theta_o} = \frac{x_{np}}{f}$$ | 32-20 |

**8. *The Compound Microscope**

The compound microscope is used to look at very small objects that are nearby. It consists of two converging lenses (or lens systems), an objective, and an ocular or eyepiece. The object to be viewed is placed just outside the focal point of the objective, which forms an enlarged image of the object at the focal plane of the eyepiece. The eyepiece acts as a simple magnifier to view the final image.

| | | |
|---|---|---|
| Magnifying power (angular magnification) | $$M = m_o M_e = -\frac{L}{f_o}\frac{x_{np}}{f_e}$$ | 32-22 |

where $L$ is the tube length, the distance between the second focal point of the objective and the first focal point of the eyepiece.

**9. *The Telescope**

The telescope is used to view objects that are far away. The objective of the telescope forms a real image of the object that is much smaller than the object but much closer. The eyepiece is then used as a simple magnifier to view the image. A reflecting telescope uses a mirror for its objective.

| | | |
|---|---|---|
| Magnifying power (angular magnification) | $$M = \frac{\theta_e}{\theta_o} = -\frac{f_o}{f_e}$$ | 32-23 |

# PROBLEMS

- Single-concept, single-step, relatively easy
- •• Intermediate-level, may require synthesis of concepts
- ••• Challenging
- SSM Solution is in the *Student Solutions Manual*
- iSOLVE Problems available on iSOLVE online homework service
- iSOLVE✔ These "Checkpoint" online homework service problems ask students additional questions about their confidence level, and how they arrived at their answer.

In a few problems, you are given more data than you actually need; in a few other problems, you are required to supply data from your general knowledge, outside sources, or informed estimates.

## Conceptual Problems

**1** • Can a virtual image be photographed?

**2** • Suppose each axis of a coordinate system, like the one shown in Figure 32-4, is painted a different color. One photograph is taken of the coordinate system and another is taken of its image in a plane mirror. Is it possible to tell that one of the photographs is of a mirror image, rather than both being photographs of the real coordinate system from different angles?

**3** •• iSOLVE True or False

(a) The virtual image formed by a concave mirror is always smaller than the object.

(b) A concave mirror always forms a virtual image.

(c) A convex mirror never forms a real image of a real object.

(d) A concave mirror never forms an enlarged real image of an object.

**4** •• SSM Under what condition will a concave mirror produce (a) an upright image, (b) a virtual image, (c) an image smaller than the object, and (d) an image larger than the object?

**5** •• Answer Problem 4 for a convex mirror.

**6** •• Convex mirrors are often used for rearview mirrors on cars and trucks to give a wide-angle view. Below the mirror is written, "Warning, objects are closer than they appear." Yet, according to a ray diagram, such as the diagram shown in Figure 32-19, the image distance for distant objects is much smaller than the object distance. Why then do they appear more distant?

**7** •• As an object is moved from a great distance toward the focal point of a concave mirror, the image moves from (a) a great distance toward the focal point and is always real. (b) the focal point to a great distance from the mirror and is always real. (c) the focal point toward the center of curvature of the mirror and is always real. (d) the focal point to a great distance from the mirror and changes from a real image to a virtual image.

**8** • A bird above the water is viewed by a scuba diver submerged beneath the water's surface directly below the bird. Does the bird appear to the diver to be closer to or farther from the surface than it actually is?

**9** • [SSM] Under what conditions will the focal length of a thin lens be (a) positive and (b) negative? Consider both the case where the index of refraction of the lens is greater than and less than the surrounding medium.

**10** • The focal length of a simple lens is different for different colors of light. Why?

**11** •• An object is placed 40 cm from a lens of focal length −10 cm. The image is (a) real, inverted, and diminished. (b) real, inverted, and enlarged. (c) virtual, inverted, and diminished. (d) virtual, upright, and diminished. (e) virtual, upright, and enlarged.

**12** •• If a real object is placed just inside the focal point of a converging lens, the image is (a) real, inverted, and enlarged. (b) virtual, upright, and diminished. (c) virtual, upright, and enlarged. (d) real, inverted, and diminished.

**13** • Both the eye and the camera work by forming real images, the eye's image forming on the retina and the camera's image forming on the film. Explain the difference between the ways in which these two systems accommodate objects located at different object distances and still keep a focused image.

**14** • [SSM] If an object is placed 25 cm from the eye of a farsighted person who does not wear corrective lenses, a sharp image is formed (a) behind the retina, and the corrective lens should be convex. (b) behind the retina, and the corrective lens should be concave. (c) in front of the retina, and the corrective lens should be convex. (d) in front of the retina, and the corrective lens should be concave.

**15** •• Myopic (nearsighted) persons sometimes claim to see better under water without corrective lenses. Why? (a) The accommodation of the eye's lens is better under water. (b) Refraction at the water–cornea interface is less than at the air–cornea interface. (c) Refraction at the water–cornea interface is greater than at the air–cornea interface. (d) No reason; the effect is only an illusion and not really true.

**16** •• A nearsighted person who wears corrective lenses would like to examine an object at close distance. Identify the correct statement. (a) The corrective lenses give an enlarged image and should be worn while examining the object. (b) The corrective lenses give a reduced image of the object and should be removed. (c) The corrective lenses result in a magnification of unity; it does not matter whether they are worn or removed.

**17** • [SSM] The image of a real object formed by a convex mirror (a) is always real and inverted. (b) is always virtual and enlarged. (c) may be real. (d) is always virtual and diminished.

**18** • The image of a real object formed by a converging lens (a) is always real and inverted. (b) is always virtual and enlarged. (c) may be real. (d) is always virtual and diminished.

**19** • The glass of a converging lens has an index of refraction of 1.6. When the lens is in air, its focal length is 30 cm. If the lens is immersed in water, its focal length will be (a) greater than 30 cm. (b) less than 30 cm. (c) the same as before, 30 cm. (d) negative.

**20** •• True or false:
(a) A virtual image cannot be displayed on a screen.
(b) A negative image distance implies that the image is virtual.
(c) All rays parallel to the axis of a spherical mirror are reflected through a single point.
(d) A diverging lens cannot form a real image from a real object.
(e) The image distance for a positive lens is always positive.

**21** • [SSM] Explain the following statement: A microscope is an object magnifier, but a telescope is an angle magnifier.

## Estimation and Approximation

**22** •• The lens-maker's equation contains three design parameters, the index of refraction of the lens and the radius of curvature of its two surfaces. Thus, there are many ways to design a lens with a particular focal length. Use the lens-maker's equation to design three different thin converging lenses, each with a focal length of 27 cm and each made from glass with an index of refraction of 1.6. Draw a sketch of each of your designs.

**23** •• Repeat Problem 22, but for a diverging lens of focal length −27 cm.

**24** •• [SSM] Estimate the maximum value that could be usefully obtained for the magnifying power of a simple magnifier, using Equation 32-20. (*Hint: Think about the smallest focal length lens that could be made from glass and still be used as a magnifier.*)

## Plane Mirrors

**25** • The image of the point object P in Figure 32-56 is viewed by an eye, as shown. Draw a bundle of rays from the object that reflect from the mirror and enter the eye. For this object position and mirror, indicate the region of space in which the eye can be positioned and still see the image.

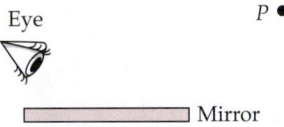

Eye        P •

◻ Mirror

**FIGURE 32-56** Problem 25

**26** • A person 1.62 m tall wants to be able to see her full image in a plane mirror. (a) What must be the minimum height of the mirror? (b) How far above the floor should the mirror be placed, assuming that the top of the person's head is 15 cm above her eye level? Draw a ray diagram.

**27** • [SSM] Two plane mirrors make an angle of 90°. The light from an object point that is arbitrarily positioned in front of the mirrors produces images at three locations. For each image location, draw two rays from the object that, upon one or two reflections, appear to come from the image location.

**28** • (a) Two plane mirrors make an angle of 60° with each other. Draw a sketch to show the location of all the images formed of a point object on the bisector of the angle between the mirrors. (b) Repeat for an angle of 120°.

**29** •• When two plane mirrors are parallel, such as on opposite walls in a barber shop, multiple images arise because each image in one mirror serves as an object for the other mirror. A point object is placed between parallel mirrors separated by 30 cm. The object is 10 cm in front of the left mirror and 20 cm in front of the right mirror. (a) Find the distance from the left mirror to the first four images in that mirror. (b) Find the distance from the right mirror to the first four images in that mirror.

## Spherical Mirrors

**30** •• **SSM** A concave spherical mirror has a radius of curvature of 24 cm. Draw ray diagrams to locate the image (if one is formed) for an object at a distance of (a) 55 cm, (b) 24 cm, (c) 12 cm, and (d) 8 cm from the mirror. For each case, state whether the image is real or virtual; upright or inverted; and enlarged, reduced, or the same size as the object.

**31** • Use the mirror equation (Equation 32-4) to locate and describe the images for the object distances and mirror of Problem 30.

**32** •• Repeat Problem 30 for a convex mirror with the same radius of curvature.

**33** • Use the mirror equation (Equation 32-4) to locate and describe the images for the object distances and convex mirror of Problem 32.

**34** • Show that a convex mirror cannot form a real image of a real object, no matter where the object is placed, by showing that $s'$ is always negative for a positive $s$.

**35** • **SSM** **iSOLVE**✓ A dentist wants a small mirror that will produce an upright image with a magnification of 5.5 when the mirror is located 2.1 cm from a tooth. (a) What should the radius of curvature of the mirror be? (b) Should the mirror be concave or convex?

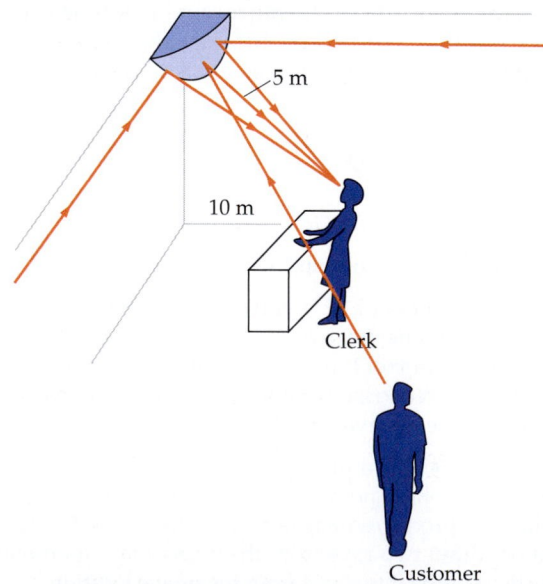

**FIGURE 32-57** Problem 36

**36** •• **iSOLVE**✓ Convex mirrors are used in stores to provide a wide angle of surveillance for a reasonable mirror size. The mirror shown in Figure 32-57 allows a clerk 5 m away

from the mirror to survey the entire store. It has a radius of curvature of 1.2 m. (a) If a customer is 10 m from the mirror, how far from the mirror surface is his image? (b) Is the image in front of or behind the mirror? (c) If the customer is 2 m tall, how tall is his image?

**37** •• **iSOLVE** A certain telescope uses a concave spherical mirror of radius 8 m. Find the location and diameter of the image of the moon formed by this mirror. The moon has a diameter of $3.5 \times 10^6$ m and is $3.8 \times 10^8$ m from the earth.

**38** •• A concave spherical mirror has a radius of curvature of 6 cm. A point object is on the axis 9 cm from the mirror. Construct a precise ray diagram showing rays from the object that make angles of 5°, 10°, 30°, and 60° with the axis, which strike the mirror and are reflected back across the axis. (Use a compass to draw the mirror, and use a protractor to measure the angles needed to find the reflected rays.) What is the spread $\delta x$ of the points where these rays cross the axis?

**39** •• **SSM** A concave mirror has a radius of curvature 6 cm. Draw rays parallel to the axis at 0.5 cm, 1 cm, 2 cm, and 4 cm above the axis, and find the points at which the reflected rays cross the axis. (Use a compass to draw the mirror and a protractor to find the angle of reflection for each ray.) (a) What is the spread $\delta x$ of the points where these rays cross the axis? (b) By what percentage could this spread be reduced if the edge of the mirror were blocked off so that parallel rays more than 2 cm from the axis could not strike the mirror?

**40** •• **iSOLVE** An object located 100 cm from a concave mirror forms a real image 75 cm from the mirror. The mirror is then turned around so that its convex side faces the object. The mirror is moved so that the image is now 35 cm behind the mirror. How far was the mirror moved? Was it moved toward the object or away from the object?

**41** •• Parallel light from a distant object strikes the large mirror shown in Figure 32-58 at $r = 5$ m and is reflected by the small mirror that is 2 m from the large mirror. The small mirror is actually spherical, not planar as shown. The light is focused at the vertex of the large mirror. (a) What is the radius of curvature of the small mirror? (b) Is the mirror convex or concave?

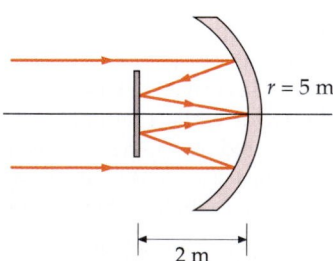

**FIGURE 32-58** Problem 41

## Images Formed by Refraction

**42** • A sheet of paper with writing on it is protected by a thick glass plate having an index of refraction of 1.5. If the plate is 2 cm thick, at what distance beneath the top of the plate does the writing appear when it is viewed from directly overhead?

**43** • ISOLVE✓ A fish is 10 cm from the front surface of a fish bowl of radius 20 cm. (a) Where does the fish appear to be to someone viewing the fish from in front of the bowl? (b) Where does the fish appear to be when it is 30 cm from the front surface of the bowl?

**44** •• SSM A very long glass rod of 3.5-cm diameter has one end ground to a convex spherical surface of radius 7.2 cm. Its index of refraction is 1.5. (a) A point object in air is on the axis of the rod 35 cm from the surface. Find the image and state whether the image is real or virtual. Repeat (b) for an object 6.5 cm from the surface and (c) for an object very far from the surface. Draw a ray diagram for each case.

**45** •• At what distance from the glass rod of Problem 44 should the object be placed, so that the light rays in the rod are parallel? Draw a ray diagram for this situation.

**46** •• Repeat Problem 44 for a glass rod with a concave hemispherical surface of radius −7.2 cm.

**47** •• Repeat Problem 44 when the glass rod and the objects are immersed in water.

**48** •• Repeat Problem 44 for a glass rod with a concave hemispherical surface of radius −7.5 cm when the glass rod and the objects are immersed in water.

**49** •• SSM ISOLVE A glass rod 96 cm long with an index of refraction of 1.6 has its ends ground to convex spherical surfaces of radii 8 cm and 16 cm. A point object is in air on the axis of the rod 20 cm from the end with the 8-cm radius. (a) Find the image distance due to refraction at the first surface. (b) Find the final image due to refraction at both surfaces. (c) Is the final image real or virtual?

**50** •• Repeat Problem 49 for a point object in air on the axis of the glass rod 20 cm from the end with the 16-cm radius.

## Thin Lenses

**51** • The following thin lenses are made of glass with an index of refraction of 1.5. Make a sketch of each lens, and find its focal length in air: (a) double convex, $r_1 = 15$ cm and $r_2 = -26$ cm; (b) plano-convex, $r_1 = \infty$ and $r_2 = -15$ cm; (c) double concave, $r_1 = -15$ cm and $r_2 = +15$ cm; and (d) plano-concave, $r_1 = \infty$ and $r_2 = +26$ cm.

**52** • Find the focal length of a glass lens of index of refraction 1.62 that has a concave surface with radius of magnitude 100 cm and a convex surface with a radius of magnitude 40 cm.

**53** • SSM A double concave lens of index of refraction 1.45 has radii of magnitudes 30 cm and 25 cm. An object is located 80 cm to the left of the lens. Find (a) the focal length of the lens, (b) the location of the image, and (c) the magnification of the image. (d) Is the image real or virtual? Is the image upright or inverted?

**54** • ISOLVE The following thin lenses are made of glass of index of refraction 1.6. Make a sketch of each lens, and find its focal length in air: (a) $r_1 = 20$ cm and $r_2 = 10$ cm, (b) $r_1 = 10$ cm and $r_2 = 20$ cm, and (c) $r_1 = -10$ cm and $r_2 = -20$ cm.

**55** • SSM ISOLVE✓ An object 3 cm high is placed 25 cm in front of a thin lens of power 10 D. Draw a precise ray diagram to find the position and the size of the image, and check your results using the thin-lens equation.

**56** • Repeat Problem 55 for an object 1.5 cm high that is placed 20 cm in front of a thin lens of power 10 D.

**57** • Repeat Problem 55 for an object 1.5 cm high that is placed 20 cm in front of a thin lens of power −10 D.

**58** •• (a) What is meant by a negative object distance? How can a negative object distance occur? Find the image distance and the magnification and state whether the image is virtual or real and upright or inverted for a thin lens in air when (b) $s = -20$ cm, $f = +20$ cm and (c) $s = -10$ cm, $f = -30$ cm. Draw a ray diagram for each of these cases.

**59** •• SSM ISOLVE✓ Two converging lenses, each of focal length 10 cm, are separated by 35 cm. An object is 20 cm to the left of the first lens. (a) Find the position of the final image using both a ray diagram and the thin-lens equation. (b) Is the image real or virtual? Is the image upright or inverted? (c) What is the overall lateral magnification of the image?

**60** •• Rework Problem 59 for a second lens that is a diverging lens of focal length −15 cm.

**61** •• (a) Show that to obtain a magnification of magnitude $m$ with a converging thin lens of focal length $f$, the object distance must be given by $s = (m - 1)f/m$. (b) A camera lens with a 50-mm focal length is used to take a picture of a person 1.75 m tall. How far from the camera should the person stand so that the image size is 24 mm?

**62** •• A converging lens has a focal length of $f = 12$ cm. (a) Using a spreadsheet program or graphing calculator, plot the image distance $s'$ as a function of the object distance $s$, for values of $s$ ranging from $s = 1.1f$ to $s = 10f$. (b) On the same graph used in Part (a), but using a different $y$ axis, plot the magnification of the lens as a function of the object distance $s$. (c) What type of image is produced for this range of object distances, real or virtual, upright or inverted? (d) Discuss the significance of any asymptotic limits your graph has.

**63** •• A converging lens has a focal length of $f = 12$ cm. (a) Using a spreadsheet program or graphing calculator, plot the image distance $s'$ as a function of the object distance $s$, for values of $s$ ranging from $s = 0.01f$ to $s = 0.9f$. (b) On the same graph used in Part (a), but using a different $y$ axis, plot the magnification of the lens as a function of the object distance $s$. (c) What type of image is produced for this range of object distances, real or virtual, upright or inverted? (d) Discuss the significance of any asymptotic limits your graph has.

**64** •• SSM An object is 15 cm in front of a positive lens of focal length 15 cm. A second positive lens of focal length 15 cm is 20 cm from the first lens. Find the final image and draw a ray diagram.

**65** •• Rework Problem 64 for a second lens with a focal length of −15 cm.

**66 •••** In a convenient form of the thin-lens equation used by Newton, the object and image distances are measured from the focal points. Show that if $x = s - f$ and $x' = s' - f$, the thin-lens equation can be written as $xx' = f^2$, and the lateral magnification is given by $m = -x'/f = -f/x$. Indicate $x$ and $x'$ on a sketch of a lens.

**67 •••** In *Bessel's method* for finding the focal length $f$ of a lens, an object and a screen are separated by distance $D$, where $D > 4f$. It is then possible to place the lens at either of two locations, both between the object and the screen, so that there is an image of the object on the screen, in one case magnified and in the other case reduced. Show that if the distance between the two lens locations is given by $L$, that

$$f = \frac{D^2 - L^2}{4D}$$

(*Hint: Refer to Figure 32-59.*) The two lens locations are such that the object distance with the lens in the one setting is equal to the image distance with the lens in the other setting and vice versa.

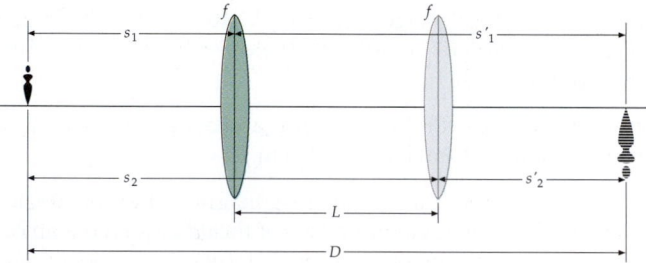

**FIGURE 32-59** Problems 67 and 68

**68 ••** An optician uses *Bessel's method* to find the focal length of a lens, as described in Problem 67. The object-to-image distance was set at 1.7 m. The position of the lens was then adjusted to get a sharp image on the screen. A second image was found when the lens was moved a distance of 72 cm. (*a*) Using the result from Problem 67, find the focal length of the lens. (*b*) What were the two locations of the lens with respect to the object?

**69 •••** An object is 17.5 cm to the left of a lens of focal length 8.5 cm. A second lens of focal length $-30$ cm is 5 cm to the right of the first lens. (*a*) Find the distance between the object and the final image formed by the second lens. (*b*) What is the overall magnification? (*c*) Is the final image real or virtual? Is the final image upright or inverted?

# *Aberrations

**70 •** **SSM** Chromatic aberration is a common defect of (*a*) concave and convex lenses. (*b*) concave lenses only. (*c*) concave and convex mirrors. (*d*) all lenses and mirrors.

**71 •** True or false:

(*a*) Aberrations occur only for real images.
(*b*) Chromatic aberration does not occur with mirrors.

**72 •** A double convex lens of radii $r_1 = +10$ cm and $r_2 = -10$ cm is made from glass with indexes of refraction of 1.53 for blue light and 1.47 for red light. Find the focal length of this lens for (*a*) red light and (*b*) blue light.

# *The Eye

**73 ••** **SSM** The Model Eye I: A simple model for the eye is a lens with variable power $P$ located a fixed distance $d$ in front of a screen, with the space between the lens and the screen filled by air. Refer to Figure 32-60. The "eye" can focus for all values of $s$ such that $x_{np} \leq s \leq x_{fp}$. This "eye" is said to be normal if it can focus on very distant objects. (*a*) Show that for a normal "eye," the minimum value of $P$ is

$$P_{min} = \frac{1}{d}$$

(*b*) Show that the maximum value of $P$ is

$$P_{max} = \frac{1}{x_{np}} + \frac{1}{d}$$

(*c*) The difference $A = P_{max} - P_{min}$ is called the accommodation. Find the minimum power and accommodation for a model eye with $d = 2.5$ cm and $x_{np} = 25$ cm.

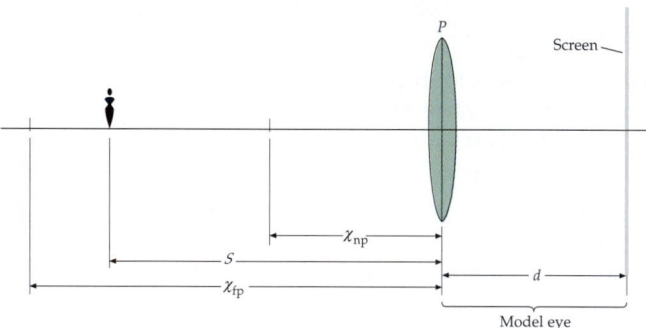

**FIGURE 32-60** Problems 73, 74, and 75

**74 ••** The Model Eye II: In an eye that exhibits nearsightedness, the eye cannot focus on distant objects. Refer to Figure 32-60 and to Problem 73. (*a*) Show that for a nearsighted model eye capable of focusing out to a maximum distance $x_{fp}$, the minimum value of $P$ is greater than that of a normal eye and is given by

$$P_{min} = \frac{1}{x_{fp}} + \frac{1}{d}$$

(*b*) To correct for nearsightedness, a contact lens may be placed directly in front of the model-eye's lens. What power contact lens would be needed to correct the vision of a nearsighted model eye with $x_{fp} = 50$ cm?

**75 ••** The Model Eye III: In an eye that exhibits farsightedness, the eye may be able to focus on distant objects but cannot focus on close objects. Refer to Figure 32-60 and to Problem 73. (*a*) Show that for a farsighted model eye capable of focusing only as close as a distance $x'_{np}$, the maximum value of $P$ is given by

$$P_{max} = \frac{1}{x'_{np}} + \frac{1}{d}$$

(*b*) Show that compared to a model eye capable of focusing as close as a distance $x_{np}$ (where $x_{np} < x'_{np}$), the maximum power of the farsighted lens is too small by

$$\frac{1}{x_{np}} - \frac{1}{x'_{np}}$$

(c) What power contact lens would be needed to correct the vision of a farsighted model eye, with $x_{np}$ = 150 cm, so that the eye may focus on objects as close as 15 cm?

**76** • **ISOLVE✔** Suppose the eye were designed like a camera with a lens of fixed focal length $f$ = 2.5 cm that could move toward or away from the retina. Approximately how far would the lens have to move to focus the image of an object 25 cm from the eye onto the retina? (*Hint:* Find the distance from the retina to the image behind it for an object at 25 cm.)

**77** • **ISOLVE✔** Find the change in the focal length of the eye when an object originally at 3 m is brought to 30 cm from the eye.

**78** • A farsighted person requires lenses with a power of 1.75 D to read comfortably from a book that is 25 cm from the eye. What is that person's near point without the lenses?

**79** • **SSM** If two point objects close together are to be seen as two distinct objects, the images must fall on the retina on two different cones that are not adjacent. That is, there must be an unactivated cone between them. The separation of the cones is about 1 $\mu$m. Model the eye as a uniform 2.5-cm-diameter sphere with a refractive index of 1.34. (a) What is the smallest angle the two points can subtend? (See Figure 32-61.) (b) How close together can two points be if they are 20 m from the eye?

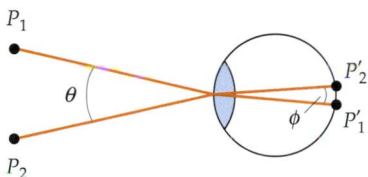

**FIGURE 32-61** Problem 79

**80** •• A person with a near point of 80 cm needs to read from a computer screen that is 45 cm from her eye. (a) Find the focal length of the lenses in reading glasses that will produce an image of the screen at 80 cm from her eye. (b) What is the power of the lenses?

**81** •• A nearsighted person cannot focus clearly on objects that are more distant than 225 cm from her eye. What power lenses are required for her to see distant objects clearly?

**82** •• Since the index of refraction of the lens of the eye is not very different from that of the surrounding material, most of the refraction takes place at the cornea, where $n$ changes abruptly from 1.0 in air to approximately 1.4. Assuming the cornea to be a homogeneous sphere with an index of refraction of 1.4, calculate the cornea's radius if it focuses parallel light on the retina a distance 2.5 cm away. Do you expect your result to be larger or smaller than the actual radius of the cornea?

**83** •• The near point of a certain person is 80 cm. Reading glasses are prescribed so that he can read a book at 25 cm from his eye. The glasses are 2 cm from the eye. What diopter lens should be used in the glasses?

**84** ••• At age 45, a person is fitted for reading glasses of power 2.1 D in order to read at 25 cm. By the time she

reaches 55, she discovers herself holding her newspaper at a distance of 40 cm in order to see it clearly with her glasses on. (a) Where was her near point at age 45? (b) Where is her near point at age 55? (c) What power is now required for the lenses of her reading glasses so that she can again read at 25 cm? (Assume the glasses are 2.2 cm from her eyes.)

## *The Simple Magnifier

**85** • **SSM** **ISOLVE** A person with a near-point distance of 30 cm uses a simple magnifier of power 20 D. What is the magnification obtained if the final image is at infinity?

**86** • A person with a near-point distance of 25 cm wishes to obtain a magnifying power of 5 with a simple magnifier. What should be the focal length of the lens used?

**87** • What is the magnifying power of a lens of focal length 7 cm when the image is viewed at infinity by a person whose near point is at 35 cm?

**88** •• A lens of focal length 6 cm is used as a simple magnifier with the image at infinity by one person whose near point is 25 cm and by another person whose near point is 40 cm. What is the effective magnifying power of the lens for each person? Compare the size of the image on the retina when each person looks at the same object with the magnifier.

**89** •• A botanist examines a leaf using a convex lens of power 12 D as a simple magnifier. What is the expected angular magnification if (a) the final image is at infinity and (b) the final image is at 25 cm?

**90** •• **SSM** (a) Show that if the final image of a simple magnifier is to be at the near point of the eye rather than at infinity, the angular magnification is given by

$$M = \frac{x_{np}}{f} + 1$$

(b) Find the magnification of a 20-D lens for a person with a near point of 30 cm if the final image is at the near point. Draw a ray diagram for this situation.

**91** •• Show that when the image of a simple magnifier is viewed at the near point, the lateral and angular magnification of the magnifier are equal.

## *The Microscope

**92** •• **ISOLVE✔** A microscope objective has a focal length of 17 mm. It forms an image at 16 cm from its second focal point. (a) How far from the objective is the object located? (b) What is the magnifying power for a person whose near point is at 25 cm if the focal length of the eyepiece is 51 mm?

**93** •• **SSM** A microscope has an objective of focal length 8.5 mm and an eyepiece that gives an angular magnification of 10 for a person whose near point is 25 cm. The tube length is 16 cm. (a) What is the lateral magnification of the objective? (b) What is the magnifying power of the microscope?

**94** •• [ISOLVE]✓ A crude, symmetric handheld microscope consists of two converging 20-D lenses fastened in the ends of a tube 30 cm long. (a) What is the tube length of this microscope? (b) What is the lateral magnification of the objective? (c) What is the magnifying power of the microscope? (d) How far from the objective should the object be placed?

**95** •• [SSM] A compound microscope has an objective lens with a power of 45 D and an eyepiece with a power of 80 D. The lenses are separated by 28 cm. Assuming that the final image is formed 25 cm from the eye, what is the magnifying power?

**96** ••• A microscope has a magnifying power of 600, and an eyepiece of angular magnification of 15. The objective lens is 22 cm from the eyepiece. Without making any approximations, calculate (a) the focal length of the eyepiece, (b) the location of the object so that it is in focus for a normal relaxed eye, and (c) the focal length of the objective lens.

## *The Telescope

**97** • [ISOLVE]✓ A simple telescope has an objective with a focal length of 100 cm and an eyepiece of focal length 5 cm. It is used to look at the moon, which subtends an angle of about 0.009 rad. (a) What is the diameter of the image formed by the objective? (b) What angle is subtended by the final image at infinity? (c) What is the magnifying power of the telescope?

**98** • The objective lens of the refracting telescope at the Yerkes Observatory has a focal length of 19.5 m. When it is used to look at the moon, which subtends an angle of about 0.009 rad, what is the diameter of the image of the moon formed by the objective?

**99** •• [SSM] The 200-in (5.1-m) mirror of the reflecting telescope at Mt. Palomar has a focal length of 1.68 m. (a) By what factor is the light-gathering power increased over the 40-in (1.016-m) diameter refracting lens of the Yerkes Observatory telescope? (b) If the focal length of the eyepiece is 1.25 cm, what is the magnifying power of this telescope?

**100** •• [ISOLVE] An astronomical telescope has a magnifying power of 7. The two lenses are 32 cm apart. Find the focal length of each lens.

**101** •• A disadvantage of the astronomical telescope for terrestrial use (e.g., at a football game) is that the image is inverted. A Galilean telescope uses a converging lens as its objective, but a diverging lens as its eyepiece. The image formed by the objective is behind the eyepiece at its focal point so that the final image is virtual, upright, and at infinity. (a) Show that the magnifying power is $M = -f_o/f_e$, where $f_o$ is the focal length of the objective and $f_e$ is that of the eyepiece (which is negative). (b) Draw a ray diagram to show that the final image is indeed virtual, upright, and at infinity.

**102** •• A Galilean telescope (see Problem 101) is designed so that the final image is at the near point, which is 25 cm (rather than at infinity). The focal length of the objective is 100 cm and that of the eyepiece is −5 cm. (a) If the object distance is 30 m, where is the image of the objective? (b) What is the object distance for the eyepiece so that the final image is at the near point? (c) How far apart are the lenses? (d) If the object height is 1.5 m, what is the height of the final image? What is the angular magnification?

**103** ••• If you look into the wrong end of a telescope, that is, into the objective, you will see distant objects reduced in size. For a refracting telescope with an objective of focal length 2.25 m and an eyepiece of focal length 1.5 cm, by what factor is the angular size of the object reduced?

## General Problems

**104** • Show that a diverging lens can never form a real image from a real object. (*Hint: Show that s' is always negative.*)

**105** • [SSM] A camera uses a positive lens to focus light from an object onto film. Unlike the eye, the camera lens has a fixed focal length, but the lens itself can be moved slightly to vary the image distance to the image on the film. A telephoto lens has a focal length of 200 mm. By how much must the lens move to change from focusing on an object at infinity to an object at a distance of 30 m?

**106** • A wide-angle lens of a camera has a focal length of 28 mm. By how much must the lens move to change from focusing on an object at infinity to an object at a distance of 5 m? (See Problem 105.)

**107** • A thin converging lens of focal length 10 cm is used to obtain an image that is twice as large as a small object. Find the object and image distances if (a) the image is to be upright and (b) the image is to be inverted. Draw a ray diagram for each case.

**108** •• You are given two converging lenses with focal lengths of 75 mm and 25 mm. (a) Show how the lenses should be arranged to form an astronomical telescope. State which lens to use as the objective, which lens to use as the eyepiece, how far apart to place the lenses and, what angular magnification you expect. (b) Draw a ray diagram to show how rays from a distant object are magnified by the telescope.

**109** •• (a) Show how the same two lenses in Problem 108 should be arranged as a compound microscope with a tube length of 160 mm. State which lens to use as the objective, which lens to use as the eyepiece, how far apart to place the lenses, and what overall magnification you expect to get, assuming the user has a near point of 25 cm. (b) Draw a ray diagram to show how rays from a close object are magnified into a larger image.

**110** •• [SSM] A scuba diver wears a diving mask with a face plate that bulges outward with a radius of curvature of 0.5 m. There is thus a convex spherical surface between the water and the air in the mask. A fish is 2.5 m in front of the diving mask. (a) Where does the fish appear to be? (b) What is the magnification of the image of the fish?

**111** •• [ISOLVE] A 35-mm camera has a picture size of 24 mm by 36 mm. It is used to take a picture of a person 175-cm tall so that the image just fills the height (24 mm) of the film. How far should the person stand from the camera if the focal length of the lens is 50 mm?

**112** •• A 35-mm camera with interchangeable lenses is used to take a picture of a hawk that has a wing span of 2 m. The hawk is 30 m away. What would be the ideal focal length of the lens used so that the image of the wings just fills the width of the film, which is 36 mm?

**113 ••** An object is placed 12 cm to the left of a lens of focal length 10 cm. A second lens of focal length 12.5 cm is placed 20 cm to the right of the first lens. (a) Find the position of the final image. (b) What is the magnification of the image? (c) Sketch a ray diagram showing the final image.

**114 ••** (a) Show that if $f$ is the focal length of a thin lens in air, its focal length in water is

$$f' = \frac{n_w(n-1)}{n - n_w}$$

where $n_w$ is the index of refraction of water and $n$ is that of the lens. (b) Calculate the focal length in air and in water of a double concave lens of index of refraction $n = 1.5$ that has radii of magnitude 30 cm and 35 cm.

**115 ••** **SSM** (a) Find the focal length of a *thick*, double convex lens with an index of refraction of 1.5, a thickness of 4 cm, and radii of +20 cm and −20 cm. (b) Find the focal length of this lens in water.

**116 ••** A 2-cm-thick layer of water ($n = 1.33$) floats on top of a 4-cm-thick layer of carbon tetrachloride ($n = 1.46$) in a tank. How far below the top surface of the water does the bottom of the tank appear to be to an observer looking from above at normal incidence?

**117 ••** While sitting in your car, you see a jogger in your side mirror, which is convex with a radius of curvature of magnitude 2 m. The jogger is 5 m from the mirror and is approaching at 3.5 m/s. How fast does the jogger appear to be running when viewed in the mirror?

**118 ••** In the seventeenth century, Antonie van Leeuwenhoek, the first great microscopist, used simple spherical lenses made first of water droplets and then of glass for his first instruments. He made staggering discoveries with these simple lenses. Consider a glass sphere of radius 2.0 mm with an index of refraction of 1.50. Find the focal length of this lens. (*Hint:* Use the equation for refraction at a single spherical surface to find the image distance for an infinite object distance for the first surface. Then use this image point as the object point for the second surface.)

**119 •••** An object is 15 cm to the left of a thin convex lens of focal length 10 cm. A concave mirror of radius 10 cm is 25 cm to the right of the lens. (a) Find the position of the final image formed by the mirror and lens. (b) Is the image real or virtual?

Is the image upright or inverted? (c) On a diagram, show where your eye must be to see this image.

**120 •••** **SSM** When a bright light source is placed 30 cm in front of a lens, there is an upright image 7.5 cm from the lens. There is also a faint inverted image 6 cm in front of the lens due to reflection from the front surface of the lens. When the lens is turned around, this weaker, inverted image is 10 cm in front of the lens. Find the index of refraction of the lens.

**121 •••** A horizontal concave mirror with radius of curvature of 50 cm holds a layer of water with an index of refraction of 1.33 and a maximum depth of 1 cm. At what height above the mirror must an object be placed so that its image is at the same position as the object?

**122 •••** A lens with one concave side with a radius of magnitude 17 cm and one convex side with a radius of magnitude 8 cm has a focal length in air of 27.5 cm. When placed in a liquid with an unknown index of refraction, the focal length increases to 109 cm. What is the index of refraction of the liquid?

**123 •••** A glass ball of radius 10 cm has an index of refraction of 1.5. The back half of the ball is silvered so that it acts as a concave mirror (Figure 32-62). Find the position of the final image seen by an eye positioned to the left of the object and ball, for an object at (a) 30 cm and (b) 20 cm to the left of the front surface of the ball.

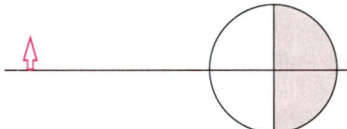

**FIGURE 32-62** Problem 123

**124 •••** (a) Show that a small change $dn$ in the index of refraction of a lens material produces a small change in the focal length $df$ given approximately by $df/f = -dn/(n-1)$. (b) Use this result to find the focal length of a thin lens for blue light, for which $n = 1.53$, if the focal length for red light, for which $n = 1.47$, is 20 cm.

**125 •••** **SSM** The lateral magnification of a spherical mirror or a thin lens is given by $m = -s'/s$. Show that for objects of small horizontal extent, the longitudinal magnification is approximately $-m^2$. (*Hint:* Show that $ds'/ds = -s'^2/s^2$.)

# Interference and Diffraction

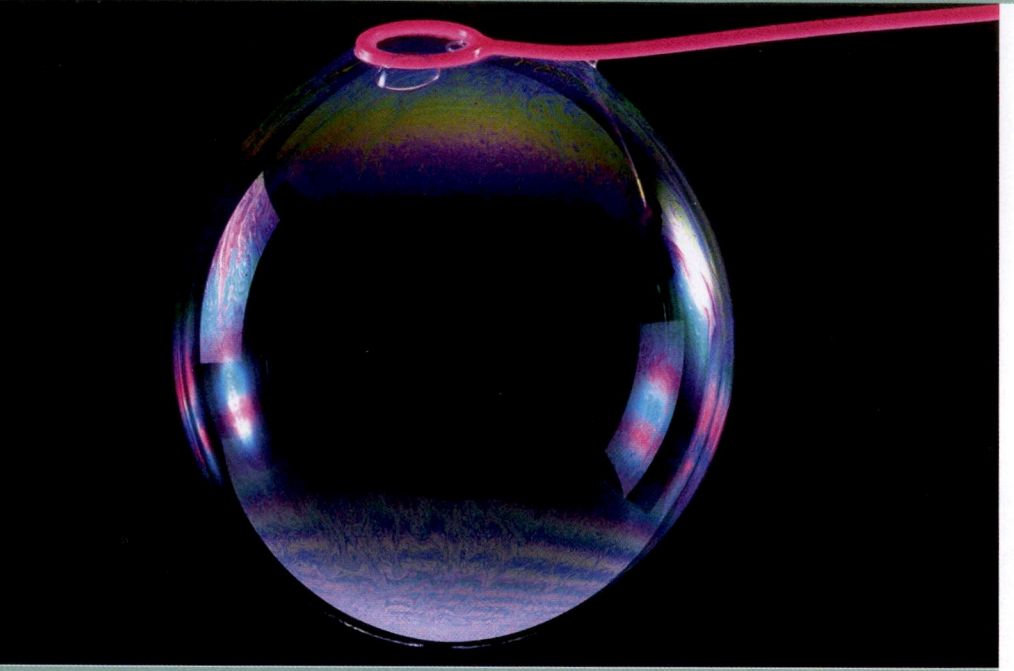

33-1  Phase Difference and Coherence

33-2  Interference in Thin Films

33-3  Two-Slit Interference Pattern

33-4  Diffraction Pattern of a Single Slit

*33-5  Using Phasors to Add Harmonic Waves

33-6  Fraunhofer and Fresnel Diffraction

33-7  Diffraction and Resolution

*33-8  Diffraction Gratings

WHITE LIGHT IS REFLECTED OFF A SOAP BUBBLE. WHEN LIGHT OF ONE WAVELENGTH IS INCIDENT ON A THIN SOAP-AND-WATER FILM, LIGHT IS REFLECTED BOTH OFF THE FRONT SURFACE AND OFF THE BACK SURFACE OF THE FILM. IF THE ORDER OF MAGNITUDE OF THE THICKNESS OF THE FILM IS THAT OF THE WAVELENGTH OF THE LIGHT, THE TWO REFLECTED LIGHT WAVES INTERFERE. IF THE TWO REFLECTED WAVES ARE 180° OUT OF PHASE, THE REFLECTED WAVE INTERFERES DESTRUCTIVELY, SO THE NET RESULT IS THAT NO LIGHT IS REFLECTED. IF WHITE LIGHT, WHICH CONTAINS MANY WAVELENGTHS, IS INCIDENT ON THE THIN FILM, THEN THE REFLECTED WAVES WILL INTERFERE DESTRUCTIVELY ONLY FOR CERTAIN WAVELENGTHS, AND FOR OTHER WAVELENGTHS THEY WILL INTERFERE CONSTRUCTIVELY. THIS PROCESS PRODUCES THE COLORED FRINGES THAT YOU SEE IN THE SOAP BUBBLE.

**?** **Have you ever wondered if the phenomena that produces the bands that you see in the light reflected off a soap bubble has any practical applications? Example 33-2 and Problem 21 reveal how the density of the bands relates to the difference in the thickness of the film for a given distance along the film.**

nterference and diffraction are the important phenomena that distinguish waves from particles.[†] Interference is the combining by superposition of two or more waves that meet at one point in space. Diffraction is the bending of waves around corners that occurs when a portion of a wavefront is cut off by a barrier or obstacle.

➤ In this chapter, we will see how the pattern of the resulting wave can be calculated by treating each point on the original wavefront as a point source, according to Huygens's principle, and calculating the interference pattern resulting from these sources.

## 33-1 Phase Difference and Coherence

When two harmonic waves of the same frequency and wavelength but differing in phase combine, the resultant wave is a harmonic wave whose amplitude depends on the phase difference. If the phase difference is zero or an integer times 360°,

[†] Before you study this chapter, you may wish to review Chapter 15 and Chapter 16, where the general topics of interference and diffraction of waves are first discussed.

the waves are in phase and interfere constructively. The resultant amplitude equals the sum of the individual amplitudes, and the intensity (which is proportional to the square of the amplitude) is maximum. If the phase difference is 180° or any odd integer times 180°, the waves are out of phase and interfere destructively. The resultant amplitude is then the difference between the individual amplitudes, and the intensity is a minimum. If the amplitudes are equal, the maximum intensity is four times that of either source and the minimum intensity is zero.

A phase difference between two waves is often the result of a difference in path length. A path difference of one wavelength produces a phase difference of 360°, which is equivalent to no phase difference at all. A path difference of one-half wavelength produces a 180° phase difference. In general, a path difference of $\Delta r$ contributes a phase difference $\delta$ given by

$$\delta = \frac{\Delta r}{\lambda} 2\pi = \frac{\Delta r}{\lambda} 360° \qquad\qquad 33\text{-}1$$

PHASE DIFFERENCE DUE TO A PATH DIFFERENCE

---

*PHASE DIFFERENCE*                          **EXAMPLE 33-1**

(*a*) **What is the minimum path difference that will produce a phase difference of 180° for light of wavelength 800 nm?** (*b*) **What phase difference will that path difference produce in light of wavelength 700 nm?**

**PICTURE THE PROBLEM** The phase difference is to 360° as the path length difference is to the wavelength.

(a) The phase difference $\delta$ is to 360° as the path length difference $\Delta r$ is to the wavelength $\lambda$. We know that $\lambda = 800$ nm and $\delta = 180°$:

$$\frac{\delta}{360°} = \frac{\Delta r}{\lambda}$$

$$\Delta r = \frac{\delta}{360°}\lambda = \frac{180°}{360°}(800 \text{ nm}) = \boxed{400 \text{ nm}}$$

(b) Set $\lambda = 700$ nm, $\Delta r = 400$ nm, and solve for $\delta$:

$$\delta = \frac{\Delta r}{\lambda} 360° = \frac{400 \text{ nm}}{700 \text{ nm}} 360°$$

$$= \boxed{206° = 3.59 \text{ rad}}$$

---

Another cause of phase difference is the 180° phase change a wave sometimes undergoes upon reflection from a boundary surface. This phase change is analogous to the inversion of a pulse on a string when it reflects from a point where the density suddenly increases, such as when a light string is attached to a heavier string or rope. The inversion of the reflected pulse is equivalent to a phase change of 180° for a sinusoidal wave (which can be thought of as a series of pulses). When light traveling in air strikes the surface of a medium in which light travels more slowly, such as glass or water, there is a 180° phase change in the reflected light. When light is originally traveling in glass or water, there is no phase change in the light reflected from the glass–air or water–air interface. This is analogous to the reflection without inversion of a pulse on a heavy string at a point where the heavy string is attached to a lighter string.

> If light traveling in one medium strikes the surface of a medium in which light travels more slowly, there is a 180° phase change in the reflected light.

PHASE DIFFERENCE DUE TO REFLECTION

As we saw in Chapter 16, interference of waves is observed when two or more coherent waves overlap. Interference of overlapping waves from two sources is not observed unless the sources are coherent. Because the light from each source is usually the result of millions of atoms radiating independently, the phase difference between the waves from such sources fluctuates randomly many times per second, so two light sources are usually not coherent. Coherence in optics is often achieved by splitting the light beam from a single source into two or more beams that can then be combined to produce an interference pattern. The light beam can be split by reflecting the light from the two closely spaced surfaces of a thin film (Section 33-2), by diffracting the beam through two small openings or slits in an opaque barrier (Section 33-3), or by using a single point source and its image in a plane mirror for the two sources (Section 33-3). Today, lasers are the most important sources of coherent light in the laboratory.

Light from an ideal monochromatic source is an infinitely long sinusoidal wave, and light from certain lasers approach this ideal. However, light from conventional *monochromatic* sources, such as gas discharge tubes designed for this purpose, consists of packets of sinusoidal light that are only a few million wavelengths long. The light from such a source consists of many such packets, each approximately the same length. The packets have essentially the same wavelength, but the packets differ in phase in a random manner. The length of one of these packets is called the **coherence length** of the light, and the time it takes one of the packets to pass a point in space is the **coherence time.** The light emitted by a gas discharge tube designed to produce monochromatic light has a coherence length of only a few millimeters. By comparison, some highly stable lasers produce light with a coherence length many kilometers long.

# 33-2 Interference in Thin Films

You have probably noticed the colored bands in a soap bubble or in the film on the surface of oily water. These bands are due to the interference of light reflected from the top and bottom surfaces of the film. The different colors arise because of variations in the thickness of the film, causing interference for different wavelengths at different points.

Consider a thin film of water (such as a small section of a soap bubble) of uniform thickness viewed at small angles with the normal, as shown in Figure 33-1. Part of the light is reflected from the upper air–water interface where it undergoes a 180° phase change. Most of the light enters the film and part of it is reflected by the bottom water–air interface. There is no phase change in this reflected light. If the light is nearly perpendicular to the surfaces, both the ray reflected from the top surface and the ray reflected from the bottom surface can enter the eye at point $P$ in the figure. The path difference between these two rays is $2t$, where $t$ is the thickness of the film. This path difference produces a phase difference of $(2t/\lambda')360°$, where $\lambda' = \lambda/n$ is the wavelength of the light in the film, and $n$ is the index of refraction of the film. The total phase difference between these two rays is thus 180° plus the phase difference due to the path difference. Destructive interference occurs when the path difference $2t$ is zero or a whole number of wavelengths $\lambda'$ (in the film). Constructive interference occurs when the path difference is an odd number of half-wavelengths.

When a thin film of water lies on a glass surface, as in Figure 33-2, the ray that reflects from the lower water–glass interface also undergoes a 180° phase change, because the index of refraction of glass (approximately 1.50) is greater than that of water (approximately 1.33). Thus, both the rays shown in the figure have undergone a 180° phase change upon reflection. The phase difference between these rays is due solely to the path difference and is given by $\delta = (2t/\lambda')360°$.

When a thin film of varying thickness is viewed with monochromatic light, such as the yellow light from a sodium lamp, alternating bright and dark bands

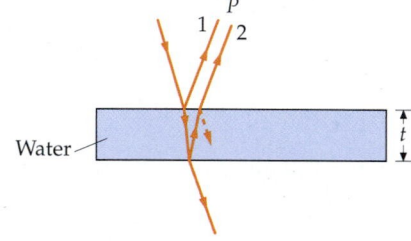

**FIGURE 33-1** Light rays reflected from the top and bottom surfaces of a thin film are coherent because both rays come from the same source. If the light is incident almost normally, the two reflected rays will be very close to each other and will produce interference.

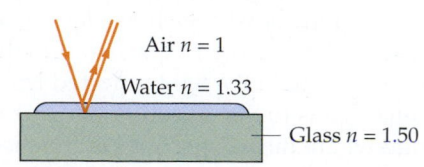

**FIGURE 33-2** The interference of light reflected from a thin film of water resting on a glass surface. In this case, both rays undergo a change in phase of 180° upon reflection.

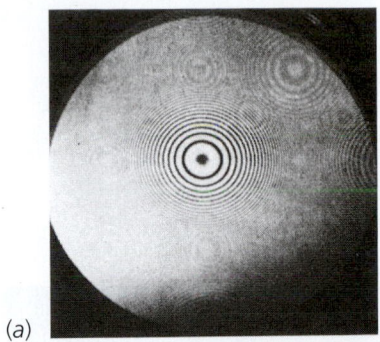

(a)

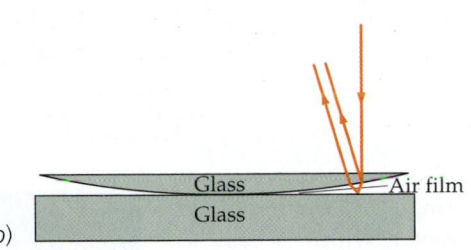

(b)

**FIGURE 33-3** (a) Newton's rings observed with light reflected from a thin film of air between a plane glass surface and a spherical glass surface. At the center, the thickness of the air film is negligible and the interference is destructive because of the 180° phase change of one of the rays upon reflection. (b) Glass surfaces for the observation of Newton's rings shown in Figure 33-3a. The thin film in this case is the film of air between the glass surfaces.

or lines called **fringes** are observed. The distance between a bright fringe and a dark fringe is that distance over which the film's thickness changes so that the path difference $2t$ is $\lambda'/2$. Figure 33-3a shows the interference pattern observed when light is reflected from an air film between a spherical glass surface and a plane glass surface in contact. These circular interference fringes are known as **Newton's rings**. Typical rays reflected at the top and bottom of the air film are shown in Figure 33-3b. Near the point of contact of the surfaces, where the path difference between the ray reflected from the upper glass–air interface and the ray reflected from the lower air–glass interface is essentially zero or is at least small compared with the wavelength of light, the interference is perfectly destructive because of the 180° phase shift of the ray reflected from the lower air–glass interface. This central region in Figure 33-3a is therefore dark. The first bright fringe occurs at the radius at which the path difference is $\lambda/2$, which contributes a phase difference of 180°. This adds to the phase shift due to reflection to produce a total phase difference of 360°, which is equivalent to a zero phase difference. The second dark region occurs at the radius at which the path difference is $\lambda$, and so on.

**FIGURE 33-4** The angle $\theta$, which is less than 0.02°, is exaggerated. The incoming and outgoing rays are essentially perpendicular to all air–glass interfaces.

---

| A WEDGE OF AIR | E X A M P L E   3 3 - 2 |
| --- | --- |

A wedge-shaped film of air is made by placing a small slip of paper between the edges of two flat pieces of glass, as shown in Figure 33-4. Light of wavelength 500 nm is incident normally on the glass, and interference fringes are observed by reflection. If the angle $\theta$ made by the plates is $3 \times 10^{-4}$ rad, how many dark interference fringes per centimeter are observed?

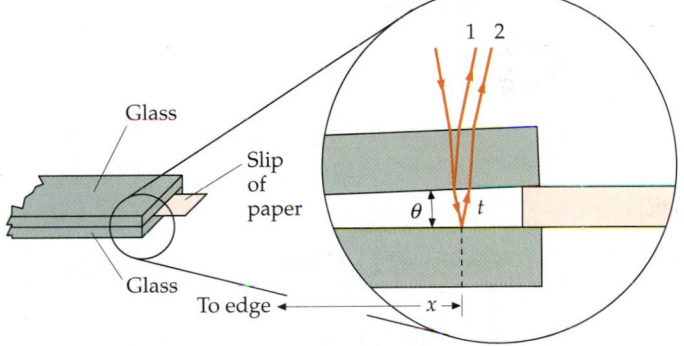

**PICTURE THE PROBLEM** We find the number of fringes per centimeter by finding the horizontal distance $x$ to the $m$th fringe and solving for $m/x$. Because the ray reflected from the bottom plate undergoes a 180° phase shift, the point of contact (where the path difference is zero) will be dark. The first dark fringe after this point occurs when $2t = \lambda'$, where $\lambda' = \lambda$ is the wavelength in the air film, and $t$ is the plate separation at $x$, as shown in Figure 33-4. Since the angle $\theta$ is small, we can use the small-angle approximation $\theta \approx t/x$.

1. The $m$th dark fringe occurs when the path difference $2t$ equals $m$ wavelengths:

$$2t = m\lambda' = m\lambda$$

$$m = \frac{2t}{\lambda}$$

2. The thickness $t$ is related to the angle $\theta$:

$$\theta = \frac{t}{x}$$

3. Substitute $t = x\theta$ into the equation for $m$:

$$m = \frac{2x\theta}{\lambda}$$

4. Calculate $m/x$:

$$\frac{m}{x} = \frac{2\theta}{\lambda} = \frac{2(3 \times 10^{-4})}{5 \times 10^{-7}\,\text{m}} = 1200\,\text{m}^{-1}$$

$$= \boxed{12\,\text{cm}^{-1}}$$

**REMARKS** We therefore observe 12 dark fringes per centimeter. In practice, the number of fringes per centimeter, which is easy to count, can be used to determine the angle. Note that if the angle of the wedge is increased, the fringes become more closely spaced.

**EXERCISE** How many dark fringes per centimeter are observed if light of wavelength 650 nm is used? (*Answer* 9.2 cm$^{-1}$)

Figure 33-5*a* shows interference fringes produced by a wedge-shaped air film between two flat glass plates, as in Example 33-2. Plates that produce straight fringes, such as those in Figure 33-5*a*, are said to be **optically flat**. To be optically flat, a surface must be flat to within a small fraction of a wavelength. A similar wedge-shaped air film formed by two ordinary glass plates yields the irregular fringe pattern in Figure 33-5*b*, which indicates that these plates are not optically flat.

One application of interference effects in thin films is in nonreflecting lenses, which are made by covering a lens with a thin film of a material that has an index of refraction of approximately 1.38, which is between that of glass and air. Then the intensities of the light reflected from the top and bottom surfaces of the film are approximately equal, and since both rays undergo a 180° phase change, there is no phase difference between the rays due to reflection. The thickness of the film is chosen to be $\lambda'/4 = \lambda/4n$, where $\lambda$ is in the middle of the visible spectrum, so that there is a phase change of 180° due to the path difference of $\lambda'/2$. Reflection from the coated surface is thus minimized, whereas transmission through the surface is maximized.

*(a)*

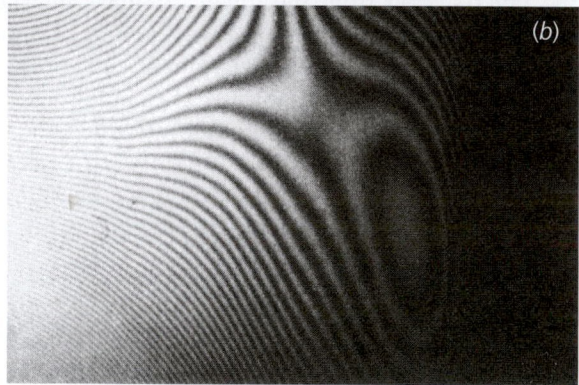

*(b)*

**FIGURE 33-5** (*a*) Straight-line fringes from a wedge-shaped film of air, like that shown in Figure 33-4. The straightness of the fringes indicates that the glass plates are optically flat. (*b*) Fringes from a wedge-shaped film of air between glass plates that are not optically flat.

# 33-3 Two-Slit Interference Pattern

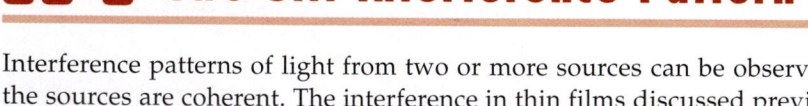

Interference patterns of light from two or more sources can be observed only if the sources are coherent. The interference in thin films discussed previously can be observed because the two beams come from the same light source but are separated by reflection. In Thomas Young's famous experiment, in which he demonstrated the wave nature of light, two coherent light sources are produced by illuminating two very narrow parallel slits with a single light source. We saw in Chapter 15 that when a wave encounters a barrier with a very small opening, the opening acts as a point source of waves (Figure 33-6). In Young's experiment, diffraction causes each slit to act as a line source (which is equivalent to a point source in two dimensions). The interference pattern is observed on a screen far from the slits (Figure 33-7*a*). At very large distances from the slits, the lines from

**FIGURE 33-6** Plane water waves in a ripple tank encountering a barrier with a small opening. The waves to the right of the barrier are circular waves that are concentric about the opening, just as if there were a point source at the opening.

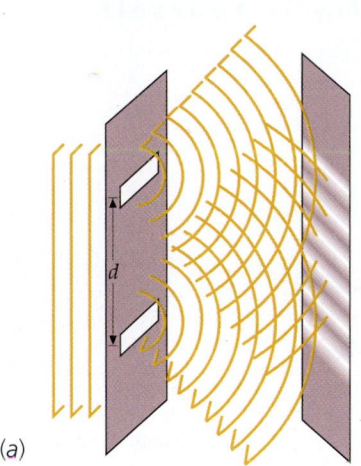

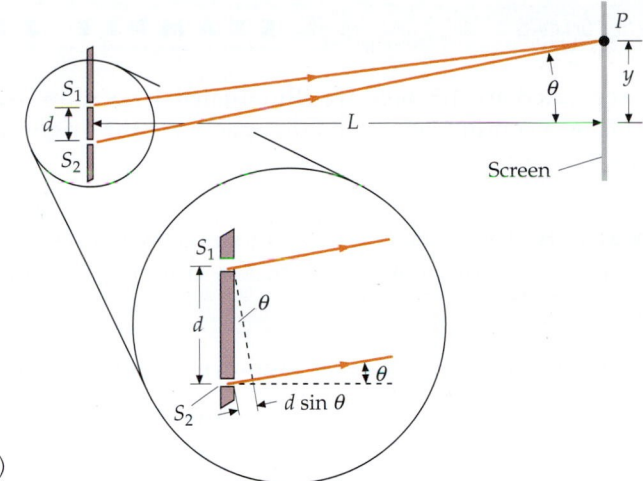

(a)                                                                      (b)

the two slits to some point $P$ on the screen are approximately parallel, and the path difference is approximately $d \sin \theta$, where $d$ is the separation of the slits, as shown in Figure 33-7b. When the path difference is equal to an integral number of wavelengths, the interference is constructive. We thus have interference maxima at an angle $\theta_m$ given by

$$d \sin \theta_m = m\lambda, \qquad m = 0, 1, 2, \ldots \tag{33-2}$$

TWO-SLIT INTERFERENCE MAXIMA

where $m$ is called the **order number.** The interference minima occur at

$$d \sin \theta_m = (m - \tfrac{1}{2})\lambda, \qquad m = 1, 2, 3, \ldots \tag{33-3}$$

TWO-SLIT INTERFERENCE MINIMA

The phase difference $\delta$ at a point $P$ is related to the path difference $d \sin \theta$ by

$$\frac{\delta}{2\pi} = \frac{d \sin \theta}{\lambda} \tag{33-4}$$

We can relate the distance $y_m$ measured along the screen from the central point to the $m$th bright fringe (see Figure 33-7b) to the distance $L$ from the slits to the screen:

$$\tan \theta_m = \frac{y_m}{L}$$

For small angles, $\tan \theta \approx \sin \theta$. Substituting $y_m/L$ for $\sin \theta_m$ in Equation 33-2 and solving for $y_m$ gives

$$y_m = m\frac{\lambda L}{d} \tag{33-5}$$

DISTANCE ON SCREEN TO THE MTH BRIGHT FRINGE

From this result, we see that for small angles the fringes are equally spaced on the screen.

**FIGURE 33-7** (a) Two slits act as coherent sources of light for the observation of interference in Young's experiment. Cylindrical waves from the slits overlap and produce an interference pattern on a screen. (b) Geometry for relating the distance $y$ measured along the screen to $L$ and $\theta$. When the screen is very far away compared with the slit separation, the rays from the slits to a point on the screen are approximately parallel, and the path difference between the two rays is $d \sin \theta$.

---

Two narrow slits separated by 1.5 mm are illuminated by yellow light of wavelength 589 nm from a sodium lamp. Find the spacing of the bright fringes observed on a screen 3 m away.

**PICTURE THE PROBLEM** The distance $y_m$ measured along the screen to the $m$th bright fringe is given by Equation 33-2, with $L = 3$ m, $d = 1.5$ mm, and $\lambda = 589$ nm.

**Cover the column to the right and try these on your own before looking at the answers.**

| Steps | Answers |
|---|---|
| 1. Make a sketch of the situation (Figure 33-8). | |
| 2. Using the sketch, obtain an expression for the spacing between fringes. | fringe spacing $= \dfrac{y_3}{3}$ |
| 3. Apply Equation 33-2 to the $m = 3$ fringe. | $d \sin \theta_3 = 3\lambda$ |
| 4. Using trig, relate $y_3$ and $\theta_3$. | $\sin \theta_3 \approx \tan \theta_3 = \dfrac{y_3}{L}$ |
| 5. Substitute into the step 3 result and solve for the fringe spacing. | $\dfrac{y_3}{3} = \boxed{1.18 \text{ mm}}$ |

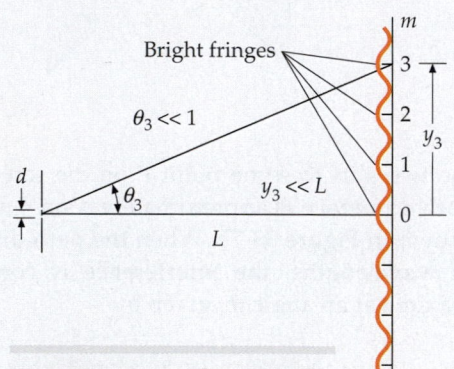

**FIGURE 33-8** The vertical scale of the figure is expanded.

**REMARKS** The fringes are uniformly spaced only to the degree that the small-angle approximation is valid. That is, to the degree that $\lambda/d \ll 1$. In this example, $\lambda/d = (589 \text{ nm})/(1.5 \text{ mm}) = 0.0004$.

**EXERCISE** A point source of light ($\lambda = 589$ nm) is placed 0.4 mm above the surface of a glass mirror. Interference fringes are observed on a screen 6 m away, and the interference is between the light reflected off the front surface of the glass and the light traveling from the source directly to the screen. Find the spacing of the fringes. (*Answer*  4.42 mm)

## Calculation of Intensity

To calculate the intensity of the light on the screen at a general point $P$, we need to add two harmonic wave functions that differ in phase.[†] The wave functions for electromagnetic waves are the electric field vectors. Let $E_1$ be the electric field at some point $P$ on the screen due to the waves from slit 1, and let $E_2$ be the electric field at that point due to waves from slit 2. Since the angles of interest are small, we can assume that these fields are parallel. Both electric fields oscillate with the same frequency (they result from a single source that illuminates both slits) and they have the same amplitude. (The path difference is only of the order of a few wavelengths of light at most.) They have a phase difference $\delta$ given by Equation 33-4. If we represent these wave functions by

$$E_1 = A_0 \sin \omega t$$

and

$$E_2 = A_0 \sin(\omega t + \delta)$$

the resultant wave function is

---

[†] We did this in Chapter 16 where we first discussed the superposition of two waves.

$$E = E_1 + E_2 = A_0 \sin \omega t + A_0 \sin(\omega t + \delta)$$

$$= 2A_0 \cos \tfrac{1}{2}\delta \sin(\omega t + \tfrac{1}{2}\delta) \qquad \text{33-6}$$

where we used the identity

$$\sin \alpha + \sin \beta = 2 \cos \tfrac{1}{2}(\alpha - \beta) \sin \tfrac{1}{2}(\alpha + \beta) \qquad \text{33-7}$$

The amplitude of the resultant wave is thus $2A_0 \cos \tfrac{1}{2}\delta$. It has its maximum value of $2A_0$ when the waves are in phase and is zero when they are $180°$ out of phase. Since the intensity is proportional to the square of the amplitude, the intensity at any point $P$ is

$$I = 4I_0 \cos^2 \tfrac{1}{2}\delta \qquad \text{33-8}$$

INTENSITY IN TERMS OF PHASE DIFFERENCE

where $I_0$ is the intensity of the light on the screen from either slit separately. The phase angle $\delta$ is related to the position on the screen by Equation 33-4.

Figure 33-9a shows the intensity pattern as seen on a screen. A graph of the intensity as a function of $\sin \theta$ is shown in Figure 33-9b. For small $\theta$, this is equivalent to a plot of intensity versus $y$ (since $y = L \tan \theta \approx L \sin \theta$). The intensity $I_0$ is that from each slit separately. The dashed line in Figure 33-9b shows the average intensity $2I_0$, which is the result of averaging over a distance containing many interference maxima and minima. This is the intensity that would arise from the two sources if they acted independently without interference, that is, if they were not coherent. Then the phase difference between the two sources would fluctuate randomly, so that only the average intensity would be observed.

Figure 33-10 shows another method of producing the two-slit interference pattern, an arrangement known as **Lloyd's mirror.** A single slit is placed at a distance $\tfrac{1}{2}d$ above the plane of a mirror. Light striking the screen directly from the source interferes with the light that is reflected from the mirror. The reflected light can be considered to come from the virtual image of the slit formed by the mirror. Because of the $180°$ change in phase upon reflection at the mirror, the interference pattern is that of two coherent line sources that differ in phase by $180°$. The pattern is the same as that shown in Figure 33-9 for two slits, except that the maxima and minima are interchanged. Constructive interference occurs at points for which the path difference is a half-wavelength or any odd number of half-wavelengths. At these points, the $180°$ phase difference due to the path difference combines with the $180°$ phase difference of the sources to produce constructive interference.

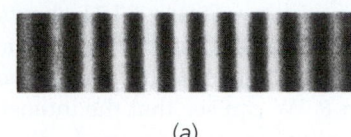

(a)

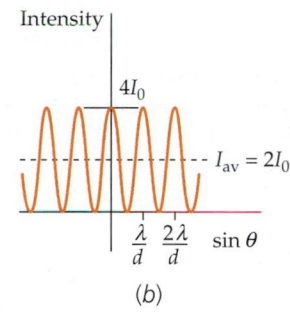

(b)

**FIGURE 33-9** (a) The interference pattern observed on a screen far away from the two slits shown in Figure 33-7. (b) Plot of intensity versus $\sin \theta$. The maximum intensity is $4I_0$, where $I_0$ is the intensity due to each slit separately. The average intensity (dashed line) is $2I_0$.

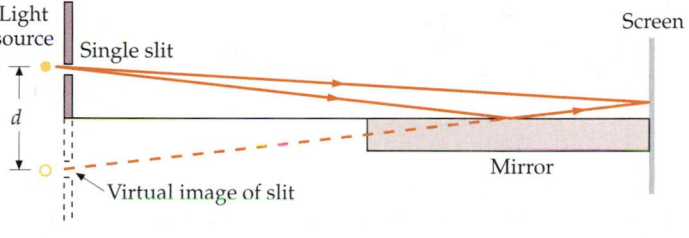

**FIGURE 33-10** Lloyd's mirror for producing a two-slit interference pattern. The two sources (the slit and its image) are coherent and are $180°$ out of phase. The central interference band at the point equidistant from the sources is dark.

# 33-4 Diffraction Pattern of a Single Slit

In our discussion of the interference patterns produced by two or more slits, we assumed that the slits were very narrow so that we could consider the slits to be line sources of cylindrical waves, which in our two-dimensional diagrams are point sources of circular waves. We could therefore assume that the intensity due to one slit acting alone was the same ($I_0$) at any point $P$ on the screen, independent of the angle $\theta$ made between the ray to point $P$ and the normal line between the slit and the screen. When the slit is not narrow, the intensity on a screen far away is not independent of angle but decreases as the angle increases. Consider a slit of width $a$. Figure 33-11 shows the intensity pattern on a screen far away from the slit of width $a$ as a function of $\sin \theta$. We can see that the intensity is maximum in the forward direction ($\sin \theta = 0$) and decreases to zero at an angle that depends on the slit width $a$ and the wavelength $\lambda$.

Most of the light intensity is concentrated in the broad **central diffraction maximum,** although there are minor secondary maxima bands on either side of the central maximum. The first zeroes in the intensity occur at angles specified by

$$\sin \theta_1 = \lambda/a \qquad\qquad 33\text{-}9$$

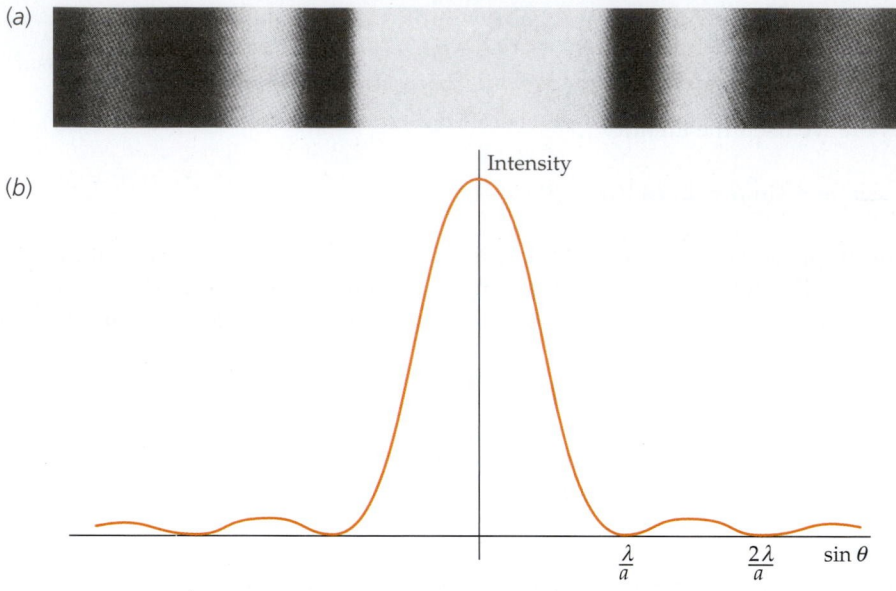

(a)

(b)

**FIGURE 33-11** (*a*) Diffraction pattern of a single slit as observed on a screen far away. (*b*) Plot of intensity versus $\sin \theta$ for the pattern in Figure 33-11*a*.

Note that for a given wavelength $\lambda$, Equation 33-9 describes how variations in the slit width result in variations in the angular width of the central maximum. If we *increase* the slit width $a$, the angle $\theta_1$ at which the intensity first becomes zero *decreases,* giving a more narrow central diffraction maximum. Conversely, if we *decrease* the slit width, the angle of the first zero *increases,* giving a wider central diffraction maximum. When $a$ is smaller than $\lambda$, then $\sin \theta_1$ would have to exceed 1 to satisfy Equation 33-9. Thus, for $a$ less than $\lambda$, there are no points of zero intensity in the pattern, and the slit acts as a line source (a point source in two dimensions) radiating light energy essentially equal in all directions.

Multiplying both sides of Equation 33-9 by $a/2$ gives

$$\tfrac{1}{2}a \sin \theta_1 = \tfrac{1}{2}\lambda \qquad\qquad 33\text{-}10$$

The quantity $\tfrac{1}{2}a \sin \theta_1$ is the path difference between a light ray leaving the middle of the upper half of the slit and one leaving the middle of the lower half of the slit. We see that the first diffraction *minimum* occurs when these two rays are 180° out of phase, that is, when their path difference equals a half-wavelength. We can understand this result by considering each point on a wavefront to be a point source of light in accordance with Huygens's principle. In Figure 33-12, we have placed a line of dots on the wavefront at the slit to represent these point sources schematically. Suppose, for example, that we have 100 such dots and that we look at an angle $\theta_1$ for which $a \sin \theta_1 = \lambda$. Let us consider the slit to be divided into two halves, with the first 50 sources in the upper half and sources 51 through 100 in the lower half. When the path difference between the middle of the upper half and the middle of the lower half of the slit equals a half-wavelength, the path difference between source 1 (the first source in the upper half) and source 51 (the first source in the lower half) is $\tfrac{1}{2}\lambda$. The waves from these two sources will be out of phase by 180° and will thus cancel. Similarly, waves from the second source in each region (source 2 and source 52) will cancel. Continuing this argument, we can see that the waves from each pair of sources separated by $a/2$ will cancel. Thus, there will be no light energy at this angle. We can extend this argument to the second and third minima in the diffraction pattern of Figure 33-11. At an angle $\theta_2$ where $a \sin \theta_2 = 2\lambda$, we can divide the slit into four regions, two regions for the top half and two regions for the bottom half. Using this same argument, the light intensity from the top half is zero because of the cancellation of pairs of sources, and, similarly, the light intensity from the bottom half is zero. The general expression for the points of zero intensity in the diffraction pattern of a single slit is thus

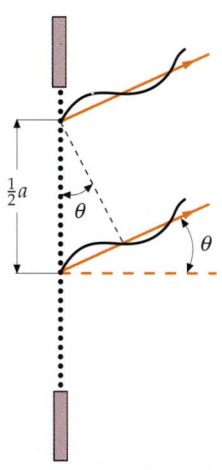

**FIGURE 33-12** A single slit is represented by a large number of point sources of equal amplitude. At the first diffraction minimum of a single slit, the waves from each point source in the upper half of the slit are 180° out of phase with the wave from the point source a distance $a/2$ lower in the slit. Thus, the interference from each such pair of point sources is destructive.

$$a \sin \theta_m = m\lambda, \qquad m = 1, 2, 3 \ldots \qquad\qquad 33\text{-}11$$

POINTS OF ZERO INTENSITY FOR A SINGLE-SLIT DIFFRACTION PATTERN

Usually, we are just interested in the first occurrence of a minimum in the light intensity because nearly all of the light energy is contained in the central diffraction maximum.

In Figure 33-13, the distance $y_1$ from the central maximum to the first diffraction minimum is related to the angle $\theta_1$ and the distance $L$ from the slit to the screen by

$$\tan \theta_1 = \frac{y_1}{L}$$

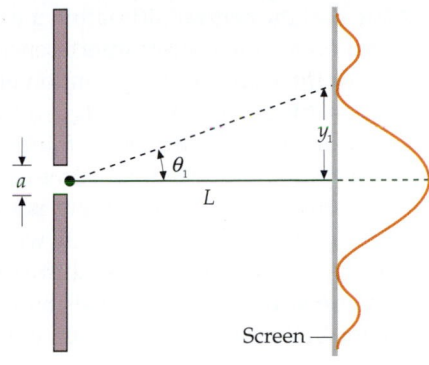

**FIGURE 33-13** The distance $y_1$ measured along the screen from the central maximum to the first diffraction minimum is related to the angle $\theta_1$ by $\tan \theta_1 = y_1/L$, where $L$ is the distance to the screen.

*WIDTH OF THE CENTRAL DIFFRACTION MAXIMUM*      **EXAMPLE 33-4**

In a lecture demonstration of single-slit diffraction, a laser beam of wavelength 700 nm passes through a vertical slit 0.2 mm wide and hits a screen 6 m away. Find the width of the central diffraction maximum on the screen; that is, find the distance between the first minimum on the left and the first minimum on the right of the central maximum.

**PICTURE THE PROBLEM** Referring to Figure 33-13, the width of the central diffraction maximum is $2y_1$.

1. The half-width of the central maxima $y_1$ is related to the angle $\theta_1$ by:

$$\tan \theta_1 = \frac{y_1}{L}$$

2. The angle $\theta_1$ is related to the slit width $a$ by Equation 33-11:

$$\sin \theta_1 = \lambda/a$$

3. Solve the step 2 result for $\theta_1$, substitute into the step 1 result, and solve for $2y_1$:

$$2y_1 = 2L \tan \theta_1 = 2L \tan\left(\sin^{-1}\frac{\lambda}{a}\right)$$

$$= 2(6 \text{ m}) \tan\left(\sin^{-1}\frac{700 \times 10^{-9} \text{ m}}{0.0002 \text{ m}}\right)$$

$$= 4.2 \times 10^{-2} \text{ m} = \boxed{4.20 \text{ cm}}$$

**REMARKS** Since $\sin \theta_1 = \lambda/a = (700 \text{ nm})/(0.2 \text{ mm}) = 0.0035$, we can use the small-angle approximation to evaluate $2y_1$. In this approximation, $\sin \theta_1 = \tan \theta_1$, so $\lambda/a = y_1/L$ and $2y_1 = 2L\lambda/a = 2(6 \text{ m})(700 \text{ nm})/(0.2 \text{ mm}) = 4.20 \text{ cm}$. (This approximate value is in agreement with the exact value to within 0.0006 percent.)

## Interference–Diffraction Pattern of Two Slits

When there are two or more slits, the intensity pattern on a screen far away is a combination of the single-slit diffraction pattern and the multiple-slit interference pattern we have studied. Figure 33-14 shows the intensity pattern on a screen far from two slits whose separation $d$ is 10 times the width $a$ of each slit. The pattern is the same as the two-slit pattern with very narrow slits (Figure 33-11) except that it is modulated by the single-slit diffraction pattern; that is, the intensity due to each slit separately is now not constant but decreases with angle, as shown in Figure 33-14b.

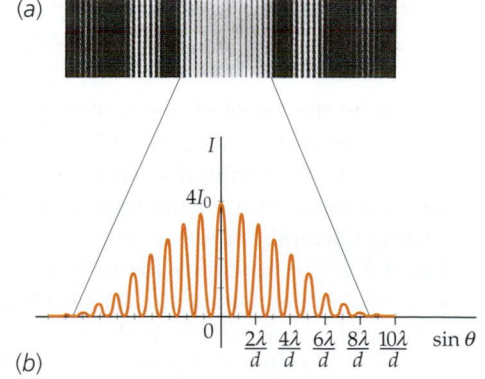

(a)

(b)

**FIGURE 33-14** (a) Interference–diffraction pattern for two slits whose separation $d$ is equal to 10 times their width $a$. The tenth interference maximum on either side of the central interference maximum is missing because it falls at the first diffraction minimum. (b) Plot of intensity versus $\sin \theta$ for the central band of the pattern in Figure 33-14a.

Note that the central diffraction maximum in Figure 33-14 contains 19 interference maxima—the central interference maximum and 9 maxima on either side. The tenth interference maximum on either side of the central one is at the angle $\theta_{10}$, given by $\sin \theta_{10} = 10\lambda/d = \lambda/a$, since $d = 10a$. This coincides with the position of the first diffraction minimum, so this interference maximum is not seen. At these points, the light from the two slits would be in phase and would interfere constructively, but there is no light from either slit because the points are diffraction minima. In general, we can see that if $m = d/a$, the $m$th interference maximum will fall at the first diffraction minimum. Since the $m$th fringe is not seen, there will be $m - 1$ fringes on each side of the central fringe for a total of $N$ fringes in the central maximum, where $N$ is given by

$$N = 2(m - 1) + 1 = 2m - 1 \qquad\qquad\qquad 33\text{-}12$$

---

*INTERFERENCE AND DIFFRACTION* **EXAMPLE 33-5**

Two slits of width $a = 0.015$ mm are separated by a distance $d = 0.06$ mm and are illuminated by light of wavelength $\lambda = 650$ nm. How many bright fringes are seen in the central diffraction maximum?

**PICTURE THE PROBLEM** We need to find the value of $m$ for which the $m$th interference maximum coincides with the first diffraction minimum. Then there will be $N = 2m - 1$ fringes in the central maximum.

1. Find the angle $\theta_1$ of the first diffraction minimum:

$$\sin \theta_1 = \frac{\lambda}{a} \text{ (first diffraction minimum)}$$

2. Find the angle $\theta_m$ of the $m$th interference maxima:

$$\sin \theta_m = \frac{m\lambda}{d} \text{ ($m$th interference maxima)}$$

3. Set these angles equal and solve for $m$:

$$\frac{m\lambda}{d} = \frac{\lambda}{a}$$

$$m = \frac{d}{a} = \frac{0.06 \text{ mm}}{0.015 \text{ mm}} = 4$$

4. The first diffraction minimum coincides with the fourth bright fringe. Therefore, there are 3 bright fringes visible on either side of the central diffraction maximum. These 6 maxima, plus the central interference maximum, combine for a total of 7 bright fringes in the central diffraction maximum:

$$N = \boxed{7 \text{ bright fringes}}$$

---

# *33-5 Using Phasors to Add Harmonic Waves

To calculate the interference pattern produced by three, four, or more coherent light sources and to calculate the diffraction pattern of a single slit, we need to combine several harmonic waves of the same frequency that differ in phase. A simple geometric interpretation of harmonic wave functions leads to a method of adding harmonic waves of the same frequency by geometric construction.

Let the wave functions for two waves at some point be $E_1 = A_1 \sin \alpha$ and $E_2 = A_2 \sin(\alpha + \delta)$, where $\alpha = \omega t$. Our problem is then to find the sum:

$$E_1 + E_2 = A_1 \sin \alpha + A_2 \sin(\alpha + \delta)$$

We can represent each wave function by a two-dimensional vector, as shown in Figure 33-15. The geometric method of addition is based on the fact that the $y$ (or $x$) component of the resultant of two vectors equals the sum of the $y$ (or $x$) components of the vectors, as illustrated in the figure. The wave function $E_1$ is represented by the vector $\vec{A}_1$. As the time varies, this vector rotates in the $xy$ plane with angular frequency $\omega$. Such a vector is called a **phasor**. (We encountered phasors in our study of ac circuits in Section 29-4.) The wave function $E_2$ is the $y$ component of a phasor of magnitude $A_2$ that makes an angle $\alpha + \delta$ with the $x$ axis. By the laws of vector addition, the sum of these components equals the $y$ component of the resultant phasor $\vec{A}$, as shown in Figure 33-15. The $y$ component of the resultant phasor, $A \sin(\alpha + \delta')$, is a harmonic wave function that is the sum of the two original wave functions:

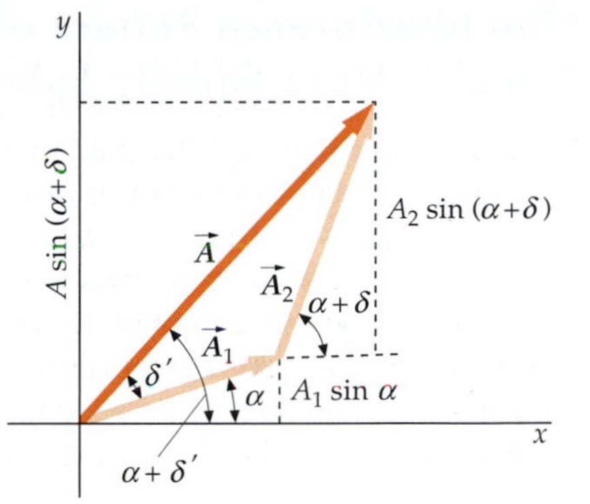

$$A_1 \sin \alpha + A_2 \sin(\alpha + \delta) = A \sin(\alpha + \delta')$$   33-13

where $A$ (the amplitude of the resultant wave) and $\delta'$ (the phase of the resultant wave relative to the first wave) are found by adding the phasors representing the waves. As time varies, $\alpha$ varies. The phasors representing the two wave functions and the resultant phasor representing the resultant wave function rotate in space, but their relative positions do not change because they all rotate with the same angular velocity $\omega$.

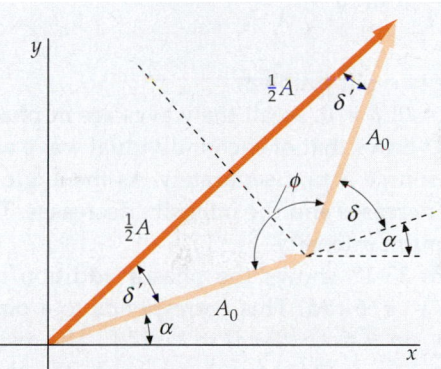

**FIGURE 33-15** Phasor representation of wave functions.

---

*WAVE SUPERPOSITION USING PHASORS*   **EXAMPLE 33-6 Try It Yourself**

Use the phasor method of addition to derive Equation 33-16 for the superposition of two waves of the same amplitude.

**PICTURE THE PROBLEM** Represent the waves $y_1 = A_0 \sin \alpha$ and $y_2 = A_0 \sin(\alpha + \delta)$ by vectors (phasors) of length $A_0$ making an angle $\delta$ with one another. The resultant wave $y_r = A \sin(\alpha + \delta')$ is represented by the sum of these vectors, which form an isosceles triangle, as shown in Figure 33-16.

**FIGURE 33-16**

Cover the column to the right and try these on your own before looking at the answers.

| Steps | Answers |
|---|---|
| 1. Find the phase angle $\delta'$ in terms of $\phi$ from the fact that the three angles in the triangle must sum to 180°. | $\delta' + \delta' + \phi = 180°$ |
| 2. Relate $\phi$ to $\delta$. | $\delta + \phi = 180°$ |
| 3. Eliminate $\phi$ from the step 1 and step 2 results and solve for $\delta'$. | $\delta' = \frac{1}{2}\delta$ |
| 4. Write $\cos \delta'$ in terms of $A$ and $A_0$. | $\cos \delta' = \dfrac{\frac{1}{2}A}{A_0}$ |
| 5. Solve for $A$ in terms of $\delta$. | $A = 2A_0 \cos \delta' = 2A_0 \cos \frac{1}{2}\delta$ |
| 6. Use your results for $A$ and $\delta'$ to write the resultant wave function. | $y_r = A \sin(\alpha + \delta')$ |
| | $= \boxed{2A_0 \cos(\tfrac{1}{2}\delta)\sin(\alpha + \tfrac{1}{2}\delta)}$ |

**EXERCISE** Find the resultant wave function of the two waves $E_1 = 4\sin(\omega t)$ and $E_2 = 3\sin(\omega t + 90°)$ [*Answer* $E_1 + E_2 = 5\sin(\omega t + 37°)$]

## *The Interference Pattern of Three or More Equally Spaced Sources

We can apply the phasor method of addition to calculate the interference pattern of three or more equally spaced, coherent sources in phase. We are most interested in the interference maxima and minima. Figure 33-17 illustrates the case of three sources. The geometry is the same as for two sources. At a great distance from the sources, the rays from the sources to a point $P$ on the screen are approximately parallel. The path difference between the first and second source is then $d \sin \theta$, as before, and the path difference between the first and third source is $2d \sin \theta$. The wave at point $P$ is the sum of three waves. Let $\alpha = \omega t$ be the phase of the first wave at point $P$. We thus have the problem of adding three waves of the form

$$E_1 = A_0 \sin \alpha$$

$$E_2 = A_0 \sin(\alpha + \delta)$$

$$E_3 = A_0 \sin(\alpha + 2\delta) \qquad \text{33-14}$$

where

$$\delta = \frac{2\pi}{\lambda} d \sin \theta \approx \frac{2\pi yd}{\lambda \ L} \qquad \text{33-15}$$

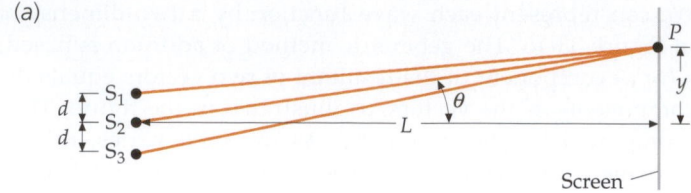

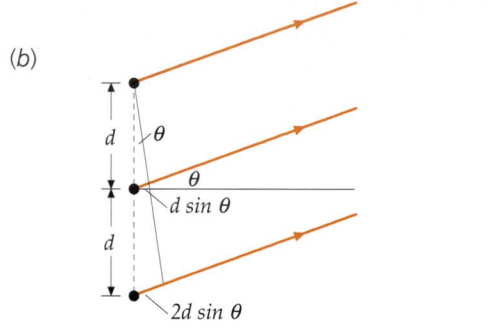

**FIGURE 33-17** Geometry for calculating the intensity pattern far away from three equally spaced, coherent sources that are in phase.

as in the two-slit problem.

At $\theta = 0$, $\delta = 0$, so all the waves are in phase. The amplitude of the resultant wave is 3 times that of each individual wave and the intensity is 9 times that due to each source acting separately. As the angle $\delta$ increases from $\theta = 0$, the phase angle $\delta$ increases and the intensity decreases. The position $\theta = 0$ is thus a position of maximum intensity.

Figure 33-18 shows the phasor addition of three waves for a phase angle $\delta = 30° = \pi/6$ rad. This corresponds to a point $P$ on the screen for which $\theta$ is given by $\sin \theta = \lambda\delta/(2\pi d) = \lambda/(12d)$. The resultant amplitude $A$ is considerably less than 3 times that of each source. As the phase angle $\delta$ increases, the resultant amplitude decreases until the amplitude is zero at $\delta = 120°$. For this phase difference, the three phasors form an equilateral triangle (Figure 33-19). This first interference minimum for three sources occurs at a smaller phase angle $\delta$ (and therefore at a smaller space angle $\theta$) than it does for only two sources (for which the first minimum occurs at $\delta = 180°$). As $\delta$ increases from 120°, the resultant amplitude increases, reaching a secondary maximum near $\delta = 180°$. At the phase angle $\delta = 180°$, the amplitude is the same as that from a single source, since the waves from the first two sources cancel each other, leaving only the third. The intensity of the secondary maximum is one-ninth that of the maximum at $\theta = 0$. As $\delta$ increases beyond 180°, the amplitude again decreases and is zero at $\delta = 180° + 60° = 240°$. For $\delta$ greater than 240°, the amplitude increases and is again 3 times that of each source when $\delta = 360°$. This phase angle corresponds to a path difference of 1 wavelength for the waves from the first two sources and 2 wavelengths for the waves from the first and third sources. Hence, the

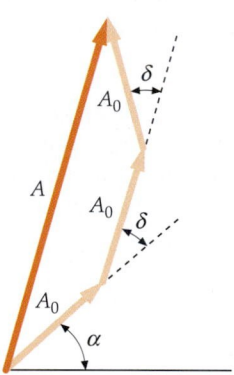

**FIGURE 33-18** Phasor diagram for determining the resultant amplitude $A$ due to three waves, each of amplitude $A_0$, that have phase differences of $\delta$ and $2\delta$ due to path differences of $d \sin \theta$ and $2d \sin \theta$. The angle $\alpha = \omega t$ varies with time, but this does not affect the calculation of $A$.

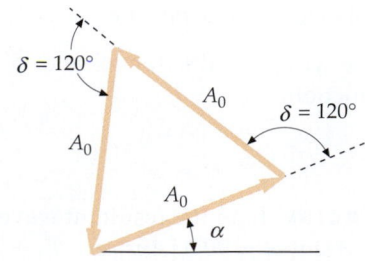

**FIGURE 33-19** The resultant amplitude for the waves from three sources is zero when $\delta$ is 120°. This interference minimum occurs at a smaller angle $\theta$ than does the first minimum for two sources, which occurs when $\delta$ is 180°.

three waves are in phase at this point. The largest maxima, called the principal maxima, are at the same positions as for just two sources, which are those points corresponding to the angles $\theta$ given by

$$d \sin \theta_m = m\lambda, \qquad m = 0, 1, 2, \ldots \qquad \text{33-16}$$

These maxima are stronger and narrower than those for two sources. They occur at points for which the path difference between adjacent sources is zero or an integral number of wavelengths.

These results can be generalized to more than three sources. For four equally spaced sources that are in phase, the principal interference maxima are again given by Equation 33-16, but these maxima are even more intense, they are narrower, and there are two small secondary maxima between each pair of principal maxima. At $\theta = 0$, the intensity is 16 times that due to a single source. The first interference minimum occurs when $\delta$ is 90°, as can be seen from the phasor diagram of Figure 33-20. The first secondary maximum is near $\delta = 132°$, leaving only the wave from the fourth source. The intensity of the secondary maximum is approximately one-sixteenth that of the central maximum. There is another minimum at $\delta = 180°$, another secondary maximum near $\delta = 228°$, and another minimum at $\delta = 270°$ before the next principal maximum at $\delta = 360°$.

Figure 33-21 shows the intensity patterns for two, three, and four equally spaced coherent sources. Figure 33-22 shows a graph of $I/I_0$, where $I_0$ is the intensity due to each source acting separately. For three sources, there is a very small secondary maximum between each pair of principal maxima, and the principal maxima are sharper and more intense than those due to just two sources. For four sources, there are two small secondary maxima between each pair of principal maxima, and the principal maxima are even more narrow and intense.

From this discussion, we can see that as we increase the number of sources, the intensity becomes more and more concentrated in the principal maxima given by Equation 33-16, and these maxima become narrower. For $N$ sources, the intensity of the principal maxima is $N^2$ times that due to a single source. The first minimum occurs at a phase angle of $\delta = 360°/N$, for which the $N$ phasors form a closed polygon of $N$ sides. There are $N - 2$ secondary maxima between each pair of principal maxima. These secondary maxima are very weak compared with the principal maxima. As the number of sources is increased, the principal maxima become sharper and more intense, and the intensities of the secondary maxima become negligible compared to those of the principal maxima.

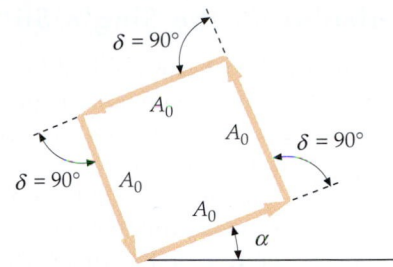

**FIGURE 33-20** Phasor diagram for the first minimum for four equally spaced in-phase sources. The amplitude is zero when the phase difference of the waves from adjacent sources is 90°.

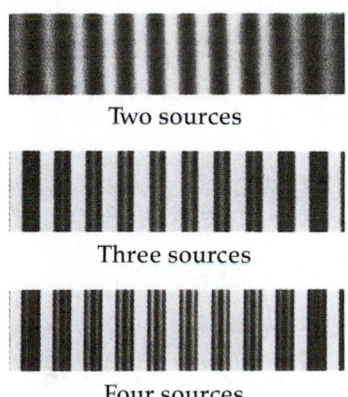

**FIGURE 33-21** Intensity patterns for two, three, and four equally spaced coherent sources. There is a secondary maximum between each pair of principal maxima for three sources, and two secondary maxima for four sources.

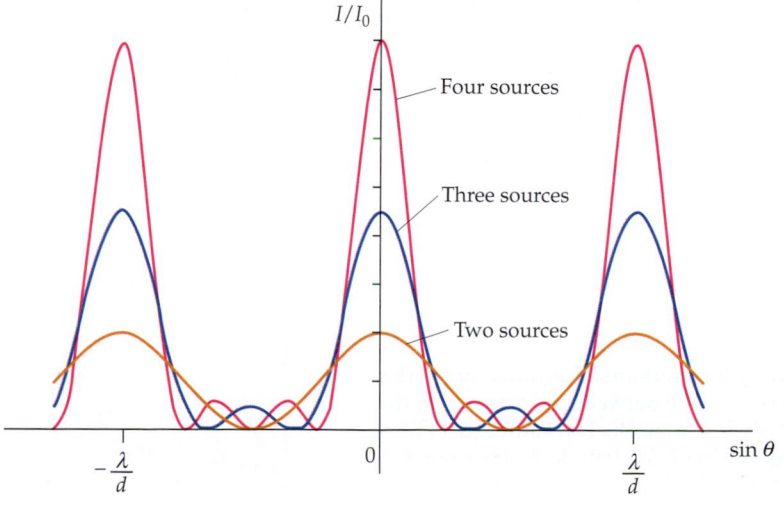

**FIGURE 33-22** Plot of relative intensity versus $\sin \theta$ for two, three, and four equally spaced coherent sources.

## *Calculating the Single-Slit Diffraction Pattern

We now use the phasor method for the addition of harmonic waves to calculate the intensity pattern shown in Figure 33-11. We assume that the slit of width $a$ is divided into $N$ equal intervals and that there is a point source of waves at the midpoint of each interval (Figure 33-23). If $d$ is the distance between two adjacent sources and $a$ is the width of the opening, we have $d = a/N$. Since the screen on which we are calculating the intensity is far from the sources, the rays from the sources to a point $P$ on the screen are approximately parallel. The path difference between any two adjacent sources is $d \sin \theta$, and the phase difference $\delta$ is related to the path difference by

$$\frac{\delta}{2\pi} = \frac{d \sin \theta}{\lambda}$$

If $A_0$ is the amplitude due to a single source, the amplitude at the central maximum, where $\theta = 0$ and all the waves are in phase, is $A_{max} = NA_0$ (Figure 33-24).

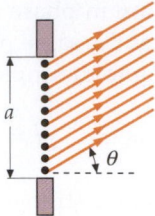

**FIGURE 33-23** Diagram for calculating the diffraction pattern far away from a narrow slit. The slit width $a$ is assumed to contain a large number of in-phase point sources separated by a distance $d$. The rays from these sources to a point far away are approximately parallel. The path difference for the waves from adjacent sources is $d \sin \theta$.

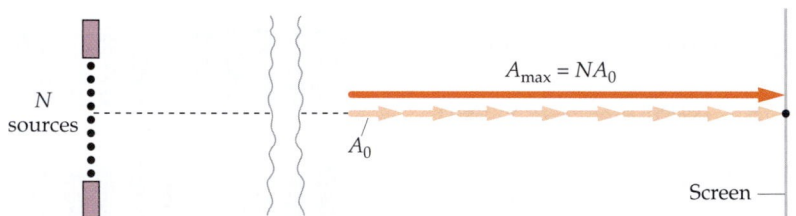

**FIGURE 33-24** A single slit is represented by $N$ sources, each of amplitude $A_0$. At the central maximum point, where $\theta = 0$, the waves from the sources add in phase, giving a resultant amplitude $A_{max} = NA_0$.

We can find the amplitude at some other point at an angle $\theta$ by using the phasor method for the addition of harmonic waves. As in the addition of two, three, or four waves, the intensity is zero at any point where the phasors representing the waves form a closed polygon. In this case, the polygon has $N$ sides (Figure 33-25). At the first minimum, the wave from the first source just below the top of the opening and the wave from the source just below the middle of the opening are 180° out of phase. In this case, the waves from the source near the top of the opening differ from those from the bottom of the opening by nearly 360°. [The phase difference is, in fact, 360° − (360°/N).] Thus, if the number of sources is very large, 360°/N is negligible and we get complete cancellation if the waves from the first and last sources are out of phase by 360°, corresponding to a path difference of 1 wavelength, in agreement with Equation 33-11.

We will now calculate the amplitude at a general point at which the waves from two adjacent sources differ in phase by $\delta$. Figure 33-26 shows the phasor diagram for the addition of $N$ waves, where the subsequent waves differ in phase from the first wave by $\delta, 2\delta, \ldots, (N-1)\delta$. When $N$ is very large and $\delta$ is very small, the phasor diagram approximates the arc of a circle. The resultant amplitude $A$ is the length of the chord of this arc. We will calculate this resultant amplitude in terms of the phase difference $\phi$ between the first wave and the last wave. From Figure 33-26, we have

$$\sin \tfrac{1}{2}\phi = \frac{A/2}{r}$$

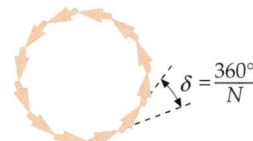

**FIGURE 33-25** Phasor diagram for calculating the first minimum in the single-slit diffraction pattern. When the waves from the $N$ sources completely cancel, the $N$ phasors form a closed polygon. The phase difference between the waves from adjacent sources is then $\delta = 360°/N$. When $N$ is very large, the waves from the first and last sources are approximately in phase.

**FIGURE 33-26** Phasor diagram for calculating the resultant amplitude due to the waves from $N$ sources in terms of the phase difference $\phi$ between the wave from the first source just below the top of the slit and the wave from the last source just above the bottom of the slit. When $N$ is very large, the resultant amplitude $A$ is the chord of a circular arc of length $NA_0 = A_{max}$.

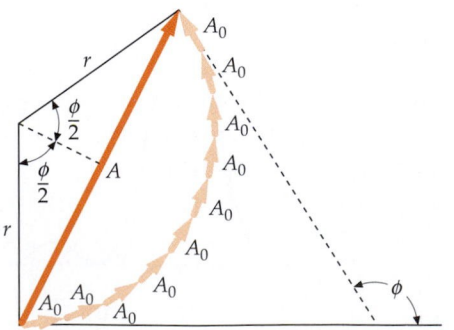

or

$$A = 2r \sin \tfrac{1}{2}\phi \qquad\qquad 33\text{-}17$$

where $r$ is the radius of the arc. Since the length of the arc is $A_{max} = NA_0$ and the angle subtended is $\phi$, we have

$$\phi = \frac{A_{max}}{r} \qquad\qquad 33\text{-}18$$

or

$$r = \frac{A_{max}}{\phi}$$

Substituting this into Equation 33-17 gives

$$A = \frac{2A_{max}}{\phi} \sin \frac{1}{2}\phi = A_{max} \frac{\sin \tfrac{1}{2}\phi}{\tfrac{1}{2}\phi}$$

Since the amplitude at the center of the central maximum ($\theta = 0$) is $A_{max}$, the ratio of the intensity at any other point to that at the center of the central maximum is

$$\frac{I}{I_0} = \frac{A^2}{A^2_{max}} = \left(\frac{\sin \tfrac{1}{2}\phi}{\tfrac{1}{2}\phi}\right)^2$$

or

$$I = I_0\left(\frac{\sin \tfrac{1}{2}\phi}{\tfrac{1}{2}\phi}\right)^2 \qquad\qquad 33\text{-}19$$

INTENSITY FOR A SINGLE-SLIT DIFFRACTION PATTERN

The phase difference $\phi$ between the first and last waves is related to the path difference $a \sin \theta$ between the top and bottom of the opening by:

$$\frac{\phi}{2\pi} = \frac{a \sin \theta}{\lambda} \qquad\qquad 33\text{-}20$$

Equation 33-19 and Equation 33-20 describe the intensity pattern shown in Figure 33-11. The first minimum occurs at $a \sin \theta = \lambda$, which is the point where the waves from the middle of the upper half and the middle of the lower half of the slit have a path difference of $\lambda/2$ and are 180° out of phase. The second minimum occurs at $a \sin \theta = 2\lambda$, where the waves from the upper half of the upper half of the slit and those from the lower half of the upper half of the slit have a path difference of $\lambda/2$ and are 180° out of phase.

There is a secondary maximum approximately midway between the first and second minima at $a \sin \theta \approx \tfrac{3}{2}\lambda$. Figure 33-27 shows the phasor diagram for determining the approximate intensity of this secondary maximum. The phase difference between the first and last waves is approximately 360° + 180°. The phasors thus complete $1\tfrac{1}{2}$ circles. The resultant amplitude is the diameter of a circle with a circumference that is two-thirds the total length $A_{max}$. If $C = \tfrac{2}{3}A_{max}$ is the circumference, the diameter $A$ is

$$A = \frac{C}{\pi} = \frac{\tfrac{2}{3}A_{max}}{\pi} = \frac{2}{3\pi}A_{max}$$

and

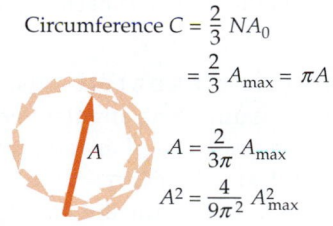

Circumference $C = \tfrac{2}{3} NA_0$

$\qquad\qquad = \tfrac{2}{3} A_{max} = \pi A$

$A = \dfrac{2}{3\pi} A_{max}$

$A^2 = \dfrac{4}{9\pi^2} A^2_{max}$

**FIGURE 33-27** Phasor diagram for calculating the approximate amplitude of the first secondary maximum of the single-slit diffraction pattern. This secondary maximum occurs near the midpoint between the first and second minima when the $N$ phasors complete $1\tfrac{1}{2}$ circles.

$$A^2 = \frac{4}{9\pi^2} A^2_{max}$$

The intensity at this point is

$$I = \frac{4}{9\pi^2} I_0 \approx \frac{1}{22.2} I_0 \qquad\qquad 33\text{-}21$$

## *Calculating the Interference–Diffraction Pattern of Multiple Slits

The intensity of the two-slit interference–diffraction pattern can be calculated from the two-slit pattern (Equation 33-8) with the intensity of each slit ($I_0$ in that equation) replaced by the diffraction pattern intensity due to each slit, $I$, given by Equation 33-19. The intensity for the two-slit interference–diffraction pattern is thus

$$I = 4I_0 \left( \frac{\sin\frac{1}{2}\phi}{\frac{1}{2}\phi} \right)^2 \cos^2 \frac{1}{2}\delta \qquad\qquad 33\text{-}22$$

INTERFERENCE–DIFFRACTION INTENSITY FOR TWO SLITS

where $\phi$ is the difference in phase between rays from the top and bottom of each slit, which is related to the width of each slit by

$$\phi = \frac{2\pi}{\lambda} a \sin \theta$$

and $\delta$ is the difference in phase between rays from the centers of two adjacent slits, which is related to the slit separation by

$$\delta = \frac{2\pi}{\lambda} d \sin \theta$$

In Equation 33-22, the intensity $I_0$ is the intensity at $\theta = 0$ due to one slit alone.

*FIVE-SLIT INTERFERENCE–DIFFRACTION PATTERN*    **E X A M P L E   3 3 - 7**

**Find the interference–diffraction intensity pattern for five equally spaced slits, where $a$ is the width of each slit and $d$ is the distance between adjacent slits.**

**PICTURE THE PROBLEM** First, find the interference intensity pattern for the five slits, assuming no angular variations in the intensity due to diffraction. To do this, first construct a phasor diagram to find the amplitude of the resultant wave in an arbitrary direction $\theta$. Intensity is proportional to the square of the amplitude. Next, correct for the variation of intensity with $\theta$ by using the single-slit diffraction pattern intensity relation (Equation 33-19 and Equation 33-20).

1. The diffraction pattern intensity $I'$ due to a slit of width $a$ is given by Equation 33-19 and Equation 33-20:

$$I' = I_0 \left( \frac{\sin\frac{1}{2}\phi}{\frac{1}{2}\phi} \right)^2$$

where

$$\phi = \frac{2\pi}{\lambda} a \sin \theta$$

2. The interference pattern intensity $I$ is proportional to the square of the amplitude $A$ of the superposition of the wave functions for the light from the five slits:

$I \propto A^2$

where

$$A \sin(\alpha + \delta') = A_0 \sin \alpha + A_0 \sin(\alpha + \delta) + A_0 \sin(\alpha + 2\delta)$$
$$+ A_0 \sin(\alpha + 3\delta) + A_0 \sin(\alpha + 4\delta)$$

with $\alpha = \omega t$ and $\delta = \dfrac{2\pi}{\lambda} d \sin \theta$

3. To solve for $A$, we construct a phasor diagram (Figure 33-28). The amplitude $A$ equals the sum of the projections of the individual phasors onto the resultant phasor:

$\delta' = \beta + \delta$

so

$\beta = \delta' - \delta = 2\delta - \delta = \delta$

**FIGURE 33-28**

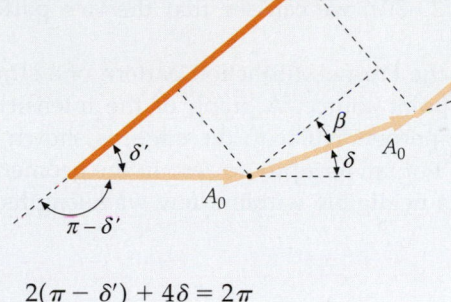

4. To find $\delta'$, we add the exterior angles. The sum of the exterior angles equals $2\pi$. (If you walk the perimeter of a polygon you rotate through the sum of the exterior angles, and you rotate through $2\pi$ radians):

$2(\pi - \delta') + 4\delta = 2\pi$

so

$\delta' = 2\delta$

5. Solve for $A$ from the figure:

$A = 2A_0 \cos \delta' + 2A_0 \cos \beta + A_0$

6. Substitute for $\delta'$ using the step 4 result, and substitute for $\beta$ using the relation $\beta = \delta$:

$A = A_0(2 \cos 2\delta + 2 \cos \delta + 1)$

7. Square both sides to relate the intensities. Recall, $I'$ is the intensity from a single slit, and $A_0$ is the amplitude from a single slit:

$A^2 = A_0^2(2 \cos 2\delta + 2 \cos \delta + 1)^2$

so

$I = I'(2 \cos 2\delta + 2 \cos \delta + 1)^2$

8. Substitute for $I'$ using the step 1 result:

$$I = I_0 \left( \frac{\sin \frac{1}{2}\phi}{\frac{1}{2}\phi} \right)^2 (2 \cos 2\delta + 2 \cos \delta + 1)^2$$

where $\phi = \dfrac{2\pi}{\lambda} a \sin \theta$ and $\delta = \dfrac{2\pi}{\lambda} d \sin \theta$

**PLAUSIBILITY CHECK** If $\theta = 0$, both $\phi = 0$ and $\delta = 0$. So, for $\theta = 0$, step 5 becomes $A = 5A_0$ and step 8 becomes $I = 5^2 I_0 = 25 I_0$ as expected.

## 33-6   Fraunhofer and Fresnel Diffraction

Diffraction patterns, like the single-slit pattern shown in Figure 33-11, that are observed at points for which the rays from an aperture or an obstacle are nearly parallel are called **Fraunhofer diffraction patterns.** Fraunhofer patterns can be observed at great distances from the obstacle or the aperture so that the rays reaching any point are approximately parallel, or they can be observed using a lens to focus parallel rays on a viewing screen placed in the focal plane of the lens.

The diffraction pattern observed near an aperture or an obstacle is called a **Fresnel diffraction pattern.** Because the rays from an aperture or an obstacle close to a screen cannot be considered parallel, Fresnel diffraction is much more difficult to analyze. Figure 33-29 illustrates the difference between the Fresnel and the Fraunhofer patterns for a single slit.[†]

Figure 33-30a shows the Fresnel diffraction pattern of an opaque disk. Note the bright spot at the center of the pattern caused by the constructive interference of the light waves diffracted from the edge of the disk. This pattern is of some historical interest. In an attempt to discredit Augustin Fresnel's wave theory of light, Siméon Poisson pointed out that it predicted a bright spot at the center of the shadow, which he assumed was a ridiculous contradiction of fact. However, Fresnel immediately demonstrated experimentally that such a spot does, in fact, exist. This demonstration convinced many doubters of the validity of the wave theory of light. The Fresnel diffraction pattern of a circular aperture is shown in Figure 33-30b. Comparing this with the pattern of the opaque disk in Figure 33-30a, we can see that the two patterns are complements of each other.

Figure 33-31a shows the Fresnel diffraction pattern of a straight edge illuminated by light from a point source. A graph of the intensity versus distance (measured along a line perpendicular to the edge) is shown in Figure 33-31b. The light intensity does not fall abruptly to zero in the geometric shadow, but it decreases rapidly and is negligible within a few wavelengths of the edge. The

As the screen is moved closer,

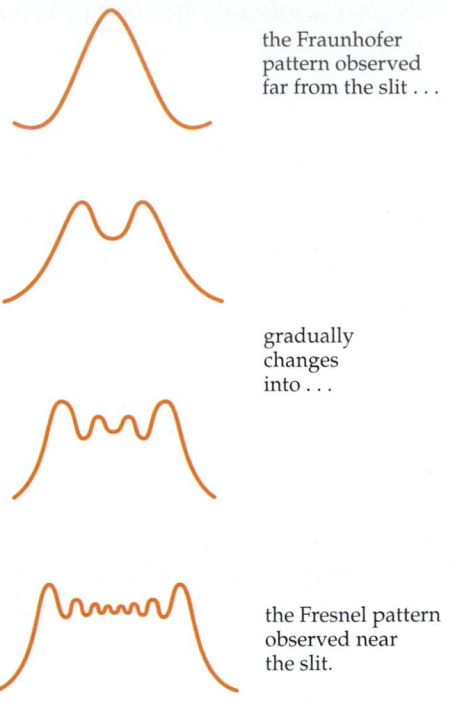

the Fraunhofer pattern observed far from the slit . . .

gradually changes into . . .

the Fresnel pattern observed near the slit.

**FIGURE 33-29** Diffraction patterns for a single slit at various screen distances.

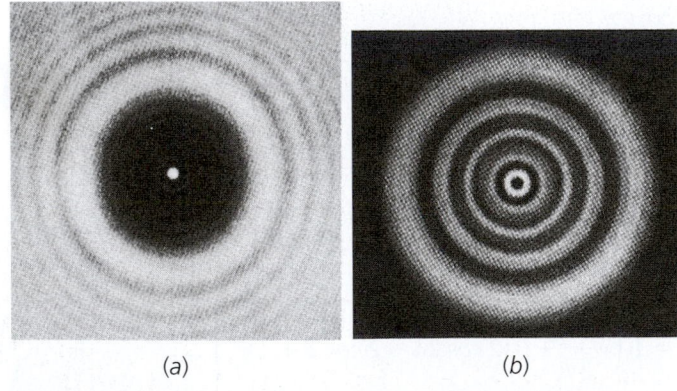

(a)                    (b)

**FIGURE 33-30** (a) The Fresnel diffraction pattern of an opaque disk. At the center of the shadow, the light waves diffracted from the edge of the disk are in phase and produce a bright spot called the *Poisson spot*. (b) The Fresnel diffraction pattern of a circular aperture. Compare this with Figure 33-30a.

(a)

(b)

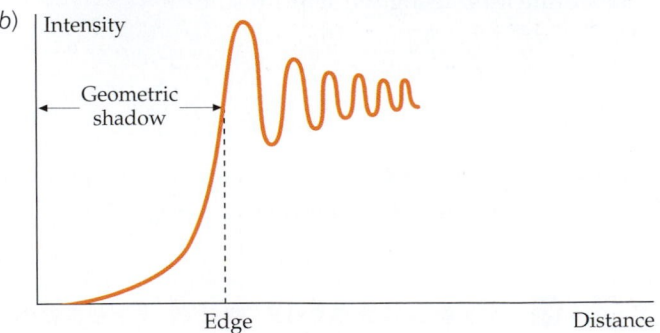

**FIGURE 33-31** (a) The Fresnel diffraction of a straightedge. (b) A graph of intensity versus distance along a line perpendicular to the edge.

---

[†] See Richard E. Haskel, "A Simple Experiment on Fresnel Diffraction," *American Journal of Physics* 38 (1970): 1039.

Fresnel diffraction pattern of a rectangular aperture is shown in Figure 33-32. These patterns cannot be seen with extended light sources like an ordinary lightbulb, because the dark fringes of the pattern produced by light from one point on the source overlap the bright fringes of the pattern produced by light from another point.

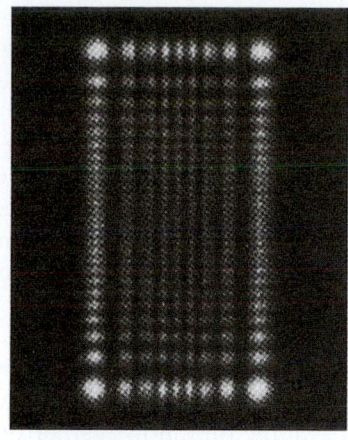

**FIGURE 33-32** The Fresnel diffraction pattern of a rectangular aperture.

## 33-7 Diffraction and Resolution

Diffraction due to a circular aperture has important implications for the resolution of many optical instruments. Figure 33-33 shows the Fraunhofer diffraction pattern of a circular aperture. The angle $\theta$ subtended by the first diffraction minimum is related to the wavelength and the diameter of the opening $D$ by

$$\sin \theta = 1.22 \frac{\lambda}{D} \qquad \text{33-23}$$

Equation 33-23 is similar to Equation 33-9 except for the factor 1.22, which arises from the mathematical analysis, and is similar to the equation for a single slit but more complicated because of the circular geometry. In many applications, the angle $\theta$ is small, so $\sin \theta$ can be replaced by $\theta$. The first diffraction minimum is then at an angle $\theta$ given by

$$\theta \approx 1.22 \frac{\lambda}{D} \qquad \text{33-24}$$

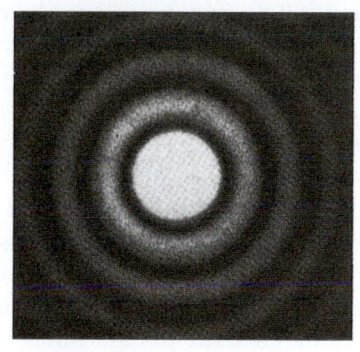

**FIGURE 33-33** The Fraunhofer diffraction pattern of a circular aperture.

Figure 33-34 shows two point sources that subtend an angle $\alpha$ at a circular aperture far from the sources. The intensities of the Fraunhofer diffraction pattern are also indicated in this figure. If $\alpha$ is much greater than $1.22\lambda/D$, the sources will be seen as two sources. However, as $\alpha$ is decreased, the overlap of the diffraction patterns increases, and it becomes difficult to distinguish the two sources from one source. At the critical angular separation, $\alpha_c$, given by

$$\alpha_c = 1.22 \frac{\lambda}{D} \qquad \text{33-25}$$

(a)

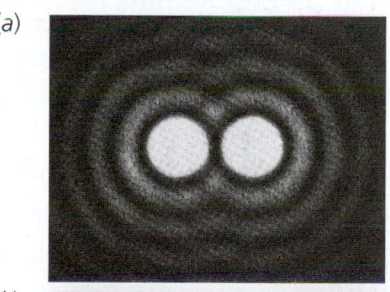

the first minimum of the diffraction pattern of one source falls on the central maximum of the other source. These objects are said to be just resolved by **Rayleigh's criterion for resolution.** Figure 33-35 shows the diffraction patterns for two sources when $\alpha$ is greater than the critical angle for resolution and when $\alpha$ is just equal to the critical angle for resolution.

Equation 33-25 has many applications. The *resolving power* of an optical instrument, such as a microscope or telescope, is the ability of the instrument to resolve

(b)

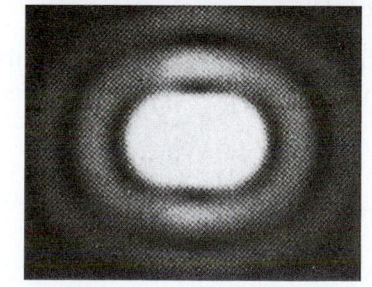

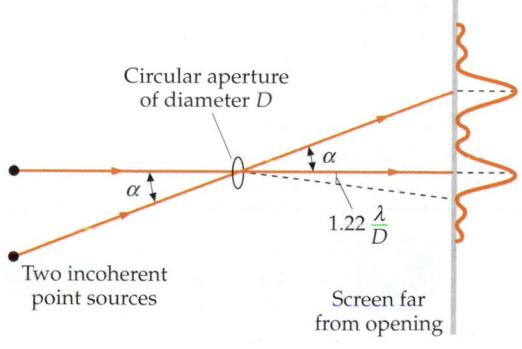

**FIGURE 33-34** Two distant sources that subtend an angle $\alpha$. If $\alpha$ is much greater than $1.22\lambda/D$, where $\lambda$ is the wavelength of light and $D$ is the diameter of the aperture, the diffraction patterns have little overlap and the sources are easily seen as two sources. If $\alpha$ is not much greater than $1.22\lambda/D$, the overlap of the diffraction patterns makes it difficult to distinguish two sources from one.

**FIGURE 33-35** The diffraction patterns for a circular aperture and two incoherent point sources when (*a*) $\alpha$ is much greater than $1.22\lambda/D$ and (*b*) when $\alpha$ is at the limit of resolution, $\alpha_c = 1.22\lambda/D$.

Circular aperture of diameter $D$

$\alpha$

$\alpha$

$1.22 \dfrac{\lambda}{D}$

Two incoherent point sources

Screen far from opening

two objects that are close together. The images of the objects tend to overlap because of diffraction at the entrance aperture of the instrument. We can see from Equation 33-25 that the resolving power can be increased either by increasing the diameter $D$ of the lens (or mirror) or by decreasing the wavelength $\lambda$. Astronomical telescopes use large objective lenses or mirrors to increase their resolution as well as to increase their light-gathering power. An array of 27 radio antennas (Figure 33-36) mounted on rails can be configured to form a single telescope with a resolution distance of 36 km (22 mi). In a microscope, a film of transparent oil with index of refraction of approximately 1.55 is sometimes used under the objective to decrease the wavelength of the light ($\lambda' = \lambda/n$). The wavelength can be reduced further by using ultraviolet light and photographic film; however, ordinary glass is opaque to ultraviolet light, so the lenses in an ultraviolet microscope must be made from quartz or fluorite. To obtain very high resolutions, electron microscopes are used—microscopes that use electrons rather than light. The wavelengths of electrons vary inversely with the square root of their kinetic energy and can be made as small as desired.[†]

**FIGURE 33-36** The very large array (VLA) of radio antennas is located near Socorro, New Mexico. The 25-m-diameter antennas are mounted on rails, which can be arranged in several configurations, and can be extended over a diameter of 36 km. The data from the antennas are combined electronically, so the array is really a single high-resolution telescope.

---

[†] The wave properties of electrons are discussed in Chapter 34.

---

*PHYSICS IN THE LIBRARY*   **EXAMPLE 33-8**   **Put It in Context**

While studying in the library, you lean back in your chair and ponder the small holes you notice in the ceiling tiles. You notice that the holes are approximately 5 mm apart. You can clearly see the holes directly above you, about 2 m up, but the tiles far away do not appear to have these holes. You wonder if the reason you cannot see the distant holes is because they are not within the criteria for resolution established by Rayleigh. Is this a feasible explanation for the disappearance of the holes? You notice the holes disappear about 20 m from you.

**PICTURE THE PROBLEM** We will need to make assumptions about the situation. If we use Equation 33-25, we will need to know the wavelength of light and the aperture diameter. Assuming our pupil is the aperture, we can assume approximately 5 mm for the diameter. (This is the number used in our physics textbook.) The light is probably centered around 500 nm or so.

1. The angular limit for resolution by the eye depends on the ratio of the wavelength and the diameter of the pupil:

$$\theta_c \approx 1.22 \frac{\lambda}{D}$$

2. The angle subtended by two holes depends on their separation distance $d$ and their distance $L$ from your eye:

$$\theta \approx \frac{d}{L}$$

3. Equating the two angles and putting in the numbers gives:

$$\frac{d}{L} \approx 1.22 \frac{\lambda}{D}$$

$$\frac{5 \text{ mm}}{L} \approx 1.22 \frac{500 \text{ nm}}{5 \text{ mm}}$$

4. Solving for $L$ gives:

$$L = 40 \text{ m}$$

5. By a factor of 2, 40 m is too large. However, you are suspect of the value given for the pupil diameter in your physics textbook. You know the pupil is smaller when the light is bright, and the library ceiling is very bright and colored white. An online search for eye pupil diameter soon turns up the information you need. The pupil diameter ranges from 2 to 3 mm up to 7 mm:

Success. If the pupil diameter is 2.5 mm, the value of $L$ is 20 m.

MASTER the CONCEPT
WEB

It is instructive to compare the limitation on resolution of the eye due to diffraction, as seen in Example 33-8, with the limitation on resolution due to the separation of the receptors (cones) on the retina. To be seen as two distinct objects, the images of the objects must fall on the retina on two nonadjacent cones. (See Problem 79 in Chapter 32.) Because the retina is about 2.5 cm from the eye lens, the distance $y$ on the retina corresponding to an angular separation of $1.5 \times 10^{-4}$ rad is found from

$$\alpha_c = 1.5 \times 10^{-4} \, \text{rad} = \frac{y}{2.5 \, \text{cm}}$$

or

$$y = 3.75 \times 10^{-4} \, \text{cm} = 3.75 \times 10^{-6} \, \text{m} = 3.75 \, \mu\text{m}$$

The actual separation of the cones in the fovea centralis, where the cones are the most tightly packed, is about 1 $\mu$m. Outside this region, they are about 3 $\mu$m to 5 $\mu$m apart.

# *33-8 Diffraction Gratings

A useful tool for measuring the wavelength of light is the **diffraction grating,** which consists of a large number of equally spaced lines or slits on a flat surface. Such a grating can be made by cutting parallel, equally spaced grooves on a glass or metal plate with a precision ruling machine. With a reflection grating, light is reflected from the ridges between the lines or grooves. Phonograph records and compact disks exhibit some of the properties of reflection gratings. In a transmission grating, the light passes through the clear gaps between the rulings. Inexpensive, optically produced plastic gratings with 10,000 or more slits per centimeter are common items. The spacing of the slits in a grating with 10,000 slits per centimeter is $d = (1 \, \text{cm})/10{,}000 = 10^{-4}$ cm.

Consider a plane light wave incident normally on a transmission grating (Figure 33-37). Assume that the width of each slit is very small so that it produces a widely diffracted beam. The interference pattern produced on a screen a large distance from the grating is due to a large number of equally spaced light sources. Suppose we have $N$ slits with separation $d$ between adjacent slits. At $\theta = 0$, the light from each slit is in phase with that from all the other slits, so the

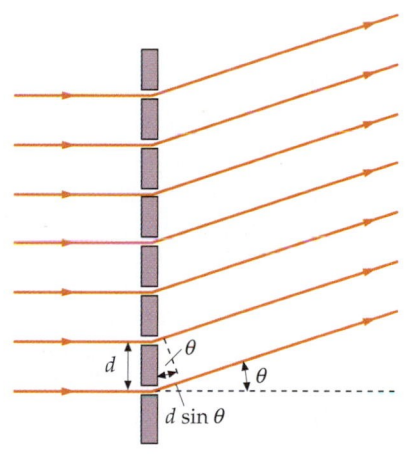

**FIGURE 33-37** Light incident normally on a diffraction grating. At an angle $\theta$, the path difference between rays from adjacent slits is $d \sin \theta$.

Compact disks act as reflection gratings.

amplitude of the wave is $NA_0$, where $A_0$ is the amplitude from each slit, and the intensity is $N^2 I_0$, where $I_0$ is the intensity due to a single slit alone. At an angle $\theta$ such that $d \sin \theta = \lambda$, the path difference between any two successive slits is $\lambda$, so again the light from each slit is in phase with that from all the other slits and the intensity is $N^2 I_0$. The interference maxima are thus at angles $\theta$ given by

$$d \sin \theta_m = m\lambda, \qquad m = 0, 1, 2, \ldots \qquad \text{33-26}$$

The position of an interference maximum does not depend on the number of sources, but the more sources there are, the sharper and more intense the maximum will be.

To see that the interference maxima will be sharper when there are many slits, consider the case of $N$ illuminated slits, where $N$ is large ($N \gg 1$). The distance from the first slit to the $N$th slit is $(N - 1)d \approx Nd$. When the path difference for the light from the first slit and that from the $N$th slit is $\lambda$, the resulting intensity will be zero. (We saw this in our discussion of single-slit diffraction.) Since the first and $N$th slits are separated by approximately $Nd$, the intensity will be zero at angle $\theta_{\min}$ given by

$$Nd \sin \theta_{\min} = \lambda$$

so

$$\theta_{\min} \approx \sin \theta_{\min} = \frac{\lambda}{Nd}$$

The width of the interference maximum $2\theta_{\min}$ is thus inversely proportional to $N$. Therefore, the greater the number of illuminated slits $N$, the sharper the maximum. Since the intensity in the maximum is proportional to $N^2 I_0$, the intensity in the maximum times the width of the maximum is proportional to $NI_0$. The intensity times the width is a measure of power per unit length in the maximum.

Figure 33-38a shows a student spectroscope that uses a diffraction grating to analyze light. In student laboratories, the light source is typically a glass tube containing atoms of a gas (e.g., helium or sodium vapor) that are excited by a bombardment of electrons accelerated by high voltage across the tube. The light emitted by such a source contains only certain wavelengths that are characteristic of the atoms in the source. Light from the source passes through a narrow collimating slit and is made parallel by a lens. Parallel light from the lens is incident on the grating. Instead of falling on a screen a large distance away, the parallel light from the grating is focused by a telescope and viewed by the eye. The telescope is mounted on a rotating platform that has been calibrated so that

**FIGURE 33-38** (*a*) A typical student spectroscope. Light from a collimating slit near the source is made parallel by a lens and falls on a grating. The diffracted light is viewed with a telescope at an angle that can be accurately measured. (*b*) Aerial view of the very large array (VLA) radio telescope in New Mexico. Radio signals from distant galaxies add constructively when Equation 33-26 is satisfied, where $d$ is the distance between two adjacent telescopes.

(a)

(b)

the angle $\theta$ can be measured. In the forward direction ($\theta = 0$), the central maximum for all wavelengths is seen. If light of a particular wavelength $\lambda$ is emitted by the source, the first interference maximum is seen at the angle $\theta$ given by Equation 33-26 with $m = 1$. Each wavelength emitted by the source produces a separate image of the collimating slit in the spectroscope called a **spectral line.** The set of lines corresponding to $m = 1$ is called the **first-order spectrum.** The **second-order spectrum** corresponds to $m = 2$ for each wavelength. Higher orders may be seen, providing the angle $\theta$ given by Equation 33-26 is less than 90°. Depending on the wavelengths, the orders may be mixed; that is, the third-order line for one wavelength may occur before the second-order line for another wavelength. If the spacing of the slits in the grating is known, the wavelengths emitted by the source can be determined by measuring the angle $\theta$.

---

*RESOLVING THE SODIUM D LINES* **E X A M P L E 3 3 - 9**

**Sodium light is incident on a diffraction grating with 12,000 lines per centimeter. At what angles will the two yellow lines (called the sodium D lines) of wavelengths 589.00 nm and 589.59 nm be seen in the first order?**

**PICTURE THE PROBLEM** Apply $d \sin \theta_m = m\lambda$ to each wavelength, with $m = 1$ and $d = 1 \text{ cm}/12{,}000$.

1. The angle $\theta_m$ is given by $d \sin \theta_m = m\lambda$ with $m = 1$:

$$\sin \theta_1 = \frac{\lambda}{d}$$

2. Calculate $\theta_1$ for $\lambda = 589.00$ nm:

$$\theta_1 = \sin^{-1}\left[ \frac{589.00 \times 10^{-9} \text{ m}}{(1 \text{ cm}/12{,}000)} \times \left( \frac{100 \text{ cm}}{1 \text{ m}} \right) \right]$$

$$= \boxed{44.98°}$$

3. Repeat the calculation for $\lambda = 589.59$ nm:

$$\theta_1 = \sin^{-1}\left[ \frac{589.59 \times 10^{-9} \text{ m}}{(1 \text{ cm}/12{,}000)} \times \left( \frac{100 \text{ cm}}{1 \text{ m}} \right) \right]$$

$$= \boxed{45.03°}$$

**REMARKS** Note that light of longer wavelength is diffracted through larger angles.

**EXERCISE** Find the angles for the two yellow lines if the grating has 15,000 lines per centimeter. (*Answer* 62.07° and 62.18°)

An important feature of a spectroscope is its ability to resolve spectral lines of two nearly equal wavelengths $\lambda_1$ and $\lambda_2$. For example, the two prominent yellow lines in the spectrum of sodium have wavelengths 589.00 and 589.59 nm. These can be seen as two separate wavelengths if their interference maxima do not overlap. According to Rayleigh's criterion for resolution, these wavelengths are resolved if the angular separation of their interference maxima is greater than the angular separation between an interference maximum and the first interference minimum on either side of it. The **resolving power** of a diffraction grating is defined to be $\lambda/|\Delta\lambda|$, where $|\Delta\lambda|$ is the smallest difference between two nearby wavelengths, each approximately equal to $\lambda$, that may be resolved. The resolving power is proportional to the number of slits illuminated because the more slits illuminated, the sharper the interference maxima. The resolving power $R$ can be shown to be

$$R = \frac{\lambda}{|\Delta\lambda|} = mN \qquad\qquad 33\text{-}27$$

where $N$ is the number of slits and $m$ is the order number (see Problem 76). We can see from Equation 33-27 that to resolve the two yellow lines in the sodium spectrum the resolving power must be

$$R = \frac{589.00 \text{ nm}}{589.59 \text{ nm} - 589.00 \text{ nm}} = 998$$

Thus, to resolve the two yellow sodium lines in the first order ($m = 1$), we need a grating containing 998 or more slits in the area illuminated by the light.

## *Holograms

An interesting application of diffraction gratings is the production of a three-dimensional photograph called a **hologram** (Figure 33-39). In an ordinary photograph, the intensity of reflected light from an object is recorded on a film. When the film is viewed by transmitted light, a two-dimensional image is produced. In a hologram, a beam from a laser is split into two beams, a reference beam and an object beam. The object beam reflects from the object to be photographed and the interference pattern between it, and the reference beam is recorded on a photographic film. This can be done because the laser beam is coherent so that the relative phase difference between the reference beam and the object beam can be kept constant during the exposure. The interference fringes on the film act as a diffraction grating. When the film is illuminated with a laser, a three-dimensional image of the object is produced.

Holograms that you see on credit cards or postage stamps, called rainbow holograms, are more complicated. A horizontal strip of the original hologram is used to make a second hologram. The three-dimensional image can be seen as the viewer moves from side to side, but if viewed with laser light, the image disappears when the viewer's eyes move above or below the slit image. When viewed with white light, the image is seen in different colors as the viewer moves in the vertical direction.

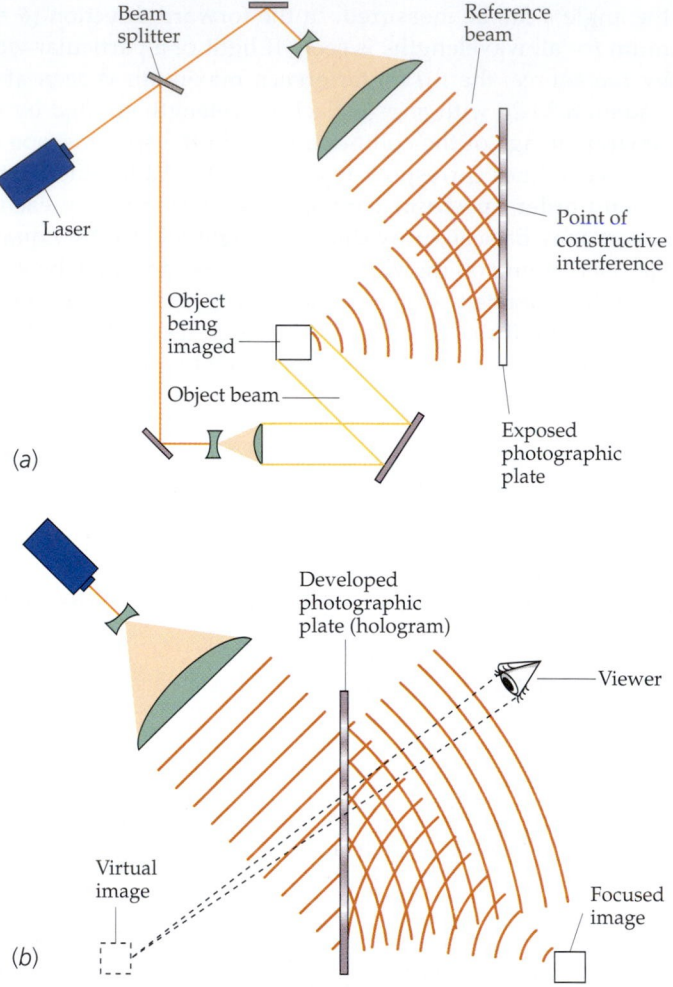

**FIGURE 33-39** (a) The production of a hologram. The interference pattern produced by the reference beam and object beam is recorded on a photographic film. (b) When the film is developed and illuminated by coherent laser light, a three-dimensional image is seen.

(a)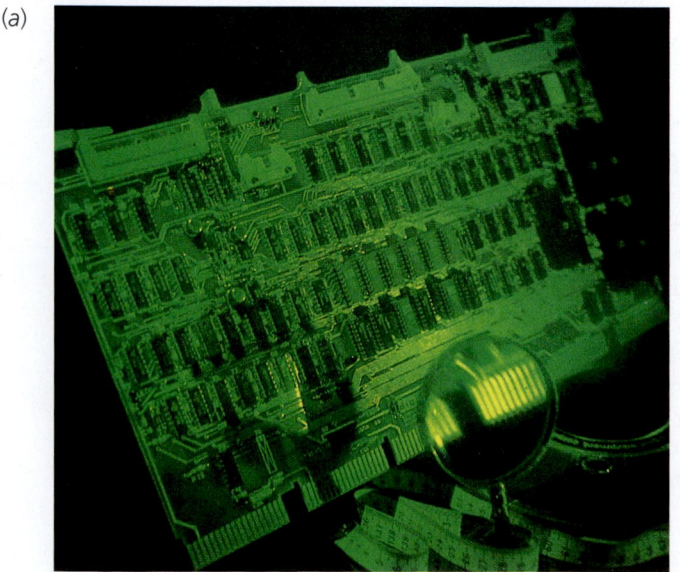

(b)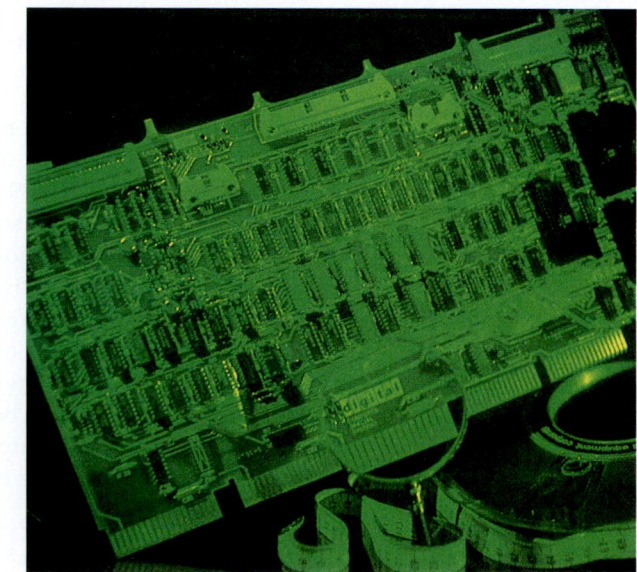

**A hologram viewed from two different angles. Note that different parts of the circuit board appear behind the front magnifying lens.**

| Topic | Relevant Equations and Remarks |
|---|---|
| **1. Interference** | Two superposing light waves interfere if their phase difference remains constant for a time long enough to observe. They interfere constructively if their phase difference is zero or an integer times 360°. They interfere destructively if their phase difference is 180° or an odd integer times 180°. |
| Phase difference due to a path difference | $$\frac{\delta}{2\pi} = \frac{\Delta r}{\lambda} \qquad\qquad \text{33-1}$$ |
| Phase difference due to reflection | A phase difference of 180° is introduced when a light wave is reflected from a boundary between two media for which the wave speed is greater on the incident-wave side of the boundary. |
| Thin films | The interference of light waves reflected from the front and back surfaces of a thin film produces interference fringes, commonly observed in soap films or oil films. The difference in phase between the two reflected waves results from the path difference of twice the thickness of the film plus any phase change due to reflection of one or both of the rays. |
| Two slits | The path difference at an angle $\theta$ on a screen far away from two narrow slits separated by a distance $d$ is $d \sin \theta$. If the intensity due to each slit separately is $I_0$, the intensity at points of constructive interference is $4I_0$, and the intensity at points of destructive interference is zero. |
| Interference maxima (sources in phase) | $$d \sin \theta_m = m\lambda, \qquad m = 0, 1, 2, \ldots \qquad\qquad \text{33-2}$$ |
| Interference minima (sources 180° out of phase) | $$d \sin \theta_m = (m - \tfrac{1}{2})\lambda, \qquad m = 1, 2, 3, \ldots \qquad\qquad \text{33-3}$$ |
| **2. Diffraction** | Diffraction occurs whenever a portion of a wavefront is limited by an obstacle or aperture. The intensity of light at any point in space can be computed using Huygens's principle by taking each point on the wavefront to be a point source and computing the resulting interference pattern. |
| Fraunhofer patterns | Fraunhofer patterns are observed at great distances from the obstacle or aperture so that the rays reaching any point are approximately parallel, or they can be observed using a lens to focus parallel rays on a viewing screen placed in the focal plane of the lens. |
| Fresnel patterns | Fresnel patterns are observed at points close to the source. |
| Single slit | When light is incident on a single slit of width $a$, the intensity pattern on a screen far away shows a broad central diffraction maximum that decreases to zero at an angle $\theta_1$ given by $$\sin \theta_1 = \frac{\lambda}{a} \qquad\qquad \text{33-9}$$ The width of the central maximum is inversely proportional to the width of the slit. The zeros in the single-slit diffraction pattern occur at angles given by $$a \sin \theta_m = m\lambda, \qquad m = 1, 2, 3, \ldots \qquad\qquad \text{33-11}$$ The maxima on either side of the central maxima have intensities that are much smaller than the intensity of the central maxima. |

| | |
|---|---|
| Two slits | The interference–diffraction pattern of two slits is the two-slit interference pattern modulated by the single-slit diffraction pattern. |
| Resolution of two sources | When light from two point sources that are close together passes through an aperture, the diffraction patterns of the sources may overlap. If the overlap is too great, the two sources cannot be resolved as two separate sources. When the central diffraction maximum of one source falls at the diffraction minimum of the other source, the two sources are said to be just resolved by Rayleigh's criterion for resolution. For a circular aperture of diameter $D$, the critical angular separation of two sources for resolution by Rayleigh's criterion is |
| Rayleigh's criterion | $$\alpha_c = 1.22 \frac{\lambda}{D} \qquad \text{33-25}$$ |
| *Gratings | A diffraction grating consisting of a large number of equally spaced lines or slits is used to measure the wavelength of light emitted by a source. The positions of the $m$th order interference maxima from a grating are at angles given by $$d \sin \theta_m = m\lambda, \quad m = 0, 1, 2 \ldots \qquad \text{33-26}$$ The resolving power of a grating is $$R = \frac{\lambda}{|\Delta\lambda|} = mN \qquad \text{33-27}$$ where $N$ is the number of slits of the grating that are illuminated and $m$ is the order number. |
| 3. *Phasors | Two or more harmonic waves can be added by representing each wave as a two-dimensional vector called a phasor. The phase difference between two harmonic waves is represented as the angle between the phasors. |

## PROBLEMS

- • Single-concept, single-step, relatively easy
- •• Intermediate-level, may require synthesis of concepts
- ••• Challenging
- SSM Solution is in the *Student Solutions Manual*
- iSOLVE Problems available on iSOLVE online homework service
- iSOLVE ✔ These "Checkpoint" online homework service problems ask students additional questions about their confidence level, and how they arrived at their answer.

In a few problems, you are given more data than you actually need; in a few other problems, you are required to supply data from your general knowledge, outside sources, or informed estimates.

### Conceptual Problems

**1** • SSM When destructive interference occurs, what happens to the energy in the light waves?

**2** • Which of the following pairs of light sources are coherent: (*a*) two candles, (*b*) one point source and its image in a plane mirror, (*c*) two pinholes uniformly illuminated by the same point source, (*d*) two headlights of a car, (*e*) two images of a point source due to reflection from the front and back surfaces of a soap film.

**3** • The spacing between Newton's rings decreases rapidly as the diameter of the rings increases. Explain qualitatively why this occurs.

**4** •• If the angle of a wedge-shaped air film, such as that in Example 33-2, is too large, fringes are not observed. Why?

**5** •• Why must a film that is used to observe interference colors be thin?

**6** • SSM A loop of wire is dipped in soapy water and held so that the soap film is vertical. (*a*) Viewed by reflection with white light, the top of the film appears black. Explain why. (*b*) Below the black region are colored bands. Is the first band red or violet? (*c*) Describe the appearance of the film when it is viewed by *transmitted* light.

**7** • As the width of a slit producing a single-slit diffraction pattern is slowly and steadily reduced, how will the diffraction pattern change?

**8** • Equation 33-2, which is $d \sin \theta_m = m\lambda$, and Equation 33-11, which is $a \sin \theta_m = m\lambda$, are sometimes confused. For each equation, define the symbols and explain the equation's application.

**9** • When a diffraction grating is illuminated with white light, the first-order maximum of green light (a) is closer to the central maximum than that of red light. (b) is closer to the central maximum than that of blue light. (c) overlaps the second-order maximum of red light. (d) overlaps the second-order maximum of blue light.

**10** • [SSM] A double-slit interference experiment is set up in a chamber that can be evacuated. Using monochromatic light, an interference pattern is observed when the chamber is open to air. As the chamber is evacuated, one will note that (a) the interference fringes remain fixed. (b) the interference fringes move closer together. (c) the interference fringes move farther apart. (d) the interference fringes disappear completely.

**11** • True or false:

(a) When waves interfere destructively, the energy is converted into heat energy.
(b) Interference is observed only for waves from coherent sources.
(c) In the Fraunhofer diffraction pattern for a single slit, the narrower the slit, the wider the central maximum of the diffraction pattern.
(d) A circular aperture can produce both a Fraunhofer diffraction pattern and a Fresnel diffraction pattern.
(e) The ability to resolve two point sources depends on the wavelength of the light.

## Estimation and Approximation

**12** • [SSM] It is claimed that the Great Wall of China is the only human object that can be seen from space with the naked eye. Make an argument in support of this claim based on the resolving power of the human eye. Evaluate the validity of your argument for observers both in low-earth orbit ($\sim$ 400 km altitude) and on the moon.

**13** • Naturally occuring coronas (brightly colored rings) are sometimes seen around the moon or the sun when viewed through a thin cloud. (Warning: When viewing a sun corona, be sure that the entire sun is blocked by the edge of a building, a tree, or a traffic pole to safeguard your eyes.) These coronas are due to diffraction of light by small water droplets in the cloud. A typical angular diameter for a coronal ring is about 10°. From this, estimate the size of the water droplets in the cloud. Assume that the water droplets can be modeled as opaque disks with the same radius as the droplet, and that the Fraunhofer diffraction pattern from an opaque disk is the same as the pattern from an aperture of the same diameter. (This statement is known as *Babinet's principle.*)

**14** • An artificial corona (see Problem 13) can be made by placing a suspension of polystyrene microspheres in water. Polystyrene microspheres are small, uniform spheres made of plastic with an index of refraction of 1.59. Assuming that the water has a refractive index of 1.33, what is the angular diameter of such an artificial corona if 5 $\mu$m diameter particles are illuminated by a helium–neon laser with wavelength in air $\lambda = 632.8$ nm?

**15** • Coronas (see Problem 13) can be caused by pollen grains, typically of birch or pine. Such grains are irregular in shape, but they can be treated as if they had an average diameter of approximately 25 $\mu$m. What is the coronal radius (in degrees) for blue light? What is the coronal radius (in degrees) for red light?

**16** •• [SSM] Human hair has a diameter of approximately 70 $\mu$m. If we illuminate a hair using a helium–neon laser with wavelength $\lambda = 632.8$ nm and intercept the light scattered from the hair on a screen 10 m away, what will be the separation of the first diffraction peak from the center? (The diffraction pattern of a hair with diameter $d$ is the same as the diffraction pattern of a single slit with width $a = d$.)

## Phase Difference and Coherence

**17** • [ISOLVE] (a) What minimum path difference is needed to introduce a phase shift of 180° in light of wavelength 600 nm? (b) What phase shift will that path difference introduce in light of wavelength 800 nm?

**18** • [ISOLVE] Light of wavelength 500 nm is incident normally on a film of water $10^{-4}$ cm thick. The index of refraction of water is 1.33. (a) What is the wavelength of the light in the water? (b) How many wavelengths are contained in the distance $2t$, where $t$ is the thickness of the film? (c) What is the phase difference between the wave reflected from the top of the air–water interface and the wave reflected from the bottom of the water–air interface after it has traveled this distance?

**19** •• [SSM] [ISOLVE ✓] Two coherent microwave sources that produce waves of wavelength 1.5 cm are in the $xy$ plane, one on the $y$ axis at $y = 15$ cm and the other at $x = 3$ cm, $y = 14$ cm. If the sources are in phase, find the difference in phase between the two waves from these sources at the origin.

## Interference in Thin Films

**20** • A wedge-shaped film of air is made by placing a small slip of paper between the edges of two flat plates of glass. Light of wavelength 700 nm is incident normally on the glass plates, and interference bands are observed by reflection. (a) Is the first band near the point of contact of the plates dark or bright? Why? (b) If there are five dark bands per centimeter, what is the angle of the wedge?

**21** •• [SSM] The diameters of fine fibers can be accurately measured using interference patterns. Two optically flat pieces of glass of length $L$ are arranged with the wire between them, as shown in Figure 33-40. The setup is illuminated by monochromatic light, and the resulting interference fringes are detected. Suppose that $L = 20$ cm and that yellow sodium light ($\lambda \approx 590$ nm) is used for illumination. If 19 bright fringes are seen along this 20-cm distance, what are the limits on the diameter of the wire? (*Hint:* The nineteenth fringe might not be right at the end, but you do not see a twentieth fringe at all.)

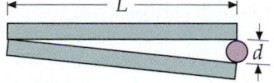

**FIGURE 33-40** Problem 21

**22** •• [SOLVE✓] Light of wavelength 600 nm is used to illuminate two glass plates at normal incidence. The plates are 22 cm in length, touch at one end, and are separated at the other end by a wire of radius 0.025 mm. How many bright fringes appear along the total length of the plates?

**23** •• A thin film having an index of refraction of 1.5 is surrounded by air. It is illuminated normally by white light and is viewed by reflection. Analysis of the resulting reflected light shows that the wavelengths 360, 450, and 602 nm are the only missing wavelengths in or near the visible portion of the spectrum. That is, for these wavelengths, there is destructive interference. (a) What is the thickness of the film? (b) What visible wavelengths are brightest in the reflected interference pattern? (c) If this film were resting on glass with an index of refraction of 1.6, what wavelengths in the visible spectrum would be missing from the reflected light?

**24** •• [SOLVE] A drop of oil ($n = 1.22$) floats on water ($n = 1.33$). When reflected light is observed from above, as shown in Figure 33-41, what is the thickness of the drop at the point where the second red fringe, counting from the edge of the drop, is observed? Assume red light has a wavelength of 650 nm.

**FIGURE 33-41**
**Problem 24**

**25** •• A film of oil of index of refraction $n = 1.45$ rests on an optically flat piece of glass of index of refraction $n = 1.6$. When illuminated with white light at normal incidence, light of wavelengths 690 nm and 460 nm is predominant in the reflected light. Determine the thickness of the oil film.

**26** •• [SSM] [SOLVE✓] A film of oil of index of refraction $n = 1.45$ floats on water ($n = 1.33$). When illuminated with white light at normal incidence, light of wavelengths 700 nm and 500 nm is predominant in the reflected light. Determine the thickness of the oil film.

## Newton's Rings

**27** •• [SSM] A Newton's ring apparatus consists of a plano-convex glass lens with radius of curvature $R$ that rests on a flat glass plate, as shown in Figure 33-42. The thin film is air of variable thickness. The pattern is viewed by reflected light. (a) Show that for a thickness $t$ the condition for a bright (constructive) interference ring is

$$t = \left(m + \frac{1}{2}\right)\frac{\lambda}{2}, \qquad m = 0, 1, 2, \ldots$$

(b) Apply the Pythagorean theorem to the triangle of sides $r$, $R - t$, and hypotenuse $R$ to show that for $t \ll R$, the radius of a fringe is related to $t$ by

$$r = \sqrt{2tR}$$

(c) How would the transmitted pattern look in comparison with the reflected one? (d) Use $R = 10$ m and a lens diameter of 4 cm. How many bright fringes would you see if the apparatus was illuminated by yellow sodium light ($\lambda \approx 590$ nm) and viewed by reflection? (e) What would be the diameter of the sixth bright fringe? (f) If the glass used in the apparatus has an index of refraction $n = 1.5$ and water ($n_w = 1.33$) is placed between the two pieces of glass, what change will take place in the bright-fringe pattern?

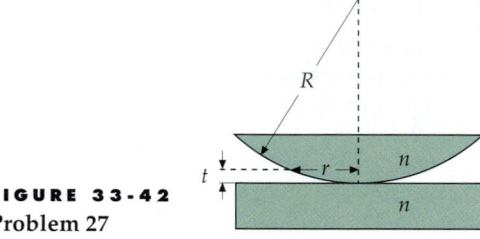

**FIGURE 33-42**
**Problem 27**

**28** •• [SOLVE✓] A plano-convex glass lens of radius of curvature 2.0 m rests on an optically flat glass plate. The arrangement is illuminated from above with monochromatic light of 520-nm wavelength. The indexes of refraction of the lens and plate are 1.6. Determine the radii of the first and second bright fringe in the reflected light. (Use the results from Problem 27b to relate $r$ to $t$.)

**29** •• Suppose that before the lens of Problem 28 is placed on the plate, a film of oil of refractive index 1.82 is deposited on the plate. What will then be the radii of the first and second bright fringes? (Use the results from Problem 27b to relate $r$ to $t$.)

## Two-Slit Interference Pattern

**30** • [SSM] Two narrow slits separated by 1 mm are illuminated by light of wavelength 600 nm, and the interference pattern is viewed on a screen 2 m away. Calculate the number of bright fringes per centimeter on the screen.

**31** • [SOLVE✓] Using a conventional two-slit apparatus with light of wavelength 589 nm, 28 bright fringes per centimeter are observed on a screen 3 m away. What is the slit separation?

**32** • Light of wavelength 633 nm from a helium–neon laser is shone normally on a plane containing two slits. The first interference maximum is 82 cm from the central maximum on a screen 12 m away. (a) Find the separation of the slits. (b) How many interference maxima can be observed?

**33** •• [SOLVE] Two narrow slits are separated by a distance $d$. Their interference pattern is to be observed on a screen a large distance $L$ away. (a) Calculate the spacing $\Delta y$ of the maxima on the screen for light of wavelength 500 nm, when $L = 1$ m and $d = 1$ cm. (b) Would you expect to be able to observe the interference of light on the screen for this situation? (c) How close together should the slits be placed for the maxima to be separated by 1 mm for this wavelength and screen distance?

**34** •• Light is incident at an angle $\phi$ with the normal to a vertical plane containing two slits of separation $d$ (Figure 33-43). Show that the interference maxima are located at angles $\theta_m$ given by $\sin\theta_m + \sin\phi = m\lambda/d$.

**FIGURE 33-43** Problems 34 and 35

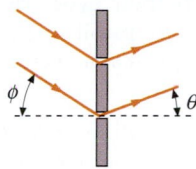

**35** •• **SSM** White light falls at an angle of 30° to the normal of a plane containing a pair of slits separated by 2.5 $\mu$m. What visible wavelengths give a bright interference maximum in the transmitted light in the direction normal to the plane? (See Problem 34.)

**36** •• Two small speakers are separated by a distance of 5 cm, as shown in Figure 33-44. The speakers are driven in phase with a sine wave signal of frequency 10 kHz. A small microphone is placed a distance 1 m away from the speakers on the axis running through the middle of the two speakers, and the microphone is then moved perpendicular to the axis. Where does the microphone record the first minimum and the first maximum of the interference pattern from the speakers? The speed of sound in air is 343 m/s.

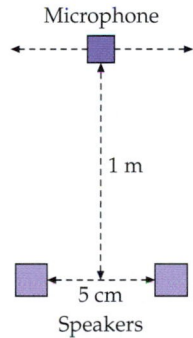

Microphone

1 m

5 cm

Speakers

**FIGURE 33-44** Problem 36

## Diffraction Pattern of a Single Slit

**37** • **iSOLVE** Light of wavelength 600 nm is incident on a long narrow slit. Find the angle of the first diffraction minimum if the width of the slit is (a) 1 mm, (b) 0.1 mm, and (c) 0.01 mm.

**38** • **iSOLVE** Plane microwaves are incident on a thin metal sheet with a long, narrow slit of width 5 cm in it. The first diffraction minimum is observed at $\theta = 37°$. What is the wavelength of the microwaves?

**39** •• **SSM** Measuring the distance to the moon (lunar ranging) is routinely done by firing short-pulse lasers and measuring the time it takes for the pulses to reflect back from the moon. A pulse is fired from the earth; to send it out, the pulse is expanded so that it fills the aperture of a 6-in-diameter telescope. (a) Assuming the only thing spreading the beam out to be diffraction, how large will the beam be when it reaches the moon, 382,000 km away? (b) The pulse is reflected off a retroreflecting mirror left by the Apollo 11 astronauts. If the diameter of the mirror is 20 in, how large will the beam be when it gets back to the earth? (c) What fraction of the power of the beam is reflected back to the earth? (d) If the beam is refocused

on return by the same 6 in telescope, what fraction of the original beam energy is recaptured? Ignore any atmospheric losses.

## Interference–Diffraction Pattern of Two Slits

**40** • How many interference maxima will be contained in the central diffraction maximum in the interference–diffraction pattern of two slits if the separation $d$ of the slits is 5 times their width $a$? How many will there be if $d = na$ for any value of $n$?

**41** •• A two-slit Fraunhofer interference–diffraction pattern is observed with light of wavelength 500 nm. The slits have a separation of 0.1 mm and a width of $a$. (a) Find the width $a$ if the fifth interference maximum is at the same angle as the first diffraction minimum. (b) For this case, how many bright interference fringes will be seen in the central diffraction maximum?

**42** •• **iSOLVE** A two-slit Fraunhofer interference–diffraction pattern is observed with light of wavelength 700 nm. The slits have widths of 0.01 mm and are separated by 0.2 mm. How many bright fringes will be seen in the central diffraction maximum?

**43** •• **SSM** Suppose that the *central* diffraction maximum for two slits contains 17 interference fringes for some wavelength of light. How many interference fringes would you expect in the first *secondary* diffraction maximum?

**44** •• **iSOLVE** Light of wavelength 550 nm illuminates two slits of width 0.03 mm and separation 0.15 mm. (a) How many interference maxima fall within the full width of the central diffraction maximum? (b) What is the ratio of the intensity of the third interference maximum to the side of the centerline (not counting the center interference maximum) to the intensity of the center interference maximum?

## *Using Phasors to Add Harmonic Waves

**45** • Find the resultant of the two waves $E_1 = 2 \sin \omega t$ and $E_2 = 3 \sin(\omega t + 270°)$.

**46** • **SSM** Find the resultant of the two waves $E_1 = 4 \sin \omega t$ and $E_2 = 3 \sin(\omega t + 60°)$.

**47** •• At the second secondary maximum of the diffraction pattern of a single slit, the phase difference between the waves from the top and bottom of the slit is approximately $5\pi$. The phasors used to calculate the amplitude at this point complete 2.5 circles. If $I_0$ is the intensity at the central maximum, find the intensity $I$ at this second secondary maximum.

**48** •• (a) Show that the positions of the interference minima on a screen a large distance $L$ away from three equally spaced sources (spacing $d$, with $d \gg \lambda$) are given approximately by

$$y = \frac{n\lambda L}{3d}, \text{where } n = 1, 2, 4, 5, 7, 8, 10, \ldots$$

that is, $n$ is not a multiple of 3. (b) For $L = 1$ m, $\lambda = 5 \times 10^{-7}$ m, and $d = 0.1$ mm, calculate the width of the principal interference maxima (the distance between successive minima) for three sources.

**49** •• (a) Show that the positions of the interference minima on a screen a large distance $L$ away from four equally spaced sources (spacing $d$, with $d \gg \lambda$) are given approximately by

$$y = \frac{n\lambda L}{4d}, \text{ where } n = 1, 2, 3, 5, 6, 7, 9, 10, \ldots$$

that is, $n$ is not a multiple of 4. (b) For $L = 2$ m, $\lambda = 6 \times 10^{-7}$ m, and $d = 0.1$ mm, calculate the width of the principal interference maxima (the distance between successive minima) for four sources. Compare this width with that for two sources with the same spacing.

**50** •• Light of wavelength 480 nm falls normally on four slits. Each slit is 2 $\mu$m wide and is separated from the next slit by 6 $\mu$m. (a) Find the angle from the center to the first point of zero intensity of the single-slit diffraction pattern on a distant screen. (b) Find the angles of any bright interference maxima that lie inside the central diffraction maximum. (c) Find the angular spread between the central interference maximum and the first interference minimum on either side of it. (d) Sketch the intensity as a function of angle.

**51** ••• Three slits, each separated from its neighbor by 0.06 mm, are illuminated by a coherent light source of wavelength 550 nm. The slits are extremely narrow. A screen is located 2.5 m from the slits. The intensity on the centerline is 0.05 W/m². Consider a location 1.72 cm from the centerline. (a) Draw the phasors, according to the phasor model for the addition of harmonic waves, appropriate for this location. (b) From the phasor diagram, calculate the intensity of light at this location.

**52** ••• **SSM** For single-slit diffraction, calculate the first three values of $\phi$ (the total phase difference between rays from each edge of the slit) that produce subsidiary maxima by (a) using the phasor model and (b) setting $dI/d\phi = 0$, where $I$ is given by Equation 33-19.

## Diffraction and Resolution

**53** • **ISOLVE✓** Light of wavelength 700 nm is incident on a pinhole of diameter 0.1 mm. (a) What is the angle between the central maximum and the first diffraction minimum for a Fraunhofer diffraction pattern? (b) What is the distance between the central maximum and the first diffraction minimum on a screen 8 m away?

**54** • Two sources of light of wavelength 700 nm are 10 m away from the pinhole of Problem 53. How far apart must the sources be for their diffraction patterns to be resolved by Rayleigh's criterion?

**55** • **SSM** Two sources of light of wavelength 700 nm are separated by a horizontal distance $x$. They are 5 m from a vertical slit of width 0.5 mm. What is the least value of $x$ for which the diffraction pattern of the sources can be resolved by Rayleigh's criterion?

**56** • The headlights on a small car are separated by 112 cm. At what maximum distance could you resolve the headlights if the diameter of your pupil is 5 mm and the effective wavelength of the light is 550 nm?

**57** • You are told not to shoot until you see the whites of their eyes. If their eyes are separated by 6.5 cm and the diameter of your pupil is 5 mm, at what distance can you resolve the two eyes using light of wavelength 550 nm?

**58** •• **ISOLVE✓** The ceiling of your lecture hall is probably covered with acoustic tile, which has small holes separated by about 6 mm. (a) Using light with a wavelength of 500 nm, how far could you be from this tile and still resolve these holes? The diameter of the pupil of your eye is about 5 mm. (b) Could you resolve these holes better with red light or with violet light?

**59** •• **ISOLVE** The telescope on Mount Palomar has a diameter of 200 in. Suppose a double star were 4 light-years away. Under ideal conditions, what must be the minimum separation of the two stars for their images to be resolved using light of wavelength 550 nm?

**60** •• **SSM** The star Mizar in Ursa Major is a binary system of stars of nearly equal magnitudes. The angular separation between the two stars is 14 seconds of arc. What is the minimum diameter of the pupil that allows resolution of the two stars using light of wavelength 550 nm?

## *Diffraction Gratings

**61** • **ISOLVE** A diffraction grating with 2000 slits per centimeter is used to measure the wavelengths emitted by hydrogen gas. At what angles $\theta$ in the first-order spectrum would you expect to find the two violet lines of wavelengths 434 nm and 410 nm?

**62** • **SSM** With the diffraction grating used in Problem 61, two other lines in the first-order hydrogen spectrum are found at angles $\theta_1 = 9.72 \times 10^{-2}$ rad and $\theta_2 = 1.32 \times 10^{-1}$ rad. Find the wavelengths of these lines.

**63** • Repeat Problem 61 for a diffraction grating with 15,000 slits per centimeter.

**64** • **ISOLVE** What is the longest wavelength that can be observed in the fifth-order spectrum using a diffraction grating with 4000 slits per centimeter?

**65** • The colors of many butterfly wings and beetle carapaces are due to effects of diffraction. The *Morpho* butterfly has structural elements on its wings that effectively act as a diffraction grating with spacing 880 nm. At what angle $\theta_1$ will normally incident blue light of wavelength $\lambda = 440$ nm be diffracted by the *Morpho's* wings?

**66** •• A diffraction grating of 2000 slits per centimeter is used to analyze the spectrum of mercury. (a) Find the angular separation in the first-order spectrum of the two lines of wavelength 579 nm and 577 nm. (b) How wide must the beam on the grating be for these lines to be resolved?

**67** •• **SSM** A diffraction grating with 4800 lines per centimeter is illuminated at normal incidence with white light (wavelength range of 400 nm to 700 nm). For how many orders can one observe the complete spectrum in the transmitted light? Do any of these orders overlap? If so, describe the overlapping regions.

**68** •• **iSOLVE** A square diffraction grating with an area of 25 cm² has a resolution of 22,000 in the fourth order. At what angle should you look to see a wavelength of 510 nm in the fourth order?

**69** •• Sodium light of wavelength 589 nm falls normally on a 2-cm-square diffraction grating ruled with 4000 lines per centimeter. The Fraunhofer diffraction pattern is projected onto a screen at 1.5 m by a lens of focal length 1.5 m placed immediately in front of the grating. Find (a) the positions of the first two intensity maxima on one side of the central maximum, (b) the width of the central maximum, and (c) the resolution in the first order.

**70** •• The spectrum of neon is exceptionally rich in the visible region. Among the many lines are two lines at wavelengths of 519.313 nm and 519.322 nm. If light from a neon discharge tube is normally incident on a transmission grating with 8400 lines per centimeter and the spectrum is observed in second order, what must be the width of the grating that is illuminated, so that these two lines can be resolved?

**71** •• **SSM** Mercury has several stable isotopes, among them ¹⁹⁸Hg and ²⁰²Hg. The strong spectral line of mercury, at about 546.07 nm, is a composite of spectral lines from the various mercury isotopes. The wavelengths of this line for ¹⁹⁸Hg and ²⁰²Hg are 546.07532 nm and 546.07355 nm, respectively. What must be the resolving power of a grating capable of resolving these two isotopic lines in the third-order spectrum? If the grating is illuminated over a 2-cm-wide region, what must be the number of lines per centimeter of the grating?

**72** •• **iSOLVE** A transmission grating is used to study the spectral region extending from 480 nm to 500 nm. The angular spread of this region is 12° in the third order. (a) Find the number of lines per centimeter. (b) How many orders are visible?

**73** •• **iSOLVE** White light is incident normally on a transmission grating and the spectrum is observed on a screen 8.0 m from the grating. In the second-order spectrum, the separation between light of 520-nm wavelength and 590-nm wavelength is 8.4 cm. (a) Determine the number of lines per centimeter of the grating. (b) What is the separation between these two wavelengths in the first-order spectrum and the third-order spectrum?

**74** ••• A diffraction grating has $n$ lines per unit length. Show that the angular separation of two lines of wavelengths $\lambda$ and $\lambda + \Delta\lambda$ is approximately

$$\Delta\theta = \Delta\lambda \Big/ \sqrt{\frac{1}{n^2 m^2} - \lambda^2}$$

where $m$ is the order number.

**75** ••• For a diffraction grating in which all the surfaces are normal to the incident radiation, most of the energy goes into the zeroth order, which is useless from a spectroscopic point of view, since in zeroth order all the wavelengths are at 0°. Therefore, modern reflection gratings have shaped, or *blazed*, grooves, as shown in Figure 33-45. This shifts the specular reflection, which contains most of the energy, from the zeroth order to some higher order. (a) Calculate the blaze

angle $\phi_m$ in terms of the groove separation $d$, the wavelength $\lambda$, and the order number $m$ in which specular reflection is to occur for $m = 1, 2, \ldots$. (b) Calculate the proper blaze angle for the specular reflection to occur in the second order for light of wavelength 450 nm incident on a grating with 10,000 lines per centimeter.

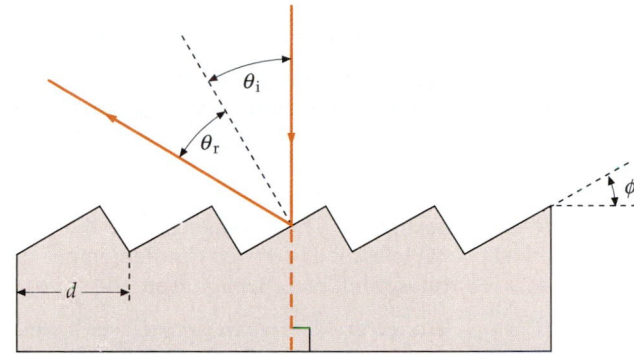

**FIGURE 33-45** Problem 75

**76** ••• In this problem, you will derive Equation 33-27 for the resolving power of a diffraction grating containing $N$ slits separated by a distance $d$. To do this, you will calculate the angular separation between the maximum and minimum for some wavelength $\lambda$ and set it equal to the angular separation of the $m$th-order maximum for two nearby wavelengths. (a) Show that the phase difference $\phi$ between the light from two adjacent slits is given by

$$\phi = \frac{2\pi d}{\lambda} \sin\theta$$

(b) Differentiate this expression to show that a small change in angle $d\theta$ results in a change in phase of $d\phi$ given by

$$d\phi = \frac{2\pi d}{\lambda} \cos\theta\, d\theta$$

(c) For $N$ slits, the angular separation between an interference maximum and an interference minimum corresponds to a phase change of $d\phi = 2\pi/N$. Use this to show that the angular separation $d\theta$ between the maximum and minimum for some wavelength $\lambda$ is given by

$$d\theta = \frac{\lambda}{Nd \cos\theta}$$

(d) The angle of the $m$th-order interference maximum for wavelength $\lambda$ is given by Equation 33-26. Compute the differential of each side of this equation to show that angular separation of the $m$th-order maximum for two nearly equal wavelengths differing by $d\lambda$ is given by

$$d\theta \approx \frac{m\, d\lambda}{d \cos\theta}$$

(e) According to Rayleigh's criterion, two wavelengths will be resolved in the $m$th order if the angular separation of the wavelengths, given by the Part (d) result, equals the angular separation of the interference maximum and the interference minimum given by the Part (c) result. Use this to derive Equation 33-27 for the resolving power of a grating.

## General Problems

**77** • [SSM] A long, narrow horizontal slit lies 1 mm above a plane mirror, which is in the horizontal plane. The interference pattern produced by the slit and its image is viewed on a screen 1 m from the slit. The wavelength of the light is 600 nm. (*a*) Find the distance from the mirror to the first maximum. (*b*) How many dark bands per centimeter are seen on the screen?

**78** •• A radio telescope is situated at the edge of a lake. The telescope is looking at light from a radio galaxy that is just rising over the horizon. If the height of the antenna is 20 m above the surface of the lake, at what angle above the horizon will the light from the radio galaxy go through its first interference maximum? The wavelength of the radio waves received by the telescope is $\lambda = 20$ cm. Remember that the light has a 180° phase shift on reflection from the water.

**79** • [iSOLVE] In a lecture demonstration, a laser beam of wavelength 700 nm passes through a vertical slit 0.5 mm wide and hits a screen 6 m away. Find the horizontal width of the principal diffraction maximum on the screen; that is, find the distance between the first minimum on the left and the first minimum on the right of the central maximum.

**80** • What minimum aperture, in millimeters, is required for opera glasses (binoculars) if an observer is to be able to distinguish the soprano's individual eyelashes (separated by 0.5 mm) at an observation distance of 25 m? Assume the effective wavelength of the light to be 550 nm.

**81** • The diameter of the aperture of the radio telescope at Arecibo, Puerto Rico, is 300 m. What is the resolving power of the telescope when tuned to detect microwaves of 3.2-cm wavelength?

**82** •• [SSM] [iSOLVE] A thin layer of a transparent material with an index of refraction of 1.30 is used as a nonreflective coating on the surface of glass with an index of refraction of 1.50. What should the thickness of the material be for the material to be nonreflecting for light of wavelength 600 nm?

**83** •• A *Fabry–Perot interferometer* consists of two parallel, half-silvered mirrors separated by a small distance *a*. Show that when light is incident on the interferometer with an angle of incidence $\theta$, the transmitted light will have maximum intensity when $a = (m\lambda/2) \cos \theta$.

**84** •• A mica sheet 1.2 $\mu$m thick is suspended in air. In reflected light, there are gaps in the visible spectrum at 421, 474, 542, and 633 nm. Find the index of refraction of the mica sheet.

**85** •• A camera lens is made of glass with an index of refraction of 1.6. This lens is coated with a magnesium fluoride film ($n = 1.38$) to enhance its light transmission. This film is to produce zero reflection for light of wavelength 540 nm.

Treat the lens surface as a flat plane and the film as a uniformly thick flat film. (*a*) How thick must the film be to accomplish its objective? (*b*) Would there be destructive interference for any other visible wavelengths? (*c*) By what factor would the reflection for light of wavelengths 400 nm and 700 nm be reduced by this film? Neglect the variation in the reflected light amplitudes from the two surfaces.

**86** •• In a pinhole camera, the image is fuzzy because of geometry (rays arrive at the film after passing through different parts of the pinhole) and because of diffraction. As the pinhole is made smaller, the fuzziness due to geometry is reduced, but the fuzziness due to diffraction is increased. The optimum size of the pinhole for the sharpest possible image occurs when the spread due to diffraction equals the spread due to the geometric effects of the pinhole. Estimate the optimum size of the pinhole if the distance from the pinhole to the film is 10 cm and the wavelength of the light is 550 nm.

**87** •• [SSM] [iSOLVE✓] The Impressionist painter Georges Seurat used a technique called *pointillism*, in which his paintings are composed of small, closely spaced dots of pure color, each about 2 mm in diameter. The illusion of the colors blending together smoothly is produced in the eye of the viewer by diffraction effects. Calculate the minimum viewing distance for this effect to work properly. Use the wavelength of visible light that requires the *greatest* distance, so that you are sure the effect will work for *all* visible wavelengths. Assume the pupil of the eye has a diameter of 3 mm.

**88** ••• [SSM] A *Jamin refractometer* is a device for measuring or for comparing the indexes of refraction of gases. A beam of monochromatic light is split into two parts, each of which is directed along the axis of a separate cylindrical tube before being recombined into a single beam that is viewed through a telescope. Suppose that each tube is 0.4 m long and that sodium light of wavelength 589 nm is used. Both tubes are initially evacuated, and constructive interference is observed in the center of the field of view. As air is slowly allowed to enter one of the tubes, the central field of view changes to dark and back to bright a total of 198 times. (*a*) What is the index of refraction of air? (*b*) If the fringes can be counted to ± 0.25 fringe, where one fringe is equivalent to one complete cycle of intensity variation at the center of the field of view, to what accuracy can the index of refraction of air be determined by this experiment?

**89** ••• Light of wavelength $\lambda$ is diffracted through a single slit of width *a*, and the resulting pattern is viewed on a screen a long distance *L* away from the slit. (*a*) Show that the width of the central maximum on the screen is approximately $2L\lambda/a$. (*b*) If a slit of width $2L\lambda/a$ is cut in the screen and is illuminated by light of the same wavelength, show that the width of its central diffraction maximum at the same distance *L* is *a* to the same approximation.

# Wave–Particle Duality and Quantum Physics

CHAPTER
# 34

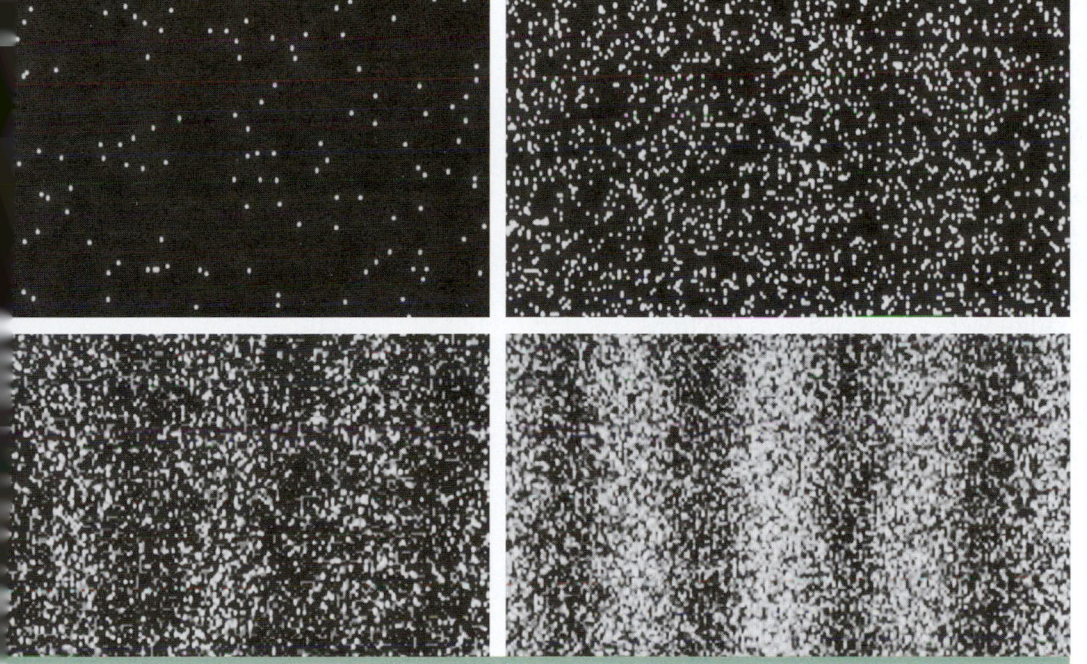

ELECTRON INTERFERENCE PATTERN PRODUCED BY ELECTRONS INCIDENT ON A BARRIER CONTAINING TWO SLITS: (A) 100 ELECTRONS, (B) 3000 ELECTRONS, (C) 20,000 ELECTRONS, AND (D) 70,000 ELECTRONS. THE MAXIMA AND MINIMA DEMONSTRATE THE WAVE NATURE OF THE ELECTRON AS IT TRAVERSES THE SLITS. INDIVIDUAL DOTS ON THE SCREEN INDICATE THE PARTICLE NATURE OF THE ELECTRON AS IT EXCHANGES ENERGY WITH THE DETECTOR. THE PATTERN IS THE SAME WHETHER ELECTRONS OR PHOTONS (PARTICLES OF LIGHT) ARE USED.

**?** **How do you calculate the wavelength of an electron? This is revealed in Example 34-4.**

34-1    Light

34-2    The Particle Nature of Light: Photons

34-3    Energy Quantization in Atoms

34-4    Electrons and Matter Waves

34-5    The Interpretation of the Wave Function

34-6    Wave–Particle Duality

34-7    A Particle in a Box

34-8    Expectation Values

34-9    Energy Quantization in Other Systems

**W**e have seen that the propagation of waves through space is quite different from the propagation of particles. Waves bend around corners (diffraction) and interfere with one another, producing interference patterns. If a wave encounters a small aperture, the wave spreads out on

1117

the other side as if the aperture were a point source. If two coherent waves of equal intensity $I_0$ meet in space, the result can be a wave of intensity $4I_0$ (constructive interference), an intensity of zero (destructive interference), or a wave of intensity between zero and $4I_0$, depending on the phase difference between the waves at their meeting point.

The propagation of particles is quite unlike the propagation of waves. Particles travel in straight lines until they collide with something, after which the particles again travel in straight lines. If two particle beams meet in space, they never produce an interference pattern.

Particles and waves also exchange energy differently. Particles exchange energy in collisions that occur at specific points in space and in time. The energy of waves, on the other hand, is spread out in space and deposited continuously as the wave fronts interact with matter.

Sometimes the propagation of a wave cannot be distinguished from the propagation of a beam of particles. If the wavelength $\lambda$ is very small compared to distances from the edges of objects, diffraction effects are negligible and the wave travels in straight lines. Also, interference maxima and minima are so close together in space as to be unobservable. The result is that the wave interacts with a detector, like a beam of numerous small particles each exchanging a small amount of energy; the exchange cannot distinguish particles from waves. For example, you do not observe the individual air molecules bouncing off your face if the wind blows on it. Instead, the interaction of billions of particles appears to be continuous, like that of a wave.

At the beginning of the twentieth century, it was thought that sound, light, and other electromagnetic radiation (e.g., radio) were waves; whereas electrons, protons, atoms, and similar constituents of nature were understood to be particles. The first 30 years of the century revealed startling developments in theoretical and experimental physics, such as the finding that light, thought to be a wave, actually exchanges energy in discrete lumps or quanta, just like particles; and the finding that an electron, thought to be a particle, exhibits diffraction and interference as it propagates through space, just like a wave.

The fact that light exchanges energy like a particle implies that light energy is not continuous but is *quantized.* Similarly, the wave nature of the electron, along with the fact that the standing wave condition requires a discrete set of frequencies, implies that the energy of an electron in a confined region of space is not continuous, but is quantized to a discrete set of values.

➤ **In this chapter, we begin by discussing some basic properties of light and electrons, examining their wave and particle characteristics. We then consider some of the detailed properties of matter waves, showing, in particular, how standing waves imply the quantization of energy. Finally, we discuss some of the important features of the theory of quantum physics, which was developed in the 1920s and which has been extremely successful in describing nature. Quantum physics is now the basis of our understanding of both the microscopic and very low temperature worlds.**

## 34-1 Light

The question of whether light consists of a beam of particles or waves in motion is one of the most interesting in the history of science. Isaac Newton used a particle theory of light to explain the laws of reflection and refraction; however, for refraction, Newton needed to assume that light travels faster in water or glass than in air, an assumption later shown to be false. The chief early proponents of the wave theory were Robert Hooke and Christian Huygens, who explained refraction by assuming that light travels more slowly in glass or water than it does in air. Newton rejected the wave theory because, in his time, light was believed to travel through a medium only in straight lines—diffraction had not yet been observed.

Because of Newton's great reputation and authority, his particle theory of light was accepted for more than a century. Then, in 1801, Thomas Young demonstrated the wave nature of light in a famous experiment in which two coherent light sources are produced by illuminating a pair of narrow, parallel slits with a single source (Figure 34-1). In Chapter 33, we saw that when light encounters a small opening, the opening acts as a point source of waves (Figure 33-7). In Young's experiment, each slit acts as a line source, which is equivalent to a point source in two dimensions. The interference pattern is observed on a screen placed behind the slits. Interference maxima occur at angles so that the path difference is an integral number of wavelengths. Similarly, interference minima occur if the path difference is one-half wavelength or any odd number of half wavelengths. Figure 34-1b shows the intensity pattern as seen on the screen. Young's experiment and many other experiments demonstrate that light propagates like a wave.

In the early nineteenth century, the French physicist Augustin Fresnel (1788–1827) performed extensive experiments on interference and diffraction and put the wave theory on a rigorous mathematical basis. Fresnel showed that the observed straight-line propagation of light is a result of the very short wavelengths of visible light.

The classical wave theory of light culminated in 1860 when James Clerk Maxwell published his mathematical theory of electromagnetism. This theory yielded a wave equation that predicted the existence of electromagnetic waves that propagate with a speed that can be calculated from the laws of electricity and magnetism. The fact that the result of this calculation was $c \approx 3 \times 10^8$ m/s, the same as the speed of light, suggested to Maxwell that light is an electromagnetic wave. The eye is sensitive to electromagnetic waves with wavelengths in the range from approximately 400 nm (1 nm = $10^{-9}$ m) to approximately 700 nm. This range is called *visible light*. Other electromagnetic waves (e.g., microwaves, radio, television, and X rays) differ from light only in wavelength and in frequency.

(a)

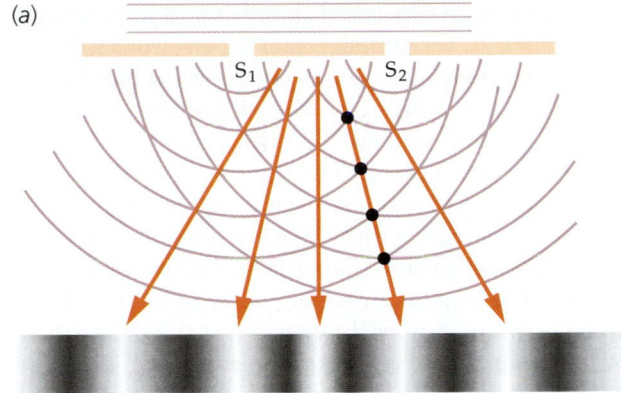

(b)

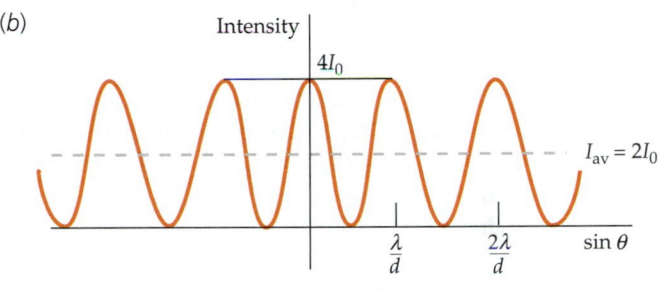

**FIGURE 34-1** (*a*) Two slits act as coherent sources of light for the observation of interference in Young's experiment. Cylindrical waves from the slits overlap and produce an interference pattern on a screen far away. (*b*) The intensity pattern produced in Figure 34-1*a*. The intensity is maximum at points where the path difference is an integral number of wavelengths, and the intensity is zero where the path difference is an odd number of half wavelengths.

# 34-2    The Particle Nature of Light: Photons

The diffraction of light and the existence of an interference pattern in the two-slit experiment give clear evidence that light has wave properties. However, early in the twentieth century, it was found that light energy comes in discrete amounts.

## The Photoelectric Effect

The quantum nature of light and the quantization of energy were suggested by Albert Einstein in 1905 in his explanation of the photoelectric effect. Einstein's work marked the beginning of quantum theory, and for his work, Einstein received the Nobel Prize for physics. Figure 34-2 shows a schematic diagram of the basic apparatus for studying the photoelectric effect. Light of a single frequency enters an evacuated chamber and falls on a clean metal surface C (C for cathode), causing electrons to be emitted. Some of these electrons strike the second metal plate A (A for anode), constituting an electric current between the plates. Plate A is negatively charged, so the electrons are repelled by it, with only the most energetic electrons reaching the plate. The maximum kinetic energy of the emitted electrons is measured by slowly increasing the voltage until the current becomes zero. Experiments give the surprising result that the maximum kinetic energy of the emitted electrons is *independent of the intensity* of the incident light. Classically, we would expect that increasing the rate at which light energy falls on the metal surface would increase the energy absorbed by individual electrons and, therefore, would increase the maximum kinetic energy of the electrons emitted. Experimentally, this is not what happens. The maximum kinetic energy of the emitted electrons is the same for a given wavelength of incident light, no matter how intense the light. Einstein demonstrated that this experimental result can be explained if light energy is quantized in small bundles called **photons**. The energy $E$ of each photon is given by

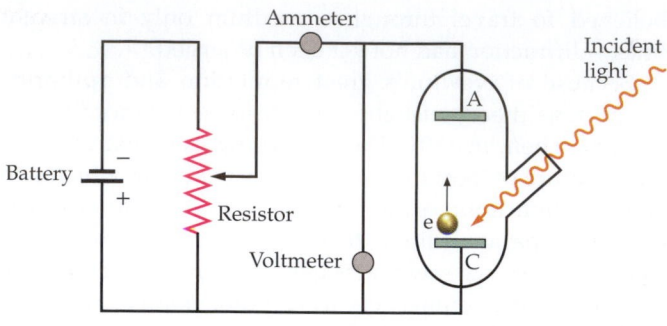

**FIGURE 34-2** A schematic drawing of the apparatus for studying the photoelectric effect. Light of a single frequency enters an evacuated chamber and strikes the cathode C, which then ejects electrons. The current in the ammeter measures the number of these electrons that reach the anode A per unit time. The anode is made electrically negative with respect to the cathode to repel the electrons. Only those electrons with enough initial kinetic energy to overcome the repulsion can reach the anode. The voltage between the two plates is slowly increased until the current becomes zero, which happens when even the most energetic electrons do not make it to plate A.

$$E = hf = \frac{hc}{\lambda} \qquad\qquad 34\text{-}1$$

EINSTEIN EQUATION FOR PHOTON ENERGY

where $f$ is the frequency, and $h$ is a constant now known as **Planck's constant.**[†] The measured value of this constant is

$$h = 6.626 \times 10^{-34} \,\text{J·s} = 4.136 \times 10^{-15} \,\text{eV·s} \qquad\qquad 34\text{-}2$$

PLANCK'S CONSTANT

Equation 34-1 is sometimes called the **Einstein equation.**

At the fundamental level, a light beam consists of a beam of particles—photons—each with energy $hf$. The intensity (power per unit area) of a monochromatic light beam is the number of photons per unit area per unit of time, times the energy per photon. The interaction of the light beam with the metal surface consists of collisions between photons and electrons. In these collisions, the photons disappear, with each photon giving all its energy to an electron, and the electron emitted from the surface thus receives its energy from a single photon. If the intensity of light is increased, more photons fall on the surface per unit time, and more electrons are emitted. However, each photon still has the same energy $hf$, so the energy absorbed by each electron is unchanged.

If $\phi$ is the minimum energy necessary to remove an electron from a metal surface, the maximum kinetic energy of the electrons emitted is given by

$$K_{max} = \left(\tfrac{1}{2}mv^2\right)_{max} = hf - \phi \qquad\qquad 34\text{-}3$$

EINSTEIN'S PHOTOELECTRIC EQUATION

The quantity $\phi$, called the **work function,** is a characteristic of the particular metal. (Some electrons will have kinetic energies less than $hf - \phi$, because of the loss of energy from traveling through the metal.)

† In 1900, the German physicist Max Planck introduced this constant to explain discrepancies between the theoretical curves and experimental data on the spectrum of blackbody radiation. Planck also assumed that the radiation was emitted and absorbed by a blackbody in quanta of energy $hf$, but he considered his assumption to be just a calculational device rather than a fundamental property of electromagnetic radiation. Blackbody radiation was discussed in Chapter 20.

According to Einstein's photoelectric equation, a plot of $K_{max}$ versus frequency $f$ should be a straight line with the slope $h$. This was a bold prediction, because, at the time, there was no evidence that Planck's constant had any application outside of blackbody radiation. In addition, there was no experimental data on $K_{max}$ versus frequency $f$, because no one before had even suspected that the frequency of the light was related to $K_{max}$. This prediction was difficult to verify experimentally, but careful experiments by R. A. Millikan approximately 10 years later showed that Einstein's equation was correct. Figure 34-3 shows a plot of Millikan's data.

Photons with frequencies less than a **threshold frequency** $f_t$, and therefore with wavelengths greater than a **threshold wavelength** $\lambda_t = c/f_t$, do not have enough energy to eject an electron from a particular metal. The threshold frequency and the corresponding threshold wavelength can be related to the work function $\phi$ by setting the maximum kinetic energy of the electrons equal to zero in Equation 34-3. Then

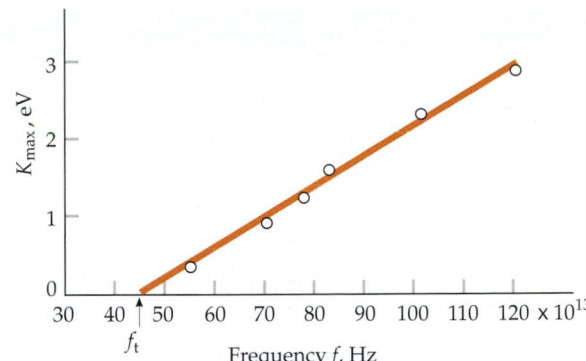

**FIGURE 34-3** Millikan's data for the maximum kinetic energy $K_{max}$ versus frequency $f$ for the photoelectric effect. The data fall on a straight line that has a slope $h$, as predicted by Einstein approximately a decade before the experiment was performed.

$$\phi = hf_t = \frac{hc}{\lambda_t} \qquad 34\text{-}4$$

Work functions for metals are typically a few electron volts. Since wavelengths are usually given in nanometers and energies in electron volts, it is useful to have the value of $hc$ in electron volt–nanometers:

$$hc = (4.1357 \times 10^{-15}\ \text{eV·s})(2.9979 \times 10^8\ \text{m/s}) = 1.240 \times 10^{-6}\ \text{eV·m}$$

or

$$hc = 1240\ \text{eV·nm} \qquad 34\text{-}5$$

---

*PHOTON ENERGIES FOR VISIBLE LIGHT*                    **EXAMPLE 34-1**

Calculate the photon energies for light of wavelengths 400 nm (violet) and 700 nm (red). (These are the approximate wavelengths at the two extremes of the visible spectrum.)

1. The energy is related to the wavelength by Equation 34-1:     $E = hf = \dfrac{hc}{\lambda}$

2. For $\lambda = 400$ nm, the energy is:     $E = \dfrac{hc}{\lambda} = \dfrac{1240\ \text{eV·nm}}{400\ \text{nm}} = \boxed{3.10\ \text{eV}}$

3. For $\lambda = 700$ nm, the energy is:     $E = \dfrac{hc}{\lambda} = \dfrac{1240\ \text{eV·nm}}{700\ \text{nm}} = \boxed{1.77\ \text{eV}}$

**REMARKS** We can see from these calculations that visible light contains photons with energies that range from approximately 1.8 eV to 3.1 eV. X rays, which have much shorter wavelengths, contain photons with energies of the order of keV. Gamma rays emitted by nuclei have even shorter wavelengths and photons of energy of the order of MeV.

**EXERCISE** Find the energy of a photon corresponding to electromagnetic radiation in the FM radio band of wavelength 3 m. (*Answer* $4.13 \times 10^{-7}$ eV)

**EXERCISE** Find the wavelength of a photon whose energy is (*a*) 0.1 eV, (*b*) 1 keV, and (*c*) 1 MeV. (*Answer* (*a*) 12.4 $\mu$m, (*b*) 1.24 nm, (*c*) 1.24 pm)

---

*THE NUMBER OF PHOTONS PER SECOND IN SUNLIGHT*     **EXAMPLE 34-2**     **Try It Yourself**

The intensity of sunlight at the earth's surface is approximately 1400 W/m². Assuming the average photon energy is 2 eV (corresponding to a wavelength of approximately 600 nm), calculate the number of photons that strike an area of 1 cm² each second.

**PICTURE THE PROBLEM** Intensity (power per unit area) is given as is the area. From this, we can calculate the power, which is the energy per unit time.

**Cover the column to the right and try these on your own before looking at the answers.**

| Steps | Answers |
|---|---|
| 1. The energy $\Delta E$ is related to the number $N$ of photons and the energy per photon $hf = 2$ eV. | $\Delta E = Nhf$ |
| 2. The intensity $I$ (power per unit area) and the area $A$ are given, so we can find the power. | $I = \dfrac{P}{A}$ |
| 3. Knowing the power (energy per unit time) and the time, we can find the energy. | $\Delta E = P\Delta t$ |
| 4. Substitute the results from steps 1–3 and solve for $N$. Take care to get the units correct. | $N = \dfrac{IA\Delta t}{hf} = \boxed{4.38 \times 10^{17}}$ |

**REMARKS** This is an enormous number of photons. In most everyday situations, the number of photons is so great that the quantization of light is not noticeable.

**EXERCISE** Calculate the photon density (in photons per cubic centimeter) of the sunlight in Example 34-2. The number arriving on an area of 1 cm² in one second is the number in a column whose cross section is 1 cm² and whose height is the distance light travels in one second. (*Answer* $1.46 \times 10^7$ cm⁻³)

## Compton Scattering

The first use of the photon concept was to explain the results of photoelectric-effect experiments. The photon concept was used by Arthur H. Compton to explain the results of his measurements of the scattering of X rays by free electrons in 1923. According to classical theory, if an electromagnetic wave of frequency $f_1$ is incident on material containing free charges, the charges will oscillate with this frequency and reradiate electromagnetic waves of the same frequency. Compton considered these reradiated waves as scattered photons, and he pointed out that if the scattering process were a collision between a photon and an electron (Figure 34-4), the electron would recoil and thus absorb energy. The scattered photon would then have less energy, and therefore a lower frequency and longer wavelength, than the incident photon.

According to classical electromagnetic theory (see Section 30-3), the energy and momentum of an electromagnetic wave are related by

$$E = pc \qquad\qquad 34\text{-}6$$

The momentum of a photon is thus related to its wavelength $\lambda$ by $p = E/c = hf/c = h/\lambda$.

$$p = \frac{h}{\lambda} \qquad\qquad 34\text{-}7$$

*MOMENTUM OF A PHOTON*

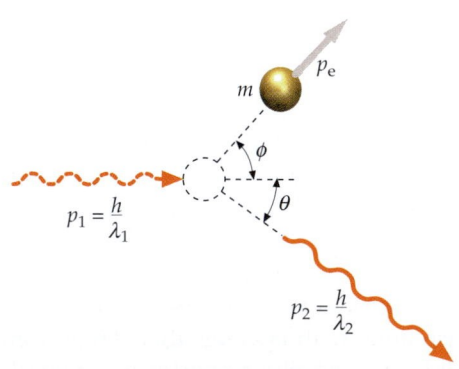

**FIGURE 34-4** The scattering of light by an electron is considered as a collision of a photon of momentum $h/\lambda_1$ and a stationary electron. The scattered photon has less energy and therefore has a greater wavelength than does the incident electron.

Compton applied the laws of conservation of momentum and energy to the collision of a photon and an electron to calculate the momentum $p_2$ and thus the wavelength $\lambda_2 = h/p_2$ of the scattered photon (see Figure 34-4). Applying conservation of momentum to the collision gives

$$\vec{p}_1 = \vec{p}_2 + \vec{p}_e \qquad\qquad 34\text{-}8$$

where $\vec{p}_1$ is the momentum of the incident photon and $\vec{p}_e$ is the momentum of the electron after the collision. The initial momentum of the electron is zero. Rearranging Equation 34-8, we have $\vec{p}_e = \vec{p}_1 - \vec{p}_2$. Taking the dot product of each side with itself gives

$$p_e^2 = p_1^2 + p_2^2 - 2p_1 p_2 \cos\theta \qquad\qquad 34\text{-}9$$

where $\theta$ is the angle the scattered photon makes with the direction of the incident photon. Because the kinetic energy of the electron after the collision can be a significant fraction of the rest energy of an electron, the relativistic expression relating the energy $E$ of the electron to its momentum is used. This expression (Equation R-17) is

$$E = \sqrt{p_e^2 c^2 + (m_e c^2)^2}$$

where $m_e$ is the rest mass of the electron. Applying conservation of energy to the collision gives

$$p_1 c + m_e c^2 = p_2 c + \sqrt{p_e^2 c^2 + (m_e c^2)^2} \qquad\qquad 34\text{-}10$$

where $pc$ (Equation 34-6) has been used to express the energies of the photons. Eliminating $p_e^2$ from Equation 34-9 and Equation 34-10 gives

$$\frac{1}{p_2} - \frac{1}{p_1} = \frac{1}{m_e c}(1 - \cos\theta)$$

and substituting for $p_1$ and $p_2$, using Equation 34-7 gives

$$\lambda_2 - \lambda_1 = \frac{h}{m_e c}(1 - \cos\theta) \qquad\qquad 34\text{-}11$$

COMPTON EQUATION

The increase in wavelengths is independent of the wavelength $\lambda_1$ of the incident photon. The quantity $h/m_e c$ has dimensions of length and is called the Compton wavelength. Its value is

$$\lambda_C = \frac{h}{m_e c} = \frac{hc}{m_e c^2} = \frac{1240\ \text{eV·nm}}{5.11 \times 10^5\ \text{eV}} = 2.43 \times 10^{-12}\ \text{m} = 2.43\ \text{pm} \qquad 34\text{-}12$$

Because $\lambda_2 - \lambda_1$ is small, it is difficult to observe unless $\lambda_1$ is so small that the fractional change $(\lambda_2 - \lambda_1)/\lambda_1$ is appreciable. Compton used X rays of wavelength 71.1 pm (1 pm $= 10^{-12}$ m $= 10^{-3}$ nm). The energy of a photon of this wavelength is $E = hc/\lambda = (1240\ \text{eV·nm})/(0.0711\ \text{nm}) = 17.4$ keV. Since this is much greater than the binding energy of the valence electrons in most atoms (which is of the order of a few eV), these electrons can be considered to be essentially free. Compton's measurements of $\lambda_2 - \lambda_1$ as a function of scattering angle $\theta$ agreed with Equation 34-11, thereby confirming the correctness of the photon concept (i.e., of the particle nature of light).

---

*FINDING THE INCREASE IN WAVELENGTH*　　　　　　　　**EXAMPLE 34-3**

An X-ray photon of wavelength 6 pm makes a head-on collision with an electron, so that the scattered photon goes in a direction opposite to that of the incident photon. The electron is initially at rest. (*a*) How much longer is the wavelength of the scattered photon than that of the incident photon? (*b*) What is the kinetic energy of the recoiling electron?

**PICTURE THE PROBLEM** We can calculate the increase in wavelength, and thus the new wavelength, from Equation 34-11. We then use the new wavelength to find the energy of the scattered photon and then to find the kinetic energy of the recoiling electron from conservation of energy (Figure 34-5).

**FIGURE 34-5**

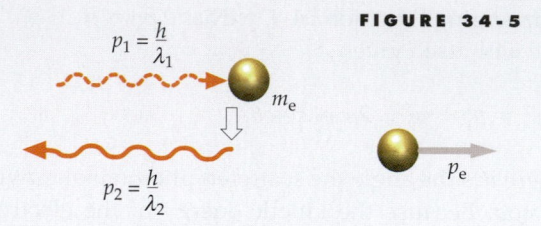

(*a*) Use Equation 34-11 to calculate the increase in wavelength:

$$\Delta\lambda = \lambda_2 - \lambda_1$$

$$= \frac{h}{m_e c}(1 - \cos\theta)$$

$$= (2.43 \text{ pm})(1 - \cos 180°) = \boxed{4.86 \text{ pm}}$$

(*b*) 1. The kinetic energy of the recoiling electron equals the energy of the incident photon $E_1$ minus the energy of the scattered photon $E_2$:

$$K_e = E_1 - E_2 = \frac{hc}{\lambda_1} - \frac{hc}{\lambda_2}$$

2. Calculate $\lambda_2$ from the given wavelength of the incident photon and the change found in Part (*a*):

$$\lambda_2 = \lambda_1 + \Delta\lambda = 6 \text{ pm} + 4.86 \text{ pm}$$

$$= 10.86 \text{ pm}$$

3. Substitute the calculated values of $E_1$ and $E_2$ to find the energy of the recoiling electron:

$$K_e = \frac{hc}{\lambda_1} - \frac{hc}{\lambda_2}$$

$$= \frac{1240 \text{ eV·nm}}{6.0 \text{ pm}} - \frac{1240 \text{ eV·nm}}{10.86 \text{ pm}}$$

$$= \frac{1.24 \text{ keV·nm}}{6.0 \times 10^{-3} \text{ nm}} - \frac{1.24 \text{ keV·nm}}{10.86 \times 10^{-3} \text{ nm}}$$

$$= 207 \text{ keV} - 114 \text{ keV}$$

$$= \boxed{93 \text{ keV}}$$

**REMARKS** The kinetic energy of the scattered electron is 93 keV and the rest energy of an electron is 511 keV, so the kinetic energy is 18 percent of the rest energy. Thus, the nonrelativistic formula for the kinetic energy ($\frac{1}{2}m_e v^2$) is not valid.

**EXERCISE** What is the speed of the scattered electron given by the nonrelativistic formula for the kinetic energy ($\frac{1}{2}m_e v^2$)? (*Answer* 0.6c)

## 34-3 Energy Quantization in Atoms

Ordinary white light has a continuous spectrum; that is, it contains *all* the wavelengths in the visible spectrum. But if atoms in a gas at low pressure are excited by an electric discharge, they emit light of specific wavelengths that are characteristic of the element or the compound. Since the energy of a photon is related to its wavelength by $E = hf = hc/\lambda$, a discrete set of wavelengths implies a discrete set of energies. Conservation of energy then implies that if an atom absorbs or emits a photon, its internal energy changes by a discrete amount, which is ± the energy of the photon. In 1913, this led Niels Bohr to postulate that the internal energy of an atom can have only a discrete set of values. That is, the internal energy of an atom is **quantized.** If an atom radiates light of frequency $f$, the atom makes a transition from one allowed level to another level that is lower in energy by $\Delta E = hf$. Bohr was able to construct a semiclassical model of the hydrogen atom that had a discrete set of energy levels consistent with the observed spectrum of emitted light.[†] However, the *reason* for the quantization of energy levels in atoms and other systems remained a mystery until the wave nature of electrons was discovered a decade later.

# 34-4 Electrons and Matter Waves

In 1897, J. J. Thomson showed that the rays of a cathode-ray tube (Figure 34-6) can be deflected by electric and magnetic fields and therefore must consist of electrically charged particles. By measuring the deflections of these particles, Thomson showed that all the particles have the same charge-to-mass ratio $q/m$. He also showed that particles with this charge-to-mass ratio can be obtained using any material for the cathode, which means that these particles, now called **electrons,** are a fundamental constituent of all matter.

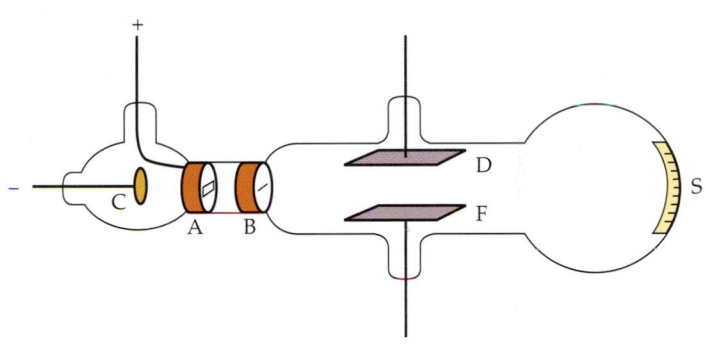

## The de Broglie Hypothesis

Since light seems to have both wave and particle properties, it is natural to ask whether matter (e.g., electrons, protons) might also have both wave and particle characteristics. In 1924, a French physics student, Louis de Broglie, suggested this idea in his doctoral dissertation. de Broglie's work was highly speculative, because there was no evidence at that time of any wave aspects of matter.

For the wavelength of electron waves, de Broglie chose

**FIGURE 34-6** Schematic diagram of the cathode-ray tube Thomson used to measure $q/m$ for the particles that comprise cathode rays (electrons). Electrons from the cathode C pass through the slits at A and B and strike a phosphorescent screen S. The beam can be deflected by an electric field between plates D and F or by a magnetic field (not shown).

$$\lambda = \frac{h}{p}$$

34-13

DE BROGLIE RELATION FOR THE WAVELENGTH OF ELECTRON WAVES

where $p$ is the momentum of the electron. Note that this is the same as Equation 34-7 for a photon. For the frequency of electron waves, de Broglie chose the Einstein equation relating the frequency and energy of a photon.

$$f = \frac{E}{h}$$

34-14

DE BROGLIE RELATION FOR THE FREQUENCY OF ELECTRON WAVES

These equations are thought to apply to all matter. However, for macroscopic objects, the wavelengths calculated from Equation 34-13 are so small that it is impossible to observe the usual wave properties of interference or diffraction. Even a dust particle with a mass as small as 1 $\mu$g is much too massive for any wave characteristics to be noticed, as we see in the following example.

[†] We will study the Bohr model in Chapter 36.

Find the de Broglie wavelength of a $10^{-6}$ g particle moving with a speed of $10^{-6}$ m/s.

**Cover the column to the right and try this on your own before looking at the answers.**

| Steps | Answers |
|---|---|
| Write the definition of the de Broglie wavelength and substitute the given data. | $\lambda = \dfrac{h}{p} = \dfrac{h}{mv} = \dfrac{6.63 \times 10^{-34}\ \text{J·s}}{(10^{-9}\ \text{kg})(10^{-6}\ \text{m/s})}$ |
| | $= \boxed{6.63 \times 10^{-19}\ \text{m}}$ |

**REMARKS** This wavelength is several orders of magnitude smaller than the diameter of an atomic nucleus, which is about $10^{-15}$ m.

Since the wavelength found in Example 34-4 is so small, much smaller than any possible apertures or obstacles, diffraction or interference of such waves cannot be observed. In fact, the propagation of waves of very small wavelengths is indistinguishable from the propagation of particles. The momentum of the particle in Example 34-4 was only $10^{-15}$ kg·m/s. A macroscopic particle with a greater momentum would have an even smaller de Broglie wavelength. We therefore do not observe the wave properties of such macroscopic objects as baseballs and billiard balls.

**EXERCISE** Find the de Broglie wavelength of a baseball of mass 0.17 kg moving at 100 km/h. (*Answer*  $1.4 \times 10^{-34}$ m)

The situation is different for low-energy electrons and other microscopic particles. Consider a particle with kinetic energy $K$. Its momentum is found from

$$K = \frac{p^2}{2m}$$

or

$$p = \sqrt{2mK}$$

Its wavelength is then

$$\lambda = \frac{h}{p} = \frac{h}{\sqrt{2mK}}$$

If we multiply the numerator and the denominator by $c$, we obtain

$$\lambda = \frac{hc}{\sqrt{2mc^2K}} = \frac{1240\ \text{eV·nm}}{\sqrt{2mc^2K}} \qquad\qquad 34\text{-}15$$

WAVELENGTH ASSOCIATED WITH A PARTICLE OF MASS $M$

where we have used $hc = 1240$ eV·nm. For electrons, $mc^2 = 0.511$ MeV. Then,

$$\lambda = \frac{1240\ \text{eV·nm}}{\sqrt{2mc^2K}} = \frac{1240\ \text{eV·nm}}{\sqrt{2(0.511 \times 10^6\ \text{eV})K}}$$

or

$$\lambda = \frac{1.226}{\sqrt{K}} \text{ nm}, \quad K \text{ in electron volts} \qquad\qquad 34\text{-}16$$

ELECTRON WAVELENGTH

Equation 34-15 and Equation 34-16 do not hold for relativistic particles whose kinetic energies are a significant fraction of their rest energies $mc^2$. (Rest energies were discussed in Chapter 7 and in Chapter R.)

**EXERCISE** Find the wavelength of an electron whose kinetic energy is 10 eV. (*Answer* 0.388 nm. From this result, we see that a 10-eV electron has a de Broglie wavelength of about 0.4 nm. This is on the same order of magnitude as the size of the atom and the spacing of atoms in a crystal.)

## Electron Interference and Diffraction

The observation of diffraction and interference of electron waves would provide the crucial test of the existence of wave properties of electrons. This was first discovered serendipitously in 1927 by C. J. Davisson and L. H. Germer as they were studying electron scattering from a nickel target at the Bell Telephone Laboratories. After heating the target to remove an oxide coating that had accumulated during an accidental break in the vacuum system, they found that the scattered electron intensity as a function of the scattering angle showed maxima and minima. Their target had crystallized, and by accident they had observed electron diffraction. Davisson and Germer then prepared a target consisting of a single crystal of nickel and investigated this phenomenon extensively. Figure 34-7a illustrates their experiment. Electrons from an electron gun are directed at a crystal and detected at some angle $\phi$ that can be varied. Figure 34-7b shows a typical pattern observed. There is a strong scattering maximum at an angle of 50°. The angle for maximum scattering of waves from a crystal depends on the wavelength of the waves and the spacing of the atoms in the crystal. Using the known spacing of atoms in their crystal, Davisson and Germer calculated the wavelength that could produce such a maximum and found that it agreed with the de Broglie equation (Equation 34-16) for the electron energy they were using. By varying the energy of the incident electrons, they could vary the electron wavelengths and produce maxima and minima at different locations in the diffraction patterns. In all cases, the measured wavelengths agreed with de Broglie's hypothesis.

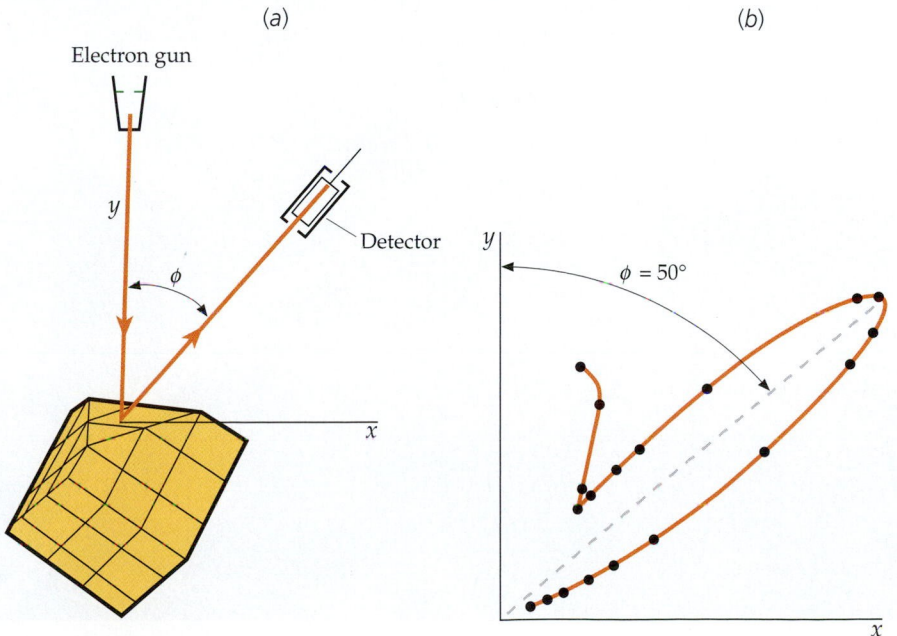

(a)  (b)

**FIGURE 34-7** The Davisson–Germer experiment. (*a*) Electrons are scattered from a nickel crystal into a detector. (*b*) Intensity of scattered electrons versus scattering angle. The maximum is at the angle predicted by the diffraction of waves of wavelength $\lambda$ given by the de Broglie formula.

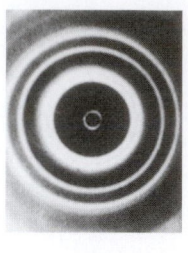
(a)

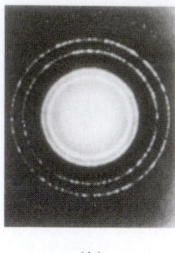
(b)

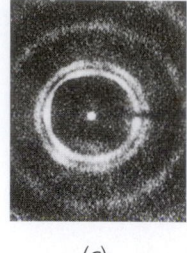
(c)

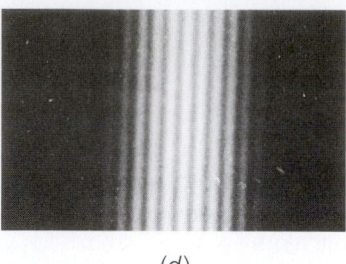
(d)

**FIGURE 34-8** (*a*) The diffraction pattern produced by X rays of wavelength 0.071 nm on an aluminum foil target. (*b*) The diffraction pattern produced by 600-eV electrons ($\lambda = 0.050$ nm) on an aluminum foil target. (*c*) The diffraction of 0.0568 eV neutrons ($\lambda = 0.12$ nm) incident on a copper foil. (*d*) A two-slit electron diffraction-interference pattern.

Another demonstration of the wave nature of electrons was provided in the same year by G. P. Thomson (son of J. J. Thomson) who observed electron diffraction in the transmission of electrons through thin metal foils. A metal foil consists of tiny, randomly oriented crystals. The diffraction pattern resulting from such a foil is a set of concentric circles. Figure 34-8*a* and Figure 34-8*b* show the diffraction pattern observed using X rays and electrons on an aluminum foil target. Figure 34-8*c* shows the diffraction patterns of neutrons on a copper foil target. Note the similarity of the patterns. The diffraction of hydrogen and helium atoms was observed in 1930. In all cases, the measured wavelengths agree with the de Broglie predictions. Figure 34-8*d* shows a diffraction pattern produced by electrons incident on two narrow slits. This experiment is equivalent to Young's famous double-slit experiment with light. The pattern is identical to the pattern observed with photons of the same wavelength. (Compare with Figure 34-1*b*.)

Shortly after the wave properties of the electron were demonstrated, it was suggested that electrons rather than light might be used to *see* small objects. As discussed in Chapter 33, reflected waves or transmitted waves can resolve details of objects only if the details are larger than the wavelength of the reflected wave. Beams of electrons, which can be focused electrically, can have very small wavelengths—much shorter than visible light. Today, the electron microscope (Figure 34-9) is an important research tool used to visualize specimens at scales far smaller than those possible with a light microscope.

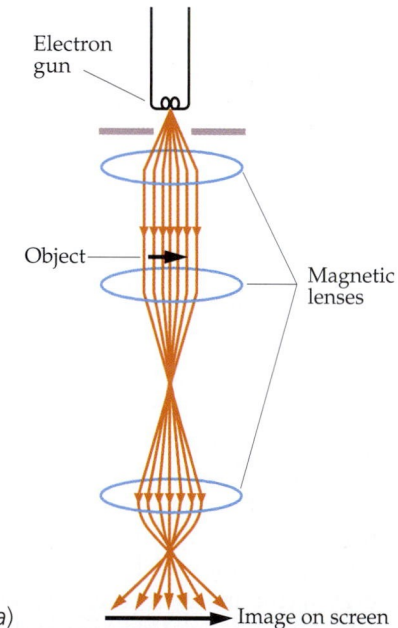

Electron gun

Object

Magnetic lenses

Image on screen

(a)

(b)

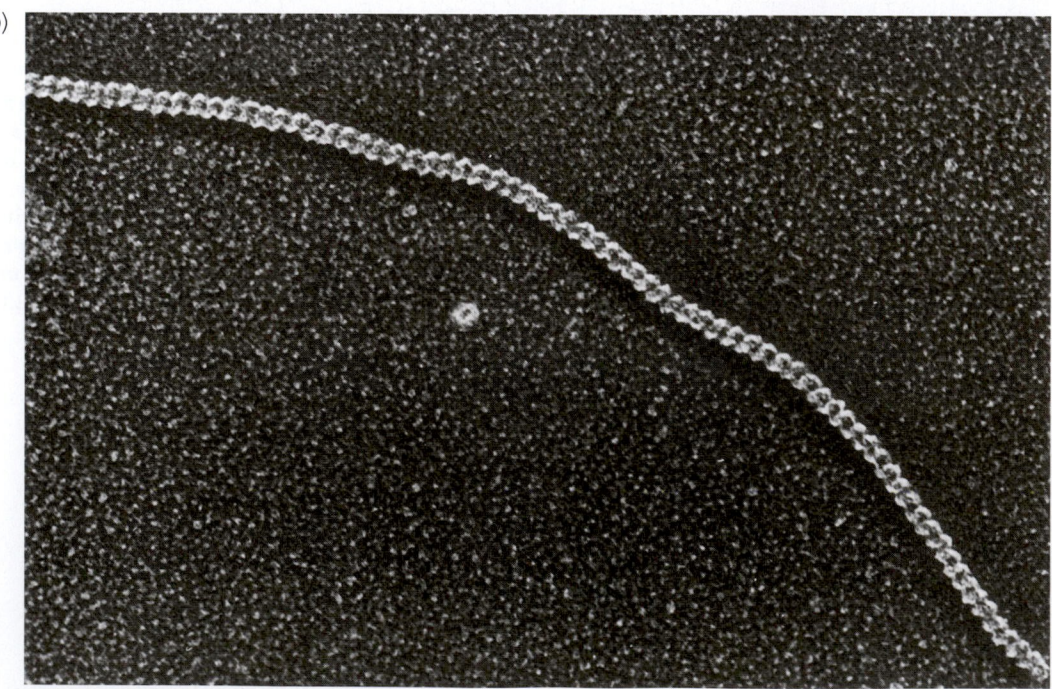

**FIGURE 34-9**
(*a*) An electron microscope. Electrons from a heated filament (the electron gun) are accelerated by a large potential difference. The electron beam is made parallel by a magnetic focusing lens. The electrons strike a thin target and are then focused by a second magnetic lens. The third magnetic lens projects the electron beam onto a fluorescent screen to produce the image.
(*b*) An electron micrograph of a DNA molecule.

## Standing Waves and Energy Quantization

Given that electrons have wave-like properties, it should be possible to produce standing electron waves. If energy is associated with the frequency of a standing wave, as in $E = hf$ (Equation 34-14), then standing waves imply quantized energies.

The idea that the discrete energy states in atoms could be explained by standing waves led to the development by Erwin Schrödinger and others in 1926 of a detailed mathematical theory known as quantum theory, quantum mechanics, or wave mechanics. In this theory, the electron is described by a wave function $\psi$ that obeys a wave equation called the Schrödinger equation. The form of the Schrödinger equation of a particular system depends on the forces acting on the particle, which are described by the potential energy functions associated with those forces. In Chapter 35 we discuss this equation, which is somewhat similar to the classical wave equations for sound or for light. Schrödinger solved the standing wave problem for the hydrogen atom, the simple harmonic oscillator, and other systems of interest. He found that the allowed frequencies, combined with $E = hf$, resulted in the set of energy levels found experimentally for the hydrogen atom, thereby demonstrating that quantum theory provides a general method of finding the quantized energy levels for a given system. Quantum theory is the basis for our understanding of the modern world, from the inner workings of the atomic nucleus to the radiation spectra of distant galaxies.

## 34-5  The Interpretation of the Wave Function

The wave function for waves on a string is the string displacement $y$. The wave function for sound waves can be either the displacement of the air molecules $s$, or the pressure $P$. The wave function for electromagnetic waves is the electric field $\vec{E}$ and the magnetic field $\vec{B}$. What is the wave function for electron waves? The symbol we use for this wave function is $\psi$ (the Greek letter psi). When Schrödinger published his wave equation, neither he nor anyone else knew just how to interpret the wave function $\psi$. We can get a hint about how to interpret $\psi$ by considering the quantization of light waves. For classical waves, such as sound or light, the energy per unit volume in the wave is proportional to the square of the wave function. Since the energy of a light wave is quantized, the energy per unit volume is proportional to the number of photons per unit volume. We might therefore expect the square of the photon's wave function to be proportional to the number of photons per unit volume in a light wave. But suppose we have a very low-energy source of light that emits just one photon at a time. In any unit volume, there is either one photon or none. The square of the wave function must then describe the *probability* of finding a photon in some unit volume.

The Schrödinger equation describes a single particle. The square of the wave function for a particle must then describe the *probability density*, which is the probability per unit volume, of finding the particle at a location. The probability of finding the particle in some volume element must also be proportional to the size of the volume element $dV$. Thus, in one dimension, the probability of finding a particle in a region $dx$ at the position $x$ is $\psi^2(x)\,dx$. If we call this probability $P(x)\,dx$, where $P(x)$ is the **probability density,** we have

$$P(x) = \psi^2(x)$$

34-17

PROBABILITY DENSITY

Generally the wave function depends on time as well as position, and is written $\psi(x,t)$. However, for standing waves, the probability density is independent of time. Since we will be concerned mostly with standing waves in this chapter, we omit the time dependence of the wave function and write it $\psi(x)$ or just $\psi$.

The probability of finding the particle in $dx$ at point $x_1$ or at point $x_2$ is the sum of the separate probabilities $P(x_1)\,dx + P(x_2)\,dx$. If we have a particle at all, the probability of finding the particle somewhere must be 1. Then the sum of the probabilities over all the possible values of $x$ must equal 1. That is,

$$\int_{-\infty}^{\infty} \psi^2\, dx = 1 \qquad\qquad 34\text{-}18$$

NORMALIZATION CONDITION

Equation 34-18 is called the **normalization condition**. If $\psi$ is to satisfy the normalization condition, it must approach zero as $x$ approaches infinity. This places a restriction on the possible solutions of the Schrödinger equation. There are solutions to the Schrödinger equation that do not approach zero as $x$ approaches infinity. However, these are not acceptable as wave functions.

---

*PROBABILITY CALCULATION FOR A CLASSICAL PARTICLE*     **EXAMPLE 34-5**

**It is known that a classical point particle moves back and forth with constant speed between two walls at $x = 0$ and $x = 8$ cm (Figure 34-10). No additional information about the location of the particle is known. (a) What is the probability density $P(x)$? (b) What is the probability of finding the particle at $x = 2$ cm? (c) What is the probability of finding the particle between $x = 3.0$ cm and $x = 3.4$ cm?**

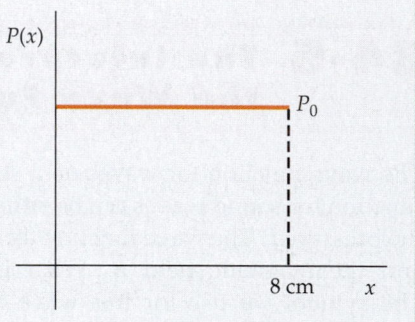

**PICTURE THE PROBLEM** We do not know the initial position of the particle. Since the particle moves with constant speed, it is equally likely to be anywhere in the region $0 < x < 8$ cm. The probability density $P(x)$ is therefore independent of $x$, for $0 < x < 8$ cm, and zero outside of this range. We can find $P(x)$, for $0 < x < 8$ cm, by normalization, that is, by requiring that the probability that the particle is somewhere between $x = 0$ and $x = 8$ cm is 1.

**FIGURE 34-10** The probability function $P(x)$.

(a) 1. The probability density $P(x)$ is uniform between the walls and zero elsewhere:

$$P(x) = P_0, \quad 0 < x < 8 \text{ cm}$$
$$P(x) = 0, \quad x < 0 \quad \text{or} \quad x > 8 \text{ cm}$$

2. Apply the normalization condition:

$$\int_{-\infty}^{+\infty} P(x)\,dx = \int_0^{8\,\text{cm}} P_0\,dx = P_0\,(8\text{ cm}) = 1$$

3. Solve for $P_0$:

$$P_0 = \boxed{\dfrac{1}{8\text{ cm}}}$$

(b) The probability of finding the particle in some range $\Delta x$ is proportional to $P_0\Delta x = \Delta x/(8\text{ cm})$. Since it is given that $\Delta x = 0$, the probability of finding the particle at the point $x = 2$ cm is 0. Alternatively, since there is an infinite number of points between $x = 0$ and $x = 8$ cm, and the particle is equally likely to be at any point, the chance that the particle will be at any one particular point must be zero.

> The probability of finding the particle at the point $x = 2$ cm is 0.

(c) Since the probability density is constant, the probability of a particle being in some range $\Delta x$ in the region $0 < x < 8$ cm is $P_0 \Delta x$. The probability of the particle being in the region 3.0 cm $< x <$ 3.4 cm is thus:

$$P_0 \Delta x = \left(\frac{1}{8 \text{ cm}}\right) 0.4 \text{ cm} = \boxed{0.05}$$

**REMARKS** Note in step 2 of Part (a) that we need only integrate from 0 to 8 cm, because $P(x)$ is zero outside this range.

# 34-6  Wave–Particle Duality

We have seen that light, which we ordinarily think of as a wave, exhibits particle properties when it interacts with matter, as in the photoelectric effect or in Compton scattering. Electrons, which we usually think of as particles, exhibit the wave properties of interference and diffraction when they pass near the edges of obstacles. All carriers of momentum and energy (e.g., electrons, atoms, light, or sound), exhibit both wave and particle characteristics. It might be tempting to say that an electron, for example, is both a wave and a particle, but what does this mean? In classical physics, the concepts of waves and particles are mutually exclusive. A **classical particle** behaves like a piece of shot; it can be localized and scattered, it exchanges energy suddenly at a point in space, and it obeys the laws of conservation of energy and momentum in collisions. It does *not* exhibit interference or diffraction. A **classical wave,** on the other hand, behaves like a water wave; it exhibits diffraction and interference, and its energy is spread out continuously in space and time. They are mutually exclusive. Nothing can be both a classical particle and a classical wave at the same time.

After Thomas Young observed the two-slit interference pattern with light in 1801, light was thought to be a classical wave. On the other hand, the electrons discovered by J. J. Thomson were thought to be classical particles. We now know that these classical concepts of waves and particles do not adequately describe the complete behavior of any phenomenon.

> Everything propagates like a wave and exchanges energy like a particle.

Often the concepts of the classical particle and the classical wave give the same results. If the wavelength is very small, diffraction effects are negligible, so the waves travel in straight lines like classical particles. Also, interference is not seen for waves of very short wavelength, because the interference fringes are too closely spaced to be observed. It then makes no difference which concept we use. If diffraction is negligible, we can think of light as a wave propagating along rays, as in geometrical optics, or as a beam of photon particles. Similarly, we can think of an electron as a wave propagating in straight lines along rays or, more commonly, as a particle.

We can also use either the wave or particle concept to describe exchanges of energy if we have a large number of particles and we are interested only in the average values of energy and momentum exchanges.

## The Two-Slit Experiment Revisited

The wave–particle duality of nature is illustrated by the analysis of the experiment in which a single electron is incident on a barrier with two slits. The analysis is the same whether we use an electron or a photon (light). To describe the propagation of an electron, we must use wave theory. Let us assume the source is a point source, such as a needlepoint, so we have spherical waves spreading out from the source. After passing through the two slits, the wave-fronts spread out—as if each slit was a source of wavefronts. The wave function $\psi$ at a point on a screen or film far from the slits depends on the difference in path lengths from the source to the point, one path through one slit, and the other path through the other slit. At points on the screen for which the difference in path lengths is either zero or an integral number of wavelengths, the amplitude of the wave function is a maximum. Since the probability of detecting the electron is proportional to $\psi^2$, the electron is very likely to arrive near these points. At points for which the path difference is an odd number of half wave-lengths, the wave function $\psi$ is zero, so the electron is very unlikely to arrive near these points. The chapter opening photos show the interference pattern produced by 10 electrons, 100 electrons, 3000 electrons, and 70,000 electrons. Note that, although the electron propagates through the slits like a wave, the electron interacts with the screen at a single point like a particle.

## The Uncertainty Principle

An important principle consistent with the wave–particle duality of nature is the uncertainty principle. It states that, in principle, it is impossible to simultane-ously measure both the position and the momentum of a particle with unlimited precision. A common way to measure the position of an object is to look at the object with light. If we do this, we scatter light from the object and determine the position by the direction of the scattered light. If we use light of wavelength $\lambda$, we can measure the position $x$ only to an uncertainty $\Delta x$ of the order of $\lambda$ because of diffraction effects.

$$\Delta x \sim \lambda$$

To reduce the uncertainty in position, we therefore use light of very short wave-length, perhaps even X rays. In principle, there is no limit to the accuracy of such a position measurement, because there is no limit on how small the wavelength $\lambda$ can be.

We can determine the momentum $p_x$ of the object if we know the mass and can determine its velocity. The momentum of the object can be found by measuring the object's position at two nearby times and computing its velocity. If we use light of wavelength $\lambda$, the photons carry momentum $h/p_x$. If these photons are scattered by the object we are looking at, the scattering changes the momentum of the object in an uncontrollable way. Each photon carries momentum $h/\lambda$, so the uncertainty in the momentum $\Delta p_x$ of the object, introduced by looking at it, is of the order of $h/\lambda$:

$$\Delta p_x \sim \frac{h}{\lambda}$$

If the wavelength of the radiation is small, the momentum of each photon will be large and the momentum measurement will have a large uncertainty. Reducing the intensity of light cannot eliminate this uncertainty; such a reduction merely reduces the number of photons in the beam. To *see* the object, we must scatter at least one photon. Therefore, the uncertainty in the momentum measurement of the object will be large if $\lambda$ is small, and the uncertainty in the position measure-ment of the object will be large if $\lambda$ is large.

Of course we could always *look at* the objects by scattering electrons instead of photons, but the same difficulty remains. If we use low-momentum electrons to reduce the uncertainty in the momentum measurement, we have a large uncertainty in the position measurement because of diffraction of the electrons. The relation between the wavelength and momentum $\lambda = h/p_x$ is the same for electrons as it is for photons.

The product of the intrinsic uncertainties in position and momentum is

$$\Delta x \, \Delta p_x \sim \lambda \times \frac{h}{\lambda} = h$$

If we define precisely what we mean by uncertainties in measurement, we can give a precise statement of the uncertainty principle. If $\Delta x$ and $\Delta p$ are defined to be the standard deviations in the measurements of position and momentum, it can be shown that their product must be greater than or equal to $\hbar/2$.

$$\Delta x \, \Delta p_x \geq \tfrac{1}{2}\hbar \qquad\qquad 34\text{-}19$$

where $\hbar = h/2\pi$.[†]

Equation 34-19 provides a statement of the uncertainty principle first enunciated by Werner Heisenberg in 1927. In practice, the experimental uncertainties are usually much greater than the intrinsic lower limit that results from wave–particle duality.

# 34-7  A Particle in a Box

We can illustrate many of the important features of quantum physics without solving the Schrödinger equation by considering a simple problem of a particle of mass $m$ confined to a one-dimensional box of length $L$, like the particle in Example 34-5. This can be considered a crude description of an electron confined within an atom or a proton confined within a nucleus. If a classical particle bounces back and forth between the walls of the box, the particle's energy and momentum can have any values. However, according to quantum theory, the particle is described by a wave function $\psi$, whose square describes the probability of finding the particle in some region. Since we are assuming that the particle is indeed inside the box, the wave function must be zero everywhere outside the box. If the box is between $x = 0$ and $x = L$, we have

$$\psi = 0, \text{ for } x \leq 0 \text{ and for } x \geq L$$

In particular, if we assume the wave function to be continuous everywhere, it must be zero at the end points of the box $x = 0$ and $x = L$. This is the same condition as the condition for standing waves on a string fixed at $x = 0$ and $x = L$, and the results are the same. The allowed wavelengths for a particle in the box are those where the length $L$ equals an integral number of half wavelengths (Figure 34-11).

$$L = n\frac{\lambda_n}{2}, \qquad n = 1, 2, 3, \ldots \qquad\qquad 34\text{-}20$$

STANDING-WAVE CONDITION FOR A PARTICLE IN A BOX OF LENGTH $L$

The total energy of the particle is its kinetic energy

$$E = \frac{1}{2}mv^2 = \frac{p^2}{2m}$$

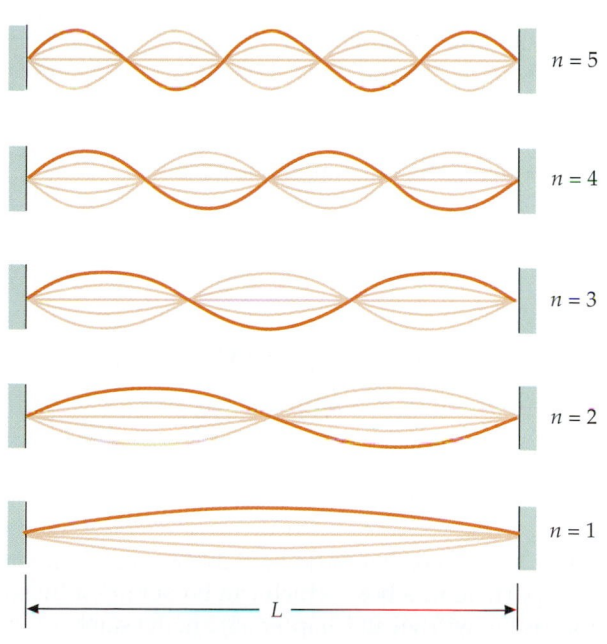

**FIGURE 34-11** Standing waves on a string fixed at both ends. The standing-wave condition is the same as for standing electron waves in a box.

---

[†] The combination $h/2\pi$ occurs so often it is given a special symbol, somewhat analogous to giving the special symbol $\omega$ for $2\pi f$, which occurs often in oscillations.

Substituting the de Broglie relation $p_n = h/\lambda_n$,

$$E_n = \frac{p_n^2}{2m} = \frac{(h/\lambda_n)^2}{2m} = \frac{h^2}{2m\lambda_n^2}$$

Then the standing-wave condition $\lambda_n = 2L/n$ gives the allowed energies.

$$E_n = n^2 \frac{h^2}{8mL^2} = n^2 E_1 \qquad\qquad 34\text{-}21$$

ALLOWED ENERGIES FOR A PARTICLE IN A BOX

where

$$E_1 = \frac{h^2}{8mL^2} \qquad\qquad 34\text{-}22$$

GROUND-STATE ENERGY FOR A PARTICLE IN A BOX

is the energy of the lowest state, which is the ground state.

The condition $\psi = 0$ at $x = 0$ and $x = L$ is called a **boundary condition.** Boundary conditions in quantum theory lead to energy quantization. Figure 34-12 shows the energy-level diagram for a particle in a box. Note that the lowest energy is not zero. This result is a general feature of quantum theory. If a particle is confined to some region of space, the particle has a minimum kinetic energy, which is called the **zero-point energy.** The smaller the region of space the particle is confined to, the greater its zero-point energy. In Equation 34-22, this is indicated by the fact that $E_1$ varies as $1/L^2$.

If an electron is confined (i.e., bound to an atom) in some energy state $E_i$, the electron can make a transition to another energy state $E_f$ with the emission of a photon (if $E_f < E_i$; if $E_f$ is greater than $E_i$, the system absorbs a photon). The transition from state 3 to the ground state is indicated in Figure 34-12 by the vertical arrow. The frequency of the emitted photon is found from the conservation of energy[†]

$$hf = E_i - E_f \qquad\qquad 34\text{-}23$$

The wavelength of the photon is then

$$\lambda = \frac{c}{f} = \frac{hc}{E_i - E_f} \qquad\qquad 34\text{-}24$$

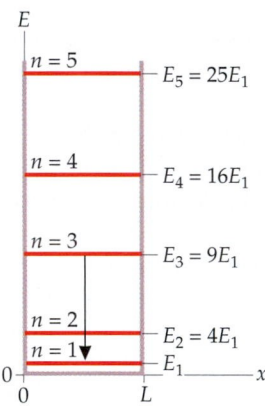

**FIGURE 34-12** Energy-level diagram for a particle in a box. Classically, a particle can have any energy value. Quantum mechanically, only those energy values given by Equation 34-22 are allowed. A transition between the state $n = 3$ and the ground state $n = 1$ is indicated by the vertical arrow.

## Standing-Wave Functions

The amplitude of a vibrating string fixed at $x = 0$ and $x = L$ is given by Equation 16-15:

$$y_n = A_n \sin k_n x$$

where $A_n$ is a constant and $k_n = 2\pi/\lambda_n$ is the wave number. The wave functions for a particle in a box (which can be obtained by solving the Schrödinger equation, as we will see in Chapter 35) are the same

---

[†] This equation was first proposed by Niels Bohr in his semiclassical model of the hydrogen atom in 1913, about 10 years before de Broglie's suggestion that electrons have wave properties. We will study the Bohr model in Chapter 36.

$$\psi_n(x) = A_n \sin k_n x$$

where $k_n = 2\pi/\lambda_n$. Using $\lambda_n = 2L/n$, we have

$$k_n = \frac{2\pi}{\lambda_n} = \frac{2\pi}{2L/n} = \frac{n\pi}{L}$$

The wave functions can thus be written

$$\psi_n(x) = A_n \sin\left(n\pi\frac{x}{L}\right)$$

The constant $A_n$ is determined by the normalization condition (Equation 34-18):

$$\int_{-\infty}^{\infty} \psi^2\,dx = \int_0^L A_n^2 \sin^2\left(n\pi\frac{x}{L}\right)dx = 1$$

Note that we need integrate only from $x = 0$ to $x = L$ because $\psi(x)$ is zero every-where else. The result of evaluating the integral and solving for $A_n$ is

$$A_n = \sqrt{\frac{2}{L}}$$

independent of $n$. The normalized wave functions for a particle in a box are thus

$$\psi_n(x) = \sqrt{\frac{2}{L}}\sin\left(n\pi\frac{x}{L}\right) \qquad\qquad 34\text{-}25$$

WAVE FUNCTIONS FOR A PARTICLE IN A BOX

These standing-wave functions for $n = 1$, $n = 2$, and $n = 3$ are shown in Fig-ure 34-13.

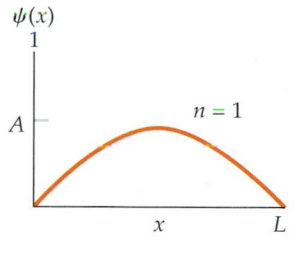
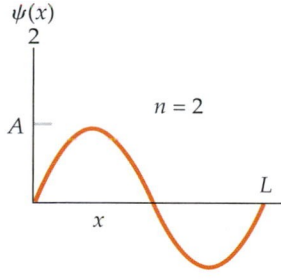
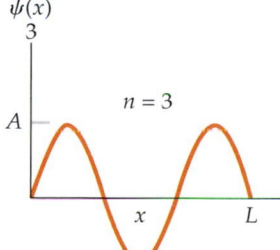

**FIGURE 34-13** Standing-wave functions for $n = 1$, $n = 2$, and $n = 3$.

The number $n$ is called a **quantum number.** It characterizes the wave function for a particular state and for the energy of that state. In our one-dimensional problem, a quantum number arises from the boundary condition on the wave function that it must be zero at $x = 0$ and $x = L$. In three-dimensional problems, three quantum numbers arise, one associated with a boundary condition in each dimension.

Figure 34-14 shows plots of $\psi^2$ for the ground state $n = 1$, the first excited state $n = 2$, the second excited state $n = 3$, and the state $n = 10$. In the ground state, the particle is most likely to be found near the center of the box, as indicated by the maximum value of $\psi^2$ at $x = L/2$. In the first excited state, the particle is least likely to be found near the center of the box because $\psi^2$ is small near $x = L/2$.

For very large values of $n$, the maxima and minima of $\psi^2$ are very close together, as illustrated for $n = 10$. The average value of $\psi^2$ is indicated in this figure by the dashed line. For very large values of $n$, the maxima are so closely spaced that $\psi^2$ cannot be distinguished from its average value. The fact that $(\psi^2)_{av}$ is constant across the whole box means that the particle is equally likely to be found anywhere in the box—the same as in the classical result. This is an example of **Bohr's correspondence principle:**

> In the limit of very large quantum numbers, the classical calculation and the quantum calculation must yield the same results.

BOHR'S CORRESPONDENCE PRINCIPLE

The region of very large quantum numbers is also the region of very large energies. For large energies, the percentage change in energy between adjacent quantum states is very small, so energy quantization is not important (see Problem 83).

We are so accustomed to thinking of the electron as a classical particle that we tend to think of an electron in a box as a particle bouncing back and forth between the walls. But the probability distributions shown in Figure 34-14 are stationary; that is, they do not depend on time. A better picture for an electron in a bound state is a cloud of charge with the charge density proportional to $\psi^2$. Figure 34-14 can then be thought of as plots of the charge density versus $x$ for the various states. In the ground state, $n = 1$, the electron cloud is centered in the middle of the box and is spread out over most of the box, as indicated in Figure 34-14a. In the first excited state, $n = 2$, the charge density of the electron cloud has two maxima, as indicated in Figure 34-14b. For very large values of $n$, there are many closely spaced maxima and minima in the charge density resulting in an average charge density that is approximately uniform throughout the box. This electron-cloud picture of an electron is very useful in understanding the structure of atoms and molecules. However, it should be noted that whenever an electron is observed to interact with matter or radiation, it is always observed as a whole unit charge.

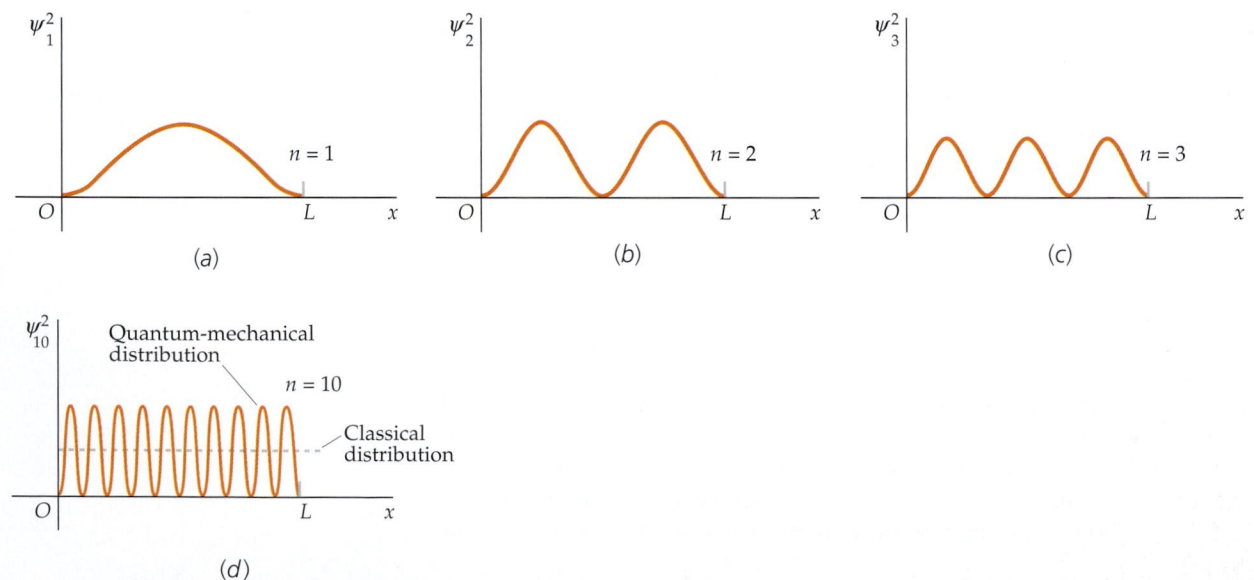

**FIGURE 34-14** $\psi^2$ versus $x$ for a particle in a box of length $L$ for (a) the ground state, $n = 1$; (b) the first excited state, $n = 2$; (c) the second excited state, $n = 3$; and (d) the state $n = 10$. For $n = 10$, the maxima and minima of $\psi^2$ are so close together that individual maxima may be hard to distinguish. The dashed line indicates the average value of $\psi^2$. It gives the classical prediction that the particle is equally likely to be found near any point in the box.

*PHOTON EMISSION BY A PARTICLE IN A BOX* | **EXAMPLE 34-6**

An electron is in a one-dimensional box of length 0.1 nm. (*a*) Find the ground-state energy. (*b*) Find the energy in electron volts of the five lowest states, and then sketch an energy-level diagram. (*c*) Find the wavelength of the photon emitted for each transition from the state $n = 3$ to a lower-energy state.

**PICTURE THE PROBLEM** For Part (*a*) and Part (*b*), the energies are given by $E_n = n^2 E_1$ (Equation 34-21), where the ground-state energy $E_1 = h^2/8mL^2$ (Equation 34-22). For Part (*c*), the photon wavelengths are given by $\lambda = hc/(E_i - E_f)$ (Equation 34-24).

(*a*) Use $hc = 1240$ eV·nm and $mc^2 = 5.11 \times 10^5$ eV to calculate $E_1$:

$$E_1 = \frac{(hc)^2}{8(mc^2)L^2}$$

$$= \frac{(1240 \text{ eV·nm})^2}{8(5.11 \times 10^5 \text{ eV})(0.1 \text{ nm})^2} = \boxed{37.6 \text{ eV}}$$

(*b*) Calculate $E_n = n^2 E_1$ for $n = 2, 3, 4,$ and $5$:

$$E_2 = (2)^2(37.6 \text{ eV}) = \boxed{150 \text{ eV}}$$

$$E_3 = (3)^2(37.6 \text{ eV}) = \boxed{338 \text{ eV}}$$

$$E_4 = (4)^2(37.6 \text{ eV}) = \boxed{602 \text{ eV}}$$

$$E_5 = (5)^2(37.6 \text{ eV}) = \boxed{940 \text{ eV}}$$

(*c*) 1. Use the energies found in Part (*b*) to calculate the wavelength for a transition from state 3 to state 2:

$$\lambda = \frac{hc}{E_3 - E_2}$$

$$= \frac{1240 \text{ eV·nm}}{338 \text{ eV} - 150 \text{ eV}} = \boxed{6.60 \text{ nm}}$$

2. Then use the energies in Part (*a*) and Part (*b*) to calculate the wavelength for a transition from state 3 to state 1:

$$\lambda = \frac{hc}{E_3 - E_1}$$

$$= \frac{1240 \text{ eV·nm}}{338 \text{ eV} - 37.6 \text{ eV}} = \boxed{4.13 \text{ nm}}$$

**REMARKS** The energy-level diagram is shown in Figure 34-15. The transitions from $n = 3$ to $n = 2$ and from $n = 3$ to $n = 1$ are indicated by the vertical arrows. The ground-state energy of 37.6 eV is on the same order of magnitude as the kinetic energy of the electron in the ground state of the hydrogen atom, which is 13.6 eV. In the hydrogen atom, the electron also has potential energy of $-27.2$ eV in the ground state, giving it a total ground-state energy of $-13.6$ eV.

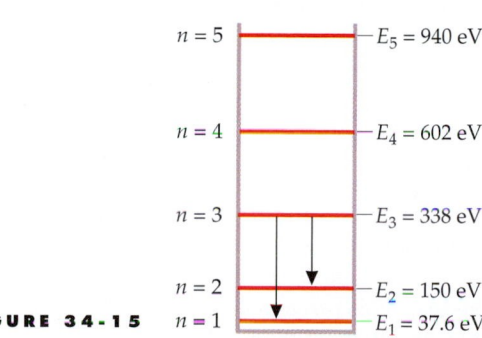

**FIGURE 34-15**

$n = 5$ — $E_5 = 940$ eV
$n = 4$ — $E_4 = 602$ eV
$n = 3$ — $E_3 = 338$ eV
$n = 2$ — $E_2 = 150$ eV
$n = 1$ — $E_1 = 37.6$ eV

**EXERCISE** Calculate the wavelength of the photon emitted if the electron in the box makes a transition from $n = 4$ to $n = 3$. (*Answer* 4.69 nm)

## 34-8 Expectation Values

The solution of a classical mechanics problem is typically specified by giving the position of a particle as a function of time. But the wave nature of matter prevents us from doing this for microscopic systems. The most that we can know is the probability of measuring a certain value of the position $x$. If we measure the position for a large number of identical systems, we get a range of values corresponding to the probability distribution. The average value of $x$ obtained from such measurements is called the **expectation value** and is written $<x>$. The expectation value of $x$ is the same as the average value of $x$ that we would expect to obtain from a measurement of the positions of a large number of particles with the same wave function $\psi(x)$.

Since $\psi^2(x)\, dx$ is the probability of finding a particle in the region $dx$, the expectation value of $x$ is

$$< x > = \int_{-\infty}^{+\infty} x\psi^2(x)\, dx \qquad\qquad 34\text{-}26$$

EXPECTATION VALUE OF $X$ DEFINED

The expectation value of any function $f(x)$ is given by

$$< f(x) > = \int_{-\infty}^{+\infty} f(x)\psi^2(x)\, dx \qquad\qquad 34\text{-}27$$

EXPECTATION VALUE OF $F(X)$ DEFINED

## *Calculating Probabilities and Expectation Values

The problem of a particle in a box allows us to illustrate the calculation of the probability of finding the particle in various regions of the box and the expectation values for various energy states. We give two examples, using the wave functions given by Equation 34-25.

---

*THE PROBABILITY OF THE PARTICLE BEING FOUND IN A SPECIFIED REGION OF A BOX*

**EXAMPLE 34-7**

**A particle in a one-dimensional box of length $L$ is in the ground state. Find the probability of finding the particle (a) anywhere in the region of length $\Delta x = 0.01L$, centered at $x = \frac{1}{2}L$ and (b) in the region $0 < x < \frac{1}{4}L$.**

**PICTURE THE PROBLEM** The probability $P$ of finding the particle in some infinitesimal range $dx$ is $\psi^2\, dx$. For a particle in the ground state, $\psi$ is given by Equation 34-25, with $n = 1$; $\psi^2$ is illustrated in Figure 34-14. The probability of finding $x$ in some region is just the area under this curve for the region. For Part (a), the region is $\Delta x = 0.01L$, centered at $x = L/2$, and the area under the $\psi^2$ versus $x$ curve is shown in Figure 34-16a. This area is $\approx \psi^2\, \Delta x$. For Part (b), the region is $0 < x < L/4$, and the area under the curve is shown in Figure 34-16b. To calculate this area, we must integrate $\psi^2$ from $x = 0$ to $x = L/4$.

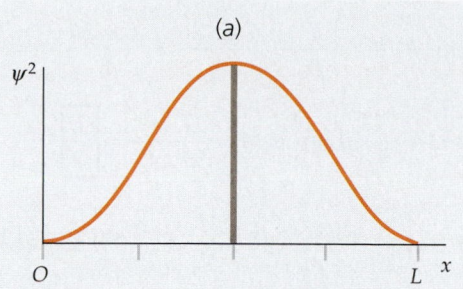

**FIGURE 34-16**

(a) 1. The probability of finding the particle is the area under the curve shown in Figure 34-16a. To calculate this area, we need to calculate the height of curve at $x = L/2$:

$$\psi(x) = \sqrt{\frac{2}{L}} \sin\left(\pi \frac{x}{L}\right)$$

so

$$\psi^2(L/2) = \frac{2}{L} \sin^2 \frac{\pi}{2} = \frac{2}{L}$$

2. The area is the height times the width, and the width is $\Delta x = 0.01L$:

$$P = \psi^2(L/2)\Delta x = \frac{2}{L} \times 0.01L = \boxed{0.02}$$

(b) 1. The probability of finding the particle is the area under the curve shown in Figure 34-16b. To calculate this area, we need to integrate from $x = 0$ to $x = L/4$:

$$P = \int_0^{L/4} \psi^2(x)\, dx = \int_0^{L/4} \frac{2}{L} \sin^2 \frac{\pi x}{L}\, dx$$

2. The integral can be evaluated a number of ways. If a table of integrals is used, a change in the integration variable in required. Changing the integration variable to $\theta = \pi x/L$ gives:

$$P = \frac{2}{\pi} \int_0^{\pi/4} \sin^2 \theta\, d\theta$$

3. The integral can be found in tables:

$$\int_0^{\pi/4} \sin^2 \theta\, d\theta = \left(\frac{\theta}{2} - \frac{\sin 2\theta}{4}\right)\Bigg|_0^{\pi/4} = \left(\frac{\pi}{8} - \frac{1}{4}\right)$$

4. Use the result from Part (b), step 3 to calculate the probability:

$$P = \frac{2}{\pi}\left(\frac{\pi}{8} - \frac{1}{4}\right) = \boxed{0.091}$$

**REMARKS** An integral was not necessary for Part (a) because the area of interest could be well approximated by a rectangle of height $\psi^2$ and width $x$. The chance of finding the particle in the region $\Delta x = 0.01L$ at $x = \frac{1}{2}L$ is approximately 2 percent. The chance of finding the particle in the region $0 < x < L/4$ is about 9.1 percent.

*CALCULATING EXPECTATION VALUES*  **E X A M P L E   3 4 - 8**

Find (a) $<x>$ and (b) $<x^2>$ for a particle in its ground state in a box of length $L$.

**PICTURE THE PROBLEM** We use $<f(x)> = \int f(x)\psi^2(x)\, dx$, with
$$\psi_n(x) = \sqrt{\frac{2}{L}} \sin \frac{n\pi x}{L}.$$

(a) 1. Write $<x>$ using the ground-state wave function given by Equation 34-25, with $n = 1$:

$$<x> = \int_{-\infty}^{+\infty} x\psi^2(x)\, dx = \frac{2}{L} \int_0^L x \sin^2\left(\frac{\pi x}{L}\right) dx$$

2. To evaluate this integral by using a table of integrals, first change the integration variable to $\theta = \pi x/L$:

$$<x> = \frac{2}{L}\left(\frac{L}{\pi}\right)^2 \int_0^\pi \theta \sin^2 \theta\, d\theta$$

$$= \frac{2L}{\pi^2} \int_0^\pi \theta \sin^2 \theta\, d\theta$$

3. The table of integrals gives:

$$\int_0^\pi \theta \sin^2 \theta \, d\theta = \left[ \frac{\theta^2}{4} - \frac{\theta \sin 2\theta}{4} - \frac{\cos 2\theta}{8} \right]_0^\pi = \frac{\pi^2}{4}$$

4. Substitute this value into the expression in step 2:

$$<x> = \frac{2L}{\pi^2} \int_0^\pi \theta \sin^2 \theta \, d\theta = \frac{2L}{\pi^2} \frac{\pi^2}{4} = \boxed{\frac{L}{2}}$$

(b) 1. Repeat step 1 and step 2 of Part (a) for $<x^2>$:

$$<x^2> = \int_{-\infty}^{+\infty} x^2 \psi^2(x) \, dx = \int_0^L x^2 \frac{2}{L} \sin^2(\pi x/L) \, dx$$

$$= \frac{2}{L} \left( \frac{L}{\pi} \right)^3 \int_0^\pi \theta^2 \sin^2 \theta \, d\theta = \frac{2L^2}{\pi^3} \int_0^\pi \theta^2 \sin^2 \theta \, d\theta$$

2. Evaluating the integral using a table of integrals gives:

$$\int_0^\pi \theta^2 \sin^2 \theta \, d\theta = \left[ \frac{\theta^3}{6} - \left( \frac{\theta^2}{4} - \frac{1}{8} \right) \sin 2\theta - \frac{\theta \cos 2\theta}{4} \right]\Big|_0^\pi$$

$$= \frac{\pi^3}{6} - \frac{\pi}{4}$$

3. Substitute this value into the expression in step 1 of Part (b):

$$<x^2> = \frac{2L^2}{\pi^3} \left( \frac{\pi^3}{6} - \frac{\pi}{4} \right) = \left( \frac{1}{3} - \frac{1}{2\pi^2} \right) L^2 = \boxed{0.283L^2}$$

**REMARKS** The expectation value of $x$ is $L/2$, as we would expect, because the probability distribution is symmetric about the midpoint of the box. Note that $<x^2>$ is not equal to $<x>^2$.

## 34-9 Energy Quantization in Other Systems

The quantized energies of a system are generally determined by solving the Schrödinger equation for that system. The form of the Schrödinger equation depends on the potential energy of the particle. The potential energy for a one-dimensional box from $x = 0$ to $x = L$ is shown in Figure 34-17. This potential energy function is called an **infinite square-well potential,** and it is described mathematically by

$$U(x) = 0, \quad 0 < x < L$$

$$U(x) = \infty, \quad x < 0 \quad \text{or} \quad x > L \qquad \text{34-28}$$

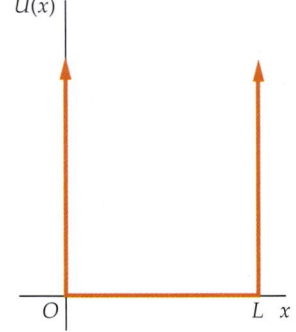

**FIGURE 34-17** The infinite square-well potential energy. For $x < 0$ and $x > L$, the potential energy $U(x)$ is infinite. The particle is confined to the region in the well $0 < x < L$.

Inside the box the particle moves freely, so the potential energy is uniform. For convenience, we choose the value of this potential energy to be zero. Outside the box the potential energy is infinite, so the particle cannot exist outside the box no matter what its energy. We did not need to solve the Schrödinger equation for this potential because the wave functions and quantized frequencies are the same as for a string fixed at both ends, which we studied in Chapter 16. Although this problem seems artificial, actually it is useful for some physical problems, such as a neutron inside a nucleus.

### The Harmonic Oscillator

More realistic than the particle in a box is the harmonic oscillator, which applies to an object of mass $m$ on a spring of force constant $k$ or to any system undergoing small oscillations about a stable equilibrium. Figure 34-18 shows the potential energy function

$$U(x) = \tfrac{1}{2}kx^2 = \tfrac{1}{2}m\omega_0^2 x^2$$

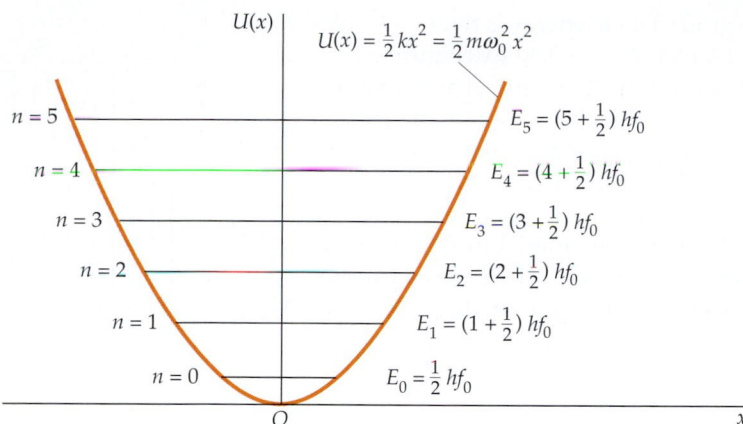

$$U(x) = \frac{1}{2}kx^2 = \frac{1}{2}m\omega_0^2 x^2$$

$n = 5$      $E_5 = (5 + \frac{1}{2})\, hf_0$

$n = 4$      $E_4 = (4 + \frac{1}{2})\, hf_0$

$n = 3$      $E_3 = (3 + \frac{1}{2})\, hf_0$

$n = 2$      $E_2 = (2 + \frac{1}{2})\, hf_0$

$n = 1$      $E_1 = (1 + \frac{1}{2})\, hf_0$

$n = 0$      $E_0 = \frac{1}{2}\, hf_0$

**FIGURE 34-18** Harmonic oscillator potential energy function. The allowed energy levels are indicated by the equally spaced horizontal lines. Also, $\omega_0 = 2\pi f_0$.

where $\omega_0 = \sqrt{k/m}$ is the natural frequency of the oscillator. Classically, the object oscillates between $x = +A$ and $x = -A$. Its total energy is $E = \frac{1}{2}m\omega_0^2 A^2$, which can have any nonnegative value, including zero.

In quantum theory, the particle is represented by the wave function $\psi(x)$, which is determined by solving the Schrödinger equation for this potential. Normalizable wave functions $\psi_n(x)$ occur only for discrete values of the energy $E_n$ given by

$$E_n = (n + \tfrac{1}{2})hf_0, \quad n = 0, 1, 2, 3, \ldots \tag{34-29}$$

where $f_0 = \omega_0/2\pi$ is the classical frequency of the oscillator. Note that the energy levels of a harmonic oscillator are evenly spaced with separation $hf$, as indicated in Figure 34-18. Compare this with the uneven spacing of the energy levels for the particle in a box, as shown in Figure 34-12. If a harmonic oscillator makes a transition from energy level $n$ to the next lowest energy level $n - 1$, the frequency $f$ of the photon emitted is given by $hf = E_i - E_f$ (Equation 34-23). Applying this equation gives

$$hf = E_n - E_{n-1} = (n + \tfrac{1}{2})hf_0 - (n - 1 + \tfrac{1}{2})hf_0 = hf_0$$

The frequency $f$ of the emitted photon is therefore equal to the classical frequency $f_0$ of the oscillator.

## The Hydrogen Atom

In the hydrogen atom, an electron is bound to a proton by the electrostatic force of attraction (discussed in Chapter 21). This force varies inversely as the square of the separation distance (exactly like the gravitational attraction of the earth and sun). The potential energy of the electron–proton system therefore varies inversely with separation distance (Equation 23-9). As in the case of gravitational potential energy, the potential energy of the electron–proton system is chosen to be zero if the electron is an infinite distance from the proton. Then for all finite distances, the potential energy is negative. Like the case of an object orbiting the earth, the electron–proton system is a bound system if its total energy is negative. Like the energies of a particle in a box and of a harmonic oscillator, the energies are described by a quantum number $n$. As we will see in Chapter 36, the allowed energies of the hydrogen atom are given by

$$E_n = -\frac{13.6 \text{ eV}}{n^2}, \quad n = 1, 2, 3, \ldots \tag{34-30}$$

The lowest energy corresponds to $n = 1$. The ground-state energy is thus $-13.6$ eV. The energy of the first excited state is $-(13.6 \text{ eV})/2^2 = -3.40$ eV. Figure 34-19 shows the energy-level diagram for the hydrogen atom. The vertical arrows indicate transitions from a higher state to a lower state with the emission of electromagnetic radiation. Only those transitions ending at the first excited state ($n = 2$) involve energy differences in the range of visible light of 1.77 eV to 3.10 eV, as calculated in Example 34-1.

Other atoms are more complicated than the hydrogen atom, but their energy levels are similar in many ways to those of hydrogen. Their ground-state energies are of the order of $-1$ eV to $-10$ eV, and many transitions involve energies corresponding to photons in the visible range.

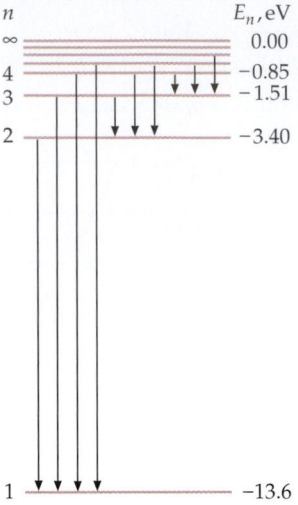

**FIGURE 34-19** Energy-level diagram for the hydrogen atom. The energy of the ground state is $-13.6$ eV. As $n$ approaches $\infty$ the energy approaches 0, which is the highest energy state for an electron bound to the atom.

# SUMMARY

1. All phenomenon propagate like waves and interact like particles.
2. The quantum of light is called a photon and has energy $E = hf$, where $h$ is Planck's constant.
3. The relation between wavelength and momentum of electrons, photons, and other particles is given by the de Broglie relation $\lambda = h/p$.
4. Energy quantization in bound systems arises from standing-wave conditions, which are equivalent to boundary conditions on the wave function.
5. The uncertainty principle is a fundamental law of nature that places theoretical restrictions on the precision of a simultaneous measurement of the position and momentum of a particle. It follows from the general properties of waves.

| Topic | Relevant Equations and Remarks | |
|---|---|---|
| **1. Constants and Values** | | |
| Planck's constant | $h = 6.626 \times 10^{-34}$ J·s $= 4.136 \times 10^{-15}$ eV·s | **34-2** |
| $hc$ | $hc = 1240$ eV·nm | **34-5** |
| Compton Wavelength | $\lambda_C = \dfrac{h}{m_e c} = 2.43$ pm | **34-12** |
| **2. The Particle Nature of Light: Photons** | Energy is quantized. | |
| Photon energy | $E = hf$ | **34-1** |
| Momentum of a photon | $p = \dfrac{h}{\lambda}$ | **34-7** |

3. **Frequency–Wavelength (energy–momentum) Relations**

| | | |
|---|---|---|
| Photons | $E = pc, \quad \text{so} \quad \lambda f = c$ | R-17 |
| Nonrelativistic particles | $K = \dfrac{p^2}{2m}, \quad \text{so} \quad \lambda = \dfrac{hc}{\sqrt{2mc^2K}}$ | 34-15 |
| Electrons | $\lambda = \dfrac{1.226}{\sqrt{K}} \text{ nm}, K \text{ in electron volts}$ | 34-16 |
| Photoelectric effect | $K_{\max} = (\tfrac{1}{2}mv^2)_{\max} = hf - \phi$ <br><br> where $\phi$ is the work function of the cathode. | 34-3 |
| Compton scattering | $\lambda_2 - \lambda_1 = \dfrac{h}{m_e c}(1 - \cos\theta) = \lambda_C(1 - \cos\theta)$ | 34-11 |

4. **Quantum Mechanics**

The state of a particle, such as an electron, is described by its wave function $\psi$, which is the solution of the Schrödinger wave equation.

| | | |
|---|---|---|
| Probability density | The probability of finding the particle in some region of space $dx$ is given by <br><br> $P(x) = \psi^2(x)\, dx$ | 34-17 |
| Normalization condition | $\displaystyle \int_{-\infty}^{\infty} \psi^2\, dx = 1$ | 34-18 |
| Quantum number | The wave function for a particular energy state is characterized by a quantum number $n$. In three dimensions there are three quantum numbers, one associated with a boundary condition in each dimension. | |
| Expectation value | The expectation value of $x$ is the same as the average value of $x$ that we would expect to obtain from a measurement of the positions of a large number of particles with the same wave function $\psi(x)$. <br><br> $\displaystyle <x> = \int_{-\infty}^{+\infty} x\psi^2(x)\, dx$ | 34-26 |
| | $\displaystyle <f(x)> = \int_{-\infty}^{+\infty} f(x)\psi^2(x)\, dx$ | 34-27 |

5. **Wave–Particle Duality**

Light, electrons, neutrons, and all carriers of momentum and energy exhibit both wave and particle properties. Everything propagates like a classical wave exhibiting diffraction and interference, but exchanges energy in discrete lumps like a classical particle. Because the wavelength of macroscopic objects is so small, diffraction and interference are not observed. Also, if a macroscopic amount of energy is exchanged, so many quanta are involved that the particle nature of the energy is not evident.

6. **Uncertainty Principle**

The wave–particle duality of nature leads to the uncertainty principle, which states that the product of the uncertainty in a measurement of position and the uncertainty in a measurement of momentum must be greater than or equal to $\tfrac{1}{2}\hbar$, where $\hbar$ is Planck's constant divided by $2\pi$.

$\Delta x\, \Delta p \geq \tfrac{1}{2}\hbar$        34-19

## Conceptual Problems

**1** • **SSM** The quantized character of electromagnetic radiation is revealed by (a) the Young double-slit experiment. (b) diffraction of light by a small aperture. (c) the photoelectric effect. (d) the J. J. Thomson cathode-ray experiment.

**2** •• Two monochromatic light sources, A and B, emit the same number of photons per second. The wavelength of A is $\lambda_A = 400$ nm, and the wavelength of B is $\lambda_B = 600$ nm. The power radiated by source B is (a) equal to the power of source A. (b) less than the power of source A. (c) greater than the power of source A. (d) cannot be compared to the power from source A using the available data.

**3** • True or false:

(a) In the photoelectric effect, the current is proportional to the intensity of the incident light.
(b) In the photoelectric effect, the work function of a metal depends on the frequency of the incident light.
(c) In the photoelectric effect, the maximum kinetic energy of electrons emitted varies linearly with the frequency of the incident light.
(d) In the photoelectric effect, the energy of a photon is proportional to its frequency.

**4** • In the photoelectric effect, the number of electrons emitted per second is (a) independent of the light intensity. (b) proportional to the light intensity. (c) proportional to the work function of the emitting surface. (d) proportional to the frequency of the light.

**5** • **SSM** The work function of a surface is $\phi$. The threshold wavelength for emission of photoelectrons from the surface is (a) $hc/\phi$. (b) $\phi/hf$. (c) $hf/\phi$. (d) none of the answers are correct.

**6** •• When light of wavelength $\lambda_1$ is incident on a certain photoelectric cathode, no electrons are emitted, no matter how intense the incident light is. Yet, when light of wavelength $\lambda_2 < \lambda_1$ is incident, electrons are emitted, even when the incident light has low intensity. Explain.

**7** • True or false:

(a) The de Broglie wavelength of an electron varies inversely with its momentum.
(b) Electrons can be diffracted.
(c) Neutrons can be diffracted.
(d) An electron microscope is used to look at electrons.

**8** • If the de Broglie wavelength of an electron and a proton are equal, then (a) the velocity of the proton is greater than the velocity of the electron. (b) the velocity of the proton and the electron are equal. (c) the velocity of the proton is less than the velocity of the electron. (d) the energy of the proton is greater than the energy of the electron. (e) both (a) and (d) are correct.

**9** • A proton and an electron have equal kinetic energies. It follows that the de Broglie wavelength of the proton is (a) greater than that of the electron. (b) equal to that of the electron. (c) less than that of the electron.

**10** • Can the expectation value of $x$ ever equal a value for which the probability function $P(x)$ is zero? Give a specific example.

**11** • **SSM** Explain why the maximum kinetic energy of electrons emitted in the photoelectric effect does not depend on the intensity of the incident light, but the total number of electrons emitted does depend on the intensity of the incident light.

**12** •• A six-sided die has the numeral 1 painted on three sides and the numeral 2 painted on the other three sides. (a) What is the probability of a 1 coming up when the die is thrown? (b) What is the expectation value of the numeral that comes up when the die is thrown?

**13** •• It was once believed that if two identical experiments are done on identical systems under the same conditions, the results must be identical. Explain why this is not true and how it can be modified, so that it is consistent with quantum physics.

## Estimation and Approximation

**14** •• Students in a physics lab are trying to determine the value of Planck's constant $h$, using a photoelectric apparatus similar to the one shown in Figure 34-2. For a light source, the students use a helium–neon laser with tunable wavelength. The data that the students obtain for the maximum electron kinetic energy are

| $\lambda$ | 544 nm | 594 nm | 604 nm | 612 nm | 633 nm |
|---|---|---|---|---|---|
| $K_{max}$ | 0.360 eV | 0.199 eV | 0.156 eV | 0.117 eV | 0.062 eV |

(a) Using a spreadsheet program or graphing calculator, convert the wavelengths to frequencies and plot $K_{max}$ versus frequency. (b) Use the graph to estimate the value of Planck's constant implied by the students' data. (*Note:* You may wish to use the linear regression function of your spreadsheet program or graphing calculator.) (c) Compare your result with the accepted value for Planck's constant.

**15 • •** The cathode that was used by the students in the experiment described in Problem 14 is known to be constructed from one of the following metals, with the given work function

| Metals | Tungsten | Silver | Potassium | Cesium |
|---|---|---|---|---|
| Work function | 4.58 eV | 2.4 eV | 2.1 eV | 1.9 eV |

Solve this problem using the same data as given in Problem 14. (a) Using a spreadsheet program or graphing calculator, convert the wavelengths to frequencies, and plot $K_{max}$ versus frequency. (b) Use the graph to estimate the value of the work function implied by the students' data. (*Note:* You may wish to use the linear regression function of your spreadsheet program or graphing calculator.) (c) Which metal was most likely used for the cathode in their experiment?

**16 • •** SSM Students in an advanced physics lab use X rays to measure the Compton wavelength, $\lambda_C$. The students obtain the following wavelength shifts $\lambda_2 - \lambda_1$ as a function of scattering angle $\theta$

| $\theta$ | 45° | 75° | 90° | 135° | 180° |
|---|---|---|---|---|---|
| $\lambda_2 - \lambda_1$ | 0.647 pm | 1.67 pm | 2.45 pm | 3.98 pm | 4.95 pm |

Use their data to estimate the value for the Compton wavelength. Compare this number with the accepted value.

**17 • •** SSM Baseball, tennis, golf, and soccer are sports that involve placing a ball in play with a certain speed. Estimate which of these sports has a ball with the smallest de Broglie wavelength when the ball is moving with the highest speed typically created by a professional athlete.

## The Particle Nature of Light: Photons

**18 •** ISOLVE✔ Find the photon energy in joules and in electron volts for an electromagnetic wave of frequency (a) 100 MHz in the FM radio band and (b) 900 kHz in the AM radio band.

**19 •** What are the frequencies of photons that have the following energies? (a) 1 eV, (b) 1 keV, and (c) 1 MeV.

**20 •** SSM Find the photon energy for light of wavelength (a) 450 nm, (b) 550 nm, and (c) 650 nm.

**21 •** Find the photon energy if the wavelength is (a) 0.1 nm (about 1 atomic diameter) and (b) 1 fm (1 fm = $10^{-15}$ m, about 1 nuclear diameter).

**22 • •** ISOLVE The wavelength of light emitted by a 3-mW helium–neon laser is 632 nm. If the diameter of the laser beam is 1 mm, what is the density of photons in the beam?

**23 •** SSM Lasers used in the telecommunications network typically have a wavelength near 1.55 $\mu$m. How many photons per second are being transmitted if such a laser has an output power of 2.5 mW?

## The Photoelectric Effect

**24 •** The work function for tungsten is 4.58 eV. (a) Find the threshold frequency and wavelength for the photoelectric effect. Find the maximum kinetic energy of the electrons if the wavelength of the incident light is (b) 200 nm and (c) 250 nm.

**25 •** ISOLVE✔ When light of wavelength 300 nm is incident on potassium, the emitted electrons have maximum kinetic energy of 2.03 eV. (a) What is the energy of an incident photon? (b) What is the work function for potassium? (c) What would be the maximum kinetic energy of the electrons if the incident light had a wavelength of 430 nm? (d) What is the threshold wavelength for the photoelectric effect with potassium?

**26 •** The threshold wavelength for the photoelectric effect for silver is 262 nm. (a) Find the work function for silver. (b) Find the maximum kinetic energy of the electrons if the incident radiation has a wavelength of 175 nm.

**27 •** The work function for cesium is 1.9 eV. (a) Find the threshold frequency and threshold wavelength for the photoelectric effect. Find the maximum kinetic energy of the electrons if the wavelength of the incident light is (b) 250 nm and (c) 350 nm.

**28 • •** SSM ISOLVE When a surface is illuminated with light of wavelength 780 nm, the maximum kinetic energy of the emitted electrons is 0.37 eV. What is the maximum kinetic energy if the surface is illuminated with light of wavelength 410 nm?

## Compton Scattering

**29 •** Find the shift in wavelength of photons scattered by electrons at $\theta = 60°$.

**30 •** When photons are scattered by electrons in carbon, the shift in wavelength is 0.33 pm. Find the scattering angle.

**31 •** The wavelength of Compton-scattered photons is measured at $\theta = 135°$. If $\Delta\lambda / \lambda$ is to be 2.3 percent, what should the wavelength of the incident photons be?

**32 •** SSM ISOLVE Compton used photons of wavelength 0.0711 nm. (a) What is the energy of these photons? (b) What is the wavelength of the photon scattered at $\theta = 180°$? (c) What is the energy of the photon scattered at this angle?

**33 •** For the photons used by Compton (see Problem 32), find the momentum of the incident photon and the momentum of the photon scattered at 180°; use the conservation of momentum to find the momentum of the recoil electron in this case.

**34 • •** An X-ray photon of wavelength 6 pm that collides with an electron is scattered by an angle of 90°. (a) What is the change in wavelength of the photon? (b) What is the kinetic energy of the scattered electron?

**35 • •** How many head-on, Compton-scattering events are necessary to double the wavelength of a photon with an initial wavelength of 200 pm?

## Electrons and Matter Waves

**36** • Use Equation 34-16 to calculate the de Broglie wavelength for an electron of kinetic energy (a) 2.5 eV, (b) 250 eV, (c) 2.5 keV, and (d) 25 keV.

**37** • An electron is moving at $v = 2.5 \times 10^5$ m/s. Find the electron's de Broglie wavelength.

**38** • **ISOLVE** An electron has a wavelength of 200 nm. Find (a) its momentum and (b) its kinetic energy.

**39** •• **SSM** An electron, a proton, and an alpha particle (the nucleus of a helium atom) each have a kinetic energy of 150 keV. Find (a) their momenta and (b) their de Broglie wavelengths.

**40** • **ISOLVE** A neutron in a reactor has kinetic energy of approximately 0.02 eV. Calculate the de Broglie wavelength of this neutron from Equation 34-15, where $mc^2 = 940$ MeV is the rest energy of the neutron.

**41** • Use Equation 34-15 to find the de Broglie wavelength of a proton (rest energy $mc^2 = 938$ MeV) that has a kinetic energy of 2 MeV.

**42** • **SSM** A proton is moving at $v = 0.003c$, where $c$ is the speed of light. Find the electron's de Broglie wavelength.

**43** • What is the kinetic energy of a proton whose de Broglie wavelength is (a) 1 nm and (b) 1 fm?

**44** • Which sport has a ball with the longest de Broglie wavelength; baseball, with a ball weighing 5 oz and moving at 95 mi/h, or tennis, with a ball weighing 2 oz and moving at 130 mi/h?

**45** • The energy of the electron beam in Davisson and Germer's experiment was 54 eV. Calculate the wavelength for these electrons.

**46** • The distance between $Li^+$ and $Cl^+$ ions in a LiCl crystal is 0.257 nm. Find the energy of electrons that have a wavelength equal to this spacing.

**47** • **SSM** An electron microscope uses electrons of energy 70 keV. Find the wavelength of these electrons.

**48** • What is the de Broglie wavelength of a neutron with speed $10^6$ m/s?

## Wave–Particle Duality

**49** • Suppose you have a spherical object of mass 4 g moving at 100 m/s. What size aperture is necessary for the object to show diffraction? Show that no common objects would be small enough to squeeze through such an aperture.

**50** • A neutron has a kinetic energy of 10 MeV. What size object is necessary to observe neutron diffraction effects? Is there anything in nature of this size that could serve as a target to demonstrate the wave nature of 10-MeV neutrons?

**51** • What is the de Broglie wavelength of an electron of kinetic energy 200 eV? What are some common targets that could demonstrate the wave nature of such an electron?

## A Particle in a Box

**52** •• **SSM** Use a spreadsheet program or graphing calculator to plot $\psi(x)$ and the probability distribution $\psi^2(x)$ of a particle in a box for the states $n = 1$, $n = 2$, and $n = 3$.

**53** •• (a) Find the energy of the ground state ($n = 1$) and the first two excited states of a proton in a one-dimensional box of length $L = 10^{-15}$ m = 1 fm. (These are the order of magnitude of nuclear energies.) Make an energy-level diagram for this system, and calculate the wavelength of electromagnetic radiation emitted when the proton makes a transition from (b) $n = 2$ to $n = 1$, (c) $n = 3$ to $n = 2$, and (d) $n = 3$ to $n = 1$.

**54** •• (a) Find the energy of the ground state ($n = 1$) and the first two excited states of a proton in a one-dimensional box of length 0.2 nm (about the diameter of a $H_2$ molecule). Calculate the wavelength of electromagnetic radiation emitted when the proton makes a transition from (b) $n = 2$ to $n = 1$, (c) $n = 3$ to $n = 2$, and (d) $n = 3$ to $n = 1$.

## *Calculating Probabilities and Expectation Values

**55** •• **ISOLVE** A particle is in the ground state of a box of length $L$. Find the probability of finding the particle in the interval $\Delta x = 0.002L$ at (a) $x = L/2$, (b) $x = 2L/3$, and (c) $x = L$. (Since $\Delta x$ is very small, you need not do any integration because the wave function is slowly varying.)

**56** •• **SSM** **ISOLVE✓** Repeat Problem 55 for a particle in the first excited state ($n = 2$).

**57** •• (a) Find $<x>$ for the second excited state ($n = 2$) for a particle in a box of length $L$ and (b) find $<x^2>$.

**58** •• A particle in a one-dimensional box is in the first excited state ($n = 2$). (a) Sketch $\psi^2(x)$ versus $x$ for this state. (b) What is the expectation value $<x>$ for this state? (c) What is the probability of finding the particle in some small region $dx$ centered at $x = L/2$? (d) Are your answers for Part (b) and Part (c) contradictory? If not, explain.

**59** •• A particle of mass $m$ has a wave function given by $\psi(x) = Ae^{-|x|/a}$, where $A$ and $a$ are constants. (a) Find the normalization constant $A$. (b) Calculate the probability of finding the particle in the region $-a \le x \le a$.

**60** •• A one-dimensional box is on the $x$-axis in the region of $0 \le x \le L$. A particle in this box is in its ground state. Calculate the probability that the particle will be found in the region (a) $0 < x < \frac{1}{2}L$, (b) $0 < x < L/3$, and (c) $0 < x < 3L/4$.

**61** •• Repeat Problem 60 for a particle in the first excited state of the box.

**62** •• The classical probability distribution function for a particle in a box of length $L$ is given by $P(x) = 1/L$. Use this to show that $<x> = L/2$ and $<x^2> = L^2/3$ for a classical particle in the box described in Problem 60.

**63** •• (a) For the wave functions

$$\psi_n(x) = \sqrt{\frac{2}{L}} \sin \frac{n\pi x}{L}, \quad n = 1, 2, 3, \ldots$$

corresponding to a particle in the $n$th state of the box described in Problem 60, show that $<x> = L/2$ and $<x^2> = L^2/3 - L^2/2n^2\pi^2$. (b) Compare this result for $n >> 1$ with your answer for the classical distribution of Problem 62.

**64** •• **SSM** (a) Use a spreadsheet program or graphing calculator to plot the expectation value for position $<x>$ and the square of the position $<x^2>$ as a function of the quantum number $n$ for the particle in the box described in Problem 60, for values of $n$ from 1 to 100. Assume $L = 1$ m for your graph. Refer to Problem 63. (b) Comment on the significance of any asymptotic limits that your graph shows.

**65** •• The wave functions for a particle of mass $m$ in a one-dimensional box of length $L$ centered at the origin (so that the ends are at $x = \pm L/2$) are given by

$$\psi(x) = \sqrt{\frac{2}{L}} \cos \frac{n\pi x}{L}, \qquad n = 1, 3, 5, 7, \ldots$$

and

$$\psi(x) = \sqrt{\frac{2}{L}} \sin \frac{n\pi x}{L}, \qquad n = 2, 4, 6, 8, \ldots$$

Calculate $<x>$ and $<x^2>$ for the ground state.

**66** •• Calculate $<x>$ and $<x^2>$ for the first excited state of the box described in Problem 65.

## General Problems

**67** • **SSM** **iSOLVE** A light beam of wavelength 400 nm has an intensity of 100 W/m². (a) What is the energy of each photon in the beam? (b) How much energy strikes an area of 1 cm² perpendicular to the beam in 1 s? (c) How many photons strike this area in 1 s?

**68** • **iSOLVE** A 1-$\mu$g particle is moving with a speed of approximately $10^{-1}$ cm/s in a box of length 1 cm. Treating this as a one-dimensional particle in a box, calculate the approximate value of the quantum number $n$.

**69** • (a) For the classical particle of Problem 68, find $\Delta x$ and $\Delta p$, assuming that these uncertainties are given by $\Delta x/L = 0.01$ percent and $\Delta p/p = 0.01$ percent. (b) What is $(\Delta x \Delta p)/\hbar$?

**70** • **iSOLVE** In 1987, a laser at Los Alamos National Laboratory produced a flash that lasted $1 \times 10^{-12}$ s and that had a power of $5 \times 10^{15}$ W. Estimate the number of emitted photons if their wavelength was 400 nm.

**71** • **iSOLVE** You cannot "see" anything smaller than the wavelength $\lambda$ used. What is the minimum energy of an electron needed in an electron microscope to "see" an atom, which has a diameter of about 0.1 nm?

**72** • **iSOLVE** A common flea that has a mass of 0.008 g can jump vertically as high as 20 cm. Estimate the de Broglie wavelength for the flea immediately after takeoff.

**73** •• **SSM** **iSOLVE**✓ Suppose that a 100-W source radiates light of wavelength 600 nm uniformly in all directions and that the eye can detect this light if only 20 photons per second enter a dark-adapted eye with a pupil 7 mm in diameter. How far from the source can the light be detected under these rather extreme conditions?

**74** •• **iSOLVE** The diameter of the pupil of the eye under room-light conditions is approximately 5 mm. (It can vary from approximately 1 mm to 8 mm.) Find the intensity of light of wavelength 600 nm so that 1 photon per second passes through the pupil.

**75** •• **iSOLVE** A lightbulb radiates 90 W uniformly in all directions. (a) Find the intensity at a distance of 1.5 m. (b) If the wavelength is 650 nm, find the number of photons per second that strike a surface of area 1 cm², which is oriented so that the line to the bulb is perpendicular to the surface.

**76** •• When light of wavelength $\lambda_1$ is incident on the cathode of a photoelectric tube, the maximum kinetic energy of the emitted electrons is 1.8 eV. If the wavelength is reduced to $\lambda_1/2$, the maximum kinetic energy of the emitted electrons is 5.5 eV. Find the work function $\phi$ of the cathode material.

**77** •• A photon of energy $E$ undergoes Compton scattering at an angle of $\theta$. Show that the energy $E'$ of the scattered photon is given by

$$E' = \frac{E}{(E/m_e c^2)(1 - \cos \theta) + 1}$$

**78** •• **iSOLVE**✓ A particle is confined to a one-dimensional box. In making a transition from the state $n$ to the state $n - 1$, radiation of 114.8 nm is emitted; in the transition from the state $n - 1$ to the state $n - 2$, radiation of wavelength 147 nm is emitted. The ground-state energy of the particle is 1.2 eV. Determine $n$.

**79** •• **SSM** The Pauli exclusion principle states that no more than one electron may occupy a particular quantum state at a time. Therefore, if we wish to model an atom as a collection of electrons trapped in a one-dimensional box, each electron in the box must have a unique value of the quantum number $n$. Calculate the energy that the most energetic electron would have for the uranium atom with atomic number 92, assuming the box has a width of 0.05 nm. How does this energy compare to the rest-mass energy of the electron?

**80** •• **iSOLVE** A beam of electrons, each with the same kinetic energy, illuminates a pair of slits separated by a distance $d = 54$ nm. The beam forms bright and dark fringes on a screen located a distance $L = 1.5$ m beyond the two slits. The arrangement is otherwise identical to that used in the optical two-slit interference experiment (described in Chapter 33 and in Figure 33-7) and the fringes have the appearance shown in Figure 34-18(d). The bright fringes are found to be separated by a distance of 0.68 mm. What is the kinetic energy of the electrons in the beam?

**81** •• **iSOLVE** When a surface is illuminated with light of wavelength $\lambda$, the maximum kinetic energy of the emitted electrons is 1.2 eV. If the wavelength $\lambda' = 0.8\lambda$ is used, the maximum kinetic energy increases to 1.76 eV. For wavelength $\lambda' = 0.6\lambda$, the maximum kinetic energy of the emitted electrons is 2.676 eV. Determine the work function of the surface and the wavelength $\lambda$.

**82** •• A simple pendulum of length 1 m has a bob of mass 0.3 kg. The energy of this oscillator is quantized to the values $E_n = (n + \frac{1}{2})hf_0$, where $n$ is an integer and $f_0$ is the frequency of the pendulum. (a) Find $n$ if the angular amplitude is 10°. (b) Find $n$ if the energy changes by 0.01 percent.

**83** •• **SSM** (a) Show that for large $n$, the fractional difference in energy between state $n$ and state $n + 1$ for a particle in a one-dimensional box is given approximately by

$$(E_{n+1} - E_n)/E_n \approx 2/n$$

(b) What is the approximate percentage energy difference between the states $n_1 = 1000$ and $n_2 = 1001$? (c) Comment on how this result is related to Bohr's correspondence principle.

**84** •• A mode-locked, titanium–sapphire laser has a wavelength of 850 nm and produces 100 million pulses of light each second. Each pulse has a duration of 125 femtoseconds (1 fs $= 10^{-15}$ s) and contains $5 \times 10^9$ photons. What is the average power produced by the laser?

**85** •• **iSOLVE** This problem is one of estimating the time lag (expected classically but not observed) in the photoelectric effect. Let the intensity of the incident radiation be 0.01 W/m². (a) If the area of the atom is 0.01 nm², find the energy per second falling on an atom. (b) If the work function is 2 eV, how long would it take classically for this much energy to fall on one atom?

# Applications of the Schrödinger Equation

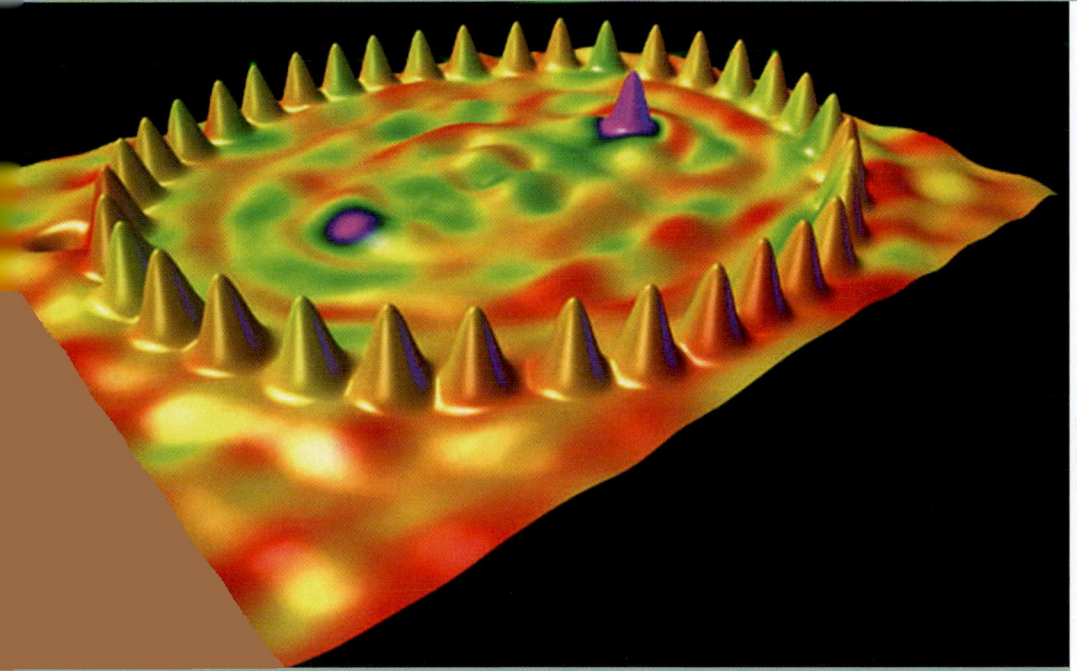

A QUANTUM MIRAGE. THE SCANNING TUNNELING MICROSCOPE (STM) ALLOWS ONE TO PUSH INDIVIDUAL ATOMS AROUND ON A SURFACE AND TO IMAGE THEM. ESPECIALLY INTRIGUING ARE IMAGES OF *QUANTUM CORRALS*, WHICH ARE CIRCULAR OR ELLIPTICAL ARRANGEMENTS ON A SURFACE INSIDE OF WHICH THE WAVES CORRESPONDING TO ELECTRONS NEAR THE SUBSTRATE SURFACE CAN BE REVEALED. THIS IMAGE COMES FROM IBM, WHERE PHYSICISTS PLACED THIRTY-SIX COBALT ATOMS IN AN ELLIPTICAL "STONEHENGE" PATTERN ON A COPPER SURFACE. AN EXTRA MAGNETIC COBALT ATOM WAS PLACED AT ONE OF THE TWO FOCI OF THE ELLIPSE, CAUSING VISIBLE INTERACTIONS WITH THE SURFACE ELECTRON WAVES. BUT THE WAVES ALSO SEEM TO BE INTERACTING WITH A PHANTOM COBALT ATOM AT THE OTHER FOCUS, AN ATOM THAT IS NOT REALLY THERE.

 **Could the phantom cobalt atom described above be caused by reflections of waves from the corral of cobalt atoms? The reflection and transmissions of one-dimensional waves is discussed in Section 35-4.**

35-1  The Schrödinger Equation

35-2  A Particle in a Finite Square Well

35-3  The Harmonic Oscillator

35-4  Reflection and Transmission of Electron Waves: Barrier Penetration

35-5  The Schrödinger Equation in Three Dimensions

35-6  The Schrödinger Equation for Two Identical Particles

In Chapter 34, we found that electrons and other particles have wave properties and are described by a wave function $\Psi(x, t)$. The probability of finding the particle in some region of space is proportional to the square of the wave function. We mentioned that the wave function is a solution of the Schrödinger equation, and we discussed some solutions qualitatively without reference to the equation itself. In particular, we showed how the standing-wave conditions lead to quantization of energy for a particle confined to a one-dimensional box.

➤ This chapter is a continuation of the material introduced in Chapter 34. We discuss the Schrödinger equation and apply the equation to the particle in the box problem and to several other situations in which a particle is confined to a region of space to illustrate how boundary conditions lead to energy quantization. We then show how the Schrödinger equation leads to barrier penetration and discuss the extension of the Schrödinger equation to more than one dimension and to more than one particle.

# 35-1 The Schrödinger Equation

Like the classical wave equation (Equation 15-9*b*), the Schrödinger equation is a partial differential equation in space and time. Like Newton's laws of motion, the Schrödinger equation cannot be derived. Its validity, like that of Newton's laws, lies in its agreement with experiment. In one dimension, the Schrödinger equation is[†]

$$-\frac{\hbar^2}{2m}\frac{\partial^2\Psi(x,t)}{\partial x^2} + U\Psi(x,t) = i\hbar\frac{\partial\Psi(x,t)}{\partial t} \qquad 35\text{-}1$$

TIME-DEPENDENT SCHRÖDINGER EQUATION

where $U$ is the potential energy function. Equation 35-1 is called the **time-dependent Schrödinger equation.** Unlike the classical wave equation, it relates the second space derivative of the wave function to the *first* time derivative of the wave function, and it contains the imaginary number $i = \sqrt{-1}$. The wave functions that are solutions of this equation are not necessarily real. $\Psi(x, t)$ is not a measurable function like the classical wave functions for sound or electromagnetic waves. The probability of finding a particle in some region of space $dx$ is certainly real though, so we must modify slightly the equation for probability density given in Chapter 34 (Equation 34-17). We take for the probability of finding a particle in some region $dx$

$$P(x, t)\, dx = |\Psi(x, t)|^2\, dx = \Psi^*\Psi\, dx \qquad 35\text{-}2$$

where $\Psi^*$, the complex conjugate of $\Psi$, is obtained from $\Psi$ by replacing $i$ by $-i$ wherever it appears.[‡]

In classical mechanics, the standing-wave solutions to the wave equation (Equation 16-16) are of great interest and value. Not surprisingly, standing-wave solutions to the Schrödinger wave equation are also of great interest and value. The wave function for the standing-wave motion of a uniform taut string is $A \sin(kx) \cos(\omega t + \delta)$, and this is representative of all standing waves. A standing wave function can always be expressed as a function of position multiplied by a function of time, where the function of time is one that varies sinusoidally with time. Standing-wave solutions to the one-dimensional Schrödinger wave equation are thus expressed

$$\Psi(x, t) = \psi(x)e^{-i\omega t} \qquad 35\text{-}3$$

where $e^{-i\omega t} = \cos(\omega t) - i \sin(\omega t)$. [In Appendix D, it is shown that $e^{-i\omega t} = \cos(\omega t) - i \sin(\omega t)$.] The right side of Equation 35-1 is then

$$i\hbar\frac{\partial\Psi(x,t)}{\partial t} = i\hbar(-i\omega)\psi(x)e^{-i\omega t} = \hbar\omega\psi(x)e^{-i\omega t} = E\psi(x)e^{-i\omega t}$$

where $E = \hbar\omega$ is the energy of the particle.

The Schrödinger wave equation has standing-wave solutions only if the potential energy function depends upon position alone. Substituting $\psi(x)e^{-i\omega t}$ into Equation 35-1 and canceling the common factor $e^{-i\omega t}$, we obtain an equation for $\psi(x)$, called the **time-independent Schrödinger equation:**

---

[†] Although we simply state the Schrödinger equation, Schrödinger himself had a vast knowledge of classical wave theory that led him to this equation.

[‡] Every complex number can be written in the form $z = a + bi$, where $a$ and $b$ are real numbers and $i = \sqrt{-1}$. The complex conjugate of $z$ is $z^* = a - bi$, so $z^*z = (a + bi)(a - bi) = a^2 + b^2 = |z|^2$. Complex numbers are discussed more fully in Appendix D.

$$-\frac{\hbar^2}{2m}\frac{d^2\psi(x)}{dx^2} + U(x)\psi(x) = E\psi(x) \qquad \text{35-4}$$

TIME-INDEPENDENT SCHRÖDINGER EQUATION

where we have written $U$ as $U(x)$ to emphasize that while $U$ may depend on position, $U$ does not depend on time. The function $U(x)$ represents the environment of the particle being described. It is this potential energy function in the Schrödinger equation that establishes the difference between different problems, just as the expressions for forces acting on a particle play in classical physics.

The calculation of the allowed energy levels in a system involves only the time-independent Schrödinger equation, whereas finding the probabilities of transition between these levels requires the solution of the time-dependent equation. In this book, we will be concerned only with the time-independent Schrödinger equation.

The solution of Equation 35-4 depends on the form of the potential energy function $U(x)$. When $U(x)$ is such that the particle is confined to some region of space, only certain discrete energies $E_n$ give solutions $\psi_n$ that can satisfy the normalization condition (Equation 34-18):

$$\int_{-\infty}^{\infty} |\psi|^2\,dx = 1$$

The complete time-dependent wave functions are then given, from Equation 35-3, by

$$\Psi_n(x, t) = \psi_n(x)e^{-i\omega_n t} = \psi_n(x)e^{-i(E_n/\hbar)t} \qquad \text{35-5}$$

## A Particle in an Infinite Square-Well Potential

We will illustrate the use of the time-independent Schrödinger equation by solving it for the problem of a particle in a box. The potential energy for a one-dimensional box from $x = 0$ to $x = L$ is shown in Figure 35-1. It is called an **infinite square-well potential** and is described mathematically by

$$U(x) = 0, \quad 0 < x < L$$

$$U(x) = \infty, \quad x < 0 \text{ or } x > L \qquad \text{35-6}$$

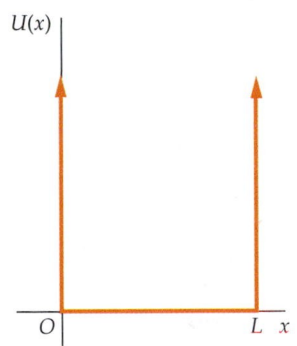

**FIGURE 35-1** The infinite square-well potential energy. For $x < 0$ and $x > L$, the potential energy $U(x)$ is infinite. The particle is confined to the region in the well $0 < x < L$.

Inside the box, the potential energy is zero, whereas outside the box it is infinite. Since we require the particle to be in the box, we have $\psi(x) = 0$ everywhere outside the box. We then need to solve the Schrödinger equation inside the box subject to the condition that $\psi(x)$ must be zero at $x = 0$ and at $x = L$.

Inside the box $U(x) = 0$, so the Schrödinger equation is written

$$-\frac{\hbar^2}{2m}\frac{d^2\psi(x)}{dx^2} = E\psi(x)$$

or

$$\frac{d^2\psi(x)}{dx^2} + k^2\psi(x) = 0 \qquad \text{35-7}$$

where

$$k^2 = \frac{2mE}{\hbar^2} \qquad \text{35-8}$$

The general solution of Equation 35-7 can be written as

$$\psi(x) = A \sin kx + B \cos kx \tag{35-9}$$

where $A$ and $B$ are constants. At $x = 0$, we have

$$\psi(0) = A \sin(k0) + B \cos(k0) = 0 + B$$

The boundary condition $\psi(x) = 0$ at $x = 0$ thus gives $B = 0$, and Equation 35-9 becomes

$$\psi(x) = A \sin kx \tag{35-10}$$

The wave function is thus a sine wave with the wavelength $\lambda$ related to the wave number $k$ in the usual way, $\lambda = 2\pi/k$. The boundary condition $\psi(x) = 0$ at $x = L$ restricts the possible values of $k$ and therefore the values of the wavelength $\lambda$, and (from Equation 35-8) the energy $E = \hbar^2 k^2/2m$. We have

$$\psi(L) = A \sin kL = 0 \tag{35-11}$$

This condition is satisfied if $kL$ is $\pi$ or any integer times $\pi$, that is, if $k$ is restricted to the values $k_n$ given by

$$k_n = n\frac{\pi}{L}, \quad n = 1, 2, 3, \ldots \tag{35-12}$$

The condition (Equation 35-11) is also satisfied for $n = 0$. The function $\psi(x) = A \sin 0 = 0$ for all values of $x$ is a solution to the wave equation. However, this solution is rejected as a wave function on physical grounds. It cannot be normalized and cannot be a wave function for a particle. Substituting this result into Equation 35-8 and solving for $E$ gives us the allowed energy values:

$$E_n = \frac{\hbar^2 k_n^2}{2m} = \frac{\hbar^2}{2m}\left(n\frac{\pi}{L}\right)^2 = n^2\left(\frac{h^2}{8mL^2}\right) = n^2 E_1 \tag{35-13}$$

where

$$E_1 = \frac{h^2}{8mL^2} \tag{35-14}$$

Equation 35-14 is the same as Equation 34-22, which we obtained by fitting an integral number of half-wavelengths into the box.

For each value of $n$, there is wave function $\psi_n(x)$ given by

$$\psi_n(x) = A_n \sin \frac{n\pi x}{L} \tag{35-15}$$

which is the same as Equation 34-25 with the constant $A_n = \sqrt{2/L}$ determined by normalization.[†]

# 35-2 A Particle in a Finite Square Well

The quantization of energy that we found for a particle in an infinite square well is a result that follows from the general solution of the Schrödinger equation for any particle confined to some region of space. We will illustrate this by considering the qualitative behavior of the wave function for a slightly more general potential energy function, the finite square well, which is shown in Figure 35-2.

---

† See Equation 34-18.

This potential energy function is described mathematically by

$$U(x) = U_0, \quad x < 0$$

$$U(x) = 0, \quad 0 < x < L \qquad\qquad 35\text{-}16$$

$$U(x) = U_0, \quad x > L$$

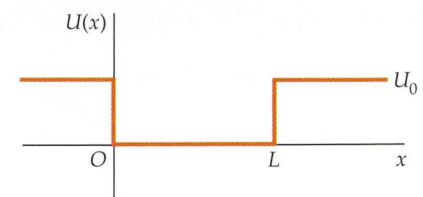

**FIGURE 35-2** The finite square-well potential energy.

This potential energy function is discontinuous at $x = 0$ and $x = L$, but it is finite everywhere. The solutions of the Schrödinger equation for this type of potential energy function depend on whether the total energy $E$ is greater or less than $U_0$. We will not discuss the case of $E > U_0$, except to remark that in that case the particle is not confined and any value of the energy is allowed; that is, there is no energy quantization. Here we assume that $0 \leq E < U_0$.

Inside the well, $U(x) = 0$, and the time-independent Schrödinger equation is the same as for the infinite well (Equation 35-7):

$$-\frac{\hbar^2}{2m}\frac{d^2\psi(x)}{dx^2} = E\psi(x)$$

or

$$\frac{d^2\psi(x)}{dx^2} + k^2\psi(x) = 0$$

where $k^2 = 2mE/\hbar^2$. The general solution is of the form

$$\psi(x) = A \sin kx + B \cos kx$$

In this case, $\psi(x)$ is not required to be zero at $x = 0$, so $B$ is not zero. Outside the well, the time-independent Schrödinger equation is

$$-\frac{\hbar^2}{2m}\frac{d^2\psi(x)}{dx^2} + U_0\psi(x) = E\psi(x)$$

or

$$\frac{d^2\psi(x)}{dx^2} - \alpha^2\psi(x) = 0 \qquad\qquad 35\text{-}17$$

where

$$\alpha^2 = \frac{2m}{\hbar^2}(U_0 - E) > 0 \qquad\qquad 35\text{-}18$$

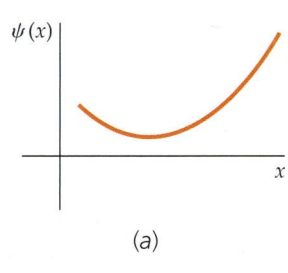

(a)

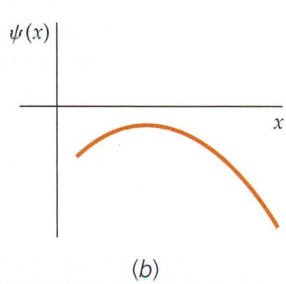

(b)

The wave functions and allowed energies for the particle can be found by solving Equation 35-17 for $\psi(x)$ outside the well and then requiring that $\psi(x)$ and $d\psi(x)/dx$ be continuous at the boundaries $x = 0$ and $x = L$. The solution of Equation 35-17 is not difficult (for positive values of $x$, it is of the form $\psi(x) = Ce^{-\alpha x}$), but applying the boundary conditions involves much tedious algebra and is not important for our purpose. The important feature of Equation 35-17 is that the second derivative of $\psi(x)$, which is related to the concavity of the wave function, has the same sign as the wave function $\psi$. If $\psi$ is positive, $d^2\psi/dx^2$ is also positive and the wave function curves away from the axis, as shown in Figure 35-3a. Similarly, if $\psi$ is negative, $d^2\psi/dx^2$ is negative and $\psi$ again curves away from the axis, as shown in Figure 35-3b. This behavior is

**FIGURE 35-3** (a) A positive function with positive concavity. (b) A negative function with negative concavity.

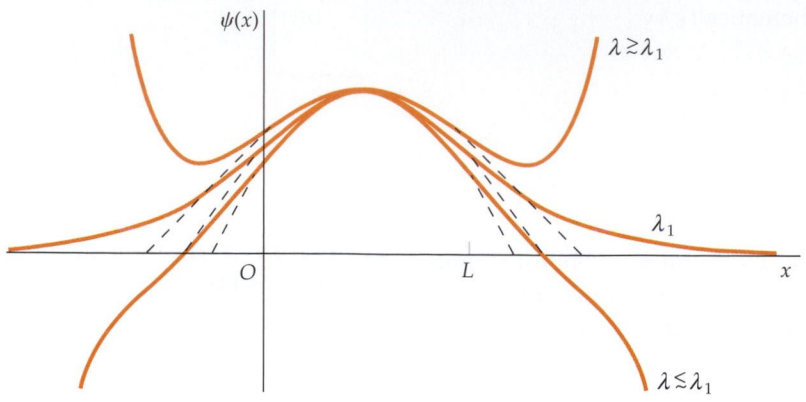

**FIGURE 35-4** Functions satisfying the Schrödinger equation with wavelengths near the wavelength $\lambda_1$ corresponding to the ground-state energy $E_1 = \hbar^2/2m\lambda_1^2$ in the finite well. If $\lambda$ is slightly greater than $\lambda_1$, the function approaches infinity, like the function in Figure 35-3a. At the critical wavelength $\lambda_1$, the function and its slope approach zero together. If $\lambda$ is slightly less than $\lambda_1$, the function crosses the $x$ axis while the slope is still negative. The slope then becomes more negative because its rate of change $d^2\psi/dx^2$ is now negative. This function approaches negative infinity as $x$ approaches infinity.

very different from the behavior inside the well, where $\psi$ and $d^2\psi/dx^2$ have opposite signs so that $\psi$ always curves toward the axis like a sine or cosine function. Because of this behavior outside the well, for most values of the energy $E$ in Equation 35-17, $\psi(x)$ becomes infinite as $x$ approaches $\pm\infty$; that is, most wave functions $\psi(x)$ are not well behaved outside the well. Though they satisfy the Schrödinger equation, such functions are not proper wave functions because they cannot be normalized. The solutions of the Schrödinger equation are well behaved (i.e., they approach 0 as $|x|$ becomes very large) only for certain values of the energy. These energy values are the allowed energies for the finite square well.

Figure 35-4 shows a well-behaved wave function, with a wavelength $\lambda_1$ inside the well corresponding to the ground-state energy. The behavior of the wave functions corresponding to nearby wavelengths and energies is also shown. Figure 35-5 shows the wave functions and probability distributions for the ground state and first two excited states. From this figure, we can see that the wavelengths inside the well are slightly longer than the corresponding wavelengths for the infinite well (Figure 34-14), so the corresponding energies are slightly less than those for the infinite well. Another feature of the finite-well problem is that there are only a finite number of allowed energies. For very small values of $U_0$, there is only one allowed energy.

Note that the wave function penetrates beyond the edges of the well at $x = L$ and $x = 0$, indicating that there is some small probability of finding the particle in the region in which its total energy $E$ is less than its potential energy $U_0$. This region is called the *classically forbidden region* because the kinetic energy, $E - U_0$, would be negative when $U_0 > E$. Since negative kinetic energy has no meaning in classical physics, it is interesting to speculate on the result of an attempt to observe the particle in the classically forbidden region. It can be shown from the uncertainty principle that if an attempt is made to localize the particle in the classically forbidden region, such a measurement introduces an uncertainty in the momentum of the particle corresponding to a minimum kinetic energy that is greater than $U_0 - E$. This is just great enough to prevent us from measuring a negative kinetic energy. The penetration of the wave function into a classically forbidden region does have important consequences in barrier penetration, which will be discussed in Section 35-4.

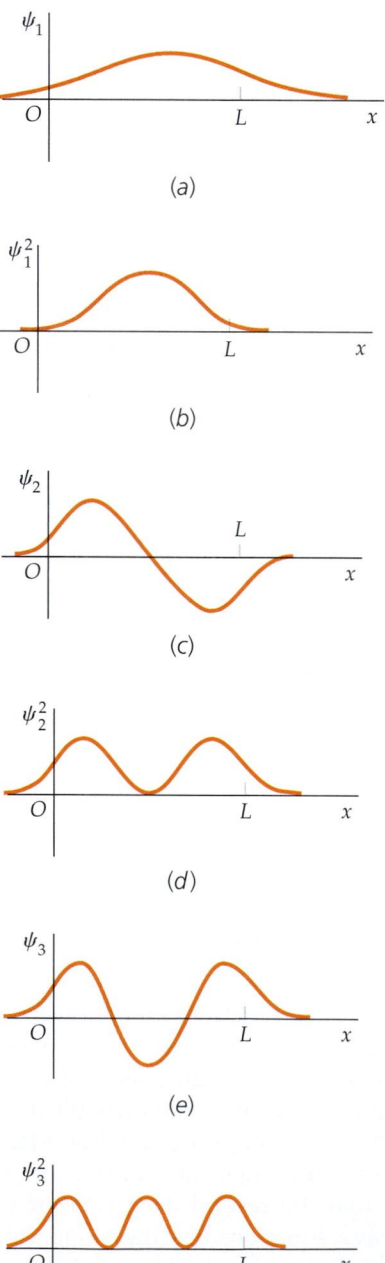

**FIGURE 35-5** Graphs of the wave functions $\psi_n(x)$ and probability distributions $\psi_n^2(x)$ for $n = 1$, $n = 2$, and $n = 3$ for the finite square well. Compare these graphs with those of Figure 34-14 for the infinite square well, where the wave functions are zero at $x = 0$ and $x = L$. The wavelengths here are slightly longer than the corresponding wavelengths for the infinite well, so the allowed energies are somewhat smaller.

Much of our discussion of the finite-well problem applies to any problem in which $E > U(x)$ in some region and $E < U(x)$ outside that region, as we see in the next section.

# 35-3 The Harmonic Oscillator

The potential energy for a particle of mass $m$ attached to a spring of force constant $k$ is

$$U(x) = \tfrac{1}{2}kx^2 = \tfrac{1}{2}m\omega_0^2 x^2 \qquad\qquad 35\text{-}19$$

where $\omega_0 = \sqrt{k/m}$ is the natural frequency of the oscillator. Classically, the object oscillates between $x = +A$ and $x = -A$. The object's total energy is $E = \tfrac{1}{2}m\omega_0^2 A^2$, which can have any positive value or zero.

This potential energy function, shown in Figure 35-6, applies to virtually any system undergoing small oscillations about a position of stable equilibrium. For example, it could apply to the oscillations of the atoms of a diatomic molecule, such as $H_2$ or HCl, oscillating about their equilibrium separation. Between the classical turning points ($|x| < A$), the total energy is greater than the potential energy, and the Schrödinger equation can be written

$$\frac{d^2\psi(x)}{dx^2} = -k^2\psi(x) \qquad\qquad 35\text{-}20$$

where $k^2 = (2m/\hbar^2)[E - U(x)]$ now depends on $x$. The solutions of this equation are no longer simple sine or cosine functions because the wave number $k = 2\pi/\lambda$ now varies with $x$; but since $d^2\psi/dx^2$ and $\psi$ have opposite signs, $\psi$ will always curve toward the axis and the solutions will oscillate.

Outside the classical turning points ($|x| > A$), the potential energy is greater than the total energy and the Schrödinger equation is similar to Equation 35-17:

$$\frac{d^2\psi(x)}{dx^2} = +\alpha^2\psi(x) \qquad\qquad 35\text{-}21$$

except that here $\alpha^2 = (2m/\hbar^2)[U(x) - E] > 0$ depends on $x$. For $|x| > A$, $d^2\psi/dx^2$ and $\psi$ have the same sign, so $\psi$ will curve away from the axis and there will be only certain values of $E$ for which solutions exist that approach zero as $x$ approaches infinity.

For the harmonic oscillator potential energy function, the Schrödinger equation is

$$-\frac{\hbar^2}{2m}\frac{d^2\psi(x)}{dx^2} + \tfrac{1}{2}m\omega_0^2 x^2\psi(x) = E\psi(x) \qquad\qquad 35\text{-}22$$

## Wave Functions and Energy Levels

Rather than pursue a general solution to the Schrödinger equation for this system, we simply present the solution for the ground state and the first excited state.

The ground-state wave function $\psi_0(x)$ is found to be a Gaussian function centered at the origin:

$$\psi_0(x) = A_0 e^{-ax^2} \qquad\qquad 35\text{-}23$$

where $A_0$ and $a$ are constants. This function and the wave function for the first excited state are shown in Figure 35-7.

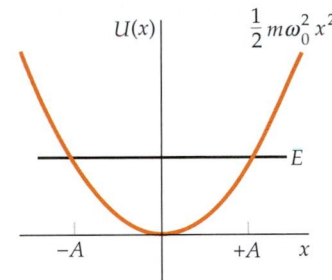

**FIGURE 35-6** Harmonic oscillator potential.

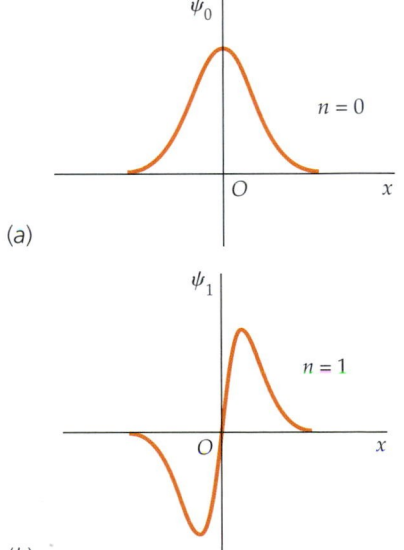

(a)

(b)

**FIGURE 35-7** (a) The ground-state wave function for the harmonic oscillator potential. (b) The wave function for the first excited state of the harmonic oscillator potential.

VERIFYING THE GROUND-STATE WAVE FUNCTION     **E X A M P L E   3 5 - 1**

Verify that $\psi_0(x) = A_0 e^{-ax^2}$, where $a$ is a positive constant, is a solution of the Schrödinger equation for the harmonic oscillator.

**PICTURE THE PROBLEM** We take the first and second derivative of $\psi$ with respect to $x$ and substitute into Equation 35-22. Since this is the ground-state wave function, we write $E_0$ for the energy $E$.

1. Compute $d\psi_0/dx$:

$$\frac{d\psi_0(x)}{dx} = \frac{d}{dx}(A_0 e^{-ax^2}) = -2ax A_0 e^{-ax^2}$$

2. Compute $d^2\psi_0/dx^2$:

$$\frac{d^2\psi_0(x)}{dx^2} = -2a A_0 e^{-ax^2} + 4a^2 x^2 A_0 e^{-ax^2}$$

3. Substitute these derivatives into the Schrödinger equation:

$$-\frac{\hbar^2}{2m}\frac{d^2\psi(x)}{dx^2} + \frac{1}{2}m\omega_0^2 x^2 \psi(x) = E\psi(x)$$

$$-\frac{\hbar^2}{2m}(-2a A_0 e^{-ax^2} + 4a^2 x^2 A_0 e^{-ax^2}) + \frac{1}{2}m\omega_0^2 x^2 A_0 e^{-ax^2} = E_0 A_0 e^{-ax^2}$$

4. Cancel the common factor $A_0 e^{-ax^2}$ and show the result in standard polynomial form:

$$-\frac{\hbar^2}{2m}(-2a + 4a^2 x^2) + \frac{1}{2}m\omega_0^2 x^2 = E_0$$

so

$$\left(-\frac{2\hbar^2 a^2}{m} + \frac{1}{2}m\omega_0^2\right)x^2 + \left(\frac{\hbar^2 a}{m} - E_0\right) = 0$$

5. The equation in step 4 must hold for all $x$. Set $x = 0$ and solve for $E_0$:

$$0 + \left(\frac{\hbar^2 a}{m} - E_0\right) = 0$$

so

$$E_0 = \frac{\hbar^2 a}{m}$$

6. Substitute this result for $E_0$ into the equation in step 4 and simplify:

$$\left(-\frac{2\hbar^2 a^2}{m} + \frac{1}{2}m\omega_0^2\right)x^2 + 0 = 0$$

7. It follows that the coefficient of $x^2$ must equal zero:

$$-\frac{2\hbar^2 a^2}{m} + \frac{1}{2}m\omega_0^2 = 0$$

8. Solve for $a$:

$$a = \frac{m\omega_0}{2\hbar}$$

9. Substitute this result into the equation for $E_0$ in step 5:

$$E_0 = \frac{\hbar^2 a}{m} = \frac{1}{2}\hbar\omega_0$$

We have shown that the given function satisfies the Schrödinger equation for any value of $A_0$, as long as the energy is given by $E_0 = \frac{1}{2}\hbar\omega_0$.

**REMARKS** The step 4 equation is a polynomial that is equal to zero. A theorem that would have simplified the solution is: If a polynomial is equal to zero over a continuous range of values of $x$, then each of the polynomial coefficients is equal to zero. For example, if $Ax^3 + Bx^2 + Cx + D = 0$ on the interval $1 < x < 2$, then $A = B = C = D = 0$. The proof of this result is the topic of Problem 43.

We see from this example that the ground-state energy is given by

$$E_0 = \frac{\hbar^2 a}{m} = \frac{1}{2}\hbar\omega_0 \qquad\qquad 35\text{-}24$$

The first excited state has a node in the center of the potential well, just as with the particle in a box.[†] The wave function $\psi_1(x)$ is

$$\psi_1(x) = A_1 x e^{-ax^2} \qquad\qquad 35\text{-}25$$

where $a = m\omega_0/2\hbar$, as in Example 35-1. This function is also shown in Figure 35-7. Substituting $\psi_1(x)$ into the Schrödinger equation, as was done for $\psi_0(x)$ in Example 35-1, yields the energy of the first excited state,

$$E_1 = \frac{3}{2}\hbar\omega_0$$

In general, the energy of the $n$th excited state of the harmonic oscillator is

$$E_n = (n + \tfrac{1}{2})\hbar\omega_0, \quad n = 0, 1, 2, \dots \qquad 35\text{-}26$$

as indicated in Figure 35-8. The fact that the energy levels are evenly spaced by the amount $\hbar\omega_0$ is a peculiarity of the harmonic oscillator potential. As we saw in Chapter 34, the energy levels for a particle in a box, or for the hydrogen atom, are not evenly spaced. The precise spacing of energy levels is closely tied to the particular form of the potential energy function.

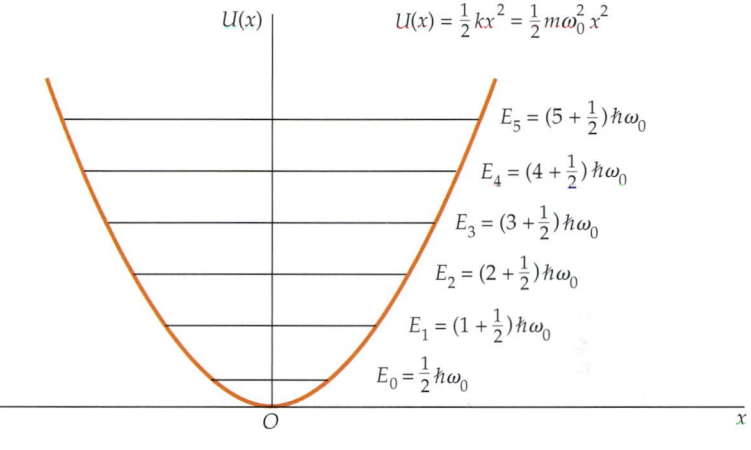

**FIGURE 35-8**  Energy levels in the harmonic oscillator potential.

# 35-4 Reflection and Transmission of Electron Waves: Barrier Penetration

In Sections 35-2 and 35-3, we were concerned with bound-state problems in which the potential energy is larger than the total energy for large values of $|x|$. In this section, we consider some simple examples of unbound states for which $E$ is greater than $U(x)$. For these problems, $d^2\psi/dx^2$ and $\psi$ have opposite signs, so $\psi(x)$ curves toward the axis and does not become infinite at large values of $|x|$.

## Step Potential

Consider a particle of energy $E$ moving in a region in which the potential energy is the step function

$$U(x) = 0, \quad x < 0$$

$$U(x) = U_0, \quad x > 0$$

as shown in Figure 35-9. We are interested in what happens when a particle moving from left to right encounters the step.

The classical answer is simple. To the left of the step, the particle moves with a speed $v = \sqrt{2E/m}$. At $x = 0$, an impulsive force acts on the particle. If the initial energy $E$ is less than $U_0$, the particle will be turned around and will then move to the left at its original speed; that is, the particle will be reflected by the step. If $E$ is greater than $U_0$, the particle will continue to move to the right but with reduced speed given by $v = \sqrt{2(E - U_0)/m}$. We can picture this classical problem as a ball rolling along a level surface and coming to a steep hill of height $h$ given by $mgh = U_0$. If the initial kinetic energy of the ball is less than

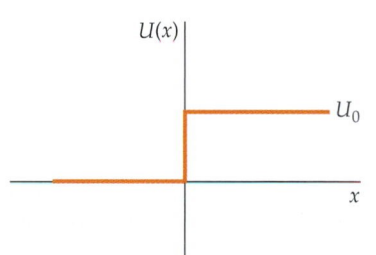

**FIGURE 35-9**  Step potential. A classical particle incident from the left, with total energy $E > U_0$, is always transmitted. The change in potential energy at $x = 0$ merely provides an impulsive force that reduces the speed of the particle. A wave incident from the left is partially transmitted and partially reflected because the wavelength changes abruptly at $x = 0$.

---

[†] Each higher-energy state has one additional node in the wave function.

*mgh*, the ball will roll part way up the hill and then back down and to the left along the lower surface at its original speed. If $E$ is greater than *mgh*, the ball will roll up the hill and proceed to the right at a lesser speed.

The quantum-mechanical result is similar when $E$ is less than $U_0$. Figure 35-10 shows the wave function for the case $E < U_0$. The wave function does not go to zero at $x = 0$ but rather decays exponentially, like the wave function for the bound state in a finite square-well problem. The wave penetrates slightly into the classically forbidden region $x > 0$, but it is eventually completely reflected. This problem is somewhat similar to that of total internal reflection in optics.

For $E > U_0$, the quantum-mechanical result differs markedly from the classical result. At $x = 0$, the wavelength changes abruptly from $\lambda_1 = h/p_1 = h/\sqrt{2mE}$ to $\lambda_2 = h/p_2 = h/\sqrt{2m(E - U_0)}$. We know from our study of waves that when the wavelength changes suddenly, part of the wave is reflected and part of the wave is transmitted. Since the motion of an electron (or other particle) is governed by a wave equation, the electron sometimes will be transmitted and sometimes will be reflected. The probabilities of reflection and transmission can be calculated by solving the Schrödinger equation in each region of space and comparing the amplitudes of the transmitted waves and reflected waves with that of the incident wave. This calculation and its result are similar to finding the fraction of light reflected from an air–glass interface. If $R$ is the probability of reflection, called the reflection coefficient, this calculation gives

$$R = \frac{(k_1 - k_2)^2}{(k_1 + k_2)^2} \qquad \text{35-27}$$

where $k_1$ is the wave number for the incident wave and $k_2$ is the wave number for the transmitted wave. This result is the same as the result in optics for the reflection of light at normal incidence from the boundary between two media having different indexes of refraction $n$ (Equation 31-11). The probability of transmission $T$, called the **transmission coefficient,** can be calculated from the reflection coefficient, since the probability of transmission plus the probability of reflection must equal 1:

$$T + R = 1 \qquad \text{35-28}$$

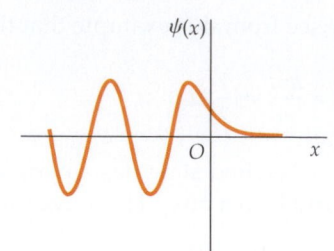

**FIGURE 35-10** When the total energy $E$ is less than $U_0$, the wave function penetrates slightly into the region $x > 0$. However, the probability of reflection for this case is 1, so no energy is transmitted.

---

*REFLECTION AND TRANSMISSION AT A STEP BARRIER*     **EXAMPLE 35-2**

A particle of energy $E_0$ traveling in a region in which the potential energy is zero is incident on a potential barrier of height $U_0 = 0.2E_0$. Find the probability that the particle will be reflected.

**PICTURE THE PROBLEM** We need to calculate the wave numbers $k_1$ and $k_2$ and use them to calculate the reflection coefficient $R$ from Equation 35-27. The wave numbers are related to the kinetic energy $K$ by $K = p^2/2m = \hbar^2 k^2/2m$.

1. The probability of reflection is the reflection coefficient:
$$R = \frac{(k_1 - k_2)^2}{(k_1 + k_2)^2}$$

2. Calculate $k_1$ from the initial kinetic energy $E_0$:
$$E_0 = \frac{\hbar^2 k_1^2}{2m}$$
$$k_1 = \sqrt{2mE_0}/\hbar^2 = 1.41\sqrt{mE_0/\hbar^2}$$

3. Relate $k_2$ to the final kinetic energy $K_2$:
$$\frac{\hbar k_2^2}{2m} = K_2 = E_0 - U_0 = E_0 - 0.2E_0 = 0.8E_0$$

4. Solve for $k_2$:

$$k_2 = \sqrt{2m(0.8E_0)/\hbar^2} = 1.26\sqrt{mE_0}/\hbar^2$$

5. Substitute these values into Equation 35-27 to calculate $R$:

$$R = \frac{(k_1 - k_2)^2}{(k_1 + k_2)^2} = \frac{(1.41 - 1.26)^2}{(1.41 + 1.26)^2} = \boxed{0.00316}$$

**REMARKS**  The probability of reflection is only 0.3 percent. This probability is small because the barrier height reduces the kinetic energy by only 20 percent. Since $k$ is proportional to the square root of the kinetic energy, the wave number and therefore the wavelength is changed by only 10 percent.

 **EXERCISE**  Express the index of refraction $n$ of light in terms of the wave number $k$, and show that Equation 31-11 for the reflection of light at normal incidence is the same as Equation 35-27.

In quantum mechanics, a localized particle is represented by a wave packet, which has a maximum at the most probable position of the particle. Figure 35-11 shows a wave packet representing a particle of energy $E$ incident on a step potential of height $U_0$, which is less than $E$. After the encounter, there are two wave packets. The relative heights of the transmitted packet and reflected packet indicate the relative probabilities of transmission and reflection. For the situation shown here, $E$ is much greater than $U_0$, and the probability of transmission is much greater than that of reflection.

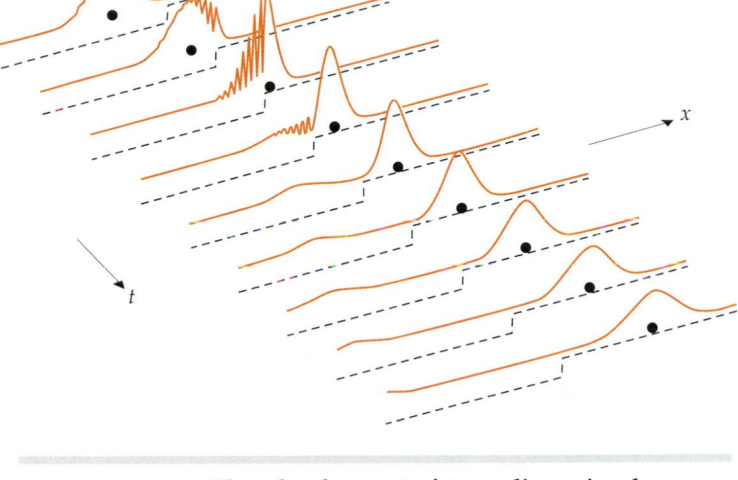

**FIGURE 35-11**  Time development of a one-dimensional wave packet representing a particle incident on a step potential for $E > U_0$. The position of a classical particle is indicated by the dot. Note that part of the packet is transmitted and part is reflected.

## Barrier Penetration

Figure 35-12a shows a rectangular potential barrier of height $U_0$ and width $a$ given by

$$U(x) = 0, \quad x < 0$$

$$U(x) = U_0, \quad 0 < x < a$$

$$U(x) = 0, \quad x > a$$

We consider a particle of energy $E$, which is slightly less than $U_0$, that is incident on the barrier from the left. Classically, the particle would always be reflected. However, a wave incident from the left does not decrease immediately to zero at the barrier, but it will instead decay exponentially in the classically forbidden region $0 < x < a$. Upon reaching the far wall of the barrier ($x = a$), the wave function must join smoothly to a sinusoidal wave function to the right of the barrier, as shown in Figure 35-12b. This implies that there is some probability of the particle (which is represented by the wave function) being found on the far side of the barrier even though, classically, it should never pass through the barrier. For  the case in which the quantity $\alpha a = \sqrt{2ma^2(U_0 - E)/\hbar^2}$ is much greater than 1, the transmission coefficient is proportional to $e^{-2\alpha a}$:

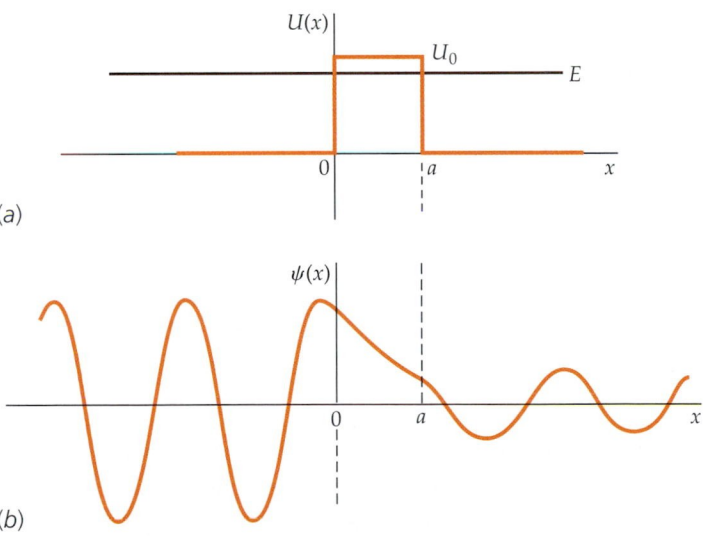

**FIGURE 35-12**  (a) A rectangular potential barrier. (b) The penetration of the barrier by a wave with total energy less than the barrier energy. Part of the wave is transmitted by the barrier even though, classically, the particle cannot enter the region $0 < x < a$ in which the potential energy is greater than the total energy. To the left of the barrier, there is both an incident and a reflected wave. These waves form a resultant wave so that $\psi$ is a superposition of a standing wave and a traveling wave (traveling toward the barrier). To the right of the barrier is only the transmitted wave that is traveling away from the barrier.

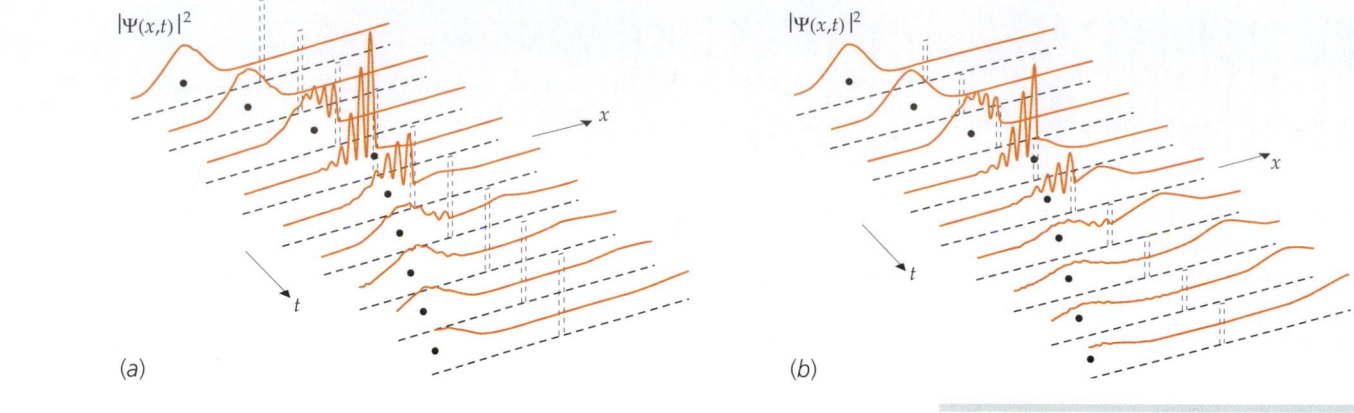

(a)                                                    (b)

$$T \propto e^{-2\alpha a} \qquad\qquad 35\text{-}29$$

TRANSMISSION THROUGH A BARRIER

**FIGURE 35-13** Barrier penetration. (a) The same particle incident on a barrier of height much greater than the energy of the particle. A very small part of the packet tunnels through the barrier. In both drawings, the position of a classical particle is indicated by a dot. (b) A wave packet representing a particle incident on a barrier of height just slightly greater than the energy of the particle. For this particular choice of energies, the probability of transmission is approximately equal to the probability of reflection, as indicated by the relative sizes of the transmitted and reflected packets.

with $\alpha = \sqrt{2m(U_0 - E)}/\hbar^2$. The probability of penetration of the barrier thus decreases exponentially with the barrier thickness $a$ and with the square root of the relative barrier height $(U_0 - E)$.

Figure 35-13a shows a wave packet incident on a potential barrier of height $U_0$ that is considerably greater than the energy of the particle. The probability of penetration is very small, as indicated by the relative sizes of the reflected and transmitted packets. In Figure 35-13b, the barrier is just slightly greater than the energy of the particle. In this case, the probability of penetration is about the same as the probability of reflection. Figure 35-14 shows a particle incident on two potential barriers of height just slightly greater than the energy of the particle.

As we have mentioned, the penetration of a barrier is not unique to quantum mechanics. When light is totally reflected from a glass–air interface, the light wave can penetrate the air barrier if a second piece of glass is brought within a few wavelengths of the first. This effect can be demonstrated with a laser beam and two 45° prisms (Figure 35-15). Similarly, water waves in a ripple tank can penetrate a gap of deep water (Figure 35-16).

The theory of barrier penetration was used by George Gamow in 1928 to explain the enormous variation in the half-lives for $\alpha$ decay of radioactive nuclei.

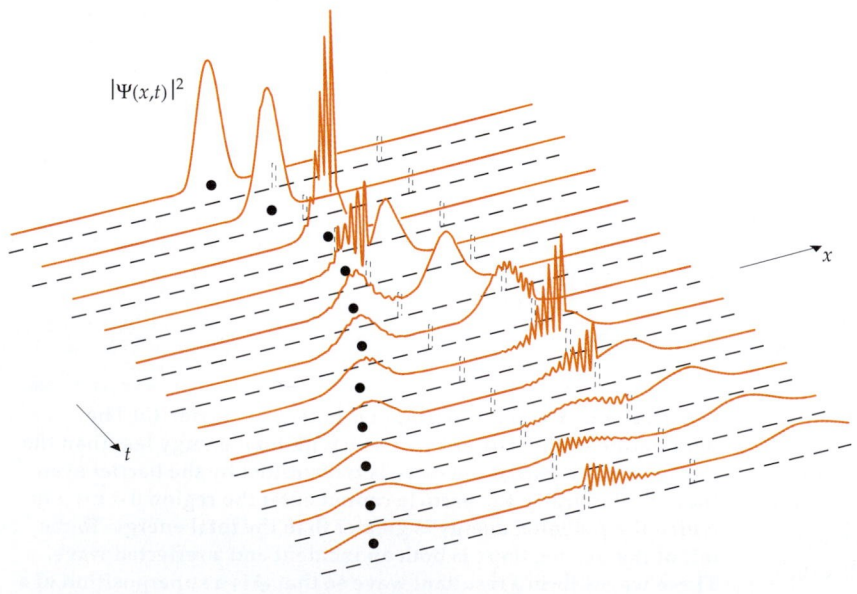

**FIGURE 35-14** A wave packet representing a particle incident on two barriers. At each encounter, part of the packet is transmitted and part reflected, resulting in part of the packet being trapped between the barriers for some time.

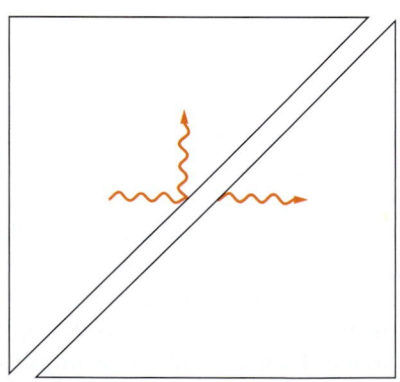

**FIGURE 35-15** The penetration of an optical barrier. If the second prism is close enough to the first, part of the wave penetrates the air barrier even when the angle of incidence in the first prism is greater than the critical angle.

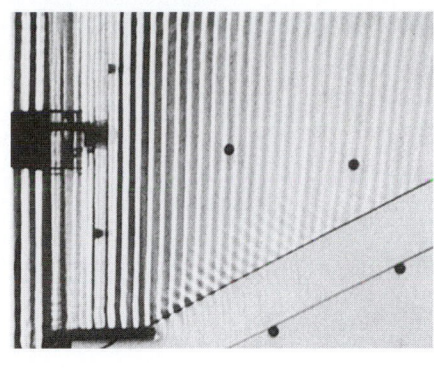 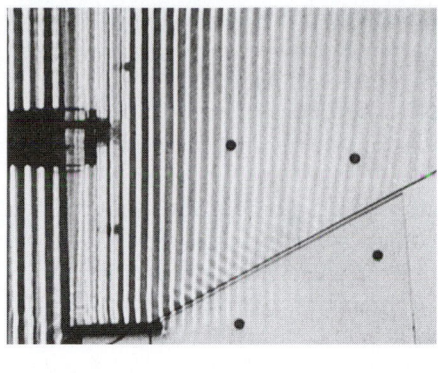

(a)                    (b)

**FIGURE 35-16** The penetration of a barrier by water waves in a ripple tank. In Figure 35-16a, the waves are totally reflected from a gap of deeper water. When the gap is very narrow, as in Figure 35-16b, a transmitted wave appears. The dark circles are spacers that are used to support the prisms from below.

(Alpha particles are helium nuclei emitted from larger atoms in radioactive decay; they consist of two protons and two neutrons tightly bound together.) In general, the smaller the energy of the emitted $\alpha$ particle, the longer the half-life. The energies of $\alpha$ particles from natural radioactive sources range from approximately 4 MeV to 7 MeV, whereas the half-lives range from approximately $10^{-5}$ second to $10^{10}$ years. Gamow represented a radioactive nucleus by a potential well containing an $\alpha$ particle, as shown in Figure 35-17. Without knowing very much about the nuclear force that is exerted on the $\alpha$ particle within the nucleus, Gamow represented it by a square well. Just outside the well, the $\alpha$ particle with its charge of $+2e$ is repelled by the nucleus with its charge $+Ze$, where $Ze$ is the remaining nuclear charge. This force is represented by the Coulomb potential energy $+k(2e)(Ze)/r$. The energy $E$ is the measured kinetic energy of the emitted $\alpha$ particle, because when it is far from the nucleus its potential energy is zero. After the $\alpha$ particle is formed inside the radioactive nucleus, it bounces back and forth inside the nucleus, hitting the barrier at the nuclear radius $R$. Each time the $\alpha$ particle strikes the barrier, there is some small probability of the particle penetrating the barrier and appearing outside the nucleus. We can see from Figure 35-17 that a small increase in $E$ reduces the relative height of the barrier $U - E$ and also the barrier's thickness. Because the probability of penetration is so sensitive to the barrier thickness and relative height, a small increase in $E$ leads to a large increase in the probability of transmission and therefore to a shorter lifetime. Gamow was able to derive an expression for the half-life as a function of $E$ that is in excellent agreement with experimental results.

In the **scanning tunneling electron microscope** developed in the 1980s, a thin space between a material specimen and a tiny probe acts as a barrier to electrons bound in the specimen. A small voltage applied between the probe and the specimen causes the electrons to *tunnel* through the vacuum separating the two surfaces if the surfaces are close enough together. The tunneling current is extremely sensitive to the size of the gap between the probe and specimen. If a constant tunneling current is maintained as the probe scans the specimen, the surface of the specimen can be mapped out by the motions of the probe. In this way, the surface features of a specimen can be measured with a resolution of the order of the size of an atom.

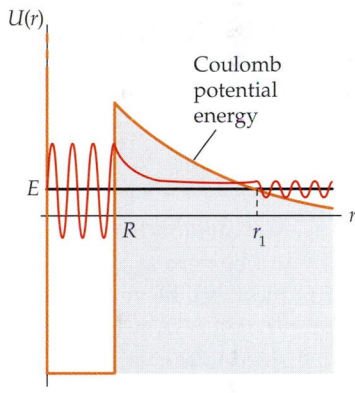

**FIGURE 35-17** Model of a potential energy function for an $\alpha$ particle in a radioactive nucleus. The strong attractive nuclear force when $r$ is less than the nuclear radius $R$ can be approximately described by the potential well shown. Outside the nucleus the nuclear force is negligible, and the potential is given by Coulomb's law, $U(r) = +k(2e)(Ze)/r$, where $Ze$ is the nuclear charge and $2e$ is the charge of the $\alpha$ particle. The wave function of the alpha particle, shown in red, is placed on the graph.

## 35-5 The Schrödinger Equation in Three Dimensions

The one-dimensional time-independent Schrödinger equation is easily extended to three dimensions. In rectangular coordinates, it is

$$-\frac{\hbar^2}{2m}\left(\frac{\partial^2\psi}{\partial x^2} + \frac{\partial^2\psi}{\partial y^2} + \frac{\partial^2\psi}{\partial z^2}\right) + U\psi = E\psi \qquad 35\text{-}30$$

where the wave function $\psi$ and the potential energy $U$ are generally functions of all three coordinates, $x$, $y$, and $z$. To illustrate some of the features of problems in three dimensions, we consider a particle in a three-dimensional infinite square well given by $U(x, y, z) = 0$ for $0 < x < L$, $0 < y < L$, and $0 < z < L$. Outside this cubical region, $U(x, y, z) = \infty$. For this problem, the wave function must be zero at the edges of the well.

There are standard methods in partial differential equations for solving Equation 35-30. We can guess the form of the solution from our knowledge of probability. For a one-dimensional box along the $x$ axis, we have found the probability that a particle is in the region $dx$ at $x$ to be $A_1^2 \sin^2(k_1 x)\, dx$ (from Equation 35-10), where $A_1$ is a normalization constant and $k_1 = n\pi/L$ is the wave number. Similarly, for a box along the $y$ axis, the probability of a particle being in a region $dy$ at $y$ is $A_2^2 \sin^2(k_2 y)\, dy$. The probability of two independent events occurring is the product of the probabilities of each event occurring.[†] So the probability of a particle being in region $dx$ at $x$ *and* in region $dy$ at $y$ is $A_1^2 \sin^2(k_1 x)\, dx\, A_2^2 \sin^2(k_2 y)\, dy = A_1^2 \sin^2(k_1 x) A_2^2 \sin^2(k_2 y)\, dx\, dy$. The probability of a particle being in the region $dx$, $dy$, and $dz$ is $\psi(x, y, z)\, dx\, dy\, dz$, where $\psi(x, y, z)$ is the solution of Equation 35-30. This solution is of the form

$$\psi(x, y, z) = A \sin^2(k_1 x) \sin^2(k_2 y) \sin^2(k_3 z)\, dx\, dy\, dz \qquad \text{35-31}$$

where the constant $A$ is determined by normalization. Inserting this solution into Equation 35-30, we obtain for the energy

$$E = \frac{\hbar^2}{2m}(k_1^2 + k_2^2 + k_3^2)$$

which is equivalent to $E = (p_x^2 + p_y^2 + p_z^2)/(2m)$, with $p_x = \hbar k_1$, and so on. The wave function will be zero at $x = L$ if $k_1 = n_1\pi/L$, where $n_1$ is an integer. Similarly, the wave function will be zero at $y = L$ if $k_2 = n_2\pi/L$, and the wave function will be zero at $z = L$ if $k_3 = n_3\pi/L$. (It is also zero at $x = 0$, $y = 0$, and $z = 0$.) The energy is thus quantized to the values

$$E_{n_1, n_2, n_3} = \frac{\hbar^2 \pi^2}{2mL^2}(n_1^2 + n_2^2 + n_3^2) = E_1\,(n_1^2 + n_2^2 + n_3^2) \qquad \text{35-32}$$

where $n_1$, $n_2$, and $n_3$ are integers and $E_1$ is the ground-state energy of the one-dimensional well. Note that the energy and wave function are characterized by three quantum numbers, each arising from the boundary conditions for one of the coordinates.

The lowest energy state (the ground state) for the cubical well occurs when $n_1 = n_2 = n_3 = 1$ and has the value

$$E_{1,1,1} = \frac{3\hbar^2 \pi^2}{2mL^2} = 3E_1$$

The first excited energy level can be obtained in three different ways: $n_1 = 2$, $n_2 = n_3 = 1$; $n_2 = 2$, $n_1 = n_3 = 1$; or $n_3 = 2$, $n_1 = n_2 = 1$. Each has a different wave function. For example, the wave function for $n_1 = 2$ and $n_2 = n_3 = 1$ is

$$\psi_{2,1,1} = A \sin\frac{2\pi x}{L} \sin\frac{\pi y}{L} \sin\frac{\pi z}{L} \qquad \text{35-33}$$

(where the value of the normalization constant $A$ is different than the value of the normalization constant in Equation 35-31). There are thus three different quantum states as described by the three different wave functions corresponding to the same energy level. An energy level with which more than one wave function is associated is said to be **degenerate.** In this case, there is threefold degeneracy.

---

[†] For example, if you throw two dice, the probability of the first die coming up 6 is $1/6$ and the probability of the second die coming up an odd number is $1/2$. The probability of the first die coming up 6 *and* the second die coming up an odd number is $(1/6)(1/2) = 1/12$.

Degeneracy is related to the spatial symmetry of the system. If, for example, we consider a noncubic well, where $U = 0$ for $0 < x < L_1$, $0 < y < L_2$, and $0 < z < L_3$, the boundary conditions at the edges would lead to the quantum conditions $k_1L_1 = n_1\pi$, $k_2L_2 = n_2\pi$, and $k_3L_3 = n_3\pi$, and the total energy would be

$$E_{n_1,n_2,n_3} = \frac{\hbar^2\pi^2}{2m}\left(\frac{n_1^2}{L_1^2} + \frac{n_2^2}{L_2^2} + \frac{n_3^2}{L_3^2}\right) \qquad 35\text{-}34$$

These energy levels are not degenerate if $L_1$, $L_2$, and $L_3$ are all different. Figure 35-18 shows the energy levels for the ground state and first two excited states for an infinite cubic well in which the excited states are degenerate and for a noncubic infinite well in which $L_1$, $L_2$, and $L_3$ are all slightly different so that the excited levels are slightly split apart and the degeneracy is removed. The ground state is the state where the quantum numbers $n_1$, $n_2$, and $n_3$ all equal 1. None of the three quantum numbers can be zero. If any one of $n_1$, $n_2$, and $n_3$ were zero, the corresponding wave number $k$ would also equal zero and the corresponding wave function (Equation 35-31) would equal zero for all values of $x$, $y$, and $z$.

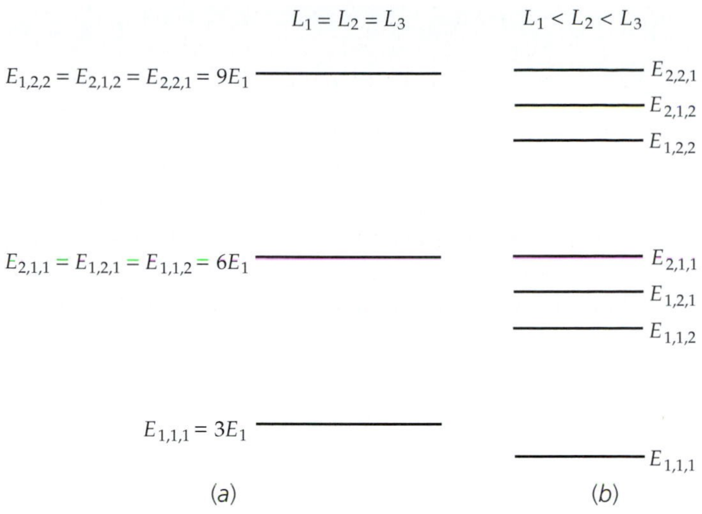

$L_1 = L_2 = L_3$ — — — $L_1 < L_2 < L_3$

$E_{1,2,2} = E_{2,1,2} = E_{2,2,1} = 9E_1$ ——— ——— $E_{2,2,1}$
——— $E_{2,1,2}$
——— $E_{1,2,2}$

$E_{2,1,1} = E_{1,2,1} = E_{1,1,2} = 6E_1$ ——— ——— $E_{2,1,1}$
——— $E_{1,2,1}$
——— $E_{1,1,2}$

$E_{1,1,1} = 3E_1$ ———

——— $E_{1,1,1}$

(a)    (b)

**FIGURE 35-18** Energy-level diagrams for (a) a cubic infinite well and (b) a noncubic infinite well. In Figure 35-18a the energy levels are degenerate; that is, there are two or more wave functions having the same energy. The degeneracy is removed when the symmetry of the potential is removed, as in Figure 35-18b.

*ENERGY LEVELS FOR A PARTICLE IN A THREE-DIMENSIONAL BOX*    **E X A M P L E    3 5 - 3**

A particle is in a three-dimensional box with $L_3 = L_2 = 2L_1$. Give the quantum numbers $n_1$, $n_2$, and $n_3$ that correspond to the thirteen quantum states of this box that have the lowest energies.

**PICTURE THE PROBLEM** We can use Equation 35-34 to write the energies in terms of the ratios $L_2/L_1 = 2$ and $L_3/L_1 = 2$, then find by inspection the values of the quantum numbers that give the lowest energies.

1. The energy of a level is given by Equation 35-34:

$$E_{n_1,n_2,n_3} = \frac{\hbar^2\pi^2}{2m}\left(\frac{n_1^2}{L_1^2} + \frac{n_2^2}{L_2^2} + \frac{n_3^2}{L_3^2}\right)$$

2. Factor out $1/L_1^2$:

$$E_{n_1,n_2,n_3} = \frac{\hbar^2\pi^2}{2mL_1^2}\left(n_1^2 + n_2^2\frac{L_1^2}{L_2^2} + n_3^2\frac{L_1^2}{L_3^2}\right)$$

$$= E_1(n_1^2 + n_2^2/4 + n_3^2/4)$$

3. The lowest energy is $E_{1,1,1}$:

$$E_{1,1,1} = E_1(1^2 + 1^2\tfrac{1}{4} + 1^2\tfrac{1}{4}) = 1.5E_1 \qquad \text{(1st)}$$

4. The energy increases the least when we increase $n_2$ or $n_3$. Try various values of the quantum numbers:

$$E_{1,2,1} = E_{1,1,2} = E_1(1^2 + 2^2\tfrac{1}{4} + 1\tfrac{1}{4}) = 2.25E_1 \qquad \text{(2nd and 3rd)}$$

$$E_{1,2,2} = E_1(1^2 + 2^2\tfrac{1}{4} + 2^2\tfrac{1}{4}) = 3.0E_1 \qquad \text{(4th)}$$

$$E_{1,3,1} = E_{1,1,3} = E_1(1^2 + 3^2\tfrac{1}{4} + 1^2\tfrac{1}{4}) = 3.50E_1 \qquad \text{(5th and 6th)}$$

$$E_{1,3,2} = E_{1,2,3} = E_1(1^2 + 3^2\tfrac{1}{4} + 2^2\tfrac{1}{4}) = 4.25E_1 \qquad \text{(7th and 8th)}$$

$$E_{2,1,1} = E_1(2^2 + 1^2\tfrac{1}{4} + 1^2\tfrac{1}{4}) = 4.5E_1 \qquad \text{(9th)}$$

$$E_{2,2,1} = E_{2,1,2} = E_1(2^2 + 2^2\tfrac{1}{4} + 1^2\tfrac{1}{4}) = 5.25E_1$$
$$E_{1,4,1} = E_{1,1,4} = E_1(1^2 + 4^2\tfrac{1}{4} + 1^2\tfrac{1}{4}) = 5.25E_1 \qquad \text{(10th, 11th, 12th, and 13th)}$$

**REMARKS** Note the degeneracy of the levels.

**EXERCISE** Find the quantum numbers and energies of the next two energy levels in step 4. (*Answer* $E_{1,3,3} = 5.5E_1$, $E_{1,4,2} = E_{1,2,4} = E_{2,2,2} = 6.0E_1$)

Write the degenerate wave functions for the fourth and fifth excited states (levels 5 and 6) of the results in step 4 of Example 35-3.

**PICTURE THE PROBLEM** Use Equation 35-33 with $k_i = n_i \pi / L$.

**Cover the column to the right and try these on your own before looking at the answers.**

| Steps | Answers |
|---|---|
| Write the wave functions corresponding to the energies $E_{1,3,1}$ and $E_{1,1,3}$. | $\psi_{1,3,1} = A \sin \dfrac{\pi x}{L} \sin \dfrac{3\pi y}{L} \sin \dfrac{\pi z}{L}$ |
| | $\psi_{1,1,3} = A \sin \dfrac{\pi x}{L} \sin \dfrac{\pi y}{L} \sin \dfrac{3\pi z}{L}$ |

## 35-6 The Schrödinger Equation for Two Identical Particles

Our discussion of quantum mechanics has thus far been limited to situations in which a single particle moves in some force field characterized by a potential energy function $U$. The most important physical problem of this type is the hydrogen atom, in which a single electron moves in the Coulomb potential of the proton nucleus. This problem is actually a two-body problem, since the proton also moves in the field of the electron. However, the motion of the much more massive proton requires only a very small correction to the energy of the atom that is easily made in both classical and quantum mechanics. When we consider more complicated problems, such as the helium atom, we must apply quantum mechanics to two or more electrons moving in an external field. Such problems are complicated by the interaction of the electrons with each other and also by the fact that the electrons are identical.

The interaction of two electrons with each other is electromagnetic and is essentially the same as the classical interaction of two charged particles. The Schrödinger equation for an atom with two or more electrons cannot be solved exactly, so approximation methods must be used. This is not very different from the situation in classical problems with three or more particles. However, the complications arising from the identity of electrons are purely quantum mechanical and have no classical counterpart. They are due to the fact that it is impossible to keep track of which electron is which. Classically, identical particles can be identified by their positions, which in principle can be determined with unlimited accuracy. This is impossible quantum mechanically because of the uncertainty principle. Figure 35-19 offers a schematic illustration of the problem.

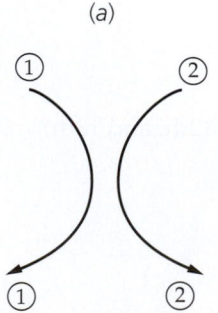
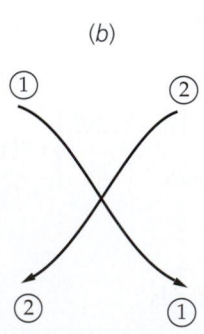
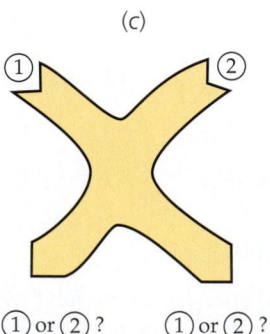

**FIGURE 35-19** Two possible classical electron paths (Figure 35-19*a* and Figure 35-19*b*). If electrons were classical particles, they could be distinguished by the paths followed. However, because of the quantum-mechanical wave properties of electrons, the paths are spread out, as indicated by the shaded region in Figure 35-19*c*. It is impossible to distinguish which electron is which after they separate.

The indistinguishability of identical particles has important consequences. For instance, consider the very simple case of two identical, noninteracting particles in a one-dimensional infinite square well. The time-independent Schrödinger equation for two particles, each of mass $m$, is

$$-\frac{\hbar^2}{2m}\frac{\partial^2\psi(x_1, x_2)}{\partial x_1^2} - \frac{\hbar^2}{2m}\frac{\partial^2\psi(x_1, x_2)}{\partial x_2^2} + U\psi(x_1, x_2) = E\psi(x_1, x_2) \qquad 35\text{-}35$$

where $x_1$ and $x_2$ are the coordinates of the two particles. If the particles interact, the potential energy $U$ contains terms with both $x_1$ and $x_2$ that cannot be separated into separate terms containing only $x_1$ or $x_2$. For example, the electrostatic repulsion of two electrons in one dimension is represented by the potential energy $ke^2/|x_2 - x_1|$. However, if the particles do not interact (as we are assuming here), we can write $U = U_1(x_1) + U_2(x_2)$. For the infinite square well, we need only solve the Schrödinger equation inside the well where $U = 0$, and require that the wave function be zero at the walls of the well. With $U = 0$, Equation 35-35 looks just like the expression for a particle in a two-dimensional well (Equation 35-30, with no $z$ and with $y$ replaced by $x_2$).

Solutions of this equation can be written in the form[†]

$$\psi_{n,m} = \psi_n(x_1)\psi_m(x_2) \qquad 35\text{-}36$$

where $\psi_n$ and $\psi_m$ are the single-particle wave functions for a particle in an infinite well and $n$ and $m$ are the quantum numbers of particles 1 and 2, respectively. For example, for $n = 1$ and $m = 2$, the wave function is

$$\psi_{1,2} = A \sin\frac{\pi x_1}{L}\sin\frac{2\pi x_2}{L} \qquad 35\text{-}37$$

The probability of finding particle 1 in $dx_1$ and particle 2 in $dx_2$ is $\psi_{n,m}^2(x_1, x_2)\,dx_1\,dx_2$, which is just the product of the separate probabilities $\psi_n^2(x_1)\,dx_1$ and $\psi_m^2(x_2)\,dx_2$. However, even though we have labeled the particles 1 and 2, we cannot distinguish which is in $dx_1$ and which is in $dx_2$ if they are identical. The mathematical descriptions of identical particles must be the same if we interchange the labels. The probability density $\psi^2(x_1, x_2)$ must therefore be the same as $\psi^2(x_2, x_1)$:

$$\psi^2(x_2, x_1) = \psi^2(x_1, x_2) \qquad 35\text{-}38$$

Equation 35-38 is satisfied if $\psi(x_2, x_1)$ is either **symmetric** or **antisymmetric** on the exchange of particles—that is, if either

$$\psi(x_2, x_1) = \psi(x_1, x_2), \quad \text{symmetric} \qquad 35\text{-}39$$

or

$$\psi(x_2, x_1) = -\psi(x_1, x_2), \quad \text{antisymmetric} \qquad 35\text{-}40$$

Note that the wave functions given by Equations 35-36 and 35-37 are neither symmetric nor antisymmetric. If we interchange $x_1$ and $x_2$ in these wave functions, we get a different wave function, which implies that the particles can be distinguished.

We can find symmetric and antisymmetric wave functions that are solutions of the Schrödinger equation by adding or subtracting $\psi_{n,m}$ and $\psi_{m,n}$. Adding them, we obtain

---

[†] Again, this result can be obtained by solving Equation 35-35, but it also can be understood in terms of our knowledge of probability. The probability of electron 1 being in region $dx_1$ and electron 2 being in region $dx_2$ is the product of the individual probabilities.

$$\psi_S = A'[\psi_n(x_1)\psi_m(x_2) + \psi_n(x_2)\psi_m(x_1)], \quad \text{symmetric} \qquad 35\text{-}41$$

and subtracting them, we obtain

$$\psi_A = A'[\psi_n(x_1)\psi_m(x_2) - \psi_n(x_2)\psi_m(x_1)], \quad \text{antisymmetric} \qquad 35\text{-}42$$

For example, the symmetric and antisymmetric wave functions for the first excited state of two identical particles in an infinite square well would be

$$\psi_S = A'\left(\sin\frac{\pi x_1}{L}\sin\frac{2\pi x_2}{L} + \sin\frac{\pi x_2}{L}\sin\frac{2\pi x_1}{L}\right) \qquad 35\text{-}43$$

and

$$\psi_A = A'\left(\sin\frac{\pi x_1}{L}\sin\frac{2\pi x_2}{L} - \sin\frac{\pi x_2}{L}\sin\frac{2\pi x_1}{L}\right) \qquad 35\text{-}44$$

There is an important difference between antisymmetric and symmetric wave functions. If $n = m$, the antisymmetric wave function is identically zero for all values of $x_1$ and $x_2$, whereas the symmetric wave function is not. Thus, if the wave function describing two identical particles is antisymmetric, the quantum numbers $n$ and $m$ of two particles cannot be the same. This is an example of the **Pauli exclusion principle,** which was first stated by Wolfgang Pauli for electrons in an atom:

No two electrons in an atom can have the same quantum numbers.

PAULI EXCLUSION PRINCIPLE

It is found that electrons, protons, neutrons, and some other particles have antisymmetric wave functions and obey the Pauli exclusion principle. These particles are called **fermions.** Other particles (e.g., $\alpha$ particles, deuterons, photons, and mesons) have symmetric wave functions and do not obey the Pauli exclusion principle. These particles are called **bosons.**

# SUMMARY

1. The Schrödinger equation is a differential equation that relates the second spatial derivative of a wave function to its first time derivative. Wave functions that describe physical situations are solutions of this differential equation.

2. Because a wave function must be normalizable, it must be well behaved; that is, it must approach zero as $x$ approaches infinity. For bound systems such as a particle in a box, a simple harmonic oscillator, or an electron in an atom, this requirement leads to energy quantization.

3. The well-behaved wave functions for bound systems describe standing waves.

| Topic | Relevant Equations and Remarks | |
|---|---|---|
| 1. **Time-Independent Schrödinger Equation** | $-\dfrac{\hbar^2}{2m}\dfrac{d^2\psi(x)}{dx^2} + U(x)\psi(x) = E\psi(x)$ | 35-4 |

| | |
|---|---|
| Allowable solutions | In addition to satisfying the Schrödinger equation, a wave function $\psi(x)$ must be continuous and (if $U$ is not infinite) must have a continuous first derivative $d\psi/dx$. Because the probability of finding an electron somewhere must be 1, the wave function must obey the normalization condition $$\int_{-\infty}^{\infty} |\psi|^2 \, dx = 1$$ This condition implies the boundary condition that $\psi$ must approach 0 as $x$ approaches $\pm\infty$. Such boundary conditions lead to the quantization of energy. |

**2. Confined Particles**

When the total energy $E$ is greater than the potential energy $U(x)$ in some region (the classically allowed region) and less than $U(x)$ outside that region, the wave function oscillates within the classically allowed region and increases or decreases exponentially outside that region. The wave function approaches zero as $x$ approaches $\infty$ only for certain values of the total energy $E$. The energy is thus quantized.

| | |
|---|---|
| In a finite square well | In a finite well of height $U_0$, there are only a finite number of allowed energies, and these are slightly less than the corresponding energies in an infinite well. |
| In the simple harmonic oscillator | In the oscillator potential energy function $U(x) = \frac{1}{2}m\omega_0^2 x^2$, the allowed energies are equally spaced and given by |

$$E_n = (n + \tfrac{1}{2})\hbar\omega_0 \qquad\qquad \textbf{35-26}$$

The ground-state wave function is given by

$$\psi_0(x) = A_0 e^{-ax^2} \qquad\qquad \textbf{35-23}$$

where $A_0$ is the normalization constant and $a = m\omega_0/(2\hbar)$.

**3. Reflection and Barrier Penetration**

When the potential changes abruptly over a small distance, a particle may be reflected even though $E > U(x)$. A particle may penetrate a region in which $E < U(x)$. Reflection and penetration of electron waves are similar to those for other kinds of waves.

**4. The Schrödinger Equation in Three Dimensions**

The wave function for a particle in a three-dimensional box can be written

$$\psi(x, y, z) = \psi_1(x)\psi_2(y)\psi_3(z)$$

where $\psi_1$, $\psi_2$, and $\psi_3$ are wave functions for a one-dimensional box.

| | |
|---|---|
| Degeneracy | When more than one wave function is associated with the same energy level, the energy level is said to be degenerate. Degeneracy arises because of spatial symmetry. |

**5. The Schrödinger Equation for Two Identical Particles**

A wave function that describes two identical particles must be either symmetric or antisymmetric when the coordinates of the particles are exchanged. Fermions (which include electrons, protons, and neutrons) are described by antisymmetric wave functions and obey the Pauli exclusion principle, which states that no two particles can have the same values for their quantum number. Bosons (which include $\alpha$ particles, deuterons, photons, and mesons) have symmetric wave functions and do not obey the Pauli exclusion principle.

# PROBLEMS

- Single-concept, single-step, relatively easy
- Intermediate-level, may require synthesis of concepts
- Challenging
- **SSM** Solution is in the *Student Solutions Manual*
- **iSOLVE** Problems available on iSOLVE online homework service
- **iSOLVE✓** These "Checkpoint" online homework service problems ask students additional questions about their confidence level, and how they arrived at their answer.

In a few problems, you are given more data than you actually need; in a few other problems, you are required to supply data from your general knowledge, outside sources, or informed estimates.

## Conceptual Problems

**1** • True or false: Boundary conditions on the wave function lead to energy quantization.

**2** • Sketch (a) the wave function and (b) the probability distribution for the $n = 4$ state for the finite square-well potential.

**3** • Sketch (a) the wave function and (b) the probability distribution for the $n = 5$ state for the finite square-well potential.

## Estimation and Approximation

**4** • **SSM** The Schrödinger equation could be applied equally well to baseballs as to electrons; yet, we would never analyze the motion of a baseball with a wave function. Explain why this is the case by estimating the quantum mechanically predicted lowest energy level of a baseball trapped inside a locker. You can treat the locker as if it were a one-dimensional infinite potential well. What value of the quantum number $n$ would you need for a ball rolling around in a locker, after you toss the ball in, so that the kinetic energy is approximately equal to the quantum mechanically calculated energy?

## The Schrödinger Equation

**5** •• Show that if $\psi_1(x)$ and $\psi_2(x)$ are each solutions to the time-independent Schrödinger equation (Equation 35-4), then $\psi_3(x) = \psi_1(x) + \psi_2(x)$ is also a solution. This is known as the superposition principle and it applies to the solutions of all linear differential equations.

## The Harmonic Oscillator

**6** •• The harmonic oscillator problem may be used to describe molecules. For example, the hydrogen molecule $H_2$ is found to have equally spaced energy levels separated by $8.7 \times 10^{-20}$ J. What value of spring constant would be needed to get this energy spacing, assuming that the molecule can be modeled as a single hydrogen atom attached to a spring?

**7** •• Show that the expectation value $<x> = \int x|\psi|^2\,dx$ is zero for both the ground state and the first excited state of the harmonic oscillator.

**8** •• **SSM** Use the procedure of Example 35-1 to verify that the energy of the first excited state of the harmonic oscillator is $E_1 = \frac{3}{2}\hbar\omega_0$. (Note: *Rather than solve for a again, use the result $a = m\omega_0/(2\hbar)$ obtained in Example 35-1.*)

**9** •• Show that the normalization constant $A_0$ of Equation 35-23 is $A_0 = (2m\omega_0/h)^{1/4}$.

**10** •• Show that for the ground state of the harmonic oscillator $<x^2> = \int x^2|\psi|^2\,dx = \hbar/(2m\omega_0) = 1/(4a)$. Use this result to show that the average potential energy equals half the total energy.

**11** •• The quantity $\sqrt{<x^2> - <x>^2}$ is a measure of the average spread in the location of a particle. (a) Consider an electron trapped in a harmonic oscillator potential. Its lowest energy level is found to be $2.1 \times 10^{-4}$ eV. Calculate $\sqrt{<x^2> - <x>^2}$ for this electron. (See Problems 7 and 10.) (b) Now consider an electron trapped in an infinite square-well potential. If the width of the well is equal to $\sqrt{<x^2> - <x>^2}$, what would be the lowest energy level for this electron?

**12** ••• Classically, the average kinetic energy of the harmonic oscillator equals the average potential energy. We may assume that this is also true for the quantum mechanical harmonic oscillator. Use this condition to determine the expectation value of $p^2$ for the ground state of the harmonic oscillator.

**13** ••• We know that for the classical harmonic oscillator, $p_{av} = 0$. It can be shown that for the quantum mechanical harmonic oscillator, $<p> = 0$. Use the results of Problem 7, Problem 10, and Problem 12 to determine the uncertainty product $\Delta x\,\Delta p$ for the ground state of the harmonic oscillator.

## Reflection and Transmission of Electron Waves: Barrier Penetration

**14** •• **SSM** **iSOLVE✓** A particle of mass $m$ with wave number $k_1$ is traveling to the right along the negative $x$ axis. The potential energy of the particle is equal to zero everywhere on the negative $x$ axis and is equal to $U_0$ everywhere on the positive $x$ axis, $U_0 > 0$. (a) Show that if the total energy is $E = \alpha U_0$, where $\alpha \geq 1$, wave number $k_2$ in the region $x > 0$ is given by

$$k_2 = \sqrt{\frac{\alpha - 1}{\alpha}}\,k_1$$

(b) Using a spreadsheet program or graphing calculator, graph the reflection coefficient $R$ and the transmission coefficient $T$ for $1 \leq \alpha \leq 5$.

**15** •• [SOLVE]✓ Suppose that the potential in Problem 14 drops from zero to $-U_0$ at $x = 0$ so that the particle speeds up instead of slowing down. The wave number for the incident particle is again $k_1$, and the total energy is $2U_0$. (a) What is the wave number for the particle in the region of positive $x$? (b) Calculate the reflection coefficient $R$ at $x = 0$. (c) What is the transmission coefficient $T$? (d) If one million particles with wave number $k_1$ are incident upon the potential drop, how many particles are expected to continue along in the positive $x$ direction? How does this compare with the classical prediction?

**16** •• [SOLVE]✓ A particle of energy $E$ approaches a step barrier of height $U_0$. What should be the ratio $E/U_0$ so that the reflection coefficient is $1/2$?

**17** •• [SOLVE] Use Equation 35-29 to calculate the order of magnitude of the probability that a proton will tunnel out of a nucleus in one collision with the nuclear barrier if it has energy 6 MeV below the top of the potential barrier and the barrier thickness is $10^{-15}$ m.

**18** •• [SSM] A 10-eV electron is incident on a potential barrier of height 25 eV and width of 1 nm. (a) Use Equation 35-29 to calculate the order of magnitude of the probability that the electron will tunnel through the barrier. (b) Repeat your calculation for a width of 0.1 nm.

**19** ••• To understand how a small change in $\alpha$-particle energy can dramatically change the tunneling probability from a nucleus, consider an $\alpha$ particle emitted by a uranium nucleus ($Z = 92$). (a) Referring to Figure 35-17, calculate the distance of closest approach $r_1$ that $\alpha$ particles with kinetic energies of 4 MeV and 7 MeV could make to the uranium nucleus. (b) Use the result from Part (a) to calculate the relative transmission coefficient $e^{-2\alpha a}$ for the same $\alpha$ particles. (Note: *The actual half-lives of uranium nuclei vary over nine orders of magnitude. Your calculation will show a smaller range than this; however, to find half-life, you must also include the frequency with which the $\alpha$ particle strikes the barrier.*)

### The Schrödinger Equation in Three Dimensions

**20** • A particle is confined to a three-dimensional box that has sides $L_1$, $L_2 = 2L_1$, and $L_3 = 3L_1$. Give the quantum numbers $n_1$, $n_2$, $n_3$ that correspond to the lowest ten quantum states of this box.

**21** • Give the wave functions for the lowest ten quantum states of the particle in Problem 20.

**22** • (a) Repeat Problem 20 for the case $L_2 = 2L_1$ and $L_3 = 4L_1$. (b) What quantum numbers correspond to degenerate energy levels?

**23** • [SSM] Give the wave functions for the lowest ten quantum states of the particle in Problem 22.

**24** • A particle moves in a potential well given by $U(x, y, z) = 0$ for $-L/2 < x < L/2, 0 < y < L$, and $0 < z < L$; and $U = \infty$ outside these ranges. (a) Write an expression for the ground-state wave function for this particle. (b) How do the allowed energies compare with those for a well having $U = 0$ for $0 < x < L$, rather than for $-L/2 < x < L/2$?

**25** •• A particle moves freely in the two-dimensional region defined by $0 \le x \le L$ and $0 \le y \le L$. (a) Find the wave functions satisfying the Schrödinger equation. (b) Find the

corresponding energies. (c) Find the lowest two states that are degenerate. Give the quantum numbers for this case. (d) Find the lowest three states that have the same energy. Give the quantum numbers for the three states having the same energy.

### The Schrödinger Equation for Two Identical Particles

**26** • Show that Equation 35-37 satisfies Equation 35-35 with $U = 0$, and find the energy of this state.

**27** • [SOLVE] What is the ground-state energy of ten noninteracting bosons in a one-dimensional box of length $L$?

**28** • [SSM] [SOLVE] What is the ground-state energy of ten noninteracting fermions, such as neutrons, in a one-dimensional box of length $L$? (Because the quantum number associated with spin can have two values, each spatial state can hold two neutrons.)

### Orthogonality of Wave Functions

*The integral of two functions over some space interval is somewhat analogous to the dot product of two vectors. If this integral is zero, the functions are said to be orthogonal, which is analogous to two vectors being perpendicular. The following problems illustrate the general principle that any two wave functions corresponding to different energy levels in the same potential are orthogonal. A general hint for all these problems is that the integral of an antisymmetric integrand over symmetric limits is equal to zero.*

**29** •• Show that the ground-state wave function and the wave function of the first excited state of the harmonic oscillator are orthogonal; that is, show that $\int \psi_0(x) \psi_1(x) dx = 0$.

**30** •• The wave function for the state $n = 2$ of the harmonic oscillator is $\psi_2(x) = A_2(2ax^2 - \frac{1}{2})e^{-ax^2}$, where $A_2$ is the normalization constant for this wave function. Show that the wave functions for the states $n = 1$ and $n = 2$ of the harmonic oscillator are orthogonal.

**31** •• For the wave functions $\psi_n(x) = \sqrt{2/L} \sin(n\pi x/L)$ corresponding to a particle in an infinite square-well potential from 0 to $L$, show that $\int \psi_n(x) \psi_m(x) dx = 0$; that is, show that $\psi_n$ and $\psi_m$ are orthogonal.

### General Problems

**32** •• Consider a particle in a one-dimensional box of length $L$ that is centered at the origin (see Problem 65 in Chapter 34). (a) What are the values of $\psi_1(0)$ and $\psi_2(0)$? (b) What are the values of $<x>$ for the states $n = 1$ and $n = 2$? (c) Evaluate $<x^2>$ for the states $n = 1$ and $n = 2$.

**33** •• [SSM] Eight identical noninteracting fermions (e.g., neutrons) are confined to a two-dimensional square box of side length $L$. Determine the energies of the three lowest states. (See Problem 26.)

**34** •• [SOLVE] A particle is confined to a two-dimensional box defined by the following boundary conditions: $U(x, y) = 0$, for $-L/2 \le x \le L/2$ and $-3L/2 \le y \le 3L/2$; and $U(x, y) = \infty$ elsewhere. (a) Determine the energies of the lowest three bound states. Are any of these states degenerate? (b) Identify the quantum numbers of the lowest doubly degenerate bound state and determine its energy.

**35** ••• The classical probability distribution function for a particle in a one-dimensional box of length $L$ is $P = 1/L$. (See Example 34-5.) (a) Show that the classical expectation value of $x^2$ for a particle in a one-dimensional box of length $L$ centered at the origin (Problem 32) is $L^2/12$. (b) Find the quantum expectation value of $x^2$ for the $n$th state of a particle in the one-dimensional box of Problem 32, and show that it approaches the classical limit $L^2/12$ for $n \gg 1$.

**36** •• Show that Equation 35-27 and Equation 35-28 imply that the transmission coefficient for particles of energy $E$ incident on a step barrier $U_0 < E$ is given by

$$T = \frac{4k_1 k_2}{(k_1 + k_2)^2} = \frac{4r}{(1 + r)^2}$$

where $r = k_2/k_1$.

**37** •• (a) Show that for the case of a particle of energy $E$ incident on a step barrier $U_0 < E$, the wave numbers $k_1$ and $k_2$ are related by

$$\frac{k_2}{k_1} = r = \sqrt{1 - \frac{U_0}{E}}$$

(b) Use this result to show that $R = (1 - r)^2/(1 + r)^2$

**38** •• (a) Using a spreadsheet program or graphing calculator and the results of Problem 36 and Problem 37, graph the transmission coefficient $T$ and reflection coefficient $R$ as a function of incident energy $E$ for values of $E$ ranging from $E = U_0$ to $E = 10.0U_0$. (b) What limiting values do your graphs indicate?

**39** ••• Determine the normalization constant $A_2$ in Problem 30.

**40** ••• Consider the time-independent, one-dimensional Schrödinger equation when the potential function is symmetric about the origin, that is, when $U(x)$ is even.[†] (a) Show that if $\psi(x)$ is a solution of the Schrödinger equation with energy $E$, then $\psi(-x)$ is also a solution with the same energy $E$, and that, therefore, $\psi(x)$ and $\psi(-x)$ can differ by only a multiplicative constant. (b) Write $\psi(x) = C\psi(-x)$, and show that $C = \pm 1$. Note that $C = +1$ means that $\psi(x)$ is an even function of $x$, and $C = -1$ means that $\psi(x)$ is an odd function of $x$.

**41** ••• SSM In this problem you will derive the ground-state energy of the harmonic oscillator using the precise form of the uncertainty principle, $\Delta x\, \Delta p \geq \hbar/2$, where $\Delta x$ and $\Delta p$ are defined to be the standard deviations $(\Delta x)^2 = [(x - x_{av})^2]_{av}$ and $(\Delta p)^2 = [(p - p_{av})^2]_{av}$ (see Equation 17-35a). Proceed as follows:

1. Write the total classical energy in terms of the position $x$ and momentum $p$ using $U(x) = m\omega_0^2 x^2$ and $K = p^2/2m$.

2. Use the result of Equation 17-35 to write $(\Delta x)^2 = [(x - x_{av})^2]_{av} = (x^2)_{av} - x_{av}^2$ and $(\Delta p)^2 = [(p - p_{av})^2]_{av} = (p^2)_{av} - p_{av}^2$.

3. Use the symmetry of the potential energy function to argue that $x_{av}$ and $p_{av}$ must be zero, so that $(\Delta x)^2 = (x^2)_{av}$ and $(\Delta p)^2 = (p^2)_{av}$.

4. Assume that $\Delta p = \hbar/(2\Delta x)$ to eliminate $(p^2)_{av}$ from the average energy $E_{av} = (p^2)_{av}/(2m) + \frac{1}{2}m\omega^2(x^2)_{av}$ and write $E_{av}$ as $E_{av} = \hbar^2/(8mZ) + \frac{1}{2}m\omega^2 Z$, where $Z = (x^2)_{av}$.

5. Set $dE/dZ = 0$ to find the value of $Z$ for which $E$ is a minimum.

6. Show that the minimum energy is given by $(E_{av})_{min} = +\frac{1}{2}\hbar\omega_0$.

**42** ••• A particle of mass $m$ near the earth's surface at $z = 0$ can be described by the potential energy

$$U = mgz, \quad z > 0$$
$$U = \infty, \quad z < 0$$

For some positive value of total energy $E$, indicate the classically allowed region on a sketch of $U(z)$ versus $z$. Sketch also the classical kinetic energy versus $z$. The Schrödinger equation for this problem is quite difficult to solve. Using arguments similar to those in Section 35-2 about the curvature of the wave function as given by the Schrödinger equation, sketch your guesses for the shape of the wave function for the ground state and for the first two excited states.

---

[†] A function $f(x)$ is even if $f(x) = f(-x)$ for all $x$.

# Atoms

AT A DISTANCE OF 6,000 LIGHT YEARS FROM EARTH, THE STAR CLUSTER RCW 38 IS A RELATIVELY CLOSE STAR-FORMING REGION. THIS IMAGE COVERS AN AREA ABOUT 5 LIGHT YEARS ACROSS AND CONTAINS THOUSANDS OF HOT, VERY YOUNG STARS FORMED LESS THAN A MILLION YEARS AGO. X RAYS FROM THE HOT UPPER ATMOSPHERES OF 190 OF THESE STARS WERE DETECTED BY CHANDRA, AN X-RAY OBSERVATORY ORBITING EARTH. THE MECHANISMS GENERATING THESE X RAYS IS NOT KNOWN.

ON EARTH, X-RAY MACHINES PRODUCE X RAYS BY BOMBARDING A TARGET WITH HIGH-ENERGY ELECTRONS. THE ATOMIC NUMBER OF THE ATOMS THAT MAKE UP THE TARGET CAN BE DETERMINED BY ANALYZING THE RESULTING X-RAY SPECTRA.

**?** **How is the atomic number obtained from the spectral analysis? Example 36-8 shows one way to accomplish this task.**

36-1    The Nuclear Atom

36-2    The Bohr Model of the Hydrogen Atom

36-3    Quantum Theory of Atoms

36-4    Quantum Theory of the Hydrogen Atom

36-5    The Spin–Orbit Effect and Fine Structure

36-6    The Periodic Table

36-7    Optical Spectra and X-Ray Spectra

There are 113 chemical elements that have been discovered, and there are a couple of additional chemical elements that recently have been reported. Each element is characterized by an atom that contains a number of protons $Z$, an equal number of electrons, and a number of neutrons $N$. The number of protons $Z$ is called the **atomic number.** The lightest atom, hydrogen (H), has $Z = 1$; the next lightest atom, helium (He), has $Z = 2$; the third lightest, lithium (Li), has $Z = 3$; and so forth. Nearly all the mass of the atom is concentrated in a tiny nucleus, which contains the protons and neutrons. Typically, the nuclear radius is approximately 1 fm to 10 fm (1 fm = $10^{-15}$ m). The distance between the nucleus and the electrons is approximately 0.1 nm = 100,000 fm. This distance determines the *size* of the atom.

The chemical properties and physical properties of an element are determined by the number and arrangement of the electrons in the atom. Because each proton has a positive charge $+e$, the nucleus has a total positive charge $+Ze$. The electrons are negatively charged $(-e)$, so they are attracted to the nucleus and repelled by each other. Since electrons and protons have equal but opposite charges, and there are an equal number of electrons and protons in an atom, atoms are electrically neutral. Atoms that lose or gain one or more electrons are then electrically charged and are called *ions*.

➤ We will begin our study of atoms by discussing the Bohr model, a semi-classical model developed by Niels Bohr in 1913 to explain the spectra emitted by hydrogen atoms. Although this *prequantum mechanics* model has many shortcomings, it provides a useful framework for the discussion of atomic phenomena. For example, we now know that the electron does not circle the nucleus in well-defined orbits, as in the Bohr model, but instead is described by a wave function that satisfies the Schrödinger equation. However, the probability distributions that follow from the full quantum theory do in fact have maxima at the positions of the Bohr orbits. After discussing the Bohr model, we will apply our knowledge of quantum mechanics from Chapter 35 to give a qualitative description of the hydrogen atom. We will then discuss the structure of other atoms and the periodic table of the elements. Finally, we will discuss both optical and X-ray spectra.

## 36-1 The Nuclear Atom

### Atomic Spectra

By the beginning of the twentieth century, a large body of data had been collected on the emission of light by atoms in a gas when the atoms are excited by an electric discharge. Viewed through a spectroscope with a narrow-slit aperture, this light appears as a discrete set of lines of different colors or wavelengths; the spacing and intensities of the lines are characteristic of the element. The wavelengths of these spectral lines could be accurately determined, and much effort went into finding regularities in the spectra. Figure 36-1 shows line spectra for hydrogen and for mercury.

In 1885 a Swiss schoolteacher, Johann Balmer, found that the wavelengths of the lines in the visible spectrum of hydrogen can be represented by the formula

$$\lambda = (364.6 \text{ nm}) \frac{m^2}{m^2 - 4}, \quad m = 3, 4, 5, \ldots \qquad 36\text{-}1$$

Balmer suggested that this might be a special case of a more general expression that would be applicable to the spectra of other elements. Such an expression, found by Johannes R. Rydberg and Walter Ritz and known as the **Rydberg–Ritz formula,** gives the reciprocal wavelength as

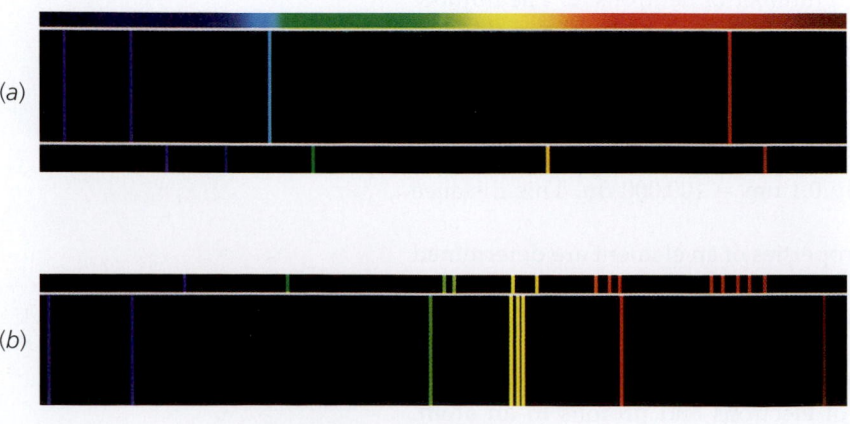

(a)

(b)

**FIGURE 36-1** (*a*) Line spectrum of hydrogen and (*b*) line spectrum of mercury.

$$\frac{1}{\lambda} = R\left(\frac{1}{n_2^2} - \frac{1}{n_1^2}\right)$$    36-2

where $n_1$ and $n_2$ are integers, with $n_1 > n_2$, and $R$ is the Rydberg constant. The Rydberg constant is the same for all spectral series of the same element and varies only slightly in a regular way from element to element. For hydrogen, $R$ has the value

$$R_H = 1.097776 \times 10^7 \, \text{m}^{-1}$$

The Rydberg–Ritz formula gives the wavelengths for all the lines in the spectra of hydrogen, as well as alkali elements such as lithium and sodium. The hydrogen Balmer series given by Equation 36-1 is also given by Equation 36-2, with $R = R_H$, $n_2 = 2$, and $n_1 = m$.

Many attempts were made to construct a model of the atom that would yield these formulas for its radiation spectrum. The most popular model, due to J. J. Thomson, considered various arrangements of electrons embedded in some kind of fluid that contained most of the mass of the atom and had enough positive charge to make the atom electrically neutral. Thomson's model, called the "plum pudding" model, is illustrated in Figure 36-2. Since classical electromagnetic theory predicted that a charge oscillating with frequency $f$ would radiate electromagnetic energy of that frequency, Thomson searched for configurations that were stable and that had normal modes of vibration of frequencies equal to those of the spectrum of the atom. A difficulty of this model and all other models was that, according to classical physics, electric forces alone cannot produce stable equilibrium. Thomson was unsuccessful in finding a model that predicted the observed frequencies for any atom.

The Thomson model was essentially ruled out by a set of experiments by H. W. Geiger and E. Marsden, under the supervision of E. Rutherford in approximately 1911, in which alpha particles from radioactive radium were scattered by atoms in a gold foil. Rutherford showed that the number of alpha particles scattered at large angles could not be accounted for by an atom in which the positive charge was distributed throughout the atomic size (known to be about 0.1 nm in diameter) but required that the positive charge and most of the mass of the atom be concentrated in a very small region, now called the nucleus, of diameter of the order of $10^{-6}$ nm = 1 fm.

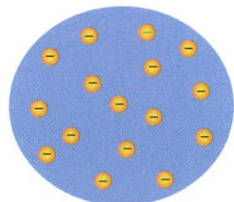

**FIGURE 36-2** J. J. Thomson's plum pudding model of the atom. In this model, the negative electrons are embedded in a fluid of positive charge. For a given configuration in such a system, the resonance frequencies of oscillations of the electrons can be calculated. According to classical theory, the atom should radiate light of frequency equal to the frequency of oscillation of the electrons. Thomson could not find any configuration that would give frequencies in agreement with the measured frequencies of the spectrum of any atom.

## 36-2  The Bohr Model of the Hydrogen Atom

Niels Bohr, working in the Rutherford laboratory in 1912, proposed a model of the hydrogen atom that extended the work of Planck, Einstein, and Rutherford and successfully predicted the observed spectra. According to Bohr's model, the electron of the hydrogen atom moves under the influence of the Coulomb attraction to the positive nucleus according to classical mechanics, which predicts circular or elliptical orbits with the force center at one focus, as in the motion of the planets around the sun. For simplicity, Bohr chose a circular orbit, as shown in Figure 36-3.

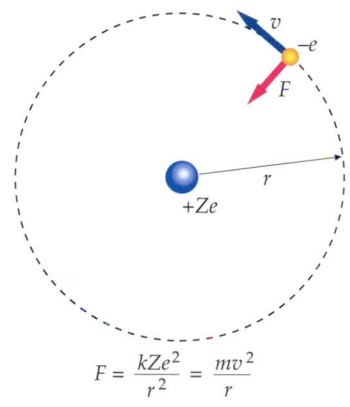

$$F = \frac{kZe^2}{r^2} = \frac{mv^2}{r}$$

**FIGURE 36-3** Electron of charge $-e$ traveling in a circular orbit of radius $r$ around the nuclear charge $+Ze$. The attractive electrical force $kZe^2/r^2$ keeps the electron in its orbit.

### Energy for a Circular Orbit

Consider an electron of charge $-e$ moving in a circular orbit of radius $r$ about a positive charge $Ze$ such as the nucleus of a hydrogen atom ($Z = 1$) or of a singly ionized helium atom ($Z = 2$). The total energy of the electron can be related to the radius of the orbit. The potential energy of the electron of charge $-e$ at a distance $r$ from a positive charge $Ze$ is

$$U = \frac{kq_1q_2}{r} = \frac{k(Ze)(-e)}{r} = -\frac{kZe^2}{r}$$  36-3

where $k$ is the Coulomb constant. The kinetic energy $K$ can be obtained as a function of $r$ by using Newton's second law, $F_{net} = ma$. Setting the Coulomb attractive force equal to the mass times the centripetal acceleration gives

$$\frac{kZe^2}{r^2} = m\frac{v^2}{r}$$  36-4a

Multiplying both sides by $r/2$ gives

$$K = \frac{1}{2}mv^2 = \frac{1}{2}\frac{kZe^2}{r}$$  36-4b

Thus, the kinetic energy and the potential energy vary inversely with $r$. Note that the magnitude of the potential energy is twice that of the kinetic energy:

$$U = -2K$$  36-5

This is a general result in $1/r^2$ force fields. It also holds for circular orbits in a gravitational field (see Example 11-6 in Section 11-3). The total energy is the sum of the kinetic energy and the potential energy:

$$E = K + U = \frac{1}{2}\frac{kZe^2}{r} - \frac{kZe^2}{r}$$

or

$$E = -\frac{1}{2}\frac{kZe^2}{r}$$  36-6

ENERGY IN A CIRCULAR ORBIT FOR A $1/r^2$ FORCE

Although mechanical stability is achieved because the Coulomb attractive force provides the centripetal force necessary for the electron to remain in orbit, classical *electromagnetic* theory says that such an atom would be unstable electrically. The atom would be unstable because the electron must accelerate when moving in a circle and therefore radiate electromagnetic energy of frequency equal to that of its motion. According to the classical theory, such an atom would quickly collapse, with the electron spiraling into the nucleus as it radiates away its energy.

## Bohr's Postulates

Bohr circumvented the difficulty of the collapsing atom by *postulating* that only certain orbits, called stationary states, are allowed, and that in these orbits the electron does not radiate. An atom radiates only when the electron makes a transition from one allowed orbit (stationary state) to another.

The electron in the hydrogen atom can move only in certain nonradiating, circular orbits called stationary states.

The second postulate relates the frequency of radiation to the energies of the stationary states. If $E_i$ and $E_f$ are the initial and final energies of the atom, the frequency of the emitted radiation during a transition is given by

$$f = \frac{E_i - E_f}{h} \qquad\qquad 36\text{-}7$$

where $h$ is Planck's constant. This postulate is equivalent to the assumption of conservation of energy with the emission of a photon of energy $hf$. Combining Equation 36-6 and Equation 36-7, we obtain for the frequency

$$f = \frac{E_1 - E_2}{h} = \frac{1}{2}\frac{kZe^2}{h}\left(\frac{1}{r_2} - \frac{1}{r_1}\right) \qquad\qquad 36\text{-}8$$

where $r_1$ and $r_2$ are the radii of the initial and final orbits.

To obtain the frequencies implied by the Rydberg–Ritz formula, $f = c/\lambda = cR(1/n_2^2 - 1/n_1^2)$, it is evident that the radii of stable orbits must be proportional to the squares of integers. Bohr searched for a quantum condition for the radii of the stable orbits that would yield this result. After much trial and error, Bohr found that he could obtain it if he postulated that the angular momentum of the electron in a stable orbit equals an integer times $\hbar$ ("bar," Planck's constant divided by $2\pi$). Since the angular momentum of a circular orbit is just $mvr$, this postulate is

$$mvr = \frac{nh}{2\pi} = n\hbar, \quad n = 1, 2, 3, \ldots \qquad\qquad 36\text{-}9$$

where $\hbar = h/2\pi = 1.055 \times 10^{-34}\,\text{J·s} = 6.582 \times 10^{-16}\,\text{eV·s}$.

Equation 36-9 relates the speed $v$ to the radius $r$. Equation 36-4$a$, from Newton's second law, gives us another equation relating the speed to the radius:

$$\frac{kZe^2}{r^2} = m\frac{v^2}{r}$$

or

$$v^2 = \frac{kZe^2}{mr} \qquad\qquad 36\text{-}10$$

We can determine $r$ by eliminating $v$ between Equations 36-9 and 36-10. Solving Equation 36-9 for $v$ and squaring gives

$$v^2 = n^2\frac{\hbar^2}{m^2 r^2}$$

Equating this expression for $v^2$ with the expression given by Equation 36-10, we get

$$n^2 \frac{\hbar^2}{m^2 r^2} = \frac{kZe^2}{mr}$$

Solving for $r$, we obtain

$$r = n^2 \frac{\hbar^2}{mkZe^2} = n^2 \frac{a_0}{Z} \qquad \text{36-11}$$

RADIUS OF THE BOHR ORBITS

where $a_0$ is called the **first Bohr radius.**

$$a_0 = \frac{\hbar^2}{mke^2} \approx 0.0529 \text{ nm} \qquad \text{36-12}$$

FIRST BOHR RADIUS

Substituting the expressions for $r$ in Equation 36-11 into Equation 36-8 for the frequency gives

$$f = \frac{1}{2} \frac{kZe^2}{h} \left( \frac{1}{r_2} - \frac{1}{r_1} \right) = Z^2 \frac{mk^2e^4}{4\pi\hbar^3} \left( \frac{1}{n_2^2} - \frac{1}{n_1^2} \right) \qquad \text{36-13}$$

If we compare this expression with $Z = 1$ for $f = c/\lambda$ with the empirical Rydberg–Ritz formula (Equation 36-2), we obtain for the Rydberg constant

$$R = \frac{mk^2e^4}{4\pi c\hbar^3} \qquad \text{36-14}$$

Using the values of $m$, $e$, and $\hbar$ known in 1913, Bohr calculated $R$ and found his result to agree (within the limits of the uncertainties of the constants) with the value obtained from spectroscopy.

---

*STANDING-WAVE CONDITION IMPLIES*
*QUANTIZATION OF ANGULAR MOMENTUM*
**E X A M P L E   3 6 - 1**

**For waves in a circle, the standing-wave condition is that there is an integral number of wavelengths in the circumference. That is, $n\lambda = 2\pi r$, where $n = 1, 2, 3,$ and so on. Show that this condition for electron waves implies quantization of angular momentum.**

1. Write the standing-wave condition:
$$n\lambda = 2\pi r$$

2. Use the de Broglie relation (Equation 34-10) to relate the momentum $p$ to $\lambda$:
$$p = \frac{h}{\lambda} = \frac{nh}{2\pi r} = n\frac{\hbar}{r}$$

3. The angular momentum of an electron in a circular orbit is $mvr = pr$, where $p = mv$:
$$L = mvr = pr = \boxed{n\hbar}$$

## Energy Levels

The total mechanical energy of the electron in the hydrogen atom is related to the radius of the circular orbit by Equation 36-6. If we substitute the quantized values of $r$ as given by Equation 36-11, we obtain

$$E_n = -\frac{1}{2}\frac{kZe^2}{r} = -\frac{1}{2}\frac{kZ^2e^2}{n^2a_0} = -\frac{1}{2}\frac{mk^2Z^2e^4}{n^2\hbar^2}$$

or

$$E_n = -Z^2\frac{E_0}{n^2} \qquad\qquad 36\text{-}15$$

ENERGY LEVELS IN THE HYDROGEN ATOM

where

$$E_0 = \frac{mk^2e^4}{2\hbar^2} = \frac{1}{2}\frac{ke^2}{a_0} \approx 13.6\ \text{eV} \qquad\qquad 36\text{-}16$$

The energies $E_n$ with $Z = 1$ are the quantized allowed energies for the hydrogen atom.

Transitions between these allowed energies result in the emission or absorption of a photon whose frequency is given by $f = (E_i - E_f)/h$, and whose wavelength is

$$\lambda = \frac{c}{f} = \frac{hc}{E_i - E_f} \qquad\qquad 36\text{-}17$$

As we found in Chapter 34, it is convenient to have the value of $hc$ in electron-volt nanometers:

$$hc = 1240\ \text{eV·nm} \qquad\qquad 36\text{-}18$$

Since the energies are quantized, the frequencies and wavelengths of the radiation emitted by the hydrogen atom are quantized in agreement with the observed line spectrum.

Figure 36-4 shows the energy-level diagram for hydrogen. The energy of the hydrogen atom in the ground state is $E_1 = -13.6$ eV. As $n$ approaches infinity the energy approaches zero. The process of removing an electron from an atom is called ionization, and the energy required to remove the electron is the **ionization energy**. The ionization energy of the ground-state hydrogen atom, which is also its binding energy, is 13.6 eV. A few transitions from a higher state to a lower state are indicated in Figure 36-4. When Bohr published his model of the hydrogen atom, the Balmer series, corresponding to $n_2 = 2$ and $n_1 = 3, 4, 5$, and so on; and the Paschen series, corresponding to $n_2 = 3$ and $n_1 = 4, 5, 6$, and so on, were known. In 1916, T. Lyman found the series corresponding to $n_2 = 1$, and in 1922 and 1924, F. Brackett and H. A. Pfund, respectively, found the series corresponding to $n_2 = 4$ and $n_2 = 5$. Only the Balmer series lies in the visible portion of the electromagnetic spectrum.

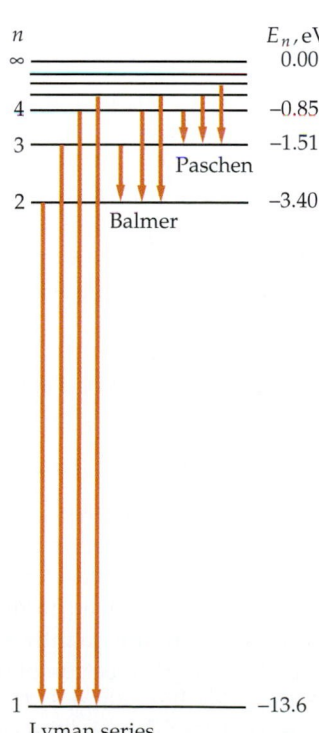

**FIGURE 36-4** Energy-level diagram for hydrogen showing the first few transitions in each of the Lyman, Balmer, and Paschen series. The energies of the levels are given by Equation 36-15.

**EXAMPLE 36-2**

Find (*a*) the energy and (*b*) the wavelength of the line with the longest wavelength in the Lyman series.

**PICTURE THE PROBLEM** From Figure 36-4, we can see that the Lyman series corresponds to transitions ending at the ground-state energy, $E_f = E_1 = -13.6$ eV. Since $\lambda$ varies inversely with energy, the transition with the longest wavelength is the transition with the lowest energy, which is that from the first excited state $n = 2$ to the ground state $n = 1$.

1. The energy of the photon is the difference in the energies of the initial and final atomic state:

$$E_{photon} = \Delta E_{atom} = E_i - E_f$$

$$= E_2 - E_1 = \frac{-13.6 \text{ eV}}{2^2} - \frac{-13.6 \text{ eV}}{1^2}$$

$$= -3.40 \text{ eV} + 13.6 \text{ eV} = \boxed{10.2 \text{ eV}}$$

2. The wavelength of the photon is:

$$\lambda = \frac{hc}{E_2 - E_1} = \frac{1240 \text{ eV·nm}}{10.2 \text{ eV}} = \boxed{121.6 \text{ nm}}$$

**REMARKS** This photon is outside the visible spectrum, in the ultraviolet region. Since all the other lines in the Lyman series have even greater energies and shorter wavelengths, the Lyman series is completely in the ultraviolet region.

**EXERCISE** Find the shortest wavelength for a line in the Lyman series. (*Answer* 91.2 nm)

Despite its spectacular successes, the Bohr model of the hydrogen atom had many shortcomings. There was no justification for the postulates of stationary states or for the quantization of angular momentum other than the fact that these postulates led to energy levels that agreed with spectroscopic data. Furthermore, attempts to apply the model to more complicated atoms had little success. The quantum-mechanical theory resolves these difficulties. The stationary states of the Bohr model correspond to the standing-wave solutions of the Schrödinger equation analogous to the standing electron waves for a particle in a box discussed in Chapter 34 and Chapter 35. Energy quantization is a direct consequence of the standing-wave solutions of the Schrödinger equation. For hydrogen, these quantized energies agree with those obtained from the Bohr model and with experiment. The quantization of angular momentum that had to be postulated in the Bohr model is predicted by the quantum theory.

## 36-3 Quantum Theory of Atoms

### The Schrödinger Equation in Spherical Coordinates

In quantum theory, the electron is described by its wave function $\psi$. The probability of finding the electron in some volume $dV$ of space equals the product of the absolute square of the electron wave function $|\psi|^2$ and $dV$. Boundary conditions on the wave function lead to the quantization of the wavelengths and frequencies and thereby to the quantization of the electron energy.

Consider a single electron of mass $m$ moving in three dimensions in a region in which the potential energy is $U$. The time-independent Schrödinger equation for such a particle is given by Equation 35-30:

$$-\frac{\hbar^2}{2m}\left(\frac{\partial^2\psi}{\partial x^2}+\frac{\partial^2\psi}{\partial y^2}+\frac{\partial^2\psi}{\partial z^2}\right)+U\psi=E\psi \qquad 36\text{-}19$$

For a single isolated atom, the potential energy $U$ depends only on the radial distance $r=\sqrt{x^2+y^2+z^2}$. The problem is then most conveniently treated using the spherical coordinates $r$, $\theta$, and $\phi$, which are related to the rectangular coordinates $x$, $y$, and $z$ by

$$z=r\cos\theta$$

$$x=r\sin\theta\cos\phi$$

$$y=r\sin\theta\sin\phi \qquad 36\text{-}20$$

These relations are shown in Figure 36-5. The transformation of the bracketed term in Equation 36-19 is straightforward but involves much tedious calculation, which we will omit. The result is

$$\frac{\partial^2\psi}{\partial x^2}+\frac{\partial^2\psi}{\partial y^2}+\frac{\partial^2\psi}{\partial z^2}=\frac{1}{r^2}\frac{\partial}{\partial r}\left(r^2\frac{\partial\psi}{\partial r}\right)+\frac{1}{r^2}\left[\frac{1}{\sin\theta}\frac{\partial}{\partial\theta}\left(\sin\theta\frac{\partial\psi}{\partial\theta}\right)+\frac{1}{\sin^2\theta}\frac{\partial^2\psi}{\partial\phi^2}\right]$$

Substituting into Equation 36-19 gives

$$-\frac{\hbar^2}{2mr^2}\frac{\partial}{\partial r}\left(r^2\frac{\partial\psi}{\partial r}\right)-\frac{\hbar^2}{2mr^2}\left[\frac{1}{\sin\theta}\frac{\partial}{\partial\theta}\left(\sin\theta\frac{\partial\psi}{\partial\theta}\right)+\frac{1}{\sin^2\theta}\frac{\partial^2\psi}{\partial\phi^2}\right]+U(r)\psi=E\psi$$

$$36\text{-}21$$

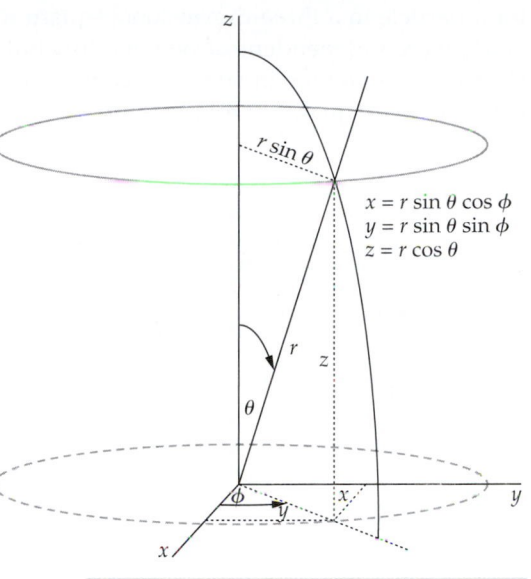

**FIGURE 36-5** Geometric relations between spherical coordinates and rectangular coordinates.

Despite the formidable appearance of this equation, it was not difficult for Schrödinger to solve because it is similar to other partial differential equations in classical physics that had been thoroughly studied. We will not solve this equation but merely discuss qualitatively some of the interesting features of the wave functions that satisfy it.

The first step in the solution of a partial differential equation, such as Equation 36-21, is to separate the variables by writing the wave function $\psi(r,\theta,\phi)$ as a product of functions of each single variable:

$$\psi(r,\theta,\phi)=R(r)f(\theta)g(\phi) \qquad 36\text{-}22$$

where $R$ depends only on the radial coordinate $r$, $f$ depends only on $\theta$, and $g$ depends only on $\phi$. When this form of $\psi(r,\theta,\phi)$ is substituted into Equation 36-21, the partial differential equation can be transformed into three ordinary differential equations, one for $R(r)$, one for $f(\theta)$, and one for $g(\phi)$. The potential energy $U(r)$ appears only in the equation for $R(r)$, which is called the **radial equation.** The particular form of $U(r)$ given in Equation 36-19 therefore has no effect on the solutions of the equations for $f(\theta)$ and $g(\phi)$, and therefore has no effect on the angular dependence of the wave function $\psi(r,\theta,\phi)$. These solutions are applicable to any problem in which the potential energy depends only on $r$.

## Quantum Numbers in Spherical Coordinates

In three dimensions, the requirement that the wave function be continuous and normalizable introduces three quantum numbers, one associated with each spatial dimension. In spherical coordinates the quantum number associated with $r$ is labeled $n$, that associated with $\theta$ is labeled $\ell$, and that associated with $\phi$ is labeled $m_\ell$.[†] The quantum numbers $n_1$, $n_2$, and $n_3$ that we found in Chapter 35

---

[†] For simplicity, $m_\ell$ is sometimes written as $m$.

for a particle in a three-dimensional square well in rectangular coordinates $x$, $y$, and $z$ were independent of one another, but the quantum numbers associated with wave functions in spherical coordinates are interdependent. The possible values of these quantum numbers are

$$n = 1, 2, 3, \ldots$$

$$\ell = 0, 1, 2, 3, \ldots, n - 1$$

$$m_\ell = -\ell, (-\ell + 1), \ldots, -2, -1, 0, 1, 2, \ldots, (\ell + 1), \ell \qquad 36\text{-}23$$

QUANTUM NUMBERS IN SPHERICAL COORDINATES

That is, $n$ can be any positive integer; $\ell$ can be 0 or any positive integer up to $n - 1$; and $m_\ell$ can have $2\ell + 1$ possible values, ranging from $-\ell$ to $+\ell$ in integral steps.

The number $n$ is called the **principal quantum number.** It is associated with the dependence of the wave function on the distance $r$ and therefore with the probability of finding the electron at various distances from the nucleus. The quantum numbers $\ell$ and $m_\ell$ are associated with the angular momentum of the electron and with the angular dependence of the electron wave function. The quantum number $\ell$ is called the **orbital quantum number.** The magnitude $L$ of the orbital angular momentum $\vec{L}$ is related to the orbital quantum number $\ell$ by

$$L = \sqrt{\ell(\ell + 1)}\hbar \qquad 36\text{-}24$$

The quantum number $m_\ell$ is called the **magnetic quantum number.** It is related to the component of the angular momentum along some direction in space. All spatial directions are equivalent for an isolated atom, but placing the atom in a magnetic field results in the direction of the magnetic field being separated out from the other directions. The convention is that the $z$ direction is chosen for the magnetic-field direction. Then the $z$ component of the angular momentum of the electron is given by the quantum condition

$$L_z = m_\ell \hbar \qquad 36\text{-}25$$

This quantum condition arises from the boundary condition on the azimuth coordinate $\phi$ that the probability of finding the electron at some arbitrary angle $\phi_1$ must be the same as that of finding the electron at angle $\phi_1 + 2\pi$ because these are the same points in space.

If we measure the angular momentum of the electron in units of $\hbar$, we see that the angular-momentum magnitude is quantized to the value $\sqrt{\ell(\ell + 1)}$ units and that its component along any direction can have only the $2\ell + 1$ values ranging from $-\ell$ to $+\ell$ units. Figure 36-6 shows a vector-model diagram illustrating the possible orientations of the angular-momentum vector for $\ell = 2$. Note that only specific values of $\theta$ are allowed; that is, the directions in space are quantized.

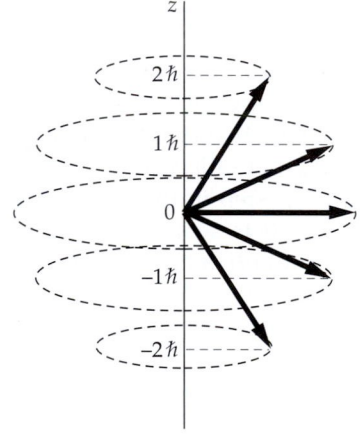

**FIGURE 36-6** Vector-model diagram illustrating the possible values of the $z$ component of the angular-momentum vector for the case $\ell = 2$. The magnitude of the angular momentum is $L = \hbar\sqrt{\ell(\ell + 1)} = \hbar\sqrt{2(2 + 1)} = \hbar\sqrt{6}$.

*THE DIRECTIONS OF THE ANGULAR MOMENTUM* **EXAMPLE 36-3**

If the angular momentum is characterized by the quantum number $\ell = 2$, what are the possible values of $L_z$, and what is the smallest possible angle between $\vec{L}$ and the $z$ axis?

**PICTURE THE PROBLEM**  The possible orientations of $\vec{L}$ and the z axis are shown in Figure 36-6. The z-axis direction is parallel with that of the external magnetic field in the vicinity of the atom.

1. Write the possible values of $L_z$:

$$L_z = m_\ell \hbar, \text{ where } m_\ell = -2, -1, 0, 1, 2$$

2. Express the angle $\theta$ between $\vec{L}$ and the z axis in terms of $L$ and $L_z$:

$$\cos \theta = \frac{L_z}{L} = \frac{m\hbar}{\sqrt{\ell(\ell+1)}\hbar} = \frac{m}{\sqrt{\ell(\ell+1)}}$$

3. The smallest angle occurs when $m_\ell = \ell = 2$:

$$\cos \theta_{\min} = \frac{2}{\sqrt{2(2+1)}} = \frac{2}{\sqrt{6}} = 0.816$$

$$\theta_{\min} = \boxed{35.3°}$$

**REMARKS**  We note the somewhat strange result that the angular-momentum vector cannot lie along the z axis.

**EXERCISE**  An atom in a region with a magnetic field has an angular momentum characterized by the quantum number $\ell = 4$. What are the possible values of $m_\ell$? (*Answer*  $-4, -3, -2, -1, 0, 1, 2, 3, 4$)

# 36-4 Quantum Theory of the Hydrogen Atom

We can treat the simplest atom, the hydrogen atom, as a stationary nucleus, a proton, that has a single moving particle, an electron, with kinetic energy $p^2/2m$. The potential energy $U(r)$ due to the electrostatic attraction between the electron and the proton[†] is

$$U(r) = -\frac{kZe^2}{r} \qquad\qquad 36\text{-}26$$

For this potential-energy function, the Schrödinger equation can be solved exactly. In the lowest energy state, which is the ground state, the principal quantum number $n$ has the value 1, $\ell$ is 0, and $m_\ell$ is 0.

## Energy Levels

The allowed energies of the hydrogen atom that result from the solution of the Schrödinger equation are

$$E_n = -\frac{mk^2e^4}{2\hbar^2 n^2} = -Z^2\frac{E_0}{n^2}, \quad n = 1, 2, 3, \ldots \qquad\qquad 36\text{-}27$$

ENERGY LEVELS FOR HYDROGEN

where

$$E_0 = -\frac{mk^2e^4}{2\hbar^2} \approx 13.6 \text{ eV} \qquad\qquad 36\text{-}28$$

---

† We include the factor Z, which is 1 for hydrogen, so that we can apply our results to other one-electron atoms, such as singly ionized helium He⁺, for which Z = 2.

These energies are the same as in the Bohr model. Note that the energy is negative, indicating that the electron is bound to the nucleus (thus the term *bound state*), and that the energy depends only on the principal quantum number $n$. The fact that the energy does not depend on the orbital quantum number $\ell$ is a peculiarity of the inverse-square force and holds only for an inverse $r$ potential such as Equation 36-26. For more complicated atoms having several electrons, the interaction of the electrons leads to a dependence of the energy on $\ell$. In general, the lower the value of $\ell$, the lower the energy for such atoms. Since there is usually no preferred direction in space, the energy for any atom does not ordinarily depend on the magnetic quantum number $m_\ell$, which is related to the $z$ component of the angular momentum. The energy does depend on $m_\ell$ if the atom is in a magnetic field.

Figure 36-7 shows an energy-level diagram for hydrogen. This diagram is similar to Figure 36-4, except that the states with the same value of $n$ but with different values of $\ell$ are shown separately. These states (called *terms*) are referred to by giving the value of $n$ along with a code letter: s for $\ell = 0$, p for $\ell = 1$, d for $\ell = 2$, and f for $\ell = 3$.[†] (Lowercase letters s, p, d, f, and so on, are used to identify the orbital angular momentum of an individual electron; uppercase letters S, P, D, F, and so on, are used to identify the orbital angular momentum for the entire multielectron atom. For hydrogen, either uppercase or lowercase letters will do, but most people use lowercase as we have done.) When an atom makes a transition from one allowed energy state to another, electromagnetic radiation in the form of a photon is emitted or absorbed. Such transitions result in spectral lines that are characteristic of the atom. The transitions obey the **selection rules:**

$$\Delta m_\ell = 0 \text{ or } \pm 1$$

$$\Delta\ell = \pm 1 \qquad\qquad\qquad 36\text{-}29$$

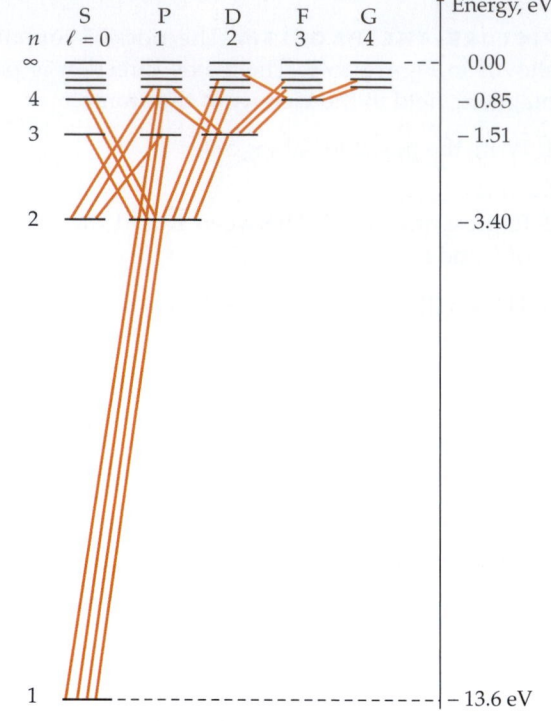

**FIGURE 36-7** Energy-level diagram for hydrogen. The diagonal lines show transitions that involve emission or absorption of radiation and obey the selection rule $\Delta\ell = \pm 1$. States with the same value of $n$ but with different values of $\ell$ have the same energy $-E_0/n^2$, where $E_0 = 13.6$ eV as in the Bohr model.

These selection rules are related to the conservation of angular momentum and to the fact that the photon itself has an intrinsic angular momentum that has a maximum component along any axis of $1\hbar$. The wavelengths of the light emitted by hydrogen (and by other atoms) are related to the energy levels by

$$hf = \frac{hc}{\lambda} = E_i - E_f \qquad\qquad\qquad 36\text{-}30$$

where $E_i$ and $E_f$ are the energies of the initial and final states.

## Wave Functions and Probability Densities

The solutions of the Schrödinger equation in spherical coordinates are characterized by the quantum numbers $n$, $\ell$, and $m_\ell$, and are written $\psi_{n\ell m_\ell}$. For any given value of $n$, there are $n$ possible values of $\ell$ ($\ell = 0, 1, \ldots, n - 1$), and for each value of $\ell$, there are $2\ell + 1$ possible values of $m_\ell$. For hydrogen, the energy depends only on $n$, so there are generally many different wave functions that correspond to the same energy (except at the lowest energy level, for which $n = 1$ and therefore $\ell$ and $m_\ell$ must be 0). These energy levels are therefore degenerate (see Section 35-5). The origins of this degeneracy are the $1/r$ dependence of the potential energy and the fact that, in the absence of any external fields, there is no preferred direction in space.[‡]

---

† These code letters are remnants of spectroscopists descriptions of various spectral lines as *sharp, principal, diffuse,* and *fundamental.* For values greater than 3, the letters follow alphabetically; thus, g is used for $\ell = 4$, and so forth.
‡ If spin, relativistic effects, the spin of the nucleus, and quantum electrodynamics are considered, the degeneracy is broken.

**The Ground State** In the lowest energy state, the ground state of hydrogen, the principal quantum number $n$ has the value 1, $\ell$ is 0, and $m_\ell$ is 0. The energy is $-13.6$ eV, and the angular momentum is zero. (In the Bohr model of the atom the angular momentum in the ground state is equal to $\hbar$, not zero.) The wave function for the ground state is

$$\psi_{1,0,0} = C_{1,0,0}e^{-Zr/a_0} \qquad\qquad 36\text{-}31$$

where

$$a_0 = \frac{\hbar^2}{mke^2} = 0.0529 \text{ nm}$$

is the first Bohr radius and $C_{1,\,0,\,0}$ is a constant that is determined by normalization. In three dimensions, the normalization condition is

$$\int |\psi|^2 \, dV = 1$$

where $dV$ is a volume element and the integration is performed over all space. In spherical coordinates, the volume element (Figure 36-8) is

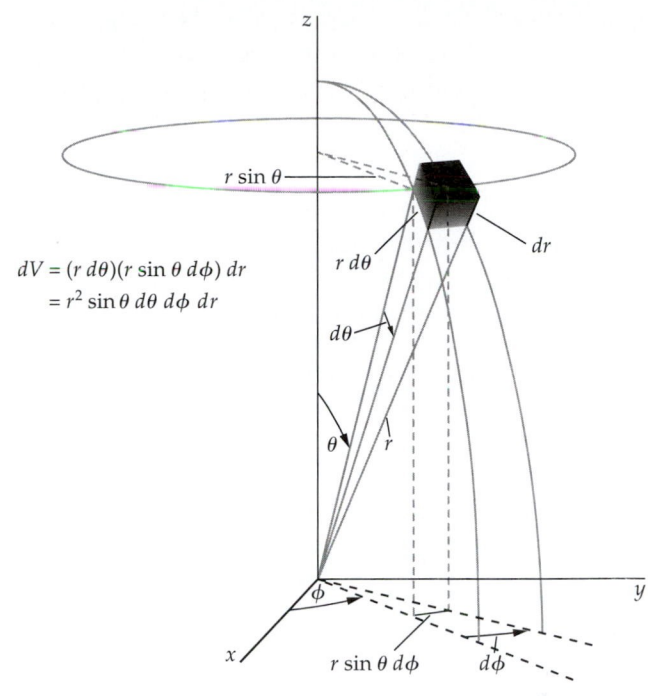

$$dV = (r\,d\theta)(r\sin\theta\,d\phi)\,dr$$
$$= r^2 \sin\theta\,d\theta\,d\phi\,dr$$

**FIGURE 36-8** Volume element in spherical coordinates.

$$dV = (r\,d\theta)(r\sin\theta\,d\phi)\,dr = r^2\sin\theta\,d\theta\,d\phi\,dr$$

We integrate over all space by integrating over $\phi$, from $\phi = 0$ to $\phi = 2\pi$, over $\theta$, from $\theta = 0$ to $\theta = \pi$; and over $r$, from $r = 0$ to $r = \infty$. The normalization condition is thus

$$\int |\psi|^2 \, dV = \int_0^\infty \left[ \int_0^\pi \left( \int_0^{2\pi} |\psi|^2\, r^2 \sin\theta\,d\phi \right) d\theta \right] dr$$

$$= \int_0^\infty \left[ \int_0^\pi \left( \int_0^{2\pi} C_{1,0,0}^2 e^{-2Zr/a_0} r^2 \sin\theta\,d\phi \right) d\theta \right] dr = 1$$

Since there is no $\theta$ or $\phi$ dependence in $\psi_{1,0,0}$, the triple integral can be factored into the product of three integrals. This gives

$$\int |\psi|^2 \, dV = \left( \int_0^{2\pi} d\phi \right) \left( \int_0^\pi \sin\theta\,d\theta \right) \left( \int_0^\infty C_{1,0,0}^2 e^{-2Zr/a_0} r^2\,dr \right)$$

$$= 2\pi \cdot 2 \cdot C_{1,0,0}^2 \left( \int_0^\infty r^2 e^{-2Zr/a_0}\,dr \right) = 1$$

The remaining integral is of the form $\int_0^\infty x^n e^{-ax}\,dx$, with $n$ a positive integer and with $a > 0$. Using successive integration-by-parts operations[†] yields the result

$$\int_0^\infty x^n e^{-ax}\,dx = \frac{n!}{a^{n+1}}$$

---

† This integral can also be looked up in a table of integrals.

so

$$\int_0^\infty r^2 e^{-2Zr/a_0}\, dr = \frac{a_0^3}{4Z^3}$$

Then

$$4\pi C_{1,0,0}^2\left(\frac{a_0^3}{4Z^3}\right) = 1$$

so

$$C_{1,0,0} = \frac{1}{\sqrt{\pi}}\left(\frac{Z}{a_0}\right)^{3/2} \qquad\qquad 36\text{-}32$$

The normalized ground-state wave function is thus

$$\psi_{1,0,0} = \frac{1}{\sqrt{\pi}}\left(\frac{Z}{a_0}\right)^{3/2} e^{-Zr/a_0} \qquad\qquad 36\text{-}33$$

The probability of finding the electron in a volume $dV$ is $|\psi|^2\, dV$. The probability density $|\psi|^2$ is illustrated in Figure 36-9. Note that this probability density is spherically symmetric; that is, the probability density depends only on $r$, and is independent of $\theta$ or $\phi$. The probability density is maximum at the origin.

We are more often interested in the probability of finding the electron at some radial distance $r$ between $r$ and $r + dr$. This radial probability $P(r)\, dr$ is the probability density $|\psi|^2$ times the volume of the spherical shell of thickness $dr$, which is $dV = 4\pi r^2\, dr$. The probability of finding the electron in the range from $r$ to $r + dr$ is thus $P(r)\, dr = |\psi|^2 4\pi r^2\, dr$, and the **radial probability density** is

$$P(r) = 4\pi r^2 |\psi|^2 \qquad\qquad 36\text{-}34$$

RADIAL PROBABILITY DENSITY

For the hydrogen atom in the ground state, the radial probability density is

$$P(r) = 4\pi r^2 |\psi|^2 = 4\pi C_{1,0,0}^2 r^2 e^{-2Zr/a_0} = 4\left(\frac{Z}{a_0}\right)^3 r^2 e^{-2Zr/a_0} \qquad\qquad 36\text{-}35$$

Figure 36-10 shows the radial probability density $P(r)$ as a function of $r$. The maximum value of $P(r)$ occurs at $r = a_0/Z$, which for $Z = 1$ is the first Bohr radius. In contrast to the Bohr model, in which the electron stays in a well-defined orbit at $r = a_0$, we see that it is possible for the electron to be found at any distance from the nucleus. However, the most probable distance is $a_0$ (assuming $Z = 1$), and the chance of finding the electron at a much different distance is small. It is often useful to think of the electron in an atom as a charged cloud of charge density $\rho = -e|\psi|^2$, but we should remember that when it interacts with matter, an electron is always observed as a single charge.

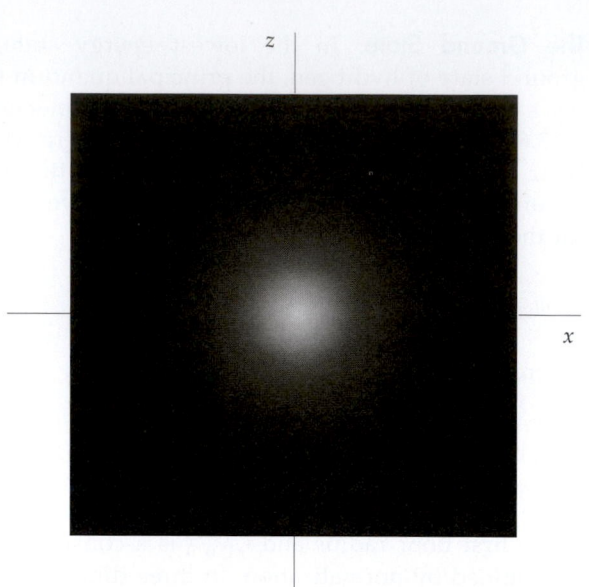

**FIGURE 36-9** Computer-generated picture of the probability density $|\psi|^2$ for the ground state of hydrogen. The quantity $-e|\psi|^2$ can be thought of as the electron charge density in the atom. The density is spherically symmetric, is greatest at the origin, and decreases exponentially with $r$.

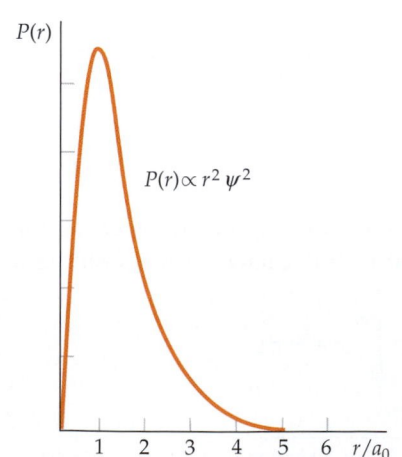

**FIGURE 36-10** Radial probability density $P(r)$ versus $r/a_0$ for the ground state of the hydrogen atom. $P(r)$ is proportional to $r^2\psi^2$. The value of $r$ for which $P(r)$ is maximum is the most probable distance $r = a_0$.

PROBABILITY THAT THE ELECTRON IS IN A THIN SPHERICAL SHELL    **EXAMPLE 36-4**

Find the probability of finding the electron in a thin spherical shell of radius $r$ and thickness $\Delta r = 0.06a_0$ at (a) $r = a_0$ and (b) $r = 2a_0$ for the ground state of the hydrogen atom.

**PICTURE THE PROBLEM**   Because the range $\Delta r$ is so small compared to $r$, the variation in the radial probability density $P(r)$ in the shell can be neglected. The probability of finding the electron in some small range $\Delta r$ is then $P(r) \, \Delta r$.

1. Use Equation 36-35 with $Z = 1$ and $r = a_0$:
$$P(r) \, \Delta r = \left[ 4\left(\frac{1}{a_0}\right)^3 r^2 e^{-2r/a_0} \right] \Delta r = \left[ 4\left(\frac{1}{a_0}\right)^3 a_0^2 e^{-2} \right](0.06a_0) = \boxed{0.0325}$$

2. Use Equation 36-35 with $Z = 1$ and $r = 2a_0$:
$$P(r) \, \Delta r = \left[ 4\left(\frac{1}{a_0}\right)^3 r^2 e^{-2r/a_0} \right] \Delta r = \left[ 4\left(\frac{1}{a_0}\right)^3 4a_0^2 e^{-4} \right](0.06a_0) = \boxed{0.0176}$$

**REMARKS**   There is approximately a 3 percent chance of finding the electron in this range at $r = a_0$, but at $r = 2a_0$ the chance is slightly less than 2 percent.

**The First Excited State**   In the first excited state, $n = 2$ and $\ell$ can be either 0 or 1. For $\ell = 0$, $m_\ell = 0$, and we again have a spherically symmetric wave function, this time given by

$$\psi_{2,0,0} = C_{2,0,0}\left( 2 - \frac{Zr}{a_0} \right)e^{-Zr/(2a_0)} \qquad\qquad 36\text{-}36$$

For $\ell = 1$, $m_\ell$ can be $+1$, 0, or $-1$. The corresponding wave functions are

$$\psi_{2,1,0} = C_{2,1,0}\frac{Zr}{a_0}e^{-Zr/(2a_0)}\cos\theta \qquad\qquad 36\text{-}37$$

$$\psi_{2,1,\pm1} = C_{2,1,1}\frac{Zr}{a_0}e^{-Zr/(2a_0)}\sin\theta\, e^{\pm i\phi} \qquad\qquad 36\text{-}38$$

where $C_{2,0,0}$, $C_{2,1,0}$, and $C_{2,1,1}$ are normalization constants. The probability densities are given by

$$\psi_{2,0,0}^2 = C_{2,0,0}^2\left( 2 - \frac{Zr}{a_0} \right)^2 e^{-Zr/a_0} \qquad\qquad 36\text{-}39$$

$$\psi_{2,1,0}^2 = C_{2,1,0}^2\left( \frac{Zr}{a_0} \right)^2 e^{-Zr/a_0}\cos^2\theta \qquad\qquad 36\text{-}40$$

$$|\psi_{2,1,\pm1}|^2 = C_{2,1,1}^2\left( \frac{Zr}{a_0} \right)^2 e^{-Zr/a_0}\sin^2\theta \qquad\qquad 36\text{-}41$$

The wave functions and probability densities for $\ell \neq 0$ are not spherically symmetric, but instead depend on the angle $\theta$. The probability densities do not depend on $\phi$. Figure 36-11 shows the probability density $|\psi|^2$ for $n = 2$, $\ell = 0$, and $m_\ell = 0$ (Figure 36-11a); for $n = 2$, $\ell = 1$, and $m_\ell = 0$ (Figure 36-11b); and for $n = 2$, $\ell = 1$, and $m_\ell = \pm1$ (Figure 36-11c). An important feature of these plots is that the electron cloud is spherically symmetric for $\ell = 0$ and is not spherically symmetric for $\ell \neq 0$. These angular distributions of the electron charge density depend only on the values of $\ell$ and $m_\ell$ and not on the radial part of the wave function.

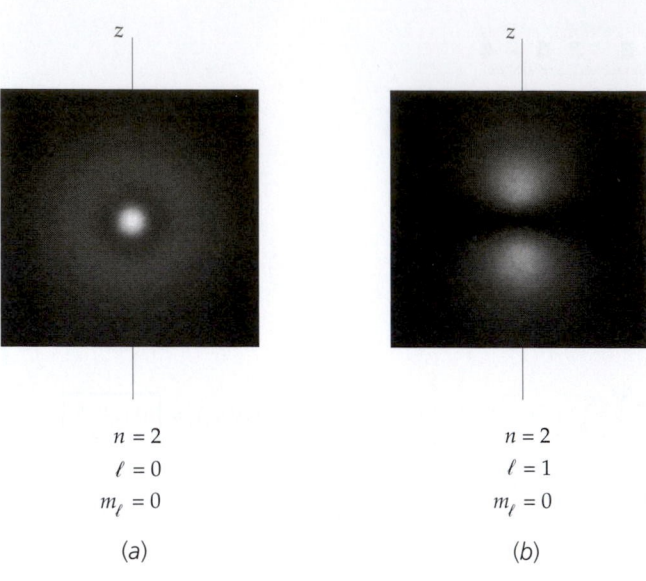
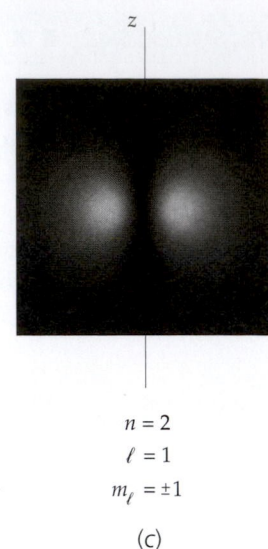

| $n = 2$ | $n = 2$ | $n = 2$ |
| $\ell = 0$ | $\ell = 1$ | $\ell = 1$ |
| $m_\ell = 0$ | $m_\ell = 0$ | $m_\ell = \pm 1$ |
| (a) | (b) | (c) |

**FIGURE 36-11**
Computer-generated picture of the probability densities $|\psi|^2$ for the electron in the $n = 2$ states of hydrogen. All three images represent figures of revolution about the z axis. (a) For $\ell = 0$, $|\psi|^2$ is spherically symmetric. (b) For $\ell = 1$ and $m_\ell = 0$, $|\psi|^2$ is proportional to $\cos^2 \theta$. (c) For $\ell = 1$ and $m_\ell = +1$ or $-1$, $|\psi|^2$ is proportional to $\sin^2 \theta$.

Similar charge distributions for the valence electrons of more complicated atoms play an important role in the chemistry of molecular bonding.

Figure 36-12 shows the probability of finding the electron at a distance $r$ as a function of $r$ for $n = 2$, when $\ell = 1$ and when $\ell = 0$. We can see from the figure that the probability distribution depends on $\ell$ as well as on $n$.

For $n = 1$, we found that the most likely distance between the electron and the nucleus is $a_0$, which is the first Bohr radius, whereas for $n = 2$ and $\ell = 1$, the most likely distance between the electron and the nucleus is $4a_0$. These are the orbital radii for the first and second Bohr orbits (Equation 36-11). For $n = 3$ (and $\ell = 2$),[†] the most likely distance between the electron and nucleus is $9a_0$, which is the radius of the third Bohr orbit.

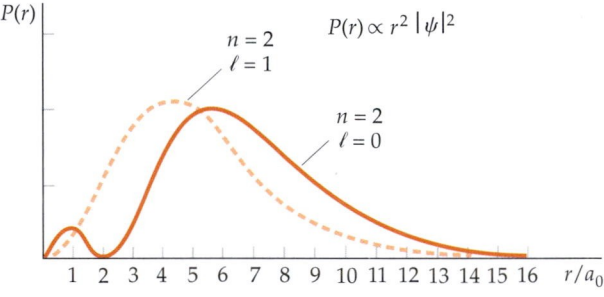

**FIGURE 36-12** Radial probability density $P(r)$ versus $r/a_0$ for the $n = 2$ states of hydrogen. For $\ell = 1$, $P(r)$ is maximum at the Bohr value $r = 2^2 a_0$. For $\ell = 0$, there is a maximum near this value and a much smaller maximum near the origin.

# 36-5 The Spin-Orbit Effect and Fine Structure

The orbital magnetic moment of an atomic electron can be derived semiclassically, even though it is quantum mechanical in origin.[‡] Consider a particle of mass $m$ and charge $q$ moving with speed $v$ in a circle of radius $r$. The magnitude of the angular momentum of the particle is $L = mvr$, and the magnitude of the magnetic moment is the product of the current and the area of the circle $\mu = IA = I\pi r^2$. If $T$ is the time for the charge to complete one revolution, the current (charge passing a point per unit time) is $q/T$. Since the period $T$ is the distance $2\pi r$ divided by the velocity $v$, the current is $I = q/T = qv/(2\pi r)$. The magnetic moment is then

$$\mu = IA = \frac{qv}{2\pi r} \pi r^2 = \frac{1}{2} qvr = \frac{q}{2m} L$$

where we have substituted $L/m$ for $vr$. If the charge $q$ is positive, the angular momentum and magnetic moment are in the same direction. We can therefore write

$$\vec{\mu} = \frac{q}{2m} \vec{L} \qquad\qquad 36\text{-}42$$

---

[†] The correspondence with the Bohr model is closest for the maximum value of $\ell$, which is $n - 1$.
[‡] This topic was first presented in Section 27-5.

Equation 36-42 is the general classical relation between magnetic moment and angular momentum. It also holds in the quantum theory of the atom for orbital angular momentum, but not for the intrinsic spin angular momentum of the electron. For electron spin, the magnetic moment is twice that predicted by Equation 36-42.[†] The extra factor of 2 is a result from quantum theory that has no analog in classical mechanics.

The quantum of angular momentum is $\hbar$, so we express the magnetic moment in terms of $\vec{L}/\hbar$:

$$\vec{\mu} = \frac{q\hbar}{2m} \frac{\vec{L}}{\hbar}$$

For an electron, $m = m_e$ and $q = -e$, so the magnetic moment of the electron due to its orbital motion is

$$\vec{\mu}_\ell = -\frac{e\hbar}{2m_e} \frac{\vec{L}}{\hbar} = -\mu_B \frac{\vec{L}}{\hbar}$$

where $\mu_B = e\hbar/(2m_e) = 5.79 \times 10^{-5}$ eV/T is the quantum unit of magnetic moment called a Bohr magneton. The magnetic moment of an electron due to its intrinsic spin angular momentum $\vec{S}$ is

$$\vec{\mu}_S = -2 \times \frac{e\hbar}{2m_e} \frac{\vec{S}}{\hbar} = -2\mu_B \frac{\vec{S}}{\hbar}$$

In general, an electron in an atom has both orbital angular momentum characterized by the quantum number $\ell$ and spin angular momentum characterized by the quantum number $s$. Analogous classical systems that have two kinds of angular momentum are the earth, which is spinning about its axis of rotation in addition to revolving about the sun, and a precessing gyroscope that has angular momentum of precession in addition to its spin. The total angular momentum $\vec{J}$ is the sum of the orbital angular momentum $\vec{L}$ and the spin angular momentum $\vec{S}$, where

$$\vec{J} = \vec{L} + \vec{S} \qquad\qquad\qquad 36\text{-}43$$

Classically $\vec{J}$ is an important quantity because the resultant torque on a system equals the rate of change of the total angular momentum, and in the case of only central forces, the total angular momentum is conserved. For a classical system, the direction of the total angular momentum $\vec{J}$ is without restrictions and the magnitude of $\vec{J}$ can take on any value between $J_{max} = L + S$ and $J_{min} = |L - S|$. However, in quantum mechanics, the directions of both $\vec{L}$ and $\vec{S}$ are more restricted and the magnitudes $L$ and $S$ are both quantized. Furthermore, like $\vec{L}$ and $\vec{S}$, the direction of the total angular momentum $\vec{J}$ is restricted and the magnitude of $\vec{J}$ is quantized. For an electron with orbital angular momentum characterized by the quantum number $\ell$ and spin $s = \frac{1}{2}$, the total angular-momentum magnitude $J$ is equal to $\sqrt{j(j + 1)}\hbar$, where the quantum number $j$ is given by

$$j = +\tfrac{1}{2}, \quad \ell = 0$$

---

[†] This result, and the phenomenon of electron spin itself, was predicted in 1927 by Paul Dirac, who combined special relativity and quantum mechanics into a relativistic wave equation called the Dirac equation. Precise measurements indicate that the magnetic moment of the electron due to its spin is 2.00232 times that predicted by Equation 36-42. The fact that the intrinsic magnetic moment of the electron is approximately twice what we would expect makes it clear that the simple model of the electron as a spinning ball is not to be taken literally.

and either

$$j = \ell + \tfrac{1}{2} \quad \text{or} \quad j = \ell - \tfrac{1}{2}, \quad \ell > 0 \qquad\qquad 36\text{-}44$$

Figure 36-13 is a vector model illustrating the two possible combinations $j = \tfrac{3}{2}$ and $j = \tfrac{1}{2}$ for the case of $\ell = 1$. The lengths of the vectors are proportional to $\sqrt{\ell(\ell+1)}\hbar$, $\sqrt{s(s+1)}\hbar$, and $\sqrt{j(j+1)}\hbar$. The spin angular momentum and the orbital angular momentum are said to be *parallel* when $j = \ell + s$ and *antiparallel* when $j = \ell - s$.

Atomic states with the same $n$ and $\ell$ values but with different $j$ values have slightly different energies because of the interaction of the spin of the electron with its orbital motion. This effect is called the **spin–orbit effect**. The resulting splitting of spectral lines is called **fine-structure splitting**.

In the notation $n\ell_j$, the ground state of the hydrogen atom is written $1s_{1/2}$, where the 1 indicates that $n = 1$, the s indicates that $\ell = 0$, and the 1/2 indicates that $j = \tfrac{1}{2}$. The $n = 2$ states can have either $\ell = 0$ or $\ell = 1$, and the $\ell = 1$ state can have either $j = \tfrac{3}{2}$ or $j = \tfrac{1}{2}$. These states are thus denoted by $2s_{1/2}$, $2p_{3/2}$, and $2p_{1/2}$. Because of the spin–orbit effect, the $2p_{3/2}$ and $2p_{1/2}$ states have slightly different energies resulting in the fine-structure splitting of the transitions $2p_{3/2} \to 2p_{1/2}$ and $2p_{1/2} \to 2s_{1/2}$.

We can understand the spin–orbit effect qualitatively from a simple Bohr-model picture, as shown in Figure 36-14. In this figure, the electron moves in a circular orbit around a fixed proton. In Figure 36-14a, the orbital angular momentum $\vec{L}$ is up. In an inertial reference frame in which the electron is momentarily at rest (see Figure 36-14b), the proton is moving at right angles to the line connecting the proton and the electron. The moving proton produces a magnetic field $\vec{B}$ at the position of the electron. The direction of $\vec{B}$ is up, parallel to $\vec{L}$. The energy of the electron depends on its spin because of the magnetic moment $\vec{\mu}_s$ associated with the electron's spin. The energy is lowest when $\vec{\mu}_s$ is parallel to $\vec{B}$ and the energy is highest when it is antiparallel. This energy is given by (Equation 36-16)

$$U = -\vec{\mu}_s \cdot \vec{B} = -\mu_{s_z}B \approx -\mu_B B \qquad\qquad 36\text{-}45^\dagger$$

Since $\vec{\mu}_s$ is directed opposite to its spin (because the electron has a negative charge), the energy is lowest when the spin $\vec{S}$ is antiparallel to $\vec{B}$ and thus to $\vec{L}$. The energy of the $2p_{1/2}$ state in hydrogen, in which $\vec{L}$ and $\vec{S}$ are antiparallel (Figure 36-15), is therefore slightly lower than that of the $2p_{3/2}$ state, in which $\vec{L}$ and $\vec{S}$ are parallel.

---

† Transferring the energy of the dipole to the frame of the proton gives a factor of 2, which is included in this result.

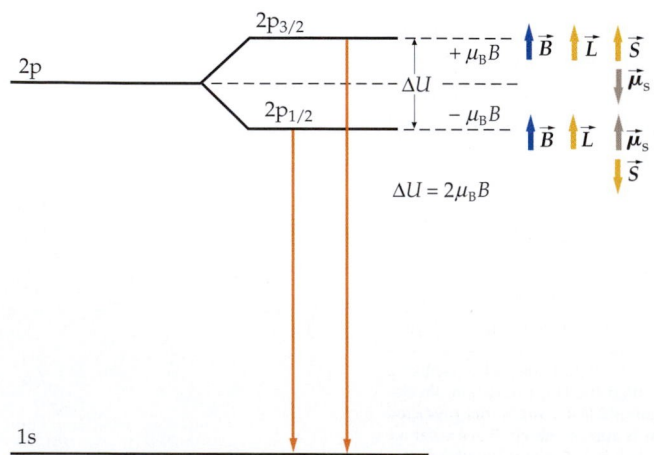

$$\Delta U = 2\mu_B B$$

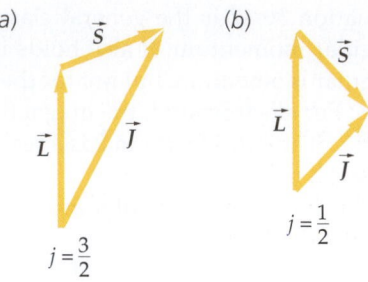

**FIGURE 36-13** Vector diagrams illustrating the addition of orbital angular momentum and spin angular momentum for the case $\ell = 1$ and $s = \tfrac{1}{2}$. There are two possible values of the quantum number for the total angular momentum: $j = \ell + s = \tfrac{3}{2}$ and $j = \ell - s = \tfrac{1}{2}$.

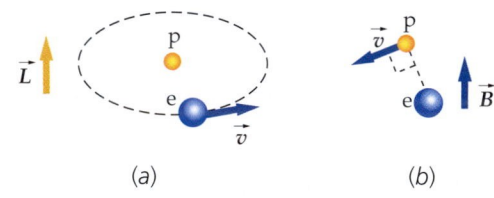

**FIGURE 36-14** (a) An electron moving about a proton in a circular orbit in the horizontal plane with angular momentum $\vec{L}$ up. (b) In an inertial reference frame in which the electron is momentarily at rest there is, at the location of the electron, a magnetic field $\vec{B}$ due to the motion of the proton that is also directed up. When the electron spin $\vec{S}$ is parallel to $\vec{L}$, its magnetic moment $\vec{\mu}_s$ is antiparallel to $\vec{L}$ and $\vec{B}$, so the spin–orbit energy is at its greatest.

**FIGURE 36-15** Fine-structure energy-level diagram. On the left, the levels in the absence of a magnetic field are shown. The effect of the field is shown on the right. Because of the spin–orbit interaction, the magnetic field splits the 2p level into two energy levels, with the $j = \tfrac{3}{2}$ level having slightly greater energy than the $j = \tfrac{1}{2}$ level. The spectral line due to the transition 2p → 1s is therefore split into two lines of slightly different wavelengths.

**E X A M P L E   3 6 - 5**

As a consequence of fine-structure splitting, the energies of the $2p_{3/2}$ and $2p_{1/2}$ levels in hydrogen differ by $4.5 \times 10^{-5}$ eV. If the 2p electron sees an internal magnetic field of magnitude $B$, the spin–orbit energy splitting will be of the order of $\Delta E = 2\mu_B B$, where $\mu_B$ is the Bohr magneton. From this, estimate the magnetic field that the 2p electron in hydrogen experiences.

1. Write the spin–orbit energy splitting in terms of the magnetic moment:

$$\Delta E = 2\mu_B B = 4.5 \times 10^{-5}\,\text{eV}$$

2. Solve for the magnetic field $B$:

$$B = \frac{4.5 \times 10^{-5}\,\text{eV}}{2\mu_B} = \frac{4.5 \times 10^{-5}\,\text{eV}}{2(5.79 \times 10^{-5}\,\text{eV/T})}$$

$$= \boxed{0.389\,\text{T}}$$

# 36-6  The Periodic Table

For atoms with more than one electron, the Schrödinger equation cannot be solved exactly. However, powerful approximation methods allow us to determine the energy levels of the atoms and wave functions of the electrons to a high degree of accuracy. As a first approximation, the $Z$ electrons in an atom are assumed to be noninteracting. The Schrödinger equation can then be solved, and the resulting wave functions used to calculate the interaction of the electrons, which in turn can be used to better approximate the wave functions. Because the spin of an electron can have two possible components along an axis, there is an additional quantum number $m_s$, which can have the possible values $+\frac{1}{2}$ or $-\frac{1}{2}$. The state of each electron is thus described by the four quantum numbers $n$, $\ell$, $m$, and $m_s$. The energy of the electron is determined mainly by the principal quantum number $n$ (which is related to the radial dependence of the wave function) and by the orbital angular-momentum quantum number $\ell$. Generally, the lower the values of $n$, the lower the energy; and for a given value of $n$, the lower the value of $\ell$, the lower the energy. The dependence of the energy on $\ell$ is due to the interaction of the electrons in the atom with each other. In hydrogen, of course, there is only one electron, and the energy is independent of $\ell$. The specification of $n$ and $\ell$ for each electron in an atom is called the **electron configuration.** Customarily, $\ell$ is specified according to the same code used to label the states of the hydrogen atom rather than by its numerical value. The code is

|         | s | p | d | f | g | h |
|---------|---|---|---|---|---|---|
| $\ell$ value | 0 | 1 | 2 | 3 | 4 | 5 |

The $n$ values are sometimes referred to as shells, which are identified by another letter code: $n = 1$ denotes the $K$ shell;[†] $n = 2$, the $L$ shell; and so on.

The electron configuration of atoms is constrained by the Pauli exclusion principle, which states that no two electrons in an atom can be in the same quantum state; that is, no two electrons can have the same set of values for the quantum numbers $n$, $\ell$, $m_\ell$, and $m_s$. Using the exclusion principle and the restrictions on the quantum numbers discussed in the previous sections ($n$ is a positive integer, $\ell$ is an integer that ranges from 0 to $n - 1$, $m_\ell$ can have $2\ell + 1$ values from $-\ell$ to $\ell$ in integral steps, and $m_s$ can be either $+\frac{1}{2}$ or $-\frac{1}{2}$), we can understand much of the structure of the periodic table.

---

† The designation of the $n = 1$ shell as $K$ is usually found when dealing with X-ray levels where the final shell in an inner electron transition is labeled as $K$, $L$, $M$, and so on.

We have already discussed the lightest element, hydrogen, which has just one electron. In the ground (lowest energy) state, the electron has $n = 1$ and $\ell = 0$, with $m_\ell = 0$ and $m_s = +\frac{1}{2}$ or $-\frac{1}{2}$. We call this a 1s electron. The 1 signifies that $n = 1$, and the s signifies that $\ell = 0$.

As electrons are added to make the heavier atoms, the electrons go into those states that will give the lowest total energy consistent with the Pauli exclusion principle.

## Helium ($Z = 2$)

The next element after hydrogen is helium ($Z = 2$), which has two electrons. In the ground state, both electrons are in the K shell with $n = 1$, $\ell = 0$, and $m_\ell = 0$; one electron has $m_s = +\frac{1}{2}$ and the other has $m_s = -\frac{1}{2}$. This configuration is lower in energy than any other two-electron configuration. The resultant spin of the two electrons is zero. Since the orbital angular momentum is also zero, the total angular momentum is zero. The electron configuration for helium is written 1s². The 1 signifies that $n = 1$, the s signifies that $\ell = 0$, and the superscript 2 signifies that there are two electrons in this state. Since $\ell$ can be only 0 for $n = 1$, these two electrons fill the K ($n = 1$) shell. The energy required to remove the most loosely bound electron from an atom in the ground state is called the **ionization energy.** This energy is the binding energy of the last electron placed in the atom. For helium, the ionization energy is 24.6 eV, which is relatively large. Helium is therefore basically inert.

---

*ELECTRON INTERACTION ENERGY IN HELIUM* **EXAMPLE 3 6 - 6**

(*a*) **Use the measured ionization energy to calculate the energy of interaction of the two electrons in the ground state of the helium atom. (*b*) Use your result to estimate the average separation of the two electrons.**

**PICTURE THE PROBLEM** The energy of one electron in the ground state of helium is $E_1$ (which is negative) given by Equation 36-27, with $n = 1$ and $Z = 2$. If the electrons did not interact, the energy of the second electron would also be $E_1$, the same as that of the first electron. Thus, for an atom with noninteracting electrons, the ionization energy would be $|E_1|$ and the ground-state energy would be $E_{non} = 2E_1$. This is represented by the lowest level in Figure 36-16. Because of the interaction energy, the ground-state energy is greater than $2E_1$. This is represented by the higher level labeled $E_g$ in the figure. When we add $E_{ion} = 24.6$ eV to ionize He, we obtain ionized helium, written He⁺, which has just one electron and therefore energy $E_1$.

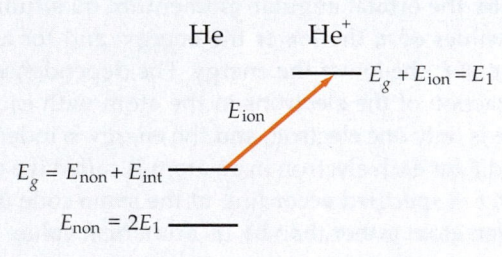

**FIGURE 36-16**

(*a*) 1. The energy of interaction plus the energy of two noninteracting electrons equals the ground-state energy of helium:

$$E_{int} + E_{non} = E_g$$

2. Solve for $E_{int}$ and substitute $E_{non} = 2E_1$:

$$E_{int} = E_g - E_{non} = E_g - 2E_1$$

3. Use Equation 36-27 to calculate the energy $E_1$ of one electron in the ground state:

$$E_n = -Z^2 \frac{E_0}{n^2}$$

so

$$E_1 = -(2)^2 \frac{13.6 \text{ eV}}{1^2} = -54.4 \text{ eV}$$

4. Substitute this value for $E_1$:

$$E_{int} = E_g - 2E_1 = E_g - 2(-54.4 \text{ eV})$$
$$= E_g + 108.8 \text{ eV}$$

5. The ground-state energy of He, $E_g$, plus the ionization energy equals the ground-state energy of He$^+$, which is $E_1$:

$$E_g + E_{ion} = E_1 = -54.4 \text{ eV}$$

6. Substitute $E_{ion} = 24.6$ eV to calculate $E_g$:

$$E_g = E_1 - E_{ion} = -54.4 \text{ eV} - 24.6 \text{ eV}$$
$$= -79 \text{ eV}$$

7. Substitute this result for $E_g$ to obtain $E_{int}$:

$$E_{int} = E_g + 108.8 \text{ eV} = -79 \text{ eV} + 108.8 \text{ eV}$$
$$= \boxed{29.8 \text{ eV}}$$

(b) 1. The energy of interaction of two electrons separated by distance $r_s$ apart is the potential energy:

$$U = +\frac{ke^2}{r_s}$$

2. Set $U$ equal to 29.8 eV, and solve for $r$. It is convenient to express $r$ in terms of $a_0$, the radius of the first Bohr orbit in hydrogen, and to use Equation 36-16:

$$r_s = \frac{ke^2}{U} = \frac{ke^2}{a_0}\frac{a_0}{U} = 2\frac{ke^2}{2a_0}\frac{a_0}{U} = 2\frac{E_0}{U}a_0$$
$$= 2\frac{13.6 \text{ eV}}{29.8 \text{ eV}}a_0 = \boxed{0.913a_0}$$

**PLAUSIBILITY   CHECK**   This   separation   is   approximately   the   size   of   the diameter $d_1$ of the first Bohr orbit for an electron in helium, which is $d_1 = 2r_1 = 2a_0/Z = a_0$.

## Lithium ($Z = 3$)

The next element, lithium, has three electrons. Since the $K$ shell ($n = 1$) is completely filled with two electrons, the third electron must go into a higher energy shell. The next lowest energy shell after $n = 1$ is the $n = 2$ or $L$ shell. The outer electron is much farther from the nucleus than are the two inner $n = 1$ electrons. It is most likely to be found at a radius near that of the second Bohr orbit, which is four times the radius of the first Bohr orbit.

The nuclear charge is partially screened from the outer electron by the two inner electrons. Recall that the electric field outside a spherically symmetric charge density is the same as if all the charge were at the center of the sphere. If the outer electron were completely outside the charge cloud of the two inner electrons, the electric field the outer electron would see would be that of a single charge $+e$ at the center due to the nuclear charge of $+3e$ and the charge $-2e$ of the inner electron cloud. However, the outer electron does not have a well-defined orbit; instead, it is itself a charge cloud that penetrates the charge cloud of the inner electrons to some extent. Because of this penetration, the effective nuclear charge $Z'e$ is somewhat greater than $+1e$. The energy of the outer electron at a distance $r$ from a point charge $+Z'e$ is given by Equation 36-6, with the nuclear charge $+Z$ replaced by $+Z'$.

$$E = -\frac{1}{2}\frac{kZ'e^2}{r} \qquad\qquad 36-46$$

The greater the penetration of the inner electron cloud, the greater the effective nuclear charge $Z'e$ and the lower the energy. Because the penetration is greater for $\ell$ values closer to zero (see Figure 36-12), the energy of the outer electron in lithium is lower for the s state ($\ell = 0$) than for the p state ($\ell = 1$). The electron configuration of lithium in the ground state is therefore $1s^2 2s$. The ionization energy of lithium is only 5.39 eV. Because its outer electron is so loosely bound to the atom, lithium is very active chemically. It behaves like a one-electron atom, similar to hydrogen.

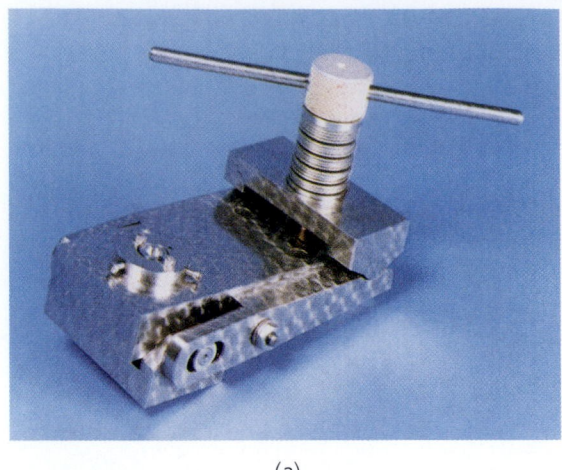
(a)

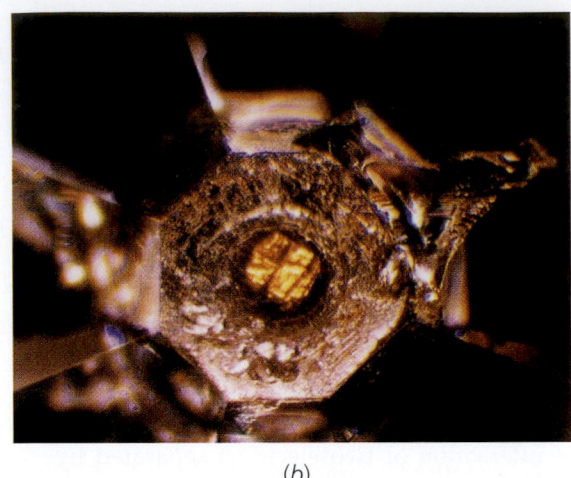
(b)

(a) A diamond anvil cell, in which the facets of two diamonds (approximately 1 mm² each) are used to compress a sample substance, subjecting it to very high pressure. (b) Samarium monosulfide (SmS) is normally a black, dull-looking semiconductor. When it is subjected to pressure above 7000 atm, an electron from the 4f state is dislocated into the 5d state. The resulting compound glitters like gold and behaves like a metal.

---

*EFFECTIVE NUCLEAR CHARGE FOR AN OUTER ELECTRON*     **EXAMPLE 36-7**

Suppose the electron cloud of the outer electron in the lithium atom in the ground state were completely outside the electron clouds of the two inner electrons, the nuclear charge would be shielded by the two inner electrons and the effective nuclear charge would be $Z'e = 1e$. Then the energy of the outer electron would be $-(13.6 \text{ eV})/2^2 = -3.4$ eV. However, the ionization energy of lithium is 5.39 eV, not 3.4 eV. Use this fact to calculate the effective nuclear charge $Z'$ seen by the outer electron in lithium.

**PICTURE THE PROBLEM** Because the outer electron is in the $n = 2$ shell, we will take $r = 4a_0$ for its average distance from the nucleus. We can then calculate $Z'$ from Equation 36-46. Since $r$ is given in terms of $a_0$, it will be convenient to use the fact that $E_0 = ke^2/(2a_0) = 13.6$ eV (Equation 36-16).

1. Equation 36-46 relates the energy of the outer electron to its average distance $r$ and the effective nuclear charge $Z'$:

$$E = \frac{1}{2}\frac{kZ'e^2}{r}$$

2. Substitute the given values $r = 4a_0$ and $E = -5.39$ eV:

$$-5.39 \text{ eV} = -\frac{1}{2}\frac{kZ'e^2}{4a_0}$$

3. Use $ke^2/(2a_0) = E_0 = 13.6$ eV and solve for $Z'$:

$$-5.39 \text{ eV} = -\frac{Z'}{4}\frac{ke^2}{2a_0} = -\frac{Z'}{4}(13.6 \text{ eV})$$

so

$$Z' = 4\frac{5.39 \text{ eV}}{13.6 \text{ eV}} = \boxed{1.59}$$

**REMARKS** This calculation is interesting but not very rigorous. We essentially used the radius ($r = 4a_0$) for the circular orbit from the semiclassical Bohr model and the measured ionization energy to calculate the effective inner charge seen by the outer electron. We know, of course, that this outer electron does not move in a circular orbit of constant radius, but is better represented by a stationary charged cloud of charge density $|\psi|^2$ that penetrates the charged clouds of the inner electrons.

## Beryllium (Z = 4)

The energy of the beryllium atom is a minimum if both outer electrons are in the 2s state. There can be two electrons with $n = 2$, $\ell = 0$, and $m_\ell = 0$ because of the two possible values for the spin quantum number $m_s$. The configuration of beryllium is thus $1s^2 2s^2$.

Hydrogen

## Boron to Neon (Z = 5 to Z = 10)

Since the 2s subshell is filled, the fifth electron must go into the next available (lowest energy) subshell, which is the 2p subshell, with $n = 2$ and $\ell = 1$. Since there are three possible values of $m$ (+1, 0, and −1) and two values of $m_s$ for each value of $m_\ell$, there can be six electrons in this subshell. The electron configuration for boron is $1s^2 2s^2 2p$. The electron configurations for the elements carbon ($Z = 6$) to neon ($Z = 10$) differ from that for boron only in the number of electrons in the 2p subshell. The ionization energy increases with $Z$ for these elements, reaching the value of 21.6 eV for the last element in the group, neon. Neon has the maximum number of electrons allowed in the $n = 2$ shell. The electron configuration for neon is $1s^2 2s^2 2p^6$. Because of its very high ionization energy, neon, like helium, basically is chemically inert. The element just before neon, fluorine, has a hole in the 2p subshell; that is, it has room for one more electron. It readily combines with elements such as lithium that have one outer electron. Lithium, for example, will donate its single outer electron to the fluorine atom to make an $F^-$ ion and a $Li^+$ ion. These ions then bond together to form a molecule of lithium fluoride.

Carbon

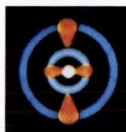

Silicon

## Sodium to Argon (Z = 11 to Z = 18)

The eleventh electron must go into the $n = 3$ shell. Since this electron is very far from the nucleus and from the inner electrons, it is weakly bound in the sodium ($Z = 11$) atom. The ionization energy of sodium is only 5.14 eV. Sodium therefore combines readily with atoms such as fluorine. With $n = 3$, the value of $\ell$ can be 0, 1, or 2. Because of the lowering of the energy due to penetration of the electron shield formed by the other ten electrons (similar to that discussed for lithium) the 3s state is lower than the 3p or 3d states. This energy difference between subshells of the same $n$ value becomes greater as the number of electrons increases. The electron configuration of sodium is $1s^2 2s^2 2p^6 3s^1$. As we move to elements with higher values of $Z$, the 3s subshell and then the 3p subshell fill. These two subshells can accommodate $2 + 6 = 8$ electrons. The configuration of argon ($Z = 18$) is $1s^2 2s^2 2p^6 3s^2 3p^6$. One might expect the nineteenth electron to go into the third subshell (the d subshell with $\ell = 2$), but the penetration effect is now so strong that the energy of the next electron is lower in the 4s subshell than in the 3d subshell. There is thus another large energy difference between the eighteenth and nineteenth electrons, and so argon, with its full 3p subshell, is basically stable and inert.

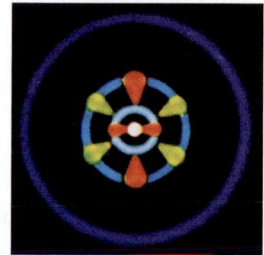

Iron

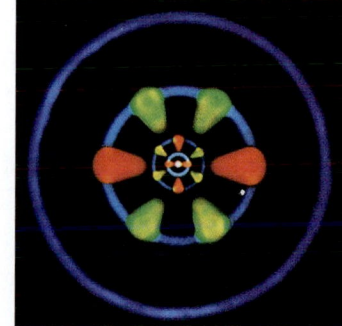

Silver

## Elements With Z > 18

The nineteenth electron in potassium ($Z = 19$) and the twentieth electron in calcium ($Z = 20$) go into the 4s subshell rather than the 3d subshell. The electron configurations of the next ten elements, scandium ($Z = 21$) through zinc ($Z = 30$),

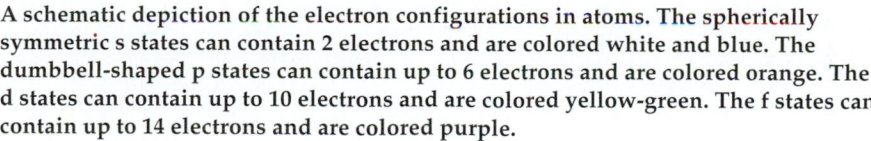

A schematic depiction of the electron configurations in atoms. The spherically symmetric s states can contain 2 electrons and are colored white and blue. The dumbbell-shaped p states can contain up to 6 electrons and are colored orange. The d states can contain up to 10 electrons and are colored yellow-green. The f states can contain up to 14 electrons and are colored purple.

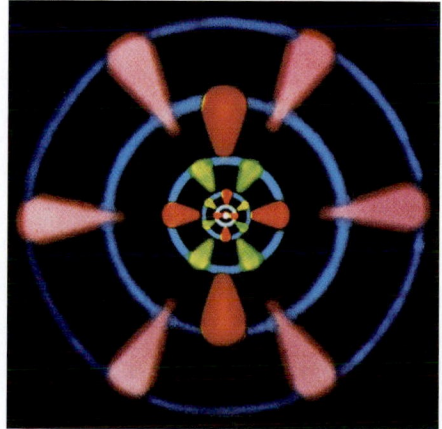

Europium

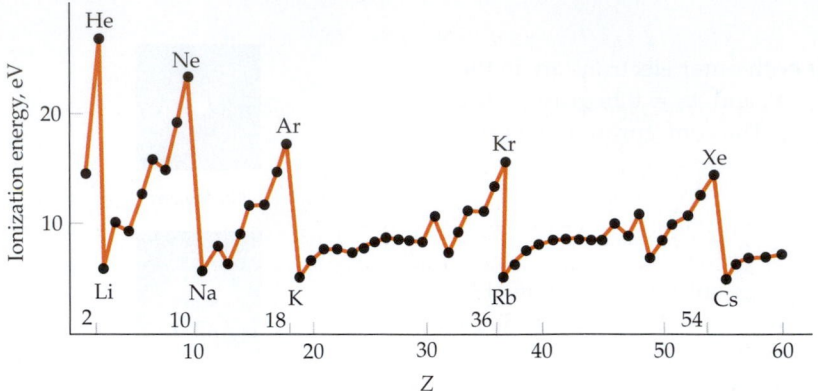

**FIGURE 36-17** Ionization energy versus Z for Z = 1 to Z = 60. This energy is the binding energy of the last electron in the atom. The binding energy increases with Z until a shell is closed at Z = 2, 10, 18, 36, and 54. Elements with a closed shell plus one outer electron, such as sodium (Z = 11), have very low binding energies because the outer electron is very far from the nucleus and is shielded by the inner core electrons.

differ only in the number of electrons in the 3d shell, except for chromium (Z = 24) and copper (Z = 29), each of which has only one 4s electron. These ten elements are called **transition elements.**

Figure 36-17 shows a plot of the ionization energy versus Z for Z = 1 to Z = 60. The peaks in ionization energy at Z = 2, 10, 18, 36, and 54 mark the closing of a shell or subshell. Table 36-1 gives the ground-state electron configurations of the elements up to atomic number 109.

## TABLE 36-1

**Electron Configurations of the Atoms in Their Ground States**

**For some of the rare-earth elements (Z = 57 to 71) and the heavy elements (Z > 89) the configurations are not firmly established.**

| | | Shell (n): | K (1) | L (2) | | M (3) | | | N (4) | | | | O (5) | | | | P (6) | | | Q (7) |
|---|---|---|---|---|---|---|---|---|---|---|---|---|---|---|---|---|---|---|---|---|
| | | | s | s | p | s | p | d | s | p | d | f | s | p | d | f | s | p | d | s |
| Z | Element | Subshell (ℓ): | (0) | (0) | (1) | (0) | (1) | (2) | (0) | (1) | (2) | (3) | (0) | (1) | (2) | (3) | (0) | (1) | (2) | (1) |
| 1 | H hydrogen | | 1 | | | | | | | | | | | | | | | | | |
| 2 | He helium | | 2 | | | | | | | | | | | | | | | | | |
| 3 | Li lithium | | 2 | 1 | | | | | | | | | | | | | | | | |
| 4 | Be beryllium | | 2 | 2 | | | | | | | | | | | | | | | | |
| 5 | B boron | | 2 | 2 | 1 | | | | | | | | | | | | | | | |
| 6 | C carbon | | 2 | 2 | 2 | | | | | | | | | | | | | | | |
| 7 | N nitrogen | | 2 | 2 | 3 | | | | | | | | | | | | | | | |
| 8 | O oxygen | | 2 | 2 | 4 | | | | | | | | | | | | | | | |
| 9 | F fluorine | | 2 | 2 | 5 | | | | | | | | | | | | | | | |
| 10 | Ne neon | | 2 | 2 | 6 | | | | | | | | | | | | | | | |
| 11 | Na sodium | | 2 | 2 | 6 | 1 | | | | | | | | | | | | | | |
| 12 | Mg magnesium | | 2 | 2 | 6 | 2 | | | | | | | | | | | | | | |
| 13 | Al aluminum | | 2 | 2 | 6 | 2 | 1 | | | | | | | | | | | | | |
| 14 | Si silicon | | 2 | 2 | 6 | 2 | 2 | | | | | | | | | | | | | |
| 15 | P phosphorus | | 2 | 2 | 6 | 2 | 3 | | | | | | | | | | | | | |
| 16 | S sulfur | | 2 | 2 | 6 | 2 | 4 | | | | | | | | | | | | | |
| 17 | Cl chlorine | | 2 | 2 | 6 | 2 | 5 | | | | | | | | | | | | | |
| 18 | Ar argon | | 2 | 2 | 6 | 2 | 6 | | | | | | | | | | | | | |
| 19 | K potassium | | 2 | 2 | 6 | 2 | 6 | | 1 | | | | | | | | | | | |

## TABLE 36-1 (continued)

**Electron Configurations of the Atoms in Their Ground States**

For some of the rare-earth elements ($Z = 57$ to $71$) and the heavy elements ($Z > 89$) the configurations are not firmly established.

| Z | Element | K(1) s(0) | L(2) s(0) | L(2) p(1) | M(3) s(0) | M(3) p(1) | M(3) d(2) | N(4) s(0) | N(4) p(1) | N(4) d(2) | N(4) f(3) | O(5) s(0) | O(5) p(1) | O(5) d(2) | O(5) f(3) | P(6) s(0) | P(6) p(1) | P(6) d(2) | Q(7) s(1) |
|---|---|---|---|---|---|---|---|---|---|---|---|---|---|---|---|---|---|---|---|
| 20 | Ca calcium | 2 | 2 | 6 | 2 | 6 | . | 2 | | | | | | | | | | | |
| 21 | Sc scandium | 2 | 2 | 6 | 2 | 6 | 1 | 2 | | | | | | | | | | | |
| 22 | Ti titanium | 2 | 2 | 6 | 2 | 6 | 2 | 2 | | | | | | | | | | | |
| 23 | V vanadium | 2 | 2 | 6 | 2 | 6 | 3 | 2 | | | | | | | | | | | |
| 24 | Cr chromium | 2 | 2 | 6 | 2 | 6 | 5 | 1 | | | | | | | | | | | |
| 25 | Mn manganese | 2 | 2 | 6 | 2 | 6 | 5 | 2 | | | | | | | | | | | |
| 26 | Fe iron | 2 | 2 | 6 | 2 | 6 | 6 | 2 | | | | | | | | | | | |
| 27 | Co cobalt | 2 | 2 | 6 | 2 | 6 | 7 | 2 | | | | | | | | | | | |
| 28 | Ni nickel | 2 | 2 | 6 | 2 | 6 | 8 | 2 | | | | | | | | | | | |
| 29 | Cu copper | 2 | 2 | 6 | 2 | 6 | 10 | 1 | | | | | | | | | | | |
| 30 | Zn zinc | 2 | 2 | 6 | 2 | 6 | 10 | 2 | | | | | | | | | | | |
| 31 | Ga gallium | 2 | 2 | 6 | 2 | 6 | 10 | 2 | 1 | | | | | | | | | | |
| 32 | Ge germanium | 2 | 2 | 6 | 2 | 6 | 10 | 2 | 2 | | | | | | | | | | |
| 33 | As arsenic | 2 | 2 | 6 | 2 | 6 | 10 | 2 | 3 | | | | | | | | | | |
| 34 | Se selenium | 2 | 2 | 6 | 2 | 6 | 10 | 2 | 4 | | | | | | | | | | |
| 35 | Br bromine | 2 | 2 | 6 | 2 | 6 | 10 | 2 | 5 | | | | | | | | | | |
| 36 | Kr krypton | 2 | 2 | 6 | 2 | 6 | 10 | 2 | 6 | | | | | | | | | | |
| 37 | Rb rubidium | 2 | 2 | 6 | 2 | 6 | 10 | 2 | 6 | . | . | 1 | | | | | | | |
| 38 | Sr strontium | 2 | 2 | 6 | 2 | 6 | 10 | 2 | 6 | . | . | 2 | | | | | | | |
| 39 | Y yttrium | 2 | 2 | 6 | 2 | 6 | 10 | 2 | 6 | 1 | . | 2 | | | | | | | |
| 40 | Zr zirconium | 2 | 2 | 6 | 2 | 6 | 10 | 2 | 6 | 2 | . | 2 | | | | | | | |
| 41 | Nb niobium | 2 | 2 | 6 | 2 | 6 | 10 | 2 | 6 | 4 | . | 1 | | | | | | | |
| 42 | Mo molybdenum | 2 | 2 | 6 | 2 | 6 | 10 | 2 | 6 | 5 | . | 1 | | | | | | | |
| 43 | Tc technetium | 2 | 2 | 6 | 2 | 6 | 10 | 2 | 6 | 6 | . | 1 | | | | | | | |
| 44 | Ru ruthenium | 2 | 2 | 6 | 2 | 6 | 10 | 2 | 6 | 7 | . | 1 | | | | | | | |
| 45 | Rh rhodium | 2 | 2 | 6 | 2 | 6 | 10 | 2 | 6 | 8 | . | 1 | | | | | | | |
| 46 | Pd palladium | 2 | 2 | 6 | 2 | 6 | 10 | 2 | 6 | 10 | . | . | | | | | | | |
| 47 | Ag silver | 2 | 2 | 6 | 2 | 6 | 10 | 2 | 6 | 10 | . | 1 | | | | | | | |
| 48 | Cd cadmium | 2 | 2 | 6 | 2 | 6 | 10 | 2 | 6 | 10 | . | 2 | | | | | | | |
| 49 | In indium | 2 | 2 | 6 | 2 | 6 | 10 | 2 | 6 | 10 | . | 2 | 1 | | | | | | |
| 50 | Sn tin | 2 | 2 | 6 | 2 | 6 | 10 | 2 | 6 | 10 | . | 2 | 2 | | | | | | |
| 51 | Sb antimony | 2 | 2 | 6 | 2 | 6 | 10 | 2 | 6 | 10 | . | 2 | 3 | | | | | | |
| 52 | Te tellurium | 2 | 2 | 6 | 2 | 6 | 10 | 2 | 6 | 10 | . | 2 | 4 | | | | | | |
| 53 | I iodine | 2 | 2 | 6 | 2 | 6 | 10 | 2 | 6 | 10 | . | 2 | 5 | | | | | | |
| 54 | Xe xenon | 2 | 2 | 6 | 2 | 6 | 10 | 2 | 6 | 10 | . | 2 | 6 | | | | | | |
| 55 | Cs cesium | 2 | 2 | 6 | 2 | 6 | 10 | 2 | 6 | 10 | . | 2 | 6 | . | . | 1 | | | |
| 56 | Ba barium | 2 | 2 | 6 | 2 | 6 | 10 | 2 | 6 | 10 | . | 2 | 6 | . | . | 2 | | | |
| 57 | La lanthanum | 2 | 2 | 6 | 2 | 6 | 10 | 2 | 6 | 10 | . | 2 | 6 | 1 | . | 2 | | | |

## TABLE 36-1 (continued)

Electron Configurations of the Atoms in Their Ground States

For some of the rare-earth elements ($Z$ = 57 to 71) and the heavy elements ($Z > 89$) the configurations are not firmly established.

| | | Shell ($n$): | K (1) | L (2) | | M (3) | | | N (4) | | | | O (5) | | | | P (6) | | | Q (7) |
|---|---|---|---|---|---|---|---|---|---|---|---|---|---|---|---|---|---|---|---|---|
| | | | s | s | p | s | p | d | s | p | d | f | s | p | d | f | s | p | d | s |
| $Z$ | Element | Subshell ($\ell$): | (0) | (0) | (1) | (0) | (1) | (2) | (0) | (1) | (2) | (3) | (0) | (1) | (2) | (3) | (0) | (1) | (2) | (1) |
| 58 | Ce | cerium | 2 | 2 | 6 | 2 | 6 | 10 | 2 | 6 | 10 | 1 | 2 | 6 | 1 | . | 2 | | | |
| 59 | Pr | praseodymium | 2 | 2 | 6 | 2 | 6 | 10 | 2 | 6 | 10 | 3 | 2 | 6 | . | . | 2 | | | |
| 60 | Nd | neodymium | 2 | 2 | 6 | 2 | 6 | 10 | 2 | 6 | 10 | 4 | 2 | 6 | . | . | 2 | | | |
| 61 | Pm | promethium | 2 | 2 | 6 | 2 | 6 | 10 | 2 | 6 | 10 | 5 | 2 | 6 | . | . | 2 | | | |
| 62 | Sm | samarium | 2 | 2 | 6 | 2 | 6 | 10 | 2 | 6 | 10 | 6 | 2 | 6 | . | . | 2 | | | |
| 63 | Eu | europium | 2 | 2 | 6 | 2 | 6 | 10 | 2 | 6 | 10 | 7 | 2 | 6 | . | . | 2 | | | |
| 64 | Gd | gadolinium | 2 | 2 | 6 | 2 | 6 | 10 | 2 | 6 | 10 | 7 | 2 | 6 | 1 | . | 2 | | | |
| 65 | Tb | terbium | 2 | 2 | 6 | 2 | 6 | 10 | 2 | 6 | 10 | 9 | 2 | 6 | . | . | 2 | | | |
| 66 | Dy | dysprosium | 2 | 2 | 6 | 2 | 6 | 10 | 2 | 6 | 10 | 10 | 2 | 6 | . | . | 2 | | | |
| 67 | Ho | holmium | 2 | 2 | 6 | 2 | 6 | 10 | 2 | 6 | 10 | 11 | 2 | 6 | . | . | 2 | | | |
| 68 | Er | erbium | 2 | 2 | 6 | 2 | 6 | 10 | 2 | 6 | 10 | 12 | 2 | 6 | . | . | 2 | | | |
| 69 | Tm | thulium | 2 | 2 | 6 | 2 | 6 | 10 | 2 | 6 | 10 | 13 | 2 | 6 | . | . | 2 | | | |
| 70 | Yb | ytterbium | 2 | 2 | 6 | 2 | 6 | 10 | 2 | 6 | 10 | 14 | 2 | 6 | . | . | 2 | | | |
| 71 | Lu | lutetium | 2 | 2 | 6 | 2 | 6 | 10 | 2 | 6 | 10 | 14 | 2 | 6 | 1 | . | 2 | | | |
| 72 | Hf | hafnium | 2 | 2 | 6 | 2 | 6 | 10 | 2 | 6 | 10 | 14 | 2 | 6 | 2 | . | 2 | | | |
| 73 | Ta | tantalum | 2 | 2 | 6 | 2 | 6 | 10 | 2 | 6 | 10 | 14 | 2 | 6 | 3 | . | 2 | | | |
| 74 | W | tungsten (wolfram) | 2 | 2 | 6 | 2 | 6 | 10 | 2 | 6 | 10 | 14 | 2 | 6 | 4 | . | 2 | | | |
| 75 | Re | rhenium | 2 | 2 | 6 | 2 | 6 | 10 | 2 | 6 | 10 | 14 | 2 | 6 | 5 | . | 2 | | | |
| 76 | Os | osmium | 2 | 2 | 6 | 2 | 6 | 10 | 2 | 6 | 10 | 14 | 2 | 6 | 6 | . | 2 | | | |
| 77 | Ir | iridium | 2 | 2 | 6 | 2 | 6 | 10 | 2 | 6 | 10 | 14 | 2 | 6 | 7 | . | 2 | | | |
| 78 | Pt | platinum | 2 | 2 | 6 | 2 | 6 | 10 | 2 | 6 | 10 | 14 | 2 | 6 | 9 | . | 1 | | | |
| 79 | Au | gold | 2 | 2 | 6 | 2 | 6 | 10 | 2 | 6 | 10 | 14 | 2 | 6 | 10 | . | 1 | | | |
| 80 | Hg | mercury | 2 | 2 | 6 | 2 | 6 | 10 | 2 | 6 | 10 | 14 | 2 | 6 | 10 | . | 2 | | | |
| 81 | Tl | thallium | 2 | 2 | 6 | 2 | 6 | 10 | 2 | 6 | 10 | 14 | 2 | 6 | 10 | . | 2 | 1 | | |
| 82 | Pb | lead | 2 | 2 | 6 | 2 | 6 | 10 | 2 | 6 | 10 | 14 | 2 | 6 | 10 | . | 2 | 2 | | |
| 83 | Bi | bismuth | 2 | 2 | 6 | 2 | 6 | 10 | 2 | 6 | 10 | 14 | 2 | 6 | 10 | . | 2 | 3 | | |
| 84 | Po | polonium | 2 | 2 | 6 | 2 | 6 | 10 | 2 | 6 | 10 | 14 | 2 | 6 | 10 | . | 2 | 4 | | |
| 85 | At | astatine | 2 | 2 | 6 | 2 | 6 | 10 | 2 | 6 | 10 | 14 | 2 | 6 | 10 | . | 2 | 5 | | |
| 86 | Rn | radon | 2 | 2 | 6 | 2 | 6 | 10 | 2 | 6 | 10 | 14 | 2 | 6 | 10 | . | 2 | 6 | | |
| 87 | Fr | francium | 2 | 2 | 6 | 2 | 6 | 10 | 2 | 6 | 10 | 14 | 2 | 6 | 10 | . | 2 | 6 | . | 1 |
| 88 | Ra | radium | 2 | 2 | 6 | 2 | 6 | 10 | 2 | 6 | 10 | 14 | 2 | 6 | 10 | . | 2 | 6 | . | 2 |
| 89 | Ac | actinium | 2 | 2 | 6 | 2 | 6 | 10 | 2 | 6 | 10 | 14 | 2 | 6 | 10 | . | 2 | 6 | 1 | 2 |
| 90 | Th | thorium | 2 | 2 | 6 | 2 | 6 | 10 | 2 | 6 | 10 | 14 | 2 | 6 | 10 | . | 2 | 6 | 2 | 2 |
| 91 | Pa | protactinium | 2 | 2 | 6 | 2 | 6 | 10 | 2 | 6 | 10 | 14 | 2 | 6 | 10 | 1 | 2 | 6 | 2 | 2 |
| 92 | U | uranium | 2 | 2 | 6 | 2 | 6 | 10 | 2 | 6 | 10 | 14 | 2 | 6 | 10 | 3 | 2 | 6 | 1 | 2 |
| 93 | Np | neptunium | 2 | 2 | 6 | 2 | 6 | 10 | 2 | 6 | 10 | 14 | 2 | 6 | 10 | 4 | 2 | 6 | 1 | 2 |
| 94 | Pu | plutonium | 2 | 2 | 6 | 2 | 6 | 10 | 2 | 6 | 10 | 14 | 2 | 6 | 10 | 6 | 2 | 6 | . | 2 |
| 95 | Am | americium | 2 | 2 | 6 | 2 | 6 | 10 | 2 | 6 | 10 | 14 | 2 | 6 | 10 | 7 | 2 | 6 | . | 2 |

## TABLE 36-1 (continued)

**Electron Configurations of the Atoms in Their Ground States**
**For some of the rare-earth elements ($Z = 57$ to $71$) and the heavy elements ($Z > 89$) the configurations are not firmly established.**

| Z | Element | | K(1) s (0) | L(2) s (0) | L(2) p (1) | M(3) s (0) | M(3) p (1) | M(3) d (2) | N(4) s (0) | N(4) p (1) | N(4) d (2) | N(4) f (3) | O(5) s (0) | O(5) p (1) | O(5) d (2) | O(5) f (3) | P(6) s (0) | P(6) p (1) | P(6) d (2) | Q(7) s (1) |
|----|----|----|----|----|----|----|----|----|----|----|----|----|----|----|----|----|----|----|----|----|
| 96 | Cm | curium | 2 | 2 | 6 | 2 | 6 | 10 | 2 | 6 | 10 | 14 | 2 | 6 | 10 | 7 | 2 | 6 | 1 | 2 |
| 97 | Bk | berkelium | 2 | 2 | 6 | 2 | 6 | 10 | 2 | 6 | 10 | 14 | 2 | 6 | 10 | 8 | 2 | 6 | 1 | 2 |
| 98 | Cf | californium | 2 | 2 | 6 | 2 | 6 | 10 | 2 | 6 | 10 | 14 | 2 | 6 | 10 | 10 | 2 | 6 | . | 2 |
| 99 | Es | einsteinium | 2 | 2 | 6 | 2 | 6 | 10 | 2 | 6 | 10 | 14 | 2 | 6 | 10 | 11 | 2 | 6 | . | 2 |
| 100 | Fm | fermium | 2 | 2 | 6 | 2 | 6 | 10 | 2 | 6 | 10 | 14 | 2 | 6 | 10 | 12 | 2 | 6 | . | 2 |
| 101 | Md | mendelevium | 2 | 2 | 6 | 2 | 6 | 10 | 2 | 6 | 10 | 14 | 2 | 6 | 10 | 13 | 2 | 6 | . | 2 |
| 102 | No | nobelium | 2 | 2 | 6 | 2 | 6 | 10 | 2 | 6 | 10 | 14 | 2 | 6 | 10 | 14 | 2 | 6 | . | 2 |
| 103 | Lr | lawrencium | 2 | 2 | 6 | 2 | 6 | 10 | 2 | 6 | 10 | 14 | 2 | 6 | 10 | 14 | 2 | 6 | 1 | 2 |
| 104 | Rf | rutherfordium | 2 | 2 | 6 | 2 | 6 | 10 | 2 | 6 | 10 | 14 | 2 | 6 | 10 | 14 | 2 | 6 | 2 | 2 |
| 105 | Db | dubnium | 2 | 2 | 6 | 2 | 6 | 10 | 2 | 6 | 10 | 14 | 2 | 6 | 10 | 14 | 2 | 6 | 3 | 2 |
| 106 | Sg | seaborgium | 2 | 2 | 6 | 2 | 6 | 10 | 2 | 6 | 10 | 14 | 2 | 6 | 10 | 14 | 2 | 6 | 4 | 2 |
| 107 | Bh | bohrium | 2 | 2 | 6 | 2 | 6 | 10 | 2 | 6 | 10 | 14 | 2 | 6 | 10 | 14 | 2 | 6 | 5 | 2 |
| 108 | Hs | hassium | 2 | 2 | 6 | 2 | 6 | 10 | 2 | 6 | 10 | 14 | 2 | 6 | 10 | 14 | 2 | 6 | 6 | 2 |
| 109 | Mt | meitnerium | 2 | 2 | 6 | 2 | 6 | 10 | 2 | 6 | 10 | 14 | 2 | 6 | 10 | 14 | 2 | 6 | 7 | 2 |

## 36-7 Optical Spectra and X-Ray Spectra

When an atom is in an excited state (i.e., when it is in an energy state above the ground state), it makes transitions to lower energy states, and in doing so emits electromagnetic radiation. The wavelength of the electromagnetic radiation emitted is related to the initial and final states by the Bohr formula (Equation 36-17), $\lambda = hc/(E_i - E_f)$, where $E_i$ and $E_f$ are the initial and final energies and $h$ is Planck's constant. The atom can be excited to a higher energy state by bombarding the atom with a beam of electrons, as in a spectral tube with a high voltage across it. Since the excited energy states of an atom form a discrete (rather than continuous) set, only certain wavelengths are emitted. These wavelengths of the emitted radiation constitute the emission spectrum of the atom.

### Optical Spectra

To understand atomic spectra we need to understand the excited states of the atom. The situation for an atom with many electrons is, in general, much more complicated than that of hydrogen with just one electron. An excited state of the atom may involve a change in the state of any one of the electrons, or even two or more electrons. Fortunately, in most cases, an excited state of an atom involves the excitation of just one of the electrons in the atom. The energies of excitation of the outer, valence electrons of an atom are of the order of a few electron volts. Transitions involving these electrons result in photons in or near the visible or **optical spectrum.** (Recall that the energies of visible photons range from approximately 1.5 eV to 3 eV.) The excitation energies can often be calculated from a simple model in which the atom is pictured as a single electron plus a stable core consisting of the nucleus plus the other inner electrons. This model works particularly well for the alkali metals: Li, Na, K, Rb, and Cs. These elements are in the first column of the periodic table. The optical spectra of these elements are similar to the optical spectra of hydrogen.

Figure 36-18 shows an energy-level diagram for the optical transitions in sodium, whose electrons form a neon core plus one outer electron. Since the total spin angular momentum of the core adds up to zero, the spin of each state of sodium is $\frac{1}{2}$. Because of the spin–orbit effect, the states with $J = L - \frac{1}{2}$ have a slightly different energy than those with $J = L + \frac{1}{2}$ (except for states with $L = 0$). Each state (except for the $L = 0$ states) is therefore split into two states, called a doublet. The doublet splitting is very small and not evident on the energy scale of this diagram. The usual spectroscopic notation is that these states are labeled with a superscript given by $2S + 1$, followed by a letter denoting the orbital angular momentum, followed by a subscript denoting the total angular momentum $J$. For states with a total spin angular momentum $S = \frac{1}{2}$ the superscript is 2, indicating the state is a doublet. Thus $^2P_{3/2}$, read as "doublet P three halves," denotes a state in which $L = 1$ and $J = \frac{3}{2}$. (The $L = 0$, or $S$, states are customarily labeled as if they were doublets even though they are not.) In the first excited state, the outer electron is excited from the 3s level to the 3p level, which is approximately 2.1 eV above the ground state. The energy difference between the $P_{3/2}$ and $P_{1/2}$

A neon sign outside a Chinatown restaurant in Paris. Neon atoms in the tube are excited by an electron current passing through the tube. The excited neon atoms emit light in the visible range as they decay toward their ground states. The colors of neon signs result from the characteristic red-orange spectrum of neon plus the color of the glass tube itself.

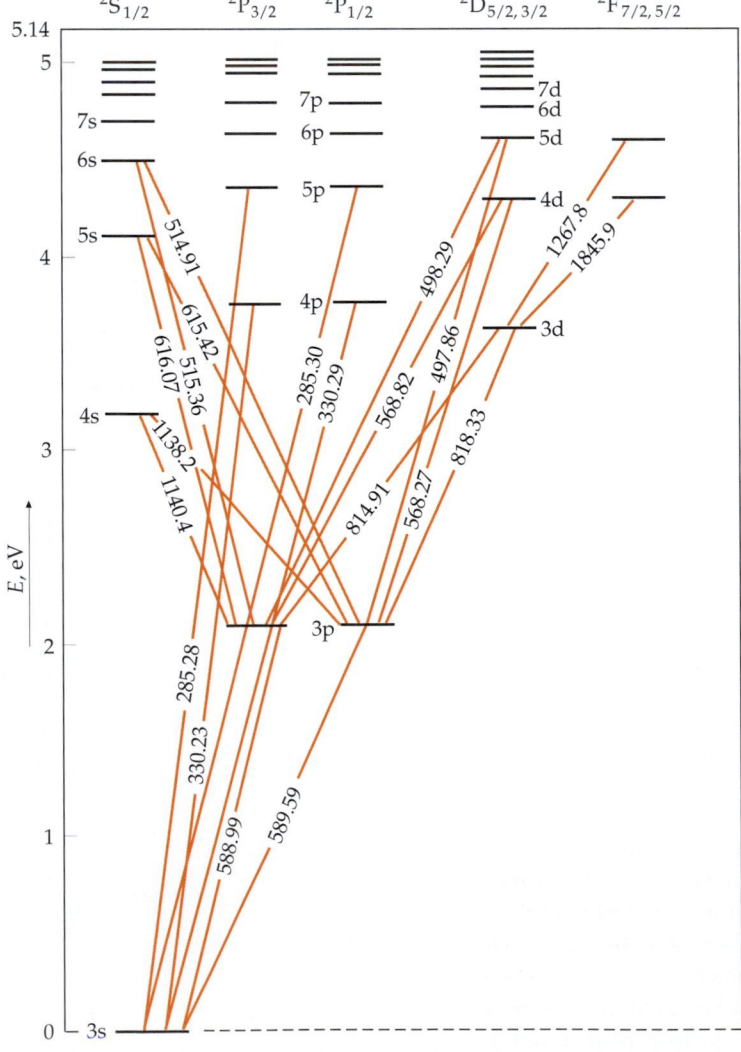

**FIGURE 36-18** Energy-level diagram for sodium. The diagonal lines show observed optical transitions, with wavelengths given in nanometers. The energy of the ground state has been chosen as the zero point for the scale on the left.

states due to the spin–orbit effect is about 0.002 eV. Transitions from these states to the ground state give the familiar sodium yellow doublet:

$$3p(^2P_{1/2}) \rightarrow 3s(^2S_{1/2}), \quad \lambda = 589.6 \text{ nm}$$

$$3p(^2P_{3/2}) \rightarrow 3s(^2S_{1/2}), \quad \lambda = 589.0 \text{ nm}$$

The energy levels and spectra of other alkali metal atoms are similar to those for sodium. The optical spectrum for atoms such as helium, beryllium, and magnesium that have two outer electrons is considerably more complex because of the interaction of the two outer electrons.

## X-Ray Spectra

X rays are usually produced in the laboratory by bombarding a target element with a high-energy beam of electrons in an X-ray tube. The result (Figure 36-19) consists of a continuous spectrum that depends only on the energy of the bombarding electrons and a line spectrum that is characteristic of the target element. The characteristic spectrum results from excitation of the inner core electrons in the target element.

The energy needed to excite an inner core electron—for example, an electron in the $n = 1$ state ($K$ shell)—is much greater than the energy required to excite an outer, valence electron. An inner electron cannot be excited to any of the filled states (e.g., the $n = 2$ states in sodium) because of the exclusion principle. The energy required to excite an inner core electron to an unoccupied state is typically of the order of several kilo-electron volts. If an electron is knocked out of the $n = 1$ state ($K$ shell), there is a vacancy left in this shell. This vacancy can be filled if an electron in the $L$ shell (or in a higher shell) makes a transition into the $K$ shell. The photons emitted by electrons making such transitions also have energies of the order of kilo-electron volts and produce the sharp peaks in the X-ray spectrum, as shown in Figure 36-19. The $K_\alpha$ line arises from transitions from the $n = 2$ ($L$) shell to the $n = 1$ ($K$) shell. The $K_\beta$ line arises from transitions from the $n = 3$ shell to the $n = 1$ shell. These and other lines arising from transitions ending at the $n = 1$ shell make up the $K$ series of the characteristic X-ray spectrum of the target element. Similarly, a second series, the $L$ series, is produced by transitions from higher energy states to a vacated place in the $n = 2$ ($L$) shell. The letters $K, L, M,$ and so on, designate the final shell of the electron making the transition and the series $\alpha, \beta,$ and so on, designates the number of shells above the final shell for the initial state of the electron.

In 1913, the English physicist Henry Moseley measured the wavelengths of the characteristic $K_\alpha$ X-ray spectra for approximately forty elements. Using this data, Moseley showed that a plot of $\lambda^{-1/2}$ versus the order in which the elements appeared in the periodic table resulted in a straight line (with a few gaps and a few outliers). From his data, Moseley was able to accurately determine the atomic number $Z$ for each known element, and to predict the existence of some elements that were later discovered. The equation of the straight line of his plot is given by

$$\frac{1}{\sqrt{\lambda_{K_\alpha}}} = a(Z - 1)$$

The work of Bohr and Moseley can be combined to obtain an equation relating the wavelength of the emitted photon and the atomic number. According to the Bohr model of a single-electron atom (see Equation 36-13), the wavelength of the emitted photon when the electron makes the transition from $n = 2$ to $n = 1$ is given by

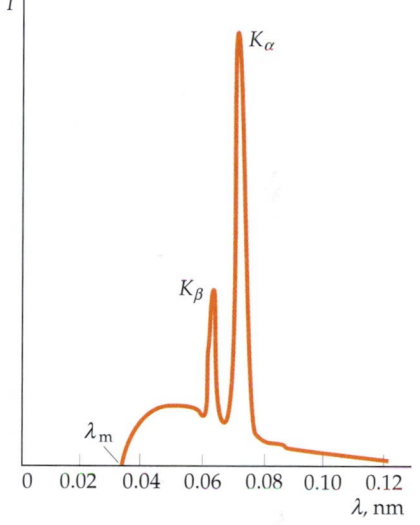

**FIGURE 36-19** X-ray spectrum of molybdenum. The sharp peaks labeled $K_\alpha$ and $K_\beta$ are characteristic of the element. The cutoff wavelength $\lambda_m$ is independent of the target element and is related to the voltage $V$ of the X-ray tube by $\lambda_m = hc/eV$.

$$\frac{1}{\lambda} = Z^2 \frac{E_0}{hc}\left(1 - \frac{1}{2^2}\right)$$

where $E_0 = 13.6$ eV is the binding energy of the ground-state hydrogen atom. Taking the square root of both sides gives

$$\frac{1}{\sqrt{\lambda_{K_\alpha}}} = \left[\frac{E_0}{hc}\left(1 - \frac{1}{2^2}\right)\right]^{1/2} Z$$

Moseley's equation and this equation are in agreement if $Z - 1$ is substituted for $Z$ in Moseley's equation and if $a = 3E_0/(4hc)$. This raises the question, why a factor of $Z - 1$ instead of a factor of $Z$? Part of the explanation is that the formula from the Bohr theory ignores the shielding of the nuclear charge. In a multi-electron atom, electrons in the $n = 2$ states are electrically shielded from the nuclear charge by the two electrons in the $n = 1$ state, so the $n = 2$ state electrons are attracted by an effective nuclear charge of about $(Z - 2)e$. However, when there is only one electron in the K shell, the $n = 2$ electrons are attracted by an effective nuclear charge of about $(Z - 1)e$. When an electron from state $n$ drops into the vacated state in the $n = 1$ shell, a photon of energy $E_2 - E_1$ is emitted. The wavelength of this photon is

$$\lambda_{K_\alpha} = \frac{hc}{(Z - 1)^2 E_0 \left(1 - \frac{1}{2^2}\right)} \qquad \text{36-47}$$

which is obtained from the previous equation with $Z - 1$ substituted for $Z$.

---

*IDENTIFYING THE ELEMENT FROM THE $K_\alpha$ X-RAY LINE*     **E X A M P L E 3 6 - 8**

**The wavelength of the $K_\alpha$ X-ray line for a certain element is $\lambda = 0.0721$ nm. What is the element?**

**PICTURE THE PROBLEM** The $K_\alpha$ line corresponds to a transition from $n = 2$ to $n = 1$. The wavelength is related to the atomic number $Z$ by Equation 36-47.

1. Solve Equation 36-47 for $(Z - 1)^2$:

$$\lambda_{K_\alpha} = \frac{hc}{(Z - 1)^2 E_0 \left(1 - \frac{1}{2^2}\right)}$$

so

$$(Z - 1)^2 = \frac{4hc}{3\lambda_{K_\alpha} E_0}$$

2. Substitute the given data and solve for $Z$:

$$(Z - 1)^2 = \frac{4(1240 \text{ eV·nm})}{3(0.0721 \text{ nm})(13.6 \text{ eV})} = 1686$$

so

$$Z = 1 + \sqrt{1686} = 42.06$$

3. Since $Z$ is an integer, we round to the nearest integer:

$$Z = 42$$

The element is molybdenum.

# SUMMARY

1.  The Bohr model is important historically because it was the first model to succeed at explaining the discrete optical spectrum of atoms in terms of the quantization of energy. It has been superceded by the quantum-mechanical model.

2.  The quantum theory of atoms results from the application of the Schrödinger equation to a bound system consisting of nucleus of charge $+Ze$ and $Z$ electrons of charge $-e$.

3.  For the simplest atom, hydrogen, consisting of one proton and one electron, the time independent Schrödinger equation can be solved exactly to obtain the wave functions $\psi$, which depend on the quantum numbers $n$, $\ell$, $m_\ell$, and $m_s$.

4.  The electron configuration of atoms is governed by the Pauli exclusion principle, which states that no two electrons in an atom can have the same set of values for the quantum numbers $n$, $\ell$, $m_\ell$, and $m_s$. Using the exclusion principle and the restrictions on the quantum numbers, we can understand much of the structure of the periodic table.

| Topic | Relevant Equations and Remarks | |
|---|---|---|
| **1. The Bohr Model of the Hydrogen Atom** | | |
| Postulates for the hydrogen atom | | |
| Nonradiating orbits | The electron moves in a circular nonradiating orbit around the proton. | |
| Photon frequency from energy conservation | $f = \dfrac{E_i - E_f}{h}$ | 36-7 |
| Quantized angular momentum | $mvr = n\hbar, \qquad n = 1, 2, 3, \ldots$ | 36-9 |
| First Bohr radius | $a_0 = \dfrac{\hbar^2}{mke^2} \approx 0.0529 \text{ nm}$ | 36-12 |
| Radius of the Bohr orbits | $r = n^2 \dfrac{a_0}{Z}$ | 36-11 |
| Energy levels in the hydrogen atom | $E_n = -\dfrac{mk^2e^4}{2\hbar^2}\dfrac{Z^2}{n^2} = -Z^2\dfrac{E_0}{n^2}$ | 36-15 |
| | where | |
| | $E_0 = \dfrac{mk^2e^4}{2\hbar^2} = \dfrac{1}{2}\dfrac{ke^2}{a_0} \approx 13.6 \text{ eV}$ | 36-16 |
| Wavelengths emitted by the hydrogen atom | $\lambda = \dfrac{c}{f} = \dfrac{hc}{E_i - E_f} = \dfrac{1240 \text{ eV·nm}}{E_i - E_f}$ | 36-17, 36-18 |
| **2. Quantum Theory of Atoms** | The electron is described by a wave function $\psi$ that is a solution of the Schrödinger equation. Energy quantization arises from standing-wave conditions. $\psi$ is described by the quantum numbers $n$, $\ell$, and $m_\ell$, and the spin quantum number $m_s = \pm\frac{1}{2}$. | |
| Schrödinger equation in spherical coordinates | $-\dfrac{\hbar^2}{2mr^2}\dfrac{\partial}{\partial r}\left(r^2\dfrac{\partial\psi}{\partial r}\right) - \dfrac{\hbar^2}{2mr^2}\left[\dfrac{1}{\sin\theta}\dfrac{\partial}{\partial\theta}\left(\sin\theta\dfrac{\partial\psi}{\partial\theta}\right) + \dfrac{1}{\sin^2\theta}\dfrac{\partial^2\psi}{\partial\phi^2}\right] + U(r)\psi = E\psi$ | 36-21 |

| | | |
|---|---|---|
| The solutions can be written as products of functions of $r$, $\theta$, and $\phi$ separately. | $\psi(r, \theta, \phi) = R(r)f(\theta)g(\phi)$ | 36-22 |

**Quantum numbers in spherical coordinates**

| | | |
|---|---|---|
| Principal quantum number | $n = 1, 2, 3, \ldots$ | |
| Orbital quantum number | $\ell = 0, 1, 2, 3, \ldots, n - 1$ | |
| Magnetic quantum number | $m_\ell = -\ell, (-\ell + 1), \ldots, -2, -1, 0, 1, 2, \ldots, (\ell + 1), \ell$ | 36-23 |
| Orbital angular momentum | $L = \sqrt{\ell(\ell + 1)}\hbar$ | 36-24 |
| $z$ component of angular momentum | $L_z = m_\ell \hbar$ | 36-25 |

## 3. Quantum Theory of the Hydrogen Atom

| | | |
|---|---|---|
| Energy levels for hydrogen (same as for the Bohr model) | $E_n = -\dfrac{mk^2e^4}{2\hbar^2} = -Z^2\dfrac{E_0}{n^2}, n = 1, 2, 3, \ldots$ | 36-27 |

where

$$E_0 = \frac{mk^2e^4}{2\hbar^2} \approx 13.6 \text{ eV} \qquad \text{36-28}$$

| | | |
|---|---|---|
| Wavelengths emitted by the hydrogen atom (same as for Bohr model) | $\lambda = \dfrac{c}{f} = \dfrac{hc}{E_i - E_f} = \dfrac{1240 \text{ eV·nm}}{E_i - E_f}$ | 36-17, 36-18 |

**Wave functions**

| | | |
|---|---|---|
| The ground state | $\psi_{1,0,0} = C_{1,0,0}e^{-Zr/a_0} = \dfrac{1}{\sqrt{\pi}}\left(\dfrac{Z}{a_0}\right)^{3/2} e^{-Zr/a_0}$ | 36-31, 36-33 |
| The first excited state | $\psi_{2,0,0} = C_{2,0,0}\left(2 - \dfrac{Zr}{a_0}\right)e^{-Zr/(2a_0)}$ | 36-36 |
| | $\psi_{2,1,0} = C_{2,1,0}\dfrac{Zr}{a_0} e^{-Zr/(2a_0)} \cos\theta$ | 36-37 |
| | $\psi_{2,1,\pm 1} = C_{2,1,1}\dfrac{Zr}{a_0} e^{-Zr/(2a_0)} \sin\theta\, e^{\pm i\phi}$ | 36-38 |

| | | |
|---|---|---|
| Probability densities | For $\ell = 0$, $|\psi|^2$ is spherically symmetric. For $\ell \neq 0$, $|\psi|^2$ depends on the angle $\theta$. | |
| Radial probability density | $P(r) = 4\pi r^2|\psi|^2$ | 36-34 |

The radial probability density is maximum at the distances corresponding roughly to the Bohr orbits.

## 4. The Spin–Orbit Effect and Fine Structure

The total angular momentum of an electron in an atom is a combination of the orbital angular momentum and spin angular momentum. It is characterized by the quantum number $j$, which can be either $|\ell - \frac{1}{2}|$ or $\ell + \frac{1}{2}$. Because of the interaction of the orbital and spin magnetic moments, the state $j = |\ell - \frac{1}{2}|$ has lower energy than the state $j = \ell + \frac{1}{2}$, for $\ell \geq 1$. This small splitting of the energy states gives rise to a small splitting of the spectral lines called fine structure.

## 5. The Periodic Table

Beginning with hydrogen, each larger neutral atom adds one electron. The electrons go into those states that will give the lowest energy consistent with the Pauli exclusion principle.

The state of an atom is described by its electron configuration, which gives the values of $n$ and $\ell$ for each electron. The $\ell$ values are specified by a code:

|            | s | p | d | f | g | h |
|------------|---|---|---|---|---|---|
| $\ell$ values | 0 | 1 | 2 | 3 | 4 | 5 |

**Pauli exclusion principle**

No two electrons in an atom can have the same set of values for the quantum numbers $n$, $\ell$, $m_\ell$, and $m_s$.

## 6. Atomic Spectra

Atomic spectra include optical spectra and X-ray spectra. Optical spectra result from transitions between energy levels of a single outer electron moving in the field of the nucleus and core electrons of the atom. Characteristic X-ray spectra result from the excitation of an inner core electron and the subsequent filling of the vacancy by other electrons in the atom.

**Selection rules**

Transitions between energy states with the emission of a photon are governed by the following selection rules

$$\Delta m_\ell = 0 \text{ or } \pm 1 \qquad\qquad 36\text{-}29$$

$$\Delta \ell = \pm 1$$

# PROBLEMS

- • Single-concept, single-step, relatively easy
- •• Intermediate-level, may require synthesis of concepts
- ••• Challenging
- SSM Solution is in the *Student Solutions Manual*
- iSOLVE Problems available on iSOLVE online homework service
- iSOLVE✓ These "Checkpoint" online homework service problems ask students additional questions about their confidence level, and how they arrived at their answer.

In a few problems, you are given more data than you actually need; in a few other problems, you are required to supply data from your general knowledge, outside sources, or informed estimates.

## Conceptual Problems

**1** • SSM As $n$ increases, does the spacing of adjacent energy levels increase or decrease?

**2** • iSOLVE The energy of the ground state of doubly ionized lithium ($Z = 3$) is _____, where $E_0 = 13.6$ eV. (a) $-9E_0$, (b) $-3E_0$, (c) $-E_0/3$, (d) $-E_0/9$.

**3** • Bohr's quantum condition on electron orbits requires (a) that the angular momentum of the electron about the hydrogen nucleus equals $n\hbar$. (b) that no more than one electron occupy a given stationary state. (c) that the electrons spiral into the nucleus while radiating electromagnetic waves. (d) that the energies of an electron in a hydrogen atom be equal to $nE_0$, where $E_0$ is a constant and $n$ is an integer. (e) none of the above.

**4** • According to the Bohr model, if an electron moves to a larger orbit, does the electron's total energy increase or decrease? Does the electron's kinetic energy increase or decrease?

**5** • According to the Bohr model, the kinetic energy of the electron in the ground state of hydrogen is 13.6 eV $= E_0$. The kinetic energy of the electron in the state $n = 2$ is (a) $4E_0$. (b) $2E_0$. (c) $E_0/2$. (d) $E_0/4$.

**6** • According to the Bohr model, the radius of the $n = 1$ orbit in the hydrogen atom is $a_0 = 0.053$ nm. What is the radius of the $n = 5$ orbit? (a) $5a_0$, (b) $25a_0$, (c) $a_0$, (d) $a_0/5$, or (e) $a_0/25$.

**7** • SSM For the principal quantum number $n = 4$, how many different values can the orbital quantum number $\ell$ have? (a) 4, (b) 3, (c) 7, (d) 16, or (e) 6.

**8** • For the principal quantum number $n = 4$, how many different values can the magnetic quantum number $m_\ell$ have? (a) 4, (b) 3, (c) 7, (d) 16, or (e) 6.

**9** • The p state of an electron configuration corresponds to (a) $n = 2$, (b) $\ell = 2$, (c) $\ell = 1$, (d) $n = 0$, or (e) $\ell = 0$.

**10** •• [SSM] Why is the energy of the 3s state considerably lower than the energy of the 3p state for sodium, whereas in hydrogen these states have essentially the same energy?

**11** •• Discuss the evidence from the periodic table of the need for a fourth quantum number. How would the properties of helium differ if there were only three quantum numbers, $n$, $\ell$, and $m_\ell$?

**12** •• Separate the following six elements—potassium, calcium, titanium, chromium, manganese, and copper—into two groups of three each, so that those in a group have similar properties.

**13** • What element has the electron configuration (a) $1s^2 2s^2 2p^6 3s^2 3p^3$ and (b) $1s^2 2s^2 2p^6 3s^2 3p^6 3d^5 4s^1$?

**14** • [SSM] For the principal quantum number $n = 3$, what are the possible values of the quantum numbers $\ell$ and $m_\ell$?

**15** • [iSOLVE✓] An electron in the $L$ shell means that (a) $\ell = 0$, (b) $\ell = 1$, (c) $n = 1$, (d) $n = 2$, or (e) $m_\ell = 2$.

**16** •• The Bohr theory and the Schrödinger theory of the hydrogen atom give the same results for the energy levels. Discuss the advantages and disadvantages of each model.

**17** •• The Sommerfeld–Hosser displacement theorem states that the optical spectrum of any neutral atom is very similar to the spectrum of the singly charged positive ion of the element immediately following it in the periodic table. Discuss why this is true.

**18** •• [SSM] The Ritz combination principle states that for any atom, one can find different spectral lines $\lambda_1$, $\lambda_2$, $\lambda_3$, and $\lambda_4$, so that $1/\lambda_1 + 1/\lambda_2 = 1/\lambda_3 + 1/\lambda_4$. Show why this is true using an energy-level diagram.

**19** • Using the triplet of numbers $(n, \ell, m_\ell)$ to represent an electron with principal quantum number $n$, orbital quantum number $\ell$, and magnetic quantum number $m_\ell$, which of the following transitions is allowed? (a) $(5, 2, 2) \rightarrow (3, 1, 2)$; (b) $(2, 0, 0) \rightarrow (3, 0, 1)$; (c) $(4, 3, -2) \rightarrow (3, 2, 0)$; (d) $(1, 0, 0) \rightarrow (2, 1, -1)$; or (e) $(2, 1, 0) \rightarrow (3, 0, 0)$.

## Estimation and Approximation

**20** •• [SSM] In laser cooling and trapping, atoms in a beam traveling in one direction are slowed by interaction with an intense laser beam in the opposite direction. The photons scatter off the atoms via resonance absorption, a process by which the incident photon is absorbed by the atom, and a short time later a photon of equal energy is emitted in a random direction. The net result of a single such scattering event is a transfer of momentum to the atom in a direction opposite to the motion of the atom, followed by a second transfer of momentum to the atom in a random direction. Thus, during photon absorption the atom loses speed, but during photon emission the change in speed of the atom is, on average, zero (because the directions of the emitted photons are random). An analogy often made to this process is that of slowing down a bowling ball by bouncing ping-pong balls off of it. (a) Given a typical photon energy used in these experiments of about 1 eV and a momentum typical for an atom with a

thermal speed appropriate to a temperature of about 500 K (a typical temperature for an atomic beam), estimate the number of photon-atom collisions that are required to bring an atom to rest. (The average kinetic energy of an atom is equal to $\frac{3}{2}kT$, where $k$ is the Boltzmann constant. Use this to estimate the speed of the atoms.) (b) Compare this with the number of ping-pong ball–bowling ball collisions that are required to bring the bowling ball to rest. (Assume the speeds of the incident ping-pong balls are all equal to the initial speed of the bowling ball.) (c) $^{85}$Rb is a type of atom often used in cooling experiments. The wavelength of the light resonant with the cooling transition is $\lambda = 780.24$ nm. Estimate the number of photons needed to slow down an $^{85}$Rb atom from a typical thermal velocity of 300 m/s to a stop.

**21** •• (a) We can define a thermal de Broglie wavelength $\lambda_T$ for an atom in a gas at temperature $T$ as being the de Broglie wavelength for an atom moving at the rms speed appropriate to that temperature. (The average kinetic energy of an atom is equal to $\frac{3}{2}kT$, where $k$ is the Boltzman constant. Use this to calculate the rms speed of the atoms.) Show that $\lambda_T = \sqrt{\dfrac{h^2}{3mkT}}$, where $m$ is the mass of the atom. (b) Cooled neutral atoms can form a *Bose condensate* (a new state of matter) when their thermal de Broglie wavelength becomes larger than the average interatomic spacing. From this criterion, estimate the temperature needed to create a Bose condensate in a gas of $^{85}$Rb atoms whose number density is $10^{12}$ atoms/cm$^3$.

## The Bohr Model of the Hydrogen Atom

**22** • Use the known values of the constants in Equation 36-12 to show that $a_0$ is approximately 0.0529 nm.

**23** • [iSOLVE✓] The longest wavelength in the Lyman series was calculated in Example 36-2. Find the wavelengths for the transitions (a) $n_1 = 3$ to $n_2 = 1$ and (b) $n_1 = 4$ to $n_2 = 1$.

**24** • Find the photon energy for the three longest wavelengths in the Balmer series and calculate the wavelengths.

**25** •• (a) Find the photon energy and wavelength for the series limit (shortest wavelength) in the Paschen series ($n_2 = 3$). (b) Calculate the wavelength for the three longest wavelengths in this series and indicate their positions on a horizontal linear scale.

**26** •• [SSM] Repeat Problem 25 for the Brackett series, $n_2 = 4$.

**27** •• [iSOLVE✓] The hydrogen spectrum is found by collimating the light from a hydrogen discharge tube and shining it on a grating to disperse the light into its various colors. The grating spacing is $d = 3.377$ $\mu$m. A bright red line ($m = 1$) is seen at an angle of 11.233° from the center of the spectroscope (see Figure 36-1). (a) What is the wavelength of this spectral line? (b) Assuming this line is from a transition from level $n_1 = 3$ to level $n_2 = 2$ (i.e., the longest wavelength Balmer series transition), what do you calculate for the value of the Rydberg constant?

**28** ••• In this problem, you will estimate the radius and the energy of the lowest stationary state of the hydrogen atom using the uncertainty principle. The total energy of the electron with momentum $p$ and mass $m$ a distance $r$ from the proton in the hydrogen atom is given by

$$E = p^2/2m - ke^2/r$$

where $k$ is the Coulomb constant. Assume that the minimum value of $p^2 \approx (\Delta p)^2 = \hbar^2/r^2$, where $\Delta p$ is the uncertainty in $p$ and we have taken $\Delta r \sim r$ for the order of magnitude of the uncertainty in position; the energy is then

$$E = \hbar^2/2mr^2 - ke^2/r$$

Find the radius $r$ for which this energy is a minimum, and calculate the minimum value of $E$ in electron volts.

**29** ••• **SSM** In the center-of-mass reference frame of a hydrogen atom, the electron and nucleus have equal and opposite momenta of magnitude $p$. (a) Show that the total kinetic energy of the electron and nucleus can be written $K = p^2/(2\mu)$ where $\mu = m_e M/(M + m_e)$ is called the reduced mass, $m_e$ is the mass of the electron, and $M$ is the mass of the nucleus. (b) For the equations for the Bohr model of the atom, the motion of the nucleus can be taken into account by replacing the mass of the electron with the reduced mass. Use Equation 36-14 to calculate the Rydberg constant for a hydrogen atom with a nucleus of mass $M = m_p$. Find the approximate value of the Rydberg constant by letting $M$ go to infinity in the reduced mass formula. To how many figures does this approximate value agree with the actual value? (c) Find the percentage correction for the ground-state energy of the hydrogen atom by using the reduced mass in Equation 36-16. *Remark: In general, the reduced mass for a two-body problem with masses $m_1$ and $m_2$ is given by*

$$\mu = \frac{m_1 m_2}{m_1 + m_2}$$

**30** •• **SSM** **iSOLVE**✔ The Pickering series of the spectrum of He$^+$ (singly-ionized helium) consists of spectral lines due to transitions to the $n = 4$ state of He$^+$. Experimentally, every other line of the Pickering series is very close to a spectral line in the Balmer series for hydrogen transitions to $n = 2$. (a) Show that this is true. (b) Calculate the wavelength of a transition from the $n = 6$ level to the $n = 4$ level of He$^+$, and show that it corresponds to one of the Balmer lines.

## Quantum Numbers in Spherical Coordinates

**31** • For $\ell = 1$, find (a) the magnitude of the angular momentum $L$ and (b) the possible values of $m_\ell$. (c) Draw a vector diagram to scale showing the possible orientations of $\vec{L}$ with the $z$ axis.

**32** • **iSOLVE**✔ Repeat Problem 31 for $\ell = 3$.

**33** • If $n = 3$, (a) what are the possible values of $\ell$? (b) For each value of $\ell$ in Part (a), list the possible values of $m_\ell$. (c) Using the fact that there are two quantum states for each value of $\ell$ and $m_\ell$ because of electron spin, find the total number of electron states with $n = 3$.

**34** • **iSOLVE**✔ Find the total number of electron states with (a) $n = 4$ and (b) $n = 2$. (See Problem 33.)

**35** •• **SSM** **iSOLVE** Find the minimum value of the angle $\theta$ between $\vec{L}$ and the $z$ axis for (a) $\ell = 1$, (b) $\ell = 4$, and (c) $\ell = 50$.

**36** •• What are the possible values of $n$ and $m_\ell$ if (a) $\ell = 3$, (b) $\ell = 4$, and (c) $\ell = 0$.

**37** •• For an $\ell = 2$ state, find (a) the square magnitude of the angular momentum $\vec{L}^2$, (b) the maximum value of $L_z^2$, and (c) the smallest value of $L_x^2 + L_y^2$.

## Quantum Theory of the Hydrogen Atom

**38** • For the ground state of the hydrogen atom, find the values of (a) $\psi$, (b) $\psi^2$, and (c) the radial probability density $P(r)$ at $r = a_0$. Give your answers in terms of $a_0$.

**39** • **SSM** (a) If spin is not included, how many different wave functions are there corresponding to the first excited energy level $n = 2$ for hydrogen? (b) List these functions by giving the quantum numbers for each state.

**40** •• For the ground state of the hydrogen atom, calculate the probability of finding the electron in the range $\Delta r = 0.03a_0$ at (a) $r = a_0$ and (b) $r = 2a_0$.

**41** •• The value of the constant $C_{2,0,0}$ in Equation 36-36 is given by

$$C_{2,0,0} = \frac{1}{4\sqrt{2\pi}} \left(\frac{Z}{a_0}\right)^{3/2}$$

Find the values of (a) $\psi$, (b) $\psi^2$, and (c) the radial probability density $P(r)$ at $r = a_0$ for the state $n = 2$, $\ell = 0$, and $m_\ell = 0$ in hydrogen. Give your answers in terms of $a_0$.

**42** ••• Show that the radial probability density for the $n = 2$, $\ell = 1$, and $m_\ell = 0$ state of a one-electron atom can be written as $P(r) = A \cos^2 \theta\, r^4 e^{-Zr/a_0}$, where $A$ is a constant.

**43** ••• Calculate the probability of finding the electron in the range $\Delta r = 0.02a_0$ for (a) $r = a_0$ and (b) $r = 2a_0$ for the state $n = 2$, $\ell = 0$, and $m_\ell = 0$ in hydrogen. (See Problem 41 for the value of $C_{2,0,0}$.)

**44** •• **SSM** Show that the ground-state hydrogen wave function (Equation 36-33) is a solution to Schödinger's equation (Equation 36-21) and the potential energy function equation (Equation 36-26).

**45** •• Show by unit cancellation that the expression for the hydrogen ground-state energy given by Equation 36-28 has the dimensions of energy.

**46** •• By dimensional analysis, show that the expression for the first Bohr radius given by Equation 36-12 has the dimensions of length.

**47** •• The radial probability distribution function for a one-electron atom in its ground state can be written $P(r) = Cr^2 e^{-2Zr/a_0}$, where $C$ is a constant. Show that $P(r)$ has its maximum value at $r = a_0/Z$.

**48** ••• Show that the number of states in the hydrogen atom for a given $n$ is $2n^2$.

**49** ••• Calculate the probability that the electron in the ground state of a hydrogen atom is in the region $0 < r < a_0$.

## The Spin–Orbit Effect and Fine Structure

**50** • **SSM** **iSOLVE** The potential energy of a magnetic moment in an external magnetic field is given by $U = -\boldsymbol{\mu} \cdot \vec{B}$. (a) Calculate the difference in energy between the two possible orientations of an electron in a magnetic field $\vec{B} = 1.50\ \text{T}\hat{k}$. (b) If these electrons are bombarded with photons of energy equal to this energy difference, "spin flip" transitions can be induced. Find the wavelength of the photons needed for such transitions. This phenomenon is called *electron spin resonance*.

**51** • The total angular momentum of a hydrogen atom in a certain excited state has the quantum number $j = \frac{1}{2}$. What can you say about the orbital angular-momentum quantum number $\ell$?

**52** • A hydrogen atom is in the state $n = 3$, $\ell = 2$. What are the possible values of $j$?

**53** • Using a scaled vector diagram, show how the orbital angular momentum $\vec{L}$ combines with the spin orbital angular momentum $\vec{S}$ to produce the two possible values of total angular momentum $\vec{J}$ for the $\ell = 3$ state of the hydrogen atom.

## The Periodic Table

**54** • The total number of quantum states of hydrogen with quantum number $n = 4$ is (a) 4, (b) 16, (c) 32, (d) 36, or (e) 48.

**55** • How many of oxygen's eight electrons are found in the p state? (a) 0, (b) 2, (c) 4, (d) 6, or (e) 8.

**56** • **SSM** Write the electron configuration of (a) carbon and (b) oxygen.

**57** • Give the possible values of the z component of the orbital angular momentum of (a) a d electron, and (b) an f electron.

## Optical Spectra and X-Ray Spectra

**58** • The optical spectra of atoms with two electrons in the same outer shell are similar, but they are quite different from the spectra of atoms with just one outer electron because of the interaction of the two electrons. Separate the following elements into two groups so that those in each group have similar spectra: lithium, beryllium, sodium, magnesium, potassium, calcium, chromium, nickel, cesium, barium.

**59** • Write down the possible electron configurations for the first excited state of (a) hydrogen, (b) sodium, and (c) helium.

**60** • **iSOLVE** Indicate which of the following elements should have optical spectra similar to hydrogen and which should be similar to helium: Li, Ca, Ti, Rb, Hg, Ag, Cd, Ba, Fr, and Ra.

**61** • **SSM** (a) Calculate the next two longest wavelengths in the K series (after the $K_\alpha$ line) of molybdenum. (b) What is the wavelength of the shortest wavelength in this series?

**62** • The wavelength of the $K_\alpha$ line for a certain element is 0.3368 nm. What is the element?

**63** • Calculate the wavelength of the $K_\alpha$ line in (a) magnesium ($Z = 12$) and (b) copper ($Z = 29$).

## General Problems

**64** • What is the energy of the shortest wavelength photon emitted by the hydrogen atom?

**65** • The wavelength of a spectral line of hydrogen is 97.254 nm. Identify the transition that results in this line under the assumption that the transition is to the ground state.

**66** • The wavelength of a spectral line of hydrogen is 1093.8 nm. Identify the transition that results in this line.

**67** • Spectral lines of the following wavelengths are emitted by singly ionized helium: 164 nm, 230.6 nm, and 541 nm. Identify the transitions that result in these spectral lines.

**68** •• **SSM** We are often interested in finding the quantity $ke^2/r$ in electron volts when $r$ is given in nanometers. Show that $ke^2 = 1.44\ \text{eV·nm}$.

**69** •• The wavelengths of the photons emitted by potassium corresponding to transitions from the $4P_{3/2}$ and $4P_{1/2}$ states to the ground state are 766.41 nm and 769.90 nm. (a) Calculate the energies of these photons in electron volts. (b) The difference in the energies of these photons equals the difference in energy $\Delta E$ between the $4P_{3/2}$ and $4P_{1/2}$ states in potassium. Calculate $\Delta E$. (c) Estimate the magnetic field that the 4p electron in potassium experiences.

**70** •• To observe the characteristic K lines of the X-ray spectrum, one of the $n = 1$ electrons must be ejected from the atom. This is generally accomplished by bombarding the target material with electrons of sufficient energy to eject this tightly bound electron. What is the minimum energy required to observe the K lines of (a) tungsten, (b) molybdenum, and (c) copper?

**71** •• **SSM** The combination of physical constants $\alpha = e^2 k/\hbar c$, where $k$ is the Coulomb constant, is known as the *fine-structure constant*. It appears in numerous relations in atomic physics. (a) Show that $\alpha$ is dimensionless. (b) Show that in the Bohr model of hydrogen $v_n = c\alpha/n$, where $v_n$ is the speed of the electron in the stationary state of quantum number $n$.

**72** •• The *positron* is a particle that is identical to the electron except that it carries a positive charge of $e$. *Positronium* is the bound state of an electron and positron. (a) Calculate the energies of the five lowest energy states of positronium using the reduced mass, as given in Problem 29. (b) Do transitions between any of the levels found in Part (a) fall in the visible range of wavelengths? If so, which transitions are these?

**73** • In 1947, Lamb and Retherford showed that there was a very small energy difference between the $2S_{1/2}$ and the $2P_{1/2}$ states of the hydrogen atom. They measured this difference essentially by causing transitions between the two states using very long wavelength electromagnetic radiation. The energy difference (the Lamb shift) is $4.372 \times 10^{-6}$ eV and is explained by quantum electrodynamics as being due to fluctuations in the energy level of the vacuum. (a) What is the frequency of a photon whose energy is equal to the Lamb shift energy? (b) What is the wavelength of this photon? In what spectral region does it belong?

**74** • [SSM] A Rydberg atom is one in which an outer shell electron is placed into a *very* high excited state (n ≈ 40 or higher). Such atoms are useful for experiments that probe the transition from quantum-mechanical behavior to classical. Furthermore, these excited states have extremely long lifetimes (i.e., the electron will stay in this high excited state for a very long time). A hydrogen atom is in the $n = 45$ state. (a) What is the ionization energy of the atom when it is in this state? (b) What is the energy level separation (in eV) between this state and the $n = 44$ state? (c) What is the wavelength of a photon resonant with this transition? (d) How large is the atom in the $n = 45$ state?

**75** •• The deuteron, the nucleus of deuterium (heavy hydrogen), was first recognized from the spectrum of hydrogen. The deuteron has a mass twice that of the proton. (a) Calculate the Rydberg constant for hydrogen and for deuterium using the reduced mass as given in Problem 29. (b) Using the result

obtained in Part (a), determine the wavelength difference between the longest wavelength Balmer lines of hydrogen and deuterium.

**76** •• The muonium atom is a hydrogen atom with the electron replaced by a $\mu^-$ particle. The $\mu^-$ is identical to an electron but has a mass 207 times as great as the electron. (a) Calculate the energies of the five lowest energy levels of muonium using the reduced mass as given in Problem 29. (b) Do transitions between any of the levels found in Part (a) fall in the visible range of wavelengths, (i.e., between $\lambda = 700$ nm and 400 nm)? If so, which transitions are these?

**77** •• The triton, a nucleus consisting of a proton and two neutrons, is unstable with a fairly long half-life of approximately 12 years. Tritium is the bound state of an electron and a triton. (a) Calculate the Rydberg constant of tritium using the reduced mass as given in Problem 29. (b) Using the result obtained in Part (a) and the result obtained in Part (b) of Problem 76, determine the wavelength difference between the longest wavelength Balmer lines of tritium and deuterium and between tritium and hydrogen.

# Molecules

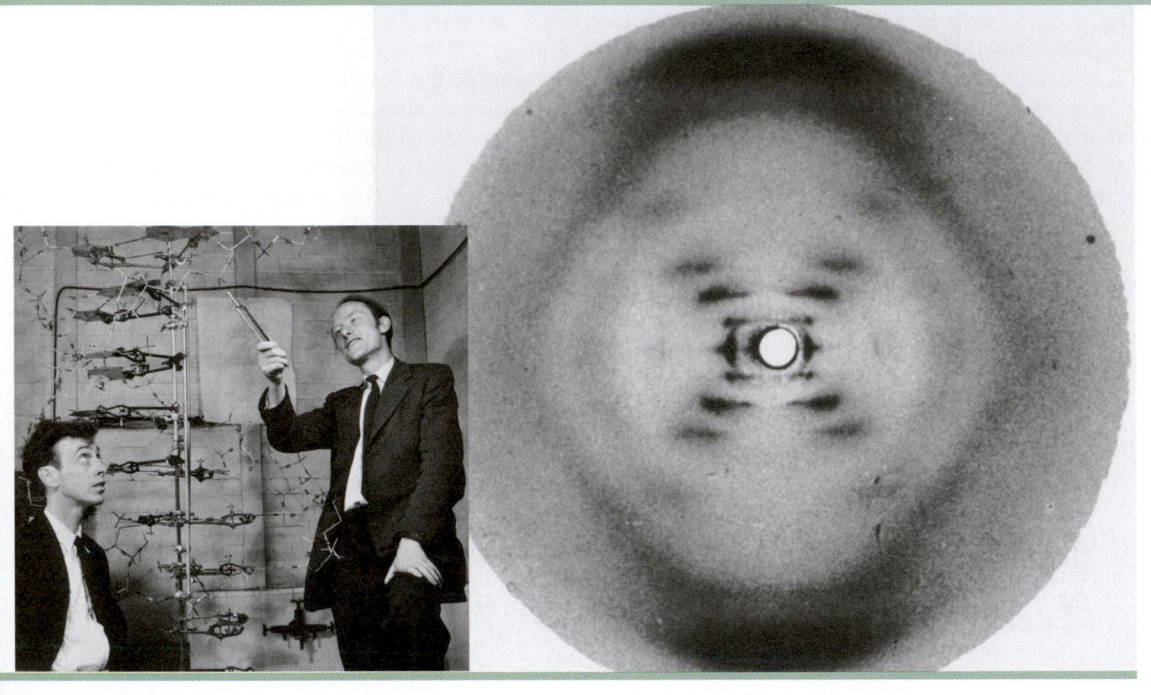

37-1    Molecular Bonding

*37-2    Polyatomic Molecules

37-3    Energy Levels and Spectra of Diatomic Molecules

**M**ost atoms bond together to form molecules or solids. Molecules may exist as separate entities, as in gaseous $O_2$ or $N_2$, or they may bond together to form liquids or solids. A molecule is the smallest constituent of a substance that retains its chemical properties.

➤ **In this chapter, we use our understanding of quantum mechanics to discuss molecular bonding and the energy levels and spectra of diatomic molecules. Much of our discussion will be qualitative because, as in atomic physics, the quantum-mechanical calculations are very difficult.**

## 37-1 Molecular Bonding

There are two extreme views that we can take of a molecule. Consider, for example, $H_2$. We can think of $H_2$ either as two H atoms joined together or as a quantum-mechanical system of two protons and two electrons. The latter picture is more fruitful in this case because neither of the electrons in the $H_2$ molecule can be identified as belonging to either proton. Instead, the wave function for each electron is spread out in space throughout the whole molecule. For more complicated molecules, however, an intermediate picture is useful. For example, the nitrogen molecule $N_2$ consists of 14 protons and 14 electrons, but only two of the electrons take part in the bonding. We therefore can consider this molecule as two $N^+$ ions and two electrons that belong to the molecule as a whole. The molecular wave functions for these bonding electrons are called **molecular**

**THE DISCOVERERS OF THE STRUCTURE OF DNA** JAMES WATSON AT LEFT AND FRANCIS CRICK ARE SHOWN WITH THEIR MODEL OF PART OF A DNA MOLECULE IN 1953. CRICK AND WATSON MET AT THE CAVENDISH LABORATORY, CAMBRIDGE, IN 1951. THEIR WORK ON THE STRUCTURE OF DNA WAS PERFORMED WITH A KNOWLEDGE OF CHARGAFF'S RATIOS OF THE BASES IN DNA AND SOME ACCESS TO THE X-RAY CRYSTALLOGRAPHY OF MAURICE WILKINS AND ROSALIND FRANKLIN AT KING'S COLLEGE LONDON. COMBINING ALL OF THIS WORK LED TO THE DEDUCTION THAT DNA EXISTS AS A DOUBLE HELIX, THUS TO ITS STRUCTURE. CRICK, WATSON, AND WILKINS SHARED THE 1962 NOBEL PRIZE FOR PHYSIOLOGY OR MEDICINE; FRANKLIN DIED FROM CANCER IN 1958.

**X-RAY DIFFRACTION PATTERN OF THE B FORM OF DNA** ROSALIND FRANKLIN'S COLLEAGUE MAURICE WILKINS, WITHOUT OBTAINING HER PERMISSION, MADE AVAILABLE TO WATSON AND CRICK HER THEN UNPUBLISHED X-RAY DIFFRACTION PATTERN OF THE B FORM OF DNA, WHICH WAS CRUCIAL EVIDENCE FOR THE HELICAL STRUCTURE. IN HIS ACCOUNT OF THIS DISCOVERY, WATSON WROTE: "THE INSTANT I SAW THE PICTURE, MY MOUTH FELL OPEN AND PULSE BEGAN TO RACE. . . . THE BLACK CROSS OF REFLECTIONS WHICH DOMINATED THE PICTURE COULD ARISE ONLY FROM A HELICAL STRUCTURE. . . . MERE INSPECTION OF THE X-RAY PICTURE GAVE SEVERAL OF THE VITAL HELICAL PERAMETERS" (FROM STENT, GUNTHER, *THE DOUBLE HELIX*, NEW YORK: NORTON, 1980).

**orbitals.** In many cases, these molecular wave functions can be constructed from combinations of the atomic wave functions with which we are familiar.

The two principal types of bonds responsible for the formation of molecules are the ionic bond and the covalent bond. Other types of bonds that are important in the bonding of liquids and solids are van der Waals bonds, metallic bonds, and hydrogen bonds. In many cases, bonding is a mixture of these mechanisms.

## The Ionic Bond

The simplest type of bond is the **ionic bond,** which is found in salts such as sodium chloride (NaCl). The sodium atom has one 3s electron outside a stable core. The ionization energy of sodium is the energy needed to remove this electron from an isolated sodium atom. This energy is just 5.14 eV (see Figure 36-18). The removal of this electron leaves an isolated positive ion with a spherically symmetric, closed-shell electron core. Chlorine, on the other hand, is one electron short of having a closed shell. The energy released by an isolated atom's acquisition of one electron is called its **electron affinity,** which in the case of chlorine is 3.62 eV. The acquisition of one electron by chlorine results in a negative ion with a spherically symmetric, closed-shell electron core. Thus, the formation of a $Na^+$ ion and a $Cl^-$ ion by the donation of one electron of sodium to chlorine requires only 5.14 eV − 3.62 eV = 1.52 eV at infinite separation. The electrostatic potential energy $U_e$ of the two ions when they are a distance $r$ apart is $-ke^2/r$. When the separation of the ions is less than approximately 0.95 nm, the negative potential energy of attraction is of greater magnitude than the 1.52 eV of energy needed to create the ions. Thus, at separation distances less than 0.95 nm, it is energetically favorable (i.e., the total energy of the system is reduced) for the sodium atom to donate an electron to the chlorine atom to form NaCl.

Since the electrostatic attraction increases as the ions get closer together, it might seem that equilibrium could not exist. However, when the separation of the ions is very small, there is a strong repulsion that is quantum mechanical in nature and is related to the exclusion principle. This **exclusion-principle repulsion** is responsible for the repulsion of the atoms in all molecules (except $H_2)^†$ for all bonding mechanisms. We can understand it qualitatively as follows. When the ions are very far apart, the wave function for a core electron in one of the ions does not overlap that of any electron in the other ion. We can distinguish the electrons by the ion to which they belong. This means that electrons in the two ions can have the same quantum numbers because they occupy different regions of space. However, as the distance between the ions decreases, the wave functions of the core electrons begin to overlap; that is, the electrons in the two ions begin to occupy the same region of space. Because of the exclusion principle, some of these electrons must go into higher energy quantum states.$^‡$ But energy is required to shift the electrons into higher energy quantum states. This increase in energy when the ions are pushed closer together is equivalent to a repulsion of the ions. It is not a sudden process. The energy states of the electrons change gradually as the ions are brought together. A sketch of the potential energy $U(r)$ of the $Na^+$ and $Cl^-$ ions versus separation distance $r$ is shown in Figure 37-1. The energy is lowest at an equilibrium separation $r_0$ of approximately 0.236 nm. At smaller separations, the energy rises steeply

**FIGURE 37-1** Potential energy for $Na^+$ and $Cl^-$ ions as a function of separation distance $r$. The energy at infinite separation is chosen to be 1.52 eV, corresponding to the energy $-\Delta E$ needed to form the ions from neutral atoms. The minimum energy is at the equilibrium separation $r_0 = 0.236$ nm for the ions in the molecule.

---

$^†$ In $H_2$, the repulsion is simply that of the two positively charged protons.

$^‡$ Recall from our discussion in Chapter 35 that the exclusion principle is related to the fact that the wave function for two identical electrons is antisymmetric on the exchange of the electrons and that an antisymmetric wave function for two electrons with the same quantum numbers is zero if the space coordinates of the electrons are the same.

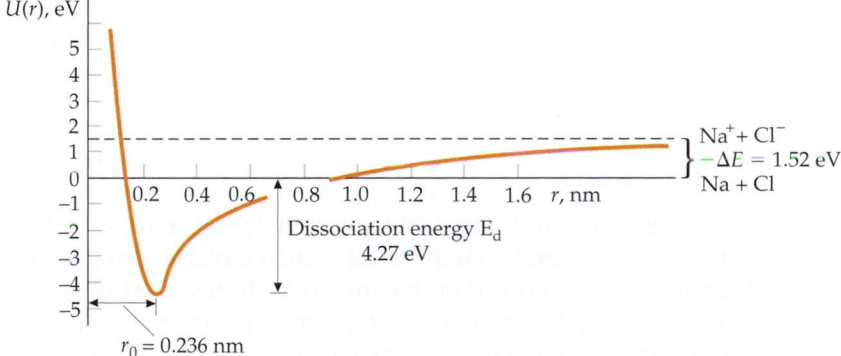

as a result of the exclusion principle. The energy required to separate the ions and form neutral sodium and chlorine atoms is called the **dissociation energy** $E_d$, which is approximately 4.27 eV for NaCl.

The equilibrium separation distance of 0.236 nm is for gaseous diatomic NaCl, which can be obtained by evaporating solid NaCl. Normally, NaCl exists in a cubic crystal structure, with the $Na^+$ and $Cl^-$ ions at the alternate corners of a cube. The separation of the ions in a crystal is somewhat larger, approximately 0.28 nm. Because of the presence of neighboring ions of opposite charge, the electrostatic energy per ion pair is lower when the ions are in a crystal.

---

*THE ENERGY OF A SODIUM-FLUORIDE MOLECULE*        **EXAMPLE 37-1**

The electron affinity of fluorine is 3.40 eV, and the equilibrium separation of sodium fluoride (NaF) is 0.193 nm. (*a*) How much energy is needed to form $Na^+$ and $F^-$ ions from neutral sodium and fluorine atoms? (*b*) What is the electrostatic potential energy of the $Na^+$ and $F^-$ ions at their equilibrium separation? (*c*) The dissociation energy of NaF is 5.38 eV. What is the energy due to repulsion of the ions at the equilibrium separation?

**PICTURE THE PROBLEM** (*a*) The energy $\Delta E$ needed to form $Na^+$ and $F^-$ ions from the neutral sodium and fluorine atoms is the difference between the ionization energy of sodium (5.14 eV) and the electron affinity of fluorine. (*b*) The electrostatic potential energy with $U = 0$ at infinity is $U_e = -ke^2/r$. (*c*) If we choose the potential energy at infinity to be $\Delta E$, the total potential energy is $U_{tot} = U_e + \Delta E + U_{rep}$, where $U_{rep}$ is the energy of repulsion, which is found by setting the dissociation energy equal to $-U_{tot}$.

(*a*) Calculate the energy needed to form $Na^+$ and $F^-$ ions from the neutral sodium and fluorine atoms (see Picture the Problem):

$$\Delta E = 5.14 \text{ eV} - 3.40 \text{ eV} = \boxed{1.74 \text{ eV}}$$

(*b*) 1. Calculate the electrostatic potential energy at the equilibrium separation of $r = 0.193$ nm:

$$U_e = -\frac{ke^2}{r}$$

$$= -\frac{(8.99 \times 10^9 \text{ N·m}^2/\text{C}^2)(1.60 \times 10^{-19} \text{ C})^2}{1.93 \times 10^{-10} \text{ m}}$$

$$= -1.19 \times 10^{-18} \text{ J}$$

2. Convert from joules to electron volts:

$$U_e = -1.19 \times 10^{-18} \text{ J}\left(\frac{1 \text{ eV}}{1.60 \times 10^{-19} \text{ J}}\right) = \boxed{-7.45 \text{ eV}}$$

(*c*) The dissociation energy equals the negative of the total potential energy:

$$E_d = -U_{tot} = -(U_e + \Delta E + U_{rep})$$

so

$$U_{rep} = -(E_d + \Delta E + U_e)$$

$$= -(5.38 \text{ eV} + 1.74 \text{ eV} - 7.45 \text{ eV})$$

$$= \boxed{0.33 \text{ eV}}$$

---

## The Covalent Bond

A completely different mechanism, the **covalent bond,** is responsible for the bonding of identical or similar atoms to form such molecules as gaseous hydrogen ($H_2$), nitrogen ($N_2$), and carbon monoxide (CO). If we calculate the energy needed to form $H^+$ and $H^-$ ions by the transfer of an electron from one atom to the other and then add this energy to the electrostatic potential energy, we find that there is no separation distance for which the total energy is negative. The bond thus cannot be ionic. Instead, the attraction of two hydrogen atoms is an entirely quantum-mechanical effect. The decrease in energy when two hydrogen atoms approach each other is due to the sharing of the two electrons by both atoms. It is intimately connected with the symmetry properties of the wave functions of electrons.

We can gain some insight into covalent bonding by considering a simple, one-dimensional quantum-mechanics problem of two finite square wells. We first consider a single electron that is equally likely to be in either well. Because the wells are identical, the probability distribution, which is proportional to $|\psi^2|$, must be symmetric about the midpoint between the wells. Then $\psi$ must be either symmetric or antisymmetric with respect to the two wells. The two possibilities for the ground state are shown in Figure 37-2$a$ for the case in which the wells are far apart and in Figure 37-2$b$ for the case in which the wells are close together. An important feature of Figure 37-2$b$ is that in the region between the wells the symmetric wave function is large and the antisymmetric wave function is small.

Now consider adding a second electron to the two wells. We saw in Chapter 35 that the wave functions for particles that obey the exclusion principle are antisymmetric on exchange of the particles. Thus the total wave function for the two electrons must be antisymmetric on exchange of the electrons. Note that exchanging the electrons in the wells here is the same as exchanging the wells. The total wave function for two electrons can be written as a product of a space part and a spin part. So, an antisymmetric wave function can be the product of a symmetric space part and an antisymmetric spin part or of a symmetric spin part and an antisymmetric space part.

To understand the symmetry of the total wave function, we must therefore understand the symmetry of the spin part of the wave function. The spin of a single electron can have two possible values for its quantum number $m_S$: $m_S = +\frac{1}{2}$, which we call spin up, or $m_S = -\frac{1}{2}$, which we call spin down. We will use arrows to designate the spin wave function for a single electron: $\uparrow_1$ or $\uparrow_2$ for electron 1 or electron 2 with spin up and $\downarrow_1$ or $\downarrow_2$ for electron 1 or electron 2 with spin down. The total spin quantum number for two electrons can be $S = 1$, with $m_S = +1, 0$, or $-1$; or $S = 0$, with $m_S = 0$. We use $\phi_{S, m_S}$ to denote the spin wave function for two electrons. The spin state $\phi_{1, +1}$, corresponding to $S = 1$ and $m_S = +1$, can be written

$$\phi_{1,+1} = \uparrow_1\uparrow_2, \quad S = 1, m_S = +1 \qquad 37\text{-}1$$

Similarly, the spin state for $S = 1$, $m_S = -1$ is

$$\phi_{1,-1} = \downarrow_1\downarrow_2, \quad S = 1, m_S = -1 \qquad 37\text{-}2$$

Note that both of these states are symmetric upon exchange of the electrons. The spin state corresponding to $S = 1$ and $m_S = 0$ is not quite so obvious. It turns out to be proportional to

$$\phi_{1,0} = \uparrow_1\downarrow_2 + \uparrow_2\downarrow_1, \quad S = 1, m_S = 0 \qquad 37\text{-}3$$

This spin state is also symmetric upon exchange of the electrons. The spin state for two electrons with antiparallel spins ($S = 0$) is

$$\phi_{0,0} = \uparrow_1\downarrow_2 - \uparrow_2\downarrow_1, \quad S = 0, m_S = 0 \qquad 37\text{-}4$$

This spin state is antisymmetric upon exchange of electrons.

We thus have the important result that the *spin* part of the wave function is symmetric for parallel spins ($S = 1$) and antisymmetric for antiparallel spins

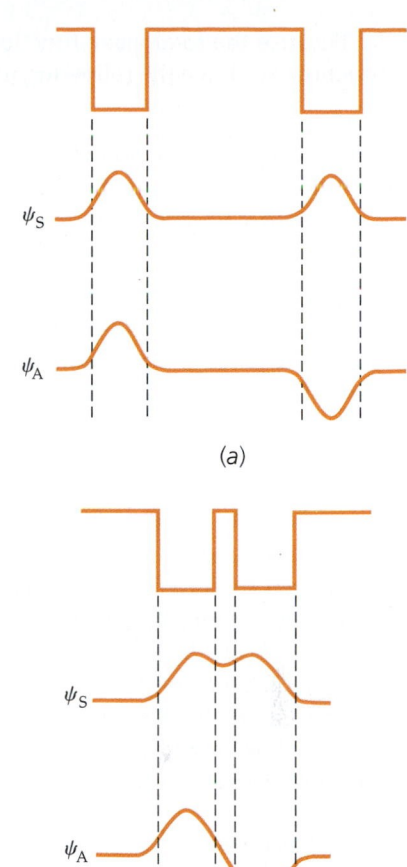

(a)

(b)

**FIGURE 37-2** (*a*) Two square wells far apart. The electron wave function can be either symmetric ($\psi_S$) or antisymmetric ($\psi_A$) in space. The probability distributions and energies are the same for the two wave functions when the wells are far apart. (*b*) Two square wells that are close together. Between the wells, the antisymmetric space wave function is approximately zero, whereas the symmetric space wave function is quite large.

($S = 0$). Because the total wave function is the product of the space function and spin function, we have the following important result:

For the total wave function of two electrons to be antisymmetric, the space part of the wave function must be antisymmetric for parallel spins ($S = 1$) and symmetric for antiparallel spins ($S = 0$).

SPIN ALIGNMENT AND WAVE-FUNCTION SYMMETRY

We can now consider the problem of two hydrogen atoms. Figure 37-3a shows a spatially symmetric wave function $\psi_S$ and a spatially antisymmetric wave function $\psi_A$ for two hydrogen atoms that are far apart, and Figure 37-3b shows the same two wave functions for two hydrogen atoms that are close together. The squares of these two wave functions are shown in Figure 37-3c. Note that the probability distribution $|\psi|^2$ in the region between the protons is large for the symmetric wave function and small for the antisymmetric wave function. Thus, when the space part of the wave function is symmetric ($S = 0$), the electrons are often found in the region between the protons. The negatively charged electron cloud representing these electrons is concentrated in the space between the protons, as shown in the upper part of Figure 37-3c, and the protons are bound together by this negatively charged cloud. Conversely, when the space part of the wave function is antisymmetric ($S = 1$), the electrons spend little time between the protons, and the atoms do not bind together to form a molecule. In this case, the electron cloud is not concentrated in the space between the protons, as shown in the lower part of Figure 37-3c.

The total electrostatic potential energy for the $H_2$ molecule consists of the positive energy of repulsion of the two electrons and the negative potential energy of attraction of each electron for each proton. Figure 37-4 shows the electrostatic potential energy for two hydrogen atoms versus separation for the case in which the space part of the electron wave function is symmetric ($U_S$) and for the case in which it is antisymmetric ($U_A$). We can see that the potential energy for the symmetric state is the lower of the two and that the shape of this potential energy curve is similar to that for ionic bonding. The equilibrium separation for $H_2$ is $r_0 = 0.074$ nm, and the binding energy is 4.52 eV. For the antisymmetric state, the potential energy is never negative and there is no bonding.

We can now see why three hydrogen atoms do not bond to form $H_3$. If a third hydrogen atom is brought near an $H_2$ molecule, the third electron cannot be in a 1s state and have its spin antiparallel to the spin of both of the other electrons. If this electron is in an antisymmetric space state with respect to exchange with one

**FIGURE 37-3** One-dimensional symmetric and antisymmetric wave functions for two hydrogen atoms (a) far apart and (b) close together. (c) Electron probability distributions ($|\psi|^2$) for the wave functions in Figure 37-3b. For the symmetric wave function, the electron charge density is large between the protons. This negative charge density holds the protons together in the hydrogen molecule $H_2$. For the antisymmetric wave function, the electron charge density is not large between the protons.

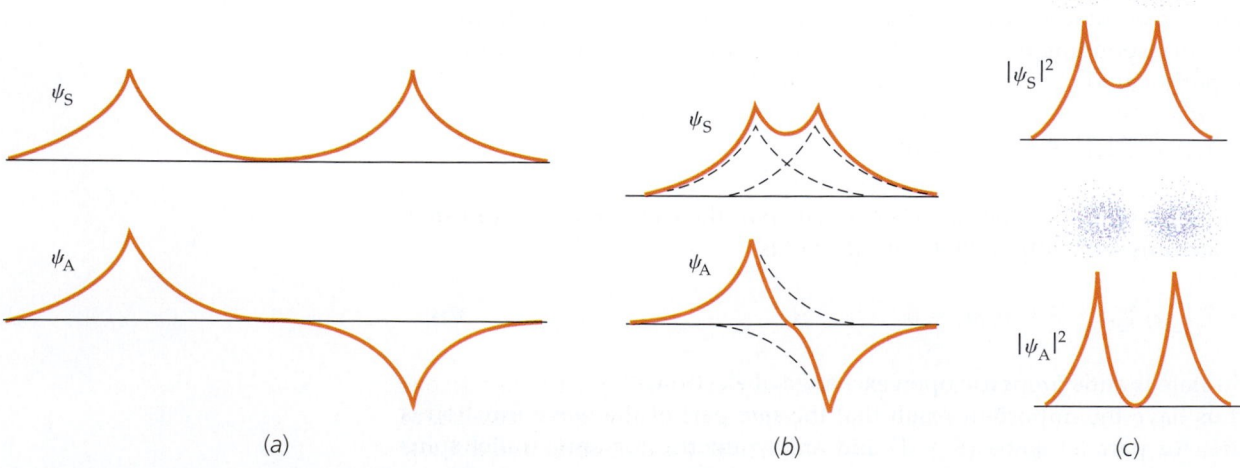

(a)        (b)        (c)

of the electrons, the repulsion of this atom is greater than the attraction of the other. As the three atoms are pushed together, the third electron is, in effect, forced into a higher quantum-energy state by the exclusion principle. The bond between two hydrogen atoms is called a **saturated bond** because there is no room for another electron. The two shared electrons essentially fill the 1s states of both atoms.

We can also see why two helium atoms do not normally bond together to form the He$_2$ molecule. There are no valence electrons that can be shared. The electrons in the closed shells are forced into higher energy states when the two atoms are brought together. At low temperatures or high pressures, helium atoms do bond together due to van der Waals forces, which we will discuss next. This bonding is so weak that at atmospheric pressure helium boils at 4 K, and it does not form a solid at any temperature unless the pressure is greater than about 20 atm.

When two identical atoms bond, as in O$_2$ or N$_2$, the bonding is purely covalent. However, the bonding of two dissimilar atoms is often a mixture of covalent and ionic bonding. Even in NaCl, the electron donated by sodium to chlorine has some probability of being at the sodium atom because its wave function in the vicinity of the sodium atom, while small, is not zero. Thus, this electron is partially shared in a covalent bond, although this bonding is only a small part of the total bond, which is mainly ionic.

A measure of the degree to which a bond is ionic or covalent can be obtained from the electric dipole moment of the molecule. For example, if the bonding in NaCl were purely ionic, the center of positive charge would be at the Na$^+$ ion and the center of negative charge would be at the Cl$^-$ ion. The electric dipole moment would have the magnitude

$$p_{ionic} = er_0 \qquad\qquad 37\text{-}5$$

where $r_0 = 2.36 \times 10^{-10}$ m is the equilibrium separation of the ions. Thus, the dipole moment of NaCl would be (from Figure 37-1)

$$p_{ionic} = er_0$$
$$= (1.60 \times 10^{-19}\,\text{C})(2.36 \times 10^{-10}\,\text{m}) = 3.78 \times 10^{-29}\,\text{C·m}$$

The actual measured electric dipole moment of NaCl is

$$p_{measured} = 3.00 \times 10^{-29}\,\text{C·m}$$

We can define the ratio of $p_{measured}$ to $p_{ionic}$ as the fractional amount of ionic bonding. For NaCl, this ratio is $3.00/3.78 = 0.79$. Thus, the bonding in NaCl is about 79 percent ionic.

**EXERCISE** The equilibrium separation of HCl is 0.128 nm and its measured electric dipole moment is $3.60 \times 10^{-30}$ C·m. What is the percentage of ionic bonding in HCl? (*Answer* 18 percent)

## Other Bonding Types

**The van der Waals Bond** Any two separated molecules will be attracted to one another by electrostatic forces called van der Waals forces. So will any two atoms that do not form ionic or covalent bonds. The **van der Waals bonds** due to these

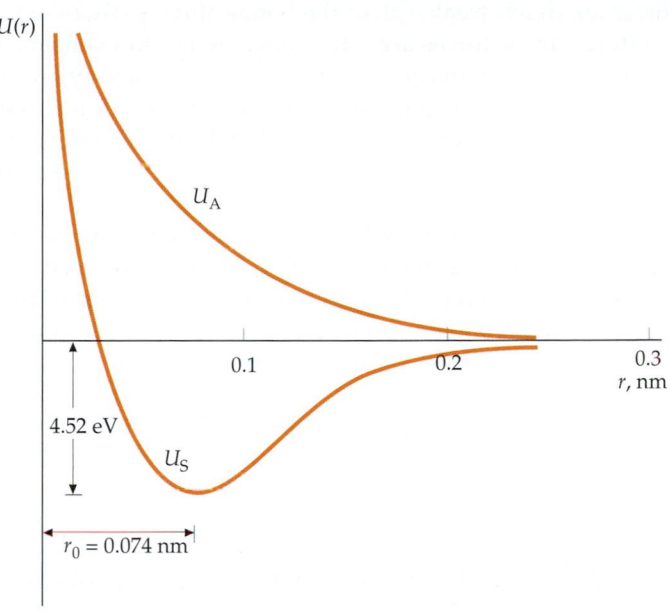

**FIGURE 37-4** Potential energy versus separation for two hydrogen atoms. The curve labeled $U_S$ is for a wave function with a symmetric space part, and the curve labeled $U_A$ is for a wave function with an antisymmetric space part.

forces are much weaker than the bonds already discussed. At high enough temperatures, these forces are not strong enough to overcome the ordinary thermal agitation of the atoms or molecules, but at sufficiently low temperatures, thermal agitation becomes negligible, and the van der Waals forces will cause virtually all substances to condense into a liquid and then a solid form.[†] The van der Waals forces arise from the interaction of the instantaneous electric dipole moments of the molecules.

Figure 37-5 shows how two polar molecules—molecules with *permanent* electric dipole moments, such as $H_2O$—can bond. The electric field due to the dipole moment of one molecule orients the other molecule so that the two dipole moments attract. Nonpolar molecules also attract other nonpolar molecules via the van der Waals forces. Although nonpolar molecules have zero electric dipole moments on the average, they have instantaneous dipole moments that are generally not zero because of fluctuations in the positions of the charges. When two nonpolar molecules are near each other, the fluctuations in the instantaneous dipole moments tend to become correlated so as to produce attraction. This is illustrated in Figure 37-6.

**The Hydrogen Bond** Another bonding mechanism of great importance is the hydrogen bond, which is formed by the sharing of a proton (the nucleus of the hydrogen atom) between two atoms, frequently two oxygen atoms. This sharing of a proton is similar to the sharing of electrons responsible for the covalent bond already discussed. It is facilitated by the small mass of the proton and by the absence of inner core electrons in hydrogen. The hydrogen bond often holds groups of molecules together and is responsible for the cross-linking that allows giant biological molecules and polymers to hold their fixed shapes. The well-known helical structure of DNA is due to hydrogen-bond linkages across turns of the helix (Figure 37-7).

**The Metallic Bond** In a metal, two atoms do not bond together by exchanging or sharing an electron to form a molecule. Instead, each valence electron is shared by many atoms. The bonding is thus distributed throughout the entire metal. A metal can be thought of as a lattice of positive ions held together by a *gas* of essentially free electrons that roam throughout the solid. In the quantum-mechanical picture, these free electrons form a cloud of negative charge density between the positively charged lattice ions that holds the ions together. In this

---

[†] Helium is the only element that does not solidify at any temperature at atmospheric pressure.

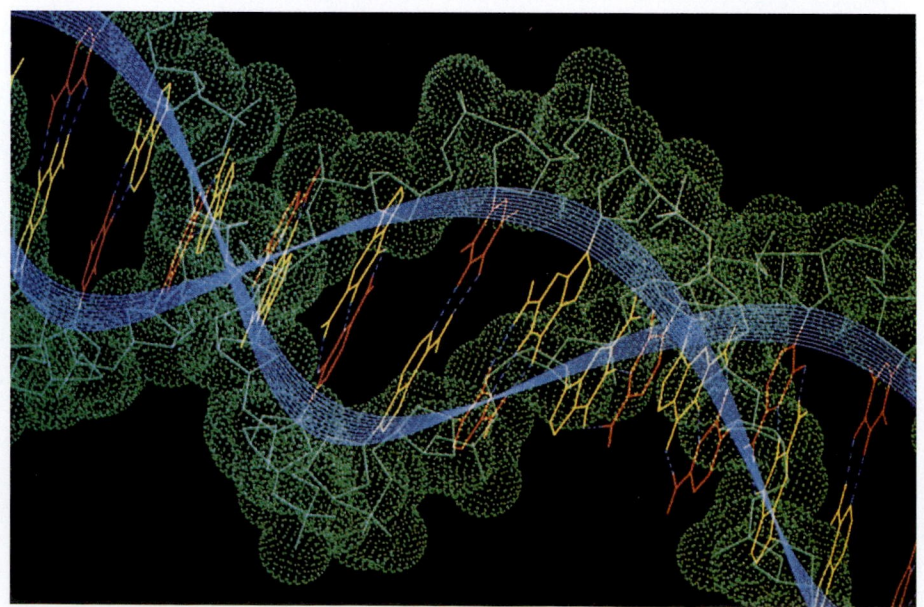

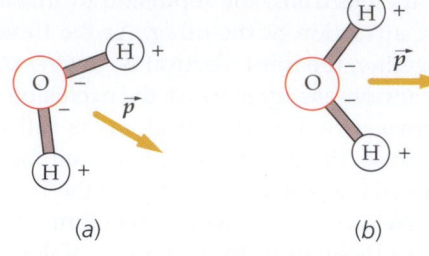

(a)                    (b)

**FIGURE 37-5** Bonding of $H_2O$ molecules because of the attraction of the electric dipoles. The dipole moment of each molecule is indicated by $\vec{p}$. The field of one dipole orients the other dipole so the moments tend to be parallel. When the dipole moments are approximately parallel, the center of negative charge of one molecule is close to the center of positive charge of the other molecule, and the molecules attract.

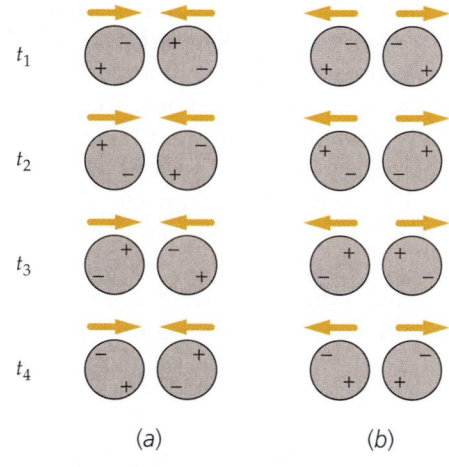

(a)                    (b)

**FIGURE 37-6** van der Waals attraction of molecules with zero average dipole moments. (*a*) Possible orientations of instantaneous dipole moments at different times leading to attraction. (*b*) Possible orientations leading to repulsion. The electric field of the instantaneous dipole moment of one molecule tends to polarize the other molecule; thus the orientations leading to attraction (Figure 37-6*a*) are much more likely than those leading to repulsion (Figure 37-6*b*).

**FIGURE 37-7** The DNA molecule.

respect, the metallic bond is somewhat similar to the covalent bond. However, with the metallic bond, there are far more than just two atoms involved, and the negative charge is distributed uniformly throughout the volume of the metal. The number of free electrons varies from metal to metal but is of the order of one per atom.

## *37-2  Polyatomic Molecules

Molecules with more than two atoms range from such relatively simple molecules as water, which has a molecular mass number of 18, to such giants as proteins and DNA, which can have molecular masses of hundreds of thousands up to many millions. As with diatomic molecules, the structure of polyatomic molecules can be understood by applying basic quantum mechanics to the bonding of individual atoms. The bonding mechanisms for most polyatomic molecules are the covalent bond and the hydrogen bond. We will discuss only some of the simplest polyatomic molecules—$H_2O$, $NH_3$, and $CH_4$—to illustrate both the simplicity and complexity of the application of quantum mechanics to molecular bonding.

The basic requirement for the sharing of electrons in a covalent bond is that the wave functions of the valence electrons in the individual atoms must overlap as much as possible. As our first example, we will consider the water molecule. The ground-state configuration of the oxygen atom is $1s^2 2s^2 2p^4$. The 1s and 2s electrons are in closed-shell states and do not contribute to the bonding. The 2p shell has room for six electrons, two in each of the three space states (orbitals) corresponding to $\ell = 1$. In an isolated atom, we describe these space states by the hydrogen-like wave functions corresponding to $\ell = 1$ and $m = +1$, 0, and $-1$. Since the energy is the same for these three space states, we could equally well use any linear combination of these wave functions. When an atom participates in molecular bonding, certain combinations of these atomic wave functions are important. These combinations are called the $p_x$, $p_y$, and $p_z$ **atomic orbitals.** The angular dependence of these orbitals is

$$p_x \propto \sin\theta\cos\phi \qquad\qquad 37\text{-}6$$

$$p_y \propto \sin\theta\sin\phi \qquad\qquad 37\text{-}7$$

$$p_z \propto \cos\phi \qquad\qquad 37\text{-}8$$

The electron charge distribution for these orbitals is maximum along the $x$, $y$, or $z$ axis, respectively, as shown in Figure 37-8.

**FIGURE 37-8** Computer-generated dot plot illustrating the spatial dependence of the electron charge distribution in the $p_x$, $p_y$, and $p_z$ atomic orbitals.

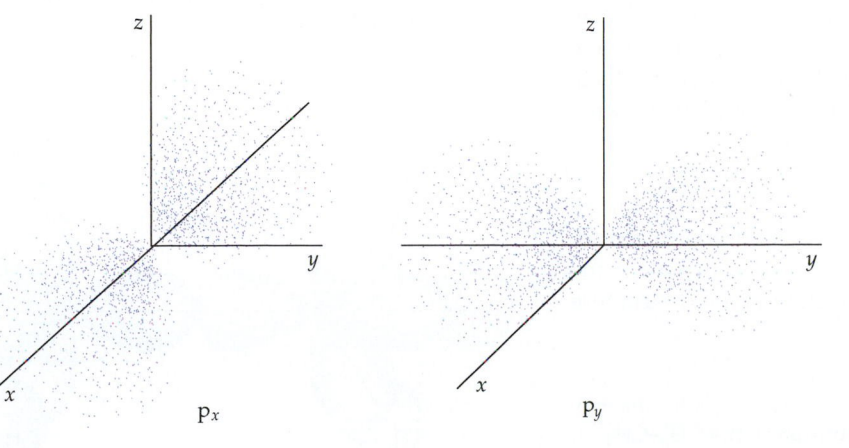

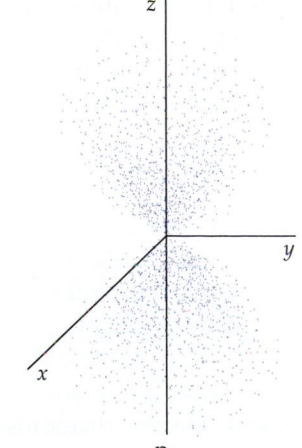

For the oxygen in an $H_2O$ molecule, maximum overlap of the electron wave functions occurs when two of the four 2p electrons are paired with their spins antiparallel in one of the atomic orbitals (for this example, assume the $p_z$ orbital), one of the other electrons is in a second orbital (the $p_x$ orbital), and the other electron is in the third orbital (the $p_y$ orbital). Each of the unpaired electrons (in the $p_x$ and $p_y$ orbitals, in this illustration) forms a bond with the electron of a hydrogen atom, as shown in Figure 37-9. Because of the repulsion of the two hydrogen atoms, the angle between the O–H bonds is actually greater than 90°. The effect of this repulsion can be calculated, and the result is in agreement with the measured angle of 104.5°.

Similar reasoning leads to an understanding of the bonding in $NH_3$. In the ground state, nitrogen has three electrons in the 2p state. When these three electrons are in the $p_x$, $p_y$, and $p_z$ atomic orbitals, they bond to the electrons of hydrogen atoms. Again, because of the repulsion of the hydrogen atoms, the angles between the bonds are somewhat larger than 90°.

The bonding of carbon atoms is somewhat more complicated. Carbon forms a wide variety of different types of molecular bonds, leading to a great diversity in the kinds of organic molecules. The ground-state configuration of carbon is $1s^2 2s^2 2p^2$. From our previous discussion, we might expect carbon to be divalent—that is, bonding only through its two 2p electrons—with the two bonds forming at approximately 90°. However, one of the most important features of the chemistry of carbon is that tetravalent carbon compounds, such as $CH_4$, are overwhelmingly favored.

The observed valence of 4 for carbon comes about in an interesting way. One of the first excited states of carbon occurs when a 2s electron is excited to a 2p state, giving a configuration of $1s^2 2s^1 2p^3$. In this excited state, we can have four unpaired electrons, one each in the 2s, $2p_x$, $2p_y$, and $2p_z$ atomic orbitals. We might expect there to be three similar bonds corresponding to the three p orbitals and one different bond corresponding to the s orbital. However, when carbon forms tetravalent bonds, these four atomic orbitals become mixed and form four new *equivalent* molecular orbitals called **hybrid orbitals.** This mixing of atomic orbitals, called **hybridization,** is among the most important features involved in the physics of complex molecular bonds. Figure 37-10 shows the tetrahedral structure of the methane molecule ($CH_4$), and Figure 37-11 shows the structure of the ethane molecule ($CH_3$–$CH_3$), which is similar to two joined methane molecules in which one of the C–H bonds is replaced with a C–C bond.

Carbon orbitals can also hybridize, with the s, $p_x$, and $p_y$ orbitals combining to form three hybrid orbitals in the $xy$ plane with 120° bonds and the $p_z$ orbital remaining unmixed. An example of this configuration is graphite, in which the bonds in the $xy$ plane provide the strongly layered structure characteristic of the material.

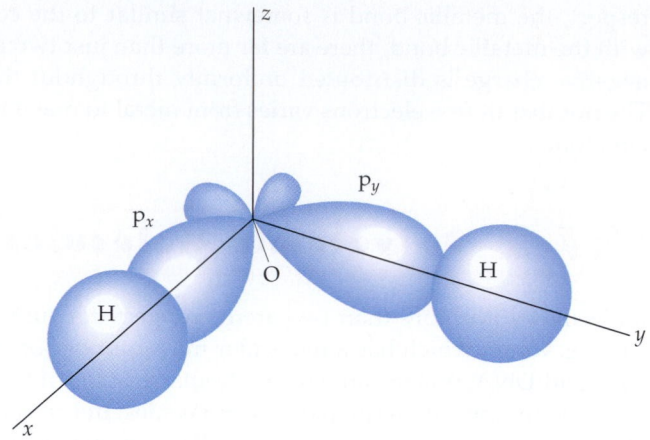

**FIGURE 37-9** Electron charge distribution in the $H_2O$ molecule.

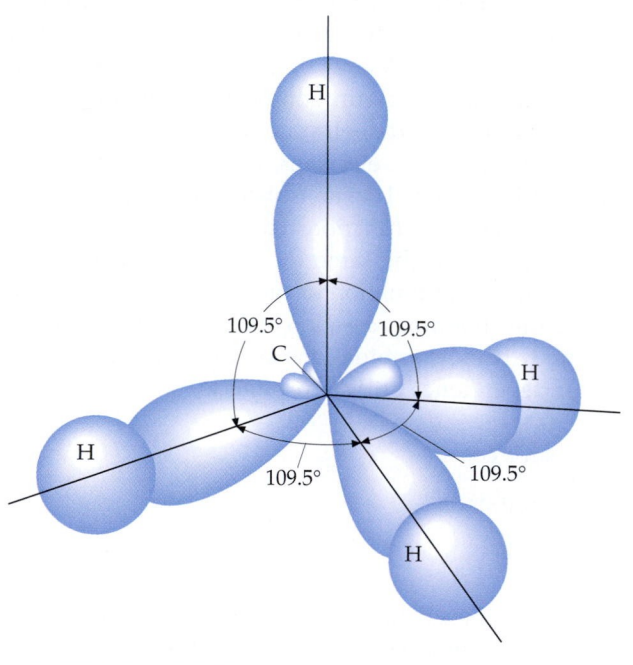

**FIGURE 37-10** Electron charge distribution in the $CH_4$ (methane) molecule.

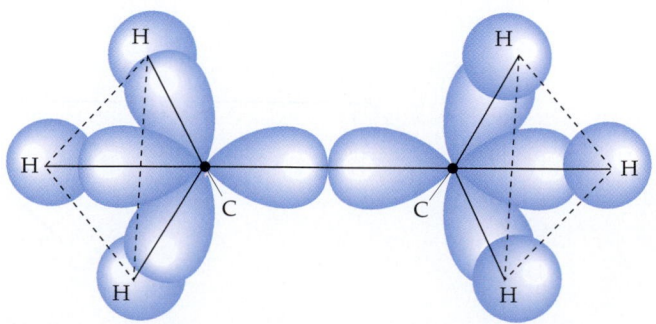

**FIGURE 37-11** Electron charge distribution in the $CH_3$–$CH_3$ (ethane) molecule.

# 37-3 Energy Levels and Spectra of Diatomic Molecules

As is the case with an atom, a molecule often emits electromagnetic radiation when it makes a transition from an excited energy state to a state of lower energy. Conversely, a molecule can absorb radiation and make a transition from a lower energy state to a higher energy state. The study of molecular emission and absorption spectra thus provides us with information about the energy states of molecules. For simplicity, we will consider only diatomic molecules here.

The energy of a molecule can be conveniently separated into three parts: electronic, due to the excitation of the electrons of the molecule; vibrational, due to the oscillations of the atoms of the molecule; and rotational, due to the rotation of the molecule about its center of mass. The magnitudes of these energies are sufficiently different that they can be treated separately. The energies due to the electronic excitations of a molecule are of the order of magnitude of 1 eV, the same as for the excitation of an atom. The energies of vibration and rotation are much smaller than this.

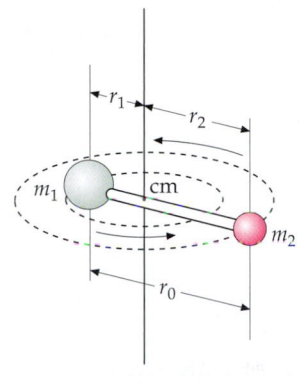

**FIGURE 37-12** Diatomic molecule rotating about an axis through its center of mass.

## Rotational Energy Levels

Figure 37-12 shows a simple schematic model of a diatomic molecule consisting of a mass $m_1$ and a mass $m_2$ separated by a distance $r$ and rotating about its center of mass. Classically, the kinetic energy of rotation (see Equation 9-11) is

$$E = \tfrac{1}{2} I\omega^2 \qquad\qquad 37\text{-}9$$

where $I$ is the moment of inertia and $\omega$ is the angular frequency of rotation. If we write this in terms of the angular momentum $L = I\omega$, we have

$$E = \frac{(I\omega)^2}{2I} = \frac{L^2}{2I} \qquad\qquad 37\text{-}10$$

The solution of the Schrödinger equation for rotation leads to quantization of the angular momentum with values given by

$$L^2 = \ell(\ell + 1)\hbar^2, \quad \ell = 0, 1, 2, \ldots \qquad\qquad 37\text{-}11$$

where $\ell$ is the **rotational quantum number.** This is the same quantum condition on angular momentum that holds for the orbital angular momentum of an electron in an atom. Note, however, that $L$ in Equation 37-10 refers to the angular momentum of the entire molecule rotating about its center of mass. The energy levels of a rotating molecule are therefore given by

$$E_\ell = \frac{\ell(\ell + 1)\hbar^2}{2I} = \ell(\ell + 1)E_{0r}, \quad \ell = 0, 1, 2, \ldots \qquad\qquad 37\text{-}12$$

ROTATIONAL ENERGY LEVELS

where $E_{0r}$ is the characteristic rotational energy of a particular molecule, which is inversely proportional to its moment of inertia:

$$E_{0r} = \frac{\hbar^2}{2I} \qquad\qquad 37\text{-}13$$

CHARACTERISTIC ROTATIONAL ENERGY

A measurement of the rotational energy of a molecule from its rotational spectrum can be used to determine the moment of inertia of the molecule, which can then be used to find the separation of the atoms in the molecule. The moment of inertia about an axis through the center of mass of a diatomic molecule (see Figure 37-12) is

$$I = m_1 r_1^2 + m_2 r_2^2$$

Using $m_1 r_1 = m_2 r_2$, where $r_1$ is the distance of atom 1 from the center of mass, $r_2$ is the distance of atom 2 from the center of mass, and $r_0 = r_1 + r_2$, we can write the moment of inertia (see Problem 31) as

$$I = \mu r_0^2 \qquad\qquad 37\text{-}14$$

where $\mu$, called the **reduced mass,** is

$$\mu = \frac{m_1 m_2}{m_1 + m_2} \qquad\qquad 37\text{-}15$$

DEFINITION—REDUCED MASS

If the masses are equal ($m_1 = m_2 = m$), as in $H_2$ and $O_2$, the reduced mass is $\mu = \frac{1}{2}m$ and

$$I = \frac{1}{2} m r_0^2 \qquad\qquad 37\text{-}16$$

A unit of mass convenient for discussing atomic and molecular masses is the **unified mass unit,** u, which is defined as one-twelfth the mass of the neutral carbon-12 ($^{12}C$) atom. The mass of one $^{12}C$ atom is thus 12 u. The mass of an atom in unified mass units is therefore numerically equal to the molar mass of the atom in grams. The unified mass unit is related to the gram and kilogram by

$$1\,u = \frac{1\,g}{N_A} = \frac{10^{-3}\,kg}{6.0221 \times 10^{23}} = 1.6606 \times 10^{-27}\,kg \qquad 37\text{-}17$$

where $N_A$ is Avogadro's number.

---

*THE REDUCED MASS OF A DIATOMIC MOLECULE*     **E X A M P L E   3 7 - 2**

**Find the reduced mass of the HCl molecule.**

**PICTURE THE PROBLEM** We find the masses of the hydrogen and chlorine atoms in the periodic table[†] in Appendix C and use the definition in Equation 37-15.

1. The reduced mass $\mu$ is related to the individual masses $m_H$ and $m_{Cl}$:

$$\mu = \frac{m_H m_{Cl}}{m_H + m_{Cl}}$$

2. Find the masses in the periodic table:

$$m_H = 1.01\,u, \quad m_{Cl} = 35.5\,u$$

3. Substitute to calculate the reduced mass:

$$\mu = \frac{m_H m_{Cl}}{m_H + m_{Cl}} = \frac{(1.01\,u)(35.5\,u)}{1.01\,u + 35.5\,u}$$

$$= \boxed{0.982\,u}$$

---

† The masses in these tables are weighted according to the natural isotopic distribution. Thus, the mass of carbon is given as 12.011 rather than 12.000 because natural carbon consists of about 98.9 percent $^{12}C$ and 1.1 percent $^{13}C$. Similarly, natural chlorine consists of about 76 percent $^{35}Cl$ and 24 percent $^{37}Cl$.

**REMARKS** Note that the reduced mass is less than the mass of either atom in the molecule, and that it is approximately equal to the mass of the hydrogen atom. When one atom of a diatomic molecule is much more massive than the other, the center of mass of the molecule is approximately at the center of the more massive atom, and the reduced mass is approximately equal to the mass of the lighter atom.

---

*ROTATIONAL KINETIC ENERGY OF A MOLECULE*        **EXAMPLE 37-3**

**Estimate the characteristic rotational energy of an $O_2$ molecule, assuming that the separation of the atoms is 0.1 nm.**

1. The characteristic rotational energy is inversely propor-    $E_{0r} = \dfrac{\hbar^2}{2I}$
   tional to the moment of inertia:

2. Calculate the moment of inertia:    $I = \mu r_0^2 = \frac{1}{2} m r_0^2$

3. Substitute this expression for $I$ into the expression for    $E_{0r} = \dfrac{\hbar^2}{m r_0^2}$
   $E_{0r}$:

4. Use $m = 16$ u for the mass of oxygen and the given val-    $E_{0r} = \dfrac{\hbar^2}{m r_0^2}$
   ues of the constants to calculate $E_{0r}$:

   $$= \frac{(1.055 \times 10^{-34}\ \text{J·s})^2}{(16\ \text{u})(10^{-10}\ \text{m})^2} \times \left( \frac{1\ \text{u}}{1.66 \times 10^{-27}\ \text{kg}} \right)$$

   $$= 4.19 \times 10^{-23}\ \text{J} = \boxed{2.62 \times 10^{-4}\ \text{eV}}$$

---

We can see from Example 37-3 that the rotational energy levels are several orders of magnitude smaller than energy levels due to electron excitation, which have energies of the order of 1 eV or higher. Transitions within a given set of rotational energy levels yield photons in the microwave region of the electromagnetic spectrum. The rotational energies are also small compared with the typical thermal energy $kT$ at normal temperatures. For $T = 300$ K, for example, $kT$ is about $2.6 \times 10^{-2}$ eV, which is approximately 100 times the characteristic rotational energy as calculated in Example 37-3, and approximately 1 percent of the typical electronic energy. Thus, at ordinary temperatures, a molecule can be easily excited to the lower rotational energy levels by collisions with other molecules. But such collisions cannot excite the molecule to its electronic energy levels above the ground state.

## Vibrational Energy Levels

The quantization of energy in a simple harmonic oscillator was one of the first problems solved by Schrödinger in his paper proposing his wave equation. Solving the Schrödinger equation for a simple harmonic oscillator gives

$$E_\nu = (\nu + \tfrac{1}{2})hf, \quad \nu = 0, 1, 2, \ldots \qquad \text{37-18}$$

VIBRATIONAL ENERGY LEVELS

where $f$ is the frequency of the oscillator and $\nu$ (lowercase Greek nu) is the **vibrational quantum number.**[†] An interesting feature of this result is that the energy levels are equally spaced with intervals equal to $hf$. The frequency of vibration

---

[†] We use $\nu$ here rather than $n$ so as not to confuse the vibrational quantum number with the principal quantum number $n$ for electronic energy levels.

of a diatomic molecule can be related to the force exerted by one atom on the other. Consider two objects of mass $m_1$ and $m_2$ connected by a spring of force constant $k_F$. The frequency of oscillation of this system (see Problem 36) can be shown to be

$$f = \frac{1}{2\pi}\sqrt{\frac{k_F}{\mu}} \qquad\qquad 37\text{-}19$$

where $\mu$ is the reduced mass given by Equation 37-15. The effective force constant $k_F$ of a diatomic molecule can thus be determined from a measurement of the frequency of oscillation of the molecule.

A selection rule on transitions between vibrational states (of the same electronic state) requires that the vibrational quantum number $\nu$ can change only by $\pm 1$, so the energy of a photon emitted by such a transition is $hf$ and the frequency is $f$, the same as the frequency of vibration. There is a similar selection rule that $\ell$ must change by $\pm 1$ for transitions between rotational states.

A typical measured frequency of a transition between vibrational states is $5 \times 10^{13}$ Hz, which gives

$$E \approx hf = (4.14 \times 10^{-15}\ \text{eV·s})(5 \times 10^{13}\ \text{s}^{-1}) = 0.2\ \text{eV}$$

for the order of magnitude of vibrational energies. This typical vibrational energy is approximately 1000 times greater than the typical rotational energy $E_{0r}$ of the $O_2$ molecule we found in Example 37-3 and about 8 times greater than the typical thermal energy $kT = 0.026$ eV at $T = 300$ K. Thus, the vibrational levels are almost never excited by molecular collisions at ordinary temperatures.

---

*DETERMINING THE FORCE CONSTANT* **EXAMPLE 37-4**

**The frequency of vibration of the CO molecule is $6.42 \times 10^{13}$ Hz. What is the effective force constant for this molecule?**

**PICTURE THE PROBLEM** We use Equation 37-19 to relate $k_F$ to the frequency and the reduced mass, and calculate $\mu$ from its definition.

1. The effective force constant is related to the frequency and reduced mass by Equation 37-19:

$$f = \frac{1}{2\pi}\sqrt{\frac{k_F}{\mu}}$$

$$k_F = (2\pi f)^2 \mu$$

2. Calculate the reduced mass using 12 u for the mass of the carbon atom and 16 u for the mass of the oxygen atom:

$$\mu = \frac{m_1 m_2}{m_1 + m_2} = \frac{(12\ \text{u})(16\ \text{u})}{12\ \text{u} + 16\ \text{u}} = 6.86\ \text{u}$$

3. Substitute this value of $\mu$ into the equation for $k_F$ in Step 1 and convert to SI units:

$$k_F = (2\pi f)^2 \mu$$

$$= 4\pi^2\,(6.42 \times 10^{13}\ \text{Hz})^2\,(6.86\ \text{u})$$

$$= 1.12 \times 10^{30}\ \text{u/s}^2 \times \left(\frac{1.66 \times 10^{-27}\ \text{kg}}{1\ \text{u}}\right)$$

$$= \boxed{1.85 \times 10^3\ \text{N/m}}$$

MASTER the CONCEPT
WEB

## Emission Spectra

Figure 37-13 shows schematically some electronic, vibrational, and rotational energy levels of a diatomic molecule. The vibrational levels are labeled with the quantum number $\nu$ and the rotational levels are labeled with $\ell$. The lower vibrational levels are evenly spaced, with $\Delta E = hf$. For higher vibrational levels, the approximation that the vibration is simple harmonic is not valid and the levels are not quite evenly spaced. Note that the potential energy curves representing the force between the two atoms in the molecule do not have exactly the same shape for the electronic ground and excited states. This implies that the fundamental frequency of vibration $f$ is different for different electronic states. For transitions between vibrational states of different electronic states, the selection rule $\Delta \nu = \pm 1$ does not hold. Such transitions result in the emission of photons of wavelengths in or near the visible spectrum, so the emission spectrum of a molecule for electronic transitions is also sometimes called the optical spectrum.

**FIGURE 37-13** Electronic, vibrational, and rotational energy levels of a diatomic molecule. The rotational levels are shown in an enlargement of the $\nu = 0$ and $\nu = 1$ vibrational levels of the electronic ground state.

The spacing of the rotational levels increases with increasing values of $\ell$. Since the energies of rotation are so much smaller than those of vibrational excitation or electronic excitation of a molecule, molecular rotation shows up in optical spectra as a fine splitting of the spectral lines. When the fine structure is not resolved, the spectrum appears as bands, as shown in Figure 37-14a. Close inspection of these bands reveals that they have a fine structure due to the rotational energy levels, as shown in the enlargement in Figure 37-14c.

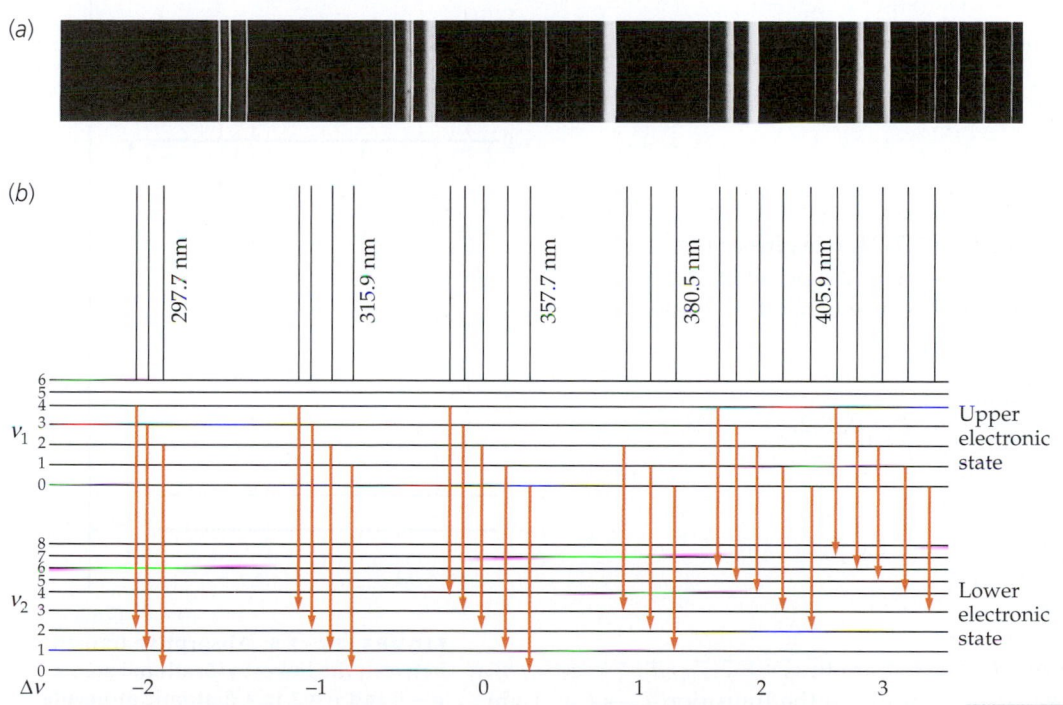

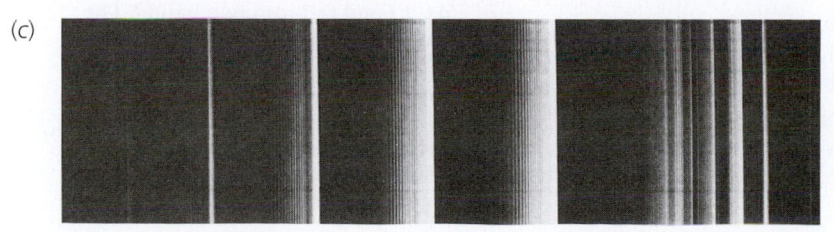

**FIGURE 37-14** (a) Part of the emission spectrum of $N_2$. The spectral lines are due to transitions between the vibrational levels of two electronic states, as indicated in the energy level diagram (b). (c) An enlargement of part of Figure 37-14a shows that the apparent lines are in fact bands with structure caused by rotational levels.

## Absorption Spectra

Much molecular spectroscopy is done using infrared absorption techniques in which only the vibrational and rotational energy levels of the ground-state electronic level are excited. For ordinary temperatures, the vibrational energies are sufficiently large in comparison with the thermal energy $kT$ that most of the molecules are in the lowest vibrational state $\nu = 0$, for which the energy is $E_0 = \frac{1}{2}hf$. The transition from $\nu = 0$ to $\nu = 1$ is the predominant transition in absorption. The rotational energies, however, are sufficiently less than the thermal energy $kT$ that the molecules are distributed among several rotational energy states. If the molecule is originally in a vibrational state characterized by $\nu = 0$ and a rotational state characterized by the quantum number $\ell$, the molecule's initial energy is

$$E_\ell = \tfrac{1}{2}hf + \ell(\ell + 1)E_{0r} \qquad\qquad 37\text{-}20$$

where $E_{0r}$ is given by Equation 37-13. From this state, two transitions are permitted by the selection rules. For a transition to the next higher vibrational state $\nu = 1$ and a rotational state characterized by $\ell + 1$, the final energy is

$$E_{\ell+1} = \tfrac{3}{2}hf + (\ell + 1)(\ell + 2)E_{0r} \qquad\qquad 37\text{-}21$$

For a transition to the next higher vibrational state and to a rotational state characterized by $\ell - 1$, the final energy is

$$E_{\ell-1} = \tfrac{3}{2}hf + (\ell - 1)\ell E_{0r} \qquad\qquad 37\text{-}22$$

The energy differences are

$$\Delta E_{\ell \to \ell+1} = E_{\ell+1} - E_\ell = hf + 2(\ell + 1)E_{0r} \qquad\qquad 37\text{-}23$$

where $\ell = 0, 1, 2,$ and so on, and

$$\Delta E_{\ell \to \ell-1} = E_{\ell-1} - E_\ell = hf - 2\ell E_{0r} \qquad\qquad 37\text{-}24$$

where $\ell = 1, 2, 3,$ and so on. (In Equation 37-24, $\ell$ begins at $\ell = 1$ because from $\ell = 0$ only the transition $\ell \to \ell + 1$ is possible.) Figure 37-15 illustrates these transitions. The frequencies of these transitions are given by

$$f_{\ell \to \ell+1} = \frac{\Delta E_{\ell \to \ell+1}}{h} = f + \frac{2(\ell + 1)E_{0r}}{h}, \quad \ell = 0, 1, 2, \ldots \quad 37\text{-}25$$

and

$$f_{\ell \to \ell-1} = \frac{\Delta E_{\ell \to \ell-1}}{h} = f - \frac{2\ell E_{0r}}{h}, \quad \ell = 1, 2, 3, \ldots \quad 37\text{-}26$$

The frequencies for the transitions $\ell \to \ell + 1$ are thus $f + 2(E_{0r}/h), f + 4(E_{0r}/h), f + 6(E_{0r}/h)$, and so forth; those corresponding to the transition $\ell \to \ell - 1$ are $f - 2(E_{0r}/h), f - 4(E_{0r}/h), f - 6(E_{0r}/h)$, and so forth. We thus expect the absorption spectrum to contain frequencies equally spaced by $2E_{0r}/h$ except for a gap of $4E_{0r}/h$ at the vibrational frequency $f$, as shown in Figure 37-16. A measurement of the position of the gap gives $f$ and a measurement of the spacing of the absorption peaks gives $E_{0r}$, which is inversely proportional to the moment of inertia of the molecule.

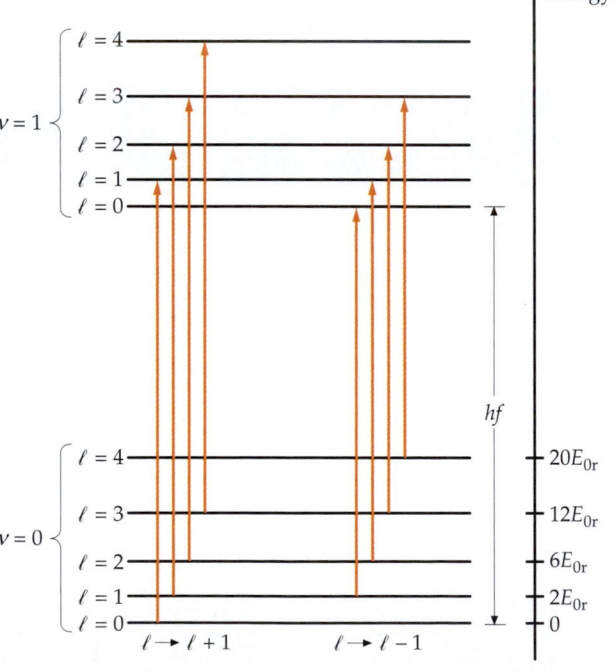

**FIGURE 37-15** Absorptive transitions between the lowest vibrational states $\nu = 0$ and $\nu = 1$ in a diatomic molecule. These transitions obey the selection rule $\Delta \ell \pm 1$ and fall into two bands. The energies of the $\ell \to \ell + 1$ band are $hf + 2E_{0r}, hf + 4E_{0r}, hf + 6E_{0r}$, and so forth; whereas the energies of the $\ell \to \ell - 1$ band are $hf - 2\,E_{0r}, hf - 4\,E_{0r}, hf - 6E_{0r}$, and so forth.

Figure 37-17 shows the absorption spectrum of HCl. The double-peak structure results from the fact that chlorine occurs naturally in two isotopes, $^{35}Cl$ and $^{37}Cl$, which gives HCl with two different moments of inertia. If all the rotational levels were equally populated initially, we would expect the intensities of each absorption line to be equal. However, the population of a rotational level is proportional to the degeneracy of the level, that is, to the number of states with the same value of $\ell$, which is $2\ell + 1$, and to the Boltzmann factor $e^{-E/kT}$, where $E$ is the energy of the state. For low values of $\ell$, the population increases slightly because of the degeneracy factor, whereas for higher values of $\ell$, the population decreases because of the Boltzmann factor. The intensities of the absorption lines therefore increase with $\ell$ for low values of $\ell$ and then decrease with $\ell$ for high values of $\ell$, as can be seen from the figure.

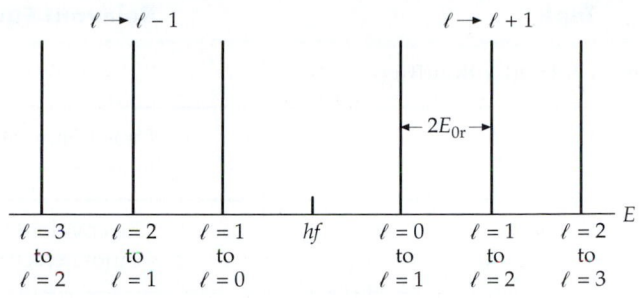

**FIGURE 37-16** Expected absorption spectrum of a diatomic molecule. The right branch corresponds to transitions $\ell \to \ell + 1$ and the left branch corresponds to the transitions $\ell \to \ell - 1$. The lines are equally spaced by $2E_{0r}$. The energy midway between the branches is $hf$, where $f$ is the frequency of vibration of the molecule.

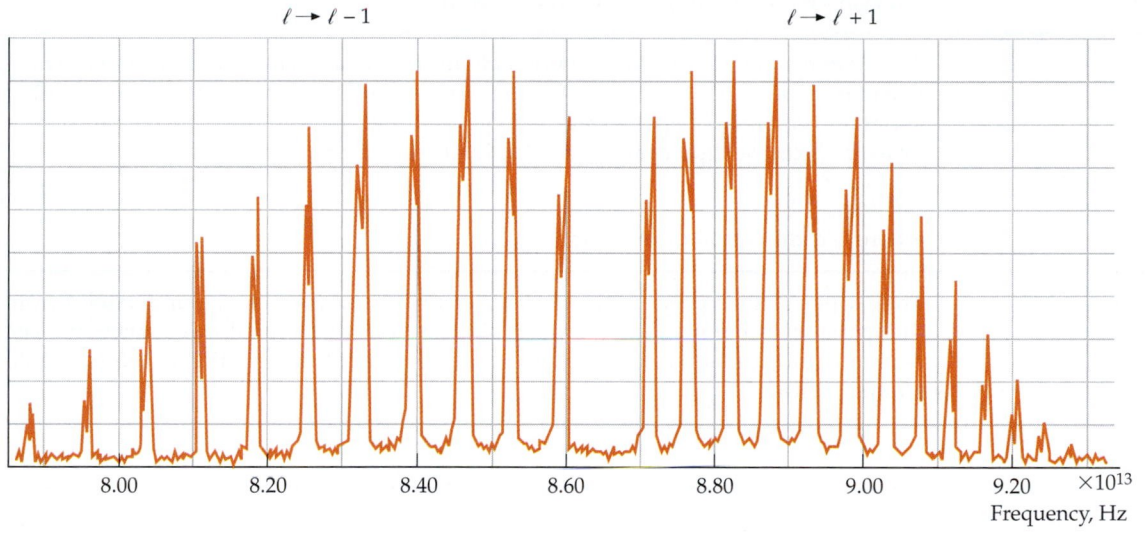

**FIGURE 37-17** Absorption spectrum of the diatomic molecule HCl. The double-peak structure results from the two isotopes of chlorine, $^{35}Cl$ (abundance 75.5 percent) and $^{37}Cl$ (abundance 24.5 percent). The intensities of the peaks vary because the population of the initial state depends on $\ell$.

# SUMMARY

1. Atoms are usually found in nature bonded to form molecules or in the lattices of crystalline solids.

2. Ionic bonds and covalent bonds are the principal mechanisms responsible for forming molecules. van der Waals bonds and metallic bonds are important in the formation of liquids and solids. Hydrogen bonds enable large biological molecules to maintain their shape.

3. Like atoms, molecules emit electromagnetic radiation when making a transition from a higher energy state to a lower energy state. The internal energy of a molecule can be separated into three parts: electronic, vibrational, and rotational energy.

4. The molecules in liquids are characterized by a temporary short-range order. The molecules or ions in solids have a more lasting order. Amorphous solids maintain a short-range order similar to the short-range order of a liquid. Crystalline solids display a long-range order determined by their minimum potential energy state.

| Topic | Relevant Equations and Remarks |
|-------|--------------------------------|

### 1. Molecular Bonding

| Ionic | Ionic bonds result when an electron is transferred from one atom to another, resulting in a positive ion and a negative ion that bond together. |
|-------|--------------------------------|
| Covalent | The covalent bond is a quantum-mechanical effect that arises from the sharing of one or more electrons by atoms. |
| van der Waals | The van der Waals bonds are weak bonds that result from the interaction of the instantaneous electric dipole moments of molecules. |
| Hydrogen | The hydrogen bond results from the sharing of a proton of the hydrogen atom by other atoms. |
| Metallic | In the metallic bond, the positive lattice ions of the metal are held together by a cloud of negative charge comprised of free electrons. |
| Mixed | A diatomic molecule formed from two identical atoms, such as $O_2$, must bond by covalent bonding. The bonding of two nonidentical atoms is often a mixture of covalent and ionic bonding. The percentage of ionic bonding can be found from the ratio of the measured electric dipole moment to the ionic electric dipole moment defined by |

$$p_{ionic} = er_0 \qquad \text{37-5}$$

where $r_0$ is the equilibrium separation of the ions.

### 2. *Polyatomic Molecules

The shapes of such polyatomic molecules as $H_2O$ and $NH_3$ can be understood from the spatial distribution of the atomic-orbital or molecular-orbital wave functions. The tetravalent nature of the carbon atom is a result of the hybridization of the 2s and 2p atomic orbitals.

### 3. Diatomic Molecules

| Moment of inertia | $$I = \mu r_0^2 \qquad \text{37-14}$$ where $r_0$ is the equilibrium separation, and $\mu$ is the reduced mass. |
|-------|--------------------------------|
| Reduced mass | $$\mu = \frac{m_1 m_2}{m_1 + m_2} \qquad \text{37-15}$$ |
| Rotational energy levels | $$E_\ell = \frac{\ell(\ell + 1)\hbar^2}{2I} = \ell(\ell + 1)E_{0r}, \quad \ell = 0, 1, 2, \ldots \qquad \text{37-12}$$ where $$E_{0r} = \frac{\hbar^2}{2I} \qquad \text{37-13}$$ |
| Vibrational energy levels | $$E_\nu = (\nu + \tfrac{1}{2})hf, \quad \nu = 0, 1, 2, \ldots \qquad \text{37-18}$$ |
| Effective force constant $k_F$ | $$f = \frac{1}{2\pi}\sqrt{\frac{k_F}{\mu}} \qquad \text{37-19}$$ |

### 4. Molecular Spectra

The optical spectra of molecules have a band structure due to transitions between rotational levels. Information about the structure and bonding of a molecule can be found from its rotational and vibrational absorption spectrum involving transitions from one vibrational–rotational level to another. These transitions obey the selection rules

$$\Delta\nu = \pm 1, \quad \Delta\ell = \pm 1$$

- Single-concept, single-step, relatively easy
- • Intermediate-level, may require synthesis of concepts
- • • Challenging
- SSM Solution is in the *Student Solutions Manual*
- iSOLVE Problems available on iSOLVE online homework service
- iSOLVE✓ These "Checkpoint" online homework service problems ask students additional questions about their confidence level, and how they arrived at their answer.

> **In a few problems, you are given more data than you actually need; in a few other problems, you are required to supply data from your general knowledge, outside sources, or informed estimates.**

## Conceptual Problems

**1** • **SSM** Would you expect the NaCl molecule to be polar or nonpolar?

**2** • Would you expect the $N_2$ molecule to be polar or nonpolar?

**3** • Does neon occur naturally as Ne or $Ne_2$? Why?

**4** • What type of bonding mechanism would you expect for (a) the HF molecule, (b) the KBr molecule, (c) the $N_2$ molecule? (d) Ag atoms in a solid?

**5** • • **SSM** The elements on the far right column of the periodic table are sometimes called noble gases because they virtually never react with other atoms to form molecules. However, this behavior is sometimes modified if the resulting molecule is formed in an electronic excited state. An example is ArF. When it is formed in the excited state, it is written ArF* and is called an excimer (for excited dimer). Refer to Figure 37-13 and discuss how this diagram would look for ArF in which the ArF ground state is unstable but the ArF* excited state is stable. *Remark: Excimers are used in certain kinds of lasers.*

**6** • Find other elements with the same subshell electron configuration in the two outermost orbitals as carbon. Would you expect the same type of hybridization for these elements as for carbon?

**7** • How does the value of the effective force constant calculated for the CO molecule in Example 37-4 compare with the value of the force constant of the suspension springs on a typical automobile, which is about 1.5 kN/m?

**8** • Explain why the moment of inertia of a diatomic molecule increases slightly with increasing angular momentum.

**9** • Why would you expect the separation distance between the two protons to be larger in the $H_2^+$ ion than in the $H_2$ molecule?

**10** • Why does an atom usually absorb radiation only from the ground state, whereas a diatomic molecule can absorb radiation from many different rotational states?

**11** • • The vibrational energy levels of diatomic molecules are described by a single vibrational frequency $f$ that is the frequency of vibration between the two atoms of the molecule. What would you expect to see in the case of polyatomic molecules? Consider in particular the water molecule $H_2O$ (Figure 37-9).

## Estimation and Approximation

**12** • • The Anharmonic Oscillator: The potential energy between the atoms in a diatomic molecule has a minimum as shown in Figure 37-13. Near this minimum the graph for the energy as a function of distance between the atoms may be approximated as a parabola, leading to the harmonic oscillator model for the vibrating molecule. An improved approximation is called the anharmonic oscillator and leads to a modification of the energy formula (Equation 37-18). The improved formula for energy is

$$E_\nu = (\nu + \tfrac{1}{2})hf - (\nu + \tfrac{1}{2})^2 hf\alpha$$

For the $O_2$ molecule, the constants have the values $f = 4.74 \times 10^{13}$ $s^{-1}$ and $\alpha = 7.6 \times 10^{-3}$. Use this formula to estimate the value of the quantum number $\nu$ for which the improved formula corrects the original formula by 10 percent.

**13** • • To understand why quantum mechanics is not needed to describe many macroscopic systems, estimate the quantum number $\ell$ and spacing between adjacent energy levels for a baseball ($m \sim 300$ g, $r \sim 3$ cm) spinning about its own axis at 20 rev/min. *Hint: Pick $\ell$ so the quantum energy formula (Equation 37-12) gives the correct energy for the given system. Then find the energy increase for the next highest energy level.*

**14** • • **SSM** Repeat Problem 13, finding the quantum number $\nu$ and spacing between adjacent energy levels for a 5-kg mass attached to 1500-N/m spring vibrating with an amplitude of 2 cm. *Hint: Pick $\nu$ so that the quantum energy formula (Equation 37-18) gives the correct energy for the given system. Then find the energy increase for the next highest energy level.*

## Molecular Bonding

**15** • iSOLVE✓ Calculate the separation of $Na^+$ and $Cl^-$ ions, for which the potential energy is $-1.52$ eV.

**16** • The dissociation energy of $Cl_2$ is 2.48 eV. Consider the formation of NaCl according to the reaction $2Na + Cl \rightarrow 2NaCl$. Does this reaction absorb energy or release energy? How much energy per molecule is absorbed or released?

**17** • The dissociation energy is sometimes expressed in kilocalories per mole (kcal/mol). (a) Find the relation between the units eV/molecule and kcal/mol. (b) Find the dissociation energy of molecular NaCl in kcal/mol.

**18** • SSM ISOLVE✓ The equilibrium separation of the HF molecule is 0.0917 nm, and its measured electric dipole moment is $6.40 \times 10^{-30}$ C·m. What percentage of the bonding is ionic?

**19** •• The dissociation energy of RbF is 5.12 eV, and the equilibrium separation is 0.227 nm. The electron affinity of fluorine is 3.40 eV, and the ionization energy of rubidium is 4.18 eV. Determine the core-repulsion energy of RbF.

**20** •• ISOLVE The equilibrium separation of the $K^+$ and $Cl^-$ ions in KCl is about 0.267 nm. (a) Calculate the potential energy of attraction of the ions, assuming them to be point charges at this separation. (b) The ionization energy of potassium is 4.34 eV, and the electron affinity of chlorine is 3.62 eV. Find the dissociation energy neglecting any energy of repulsion. (See Figure 37-1.) The measured dissociation energy is 4.49 eV. What is the energy due to repulsion of the ions at the equilibrium separation?

**21** •• Indicate the mean value of $r$ for two vibration levels in the potential energy curve for a diatomic molecule. Show that because of the asymmetry in the curve, $r_{av}$ increases with increasing vibration energy, and therefore solids expand when heated.

**22** •• ISOLVE Calculate the potential energy of attraction between the $Na^+$ and $Cl^-$ ions at the equilibrium separation $r_0 = 0.236$ nm. Compare this result with the dissociation energy given in Figure 37-1. What is the energy due to repulsion of the ions at the equilibrium separation?

**23** •• The equilibrium separation of the $K^+$ and $F^-$ ions in KF is about 0.217 nm. (a) Calculate the potential energy of attraction of the ions, assuming them to be point charges at this separation. (b) The ionization energy of potassium is 4.34 eV, and the electron affinity of fluorine is 3.40 eV. Find the dissociation energy neglecting any energy of repulsion. (c) The measured dissociation energy is 5.07 eV. Calculate the energy due to repulsion of the ions at the equilibrium separation.

**24** ••• SSM Assume that the potential energy associated with the core repulsion of the two ions of a diatomic molecule with ionic bonding can be represented by a potential energy of the form $U_{rep} = C/r^n$, so the total potential energy is $U = U_e + U_{rep} + \Delta E$, where $U_e = -ke^2/r$. $\Delta E$ is the energy of the two ions at infinite separation less the energy of the two neutral atoms at infinite separation (see Figure 37-1). Use $\dfrac{dU}{dr} = 0$ at $r = r_0$ to show that $n = \dfrac{|U_e(r_0)|}{U_{rep}(r_0)}$.

**25** ••• (a) Find $U_{rep}$ at $r = r_0$ for NaCl. (b) Assume $U_{rep} = C/r^n$ and find $C$ and $n$ for NaCl. (See Problem 24.)

## Energy Levels of Spectra of Diatomic Molecules

**26** • ISOLVE✓ The characteristic rotational energy $E_{0r}$ for the rotation of the $N_2$ molecule is $2.48 \times 10^{-4}$ eV. From this, find the separation distance of the 2 nitrogen atoms.

**27** • SSM The separation of the two oxygen atoms in a molecule of $O_2$ is actually slightly greater than the 0.1 nm used in Example 37-3, and the characteristic energy of rota-

tion $E_{0r}$ is $1.78 \times 10^{-4}$ eV rather than the result obtained in that example. Use this value to calculate the separation distance of the two oxygen atoms.

**28** •• Show that the reduced mass is smaller than either mass in a diatomic molecule and calculate it for (a) $H_2$, (b) $N_2$, (c) CO, and (d) HCl. Express your answers in unified mass units.

**29** •• The CO molecule has a binding energy of approximately 11 eV. Find the vibrational quantum number $v$ that would cause the molecule to have this much energy, and thus cause it to "shake" apart.

**30** •• SSM The equilibrium separation between the nuclei of the LiH molecule is 0.16 nm. Determine the energy separation between the $\ell = 3$ and $\ell = 2$ rotational levels of this diatomic molecule.

**31** •• SSM Derive Equations 37-14 and 37-15 for the moment of inertia in terms of the reduced mass of a diatomic molecule.

**32** •• ISOLVE✓ Use the separation of the $K^+$ and $Cl^-$ ions given in Problem 20 and the reduced mass of KCl to calculate the characteristic rotational energy $E_{0r}$.

**33** •• ISOLVE The central frequency for the absorption band of HCl shown in Figure 37-17 is at $f = 8.66 \times 10^{13}$ Hz, and the absorption peaks are separated by about $f = 6 \times 10^{11}$ Hz. Use this information to find (a) the lowest (zero-point) vibrational energy for HCl, (b) the moment of inertia of HCl, and (c) the equilibrium separation of the atoms.

**34** •• ISOLVE✓ Calculate the effective force constant for HCl from its reduced mass and the fundamental vibrational frequency obtained from Figure 37-17.

**35** •• To see how the population of rotational states of the oxygen molecule depends on the angular momentum quantum number $\ell$, use a spreadsheet program or graphing calculator to graph the function $(2\ell + 1)e^{-E_\ell/kT}$, where $E_\ell = \ell(\ell + 1)E_{0r}$ for values of $0 \le \ell \le 10$ at $T = 100$K, 200K, 300K, and 500K.

**36** •• SSM Two objects of mass $m_1$ and $m_2$ are attached to a spring of force constant $k$ and equilibrium length $r_0$. (a) Show that when $m_1$ is moved a distance $\Delta r_1$ from the center of mass, the force exerted by the spring is

$$F = -k\left(\frac{m_1 + m_2}{m_2}\right)\Delta r_1$$

(b) Show that the frequency of oscillation is $f = \left(\frac{1}{2\pi}\right)\sqrt{k/\mu}$, where $\mu$ is the reduced mass.

**37** ••• Calculate the reduced mass for the $H^{35}Cl$ and $H^{37}Cl$ molecules and the fractional difference $\Delta\mu/\mu$. Show that the mixture of isotopes in HCl leads to a fractional difference in the frequency of a transition from one rotational state to another given by $\Delta f/f = -\Delta\mu/\mu$. Compute $\Delta f/f$ and compare your result with Figure 37-17.

## General Problems

**38** • Show that when one atom in a diatomic molecule is much more massive than the other the reduced mass is approximately equal to the mass of the lighter atom.

**39** •• The equilibrium separation between the nuclei of the CO molecule is 0.113 nm. Determine the energy difference between the $\ell = 2$ and $\ell = 1$ rotational energy levels of this molecule.

**40** •• SSM The effective force constant for the HF molecule is 970 N/m. Find the frequency of vibration for this molecule.

**41** •• ISOLVE The frequency of vibration of the NO molecule is $5.63 \times 10^{13}$ Hz. Find the effective force constant for NO.

**42** •• The effective force constant of the hydrogen bond in the $H_2$ molecule is 580 N/m. Obtain the energies of the four lowest vibrational levels of the $H_2$, HD, and $D_2$ molecules and the wavelengths of photons resulting from transitions between adjacent vibrational levels of these molecules.

**43** •• The potential energy between two atoms in a molecule can often be described rather well by the Lenard–Jones potential, which can be written as

$$U = U_0\left[\left(\frac{a}{r}\right)^{12} - 2\left(\frac{a}{r}\right)^6\right],$$

where $U_0$ and $a$ are constants. Find the interatomic separation $r_0$ in terms of $a$ for which the potential energy is a minimum. Find the corresponding value of $U_{min}$. Use Figure 37-4 to obtain numerical values of $r_0$ and $U_0$ for the $H_2$ molecule, and express your answers in nanometers and electron volts.

**44** •• In this problem, you are to find how the van der Waals force between a polar molecule and a nonpolar molecule depends on the distance between the molecules. Let the dipole moment of the polar molecule be in the $x$ direction and the nonpolar molecule be a distance $x$ away. (a) How does the electric field due to an electric dipole depend on distance $x$? (b) Use the facts that (1) the potential energy of an electric dipole of moment $\vec{p}$ in an electric field $\vec{E}$ is $U = -\vec{p} \cdot \vec{E}$, and (2) the induced dipole moment of the nonpolar molecule is

proportional to $E$, to find how the potential energy of interaction of the two molecules depends on separation distance. (c) Using $F_x = -dU/dx$, find the $x$ dependence of the force between the two molecules.

**45** •• Find the dependence of the force on separation distance between two polar molecules. (See Problem 44.)

**46** •• Use the infrared absorption spectrum of HCl in Figure 37-17 to obtain (a) the characteristic rotational energy $E_{0r}$ (in eV) and (b) the vibrational frequency $f$ and the vibrational energy $hf$ (in eV).

**47** •• SSM For a molecule such as CO, which has a permanent electric dipole moment, radiative transitions obeying the selection rule $\Delta\ell = \pm 1$ between two rotational energy levels of the same vibrational level are allowed. (That is, the selection rule $\Delta\nu = \pm 1$ does not hold.) (a) Find the moment of inertia of CO and calculate the characteristic rotational energy $E_{0r}$ (in eV). (b) Make an energy-level diagram for the rotational levels for $\ell = 0$ to $\ell = 5$ for some vibrational level. Label the energies in electron volts, starting with $E = 0$ for $\ell = 0$. Indicate on your diagram the transitions that obey $\Delta\ell = -1$, and calculate the energy of the photon emitted. (c) Find the wavelength of the photons emitted for each transition in (b). In what region of the electromagnetic spectrum are these photons?

**48** ••• SSM Use the results of Problem 24 to calculate the vibrational frequency of the LiCl molecule. The dissociation energy of LiCl is 4.86 eV, and the equilibrium separation is 0.202 nm. The electron affinity of chlorine is 3.62 eV, and the ionization energy of lithium is 5.39 eV. To do this, expand the potential about $r = r_0$, where $r_0$ is the equilibrium separation, in a Taylor series. Retain only the term proportional to $(r - r_0)^2$. Recall that the potential energy of a simple harmonic oscillator is given by $U_{SHO} = \frac{1}{2}m\omega^2 x^2$. What is the wavelength resulting from transitions between adjacent harmonic oscillator levels of this molecule?

# 38

# Solids

**SILICON INGOT** SILICON IS A
SEMICONDUCTOR, AND SLICES OF
INGOTS LIKE THIS ARE USED TO
PRODUCE TRANSISTORS AND OTHER
ELECTRONIC DEVICES. TO MAKE A
TRANSISTOR, ATOMS OF ARSENIC
AND GALLIUM ARE INJECTED INTO
THE SILICON.

**?** **Do you know how many
atoms of arsenic it takes to
increase the charge-carrier density
by a factor of 5 million? For more
on this topic, see Example 38-7.**

38-1    The Structure of Solids

38-2    A Microscopic Picture of Conduction

38-3    The Fermi Electron Gas

38-4    Quantum Theory of Electrical Conduction

38-5    Band Theory of Solids

38-6    Semiconductors

*38-7   Semiconductor Junctions and Devices

38-8    Superconductivity

38-9    The Fermi–Dirac Distribution

**T**he first microscopic model of electric conduction in metals was proposed by Paul K. Drude in 1900 and developed by Hendrik A. Lorentz about 1909. This model successfully predicts that the current is proportional to the potential drop (Ohm's law) and relates the resistivity $\rho$ of conductors to the mean speed $v_{av}$ and the mean free path[†] $\lambda$ of the free electrons within the conductor. However, when $v_{av}$ and $\lambda$ are interpreted classically, there is a disagreement between the calculated values and the measured values of the resistivity, and a similar disagreement between the predicted temperature dependence and the observed temperature dependence. Thus, the classical theory fails to adequately describe the resistivity of metals. Furthermore, the classical theory says nothing about the most striking property of solids, namely that some materials are conductors, others are insulators, and still others are semiconductors, which are materials whose resistivity falls between that of conductors and insulators.

† The mean free path is the average distance traveled between collisions.

When $v_{av}$ and $\lambda$ are interpreted using quantum theory, both the magnitude and the temperature dependence of the resistivity are correctly predicted. In addition, quantum theory allows us to determine if a material will be a conductor, an insulator, or a semiconductor.

➤ **In this chapter, we use our understanding of quantum mechanics to discuss the structure of solids and solid-state semiconducting devices. Much of our discussion will be qualitative because, as in atomic physics, the quantum-mechanical calculations are very difficult.**

## 38-1 The Structure of Solids

The three phases of matter we observe—gas, liquid, and solid—result from the relative strengths of the attractive forces between molecules and the thermal energy of the molecules. Molecules in the gas phase have a relatively high thermal kinetic energy, and such molecules have little influence on one another except during their frequent but brief collisions. At sufficiently low temperatures (depending on the type of molecule), van der Waals forces will cause practically every substance to condense into a liquid and then into a solid. In liquids, the molecules are close enough—and their kinetic energy is low enough—that they can develop a temporary **short-range order.** As thermal kinetic energy is further reduced, the molecules form solids, which are characterized by a lasting order.

If a liquid is cooled slowly so that the kinetic energy of its molecules is reduced slowly, the molecules (or atoms or ions) may arrange themselves in a regular crystalline array, producing the maximum number of bonds and leading to a minimum potential energy. However, if the liquid is cooled rapidly so that its internal energy is removed before the molecules have a chance to arrange themselves, the solid formed is often not crystalline but instead resembles a snapshot of the liquid. Such a solid is called an **amorphous solid.** It displays short-range order but not the **long-range order** (over many molecular diameters) that is characteristic of a crystal. Glass is a typical amorphous solid. A characteristic result of the long-range ordering of a crystal is that it has a well-defined melting point, whereas an amorphous solid merely softens as its temperature is increased. Many materials may solidify into either an amorphous state or a crystalline state depending on how the materials are prepared; others exist only in one form or the other.

Most common solids are polycrystalline; that is, they consist of many single crystals that meet at grain boundaries. The size of a single crystal is typically a fraction of a millimeter. However, large single crystals do occur naturally and can be produced artificially. The most important property of a single crystal is the symmetry and regularity of its structure. It can be thought of as having a single unit structure that is repeated throughout the crystal. This smallest unit of a crystal is called the **unit cell;** its structure depends on the type of bonding—ionic, covalent, metallic, hydrogen, van der Waals—between the atoms, ions, or molecules. If more than one kind of atom is present, the structure will also depend on the relative sizes of the atoms.

Figure 38-1 shows the structure of the ionic crystal sodium chloride (NaCl). The $Na^+$ and $Cl^-$ ions are spherically symmetric, and the $Cl^-$ ion is approximately twice as large as the $Na^+$ ion. The minimum potential energy for this crystal occurs when an ion of either kind has six nearest neighbors of the other kind. This structure is called *face-centered-cubic* (fcc). Note that the $Na^+$ and $Cl^-$ ions in solid NaCl are *not* paired into NaCl molecules.

The net attractive part of the potential energy of an ion in a crystal can be written

$$U_{att} = -\alpha \frac{ke^2}{r}$$

38-1

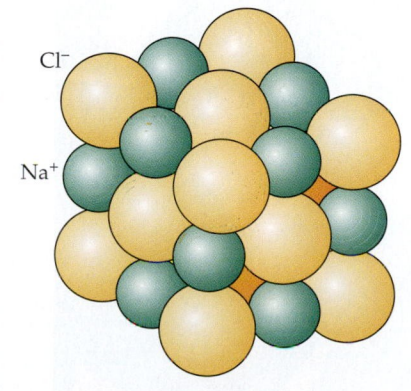

$Cl^-$

$Na^+$

**FIGURE 38-1** Face-centered-cubic structure of the NaCl crystal.

where $r$ is the separation distance between neighboring ions (0.281 nm for the $Na^+$ and $Cl^-$ ions in crystalline NaCl), and $\alpha$, called the **Madelung constant**, depends on the geometry of the crystal. If only the six nearest neighbors of each ion were important, $\alpha$ would be six. However, in addition to the six neighbors of the opposite charge at a distance $r$, there are twelve ions of the same charge at a distance $\sqrt{2}r$, eight ions of opposite charge at a distance $\sqrt{3}r$, and so on. The Madelung constant is thus an infinite sum:

$$\alpha = 6 - \frac{12}{\sqrt{2}} + \frac{8}{\sqrt{3}} - \dots \qquad \text{38-2}$$

The result for face-centered-cubic structures is $\alpha = 1.7476.$[†]

[†] A large number of terms are needed to calculate the Madelung constant accurately because the sum converges very slowly.

**Crystal structure.** (*a*) The hexagonal symmetry of a snowflake arises from a hexagonal symmetry in its lattice of hydrogen atoms and oxygen atoms. (*b*) NaCl (salt) crystals, magnified approximately thirty times. The crystals are built up from a cubic lattice of sodium and chloride ions. In the absence of impurities, an exact cubic crystal is formed. This (false-color) scanning electron micrograph shows that in practice the basic cube is often disrupted by dislocations, giving rise to crystals with a wide variety of shapes. The underlying cubic symmetry, though, remains evident. (*c*) A crystal of quartz ($SiO_2$, silicon dioxide), the most abundant and widespread mineral on the earth. If molten quartz solidifies without crystallizing, glass is formed. (*d*) A soldering iron tip, ground down to reveal the copper core within its iron sheath. Visible in the iron is its underlying microcrystalline structure.

(*a*)

(*b*)

(*c*)

(*d*)

When $Na^+$ and $Cl^-$ ions are very close together, they repel each other because of the overlap of their electrons and the exclusion-principle repulsion discussed in Section 37-1. A simple empirical expression for the potential energy associated with this repulsion that works fairly well is

$$U_{rep} = \frac{A}{r^n}$$

where $A$ and $n$ are constants. The total potential energy of an ion is then

$$U = -\alpha\frac{ke^2}{r} + \frac{A}{r^n} \qquad\qquad 38\text{-}3$$

The equilibrium separation $r = r_0$ is that at which the force $F = -dU/dr$ is zero. Differentiating and setting $dU/dr = 0$ at $r = r_0$, we obtain

$$A = \frac{\alpha ke^2 r_0^{n-1}}{n} \qquad\qquad 38\text{-}4$$

The total potential energy can thus be written

$$U = -\alpha\frac{ke^2}{r_0}\left[\frac{r_0}{r} - \frac{1}{n}\left(\frac{r_0}{r}\right)^n\right] \qquad\qquad 38\text{-}5$$

At $r = r_0$, we have

$$U(r_0) = -\alpha\frac{ke^2}{r_0}\left(1 - \frac{1}{n}\right) \qquad\qquad 38\text{-}6$$

If we know the equilibrium separation $r_0$, the value of $n$ can be found approximately from the *dissociation energy* of the crystal, which is the energy needed to break up the crystal into atoms.

---

*SEPARATION DISTANCE BETWEEN $Na^+$ AND $Cl^-$ IN NACL*   **EXAMPLE  38-1**

**Calculate the equilibrium spacing $r_0$ for NaCl from the measured density of NaCl, which is $\rho = 2.16$ g/cm³.**

**PICTURE THE PROBLEM** We consider each ion to occupy a cubic volume of side $r_0$. The mass of 1 mol of NaCl is 58.4 g, which is the sum of the atomic masses of sodium and chlorine. There are $2N_A$ ions in 1 mol of NaCl, where $N_A = 6.02 \times 10^{23}$ is Avogadro's number.

1. The volume $v$ per mol of NaCl equals the number of ions times the volume per ion:

$$v = 2N_A r_0^3$$

2. Relate $r_0$ to the density $\rho$:

$$\rho = \frac{M}{v} = \frac{M}{2N_A r_0^3}$$

3. Solve for $r_0^3$ and substitute the known values:

$$r_0^3 = \frac{M}{2N_A\rho} = \frac{58.4 \text{ g}}{2(6.02 \times 10^{23})(2.16 \text{ g/cm}^3)}$$

$$= 2.25 \times 10^{-23} \text{ cm}^3$$

so

$$r_0 = 2.82 \times 10^{-8} \text{ cm} = \boxed{0.282 \text{ nm}}$$

The measured dissociation energy of NaCl is 770 kJ/mol. Using 1 eV = $1.602 \times 10^{-19}$ J, and the fact that 1 mol of NaCl contains $N_A$ pairs of ions, we can express the dissociation energy in electron volts per ion pair. The conversion between electron volts per ion pair and kilojoules per mole is

$$1\frac{\text{eV}}{\text{ion pair}} \times \frac{6.022 \times 10^{23} \text{ ion pairs}}{1 \text{ mol}} \times \frac{1.602 \times 10^{-19} \text{ J}}{1 \text{ eV}}$$

The result is

$$1\frac{\text{eV}}{\text{ion pair}} = 96.47 \frac{\text{kJ}}{\text{mol}} \qquad\qquad 38\text{-}7$$

Thus, 770 kJ/mol = 7.98 eV per ion pair. Substituting −7.98 eV for $U(r_0)$, 0.282 nm for $r_0$, and 1.75 for $\alpha$ in Equation 38-6, we can solve for $n$. The result is $n = 9.35 \approx 9$.

Most ionic crystals, such as LiF, KF, KCl, KI, and AgCl, have a face-centered-cubic structure. Some elemental solids that have fcc structure are silver, aluminum, gold, calcium, copper, nickel, and lead.

Figure 38-2 shows the structure of CsCl, which is called the *body-centered-cubic* (bcc) structure. In this structure, each ion has eight nearest neighbor ions of the opposite charge. The Madelung constant for these crystals is 1.7627. Elemental solids with bcc structure include barium, cesium, iron, potassium, lithium, molybdenum, and sodium.

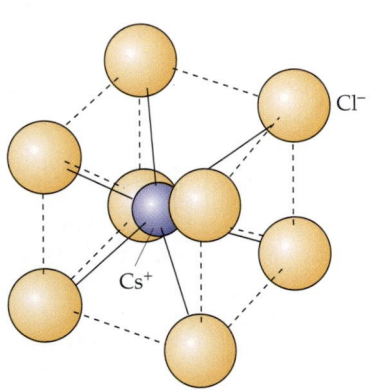

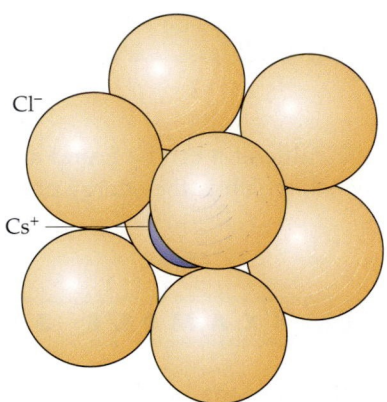

**FIGURE 38-2** Body-centered-cubic structure of the CsCl crystal.

Figure 38-3 shows another important crystal structure: the *hexagonal close-packed* (hcp) structure. This structure is obtained by stacking identical spheres, such as bowling balls. In the first layer, each ball touches six others; thus, the name *hexagonal*. In the next layer, each ball fits into a triangular depression of the first layer. In the third layer, each ball fits into a triangular depression of the second layer, so it lies directly over a ball in the first layer. Elemental solids with hcp structure include beryllium, cadmium, cerium, magnesium, osmium, and zinc.

In some solids with covalent bonding, the crystal structure is determined by the directional nature of the bonds. Figure 38-4 illustrates the diamond structure of carbon, in which each atom is bonded to four other atoms as a result of hybridization, which is discussed in Section 38-2. This is also the structure of germanium and silicon.

**FIGURE 38-3** Hexagonal close-packed crystal structure.

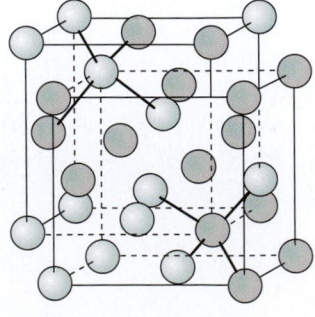

**FIGURE 38-4** Diamond crystal structure. This structure can be considered to be a combination of two interpenetrating face-centered-cubic structures.

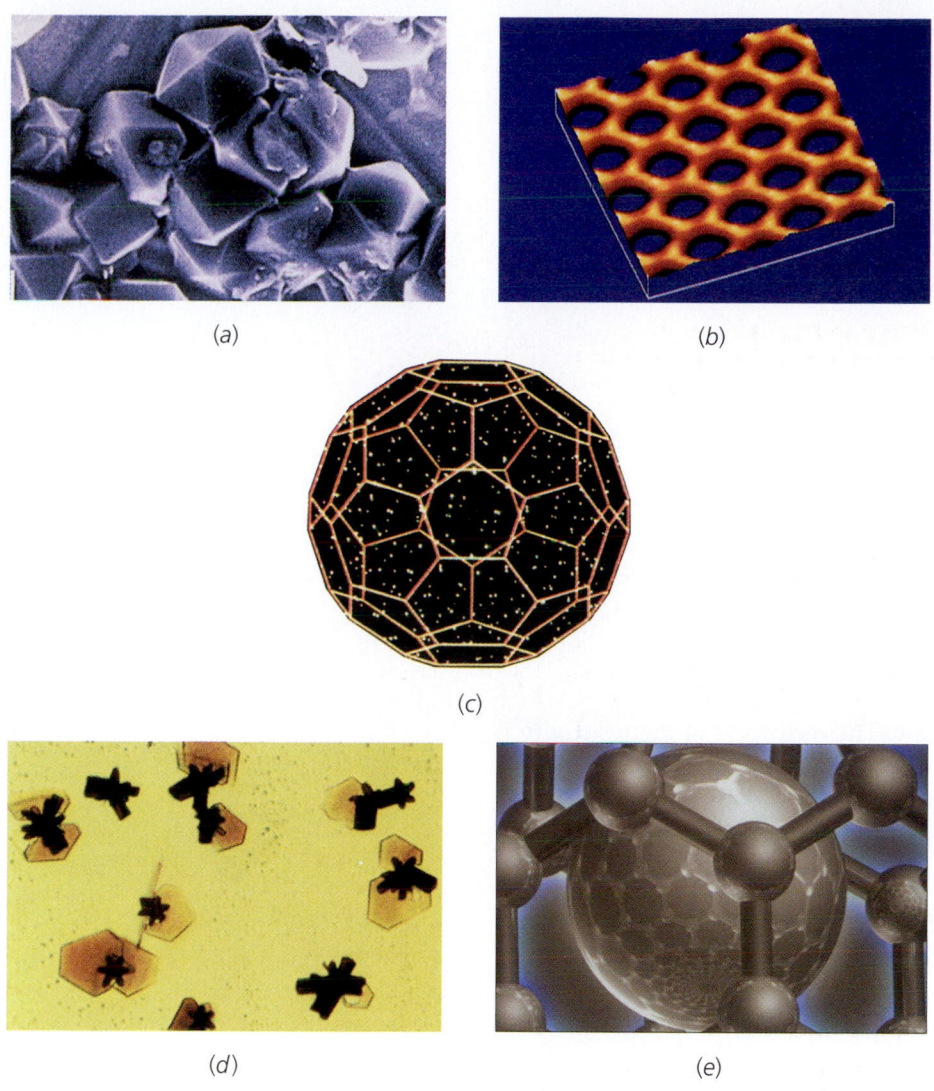

(a)

(b)

(c)

(d)

(e)

Carbon exists in three well-defined crystalline forms: diamond, graphite, and fullerenes (short for "buckminsterfullerenes"), the third of which was predicted and discovered less than two decades ago. The forms differ in how the carbon atoms are packed together in a lattice. A fourth form of carbon, in which no well-defined crystalline form exists, is common charcoal. (a) Synthetic diamonds, magnified approximately 75,000 times. In diamond, each carbon atom is centered in a tetrahedron of four other carbon atoms. The strength of these bonds accounts for the hardness of a diamond. (b) An atomic-force micrograph of graphite. In graphite, carbon atoms are arranged in sheets, with each sheet made up of atoms in hexagonal rings. The sheets slide easily across one another, a property that allows graphite to function as a lubricant. (c) A single sheet of carbon rings can be closed on itself if certain rings are allowed to be pentagonal, instead of hexagonal. A computer-generated image of the smallest such structure, $C_{60}$, is shown here. Each of the sixty vertices corresponds to a carbon atom; twenty of the faces are hexagons and twelve of the faces are pentagons. The same geometric pattern is encountered in a soccer ball. (d) Fullerene crystals, in which $C_{60}$ molecules are close-packed. The smaller crystals tend to form thin brownish platelets; larger crystals are usually rod-like in shape. Fullerenes exist in which more than sixty carbon atoms appear. In the crystals shown here, about one-sixth of the molecules are $C_{70}$. (e) Carbon nanotubes have very interesting electrical properties. A single graphite sheet is a semimetal, which means that it has properties intermediate between those of semiconductors and those of metals. When a graphite sheet is rolled into a nanotube, not only do the carbon atoms have to line up around the circumference of the tube, but the wavefunctions of the electrons must also match up. This boundary-matching requirement places restrictions on these wavefunctions, which affects the motion of the electrons. Depending on exactly how the tube is rolled up, the nanotube can be either a semiconductor or a metal.

# 38-2 A Microscopic Picture of Conduction

We consider a metal as a regular three-dimensional lattice of ions filling some volume $V$ and containing a large number $N$ of electrons that are free to move throughout the whole metal. Experimentally, the number of free electrons in a metal is approximately one electron to four electrons per atom. In the absence of an electric field, the free electrons move about the metal randomly, much the way gas molecules move about in a container. We will often refer to these free electrons in a metal as an electron gas.

The current in a conducting wire segment is proportional to the voltage drop across the segment:

$$I = \frac{V}{R}, \quad \text{or } (V = IR)$$

The resistance $R$ is proportional to the length $L$ of the wire segment and inversely proportional to the cross-sectional area $A$:

$$R = \rho \frac{L}{A}$$

where $\rho$ is the resistivity. Substituting $\rho L/A$ for $R$, and $EL$ for $V$, we can write the current in terms of the electric field strength $E$ and the resistivity. We have

$$I = \frac{V}{R} = \frac{EL}{\rho L/A} = \frac{1}{\rho} EA$$

Dividing both sides by the area $A$ gives $I/A = (1/\rho)E$, or $J = (1/\rho)E$, where $J = I/A$ is the magnitude of the **current density** vector $\vec{J}$. The current density vector is defined as

$$\vec{J} = qn\vec{v}_{\text{d}} \qquad \qquad 38\text{-}8$$

DEFINITION—CURRENT DENSITY

where $q$, $n$, and $\vec{v}_{\text{d}}$ are the charge, the number density, and the drift velocity of the charge carrier. (This follows from Equation 25-3.) In vector form, the relation between the current density and the electric field is

$$\vec{J} = \frac{1}{\rho} \vec{E} \qquad \qquad 38\text{-}9$$

This relation is the point form of Ohm's law. The reciprocal of the resistivity is called the **conductivity.**

According to Ohm's law, the resistivity is independent of both the current density and the electric field $\vec{E}$. Combining Equation 38-8 and Equation 38-9 gives

$$-en_{\text{e}}\vec{v}_{\text{d}} = \frac{1}{\rho} \vec{E} \qquad \qquad 38\text{-}10$$

where $-e$ has been substituted for $q$. According to Equation 38-10, the drift velocity $\vec{v}_{\text{d}}$ is proportional to $\vec{E}$.

In the presence of an electric field, a free electron experiences a force $-e\vec{E}$. If this were the only force acting, the electron would have a constant acceleration $-e\vec{E}/m_{\text{e}}$. However, Equation 38-10 implies a steady-state situation with a constant drift velocity that is proportional to the field $\vec{E}$. In the microscopic model, it is assumed that a free electron is accelerated for a short time and then makes a collision with a lattice ion. The velocity of the electron immediately after the collision is completely unrelated to the drift velocity. The justification for this assumption is that the magnitude of the drift velocity is extremely small compared with the random thermal speeds of the electrons.

For a typical electron, its velocity a time $t$ after its last collision is $\vec{v}_0 - (e\vec{E}/m_{\text{e}})t$, where $\vec{v}_0$ is its velocity immediately after that collision. Since the direction of $\vec{v}_0$ is random, it does not contribute to the average velocity of the electrons. Thus, the average velocity or drift velocity of the electrons is

$$\vec{v}_{\text{d}} = -\frac{e\vec{E}}{m_{\text{e}}} \tau \qquad \qquad 38\text{-}11$$

where $\tau$ is the average time since the last collision. Substituting for $\vec{v}_{\text{d}}$ in Equation 38-10, we obtain

$$-n_{\text{e}}e\left(-\frac{e\vec{E}}{m_{\text{e}}}\tau\right) = \frac{1}{\rho}\vec{E}$$

so

$$\rho = \frac{m_e}{n_e e^2 \tau}$$  38-12

The time $\tau$, called the **collision time**, is also the average time between collisions.[†] The average distance an electron travels between collisions is $v_{av}\tau$, which is called the mean free path $\lambda$:

$$\lambda = v_{av}\tau$$  38-13

where $v_{av}$ is the mean speed of the electrons. (The mean speed is many orders of magnitude greater than the drift speed.) In terms of the mean free path and the mean speed, the resistivity is

$$\rho = \frac{m_e v_{av}}{n_e e^2 \lambda}$$  38-14

RESISTIVITY IN TERMS OF $v_{AV}$ AND $\lambda$

According to Ohm's law, the resistivity $\rho$ is independent of the electric field $\vec{E}$. Since $m_e$, $n_e$, and $e$ are constants, the only quantities that could possibly depend on $\vec{E}$ are the mean speed $v_{av}$ and the mean free path $\lambda$. Let us examine these quantities to see if they can possibly depend on the applied field $\vec{E}$.

## Classical Interpretation of $v_{av}$ and $\lambda$

Classically, at $T = 0$ all the free electrons in a conductor should have zero kinetic energy. As the conductor is heated, the lattice ions acquire an average kinetic energy of $\frac{3}{2}kT$, which is imparted to the electron gas by the collisions between the electrons and the ions. (This is a result of the equipartition theorem studied in Chapters 17 and 18.) The electron gas would then have a Maxwell–Boltzmann distribution just like a gas of molecules. In equilibrium, the electrons would be expected to have a mean kinetic energy of $\frac{3}{2}kT$, which at ordinary temperatures ($\sim$300 K) is approximately 0.04 eV. At $T = 300$ K, their root-mean-square (rms) speed,[‡] which is slightly greater than the mean speed, is

$$v_{av} \approx v_{rms} = \sqrt{\frac{3kT}{m_e}} = \sqrt{\frac{3(1.38 \times 10^{-23}\,\text{J/K})(300\,\text{K})}{9.11 \times 10^{-31}\,\text{kg}}}$$  38-15

$$= 1.17 \times 10^5\,\text{m/s}$$

Note that this is about nine orders of magnitude greater than the typical drift speed of $3.5 \times 10^{-5}$ m/s, which was calculated in Example 25-1. The very small drift speed caused by the electric field therefore has essentially no effect on the very large mean speed of the electrons, so $v_{av}$ in Equation 38-14 cannot depend on the electric field $\vec{E}$.

The mean free path is related classically to the size of the lattice ions in the conductor and to the number of ions per unit volume. Consider one electron moving with speed $v$ through a region of stationary ions that are assumed to be hard spheres (Figure 38-5). The size of the electron is assumed to be negligible.

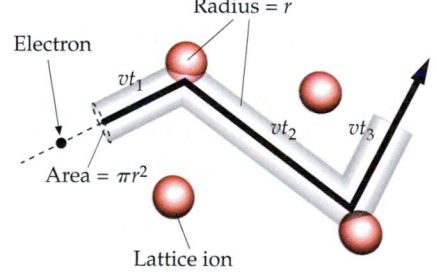

**FIGURE 38-5** Model of an electron moving through the lattice ions of a conductor. The electron, which is considered to be a point particle, collides with an ion if it comes within a distance $r$ of the center of the ion, where $r$ is the radius of the ion. If the electron speed is $v$, it collides in time $t$ with all the ions whose centers are in the volume $\pi r^2 vt$. While this picture is in accord with the classical Drude model for conduction in metals, it is in conflict with the current quantum-mechanical model presented later in this chapter.

---

[†] It is tempting but incorrect to think that if $\tau$ is the average time between collisions, the average time since its last collision is $\frac{1}{2}\tau$ rather than $\tau$. (If you find this confusing, you may take comfort in the fact that Drude used the incorrect result $\frac{1}{2}\tau$ in his original work.)
[‡] See Equation 17-23.

The electron will collide with an ion if it comes within a distance $r$ from the center of the ion, where $r$ is the radius of the ion. In some time $t_1$, the electron moves a distance $vt_1$. If there is an ion whose center is in the cylindrical volume $\pi r^2 vt_1$, the electron will collide with the ion. The electron will then change directions and collide with another ion in time $t_2$ if the center of the ion is in the volume $\pi r^2 vt_2$. Thus, in the total time $t = t_1 + t_2 + \ldots$, the electron will collide with the number of ions whose centers are in the volume $\pi r^2 vt$. The number of ions in this volume is $n_{ion} \pi r^2 vt$, where $n_{ion}$ is the number of ions per unit volume. The total path length divided by the number of collisions is the mean free path:

$$\lambda = \frac{vt}{n_{ion} \pi r^2 vt} = \frac{1}{n_{ion} \pi r^2} = \frac{1}{n_{ion} A} \qquad \text{38-16}$$

where $A = \pi r^2$ is the cross-sectional area of a lattice ion.

## Successes and Failures of the Classical Model

Neither $n_{ion}$ nor $r$ depends on the electric field $\vec{E}$, so $\lambda$ also does not depend on $\vec{E}$. Thus, according to the classical interpretation of $v_{av}$ and $\lambda$, neither depend on $\vec{E}$, so the resistivity $\rho$ does not depend on $\vec{E}$, in accordance with Ohm's law. However, the classical theory gives an incorrect temperature dependence for the resistivity. Because $\lambda$ depends only on the radius and the number density of the lattice ions, the only quantity in Equation 38-14 that depends on temperature in the classical theory is $v_{av}$, which is proportional to $\sqrt{T}$. But experimentally, $\rho$ varies linearly with temperature. Furthermore, when $\rho$ is calculated at $T = 300$ K using the Maxwell–Boltzmann distribution for $v_{av}$ and Equation 38-16 for $\lambda$, the numerical result is about six times greater than the measured value.

The classical theory of conduction fails because electrons are not classical particles. The wave nature of the electrons must be considered. Because of the wave properties of electrons and the exclusion principle (to be discussed in the following section), the energy distribution of the free electrons in a metal is not even approximately given by the Maxwell–Boltzmann distribution. Furthermore, the collision of an electron with a lattice ion is not similar to the collision of a baseball with a tree. Instead, it involves the scattering of electron waves by the lattice. To understand the quantum theory of conduction, we need a qualitative understanding of the energy distribution of free electrons in a metal. This will also help us understand the origin of contact potentials between two dissimilar metals in contact and the contribution of free electrons to the heat capacity of metals.

## 38-3 The Fermi Electron Gas

We have used the term *electron gas* to describe the free electrons in a metal. Whereas the molecules in an ordinary gas, such as air, obey the classical Maxwell–Boltzmann energy distribution, the free electrons in a metal do not. Instead, they obey a quantum energy distribution called the Fermi–Dirac distribution. Because the behavior of this electron gas is so different from a gas of molecules, the electron gas is often called a **Fermi electron gas.** The main features of a Fermi electron gas can be understood by considering an electron in a metal to be a particle in a box, a problem whose one-dimensional version we studied extensively in Chapter 34. We discuss the main features of a Fermi electron gas semiquantitatively in this section and leave the details of the Fermi–Dirac distribution to Section 38-9.

## Energy Quantization in a Box

In Chapter 34, we found that the wavelength associated with an electron of momentum $p$ is given by the de Broglie relation:

$$\lambda = \frac{h}{p} \qquad\qquad 38\text{-}17$$

where $h$ is Planck's constant. When a particle is confined to a finite region of space, such as a box, only certain wavelengths $\lambda_n$ given by standing-wave conditions are allowed. For a one-dimensional box of length $L$, the standing-wave condition is

$$n\frac{\lambda_n}{2} = L \qquad\qquad 38\text{-}18$$

This results in the quantization of energy:

$$E_n = \frac{p_n^2}{2m} = \frac{(h/\lambda_n)^2}{2m} = \frac{h^2}{2m}\frac{1}{\lambda_n^2} = \frac{h^2}{2m}\frac{1}{(2L/n)^2}$$

or

$$E_n = n^2\frac{h^2}{8mL^2} \qquad\qquad 38\text{-}19$$

The wave function for the $n$th state is given by

$$\psi_n(x) = \sqrt{\frac{2}{L}}\sin\frac{n\pi x}{L} \qquad\qquad 38\text{-}20$$

The quantum number $n$ characterizes the wave function for a particular state and the energy of that state. In three-dimensional problems, three quantum numbers arise, one associated with each dimension.

## The Exclusion Principle

The distribution of electrons among the possible energy states is dominated by the exclusion principle, which states that no two electrons in an atom can be in the same quantum state; that is, they cannot have the same set of values for their quantum numbers. The exclusion principle applies to all "spin one-half" particles, which include electrons, protons, and neutrons. These particles have a *spin* quantum number $m_s$ which has two possible values, $+\frac{1}{2}$ and $-\frac{1}{2}$. The quantum state of a particle is characterized by the spin quantum number $m_s$ plus the quantum numbers associated with the spatial part of the wave function. Because the spin quantum numbers have just two possible values, the exclusion principle can be stated in terms of the spatial states:

There can be at most two electrons with the same set of values for their *spatial* quantum numbers.

EXCLUSION PRINCIPLE IN TERMS OF SPATIAL STATES

When there are more than two electrons in a system, such as an atom or metal, only two can be in the lowest energy state. The third and fourth electrons must go into the second-lowest state, and so on.

**EXAMPLE 38-2**

**FIGURE 38-6**

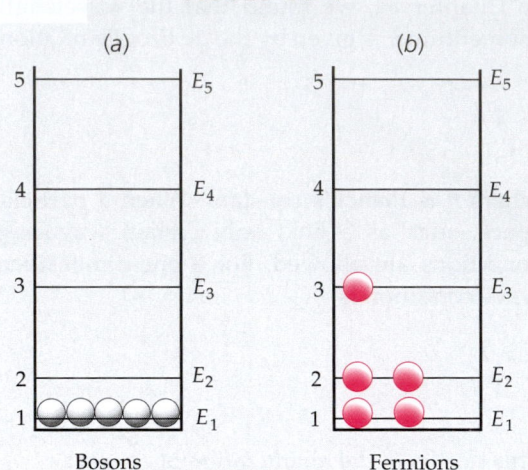

Compare the total energy of the ground state of five identical bosons of mass $m$ in a one-dimensional box with that of five identical fermions of mass $m$ in the same box.

**PICTURE THE PROBLEM** The ground state is the lowest possible energy state. The energy levels in a one-dimensional box are given by $E_n = n^2 E_1$, where $E_1 = h^2/(8mL^2)$. The lowest energy for five bosons occurs when all the bosons are in the state $n = 1$, as shown in Figure 38-6a. For fermions, the lowest state occurs with two fermions in the state $n = 1$, two fermions in the state $n = 2$, and one fermion in the state $n = 3$, as shown in Figure 38-6b.

1. The energy of five bosons in the state $n = 1$ is:

$$E = 5E_1$$

2. The energy of two fermions in the state $n = 1$, two fermions in the state $n = 2$, and one fermion in the state $n = 3$ is:

$$E = 2E_1 + 2E_2 + 1E_3 = 2E_1 + 2(2)^2E_1 + 1(3)^2E_1$$
$$= 2E_1 + 8E_1 + 9E_1 = 19E_1$$

3. Compare the total energies:

> The five identical fermions have 3.8 times the total energy of the five identical bosons.

**REMARKS** We see that the exclusion principle has a large effect on the total energy of a multiple-particle system.

## The Fermi Energy

When there are many electrons in a box, at $T = 0$ the electrons will occupy the lowest energy states consistent with the exclusion principle. If we have $N$ electrons, we can put two electrons in the lowest energy level, two electrons in the next lowest energy level, and so on. The $N$ electrons thus fill the lowest $N/2$ energy levels (Figure 38-7). The energy of the last filled (or half-filled) level at $T = 0$ is called the Fermi energy $E_F$. If the electrons moved in a one-dimensional box, the Fermi energy would be given by Equation 38-19, with $n = N/2$:

$$E_F = \left(\frac{N}{2}\right)^2 \frac{h^2}{8m_e L^2} = \frac{h^2}{32m_e}\left(\frac{N}{L}\right)^2 \qquad \text{38-21}$$

FERMI ENERGY AT $T = 0$ IN ONE DIMENSION

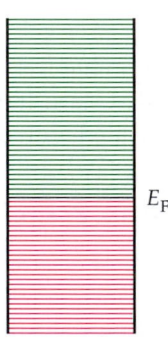

**FIGURE 38-7** At $T = 0$ the electrons fill up the allowed energy states to the Fermi energy $E_F$. The levels are so closely spaced that they can be assumed to be continuous.

In a one-dimensional box, the Fermi energy depends on the number of free electrons per unit length of the box.

**EXERCISE** Suppose there is an ion, and therefore a free electron, every 0.1 nm in a one-dimensional box. Calculate the Fermi energy. *Hint:* Write Equation 38-21 as

$$E_F = \frac{(hc)^2}{32m_e c^2}\left(\frac{N}{L}\right)^2 = \frac{(1240 \text{ eV·nm})^2}{32(0.511 \text{ MeV})}\left(\frac{N}{L}\right)^2$$

(*Answer*   $E_F = 9.4$ eV)

In our model of conduction, the free electrons move in a *three-dimensional* box of volume $V$. The derivation of the Fermi energy in three dimensions is somewhat difficult, so we will just give the result. In three dimensions, the Fermi energy at $T = 0$ is

$$E_F = \frac{h^2}{8m_e}\left(\frac{3N}{\pi V}\right)^{2/3}$$   38-22a

FERMI ENERGY AT $T = 0$ IN THREE DIMENSIONS

The Fermi energy depends on the number density of electrons $N/V$. Substituting numerical values for the constants gives

$$E_F = (0.365 \text{ eV·nm}^2)\left(\frac{N}{V}\right)^{2/3}$$   38-22b

FERMI ENERGY AT $T = 0$ IN THREE DIMENSIONS

---

*THE FERMI ENERGY FOR COPPER*                    **EXAMPLE   38-3**

The number density for electrons in copper was calculated in Example 25-1 and found to be 84.7/nm³. Calculate the Fermi energy at $T = 0$ for copper.

1. The Fermi energy is given by Equation 38-22:

$$E_F = (0.365 \text{ eV·nm}^2)\left(\frac{N}{V}\right)^{2/3}$$

2. Substitute the given number density for copper:

$$E_F = (0.365 \text{ eV·nm}^2)(84.7/\text{nm}^3)^{2/3}$$

$$= \boxed{7.04 \text{ eV}}$$

**REMARKS** Note that the Fermi energy is much greater than $kT$ at ordinary temperatures. For example, at $T = 300$ K, $kT$ is only about 0.026 eV.

**EXERCISE** Use Equation 38-22b to calculate the Fermi energy at $T = 0$ for gold, which has a free-electron number density of 59.0/nm³. (*Answer*   5.53 eV)

---

Table 38-1 lists the free-electron number densities and Fermi energies at $T = 0$ for several metals.

The average energy of a free electron can be calculated from the complete energy distribution of the electrons, which is discussed in Section 38-9. At $T = 0$, the average energy turns out to be

$$E_{av} = \tfrac{3}{5}E_F$$   38-23

AVERAGE ENERGY OF ELECTRONS
IN A FERMI GAS AT $T = 0$

For copper, $E_{av}$ is approximately 4 eV. This average energy is huge compared with typical thermal energies of about $kT \approx 0.026$ eV at a normal temperature of $T = 300$ K. This result is very different from the classical Maxwell–Boltzmann distribution result that at $T = 0$, $E = 0$, and that at some temperature $T$, $E$ is of the order of $kT$.

## TABLE 38-1

**Free-Electron Number Densities[†] and Fermi Energies at $T = 0$ for Selected Elements**

|    | Element | N/V, electrons/nm³ | $E_F$, eV |
|----|---------|--------------------|-----------|
| Al | Aluminum | 181 | 11.7 |
| Ag | Silver | 58.6 | 5.50 |
| Au | Gold | 59.0 | 5.53 |
| Cu | Copper | 84.7 | 7.04 |
| Fe | Iron | 170 | 11.2 |
| K | Potassium | 14.0 | 2.11 |
| Li | Lithium | 47.0 | 4.75 |
| Mg | Magnesium | 86.0 | 7.11 |
| Mn | Manganese | 165 | 11.0 |
| Na | Sodium | 26.5 | 3.24 |
| Sn | Tin | 148 | 10.2 |
| Zn | Zinc | 132 | 9.46 |

† Number densities are measured using the Hall effect, discussed in Section 26-4.

## The Fermi Factor at $T = 0$

The probability of an energy state being occupied is called the **Fermi factor,** $f(E)$. At $T = 0$ all the states below $E_F$ are filled, whereas all those above this energy are empty, as shown in Figure 38-8. Thus, at $T = 0$ the Fermi factor is simply

$$f(E) = 1, \qquad E < E_F$$
$$f(E) = 0, \qquad E > E_F \qquad\qquad 38\text{-}24$$

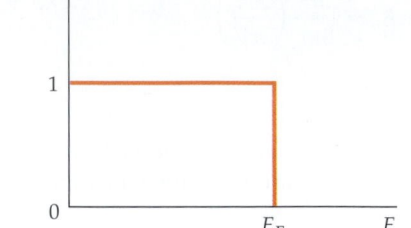

**FIGURE 38-8** Fermi factor versus energy at $T = 0$.

## The Fermi Factor for $T > 0$

At temperatures greater than $T = 0$, some electrons will occupy higher energy states because of thermal energy gained during collisions with the lattice. However, an electron cannot move to a higher or lower state unless it is unoccupied. Since the kinetic energy of the lattice ions is of the order of $kT$, electrons cannot gain much more energy than $kT$ in collisions with the lattice ions. Therefore, only those electrons with energies within about $kT$ of the Fermi energy can gain energy as the temperature is increased. At 300 K, $kT$ is only 0.026 eV, so the exclusion principle prevents all but a very few electrons near the top of the energy distribution from gaining energy through random collisions with the lattice ions. Figure 38-9 shows the Fermi factor for some temperature $T$. Since for $T > 0$ there is no distinct energy that separates filled levels from unfilled levels, the definition of the Fermi energy must be slightly modified. At temperature $T$, the Fermi energy is defined to be that energy for which the probability of being occupied is $\frac{1}{2}$. For all but extremely high temperatures, the difference between the Fermi energy at temperature $T$ and the Fermi energy at temperature $T = 0$ is very small.

The **Fermi temperature** $T_F$ is defined by

$$kT_F = E_F \qquad\qquad 38\text{-}25$$

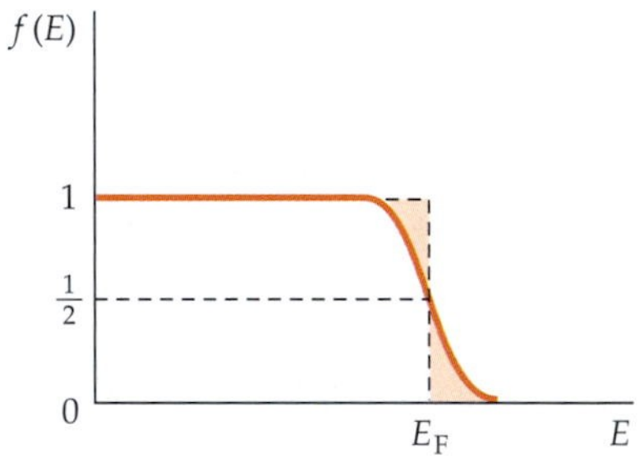

**FIGURE 38-9** The Fermi factor for some temperature $T$. Some electrons with energies near the Fermi energy are excited, as indicated by the shaded regions. The Fermi energy is that value of $E$ for which $f(E) = \frac{1}{2}$.

For temperatures much lower than the Fermi temperature, the average energy of the lattice ions will be much less than the Fermi energy, and the electron energy distribution will not differ greatly from that at $T = 0$.

---

*THE FERMI TEMPERATURE FOR COPPER* **EXAMPLE 38-4**

**Find the Fermi temperature for copper.**

Use $E_F = 7.04$ eV and $k = 8.62 \times 10^{-5}$ eV/K in Equation 38-25: $\quad T_F = \dfrac{E_F}{k} = \dfrac{7.04 \text{ eV}}{8.62 \times 10^{-5} \text{ eV/K}} = \boxed{81{,}700 \text{ K}}$

**REMARKS** We can see from this example that the Fermi temperature of copper is much greater than any temperature $T$ for which copper remains a solid.

Because an electric field in a conductor accelerates all of the conduction electrons together, the exclusion principle does not prevent the free electrons in filled states from participating in conduction. Figure 38-10 shows the Fermi factor in one dimension versus *velocity* for an ordinary temperature. The factor is

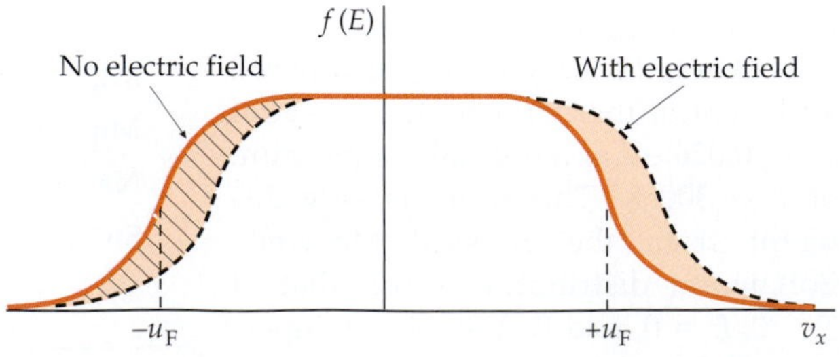

**FIGURE 38-10** Fermi factor versus velocity in one dimension with no electric field (solid) and with an electric field in the $-x$ direction (dashed). The difference is greatly exaggerated.

approximately 1 for speeds $v_x$ in the range $-u_F < v_x < u_F$, where the Fermi speed $u_F$ is related to the Fermi energy by $E_F = \frac{1}{2}m_e u_F^2$. Then

$$u_F = \sqrt{\frac{2E_F}{m_e}}$$
38-26

---

*THE FERMI SPEED FOR COPPER*                                         **EXAMPLE 38-5**

**Calculate the Fermi speed for copper.**

Use Equation 38-26 with $E_F = 7.04$ eV:
$$u_F = \sqrt{\frac{2(7.04\ \text{eV})}{9.11 \times 10^{-31}\ \text{kg}}\left(\frac{1.60 \times 10^{-19}\ \text{J}}{1\ \text{eV}}\right)} = \boxed{1.57 \times 10^6\ \text{m/s}}$$

---

The dashed curve in Figure 38-10 shows the Fermi factor after the electric field has been acting for some time $t$. Although all of the free electrons have their velocities shifted in the direction opposite to the electric field, the net effect is equivalent to shifting only the electrons near the Fermi energy.

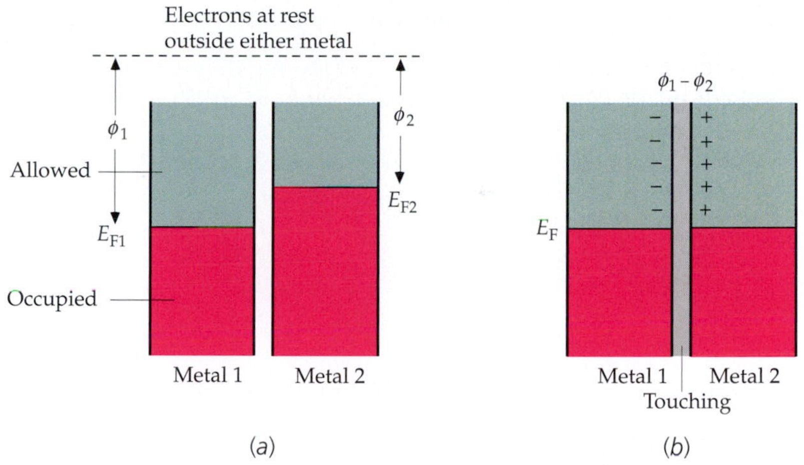

## Contact Potential

When two different metals are placed in contact, a potential difference $V_{contact}$ called the **contact potential** develops between them. The contact potential depends on the work functions of the two metals, $\phi_1$ and $\phi_2$ (we encountered work functions when the photoelectric effect was introduced in Chapter 34), and the Fermi energies of the two metals. When the metals are in contact, the total energy of the system is lowered if electrons near the boundary move from the metal with the higher Fermi energy into the metal with the lower Fermi energy until the Fermi energies of the two metals are the same, as shown in Figure 38-11. When equilibrium is established, the metal with the lower initial Fermi energy is negatively charged and the other is positively charged, so that between them there is a potential difference $V_{contact}$ given by

$$V_{contact} = \frac{\phi_1 - \phi_2}{e}$$
38-27

Table 38-2 lists the work functions for several metals.

**FIGURE 38-11** (*a*) Energy levels for two different metals with different Fermi energies and work functions. (*b*) When the metals are in contact, electrons flow from the metal that initially has the higher Fermi energy to the metal that initially has the lower Fermi energy until the Fermi energies are equal.

## TABLE 38-2

**Work Functions for Some Metals**

| | Metal | $\phi$, eV | | Metal | $\phi$, eV |
|---|---|---|---|---|---|
| Ag | Silver | 4.7 | K | Potassium | 2.1 |
| Au | Gold | 4.8 | Mn | Manganese | 3.8 |
| Ca | Calcium | 3.2 | Na | Sodium | 2.3 |
| Cu | Copper | 4.1 | Ni | Nickel | 5.2 |

---

The threshold wavelength for the photoelectric effect is 271 nm for tungsten and 262 nm for silver. What is the contact potential developed when silver and tungsten are placed in contact?

**PICTURE THE PROBLEM**  The contact potential is proportional to the difference in the work functions for the two metals. The work function $\phi$ can be found from the given threshold wavelengths using $\phi = hc/\lambda_t$ (Equation 34-4).

1. The contact potential is given by Equation 38-27:

$$V_{contact} = \frac{\phi_1 - \phi_2}{e}$$

2. The work function is related to the threshold wavelength (Equation 34-4):

$$\phi = \frac{hc}{\lambda_t}$$

3. Substitute $\lambda_t = 271$ nm for tungsten:

$$\phi_W = \frac{hc}{\lambda_t} = \frac{1240 \text{ eV·nm}}{271 \text{ nm}} = 4.58 \text{ eV}$$

4. Substitute $\lambda_t = 262$ nm for silver:

$$\phi_{Ag} = \frac{1240 \text{ eV·nm}}{262 \text{ nm}} = 4.73 \text{ eV}$$

5. The contact potential is thus:

$$V_{contact} = \frac{\phi_{Ag} - \phi_W}{e} = 4.73 \text{ V} - 4.58 \text{ V}$$

$$= \boxed{0.15 \text{ V}}$$

## Heat Capacity Due to Electrons in a Metal

The quantum-mechanical modification of the electron distribution in metals allows us to understand why the contribution of the electron gas to the heat capacity of a metal is much less that of the ions. According to the classical equipartition theorem, the energy of the lattice ions in $n$ moles of a solid is $3nRT$, and thus the molar specific heat is $c' = 3R$, where $R$ is the universal gas constant (see Section 18-7). In a metal, there is a free electron gas containing a number of electrons approximately equal to the number of lattice ions. If these electrons obey the classical equipartition theorem, they should have an energy of $\frac{3}{2}nRT$ and contribute an additional $\frac{3}{2}R$ to the molar specific heat. But measured heat capacities of metals are just slightly greater than those of insulators. We can understand this because at some temperature $T$, only those electrons with energies near the Fermi energy can be excited by random collisions with the lattice ions. The number of these electrons is of the order of $(kT/E_F)N$, where $N$ is the total number of electrons. The energy of these electrons is increased from that at $T = 0$ by an amount that is of the order of $kT$. So the total increase in thermal energy is of the order of $(kT/E_F)N \times kT$. We can thus express the energy of $N$ electrons at temperature $T$ as

$$E = NE_{av}(0) + \alpha N \frac{kT}{E_F} kT \qquad\qquad 38\text{-}28$$

where $\alpha$ is some constant that we expect to be of the order of 1 if our reasoning is correct. The calculation of $\alpha$ is quite difficult. The result is $\alpha = \pi^2/4$. Using this result and writing $E_F$ in terms of the Fermi temperature, $E_F = kT_F$, we obtain the following for the contribution of the electron gas to the heat capacity at constant volume:

$$C_V = \frac{dE}{dT} = 2\alpha Nk\frac{kT}{E_F} = \frac{\pi^2}{2} nR\frac{T}{T_F}$$

where we have written $Nk$ in terms of the gas constant $R$ ($Nk = nR$). The molar specific heat at constant volume is then

$$c'_V = \frac{\pi^2}{2} R\frac{T}{T_F} \qquad\qquad 38\text{-}29$$

We can see that because of the large value of $T_F$, the contribution of the electron gas is a small fraction of $R$ at ordinary temperatures. Because $T_F = 81,700$ K for copper, the molar specific heat of the electron gas at $T = 300$ K is

$$c'_V = \frac{\pi^2}{2}\left(\frac{300\ \text{K}}{81,700}\right)R \approx 0.02\ R$$

which is in good agreement with experiment.

## 38-4 Quantum Theory of Electrical Conduction

We can use Equation 38-14 for the resistivity if we use the Fermi speed $u_F$ in place of $v_{av}$:

$$\rho = \frac{m_e u_F}{n_e e^2 \lambda} \qquad\qquad 38\text{-}30$$

We now have two problems. First, since the Fermi speed $u_F$ is approximately independent of temperature, the resistivity given by Equation 38-30 is independent of temperature unless the mean free path depends on it. The second problem concerns magnitudes. As mentioned earlier, the classical expression for resistivity using $v_{av}$ calculated from the Maxwell–Boltzmann distribution gives values that are about 6 times too large at $T = 300$ K. Since the Fermi speed $u_F$ is about 16 times the Maxwell–Boltzmann value of $v_{av}$, the magnitude of $\rho$ predicted by Equation 38-30 will be approximately 100 times greater than the experimentally determined value. The resolution of both of these problems lies in the calculation of the mean free path $\lambda$.

### The Scattering of Electron Waves

In Equation 38-16 for the classical mean free path $\lambda = 1/(n_{ion}A)$, the quantity $A = \pi r^2$ is the area of the lattice ion as seen by an electron. In the quantum calculation, the mean free path is related to the scattering of electron waves by the crystal lattice. Detailed calculations show that, for a *perfectly* ordered crystal, $\lambda = \infty$; that is, there is no scattering of the electron waves. The scattering of electron waves arises because of *imperfections* in the crystal lattice, which have nothing to do with the actual cross-sectional area $A$ of the lattice ions. According to the quantum theory of electron scattering, $A$ depends merely on *deviations* of the lattice ions from a perfectly ordered array and not on the size of the ions. The most common causes of such deviations are thermal vibrations of the lattice ions or impurities.

We can use $\lambda = 1/(n_{ion}A)$ for the mean free path if we reinterpret the area $A$. Figure 38-12 compares the classical picture and the quantum picture of this area. In the quantum picture, the lattice ions are points that have no size but present an

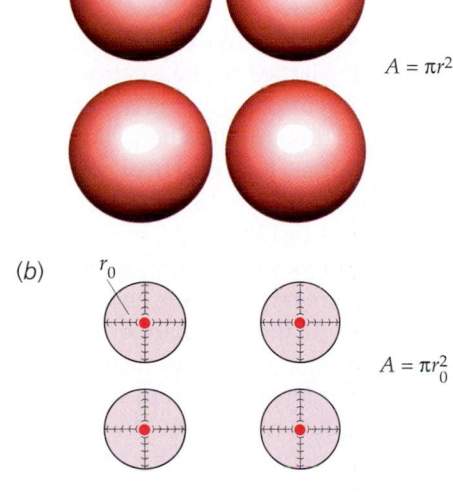

*(a)*

$A = \pi r^2$

*(b)*

$A = \pi r_0^2$

**FIGURE 38-12** (*a*) Classical picture of the lattice ions as spherical balls of radius $r$ that present an area $\pi r^2$ to the electrons. (*b*) Quantum-mechanical picture of the lattice ions as points that are vibrating in three dimensions. The area presented to the electrons is $\pi r_0^2$, where $r_0$ is the amplitude of oscillation of the ions.

area $A = \pi r_0^2$, where $r_0$ is the amplitude of thermal vibrations. In Chapter 14, we saw that the energy of vibration in simple harmonic motion is proportional to the square of the amplitude, which is $\pi r_0^2$. Thus, the effective area $A$ is proportional to the energy of vibration of the lattice ions. From the equipartition theorem,[†] we know that the average energy of vibration is proportional to $kT$. Thus, $A$ is proportional to $T$, and $\lambda$ is proportional to $1/T$. Then the resistivity given by Equation 38-14 is proportional to $T$, in agreement with experiment.

The effective area $A$ due to thermal vibrations can be calculated, and the results give values for the resistivity that are in agreement with experiment. At $T = 300$ K, for example, the effective area turns out to be about 100 times smaller than the actual area of a lattice ion. We see, therefore, that the free-electron model of metals gives a good account of electrical conduction if the classical mean speed $v_{av}$ is replaced by the Fermi speed $u_F$ and if the collisions between electrons and the lattice ions are interpreted in terms of the scattering of electron waves, for which only deviations from a perfectly ordered lattice are important.

The presence of impurities in a metal also causes deviations from perfect regularity in the crystal lattice. The effects of impurities on resistivity are approximately independent of temperature. The resistivity of a metal containing impurities can be written $\rho = \rho_t + \rho_i$, where $\rho_t$ is the resistivity due to the thermal motion of the lattice ions and $\rho_i$ is the resistivity due to impurities. Figure 38-13 shows typical resistance curves versus temperature curves for metals with impurities. As the absolute temperature approaches zero, $\rho_t$ approaches zero, and the resistivity approaches the constant $\rho_i$ due to impurities.

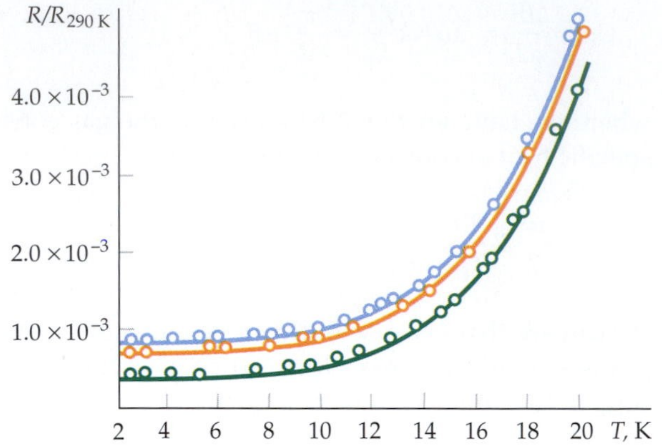

**FIGURE 38-13** Relative resistance versus temperature for three samples of sodium. The three curves have the same temperature dependence but different magnitudes because of differing amounts of impurities in the samples.

# 38-5 Band Theory of Solids

Resistivities vary enormously between insulators and conductors. For a typical insulator, such as quartz, $\rho \sim 10^{16}$ $\Omega \cdot$m, whereas for a typical conductor, $\rho \sim 10^{-8}$ $\Omega \cdot$m. The reason for this enormous variation is the variation in the number density of free electrons $n_e$. To understand this variation, we consider the effect of the lattice on the electron energy levels.

We begin by considering the energy levels of the individual atoms as they are brought together. The allowed energy levels in an isolated atom are often far apart. For example, in hydrogen, the lowest allowed energy $E_1 = -13.6$ eV is 10.2 eV below the next lowest allowed energy $E_2 = (-13.6 \text{ eV})/4 = -3.4$ eV.[‡] Let us consider two identical atoms and focus our attention on one particular energy level. When the atoms are far apart, the energy of a particular level is the same for each atom. As the atoms are brought closer together, the energy level for each atom changes because of the influence of the other atom. As a result, the level splits into two levels of slightly different energies for the two-atom system. If we bring three atoms close together, a particular energy level splits into three separate levels of slightly different energies. Figure 38-14 shows the energy splitting of two energy levels for six atoms as a function of the separation of the atoms.

If we have $N$ identical atoms, a particular energy level in the isolated atom splits into $N$ different, closely spaced energy levels when the atoms are close together. In a macroscopic solid, $N$ is very large—of the order of $10^{23}$—so each energy level splits into a very large

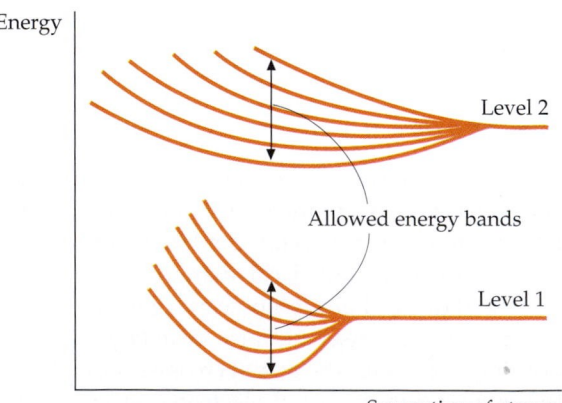

**FIGURE 38-14** Energy splitting of two energy levels for six atoms as a function of the separation of the atoms. When there are many atoms, each level splits into a near-continuum of levels called a band.

---

† The equipartition theorem does hold for the lattice ions, which obey the Maxwell–Boltzmann energy distribution.
‡ The energy levels in hydrogen are discussed in Chapter 36.

number of levels called a **band.** The levels are spaced almost continuously within the band. There is a separate band of levels for each particular energy level of the isolated atom. The bands may be widely separated in energy, they may be close together, or they may even overlap, depending on the kind of atom and the type of bonding in the solid.

The lowest energy bands, corresponding to the lowest energy levels of the atoms in the lattice, are filled with electrons that are bound to the individual atoms. The electrons that can take part in conduction occupy the higher energy bands. The highest energy band that contains electrons is called the **valence band.** The valence band may be completely filled with electrons or only partially filled, depending on the kind of atom and the type of bonding in the solid.

We can now understand why some solids are conductors and why others are insulators. If the valence band is only partially full, there are many available empty energy states in the band, and the electrons in the band can easily be raised to a higher energy state by an electric field. Accordingly, this material is a good conductor. If the valence band is full and there is a large energy gap between it and the next available band, a typical applied electric field will be too weak to excite an electron from the upper energy levels of the filled band across the large gap into the energy levels of the empty band, so the material is an insulator. The lowest band in which there are unoccupied states is called the **conduction band.** In a conductor, the valence band is only partially filled, so the valence band is also the conduction band. An energy gap between allowed bands is called a **forbidden energy band.**

The band structure for a conductor, such as copper, is shown in Figure 38-15a. The lower bands (not shown) are filled with the inner electrons of the atoms. The valence band is only about half full. When an electric field is established in the conductor, the electrons in the conduction band are accelerated, which means that their energy is increased. This is consistent with the Pauli exclusion principle because there are many empty energy states just above those occupied by electrons in this band. These electrons are thus the conduction electrons.

Figure 38-15b shows the band structure for magnesium, which is also a conductor. In this case, the highest occupied band is full, but there is an empty band above it that overlaps it. The two bands thus form a combined valence–conduction band that is only partially filled.

Figure 38-15c shows the band structure for a typical insulator. At $T = 0$ K, the valence band is completely full. The next energy band containing empty energy states, the conduction band, is separated from the valence band by a large energy gap. At $T = 0$, the conduction band is empty. At ordinary temperatures, a few electrons can be excited to states in this band, but most cannot be excited to states because the energy gap is large compared with the energy an electron might obtain by thermal excitation. Very few electrons can be thermally excited to the nearly empty conduction band, even at fairly high temperatures. When an electric field of ordinary magnitude is established in the solid, electrons cannot be

**FIGURE 38-15** Four possible band structures for a solid. (*a*) A typical conductor. The valence band is only partially full, so electrons can be easily excited to nearby energy states. (*b*) A conductor in which the allowed energy bands overlap. (*c*) A typical insulator. There is a forbidden band with a large energy gap between the filled valence band and the conduction band. (*d*) A semiconductor. The energy gap between the filled valence band and the conduction band is very small, so some electrons are excited to the conduction band at normal temperatures, leaving holes in the valence band.

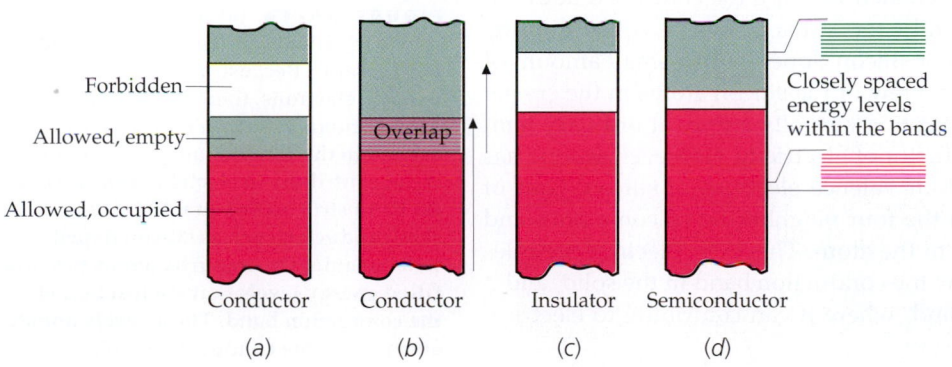

Forbidden

Allowed, empty

Allowed, occupied

Overlap

Closely spaced energy levels within the bands

Conductor        Conductor        Insulator    Semiconductor

(a)              (b)              (c)              (d)

accelerated because there are no empty energy states at nearby energies. We describe this by saying that there are no free electrons. The small conductivity that is observed is due to the very few electrons that are thermally excited into the nearly empty conduction band. When an electric field applied to an insulator is sufficiently strong to cause an electron to be excited across the energy gap to the empty band, dielectric breakdown occurs.

In some materials, the energy gap between the filled valence band and the empty conduction band is very small, as shown in Figure 38-15d. At $T = 0$, there are no electrons in the conduction band and the material is an insulator. However, at ordinary temperatures, there are an appreciable number of electrons in the conduction band due to thermal excitation. Such a material is called an **intrinsic semiconductor.** For typical intrinsic semiconductors, such as silicon and germanium, the energy gap is only about 1 eV. In the presence of an electric field, the electrons in the conduction band can be accelerated because there are empty states nearby. Also, for each electron in the conduction band there is a vacancy, or hole, in the nearly filled valence band. In the presence of an electric field, electrons in this band can also be excited to a vacant energy level. This contributes to the electric current and is most easily described as the motion of a hole in the direction of the field and opposite to the motion of the electrons. The hole thus acts like a positive charge. To visualize the conduction of holes, think of a two-lane, one-way road with one lane full of parked cars and the other lane empty. If a car moves out of the filled lane into the empty lane, it can move ahead freely. As the other cars move up to occupy the space left, the empty space propagates backward in the direction opposite the motion of the cars. Both the forward motion of the car in the nearly empty lane and the backward propagation of the empty space contribute to a net forward propagation of the cars.

An interesting characteristic of semiconductors is that the resistivity of the material decreases as the temperature increases, which is contrary to the case for normal conductors. The reason is that as the temperature increases, the number of free electrons increases because there are more electrons in the conduction band. The number of holes in the valence band also increases, of course. In semiconductors, the effect of the increase in the number of charge carriers, both electrons and holes, exceeds the effect of the increase in resistivity due to the increased scattering of the electrons by the lattice ions due to thermal vibrations. Semiconductors therefore have a negative temperature coefficient of resistivity.

# 38-6 Semiconductors

The semiconducting property of intrinsic semiconductors materials makes them useful as a basis for electronic circuit components whose resistivity can be controlled by application of an external voltage or current. Most such *solid-state devices,* however, such as the semiconductor diode and the transistor, make use of **impurity semiconductors,** which are created through the controlled addition of certain impurities to intrinsic semiconductors. This process is called **doping.** Figure 38-16a is a schematic illustration of silicon doped with a small amount of arsenic so that the arsenic atoms replace a few of the silicon atoms in the crystal lattice. The conduction band of pure silicon is virtually empty at ordinary temperatures, so pure silicon is a poor conductor of electricity. However, arsenic has five valence electrons rather than the four valence electrons of silicon. Four of these electrons take part in bonds with the four neighboring silicon atoms, and the fifth electron is very loosely bound to the atom. This extra electron occupies an energy level that is just slightly below the conduction band in the solid, and it is easily excited into the conduction band, where it can contribute to electrical conduction.

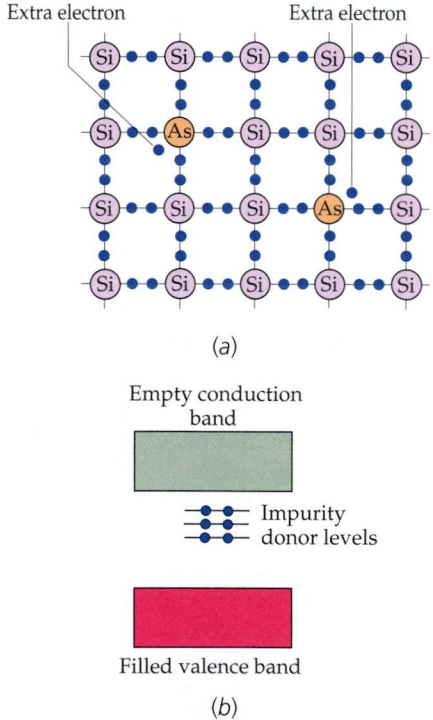

(a)

(b)

**FIGURE 38-16** (*a*) A two-dimensional schematic illustration of silicon doped with arsenic. Because arsenic has five valence electrons, there is an extra, weakly bound electron that is easily excited to the conduction band, where it can contribute to electrical conduction. (*b*) Band structure of an *n*-type semiconductor, such as silicon doped with arsenic. The impurity atoms provide filled energy levels that are just below the conduction band. These levels donate electrons to the conduction band.

The effect on the band structure of a silicon crystal achieved by doping it with arsenic is shown in Figure 38-16b. The levels shown just below the conduction band are due to the extra electrons of the arsenic atoms. These levels are called **donor levels** because they donate electrons to the conduction band without leaving holes in the valence band. Such a semiconductor is called an **n-type semiconductor** because the major charge carriers are negative electrons. The conductivity of a doped semiconductor can be controlled by controlling the amount of impurity added. The addition of just one part per million can increase the conductivity by several orders of magnitude.

Another type of impurity semiconductor can be made by replacing a silicon atom with a gallium atom, which has three valence electrons (Figure 38-17a). The gallium atom accepts electrons from the valence band to complete its four covalent bonds, thus creating a hole in the valence band. The effect on the band structure of silicon achieved by doping it with gallium is shown in Figure 38-17b. The empty levels shown just above the valence band are due to the holes from the ionized gallium atoms. These levels are called **acceptor levels** because they accept electrons from the filled valence band when these electrons are thermally excited to a higher energy state. This creates holes in the valence band that are free to propagate in the direction of an electric field. Such a semiconductor is called a **p-type semiconductor** because the charge carriers are positive holes. The fact that conduction is due to the motion of positive holes can be verified by the Hall effect. (The Hall effect is discussed in chapter 26.)

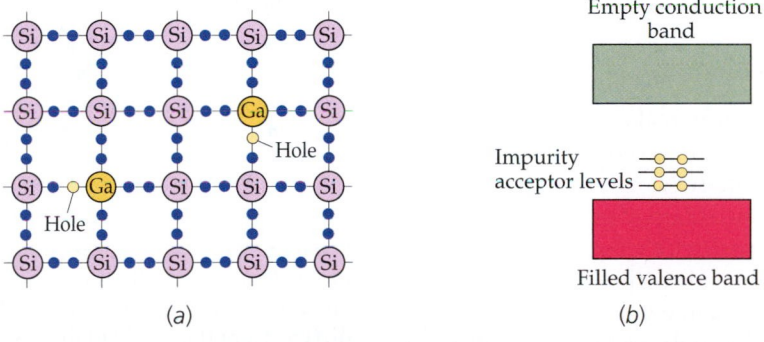

(a)                                        (b)

**FIGURE 38-17** (a) A two-dimensional schematic illustration of silicon doped with gallium. Because gallium has only three valence electrons, there is a hole in one of its bonds. As electrons move into the hole the hole moves about, contributing to the conduction of electrical current. (b) Band structure of a p-type semiconductor, such as silicon doped with gallium. The impurity atoms provide empty energy levels just above the filled valence band that accept electrons from the valence band.

Synthetic crystal silicon is produced beginning with a raw material containing silicon (for instance, common beach sand), separating out the silicon, and melting it. From a seed crystal, the molten silicon grows into a cylindrical crystal, such as the one shown here. The crystals (typically about 1.3 m long) are formed under highly controlled conditions to ensure that they are flawless and the crystals are then sliced into thousands of thin wafers onto which the layers of an integrated circuit are etched.

---

**NUMBER DENSITY OF FREE ELECTRONS IN ARSENIC-DOPED SILICON**

**EXAMPLE 38-7**   **Try It Yourself**

The number of free electrons in pure silicon is approximately $10^{10}$ electrons/cm³ at ordinary temperatures. If one silicon atom out of every million atoms is replaced by an arsenic atom, how many free electrons per cubic centimeter are there? (The density of silicon is 2.33 g/cm³ and its molar mass is 28.1 g/mol.)

**PICTURE THE PROBLEM** The number of silicon atoms per cubic centimeter, $n_{Si}$ can be found from $n_{Si} = N_A \rho / M$. Then, since each arsenic atom contributes one free electron, the number of electrons contributed by the arsenic atoms is $10^{-6} n_{Si}$.

**Cover the column to the right and try these on your own before looking at the answers.**

| Steps | Answers |
|---|---|
| 1. Calculate the number of silicon atoms per cubic centimeter. | $n_{Si} = \dfrac{\rho N_A}{M}$ $= \dfrac{(2.33 \text{ g/cm}^3)(6.02 \times 10^{23} \text{ atoms/mol})}{28.1 \text{ g/mol}}$ $= 4.99 \times 10^{22} \text{ atoms/cm}^3$ |
| 2. Multiply by $10^{-6}$ to obtain the number of arsenic atoms per cubic centimeter, which equals the added number of free electrons per cubic centimeter. | $n_e = 10^{-6} n_{Si} = \boxed{4.99 \times 10^{16} \text{ electrons/cm}^3}$ |

**REMARKS** Because silicon has so few free electrons per atom, the number of conduction electrons is increased by a factor of approximately 5 million by doping silicon with just one arsenic atom per million silicon atoms.

**EXERCISE** How many free electrons are there per silicon atom in pure silicon? (*Answer*  $2 \times 10^{-13}$)

## *38-7  Semiconductor Junctions and Devices

Semiconductor devices such as diodes and transistors make use of *n*-type semiconductors and *p*-type semiconductors joined together, as shown in Figure 38-18. In practice, the two types of semiconductors are often incorporated into a single silicon crystal doped with donor impurities on one side and acceptor impurities on the other side. The region in which the semiconductor changes from a *p*-type semiconductor to an *n*-type semiconductor is called a **junction.**

When an *n*-type semiconductor and a *p*-type semiconductor are placed in contact, the initially unequal concentrations of electrons and holes result in the diffusion of electrons across the junction from the *n* side to the *p* side and holes from the *p* side to the *n* side until equilibrium is established. The result of this diffusion is a net transport of positive charge from the *p* side to the *n* side. Unlike the case when two different metals are in contact, the electrons cannot travel very far from the junction region because the semiconductor is not a particularly good conductor. The diffusion of electrons and holes therefore creates a double layer of charge at the junction similar to that on a parallel-plate capacitor. There is, thus, a potential difference $V$ across the junction, which tends to inhibit further diffusion. In equilibrium, the *n* side with its net positive charge will be at a higher potential than the *p* side with its net negative charge. In the junction region, between the charge layers, there will be very few charge carriers of either type, so the junction region has a high resistance. Figure 38-19 shows the energy-level diagram for a *pn* junction. The junction region is also called the **depletion region** because it has been depleted of charge carriers.

### *Diodes

In Figure 38-20, an external potential difference has been applied across a *pn* junction by connecting a battery and a resistor to the semiconductor. When the positive terminal of the battery is connected to the *p* side of the junction, as shown in Figure 38-20*a*, the junction is said to be **forward biased.** Forward

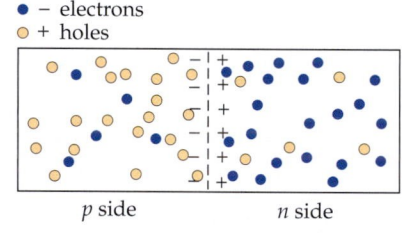

**FIGURE 38-18** A *pn* junction. Because of the difference in their concentrations on either side of the *pn* junction, holes diffuse from the *p* side to the *n* side, and electrons diffuse from the *n* side to the *p* side. As a result, there is a double layer of charge at the junction, with the *p* side being negative and the *n* side being positive.

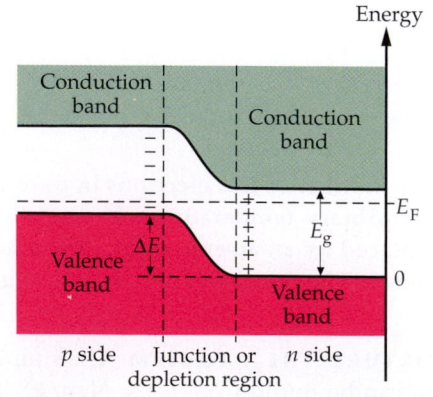

**FIGURE 38-19** Electron energy levels for a *pn* junction.

(a)

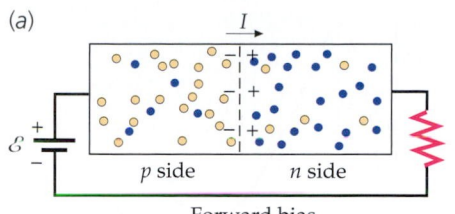

Forward bias

(b)

No current

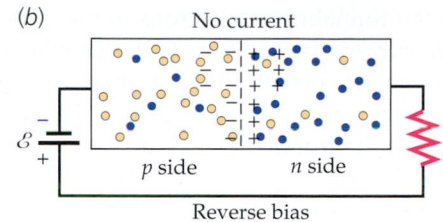

Reverse bias

**FIGURE 38-20** A *pn*-junction diode. (a) Forward-biased *pn* junction. The applied potential difference enhances the diffusion of holes from the *p* side to the *n* side and of electrons from the *n* side to the *p* side, resulting in a current *I*. (b) Reverse-biased *pn* junction. The applied potential difference inhibits the further diffusion of holes and electrons across the junction, so there is no current.

biasing lowers the potential across the junction. The diffusion of electrons and holes is thereby increased as they attempt to reestablish equilibrium, resulting in a current in the circuit.

If the positive terminal of the battery is connected to the *n* side of the junction, as shown in Figure 38-20*b*, the junction is said to be **reverse biased.** Reverse biasing tends to increase the potential difference across the junction, thereby further inhibiting diffusion. Figure 38-21 shows a plot of current versus voltage for a typical semiconductor junction. Essentially, the junction conducts only in one direction. A single-junction semiconductor device is called a **diode.**[†] Diodes have many uses. One is to convert alternating current into direct current, a process called rectification.

Note that the current in Figure 38-21 suddenly increases in magnitude at extreme values of reverse bias. In such large electric fields, electrons are stripped from their atomic bonds and accelerated across the junction. These electrons, in turn, cause others to break loose. This effect is called **avalanche breakdown.** Although such a breakdown can be disastrous in a circuit where it is not intended, the fact that it occurs at a sharply defined voltage makes it of use in a special voltage reference standard known as a **Zener diode.** Zener diodes are also used to protect devices from excessively high voltages.

An interesting effect, one that we discuss only qualitatively, occurs if both the *n* side and the *p* side of a *pn*-junction diode are so heavily doped that the donors on the *n* side provide so many electrons that the lower part of the conduction band is practically filled, and the acceptors on the *p* side accept so many electrons that the upper part of the valence band is nearly empty. Figure 38-22*a* shows the energy-level diagram for this situation. Because the depletion region is now so narrow, electrons can easily penetrate the potential barrier across the junction and tunnel to the other side. The flow of electrons through the barrier is called a **tunneling current,** and such a heavily doped diode is called a **tunnel diode.**

At equilibrium with no bias, there is an equal tunneling current in each direction. When a small bias voltage is applied across the junction, the energy-level diagram is as shown in Figure 38-22*b*, and the tunneling of electrons from the

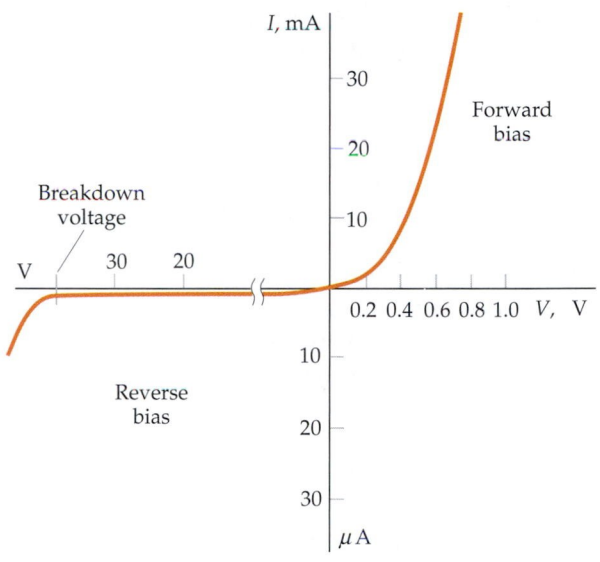

**FIGURE 38-21** Plot of current versus applied voltage across a *pn* junction. Note the different scales on both axes for the forward and reverse bias conditions.

**FIGURE 38-22** Electron energy levels for a heavily doped *pn*-junction tunnel diode. (a) With no bias voltage, some electrons tunnel in each direction. (b) With a small bias voltage, the tunneling current is enhanced in one direction, making a sizable contribution to the net current. (c) With further increases in the bias voltage, the tunneling current decreases dramatically.

(a)

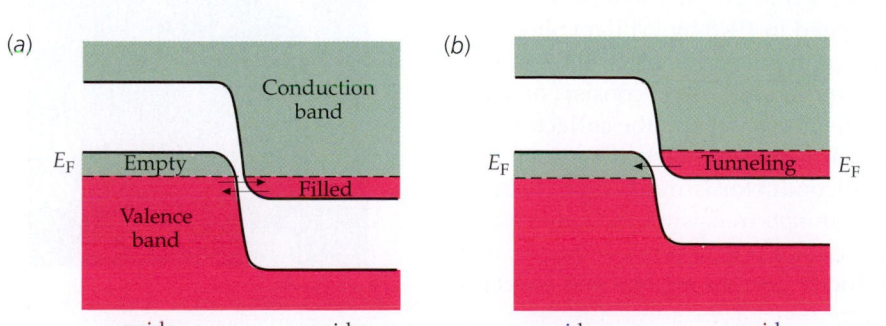

(b)

(c)

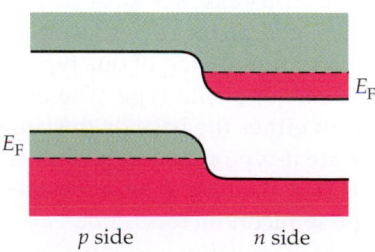

---

[†] The name *diode* originates from a vacuum tube device consisting of just two electrodes that also conducts electric current in one direction only.

*n* side to the *p* side is increased, whereas the tunneling of electrons in the opposite direction is decreased. This tunneling current, in addition to the usual current due to diffusion, results in a considerable net current. When the bias voltage is increased slightly, the energy-level diagram is as shown in Figure 38-22*c*, and the tunneling current is decreased. Although the diffusion current is increased, the net current is decreased. At large bias voltages, the tunneling current is completely negligible, and the total current increases with increasing bias voltage due to diffusion, as in an ordinary *pn*-junction diode. Figure 38-23 shows the current curve versus the voltage curve for a tunnel diode. Such diodes are used in electric circuits because of their very fast response time. When operated near the peak in the current curve versus the voltage curve, a small change in bias voltage results in a large change in the current.

Another use for the *pn*-junction semiconductor is the **solar cell,** which is illustrated schematically in Figure 38-24. When a photon of energy greater than the gap energy (1.1 eV in silicon) strikes the *p*-type region, it can excite an electron from the valence band into the conduction band, leaving a hole in the valence band. This region is already rich in holes. Some of the electrons created by the photons will recombine with holes, but some will migrate to the junction. From there, they are accelerated into the *n*-type region by the electric field between the double layer of charge. This creates an excess negative charge in the *n*-type region and an excess positive charge in the *p*-type region. The result is a potential difference between the two regions, which in practice is approximately 0.6 V. If a load resistance is connected across the two regions, a charge flows through the resistance. Some of the incident light energy is thus converted into electrical energy. The current in the resistor is proportional to the number of incident photons, which is in turn proportional to the intensity of the incident light.

There are many other applications of semiconductors with *pn* junctions. Particle detectors, called **surface-barrier detectors,** consist of a *pn*-junction semiconductor with a large reverse bias so that there is ordinarily no current. When a high-energy particle, such as an electron, passes through the semiconductor, it creates many electron–hole pairs as it loses energy. The resulting current pulse signals the passage of the particle. **Light-emitting diodes** (LEDs) are *pn*-junction semiconductors with a large forward bias that produces a large excess concentration of electrons on the *p* side and holes on the *n* side of the junction. Under these conditions, the diode emits light as the electrons and holes recombine. This is essentially the reverse of the process that occurs in a solar cell, in which electron–hole pairs are created by the absorption of light. LEDs are commonly used as warning indicators and as sources of infrared light beams.

## *Transistors

The transistor, a semiconducting device that is used to produce a desired output signal in response to an input signal, was invented in 1948 by William Shockley, John Bardeen, and Walter Brattain and has revolutionized the electronics industry and our everyday world. A *simple bipolar junction transistor*[†] consists of three distinct semiconductor regions called the **emitter,** the **base,** and the **collector.** The base is a very thin region of one type of semiconductor sandwiched between two regions of the opposite type. The emitter semiconductor is much more heavily doped than either the base or the collector. In an *npn* transistor, the emitter and collector are *n*-type semiconductors and the base is a *p*-type semiconductor; in a *pnp* transistor, the base is an *n*-type semiconductor and the emitter and collector are *p*-type semiconductors.

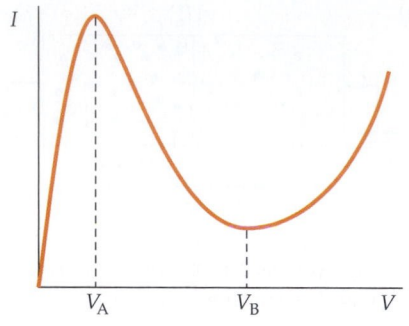

**FIGURE 38-23** Current versus applied (bias) voltage *V* for a tunnel diode. For $V < V_A$, an increase in the bias voltage *V* enhances tunneling. For $V_A < V < V_B$, an increase in the bias voltage inhibits tunneling. For $V > V_B$, the tunneling is negligible, and the diode behaves like an ordinary *pn*-junction diode.

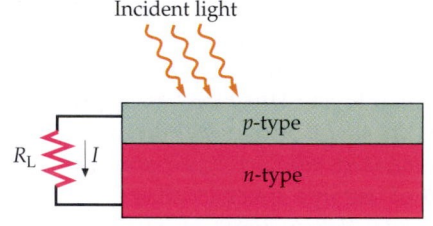

**FIGURE 38-24** A *pn*-junction semiconductor as a solar cell. When light strikes the *p*-type region, electron–hole pairs are created, resulting in a current through the load resistance $R_L$.

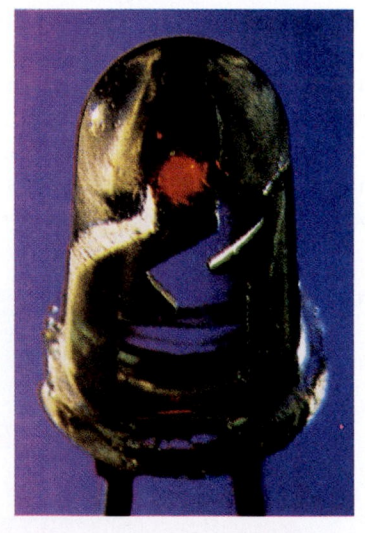

A light-emitting diode (LED).

---

[†] Besides the bipolar junction transistor, there are other categories of transistors, notably, the field-effect transistor.

Figure 38-25 and Figure 38-26 show, respectively, a *pnp* transistor and an *npn* transistor with the symbols used to represent each transistor in circuit diagrams. We see that either transistor consists of two *pn* junctions. We will discuss the operation of a *pnp* transistor. The operation of an *npn* transistor is similar.

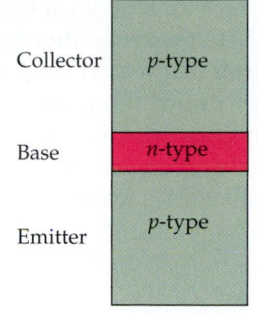

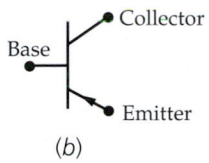

(b)

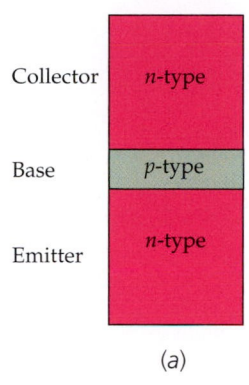

(a)

**FIGURE 38-25** A *pnp* transistor.
(*a*) The heavily doped emitter emits holes that pass through the thin base to the collector. (*b*) Symbol for a *pnp* transistor in a circuit. The arrow points in the direction of the conventional current, which is the same as that of the emitted holes.

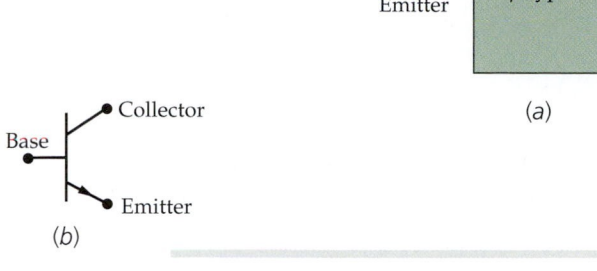

**FIGURE 38-26** An *npn* transistor.
(*a*) The heavily doped emitter emits electrons that pass through the thin base to the collector. (*b*) Symbol for an *npn* transistor. The arrow points in the direction of the conventional current, which is opposite the direction of the emitted electrons.

In normal operation, the emitter–base junction is forward biased, and the base–collector junction is reverse biased, as shown in Figure 38-27. The heavily doped *p*-type emitter emits holes that flow toward the emitter–base junction. This flow constitutes the emitter current $I_e$. Because the base is very thin, most of these holes flow across the base into the collector. This flow in the collector constitutes a current $I_c$. However, some of the holes recombine in the base producing a positive charge that inhibits the further flow of current. To prevent this, some of the holes that do not reach the collector are drawn off the base as a base current $I_b$ in a wire connected to the base. In Figure 38-27, therefore, $I_c$ is almost but not quite equal to $I_e$, and $I_b$ is much smaller than either $I_c$ or $I_e$. It is customary to express $I_c$ as

$$I_c = \beta I_b \qquad\qquad 38\text{-}31$$

where $\beta$ is called the **current gain** of the transistor. Transistors can be designed to have values of $\beta$ as low as ten or as high as several hundred.

Figure 38-28 shows a simple *pnp* transistor used as an amplifier. A small, time-varying input voltage $v_s$ is connected in series with a bias voltage $V_{eb}$. The base current is then the sum of a steady current $I_b$ produced by the bias voltage $V_{eb}$ and a varying current $i_b$ due to the signal voltage $v_s$. Because $v_s$ may at any

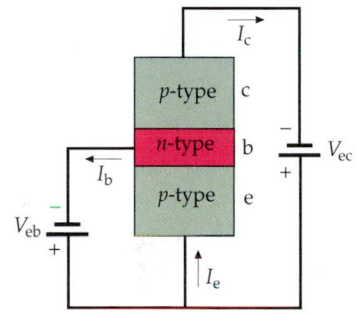

**FIGURE 38-27** A *pnp* transistor biased for normal operation. Holes from the emitter can easily diffuse across the base, which is only tens of nanometers thick. Most of the holes flow to the collector, producing the current $I_c$.

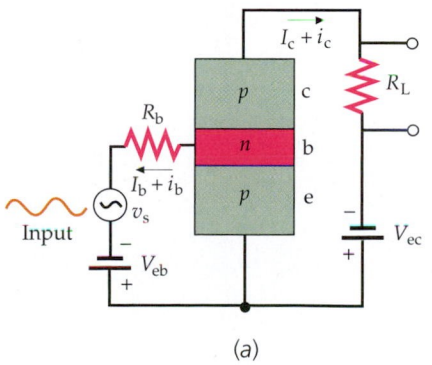

(a)

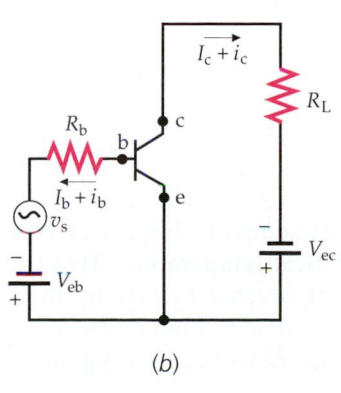

(b)

**FIGURE 38-28** (*a*) A *pnp* transistor used as an amplifier. A small change $i_b$ in the base current results in a large change $i_c$ in the collector current. Thus, a small signal in the base circuit results in a large signal across the load resistor $R_L$ in the collector circuit. (*b*) The same circuit as in Figure 38-28*a* with the conventional symbol for the transistor.

instant be either positive or negative, the bias voltage $V_{eb}$ must be large enough to ensure that there is always a forward bias on the emitter–base junction. The collector current will consist of two parts: a direct current $I_c = \beta I_b$ and an alternating current $i_c = \beta i_b$. We thus have a current amplifier in which the time-varying output current $i_c$ is $\beta$ times the input current $i_b$. In such an amplifier, the steady currents $I_c$ and $I_b$, although essential to the operation of the transistor, are usually not of interest. The input signal voltage $v_s$ is related to the base current by Ohm's law:

$$i_b = \frac{v_s}{R_b + r_b} \qquad\qquad 38\text{-}32$$

where $r_b$ is the internal resistance of the transistor between the base and emitter. Similarly, the collector current $i_c$ produces a voltage $v_L$ across the output or load resistance $R_L$ given by

$$v_L = i_c R_L \qquad\qquad 38\text{-}33$$

Using Equation 38-31 and Equation 38-32, we have

$$i_c = \beta i_b = \beta \frac{v_s}{R_b + r_b} \qquad\qquad 38\text{-}34$$

The output voltage is thus related to the input voltage by

$$v_L = \beta \frac{v_s}{R_b + r_b} R_L = \beta \frac{R_L}{R_b + r_b} v_s \qquad\qquad 38\text{-}35$$

The ratio of the output voltage to the input voltage is the **voltage gain** of the amplifier:

$$\text{Voltage gain} = \frac{v_L}{v_s} = \beta \frac{R_L}{R_b + r_b} \qquad\qquad 38\text{-}36$$

A typical amplifier (e.g., in a tape player) has several transistors, similar to the one shown in Figure 38-28, connected in series so that the output of one transistor serves as the input for the next. Thus, the very small voltage fluctuations produced by the motion of the magnetic tape past the pickup heads controls the large amounts of power required to drive the loudspeakers. The power delivered to the speakers is supplied by the dc sources connected to each transistor.

The technology of semiconductors extends well beyond individual transistors and diodes. Many of the electronic devices we now take for granted, such as laptop computers and the processors that govern the operation of vehicles and appliances, rely on large-scale integration of many transistors and other circuit components on a single chip. Large-scale integration combined with advanced concepts in semiconductor theory has created remarkable new instruments for scientific research.

## 38-8 Superconductivity

There are some materials for which the resistivity suddenly drops to zero below a certain temperature $T_c$, which is called the **critical temperature.** This amazing phenomenon, called **superconductivity,** was discovered in 1911 by the Dutch physicist H. Kamerlingh Onnes, who developed a technique for liquefying helium (boiling point 4.2 K) and put his technique to work exploring the properties of materials at temperatures in this range. Figure 38-29 shows Onnes's plot of the resistance of mercury versus temperature. The critical temperature for mercury is approximately the same as the boiling point of helium, which is 4.2 K. Critical temperatures for other superconducting elements range from less than 0.1 K for hafnium and iridium to 9.2 K for niobium. The temperature range for superconductors goes much higher for a number of metallic compounds. For example, the superconducting alloy $Nb_3Ge$, discovered in 1973, has a critical temperature of 25 K, which was the highest known until 1986, when the discoveries of J. Georg Bednorz and K. Alexander Müller launched the era of high-temperature superconductors, now defined as materials that exhibit superconductivity at temperatures above 77 K (the temperature at which nitrogen boils). To date (April 2003), the highest temperature at which superconductivity has been demonstrated, using thallium doped $HgBa_2Ca_2Cu_3O_8+delta$, is 138 K at atmospheric pressure. At extremely high pressures, some materials exhibit superconductivity at temperatures as high as 164 K.

The resistivity of a superconductor is zero. There can be a current in a superconductor even when there is no emf in the superconducting circuit. Indeed, in superconducting rings in which there was no electric field, steady currents have been observed to persist for years without apparent loss. Despite the cost and inconvenience of refrigeration with expensive liquid helium, many superconducting magnets have been built using superconducting materials, because such magnets require no power expenditure to maintain the large current needed to produce a large magnetic field.

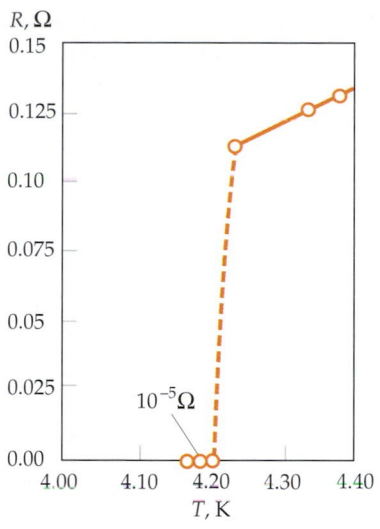

**FIGURE 38-29** Plot by H. Kamerlingh Onnes of the resistance of mercury versus temperature, showing the sudden decrease at the critical temperature of $T = 4.2$ K.

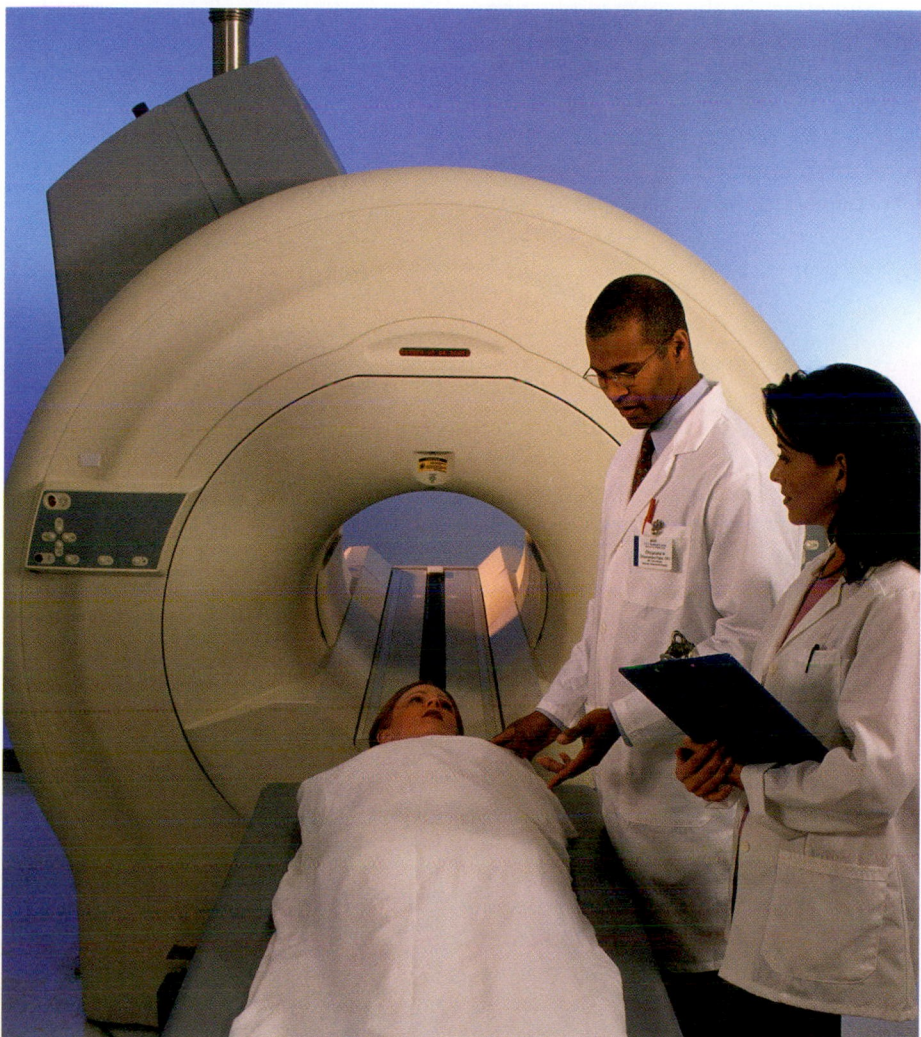

The wires for the magnetic field of a magnetic resonance imaging (MRI) machine carry large currents. To keep the wires from overheating, they are maintained at superconducting temperatures. To accomplish this, they are immersed in liquid helium.

The discovery of high-temperature superconductors has revolutionized the study of superconductivity because relatively inexpensive liquid nitrogen, which boils at 77 K, can be used for a coolant. However, many problems, such as brittleness and the toxicity of the materials, make these new superconductors difficult to use. The search continues for new materials that will superconduct at even higher temperatures.

## The BCS Theory

It had been recognized for some time that superconductivity is due to a collective action of the conducting electrons. In 1957, John Bardeen, Leon Cooper, and Robert Schrieffer published a successful theory of superconductivity now known by the initials of the inventors as the **BCS theory.** According to this theory, the electrons in a superconductor are coupled in pairs at low temperatures. The coupling comes about because of the interaction between electrons and the crystal lattice. One electron interacts with the lattice and perturbs it. The perturbed lattice interacts with another electron in such a way that there is an attraction between the two electrons that at low temperatures can exceed the Coulomb repulsion between them. The electrons form a bound state called a **Cooper pair.** The electrons in a Cooper pair have equal and opposite spins, so they form a system with zero spin. Each Cooper pair acts as a *single particle* with zero spin, in other words, as a boson. Bosons do not obey the exclusion principle. Any number of Cooper pairs may be in the same quantum state with the same energy. In the ground state of a superconductor (at $T = 0$), all the electrons are in Cooper pairs and all the Cooper pairs are in the same energy state. In the superconducting state, the Cooper pairs are correlated so that they all act together. An electric current can be produced in a superconductor because all of the electrons in this collective state move together. But energy cannot be dissipated by individual collisions of electron and lattice ions unless the temperature is high enough to break the binding of the Cooper pairs. The required energy is called the **superconducting energy gap** $E_g$. In the BCS theory, this energy at absolute zero is related to the critical temperature by

$$E_g = 3.5 \, kT_c \qquad\qquad 38\text{-}37$$

The energy gap can be determined by measuring the current across a junction between a normal metal and a superconductor as a function of voltage. Consider two metals separated by a layer of insulating material, such as aluminum oxide, that is only a few nanometers thick. The insulating material between the metals forms a barrier that prevents most electrons from traversing the junction. However, waves can tunnel through a barrier if the barrier is not too thick, even if the energy of the wave is less than that of the barrier.

When the materials on either side of the gap are normal nonsuperconducting metals, the current resulting from the tunneling of electrons through the insulating layer obeys Ohm's law for low applied voltages (Figure 38-30a). When one of the metals is a normal metal and the other is a superconductor, there is no current (at absolute zero) unless the applied voltage $V$ is greater than a critical voltage $V_c = E_g/(2e)$, where $E_g$ is the superconductor energy gap. Figure 38-30b shows the plot of current versus voltage for this situation. The current escalates rapidly when the energy $2eV$ absorbed by a Cooper pair traversing the barrier approaches $E_g = 2eV_c$, the minimum energy needed to break up the pair. (The small current visible in Figure 38-30b before the critical voltage is reached is present because at any temperature above absolute zero some of the electrons in the superconductor are thermally excited above the energy gap and are therefore not paired.) At voltages slightly above $V_c$, the current versus voltage curve becomes that for a normal metal. The superconducting energy gap can thus be measured by measuring the average voltage for the transition region.

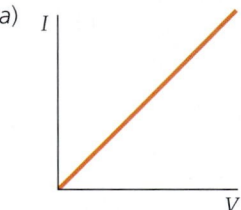

(a)

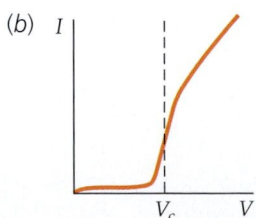

(b)

**FIGURE 38-30** Tunneling current versus voltage for a junction of two metals separated by a thin oxide layer. (*a*) When both metals are normal metals, the current is proportional to the voltage, as predicted by Ohm's law. (*b*) When one metal is a normal metal and another metal is a superconductor, the current is approximately zero until the applied voltage $V$ approaches the critical voltage $V_c = E_g/(2e)$.

**EXAMPLE 38-8**

Calculate the superconducting energy gap for mercury ($T_c = 4.2$ K) predicted by the BCS theory.

1. The BCS prediction for the energy gap is:

$$E_g = 3.5kT_c$$

2. Substitute $T_c = 4.2$ K:

$$E_g = 3.5kT_c$$

$$= 3.5(1.38 \times 10^{-23} \text{ J/K})(4.2 \text{ K})\left(\frac{1 \text{ ev}}{1.6 \times 10^{-19} \text{ J}}\right)$$

$$= \boxed{1.27 \times 10^{-3} \text{ eV}}$$

Note that the energy gap for a typical superconductor is much smaller than the energy gap for a typical semiconductor, which is of the order of 1 eV. As the temperature is increased from $T = 0$, some of the Cooper pairs are broken. Then there are fewer pairs available for each pair to interact with, and the energy gap is reduced until at $T = T_c$ the energy gap is zero (Figure 38-31).

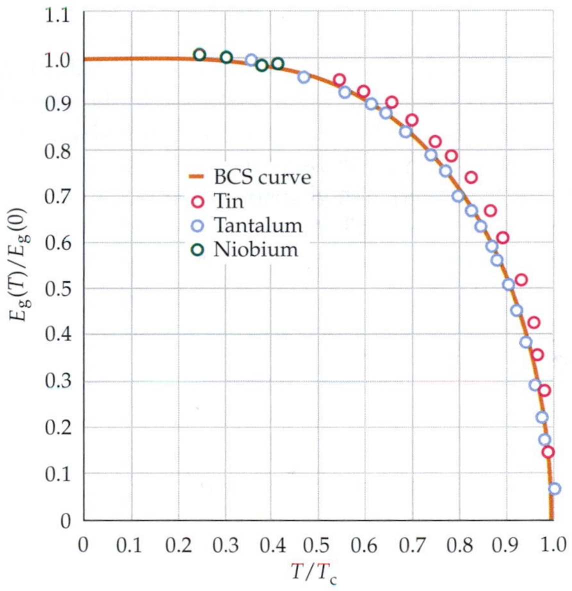

## The Josephson Effect

When two superconductors are separated by a thin nonsuperconducting barrier (e.g., a layer of aluminum oxide a few nanometers thick), the junction is called a **Josephson junction,** based on the prediction in 1962 by Brian Josephson that Cooper pairs could tunnel across such a junction from one superconductor to the other with no resistance. The tunneling of Cooper pairs constitutes a current, which does not require a voltage to be applied across the junction. The current depends on the difference in phase of the wave functions that describe the Cooper pairs. Let $\phi_1$ be the phase constant for the wave function of a Cooper pair in one superconductor. All the Cooper pairs in a superconductor act coherently and have the same phase constant. If $\phi_2$ is the phase constant for the Cooper pairs in the second superconductor, the current across the junction is given by

$$I = I_{max} \sin(\phi_2 - \phi_1) \qquad 38\text{-}38$$

where $I_{max}$ is the maximum current, which depends on the thickness of the barrier. This result has been observed experimentally and is known as the **dc Josephson effect.**

Josephson also predicted that if a dc voltage $V$ were applied across a Josephson junction, there would be a current that alternates with frequency $f$ given by

$$f = \frac{2e}{h} V \qquad 38\text{-}39$$

This result, known as the **ac Josephson effect,** has been observed experimentally, and careful measurement of the frequency allows a precise determination of the ratio $e/h$. Because frequency can be measured very accurately, the ac Josephson effect is also used to establish precise voltage standards. The inverse effect, in which the application of an alternating voltage across a Josephson junction results in a dc current, has also been observed.

**FIGURE 38-31** Ratio of the energy gap at temperature $T$ to that at temperature $T = 0$ as a function of the relative temperature $T/T_c$. The solid curve is that predicted by the BCS theory.

**EXAMPLE 38-9**

Using $e = 1.602 \times 10^{-19}$ C and $h = 6.626 \times 10^{-34}$ J·s, calculate the frequency of the Josephson current if the applied voltage is 1 $\mu$V.

Substitute the given values into Equation 38-39 to calculate $f$:   $f = \dfrac{2e}{h} V = \dfrac{2(1.602 \times 10^{-19}\,\text{C})}{6.626 \times 10^{-34}\,\text{J·s}} (10^{-6}\,\text{V})$

$$= 4.835 \times 10^8\,\text{Hz} = \boxed{483.5\,\text{MHz}}$$

## 38-9 The Fermi–Dirac Distribution†

The classical Maxwell–Boltzmann distribution (Equation 17-39) gives the number of molecules with energy $E$ in the range between $E$ and $E + dE$. It is the product of $g(E)\,dE$ where $g(E)$ is the **density of states** (number of energy states in the range $dE$) and the Boltzmann factor $e^{-E/(kT)}$, which is the probability of a state being occupied. The distribution function for free electrons in a metal is called the **Fermi–Dirac distribution.** The Fermi–Dirac distribution can be written in the same form as the Maxwell–Boltzmann distribution, with the density of states calculated from quantum theory and the Boltzmann factor replaced by the Fermi factor. Let $n(E)\,dE$ be the number of electrons with energies between $E$ and $E + dE$. This number is written

$$n(E)\,dE = g(E)\,dE\,f(E) \tag{38-40}$$

ENERGY DISTRIBUTION FUNCTION

where $g(E)\,dE$ is the number of states between $E$ and $E + dE$ and $f(E)$ is the probability of a state being occupied, which is the Fermi factor. The density of states in three dimensions is somewhat difficult to calculate, so we just give the result. For electrons in a metal of volume $V$, the density of states is

$$g(E) = \frac{8\sqrt{2}\pi m_e^{3/2} V}{h^3} E^{1/2} \tag{38-41}$$

DENSITY OF STATES

As in the classical Maxwell–Boltzmann distribution, the density of states is proportional to $E^{1/2}$.

At $T = 0$, the Fermi factor is given by Equation 38-24:

$$f(E) = 1, \qquad E < E_F$$
$$f(E) = 0, \qquad E > E_F$$

The integral of $n(E)\,dE$ over all energies gives the total number of electrons $N$. We can derive Equation 38-22a for the Fermi energy at $T = 0$ by integrating $n(E)\,dE$ from $E = 0$ to $E = \infty$. We obtain

---

† This material is somewhat complicated. You may wish to read it in two passes, the first to gain an overview of the topic, and the second to understand it in depth.

$$N = \int_0^\infty n(E)\, dE = \int_0^{E_F} n(E)\, dE + \int_{E_F}^\infty n(E)\, dE = \frac{8\sqrt{2}\pi m_e^{3/2} V}{h^3} \int_0^{E_F} E^{1/2}\, dE + 0$$

$$= \frac{16\sqrt{2}\pi m_e^{3/2} V}{3h^3} E_F^{3/2}$$

Note that at $T = 0$, $n(E)$ is zero for $E > E_F$. Solving for $E_F$ gives the Fermi energy at $T = 0$:

$$E_F = \frac{h^2}{8m_e} \left( \frac{3N}{\pi V} \right)^{2/3} \tag{38-42}$$

which is Equation 38-22$a$. In terms of the Fermi energy, the density of states (Equation 38-41) is

$$g(E) = \frac{8\sqrt{2}\pi m_e^{3/2} V}{h^3} E^{1/2} = \frac{3}{2} N E_F^{-3/2} E^{1/2} \tag{38-43}$$

DENSITY OF STATES IN TERMS OF $E_F$

which is obtained by solving Equation 38-42 for $m_e$, and then substituting for $m_e$ in Equation 38-41. The average energy at $T = 0$ is calculated from

$$E_{av} = \frac{\displaystyle\int_0^{E_F} E g(E)\, dE}{\displaystyle\int_0^{E_F} g(E)\, dE} = \frac{1}{N} \int_0^{E_F} E g(E)\, dE$$

where $N = \int_0^{E_F} g(E)\, dE$ is the total number of electrons. Performing the integration, we obtain Equation 38-23:

$$E_{av} = \tfrac{3}{5} E_F \tag{38-44}$$

AVERAGE ENERGY AT $T = 0$

At $T > 0$, the Fermi factor is more complicated. It can be shown to be

$$f(E) = \frac{1}{e^{(E-E_F)/(kT)} + 1} \tag{38-45}$$

FERMI FACTOR

We can see from this equation that for $E$ greater than $E_F$, $e^{(E-E_F)/(kT)}$ becomes very large as $T$ approaches zero, so at $T = 0$, the Fermi factor is zero for $E > E_F$. On the other hand, for $E$ less than $E_F$, $e^{(E-E_F)/(kT)}$ approaches 0 as $T$ approaches zero, so at $T = 0$, $f(E) = 1$ for $E < E_F$. Thus, the Fermi factor given by Equation 38-45 holds for all temperatures. Note also that for any nonzero value of $T$, $f(E) = \tfrac{1}{2}$ at $E = E_F$.
The complete Fermi–Dirac distribution function is thus

$$n(E)\, dE = g(E)f(E)\, dE = \frac{8\sqrt{2}\pi m_e^{3/2} V}{h^3} E^{1/2} \frac{1}{e^{(E-E_F)/(kT)} + 1}\, dE \tag{38-46}$$

FERMI–DIRAC DISTRIBUTION

We can see that for those few electrons with energies much greater than the Fermi energy, the Fermi factor approaches $1/e^{(E-E_F)/(kT)} = e^{(E_F-E)/(kT)} = e^{E_F/(kT)}e^{-E/(kT)}$, which is proportional to $e^{-E/(kT)}$. Thus, the high-energy tail of the Fermi–Dirac energy distribution decreases as $e^{-E/(kT)}$, just like the classical Maxwell–Boltzmann energy distribution. The reason is that in this high-energy region, there are many unoccupied energy states and few electrons, so the Pauli exclusion principle is not important, and the distribution approaches the classical distribution. This result has practical importance because it applies to the conduction electrons in semiconductors.

---

*FERMI FACTOR FOR COPPER AT 300 K*      **EXAMPLE 38-10**

**At what energy is the Fermi factor equal to 0.1 for copper at $T = 300$ K?**

**PICTURE THE PROBLEM** We set $f(E) = 0.1$ in Equation 38-45, using $T = 300$ K and $E_F = 7.04$ eV from Table 38-1 and solve for $E$.

1. Solve Equation 38-45 for $e^{(E-E_F)/(kT)}$:

$$f(E) = \frac{1}{e^{(E-E_F)/(kT)} + 1}$$

so

$$e^{(E-E_F)/(kT)} = \frac{1}{f(E)} - 1 = \frac{1}{0.1} - 1 = 9$$

2. Take the logarithm of both sides:

$$\frac{E - E_F}{kT} = \ln 9$$

3. Solve for $E$. For $E_F$, use the value for $E_F$ at $T = 0$ K listed in Table 38-1:

$$E = E_F + (\ln 9)kT$$

$$= 7.04 \text{ eV} + (\ln 9)(8.62 \times 10^{-5} \text{ eV/K})(300 \text{ K})$$

$$= \boxed{7.10 \text{ eV}}$$

**REMARKS** The Fermi factor drops from about 1 to 0.1 at just 0.06 eV above the Fermi energy of approximately 7 eV.

---

*PROBABILITY OF A HIGHER ENERGY STATE BEING OCCUPIED*    **EXAMPLE 38-11**

**Find the probability that an energy state in copper 0.1 eV above the Fermi energy is occupied at $T = 300$ K.**

**PICTURE THE PROBLEM** The probability is the Fermi factor given in Equation 38-45, with $E_F = 7.04$ eV, and $E = 7.14$ eV.

1. The probability of a state being occupied equals the Fermi factor:

$$P = f(E) = \frac{1}{e^{(E-E_F)/(kT)} + 1}$$

2. Calculate the exponent in the Fermi factor (exponents are always dimensionless):

$$\frac{E - E_F}{kT} = \frac{7.14 \text{ eV} - 7.04 \text{ eV}}{(8.62 \times 10^{-5} \text{ eV/K})(300 \text{ K})} = 3.87$$

3. Use this result to calculate the Fermi factor:

$$f(E) = \frac{1}{e^{(E-E_F)/(kT)} + 1} = \frac{1}{e^{3.87} + 1}$$

$$= \frac{1}{48 + 1} = \boxed{0.0204}$$

**REMARKS** The probability of an electron having an energy 0.1 eV above the Fermi energy at 300 K is only about 2 percent.

PROBABILITY OF A LOWER ENERGY STATE BEING OCCUPIED **EXAMPLE 38-12 Try It Yourself**

Find the probability that an energy state in copper 0.1 eV *below* the Fermi energy is occupied at $T = 300$ K.

**PICTURE THE PROBLEM** The probability is the Fermi factor given in Equation 38-45, with $E_F = 7.04$ eV, and $E = 6.94$ eV.

**Cover the column to the right and try these on your own before looking at the answers.**

| Steps | Answers |
|---|---|
| 1. Write the Fermi factor. | $f(E) = \dfrac{1}{e^{(E-E_F)/(kT)} + 1}$ |
| 2. Calculate the exponent in the Fermi factor. | $\dfrac{E - E_F}{kT} = \dfrac{6.94 \text{ eV} - 7.04 \text{ eV}}{(8.62 \times 10^{-5} \text{ eV/K})(300 \text{ K})} = -3.87$ |
| 3. Use your result from step 2 to calculate the Fermi factor. | $f(E) = \dfrac{1}{e^{(E-E_F)/(kT)} + 1} = \dfrac{1}{e^{-3.87} + 1}$ |
| | $= \dfrac{1}{0.021 + 1} = \boxed{0.979}$ |

**REMARKS** The probability of an electron having an energy of 0.1 eV *below* the Fermi energy at 300 K is approximately 98 percent.

**EXERCISE** What is the probability of an energy state 0.1 eV below the Fermi energy being unoccupied at 300 K? (*Answer* $1 - 0.98 = 0.02$ or 2 percent. This is the probability of there being a hole at this energy.)

# SUMMARY

| Topic | Relevant Equations and Remarks |
|---|---|
| 1. **The Structure of Solids** | Solids are often found in crystalline form in which a small structure, which is called the unit cell, is repeated over and over. A crystal may have a face-centered-cubic, body-centered-cubic, hexagonal close-packed, or other structure depending on the type of bonding between the atoms, ions, or molecules in the crystal and on the relative sizes of the atoms. |

Potential energy

$$U = -\alpha \frac{ke^2}{r} + \frac{A}{r^n} \qquad 38\text{-}3$$

where $r$ is the separation distance between neighboring ions and $\alpha$ is the Madelung constant, which depends on the geometry of the crystal and is of the order of 1.8, and $n$ is approximately 9.

2. **A Microscopic Picture of Conduction**

Resistivity

$$\rho = \frac{m_e v_{av}}{n_e e^2 \lambda} \qquad 38\text{-}14$$

where $v_{av}$ is the average speed of the electrons and $\lambda$ is their mean free path between collisions with the lattice ions.

| | | |
|---|---|---|
| Mean free path | $$\lambda = \frac{vt}{n_{ion}\pi r^2 vt} = \frac{1}{n_{ion}\pi r^2} = \frac{1}{n_{ion}A}$$ | 38-16 |

where $n_{ion}$ is the number of lattice ions per unit volume, $r$ is their effective radius, and $A$ is their effective cross-sectional area.

**3. Classical Interpretation of $v_{av}$ and $\lambda$**

$v_{av}$ is determined from the Maxwell–Boltzmann distribution, and $r$ is the actual radius of a lattice ion.

**4. Quantum Interpretation of $v_{av}$ and $\lambda$**

$v_{av}$ is determined from the Fermi–Dirac distribution and is approximately constant independent of temperature. The mean free path is determined from the scattering of electron waves, which occurs only because of deviations from a perfectly ordered array. The radius $r$ is the amplitude of vibration of the lattice ion, which is proportional to $\sqrt{T}$, so $A$ is proportional to $T$.

**5. The Fermi Electron Gas**

| | |
|---|---|
| Fermi energy $E_F$ at $T = 0$ | $E_F$ is the energy of the last filled (or half-filled) energy state. |

| | |
|---|---|
| $E_F$ at $T > 0$ | $E_F$ is the energy at which the probability of being occupied is $\frac{1}{2}$. |

| | |
|---|---|
| Approximate magnitude of $E_F$ | $E_F$ is between 5 eV and 10 eV for most metals. |

| | | |
|---|---|---|
| Dependence of $E_F$ on the number density of free electrons | $$E_F = \frac{h^2}{8m_e}\left(\frac{3N}{\pi V}\right)^{2/3} = (0.365 \text{ eV·nm}^2)\left(\frac{N}{V}\right)^{2/3}$$ | 38-22a, 38-22b |

| | | |
|---|---|---|
| Average energy at $T = 0$ | $E_{av} = \frac{3}{5}E_F$ | 38-23 |

| | | |
|---|---|---|
| Fermi factor at $T = 0$ | The Fermi factor $f(E)$ is the probability of a state being occupied.<br><br>$f(E) = 1, \quad E < E_F$<br><br>$f(E) = 0, \quad E > E_F$ | <br><br><br><br>38-24 |

| | | |
|---|---|---|
| Fermi temperature | $$T_F = \frac{E_F}{k}$$ | 38-25 |

| | | |
|---|---|---|
| Fermi speed | $$u_F = \sqrt{\frac{2E_F}{m_e}}$$ | 38-26 |

| | |
|---|---|
| Contact potential | When two different metals are placed in contact, electrons flow from the metal with the higher Fermi energy to the metal with the lower Fermi energy until the Fermi energies of the two metals are equal. In equilibrium, there is a potential difference between the metals that is equal to the difference in the work function of the two metals divided by the electronic charge $e$: |

$$V_{contact} = \frac{\phi_1 - \phi_2}{e} \qquad \text{38-27}$$

| | | |
|---|---|---|
| Specific heat due to conduction electrons | $$c_V' = \frac{\pi^2}{2}R\frac{T}{T_F}$$ | 38-29 |

**6. Band Theory of Solids**

When many atoms are brought together to form a solid, the individual energy levels are split into bands of allowed energies. The splitting depends on the type of bonding and the lattice separation. The highest energy band that contains electrons is called the valence band. In a conductor, the valence band is only partially full, so there are many available empty energy states for excited electrons. In an insulator, the valence band is completely full and there is a large energy gap between it and the next allowed band, the conduction band. In a semiconductor, the energy gap between the filled valence band and the empty conduction band is small; so, at ordinary temperatures, an appreciable number of electrons are thermally excited into the conduction band.

| | |
|---|---|
| **7. Semiconductors** | The conductivity of a semiconductor can be greatly increased by doping. In an $n$-type semiconductor, the doping adds electrons at energies just below that of the conduction band. In a $p$-type semiconductor, holes are added at energies just above that of the valence band. |
| **8. *Semiconductor Junctions and Devices** | |
| *Junctions | Semiconductor devices such as diodes and transistors make use of $n$-type semiconductors and $p$-type semiconductors. The two types of semiconductors are typically a single silicon crystal doped with donor impurities on one side and acceptor impurities on the other side. The region in which the semiconductor changes from a $p$-type semiconductor to an $n$-type semiconductor is called a junction. Junctions are used in diodes, solar cells, surface barrier detectors, LEDs, and transistors. |
| *Diodes | A diode is a single-junction device that carries current in one direction only. |
| *Zener diodes | A Zener diode is a diode with a very high reverse bias. It breaks down suddenly at a distinct voltage and can therefore be used as a voltage reference standard. |
| *Tunnel diodes | A tunnel diode is a diode that is heavily doped so that electrons tunnel through the depletion barrier. At normal operation, a small change in bias voltage results in a large change in current. |
| *Transistors | A transistor consists of a very thin semiconductor of one type sandwiched between two semiconductors of the opposite type. Transistors are used in amplifiers because a small variation in the base current results in a large variation in the collector current. |
| **9. Superconductivity** | In a superconductor, the resistance drops suddenly to zero below a critical temperature $T_c$. Superconductors with critical temperatures as high as 138 K have been discovered. |
| The BCS theory | Superconductivity is described by a theory of quantum mechanics called the BCS theory in which the free electrons form Cooper pairs. The energy needed to break up a Cooper pair is called the superconducting energy gap $E_g$. When all the electrons are paired, individual electrons cannot be scattered by a lattice ion, so the resistance is zero. |
| Tunneling | When a normal conductor is separated from a superconductor by a thin layer of oxide, electrons can tunnel through the energy barrier if the applied voltage across the layer is $E_g/(2e)$, where $E_g$ is the energy needed to break up a Cooper pair. The energy gap $E_g$ can be determined by a measurement of the tunneling current versus the applied voltage. |
| Josephson junction | A system of two superconductors separated by a thin layer of nonconducting material is called a Josephson junction. |
| dc Josephson effect | A dc current is observed to tunnel through a Josephson junction even in the absence of voltage across the junction. |
| ac Josephson effect | When a dc voltage $V$ is applied across a Josephson junction, an ac current is observed with a frequency |

$$f = \frac{2e}{h} V \qquad \text{38-39}$$

Measurement of the frequency of this current allows a precise determination of the ratio $e/h$.

**10. The Fermi–Dirac Distribution**   The number of electrons with energies between $E$ and $E + dE$ is given by

$$n(E)\, dE = g(E)\, dE\, f(E) \qquad\qquad 38\text{-}40$$

where $g(E)$ is the density of states and $f(E)$ is the Fermi factor.

| | | |
|---|---|---|
| Density of states | $g(E) = \dfrac{8\sqrt{2}\,\pi m_e^{3/2} V}{h^3} E^{1/2}$ | 38-41 |
| Fermi factor at temperature | $f(E) = \dfrac{1}{e^{(E-E_F)/(kT)} + 1}$ | 38-45 |

# PROBLEMS

- Single-concept, single-step, relatively easy
- •• Intermediate-level, may require synthesis of concepts
- ••• Challenging
- SSM Solution is in the *Student Solutions Manual*
- iSOLVE Problems available on iSOLVE online homework service
- iSOLVE✔ These "Checkpoint" online homework service problems ask students additional questions about their confidence level, and how they arrived at their answer.

In a few problems, you are given more data than you actually need; in a few other problems, you are required to supply data from your general knowledge, outside sources, or informed estimates.

## Conceptual Problems

**1**   • In the classical model of conduction, the electron loses energy on average in a collision because it loses the drift velocity it had picked up since the last collision. Where does this energy appear?

**2**   • SSM   When the temperature of pure copper is lowered from 300 K to 4 K, its resistivity drops by a much greater factor than that of brass when it is cooled the same way. Why?

**3**   • Thomas refuses to believe that a potential difference can be created simply by bringing two different metals into contact with each other. John talks him into making a small wager, and is about to cash in. (a) Which two metals from Table 38-2 would demonstrate his point most effectively? (b) What is the value of that contact potential?

**4**   • (a) In Problem 3, which choices of different metals would make the least impressive demonstration? (b) What is the value of that contact potential?

**5**   • A metal is a good conductor because the valence energy band for electrons is (a) completely full. (b) full, but there is only a small gap to a higher empty band. (c) partly full. (d) empty. (e) None of these is correct.

**6**   • Insulators are poor conductors of electricity because (a) there is a small energy gap between the valence band and the next higher band where electrons can exist. (b) there is a large energy gap between the full valence band and the next higher band where electrons can exist. (c) the valence band has a few vacancies for electrons. (d) the valence band is only partly full. (e) None of these is correct.

**7**   • True or false:

(a) Solids that are good electrical conductors are usually good heat conductors.
(b) The classical free-electron theory adequately explains the heat capacity of metals.
(c) At $T = 0$, the Fermi factor is either 1 or 0.
(d) The Fermi energy is the average energy of an electron in a solid.
(e) The contact potential between two metals is proportional to the difference in the work functions of the two metals.
(f) At $T = 0$, an intrinsic semiconductor is an insulator.
(g) Semiconductors conduct current in one direction only.

**8**   • SSM   How does the change in the resistivity of copper compare with that of silicon when the temperature increases?

**9**   • Which of the following elements are most likely to act as acceptor impurities in germanium? (a) bromine (b) gallium (c) silicon (d) phosphorus (e) magnesium

**10**   • Which of the following elements are most likely to serve as donor impurities in germanium? (a) bromine (b) gallium (c) silicon (d) phosphorus (e) magnesium

**11**   • An electron hole is created when a photon is absorbed by a semiconductor. How does this enable the semiconductor to conduct electricity?

**12** • Examine the positions of phosphorus, P; boron, B; thallium, Tl; and antimony, Sb in Table 36-1. (a) Which of these elements can be used to dope silicon to create an n-type semiconductor? (b) Which of these elements can be used to dope silicon to create a p-type semiconductor?

**13** • When light strikes the p-type semiconductor in a pn junction solar cell, (a) only free electrons are created. (b) only positive holes are created. (c) both electrons and holes are created. (d) positive protons are created. (e) None of these is correct.

## Estimation and Approximation

**14** • The ratio of the resistivity of the most resistive (least conductive) material to that of the least resistive material (excluding superconductors) is approximately $10^{24}$. You can develop a feeling for how remarkable this range is by considering what the ratio is of the largest to smallest values of other material properties. Choose any three properties of matter, and using tables in this book or some other resource, calculate the ratio of the largest instance of the property to the smallest instance of that property (other than zero) and rank these in decreasing order. Can you find any other property that shows a range as large as that of electrical resistivity?

**15** • A device is said to be "ohmic" if a graph of current versus applied voltage is a straight line, and the resistance $R$ of the device is the slope of this line. A pn junction is an example of a nonohmic device, as may be seen from Figure 38-21. For nonohmic devices, it is sometimes convenient to define the *differential resistance* as the reciprocal of the slope of the $I$ versus $V$ curve. Using the curve in Figure 38-21, estimate the differential resistance of the pn junction at applied voltages of –20 V, +0.2 V, +0.4 V, +0.6 V, and +0.8 V.

## The Structure of Solids

**16** • **iSOLVE** ✔ Calculate the distance $r_0$ between the $K^+$ and the $Cl^-$ ions in KCl, assuming that each ion occupies a cubic volume of side $r_0$. The molar mass of KCl is 74.55 g/mol and its density is 1.984 g/cm³.

**17** • **iSOLVE** ✔ The distance between the $Li^+$ and $Cl^-$ ions in LiCl is 0.257 nm. Use this and the molar mass of LiCl (42.4 g/mol) to compute the density of LiCl.

**18** • **SSM** **iSOLVE** Find the value of $n$ in Equation 38-6 that gives the measured dissociation energy of 741 kJ/mol for LiCl, which has the same structure as NaCl and for which $r_0 = 0.257$ nm.

**19** •• (a) Use Equation 38-6 and calculate $U(r_0)$ for calcium oxide, CaO, where $r_0 = 0.208$ nm. Assume $n = 8$. (b) If $n = 10$, what is the fractional change in $U(r_0)$?

## A Microscopic Picture of Conduction

**20** • A measure of the density of the free-electron gas in a metal is the distance $r_s$, which is defined as the radius of the sphere whose volume equals the volume per conduction electron. (a) Show that $r_s = [3/(4\pi n)]^{1/3}$, where $n$ is the free-electron number density. (b) Calculate $r_s$ for copper in nanometers.

**21** • (a) Given a mean free path $\lambda = 0.4$ nm and a mean speed $v_{av} = 1.17 \times 10^5$ m/s for the current flow in copper at a temperature of 300 K, calculate the classical value for the resistivity $\rho$ of copper. (b) The classical model suggests that the mean free path is temperature independent and that $v_{av}$ depends on temperature. From this model, what would $\rho$ be at 100 K?

## The Fermi Electron Gas

**22** •• **SSM** Silicon has an atomic weight of 28.09 and a density of $2.41 \times 10^3$ kg/m³. Each atom of silicon has two valence electrons and the Fermi energy of the material is 4.88 eV. (a) Given that the electron mean free path at room temperature is $\lambda = 27.0$ nm, estimate the resistivity. (b) The accepted value for the resistivity of silicon is 640 $\Omega \cdot$m (at room temperature). How does this accepted value compare to the value calculated in part (a)?

**23** • Calculate the number density of free electrons in (a) Ag ($\rho = 10.5$ g/cm³) and (b) Au ($\rho = 19.3$ g/cm³), assuming one free electron per atom, and compare your results with the values listed in Table 38-1.

**24** • The density of aluminum is 2.7 g/cm³. How many free electrons are present per aluminum atom?

**25** • The density of tin is 7.3 g/cm³. How many free electrons are present per tin atom?

**26** • **SSM** **iSOLVE** ✔ Calculate the Fermi temperature for (a) Al, (b) K, and (c) Sn.

**27** • **iSOLVE** What is the speed of a conduction electron whose energy is equal to the Fermi energy $E_F$ for (a) Na, (b) Au, and (c) Sn?

**28** • **iSOLVE** ✔ Calculate the Fermi energy for (a) Al, (b) K, and (c) Sn using the number densities given in Table 38-1.

**29** • **iSOLVE** Find the average energy of the conduction electrons at $T = 0$ in (a) copper and (b) lithium.

**30** • **iSOLVE** Calculate (a) the Fermi temperature and (b) the Fermi energy at $T = 0$ for iron.

**31** •• **SSM** (a) Assuming that gold contributes one free electron per atom to the metal, calculate the electron density in gold knowing that its atomic weight is 196.97 and its mass density is $19.3 \times 10^3$ kg/m³. (b) If the Fermi speed for gold is $1.39 \times 10^6$ m/s, what is the Fermi energy in electron volts? (c) By what factor is the Fermi energy higher than the $kT$ energy at room temperature? (d) Explain the difference between the Fermi energy and the $kT$ energy.

**32** •• **SSM** The pressure of an ideal gas is related to the average energy of the gas particles by $PV = \frac{2}{3}NE_{av}$, where $N$ is the number of particles and $E_{av}$ is the average energy. Use this to calculate the pressure of the Fermi electron gas in copper in newtons per square meter, and compare your result with atmospheric pressure, which is about $10^5$ N/m². (*Note:* The units are most easily handled by using the conversion factors $1$ N/m² $= 1$ J/m³ and $1$ eV $= 1.6 \times 10^{-19}$ J.)

**33** •• The bulk modulus $B$ of a material can be defined by

$$B = -V\frac{\partial P}{\partial V}$$

(a) Use the ideal-gas relation $PV = \frac{2}{3}NE_{av}$ and Equation 38-22 and Equation 38-23 to show that

$$P = \frac{2NE_F}{5V} = CV^{-5/3}$$

where $C$ is a constant independent of $V$. (b) Show that the bulk modulus of the Fermi electron gas is therefore

$$B = \frac{5}{3}P = \frac{2NE_F}{5V}$$

(c) Compute the bulk modulus in newtons per square meter for the Fermi electron gas in copper and compare your result with the measured value of $140 \times 10^9$ N/m$^2$.

**34** • Calculate the contact potential between (a) Ag and Cu, (b) Ag and Ni, and (c) Ca and Cu.

## Heat Capacity Due to Electrons in a Metal

**35** •• [SSM] Gold has a Fermi energy of 5.53 eV. Determine the molar specific heat at constant volume and room temperature for gold.

## Quantum Theory of Electrical Conduction

**36** • The resistivities of Na, Au, and Sn at $T = 273$ K are 4.2 $\mu\Omega\cdot$cm, 2.04 $\mu\Omega\cdot$cm, and 10.6 $\mu\Omega\cdot$cm, respectively. Use these values and the Fermi speeds calculated in Problem 27 to find the mean free paths $\lambda$ for the conduction electrons in these elements.

**37** •• [SSM] The resistivity of pure copper is increased approximately $1 \times 10^{-8}$ $\Omega\cdot$m by the addition of 1 percent (by number of atoms) of an impurity throughout the metal. The mean free path depends on both the impurity and the oscillations of the lattice ions according to the equation $1/\lambda = 1/\lambda_t + 1/\lambda_i$. (a) Estimate $\lambda_i$ from the data given in Table 38-1. (b) If $r$ is the effective radius of an impurity lattice ion seen by an electron, the scattering cross section is $\pi r^2$. Estimate this area, using the fact that $r$ is related to $\lambda_i$ by Equation 38-16.

## Band Theory of Solids

**38** • When light of wavelength 380.0 nm falls on a semiconductor, electrons are promoted from the valence band to the conduction band. Calculate the energy gap, in electron volts, for this semiconductor.

**39** • [SSM] [iSOLVE✓] You are an electron sitting at the top of the valence band in a silicon atom, longing to jump across the 1.14-eV energy gap that separates you from the bottom of the conduction band and all of the adventures that it may contain. What you need, of course, is a photon. What is the maximum photon wavelength that will get you across the gap?

**40** • Repeat Problem 39 for germanium, for which the energy gap is 0.74 eV.

**41** • Repeat Problem 39 for diamond, for which the energy gap is 7.0 eV.

**42** •• A photon of wavelength 3.35 $\mu$m has just enough energy to raise an electron from the valence band to the conduction band in a lead sulfide crystal. (a) Find the energy gap between these bands in lead sulfide. (b) Find the temperature $T$ for which $kT$ equals this energy gap.

## Semiconductors

**43** • The donor energy levels in an $n$-type semiconductor are 0.01 eV below the conduction band. Find the temperature for which $kT = 0.01$ eV.

**44** •• When a thin slab of semiconducting material is illuminated with monochromatic electromagnetic radiation, most of the radiation is transmitted through the slab if the wavelength is greater than 1.85 mm. For wavelengths less than 1.85 mm, most of the incident radiation is absorbed. Determine the energy gap of this semiconductor.

**45** •• The relative binding of the extra electron in the arsenic atom that replaces an atom in silicon or germanium can be understood from a calculation of the first Bohr orbit of this electron in these materials. Four of arsenic's outer electrons form covalent bonds, so the fifth electron sees a singly charged center of attraction. This model is a modified hydrogen atom. In the Bohr model of the hydrogen atom, the electron moves in free space at a radius $a_0$ given by Equation 36-12, which is

$$a_0 = \frac{4\pi\epsilon_0\hbar^2}{m_e e^2}$$

When an electron moves in a crystal, we can approximate the effect of the other atoms by replacing $\epsilon_0$ with $\kappa\epsilon_0$ and $m_e$ with an effective mass for the electron. For silicon, $\kappa$ is 12 and the effective mass is approximately $0.2m_e$. For germanium, $\kappa$ is 16 and the effective mass is approximately $0.1m_e$. Estimate the Bohr radii for the outer electron as it orbits the impurity arsenic atom in silicon and germanium.

**46** •• [SSM] The ground-state energy of the hydrogen atom is given by

$$E_1 = -\frac{mk^2e^4}{2\hbar^2} = -\frac{e^2m_e}{8\epsilon_0^2h^2}$$

Modify this equation in the spirit of Problem 45 by replacing $\epsilon_0$ by $\kappa\epsilon_0$ and $m_e$ by an effective mass for the electron to estimate the binding energy of the extra electron of an impurity arsenic atom in (a) silicon and (b) germanium.

**47** •• A doped $n$-type silicon sample with $10^{16}$ electrons per cubic centimeter in the conduction band has a resistivity of $5 \times 10^{-3}$ $\Omega\cdot$m at 300 K. Find the mean free path of the electrons. Use the effective mass of $0.2m_e$ for the mass of the electrons. (See Problem 45.) Compare this mean free path with that of conduction electrons in copper at 300 K.

**48** •• The measured Hall coefficient of a doped silicon sample is 0.04 V·m/A·T at room temperature. If all the doping impurities have contributed to the total charge carriers of the sample, find (a) the type of impurity (donor or acceptor) used to dope the sample and (b) the concentration of these impurities.

## *Semiconductor Junctions and Devices

**49** •• Simple theory for the current versus the bias voltage across a $pn$ junction yields the equation $I = I_0(e^{eV_b/kT} - 1)$.

Sketch $I$ versus $V_b$ for both positive and negative values of $V_b$ using this equation.

**50** • The plate current in an *npn* transistor circuit is 25.0 mA. If 88 percent of the electrons emitted reach the collector, what is the base current?

**51** •• SSM ISOLVE In Figure 38-27 for the *pnp*-transistor amplifier, suppose $R_b = 2$ k$\Omega$ and $R_L = 10$ k$\Omega$. Suppose further that a 10-$\mu$A ac base current generates a 0.5-mA ac collector current. What is the voltage gain of the amplifier?

**52** •• Germanium can be used to measure the energy of incident particles. Consider a 660-keV gamma ray emitted from $^{137}$Cs. (a) Given that the band gap in germanium is 0.72 eV, how many electron–hole pairs can be generated as this gamma ray travels through germanium? (b) The number of pairs $N$ in Part (a) will have statistical fluctuations given by $\pm\sqrt{N}$. What then is the energy resolution of this detector in this photon energy region?

**53** •• Make a sketch showing the valence and conduction band edges and Fermi energy of a *pn*-junction diode when biased (a) in the forward direction and (b) in the reverse direction.

**54** •• SSM A "good" silicon diode has the current–voltage characteristic given in Problem 49. Let $kT = 0.025$ eV (room temperature) and the saturation current $I_0 = 1$ nA. (a) Show that for small reverse-bias voltages, the resistance is 25 M$\Omega$. (*Hint: Do a Taylor expansion of the exponential function or use your calculator and enter small values for $V_b$.*) (b) Find the dc resistance for a reverse bias of 0.5 V. (c) Find the dc resistance for a 0.5-V forward bias. What is the current in this case? (d) Calculate the ac resistance $dV/dI$ for a 0.5-V forward bias.

**55** •• A slab of silicon of thickness $t = 1.0$ mm and width $w = 1.0$ cm is placed in a magnetic field $B = 0.4$ T. The slab is in the $xy$ plane, and the magnetic field points in the positive $z$ direction. When a current of 0.2 A flows through the sample in the positive $x$ direction, a voltage difference of 5 mV develops across the width of the sample with the electric field in the sample pointing in the positive $y$ direction. Determine the semiconductor type ($n$ or $p$) and the concentration of charge carriers.

## The BCS Theory

**56** • (a) Use Equation 38-37 to calculate the superconducting energy gap for tin, and compare your result with the measured value of $6 \times 10^{-4}$ eV. (b) Use the measured value to calculate the wavelength of a photon having sufficient energy to break up Cooper pairs in tin ($T_c = 3.72$ K) at $T = 0$.

**57** • SSM Repeat Problem 56 for lead ($T_c = 7.19$ K), which has a measured energy gap of $2.73 \times 10^{-3}$ eV.

## The Fermi–Dirac Distribution

**58** •• The number of electrons in the conduction band of an insulator or intrinsic semiconductor is governed chiefly by the Fermi factor. Since the valence band in these materials is nearly filled and the conduction band is nearly empty, the Fermi energy $E_F$ is generally midway between the top of the valence band and the bottom of the conduction band, that is,

at $E_g/2$, where $E_g$ is the band gap between the two bands and the energy is measured from the top of the valence band. (a) In silicon, $E_g \approx 1.0$ eV. Show that in this case the Fermi factor for electrons at the bottom of the conduction band is given by $\exp(-E_g/2kT)$ and evaluate this factor. Discuss the significance of this result if there are $10^{22}$ valence electrons per cubic centimeter and the probability of finding an electron in the conduction band is given by the Fermi factor. (b) Repeat the calculation in Part (a) for an insulator with a band gap of 6.0 eV.

**59** •• Approximately how many energy states, with energies between 2.00 eV and 2.20 eV, are available to electrons in a cube of silver measuring 1.00 mm on a side?

**60** •• SSM (a) Use Equation 38-22a to calculate the Fermi energy for silver. (b) Determine the average energy of a free electron and (c) find the Fermi speed for silver.

**61** •• ISOLVE Show that at $E = E_F$, the Fermi factor is $F = 0.5$.

**62** •• What is the difference between the energies at which the Fermi factor is 0.9 and 0.1 at 300 K in (a) copper, (b) potassium, and (c) aluminum.

**63** •• SSM What is the probability that a conduction electron in silver will have a kinetic energy of 4.9 eV at $T = 300$ K?

**64** •• Show that $g(E) = (3N/2)E_F^{-3/2}E^{1/2}$ (Equation 38-43) follows from Equation 38-41 for $g(E)$, and from Equation 38-22a for $E_F$.

**65** •• Carry out the integration $E_{av} = (1/N)\int_0^{E_F} Eg(e)\,dE$ to show that the average energy at $T = 0$ is $\frac{3}{5}E_F$.

**66** •• The density of the electron states in a metal can be written $g(E) = AE^{1/2}$, where $A$ is a constant and $E$ is measured from the bottom of the conduction band. (a) Show that the total number of states is $\frac{2}{3}AE_F^{3/2}$. (b) Approximately what fraction of the conduction electrons are within $kT$ of the Fermi energy? (c) Evaluate this fraction for copper at $T = 300$ K.

**67** •• What is the probability that a conduction electron in silver will have a kinetic energy of 5.49 eV at $T = 300$ K?

**68** •• ISOLVE Use Equation 38-41 for the density of states to estimate the fraction of the conduction electrons in copper that can absorb energy from collisions with the vibrating lattice ions at (a) 77 K and (b) 300 K.

**69** •• In an intrinsic semiconductor, the Fermi energy is about midway between the top of the valence band and the bottom of the conduction band. In germanium, the forbidden energy band has a width of 0.7 eV. Show that at room temperature the distribution function of electrons in the conduction band is given by the Maxwell–Boltzmann distribution function.

**70** ••• SSM (a) Show that for $E \geq 0$, the Fermi factor may be written as $f(E) = 1/(Ce^{E/kT} + 1)$. (b) Show that if $C \gg e^{-E/(kT)}$, $f(E) = Ae^{-E/(kT)} \ll 1$; in other words, show that the Fermi factor is a constant times the classical Boltzmann factor if $A \ll 1$. (c) Use $\int n(E)\,dE = N$ and Equation 38-41 to determine the constant $A$. (d) Using the result obtained in Part (c), show that the classical approximation is applicable when the electron concentration is very small and/or the temperature is very high. (*e*) Most semiconductors have impurities added in a process called doping, which increases the free electron concentration so that it is about $10^{17}/cm^3$ at room temperature. Show that for these systems, the classical distribution function is applicable.

**71** ••• Show that the condition for the applicability of the classical distribution function for an electron gas ($A \ll 1$ in Problem 70) is equivalent to the requirement that the average separation between electrons is much greater than their de Broglie wavelength.

**72** ••• The root-mean-square (rms) value of a variable is obtained by calculating the average value of the square of that variable and then taking the square root of the result. Use this procedure to determine the rms energy of a Fermi distribution. Express your result in terms of $E_F$ and compare it to the average energy. Why do $E_{av}$ and $E_{rms}$ differ?

## General Problems

**73** • The density of potassium is 0.851 g/cm³. How many free electrons are there per potassium atom?

**74** • Calculate the number density of free electrons for (*a*) Mg ($\rho = 1.74$ g/cm³) and (*b*) Zn ($\rho = 7.1$ g/cm³), assuming two free electrons per atom, and compare your results with the values listed in Table 38-1.

**75** •• **iSOLVE** Estimate the fraction of free electrons in copper that are in excited states above the Fermi energy at (*a*) 300 K (about room temperature) and (*b*) at 1000 K.

**76** •• **SSM** Determine the energy that has 10 percent free electron occupancy probability for manganese at T = 1300 K.

**77** •• The semiconducting compound CdSe is widely used for light-emitting diodes (LEDs). The energy gap in CdSe is 1.8 eV. What is the frequency of the light emitted by a CdSe LED?

**78** ••• **SSM** A 2-cm² wafer of pure silicon is irradiated with light having a wavelength of 775 nm. The intensity of the light beam is 4.0 W/m² and every photon that strikes the sample is absorbed and creates an electron–hole pair. (*a*) How many electron–hole pairs are produced in one second? (*b*) If the number of electron–hole pairs in the sample is $6.25 \times 10^{11}$ in the steady state, at what rate do the electron–hole pairs recombine? (*c*) If every recombination event results in the radiation of one photon, at what rate is energy radiated by the sample?

# Relativity

**THE ANDROMEDA GALAXY** BY MEASURING THE FREQUENCY OF THE LIGHT COMING TO US FROM DISTANT OBJECTS, WE ARE ABLE TO DETERMINE HOW FAST THESE OBJECTS ARE APPROACHING TOWARD US OR RECEDING FROM US.

**?** **Have you wondered how the frequency of the light enables us to determine the speed of recession of a distant galaxy? This is discussed in Example 39-5.**

39-1 Newtonian Relativity

39-2 Einstein's Postulates

39-3 The Lorentz Transformation

39-4 Clock Synchronization and Simultaneity

39-5 The Velocity Transformation

39-6 Relativistic Momentum

39-7 Relativistic Energy

39-8 General Relativity

The theory of relativity consists of two rather different theories, the special theory and the general theory. The special theory, developed by Albert Einstein and others in 1905, concerns the comparison of measurements made in different inertial reference frames moving with constant velocity relative to one another. Its consequences, which can be derived with a minimum of mathematics, are applicable in a wide variety of situations encountered in physics and in engineering. On the other hand, the general theory, also developed by Einstein and others around 1916, is concerned with accelerated reference frames and gravity. A thorough understanding of the general theory requires sophisticated mathematics, and the applications of this theory are chiefly in the area of gravitation. The general theory is of great importance in cosmology, but it is rarely encountered in other areas of physics or in engineering. The general theory is used, however, in the engineering of the Global Positioning System (GPS).[†]

---

† The satellites used in GPS contain atomic clocks.

➤In this chapter, we concentrate on the special theory (often referred to as *special relativity*). General relativity will be discussed briefly near the end of the chapter.

# 39-1 Newtonian Relativity

Newton's first law does not distinguish between a particle at rest and a particle moving with constant velocity. If there is no net external force acting, the particle will remain in its initial state, either at rest or moving with its initial velocity. A particle at rest relative to you is moving with constant velocity relative to an observer who is moving with constant velocity relative to you. How might we distinguish whether you and the particle are at rest and the second observer is moving with constant velocity, or the second observer is at rest and you and the particle are moving?

Let us consider some simple experiments. Suppose we have a railway boxcar moving along a straight, flat track with a constant velocity $v$. We note that a ball at rest in the boxcar remains at rest. If we drop the ball, it falls straight down, relative to the boxcar, with an acceleration $g$ due to gravity. Of course, when viewed from the track the ball moves along a parabolic path because it has an initial velocity $v$ to the right. No mechanics experiment that we can do— measuring the period of a pendulum, observing the collisions between two objects, or whatever—will tell us whether the boxcar is moving and the track is at rest or the track is moving and the boxcar is at rest. If we have a coordinate system attached to the track and another attached to the boxcar, Newton's laws hold in either system.

A set of coordinate systems at rest relative to each other is called a *reference frame*. A reference frame in which Newton's laws hold is called an *inertial reference frame*.[†] All reference frames moving at constant velocity relative to an inertial reference frame are also inertial reference frames. If we have two inertial reference frames moving with constant velocity relative to each other, there are no mechanics experiments that can tell us which is at rest and which is moving or if they are both moving. This result is known as the principle of **Newtonian relativity**:

Absolute motion cannot be detected.

right

PRINCIPLE OF NEWTONIAN RELATIVITY

This principle was well known by Galileo, Newton, and others in the seventeenth century. By the late nineteenth century, however, this view had changed. It was then generally thought that Newtonian relativity was not valid and that absolute motion could be detected in principle by a measurement of the speed of light.

## Ether and the Speed of Light

We saw in Chapter 15 that the velocity of a wave depends on the properties of the medium in which the wave travels and not on the velocity of the source of the waves. For example, the velocity of sound relative to still air depends on the temperature of the air. Light and other electromagnetic waves (radio, X rays, etc.) travel through a vacuum with a speed $c \approx 3 \times 10^8$ m/s that is predicted by James Clerk Maxwell's equations for electricity and magnetism. But what is this speed

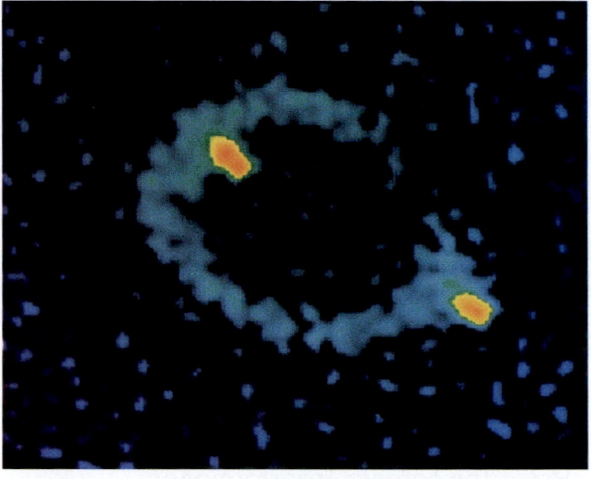

This ring-like structure of the radio source MG1131 + 0456 is thought to be due to *gravitational lensing*, first proposed by Albert Einstein in 1936, in which a source is imaged into a ring by a large, massive object in the foreground.

---

[†] Reference frames were first discussed in Section 2-1. Inertial reference frames were also discussed in Section 4-1.

relative to? What is the equivalent of still air for a vacuum? A proposed medium for the propagation of light was called the *ether*; it was thought to pervade all space. The velocity of light relative to the ether was assumed to be $c$, as predicted by Maxwell's equations. The velocity of any object relative to the ether was considered its absolute velocity.

Albert Michelson, first in 1881 and then again with Edward Morley in 1887, set out to measure the velocity of the earth relative to the ether by an ingenious experiment in which the velocity of light relative to the earth was compared for two light beams, one in the direction of the earth's motion relative to the sun and the other perpendicular to the direction of the earth's motion. Despite painstakingly careful measurements, they could detect no difference. The experiment has since been repeated under various conditions by a number of people, and no difference has ever been found. The absolute motion of the earth relative to the ether cannot be detected.

## 39-2 Einstein's Postulates

In 1905, at the age of 26, Albert Einstein published a paper on the electrodynamics of moving bodies.[†] In this paper, he postulated that absolute motion cannot be detected by any experiment. That is, there is no ether. The earth can be considered to be at rest and the velocity of light will be the same in any direction.[‡] His theory of special relativity can be derived from two postulates. Simply stated, these postulates are as follows:

> Postulate 1: Absolute uniform motion cannot be detected.
>
> Postulate 2: The speed of light is independent of the motion of the source.

EINSTEIN'S POSTULATES

Postulate 1 is merely an extension of the Newtonian principle of relativity to include all types of physical measurements (not just those that are mechanical). Postulate 2 describes a common property of all waves. For example, the speed of sound waves does not depend on the motion of the sound source. The sound waves from a car horn travel through the air with the same velocity independent of whether the car is moving or not. The speed of the waves depends only on the properties of the air, such as its temperature.

Although each postulate seems quite reasonable, many of the implications of the two postulates together are quite surprising and contradict what is often called common sense. For example, one important implication of these postulates is that every observer measures the same value for the speed of light independent of the relative motion of the source and the observer. Consider a light source $S$ and two observers, $R_1$ at rest relative to $S$ and $R_2$ moving toward $S$ with speed $v$, as shown in Figure 39-1$a$. The speed of light measured by $R_1$ is $c = 3 \times 10^8$ m/s. What is the speed measured by $R_2$? The answer is *not* $c + v$. By postulate 1, Figure 39-1$a$ is equivalent to Figure 39-1$b$, in which $R_2$ is at rest and the source $S$ and $R_1$ are moving with speed $v$. That is, since absolute motion cannot be detected, it is not possible to say which is really moving and which is at rest. By postulate 2, the speed of light from a moving source is independent of the

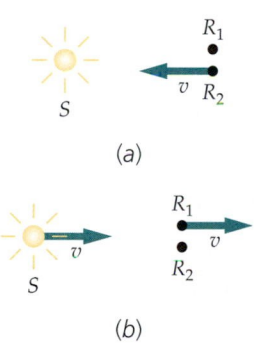

**FIGURE 39-1** (*a*) A stationary light source $S$ and a stationary observer $R_1$, with a second observer $R_2$ moving toward the source with speed $v$. (*b*) In the reference frame in which the observer $R_2$ is at rest, the light source $S$ and observer $R_1$ move to the right with speed $v$. If absolute motion cannot be detected, the two views are equivalent. Since the speed of light does not depend on the motion of the source, observer $R_2$ measures the same value for that speed as observer $R_1$.

---

[†] *Annalen der Physik*, vol. 17, 1905, p. 841. For a translation from the original German, see W. Perrett and G. B. Jeffery (trans.), *The Principle of Relativity: A Collection of Original Memoirs on the Special and General Theory of Relativity* by H. A. Lorentz, A. Einstein, H. Minkowski, and W. Weyl, Dover, New York, 1923.

[‡] Einstein did not set out to explain the results of the Michelson–Morley experiment. His theory arose from his considerations of the theory of electricity and magnetism and the unusual property of electromagnetic waves that they propagate in a vacuum. In his first paper, which contains the complete theory of special relativity, he made only a passing reference to the Michelson–Morley experiment, and in later years he could not recall whether he was aware of the details of this experiment before he published his theory.

motion of the source. Thus, looking at Figure 39-1b, we see that $R_2$ measures the speed of light to be $c$, just as $R_1$ does. This result is often considered as an alternative to Einstein's second postulate:

> Postulate 2 (alternate): Every observer measures the same value $c$ for the speed of light.

This result contradicts our intuitive ideas about relative velocities. If a car moves at 50 km/h away from an observer and another car moves at 80 km/h in the same direction, the velocity of the second car relative to the first car is 30 km/h. This result is easily measured and conforms to our intuition. However, according to Einstein's postulates, if a light beam is moving in the direction of the cars, observers in both cars will measure the same speed for the light beam. Our intuitive ideas about the combination of velocities are approximations that hold only when the speeds are very small compared with the speed of light. Even in an airplane moving with the speed of sound, to measure the speed of light accurately enough to distinguish the difference between the results $c$ and $c + v$, where $v$ is the speed of the plane, would require a measurement with six-digit accuracy.

## 39-3 The Lorentz Transformation

Einstein's postulates have important consequences for measuring time intervals and space intervals, as well as relative velocities. Throughout this chapter, we will be comparing measurements of the positions and times of events (such as lightning flashes) made by observers who are moving relative to each other. We will use a rectangular coordinate system $xyz$ with origin $O$, called the $S$ reference frame, and another system $x'y'z'$ with origin $O'$, called the $S'$ frame, that is moving with a constant velocity $\vec{v}$ relative to the $S$ frame. Relative to the $S'$ frame, the $S$ frame is moving with a constant velocity $-\vec{v}$. For simplicity, we will consider the $S'$ frame to be moving along the $x$ axis in the positive $x$ direction relative to $S$. In each frame, we will assume that there are as many observers as are needed who are equipped with measuring devices, such as clocks and metersticks, that are identical when compared at rest (see Figure 39-2).

We will use Einstein's postulates to find the general relation between the coordinates $x$, $y$, and $z$ and the time $t$ of an event as seen in reference frame $S$ and the coordinates $x'$, $y'$, and $z'$ and the time $t'$ of the same event as seen in reference frame $S'$, which is moving with uniform velocity relative to $S$. We assume that the origins are coincident at time $t = t' = 0$. The classical relation, called the **Galilean transformation**, is

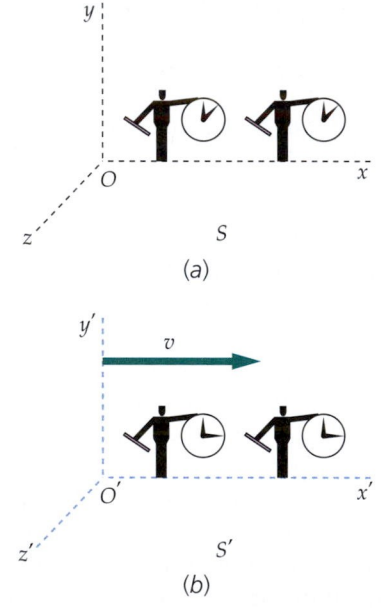

**FIGURE 39-2** Coordinate reference frames $S$ and $S'$ moving with relative speed $v$. In each frame, there are observers with metersticks and clocks that are identical when compared at rest.

$$x = x' + vt', \qquad y = y', \qquad z = z', \qquad t = t' \qquad \text{39-1}a$$

GALILEAN TRANSFORMATION

The inverse transformation is

$$x' = x - vt, \qquad y' = y, \qquad z' = z, \qquad t' = t \qquad \text{39-1}b$$

These equations are consistent with experimental observations as long as $v$ is much less than $c$. They lead to the familiar classical addition law for velocities. If a particle has velocity $u_x = dx/dt$ in frame $S$, its velocity in frame $S'$ is

$$u_x' = \frac{dx'}{dt'} = \frac{dx'}{dt} = \frac{dx}{dt} - v = u_x - v \qquad \text{39-2}$$

If we differentiate this equation again, we find that the acceleration of the particle is the same in both frames:

$$a_x = \frac{du_x}{dt} = \frac{du_x'}{dt'} = a_x'$$

It should be clear that the Galilean transformation is not consistent with Einstein's postulates of special relativity. If light moves along the $x$ axis with speed $u_x' = c$ in $S'$, these equations imply that the speed in $S'$ is $u_x = c + v$ rather than $u_x = c$, which is consistent with Einstein's postulates and with experiment. The classical transformation equations must therefore be modified to make them consistent with Einstein's postulates. We will give a brief outline of one method of obtaining the relativistic transformation.

We assume that the relativistic transformation equation for $x$ is the same as the classical equation (Equation 39-1$a$) except for a constant multiplier on the right side. That is, we assume the equation is of the form

$$x = \gamma(x' + vt') \qquad\qquad 39\text{-}3$$

where $\gamma$ is a constant that can depend on $v$ and $c$ but not on the coordinates. The inverse transformation must look the same except for the sign of the velocity:

$$x' = \gamma(x - vt) \qquad\qquad 39\text{-}4$$

Let us consider a light pulse that starts at the origin of $S$ at $t = 0$. Since we have assumed that the origins are coincident at $t = t' = 0$, the pulse also starts at the origin of $S'$ at $t' = 0$. Einstein's postulates require that the equation for the $x$ component of the wave front of the light pulse is $x = ct$ in frame $S$ and $x' = ct'$ in frame $S'$. Substituting $ct$ for $x$ and $ct'$ for $x'$ in Equation 39-3 and Equation 39-4, we obtain

$$ct = \gamma(ct' + vt') = \gamma(c + v)t' \qquad\qquad 39\text{-}5$$

and

$$ct' = \gamma(ct - vt) = \gamma(c - v)t \qquad\qquad 39\text{-}6$$

We can eliminate the ratio $t'/t$ from these two equations and determine $\gamma$. Thus,

$$\gamma = \frac{1}{\sqrt{1 - \dfrac{v^2}{c^2}}} \qquad\qquad 39\text{-}7$$

Note that $\gamma$ is always greater than 1, and that when $v$ is much less than $c$, $\gamma \approx 1$. The relativistic transformation for $x$ and $x'$ is therefore given by Equation 39-3 and Equation 39-4, with $\gamma$ given by Equation 39-7. We can obtain equations for $t$ and $t'$ by combining Equation 39-3 with the inverse transformation given by Equation 39-4. Substituting $x = \gamma(x' + vt')$ for $x$ in Equation 39-4, we obtain

$$x' = \gamma\big[\gamma(x' + vt') - vt\big] \qquad\qquad 39\text{-}8$$

which can be solved for $t$ in terms of $x'$ and $t'$. The complete relativistic transformation is

$$x = \gamma(x' + vt'), \qquad y = y', \qquad z = z' \qquad\qquad\qquad 39\text{-}9$$

$$t = \gamma\left(t' + \frac{vx'}{c^2}\right) \qquad\qquad\qquad 39\text{-}10$$

LORENTZ TRANSFORMATION

The inverse transformation is

$$x' = \gamma(x - vt), \qquad y' = y, \qquad z' = z \qquad\qquad\qquad 39\text{-}11$$

$$t' = \gamma\left(t - \frac{vx}{c^2}\right) \qquad\qquad\qquad 39\text{-}12$$

The transformation described by Equation 39-9 through Equation 39-12 is called the **Lorentz transformation.** It relates the space and time coordinates $x, y, z$, and $t$ of an event in frame $S$ to the coordinates $x', y', z'$, and $t'$ of the same event as seen in frame $S'$, which is moving along the $x$ axis with speed $v$ relative to frame $S$.

We will now look at some applications of the Lorentz transformation.

## Time Dilation

Consider two events that occur at a single point $x'_0$ at times $t'_1$ and $t'_2$ in frame $S'$. We can find the times $t_1$ and $t_2$ for these events in $S$ from Equation 39-10. We have

$$t_1 = \gamma\left(t'_1 + \frac{vx'_0}{c^2}\right)$$

and

$$t_2 = \gamma\left(t'_2 + \frac{vx'_0}{c^2}\right)$$

so

$$t_2 - t_1 = \gamma(t'_2 - t'_1)$$

The time between events that happen at the *same place* in a reference frame is called **proper time** $t_p$. In this case, the time interval $t'_2 - t'_1$ measured in frame $S'$ is proper time. The time interval $\Delta t$ measured in any other reference frame is always longer than the proper time. This expansion is called **time dilation:**

$$\Delta t = \gamma \, \Delta t_p \qquad\qquad\qquad 39\text{-}13$$

TIME DILATION

*SPATIAL SEPARATION AND TEMPORAL SEPARATION OF TWO EVENTS*

**E X A M P L E   3 9 - 1**

Two events occur at the same point $x'_0$ at times $t'_1$ and $t'_2$ in frame $S'$, which is traveling at speed $v$ relative to frame $S$. (*a*) What is the spatial separation of these events in frame $S$? (*b*) What is the temporal separation of these events in frame $S$?

**PICTURE THE PROBLEM** The spatial separation in $S$ is $x_2 - x_1$, where $x_2$ and $x_1$ are the coordinates of the events in $S$, which are found using Equation 39-9.

(a) 1. The position $x_1$ in $S$ is given by Equation 39-9 with $x_1' = x_0'$:

$$x_1 = \gamma(x_0' + vt_1')$$

2. Similarly, the position $x_2$ in $S$ is given by:

$$x_2 = \gamma(x_0' + vt_2')$$

3. Subtract to find the spatial separation:

$$\Delta x = x_2 - x_1 = \gamma v(t_2' - t_1') = \boxed{\dfrac{v(t_2' - t_1')}{\sqrt{1 - (v^2/c^2)}}}$$

(b) Using the time dilation formula, relate the two time intervals. The two events occur at the same place in $S'$, so the proper time between the two events is $\Delta t_p = t_2' - t_1'$:

$$\Delta t = t_2 - t_1 = \gamma(t_2' - t_1') = \boxed{\dfrac{(t_2' - t_1')}{\sqrt{1 - (v^2/c^2)}}}$$

**REMARKS** Dividing the Part (a) result by the Part (b) result gives $\Delta x/\Delta t = v$. The spatial separation of these two events in $S$ is the distance a fixed point, such as $x_0'$ in $S'$, moves in $S$ during the time interval between the events in $S$.

We can understand time dilation directly from Einstein's postulates without using the Lorentz transformation. Figure 39-3a shows an observer $A'$ a distance $D$ from a mirror. The observer and the mirror are in a spaceship that is at rest in frame $S'$. The observer explodes a flash gun and measures the time interval $\Delta t'$ between the original flash and his seeing the return flash from the mirror. Because light travels with speed $c$, this time is

$$\Delta t' = \frac{2D}{c}$$

We now consider these same two events, the original flash of light and the receiving of the return flash, as observed in reference frame $S$, in which observer $A'$ and the mirror are moving to the right with speed $v$, as shown in Figure 39-3b.

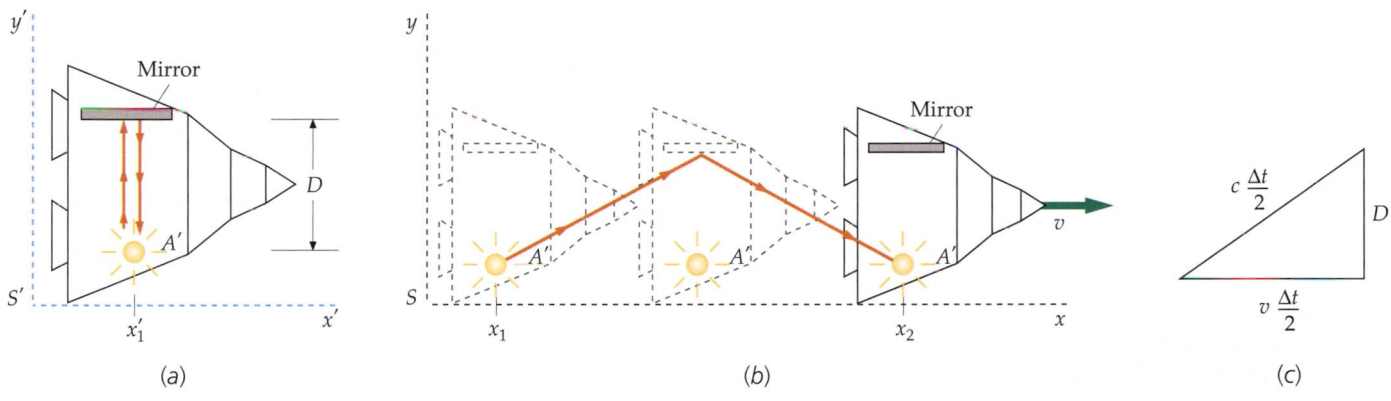

(a)    (b)    (c)

The events happen at two different places $x_1$ and $x_2$ in frame $S$. During the time interval $\Delta t$ (as measured in $S$) between the original flash and the return flash, observer $A'$ and his spaceship have moved a horizontal distance $v \Delta t$. In Figure 39-3b, we can see that the path traveled by the light is longer in $S$ than in $S'$. However, by Einstein's postulates, light travels with the same speed $c$ in frame $S$ as it does in frame $S'$. Because light travels farther in $S$ at the same speed, it takes longer in $S$ to reach the mirror and return. The time interval in $S$ is thus longer than it is in $S'$. From the triangle in Figure 39-3c, we have

$$\left(\frac{c \, \Delta t}{2}\right)^2 = D^2 + \left(\frac{v \, \Delta t}{2}\right)^2$$

**FIGURE 39-3** (a) Observer $A'$ and the mirror are in a spaceship at rest in frame $S'$. The time it takes for the light pulse to reach the mirror and return is measured by $A'$ to be $2D/c$. (b) In frame $S$, the spaceship is moving to the right with speed $v$. If the speed of light is the same in both frames, the time it takes for the light to reach the mirror and return is longer than $2D/c$ in $S$ because the distance traveled is greater than $2D$. (c) A right triangle for computing the time $\Delta t$ in frame $S$.

or

$$\Delta t = \frac{2D}{\sqrt{c^2 - v^2}} = \frac{2D}{c}\frac{1}{\sqrt{1 - (v^2/c^2)}}$$

Using $\Delta t' = 2D/c$, we obtain

$$\Delta t = \frac{\Delta t'}{\sqrt{1 - (v^2/c^2)}} = \gamma \,\Delta t'$$

---

*How Long Is a One-Hour Nap?*  **EXAMPLE 39-2**  **Try It Yourself**

**Astronauts in a spaceship traveling at $v = 0.6c$ relative to the earth sign off from space control, saying that they are going to nap for 1 h and then call back. How long does their nap last as measured on the earth?**

**PICTURE THE PROBLEM** Because the astronauts go to sleep and wake up at the same place in their reference frame, the time interval for their nap of 1 h as measured by them is proper time. In the earth's reference frame, they move a considerable distance between these two events. The time interval measured in the earth's frame (using two clocks located at those events) is longer by the factor $\gamma$.

**Cover the column to the right and try these on your own before looking at the answers.**

| Steps | Answers |
|---|---|
| 1. Relate the time interval measured on the earth $\Delta t$ to the proper time $\Delta t_p$. | $\Delta t = \gamma \,\Delta t_p$ |
| 2. Calculate $\gamma$ for $v = 0.6c$. | $\gamma = 1.25$ |
| 3. Substitute to calculate the time of the nap in the earth's frame. | $\Delta t = \gamma \,\Delta t_p = \boxed{1.25\ \text{h}}$ |

**EXERCISE** If the spaceship is moving at $v = 0.8c$, how long would a 1 h nap last as measured on the earth? (*Answer* 1.67 h)

## Length Contraction

A phenomenon closely related to time dilation is **length contraction.** The length of an object measured in the reference frame in which the object is at rest is called its **proper length** $L_p$. In a reference frame in which the object is moving, the measured length is shorter than its proper length. Consider a rod at rest in frame $S'$ with one end at $x_2'$ and the other end at $x_1'$. The length of the rod in this frame is its proper length $L_p = x_2' - x_1'$. Some care must be taken to find the length of the rod in frame $S$. In this frame, the rod is moving to the right with speed $v$, the speed of frame $S'$. The length of the rod in frame $S$ is defined as $L = x_2 - x_1$, where $x_2$ is the position of one end at some time $t_2$, and $x_1$ is the position of the other end *at the same time* $t_1 = t_2$ as measured in frame $S$. Equation 39-11 is convenient to use to calculate $x_2 - x_1$ at some time $t$ because it relates $x$ and $x'$ to $t$, whereas Equation 39-9 is not convenient because it relates $x$ and $x'$ to $t'$:

$$x_2' = \gamma(x_2 - vt_2)$$

and

$$x_1' = \gamma(x_1 - vt_1)$$

Since $t_2 = t_1$, we obtain

$$x_2' - x_1' = \gamma(x_2 - x_1)$$

$$x_2 - x_1 = \frac{1}{\gamma}(x_2' - x_1') = (x_2' - x_1')\sqrt{1 - \frac{v^2}{c^2}}$$

or

$$L = \frac{1}{\gamma}L_\mathrm{p} = L_\mathrm{p}\sqrt{1 - \frac{v^2}{c^2}} \qquad\qquad 39\text{-}14$$

LENGTH CONTRACTION

Thus, the length of a rod is smaller when it is measured in a frame in which it is moving. Before Einstein's paper was published, Hendrik A. Lorentz and George F. FitzGerald tried to explain the null result of the Michelson–Morley experiment by assuming that distances in the direction of motion contracted by the amount given in Equation 39-14. This length contraction is now known as the **Lorentz–FitzGerald contraction.**

*THE LENGTH OF A MOVING METERSTICK*                **E X A M P L E   3 9 - 3**

**A stick that has a proper length of 1 m moves in a direction along its length with speed $v$ relative to you. The length of the stick as measured by you is 0.914 m. What is the speed $v$?**

**PICTURE THE PROBLEM** Since both $L$ and $L_\mathrm{p}$ are given, we can find $v$ directly from Equation 39-14.

1. Equation 39-14 relates the lengths $L$ and $L_\mathrm{p}$ and the speed $v$:

$$L = L_\mathrm{p}\sqrt{1 - \frac{v^2}{c^2}}$$

2. Solve for $v$:

$$v = c\sqrt{1 - \frac{L^2}{L_\mathrm{p}^2}} = c\sqrt{1 - \frac{(0.914\ \mathrm{m})^2}{(1\ \mathrm{m})^2}} = \boxed{0.406c}$$

An interesting example of time dilation or length contraction is afforded by the appearance of muons as secondary radiation from cosmic rays. Muons decay according to the statistical law of radioactivity:

$$N(t) = N_0 e^{-t/\tau} \qquad\qquad 39\text{-}15$$

where $N_0$ is the original number of muons at time $t = 0$, $N(t)$ is the number remaining at time $t$, and $\tau$ is the mean lifetime, which is approximately 2 $\mu$s for muons at rest. Since muons are created (from the decay of pions) high in the atmosphere, usually several thousand meters above sea level, few muons should reach sea level. A typical muon moving with speed $0.9978c$ would travel only about 600 m in 2 $\mu$s. However, the lifetime of the muon measured in the earth's reference frame is increased by the factor $1/\sqrt{1 - (v^2/c^2)}$, which is 15 for this particular speed. The mean lifetime measured in the earth's reference frame is therefore 30 $\mu$s, and a muon with speed $0.9978c$ travels approximately 9000 m in this time. From the muon's point of view, it lives only 2 $\mu$s, but the atmosphere is rushing past it with a speed of $0.9978c$. The distance of 9000 m in the earth's frame is thus contracted to only 600 m in the muon's frame, as indicated in Figure 39-4.

It is easy to distinguish experimentally between the classical and relativistic predictions of the observation of muons at sea level. Suppose that we observe $10^8$ muons at an altitude of 9000 m in some time interval with a muon detector. How many would we expect to observe at sea level in the same time interval? According to the nonrelativistic prediction, the time it takes for these muons to travel 9000 m is $(9000 \text{ m})/(0.998c) \approx 30 \text{ } \mu s$, which is 15 lifetimes. Substituting $N_0 = 10^8$ and $t = 15\tau$ into Equation 39-15, we obtain

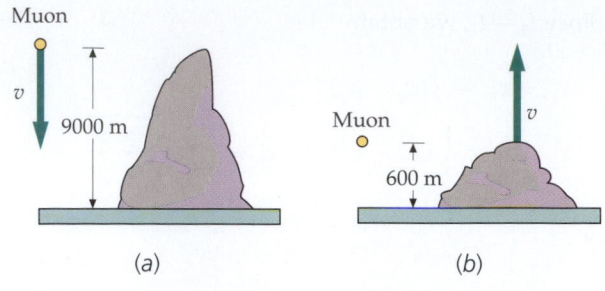

$$N = 10^8 e^{-15} = 30.6$$

We would thus expect all but about 31 of the original 100 million muons to decay before reaching sea level.

According to the relativistic prediction, the earth must travel only the contracted distance of 600 m in the rest frame of the muon. This takes only $2 \text{ } \mu s = 1\tau$. Therefore, the number of muons expected at sea level is

$$N = 10^8 e^{-1} = 3.68 \times 10^7$$

Thus, relativity predicts that we would observe 36.8 million muons in the same time interval. Experiments of this type have confirmed the relativistic predictions.

**FIGURE 39-4** Although muons are created high above the earth and their mean lifetime is only about 2 $\mu$s when at rest, many appear at the earth's surface. (*a*) In the earth's reference frame, a typical muon moving at 0.998*c* has a mean lifetime of 30 $\mu$s and travels 9000 m in this time. (*b*) In the reference frame of the muon, the distance traveled by the earth is only 600 m in the muon's lifetime of 2 $\mu$s.

## The Relativistic Doppler Effect

For light or other electromagnetic waves in a vacuum, a distinction between motion of source and receiver cannot be made. Therefore, the expressions we derived in Chapter 15 for the Doppler effect cannot be correct for light. The reason is that in that derivation, we assumed the time intervals in the reference frames of the source and receiver to be the same.

Consider a source moving toward a receiver with velocity $v$, relative to the receiver. If the source emits $N$ electromagnetic waves in a time $\Delta t_R$ (measured in the frame of the receiver), the first wave will travel a distance $c \Delta t_R$ and the source will travel a distance $v \Delta t_R$ measured in the frame of the receiver. The wavelength will be

$$\lambda' = \frac{c \Delta t_R - v \Delta t_R}{N}$$

The frequency $f'$ observed by the receiver will therefore be

$$f' = \frac{c}{\lambda'} = \frac{c}{c - v} \frac{N}{\Delta t_R} = \frac{1}{1 - (v/c)} \frac{N}{\Delta t_R}$$

If the frequency of the source is $f_0$, it will emit $N = f_0 \Delta t_S$ waves in the time $\Delta t_S$ measured by the source. Then

$$f' = \frac{1}{1 - (v/c)} \frac{N}{\Delta t_R} = \frac{1}{1 - (v/c)} \frac{f_0 \Delta t_S}{\Delta t_R} = \frac{f_0}{1 - (v/c)} \frac{\Delta t_S}{\Delta t_R}$$

Here $\Delta t_S$ is the proper time interval (the first wave and the $N$th wave are emitted at the same place in the source's reference frame). Times $\Delta t_S$ and $\Delta t_R$ are related by Equation 39-13 for time dilation:

$$\Delta t_R = \gamma \Delta t_S = \frac{\Delta t_S}{\sqrt{1 - (v^2/c^2)}}$$

Thus, when the source and the receiver are moving toward one another we obtain

$$f' = \frac{f_0}{1 - (v/c)}\frac{1}{\gamma} = \frac{\sqrt{1 - (v/c)^2}}{1 - (v/c)}f_0 = \sqrt{\frac{1 + (v/c)}{1 - (v/c)}}f_0, \quad \text{approaching} \quad 39\text{-}16a$$

This differs from our classical equation only in the time-dilation factor. It is left as a problem (Problem 27) for you to show that the same results are obtained if the calculations are done in the reference frame of the source.

When the source and the receiver are moving away from one another, the same analysis shows that the observed frequency is given by

$$f' = \frac{\sqrt{1 - (v/c)^2}}{1 + (v/c)}f_0 = \sqrt{\frac{1 - (v/c)}{1 + (v/c)}}f_0, \quad \text{receding} \qquad\qquad 39\text{-}16b$$

An application of the relativistic Doppler effect is the **redshift** observed in the light from distant galaxies. Because the galaxies are moving away from us, the light they emit is shifted toward the longer red wavelengths. The speed of the galaxies relative to us can be determined by measuring this shift.

---

*CONVINCING THE JUDGE*　　　　　　　　**EXAMPLE　39-4**　　**Put It in Context**

As part of a community volunteering option on your campus, you are spending the day shadowing two police officers. You have just had the excitement of pulling over a car that went through a red light. The driver claims that the red light looked green because the car was moving toward the stoplight, which shifted the wavelength of the observed light. You quickly do some calculations to see if the driver has a reasonable case or not.

**PICTURE THE PROBLEM** We can use the Doppler shift formula for approaching objects in Equation 39-16a. This will tell us the velocity, but we need to know the frequencies of the light. We can make good guesses for the wavelengths of red light and green light and use the definition of the speed of a wave $c = f\lambda$ to determine the frequencies.

1. The observer is approaching the light source, so we use the Doppler formula (Equation 39-16a) for approaching sources:

$$f' = \sqrt{\frac{1 + (v/c)}{1 - (v/c)}}f_0$$

2. Substitute $c/\lambda$ for $f$, then simplify:

$$\frac{c}{\lambda'} = \sqrt{\frac{1 + (v/c)}{1 - (v/c)}}\frac{c}{\lambda_0}$$

$$\left(\frac{\lambda_0}{\lambda'}\right)^2 = \frac{1 + (v/c)}{1 - (v/c)}$$

3. Cross multiply and solve for $v/c$:

$$(\lambda_0)^2\left(1 - \frac{v}{c}\right) = (\lambda')^2\left(1 + \frac{v}{c}\right)$$

$$(\lambda_0)^2 - (\lambda')^2 = \left[(\lambda_0)^2 + (\lambda')^2\right]\left(\frac{v}{c}\right)$$

$$\frac{v}{c} = \frac{(\lambda_0)^2 - (\lambda')^2}{(\lambda_0)^2 + (\lambda')^2} = \frac{1 - (\lambda'/\lambda_0)^2}{1 + (\lambda'/\lambda_0)^2}$$

4. The values for the wavelengths for the colors of the visible spectrum can be found in Table 30-1. The wavelengths for red are 725 nm or longer, and the wavelengths for green are 675 nm or shorter. Solve for the speed needed to shift the wavelength from 725 nm to 675 nm:

$$\frac{\lambda'}{\lambda_0} = \frac{675 \text{ nm}}{725 \text{ nm}} = 0.931$$

$$\frac{v}{c} = \frac{1 - 0.931^2}{1 + 0.931^2} = 0.0713$$

$$v = 0.0713c = 2.14 \times 10^7 \text{ m/s} = 4.79 \times 10^7 \text{ mi/h}$$

5. This speed is beyond any possible speed for a car:

The driver does not have a plausible case.

---

*FINDING SPEED FROM THE DOPPLER SHIFT*     **EXAMPLE 39-5**   **Try It Yourself**

The longest wavelength of light emitted by hydrogen in the Balmer series is $\lambda_0 = 656$ nm. In light from a distant galaxy, this wavelength is measured to be $\lambda' = 1458$ nm. Find the speed at which the distant galaxy is receding from the earth.

**Cover the column to the right and try these on your own before looking at the answers.**

**Steps**

**Answers**

1. Use Equation 39-16b to relate the speed $v$ to the received frequency $f'$ and the emitted frequency $f_0$.

$$f' = \sqrt{\frac{1 - (v/c)}{1 + (v/c)}} f_0$$

2. Substitute $f' = c/\lambda'$ and $f_0 = c/\lambda_0$ and solve for $v/c$.

$$\frac{v}{c} = \frac{1 - (\lambda_0/\lambda')^2}{1 + (\lambda_0/\lambda')^2} = 0.664$$

$$v = \boxed{0.664c}$$

---

# 39-4 Clock Synchronization and Simultaneity

We saw in Section 39-3 that proper time is the time interval between two events that occur at the same point in some reference frame. It can therefore be measured on a single clock. (Remember, in each frame there is a clock at each point in space, and the time of an event in a given frame is measured by the clock at that point.) However, in another reference frame moving relative to the first, the same two events occur at different places, so two clocks are needed to record the times. The time of each event is measured on a different clock, and the interval is found by subtraction. This procedure requires that the clocks be **synchronized.** We will show in this section that

Two clocks that are synchronized in one reference frame are typically not synchronized in any other frame moving relative to the first frame.

SYNCHRONIZED CLOCKS

Here is a corollary to this result:

Two events that are simultaneous in one reference frame typically are not simultaneous in another frame that is moving relative to the first.[†]

SIMULTANEOUS EVENTS

[†] This is true *unless* the $x$ coordinates of the two events are equal, where the $x$ axis is parallel with the relative velocity of the two frames.

Comprehension of these facts usually resolves all relativity paradoxes. Unfortunately, the intuitive (and incorrect) belief that simultaneity is an absolute relation is difficult to overcome.

Suppose we have two clocks at rest at point $A$ and point $B$ a distance $L$ apart in frame $S$. How can we synchronize these two clocks? If an observer at $A$ looks at the clock at $B$ and sets her clock to read the same time, the clocks will not be synchronized because of the time $L/c$ it takes light to travel from one clock to another. To synchronize the clocks, the observer at $A$ must set her clock ahead by the time $L/c$. Then she will see that the clock at $B$ reads a time that is $L/c$ behind the time on her clock, but she will calculate that the clocks are synchronized when she allows for the time $L/c$ for the light to reach her. Any other observers in $S$ (except those equidistant from the clocks) will see the clocks reading different times, but they will also calculate that the clocks are synchronized when they correct for the time it takes the light to reach them. An equivalent method for synchronizing two clocks would be for an observer $C$ at a point midway between the clocks to send a light signal and for the observers at $A$ and $B$ to set their clocks to some prearranged time when they receive the signal.

We now examine the question of **simultaneity.** Suppose $A$ and $B$ agree to explode flashguns at $t_0$ (having previously synchronized their clocks). Observer $C$ will see the light from the two flashes at the same time, and because he is equidistant from $A$ and $B$, he will conclude that the flashes were simultaneous. Other observers in frame $S$ will see the light from $A$ or $B$ first, depending on their location, but after correcting for the time the light takes to reach them, they also will conclude that the flashes were simultaneous. We can thus define simultaneity as follows:

> Two events in a reference frame are simultaneous if light signals from the events reach an observer halfway between the events at the same time.

DEFINITION—SIMULTANEITY

To show that two events that are simultaneous in frame $S$ are not simultaneous in another frame $S'$ moving relative to $S$, we will use an example introduced by Einstein. A train is moving with speed $v$ past a station platform. We will consider the train to be at rest in $S'$ and the platform to be at rest in $S$. We have observers $A'$, $B'$, and $C'$ at the front, back, and middle of the train. We now suppose that the train and platform are struck by lightning at the front and back of the train and that the lightning bolts are simultaneous in the frame of the platform $S$ (Figure 39-5). That is, an observer $C$ on the platform halfway between the positions $A$ and $B$, where the lightning strikes, sees the two flashes at the same time. It is convenient to suppose that the lightning scorches the train and platform so that the events can be easily located. Because $C'$ is in the middle of the train, halfway between the places on the train that are scorched, the events are simultaneous in $S'$ only if $C'$ sees the flashes at the same time. However, the flash from the front of the train is seen by $C'$ before the flash from the back of the

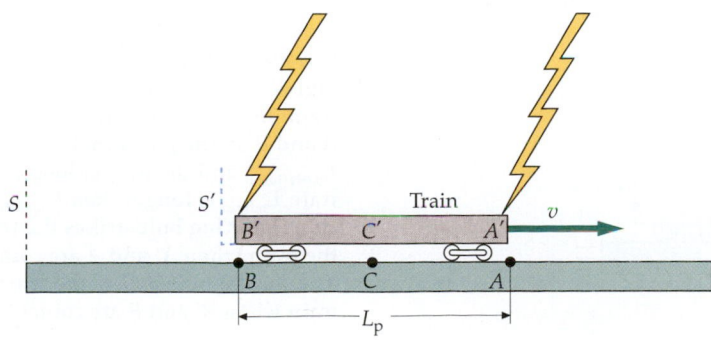

**FIGURE 39-5** In frame $S$ attached to the platform, simultaneous lightning bolts strike the ends of a train traveling with speed $v$. The light from these simultaneous events reaches observer $C$, standing midway between the events, at the same time. The distance between the bolts is $L_{p,\text{platform}}$.

train. We can understand this by considering the motion of $C'$ as seen in frame $S$ (Figure 39-6). By the time the light from the front flash reaches $C'$, $C'$ has moved some distance toward the front flash and some distance away from the back flash. Thus, the light from the back flash has not yet reached $C'$, as indicated in the figure. Observer $C'$ must therefore conclude that the events are not simultaneous and that the front of the train was struck before the back. Furthermore, all observers in $S'$ on the train will agree with $C'$ when they have corrected for the time it takes the light to reach them.

Figure 39-7 shows the events of the lightning bolts as seen in the reference frame of the train ($S'$). In this frame the platform is moving, so the distance between the burns on the platform is contracted. The platform is shorter than it is in $S$, and, since the train is at rest, the train is longer than its contracted length in $S$. When the lightning bolt strikes the front of the train at $A'$, the front of the train is at point $A$, and the back of the train has not yet reached point $B$. Later, when the lightning bolt strikes the back of the train at $B'$, the back has reached point $B$ on the platform.

The time discrepancy of two clocks that are synchronized in frame $S$ as seen in frame $S'$ can be found from the Lorentz transformation equations. Suppose we have clocks at points $x_1$ and $x_2$ that are synchronized in $S$. What are the times $t_1$ and $t_2$ on these clocks as observed from frame $S'$ at a time $t'_0$? From Equation 39-12, we have

$$t'_0 = \gamma\left(t_1 - \frac{vx_1}{c^2}\right)$$

and

$$t'_0 = \gamma\left(t_2 - \frac{vx_2}{c^2}\right)$$

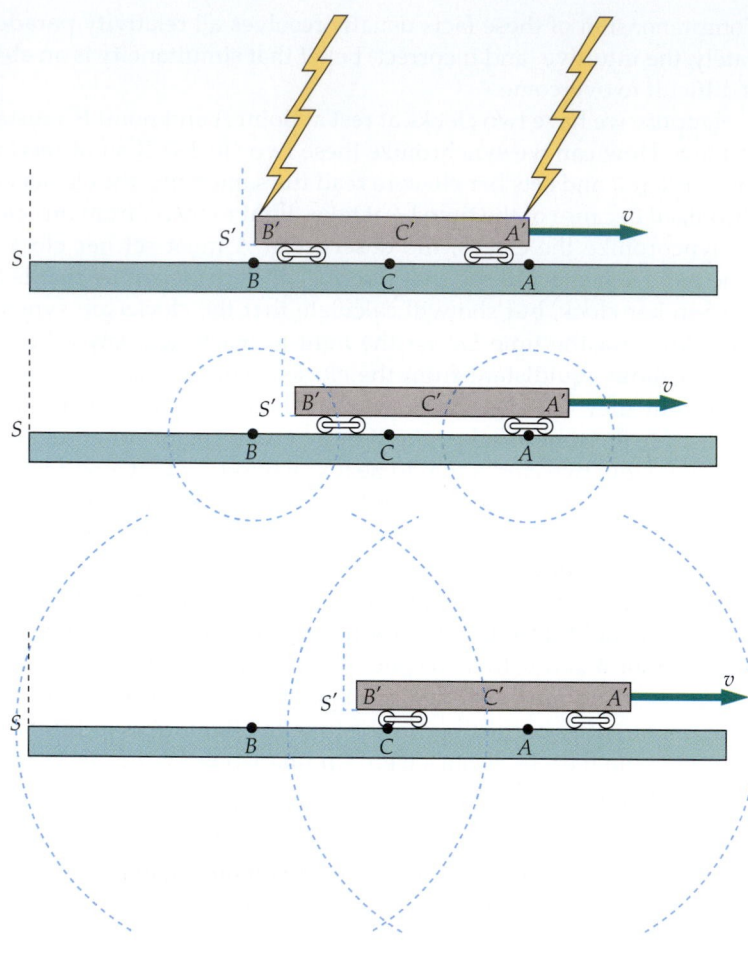

**FIGURE 39-6** In frame $S$ attached to the platform, the light from the lightning bolt at the front of the train reaches observer $C'$, standing on the train at its midpoint, before the light from the bolt at the back of the train. Since $C'$ is midway between the events (which occur at the front and rear of the train), these events are not simultaneous for him.

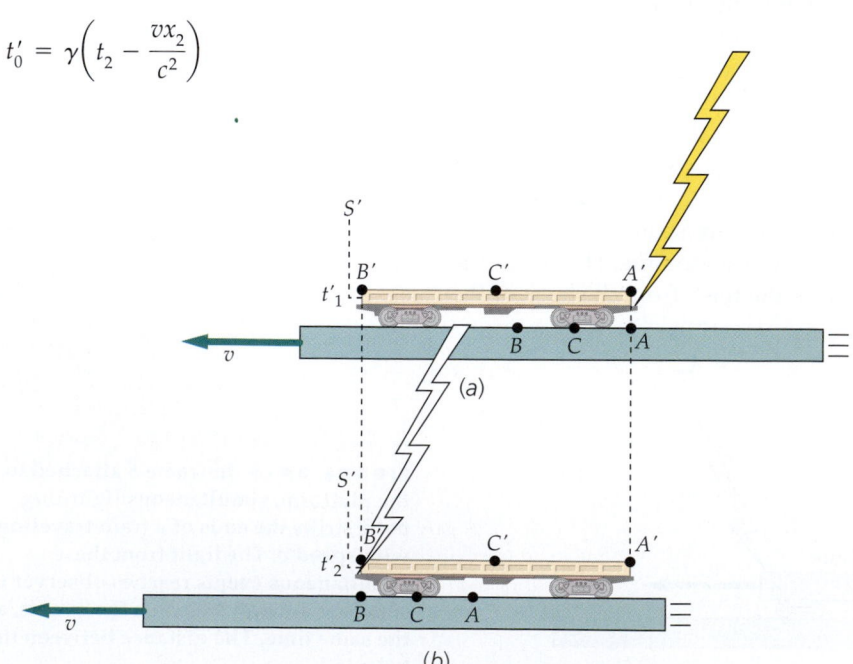

**FIGURE 39-7** The lightning bolts of Figure 39-5 as seen in frame $S'$ of the train. In this frame, the distance between $A$ and $B$ on the platform is less than $L_{p,platform}$, and the proper length of the train $L_{p,train}$ is longer than $L_{p,platform}$. The first lightning bolt strikes the front of the train when $A'$ and $A$ are coincident. The second bolt strikes the rear of the train when $B'$ and $B$ are coincident.

Then

$$t_2 - t_1 = \frac{v}{c^2}(x_2 - x_1)$$

Note that the chasing clock (at $x_2$) leads the other (at $x_1$) by an amount that is proportional to their proper separation $L_p = x_2 - x_1$.

If two clocks are synchronized in the frame in which they are both at rest, in a frame in which they are moving along the line through both clocks, the chasing clock leads (shows a later time) by an amount

$$\Delta t_S = L_p \frac{v}{c^2} \qquad\qquad 39\text{-}17$$

where $L_p$ is the proper distance between the clocks.

CHASING CLOCK SHOWS LATER TIME

A numerical example should help clarify time dilation, clock synchronization, and the internal consistency of these results.

---

*SYNCHRONIZING CLOCKS*                    **EXAMPLE  39-6**

An observer in a spaceship has a flashgun and a mirror, as shown in Figure 39-3. The distance from the gun to the mirror is 15 light-minutes (written $15c\cdot$min) and the spaceship, at rest in frame $S'$, travels with speed $v = 0.8c$ relative to a very long space platform that is at rest in frame $S$. The platform has two synchronized clocks, one clock at the position $x_1$ of the spaceship when the observer explodes the flashgun, and the other clock at the position $x_2$ of the spaceship when the light returns to the gun from the mirror. Find the time intervals between the events (exploding the flashgun and receiving the return flash from the mirror) (*a*) in the frame of the spaceship and (*b*) in the frame of the platform. (*c*) Find the distance traveled by the spaceship and (*d*) the amount by which the clocks on the platform are out of synchronization according to observers on the spaceship.

(*a*) 1. In the spaceship, the light travels from the gun to the mirror and back, a total distance $D = 30\ c\cdot$min. The time required is $D/c$:

$$\Delta t' = \frac{D}{c} = \frac{30\ c\cdot\text{min}}{c} = 30\ \text{min}$$

2. Since these events happen at the same place in the spaceship, the time interval is proper time:

$$\Delta t_p = \boxed{30\ \text{min}}$$

(*b*) 1. In frame $S$, the time between the events is longer by the factor $\gamma$:

$$\Delta t = \gamma\, \Delta t_p = \gamma(30\ \text{min})$$

2. Calculate $\gamma$:

$$\gamma = \frac{1}{\sqrt{1 - (v^2/c^2)}} = \frac{1}{\sqrt{1 - (0.8)^2}} = \frac{1}{\sqrt{0.36}} = \frac{5}{3}$$

3. Use this value of $\gamma$ to calculate the time between the events as observed in frame $S$:

$$\Delta t = \gamma\, \Delta t_p = \tfrac{5}{3}(30\ \text{min}) = \boxed{50\ \text{min}}$$

(*c*) In frame $S$, the distance traveled by the spaceship is $v\,\Delta t$:

$$x_2 - x_1 = v\,\Delta t = (0.8c)(50\ \text{min}) = \boxed{40\ c\cdot\text{min}}$$

(d) 1. The amount that the clocks on the platform are out of synchronization is related to the proper distance between the clocks $L_p$:

$$\Delta t_s = L_p \frac{v}{c^2}$$

2. The Part (c) result is the proper distance between the clocks on the platform:

$$L_p = x_2 - x_1 = 40\ c\cdot\text{min}$$

so

$$\Delta t_s = L_p \frac{v}{c^2} = (40\ c\cdot\text{min})\frac{(0.8c)}{c^2} = \boxed{32\ \text{min}}$$

**REMARKS** Observers on the platform would say that the spaceship's clock is running slow because it records a time of only 30 min between the events, whereas the time measured by observers on the platform is 50 min.

Figure 39-8 shows the situation viewed from the spaceship in $S'$. The platform is traveling past the ship with speed $0.8c$. There is a clock at point $x_1$, which coincides with the ship when the flashgun is exploded, and another at point $x_2$, which coincides with the ship when the return flash is received from the mirror. We assume that the clock at $x_1$ reads 12:00 noon at the time of the light flash. The clocks at $x_1$ and $x_2$ are synchronized in $S$ but not in $S'$. In $S'$, the clock at $x_2$, which is chasing the one at $x_1$, leads by 32 min; it would thus read 12:32 to an observer in $S'$. When the spaceship coincides with $x_2$, the clock there reads 12:50. The time between the events is therefore 50 min in $S$. Note that according to observers in $S'$, this clock ticks off 50 min − 32 min = 18 min for a trip that takes 30 min in $S'$. Thus, observers in $S'$ see this clock run slow by the factor 30/18 = 5/3.

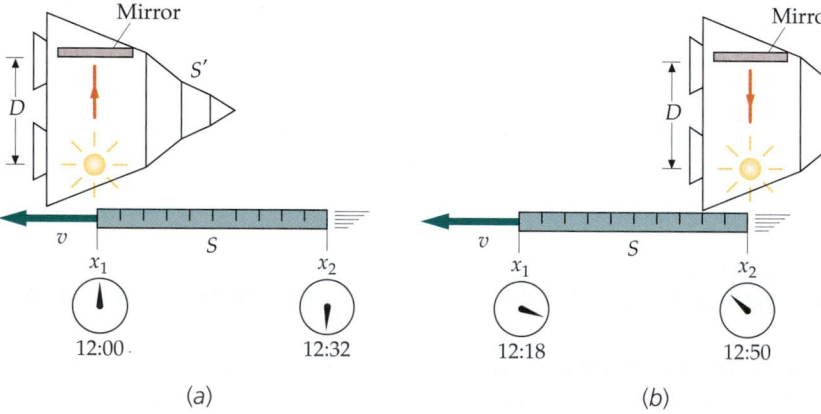

(a)          (b)

**FIGURE 39-8** Clocks on a platform as observed from the spaceship's frame of reference $S'$. During the time $\Delta t' = 30$ min it takes for the platform to pass the spaceship, the clocks on the platform run slow and tick off $(30\ \text{min})/\gamma = 18$ min. But the clocks are unsynchronized, with the chasing clock leading by $L_p v/c^2$, which for this case is 32 min. The time it takes for the spaceship to go from $x_1$ to $x_2$, as measured on the platform, is therefore 32 min + 18 min = 50 min.

Every observer in one frame sees the clocks in the other frame run slow. According to observers in $S$, who measure 50 min for the time interval, the time interval in $S'$ (30 min) is too small, so they see the single clock in $S'$ run too slow by the factor 5/3. According to the observers in $S'$, the observers in $S$ measure a time that is too *long* despite the fact that their clocks run too slow because the clocks in $S$ are out of synchronization. The clocks tick off only 18 min, but the second clock leads the first clock by 32 min, so the time interval is 50 min.

## The Twin Paradox

Homer and Ulysses are identical twins. Ulysses travels at high speed to a planet beyond the solar system and returns while Homer remains at home. When they are together again, which twin is older, or are they the same age? The correct answer is that Homer, the twin who stays at home, is older. This problem, with variations, has been the subject of spirited debate for decades, though there are very few who disagree with the answer. The problem appears to be a paradox because of the seemingly symmetric roles played by the twins with the asymmetric result in their aging. The paradox is resolved when the asymmetry of the twins' roles is noted. The relativistic result conflicts with common sense based on our strong but incorrect belief in absolute simultaneity. We will consider a particular case with some numerical magnitudes that, though impractical, make the calculations easy.

Let planet $P$ and Homer on the earth be at rest in reference frame $S$ a distance $L_p$ apart, as illustrated in Figure 39-9. We neglect the motion of the earth. Reference frames $S'$ and $S''$ are moving with speed $v$ toward and away from the planet, respectively. Ulysses quickly accelerates to speed $v$, then coasts in $S'$ until he reaches the planet, where he quickly decelerates to a stop and is momentarily at rest in $S$. To return, Ulysses quickly accelerates to speed $v$ toward the earth and then coasts in $S''$ until he reaches the earth, where he quickly decelerates to a stop. We can assume that the acceleration (and deceleration) times are negligible compared with the coasting times. We use the following values for illustration: $L_p$ = 8 light-years (8 $c \cdot$y) and $v$ = 0.8$c$. Then $\sqrt{1 - (v^2/c^2)}$ = 3/5 and $\gamma$ = 5/3.

It is easy to analyze the problem from Homer's point of view on the earth. According to Homer's clock, Ulysses coasts in $S'$ for a time $L_p/v$ = 10 y and in $S''$ for an equal time. Thus, Homer is 20 y older when Ulysses returns. The time interval in $S'$ between Ulysses's leaving the earth and his arriving at the planet is shorter because it is proper time. The time it takes to reach the planet by Ulysses's clock is

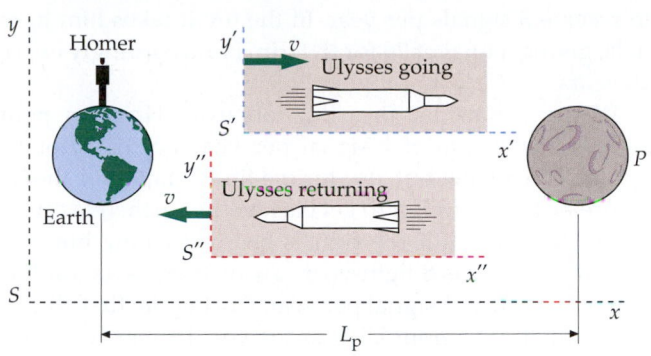

$$\Delta t' = \frac{\Delta t}{\gamma} = \frac{10 \text{ y}}{5/3} = 6 \text{ y}$$

Since the same time is required for the return trip, Ulysses will have recorded 12 y for the round trip and will be 8 y younger than Homer upon his return.

From Ulysses's point of view, the distance from the earth to the planet is contracted and is only

$$L' = \frac{L_p}{\gamma} = \frac{8 \text{ } c \cdot \text{y}}{5/3} = 4.8 \text{ } c \cdot \text{y}$$

At $v$ = 0.8$c$, it takes only 6 y each way.

The real difficulty in this problem is for Ulysses to understand why his twin aged 20 y during his absence. If we consider Ulysses as being at rest and Homer as moving away, Homer's clock should run slow and measure only 3/5(6 y) = 3.6 y. Then why shouldn't Homer age only 7.2 y during the round trip? This, of course, is the paradox. The difficulty with the analysis from the point of view of Ulysses is that he does not remain in an inertial frame. What happens while Ulysses is stopping and starting? To investigate this problem in detail, we would need to treat accelerated reference frames, a subject dealt with in the study of general relativity and beyond the scope of this book. However, we can get some insight into the problem by having the twins send regular signals to each other so that they can record the other's age continuously. If they arrange to send a signal once a year, each can determine the age of the other merely by counting the signals received. The arrival frequency of the signals will not be 1 per year because of the Doppler shift. The frequency observed will be given by Equation 39-16a and Equation 39-16b. Using $v/c$ = 0.8 and $v^2/c^2$ = 0.64, we have for the case in which the twins are receding from each other

$$f' = \frac{\sqrt{1 - (v^2/c^2)}}{1 + (v/c)} f_0 = \frac{\sqrt{1 - 0.64}}{1 + 0.8} f_0 = \frac{1}{3} f_0$$

When they are approaching, Equation 39-16a gives $f'$ = 3$f_0$.

Consider the situation first from the point of view of Ulysses. During the 6 y it takes him to reach the planet (remember that the distance is contracted in his frame), he receives signals at the rate of $\frac{1}{3}$ signal per year, and so he receives 2 signals. As soon as Ulysses turns around and starts back to the earth, he begins to receive 3 signals per year. In the 6 y it takes him to return he receives 18 signals, giving a total of 20 for the trip. He accordingly expects his twin to have aged 20 years.

We now consider the situation from Homer's point of view. He receives signals at the rate of $\frac{1}{3}$ signal per year not only for the 10 y it takes Ulysses to reach the planet but also for the time it takes for the last signal sent by Ulysses before he turns around to get back to the earth. (He cannot know that Ulysses has turned around until the signals begin reaching him with increased frequency.) Since the planet is 8 light-years away, there is an additional 8 y of receiving signals at the rate of $\frac{1}{3}$ signal per year. During the first 18 y, Homer receives 6 signals. In the final 2 y before Ulysses arrives, Homer receives 6 signals, or 3 per year. (The first signal sent after Ulysses turns around takes 8 y to reach the earth, whereas Ulysses, traveling at $0.8c$, takes 10 y to return and therefore arrives just 2 y after Homer begins to receive signals at the faster rate.) Thus, Homer expects Ulysses to have aged 12 y. In this analysis, the asymmetry of the twins' roles is apparent. When they are together again, both twins agree that the one who has been accelerated will be younger than the one who stayed home.

The predictions of the special theory of relativity concerning the twin paradox have been tested using small particles that can be accelerated to such large speeds that $\gamma$ is appreciably greater than 1. Unstable particles can be accelerated and trapped in circular orbits in a magnetic field, for example, and their lifetimes can then be compared with those of identical particles at rest. In all such experiments, the accelerated particles live longer on the average than the particles at rest, as predicted. These predictions have also been confirmed by the results of an experiment in which high-precision atomic clocks were flown around the world in commercial airplanes, but the analysis of this experiment is complicated due to the necessity of including gravitational effects treated in the general theory of relativity.

# 39-5 The Velocity Transformation

We can find how velocities transform from one reference frame to another by differentiating the Lorentz transformation equations. Suppose a particle has velocity $u'_x = dx'/dt'$ in frame $S'$, which is moving to the right with speed $v$ relative to frame $S$. The particle's velocity in frame $S$ is

$$u_x = \frac{dx}{dt}$$

From the Lorentz transformation equations (Equation 39-9 and Equation 39-10), we have

$$dx = \gamma(dx' + v\,dt')$$

and

$$dt = \gamma\left(dt' + \frac{v\,dx'}{c^2}\right)$$

The velocity in $S$ is thus

$$u_x = \frac{dx}{dt} = \frac{\gamma(dx' + v\,dt')}{\gamma\left(dt' + \dfrac{v\,dx'}{c^2}\right)} = \frac{\dfrac{dx'}{dt'} + v}{1 + \dfrac{v}{c^2}\dfrac{dx'}{dt'}} = \frac{u'_x + v}{1 + \dfrac{v\,u'_x}{c^2}}$$

If a particle has components of velocity along the $y$ or $z$ axes, we can use the same relation between $dt$ and $dt'$, with $dy = dy'$ and $dz = dz'$, to obtain

$$u_y = \frac{dy}{dt} = \frac{dy'}{\gamma\left(dt' + \dfrac{v\,dx'}{c^2}\right)} = \frac{\dfrac{dy'}{dt'}}{\gamma\left(1 + \dfrac{v}{c^2}\dfrac{dx'}{dt'}\right)} = \frac{u'_y}{\gamma\left(1 + \dfrac{vu'_x}{c^2}\right)}$$

and

$$u_z = \frac{u'_z}{\gamma\left(1 + \dfrac{vu'_x}{c^2}\right)}$$

The complete relativistic velocity transformation is

$$u_x = \frac{u'_x + v}{1 + \dfrac{vu'_x}{c^2}} \qquad\qquad 39\text{-}18a$$

$$u_y = \frac{u'_y}{\gamma\left(1 + \dfrac{vu'_x}{c^2}\right)} \qquad\qquad 39\text{-}18b$$

$$u_z = \frac{u'_z}{\gamma\left(1 + \dfrac{vu'_x}{c^2}\right)} \qquad\qquad 39\text{-}18c$$

RELATIVISTIC VELOCITY TRANSFORMATION

The inverse velocity transformation equations are

$$u'_x = \frac{u_x - v}{1 - \dfrac{vu_x}{c^2}} \qquad\qquad 39\text{-}19a$$

$$u'_y = \frac{u_y}{\gamma\left(1 - \dfrac{vu_x}{c^2}\right)} \qquad\qquad 39\text{-}19b$$

$$u'_z = \frac{u_z}{\gamma\left(1 - \dfrac{vu_x}{c^2}\right)} \qquad\qquad 39\text{-}19c$$

These equations differ from the classical and intuitive result $u_x = u'_x + v$, $u_y = u'_y$, and $u_z = u'_z$ because the denominators in the equations are not equal to 1. When $v$ and $u'_x$ are small compared with the speed of light $c$, $\gamma \approx 1$ and $vu'_x/c^2 \ll 1$. Then the relativistic and classical expressions are the same.

*RELATIVE VELOCITY AT NONRELATIVISTIC SPEEDS*          **E X A M P L E   3 9 - 7**

**A supersonic plane moves away from you along the $x$ axis with speed 1000 m/s (about 3 times the speed of sound) relative to you. A second plane moves along the $x$ axis away from you, and away from the first plane, at speed 500 m/s relative to the first plane. How fast is the second plane moving relative to you?**

**PICTURE THE PROBLEM** These speeds are so small compared with $c$ that we expect the classical equations for combining velocities to be accurate. We show this by calculating the correction term in the denominator of Equation 39-18a. Let frame $S$ be your rest frame and frame $S'$ be moving with velocity $v = 1000$ m/s. The first plane is then at rest in frame $S'$, and the second plane has velocity $u_x' = 500$ m/s in $S'$.

1. Let $S$ and $S'$ be the reference frames of you and the first plane, respectively. Also, let $u_x$ and $u_x'$ be the velocities of the second plane relative to $S$ and $S'$, respectively. Equation 39-18a can be used to find $u_x$. The velocity of the second plane relative to you is $v$:

$$u_x = \frac{u_x' + v}{1 + \frac{vu_x'}{c^2}}$$

2. If the correction term in the denominator is neglgible, Equation 39-18a gives the classical formula for combining velocities. Calculate the value of this correction term:

$$\frac{vu_x'}{c^2} = \frac{(1000)(500)}{(3 \times 10^8)^2} \approx 5.6 \times 10^{-12}$$

3. This correction term is so small that the classical and relativistic results are essentially the same:

$$u_x \approx u_x' + v$$

$$= 500 \text{ m/s} + 1000 \text{ m/s} = \boxed{1500 \text{ m/s}}$$

---

*RELATIVE VELOCITY AT RELATIVISTIC SPEEDS*      **EXAMPLE 39-8**

**Work Example 39-7 if the first plane moves with speed $v = 0.8c$ relative to you and the second plane moves with the same speed $0.8c$ relative to the first plane.**

**PICTURE THE PROBLEM** These speeds are not small compared with $c$, so we use the relativistic expression (Equation 39-18a). We again assume that you are at rest in frame $S$ and the first plane is at rest in frame $S'$ that is moving at $v = 0.8c$ relative to you. The velocity of the second plane in $S'$ is $u_x' = 0.8c$.

Use Equation 39-18a to calculate the speed of the second plane relative to you:

$$u_x = \frac{u_x' + v}{1 + \frac{vu_x'}{c^2}} = \frac{0.8c + 0.8c}{1 + \frac{(0.8c)(0.8c)}{c^2}} = \frac{1.6c}{1.64} = \boxed{0.98c}$$

---

The result in Example 39-8 is quite different from the classically expected result of $0.8c + 0.8c = 1.6c$. In fact, it can be shown from Equations 39-18 that if the speed of an object is less than $c$ in one frame, it is less than $c$ in all other frames moving relative to that frame with a speed less than $c$. (See Problem 23.) We will see in Section 39-7 that it takes an infinite amount of energy to accelerate a particle to the speed of light. The speed of light $c$ is thus an upper, unattainable limit for the speed of a particle with mass. (There are massless particles, such as photons, that always move at the speed of light.)

---

*RELATIVE SPEED OF A PHOTON*      **EXAMPLE 39-9**

**A photon moves along the $x$ axis in frame $S'$, with speed $u_x' = c$. What is its speed in frame $S$?**

The speed in $S$ is given by Equation 39-18a:

$$u_x = \frac{u_x' + v}{1 + \frac{vu_x'}{c^2}} = \frac{c + v}{1 + \frac{vc}{c^2}} = \frac{c + v}{1 + \frac{v}{c}} = \frac{c + v}{\frac{1}{c}(c + v)} = \boxed{c}$$

**REMARKS** The speed in both frames is $c$, independent of $v$. This is in accord with Einstein's postulates.

---

**Two spaceships, each 100 m long when measured at rest, travel toward each other with speeds of 0.85$c$ relative to the earth. (a) How long is each spaceship as measured by someone on the earth? (b) How fast is each spaceship traveling as measured by an observer on one of the spaceships? (c) How long is one spaceship when measured by an observer on one of the spaceships? (d) At time $t = 0$ on the earth, the front ends of the ships are together as they just begin to pass each other. At what time on the earth are their back ends together?**

**PICTURE THE PROBLEM** (a) The length of each spaceship as measured on the earth is the contracted length $\sqrt{1 - (v_1^2/c^2)}\ L_p$ (Equation 39-14), where $v_1$ is the speed of either spaceship. To solve Part (b), let the earth be in frame $S$, and the spaceship on the left be in frame $S'$ moving with velocity $v = 0.85c$ relative to $S$. Then the spaceship on the right moves with velocity $u_x = -0.85c$, as shown in Figure 39-10. (c) The length of one spaceship as seen by the other is $\sqrt{1 - (v_2^2/c^2)}\ L_p$, where $v_2$ is the speed of one spaceship relative to the other.

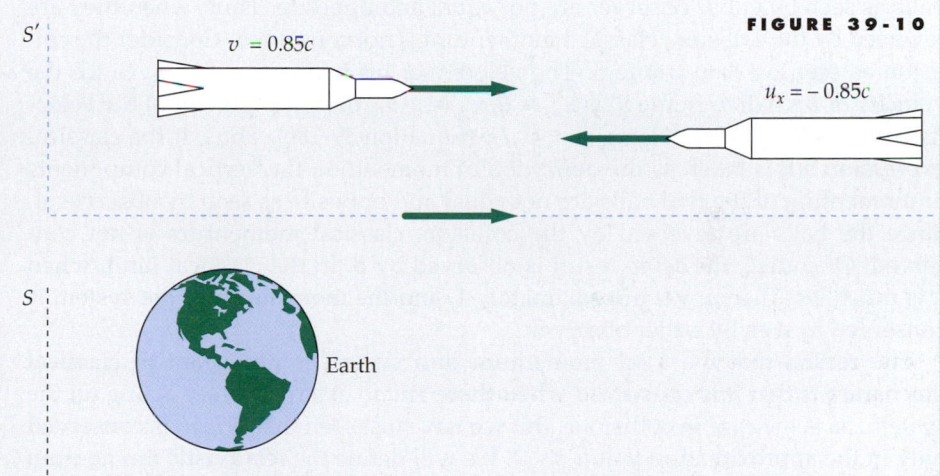

**FIGURE 39-10**

$v = 0.85c$

$u_x = -0.85c$

$S$

Earth

(a) The length of each spaceship in the earth's frame is the proper length divided by $\gamma$:

$$L = \sqrt{1 - \frac{v_1^2}{c^2}}\ L_p = \sqrt{1 - \frac{(0.85c)^2}{c^2}}\ (100\ \text{m}) = \boxed{52.7\ \text{m}}$$

(b) Use the velocity transformation formula (Equation 39-19a) to find the velocity $u'_x$ of the spaceship on the right as seen in frame $S'$:

$$u'_x = \frac{u_x - v}{1 - \dfrac{vu_x}{c^2}} = \frac{-0.85c - 0.85c}{1 - \dfrac{(0.85c)(-0.85c)}{c^2}} = \frac{-1.70c}{1.7225} = \boxed{-0.987c}$$

(c) In the frame of the left spaceship, the right spaceship is moving with speed $v_2 = |u'_x| = 0.987c$. Use this to calculate the contracted length of the spaceship on the right:

$$L = \sqrt{1 - \frac{v_1^2}{c^2}}\ L_p = \sqrt{1 - \frac{(0.987c)^2}{c^2}}\ (100\ \text{m}) = \boxed{16.1\ \text{m}}$$

(d) If the front ends of the spaceships are together at $t = 0$ on the earth, their back ends will be together after the time it takes either spaceship to move the length of the spaceship in the earth's frame:

$$t = \frac{L}{v_1} = \frac{52.7\ \text{m}}{0.85c} = \frac{52.7\ \text{m}}{(0.85)(3 \times 10^8\ \text{m/s})} = \boxed{2.07 \times 10^{-7}\ \text{s}}$$

---

# 39-6 Relativistic Momentum

We have seen in previous sections that Einstein's postulates require important modifications in our ideas of simultaneity and in our measurements of time and length. Einstein's postulates also require modifications in our concepts of mass, momentum, and energy. In classical mechanics, the momentum of a particle is defined as the product of its mass and its velocity, $m\vec{u}$, where $\vec{u}$ is the velocity. In an isolated system of particles, with no net force acting on the system, the total momentum of the system remains constant.

We can see from a simple thought experiment that the quantity $\Sigma m_i \vec{u}_i$ is not conserved in an isolated system. We consider two observers: observer $A$ in reference frame $S$ and observer $B$ in frame $S'$, which is moving to the right in the $x$ direction with speed $v$ with respect to frame $S$. Each has a ball of mass $m$. The two balls are identical when compared at rest. One observer throws his ball up with a speed $u_0$ relative to him and the other throws his ball down with a speed $u_0$ relative to him, so that each ball travels a distance $L$, makes an elastic collision with the other ball, and returns. Figure 39-11 shows how the collision looks in each reference frame. Classically, each ball has vertical momentum of magnitude $mu_0$. Since the vertical components of the momenta are equal and opposite, the total vertical component of momentum is zero before the collision. The collision merely reverses the momentum of each ball, so the total vertical momentum is zero after the collision.

Relativistically, however, the vertical components of the velocities of the two balls as seen by either observer are not equal and opposite. Thus, when they are reversed by the collision, classical momentum is not conserved. Consider the collision as seen by $A$ in frame $S$. The velocity of his ball is $u_{Ay} = +u_0$. Since the velocity of $B$'s ball in frame $S'$ is $u'_{Bx} = 0$, $u'_{By} = -u_0$, the $y$ component of the velocity of $B$'s ball in frame $S$ is $u_{By} = -u_0/\gamma$ (Equation 39-18b). Thus, if the classical expression $m\vec{u}$ is taken as the definition of momentum, the vertical components of momentum of the two balls are not equal and opposite as seen by observer $A$. Since the balls are reversed by the collision, classical momentum is not conserved. Of course, the same result is observed by $B$. In the classical limit, when $u$ is much less than $c$, $\gamma$ is approximately 1, and the momentum of the system is conserved as seen by either observer.

The reason that the total momentum of a system is important in classical mechanics is that it is conserved when there are no external forces acting on the system, as is the case in collisions. But we have just seen that $\Sigma\, m_i \vec{u}_i$ is conserved only in the approximation that $u \ll c$. We will define the relativistic momentum $\vec{p}$ of a particle to have the following properties:

1. In collisions, $\vec{p}$ is conserved.

2. As $u/c$ approaches zero, $\vec{p}$ approaches $m\vec{u}$.

We will show that the quantity

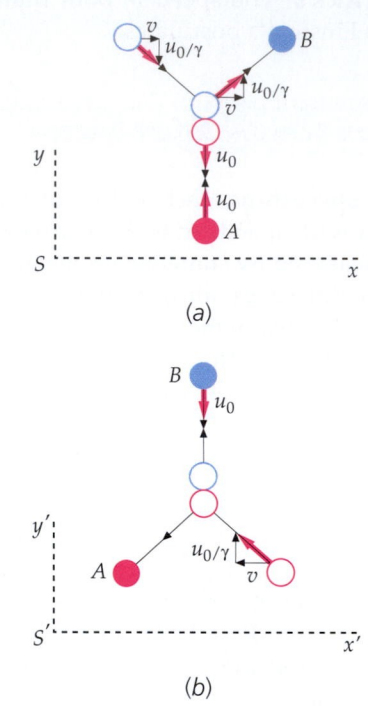

(a)

(b)

**FIGURE 39-11** (a) Elastic collision of two identical balls as seen in frame $S$. The vertical component of the velocity of ball $B$ is $u_0/\gamma$ in $S$ if it is $u_0$ in $S'$. (b) The same collision as seen in $S'$. In this frame, ball $A$ has a vertical component of velocity equal to $u_0/\gamma$.

$$\vec{p} = \frac{m\vec{u}}{\sqrt{1 - \dfrac{u^2}{c^2}}} \qquad\qquad 39\text{-}20$$

RELATIVISTIC MOMENTUM

is conserved in the elastic collision shown in Figure 39-11. Since this quantity also approaches $m\vec{u}$ as $u/c$ approaches zero, we take this equation for the definition of the **relativistic momentum** of a particle.

One interpretation of Equation 39-20 is that the mass of an object increases with speed. Then the quantity $m_{rel} = m/\sqrt{1 - (u^2/c^2)}$ is called the *relativistic mass*. The relativistic mass of a particle when it is at rest in some reference frame is then called its *rest mass* $m$. In this chapter, we will treat the terms mass and rest mass as synonymous, and both terms will be labeled $m$.

## Illustration of Conservation of the Relativistic Momentum

We will compute the $y$ component of the relativistic momentum of each particle in the reference frame $S$ for the collision of Figure 39-11 and show that the $y$ component of the total relativistic momentum is zero. The speed of ball $A$ in $S$ is $u_0$, so the $y$ component of its relativistic momentum is

$$p_{Ay} = \frac{mu_0}{\sqrt{1 - (u_0^2/c^2)}}$$

The speed of ball $B$ in $S$ is more complicated. Its $x$ component is $v$ and its $y$ component is $-u_0/\gamma$. Thus,

$$u_B^2 = u_{Bx}^2 + u_{By}^2 = v^2 + \left[-u_0\sqrt{1 - (v^2/c^2)}\right]^2 = v^2 + u_0^2 - \frac{u_0^2 v^2}{c^2}$$

Using this result to compute $\sqrt{1 - (u_B^2/c^2)}$, we obtain

$$1 - \frac{u_B^2}{c^2} = 1 - \frac{v^2}{c^2} - \frac{u_0^2}{c^2} + \frac{u_0^2 v^2}{c^4} = \left(1 - \frac{v^2}{c^2}\right)\left(1 - \frac{u_0^2}{c^2}\right)$$

and

$$\sqrt{1 - (u_B^2/c^2)} = \sqrt{1 - (v^2/c^2)}\,\sqrt{1 - (u_0^2/c^2)} = (1/\gamma)\,\sqrt{1 - (u_0^2/c^2)}$$

The $y$ component of the relativistic momentum of ball $B$ as seen in $S$ is therefore

$$p_{By} = \frac{mu_{By}}{\sqrt{1 - (u_B^2/c^2)}} = \frac{-mu_0/\gamma}{(1/\gamma)\,\sqrt{1 - (u_0^2/c^2)}} = \frac{-mu_0}{\sqrt{1 - (u_0^2/c^2)}}$$

Since $p_{By} = -p_{Ay}$, the $y$ component of the total momentum of the two balls is zero. If the speed of each ball is reversed by the collision, the total momentum will remain zero and momentum will be conserved.

## 39-7 Relativistic Energy

In classical mechanics, the work done by the net force acting on a particle equals the change in the kinetic energy of the particle. In relativistic mechanics, we equate the net force to the rate of change of the relativistic momentum. The work done by the net force can then be calculated and set equal to the change in kinetic energy.

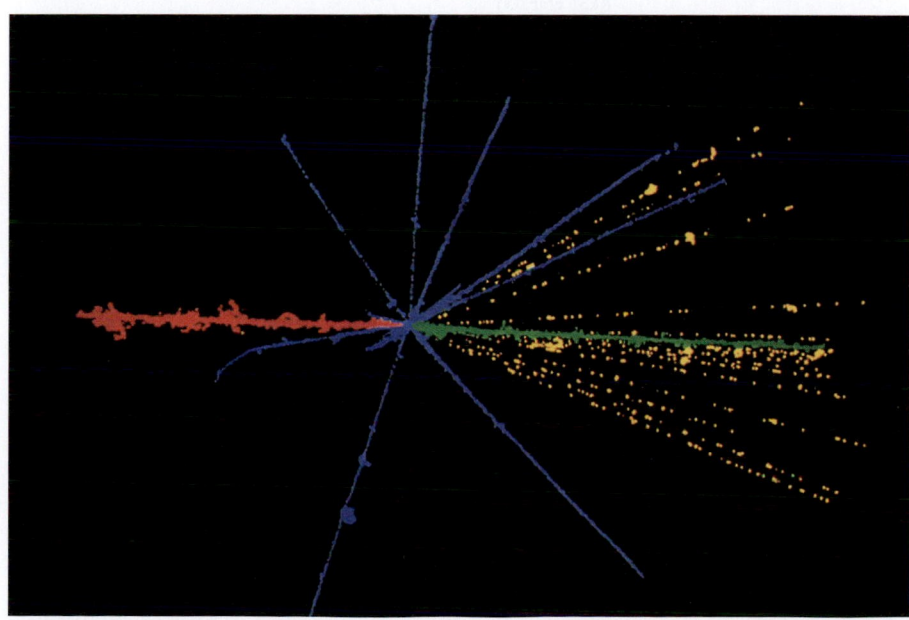

The creation of elementary particles demonstrates the conversion of kinetic energy to rest energy. In this 1950 photograph of a cosmic ray shower, a high-energy sulfur nucleus (red) collides with a nucleus in a photographic emulsion and produces a spray of particles, including a fluorine nucleus (green), other nuclear fragments (blue), and approximately 16 pions (yellow).

As in classical mechanics, we will define kinetic energy as the work done by the net force in accelerating a particle from rest to some final velocity $u_f$. Considering one dimension only, we have

$$K = \int_{u=0}^{u=u_f} F_{net}\, ds = \int_0^{u_f} \frac{dp}{dt}\, ds = \int_0^{u_f} u\, dp = \int_0^{u_f} u\, d\left(\frac{mu}{\sqrt{1-(u^2/c^2)}}\right) \qquad 39\text{-}21$$

where we have used $u = ds/dt$. It is left as a problem (Problem 37) for you to show that

$$d\left(\frac{mu}{\sqrt{1-(u^2/c^2)}}\right) = m\left(1 - \frac{u^2}{c^2}\right)^{-3/2} du$$

If we substitute this expression into the integrand in Equation 39-21, we obtain

$$K = \int_0^{u_f} u\, d\left(\frac{mu}{\sqrt{1-(u^2/c^2)}}\right) = \int_0^{u_f} m\left(1 - \frac{u^2}{c^2}\right)^{-3/2} u\, du$$

$$= mc^2\left(\frac{1}{\sqrt{1-(u_f^2/c^2)}} - 1\right)$$

or

$$K = \frac{mc^2}{\sqrt{1-(u^2/c^2)}} - mc^2 \qquad 39\text{-}22$$

RELATIVISTIC KINETIC ENERGY

(In this expression the final speed $u_f$ is arbitrary, so the subscript f is not needed.)

The expression for kinetic energy consists of two terms. The first term depends on the speed of the particle. The second, $mc^2$, is independent of the speed. The quantity $mc^2$ is called the **rest energy** $E_0$ of the particle. The rest energy is the product of the mass and $c^2$:

$$E_0 = mc^2 \qquad 39\text{-}23$$

REST ENERGY

The total **relativistic energy** $E$ is then defined to be the sum of the kinetic energy and the rest energy:

$$E = K + mc^2 = \frac{mc^2}{\sqrt{1-(u^2/c^2)}} \qquad 39\text{-}24$$

RELATIVISTIC ENERGY

Thus, the work done by an unbalanced force increases the energy from the rest energy $mc^2$ to the final energy $mc^2/\sqrt{1-(u^2/c^2)} = m_{rel}c^2$, where $m_{rel} = m/\sqrt{1-(u^2/c^2)}$ is the relativistic mass. We can obtain a useful expression for the velocity of a particle by multiplying Equation 39-20 for the relativistic momentum by $c^2$ and comparing the result with Equation 39-24 for the relativistic energy. We have

$$pc^2 = \frac{mc^2 u}{\sqrt{1-(u^2/c^2)}} = Eu$$

or

$$\frac{u}{c} = \frac{pc}{E}$$ 39-25

Energies in atomic and nuclear physics are usually expressed in units of electron volts (eV) or mega-electron volts (MeV):

$$1 \text{ eV} = 1.602 \times 10^{-19} \text{ J}$$

A convenient unit for the masses of atomic particles is $\text{eV}/c^2$ or $\text{MeV}/c^2$, which is the rest energy of the particle divided by $c^2$. The rest energies of some elementary particles and light nuclei are given in Table 39-1.

## TABLE 39-1

**Rest Energies of Some Elementary Particles and Light Nuclei**

| Particle | Symbol | Rest energy, MeV |
|----------|--------|------------------|
| Photon | $\gamma$ | 0 |
| Electron (positron) | $e$ or $e^-$ ($e^+$) | 0.5110 |
| Muon | $\mu^\pm$ | 105.7 |
| Pion | $\pi^0$ | 135 |
| | $\pi^\pm$ | 139.6 |
| Proton | p | 938.280 |
| Neutron | n | 939.573 |
| Deuteron | $^2$H or d | 1875.628 |
| Triton | $^3$H or t | 2808.944 |
| Helium-3 | $^3$He | 2808.41 |
| Alpha particle | $^4$He or $\alpha$ | 3727.409 |

---

*TOTAL ENERGY, KINETIC ENERGY, AND MOMENTUM*     **EXAMPLE 39-11**

An electron (rest energy 0.511 MeV) moves with speed $u = 0.8c$. Find (a) its total energy, (b) its kinetic energy, and (c) the magnitude of its momentum.

(a) The total energy is given by Equation 39-24:

$$E = \frac{mc^2}{\sqrt{1 - (u^2/c^2)}} = \frac{0.511 \text{ MeV}}{\sqrt{1 - 0.64}} = \frac{0.511 \text{ MeV}}{0.6} = \boxed{0.852 \text{ MeV}}$$

(b) The kinetic energy is the total energy minus the rest energy:

$$K = E - mc^2 = 0.852 \text{ MeV} - 0.511 \text{ MeV} = \boxed{0.341 \text{MeV}}$$

(c) The magnitude of the momentum is found from Equation 39-20. We can simplify by multiplying both numerator and denominator by $c^2$ and using the Part (a) result:

$$p = \frac{mu}{\sqrt{1 - (u^2/c^2)}}$$

$$= \frac{mc^2}{\sqrt{1 - (u^2/c^2)}} \frac{u}{c^2} = (0.852 \text{ MeV}) \frac{0.8c}{c^2} = \boxed{0.681 \text{ MeV}/c}$$

**REMARKS** The technique used to solve Part (c) (multiplying numerator and denominator by $c^2$) is equivalent to using Equation 39-25.

The expression for kinetic energy given by Equation 39-22 does not look much like the classical expression $\frac{1}{2}mu^2$. However, when $u$ is much less than $c$, we can approximate $1/\sqrt{1 - (u^2/c^2)}$ using the binomial expansion

$$(1 + x)^n = 1 + nx + n(n - 1)\frac{x^2}{2} + \cdots \approx 1 + nx \qquad\qquad 39\text{-}26$$

Then

$$\frac{1}{\sqrt{1 - (u^2/c^2)}} = \left(1 - \frac{u^2}{c^2}\right)^{-1/2} \approx 1 + \frac{1}{2}\frac{u^2}{c^2}$$

From this result, when $u$ is much less than $c$, the expression for relativistic kinetic energy becomes

$$K = mc^2\left[\frac{1}{\sqrt{1 - (u^2/c^2)}} - 1\right] \approx mc^2\left(1 + \frac{1}{2}\frac{u^2}{c^2} - 1\right) = \frac{1}{2}mu^2$$

Thus, at low speeds, the relativistic expression is the same as the classical expression.

We note from Equation 39-24 that as the speed $u$ approaches the speed of light $c$, the energy of the particle becomes very large because $1/\sqrt{1 - (u^2/c^2)}$ becomes very large. At $u = c$, the energy becomes infinite. For $u$ greater than $c$, $\sqrt{1 - (u^2/c^2)}$ is the square root of a negative number and is therefore imaginary. A simple interpretation of the result that it takes an infinite amount of energy to accelerate a particle to the speed of light is that no particle that is ever at rest in any inertial reference frame can travel as fast or faster than the speed of light $c$. As we noted in Example 39-8, if the speed of a particle is less than $c$ in one reference frame, it is less than $c$ in all other reference frames moving relative to that frame at speeds less than $c$.

In practical applications, the momentum or energy of a particle is often known rather than the speed. Equation 39-20 for the relativistic momentum and Equation 39-24 for the relativistic energy can be combined to eliminate the speed $u$. The result is

$$E^2 = p^2c^2 + (mc^2)^2 \qquad\qquad 39\text{-}27$$

RELATION FOR TOTAL ENERGY, MOMENTUM, AND REST ENERGY

This useful equation can be conveniently remembered from the right triangle shown in Figure 39-12. If the energy of a particle is much greater than its rest energy $mc^2$, the second term on the right side of Equation 39-27 can be neglected, giving the useful approximation

$$E \approx pc, \qquad \text{for } E \gg mc^2 \qquad\qquad 39\text{-}28$$

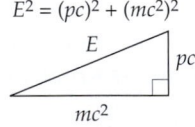

$E^2 = (pc)^2 + (mc^2)^2$

**FIGURE 39-12** Right triangle to remember Equation 39-27.

Equation 39-28 is an exact relation between energy and momentum for particles with no mass, such as photons.

**EXERCISE** A proton (mass 938 MeV/$c^2$) has a total energy of 1400 MeV. Find (a) $1/\sqrt{1 - (u^2/c^2)}$, (b) the momentum of the proton, and (c) the speed $u$ of the proton. (*Answer* (a) 1.49, (b) $p = 1.04 \times 10^3$ MeV/$c$, and (c) $u = 0.74c$)

## Mass and Energy

Einstein considered Equation 39-23 relating the energy of a particle to its mass to be the most significant result of the theory of relativity. Energy and inertia, which were formerly two distinct concepts, are related through this famous equation. As discussed in Chapter 7, the conversion of rest energy to kinetic energy with a corresponding decrease in mass is a common occurrence in radioactive decay and nuclear reactions, including nuclear fission and nuclear fusion. We illustrated this in Section 7-3 with the deuteron, whose mass is 2.22 $MeV/c^2$ less than the mass of its parts, a proton and a neutron. When a neutron and a proton combine to form a deuteron, 2.22 MeV of energy is released. The breaking up of a deuteron into a neutron and a proton requires 2.22 MeV of energy input. The proton and the neutron are thus bound together in a deuteron by a binding energy of 2.22 MeV. Any stable composite particle, such as a deuteron or a helium nucleus (2 neutrons plus 2 protons), that is made up of other particles has a mass and rest energy that are less than the sum of the masses and rest energies of its parts. The difference in rest energy is the binding energy of the composite particle. The binding energies of atoms and molecules are of the order of a few electron volts, which leads to a negligible difference in mass between the composite particle and its parts. The binding energies of nuclei are of the order of several MeV, which leads to a noticeable difference in mass. Some very heavy nuclei, such as radium, are radioactive and decay into a lighter nucleus plus an alpha particle. In this case, the original nucleus has a rest energy greater than that of the decay particles. The excess energy appears as the kinetic energy of the decay products.

To further illustrate the interrelation of mass and energy, we consider a perfectly inelastic collision of two particles. Classically, kinetic energy is lost in such a collision. Relativistically, this loss in kinetic energy shows up as an increase in rest energy of the system; that is, the total energy of the system is conserved. Consider a particle of mass $m_1$ moving with initial speed $u_1$ that collides with a particle of mass $m_2$ moving with initial speed $u_2$. The particles collide and stick together, forming a particle of mass $M$ that moves with speed $u_f$, as shown in Figure 39-13. The initial total energy of particle 1 is

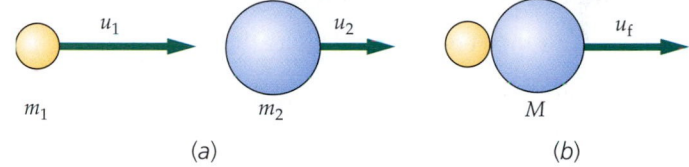

(a)          (b)

**FIGURE 39-13** A perfectly inelastic collision between two particles. One particle of mass $m_1$ collides with another particle of mass $m_2$. After the collision, the particles stick together, forming a composite particle of mass $M$ that moves with speed $u_f$ so that relativistic momentum is conserved. Kinetic energy is lost in this process. If we assume that the total energy is conserved, the loss in kinetic energy must equal $c^2$ times the increase in the mass of the system.

$$E_1 = K_1 + m_1 c^2$$

where $K_1$ is its initial kinetic energy. Similarly the initial total energy of particle 2 is

$$E_2 = K_2 + m_2 c^2$$

The total initial energy of the system is

$$E_i = E_1 + E_2 = K_1 + m_1 c^2 + K_2 + m_2 c^2 = K_i + M_i c^2$$

where $K_i = K_1 + K_2$ and $M_i = m_1 + m_2$ are the initial kinetic energy and initial mass of the system. The final total energy of the system is

$$E_f = K_f + M_f c^2$$

If we set the final total energy equal to the initial total energy, we obtain

$$K_f + M_f c^2 = K_i + M_i c^2$$

Rearranging gives $K_f - K_i = -(M_f - M_i)c^2$, which can be expressed

$$\Delta K + (\Delta M)c^2 = 0 \qquad\qquad 39\text{-}29$$

where $\Delta M = M_f - M_i$ is the change in mass of the system.

A particle of mass 2 MeV/$c^2$ and kinetic energy 3 MeV collides with a stationary particle of mass 4 MeV/$c^2$. After the collision, the two particles stick together. Find (*a*) the initial momentum of the system, (*b*) the final velocity of the two-particle system, and (*c*) the mass of the two-particle system.

**PICTURE THE PROBLEM** (*a*) The initial momentum of the system is the initial momentum of the incoming particle, which can be found from the total energy of the particle. (*b*) The final velocity of the system can be found from its total energy and momentum using $u/c = pc/E$ (Equation 39-25). The energy is found from conservation of energy, and the momentum from conservation of momentum. (*c*) Since the final energy and momentum are known, the final mass can be found from $E^2 = p^2c^2 + (Mc^2)^2$.

(*a*) 1. The initial momentum of the system is the initial momentum of the incoming particle. The momentum of a particle is related to its energy and mass (Equation 39-27):

$$E_1^2 = p_1^2c^2 + (m_1c^2)^2$$
$$p_1c = \sqrt{E_1^2 - (m_1c^2)^2}$$

2. The total energy of the moving particle is the sum of its kinetic energy and its rest energy:

$$E_1 = 3\,\text{MeV} + 2\,\text{MeV} = 5\,\text{MeV}$$

3. Use this total energy to calculate the momentum:

$$p_1c = \sqrt{E_1^2 - (m_1c^2)^2} = \sqrt{(5\,\text{MeV})^2 - (2\,\text{MeV})^2} = \sqrt{21}\,\text{MeV}$$

$$\boxed{p_1 = 4.58\,\text{MeV}/c}$$

(*b*) 1. We can find the final velocity of the system from its total energy $E_f$ and its momentum $p_f$ using Equation 39-25:

$$\frac{u_f}{c} = \frac{p_f c}{E_f}$$

2. By the conservation of total energy, the final energy of the system equals the initial total energy of the two particles:

$$E_f = E_i = E_1 + E_2 = 5\,\text{MeV} + 4\,\text{MeV} = 9\,\text{MeV}$$

3. By the conservation of momentum, the final momentum of the two-particle system equals the initial momentum:

$$p_f = 4.58\,\text{MeV}/c$$

4. Calculate the velocity of the two-particle system from its total energy and momentum using $u/c = pc/E$:

$$\frac{u_f}{c} = \frac{p_f c}{E_f} = \frac{4.58\,\text{MeV}}{9\,\text{MeV}} = 0.509$$

$$\boxed{u_f = 0.509c}$$

(*c*) We can find the mass $M_f$ of the final two-particle system from Equation 39-27 using $pc = 4.58$ MeV and $E = 9$ MeV:

$$E_f^2 = (p_f c)^2 + (M_f c^2)^2$$
$$(9\,\text{MeV})^2 = (4.58\,\text{MeV})^2 + (M_f c^2)^2$$

$$\boxed{M_f = 7.75\,\text{MeV}/c^2}$$

**REMARKS** Note that the mass of the system increased from 6 MeV/$c^2$ to 7.75 MeV/$c^2$. This increase times $c^2$ equals the loss in kinetic energy of the system, as you will show in the following exercise.

**EXERCISE** (*a*) Find the final kinetic energy of the two-particle system in Example 39-12. (*b*) Find the loss in kinetic energy, $K_{\text{loss}}$, in the collision. (*c*) Show that $K_{\text{loss}} = (\Delta M)c^2$, where $\Delta M$ is the change in mass of the system. [*Answer* (*a*) $K_f = E_f - M_f c^2 = 9\,\text{MeV} - 7.75\,\text{MeV} = 1.25\,\text{MeV}$, (*b*) $K_{\text{loss}} = K_i - K_f = 3\,\text{MeV} - 1.25\,\text{MeV} = 1.75\,\text{MeV}$, and (*c*) $(\Delta M)c^2 = (M_f - M_i)c^2 = 7.75\,\text{MeV} - (2\,\text{MeV} + 4\,\text{MeV}) = 1.75\,\text{MeV} = K_{\text{loss}}$]

| MOMENTUM AND TOTAL-ENERGY CONSERVATION | **EXAMPLE  39-13** |
| --- | --- |

A $1 \times 10^6$-kg rocket has $1 \times 10^3$ kg of fuel on board. The rocket is parked in space when it suddenly becomes necessary to accelerate. The rocket engines ignite, and the $1 \times 10^3$ kg of fuel are consumed. The exhaust (spent fuel) is ejected during a very short time interval at a speed of $c/2$ relative to $S$—the inertial reference frame in which the rocket is initially at rest. (*a*) Calculate the change in the mass of the rocket-fuel system. (*b*) Calculate the final speed of the rocket $u_R$ relative to $S$. (*c*) Again, calculate the final speed of the rocket relative to $S$, this time using classical (newtonian) mechanics.

**PICTURE THE PROBLEM**  The speed of the rocket and the change in the mass of the system can be calculated via conservation of momentum and conservation of energy. In reference frame $S$, the total momentum of the rocket plus fuel is zero. After the burn, the magnitude of the momentum of the rocket equals that of the ejected fuel. Let $m_R = 1 \times 10^6$ kg be the mass of the rocket, not including the mass of the fuel, let $m_{F,i} = 1 \times 10^3$ kg be the mass of the fuel *before* the burn, and let $m_{F,f}$ be the mass of the fuel *after* the burn. The mass of the rocket, $m_R$, remains fixed, but during the burn the mass of the fuel decreases. (The fuel has less chemical energy after the burn, and so has less mass as well.)

(*a*) 1. The magnitudes of the momentum of the rocket and the momentum of the ejected fuel are equal.  For the reasons stated above, the mass of the rocket, not including the $1 \times 10^3$ kg of fuel, does not change during the burn:

$$p_R = p_F$$

$$\frac{m_R u_R}{\sqrt{1 - (u_R^2/c^2)}} = \frac{m_{F,f} u_F}{\sqrt{1 - (u_F^2/c^2)}} = p$$

where

$p = p_R = p_F$, $m_R = 1 \times 10^6$ kg, $u_F = 0.5c$, and $u_R$ is the final speed of the rocket.

2. The total energy of the system does not change:

$$E_f = E_i$$

3. The initial energy is the rest energy of the rocket and fuel before the burn.  The final energy is the energy of the rocket plus energy of the fuel. The energy of each is related to its momentum by Equation 39-27:

$$E_i = m_R c^2 + m_{F,i} c^2 = (m_R + m_{F,i})c^2$$

$$E_{R,f}^2 = p^2 c^2 + (m_R c^2)^2$$

$$E_{F,f}^2 = p^2 c^2 + (m_{F,f} c^2)^2$$

so

$$E_f = E_{R,f} + E_{F,f}$$

$$E_f = \sqrt{p^2 c^2 + (m_R c^2)^2} + \sqrt{p^2 c^2 + (m_{F,f} c^2)^2}$$

4. Equate the initial and final energies:

$$\sqrt{p^2 c^2 + (m_R c^2)^2} + \sqrt{p^2 c^2 + (m_{F,f} c^2)^2} = (m_R + m_{F,i})c^2$$

5. The step 4 result and the step 1 result,

$$p = \frac{m_{F,f} u_F}{\sqrt{1 - (u_F^2/c^2)}}, \text{ constitute two simultaneous equa-}$$

tions with unknowns $p$ and $m_{F,f}$. Solving for $m_{F,f}$ gives:

$$m_{F,f} = 866 \text{ kg}$$

so

$$m_{loss} = m_{F,i} - m_{F,f} = 1000 \text{ kg} - 866 \text{ kg} = \boxed{134 \text{ kg}}$$

(*b*) 1. To solve for $u_R$, we use Equation 39-25:

$$\frac{u_R}{c} = \frac{pc}{E_{R,f}}$$

2. To solve for $p$, we substitute the value for $m_{F,f}$ into the Part (*a*), step 1 result:

$$p = \frac{m_{F,f} u_F}{\sqrt{1 - (u_F^2/c^2)}} = \frac{(866 \text{ kg})\frac{1}{2}c}{\sqrt{1 - \frac{1}{4}}} = (5.00 \times 10^2 \text{ kg})c$$

3. We use the value for $p$ to solve for $E_{R,f}$:

$$E_{R,f}^2 = p^2c^2 + (m_Rc^2)^2 = (5.00 \times 10^2 \text{ kg})^2c^4 + (10^6 \text{ kg})^2c^4 = (1.00 \times 10^{12} \text{ kg}^2)c^4$$

so

$$E_{R,f} = (1.00 \times 10^6 \text{ kg})c^2$$

4. Using our Part (b), step 1 result, we solve for $u_R$:

$$u_R = \frac{pc^2}{E_{R,f}} = \frac{(5.00 \times 10^2 \text{ kg})c^3}{(1.00 \times 10^6 \text{ kg})c^2} = \boxed{5.00 \times 10^{-4}c = 1.50 \times 10^5 \text{ m/s}}$$

(c) Equate the magnitude of the classical expressions for the momentum of the rocket and burned fuel and solve for $u_R$:

$$m_R u_R = m_F u_F$$

$$u_R = \frac{m_F}{m_R} u_F = \frac{10^3 \text{ kg}}{10^6 \text{ kg}} 0.5c = \boxed{1.5 \times 10^5 \text{ m/s}}$$

**REMARK** If carried out to five figures, the relativistic calculation gives $u_R = 4.9994 \times 10^4 \, c$ for the final speed of the rocket. However, the classical calculation gives $u_R = 5.0000 \times 10^4 \, c$. These two values differ by less than one part in 8000.

 **EXERCISE** If the matter being ejected were a $1 \times 10^3$-kg rigid block launched by a spring with one end attached to the rocket, would the rest mass of the block change or would the rest mass of the spring change? (*Answer* Only the rest mass of the spring would change.)

# 39-8 General Relativity

The generalization of the theory of relativity to noninertial reference frames by Einstein in 1916 is known as the general theory of relativity. It is much more difficult mathematically than the special theory of relativity, and there are fewer situations in which it can be tested. Nevertheless, its importance calls for a brief qualitative discussion.

The basis of the general theory of relativity is the **principle of equivalence:**

> A homogeneous gravitational field is completely equivalent to a uniformly accelerated reference frame.

PRINCIPLE OF EQUIVALENCE

This principle arises in Newtonian mechanics because of the apparent identity of gravitational mass and inertial mass. In a uniform gravitational field, all objects fall with the same acceleration $\vec{g}$ independent of their mass because the gravitational force is proportional to the (gravitational) mass, whereas the acceleration varies inversely with the (inertial) mass. Consider a compartment in space undergoing a uniform acceleration $\vec{a}$, as shown in Figure 39-14a. No mechanics experiment can be performed *inside* the compartment that will distinguish whether the compartment is actually accelerating in space or is at rest (or is moving with uniform velocity) in the presence of a uniform gravitational field $\vec{g} = -\vec{a}$, as shown in Figure 39-14b. If objects are dropped in the compartment, they will fall to the floor with an acceleration $\vec{g} = -\vec{a}$. If people stand on a spring scale, it will read their weight of magnitude $ma$.

Einstein assumed that the principle of equivalence applies to all physics and not just to mechanics. In effect, he assumed that there is no experiment of any kind that can distinguish uniformly accelerated motion from the presence of a gravitational field.

One consequence of the principle of equivalence—the deflection of a light beam in a gravitational field—was one of the first to be tested experimentally. In a region

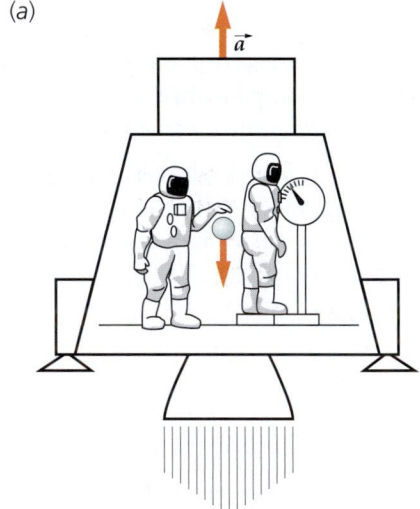

(a)

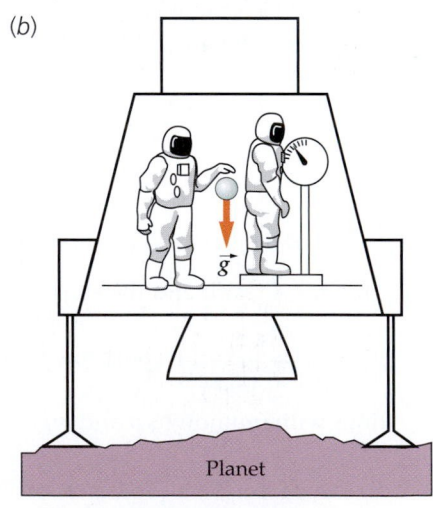

(b)

Planet

**FIGURE 39-14** The results of experiments in a uniformly accelerated reference frame (*a*) cannot be distinguished from those in a uniform gravitational field (*b*) if the acceleration $\vec{a}$ and the gravitational field $\vec{g}$ have the same magnitude.

with no gravitational field, a light beam will travel in a straight line at speed $c$. The principle of equivalence tells us that a region with no gravitational field exists only in a compartment that is in free fall. Figure 39-15 shows a beam of light entering a compartment that is accelerating relative to a nearby reference frame in free fall. Successive positions of the compartment at equal time intervals are shown in Figure 39-15$a$. Because the compartment is accelerating, the distance it moves in each time interval increases with time. The path of the beam of light as observed from inside the compartment is therefore a parabola, as shown in Figure 39-15$b$. But according to the principle of equivalence, there is no way to distinguish between an accelerating compartment and one moving with uniform velocity in a uniform gravitational field. We conclude, therefore, that a beam of light will accelerate in a gravitational field, just like objects that have mass. For example, near the surface of the earth, light will fall with an acceleration of $9.81 \text{ m/s}^2$. This is difficult to observe because of the enormous speed of light. In a distance of 3000 km, which takes light about 0.01 s to traverse, a beam of light should fall approximately 0.5 mm. Einstein pointed out that the deflection of a light beam in a gravitational field might be observed when light from a distant star passes close to the sun, as illustrated in Figure 39-16. Because of the brightness of the sun, this cannot ordinarily be seen. Such a deflection was first observed in 1919 during an eclipse of the sun. This well-publicized observation brought instant worldwide fame to Einstein.

A second prediction from Einstein's theory of general relativity, which we will not discuss in detail, is the excess precession of the perihelion of the orbit of Mercury of about 0.01° per century. This effect had been known and unexplained for some time, so, in a sense, explaining it constituted an immediate success of the theory.

A third prediction of general relativity concerns the change in time intervals and frequencies of light in a gravitational field. In Chapter 11, we found that the gravitational potential energy between two masses $M$ and $m$ a distance $r$ apart is

$$U = -\frac{GMm}{r}$$

where $G$ is the universal gravitational constant, and the point of zero potential energy has been chosen to be when the separation of the masses is infinite. The potential energy per unit mass near a mass $M$ is called the *gravitational potential* $\phi$:

$$\phi = -\frac{GM}{r}$$

39-30

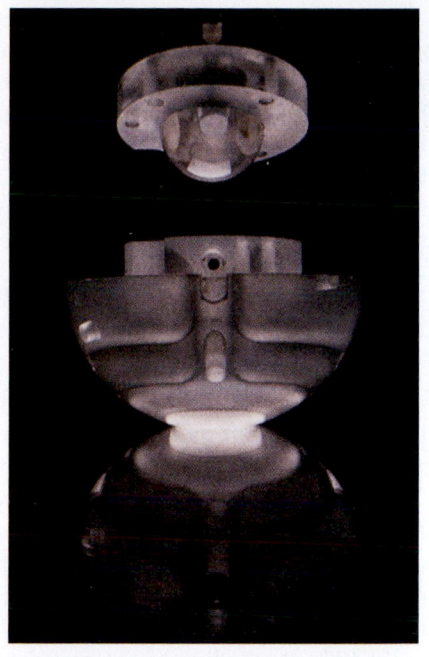

The quartz sphere in the top part of the container is probably the world's most perfectly round object. It is designed to spin as a gyroscope in a satellite orbiting the earth. General relativity predicts that the rotation of the earth will cause the axis of rotation of the gyroscope to precess in a circle at a rate of approximately 1 revolution in 100,000 years.

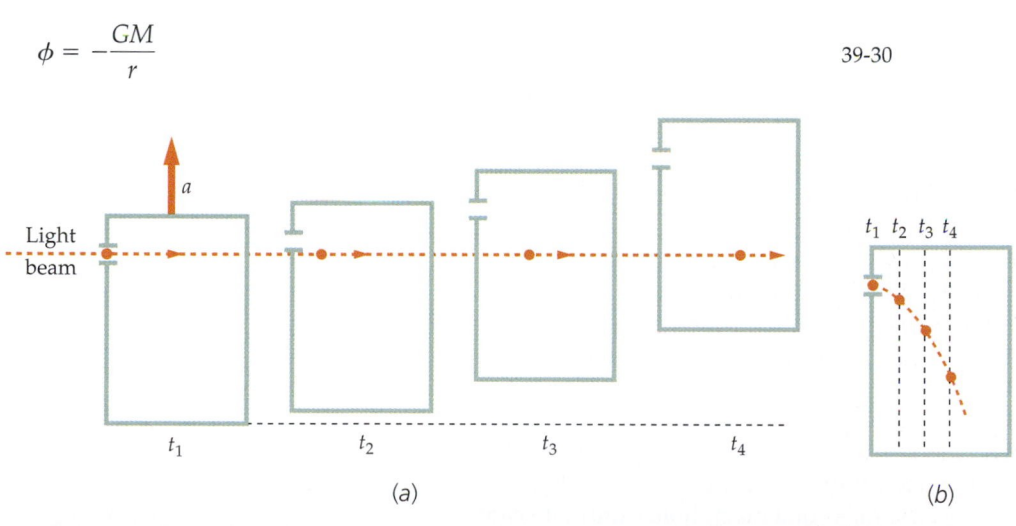

(a)

(b)

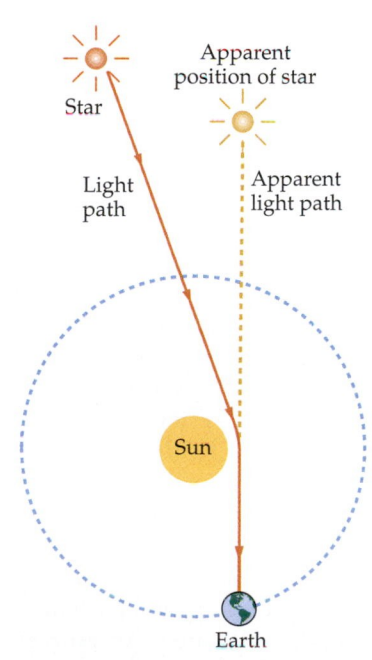

**FIGURE 39-15** (*a*) A light beam moving in a straight line through a compartment that is undergoing uniform acceleration relative to a nearby reference frame in free fall. The position of the beam is shown at equally spaced times $t_1$, $t_2$, $t_3$, and $t_4$. (*b*) In the reference frame of the compartment, the light travels in a parabolic path as a ball would if it were projected horizontally. The vertical displacements are greatly exaggerated in Figure 39-15$a$ and Figure 39-15$b$ for emphasis.

**FIGURE 39-16** The deflection (greatly exaggerated) of a beam of light due to the gravitational attraction of the sun.

According to the general theory of relativity, clocks run more slowly in regions of lower gravitational potential. (Since the gravitational potential is negative, as can be seen from Equation 39-30, the nearer the mass the more negative, and therefore the lower the gravitational potential.) If $\Delta t_1$ is a time interval between two events measured by a clock where the gravitational potential is $\phi_1$ and $\Delta t_2$ is the interval between the same events as measured by a clock where the gravitational potential is $\phi_2$, general relativity predicts that the fractional difference between these times will be approximately[†]

$$\frac{\Delta t_2 - \Delta t_1}{\Delta t} = \frac{1}{c^2}(\phi_2 - \phi_1)$$
39-31

A clock in a region of low gravitational potential will therefore run slower than a clock in a region of high potential. Since a vibrating atom can be considered to be a clock, the frequency of vibration of an atom in a region of low potential, such as near the sun, will be lower than the frequency of vibration of the same atom on the earth. This shift toward a lower frequency, and therefore a longer wavelength, is called the **gravitational redshift.**

As our final example of the predictions of general relativity, we mention **black holes,** which were first predicted by J. Robert Oppenheimer and Hartland Snyder in 1939. According to the general theory of relativity, if the density of an object such as a star is great enough, its gravitational attraction will be so great that once inside a critical radius, nothing can escape, not even light or other electromagnetic radiation. (The effect of a black hole on objects outside the critical radius is the same as that of any other mass.) A remarkable property of such an object is that nothing that happens inside it can be communicated to the outside. As sometimes occurs in physics, a simple but incorrect calculation gives the correct results for the relation between the mass and the critical radius of a black hole. In Newtonian mechanics, the speed needed for a particle to escape from the surface of a planet or a star of mass $M$ and radius $R$ is given by Equation 11-21:

$$v_e = \sqrt{\frac{2GM}{R}}$$

If we set the escape speed equal to the speed of light and solve for the radius, we obtain the critical radius $R_S$, called the **Schwarzschild radius:**

$$R_S = \frac{2GM}{c^2}$$
39-32

For an object with a mass equal to five times that of our sun (theoretically the minimum mass for a black hole) to be a black hole, its radius would have to be approximately 15 km. Since no radiation is emitted from a black hole and its radius is expected to be small, the detection of a black hole is not easy. The best chance of detection occurs if a black hole is a close companion to a normal star in a binary star system. Then both stars revolve around their center of mass and the gravitational field of the black hole will pull gas from the normal star into the black hole. However, to conserve angular momentum, the gas does not go straight into the black hole. Instead, the gas orbits around the black hole in a disk, called an accretion disk, while slowly being pulled closer to the black hole. The gas in this disk emits X rays because the temperature of the gas being pulled inward reaches several millions of kelvins. The mass of a black-hole candidate can often be estimated. An estimated mass of at least five solar masses, along with the emission of X rays, establishes a strong inference that the candidate is, in fact, a black hole. In addition to the black holes just described, there are supermassive black holes that exist at the centers of galaxies. At the center of the Milky Way is a supermassive black hole with a mass of about two million solar masses.

This extremely accurate hydrogen maser clock was launched in a satellite in 1976, and its time was compared to that of an identical clock on the earth. In accordance with the prediction of general relativity, the clock on the earth, where the gravitational potential was lower, *lost* approximately $4.3 \times 10^{-10}$ s each second compared with the clock orbiting the earth at an altitude of approximately 10,000 km.

---

[†] Since this shift is usually very small, it does not matter by which interval we divide on the left side of the equation.

| Topic | Relevant Equations and Remarks | |
|---|---|---|
| **1. Einstein's Postulates** | The special theory of relativity is based on two postulates of Albert Einstein. All of the results of special relativity can be derived from these postulates. | |
| | Postulate 1: Absolute uniform motion cannot be detected. | |
| | Postulate 2: The speed of light is independent of the motion of the source. | |
| | An important implication of these postulates is | |
| | Postulate 2 (alternate): Every observer measures the same value $c$ for the speed of light. | |
| **2. The Lorentz Transformation** | $x = \gamma(x' + vt'), \qquad y = y', \quad z = z'$ | 39-9 |
| | $t = \gamma\left(t' + \dfrac{vx'}{c^2}\right)$ | 39-10 |
| | $\gamma = \dfrac{1}{\sqrt{1 - (v^2/c^2)}}$ | 39-7 |
| Inverse transformation | $x' = \gamma(x - vt), \qquad y' = y, \quad z' = z$ | 39-11 |
| | $t' = \gamma\left(t - \dfrac{vx}{c^2}\right)$ | 39-12 |
| **3. Time Dilation** | The time interval measured between two events that occur at the same point in space in some reference frame is called the proper time $t_p$. In another reference frame in which the events occur at different places, the time interval between the events is longer by the factor $\gamma$. | |
| | $\Delta t = \gamma\,\Delta t_p$ | 39-13 |
| **4. Length Contraction** | The length of an object measured in the reference frame in which the object is at rest is called its proper length $L_p$. When measured in another reference frame, the length of the object is | |
| | $L = \dfrac{L_p}{\gamma}$ | 39-14 |
| **5. The Relativistic Doppler Effect** | $f' = \dfrac{\sqrt{1 - (v^2/c^2)}}{1 - (v/c)}\,f_0, \qquad$ approaching | 39-16a |
| | $f' = \dfrac{\sqrt{1 - (v^2/c^2)}}{1 + (v/c)}\,f_0, \qquad$ receding | 39-16b |
| **6. Clock Synchronization and Simultaneity** | Two events that are simultaneous in one reference frame typically are not simultaneous in another frame that is moving relative to the first. If two clocks are synchronized in the frame in which they are at rest, they will be out of synchronization in another frame. In the frame in which they are moving, the chasing clock leads by an amount | |
| | $\Delta t_S = L_p\dfrac{v}{c^2}$ | 39-17 |
| | where $L_p$ is the proper distance between the clocks. | |
| **7. The Velocity Transformation** | $u_x = \dfrac{u'_x + v}{1 + (vu'_x/c^2)}$ | 39-18a |

$$u_y = \frac{u'_y}{\gamma[1 + (vu'_x/c^2)]} \qquad\qquad 39\text{-}18b$$

$$u_z = \frac{u'_z}{\gamma[1 + (vu'_x/c^2)]} \qquad\qquad 39\text{-}18c$$

Inverse velocity transformation
$$u'_x = \frac{u_x - v}{1 - (vu_x/c^2)} \qquad\qquad 39\text{-}19a$$

$$u'_y = \frac{u_y}{\gamma[1 - (vu_x/c^2)]} \qquad\qquad 39\text{-}19b$$

$$u'_z = \frac{u_z}{\gamma[1 - (vu_x/c^2)]} \qquad\qquad 39\text{-}19c$$

**8. Relativistic Momentum**
$$\vec{p} = \frac{m\vec{u}}{\sqrt{1 - (u^2/c^2)}} \qquad\qquad 39\text{-}20$$

where $m$ is the mass of the particle.

**9. Relativistic Energy**

Kinetic energy
$$K = \frac{mc^2}{\sqrt{1 - (u^2/c^2)}} - mc^2 = \frac{mc^2}{\sqrt{1 - (u^2/c^2)}} - E_0 \qquad\qquad 39\text{-}22$$

Rest energy
$$E_0 = mc^2 \qquad\qquad 39\text{-}23$$

Total energy
$$E = K + E_0 = \frac{mc^2}{\sqrt{1 - (u^2/c^2)}} \qquad\qquad 39\text{-}24$$

**10. Useful Formulas for Speed, Energy, and Momentum**
$$\frac{u}{c} = \frac{pc}{E} \qquad\qquad 39\text{-}25$$

$$E^2 = p^2c^2 + (mc^2)^2 \qquad\qquad 39\text{-}27$$

$$E \approx pc, \qquad \text{for } E \gg mc^2 \qquad\qquad 39\text{-}28$$

# PROBLEMS

- Single-concept, single-step, relatively easy
- •• Intermediate-level, may require synthesis of concepts
- ••• Challenging
- **SSM** Solution is in the *Student Solutions Manual*
- **iSOLVE** Problems available on iSOLVE online homework service
- **iSOLVE✓** These "Checkpoint" online homework service problems ask students additional questions about their confidence level, and how they arrived at their answer.

In a few problems, you are given more data than you actually need; in a few other problems, you are required to supply data from your general knowledge, outside sources, or informed estimates.

## Conceptual Problems

**1** • **SSM** The approximate total energy of a particle of mass $m$ moving at speed $u \ll c$ is (a) $mc^2 + \frac{1}{2}mu^2$. (b) $\frac{1}{2}mu^2$. (c) $cmu$. (d) $mc^2$. (e) $\frac{1}{2}cmu$.

**2** • **SSM** A set of twins work in an office building. One twin works on the top floor and the other twin works in the basement. Considering general relativity, which twin will age more quickly? (a) They will age at the same rate. (b) The twin who works on the top floor will age more quickly. (c) The twin who works in the basement will age more quickly. (d) It depends on the speed of the office building. (e) None of these is correct.

**3** • True or false:

(a) The speed of light is the same in all reference frames.

(b) Proper time is the shortest time interval between two events.

(c) Absolute motion can be determined by means of length contraction.

(d) The light-year is a unit of distance.

(e) Simultaneous events must occur at the same place.

(f) If two events are not simultaneous in one frame, they cannot be simultaneous in any other frame.

(g) If two particles are tightly bound together by strong attractive forces, the mass of the system is less than the sum of the masses of the individual particles when separated.

**4** • An observer sees a system consisting of a mass oscillating on the end of a spring moving past at a speed $u$ and notes that the period of the system is $T$. Another observer, who is moving with the mass–spring system, also measures its period. The second observer will find a period that is (a) equal to $T$. (b) less than $T$. (c) greater than $T$. (d) either (a) or (b) depending on whether the system was approaching or receding from the first observer. (e) There is not sufficient information to answer the question.

**5** • The Lorentz transformation for $y$ and $z$ is the same as the classical result: $y = y'$ and $z = z'$. Yet the relativistic velocity transformation does not give the classical result $u_y = u'_y$ and $u_z = u'_z$. Explain.

## Estimation and Approximation

**6** •• The sun radiates energy at the rate of approximately $4 \times 10^{26}$ W. Assume that this energy is produced by a reaction whose net result is the fusion of 4 H nuclei to form 1 He nucleus, with the release of 25 MeV for each He nucleus formed. Calculate the sun's loss of mass per day.

**7** •• SSM The most distant galaxies that can be seen by the Hubble telescope are moving away from us with a redshift parameter of about $z = 5$. (See Problem 30 for a definition of $z$.) (a) What is the velocity of these galaxies relative to us (expressed as a fraction of the speed of light)? (b) *Hubble's law* states that the recession velocity is given by the expression $v = Hx$, where $v$ is the velocity of recession, $x$ is the distance, and $H$ is the Hubble constant, $H = 75$ km/s/Mpc. (1 pc = 3.26 $c \cdot$y.) Estimate the distance of such a galaxy using the information given.

## Time Dilation and Length Contraction

**8** • The proper mean lifetime of a muon is 2 $\mu$s. Muons in a beam are traveling through a laboratory at $0.95c$. (a) What is their mean lifetime as measured in the laboratory? (b) How far do they travel, on average, before they decay?

**9** •• In the Stanford linear collider, small bundles of electrons and positrons are fired at each other. In the laboratory's frame of reference, each bundle is approximately 1 cm long and 10 $\mu$m in diameter. In the collision region, each particle has an energy of 50 GeV, and the electrons and the positrons are moving in opposite directions. (a) How long and how wide is each bundle in its own reference frame? (b) What must be the minimum proper length of the accelerator for a bundle to have both its ends simultaneously in the accelerator in its own reference frame? (The ac-

tual length of the accelerator is less than 1000 m.) (c) What is the length of a positron bundle in the reference frame of the electron bundle?

**10** •• SSM Unobtainium (Un) is an unstable particle that decays into normalium (Nr) and standardium (St) particles. (a) An accelerator produces a beam of Un that travels to a detector located 100 m away from the accelerator. The particles travel with a velocity of $v = 0.866c$. How long do the particles take (in the laboratory frame) to get to the detector? (b) By the time the particles get to the detector, half of the particles have decayed. What is the half-life of Un? (*Note:* Half-life as it would be measured in a frame moving with the particles.) (c) A new detector is going to be used, which is located 1000 m away from the accelerator. How fast should the particles be moving if half of the particles are to make it to the new detector?

**11** •• Star A and Star B are at rest relative to the earth. Star A is 27 $c \cdot$y from earth, and Star B is located beyond (behind) Star A as viewed from earth. (a) A spaceship is making a trip from earth to Star A at a speed such that the trip from earth to Star A takes 12 y according to clocks on the spaceship. At what speed, relative to earth, must the ship travel? (Assume that the times for acceleration are very short compared to the overall trip time.) (b) Upon reaching Star A, the ship speeds up and departs for Star B at a speed such that the gamma factor, $\gamma$, is twice that of Part (a). The trip from Star A to Star B takes 5 y (ship's time). How far, in $c \cdot$y, is Star B from Star A in the rest frame of the earth and the two stars? (c) Upon reaching Star B, the ship departs for earth at the same speed as in Part (b). It takes it 10 y (ship's time) to return to earth. If you were born on earth the day the ship left earth (and you remain on earth), how old are you on the day the ship returns to earth?

**12** • A spaceship travels to a star 35 $c \cdot$y away at a speed of $2.7 \times 10^8$ m/s. How long does the spaceship take to get to the star (a) as measured on the earth and (b) as measured by a passenger on the spaceship?

**13** • Use the binomial expansion equation

$$(1 + x)^n = 1 + nx + \frac{n(n-1)}{2} x^2 + \ldots \approx 1 + nx, \quad \text{for } x \ll 1$$

to derive the following results for the case when $v$ is much less than $c$.

(a) $\gamma \approx 1 + \dfrac{1}{2}\dfrac{v^2}{c^2}$

(b) $\dfrac{1}{\gamma} \approx 1 - \dfrac{1}{2}\dfrac{v^2}{c^2}$

(c) $\gamma - 1 \approx 1 - \dfrac{1}{\gamma} \approx \dfrac{1}{2}\dfrac{v^2}{c^2}$

**14** •• A clock on Spaceship A measures the time interval between two events, both of which occur at the location of the clock. You are on Spaceship B. According to your careful measurements, the time interval between the two events is 1 percent longer than that measured by the two clocks on Spaceship A. How fast is Ship A moving relative to Ship B. (Use one or more of the results of Problem 13.)

**15** •• If a plane flies at a speed of 2000 km/h, how long must the plane fly before its clock loses 1 s because of time dilation? (Use one or more of the results of Problem 13.)

## The Lorentz Transformation, Clock Synchronization, and Simultaneity

**16** •• Show that when $v \ll c$ the transformation equations for $x$, $t$, and $u$ reduce to the Galilean equations.

**17** •• **SSM** **iSOLVE** A spaceship of proper length $L_p = 400$ m moves past a transmitting station at a speed of $0.76c$. At the instant that the nose of the spaceship passes the transmitter, clocks at the transmitter and in the nose of the spaceship are synchronized to $t = t' = 0$. The instant that the tail of the spaceship passes the transmitter a signal is sent and subsequently detected by the receiver in the nose of the spaceship. (a) When, according to the clock in the spaceship, is the signal sent? (b) When, according to the clock at the transmitter, is the signal received by the spaceship? (c) When, according to the clock in the spaceship, is the signal received? (d) Where, according to an observer at the transmitter, is the nose of the spaceship when the signal is received?

**18** •• In frame $S$, event $B$ occurs 2 $\mu$s after event $A$, which occurs at $x = 1.5$ km from event $A$. How fast must an observer be moving along the $+x$ axis so that events $A$ and $B$ occur simultaneously? Is it possible for event $B$ to precede event $A$ for some observer?

**19** •• Observers in reference frame $S$ see an explosion located at $x_1 = 480$ m. A second explosion occurs 5 $\mu$s later at $x_2 = 1200$ m. In reference frame $S'$, which is moving along the $+x$ axis at speed $v$, the explosions occur at the same point in space. What is the separation in time between the two explosions as measured in $S'$?

**20** ••• Two events in $S$ are separated by a distance $D = x_2 - x_1$ and a time $T = t_2 - t_1$. (a) Use the Lorentz transformation to show that in frame $S'$, which is moving with speed $v$ relative to $S$, the time separation is $t_2' - t_1' = \gamma(T - vD/c^2)$. (b) Show that the events can be simultaneous in frame $S'$ only if $D$ is greater than $cT$. (c) If one of the events is the *cause* of the other, the separation $D$ must be less than $cT$, since $D/c$ is the smallest time that a signal can take to travel from $x_1$ to $x_2$ in frame $S$. Show that if $D$ is less than $cT$, $t_2'$ is greater than $t_1'$ in all reference frames. This shows that if the cause precedes the effect in one frame, it must precede it in all reference frames. (d) Suppose that a signal could be sent with speed $c' > c$ so that in frame $S$ the cause precedes the effect by the time $T = D/c'$. Show that there is then a reference frame moving with speed $v$ less than $c$ in which the effect precedes the cause.

**21** ••• A rocket with a proper length of 700 m is moving to the right at a speed of $0.9c$. It has two clocks, one in the nose and one in the tail, that have been synchronized in the frame of the rocket. A clock on the ground and the nose clock on the rocket both read $t = 0$ as they pass. (a) At $t = 0$, what does the tail clock on the rocket read as seen by an observer on the ground? When the tail clock on the rocket passes the ground clock, (b) what does the tail clock read as seen by an observer on the ground, (c) what does the nose clock read as seen by an observer on the ground, and (d) what does the nose clock read as seen by an observer on the rocket? (e) At $t = 1$ h, as measured on the rocket, a light signal is sent from the nose of the rocket to an observer standing by the ground clock. What does the ground clock read when the observer receives this signal? (f) When the observer on the ground receives the signal, he sends a return signal to the nose of the rocket. When is this signal received at the nose of the rocket as seen on the rocket?

**22** ••• **SSM** An observer in frame $S$ standing at the origin observes two flashes of colored light separated spatially by $\Delta x = 2400$ m. A blue flash occurs first, followed by a red flash 5 $\mu$s later. An observer in $S'$ moving along the $x$ axis at speed $v$ relative to $S$ also observes the flashes 5 $\mu$s apart and with a separation of 2400 m, but the red flash is observed first. Find the magnitude and direction of $v$.

## The Velocity Transformation

**23** •• Show that if $u_x'$ and $v$ in Equation 39-18a are both positive and less than $c$, then $u_x$ is positive and less than $c$. [*Hint: Let $u_x' = (1 - \varepsilon_1)c$ and $v = (1 - \varepsilon_2)c$, where $\varepsilon_1$ and $\varepsilon_2$ are positive numbers that are less than 1.*]

**24** •• **SSM** A spaceship, at rest in a certain reference frame $S$, is given a speed increment of $0.50c$ (call this boost 1). Relative to its new rest frame, the spaceship is given a further $0.50c$ increment 10 seconds later (as measured in its new rest frame; call this boost 2). This process is continued indefinitely, at 10-s intervals, as measured in the rest frame of the ship. (Assume that the boost itself takes a very short time compared to 10 s.) (a) Using a spreadsheet program, calculate and graph the velocity of the spaceship in reference frame $S$ as a function of the boost number for boost 1 to boost 10. (b) Graph the gamma factor the same way. (c) How many boosts does it take until the velocity of the ship in $S$ is greater than $0.999c$? (d) How far has the spaceship moved after 5 boosts, as measured in reference frame $S$? What is the average speed of the spaceship (between boost 1 and boost 5) as measured in $S$?

## The Relativistic Doppler Effect

**25** • Sodium light of wavelength 589 nm is emitted by a source that is moving toward the earth with speed $v$. The wavelength measured in the frame of the earth is 547 nm. Find $v$.

**26** • A distant galaxy is moving away from us at a speed of $1.85 \times 10^7$ m/s. Calculate the fractional redshift $(\lambda' - \lambda_0)/\lambda_0$ in the light from this galaxy.

**27** •• Derive Equation 39-16a for the frequency received by an observer moving with speed $v$ toward a stationary source of electromagnetic waves.

**28** • Show that if $v$ is much less than $c$, the Doppler shift is given approximately by

$$\Delta f/f \approx \pm v/c$$

**29** •• **SSM** A clock is placed in a satellite that orbits the earth with a period of 90 min. By what time interval will this clock differ from an identical clock on the earth after 1 y? (Assume that special relativity applies and neglect general relativity.)

**30** •• For light that is Doppler-shifted with respect to an observer, define the redshift parameter

$$z = \frac{f - f'}{f'}$$

where $f$ is the frequency of the light measured in the rest frame of the emitter, and $f'$ is the frequency measured in the rest frame of the observer. If the emitter is moving directly away from the observer, show that the relative velocity between the emitter and the observer is

$$v = c\left(\frac{u^2 - 1}{u^2 + 1}\right)$$

where $u = z + 1$.

**31** • A light beam moves along the $y'$ axis with speed $c$ in frame $S'$, which is moving to the right with speed $v$ relative to frame $S$. (a) Find the $x$ and $y$ components of the velocity of the light beam in frame $S$. (b) Show that the magnitude of the velocity of the light beam in $S$ is $c$.

**32** • A spaceship is moving east at speed $0.90c$ relative to the earth. A second spaceship is moving west at speed $0.90c$ relative to the earth. What is the speed of one spaceship relative to the other spaceship?

**33** •• SSM A particle moves with speed $0.8c$ along the $x''$ axis of frame $S''$, which moves with speed $0.8c$ along the $x'$ axis relative to frame $S'$. Frame $S'$ moves with speed $0.8c$ along the $x$ axis relative to frame $S$. (a) Find the speed of the particle relative to frame $S'$. (b) Find the speed of the particle relative to frame $S$.

## Relativistic Momentum and Relativistic Energy

**34** • SSM A proton (rest energy 938 MeV) has a total energy of 2200 MeV. (a) What is its speed? (b) What is its momentum?

**35** • If the kinetic energy of a particle equals twice its rest energy, what percentage error is made by using $p = mu$ for its momentum?

**36** •• iSOLVE✔ A particle with momentum of 6 MeV/$c$ has total energy of 8 MeV. (a) Determine the mass of the particle. (b) What is the energy of the particle in a reference frame in which its momentum is 4 MeV/$c$? (c) What are the relative velocities of the two reference frames?

**37** •• Show that

$$d\left(\frac{mu}{\sqrt{1 - (u^2/c^2)}}\right) = m\left(1 - \frac{u^2}{c^2}\right)^{-3/2} du$$

**38** •• iSOLVE✔ The $K^0$ particle has a mass of 497.7 MeV/$c^2$. It decays into a $\pi^-$ and $\pi^+$, each with mass 139.6 MeV/$c^2$. Following the decay of a $K^0$, one of the pions is at rest in the laboratory. Determine the kinetic energy of the other pion and of the $K^0$ prior to the decay.

**39** •• SSM Two protons approach each other head-on at $0.5c$ relative to reference frame $S'$. (a) Calculate the total kinetic energy of the two protons as seen in frame $S'$. (b) Calculate the total kinetic energy of the protons as seen in reference frame $S$, which is moving with speed $0.5c$ relative to $S'$ so that one of the protons is at rest.

**40** •• An antiproton has the same rest energy as a proton. It is created in the reaction $p + p \rightarrow p + p + p + \bar{p}$. In an experiment, protons at rest in the laboratory are bombarded with protons of kinetic energy $K_L$, which must be great enough so that kinetic energy equal to $2mc^2$ can be converted into the rest energy of the two particles. In the frame of the laboratory, the total kinetic energy cannot be converted into rest energy because of conservation of momentum. However, in the zero-momentum reference frame in which the two initial protons are moving toward each other with equal speed $u$, the total kinetic energy can be converted into rest energy. (a) Find the speed of each proton $u$ so that the total kinetic energy in the zero-momentum frame is $2mc^2$. (b) Transform to the laboratory's frame in which one proton is at rest, and find the speed $u'$ of the other proton. (c) Show that the kinetic energy of the moving proton in the laboratory's frame is $K_L = 6mc^2$.

**41** ••• iSOLVE A particle of mass 1 MeV/$c^2$ and kinetic energy 2 MeV collides with a stationary particle of mass 2 MeV/$c^2$. After the collision, the particles stick together. Find (a) the speed of the first particle before the collision, (b) the total energy of the first particle before the collision, (c) the initial total momentum of the system, (d) the total kinetic energy after the collision, and (e) the mass of the system after the collision.

## General Relativity

**42** •• SSM Light traveling in the direction of increasing gravitational potential undergoes a frequency redshift. Calculate the shift in wavelength if a beam of light of wavelength $\lambda = 632.8$ nm is sent up a vertical shaft of height $L = 100$ m.

**43** •• Let us revisit a problem from Chapter 3: Two cannons are pointed directly toward each other, as shown in Figure 39-17. When fired, the cannonballs will follow the trajectories shown. Point $P$ is the point where the trajectories cross each other. Ignore the effects of air resistance. Using the principle of equivalence, show that if the cannons are fired simultaneously, the cannonballs will hit each other at point $P$.

**FIGURE 39-17** Problem 43

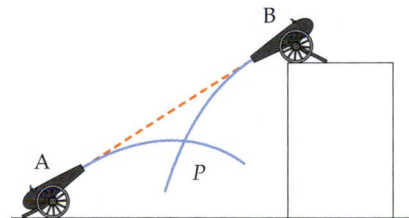

**44** ••• A horizontal turntable rotates with angular speed $\omega$. There is a clock at the center of the turntable and one at a distance $r$ from the center. In an inertial reference frame, the clock at distance $r$ is moving with speed $u = r\omega$. (a) Show that from time dilation according to special relativity, time intervals $\Delta t_0$ for the clock at rest and $\Delta t_r$ for the moving clock are related by

$$\frac{\Delta t_r - \Delta t_0}{\Delta t_0} = -\frac{r^2\omega^2}{2c^2}, \qquad \text{if } r\omega \ll c$$

(b) In a reference frame rotating with the table, both clocks are at rest. Show that the clock at distance $r$ experiences a pseudo-force $F_r = mr\omega^2$ in this accelerated frame and that this is equivalent to a difference in gravitational potential between $r$ and the origin of $\phi_r - \phi_0 = -\frac{1}{2}r^2\omega^2$. Use this potential difference given in Part (b) to show that in this frame the difference in time intervals is the same as in the inertial frame.

## General Problems

**45** • How fast must a muon travel so that its mean lifetime is 46 $\mu$s if its mean lifetime at rest is 2 $\mu$s?

**46** • **iSOLVE** A distant galaxy is moving away from the earth with a speed that results in each wavelength received on the earth being shifted so that $\lambda' = 2\lambda_0$. Find the speed of the galaxy relative to the earth.

**47** •• **SSM** Frames $S$ and $S'$ are moving relative to each other along the $x$ and $x'$ axes. Observers in the two frames set their clocks to $t = 0$ when the origins coincide. In frame $S$, event 1 occurs at $x_1 = 1.0\ c\cdot y$ and $t_1 = 1$ y and event 2 occurs at $x_2 = 2.0\ c\cdot y$ and $t_2 = 0.5$ y. These events occur simultaneously in frame $S'$. (a) Find the magnitude and direction of the velocity of $S'$ relative to $S$. (b) At what time do both these events occur as measured in $S'$?

**48** •• An interstellar spaceship travels from the earth to a distant star system 12 light-years away (as measured in the earth's frame). The trip takes 15 y as measured on the spaceship. (a) What is the speed of the spaceship relative to the earth? (b) When the ship arrives, it sends a signal to the earth. How long after the ship leaves the earth will it be before the earth receives the signal?

**49** •• The neutral pion $\pi^0$ has a mass of 135 MeV/$c^2$. This particle can be created in a proton–proton collision:

$$p + p \rightarrow p + p + \pi^0$$

Determine the threshold kinetic energy for the creation of a $\pi^0$ in a collision of a moving proton and a stationary proton. (See Problem 40.)

**50** •• A rocket with a proper length of 1000 m moves in the $+x$ direction at 0.6$c$ with respect to an observer on the ground. An astronaut stands at the rear of the rocket and fires a bullet toward the front of the rocket at 0.8$c$ relative to the rocket. How long does it take the bullet to reach the front of the rocket (a) as measured in the frame of the rocket, (b) as measured in the frame of the ground, and (c) as measured in the frame of the bullet?

**51** ••• **SSM** In a simple thought experiment, Einstein showed that there is mass associated with electromagnetic radiation. Consider a box of length $L$ and mass $M$ resting on a frictionless surface. At the left wall of the box is a light source that emits radiation of energy $E$, which is absorbed at the right wall of the box. According to classical electromagnetic theory, this radiation carries momentum of magnitude $p = E/c$ (Equation 32-13). (a) Find the recoil velocity of the box so that momentum is conserved when the light is emitted. (Since $p$ is small and $M$ is large, you may use classical mechanics.) (b) When the light is absorbed at the right wall of the box the box stops, so the total momentum remains zero. If we neglect the very small velocity of the box, the time it takes for the radiation to travel across the box is $\Delta t = L/c$. Find the distance moved by the box in this time. (c) Show that if the center of

mass of the system is to remain at the same place, the radiation must carry mass $m = E/c^2$.

**52** ••• Reference frame $S'$ is moving along the $x'$ axis at 0.6$c$ relative to frame $S$. A particle that is originally at $x' = 10$ m at $t_1' = 0$ is suddenly accelerated and then moves at a constant speed of $c/3$ in the $-x'$ direction until time $t_2' = 60$ m/$c$, when it is suddenly brought to rest. As observed in frame $S$, find (a) the speed of the particle, (b) the distance and the direction that the particle traveled from $t_1'$ to $t_2'$, and (c) the time the particle traveled.

**53** ••• In reference frame $S$, the acceleration of a particle is $\vec{a} = a_x\hat{i} + a_y\hat{j} + a_z\hat{k}$. Derive expressions for the acceleration components $a_x'$, $a_y'$, and $a_z'$ of the particle in reference frame $S'$ that is moving relative to $S$ in the $x$ direction with velocity $v$.

**54** ••• Using the relativistic conservation of momentum and energy and the relation between energy and momentum for a photon $E = pc$, prove that a free electron (i.e., one not bound to an atomic nucleus) cannot absorb or emit a photon.

**55** ••• **SSM** When a projectile particle with kinetic energy greater than the threshold kinetic energy $K_{th}$ strikes a stationary target particle, one or more particles may be created in the inelastic collision. Show that the threshold kinetic energy of the projectile is given by

$$K_{th} = \frac{(\Sigma m_{in} + \Sigma m_{fin})(\Sigma m_{fin} - \Sigma m_{in})c^2}{2m_{target}}$$

Here $\Sigma m_{in}$ is the sum of the masses of the projectile and target particles, $\Sigma m_{fin}$ is the sum of the masses of the final particles, and $m_{target}$ is the mass of the target particle. Use this expression to determine the threshold kinetic energy of protons incident on a stationary proton target for the production of a proton–antiproton pair; compare your result with the result of Problem 40.

**56** ••• A particle of mass $M$ decays into two identical particles of mass $m$, where $m = 0.3M$. Prior to the decay, the particle of mass $M$ has an energy of $4Mc^2$ in the laboratory. The velocities of the decay products are along the direction of motion of $M$. Find the velocities of the decay products in the laboratory.

**57** ••• A stick of proper length $L_p$ makes an angle $\theta$ with the $x$ axis in frame $S$. Show that the angle $\theta'$ made with the $x'$ axis in frame $S'$, which is moving along the $+x$ axis with speed $v$, is given by $\tan\theta' = \gamma\tan\theta$ and that the length of the stick in $S'$ is

$$L' = L_p\left(\frac{1}{\gamma^2}\cos^2\theta + \sin^2\theta\right)^{1/2}$$

**58** ••• Show that if a particle moves at an angle $\theta$ with the $x$ axis with speed $u$ in frame $S$, it moves at an angle $\theta'$ with the $x'$ axis in $S'$ given by

$$\tan\theta' = \frac{\sin\theta}{\gamma[\cos\theta - (v/u)]}$$

**59** ••• **SSM** For the special case of a particle moving with speed $u$ along the $y$ axis in frame $S$, show that its momentum and energy in frame $S'$, a frame that is moving along the $x$ axis with velocity $v$, are related to its momentum and energy in $S$ by the transformation equations

$$p'_x = \gamma\left(p_x - \frac{vE}{c^2}\right), \qquad p'_y = p_y, \qquad p'_z = p_z$$

$$\frac{E'}{c} = \gamma\left(\frac{E}{c} - \frac{vp_x}{c}\right)$$

Compare these equations with the Lorentz transformation for $x'$, $y'$, $z'$, and $t'$. These equations show that the quantities $p_x$, $p_y$, $p_z$, and $E/c$ transform in the same way as do $x$, $y$, $z$, and $ct$.

**60** ••• The equation for the spherical wavefront of a light pulse that begins at the origin at time $t = 0$ is $x^2 + y^2 + z^2 - (ct)^2 = 0$. Using the Lorentz transformation, show that such a light pulse also has a spherical wavefront in frame $S'$ by showing that $x'^2 + y'^2 + z'^2 - (ct')^2 = 0$ in $S'$.

**61** ••• In Problem 60, you showed that the quantity $x^2 + y^2 + z^2 - (ct)^2$ has the same value (0) in both $S$ and $S'$. Such a quantity is called an *invariant*. From the results of Problem 59, the quantity $p_x^2 + p_y^2 + p_z^2 - E^2/c^2$ must also be an invariant. Show that this quantity has the value $-m^2c^2$ in both the $S$ and $S'$ reference frames.

**62** ••• **SSM** A long rod that is parallel to the $x$ axis is in free fall with acceleration $g$ parallel to the $-y$ axis. An observer in a rocket moving with speed $v$ parallel to the $x$ axis passes by and watches the rod falling. Using the Lorentz transformations, show that the observer will measure the rod to be bent into a parabolic shape. Is the parabola concave upward or concave downward?

# Nuclear Physics

**A NUCLEAR POWER PLANT IN GERMANY** THE FISSION REACTOR CORE IS HOUSED IN A HEMISPHERICAL CONTAINMENT STRUCTURE (CENTER). TWO LARGE COOLING TOWERS ARE TO ITS LEFT.

**?** How much energy is released in the fission of 1 g of $^{235}$U? This calculation is the subject of Example 40-6.

40-1 Properties of Nuclei

40-2 Radioactivity

40-3 Nuclear Reactions

40-4 Fission and Fusion

To the chemist, the atomic nucleus is essentially a point charge that contains most of the mass of the atom. It plays a negligible role in the structure of atoms and molecules.

> In this chapter, we will look at the nucleus from the physicist's perspective and see how the protons and neutrons that make up the nucleus have played important roles in our everyday life as well as in the history and structure of the universe. The fission of very heavy nuclei, such as uranium, is a major source of power today, while the fusion of very light nuclei is the energy source that powers the stars, including our sun, and may hold the key to our energy needs of the future.

## 40-1 Properties of Nuclei

The nucleus of an atom contains just two kinds of particles, protons and neutrons,[†] which have approximately the same mass (the neutron is approximately 0.2 percent more massive). The proton has a charge of $+e$ and the neutron is uncharged. The number of protons, $Z$, is the atomic number of the atom, which

---

† The most prevalent hydrogen nucleus contains a single proton.

also equals the number of electrons in the atom. The number of neutrons, $N$, is approximately equal to $Z$ for light nuclei, and for heavier nuclei the number of neutrons is increasingly greater than $Z$. The total number of nucleons[†] $A = N + Z$ is called the mass number of the nucleus. A particular nuclear species is called a **nuclide.** Two or more nuclides with the same atomic number $Z$ but with different $N$ and $A$ numbers are called **isotopes.** A particular nuclide is designated by its atomic symbol (H for hydrogen, He for helium, etc.), with the mass number $A$ as a presuperscript. The lightest element, hydrogen, has three isotopes: protium, $^1$H, whose nucleus is just a single proton; deuterium, $^2$H, whose nucleus contains one proton and one neutron; and tritium, $^3$H, whose nucleus contains one proton and two neutrons. Although the mass of the deuterium atom is about twice the mass of the protium atom and the mass of the tritium atom is about three times the mass of protium, these three atoms have nearly identical chemical properties because they each have one electron. On the average, there are about three stable isotopes for each element, although some atoms have only one stable isotope while others have five or six. The most common isotope of the second lightest element, helium, is $^4$He. The $^4$He nucleus is also known as an $\alpha$ particle. Another isotope of helium is $^3$He.

Inside the nucleus, the nucleons exert a strong attractive force on their nearby neighbors. This force, called the **strong nuclear force** or the **hadronic force,** is much stronger than the electrostatic force of repulsion between the protons and is very much stronger than the gravitational forces between the nucleons. (Gravity is so weak that it can always be neglected in nuclear physics.) The strong nuclear force is roughly the same between two neutrons, two protons, or a neutron and a proton. Two protons, of course, also exert a repulsive electrostatic force on each other due to their charges, which tends to weaken the attraction between them somewhat. The strong nuclear force decreases rapidly with distance, and it is negligible when two nucleons are more than a few femtometers apart.

## Size and Shape

The size and shape of the nucleus can be determined by bombarding it with high-energy particles and observing the scattering. The results depend somewhat on the kind of experiment. For example, a scattering experiment using electrons measures the charge distribution of the nucleus, whereas a scattering experiment using neutrons determines the region of influence of the strong nuclear force. A wide variety of experiments suggest that most nuclei are approximately spherical, with radii given approximately by

$$R = R_0 A^{1/3} \qquad \qquad \text{40-1}$$

NUCLEAR RADIUS

where $R_0$ is approximately 1.2 fm. The fact that the radius of a spherical nucleus is proportional to $A^{1/3}$ implies that the volume of the nucleus is proportional to $A$. Since the mass of the nucleus is also approximately proportional to $A$, the densities of all nuclei are approximately the same. This is analogous to a drop of liquid, which also has constant density independent of its size. The **liquid-drop model** of the nucleus has proved quite successful in explaining nuclear behavior, especially the fission of heavy nuclei.

---

† The word *nucleon* refers to either a neutron or a proton.

## N and Z Numbers

For light nuclei, the greatest stability is achieved when the numbers of protons and neutrons are approximately equal, $N \approx Z$. For heavier nuclei, instability caused by the electrostatic repulsion between the protons is minimized when there are more neutrons than protons. We can see this by looking at the $N$ and $Z$ numbers for the most abundant isotopes of some representative elements: for $^{16}_{8}O$, $N = 8$ and $Z = 8$; for $^{40}_{20}Ca$, $N = 20$ and $Z = 20$; for $^{56}_{26}Fe$, $N = 30$ and $Z = 26$; for $^{207}_{82}Pb$, $N = 125$ and $Z = 82$; and for $^{238}_{92}U$, $N = 146$ and $Z = 92$. (The atomic number $Z$ has been included here as a presubscript of the atomic symbol for emphasis. It is not actually needed because the atomic number is implied by the atomic symbol.)

Figure 40-1 shows a plot of $N$ versus $Z$ for the known stable nuclei. The curve follows the straight line $N = Z$ for small values of $N$ and $Z$. We can understand this tendency for $N$ and $Z$ to be equal by considering the total energy of $A$ particles in a one-dimensional box. For $A = 8$, Figure 40-2 shows the energy levels for eight neutrons and for four neutrons and four protons. Because of the exclusion principle, only two identical particles (with opposite spins) can be in the same space state. Since protons and neutrons are not identical, we can put two each in a state, as shown in Figure 40-2b. Thus, the total energy for four protons and four neutrons is less than the total energy for eight neutrons (or eight protons), as shown in Figure 40-2a. When the Coulomb energy of repulsion, which is proportional to $Z^2$, is included, this result changes somewhat. For large values of $A$ and $Z$, the total energy may be increased less by adding two neutrons than by adding one neutron and one proton because of the electrostatic repulsion involved in the latter case. This explains why $N > Z$ for the larger values of $A$ (for the heavier nuclei).

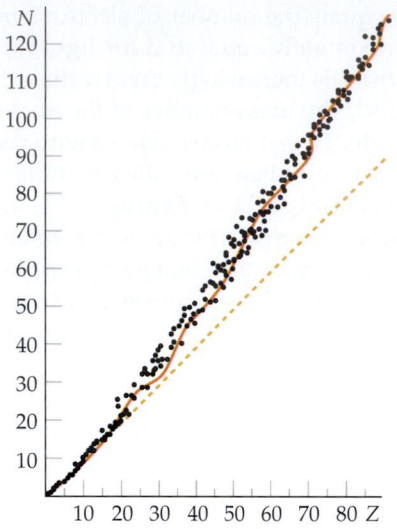

**FIGURE 40-1** Plot of number of neutrons $N$ versus number of protons $Z$ for the stable nuclides. The dashed line is $N = Z$.

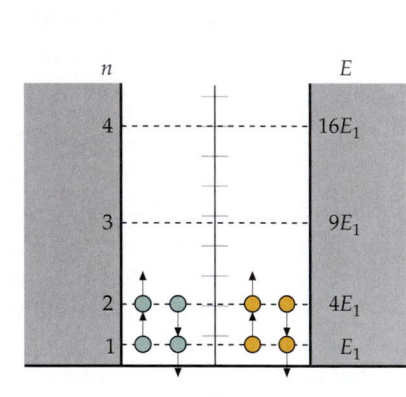

(a)                    (b)

**FIGURE 40-2** (a) Eight neutrons in a one-dimensional box. In accordance with the exclusion principle, only two neutrons (with opposite spins) can be in a given energy level. (b) Four neutrons and four protons in a one-dimensional box. Because protons and neutrons are not identical particles, two of each can be in the same energy level. The total energy is much less for this case than for that in Figure 40-2a.

**EXERCISE** (a) Calculate the total energy of the eight neutrons in the one-dimensional box shown in Figure 40-2a. (b) Calculate the total energy of the four neutrons and four protons in the one-dimensional box shown in Figure 40-2b. (*Answer* (a) $60E_1$ (b) $20E_1$)

## Mass and Binding Energy

The mass of a nucleus is less than the mass of its parts by $E_b/c^2$, where $E_b$ is the binding energy and $c$ is the speed of light. When two or more nucleons fuse together to form a nucleus, the total mass decreases and energy is given off. Conversely, to break up a nucleus into its parts, energy must be put into the system to produce the increase in mass.

Atomic masses and nuclear masses are often given in unified mass units (u), defined as one-twelfth the mass of the neutral $^{12}C$ atom. The rest energy of one unified mass unit is

$$(1 \text{ u})c^2 = 931.5 \text{ MeV}$$

40-2

Consider $^4$He, for example, which consists of two protons and two neutrons. The mass of an atom can be accurately measured in a mass spectrometer. The mass of the $^4$He atom is 4.002 603 u. This includes the masses of the two electrons in the atom. The mass of the $^1$H atom is 1.007 825 u, and the mass of the neutron is 1.008 665 u. The sum of the masses of two $^1$H atoms plus two neutrons is 2(1.007 825 u) + 2(1.008 665 u) = 4.032 980 u, which is greater than the mass of the $^4$He atom by 0.030 377 u.[†] We can find the binding energy of the $^4$He nucleus from this mass difference of 0.030 377 u by using the mass conversion factor (1 u)$c^2$ = 931.5 MeV from Equation 40-2. Then

$$(0.030\ 377\ \text{u})c^2 = (0.030\ 377\ \text{u})c^2 \times \frac{931.5\ \text{MeV}/c^2}{1\ \text{u}}$$
$$= 28.30\ \text{MeV}$$

The total binding energy of $^4$He is thus 28.30 MeV. In general, the binding energy of a nucleus of an atom of atomic mass $M_A$ containing $Z$ protons and $N$ neutrons is found by calculating the difference between the mass of the parts and the mass of the nucleus and then multiplying by $c^2$:

$$E_b = (ZM_H + Nm_n - M_A)c^2 \qquad\qquad 40\text{-}3$$

TOTAL NUCLEAR BINDING ENERGY

where $M_H$ is the mass of the $^1$H atom and $m_n$ is the mass of the neutron. (Note that the mass of the $Z$ electrons in the term $ZM_H$ is canceled by the mass of the $Z$ electrons in the term $M_A$.[‡]) The atomic masses of the neutron and of some selected isotopes are listed in Table 40-1.

[†] Note that by using the masses of two $^1$H atoms rather than two protons, the masses of the electrons in the atom are accounted for. We do this because it is atomic masses, not nuclear masses, that are measured directly and listed in mass tables.
[‡] The mass associated with the binding energies of the electrons are not accounted for in this calculation.

## TABLE 40-1

**Atomic Masses of the Neutron and Selected Isotopes[†]**

| Element | Symbol | Z | Atomic mass, u | Element | Symbol | Z | Atomic mass, u |
|---|---|---|---|---|---|---|---|
| Neutron | n | 0 | 1.008 665 | Oxygen | $^{16}$O | 8 | 15.994 915 |
| Hydrogen | $^1$H | 1 | 1.007 825 | Sodium | $^{23}$Na | 11 | 22.989 770 |
| Deuterium | $^2$H or D | 1 | 2.013 553 | Potassium | $^{39}$K | 19 | 38.963 707 |
| Tritium | $^3$H or T | 1 | 3.016 049 | Iron | $^{56}$Fe | 26 | 55.934 942 |
| Helium | $^3$He | 2 | 3.016 029 | Copper | $^{63}$Cu | 29 | 62.929 601 |
| | $^4$He | 2 | 4.002 603 | Silver | $^{107}$Ag | 47 | 106.905 093 |
| Lithium | $^6$Li | 3 | 6.015 122 | Gold | $^{197}$Au | 79 | 196.966 552 |
| | $^7$Li | 3 | 7.016 004 | Lead | $^{208}$Pb | 82 | 207.976 636 |
| Boron | $^{10}$B | 5 | 10.012 937 | Polonium | $^{212}$Po | 84 | 211.988 852 |
| Carbon | $^{12}$C | 6 | 12.000 000 | Radon | $^{222}$Rn | 86 | 222.017 571 |
| | $^{13}$C | 6 | 13.003 354 | Radium | $^{226}$Ra | 88 | 226.025 403 |
| | $^{14}$C | 6 | 14.003 242 | Uranium | $^{238}$U | 92 | 238.050 783 |
| Nitrogen | $^{13}$N | 7 | 13.005 739 | Plutonium | $^{242}$Pu | 94 | 242.058 737 |
| | $^{14}$N | 7 | 14.003 074 | | | | |

[†]Mass values obtained at <http://physics.nist.gov/PhysRefData/Compositions/index.html>.

---

*BINDING ENERGY OF THE LAST NEUTRON*          **E X A M P L E   4 0 - 1**

**Find the binding energy of the last neutron in ⁴He.**

**PICTURE THE PROBLEM** The binding energy is $c^2$ times the difference in mass of ³He plus a neutron and ⁴He. We find these masses from Table 40-1 and convert to energy using Equation 40-3.

1. Add the mass of the neutron to that of ³He:

$$m_{^3\text{He}} + m_n = 3.016\ 029\ u + 1.008\ 665\ u$$

$$= 4.024\ 694\ u$$

2. Subtract the mass of ⁴He from the result:

$$\Delta m = (m_{^3\text{He}} + m_n) - m_{^4\text{He}}$$

$$= 4.024\ 694\ u - 4.002\ 603\ u = 0.022\ 091\ u$$

3. Multiply this mass difference by $c^2$ and convert to MeV:

$$E_b = (\Delta m)c^2$$

$$= (0.022\ 091\ u)c^2 \times \frac{931.5\ \text{MeV}/c^2}{1\ u}$$

$$= \boxed{20.58\ \text{MeV}}$$

---

Figure 40-3 shows the binding energy per nucleon $E_b/A$ versus $A$. The mean value is approximately 8.3 MeV. The flatness of this curve for $A > 50$ shows that $E_b$ is approximately proportional to $A$. This indicates that there is saturation of nuclear forces in the nucleus as would be the case if each nucleon were attracted only to its nearest neighbors. Such a situation also leads to a constant nuclear density consistent with the measurements of the radius. If, for example, there were no saturation and each nucleon bonded to each other nucleon, there would be $A - 1$ bonds for each nucleon and a total of $A(A - 1)$ bonds altogether. The total binding energy, which is a measure of the energy needed to break all these bonds, would then be proportional to $A(A - 1)$, and $E_b/A$ would not be approximately constant. The steep rise in the curve for low $A$ is due to the increase in the number of nearest neighbors and therefore to the increased number of bonds per nucleon. The gradual decrease at high $A$ is due to the Coulomb repulsion of the protons, which increases as $Z^2$ and decreases the binding energy. Eventually, for very large $A$, this Coulomb repulsion becomes so great that a nucleus with $A$ greater than approximately 300 is unstable and undergoes spontaneous fission.

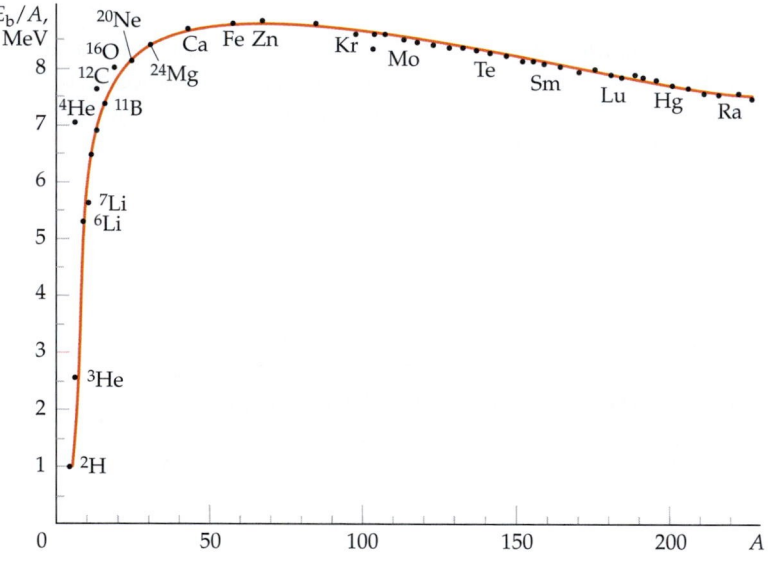

**FIGURE 40-3** The binding energy per nucleon versus the mass number $A$. For nuclei with values of $A$ greater than 50, the curve is approximately constant, indicating that the total binding energy is approximately proportional to $A$.

# 40-2 Radioactivity

Many nuclei are radioactive; that is, they decay into other nuclei by the emission of particles such as photons, electrons, neutrons, or $\alpha$ particles. The terms $\alpha$ decay, $\beta$ decay, and $\gamma$ decay were used before it was discovered that $\alpha$ particles are

$^4$He nuclei, $\beta$ particles are either electrons ($\beta^-$) or positrons[†] ($\beta^+$), and $\gamma$-rays are photons. The rate of decay is not constant over time, but decreases exponentially. *This exponential time dependence is characteristic of all radioactivity and indicates that radioactive decay is a statistical process.* Because each nucleus is well shielded from others by the atomic electrons, pressure and temperature changes have little or no effect on the rate of radioactive decay or other nuclear properties.

Let $N$ be the number of radioactive nuclei at some time $t$. If the decay of an individual nucleus is a random event, we expect the number of nuclei that decay in some time interval $dt$ to be proportional to $N$ and to $dt$. Because of these decays, the number $N$ will decrease. The change in $N$ is given by

$$dN = -\lambda N\, dt \qquad\qquad 40\text{-}4$$

where $\lambda$ is a constant of proportionality called the **decay constant.** The rate of change of $N$, $dN/dt$, is proportional to $N$. This is characteristic of exponential decay. To solve Equation 40-4 for $N$, we first divide each side by $N$, thus separating the variables $N$ and $t$:

$$\frac{dN}{N} = -\lambda\, dt$$

Integrating, we obtain

$$\int_{N_0}^{N'} \frac{dN}{N} = -\lambda \int_0^{t'} dt$$

or

$$\ln\frac{N'}{N_0} = -\lambda t' \qquad\qquad 40\text{-}5$$

where $N'$ is the number of nuclei that remain at time $t'$. For convenience, we drop the primes from $N'$ and $t'$. This introduces no ambiguity because the parameters $N$ and $t$ have been integrated out of the equation. Taking the exponential of each side, we obtain

$$\frac{N}{N_0} = e^{-\lambda t}$$

or

$$N = N_0 e^{-\lambda t} \qquad\qquad 40\text{-}6$$

The number of radioactive decays per second is called the decay rate $R$:

$$R = -\frac{dN}{dt} = \lambda N = \lambda N_0 e^{-\lambda t} = R_0 e^{-\lambda t} \qquad\qquad 40\text{-}7$$

DECAY RATE

where

$$R_0 = \lambda N_0 \qquad\qquad 40\text{-}8$$

---

[†] The positron is identical to an electron except it has a charge of $+e$.

is the rate of decay at time $t = 0$. The decay rate $R$ is the quantity that is determined experimentally.

The average or **mean lifetime** $\tau$ is equal to the reciprocal of the decay constant (see Problem 42):

$$\tau = \frac{1}{\lambda} \qquad\qquad 40\text{-}9$$

The mean lifetime is analogous to the time constant in the exponential decrease in the charge on a capacitor in an $RC$ circuit that we discussed in Section 25-6. After a time equal to the mean lifetime, the number of radioactive nuclei and the decay rate are each equal to $e^{-1} = 37$ percent of their original values. The **half-life** $t_{1/2}$ is defined as the time it takes for the number of nuclei and the decay rate to decrease by half. Setting $t = t_{1/2}$ and $N = N_0/2$ in Equation 40-6 gives

$$\frac{N_0}{2} = N_0 e^{-\lambda t_{1/2}} \qquad\qquad 40\text{-}10$$

or

$$e^{+\lambda t_{1/2}} = 2$$

Solving for $t_{1/2}$ gives

$$t_{1/2} = \frac{\ln 2}{\lambda} = \frac{0.693}{\lambda} = 0.693\tau \qquad\qquad 40\text{-}11$$

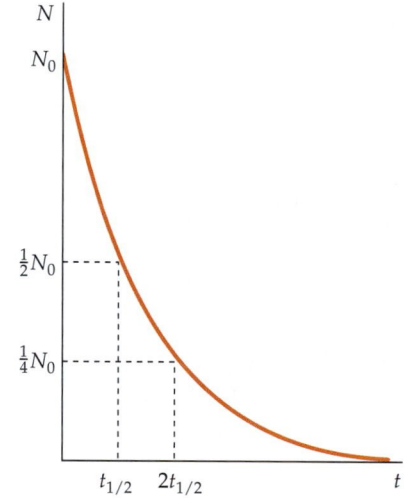

**FIGURE 40-4** Exponential radioactive decay. After each half-life $t_{1/2}$, the number of nuclei remaining has decreased by one-half. The decay rate $R = \lambda N$ has the same time dependence, as does $N$.

Figure 40-4 shows a plot of $N$ versus $t$. If we multiply the numbers on the $N$ axis by $\lambda$, this graph becomes a plot of $R$ versus $t$. After each time interval of one half-life, both the number of nuclei left and the decay rate have decreased to half of their previous values. For example, if the decay rate is $R_0$ initially, it will be $\frac{1}{2}R_0$ after one half-life, $(\frac{1}{2})(\frac{1}{2})R_0$ after two half-lives, and so forth. After $n$ half-lives, the decay rate will be

$$R = (\tfrac{1}{2})^n R_0 \qquad\qquad 40\text{-}12$$

The half-lives of radioactive nuclei vary from very small times (less than 1 $\mu$s) to very large times (up to $10^{16}$ y).

---

*COUNTING RATE FOR RADIOACTIVE DECAY*      **EXAMPLE 40-2**

A radioactive source has a half-life of 1 min. At time $t = 0$, the radioactive source is placed near a detector, and the counting rate (the number of decay particles detected per unit time) is observed to be 2000 counts/s. Find the counting rate at times $t = 1$ min, $t = 2$ min, $t = 3$ min, and $t = 10$ min.

**PICTURE THE PROBLEM** The counting rate decreases by a factor of 2 each minute.

1. Since the half-life is 1 min, the counting rate will be half    $r_1 = \frac{1}{2}r_0 = \frac{1}{2}(2000 \text{ counts/s})$
as great at $t = 1$ min as at $t = 0$:

$$= \boxed{1000 \text{ counts/s at 1 min}}$$

2. At $t = 2$ min, the rate is half that at 1 min. It decreases by one-half each minute:

$r_2 = \frac{1}{2}r_1 = \frac{1}{2}(1000 \text{ counts/s})$

$= \boxed{500 \text{ counts/s at 2 min}}$

$r_3 = \frac{1}{2}r_2 = \frac{1}{2}(500 \text{ counts/s})$

$= \boxed{250 \text{ counts/s at 3 min}}$

3. At $t = 10$ min, the rate will be $(\frac{1}{2})^{10}$ times the initial rate:

$r_{10} = (\frac{1}{2})^{10}r_0 = (\frac{1}{2})^{10} (2000 \text{ counts/s})$

$= 1.95 \text{ counts/s}$

$\approx \boxed{2 \text{ counts/s at 10 min}}$

---

*DETECTION-EFFICIENCY CONSIDERATIONS*        **EXAMPLE 40-3**

If the detection efficiency in Example 40-2 is 20 percent, (*a*) how many radioactive nuclei are there at time $t = 0$ and (*b*) at time $t = 1$ min? (*c*) How many nuclei decay in the first minute?

**PICTURE THE PROBLEM** The detection efficiency depends on the probability that a radioactive decay particle will enter the detector and the probability that upon entering the detector it will produce a count. If the efficiency is 20 percent, the decay rate must be five times the counting rate.

(*a*) 1. The number of radioactive nuclei is related to the decay rate $R$, and the decay constant $\lambda$:

$R = \lambda N$

2. The decay constant is related to the half-life:

$\lambda = \dfrac{0.693}{t_{1/2}} = \dfrac{0.693}{1 \text{ min}} = 0.693 \text{ min}^{-1}$

3. Because the detection efficiency is 20 percent, the decay rate is five times the counting rate. Calculate the initial decay rate:

$R_0 = 5 \times 2000 \text{ counts/s}$

$= 10,000 \text{ s}^{-1}$

4. Substitute to calculate the initial number of radioactive nuclei $N_0$ at $t = 0$:

$N_0 = \dfrac{R_0}{\lambda} = \dfrac{10,000 \text{ s}^{-1}}{0.693 \text{ min}^{-1}} \times \dfrac{60 \text{ s}}{1 \text{ min}}$

$= \boxed{8.66 \times 10^5}$

(*b*) At time $t = 1$ min $= t_{1/2}$, there are half as many radioactive nuclei as at $t = 0$:

$N_1 = \frac{1}{2}(8.66 \times 10^5) = \boxed{4.33 \times 10^5}$

(*c*) The number of nuclei that decay in the first minute is $N_0 - N_1$:

$\Delta N = N_0 - N_1$

$= 8.66 \times 10^5 - 4.33 \times 10^5$

$= \boxed{4.33 \times 10^5}$

---

The SI unit of radioactive decay is the **becquerel** (Bq), which is defined as one decay per second:

$1 \text{ Bq} = 1 \text{ decay/s}$                    40-13

A historical unit that applies to all types of radioactivity is the **curie** (Ci), which is defined as

$$1 \text{ Ci} = 3.7 \times 10^{10} \text{ decays/s} = 3.7 \times 10^{10} \text{ Bq} \qquad 40\text{-}14$$

The curie is the rate at which radiation is emitted by 1 g of radium. Since this is a very large unit, the millicurie (mCi) or microcurie ($\mu$Ci) are often used.

## Beta Decay

Beta decay occurs in nuclei that have too many neutrons or too few neutrons for stability. In $\beta$ decay, $A$ remains the same while $Z$ either increases by 1 ($\beta^-$ decay) or decreases by 1 ($\beta^+$ decay).

The simplest example of $\beta$ decay is the decay of the free neutron into a proton plus an electron. (The half-life of a free neutron is about 10.8 min.) The energy of decay is 0.782 MeV, which is the difference between the rest energy of the neutron and the rest energy of the proton plus electron. More generally, in $\beta^-$ decay, a nucleus of mass number $A$ and atomic number $Z$ decays into a nucleus, referred to as the **daughter nucleus,** of mass number $A$ and atomic number $Z' = Z + 1$ with the emission of an electron. If the decay energy were shared by only the daughter nucleus and the emitted electron, the energy of the electron would be uniquely determined by the conservation of energy and momentum. Experimentally, however, the energies of the electrons emitted in the $\beta^-$ decay of a nucleus are observed to vary from zero to the maximum energy available. A typical energy spectrum for these electrons is shown in Figure 40-5.

To explain the apparent nonconservation of energy in $\beta$ decay, Wolfgang Pauli in 1930 suggested that a third particle, which he called the **neutrino,** is also emitted. Because the measured maximum energy of the emitted electrons is equal to the total available for the decay, the rest energy and therefore the mass of the neutrino was assumed to be zero. (It is now believed that the mass of the neutrino is very small but not zero.) In 1948, measurements of the momenta of the emitted electron and the recoiling nucleus showed that the neutrino was also needed for the conservation of linear momentum in $\beta$ decay. The neutrino was first observed experimentally in 1957. It is now known that there are at least three kinds of neutrinos, one ($v_e$) associated with electrons, one ($v_\mu$) associated with muons, and one ($v_\tau$), which was first observed at Fermi National Laboratory in 2000, associated with the tau particle, $\tau$. Moreover, each neutrino has an antiparticle, written $\bar{v}_e$, $\bar{v}_\mu$, and $\bar{v}_\tau$. It is the electron antineutrino that is emitted in the decay of a neutron, which is written[†]

$$\text{n} \rightarrow \text{p} + \beta^- + \bar{v}_e \qquad 40\text{-}15$$

In $\beta^+$ decay, a proton changes into a neutron with the emission of a positron (and a neutrino). A free proton cannot decay by positron emission because of conservation of energy (the mass of the neutron plus the positron is greater than that of the proton); however, because of binding-energy effects, a proton inside a nucleus can decay. A typical $\beta^+$ decay is

$$^{13}_{7}\text{N} \rightarrow \, ^{13}_{6}\text{C} + \beta^+ + v_e \qquad 40\text{-}16$$

The electrons or the positrons emitted in $\beta$ decay do not exist inside the nucleus. They are created in the process of decay, just as photons are created when an atom makes a transition from a higher energy state to a lower energy state.

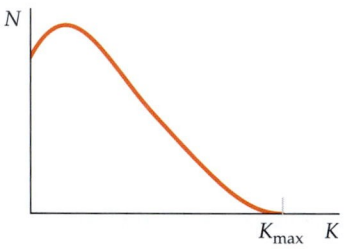

**FIGURE 40-5** Number of electrons emitted in $\beta^-$ decay versus kinetic energy. The fact that all the electrons do not have the same energy $K_{max}$ suggests that another particle, one that shares the energy available for decay, is emitted.

---

† This reaction is also written $\text{n} \rightarrow \text{p} + \text{e}^- + \bar{v}_e$.

SECTION 40-2  **Radioactivity**  |  **1315**

An important example of $\beta$ decay is that of $^{14}C$, which is used in radioactive carbon dating:

$$^{14}C \rightarrow {}^{14}N + \beta^- + \bar{v}_e \qquad\qquad 40\text{-}17$$

The half-life for this decay is 5730 y. The radioactive isotope $^{14}C$ is produced in the upper atmosphere in nuclear reactions caused by cosmic rays. The chemical behavior of carbon atoms with $^{14}C$ nuclei is the same as those with ordinary $^{12}C$ nuclei. For example, atoms with these nuclei combine with oxygen to form $CO_2$ molecules. Since living organisms continually exchange $CO_2$ with the atmosphere, the ratio of $^{14}C$ to $^{12}C$ in a living organism is the same as the equilibrium ratio in the atmosphere, which is about $1.3 \times 10^{-12}$. After an organism dies, it no longer absorbs $^{14}C$ from the atmosphere, so the ratio of $^{14}C$ to $^{12}C$ continually decreases due to the radioactive decay of $^{14}C$. The number of $^{14}C$ decays per minute per gram of carbon in a living organism can be calculated from the known half-life of $^{14}C$ and the number of $^{14}C$ nuclei in a gram of carbon. The result is that there are approximately 15.0 decays per minute per gram of carbon in a living organism. Using this result and the measured number of decays per minute per gram of carbon in a nonliving sample of bone, wood, or other object containing carbon, we can determine the age of the sample. For example, if the measured rate were 7.5 decays per minute per gram, the age of the sample would be one half-life = 5730 years.

---

*How Old Is the Artifact?*                           **EXAMPLE 40-4**   **Put It in Context**

**You have a summer job working in an archeological research lab. Your supervisor calls to tell you that they found a new bone at their current dig and asks you to determine the age of the bone from a sample that is in the mail. When the bone arrives, you take a sample that contains 200 grams of carbon and you find a beta decay rate of 400 decays/min.**

**PICTURE THE PROBLEM** We first obtain a rough estimate of the age of the bone. If the bone were from a living organism, we would expect the decay rate to be $[(15 \text{ decays/min})/g](200 \text{ g}) = 3000$ decays/min. Since 400/3000 is roughly 1/8 (actually 1/7.5), the sample must be approximately three half-lives old, which is about $3(5730 \text{ y}) = 17{,}190$ y. To find the age of the bone more accurately, we need to determine the number of half-lives of the bone. We can do this by using the equality $R_n = (1/2)^n R_0$ where $R_n$ is the current decay rate, $R_0$ is the initial decay rate, and $n$ is the number of half-lives. We can determine the initial decay rate by multiplying the decay rate per gram times the mass of the carbon of the sample.

1. Write the decay rate after $n$ half-lives in terms of the initial decay rate:

$$R_n = (\tfrac{1}{2})^n R_0$$

2. Calculate the initial decay rate (the decay for 200 g of carbon when the organism stopped breathing):

$$R_0 = [(15 \text{ decays/min})/g](200 \text{ g})$$
$$= 3000 \text{ decays/min}$$

3. Substitute the values for $R_0$ and $R_n$ into the step 1 equation and solve for $n$:

$$R_n = (\tfrac{1}{2})^n R_0$$

$$400 \frac{\text{decays}}{\text{min}} = (\tfrac{1}{2})^n 3000 \frac{\text{decays}}{\text{min}}$$

$$(\tfrac{1}{2})^n = \frac{400}{3000}$$

$$2^n = \frac{3000}{400} = 7.5$$

4. We solve for $n$ by taking the logarithm of each side:

$$n \ln 2 = \ln 7.5$$

$$n = \frac{\ln 7.5}{\ln 2} = 2.91$$

5. The age of the bone is $nt_{1/2}$:

$$t = nt_{1/2} = 2.91(5730 \text{ y}) = \boxed{1.67 \times 10^4 \text{ y}}$$

**EXERCISE** Picture the Problem of Example 40-4 states "Since 400/3000 is roughly 1/8 (actually 1/7.5), the sample must be approximately three half-lives old, . . .." Explain. [*Answer* It is because $\frac{1}{7.5} \approx \frac{1}{8} = (\frac{1}{2})^3$, so $n = 3$.]

## Gamma Decay

In $\gamma$ decay, a nucleus in an excited state decays to a lower-energy state by the emission of a photon. This is the nuclear counterpart of spontaneous emission of photons by atoms and molecules. Unlike $\beta$ decay or $\alpha$ decay, neither the mass number $A$ nor the atomic number $Z$ change during $\gamma$ decay. Since the spacing of the nuclear energy levels is of the order of 1 MeV (as compared with spacing of the order of 1 eV in atoms), the wavelengths of the emitted photons are of the order of 1 pm (1 pm = $10^{-12}$ m):

$$\lambda = \frac{hc}{E} \approx \frac{1240 \text{ eV·nm}}{1 \text{ MeV}} = 0.00124 \text{ nm} = 1.24 \text{ pm}$$

The mean lifetime for $\gamma$ decay is often very short. Usually it is observed only because it follows either $\alpha$ decay or $\beta$ decay. For example, if a radioactive parent nucleus decays by $\beta$ decay to an excited state of the daughter nucleus, the daughter nucleus then decays to its ground state by $\gamma$ emission. Direct measurements of mean lifetimes as short as approximately $10^{-11}$ s are possible. Measurements of mean lifetimes shorter than $10^{-11}$ s are difficult, but they can sometimes be made by indirect methods.

A few $\gamma$ emitters have very long lifetimes, of the order of hours. Nuclear energy states that have such long lifetimes are called **metastable states.**

## Alpha Decay

All very heavy nuclei ($Z > 83$) are theoretically unstable via $\alpha$ decay because the mass of the original radioactive nucleus is greater than the sum of the masses of the decay products—an $\alpha$ particle and the daughter nucleus. Consider the decay of $^{232}$Th ($Z = 90$) into $^{228}$Ra ($Z = 88$) plus an $\alpha$ particle. This is written as

$$^{232}\text{Th} \rightarrow {}^{228}\text{Ra} + \alpha = {}^{228}\text{Ra} + {}^{4}\text{He} \qquad\qquad 40\text{-}18$$

The mass of the $^{232}$Th atom is 232.038 050 u. The mass of the daughter atom $^{228}$Ra is 228.031 064 u. Adding 4.002 603 u to this for the mass of $^4$He, we get 232.033 667 u for the total mass of the decay products. This is less than the mass of $^{232}$Th by 0.004 382 u, which multiplied by 931.5 MeV/$c^2$ gives 4.08 MeV/$c^2$ for the excess mass of $^{232}$Th over that of the decay products. The isotope $^{232}$Th is therefore theoretically unstable to $\alpha$ decay. This decay does in fact occur in nature with the emission of an $\alpha$ particle of kinetic energy 4.08 MeV. (The kinetic energy of the $\alpha$ particle is actually somewhat less than 4.08 MeV because some of the decay energy is shared by the recoiling $^{228}$Ra nucleus.)

When a nucleus emits an $\alpha$ particle, both $N$ and $Z$ decrease by 2 and $A$ decreases by 4. The daughter of a radioactive nucleus is often itself radioactive and decays by either $\alpha$ decay or $\beta$ decay or both. If the original nucleus has a mass number $A$ that is 4 times an integer, the daughter nucleus and all those in the chain will also have mass numbers equal to 4 times an integer. Similarly, if the mass number of the original nucleus is $4n + 1$, where $n$ is an integer, all the nuclei in the decay chain will have mass numbers given by $4n + 1$, with $n$ decreasing by one at each decay. We can see, therefore, that there are four possible $\alpha$-decay chains, depending on whether $A$ equals $4n$, $4n + 1$, $4n + 2$, or $4n + 3$, where $n$ is an integer. All but one of these decay chains are found on the earth. The $4n + 1$ series is not found because its longest-lived member (other than the stable end product $^{209}$Bi) is $^{237}$Np, which has a half-life of only $2 \times 10^6$ y. Because this is much less than the age of the earth, this series has disappeared.

Figure 40-6 shows the thorium series, for which $A = 4n$. It begins with an $\alpha$ decay from $^{232}$Th to $^{228}$Ra. The daughter nuclide of an $\alpha$ decay is on the left or neutron-rich side of the stability curve (the dashed line in the figure), so it often decays by $\beta^-$ decay. In the thorium series, $^{228}$Ra decays by $\beta^-$ decay to $^{228}$Ac, which in turn decays by $\beta^-$ decay to $^{228}$Th. There are then four $\alpha$ decays to $^{212}$Pb, which decays by $\beta^-$ decay to $^{212}$Bi. The series branches at $^{212}$Bi, which decays either by $\alpha$ decay to $^{208}$Tl or by $\beta^-$ decay to $^{212}$Po. The branches meet at the stable lead isotope $^{208}$Pb.

The energies of $\alpha$ particles from natural radioactive sources range from approximately 4 MeV to 7 MeV, and the half-lives of the sources range from approximately $10^{-5}$ s to $10^{10}$ y. In general, the smaller the energy of the emitted $\alpha$ particle, the longer the half-life. As we discussed in Section 35-4, the enormous variation in half-lives was explained by George Gamow in 1928. He considered $\alpha$ decay to be a process in which an $\alpha$ particle is first formed inside a nucleus and then tunnels through the Coulomb barrier (Figure 40-7). A slight increase in the energy of the $\alpha$ particle reduces the relative height $U - E$ of the barrier and also the thickness. Because the probability of penetration is so sensitive to the relative height and thickness of the barrier, a small increase in $E$ leads to a large increase in the probability of barrier penetration and therefore to a shorter lifetime. Gamow was able to derive an expression for the half-life as a function of $E$ that is in excellent agreement with experimental results.

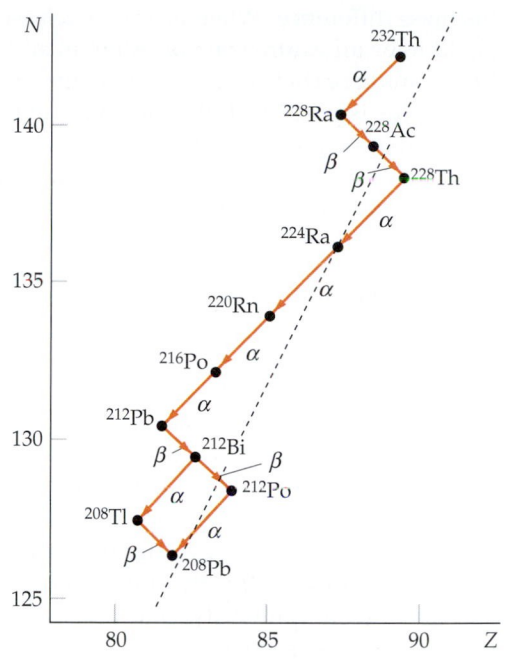

**FIGURE 40-6** The thorium ($4n$) $\alpha$ decay series. The dashed line is the curve of stability.

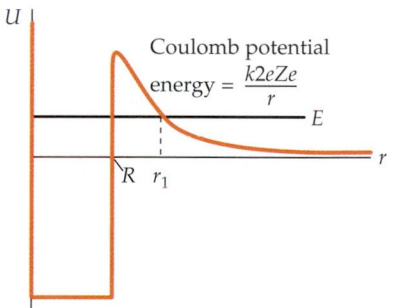

**FIGURE 40-7** A model of the potential energy for an $\alpha$ particle and a nucleus. The strong attractive nuclear force that exists for values of $r$ less than the nuclear radius $R$ is indicated by the potential well. Outside the nucleus, the nuclear force is negligible, and the potential energy is given by Coulomb's law $U = +k2eZe/r$, where $Ze$ is the nuclear charge and $2e$ is the charge of the $\alpha$ particle. The energy $E$ is the kinetic energy of the $\alpha$ particle when it is far away from the nucleus. A small increase in $E$ reduces the relative height $U - E$ of the barrier and also reduces its thickness, leading to a much greater chance of penetration. An increase in the energy of the emitted $\alpha$ particles by a factor of 2 results in a reduction of the half-life by a factor of more than $10^{20}$.

# 40-3    Nuclear Reactions

Information about nuclei is typically obtained by bombarding the nuclei with various particles and observing the results. Although the first experiments of this type were limited by the need to use naturally occurring radiation, they produced many important discoveries. In 1932, J. D. Cockcroft and E. T. S. Walton succeeded in producing the reaction

$$p + {}^7\text{Li} \rightarrow {}^8\text{Be} \rightarrow {}^4\text{He} + {}^4\text{He}$$

using artifically accelerated protons. At about the same time, the Van de Graaff electrostatic generator was built (by R. Van de Graaff in 1931) as was the first cyclotron (by E. O. Lawrence and M. S. Livingston in 1932). Since then, enormous advances in the technology for accelerating and detecting particles have been made, and many nuclear reactions have been studied.

When a particle is incident on a nucleus, several different things can happen. The incident particle may be scattered elastically or inelastically, or the incident particle may be absorbed by the nucleus, and another particle or particles may be emitted. In inelastic scattering, the nucleus is left in an excited state and subsequently decays by emitting photons (or other particles).

The amount of energy released or absorbed in a reaction (in the center of mass reference frame) is called the **Q value** of the reaction. The Q value equals $c^2$ times this mass difference. When energy is released by a nuclear reaction, the reaction is said to be an **exothermic reaction.** In an exothermic reaction, the total mass of the incoming particles is greater than the total mass of the outgoing particles, and the Q value is positive. If the total mass of the incoming particles is less than that of the outgoing particles, energy is required for the reaction to take place, and the reaction is said to be an **endothermic reaction.** The Q value of an endothermic reaction is negative. In general, if $\Delta m$ is the increase in mass, the Q value is

$$Q = -(\Delta m)c^2 \qquad \text{40-19}$$

Q VALUE

An endothermic reaction cannot take place below a specific threshold energy. In the laboratory reference frame in which stationary particles are bombarded by incoming particles, the threshold energy is somewhat greater than $|Q|$ because the outgoing particles must have some kinetic energy to conserve momentum.

A measure of the effective size of a nucleus for a particular nuclear reaction is the **cross section** $\sigma$. If $I$ is the number of incident particles per unit time per unit area (the incident intensity) and $R$ is the number of reactions per unit time per nucleus, the cross section is

$$\sigma = \frac{R}{I} \qquad \text{40-20}$$

The cross section $\sigma$ has the dimensions of area. Since nuclear cross sections are of the order of the square of the nuclear radius, a convenient unit for them is the **barn,** which is defined as

$$1 \text{ barn} = 10^{-28} \text{ m}^2 \qquad \text{40-21}$$

The cross section for a particular reaction is a function of energy. For an endothermic reaction, it is zero for energies below the threshold energy.

**Find the Q value of the reaction p + $^7$Li → $^4$He + $^4$He and state whether the reaction is exothermic or endothermic.**

**PICTURE THE PROBLEM** We find the masses of the atoms from Table 40-1 and calculate the difference in the total mass of the outgoing particles and the incoming particles. The Q value is $-(\Delta m)c^2$. If we use the mass of hydrogen rather than the mass of the proton, there will be four electrons on each side of the reaction, so the electron masses will cancel.

1. Find the mass of each atom from Table 40-1:

| | |
|---|---|
| $^1$H | 1.007 825 u |
| $^7$Li | 7.016 004 u |
| $^4$He | 4.002 603 u |

2. Calculate the initial mass $m_i$ of the incoming particles:

$$m_i = 1.007\ 825 \text{ u} + 7.016\ 004 \text{ u} = 8.023\ 829 \text{ u}$$

3. Calculate the final mass $m_f$:

$$m_f = 2(4.002\ 603 \text{ u}) = 8.005\ 206 \text{ u}$$

4. Calculate the increase in mass:

$$\Delta m = m_f - m_i = 8.005\ 206 \text{ u} - 8.023\ 829 \text{ u}$$
$$= -0.018\ 623 \text{ u}$$

5. Calculate the $Q$ value:

$$Q = -(\Delta m)c^2 = (+0.018\ 623\ \text{u})c^2 \times \left[931.5\ \text{MeV}/(\text{u}\cdot c^2)\right]$$

$$= \boxed{17.35\ \text{MeV}}$$

$\boxed{Q \text{ is positive, so the reaction is exothermic.}}$

**REMARKS** Since the initial mass is greater than the final mass, the initial energy is greater than the final energy and the reaction is exothermic, yielding 17.35 MeV.

## Reactions With Neutrons

Nuclear reactions that involve neutrons are important for understanding nuclear reactors. The most likely reaction between a nucleus and a neutron having an energy of more than about 1 MeV is scattering. However, even if the scattering is elastic, the neutron loses some energy to the nucleus because the nucleus recoils. If a neutron is scattered many times in a material, its energy decreases until the neutron is of the order of the energy of thermal motion $kT$, where $k$ is Boltzmann's constant and $T$ is the absolute temperature. (At ordinary room temperatures, $kT$ is approximately 0.025 eV.) The neutron is then equally likely to gain or lose energy from a nucleus when it is elastically scattered. A neutron with energy of the order of $kT$ is called a **thermal neutron.**

At low energies, a neutron is likely to be captured, with the emission of a $\gamma$ ray from the excited nucleus. Figure 40-8 shows the neutron-capture cross section for silver as a function of the energy of the neutron. The large peak in this curve is called a **resonance.** Except for the resonance, the cross section varies fairly smoothly with energy, decreasing with increasing energy roughly as $1/v$, where $v$ is the speed of the neutron. We can understand this energy dependence as follows: Consider a neutron moving with speed $v$ near a nucleus of diameter $2R$. The time it takes the neutron to pass the nucleus is $2R/v$. Thus, the neutron-capture cross section is proportional to the time spent by the neutron in the vicinity of the silver nucleus. The dashed line in Figure 40-8 indicates this $1/v$ dependence. At the maximum of the resonance, the value of the cross section is very large ($\sigma > 5000$ barns) compared with a value of only about 10 barns just past the resonance. Many elements show similar resonances in their neutron-capture cross sections. For example, the maximum cross section for $^{113}\text{Cd}$ is approximately 57,000 barns. This material is thus very useful for shielding against low-energy neutrons.

An important nuclear reaction that involves neutrons is fission, which is discussed in the next section.

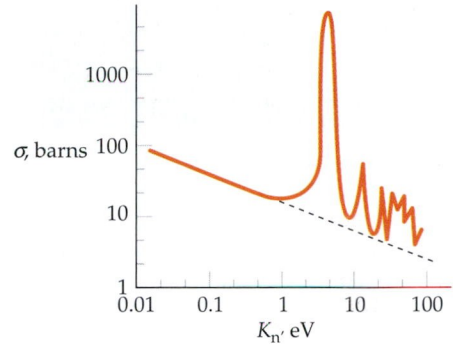

**FIGURE 40-8** Neutron-capture cross section for silver as a function of the energy of the neutron. The straight line indicates the $1/v$ dependence of the cross section, which is proportional to the time spent by the neutron in the vicinity of the silver nucleus. Superimposed on this dependence are a large resonance and several smaller resonances.

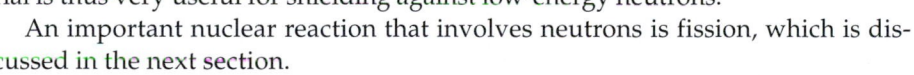

## 40-4 Fission and Fusion

Figure 40-9 shows a plot of the nuclear mass difference per nucleon $(M - Zm_p - Nm_n)/A$ in units of MeV/$c^2$ versus $A$. This is just the negative of the binding-energy curve shown in Figure 40-3. From Figure 40-9, we can see that the mass per nucleon for both very heavy ($A \approx 200$) and very light ($A \leq 20$) nuclides is more than that for nuclides of intermediate mass. Thus, energy is released when a very heavy nucleus, such as $^{235}\text{U}$, breaks up into two lighter nuclei—a process called **fission**—or when two very light nuclei, such as $^2\text{H}$ and $^3\text{H}$, fuse together to form a nucleus of greater mass—a process called **fusion.**

The application of both fission and fusion to the development of nuclear weapons has had a profound effect on our lives during the past 58 years. The peaceful application of these reactions to the development of energy resources

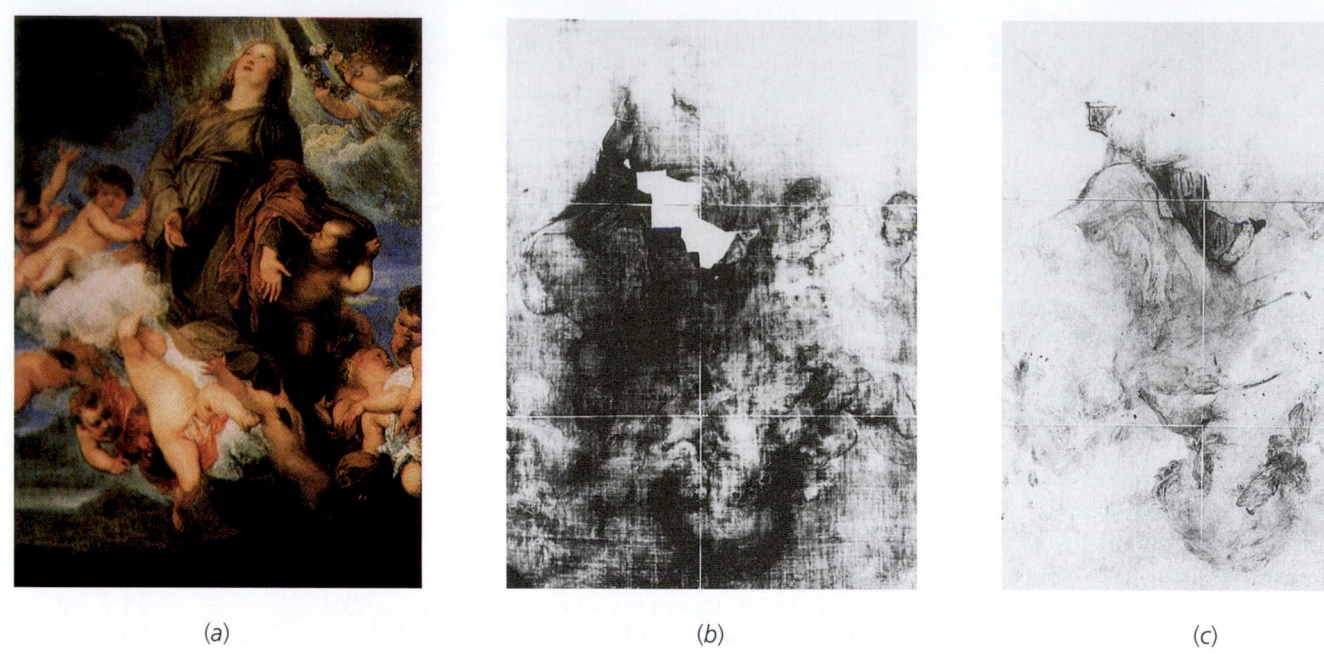

(a)  (b)  (c)

Hidden layers in paintings are analyzed by bombarding the painting with neutrons and observing the radiative emissions from nuclei that have captured a neutron. Different elements used in the painting have different half-lives. (*a*) Van Dyck's painting *Saint Rosalie Interceding for the Plague-Stricken of Palermo*. The black-and-white images in (*b*) and (*c*) were formed using a special film sensitive to electrons emitted by the radioactively decaying elements. Image (*b*), taken a few hours after the neutron irradiation, reveals the presence of manganese, found in umber, which is a dark earth pigment used for the painting's base layer. (Blank areas show where modern repairs, free of manganese, have been made.) The image in (*c*) was taken 4 days later, after the umber emissions had died away and when phosphorus, found in charcoal and boneblack, was the main radiating element. Upside down is revealed a sketch of Van Dyck himself. The self-portrait, executed in charcoal, had been overpainted by the artist.

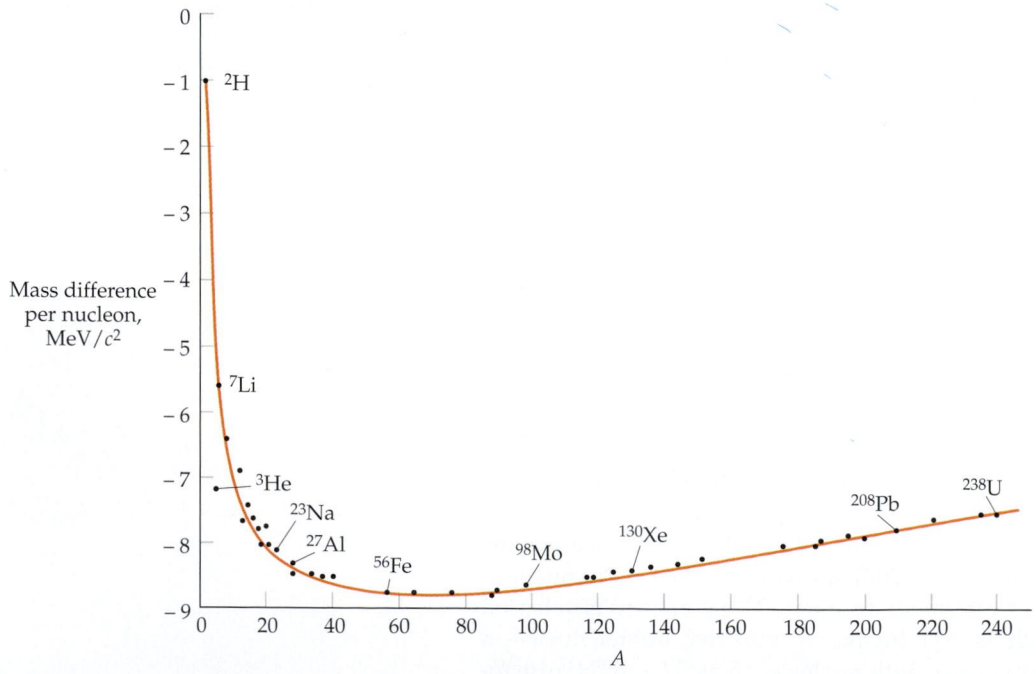

**FIGURE 40-9** Plot of mass difference per nucleon $(M - Zm_p - Nm_n)/A$ in units of $MeV/c^2$ versus $A$. The rest mass per nucleon is less for nuclei of intermediate mass than for very light nuclei or very heavy nuclei.

may have an even greater effect in the future. We will look at some of the features of fission and fusion that are important for their application in reactors to generate power.

## Fission

Very heavy nuclei ($Z > 92$) are subject to spontaneous fission. They break apart into two nuclei even if the nuclei are left to themselves with no outside disturbance. We can understand this by considering the analogy of a charged liquid drop. If the drop is not too large, surface tension can overcome the repulsive forces of the charges and hold the drop together. There is, however, a certain maximum size beyond which the drop will be unstable and will spontaneously break apart. Spontaneous fission puts an upper limit on the size of a nucleus and therefore on the number of elements that are possible.

Some heavy nuclei—uranium and plutonium, in particular—can be induced to fission by the capture of a neutron. In the fission of $^{235}$U, for example, the uranium nucleus is excited by the capture of a neutron, causing it to split into two nuclei and emit several neutrons. The Coulomb force of repulsion drives the fission fragments apart, with the energy eventually showing up as thermal energy. Consider, for example, the fission of a nucleus of mass number $A = 200$ into two nuclei of mass number $A = 100$. Since the rest energy for $A = 200$ is about 1 MeV per nucleon greater than that for $A = 100$, approximately 200 MeV per nucleus is released in such a fission. This is a large amount of energy. By contrast, in the chemical reaction of combustion, only about 4 eV of energy is released per molecule of oxygen consumed.

---

*ENERGY RELEASED IN THE FISSION OF $^{235}$U*                      **EXAMPLE 40-6**

**Calculate the total energy in kilowatt-hours released in the fission of 1 g of $^{235}$U, assuming that 200 MeV is released per fission.**

**PICTURE THE PROBLEM** We need to find the number of uranium nuclei in one gram of $^{235}$U, which we find using the fact that there are Avogadro's number ($N_A = 6.02 \times 10^{23}$) of nuclei in 235 grams.

1. The total energy is the number of nuclei times the energy per nucleus:

$$E = NE_{nucleus} = N(200 \text{ MeV/nucleus})$$

2. Calculate $N$:

$$N = \frac{6.02 \times 10^{23} \text{ nuclei/mol}}{235 \text{ g/mol}} \times 1 \text{ g}$$

$$= 2.56 \times 10^{21} \text{ nuclei}$$

3. Calculate the energy per gram in eV and convert to kW·h:

$$E = \frac{200 \times 10^6 \text{ eV}}{1 \text{ nucleus}} \times 2.56 \times 10^{21} \text{ nuclei}$$

$$= 5.12 \times 10^{29} \text{ eV} = 8.19 \times 10^{10} \text{ J}$$

$$= 8.19 \times 10^7 \text{ kW·s} = \boxed{2.28 \times 10^4 \text{ kW·h}}$$

---

The fission of uranium was discovered in 1938 by Otto Hahn and Fritz Strassmann, who found, by careful chemical analysis, that medium-mass elements (e.g., barium and lanthanum) were produced in the bombardment of uranium with neutrons. The discovery that several neutrons were emitted in the fission process led to speculation concerning the possibility of using these neutrons to cause further fissions, thereby producing a chain reaction. When $^{235}$U captures a neutron, the resulting $^{236}$U nucleus emits $\gamma$-rays as it deexcites to the ground state approximately 15 percent of the time and undergoes fission approximately 85 percent of the time. The fission process is somewhat analogous to the oscillation of a liquid drop, as shown in Figure 40-10. If the oscillations are violent enough, the drop splits in two. Using the liquid-drop model, Niels Bohr and John Wheeler calculated the critical energy $E_c$ needed by the $^{236}$U nucleus to undergo fission. ($^{236}$U is the nucleus formed momentarily by the capture of a neutron by $^{235}$U.) For this nucleus, the critical energy is 5.3 MeV, which is less than the 6.4 MeV of excitation energy produced when $^{235}$U captures a neutron. The capture of a neutron by $^{235}$U therefore produces an excited state of the $^{236}$U nucleus that has more than enough energy to break apart. On the other hand, the critical energy for fission of the $^{239}$U nucleus is 5.9 MeV. The capture of a neutron by a $^{238}$U nucleus produces an excitation energy of only 5.2 MeV. Therefore, when a neutron is captured by $^{238}$U to form $^{239}$U, the excitation energy is not great enough for fission to occur. In this case, the excited $^{239}$U nucleus deexcites by $\gamma$ emission and then decays to $^{239}$N$_p$ by $\beta$ decay, and then again to $^{239}$Pu by $\beta$ decay.

A fissioning nucleus can break into two medium-mass fragments in many different ways, as shown in Figure 40-11. Depending on the particular reaction, 1, 2, or 3 neutrons may be emitted. The average number of neutrons emitted in the fission of $^{235}$U is approximately 2.5. A typical fission reaction is

$$n + {}^{235}\text{U} \rightarrow {}^{141}\text{Ba} + {}^{92}\text{Kr} + 3n$$

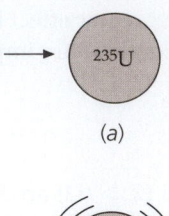

(a)

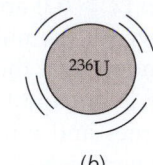

(b)

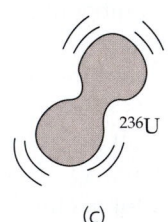

(c)

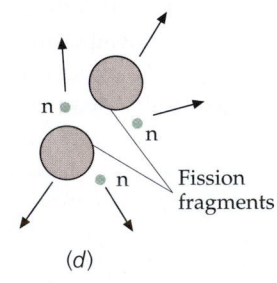

Fission
fragments

(d)

**FIGURE 40-10** Schematic illustration of nuclear fission. (*a*) The absorption of a neutron by $^{235}$U leads to (*b*) $^{236}$U in an excited state. (*c*) The oscillation of $^{236}$U has become unstable. (*d*) The nucleus splits apart into two nuclei of medium mass and emits several neutrons that can produce fission in other nuclei.

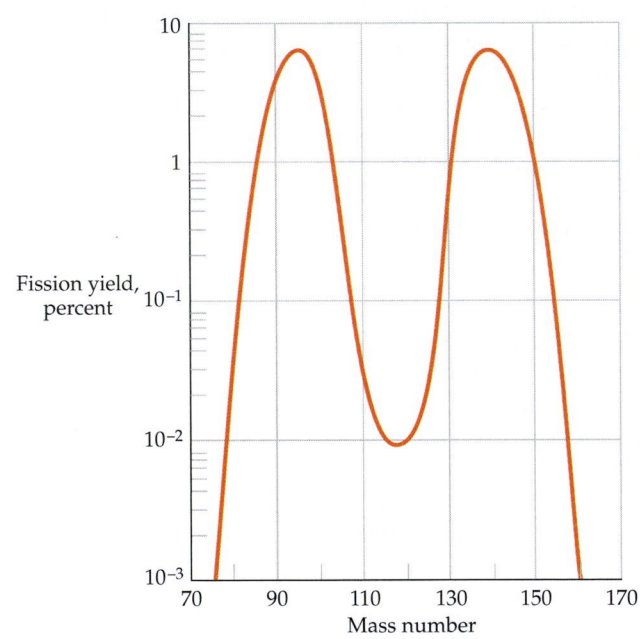

**FIGURE 40-11** Distribution of the possible fission fragments of $^{235}$U. The splitting of $^{235}$U into two fragments of unequal mass is more likely than its splitting into fragments of equal mass.

## Nuclear Fission Reactors

To sustain a chain reaction in a fission reactor, one of the neutrons (on the average) emitted in the fission of $^{235}$U must be captured by another $^{235}$U nucleus and cause it to fission. The **reproduction constant** $k$ of a reactor is defined as the average number of neutrons from each fission that cause a subsequent fission. The maximum possible value of $k$ is 2.5, but it is normally less than this for two important reasons: (1) Some of the neutrons may escape from the region containing fissionable nuclei and (2) some of the neutrons may be captured by nonfissioning nuclei in the reactor. If $k$ is exactly 1, the reaction will be self-sustaining. If $k$ is less than 1, the reaction will die out. If $k$ is significantly greater than 1, the reaction

The inside of a nuclear power plant in Kent, England. A technician is standing on the reactor charge transfer plate, into which uranium fuel rods fit.

rate will increase rapidly and *run away*. In the design of nuclear bombs, such a runaway reaction is desired. In power reactors, the value of $k$ must be kept very nearly equal to 1.

Since the neutrons emitted in fission have energies of the order of 1 MeV, whereas the chance for neutron capture leading to fission in $^{235}$U is largest at small energies, the chain reaction can be sustained only if the neutrons are slowed down before they escape from the reactor. At high energies (1 MeV to 2 MeV), neutrons lose energy rapidly by inelastic scattering from $^{238}$U, the principal constituent of natural uranium. (Natural uranium contains 99.3 percent $^{238}$U and only 0.7 percent fissionable $^{235}$U.) Once the neutron energy is below the excitation energies of the nuclei in the reactor (about 1 MeV), the main process of energy loss is by elastic scattering, in which a fast neutron collides with a nucleus at rest and transfers some of its kinetic energy to that nucleus. Such energy transfers are efficient only if the masses of the two bodies are comparable. A neutron will not transfer much energy in an elastic collision with a heavy uranium nucleus. Such a collision is like one between a marble and a billiard ball. The marble will be deflected by the much more massive billiard ball, and very little of its kinetic energy will be transferred to the billiard ball. A **moderator** consisting of material, such as water or carbon, that contains light nuclei is therefore placed around the fissionable material in the core of the reactor to slow down the neutrons. The neutrons are slowed down by elastic collisions with the nuclei of the moderator until they are in thermal equilibrium with the moderator. Because of the relatively large neutron-capture cross section of the hydrogen nucleus in water, reactors that use ordinary water as a moderator cannot easily achieve $k \approx 1$ unless they use enriched uranium, in which the $^{235}$U content has been increased from 0.7 percent to between 1 percent and 4 percent. Natural uranium can be used if heavy water ($D_2O$) is used instead of ordinary (light) water ($H_2O$) as the moderator. Although heavy water is expensive, most Canadian reactors use heavy water for a moderator to avoid the cost of constructing uranium-enrichment facilities.

Figure 40-12 shows some of the features of a pressurized-water reactor commonly used in the United States to generate electricity. Fission in the core heats the water to a high temperature in the primary loop, which is closed. This water, which also serves as the moderator, is under high pressure to prevent the water from boiling. The hot water is pumped to a heat exchanger, where it heats the water in the secondary loop and converts the water to steam, which is then used to drive the turbines that produce electrical power. Note that the water in the secondary loop is isolated from the water in the primary loop to prevent its contamination by the radioactive nuclei in the reactor core.

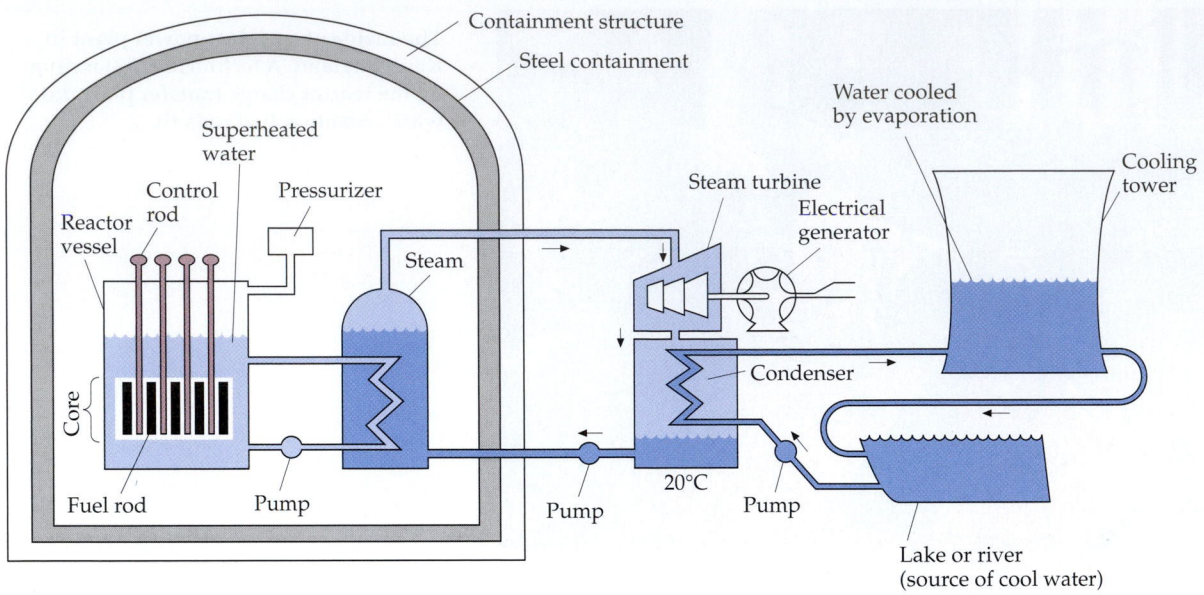

The ability to control the reproduction factor $k$ precisely is important if a power reactor is to be operated safely. Both natural negative-feedback mechanisms and mechanical methods of control are used. If $k$ is greater than 1, the reaction rate increases and the temperature of the reactor increases. If water is used as a moderator, its density decreases with increasing temperature and the water becomes a less effective moderator. A second important control method is the use of control rods made of a material, such as cadmium, that has a very large neutron-capture cross section. When the reactor is started up, the control rods are inserted so that $k$ is less than 1. As the rods are gradually withdrawn from the reactor, fewer neutrons are captured by the control rods and $k$ increases to 1. If $k$ becomes greater than 1, the rods are inserted further.

Mechanical control of the reaction rate of a nuclear reactor using control rods is possible only because some of the neutrons emitted in the fission process are **delayed neutrons.** The time needed for a neutron to slow down from 1 MeV or 2 MeV to the thermal-energy level and then be captured is only of the order of a millisecond. If all the neutrons emitted in fission were prompt neutrons, that is, emitted immediately in the fission process, mechanical control would not be possible because the reactor would run away before the rods could be further inserted. However, approximately 0.65 percent of the neutrons emitted are delayed by an average time of about 14 s. These neutrons are emitted not in the fission process itself but in the decay of the fission fragments. The effect of the delayed neutrons can be seen in the following examples.

---

*DOUBLING TIME*                                    **EXAMPLE 40-7**

**If the average time between fission generations (the time it takes for a neutron emitted in one fission to cause another) is $t_1 = 1$ ms $= 0.001$ s and if the average number of neutrons from each fission that cause a subsequent fission is 1.001, how long will it take for the reaction rate to double?**

**PICTURE THE PROBLEM** The reaction rate is the number of nuclei that fission per unit time. The time to double the reaction rate is the product of the number of generations $N$ needed to double the reaction rate and the generation time. If $k = 1.001$, the reaction rate after $N$ generations is $1.001^N$. We find the number of generations by setting $1.001^N$ equal to 2 and solving for $N$.

1. Set $1.001^N$ equal to 2 and solve for $N$:

$$(1.001)^N = 2$$

$$N \ln 1.001 = \ln 2$$

$$N = \frac{\ln 2}{\ln 1.001} = 693$$

2. Multiply the number of generations by the generation time: $\quad t = Nt_1 = 693(0.001 \text{ s}) = \boxed{0.693 \text{ s}}$

**REMARKS** The doubling time of about 0.7 s is not enough time for insertion of control rods.

---

*DELAYED NEUTRONS AND CONTROL-ROD INSERTION*  **EXAMPLE 40-8** **Try It Yourself**

**Assuming that 0.65 percent of the neutrons emitted are delayed by 14 s, find the average generation time and the doubling time if $k = 1.001$.**

**PICTURE THE PROBLEM** The doubling time is $Nt_{av}$, where $t_{av}$ is the average time between generations. Since 99.35 percent of the generation times are 0.001 s and 0.65 percent are 14 s, the average generation time is $0.9935(0.001 \text{ s}) + 0.0065(14 \text{ s})$.

**Cover the column to the right and try these on your own before looking at the answers.**

| Steps | Answers |
|---|---|
| 1. Compute the average generation time. | $t_{av} = 0.9935(0.001 \text{ s}) + 0.0065(14 \text{ s})$ |
| | $= 0.092 \text{ s}$ |
| 2. Use your result to find the time for 693 generations. | $t = \boxed{63.8 \text{ s}}$ |

**REMARKS** Even though the number of delayed neutrons is less than 1 percent, they have a large effect on the doubling time. Here they increase the generation time by a factor of 92, resulting in a doubling time of about 64 s, which is plenty of time for mechanical insertion of control rods.

Because of the limited supply of natural uranium, the small fraction of $^{235}$U in natural uranium, and the limited capacity of enrichment facilities, reactors based on the fission of $^{235}$U cannot meet our energy needs for very long. A promising alternative is the **breeder reactor.** When the relatively plentiful but nonfissionable $^{238}$U nucleus captures a neutron, it decays by $\beta$ decay (with a half-life of 20 min) to $^{239}$Np, which in turn decays by $\beta$ decay (with a half-life of 2.35 days) to the fissionable nuclide $^{239}$Pu. Since $^{239}$Pu fissions with fast neutrons, no moderator is needed. A reactor initially fueled with a mixture of $^{238}$U and $^{239}$Pu will breed as much fuel as it uses or more if one or more of the neutrons emitted in the fission of $^{239}$Pu is captured by $^{238}$U. Practical studies indicate that a typical breeder reactor can be expected to double its fuel supply in 7 to 10 years.

There are two major safety problems inherent with breeder reactors. The fraction of delayed neutrons is only 0.3 percent for the fission of $^{239}$Pu, so the time between generations is much less than that for ordinary reactors. Mechanical control is therefore much more difficult. Also, because the operating temperature of a breeder reactor is relatively high and a moderator is not desired, a heat-transfer material, such as liquid sodium metal, is used rather than water (which is the moderator as well as the heat-transfer material in an ordinary reactor). If the temperature of the reactor increases, the resulting decrease in the density of the heat-transfer material leads to positive feedback, since it will absorb fewer neutrons than before. Because of these safety considerations, breeder reactors are not yet in commercial use in the United States. There are, however, several in operation in France, Great Britain, and the former Soviet Union.

## Fusion

In fusion, two light nuclei, such as deuterium ($^2$H) and tritium ($^3$H), fuse together to form a heavier nucleus. A typical fusion reaction is

$$^2H + {}^3H \rightarrow {}^4He + n + 17.6 \text{ MeV}$$

The energy released in fusion depends on the particular reaction. For the $^2$H + $^3$H reaction, the energy released is 17.6 MeV. Although this is less than the energy released in a fission reaction, it is a greater amount of energy per unit mass. The energy released in this fusion reaction is (17.6 MeV)/(5 nucleons) = 3.52 MeV per nucleon. This is approximately 3.5 times as great as the 1 MeV per nucleon released in fission.

The production of power from the fusion of light nuclei holds great promise because of the relative abundance of the fuel and the absence of some of the dangers inherent in fission reactors. Unfortunately, the technology necessary to make fusion a practical source of energy has not yet been developed. We will consider the $^2$H + $^3$H reaction; other reactions present similar problems.

Because of the Coulomb repulsion between the $^2$H and $^3$H nuclei, very large kinetic energies, of the order of 1 MeV, are needed to get the nuclei close enough together for the attractive nuclear forces to become effective and to cause fusion. Such energies can be obtained in an accelerator, but since the scattering of one nucleus by the other is much more probable than fusion, the bombardment of one nucleus by another in an accelerator requires the input of more energy than is recovered. To obtain energy from fusion, the particles must be heated to a temperature great enough for the fusion reaction to occur as the result of random thermal collisions. Because a significant number of particles have kinetic energies greater than the mean kinetic energy, $\frac{3}{2}kT$, and because some particles can tunnel through the Coulomb barrier, a temperature $T$ corresponding to $kT \approx 10$ keV is adequate to ensure that a reasonable number of fusion reactions will occur if the density of the particles is sufficiently high. The temperature corresponding to $kT = 10$ keV is of the order of $10^8$ K. These temperatures occur in the interiors of stars, where such reactions are common. At these temperatures, a gas consists of positive ions and negative electrons and is called a **plasma.** One of the problems arising in attempts to produce controlled fusion reactions is the problem of confining the plasma long enough for the reactions to take place. In the interior of the sun, the plasma is confined by the enormous gravitational field of the sun. In a laboratory on the earth, confinement is a difficult problem.

The energy required to heat a plasma is proportional to the number density of its ions, $n$, whereas the collision rate is proportional to $n^2$, the square of the number density. If $\tau$ is the confinement time, the output energy is proportional to $n^2\tau$. If the output energy is to exceed the input energy, we must have

$$C_1 n^2 \tau > C_2 n$$

where $C_1$ and $C_2$ are constants. In 1957, the British physicist J. D. Lawson evaluated these constants from estimates of the efficiencies of various hypothetical fusion reactors and derived the following relation between density and confinement time, known as **Lawson's criterion:**

$$n\tau > 10^{20} \text{ s·particles/m}^3 \qquad \text{40-22}$$

LAWSON'S CRITERION

If Lawson's criterion is met and the thermal energy of the ions is great enough ($kT \sim 10$ keV), the energy released by a fusion reactor will just equal the energy input; that is, the reactor will just break even. For the reactor to be practical, much more energy must be released.

Two schemes for achieving Lawson's criterion are currently under investigation. In one scheme, **magnetic confinement,** a magnetic field is used to confine the plasma (see Section 26-2). In the most common arrangement, first developed in the former USSR and called the Tokamak, the plasma is confined in a large toroid. The magnetic field is a combination of the doughnut-shaped magnetic field due to the windings of the toroid and the self-field due to the current of the circulating plasma. The break-even point has been achieved recently using magnetic confinement, but we are still a long way from building a practical fusion reactor.

In a second scheme, called **inertial confinement,** a pellet of solid deuterium and tritium is bombarded from all sides by intense pulsed laser beams of energies of the order of $10^4$ J lasting about $10^{-8}$ s. (Intense beams of ions are also used.) Computer simulation studies indicate that the pellet should be compressed to approximately $10^4$ times its normal density and heated to a temperature greater than $10^8$ K. This should produce approximately $10^6$ J of fusion energy in $10^{-11}$ s, which is so brief that confinement is achieved by inertia alone.

Because the break-even point is just barely being achieved in magnetic-confinement fusion, and because the building of a fusion reactor involves many practical problems that have not yet been solved, the availability of fusion to meet our energy needs is not expected for at least several decades. However, fusion holds great promise as an energy source for the future.

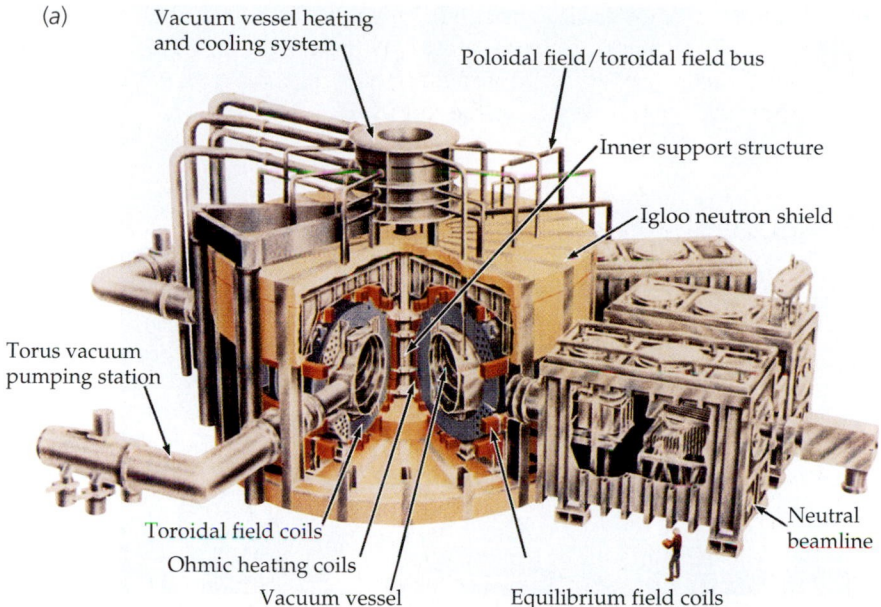

(a)

Vacuum vessel heating and cooling system

Poloidal field/toroidal field bus

Inner support structure

Igloo neutron shield

Torus vacuum pumping station

Toroidal field coils

Ohmic heating coils

Vacuum vessel

Equilibrium field coils

Neutral beamline

(b)

(c)

(*a*) Schematic of the Tokamak Fusion Test Reactor (TFTR). The toroidal coils, surrounding the doughnut-shaped vacuum vessel, are designed to conduct current for 3-s pulses, separated by waiting times of 5 min. Pulses peak at 73,000 A, producing a magnetic field of 5.2 T. This magnetic field is the principal means of confining the deuterium–tritium plasma that circulates within the vacuum vessel. Current for the pulses is delivered by converting the rotational energy of two 600-ton flywheels. Sets of poloidal coils, perpendicular to the toroidal coils, carry an oscillating current that generates a current through the confined plasma itself, heating it ohmically. Additional poloidal fields help stabilize the confined plasma. Between four and six neutral-beam injection systems (only one of which is shown in the schematic) are used to inject high-energy deuterium atoms into the deuterium–tritium plasma, heating beyond what could be obtained ohmically, ultimately to the point of fusion. (*b*) The TFTR itself. The diameter of the vacuum vessel is 7.7 m. (*c*) An 800-*kA* plasma, lasting 1.6 s, as it discharges within the vacuum vessel.

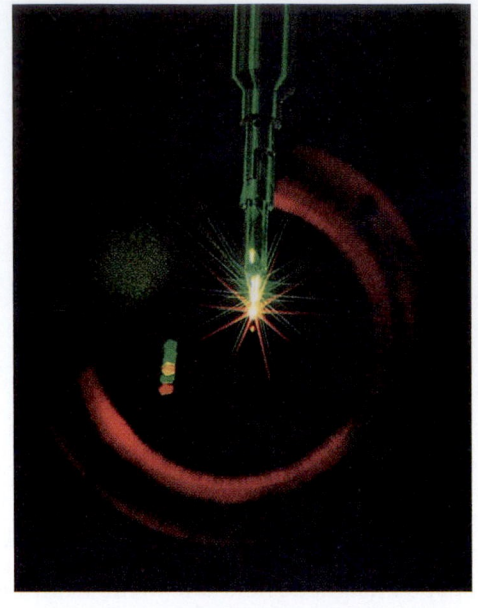

(a)    (b)

(a) The Nova target chamber, an aluminum sphere approximately 5 m in diameter, inside which 10 beams from the world's most powerful laser converge onto a hydrogen-containing pellet 0.5 mm in diameter. (b) The resulting fusion reaction is visible as a tiny star, lasting $10^{-10}$ s, releasing $10^{13}$ neutrons.

# SUMMARY

| Topic | Relevant Equations and Remarks |
|---|---|
| 1. **Properties of Nuclei** | Nuclei have $N$ neutrons, $Z$ protons, and a mass number $A = N + Z$. For light nuclei, $N$ and $Z$ are approximately equal, whereas for heavy nuclei, $N$ is greater than $Z$. |
| Isotopes | Isotopes consist of two or more nuclei with the same atomic number $Z$ but with different values of $N$ and $A$. |
| Size and shape | Most nuclei are approximately spherical in shape and have a volume that is proportional to $A$. Because the mass is proportional to $A$, nuclear density is independent of $A$. |
| Radius | $R = R_0 A^{1/3} \approx (1.2 \text{ fm}) A^{1/3}$     **40-1** |
| Mass and binding energy | The mass of a stable nucleus is less than the sum of the masses of its nucleons. The mass difference $\Delta m$ times $c^2$ equals the binding energy $E_b$ of the nucleus. The binding energy is approximately proportional to the mass number $A$. |
| 2. **Radioactivity** | Unstable nuclei are radioactive and decay by emitting $\alpha$ particles ($^4$He nuclei), $\beta$ particles (electrons or positrons), or $\gamma$-rays (photons). All radioactivity is statistical in nature and follows an exponential decay law: |
| | $N = N_0 e^{-\lambda t}$     **40-6** |

| | | |
|---|---|---|
| Decay law | $R = \lambda N = R_0 e^{-\lambda t}$ | **40-7** |
| Mean life | $\tau = \dfrac{1}{\lambda}$ | **40-9** |
| Half-life | $t_{1/2} = 0.693\tau$ | **40-11** |

The half-lives of $\alpha$ decay range from a fraction of a second to millions of years. For $\beta$ decay, the half-lives range up to hours or days. For $\gamma$ decay, the half-lives are usually less than a microsecond.

**Decay-rate units**

The number of decays per second of 1 g of radium is the curie (Ci).

$1\ \text{Ci} = 3.7 \times 10^{10}\ \text{decays/s} = 3.7 \times 10^{10}\ \text{Bq}$

$(1\ \text{Bq} = 1\ \text{decay/s})$

## 3. Nuclear Reactions

**Q value**

The $Q$ value equals $c^2$ times the total mass of the incoming particles less the total mass of the outgoing particles in the center of mass reference frame. If the net mass change is $\Delta m$, the $Q$ value is

$$Q = -(\Delta m)c^2 \qquad\qquad \textbf{40-19}$$

**Exothermic reaction**

The mass decreases, $Q$ is positive and measures the energy released.

**Endothermic reaction**

The mass increases, $Q$ is negative. Then $|Q|$ is the threshold energy for the reaction in the center of mass reference frame.

## 4. Fission

Fission occurs when some heavy elements, such as $^{235}$U or $^{239}$Pu, capture a neutron and split apart into two medium-mass nuclei. The two nuclei then fly apart because of electrostatic repulsion, releasing a large amount of energy. A chain reaction is possible because several neutrons are emitted by a nucleus when it undergoes fission. A chain reaction can be sustained in a reactor if, on the average, one of the emitted neutrons is slowed down by scattering in the reactor and is then captured by another fissionable nucleus. Very heavy nuclei ($Z > 92$) are subject to spontaneous fission.

## 5. Fusion

A large amount of energy is released when two light nuclei, such as $^2$H and $^3$H, fuse together. Fusion takes place spontaneously inside the sun and other stars, where the temperature is great enough (about $10^8$ K) for thermal motion to bring the charged hydrogen ions close enough together to fuse. Although controlled fusion holds great promise as a future energy source, practical difficulties have thus far prevented its development.

**Lawson criterion**

The minimum product of particle density $n$ and confinement time $\tau$ to get more energy out of a fusion reactor than is put in is $n\tau > 10^{20}$ s·particles/m$^3$.

# PROBLEMS

- • Single-concept, single-step, relatively easy
- •• Intermediate-level, may require synthesis of concepts
- ••• Challenging
- SSM Solution is in the *Student Solutions Manual*
- ISOLVE Problems available on iSOLVE online homework service
- ISOLVE ✓ These "Checkpoint" online homework service problems ask students additional questions about their confidence level, and how they arrived at their answer.

In a few problems, you are given more data than you actually need; in a few other problems, you are required to supply data from your general knowledge, outside sources, or informed estimates.

## Conceptual Problems

**1** • Give the symbols for two other isotopes of (a) $^{14}$N, (b) $^{56}$Fe, and (c) $^{118}$Sn.

**2** • Why is the decay series $A = 4n + 1$ not found in nature?

**3** • A decay by $\alpha$ emission is often followed by $\beta$ decay. When this occurs, it is by $\beta^-$ and not $\beta^+$ decay. Why?

**4** • SSM The half-life of $^{14}$C is much less than the age of the universe, yet $^{14}$C is found in nature. Why?

**5** • What effect would a long-term variation in cosmic-ray activity have on the accuracy of $^{14}$C dating?

**6** • Why is there not an element with $Z = 130$?

**7** • Why is a moderator needed in an ordinary nuclear fission reactor?

**8** • Explain why water is more effective than lead in slowing down fast neutrons.

**9** • The stable isotope of sodium is $^{23}$Na. What kind of radioactivity would you expect of (a) $^{22}$Na and (b) $^{24}$Na?

**10** • What is the advantage of a breeder reactor over an ordinary reactor? What are the disadvantages?

**11** • True or false:

(a) The atomic nucleus contains protons, neutrons, and electrons.
(b) The mass of $^2$H is less than the mass of a proton plus a neutron.
(c) After two half-lives, all the radioactive nuclei in a given sample have decayed.
(d) In a breeder reactor, fuel can be produced as fast as it is consumed.

**12** • Why do extreme changes in the temperature or the pressure of a radioactive sample have little or no effect on the radioactivity?

**13** • SSM Write balanced reactions for each of the following nuclear decays: (a) beta decay of $^{16}$N, (b) alpha decay of $^{248}$Fm, (c) positron decay of $^{12}$N, (d) beta decay of $^{81}$Se, (e) positron decay of $^{61}$Cu, and (f) alpha decay of $^{228}$Th.

**14** • SSM Write and balance reaction equations for each of the following: (a) $^{240}$Pu undergoes spontaneous fission to form two fission fragments and three neutrons. One of the fission fragments is a $^{90}$Sr nucleus. (b) A $^{72}$Ge nucleus absorbs an alpha particle and ejects a photon. (c) An $^{127}$I nucleus absorbs a deuteron and ejects a neutron. (d) A $^{235}$U nucleus absorbs a slow neutron and fissions forming a $^{113}$Ag nucleus, two neutrons, and another fission fragment. (e) A $^{55}$Mn nucleus is struck with a high-energy $^7$Li nucleus resulting in a triton, $^3$H, and a new nucleus. (f) $^{238}$U absorbs a slow neutron, resulting in a compound nucleus that emits a beta particle, followed a short time later by a second beta particle. What is the resulting nucleus?

## Estimation and Approximation

**15** • We found in Chapter 25 that the ratio of the resistivity of the most insulating material to that of the least resistive material (excluding superconductors) is approximately $10^{22}$. There are very few properties of materials that show such a wide range of values. Using information in the textbook or other resources, find the ratio of largest to smallest for some nuclear related properties of matter. Some examples might be the range of mass densities found in an atom, the half-life of radioactive nuclei, or the range of nuclear masses.

**16** •• According to the United States Department of Energy, the U.S. population consumes approximately $10^{20}$ joules of energy each year. Estimate the mass (in kg) of (a) uranium that would be needed to produce this much energy using nuclear fission and (b) deuterium and tritium that would be needed to produce this much energy using nuclear fusion.

## Properties of Nuclei

**17** • SSM Calculate the binding energy and the binding energy per nucleon from the masses given in Table 40-1 for (a) $^{12}$C, (b) $^{56}$Fe, and (c) $^{238}$U.

**18** • Repeat Problem 17 for (a) $^6$Li, (b) $^{39}$K, and (c) $^{208}$Pb.

**19** • Use Equation 40-1 to compute the radii of the following nuclei: (a) $^{16}$O, (b) $^{56}$Fe, and (c) $^{197}$Au.

**20** • In a fission process, a $^{239}$Pu nucleus splits into two nuclei whose mass number ratio is 3 to 1. Calculate the radii of the nuclei formed in this process.

**21** •• SSM The neutron, when isolated from an atomic nucleus, decays into a proton, an electron, and an antineutrino as follows: $^1_0 n \rightarrow ^1_1 H + ^{\,0}_{-1} e + ^0_0 \bar{v}$. The thermal energy of a neutron is of the order of $kT$, where $k$ is the Boltzmann constant. (*a*) In both joules and electron volts, calculate the energy of a thermal neutron at 25°C. (*b*) What is the speed of this thermal neutron? (*c*) A beam of monoenergetic thermal neutrons is produced at 25°C with an intensity $I$. After traveling 1350 km, the beam has an intensity of $I/2$. Using this information, estimate the half-life of the neutron. Express your answer in minutes.

**22** • Use Equation 40-1 for the radius of a spherical nucleus and the approximation that the mass of a nucleus of mass number $A$ is $A \times (1u)$ to calculate the density of nuclear matter in grams per cubic centimeter.

**23** •• Consider the following fission process: $^{235}_{92}U + ^1_0n \rightarrow ^{95}_{42}Mo + ^{139}_{57}La + 2^1_0n$. Determine the electrostatic potential energy, in MeV, of the reaction products when the $^{95}$Mo nucleus and the $^{139}$La nucleus are just touching immediately after being formed in the fission process.

**24** •• **SSM** In 1920, 12 years before the discovery of the neutron, Ernest Rutherford argued that proton–electron pairs might exist in the confines of the nucleus in order to explain the mass number, $A$, being greater than the nuclear charge, $Z$. He also used this argument to account for the source of beta particles in radioactive decay. Rutherford's scattering experiments in 1910 showed that the nucleus had a diameter of approximately 10 fm. Using this nuclear diameter, the uncertainty principle, and given that beta particles have an energy range of 0.02 MeV to 3.40 MeV, show why electrons cannot be contained within the nucleus.

## Radioactivity

**25** • **SOLVE** Homer enters the visitors' chambers, and his Geiger beeper goes off. He shuts off the beep, removes the device from his shoulder patch, and holds it near the only new object in the room—an orb that is to be presented as a gift from the visiting Cartesians. Pushing a button marked "monitor," Homer reads that the orb is a radioactive source with a counting rate of 4000 counts/s. After 10 min, the counting rate has dropped to 1000 counts/s. The source's half-life appears on the Geiger-beeper display. (*a*) What is the half-life of the source? (*b*) What will the counting rate be 20 min after the monitoring device was switched on?

**26** • **SOLVE** A certain source gives 2000 counts/s at time $t = 0$. Its half-life is 2 min. What is the counting rate after (*a*) 4 min, (*b*) 6 min, and (*c*) 8 min?

**27** • **SOLVE** The counting rate from a radioactive source is 8000 counts/s at time $t = 0$, and 10 min later the rate is 1000 counts/s. (*a*) What is the half-life? (*b*) What is the decay constant? (*c*) What is the counting rate after 20 min?

**28** • **SOLVE** The half-life of radium is 1620 y. Calculate the number of disintegrations per second of 1 g of radium, and show that the disintegration rate is approximately 1 Ci.

**29** • **SOLVE** A radioactive silver foil ($t_{1/2} = 2.4$ min) is placed near a Geiger counter and 1000 counts/s are observed at time $t = 0$. (*a*) What is the counting rate at $t = 2.4$ min and at $t = 4.8$ min? (*b*) If the counting efficiency is 20 percent, how many radioactive nuclei are there at time $t = 0$? At time

$t = 2.4$ min? (*c*) At what time will the counting rate be about 30 counts/s?

**30** • Use Table 40-1 to calculate the energy in MeV for the $\alpha$ decay of (*a*) $^{226}$Ra and (*b*) $^{242}$Pu.

**31** •• **SSM** Plutonium is a highly hazardous and toxic material to the human body. Once it enters the body it collects primarily in the bones, although it also can be found in other organs. Red blood cells are synthesized within the marrow of the bones. The isotope $^{239}$Pu is an alpha emitter with a half-life of 24,360 years. Since alpha particles are an ionizing radiation, the blood-making ability of the marrow is, in time, destroyed by the presence of $^{239}$Pu. In addition, many kinds of cancers will also be initiated in the surrounding tissues by the ionizing effects of the alpha particles. (*a*) If a person accidentally ingested 2.0 µg of $^{239}$Pu and it is absorbed by the bones of the victim, how many alpha particles are produced per second within the skeleton? (*b*) When, in years, will the activity be 1000 alpha particles per second?

**32** •• Consider a parent, $^A_ZX$, alpha-emitting nucleus initially at rest. The nucleus decays into a daughter nucleus, $^{A-4}_{Z-2}Y$, and an alpha particle as follows: $^A_ZX \rightarrow ^{A-4}_{Z-2}Y + ^4_2\alpha + Q$. (*a*) Show that the alpha particle has a kinetic energy of $(A-4)Q/A$. (*b*) Show that the kinetic energy of the recoiling daughter nucleus is given by $K_Y = 4Q/A$.

**33** •• **SSM** The fissile material $^{239}$Pu is an alpha emitter. Write the equation of this reaction. Given that $^{239}$Pu, $^{235}$U, and an alpha particle have respective masses of 239.052 156 u, 235.043 923 u, and 4.002 603 u, use the relations appearing in Problem 32 to calculate the kinetic energies of the alpha particle and the recoiling daughter nucleus.

**34** • Through a friend in the security department at the museum, Angela obtained a sample with 175 g of carbon. The decay rate of $^{14}$C was 8.1 Bq. How old is the sample?

**35** • **SOLVE** A sample of a radioactive isotope is found to have an activity of 115.0 Bq immediately after it is pulled from the reactor that formed the isotope. Its activity 2 h 15 min later is measured to be 85.2 Bq. (*a*) Calculate the decay constant and the half-life of the sample. (*b*) How many radioactive nuclei were there in the sample initially?

**36** •• **SSM** Radiation has long been used in medical therapy to control the development and growth of cancer cells. Cobalt-60, a gamma emitter of 1.17 MeV and 1.33 MeV energies, is used to irradiate and destroy deep-seated cancers. Small needles made of $^{60}$Co of a specified activity are encased in gold and used as body implants in tumors for time periods that are related to tumor size, tumor cell reproductive rate, and the activity of the needle. (*a*) A 1.00 µg sample of $^{60}$Co, with a half-life of 5.27 y, is prepared in the cyclotron of a medical center to irradiate a small internal tumor with gamma rays. In curies, determine the initial activity of the sample. (*b*) What is the activity of the sample after 1.75 y?

**37** •• **SOLVE** Measurements of the activity of a radioactive sample have yielded the results shown in the following table. Plot the activity as a function of time, using semilogarithmic paper, and determine the decay constant and the half-life of the radioisotope.

| Time, min | Activity | Time, min | Activity |
|-----------|----------|-----------|----------|
| 0 | 4287 | 20 | 880 |
| 5 | 2800 | 30 | 412 |
| 10 | 1960 | 40 | 188 |
| 15 | 1326 | 60 | 42 |

**38** •• (a) Show that if the decay rate is $R_0$ at time $t = 0$ and $R_1$ at some later time $t_1$, the decay constant is given by $\lambda = t_1^{-1} \ln(R_0/R_1)$ and the half-life is given by $t_{1/2} = 0.693 t_1 / \ln(R_0/R_1)$. (b) Use these results to find the decay constant and the half-life if the decay rate is 1200 Bq at $t = 0$ and 800 Bq at $t_1 = 60$ s.

**39** •• **ISOLVE** A wooden casket is thought to be 18,000 years old. How much carbon would have to be recovered from this object to yield a $^{14}$C counting rate of no less than 5 counts/min?

**40** •• **ISOLVE** A 1.00-mg sample of substance of atomic mass 59.934 u emits $\beta$ particles with an activity of 1.131 Ci. Find the decay constant for this substance in s$^{-1}$ and its half-life in years.

**41** •• **SSM** The counting rate from a radioactive source is measured every minute. The resulting counts per second are 1000, 820, 673, 552, 453, 371, 305, and 250. Plot the counting rate versus time on semilogarithmic graph paper, and use your graph to find the half-life of the source.

**42** •• **ISOLVE** A sample of radioactive material is initially found to have an activity of 115.0 decays/min. After 4 d 5 h, its activity is measured to be 73.5 decays/min. (a) Calculate the half-life of the material. (b) How long (from the initial time) will it take for the sample to reach an activity level of 10.0 decays/min?

**43** •• **ISOLVE**✓ The rubidium isotope $^{87}$Rb is a $\beta$ emitter with a half-life of $4.9 \times 10^{10}$ y that decays into $^{87}$Sr. It is used to determine the age of rocks and fossils. Rocks containing the fossils of early animals contain a ratio of $^{87}$Sr to $^{87}$Rb of 0.0100. Assuming that there was no $^{87}$Sr present when the rocks were formed, calculate the age of these fossils.

**44** ••• If there are $N_0$ radioactive nuclei at time $t = 0$, the number that decay in some time interval $dt$ at time $t$ is $-dN = \lambda N_0 e^{-\lambda t}\, dt$. If we multiply this number by the lifetime $t$ of these nuclei, sum over all the possible lifetimes from $t = 0$ to $t = \infty$, and divide by the total number of nuclei, we get the mean lifetime $\tau$.

$$\tau = \frac{1}{N_0} \int_0^\infty t |dN| = \int_0^\infty t \lambda e^{-\lambda t}\, dt$$

Show that $\tau = 1/\lambda$.

## Nuclear Reactions

**45** • Using Table 40-1, find the $Q$ values for the following reactions: (a) $^1$H + $^3$H → $^3$He + n + $Q$ and (b) $^2$H + $^2$H → $^3$He + n + $Q$.

**46** • Using Table 40-1, find the $Q$ values for the following reactions: (a) $^2$H + $^2$H → $^3$H + $^1$H + $Q$, (b) $^2$H + $^3$He → $^4$He + $^1$H + $Q$, and (c) $^6$Li + n → $^3$H + $^4$He + $Q$.

**47** •• **SSM** (a) Use the atomic masses $m = 14.003\,242$ u for $^{14}_{6}$C and $m = 14.003\,074$ u for $^{14}_{7}$N to calculate the $Q$ value (in MeV) for the $\beta$ decay

$$^{14}_{6}\text{C} \rightarrow {}^{14}_{7}\text{N} + \beta^- + \overline{v}_e$$

(b) Explain why you do not need to add the mass of the $\beta^-$ to that of atomic $^{14}_{7}$N for this calculation.

**48** •• (a) Use the atomic masses $m = 13.005\,738$ u for $^{13}_{7}$N and $m = 13.003\,354$ u for $^{13}_{6}$C to calculate the $Q$ value (in MeV) for the $\beta$ decay

$$^{13}_{7}\text{N} \rightarrow {}^{13}_{6}\text{C} + \beta^+ + v_e$$

(b) Explain why you need to add two electron masses to the mass of $^{13}_{6}$C in the calculation of the $Q$ value for this reaction.

## Fission and Fusion

**49** • **SSM** **ISOLVE**✓ Assuming an average energy of 200 MeV per fission, calculate the number of fissions per second needed for a 500-MW reactor.

**50** • **ISOLVE** If the reproduction factor in a reactor is $k = 1.1$, find the number of generations needed for the power level to (a) double, (b) increase by a factor of 10, and (c) increase by a factor of 100. Find the time needed in each case if (d) there are no delayed neutrons, so that the time between generations is 1 ms and (e) there are delayed neutrons that make the average time between generations 100 ms.

**51** •• **SSM** Consider the following fission reaction: $^{235}_{92}$U + $^1_0$n → $^{95}_{42}$Mo + $^{139}_{57}$La + $2^1_0$n + $Q$. The masses of the neutron, U, Mo, and La are 1.008 665 u, 235.043 923 u, 94.905 842 u, and 138.906 348 u, respectively. Calculate the $Q$ value, in MeV, for this fission reaction. Compare the result to the result obtained in Problem 23.

**52** •• **ISOLVE** In 1989, researchers claimed to have achieved fusion in an electrochemical cell at room temperature. They claimed a power output of 4 W from deuterium fusion reactions in the palladium electrode of their apparatus. If the two most likely reactions are

$$^2\text{H} + {}^2\text{H} \rightarrow {}^3\text{He} + \text{n} + 3.27 \text{ MeV}$$

and

$$^2\text{H} + {}^2\text{H} \rightarrow {}^3\text{He} + {}^1\text{H} + 4.03 \text{ MeV}$$

with 50 percent of the reactions going by each branch, how many neutrons per second would we expect to be emitted in the generation of 4 W of power?

**53** •• **ISOLVE** A fusion reactor that uses only deuterium for fuel would have the two reactions in Problem 52 taking place in the reactor. The $^3$H produced in the second reaction reacts immediately with another $^2$H to produce

$$^3\text{H} + {}^2\text{H} \rightarrow {}^4\text{He} + \text{n} + 17.6 \text{ MeV}$$

The ratio of $^2$H to $^1$H atoms in naturally occurring hydrogen is $1.5 \times 10^{-4}$. How much energy would be produced from 4 L of water if all of the $^2$H nuclei undergo fusion?

**54** ••• SSM iSOLVE The fusion reaction between $^2$H and $^3$H is

$$^3\text{H} + {}^2\text{H} \rightarrow {}^4\text{He} + \text{n} + 17.6 \text{ MeV}$$

Using the conservation of momentum and the given $Q$ value, find the final energies of both the $^4$He nucleus and the neutron, assuming the initial kinetic energy of the system is 1.00 MeV and the initial momentum of the system is zero.

**55** ••• Energy is generated in the sun and other stars by fusion. One of the fusion cycles, the proton–proton cycle, consists of the following reactions:

$$^1\text{H} + {}^1\text{H} \rightarrow {}^2\text{H} + \beta^+ + v_e$$

$$^1\text{H} + {}^2\text{H} \rightarrow {}^3\text{He} + \gamma$$

followed by

$$^1\text{H} + {}^3\text{H} \rightarrow {}^4\text{He} + \beta^+ + v_e$$

(a) Show that the net effect of these reactions is

$$4{}^1\text{H} \rightarrow {}^4\text{He} + 2\beta^+ + 2v_e + \gamma$$

(b) Show that rest energy of 24.7 MeV is released in this cycle (not counting the energy of 1.02 MeV released when each positron meets an electron and the two annihilate). (c) The sun radiates energy at the rate of approximately $4 \times 10^{26}$ W. Assuming that this is due to the conversion of four protons into helium plus $\gamma$-rays and neutrinos, which releases 26.7 MeV, what is the rate of proton consumption in the sun? How long will the sun last if it continues to radiate at its present level? (Assume that protons constitute about half of the total mass, $2 \times 10^{30}$ kg, of the sun.)

## General Problems

**56** • (a) Show that $ke^2 = 1.44$ MeV·fm, where $k$ is the Coulomb constant and $e$ is the electron charge. (b) Show that $hc = 1240$ MeV·fm.

**57** • SSM iSOLVE ✔ The counting rate from a radioactive source is 6400 counts/s. The half-life of the source is 10 s. Make a plot of the counting rate as a function of time for times up to 1 min. What is the decay constant for this source?

**58** • Find the energy needed to remove a neutron from (a) $^4$He and (b) $^7$Li.

**59** • The isotope $^{14}$C decays according to $^{14}\text{C} \rightarrow {}^{14}\text{N} + e^- + \bar{v}_e$. The atomic mass of $^{14}$N is 14.003 074 u. Determine the maximum kinetic energy of the electron. (Neglect recoil of the nitrogen atom.)

**60** • iSOLVE A neutron star is an object of nuclear density. If our sun were to collapse to a neutron star, what would be the radius of that object?

**61** •• SSM Show that the $^{109}$Ag nucleus is stable against alpha decay, $^{109}_{47}\text{Ag} \rightarrow {}^4_2\text{He} + {}^{105}_{45}\text{Rh} + Q$. The mass of the $^{109}$Ag nucleus is 108.904 756 u, and the products of the decay are 4.002 603 u and 104.905 250 u, respectively.

**62** •• Gamma rays can be used to induce photofission (fission triggered by the absorption of a photon) in nuclei. Calculate the threshold photon wavelength for the following nuclear reaction: $^2\text{H} + \gamma \rightarrow {}^1\text{H} + {}^1\text{n}$. Use Table 40-1 for the masses of the interacting particles.

**63** • The relative abundance of $^{40}$K (molecular mass 40.0 g/mol) is $1.2 \times 10^{-4}$. The isotope $^{40}$K is radioactive with a half-life of $1.3 \times 10^9$ y. Potassium is an essential element of every living cell. In the human body the mass of potassium constitutes approximately 0.36 percent of the total mass. Determine the activity of this radioactive source in a student whose mass is 60 kg.

**64** •• When a positron makes contact with an electron, the electron–positron pair annihilate via the reaction $\beta^+ + \beta^- \rightarrow 2\gamma$. Calculate the minimum total energy, in MeV, of the two photons created when a positron–electron pair annihilate.

**65** •• The isotope $^{24}$Na is a $\beta$ emitter with a half-life of 15 h. A saline solution containing this radioactive isotope with an activity of 600 kBq is injected into the bloodstream of a patient. Ten hours later, the activity of 1 mL of blood from this individual yields a counting rate of 60 Bq. Determine the volume of blood in this patient.

**66** •• SSM (a) Determine the closest distance of approach of an 8-MeV $\alpha$ particle in a head-on collision with a nucleus of $^{197}$Au and a nucleus of $^{10}$B, neglecting the recoil of the struck nuclei. (b) Repeat the calculation taking into account the recoil of the struck nuclei.

**67** •• Twelve nucleons are in a one-dimensional infinite square well of length $L = 3$ fm. (a) Using the approximation that the mass of a nucleon is 1 u, find the lowest energy of a nucleon in the well. Express your answer in MeV. What is the ground-state energy of the system of 12 nucleons in the well if (b) all the nucleons are neutrons so that there can be only 2 in each state and (c) 6 of the nucleons are neutrons and 6 are protons so that there can be 4 nucleons in each state? (Neglect the energy of Coulomb repulsion of the protons.)

**68** •• The helium nucleus or $\alpha$ particle is a very tightly bound system. Nuclei with $N = Z = 2n$, where $n$ is an integer (e.g., $^{12}$C, $^{16}$O, $^{20}$Ne, and $^{24}$Mg), may be thought of as agglomerates of $\alpha$ particles. (a) Use this model to estimate the binding energy of a pair of $\alpha$ particles from the atomic masses of $^4$He and $^{16}$O. Assume that the four $\alpha$ particles in $^{16}$O form a regular tetrahedron with one $\alpha$ particle at each vertex. (b) From the result obtained in Part (a) determine, on the basis of this model, the binding energy of $^{12}$C and compare your result with the result obtained from the atomic mass of $^{12}$C.

**69** •• Radioactive nuclei with a decay constant of $\lambda$ are produced in an accelerator at a constant rate $R_p$. The number of radioactive nuclei $N$ then obeys the equation $dN/dt = R_p - \lambda N$. (a) If $N$ is zero at $t = 0$, sketch $N$ versus $t$ for this situation. (b) The isotope $^{62}$Cu is produced at a rate of 100 per second by placing ordinary copper ($^{63}$Cu) in a beam of high-energy photons. The reaction is

$$\lambda + {}^{63}\text{Cu} \rightarrow {}^{62}\text{Cu} + \text{n}$$

$^{62}$Cu decays by $\beta$ decay with a half-life of 10 min. After a time long enough so that $dN/dt \approx 0$, how many $^{62}$Cu nuclei are there?

**70** •• SSM iSOLVE The total energy consumed in the United States in 1 y is approximately $7.0 \times 10^{19}$ J. How many kilograms of $^{235}$U would be needed to provide this amount of energy if we assume that 200 MeV of energy is released by each fissioning uranium nucleus, that all of the uranium atoms undergo fission, and that all of the energy-conversion mechanisms used are 100 percent efficient?

**71** •• (a) Find the wavelength of a particle in the ground state of a one-dimensional infinite square well of length $L = 2$ fm. (b) Find the momentum in units of MeV/$c$ for a particle with this wavelength. (c) Show that the total energy of an electron with this wavelength is approximately $E \approx pc$. (d) What is the kinetic energy of an electron in the ground state of this well? This calculation shows that if an electron were confined in a region of space as small as a nucleus, it would have a very large kinetic energy.

**72** •• If $^{12}$C, $^{11}$B, and $^{1}$H have respective masses of 12.000 000 u, 11.009 306 u, and 1.007 825 u, determine the minimum energy, $Q$, in MeV, required to remove a proton from a $^{12}$C nucleus.

**73** ••• SSM Assume that a neutron decays into a proton plus an electron without the emission of a neutrino. The energy shared by the proton and the electron is then 0.782 MeV. In the rest frame of the neutron, the total momentum is zero, so the momentum of the proton must be equal and opposite the momentum of the electron. This determines the relative energies of the two particles, but because the electron is relativistic, the exact calculation of these relative energies is somewhat difficult. (a) Assume that the kinetic energy of the electron is 0.782 MeV and calculate the momentum $p$ of the electron in units of MeV/$c$. (Hint: Use Equation 39-28.) (b) Using your result from Part (a), calculate the kinetic energy $p^2/2m_p$ of the proton. (c) Since the total energy of the electron plus the proton is 0.782 MeV, the calculation in Part (b) gives a correction to the assumption that the energy of the electron is 0.782 MeV. What percentage of 0.782 MeV is this correction?

**74** ••• Consider a neutron of mass $m$ moving with speed $v_L$ and making an elastic head-on collision with a nucleus of mass $M$ that is at rest in the laboratory frame of reference. (a) Show that the speed of the center of mass in the lab frame is $V = mv_L/(m + M)$. (b) What is the speed of the nucleus in the center-of-mass frame before the collision and after the collision? (c) What is the speed of the nucleus in the lab frame after the collision? (d) Show that the energy of the nucleus after the collision in the lab frame is

$$\frac{1}{2}M(2V)^2 = \frac{4mM}{(m + M)^2}\left(\frac{1}{2}mv_L^2\right)$$

(e) Show that the fraction of the energy lost by the neutron in this elastic collision is

$$\frac{-\Delta E}{E} = \frac{4mM}{(m + M)^2} = \frac{4(m/M)}{(1 + m/M)^2}$$

**75** ••• (a) Use the result from the Part (e) equation of Problem 74 to show that after $N$ head-on collisions of a neutron with carbon nuclei at rest, the energy of the neutron is approximately $(0.714)^N E_0$, where $E_0$ is its original energy. (b) How many head-on collisions are required to reduce the energy of the neutron from 2 MeV to 0.02 eV, assuming stationary carbon nuclei?

**76** ••• On the average, a neutron loses 63 percent of its energy in a collision with a hydrogen atom and 11 percent of its energy in a collision with a carbon atom. Calculate the number of collisions needed to reduce the energy of a neutron from 2 MeV to 0.02 eV if the neutron collides with (a) hydrogen atoms and (b) carbon atoms. (See Problem 75.)

**77** ••• SSM Frequently, the daughter of a radioactive parent is itself radioactive. Suppose the parent, designated by A, has a decay constant $\lambda_A$; while the daughter, designated by B, has a decay constant $\lambda_B$. The number of nuclei of B are then given by the solution to the differential equation

$$dN_B/dt = \lambda_A N_A - \lambda_B N_B$$

(a) Justify this differential equation. (b) Show that the solution for this equation is

$$N_B(t) = \frac{N_{A0}\lambda_A}{\lambda_B - \lambda_A}(e^{-\lambda_A t} - e^{-\lambda_B t})$$

where $N_{A0}$ is the number of A nuclei present at $t = 0$ when there are no B nuclei. (c) Show that $N_B(t) > 0$ whether $\lambda_A > \lambda_B$ or $\lambda_B > \lambda_A$. (d) Make a plot of $N_A(t)$ and $N_B(t)$ as a function of time when $\tau_B = 3\tau_A$.

**78** ••• Suppose isotope A decays to isotope B with a decay constant $\lambda_A$, and isotope B in turn decays with a decay constant $\lambda_B$. Suppose a sample contains, at $t = 0$, only isotope A. Derive an expression for the time at which the number of isotope B nuclei will be a maximum. (See Problem 77.)

**79** ••• An example of the situation discussed in Problem 77 is the radioactive isotope $^{229}$Th, an $\alpha$ emitter with a half-life of 7300 y. Its daughter, $^{225}$Ra, is a $\beta$ emitter with a half-life of 14.8 d. In this instance, as in many instances, the half-life of the parent is much longer than the half-life of the daughter. Using the expression given in Problem 77, Part (b), starting with a sample of pure $^{229}$Th containing $N_{A0}$ nuclei, show that the number, $N_B$, of $^{225}$Ra nuclei will, after several years, be a constant given by

$$N_B = \frac{\lambda_A}{\lambda_B}N_A$$

The number of daughter nuclei are said to be in *secular equilibrium*.

# Elementary Particles and the Beginning of the Universe

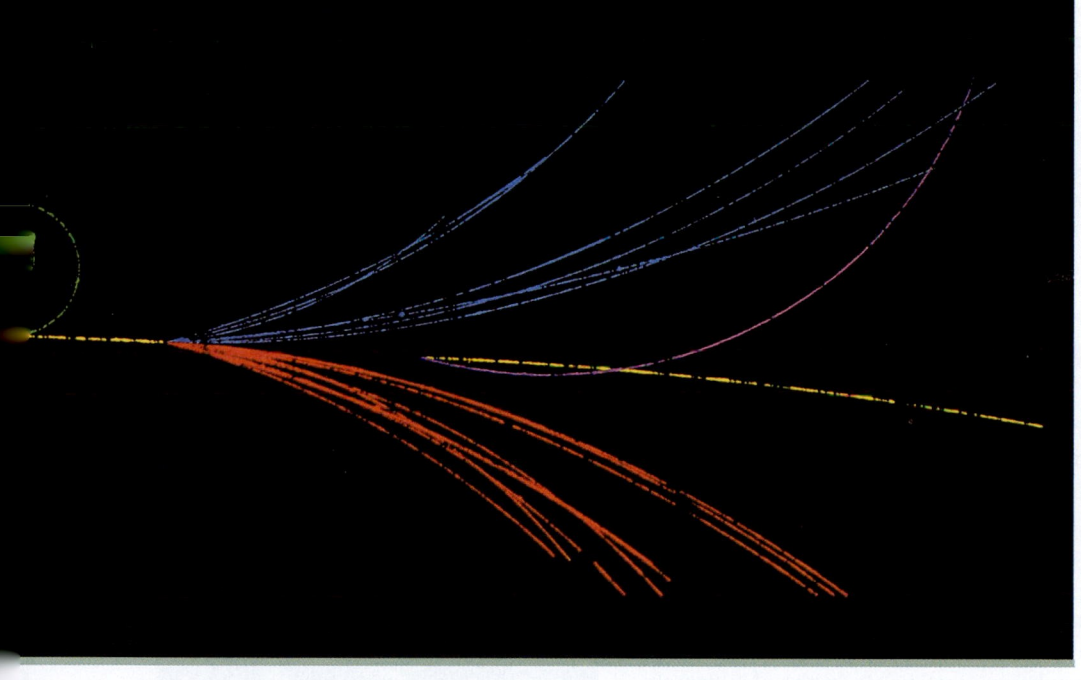

TRACKS IN A BUBBLE CHAMBER PRODUCED BY AN INCOMING HIGH-ENERGY PROTON (YELLOW) INCIDENT FROM THE LEFT, COLLIDING WITH A PROTON AT REST. THE SMALL GREEN SPIRAL IS AN ELECTRON KNOCKED OUT OF AN ATOM. IT CURVES TO THE LEFT BECAUSE OF AN EXTERNAL MAGNETIC FIELD IN THE CHAMBER. THE COLLISION PRODUCES SEVEN NEGATIVE PARTICLES ($\pi^-$)(BLUE); A NEUTRAL PARTICLE $\Lambda^0$ THAT LEAVES NO TRACK; AND NINE POSITIVE PARTICLES (RED) INCLUDING SEVEN $\pi^+$, A $K^+$, AND A PROTON. THE $\Lambda^0$ TRAVELS IN THE ORIGINAL DIRECTION OF THE INCOMING PROTON BEFORE DECAYING INTO A PROTON (YELLOW) AND A $\pi^-$ (PURPLE).

41-1   Hadrons and Leptons

41-2   Spin and Antiparticles

41-3   The Conservation Laws

41-4   Quarks

41-5   Field Particles

41-6   The Electroweak Theory

41-7   The Standard Model

41-8   The Evolution of the Universe

In John Dalton's atomic theory of matter (1808), the atom was considered to be the smallest indivisible constituent of matter, that is, an elementary particle. Then, with the discovery of the electron by J. J. Thomson (1897), the Bohr theory of the nuclear atom (1913), and the discovery of the neutron (1932), it became clear that atoms and even nuclei have considerable structure. For a time, it was thought that there were just four "elementary" particles: proton, neutron, electron, and photon. However, the positron or antielectron was discovered in 1932, and shortly thereafter the muon, the pion, and many other particles were predicted and discovered.

Since the 1950s, enormous sums of money have been spent constructing particle accelerators of greater and greater energies in hopes of finding particles predicted by various theories. At present, we know of several hundred particles that at one time or another have been considered to be elementary, and research teams at the giant accelerator laboratories around the world are searching for and finding new particles. Some of these particles have such short lifetimes (of the order of $10^{-23}$ s) that they can be detected only indirectly. Many particles are observed only in nuclear reactions with high-energy accelerators. In addition to the usual particle properties of mass, charge, and spin, new properties have been found and given whimsical names such as strangeness, charm, color, topness, and bottomness.

➤ In this chapter, we will first look at the various ways of classifying the multitude of particles that have been found. We will then describe the current theory of elementary particles, called the *standard model,* in which all matter in nature—from the exotic particles produced in the giant accelerator laboratories to ordinary grains of sand—is considered to be constructed from just two families of elementary particles, leptons and quarks. In the final section, we will use our knowledge of elementary particles to discuss the big bang theory of the origin of the universe.

# 41-1 Hadrons and Leptons

All the different forces observed in nature, from ordinary friction to the tremendous forces involved in supernova explosions, can be understood in terms of the four basic interactions: (1) the strong nuclear interaction (also called the hadronic interaction), (2) the electromagnetic interaction, (3) the weak (nuclear) interaction, and (4) the gravitational interaction. The four basic interactions provide a convenient structure for the classification of particles. Some particles participate in all four interactions, whereas other particles participate in only some of the interactions. For example, all particles participate in gravity, the weakest of the interactions. All particles that carry electric charge participate in the electromagnetic interaction.

Particles that interact via the strong interaction are called **hadrons.** There are two kinds of hadrons: **baryons,** which have spin $\frac{1}{2}$ (or $\frac{3}{2}$, $\frac{5}{2}$, etc.), and **mesons,** which have zero or integral spin. Baryons, which include nucleons, are the most massive of the elementary particles. Mesons have intermediate masses between the mass of the electron and the mass of the proton. Particles that decay via the strong interaction have very short lifetimes of the order of $10^{-23}$ s, which is about the time it takes light to travel a distance equal to the diameter of a nucleus. On the other hand, particles that decay via the weak interaction have much longer lifetimes of the order of $10^{-10}$ s. Table 41-1 lists some of the properties of those hadrons that are stable against decay via the strong interaction.

Hadrons are rather complicated entities with complex structures. If we use the term *elementary particle* to mean a point particle without structure that is not constructed from some more elementary entities, hadrons do not fit the bill. It is now believed that all hadrons are composed of more fundamental entities called *quarks,* which are truly elementary particles.

Particles that participate in the weak interaction but not in the strong interaction are called **leptons.** These include electrons, muons, and neutrinos, which are all less massive than the lightest hadron. The word *lepton,* meaning "light particle," was chosen to reflect the relatively small mass of these particles. However, the most recently discovered lepton, the *tau,* found by Martin Lewis Perl in 1975, has a mass of 1784 MeV/$c^2$, nearly twice the mass of the proton (938 MeV/$c^2$), so we now have a "heavy lepton." As far as we know, leptons are point particles with no structure and can be considered to be truly elementary in the sense that they are not composed of other particles.

There are six leptons. They are the electron and the electron neutrino, the muon and the muon neutrino, and the tau and the tau neutrino. (Each of these leptons has an antiparticle.) The masses of the electron, the muon, and the tau are quite different. The mass of the electron is 0.511 MeV/$c^2$, the mass of the muon is 106 MeV/$c^2$, and the mass of the tau is 1784 MeV/$c^2$. The standard model predicts that neutrinos, like photons, are without mass. Neutrinos were originally

The Super-Kamiokande detector, built in Japan in 1996 as a joint Japanese–American experiment, is essentially a water tank the size of a large cathedral installed in a deep zinc mine 1 mile inside a mountain. When neutrinos pass through the tank, one of the nutrinos occasionally collides with an atom, sending blue light through the water to an array of detectors. This photograph shows the detector wall and top with approximately 9000 photomultiplier tubes that help detect the neutrinos. Experimental results reported in June 1998 indicate that the mass of the neutrino cannot be zero.

## TABLE 41-1

**Hadrons That Are Stable Against Decay via the Strong Nuclear Interaction**

| Name | Symbol | Mass, MeV/$c^2$ | Spin, $\hbar$ | Charge, $e$ | Antiparticle | Mean Lifetime, s | Typical Decay Products[†] |
|------|--------|------|------|------|------|------|------|
| **Baryons** | | | | | | | |
| Nucleon | p (proton) | 938.3 | $\frac{1}{2}$ | +1 | $\mathrm{p}^-$ | Infinite | |
| | n (neutron) | 939.6 | $\frac{1}{2}$ | 0 | $\bar{\mathrm{n}}$ | 930 | $\mathrm{p} + \mathrm{e}^- + \bar{\nu}_e$ |
| Lambda | $\Lambda^0$ | 1116 | $\frac{1}{2}$ | 0 | $\bar{\Lambda}^0$ | $2.5 \times 10^{-10}$ | $\mathrm{p} + \pi^-$ |
| Sigma[‡] | $\Sigma^+$ | 1189 | $\frac{1}{2}$ | +1 | $\bar{\Sigma}^-$ | $0.8 \times 10^{-10}$ | $\mathrm{n} + \pi^+$ |
| | $\Sigma^0$ | 1193 | $\frac{1}{2}$ | 0 | $\bar{\Sigma}^0$ | $10^{-20}$ | $\Lambda^0 + \gamma$ |
| | $\Sigma^-$ | 1197 | $\frac{1}{2}$ | −1 | $\bar{\Sigma}^+$ | $1.7 \times 10^{-10}$ | $\mathrm{n} + \pi^-$ |
| Xi | $\Xi^0$ | 1315 | $\frac{1}{2}$ | 0 | $\bar{\Xi}^0$ | $3.0 \times 10^{-10}$ | $\Lambda^0 + \pi^0$ |
| | $\Xi^-$ | 1321 | $\frac{1}{2}$ | −1 | $\Xi^+$ | $1.7 \times 10^{-10}$ | $\Lambda^0 + \pi^-$ |
| Omega | $\Omega^-$ | 1672 | $\frac{3}{2}$ | −1 | $\Omega^+$ | $1.3 \times 10^{-10}$ | $\Xi^0 + \pi^-$ |
| **Mesons** | | | | | | | |
| Pion | $\pi^+$ | 139.6 | 0 | +1 | $\pi^-$ | $2.6 \times 10^{-8}$ | $\mu^+ + \nu_\mu$ |
| | $\pi^0$ | 135 | 0 | 0 | $\pi^0$ | $0.8 \times 10^{-16}$ | $\gamma + \gamma$ |
| | $\pi^-$ | 139.6 | 0 | −1 | $\pi^+$ | $2.6 \times 10^{-8}$ | $\mu^- + \bar{\nu}_\mu$ |
| Kaon[§] | $\mathrm{K}^+$ | 493.7 | 0 | +1 | $\mathrm{K}^-$ | $1.24 \times 10^{-8}$ | $\pi^+ + \pi^0$ |
| | $\mathrm{K}^0$ | 497.7 | 0 | 0 | $\bar{\mathrm{K}}^0$ | $0.88 \times 10^{-10}$ and | $\pi^+ + \pi^-$ |
| | | | | | | $5.2 \times 10^{-8}$ | $\pi^+ + \mathrm{e}^- + \bar{\nu}_e$ |
| Eta | $\eta^0$ | 549 | 0 | 0 | | $2 \times 10^{-19}$ | $\gamma + \gamma$ |

† Other decay modes also occur for most particles.
‡ The $\Sigma^0$ is included here for completeness even though it does decay via the strong interaction.
§ The $\mathrm{K}^0$ has two distinct lifetimes, sometimes referred to as $\mathrm{K}^0_{\text{short}}$ and $\mathrm{K}^0_{\text{long}}$. All other particles have a unique lifetime.

thought to be massless. However, there is now strong evidence that their mass, though very small, is greater than zero. In the late 1990s, experiments using a detector in Japan called the Super-Kamiokande (Super-K) found that neutrinos emitted from the sun arrived on the earth in much smaller numbers than the numbers predicted from the fusion processes in the sun. This can be explained if the mass of the neutrino were not zero.[†] In addition, a neutrino mass as small as a few eV/$c^2$ would have great cosmological significance. The answer to the question of whether the universe will continue to expand indefinitely or will reach a maximum size and begin to contract depends on the total mass in the universe. Thus, the answer could depend on whether the mass of the neutrino is actually zero, or is merely small, since the cosmic density of each species of neutrino is $\sim 100$ per cm$^3$. The observation of electron neutrinos from the supernova 1987A puts an upper limit on the mass of these neutrinos. Since the velocity of a particle with mass depends on its energy, the arrival time of a burst of neutrinos with mass from a supernova would be spread out in time. The fact that the electron neutrinos from the 1987 supernova all arrived at the earth within 13 s of one another results in an upper limit of about 16 eV/$c^2$ for their mass. Note that

† The connection between the shortfall of solar-neutrino detections and the mass of the neutrino is elucidated in "On Morphing Neutrinos and Why They Must Have Mass" by Eugene Hecht, *The Physics Teacher* 41 (2003): 164–168.

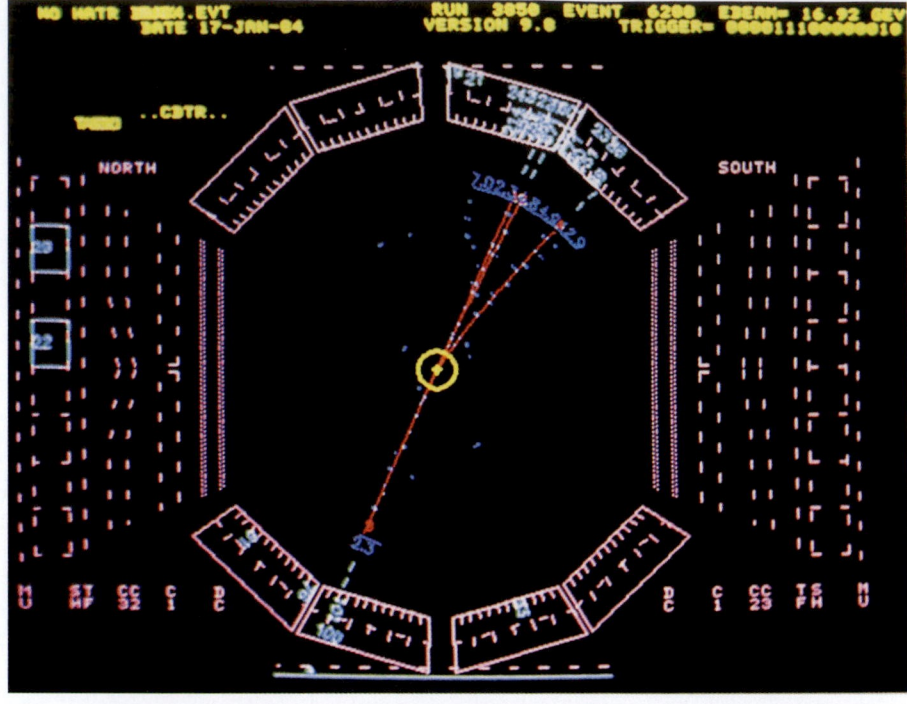

(a)

(b)

(a) A computer display of the production and decay of a $\tau^+$ and $\tau^-$ pair. An electron and a positron annihilate at the center marked by the yellow cross, producing a $\tau^+$ and $\tau^-$ pair, which travel in opposite directions, but quickly decay while still inside the beam pipe (yellow circle). The $\tau^+$ decays into two invisible neutrinos and a $\mu^+$, which travels toward the bottom left. Its track in the drift chamber is calculated by a computer and indicated in red. It penetrates the lead–argon counters outlined in purple and is detected at the blue dot near the bottom blue line that marks the end of a muon detector. The $\tau^-$ decays into three charged pions (red tracks moving upward) plus invisible neutrinos. (b) The Mark I detector, built by a team from the Stanford Linear Accelerator Center (SLAC) and the Lawrence Berkeley Laboratory, became famous for many discoveries, including the $\psi$/J meson and the $\tau$ lepton. Tracks of particles are recorded by wire spark chambers wrapped in concentric cylinders around the beam pipe extending out to the ring where physicist Carl Friedberg has his right foot. Beyond this are two rings of protruding tubes, housing photomultipliers that view various scintillation counters. The rectangular magnets at the left guide the counterrotating beams that collide in the center of the detector.

an upper limit does not imply that the mass is not zero. Measurements of the relative number of muon neutrinos and electron neutrinos entering the huge, underground Super-K detector suggest that at least one type of neutrino can oscillate between types (e.g., between a mu neutrino and a tau neutrino). Further measurements of antineutrinos from nuclear reactors strongly shows that all three types of neutrinos oscillate between types and thus have mass. Measurements made in Japan, using the *Kamioka Liquid Scintillator Anti-Neutrino Detector* (KamLAND), show that oscillations from one species of neutrino to another species of neutrino can be observed over path lengths as short as 180 km (Figure 41-1).

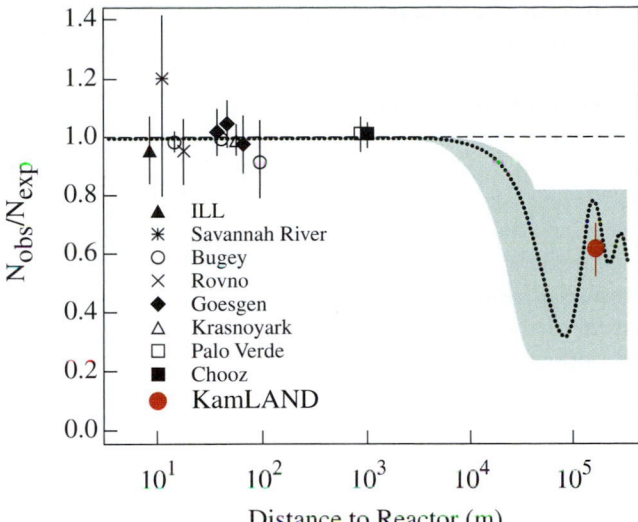

**FIGURE 41-1** First evidence for antineutrino disappearance. The ratio of the number of antineutrinos observed $N_{obs}$ to the number that one would expect to observe $N_{exp}$ (assuming no neutrino oscillations) is plotted versus distance to the nearest antineutrino sources. The KamLAND site is 180,000 m (180 km) from nearby antineutrino sources (nuclear reactors), while the other eight detector sites are less than 1000 m from nearby nuclear reactors. For these eight sites, $N_{obs}/N_{exp} = 1.0$, which is what is expected assuming no neutrino oscillations. However, the KamLAND detector found $N_{obs}/N_{exp} = 0.6$. This result is strong evidence that while neutrinos do not oscillate in significant numbers while traveling over path lengths of less than 1.0 km, they do oscillate in significant numbers while traveling over path lengths only a few orders of magnitude longer than that.

# 41-2 Spin and Antiparticles

One important characteristic of a particle is its intrinsic spin angular momentum. We have already discussed the fact that the electron has a quantum number $m_s$ that corresponds to the $z$ component of its intrinsic spin characterized by the quantum number $s = \frac{1}{2}$. Protons, neutrons, neutrinos, and the various other particles that also have an intrinsic spin characterized by the quantum number $s = \frac{1}{2}$ are called **spin-$\frac{1}{2}$ particles.** Particles that have spin $\frac{1}{2}$ (or $\frac{3}{2}, \frac{5}{2}$, etc.) are called fermions and obey the Pauli exclusion principle. Particles such as pions and other mesons have zero spin or integral spin ($s = 0, 1, 2$, etc.). These particles are called bosons and do not obey the Pauli exclusion principle. That is, any number of these particles can be in the same quantum state.

Spin-$\frac{1}{2}$ particles are described by the Dirac equation, which is an extension of the Schrödinger equation that includes special relativity. One feature of Paul Dirac's theory, proposed in 1927, is the prediction of the existence of antiparticles. In special relativity, the energy of a particle is related to the mass and the momentum of the particle by $E = \pm\sqrt{p^2c^2 + m^2c^4}$ (Equation 39-28). We usually choose the positive solution and dismiss the negative-energy solution with a physical argument. However, the Dirac equation requires the existence of wave functions that correspond to the negative-energy states. Dirac got around this difficulty by postulating that all the negative-energy states were filled and would therefore not be observable. Only holes in the "infinite sea" of negative-energy states would be observed. For example, a hole in the negative sea of electron energy states would appear as a particle identical to the electron except with positive charge. When such a particle came in the vicinity of an electron the two particles would annihilate, releasing two photons with a total energy of $2m_ec^2$, where $m_e$ is the mass of the electron. This interpretation received little attention until a particle with just these properties, called the positron, was discovered in 1932 by Carl Anderson.

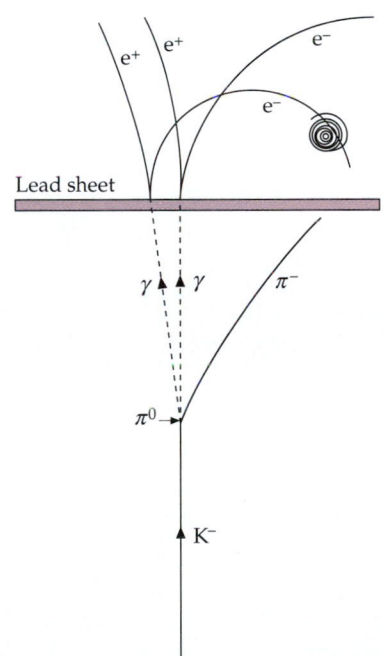

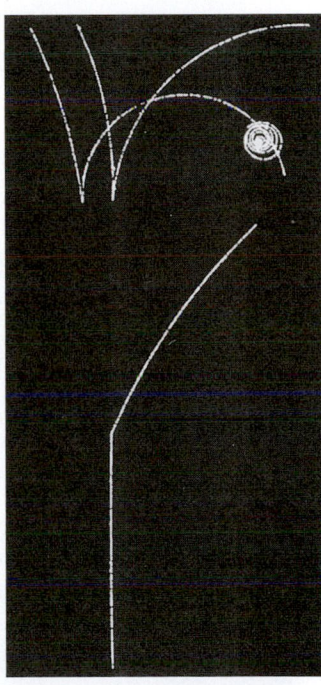

A negative kaon ($K^-$) enters a bubble chamber from the bottom and decays into a $\pi^-$, which moves off to the right, and a $\pi^0$, which immediately decays into two photons whose paths are indicated by the dashed lines in the drawing. Each photon interacts in the lead sheet, producing an electron–positron pair. The spiral at the right is another electron that has been knocked out of an atom in the chamber. (Other extraneous tracks have been removed from the photograph.)

Antiparticles are never created alone but always in particle–antiparticle pairs. In the creation of an electron–positron pair by a photon, the energy of the photon must be greater than the rest energy of the electron plus that of the positron, which is $2m_e c^2 \approx 1.02$ MeV. Although the positron is stable, it has only a short-term existence in our universe because of the large supply of electrons in matter. The fate of a positron is annihilation according to the reaction

$$e^+ + e^- \rightarrow \gamma + \gamma \tag{41-1}$$

The probability of this reaction is large only if the positron is at rest or nearly at rest. In the center-of-mass reference frame, the momentum of the two particles prior to annihilation is zero, so two photons moving in opposite directions are needed to conserve linear momentum.

The fact that we call electrons *particles* and positrons *antiparticles* does not imply that positrons are less fundamental than electrons. It merely reflects the nature of our universe. If our matter were made up of negative protons and positive electrons, then positive protons and negative electrons would suffer quick annihilation and would be called antiparticles.

The antiproton ($p^-$) was discovered in 1955 by Emilio Segrè and Owen Chamberlain using a beam of protons in the Bevatron at Berkeley to produce the reaction[†]

$$p^+ + p^+ \rightarrow p^+ + p^+ + p^+ + p^- \tag{41-2}$$

---

[†] The antiproton is sometimes denoted by $\bar{p}$ rather than $p^-$. For neutral particles, such as the neutron, the bar must be used to denote the antiparticle. Thus, the antineutron is denoted by $\bar{n}$. The normal electron and proton are often denoted by e and p without the minus or plus superscripts.

An aerial view of the European Laboratory for Particle Physics (CERN) just outside of Geneva, Switzerland. The large circle shows the Large Electron–Positron collider (LEP) tunnel, which is 27 km in circumference. The irregular dashed line is the border between France and Switzerland.

The creation of a proton–antiproton pair (Figure 41-2) requires kinetic energy of at least $2m_pc^2 = 1877$ MeV $= 1.877$ GeV in the zero-momentum reference frame in which the two protons approach each other with equal and opposite momenta. In the laboratory frame in which one of the protons is initially at rest, the kinetic energy of the incoming proton must be at least $6m_pc^2 = 5.63$ GeV (see Problem 40 of Chapter 39). This energy was not available in laboratories before the development of high-energy accelerators in the 1950s. Antiprotons annihilate with protons to produce two gamma rays in a reaction similar to the reaction in Equation 41-1.

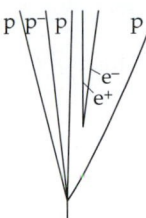

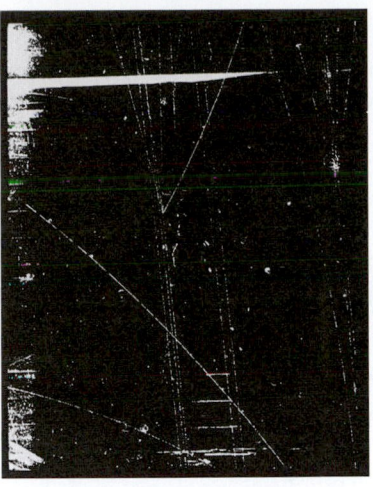

**FIGURE 41-2** Bubble-chamber tracks that show the creation of a proton–antiproton pair in the collision of an incident 25-GeV proton, with a stationary proton in liquid hydrogen.

The tunnel of the proton–antiproton collider at CERN. The same bending magnets and focusing magnets can be used for protons or antiprotons moving in opposite directions. The rectangular box in the foreground is a focusing magnet, and the next four boxes are the bending magnets.

*PROTON–ANTIPROTON ANNIHILATION*  **EXAMPLE 41-1**  **Put It in Context**

You have been reading about nuclear physics and particle interactions. In particular, you have been looking at the reaction $p^+ + p^- \rightarrow \gamma + \gamma$ (proton–antiproton annihilation). You wonder if the photons produced are visible to the human eye if the two protons are initially at rest. Are the photons visible to the human eye?

**PICTURE THE PROBLEM** If the photons are visible, they should have wavelengths in the visible range (400 nm to 800 nm). Because the proton and the antiproton are at rest, conservation of momentum requires that the two photons created in their annihilation have equal and opposite momenta and therefore equal energies, frequencies, and wavelengths. Conservation of energy implies that the photons have a combined energy equal to the rest energy of the proton plus the rest energy of the antiproton (approximately 938 MeV each).

1. Set the total energy of the two photons, $2E_\gamma$, equal to the rest energy of the proton plus antiproton and solve for $E_\gamma$:

$$2E_\gamma = 2m_pc^2$$

so

$$E_\gamma = m_pc^2 = 938 \text{ MeV}$$

2. Set the energy of the photon equal to $hf = hc/\lambda$ and solve for the wavelength $\lambda$:

$$E_\gamma = hf = \frac{hc}{\lambda}$$

$$\lambda = \frac{hc}{E_\gamma} = \frac{1240 \text{ eV·nm}}{938 \text{ MeV}}$$

$$= 1.32 \times 10^{-6} \text{ nm} = 1.32 \text{ fm}$$

3. Compare this wavelength with the wavelengths of visible light:

| The photons are *not* in the visible spectrum. |

**REMARKS** The wavelength of the photons is more than eight orders of magnitude less than 400 nm—the shortest wavelength in the visible spectrum.

# 41-3 The Conservation Laws

One of the maxims of nature is "anything that can happen does." If a conceivable decay or reaction does not occur, there must be a reason. The reason is usually expressed in terms of a conservation law. The conservation of energy rules out the decay of any particle for which the total mass of the decay products would be greater than the initial mass of the particle before decay. The conservation of linear momentum requires that when an electron and a positron at rest annihilate, two photons must be emitted. Angular momentum must also be conserved in a reaction or a decay. A fourth conservation law that restricts the possible particle decays and reactions is that of electric charge. The net electric charge before a decay or a reaction must equal the net charge after the decay or the reaction.

There are two additional conservation laws that are important in the reactions and the decays of elementary particles: the conservation of baryon number and the conservation of lepton number. Consider the proposed decay

$$p \rightarrow \pi^0 + e^+$$

This decay would conserve charge, energy, angular momentum, and linear momentum, but it does not occur. It does not conserve either lepton number or baryon number. The conservation of lepton number and baryon number implies that whenever a lepton or a baryon is created, an antiparticle of the same type is also created. We assign the **lepton number** $L = +1$ to all leptons, $L = -1$ to all antileptons, and $L = 0$ to all other particles. Similarly, the **baryon number** $B = +1$ is assigned to all baryons, $B = -1$ to all antibaryons, and $B = 0$ to all other particles. The baryon numbers and the lepton numbers cannot change in a reaction or a decay. The conservation of baryon number along with the conservation of energy implies that the least massive baryon, the proton, must be stable.

The conservation of lepton number implies that the neutrino emitted in the $\beta$ decay of the free neutron is an antineutrino:

$$n \rightarrow p^+ + e^- + \bar{\nu}_e \qquad\qquad 41\text{-}3$$

The fact that neutrinos and antineutrinos are different is illustrated by an experiment in which $^{37}$Cl is bombarded with an intense antineutrino beam from the decay of reactor neutrons. If neutrinos and antineutrinos were the same, we would expect the following reaction:

$$^{37}\text{Cl} + \bar{\nu}_e \rightarrow {}^{37}\text{Ar} + e^- \qquad\qquad 41\text{-}4$$

This reaction is not observed. However, if protons are bombarded with anti-neutrinos, the reaction

$$p + \bar{\nu}_e \rightarrow n + e^+ \tag{41-5}$$

*is* observed. Note that the lepton number is $-1$ on the left side of the reaction in Equation 41-4 and $+1$ on the right side of the reaction. But the lepton number is $-1$ on both sides of the reaction in Equation 41-5.

Not only are neutrinos and antineutrinos distinct particles, but the neutrinos associated with electrons are distinct from the neutrinos associated with muons. Electron-like leptons (e and $\nu_e$), muon-like leptons ($\mu$ and $\nu_\mu$), and tau-like leptons ($\tau$ and $\nu_\tau$) are each separately conserved, so we assign separate lepton numbers $L_e$, $L_\mu$, and $L_\tau$ to the particles. For e and $\nu_e$, $L_e = +1$; for their anti-particles, $L_e = -1$; and for all other particles, $L_e = 0$. The lepton numbers $L_\mu$ and $L_\tau$ are similarly assigned.

---

*What Laws Are Being Violated?*                    **EXAMPLE    41-2**

**What conservation laws (if any) are violated by the following proposed decays: (a) n $\rightarrow$ p + $\pi$, (b) $\Lambda^0 \rightarrow$ p$^-$ + $\pi^+$, and (c) $\mu^- \rightarrow$ e$^-$ + $\gamma$?**

(a) There are no leptons in this decay, so there is no problem with the conservation of lepton number. The net charge is zero before the decay and after the decay, so charge is conserved. Also, the baryon number is $+1$ before the decay and after the decay. However, the rest energy of the proton (938.3 MeV) plus the rest energy of the pion (139.6 MeV) is greater than the rest energy of the neutron (939.6 MeV). Thus, this decay violates the conservation of energy.

(b) Again, there are no leptons involved, and the net charge is zero before the decay and after the decay. Also, the rest energy of the $\Lambda^0$ (1116 MeV) is greater than the rest energy of the antiproton (938.3 MeV) plus the rest energy of the pion (139.6 MeV), so energy is conserved, with the loss in rest energy equaling the gain in kinetic energy of the decay products. However, this decay does not conserve baryon number, which is $+1$ for the $\Lambda^0$ and $-1$ for the antiproton.

(c) This reaction does not conserve muon lepton number or electron lepton number. The muon does decay via $\mu^- \rightarrow$ e$^-$ + $\bar{\nu}_e$ + $\nu_\mu$, which does conserve both muon lepton numbers and electron lepton numbers.

---

There are some conservation laws that are not universal but apply only to certain kinds of interactions. In particular, there are quantities that are conserved in decays and reactions that occur via the strong interaction but not in decays or reactions that occur via the weak interaction. One of these quantities that is particularly important is **strangeness,** introduced by M. Gell-Mann and K. Nishijima in 1952 to explain the strange behavior of some of the heavy baryons and mesons. Consider the reaction

$$p + \pi^- \rightarrow \Lambda^0 + K^0 \tag{41-6}$$

The proton and the pion interact via the strong interaction. Both the $\Lambda^0$ and $K^0$ decay into hadrons

$$\Lambda^0 \rightarrow p + \pi^- \tag{41-7}$$

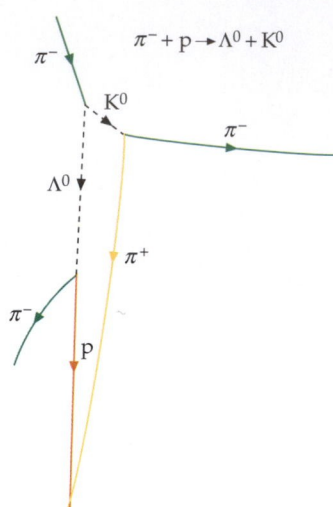

$$\pi^- + p \rightarrow \Lambda^0 + K^0$$

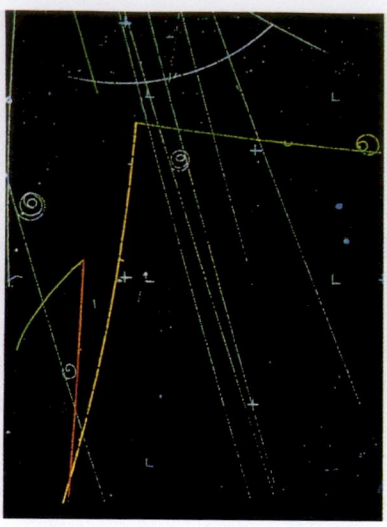

An early photograph of bubble-chamber tracks at the Lawrence Berkeley Laboratory, showing the production and the decay of two strange particles, the $K^0$ and the $\Lambda^0$. These neutral particles are identified by the tracks of their decay particles. The lambda particle was named because of the similarity of the tracks of its decay particles to the Greek letter $\Lambda$. (The blue tracks are particles not involved in the reaction of Equation 41-6.)

and

$$K^0 \rightarrow \pi^+ + \pi^- \qquad\qquad 41\text{-}8$$

However, the decay times for both the $\Lambda^0$ and $K^0$ are of the order of $10^{-10}$ s, which is characteristic of the weak interaction, rather than $10^{-23}$ s, which would be expected for the strong interaction. Other particles showing similar behavior were called **strange particles.** These particles are always produced in pairs and never singly, even when all other conservation laws are met. This behavior is described by assigning a new property called strangeness to these particles. In reactions and decays that occur via the strong interaction, strangeness is conserved. In reactions and decays that occur via the weak interaction, the strangeness can change by $\pm 1$. The strangeness of the ordinary hadrons—the nucleons and pions—was arbitrarily taken to be zero. The strangeness of the $K^0$ was arbitrarily chosen to be $+1$. The strangeness of the $\Lambda^0$ particle must then be $-1$ so that strangeness is conserved in the reaction of Equation 41-6. The strangeness of other particles could then be assigned by looking at their various reactions and decays. In reactions and decays that occur via the weak interaction, the strangeness can change by $\pm 1$.

Figure 41-3 shows the masses of the baryons and the mesons that are stable against decay via the strong interaction versus strangeness. We can see from this figure that these particles cluster in multiplets of one, two, or three particles of approximately equal mass, and that the strangeness of a multiplet of particles is related to the *center of charge* of the multiplet.

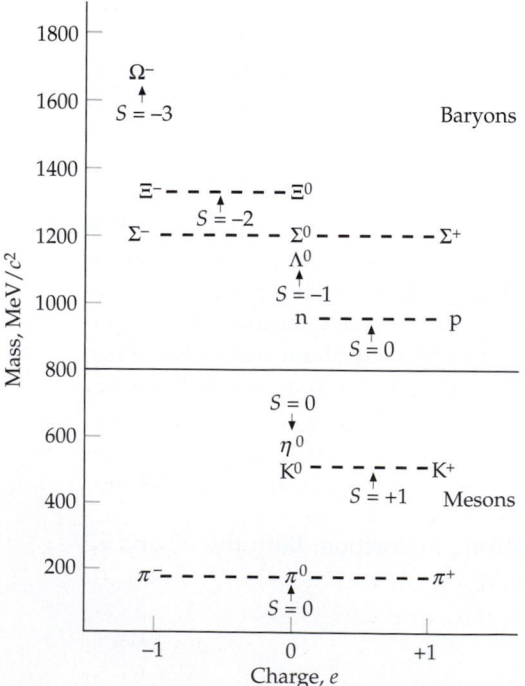

**FIGURE 41-3** The strangeness of hadrons shown on a plot of mass versus charge. The strangeness of a baryon-charge multiplet is related to the number of places the center of charge of the multiplet is displaced from that of the nucleon doublet. For each displacement of $e$, the strangeness changes by $\pm 1$. For mesons, the strangeness is related to the number of places the center of charge is displaced from that of the pion triplet. Because of the unfortunate original assignment of $+1$ for the strangeness of kaons, all of the baryons that are stable against decay via the strong interaction have negative or zero strangeness.

**EXAMPLE    41-3**

State whether the following decays can occur via the strong interaction, via the weak interaction, or not at all: (*a*) $\Sigma^+ \rightarrow p + \pi^0$, (*b*) $\Sigma^0 \rightarrow \Lambda^0 + \gamma$, and (*c*) $\Xi^0 \rightarrow n + \pi^0$.

**PICTURE THE PROBLEM**  We first note that the mass of each decaying particle is greater than the mass of the decay products, so there is no problem with energy conservation in any of the decays. In addition, there are no leptons involved in any of the decays, and charge and baryon number are both conserved in all the decays. The decay will occur via the strong interaction if strangeness is conserved. If $\Delta S = \pm 1$, the decay will occur via the weak interaction. If $|S|$ changes by more than 1, the decay will not occur.

(*a*) From Figure 41-3, we can see that the strangeness of the $\Sigma^+$ is $-1$, whereas the strangeness of both the proton and the pion is zero. This decay is possible via the weak interaction but not the strong interaction. It is, in fact, one of the decay modes of the $\Sigma^+$ particle with a lifetime of the order of $10^{-10}$ s.

(*b*) Since the strangeness of both the $\Sigma^0$ and $\Lambda^0$ is $-1$, this decay can proceed via the strong interaction. It is, in fact, the dominant mode of decay of the $\Sigma^0$ particle with a lifetime of approximately $10^{-20}$ s.

(*c*) The strangeness of the $\Xi^0$ is $-2$, whereas the strangeness of both the neutron and the pion is zero. Since strangeness cannot change by 2 in a decay or in a reaction, this decay cannot occur.

## 41-4    Quarks

Leptons appear to be truly elementary particles in that they do not break down into smaller entities and they seem to have no measurable size or structure. Hadrons, on the other hand, are complex particles with size and structure, and they decay into other hadrons. Furthermore, at the present time, there are only six known leptons, whereas there are many more hadrons. Except for the $\Sigma^0$ particle, Table 41-1 includes only hadrons that are stable against decay via the strong interaction. Hundreds of other hadrons have been discovered; their properties, such as charge, spin, mass, strangeness, and decay schemes, have been measured.

The most important advance in our understanding of elementary particles was the quark model proposed by M. Gell-Mann and G. Zweig in 1963 in which all hadrons consist of combinations of two or three truly elementary particles called **quarks.**[†] In the original model, quarks came in three types, called **flavors,** labeled *u*, *d*, and *s* (for *up*, *down*, and *strange*). An unusual property of quarks is that they carry fractional electron charges. The charge of the *u* quark is $+\frac{2}{3}e$ and the charge of the *d* and *s* quarks is $\frac{1}{3}e$. Each quark has spin $\frac{1}{2}$ and a baryon number of $\frac{1}{3}$. The strangeness of the *u* and *d* quark is 0, and the strangeness of the *s* quark is $-1$. Each quark has an antiquark with the opposite electric charge, baryon number, and strangeness. Baryons consist of three quarks (or three antiquarks for antiparticles), whereas mesons consist of a quark and an antiquark, giving mesons a baryon number $B = 0$, as required. The proton consists of the combination *uud* and the neutron consists of the combination *udd*. Baryons with a strangeness $S = -1$ contain one *s* quark. All the particles listed in Table 41-1 can be constructed from these three quarks and three antiquarks.[‡] The great strength of the quark model is that all the allowed combinations of three quarks or

---

[†] The name *quark* was chosen by M. Gell-Mann from a quotation from *Finnegan's Wake* by James Joyce.
[‡] The correct quark combinations of hadrons are not always obvious, because of the symmetry requirements on the total wave function. For example, the $\pi^0$ meson is represented by a linear combination of $u\bar{u}$ and $d\bar{d}$.

quark–antiquark pairs result in known hadrons. Strong evidence for the existence of quarks inside a nucleon is provided by high-energy scattering experiments called *deep inelastic scattering*. In these experiments, a nucleon is bombarded with electrons, muons, or neutrinos of energies from 15 GeV to 200 GeV. Analyses of particles scattered at large angles indicate that inside the nucleon are three spin-$\frac{1}{2}$ particles of sizes much smaller than that of the nucleon. These experiments are analogous to Rutherford's scattering of $\alpha$ particles by atoms in which the presence of a tiny nucleus in the atom was inferred from the large-angle scattering of the $\alpha$ particles.

---

GIVEN THE CONSTITUENT QUARK SPECIES, IDENTIFY THE PARTICLE

**EXAMPLE 41-4**

**What are the properties of the particles made up of the following quarks: (a) $u\bar{d}$, (b) $\bar{u}d$, (c) $dds$, and (d) $uss$?**

**PICTURE THE PROBLEM** Baryons are made up of three quarks, whereas mesons consist of a quark and an antiquark. We add the electric charges of the quarks to find the total charge of the hadron. We also find the strangeness of the hadron by adding the strangeness of the quarks.

(a) Because $u\bar{d}$ is a quark–antiquark combination, it has baryon number 0 and is therefore a meson. There is no strange quark here, so the strangeness of the meson is zero. The charge of the up quark is $+\frac{2}{3}e$ and the charge of the antidown quark is $+\frac{1}{3}e$, so the charge of the meson is $+1e$. This is the quark combination of the $\pi^+$ meson.

(b) The particle $\bar{u}d$ is also a meson with zero strangeness. Its electric charge is $-\frac{2}{3}e + (-\frac{1}{3}e) = -1e$. This is the quark combination of the $\pi^-$ meson.

(c) The particle $dds$ is a baryon with strangeness $-1$ because it contains one strange quark. Its electric charge is $-\frac{1}{3}e - \frac{1}{3}e - \frac{1}{3}e = -1e$. This is the quark combination for the $\Sigma^-$ particle.

(d) The particle $uss$ is a baryon with strangeness $-2$. Its electric charge is $+\frac{2}{3}e - \frac{1}{3}e - \frac{1}{3}e = 0$. This is the quark combination for the $\Xi^0$ particle.

MASTER the CONCEPT WEB

---

In 1967, a fourth quark was proposed to explain some discrepancies between experimental determinations of certain decay rates and calculations based on the quark model. The fourth quark is labeled $c$ for a new property called **charm.** Like strangeness, charm is conserved in strong interactions but changes by $\pm 1$ in weak interactions. In 1975, a new heavy meson called the $\psi/J$ particle (or simply the $\psi$ particle) was discovered that has the properties expected of a $c\bar{c}$ combination. Since then, other mesons with combinations such as $c\bar{d}$ and $\bar{c}d$ as well as baryons containing the charmed quark, have been discovered. Two more quarks labeled $t$ and $b$ (for *top* and *bottom*) were proposed in the 1970s. In 1977, a massive new meson called the $\Upsilon$ meson or **bottomonium,** which is considered to have the quark combination $b\bar{b}$, was discovered. The top quark was observed in 1995. The properties of the six quarks are listed in Table 41-2.

The six quarks and six leptons (and their antiparticles) are thought to be the fundamental elementary particles of which all matter is composed. Table 41-3 lists the masses of the fundamental particles. In this table, the masses given for neutrinos are upper limits. The masses given for quarks are educated guesses. There is experimental evidence for the existence of each of these particles.

## TABLE 41-2

**Properties of Quarks and Antiquarks**

| Flavor | Spin | Charge | Baryon Number | Strangeness | Charm | Topness | Bottomness |
|---|---|---|---|---|---|---|---|
| **Quarks** | | | | | | | |
| $u$ (up) | $\frac{1}{2}\hbar$ | $+\frac{2}{3}e$ | $+\frac{1}{3}$ | 0 | 0 | 0 | 0 |
| $d$ (down) | $\frac{1}{2}\hbar$ | $-\frac{1}{3}e$ | $+\frac{1}{3}$ | 0 | 0 | 0 | 0 |
| $s$ (strange) | $\frac{1}{2}\hbar$ | $-\frac{1}{3}e$ | $+\frac{1}{3}$ | $-1$ | 0 | 0 | 0 |
| $c$ (charmed) | $\frac{1}{2}\hbar$ | $+\frac{2}{3}e$ | $+\frac{1}{3}$ | 0 | $+1$ | 0 | 0 |
| $t$ (top) | $\frac{1}{2}\hbar$ | $+\frac{2}{3}e$ | $+\frac{1}{3}$ | 0 | 0 | $+1$ | 0 |
| $b$ (bottom) | $\frac{1}{2}\hbar$ | $-\frac{1}{3}e$ | $+\frac{1}{3}$ | 0 | 0 | 0 | $+1$ |
| **Antiquarks** | | | | | | | |
| $\overline{u}$ | $\frac{1}{2}\hbar$ | $-\frac{2}{3}e$ | $-\frac{1}{3}$ | 0 | 0 | 0 | 0 |
| $\overline{d}$ | $\frac{1}{2}\hbar$ | $+\frac{1}{3}e$ | $-\frac{1}{3}$ | 0 | 0 | 0 | 0 |
| $\overline{s}$ | $\frac{1}{2}\hbar$ | $+\frac{1}{3}e$ | $-\frac{1}{3}$ | $+1$ | 0 | 0 | 0 |
| $\overline{c}$ | $\frac{1}{2}\hbar$ | $-\frac{2}{3}e$ | $-\frac{1}{3}$ | 0 | $-1$ | 0 | 0 |
| $\overline{t}$ | $\frac{1}{2}\hbar$ | $-\frac{2}{3}e$ | $-\frac{1}{3}$ | 0 | 0 | $-1$ | 0 |
| $\overline{b}$ | $\frac{1}{2}\hbar$ | $+\frac{1}{3}e$ | $-\frac{1}{3}$ | 0 | 0 | 0 | $-1$ |

## Quark Confinement

Despite considerable experimental effort, no isolated quark has ever been observed. It is now believed that it is impossible to obtain an isolated quark. Although the force between quarks is not known, it is believed that the potential energy of two quarks increases with increasing separation distance so that an infinite amount of energy would be needed to separate the quarks completely. This would be true, for example, if the force of attraction between two quarks remains constant or increases with separation distance, rather than decreasing with increasing separation distance as is the case for other fundamental forces, such as the electric force between two charges, the gravitational force between two masses, and the strong nuclear force between two hadrons.

When a large amount of energy is added to a quark system, such as a nucleon, a quark–antiquark pair is created and the original quarks remain confined within the original system. Because quarks cannot be isolated, but are always bound in a baryon or a meson, the mass of a quark cannot be accurately known, which is why the masses listed in Table 41-3 are merely educated guesses.

## TABLE 41-3

**Masses of Fundamental Particles**

| Particle | Mass |
|---|---|
| **Quarks** | |
| $u$ (up) | 336 MeV/$c^2$ |
| $d$ (down) | 338 MeV/$c^2$ |
| $s$ (strange) | 540 MeV/$c^2$ |
| $c$ (charmed) | 1,500 MeV/$c^2$ |
| $t$ (top) | 174,000 MeV/$c^2$ |
| $b$ (bottom) | 4,500 MeV/$c^2$ |
| **Leptons** | |
| $e^-$ (electron) | 0.511 MeV/$c^2$ |
| $\nu_e$ (electron neutrino) | $< 7$ eV/$c^2$ |
| $\mu^-$ (muon) | 105.659 MeV/$c^2$ |
| $\nu_\mu$ (muon neutrino) | $< 0.27$ MeV/$c^2$ |
| $\tau^-$ (tau) | 1,784 MeV/$c^2$ |
| $\nu_\tau$ (tau neutrino) | $< 31$ MeV/$c^2$ |

## 41-5 Field Particles

In addition to the six fundamental leptons and six fundamental quarks, there are other particles, called *field particles* or *field quanta*, that are associated with the forces exerted by one elementary particle on another. In **quantum electrodynamics,** the electromagnetic field of a single charged particle is described by **virtual photons** that are continuously being emitted and reabsorbed by the particle. If we put energy into the system by accelerating the charge, some of these virtual photons are shaken off and become real, observable photons. The photon is said to mediate the electromagnetic interaction. Each of the four basic interactions can be described via mediating field particles.

The field quantum associated with the gravitational interaction, called the **graviton,** has not yet been observed. The gravitational *charge* analogous to electric charge is mass.

The weak interaction is thought to be mediated by three field quanta called **vector bosons:** $W^+$, $W^-$, and $Z^0$. These particles were predicted by Sheldon Glashow, Abdus Salam, and Steven Weinberg in a theory called the *electroweak theory*, which we discuss in the next section. The W and Z particles were first observed in 1983 by a group of over a hundred scientists led by Carlo Rubbia using the high-energy accelerator at CERN in Geneva, Switzerland. The masses of the $W^{\pm}$ particles (about 80 GeV/$c^2$) and the Z particle (about 91 GeV/$c^2$) measured in this experiment were in excellent agreement with those predicted by the electroweak theory. (The $W^-$ particle is the antiparticle of the $W^+$ particle, so they must have identical masses.)

The field quanta associated with the strong force between quarks are called **gluons.** Isolated gluons have not been observed experimentally. The *charge* responsible for the strong interactions comes in three varieties, labeled *red, green,* and *blue* (analogous with the three primary colors), and the strong charge is called the **color charge.** The field theory for strong interactions, analogous to quantum electrodynamics for electromagnetic interactions, is called **quantum chromodynamics (QCD).**

Table 41-4 lists the bosons responsible for mediating the basic interactions.

## TABLE 41-4

**Bosons That Mediate the Basic Interactions**

| Interaction | Boson | Spin | Mass | Electric Charge |
|---|---|---|---|---|
| Strong | $g$ (gluon)[†] | 1 | 0 | 0 |
| Weak | $W^{\pm}$ | 1 | 80.22 GeV/$c^2$ | $\pm 1e$ |
| | $Z^0$ | 1 | 91.19 GeV/$c^2$ | 0 |
| Electromagnetic | $\gamma$ (photon) | 1 | 0 | 0 |
| Gravitational | Graviton[†] | 2 | 0 | 0 |

† Not yet observed.

## 41-6 The Electroweak Theory

In the **electroweak theory,** the electromagnetic and weak interactions are considered to be two different manifestations of a more fundamental electroweak interaction. At very high energies ($\gg$ 100 GeV), the electroweak interaction would be mediated by four bosons. From symmetry considerations, these would be a triplet consisting of $W^+$, $W^0$, and $W^-$, all of equal mass, and a singlet $B^0$ of some other mass. Neither the $W^0$ nor the $B^0$ would be observed directly, but one linear combination of the $W^0$ and the $B^0$ would be the $Z^0$ and another would be the photon. At ordinary energies, the symmetry is broken. This leads to the separation of the electromagnetic interaction mediated by the massless photon and the weak interaction mediated by the $W^+$, $W^-$, and $Z^0$ particles. The fact that the photon is massless and that the W and Z particles have masses of the order of 100 GeV/$c^2$ shows that the symmetry assumed in the electroweak theory does not exist at lower energies.

The symmetry-breaking mechanism is called a **Higgs field,** which requires a new boson, the **Higgs boson,** whose rest energy is expected to be of the order of 1 TeV (1 TeV = $10^{12}$ eV). The Higgs boson has not yet been observed. Calculations show that Higgs bosons (if they exist) should be produced in head-on collisions between protons of energies of the order of 20 TeV. Such energies are not presently available.

# 41-7 The Standard Model

The combination of the quark model, electroweak theory, and quantum chromo-dynamics is called the **standard model.** In this model, the fundamental particles are the leptons and quarks, each of which comes in six flavors, as shown in Table 41-3; the force carriers are the photon, the $W^{\pm}$ and $Z$ particles, and the gluons (of which there are eight types). The leptons and quarks are all spin-$\frac{1}{2}$ fermions, which obey the Pauli exclusion principle, and the force carriers are integral-spin bosons, which do not obey the Pauli exclusion principle. Every force in nature is due to one of the four basic interactions: strong, electromagnetic, weak, and gravitational. A particle experiences one of the basic interactions if it carries a charge associated with that interaction. Electric charge is the familiar charge that we have studied previously. Weak charge, also called flavor charge, is carried by leptons and quarks. The charge associated with the strong interaction is called color charge and is carried by quarks and gluons but not by leptons. The charge associated with the gravitational force is mass. It is important to note that the photon, which mediates the electromagnetic interaction, does not carry electric charge. Similarly, the $W^{\pm}$ and $Z$ particles, which mediate the weak interaction, do not carry weak charge. However, the gluons, which mediate the strong interaction, do carry color charge. This fact is related to the confinement of quarks as discussed in Section 41-4.

All matter is made up of leptons or quarks. There are no known composite particles consisting of leptons bound together by the weak force. Leptons exist only as isolated particles. Hadrons (baryons and mesons) are composite particles consisting of quarks bound together by the color charge. A result of QCD theory is that only color-neutral combinations of quarks are allowed. Three quarks of different colors can combine to form color-neutral baryons, such as the neutron and the proton. Mesons contain a quark and an antiquark and are also color-neutral. Excited states of hadrons are considered to be different particles. For example, the $\Delta^{+}$ particle is an excited state of the proton. Both are made up of the *uud* quarks, but the proton is in the ground state with spin $\frac{1}{2}$ and a rest energy of 938 MeV, whereas the $\Delta^{+}$ particle is in the first excited state with spin $\frac{3}{2}$ and a rest energy of 1232 MeV. The two $u$ quarks can be in the same spin state in the $\Delta^{+}$ without violating the exclusion principle, because they have different color. All baryons eventually decay to the lightest baryon, the proton. The proton cannot decay because of conservation of energy and conservation of baryon number.

The strong interaction has two parts, the fundamental interaction or color interaction and what is called the *residual strong interaction*. The fundamental interaction is responsible for the force exerted by one quark on another quark and is mediated by gluons. The residual strong interaction is responsible for the force between color-neutral nucleons, such as the neutron and the proton. This force is due to the residual strong interactions between the color-charged quarks that make up the nucleons and can be viewed as being mediated by the exchange of mesons. The residual strong interaction between color-neutral nucleons can be thought of as analogous to the residual electromagnetic interaction between neutral atoms that bind them together to form molecules. Table 41-5 lists some of the properties of the basic interactions.

## TABLE 41-5

**Properties of the Basic Interactions**

| | Gravitational | Weak | Electromagnetic | Strong Fundamental | Strong Residual |
|---|---|---|---|---|---|
| Acts on | Mass | Flavor | Electric charge | Color charge | |
| Particles experiencing | All | Quarks, leptons | Electrically charged | Quarks, gluons | Hadrons |
| Particles mediating | Graviton | $W^{\pm}, Z$ | $\gamma$ | Gluons | Mesons |
| Strength for two quarks at $10^{-18}$ m† | $10^{-41}$ | 0.8 | 1 | 25 | (not applicable) |
| Strength for two protons in nucleus† | $10^{-36}$ | $10^{-7}$ | 1 | (not applicable) | 20 |

† Strengths are relative to electromagnetic strength.

For each particle there is an antiparticle. A particle and its antiparticle have identical mass and spin but opposite electric charge. For leptons, the lepton numbers $L_e$, $L_\mu$, and $L_\tau$ of the antiparticles are the negatives of the corresponding numbers for the particles. For example, the lepton number for the electron is $L_e = +1$, and the lepton number for the positron is $L_e = -1$. For hadrons, the baryon number, strangeness, charm, topness, and bottomness are the sums of those quantities for the quarks that make up the hadron. The number of each antiparticle is the negative of the number for the corresponding particle. For example, the lambda particle $\Lambda^0$, which is made up of the *uds* quarks, has $B = 1$ and $S = -1$, whereas its antiparticle $\overline{\Lambda}^0$, which is made up of the $\overline{u}\overline{d}\overline{s}$ quarks, has $B = -1$ and $S = +1$. A particle such as the photon $\gamma$ or the $Z^0$ particle that has zero electric charge, $B = 0$, $L = 0$, $S = 0$; and zero charm, topness, and bottomness is its own antiparticle. Note that the $K^0$ meson ($d\overline{s}$) has a zero value for all of these quantities except strangeness, which is $+1$. Its antiparticle, the $\overline{K}^0$ meson ($\overline{d}s$), has strangeness $-1$, which makes it distinct from the $K^0$. The $\pi^+$ ($u\overline{d}$) and $\pi^-$ ($\overline{u}d$) are somewhat special in that they have electric charge but zero values for $L$, $B$, and $S$. They are antiparticles of each other, but since there is no conservation law for mesons, it is impossible to say which is the particle and which is the antiparticle. Similarly, the $W^+$ and $W^-$ are antiparticles of each other.

## Grand Unification Theories

With the success of the electroweak theory, attempts have been made to combine the strong, electromagnetic, and weak interactions in various **grand unification theories (GUTs).** In one of these theories, leptons and quarks are considered to be two aspects of a single class of particles. Under certain conditions, a quark could change into a lepton and vice versa, even though this would appear to violate the conservation of lepton number and baryon number. One of the exciting predictions of this theory is that the proton is not stable but merely has a very long lifetime of the order of $10^{32}$ y. Such a long lifetime makes proton decay difficult to observe. However, projects are ongoing in which detectors monitor very large numbers of protons in search of an event indicating the decay of a proton.

## 41-8 The Evolution of the Universe

In the presently accepted model, the universe began with a singular cataclysmic event called the **big bang** and is expanding. The first evidence that the universe is expanding was the astronomer Edwin Powell Hubble's discovery of the relation between the redshifts in the spectra of galaxies and their distances from us. This relation is illustrated in Figure 41-4 for a group of spiral galaxies used by astronomers for calibrating distances. Provided that the redshift is due to the Doppler effect, the recession velocity $v$ of a galaxy is related to its distance $r$ from us by Hubble's law,

$$v = Hr \qquad 41\text{-}9$$

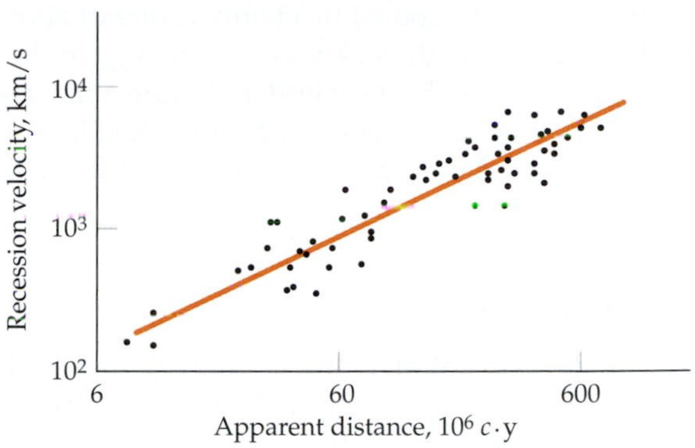

where $H$ is the **Hubble constant.** In principle, the value of $H$ is easy to obtain since it relies on the direct calculation of $v$ from redshift measurements. However, astronomical distances are very difficult to obtain, and they have been computed for only a fraction of the $10^{10}$ or so galaxies in the observable universe. Thus, the value of $H$ changes as distance calibration data are refined. The currently accepted value of the Hubble constant is

$$H = \frac{23 \text{ km/s}}{10^6 \, c \cdot y} \qquad 41\text{-}10$$

Hubble's law tells us that the galaxies are all rushing away from us, with those the farthest away moving the fastest. However, there is no reason why our location should be special. An observer in any galaxy would make the same observations and compute the same Hubble constant. Thus, Hubble's law suggests that all of the galaxies are receding from each other at an average speed of 23 km/s per $10^6 \, c \cdot y$ of separation. In other words, the universe is expanding. Notice that the basic dimension of $H$ is reciprocal time. The quantity $1/H$ is called the **Hubble age** and equals about $1.3 \times 10^{10}$ y. This would correspond to the age of the universe if the gravitational pull on the receding galaxies were ignored.

**FIGURE 41-4** A plot of the recession velocities of individual galaxies versus apparent distance.

---

*USING HUBBLE'S LAW*                    **EXAMPLE 41-5**

**Redshift measurements of a galaxy in the constellation Virgo yield a recession velocity of 1200 km/s. How far is it to that galaxy?**

**PICTURE THE PROBLEM** We calculate the distance from Hubble's law.

Use Hubble's law to find $r$:
$$r = \frac{v}{H} = (1200 \text{ km/s}) \frac{10^6 \, c \cdot y}{23 \text{ km/s}} = \boxed{52 \times 10^6 \, c \cdot y}$$

■ **EXERCISE** Show that $1/H = 1.3 \times 10^{10}$ y.

## The 2.7-K Background Radiation

In investigating ways of accounting for the cosmic abundance of elements heavier than hydrogen, cosmologists recognized that nucleosynthesis in stars could explain the abundance of elements heavier than helium but could not by itself explain that of helium. Helium must therefore have been formed during the big bang. To synthesize an amount of helium sufficient to account for its present abundance, the big bang would have to have occurred at an extremely high initial temperature to provide the necessary reaction rate before fusion was shut down by the decreasing density of the very rapid initial expansion. The high temperature implies a corresponding thermal (blackbody) radiation field that would cool as the expansion progressed. Theoretical analysis predicted that from the estimated time of the big bang to the present, the remnants of the radiation field should have cooled to a temperature of about 3 K, corresponding to a black-body spectrum with peak wavelength $\lambda_{max}$ in the microwave region. In 1965, the predicted cosmic background radiation was discovered by Arno Penzias and Robert Wilson at the Bell Labs. Since this landmark discovery, careful analysis has established that the temperature of the background field is $2.7 \pm 0.1$ K and has shown that it has an isotropic distribution in space.

## The Big Bang

The singular event that initiated the expansion of the universe is thought to have been a huge explosion. Initially, the four forces of nature (strong, electromagnetic, weak, and gravitational) were unified into a single force. Physicists have been successful in developing theoretical descriptions that unify the first three forces, but a theory of quantum gravity, needed for the extreme densities of the single-force period, does not yet exist. Consequently, until the cooling universe "froze" or "condensed out" the gravitational force at approximately $10^{-43}$ s after the big bang, when the temperature was still $10^{32}$ K, we have no means of describing what was occurring. At this point, the average energy of the particles created would have been about $10^{19}$ GeV. As the universe continued to cool below $10^{32}$ K, the three forces other than gravity remained unified and are described by the grand unification theories (GUTs). Quarks and leptons were indistinguishable and particle quantum numbers were not conserved. It was during this period that a slight excess of quarks over antiquarks occurred, roughly 1 part in $10^9$, that ultimately resulted in the predominance of matter over antimatter that we now observe in the universe.

At $10^{-35}$ s, the universe had expanded sufficiently to cool to approximately $10^{27}$ K, at which point another phase transition occurred as the strong force condensed out of the GUTs group, leaving only the electromagnetic and weak forces still unified as the **electroweak force.** During this period, the previously free quarks in the dense mixture of roughly equal numbers of quarks, leptons, their antiparticles, and photons began to combine into hadrons and their antiparticles, including the nucleons. By the time the universe had cooled to approximately $10^{13}$ K, at about $t = 10^{-6}$ s, the hadrons had mostly disappeared. This is because $10^{13}$ K corresponds to $kT \sim 1$ GeV, which is the minimum energy needed to create nucleons and antinucleons from the photons present via the reactions

$$\gamma \rightarrow p^+ + p^- \qquad\qquad\qquad 41\text{-}11a$$

and

$$\gamma \rightarrow n^+ + \overline{n} \qquad\qquad\qquad 41\text{-}11b$$

The particle–antiparticle pairs annihilated and there was no new production to replace them. Only the slight earlier excess of quarks over antiquarks led to a slight excess of protons and neutrons over their antiparticles. The annihilations resulted in photons and leptons, and after about $t = 10^{-4}$ s, those particles in roughly equal numbers dominated the universe. This was the **lepton era.** At about $t = 10$ s, the temperature had fallen to $10^{10}$ K ($kT \sim 1$ MeV). Further expansion and cooling dropped the average photon energy below the energy needed to form an electron–positron pair. Annihilation then removed all of the positrons as it had the antiprotons and antineutrons earlier, leaving only the small excess of electrons arising from charge conservation, and the **radiation era** began. The particles present were primarily photons and neutrinos.

Within a few more minutes, the temperature dropped sufficiently to enable fusing protons and neutrons to form nuclei that were not immediately photodisintegrated. The nuclei of deuterium, helium, and lithium were produced in this **nucleosynthesis period,** but the rapid expansion soon dropped the temperature too low for the fusion to continue and the formation of heavier elements had to await the birth of stars.

A long time later, when the temperature had dropped to about 3000 K as the universe grew to about 1/1000 of its present size, $kT$ dropped below typical atomic ionization energies and atoms were formed. By then, the expansion had redshifted the radiation field so that the total radiation energy was about equal to the energy represented by the remaining mass. As expansion and cooling continued, the energy of the steadily redshifting radiation declined at a steady rate until, at $t = 10^{10}$ y (now), matter came to dominate the universe, with its energy density exceeding that of the 2.7-K radiation remaining from the big bang by a factor of about 1000.

## SUMMARY

| Topic | Relevant Equations and Remarks |
|---|---|
| **1. Basic Interactions** | There are four basic interactions: strong, electromagnetic, weak, and gravitational. |
| Strong | The *charge* associated with the strong interaction is called color. Quarks and gluons have color and experience the strong interaction. Hadrons (baryons and mesons) experience a residual strong interaction resulting from the fundamental strong interaction between the quarks that make up the hadrons. Decay times via strong interaction are typically $10^{-23}$ s. |
| Electromagnetic | All particles with electric charge experience the force due to the electromagnetic interaction. |
| Weak | The *charge* associated with the weak interaction is called flavor. Quarks and leptons have flavor and experience the weak interaction. Decay times via weak interaction are typically $10^{-10}$ s. |
| Gravitational | All particles with mass experience the force due to the gravitational interaction. |
| **2. Fundamental Particles** | There are two families of fundamental particles, leptons and quarks, each containing six members. It is thought that these particles have no size and no internal structure. |
| Leptons | Leptons are spin-$\frac{1}{2}$ fermions: the electron e and its neutrino $\nu_e$, the muon $\mu$ and its neutrino $\nu_\mu$, and the tau $\tau$ and its neutrino $\nu_\tau$. The electron, muon, and tau have mass, electric charge, and flavor, but not color; so they participate in the gravitational, electromagnetic, and weak interactions, but not the strong interaction. The neutrinos have flavor but no electric charge and no color. They appear to have a very small mass. |
| Quarks | There are six quarks, called up $u$, down $d$, strange $s$, charmed $c$, top $t$, and bottom $b$. Each is a spin-$\frac{1}{2}$ fermion. The quarks participate in all of the basic interactions. Because they are always confined in mesons or baryons, their masses can only be estimated. |
| **3. Hadrons** | Hadrons are composite particles that are made up of quarks. There are two types of hadrons, baryons and mesons. Baryons, which include the neutron and proton, are fermions of half-integral spin consisting of three quarks. Mesons, which include pions and kaons, have zero or integral spin. Hadrons interact with each other via the residual strong interaction. |

**4. Field Particles**

In addition to the six fundamental leptons and six fundamental quarks, there are field particles that are associated with the basic interactions.

| Interaction | Field Particle |
|---|---|
| Gravitational | Graviton |
| Electromagnetic | Photon |
| Weak | $W^+, W^-, Z^0$ |
| Strong | Gluons |

**5. The Conservation Laws**

Some quantities, such as energy, linear momentum, electric charge, angular momentum, baryon number, and each of the three lepton numbers, are strictly conserved in all reactions and decays. Others, such as strangeness and charm, are conserved in reactions and decays that proceed via the strong interaction but not in those that proceed via the weak interaction.

**6. Particles and Antiparticles**

Particles and their antiparticles have identical masses but opposite values for their other properties, such as charge, lepton number, baryon number, and strangeness. Particle–antiparticle pairs can be produced in various nuclear reactions if the energy available is greater than $2mc^2$, where $m$ is the mass of the particle.

**7. Hubble's Law**

Hubble's law relates the recession velocity of a galaxy, determined from the redshift of its spectrum, to the distance of the galaxy from us:

$$v = Hr \qquad\qquad\qquad\qquad \textbf{41-9}$$

where the Hubble constant $H = 23$ km/s per million light-years. From Hubble's law, we conclude that the universe is expanding and that the expansion began approximately $1/H$ years ago.

**8. The Big Bang**

According to the model currently used to describe the evolution of the universe, the universe began with a big bang approximately $10^{10}$ years ago. The big bang model is supported by substantial experimental observations, including the isotropic, 2.7-K, background blackbody radiation spectrum.

# PROBLEMS

- • Single-concept, single-step, relatively easy
- •• Intermediate-level, may require synthesis of concepts
- ••• Challenging
- **SSM** Solution is in the *Student Solutions Manual*
- **iSOLVE** Problems available on iSOLVE online homework service
- **iSOLVE** ✔ These "Checkpoint" online homework service problems ask students additional questions about their confidence level, and how they arrived at their answer.

In a few problems, you are given more data than you actually need; in a few other problems, you are required to supply data from your general knowledge, outside sources, or informed estimates.

## Conceptual Problems

**1** • How are baryons and mesons similar? How are they different?

**2** • The muon and the pion have nearly the same mass. How do these particles differ?

**3** • **SSM** How can you tell whether a decay proceeds via the strong interaction or via the weak interaction?

**4** • True or false:

(a) All baryons are hadrons.
(b) All hadrons are baryons.

**5** • True or false: Mesons are spin-$\frac{1}{2}$ particles.

**6** • How can you tell whether a particle is a meson or a baryon by looking at its quark content?

**7** • Are there any quark–antiquark combinations that result in a nonintegral electric charge?

**8** • True or false:

(a) Leptons consist of three quarks.
(b) The times for decays via the weak interaction are typically longer than the times for decays via the strong interaction.
(c) Electrons interact with protons via the strong interaction.
(d) Strangeness is not conserved in weak interactions.
(e) Neutrons have no charm.

**9** • **SSM** Based on the assumption that a pion, $\pi^+$, interacts with an antiproton, $\bar{p}$, is it possible that a proton, p, could be produced by such an interaction?

## Estimation and Approximation

**10** •• Grand unification theories predict that the proton has a long but finite lifetime. Current experiments based on detecting the decay of protons in water infer that this lifetime is at least $10^{32}$ years. Assume $10^{32}$ years is, in fact, the lifetime of the proton. Estimate the expected time between proton-decays that occur in the water of a filled Olympic-size swimming pool. An Olympic-size swimming pool is 100 m × 25 m × 2 m. Give your answer in days.

**11** • **ISOLVE** Table 41-5 lists some properties of the four fundamental interactions. To gain a better appreciation for the significance of this table, confirm the numerical entries in the second and fourth column of the last row of the table by estimating the ratio of the electromagnetic force to the gravitational force between two protons located in a nucleus.

## Spin and Antiparticles

**12** • **SSM** Two pions at rest annihilate according to the reaction $\pi^+ + \pi^- \rightarrow \gamma + \gamma$. (a) Why must the energies of the two $\gamma$-rays be equal? (b) Find the energy of each $\gamma$-ray. (c) Find the wavelength of each $\gamma$-ray.

**13** • Find the minimum energy of the photon needed for the following pair-production reactions: (a) $\gamma \rightarrow \pi^+ + \pi^-$, (b) $\gamma \rightarrow p + p^-$, and (c) $\gamma \rightarrow \mu^- + \mu^+$.

## The Conservation Laws

**14** • State which of the following decays or reactions violate one or more of the conservation laws, and give the law or laws violated in each case: (a) $p^+ \rightarrow n + e^+ + \bar{\nu}_e$, (b) $n \rightarrow p^+ + \pi^-$, (c) $e^+ + e^- \rightarrow \gamma$, (d) $p + p^- \rightarrow \gamma + \gamma$, and (e) $\bar{\nu}_e + p \rightarrow n + e^+$.

**15** • Determine the change in strangeness in each reaction that follows, and state whether the reaction can proceed via the strong interaction, via the weak interaction, or not at all: (a) $\Omega^- \rightarrow \Xi^0 + \pi^-$, (b) $\Xi^0 \rightarrow p + \pi^- + \pi^0$, and (c) $\Lambda^0 \rightarrow p^+ + \pi^-$.

**16** • Determine the change in strangeness for each decay, and state whether the decay can proceed via the strong interaction, via the weak interaction, or not at all: (a) $\Omega^- \rightarrow \Lambda^0 + K^-$ and (b) $\Xi^0 \rightarrow p + \pi^-$.

**17** • Determine the change in strangeness for each decay, and state whether the decay can proceed via the strong interaction, via the weak interaction, or not at all: (a) $\Omega^- \rightarrow \Lambda^0 + \bar{\nu}_e + e^-$ and (b) $\Sigma^+ \rightarrow p + \pi^0$.

**18** • (a) Which of the following decays of the $\tau$ particle is possible?

$$\tau \rightarrow \mu^- + \bar{\nu}_\mu + \nu_\tau$$
$$\tau \rightarrow \mu^- + \nu_\mu + \bar{\nu}_\tau$$

(b) Explain why the other decay is not possible. (c) Calculate the kinetic energy of the decay products for the decay that is possible.

**19** •• **ISOLVE** Consider the following decay chain:

$$\Omega^- \rightarrow \Xi^0 + \pi^-$$
$$\Xi^0 \rightarrow \Sigma^+ + e^- + \bar{\nu}_e$$
$$\pi^- \rightarrow \mu^- + \bar{\nu}_\mu$$
$$\Sigma^+ \rightarrow n + \pi^+$$
$$\pi^+ \rightarrow \mu^+ + \nu_\mu$$
$$\mu^+ \rightarrow e^+ + \bar{\nu}_\mu + \nu_e$$
$$\mu^- \rightarrow e^- + \bar{\nu}_e + \nu_\mu$$

(a) Are all the final products shown stable? If not, finish the decay chain. (b) Write the overall decay reaction for $\Omega^-$ to the final products. (c) Check the overall decay reaction for the conservation of electric charge, baryon number, lepton number, and strangeness.

**20** •• **SSM** **ISOLVE** Test the following decays for violation of the conservation of energy, electric charge, baryon number, and lepton number: (a) $n \rightarrow \pi^+ + \pi^- + \mu^+ + \mu^-$ and (b) $\pi^0 \rightarrow e^+ + e^- + \gamma$. Assume that linear momentum and angular momentum are conserved. State which conservation laws (if any) are violated in each decay.

## Quarks

**21** • Find the baryon number, charge, and strangeness for the following quark combinations and identify the hadron: (a) $uud$, (b) $udd$, (c) $uus$, (d) $dds$, (e) $uss$, and (f) $dss$.

**22** • Repeat Problem 21 for the following quark combinations: (a) $u\bar{d}$, (b) $\bar{u}d$, (c) $u\bar{s}$, and (d) $\bar{u}s$.

**23** • The $\Delta^{++}$ particle is a baryon that decays via the strong interaction. Its strangeness, charm, topness, and bottomness are all zero. What combination of quarks gives a particle with these properties?

**24** • Find a possible combination of quarks that gives the correct values for electric charge, baryon number, and strangeness for (a) $K^+$ and (b) $K^0$.

**25** • The $D^+$ meson has no strangeness, but it has charm of +1. (a) What is a possible quark combination that will give the correct properties for this particle? (b) Repeat Part (a) for the $D^-$ meson, which is the antiparticle of the $D^+$ meson.

**26** • Find a possible combination of quarks that gives the correct values for electric charge, baryon number, and strangeness for (a) $K^-$ (the $K^-$ is the antiparticle of the $K^+$) and (b) $\bar{K}^0$.

**27** •• **SSM** Find a possible quark combination for the following particles: (a) $\Lambda^0$, (b) $p^-$, and (c) $\Sigma^-$.

**28** •• Find a possible quark combination for the following particles: (a) $\bar{n}$, (b) $\Xi^0$, and (c) $\Sigma^+$.

**29** •• Find a possible quark combination for the following particles: (a) $\Omega^-$ and (b) $\Xi^-$.

**30** •• State the properties of the particles made up of the following quarks: (a) $ddd$, (b) $u\bar{c}$, (c) $u\bar{b}$, and (d) $\overline{sss}$.

## The Evolution of the Universe

**31** • SSM A galaxy is receding from the earth at 2.5 percent the speed of light. Estimate the distance from the earth to this galaxy.

**32** • Estimate the speed of a galaxy that is $12 \times 10^9$ $c\cdot y$ away from us.

**33** •• Equation 39-16b deals with the relativistic Doppler frequency shift for a light from a source that is receding from an observer. Show that the relativistic Doppler wavelength shift is

$$\lambda' = \lambda_0 \sqrt{\frac{1 + v/c}{1 - v/c}}$$

**34** •• SSM The red line in the spectrum of atomic hydrogen is frequently referred to as the $H\alpha$ line, and it has a wavelength of 656.3 nm. Using Hubble's law and the relativistic Doppler equation from Problem 33, determine the wavelength of the $H\alpha$ line in the spectrum emitted from galaxies at distances of (a) $5 \times 10^6$ $c\cdot y$, (b) $50 \times 10^6$ $c\cdot y$, (c) $500 \times 10^6$ $c\cdot y$, and (d) $5 \times 10^9$ $c\cdot y$ from the earth.

## General Problems

**35** • (a) What conditions are necessary for a particle and its antiparticle to be identical? Find the antiparticle for (b) $\pi^0$ and (c) $\Xi^0$.

**36** •• Consider the following decay chain:

$$\Xi^0 \to \Lambda^0 + \pi^0$$
$$\Lambda^0 \to p + \pi^-$$
$$\pi^0 \to \gamma + \gamma$$
$$\pi^- \to \mu^- + \bar{\nu}_\mu$$
$$\mu^- \to e^- + \bar{\nu}_e + \nu_\mu$$

(a) Are all the final products shown stable? If not, finish the decay chain. (b) Write the overall decay reaction for $\Xi^0$ to the final products. (c) Check the overall decay reaction for the conservation of electric charge, baryon number, lepton number, and strangeness. (d) In the first step of the chain, could the $\Lambda^0$ have been a $\Sigma^0$?

**37** •• SSM In Problem 36, one of the reactions is $\pi^0 \to \gamma + \gamma$. (a) In terms of the quark model, show how this reaction can take place. (b) Why is it that the number of photons produced must be at least two?

**38** •• Test the following decays for violation of the conservation of energy, electric charge, baryon number, and lepton number: (a) $\Lambda^0 \to p + \pi^-$, (b) $\Sigma^- \to n + p^-$, and (c) $\mu^- \to e^- + \bar{\nu}_e + \nu_\mu$. Assume that linear momentum and angular momentum are conserved. State which conservation laws (if any) are violated in each decay.

**39** •• SSM Using Figure 41-2 and the laws of conservation of charge number, baryon number, strangeness, and spin, identify the unknown particle in each of the following strong reactions: (a) $p + \pi^- \to \Sigma^0 + ?$, (b) $p + p \to \pi^+ + n + K^+ + ?$, and (c) $p + K^- \to \Xi^- + ?$

**40** •• ISOLVE Consider the following high-energy particle reaction: $p + p \to \Lambda^0 + K^0 + p + ?$ In this reaction, stationary protons are bombarded with a beam of high-energy protons. (a) Use the laws of conservation of charge number, baryon number, strangeness (Figure 41-2), and spin to determine the unknown particle. (b) Calculate the $Q$ value for the reaction. (c) The threshold energy $K_{th}$ for this reaction is given by

$$K_{th} = -\frac{Q}{2m_p}(m_p + m_p + M_1 + M_2 + M_3 + M_4)$$

where $M_1, M_2, M_3$, and $M_4$ are the masses of the reaction products. Find $K_{th}$.

**41** •• ISOLVE Light from a distant galaxy shows a redshift of the $H\alpha$ line of hydrogen from 656.3 nm to 1458 nm. (a) What is the recessional velocity of the galaxy? (b) Estimate the distance to this galaxy.

**42** ••• ISOLVE (a) Calculate the total kinetic energy of the decay products for the decay $\Lambda^0 \to p + \pi^-$. Assume the $\Lambda^0$ is initially at rest. (b) Find the ratio of the kinetic energy of the pion to the kinetic energy of the proton. (c) Find the kinetic energies of the proton and the pion for this decay.

**43** ••• SSM A $\Sigma^0$ particle at rest decays into a $\Lambda^0$ plus a photon. (a) What is the total energy of the decay products? (b) Assuming that the kinetic energy of the $\Lambda^0$ is negligible compared with the energy of the photon, calculate the approximate momentum of the photon. (c) Use your result from Part (b) to calculate the kinetic energy of the $\Lambda^0$. (d) Use your result from Part (c) to obtain a better estimate of the momentum and the energy of the photon.

**44** ••• In this problem, you will calculate the difference in the time of arrival of two neutrinos of different energy from a supernova that is 170,000 light-years away. Let the energies of the neutrinos be $E_1 = 20$ MeV and $E_2 = 5$ MeV, and assume that the mass of a neutrino is 20 eV/$c^2$. Because the total energy of the neutrinos is so much greater than their rest energy, the neutrinos have speeds that are very nearly equal to $c$ and energies that are approximately $E \approx pc$. (a) If $t_1$ and $t_2$ are the times it takes for neutrinos of speeds $u_1$ and $u_2$ to travel a distance $x$, show that

$$\Delta t = t_2 - t_1 = x\frac{u_1 - u_2}{u_1 u_2} \approx \frac{x \, \Delta u}{c^2}$$

(b) The speed of a neutrino of mass $m$ and total energy $E$ can be found from Equation 39-25. Show that when $E >> mc^2$, the speed $u$ is given approximately by

$$\frac{u}{c} \approx 1 - \frac{1}{2}\left(\frac{mc^2}{E}\right)^2$$

(c) Use the results from Part (b) to calculate $u_1 - u_2$ for the energies and mass given, and calculate $\Delta t$ from the result from Part (a) for $x = 170{,}000$ $c\cdot y$. (d) Repeat the calculation in Part (c) using $mc^2 = 40$ eV for the rest energy of a neutrino.

# APPENDIX A

## SI Units and Conversion Factors

### Basic Units

| | |
|---|---|
| Length | The *meter* (m) is the distance traveled by light in a vacuum in 1/299,792,458 s. |
| Time | The *second* (s) is the duration of 9,192,631,770 periods of the radiation corresponding to the transition between the two hyperfine levels of the ground state of the $^{133}$Cs atom. |
| Mass | The *kilogram* (kg) is the mass of the international standard body preserved at Sèvres, France. |
| Current | The *ampere* (A) is that current in two very long parallel wires 1 m apart that gives rise to a magnetic force per unit length of $2 \times 10^{-7}$ N/m. |
| Temperature | The *kelvin* (K) is 1/273.16 of the thermodynamic temperature of the triple point of water. |
| Luminous intensity | The *candela* (cd) is the luminous intensity, in the perpendicular direction, of a surface of area 1/600,000 m² of a blackbody at the temperature of freezing platinum at a pressure of 1 atm. |

### Derived Units

| | | |
|---|---|---|
| Force | newton (N) | $1\,\text{N} = 1\,\text{kg·m/s}^2$ |
| Work, energy | joule (J) | $1\,\text{J} = 1\,\text{N·m}$ |
| Power | watt (W) | $1\,\text{W} = 1\,\text{J/s}$ |
| Frequency | hertz (Hz) | $1\,\text{Hz} = \text{cy/s}$ |
| Charge | coulomb (C) | $1\,\text{C} = 1\,\text{A·s}$ |
| Potential | volt (V) | $1\,\text{V} = 1\,\text{J/C}$ |
| Resistance | ohm ($\Omega$) | $1\,\Omega = 1\,\text{V/A}$ |
| Capacitance | farad (F) | $1\,\text{F} = 1\,\text{C/V}$ |
| Magnetic field | tesla (T) | $1\,\text{T} = 1\,\text{N/(A·m)}$ |
| Magnetic flux | weber (Wb) | $1\,\text{Wb} = 1\,\text{T·m}^2$ |
| Inductance | henry (H) | $1\,\text{H} = 1\,\text{J/A}^2$ |

## Conversion Factors

Conversion factors are written as equations for simplicity;
relations marked with an asterisk are exact.

### Length

1 km = 0.6215 mi

1 mi = 1.609 km

1 m = 1.0936 yd = 3.281 ft = 39.37 in.

*1 in. = 2.54 cm

*1 ft = 12 in. = 30.48 cm

*1 yd = 3 ft = 91.44 cm

1 lightyear = 1 $c \cdot y$ = 9.461 × 10$^{15}$ m

*1 Å = 0.1 nm

### Area

*1 m$^2$ = 10$^4$ cm$^2$

1 km$^2$ = 0.3861 mi$^2$ = 247.1 acres

*1 in.$^2$ = 6.4516 cm$^2$

1 ft$^2$ = 9.29 × 10$^{-2}$ m$^2$

1 m$^2$ = 10.76 ft$^2$

*1 acre = 43,560 ft$^2$

1 mi$^2$ = 640 acres = 2.590 km$^2$

### Volume

*1 m$^3$ = 10$^6$ cm$^3$

*1 L = 1000 cm$^3$ = 10$^{-3}$ m$^3$

1 gal = 3.786 L

1 gal = 4 qt = 8 pt = 128 oz = 231 in$^3$

1 in$^3$ = 16.39 cm$^3$

1 ft$^3$ = 1728 in.$^3$ = 28.32 L
$\quad$ = 2.832 × 10$^4$ cm$^3$

### Time

*1 h = 60 min = 3.6 ks

*1 d = 24 h = 1440 min = 86.4 ks

1 y = 365.24 d = 3.156 × 10$^7$ s

### Speed

*1 m/s = 3.6 km/h

1 km/h = 0.2778 m/s = 0.6215 mi/h

1 mi/h = 0.4470 m/s = 1.609 km/h

1 mi/h = 1.467 ft/s

### Angle and Angular Speed

*$\pi$ rad = 180°

1 rad = 57.30°

1° = 1.745 × 10$^{-2}$ rad

1 rev/min = 0.1047 rad/s

1 rad/s = 9.549 rev/min

### Mass

*1 kg = 1000 g

*1 tonne = 1000 kg = 1 Mg

1 u = 1.6606 × 10$^{-27}$ kg

1 kg = 6.022 × 10$^{26}$ u

1 slug = 14.59 kg

1 kg = 6.852 × 10$^{-2}$ slug

1 u = 931.50 MeV/$c^2$

### Density

*1 g/cm$^3$ = 1000 kg/m$^3$ = 1 kg/L

(1 g/cm$^3$)$g$ = 62.4 lb/ft$^3$

### Force

1 N = 0.2248 lb = 10$^5$ dyn

*1 lb = 4.448222 N

(1 kg)$g$ = 2.2046 lb

### Pressure

*1 Pa = 1 N/m$^2$

*1 atm = 101.325 kPa = 1.01325 bars

1 atm = 14.7 lb/in.$^2$ = 760 mmHg
$\quad$ = 29.9 in.Hg = 33.8 ftH$_2$O

1 lb/in.$^2$ = 6.895 kPa

1 torr = 1 mmHg = 133.32 Pa

1 bar = 100 kPa

### Energy

*1 kW·h = 3.6 MJ

*1 cal = 4.1840 J

1 ft·lb = 1.356 J = 1.286 × 10$^{-3}$ Btu

*1 L·atm = 101.325 J

1 L·atm = 24.217 cal

1 Btu = 778 ft·lb = 252 cal = 1054.35 J

1 eV = 1.602 × 10$^{-19}$ J

1 u·$c^2$ = 931.50 MeV

*1 erg = 10$^{-7}$ J

### Power

1 horsepower = 550 ft·lb/s = 745.7 W

1 Btu/h = 1.055 kW

1 W = 1.341 × 10$^{-3}$ horsepower
$\quad$ = 0.7376 ft·lb/s

### Magnetic Field

*1 T = 10$^4$ G

### Thermal Conductivity

1 W/(m·K) = 6.938 Btu·in./(h·ft$^2$·F°)

1 Btu·in./(h·ft$^2$·F°) = 0.1441 W/(m·K)

# APPENDIX B

## Numerical Data

### Terrestrial Data

| | |
|---|---|
| Free-fall acceleration $g$ | 9.80665 m/s$^2$; 32.1740 ft/s$^2$ |
| (Standard value at sea level at 45° latitude)[†] | |
|   Standard value | |
|     At sea level, at equator[†] | 9.7804 m/s$^2$ |
|     At sea level, at poles[†] | 9.8322 m/s$^2$ |
| Mass of earth $M_E$ | $5.98 \times 10^{24}$ kg |
| Radius of earth $R_E$, mean | $6.37 \times 10^6$ m; 3960 mi |
| Escape speed $\sqrt{2R_E g}$ | $1.12 \times 10^4$ m/s; 6.95 mi/s |
| Solar constant[‡] | 1.35 kW/m$^2$ |
| Standard temperature and pressure (STP): | |
|   Temperature | 273.15 K |
|   Pressure | 101.325 kPa (1.00 atm) |
| Molar mass of air | 28.97 g/mol |
| Density of air (STP), $\rho_{air}$ | 1.293 kg/m$^3$ |
| Speed of sound (STP) | 331 m/s |
| Heat of fusion of $H_2O$ (0°C, 1 atm) | 333.5 kJ/kg |
| Heat of vaporization of $H_2O$ (100°C, 1 atm) | 2.257 MJ/kg. |

† Measured relative to the earth's surface.
‡ Average power incident normally on 1 m$^2$ outside the earth's atmosphere at the mean distance from the earth to the sun.

### Astronomical Data[†]

| | |
|---|---|
| **Earth** | |
|   Distance to moon[‡] | $3.844 \times 10^8$ m; $2.389 \times 10^5$ mi |
|   Distance to sun, mean[†] | $1.496 \times 10^{11}$ m; $9.30 \times 10^7$ mi; 1.00 AU |
|   Orbital speed, mean | $2.98 \times 10^4$ m/s |
| **Moon** | |
|   Mass | $7.35 \times 10^{22}$ kg |
|   Radius | $1.738 \times 10^6$ m |
|   Period | 27.32 d |
|   Acceleration of gravity at surface | 1.62 m/s$^2$ |
| **Sun** | |
|   Mass | $1.99 \times 10^{30}$ kg |
|   Radius | $6.96 \times 10^8$ m |

† Additional solar-system data is available from NASA at <http://nssdc.gsfc.nasa.gov/planetary/planetfact.html>.
‡ Center to center.

## Physical Constants[†]

| | | |
|---|---|---|
| Gravitational constant | $G$ | $6.673(10) \times 10^{-11}$ N·m$^2$/kg$^2$ |
| Speed of light | $c$ | $2.997\ 924\ 58 \times 10^8$ m/s |
| Fundamental charge | $e$ | $1.602\ 1764\ 62(63) \times 10^{-19}$ C |
| Avogadro's number | $N_A$ | $6.022\ 141\ 99(47) \times 10^{23}$ particles/mol |
| Gas constant | $R$ | $8.314\ 472(15)$ J/(mol·K) |
| | | $1.987\ 2065(36)$ cal/(mol·K) |
| | | $8.205\ 746(15) \times 10^{-2}$ L·atm/(mol·K) |
| Boltzmann constant | $k = R/N_A$ | $1.380\ 6503(24) \times 10^{-23}$ J/K |
| | | $8.617\ 342(15) \times 10^{-5}$ eV/K |
| Stefan-Boltzmann constant | $\sigma = (\pi^2/60)k^4/(\hbar^3 c^2)$ | $5.670\ 400(40) \times 10^{-8}$ W/(m$^2$k$^4$) |
| Atomic mass constant | $m_u = \frac{1}{12}m(^{12}C)$ | $1.660\ 538\ 73(13) \times 10^{-27}$ kg = 1u |
| Coulomb constant | $k = 1/(4\pi\epsilon_0)$ | $8.987\ 551\ 788 \ldots \times 10^9$ N·m$^2$/C$^2$ |
| Permittivity of free space | $\epsilon_0$ | $8.854\ 187\ 817 \ldots \times 10^{-12}$ C$^2$/(N·m$^2$) |
| Permeability of free space | $\mu_0$ | $4\pi \times 10^{-7}$ N/A$^2$ |
| | | $1.256\ 637 \times 10^{-6}$ N/A$^2$ |
| Planck's constant | $h$ | $6.626\ 068\ 76(52) \times 10^{-34}$ J·s |
| | | $4.135\ 667\ 27(16) \times 10^{-15}$ eV·s |
| | $\hbar = h/2\pi$ | $1.054\ 571\ 596(82) \times 10^{-34}$ J·s |
| | | $6.582\ 118\ 89(26) \times 10^{-16}$ eV·s |
| Mass of electron | $m_e$ | $9.109\ 381\ 88(72) \times 10^{-31}$ kg |
| | | $0.510\ 998\ 902(21)$ MeV/$c^2$ |
| Mass of proton | $m_p$ | $1.672\ 621\ 58(13) \times 10^{-27}$ kg |
| | | $938.271\ 998(38) \times$ MeV/$c^2$ |
| Mass of neutron | $m_n$ | $1.674\ 927\ 16(13) \times 10^{-27}$ kg |
| | | $939.565\ 330(38)$ MeV/$c^2$ |
| Bohr magneton | $m_B = eh/2m_e$ | $9.274\ 0008\ 99(37) \times 10^{-24}$ J/T |
| | | $5.788\ 381\ 749(43) \times 10^{-5}$ eV/T |
| Nuclear magneton | $m_n = eh/2m_p$ | $5.050\ 783\ 17(20) \times 10^{-27}$ J/T |
| | | $3.152\ 451\ 238(24) \times 10^{-8}$ eV/T |
| Magnetic flux quantum | $\phi_0 = h/2e$ | $2.067\ 833\ 636(81) \times 10^{-15}$ T·m$^2$ |
| Quantized Hall resistance | $R_K = h/e^2$ | $2.581\ 280\ 7572(95) \times 10^4$ $\Omega$ |
| Rydberg constant | $R_H$ | $1.097\ 373\ 156\ 8549(83) \times 10^7$ m$^{-1}$ |
| Josephson frequency-voltage quotient | $K_J = 2e/h$ | $4.835\ 978\ 98(19) \times 10^{14}$ Hz/V |
| Compton wavelength | $\lambda_C = h/m_e c$ | $2.426\ 310\ 215(18) \times 10^{-12}$ m |

[†] The values for these and other constants may be found on the Internet at http://physics.nist.gov/cuu/Constants/index.html. The numbers in parentheses represent the uncertainties in the last two digits. (For example, 2.044 43(13) stands for 2.044 43 ± 0.000 13.) Values with without uncertainties are exact, including those values with ellipses (like the value of pi is exactly 3.1415. . .).

**For additional data, see the following tables in the text.**

**1-1** Prefixes for Powers of 10, p. 5

**1-2** Dimensions of Physical Quantities, p. 7

**1-3** The Universe by Orders of Magnitude, p. 12

**3-1** Properties of Vectors, p. 58

**5-1** Approximate Values of Frictional Coefficients, p. 120

**6-1** Properties of Dot Products, p. 159

**7-1** Rest Energies of Some Elementary Particles and Light Nuclei, p. 201

**9-1** Moments of Inertia of Uniform Bodies of Various Shapes, p. 274

**9-2** Analogs in Rotational and Linear Motion, p. 287

**11-1** Mean Orbital Radii and Orbital Periods for the Planets, p. 340

**12-1** Young's Modulus $Y$ and Strengths of Various Materials, p. 381

**12-2** Approximate Values of the Shear Modulus $M_s$ of Various Materials, p. 382

**13-1** Densities of Selected Substances, p. 396

**13-2** Approximate Values for the Bulk Modulus $B$ of Various Materials, p. 398

**13-3** Coefficients of Viscosity for Various Fluids, p. 414

**15-1** Intensity and Intensity Level of Some Common Sounds $(I_0 = 10^{-12}\ \text{W/m}^2)$, p. 481

**17-1** The Temperatures of Various Places and Phenomena, p. 537

**18-1** Specific Heats and Molar Specific Heats of Some Solids and Liquids, p. 560

**18-2** Normal Melting Point (MP), Latent Heat of Fusion ($L_f$), Normal Boiling Point (BP), and Latent Heat of Vaporization ($L_v$) for Various Substances at 1 atm, p. 563

**18-3** Molar Heat Capacities J/mol·K of Various Gases at 25°C, p. 574

**20-1** Approximate Values of the Coefficients of Thermal Expansion for Various Substances, p. 629

**20-2** Critical Temperatures $T_c$ for Various Substances, p. 634

**20-3** Thermal Conductivities $k$ for Various Materials, p. 636

**20-4** $R$ Factors $\Delta x/k$ for Various Building Materials, p. 639

**21-1** The Triboelectric Series, p. 652

**21-2** Some Electric Fields in Nature, p. 661

**24-1** Dielectric Constants and Dielectric Strengths of Various Materials, p. 768

**25-1** Resistivities and Temperature Coefficients, p. 792

**25-2** Wire Diameters and Cross-Sectional Areas for Commonly Used Copper Wires, p. 792

**27-1** Magnetic Susceptibility of Various Materials at 20 C, p. 876

**27-2** Maximum Values of $\mu_0 M_s$ and $K_m$ for Some Ferromagnetic Materials, p. 882

**30-1** The Electromagnetic Spectrum, p. 977

**36-1** Electron Configurations of the Atoms in Their Ground States, p. 1196

**38-1** Free-Electron Number Densities and Fermi Energies at $T = 0$ for Selected Elements, p. 1239

**38-2** Work Functions for Some Metals, p. 1241

**39-1** Rest Energies of Some Elementary Particles and Light Nuclei, p. 1291

**40-1** Atomic Masses of the Neutron and Selected Isotopes, p. 1309

**41-1** Hadrons That Are Stable Against Decay via the Strong Nuclear Interaction, p. 1337

**41-2** Properties of Quarks and Antiquarks, p. 1347

**41-3** Masses of Fundamental Particles, p. 1347

**41-4** Bosons That Mediate the Basic Interactions, p. 1348

**41-5** Properties of the Basic Interactions, p. 1350

## Geometry and Trigonometry

| | |
|---|---|
| $C = \pi d = 2\pi r$ | definition of $\pi$ |
| $A = \pi r^2$ | area of circle |
| $V = \frac{4}{3}\pi r^3$ | spherical volume |
| $A = dV/dr = 4\pi r^2$ | spherical surface area |
| $V = A_{\text{base}}L = \pi r^2 L$ | cylindrical volume |
| $A = dV/dr = 2\pi rL$ | cylindrical surface area |

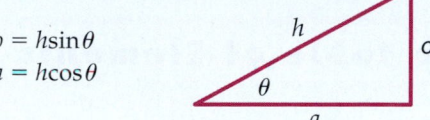

$o = h\sin\theta$
$a = h\cos\theta$

$\sin^2\theta + \cos^2\theta = 1$

$\sin(A \pm B) = \sin A \cos B \pm \cos A \sin B$

$\cos(A \pm B) = \cos A \cos B \mp \sin A \sin B$

$\sin A \pm \sin B = 2\sin[\frac{1}{2}(A \pm B)]\cos[\frac{1}{2}(A \mp B)]$

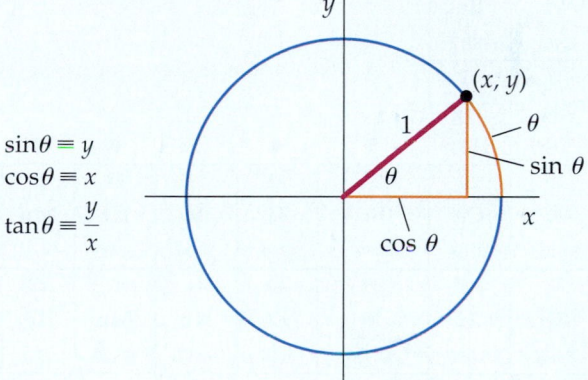

$\sin\theta \equiv y$
$\cos\theta \equiv x$
$\tan\theta \equiv \dfrac{y}{x}$

IF $|\theta| << 1$, THEN
$\quad \cos\theta \approx 1$ AND $\tan\theta \approx \sin\theta \approx \theta \qquad$ ($\theta$ in radians)

## Quadratic Formula

If $ax^2 + bx + c = 0$, then $x = \dfrac{-b \pm \sqrt{b^2 - 4ac}}{2a}$

## Binomial Expansion

If $|x| < 1$, then $(1 + x)^n =$

$$1 + nx + \frac{n(n-1)}{2!}x^2 + \frac{n(n-1)(n-2)}{3!}x^3 + \ldots$$

If $|x| << 1$, then $(1 + x)^n \approx 1 + nx$

If $|\Delta x|$ is small, then $\Delta F \approx \dfrac{dF}{dx}\Delta x$

# APPENDIX C

## Periodic Table of Elements

| 1 | | | | | | | | | | | | | | | | | 18 |
|---|---|---|---|---|---|---|---|---|---|---|---|---|---|---|---|---|---|
| **1**<br>**H**<br>1.00797 | **2** | | | | | | | | | | | | **13** | **14** | **15** | **16** | **17** | **2**<br>**He**<br>4.003 |
| **3**<br>**Li**<br>6.941 | **4**<br>**Be**<br>9.012 | | | | | | | | | | | | **5**<br>**B**<br>10.81 | **6**<br>**C**<br>12.011 | **7**<br>**N**<br>14.007 | **8**<br>**O**<br>15.9994 | **9**<br>**F**<br>19.00 | **10**<br>**Ne**<br>20.179 |
| **11**<br>**Na**<br>22.990 | **12**<br>**Mg**<br>24.31 | **3** | **4** | **5** | **6** | **7** | **8** | **9** | **10** | **11** | **12** | | **13**<br>**Al**<br>26.98 | **14**<br>**Si**<br>28.09 | **15**<br>**P**<br>30.974 | **16**<br>**S**<br>32.064 | **17**<br>**Cl**<br>35.453 | **18**<br>**Ar**<br>39.948 |
| **19**<br>**K**<br>39.102 | **20**<br>**Ca**<br>40.08 | **21**<br>**Sc**<br>44.96 | **22**<br>**Ti**<br>47.88 | **23**<br>**V**<br>50.94 | **24**<br>**Cr**<br>52.00 | **25**<br>**Mn**<br>54.94 | **26**<br>**Fe**<br>55.85 | **27**<br>**Co**<br>58.93 | **28**<br>**Ni**<br>58.69 | **29**<br>**Cu**<br>63.55 | **30**<br>**Zn**<br>65.38 | | **31**<br>**Ga**<br>69.72 | **32**<br>**Ge**<br>72.59 | **33**<br>**As**<br>74.92 | **34**<br>**Se**<br>78.96 | **35**<br>**Br**<br>79.90 | **36**<br>**Kr**<br>83.80 |
| **37**<br>**Rb**<br>85.47 | **38**<br>**Sr**<br>87.62 | **39**<br>**Y**<br>88.906 | **40**<br>**Zr**<br>91.22 | **41**<br>**Nb**<br>92.91 | **42**<br>**Mo**<br>95.94 | **43**<br>**Tc**<br>(98) | **44**<br>**Ru**<br>101.1 | **45**<br>**Rh**<br>102.905 | **46**<br>**Pd**<br>106.4 | **47**<br>**Ag**<br>107.870 | **48**<br>**Cd**<br>112.41 | | **49**<br>**In**<br>114.82 | **50**<br>**Sn**<br>118.69 | **51**<br>**Sb**<br>121.75 | **52**<br>**Te**<br>127.60 | **53**<br>**I**<br>126.90 | **54**<br>**Xe**<br>131.29 |
| **55**<br>**Cs**<br>132.905 | **56**<br>**Ba**<br>137.33 | **57–71**<br>**Rare**<br>**Earths** | **72**<br>**Hf**<br>178.49 | **73**<br>**Ta**<br>180.95 | **74**<br>**W**<br>183.85 | **75**<br>**Re**<br>186.2 | **76**<br>**Os**<br>190.2 | **77**<br>**Ir**<br>192.2 | **78**<br>**Pt**<br>195.09 | **79**<br>**Au**<br>196.97 | **80**<br>**Hg**<br>200.59 | | **81**<br>**Tl**<br>204.37 | **82**<br>**Pb**<br>207.19 | **83**<br>**Bi**<br>208.98 | **84**<br>**Po**<br>(210) | **85**<br>**At**<br>(210) | **86**<br>**Rn**<br>(222) |
| **87**<br>**Fr**<br>(223) | **88**<br>**Ra**<br>(226) | **89–103**<br>Actinides | **104**<br>**Rf**<br>(261) | **105**<br>**Ha**<br>(260) | **106**<br>(263) | **107**<br>(262) | **108**<br>(265) | **109**<br>(266) | | | | | | | | | | |

| | 57 | 58 | 59 | 60 | 61 | 62 | 63 | 64 | 65 | 66 | 67 | 68 | 69 | 70 | 71 |
|---|---|---|---|---|---|---|---|---|---|---|---|---|---|---|---|
| **Rare Earths**<br>**(Lanthanides)** | **La**<br>138.91 | **Ce**<br>140.12 | **Pr**<br>140.91 | **Nd**<br>144.24 | **Pm**<br>(147) | **Sm**<br>150.36 | **Eu**<br>152.0 | **Gd**<br>157.25 | **Tb**<br>158.92 | **Dy**<br>162.50 | **Ho**<br>164.93 | **Er**<br>167.26 | **Tm**<br>168.93 | **Yb**<br>173.04 | **Lu**<br>174.97 |
| | 89 | 90 | 91 | 92 | 93 | 94 | 95 | 96 | 97 | 98 | 99 | 100 | 101 | 102 | 103 |
| **Actinides** | **Ac**<br>227.03 | **Th**<br>232.04 | **Pa**<br>231.04 | **U**<br>238.03 | **Np**<br>237.05 | **Pu**<br>(244) | **Am**<br>(243) | **Cm**<br>(247) | **Bk**<br>(247) | **Cf**<br>(251) | **Es**<br>(252) | **Fm**<br>(257) | **Md**<br>(258) | **No**<br>(259) | **Lr**<br>(260) |

The 1–18 group designation has been recommended by the International Union of Pure and Applied Chemistry (IUPAC).

## Atomic Numbers and Atomic Masses†

| Name | Symbol | Atomic Number | Mass | Name | Symbol | Atomic Number | Mass |
|------|--------|---------------|------|------|--------|---------------|------|
| Actinium | Ac | 89 | 227.03 | Mercury | Hg | 80 | 200.59 |
| Aluminum | Al | 13 | 26.98 | Molybdenum | Mo | 42 | 95.94 |
| Americium | Am | 95 | (243) | Neodymium | Nd | 60 | 144.24 |
| Antimony | Sb | 51 | 121.75 | Neon | Ne | 10 | 20.179 |
| Argon | Ar | 18 | 39.948 | Neptunium | Np | 93 | 237.05 |
| Arsenic | As | 33 | 74.92 | Nickel | Ni | 28 | 58.69 |
| Astatine | At | 85 | (210) | Niobium | Nb | 41 | 92.91 |
| Barium | Ba | 56 | 137.3 | Nitrogen | N | 7 | 14.007 |
| Berkelium | Bk | 97 | (247) | Nobelium | No | 102 | (259) |
| Beryllium | Be | 4 | 9.012 | Osmium | Os | 76 | 190.2 |
| Bismuth | Bi | 83 | 208.98 | Oxygen | O | 8 | 15.9994 |
| Boron | B | 5 | 10.81 | Palladium | Pd | 46 | 106.4 |
| Bromine | Br | 35 | 79.90 | Phosphorus | P | 15 | 30.974 |
| Cadmium | Cd | 48 | 112.41 | Platinum | Pt | 78 | 195.09 |
| Calcium | Ca | 20 | 40.08 | Plutonium | Pu | 94 | (244) |
| Californium | Cf | 98 | (251) | Polonium | Po | 84 | (210) |
| Carbon | C | 6 | 12.011 | Potassium | K | 19 | 39.098 |
| Cerium | Ce | 58 | 140.12 | Praseodymium | Pr | 59 | 140.91 |
| Cesium | Cs | 55 | 132.905 | Promethium | Pm | 61 | (147) |
| Chlorine | Cl | 17 | 35.453 | Protactinium | Pa | 91 | 231.04 |
| Chromium | Cr | 24 | 52.00 | Radium | Ra | 88 | (226) |
| Cobalt | Co | 27 | 58.93 | Radon | Rn | 86 | (222) |
| Copper | Cu | 29 | 63.55 | Rhenium | Re | 75 | 186.2 |
| Curium | Cm | 96 | (247) | Rhodium | Rh | 45 | 102.905 |
| Dysprosium | Dy | 66 | 162.50 | Rubidium | Rb | 37 | 85.47 |
| Einsteinium | Es | 99 | (252) | Ruthenium | Ru | 44 | 101.1 |
| Erbium | Er | 68 | 167.26 | Rutherfordium | Rf | 104 | (261) |
| Europium | Eu | 63 | 152.0 | Samarium | Sm | 62 | 150.36 |
| Fermium | Fm | 100 | (257) | Scandium | Sc | 21 | 44.96 |
| Fluorine | F | 9 | 19.00 | Selenium | Se | 34 | 78.96 |
| Francium | Fr | 87 | (223) | Silicon | Si | 14 | 28.09 |
| Gadolinium | Gd | 64 | 157.25 | Silver | Ag | 47 | 107.870 |
| Gallium | Ga | 31 | 69.72 | Sodium | Na | 11 | 22.990 |
| Germanium | Ge | 32 | 72.59 | Strontium | Sr | 38 | 87.62 |
| Gold | Au | 79 | 196.97 | Sulfur | S | 16 | 32.064 |
| Hafnium | Hf | 72 | 178.49 | Tantalum | Ta | 73 | 180.95 |
| Hahnium | Ha | 105 | (260) | Technetium | Tc | 43 | (98) |
| Helium | He | 2 | 4.003 | Tellurium | Te | 52 | 127.60 |
| Holmium | Ho | 67 | 164.93 | Terbium | Tb | 65 | 158.92 |
| Hydrogen | H | 1 | 1.0079 | Thallium | Tl | 81 | 204.37 |
| Indium | In | 49 | 114.82 | Thorium | Th | 90 | 232.04 |
| Iodine | I | 53 | 126.90 | Thulium | Tm | 69 | 168.93 |
| Iridium | Ir | 77 | 192.2 | Tin | Sn | 50 | 118.69 |
| Iron | Fe | 26 | 55.85 | Titanium | Ti | 22 | 47.88 |
| Krypton | Kr | 36 | 83.80 | Tungsten | W | 74 | 183.85 |
| Lanthanum | La | 57 | 138.91 | Uranium | U | 92 | 238.03 |
| Lawrencium | Lr | 103 | (260) | Vanadium | V | 23 | 50.94 |
| Lead | Pb | 82 | 207.2 | Xenon | Xe | 54 | 131.29 |
| Lithium | Li | 3 | 6.941 | Ytterbium | Yb | 70 | 173.04 |
| Lutetium | Lu | 71 | 174.97 | Yttrium | Y | 39 | 88.906 |
| Magnesium | Mg | 12 | 24.31 | Zinc | Zn | 30 | 65.38 |
| Manganese | Mn | 25 | 54.94 | Zirconium | Zr | 40 | 91.22 |
| Mendelevium | Md | 101 | (258) | | | | |

† More precise values for the atomic masses, along with the uncertainties in the masses, can be found at http://physics.nist.gov/PhysRefData/.

# APPENDIX D

## Review of Mathematics

In this appendix, we will review some of the basic results of algebra, geometry, trigonometry, and calculus. In many cases, we will merely state results without proof. Table D-1 lists some mathematical symbols.

### Equations

The following operations can be performed on mathematical equations to facilitate their solution:

1. The same quantity can be added to or subtracted from each side of the equation.

2. Each side of the equation can be multiplied or divided by the same quantity.

3. Each side of the equation can be raised to the same power.

It is important to understand that the preceding rules apply to each *side* of the equation and not to each term in the equation.

### EXAMPLE D-1

Solve the following equation for $x$: $(x - 3)^2 + 7 = 23$.

| | |
|---|---|
| 1. Subtract 7 from each side: | $(x - 3)^2 = 16$ |
| 2. Take the square root of each side: | $x - 3 = \pm 4$ |
| 3. Add 3 to each side: | $x = 4 + 3 = 7$ |
| | or |
| | $x = -4 + 3 = -1$ |

**REMARKS** Note that in step 2 we do not need to write $-(x - 3) = \pm 4$ because all possibilities are included in

$$x - 3 = \pm 4.$$

**CHECK THE RESULT** We check our result by substituting each value into the original equation: $(7 - 3)^2 + 7 = 16 + 7 = 23$ and $(-1 - 3)^2 + 7 = 16 + 7 = 23$.

### TABLE D-1

**Mathematical Symbols**

| | |
|---|---|
| $=$ | is equal to |
| $\neq$ | is not equal to |
| $\approx$ | is approximately equal to |
| $\sim$ | is of the order of |
| $\propto$ | is proportional to |
| $>$ | is greater than |
| $\geq$ | is greater than or equal to |
| $\gg$ | is much greater than |
| $<$ | is less than |
| $\leq$ | is less than or equal to |
| $\ll$ | is much less than |
| $\Delta x$ | change in $x$ |
| $|x|$ | absolute value of $x$ |
| $n!$ | $n(n - 1)(n - 2) \ldots 1$ |
| $\Sigma$ | sum |
| lim | limit |
| $\Delta t \to 0$ | $\Delta t$ approaches zero |
| $\dfrac{dx}{dt}$ | derivative of $x$ with respect to $t$ |
| $\dfrac{\partial x}{\partial t}$ | partial derivative of $x$ with respect to $t$ |
| $\int$ | integral |

**EXAMPLE D - 2**

Solve the following equation for *x:*

$$\frac{1}{x} + \frac{1}{4} = \frac{1}{3}$$

1. Subtract from each side:

$$\frac{1}{x} = \frac{1}{3} - \frac{1}{4} = \frac{4}{12} - \frac{3}{12} = \frac{1}{12}$$

2. Multiply each side by 12*x:*

$$x = 12$$

**REMARKS** This type of equation occurs both in geometric optics and in analyses of electric circuits. Although it is easy to solve, errors are often made. A typical mistake is to take the reciprocal of each *term,* obtaining $x + 4 = 3$. Taking the reciprocal of each term is not allowed; taking the reciprocal of each *side* of an equation is allowed. Note that multiplying each side by 12*x* in step 2 is equivalent to taking the reciprocal of each side of the equation.

## Direct and Inverse Proportion

The relationships of direct proportion and inverse proportion are so important in physics that they deserve special consideration. Often much algebraic manipulation can be avoided through a simple knowledge of these relationships. Suppose, for example, that you work for 5 days at a certain pay rate and earn $400. How much would you earn at the same pay rate if you worked 8 days? In this problem, the money earned is *directly proportional* to the time worked. We can write an equation relating the money earned $M$ to the time worked $t$ using a constant of proportionality $R$:

$$M = Rt$$

The constant of proportionality in this case is the pay rate. We can express $R$ in dollars per day. Since $400 was earned in 5 d, the value of $R$ is $400/(5 d) = $80/d. In 8 d, the amount earned is therefore

$$M = (\$80/d)(8\ d) = \$640$$

However, we do not have to find the pay rate explicitly to work the problem. Since the amount earned in 8 d is $\frac{8}{5}$ times that earned in 5 d, this amount is

$$M = \tfrac{8}{5}(\$400) = \$640$$

We can use a similar example to illustrate inverse proportion. If you get a 25% raise, how long would you need to work to earn $400? Here we consider $R$ to be a variable and we wish to solve for $t$:

$$t = \frac{M}{R}$$

In this equation, the time $t$ is *inversely proportional* to the pay rate $R$. Thus, if the new rate is $\frac{5}{4}$ times the old rate, the new time will be $\frac{4}{5}$ times the old time or 4 d.

There are some situations in which one quantity varies as the square or some other power of another quantity where the ideas of proportionality are also very useful. Suppose, for example, that a 10-in. diameter pizza costs $8.50. How much would you expect a 12-in. diameter pizza to cost? We expect the cost of a pizza to be approximately proportional to the amount of its contents, which is proportional to the area of the pizza. Since the area is in turn proportional to the square of the diameter, the cost should be proportional to the square of the diameter. If we increase the diameter by a factor of $12/10$, the area increases by a factor of $(12/10)^2 = 1.44$, so we should expect the cost to be $(1.44)(\$8.50) = \$12.24$.

## EXAMPLE D-3

**The intensity of light from a point source varies inversely with the square of the distance from the source. If the intensity is 3.20 W/m² at 5 m from a source, what is it at 6 m from the source?**

1. Write an equation expressing the fact that the intensity varies inversely with the square of the distance:

$$I = \frac{C}{r^2}$$

where $C$ is some constant.

2. Let $I_1$ be the intensity at $r_1 = 5$ m and $I_2$ be the intensity at $r_2 = 6$ m, and express the ratio $I_2/I_1$ in terms of $r_1$ and $r_2$:

$$\frac{I_2}{I_1} = \frac{C/r_2^2}{C/r_1^2} = \frac{r_1^2}{r_2^2} = \left(\frac{r_1}{r_2}\right)^2 = \left(\frac{5}{6}\right)^2 = 0.694$$

3. Solve for $I_2$:

$$I_2 = 0.694 I_1 = (0.694)(3.20 \text{ W/m}^2) = 2.22 \text{ W/m}^2$$

## Linear Equations

An equation in which the variables occur only to the first power is said to be linear. A linear equation relating $y$ and $x$ can always be put into the standard form

$$y = mx + b \qquad\qquad \text{D-1}$$

where $m$ and $b$ are constants that may be either positive or negative. Figure D-1 shows a graph of the values of $x$ and $y$ that satisfy (Equation D-1). The constant $b$, called the **intercept,** is the value of $y$ at $x = 0$. The constant $m$ is the **slope** of the line, which equals the ratio of the change in $y$ to the corresponding change in $x$. In the figure, we have indicated two points on the line, $x_1, y_1$ and $x_2, y_2$, and the changes $\Delta x = x_2 - x_1$ and $\Delta y = y_2 - y_1$. The slope $m$ is then

$$m = \frac{y_2 - y_1}{x_2 - x_1} = \frac{\Delta y}{\Delta x}$$

If $x$ and $y$ are both unknown, there is no unique solution for their values. Any pair of values $x_1, y_1$ on the line in Figure D-1 will satisfy the equation. If we have two equations, each with the same two unknowns $x$ and $y$, the equations can be solved simultaneously for the unknowns.

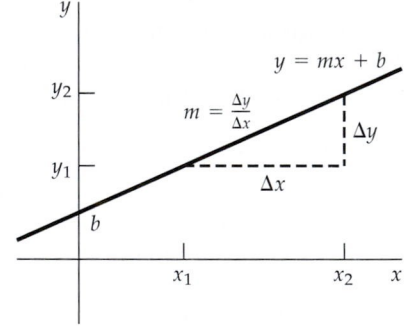

**FIGURE D-1** Graph of the linear equation $y = mx + b$, where $b$ is the intercept and $m = \Delta y/\Delta x$ is the slope.

## EXAMPLE D-4

**Find the values of $x$ and $y$ that satisfy**

$$3x - 2y = 8 \qquad\qquad \text{D-2}$$

**and**

$$y - x = 2 \qquad\qquad \text{D-3}$$

**PICTURE THE PROBLEM** Figure D-2 shows a graph of each of these equations. At the point where the lines intersect, the values of $x$ and $y$ satisfy both equations. We can solve two simultaneous equations by first solving either equation for one variable in terms of the other variable and then substituting the result into the other equation. An alternative method is to multiply one equation by a constant such that one of the unknown terms is eliminated when the equations are added or subtracted.

1. Solve (Equation D-3) for $y$:

$$y = x + 2$$

2. Substitute this value for $y$ into (Equation D-2):

$$3x - 2(x + 2) = 8$$

3. Simplify and solve for $x$:

$$3x - 2x - 4 = 8$$
$$x = 12$$

4. To solve these equations using the alternative method, we first multiply (Equation D-3) by 2:

$$2y - 2x = 4$$

5. Add this equation to (Equation D-2):

$$2y - 2x = 4$$
$$\underline{3x - 2y = 8}$$
$$3x - 2x = 12$$
$$x = 12$$

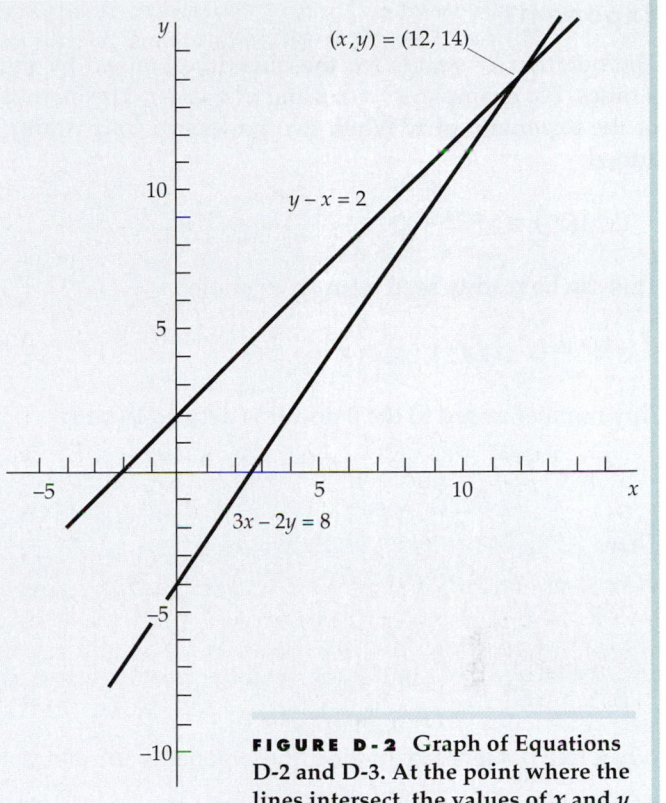

**FIGURE D-2** Graph of Equations D-2 and D-3. At the point where the lines intersect, the values of $x$ and $y$ satisfy both equations.

## Factoring

Equations can often be simplified by factoring. Three important examples are

1. Common factor:  $2ax + 3ay = a(2x + 3y)$
2. Perfect square:  $x^2 - 2xy + y^2 = (x - y)^2$
3. Difference of squares:  $x^2 - y^2 = (x + y)(x - y)$

## The Quadratic Formula

An equation that contains a variable to the second power is called a *quadratic equation*. The standard form for a quadratic equation is

$$ax^2 + bx + c = 0 \qquad \text{D-4}$$

where $a$, $b$, and $c$ are constants. The general solution of this equation is

$$x = -\frac{b}{2a} \pm \frac{1}{2a}\sqrt{b^2 - 4ac} \qquad \text{D-5}$$

When $b^2$ is greater than $4ac$, there are two solutions corresponding to the + and − signs. Figure D-3 shows a graph of $y$ versus $x$ where $y = ax^2 + bx + c$. The curve, called a **parabola**, crosses the $x$ axis twice. The values of $x$ for which $y = 0$ are the solutions to (Equation D-4). When $b^2 < 4ac$, the graph of $y$ versus $x$ does not intersect the $x$ axis, as is shown in Figure D-4, and there are no real solutions to (Equation D-4). When $b^2 = 4ac$, the graph of $y$ versus $x$ is tangent to the $x$ axis at the point $x = -b/2a$.

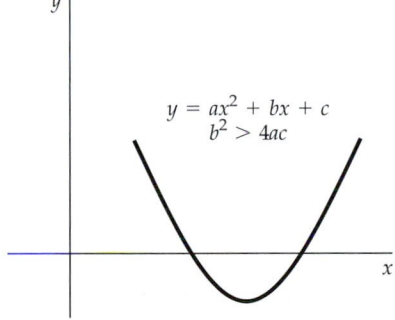

**FIGURE D-3** Graph of $y$ versus $x$ when $y = ax^2 + bx + c$ for the case $b^2 > 4ac$. The two values of $x$ for which $y = 0$ satisfy the quadratic equation (Equation D-4).

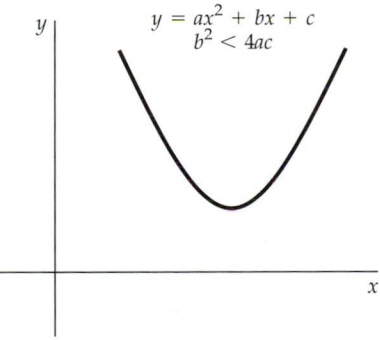

**FIGURE D-4** Graph of $y$ versus $x$ when $y = ax^2 + bx + c$ for the case $b^2 < 4ac$. In this case, there are no (real) values of $x$ for which $y = 0$.

## Exponents

The notation $x^n$ stands for the quantity obtained by multiplying $x$ times itself $n$ times. For example, $x^2 = x \cdot x$ and $x^3 = x \cdot x \cdot x$. The quantity $n$ is called the **power,** or the **exponent,** of $x$. When two powers of $x$ are multiplied, the exponents are added:

$$(x^m)(x^n) = x^{m+n} \qquad \text{D-6}$$

This can be readily seen from an example:

$$x^2 x^3 = (x \cdot x)(x \cdot x \cdot x) = x^5$$

Any number raised to the 0 power is defined to be 1:

$$x^0 = 1 \qquad \text{D-7}$$

Then

$$x^n x^{-n} = x^0 = 1$$
$$\qquad \text{D-8}$$
$$x^{-n} = \frac{1}{x^n}$$

When two powers are divided, the exponents are subtracted:

$$\frac{x^n}{x^m} = x^n x^{-m} = n^{n-m} \qquad \text{D-9}$$

Using these rules, we have

$$x^{1/2} \cdot x^{1/2} = x$$

so

$$x^{1/2} = \sqrt{x}$$

When a power is raised to another power, the exponents are multiplied:

$$(x^n)^m = x^{nm} \qquad \text{D-10}$$

## Logarithms

If $y$ is related to $x$ by $y = a^x$, the number $x$ is said to be the logarithm of $y$ to the base $a$ and is written

$$x = \log_a y$$

then if $x = 1$ then $y = a^1 = a$ and

$$\log_a a = 1 \qquad \text{D-11}$$

and, if $x = 0$ then $y = a^0 = 1$ and

$$\log_a 1 = 0 \qquad \text{D-12}$$

Also, if $y_1 = a^n$ and $y_2 = a^m$, then

$$y_1 y_2 = a^n a^m = a^{n+m}$$

and

$$\log_a y_1 = n, \log_a y_2 = m, \text{ and } \log_a y_1 y_2 = n + m$$

so

$$\log_a y_1 y_2 = \log_a y_1 + \log_a y_2 \qquad \text{D-13}$$

It immediately follows that

$$\log_a y^n = n \log_a y \qquad \text{D-14}$$

There are two bases in common use: base 10, called **common logarithms,** and base $e(e = 2.728 \dots)$, called **natural logarithms.** When no base is specified, the base is usually understood to be 10. Thus, $\log 100 = \log_{10} 100 = 2$ since $100 = 10^2$.

The symbol ln is used for natural logarithms. Thus,

$$\log_e x = \ln x \qquad \text{D-15}$$

and $y = \ln x$ implies

$$x = e^y \qquad \text{D-16}$$

Logarithms can be changed from one base to another. Suppose that

$$z = \log x \qquad \text{D-17}$$

Then

$$10^z = 10^{\log x} = x \qquad \text{D-18}$$

Taking the natural logarithm of both sides of (Equation D-18), we obtain

$$z \ln 10 = \ln x$$

Substituting $\log x$ for $z$ (see Equation D-17) gives

$$\ln x = (\ln 10)\log x \qquad \text{D-19}$$

## The Exponential Function

When the rate of change of a quantity is proportional to the quantity itself, the quantity increases or decreases exponentially. An example of *exponential decrease* is nuclear decay. If $N$ is the number of radioactive nuclei at some time, then the change $dN$ in some very small time interval $dt$ will be proportional to $N$ and to $dt$:

$$dN = -\lambda N \, dt$$

where the constant of proportionality $\lambda$ is the decay rate. The function $N$ satisfying this equation is

$$N = N_0 e^{-\lambda t} \qquad \text{D-20}$$

where $N_0$ is the number at time $t = 0$. Figure D-5 shows $N$ versus $t$. A characteristic of exponential decay is that $N$ decreases by a constant factor in a given time interval. The time interval for $N$ to decrease to half its original value is its half-life $t_{1/2}$. The half life is obtained from (Equation D-20) by setting $N = \frac{1}{2}N_0$ and solving for the time. This gives

$$t_{1/2} = \frac{\ln 2}{\lambda} = \frac{0.693}{\lambda} \qquad \text{D-21}$$

An example of *exponential increase* is population growth. If the number of organisms is $N$, the change in $N$ after a small time interval $dt$ is given by

$$dN = +\lambda N \, dt$$

where $\lambda$ is a constant that characterizes the rate of increase. The function $N$ satisfying this equation is

$$N = N_0 e^{\lambda t} \qquad \text{D-22}$$

A graph of this function is shown in Figure D-6. An exponential increase is characterized by a doubling time $T_2$, which is related to $\lambda$ by

$$T_2 = \frac{\ln 2}{\lambda} = \frac{0.693}{\lambda} \qquad \text{D-23}$$

If the rate of increase $\lambda$ is expressed as a percentage, $r = \lambda/100\%$, the doubling time is

$$T_2 = \frac{69.3}{r} \qquad \text{D-24}$$

For example, if the population increases by 2 percent per year, the population will double every $69.3/2 \approx 35$ years. Table D-2 lists some useful relations for exponential and logarithmic functions.

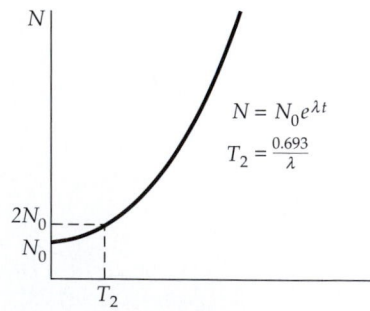

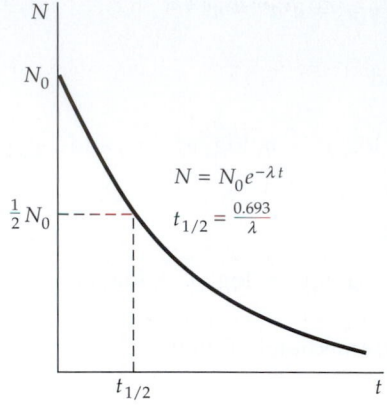

**FIGURE D-5** Graph of $N$ versus $t$ when $N$ decreases exponentially. The time $t_{1/2}$ is the time it takes for $N$ to decrease by one-half.

## TABLE D-2

**Exponential and Logarithmic Functions**

$e = 2.71828;$

$e^0 = 1$

If $y = e^x$, then $x = \ln y$.

$e^{\ln x} = x$

$e^x e^y = e^{(x+y)}$

$(e^x)^y = e^{xy} = (e^y)^x$

$\ln e = 1; \quad \ln 1 = 0$

$\ln xy = \ln x + \ln y$

$\ln \dfrac{x}{y} = \ln x - \ln y$

$\ln e^x = x; \quad \ln a^x = x \ln a$

$\ln x = (\ln 10) \log x$

$\qquad = 2.3026 \log x$

$\log x = (\log e) \ln x = 0.43429 \ln x$

$e^x = 1 + x + \dfrac{x^2}{2!} + \dfrac{x^3}{3!} + \ldots$

$\ln(1 + x) = x \pm \dfrac{x^2}{2} + \dfrac{x^3}{3} \pm \dfrac{x^4}{4} + \ldots$

**FIGURE D-6** Graph of $N$ versus $t$ when $N$ increases exponentially. The time $T_2$ is the time it takes for $N$ to double.

## Geometry

The ratio of the circumference of a circle to its diameter is a natural number $\pi$, which has the approximate value

$$\pi = 3.141592$$

The circumference $C$ of a circle is thus related to its diameter $d$ and its radius $r$ by

$$C = \pi d = 2\pi r \qquad \text{circumference of circle} \qquad \text{D-25}$$

The area of a circle is

$$A = \pi r^2 \qquad \text{area of circle} \qquad \text{D-26}$$

The area of a parallelogram is the base $b$ times the height $h$ (Figure D-7) and that of a triangle is one-half the base times the height (Figure D-8). A sphere of radius $r$ (Figure D-9) has a surface area given by

$$A = 4\pi r^2 \qquad \text{spherical surface area} \qquad \text{D-27}$$

and a volume given by

$$V = \tfrac{4}{3}\pi r^3 \qquad \text{spherical volume} \qquad \text{D-28}$$

A cylinder of radius $r$ and length $L$ (Figure D-10) has surface area (not including the end faces) of

$$A = 2\pi rL \qquad \text{cylindrical surface} \qquad \text{D-29}$$

and volume of

$$V = \pi r^2 L \qquad \text{cylindrical volume} \qquad \text{D-30}$$

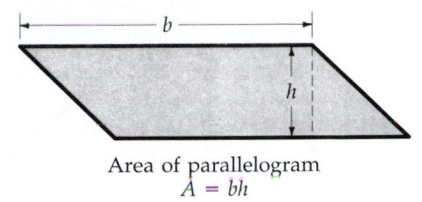

Area of parallelogram
$A = bh$

**FIGURE D-7** Area of a parallelogram.

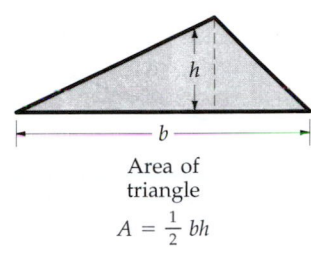

Area of
triangle
$A = \frac{1}{2} bh$

**FIGURE D-8** Area of a triangle.

## Trigonometry

The angle between two intersecting straight lines is measured as follows. A circle is drawn with its center at the intersection of the lines, and the circular arc is divided into 360 parts called **degrees**. The number of degrees in the arc between the lines is the measure of the angle between the lines. For very small angles, the degree is divided into minutes (') and seconds (") with $1' = 1°/60$ and $1'' = 1'/60 = 1°/3600$. For scientific work, a more useful measure of an angle is the radian (rad). In radian measure, the angle between two intersecting straight lines is found by again drawing a circle with its center at the intersection of the lines. The measure of the angle in radians is then defined as the length of the circular arc between the lines divided by the radius of the circle (Figure D-11). If $s$ is the arc length and $r$ is the radius of the circle, the angle $\theta$ measured in radians is

$$\theta = \frac{s}{r} \qquad \text{D-31}$$

Since the angle measured in radians is the ratio of two lengths, it is dimensionless. The relation between radians and degrees is

$$360° = 2\pi \text{ rad}$$

or

$$1 \text{ rad} = \frac{360°}{2\pi} = 57.3°$$

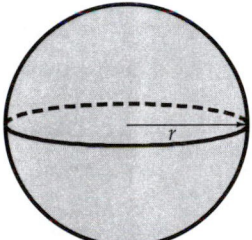

Spherical surface area
$A = 4\pi r^2$
Spherical volume
$V = \frac{4}{3}\pi r^3$

**FIGURE D-9** Surface area and volume of a sphere.

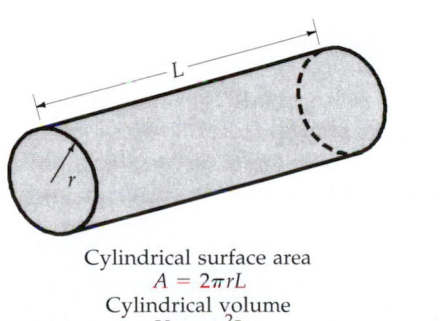

Cylindrical surface area
$A = 2\pi rL$
Cylindrical volume
$V = \pi r^2 L$

**FIGURE D-10** Surface area (not including the end faces) and volume of a cylinder.

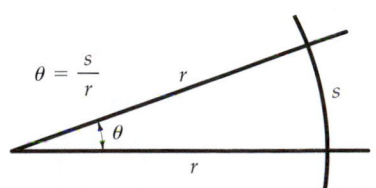

$\theta = \dfrac{s}{r}$

**FIGURE D-11** The angle $\theta$ in radians is defined to be the ratio $s/r$, where $s$ is the arc length intercepted on a circle of radius $r$.

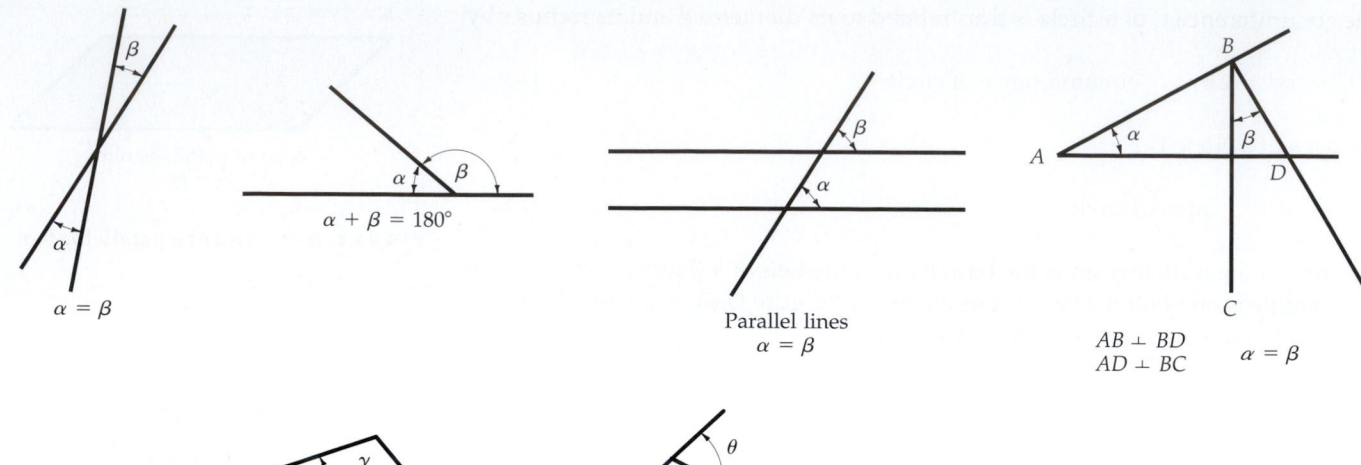

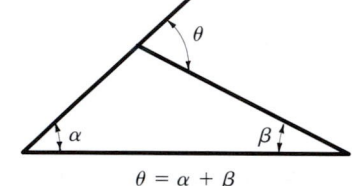

**FIGURE D-12** Some useful relations for angles.

Figure D-12 shows some useful relations for angles.

Figure D-13 shows a right triangle formed by drawing the line $BC$ perpendicular to $AC$. The lengths of the sides are labeled $a$, $b$, and $c$. The trigonometric functions $\sin \theta$, $\cos \theta$, and $\tan \theta$ for an acute angle $\theta$ are defined as

$$\sin \theta = \frac{a}{c} = \frac{\text{opposite side}}{\text{hypotenuse}} \qquad \text{D-33}$$

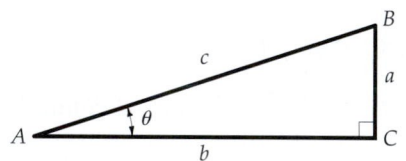

**FIGURE D-13** A right triangle with sides of length $a$ and $b$ and a hypotenuse of length $c$.

$$\cos \theta = \frac{b}{c} = \frac{\text{adjacent side}}{\text{hypotenuse}} \qquad \text{D-34}$$

$$\tan \theta = \frac{a}{b} = \frac{\text{opposite side}}{\text{adjacent side}} = \frac{\sin \theta}{\cos \theta} \qquad \text{D-35}$$

Three other trigonometric functions, defined as the reciprocals of these functions, are

$$\sec \theta = \frac{c}{b} = \frac{1}{\cos \theta} \qquad \text{D-36}$$

$$\csc \theta = \frac{c}{a} = \frac{1}{\sin \theta} \qquad \text{D-37}$$

$$\cot \theta = \frac{b}{a} = \frac{1}{\tan \theta} = \frac{\cos \theta}{\sin \theta} \qquad \text{D-38}$$

The angle $\theta$ whose sine is $x$ is called the arcsine of $x$, and is written $\sin^{-1} x$. That is, if

$$\sin \theta = x$$

then

$$\theta = \arcsin x = \sin^{-1} x \qquad \text{D-39}$$

The arcsine is the inverse of the sine. The inverse of the cosine and tangent are defined similarly. The angle whose cosine is $y$ is the arccosine of $y$. That is, if

$$\cos \theta = y$$

then

$$\theta = \arccos y = \cos^{-1} y \qquad \text{D-40}$$

The angle whose tangent is $z$ is the arctangent of $z$. That is, if

$$\tan \theta = z$$

$$\theta = \arctan z = \tan^{-1} z \qquad \text{D-41}$$

The Pythagorean theorem

$$a^2 + b^2 = c^2 \qquad \text{D-42}$$

gives some useful identities. If we divide each term in this equation by $c^2$, we obtain

$$\frac{a^2}{c^2} + \frac{b^2}{c^2} = 1$$

or, from the definitions of $\sin \theta$ and $\cos \theta$,

$$\sin^2 \theta + \cos^2 \theta = 1 \qquad \text{D-43}$$

Similarly, we can divide each term in (Equation D-42) by $a^2$ or $b^2$ and obtain

$$1 + \cot^2 \theta = \csc^2 \theta \qquad \text{D-44}$$

and

$$1 + \tan^2 \theta = \sec^2 \theta \qquad \text{D-45}$$

These and other useful trigonometric formulas are listed in Table D-3.

The following equation is actually two equations. For one of the equations the two upper signs are taken, and for the other equation the two lower signs are taken. The same sign rule applies to others of the following "equations":

$$\sin(A \pm B) = \sin A \cos B \pm \cos A \sin B$$

$$\cos(A \pm B) = \cos A \cos B \mp \sin A \sin B$$

$$\tan(A \pm B) = \frac{\tan A \pm \tan B}{1 \mp \tan A \tan B}$$

$$\sin A \pm \sin B = 2 \sin\left[\tfrac{1}{2}(A \pm B)\right] \cos\left[\tfrac{1}{2}(A \pm B)\right]$$

$$\cos A + \cos B = 2 \cos\left[\tfrac{1}{2}(A + B)\right] \cos\left[\tfrac{1}{2}(A - B)\right]$$

$$\cos A - \cos B = 2 \sin\left[\tfrac{1}{2}(A + B)\right] \sin\left[\tfrac{1}{2}(B - A)\right]$$

$$\tan A \pm \tan B = \frac{\sin(A \pm B)}{\cos A \cos B}$$

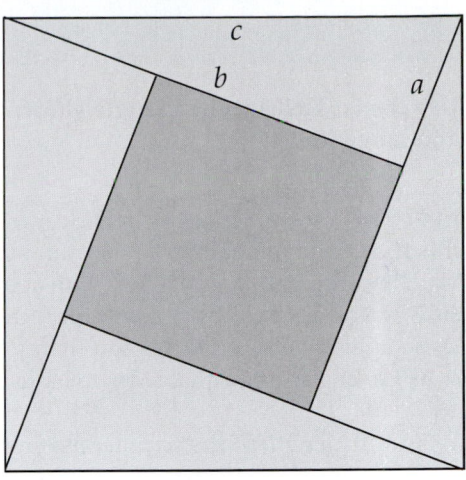

When this figure was first published the letters were absent and it was accompanied by the single word "Behold!" Using the drawing, establish the Pythagorean theorem ($a^2 + b^2 = c^2$).

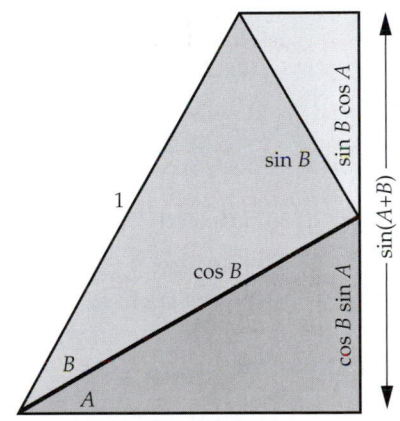

Using this drawing, establish the identity $\sin(A + B) = \sin A \cos B + \cos A \sin B$. You can also use it to establish the identity $\cos(A + B) = \cos A \cos B - \sin A \sin B$. Try it.

## TABLE D-3

**Trigonometric Formulas**

$$\sin^2 \theta + \cos^2 \theta = 1; \quad \sec^2 \theta - \tan^2 \theta = 1; \quad \csc^2 \theta - \cot^2 \theta = 1$$

$$\sin 2\theta = 2 \sin \theta \cos \theta$$

$$\cos 2\theta = \cos^2 \theta - \sin^2 \theta = 2 \cos^2 \theta - 1 = 1 - 2 \sin^2 \theta$$

$$\tan 2\theta = \frac{2 \tan \theta}{1 - \tan^2 \theta}$$

$$\sin \tfrac{1}{2}\theta = \sqrt{\frac{1 - \cos \theta}{2}}; \quad \cos \tfrac{1}{2}\theta = \sqrt{\frac{1 + \cos \theta}{2}}; \quad \tan \tfrac{1}{2}\theta = \sqrt{\frac{1 - \cos \theta}{1 + \cos \theta}}$$

## EXAMPLE D-5

Use the isosceles right triangle shown in Figure D-14 to find the sine, cosine, and tangent of 45°.

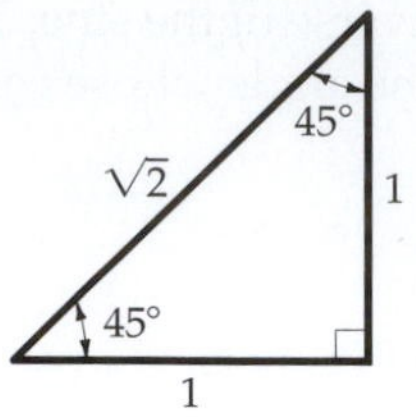

**FIGURE D-14**
An isosceles right triangle.

**PICTURE THE PROBLEM**  It is clear from the figure that the two acute angles of this triangle are equal. Since the sum of the three angles in a triangle must equal 180°, and the right angle is 90°, each acute angle must be 45°. If we multiply each side of any triangle by a common factor, we obtain a similar triangle with the same angles as the first. We can therefore choose any convenient length for one side. We choose the equal sides to have a length of 1 unit.

1. Find the length of the hypotenuse from the Pythagorean theorem:

$$c = \sqrt{a^2 + b^2} = \sqrt{1^2 + 1^2} = \sqrt{2}\ \text{units}$$

2. Calculate sin 45° from its definition:

$$\sin 45° = \frac{a}{c} = \frac{1}{\sqrt{2}} = 0.707$$

3. Calculate cos 45° from its definition:

$$\cos 45° = \frac{b}{c} = \frac{1}{\sqrt{2}} = 0.707$$

4. Calculate tan 45° from its definition:

$$\tan 45° = \frac{a}{b} = \frac{1}{1} = 1$$

## EXAMPLE D-6

The sine of 30° is exactly 0.5. Find the ratios of the sides of a 30–60° right triangle.

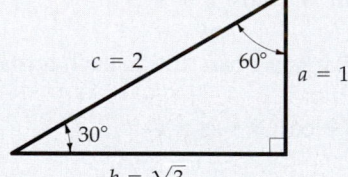

**PICTURE THE PROBLEM** This common triangle is shown in Figure D-15. We choose a length of 1 unit for the side opposite the 30° angle.

**FIGURE D-15** A 30–60° right triangle.

1. Calculate the hypotenuse from the definition of the sine and the choice of 1 unit for the opposite side:

$$\sin 30° = \frac{a}{c} = \frac{1}{c} = 0.5$$

2. Use the Pythagorean theorem to find the length $b$ of the side opposite the 60° angle:

$$c = \frac{1}{0.5} = 2$$

$$b = \sqrt{c^2 - a^2} = \sqrt{2^2 - 1^2} = \sqrt{3}$$

3. Use these results to calculate cos 30°, tan 30°, sin 60°, cos 60°, and tan 60°:

$$\cos 30° = \frac{b}{c} = \frac{\sqrt{3}}{2} = 0.866$$

$$\tan 30° = \frac{a}{b} = \frac{1}{\sqrt{3}} = 0.577$$

$$\sin 60° = \frac{b}{c} = \cos 30° = 0.866$$

$$\cos 60° = \frac{a}{c} = \sin 30° = 0.500$$

$$\tan 60° = \frac{b}{a} = \frac{\sqrt{3}}{1} = 1.732$$

For small angles, the length $a$ is nearly equal to the arc length $s$, as can be seen in Figure D-16. The angle $\theta = s/c$ is therefore nearly equal to $\sin \theta = a/c$:

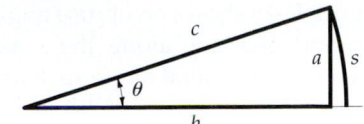

$$\sin \theta \approx \theta \qquad \text{for small values of } \theta \qquad\qquad \text{D-46}$$

Similarly, the lengths $c$ and $b$ are nearly equal, so $\tan \theta = a/b$ is nearly equal to both $\theta$ and $\sin \theta$ for small values of $\theta$:

$$\tan \theta \approx \sin \theta \approx \theta \qquad \text{for small values of } \theta \qquad\qquad \text{D-47}$$

Equations D-46 and D-47 hold only if $\theta$ is measured in radians. Since $\cos \theta = b/c$ and these lengths are nearly equal for small values of $\theta$, we have

$$\cos \theta \approx 1 \qquad \text{for small values of } \theta \qquad\qquad \text{D-48}$$

## EXAMPLE D-7

**By how much do sin $\theta$, tan $\theta$, and $\theta$ differ when $\theta = 15°$?**

1. Convert 15° to radians:

$$\theta = 15° \frac{2\pi \text{ rad}}{360°} = 0.262 \text{ rad}$$

2. Find sin 15° and tan 15° using a calculator:

$$\sin 15° = 0.259$$

$$\tan 15° = 0.268$$

3. Compute the percentage difference between $\theta$ and sin $\theta$:

$$\frac{|\sin \theta - \theta|}{\theta} = \frac{|0.259 - 0.262|}{0.262} = \frac{0.003}{0.262} = 0.011 \approx 1\%$$

4. Compute the percentage difference between $\theta$ and tan $\theta$:

$$\frac{|\tan \theta - \theta|}{\theta} = \frac{|0.268 - 0.262|}{0.262} = \frac{0.006}{0.262} = 0.023 \approx 2\%$$

**REMARKS** For smaller angles, the approximation $\theta < \sin \theta < \tan \theta$ is even more accurate.

Example D-7 shows that if accuracy of a few percent is needed, small angle approximations can be used only for angles of about 15° or less. Figure D-17 shows graphs of $\theta$, sin $\theta$, and tan $\theta$ versus $\theta$, for small values of $\theta$.

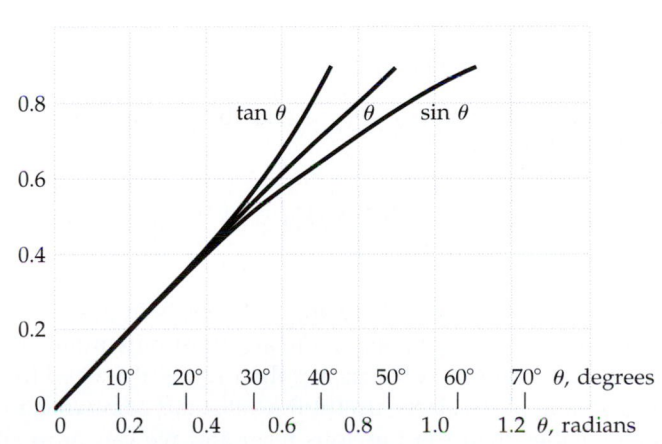

**FIGURE D-17** Graphs of tan $\theta$, $\theta$, and sin $\theta$ versus $\theta$ for small values of $\theta$.

Figure D-18 shows an obtuse angle with its vertex at the origin and one side along the $x$ axis. The trigonometric functions for a general angle such as this are defined by

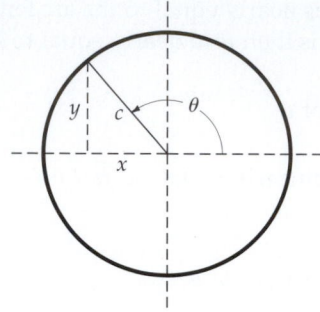

$$\sin \theta = \frac{y}{c} \qquad\qquad\qquad\qquad \text{D-49}$$

$$\cos \theta = \frac{x}{c} \qquad\qquad\qquad\qquad \text{D-50}$$

**FIGURE D-18** Diagram for defining the trigonometric functions for an obtuse angle.

$$\tan \theta = \frac{y}{x} \qquad\qquad\qquad\qquad \text{D-51}$$

Figure D-19 shows plots of these functions versus $\theta$. The sine function has a period of $2\pi$ rad. Thus, for any value of $\theta$ $\sin(\theta + 2\pi) = \sin \theta$ and so forth. That is, when an angle changes by $2\pi$ rad, the function returns to its original value. The tangent function has a period of $\pi$ rad. Thus, $\sin(\theta + 2\pi) = \sin \theta$ and so forth. Some other useful relations are

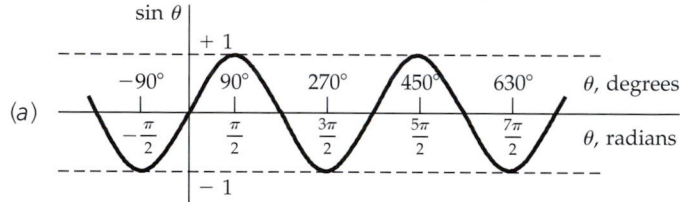

$$\sin(\pi - \theta) = \sin \theta \qquad\qquad\qquad\qquad \text{D-52}$$

$$\cos(\pi - \theta) = -\cos \theta \qquad\qquad\qquad\qquad \text{D-53}$$

$$\sin(\pi/2 - \theta) = \cos \theta \qquad\qquad\qquad\qquad \text{D-54}$$

$$\cos(\pi/2 - \theta) = \sin \theta \qquad\qquad\qquad\qquad \text{D-55}$$

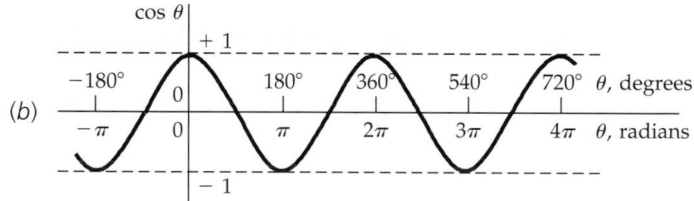

The trigonometric functions can be expressed as power series in $\theta$. The series for $\sin \theta$ and $\cos \theta$ are

$$\sin \theta = \theta - \frac{\theta^3}{3!} + \frac{\theta^5}{5!} - \frac{\theta^7}{7!} + \;\dots \qquad\qquad \text{D-56}$$

$$\cos \theta = 1 - \frac{\theta^2}{2!} + \frac{\theta^4}{4!} - \frac{\theta^6}{6!} + \;\dots \qquad\qquad \text{D-57}$$

When $\theta$ is small, good approximations are obtained using only the first few terms in the series.

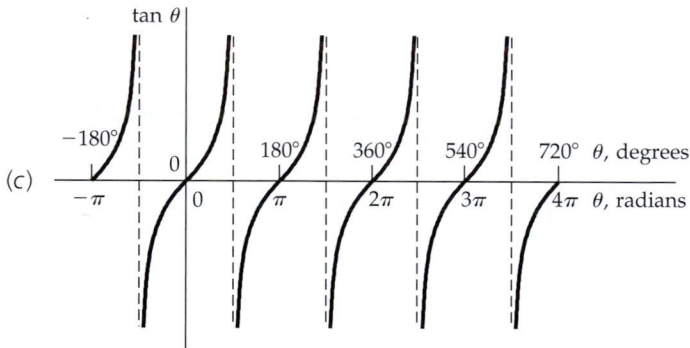

**FIGURE D-19** The trigonometric functions $\sin \theta$, $\cos \theta$, and $\tan \theta$ versus $\theta$.

## The Binomial Expansion

The binomial theorem is very useful for making approximations. One form of this theorem is

$$(1 + x)^n = 1 + nx + \frac{n(n-1)}{2!}x^2 + \frac{n(n-1)(n-2)}{3!}x^3 + \;\dots \qquad \text{D-58}$$

If $n$ is a positive integer, there are $n + 1$ terms in this series. If $n$ is a real number other than a positive integer, there are an infinite number of terms. The series is valid for any value of $n$ if $x^2$ is less than 1. It is also valid for $x^2 = 1$ if $n$ is positive. The series is particularly useful if $x^2$ is small compared to 1. Then each term is much smaller than the previous term and we can drop all but the first two or three terms in the equation. If $x^2$ is much less compared to 1, we have

$$(1 + x)^n \approx 1 + nx, \qquad |x| \ll 1 \qquad\qquad\qquad \text{D-59}$$

**EXAMPLE D-8**

Use (Equation D-59) to find an approximate value for the square root of 101.

1. Write $(101)^{1/2}$ so it is in the form $(1 + x)^n$ with $x$ much less than 1:

$(101)^{1/2} = (100 + 1)^{1/2} = (100)^{1/2}(1 + 0.01)^{1/2} = 10(1 + 0.01)^{1/2}$

2. Use Equation D-59 with $n = \frac{1}{2}$ and $x = 0.1$ to expand $(1 + 0.01)^{1/2}$.

$(1 + 0.01)^{1/2} \approx 1 + \frac{1}{2}(0.01) = 1.005$

3. Substitute this result into the equation in step 1:

$(101)^{1/2} = 10(1 + 0.01)^{1/2} \approx \boxed{10.05}$

**REMARKS** We can assess the accuracy of this result by computing the first term in Equation D-58 that was neglected. This term is

$$\frac{n(n-1)}{2}x^2 = \frac{\frac{1}{2}(-\frac{1}{2})}{2}(0.01)^2 = -\frac{0.0001}{8} \approx -0.00001 = -0.001$$

We therefore expect our answer to be correct to within about 0.001%. The value of $(101)^{1/2}$ to eight significant figures is 10.049875, which differs from 10.05 by 0.000124 or about 0.001% of 10.05.

## Complex Numbers

A general complex number $z$ can be written

$$z = a + bi \qquad\qquad \text{D-60}$$

where $a$ and $b$ are real numbers and $i = \sqrt{-1}$. The quantity $a$ is called the real part and the quantity $ib$ is called the imaginary part of $z$. We can represent a complex number in a plane as shown in Figure D-20, where the $x$ axis is the real axis and the $y$ axis is the imaginary axis. We can use the relations $a = r\cos\theta$ and $b = r\sin\theta$ from Figure D-20 to write the complex number $z$ in polar coordinates:

$$z = r\cos\theta + ir\sin\theta \qquad\qquad \text{D-61}$$

where $r = \sqrt{a^2 + b^2}$ is called the magnitude of $z$.

When complex numbers are added or subtracted, the real and imaginary parts are added or subtracted separately:

$$z_1 + z_2 = (a_1 + ib_1) + (a_2 + ib_2) = (a_1 + a_2) + i(b_1 + b_2) \qquad\qquad \text{D-62}$$

However, when two complex numbers are multiplied, each part of one number is multiplied by each part of the other number:

$$z_1 z_2 = (a_1 + ib_1)(a_2 + ib_2) = a_1 a_2 + i^2 b_1 b_2 + i(a_1 b_2 + a_2 b_1) \qquad\qquad \text{D-63}$$

$$= a_1 a_2 - b_1 b_2 + i(a_1 b_2 + a_2 b_1)$$

where we have used $i^2 = -1$.

The complex conjugate $z^*$ of the complex number $z$ is that number obtained by replacing $i$ with $-i$. If $z = a + ib$ then

$$z^* = (a + ib)^* = a - ib \qquad\qquad \text{D-64}$$

The product of a complex number and its complex conjugate equals the square of the magnitude of the number:

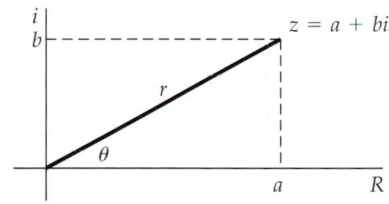

$z = a + bi$
$= r\cos\theta + (r\sin\theta)i$
$= r(\cos\theta + i\sin\theta)$

**FIGURE D-20** Representation of a complex number in a plane. The real part of the complex number is plotted along the horizontal axis, and the imaginary part is plotted along the vertical axis.

$$zz^* = (a + ib)(a - ib) = a^2 + b^2 \qquad \text{D-65}$$

A particularly useful function of a complex number is the exponential $e^{i\theta}$. Using the expansion for $e^x$ given in Table D-2, we have

$$e^{i\theta} = 1 + i\theta + \frac{(i\theta)^2}{2!} + \frac{(i\theta)^3}{3!} + \frac{(i\theta)^4}{4!} + \dots$$

Using $i^2 = -1$, $i^3 = -i$, $i^4 = +1$, and so forth and separating the real parts from the imaginary parts, this expansion can be written

$$e^{i\theta} = \left(1 - \frac{\theta^2}{2!} + \frac{\theta^4}{4!} - \dots\right) + i\left(\theta - \frac{\theta^3}{3!} + \dots\right)$$

Comparing this result with Equations D-56 and D-57, we can see that

$$e^{i\theta} = \cos\theta + i\sin\theta \qquad \text{D-66}$$

Using this result, we can express a general complex number as an exponential:

$$z = a + ib = r\cos\theta + ir\sin\theta = re^{i\theta} \qquad \text{D-67}$$

where $r = \sqrt{a^2 + b^2}$.

### Solving differential equations using complex numbers

Consider an equation of the form

$$a\frac{d^2x}{dt^2} + b\frac{dx}{dt} + cx = A\cos\omega t \qquad \text{D-68}$$

that represents a physical process, such as a damped harmonic oscillator driven by a sinusoidal force, or a series $RLC$ combination being driven by a sunusoidal potential drop. Each of the parameters in (Equation D-68) is a real number. We wish to obtain the steady-state solution to this equation using complex numbers. To do so, we first construct the equation:

$$a\frac{d^2y}{dt^2} + b\frac{dy}{dt} + cy = A\sin\omega t \qquad \text{D-69}$$

Equation D-69 has no physical meaning and we have no interest in solving it. However, it is of use in solving Equation D-68. After multiplying both sides of (Equation D-69) by $i$ we add it and (Equation D-68) to obtain

$$\left(a\frac{d^2x}{dt^2} + ai\frac{d^2y}{dt^2}\right) + \left(b\frac{dx}{dt} + bi\frac{dy}{dt}\right) + (cx + ciy) = A\cos\omega t + Ai\sin\omega t$$

We next combine terms to get

$$a\frac{d^2(x + iy)}{dt^2} + b\frac{d(x + iy)}{dt} + c(x + iy) = A(\cos\omega t + i\sin\omega t) \qquad \text{D-70}$$

whose validity depends on the derivative of a sum being equal to the sum of the derivatives. We simplify our result by defining $z = x + iy$ and by using the identity $e^{i\omega t} = \cos\omega t + i\sin\omega t$. Substituting these into (Equation D-70) we obtain

$$a\frac{d^2z}{dt^2} + b\frac{dz}{dt} + cz = Ae^{i\omega t} \qquad\text{D-71}$$

which we now solve for $z$. Once $z$ is obtained we can solve for $x$ using $x = \text{Re}(z)$.

Since we are looking only for the steady state solution for (Equation D-68) we can assume its solution is of the form $x = x_0 \cos(\omega t - \phi)$, where $\phi$ is a constant. This is equivalent to assuming that the solution to (Equation D-71) is of the form $z = z_0 e^{i\omega t}$, where $z_0$ is a complex number. Then $dz/dt = i\omega z$, $d^2z/dt^2 = -\omega^2 z$, and $e^{i\omega t} = z/z_0$.

Substituting these into (Equation D-70) gives $-a\omega^2 z + i\omega bz + cz = A\dfrac{z}{z_0}$

Dividing both sides of this equation by $z$ and solving for $z_0$ gives

$$z_0 = \frac{A}{-a\omega^2 + i\omega b + c}$$

Expressing the denominator in polar form gives $(-a\omega^2 + c) + i\omega b = \sqrt{(-a\omega^2 + c)^2 + \omega^2 b^2}\, e^{i\phi}$, where $\tan\phi = \omega^2 b^2/(-a\omega^2 + c)$. Thus,

$$z_0 = \frac{A}{\sqrt{(-a\omega^2 + c)^2 + \omega^2 b^2}}\, e^{-i\phi}$$

so

$$z = z_0 e^{i\omega t} = \frac{A}{\sqrt{(-a\omega^2 + c)^2 + \omega^2 b^2}}\, e^{i(\omega t - \phi)}$$

$$= \frac{A}{\sqrt{(-a\omega^2 + c)^2 + \omega^2 b^2}}(\cos(\omega t - \delta) + i\sin(\omega t - \delta)) \qquad\text{D-72}$$

It follows that

$$x = \text{Re}(z) = \frac{A}{\sqrt{(-a\omega^2 + c)^2 + \omega^2 b^2}}\cos(\omega t - \phi) \qquad\text{D-73}$$

## Differential Calculus

When we say that $x$ is a function of $t$, we mean that for each value of $t$ there is a single corresponding value of $x$. An example is $x = At^2$, where $A$ is a constant. To indicate that $x$ is a function of $t$, we sometimes write $x(t)$ for $x$. Figure D-21 is a graph of $x$ versus $t$ for a typical function $x(t)$. At a particular value $t = t_1$, $x$ has the value of $x_1$ as indicated. At another value $t_2$, $x$ has the value $x_2$. The change in $t$, $t_2 - t_1$, is written $\Delta t = t_2 - t_1$ and the corresponding change in $x$ is written $\Delta x = x_2 - x_1$. The ratio $\Delta x/\Delta t$ is the slope of the straight line connecting $(x_1, t_1)$ and $(x_2, t_2)$. If we make $\Delta t$ smaller and smaller, the line connecting $(x_1, t_1)$ and $(x_2, t_2)$ approaches the line that is tangent to the curve at the point $(x_1, t_1)$. The slope of this tangent line is called the derivative of $x$ with respect to $t$ and is written $dx/dt$:

$$\frac{dx}{dt} = \lim_{\Delta t \to 0} \frac{\Delta x}{\Delta t} \qquad\text{D-74}$$

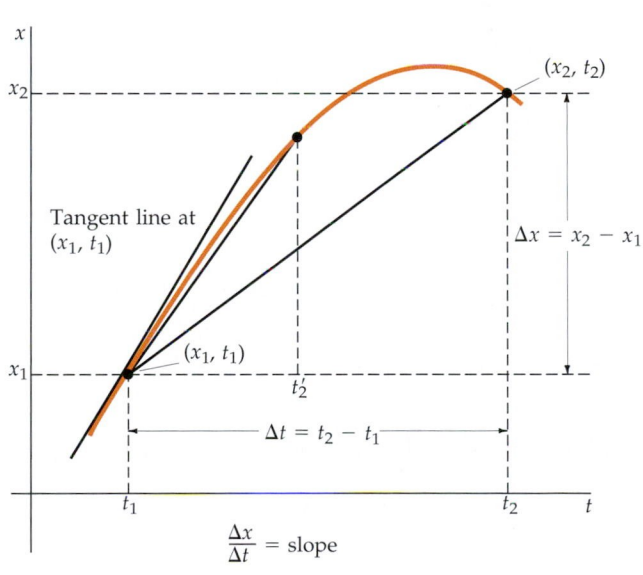

**FIGURE D-21** Graph of a typical function $x(t)$. The points $(x_1, t_1)$ and $(x_2, t_2)$ are connected by a straight line. The slope of this line is $\Delta x/\Delta t$. As the time interval beginning at $t_1$ is decreased, the slope for that interval approaches the slope of the line tangent to the curve at time $t_1$, which is the derivative of $x$ with respect to $t$.

The derivative of a function of $t$ is generally another function of $t$. If $x$ is a constant, the graph of $x$ versus $t$ is a horizontal line with zero slope. The derivative of a constant is thus zero. In Figure D-22, $x$ is proportional to $t$:

$$x = Ct$$

This function has a constant slope equal to $C$. Thus the derivative of $Ct$ is $C$. Table D-4 lists some properties of derivatives and the derivatives of some particular functions that occur often in physics. It is followed by comments aimed at making these properties and rules clearer. More detailed discussion can be found in most calculus books.

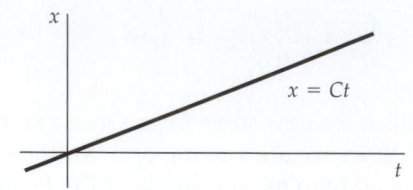

**FIGURE D-22** Graph of the linear function $x = Ct$. This function has a constant slope $C$.

## TABLE D-4

**Properties of Derivatives and Derivatives of Particular Functions**

Linearity

1. The derivative of a constant times a function equals the constant times the derivative of the function:

$$\frac{d}{dt}[Cf(t)] = C\frac{df(t)}{dt}$$

2. The derivative of a sum of functions equals the sum of the derivatives of the functions:

$$\frac{d}{dt}[f(t) + g(t)] = \frac{df(t)}{dt} + \frac{dg(t)}{dt}$$

Chain rule

3. If $f$ is a function of $x$ and $x$ is in turn a function of $t$, the derivative of $f$ with respect to $t$ equals the product of the derivative of $f$ with respect to $x$ and the derivative of $x$ with respect to $t$:

$$\frac{d}{dt}f(t) = \frac{df}{dx}\frac{dx}{dt}$$

Derivative of a product

4. The derivative of a product of functions $f(t)g(t)$ equals the first function times the derivative of the second plus the second function times the derivative of the first:

$$\frac{d}{dt}[f(t)g(t)] = f(t)\frac{dg(t)}{dt} + \frac{df(t)}{dt}g(t)$$

Reciprocal derivative

5. The derivative of $t$ with respect to $x$ is the reciprocal of the derivative of $x$ with respect to $t$, assuming that neither derivative is zero:

$$\frac{dx}{dt} = \left(\frac{dt}{dx}\right)^{-1} \quad \text{if} \quad \frac{dt}{dx} \neq 0$$

Derivatives of particular functions

6. $\dfrac{dC}{dt} = 0$ where $C$ is a constant

7. $\dfrac{d(t^n)}{dt} = nt^{n-1}$

8. $\dfrac{d}{dt}\sin \omega t = \omega \cos \omega t$

9. $\dfrac{d}{dt}\cos \omega t = -\omega \sin \omega t$

10. $\dfrac{d}{dt}\tan \omega t = \omega \sec^2 \omega t$

11. $\dfrac{d}{dt}e^{bt} = be^{bt}$

12. $\dfrac{d}{dt}\ln bt = \dfrac{1}{t}$

## EXAMPLE D-9

Find the derivative of $x = at^2 + bt + c$, where $a$, $b$, and $c$ are constants.

**PICTURE THE PROBLEM** From rule 2, we can differentiate each term separately and add the results.

1. Use rules 1 and 7 to find the derivative of the first term:

$$\frac{d(at^2)}{dt} = 2at^1 = 2at$$

2. Compute the derivatives of the second and third terms:

$$\frac{d(bt)}{dt} = b, \quad \frac{d(c)}{dt} = 0$$

3. Add these results:

$$\frac{dx}{dt} = 2at + b$$

## Comments on Rules 1 Through 5

Rules 1 and 2 follow from the fact that the limiting process is linear. We can understand rule 3, the chain rule, by multiplying $\Delta f / \Delta t$ by $\Delta x / \Delta x$ and noting that, since $x$ is a function of $t$, both $\Delta x$ and $\Delta f$ approach zero as $\Delta t$ approaches zero. Since the limit of a product of two functions equals the product of their limits, we have

$$\lim_{\Delta t \to 0} \frac{\Delta f}{\Delta t} = \lim_{\Delta t \to 0} \frac{\Delta f \, \Delta x}{\Delta x \, \Delta t} = \left( \lim_{\Delta t \to 0} \frac{\Delta f}{\Delta x} \right)\left( \lim_{\Delta t \to 0} \frac{\Delta x}{\Delta t} \right) = \frac{df}{dx}\frac{dx}{dt}$$

Rule 4 is not immediately apparent. The derivative of a product of functions is the limit of the ratio

$$\frac{f(t + \Delta t)g(t + \Delta t) - f(t)g(t)}{\Delta t}$$

If we add and subtract the quantity $f(t + \Delta t)g(t)$ in the numerator, we can write this ratio as

$$\frac{f(t + \Delta t)g(t + \Delta t) - f(t + \Delta t)g(t) + f(t + \Delta t)g(t) - f(t)g(t)}{\Delta t}$$

$$= f(t + \Delta t)\left[ \frac{g(t + \Delta t) - g(t)}{\Delta t} \right] + g(t)\left[ \frac{f(t + \Delta t) - f(t)}{\Delta t} \right]$$

As $\Delta t$ approaches zero, the terms in square brackets become $dg(t)/dt$ and $df(t)/dt$, respectively, and the limit of the expression is

$$f(t)\frac{dg(t)}{dt} + g(t)\frac{df(t)}{dt}$$

Rule 5 follows directly from the definition:

$$\frac{dx}{dt} = \lim_{\Delta t \to 0} \frac{\Delta x}{\Delta t} = \lim_{\Delta x \to 0} \left( \frac{\Delta t}{\Delta x} \right)^{-1} = \left( \frac{dt}{dx} \right)^{-1}$$

## Comments on Rule 7

We can obtain this important result using the binomial expansion. We have

$$f(t) = t^n$$

$$f(t + \Delta t) = (t + \Delta t)^n = t^n\left(1 + \frac{\Delta t}{t}\right)^n$$

$$= t^n\left[1 + n\frac{\Delta t}{t} + \frac{n(n-1)}{2!}\left(\frac{\Delta t}{t}\right)^2 + \frac{n(n-1)(n-2)}{3!}\left(\frac{\Delta t}{t}\right)^3 + \cdots\right]$$

Then

$$f(t - \Delta t) - f(t) = t^n\left[n\frac{\Delta t}{t} + \frac{n(n-1)}{2!}\left(\frac{\Delta t}{t}\right)^2 + \cdots\right]$$

and

$$\frac{f(t - \Delta t) - f(t)}{\Delta t} = nt^{n-1} + \frac{n(n-1)}{2!}t^{n-2}\,\Delta t + \cdots$$

The next term omitted from the last sum is proportional to $(\Delta t)^2$, the following to $(\Delta t)^3$, and so on. Each term except the first approaches zero as $\Delta t$ approaches zero. Thus

$$\frac{df}{dt} = \lim_{\Delta x \to 0}\frac{f(t + \Delta t) - f(t)}{\Delta t} = nt^{n\pm 1}$$

## Comments on Rules 8 to 10

We first write $\sin \omega t = \sin \theta$ with $\theta = \omega t$ and use the chain rule,

$$\frac{d\sin\theta}{dt} = \frac{d\sin\theta}{d\theta}\frac{d\theta}{dt} = \omega\frac{d\sin\theta}{d\theta}$$

We then use the trigonometric formula for the sine of the sum of two angles $\theta$ and $\Delta\theta$:

$$\sin(\theta + \Delta\theta) = \sin\Delta\theta\cos\theta + \cos\Delta\theta\sin\theta$$

Since $\Delta\theta$ is to approach zero, we can use the small-angle approximations

$$\sin\Delta\theta \approx \Delta\theta \qquad \text{and} \qquad \cos\Delta\theta \approx 1$$

Then

$$\sin(\theta + \Delta\theta) \approx \Delta\theta\cos\theta + \sin\theta$$

and

$$\frac{\sin(\theta + \Delta\theta) - \sin\theta}{\Delta\theta} \approx \cos\theta$$

Similar reasoning can be applied to the cosine function to obtain rule 9.

Rule 10 is obtained by writing $\tan\theta = \sin\theta/\cos\theta$ and applying rule 4 along with rules 8 and 9.

$$\frac{d}{dt}(\tan\theta) = \frac{d}{dt}(\sin\theta)(\cos\theta)^{-1} = \sin\theta\frac{d}{dt}(\cos\theta)^{-1} + \frac{d(\sin\theta)}{dt}(\cos\theta)^{-1}$$

$$= \sin\theta(-1)(\cos\theta)^{-2}(-\sin\theta) + (\cos\theta)(\cos\theta)^{-1}$$

$$= \frac{\sin^2\theta}{\cos^2\theta} + 1 = \tan^2\theta + 1 = \sec^2\theta$$

## Comments on Rule 11

Again we use the chain rule

$$\frac{de^\theta}{dt} = \frac{bde^\theta}{bdt} = b\frac{de^\theta}{d(bt)} = b\frac{de^\theta}{d\theta} \qquad \text{with} \qquad \theta = bt$$

and the series expansion for the exponential function:

$$e^{\theta+\Delta\theta} = e^\theta e^{\Delta\theta} = e^\theta\left[1 + \Delta\theta + \frac{(\Delta\theta)^2}{2!} + \frac{(\Delta\theta)^3}{3!} + \cdots\right]$$

Then

$$\frac{e^{\theta+\Delta\theta} - e^\theta}{\Delta\theta} = e^\theta + e^\theta\frac{\Delta\theta}{2!} + e^\theta\frac{(\Delta\theta)^2}{3!} + \cdots$$

As $\Delta\theta$ approaches zero, the right side of the equation above approaches $e^\theta$.

## Comments on Rule 12

Let

$$y = \ln bt$$

Then

$$e^y = bt \qquad \text{and} \qquad \frac{dt}{dy} = \frac{1}{b}e^y = t$$

Then using rule 5, we obtain

$$\frac{dy}{dt} = \left(\frac{dt}{dy}\right)^{-1} = \frac{1}{t}$$

## Integral Calculus

Integration is related to the problem of finding the area under a curve. It is also the inverse of differentiation. Figure D-23 shows a function $f(t)$. The area of the shaded element is approximately $f_i\Delta t_i$, where $f_i$ is evaluated anywhere in the interval $\Delta t_i$. This approximation improves if $\Delta t_i$ is very small. The total area from $t_1$ to $t_2$ is found by summing all the area elements from $t_1$ to $t_2$ and taking the limit as each $\Delta t_i$ approaches zero. This limit is called the integral of $f$ over $t$ and is written

$$\int_{t_1}^{t_2} f\,dt = \text{Area} = \lim_{\Delta t_i \to 0} \sum_i f_i\,\Delta t_i$$

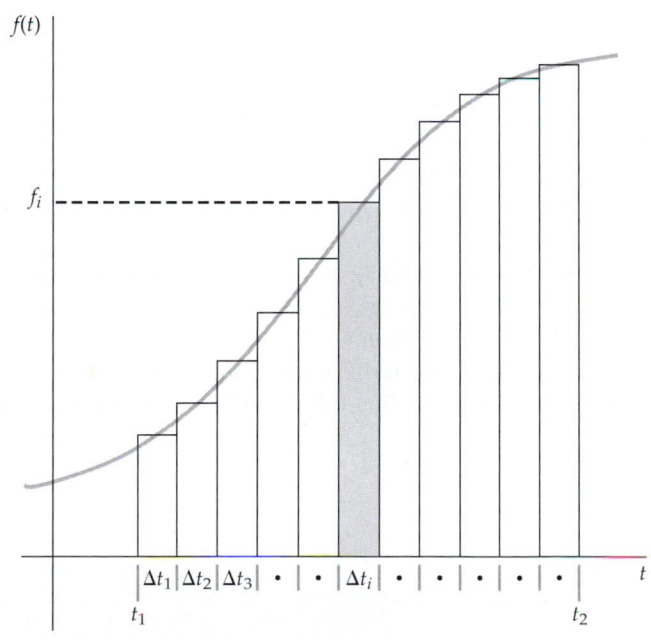

**FIGURE D-23** A general function $f(t)$. The area of the shaded element is approximately $f_i\Delta t_i$, where $f_i$ is evaluated anywhere in the interval.

If we integrate some function $f(t)$ from $t_1$ to some general value of $t$, we obtain another function of $t$. Let us call this function $y$:

$$y = \int_{t_1}^{t} f \, dt$$

The function $y$ is the area under the $f$-versus-$t$ curve from $t_1$ to a general value $t$. For a small interval $\Delta t$, the change in the area $\Delta y$ is approximately $f \Delta t$.

$$\Delta y \approx f \Delta t$$

$$f \approx \frac{\Delta y}{\Delta t}$$

If we take the limit as $\Delta t$ approaches 0, we can see that $f$ is the derivative of $y$:

$$f = \frac{dy}{dt}$$

The relation between $y$ and $f$ is often written

$$y = \int f \, dt$$

where $\int f \, dt$ is called an **indefinite integral.** To evaluate an indefinite integral, we find the function $y$ whose derivative is $f$. The definite integral of $f$ from $t_1$ to $t_2$ is $y(t_1) - y(t_2)$, where $df/dt = y$:

$$\int_{t_1}^{t_2} f \, dt = y(t_2) - y(t_1)$$

**EXAMPLE D-10**

**Find the indefinite integral of $f(t) = t$.**

The function whose derivative is $t$ is $\frac{1}{2}t^2$ plus any constant:

$$\int t \, dt = \frac{1}{2}t^2 + C$$

where $C$ is any constant

Table D-5 lists some important integration formulas. More extensive lists of differentiation and integration formulas can be found on the Internet. They can be found by using a search engine and searching for *table of integrals,* and also in handbooks such as Herbert Dwight's *Tables of Integrals and Other Mathematical Data,* fourth edition, Macmillan Publishing Company, Inc., New York, 1961.

## TABLE D-5

**Integration Formulas[†]**

1. $\displaystyle\int A \, dt = At$

2. $\displaystyle\int At \, dt = \frac{1}{2}At^2$

3. $\displaystyle\int At^n \, dt = A\frac{t^{n+1}}{n+1} \quad n \neq -1$

4. $\displaystyle\int At^{-1} \, dt = A \ln t$

5. $\displaystyle\int e^{bt} \, dt = \frac{1}{b}e^{bt}$

6. $\displaystyle\int \cos \omega t \, dt = \frac{1}{\omega}\sin \omega t$

7. $\displaystyle\int \sin \omega t \, dt = -\frac{1}{\omega}\cos \omega t$

8. $\displaystyle\int_0^\infty e^{-ax} \, dx = \frac{1}{a}$

9. $\displaystyle\int_0^\infty e^{-ax^2} \, dx = \frac{1}{2}\sqrt{\frac{\pi}{a}}$

10. $\displaystyle\int_0^\infty xe^{-ax^2} \, dx = \frac{2}{a}$

11. $\displaystyle\int_0^\infty x^2 e^{-ax^2} \, dx = \frac{1}{4}\sqrt{\frac{\pi}{a^3}}$

12. $\displaystyle\int_0^\infty x^3 e^{-ax^2} \, dx = \frac{4}{a^2}$

13. $\displaystyle\int_0^\infty x^4 e^{-ax^2} \, dx = \frac{3}{8}\sqrt{\frac{\pi}{a^5}}$

[†] In these formulas, $A$, $b$, and $\omega$ are constants. In formulas 1 through 7 an arbitrary constant $C$ can be added to the right side of each equation. The constant $a$ is greater than zero.

# ILLUSTRATION CREDITS

## Chapter 11

**Opener p. 339** Stocktrek/Corbis; **p. 340** Collection of Historical Scientific Instruments, Harvard University; **p. 344 (top)** NASA; **(bottom)** NASA; **p. 345** Courtesy Central Scientific Company; **p. 353** NASA.

## Chapter 12

**Opener p. 370** Courtesy of Department of Physics, Purdue University; **p. 372** © 2002 Estate of Alexander Calder/Artists Rights Society (ARS), New York; **p. 379** Photodisk.

## Chapter 13

**Opener p. 395** Andy Pernick/Bureau of Reclamation; **p. 401** Vanessa Vick/Photo Researchers, Inc.; **p. 403** Chuck O'Rear/Woodfin Camp and Assoc.; **p. 405** David Burnett/Woodfin Camp and Assoc.; **p. 407 (top)** Estate of Harold E. Edgerton; **(bottom)** Takeski Takahara/Photo Researchers, Inc.; **p. 408** P. Motta/Photo Researchers, Inc.; **p. 412** Michael Dunn/The Stock Market; **p. 415** Picker International.

## Chapter 14

**Opener p. 425** Barry Slaven/Visuals Unlimited; **p. 427** Citibank; **p. 429** NASA; **p. 433** Institute for Marine Dynamics; **p. 442** Richard Menga/Fundamental Photographers; **p. 447 (top)** Monroe Auto Equipment; **(bottom)** David Wrobel/Visuals Unlimited; **p. 449** Eye Wire/Getty; **p. 451** Royal Swedish Academy of Music.

## Chapter 15

**Opener p. 465** John Cetrino/Check Six/Picture Quest; **p. 466 Figure 15-1** Richard Menga/Fundamental Photographs; **Figure 15-2** Richard Menga/Fundamental Photographs; **p. 478 (top right)** David Sacks/The Image Bank/Getty; **(left)** Maynard and Boucher/Visuals Unlimited; **p. 480** From Winston E. Cock, *Lasers and Holography*, Dover Publications, New York, 1981; **p. 484** Courtesy of Davies Symphony Hall; **p. 485 (top and bottom)** Fundamental Photographs; **p. 486 (top)** Bernard Benoit/Photo Researchers, Inc.; **(a)** Education Development Center; **p. 491 (top, a)** Sandia National Laboratory; **(top, b)** Robert de Gast/Photo Researchers; **(c)** Estate of Harold E. Edgerton/Palm Press Inc.; **(bottom, b)** Education Development Center; **p. 496 Figure 15-30** Estate of Harold E. Edgerton/Palm Press Inc.

## Chapter 16

**Opener p. 503** David Yost/Steinway & Sons; **p. 507** Rubberball Productions; **p. 510 (a)** Berenice Abbott (8J 1328)/Photo Researchers; **p. 513 (left)** University of Washington; **(center)** University of Washington; **(right)** University of Washington; **p. 519** Professor Thomas D. Rossing, Northern Illinois University, DeKalb; **p. 525** Courtesy of Chuck Adler.

## Chapter 17

**Opener p. 532** Hoby Finn/PhotoDisk/Getty; **p. 535 (a)** Courtesy of Taylor Precision Products; **(b)** Courtesy Honeywell, Inc.; **p. 536** Richard Menga/Fundamental Photographs; **p. 538** NASA; **p. 554** Jet Propulsion Laboratory/NASA.

## Chapter 18

**Opener p. 558** Donna Day/PhotoDisk/Getty; **p. 559** Phoenix Pipe & Tube/Lana Berkovich; **p. 561** From Frank Press and Raymond Sievert, *Understanding Earth*, 3rd ed., W.H. Freeman and Co., 2001; **p. 562** From Donald Wink, Sharon Gislason, and Sheila McNicholas, *The Practice of Chemistry*, W.H. Freeman and Co., 2002; **p. 582** Will and Deni McIntyre/Photo Researchers.

## Chapter 19

**Opener p. 595 (top)** Paul Chesley/National Geographic/Getty; **p. 595 (bottom)** Sandia National Laboratory; **p. 598** © 2002 Robert Briggs; **p. 601** Anderson Ross/PhotoDisk/Getty; **p. 605 (right)** Michael Collier/Stock, Boston; **(left)** Jean-Pierre Horlin/The Image Bank; **p. 606 (top left)** Sandia National Laboratory; **(top right)** Peter Miller/The Image Bank; **(bottom)** Sandia National Laboratory.

## Chapter 20

**Opener p. 628** Frank Siteman/Stock Boston, Inc./PictureQuest; **p. 637** Alfred Pasieka/Photo Researchers, Inc.; **p. 639** Courtesy of Eugene Mosca; **p. 643** Science Photo Library/Photo Researchers, Inc.

## Chapter 21

**p. 652 Figure 21-1** From *PSSC Physics*, 2nd Edition, 1965. D. C. Heath & Co. and Education Development Center, Inc., Newton, MA; **p. 653** Bruce Terris/IBM Almaden Research Center; **p. 656 (left)** © Grant Heilman; **(right)** Ann Roman Picture Library; **(bottom)** Bundy Library, Norwalk, CT; **p. 666 Figure 21-19 (b)** and **Figure 21-20 (b)** Harold M. Waage; **p. 667 Figure 21-21 (b)** Harold M. Waage; **p. 668** Courtesy of Hulon Forrester/Video Display Corporation, Tucker, Georgia; **p. 670 Figure 21-26** Courtesy of Videojet Systems International.

## Chapter 22

**Opener p. 682** © Kent Wood/Photo Researchers, Inc.; **p. 685** Ben Damsky/Electric Power Research Institute; **p. 697** Runk/Schoenberger from Grant Heilman; **p. 705** Harold M. Waage.

## Chapter 23

**Opener p. 717** Courtesy of the U.S. Department of Energy; **p. 724** © 1990 Richard Megna/Fundamental Photographs; **p. 736 Figure 23-23 (b)** © Karen R. Preuss; **p. 739 Figure 23-25 (b)** Harold M. Waage.

## Chapter 24

**Opener p. 748** © David Young-Wolff/PhotoEdit; **p. 749** © tom pantages images; **p. 750** © Steve Allen/The Image Bank/Getty Images; **p. 753 Figure 24-2 (b)** Harold M. Waage; **p. 755 (top)** Courtesy of Gepco, International, Inc., Des Plaines, IL; **(bottom)** © Bruce Iverson; **p. 756 (top left)** © Bruce Iverson; **(top right)** © Paul Brierly; **(middle)** Courtesy Tusonix, Tucson, AZ; **p. 764** Lawrence Livermore National Laboratory; **p. 767** © Manfred Kage/Peter Arnold, Inc.

## Chapter 25

**Opener p. 786** © Tom Stewart/CORBIS; **p. 793** © Chris Rogers/ The Stock Market; **p. 795** © Stephen Frink/CORBIS; **p. 796** © Paul Silverman/Fundamental Photographs.

## Chapter 26

**Opener p. 829** © B. Sidney/TAXI/Getty Images; **p. 832 Figure 26-8 (b)** © 1995 tom pantages; **p. 834 (a)** Larry Langrill; **(b)** © Lawrence Berkeley Laboratory/Science Photo Library; **p. 835 Figure 26-13 (b)** Carl E. Nielsen.

## Chapter 27

**Opener p. 856** Bob Williamson, Oakland University, Rochester, Michigan; **p. 858 (a and b)** © 1990 Richard Megna/Fundamental Photographs; **p. 860 Figure 27-6 (b)** © 1990 Richard Megna/ Fundamental Photographs; **p. 863 Figure 27-10 (c)** © 1990 Richard Megna/Fundamental Photographs; **p. 865** © Bruce Iverson; **p. 866 Figure 27-15 (b)** © 1990 Richard Megna/Fundamental Photographs; **p. 867** Courtesy of F. W. Bell; **p. 869 Figure 27-20 (a)** Photo by Gene Mosca; **p. 873 (a and b)** Courtesy Princeton University Plasma Physics Laboratory; **p. 880 (top)** J. F. Allen, St. Andrews University, Scotland; **(bottom)** Photo by Gene Mosca; **p. 881 (top)** Robert J. Celotta, National Institute of Standards and Technology; **(middle)** © Paul Silverman/Fundamental Photographs, **(bottom a)** Akira Tonomura, Hitachi Advanced Research Library, Hatomaya, Japan; **(bottom b)** © Bruce Iverson; **p. 883 (a)** © 2003 Western Digital Corporation. All rights reserved; **(b)** Tom Chang/IBM Storage Systems Division, San Jose, CA; **p. 885** © Bill Pierce/*Time* Magazines, Inc.

## Chapter 28

**Opener p. 897** © 1990 Richard Megna/Fundamental Photographs; **p. 915 Figure 28-28 (b)** © Michael Holford, Collection of the Science Museum, London; **p. 922 (top)** A. Leitner/Renselaer Polytechnic Institute; **(bottom)** © Palmer/ Kane, Inc./CORBIS; **p. 926 Figures 28-42 (a and b)** Courtesy PASCO Scientific Co.

## Chapter 29

**Opener p. 922** © Roger Ressmeyer/CORBIS; **p. 936 (a)** Courtesy of U.S. Department of the Interior, Department of Reclamation; **(b)** © Lee Langum/Photo Researchers, Inc.; **p. 956** © George H. Clark Radioana Collection-Archive Center, National Museum of American History; **p. 957 (a)** © Yoav/Phototake; **(b)** © Daniel S. Brody/Stock Boston.

## Chapter 30

**Opener p. 971** Courtesy of NASA.

## Chapter 31

**Opener p. 997** © James L. Amos/CORBIS; **p. 998 Figure 31-1** CORBIS-Bettmann; **p. 999 Figure 31-2** CORBIS-Bettmann; **(bottom)** Adapted from Eastman Kodak and Wabash Instrument Corporation; **p. 1002 (a and b)** © 1991 Paul Silverman/Fundamental Photographs; **p. 1007 (a)** © Chuck O'Rear/West Light; **(b)** Courtesy of Ahmed H. Zewail, California Institute of Technology; **(c)** © Chuck O'Rear/West Light; **(d)** © Michael W. Berns/*Scientific American*; **(e)** © 1988 by David Scharf. All rights reserved; **p. 1012 Figure 31-19 (a)** © 1987 Ken Kay/Fundamental Photographs; **(b)** Courtesy Battelle-Northwest Laboratories; **(bottom)** © 1990 Richard Magna/Fundamental Photographs; **p. 1013 Figure 31-20 (b)** © Macduff Everton/CORBIS; **p. 1014** © 1987 Pete Saloutos/ The Stock Market; **p. 1015 Figure 31-23 (b)** © 1987 Ken Kay/ Fundamental Photographs; **p. 1017 Figure 31-27 (c)** © Ted Horowitz/The Stock Market; **(bottom a)** © Dan Boyd/Courtesy Naval Research Laboratory; **(bottom b)** Courtesy AT&T Archives; **p. 1018 Figure 31-28 (c)** © Robert Greenler; **p. 1019** © David Parker/Science Photo Library/Photo Researchers; **p. 1020 (a)** © Robert Greenler; **(b)** Giovanni DeAmici, NSF, Lawrence Berkeley Laboratory; **p. 1021 (a and b)** Larry Langrill; **p. 1023 (a)** © 1970 Fundamental Photographs; **(b)** 1990 PAR/NYC, Inc./ Photo by Elizabeth Algieri; **p. 1025 Figure 31-43 (b)** © 1987 Paul Silverman Photographs; **p. 1026 (a)** Glen A. Izett, U.S. Geological Survey, Denver, Colorado; **(b)** Glen A. Izett, U.S. Geological Survey, Denver, Colorado; **(c)** Dr. Anthony J. Gow/Cold Regions Research and Engineering Laboratory, Hanover, New Hampshire; **(d)** Dr. Anthony J. Gow/Cold Regions Research and Engineering Laboratory, Hanover, New Hampshire; **(e)** © Sepp Seitz/ Woodfin Camp and Associates.

## Chapter 32

**Opener p. 1038** Photo by Gene Mosca; **p. 1039 Figure 32-2** Photo by Demetrios Zangos; **p. 1048 (a and b)** © 1990 Richard Megna/ Fundamental Photographs; **p. 1054 Figure 32-29 (a, bottom)** Nils Abramson; **(b, bottom)** © 1974 Fundamental Photographs; **Figure 32.30 (b)** © Fundamental Photographs; **p. 1057** © Bohdan Hrynewych/Stock Boston; **p. 1068 (a, b, and c)** Lennart Nilsson, **(d)** Courtesy IMEC and University of Pennsylvania Department of Electrical Engineering; **p. 1072 (a)** © Scala/Art Resource; **(b)** © Royal Astronomical Society Library; **(c)** Lick Observatory, courtesy of the University of California Regents; **(d)** California Institute of Technology; **(e)** © 1980 Gary Ladd; **p. 1073 (a, b and c)** © California Association for Research in Astronomy; **(bottom)** Courtesy of NASA.

## Chapter 33

**Opener p. 1084** © Aaron Haupt/Photo Researchers, NY; **p. 1087 Figure 33-3 (a)** Courtesy of Bausch & Lomb; **p. 1088 Figure 33-5 (a and b)** Courtesy T. A. Wiggins; **Figure 33-6** From *PSSC Physics*, 2nd Edition, 1965. D.C. Heath & Co. and Education Development Center, Newton, MA; **p. 1091 Figure 33-9 (a)** Courtesy of Michael Cagnet; **p. 1092 Figure 33-11 (a)** Courtesy of Michael Cagnet; **p. 1093 Figure 33-14 (a)** Courtesy of Michael Cagnet; **p. 1097 Figure 33-21** Courtesy Michael Cagnet; **p. 1102 Figure 33-30 (a and b)** M. Cagnet, M. Fraçon, J.C. Thrierr, *Atlas of Optical Phenomena*; **Figure 33-31 (a)** Courtesy Battelle-Northwest Laboratories; **p. 1103 Figures 33-32, 33-33,** and **33-35 (a and b)** Courtesy of Michael Cagnet; **p. 1104 Figure 33-36** Courtesy of National Radio Astronomy Observatory/ Associated Universities, Inc./National Science Foundation. Photographer: Kelly Gatlin. Digital composite: Patricia Smiley; **p. 1105 (bottom)** © Kevin R. Morris/CORBIS; **p. 1106 Figure 33-38 (a)** Clarence Bennett/Oakland University, Rochester, Michigan; **(b)** NRAO/AUI/Science Photo Library/Photo Researchers; **p. 1108 (a and b)** © 1981 by Ronald R. Erickson, Hologram by Nicklaus Phillips, 1978, for Digital Equipment Corporation.

## Chapter 34

**Opener p. 1117** Courtesy of Akira Tononmura, Advanced Research Laboratory, Hitachi, Ltd.; **p. 1128 (a & b)** *PSSC Physics,* 2nd ed., 1965. D.C. Heath & Co., and Education Development Center, Inc., Newton, MA; **(c)** C. G. Shull; **(d)** Claus Jönsson; **p. 1128 (bottom)** Jack Griffith/ University of North Carolina

## Chapter 35

**Opener p. 1149** Courtesy of IBM and the IBM Almadin Laboratories; **p. 1161** Education Development Center

## Chapter 36

**Opener p. 1171** Courtesy of NASA and the Harvard-Smithsonian Center for Astrophysics; **p. 1172** Adapted from Eastman Kodak and Wabash Instrument Corporation; **p. 1192 (a & b)** A Jayaraman/AT&T Bell Labs; **p. 1193** © David Parker/Photo Researchers; **p. 1198** © Robert Landau/Westlight.

## Chapter 37

**Opener p. 1208 (a)** Norman Collection for the History of Molecular Biology, **(b)** © A. Barrington Brown/Photo Researchers, NY; **p. 1214** © Will and Demi McIntire/Photo Researchers; **p. 1221** Courtesy of Dr. J.A. Marquissee.

## Chapter 38

**Opener p. 1228** © 2003 John Alves/Mystic Wanderer Images; **p. 1230 (a)** Richard Walters, 2/89, p. 52/*Discover;* **(b)** © Dr. Jeremy Burgess/ Science Photo Library/Photo Researchers; **(c)** © Thomas R. Taylor/ Photo Researchers; **(d)** Courtesy the AT&T Archives; **p. 1233 (a)** Chris Kovach, 3/91, p. 69/*Discover;* **(b)** Srinivas Manne, University of California, Santa Barbara; **(c)** Dr. F.A. Quiocho and J.S. Spurlino/

Howard Hughes Medical Institute, Baylor College of Medicine; **(d)** W. Krätschmer/Max-Planck Institute for Nuclear Physics; **(e)** © Kenneth Weard/BioGrafx/Science Source/Photo Researchers; **p. 1247** Museum of Modern Art; **p. 1250** © C. Falco/Photo Researchers; **p. 1253** © Royalty Free/CORBIS

## Chapter 39

**Opener p. 1267** Courtesy of NASA; **p. 1268** Courtesy NRAO/AUI; **p. 1289** © C. Powell, P. Fowler, & D. Perkins/Science Photo Library/ Photo Researchers; **p. 1297** © Michael Freeman

## Chapter 40

**Opener p. 1306** © Hans Wolf/The Image Bank; **p. 1320(a)** © 1991 by the Metropolitan Museum of Art, **(b & c)** Courtesy of Paintings Conservation Dept., Metropolitan Museum of Art; **p. 1323** © Jerry Mason/Photo Researchers; **p. 1327 (all)** Courtesy of the Princeton Plasma Physics Laboratory; **p. 1328 (a & b)** Courtesy of the Lawrence Livermore National Laboratory/U.S. Department of Energy

## Chapter 41

**Opener p. 1335** © Lawerence Livermore Lab/Science Photo Library/Photo Researchers; **p. 1336** ICCR (Institute for Cosmic Ray Research), The University of Tokyo; **p. 1338 (top)** © Science Photo Library/Photo Researchers, **(bot)**, © Lawrence Berkeley Laboratory/ Science Photo Library/Photo Researchers; **p. 1339 (top)** © Lawrence Berkeley Laboratory/Science Photo Library/Photo Researchers, **(bot)** Fig. 4 from "First Results from KamLAND: Evidence for Reactor Antineutrino Disappearance" by the KamLAND Collaboration, *Physical Review Letters,* Vol. 90, No. 2, December 17, 2003. Copyright © 2003 The American Physical Society. Reprinted with permission; **p. 1340** Richard Ehrlich; **p. 1341 (both)** CERN; **p. 1344** © Lawrence Berkeley Laboratory/Science Photo Library/ Photo Researchers.

# ANSWERS

Problem answers are calculated using $g = 9.81$ m/s$^2$ unless otherwise specified in the Problem. Differences in the last figure can easily result from differences in rounding the input data and are not important.

# ANSWERS

## Chapter 1

1. (c)
3. (c)
5. (a)
7. (e)
9. (a) True
   (b) False
   (c) True
11. $1.19 \times 10^{57}$
13. (a) $10^{-8}$ m
    (b) 20 atoms
15. (a) $3 \times 10^{10}$ diapers
    (b) $1.5 \times 10^7$ m$^3$
    (c) 0.6 mi$^2$
17. \$177 M
19. (a) 0.000040 W
    (b) 0.000000004 s
    (c) 3,000,000 W
    (d) 25,000 m
21. (a) $C_1$ is in m; $C_2$ is in m/s
    (b) $C_1$ is in m/s$^2$
    (c) $C_1$ is in m/s$^2$
    (d) $C_1$ is in m; $C_2$ is in s$^{-1}$
    (e) $C_1$ is in m/s; $C_2$ is in s$^{-1}$
23. (a) $4 \times 10^7$ m
    (b) $6.37 \times 10^6$ m
    (c) $2.48 \times 10^4$ mi; $3.96 \times 10^3$ mi
25. 210 cm
27. 1.28 km
29. (a) 36.0 km/h·s
    (b) 10.0 m/s$^2$
    (c) 88.0 ft/s
    (d) 26.8 m/s
31. 4050 m$^2$
33. (a) m/s$^2$
    (b) s
    (c) m
35. $T^{-1}$
37. $mv^2/r$
41. $M/L^3$
43. (a) 30,000
    (b) 0.0062
    (c) 0.000004
    (d) 217,000
45. (a) $1.44 \times 10^5$
    (b) $2.55 \times 10^{-8}$
    (c) $8.27 \times 10^3$
    (d) $6.27 \times 10^2$
47. $4 \times 10^6$
49. (a) $1.69 \times 10^3$
    (b) 4.8
    (c) 5.6
    (d) 10
51. 31.7 y
53. $2.0 \times 10^{23}$
55. (a) $1.41 \times 10^{17}$ kg/m$^3$
    (b) 216 m
57. (a) $4.85 \times 10^{-6}$ parsec
    (b) $3.08 \times 10^6$ m
    (c) $9.47 \times 10^{15}$ m
    (d) $6.33 \times 10^4$ AU
    (e) $3.25$ c·y
59. The claim is conservative as the actual weight of water used is closer to 55,000 tons.
61. (a) $n = 3/2$; $C = 17.0$ y/(Gm)$^{3/2}$
    (b) 0.510 Gm
63. $1.16 \times 10^{19}$ lb

## Chapter 2

1. 0
3. It is safer to land against the wind.
5. (a) Negative
   (b) During the last five steps, gradually slow the speed of walking, until the wall is reached.
   (c)

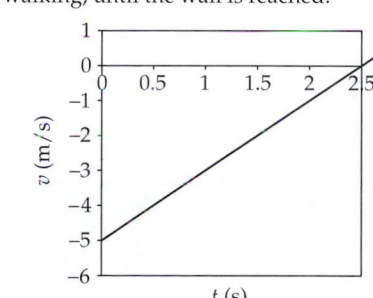

7. *(a)* True

  *(b)* True in one dimension

9. False

11. *(a)*

13. *(b)*

15. Yes. In any round-trip, A to B, and back to A, the average velocity is zero.

17. No. If the velocity is constant, a graph of position as a function of time is linear with a constant slope equal to the velocity.

19. *(b)*

21. *(a)* False

  *(b)* True

23. $v_{\text{top of flight}} = 0; a_{\text{top of flight}} = -g$

25. *(b)*

27. *(c)*

29. *(c)*

31. *(c)*

33. *(a)*

35. *(d)*

37. *(d)*

39. Velocity: *(a)* negative at $t_0$ and $t_1$; *(b)* positive at $t_3$, $t_4$, $t_6$, and $t_7$; *(c)* zero at $t_2$ and $t_5$

  Acceleration: *(a)* negative at $t_4$; *(b)* positive at $t_2$ and $t_6$; *(c)* zero at $t_0, t_1, t_3, t_5,$ and $t_7$

41. *(a)* Graphs *(a)*, *(f)*, and *(i)*

  *(b)* Graphs *(c)* and *(d)*

  *(c)* Graphs *(a)*, *(d)*, *(e)*, *(f)*, *(h)*, and *(i)*

  *(d)* Graphs *(b)*, *(c)*, and *(g)*

  *(e)* Graphs *(a)* and *(i)* are mutually consistent. Graphs *(d)* and *(h)* are mutually consistent. Graphs *(f)* and *(i)* are also mutually consistent.

43. *(a)* 54.2 m/s; *(b)* $-123g$

45. 4.02 m/s²

47. 14.2 ms

49. *(a)* 0.278 km/min

  *(b)* −0.0833 km/min

  *(c)* 0

  *(d)* 0.128 km/min

51. *(a)* 2.25 h

  *(b)* 4.99 h

  *(c)* 880 km/h

  *(d)* 611 km/h

53. *(a)* 4.33 y

  *(b)* $4.33 \times 10^6$ y; No

55. 35.8 m

57. *(a)* 0

  *(b)* 0.333 m/s

  *(c)* −2.00 m/s

  *(d)* 1.00 m/s

59. 122 km/h; $1.04v_{\text{av}}$

61. *(a)*

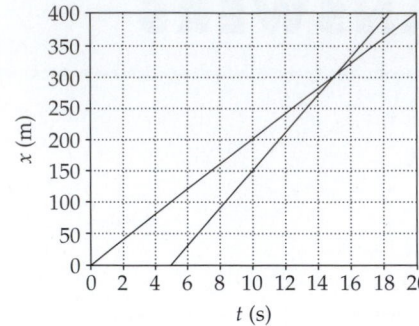

  *(b)* 15 s

  *(c)* 300 m

  *(d)* 100 m

63. 6 h

65. −2.00 m/s²

67. *(a)* 2 m

  *(b)* $\Delta x = (2t - 5)\Delta t + (\Delta t)^2$, where $x$ is in meters if $t$ is in seconds.

  *(c)* $v = 2t - 5$, where $v$ is in meters per second if $t$ is in seconds.

69. *(a)* $a_{\text{av,AB}} = 3.33$ m/s²; $a_{\text{av,BC}} = 0$; $a_{\text{av,CE}} = -7.50$ m/s²

  *(b)* 75.0 m

  *(c)*

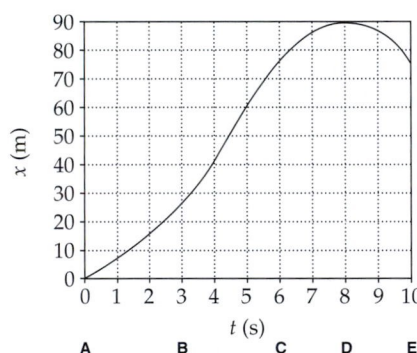

  *(d)* At point $D$, $t = 8$ s, the graph crosses the time axis and so $v = 0$.

71. *(a)* 80.0 m/s

  *(b)* 400 m

  *(c)* 40.0 m/s

73. 15.6 m/s²

75. *(a)* 4.68 s

  *(b)* 20.4 m

  *(c)* 0.991 s and 3.09 s

77. *(a)*

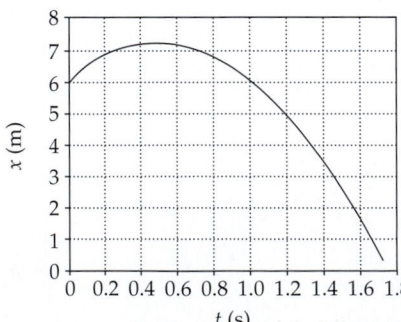

  *(b)* 7.27 m

  *(c)* 1.73 s

  *(d)* 11.9 m/s

79.  43.6 m

81.  68.0 m/s

83.  (a)  666 m

     (b)  13.6 m/s

85.  (a)  10.4 s

     (b)  27.4 s

     (c)  12.8 s

87.  (a)  19.0 km

     (b)  2 min 18 s

     (c)  610 m/s

89.  40.0 cm/s; $-6.88$ cm/s$^2$

91.  (a)  4.76 m/s or 10.7 mi/h

     (b)  0.595

93.  10.9 m

95.  27.6 m

97.  4.59 km

99.  2.40 m; 1.40 s

103. (a)  $-25.7$ m/s$^2$

     (b)  2.33 s

105. (a)  1.03 $c \cdot y/y^2 = 1.03\ c/y$

     (b)  $\approx 2$ d

107. 4.80 m/s

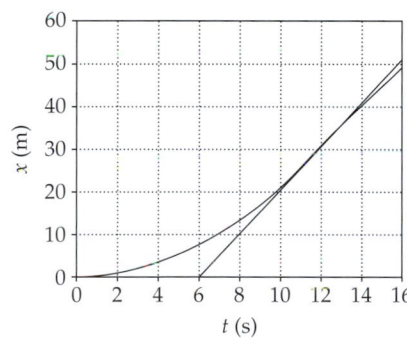

109. $\frac{1}{3} h$

111. (a)  34.7 s

     (b)  1.20 km

     (c)

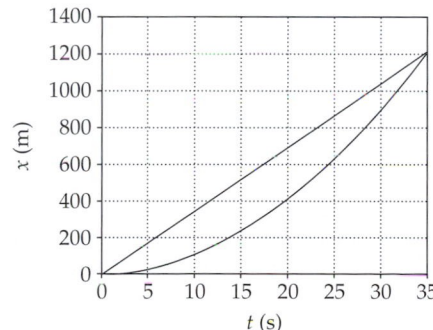

113. (a)  $L_1 = \frac{2}{3} L$

     (b)  $t = \frac{2}{3} t_{\text{fin}}$

115. (a)  $A = 90$ m

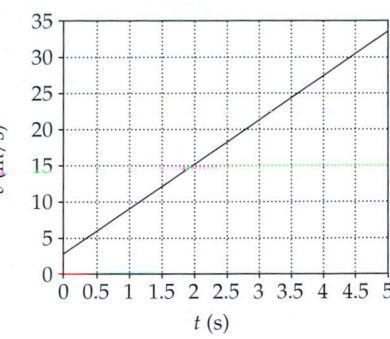

     (b)  $x(t) = (3\ \text{m/s}^2)t^2 + (3\ \text{m/s})t;\ \Delta x = 90.0$ m

117. $x(t) = (\frac{7}{3}\ \text{m/s}^3)t^3 - (5\ \text{m/s})t$

119. (a)  0.250 m/s per box

     (b)  $v(1\ \text{s}) = 0.930$ m/s; $v(2\ \text{s}) = 3.20$ m/s; $v(3\ \text{s}) = 6.20$ m/s

     (c)  $x(3\ \text{s}) = 7.00$ m

$v$ vs. $t$

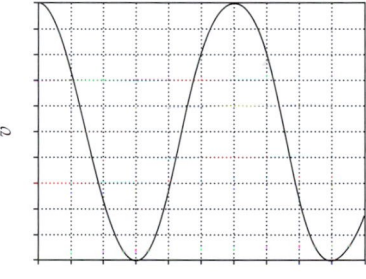

121.

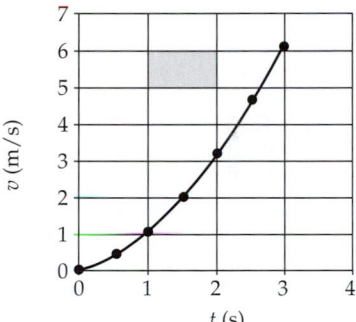

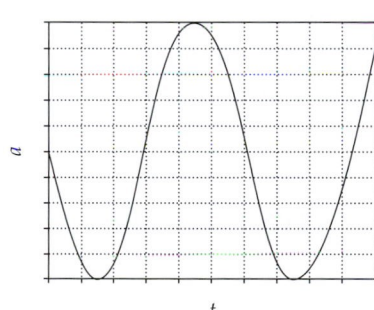

123. (a)  $v(t) = (0.1\ \text{m/s}^3)t^2$

     (b)  2.23 m/s

125. 12.8 m/s$^2$; 30.5%

127. (a)  $a = \omega v_{\text{max}}\cos(\omega t)$; Because $a$ varies sinusoidally with time, it is *not* constant.

     (b)  $x = x_0 + \dfrac{v_{\text{max}}}{\omega}[1 - \cos(\omega t)]$

129. (a) $x(t) = x_0 e^{(t - t_0)/b}$    (b = 1 s)

(b) Because the numerical value of $b$, expressed in SI units, is one, the numerical values of $a$, $v$, and $x$ are the same at each instant in time.

# Chapter 3

1. The magnitude of the displacement of a particle is less than or equal to the distance it travels along its path.

3. The displacement for any trip around the track is ZERO. Thus we see that no matter how fast the race car travels, the average velocity is always ZERO at the end of each complete circuit.

5. No. The magnitude of a component of a vector must be less than or equal to the magnitude of the vector. If the angle $\theta$ shown in the figure is equal to 0° or multiples of 90°, then the magnitude of the vector and its component are equal.

7. No

9. (e)

11. (c)

13. (a) The velocity vector, as a consequence of always being in the direction of motion, is tangent to the path.

(b)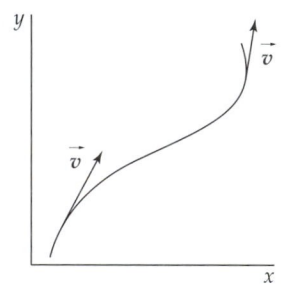

15. (a) A car moving along a straight road while braking

(b) A car moving along a straight road while speeding up

(c) A particle moving around a circular track at constant speed

17. (a) $\vec{v}_2$ ↑  $-\vec{v}_1$ ↓ | ↑ $\vec{v}_2$
    ↓↓ $\Delta\vec{v} = \vec{v}_2 - \vec{v}_1$
    $\vec{v}_1$ ↑    ↓ $\vec{a} = \Delta\vec{v}/\Delta t$

(b) $\vec{v}_1$ ↓  $-\vec{v}_2$ | ↑ $\Delta\vec{v} = \vec{v}_2 - \vec{v}_1$
    ↑ $-\vec{v}_2$
    $\vec{v}_2$ ↓    ↓ $\vec{a} = \Delta\vec{v}/\Delta t$

(c)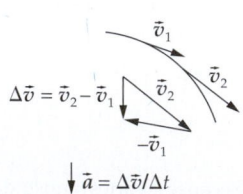

19.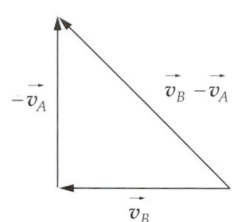

$$\vec{v}_1 \downarrow \quad \uparrow \vec{v}_2 \qquad {-\vec{v}_1} \uparrow \uparrow \Big| \Delta\vec{v} = \vec{v}_2 - \vec{v}_1$$
$$-\vec{v}_2$$
$$\uparrow \vec{a} = \Delta\vec{v}/\Delta t$$

21. True

23. (d)

25. (a) False

(b) True

27.

$$-\vec{v}_A \qquad \vec{v}_B - \vec{v}_A$$
$$\vec{v}_B$$

29. (a)

| Path | Direction of velocity vector |
|------|------------------------------|
| AB   | north                        |
| BC   | northeast                    |
| CD   | east                         |
| DE   | southeast                    |
| EF   | south                        |

(b)

| Path | Direction of acceleration vector |
|------|----------------------------------|
| AB   | north                            |
| BC   | southeast                        |
| CD   | 0                                |
| DE   | southwest                        |
| EF   | north                            |

(c) The magnitudes are approximately equal.

31. The droplet leaving the bottle has the same horizontal velocity as the ship. During the time the droplet is in the air, it is also moving horizontally with the same velocity as the rest of the ship. Because of this, it falls into the vessel, which has the same horizontal velocity. Because you have the same horizontal velocity as the ship, you see the same thing as if the ship were standing still.

33. True

35. The principal reason is aerodynamic drag; when moving through a fluid such as the atmosphere, the ball's acceleration will depend strongly on its velocity.

37. 14.8 m/s

39. $R = 22.2$ m; $\alpha = 22.5°$

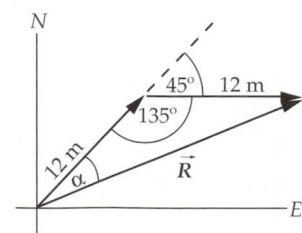

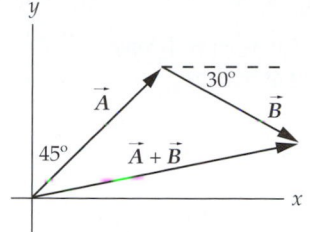

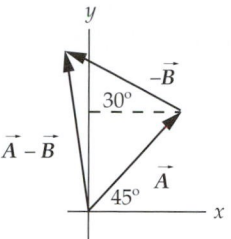

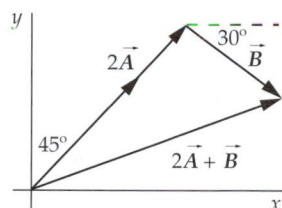

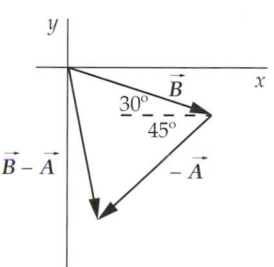

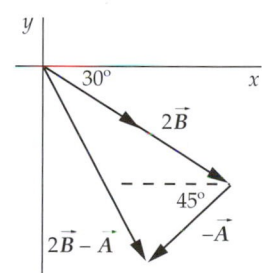

41. *(a)*

*(b)*

*(c)*

*(d)*

*(e)*

43. *(b)*

45.

| | $A$ | $\theta$ | $A_x$ | $A_y$ |
|---|---|---|---|---|
| *(a)* | 10 m | 30° | 8.66 m | 5 m |
| *(b)* | 5 m | 45° | 3.54 m | 3.54 m |
| *(c)* | 7 km | 60° | 3.50 km | 6.06 km |
| *(d)* | 5 km | 90° | 0 | 5 km |
| *(e)* | 15 km/s | 150° | −13.0 km/s | 7.50 km/s |
| *(f)* | 10 m/s | 240° | −5.00 m/s | −8.66 m/s |
| *(g)* | 8 m/s² | 270° | 0 | −8.00 m/s² |

47. *(a)* 5.83; 31.0°

    *(b)* 122; −35.0°

    *(c)* 5.39; $\theta = 42.1°$; $\phi = 236°$

49. *(a)* $\vec{v} = (5 \text{ m/s})\hat{i} + (8.66 \text{ m/s})\hat{j}$

    *(b)* $\vec{A} = (-3.54 \text{ m})\hat{i} + (-3.54 \text{ m})\hat{j}$

    *(c)* $\vec{r} = (14 \text{ m})\hat{i} - (6 \text{ m})\hat{j}$

51. $\vec{D} = (3 \text{ m})\hat{i} + (3 \text{ m})\hat{j} + (3 \text{ m})\hat{k}$; $D = 5.20 \text{ m}$

53. $\vec{v}_{av} = (14.1 \text{ km/h})\hat{i} + (-4.1 \text{ km/h})\hat{j}$

55. *(b)*

57. *(a)* $\vec{v}_{av} = (33.3 \text{ m/s})\hat{i} + (26.7 \text{ m/s})\hat{j}$

    *(b)* $\vec{a}_{av} = (-3.00 \text{ m/s}^2)\hat{i} + (-1.77 \text{ m/s}^2)\hat{j}$

59. $\vec{v} = (30 \text{ m/s})\hat{i} + [40 \text{ m/s} - (10 \text{ m/s}^2)t]\hat{j}$; $\vec{a} = (-10 \text{ m/s}^2)\hat{j}$

61. *(a)* $\vec{v}_{av} = (20 \text{ m/s})(-\hat{i} + \hat{j})$

    *(b)* $\vec{a}_{av} = (-2 \text{ m/s}^2)\hat{i}$

    *(c)* $\Delta\vec{r} = (600 \text{ m})(-\hat{i} + \hat{j})$

63. *(a)* 13.1° west of north

    *(b)* 300 km/h

65. 8.47°; 2.57 h

67. You should fly your plane across the wind.

69. *(a)* $\vec{r}_{AB} (6 \text{ s}) = (120 \text{ m})\hat{i} + (4 \text{ m})\hat{j}$

    *(b)* $\vec{v}_{AB} (6 \text{ s}) = (-20 \text{ m/s})\hat{i} - (12 \text{ m/s})\hat{j}$

    *(c)* $\vec{a}_{AB} = (-2 \text{ m/s}^2)\hat{j}$

71. $1.52 \times 10^{-6} \text{ m/s}^2$; $1.55 \times 10^{-7}g$

73. $3.44 \times 10^{-3}g$; $6.07 \times 10^{-4}g$

75. 33.4 min⁻¹

77. $h = \dfrac{(v_0 \sin \theta_0)^2}{2g}$

79. 33.8 m/s

81. 20.3 m/s; 36.2°

83. 69.3°

85. *(a)* 18.0 m/s

    *(b)* 14.0°

87. *(a)* 8.14 m/s

    *(b)* 23.2 m/s

89. −63.4°

91. 209 m

93. *(a)* 0.452 s

    *(b)* 22.6 m

95. *(a)* 485 km

    *(b)* 1.70 km/s

101. $L = \dfrac{2v_0^2 \tan \theta}{g \cos \theta}$

103. 10.8 m/s; $v = (6.50 \text{ m/s})\hat{i} + (-21.6 \text{ m/s})\hat{j}$

105. 40.5 m/s; 0.994 s

107. 7.41 m/s; 0.756 s; 15.9 m/s; 17.5 m/s; 25.0°

109. 0.785 m

111. 4.91 m/s²; 8.50 m/s²

113. *(a)*

$y$ (m)

[graph with y-axis labeled from 0 to 25, x-axis labeled $x$ (m) from 0 to 15]

    *(b)* $\vec{v} = (5 \text{ m/s})\hat{i} + (10 \text{ m/s})\hat{j}$; 11.2 m/s

115. $31.3°$; 8.06 m

117. Fourth step

119. (a) $v_{min} = \dfrac{x}{\cos\theta}\sqrt{\dfrac{g}{2(x\tan\theta - h)}}$

   (b) $v_{min} > 26.0$ m/s $= 58.0$ mi/h

   (c) $h_{max} < x\tan\theta$

# Chapter 4

1. If an object with no net force acting on it is at rest or is moving with a constant speed in a straight line (i.e., with constant velocity) relative to the reference frame, then the reference frame is an inertial reference frame.

3. No. If the net force acting on an object is zero, its acceleration is zero. The only conclusion one can draw is that the *net* force acting on the object is zero.

5. No. Correctly predicting the direction of the subsequent motion requires knowledge of the initial velocity as well as the acceleration.

7. The mass of an object is an intrinsic property of the object whereas the weight of an object depends directly on the local gravitational field. Therefore, the mass of the object would not change and $w_{grav} = mg_{local}$. Note that if the gravitational field is zero then the gravitational force is also zero.

9. Your apparent weight would be greater than your true weight when observed from a reference frame that is accelerating upward. That is, when the surface on which you are standing has an acceleration $a$ such that $a_y$ is positive.

11. (a) $F_{n21} = m_1 g$

   (b) $F_{n12} = m_1 g$

   (c) $F_{nT2} = (m_1 + m_2)g$

   (d) $F_{n2T} = (m_1 + m_2)g$

13. (b)

15. (c)

17. (a)          (b)

19. (a) True

   (b) False

   (c) False

   (d) False

21. (d)

23. The velocity of the elevator has no effect on the person's apparent weight.

25. (a) 782 N; 62.6 N

   (b) Because there is no acceleration, the forces are the same going up and going down the incline.

27. (a) 6.00 m/s²

   (b) 1/3

   (c) 2.25 m/s²

29. $-3.75$ kN

31. (a) 4.24 m/s² @ 45.0° from each force

   (b) 8.40 m/s² @ 14.6° from $2\vec{F}_0$

33. 12.0 kg

35. (a) 4.00 m/s²

   (b) 2.40 m/s²

37. (a) $\vec{a} = (1.50\text{ m/s}^2)\hat{\imath} + (-3.50\text{ m/s}^2)\hat{\jmath}$

   (b) $\vec{v} = (4.50\text{ m/s})\hat{\imath} + (-10.5\text{ m/s})\hat{\jmath}$

   (c) $\vec{r} = (6.75\text{ m})\hat{\imath} + (-15.8\text{ m})\hat{\jmath}$

39. (a) 530 N

   (b) 119 lb

41. (a) 60.0 N

   (b) 57.7 N

43. $T_2 > T_1$

45. (a) 36.9°

   (b) 4.08 N

   (c) 3.43 N; 2.40 N; 3.43 N

47. (a) $\vec{a} = (0.500\text{ m/s}^2)\hat{\imath} + (2.60\text{ m/s}^2)\hat{\jmath}$

   (b) $\vec{F}_3 = (-5.00\text{ N})\hat{\imath} + (-26.0\text{ N})\hat{\jmath}$

49. (a) $T = \dfrac{w}{2\sin\theta}$; $\theta = 90°$; $T \to T_{max}$ as $\theta \to 0°$

   (b) 19.6 N

51. (a) 11.8 kN

   (b) 9.81 kN

   (c) 7.81 kN

53. (a) 3.82 kN

   (b) 4.30 kN

55. 56.0 N

57. (d)

59. (a) 508 N; 508 N

   (b) $mg$; 0

61. 552 N

63. (a)

65. (e)

67. (a) 19.6 N

   (b) 19.6 N

   (c) 25.6 N

   (d) 14.6 N

69. (a) 1.31 m/s²

   (b) 16.7 N; 21.3 N

71. (a) $a = \dfrac{F}{m_1 + m_2}$; $F_{2,1} = \dfrac{Fm_1}{m_1 + m_2}$

   (b) 0.400 m/s²; 0.800 N

75. (a)

77. (a) $a = \dfrac{g(m_2 - m_1\sin\theta)}{m_1 + m_2}$; $T = \dfrac{gm_1m_2(1 + \sin\theta)}{m_1 + m_2}$

   (b) 2.45 m/s²; 36.8 N

79. (a) 1.37 m/s²; 61.4 N

   (b) 1.19; The answer is the ratio of two quantities with the same units and so has no units.

81. (a) 398 N

   (b) 368 N

83. (a) 5.00 cm

   (b) $a_{5kg} = 4.91$ m/s²; $a_{20kg} = 2.45$ m/s²; $T = 24.5$ N

85. 1.36 kg or 1.06 kg

87. $F = 2T = \dfrac{4m_1m_2g}{m_1 + m_2}$

91. (a) $-100$ m/s$^2$

(b) 6.13 cm

(c) 35.0 ms

93. 305 N; 1.55 kN

95. (a) 1.50 m/s

(b) 1.50 m

(c) 0.500 m/s

(d) 12.0 N

97. (a) $a = \dfrac{F}{m_1 + m_2}$

(b) $F_{net} = \dfrac{Fm_2}{m_1 + m_2}$

(c) $T = \dfrac{Fm_1}{m_1 + m_2}$

(d)

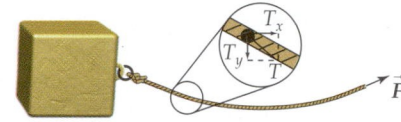

Yes . . . correct answers appear above.

99. (a) 55.0 g

(b) 2.45 m/s$^2$; 2.03 N

101. (a) $\frac{1}{3}(F_2 + 2F_1)$

(b) $\dfrac{3T_0}{4C}$

# Chapter 5

1. The force of friction between the object and the floor of the truck must be the force that causes the object to accelerate.

3. (d)

5. (b)

7. As the spring is extended, the force exerted by the spring on the block increases. Once that force is greater than the maximum value of the force of static friction on the block, the block will begin to move. However, as it accelerates, it will shorten the length of the spring, decreasing the force that the spring exerts on the block. As this happens, the force of kinetic friction can then slow the block to a stop, which starts the cycle over again. One interesting application of this to the real world is the bowing of a violin string: the string under tension acts like the spring, while the bow acts as the block, so as the bow is dragged across the string, the string periodically sticks and frees itself from the bow.

9. (e)

11. Block 1 will hit the pulley before block 2 hits the wall.

13. (d)

15. (d)

17. For a rock, which has a relatively small surface area compared to its mass, the terminal speed will be relatively high; for a lightweight, spread-out object like a feather, the opposite is true.

   Another issue is that the higher the terminal velocity is, the longer it takes for a falling object to reach terminal velocity: from this, the feather will reach its terminal velocity quickly,

and fall at an almost constant speed very soon after being dropped; a rock, if not dropped from a great height, will have almost the same acceleration as if it were in free-fall for the duration of its fall, and thus be continually speeding up as it falls.

19. (a) M/T; kg/s

(b) M/L; kg/m

(c) ML/T$^2$

(d) 56.9 m/s

(e) 86.9 m/s

21. (b)

23. (a) 15.0 N

(b) 12.0 N

25. 500 N

27. (a) $-5.89$ m/s$^2$

(b) 76.4 m

29. (a) 49.1 N

(b) 123 N

31. 4.57°

33. (a) 0.667

(b) 2.16 m/s$^2$; 1.36 s

35. (a)

37. 2.36 m/s$^2$; 37.2 N

39. (a) 0.599

(b) 9.25 m

(c) 4.73 m/s

41. (a) 2.75 m/s$^2$

(b) 10.1 s

43. (a) 0.965 m/s$^2$ directed up the incline

(b) 0.184 N

45. (a) 25.0°

(b) 0.118 N

47. (a) The static-frictional force opposes the motion of the object, and the maximum value of the static-frictional force is proportional to the normal force $F_N$. The normal force is equal to the weight minus the vertical component $F_V$ of the force $F$. Keeping the magnitude $F$ constant while increasing $\theta$ from 0 results in a decrease in $F_v$ and thus a corresponding decrease in the maximum static-frictional force $f_{max}$. The object will begin to move if the horizontal component $F_H$ of the force $F$ exceeds $f_{max}$. An increase in $\theta$ results in a decrease in $F_H$. As $\theta$ increases from 0, the decrease in $F_N$ is larger than the decrease in $F_H$, so the object is more and more likely to slip. However, as $\theta$ approaches 90°, $F_H$ approaches 0, and no movement will be initiated. If $F$ is large enough and if $\theta$ increases from 0, then at some value of $\theta$, the block will start to move.

(b)

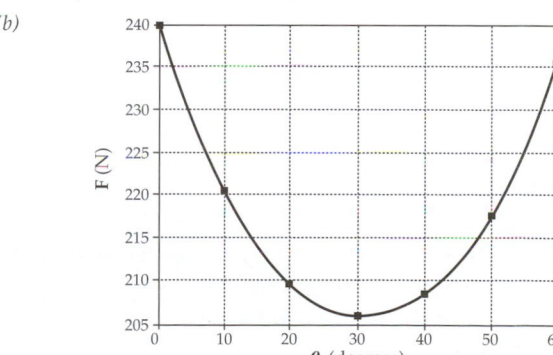

49. *(b)*

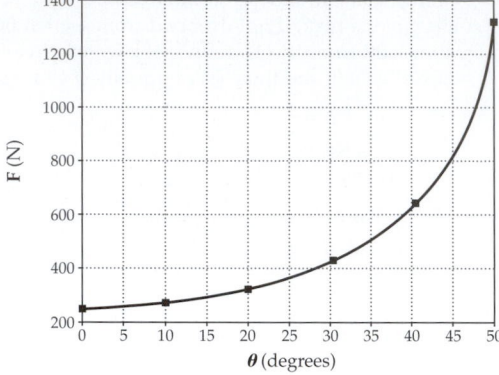

51. *(a)* 0.238
    *(b)* 1.40 m/s²

53. *(a)* 17.7 N
    *(b)* 1.47 m/s²; 5.88 N
    *(c)* 1.96 m/s²; 7.87 m/s²

55. *(a)* 0.163 m/s²
    *(b)* 0.0381 m
    *(c)* −0.254 m/s²

57. −8.41 glapp/plipp²; 0.191

59. *(a)* −1.57 N; 83.8 N
    *(b)* 6.49 N; 37.5 N

61. *(a)* −2.60 m/s²; 19.2 m
    *(b)* −2.11 m/s²; 23.7 m

63. *(a)* 0.297
    *(b)* 2.82 m/s

65. *(c)*

67. *(a)* 1.41 m/s
    *(b)* 8.50 N

69. *(a)* 8.33 m/s²; upward
    *(b)* 542 N; upward
    *(c)* 1.18 kN; upward

71. $T_2 = [m_2(L_1 + L_2)]\left(\dfrac{2\pi}{T}\right)^2$; $T_1 = [m_2(L_1 + L_2) + m_1 L_1]\left(\dfrac{2\pi}{T}\right)^2$

73. 53.3°; 410 N

75. *(a)* 0.395 N
    *(b)* 0.644

77. $3.44 \times 10^{-3}g$; $6.07 \times 10^{-4}g$

79. 51.6°

81. *(a)* $a_c = \dfrac{v_0^2}{r}\left(\dfrac{1}{1 + \left(\dfrac{\mu_k v_0}{r}\right)t}\right)^2$
    *(b)* $a_t = -\mu_k a_c$
    *(c)* $a = a_c \sqrt{1 + \mu_k^2}$

83. 12.8 m/s

85. *(a)* 7.25 m/s
    *(b)* 0.536

87. 21.7°

89. *(a)* 7.832 kN
    *(b)* −766 kN

91. $v_{min} = 20.1$ km/h; $v_{max} = 56.1$ km/h

93. $2.79 \times 10^{-4}$ kg/m

95. 88.2 km/h

97. 3.31 s; 100

99. $y(3.5 \text{ s}) \approx 60.4$ m; $y_{max} \approx 60.6$ m @ $t = 3.3$ s; $t_{flight} \approx 7$ s; The ball spends a little longer coming down than it does going up.

101. 0.511

103. *(a)* 0.289
     *(b)* 600 N

105. 1.49 kN

107. $a = g(\sin \theta_1 - \tan \theta_0 \cos \theta_1)$

109. *(a)* 49.4 m/s²
     *(b)* 4.49 s

111. *(a)* 193 N
     *(b)* 51.8 N
     *(c)* The sled does not move.
     *(d)* $\mu_k$ is undetermined.
     *(e)* 536 N

113. 0.433

115. 23.6 rev/min

117. *(a)* Toward the earth's axis.
     *(b)* A stone dropped from a hand at a location on the earth. The effective weight of the stone is equal to $m\vec{a}_{st,\,surf}$, where $\vec{a}_{st,\,surf}$ is the acceleration of the falling stone (neglecting air resistance) relative to the local surface of the earth. The gravitational force on the stone is equal to $m\vec{a}_{st,\,iner}$, where $\vec{a}_{st,\,iner}$ is the acceleration of the stone relative to an inertial reference frame. These accelerations are related by $\vec{a}_{st,\,surf} + \vec{a}_{surf,\,iner} = \vec{a}_{st,\,iner}$, where $\vec{a}_{surf,\,iner}$ is the acceleration of the local surface of the earth relative to the inertial frame (the acceleration of the surface due to the rotation of the earth). Multiplying through this equation by $m$ and rearranging gives $m\vec{a}_{st,\,surf} = m\vec{a}_{st,\,iner} - m\vec{a}_{surf,\,iner}$, which relates the apparent weight to the acceleration due to gravity and the acceleration due to the earth's rotation. A vector addition diagram can be used to show that the magnitude of $m\vec{a}_{st,\,surf}$ is slightly less than that of $m\vec{a}_{st,\,iner}$.
     *(c)* 983 cm/s²

# Chapter 6

1. *(a)* False
   *(b)* True
   *(c)* True

3. False

5. No. The work done on any object by any force $\vec{F}$ is defined as $dW \equiv \vec{F} \cdot d\vec{r}$. The direction of $\vec{F}_{net}$ is toward the center of the circle in which the object is traveling, and $d\vec{r}$ is tangent to the circle. No work is done by the net force because $\vec{F}_{net}$ and $d\vec{r}$ are perpendicular, so the dot product is zero.

7. Because $W \propto x^2$, doubling the distance the spring is stretched will require four times as much work.

9. *(d)*

11. *(a)* False
    *(b)* False
    *(c)* True

13. *(a)* False
    *(b)* False

15. *(a)* 0.245 m
    *(b)* 120 J

17. ≈ 1%

19. 20.8 kJ

21. (a) 147 J

   (b) 266 J

23. 10.6 kJ

25. (a) 6.00 J

   (b) 12.0 J

   (c) 3.46 m/s

27. $W = -\frac{1}{2}kx_1^2 - \frac{1}{3}ax_1^3$

29. (a) $m(y) = 40\ \text{kg} - (1\ \text{kg/m})y$

   (b) 5.89 kJ

31. (a) 4.17 N

   (b) $\vec{T}, \vec{F}_g$, and $\vec{F}_n$; Because all of these forces act perpendicularly to the direction of motion of the object, none of them do any work.

35. 180°

37. (a) −24

   (b) −10

   (c) 0

39. (a) 1.00 J

   (b) 0.213 N

43. No. Let $\vec{A} = \hat{i}, \vec{B} = 3\hat{i} + 4\hat{j}$ and $\vec{C} = 3\hat{i} - 4\hat{j}$ and form $\vec{A} \cdot \vec{B}$ and $\vec{A} \cdot \vec{C}$.

45. (b) The results of (a) and (b) tell us that $\vec{a}$ is perpendicular to $\vec{v}$ and parallel (or antiparallel) to $\vec{r}$.

47. (a) 98.1 W

   (b) 392 J

49. (a) $v = (\frac{5}{8}\ \text{m/s}^2)t$

   (b) $P = 3.13t\ \text{W/s}$

   (c) 9.38 W

51. 445 W

53. $v = \sqrt{v_0^2 + 2gH}$

55. 4.71 kJ

57. (a) 392 J

   (b) 2.45 m; 4.91 m/s

   (c) 24.1 J; 368 J

   (d) 392 J; 19.8 m/s

59. (a) 0.100 m

   (b) 0.141 m

61. (a) $U(\theta) = (m_2\ell_2 - m_1\ell_1)g \sin\theta$

   (b) $U$ is a minimum at $\theta = -\pi/2$ and a maximum at $\theta = \pi/2$

   (c) $U = 0$ independently of $\theta$

63. (a) $F_x = \dfrac{C}{x^2}$

   (b) $F_x$ is positive for $x \neq 0$ and therefore $\vec{F}$ is directed away from the origin.

   (c) $U(x)$ decreases with increasing $x$.

   (d) $F_x$ is negative for $x \neq 0$ and therefore $\vec{F}$ is directed toward the origin. $U(x)$ increases with increasing $x$.

65. $U(x) = \dfrac{a}{x} + U_0$

67. (a) $F_x = 4x(x + 2)(x - 2)$

   (b) −2 m, 0, 2 m

   (c) Unstable equilibrium at $x = -2$ m; stable equilibrium at $x = 0$; unstable equilibrium at $x = 2$ m

69. (a) 0 and 2 m; neutral equilibrium for $x > 3$ m

   (b)

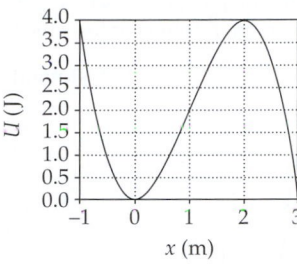

   (c) Stable equilibrium at $x = 0$; unstable equilibrium at $x = 2$ m

   (d) 2.00 m/s

71. (a) $U(y) = -mgy - 2Mg\left(L - \sqrt{y^2 + d^2}\right)$

   (b) $y = d\sqrt{\dfrac{m^2}{4M^2 - m^2}}$

   (c) Stable equilibrium

73. (a) 706 MJ

   (b) 11.8 MW

75. 0.500 m

77. (a) 34.4 N

   (b)

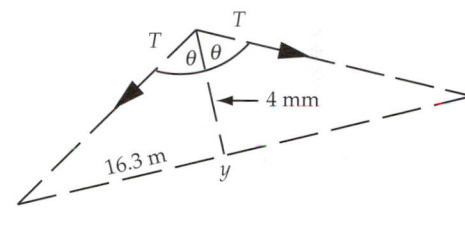

   1.68 N

   (c) 3.38 mJ

79. (a) $F(x) = mC^2x$

   (b) $W = \frac{1}{2}mC^2x_1^2$

81. In the following, if $t$ is in seconds and $m$ is in kilograms, then $v$ is in m/s, $a$ is in m/s², $P$ is in W, and $W$ is in J.

   (a) $v = (6t^2 - 8t); a = (12t - 8)$

   (b) $P = 8mt(9t^2 - 18t + 8)$

   (c) $W = 2mt_1^2(3t_1 - 4)^2$

83. 5.74 km

85. (a)

| $x$ | $W$ |
|---|---|
| (m) | (J) |
| −4 | 6 |
| −3 | 4 |
| −2 | 2 |
| −1 | 0.5 |
| 0 | 0 |
| 1 | 0.5 |
| 2 | 1.5 |
| 3 | 2.5 |
| 4 | 3 |

   (b)

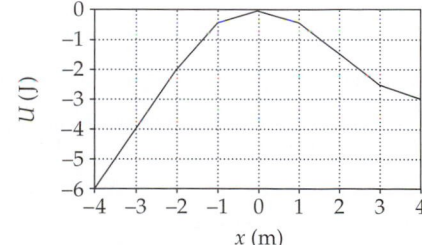

87. (b) $W = (10\pi \text{ m})F_0$ if the rotation is clockwise; $-(10\pi \text{ m})F_0$ if the rotation is counterclockwise. Because $W \neq 0$ for a complete circuit, $\vec{F}$ is not conservative.

89. (a)

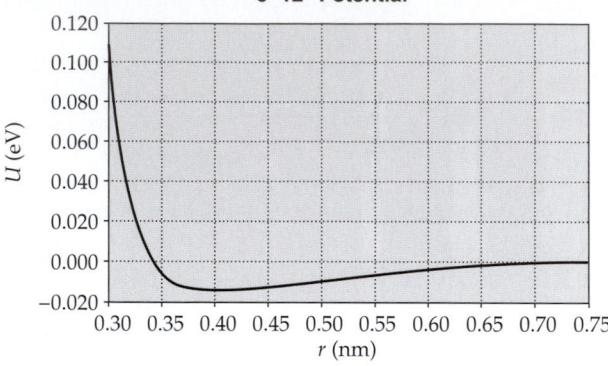

**"6–12" Potential**

(b) The minimum value is about $-0.0107$ eV, occurring at a separation of approximately 0.380 nm. Because the function is concave upward at this separation, this separation is one of stable equilibrium, although very shallow.

(c) $-6.69 \times 10^{-12}$ N; $7.49 \times 10^{-11}$ N

# Chapter 7

1. (a)
3. (a) False
   (b) False
5. As she starts pedaling, chemical energy inside her body is converted into kinetic energy as the bike picks up speed. As she rides it up the hill, chemical energy is converted into gravitational potential energy. While freewheeling down the hill, potential energy is converted to kinetic energy, and while braking to a stop, kinetic energy is converted into thermal energy (a more random form of kinetic energy) by the frictional forces acting on the bike.
7. (d)
9. No. From the work–kinetic energy theorem, no total work is being done on the rock, as its kinetic energy is constant. However, the rod must exert a tangential force on the rock to keep the speed constant. The effect of this force is to cancel the component of the force of gravity that is tangential to the trajectory of the rock.
11. 33.6 s
13. $3.04 \times 10^{19}$ J/y; $\approx 6\%$
15. $1.10 \times 10^6$ L/s
17. (c)
19. 3.89 m
21. 5.05 m
23. 25.6°
25. $U = \dfrac{[mg(\sin\theta + \mu_s \cos\theta)]^2}{2k}$
27. $6mg$
29. (c)
31. $6mg$
33. (a) 31.0 m
    (b) $-31.7$ J
    (c) 33.7 m/s

35. (a) 151 m
    (b) 45.3 m/s
37. (a) $\frac{5}{2}mgL$
    (b) $6mg$
39. (a) 20.2°
    (b) 6.39 m/s
41. $v = L\sqrt{2\dfrac{g}{L}(1-\cos\theta) + \dfrac{k}{m}\left(\sqrt{\frac{13}{4} - 3\cos\theta} - \frac{1}{2}\right)^2}$
43. (a) 94.2 kJ
    (b) The energy required to do this work comes from chemical energy stored in the body.
    (c) 471 kJ
45. (a) 104 J
    (b) 70.2 J
    (c) 33.8 J
    (d) 2.91 m/s
47. (a) 7.67 m/s
    (b) 58.9 J
    (c) 0.333
49. (a) $W_f = (13.7 \text{ N})y$
    (b) $E_{\text{mech}} = -(13.7 \text{ N})y$
    (c) 1.98 m/s
51. 0.875 m; 2.49 m/s
53. (a) $9.00 \times 10^{13}$ J
    (b) \$$2.5 \times 10^6$
    (c) 28,400 y
55. $1.88 \times 10^{-28}$ kg
57. $3.56 \times 10^{14}$ reactions
59. 0.782 MeV
61. (a) 3.16 kg
    (b) $8.04 \times 10^9$ kg
63. (b)
65. 57.6 MJ
67. (a) 0.208
    (b) 3.45 MJ
69. (a) From the FBD we can see that the forces acting on the box are the normal force exerted by the inclined plane, kinetic friction force, and the gravitational force (the weight of the box) exerted by the earth.
    (b) 0.451 m
    (c) 1.33 J
    (d) 2.52 m/s
71. 11.3 kW; $-6.77$ kW
73. (a) 1.60 kJ
    (b) 619 J
    (c) 23.4 m/s
75. (a) 147 J
    (b) The energy is transferred to the girder from its surroundings, which are warmer than the girder. As the temperature of the girder rises, the atoms in the girder vibrate with a greater average kinetic energy, leading to a larger average separation, which causes the girder's expansion.

77. (a)

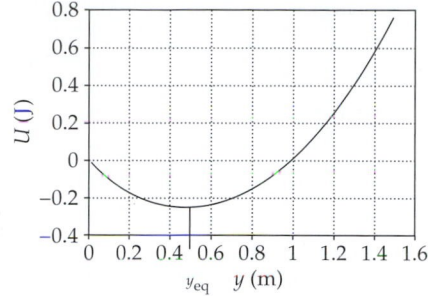

(b) $F = -ky + mg$

(c) $y_{max} = \dfrac{2mg}{k}$

(d) $y_{eq} = \dfrac{mg}{k}$

(e) $W_f = \dfrac{m^2g^2}{2k}$

79. (a) 17.3 m

(b) 4.91 kN

(c) 4.91 m/s²

(d) 13.4 kN, upward

(e) 5.46 kN; 63.9°

(f) 1.44 kN

81. (a) 491 N; 981 N

(b) 9.82 kW; 29.4 kW

(c) 8.85°

(d) 6.36 km/L

83. (a) 17.4 MJ

(b) $1.39 \times 10^{10}$ J

(c) $9.73 \times 10^9$ J

(d) 1.59 MW

87. (a) $v = \sqrt{\dfrac{2mgY}{M+m}}$

(b) $v = \sqrt{\dfrac{2mgY}{M+m}}$

89. $D = 2\sqrt{HL(1 - \cos\theta)}$

91. (a) $K_{max} = mgh + \dfrac{m^2g^2}{2k}$

(b) $x_{max} = \dfrac{mg}{k} + \sqrt{\dfrac{m^2g^2}{k^2} + \dfrac{2mgh}{k}}$

(c) $x = \dfrac{mg}{k} + \sqrt{\dfrac{2m^2g^2}{k^2} + \dfrac{4mgh}{k}}$

93. (a)

**Potential Energy**

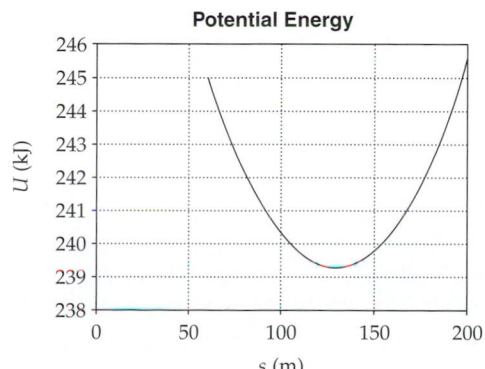

$s$ (m)

(b) 5.39 kJ

## Chapter 8

1. A doughnut.

3. (b)

5. No. Consider a 1-kg block with a speed of 1 m/s and a 2-kg block with a speed of 0.707 m/s. The blocks have equal kinetic energies but momenta of magnitude 1 kg·m/s and 1.414 kg·m/s, respectively.

7. $\vec{p}_{recoil} = \vec{p}_{rifle} = -\vec{p}_{bullet}$ or $\vec{p}_{rifle} + \vec{p}_{bullet} = 0$

9. Conservation of momentum requires only that the net external force acting on the system be zero. It does not require the presence of a medium such as air.

11. Think of someone pushing a box across a floor. Her push on the box is equal but opposite to the push of the box on her, but the action and reaction forces act on *different objects*. You can only add forces when they act on the same object.

13. The problem is that the comic situations violate the conservation of momentum! To move forward requires pushing something backward, which Superman doesn't appear to be doing when flying around. In a similar manner, if Superman picks up a train and throws it at Lex Luthor, he (Superman) ought to be tossed backward at a pretty high speed to satisfy the conservation of momentum.

15. The friction of the tire against the road causes the car to slow down. This is rather subtle, as the tire is in contact with the ground without slipping at all times, so as you push on the brakes harder, the force of static friction of the road against the tires must increase. Also, of course, the brakes heat up, and not the tires.

17. Assume that the ball travels at 80 mi/h = 35 m/s. The ball stops in a distance of about 1 cm, so the distance traveled is about 2 cm at an average speed of about 18 m/s. The collision time is $\dfrac{0.02 \text{ m}}{18 \text{ m/s}} \approx 1$ ms.

19. (a) False

(b) True

(c) True

21. (a) The loss of kinetic energy is the same in both cases.

(b) The percentage loss is greatest for the case in which the two objects have oppositely directed velocities of magnitude $\frac{1}{2}v$.

23. (b) is correct because all of 1's kinetic energy is transferred to 2 when $m_2 = m_1$.

25. The water is changing direction when it rounds the corner in the nozzle. Therefore, the nozzle must exert a force on the stream of water to change its direction, and, from Newton's 3rd law, the water exerts an equal but opposite force on the nozzle.

27. No. $\vec{F}_{ext,net} = d\vec{p}/dt$ defines the relationship between the net force acting on a system and the rate at which its momentum changes. The net external force acting on the pendulum bob is the sum of the force of gravity and the tension in the string and these forces do not add to zero.

29. Think of the stream of air molecules hitting the sail. Imagine that they bounce off the sail elastically–their net change in momentum is then roughly twice the change in momentum that they experienced going through the fan. Another way of looking at it: initially, the air is at rest, but after passing through the fan and bouncing off the sail, it is moving backward; therefore, the boat must exert a net force on the air pushing it backward, and there must be a force on the boat pushing it forward.

31. (a) 2.33 s

    (b) 6.74 m/s

33. (0.233 m, 0)

35. (2.00 m, 1.40 m)

37. (1.50 m, 1.36 m)

41. $z_{cm} = \frac{1}{2}R$

43. $\vec{v}_{cm} = (3\ \text{m/s})\hat{i} - (1.5\ \text{m/s})\hat{j}$

45. $\vec{a}_{cm} = (2.4\ \text{m/s}^2)\hat{i}$

47. (a) $F_n = (m_p + m_b)g$

    (b) $F_n = (m_p + 2m_b)g$

    (c) $F_n = m_p g$

49. (a) $F_n = (m_p + m_b)g$

    (b) $F_n = m_p g + m_b g\left(1 + \sqrt{1 + \dfrac{2kh}{m_b g}}\right)$

51. $\vec{v}_{10} = (4\ \text{m/s})\hat{i}$

53. $\vec{v}' = 2v\hat{i} - v\hat{j}$

55. $-\sqrt{\dfrac{gh}{3}}\,\hat{i}$

57. (a) 43.5 J

    (b) $\vec{v}_{cm} = (1.50\ \text{m/s})\hat{i}$

    (c) $\vec{v}_{1,rel} = (3.50\ \text{m/s})\hat{i}$ and $\vec{v}_{2,rel} = (-3.50\ \text{m/s})\hat{i}$

    (d) 36.8 J

    (e) $K_{cm} = 6.75\ \text{J} = K - K_{rel}$

59. (a) 10.8 N·s

    (b) 1.34 kN

61. 1.81 MN·s; 10.602 MN

63. 230 N

65. (a) $\vec{I} = (1.08\ \text{N·s})\hat{i}$ (directed into wall)

    (b) 360 N, into wall

    (c) 0.480 N·s, away from wall

    (d) 3.84 N, away from wall

67. (a) 20.0 m/s

    (b) 20% of the initial kinetic energy is transformed into thermal energy, sound, and the deformation of metal.

69. (a) −2.00 m/s

    (b) The collision was inelastic.

71. (a) $\vec{v}_{cm} = (23.1\ \text{m/s})\hat{i}$

    (b) −254 m/s

73. (a) 5.00 m/s

    (b) 0.250 m

    (c) $v_{1f} = 0; v_{2f} = 7.00\ \text{m/s}$

75. (a) $0.200v_0$

    (b) $0.400v_0$

77. 450 m/s

81. $h = \dfrac{v^2}{8g}\left(\dfrac{m_1}{m_2}\right)^2$

83. 0.0529

85. $1.50 \times 10^6$ m/s

87. (a) $\vec{v}_1 = (312\ \text{m/s})\hat{i} + (66.6\ \text{m/s})\hat{j}$

    (b) 5.61 km

    (c) 35.8 kJ

89. 0.913

91. (a) 20% of its mechanical energy is lost.

    (b) 0.894

93. (a) 1.70 m/s

    (b) 0.833

95. (a) 60°

    (b) 2.50 m/s; 4.33 m/s

97. (a) 1.00 m/s; 1.73 m/s

    (b) The collision was elastic.

99. $v_1 = 8.66\ \text{m/s}; v_2 = 5.00\ \text{m/s}$

101. In an elastic collision

$$K_i = K_f = \frac{p_1^2}{2}\left[\frac{m_1^2 + 6m_1 m_2 + m_2^2}{m_1^2 m_2 + m_1 m_2^2}\right] = \frac{p_1'^2}{2}\left[\frac{m_1^2 + 6m_1 m_2 + m_2^2}{m_1^2 m_2 + m_1 m_2^2}\right]$$

If $p_1' = +p_1$, the particles do not collide.

103. (a) $\vec{v}_{cm} = 0$

    (b) $\vec{u}_3 = (-5\ \text{m/s})\hat{i}; \vec{u}_5 = (3\ \text{m/s})\hat{i}$

    (c) $\vec{u}_3' = (5\ \text{m/s})\hat{i}; u_5' = 0.75\ \text{m/s}$

    (d) $\vec{v}_3' = (5\ \text{m/s})\hat{i}; \vec{v}_5' = (-3\ \text{m/s})\hat{i}$

    (e) 60.0 J; 60.0 J

105. (a) 360 kN

    (b) 120 s

    (c) 1.72 km/s

107. (a) $\tau_0 = 1 + \dfrac{a_0}{g}$

    (b) $v_f = gI_{sp}\left(\ln\dfrac{m_0}{m_f} - \dfrac{1}{\tau_0}\left(1 - \dfrac{m_f}{m_0}\right)\right)$

    (c)

    (d) 28.1

109. 0.192 m/s; 31.3 mJ; 12.0 mJ

111. 0.462 m/s

113. (a) $\vec{p} = -(1.10 \times 10^5\ \text{kg·km/h})\hat{i} + (1.05 \times 10^5\ \text{kg·km/h})\hat{j}$

    (b) 43.4 km/h; 46.3° west of north

115. (a) 6.26 m/s

    (b) 20.0 m

117. 3.72 m

119. (a) The velocity of the basketball will be equal in magnitude but opposite in direction to the velocity of the baseball.

    (b) $v_{1f} = 0$

    (c) $v_{2f} = 2v$

121. (a) 29.6 km/s

    (b) 8.10; The energy comes from an immeasurably small slowing of Saturn.

123. $3.00 \times 10^5$ m/s

125. (a) 0.600 m/s²

    (b) 960 N

127. No. The driver was traveling at 23.3 km/h.

129. 8.85 kg

131. $\frac{1}{14}r$

133. (a) $0.716^N E_0$

    (b) 55

135. (a) $y_{cm} = \dfrac{v^2}{2L}t^2$

   (b) $a_{cm} = \dfrac{v^2}{L}$

   (c) $F = \dfrac{vt}{L}\left(\dfrac{v}{gt} + 1\right)Mg$

137. $v_{2f} = \left(\dfrac{m_b}{m_2 + m_b}\right)\left[1 + \dfrac{m_1}{m_1 + m_b}\right]v;$

   $v_{1f} = -\dfrac{m_2 m_b(2m_1 + m_b)}{(m_1 + m_b)^2\,(m_2 + m_b)}v$

139. $-0.960$ m/s$^2$

141. $v = (1.70 \text{ m}^{1/2}/\text{s})\sqrt{L}$

# Chapter 9

1. Because $r$ is greater for the point on the rim, it moves the greater distance. Both turn through the same angle. Because $r$ is greater for the point on the rim, it has the greater speed. Both have the same angular velocity. Both have zero tangential acceleration. Both have zero angular acceleration. Because $r$ is greater for the point on the rim, it has the greater centripetal acceleration.

3. (c)

5. (d)

7. No. A net torque is required to *change* the rotational state of an object. A net torque may decrease the angular speed of an object. All we can say for sure is that a net torque will *change* the angular speed of an object.

9. (b)

11. (b)

13. For a given applied force, this increases the torque about the hinges of the door, which increases the door's angular acceleration, leading to the door being opened more quickly. It is clear that putting the knob far from the hinges means that the door can be opened with less effort (force). However, it also means that the hand on the knob must move through the greatest distance to open the door, so it may not be the quickest way to open the door. Also, if the knob were at the center of the door, you would have to walk around the door after opening it, assuming the door is opening toward you.

15. (b)

17. (b)

19. (a)

21. True. If the sphere is slipping, then there is kinetic friction that dissipates the mechanical energy of the sphere.

23. 10.3%

25. 6.42

27. (a) 15.6 rad/s

   (b) 46.8 rad

   (c) 7.45 rev

   (d) 73.0 m/s$^2$

29. (a) 40.0 rad/s

   (b) 0.960 m/s$^2$; 192 m/s$^2$

31. (a) 0.589 rad/s$^2$

   (b) 4.71 rad

33. (d)

35. 1.04 rad/s; 9.92 rev/min

37. (a) 1.87 N·m

(b) 124 rad/s$^2$

(c) 620 rad/s

39. (a) $g \sin\theta$

   (b) Because the line-of-action of the tension passes through the pendulum's pivot point, its lever arm is zero and it causes no torque.

   (c) $g \sin\theta$

41. (a) $d\tau_f = \dfrac{2\mu_k Mg}{R^2}r^2 dr$

   (b) $\tau_f = \frac{2}{3}MR\mu_k g$

   (c) $\Delta t = \dfrac{3R\omega}{4\mu_k g}$

43. 56.0 kg·m$^2$

45. (a) 28.0 kg·m$^2$

   (b) 28.0 kg·m$^2$

47. 2.60 kg·m$^2$

49. (b) $I_{cm} = \frac{1}{12}m(a^2 + b^2)$

51. $5.41 \times 10^{-47}$ kg·m$^2$

55. $I = \frac{3}{10}MR^2$

57. $I_x = 3M\left(\dfrac{H^2}{5} + \dfrac{R^2}{20}\right)$

59. (a) 84.6 mJ

   (b) 347 rev/min

63. (a) 19.6 kN

   (b) 5.89 kN·m

   (c) 0.267 rad/s

   (d) 1.57 kN

65. (a) 3.62 rad/s

   (b) 3.62 rad/s

67. Unless $M$, the mass of the ladder, is zero, $v_r > v_f$. It is better to let go and fall to the ground.

69. 3.11 m/s$^2$; $T_1 = 12.5$ N; $T_2 = 13.4$ N

71. 8.23 m/s

73. (a) $a = \dfrac{g}{1 + \dfrac{2M}{5m}}$

   (b) $T = \dfrac{2mMg}{5m + 2M}$

75. (a) 72.0 kg

   (b) 1.37 rad/s$^2$; $T_1 = 294$ N; $T_2 = 746$ N

77. (a) $a = \dfrac{g \sin\theta}{1 + \dfrac{m_1}{2m_2}}$

   (b) $T = \dfrac{\frac{1}{2}m_1 g \sin\theta}{1 + \dfrac{m_1}{2m_2}}$

   (c) $E = m_2 gh$

   (d) $E_{bottom} = m_2 gh$

   (e) $v = \sqrt{\dfrac{2gh}{1 + \dfrac{m_1}{2m_2}}}$

   (f) For $\theta = 0$: $a = T = 0$

   For $\theta = 90°$: $a = \dfrac{g}{1 + \dfrac{m_1}{2m_2}}$, $T = \frac{1}{2}m_1 a$, and $v = \sqrt{\dfrac{2gh}{1 + \dfrac{m_1}{2m_2}}}$

   For $m_1 = 0$: $a = g \sin\theta$, $T = 0$, and $v = \sqrt{2gh}$

79. $0.0864 \text{ m/s}^2$; $3.14 \text{ m/s}$

81. $0.192 \text{ m/s}^2$; $0.962 \text{ N}$

83. $1.13 \text{ kJ}$

85. $45.9 \text{ m}$

87. $19.5°$

89. (a) $a = \frac{2}{3}g \sin \theta$

    (b) $f_s = \frac{1}{3}mg \sin \theta$

    (c) $\theta_{max} = \tan^{-1}(3\mu_s)$

91. $v' = \sqrt{\frac{4}{3}v}$

93. $223 \text{ J}$

97. (a) $\alpha = \dfrac{2F}{R(M + 3m)}$; counterclockwise

    (b) $a_c = \dfrac{F}{M + 3m}$

    (c) $a_{CB} = -\dfrac{2F}{M + 3m}$

99. (a) $0.400 \text{ rad/s}^2$; $0.200 \text{ rad/s}^2$

    (b) $4.00 \text{ N}$

101. (a) $s_1 = \dfrac{12}{49}\dfrac{v_0^2}{\mu_k g}$, $t_1 = \dfrac{2}{7}\dfrac{v_0}{\mu_k g}$, and $v_1 = \dfrac{5}{7}v_0$

     (b) $5/7$

     (c) $26.6 \text{ m}$; $3.88 \text{ s}$; $5.71 \text{ m/s}$

103. $v = \dfrac{2r\omega_0}{7}$

105. (a) $360 \text{ kN}$

     (b) $120 \text{ s}$

     (c) $1.72 \text{ km/s}$

107. (a) $v = 1.57\,v_0$

     (b) $\Delta t = \dfrac{4}{7}\dfrac{v_0}{\mu_k g}$

     (c) $\Delta x = 0.735\dfrac{v_0^2}{\mu_k g}$

111. $I = 2mR^2$

113. $0.134 \text{ m}$

115. (a) $7.36 \text{ m/s}^2$

     (b) $14.7 \text{ m/s}^2$

     (c) $2.43 \text{ m/s}$

117. (a) $780 \text{ kJ}$

     (b) $90.3 \text{ N·m}$; $150 \text{ N}$

     (c) $1380 \text{ rev}$

119. (a) $15.0 \text{ m}$

     (b) $15.4 \text{ rad/s}$

121. (a) $S^2$

     (b) $S^3$

     (c) $S^5$

123. (a) $\omega = \sqrt{\dfrac{4g}{3r}}$

     (b) $F = \frac{7}{3}Mg$

125. (a) $v = \sqrt{\dfrac{2MgD \sin \theta}{M + \dfrac{I}{r^2}}}$

     (b) $f_s = \dfrac{Mg \sin \theta}{1 + \dfrac{R}{r}}$

127. (a) $14.7 \text{ m/s}^2$

     (b) $66.7 \text{ cm}$

129. $41.7 \text{ J}$

131. The solid line on the graph shown below shows the position $y$ of the bucket when it is in free fall and the dashed line shows $y$ under the conditions modeled in this problem.

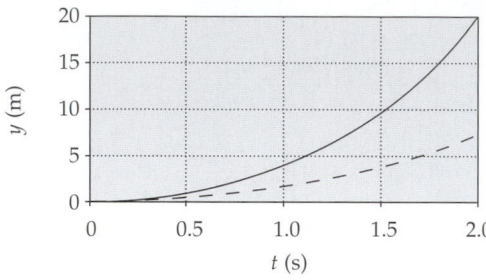

133. (a) $25.7 \text{ N}$

     (b) $3.21 \text{ kg}$

     (c) $1.10 \text{ m/s}^2$

# Chapter 10

1. (a) True

   (b) True

   (c) True

3. $90°$

5. (a) Doubling $\vec{p}$ doubles $\vec{L}$.

   (b) Doubling $\vec{r}$ doubles $\vec{L}$.

7. False

9. (e)

11. It is easier to crawl radially outward. In fact, a radially inward force is required just to prevent you from sliding outward.

13. The hardboiled egg is solid inside, so everything rotates with a uniform velocity. By contrast, it is difficult to get the viscous fluid inside a raw egg to start rotating; however, once it is rotating, stopping the shell will not stop the motion of the interior fluid, and, for this reason, the egg may start rotating again after momentarily stopping.

15. (b)

17. (b)

19. (a) The lifting of the nose of the plane rotates the angular momentum vector upward. It veers to the right in response to the torque associated with the lifting of the nose.

    (b) The angular momentum vector is rotated to the right when the plane turns to the right. In turning to the right, the torque points down. The nose will move downward.

21. (b)

23. The center of mass of the rod-and-putty system moves in a straight line, and the system rotates about its center of mass.

25. $4.17 \text{ rev/s}$

27. (a) $2.40 \times 10^{-8} \text{ kg·m}^2\text{/s}$

    (b) $5.22 \times 10^{52}$; $2.29 \times 10^{26}$

    (c) The quantization of angular momentum is not noticed in macroscopic physics because no experiment can differentiate between $\ell = 2 \times 10^{26}$ and $\ell = 2 \times 10^{26} + 1$.

29. (a) 0.331

   (b) Because experimentally $C < 2/5 = 0.4$, the mass density must be greater near the center of the earth.

31. 10.1 rad/s

33. $\vec{\tau} = FR\hat{k}$

35. (a) $24\hat{k}$

   (b) $-24\hat{j}$

   (c) $-5\hat{k}$

39. $\vec{B} = 4\hat{j} + 3\hat{k}$

45. (a) 54.0 kg·m²/s

   (b) $\omega$ increases as the particle approaches the point and decreases as it recedes.

47. (a) $1.33 \times 10^{-5}$ kg·m²/s

   (b) $1.33 \times 10^{-5}$ kg·m²/s

   (c) $1.33 \times 10^{-5}$ kg·m²/s

   (d) $8.83 \times 10^{-5}$ kg·m²/s; $-6.17 \times 10^{-5}$ kg·m²/s

49. (a) 4.00 N·m

   (b) $(0.192 \text{ rad/s}^2)t$

51. (a) $\tau_{\text{net}} = Rg(m_2 \sin \theta - m_1)$

   (b) $L = vR\left(\dfrac{I}{R^2} + m_1 + m_2\right)$

   (c) $a = \dfrac{g(m_2 \sin \theta - m_1)}{\dfrac{I}{R_2} + m_1 + m_2}$

55. (a) 5.00 rev/s

   (b) 622 J

   (c) The energy comes from your internal energy.

57. 9.67 mm/s

59. (a) $L_0 = r_0 m v_0$

   (b) $K_0 = \frac{1}{2}mv_0^2$

   (c) $T = F_c = m\dfrac{v_0^2}{r_0}$; $W = -\dfrac{3}{2}mv_0^2$

61. 54.7°

63. (a) $3.46 \times 10^{-47}$ kg·m²

   (b) 1.99 meV; 5.98 meV; 12.0 meV

65. 82.5 m/s

67. $v_{\text{cm}} = \dfrac{mv}{M + m}$; $\omega = \dfrac{mMvd}{\frac{1}{12}ML^2(M + m) + Mmd^2}$

69. $v = \sqrt{\dfrac{(0.5\,M + 0.8m)(\frac{1}{3}Md^2 + 0.64md^2)g}{0.32dm^2}}$

71. (a) $v_{\text{cm}} = \dfrac{K}{M}$

   (b) $\dfrac{4K}{M}$

   (c) $-\dfrac{2K}{M}$

   (d) $x = \frac{1}{6}\ell$

73. 0.349

75. 12 rad/s; 10.8 J

77. (a) 18.1 J·s

   (b) 0.414 rad/s

   (c) 15.2 s

   (d) 0.0791 J·s

79. (a) $\vec{L} = -(47.7 \text{ kg·m}^2/\text{s})\hat{k}$

   (b) $\vec{\tau} = (15.9 \text{ N·m})\hat{k}$

81. (a) 243 J·s

   (b) 306 J

83. (a) No, $L$ decreases.

   (b) Its kinetic energy is constant.

   (c) $v_0$ (The kinetic energy remains constant.)

85. Yes.

87. $v_r = \dfrac{\ell\omega}{2L}\sqrt{(L^2 - \ell^2)}$

91. (a) 0.228 rad/s

   (b) 0.192 rad/s

93. $4.47 \times 10^{22}$ N·m

95. 12.5 rad/s

97. (a) 26.5 rad/s

   (b) $\vec{L} = (0.303 \text{ kg·m}^2/\text{s})e^{(1.41\text{s}^{-1})t}$

# Chapter R

1. The friend in the car.

3. Yes. If two events occur at the same time *and* place in one reference frame they occur at the same time *and* place in all reference frames. (Any pair of events that occur at the same time *and* at the same place in one reference frame are called a space-time coincidence.)

5. (a)

7. $1 + (8.61 \times 10^{-11})$

9. 6.00 ns

11. (a) 6.63 m

   (b) 12.6 m

13. (a) 599 m

   (b) 13.4 km

15. (a) 129 y

   (b) 87.6 y

17. (a) 0.600 m

   (b) 2.50 ns

19. 0.800c

21. (a) $4.50 \times 10^{-10}\%$

   (b) 0.142 ms

23. 25.0 min; 25.0 min

25. 60.0 min

27. 0.400c; event $B$ can precede event $A$ provided $v > 0.400c$

29. (a) 11.3 y

   (b) 40.0 y

31. (a) 1.005

   (b) 1.155

   (c) 1.667

   (d) 7.089

33. (a) $0.155E_0$

   (b) $1.29E_0$

   (c) $6.09E_0$

35. 2.97 GeV

39. (b) 0.866c

   (c) 0.999c

41.  (a)   0.794%

(b)   68.7%

43.  (a)   0.943

45.  (a)   617 eV

(b)   79.6 eV

(c)   7.96 eV

47.  (a)   0.745

(b)   5.00 ft

(c)   No. In Keisha's rest frame, the back end of the ladder will clear the door before the front end hits the wall of the shed, while in Ernie's rest frame, the front end will hit the wall of the shed while the back end has yet to reach the door.

# Chapter 11

1.  (a)   False

(b)   True

3.  (d)

5.  (a)

7.   The gravitational field is proportional to the mass within the sphere of radius $r$ *and* inversely proportional to the square of $r$, i.e., proportional to $r^3/r^2 = r$.

9.  (d)

11.  $1.08 \times 10^{11} M_s$

13.  (a)   2.78 h

(b)   $19.3 \times 10^{42}$ kg·m$^2$/s; $7.85 \times 10^{42}$ kg·m$^2$/s; 0.703%

(c)   $4.80 \times 10^{-4}$ rad/s

15.  84.0 y

17.  (a)   $1.59 \times 10^{11}$ m

(b)   $2.71 \times 10^{10}$ m; $2.91 \times 10^{11}$ m

19.  (a)   90°

(b)   0.731 AU

21.  (a)   $1.90 \times 10^{27}$ kg

(b)   0.282 m/s$^2$; 0.0356 m/s$^2$

23.  (a)   $8.18 \times 10^4$ s

(b)   $1.22 \times 10^9$ m

25.  $1.99 \times 10^{30}$ kg

27.  $10w$, where $w$ is your weight on earth.

29.  $2.27 \times 10^4$ m/s

31.  1.43

33.  (a)   7.37 m

(b)   0.0319 mm

35.  0.605

37.  (a)   2.27 kg

(b)   It is the *inertial* mass of $m_2$.

39.  $10^9$ m

41.  $W = \dfrac{GMm_0}{R}$

43.  6.94 km/s

45.  (a)   $\vec{F}_{\text{outside}} = -\dfrac{GMm_0}{r^2}\,\hat{r}$

(b)   $U(r) = -\dfrac{GMm_0}{r}$; $U(R) = -\dfrac{GMm_0}{R}$

(d)   $U(r) = U(R) = -\dfrac{GMm_0}{R}$

(e)

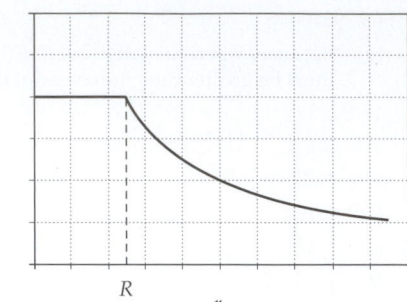

47.   2.38 km/s

49.   19.4 km/s

51.  (a)   62.7 MJ

(b)   17.4 kW·h

(c)   $139

53.  (a)   7.31 h

(b)   1.04 GJ

(c)   $8.72 \times 10^{12}$ J·s

55.   11.1 GJ

57.   $\vec{g} = (4\text{ N/kg})\hat{i}$

59.  (a)   $\vec{g} = \dfrac{Gm}{L^2}\hat{i} + \dfrac{Gm}{L^2}\hat{j}$

(b)   $g = \sqrt{2}\,\dfrac{Gm}{L^2}$

61.  (a)   $\vec{g} = (-1.67 \times 10^{-11}\text{ N/kg})\hat{i}$

(b)   $\vec{g} = (-8.34 \times 10^{-12}\text{ N/kg})\hat{i}$

(c)   2.48 m

63.  (a)   $M = \frac{1}{2}CL^2$

(b)   $\vec{g} = \dfrac{2GM}{L^2}\left[\ln\left(\dfrac{x_0}{x_0 - L}\right) - \left(\dfrac{L}{x_0 - L}\right)\right]\hat{i}$

65.  (a)   0

(b)   0

(c)   $3.20 \times 10^{-9}$ N/kg

67.   $g_1 = g_2$

69.  (a)   $F = \dfrac{Gm(M_1 + M_2)}{9a^2}$

(b)   $F = \dfrac{GmM_1}{3.61a^2}$

(c)   0

71.  (a)   $F_g = \dfrac{mg}{R}r$

(b)   $F_N = \left(\dfrac{mg}{R} - m\omega^2\right)r$

(c)   The change in mass between you and the center of the earth as you move away from the center is more important than the rotational effect.

73.   $g(x) = G\left(\dfrac{4\pi\rho_0 R^3}{3}\right)\left[\dfrac{1}{x^2} - \dfrac{1}{8(x - \frac{1}{2}R)^2}\right]$

75.   $\omega = \sqrt{\dfrac{4\pi\rho_0 G}{3}}$

77.   0.104 mm/s

79.  (a)   $\vec{F} = -\dfrac{GMm}{d^2}\left[1 - \dfrac{\dfrac{d^3}{4}}{\left\{d^2 + \dfrac{R^2}{4}\right\}^{3/2}}\right]\hat{i}$

(b)   $\vec{F}(R) = -0.821\dfrac{GMm}{R^2}\hat{i}$

81. 249 y

83. (a) $W = GM_Em\left(\dfrac{1}{r_1} - \dfrac{1}{r_2}\right)$

    (b) $W = mgR_E^2\left(\dfrac{1}{R_E} - \dfrac{1}{R_E + h}\right)$

85. $8.96 \times 10^7$ m

87. 1.70 Mm

91. $v = 1.64\sqrt{\dfrac{GM}{a}}$

93. For $r < R_1$, $g(r) = 0$; For $r > R_2$, $g(r) = \dfrac{GM}{r^2}$;

    For $R_1 < r < R_2$, $g(r) = \dfrac{GM(r^3 - R_1^3)}{r^2(R_2^3 - R_1^3)}$

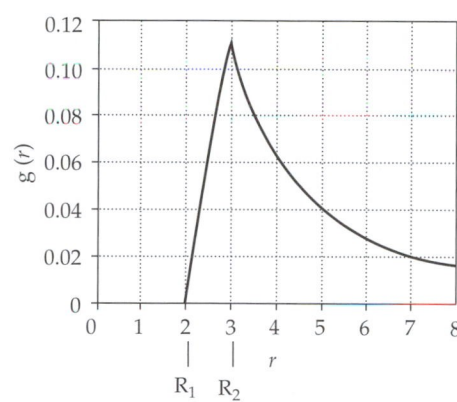

95. $g = \dfrac{2G\lambda}{r}$

97. (b) $U = -\dfrac{GMm_0}{L}\ln\left(\dfrac{x_0 + L/2}{x_0 - L/2}\right)$

99. 33.5 pN

101. (a) The gravitational force is greater on the lower robot, so if it were not for the cable its acceleration would be greater than that of the upper robot, and they would separate. In opposing this separation the cable is stressed.

    (b) 220 km

# Chapter 12

1. (a) False

   (b) True

   (c) True

   (d) False

3. No. The definition of the center of gravity does not require that there be any material at its location.

5. This technique works because the center of mass must be directly under the balance point. Hence the intersection of the two lines must be at the center of mass.

7. (b)

9. (c)

11. The tensile strengths of stone and concrete are at least an order of magnitude lower than their compressive strengths, so you want to build compressive structures to match their properties.

13. (b) 200 N/m

15. 318 N

17. (b) Taking long strides requires a larger coefficient of static friction because $\theta$ is then large.

    (c) If $\mu_s$ is small, that is, there is ice on the surface, $\theta$ must be small to avoid slipping.

19.

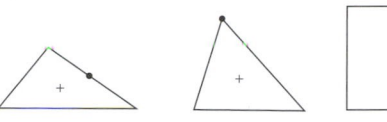
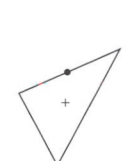

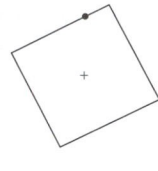

21. $(x_{cg}, y_{cg}) = \left(\dfrac{\frac{1}{2}a^2b - \pi aR^2 + \pi R^3}{ab - \pi R^2}, \dfrac{1}{2}b\right)$

23. 692 N; 90°; 2.54 kN; No block is required to prevent the mast from moving.

25. 0.728 m

27. $F_2 = \dfrac{1}{2}W$; $F_1 = \dfrac{\sqrt{3}}{2}W$

29. (a) 5.00 m

    (b) 4.87 m

31. $\vec{F}_1 = \dfrac{Mg\sqrt{h(2R - h)}}{h - R}\hat{i} + Mg\hat{j}$

33. (a) $\vec{F} = (30.0\text{ N})\hat{i} + (30.0\text{ N})\hat{j}$

    (b) $\vec{F} = (35.0\text{ N})\hat{i} + (45.0\text{ N})\hat{j}$

35. (a) $F_n = Mg - F\sqrt{\dfrac{2R - h}{h}}$

    (b) $F_{c,h} = F$

    (c) $F_{c,v} = F\sqrt{\dfrac{2R - h}{h}}$

37. (a) 6.87 N

    (b) 1.65 N·m

    (c) −8.26 N; 15.1 N

39. 636 N; 21.5°

41. (a) 70.7 N

    (b) 1.77 m

    (c) 3.54 m

    (d) 497 N

43. $\tau_{net} = (69.3\text{ N})b - (40.0\text{ N})a$

45. $D = \frac{1}{2}(\sqrt{3}b - a)$

47. $y = \dfrac{35.7\text{ m} - 30.4x}{3.57\text{ m} - (294\text{m}^{-1})x}$

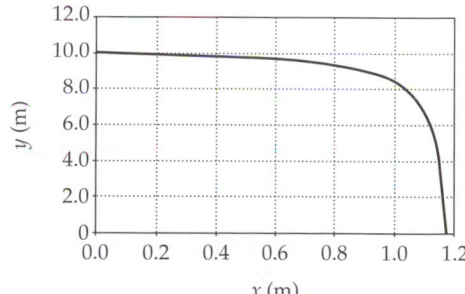

49. $h = \mu_s L \tan\theta \sin\theta$

51. $\mu_s = \dfrac{2h}{L \tan\theta \sin\theta}$

53. 59.0°

55. (a) 41.6 N

    (b) 0.136%

57. 5.01°

61. (a) $1.82 \times 10^6$ N/m²

    (b) 6.62 mJ

63. 0.686

65. It will not support the elevator.

69. $F_L = 117$ N; $F_R = 333$ N

71. $w_1 = 1.50$ N; $w_2 = 7.00$ N; $w_3 = 3.50$ N

73. 0.148

75. $\mu_s < 0.500$

77. $\mu_s = \frac{1}{2}(\cot\theta - 1)$

79. (a) 147·N

    (b) 3.62 m

81. The block will tip before it slides.

83. $\mu_s < 0.500$

85. (a) The stick remains balanced as long as the center of mass is between the two fingers. For a balanced stick the normal force exerted by the finger nearest the center of mass is greater than that exerted by the other finger. Consequently, a larger static-frictional force can be exerted by the finger closer to the center of mass, which means the slipping occurs at the other finger.

    (b) The finger farthest from the center of mass will slide inward until the normal force it exerts on the stick is sufficiently large to produce a kinetic-frictional force exceeding the maximum static-frictional force exerted by the other finger. At that point the finger that was not sliding begins to slide, the finger that was sliding stops sliding, and the process is reversed. When one finger is slipping the other is not.

87. (a) 23.0 m/s

    (b) 29.1 m/s

89. (c) $\ell_5 = 1.142$ m; $\ell_{10} = 1.464$ m; $\ell_{100} = 2.594$ m

    (d) Increasing $N$ in the spreadsheet solution suggests that the sum of the individual offsets continues to grow as $N$ increases without bound. The series is, in fact, divergent and the stack of bricks has no maximum offset or length.

91. 566 N

93. $F_n = 2mg; F = \dfrac{mg}{\cos\theta}; F_W = mg\dfrac{R-r}{\sqrt{R(2r-R)}}$

# Chapter 13

1. (e)

3. (d)

5. Nothing. The fish is in neutral buoyancy, so the upward acceleration of the fish is balanced by the downward acceleration of the displaced water.

7. (b)

9. It blows over the ball, reducing the pressure above the ball to below atmospheric pressure.

11. False

13. The buoyant force acting on the ice cubes equals the weight of the water they displace (i.e., $B = w_f = \rho_f V_f g$). When the ice melts, the volume of water displaced by the ice cubes will occupy the space previously occupied by the submerged part of the ice cubes. Therefore the water level remains constant.

15. Because the pressure increases with depth, the object will be compressed and its density will increase. Thus it will sink to the bottom.

17. The drawing shows the beaker and a strip within the water. As is readily established by a simple demonstration, the surface of the water is not level while the beaker is accelerated, showing that there is a pressure gradient. That pressure gradient results in a net force on the small element shown in the figure.

19. From Bernoulli's principle, the opening above which the air flows faster will be at a lower pressure than the other one, which will cause a circulation of air in the tunnel from opening 1 toward opening 2. It has been shown that enough air will circulate inside the tunnel even with the slightest breeze outside.

21. 0.673 kg

23. 103 kg

25. 29.8 inHg

27. 230 N

29. 198 atm

31. (a) 14.8 kN

    (b) 0.339 kg

33. 0.453 m

35. $F = \dfrac{\rho g a^3}{8}$

37. 4.36 N

39. (a) $11.1 \times 10^3$ kg/m³

    (b) lead

41. 800 kg/m³; 1.11

43. 250 kg/m³

45. 3.89 kg

47. $2.46 \times 10^7$ kg

49. 491 kN

51. (a) 9.28 cm/s

    (b) 0.331 cm

    (c) 8.31 cm, in reasonable agreement with everyday experience.

53. (a) 12.0 m/s

    (b) 133 kPa

    (c) The volume flow rates are equal.

55. (a) 4.58 L/min

    (b) 763 cm²

57. 144 kPa

59. (a) 21.2 kg/s

    (b) 636 kg·m/s

    (c) 899 kg·m/s; 899 N

61. (a) $x = 2\sqrt{h(H-h)}$

    (b) $h = \frac{1}{2}H \pm \frac{1}{2}\sqrt{H^2 - x^2}$

63. (b) $P_{top} = P_{atm} - \rho g d$

65. 1.43 mm

67. 93.4 mi/h; Since most major league pitchers can throw a fastball in the low-to-mid-90s, this drag crisis may very well play a role in the game.

69. 0.0137; 0.0115

71. The net force is zero. Neglecting the thickness of the table, the atmospheric pressure is the same above and below the surface of the table.

73. $1061 \text{ kg/m}^3$

75. 65.7%

77. If you are floating, the density (or specific gravity) of the liquid in which you are floating is immaterial as you are in translational equilibrium under the influence of your weight and the buoyant force on your body. Thus the buoyant force on your body is your weight in both (a) and (b).

79. $V = \dfrac{m}{0.96\rho_{\mathrm{W}}}$

81. 11.8 cm

83. 1 m is a reasonable diameter for the pipeline.

85. $h_{\mathrm{A}} = 12.6 \text{ m}; h_{\mathrm{B}} = 9.78 \text{ m}$

87. (a) 64.6%

    (b) 10.7 kN

    (c) $17.9 \text{ m/s}^2$

89. $3.31 \times 10^{-3} \text{ mmHg}$ or $3.31 \ \mu\text{mHg}$

91. 1.37

93. (a) $70.0 \text{ m}^3$

    (b) $7.47 \text{ m/s}^2$

95. (c) $0.126 \text{ km}^{-1}$

97. (a) 33.9 kN

    (b) 39.8 kN; 36.1 kN

99. (c) $h = \left( \sqrt{H} - \dfrac{A_2}{2A_1}\sqrt{2g}\, t \right)^2$

    (d) 1 h 46 min

# Chapter 14

1. $0; 4\pi^2 f^2 A$

3. (a) False

   (b) True

   (c) True

5. (a)

7. False

9. Assume that the first cart is given an initial velocity $v$ by the blow. After the initial blow, there are no external forces acting on the carts, so their center of mass moves at a constant velocity $v/2$. The two carts will oscillate about their center of mass in simple harmonic motion where the amplitude of their velocity is $v/2$. Therefore, when one cart has velocity $v/2$ with respect to the center of mass, the other will have velocity $-v/2$. So, the velocity with respect to the laboratory frame of reference will be $+v$ and 0, respectively. Half a period later, the situation is reversed; so, one will appear to move as the other stops, and vice-versa.

11. True

13. Examples of driven oscillators include the pendulum of a clock, a bowed violin string, and the membrane of any loudspeaker.

15. Because $f'$ varies inversely with the square root of $m$, taking into account the effective mass of the spring predicts that the frequency will be reduced.

17. (d)

19. (b)

21. $8\pi$

23. (a) 3.00 Hz

   (b) 0.333 s

   (c) 7.00 cm

   (d) 0.0833 s; Because $v < 0$, the particle is moving in the negative direction at $t = 0.0833$ s.

25. (a) $x = (25\text{ cm})\cos\left[(4.19\text{ s}^{-1})t\right]$

   (b) $v = -(105\text{ cm/s})\sin\left[(4.19\text{ s}^{-1})t\right]$

   (c) $a = -(439\text{ cm/s}^2)\cos\left[(4.19\text{ s}^{-1})t\right]$

27. (a) $x = (27.7\text{ cm})\cos\left[(4.19\text{ s}^{-1})t - 0.445\right]$

   (b) $v = -(116\text{ cm/s})\sin\left[(4.19\text{ s}^{-1})t - 0.445\right]$

   (c) $a = -(486\text{ cm/s}^2)\cos\left[(4.19\text{ s}^{-1})t - 0.445\right]$

29. (a)

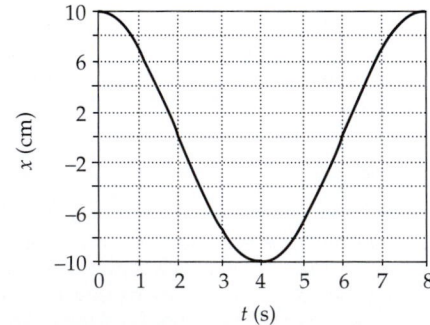

   (b)

| $t_f$ | $t_i$ | $\Delta x$ |
|---|---|---|
| (s) | (s) | (cm) |
| 1 | 0 | 2.93 |
| 2 | 1 | 7.07 |
| 3 | 2 | 7.07 |
| 4 | 3 | 2.93 |

31. (a) 7.85 m/s; 24.7 m/s$^2$

   (b) $-6.28$ m/s; $-14.8$ m/s$^2$

33. (a) 0.313 Hz

   (b) 3.14 s

   (c) $x = (40\text{ cm})\cos\left[(2\text{ s}^{-1})t + \delta\right]$

35. 22.5 J

37. (a) 0.368 J

   (b) 3.84 cm

39. 1.38 kN/m

41. (a) 6.89 Hz

   (b) 0.145 s

   (c) 0.100 m

   (d) 4.33 m/s

   (e) 187 m/s$^2$

   (f) 36.3 ms; 0

43. (a) 682 N/m

   (b) 0.417 s

   (c) 1.51 m/s

   (d) 22.7 m/s$^2$

45. (a) 3.08 kN/m

   (b) 4.16 Hz

   (c) 0.240 s

47. (a) 0.438 m/s

   (b) 0.379 m/s; 120 m/s$^2$

   (c) 95.5 ms

49. 0.262 s

51. 10.1 kJ

53. (a) 0.997 Hz

   (b) 0.502 s

   (c) 0.294 N

55. (a) 46.66 cm

   (b) 0.261 s

   (c) 0.767 m/s

57. (a) 0.270 J

   (b) $-0.736$ J

   (c) 1.01 J

   (d) 0.270 J

59. (a) 1.90 cm

   (b) 0.0542 J

   (c) $\pm 0.224$ J

   (d) 0.334 J

61. 12.2 s

63. 11.7 s

65. $T = 2\pi\sqrt{\dfrac{L}{g(1 - \sin\theta)}}$

67. 1.10 s

69. 0.504 kg·m$^2$

71. (b) 3.17 s

73. 21.1 cm from the center of the meter stick

77. (a) 1.63572 m

   (b) 14.5 mm, upward

79. 13.5°

81. 3.14%

85. (a) 0.314

(b) $-3.13 \times 10^{-2}$ percent

87. (a) 1.57%

(c) $0.430E_0$

89. (a) 1.01 Hz

(b) 2.01 Hz

(c) 0.352 Hz

91. (a) 4.98 cm

(b) 14.1 rad/s

(c) 35.4 cm

(d) 1.00 rad/s

93. (a) 0

(b) 4.00 m/s

95. (a) 14.1 cm; 0.444 s

(b) 23.1 cm; 0.363 s

(c) $(14.1 \text{ cm})\sin[(14.1 \text{ s}^{-1})t]; (23.1 \text{ cm})\sin[(17.3 \text{ s}^{-1})t]$

97. (a) $v = -(1.2 \text{ m/s})\sin\left[(3 \text{ rad/s})t + \dfrac{\pi}{4}\right]$

(b) $-0.849$ m/s

(c) 1.20 m/s

(d) 1.31 s

99. (a) The normal force is identical to the tension in a string of length $r$ that keeps the particle moving in a circular path and a component of $mg$ provides, for small displacements $\theta_0$ or $s_2$, the linear restoring force required for oscillatory motion.

(b) The particles meet at the bottom. Because $s_1$ and $s_2$ are both much smaller than $r$, the particles behave like the bobs of simple pendulums of equal length and, therefore, have the same periods.

101. 1.62 s

103. $3.86 \times 10^{-7}$ N·m/rad

105. $g'$ is closer to $g$ than is $g''$. Thus the error is greater if the clock is elevated.

107. (a) $\mu_s = \dfrac{Ak}{(m_1 + m_2)g}$

(b) $A$ is unchanged. $E$ is unchanged since $E = \frac{1}{2}kA^2$. $\omega$ is reduced by increasing the total mass of the system and $T$ is increased.

109. (b) 2.04 cm/s²

113. (a) $x = 0$

(b) $v_s = x_0\sqrt{\dfrac{k}{m_b + m_p}}$

(c) $x_f = x_0\sqrt{\dfrac{m_p}{m_b + m_p}}$

115. (a)

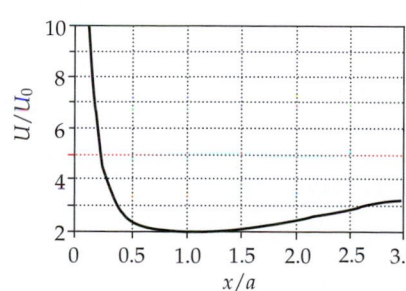

(b) $x_0 = a$ or $\alpha_0 = 1$

(c) $U(x_0 + \varepsilon) = U_0[1 + \beta + (1 + \beta)^{-1}]$

(d) $U(x_0 + \varepsilon) = \text{constant} + U_0\dfrac{\varepsilon^2}{a^2}$

119. $6.44 \times 10^{13}$ rad/s

121. $7.78\sqrt{\dfrac{R}{g}}$

123. (a) 0.0478

(b) 0.00228

127. (a)

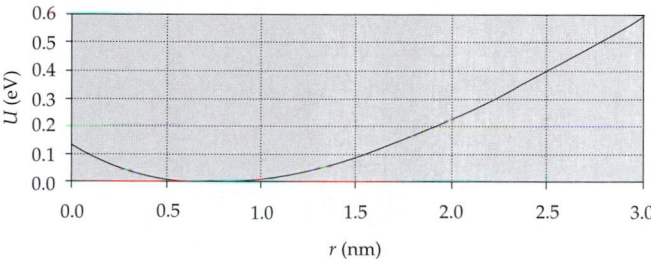

(b) $r = r_0; k = 2\beta^2 D$

(c) $\omega = 2\beta\sqrt{\dfrac{D}{m}}$

# Chapter 15

1. The speed of a transverse wave on a rope is given by $v = \sqrt{F/\mu}$ where $F$ is the tension in the rope and $\mu$ is its linear density. The waves on the rope move faster as they move up because the tension increases due to the weight of the rope below.

3. True

5. The speed of the wave $v$ on the bullwhip varies with the tension $F$ in the whip and its linear density $\mu$ according to $v = \sqrt{F/\mu}$. As the whip tapers, the wave speed in the tapered end increases due to the decrease in the mass density, so the wave travels faster.

7. No; Because the source and receiver are at rest relative to each other, there is no relative motion of the source and receiver and there will be no Doppler shift in frequency.

9. The light from the companion star will be shifted about its mean frequency periodically due to the relative approach to and recession from the earth of the companion star as it revolves about the black hole.

11. (a) True

(b) False

(c) False

13. There was only one explosion. Sound travels faster in water than air. Abel heard the sound wave in the water first, then, surfacing, heard the sound wave traveling through the air, which took longer to reach him.

15.

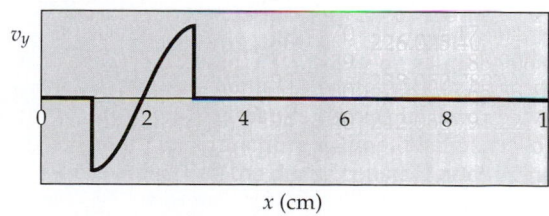

17. Path C. Because the wave speed is highest in the water, and more of path C is underwater than A or B, the sound wave will spend the least time on path C.

19. (a) 78.5 m

    (b) 69.7 m

    (c) 70.5 m . . . about 1% larger than our result in part (b) and 11% smaller than our first approximation in (a).

21. 270 m/s; 20.6%

23. 1.32 km/s

25. 19.6 g

27. (a) 265 m/s

    (b) 15.0 g

29. (b) 40.0 N

33. The lightning struck 680 m from the ball park, 58.4° W (or E) of north.

39. (a) $y(x,t) = A \sin k(x - vt)$

    (b) $y(x,t) = A \sin 2\pi\left(\dfrac{x}{\lambda} - ft\right)$

    (c) $y(x,t) = A \sin 2\pi\left(\dfrac{x}{\lambda} - \dfrac{1}{T}t\right)$

    (d) $y(x,t) = A \sin\dfrac{2\pi}{\lambda}(x - vt)$

    (e) $y(x,t) = A \sin 2\pi f\left(\dfrac{x}{v} - t\right)$

41. 9.87 W

43. (a) The wave is traveling in the $-x$ direction.; 5.00 m/s

    (b) 10.0 cm; 50.0 Hz; 0.0200 s

    (c) 0.314 m/s

45. (a) 6.82 J

    (b) 44.0 W

47. (a) 79.0 mW

    (b) Increasing $f$ by a factor of 10 would increase $P_{av}$ by a factor of 100. Increasing $A$ by a factor of 10 would increase $P_{av}$ by a factor of 100. Increasing $F$ by a factor of $10^4$ would increase $v$ by a factor of 100 and $P_{av}$ by a factor of 100.

    (c) Depending on the adjustability of the power source, increasing $f$ or $A$ would be the easiest.

49. (a) 0.750 Pa

    (b) 4.00 m

    (c) 85.0 Hz

    (d) 340 m/s

51. (a) $3.68 \times 10^{-5}$ m

    (b) $8.27 \times 10^{-2}$ Pa

53. (a) The displacement $s$ is zero.

    (b) 3.68 $\mu$m

55. (a) 138 Pa

    (b) 21.7 W/m²

    (c) 0.217 W

57. (a) 50.3 W

    (b) 2.00 m

    (c) $4.45 \times 10^{-3}$ W/m²

59. (a) 20.0 dB

    (b) 100 dB

61. 90.0 dB

65. (a) 100 m

    (b) 0.126 W

67. (a) 100 dB

    (b) 50.3 W

    (c) 2.00 m

    (d) 96.5 dB

69. (a) 81.1 dB

    (b) 80.0 dB; Eliminating the two least intense sources does not reduce the intensity level significantly.

71. 87.8 dB

73. 57.0 dB

75. (a) 260 m/s

    (b) 1.30 m

    (c) 262 Hz

77. (a) 1.70 m

    (b) 247 Hz

79. 153 Hz

81. 1021 Hz or a fraction increase of 2.06%; Because this fractional change in frequency is less than the 3% criterion for recognition of a change in frequency, it would be *impossible* to use your sense of pitch to estimate your running speed.

83. 349 mi/h

85. 7.78 kHz

87. 15.0 km west of $P$

89. (a) $f' = (1 - u_r/v)(1 - u_s/v)^{-1} f_0$

91. 1.33 m/s

93. (a) 824 Hz

    (b) 849 Hz

95. 184 m

97. $-2.07 \times 10^{-5}$ nm; 99 $2.25 \times 10^8$ m/s

99. $2.25 \times 10^{\wedge}8$ m/s . . . where the upper arrow means the 8 is an exponent.

101. 20.8 cm

103. 3.42 m/s

105. 529 Hz; 474 Hz

107. 7.99 m

109. (a) 55.1 N/m²

    (b) 3.46 W/m²

    (c) 0.109 W

111. 77.0 kN

113. 204 m

115. 24.0 cm

117. (b) $v_0 = \sqrt{\dfrac{F}{\mu}}$

    (c) As seen by an observer at rest, the pulse remains at the same position because its speed along the chain is the same as the speed of the chain. With respect to a fixed point on the chain, the pulse travels through 360°.

119. (b) 2.21 s

# Chapter 16

1.

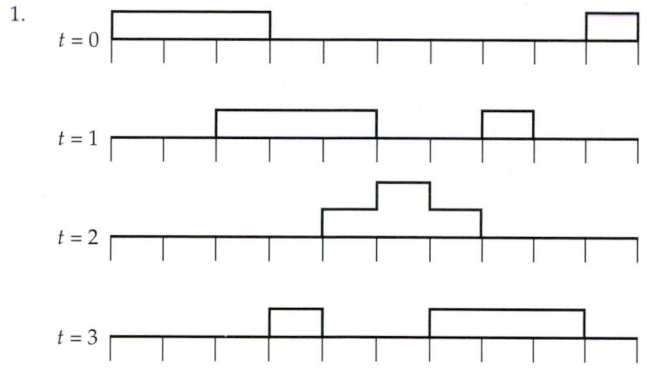

3.   (c)

5.   (b)

7.   (a)

9.   since $v \propto T$, increasing the temperature increases resonant frequencies.

11.   No; the wavelength of a wave is related to its frequency and speed of propagation ($\lambda = v/f$). The frequency of the plucked string will be the same as the wave it produces in air, but the speeds of the waves depend on the media in which they are propagating. Since the velocities of propagation differ, the wavelengths will not be the same.

13.   When the edges of the glass vibrate, sound waves are produced in the air in the glass. The resonance frequency of the air columns depends on the length of the air column, which depends on how much water is in the glass.

15.   (b)

17.   The pitch is determined mostly by the resonant cavity of the mouth, and the frequency of sounds he makes is directly proportional to their speed. Since $v_{He} > v_{air}$ (see Equation 15-5), the resonance frequency is higher if helium is the gas in the cavity.

19.   Pianos are tuned by ringing the tuning fork and the piano note simultaneously and tuning the piano string until the beats are far apart (i.e., the time between beats is very long). If we assume that 2 s is the maximum detectable period for the beats, then one should be able to tune the piano string to at least 0.5 Hz.

21.   34.0 Hz; Because $v \propto T$, the frequency will be somewhat higher in the summer.

23.   7.07 cm

25.   (a)   90.0°

   (b)   $\sqrt{2}A$

27.   (a)   0

   (b)   $2I_0$

   (c)   $4I_0$

29.   (a)   $\frac{1}{4}\lambda$

   (b)   $\frac{1}{4}\lambda$

31.   (a)   60.0 cm

   (b)   $\dfrac{2\pi}{5}$

   (c)   24.0 m/s

33.   4726 Hz; 9452 Hz

35.   (b)

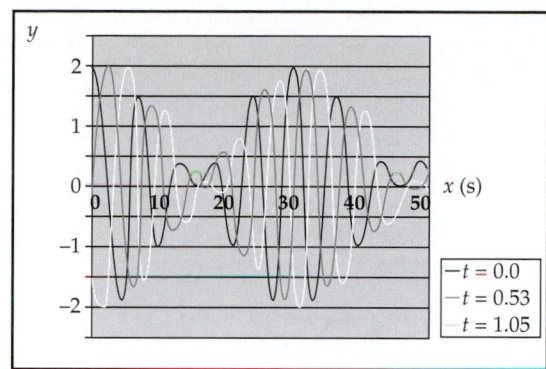

   (c)   0.500 m/s

37.   1.81; 51.5°

39.   (a)   0.279 m

   (b)   1.22 kHz

   (c)

| $m$ | $\theta_m$ |
|---|---|
|  | (rad) |
| 3 | 0.432 |
| 4 | 0.592 |
| 5 | 0.772 |
| 6 | 0.992 |
| 7 | 1.354 |
| 8 | undefined |

   (d)   0.0698 rad

41.   1.98 rad or 113°

43.   (a)   70.5 Hz

   (b)   The person on the street hears no beat frequency as the sirens of both ambulances are Doppler shifted up by the same amount (approximately 35 Hz).

45.   (a)   2.00 m; 25.0 Hz

   (b)   $y_3(x,t) = (4 \text{ mm})\sin kx \cos \omega t$, where $k = \pi\text{m}^{-1}$ and $\omega = 50\pi\text{s}^{-1}$.

47.   (a)   521 m/s

   (b)   2.80 m; 186 Hz

   (c)   372 Hz; 558 Hz

49.   141 Hz

51.   (a)   31.4 cm; 47.7 Hz

   (b)   15.0 m/s

   (c)   62.8 cm

53. (a)

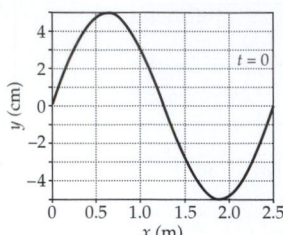

(b) 12.6 ms

(c) Since the string is moving either upward or downward when $y(x) = 0$ for all $x$, the energy of the wave is entirely kinetic energy.

55. (a) 70.8 Hz

(b) 4.89 Hz

(c) 35

57. 452 Hz; It would be better to have the pipe expand so that $v/L$, where $L$ is the length of the pipe, is independent of temperature.

59. (a) 80 cm

(b) 480 N

(c) You should place your finger 9.23 cm from the scroll bridge.

61. (a) 75.0 Hz

(b) The harmonics are the 5th and 6th.

(c) 2.00 m

63. (a) 0.574 g/m

(b) 1.29 g/m; 2.91 g/m; 6.55 g/m

65. (a) The two sounds produce a beat because the third harmonic of the A string equals the second harmonic of the E string, and the original frequency of the E string is slightly greater than 660 Hz. If $f_E = (660 + \Delta f)$Hz, a beat of $2\Delta f$ will be heard.

(b) 661.5 Hz

(c) 79.6 N

69. 76.8 N; 19.2 N; 8.53 N

71. (a) $N/f_0$

(b) $\Delta x/N$

(c) $2\pi N/\Delta x$

(d) $N$ is uncertain because the waveform dies out gradually rather than stopping abruptly at some time; hence, where the pulse starts and stops is not well defined.

73. (a) 3.40 kHz; 10.2 kHz; 17.0 kHz

(b) Frequencies near 3400 Hz will be most readily perceived.

75. $\frac{1}{3}\lambda$

77. 6.62 m

79. (a) 1.90 cm; 3.59 m/s

(b) 0; 0

(c) 1.18 cm; 2.22 m/s

(d) 0; 0

81. (a) At resonance, standing waves are set up in the tube. At a displacement antinode, the powder is moved about; at a node the powder is stationary, and so it collects at the nodes.

(b) $2fD$

(c) If we let the length $L$ of the tube be 1.2 m and assume that $v_{air} = 344$ m/s (the speed of sound in air at 20°C), then the 10th harmonic corresponds to $D = 25.3$ cm and a driving frequency of 680 Hz.

(d) If $f = 2$ kHz and $v_{He} = 1008$ m/s (the speed of sound in helium at 20°), then $D$ for the 10th harmonic in helium would be 25.3 cm, and $D$ for the 10th harmonic in air would be 8.60 cm. Hence, neglecting end effects at the driven end, a tube whose length is the least common multiple of 8.60 cm and 25.3 cm (218 cm) would work well for the measurement of the speed of sound in either air or helium.

83. (a) The pipe is closed at one end.

(b) 262 Hz

(c) 32.4 cm

85. (a) $y_1(x,t) = (0.01 \text{ m})\sin\left[\left(\frac{\pi}{2}\text{ m}^{-1}\right)x - (40\pi\text{ s}^{-1})t\right]$;

$y_2(x,t) = (0.01 \text{ m})\sin\left[\left(\frac{\pi}{2}\text{ m}^{-1}\right)x + (40\pi\text{ s}^{-1})t\right]$;

(b) 2.00 m

(c) $v_y(1\text{ m},t) = -(2.51 \text{ m/s})\sin(40\pi\text{ s}^{-1})t$

(d) $a_y(1\text{ m},t) = -(316 \text{ m/s}^2)\cos(40\pi\text{ s}^{-1})t$

87. $y_{res}(x,t) = 0.1\sin(kx - \omega t)$

89. (b) 203 Hz

91. (a) What you hear is the fundamental mode of the tube and its overtones. A more physical explanation is that the echo of the finger snap moves back and forth along the tube with a characteristic time of $2L/c$, leading to a series of clicks from each echo. Since the clicks happen with a frequency of $c/2L$, the ear interprets this as a musical note of that frequency.

(b) 38.6 cm

93. (a) Since no conditions were placed on its derivation, this expression is valid for all harmonics.

(b) 1.54%

95. (a) $v_y(x,t) = -\omega_1 A_1 \sin \omega_1 t \sin k_1 x - \omega_2 A_2 \sin \omega_2 t \sin k_2 x$

(b) $dK = \frac{1}{2}\mu[\omega_1^2 A_1^2 \sin^2 \omega_1 t \sin^2 k_1 x + 2\omega_1\omega_2 A_1 A_2 \sin \omega_1 t$ $\sin k_1 x \sin \omega_2 t \sin k_2 x + \omega_2^2 A_2^2 \sin^2 \omega_2 t \sin^2 k_2 x]dx$

(c) $K = \frac{1}{4}m\omega_1^2 A_1^2 \sin^2 \omega_1 t + \frac{1}{4}m\omega_2^2 A_2^2 \sin^2 \omega_2 t$

97. (a)

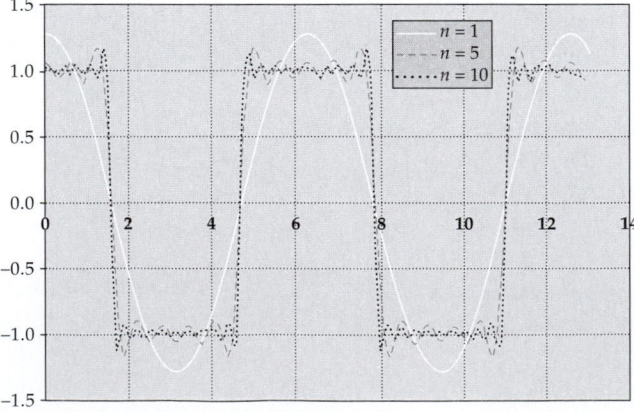

(b) $f(2\pi) = 1$ which is equivalent to the Liebnitz formula.

99. (b)

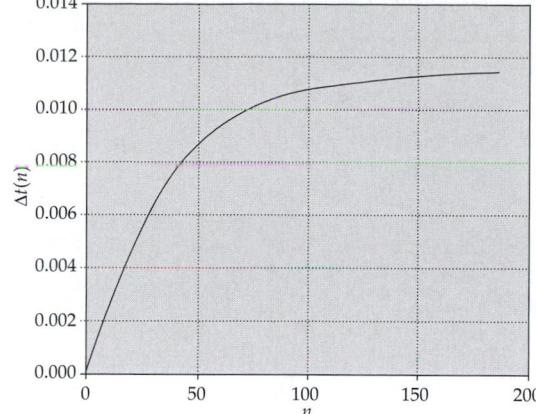

(c) The frequency heard at any time is $1/\Delta t_n$, so because $\Delta t_n$ increases over time, the frequency of the culvert whistler decreases.; 7.65 kHz

# Chapter 17

1. (a) False
   (b) False
   (c) True
   (d) False

3. Mert's room was colder.

5. From the ideal-gas law we have $P = nRT/V$. In the process depicted, both the temperature and the volume increase but the temperature increases faster than does the volume. Hence the pressure increases.

7. True

9. $K_{av}$ increases by a factor of 2; $K_{av}$ is reduced by a factor of $\frac{1}{2}$.

11. False

13. Since $10^7 \gg 273$, it does not matter.

15. (b)

17. (d)

19. The ratio of the rms speeds is inversely proportional to the square root of the ratio of the molecular masses. The kinetic energies of the molecules are the same.

21. Because the temperature remains constant, the average speed of the molecules remains constant. When the volume decreases, the molecules travel less distance between collisions, so the pressure increases because the frequency of collisions increases.

23. The average molecular speed of He gas at 300 K is about 1.4 km/s, so a significant fraction of He molecules have speeds in excess of earth's escape velocity (11.2 km/s), and thus "leak" away into space. Over time, the He content of the atmosphere decreases to almost nothing.

25. (a) $3.61 \times 10^3$ K
    (b) 225 K
    (c) If $v_{rms} > \frac{1}{5}v_e$ or $T \geq 25T_{atm}$, $H_2$ molecules escape. Therefore, the more energetic $H_2$ molecules escape from the upper atmosphere.
    (d) 164 K; 10.3 K; If we assume that the temperature on the moon with an atmosphere would have been approximately 1000 K, then all $O_2$ and $H_2$ would have escaped during the time since the formation of the moon to the present.

27. (a) 1.24 km/s
    (b) 310 m/s

(c) 264 m/s
(d) $O_2$, $CO_2$, and $H_2$ should be found on Jupiter.

29. 1063°C

31. (a) 8.40 cm
    (b) 107°C

33. −319°F

35. (a) 54.9 torr
    (b) 3704 K

37. −40°C = −40°F

39. −183°C; −297°F

41. (a) $B = 3.94 \times 10^3$ K; $R_0 = 3.97 \times 10^{-3}$ Ω
    (b) 1.31 kΩ
    (c) −389 Ω/K; −433 Ω/K
    (d) The thermistor is more sensitive (i.e., has greater sensitivity, at lower temperatures).

43. 1.79 mol; $1.08 \times 10^{24}$ molecules

45. −83.2 glips

47. (a) $3.66 \times 10^3$ mol
    (b) 60.0 mol

49. 10.0 atm

51. 1.19 kg/m³

53. 2.56 N

55. (a) 276 m/s
    (b) 872 m/s

57. 499 km/s; $2.07 \times 10^{-16}$ J

61. $K/\Delta U = 7.95 \times 10^4$

65. (a) 0.142 s
    (b) 0.143 s

67. (a) 122 K
    (b) 244 K
    (c) 1.43 atm

69. 111 mol; 55.5 mol

71. $7m_H$

73. 400.49 K

75. (a) $4.10 \times 10^{-26}$ m
    (b) 4.28 nm; The mean free path is larger by approximately a factor of 1000.

77. (a) 48.9%
    (b) 70.6%

# Chapter 18

1. $\Delta T_B = 4\Delta T_A$

3. (c)

5. Yes, if the heat adsorbed by the system is equal to the work done by the system.

7. $W_m + Q_m = \Delta E_{int}$; For an ideal gas, $\Delta E_{int}$ is a function of $T$ only. Since $W = 0$ and $Q = 0$ in a free expansion, $\Delta E_{int} = 0$ and $T$ is constant. For a real gas, $\Delta E_{int}$ depends on the density of the gas because the molecules exert weak attractive forces on each other. In a free expansion, these forces reduce the average kinetic energy of the molecules and, consequently, the temperature.

9. The temperature of the gas increases. The average kinetic energy increases with increasing volume due to the repulsive interaction between the ions.

11. (a)

13. (a) False

(b) False

(c) True

(d) True

(e) True

(f) True

(g) True

15. (d)

17. If $V$ decreases, the temperature decreases.

19. The heat capacity of a substance is proportional to the number of degrees of freedom per molecule associated with the molecule. Since there are 6 degrees of freedom per molecule in a solid, and only 3 per molecule (translational) for a monatomic liquid, you would expect the solid to have the higher heat capacity.

21. 1.63 min, an elapsed time that seems to be consistent with experience.

23. $c_p = (1.01\%)c_{water}$

25. (a) 10.5 MJ

(b) 121 W

27. 7.48 kcal

29. 48.8 mg

31. 365°C

33. 20.8°C

35. 453 kg

37. (a) 0°C

(b) 125 g

39. (a) 4.94°C

(b) No ice is left.

41. (a) 2.99°C

(b) 199.8 g

(c) The answer would be the same.

43. 618°C

45. 2.21 kJ

47. 176°C

49. 53.7 J

51. (a) 6.13 W

(b) 19.0 min

53. (a) 405 J

(b) 861 J

55. (a) 507 J

(b) 963 J

57. $\frac{3}{2}P_0V_0$

59. (a) 555 J

(b) 555 J

61. (a) 55.7 g/mol

(b) Fe

63. (a) 0; 6.24 kJ; 6.24 kJ

(b) 8.73 kJ; 6.24 kJ; 2.49 kJ

(c) 2.49 kJ

65. 59.6 L

67. $\Delta C_P = -\frac{13}{2}Nk$

69. $C_{V,water} = 5Nk$

71. (a) 465 K

(b) 387 K

73. (a) 300 K; 7.80 L; 1.14 kJ; 1.14 kJ

(b) 208 K; 5.41 L; 574 J; 0

75. (a) 263 K

(b) 10.8 L

(c) −1.48 kJ

(d) 1.48 kJ

79. −142 J

81. $Q_{D\to A} = 8.98$ kJ; $Q_{A\to B} = 13.2$ kJ; $Q_{B\to C} = -8.98$ kJ; $Q_{C\to D} = -6.56$ kJ; $W_{cycle} = 6.62$ kJ

83. (a)

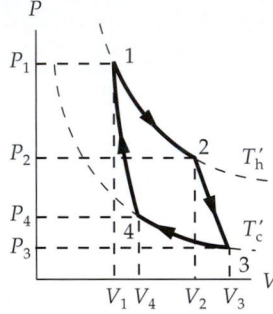

85. 180 kJ

87. (a) 65.2 K; 81.2 K

(b) 1.62 kJ

(c) 2.22 kJ

89. (a) 65.2 K; 81.2 K

(b) 2.65 kJ

(c) 3.25 kJ

91. (a) $9.20 \times 10^{-2}$ J/kg·K

(b) 0.0584 J/kg

93. 47.6 kPa; 51.5 K; 71.2 K; 148 kPa

95. (a) 2.49 kJ

(b) 3.20 kJ

97. 171 K

99. (a) $W = 0$; $Q = 3.74$ kJ

(b) $\Delta U = 3.74$ kJ; $Q = 6.24$ kJ; $W = 2.50$ kJ

101.

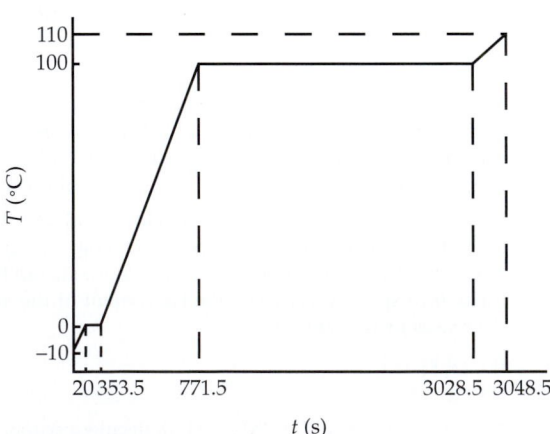

103. $4RT$

105. 396 K

107. (a) $\frac{1}{2}P_0$

(b) diatomic

(c) In the isothermal process, $T$ is constant, and the translational kinetic energy is unchanged. In the adiabatic process, $T_3 = 1.32T_0$, and the translational kinetic energy increases by a factor of 1.32.

109. (a) 93.5 kPa

   (b) 6266 K; 1.30 MPa

   (c) 56.7 kPa

111. (b) $\Delta U = 4621$ J, a result in good agreement with the result of Problem 106.

# Chapter 19

1. Friction reduces the efficiency of the engine.

3. Increasing the temperature of the steam increases the Carnot efficiency, and generally increases the efficiency of any heat engine.

5. (c)

7. (d)

9. Note that A→B is an adiabatic expansion. B→C is a constant volume process in which the entropy decreases; therefore heat is released. C→D is an adiabatic compression. D→A is a constant volume process that returns the gas to its original state. The cycle is that of the Otto engine (see Figure 19-3).

11.
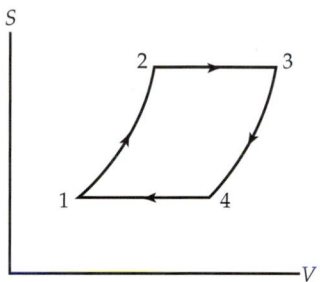

13.
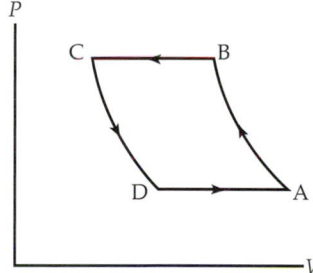

15. 56.5%

17. (a) $1.66 \times 10^{17}$ W

   (b) $5.66 \times 10^{14}$ J/K·s

   (c) $3.09 \times 10^{13}$ J/K·s

19. 29.8 kJ/K

21. (a) 500 J

   (b) 400 J

23. (a) 40.0%

   (b) 80.0 W

25. (a)
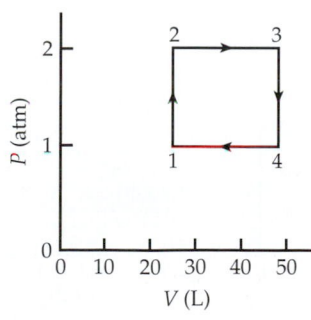

$W_{1\to2} = 0; Q_{1\to2} = 3.74$ kJ

$W_{2\to3} = 4.99$ kJ; $Q_{2\to2} = 12.5$ kJ

$W_{3\to4} = 0; Q_{3\to4} = -7.48$ kJ

$W_{4\to1} = 2.49$ kJ; $Q_{4\to1} = -6.24$ kJ

   (b) 15.4%

27.
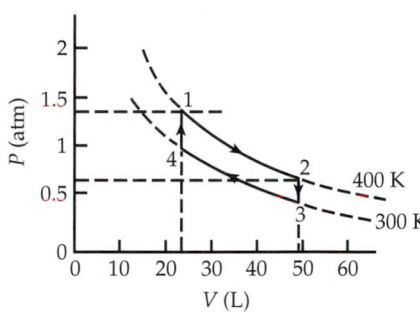

   13.1%

29. (a) 600 K; 1800 K; 600 K

   (b) 15.4%

31. (a) 5.16%; The fact that this efficiency is considerably less than the actual efficiency of a human body does not contradict the Second Law of Thermodynamics. The application of the second law to chemical reactions such as the ones that supply the body with energy have not been discussed in the text.

   (b) Most warm-blooded animals survive under roughly the same conditions as humans. To make a heat engine work with appreciable efficiency, internal body temperatures would have to be maintained at an unreasonably high level.

35. (a) 33.3%

   (b) 33.3 J

   (c) 66.7 J

   (d) 2.00

37. Let the first engine be run as a refrigerator. Then it will remove 140 J from the cold reservoir, deliver 200 J to the hot reservoir, and require 60 J of energy to operate. Now take the second engine and run it between the same reservoirs, and let it eject 140 J into the cold reservoir, thus replacing the heat removed by the refrigerator. If $\varepsilon_2$, the efficiency of this engine, is greater than 30%, then $Q_{h2}$, the heat removed from the hot reservoir by this engine, is 140 J/$(1 - \varepsilon_2) > 200$ J, and the work done by this engine is $W = \varepsilon_2 Q_{h2} > 200$ J. The end result of all this is that the second engine can run the refrigerator, replacing the heat taken from the cold reservoir, and do additional mechanical work. The two systems working together then convert heat into mechanical energy without rejecting any heat to a cold reservoir, in violation of the second law.

39. (a) 33.3%

   (b) If COP > 2, then 50 J of work will remove more than 100 J of heat from the cold reservoir and put more than 150 J of heat into the hot reservoir. So running engine (a) to operate the refrigerator with a COP > 2 will result in the transfer of heat from the cold to the hot reservoir without doing any net mechanical work in violation of the second law.

41. (a) 100°C

   (b) $Q_{1\to2} = 3.12$ kJ; $Q_{2\to3} = 0; Q_{3\to1} = -2.91$ kJ

   (c) 6.73%

   (d) 35.5%

43. (a) 5.26

   (b) 3.19 kW

   (c) 4.81 kW

45. (a) 173 kJ
    (b) 121 kJ
47. $\Delta S_u = 2.40$ J/K
49. (a) 11.5 J/K
    (b) Since the process is not quasi-static, it is nonreversible and the entropy of the universe must increase.
51. 1.22 kJ/K
53. (a) 0
    (b) 267 K
55. (a) 244 kJ/K
    (b) −244 kJ/K
    (c) $\Delta S_u > 0$
57. (a) −117 J/K
    (b) 137 J/K
    (c) 20.3 J/K
59. 1.97 kJ/K
61. (a) 0.417 J/K
    (b) 125 J
63. (a) 20.0 J
    (b) 66.7 J; 46.7 J
65. (a) 51.0%
    (b) 102 kJ
    (c) 98.0 kJ
67. 113 W/K
69. (a) Process (1) is more wasteful of *mechanical* energy. Process (2) is more wasteful of *total* energy.
    (b) 1.67 J/K; 0.833 J/K
71. 313 K
73. 10.0 W
75. (a) 253 kPa
    (b) 462 K
    (c) 6.96 kJ; 25.9%
77. (a) 253 kPa
    (b) 416 K
    (c) 6.58 kJ; 34.8%
79. 180 J
83. $\approx 10^{478}$

# Chapter 20

1. The glass bulb warms and expands first, before the mercury warms and expands.
3. (c)
5. (a) With increasing altitude $P$ decreases; from curve OF, $T$ of the liquid-gas interface diminishes, so the boiling temperature decreases. Likewise, from curve OH, the melting temperature increases with increasing altitude.
   (b) Boiling at a lower temperature means that the cooking time will have to be increased.
7. The thermal conductivity of metal and marble is much greater than that of wood; consequently, heat transfer from the hand is more rapid.
9. (c)
11. In the absence of matter to support conduction and convection, radiation is the only mechanism.

13. (a)
15. The temperature of an object is inversely proportional to the maximum wavelength at which the object radiates (Wein's displacement law). Since blue light has a shorter wavelength than red light, an object for which the wavelength of the peak of thermal emission is blue is hotter than one which is red.
17. 18.1 mW/(m·K)
19. 2.90 nm
21. (a) $\gamma \equiv \dfrac{\Delta A/A}{\Delta T}$
    (b) $\gamma \approx 2\alpha \Delta T$
23. 217°C
25. $15.4 \times 10^{-6}$ K$^{-1}$
27. 5.24 m
29. 0.255 mm
31. (a) The clock runs slow.
    (b) 8.21 s
33. $3.68 \times 10^{-12}$ N/m²
35. (a) 90°C
    (b) 82°C
    (c) 170 kPa
37. (b) $\left(P_r + \dfrac{3}{V_r^2}\right)(3V_r - 1) = 8T_r$
39. 2.07 kBtu/h
41. (a) $I_{Cu} = 962$ W; $I_{A1} = 569$ W
    (b) 1.53 kW
    (c) 0.0523 K/W
43. (a) Conservation of energy requires that the thermal current through each shell be the same.
45. $I = \dfrac{2\pi kL}{\ln(r_1/r_2)}(T_2 - T_1)$
47. $9.35 \times 10^{-3}$ m²
49. 1598°C
51. 2.10 km
53. 5767 K
55. 1.18 cm
57. (b) $\dfrac{\beta_{exp} - \beta_{th}}{\beta_{th}} < \boxed{0.3\%}$
59. $1.26 \times 10^{10}$ kW; <0.002%
61. 132 W ignoring the cylindrical insulation; 142 W taking the insulation into account.
63. $L_2 = L_1$; $\omega_2 \approx (1 - 2\alpha\Delta T)\omega_1$; $E_2 = E_1(1 - 2\alpha\Delta T)$
65. (a) 0.698 cm/h
    (b) 11.9 d
67. (b) 40.5 min

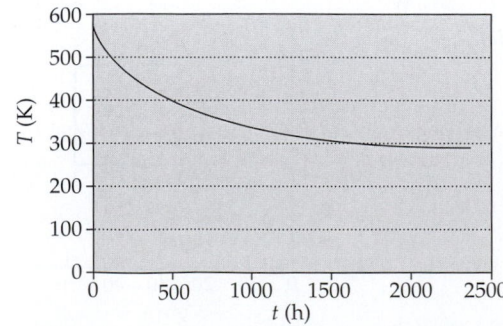

# Chapter 21

1. *Similarities:* The force between charges and masses vary as $1/r^2$.

    *Differences:* There are positive and negative charges but only positive masses. Like charges repel; like masses attract. The gravitational constant $G$ is many orders of magnitude smaller than the Coulomb constant $k$.

3. *(c)*

5. *(a)*

    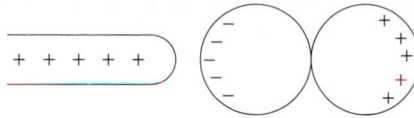

    *(b)*

    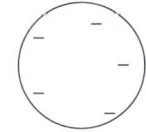 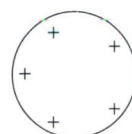

7. *(d)*

9. *(d)*

11.

    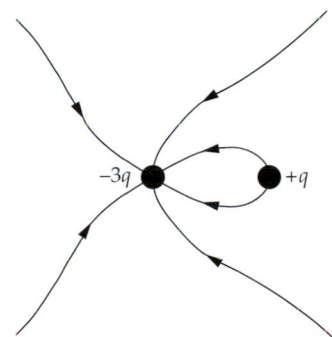

13. *(a)*

15. Because $\theta \neq 0$, a dipole in a uniform electric field will experience a restoring torque whose magnitude is $pE \sin \theta$. Hence, it will oscillate about its equilibrium orientation, $\theta = 0$. If the field is nonuniform and $dE/dx > 0$, the dipole will accelerate in the $x$ direction as it oscillates about $\theta = 0$.

17. *(a)* The force between the balls is diminished because the field produced by the two charges creates a dipolar field that opposes that of the two charges when they are out of the water.

    *(b)* The force is increased because a dipole is induced on the third metal ball. The dipole attracts each of the charged balls.

19. Assume that the wand has a negative charge. When the charged wand is brought near the tinfoil, the side nearer the wand becomes positively charged by induction, and so it swings toward the wand. When it touches the wand, some of the negative charge is transferred to the tinfoil, which thus has a net negative charge, and is now repelled by the wand.

21. *(a)* $3.46 \times 10^{10}$ N

    *(b)* $32.0 \ \mu$C

23. 141 nC

25. $5.00 \times 10^{12}$

27. $4.82 \times 10^7$ C

29. $(1.50 \times 10^{-2} \ \text{N})\hat{i}$

31. $(-8.66 \ \text{N})\hat{j}$

33. $\vec{F}_1 = (0.899 \ \text{N})\hat{i} + (1.80 \ \text{N})\hat{j}$
    $\vec{F}_2 = (-1.28 \ \text{N})\hat{i} - (1.16 \ \text{N})\hat{j}$
    $\vec{F}_3 = (0.381 \ \text{N})\hat{i} - (0.640 \ \text{N})\hat{j}$

35. $\vec{F}_q = \dfrac{kqQ}{R^2}\left(1 + \dfrac{\sqrt{2}}{2}\right)\hat{i}$

37. *(a)* $\vec{E}(6 \ \text{m}) = (999 \ \text{N/C})\hat{i}$

    *(b)* $\vec{E}(-10 \ \text{m}) = (-360 \ \text{N/C})\hat{i}$

    *(c)*

    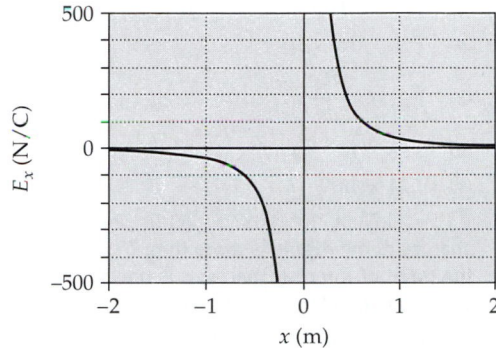

39. *(a)* $(400 \ \text{kN/C})\hat{j}$

    *(b)* $(-1.60 \ \text{mN})\hat{j}$

    *(c)* $-40.0$ nC

41. *(a)* $\vec{E}_r = (34.5 \ \text{kN/C})\hat{i}$

    *(b)* $\vec{F} = (69.0 \ \mu\text{N})\hat{i}$

43. *(a)* 12.9 kN/C, 231°

    *(b)* $2.08 \times 10^{-15}$ N, 51.3°

45. *(a)* 1.90 kN/C, 235°

    *(b)* $3.04 \times 10^{-16}$ N, 235°

47. *(a)* Because $\vec{E}$ is in the $+x$ direction, a positive test charge that is displaced from (0, 0) in the $+x$ direction will experience a force in the $x$ direction and accelerate in the $x$ direction. Consequently, the equilibrium at (0, 0) is unstable for a small displacement along the $x$ axis. If the positive test charge is displaced in the $+y$ direction, the charge at $+a$ will exert a greater force than the charge at $-a$, and the net force is then in the $-y$ direction; that is, it is a restoring force. Consequently, the equilibrium at (0, 0) is stable for small displacements along the $y$ direction.

    *(b)* Following the same arguments as in Part *(a)*, one finds that, for a negative test charge, the equilibrium is stable at (0, 0) for displacements along the $x$ direction and the equilibrium is unstable for displacements along the $y$ direction.

    *(c)* Because the two $+q$ charges repel, the charge $q_0$ at (0, 0) must be a negative charge. Because the force between charges varies as $1/r^2$, and the negative charge is midway between the two positive charges, $q_0 = -q/4$.

    *(d)* If the charge $q_0$ is displaced, the equilibrium is the same as discussed in Part *(b)*. If either of the $+q$ charges are displaced, the system is unstable.

49. *(a)* $1.76 \times 10^{11}$ C/kg

    *(b)* $1.76 \times 10^{13}$ m/s²; The direction of the acceleration is opposite the electric field.

    *(c)* $0.170 \ \mu$s

    *(d)* 25.5 cm

51. (a) $(-7.03 \times 10^{13} \text{ m/s}^2)\hat{j}$

    (b) 50.0 ns

    (c) $(-8.79 \text{ cm})\hat{j}$

53. 800 $\mu$C

55. 4.07 cm

57. (a) $8.00 \times 10^{-18}$ C·m

    (b)

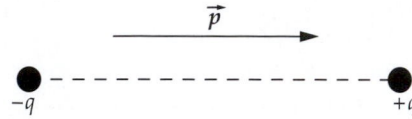

61. (a) $\vec{F}_{\text{net}} = Cp\hat{i}$

    (b) $\vec{F}_{\text{net}} = p_x \dfrac{dE_x}{dx}\hat{i}$

63. (a) $1.86 \times 10^{-9}$ kg

    (b) $1.24 \times 10^{36}$

65. $\vec{E}_{P_2} = (1.73 \times 10^6 \text{ N/C})\hat{i}$. While the separation of the two charges of the dipole is more than 10 percent of the distance to the point of interest, that is, $x$ is not much greater than $a$, this result is in excellent agreement with the result of Problem 64.

67. (a) $q_1 = 3.99 \ \mu$C, $q_2 = 2.01 \ \mu$C; or $q_1 = 2.01 \ \mu$C, $q_2 = 3.99 \ \mu$C

    (b) $q_1 = 7.12 \ \mu$C; $q_2 = -1.12 \ \mu$C

71. (a) $q_1 = 17.5 \ \mu$C, $q_2 = 183 \ \mu$C; or $q_1 = 183 \ \mu$C, $q_2 = 17.5 \ \mu$C

    (b) $q_1 = -15.0 \ \mu$C; $q_2 = 215 \ \mu$C

73. (a) 0.225 N

    (b) 0.113 N·m; counterclockwise

    (c) 0.0461 kg

    (d) $5.03 \times 10^{-7}$ C

75. (a) $q_1 = 28.0 \ \mu$C, $q_2 = 172 \ \mu$C; or $q_1 = 172 \ \mu$C, $q_2 = 28.0 \ \mu$C

    (b) 250 N

77. (a) $-97.2 \ \mu$C

    (b) $x_1 = 0.0508$ m; $x_2 = 0.169$ m

79. (a) 10.3°

    (b) 9.86°

81. $\dfrac{d^2\theta}{dt^2} = -\dfrac{2qE}{ma}\theta; \ T = 2\pi\sqrt{\dfrac{ma}{2qE}}$

83. $v = \sqrt{\dfrac{ke^2}{2mr}}$

87. (a) 8.48°

    (b) $\theta_1 = 9.42°; \theta_2 = 6.98°$

89. $\vec{E} = (-1.10 \times 10^4 \text{ N/C})\hat{j}$

91. (a) $\vec{E}_P = \dfrac{2kQy}{(a^2 + y^2)^{3/2}}\hat{j}$

    (b) $\vec{F}_y = \dfrac{2kqQy}{(a^2 + y^2)^{3/2}}\hat{j}$

    (c) The differential equation of motion is $\dfrac{d^2y}{dt^2} + \dfrac{16kqQ}{mL^3}y = 0$

    (d) 9.37 Hz

93. (b) $5.15 \times 10^{-5}$ m/s

# Chapter 22

1. (a) False. Gauss's law states that the net flux through any surface is given by $\phi_{\text{net}} = \oint_S E_n \, dA = 4\pi kQ_{\text{inside}}$. While it is true that Gauss's law is easiest to apply to symmetric charge distributions, it holds for *any* surface.

   (b) True

3. The electric field is that due to all the charges, inside and outside the surface. Gauss's law states that the net flux through any surface is given by $\phi_{\text{net}} = \oint_S E_n \, dA = 4\pi kQ_{\text{inside}}$. The lines of flux through a Gaussian surface begin on charges on one side of the surface and terminate on charges on the other side of the surface.

5. (a) False. Consider a spherical shell, in which there is no charge, in the vicinity of an infinite sheet of charge. The electric field due to the infinite sheet would be non-zero everywhere on the spherical surface.

   (b) True (assuming there are no charges inside the shell).

   (c) True

   (d) False. Consider a spherical conducting shell. Such a surface will have equal charges on its inner and outer surfaces but, because their areas differ, so will their charge densities.

7. (a)

9. (b)

11. (c)

13. False. A physical quantity is discontinuous if its value on one side of a boundary differs from that on the other. We can show that this statement is false by citing a counterexample. Consider the field of a uniformly charged sphere. $\rho$ is discontinuous at the surface, $E$ is not.

15. $3 \times 10^6$ V/m; $5.31 \times 10^{-5}$ C/m$^2$

17. (a) 17.5 nC

    (b) 26.2 N/C

    (c) 4.37 N/C

    (d) 2.57 mN/C

    (e) 2.52 mN/C

19. (a) $4.69 \times 10^5$ N/C

    (b) $1.13 \times 10^6$ N/C

    (c) $1.54 \times 10^3$ N/C

    (d) $1.55 \times 10^3$ N/C; This result agrees to within 0.1% of the result for Part (c).

21. (a) 0.300 nC

    (b) 1.43 kN/C

    (c) 183 N/C

    (d) 0.133147 N/C

    (e) 0.133149 N/C

23. (a) $E_x(0.2a) = 0.189\dfrac{kQ}{a^2}$

    (b) $E_x(0.5a) = 0.358\dfrac{kQ}{a^2}$

    (c) $E_x(0.7a) = 0.385\dfrac{kQ}{a^2}$

    (d) $E_x(a) = 0.354\dfrac{kQ}{a^2}$

(e)  $E_x(2a) = 0.179\dfrac{kQ}{a^2}$

(f)

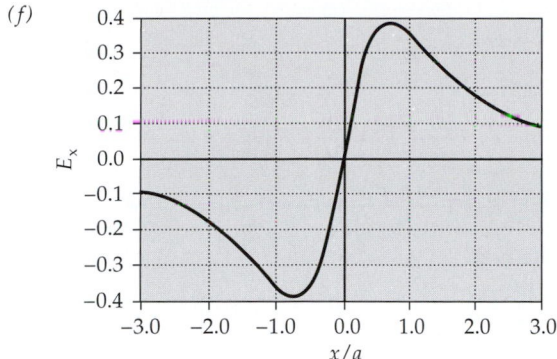

25.

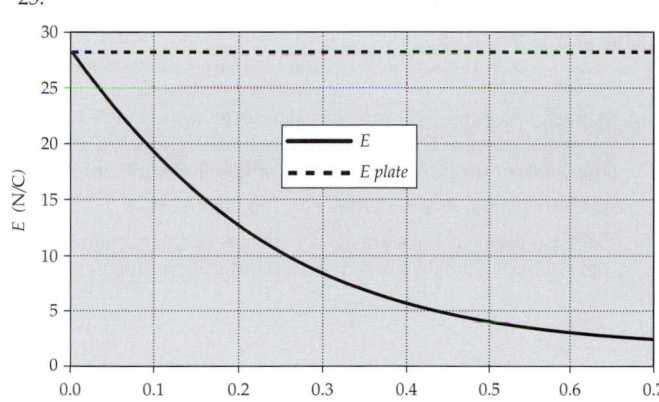

The magnitudes of the electric fields differ by more than 10% for $x = 0.03$ m.

31.  (a)  $20.0$ N·m²/C

(b)  $17.3$ N·m²/C

33.  (a)  $\phi_{\text{right}} = 1.51$ N·m²/C; $\phi_{\text{left}} = 1.51$ N·m²/C

(b)  $\phi_{\text{curved}} = 0$

(c)  $\phi_{\text{net}} = 3.02$ N·m²/C

(d)  $Q_{\text{inside}} = 2.67 \times 10^{-11}$ C

35.  (a)  $3.14$ m²

(b)  $7.19 \times 10^4$ N/C

(c)  $2.26 \times 10^5$ N·m²/C

(d)  No. The flux through the surface is independent of where the charge is located inside the sphere.

(e)  $2.26 \times 10^5$ N·m²/C

37.  $3.77 \times 10^4$ N·m²/C

39.  (a)  $E_{r<R_1} = 0$; $E_{R_1<r<R_2} = \dfrac{kq_1}{r^2}$; $E_{r>R_2} = \dfrac{k(q_1 + q_2)}{r^2}$

(b)  $-1$

(c)

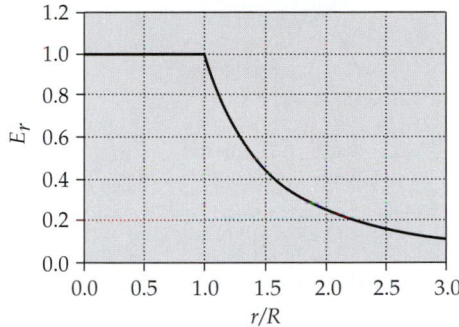

41.  (a)  $0.407$ nC

(b)  $339$ N/C

(c)  $999$ N/C

(d)  $983$ N/C

(e)  $366$ N/C

43.  $3.77$ N/C

45.  (a)  $2\pi BR^2$

(b)  $E_r(r > R) = \dfrac{BR^2}{2\,\epsilon_0 r^2}$; $E_r(r < R) = \dfrac{B}{2\epsilon_0}$

47.  (a)  $\dfrac{4}{3}\pi p(b^3 - a^3)$

(b)  $E_r(r < a) = 0$; $E_r(a < r < b) = \dfrac{\rho}{3\epsilon_0 r^2}(r^3 - a^3)$;

$E_r(r > b) = \dfrac{\rho}{3\epsilon_0 r^2}(b^3 - a^3)$

49.  (a)  $679$ nC

(b)  $E(2\text{ cm}) = 0$

(c)  $E(5.9\text{ cm}) = 0$

(d)  $E(6.1\text{ cm}) = 1.00$ kN/C

(e)  $E(10\text{ cm}) = 610$ N/C

51.  (a)  $679$ nC

(b)  $E_r(2\text{ cm}) = 339$ N/C

(c)  $E_r(5.9\text{ cm}) = 1.00$ kN/C

(d)  $E_r(6.1\text{ cm}) = 1.00$ kN/C

(e)  $E_r(10\text{ cm}) = 610$ N/C

53.  (a)  $E_n(r < 1.5\text{ cm}) = 0$; $E_n(1.5\text{ cm} < r < 4.5\text{ cm}) = \dfrac{(108\text{ N·m/C})}{r}$;

$E_n(4.5\text{ cm} < r < 6.5\text{ cm}) = 0$

(b)  $\sigma_1 = -21.2$ nC/m²; $\sigma_2 = 14.7$ nC/m²

55.  (b)  $E_n(r < R) = \dfrac{b}{4\epsilon_0}r^3$; $E_n(r > R) = \dfrac{bR^4}{4r\epsilon_0}$

57.  (a)  $\lambda_{\text{inner}} = 18.8$ nC/m

(b)  $E_n(r < 1.5\text{ cm}) = 22.6$ kN/C;

$E_n(1.5\text{ cm} < r < 4.5\text{ cm}) = \dfrac{339\text{ N·m/C}}{r}$;

$E_n(4.5\text{ cm} < r < 6.5\text{ cm}) = 0$;

$E_n(r > 6.5\text{ cm}) = \dfrac{339\text{ N·m/C}}{r}$

59.  $9.42$ kN/C

61. (a) $E_n(r < a) = \dfrac{kq}{r^2}$; $E_n(a < r < b) = 0$; $E_n(r > b) = \dfrac{kq}{r^2}$

(b)

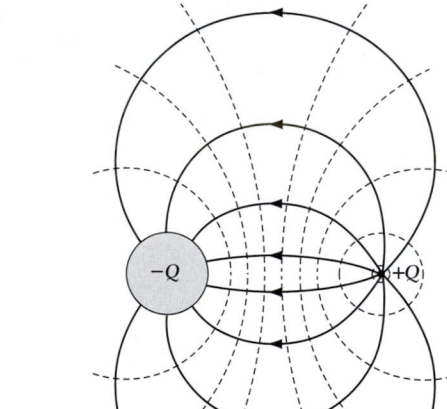

(c) $\sigma_{\text{inner}} = -\dfrac{q}{4\pi a^2}$; $\sigma_{\text{outer}} = \dfrac{q}{4\pi b^2}$

63. (a) $\sigma_{\text{inner}} = -0.553\ \mu\text{C/m}^2$; $\sigma_{\text{outer}} = 0.246\ \mu\text{C/m}^2$

(b) $E_n(r < a) = (2.25 \times 10^4\ \text{N·m}^2/\text{C})\left(\dfrac{1}{r^2}\right)$;

$E_n(0.6\ \text{m} < r < 0.9\ \text{m}) = 0$;

$E_n(r > 0.9\ \text{m}) = (2.25 \times 10^4\ \text{N·m}^2/\text{C})\left(\dfrac{1}{r^2}\right)$

(c) $\sigma_{\text{inner}} = -0.553\ \mu\text{C/m}^2$; $\sigma_{\text{outer}} = 0.589\ \mu\text{C/m}^2$;

$E_n(r < a) = (2.25 \times 10^4\ \text{N·m}^2/\text{C})\left(\dfrac{1}{r^2}\right)$;

$E_n(0.6\ \text{m} < r < 0.9\ \text{m}) = 0$;

$E_n(r > 0.9\ \text{m}) = (5.39 \times 10^4\ \text{N·m}^2/\text{C})\left(\dfrac{1}{r^2}\right)$

65. (a) $\sigma_{\text{face}} = 1.60\ \mu\text{C/m}^2$; $E_{\text{slab}} = 1.81 \times 10^5\ \text{N/C}$

(b) $\vec{E}_{\text{near}} = (-0.680 \times 10^5\ \text{N/C})\hat{r}$; $\vec{E}_{\text{far}} = (2.94 \times 10^5\ \text{N/C})\hat{r}$;
$\sigma_{\text{near}} = 0.602\ \mu\text{C/m}^2$; $\sigma_{\text{near}} = 2.60\ \mu\text{C/m}^2$

67. $1.15 \times 10^5\ \text{N/C}$

69. (a) $\dfrac{Q}{8\pi\epsilon_0 r^2}$

(b) $\dfrac{Q^2 a^2}{32\pi\epsilon_0 r^4}$

(c) $\dfrac{Q^2}{32\pi^2\epsilon_0 r^4}$

71. $1.11 \times 10^6\ \text{N/C}$

73. (a) $\vec{E}(0, 0) = (339\ \text{kN/C})\hat{i}$

(b) $\vec{E}(0.2\ \text{m}, 0.1\ \text{m}) = (1310\ \text{kN/C})\hat{i} + (-268\ \text{kN/C})\hat{j} = 1340\ \text{kN/C}$, in fourth quadrant of $xy$ plane at $\theta = -11.6°$.

(c) $\vec{E}(0.5\ \text{m}, 0.2\ \text{m}) = (203\ \text{kN/C})\hat{i}$

75. (a) $\dfrac{e}{\pi a_0^3}$

(b) $E(r) = \dfrac{ke}{r^2} e^{-2r/a}\left(1 + \dfrac{2r}{a} + \dfrac{2r^2}{a^2}\right)$

77. (a) $\dfrac{q_1}{q_2} = \dfrac{r_1}{r_2}$; $E_1 > E_2$

(b) Because $E_1 > E_2$, the resultant field points toward $s_2$.

(c) $E_1 = E_2$

(d) $\vec{E} = 0$; If $E \propto 1/r$, then $s_2$ would produce the stronger field at $P$ and $\vec{E}$ would point toward $s_1$.

79. (a) If $Q$ is positive, the field at the origin points radially outward.

(b) $E_{\text{center}} = \dfrac{kQ\ell}{2\pi R^3}$

81. (a) $E(0.4\ \text{m}, 0) = 203\ \text{kN/C}$; $\theta = 56.2°$

(b) $E(2.5\ \text{m}, 0) = 263\ \text{kN/C}$; $\theta = 153°$

83. $T = 2\pi R\sqrt{\dfrac{m}{2k\lambda q}}$

85. $7.42\ \text{rad/s}$

87. (b) $\vec{E}_1 = \dfrac{\rho b}{3\epsilon_0}\hat{r}$; $\vec{E}_2 = \dfrac{\rho b}{3\epsilon_0}\hat{r}$

89. (b) $\vec{E}_1 = \dfrac{(2\rho + \rho')b}{3\epsilon_0}\hat{r}$; $\vec{E}_2 = \dfrac{\rho'}{3\epsilon_0}b\hat{r}$

91. $200\ \text{N/C}$

95. $0.5R$

97. $4.49 \times 10^{14}\ \text{s}^{-1}$

# Chapter 23

1. A positive charge will move in whatever direction reduces its potential energy. The positive charge will reduce its potential energy if it moves toward a region of lower electric potential.

3. If $V$ is constant, its gradient is zero; consequently $\vec{E} = 0$.

5. Because the field lines are always perpendicular to equipotential surfaces, you always move perpendicular to the field.

7.

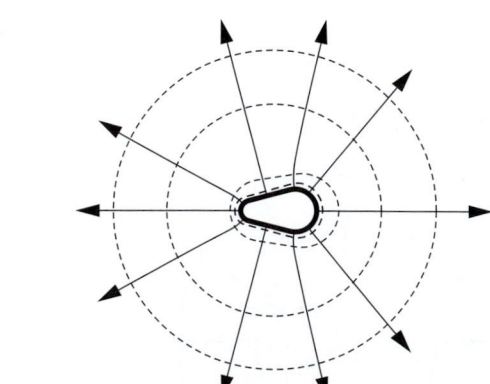

9.

11. (b)

13. (c)

15. (a) No. The potential at the surface of a conductor also depends on the local radius of the surface. Hence, $r$ and $\sigma$ can vary in such a way that $V$ is constant.

    (b) Yes; Yes.

17. $3.00 \times 10^9$ V

19. (a) $K = 0.719$ MeV

    (b) 0.0767%

21. (a) $-8.00$ kV

    (b) $-24.0$ mJ

    (c) 24.0 mJ

    (d) $-(2 \text{ kV/m})x$

    (e) $4 \text{ kV} - (2 \text{ kV/m})x$

    (f) $2 \text{ kV} - (2 \text{ kV/m})x$

23. (a) 4.50 kV

    (b) 13.5 mJ

    (c) 13.5 mJ

25. (a) $3.10 \times 10^7$ m/s

    (b) 2.50 MV/m

27. (a) $r = \dfrac{2kZe^2}{E}$

    (b) 45.4 fm; 25.3 fm

29. (a) 12.9 kV

    (b) 7.55 kV

    (c) 4.44 kV

31. (a) 270 kV

    (b) 191 kV

33. (a) $V = kq\left(\dfrac{1}{|x - a|} + \dfrac{1}{|x + a|}\right)$

    (b)

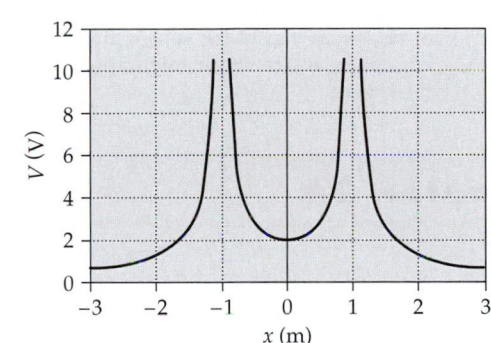

    (c) 0; 0

35. (a) $V_b - V_a$ is positive.

    (b) 25.0 kV/m

37. (a) 8.99 kV; 8.96 kV

    (b) 3.00 kV/m

    (c) 3.00 kV/m

    (d) 8.99 kV

39. (a) $V_b - V_a$ is negative.

    (b) 5.00 kV/m

41. (a) $V(x) = k\left(\dfrac{q}{|x|} + \dfrac{3q}{|x - 1|}\right)$

    (b) $-0.500$ m; 0.250 m

(c) $E_x(0.25 \text{ m}) = (21.3 \text{ m}^{-2})kq$; $E_x(-0.5 \text{ m}) = (-2.67 \text{ m}^{-2})kq$

(d)

43. (a) $V(x) = kq\left(\dfrac{2}{\sqrt{x^2 + a^2}} + \dfrac{1}{|x - a|}\right)$

    (b) $E_x(x > a) = \dfrac{2kqx}{(x^2 + a^2)^{3/2}} + \dfrac{kq}{(x - a)^2}$

45. (a) 6.02 kV

    (b) $-12.7$ kV

    (c) $-42.3$ kV

47. (a) $V(x, 0) = \dfrac{kQ}{L} \ln\left(\dfrac{\sqrt{x^2 + L^2/4} + L/2}{\sqrt{x^2 + L^2/4} - L/2}\right)$

49. (a) $Q = \frac{1}{2}\pi\sigma_0 R^2$

    (b) $V = \dfrac{2\pi k\sigma_0}{R^2}\left(\dfrac{R^2 - 2x^2}{3}\sqrt{x^2 + R^2} + \dfrac{2x^3}{3}\right)$

51. (a) $V(x) = 2\pi k\sigma_0\left(2\sqrt{x^2 + \dfrac{R^2}{2}} - \sqrt{x^2 + R^2} - x\right)$

    (b) $V(x) = \dfrac{\pi k\sigma_0 R^4}{8x^3}$

53. (a) $V(x) = \dfrac{kQ}{L} \ln\left(\dfrac{x + \dfrac{L}{2}}{x - \dfrac{L}{2}}\right)$

    (b) $V(x) = \dfrac{kQ}{x}$

55. $V_b - V_a = -\dfrac{2kq}{L} \ln\left(\dfrac{b}{a}\right)$

57. (a) $V_{\text{I}} = \dfrac{\sigma}{\epsilon_0}x$; $V_{\text{II}} = 0$; $V_{\text{III}} = \dfrac{\sigma}{\epsilon_0}(a - x)$

    (b) $V_{\text{I}} = 0$; $V_{\text{II}} = -\dfrac{\sigma}{\epsilon_0}x$; $V_{\text{III}} = -\dfrac{\sigma}{\epsilon_0}a$

59. (a) $V_1 = \dfrac{kQ}{R^3}r^2$

    (b) $dV_2 = \dfrac{3kQ}{R^3}r' \, dr'$

    (c) $V_2 = \dfrac{3kQ}{2R^3}(R^2 - r^2)$

    (d) $V = \dfrac{kQ}{2R}\left(3 - \dfrac{r^2}{R^2}\right)$

61. $r_{20\,V} = 0.499$ m; $r_{40\,V} = 0.250$ m; $r_{60\,V} = 0.166$ m; $r_{80\,V} = 0.125$ m; $r_{100\,V} = 0.0999$ m

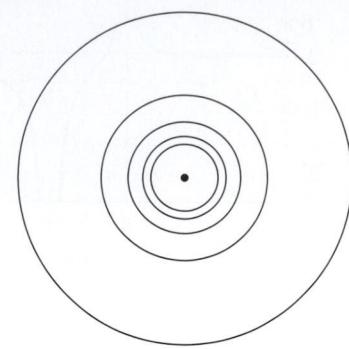

The equipotential surfaces are not equally spaced.

63. $26.6 \ \mu C/m^2$

65. $V_a - V_b = V_a = kq\left(\dfrac{1}{a} - \dfrac{1}{b}\right)$

67. (a)

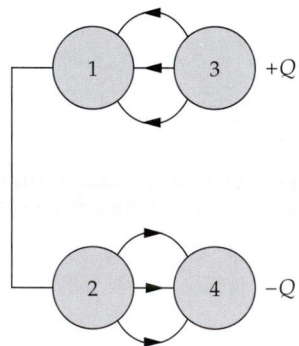

(b) $V_1 = V_2$ because the spheres are connected. From the direction of the electric field lines, it follows that $V_3 > V_1$.

(c) If sphere 3 and sphere 4 are connected, $V_3 = V_4$. The conditions of Part (b) can only be satisfied if all potentials are zero. Consequently the charge on each sphere is zero.

69. (a) $V(x) = \dfrac{2kq}{\sqrt{x^2 + a^2}}$

(b) $\vec{E}(x) = \dfrac{2kqx}{(x^2 + a^2)^{3/2}} \hat{i}$

71. (a) $V(x, y) = \dfrac{\lambda}{2\pi\epsilon_0} \ln\left[\dfrac{\sqrt{(x-a)^2 + y^2}}{\sqrt{(x+a)^2 + y^2}}\right]$; $V(0, y) = 0$

(b)

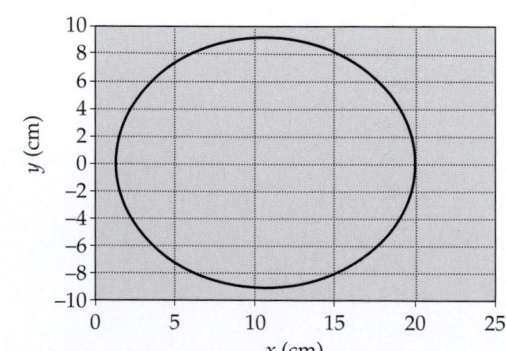

73. $\rho_0 = \dfrac{e}{\pi a^3}$

75. $-20 \ \mu C$

77. (a) $W_{+Q \rightarrow +a} = \dfrac{kQ^2}{2a}$

(b) $W_{-Q \rightarrow 0} = \dfrac{-2kQ^2}{a}$

(c) $W_{-Q \rightarrow 2a} = \dfrac{2kQ^2}{3a}$

79. $1.38 \times 10^5$ m/s

81. $R_2 = \frac{2}{3} R_1$

83. 7.12 nC

85. (a) $\Delta V = \dfrac{\sigma d}{\epsilon_0}$

(b) $\Delta V' = \dfrac{\sigma}{\epsilon_0}(d - a)$

87. (b) $\sigma = \dfrac{qd}{4\pi(d^2 + r^2)^{3/2}}$

89. (a) $V(c) = 0$; $V(b) = kQ\left(\dfrac{1}{b} - \dfrac{1}{c}\right)$; $V(a) = kQ\left(\dfrac{1}{b} - \dfrac{1}{c}\right)$

(b) $Q_b = Q$; $V(a) = V(c) = 0$; $Q_a = -Q\dfrac{a(c-b)}{b(c-a)}$;

$Q_c = -Q\dfrac{c(b-a)}{b(c-a)}$; $V(b) = kQ\dfrac{(c-b)(b-a)}{b^2(c-a)}$

91. $W = \dfrac{3Q^2}{20\pi\epsilon_0 R}$

93. (a) $R' = 0.794R$

(b) $\Delta E = 0.370E$

95. (a) $V_{av} = \dfrac{q}{4\pi\epsilon_0 R}$

(b) The superposition principle tells us that the potential at any point is the sum of the potentials due to any charge distributions in space. Because this result is independent of any properties of the sphere, this result must hold for any sphere and for any configuration of charges outside of the sphere.

# Chapter 24

1. (c)
3. True
5. (d)
7. Both statements are true.
9. True
11. (a) False
    (b) False
    (c) False
13. 0.104 nF/m $\leq C/L \leq$ 0.173 nF/m
15. $9.03 \times 10^{10}$ J
17. (a) 30 mJ
    (b) $-5.99$ mJ
    (c) $-18.0$ mJ
19. 22.2 $\mu$J
21. $v = q\sqrt{\dfrac{6\sqrt{2}k}{ma}}$

23. 75.0 nF

25. (a) 15.0 mJ

    (b) 45.0 mJ

27. (a) 0.625 J

    (b) 1.88 J

29. (a) 100 kV/m

    (b) 44.3 mJ/m$^3$

    (c) 88.6 $\mu$J

    (d) 17.7 nF

    (e) 88.5 $\mu$J; in agreement with Part (c)

31. (a) 11.1 nC

    (b) 0.553 $\mu$J

33. (a) 100

    (b) 10 V

    (c) charge: 1.00 kV; difference: 10.0 $\mu$C

35. $C_{eq} = C_2 + \dfrac{C_1 C_3}{C_1 + C_3}$

37. (a) 40.0 $\mu$C

    (b) $V_{10} = 4.00$ V; $V_{20} = 2.00$ V

39. (a) 15.2 $\mu$F

    (b) 2.40 mC; 0.632 mC

    (c) 0.304 J

41. (a) 0.242 $\mu$F

    (b) 2.42 $\mu$C; $Q_1 = 1.93$ $\mu$C; $Q_{0.25} = 0.483$ $\mu$C

    (c) 12.1 $\mu$J

43. Place four of the capacitors in series. Then the potential across each capacitor is 100 V when the potential across the combination is 400 V. The equivalent capacitance of the series is $2/4$ $\mu$F $= 0.5$ $\mu$F. If we place four such series combinations in parallel, as shown in the circuit diagram, the total capacitance between the terminals is 2 $\mu$F.

45. (a) $C_{eq} = 0.618$ $\mu$F

    (b) 1.618 $\mu$F

47. (a) 40.0 V

    (b) 4.24 m

49. (a) 0.333 mm

    (b) 3.76 m$^2$

51. (a) $E_{r<R_1} = 0; u_{r<R_1} = 0; E_{r>R_2} = 0; u_{r>R_2} = 0$

    (b) $\dfrac{kQ^2}{L} \ln\left(\dfrac{R_2}{R_1}\right)$

    (c) $\dfrac{kQ^2}{L} \ln\left(\dfrac{R_2}{R_1}\right)$

53. $C = \dfrac{\epsilon_0 (R_2^2 - R_1^2)}{2d}(\theta - \Delta\theta)$

57. $2R$

59. (a) 2.00 kV

    (b) 0

61. (a) 2.40 $\mu$F

    (b) 360 $\mu$J

63. (a) 6.00 V

    (b) 1.15 mJ

    (c) 0.288 mJ

65. (a) 200 V

    (b) $q_1 = -254$ $\mu$C; $q_2 = 146$ $\mu$C; $q_3 = 546$ $\mu$C

    (c) $V_1 = -127$ V; $V_2 = 36.5$ V; $V_3 = 91.0$ V

67. 2.71 nF

69. (a) $E_{r<R_1} = 0; u_{r<R_1} = 0; E_{r>R_2} = 0; u_{r>R_2} = 0$

    (b) $dU = \dfrac{kQ^2}{2\kappa r^2} dr$

    (c) $U = \dfrac{1}{2} Q^2 \left(\dfrac{R_2 - R_1}{4\pi\kappa\epsilon_0 R_1 R_2}\right)$

71. $C_{eq} = \left(\dfrac{4\kappa_1\kappa_2}{3\kappa_1 + \kappa_2}\right) C_0$

73. $C_{eq} = \left(\dfrac{\kappa d}{\kappa(d - t) + t}\right) C_0$

75. (a) 5.00

    (b) 1.25

    (c) 50.0

77. $C = \left(\kappa_3 + \dfrac{2\kappa_1\kappa_2}{\kappa_1 + \kappa_2}\right)\left(\dfrac{\epsilon_0 A}{2d}\right)$

79. (a) $C = \dfrac{3\epsilon_0 A}{y_0 \ln(4)}$

    (b) $\left.\dfrac{\sigma_b}{\sigma_f}\right|_{y=0} = 0; \left.\dfrac{\sigma_b}{\sigma_f}\right|_{y=y_0} = 0.750$

    (c) $\rho(y) = \dfrac{3\sigma}{[y_0(1 + 3y/y_0)^2]}$

    (d) $\rho = -\frac{3}{4}\sigma$, which is the charge per unit area in the dielectric, and just cancels out the induced surface charge density.

81. (a) 14.0 $\mu$F

    (b) 1.14 $\mu$F

83. 1.00 mm

85. $C_2 C_3 = C_1 C_4$

87. (a) $C_{new} = \dfrac{\epsilon_0 A}{2d}$

    (b) $V_{new} = 2V$

    (c) $U_{new} = \dfrac{\epsilon_0 A V^2}{d}$

    (d) $W = \dfrac{\epsilon_0 A V^2}{2d}$

89. (a) 2.22 nF

    (b) 66.6 $\mu$C

91. $Q_1 = 267$ $\mu$C; $Q_2 = 133$ $\mu$C

95. (a) $U = \dfrac{Q^2}{2\epsilon_0 A} x$

    (b) $dU = \dfrac{Q^2}{2\epsilon_0 A} dx$

    (c) $F = \dfrac{Q^2}{2\epsilon_0 A}$

    (d) $F = \frac{1}{2}QE$

97. (a) $U = \dfrac{Q^2 d}{2\epsilon_0 a[(\kappa - 1)x + a]}$

    (b) $F = \dfrac{(\kappa - 1)Q^2 d}{2a\epsilon_0 [(\kappa - 1)x + a]^2}$

    (c) $F = \dfrac{(\kappa - 1)a\epsilon_0 V^2}{2d}$

    (d) This force originates from the fringing fields around the edges of the capacitor. The effect of the force is to pull the dielectric into the space between the capacitor plates.

99. $2.55\ \mu J$

101. (a) Because $F$ increases as $\ell$ decreases, a decrease in plate separation will unbalance the system and the balance is unstable.

   (b) $V = \ell\sqrt{\dfrac{2Mg}{\epsilon_0 A}}$

103. (a) $Q_1 = (200\ \text{V})C_1;\ Q_2 = (200\ \text{V})\kappa C_1$

   (b) $U = (2 \times 10^4\ \text{V}^2)\,(1 + \kappa)C_1$

   (c) $U_f = (10^4\ \text{V}^2)C_1(1 + \kappa)^2$

   (d) $V_f = 100(1 + \kappa)\text{V}$

105. (a) $0.225\ \text{J}$

   (b) $3.50\ \text{mC};\ 1.00\ \text{mC}$

   (c) $2.25\ \text{mC}$

   (d) $0.506\ \text{J}$

107. $0.100\ \mu\text{F};\ 16.0\ \mu\text{C}$

109. (a) $1.00\ \text{mJ}$

   (b) $Q_1' = 47.6\ \mu\text{C};\ Q_2' = 152\ \mu\text{C}$

   (c) $0.476\ \text{mJ}$

111. (a) $C(V) = C_0\left(1 + \dfrac{\kappa\epsilon_0 V^2}{2Yd^2}\right)$

   (b) $7.97\ \text{kV}$

   (c) $0.209$ percent; $99.8$ percent

# Chapter 25

1. When current flows, the charges are not in equilibrium. In that case, the electric field provides the force needed for the charge flow.

3. (e)

5. (c)

7. (d)

9. You should decrease the resistance. Because the voltage across the resistor is constant, the heat out is given by $P = V^2/R$. Hence, decreasing the resistance will increase $P$.

11. (a)

13. (a)

15. (b)

17. (b)

19. (b)

21. (e)

23. A small resistance, because $P = \mathcal{E}^2/R$.

25. Yes. Kirchhoff's rules are statements of the conservation of energy and charge and therefore apply to all circuits.

27. (a) $3.12\ \text{V}$

   (b) $78.0\ \text{mV/m}$

   (c) $18.7\ \text{W}$

29. $2.03\ \text{m}$

31. $2.08\ \text{mm}$

33. $0.281\ \text{mm/s}$

35. (a) $5.93 \times 10^7\ \text{m/s}$

   (b) $37.3\ \mu\text{A}$

37. $0.210\ \text{mm/s};\ 0.531\ \text{mm/s}$

39. (a) $1.04 \times 10^8\ \text{m}^{-1}$

   (b) $1.04 \times 10^{14}\ \text{m}^{-3}$

41. (a) $33.3\ \Omega$

   (b) $0.751\ \text{A}$

43. $8.98\ \text{mm}$

45. $1.95\ \text{V}$

47. $62.2\ c\cdot y$

49. (a) $0.170$

   (b) $E$ is greater in the iron wire.

51. $1.20\ \Omega$

53. $0.0314\ \Omega$

55. (b) $90.0\ \text{mA}$

57. $\dfrac{\rho L}{\pi ab}$

59. (a) $\dfrac{\rho}{2\pi L}\ln\dfrac{b}{a}$

   (b) $2.05\ \text{A}$

61. $45.6°\text{C}$

63. (a) $15.0\ \text{A}$

   (b) $11.1\ \Omega$

   (c) $1.30\ \text{kW}$

65. (a) $\dfrac{1}{A}\,[\rho_1 L_1 + \rho_2 L_2]$

   (b) $264$

67. (a) $636\ \text{K}$

   (b) As the filament heats up its resistance increases, leading to more power being dissipated, leading to further heat, leading to a higher temperature, and so on. This thermal runaway can burn out the filament if not controlled.

69. (a) $5.00\ \text{mA}$

   (b) $50.0\ \text{V}$

71. $180\ \text{J}$

73. (a) $240\ \text{W}$

   (b) $228\ \text{W}$

   (c) $43.2\ \text{kJ}$

   (d) $2.16\ \text{kJ}$

75. (a) $6.91\ \text{MJ}$

   (b) $12.8\ \text{h}$

77. (a) $26.7\ \text{kW}$

   (b) $5.76\ \text{MC}$

   (c) $69.1\ \text{MJ}$

   (d) $57.6\ \text{km}$

   (e) $\$0.03/\text{km}$

79. (a) $1.33\ \Omega$

   (b) $I_4 = 3.00\ \text{A};\ I_3 = 4.00\ \text{A};\ I_6 = 2.00\ \text{A}$

81. (b) Because the potential difference between points $c$ and $d$ is zero, no current would flow through the resistor connected between these two points, and the addition of that resistor would not change the network.

83. $450\ \Omega$

85. (b)

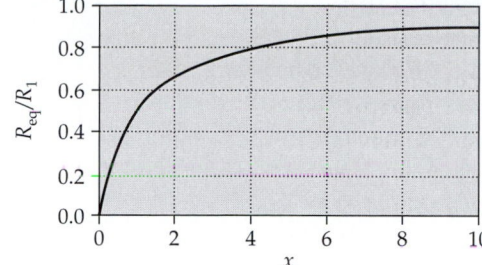

87. (a) $6.00\ \Omega$

    (b) $0.667$ A; $1.33$ A; $0.667$ A

89. 8

91. (a) $R_3 = \dfrac{R_1^2}{R_1 + R_2}$

    (b) $R_2 = 0$

    (c) $R_1 = \dfrac{R_3 + \sqrt{R_3^2 + 4R_2R_3}}{2}$

93. (a) $4.00$ A

    (b) $2.00$ V

    (c) $1.00\ \Omega$

95. (a)

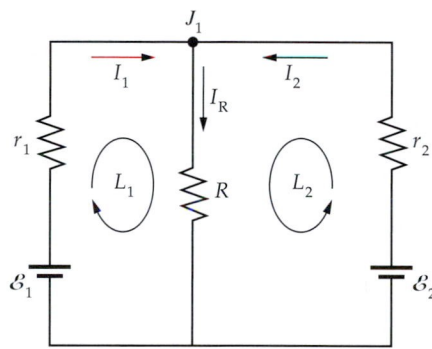

    (b) $I_1 = -57.0$ A; $I_2 = 63.0$ A; $I_R = 6.00$ A

    (c) $P_2 = 794$ W; $P_1 = 650$ W; $P_{r_1} = 32.5$ W; $P_{r_2} = 39.7$ W; $P_R = 72.0$ W

97. (a) $I_1 = 0.667$ A; $I_2 = 0.889$ A; $I_3 = 1.56$ A

    (b) $9.36$ V

    (c) $8.00$ W; $10.7$ W

99. If $r = R$, both arrangements provide the same power to the load. Examination of the second derivative of $P_p$ at $R = \frac{1}{2}r$ shows that $R = \frac{1}{2}r$ corresponds to a maximum value of $P_p$ and hence, for the parallel combination, the power delivered to the load is greater if $R < r$ and is at a maximum when $R = \frac{1}{2}r$.

103. $2.40$ V

105. (a) $3.33$ V

    (b) $3.33$ V

    (c) $3.13$ V

    (d) $2.00$ V

    (e) $0.435$ V

    (f) $1.67\ \mathrm{M}\Omega$

107. $2.50\ \Omega$

109. $195\ \mathrm{k}\Omega$

111. (a) $600\ \mu\mathrm{C}$

    (b) $0.200$ A

    (c) $3.00$ ms

    (d) $81.2\ \mu\mathrm{C}$

113. $2.18\ \mathrm{M}\Omega$

115. (a) $8.00\ \mu\mathrm{C}$

    (b) $73.7$ ms

117. (a) $5.69\ \mu\mathrm{C}$

    (b) $1.10\ \mu\mathrm{A}$

    (c) $1.10\ \mu\mathrm{A}$

    (d) $6.60\ \mu\mathrm{W}$

    (e) $2.42\ \mu\mathrm{W}$

    (f) $4.17\ \mu\mathrm{W}$

121. (a) $0.250$ A

    (b) $62.5$ mA

    (c) $I_2(t) = (62.5\ \mathrm{mA})\,(1 - e^{-t/0.750\ \mathrm{ms}})$

123. (a) $48.0\ \mu\mathrm{A}$

    (b) $0.866$ s

125. (a)

127. (b)

129. (a) $30.0$ A

    (b) $4.00$ V

131. $R = 14.0\ \Omega$; $\mathcal{E} = -7.00$ V

133. (a) $43.9\ \Omega$

    (b) $300\ \Omega$

    (c) $3.80\ \mathrm{k}\Omega$

135. (a) $2.19 \times 10^{13}/\mathrm{s}$

    (b) $210\ \mathrm{J/s}$

    (c) $27.6$ s

137. $0.164\ \mathrm{L/s}$

139. This result holds independently of the geometries of the capacitor and the resistor.

141. (a) $10.0$ ms

    (b) $V(t) = \dfrac{\mathcal{E}}{\tau}t$

    (c) $1.00\ \mathrm{G}\Omega$

    (d) $60.9$ ps

    (e) $P_1 = 6.17\ \mathrm{nW}$; $P_2 = 2.89\ \mathrm{kW}$

145. (a) $I(t) = \dfrac{V_0}{R}e^{-t/\tau}$

    (b) $P(t) = \dfrac{V_0^2}{R}e^{-2t/\tau}$

    (c) $E = \frac{1}{2}V_0^2 C_{eq}$; This is exactly the difference between the initial and final stored energies found in the preceding problem, which confirms the statement at the end of that problem that the difference in the stored energies equals the energy dissipated in the resistor.

149. (a) $10^{12}$

    (b) $0.160\ \mathrm{mA}$

    (c) $64.0$ kW

    (d) $640$ MW

    (e) $10^{-4}$

151. $R_{eq} = \dfrac{R_1 + \sqrt{R_1^2 + 4R_1R_2}}{2}$

# Chapter 26

1. (b)

3. False

5. The alternating current running through the filament is changing direction every 1/60 s, so in a magnetic field the filament experiences a force that alternates in direction at that frequency.

7. (a) True

 (b) True

 (c) True

 (d) False

 (e) True

9. Upward

11. (a)

13. From relativity; This is equivalent to the electron moving from right to left at velocity $v$ with the magnet stationary. When the electron is directly over the magnet, the field points directly up, so there is a force directed out of the page on the electron.

15. If only $\vec{F}$ and $I$ are known, one can only conclude that the magnetic field $\vec{B}$ is in the plane perpendicular to $\vec{F}$. The specific direction of $\vec{B}$ is undetermined.

17. (a) 88.6 C/kg

 (b) 26.6 nC

19. (a) $-(3.80\ \text{mN})\hat{k}$

 (b) $-(7.51\ \text{mN})\hat{k}$

 (c) 0

 (d) $(7.51\ \text{mN})\hat{j}$

21. 0.962 N

23. 0.621 pN; $\theta_x = 108°$; $\theta_y = 102°$; $\theta_z = 158°$

25. 1.48 A

29. $\vec{B} = (10\ \text{T})\hat{i} + (10\ \text{T})\hat{j} - (15\ \text{T})\hat{k}$

31. (a) 87.4 ns

 (b) $4.62 \times 10^7$ m/s

 (c) 11.4 MeV

33. (a) 142 m

 (b) 2.84 m

35. (a) $\dfrac{v_\alpha}{v_p} = \dfrac{1}{2}$

 (b) $\dfrac{K_\alpha}{K_p} = 1$

 (c) $\dfrac{L_\alpha}{L_p} = 2$

39. (a) $\phi = 24.0°$; $r_p = 0.492$ m; $v_p = 2.83 \times 10^7$ m/s

 (b) $v_d = 1.41 \times 10^7$ m/s

41. (a) $1.64 \times 10^6$ m/s

 (b) 14.0 keV

 (c) 7.66 eV

43. (a) 7.34 mm

 (b) 66.2 $\mu$T

45. (a) 63.3 cm

 (b) 2.60 cm

47. $\Delta t_{58} = 15.8\ \mu$s; $\Delta t_{60} = 16.3\ \mu$s

49. (a) 21.3 MHz

 (b) 46.0 MeV

 (c) 10.7 MHz; 23.0 MeV

53. (a) 0.302 A·m²

 (b) 0.131 N·m

55. (a) 0

 (b) $\pm(2.70 \times 10^{-3}\ \text{N·m})\hat{j}$

57. $B = \dfrac{mg}{I\pi R}$

59. (a)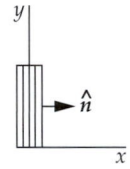

 $\vec{\tau} = (0.840\ \text{N·m})\hat{k}$

 (b)

 $\vec{\tau} = 0$

 (c)

 $\vec{\tau} = 0$

 (d)

 $\vec{\tau} = (0.594\ \text{N·m})\hat{k}$

61. 0.377 A·m²; Into the page.

67. $\mu = \frac{4}{3}\pi\sigma R^4\omega$

69. (a) $\tau = \frac{1}{4}\pi\sigma r^4\omega B \sin\theta$

 (b) $\Omega = \dfrac{\pi\sigma r^2 B}{2m}\sin\theta$

71. (a) $3.69 \times 10^{-5}$ m/s

 (b) 1.48 $\mu$V

73. 1.02 mV

75. 3.46

77. (a) 0.131 $\mu$s

 (b) $2.40 \times 10^7$ m/s

 (c) 12.0 MeV

81. (a) $\vec{B} = -\dfrac{mg}{I\ell}\tan\theta\,\hat{u}_v$

 (b) $g \sin\theta$

83. (a) $\vec{F} = (1.60 \times 10^{-18}\,\text{N})\hat{j}$

    (b) $\vec{E} = (10.0\,\text{V/m})\hat{j}$

    (c) $\Delta V = 20.0\,\text{V}$

85. 5.10 m

# Chapter 27

1. (a) The electric forces are repulsive; The magnetic forces are attractive (the two charges moving in the same direction act like two currents in the same direction).

   (b) The electric forces are again repulsive; The magnetic forces are also repulsive.

3.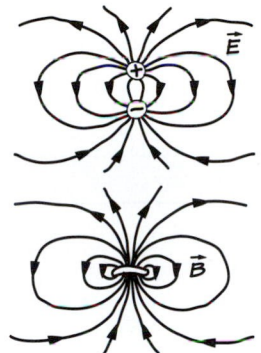

5. (a)

7. (e)

9.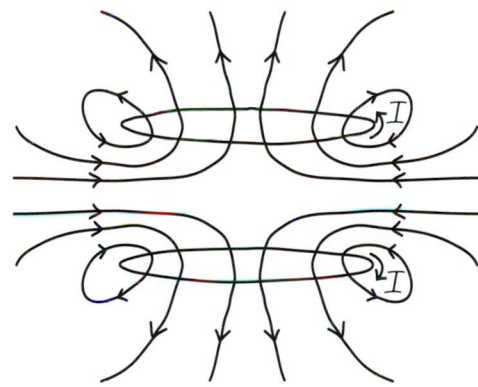

11. (a) True

    (b) True

13. (a) False

    (b) True

    (c) False

    (d) False

    (e) True

15. No. The classical relation between magnetic moment and angular momentum is $\vec{\mu} = \dfrac{q}{2m}\vec{L}$. Thus, if the angular momentum of the particle is zero, its magnetic moment will also be zero.

17. From Ampère's law, the current enclosed by a closed path within the tube is zero, and from the cylindrical symmetry it follows that $B = 0$ everywhere within the tube.

19. $H_2$, $CO_2$, and $N_2$ are diamagnetic ($\chi_m < 0$); $O_2$ is paramagnetic ($\chi_m > 0$).

21. $60.0\,\mu\text{T}$

23. (a) $\vec{B}(0, 0) = -(9.00\,\text{pT})\hat{k}$

    (b) $\vec{B}(0, 1\,\text{m}) = -(36.0\,\text{pT})\hat{k}$

    (c) $\vec{B}(0, 3\,\text{m}) = (36.0\,\text{pT})\hat{k}$

    (d) $\vec{B}(0, 4\,\text{m}) = (9.00\,\text{pT})\hat{k}$

25. (a) $\vec{B}(1\,\text{m}, 3\,\text{m}) = 0$

    (b) $\vec{B}(6\,\text{m}, 4\,\text{m}) = -(3.56 \times 10^{-23}\,\text{T})\hat{k}$

    (c) $\vec{B}(3\,\text{m}, 6\,\text{m}) = (4.00 \times 10^{-23}\,\text{T})\hat{k}$

27. $\dfrac{F_B}{F_E} = \dfrac{v^2}{c^2}$

29. $d\vec{B}(3\,\text{m}, 0, 0) = -(9.60\,\text{pT})\hat{i}$

31. (a) $B(0) = 54.5\,\mu\text{T}$

    (b) $B(0.01\,\text{m}) = 46.5\,\mu\text{T}$

    (c) $B(0.02\,\text{m}) = 31.4\,\mu\text{T}$

    (d) $B(0.35\,\text{m}) = 33.9\,\text{nT}$

33. (a) 19.1 cm

    (b) 45.3 cm

    (c) 99.5 cm

35. (a)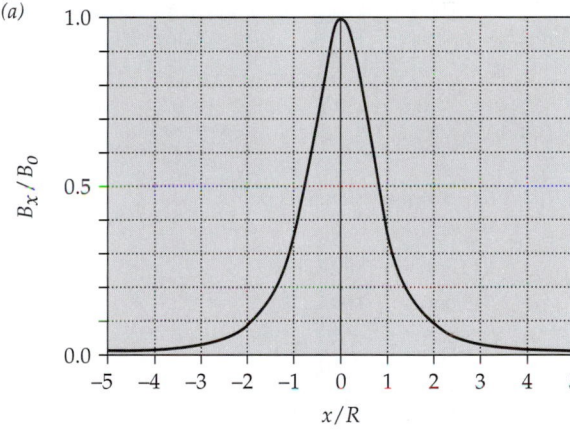

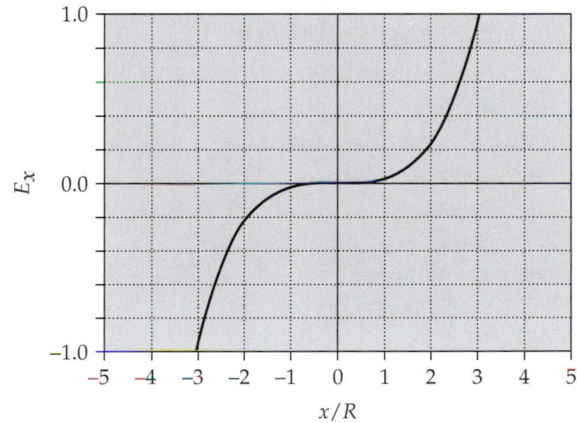

(b)

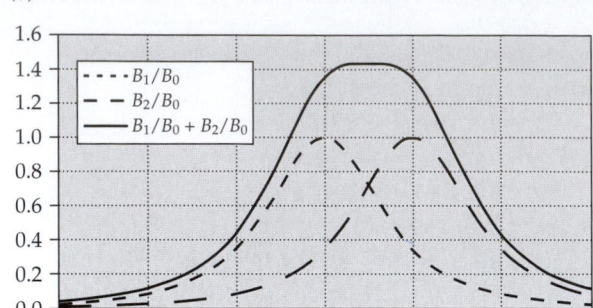

39.

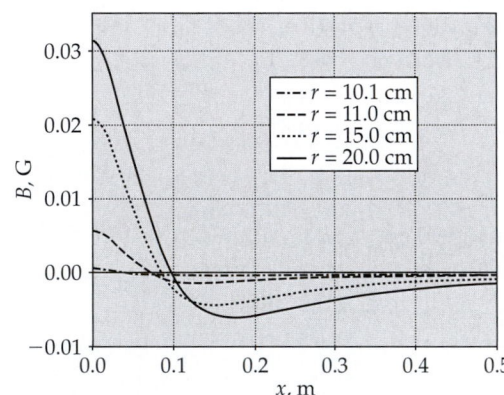

43. (a) $\vec{B}(-3\text{ cm}) = -(88.6\ \mu\text{T})\hat{k}$

(b) $\vec{B}(0) = 0$

(c) $\vec{B}(3\text{ cm}) = -(88.6\ \mu\text{T})\hat{k}$

(d) $\vec{B}(9\text{ cm}) = -(160\ \mu\text{T})\hat{k}$

45. (a) $\vec{B}(-3\text{ cm}) = -(177\ \mu\text{T})\hat{k}$

(b) $\vec{B}(0) = -(133\ \mu\text{T})\hat{k}$

(c) $\vec{B}(3\text{ cm}) = -(177\ \mu\text{T})\hat{k}$

(d) $\vec{B}(9\text{ cm}) = (106\ \mu\text{T})\hat{k}$

47. (a) $\vec{B}(z = 8\text{ cm}) = (64.0\ \mu\text{T})\hat{j}$

(b) $\vec{B}(z = 8\text{ cm}) = -(48.0\ \mu\text{T})\hat{k}$

49. (a) Because the currents repel, they are antiparallel.

(b) 39.3 mA

51. 80.2 A

53. (a) $4.50 \times 10^{-4}\text{ N/m}$

(b) 30.0 $\mu$T

55. (a) 80.0 A

(b) $-(0.240\text{ mT})\hat{j}$

57. (a) $\dfrac{3\sqrt{2}\mu_0 I^2}{4\pi a}$

(b) $\dfrac{\sqrt{2}\mu_0 I^2}{4\pi a}$

59. (a) 3.25 mT

(b) 3.25 mT

(c) 1.63 mT

63. $B_{\text{inside}} = 0;\ B_{\text{outside}} = \dfrac{\mu_0 I}{2\pi R}$

65. (a) $B_{r<R} = \dfrac{\mu_0 I}{2\pi r}$

(b) $B_{r>R} = 0$

69. (a) $B_{r<a} = 0$

(b) $B_{a<r<b} = \dfrac{\mu_0 I}{2\pi r}\dfrac{r^2 - a^2}{b^2 - a^2}$

(c) $B_{r>b} = \dfrac{\mu_0 I}{2\pi r}$

71. (a) 27.3 mT

(b) 20.0 mT

73. (a) 10.1 mT

(b) 10.1 mT; 1.52 T

75. $B_{\text{app}} = 10.053\text{ mT};\ M = 0.544\text{ A/m};\ B = 10.054\text{ mT}$

77. $-4.00 \times 10^{-5}$

79.

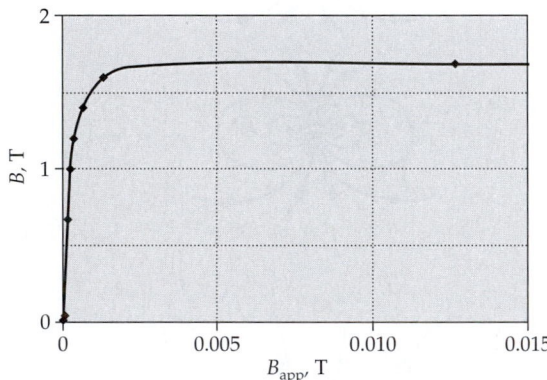

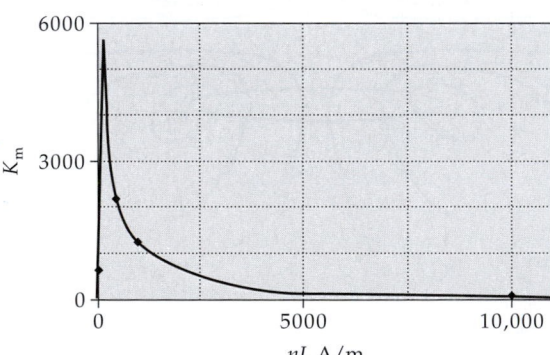

83. 1.69 $\mu_B$

85. (b) $7.46 \times 10^{-4}$

87. $B_{\text{app}} = \dfrac{\mu_0 NI}{2\pi a};\ B = \dfrac{\mu_0 NI}{2\pi a} + \mu_0 M$

89. (a) 30.2 mT

(b) 6.96 A/m

(c) 30.2 mT

91. $K_m = 11.75; \mu = 1.48 \times 10^{-5}$ N/A$^2$
93. (a) 12.6 mT
    (b) $1.36 \times 10^6$ A/m
    (c) 137
95. (a) 60.3 mT
    (b) 24.0 A
97. (a) $1.42 \times 10^6$ A/m
    (b) $K_m = 90.0; \mu = 1.13 \times 10^{-4}$ T·m/A; $\chi_m = 89.0$
99. (a) $(8.00$ T/m$)r$
    (b) $(3.20 \times 10^{-3}$ T·m$)\dfrac{1}{r}$
    (c) $(8.00 \times 10^{-6}$ T·m$)\dfrac{1}{r}$
    (d) Note that the field in the ferromagnetic region is the field that would be produced in a nonmagnetic region by a current of $400I = 1600$ A. The amperian current on the inside of the surface of the ferromagnetic material must therefore be 1600 A − 40 A = 1560 A in the direction of $I$. On the outside surface, there must then be an amperian current of 1560 A in the opposite direction.
101. $\dfrac{\mu_0 I}{4}\left(\dfrac{1}{R_1} - \dfrac{1}{R_2}\right)\hat{i}$
103. $\dfrac{\mu_0}{2\pi}\dfrac{I}{a}(1 + \sqrt{2})$
105. (a) $\vec{F}_2 = (1.00 \times 10^{-4}$ N$)\hat{i}; \vec{F}_4 = (-0.286 \times 10^{-4}$ N$)\hat{i}$
    (b) $\vec{F}_{net} = (0.714 \times 10^{-4}$ N$)\hat{i}$
107. $(7.07\ \mu$T$)\hat{i}$
109.

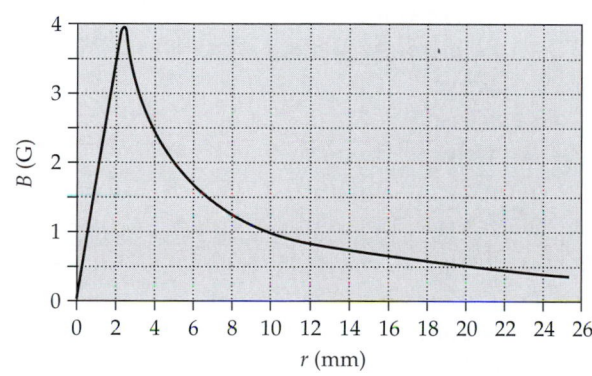

111. $\kappa = 0.246$ N·m/rad; $T = 0.523$ s
115. (a) $5.24 \times 10^{-2}$ A·m$^2$
    (b) $7.70 \times 10^5$ A/m
    (c) $2.31 \times 10^4$ A
117. (a) 70.6 A·m$^2$
    (b) 17.7 N·m
119. 3.18 cm
121. (a) 10.0 $\mu$T
    (b) 10.0 $\mu$T
    (c) 5.00 $\mu$T
123. 2.24 A
125. (c) $\dfrac{\mu_0 \omega \sigma}{2}\left(\dfrac{R^2 + 2x^2}{\sqrt{R^2 + x^2}} - 2x\right)$

# Chapter 28

1. (d)
3. (a) If the current in $B$ is clockwise, the loops repel one another.
    (b) If the current in $B$ is counterclockwise, the loops attract one another.
5. (a) and (b)

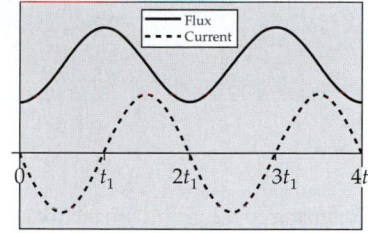

Time

7. The protection is needed because if the current is suddenly interrupted, the resulting emf generated across the inductor due to the large flux change can blow out the inductor. The diode allows the current to flow (in a loop) even when the switch is opened.
9. (a) False
    (b) True
    (c) True
    (d) False
    (e) True
11. The time-varying magnetic field of the magnet sets up eddy currents in the metal tube. The eddy currents establish a magnetic field with a magnetic moment opposite to that of the moving magnet; thus, the magnet is slowed down. If the tube is made of a nonconducting material, there are no eddy currents.
13. (a) 3.14 rad/s
    (b) 1.94 mV
    (c) No. To generate an emf of 1 V, the students would have to rotate the jump rope about 500 times faster.
    (d) The use of multiple strands of lighter wire (so that the composite wire could be rotated at the same angular speed) looped several times around would increase the induced emf.
15. 2.00 kV
17. (a) 0
    (b) $1.37 \times 10^{-5}$ Wb
    (c) 0
    (d) $1.19 \times 10^{-5}$ Wb
19. $\pi R^2 B$
21. $6.74 \times 10^{-3}$ Wb
23. (a) $\mu_0 n I N \pi R_1^2$
    (b) $\mu_0 n I N \pi R_3^2$
25. $\dfrac{\mu_0 I}{4\pi}$
27. (a) 0.314 mV
    (b) 0.785 mA
    (c) 0.247 $\mu$W

29. (a)

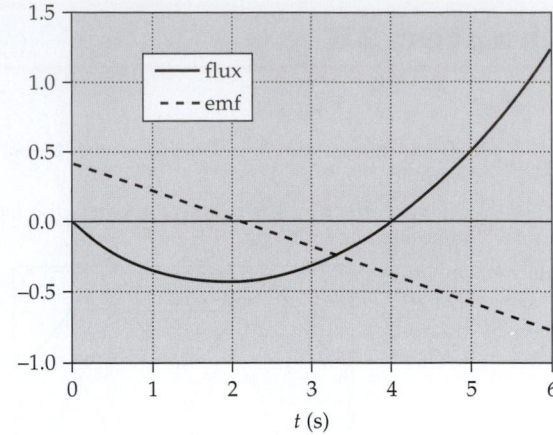

(b) Referring to the graph, we see that the flux is a minimum at $t = 2$ s and that $V = 0$ at this instant.

(c) The flux is zero at $t = 0$ and $t = 4$ s. At these times, $\mathcal{E} = 0.4$ V and $-0.4$ V, respectively.

31. (a) $-1.26$ mC

(b) 12.6 mA

(c) 630 mV

33. 79.8 $\mu$T

35. (a) $0.693 \dfrac{\mu_0 a}{\pi}$

(b) $4.16 \ \mu\Omega$.

Because the magnetic flux due to $I$ is increasing into the page, the induced current will be in such a direction that its magnetic field will oppose this increase; that is, it will be out of the page. Thus, the induced current is counter-clockwise.

37. 400 m/s

39. (a)

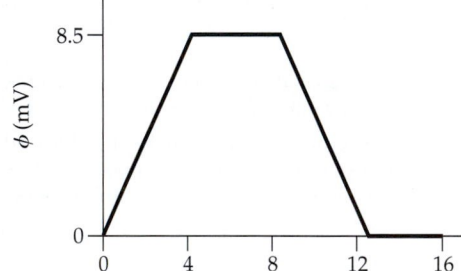

(b)

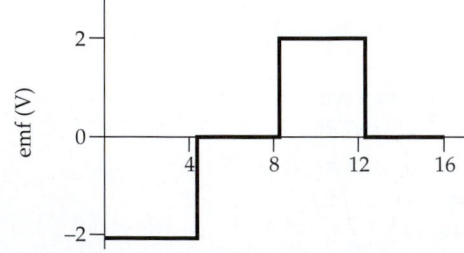

41. (a) $\dfrac{dv}{dt} = \dfrac{B\ell}{mR}(\mathcal{E} - B\ell v)$

(b) $v_t = \dfrac{\mathcal{E}}{B\ell}$

(c) 0

43. $x = \dfrac{mv_0 R}{B^2\ell^2}$

47. (d)

49. (a) (15.6 T·m) $v$

(b) 1.61 cm/s

51. (a) $\dfrac{\mu_0 Ivb}{2\pi}\left(\dfrac{1}{d + vt} - \dfrac{1}{d + a + vt}\right)$

(b) $\dfrac{\mu_0 Ivb}{2\pi}\left(\dfrac{1}{d + vt} - \dfrac{1}{d + a + vt}\right)$

53. (a) 24 Wb + (1600 H·A/s)$t$

(b) $-1.60$ kV

55. (a) 6.03 mT

(b) $7.58 \times 10^{-4}$ Wb

(c) 0.253 mH

(d) $-38.0$ mV

57. 0; 162 $\Omega$

63. $\dfrac{\mu_0 I^2}{16\pi}$

65. 0.157 $\mu$H

67. (a) $I = 0$; $dI/dt = 25.0$ A/s

(b) $I = 2.27$ A; $dI/dt = 20.5$ A/s

(c) $I = 7.90$ A; $dI/dt = 9.20$ A/s

(d) $I = 10.8$ A; $dI/dt = 3.38$ A/s

69. (a) 44.0 W

(b) 40.4 W

(c) 3.62 W

71. (a) 5.77 s

(b) 28.9 H

73. (a) 3.00 kA/s

(b) 1.50 kA/s

(c) 80.0 mA

(d) 0.123 ms

75. (a) 1.00 A; 0
    (b) 100 V; 100 V
    (c)

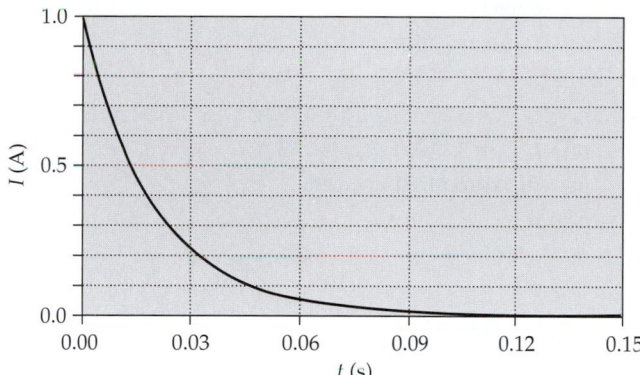

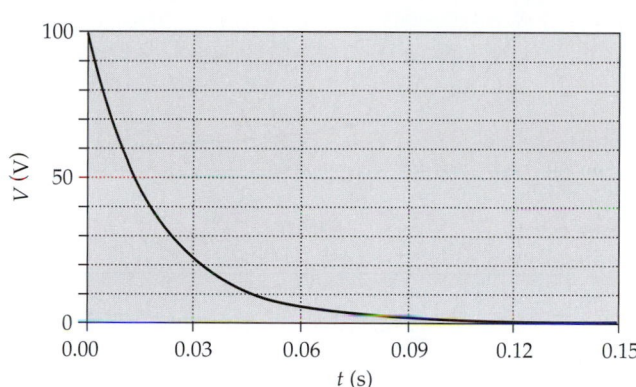

77. (a) 88.1 ms
    (b) 35.2 mH
79. (a) $\dfrac{dI_R}{dt}\Big|_{t=0} = 9.00$ kA/s; $\dfrac{dI_{8\,\text{mH}}}{dt} = 3.00$ kA/s;

    $\dfrac{dI_{4\,\text{mH}}}{dt} = 6.00$ kA/s

    (b) 1.60 A
81. (a) 3.53 J
    (b) 1.61 J
    (c) 1.92 J
83. (a) 7.07 mV
    (b) 6.64 mV
85. (b) 275 rad/s
89. (a) As the magnet passes through the coil, it induces an emf because of the changing flux through the coil. This allows the coil to sense when the magnet is passing through it.

    (b) One cannot use a cylinder made of conductive material because eddy currents induced in the material by a falling magnet would slow the magnet.

    (c) As the magnet approaches the loop, the flux increases, resulting in the increasing voltage signal. When the magnet is passing the coil, the flux goes from increasing to decreasing, so the induced emf becomes zero and then negative. The time at which the induced emf is zero is the time at which the magnet is at the center of the coil.

(d)

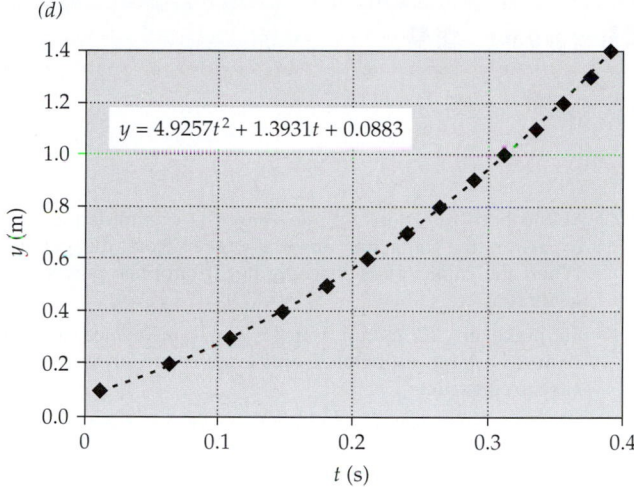

9.85 m/s$^2$

91. $I(t) = (0.350$ A$)\sin(2$ rad/s$)t$
93. (a) $-\tfrac{1}{2}r\mu_0 nI_0\omega\cos\omega t$

    (b) $-\dfrac{\mu_0 nR^2 I_0\omega}{2r}\cos\omega t$

97. (a) 30.8 N/m

    (b) $\dfrac{By_0\omega w}{R}\cos\omega t$

    (c) $\beta = \dfrac{B^2 w^2}{R}$

(d)

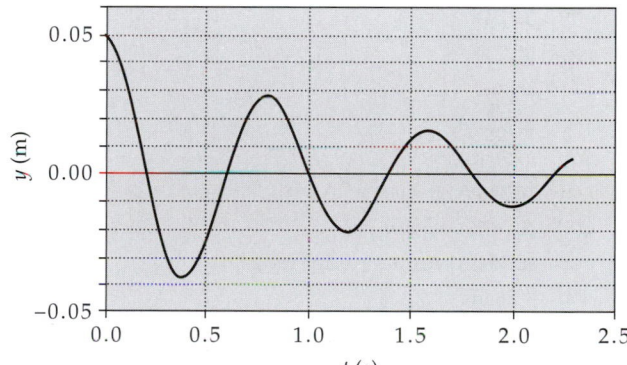

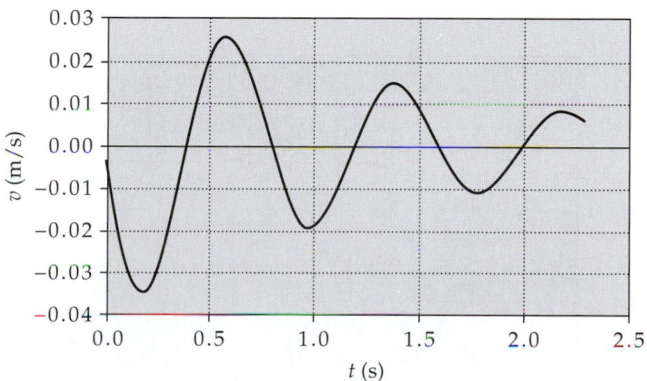

# Chapter 29

1. (b)

3. (b)

5. (c)

7. Yes to both questions. While charge is accumulating on the capacitor, the capacitor absorbs power from the generator. When the capacitor is discharging, it supplies power to the generator.

9. To make an $LC$ circuit with a small resonance frequency requires a large inductance and large capacitance. Neither is easy to construct.

11. Yes. The power factor is defined to be $\cos \delta = R/Z$ and, because $Z$ is frequency dependent, so is $\cos \delta$.

13. 0

15. True

17. (a) False

 (b) True

19. (a) 39.8 Hz

 (b) 15.1 V

21. (a) 13.6 V

 (b) 486 Hz

23. (a) 0.833 A

 (b) 1.18 A

 (c) 200 W

25. (a) 0.377 Ω

 (b) 3.77 Ω

 (c) 37.7 Ω

27. 1.59 kHz

29. (a) 25.1 mA

 (b) 17.8 mA

31. (a) $(0.346 \text{ A}) \cos \omega t$

 (b) $(0.346 \text{ A}) \cos \omega t$

 (c) $(0.344 \text{ A}) \cos(\omega t + 0.165 \text{ rad})$

33. (a) 1.26 ms

 (b) 88.0 mH

35. (a) 2.25 mJ

 (b) 712 Hz

 (c) 0.671 A

37. (a)

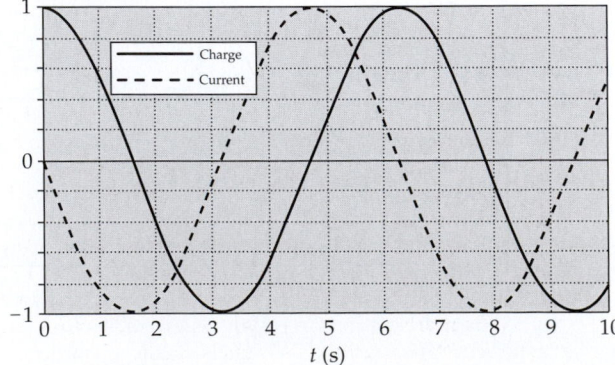

39. 29.2 mH

41. (a) 0.333

 (b) 26.7 Ω

 (c) 0.200 H

 (d) $I$ lags $\mathscr{E}$ by 70.5°

43. 0.397

45. (a) $I_{R,\text{rms}} = 6.20 \text{ A}$; $I_{R_L,\text{rms}} = 2.79 \text{ A}$; $I_{L,\text{rms}} = 5.52 \text{ A}$

 (b) $I_{R,\text{rms}} = 3.29 \text{ A}$; $I_{R_L,\text{rms}} = 2.95 \text{ A}$; $I_{L,\text{rms}} = 1.47 \text{ A}$

 (c) 50.3 percent; 80.4 percent

47. 60.0 V

49. (b) −90°

 (c) 0

57.

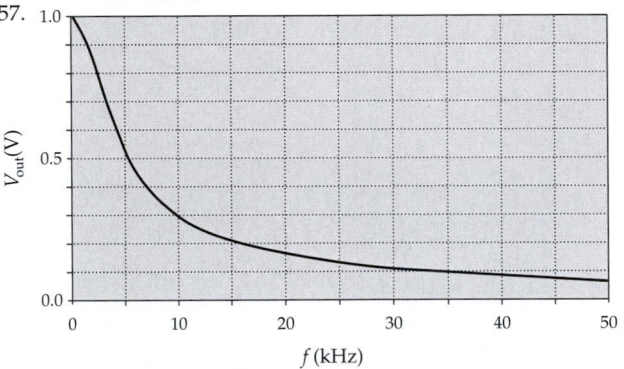

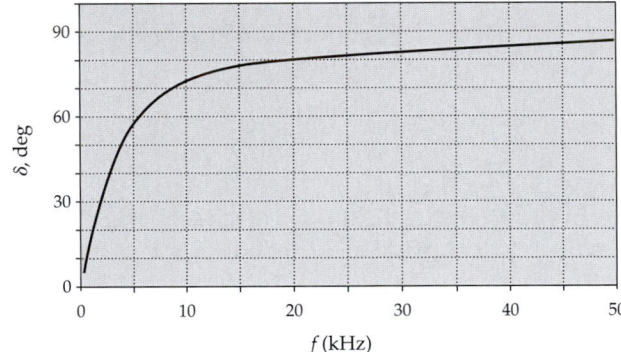

59. $\Delta \omega = R/2L$

61. 33.3 μF

63. (a) $I = -(18.75 \text{ mA}) \sin\left(1250t + \dfrac{\pi}{4}\right)$

 (b) 22.86 μF

 (c) $U_e = (4.92 \ \mu\text{J}) \cos^2\left(1250t + \dfrac{\pi}{4}\right)$

 $U_m = (4.92 \ \mu\text{J}) \sin^2\left(1250t + \dfrac{\pi}{4}\right)$

 $U = U_m + U_e = 4.92 \ \mu\text{J}$

65. (a) $5.39 \times 10^{-16}$ F

 (b) $f(x) = \dfrac{70.0 \text{ MHz}}{\sqrt{1 - (3.96 \text{ m}^{-1})x}}$

67. (a) 0.0444

 (b) 491 rad/s or 509 rad/s

71. (a) 1.13 kHz

 (b) $X_C = 79.6 \ \Omega$; $X_L = 62.8 \ \Omega$

 (c) $Z = 17.5 \ \Omega$; 4.04 A

 (d) $\delta = -73.4°$

73. (a) 14.1
   (b) 79.8 Hz
   (c) 0.275

75. (a) 10.0 A
   (b) 53.1°
   (c) 332 $\mu$F
   (d) 133 V

77. (b) $\delta = -\dfrac{\pi}{2} + \omega RC$

   (c) $\delta = \dfrac{\pi}{2} - \dfrac{R}{\omega L}$

79. (a) 80.0 V
   (b) 77.5 V
   (c) 164 V
   (d) 111 V
   (e) 181 V

81. 0.933 $\mu$F

83. (a)

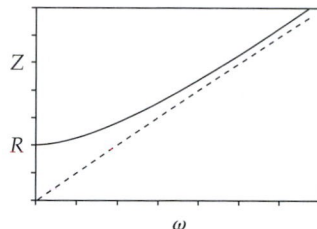

   (b)

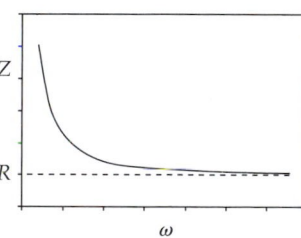

   (c)

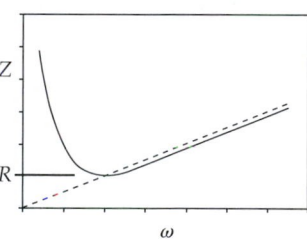

85. (a) $L = 0.800$ mH; $C = 12.5$ $\mu$F
   (b) 1.60
   (c) 2.00 A

87. (a) 12.0 $\Omega$
   (b) 7.20 $\Omega$; 9.60 $\Omega$
   (c) If the current leads the emf, the reactance is capacitive.

91. (a) $1.64 \times 10^3$ rad/s
   (b) $I_{C,rms} = 1.39$ A; $\delta_C = -84.4°$
       $I_{L,rms} = 1.41$ A; $\delta_L = 78.5°$
       $I_{rms} = 0.417$ A

93. (a) 0.396 $\mu$F
   (b) 804 $\Omega$

95. (a) 13.26 kHz
   (b) 200 mA; 600 V
   (c) $V_L = 433$ V; $V_C = 419$ V; $I = 142$ mA

103. (a) 1.67 rad/s
   (b) 3.96 W; 7.69 W

105. (a) 1/5
   (b) 50.0 A

107. (a) 1.50 A
   (b) 19

109. 3333

113. (a) 265 $\Omega$
   (b) 2.65 $\Omega$
   (c) 2.65 m$\Omega$

115. (a) 12.0 V
   (b) 8.49 V

117. (a) $Q_1 = (60 \ \mu F) \cos(120 \pi t) + 72 \ \mu F$;
       $Q_2 = (30 \ \mu F) \cos(120 \pi t) + 36 \ \mu F$
   (b) $I = -(33.9 \ \text{mA}) \sin(120 \pi t)$
   (c) 4.36 mJ
   (d) 36.0 $\mu$J

119. $I_{max} = 1.06$ A; $I_{min} = -0.0560$ A; $I_{av} = 0.500$ A; $I_{rms} = 0.636$ A

121. The inductance acts as a short circuit to the constant voltage source. The current is infinite at all times. Consequently, $I_{max} = I_{rms} = \infty$. There is no minimum current.

# Chapter 30

1. (a) False
   (b) True
   (c) True
   (d) True
   (e) False
   (f) True

3. X rays have greater frequencies, whereas light waves have longer wavelengths (see Table 30-1).

5. Consulting Table 30-1, we see that FM radio waves and television waves have wavelengths of the order of a few meters.

7. The dipole antenna should be in the horizontal plane and normal to the line from the transmitter to the receiver.

9. $I = 2.94 \times 10^7$ W/m²; $P = 5.20$ mW

11. (a) $E_{rms} = 719$ V/m; $B_{rms} = 2.40$ $\mu$T
   (b) $P_{av} = 3.87 \times 10^{26}$ W
   (c) $I = 6.36 \times 10^7$ W/m²; $P_r = 0.212$ Pa

13. $F_r = 7.09 \times 10^7$ N; Because the ratio of these forces is $1.65 \times 10^{-14}$ for the earth and $4.26 \times 10^{-14}$ for Mars, Mars has the larger ratio. The reason that the ratio is higher for Mars is that the dependence of the radiation pressure on the distance from the sun is the same for both forces ($r^{-2}$), whereas the dependence on the radii of the planets is different. Radiation pressure varies as $R^2$, whereas the gravitational force varies as $R^3$ (assuming that the two planets have the same density, an assumption that is nearly true). Consequently, the ratio of the forces goes as $R^2/R^3 = R^{-1}$. Because Mars is smaller than the earth, the ratio is larger.

15. (a)  $3.40 \times 10^{14}$ V/m·s

    (b)  5.00 A

19. (a)  10.0 A

    (b)  $2.26 \times 10^{12}$ V/m·s

    (c)  $7.90 \times 10^{-7}$ T·m

21. (a)  $I = \dfrac{A(0.01 \text{ V/s})}{\rho d}t$

    (b)  $I_d = \dfrac{(0.01 \text{ V/s})\epsilon_0 A}{d}$

    (c)  $t = \epsilon_0 \rho$

25. (a)  300 m

    (b)  3.00 m

27.  $3.00 \times 10^{18}$ Hz

29. (a)  30.0°

    (b)  7.07 m

31.  $4.15\ \mu\text{W/m}^2$; $5.22 \times 10^{17}$ W/cm²·s

33.  $0.151\ \mu\text{W/m}^2$

35. (a)  283 V/m

    (b)  $0.943\ \mu\text{T}$

    (c)  212 W/m²

    (d)  $0.707\ \mu\text{Pa}$

39. (a)  40.0 nN

    (b)  80.0 nN

41. (a)  3.46 V/m; 11.5 nT

    (b)  0.346 V/m; 1.15 nT

    (c)  0.0346 V/m; 0.115 nT

43.  $E_{\text{rms}} = 75.2$ kV/m; $B_{\text{rms}} = 0.251$ mT

45. (a)  Positive $x$ direction

    (b)  0.628 m; 477 MHz

    (c)  $\vec{E}(x, t) = (194 \text{ V/m}) \cos[10x - (3 \times 10^9)t]\hat{j}$

       $\vec{B}(x, t) = (0.647\ \mu\text{T}) \cos[10x - (3 \times 10^9)t]\hat{k}$

47.  $6.10 \times 10^{-3}$ degrees

49.  3.42 MW/m²

55.  The current induced in a loop antenna is proportional to the time-varying magnetic field. For a maximum signal, the antenna's plane should make an angle $\theta = 0°$ with the line from the antenna to the transmitter. For any other angle, the induced current is proportional to $\cos\theta$. The intensity of the signal is therefore proportional to $\cos\theta$.

57.  7.25 nV

59. (a)  $I = V_0\left(\dfrac{1}{R}\sin\omega t + \dfrac{\omega\epsilon_0\pi a^2}{d}\cos\omega t\right)$

    (b)  $B(r) = \dfrac{\mu_0 V_0}{2\pi r}\left(\dfrac{1}{R}\sin\omega t + \omega\dfrac{r^2}{a^2}\cos\omega t\right)$

    (c)  $\delta = \tan^{-1}\left(\dfrac{R\omega\epsilon_0\pi a^2}{d}\right)$

61. (a)  $\vec{S} = \dfrac{1}{\mu_0 c}[E_{1,0}^2 \cos^2(k_1 x - \omega_1 t) + 2E_{1,0}E_{2,0}\cos(k_1 x - \omega_1 t)$

       $\times \cos(k_2 x - \omega_2 t + \delta) + E_{2,0}^2 \cos^2(k_2 x - \omega_2 t + \delta)]\hat{i}$

    (b)  $\vec{S}_{\text{av}} = \dfrac{1}{2\mu_0 c}[E_{1,0}^2 + E_{2,0}^2]\hat{i}$

    (c)  $\vec{S} = \dfrac{1}{\mu_0 c}[E_{1,0}^2 \cos^2(k_1 x - \omega_1 t) - E_{2,0}^2 \cos^2(k_2 x + \omega_2 t + \delta)]\hat{i}$

    (d)  $\vec{S}_{\text{av}} = \dfrac{1}{2\mu_0 c}[E_{1,0}^2 - E_{2,0}^2]\hat{i}$

63. (a)  $9.15 \times 10^{-15}$ T

    (b)  $(1.01\ \mu\text{V})\cos(8.80 \times 10^5 \text{ s}^{-1})t$

    (c)  $(5.49\ \mu\text{V})\sin(8.80 \times 10^5 \text{ s}^{-1})t$

65. (a)  $E = \dfrac{I\rho}{\pi a^2}$

    (b)  $B = \dfrac{\mu_0 I}{2\pi a}$

    (c)  $\vec{S} = -\dfrac{I^2\rho}{2\pi^2 a^3}\hat{r}$

67.  $0.574\ \mu\text{m}$

69.  3.33 mN

# Chapter 31

1.  The population inversion between the state $E_{2,\text{Ne}}$ and the state 1.96 eV below it (see Figure 31-9) is achieved by inelastic collisions between neon atoms and helium atoms excited to the state $E_{2,\text{He}}$.

3.  The layer of water greatly reduces the light reflected back from the car's headlights, but increases the light reflected by the road of light from the headlights of oncoming cars.

5.  The change in atmospheric density results in refraction of the light from the sun, bending it toward the earth. Consequently, the sun can be seen even after it is just below the horizon. Also, the light from the lower portion of the sun is refracted more than the light from the upper portion, so the lower part appears to be slightly higher in the sky. The effect is an apparent flattening of the disk into an ellipse.

7.  She takes the path $LOS$.

9.  (d)

11.

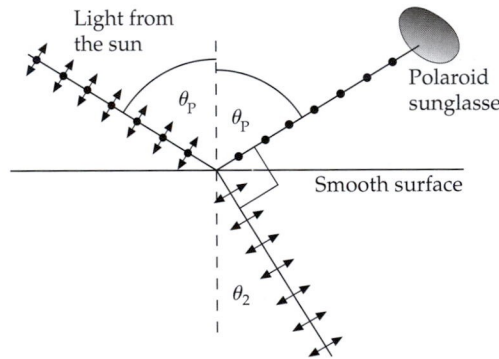

13.  (c)

15.  In resonance absorption, the molecules respond to the frequency of the light through the Einstein photon relation $E = hf$. Thus, the color appears to be the same in spite of the fact that the wavelength has changed.

17. (a)  2:00 A.M., September 1

    (b)  2:08 A.M., September 1

19. (a)  15.0 mJ

    (b)  $5.25 \times 10^{16}$

21. (a)  435 nm

    (b)  1210 nm

23. (a) 387.5 nm; $\lambda_{21} = 1138$ nm; $\lambda_{10} = 587.7$ nm

(b) $\lambda_{03} = 285.1$ nm; $\lambda_{32} = 1078$ nm; $\lambda_{21} = 1138$ nm;
$\lambda_{10} = 587.7$ nm; $\lambda_{31} = 553.6$ nm; $\lambda_{20} = 387.5$ nm

25. (a)

27. 5 min, 23 s

29. (a) 20.0 $\mu$s

(b) $\Delta t_{reaction} \approx 10^4 \Delta t$

33. $v_{water} = 2.25 \times 10^8$ m/s; $v_{glass} = 2.00 \times 10^8$ m/s

35. (a) 50.2°

(b) 38.8°

(c) 26.3°

37. 92.2 percent

41. 62.5°

43. 102 m²

45. 1.30

47. 5.43°

49. (a) 48.7°

(b) Note that $\theta_2$ equals the critical angle for a water–air interface. Therefore, the ray will not leave the water for $\theta_1 \gtrsim 41.8°$.

51. 1.02°

53. (a) 53.1°

(b) 56.3°

55. (a) $\frac{1}{8}I_0$

(b) $\frac{3}{32}I_0$

57. (a) 30.0°

(b) 1.73

59. $I_3 = \frac{1}{8}I_0 \sin^2 2\omega t$

61. 13

63. $I_4 = 0.211 I_0$

65. Right circularly polarized; $\vec{E} = E_0 \sin(kx + \omega t)\hat{j} - E_0 \cos(kx + \omega t)\hat{k}$

67. 35.3°

69. 1.45 m

71. 3.42 m

73. (a) 36.8°

(b) 38.7°

75. (a) $-1.00$ m

(b) $\theta_i = 26.6°$; $\theta_r = 26.6°$

77. For silicate flint glass: $\theta_p = 58.3°$; for borate flint glass: $\theta_p = 57.5°$; for quartz glass: $\theta_p = 57.0°$; for silicate crown glass: $\theta_p = 56.5°$

79. (b) $\theta_p > \theta_c$

81. (a) 1.33

(b) $\theta_c = 37.2°$

(c) $\theta_2 = 48.8°$

83. (a) $\dfrac{I_t}{I_0} = \left[\dfrac{4n}{(n+1)^2}\right]^{2N}$

(b) 0.783

(c) $\approx 28$

85. (a) 24.0°

(b) 4.45 km

(c) $\theta = \tan^{-1}\left(\dfrac{v_{earth}}{c}\right)$

(d) $2.99 \times 10^8$ m/s

# Chapter 32

1. Yes. Note that a virtual image is seen because the eye focuses the diverging rays to form a real image on the retina. Similarly, the camera lens can focus the diverging rays onto the film.

3. (a) False

(b) False

(c) True

(d) False

5. A convex mirror always produces a virtual erect image that is smaller than the object. It never produces an enlarged image.

7. (b)

9. (a) The lens will be positive if its index of refraction is greater than that of the surrounding medium, and the lens is thicker in the middle than at the edges. Conversely, if the index of refraction of the lens is less than that of the surrounding medium, the lens will be positive if it is thinner at its center than at the edges.

(b) The lens will be negative if its index of refraction is greater than that of the surrounding medium, and the lens is thinner at the center than at the edges. Conversely, if the index of refraction of the lens is less than that of the surrounding medium, the lens will be negative if it is thicker at the center than at the edges.

11. (d)

13. The eye accommodates by varying the focal length of a lens located a fixed distance away from the retina. A camera, on the other hand, has a fixed focal length. The distance between the lens and the film is varied.

15. (b)

17. (d)

19. (a)

21. Microscopes ordinarily produce images (either the intermediate one produced by the objective or the one viewed through the eyepiece) that are larger than the object being viewed. A telescope, on the other hand, ordinarily produces images that are much reduced compared to the object. The object is normally viewed from a great distance, and the telescope magnifies the angle subtended by the object.

23. Plano-convex lens: $r_1 = -16.2$ cm; $r_2 = \infty$

Biconvex with equal curvature: $r_1 = -32.4$ cm; $r_2 = 32.4$ cm

Biconvex with unequal curvature: $r_1 = 16.2$ cm; $r_2 = 8.10$ cm

25.
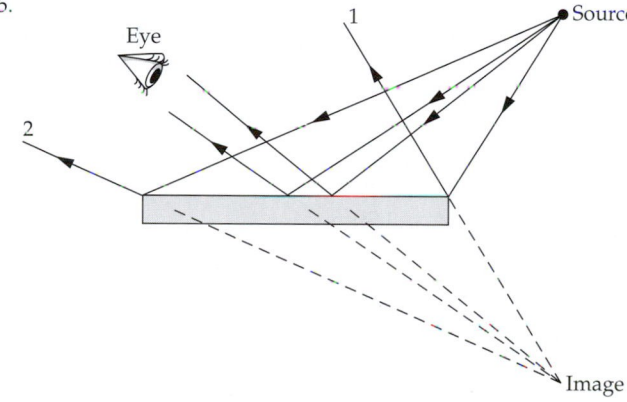

27.

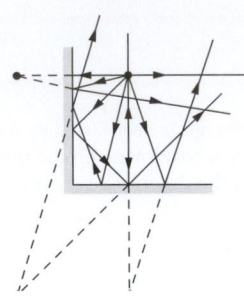

29. (a) The first image in the mirror on the left is 10 cm behind the mirror. The mirror on the right forms an image 20 cm behind that mirror or 50 cm from the left mirror. This image will result in a second image 50 cm behind the left mirror. The first image in the left mirror is 40 cm from the right mirror and forms an image 40 cm behind the right mirror or 70 cm from the left mirror. That image gives an image 70 cm behind the left mirror. The fourth image behind the left mirror is 110 cm behind that mirror.

(b) Proceeding as in Part (a) for the mirror on the right, one finds the location of the images to be 20 cm, 40 cm, 80 cm, and 100 cm behind the right-hand mirror.

31. (a) $s' = 15.8$ cm; $m = -0.316$; Because the image distance is positive and the lateral magnification is less than one and negative, the image is real, inverted, and reduced.

(b) $s' = 24.0$ cm; $m = -1$; Because the image distance is positive and the lateral magnification is one and negative, the image is real, inverted, and the same size as the object.

(c) $s' = \infty$ and there is no image.

(d) $s' = -24.0$ cm; $m = 3$; Because the image distance is negative and the lateral magnification is three and positive, the image is virtual, erect, and three times the size of the object.

33. (a) $s' = -9.85$ cm; $m = 0.179$; Because the image distance is negative and the lateral magnification is less than one in magnitude and positive, the image is virtual, erect, and reduced.

(b) $s' = -8.00$ cm; $m = 0.333$; Because the image distance is negative and the lateral magnification is less than one in magnitude and positive, the image is virtual, erect, and reduced.

(c) $s' = -6.00$ cm; $m = 0.5$; Because the image distance is negative and the lateral magnification is one-half in magnitude and positive, the image is virtual, erect, and half the size of the object.

(d) $s' = -4.80$ cm; $m = 0.600$; Because the image distance is negative and the lateral magnification is less than one and positive, the image is virtual, erect, and six-tenths the size of the object.

35. (a) 5.13 cm

(b) The mirror must be concave. A convex mirror always produces a diminished virtual image.

37. $s' = -4.00$ m; $y' = 3.68$ cm

39. (a)

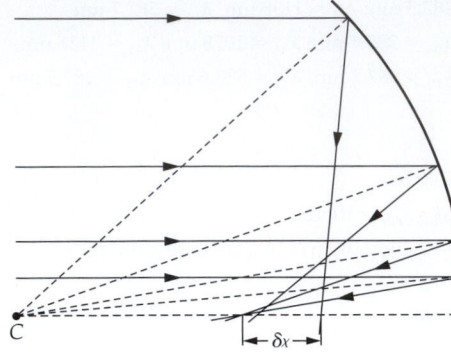

$\delta x = 1.01$ cm

(b) By blocking off the edges of the mirror so that only paraxial rays within 2 cm of the mirror axis are reflected, the spread is reduced by about 83 percent.

41. (a) $-1.33$ m

(b) Because $f_{\text{small}} < 0$, the small mirror is convex.

43. (a) $s' = -8.58$ cm, where the minus sign tells us that the image is 8.58 cm from the front surface of the bowl on the same side as the fish.

(b) $s' = -35.9$ cm, where the minus sign tells us that the image is 35.9 cm from the front surface of the bowl on the same side as the fish.

45. 14.4 cm

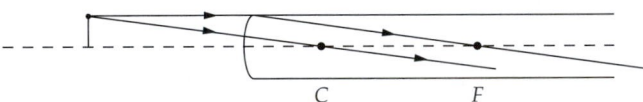

47. (a) $s' = -104$ cm, where the negative image position tells us that the image is 104 cm in front of the surface and is virtual.

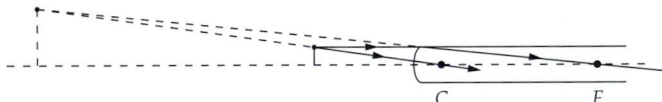

(b) $s' = -8.29$ cm, where the minus sign tells us that the image is 8.29 cm in front of the surface and is virtual.

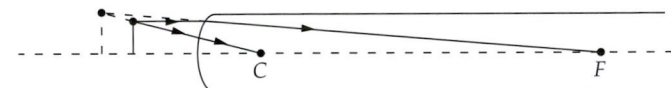

(c) $s' = 63.5$ cm, that is, the image is 63.5 cm behind (to the right of) the surface (at the focal point) and is real.

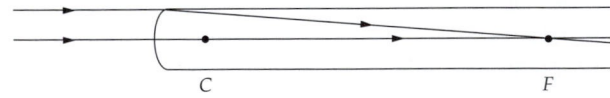

49. (a) $s' = 64.0$ cm

(b) $s' = -80.0$ cm

(c) The final image is 96 cm − 80 cm = 16 cm from the surface, the radius of the surface is 8 cm and is virtual.

51. (a) 19.0 cm

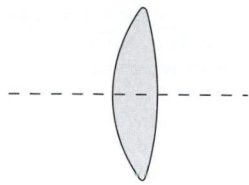

(b) 30.0 cm

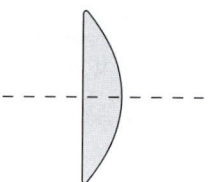

(c) −15.0 cm

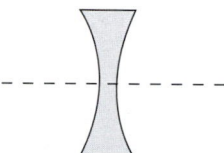

(d) −52.0 cm

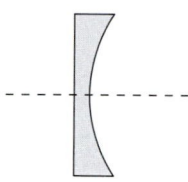

53. (a) −30.3 cm
    (b) −22.0 cm (22.0 cm to the left of the lens)
    (c) 0.275
    (d) Because $s' < 0$ and $m > 0$, the image is virtual and upright.

55.

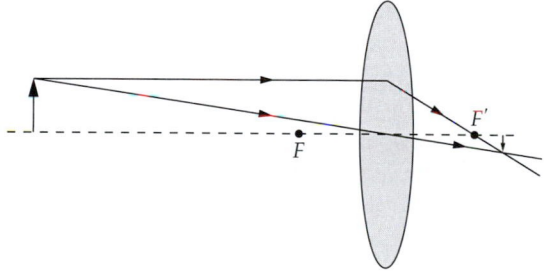

$s' = 16.7$ cm; $y' = -2.00$ cm; The image is real, inverted, and diminished. Because $s' > 0$ and $y' = -2.00$ cm, the image is real, inverted, and diminished in agreement with the ray diagram.

57.

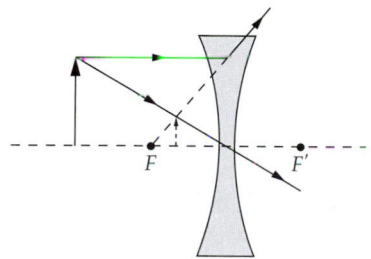

$s' = -6.67$ cm; $y' = 0.500$ cm; The image is virtual, erect, and diminished. Because $s' < 0$ and $y' = 0.500$ cm, the image is virtual, erect, and about one-third the size of the object in agreement with the ray diagram.

59. (a)

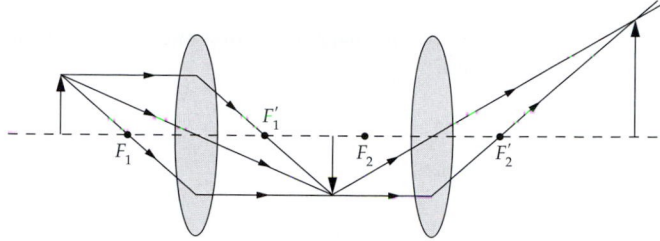

$s'_2 = 30$ cm and the final image is 85.0 cm from the object; $m_2 = -2$

(b) Because $s'_2 > 0$ and $m = m_1 m_2 = 2$, the image is real and upright.

(c) The image magnification is twice the size of the object.

61. (b) 3.70 m

63. (a) and (b)

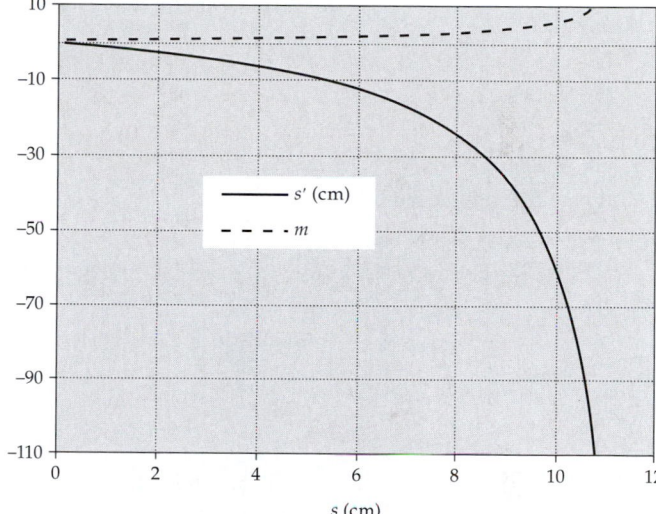

(c) The images are virtual and erect for this range of object distances.

(d) The asymptote of the graph of $s'$ versus $s$ corresponds to the image approaching infinity as the object distance approaches the focal length of the lens. The horizontal asymptote of the graph of $m$ versus $s$ indicates that, as the object moves toward the lens, the height of the image formed by the lens approaches the height of the object.

65. 15.0 cm; The final image is 50 cm from the object, real, inverted, and the same size as the object.

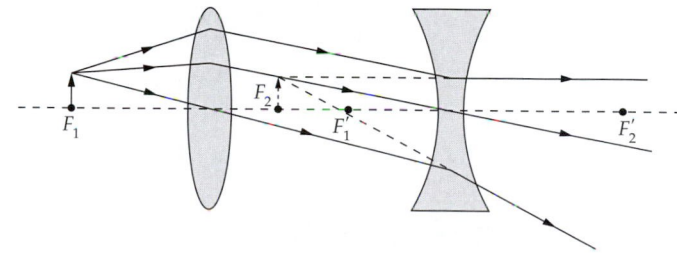

69. (a) 41.2 cm

(b) −1.53

(c) Because $m < 0$, the image is inverted. Because $s'_2 > 0$, the image is real.

71. (a) False

(b) True

73. (c) 40.0 D; 4.00 D

75. (c) 6.00 D

77. 1.72 mm

79. (a) 80.0 $\mu$rad

(b) 1.60 mm

81. 0.444 D

83. 3.07 D

85. 6.00

87. 5.00

89. (a) 3.00

(b) 4.00

93. (a) −1.88

(b) −18.8

95. −232

97. (a) 9.00 mm

(b) −20.0; −0.180 rad

99. (a) $P_{\text{Palomar}} = (25.0)P_{\text{Yerkes}}$

(b) −134

101. (b)

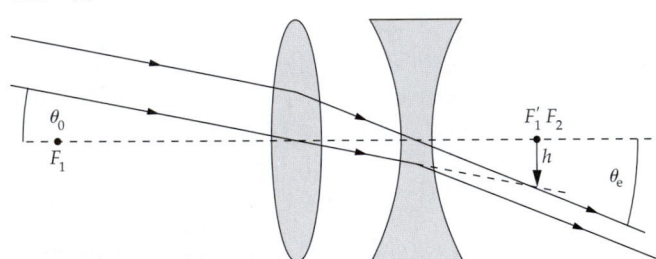

103. −1/150

105. 1.34 mm

107. (a) $s = 5.00$ cm; $s' = -10.0$ cm

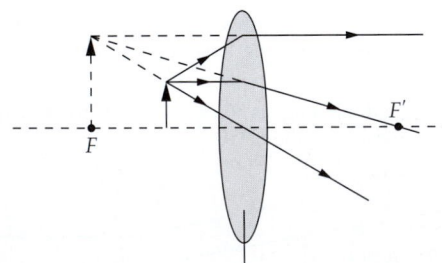

(b) $s = 15.0$ cm; $s' = 30.0$ cm

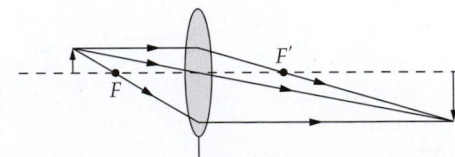

109. (a) Because the focal lengths appear in the magnification formula as a product, it would appear that it does not matter in which order we use them. The usual arrangement would be to use the shorter focal length lens as the objective, but we get the same magnification in the reverse order. What difference does it make then? It makes no difference in this problem. However, it is generally true that the smaller the focal length of a lens, the smaller its diameter. This condition makes it harder to use the shorter focal length lens, with its smaller diameter, as the eyepiece lens. If we separate the objective and eyepiece lenses by $L + f_e + f_o = 16$ cm + 7.5 cm + 2.5 cm = 26.0 cm, the overall magnification will be −21.3.

(b)

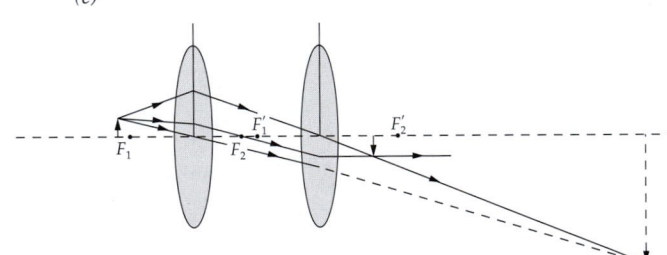

111. 3.70 m

113. (a) 9.52 cm

(b) −1.19

(c)

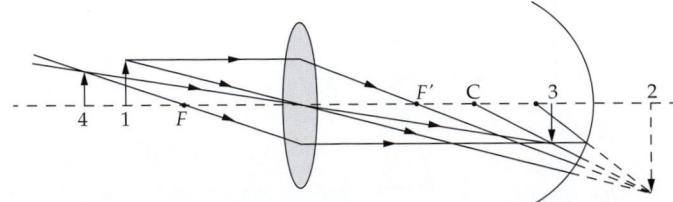

115. (a) 21.3 cm

(b) 79.2 cm

117. 0.0971 m/s

119. (a) 22.5 cm

(b) 18.0 cm

(c)

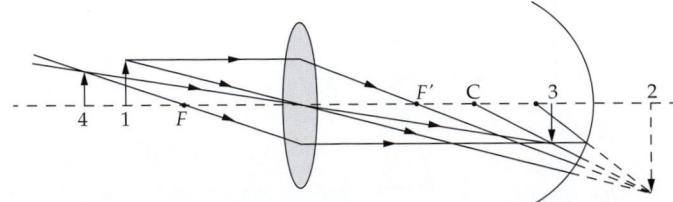

To see this image, the eye must be to the left of the image 4.

121. 36.8 cm

123. (a) The final image is 0.8 cm behind the mirror surface.

(b) The final image is at the mirror surface.

# Chapter 33

1. The energy is distributed nonuniformly in space; in some regions the energy is below average (destructive interference), in others it is higher than average (constructive interference).

3. The thickness of the air space between the flat glass and the lens is approximately proportional to the square of $d$, the diameter of the ring. Consequently, the separation between adjacent rings is proportional to $1/d$.

5. If the film is thick, the various colors (i.e., different wavelengths) will give constructive and destructive interference at that thickness. Consequently, what one observes is the reflected intensity of white light.

7. The first zeros in the intensity occur at angles given by $\sin \theta = \lambda/a$. Hence, decreasing $a$ increases $\theta$, and the diffraction pattern becomes wider.

9. (a)

11. (a) False

    (b) True

    (c) True

    (d) True

    (e) True

13. $3.50 \, \mu$m

15. $\theta_{\text{red}} \approx 1.82°$; $\theta_{\text{blue}} \approx 1.26°$.

17. (a) 300 nm

    (b) 135°

19. 164°

21. $5.46 \, \mu$m $< d < 5.75 \, \mu$m

23. (a) 600 nm

    (b) From the table, we see that the only wavelengths in the visible spectrum are 720 nm, 514 nm, and 400 nm.

    (c) From the table, we see that the missing wavelengths in the visible spectrum are 720 nm, 514 nm, and 400 nm.

25. 476 nm

27. (c) The transmitted pattern is complementary to the reflected pattern.

    (d) 68 bright fringes

    (e) 1.14 cm

    (f) The wavelength of the light in the film becomes $\lambda_{\text{air}}/n = 444$ nm. The separation between fringes is reduced and the number of fringes that will be seen is increased by the factor $n = 1.33$.

29. 0.535 nm; 0.926 nm

31. 4.95 mm

33. (a) $50.0 \, \mu$m

    (b) Not with the unaided eye. The separation is too small to be observed with the naked eye.

    (c) 0.500 mm

35. 625 nm and 417 nm

37. (a) 0.600 mrad

    (b) 6.00 mrad

    (c) 60.0 mrad

39. (a) 1.53 km

    (b) 459 m

    (c) $1.1 \times 10^{-7}$

    (d) $1.21 \times 10^{-14}$

41. (a) $20.0 \, \mu$m

    (b) 9

43. There are eight interference fringes on each side of the central maximum. The secondary diffraction maximum is half as wide as the central one. It follows that it will contain eight interference maxima.

45. $3.61 \sin(\omega t - 56.3°)$

47. $0.0162 I_0$

49. (b) 6.00 mm

    (c) The width for four sources is half the width for two sources.

51. (a)

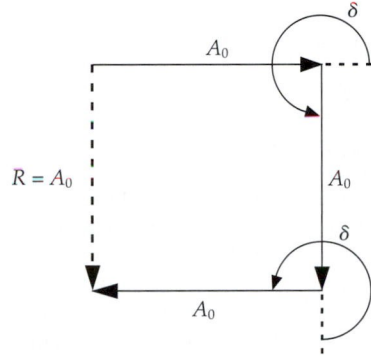

    (b) $5.56 \, \text{mW/m}^2$

53. (a) 8.54 mrad

    (b) 6.83 cm

55. 7.00 mm

57. 484 m

59. $5.00 \times 10^9$ m

61. 86.9 mrad; 82.1 mrad

63. 40.6°; 38.0°

65. 30.0°

67. Because $m_{\text{max}} = 2.98$, one can see the complete spectrum only for $m = 1$ and 2. Because 700 nm $< 2 \times 400$ nm, there is no overlap of the second-order spectrum into the first-order spectrum; however, there is overlap of long wavelengths in the second order with short wavelengths in the third-order spectrum.

69. (a) $y_1 = 0.353$ m; $y_2 = 0.706$ m

    (b) $\Delta y = 88.4 \, \mu$m

    (c) 8000

71. $R = 3.09 \times 10^5$; $n = 5.15 \times 10^4 \, \text{cm}^{-1}$

73. (a) $n = 750 \, \text{cm}^{-1}$

    (b) 4.20 cm; 12.6 cm

75. (a) $\phi = \sin^{-1}\left(\dfrac{m\lambda}{a}\right)$

    (b) 64.2°

77. (a) 0.150 mm

    (b) $3.33 \times 10^3 \, \text{m}^{-1}$

79. 1.68 cm

81. 0.130 mrad

85. (a) 97.8 nm

    (b) No, because 180 nm is not in the visible portion of the spectrum.

    (c) 0.273; 0.124

87. 12.3 m

# Chapter 34

1. (c)

3. (a) True
   (b) False
   (c) True
   (d) True

5. (a)

7. (a) True
   (b) True
   (c) True
   (d) False

9. (c)

11. In the photoelectric effect, an electron absorbs the energy of a single photon. Therefore, $K_{max} = hf - \phi$, independent of the number of photons incident on the surface. However, the number of photons incident on the surface determines the number of electrons that are emitted.

13. According to quantum theory, the average value of many measurements of the same quantity will yield the expectation value of that quantity. However, any single measurement may differ from the expectation value.

15. (a)

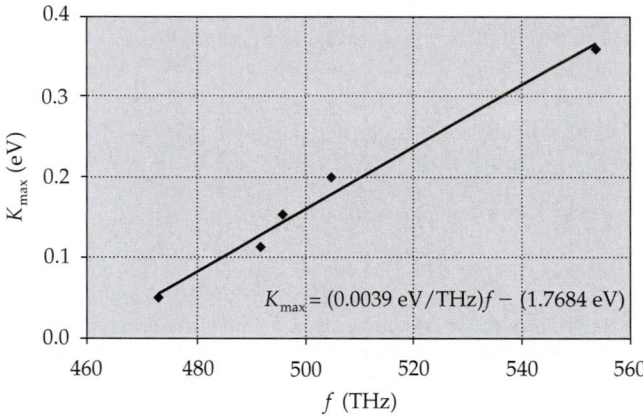

$K_{max} = (0.0039\ eV/THz)f - (1.7684\ eV)$

   (b) 1.77 eV
   (c) Cesium

17. Soccer ball

19. (a) $2.42 \times 10^{14}$ Hz
    (b) $2.42 \times 10^{17}$ Hz
    (c) $2.42 \times 10^{20}$ Hz

21. (a) 12.4 keV
    (b) 1.24 GeV

23. $1.95 \times 10^{16}\ s^{-1}$

25. (a) 4.13 eV
    (b) 2.10 eV
    (c) 0.784 eV
    (d) 590 nm

27. (a) 653 nm; $4.59 \times 10^{14}$ Hz
    (b) 3.06 eV
    (c) 1.64 eV

29. 1.21 pm

31. 180 pm

33. $p_1 = 9.32 \times 10^{-24}\ kg \cdot m/s$; $p_e = 1.80 \times 10^{-23}\ kg \cdot m/s$

35. 42

37. 2.91 nm

39. (a) $p_e = 2.09 \times 10^{-22}\ N \cdot s$; $p_p = 8.95 \times 10^{-21}\ N \cdot s$; $p_\alpha = 1.79 \times 10^{-20}\ N \cdot s$
    (b) $\lambda_p = 7.41 \times 10^{-14}\ m$; $\lambda_e = 3.17 \times 10^{-12}\ m$; $\lambda_\alpha = 3.70 \times 10^{-14}\ m$

41. 20.2 fm

43. (a) 0.820 meV
    (b) 820 MeV

45. 0.167 nm

47. 4.63 pm

49. $1.66 \times 10^{-33}$ m; This is many orders of magnitude smaller than even the diameter of a proton.

51. 0.0872 nm; This distance is of the order of the size of an atom.

53. (a) $E_1 = 206$ MeV; $E_2 = 824$ MeV; $E_3 = 1.85$ GeV

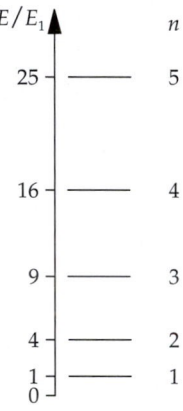

   (b) 2.01 fm
   (c) 1.20 fm
   (d) 0.752 fm

55. (a) 0.004
    (b) 0.003
    (c) 0

57. (a) $\dfrac{L}{2}$
    (b) $0.321L^2$

59. (a) $\dfrac{1}{\sqrt{a}}$
    (b) 0.865

61. (a) 0.500
    (b) 0.402
    (c) 0.750

65. $<x> = 0$; $<x^2> = L^2\left[\dfrac{1}{12} - \dfrac{1}{2\pi^2}\right]$

67. (a) 3.10 eV
    (b) $6.25 \times 10^{16}$ eV
    (c) $2.02 \times 10^{16}$

69. (a) 1.00 $\mu$m; $10^{-16}\ kg \cdot m/s$
    (b) $0.949 \times 10^{12}$

71. 1.21 eV

73. $6.80 \times 10^3$ km

75. (a) 3.18 W/m²

(b) 1.04 × 10¹⁵

79. 1.28 MeV

81. 1.04 eV; 554 nm

83. (b) 0.2 percent

(c) Classically, the energy is continuous. For very large values of $n$, the energy difference between adjacent levels is infinitesimal.

85. (a) 6.25 × 10⁻⁴ eV/s

(b) 53.3 min

# Chapter 35

1. True

3. (a) 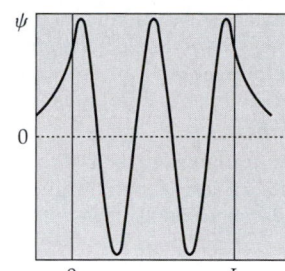 (b)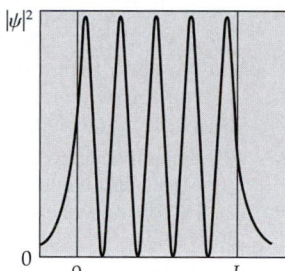

11. (a) 9.49 × 10⁻⁹ m

(b) 4.19 meV

13. $\Delta x\, \Delta p = \dfrac{\hbar}{2}$

15. (a) $\sqrt{\dfrac{3}{2}}\,k_1$

(b) 0.0102

(c) 0.990

(d) 9.90 × 10⁵

17. 0.341

19. (a) $r_{1,4\,\text{MeV}} = 6.62 \times 10^{-14}$ m; $r_{1,7\,\text{MeV}} = 3.78 \times 10^{-14}$ m

(b) $T_{4\,\text{MeV}} = 3.27 \times 10^{-51}$; $T_{7\,\text{MeV}} = 8.21 \times 10^{-39}$

21. $\psi = A \sin\!\left(\dfrac{n_1 \pi}{L_1} x\right) \sin\!\left(\dfrac{n_2 \pi}{2L_1} y\right) \sin\!\left(\dfrac{n_3 \pi}{3L_1} z\right)$

23. $\psi = A \sin\!\left(\dfrac{n_1 \pi}{L_1} x\right) \sin\!\left(\dfrac{n_2 \pi}{2L_1} y\right) \sin\!\left(\dfrac{n_3 \pi}{4L_1} z\right)$

25. (a) $\psi(x, y) = A \sin\dfrac{n\pi}{L} x \sin\dfrac{m\pi}{L} y$

(b) $E_{n,m} = \dfrac{h^2}{8mL^2}(n^2 + m^2)$

(c) $E_{1,2} = E_{2,1} = \dfrac{5h^2}{8mL^2}$

(d) $(1, 7), (7, 1), (5, 5)$; $E = \dfrac{25h^2}{4mL^2}$

27. $E_{0,10\,\text{bosons}} = \dfrac{5h^2}{4mL^2}$

33. $E_0 = \dfrac{5h^2}{mL^2}$; $E_1 = E_2 = \dfrac{21h^2}{4mL^2}$

39. $A_2 = \sqrt[4]{\dfrac{8m\omega_0}{h}}$

# Chapter 36

1. Examination of Figure 36-4 indicates that as $n$ increases, the spacing of adjacent energy levels decreases.

3. (a)

5. (d)

7. (a)

9. (c)

11. In conformity with the exclusion principle, the total number of electrons that can be accommodated in states of quantum number $n$ is $n^2$ (see Problem 48). The fact that closed shells correspond to $2n^2$ electrons indicates that there is another quantum number that can have two possible values.

13. (a) phosphorus

(b) chromium

15. (d)

17. The optical spectrum of any atom is due to the configuration of its outer-shell electrons. Ionizing the next atom in the periodic table gives you an ion with the same number of outer-shell electrons and almost the same nuclear charge. Hence, the spectra should be very similar.

19. (a) allowed

(b) not allowed

(c) not allowed

(d) allowed

(e) allowed

21. (b) 75.2 nK

23. (a) 103 nm

(b) 97.3 nm

25. (a) 1.51 eV; 821 nm

(b) 0.661 eV; 1876 nm

0.967 eV; 1282 nm

1.13 eV; 1097 nm

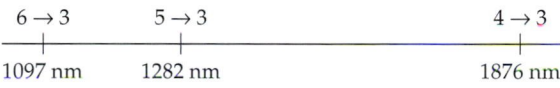

27. (a) 657.8 nm

(b) 1.0945 × 10⁷ m⁻¹

29. (b) 1.096776 × 10⁷ m⁻¹; 1.097448 × 10⁷ m⁻¹; 0.0546 percent

31. (a) 1.49 × 10⁻³⁴ J·s

(b) −1, 0, +1

(c)

33. (a) 0, 1, 2

(b) 0, −1, 0, +1; −2, −1, 0, +1, +2

(c) 18

35. (a) 45.0°

    (b) 26.6°

    (c) 8.05°

37. (a) $6\hbar^2$

    (b) $4\hbar^2$

    (c) $2\hbar^2$

39. (a) 4

    (b)

| $n$ | $\ell$ | $m$ | $(n, \ell, m)$ |
|-----|--------|-----|----------------|
| 2 | 0 | 0 | (2, 0, 0) |
| 2 | 1 | −1 | (2, 1, −1) |
| 2 | 1 | 0 | (2, 1, 0) |
| 2 | 1 | 1 | (2, 1, 1) |

41. (a) $\psi_{2,0,0}(a_0) = \dfrac{0.0605}{a_0^{3/2}}$

    (b) $[\psi_{2,0,0}(a_0)]^2 = \dfrac{0.00366}{a_0^3}$

    (c) $P(a_0) = \dfrac{0.0460}{a_0}$

43. (a) $9.20 \times 10^{-4}$

    (b) 0

49. 0.323

51. $\ell$ must equal to 0 or 1

53.

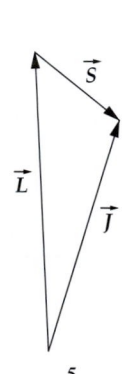
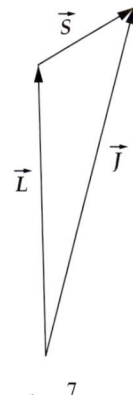

$j = \dfrac{5}{2}$     $j = \dfrac{7}{2}$

55. (c)

57. (a) $-2\hbar, -\hbar, 0, \hbar, 2\hbar$

    (b) $-4\hbar, -3\hbar, -2\hbar, -\hbar, 0, \hbar, 2\hbar, 3\hbar, 4\hbar$

59. (a) 2s or 2p

    (b) $1s^2 2s^2 2p^6 3p$

    (c) 1s2s

61. (a) 0.0610 nm; 0.0578 nm

    (b) 0.0542 nm

63. (a) 1.00 nm

    (b) 0.155 nm

65. $n_i = 4$ to $n_f = 1$

67. $n_i = 3$ to $n_f = 2$; $n_i = 9$ to $n_f = 3$; $n_i = 7$ to $n_f = 4$

69. (a) 1.6179 eV; 1.6106 eV

    (b) 0.00730 eV

    (c) 63.0 T

73. (a) 1.06 GHz

    (b) 28.4 cm; microwave

75. (a) $R_H = 1.096776 \times 10^7\,\text{m}^{-1}$; $R_D = 1.097075 \times 10^7\,\text{m}^{-1}$

    (b) 0.179 nm

77. (a) $R_T = 1.097175 \times 10^7\,\text{m}^{-1}$

    (b) 0.0600 nm; 0.238 nm

---

# Chapter 37

1. Yes. Because the center of charge of the positive Na ion does not coincide with the center of charge for the negative Cl ion, the NaCl molecule has a permanent dipole moment. Hence, it is a polar molecule.

3. Neon occurs naturally as Ne, not $Ne_2$. Neon is a rare gas atom with a closed shell electron configuration.

5. The diagram would consist of a nonbonding ground state with no vibrational or rotational states for ArF (similar to the upper curve in Figure 37-4) but for ArF* there should be a bonding excited state as with a definite minimum with respect to internuclear separation and several vibrational states as in the excited state curve of Figure 37–13.

7. The effective force constant from Example 37-4 is $1.85 \times 10^3$ N/m. This value is about 25% larger than the given value of the force constant of the suspension springs on a typical automobile.

9. For $H_2$, the concentration of negative charge between the two protons holds the protons together. In the $H_2^+$ ion, there is only one electron that is shared by the two positive charges so that most of the electronic charge is again between the two protons. However, the negative charge in the $H_2^+$ ion is not as effective as the larger charge in the $H_2$ molecule, and the protons should be farther apart. The experimental values support this argument. For $H_2$, $r_0 = 0.074$ nm, while for $H_2^+$, $r_0 = 0.106$ nm.

11. With more than two atoms in the molecule, there will be more than just one frequency of vibration because there are more possible relative motions. In advanced mechanics these are known as normal modes of vibration.

13. (a) $\ell = 2.15 \times 10^{30}$

    (b) $5.10 \times 10^{-65}$

15. 0.946 nm

17. (a) 23.0 kcal/mol

    (b) 98.2 kcal/mol

19. 0.44 eV

21.

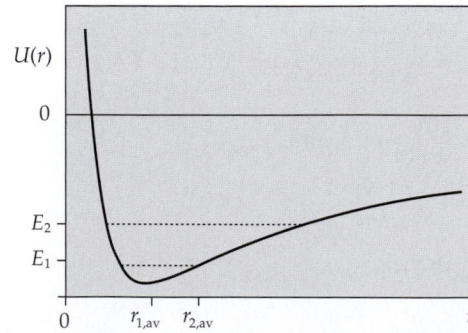

23. (a) −6.64 eV

    (b) 5.70 eV; 0.63 eV

25. (a) 0.31 eV

    (b) $n = 19.7$; $C = 1.37 \times 10^{-13}$ eV·nm

27. 0.121 nm

29. 41

33. (a)  0.179 eV

    (b)  $2.80 \times 10^{-47}$ kg·m$^2$

    (c)  0.132 nm

35.

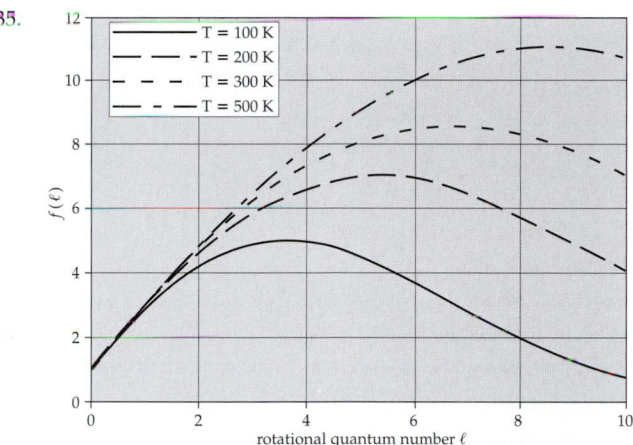

37. 0.9722 u; 0.9737 u; 0.00150; 0.00119

39. 0.955 meV

41. 1.55 kN/m

43. $r_0 = 0.074$ nm; $U_0 = 4.52$ eV

45. $1/x^4$

47. (a)  $1.45 \times 10^{-46}$ kg·m$^2$; 0.239 meV

    (b)  $\ell = 5, E = 7.14$ meV ────────

        $\ell = 4, E = 4.76$ meV ────────

        $\ell = 3, E = 2.86$ meV ────────

        $\ell = 2, E = 1.43$ meV ────────

        $\ell = 1, E = 0.476$ meV ────────

        $\ell = 0, E = 0$ ────────

    (c)  $\lambda_{1,0} = 2596$ $\mu$m; $\lambda_{2,1} = 1298$ $\mu$m; $\lambda_{3,2} = 865$ $\mu$m;
        $\lambda_{4,3} = 649$ $\mu$m; $\lambda_{5,4} = 519$ $\mu$m; microwave

## Chapter 38

1. The energy lost by the electrons in collision with the ions of the crystal lattice appears as Joule heat ($I^2R$).

3. (a)  Examining Table 38-2, we see that the greatest difference between the work functions will occur when potassium and nickel are joined.

    (b)  3.10 V

5. (c)

7. (a)  True

    (b)  False

    (c)  True

    (d)  False

    (e)  True

    (f)  True

    (g)  False

9. (b)

11. The excited electron is in the conduction band and can conduct electricity. A hole is left in the valence band allowing the positive hole to move through the band also contributing to the current.

13. (c)

15. $\infty$, 40 Ω, 20 Ω, 10 Ω, 5 Ω

17. 2.07 g/cm$^3$

19. (a)  $-10.6$ eV

    (b)  $-2.83$ percent

21. (a)  0.123 $\mu$Ω·m

    (b)  0.0708 $\mu$Ω·m

23. (a)  $n_{Ag} = 5.86 \times 10^{22}$ electrons/cm$^3$

    (b)  $n_{Ag} = 5.90 \times 10^{22}$ electrons/cm$^3$; Both these results agree with the values in Table 38-1.

25. 4.00

27. (a)  $1.07 \times 10^6$ m/s

    (b)  $1.39 \times 10^6$ m/s

    (c)  $1.89 \times 10^6$ m/s

29. (a)  4.22 eV

    (b)  2.85 eV

31. (a)  $5.90 \times 10^{28}$ e/m$^3$

    (b)  5.50 eV

    (c)  211

    (d)  $E_F$ is 211 times $kT$ at room temperature. There are so many free electrons present that most of them are crowded, as described by the Pauli exclusion principle, up to energies far higher than they would be according to the classical model.

33. (c)  $63.6 \times 10^9$ N/m$^2$; $B = 0.454B_{Cu}$

35. 0.192 J/mol·K

37. (a)  66.1 nm

    (b)  0.0179 nm$^2$

39. 1.09 $\mu$m

41. 177 nm

43. 116 K

45. 3.17 nm; 8.46 nm

47. 37.2 nm; $\lambda_{Cu} = 38.8$ nm; The mean free paths agree to within 4.02 percent.

49.

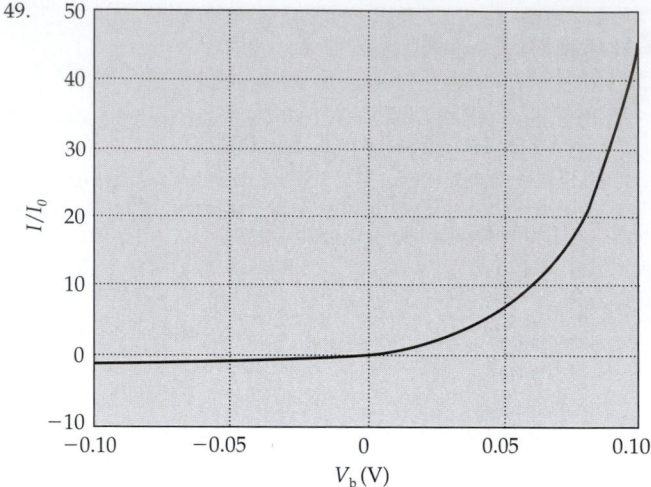

51. 250

53. (a) (b)

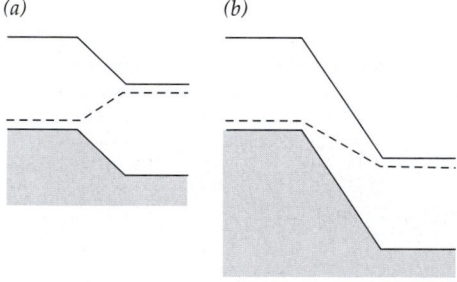

55. $n = 1.00 \times 10^{23}$ m$^{-3}$; the semiconductor is $p$-type.

57. (a) 2.17 meV; $0.8E_{g,\text{measured}}$

(b) 0.454 mm

59. $1.97 \times 10^{18}$

63. 1

67. 0.596

73. 1.07

75. (a) $5.51 \times 10^{-3}$

(b) $1.84 \times 10^{-2}$

77. 689 nm

# Chapter 39

1. (a)

3. (a) True

(b) True

(c) False

(d) True

(e) False

(f) False

(g) True

5. Although $\Delta y = \Delta y'$; $\Delta t \neq \Delta t'$. Consequently, $u_y = \Delta y/\Delta t' \neq \Delta y'/\Delta t' = u_y'$.

7. (a) 0.946

(b) 12.3 G$c \cdot$y

9. (a) $L_p = 978.5$ m; the width of the beam is unchanged

(b) $9.57 \times 10^7$ m

(c) 0.102 $\mu$m

11. (a) 0.914$c$

(b) 22.5$c \cdot$y

(c) 101 y

15. $1.85 \times 10^4$ y

17. (a) 1.76 $\mu$s

(b) 6.32 $\mu$s

(c) 3.09 $\mu$s

(d) 1.70 km

19. 4.39 $\mu$s

21. (a) 2.10 $\mu$s

(b) 2.59 $\mu$s

(c) 0.49 $\mu$s

(d) 2.59 $\mu$s

(e) 4.36 h

(f) 18.8 h

25. $2.22 \times 10^7$ m/s

29. 10.5 ms

31. (a) $u_x = v; u_y = \dfrac{c}{\gamma}$

33. (a) 0.976$c$

(b) 0.997$c$

35. 66.7 percent

39. (a) 290 MeV

(b) 629 MeV

41. (a) 0.943$c$

(b) 3.00 MeV

(c) 2.83 MeV/$c$

(d) 0.878 MeV

(e) 4.12 MeV/$c^2$

43. In a freely falling reference frame, both cannonballs travel along straight lines, so they must hit each other, as they were pointed at each other when they were fired.

45. 0.999$c$

47. (a) $S'$ moves in the negative $x$ direction

(b) 1.73 y

49. 281 MeV

51. (a) $v = -\dfrac{E}{Mc}$

(b) $d = -\dfrac{LE}{Mc^2}$

53. $a_x' = \dfrac{a_x}{\gamma^3 \delta^3}; a_y' = \dfrac{a_y}{\gamma^2 \delta^2} + \dfrac{v u_y}{\gamma^3 \delta^3 c^2} a_x$

# Chapter 40

1. (a) $^{15}$N, $^{16}$N

(b) $^{54}$Fe, $^{55}$Fe

(c) $^{54}$Fe, $^{55}$Fe

3. Generally, $\beta$-decay leaves the daughter nucleus neutron rich, that is, above the line of stability. The daughter nucleus therefore tends to decay via $\beta^-$ emission, which converts a nuclear neutron to a proton.

5. It would make the dating unreliable because the current concentration of $^{14}$C is not equal to that at some earlier time.

7. The probability for neutron capture by the fissionable nucleus is large only for slow (thermal) neutrons. The neutrons emitted in the fission process are fast (high energy) neutrons and must be slowed to thermal neutrons before they are likely to be captured by another fissionable nucleus.

9. (a) $\beta^+$

   (b) $\beta^-$

11. (a) False

    (b) True

    (c) False

    (d) True

13. (a) $^{16}_{7}\text{N} \rightarrow {}^{16}_{8}\text{O} + {}^{0}_{-1}\beta + {}^{0}_{0}\overline{v} + Q$

    (b) $^{248}_{100}\text{Fm} \rightarrow {}^{244}_{98}\text{Cf} + {}^{4}_{2}\text{He} + Q$

    (c) $^{12}_{7}\text{N} \rightarrow {}^{12}_{6}\text{C} + {}^{0}_{+1}\beta + {}^{0}_{0}v + Q$

    (d) $^{81}_{34}\text{Se} \rightarrow {}^{81}_{35}\text{Br} + {}^{0}_{-1}\beta + {}^{0}_{0}\overline{v} + Q$

    (e) $^{61}_{29}\text{Cu} \rightarrow {}^{61}_{28}\text{Ni} + {}^{0}_{+1}\beta + {}^{0}_{0}v + Q$

    (f) $^{228}_{90}\text{Th} \rightarrow {}^{224}_{88}\text{Ra} + {}^{4}_{2}\text{He} + Q$

15. Mass density $= 10^{15}$, half life $= 10^{15}$, and nuclear masses $= 200$

17. (a) 92.2 MeV; 7.68 MeV

    (b) 492 MeV; 8.79 MeV

    (c) 1802 MeV; 7.57 MeV

19. (a) 3.02 fm

    (b) 4.59 fm

    (c) 6.98 fm

21. (a) $4.11 \times 10^{-21}$ J; 25.7 MeV

    (b) 2.22 km/s

    (c) 10.1 min

23. 295 MeV

25. (a) 5 min

    (b) 250 Bq

27. (a) 200 s

    (b) $3.47 \times 10^{-3}$ s$^{-1}$

    (c) 125 Bq

29. (a) 500 Bq; 250 Bq

    (b) $1.04 \times 10^6$; $5.19 \times 10^5$

    (c) 12.1 min

31. (a) $4.55 \times 10^3$ $\alpha$/s

    (b) $5.32 \times 10^4$ y

33. $^{239}_{94}\text{Pu} \rightarrow {}^{235}_{92}\text{U} + {}^{4}_{2}\alpha + Q$; $Q = 5.24$ MeV; $K_\alpha = 5.15$ MeV; $K_U = 87.7$ keV

35. (a) 0.133 h$^{-1}$; 5.20 h

    (b) $3.11 \times 10^6$

37.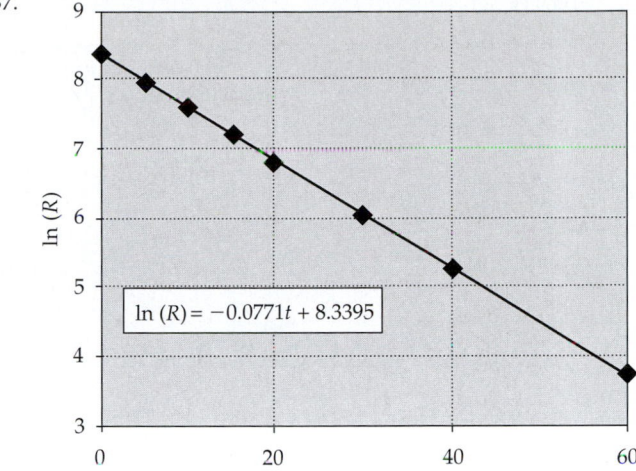

$\lambda = 0.0771$ min$^{-1}$

$t_{1/2} = 8.99$ min

39. 2.94 g

41. 3.50 min

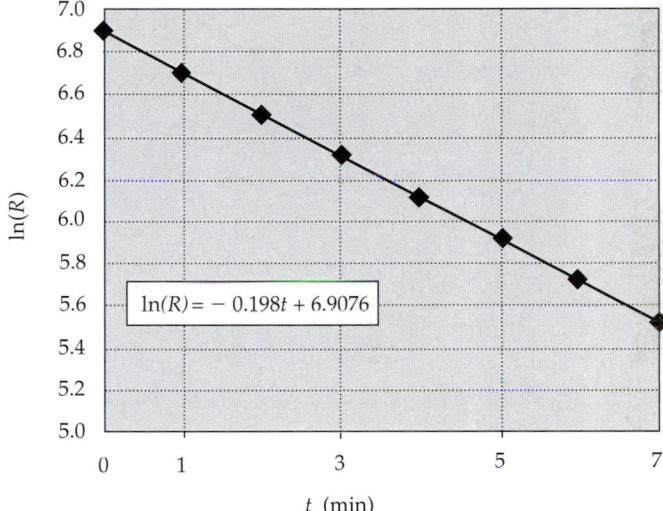

43. $7.03 \times 10^8$ y

45. (a) $-0.764$ MeV

    (b) 3.27 MeV

47. (a) 0.156 MeV

    (b) The masses given are for atoms, not nuclei, so for nuclear masses the masses are too large by the atomic number times the mass of an electron. For the given nuclear reaction, the mass of the carbon atom is too large by $6m_e$ and the mass of the nitrogen atom is too large by $7m_e$. Subtracting $6m_e$ from both sides of the reaction equation leaves an extra electron mass on the right. Not including the mass of the beta particle (electron) is mathematically equivalent to explicitly subtracting $1m_e$ from the right side of the equation.

49. $1.56 \times 10^{19}$ s$^{-1}$

51. 208 MeV; 88.1 percent

53. $3.20 \times 10^{10}$ J

55. (a) $4^1\text{H} \rightarrow {}^4\text{He} + 2\beta^+ + 2v_e + \gamma$

    (b) 24.7 MeV

    (c) $3.74 \times 10^{38}$ s$^{-1}$; $5.04 \times 10^{10}$ y

57.  $0.0693 \, s^{-1}$

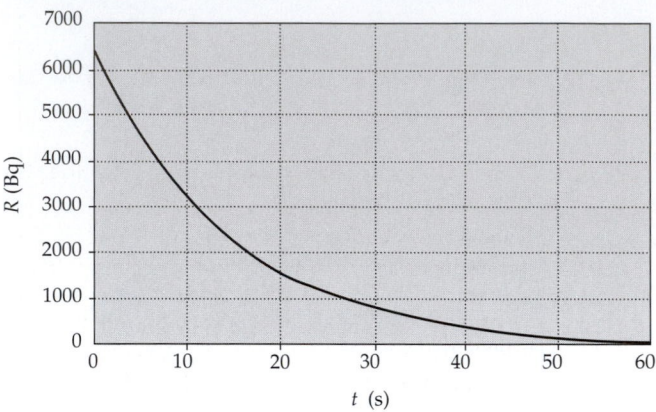

59.  156 keV

61.  $-2.88$ MeV

63.  $6.74 \times 10^3$ Bq

65.  6.30 L

67.  (a)  23.0 MeV

(b)  4.19 GeV

(c)  1.29 GeV

69.  (a)

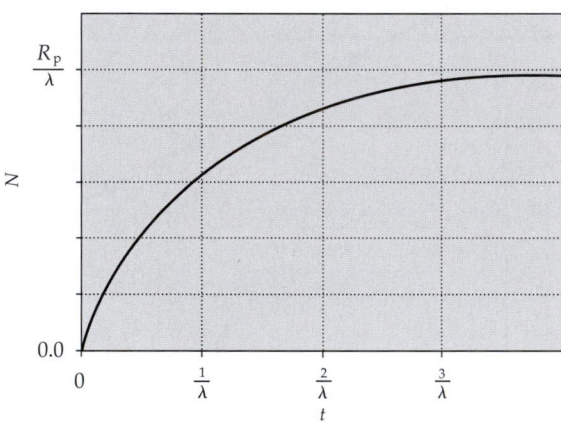

(b)  $8.66 \times 10^4$

71.  (a)  4.00 fm

(b)  310 MeV/$c$

(d)  310 MeV

73.  (a)  1.188 MeV/$c$

(b)  752 eV

(c)  0.0962 percent

75.  (b)  55

77.  (d)

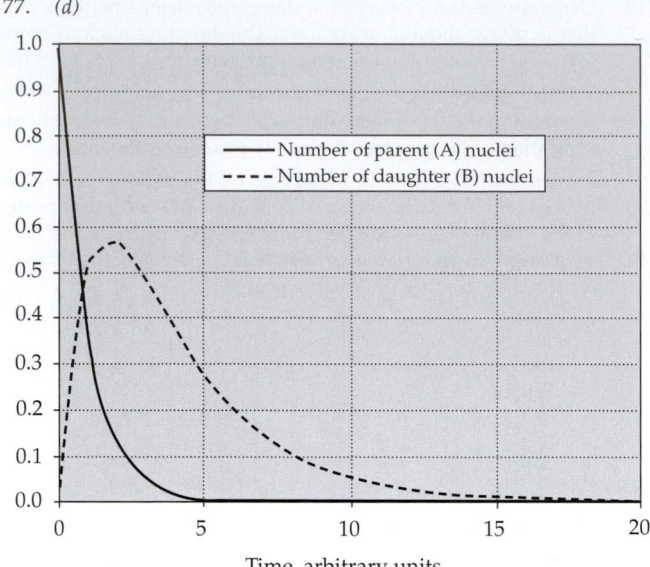

# Chapter 41

1.  *Similarities:* Baryons and mesons are hadrons; that is, they participate in the strong interaction. Both are composed of quarks.

*Differences:* Baryons consist of three quarks and are fermions. Mesons consist of two quarks and are bosons. Baryons have baryon number $+1$ or $-1$. Mesons have baryon number 0.

3.  A decay process involving the strong interaction has a very short lifetime ($\sim 10^{-23}$ s), whereas decay processes that proceed via the weak interaction have lifetimes of order $10^{-10}$ s.

5.  False

7.  No. From Table 41-2, it is evident that any quark–antiquark combination always results in an integral or zero charge.

9.  No. Such a reaction is impossible. A proton requires three quarks. Three quarks are not available because a pion is made of a quark and an antiquark and the antiproton consists of three antiquarks.

13.  (a)  279.2 MeV

(b)  1877 MeV

(c)  211.3 MeV

15.  (a)  $+1$; Because $\Delta S = +1$, the reaction can proceed via the weak interaction.

(b)  $+2$; Because $\Delta S = +2$, the reaction is not allowed.

(c)  $+1$; Because $\Delta S = +1$, the reaction can proceed via the weak interaction.

17.  (a)  $+2$; Because $\Delta S = +2$, the reaction is not allowed.

(b)  $+1$; Because $\Delta S = +1$, the reaction can proceed via the weak interaction.

19. (a) No. The neutron is not stable. $n \rightarrow p^+ + e^- + \bar{v}_e$

(b) $\Omega^- \rightarrow p^+ + e^+ + 3e^- + v_e + 3\bar{v}_e + 2\bar{v}_\mu + 2v_\mu$

(c) Because $Q = -1$ before and after the decay, charge is conserved. Because $B = 1$ before and after the decay, the baryon number is conserved. Because $L_e = 0$ before and after the decay, the lepton number for electrons is conserved. Strangeness is not conserved. However, in each baryon decay $\Delta S = +1$, and each decay is allowed via the weak interaction.

21.

| Combination | | $B$ | $Q$ | $S$ | Hadron |
|---|---|---|---|---|---|
| (a) | $uud$ | 1 | +1 | 0 | $p^+$ |
| (b) | $udd$ | 1 | 0 | 0 | $n$ |
| (c) | $uus$ | 1 | +1 | −1 | $\Sigma^+$ |
| (d) | $dds$ | 1 | −1 | −1 | $\Sigma^-$ |
| (e) | $uss$ | 1 | 0 | −2 | $\Xi^0$ |
| (f) | $dss$ | 1 | −1 | −2 | $\Xi^-$ |

23. From Table 41-2, we see that to satisfy the conditions of charge $= +2$ and zero strangeness, charm, topness, and bottomness, the quark combination must be $uuu$.

25. (a) $c\bar{d}$

(b) $\bar{c}d$

27. (a) $uds$

(b) $\bar{u}\bar{u}\bar{d}$

(c) $dds$

29. (a) $sss$

(b) $ssd$

31. $3.26 \times 10^8 \, c \cdot y$

35. (a) It must be a meson, and it must consist of a quark and its antiquark.

(b) The $\pi^0$ is its own antiparticle.

(c) The $\Xi^0$ is a baryon; it cannot be its own antiparticle. The antiparticle is the $\Xi^0 = \overline{uss}$.

37. (a) The $u$ and $\bar{u}$ annihilate, resulting in the photons.

(b) In the center-of-mass frame (center-of-momentum frame) the linear momentum is zero. A single photon does not have zero momentum, so to conserve momentum at least two photons are required.

39. (a) These properties indicate that the particle is the kaon, $K^0$.

(b) These properties indicate that the particle is either the $\Sigma^0$ or the $\Lambda^0$ baryon.

(c) These properties indicate that the particle is the kaon, $K^+$.

41. (a) $1.99 \times 10^5 \, km/s$

(b) $8.65 \times 10^9 \, c \cdot y$

43. (a) 1193 MeV

(b) 77.0 MeV/$c$

(c) 2.66 MeV

(d) 74.3 MeV; 74.3 MeV/$c$

# INDEX

Aberrations, 1062–1063

Absolute temperature scale, 537, 608–609

Absorption spectra, 1222–1223

AC (alternating current), 935–936

Acceleration
  angular, 268–269
  average, 24–25, 65–72
  examples/demonstrations of, 25–26, 85, 290–292, 344–345, 432–433
  as function of time, 26
  instantaneous, 25, 62
  integration and, 36–40
  motion with constant, 27–36
  in simple harmonic motion, 426, 428

Acceptor levels, 1247

Action
  at a distance problem, 92–93
  Newton's third law of motion on, 101–102

Adiabatic bulk modulus, 468n

Adiabatic process, 581

Air wedge, 1087–1088

Alpha decay, 1316–1317

Alternating current circuits
  capacitive reactance, 942–943
  capacitors in, 941–942
  inductive reactance in, 940–941
  inductors in, 939–941

Alternating current in a resistor
  average power delivered by generator, 938
  described, 936–937
  root-mean-square values, 937–938
  sawtooth waveform, 939

Ammeters, 809–810

Amorphous solid, 1229

Ampère, 869

Ampere (A), 4

Ampère, André-Marie, 856

Ampère's law
  Maxwell's displacement current generalized from, 972–973
  overview of, 871–874

Amplitude
  damped oscillation, 446, 448–449
  described, 427
  driven oscillation, 452
  frequency in simple harmonic motion as independent of, 430
  of pendulum's oscillation, 442–443

Analyzer, 1022

Anderson, Carl, 1339

Angle of incidence, 1011

Angular acceleration, 268–269

Angular frequency, 428–429

Angular momentum (see also Conservation of angular momentum)
  Atwood's machine and, 315–316
  fundamental unit of, 327
  Newton's second law for, 313, 324–326
  of particles, 311–312
  quantization of, 326–328
  of system rotating about symmetry axis, 312–313
  torque and, 311–317

Antiderivative of $f(t)$, 37

Antilock brake friction, 128–129

Antineutrinos, 1342–1343

Antinode, 512, 515

Antiparticles, 1339–1340, 1350

Antiproton (p-), 1340–1342

Antiquarks, 1347

Anti-Stokes Raman scattering, 1001

Aorta
  blood flow and, 408
  blood pressure and, 401–402
  calculating Reynolds number for blood flowing through, 415

Apparent weight, 90–91

Arc discharge, 737

Archimedes' principle, 402–403

Argon laser, 1107

Aristotle, 2

Aston, Francis William, 838

Astronomical telescopes, 1071–1073

Atomic magnetic moments
  Bohr magneton, 878
  due to electron spin, 878
  due to orbital motion of electron, 877–878
  overview of, 876–877
  relation between angular momentum and, 877
  saturization magnetization for iron, 878–879

Atomic masses of neutron/selected isotopes, 1309

Atomic number, 1171

Atomic orbitals, 1215

Atoms (see also Hydrogen atoms; Periodic table)
  electron configuration of, 1189–1197
  energy quantization in, 1124–1125
  exclusion-principle repulsion and, 1209
  dipole moment, 774
  magnetic moment of atomic electron, 1186–1188
  mass of, 540–541
  nuclear, 1172–1173
  optical spectra of, 1197–1199

potential energy of hydrogen, 721

properties of nuclei, 1306–1310

quantum theory of, 1178–1181

rotational energy of hydrogen, 581

spin-orbit effect and fine structure, 1186–1189

Thomson's plum pudding model of atom, 1173

X-ray spectra of, 1199–1200

Atwood's machine, 191, 315–316

Avalanche breakdown, 1249

Average acceleration, 62

Average speed, 19–20

Average velocity, 21, 59

Avogadro's number, 538


Background radiation, 1351–1352

Balmer, Johann, 1172

Band, 1245

Band theory of solids, 1244–1246

Bardeen, John, 1254

Barn, 1318

Barrier penetration, 1159–1161

Baryon number, 1342

Baryons, 1336, 1342

Base (semiconductor), 1250

Basic interactions (see also Elementary particles)
  big bang theory on evolutionary role of, 1351–1353
  bosons mediating, 1348
  charm conserved in strong, 1346
  conservation laws applying to, 1342–1345
  electromagnetic, 1336, 1348, 1350
  field particles, 1347–1348
  gravitational, 1336, 1348, 1350
  properties of, 1350
  proton-antiproton, 1340–1342
  residual strong, 1349
  in standard model, 1349–1350
  strangeness/strange particles and, 1343–1344
  strong, 1137, 1336, 1345, 1348, 1350
  weak, 1336, 1345, 1348, 1350

Batteries
  charging parallel-plate capacitor with, 757–759
  capacitors disconnected/connected to and from, 771–772
  emf (electromotive force) and, 795–797
  ideal, 795
  internal resistance of, 796–797
  jump starting car, 806

open-circuit terminal voltage of, 760
  real, 796
  terminal voltage of, 760
BCS theory, 1254–1255
Beat frequency, 507
Beats, 506–507
Becquerel, Antoine, 2
Becquerel (Bq), 1313
Bednorz, J. Georg, 1253
Benzene molecules, 11
Bernoulli's equation, 408–410
Beryllium (Z = 4), 1193
Beta decay, 1314–1316
Big bang theory, 1350–1353
Binding energy, 202, 203
Biot, Jean-Baptiste, 856
Biot-Savart law, 858–859
Birefringence, 1025, 1027
Blackbody, 642
Black holes, 1298
Black light (ultraviolet light), 1002
Bobbing cork/buoyancy, 407
Body-centered-cubic (bcc) structure, 1232
Body fat measurement, 405–406
Bohr magneton, 878
Bohr model
  correspondence between Schrödinger
    equation and, 1178
  described, 1172
  of hydrogen atom, 1173–1178
Bohr, Niels, 1125, 1134, 1172, 1199
Bohr's correspondence principle, 1136
Bohr's postulates
  first postulate: nonradiating orbits, 1175
  second postulate: photon frequency from
    energy conservation, 1175
  third postulate: quantized angular
    momentum, 1175–1176
Boiling point (BP), 533, 563, 633–634
Boltzmann's constant, 538
Boron to neon (Z = 5 to Z = 10), 1193
Bosons, 328
  defining, 1166
  Higgs, 1349
  that mediate basic interactions, 1348
Bottomonium, 1346
Boundary condition, 1134
Boyle, Robert, 538
Boyle's law, 558
BP (boiling point), 533, 563, 633–634
Breeder reactor, 1325–1326
Bremastrahlung (braking radiation), 978
Brewster, David, 1023
Btu, 559
Bulk modulus, 397–398
Buoyancy
  Archimedes' principle and, 402–403
  bobbing cork and, 407
  described, 402
  floating on a raft and, 406
  used to measure body fat, 405–406
  used to measure gold content, 404–405

Caloric theory, 558
Calorie, 559
Calorimeter, 561
Calorimetry, 561

Candela (cd), 4
Capacitance
  of the conductor, 748
  cylindrical capacitors and expression of,
    754–756
  definition of, 752
  dielectric effect on, 767
  equivalent, 761–766
  switching in computer keyboards, 758
Capacitive reactance, 942–943
Capacitors
  charging, 813–816
  charging parallel-plate capacitor with
    battery, 757–759
  combinations of, 761–766
  connected in parallel, 761–762
  connected in series, 762–764
  cylindrical, 754–756
  described, 748–749, 753
  dielectrics used in parallel-plate, 768–769
  discharging, 811–813
  electrostatic field energy produced when
    charging, 759–760
  energy conservation in charging, 815–816
  energy stored in, 756–760
  using the equivalence formula, 764–765
  homemade, 770
  Leyden jar, 749
  parallel-plate, 753–754
  in series and parallel, 766
  in series rearranged in parallel, 765–766
Carnot cycle
  described, 603, 604–606
  entropy changes during, 615–617
Carnot efficiency, 603, 606–608
Carnot engine
  conditions for reversibility, 603–604
  cycle of, 603, 604–606, 615–617
  described, 602–603
  efficiency of, 603, 606–608
Carnot, Sadi, 2
Cathode-ray tube (schematic drawing), 668
Cavendish, Henry, 345
Celsius temperature scale, 533–534
Center of mass
  described, 218–219
  examples of, 219–221
  found by integration, 222–223
  gravitational potential energy of system
    and, 221–222
Center of mass motion
  described, 223–224
  examples/demonstrations of, 224–227
Center-of-mass reference frame, 247–248
Central diffraction maximum, 1092–1093
Central ray, 1057
Centripetal direction, 73
Centripetal force, 129
Cerenkov radiation, 491
CERN (European Laboratory for Particle
  Physics), 1340, 1341
Cesium fountain clock, 4
CGS system, 5
Chamberlain, Owen, 1340
Characteristic rotational energy, 1217–1218,
  1219
Charge carrier number density in silver, 846
Charging by induction, 655–656

Charles, Jacques, 558
Charm, 1346
Chromatic aberration, 1063
Circle of least confusion, 1062
Circuits (see also Electric charge)
  alternating current, 934–959
  energy in, 794–798
  Kirchhoff's rules and multiloop, 806–809
  Kirchhoff's rules and single-loop,
    804–806
  LC, 944–948
  magnetic energy of, 915–917
  mutual inductance of, 914–915
  RC, 811–816
  RL, 917–921
  RLC, 944–956
  self-inductance, 912–914
  time constant of, 918
Circular motion
  examples/demonstrations of, 72–74
  simple harmonic motion and, 433
  uniform, 73
Circulation integral, 871
Classical particle, 1131
Classical physics
  compared to modern physics, 3
  origins and development of, 2–3
Classical wave, 1131
Clocks
  special relativity and moving, R.4-6
  special relativity, simultaneity and
    distant, R.8-9
  synchronization of, 1278, 1281–1282, R.8-9
  water, 4
Coal-fueled electric generating plant (New
  Mexico), 605
Coaxial cable, 755
Cockcroft, J. D., 1317
Coefficient of linear expansion, 628–629
Coefficient of restitution, 244
Coherence, 510–511
Coherence length, 1086
Coherence time, 1086
Cold heat reservoir, 598
Collector, 1250
Collisions
  described, 233
  entropy changes during inelastic, 614–615
  examples/demonstrations of, 234–236
  head-on, 237–244
  impulse and average force of, 234
Collisions in one dimension, 233, 234,
  237–244
Collisions in three dimensions, 244–246
Collision time, 1235
Comets
  collision movement of, 233
  displacement/velocity of, 18
Compound lenses, 1061–1062
Compound microscope, 1068–1070
Compressibility, 398
Compressive stress, 381
Compton, Arthur H., 1122, 1123
Compton equation, 1122
Compton scattering, 1122–1124
Conduction
  classical theory of, 1236
  microscopic picture of, 1233–1236

quantum theory of electrical, 1243–1244
semiconductors and, 1246–1248
verified by Hall effect, 1247
Conductivity, 1234
Conductors
charge on surface, 703
capacitance of the, 748
electrostatic equilibrium of, 702–703
electrostatic potential energy of system of, 751–752
grounded, 655–656
polarized, 655
two charged spherical, 738–739
Confined particles
in infinite square-well, 1151–1152
in simple harmonic oscillator, 1155–1157
Conical tube, 519
Conservation of angular momentum (see also Angular momentum)
ballistic pendulum and, 323
examples/demonstrations of, 318–324
gyroscope motion and, 316–317
law of, 317–318
vector nature of rotation and, 309–311
Conservation of energy (see also First law of thermodynamics)
examples/demonstrations of, 193, 196–198
law of, 184, 191–192
problems involving kinetic friction, 194–195
work-energy theorem and, 192
Conservation laws
basic interactions and, 1342–1345
of energy, 184, 191–192
of mechanical energy, 183–185, 229
Conservation of mechanical energy
applications of, 185–191
law of, 229, 183–185
Conservation of momentum, 229–232
Conservative forces, 168
Constant-volume gas thermometer, 535–537
Contact forces, 93–94
Continuity equation, 408
Continuous objects moment of inertia, 273–275
Continuous spectrum, 999–1000
Convection, 641–642
Converging lens, 1054
Conversion factor, 6
Convex mirror, 1047
Cooper, Leon, 1254
Cooper pair, 1254
Coulomb (C), 653
Coulomb, Charles, 656
Coulomb constant
described, 657
permittivity of free space and, 693
Coulomb potential, 721
Coulomb's law
calculating electric fields from, 683–690
derivation of Gauss's law from, 696, 707–708, 975
described, 656–657
for electric field due to a point charge, 662
electric force in hydrogen and, 657
net force and, 658–659
net force in two dimensions and, 660–661

potential energy function for particle given by, 1161
ratio of electric and gravitational forces and, 658
Couples, 377
Covalent bond, 1210–1213
Crab Pulsar rotation, 271
Critically damped motion, 446
Critical angle, 1015
Critical point, 634
Critical temperature, 1252–1253
Crossed fields, 836
Cross product (or vector product), 310–311
Curie (Ci), 1314
Curie, Marie, 2
Curie, Pierre, 2
Curie's law, 880
Curie temperature, 881
Current balance, 869
Current density, 1234
Current gain, 1251
Cyclotron, 839–840
Cyclotron frequency, 835
Cyclotrotion period, 835
Cylindrical capacitors, 754–756
Cylindrical symmetry, 700–701


Dalton, John, 1335
Damped oscillation
amplitude, energy, frequency, decay time and, 446–448
described, 445–446
musical example of, 448–449
Q factor and, 447–448
Daughter nucleus, 1314
Davisson, C. J., 503, 1127
De Broglie hypothesis
for frequency of electron waves, 1125
for wavelength of electron waves, 1125–1127
De Broglie, Louis, 1125
Decay constant, 1311
Decay law, 1311
Decay rates
definition of, 1311
hadrons stable against, 1337
half-life, 1312
mean lifetime, 1312
radioactivity, 1314–1317
Decay-rate units, 1314
Decay time, 446
Deep inelastic scattering, 1346
Degenerate energy levels, 1162–1163
Degree of freedom, 544
Delayed neutrons, 1324, 1325
Density
Archimedes' principle and measuring, 402–403
calculating, 397
described, 396
as specific gravity, 396
Density of states, 1256–1257
Depth inversion, 1039
Derivative, AP-23–27
finding velocity from given acceleration, 36–37
usual calculus notation of, 21

Deuteron, 202
Diamagnetic material, 874, 875, 884–885
Diatomic molecules
absorption spectra, 1222–1223
emission spectra, 1220–1221
rotational energy levels, 1217–1219
vibrational energy levels, 1219–1220
Dielectric breakdown, 737–739
Dielectrics
constants/strengths of various materials, 768
described, 767
effect on capacitance, 767
electric field inside a, 767
energy stored in presence of, 770–772
molecular view of, 772–775
used in parallel-plate capacitor, 768–769
permittivity of, 767–768
Dielectric strength, 737
Diffraction, 484–486
coherence and phase, 1084–1086
definition of, 1084
electron interference and, 1127–1128
holograms, 1108
resolution and, 1103–1105
Diffraction gratings, 1105–1108
Diffraction-interference patterns
calculating multiple slit, 1100–1101
of two slits, 1093–1094
Diffraction patterns
calculating single-slit, 1098–1100
Fraunhofer and Fresnel, 1101–1103
single slit, 1091–1093
Diffuse reflection, 1013–1014
Diodes
LEDs (light-emitting diodes), 1250
overview of, 1248–1250
tunnel, 1249
Zener, 1249
Diopter (D), 1056
Dirac equation, 1339
Discontinuity of $E_n$, 701–702
Dispersion
definition of, 1018
rainbows, 1019–1021
Displacement
defining, 17–18
dot product and, 159–165
examples/demonstrations of, 18–20, 22–23
and velocity as integrals, 37–40
Displacement vectors
addition of, 54
definition of, 53
Dissociation energy, 1210, 1231
Distortion, 1062
Distribution functions, 546–551
Donor levels, 1247
Doorbell transformer, 958–959
Doping, 1246
Doppler effect
described, 486–488
examples/demonstrations of, 486–490
frequency and, 486–492
relativistic, 1276–1278
relativity and, 490
shock waves and, 490–491
sonic boom, 492

Dot product (or scalar product)
  derivative, 160
  described, 159
  differentiating, 163
  examples/demonstrations of, 162–164
  properties of, 159–160
  vector components using, 160–163
Drag forces, 134–135, 137
Drift speed
  defining, 787
  finding the, 788–789
Drift velocity
  defining, 787
  relation between current and, 788
Driven oscillations
  amplitude for, 452
  described, 449–450, 452–453
  phase constant for, 452
  position for, 451–452
  resonance and, 450–452
Driven $RLC$ circuits
  FM tuner, 955
  low-pass filter, 954–955
  resonance, 950–952
  at resonance, 953–954
  series, 948–950, 952
  series current, phase, and power, 953
Drude, Paul K., 1228
Dulong-Petit law, 577–578

Earth
  kinetic energy of object at distance from, 350
  Newton's law of gravity and free-fall to, 344–345
  spin of, 314
Earth electric field, 704
Earth magnetic field, 829
Eddy currents, 9112
Einstein, Albert
  Postulates on speed of light by, R.2-3
  special theory of relativity equation, 201
  theory of special relativity proposed by, 2–3, 1119–1120, 1267, R.1
Einstein equation, 1120
Einstein's photoelectric equation, 1120–1121
Einstein's postulates, 1269–1270
Elastic head-on collisions
  coefficient of restitution and, 244
  exercises on, 243–244
  of neutron and nucleus, 242–243
  in one dimension, 233, 240–241
  speed of approach/speed of recession in, 241
Elastic collision in three dimension, 245–246
Electrical conduction, 1243–1244
Electric charge (*see also* Circuits)
  conservation of, 653–654
  described, 652
  fundamental unit of, 653
  quantization, 652–653
  two objects that repel/attract, 652
Electric current
  definition of, 787
  electric field driving, 793
  relation between drift velocity and, 788

Electric dipole moment, 665–666
Electric dipole radiation, 978–980
Electric dipoles
  described, 665–666
  electric field lines for, 667, 671–672
  flux through surface enclosing, 691
  potential due to, 724
Electric field lines
  described, 666–667
  due to electric dipole, 667, 671–672
  due to two positive point charges, 666
  near long wire, 685
  for oppositely charged cylinder and plate, 705
  rules for drawing, 667
  due to single positive point charge, 666
  due to two charged conducting spheres, 668
  of uniform electric field perpendicular to surface, 691
  of uniform field penetrating surface of area, 691
Electric fields
  calculated from Coulomb's law, 683–690
  calculated from Gauss's law, 694–701
  computing electric potential from, 724–726
  Coulomb's law for due to a point charge, 662
  definition of, 661–662
  discontinuity of $E_n$ at surface charge, 701–702
  driving current, 793
  due to a system of point charges, 662–665
  due to two equal and opposite charges, 665
  of the earth, 704
  electric dipoles in, 671–672
  electron moving parallel to uniform, 668–669
  electrons moving perpendicular to uniform, 669
  examples in nature, 661
  induced nonconservative, 901–903
  for infinite plane of charge, 695
  in ink-jet printer, 670–671
  inside a dielectric, 767
  on a line through two positive charges, 662–663
  motion of point charges in, 668–671
  near/far from finite line charge, 686–687
  plotting magnitude vs. distance of, 687
  potential difference and, 719–720
  produced by oscillating electric dipole, 979
  on $y$ axis due to point charges on $x$ axis, 664
Electric fields for various uniform charges
  on axis of a disk, 690
  on axis of finite line charge, 685–686
  on axis of ring charge, 688
  on axis of uniformly charged disk, 688–689
  due to an infinite line charge, 685, 700–701
  due to continuous charge distribution, 683–684
  due to infinite plane of charge, 689

due to line charge and point charge, 687
due to thin spherical shell of charge, 696–699
due to two infinite planes, 695–696
due to uniformly charged line segment, 684–685
due to uniformly charged sphere, 699–700
plane symmetry of, 695–696
spherical symmetry of, 696
on two faces of a disk, 705–707
Electric flux
  described, 691–692
  through and charge inside imaginary closed surface, 694
Electric potential
  along the $x$-axis, 723
  calculations for continuous charge distributions, 726–733
  computing electric field from, 724–726
  Coulomb potential, 721
  due to electric dipole, 724
  due to system of point charges, 720–724
  due to two point charges, 722–723
  of hydrogen atom, 721
  due to point charge, 720–721
Electric potential functions
  on the axis of charged ring, 726–727
  on the axis of uniformly charged disk, 727–729
  due to infinite line charge, 733
  due to infinite plane of charge, 729–730
  inside and outside of spherical shell of charge, 731–733
Electromagnetic energy density, 917
Electromagnetic interactions
  bosons that mediate, 1348
  definition of, 1336
  in electroweak theory, 1348–1349
  properties of, 1350
Electromagnetic spectrum, 976–977
Electromagnetic waves, 478
  electric dipole antenna/loop antenna for detecting, 979
  electric dipole radiation, 978–980
  electromagnetic spectrum, 976–977
  energy and momentum in, 981–983
  intensity of, 981–982
  laser rocket, 984–985
  Maxwell's equations on, 971–972
  polarization of, 1021–1022
  production of, 978
  radiation pressure in, 982–984
  of radio or television frequencies, 979
  speed of, 972
  wave equation for, 985–991
Electron affinity, 1209
Electron beam deflection, 838
Electron configurations
  of atoms in ground states, 1194–1197
  beryllium ($Z = 4$), 1193
  boron to neon ($Z = 5$ to $Z = 10$), 1193
  defining, 1189
  elements with $Z > 18$, 1193–1194
  helium ($Z = 2$), 1190–1191
  lithium ($Z = 3$), 1191–1192
  sodium to argon ($Z = 11$ to $Z = 18$), 1193
Electron neutrino, 1336

Electrons
  atomic magnetic moments due to orbital motion of, 877–878
  circular path moving in magnetic field, 834
  de Broglie hypothesis on matter waves and, 1125–1127
  interference and diffraction, 1127–1128
  ionization energy and, 1177, 1190, 1194
  kinetic energy of, 154–155
  as lepton, 1336–1339
  magnetic moment due to spin of, 878
  mean free path and mean speed of, 1235
  moving parallel to uniform electric field, 668–669
  moving perpendicular to uniform electric field, 669
  orbital magnetic moment of, 1186–1188
  as particles, 1340
  in quantum theory, 1129
  spin of, 327–328
  standing waves and energy quantization, 1129
  Thomson's measurement of q/m for, 837
  Thomson's study of, 1125
  two-slit experiment using, 1131–1332
Electron waves
  barrier penetration, 1159–1161
  scattering of, 1243–1244
  step potential, 1157–1159
Electroscope, 654
Electrostatic equilibrium, 702–703
Electrostatic potential energy
  described, 749–750
  of a system, 749
  of system of conductors, 751–752
  of a system of point charges, 750
Electrostatics, 651–652
Electroweak force, 1352
Electroweak theory, 1348–1349
Elementary particles (see also Basic interactions; Particles)
  big bang theory on, 1350–1353
  conservation laws and, 1342–1345
  defining, 1336
  electroweak theory, 1348–1349
  field particles, 1347–1348
  GUTs (grand unification theories) on, 1350, 1352
  hadrons and leptons, 1336–1339, 1350
  history of discovery/study of, 1335–1336
  quarks, 1336, 1345–1347, 1350
  spin and antiparticles, 1339–1342
  standard model on, 1336, 1349–1350
  strangeness/strange particles, 1343–1344
Ellipse, 340
Elliptical polarization, 1021
Emf (electromotive force)
  batteries and, 795–797
  definition of, 795–797
  induced, 897, 899–903
  motional, 907–911
Emission spectra, 1220–1221
Emissivity, 642
Emitter, 1250
Endothermic reaction, 1318
Energizing coil, 918

Energy (see also Kinetic energy; Power; Waves)
  of accelerated proton, 840–841
  association between work and, 151
  binding, 202, 203
  characteristic rotational, 581, 1217–1218, 1219
  conservation of mechanical, 183–191
  conservation of, 191–201
  contained in a pizza, 198
  conversion of chemical energy into mechanical or thermal, 199–201
  damped oscillation, 447–449
  dissociation, 1210, 1231
  Einstein equation for photon, 1120
  in electric circuits, 794–798
  electromagnetic wave, 981–983
  electrostatic potential, 749–752
  entropy and availability of, 617–618
  First law of thermodynamics, 515
  Fermi, 1238–1239
  gravitational potential, 349–353
  ground state, 205
  internal, 566
  ionization, 1177, 1190, 1194
  law of conservation of, 184, 192
  magnetic, 915–917
  mass and, 201–204
  Maxwell-Boltzmann distribution function and, 551
  net attractive part of ion in crystal potential, 1229
  nuclear, 202–204
  nucleus mass and binding, 1308–1310
  particle in a box and, 1133–1134
  potential, 167–174
  potential energy of dipole in electric field, 671
  potential energy of hydrogen atom, 721
  potential energy of nuclear fission products, 722
  potential of magnetic dipole in magnetic field, 843
  quantized, 204–205, 579
  relativistic, 1289–1296
  relativistic momentum, mass and, R.12-13
  released in fission, 1321
  rest, 1290, 1291, 1292
  selected elementary particles/light nuclei and resting, 201
  in simple harmonic motion, 434–437
  of sound waves, 477
  stored in capacitor, 756–760
  stored in inductor, 916
  stored in presence of dielectric, 770–772
  terminal voltage, power and stored, 797–798
  thermal, 638–644
  three dimensional work and, 165–167
  torque and potential, 672
  total mechanical, 184–185
  transferred via waves on string, 475–476
  two hydrogen atoms and separation vs. potential, 1213
  work and kinetic, 152–158
  zero-point, 1134
Energy for circular orbit, 1173–1174
Energy distribution function, 1256

Energy-level diagram for oscillator, 204–205
Energy levels
  diatomic molecules and rotational, 1217–1219
  ground-state wave functions and, 1155–1157
  of hydrogen, 1181
  of hydrogen atom, 1177–1178
  quantum theory on hydrogen atom, 1181–1182
  vibrational, 1219–1220
Energy quantization
  in atoms, 1124–1125
  electron standing waves and, 1129
  harmonic oscillator, 1140–1141
  hydrogen atom, 1141–1142
  infinite square-well potential, 1140
Engines
  Carnot, 602–608, 615–617
  efficiency of steam, 606–607
  heat, 596–600
  internal-combustion, 597–598
  Second law of thermodynamics and, 602, 598–599
  torques exerted by, 287
  work lost by, 607–608
Entropy
  availability of energy and, 617–618
  changes during Carnot cycle, 615–617
  changes during heat transfer, 614
  for constant-pressure processes, 613–614
  described, 610–611
  free expansion of ideal gas and, 612–613
  of an ideal gas, 611
  for inelastic collision, 614–615
  probability and, 618–620
Equation of state, 539
Equilibrium (see Static equilibrium)
Equipartition theorem
  failure of the, 578–581
  heat capacities and, 576
  kinetic theory of gases and, 544–545
Equipotential region, 733
Equipotential surfaces
  described, 733–734
  dielectric breakdown on charged sphere, 737
  hollow spherical shell, 734–735
  two charged spherical conductors, 738–739
  Van de Graaff generator, 736–737
Equivalence of heat-engine/refrigerator statements, 602
Equivalent capacitance, 761–766
Equivalent resistance, 637
Escape speed, 350–351
Ether, 1268–1269
Euler's method, 136–138
Event, R.6
Exchange interaction, 880
Exclusion-principle repulsion, 1209
Exclusion-principle in terms of spatial states, 1237
Exhaust manifold, 598
Exothermic reaction, 1318
Expectation values
  calculating probabilities and, 1138–1140
  definition of, 1138

Exploding projectile motion, 224–225
Exponents, 8, AP-12
  logarithms, AP-12–13
  exponential function, AP-13–14
Exponential decrease, 812
Eye
  focal length of cornea-lens system,
    1064–1065
  images seen through, 1063
  nearsighted, 1064


Fahrenheit temperature scale, 533–534
Falling rope (continuously varying mass),
    249–250
Faraday, Michael, 897
Faraday's law
  described, 899–901, 973
  induced emfs and, 899–903
Fermat's principle, 1011, 1028–1030
Fermi-Dirac distribution, 1256–1259,
    1257–1258
Fermi electron gas
  contact potential, 1241–1242
  described, 1236
  energy quantization in a box, 1237
  exclusion principle, 1237
  Fermi energy, 1238–1239
  heat capacity due to electrons in metal
    and, 1242–1243
Fermi factor $T = 0$, 1240, 1257, 1258, 1259
Fermi factor $T > 0$, 1240, 1257
Fermions, 327, 1166, 1339
Fermi questions, 12
Fermi speed, 1240–1241
Fermi temperature, 1240
Ferromagnetic material, 874, 875, 880–884
Fiber optics, 1017
Field particle interactions, 1347–1348
Field point, 354
Fine-structure splitting, 1188–1189
First Bohr radius, 1176
First harmonic, 511
First law of thermodynamics, 566–568
First-order spectrum, 1107
Fission
  defining, 1319
  energy released during, 1321
  nuclear fission reactors, 1322–1326
  overview of, 1319–1322
FitzGerald, George F., 1275
Fizeau, Armand, 1008, 1009
Flavors (quark), 1345
Fluid flow
  Bernoulli's equation and, 408–410
  continuity equation for incompressible
    fluid, 408
  described, 407
  leaking tank, 410
  Poiseuille's Law, 414–415
  Reynolds number, 415
  Venturi effect and, 410–412
  viscous flow, 413–415
  volume flow rate, 408
Fluids
  buoyancy and Archimedes' principle,
    402–407
  coefficients of viscosity, 414

density of, 396–397
described, 395
in motion, 407–415
pressure in, 397–402
Fluxon, 923
Flux quantization, 923
FM tuner, 955
Focal length, 1043
Focal Plane, 1043
Focal point, 1043
Focal ray, 1045, 1057
Forbidden energy band, 1245
Forces (see also Power; Spring force)
  action at a distance problem, 92–93
  balancing magnetic, 870
  centripetal, 129
  conservative, 168
  contact, 93–94
  Coulomb's law on, 656–661
  couple, 377
  drag, 134–136, 137
  due to gravity (weight), 90–91
  on elbow equilibrium, 373–374
  electric force in hydrogen, 657–658
  on electrons, 154–155
  examples/demonstrations of, 88–89
  exerted by magnetic field, 830–833
  exerted by system of charges, 658–661
  Hooke's law, 93–94
  impulse/average, 234
  in nature, 92–95
  motion in one dimension and constant,
    152–153
  Newton's second law of motion defining,
    86, 87
  Newton's third law, 101–105
  nonconservative, 172
  shear, 382
  units of, 91
  work done by variable, 156–157
Forces in nature
  action at a distance problem, 92–93
  contact forces, 93–95
  fundamental, 92
Forward biased junction, 1248–1249
Foucault, Jean, 1008
Fourier analysis, 521
Frame of reference, 23–24
Franklin, Benjamin, 652, 749
Fraunhofer diffraction patterns, 1093–1094
Free-body diagrams
  described, 95–96
  examples/demonstrations of, 86–101, 135
  guidelines for applying second law for
    rotation, 282
Free expansion, 568
Freezing point of water, 533
Frequency
  angular, 428–429
  beat, 507
  Bohr's postulates on photon, 1175
  damped oscillation, 446–447, 448–449
  de Broglie relation for electron waves,
    1125
  Doppler effect and, 486–492
  driven oscillations and resonance,
    450–451
  equation for photon, 1134

as independent of amplitude in simple
  harmonic motion, 430
Josephson current, 1256
natural, 450
photon frequencies less than threshold,
  1121
standing waves and resonance, 511
Fresnel, Augustin, 1119
Fresnel diffraction patterns, 1093–1094
Friction
  additional remarks on, 127–128
  conservation of energy and problem of
    kinetic, 194–195
  examples/demonstrations of, 120–129
  explanation of, 119
  reducing, 86
  types of, 117–118
Frictional coefficient values, 120
Friedberg, Carl, 1338
Fringes (interference)
  definition of, 1087
  distance on screen to $m$th bright, 1989
  spacing from slit spacing, 1090
  straight-line, 1088
Fundamental forces, 92
Fundamental frequency, 511
Fundamental unit of angular momentum,
  327
Fundamental unit of charge, 653
Fusion
  defining, 1319
  Lawson criterion, 1326–1327
  overview of, 1319–1321, 1326–1327
  reactors, 1327–1328


Galilean transformation, 1270–1271
Galileo, 2, 1005–1006, 1072
Galvanometer, 810
Gamma decay, 1316
Gases (see also Ideal gas)
  calculating pressure exerted by, 542
  distribution of, 546, 549–551
  expanding at constant temperature, 541
  heat capacities of, 572–577
  heating and compressing, 540
  ideal-gas law, 537–541
  kinetic theory of, 541–551
  quasi-static adiabatic compression of,
    581–583
  work and PV diagram for, 569–572
Gas thermometers, 535–537
Gauge pressure, 401
Gauss C. F., 831
Gaussian surface, 694–695
Gauss's law
  calculating electric fields from, 694–701
  derivation from Coulomb's law, 696,
    707–708, 975
  described, 690–691
  quantitative statement of, 692–694
Gauss's law of magnetism, 870–871,
  975–976
Gay-Lussac, Joseph, 558
Geiger, H. W., 1173
Gell-Mann, M., 1343, 1345
General relativity, 1296–1298
Germer, L. H., 503, 1127

GeV, 1341
Gilbert, William, 829
Gluons, 1348
Golf ball collision, 236–237
GPS (Global Positioning System), 1267
Gradient, 724
Gravitational constant $G$ measurement, 345
Gravitational field
    described, 353–354
    examples/demonstrations of, 354–360
Gravitational inertial mass, 345–346
Gravitational interactions
    bosons that mediate, 1348
    definition of, 1336
    properties of, 1350
Gravitational potential energy
    described, 349–350
    escape speed, 350–351
    examples/demonstrations of, 351–353
Gravitational redshift, 1298
Graviton, 1348
Gravity
    action at a distance problem and, 92–93
    center of mass and gravitational energy
        of system, 221–222
    as conservative force, 168
    described, 339
    equilibrium and center of, 371–372, 379
    examples/demonstrations of, 347–348
    force due to, 90–91
    Kepler's laws and, 340–342, 346–347
    Newton's law of, 342–348
    specific, 396
    torques due to, 281–282
Grounded conductors, 655–656
Ground state energy, 205
Group velocity, 522
Guitar tuning, 507
GUTs (grand unification theories), 1350, 1352
Gyroscope, 316–317

Hadronic force, 1307
Hadrons, 1336–1339
Hahn, Otto, 1321
Half-life decay, 1312
Half-wave plate, 1025
Hall effect
    von Klitzing constant, 847
    described, 845–846, 1247
    quantum, 847
Hall voltage, 846, 847
Handbell, standing waves, 519
Hanging sign equilibrium, 374–375
Hard-disk drive/magnetic storage, 883
Harmonic analysis, 520–521
Harmonic oscillator
    energy quantization in, 1140–1141
    Schrödinger equation application to,
        1155–1157
Harmonic series, 513
Harmonic sound waves, 477
Harmonic synthesis, 521
Harmonic waves
    described, 473–474
    examples/demonstrations of, 474–476
    function of, 472–473
    interference of, 505–511

Head-on collisions (see Collisions in one
    dimension)
Heat capacity
    described, 559–560
    equipartition theorem and, 576, 578–581
    of gases, 572–577
    of solids, 577–578
Heat conduction entropy changes, 615
Heat engine
    described, 596
    early history and development of,
        596–598
    efficiency of ideal internal combustion
        engine, 599–600
    efficiency of, 598–599
    Otto cycle, 598, 599–600
    second law of thermodynamics and, 599
Heat/heating
    change of phase and latent, 562–565
    and cooling, compressing ideal gas, 575
    described, 558–559
    diatomic ideal gas, 576–577
    examples/demonstrations of, 563–567,
        640–641
    measuring specific, 561–562
    mechanical equivalence of, 565
    molar specific, 560
    needed to raise temperature, 560
    specific heats/molar specific heats of
        some solids/liquids, 560
Heat loss, 640–641
Heat pumps, 609–610
Helium-neon laser, 1004–1005
Helium ($Z = 2$), 1190–1191
Henry, Joseph, 897
Herschel, Friedrich Wilhelm, 1072
Hexagonal close-packed (hcp) structure,
    1232
Higgs boson, 1349
Higgs field, 1349
Holograms, 1108
Hooke, Robert, 1118
Hooke's law, 93–94
Hubble constant, 1351
Hubble, Edwin Powell, 1072
Hubble's law, 1351
Hubble Space Telescope, 321, 1073
Huygens, Christian, 118
Huygen's principle, 1010–1011, 1027–1028
Hybridization, 1216
Hybrid orbitals, 1216
Hydraulic lift pressure, 400–401
Hydrogen atoms (see also Atoms)
    Bohr model of, 1173–1178
    energy levels of, 1177–1178
    energy quantization of, 1141–1142
    first Bohr radius on, 1176
    induced dipole moment in, 774
    mass of, 540–541
    potential energy of, 721
    potential energy vs. separation for two,
        1213
    quantum theory of, 1181–1186
    rotational energy of, 581
    symmetric and antisymmetric wave
        functions for, 1212–1213
Hydrogen bond, 1214
Hydrostatic paradox, 400

Hysteresis, 882
Hysteresis curve, 882, 883

Ice-point (normal freezing point)
    temperature, 533
Ideal battery, 795
Ideal gas (see also Gases)
    Carnot cycle for, 603, 604–606
    entropy of an, 611
    free expansion of an, 612–613
    heating and compressing, 540
    heating, cooling, and compressing, 575
    heating diatomic, 576–577
    internal energy of, 568
    temperature scale of, 536
    volume of, 539
    work done on an, 571–572
Ideal-gas law, 537–541
Ideal heat pump, 609–610
Images (see also Mirrors; Reflection;
    Refraction)
    aberrations of, 1062–1063
    in concave mirror, 1044–1045
    depth inversion of, 1039
    determining range of mirror, 1048
    formed by lens, 1058–1059
    formed by refraction, 1049–1052
    formed by two plane mirrors, 1040
    geometry calculating, 1042–1043
    lateral magnification of, 1046
    optical instruments, 1063–1073
    produced from combinations of lenses,
        1059–1062
    real, 1041
    seen from goldfish bowl, 1051
    seen from overhead branch, 1051–1052
    virtual, 1038–1039
Impulse-momentum theorem, 217
Impurity semiconductors, 1246
Incoherent sources, 511
Index of refraction, 1011
Induced currents, 897
Induced emfs
    in circular coil I, 900
    in circular coil II, 901
    described, 897
    Faraday's law and, 899–903
    in stationary circuit in changing magnetic
        field, 900
Induced nonconservative electric field,
    901–903
Induction
    charging by, 655–656
    discovery of magnetic, 897–898
    mutual, 914–915
    self-inductance, 912–914
    via grounding, 655
Inductive reactance, 940–941
Inductors
    alternating current circuits in, 939–941
    energy stored in, 916
    energy when final current attained, 919
    potential difference across, 914
Inelastic collision entropy changes, 614–615
Inelastic scattering, 1001
Inertia
    Newton's first law on, 85

reference frames of, 86–87
rotation and moment of, 272–279
Inertial confinement, 1327
Inertial reference frames, 86–87, 1268
Infinite square-well potential
mathematical description of, 1140
particle in, 1151–1152
Infrared waves, 977
Initial conditions, 37
Initial-value problem, 37
Ink-jet printer, 670–671
Instantaneous acceleration, 25, 62
Instantaneous speed, 21
Instantaneous velocity, 21, 59
Insulators, 654–656
Integrals
definition of, 37
displacement and velocity as, 37–40
Integration
definition of, 38
equations used to compute, 36–40
Euler's method and numerical, 136–138
finding center of mass by, 222–223
finding gravitational field of spherical
shell by, 358–360
review of, AP-27–28
Intensity
defined, 479
due to point source, 479–480
examples of, 481–482
soundproofing sound, 481
Interference
definition of, 1984
electron diffraction and, 1127–1128
in thin films, 1086–1088
two-slit pattern of, 1131–1132
Interference-diffraction pattern
of multiple slits, 1100–1101
of two slits, 1093–1094
Interference patterns
phasors to add harmonic
waves/calculating, 1094–1101
three or more equally spaced sources,
1096–1097
two-slit, 1088–1091
Internal-combustion engine, 597–598
International system of units
length, 3–4
mass, 4–5
time, 4
Intrinsic semiconductor, 1246
Invariance of coincidences principle, R.6
Ionic bond, 1209–1210
Ionization energy, 1177, 1190, 1194
Ions, 1172
Irreversibility, 610
Isothermal bulk modulus, 468n
Isotherms, 539
Isotopes
mass spectrometer for measuring masses
of, 838–839
separating nickel, 839

Josephson current frequency, 1256
Josephson effect, 1255–1256
Josephson junction, 1255
Joule heating, 794

Joule, James, 2, 559, 568
Joule's temperature experiment, 565–566
Jump starting cars, 806
Junction
defining, 1248
forward biased, 1248–1249
Josephson, 1255
reverse biased, 1249
Junction rule, 804
Jupiter's orbit, 341

KamLAND (Kamioka Liquid Scintillator
Anti-Neutrino Detector), 1338–1339
Karate collision, 234–235
Keck Observatory, 1073
Kelvin (K), 4
Kelvin temperature scale, 537
Kepler's laws
described, 340–342
gravity and derivation of, 346–347
Kilogram (kg)
defining unit of, 4, 88
as unit of force and mass, 91
Kinematics, 17
Kinetic energy (see also Energy; Power)
before and after two-object collision, 233
examples/demonstrations of, 154–156,
350
mass and, 1292–1294
of molecules in ideal gas, 568
power and, 164–165
of rotation, 271–272, 323
in simple harmonic motion, 434
of a system, 232–233
theorem for the kinetic energy of system,
232
total energy, momentum and, 1291–1292
work and, 152–153
work-kinetic energy theorem, 153
Kinetic friction
conservation of energy and problems of,
194–195
described, 118
Kinetic theory of gases
calculating pressure exerted by gas, 542
described, 541–542
equipartition theorem and, 544–545
mean free path and, 545–546
molecular interpretation of temperature,
542–544
Kirchhoff, Gustav, 999
Kirchhoff's rules
applied to multiloop circuits, 807–808
described, 803–804
junction rule of, 804
loop rule of, 804, 917
multiloop circuits and, 806–809
single-loop circuits and, 804–806
Klitzing, Klaus von, 847
Krypton laser, 1107

Laser rocket, 984–985
Lasers, 1003–1005, 1007
Latent heat of fusion, 563
Latent heat of vaporization, 563
Lateral magnification, 1046

Lattice, 655
Law of conservation of angular momentum
(see Conservation of angular
momentum)
Law of conservation of charge, 653
Law of conservation of energy (see
Conservation of energy)
Law of conservation of mechanical energy
(see Conservation of mechanical
energy)
Law of conservation of momentum (see
Conservation of momentum)
Law of inertia, 86–87
Law of Malus, 1022
Law of reflection, 1011–1012
Law of refraction, 1012
Lawrence, E. O., 839, 1317
Lawson, J. D., 1326
Lawson's criterion, 1326–1327
LC circuits, 944–948
LC oscillator, 947
Leaking tank fluid flow, 410
Leaning ladder equilibrium, 376–377
LEDs (light-emitting diodes), 1250
Length contraction, 1274–1276
Length units, 3–4
Lens combinations
image formed by second lens, 1059–1060
image formed by two lenses, 1060–1061
images formed by compound lenses,
1061–1062
images produced from, 1059–1062
Lenses
aberrations found when using, 1062–1063
combinations of, 1059–1062
compound microscope, 1068–1070
focal length of cornea-lens system of eye,
1064–1065
images formed by refraction, 1049–1052,
1058–1059
ray diagrams for, 1057–1059
reading glasses, 1065–1066
simple magnifier, 1066–1068
telescope, 1070–1073
thin, 1052–1057
Lens-maker's equation, 1053, 1055
Lenz's law
alternative statement of, 904
definition of, 903
induced current and, 904–905
moving coil and, 906
LEP (Large Electron-Positron) collider
[CERN], 1340, 1341
Lepton era, 1352
Lepton number, 1342
Leptons
applying conservation laws to, 1342
GUTs (grand unification theories) on,
1350
interactions of, 1336–1339
masses of fundamental, 1347
Leyden jar, 749
Lick Observatory, 1072
Light (see also Reflection; Refraction)
coherence length of, 1086
early theories/studies on, 1118–1119
energy quantization in atoms, 1124–1125
ether and speed of, 1268–1269

interference and diffraction, 1084–1108
photons, 1119–1124
polarization, 1021–1027
propagation of, 1010–1011
sources of, 999–1005
spectra of, 998–999
speed of, 1005–1010
two-slit interference pattern with, 1131–1132
wave-particle duality of, 998
Light clock, R.5
Light nuclei
    nuclear fusion and, 202
    resting energies of selected, 201
Light sources
    absorption, scattering, spontaneous emission, and stimulated emission, 1001–1002
    continuous spectra, 1000
    lasers, 1003–1005, 1107
    line spectra, 999–1000
    resonant absorption and emission, 1002
    spontaneous emission, 1000
Light spectra, 998–999
Linear charge density, 683
Linearly polarization, 1021
Linear motion analogs, 287
Line spectra, 999–1000
Liquid-drop model, 1307
Lithium (Z = 3), 1191–1192
Livingston, M. S., 839, 1317
Lloyd's mirror, 1091
Longitudinal waves, 465–466
Long-range order, 1229
Loop rule, 804
Lorentz-FitzGerald contraction, 1275
Lorentz, Hendrik A., 1228, 1275
Lorentz transformation, 1271–1272
Loudspeakers
    sound intensity of one, 480
    sound intensity of two, 509
Lyman series, 1177–1178

Macdonald Observatory, 6
Mach angle, 491
Mach number, 491
Madelung constant, 1230
Magnetically hard materials, 883
Magnetically soft materials, 883
Magnetic bottle, 386
Magnetic confinement, 1327
Magnetic dipole moment, 841–842
Magnetic dipole potential energy, 843
Magnetic domain, 880–881
Magnetic energy
    definition of, 915–916
    density of, 916–917
Magnetic field lines, 829, 979
Magnetic fields
    Ampère's law on, 871–874
    crossed electric and, 836
    on current element, 832
    force on bent wire, 833
    force exerted by, 830–833
    force on moving charge, 830–831
    force on straight wire, 833

gauss, measure of, 831
induced emfs/induced currents in changing, 897–898
motion of point charge in a, 834–841
potential energy of magnetic dipole in, 843
right-hand rule for determining direction of force from, 830
right-hand rule determining direction of, 866
on segment of current-carrying wire, 832
Magnetic field
    on axis of coil, 861
    on the axis of current loop, 860
    between parallel wires, 868–869
    calculating amount of mobile charge, 862
    at center of current loop, 859–860
    center of solenoid, 865
    at center of square current loop, 867
    due to current element (Biot-Savart law), 858–859
    due to current loop, 859
    due to current in solenoid, 863–865
    due to current in straight wire, 865–866
    due to two parallel wires, 867–868
    earliest known of, 856
    inside and outside a wire, 872
    inside tightly wound toroid, 873
    magnetic-dipole field on axis of dipole, 860
    of moving point charges, 857–858
    torque on bar magnet, 862
Magnetic flux
    definition of, 898–899
    through a solenoid, 899
Magnetic moment, 1186–1187
Magnetic quantum number, 1180
Magnetic susceptibility, 875–876
Magnetism
    applying Curie's law, 880
    atomic magnetic moments, 876–879
    diamagnetic, 874, 875, 884–885
    ferromagnetic, 874, 875, 880–884
    ferromagnetic material, 874, 875, 880–884
    Gauss's law of, 870–871, 975–976
    iron and saturization, 878–879
    magnetic susceptibility and, 875–876
    in matter, 874–885
    paramagnetic material, 874, 875, 879–880
Magnets
    torque on bar magnet, 862
    torques on current loops, 841–845
Maiman, Theodore, 1003
Malus, E. L., 1022
Maricourt, Pierre de, 829
Mark I detector, 1338
Mark, Robert, 1026
Marsden, E., 1173
Mass
    center of, 218–222
    contact forces and, 93–95
    energy and, 201–204, R12
    examples/demonstrations on, 88–89
    gravitational constant G and inertial, 345–346
    of hydrogen atom, 540–541
    international system of units for, 4–5

molar, 540
Newton's second law of motion defining, 86, 87–88
Newton's third law of motion on force between objects of, 101–105
nucleus binding energy and, 1308–1310
reduced, 1218–1219
systems with continuously varying, 248–252
units of, 91
Mass spectrometer, 838–839
Matter
    diamagnetic, 874, 875, 884–885
    ferromagnetic, 874, 875, 880–884
    magnetically hard and soft, 883
    magnetic susceptibility of, 875–876
    magnetism in, 874–885
    paramagnetic, 874, 875, 879–880
    relative permeability, 876
Maxwell-Boltzmann energy distribution function, 551
Maxwell-Boltzmann speed distribution, 549
Maxwell, James Clerk, 2, 1119, 1268, 1269
Maxwell's displacement current
    calculating, 974–975
    definition of, 972
    generalized form of Ampère's law, 972–973
Maxwell's equations
    on electromagnetic waves, 971–972
    overview of, 975–976
Mean free path, 545–546
Mean lifetime, 1312
Measurements (see Units of measurement)
Mechanical energy
    conservation of, 183–185
    conversion of chemical energy into, 199
    gravitational potential energy and, 353
    simple harmonic motion and total, 434–435
Mechanical equivalence of heat, 565
Meissner effect, 922
Melting, 562
Mesons, 1336
Metallic bond, 1214–1215
Metastable states, 1001, 1316
Meter (m), 3–4
Michelson, A. A., 1008–1009, 1269, R.2
Microwave oven electric dipole movement, 671
Midair refueling, 23
Millisecond (ms), 4
Mirages, 1017–1018
Mirror equation, 1044
Mirrors (see also Images; Reflection)
    aberrations found when using, 1062–1063
    convex, 1047
    determining range of image in, 1048
    images in concave, 1044–1045
    Lloyd's mirror, 1091
    plane, 1038–1041
    ray diagrams for, 1045–1048
    spherical, 1041–1045
Modelocking technique, 1007
Moderator, 1323
Modern physics, 3
Mode of vibration, 511
Molar heat capacities of gases, 574

Molar mass, 540
Molar specific heat, 560
Molecular bonding
    covalent bond, 1210–1213
    described, 1208–1209
    hydrogen bond, 1214
    ionic bond, 1209–1210
    metallic bond, 1214–1215
    van der Waals bond, 1213–1214
Molecular view of dielectric, 772–775
Molecules
    center of mass of a, 219–220
    energy levels and spectra of diatomic,
        1217–1223
    kinetic energy in ideal gas, 568
    polyatomic, 1215–1216
    quantized energy of, 579
    speeds distribution of, 546, 549–551
Mole (mol), 4, 538
Moment of inertia
    calculating, 272–279
    examples/demonstrations of, 273–279
    parallel-axis theorem, 275
    of uniform bodies, 274
Momentum
    angular, 311–316, 324–328
    Bohr's postulate on quantized angular,
        1175–1176
    conservation of angular, 309–311, 316–324
    conservation of linear, 227–232
    impulse-momentum theorem, 217
    law of conservation of, 228–229
    mass, energy and relativistic, R.12-13
    of a particle, 217, 227-228, R.12
    relativistic, 1287–1289
    of system of particles, 217
    total energy, kinetic energy and, 1291–1292
Morley, Edward, 1269, R.2
Moseley, Henry, 1199, 1200
Motion (see also Oscillation; Rotation)
    analogs in rotational and linear, 287
    of the center of mass, 223–227
    fluids in, 407–415
    gyroscope, 316–317
    overdamped, underdamped, critically
        damped, 445–448
    of rolling objects, 288–294
    simple harmonic, 426–433
    simple wave, 465–472
    underdamped, 446
Motional emfs
    definition of, 897, 907
    magnetic drag, 910–911
    for moving rod, 909
    potential difference across moving rod,
        909
    total charge through flipped coil, 907–909
    U-shaped conductor and sliding rod, 910
Motion along curved path
    banked curves, 132, 133–134
    centripetal force and, 129
    examples/demonstrations of, 129–134
Motion with constant acceleration
    examples with one object, 28–33
    examples with two objects, 33–36
    overview of, 27–28
    velocity vs. time, 27

Motion diagrams, 64, 72
Motion in one dimension (see also
        Newtonian mechanics)
    acceleration, 24–26
    constant forces and, 152–153
    displacement, velocity, and speed, 17–24
    integration, 36–52
    study of, 17
Motion of point charges, 668–671
    cyclotron frequency, 835
    cyclotron period, 834–835
    in magnetic field, 834–841
    mass spectrometer, 838–839
    Newton's second law and, 834
    Thomson's measurement of $q/m$, 837
    velocity selector, 836
Motion in two and three dimensions
    circular, 72–74
    displacement vector and, 53–54
    general properties of vectors in, 55–59
    position, velocity, acceleration and, 59–65
    projectile, 65–72
Moving elevator examples, 34–36
Moving sticks/special relativity, R.4, R.7-8
MP (melting point), 563
Müller, K. Alexander, 1253
Multiloop circuits
    applying Kirchhoff's rules to, 807–808
    general method for analyzing, 808
    sign rule for change in potential across
        resistor, 806
    three-branch, 808–809
Muon, 1336
Muon neutrino, 1336
Mutual inductance, 914–915


Nanolasers, 1007
National Institute of Standards and
        Technology (NIST), 4
Natural frequencies, 450, 513
Nearsighted eye, 1064
Negative-energy states, 1339
Neutral rotational equilibrium, 379
Neutrinos, 1314, 1336–1339, 1342–1343
Neutrons
    atomic masses of, 1309
    binding energy of, 1310
    delayed, 1324, 1325
    elastic head-on collision of nucleus and,
        242–243
    nuclear reactions with, 1319
    spin of, 327–328
    thermal, 1319
Newtonian mechanics
    on drag forces, 134–136
    friction and, 117–129
    law of gravity, 342–348
    on motion along a curved path, 129–134
    numerical integration using Euler's
        method, 136–138
    second law for angular momentum, 313,
        324–326
    second law, 86, 87–101
    second law for rotation, 280–287
    second law for systems, 217, 228
    theory of relativity and, 204

    third law, 101–105
    three laws defined, 85–86
Newtonian relativity
    ether and speed of light, 1268–1269
    principle of, 1268
Newton (N)
    defining, 4, 88
    as unit of force, 91
Newton, Sir Isaac, 2, 346, 998, 1118–1119
Newton's law of cooling, 635, 644
Newton's law of gravity, 342–348
Newton's rings, 1087
Nishijima, K., 1343
Nodes, 512
Nonconservative force, 172
Nonpolar molecules, 672
Normalization condition, 547
Normalization condition equation, 1130
Nova target chamber, 1328
$n$-type semiconductor, 1247, 1248
Nuclear atom, 1172–1173
Nuclear energy
    binding, 202, 203
    process of, 202
Nuclear fission products potential energy,
        722
Nuclear fission reactors, 1322–1326
Nuclear fusion, 202, 203
Nuclear physics
    fission and fusion, 1319–1328
    nuclear reactions, 1317–1319
    properties of nuclei, 1306–1310
    radioactivity, 1310–1317
Nuclear radius, 1307
Nuclear reactions
    exothermic and endothermic, 1318
    fission and fusion, 1319–1328
    with neutrons, 1319
    $Q$ value measurement of, 1317–1319
Nuclear reactor control rods, 606
Nuclei properties
    mass and binding energy of, 1308–1310
    $N$ and $Z$ numbers, 1308
    overview of, 1306–1307
    size and shape, 1307
Nucleosynthesis period, 1352–1353
Nucleus-neutrons elastic head-on collision,
        242–243
Nuclide, 1307
Numerical integration
    computing sky diver position/speed,
        137–138
    drag forces illustrating, 137
    Euler's method and, 136–137


Ohmmeters, 809–810
Ohm's law, 790–793, 1234, 1235
Onnes, H. Kamerlingh, 1252
Open-circuit terminal voltage, 760
Oppenheimer, J. Robert, 1298
Optical images
    lenses, 1049–1073
    mirrors, 1038–1048
Optical instruments
    angular magnification of simple
        magnifier, 1067–1068

compound microscope, 1068–1070
the eye, 1063–1065
reading glasses, 1065–1066
simple magnifier, 1066–1067
telescope, 1070–1073
Optically flat surface, 1088
Optical spectrum, 1197–1199
Optic axis, 1025
Orbital quantum number, 1180
Orbits
classification by energy, 351
space station, 347–348
Order of magnitude, 11–12
Orrery (solar system model), 340
Oscillating systems
examples/demonstrations of, 437–439, 440–445
potential energy of spring-earth system, 439–440
Oscillation (*see also* Motion)
damped, 445–449
described, 425
energy and speed of oscillating object, 435–436
near equilibrium, 436–437
resonance and driven, 449–453
simple harmonic motion, 426–433
Oscillator energy-level diagram, 204–205
Otto cycle, 598, 599–600
Overdamped, 445

Parallel-axis theorem, 275
Parallelogram method, 54
Parallel-plate capacitors, 753–754
Parallel ray, 1045, 1057
Parallel *RLC* circuit, 956
Paramagnetic material, 874, 875, 879–880
Paraxial rays, 1041
Partial derivative, 470
Particle in a box
allowed energies for, 1134
Bohr's correspondence principle and, 1135–1136
ground-state energy for, 1134
photon emission by, 1137
standing-wave condition for, 1133
standing-wave functions, 1134–1135
Particles (*see also* Elementary particles; System of particles)
allowed energies for particle in a box, 1134
angular momentum of, 311–312
antiparticles and, 1339–1340, 1350
behavior of classical, 1131
cyclotron to accelerate, 839–840
expectation values, 1138–1140
fermions and bosons, 1166
in finite square well, 1152–1155
ground-state energy for particle in a box, 1134
in infinite square-well potential, 1151–1152
momentum of, 217, 227–228, R.12
motion in magnetic field, 835–836
position as function of time, 22
probability density equation for, 1129, 1130–1131

resting energies of selected, 201
rotating system of, 272
Schrödinger equation for two identical, 1164–1166
simple harmonic motion and circular motion of, 433
in simple harmonic oscillator, 1155–1157
spin of, 327–328
spin-one-half, 327
standing-wave condition for, 1133
uncertainty principle and, 1132–1133
velocity transformation, 1284–1287
wave-particle duality, 1131–1133
work done on, 157
Pascal (Pa), 397
Pascal's principle, 400
Pauli exclusion principle, 1166, 1339
Pendulum clock, 443
Pendulums
in accelerated reference frame, 442
amplitude of oscillation of, 442–443
ballistic, 239–240, 323
circular motion of, 72–73
conservation of energy, 186–187
oscillating system of simple, 440–443
physical, 443–444
Perfectly inelastic collisions
described, 233
in one dimension, 237
in three dimensions, 244–245
Periodic table (*see also* Atoms)
beryllium ($Z = 4$), 1193
boron to neon ($Z = 5$ to $Z = 10$), 1193
described, 1189–1190
electron configurations of atoms in ground states, 1194–1197
elements with $Z > 18$, 1193–1194
helium ($Z = 2$), 1190–1191
lithium ($Z = 3$), 1191–1192
sodium to argon ($Z = 11$ to $Z = 18$), 1193
Periodic waves, 473–478
Perl, Martin Lewis, 1336
Permeability of free space, 857
Permeability of the material, 882
Permittivity of free space, 693
Phase
coherent and incoherent sources, 510–511
described, 474
difference between two waves, 508–509
Phase change, 562
Phase constant
for driven oscillations, 452
for harmonic waves, 427
Phase diagrams, 634
Phase difference
due to path difference, 1084–1085
due to reflection, 1085–1086
Phase of the motion, 427
Phase velocity, 522
Phasors, 943–944
Phosphorescent materials, 1001
Photoelectric effects
Einstein's explanation of, 1119–1120
work functions, 1120–1121
Photon, 205
Photon-atom interactions, 1001–1002
Photon beam in laser, 1004

Photon concept, 1122
Photons
Bohr's postulate on frequency of, 1175
Compton equation on scattering of, 1122–1124
Einstein equation for photon energy, 1120
equations for frequency and wavelength of, 1134
momentum of, 1122
photoelectric effects, 1119–1122
two-slit experiment using, 1131–1132
virtual, 1347–1348
wavelength during electron transition, 1199–1200
Physical pendulum, 443–444
Physical quantities, units of measurement for dimensions of, 7–8
Physics
classical and modern, 2–3
nuclear, 1306–1328
science of, 2
Piezoelectric effect, 775
Pitch, 520
Planck, Max, 643
Planck's constant, 1120
Plane of incidence, 1011
Plane mirrors, 1038–1041
Plane polarization, 1021
Plane symmetry, 695
Plane wave, 985
Plasma, 1326, 1327
Point charges
in the cavity at center of spherical conducting shell, 704
Coulomb's law for electric field due to, 662
electric field due to line charge and, 687
electric field lines due to two positive, 666
electric field lines of single positive, 666
electric field on $y$ axis due to point charges on $x$ axis, 664
electric potential due to system of, 720–724
electrostatic potential energy of system of, 750
magnetic field of moving, 857–858
motion in electric fields, 668–671
motion in magnetic field, 834–841
Poiseuille's Law, 414–415
Polarization
by absorption, 1022
by birefringence, 1025, 1027
intensity transmitted, 1023
overview of, 1021–1022
by reflection, 1023–1024
by scattering, 1024–1025
transmission axes of two polarizing sheets, 1026
Polarized conductor, 655
Polarized molecules, 672
Polyatomic molecules, 1215–1216
Position
described, 59–60
driven oscillation, 451–452
examples of computing, 60–61, 63, 137–138
in simple harmonic motion, 426–427

Positive lens, 1054
Positron (antiparticles), 1339–1342
Potential difference
  across inductor, 914
  continuity of $V$ and, 718–719
  definition of, 717–718
  devices measuring, 809–810
  electric fields and, 719–720
  find $V$ for constant electric field, 720
  for infinitesimal displacements, 718
  single-loop circuit, 805
Potential energy
  conservative forces and, 168
  described, 167–168
  of dipole in electric field, 671
  electrostatic, 749–752
  equilibrium and, 172–174
  examples/demonstrations of, 169–171
  gravitational, 349–353
  of hydrogen atoms, 721
  infinite square-well potential, 1140
  loss per unit time, 794
  of magnetic dipole in magnetic field, 843
  net attractive part of ion in crystal, 1229
  nonconservative forces of, 172
  of nuclear fission products, 722
  potential-energy function of, 168–169
  of spring force, 167–168, 170–171
  simple harmonic motion and, 434–435
  of spring-earth oscillating system,
    439–440
  torque and, 672
  two hydrogen atoms and separation vs.,
    1213
Potential-energy function, 168–169
Power (see also Energy; Forces; Kinetic
    energy)
  described, 163–164
  dissipated in resistor, 794–795
  kinetic energy and, 164–165
  of a motor, 164, 287
  rotation and, 286–287
  supplied by emf source, 795–797
  supplied to device or segment, 794
  terminal voltage, stored energy and,
    797–798
Power factor of $RLC$ circuit, 951
Power of lenses, 1056–1057
Precession, 316–317
Pressure
  aorta and blood, 401–402
  calculating pressure exerted by gas, 542
  described, 397–398
  force on a dam, 398–400
  gauge, 401
  hydraulic lift, 400–401
Principal quantum number, 1180
Principle of equivalence, 1296–1297
Principle of invariance of coincidences, R.6
Principle of Newtonian relativity, 1268
Principle of relativity, R.2-3
Probability density (see also Schrödinger
    equation)
  equation for, 1129, 1130–1131
  first excited state of hydrogen, 1185–1186
  ground state of hydrogen, 1183–1185
  radial, 1184, 1186
  wave functions and, 1182–1186

Projectile motion
  described, 65–66
  examples/demonstrations of, 66–72,
    224–225
  gravitational potential energy and height
    of, 351–352
  gravitational potential energy and speed
    of, 352
Propagation of light
  Fermat's principle, 1011
  Huygen's principle, 1010–1011
Proper length, 1274
Proton-antiproton pair, 1340–1341
Proton-proton collision, 246
Protons
  energy of accelerated, 840–841
  force on north moving, 831
  spin of, 327–328
$p$-type semiconductor, 1247, 1248
Pulley rotation, 285–286
Pulling through hole angular momentum,
    321–323
Push/pushing
  examples/demonstrations of, 188,
    195–196
  nonconservative force of, 172


QCD (quantum chromodynamics), 1348
$Q$ factor, 952
  damped oscillation and, 447–448
  driven oscillation and, 450
Quantization of angular momentum,
    326–328
Quantization of energy, 204–205, 579
Quantum electrodynamics, 1347
Quantum Hall effects, 847
Quantum mechanics, 3
Quantum numbers, 205
  described, 1135
  principal, orbital, and magnetic, 1180
  rotational, 1217
  in spherical coordinates, 1179–1180
  value of ground state of hydrogen, 1183
Quantum theory
  of atoms, 1178–1181
  of electrical conduction, 1243–1244
  electron described in, 1129
  of hydrogen atom, 1181–1186
Quarks
  confinement of, 1347
  definition of, 1345
  in GUTs (grand unification theories), 1350
  properties of antiquarks and, 1347
  types of, 1345–1346
Quarter-wave plate, 1025
Quasar, 205
Quasi-static adiabatic gas compression,
    581–583
Quasi-static processes, 569
$Q$ value, 1317–1319


Radial equation, 1179
Radial probability density, 1184, 1186
Radial ray, 1045
Radiation
  described, 642

  from the human body, 644
  from the sun, 643–644
  Stefan-Boltzmann law on, 642
  Wien's displacement law on, 643
Radiation era, 1352
Radioactivity
  alpha decay, 1316–1317
  beta decay, 1314–1316
  counting rate for radioactive decay,
    1312–1313
  detection-efficiency considerations,
    1313–1314
  gamma decay, 1316
  overview of, 1310–1312
Rainbows
  calculating angular radius of, 1020–1021
  as dispersion, 1019–1020
Ray approximation, 485
Ray diagrams for lenses, 1057–1059
Ray diagrams for mirrors, 1045–1048
Rayleigh, Lord, 1001
Rayleigh scattering, 1001
Rayleigh's criterion for resolution,
    1103–1104
$RC$ circuit, 811–816
Reaction, Newton's third law of motion on,
    101–102
Real battery, 796
Real image, 1041
Redshift, 1277
Reduced mass, 1218–1219
Reference frame, 1268
Reflection
  derivation of the laws of, 1027–1030
  electron waves transmission and,
    1157–1161
  Fermat's principle, 1011, 1028
  fiber optics application of, 1017
  Huygen's principle, 1010–1011, 1027–1028
  overview of, 482–484
  phase difference due to, 1085–1086
  physical mechanisms for, 1013
  of plane waves from convex mirror, 1044
  polarization by, 1023–1024
  relative intensity of reflected light, 1014
  sign conventions for, 1046
  specular and diffuse, 1013–1014
  total internal, 1015
  water-air surface, 1016
  See also Images; Light; Mirrors
Refraction (see also Images; Light)
  defining, 1011
  derivation of the laws of, 1027–1030
  dispersion, 1018–1021
  Fermat's principle, 1010, 1029–1030
  from air to water, 1014
  Huygen's principle, 1010–1011, 1028
  images formed by, 1049–1052, 1058–1059
  index of, 1011
  law of, 1011–1012
  mirages, 1017–1018
  overview of, 482–484
  physical mechanisms for, 1013
  sign conventions for, 1049–1050
  Snell's law of, 1012
  thin lenses, 1052–1057
Refrigerator statement of the second law of
    thermodynamics, 600–602

Relative permeability of material, 876
Relative velocity
  in one dimension, 23–24
  in two and three dimensions, 61–62
Relativistic Doppler effect, 1276–1278
Relativistic energy
  mass and, 1292–1296
  overview of, 1289–1292
Relativistic kinetic energy, 1290
Relativistic momentum
  conservation of, 1288–1289
  described, 1287–1288
Relativity (see Special relativity)
Relativity of simultaneity, R.8-9
Remnant field, 882
Reproduction constant $k$ of reactor,
  1322–1323
Residual strong interaction, 1349
Resistance
  calculated per unit length of wire, 792–793
  definition of, 790
  devices measuring, 809–810
  internal resistance of batteries, 796–797
  ohmic/nonohmic materials and, 790–791
  temperature coefficient of resistivity,
    791–792
Resistors
  blowing the fuse, 803
  combinations of combinations of, 802
  combinations of, 798–803
  combinations of series and parallels,
    801–802
  in parallel, 799–800
  power dissipated in resistor, 794–795
  in series, 798–799, 800–801
  sign rule for change in potential across,
    806
Resolution
  problem on, 1104–1105
  Rayleigh's criterion for, 1103–1104
Resolving power of diffraction grating,
  1107–1108
Resonance
  described, 450, 1319
  driven series $RLC$ circuit at, 953–954
  mathematical treatment of, 451–452
  overview of, 950–952
Resonance Absorption, 1001
Resonance curves, 450, 951, 952
Resonance frequency
  driven oscillations and, 450–451
  standing waves and, 511
Resonance width, 951
Resonant frequency spectrum, 511
Rest energy, 1290, 1291, 1292
Reverse biased junction, 1249
Reversibility, 1044
Reynolds number, 415
$R$ factor, 639–641
Right-hand rule, direction of
  angular velocity, 309–310
  cross product, 310–311
  magnetic field, 866
  magnetic force, 830
Ritz, Walter, 1172
$RLC$ circuits
  driven, 948–956
  $LC$ and, 944–948

parallel, 956
  power factor of, 951
  $Q$ factor for, 952
$RL$ circuits
  energy dissipated, 920
  initial currents and final currents,
    920–921
  Kirchhoff's loop rule applied to, 917
  time constant of, 918
RMS (root-mean-square) values, 937–938
Rocket propulsion
  liftoff and, 251–252
  rocket equation of, 250–251
  as system of continuously varying mass,
    248–249
Rolling friction, 118
Röntgen, Wilhelm, 2
Root mean square, 543
Rotation (see also Motion)
  analogs in linear motion and, 287
  calculating moment of inertia during,
    272–279
  examples/demonstrations of, 282–286
  kinematics of, 267–271
  kinetic energy of object in, 323
  kinetic energy of, 271–272
  Newton's second law for, 280–287
  nonslip conditions of, 283–284
  power of, 286–287
  torques, 281–282, 287
  vector nature of, 309–311
Rotational equilibrium, 378–379
Rotational kinematics
  angular acceleration, 268–269
  CD player, 269–271
  described, 267–268
Rotational quantum numbers, 1217
Ruby laser, 1003, 1004–1005
Rutherford, E., 1173
Rydberg, Johannes R., 1172
Rydberg-Ritz formula, 1172–1173

Sandia National Laboratory, 606, 133
Saturated bond, 1213
Saturation magnetization, 878–879
Savart, Félix, 856
Sawtooth waveform, 939
Scanning tunneling electron microscope,
  1161
Scattering
  deep inelastic, 1346
  inelastic, 1001
  light sources, 1001–1002
  polarization by, 1024–1025
  Rayleigh, 1001
Schrieffer, Robert, 1254
Schrödinger equation
  correspondence between Bohr model
    and, 1178
  Dirac equation extension of, 1339
  harmonic oscillator application of,
    1155–1157
  particle in finite square well application
    of, 1152–1155
  probability density described by, 1129
  quantized energies of system determined
    by, 1140

in spherical coordinates, 1178–1179
  in three dimensions, 1161–1164
  time-dependent, 1150
  time-independent, 1150–1155
  for two identical particles, 1164–1166
Schrödinger, Erwin, 1129, 1219
Schwarzschild radius, 1298
Science
  definition of, 1
  division into separate fields, 1–2
  physics, 2
Scientific model, 1
Scientific notations, 8–10
Second law of motion
  for angular motion, 313, 324–326
  for the center-of-mass motion, 223–227
  defined, 86
  force, defined by, 87–89
  free-body diagram problems using,
    95–100
  for rotation, 280–287
  written in terms of momentum, 228
Second law of thermodynamics (see also
    Thermodynamics)
  Carnot engine and, 602–610
  Clausius statement on, 596
  entropy and, 610–620
  equivalence of heat-engine and
    refrigerator statements and, 602
  heat engine and, 596–600
  heat-engine statement of the, 598–599
  heat pumps and, 609–610
  irreversibility and disorder, 610
  Kelvin statement on, 596
  refrigerator statement of, 600–602
  subject of, 595
Second-order spectrum, 1107
Second (s), 4
Segrè, Emilo, 1340
Segway (self-balancing human transporter),
  317
Self-inductance, 912–913
Semicircular hoop center of mass, 222–223
Semiconductors
  diodes, 1248–1250
  junction, 1248
  transistors, 1250–1252
  types of, 1246–1248
Series driven $RLC$ circuits, 948–955
Shear force, 382
Shear modulus, 382
Shear strain, 382
Shockley, William, 1250
Shock waves, 490–491
Short-range order, 1229
Significant figures, 10–11
Simple harmonic motion
  acceleration in, 426, 428
  angular frequency and, 428–429
  of block on a spring, 432
  circular motion and, 433
  described, 426
  energy in, 434–437
  frequency in, 430
  of oscillating object, 431
  position in, 426–427
  riding the waves, 429–430
  velocity in, 427–428

Simple magnifier, 1066–1068
Simultaneity
    clock synchronization, 1278, 1280–1282
    defining, 1279
    simultaneous events, 1278–1280
    twin paradox, 1282–1284
Simultaneity relativity, R.8-9
Single-loop circuits, 804–806
Single slit pattern of diffraction, 1091–1093
SI units
    of force and power, 4
    frequently used, 4
    prefixes for common
        multiples/submultiples of, 4–5
Snell's law of reflection, 1012
Snyder, Hartland, 1298
Sodium light, 1107
Sodium to argon ($Z = 11$ to $Z = 18$), 1193
Solar cell, 1250
Solar system
    elliptical path of planet in, 341
    Jupiter's orbit, 341
    orbits of planets around sun, 340
    plotting mean distance from sun, 342
Solenoids
    with iron core, 884
    magnetic field at center of, 865
    magnetic field due to current in, 863–865
    magnetic flux through, 899
    self-inductance of, 913–914
Solids
    amorphous, 1229
    band theory of, 1244–1246
    Fermi-Dirac distribution, 1256–1259
    Fermi electron gas, 1236–1243
    microscopic picture of conduction,
        1233–1236
    semiconductors, 1246–1248
    structure of, 1229–1232
    superconductivity, 1252–1256
Sonic boom, 492
Soundproofing, 481
Sound waves
    beats/beat frequency of, 506–507
    calculating speed of, 468–469
    Doppler effect, 488–489
    energy of, 477
    harmonic, 477
    intensity of, 479–482
    phase difference between, 508–509
    speed of, 467–468, 584–585
    standing, 517–519
    ultrasonic, 486
Source of emf, 795
Spacetime coincidence, R.6
Spacetime event, R.6
Special relativity
    applying rules of, R.10-11
    clock synchronization and simultaneity,
        1278–1284
    distant clocks, simultaneity, and, R.8-9
    Doppler effect and, 490
    Einstein's postulates, 1269–1270
    examples/demonstrations of, R.4-6, R.7-8
    Galilean transformation, 1270–1271
    general relativity, 1296–1298
    length contraction, 1274–1276
    Lorentz transformation, 1271–1272

mass and energy equation, 201
    Newtonian, 1268–1269
    Newtonian mechanics and, 204
    principle of relativity and constancy of
        speed of light, R.2-3
    proposed by Einstein, 2-3, R.1, 1267
    relativistic Doppler effect, 1276–1278
    relativistic energy, 1289–1296
    relativistic momentum, 1287–1289
    time dilation, 1272–1274
    velocity transformation, 1284–1287
Specific gravity, 396
Spectral line, 1107
Specular reflection, 1013–1014
Speed
    and acceleration of object on spring,
        432–433
    of approach in head-on elastic collision,
        241, 244
    average, 19–20
    computing sky diver, 137–138
    and energy of oscillating object, 435–436
    escape, 350–351
    gravitational potential energy and
        projectile, 352
    instantaneous, 21
    of molecular distribution, 546
    of recession in head-on elastic collision,
        241
    simple harmonic motion/circular motion
        relation with constant, 433
    of sound waves, 468–469, 584–585
    of supersonic plane, 492
    terminal, 135–136
    of waves, 467–470
Speed of light
    early understanding of, 1005–1006
    Einstein's Postulates on, R.2-3
    ether and, 1268–1269
    exercise in unit conversions measuring,
        1009
    Fizeau's method of measuring, 1008–1010
    Foucault's method of measuring, 1008
    Michelson's method of measuring,
        1008–1009
    relativity principle and constancy of,
        R.2-3
    Römer's method of measuring, 1006
Spherical aberration, 1041, 1062
Spherical coordinates
    quantum numbers in, 1179–1181
    Schrödinger equation in, 1178–1179
    volume element in, 1183
Spherical mirrors, 1041–1045
Spherical symmetry, 696
Spin 1/2 particles, 1339
Spin, angular momentum and, 327–328, 314
Spin-orbit effect, 1186–1188
Spontaneous emission, 1000
Spring
    Hooke's law and, 93–94
    potential energy of, 167–168, 170–171
    simple harmonic motion of, 432
Stable rotational equilibrium, 378–379
Standard conditions, 539
Standard model
    basic interactions in, 1349–1350
    described, 1336, 1349

    predicting neutrinos without mass,
        1336–1339
Standing sound waves, 517–519
Standing waves
    described, 511
    on string, 511–515
    superposition of, 519–520
    wave functions for, 516–517
Stanford University linear accelerator, 32
Star tracks in time exposure, 269
State variables $P$, $V$, and $T$, 539
Static equilibrium
    in an accelerated frame, 377–378
    center of gravity and, 371–372, 379
    conditions for, 371
    couples, 377
    examples/demonstrations of, 372–382
    general motion near, 436–437
    indeterminate problems of, 379–380
    stability of rotational, 378–379
    stress and strain affecting, 380–382
    Young's modulus, 380–381
Static friction, 117–118
Steady-state solution, 451
Steam engine efficiency, 606–607
Steam-point (normal boiling point)
    temperature, 533
Stefan-Boltzmann law, 642
Step potential
    described, 1157–1158
    reflection and transmission of, 1158–1159
Stimulated emission, 1001–1002
Stokes Raman scattering, 1001
Stopping distance motion, 30–31
Stored energy
    capacitors and, 756–760
    in presence of dielectrics, 770–772
    terminal voltage, power and, 797–798
Strangeness, 1343–1344
Strange particles, 1344
Strassmann, Fritz, 1321
Stress/strain
    compressive, 381
    elevator safety and, 381–382
    equilibrium and, 380–381
    Young's modulus and, 380–381
Strong interactions
    bosons that mediate, 1348
    decays occurring via, 1345
    definition of, 1336, 1348
    hadrons stable against decay via, 1337
    properties of, 1350
    residual, 1349
Strong nuclear force, 1307
Sublimation, 634
Superconducting energy gap, 1254
Superconductivity
    BCS theory, 1254–1255
    Josephson effect, 1255–1256
    overview of, 1252–1254
Superconductors
    as diamagnetic, 885
    flux quantization, 923
    magnetic properties of, 922–923
    Meissner effect, 922
    type II, 922
Super-K detector, 1338
Superposition of standing waves, 519–520

Superposition of waves
interference and, 505–511
principle of superposition, 504
wave equation and, 504–505
Surface-barrier detectors, 1250
Surface charge density, 682
Swinging pendulum circular motion, 72–73
Synchronizing clocks, 1278, 1281–1282,
R.8-9
System of particles (see also Particles)
center of mass and, 218–222
center-of-mass reference frame, 247–248
collisions in, 233–246
conservation of linear momentum,
227–232
continuously varying mass, 248–252
finding center of mass, 222–223
gravitational potential energy of, 221–222
kinetic energy of, 232–233
moment of inertia and, 273
momentum of, 217
motion of the center of mass, 223–227
theorem for the kinetic energy of a, 232

Tacoma Narrows suspension bridge, 513
Tau, 1336
Tau neutrino, 1336
Telescope
astronomical, 1071–1073
overview of, 1070–1071
schematic diagram of astronomical, 1070
Temperature
absolute temperature scale, 535–537,
608–609
big bang theory on evolutionary role of,
1352–1353
Celsius and Fahrenheit, 533–534
coefficient of linear expansion and
changes in, 628–629
critical, 1252–1253
Curie, 881
expanding gas at constant, 541
Fermi, 1240
heat needed to raise, 560
ideal-gas law and, 537–541
Joule's experiment on raising, 565–566
Kelvin, 537
molecular interpretation of, 542–544
thermal equilibrium and, 532–533
thermodynamic, 608–609
Temperature gradient, 635
Terminal speed, 135–136
Terminal voltage, 760, 796, 797
Test charge, 661
TFTR (Tokamak Fusion Test Reactor), 1327
Theory of relativity (see Special relativity)
Thermal conductivity, 635–636
Thermal contact, 533
Thermal current, 635–636
Thermal energy
conduction of, 635–638
convection transport of, 641–642
conversion of chemical energy into,
199–201
heat loss through roof, 640–641
radiation, 642–644
rearranged metal bars and, 638–639

Thermal equilibrium, 532–533
Thermal expansion, 628–632
Thermal neutron, 1319
Thermal radiation, 642–644
Thermal resistance
parallel, 637–639
series, 636–637
Thermodynamics (see also Second law of
thermodynamics)
Celsius and Fahrenheit temperature
scales, 533–535
change of phase and latent heat, 562–565
failure of equipartition theorem, 578–581
gas thermometers and absolute
temperature scale, 535–537, 608–609
heat capacities of gases, 572–577
heat capacities of solids, 577–578
heat capacity and specific heat, 559–562
ideal-gas law, 537–541
internal energy of ideal gas, 568
Joule's experiment and first law of,
565–568
kinetic theory of gases, 541–551
quasi-static adiabatic compression of a
gas, 581–583
thermal equilibrium and temperature,
532–533
work and PV diagram for gas, 569–572
Thermodynamic temperature, 608–609
Thermometric property, 533
Thin-lens equation, 1053
Thin lenses
converging (positive) lens, 1054
diopters (D), 1056
lens-maker's equation, 1053, 1055
power of, 1056–1057
refraction through, 1052–1057
thin-lens equation, 1053
Third law of motion
defined, 86
examples/demonstrations of, 101–105
Thomson, G. P., 1128
Thomson, J. J., 1125, 1173, 1335
Thomson's measurement of $q/m$, 837
Thomson's plum pudding model of atom,
1173
Three-branch multiloop circuit, 808–809
Threshold frequency, photon frequencies
less than, 1121
Threshold wavelength
finding increase in photon, 1124
photon wavelengths greater than, 1121
Time
international system of units for, 4
Time constant, 446, 812
Time constant of circuit, 918
Time-dependent Schrödinger equation,
1150
Time dilation, 1272–1274, R.6
Time-independent Schrödinger equation
described, 1150–1151
infinite square-well potential application
of, 1151–1152
particle in a finite square well application
of, 1152–1155
Time interval, R.6
Time-travel, 189–190
Tire-pressure gauge, 401

Tokamak fusion-test reactor, 873
Tone quality, 520
Toroid, 872–873
Torques
angular momentum and, 311–317
calculating, 281
due to gravity, 281–282
examples/demonstrations of, 841–845
exerted by automobile engine, 287
expressed mathematically as cross
product, 310–311
produced by couples, 377
Torsion modulus, 382
Total internal reflection, 484
Total mechanical energy, 184–185, 434–435
Total nuclear binding energy, 1309
Transformer, 956–959
Transient solution, 451
Transistors, 1250–1252
Transition elements, 1194
Transmission axis, 1022
Transmission coefficient, 1158
Transmission losses, 959
Transmitted light intensity, 1014
Transverse waves, 465–466
Triboelectric series, 652
Triple point of water, 536
Tuning fork, 518, 521
Tunnel diode, 1249
Tunneling current, 1249, 1250, 1254
Twin paradox, 1282–1284
Two slit interference-diffraction pattern,
1093–1094
Two-slit interference pattern
calculation of intensity, 1090–1091
distance on screen to $m$th bright fringe,
1089–1090
maxima and minima, 1089
overview of, 1088–1091
Type II superconductors, 922

Ultrasonic waves, 486
Ultraviolet light (black light), 1002
Ultraviolet rays, 977
Uncertainty principle, 1132–1133
Underdamped motion, 446
Unified mass unit, 1218
Uniform circular motion, 73
Units of measurement
Btu, 559
calorie, 559
cgs system of, 5
conversion of, 6–7
for dimensions of physical quantities, 7–8
force and mass, 91
international system of, 3–5
scientific notations for, 8–10
significant figures/order of magnitude,
10–12
value of using, 3
volt(V), 719
watt (W), 164
Unit vectors, 58
Universal gas constant ($R$), 558
Universal gravitational constant ($G$), 345
Universe by orders of magnitude, 12
Unstable rotational equilibrium, 379

Van Allen belts, 836
Van de Graaff generator, 736–737
Van de Graaff, R., 1317
Van der Waals bonds, 1213–1214
Van der Walls equation of state, 632–634
Vector bosons, 1348
Vectors
  acceleration, 62–65
  addition of, 54, 58
  average-velocity, 59–61
  components of, 55–56
  of cornering a turn, 63–64
  described, 62
  displacement, 53–54, 57
  examples/demonstrations of, 63–65
  multiplying by scalar, 55, 58
  negative of, 54, 58
  position and velocity, 59–61
  properties of, 58
  relative velocity, 61–62
  rotation and, 309–311
  subtracting, 55, 58
  unit, 58
  velocity, 60–62
Velocity
  average, 21, 59
  and displacement as integrals, 36–40
  examples/demonstrations of, 18, 22–23, 40
  as function of time, 26
  instantaneous, 21, 59
  phase, 522
  relative, 23–24, 61
  in simple harmonic motion, 427–428
  wave packet group, 522
Velocity transformation, 1284–1287
Velocity vectors, 60–62
Venturi effect, 410–412
Venturi meter, 411
Vibrational energy levels, 1219–1220
Virtual image, 1038–1039
Virtual photons, 1347–1348
Viscous flow, 413–415
Visible light, 976
VLA (very large array) of radio antennas (New Mexico), 1104
Voltmeters, 809–810
Volt(V), 719
Volume flow rate, 408
Von Klitzing constant, 847

Wairakei power plant (New Zealand), 605
Walton, E.T.S., 1317
Water
  making ice cubes out of, 601–602
  normal boiling point (BP) of, 533, 563, 633–634
  normal freezing point of, 533
Water clock, 4

Watt (W), 4, 164
Wave barriers
  diffraction, 484–486
  reflection and refraction, 482–484
  total internal reflection, 484
  two soldered wires, 483
Wave equation
  circular polarized plane, 990–991
  defining, 985–986
  derivation of, 986–989
  described, 470–472
  linearly polarized plane, 989–990
  superposition of waves and, 504–505
Waveforms, 520–521
Wave functions
  degenerate energy levels in, 1162–1163
  described, 466
  energy levels and ground-state, 1155–1157
  harmonic, 472–473
  hydrogen atom symmetric and antisymmetric, 1212–1213
  normalized ground-state, 1183–1184
  particle in a box standing, 1134–1135
  probability densities and, 1182–1186
  understanding spin augment and symmetry of, 1211–1212
Wave intensity
  described, 479
  due to point source, 479–480
  level and loudness of, 481–482
  loudspeaker and, 480
  soundproofing, 481
Wavelengths, 473
  de Broglie hypothesis on electron, 1125–1127
  of emitted photon during electron transition, 1199–1200
  finding increase in photon, 1124
  Lyman series, 1177–1178
  of photon, 1134
  photon wavelengths greater than threshold, 1121
  Rydberg-Ritz formula on, 1172–1173
  step potential and, 1157–1159
Wave mechanics. *See* Quantum theory
Wave number, 473
Wave packets, 521–522
Wave-particle duality, 998
  described, 1131
  two-slit experiment illustrating, 1131–1132
  uncertainty principle and, 1132–1133
Wave pulses, 466
Waves (*see also* Energy)
  described, 465
  Doppler effect and, 486–492
  electromagnetic, 478
  encountering barriers, 482–486
  harmonic, 472–476

motion by, 465–472
  periodic, 473–478
  shock, 490–491
  speed of, 467–470
  standing, 511–520
  superposition of, 504–511
  transverse and longitudinal, 465–466
Wave speed
  sound waves, 468–469
  sound waves in gas, 468
  of waves on string, 467–470
Waves in three dimensions
  described, 478–479
  wave intensity, 479–482
Wave theory, 1118–1119
Weak interactions
  bosons that mediate, 1348
  definition of, 1336
  in electroweak theory, 1348–1349
  properties of, 1350
Weber (Wb), 898
Weight
  Archimedes' principle and, 402–403
  in an elevator, 100–101
  force due to gravity, 90–91
  weightlessness and apparent, 90–91
Whipple Observatory, 1072
Wien's displacement law, 643
Work
  association between energy and, 151
  done on an ideal gas, 571–572
  done on block by spring, 158
  done by variable force, 156–157
  done on particle, 157
  kinetic energy and, 152–158
  lost between heat reservoirs, 608
  lost by an engine, 607–608
  power rate of, 163–165
  *PV* diagram for gas and, 569–572
  three dimensional energy and, 165–167
Work-energy theorem, 192
Work function, 1120–1121
Work-kinetic energy theorem, 153

X-ray spectrum, 1199–1200

Yerkes Observatory (University of Chicago), 1071
Young's modulus, 380–381
Young, Thomas, 1131

Zener diode, 1249
Zero-momentum reference frame, 247
Zero-point energy, 1134
Zeroth law of thermodynamics, 533
Zweig, G., 1345

## Physical Constants[†]

| | | |
|---|---|---|
| Atomic mass constant | $m_u = \frac{1}{12}m(^{12}C)$ | $1\text{ u} = 1.660\ 538\ 73(13) \times 10^{-27}$ kg |
| Avogadro's number | $N_A$ | $6.022\ 141\ 99(47) \times 10^{23}$ particles/mol |
| Boltzmann constant | $k = R/N_A$ | $1.380\ 6503(24) \times 10^{-23}$ J/K<br>$8.617\ 342(15) \times 10^{-5}$ eV/K |
| Bohr magneton | $m_B = e\hbar/(2m_e)$ | $9.274\ 008\ 99(37) \times 10^{-24}$ J/T $=$<br>$5.788\ 381\ 749(43) \times 10^{-5}$ eV/T |
| Coulomb constant | $k = 1/(4\pi\epsilon_0)$ | $8.987\ 551\ 788\ldots \times 10^9$ N·m²/C² |
| Compton wavelength | $\lambda_C = h/(m_e c)$ | $2.426\ 310\ 215(18) \times 10^{-12}$ m |
| Fundamental charge | $e$ | $1.602\ 176\ 462(63) \times 10^{-19}$ C |
| Gas constant | $R$ | $8.314\ 472(15)$ J/(mol·K) $=$<br>$1.987\ 2065(36)$ cal/(mol·K) $=$<br>$8.205\ 746(15) \times 10^{-2}$ L·atm/(mol·K) |
| Gravitational constant | $G$ | $6.673(10) \times 10^{-11}$ N·m²/kg² |
| Mass of electron | $m_e$ | $9.109\ 381\ 88(72) \times 10^{-31}$ kg $=$<br>$0.510\ 998\ 902(21)$ MeV/$c^2$ |
| Mass of proton | $m_p$ | $1.672\ 621\ 58(13) \times 10^{-27}$ kg $=$<br>$938.271\ 998(38)$ MeV/$c^2$ |
| Mass of neutron | $m_n$ | $1.674\ 927\ 16(13) \times 10^{-27}$ kg $=$<br>$939.565\ 330(38)$ MeV/$c^2$ |
| Permittivity of free space | $\epsilon_0$ | $8.854\ 187\ 817\ldots \times 10^{-12}$ C²/(N·m²) |
| Permeability of free space | $\mu_0$ | $4\pi \times 10^{-7}$ N/A² |
| Planck's constant | $h$ | $6.626\ 068\ 76(52) \times 10^{-34}$ J·s $=$<br>$4.135\ 667\ 27(16) \times 10^{-15}$ eV·s |
| | $\hbar = h/(2\pi)$ | $1.054\ 571\ 596(82) \times 10^{-34}$ J·s $=$<br>$6.582\ 118\ 89(26) \times 10^{-16}$ eV·s |
| Speed of light | $c$ | $2.997\ 924\ 58 \times 10^8$ m/s |
| Stefan-Boltzmann constant | $\sigma$ | $5.670\ 400(40) \times 10^{-8}$ W/(m²·K⁴) |

† The values for these and other constants can be found in Appendix B as well as on the Internet at http://physics.nist.gov/cuu/Constants/index.html. The numbers in parentheses represent the uncertainties in the last two digits. (For example, 2.044 43(13) stands for 2.044 43 ± 0.000 13.) Values without uncertainties are exact. Values with ellipses are exact (like the number $\pi = 3.1415\ldots$).

## Derivatives and Definite Integrals

$$\frac{d}{dx}\sin ax = a\cos ax \qquad \int_0^\infty e^{-ax}\,dx = \frac{1}{a} \qquad \int_0^\infty x^2 e^{-ax^2}\,dx = \frac{1}{4}\sqrt{\frac{\pi}{a^3}}$$

The $a$ in the six integrals is a positive constant.

$$\frac{d}{dx}\cos ax = -a\sin ax \qquad \int_0^\infty e^{-ax^2}\,dx = \frac{1}{2}\sqrt{\frac{\pi}{a}} \qquad \int_0^\infty x^3 e^{-ax^2}\,dx = \frac{4}{a^2}$$

$$\frac{d}{dx}e^{ax} = ae^{ax} \qquad \int_0^\infty xe^{-ax^2}\,dx = \frac{2}{a} \qquad \int_0^\infty x^4 e^{-ax^2}\,dx = \frac{3}{8}\sqrt{\frac{\pi}{a^5}}$$

## Vector Products

$$\vec{A}\cdot\vec{B} = AB\cos\theta \qquad \vec{A}\times\vec{B} = AB\sin\theta\,\hat{n} \quad (\hat{n}\text{ obtained using right-hand rule})$$

**For additional data, see the following tables in the text.**

**1-1** Prefixes for Powers of 10, p. 5

**1-2** Dimensions of Physical Quantities, p. 7

**1-3** The Universe by Orders of Magnitude, p. 12

**3-1** Properties of Vectors, p. 58

**5-1** Approximate Values of Frictional Coefficients, p. 120

**6-1** Properties of Dot Products, p. 159

**7-1** Rest Energies of Some Elementary Particles and Light Nuclei, p. 201

**9-1** Moments of Inertia of Uniform Bodies of Various Shapes, p. 274

**9-2** Analogs in Rotational and Linear Motion, p. 287

**11-1** Mean Orbital Radii and Orbital Periods for the Planets, p. 340

**12-1** Young's Modulus $Y$ and Strengths of Various Materials, p. 381

**12-2** Approximate Values of the Shear Modulus $M_s$ of Various Materials, p. 382

**13-1** Densities of Selected Substances, p. 396

**13-2** Approximate Values for the Bulk Modulus $B$ of Various Materials, p. 398

**13-3** Coefficients of Viscosity for Various Fluids, p. 414

**15-1** Intensity and Intensity Level of Some Common Sounds ($I_0 = 10^{-12}$ W/m²), p. 481

**17-1** The Temperatures of Various Places and Phenomena, p. 537

**18-1** Specific Heats and Molar Specific Heats of Some Solids and Liquids, p. 560

**18-2** Normal Melting Point (MP), Latent Heat of Fusion ($L_f$), Normal Boiling Point (BP), and Latent Heat of Vaporization ($L_V$) for Various Substances at 1 atm, p. 563

**18-3** Molar Heat Capacities J/mol·K of Various Gases at 25°C, p. 574

**20-1** Approximate Values of the Coefficients of Thermal Expansion for Various Substances, p. 629

**20-2** Critical Temperatures $T_c$ for Various Substances, p. 634

**20-3** Thermal Conductivities $k$ for Various Materials, p. 636

**20-4** $R$ Factors $\Delta x/k$ for Various Building Materials, p. 639

**21-1** The Triboelectric Series, p. 652

**21-2** Some Electric Fields in Nature, p. 661

**24-1** Dielectric Constants and Dielectric Strengths of Various Materials, p. 768

**25-1** Resistivities and Temperature Coefficients, p. 792

**25-2** Wire Diameters and Cross-Sectional Areas for Commonly Used Copper Wires, p. 792

**27-1** Magnetic Susceptibility of Various Materials at 20 C, p. 876

**27-2** Maximum Values of $\mu_0 M_s$ and $K_m$ for Some Ferromagnetic Materials, p. 882

**30-1** The Electromagnetic Spectrum, p. 977

**36-1** Electron Configurations of the Atoms in Their Ground States, p. 1196

**38-1** Free-Electron Number Densities and Fermi Energies at $T = 0$ for Selected Elements, p. 1239

**38-2** Work Functions for Some Metals, p. 1241

**39-1** Rest Energies of Some Elementary Particles and Light Nuclei, p. 1291

**40-1** Atomic Masses of the Neutron and Selected Isotopes, p. 1309

**41-1** Hadrons That Are Stable Against Decay via the Strong Nuclear Interaction, p. 1337

**41-2** Properties of Quarks and Antiquarks, p. 1347

**41-3** Masses of Fundamental Particles, p. 1347

**41-4** Bosons That Mediate the Basic Interactions, p. 1348

**41-5** Properties of the Basic Interactions, p. 1350

## Geometry and Trigonometry

$C = \pi d = 2\pi r$     definition of $\pi$

$A = \pi r^2$     area of circle

$V = \frac{4}{3}\pi r^3$     spherical volume

$A = dV/dr = 4\pi r^2$     spherical surface area

$V = A_{\text{base}}L = \pi r^2 L$     cylindrical volume

$A = dV/dr = 2\pi rL$     cylindrical surface area

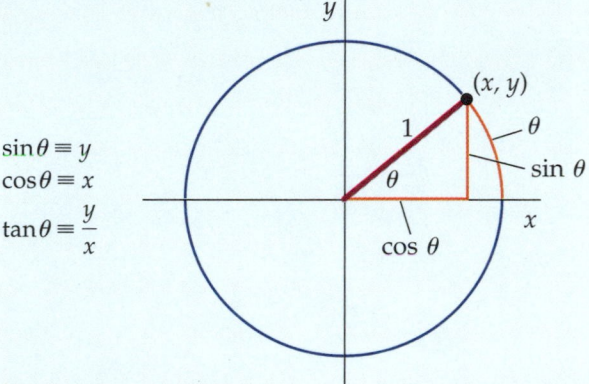

$o = h\sin\theta$

$a = h\cos\theta$

$\sin^2\theta + \cos^2\theta = 1$

$\sin(A \pm B) = \sin A \cos B \pm \cos A \sin B$

$\cos(A \pm B) = \cos A \cos B \mp \sin A \sin B$

$\sin A \pm \sin B = 2\sin[\frac{1}{2}(A \pm B)]\cos[\frac{1}{2}(A \mp B)]$

$\sin\theta \equiv y$

$\cos\theta \equiv x$

$\tan\theta \equiv \dfrac{y}{x}$

IF $|\theta| \ll 1$, THEN
   $\cos\theta \approx 1$ AND $\tan\theta \approx \sin\theta \approx \theta$     ($\theta$ in radians)

## Quadratic Formula

If $ax^2 + bx + c = 0$, then $x = \dfrac{-b \pm \sqrt{b^2 - 4ac}}{2a}$

## Binomial Expansion

If $|x| < 1$, then $(1 + x)^n =$

$$1 + nx + \frac{n(n-1)}{2!}x^2 + \frac{n(n-1)(n-2)}{3!}x^3 + \ldots$$

If $|x| \ll 1$, then $(1 + x)^n \approx 1 + nx$

If $|\Delta x|$ is small, then $\Delta F \approx \dfrac{dF}{dx}\Delta x$